INTEGRATION RULES

$$\int (Af(x) + Bg(x))\,dx = A\int f(x)\,dx + B\int g(x)\,dx$$

$$\int f'(g(x))\,g'(x)\,dx = f(g(x)) + C$$

$$\int U(x)\,dV(x) = U(x)\,V(x) - \int V(x)\,dU(x)$$

$$\int_a^b f'(x)\,dx = f(b) - f(a)$$

$$\frac{d}{dx}\int_a^x f(t)\,dt = f(x)$$

ELEMENTARY INTEGRALS

$$\int x^r\,dx = \frac{1}{r+1}x^{r+1} + C \text{ if } r \neq -1$$

$$\int \frac{dx}{x} = \ln|x| + C$$

$$\int e^x\,dx = e^x + C$$

$$\int a^x\,dx = \frac{a^x}{\ln a} + C$$

$$\int \sin x\,dx = -\cos x + C$$

$$\int \cos x\,dx = \sin x + C$$

$$\int \sec^2 x\,dx = \tan x + C$$

$$\int \csc^2 x\,dx = -\cot x + C$$

$$\int \sec x \tan x\,dx = \sec x + C$$

$$\int \csc x \cot x\,dx = -\csc x + C$$

$$\int \tan x\,dx = \ln|\sec x| + C$$

$$\int \cot x\,dx = \ln|\sin x| + C$$

$$\int \sec x\,dx = \ln|\sec x + \tan x| + C$$

$$\int \csc x\,dx = \ln|\csc x - \cot x| + C$$

$$\int \frac{dx}{\sqrt{a^2 - x^2}} = \sin^{-1}\frac{x}{a} + C$$

$$\int \frac{dx}{a^2 + x^2} = \frac{1}{a}\tan^{-1}\frac{x}{a} + C$$

$$\int \frac{dx}{a^2 - x^2} = \frac{1}{2a}\ln\left|\frac{x+a}{x-a}\right| + C$$

$$\int \frac{dx}{x\sqrt{x^2 - a^2}} = \frac{1}{a}\sec^{-1}\left|\frac{x}{a}\right| + C$$

TRIGONOMETRIC INTEGRALS

$$\int \sin^2 x\,dx = \frac{x}{2} - \frac{1}{4}\sin 2x + C$$

$$\int \cos^2 x\,dx = \frac{x}{2} + \frac{1}{4}\sin 2x + C$$

$$\int \tan^2 x\,dx = \tan x - x + C$$

$$\int \cot^2 x\,dx = -\cot x - x + C$$

$$\int \sec^3 x\,dx = \frac{1}{2}\sec x \tan x + \frac{1}{2}\ln|\sec x + \tan x| + C$$

$$\int \csc^3 x\,dx = -\frac{1}{2}\csc x \cot x + \frac{1}{2}\ln|\csc x - \cot x| + C$$

$$\int \sin ax\,\sin bx\,dx = \frac{\sin(a-b)x}{2(a-b)} - \frac{\sin(a+b)x}{2(a+b)} + C \text{ if } a^2 \neq b^2$$

$$\int \cos ax\,\cos bx\,dx = \frac{\sin(a-b)x}{2(a-b)} + \frac{\sin(a+b)x}{2(a+b)} + C \text{ if } a^2 \neq b^2$$

$$\int \sin ax\,\cos bx\,dx = -\frac{\cos(a-b)x}{2(a-b)} - \frac{\cos(a+b)x}{2(a+b)} + C \text{ if } a^2 \neq b^2$$

$$\int \sin^n x\,dx = -\frac{1}{n}\sin^{n-1} x\,\cos x + \frac{n-1}{n}\int \sin^{n-2} x\,dx$$

$$\int \cos^n x\,dx = \frac{1}{n}\cos^{n-1} x\,\sin x + \frac{n-1}{n}\int \cos^{n-2} x\,dx$$

$$\int \tan^n x\,dx = \frac{1}{n-1}\tan^{n-1} x - \int \tan^{n-2} x\,dx \text{ if } n \neq 1$$

$$\int \cot^n x\,dx = \frac{-1}{n-1}\cot^{n-1} x - \int \cot^{n-2} x\,dx \text{ if } n \neq 1$$

$$\int \sec^n x\,dx = \frac{1}{n-1}\sec^{n-2} x\,\tan x + \frac{n-2}{n-1}\int \sec^{n-2} x\,dx \text{ if } n \neq 1$$

$$\int \csc^n x\,dx = \frac{-1}{n-1}\csc^{n-2} x\,\cot x + \frac{n-2}{n-1}\int \csc^{n-2} x\,dx \text{ if } n \neq 1$$

$$\int \sin^n x\,\cos^m x\,dx = -\frac{\sin^{n-1} x\,\cos^{m+1} x}{n+m} + \frac{n-1}{n+m}\int \sin^{n-2} x\,\cos^m x\,dx \text{ if } n \neq -m$$

$$\int \sin^n x\,\cos^m x\,dx = \frac{\sin^{n+1} x\,\cos^{m-1} x}{n+m} + \frac{m-1}{n+m}\int \sin^n x\,\cos^{m-2} x\,dx \text{ if } m \neq -n$$

$$\int x\,\sin x\,dx = \sin x - x\,\cos x + C$$

$$\int x\,\cos x\,dx = \cos x + x\,\sin x + C$$

$$\int x^n\,\sin x\,dx = -x^n\,\cos x + n\int x^{n-1}\cos x\,dx$$

$$\int x^n\,\cos x\,dx = x^n\,\sin x - n\int x^{n-1}\sin x\,dx$$

(continued inside back cover)

SINGLE-VARIABLE CALCULUS

Second Edition

ROBERT A. ADAMS

Department of Mathematics
University of British Columbia

ADDISON-WESLEY
PUBLISHERS LIMITED

Don Mills, Ontario
Reading, Massachusetts
Menlo Park, California
New York • Wokingham,
England • Amsterdam
Bonn • Sydney
Singapore • Tokyo
Madrid • San Juan

SPONSORING EDITOR: Jim Grant
DESIGN: Pronk&Associates
COPY EDITOR: Valerie Adams
PRODUCTION EDITOR: Shirley Tessier
TYPE OUTPUT: Tony Gordon Limited

Canadian Cataloguing in Publication Data

Adams, Robert A. (Robert Alexander), 1940–
Single-variable calculus

2nd ed.

ISBN 0-201-50741-2

1. Calculus. I. Title.

QA303.A32 1990 515 C89-090577-0

ISBN 0-201-50741-2

Printed and bound in Canada

A B C D E F –DEY– 95 94 93 92 91 90

Dedicated to the memory of
Dr. Ronald C. Riddell

SHAJI PALAL
VILLAGE 2
EAST C
ROOM 312

CONTENTS

PREFACE

Like the first edition which preceded it, the second edition of *Single-Variable Calculus* has been designed for general calculus courses, as well as for courses for science and engineering students. As such it provides a complete introduction to the calculus of functions of a single, real variable and treats all those topics normally found in a two-semester course on the differentiation and integration of these functions. In addition to the usual "core material," it also covers a selection of optional and enrichment topics from which an instructor can select those appropriate for his or her class.

Much of the material of the first edition has been rewritten to make it more accessible to the average student with a reasonable background in high-school algebra and some previous exposure to analytic geometry. However, some optional material is more subtle and/or theoretical, and is intended mainly for stronger students. Throughout the book I have taken pains to make correct statements of results; I have tried to make the presentation as simple as possible, but no simpler.

The exercises vary greatly in both difficulty and subtlety. Numerous drill-type exercises are provided to help the student master core concepts, and many more thought-provoking ones, some theoretical, some computational, are also included to challenge the student to apply the concepts. More difficult and/or theoretical exercises are marked with an asterisk (*).

Because differential equations are used extensively to model phenomena in the sciences, this book introduces the terminology of differential equations and initial-value problems early as exercises to develop differentiation and integration skills, and to familiarize the students with the most important types. Exercises involving differential equations are marked with a dagger (†).

Principal Features of the Text

- There is an emphasis on geometry. Frequently applications of calculus are based on underlying geometric relationships among the variables involved.
- Trigonometric and exponential functions and their inverses are introduced early, and their major properties are developed before applications of differentiation are discussed. Thus these functions can be *freely* used in the applications. Students who have encountered the trigonometric functions earlier will find Section 3.1 a good review of their basic properties. Exponential functions are introduced prior to their inverses, the logarithms, but an optional section provides the alternate approach (used in the first edition) of introducing the natural logarithm first, as

the "area" under a curve. This can provide some advance motivation for the later development of the Fundamental Theorem.

- There is increased emphasis on numerical approximation of values of functions, roots of equations, and definite integrals. In particular, there is now a separate (optional) chapter on numerical integration. Topics such as the Romberg method and a discussion of the pitfalls of numerical methods help to make calculus more relevant in this age of computers and calculators. Where appropriate, students are encouraged to program calculators or computers to obtain numerical results efficiently.
- Precise statements are given for theorems. Proofs of most theorems are given or suggested immediately, but some proofs are postponed to the appendices. The three appendices develop, respectively, the technique of proof by mathematical induction, the properties of continuous functions defined on a closed, finite interval, and the properties of the Riemann integral. They provide suitable enrichment for particularly interested students and honours classes.
- The Mean-Value Theorem and its applications have been given a higher profile in a separate chapter.
- The definition of the definite integral in Chapter 6 now allows partitions with subintervals of unequal length, and a new example illustrates the added power of this improved definition.
- A chapter on plane curves now provides a (classical) introduction to the conic sections as well as the development of polar coordinates, plane parametric curves, and plane vector functions. This chapter follows that on applications of integration, so that lengths of polar and parametric curves, and areas bounded by them, can be done in proper sequence.
- The chapter on Taylor's formula and Taylor series has been reorganized so that its material can be treated either with or without having first covered most of the material in the previous chapter on numerical series.
- Like the revised first edition, this book contains a final chapter on partial differentiation, included for use in those courses which must cover a few weeks of multivariable calculus at the end of the second semester. (Of course, the author's companion volume, *Calculus of Several Variables*, takes up where *Single-Variable Calculus* leaves off, presenting a full treatment of partial differentiation, multiple integration and vector calculus.)

Core and Optional Material

Any division of material into "core" and "optional" is necessarily somewhat arbitrary. I regard most of the material of Chapters 1–6 as core, with the exception of Sections 3.6 (the alternate presentation of ln and exp), 3.8 (the hyperbolic functions), and the approximation methods of Sections 5.4 and 5.5. I also consider Sections 8.1–8.3 (basic geometric applications of integration), 9.2 (polar coordinates), 9.3 (parametric curves), 10.1 and 10.2 (the basics of infinite sequences and series), and most of Chapter 11 as core. These days, scientists and engineers require more training in the proper and effective use of numerical techniques, so Sections 5.5 (root finding), and much of Chapter 7 (numerical integration) are assuming more importance and might be regarded as "core" for some classes.

The remaining material is optional in the sense that (with minor exceptions) its prior coverage is not necessary for any of what follows. It is up to individual instructors to decide what is most appropriate for their classes.

Acknowledgments

The first edition of *Single-Variable Calculus* has been used, since its publication, for classes of general science, engineering, and mathematics majors and honours students at the University of British Columbia. I am grateful to colleagues and students at UBC, and at many other institutions where the book has been used, for their encouragement and useful comments and criticisms. Many of the changes in this edition are a result of that feedback. I am also grateful to several reviewers and proofreaders for their helpful suggestions during the preparation of this edition. Reviews of specific chapters were done by Professors T. Bisztriczky (University of Calgary), Ken Dunn (Dalhousie University), Tom Holens, (University of Manitoba), David J. Leeming, (University of Victoria), Richard Nowakowsky, (Dalhousie University), David Ryeburn (Simon Fraser University), Cedric Schubert (Queens University), and R. Grant Woods, (University of Manitoba). Painstaking reviews of the finished typescript were done by David Ryeburn and Ken Dunn, and by student Joanna Kwan (UBC). A final, thorough editorial proofreading was done by production editor Valerie Adams at Addison-Wesley.

I typeset this volume using TeX and PostScript on an AT microcomputer. I also generated most of the figures in PostScript using software developed by myself and my colleague, Professor Robert Israel. Some of the three dimensional air-brush art was prepared by Iris Ward. Prior to my starting the revision, the unrevised text of the first edition was committed to computer files in TeX format by Valerie Adams. I am very grateful to all these people for their excellent work.

I also wish to thank several people at Addison-Wesley for their assistance and encouragement. These include Sponsoring Editor Jim Grant, who guided the project and arranged for the reviews, Vice-President Andy Yull, with whom I enjoyed frequent stimulating discussions on matters of design and on numerous problems involving the TeX–PostScript interface, Executive Vice-President Joe Swan, and Editorial Director Ron Doleman, who supervised the publication of the first edition of *Single-Variable Calculus* and of *Calculus of Several Variables*, who first introduced the author to TeX and PostScript, and who assumed responsibility, along with Shirley Tessier, for the final stages of the publication of this edition.

Despite all the excellent help I have received, I am not so naïve as to believe that the text is now free of errors and obscurities, and I accept full responsibility for any that remain. Any comments, corrections, and suggestions for future revisions from readers will be much appreciated.

R.A.A.
Vancouver, Canada
September, 1989

CHAPTER 1

Functions, Limits, and Continuity

1.1 WHAT IS CALCULUS?

Much of our understanding of the world in which we live depends on our ability to describe how things change. Whether we are concerned with the motion of a pitched baseball or the path of a planet, whether the temperatures and currents of the oceans or the fluctuations of the stock market, whether the propagation of radio waves or the power produced by a chemical reaction, we are constantly forced to analyze relationships among quantities which change with time.

Algebra and geometry are useful tools for describing relationships among *static* quantities, but they do not involve concepts appropriate for describing how a quantity changes. For this we need new mathematical operations which go beyond the algebraic operations of addition, subtraction, multiplication, division and the taking of powers and roots. We require operations which measure the way related quantities change.

Calculus provides the tools for describing motion quantitatively. It introduces two new operations called *differentiation* and *integration* which, like addition and subtraction, are opposites of one another; what differentiation does, integration undoes.

For example, consider the motion of a falling rock. The height (in metres) of the rock t seconds after it is dropped from a height h_0 is a function $h(t)$ given by

$$h(t) = h_0 - 4.9t^2.$$

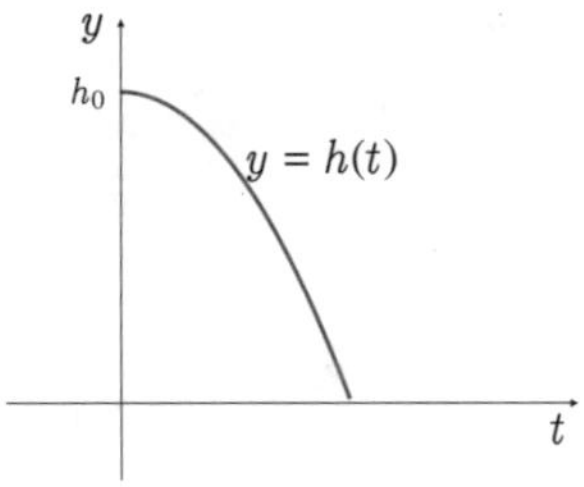

FIGURE 1.1.1

The graph of $y = h(t)$ is shown in Fig. 1.1.1. The process of differentiation enables us to find a new function, which we denote $h'(t)$ and call *the derivative* of h with respect to t:

$$h'(t) = -9.8t,$$

and which represents the *rate of change* of the height of the rock, that is, its *velocity* in metres/second.

Inversely, if we know the velocity of the rock as a function of time, integration enables us to find the height function $h(t)$.

Calculus was invented independently and in somewhat different ways by two 17th century mathematicians, Sir Isaac Newton and Gottfried Wilhelm Leibniz. Newton's motivation was a desire to analyze the motion of moving objects. Using his calculus he was able to formulate his laws of motion and gravitation, and to *calculate from them* that the planets must move around the sun in elliptical orbits, a fact that had been discovered half a century earlier by Johannes Kepler. Kepler's discovery was empirical, made from years of study of numerical data on the positions of planets.

Many of the most fundamental and important "laws of nature" are conveniently expressed as equations involving rates of change of quantities. Such equations are called *differential equations* and techniques for their study and solution are at the heart of calculus. In the falling rock example the appropriate law is Newton's second law of motion:

$$\text{Force} = \text{mass} \times \text{acceleration}.$$

The *acceleration*, -9.8 m/sec^2, is the rate of change (the *derivative*) of the velocity, which is in turn the rate of change (the *derivative*) of the height function.

Much of mathematics is related indirectly to the study of motion. We regard *lines* or *curves* as geometric objects, but the ancient Greeks thought of them as paths traced out by moving points. Nevertheless, the study of curves also involves geometric concepts such as tangency and area. The process of differentiation (Chapters 2–4) is closely tied to the geometric problem of finding tangent lines; similarly, integration (Chapters 6–8) is related to the geometric problem of finding areas of regions with curved boundaries.

Underpinning the study of calculus are the concepts of real number, coordinate system, and function. In the next three sections of this chapter we will review these concepts and set out the terminology and symbols we will use in referring to them throughout the book. The remaining sections introduce and explore the concept of **limit**, an operation on functions. The use of limits distinguishes calculus from other branches of mathematics (arithmetic, algebra, geometry) you have already encountered.

1.2 THE REAL LINE AND THE CARTESIAN PLANE

Elementary calculus depends heavily on properties of **real numbers**, that is, numbers expressible in decimal form such as

$$\begin{aligned} 5 &= 5.00000\ldots \\ -\tfrac{3}{4} &= -0.750000\ldots \\ \tfrac{1}{3} &= 0.3333\ldots \\ \sqrt{2} &= 1.4142\ldots \\ \pi &= 3.14159\ldots \end{aligned}$$

We expect that as a student of calculus you already have some familiarity with the real numbers and with the Cartesian coordinate system in the plane. Both are treated only briefly here to establish the terminology.

The real numbers can be represented geometrically as points on a number line, which we call the **real line**, shown in Fig. 1.2.1. The symbol $\mathbb{R}$ is used to denote either the real number system or, equivalently, the real line.

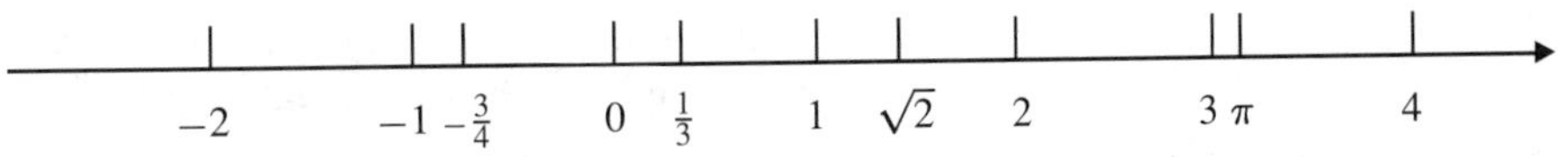

FIGURE 1.2.1

The properties of the real number system fall into three categories: algebraic properties, order properties, and completeness. The algebraic properties will already be familiar to you, and we will not dwell on them here; roughly speaking, they assert that real numbers may be added, subtracted, multiplied, and divided (except by zero) to produce more real numbers, and that the usual laws of arithmetic are satisfied.

The *order properties* refer to the order in which the numbers appear on the real line. If x lies to the left of y, then we say $x < y$ or $y > x$. Of course $x \leq y$ means that either $x < y$ or $x = y$. The order properties can be summarized as follows:

1.2.1
Order Properties of Real Numbers

i) If $x < y$ and z is any real number, then $x + z < y + z$.
ii) If $x < y$ and $z > 0$, then $xz < yz$.
iii) If $x < y$ and $z < 0$, then $xz > yz$; in particular, for $z = -1$, $-x > -y$.
iv) If $0 < x < y$ then $0 < \frac{1}{y} < \frac{1}{x}$.

Note especially the rules for multiplying an inequality by a number. If the number is positive, the inequality is preserved; if the number is negative, the inequality is reversed.

The *completeness* property of the real number system is more subtle and difficult to understand. One way to state it is as follows: If A is any set of real numbers having at least one number in it, and if there exists a real number y with the property that $x \leq y$ for every x in A, then there exists a *smallest* number y with the same property. Roughly speaking, this says that there can be no holes or gaps on the real line—every point corresponds to a real number. Certain important results in calculus require the completeness property for their proofs. Most of these results can be derived with no great difficulty from a few basic theorems, in particular Theorems 1.7.7 and 1.7.10 below. We do not prove these theorems in this chapter, but sketch their proofs in Appendix II, which is concerned with the theoretical foundations of calculus. The techniques for formal proofs involving limits in that appendix often are not studied in first courses in calculus but are deferred to subsequent courses in mathematical analysis. We will, however, make some direct use of completeness when we study infinite sequences and series in Chapter 11.

We distinguish three special subsets of the real numbers:

i) the **natural numbers**, namely the numbers 1, 2, 3, 4, ...
ii) the **integers**, namely, the numbers 0, ± 1, ± 2, ± 3, ...
iii) the **rational numbers**, that is, numbers that can be expressed in the form m/n, where m and n are integers, and $n \neq 0$.

The rational numbers are precisely those real numbers with decimal expansions that are either:

a) terminating, (that is, ending with an infinite string of zeros), or
b) repeating, (that is, ending with a string of digits that repeats over and over).

EXAMPLE 1.2.2 Show that the numbers a) $1.323232\cdots = 1.\overline{32}$, and b) $0.3405405405\cdots = 0.3\overline{405}$ are rational numbers by expressing each as a quotient of two integers. (The bars indicate the pattern of repeating digits.)

SOLUTION a) Let $x = 1.323232\cdots$ Then $x - 1 = 0.323232\cdots$ and

$$100x = 132.323232\cdots = 132 + 0.323232\cdots = 132 + x - 1.$$

Therefore, $99x = 131$ and $x = 131/99$.

b) Let $y = 0.3405405405\ldots$ Then $10y = 3.405405405\ldots$ and $10y - 3 = 0.405405405\ldots$ Also,

$$10000y = 3405.405405405\ldots = 3405 + 10y - 3.$$

Therefore, $9990y = 3402$ and $y = 3402/9990 = 63/185$.

The set of rational numbers possesses all the algebraic and order properties of the real numbers, but not the completeness property. There is, for example, no rational number whose square is 2. Hence, there is a "hole" on the "rational line" where $\sqrt{2}$ should be. To see this, suppose that there were a rational number $x = m/n$ such that $x^2 = 2$. We can assume that any common factors in the fraction m/n have been canceled, so that m and n are not both even integers. Since $m^2/n^2 = 2$, therefore $m^2 = 2n^2$ and m^2 must be an even integer. Hence m must also be even. (A product of odd numbers is always odd.) Since m is even, we can write $m = 2k$ where k is an integer. Thus $4k^2 = 2n^2$ and $n^2 = 2k^2$; n^2 is even and so is n. We have arrived at a contradiction; we assumed that m and n were not both even and then proved they were both even. Accordingly there can be no rational number whose square is 2. Because the real line has no such "holes," it is the appropriate setting for studying limits and therefore calculus.

Intervals

A subset of the real line is called an **interval** if it contains at least two numbers and also contains all real numbers between any two of its elements. For example, the set of real numbers x such that $x > 6$ is an interval, but the set of real numbers y such that $y \neq 0$ is not an interval. (Why?) It consists of two intervals.

If a and b are real numbers and $a < b$ we often refer to

i) the **closed interval** from a to b, denoted $[a, b]$, consisting of all real numbers x satisfying $a \leq x \leq b$, and

ii) the **open interval** from a to b, denoted (a, b), consisting of all real numbers x satisfying $a < x < b$. (See Fig. 1.2.2. Note the use of small solid "dots" and open "holes" to indicate that the closed interval contains its endpoints while the open open interval does not.)

We can extend the above notations to various half-open intervals and infinite intervals (see Fig. 1.2.3).

FIGURE 1.2.2

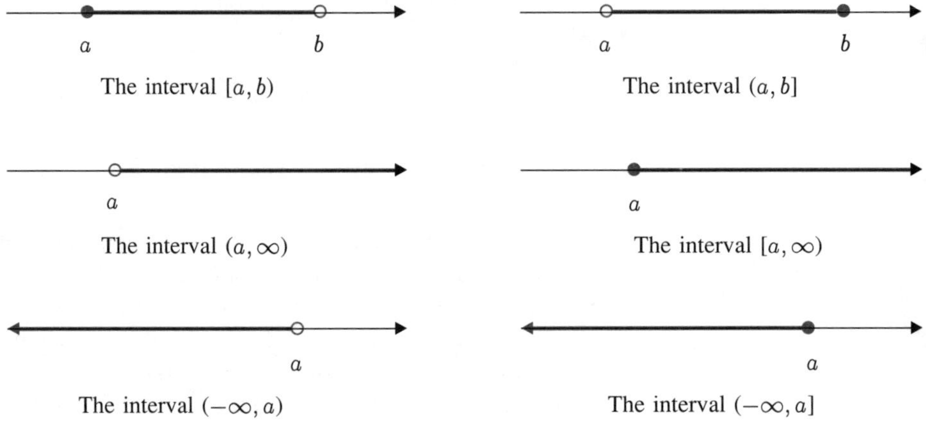

FIGURE 1.2.3

The whole real line may also be regarded as an interval: $\mathbb{R} = (-\infty, \infty)$. Infinity (denoted ∞) is *not* a real number, and intervals extending to $-\infty$ or ∞ should not be regarded as having any endpoints there.

EXAMPLE 1.2.3 Solve the following inequalities and graph their solutions on the real line:

a) $2x - 1 > x + 3$ b) $-\dfrac{x}{3} < 2x - 1$

c) $\dfrac{2}{x-1} \geq 5$ d) $\dfrac{3}{x-1} \leq \dfrac{-2}{x}$

SOLUTION a) Adding $1 - x$ to both sides we get $2x - x > 3 + 1$, so $x > 4$. (See Fig. 1.2.4a.)

b) Multiplying both sides by -3 and remembering to reverse the direction of the inequality (since we are multiplying by a negative number), we get $x > -6x+3$. Thus $7x > 3$ and $x > 3/7$. (See Fig. 1.2.4b.)

c) We would like to multiply both sides by $x - 1$, but this will necessitate reversing the inequality if $x - 1 < 0$, so we break the problem into two cases.
Case I. $x - 1 > 0$, that is, $x > 1$. Then $2 \geq 5(x - 1) = 5x - 5$. Therefore

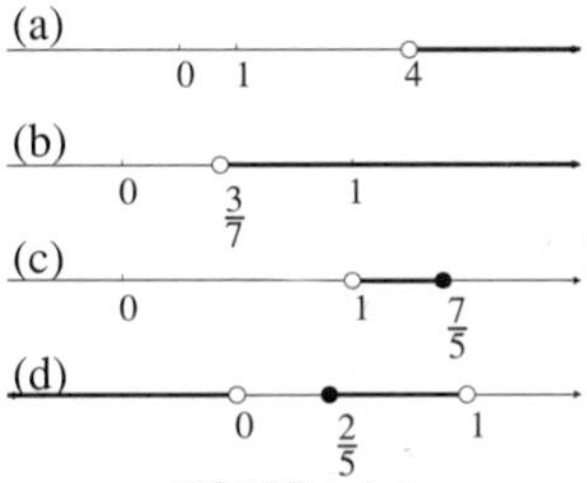

FIGURE 1.2.4

$7 \geq 5x$ and $x \leq 7/5$. This case leads to solutions $1 < x \leq 7/5$.
Case II. $x - 1 < 0$, that is, $x < 1$. Then $2 \leq 5(x-1)$ and $x \geq 7/5$. Since no real numbers x satisfy both $x < 1$ and $x \geq 7/5$, this case leads to no solutions. The only solutions of the given inequality satisfy $1 < x \leq 7/5$. (Fig. 1.2.4c.)

d) Note that one side or the other is not defined if $x = 1$ or $x = 0$. We would like to multiply the inequality by $(x-1)x$ to clear it of fractions. Accordingly, we must distinguish two cases: $(x-1)x > 0$ and $(x-1)x < 0$.
Case I. If $x > 1$ or $x < 0$, then $(x-1)x > 0$ and

$$\begin{aligned} 3x &\leq -2(x-1) \\ 5x &\leq 2, \end{aligned}$$

so $x \leq 2/5$. Since $x > 1$ or $x < 0$, this case has solutions $x < 0$.
Case II. If $0 < x < 1$, then $(x-1)x < 0$ and

$$\begin{aligned} 3x &\geq -2(x-1) \\ 5x &\geq 2, \end{aligned}$$

so $x \geq 2/5$. Since $0 < x < 1$, this case has solutions $2/5 \leq x < 1$.
The solution set can be written as a *union* of intervals: $(-\infty, 0) \cup [\frac{2}{5}, 1)$. It is shown in Fig. 1.2.4d.

Note the use of the symbol $\cup$ to denote the union of intervals. A real number is in the union of two intervals if it is in either interval. Similarly, we shall use the symbol $\cap$ to denote intersection. A real number x is in the *intersection* $I \cap J$ of two intervals I and J if x belongs to *both* of the intervals I and J. For example,

$$[1,3) \cap [2,4] = [2,3).$$

The Absolute Value

For any real number x, the **absolute value** of x, denoted $|x|$, is defined by

1.2.4
Definition of Absolute Value

$$|x| = \begin{cases} x, & \text{if } x \geq 0 \\ -x, & \text{if } x < 0 \end{cases}$$

For example, $|2| = 2$ since $2 \geq 0$, but $|-3| = -(-3) = 3$ since $-3 < 0$. Evidently $|x| \geq 0$ for every real number x, and $|x| = 0$ if and only if $x = 0$.

Geometrically, $|x|$ represents the (nonnegative) distance from x to 0 on the real number line. More generally, $|x - y|$ represents the (nonnegative) distance between the points x and y on the real line, since this distance is the same as that from the point $x - y$ to 0 (see Fig. 1.2.5):

$$|x - y| = \begin{cases} x - y, & \text{if } x \geq y \\ y - x, & \text{if } x < y. \end{cases}$$

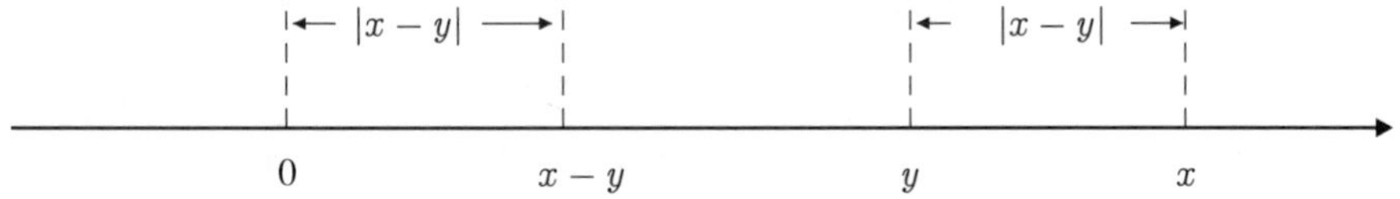

FIGURE 1.2.5

The absolute value function has the following properties:

1.2.5
Properties of Absolute Value

For any real numbers a and b,

$$|ab| = |a||b|$$
$$|a \pm b| \leq |a| + |b| \quad \text{(The Triangle Inequality).}$$

The first property is easily checked by considering the four possible cases where each of a or b is either positive or negative. The second property follows from the first because $\pm 2ab \leq |2ab| = 2|a||b|$. Therefore we have

$$\begin{aligned}|a \pm b|^2 = (a \pm b)^2 &= a^2 \pm 2ab + b^2 \\ &\leq |a|^2 + 2|a||b| + |b|^2 = (|a| + |b|)^2,\end{aligned}$$

and taking (positive) square roots of both sides we obtain $|a \pm b| \leq |a| + |b|$. If we regard 0, a and $\mp b$ as the vertices of a triangle, this inequality says that the length of the side joining a and $\mp b$ is less than the sum of the lengths of the other two sides. Of course the "triangle" is somewhat degenerate since all three points lie on a straight line.

EXAMPLE 1.2.6 Solve a) $|2x - 3| = 5$, and b) $|x + 1| < |x - 1|$.

SOLUTION a) We could break the equation $|2x - 3| = 5$ into two cases.

Case I. $2x - 3 \geq 0$, i.e., $x \geq \frac{3}{2}$. Thus $|2x - 3| = 2x - 3$ and the equation becomes $2x - 3 = 5$ and yields the solution $x = 4$.

Case II. $2x - 3 < 0$, i.e., $x < \frac{3}{2}$. Then $|2x - 3| = -(2x - 3) = 3 - 2x$. In this case the equation is $3 - 2x = 5$, and the solution is $x = -1$.

There are, therefore, two solutions, $x = 4$ and $x = -1$.

An easier way to get this answer is to interpret the equation geometrically, as represented in Fig. 1.2.6a. If $|2x - 3| = 5$, then $|2(x - \frac{3}{2})| = 5$, so $2|x - \frac{3}{2}| = 5$ and $|x - \frac{3}{2}| = \frac{5}{2}$. This says that the distance from x to $\frac{3}{2}$ is $\frac{5}{2}$. There are two such numbers, x, namely, $x = \frac{3}{2} + \frac{5}{2} = 4$ and $x = \frac{3}{2} - \frac{5}{2} = -1$.

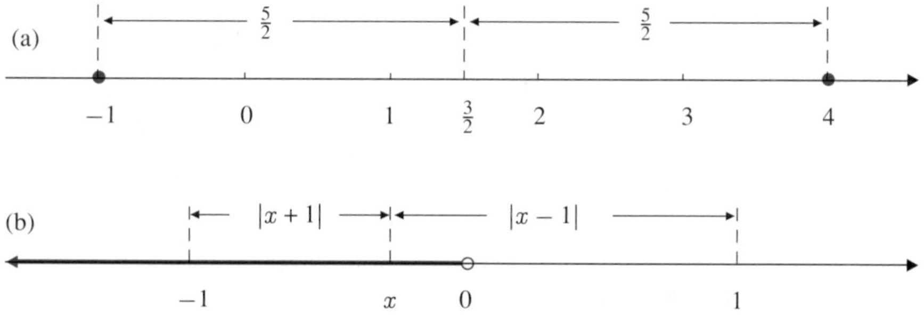

FIGURE 1.2.6

b) We could proceed as in (a) by considering all possible cases (where $x + 1$ is positive or negative and where $x - 1$ is positive or negative). However, observe that the inequality says that the distance of x from -1 is less than the distance of x from 1. Thus the solution consists of all real numbers x less than 0. This is shown in Fig. 1.2.6b.

Coordinates and Graphs

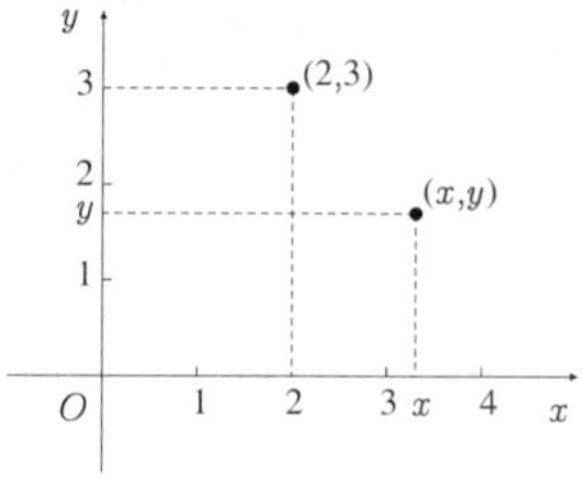

FIGURE 1.2.7

The positions of all points in a plane can be measured with respect to a pair of perpendicular real lines in the plane intersecting at the 0-point of each. It is conventional to take one line horizontal and call it the x-axis, and the other line vertical and call it the y-axis. The point of intersection of the axes is called the origin, and is denoted O. The position of any point P is then specified by an ordered pair (x, y) of real numbers, called the coordinates of P; x measures the displacement of P from O measured parallel to the x-axis, and y the displacement measured parallel to the y-axis (see Fig. 1.2.7). By the Pythagorean theorem the distance from O to P is (Fig. 1.2.8)

$$|OP| = \sqrt{x^2 + y^2}.$$

More generally, the distance between points $P_1 = (x_1, y_1)$ and $P_2 = (x_2, y_2)$ is $\sqrt{(x_2 - x_1)^2 + (y_2 - y_1)^2}$, as shown in Fig. 1.2.9.

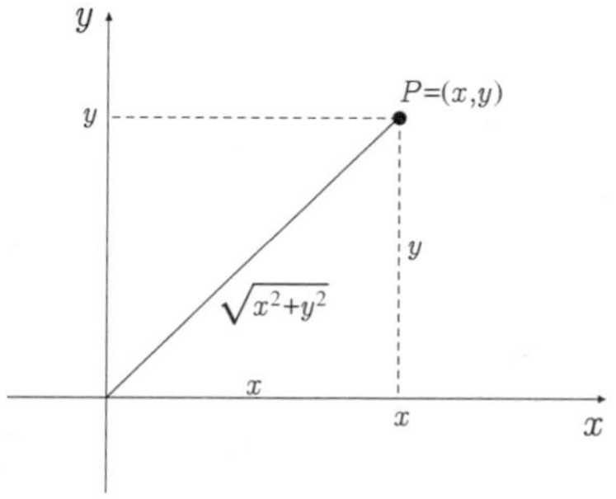

FIGURE 1.2.8

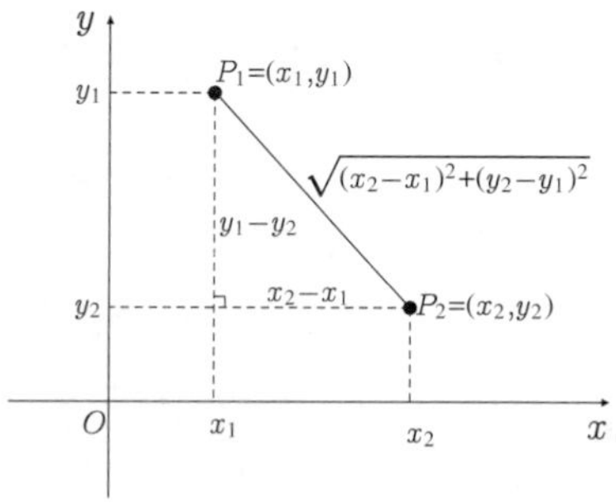

FIGURE 1.2.9

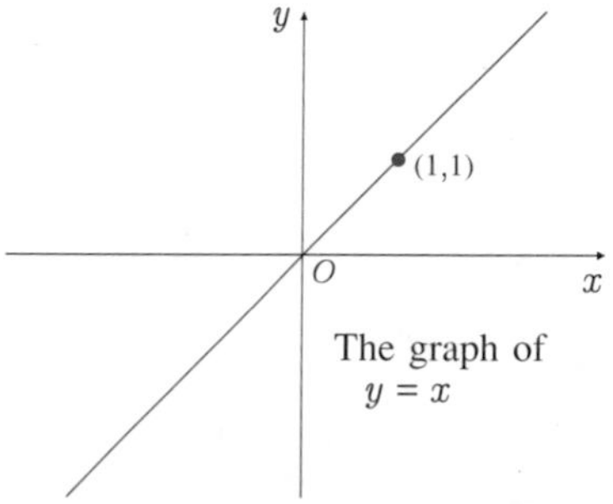

FIGURE 1.2.10

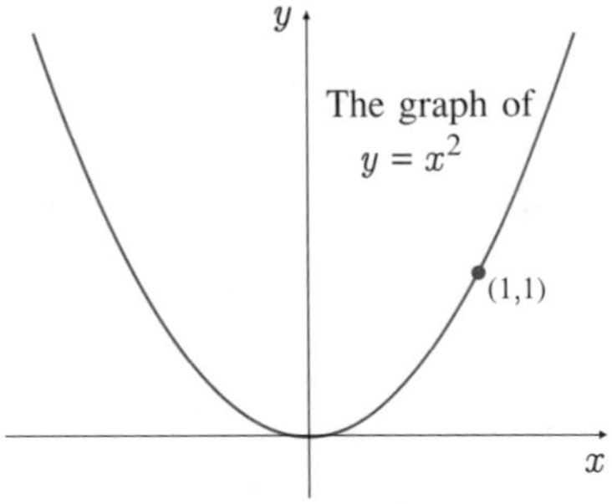

FIGURE 1.2.11

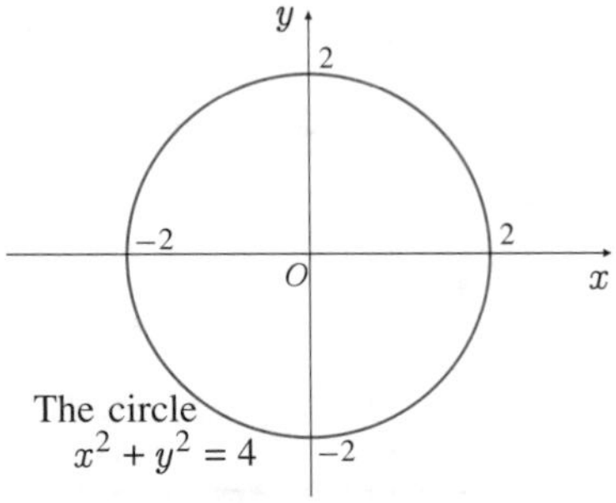

FIGURE 1.2.12

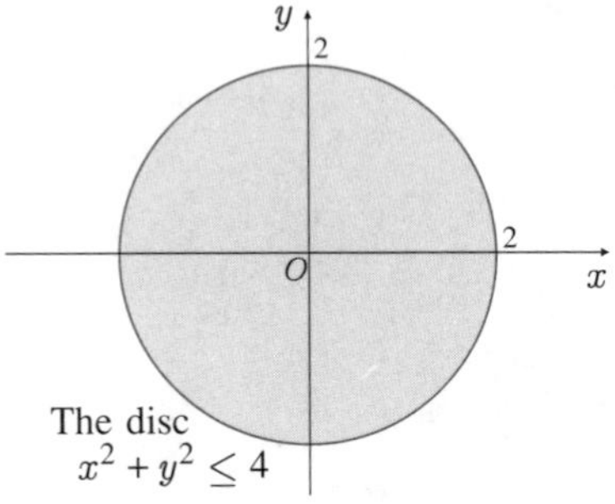

FIGURE 1.2.13

The *graph* of an equation (or inequality) involving the variables x and y is the set of all points $P = (x, y)$ whose coordinates x and y satisfy the equation (or inequality). See, for example, Figs. 1.2.10 to 1.2.13.

Any nonvertical straight line in the plane has the property that the ratio

$$m = \frac{y_2 - y_1}{x_2 - x_1}$$

has the same value for every choice of a pair of points $P_1 = (x_1, y_1)$ and $P_2 = (x_2, y_2)$ on the line. The value of m is the **slope** of the line. Observe in Figure 1.2.14, that triangles P_1AP_2 and $P_1'A'P_2'$ are similar. Thus,

$$\frac{AP_2}{P_1A} = \frac{A'P_2'}{P_1'A'} = m, \quad \text{that is,} \quad \frac{y_2 - y_1}{x_2 - x_1} = \frac{y_2' - y_1'}{x_2' - x_1'} = m.$$

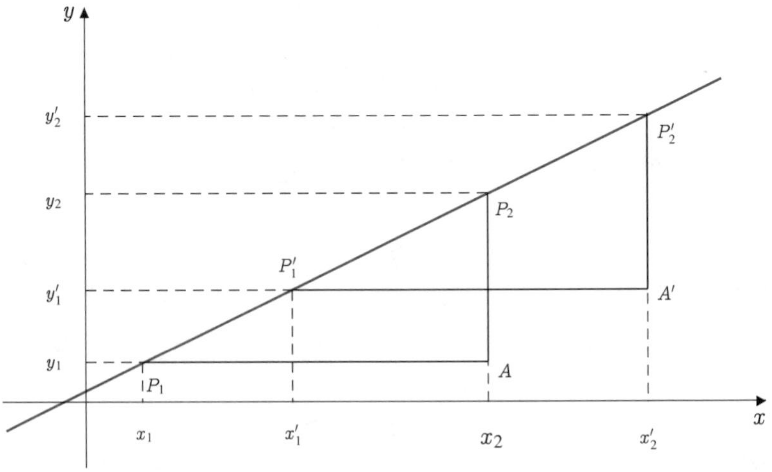

FIGURE 1.2.14

If $P = (x, y)$ is any point on the straight line with slope m passing through $P_1 = (x_1, y_1)$, then

$$\frac{y - y_1}{x - x_1} = m$$

so the coordinates x and y of P satisfy the equation

1.2.7
Point-Slope Equation of a Straight Line

$$y = y_1 + m(x - x_1).$$

This is called the **point-slope form** of the equation of a straight line. It will be very useful when we study tangent lines in Chapter 2. A horizontal line through P_1 has slope 0 and equation $y = y_1$. A vertical line through P_1 has infinite slope, so its equation is $x = x_1$. (See Fig. 1.2.15.)

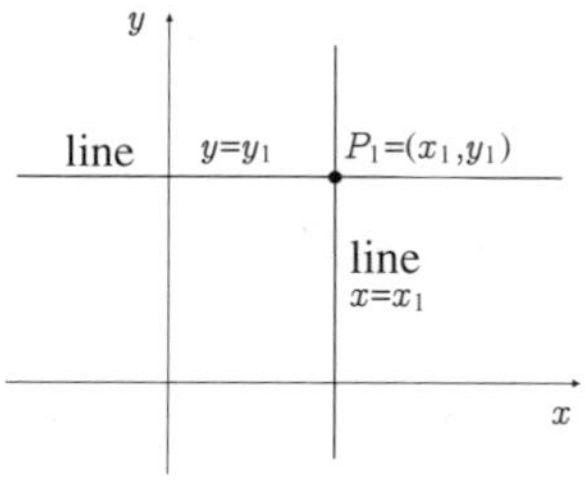

FIGURE 1.2.15

An equation of the form

$$Ax + By + C = 0$$

is called a **linear equation** provided A and B are not both zero. The graph of such an equation is always a straight line. It has infinite slope if $B = 0$ and slope $-A/B$ otherwise. Two lines are parallel if they have the same slope; they are perpendicular if the product of their slopes is -1. Some other forms of equations of straight lines are

i) The **slope–y-intercept form:** $y = mx + b$. This is a straight line with slope m and y-intercept b; that is, it passes through the point $(0, b)$.

ii) The **two-intercept form:** $\dfrac{x}{a} + \dfrac{y}{b} = 1$. This line has x-intercept a and y-intercept b; it passes through the points $(a, 0)$ and $(0, b)$.

EXAMPLE 1.2.8 Figure 1.2.16 illustrates the following examples:

a) The straight line passing through the point $(0, -3)$ and having slope 7 has equation $y = 7x - 3$. (We used the slope–y-intercept form.)

b) The straight line passing through the points $(2, 0)$ and $(0, 3)$ has equation $x/2 + y/3 = 1$ or $3x + 2y = 6$. (We used the two-intercept form.)

c) The straight line passing through the points $(1, 3)$ and $(-2, 6)$ has a slope of $(6 - 3)/(-2 - 1) = -1$. Hence it has equation $y - 3 = -1(x - 1)$ or $x + y = 4$.

d) The equation $5x + 7y + 21 = 0$ represents a straight line with slope $-5/7$, since $y = -\frac{5}{7}x - 3$. The y-intercept of this line is -3; the x-intercept is $-21/5$.

EXAMPLE 1.2.9 Identify the graph of the equation $x^2 + y^2 - 2x + 4y = 4$.

SOLUTION Observe that the expression $x^2 - 2x$ contains the first two terms of the perfect square $(x - 1)^2 = x^2 - 2x + 1$. Hence $x^2 - 2x = (x - 1)^2 - 1$. Similarly, $y^2 + 4y = (y + 2)^2 - 4$. Hence the given equation can be written

$$(x - 1)^2 + (y + 2)^2 = 1 + 4 + 4 = 9 = 3^2.$$

This says that the distance from the point (x, y) to the point $(1, -2)$ is equal to 3. Hence the equation represents a circle of radius 3 centred at the point $(1, -2)$ (see Fig. 1.2.17).

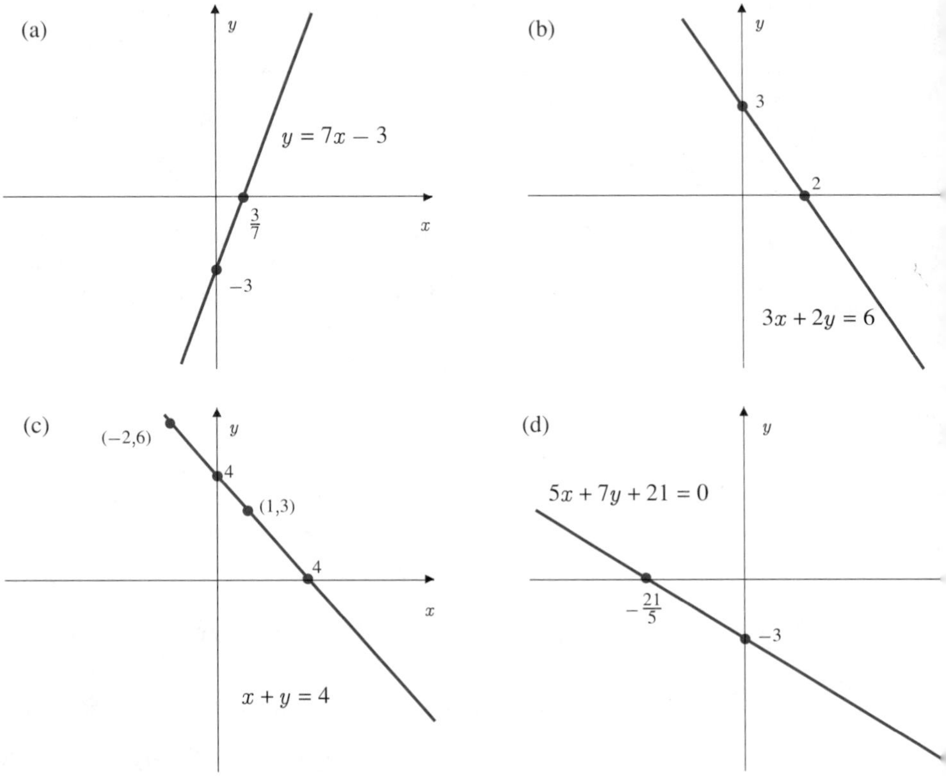

FIGURE 1.2.16

EXAMPLE 1.2.10 Sketch the graph of the equation $y = x^2 - 4x$.

SOLUTION We can make a table of values to obtain the coordinates of points on the graph.

x	-2	-1	0	1	2	3	4	5	6
y	12	5	0	-3	-4	-3	0	5	12

The graph is a parabola with vertex at $(2, -4)$ and a vertical axis, as shown in Fig. 1.2.18. (Note that different scales are used on the two axes.) The equation may be rewritten $y = (x - 2)^2 - 4$.

In the study of calculus we will encounter many kinds of curves and their equations. You are probably already familiar with some of them, such as straight lines, circles, and perhaps the other *conic sections*, parabolas, ellipses and hyperbolas. (These curves are treated in detail in Sections 9.1 and 9.2.) Other curves encountered will be less familiar. Calculus will provide useful tools to help us determine their interesting features and sketch them without having to calculate the coordinates of very many points.

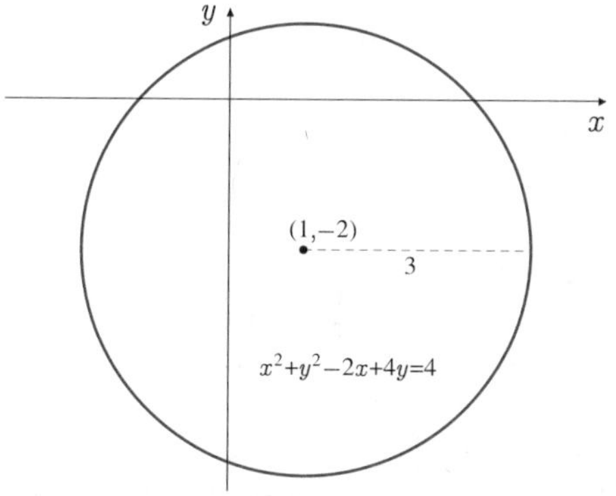

FIGURE 1.2.17

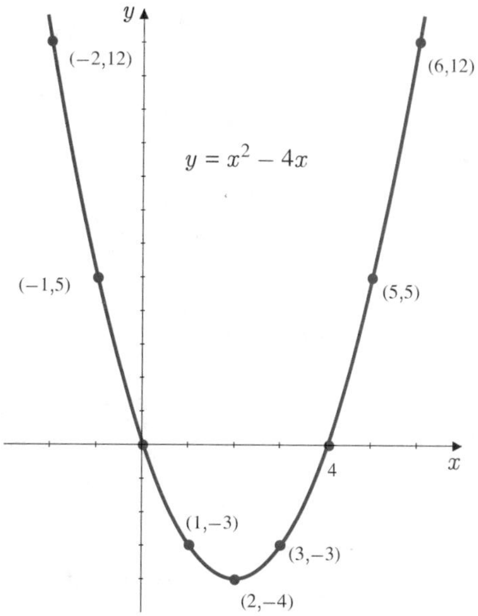

FIGURE 1.2.18

EXERCISES

1. Express the following repeating decimals as quotients of integers:
a) $9.090909\cdots = 9.\overline{09}$,
b) $3.2101010\cdots = 3.2\overline{10}$,
c) $0.00044044044\cdots = 0.00\overline{044}$,
d) $0.285714285714\cdots = 0.\overline{285714}$.

In Exercises 2–17 solve the given inequality and graph the solution set.

2. $-2x > 4$ **3.** $x^2 \le 9$

4. $5x - 3 \le 7 - 3x$ **5.** $3(2 - x) < 2(3 + x)$

6. $\dfrac{3x-4}{2} \le \dfrac{6-x}{4}$ **7.** $x^2 - 1 \ge 2x$

8. $x^3 > 4x$ **9.** $\dfrac{1}{2-x} < 3$

10. $x^2 - x \le 2$ **11.** $\dfrac{x+1}{x} > 2$

12. $\dfrac{x}{2} > 1 + \dfrac{4}{x}$ **13.** $\dfrac{3}{x-1} < \dfrac{2}{x+1}$

14. $|x - 3| > 1$ **15.** $|x + 2| \ge |x + 3|$

16. $|4 - 3x| \le 9$ **17.** $\left|\dfrac{x+1}{2}\right| \ge 1$

In Exercises 18–25 find equations of the straight lines satisfying the given conditions.

18. passing through $(2, 3)$ and having slope -4

19. passing through $(5, -3)$ and having slope 0

20. passing through $(-2, -10)$ and having slope 20

21. passing through $(2, 1)$ and $(4, 5)$

22. passing through $(0, 4)$ and $(8, 0)$

23. passing through $(0, 0)$ and parallel to $x + 3y = 17$

24. passing through $(-2, -3)$ and parallel to the line $y = 4 - 2x$

25. passing through the point $(1, 1)$ and perpendicular to the line $y = 3x$

In Exercises 26–49 sketch the graph of the given equation or inequality. When possible, identify the graph.

26. $y = 2x - 1$ **27.** $y = 2 - x$

28. $y = -1$ **29.** $x = 4$

30. $y = 2 + \dfrac{x-1}{3}$ **31.** $\dfrac{x}{4} + y = 1$

32. $2x - 5y + 5 = 0$ **33.** $\dfrac{x}{2} - \dfrac{y}{3} = 2$

34. $x^2 + y^2 = 25$ **35.** $x^2 + y^2 = 5$

36. $(x + 1)^2 + y^2 = 25$ **37.** $x^2 + (y - 4)^2 = 16$

38. $x^2 + 4x + y^2 = 12$

39. $x^2 + y^2 + 2x - 2y = 2$

40. $y = x^2$

41. $x = y^2$

42. $y = 2x^2 - 1$

43. $y = x^2 + 2x + 2$

44. $x^2 - y^2 = 0$

45. $x^2 + y^2 = 0$

46. $2x - y \geq 1$

47. $x + y < 2$

48. $x^2 + y^2 - 6y \leq 0$

49. $y \leq 1 - x^2$

50.*Show that the inequality $|a - b| \geq \big||a| - |b|\big|$ holds for any real numbers a and b.

51.* Show that the product of the slopes of perpendicular lines is -1. Use the Pythagorean theorem.

1.3 FUNCTIONS

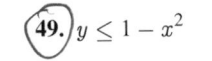

1.3.1 Definition of a Function

> A **function** f is a rule that assigns to each element x in some set $\mathcal{D}(f)$ (called the **domain** of f) a *unique* object $f(x)$ called the value of f at x.

Let us assume for the time being that the domains of our functions consist of real numbers, and that the values of the functions are also real numbers; introductory courses in calculus deal with *real-valued* functions of a single *real variable*. (In higher mathematics courses functions with other kinds of domains and values are studied.)

EXAMPLE 1.3.2 The squaring function on $\mathbb{R}$ is the function f that assigns to each real number x its square, x^2:

$$f(x) = x^2, \qquad \text{for } x \text{ in } \mathbb{R}.$$

The domain of f is the set of all real numbers: $\mathcal{D}(f) = \mathbb{R}$.

Strictly speaking, we should call the function f, and reserve the notation $f(x)$ to denote the value of f at x. However, it is common to refer to a the function loosely as $f(x)$. Thus, in the example above, the squaring function is often simply called the function x^2. Another way of denoting a function is as follows:

$$f : x \rightarrow f(x)$$

This is read as "the function f that takes x to $f(x)$." The squaring function can then be denoted

$$f : x \rightarrow x^2$$

or, even more simply, without using any f,

$$x \rightarrow x^2.$$

The definition of function given above requires that the domain be specified. If the function g is defined by

$$g(x) = x^2 \qquad \text{for } -1 \leq x < 2,$$

then g is not the same function as f in Example 1.3.2 because f and g have different domains. Most often we do not want to specify the domain of a function explicitly; since our functions are defined for real numbers and have real values, we adopt the following convention.

1.3.3
The Domain Convention

> Unless the domain of a function f is given explicitly, we assume that it is *the largest set* of real numbers for which $f(x)$ makes sense as a real number.

Thus the function x^2 has domain all real numbers, but the function $1/x$ has domain all real numbers except 0.

EXAMPLE 1.3.4 Let F be the function defined by $F(t) = 3(t - 1) + 5$. Find the values of F at the points 0, 2, -4, $t + 2$, and $F(2)$.

SOLUTION

$$
\begin{aligned}
F(0) &= 3(-1) + 5 = 2 \\
F(2) &= 3(2 - 1) + 5 = 8 \\
F(-4) &= 3(-4 - 1) + 5 = -10 \\
F(t + 2) &= 3(t + 2 - 1) + 5 = 3t + 8 \\
F(F(2)) &= F(8) = 3(7) + 5 = 26
\end{aligned}
$$

1.3.5
The Range of a Function

> The **range** of a function f is the set of all real numbers that are values of the function; that is, it is the set of all numbers y such that $y = f(x)$ for some x in $\mathcal{D}(f)$. The range of f is denoted $\mathcal{R}(f)$.

1.3.6
The Graph of a Function

> The **graph** of a *function* f is the graph of the *equation* $y = f(x)$ in the Cartesian plane. Thus it is the set of all points with coordinates $(x, f(x))$, where x belongs to $\mathcal{D}(f)$.

Figure 1.3.1 illustrates the domain, range, and graph of the function f. Note that

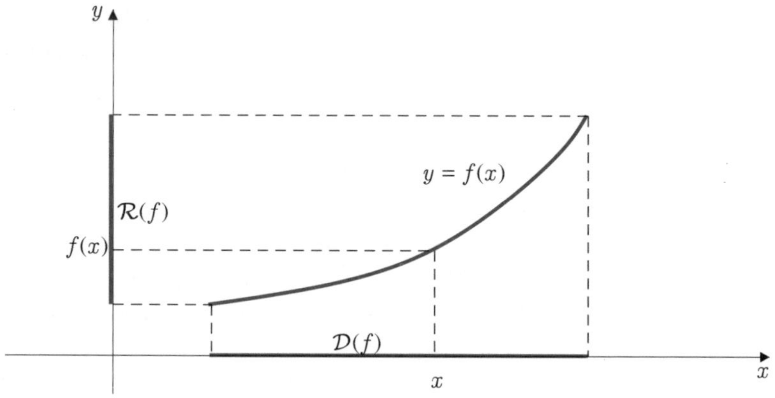

FIGURE 1.3.1

the domain of the function f can be regarded as a set of points on the x-axis, and the range as a set on the y-axis. Note that the graph of a function is such that any *vertical* line through a point in the domain meets the graph at *exactly one point* (because $f(x)$ is a *unique* number for each x). The horizontal line through this point on the graph meets the y-axis at $f(x)$. Thus $f(x)$ is the vertical displacement of that point on the graph.

EXAMPLE 1.3.7 Specify the domain and range of each of the following functions:

a) $f(x) = x^2$ b) $g(x) = x^2$ for $-1 \le x < 2$

c) $k(x) = |x|$ d) $S(x) = \dfrac{1}{x-1}$

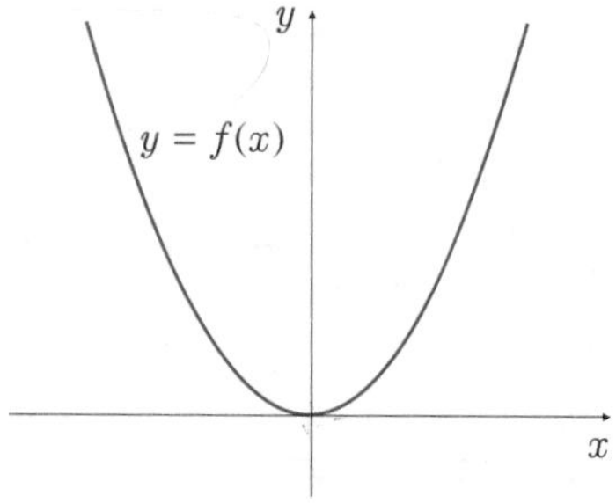

FIGURE 1.3.2

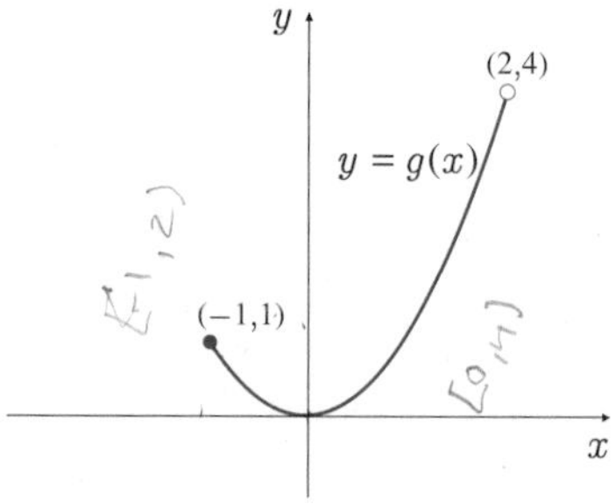

FIGURE 1.3.3

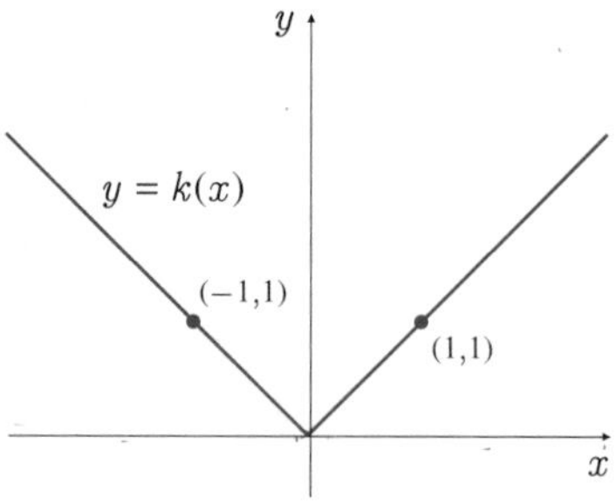

FIGURE 1.3.4

SOLUTION See Figures 1.3.2–1.3.5 for the graphs of these functions.

a) $\mathcal{D}(f) = \mathbb{R} = (-\infty, \infty)$, $\mathcal{R}(f) = [0, \infty)$.

b) $\mathcal{D}(g) = [-1, 2)$, $\mathcal{R}(g) = [0, 4)$.

c) $k(x) = x$ if $x \ge 0$ and $k(x) = -x$ if $x < 0$. Thus $k(x)$ is defined for all real numbers but takes only nonnegative values: $\mathcal{D}(k) = \mathbb{R}$, $\mathcal{R}(k) = [0, \infty)$.

d) We cannot divide by 0, so $\mathcal{D}(S) = (-\infty, 1) \cup (1, \infty)$; we cannot allow $x = 1$. Since the numerator of the fraction defining $S(x)$ is not zero, $S(x)$ can never have the value 0, but it can have any other real value y; $\dfrac{1}{x-1} = y$ if $x = \dfrac{1}{y} + 1$. Thus $\mathcal{R}(S) = (-\infty, 0) \cup (0, \infty)$.

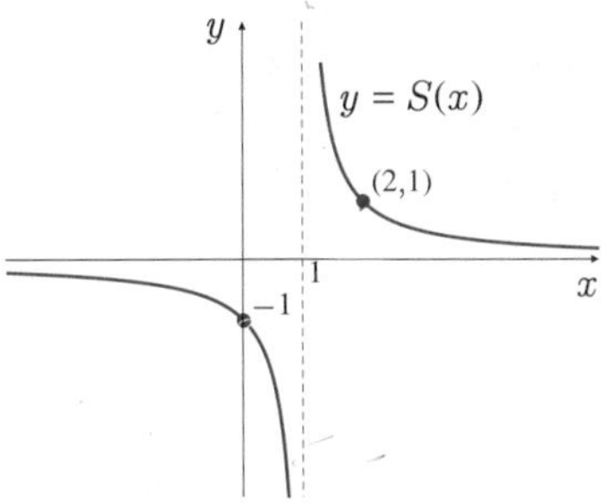

FIGURE 1.3.5

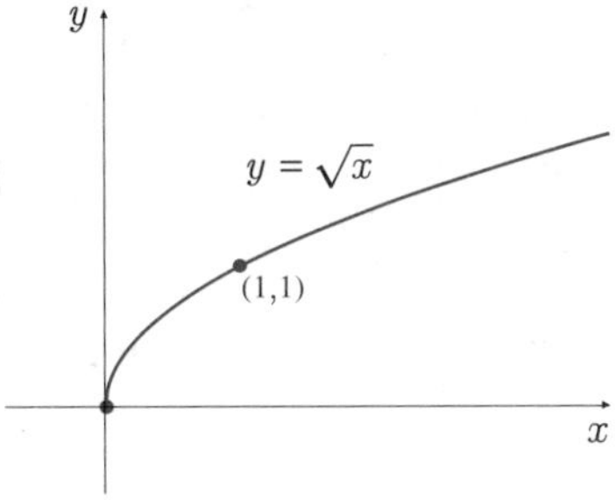

FIGURE 1.3.6

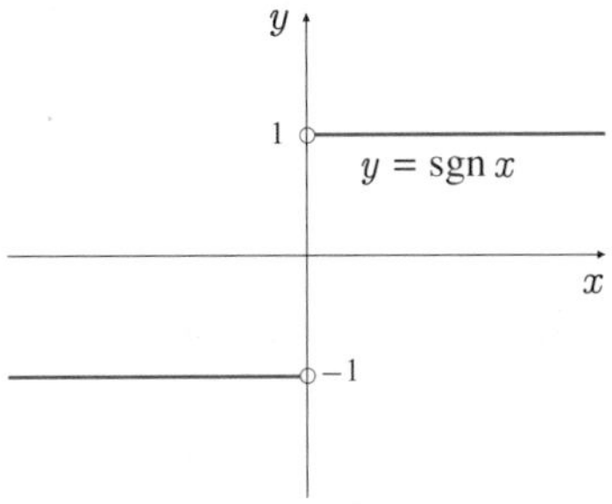

FIGURE 1.3.7

Since squares of numbers cannot be negative, the **square root function** $\sqrt{x}$ is defined for $x \ge 0$ only. The uniqueness condition in the definition of function requires that there be only *one value* of $\sqrt{x}$ for each value of x in $\mathcal{D}(\sqrt{\ }) = [0, \infty)$. By convention, the term *square root* and the symbol $\sqrt{\ }$ always refer to the *nonnegative* square root. Thus, while there are two numbers whose square is 4, only one of these is called $\sqrt{4}$, namely $\sqrt{4} = 2$. The graph of $y = \sqrt{x}$ is shown in Fig. 1.3.6. If we want to give both solutions of the equation $x^2 = a$, $(a > 0)$, we can indicate them as $x = \sqrt{a}$ and $x = -\sqrt{a}$. Students often make the mistake of writing $\sqrt{x^2} = x$, where x is some variable or algebraic expression. This is incorrect unless you know beforehand that $x \ge 0$. The following formula is always correct:

1.3.8
The Square Root of a Square

$$\sqrt{x^2} = |x|.$$

Consider the **signum function**, $\operatorname{sgn} x$, defined by

1.3.9
The Signum Function

$$\operatorname{sgn} x = \frac{x}{|x|} = \begin{cases} 1 & \text{if } x > 0, \\ -1 & \text{if } x < 0. \end{cases}$$

The name *signum* is the Latin noun meaning "sign;" the value of $\operatorname{sgn} x$ indicates whether x is positive or negative. The domain of sgn consists of all real numbers except zero, (we do not regard zero as being either positive or negative), but the range of sgn consists of only two points, -1 and 1. The graph of sgn is shown in Fig. 1.3.7.

At the moment, if we wish to sketch the graph of a function f, either we must identify the equation $y = f(x)$ beforehand as representing a standard curve such as a straight line or part of a circle, or we must calculate a number of points and connect them by a "reasonable" curve. Thus we can recognize the graph of $f(x) = 2x - 3$ as the straight line with slope 2 and y-intercept -3, and we can recognize $g(x) = \sqrt{16 - x^2}$ as the function whose graph will be the upper half of the circle with equation $x^2 + y^2 = 16$ (because $y = g(x)$ implies that $y^2 = 16 - x^2$). For more complicated functions, however, we are reduced to the laborious computation of the coordinates of many points. This approach can have more serious drawbacks as shown by the following example.

EXAMPLE 1.3.10 Use a table of values to help you sketch the graph of the function $f(x) = x^{2/3}$.

SOLUTION Using a scientific calculator we can compute the following approximate values:

x	-3	-2	-1	0	1	2	3
$f(x) = x^{2/3}$	2.08	1.59	1	0	1	1.59	2.08

If we plot these values and try to draw a curve through them, we may come up with a curve like that in Fig. 1.3.8. However, the actual graph is the curve of Fig. 1.3.9 which has a **cusp** (an infinitely sharp point) at the origin. Unless we calculate coordinates of many more points near $x = 0$ we would be likely to miss this very important feature of the graph. Using calculus we will be able to sketch

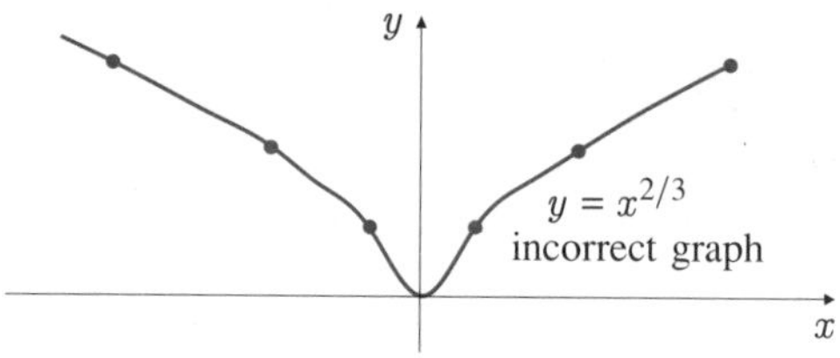

FIGURE 1.3.8

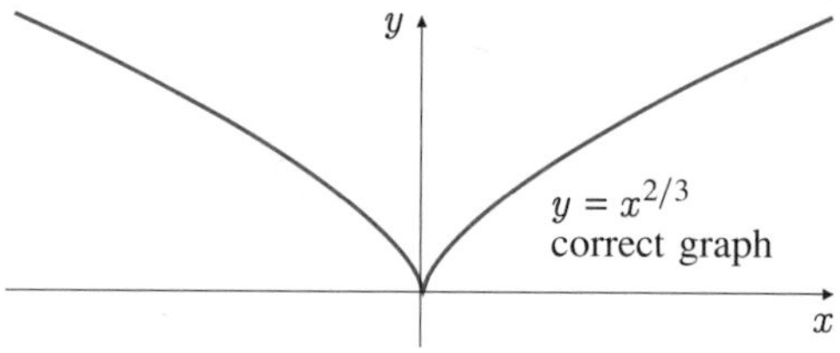

FIGURE 1.3.9

graphs of functions without missing such interesting features, and without having to calculate many coordinates.

Operations on Functions

Like numbers, functions can be added, subtracted, multiplied, and divided (except when the denominator is 0) to produce new functions. If f and g are functions, then for every x that belongs to the domains of both f and g we define:

$$\begin{aligned}(f+g)(x) &= f(x) + g(x),\\ (f-g)(x) &= f(x) - g(x),\\ (fg)(x) &= f(x)g(x),\\ \left(\frac{f}{g}\right)(x) &= \frac{f(x)}{g(x)} \qquad \text{where } g(x) \neq 0.\end{aligned}$$

Functions can also be multiplied by constants; if c is a real number then

$$(cf)(x) = cf(x).$$

EXAMPLE 1.3.11 If $f(x) = \sqrt{x+1}$ and $g(x) = \dfrac{x-1}{x}$, then

$$\begin{aligned}
(f+g)(x) &= \sqrt{x+1} + \frac{x-1}{x}, \\
(f-g)(x) &= \sqrt{x+1} - \frac{x-1}{x}, \\
(fg)(x) &= \frac{\sqrt{x+1}(x-1)}{x}, \\
\left(\frac{f}{g}\right)(x) &= \frac{x\sqrt{x+1}}{x-1}.
\end{aligned}$$

Since $\mathcal{D}(f) = [-1, \infty)$ and $\mathcal{D}(g) = (-\infty, 0) \cup (0, \infty)$, therefore we have $\mathcal{D}(f+g) = \mathcal{D}(f-g) = \mathcal{D}(fg) = [-1, 0) \cup (0, \infty)$, and $\mathcal{D}(f/g) = [-1, 0) \cup (0, 1) \cup (1, \infty)$. Note that even though $x\sqrt{x+1}/(x-1)$ makes sense when $x = 0$, $(f/g)(x)$ is not defined at $x = 0$ because $g(0)$ is not defined.

There is another method, called composition, by which functions can be combined to form new functions.

1.3.12 Composition of Functions

> The **composition** of two functions f and g is another function, denoted $f \circ g$, defined by
>
> $$f \circ g(x) = f(g(x))$$
>
> for every x for which $f(g(x))$ makes sense. Thus $\mathcal{D}(f \circ g)$ consists of all numbers x in $\mathcal{D}(g)$ such that $g(x)$ is in $\mathcal{D}(f)$.

EXAMPLE 1.3.13 Given $f(x) = \sqrt{x}$ and $g(x) = x + 1$ we calculate the following compositions:

$$\begin{gathered}
f \circ g(x) = f(g(x)) = \sqrt{g(x)} = \sqrt{x+1}, \quad \mathcal{D}(f \circ g) = [-1, \infty) \\
g \circ f(x) = g(f(x)) = f(x) + 1 = \sqrt{x} + 1, \quad \mathcal{D}(g \circ f) = [0, \infty) \\
f \circ f(x) = f(f(x)) = \sqrt{f(x)} = \sqrt{\sqrt{x}} = x^{1/4}, \quad \mathcal{D}(f \circ f) = [0, \infty) \\
g \circ g(x) = g(g(x)) = g(x) + 1 = (x+1) + 1 = x + 2, \quad \mathcal{D}(g \circ g) = I\!R \\
g \circ f \circ g(x) = g(f(g(x))) = f(g(x)) + 1 = \sqrt{x+1} + 1, \quad \mathcal{D}(g \circ f \circ g) = [-1, \infty).
\end{gathered}$$

because √ is always positive

EXAMPLE 1.3.14 If $H(x) = \dfrac{1-x}{1+x}$ calculate $H \circ H(x)$ and specify its domain.

SOLUTION

$$H \circ H(x) = H(H(x)) = \frac{1 - H(x)}{1 + H(x)} = \frac{1 - \dfrac{1-x}{1+x}}{1 + \dfrac{1-x}{1+x}} = \frac{1 + x - 1 + x}{1 + x + 1 - x} = x,$$

provided that $x \neq -1$. Note that although x is defined on $\mathbb{R}$, the domain of $H \circ H$ is $(-\infty, -1) \cup (-1, \infty)$. Because $H(x)$ is not defined for $x = -1$, neither is $H(H(x))$.

Odd and Even Functions

It often happens that the graph of a function will have certain kinds of symmetry. The simplest kinds of symmetry relate the value of f at $-x$ to its value at x.

1.3.15
Odd and Even Functions

Suppose that $-x$ belongs to $\mathcal{D}(f)$ whenever x belongs to $\mathcal{D}(f)$. We say that f is an **odd function** if

$$f(-x) = -f(x) \quad \text{for every } x \text{ in } \mathcal{D}(f).$$

We say that f is an **even function** if

$$f(-x) = f(x) \quad \text{for every } x \text{ in } \mathcal{D}(f).$$

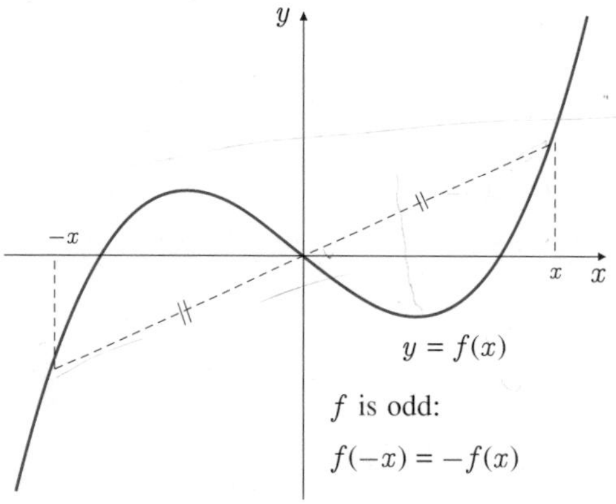

FIGURE 1.3.10

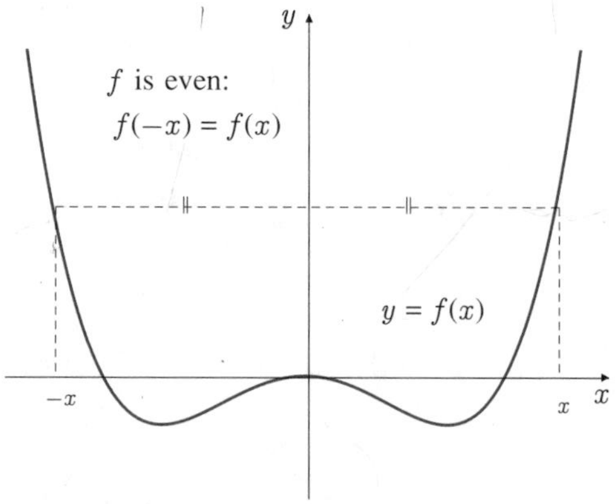

FIGURE 1.3.11

Odd powers of x such as x, x^3, x^5, ..., x^{-1}, x^{-3}, x^{-5}, ... are odd functions; even powers $x^0 = 1$, x^2, x^4, ..., x^{-2}, x^{-4}, ... are even functions. The absolute value function $|x|$ is an even function. Any function which depends only on x^2 is an even function of x.

The graph of an odd function is symmetric about the origin; a line drawn from any point on its graph to the origin will, if continued the same distance beyond the origin, come to another point on the graph. Similarly, the graph of an even function is symmetric about the y-axis; a line drawn perpendicular to the y-axis from a point on the graph will, if continued the same distance beyond the y-axis, come to another point on the graph. (See Figs. 1.3.10 and 1.3.11.) These symmetries are useful when we want to sketch the graphs of even and odd functions. Note also that an odd function defined at $x = 0$ must vanish ($= 0$) there (because $f(0) = f(-0) = -f(0)$ implies $2f(0) = 0$, and so $f(0) = 0$).

EXERCISES

Specify the domains and ranges of the functions in 1–12

1. $f(x) = 5$ **2.** $f(x) = 2 - x$

3. $F(x) = 2 - x^2$ **4.** $g(x) = \dfrac{1}{2 + x}$

5. $f(t) = \dfrac{t + 1}{t}$ **6.** $s(x) = \sqrt{x - 2}$

7. $k(x) = \sqrt{2 - x}$ **8.** $m(x) = \sqrt{1 - x^2}$

9. $g(s) = \dfrac{1}{s^2 + 1}$ **10.** $r(x) = \dfrac{x}{1 - x}$

11. $H(x) = \sqrt{x^2 - x}$ **12.** $h(x) = (x - 2)^2 - 4$

13. Which of the functions of Example 1.3.7 are even? Which are odd? Which are neither even nor odd?

Sketch the graphs of the functions in 14–21. Which of these functions are even? Which are odd? Which are neither even nor odd?

14. $f(x) = 1 - x^2$ **15.** $f(x) = x^3$.

16. $f(x) = \sqrt{|x|}$ **17.** $g(x) = \sqrt{x^2 - 2x + 1}$

18. $F(x) = \sqrt{4 - x^2}$ **19.** $G(x) = -\dfrac{1}{\sqrt{4 - x^2}}$

20. $f(x) = |x^2 - 1|$ **21.** $f(x) = |x| + |x - 2|$

22. Sketch the graph of the equation $|x| + |y| = 1$. Describe its symmetries.

For the functions f and g in Exercises 23–25 construct $f + g$, fg, g/f, $f \circ f$, $f \circ g$, $g \circ f$, and $g \circ g$. Specify the domain of each function you construct.

23. $f(x) = \dfrac{1}{x - 1}$, $g(x) = \dfrac{1}{x}$

24. $f(x) = x^4$, $g(x) = \sqrt{x - 1}$

25. $f(x) = \sqrt{1 - x^2}$, $g(x) = 2 + x$

26. Find all values of the constants A and B for which the function $F(x) = Ax + B$ satisfies the following:
a) $F \circ F(x) = F(x)$ for all x, b) $F \circ F(x) = x$ for all x.

27. The graph of a certain function f with domain $[0, 2]$ and range $[0, 1]$ is shown in Fig. 1.3.12 below. Sketch the graphs of the following functions and specify their domains and ranges.

a) $f(x) + 2$, b) $f(x) - 1$,
c) $2f(x)$, d) $-f(x)$,
e) $f(x + 2)$, f) $f(x - 1)$,
g) $f(2x)$, h) $f(-x)$,
i) $f\left(\dfrac{x}{3}\right)$, j) $1 - 2f\left(\dfrac{x - 1}{2}\right)$.

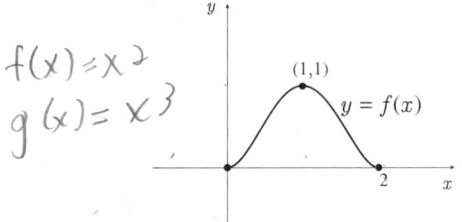

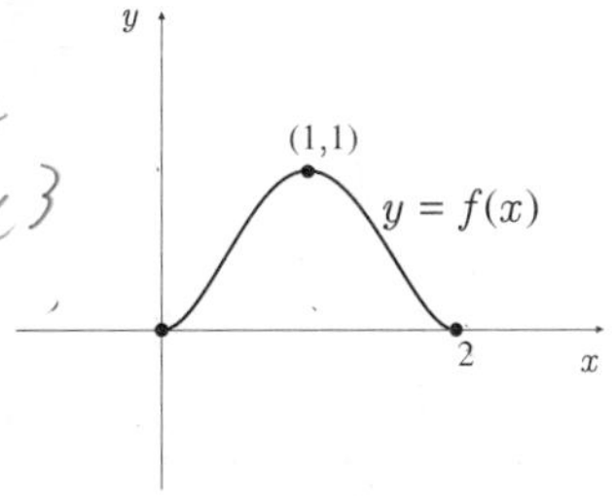

FIGURE 1.3.12

28. If f is an even function and g is an odd function, which of the functions $f \circ f$, $g \circ g$, $f \circ g$, or $g \circ f$ are even? Which are odd?

29. If f is both an even function and an odd function, show that $f(x)$ is identically zero on its domain.

30.*Suppose that the domain of a function f is symmetric about the origin, that is, $-x$ belongs to $\mathcal{D}(f)$ whenever x belongs to $\mathcal{D}(f)$. Show that

a) f is the sum of an even function and an odd function (*Hint:* Let $E(x) = [f(x) + f(-x)]/2$. Show that E is even.)

b) There is only one way of writing f as the sum of an even and an odd function. (That is, the even and odd functions in (a) are uniquely determined by f.)

1.4 INVERSE FUNCTIONS

Given any function f and any number x in its domain, there exists *exactly one* number y in its range such that $y = f(x)$. In graphical terms, a vertical line through x meets the graph of f exactly once, and y is the height of that point. Suppose that f is such that, given any y in its range, there is *exactly one* x in its domain such that $y = f(x)$. That is, a horizontal line through y meets the graph of f at exactly one point. Such a function f is said to be *one-to-one*.

If f is one-to-one, then each y in the range determines a unique x in the domain, so the equation $y = f(x)$ can be solved for x as a function of y; $x = f^{-1}(y)$. The function that we have denoted f^{-1} is called the *inverse of f*. Here is a formal definition of these concepts.

1.4.1
One-to-One Functions and Inverse Functions

A function f is **one-to-one** if $f(x_1) \neq f(x_2)$ whenever x_1 and x_2 belong to the domain of f and $x_1 \neq x_2$. A one-to-one function is also said to be **invertible**, and its **inverse function**, f^{-1}, is defined as follows: for every y in the range of f, $f^{-1}(y)$ is the unique element x in the domain of f such that $y = f(x)$.

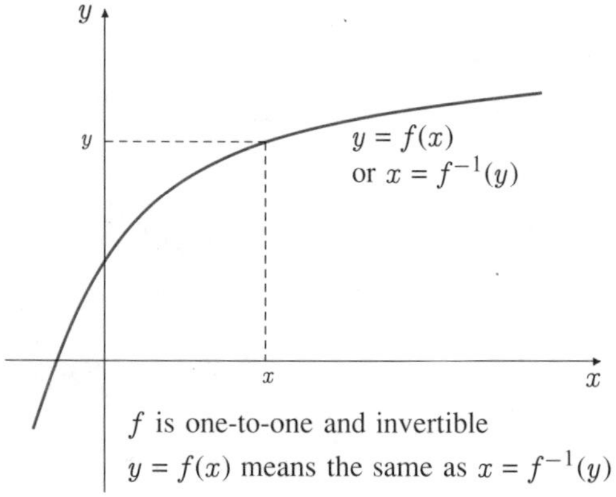

FIGURE 1.4.1

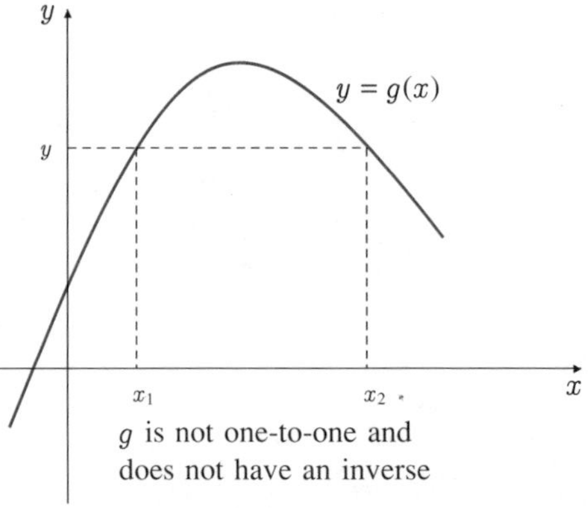

FIGURE 1.4.2

Fig. 1.4.1 shows the graph of a function f which is one-to-one; horizontal lines meet the curve $y = f(x)$ at most once. Fig. 1.4.2 shows the graph of a function g which is not one-to-one. Some horizontal lines meet this graph twice; observe there are two different points x_1 and x_2 in the domain of g such that $g(x_1) = g(x_2)$.

Do not confuse *inverse* functions with *reciprocals*. $f^{-1}(x)$ does *not* mean $1/f(x)$.

For a one-to-one function f, the two equations $y = f(x)$ and $x = f^{-1}(y)$ are equivalent; they *say the same thing*. (For example, the equations $y = x + 1$ and $x = y - 1$ are equivalent. Either of these equations can be replaced by the other.) Since we want to study f^{-1} let us reverse the roles of x and y and write

1.4.2
Equivalent Definition of Inverse Function

$$y = f^{-1}(x) \iff x = f(y).$$

The symbol $\iff$ should be read "if and only if" or, more informally, "means the same as." Observe that the domain of f^{-1} is the range of f and vice versa:

1.4.3
Domain and Range of Inverse Functions

$$\mathcal{D}(f^{-1}) = \mathcal{R}(f), \qquad \mathcal{R}(f^{-1}) = \mathcal{D}(f).$$

Also, the inverse of the inverse of a function is that function:

$$y = (f^{-1})^{-1}(x) \iff x = f^{-1}(y) \iff y = f(x).$$

We can substitute either of the equations in Box 1.4.2 into the other and obtain the following **cancellation identities**:

1.4.4
Cancellation Identities for Inverse Functions

$$f\left(f^{-1}(x)\right) = x \qquad \text{for all } x \text{ in } \mathcal{D}(f^{-1}) = \mathcal{R}(f),$$
$$f^{-1}\left(f(y)\right) = y \qquad \text{for all } y \text{ in } \mathcal{R}(f^{-1}) = \mathcal{D}(f).$$

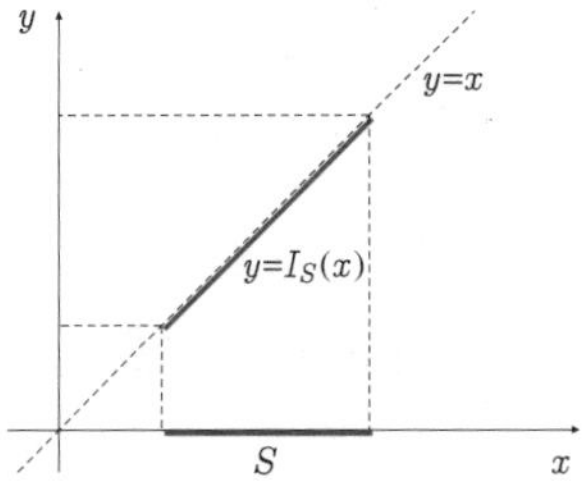

FIGURE 1.4.3

Let S be any set of real numbers. The **identity function** on S is the function I_S defined by

$$I_S(x) = x \quad \text{for all } x \text{ in } S.$$

(See Fig. 1.4.3.) The cancellation identities 1.4.4 say that the compositions $f \circ f^{-1}$ and $f^{-1} \circ f$ are the identity functions on the range of f and the domain of f respectively:

$$f \circ f^{-1}(x) = I_{\mathcal{R}(f)}, \qquad f^{-1} \circ f(x) = I_{\mathcal{D}(f)}.$$

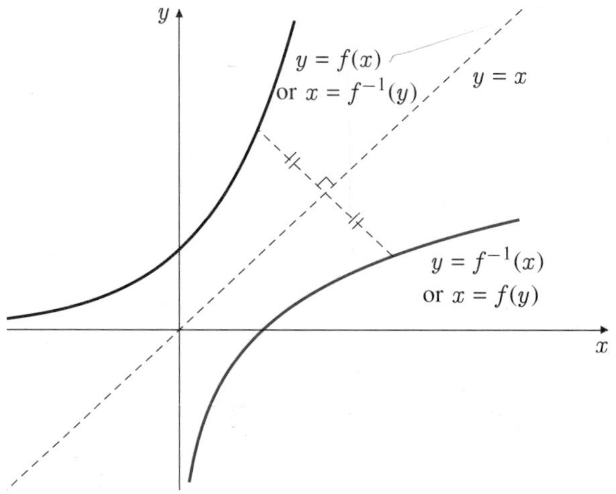

FIGURE 1.4.4

Since the points (x, y) and (y, x) in the Cartesian plane are mirror images in the line $y = x$, it follows that the curves $y = f(x)$ and $x = f(y)$ (that is, $y = f^{-1}(x)$) are also mirror images of one another in that line. See Fig. 1.4.4.

1.4.5 Graphs of Inverse Functions

> The graph of f^{-1} is the mirror image of the graph of f in the line $y = x$.

EXAMPLE 1.4.6 Show that $f(x) = 3x - 5$ is invertible and find its inverse function f^{-1}.

SOLUTION If $f(x_1) = f(x_2)$, then $3x_1 - 5 = 3x_2 - 5$, so $3(x_1 - x_2) = 0$ and $x_1 = x_2$. Therefore f in invertible. If $y = f^{-1}(x)$, then

$$x = f(y) = 3y - 5,$$

so $y = \dfrac{x+5}{3}$. Hence $f^{-1}(x) = \dfrac{x+5}{3}$. See Fig. 1.4.5.

EXAMPLE 1.4.7 Show that $g(x) = \sqrt{2x+1}$ is invertible and find its inverse.

SOLUTION If $g(x_1) = g(x_2)$, then $\sqrt{2x_1+1} = \sqrt{2x_2+1}$. Squaring both sides we get $2x_1 + 1 = 2x_2 + 1$, which implies that $x_1 = x_2$. Thus g is invertible. If $y = g^{-1}(x)$ then $x = g(y) = \sqrt{2y+1}$. It follows that $x^2 = 2y + 1$ and $y = \dfrac{x^2-1}{2}$. Thus $g^{-1}(x) = \dfrac{x^2-1}{2}$ for $x \geq 0$. (The restriction $x \geq 0$ applies since the range of g is $[0, \infty)$.) See Fig. 1.4.6 for the graphs of g and g^{-1}.

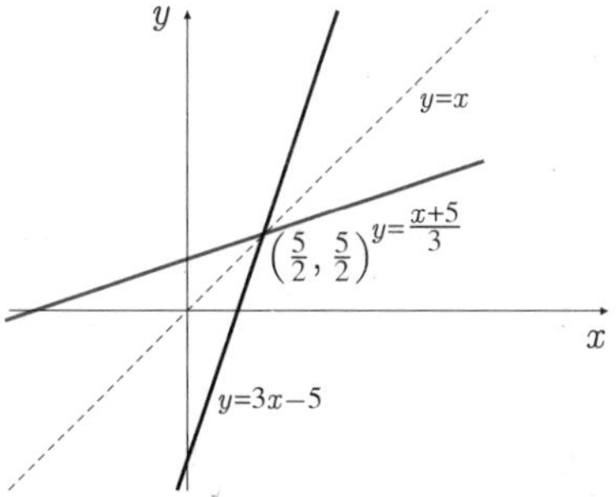

FIGURE 1.4.5

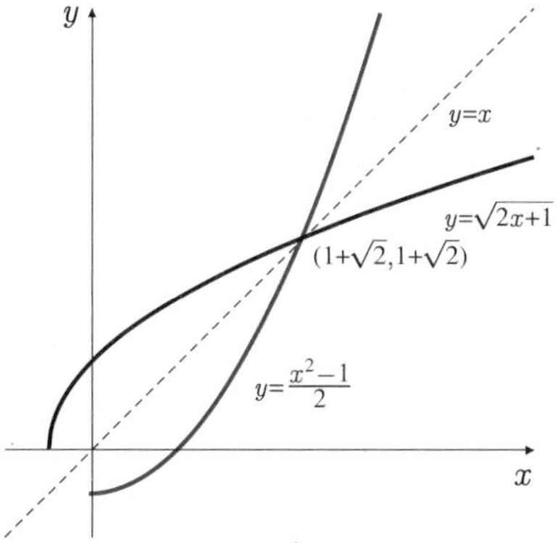

FIGURE 1.4.6

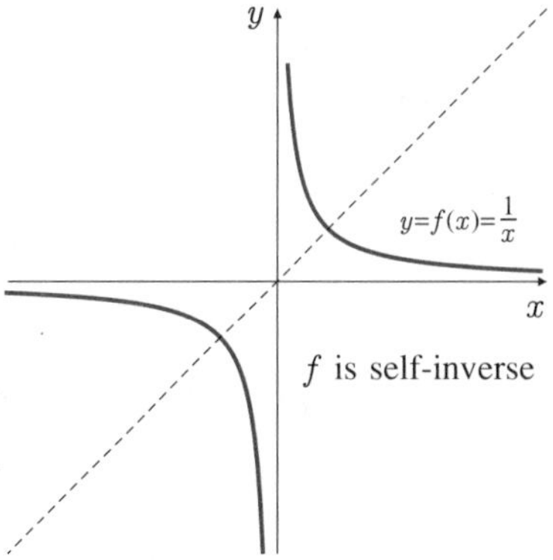

FIGURE 1.4.7

1.4.8
Self-Inverse Functions

> A function is said to be **self-inverse** if $f^{-1} = f$, that is, if $f(f(x)) = x$ for every x in the domain of f.

The function $f(x) = 1/x$ is self-inverse, as is the function H in Example 1.3.14. The graph of a self-inverse function must be its own mirror image in the line $y = x$ and so must be symmetric about that line. Conversely, any function whose graph is symmetric about the line $y = x$ is self-inverse.

Inverting Non-One-to-One Functions

Many important functions such as the trigonometric functions which we will study in Chapter 3 are not one-to-one on their whole domains. It is still possible to define an inverse for such a function, but we have to restrict the domain of the function artificially so that the restricted function is one-to-one.

As an example, consider the function $f(x) = x^2$. Unrestricted, its domain is the whole real line and it is not one-to-one since $f(-a) = f(a)$ for any a. Let us define a new function $F(x)$ equal to $f(x)$, but having a smaller domain, so that it is one-to-one. We may use the interval $[0, \infty)$ as the domain of F:

$$F(x) = x^2 \qquad \text{for} \quad 0 \leq x < \infty.$$

The graph of F is shown in Fig. 1.4.8; it is the right half of the parabola $y = x^2$, the graph of f. Evidently F is one-to-one so has an inverse F^{-1} which we calculate as follows:

If $y = F^{-1}(x)$ then $x = F(y) = y^2$ and $y \geq 0$. Thus $y = \sqrt{x}$. Hence $F^{-1}(x) = \sqrt{x}$.

The above method of restricting the domain of a non one-to-one function to make it invertible will be used extensively when we invert the trigonometric functions in Section 3.3.

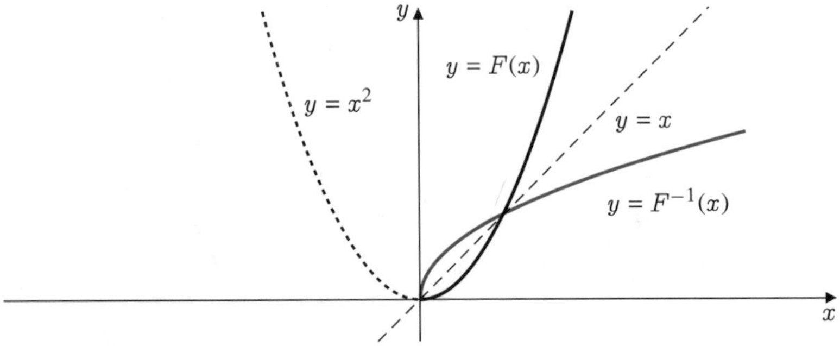

FIGURE 1.4.8

EXERCISES

Show that the functions f in Exercises 1–12 are one-to-one and calculate the inverse function f^{-1}. Specify the domain and range of f and f^{-1}.

1. $f(x) = x - 1$

2. $f(x) = 2x - 1$

3. $f(x) = \sqrt{x - 1}$

4. $f(x) = -\sqrt{x - 1}$

5. $f(x) = x^3$

6. $f(x) = 1 + \sqrt[3]{x}$

7. $f(x) = x^2, \quad x \leq 0$

8. $f(x) = (1 - 2x)^3$

9. $f(x) = \dfrac{1}{x + 1}$

10. $f(x) = \dfrac{x}{1 + x}$

11. $f(x) = \dfrac{1 - 2x}{1 + x}$

12. $f(x) = \dfrac{x}{\sqrt{x^2 + 1}}$

In Exercises 13–20 f is a one-to-one function with inverse f^{-1}. Calculate the inverses of the given functions in terms of f^{-1}.

13. $g(x) = f(x) - 2$

14. $h(x) = f(2x)$

15. $k(x) = -3f(x)$

16. $m(x) = f(x - 2)$

17. $p(x) = \dfrac{1}{1 + f(x)}$

18. $q(x) = \dfrac{f(x) - 3}{2}$

19. $r(x) = 1 - 2f(3 - 4x)$

20. $s(x) = \dfrac{1 + f(x)}{1 - f(x)}$

21. Show that $f(x) = \begin{cases} x^2 + 1 & \text{if } x \geq 0 \\ x + 1 & \text{if } x < 0 \end{cases}$ is one-to-one. What is its inverse?

22.*Verify that $f(x) = x^3 + x$ is a one-to-one function. Can you calculate its inverse?

23.*For what values of the constants a, b, and c is the function $f(x) = (x - a)/(bx - c)$ self-inverse?

24. Can an even function be self-inverse? An odd function?

1.5 LIMITS

The concept of *limit* is the cornerstone on which the development of calculus rests. Before we attempt any definition of this concept we will illustrate the idea with an example.

EXAMPLE 1.5.1 Consider the function $f(x) = \dfrac{x^2 - 1}{x - 1}$. Evidently f is defined for all real numbers x except $x = 1$ (we can't divide by zero); thus

$$\mathcal{D}(f) = (-\infty, 1) \cup (1, \infty).$$

Also, for any $x \neq 1$ we can simplify the expression for $f(x)$ by factoring the numerator and cancelling common factors:

$$f(x) = \frac{(x - 1)(x + 1)}{x - 1} = x + 1 \qquad \text{for} \quad x \neq 1.$$

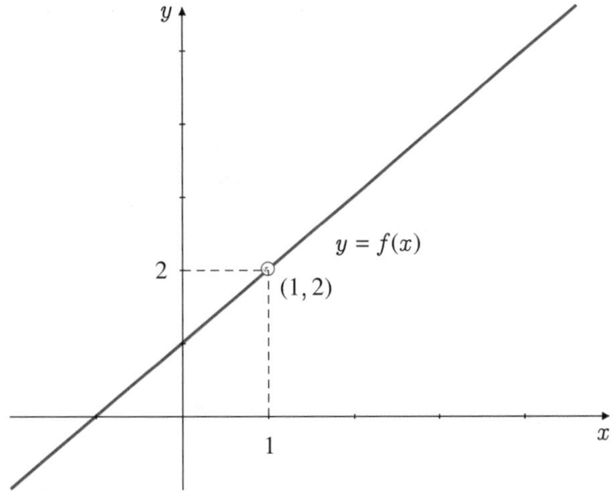

FIGURE 1.5.1

The graph of f is thus that of the straight line $y = x+1$ with one point removed, namely the point $(1, 2)$. This removed point is shown as a "hole" in the graph in Fig. 1.5.1. Even though $f(1)$ is not defined, it is clear that we can make the value of $f(x)$ *as close as we want* to 2 by choosing x *close enough* to 1. We say that $f(x)$ approaches arbitrarily close to 2 as x approaches 1, or, more simply, $f(x)$ approaches *the limit* 2 as x approaches 1. We write this as

$$\lim_{x \to 1} f(x) = \lim_{x \to 1} \frac{x^2 - 1}{x - 1} = 2.$$

1.5.2
Informal Definition of Limit

> We say that the function f approaches the **limit** L as x approaches a, and we write
>
> $$\lim_{x \to a} f(x) = L$$
>
> if $f(x)$ is defined for all x nearby (on either side of) the point a, except possibly at a itself, and if f approaches arbitrarily close to L as x approaches a.

This definition is "informal" because it is not very precise; we have not said just what we mean by such vague phrases as "nearby the point," "arbitrarily close," and "approaches." The intent of the definition, however, should be clear enough so that it can be used for most of our purposes. When we wish to prove theoretical results about limits such as those given in Theorems 1.5.6, 1.5.7 and 1.5.8 below, we must have a more precise definition of limit. We will not actually prove such results in this chapter (the proofs are in Appendix II), but we include the more formal definition here anyway, so you can see how the vague phrases from the informal definition are made precise.

1.5.3 Formal Definition of Limit

> Suppose that there exist numbers a, b, and c with $b < a < c$ such that f is defined on the interval (b, a) and on the interval (a, c). Suppose that there exists a real number L with the following property: corresponding to any positive real number ϵ, there exists another positive real number δ (which depends on ϵ) such that
>
> $$|f(x) - L| < \epsilon \quad \text{whenever} \quad 0 < |x - a| < \delta.$$
>
> Then we say that the limit of $f(x)$ as x approaches a is L, and we write
>
> $$\lim_{x \to a} f(x) = L.$$

Expressed in terms of distances, the above definition says that $f(x)$ has limit L as x approaches a if the distance from $f(x)$ to L (that is, $|f(x) - L|$) can be made as small as we please (less than ϵ) by requiring that the distance of x from a (that is, $|x - a|$) is sufficiently small (less than δ). See Fig. 1.5.2. The definition makes no demand of f at the point $x = a$. f may or may not be defined at $x = a$, and if it is defined there, $f(a)$ may or may not be equal to L.

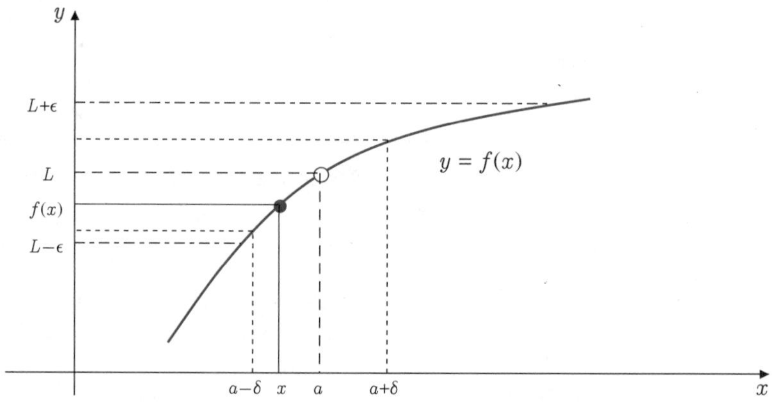

FIGURE 1.5.2

Don't be worried if you don't understand Definition 1.5.3. The less precise version 1.5.2 is quite adequate for our needs in computing values of limits. If you don't understand 1.5.2 go back and think about Example 1.5.1 again.

EXAMPLE 1.5.4 Find $\lim_{x \to 1} \dfrac{x - 3}{x + 7}$.

SOLUTION As x approaches arbitrarily close to 1, the numerator $x-3$ approaches arbitrarily close to -2, and the denominator $x+7$ approaches arbitrarily close to 8. Hence,

$$\lim_{x\to 1} \frac{x-3}{x+7} = \frac{-2}{8} = -\frac{1}{4}.$$

Example 1.5.4 is not very interesting because, unlike the fraction in Example 1.5.1, this fraction does not become "0/0" when $x=1$. If substituting a for x in the fraction $f(x)/g(x)$ gives the meaningless expression $0/0$, it is usually because $f(x)$ and $g(x)$ have a common factor which vanishes at $x=a$; we must cancel this common factor before evaluating the limit.

EXAMPLE 1.5.5 Find $\lim_{x\to 4} \dfrac{\sqrt{x}-2}{x^2-5x+4}$.

SOLUTION If we try to proceed as in Example 1.5.4 and evaluate the numerator and denominator separately at $x=4$ we will be led (as in Example 1.5.1) to the meaningless expression 0/0. In this case we must really determine whether $f(x) = \dfrac{\sqrt{x}-2}{x^2-5x+4}$ has a limit as x approaches 4 without actually letting x become equal to 4. Using a pocket calculator we can determine values of $f(x)$ for values of x close to 4:

x	$f(x)$
3.9	.0867525
3.99	.0836644
3.999	.0833663
3.9999	.0833366
3.99999	.0833337

x	$f(x)$
4.1	.0801473
4.01	.0830046
4.001	.0833004
4.0001	.0833300
4.00001	.0833330

$f(x)$ does indeed appear to be getting closer and closer to some limit as x approches very close to 4 from either side. It seems that

$$\lim_{x\to 4} f(x) \approx 0.08333\cdots.$$

The limit can be determined exactly by astute factoring and cancellation, similar to the procedure we used in Example 1.5.1: if $x \neq 4$ then

$$\frac{\sqrt{x}-2}{x^2-5x+4} = \frac{\sqrt{x}-2}{(x-4)(x-1)} = \frac{\sqrt{x}-2}{(\sqrt{x}-2)(\sqrt{x}+2)(x-1)} = \frac{1}{(\sqrt{x}+2)(x-1)}.$$

Hence,

$$\lim_{x\to 4} \frac{\sqrt{x}-2}{x^2-5x+4} = \frac{1}{(2+2)(4-1)} = \frac{1}{12} = 0.083333\cdots.$$

Because both the numerator and denominator of the original fraction contain factors that vanish at $x=4$ (the factor $\sqrt{x}-2$ in each case), the original fraction gives the meaningless 0/0 when 4 is substituted for x. Proper evaluation of the limit depends on canceling out these vanishing factors first.

The principal properties of limits that we will need hereafter are summarized in Theorems 1.5.6, 1.5.7 and 1.5.8. These properties should seem intuitively clear once you understand the concept of limit. They can be rigorously proved using Definition 1.5.3. See Appendix II for details.

THEOREM 1.5.6 *(Uniqueness of Limits)* A function cannot have two different limits at the same point; if $\lim_{x\to a} f(x) = L$ and $\lim_{x\to a} f(x) = M$, then necessarily $L = M$. □

THEOREM 1.5.7 If $\lim_{x\to a} f(x) = L$ and $\lim_{x\to a} g(x) = M$, then the following conclusions hold.

i) $\lim_{x\to a}(f(x) + g(x)) = L + M$ The limit of a sum is the sum of limits.

ii) $\lim_{x\to a}(f(x) - g(x)) = L - M$ The limit of a difference is the difference of the limits.

iii) $\lim_{x\to a}(f(x)g(x)) = LM$ The limit of a product is the product of limits.

iv) $\lim_{x\to a} \dfrac{f(x)}{g(x)} = \dfrac{L}{M}$, provided $M \neq 0$ The limit of a quotient is the quotient of the limits provided the limit of the denominator is not 0.

v) $\lim_{x\to a} cf(x) = cL$ for any constant c.

vi) If $f(x) \leq g(x)$ near a, then $L \leq M$. □

THEOREM 1.5.8 (*The Squeeze Theorem*) Suppose that $f(x) \leq g(x) \leq h(x)$ for all x near a (except possibly at $x = a$). If

$$\lim_{x\to a} f(x) = \lim_{x\to a} h(x) = L$$

then $\lim_{x\to a} g(x) = L$ also. □

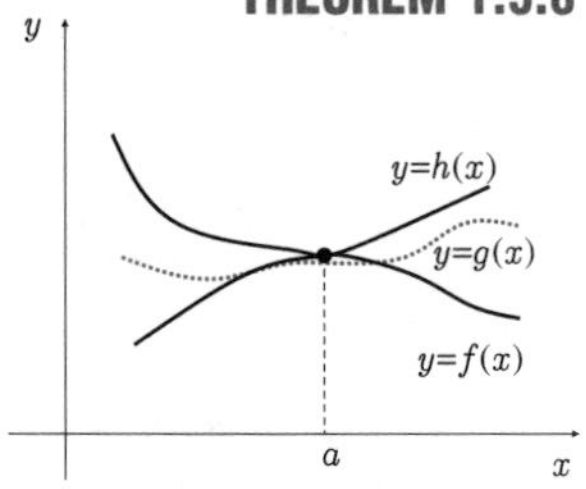

FIGURE 1.5.3

Figure 1.5.3 shows that the graph of g is trapped between those of f and h, which come together at $x = a$.

EXAMPLE 1.5.9 Determine which of the following limits exist and evaluate those that do.

a) $\lim_{x\to 2}(x^2 + 5x)$

b) $\lim_{x\to -3} \dfrac{x^2 + x - 6}{x + 3}$

c) $\lim_{x\to 0} \dfrac{1}{x}$

d) $\lim_{x\to \pi} \dfrac{x - \pi}{x^2}$

e) $\lim_{x\to 3} \dfrac{x - 3}{\sqrt{x} - \sqrt{3}}$

f) $\lim_{x\to 0} \operatorname{sgn} x = \lim_{x\to 0} \dfrac{x}{|x|}$

g) $\lim_{h\to 0} \dfrac{\dfrac{1}{x+h} - \dfrac{1}{x}}{h}$

h) $\lim_{x\to 1} f(x)$ where $f(x) = \begin{cases} x + 1 & \text{if } x \neq 1 \\ 4 & \text{if } x = 1 \end{cases}$

SOLUTION a) Evidently x^2 approaches 4, and $5x$ approaches 10 as x approaches 2. Thus $\lim_{x\to 2}(x^2 + 5x) = 4 + 10 = 14$.

b) We cannot just substitute $x = -3$ because the expression $(x^2 + x - 6)/(x + 3)$ becomes the meaningless form 0/0 in this case. This happens because $x + 3$ is a factor of both the numerator and the denominator. We have

$$\lim_{x\to -3} \frac{x^2 + x - 6}{x + 3} = \lim_{x\to -3} \frac{(x + 3)(x - 2)}{x + 3} = \lim_{x\to -3} (x - 2) = -5.$$

c) $1/x$ takes on larger and larger values as x approaches 0, (positive values if $x > 0$ and negative values if $x < 0$). Hence $\lim_{x\to 0} 1/x$ does not exist. There is no *unique real* number that $\dfrac{1}{x}$ approaches as x approaches 0.

d) In the expression $(x - \pi)/x^2$, the numerator approaches 0, and the denominator approaches π^2 as x approaches π. Hence

$$\lim_{x\to\pi} \frac{x-\pi}{x^2} = \frac{0}{\pi^2} = 0.$$

e) The expression $(x-3)/(\sqrt{x}-\sqrt{3})$ becomes 0/0 if we substitute $x = 3$. Evidently there is a common factor to cancel:

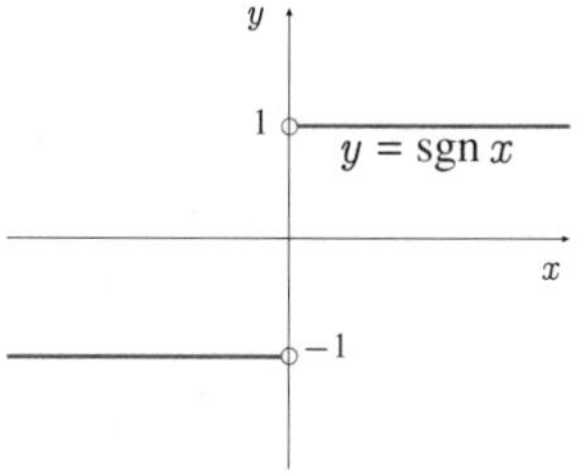

FIGURE 1.5.4

$$\lim_{x\to 3} \frac{x-3}{\sqrt{x}-\sqrt{3}} = \lim_{x\to 3} \frac{(\sqrt{x}-\sqrt{3})(\sqrt{x}+\sqrt{3})}{\sqrt{x}-\sqrt{3}} = \lim_{x\to 3}(\sqrt{x}+\sqrt{3}) = 2\sqrt{3}.$$

f) We have

$$\operatorname{sgn} x = \frac{x}{|x|} = \begin{cases} 1, & \text{if } x > 0 \\ -1, & \text{if } x < 0. \end{cases}$$

The function sgn is not defined at $x = 0$. It appears to approach different numbers (1 and -1) as x approaches 0 from the positive and negative sides respectively. (See Fig. 1.5.4.) Since limits must be unique, we conclude that $\lim_{x\to 0} x/|x|$ does not exist.

g) The expression $\left(\frac{1}{x+h} - \frac{1}{x}\right)/h$ is the meaningless 0/0 if $h = 0$ because both numerator and denominator have the factor h. To cancel this factor we need to do a little algebra:

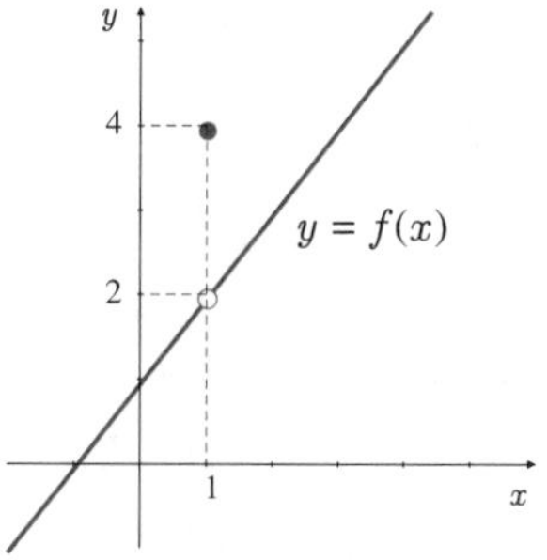

FIGURE 1.5.5

$$\begin{aligned}\lim_{h\to 0} \frac{\dfrac{1}{x+h} - \dfrac{1}{x}}{h} &= \lim_{h\to 0} \frac{x-(x+h)}{x(x+h)h} \\ &= \lim_{h\to 0} \frac{-h}{x(x+h)h} \\ &= \lim_{h\to 0} \frac{-1}{x(x+h)} = \frac{-1}{x^2}, \qquad (x \neq 0).\end{aligned}$$

h) Since $f(x) = x + 1$ for $x \neq 1$, and since $x + 1$ evidently approaches 2 as x approaches 1, we have $\lim_{x\to 1} f(x) = 2$. Note that in this case the limit is not $f(1)$. (Here $f(1) = 4$. See Fig. 1.5.5.) We say that this function f is *not continuous* at $x = 1$. (See Section 1.7 for a discussion of continuity.)

EXAMPLE 1.5.10 Find $\lim_{x\to 3} f(x)$ given that $|f(x) - 2| \le (x-3)^2$ for all $x \neq 3$.

SOLUTION Since $|f(x)-2| \le (x-3)^2$, we have $-(x-3)^2 \le f(x)-2 \le (x-3)^2$. (The given inequality says that $f(x) - 2$ is no further from 0 than is $(x-3)^2$. This implies that $f(x)-2$ lies between $-(x-3)^2$ and $(x-3)^2$.) Hence, $2-(x-3)^2 \le f(x) \le 2+(x-3)^2$. But

$$\lim_{x\to 3} \left(2-(x-3)^2\right) = 2 = \lim_{x\to 3} \left(2+(x-3)^2\right).$$

By the squeeze theorem, $\lim_{x\to 3} f(x) = 2$.

EXERCISES

In Exercises 1–28 evaluate the limit or explain why it does not exist.

1. $\lim_{x\to 4}(x^2 - 4x + 1)$

2. $\lim_{x\to -1}(1 + x + x^2 + x^3)$

3. $\lim_{x\to 1}\dfrac{x^2-1}{x+1}$

4. $\lim_{x\to -1}\dfrac{x^2-1}{x+1}$

5. $\lim_{x\to 3}\dfrac{x^2-6x+9}{x^2-9}$

6. $\lim_{x\to -2}\dfrac{x^2+2x}{x^2-4}$

7. $\lim_{t\to -2}\dfrac{t^2+t-2}{t+2}$

8.* $\lim_{h\to 0}\dfrac{\sqrt{4+h}-2}{h}$

9. $\lim_{x\to \pi}\dfrac{(x-\pi)^2}{\pi x}$

10. $\lim_{x\to -2}|x-2|$

11. $\lim_{x\to 0}\dfrac{|x-2|}{x-2}$

12. $\lim_{x\to 2}\dfrac{|x-2|}{x-2}$

13. $\lim_{t\to 1}\dfrac{t^2-1}{t^2-2t+1}$

14. $\lim_{x\to 2}\dfrac{\sqrt{4-4x+x^2}}{x-2}$

15. $\lim_{y\to 1}\dfrac{y-4\sqrt{y}+3}{y^2-1}$

16. $\lim_{x\to -1}\dfrac{x^3+1}{x+1}$

17. $\lim_{t\to 0}\dfrac{t^2+3t}{(t+2)^2-(t-2)^2}$

18. $\lim_{s\to 0}\dfrac{(s+1)^2-(s-1)^2}{s}$

19. $\lim_{x\to 2}\dfrac{x^4-16}{x^3-8}$

20. $\lim_{x\to 8}\dfrac{x^{2/3}-4}{x^{1/3}-2}$

21. $\lim_{x\to 5}\dfrac{x^2+3x-10}{3x^2+16x+5}$

22. $\lim_{x\to -5}\dfrac{x^2+3x-10}{3x^2+16x+5}$

23. $\lim_{t\to 1}\dfrac{(t^2-1)^2}{t^3-2t^2+t}$

24. $\lim_{y\to -\sqrt{2}}\dfrac{|y^2-2|}{y^2+2\sqrt{2}y+2}$

25. $\lim_{x\to 2}\left(\dfrac{1}{x-2}-\dfrac{4}{x^2-4}\right)$

26. $\lim_{x\to 2}\left(\dfrac{1}{x-2}-\dfrac{1}{x^2-4}\right)$

27. $\lim_{x\to 0}\dfrac{\sqrt{2+x^2}-\sqrt{2-x^2}}{x^2}$

28. $\lim_{x\to 0}\dfrac{|3x-1|-|3x+1|}{x}$

29. Use a calculator to compute values for the function $f(x) = \dfrac{x^3-3x^2+5x-3}{5x^2-4x-1}$ for several values of x near 1 (say $x = 1 \pm 0.1$, 1 ± 0.01, 1 ± 0.001, and so on). Guess the value of $\lim_{x\to 1} f(x)$ and then try to verify your guess.

30. Repeat Exercise 29 for $f(x) = \dfrac{2x\sqrt{x}+x-8\sqrt{x}-4}{x+\sqrt{x}-6}$ for values of x near 4.

31. Suppose that $x^4 < f(x) < x^2$ if $0 < |x| < 1$ and $x^2 < f(x) < x^4$ if $|x| > 1$. Find a) $\lim_{x\to -1} f(x)$, b) $\lim_{x\to 0} f(x)$, and c) $\lim_{x\to 1} f(x)$.

32.*Suppose $|f(x)| \le g(x)$ for all x. What can you conclude about $\lim_{x\to a} f(x)$ if $\lim_{x\to a} g(x) = 0$? What if $\lim_{x\to a} g(x) = 3$?

33.*If $\lim_{x\to a}\dfrac{f(x)-f(a)}{x-a} = 3$, find $\lim_{x\to a} f(x)$.

34.*If $\lim_{x\to 0}\dfrac{f(x)}{x^2} = -2$, find $\lim_{x\to 0} f(x)$ and $\lim_{x\to 0}\dfrac{f(x)}{x}$.

1.6 EXTENSIONS OF THE LIMIT CONCEPT

In this section we will extend the concept of limit to allow for three situations not covered by the definition of limit given in the previous section:

i) one-sided limits, that is, limits as x approaches a from one side only (the left or the right);

ii) limits at infinity, where x becomes arbitrarily large, positive or negative;

iii) infinite limits, which are not really limits at all but provide useful symbolism for describing the behavior of functions whose values become arbitrarily large, positive or negative.

One-Sided Limits

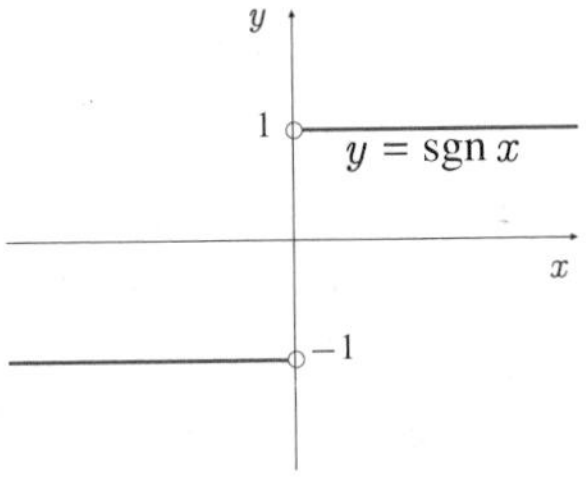

FIGURE 1.6.1

In Example 1.5.9(f) we argued that $\lim_{x\to 0} \operatorname{sgn} x$ does not exist, because $\operatorname{sgn} x$ approaches different values (1 or -1) as x approaches 0 from the right side (through positive values) or from the left side (through negative values). (See Fig. 1.6.1.) If we are only interested in letting x approach 0 from one side, we would say that the appropriate *one-sided* limit exists; specifically,

$$\lim_{x\to 0+} \operatorname{sgn} x = 1, \qquad \lim_{x\to 0-} \operatorname{sgn} x = -1.$$

We read the symbol $\lim_{x\to 0+}$ as "the limit as x approaches 0 from the right" (or "from the positive side of 0"), and the symbol $\lim_{x\to 0-}$ as "the limit as x approaches 0 from the left" (or "from the negative side of 0").

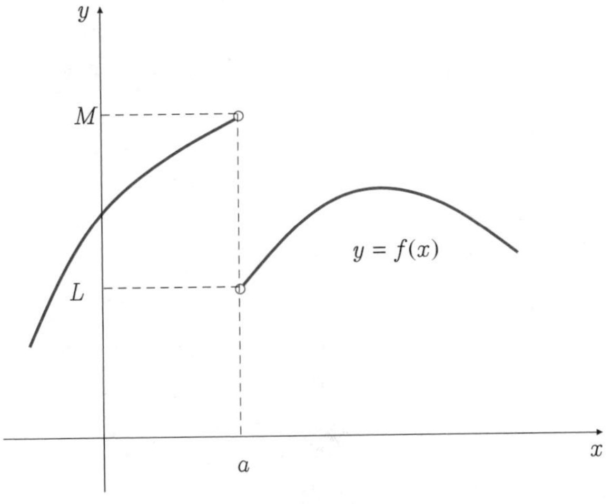

FIGURE 1.6.2

1.6.1
Left and Right Limits

We say that the function f has **right limit** L as x approaches a, or that the limit of $f(x)$ as x approaches a *from the right* is L, provided $f(x)$ is defined on the interval (a, b) for some $b > a$ and $f(x)$ approaches arbitrarily close to L as x decreases towards a. We write this as

$$\lim_{x \to a+} f(x) = L.$$

Similarly, we say that f has **left limit** M as x approaches a, or that the limit of $f(x)$ as x approaches a *from the left* is M, provided $f(x)$ is defined on the interval (b, a) for some $b < a$ and $f(x)$ approaches arbitrarily close to M as x increases towards a. We write this as

$$\lim_{x \to a-} f(x) = M.$$

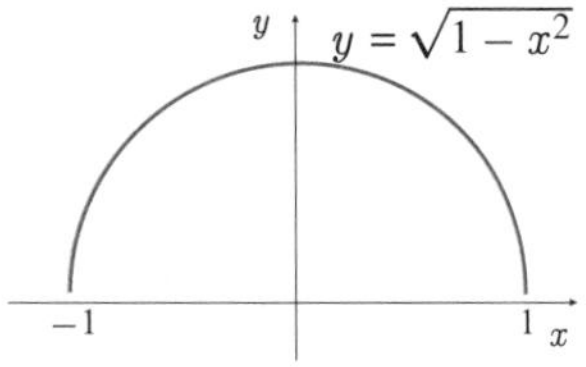

FIGURE 1.6.3

Right and left limits are illustrated in Fig. 1.6.2. As was the case for (two-sided) limits studied in the previous section, $\lim_{x \to a-} f(x)$ or $\lim_{x \to a+} f(x)$ can exist even if $f(a)$ is not defined, and if $f(a)$ is defined it need not be equal to either of these one-sided limits.

EXAMPLE 1.6.2

a) The function $f(x) = \sqrt{1 - x^2}$ is defined only on the interval $[-1, 1]$. Evidently $f(-1) = 0$ and $f(1) = 0$. If $-1 < a < 1$ then $\lim_{x \to a} f(x) = \sqrt{1 - a^2} = f(a)$. However, f has only a right limit at -1 and a left-hand limit at 1 (see Fig. 1.6.3):

$$\lim_{x \to -1+} f(x) = 0, \qquad \lim_{x \to 1-} f(x) = 0.$$

b) Let $[x]$ denote the greatest integer that does not exceed x. Thus $[3] = 3$, $[2.7] = 2$, $[-1/2] = -1$, $[0] = 0$. If n is any integer, then

$$\lim_{x \to n+} [x] = n, \qquad \lim_{x \to n-} [x] = n - 1.$$

If a is not an integer, then $n < a < n + 1$ for some integer n, and

$$\lim_{x \to a+} [x] = \lim_{x \to a-} [x] = \lim_{x \to a} [x] = n.$$

The graph of $y = [x]$ is shown in Fig. 1.6.4.

It is always true that $\lim_{x \to a} f(x) = L$ holds if and only if $\lim_{x \to a+} f(x) = L = \lim_{x \to a-} f(x)$. If $\lim_{x \to a+} f(x)$ and $\lim_{x \to a-} f(x)$ both exist but are not equal, then $\lim_{x \to a} f(x)$ does not exist.

Left limits and right limits have all the properties possessed by ordinary (two-sided) limits as stated in Theorems 1.5.6, 1.5.7 and 1.5.8.

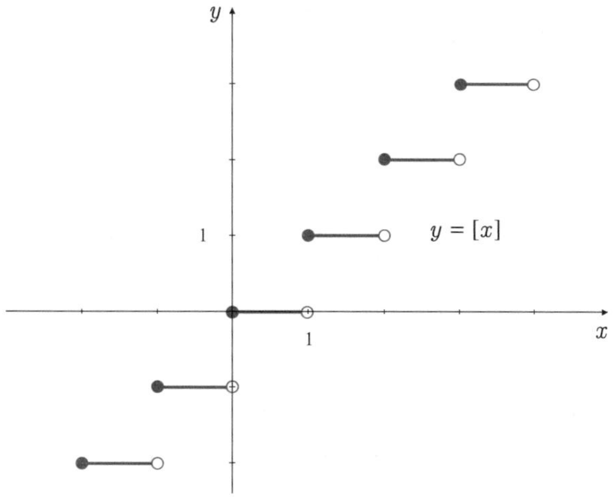

FIGURE 1.6.4

Limits at Infinity

If a function f is defined on the whole real line or on a semi-infinite interval of the form (a, ∞) or $(-\infty, a)$, it is appropriate to ask what happens to the values $f(x)$ as x becomes very large, positive (approaches infinity) or negative (approaches negative infinity).

EXAMPLE 1.6.3 Consider the function

$$f(x) = \frac{4x^2 - 1}{2x^2 + 5}.$$

As x becomes large (either positive or negative) both the numerator and the denominator of $f(x)$ become large. The quotient, however, does not become large. To see this divide both numerator and denominator by x^2 (the highest power of x occurring in the denominator), and thus rewrite $f(x)$ in the form

$$f(x) = \frac{4 - \dfrac{1}{x^2}}{2 + \dfrac{5}{x^2}}, \qquad \text{valid for } x \neq 0.$$

As x approaches infinity or negative infinity, $1/x^2$ approaches 0, so the numerator approaches 4 and the denominator approaches 2. Thus we say that

$$\lim_{x \to \infty} f(x) = 2, \qquad \lim_{x \to -\infty} f(x) = 2.$$

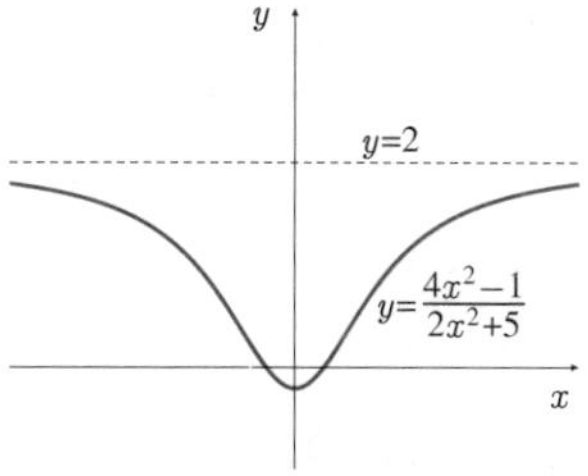

FIGURE 1.6.5

In graphical terms, as depicted in Fig. 1.6.5, the graph of f approaches the horizontal straight line $y = 2$ as x recedes far to the right or left of 0. We call the line $y = 2$, a (horizontal) **asymptote** of the graph of f. Asymptotes are straight lines to which graphs approach arbitrarily close as they recede very far from the origin.

We will study asymptotes in some detail in Section 4.4.

1.6.4
Limits at Infinity and Negative Infinity

If $f(x)$ is defined on an interval (a, ∞) and if $f(x)$ approaches arbitrarily close to L as x increases through arbitrarily large positive numbers, then we say that $f(x)$ tends to the limit L as x tends to infinity, and we write

$$\lim_{x \to \infty} f(x) = L.$$

Similarly, if $f(x)$ is defined on an interval $(-\infty, a)$ and if $f(x)$ approaches arbitrarily close to M as x decreases through arbitrarily large negative numbers, then we say that $f(x)$ tends to the limit M as x tends to negative infinity, and we write

$$\lim_{x \to -\infty} f(x) = M.$$

As did Definition 1.5.2, this definition contains somewhat vague terms ("arbitrarily close," "arbitrarily large") that would have to be made more precise if we wanted to use the definition to provide proofs of statements. (See Appendix II.) The version in this definition will serve our purposes, however. Many functions can be manipulated algebraically into a form where the limits at infinity or negative infinity can be determined by inspection. If the function is a quotient whose numerator and denominator involve sums of various powers of x, this can be achieved (as in Example 1.6.3) by dividing the numerator and denominator by the highest power of x in the denominator.

EXAMPLE 1.6.5 Evaluate the limits as x approaches ∞ and $-\infty$ for:

a) $\dfrac{1+x}{x-2x^3}$, b) $\dfrac{x}{\sqrt{x^2+1}}$, and c) $\dfrac{\sqrt{x}(x+1)(x^2+1)}{(2\sqrt{x}+3)^2(1-\sqrt{x})^5}$.

SOLUTION

a) $$\lim_{x \to \infty} \frac{1+x}{x-2x^3} = \lim_{x \to \infty} \frac{\dfrac{1}{x^3} + \dfrac{1}{x^2}}{\dfrac{1}{x^2} - 2} = \frac{0}{-2} = 0 = \lim_{x \to -\infty} \frac{1+x}{x-2x^3}.$$

b) $$\lim_{x \to \infty} \frac{x}{\sqrt{x^2+1}} = \lim_{x \to \infty} \frac{x}{\sqrt{x^2\left(1+\dfrac{1}{x^2}\right)}} = \lim_{x \to \infty} \frac{x}{x\sqrt{1+\dfrac{1}{x^2}}} = \lim_{x \to \infty} \frac{1}{\sqrt{1+\dfrac{1}{x^2}}} = 1.$$

The same calculation works as $x \to -\infty$ except that $\sqrt{x^2} = |x| = -x$ in this case:

$$\lim_{x \to -\infty} \frac{x}{\sqrt{x^2+1}} = \lim_{x \to -\infty} \frac{x}{\sqrt{x^2\left(1+\dfrac{1}{x^2}\right)}} = \lim_{x \to -\infty} \frac{x}{(-x)\sqrt{1+\dfrac{1}{x^2}}} = -1.$$

c) If we factor out the highest power of x from each factor in the numerator and denominator of the expression, these powers cancel:

$$\lim_{x\to\infty} \frac{\sqrt{x}(x+1)(x^2+1)}{(2\sqrt{x}+3)^2(1-\sqrt{x})^5} = \lim_{x\to\infty} \frac{x^{\frac{1}{2}+1+2}\left(1+\frac{1}{x}\right)\left(1+\frac{1}{x^2}\right)}{x^{\frac{2}{2}+\frac{5}{2}}\left(2+\frac{3}{\sqrt{x}}\right)^2\left(\frac{1}{\sqrt{x}}-1\right)^5}$$

$$= \lim_{x\to\infty} \frac{\left(1+\frac{1}{x}\right)\left(1+\frac{1}{x^2}\right)}{\left(2+\frac{3}{\sqrt{x}}\right)^2\left(\frac{1}{\sqrt{x}}-1\right)^5}$$

$$= \frac{1}{4(-1)} = -\frac{1}{4}$$

The expression in (c) does not have a limit at $-\infty$ because $\sqrt{x}$ is not defined for $x < 0$.

EXAMPLE 1.6.6 Find (a) $\lim_{x\to\infty}(\sqrt{x^2+x}-x)$, and (b) $\lim_{x\to-\infty}(\sqrt{x^2+x}-x)$.

SOLUTION In (a) we are trying to find the limit of the difference of two functions each of which becomes arbitrarily large as x increases to infinity. Again an algebraic trick is needed to render the limit obvious; we multiply numerator and denominator by the conjugate expression $\sqrt{x^2+x}+x$:

$$\lim_{x\to\infty}\left(\sqrt{x^2+x}-x\right) = \lim_{x\to\infty} \frac{\left(\sqrt{x^2+x}-x\right)\left(\sqrt{x^2+x}+x\right)}{\sqrt{x^2+x}+x}$$

$$= \lim_{x\to\infty} \frac{x^2+x-x^2}{\sqrt{x^2+x}+x}$$

$$= \lim_{x\to\infty} \frac{x}{x\sqrt{1+\frac{1}{x}}+x} = \lim_{x\to\infty} \frac{1}{\sqrt{1+\frac{1}{x}}+1} = \frac{1}{2}.$$

(Here $\sqrt{x^2} = x$ because $x > 0$ as $x \to \infty$.) At first glance, the situation in (b) may appear similar, but it is not. For negative x both terms in the expression are positive so we are now dealing with a sum rather than a difference of functions with large values. The limit does not exist. (The function becomes infinite as $x \to -\infty$.)

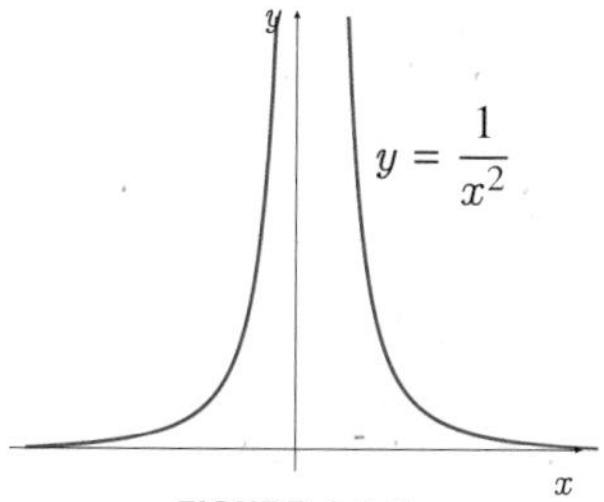

FIGURE 1.6.6

Infinite Limits

Consider the function $f(x) = \frac{1}{x^2}$. As x approaches 0 from either side the values of $f(x)$ grow larger and larger (see Fig. 1.6.6), so the limit of $f(x)$ as x approaches 0 *does not exist*. It is nevertheless convenient to describe the behavior of f near 0 by saying that $f(x)$ *approaches infinity* as x approaches zero. We write

$$\lim_{x\to 0} f(x) = \lim_{x\to 0}\frac{1}{x^2} = \infty.$$

Note that in writing this we are *not* saying that $\lim_{x\to 0} 1/x^2$ *exists*. (Infinity is not a real number.) Rather we are saying that that limit *does not exist* because $1/x^2$ becomes arbitrarily large near $x = 0$.

1.6.7
Infinite Limits

> If $f(x)$ is positive and becomes arbitrarily large as x approaches arbitrarily close to the value a then we say that $f(x)$ *tends to infinity* as x approaches a, and we write
>
> $$\lim_{x\to a} f(x) = \infty.$$
>
> Similarly, if $f(x)$ is negative and becomes arbitrarily large as x approaches a, then we say $f(x)$ *tends to negative infinity* as x approaches a, and write
>
> $$\lim_{x\to a} = -\infty.$$

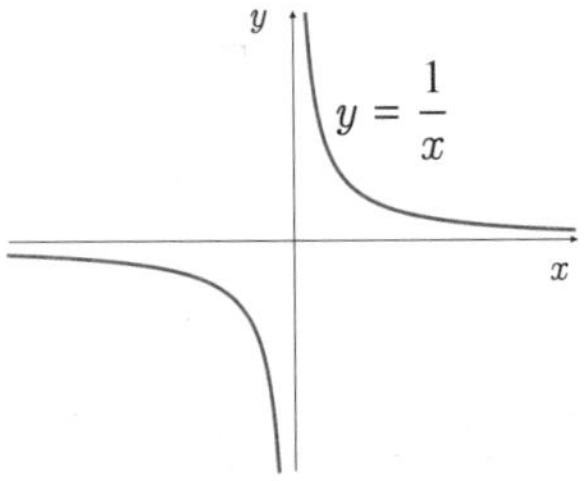

FIGURE 1.6.7

This definition can be extended to cover situations like $\lim_{x\to a+} f(x) = \infty$ (or $-\infty$), $\lim_{x\to a-} f(x) = \infty$ (or $-\infty$), $\lim_{x\to\infty} f(x) = \infty$ (or $-\infty$), and $\lim_{x\to-\infty} f(x) = \infty$ (or $-\infty$). Reconsidering Example 1.6.6(b), we can now assert that

$$\lim_{x\to-\infty} (\sqrt{x^2 + x} - x) = \infty,$$

which, we stress again, in no way contradicts our earlier assertion that this limit does not exist.

EXAMPLE 1.6.8 The function $1/x$ has no limit as x approaches 0, but in this case we cannot even say that $1/x$ approaches infinity (or negative infinity) because the values of $1/x$ are positive for $x > 0$ and negative for $x < 0$. We can only write $\lim_{x\to 0} 1/x$ does not exist. (See Fig. 1.6.7.) However, we can make assertions about one-sided limits of $1/x$ at 0:

$$\lim_{x\to 0+} \frac{1}{x} = \infty, \qquad \lim_{x\to 0-} \frac{1}{x} = -\infty.$$

These assertions do *not* imply that these one-sided limits *exist* (as real numbers); ∞ is not a number.

EXAMPLE 1.6.9 a) $\displaystyle\lim_{x\to 2} \frac{1}{|x - 2|} = \infty.$

b) $\displaystyle\lim_{x\to -5} \frac{1}{5 + x}$ does not exist; but $\displaystyle\lim_{x\to -5+} \frac{1}{5 + x} = \infty$ (the fraction is positive if $x > -5$), and $\displaystyle\lim_{x\to -5-} \frac{1}{5 + x} = -\infty$ (the fraction is negative if $x < -5$.)

c) $\displaystyle\lim_{x\to 2} \frac{x - 1}{4 - x^2}$ does not exist, but $\displaystyle\lim_{x\to 2-} \frac{x - 1}{4 - x^2} = \infty$ and $\displaystyle\lim_{x\to 2+} \frac{x - 1}{4 - x^2} = -\infty$; $\displaystyle\lim_{x\to -2} \frac{x - 1}{4 - x^2}$ does not exist, but $\displaystyle\lim_{x\to -2-} \frac{x - 1}{4 - x^2} = \infty$ and $\displaystyle\lim_{x\to -2+} \frac{x - 1}{4 - x^2} = -\infty$.

When a function $f(x)$ approaches infinity or negative infinity as x approaches some finite point a, then the graph of f approaches the vertical straight line $x = a$ as it recedes to infinity (in either direction) near a. Thus the vertical line $x = a$ is called a (vertical) **asymptote** of the graph of f. We persist in calling it an asymptote even if only a one-sided limit is infinite there. Thus the y-axis is a vertical asymptote of each of the curves of $y = 1/x$ and $y = 1/x^2$. The line $x = 2$ is a vertical asymptote of the graph $y = 1/|x-2|$; the lines $x = -2$ and $x = 2$ are vertical asymptotes of the curve $y = \dfrac{1}{\sqrt{4-x^2}}$.

EXERCISES

In Exercises 1–50 find the indicated limit or explain why it does not exist. If it does not exist, is it infinity, negative infinity, or neither? Identify any horizontal or vertical asymptotes of the functions considered.

1. $\lim_{x\to 2-} \sqrt{2-x}$

2. $\lim_{x\to 2+} \sqrt{2-x}$

3. $\lim_{x\to -2-} \sqrt{2-x}$

4. $\lim_{x\to -2+} \sqrt{2-x}$

5. $\lim_{x\to 3} \dfrac{1}{3-x}$

6. $\lim_{x\to 3} \dfrac{1}{(3-x)^2}$

7. $\lim_{x\to 3-} \dfrac{1}{3-x}$

8. $\lim_{x\to 3+} \dfrac{1}{3-x}$

9. $\lim_{x\to 0} \sqrt{x^3-x}$

10. $\lim_{x\to 0-} \sqrt{x^3-x}$

11. $\lim_{x\to 0+} \sqrt{x^3-x}$

12. $\lim_{x\to 0+} \sqrt{x^2-x^4}$

13. $\lim_{x\to -5/2} \dfrac{2x+5}{5x+2}$

14. $\lim_{x\to -2/5} \dfrac{2x+5}{5x+2}$

15. $\lim_{x\to -(2/5)-} \dfrac{2x+5}{5x+2}$

16. $\lim_{x\to -(2/5)+} \dfrac{2x+5}{5x+2}$

17. $\lim_{x\to a+} \dfrac{x-a}{x^2-a^2}$

18. $\lim_{x\to a-} \dfrac{x-a}{x^2-a^2}$

19. $\lim_{x\to a} \dfrac{|x-a|}{x^2-a^2}$

20. $\lim_{x\to a-} \dfrac{|x-a|}{x^2-a^2}$

21. $\lim_{x\to 1-} \dfrac{2x+3}{x^2+x-2}$

22. $\lim_{x\to 1+} \dfrac{2x+3}{x^2+x-2}$

23. $\lim_{x\to 2+} \dfrac{x}{(2-x)^3}$

24. $\lim_{x\to 1-} \dfrac{x}{\sqrt{1-x^2}}$

25. $\lim_{x\to 1+} \dfrac{1}{|x-1|}$

26. $\lim_{x\to 1-} \dfrac{1}{|x-1|}$

27. $\lim_{x\to 2-} \dfrac{x^2-4}{|x+2|}$

28. $\lim_{x\to 2+} \dfrac{x^2-4}{|x+2|}$

29. $\lim_{x\to -3+} \dfrac{x+3}{|x+3|}$

30. $\lim_{x\to -3-} \dfrac{x+3}{|x+3|}$

31. $\lim_{x\to -3} \dfrac{x+3}{|x+3|}$

32. $\lim_{x\to 0} \dfrac{4-5x}{x^4}$

33. $\lim_{x\to 2} \dfrac{x-3}{x^2-4x+4}$

34. $\lim_{x\to 1+} \dfrac{\sqrt{x^2-x}}{x-x^2}$

35. $\lim_{x\to\infty} \dfrac{3x^3-5x^2+7}{8+2x-5x^3}$

36. $\lim_{x\to -\infty} \dfrac{x^2-2}{x-x^2}$

37. $\lim_{x\to -\infty} \dfrac{x^2+3}{x^3+2}$

38. $\lim_{x\to\infty} \dfrac{x^3+3}{x^2+2}$

39. $\lim_{x\to\infty} \dfrac{x+x^3+x^5}{1+x^2+x^3}$

40. $\lim_{x\to\infty} \dfrac{Ax^2+Bx+C}{Dx^2+Ex+F}$

41. $\lim_{x\to\infty} \dfrac{3x+2\sqrt{x}}{1-x}$

42. $\lim_{x\to\infty} \dfrac{2x-1}{\sqrt{3x^2+x+1}}$

43. $\lim_{x\to -\infty} \dfrac{2x-1}{\sqrt{3x^2+x+1}}$

44. $\lim_{x\to -\infty} \dfrac{2x-5}{|3x+2|}$

45.* $\lim_{x\to\infty} \dfrac{x\sqrt{x+1}\left(1-\sqrt{2x+3}\right)}{7-6x+4x^2}$

46. $\lim_{x\to\infty} \left(\dfrac{x}{x+1}-\dfrac{x}{x-1}\right)$

47.* $\lim_{x\to -\infty} \left(\sqrt{x^2+2x}-\sqrt{x^2-2x}\right)$

48. $\lim_{x\to\infty} \left(\sqrt{x^2-2x}-x\right)$

49. $\lim_{x\to -\infty} \left(\sqrt{x^2+2x}+x\right)$

50. $\lim_{x\to\infty} \left(\sqrt{x^2+2x}-\sqrt{x^2-2x}\right)$

51. Parking in a certain parking lot costs \$1.50 for each hour or part of an hour. Sketch the graph of the function $C(t)$ representing the cost of parking for t hours. At what values of t does $C(t)$ have a limit? Evaluate $\lim_{t\to t_0-} C(t)$ and $\lim_{t\to t_0+} C(t)$ for an arbitrary point t_0.

52. If $\lim_{x\to 0+} f(x) = L$ find $\lim_{x\to 0-} f(x)$ if (a) f is even, (b) f is odd.

53. If $\lim_{x\to 0+} f(x) = A$ and $\lim_{x\to 0-} f(x) = B$, find
a) $\lim_{x\to 0+} f(x^3 - x)$, b) $\lim_{x\to 0-} f(x^3 - x)$,
c) $\lim_{x\to 0-} f(x^2 - x^4)$, d) $\lim_{x\to 0+} f(x^2 - x^4)$.

1.7 CONTINUITY

1.7.1 Definition of Continuity at a Point

A function f is said to be **continuous at the point** a if

$$\lim_{x\to a} f(x) = f(a).$$

According to this definition, three conditions must be satisfied if f is to be continuous at a:

i) $f(x)$ must be defined nearby $x = a$ and at $x = a$.

ii) $\lim_{x\to a} f(x)$ must exist, and

iii) the limit in (ii) must be equal to $f(a)$.

In graphical terms, f is continuous at $x = a$ if its graph extends some distance to the left and right of the point $(a, f(a))$ and has no break in it at that point. In Fig. 1.7.1 f is continuous at x_1 and at every other point between a and b except at the points x_2, x_3 and x_4 where its graph has breaks. It is *discontinuous* at x_2 and x_3 because it does not have a limit as x approaches either of these points. (It has left and right limits, but they are not equal.) It is discontinuous at x_4 because, although $\lim_{x\to x_4} f(x)$ exists, the limit is not equal to $f(x_4)$.

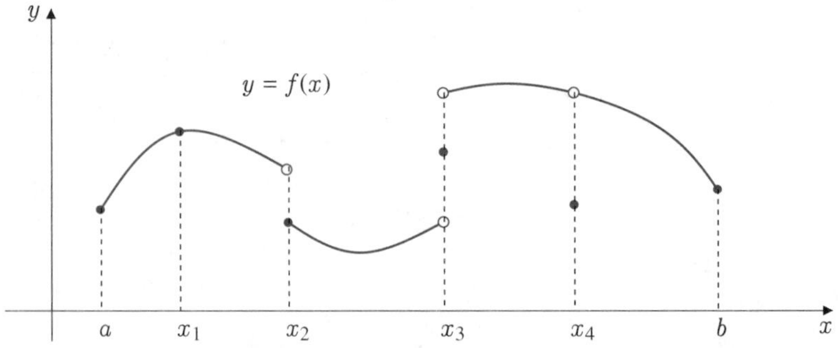

FIGURE 1.7.1

1.7.2 Definition of One-Sided Continuity

We say that f is **continuous on the right** at a if $\lim_{x\to a+} f(x) = f(a)$, and that it is **continuous on the left** at a if $\lim_{x\to a-} f(x) = f(a)$.

Evidently f is continuous at a if f is continuous on the right and on the left at a. In Fig. 1.7.1 f is continuous on the right at a and x_2, and is continuous on the left at b. It is not continuous on either side at x_3 or x_4. It is continuous on both sides at x_1 and all other points between a and b.

1.7.3 Definition of Continuity on an Interval

A function is **continuous on the interval** I if it is continuous at every point of I. If the interval is closed, the function need only be continuous on the right at the left endpoint and continuous on the left at the right endpoint (as well as continuous at every interior point).

EXAMPLE 1.7.4

a) $f(x) = 3x - 2$ on $[-1, 2]$. f is continuous on $[-1, 2]$. See Fig. 1.7.2.

b) $f(x) = 1/x$ on $[-1, 0) \cup (0, 1]$. f is continuous at every point of its domain.

c) $H(x) = \begin{cases} 1, & \text{if } x > 0 \\ 0, & \text{if } x \le 0 \end{cases}$.

H is continuous everywhere except at $x = 0$, where it is continuous on the left but not on the right (see Fig. 1.7.3).

d) The greatest integer function $[x]$ of Example 1.6.2(b) is continuous except at the integers. At each integer the function is continuous on the right but discontinuous on the left:

$$\lim_{x\to n+} [x] = n = [n], \qquad \lim_{x\to n-} [x] = n - 1 \ne n = [n].$$

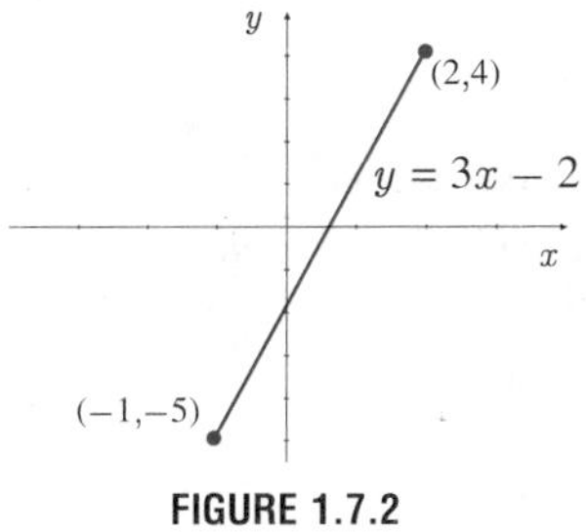

FIGURE 1.7.2

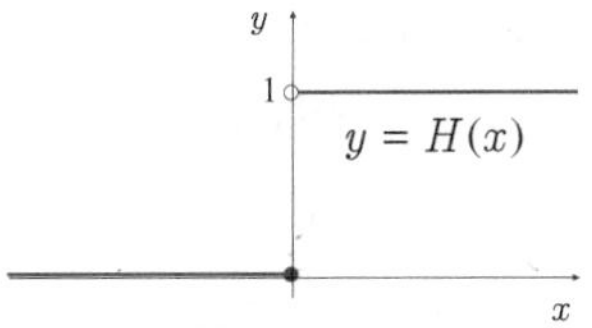

FIGURE 1.7.3

Most of the functions encountered in elementary calculus are continuous wherever they are defined. Some properties of continuous functions are collected in the following theorem. The proof involves using the definition of continuity and appropriate properties of limits given in Theorem 1.5.7. (See Appendix II for further discussion.)

THEOREM 1.7.5 a) If the functions f and g are continuous at $x = a$, then the functions $f + g$, $f - g$ and fg are also continuous at $x = a$. If $g(a) \neq 0$, then f/g is continuous at $x = a$.

b) If f is continuous at L and if $\lim_{x\to a} g(x) = L$, then

$$\lim_{x\to a} f(g(x)) = f(L) = f\left(\lim_{x\to a} g(x)\right).$$

In particular, if g is continuous at a (so $L = g(a)$), then the composition $f \circ g$ is continuous at a:

$$\lim_{x\to a} f(g(x)) = f(g(a)).$$

c) The function $f(x) = C$ (constant) is continuous on the whole real line.

d) If r is a rational number then the function $f(x) = x^r$ is continuous wherever it is defined. In particular, if n is a positive integer, then x^n is continuous on the whole real line. □

Roughly speaking, parts (a) and (b) of Theorem 1.7.5 assure us that sums, differences, products, quotients, and compositions of continuous functions are continuous wherever they are defined. Parts (c) and (d) provide us with some specific examples of continuous functions from which we may construct others by taking such combinations.

A **polynomial** is a function $P(x)$ that is the sum of a finite number of terms each of which is a constant multiple of a nonnegative integer power of x. Every polynomial $P(x)$ is of the form

$$P(x) = a_0 + a_1x + a_2x^2 + a_3x^3 + \cdots + a_nx^n$$

for some integer $n \geq 0$, where a_0, a_1, ..., a_n are constants and $a_n \neq 0$. This polynomial is said to be of degree n. For instance, $5x^2 - 3x + 1$ is a polynomial of degree 2.

Theorem 1.7.5 assures us that every polynomial is continuous everywhere on the real line. (Why?) If $P(x)$ and $Q(x)$ are two polynomials, the fraction $P(x)/Q(x)$ is called a **rational function**. A rational function is continuous everywhere on its domain, that is, everywhere on the real line except at points x where $Q(x) = 0$.

EXAMPLE 1.7.6 a) The function $5 - 2x^3 + \frac{3}{4}x^7$ is a polynomial, and so it is continuous on the whole real line $\mathbb{R}$.

b) The function $f(x) = (x^2 - x)/(x^2 + 3)$ is a rational function whose denominator does not vanish for any x. Thus $f(x)$ is continuous on $\mathbb{R}$.

c) The function $x/(x^2 - 4)$ is a rational function whose denominator vanishes at $x = 2$ and $x = -2$. It is continuous on $\mathbb{R}$ except at those two points (where it is undefined).

d) The function $\sqrt{x}$ is defined on $[0, \infty)$. It is continuous on that interval (by Theorem 1.7.5(d)); in particular, it is continuous only on the right at $x = 0$.

e) The function $f(x) = \sqrt{x^2 - 1}$ has domain $(-\infty, -1] \cup [1, \infty)$. Since $x^2 - 1$ is continuous everywhere on $\mathbb{R}$, f is continuous on its domain, by Theorem 1.7.5(b).

f) The function $1/\sqrt{x+2}$ has domain $(-2, \infty)$ and is continuous there since $x + 2$ is continuous everywhere on $\mathbb{R}$.

Theorems 1.7.7 and 1.7.10 below contain very important and useful results about continuous functions. They are more subtle than the results quoted in Theorem 1.7.5; the proofs (see Appendix II) require a careful study of the implications of the completeness property of the real numbers. Nevertheless, both results should seem obviously true on an intuitive level.

THEOREM 1.7.7 If f is continuous on the closed, finite interval $[a, b]$, then

i) there is a positive real number K such that

$$|f(x)| \le K \qquad \text{for all } x \text{ in } [a, b], \text{ and}$$

ii) there are points x_0 and x_1 in $[a, b]$ such that

$$f(x_0) \le f(x) \le f(x_1) \qquad \text{for all } x \text{ in } [a, b]. \ \square$$

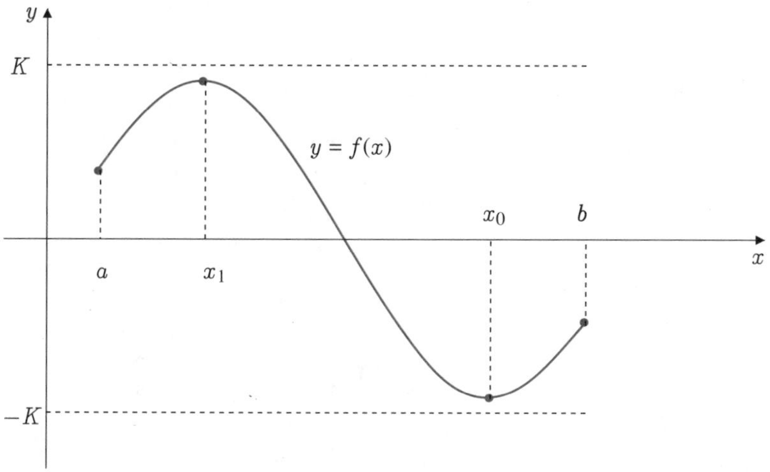

FIGURE 1.7.4

Conclusion (i) asserts that $f(x)$ is bounded on $[a, b]$; that is, it cannot take on arbitrarily large positive or negatives values. Conclusion (ii) asserts that f takes on maximum and minimum values at points of $[a, b]$. These results are illustrated in Fig. 1.7.4.

Theorem 1.7.7 will be especially useful when we study optimization problems in Section 5.1. Such problems require us to find maximum or minimum values of functions defined on intervals, and they arise frequently in applications of mathematics. Here is an example of a problem of that type which can be solved by elementary means (not requiring any calculus).

EXAMPLE 1.7.8 What is the area of the largest rectangular plot that can be enclosed by a fence of total length 200 metres?

SOLUTION If the length and width of the plot are x and y (metres), respectively, then the perimeter is $2x+2y$ (see Fig. 1.7.5). Since we have only 200 metres of fence, we want $2x+2y=200$; therefore $y=100-x$. The area of the plot can be expressed as a function of x alone:

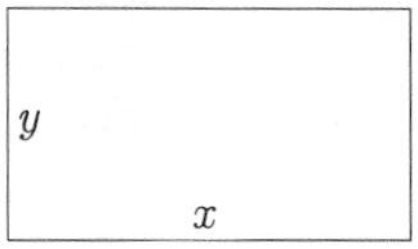

FIGURE 1.7.5

$$\text{Area } = A(x) = xy = x(100-x) = 100x - x^2.$$

Evidently x and y must be nonnegative, so $0 \le x \le 100$. We are looking for the maximum value of $A(x)$ on the interval $[0,100]$. Theorem 1.7.7 assures us that such a maximum value exists. Since $A(x) > 0$ if $0 < x < 100$ and $A(0) = 0$ and $A(100) = 0$, the maximum must occur at a point x in the open interval $(0,100)$. Calculus will provide us with useful tools for finding points where such maximum (or minimum) values occur. In this case, however, we can find the maximum value without using any calculus; we need only complete the square in the quadratic polynomial $A(x)$:

$$\begin{aligned} A(x) &= -(x^2 - 100x + 2500) + 2500 \\ &= 2500 - (x-50)^2. \end{aligned}$$

Since $(x-50)^2 \ge 0$ for all x we have $A(x) \le 2500$ for every x. Since $A(50) = 2500$, the largest rectangle with perimeter 200 metres has the area 2500 m^2. It is a square of side 50 m.

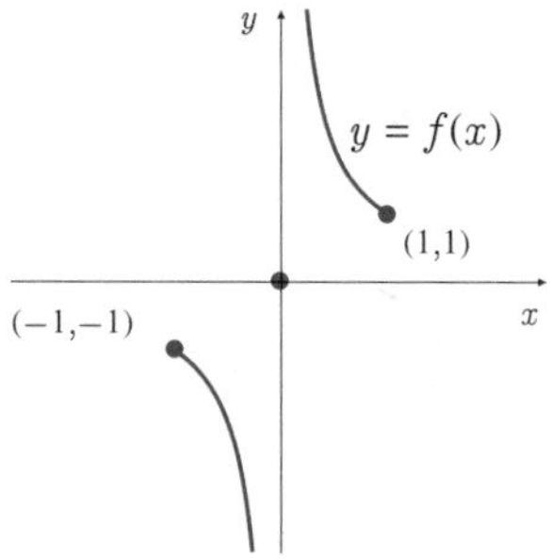

FIGURE 1.7.6

The following example shows that Theorem 1.7.7 may fail if any of its conditions are not satisfied. A function $f(x)$ may fail to be bounded, or may fail to have a maximum or minimum value, if its domain is not a finite interval, or not closed, or if f fails to be continuous anywhere in the domain.

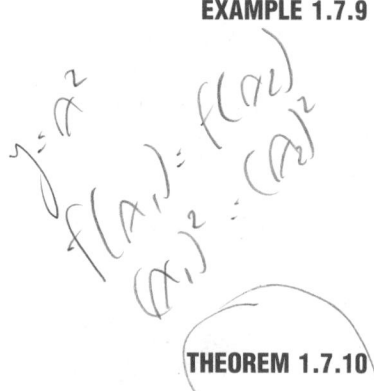

EXAMPLE 1.7.9

a) Let $f(x) = \begin{cases} 1/x, & \text{if } -1 \le x \le 1 \text{ and } x \ne 0 \\ 0, & \text{if } x = 0. \end{cases}$
   Then f is defined on the closed, finite interval $[-1,1]$ but it is not continuous at the point $x=0$ in that interval. The function is not bounded on $[-1,1]$ and does not have a maximum or minimum value there. See Fig. 1.7.6.

b) If $f(x) = x$ on $\mathbb{R}$, then f is not bounded, and so it does not have a maximum or minimum value. The domain of f is not a finite interval.

c) If $f(x) = x$ on $(1,2)$, then f, though bounded, has no maximum or minimum value. Indeed, its range is also the open interval $(1,2)$, which has no greatest or least element. The domain of f is not a closed interval.

THEOREM 1.7.10 **(The Intermediate-Value Theorem – IVT)** If f is continuous on $[a,b]$ and if s is a real number lying between $f(a)$ and $f(b)$, then there exists at least one real number c between a and b such that $f(c) = s$. □

Intuitively, we can see that since the graph of f has no breaks between a and b, and since it starts out at height $f(a)$ and ends up at height $f(b)$, it must cross the horizontal line at the height s between $f(a)$ and $f(b)$ at least once. This is depicted in Fig. 1.7.7.

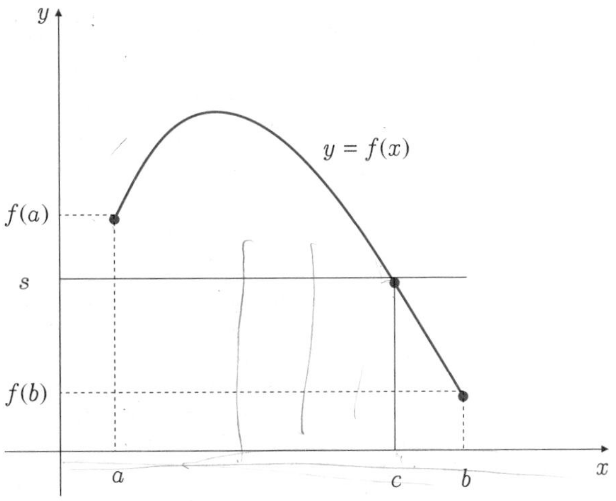

FIGURE 1.7.7

Theorem 1.7.10 is often used to show that certain equations have solutions.

EXAMPLE 1.7.11 Show that the cubic equation $x^3 + x^2 - x - 4 = 0$ has a solution in the interval $(1, 2)$.

SOLUTION If $f(x) = x^3 + x^2 - x - 4$, then $f(1) = -3$ and $f(2) = 6$. Since 0 lies between -3 and 6, there exists at least one x between 1 and 2 such that $f(x) = 0$.

EXAMPLE 1.7.12 Determine the intervals on which $f(x) = x^3 - 4x$ is positive and negative.

SOLUTION Since $f(x) = x(x^2 - 4) = x(x - 2)(x + 2)$, $f(x) = 0$ only at $x = 0$, 2, and -2. Because f is continuous on the whole real line, it must have constant sign on each of the intervals $(-\infty, -2)$, $(-2, 0)$, $(0, 2)$, and $(2, \infty)$. (This assertion follows from the Intermediate-Value Theorem. If there were points a and b in one of the intervals, say $(0, 2)$, such that $f(a) < 0$ and $f(b) > 0$, then by that theorem there would exist c between a and b, and therefore between 0 and 2, such that $f(c) = 0$. This is a contradiction.) Since $f(-3) = -15$, $f(-1) = 3$, $f(1) = -3$, and $f(3) = 15$, we must have $f(x) < 0$ on $(-\infty, -2)$ and $(0, 2)$, and $f(x) > 0$ on $(-2, 0)$ and $(2, \infty)$.

REMARK: The Theorems 1.7.7 and 1.7.10 are examples of what mathematicians call **existence theorems**. Such theorems assert the existence of something without telling you how to find it. Students sometimes complain that mathematicians worry too much about proving that a problem has a solution and not enough about how to find that solution. They argue: "If I can calculate a solution to a problem, then surely I do not need to worry about whether a solution exists." This is, however, false logic. Suppose we pose the problem "Find the largest positive integer." Of course this problem has no solution; there is no largest positive integer because we can add one to any integer and get a larger integer. Suppose, however, that we forget this and try to calculate a solution. We could proceed as follows:

Let N be the largest positive integer.
Since 1 is a positive integer, therefore $N \geq 1$.
Since N^2 is a positive integer, therefore $N^2 \leq N$.
Thus $N(N-1) \leq 0$ and we must have $N - 1 \leq 0$.
Therefore $N \leq 1$.
Since we also know $N \geq 1$, therefore $N = 1$.
Therefore 1 is the largest positive integer.

The only error we have made here is in the assumption (in the first line) that the problem has a solution. It is partly to avoid logical pitfalls like this that mathematicians prove existence theorems.

EXERCISES

In Exercises 1–6 state where in its domain the given function is continuous, where it is left or right continuous, and where it is discontinuous. Sketch its graph.

1. $f(x) = \begin{cases} x, & \text{if } x < 0 \\ x^2, & \text{if } x \geq 0 \end{cases}$

2. $f(x) = \begin{cases} -1, & \text{if } x > 1 \\ x, & \text{if } -1 \leq x \leq 1 \\ 1, & \text{if } x < -1 \end{cases}$

3. $f(x) = \begin{cases} 1 + x^2, & \text{if } x \neq 2 \\ 4.987, & \text{if } x = 2 \end{cases}$

4. $f(x) = \begin{cases} -3, & \text{if } x = -1 \\ \dfrac{x^2 - x - 2}{x + 1}, & \text{if } x \neq -1 \end{cases}$

5. $f(x) = \begin{cases} x, & \text{if } x \geq 0 \\ \dfrac{1}{x}, & \text{if } x < 0 \end{cases}$

6. $f(x) = \begin{cases} \dfrac{1}{x^4}, & \text{if } x \neq 0 \\ 0, & \text{if } x = 0 \end{cases}$

7. At what points does the parking cost function $C(t)$ of Exercise 51 in Section 1.6 fail to be continuous? Is it left or right continuous at those points?

Exercises 8–9 are are examples of optimization problems similar to Example 1.7.8. They can be solved by elementary means.

8. If the sum of two nonnegative numbers is 8, show that their product must be bounded. What is the largest possible value of their product?

9. Show that there is a point on the straight line with equation $y = (2 - 3x)/\sqrt{3}$ that is closest to the origin. Find that point. (*Hint:* This can be done geometrically (make a diagram), or else you can express the *square* of the distance from the point (x, y) to the origin as a function of x and complete the square as in Example 1.7.8.)

Find the intervals on which the functions $f(x)$ in Exercises 10–13 are positive and negative.

10. $f(x) = x^2 + 4x + 3$

11. $f(x) = \dfrac{x^2 - 1}{x}$

12. $f(x) = \dfrac{x^2 + x - 2}{x^3}$

13. $f(x) = \dfrac{x^2 - 1}{x^2 - 4}$

Exercises 14–17 involve applications of the Intermediate-Value Theorem (IVT) 1.7.10. A number r is called a zero of the function $f(x)$ if $f(r) = 0$.

14. If $f(x) = x^3 + x - 1$, show that f has a zero between $x = 0$ and $x = 1$.

15. Show that $f(x) = x^3 - 15x + 1$ has at least three zeroes in $[-4, 4]$.

16.*Show that $F(x) = (x - a)(x - b) + x$ takes on the value $\dfrac{a + b}{2}$ for some value of x.

17.*Suppose that $f(x)$ is continuous on $[0, 1]$ and satisfies $0 \leq f(x) \leq 1$ for every x in $[0, 1]$. Show that there exists c in $[0, 1]$ such that $f(c) = c$ (*Hint:* Apply IVT to $g(x) = f(x) - x$.)

18.*If an even function is continuous on the right at $x = 0$, show that it is continuous at $x = 0$.

19.*If an odd function is continuous on the right at $x = 0$, show that it is continuous at $x = 0$ and $f(0) = 0$.

CHAPTER 2

Differentiation

There are two fundamental problems considered in calculus. The **problem of slopes** is concerned with finding an equation of the straight line tangent to a given curve at a given point on that curve. The **problem of areas** is concerned with finding the area of a plane region bounded by curves and straight lines. The solution of the problem of slopes is the subject of **differential calculus**. As we will see, it has many and varied applications in mathematics and other disciplines. The problem of areas is the subject of *integral calculus*, which we begin in Chapter 6.

2.1 TANGENT LINES AND THEIR SLOPES

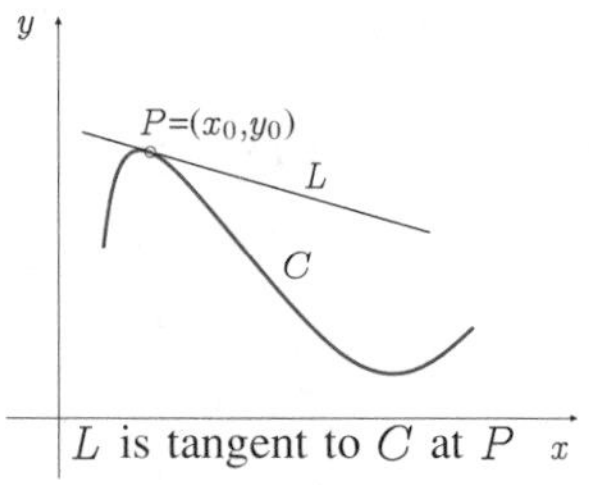

L is tangent to C at P

FIGURE 2.1.1

This section deals with the problem of finding a straight line L that is tangent to the curve C at the point P. As is often the case in mathematics, the most important step in the solution of such a fundamental problem is making a suitable definition.

For simplicity, and to avoid certain problems best postponed until later, we will not deal now with the most general kinds of curves; we will deal only with those that are the *graphs of continuous functions*. Let C be the graph of $y = f(x)$ and let P be the point (x_0, y_0) on C, so that $y_0 = f(x_0)$. We assume that P is not an endpoint of C; hence C extends some distance to the left and right of P (see Fig. 2.1.1).

What do we mean when we say that the line L is tangent to C at P? Past experience with tangent lines to circles does not help us to define tangency for more general curves. A tangent line to a circle (Fig. 2.1.2) has the following properties:

i) It meets the circle at only one point.

ii) The circle lies on only one side of the line.

iii) The tangent is perpendicular to the line joining the centre of the circle to the point of contact.

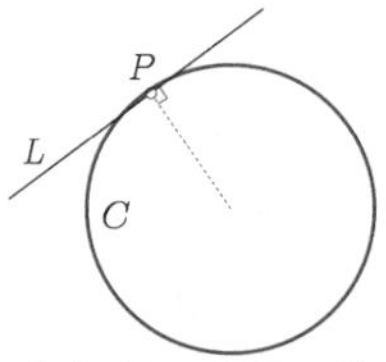

L is tangent to C at P

FIGURE 2.1.2

In general, curves do not have well-defined *centres*, so (iii) is useless for characterizing tangents to curves. The examples in Figs. 2.1.3 show that (i) and (ii) cannot be used to define tangency either. In particular, the lower right curve in Fig. 2.1.3 is not "smooth" at P so that curve should not have any tangent line there. A tangent line should have the "same direction" as the curve does at the point of tangency.

A reasonable definition of tangency can be stated in terms of limits. If Q is a point on C different from P, then the *secant line* PQ rotates around P as Q moves along the curve. If L is a line through P whose slope is the limit of the slopes of these secant lines PQ as Q approaches P along C (see Fig. 2.1.4), then L is tangent to C at P.

Since C is the graph of the *function* $y = f(x)$ vertical lines can meet C only once. Since $P = (x_0, f(x_0))$, a different point Q on the graph must have a different x-coordinate, say $x_0 + h$ where $h \neq 0$. Thus $Q = (x_0 + h, f(x_0 + h))$ and the slope of PQ is

$$\frac{f(x_0 + h) - f(x_0)}{h}.$$

This expression is called the **Newton quotient** for f at x_0. Note that h may be positive or negative depending on whether Q is to the right or left of P.

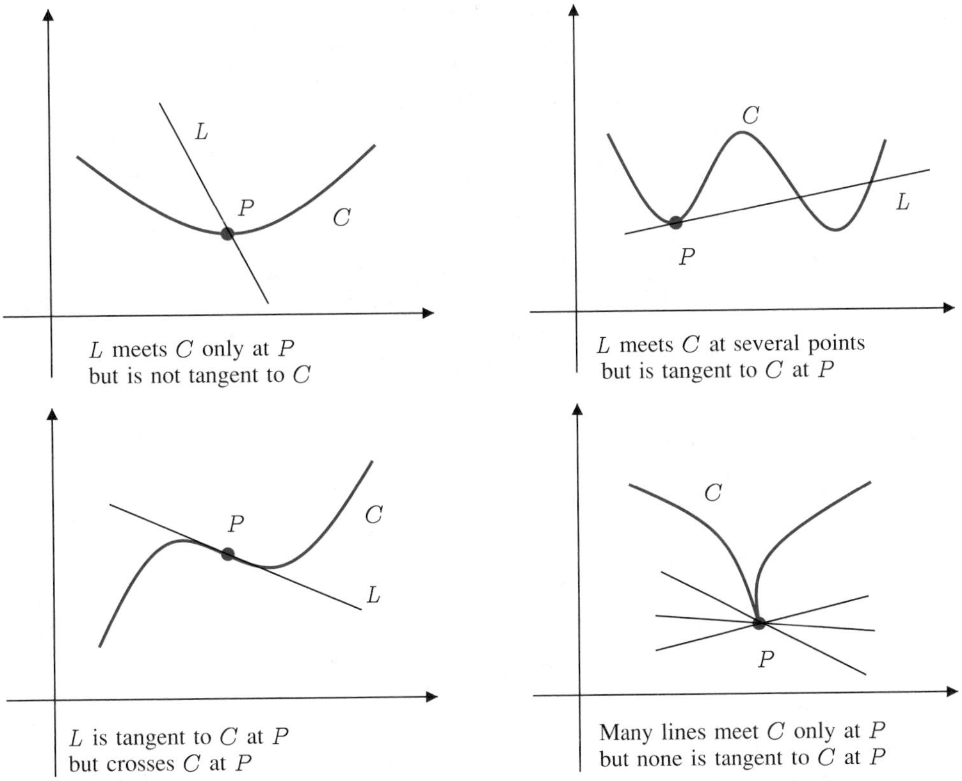
L
P
C
L meets C only at P
but is not tangent to C
C
L
P
L meets C at several points
but is tangent to C at P
P
C
L
L is tangent to C at P
but crosses C at P
C
P
Many lines meet C only at P
but none is tangent to C at P

FIGURE 2.1.3

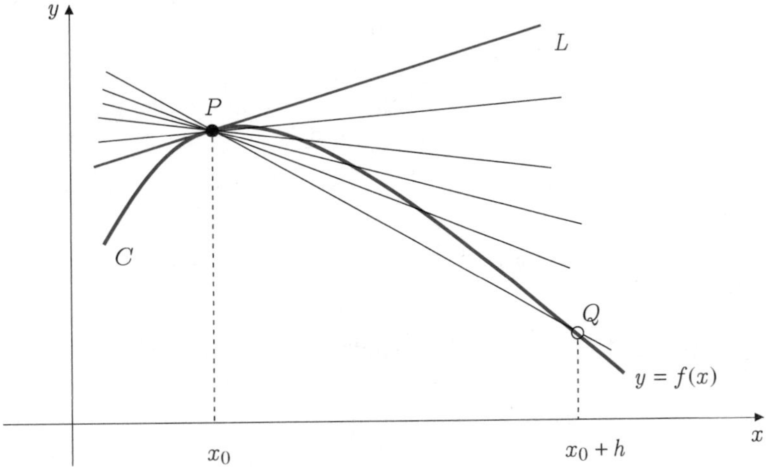
y
L
P
C
Q
y = f(x)
x
x0
x0 + h

FIGURE 2.1.4

The preceding discussion motivates us to make the following definition.

2.1.1
Non-vertical Tangent Lines

Suppose that the function f is continuous at $x = x_0$ and that

$$\lim_{h\to 0} \frac{f(x_0+h)-f(x_0)}{h} = m$$

exists. Then the straight line $y = y_0 + m(x - x_0)$, which has slope m and passes through the point $P = (x_0, f(x_0))$, is **tangent** to the graph of $y = f(x)$ at P.

EXAMPLE 2.1.2 Find an equation of the tangent line to the curve $y = x^2$ at the point $(1, 1)$.

SOLUTION Here $f(x) = x^2$, $x_0 = 1$, and $y_0 = f(1) = 1$. The required tangent has slope

$$\begin{aligned} m = \lim_{h\to 0} \frac{f(1+h)-f(1)}{h} &= \lim_{h\to 0} \frac{(1+h)^2-1}{h} \\ &= \lim_{h\to 0} \frac{2h+h^2}{h} \\ &= \lim_{h\to 0} 2+h = 2 \end{aligned}$$

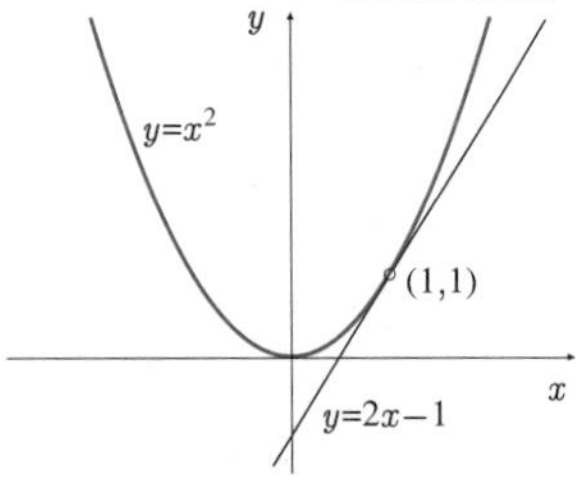

FIGURE 2.1.5

Accordingly, the equation of the tangent line at $(1, 1)$ is $y = 1 + 2(x - 1)$, or $y = 2x - 1$. See Fig. 2.1.5.

Definition 2.1.1 deals only with *non-vertical* tangent lines, that is, ones which have finite slope. It is possible for the graph of a continuous function to have a *vertical* tangent line.

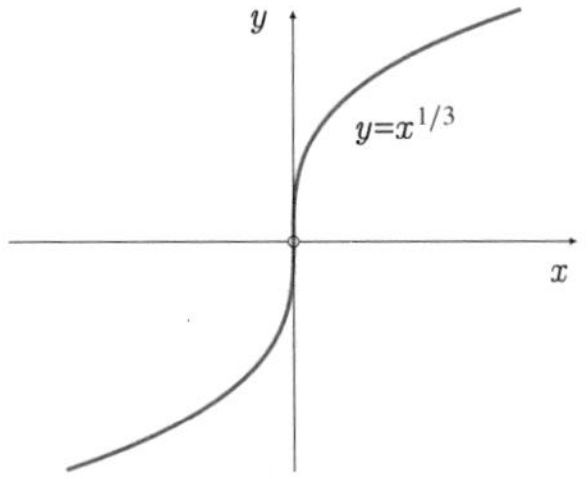

FIGURE 2.1.6

For instance, consider the graph of the function $f(x) = \sqrt[3]{x} = x^{1/3}$, which is shown in Fig. 2.1.6. The graph is a smooth curve and it seems evident that the y-axis is tangent to this curve at the origin. Let us try to calculate the limit of the Newton quotient for f at $x = 0$:

$$\lim_{h\to 0} \frac{f(0+h)-f(0)}{h} = \lim_{h\to 0} \frac{h^{1/3}}{h} = \lim_{h\to 0} h^{-2/3} = \infty.$$

Although the limit does not exist, the secant line joining the origin to another point Q on the curve has slope approaching infinity as Q approaches the origin.

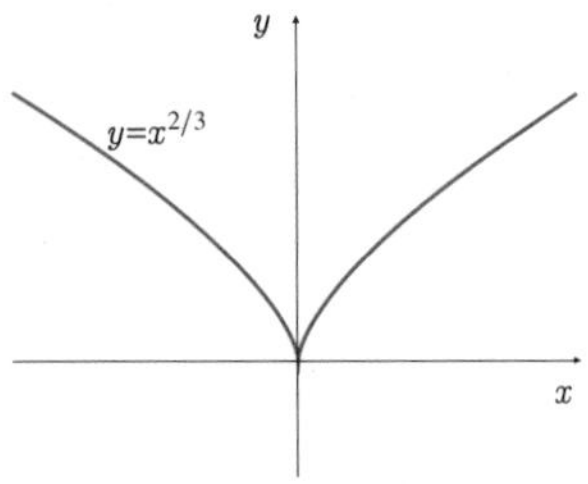

FIGURE 2.1.7

On the other hand, the function $f(x) = x^{2/3}$ whose graph is shown in Fig. 2.1.7, does not have a tangent line at the origin because it is not "smooth" there. In this case the Newton quotient is

$$\frac{f(0+h)-f(0)}{h} = \frac{h^{2/3}}{h} = h^{-1/3},$$

which has no limit as h approaches zero. (The right limit is ∞; the left limit is $-\infty$.) We say this curve has a **cusp** at the origin. A cusp is an infinitely sharp point; if you were travelling along the curve you would have to reverse your direction of travel by 180° at the origin.

The above analysis of $y = x^{1/3}$ and $y = x^{2/3}$ motivates the following extenstion to Definition 2.1.1.

2.1.3
Vertical Tangent Lines

If f is continuous at $P = (x_0, y_0)$, where $y_0 = f(x_0)$, and if either

$$\lim_{h\to 0} \frac{f(x_0+h) - f(x_0)}{h} = \infty \qquad \text{or} \qquad \lim_{h\to 0} \frac{f(x_0+h) - f(x_0)}{h} = -\infty$$

then the vertical line $y = y_0$ is tangent to the graph $y = f(x)$ at P. If the limit of the Newton quotient fails to exist in any other way than by being ∞ or $-\infty$ the graph $y = f(x)$ has no tangent line at P.

EXAMPLE 2.1.4 Does the graph of $y = |x|$ have a tangent line at $x = 0$?

SOLUTION The Newton quotient here is

$$\frac{|0+h| - |0|}{h} = \frac{|h|}{h} = \operatorname{sgn} h = \begin{cases} 1, & \text{if } h > 0 \\ -1, & \text{if } h < 0. \end{cases}$$

Since $\operatorname{sgn} h$ has different right and left limits at 0 (namely 1 and -1) the Newton quotient has no limit as $h \to 0$, so $y = |x|$ has no tangent line at $(0, 0)$. (See Fig. 2.1.8.) The graph does not have a cusp at the origin but it is kinked at that point; it suddently changes direction and is not smooth. Curves have tangent lines only at points where they are smooth. The graphs of $y = x^{2/3}$ and $y = |x|$ have tangent lines everywhere except at the origin where they are not smooth.

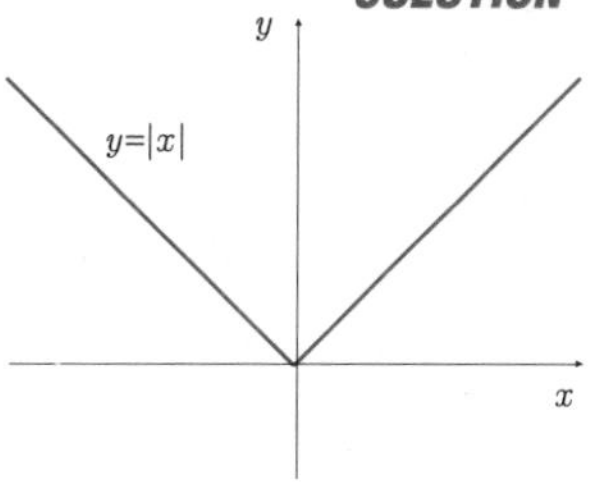

FIGURE 2.1.8

2.1.5
Definition of Slope of a Curve

The **slope** of a curve C at a point P is the slope of the tangent line to C at P if such a tangent line exists. In particular, the slope of the graph $y = f(x)$ at the point x_0 is

$$\lim_{h\to 0} \frac{f(x_0+h) - f(x_0)}{h}.$$

EXAMPLE 2.1.6 Find the slope of the curve $y = x/(3x+2)$ at the point $x = -2$.

SOLUTION If $x = -2$, then $y = 1/2$, so the required slope is

$$\lim_{h\to 0} \frac{\dfrac{-2+h}{3(-2+h)+2} - \dfrac{1}{2}}{h} = \lim_{h\to 0} \frac{-4+2h-(-4+3h)}{2(-4+3h)h}$$
$$= \lim_{h\to 0} \frac{-1}{2(-4+3h)} = \frac{1}{8}.$$

Normals

If a curve C has a tangent line L at point P, then the straight line N through P perpendicular to L is called the **normal** to C at P. If L is horizontal, then N is vertical and so it has infinite slope; if L is vertical, then N is horizontal and has 0 slope. If L is neither horizontal nor vertical, then the slope of N is the negative reciprocal of the slope of L:

$$\text{slope of the normal} = \frac{-1}{\text{slope of the tangent}}.$$

To see this, suppose that L has slope m and that N has slope p. Let P be the point (x_0, y_0). The vertical line $x = x_0 + 1$ meets L at $A = (x_0 + 1, y_0 + m)$, and it meets N at the point $B = (x_0 + 1, y_0 + p)$. (See Fig. 2.1.9.) The Pythagorean formula for the right-angled triangle APB says that $(AB)^2 = (AP)^2 + (PB)^2$, that is, $(m - p)^2 = (1 + m^2) + (1 + p^2)$. Simplifying this equation, we get $mp = -1$, which is equivalent to the assertion in the box.

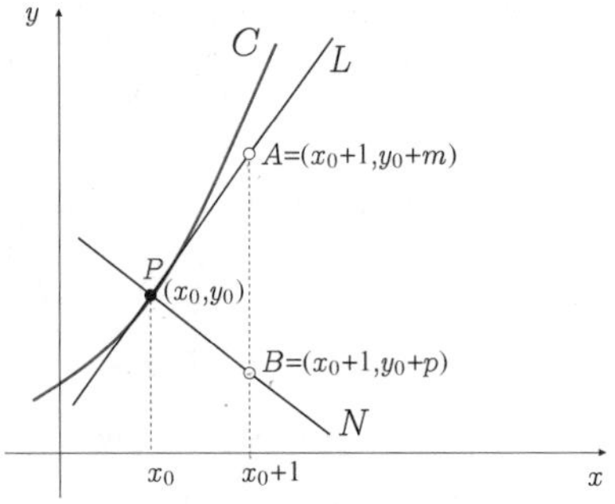

FIGURE 2.1.9

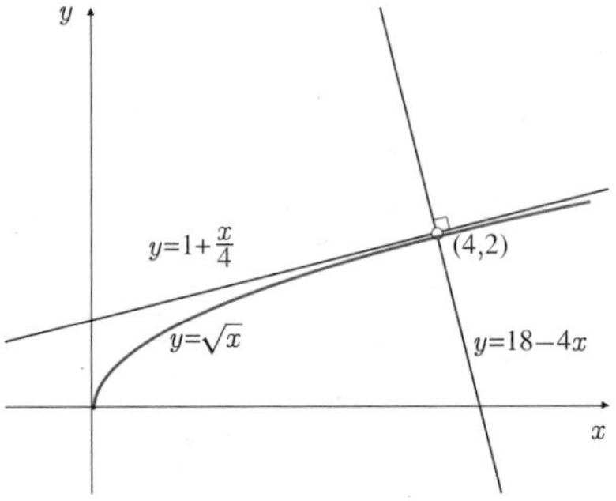

FIGURE 2.1.10

EXAMPLE 2.1.7 Find an equation of the normal to $y = x^2$ at $(1, 1)$.

SOLUTION By Example 2.1.2, the tangent line to $y = x^2$ at $(1, 1)$ has slope 2. Hence the normal has slope $-1/2$ and equation $y = 1 - \dfrac{x-1}{2}$, or $y = \dfrac{3}{2} - \dfrac{x}{2}$.

EXAMPLE 2.1.8 Find equations of the straight lines tangent and normal to the curve $y = \sqrt{x}$ at the point $(4, 2)$.

SOLUTION The slope of the tangent at (4, 2) (see Fig. 2.1.10) is

$$\begin{aligned}\lim_{h\to 0}\frac{\sqrt{4+h}-2}{h} &= \lim_{h\to 0}\frac{(\sqrt{4+h}-2)(\sqrt{4+h}+2)}{h(\sqrt{4+h}+2)}\\ &= \lim_{h\to 0}\frac{4+h-4}{h(\sqrt{4+h}+2)}\\ &= \lim_{h\to 0}\frac{1}{\sqrt{4+h}+2}=\frac{1}{4}.\end{aligned}$$

The tangent line has equation $y = 2 + \frac{1}{4}(x-4)$, or $y = \frac{x}{4} + 1$, and the normal line has the equation $y = 2 - 4(x-4)$, or $y = -4x + 18$.

EXERCISES

In Exercises 1–15 find an equation of the straight line tangent to the given curve at the point indicated.

1. $y = 2x^2 - 5$ at $(2, 3)$ **2.** $y = x^3 + 8$ at $x = 0$

3. $y = x^2 - 2x + 2$ at $(1, 1)$

4. $y = 6 - x - x^2$ at $x = -2$

✓**5.** $y = x^3 + 8$ at $x = -2$ **6.** $y = \dfrac{1}{x^2+1}$ at $(0, 1)$

7. $y = \dfrac{1}{x^2}$ at $x = 3$ **8.** $y = \dfrac{1}{x^2+1}$ at $x = -1$

✓**9.** $y = \sqrt{x+1}$ at $x = 3$ **10.** $y = \dfrac{1}{\sqrt{x}}$ at $x = 9$

11. $y = \dfrac{2x}{x+2}$ at $x = 2$ **12.** $y = \sqrt{5 - x^2}$ at $x = 1$

13. $y = x^2$ at $x = x_0$ **14.** $y = \dfrac{1}{x}$ at $\left(a, \dfrac{1}{a}\right)$

15. $y = ax^2 + bx + c$ at $x = x_0$

16. Find the slope of the curve $y = x^2 - 1$ at the point $x = x_0$. What is the equation of the tangent line to $y = x^2 - 1$ that has slope -3?

✓**17.** a) Find the slope of $y = x^3$ at the point $x = a$.
b) Find the equations of the straight lines having slope 3 that are tangent to $y = x^3$.

✓**18.** Find all points on the cirve $y = x^3 - 3x$ where the tangent line is parallel to the x-axis.

19.*If line L is tangent to curve C at point P then the smaller angle between L and the secant line PQ joining P to another point Q on C approaches 0 as Q approaches P along C. Is the converse true; if the angle between PQ and line L (which passes through P) approaches 0, must L be tangent to C?

20.*Let $P(x)$ be a polynomial. If a is a real number then $P(x)$ can be expressed in the form

$$P(x) = a_0 + a_1(x-a) + a_2(x-a)^2 + \cdots + a_n(x-a)^n$$

for some $n \geq 0$. If $\ell(x) = m(x-a) + b$, show that the straight line $y = \ell(x)$ is tangent to the graph of $y = P(x)$ at $x = a$ provided

$$P(x) - \ell(x) = (x-a)^2 Q(x)$$

where $Q(x)$ is a polynomial.

2.2 THE DERIVATIVE

A straight line has the property that its slope is the same at all points. For any other graph, however, the slope may vary from point to point. Thus the slope of the graph of $y = f(x)$ at the point x is itself a function of x. At any point x where the graph has a finite slope we say that f is differentiable, and we call the slope the derivative of f. The derivative is the limit of the Newton quotient $\big(f(x+h)-f(x)\big)/h$ as h approaches 0.

2.2.1 Definition of Derivative

> We say that the function f is **differentiable** at x if
>
> $$\lim_{h\to 0}\frac{f(x+h)-f(x)}{h}=f'(x)$$
>
> exists as a finite real number. The number $f'(x)$ is called **the derivative of f at x**. The *function* f' having domain the set of numbers x where f is differentiable, and value at x equal to $f'(x)$ is called **the derivative of f**.

The derivative of f is a function whose value at any point is the slope of the tangent line to the graph of f at that point. Thus the equation of the tangent line to $y=f(x)$ at $(x_0, f(x_0))$ is

$$y=f(x_0)+f'(x_0)(x-x_0).$$

The domain $\mathcal{D}(f')$ of f' may be smaller than the domain $\mathcal{D}(f)$ of f because it contains only those points in $\mathcal{D}(f)$ at which f is differentiable. Values of x in $\mathcal{D}(f)$ where f is not differentiable are called **singular points** of f. The graph of f' can often be sketched directly from that of f by visualizing slopes, a procedure called **graphical differentiation**. Indeed, the process of finding the derivative f' of a given function f is called **differentiation**. In Fig. 2.2.1 the graphs of f' and g' were sketched by inspecting the graphs of f and g. Note that -1 and 1 are singular points of f.

A function is differentiable on a set S if it is differentiable at every point x in S. If a function f is defined on a *closed* interval $[a,b]$ Definition 2.2.1 does not allow for the existence of a derivative at $x=a$ or $x=b$. (Why?) As we did for continuity in Section 1.7, we extend the definition to allow for a **right derivative** at $x=a$ and a **left derivative** at $x=b$:

$$f'(a+)=\lim_{h\to 0+}\frac{f(a+h)-f(a)}{h},\qquad f'(b-)=\lim_{h\to 0-}\frac{f(b+h)-f(b)}{h}.$$

We now say that f is differentiable on $[a,b]$ if $f'(x)$ exists for all x in (a,b) and $f'(a+)$ and $f'(b-)$ both exist.

We now give several examples of the computation of derivatives algebraically from the definition of derivative. Some of these are the basic building blocks from which more complicated derivatives can be calculated later. They are collected in a table at the end of the examples, and should be memorized.

EXAMPLE 2.2.2 Use the definition of derivative to calculate the derivatives of the following functions:

a) $f(x)=C$ (constant), b) $f(x)=ax+b$, c) $f(x)=x^2$,

d) $f(x)=\dfrac{1}{x}$, e) $f(x)=\sqrt{x}$.

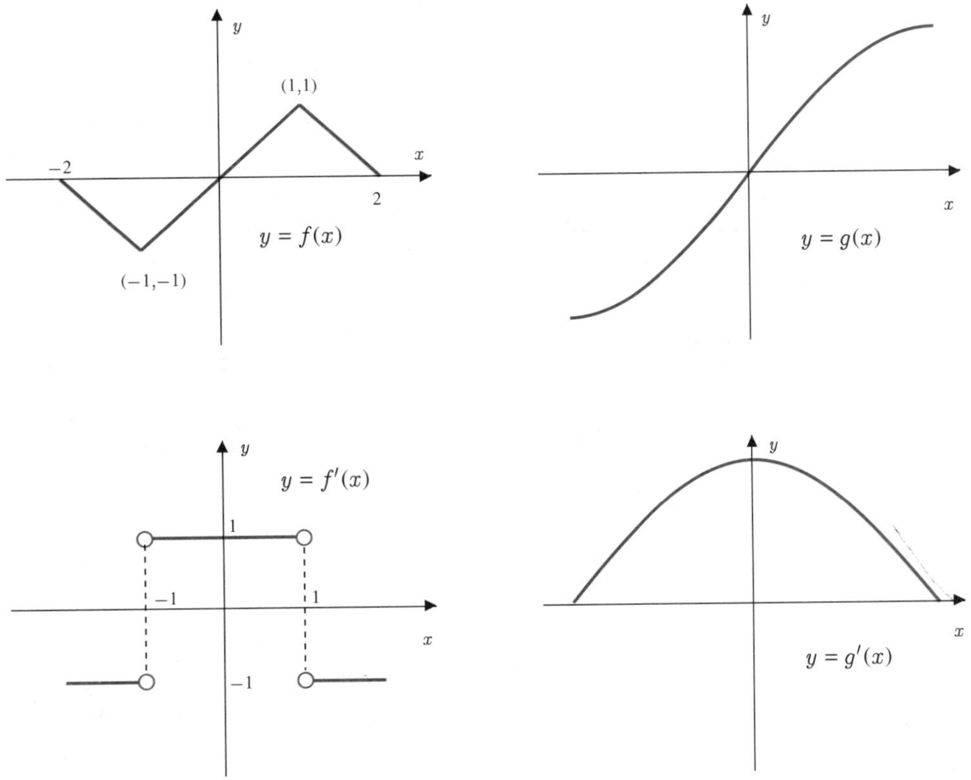

FIGURE 2.2.1

SOLUTION a) The graph $y = f(x) = C$ is a horizontal straight line, so it is evident geometrically that $f'(x) = 0$ for all x. (See Fig. 2.2.2.) Using the definition of derivative,

$$f'(x) = \lim_{h \to 0} \frac{f(x+h) - f(x)}{h} = \lim_{h \to 0} \frac{C - C}{h} = \lim_{h \to 0} 0 = 0.$$

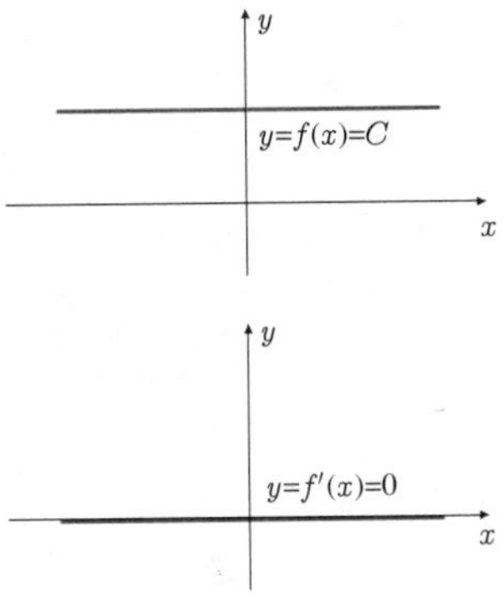

FIGURE 2.2.2

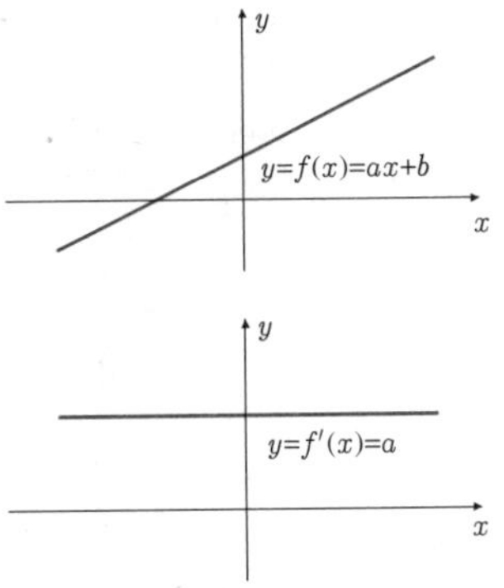

FIGURE 2.2.3

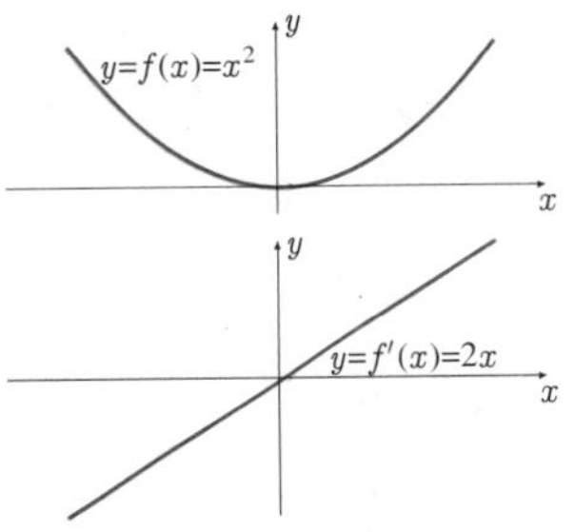

FIGURE 2.2.4

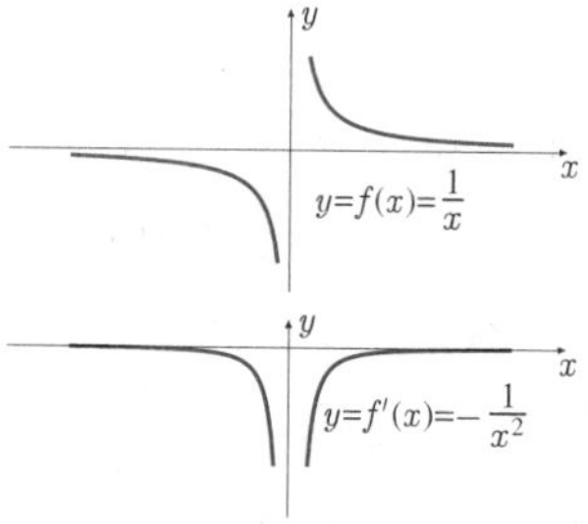

FIGURE 2.2.5

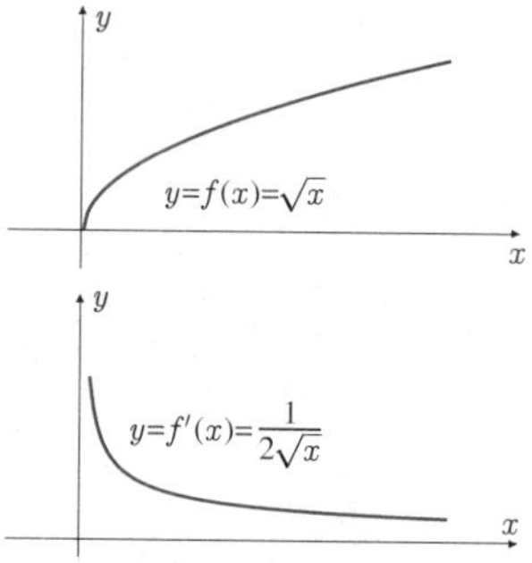

FIGURE 2.2.6

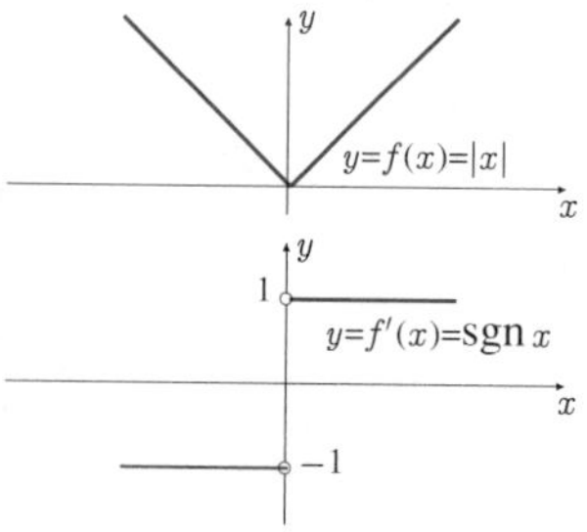

FIGURE 2.2.7

b) Again the answer is clear graphically; $y = f(x) = ax + b$ is a straight line with slope a, so $f'(x) = a$ for all x. (See Fig. 2.2.3.) By the definition,

$$\begin{aligned} f'(x) &= \lim_{h\to 0} \frac{f(x+h) - f(x)}{h} \\ &= \lim_{h\to 0} \frac{a(x+h) + b - (ax+b)}{h} \\ &= \lim_{h\to 0} \frac{ah}{h} = a. \end{aligned}$$

c) If $f(x) = x^2$, we have, for all x (see Fig. 2.2.4),

$$f'(x) = \lim_{h\to 0} \frac{(x+h)^2 - x^2}{h} = \frac{2hx + h^2}{h} = \lim_{h\to 0}(2x + h) = 2x$$

d) If $f(x) = \dfrac{1}{x}$, we have, for all $x \neq 0$ (see Fig. 2.2.5),

$$\begin{aligned} f'(x) = \lim_{h\to 0} \frac{\dfrac{1}{x+h} - \dfrac{1}{x}}{h} &= \lim_{h\to 0} \frac{\dfrac{x - (x+h)}{(x+h)x}}{h} \\ &= \lim_{h\to 0} \frac{-h}{hx(x+h)} = -\frac{1}{x^2} \end{aligned}$$

e) For $f(x) = \sqrt{x}$, $(x \geq 0)$, we calculate

$$\begin{aligned} f'(x) &= \lim_{h\to 0} \frac{\sqrt{x+h} - \sqrt{x}}{h} \\ &= \lim_{h\to 0} \frac{(\sqrt{x+h} - \sqrt{x})(\sqrt{x+h} + \sqrt{x})}{h(\sqrt{x+h} + \sqrt{x})} \\ &= \lim_{h\to 0} \frac{x + h - x}{h(\sqrt{x+h} + \sqrt{x})} \\ &= \lim_{h\to 0} \frac{1}{\sqrt{x+h} + \sqrt{x}} = \frac{1}{2\sqrt{x}} \end{aligned}$$

Note that f is not differentiable at $x = 0$. (See Fig.2.2.6.) Since 0 is in $\mathcal{D}(f)$, it is a singular point of f.

EXAMPLE 2.2.3 Find $f'(x)$ if $f(x) = |x|$.

SOLUTION We have $f(x) = \begin{cases} x, & \text{if } x \geq 0 \\ -x, & \text{if } x < 0 \end{cases}$. Thus, from part b) above, $f'(x) = 1$ if $x > 0$ and $f'(x) = -1$ if $x < 0$. Also, Example 2.1.4 shows that f is not differentiable at $x = 0$; 0 is a singular point of f.

Therefore $f'(x) = \begin{cases} 1, & \text{if } x > 0 \\ -1, & \text{if } x < 0 \end{cases} = \operatorname{sgn} x$. (See Fig. 2.2.7.)

Some Elementary Functions and Their Derivatives

$f(x)$	$f'(x)$
1	0
x	1
x^2	$2x$
$\frac{1}{x}$	$-\frac{1}{x^2}$ $(x \neq 0)$
$\sqrt{x}$	$\frac{1}{2\sqrt{x}}$ $(x > 0)$
$\|x\|$	$\operatorname{sgn} x$ $(x \neq 0)$

The first five functions $f(x)$ in the table are all of the form x^r for various values of r. The results in the table suggest the following **General Power Rule** for differentiating powers of x:

2.2.4
The General Power Rule

> If $f(x) = x^r$, then $f'(x) = rx^{r-1}$.

We will see later that the General Power Rule is true whenever it "makes sense," that is, for all real values of x and r for which x^r and x^{r-1} are well defined. This includes all real numbers x if r is a positive integer and all $x \neq 0$ if r is a negative integer. For now, let us prove the rule for $r = n$, a positive integer, by using the factorization of a difference of nth powers:

2.2.5
Factoring a Difference of n'th Powers

$$a^n - b^n = (a - b)(a^{n-1} + a^{n-2}b + a^{n-3}b^2 + \cdots + a^2b^{n-3} + ab^{n-2} + b^{n-1}).$$

(Check this formula by multiplying the two factors on the right-hand side.) We have, for $f(x) = x^n$,

$$f'(x) = \lim_{h\to 0} \frac{(x+h)^n - x^n}{h}$$

$$= \lim_{h\to 0} \frac{h\,\overbrace{[(x+h)^{n-1} + (x+h)^{n-2}x + (x+h)^{n-3}x^2 + \cdots + x^{n-1}]}^{n \text{ terms}}}{h} = nx^{n-1},$$

as required. An alternative proof using the binomial theorem is suggested in Exercise 39 at the end of this section. (Yet another proof can be based on the product rule; see Example 2.3.5(c).) The factorization method used above can also be used to demonstrate the general power rule for negative integers, $r = -n$, and reciprocals of integers, $r = 1/n$. (See Exercises 36–38 at the end of this section.)

Leibniz Notation

It is useful to have more than one notation for derivatives. If $y = f(x)$ we can use the dependent variable y to represent the function, and we can denote the derivative of the function in any of the following ways:

2.2.6
Notations for Derivatives

$$D_x y = y' = \frac{dy}{dx} = \frac{d}{dx}f(x) = f'(x) = Df(x).$$

These should be read "the derivative of y" or "y primed," or "the derivative of $f(x)$." When the independent variable x appears explicitly as it does in all the forms except y', we can say "the derivative of y with respect to x" or "the derivative of $f(x)$ with respect to x." Often the most convenient way of referring to the derivative of a function given explicitly as an expression in the variable x is to write d/dx in front of that expression. The symbol d/dx is a *differential operator*, and should be read "the derivative with respect to x of."

$$\frac{d}{dx}x^2 = 2x$$
$$\frac{d}{dx}\sqrt{x} = \frac{1}{2\sqrt{x}}$$
$$\frac{d}{dt}t^{100} = 100t^{99}$$
$$\text{if } y = u^3, \text{ then } \frac{dy}{du} = 3u^2$$

The value of the derivative of a function at a particular number x_0 in its domain can also be expressed in several ways:

$$D_x y\Big|_{x=x_0} = y'\Big|_{x=x_0} = \frac{dy}{dx}\Big|_{x=x_0} = \frac{d}{dx}f(x)\Big|_{x=x_0} = f'(x_0) = Df(x_0).$$

The symbol $\Big|_{x=x_0}$ is called an *evaluation symbol.* It signifies that the expressi preceding it should be evaluated at $x = x_0$. Thus

$$\frac{d}{dx}x^4\Big|_{x=-1} = 4x^3\Big|_{x=-1} = 4(-1)^3 = -4.$$

Here is another example in which a derivative is computed from the definition, this time for a somewhat more complicated function.

EXAMPLE 2.2.7 Use the definition of derivative to calculate $\dfrac{d}{dx}\left(\dfrac{x}{x^2+1}\right)$.

SOLUTION

$$\begin{aligned}
\frac{d}{dx}\left(\frac{x}{x^2+1}\right) &= \lim_{h\to 0}\frac{\dfrac{x+h}{(x+h)^2+1} - \dfrac{x}{x^2+1}}{h} \\
&= \lim_{h\to 0}\frac{(x+h)(x^2+1) - x(x^2+2hx+h^2+1)}{h((x+h)^2+1)(x^2+1)} \\
&= \lim_{h\to 0}\frac{x^3+x+hx^2+h-x^3-2hx^2-h^2x-x}{h((x+h)^2+1)(x^2+1)} \\
&= \lim_{h\to 0}\frac{1-x^2-hx}{((x+h)^2+1)(x^2+1)} = \frac{1-x^2}{(1+x^2)^2}
\end{aligned}$$

The notations dy/dx and $(d/dx)f(x)$ are called Leibniz notations for the derivative, after Gottfried Wilhelm Leibniz (1646–1716), one of the creators of calculus, who used such notations. The main ideas of calculus were discovered independently by Leibniz and Isaac Newton (1642–1727); the latter used notations similar to the prime (y') notations we use here. The Leibniz notation is suggested by the definition of derivative. The Newton quotient $[f(x+h) - f(x)]/h$ whose limit we take to find the derivative dy/dx can be written in the form $\Delta y/\Delta x$ where $\Delta y = f(x+h) - f(x)$ is the vertical displacement and $\Delta x = (x+h) - x = h$ is the horizontal displacement as we pass from the point $(x, f(x))$ to the point $(x+h, f(x+h))$ on the graph of f. (See Fig. 2.2.8.) Symbolically,

$$\frac{dy}{dx} = \lim_{\Delta x\to 0}\frac{\Delta y}{\Delta x}.$$

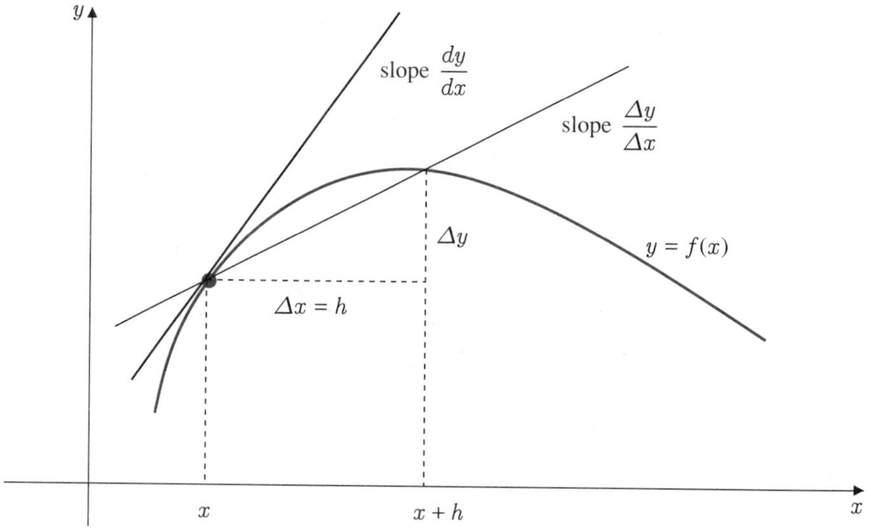

FIGURE 2.2.8

Differentials

!!DANGER!!

Many students find differentials confusing. We won't need them until Section 5.3. You can skip this for now.

The Newton quotient $\Delta y/\Delta x$ is actually the quotient of two quantities, Δy and Δx. It is not at all clear, however, that the derivative dy/dx, the limit of $\Delta y/\Delta x$ as Δx approaches zero, can be regarded as a quotient. If y is a continuous function of x, then Δy approaches zero when Δx approaches zero, so dy/dx appears to be the quotient $0/0$, which is meaningless. Nevertheless, it is sometimes useful to be able to refer to quantities dy and dx in such a way that their quotient is the derivative dy/dx. We can justify this by regarding dx as a new *independent* variable (called **the differential of** x) and defining a new *dependent* variable dy (**the differential of** y) as a function of x and dx by

2.2.8
Definition of dy in Terms of dx

$$dy = \frac{dy}{dx}dx = f'(x)\,dx.$$

For example, if $y = x^2$, we can write $dy = 2x\,dx$ to mean the same thing as $dy/dx = 2x$. This *differential formalism* will be used for the interpretation and manipulation of integrals beginning in Chapter 6. We will make little use of it until then.

Note that, defined as above, differentials are merely variables that may or may not be small in absolute value. The differentials dy and dx were originally regarded (by Leibniz and his successors) as "infinitesimals" (infinitely small but nonzero) quantities whose quotient dy/dx gave the slope of the tangent line, (a secant line meeting the graph of $y = f(x)$ at two points infinitely close together).

It can be shown that such "infinitesimal" quantities cannot exist (as real numbers). It is possible to extend the number system to contain infinitesimals and use these to develop calculus, but we will not consider this approach here.

EXERCISES

In Exercises 1–16 calculate the derivative of the given function directly from the definition of derivative.

1. $y = x^2 - 3x$

2. $f(x) = 1 + 4x - 5x^2$

3. $f(x) = x^3$

4. $y = \frac{1}{3}x^3 - x$

5. $g(x) = \frac{2 - x}{2 + x}$

6. $s = \frac{1}{3 + 4t}$

7. $F(t) = \sqrt{2t + 1}$

8. $f(x) = \frac{3}{4}\sqrt{2 - x}$

9. $y = x + \frac{1}{x}$

10. $z = \frac{s}{1 + s}$

11. $H(t) = \frac{1 + \sqrt{t}}{t}$

12. $y = \frac{1}{x^2}$

13. $y = \frac{1}{\sqrt{1 + x}}$

14. $f(t) = \frac{t^2 - 3}{t^2 + 3}$

15. $F(x) = \frac{1}{\sqrt{1 + x^2}}$

16. $f(x) = \sqrt{\frac{x - 1}{x + 1}}$

17. Where are the functions $\operatorname{sgn} x$, $x \operatorname{sgn} x$ and $x^2 \operatorname{sgn} x$ differentiable? What are their derivatives?

18. Using a calculator, find the slope of the secant line to $y = x^3 - 2x$ passing through the points corresponding to $x = 1$ and $x = 1 + \Delta x$ for several values of Δx of decreasing size, say $\Delta x = \pm 0.1, \pm 0.01, \pm 0.001, \pm 0.0001$. (Make a table.) Also calculate $\left.\frac{d}{dx}\left(x^3 - 2x\right)\right|_{x=1}$.

19. Repeat the previous exercise for the function $f(x) = \frac{1}{x}$ and the points $x = 2$ and $x = 2 + \Delta x$.

Using the definition of derivative, find equations for the tangent lines to the curves in Exercises 20–23 at the points indicated.

20. $y = 5 + 4x - x^2$ at the point where $x = 2$

21. $y = \sqrt{x + 6}$ at the point $(3, 3)$

22. $y = \frac{t}{t^2 - 2}$ at the point where $t = -2$

23. $y = \frac{2}{t^2 + t}$ at the point where $t = a$.

In Exercises 24–34 you may make use of the formulas for derivatives established in the section.

24. Find $\frac{d}{dx}x^{17}$.

25. Find $\frac{dy}{dt}$ if $y = t^{22}$.

26. Calculate $\left.\frac{d}{ds}\sqrt{s}\right|_{s=9}$.

27. Find $F'(\frac{1}{4})$ if $F(x) = \frac{1}{x}$.

28. Find an equation of the straight line tangent to the curve $y = \sqrt{x}$ at $x = x_0$.

29. Find an equation of the straight line normal to the curve $y = 1/x$ at the point where $x = a$.

30. Show that the curve $y = x^2$ and the straight line $x + 4y = 18$ intersect at right angles at one of their two intersection points. (*Hint:* find the product of their slopes at their intersection points.)

31. There are two distinct straight lines that pass through the point $(1, -3)$ and are tangent to the curve $y = x^2$. Find their equations. (*Hint:* Draw a sketch. The points of tangency are not given; let them be denoted (a, a^2).)

32. Find equations of two straight lines that have slope -2 and are tangent to the graph of $y = 1/x$.

33. Find the slope of a straight line which passes through the point $(-2, 0)$ and is tangent to the curve $y = \sqrt{x}$.

34.*Show that there are two distinct tangent lines to the curve $y = x^2$ passing through the point (a, b) provided $b < a^2$. How many tangent lines to $y = x^2$ pass through (a, b) if $b = a^2$? If $b > a^2$?

35.*Show that the derivative of an odd differentiable function is even, and that the derivative of an even differentiable function is odd.

36.*Prove the case $r = -n$, (n is a positive integer), of the General Power Rule; that is, prove that $\frac{d}{dx}x^{-n} = -n\,x^{-n-1}$. Use the factorization of a difference of nth powers given in box 2.2.5.

37. *Make use of the factoring of a difference of cubes

$$a^3 - b^3 = (a - b)(a^2 + ab + b^2)$$

to help you calculate the derivative of $f(x) = x^{1/3}$ directly from the definition of derivative.

38. *Prove the General Power Rule for $(d/dx)x^r$ where $r = 1/n$, n being a positive integer. (*Hint:*

$$\frac{d}{dx}x^{1/n} = \lim_{h \to 0} \frac{(x+h)^{1/n} - x^{1/n}}{h}$$
$$= \lim_{h \to 0} \frac{(x+h)^{1/n} - x^{1/n}}{((x+h)^{1/n})^n - (x^{1/n})^n}.$$

Apply the factorization of the difference of nth powers to the denominator of the latter quotient.)

39. Give a proof of the power rule $(d/dx)x^n = nx^{n-1}$ for positive integers n by using the Binomial Theorem

$$(x+h)^n = x^n + \frac{n}{1}x^{n-1}h + \frac{n(n-1)}{1 \times 2}x^{n-2}h^2 + \frac{n(n-1)(n-2)}{1 \times 2 \times 3}x^{n-3}h^3 + \cdots + h^n.$$

(See Exercise 13 of Appendix I.)

40. *Use right and left derivatives, $f'(a+)$ and $f'(a-)$, to define the concept of a half-line starting at $(a, f(a))$ being a right or left tangent to the graph of f at $x = a$. Show that the graph has a tangent line at $x = a$ if and only if it has right and left tangents which are opposite halves of the same straight line. What are the left and right tangents to the graphs of $y = x^{2/3}$ and $y = |x|$ at $x = 0$? (See Figures 2.1.7 and 2.1.8 in the previous section.)

2.3 DIFFERENTIATION RULES (SUMS, PRODUCTS, QUOTIENTS)

If every derivative had to be calculated directly from the definition of the derivative as in the examples of the previous section, calculus would indeed be a laborious subject. Fortunately there is an easier way. In the next two sections we present several general *differentiation rules* that enable us to calculate easily the derivatives of complicated combinations of functions if we already know the derivatives of the elementary functions from which they are constructed. For instance, we will be able to find the derivative of $\dfrac{x^2}{\sqrt{x^2+1}}$ if we know the derivatives of x^2 and $\sqrt{x}$. The rules we develop tell us how to differentiate sums, constant multiples, products, quotients and compositions of functions whose derivatives we already know.

Before developing these differentiation rules we need to establish one obvious but very important theorem which states, roughly, that the graph of a function has no break at a point where it is smooth enough to have a tangent line.

THEOREM 2.3.1 If f is differentiable at x, then f is continuous at x.

PROOF Since f is differentiable at x, we know that

$$\lim_{h \to 0} \frac{f(x+h) - f(x)}{h} = f'(x)$$

exists. In order to prove that f is continuous at x, we need to show that

$$\lim_{h \to 0} f(x+h) = f(x).$$

Using the laws of limits (Theorem 1.5.7), we have

$$\lim_{h\to 0} f(x+h) = \lim_{h\to 0}\left(f(x) + \frac{f(x+h)-f(x)}{h}h\right) = f(x) + f'(x)(0) = f(x). \square$$

Sums and Constant Multiples

2.3.2
The Derivative of a Sum or Constant Multiple

> If f and g are functions differentiable at x and if C is a constant, then the functions $f+g$ and Cf are both differentiable at x and
>
> $$(f+g)'(x) = f'(x) + g'(x), \qquad (Cf)'(x) = Cf'(x).$$

Thus the derivative of a sum of functions is the sum of their derivatives and the derivative of a constant multiple of a function is the same constant multiple of the derivative of the function.

The proofs of these rules are straightforward, using corresponding properties of limits from Theorem 1.5.7:

$$\begin{aligned}(f+g)'(x) &= \lim_{h\to 0}\frac{(f(x+h)+g(x+h))-(f(x)+g(x))}{h}\\ &= \lim_{h\to 0}\left(\frac{f(x+h)-f(x)}{h}+\frac{g(x+h)-g(x)}{h}\right)\\ &= f'(x)+g'(x)\end{aligned}$$

because the limit of a sum is the sum of the limits. Similarly,

$$(Cf)'(x) = \lim_{h\to 0}\frac{Cf(x+h)-Cf(x)}{h} = \lim_{h\to 0} C\frac{f(x+h)-f(x)}{h} = Cf'(x).$$

The above rule for differentiating sums extends to sums of any finite number of terms. In particular, therefore, the derivative of any polynomial is the sum of the derivatives of its terms.

EXAMPLE 2.3.3

a) $\dfrac{d}{dx}(x^3+x^2+x+1) = 3x^2+2x+1$

b) If $f(x) = 5\sqrt{x} + \dfrac{3}{x} - 18$, then $f'(x) = \dfrac{5}{2\sqrt{x}} - \dfrac{3}{x^2}$.

c) If $y = \frac{1}{7}t^3 - \frac{1}{3}t^7$, then $\dfrac{dy}{dt} = \frac{3}{7}t^2 - \frac{7}{3}t^6$.

d) We find the equation of the tangent line to the curve $y = 3x^2 - \dfrac{4}{x}$ at the point where $x=-2$. At this point we have $y = 14$ and the slope of the curve is

$$\left.\frac{dy}{dx}\right|_{x=-2} = \left.\left(6x+\frac{4}{x^2}\right)\right|_{x=-2} = -11.$$

Thus the tangent line has equation $y = 14 - 11(x+2)$, or $y = -11x - 8$.

The Product Rule

2.3.4
The Product Rule

> If functions f and g are differentiable at x, then their product fg is also differentiable at x, and
>
> $$(fg)'(x) = f'(x)g(x) + f(x)g'(x).$$

Note that the derivative of a product is *not* the product of the derivatives.

PROOF We set up the Newton quotient for fg and then add 0 to the numerator in a way that enables us to involve the Newton quotients for f and g separately:

$$\begin{aligned}(fg)'(x) &= \lim_{h\to 0}\frac{f(x+h)g(x+h) - f(x)g(x)}{h} \\ &= \lim_{h\to 0}\frac{f(x+h)g(x+h) - f(x+h)g(x) + f(x+h)g(x) - f(x)g(x)}{h} \\ &= \lim_{h\to 0}\left(f(x+h)\frac{g(x+h)-g(x)}{h} + g(x)\frac{f(x+h)-f(x)}{h}\right) \\ &= f(x)g'(x) + g(x)f'(x).\end{aligned}$$

To get the last line we have used the fact that g and f are differentiable and the fact that f is therefore continuous (Theorem 2.3.1), as well as properties of limits from Theorem 1.5.7. □

EXAMPLE 2.3.5 a) $\frac{d}{dx}(x^2+1)(x^3+4) = (2x)(x^3+4) + (x^2+1)(3x^2) = 5x^4 + 3x^2 + 8x.$
Of course, we could do this without the product rule by first multiplying the factors in the product:

$$\frac{d}{dx}\left((x^2+1)(x^3+4)\right) = \frac{d}{dx}(x^5 + x^3 + 4x^2 + 4) = 5x^4 + 3x^2 + 8x.$$

b) If $y = \left(2\sqrt{x} + \frac{3}{x}\right)\left(3\sqrt{x} - \frac{2}{x}\right)$, then

$$\begin{aligned}\frac{dy}{dx} &= \left(\frac{1}{\sqrt{x}} - \frac{3}{x^2}\right)\left(3\sqrt{x} - \frac{2}{x}\right) + \left(2\sqrt{x} + \frac{3}{x}\right)\left(\frac{3}{2\sqrt{x}} + \frac{2}{x^2}\right) \\ &= 6 - \frac{5}{2}x^{-3/2} + \frac{12}{x^3}.\end{aligned}$$

c) The product rule can be used to show that $(d/dx)x^n = nx^{n-1}$ for $n = 2, 3, 4, \ldots$
Since $\frac{d}{dx}x = 1$ we have

$$\frac{d}{dx}x^2 = \frac{d}{dx}(xx) = (1)(x) + (x)(1) = 2x.$$

Therefore,

$$\frac{d}{dx}x^3 = \frac{d}{dx}(x^2x) = (2x)(x) + (x^2)(1) = 3x^2,$$
$$\frac{d}{dx}x^4 = \frac{d}{dx}(x^3x) = (3x^2)(x) + (x^3)(1) = 4x^3,$$

and so on in the obvious way. For an indication of how to base a formal proof on this observed pattern, see Appendix I.

The product rule can be extended to products of any number of factors, for instance:

$$\begin{aligned}(fgh)'(x) &= f'(x)(gh)(x) + f(x)(gh)'(x)\\ &= f'(x)g(x)h(x) + f(x)g'(x)h(x) + f(x)g(x)h'(x).\end{aligned}$$

In general, the derivative of a product of n functions will have n terms; each term will be the same product but with one of the factors replaced by its derivative:

$$(f_1f_2f_3\cdots f_n)' = f_1'f_2f_3\cdots f_n + f_1f_2'f_3\cdots f_n + f_1f_2f_3'\cdots f_n + f_1f_2f_3\cdots f_n'.$$

This can be proved by *mathematical induction*. See Appendix I.

The Reciprocal Rule

2.3.6
The Reciprocal Rule

If f is differentiable at x and $f(x) \neq 0$ then $1/f$ is differentiable at x and

$$\left(\frac{1}{f}\right)'(x) = -\frac{f'(x)}{(f(x))^2}.$$

PROOF

$$\begin{aligned}\frac{d}{dx}\frac{1}{f(x)} &= \lim_{h\to 0}\frac{\dfrac{1}{f(x+h)} - \dfrac{1}{f(x)}}{h}\\ &= \lim_{h\to 0}\frac{f(x)-f(x+h)}{hf(x+h)f(x)}\\ &= \lim_{h\to 0}\left(-\frac{1}{f(x+h)f(x)}\right)\frac{f(x+h)-f(x)}{h}\\ &= -\frac{1}{(f(x))^2}f'(x)\end{aligned}$$

Again we have to use the continuity of f (Theorem 2.3.1), and our theorem on limits of products and quotients. □

EXAMPLE 2.3.7 a) $\frac{d}{dx}\left(\frac{1}{x^2+1}\right) = -\frac{2x}{(x^2+1)^2}$

b) $\frac{d}{dx}x^{-1/2} = \frac{d}{dx}\frac{1}{\sqrt{x}} = -\frac{1/(2\sqrt{x})}{(\sqrt{x})^2} = -\frac{1}{2x^{3/2}}$ Note that this result also fits the pattern of the General Power Rule.

c) We can also use the Reciprocal Rule to extend the General Power Rule to arbitrary negative integer powers $r = -n$:

$$\frac{d}{dx}x^{-n} = \frac{d}{dx}\left(\frac{1}{x^n}\right) = -\frac{nx^{n-1}}{(x^n)^2} = -nx^{-n-1}.$$

For instance,

$$\frac{d}{dx}\left(\frac{1}{x}+\frac{1}{x^2}+\frac{1}{x^3}\right) = \frac{d}{dx}(x^{-1}+x^{-2}+x^{-3})$$
$$= -x^{-2} - 2x^{-3} - 3x^{-4} = -\frac{1}{x^2} - \frac{2}{x^3} - \frac{3}{x^4}.$$

The Quotient Rule

2.3.8
The Quotient Rule

If f and g are differentiable at x, and if $g(x) \neq 0$, then the quotient f/g is differentiable at x and

$$\left(\frac{f}{g}\right)'(x) = \frac{g(x)f'(x) - f(x)g'(x)}{(g(x))^2}.$$

Sometimes students have trouble remembering this rule because of the negative sign in the numerator. Try to remember and use it in the following form:

$$(\text{quotient})' = \frac{(\text{denominator}) \times (\text{numerator})' - (\text{numerator}) \times (\text{denominator})'}{(\text{denominator})^2}$$

PROOF This is just an application of the product and reciprocal rules.

$$\frac{d}{dx}\left(\frac{f(x)}{g(x)}\right) = \frac{d}{dx}\left(f(x)\frac{1}{g(x)}\right) = f'(x)\frac{1}{g(x)} + f(x)\left(-\frac{g'(x)}{(g(x))^2}\right)$$
$$= \frac{g(x)f'(x) - f(x)g'(x)}{(g(x))^2}$$

Of course the reciprocal rule is really just a special case of the quotient rule. □

EXAMPLE 2.3.9 a) $\frac{d}{dx}\left(\frac{1-x^2}{1+x^2}\right) = \frac{(1+x^2)(-2x)-(1-x^2)(2x)}{(1+x^2)^2} = -\frac{4x}{(1+x^2)^2}$

b) If $f(t) = \frac{\sqrt{t}}{3-5t}$, then

$$f'(t) = \frac{(3-5t)\frac{1}{2\sqrt{t}} - \sqrt{t}(-5)}{(3-5t)^2} = \frac{3+5t}{2\sqrt{t}(3-5t)^2}.$$

c) If $y = \frac{a+b\theta}{m+n\theta}$, then $\frac{dy}{d\theta} = \frac{(m+n\theta)(b)-(a+b\theta)(n)}{(m+n\theta)^2} = \frac{mb-na}{(m+n\theta)^2}$.

d) If $F(r) = \frac{(1-4r^3)\sqrt{r}}{2r^2+1}$, find $F'(1)$.

$$\begin{aligned} F'(1) &= \left.\frac{(2r^2+1)\left(-12r^2\sqrt{r}+(1-4r^3)\frac{1}{2\sqrt{r}}\right)-(1-4r^3)\sqrt{r}(4r)}{(2r^2+1)^2}\right|_{r=1} \\ &= \frac{3(-12-(3/2))-(-3)(4)}{9} = -\frac{19}{6} \end{aligned}$$

Note the use of the product rule for differentiating the numerator. Note also that the evaluation takes place immediately after the derivative is calculated, before any simplification is done. It is easier to simplify a numerical expression than an algebraic one.

In Examples 2.3.9(a)–(c) above, the differentiation was accomplished immediately (after the first equal sign) and the result was then simplified by algebraic methods. The quotient rule frequently leads to fractions with numerators that look very complicated. These numerators can often be simplified algebraically (by expanding and factoring). You are well advised to attempt such simplification at all times; the usefulness of derivatives in applications of calculus usually depends on such simplifications.

EXERCISES

In Exercises 1–30 calculate the derivatives of the given functions. Simplify your answers whenever possible.

1. $y = 3x^2 - 5x - 7$

2. $y = x^8 - x^4$

3. $y = \frac{x^3}{3} - \frac{x^2}{2} + x$

4. $y = 4x^{1/2} - \frac{5}{x}$

5. $f(x) = Ax^2 + Bx + C$

6. $f(x) = \frac{6}{x^3} + \frac{2}{x^2} - 2$

7. $z = \frac{s^5 - s^3}{15}$

8. $y = x^{45} - x^{-45}$

9. $g(t) = t^{1/3} + 2t^{1/4} + 3t^{1/5}$

10. $F(x) = (3x-2)(1-5x)$

11. $y = \sqrt{x}\left(5 - x - \frac{x^2}{3}\right)$

12. $g(t) = \frac{1}{2t-3}$

13. $y = \frac{1}{x^2+5x}$

14. $y = \frac{4}{3-x}$

15. $f(t) = \frac{\pi}{2-\pi t}$

16. $g(y) = \frac{2}{1-y^2}$

17. $y = \frac{3}{x+\sqrt{x}}$

18. $y = \frac{x-2}{x+2}$

19. $f(x) = \dfrac{3 - 4x}{3 + 4x}$

20. $z = \dfrac{t^2 + 2t}{t^2 - 1}$

21. $s = \dfrac{1 + \sqrt{t}}{1 - \sqrt{t}}$

22. $f(x) = \dfrac{x^3 - 4}{x + 1}$

23. $f(x) = \dfrac{ax + b}{cx + d}$

24. $F(t) = \dfrac{t^2 + 7t - 8}{t^2 - t + 1}$

25. $f(x) = (1 + x)(1 + 2x)(1 + 3x)(1 + 4x)$

26. $f(r) = (r^{-2} + r^{-3} - 4)(r^2 + r^3 + 1)$

27. $y = (x^2 + 4)(\sqrt{x} + 1)(5x^{2/3} - 2)$

28. $y = \dfrac{(x^2 + 1)(x^3 + 2)}{(x^2 + 2)(x^3 + 1)}$

29.* $y = \dfrac{x}{2x + \dfrac{1}{3x + 1}}$

30.* $f(x) = \dfrac{(\sqrt{x} - 1)(2 - x)(1 - x^2)}{\sqrt{x}(3 + 2x)}$

31. Find $\dfrac{d}{dx}\left(\dfrac{x^2 - 4}{x^2 + 4}\right)\Bigg|_{x=-2}$

32. Find $\dfrac{d}{dt}\left(\dfrac{t(1 + \sqrt{t})}{5 - t}\right)\Bigg|_{t=4}$

33. If $f(x) = \dfrac{\sqrt{x}}{x + 1}$, find $f'(2)$.

34. Find $\dfrac{d}{dt}\Big((1 + t)(1 + 2t)(1 + 3t)(1 + 4t)\Big)\Big|_{t=0}$.

35. Find an equation of the tangent line to $y = \dfrac{2}{3 - 4\sqrt{x}}$ at the point $(1, -2)$.

36. Find equations of the tangent and normal to $y = \dfrac{x + 1}{x - 1}$ at $x = 2$.

37. Find the points on the curve $y = x + 1/x$ where the tangent line is horizontal.

38. Find the equations of all horizontal lines that are tangent to the curve $y = x^2(4 - x^2)$.

39. Find the coordinates of all points where the curve $y = \dfrac{1}{x^2 + x + 1}$ has a horizontal tangent line.

40. Find the coordinates of points on the curve $y = \dfrac{x + 1}{x + 2}$ where the tangent line is parallel to the line $y = 4x$.

41. Find the equation of the straight line that passes through the point $(0, b)$ and is tangent to the curve $y = 1/x$. Assume $b \neq 0$.

42.*Show that the curve $y = x^2$ intersects the curve $y = 1/\sqrt{x}$ at right angles.

43. Develop a "Square Root Rule" for differentiating $\sqrt{f(x)}$ and use it to find the derivative of $\sqrt{x^2 + 1}$.

44.† Show that for any constant C, the function $y = Cx^k$ $(x > 0)$ satisfies the *differential equation* $xy' = ky$. Find a function y that is a solution of the *initial-value problem*

$$\begin{cases} xy' = 3y & (x > 0) \\ y\big|_{x=2} = -4. \end{cases}$$

2.4 THE CHAIN RULE

The last and most important differentiation rule is the Chain Rule, which enables us to differentiate compositions of functions whose derivatives we already know.

2.4.1
The Chain Rule

> If the function g is differentiable at the point x and the function f is differentiable at the point $g(x)$, then the composition $f \circ g$ is differentiable at x and
>
> $$\frac{d}{dx}f(g(x)) = (f \circ g)'(x) = f'(g(x))g'(x).$$

In terms of Leibniz notation, if $y = f(u)$ and $u = g(x)$, then $y = f(g(x))$ and, since

†All exercises pertaining to differential equations are marked with this symbol.

$$\frac{dy}{du} = f'(u) = f'(g(x)) \text{ and } \frac{du}{dx} = g'(x), \text{ we have}$$

2.4.2
The Chain Rule in Leibniz Notation

$$\frac{dy}{dx} = \frac{dy}{du}\frac{du}{dx}.$$

This form of the chain rule can be interpreted as follows: as x changes, u changes du/dx times as fast as x, and y changes dy/du times as fast as u. Thus y changes $\frac{dy}{du}\frac{du}{dx}$ times as fast as x. It appears as though the symbol du cancels from the numerator and denominator, but this is not meaningful because dy/du was not defined as the quotient of two quantities, but rather as a single quantity, the derivative of y with respect to u.

Of all the differentiation rules, the Chain Rule causes the most difficulty for students initially. However it is used more frequently than the other rules, so you will get used to it quickly. We postpone the proof to the end of this section and proceed immediately to several examples illustrating how the rule is used. Think of composition as the nesting of one function inside another; the chain rule says to begin differentiating at the outside and work inward. In $f(g(x))$, f is the outside function and g is inside.

EXAMPLE 2.4.3 a) If $y = \sqrt{3x-2}$, then $\frac{dy}{dx} = \frac{1}{2\sqrt{3x-2}}\frac{d}{dx}(3x-2) = \frac{3}{2\sqrt{3x-2}}$. Here we have $y = \sqrt{u}$ where $u = 3x - 2$, so that

$$\frac{dy}{dx} = \frac{dy}{du}\frac{du}{dx} = \frac{1}{2\sqrt{u}}(3) = \frac{3}{2\sqrt{3x-2}}.$$

b) In a similar vein we have

$$\frac{d}{dx}\sqrt{x^2+1} = \frac{1}{2\sqrt{x^2+1}}(2x) = \frac{x}{\sqrt{x^2+1}}.$$

As you read this, say the following to yourself: "The derivative of the square root of something is one over twice the square root of that thing multiplied by the derivative of that thing." In (a) and (b) the outer function is square root.

c) If $f(x) = (ax+b)^n$, then $f'(x) = n(ax+b)^{n-1}(a) = an(ax+b)^{n-1}$. "The derivative of the nth power of something is n times the $(n-1)$st power of that thing, multiplied by the derivative of that thing." In general,

$$\frac{d}{dx}\big(f(x)\big)^n = n\big(f(x)\big)^{n-1} f'(x).$$

The outer function is the nth power.

d) If $h(t) = (t^3 + 3t + 4)^{1/3}$, then the outer function is the $1/3$ power:

$$h'(t) = \frac{1}{3}\left(t^3 + 3t + 4\right)^{-2/3}(3t^2 + 3) = \frac{t^2 + 1}{(t^3 + 3t + 4)^{2/3}}.$$

e)
$$\begin{aligned}\frac{d}{dx}\left(1 + \frac{1}{\sqrt{7-5x}}\right)^{50} &= \frac{d}{dx}\left(1 + (7-5x)^{-1/2}\right)^{50} \\ &= 50\left(1 + (7-5x)^{-1/2}\right)^{49}\frac{d}{dx}\left(1 + (7-5x)^{-1/2}\right) \\ &= 50\left(1 + (7-5x)^{-1/2}\right)^{49}\left(-\tfrac{1}{2}\right)(7-5x)^{-3/2}\frac{d}{dx}(7-5x) \\ &= 50\left(1 + (7-5x)^{-1/2}\right)^{49}\left(-\tfrac{1}{2}\right)(7-5x)^{-3/2}(-5) \\ &= 125(7-5x)^{-3/2}\left(1 + \frac{1}{\sqrt{7-5x}}\right)^{49}.\end{aligned}$$

Several comments should be made on part (e). Observe that the chain rule was used twice: first to differentiate the 50th power, and then to differentiate the $-1/2$ power. Although we proceeded one step at a time in the above calculation, it is usually better to try to accomplish the whole differentiation at one time as follows:

$$\begin{aligned}\frac{d}{dx}(1 + (7-5x)^{-1/2})^{50} &= 50(1 + (7-5x)^{-1/2})^{49}\left(-\tfrac{1}{2}\right)(7-5x)^{-3/2}(-5) \\ &= 125(7-5x)^{-3/2}\left(1 + \frac{1}{\sqrt{7-5x}}\right)^{49}.\end{aligned}$$

With a little practice you can learn to differentiate even very complicated expressions in one step. Note, however, that most applications of the chain rule lead to products that need to be rewritten in a different order; it is conventional to write factors in a product in order of increasing complexity.

EXAMPLE 2.4.4 Note that the reciprocal rule of the previous section is just a special case of the chain rule with outer function $1/x$.

$$\frac{d}{dx}\left(\frac{1}{f(x)}\right) = -\frac{1}{(f(x))^2}f'(x) = \frac{-f'(x)}{(f(x))^2}.$$

Since $(d/dx)(1/x) = -1/x^2$, "the derivative of one over something is -1 over that thing squared, multiplied by the derivative of that thing."

EXAMPLE 2.4.5 Recall that the derivative of the absolute value is the signum function: $\frac{d}{dx}|x| = \operatorname{sgn} x = \frac{x}{|x|}$. It follows that

$$\frac{d}{dx}f\left(|x|\right) = f'\left(|x|\right)\operatorname{sgn} x, \quad \text{and}$$
$$\frac{d}{dx}|f(x)| = (\operatorname{sgn} f(x))f'(x).$$

For instance, if $y = |1 - x^2| + \sqrt{|3x+2|}$, then

$$\begin{aligned}\frac{dy}{dx} &= \big(\operatorname{sgn}(1-x^2)\big)(-2x) + \frac{1}{2\sqrt{|3x+2|}}\big(\operatorname{sgn}(3x+2)\big)(3)\\ &= -2x\operatorname{sgn}(1-x^2) + \frac{3\operatorname{sgn}(3x+2)}{2\sqrt{|3x+2|}}\end{aligned}$$

EXAMPLE 2.4.6 Suppose that f is a differentiable function on the real line. Express the derivatives of (a) $f(3x)$, (b) $f(x^2)$, and (c) $[f(3-2f(x))]^4$ in terms of the derivative of f.

SOLUTION a) $\dfrac{d}{dx} f(3x) = \big(f'(3x)\big)(3) = 3f'(3x)$.

b) $\dfrac{d}{dx} f(x^2) = \big(f'(x^2)\big)(2x) = 2xf'(x^2)$.

c)
$$\begin{aligned}\frac{d}{dx}[f(3-2f(x))]^4 &= 4[f(3-2f(x))]^3 f'(3-2f(x))(-2f'(x))\\ &= -8f'(x)f'(3-2f(x))[f(3-2f(x))]^3.\end{aligned}$$

Proof of the Chain Rule

Suppose that f is differentiable at the point $u = g(x)$ and that g is differentiable at x. Let the function $E(k)$ be defined by

$$\begin{aligned}E(0) &= 0,\\ E(k) &= \frac{f(u+k) - f(u)}{k} - f'(u), \qquad \text{if } k \neq 0.\end{aligned}$$

By the definition of derivative, $\lim_{k\to 0} E(k) = f'(u) - f'(u) = 0 = E(0)$, so $E(k)$ is continuous at $k = 0$, Also

$$f(u+k) - f(u) = \big(f'(u) + E(k)\big)k.$$

Now put $u = g(x)$ and $k = g(x+h) - g(x)$, so that $u + k = g(x+h)$, and obtain

$$f(g(x+h)) - f(g(x)) = \big(f'(g(x)) + E(k)\big)(g(x+h) - g(x)).$$

Since g is differentiable at x, $\lim_{h\to 0}[g(x+h) - g(x)]/h = g'(x)$. Also, g is continuous at x by Theorem 2.3.1, so $\lim_{h\to 0} k = \lim_{h\to 0}(g(x+h) - g(x)) = 0$. Since E is continuous at 0, $\lim_{h\to 0} E(k) = \lim_{k\to 0} E(k) = E(0) = 0$. Hence,

$$\begin{aligned}\frac{d}{dx} f(g(x)) &= \lim_{h\to 0} \frac{f(g(x+h)) - f(g(x))}{h}\\ &= \lim_{h\to 0} \big(f'(g(x)) + E(k)\big)\,\frac{g(x+h) - g(x)}{h}\\ &= (f'g(x)) + 0)g'(x) = f'(g(x))g'(x),\end{aligned}$$

which was to be proved. □

EXERCISES

Find the derivatives of the functions in exercises 1–26.

1. $y = (2x+3)^6$

2. $y = \left(1 - \dfrac{x}{3}\right)^{99}$

3. $f(x) = (4 - x^2)^{10}$

4. $F(t) = \dfrac{(at^2+bt+c)^{12}}{3}$

5. $y = (Ax+B)^r$

6. $y = \sqrt{3x-2}$

7. $y = \dfrac{3}{5-4x}$

8. $y = (1-2x^2)^{-3/2}$

9. $y = \sqrt{5x+3}$

10. $y = \sqrt{x^2+2x+3}$

11. $F(t) = \left(2 + \dfrac{3}{t}\right)^{10}$

12. $z = (1 + x^{2/3})^{3/2}$

13. $s = \dfrac{1}{\sqrt{x^4+1}}$

14. $y = \dfrac{\sqrt{x+1}}{x+2}$

15. $y = x^3\sqrt{x^2+3}$

16. $y = \sqrt{4x + \dfrac{3}{x^2+1}}$

17. $y = \dfrac{1}{1 + \dfrac{1}{2 + \dfrac{3}{t}}}$

18. $f(t) = |2 + t^3|$

19. $y = 4x + |4x - 1|$

20. $y = (2 + |x|^3)^{1/3}$

21. $y = \dfrac{1}{2 + \sqrt{3x+4}}$

22. $f(x) = \left(1 + \sqrt{\dfrac{x-2}{3}}\right)^4$

23. $g(t) = \sqrt{(3-t)^3(2+t^2)^4}$

24. $f(y) = \left(\sqrt{y} + \dfrac{1}{1+\sqrt{y^2+2}}\right)^2$

25. $z = \left(u + \dfrac{1}{u-1}\right)^{-5/3}$

26. $y = \dfrac{x^5\sqrt{3+x^6}}{(4+x^2)^3}$

27. Sketch the graphs of the functions in Exercises 18 and 19 above.

In Exercises 28–35 express the derivative of the given function in terms of the derivative f' of the differentiable function f.

28. $f(2t+3)$

29. $f(5x - x^2)$

30. $\left[f\left(\dfrac{x}{2}\right)\right]^3$

31. $\sqrt{3 + 2f(x)}$

32. $f\left(\sqrt{3+2t}\right)$

33. $f\left(3 + 2\sqrt{x}\right)$

34. $f\big(2f(3f(x))\big)$

35. $f\big(2 - 3f(4-5t)\big)$

36. Find $\dfrac{d}{dx}\left(\dfrac{\sqrt{x^2-1}}{x^2+1}\right)\Bigg|_{x=-2}$.

37. Find $\dfrac{d}{dt}\sqrt{3t-7}\Big|_{t=3}$.

38. If $f(x) = \dfrac{1}{\sqrt{2x+1}}$, find $f'(4)$.

39. If $y = (x^3+9)^{17/2}$, find $y'\Big|_{x=-2}$.

40. Find $F'(0)$ if $F(x) = (1+x)(2+x)^2(3+x)^3(4+x)^4$.

41.*Calculate y' if $y = (x + ((3x)^5 - 2)^{-1/2})^{-6}$. Try to do it all in one step.

In exercises 42–45 find an equation of the tangent line to the given curve at the given point.

42. $y = \sqrt{1+2x^2}$ at $x = 2$

43. $y = (1 + x^{2/3})^{3/2}$ at $x = -2$

44. $y = (ax+b)^8$ at $x = b/a$

45. $y = 1/(x^2 - x + 3)^{3/2}$ at $x = -2$

46. Show that the derivative of $f(x) = (x-a)^m(x-b)^n$ vanishes at some point between a and b if m and n are positive integers.

47.*What is wrong with the following "proof" of the chain rule? Let $k = g(x+h) - g(x)$. Then $\lim_{h\to 0} k = 0$. Thus

$$\begin{aligned}
&\lim_{h\to 0}\frac{f(g(x+h)) - f(g(x))}{h}\\
&= \lim_{h\to 0}\frac{f(g(x+h)) - f(g(x))}{g(x+h)-g(x)}\,\frac{g(x+h)-g(x)}{h}\\
&= \lim_{h\to 0}\frac{f(g(x)+k) - f(g(x))}{k}\,\frac{g(x+h)-g(x)}{h}\\
&= f'(g(x))\,g'(x).
\end{aligned}$$

2.5 IMPLICIT DIFFERENTIATION

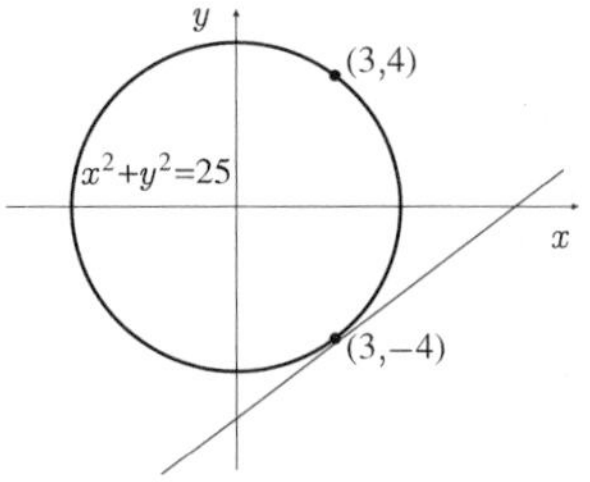

FIGURE 2.5.1

Suppose we wish to find the equation of the tangent line to the circle $x^2+y^2=25$ at the point $(3,-4)$. As presented, this curve is not the graph of a function. (Vertical lines can meet the circle more than once.) It is, in fact, the union of the graphs of two functions: the upper semicircle, $y=\sqrt{25-x^2}$ and the lower semicircle, $y=-\sqrt{25-x^2}$. As you can see in Fig. 2.5.1, $(3,-4)$ lies on the lower semicircle, so the slope of the required tangent line is

$$\frac{d}{dx}\left(-\sqrt{25-x^2}\right)\bigg|_{x=3}=\frac{-1}{2\sqrt{25-x^2}}(-2x)\bigg|_{x=3}=\frac{3}{4}.$$

Thus the tangent line has equation $y=-4+\frac{3}{4}(x-3)$.

There is another way to obtain the slope that does not require solving the given equation for y as a function of x. Part of the circle near $(3,-4)$ is certainly the graph of *some function* y. Let us differentiate the equation of the circle with respect to x regarding y as representing a function of x. Differentiating y^2 then requires the chain rule; $\frac{d}{dx}(y^2)=\frac{d}{dy}(y^2)\frac{dy}{dx}=2y\frac{dy}{dx}$. We have

$$2x+2y\frac{dy}{dx}=0.$$

Hence, $\frac{dy}{dx}=-\frac{x}{y}$ and the required tangent line has slope

$$\frac{dy}{dx}\bigg|_{(3,-4)}=-\frac{3}{-4}=\frac{3}{4}.$$

This latter technique for finding the slope is called **implicit differentiation** because we are finding the derivative of a function y defined *implicitly* by the given equation rather than presented *explicitly* as a function of x. This method has the very great advantage that it can be used even if the given equation linking x and y cannot readily be solved for y as an explicit function of x. The price we have to pay for this advantage is that the derivative dy/dx so calculated is, in general, a function of both x and y, so both coordinates of a point must be known to find the slope there. This is to be expected because an equation in x and y can define more than one function of x. In the example of the circle above, $dy/dx=-x/y$ also gives the slope at points of the upper semicircle: at $(3,4)$ the slope is $-3/4$.

There is another, more subtle, and more important disadvantage to the use of implicit differentiation. When we differentiate an equation in x and y with respect to x, we are tacitly assuming that the equation does define y as a differentiable function of x. Otherwise we would be attempting to calculate a quantity that does not exist. For example, the argument used above to show that the slope of the circle $x^2+y^2=25$ at any point on it is $-x/y$ would give the same result if applied to the equation $x^2+y^2=-1$, which is satisfied by *no points*.

It is possible to formulate and prove a theorem giving conditions under which a function defined implicitly by an equation is differentiable. We cannot attempt to do this here; such a theorem presupposes some familarity with the calculus of functions of more than one variables. *When we use implicit differentiation here we will usually be assuming, though we will not usually say so explicitly, that the required derivative exists.*

EXAMPLE 2.5.1 Find dy/dx for the equation $x^3y + xy^3 = 2$.

SOLUTION Differentiating the given equation with respect to x and remembering that y is a function of x, we get

$$3x^2y + x^3\frac{dy}{dx} + y^3 + 3xy^2\frac{dy}{dx} = 0.$$

Thus,

$$\frac{dy}{dx} = -\frac{3x^2y + y^3}{x^3 + 3xy^2}.$$

EXAMPLE 2.5.2 Find the straight line tangent at the point $(-1, 2)$ to the curve

$$\frac{x^2}{1 + xy} = \sqrt{y + 2x^2} - 3.$$

SOLUTION Observe that the given point $(-1, 2)$ does indeed satisfy the given equation, so the point lies on the curve. Differentiate the equation of the curve with respect to x to get

$$\frac{(1 + xy)(2x) - (x^2)\left(y + x\dfrac{dy}{dx}\right)}{(1 + xy)^2} = \frac{1}{2\sqrt{y + 2x^2}}\left(\frac{dy}{dx} + 4x\right) - 0.$$

Now substitute the coordinates $x = -1$, $y = 2$:

$$\frac{(-1)(-2) - (-1)^2\left(2 + (-1)\dfrac{dy}{dx}\right)}{(-1)^2} = \frac{1}{2\sqrt{2 + 2(-1)^2}}\left(\frac{dy}{dx} + 4(-1)\right).$$

We can now solve this equation for dy/dx:

$$\frac{dy}{dx} = \frac{1}{4}\frac{dy}{dx} - 1, \quad \text{so} \quad \left.\frac{dy}{dx}\right|_{(-1,2)} = -\frac{4}{3},$$

and the required tangent line has equation $y = 2 - \dfrac{4}{3}(x + 1)$.

Note that we substituted the coordinates of the given point as soon as we differentiated the given equation, and then we solved the resulting equation for the derivative. With numbers substituted for x and y it is usually much easier to solve for dy/dx than it would be with algebraic expressions. The appropriate order in which to perform the operation is as follows:

i) Differentiate the given equation with respect to x, treating y as a function of x,

ii) Substitute any numerical values for the coordinates,

iii) Solve for the required slope.

EXAMPLE 2.5.3 Prove the General Power Rule, $\frac{d}{dx}x^r = r\,x^{r-1}$, for a rational number $r = m/n$, (m and n are integers and $n \neq 0$). Assume the rule has already been proven for r an integer).

PROOF If $y = x^{m/n}$ then $y^n = x^m$. Differentiating implicitly with respect to x we obtain

$$n\,y^{n-1}\frac{dy}{dx} = m\,x^{m-1}.$$

Thus $\frac{dy}{dx} = \frac{m\,x^{m-1}}{n\,y^{n-1}} = \frac{m\,x^{m-1}}{n\,(x^{m/n})^{n-1}} = \frac{m}{n}\,x^{m-1-(m/n)(n-1)} = \frac{m}{n}\,x^{(m/n)-1}$. □

EXAMPLE 2.5.4 Show that for any constants A and B, the curves $x^2 - y^2 = A$ and $xy = B$ intersect at right angles, that is, at any point where they intersect their tangents are perpendicular.

SOLUTION The slope at any point on $x^2 - y^2 = A$ is given by $2x - 2yy' = 0$, or $y' = x/y$. The slope at any point on $xy = B$ is given by $y + xy' = 0$, or $y' = -y/x$. If the two curves (they are both hyperbolas if A and B are not 0) intersect at (x_0, y_0), then their slopes at that point are x_0/y_0 and $-y_0/x_0$, respectively. Clearly these slopes are negative reciprocals, so the tangent line to one curve is the normal line to the other at that point. Hence, the curves intersect at right angles. (See Fig. 2-5-2.)

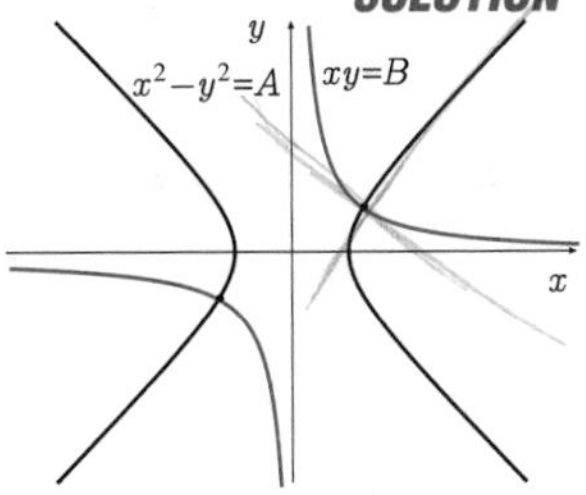

FIGURE 2.5.2

Derivatives of Inverse Functions

Suppose that the function f is differentiable on an interval (a, b) and that either $f'(x) > 0$ for $a < x < b$ or $f'(x) < 0$ for $a < x < b$. In this case, as we will prove later (see Theorem 4.1.5), f must be one-to-one on (a, b). Therefore f has an inverse, f^{-1}, defined by

$$y = f^{-1}(x) \Longleftrightarrow x = f(y), \qquad (a < y < b).$$

(Review Section 1.4 if you are uncertain about inverse functions.) Recall that the graphs of f and f^{-1} are mirror images in the line $x = y$. (See Fig. 2.5.3.) Since we are assuming that the graph $y = f(x)$ has a *non-horizontal* tangent line at any x in (a, b), its mirror image, the graph $y = f^{-1}(x)$, has a *non- vertical* tangent line at any x in the interval between $f(a)$ and $f(b)$. Therefore f^{-1} is differentiable at any such x.

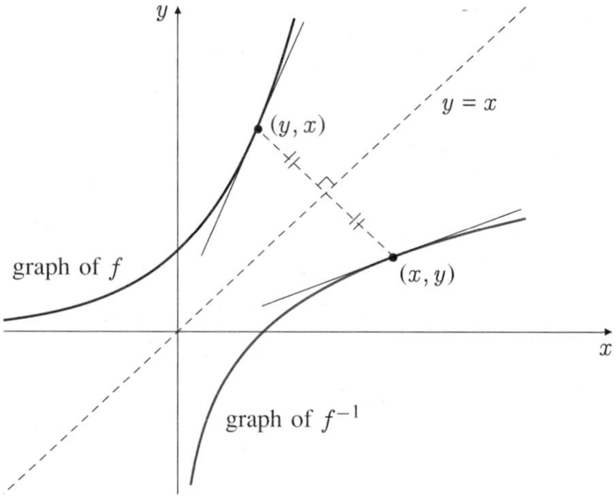

FIGURE 2.5.3

Let $y = \left(f^{-1}\right)(x)$. We want to find dy/dx. Solve the equation $y = f^{-1}(x)$ for $x = f(y)$ and differentiate implicitly with respect to x to obtain

$$1 = f'(y)\,\frac{dy}{dx}, \qquad \text{so} \qquad \frac{dy}{dx} = \frac{1}{f'(y)} = \frac{1}{f'\left(f^{-1}(x)\right)}.$$

Therefore the slope of the graph of f^{-1} at (x, y) is the reciprocal of the slope of the graph of f at (y, x), and

2.5.5
Derivative of an Inverse Function

$$\frac{d}{dx} f^{-1}(x) = \frac{1}{f'\left(f^{-1}(x)\right)}.$$

In terms of Leibniz notation we have $\left.\dfrac{dy}{dx}\right|_x = \dfrac{1}{\left.\dfrac{dx}{dy}\right|_{y=f^{-1}(x)}}$.

EXAMPLE 2.5.6 Show that $f(x) = x^3 + x$ is one-to-one on the whole real line, and, noting that $f(2) = 10$, find $\left(f^{-1}\right)'(10)$.

SOLUTION Since $f'(x) = 3x^2 + 1 > 0$ for all real numbers x, f is one-to-one and has an inverse. If $y = f^{-1}(x)$, then $x = f(y) = y^3 + y$, so $1 = (3y^2 + 1)y'$ and $y' = \dfrac{1}{3y^2 + 1}$. Now $f(2) = 10$ implies $f^{-1}(10) = 2$. Thus

$$\left(f^{-1}\right)'(10) = \left.\frac{1}{3y^2 + 1}\right|_{y=2} = \frac{1}{13}.$$

EXERCISES

In Exercises 1–8 find dy/dx in terms of x and y.

1. $xy - x + 2y = 1$

2. $x^3 + y^3 = 1$

3. $x^2 + xy = y^3$

4. $x^3y + xy^5 = 2$

5. $x^2y^3 = 2x - y$

6. $x^2 + 4(y-1)^2 = 4$

7. $\dfrac{x-y}{x+y} = \dfrac{x^2}{y} + 1$

8. $x\sqrt{x+y} = 8 - xy$

In Exercises 9–12 find an equation of the tangent line to the given curve at the given point.

9. $2x^2 + 3y^2 = 5$ at $(1, 1)$

10. $x^2y^3 - x^3y^2 = 12$ at $(-1, 2)$

11. $\dfrac{x}{y} + \left(\dfrac{y}{x}\right)^3 = 2$ at $(-1, -1)$

12. $x + 2y + 1 = \dfrac{y^2}{x-1}$ at $(2, -1)$

13.*Show that the ellipse $x^2 + 2y^2 = 2$ and the hyperbola $2x^2 - 2y^2 = 1$ intersect at right angles.

14.*Show that the ellipse $x^2/a^2 + y^2/b^2 = 1$ and the hyperbola $x^2/A^2 - y^2/B^2 = 1$ intersect at right angles if $a^2 - b^2 = A^2 + B^2$. (This condition says that the ellipse and the hyperbola have the same foci.)

15.*Use implicit differentiation to find y' if $(x-y)/(x+y) = x/y + 1$. Now show that there are, in fact, no points on that curve, so the derivative you calculated is meaningless. This is another example which demonstrates the dangers of calculating something when you don't know whether or not it exists.

16. Show that $f(x) = \dfrac{4x^3}{x^2+1}$ has an inverse and find $\left(f^{-1}\right)'(2)$.

17. Find $\left(f^{-1}\right)'(x)$ if $f(x) = 1 + 2x^3$.

18.*Find $\left(f^{-1}\right)'(-2)$ if $f(x) = x\sqrt{3 + x^2}$.

19. Assume that the function $f(x)$ satisfies $f'(x) = \dfrac{1}{x}$, and that f is one-to-one. If $y = f^{-1}(x)$, show that $dy/dx = y$.

2.6 INTERPRETATIONS OF THE DERIVATIVE

When calculus is applied to other disciplines, mathematical functions are used to represent quantities appropriate to the particular problem under consideration. In such cases the derivatives of these functions may provide useful information about the problem. The mathematical interpretation of the derivative as the slope of the graph of the function suggests a more concrete interpretation of derivative as a rate of change of the function with respect to its independent variable: the slope of a straight line is the rate of change of vertical position with respect to horizontal position along the line.

2.6.1
Average Rate of Change

> The **average rate of change** of $f(x)$ with respect to x over the interval $[a, b]$ is
> $$\frac{f(b) - f(a)}{b - a}.$$

For example, the rate of change of x^2 with respect to x over the interval $[1, 4]$ is $(4^2 - 1^2)/(4-1) = 15/3 = 5$.

The limit of the average rate of change as the interval over which the average is taken approaches zero in length is called the *instantaneous rate of change*, or simply the *rate of change*. The rate of change of $f(x)$ at $x = a$ is, letting $h = b - a$),

$$\lim_{b\to a} \frac{f(b) - f(a)}{b - a} = \lim_{h \to 0} \frac{f(a+h) - f(a)}{h} = f'(a).$$

2.6.2
Rate of Change

> The **rate of change** of $f(x)$ with respect to x at $x = x_0$ is $f'(x_0)$.

The sign of f' tells us whether $f(x)$ is increasing or decreasing as x increases. If $f'(x) > 0$ on an interval, then f is increasing as x increases on that interval; if $f'(x) < 0$, then $f(x)$ decreases as x increases. See Fig. 2.6.1. We will use these ideas extensively in Chapters 4 and 5.

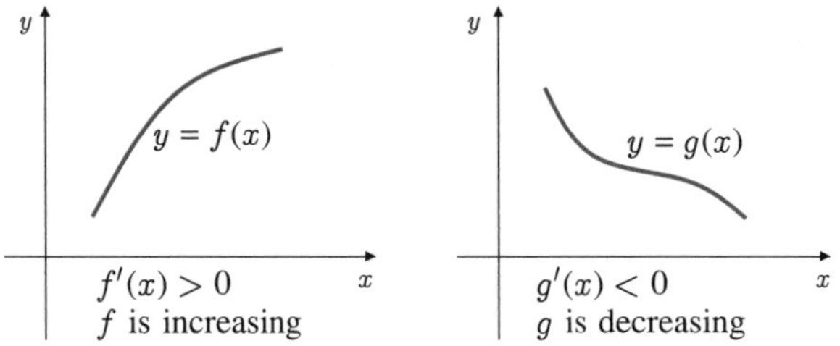

FIGURE 2.6.1

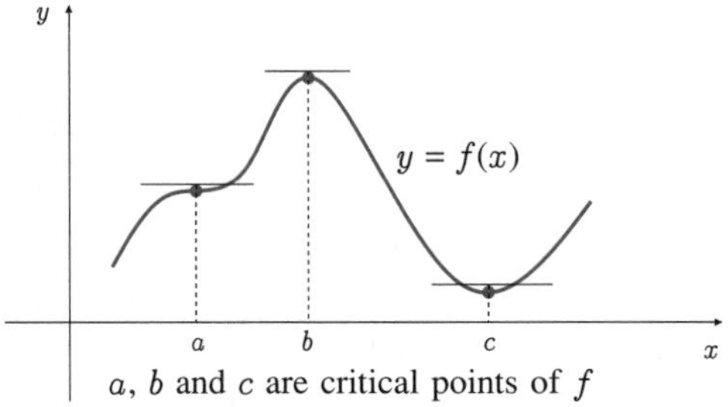

FIGURE 2.6.2

If $f'(x_0) = 0$, we say that f is **stationary** at x_0. This means that f is instantaneously neither increasing nor decreasing at that point, although it may be increasing or decreasing on intervals containing that point. Such points x_0 are called **critical points** of f. At a critical point the tangent line to the graph $y = f(x)$ is horizontal, as shown in Fig. 2.6.2. Note that f may or may not have a maximum or minimum value at a critical point. These ideas will be examined more fully in Section 4.2.

EXAMPLE 2.6.3 Find the rate of change of the area of a circle with respect to the diameter of the circle.

SOLUTION The area A and diameter D of a circle are related by the formula

$$A = \frac{\pi}{4} D^2.$$

The rate of change of the area with respect to the diameter is therefore

$$\frac{dA}{dD} = \frac{\pi D}{2}.$$

The following example illustrates the fact that if $y = f(x)$ where x and y are measured in specific units, then the derivative dy/dx, being the rate of change of y with respect to x, is measured in units of y per unit of x.

EXAMPLE 2.6.4 Suppose the temperature at a certain location t hours after noon on a certain day is T°C, (T degrees Celsius) where

$$T = \tfrac{1}{3}t^3 - 3t^2 + 8t + 10, \qquad (\text{for } 0 \le t \le 5.)$$

How fast is the temperature rising or falling at 1:00 p.m.? at 2:00 p.m.? at 3:00 p.m.? During what time interval is the temperature falling?

SOLUTION The rate of change of the temperature is given by

$$\frac{dT}{dt} = t^2 - 6t + 8 = (t-2)(t-4).$$

If $t = 1$ then $\dfrac{dT}{dt} = 3$ so the temperature is rising at rate 3°C/hour at 1:00 p.m.
If $t = 2$ then $\dfrac{dT}{dt} = 0$ so the temperature is stationary (neither rising nor falling) at 2:00 p.m.
If $t = 3$ then $\dfrac{dT}{dt} = -1$ so the temperature is falling at a rate of 1°C/hour at 3:00 p.m.
The temperature is falling during the interval when $t - 2 > 0$ and $t - 4 < 0$, that is, between 2:00 and 4:00 p.m., because $dT/dt < 0$ at these times.

Velocity and Acceleration

Suppose that an object is moving along a straight line (say the x-axis) so that at time t its position is given by $x = f(t)$. The **average velocity** of the object over the time interval $[t, t+h]$ is the change in position divided by the change in time, that is, the Newton quotient

$$v_{\text{average}} = \frac{f(t+h) - f(t)}{h}.$$

The **velocity** $v(t)$ of the object at time t is the limit of this average velocity as $h \to 0$. Thus it is the rate of change of position with respect to time, and is measured in units of distance/time;

2.6.5 Velocity

$$v(t) = \frac{dx}{dt} = f'(t).$$

If $v(t) > 0$, then x is increasing, so the object is moving to the right; if $v(t) < 0$, then x is decreasing, so the object is moving to the left. At a critical point of f, that is, a time t when $v(t) = 0$, the object is instantaneously at rest—at that instant it is not moving in either direction.

We distinguish between the term *velocity* (which involves direction of motion as well as the rate) and **speed**, which does not involve direction. The speed is the absolute value of the velocity:

2.6.6
Speed

$$s(t) = |v(t)| = \left|\frac{dx}{dt}\right|.$$

The derivative of the velocity also has a useful interpretation. The rate of change of the velocity with respect to time is the **acceleration** of the moving object. It is measured in units of distance/time^2. The value of the acceleration at time t is

2.6.7
Acceleration

$$a(t) = v'(t) = \frac{dv}{dt} = \frac{d}{dt}\frac{dx}{dt}.$$

If $a(t) > 0$, the velocity is increasing. This does not necessarily mean that the speed is increasing; if the object is moving to the left ($v(t) < 0$) and accelerating to the right ($a(t) > 0$), then it is actually slowing down. The object is speeding up only when the velocity and acceleration have the same sign.

if velocity is	and acceleration is	then object is	and speed is
positive	positive	moving right	increasing
positive	negative	moving right	decreasing
negative	positive	moving left	decreasing
negative	negative	moving left	increasing

If $a(t_0) = 0$, then the velocity and the speed are stationary at t_0. If $a(t) = 0$ during an interval of time, then the velocity is unchanging and, therefore, constant over that interval.

EXAMPLE 2.6.8 A point P moves along the x-axis in such a way that its position at time t (seconds) is given by

$$x = 2t^3 - 15t^2 + 24t \text{ cm}.$$

Find the velocity and acceleration of P at time t. In what direction is P moving at the time $t = 2$? What is its speed at that time, and is that speed increasing or decreasing? At what instant is P at rest? At what instant is the velocity stationary?

SOLUTION The velocity and acceleration of P at time t are

$$v = \frac{dx}{dt} = 6t^2 - 30t + 24 = 6(t-1)(t-4) \text{ cm/s}, \quad \text{and}$$
$$a = \frac{dv}{dt} = 12t - 30 \text{ cm/s}^2.$$

At $t = 2$ we have $v = -12$ and $a = -6$. Thus P is moving to the left with speed 12 cm/s and, since the velocity and acceleration are both negative, this speed is increasing. Since $v = 0$ if $t = 1$ or $t = 4$, P is momentarily at rest at times $t = 1$ and $t = 4$. The velocity is stationary when $a = 0$, i.e. when $t = 30/12 = 5/2$ seconds.

EXAMPLE 2.6.9 An object is hurled upward from the roof of a building 10 metres high. It rises and then falls back; its height above the ground t seconds after it is thrown is

$$y = -4.9t^2 + 8t + 10 \text{ m},$$

until it strikes the ground. What is the greatest height above the ground that the object attains? With what speed does the object strike the ground?

SOLUTION Refer to Fig. 2.6.3. The vertical velocity at time t during flight is

$$v(t) = -2(4.9)t + 8 = -9.8t + 8 \text{ m/s}.$$

The object is rising when $v > 0$, that is when $0 < t < 8/9.8$; the object is falling for $t > 8/9.8$. Thus the object is at its maximum height at time $t = 8/9.8 \approx 0.8163$ s, and this maximum height is

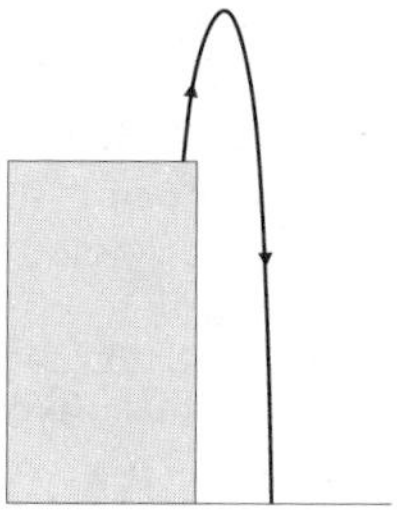

FIGURE 2.6.3

$$y_{\max} = -4.9\left(\frac{8}{9.8}\right)^2 + 8\left(\frac{8}{9.8}\right) + 10 \approx 13.27 \text{ metres}.$$

The time t at which the object strikes the ground is the positive root of the quadratic equation ($y = 0$)

$$-4.9t^2 + 8t + 10 = 0,$$

namely,

$$t = \frac{-8 - \sqrt{64 + 196}}{-9.8} \approx 2.462 \text{ s}.$$

The velocity at this time is $v = -(9.8)(2.462) + 8 \approx -16.12$. Thus the object strikes the ground with a speed of about 16.12 m/s.

Marginals

In economics the term *marginal* denotes the rate of change of a quantity with respect to a variable on which it depends. The marginal cost of production in a manufacturing operation is the rate of change of the cost of production with respect to the number of units of product produced. It is sometimes loosely defined to be the extra cost of producing one more unit when a given number of units is being produced.

EXAMPLE 2.6.10 In a manufacturing operation producing certain collectors' items, the cost of production has two components, a fixed cost of $\$A$ to set up the operation, and a cost of $\$B$ per item for the raw materials and labor. The price $\$P$ at which each item can be sold depends on the number n of items produced according to the formula $P = C - Dn$. (The more items produced, the less each is worth to a prospective buyer.) How many items should the manufacturer produce to maximize profit?

SOLUTION The total cost of producing n items is $\$A + Bn$. These can be sold for a total of $\$Pn = Cn - Dn^2$ to generate a total profit of

$$T(n) = Cn - Dn^2 - (A + Bn) = (C - B)n - Dn^2 - A.$$

The marginal profit is

$$T'(n) = C - B - 2Dn.$$

If $n < (C - B)/2D$, the marginal profit is positive, so the total profit increases as n increases. If $n > (C - B)/2D$, the marginal profit is negative, so the total profit decreases as n increases. Accordingly, the profit will be greatest if $n = (C - B)/2D$ items are produced. (Note that we are using a "real variable" to represent an integer quantity in this question. If $(C - B)/2D$ is not an integer we would use the closest integer to it for the optimal number of items to be produced.)

EXERCISES

1. Find the rate of change of the area of a square with respect to the length of its side.

2. Find the rate of change of the area of a circle with respect to its radius.

3. Find the rate of change of the volume of a sphere $\left(V = \frac{4}{3}\pi r^3\right)$ with respect to its radius r.

4. What is the rate of change of the area A of a square with respect to the length L of the diagonal of the square?

5. What is the rate of change of the circumference C of a circle with respect to the area A of the circle?

6. Find the rate of change of the side s of a cube with respect to the volume V of the cube.

In Exercises 7–10 a point moves along the x-axis so that its position x at time t is specified by the given function. In each case determine the following:

a) the time intervals on which the point is moving to the right, and b) to the left.

c) the time intervals on which the point is accelerating to the right, and d) to the left.

e) the time intervals when the particle is speeding up, and f) slowing down.

g) the acceleration at times when the velocity is zero.

h) the average velocity over the time interval $[0, 4]$

Make a sketch of the graph of x as a function of t in each case.

7. $x = t^2 - 4t + 3$

8. $x = 4 + 5t - t^2$

9. $x = t^3 - 4t + 1$

10. $x = \dfrac{t}{t^2 + 1}$

11. A ball is thrown upward from ground level with an initial speed of 9.8 m/s so that its height in metres after t seconds is given by $y = 9.8t - 4.9t^2$. What is the acceleration of the ball at any time t? How high does the ball go? How fast is it moving when it strikes the ground?

12. A ball is thrown downward from the top of a 100-metre-high tower with an initial speed of 2m/s. Its height in metres above the ground t sec later is $y = 100 - 2t - 4.9t^2$. How long does it take to reach the ground? What is its average velocity during the fall? At what instant is its velocity equal to its average velocity?

13.*The distance an aircraft travels along a runway before takeoff is given by $D = t^2$, where D is measured in metres from the starting point and t is measured in seconds from the time the brake is released. If the aircraft will become airborne when its speed reaches 200km/hr, how long will it take to become airborne, and what distance will it travel in that time?

14.*Show that if the position x of a moving point is given by a quadratic function of t, $x = At^2 + Bt + C$, then the average velocity over any time interval $[t_1, t_2]$ is equal to the instantaneous velocity at the midpoint of that time interval.

15.*The position of an object moving along the s-axis is given at time t by

$$s = \begin{cases} t^2 & \text{if } 0 \le t \le 2 \\ 4t - 4 & \text{if } 2 < t < 8 \\ -68 + 20t - t^2 & \text{if } 8 \le t \le 10. \end{cases}$$

Determine the velocity and acceleration at any time t. Is the velocity continuous? Is the acceleration continuous? What is the maximum velocity and when is it attained?

16. The cost C (in dollars) of producing n widgets per month in a widget factory is known to be given by

$$C = \frac{80,000}{n} + 4n + \frac{n^2}{100}.$$

Find the marginal cost of production if the number of widgets manufactured each month is a) 100, b) 300.

17.*In a mining operation the cost C (in dollars) of extracting each tonne of ore is given by

$$C = 10 + \frac{20}{x} + \frac{x}{1000},$$

where x is the number of tonnes extracted each day. (For small x, C decreases as x increases because of economies of scale, but for large x, C increases with x because of overloaded equipment and labour overtime.) If each tonne of ore can be sold for \$13, how many tonnes should be extracted each day to maximize the daily profit of the mine?

18.*If it costs a manufacturer $C(x)$ dollars to produce x items then his average cost of production is $C(x)/x$ dollars per item. Typically the average cost is a decreasing function of x for small x and an increasing function of x for large x. (Why?) Show that the value of x which minimizes the average cost makes the average cost equal to the marginal cost.

19. Explain why describing the marginal cost of production as "the extra cost of producing one more item" is a "loose" interpretation of the definition of marginal cost of production.

2.7 HIGHER-ORDER DERIVATIVES

If the derivative f' of a function f is itself differentiable, we can calculate *its* derivative, which we call the **second derivative** of f and denote by f'' or d^2f/dx^2:

2.7.1 The Second Derivative

$$f''(x) = \frac{d}{dx}f'(x) = \frac{d}{dx}\frac{d}{dx}f(x) = \frac{d^2}{dx^2}f(x).$$

If the function is denoted by a dependent variable, $y = f(x)$, then the second derivative can be denoted by y'' or d^2y/dx^2. Similarly one can consider third-, fourth-, and in general nth- order derivatives. The prime notation is inconvenient for derivatives of high order, so we denote the order by a superscript in parentheses (to distinguish it from an exponent): the nth derivative of $y = f(x)$ is

2.7.2 The nth Derivative

$$f^{(n)}(x) = \frac{d^n}{dx^n}f(x) = y^{(n)} = \frac{d^ny}{dx^n},$$

and it is defined to be the derivative of the $(n-1)$st derivative. For $n = 1, 2,$ and 3, primes are still normally used: $f^{(2)}(x) = f''(x)$, $f^{(3)}(x) = f'''(x)$. It is sometimes convenient to denote $f^{(0)}(x) = f(x)$, that is, to regard a function as its own zeroth-order derivative.

EXAMPLE 2.7.3 Acceleration is the second derivative of position. If $x = f(t)$, then velocity v and acceleration a are given by

$$v = \frac{dx}{dt} = f'(t), \quad \text{and} \quad a = \frac{dv}{dt} = \frac{d^2x}{dt^2} = f''(t).$$

EXAMPLE 2.7.4 Let $y = x^3$. Then $y' = 3x^2$, $y'' = 6x$, $y''' = 6$, $y^{(4)} = 0$, and all higher derivatives are zero. In general, if $f(x) = x^n$ (n is a positive integer), then

$$\begin{aligned} f^{(k)}(x) &= n(n-1)(n-2)\ldots(n-(k-1))\,x^{n-k} \\ &= \begin{cases} \dfrac{n!}{(n-k)!}\,x^{n-k} & \text{if } 0 \le k \le n \\ 0 & \text{if } k > n, \end{cases} \end{aligned}$$

where $n!$ (called n **factorial**) is defined by

2.7.5 Definition of Factorials

$$\begin{aligned} 0! &= 1 \\ 1! &= 1 \\ 2! &= 1 \times 2 = 2 \\ 3! &= 1 \times 2 \times 3 = 6 \\ 4! &= 1 \times 2 \times 3 \times 4 = 24 \\ &\vdots \\ n! &= 1 \times 2 \times 3 \times \cdots \times (n-1) \times n = (n-1)!n. \end{aligned}$$

It follows that if P is a polynomial of degree n,

$$P(x) = a_n x^n + a_{n-1}x^{n-1} + \cdots + a_1 x + a_0,$$

where a_n, a_{n-1}, $\ldots$, a_1, a_0 are constants, then $P^{(k)}(x) = 0$ for $k > n$. For $k \le n$, $P^{(k)}$ is a polynomial of degree $n - k$; in particular, $P^{(n)}(x) = n!\,a_n$, a constant function.

EXAMPLE 2.7.6 Calculate f', f'', and f''' for $f(x) = \sqrt{x^2+1}$.

SOLUTION Since $f(x) = (x^2+1)^{1/2}$ we have

$$\begin{aligned} f'(x) &= \tfrac{1}{2}(x^2+1)^{-1/2}(2x) = x(x^2+1)^{-1/2}, \\ f''(x) &= (x^2+1)^{-1/2} + x\left(-\tfrac{1}{2}\right)(x^2+1)^{-3/2}(2x) \\ &= (x^2+1)^{-1/2} - x^2(x^2+1)^{-3/2}, \\ f'''(x) &= -\tfrac{1}{2}(x^2+1)^{-3/2}(2x) - 2x(x^2+1)^{-3/2} - x^2\left(-\tfrac{3}{2}\right)(x^2+1)^{-5/2}(2x) \\ &= -3x(x^2+1)^{-3/2} + 3x^3(x^2+1)^{-5/2}. \end{aligned}$$

EXAMPLE 2.7.7 Find $y^{(n)}$ if $y = \dfrac{1}{1+x} = (1+x)^{-1}$.

SOLUTION Begin by calculating the first few derivatives:

$$\begin{aligned}
y' &= -(1+x)^{-2} \\
y'' &= -(-2)(1+x)^{-3} = 2(1+x)^{-3} \\
y''' &= 2(-3)(1+x)^{-4} = -3!(1+x)^{-4} \\
y^{(4)} &= -3!(-4)(1+x)^{-5} = 4!(1+x)^{-5}
\end{aligned}$$

The pattern here is becoming obvious. It seems that

$$y^{(n)} = (-1)^n n!(1+x)^{-n-1}.$$

Note the use of $(-1)^n$ to denote a positive sign if n is even and a negative sign if n is odd. Strictly speaking, we have not yet actually proved that the above formula is correct for every n, although it is for $n = 1$, 2, 3, and 4. To complete the proof, suppose that the formula is valid for some n. Consider $y^{(n+1)}$:

$$\begin{aligned}
y^{(n+1)} &= \frac{d}{dx}\, y^{(n)} = \frac{d}{dx}\Big((-1)^n n!(1+x)^{-n-1}\Big) \\
&= (-1)^n n!(-n-1)(1+x)^{-n-2} = (-1)^{n+1}(n+1)!(1+x)^{-(n+1)-1}.
\end{aligned}$$

Therefore, if the formula for $y^{(n)}$ is correct for one value of n, it is also correct for the next higher value of n. Since the formula is known to be true for $n = 1$ it must therefore be true for every integer $n \geq 1$; there can be no first value of n for which it is false. This technique of proof is called **mathematical induction**.

Mathematical induction is used to prove statements asserted to hold for all positive integers n from some integer n_0 onward. (Usually n_0 is 1.) The technique involves two steps:

i) Prove the statement for the lowest value of n, ($n = n_0$).

ii) Assume the statement is true for some arbitrary $n \geq n_0$ and deduce from that that it must also be true for the next integer, $n + 1$.

It follows that there can be no first value of $n > n_0$ for which the statement is false, and the statement is true for each $n \geq n_0$. See Appendix I for more discussion of mathematical induction and further examples and exercises based on it.

As a final example, let us calculate a second derivative using implicit differentiation.

EXAMPLE 2.7.8 Calculate y' and y'' given that $xy + y^2 = 2x$.

SOLUTION Differentiating implicitly with respect to x we obtain

$$y + xy' + 2yy' = 2.$$

Now differentiate a second time with respect to x:

$$y' + y' + xy'' + 2(y')^2 + 2yy'' = 0.$$

Solving the first equation for y', we get

$$y' = \frac{2-y}{2y+x}.$$

Solving the second equation for y'' and substituting the above expression for y' we finally obtain

$$y'' = -\frac{2y' + 2(y')^2}{2y+x} = -2\left(\frac{2-y}{2y+x}\right)\frac{1+\dfrac{2-y}{2y+x}}{2y+x} = -2\frac{(2-y)(y+x+2)}{(2y+x)^3}.$$

EXERCISES

Find y', y'', and y''' for the functions in Exercises 1–8.

1. $y = (3-2x)^7$

2. $y = x^2 - \dfrac{1}{x}$

3. $y = \dfrac{6}{(x-1)^2}$

4. $y = \sqrt{ax+b}$

5. $y = x^{1/3} - x^{-1/3}$

6. $y = x^{10} + 2x^8$

7. $y = (x^2+3)\sqrt{x}$

8. $y = \dfrac{x-1}{x+1}$

In Exercises 9–16 calculate enough derivatives of the given function to enable you to guess the general formula for $f^{(n)}(x)$. Then verify your guess using mathematical induction.

9. $f(x) = \dfrac{1}{x}$

10. $f(x) = \dfrac{1}{x^2}$

11. $f(x) = \dfrac{1}{2-x}$

12. $f(x) = \sqrt{x}$

13. $f(x) = \dfrac{1}{a+bx}$

14. $f(x) = x^{2/3}$

15.* $f(x) = \sqrt{1-3x}$

16.* $f(x) = \dfrac{1}{|x|}$

In Exercises 17–20 find y'' in terms of x and y.

17. $xy = x + y$

18. $x^2 + 4y^2 = 4$

19.* $x^3 - y^2 + y^3 = x$

20.* $x^3 - 3xy + y^3 = 1$

21. For $x^2 + y^2 = a^2$ show that $y'' = -\dfrac{a^2}{y^3}$.

22. For $Ax^2 + By^2 = C$ show that $y'' = -\dfrac{AC}{B^2y^3}$.

23. If f and g are twice-differentiable functions, show tha $(fg)'' = f''g + 2f'g' + fg''$.

24.* State and prove the results analogous to that of Exer cise 23 but for $(fg)^{(3)}$ and $(fg)^{(4)}$. Can you guess th formula for $(fg)^{(n)}$?

def

2.8 ANTIDERIVATIVES AND INDEFINITE INTEGRALS

2.8.1 Definition of Antiderivative

An **antiderivative** of a function f on an interval I is another function F whose derivative is equal to f on I:

$$F'(x) = f(x) \quad \text{for } x \text{ in } I.$$

For example, $F(x) = \frac{1}{2}x^2$ is an antidervative of $f(x) = x$ on any interval, since $F'(x) = x = f(x)$, and $G(t) = 2t^{3/2} - 4t - 1/t$ is an antiderivative of $g(t) = 3\sqrt{t} - 4 + 1/t^2$ on the interval $(0, \infty)$ because $G'(t) = 3\sqrt{t} - 4 + 1/t^2 = g(t)$ there.

Antiderivatives are not unique; indeed, if C is any constant, then $F(x) = \frac{1}{2}x^2 + C$ is an antiderivative of $f(x) = x$ on the real line. It should seem intuitively obvious (and will be proved later) that if the derivative of a function is identically zero on an interval I then the function must be constant on I:

$$f'(x) = 0 \quad \text{on } I \text{ implies } f(x) = C \text{ (constant) on } I.$$

It follows that if F and G are two antiderivatives of the same function f on an interval I, then the function $F(x) - G(x)$ has derivative $f(x) - f(x) = 0$ on I, so $F(x) - G(x) = C$, a constant, on I. We state this fact as a theorem now, and will provide a formal proof after we prove the Mean-Value Theorem in Section 4.1.

THEOREM 2.8.2 If F and G are both antiderivatives of a function f on an interval I, (that is, if $F'(x) = G'(x) = f(x)$ on I), then there exists a constant C such that $F(x) = G(x)+C$ for all x in I. □

It is essential that the set on which $F'(x) = G'(x)$ be an *interval* if we want to conclude that $F(x) - G(x)$ is constant on that set. For instance, $F(x) = 1$ and $G(x) = \operatorname{sgn} x$ satisfy $F'(x) = G'(x) = 0$ for all $x \neq 0$, but $F(x) - G(x) = 1 - \operatorname{sgn} x$ is 0 for $x > 0$ and 2 for $x < 0$; the difference is not constant on the set of real numbers $x \neq 0$.

Theorem 2.8.2 shows that if $F(x)$ is any one particular antiderivative of $f(x)$ on an interval I, then any antiderivative of $f(x)$ on I is of the form $F(x) + C$ where C is an arbitrary constant.

2.8.3 The Indefinite Integral

The **indefinite integral** of a function f on an interval I, denoted

$$\int f(x)\,dx,$$

is the *general* antiderivative of f on I. If F is *any* specific antiderivative of f on I then

$$\int f(x)\,dx = F(x) + C,$$

where C is an arbitrary constant.

EXAMPLE 2.8.4 a) $\displaystyle\int x\,dx = \frac{1}{2}x^2 + C$ on any interval.

b) $\int (x^3 - 5x^2 + 7)\, dx = \frac{1}{4}x^4 - \frac{5}{3}x^3 + 7x + C$ on any interval.

c) $\int \frac{dx}{x^2} = -\frac{1}{x} + C$ on any interval contained in the open interval $(0, \infty)$.

All three formulas above can be checked by differentiating the right-hand sides.

Finding antiderivatives is generally more difficult than finding derivatives; many functions do not have antiderivatives which can be expressed as combinations of finitely many elementary functions. We will develop several techniques for finding antiderivatives in Chapter 6. Until then, we must content ourselves with being able to write a few simple antiderivatives based on the known derivatives of elementary functions.

$$\int dx = \int 1\, dx = x + C \qquad \int x\, dx = \frac{x^2}{2} + C$$

$$\int x^2\, dx = \frac{x^3}{3} + C \qquad \int \frac{1}{x^2}\, dx = \int \frac{dx}{x^2} = -\frac{1}{x} + C$$

$$\int \frac{1}{\sqrt{x}}\, dx = 2\sqrt{x} + C \qquad \int x^r\, dx = \frac{x^{r+1}}{r+1} + C \quad (r \neq -1)$$

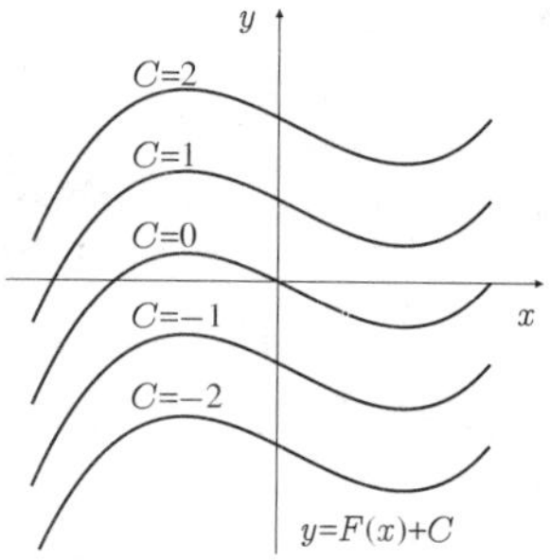

FIGURE 2.8.1

Of course, the first five of these formulas are special cases of the sixth.

The rule for differentiating sums and constant multiples of functions translates into a similar rule for antiderivatives, as reflected in Example 2.8.4(b) above.

The graphs of the different antiderivatives of the same function on the same interval are vertically displaced versions of the same curve, as shown in Fig. 2.8.1. In general only one of these curves will pass through any given point, so we can obtain a unique antiderivative of a given function on an interval by requiring the antiderivative to take on a prescribed value at a particular point x.

EXAMPLE 2.8.5 Find the function $f(x)$ whose graph passes through the point $(2, 10)$ and whose derivative is $f'(x) = 6x^2 - 1$ for all real x.

SOLUTION Since $f'(x) = 6x^2 - 1$, we have

$$f(x) = \int (6x^2 - 1)\, dx = 2x^3 - x + C$$

for some constant C. Since the graph of f passes through $(2, 10)$, we have

$$10 = f(2) = 16 - 2 + C.$$

Thus $C = -4$ and $f(x) = 2x^3 - x - 4$. (By direct calculation we verify that $f'(x) = 6x^2 - 1$ and $f(2) = 10$.)

EXAMPLE 2.8.6 Find the function $f(t)$ whose derivative is $t^{-1/2} + 5t^{-3/2} + 2t^{2/3}$ and whose graph passes through the point $(4, 1)$.

SOLUTION

$$\begin{aligned} f(t) &= \int (t^{-1/2} + 5t^{-3/2} + 2t^{2/3})dt \\ &= 2t^{1/2} - 10t^{-1/2} + \frac{6}{5}t^{5/3} + C \\ 1 = f(4) &= 4 - 5 + \frac{6}{5}(4^{5/3}) + C \end{aligned}$$

Hence, $C = 2 - \frac{6}{5}(4^{5/3})$ and

$$f(t) = 2t^{1/2} - 10t^{-1/2} + 2 + \frac{6}{5}(t^{5/3} - 4^{5/3}) \qquad \text{for } t > 0.$$

Differential Equations and Initial-Value Problems

A **differential equation** (abbreviated DE) is an equation involving one or more derivatives of an unknown function. Any function whose derivatives satisfy the differential equation is called a **solution** of the equation. For instance, $y = x^3 - x$ is a solution of the differential equation $y' = 3x^2 - 1$. This differential equation has more than one solution; in fact, $y = x^3 - x + C$ is a solution for any value of the constant C.

EXAMPLE 2.8.7 Show that for any constants A and B, the function $y = Ax^3 + B/x$ satisfies the differential equation $x^2y'' - xy' - 3y = 0$ for any $x \neq 0$.

SOLUTION If $y = Ax^3+B/x$, then for $x \neq 0$ we have $y' = 3Ax^2-B/x^2$ and $y'' = 6Ax+2B/x^3$. Therefore,

$$x^2y'' - xy' - 3y = 6Ax^3 + \frac{2B}{x} - 3Ax^3 + \frac{B}{x} - 3Ax^3 - \frac{3B}{x} = 0,$$

as required.

The **order** of a differential equation is the order of the highest order derivative appearing in the equation. The DE in the above example is a *second order* DE since it involves y''. Note that the general solution involves two arbitrary constants, A and B. The general solution of an nth order differential equation typically involves n arbitrary constants.

An **initial-value problem** (abbreviated IVP) is a problem which consists of:

i) a differential equation (to be solved for an unknown function), and

ii) prescribed values for the solution and enough of its derivatives at a particular point (the initial point) to determine values for the arbitrary constants in the solution of the DE and so provide for a unique solution.

REMARK: It is common to use the same symbol, say y, to denote both the dependent variable and the function which is the solution to a DE or an IVP; that is, we call the solution $y(x)$.

EXAMPLE 2.8.8 Use the result of Example 2.8.7 to solve the following initial-value problem.

$$\begin{cases} x^2y'' - xy' - 3y = 0 \qquad (x > 0) \\ y(1) = 2 \\ y'(1) = -6 \end{cases}$$

SOLUTION From Example 2.8.7, the DE $x^2y'' - xy' - 3y = 0$ has solution $y = Ax^3 + B/x$, which has derivative $y' = 3Ax^2 - B/x^2$. At $x = 1$ we must have $y = 2$ and $y' = -6$. Therefore

$$\begin{aligned} A+B &= 2 \\ 3A-B &= -6. \end{aligned}$$

Solving these two linear equations for A and B we get $A = -1$ and $B = 3$. Hence $y = -x^3 + 3/x$ for $x > 0$.

One of the simplest kinds of differential equation is the equation

$$\frac{dy}{dx} = f(x),$$

which is to be solved for y as a function of x. Evidently the solution is

$$y = \int f(x)\,dx.$$

Our ability to find the unknown function $y(x)$ depends on our ability to find an antiderivative of f.

EXAMPLE 2.8.9 Solve the initial-value problem

$$\begin{cases} y' = \dfrac{3}{x^2} + 2 \\ y(-2) = 1. \end{cases}$$

Where is the solution valid?

SOLUTION

$$\begin{aligned} y &= \int \left(\frac{3}{x^2} + 2\right) dx = -\frac{3}{x} + 2x + C \\ 1 &= y(-2) = \frac{3}{2} - 4 + C \end{aligned}$$

Therefore $C = \frac{7}{2}$ and

$$y = -\frac{3}{x} + 2x + \frac{7}{2}.$$

Although the solution function appears to be defined for all x except 0, it is only a solution of the given initial-value problem for $x < 0$. This is because $(-\infty, 0)$ is the largest interval which contains the initial point -2 and on which the differential equation is well defined.

EXAMPLE 2.8.10 An object falling freely near the surface of the earth is subject to a constant downward acceleration g, if the effect of air resistance is neglected. ($g \approx 9.8\text{m/s}^2$.) If we know the height y_0 of the object at an initial time $t = 0$, and if we also know the vertical velocity v_0 of the object at that time, the height $y(t)$ at other times while the object is still falling can be determined from the *second-order* initial-value problem:

$$\begin{cases} y''(t) = -g & \text{(since the acceleration } g \text{ is } \textit{downward}) \\ y(0) = y_0 \\ y'(0) = v_0. \end{cases}$$

We have

$$\begin{aligned} y'(t) &= -\int g\,dt = -gt + C_1 \\ v_0 &= y'(0) = 0 + C_1. \end{aligned}$$

Thus $C_1 = v_0$.

$$\begin{aligned} y'(t) &= -gt + v_0 \\ y(t) &= \int (-gt + v_0)dt = -\frac{1}{2}gt^2 + v_0t + C_2 \\ y_0 &= y(0) = 0 + 0 + C_2. \end{aligned}$$

Thus $C_2 = y_0$. Finally, therefore,

$$y(t) = -\frac{1}{2}gt^2 + v_0t + y_0.$$

EXAMPLE 2.8.11 A car is travelling at 72 km/h. At a certain instant its brakes are applied to produce a constant decceleration of 0.8m/s^2. How far does the car travel before coming to a stop?

SOLUTION Let $s(t)$ be the distance the car travels in the t seconds after the brakes are applied. Then $s''(t) = -0.8$ (m/s^2), so the velocity at time t is given by

$$s'(t) = \int -0.8\,dt = -0.8t + C_1.$$

Since $s'(0) = 72$ km/h $= 72 \times 1000/3600 = 20$ m/s, we have $C_1 = 20$. Thus,

$$s'(t) = 20 - 0.8t$$

and

$$s(t) = \int (20 - 0.8t)\,dt = 20t - 0.4t^2 + C_2.$$

Since $s(0) = 0$, we have $C_2 = 0$ and $s(t) = 20t - 0.4t^2$. When the car has stopped, its velocity will be 0. Hence the stopping time is the solution t of the equation

$$0 = s'(t) = 20 - 0.8t,$$

that is, $t = 25$ seconds. The distance traveled during deceleration is $s(25) = 250$ metres.

Differential equations and initial-value problems are of great importance in applications of calculus, especially for expressing in mathematical form certain laws of nature which involve rates of change of quantities. A large portion of the total mathematical endeavor of the last two hundred years has been devoted to their study. They are usually treated in separate courses on differential equations, but we mention them from time to time in this book whenever appropriate. Throughout this book, exercises about differential equations and initial-value problems are designated with the symbol †.

EXERCISES

In Exercises 1–18 find the given indefinite integrals.

1. $\int 5\,dx$

2. $\int x^2\,dx$

3. $\int \sqrt{x}\,dx$

4. $\int x^{12}\,dx$

5. $\int x^3\,dx$

6. $\int x^r\,dx \quad (r \neq -1)$

7. $\int (a^2 - x^2)\,dx$

8. $\int (A + Bx + Cx^2)\,dx$

9. $\int (2x^{1/2} + 3x^{1/3})\,dx$

10. $\int 6(x^{-1/3} - x^{-4/3})\,dx$

11. $30 \int (x^4 + x^5)\,dx$

12.* $\int \frac{dx}{(1+x)^2}$

13. $\int \left(\frac{x^3}{3} - \frac{x^2}{2} + x - 1\right) dx$

14. $105 \int (1 + t^2 + t^4 + t^6)\,dt$

15.* $\int (ax + b)^r\,dx \quad (r \neq -1)$

16.* $\int \frac{4}{\sqrt{x+1}}\,dx$

17.* $\int \sqrt{2x+3}\,dx$

18.* $\int \frac{2x}{\sqrt{x^2+1}}\,dx$

In Exercises 19–28 find the solution $y = y(x)$ to the given initial-value problem. In what interval is the solution valid?

19.† $\begin{cases} y' = x - 2 \\ y(0) = 3 \end{cases}$

20.† $\begin{cases} y' = x^{-2} - x^{-3} \\ y(-1) = 0 \end{cases}$

21.† $\begin{cases} y' = 3\sqrt{x} \\ y(4) = 1 \end{cases}$

22.† $\begin{cases} y' = x^{1/3} \\ y(0) = 5 \end{cases}$

23.† $\begin{cases} y' = Ax^2 + Bx + C \\ y(1) = 1 \end{cases}$

24.† $\begin{cases} y' = x^{-9/7} \\ y(1) = -4 \end{cases}$

25.† $\begin{cases} y'' = 2 \\ y'(0) = 5 \\ y(0) = -3 \end{cases}$

26.† $\begin{cases} y'' = x^{-4} \\ y'(1) = 2 \\ y(1) = 1 \end{cases}$

27.† $\begin{cases} y'' = x^3 - 1 \\ y'(0) = 0 \\ y(0) = 8 \end{cases}$

28.† $\begin{cases} y'' = 5x^2 - 3x^{-1/2} \\ y'(1) = 2 \\ y(1) = 0 \end{cases}$

29.† Show that for any constants A and B the function $y = y(x) = Ax + B/x$ satisfies the *second-order differential equation* $x^2y'' + xy' - y = 0$ for $x \neq 0$. Find a function y satisfying the initial value problem:

$$\begin{cases} x^2y'' + xy' - y = 0 \qquad (x > 0) \\ y(1) = 2 \\ y'(1) = 4. \end{cases}$$

30. Redo Example 2.8.11 using instead a non-constant deceleration, $s''(t) = -t$ m/s^2.

31.† Show that for any constants A and B the function $y = Ax^{r_1} + Bx^{r_2}$ satisfies, for $x > 0$, the differential equation $ax^2y'' + bxy' + cy = 0$, provided r_1 and r_2 are two distinct rational roots of the quadratic equation $ar(r-1) + br + c = 0$.

Use the result of Exercise 31 to solve the initial-value problems in Exercises 32–33 on the interval $x > 0$.

32.† $\begin{cases} x^2y'' - 6y = 0 \\ y(1) = 1 \\ y'(1) = 1 \end{cases}$

33.† $\begin{cases} 4x^2y'' + 4xy' - y = 0 \\ y(4) = 2 \\ y'(4) = -2 \end{cases}$

CHAPTER 3

The Elementary Transcendental Functions

The functions we have encountered so far in our study of calculus have been of three main types, *polynomials*, *rational functions* (quotients of polynomials), and *algebraic functions* (which involve taking fractional powers of rational functions). All of these functions are constructed using finitely many arithmetic operations (addition, subtraction, multiplication, and division) and the extraction of finitely many roots (fractional powers). Functions which cannot be so constructed are called **transcendental functions**. Much of the importance of calculus and many of its most useful applications centre on its ability to illuminate the behavior of transcendental functions that arise naturally when we attempt to model concrete problems in mathematical terms. Of special importance are the circular (or trigonometric) functions and their inverses, exponential functions, and their inverses the logarithmic functions. These functions are the subject of this chapter.

3.1 THE CIRCULAR (TRIGONOMETRIC) FUNCTIONS

Most students of calculus will be somewhat familiar with the trigonometric functions from their previous mathematical studies. Although you may have first encountered the *sine* and *cosine* of an acute angle as ratios of sides of a right-angled triangle, we will give a more general definition of them here, one which shows how to obtain their values for any real number. Later in this section we will review their use in the measurement of triangles.

3.1.1
Definition of Radian Measure, Sine and Cosine

Let C be the circle with equation $x^2+y^2=1$ and A the point $(1,0)$ on C. For any real number t, let P_t be the point on C obtained by moving a distance $|t|$ from A, measured along the circle in a counterclockwise direction if $t>0$ and in a clockwise direction if $t<0$. Then the **radian measure** of angle AOP_t is t (see Fig. 3.1.1).

$$\text{angle } AOP_t = t \text{ radians}$$

If P_t has coordinates (x_t, y_t), then **cosine** of t and **sine** of t (abbreviated $\cos t$ and $\sin t$) are defined by

$$\cos t = x_t \qquad \sin t = y_t.$$

According to this definition, any real number can be interpreted as a (signed) angle, and its cosine and sine can be determined. Since the circumference of the circle C is 2π, the relationship between radian and degree measure is 2π radians $= 360°$, or

$$\pi \text{ radians} = 180°.$$

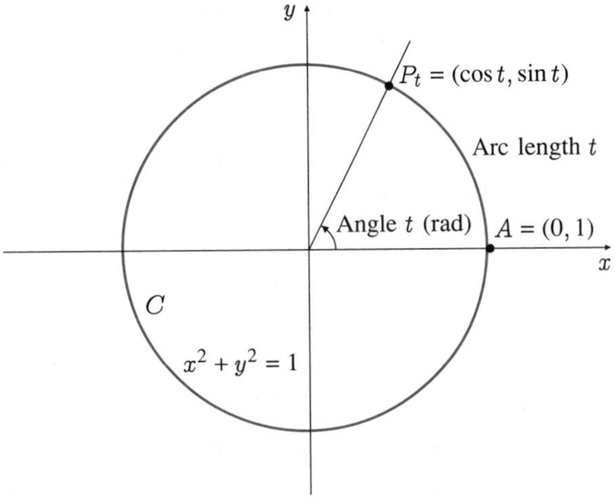

FIGURE 3.1.1

Note that the unit radians is not usually written when an angle is expressed in radians: When we refer to the angle $\frac{\pi}{2}$ we mean $\frac{\pi}{2}$ radians or 90°. **In calculus all angles are assumed to be in radians unless degrees or other units are stated explicitly.**

If an arc of length s on a circle of radius r subtends an angle t at the centre of the circle, then, because the circumference of the whole circle is $2\pi r$, we must have

$$s = \frac{t}{2\pi}(2\pi r) = rt.$$

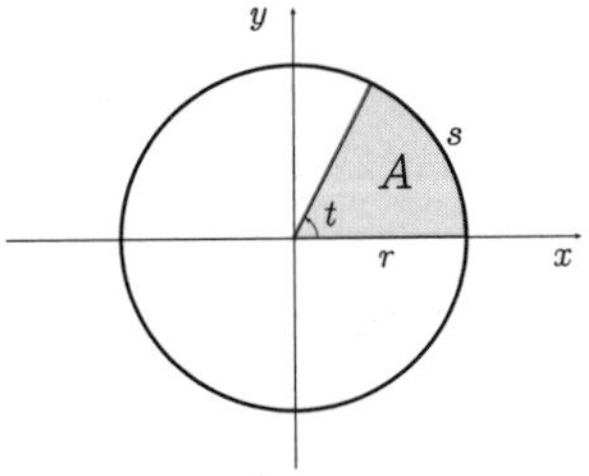

FIGURE 3.1.2

(See Fig. 3.1.2.) Similarly, the area of the circular sector with angle t is

$$A = \frac{t}{2\pi}(\pi r^2) = \frac{r^2 t}{2}.$$

REMARK: We are assuming here that we know that the area of a circle of radius r units is πr^2 square units. The number π is usually defined as the ratio of the circumference to the diameter of a circle, which is the same for all circles because all circles are "similar" geometrical figures. It is not obvious, and requires proof, that the ratio of the area of a circle to the square of its radius is the same number π. For the time being we will assume this, but will give a proof in Section 6.1 when we begin our discussion of area.

The functions cosine and sine are called **circular functions** because they are defined as coordinates of a point on a circle. Many properties of the these functions follow directly from the Definition 3.1.1. Here are the most important of these properties:

a) **The Pythagorean Identity**

$P_t = (\cos t, \sin t)$ lies on the circle $x^2 + y^2 = 1$; therefore, for any t, $x = \cos t$ and $y = \sin t$ satisfy $x^2 + y^2 = 1$. Hence,

3.1.2
The Pythagorean Identity

$$\cos^2 t + \sin^2 t = 1.$$

Note that in common usage, the symbol $\cos^2 t$ means $(\cos t)^2$ rather than $\cos(\cos t)$. Similarly $\sin^2 t$ means $(\sin t)^2$.

b) **Periodicity**

The circle C has circumference 2π; therefore, $P_{t+2\pi} = P_t$ and

3.1.3
Periodicity of cos and sin

$$\cos(t + 2\pi) = \cos t, \qquad \sin(t + 2\pi) = \sin t;$$

we say that cosine and sine are **periodic functions** with *period* 2π.

c) **Symmetry**

For any t, P_t and P_{-t} have the same x-coordinates and opposite y-coordinates. Thus cosine is an *even function* and sine is an *odd function*. See Fig. 3.1.3.

3.1.4
cos is even
sin is odd

$$\cos(-t) = \cos t, \qquad \sin(-t) = -\sin t.$$

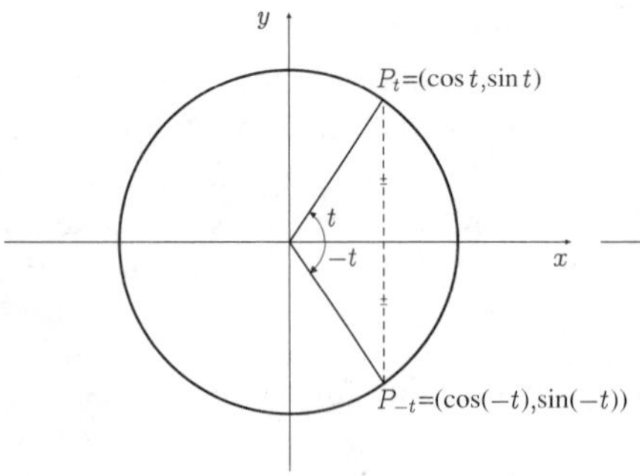

FIGURE 3.1.3

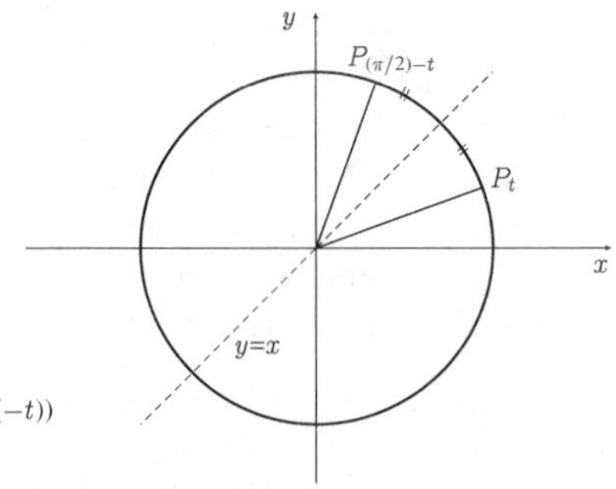

FIGURE 3.1.4

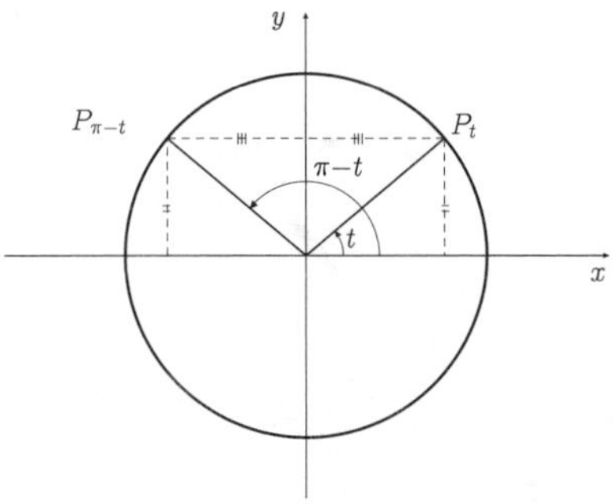

FIGURE 3.1.5

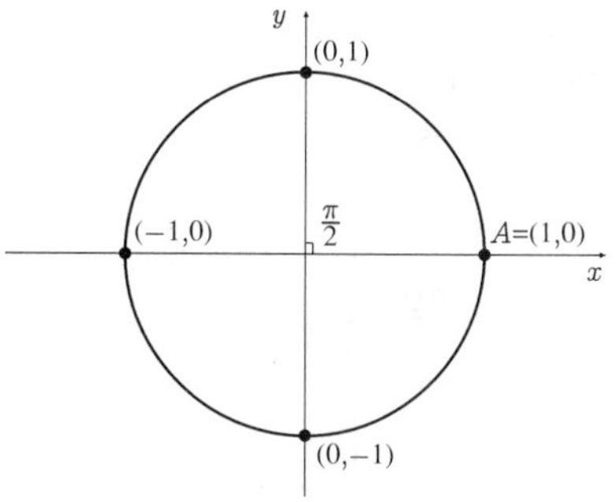

FIGURE 3.1.6

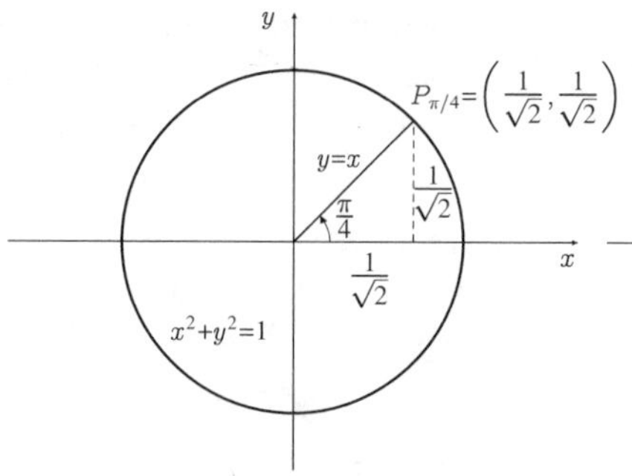

FIGURE 3.1.7

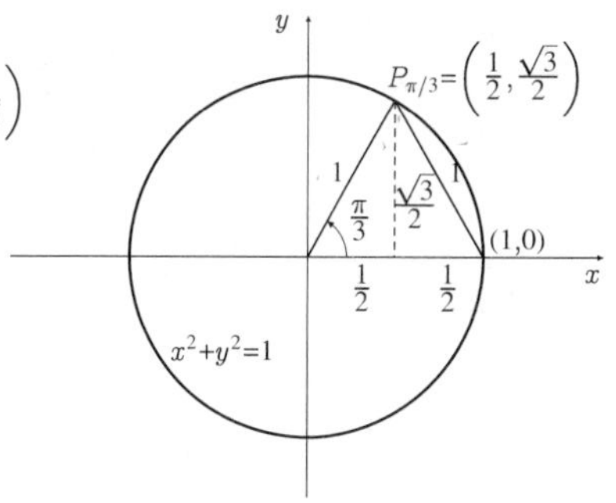

FIGURE 3.1.8

d) **Complementary Angle Identities**
The two angles t and $\frac{\pi}{2}-t$ have sum equal to $\frac{\pi}{2}$ radians (or 90°) and are therefore said to be *complementary*. The points P_t and $P_{(\pi/2)-t}$ are mirror images of each other in the line $y=x$ (see Fig 3.1.4), so their x and y coordinates are reversed:

3.1.5
Complementary Angle Identities

$$\cos\left(\frac{\pi}{2}-t\right)=\sin t, \qquad \sin\left(\frac{\pi}{2}-t\right)=\cos t.$$

e) **Supplementary Angle Identies**
The angles t and $\pi-t$ have sum π radians (or 180°) and are said to be *supplementary*. The points $P_{\pi-t}$ and P_t are mirror images of each other in the y-axis (see Fig. 3.1.5). Thus,

3.1.6
Supplementary Angle Identities

$$\cos(\pi-t)=-\cos t. \qquad \sin(\pi-t)=\sin t.$$

Cosine and Sine of Special Angles

The values of cosine and sine for multiples of $\frac{\pi}{2}$ can be read directly from Fig. 3.1.6; for example,

$$\cos 0=1, \qquad \sin 0=0, \qquad \cos\left(\frac{\pi}{2}\right)=0, \qquad \sin\left(\frac{\pi}{2}\right)=1.$$

The line $y=x$ intersects the circle $x^2+y^2=1$ at $\left(\frac{1}{\sqrt{2}},\frac{1}{\sqrt{2}}\right)$. (See Fig. 3-1-7.) Since this line makes an angle $\frac{\pi}{4}$ (or 45°) with the positive x-axis, the cosine and sine of $\frac{\pi}{4}$ must each be $\frac{1}{\sqrt{2}}$. The large triangle in Fig. 3.1.8 is equilateral with 1 unit sides. Its altitude is therefore $\frac{\sqrt{3}}{2}$. Thus the sine of $\frac{\pi}{3}$ (or 60°) is $\frac{\sqrt{3}}{2}$ and the cosine is $\frac{1}{2}$. For $\frac{\pi}{6}$, or 30°, the angle complementary to 60°, these values are reversed.

The following table summarizes the values of cosine and sine at multiples of 30° and 45° between 0° and 180°. The values for 120°, 135°, and 150° were determined by using the supplementary angle identities; for example,

$$\cos(120°) = \cos\left(\frac{2\pi}{3}\right) = \cos\left(\pi - \frac{\pi}{3}\right) = -\cos\left(\frac{\pi}{3}\right) = -\cos(60°) = -\frac{1}{2}.$$

Degrees	0°	30°	45°	60°	90°	120°	135°	150°	180°
Radians	0	$\frac{\pi}{6}$	$\frac{\pi}{4}$	$\frac{\pi}{3}$	$\frac{\pi}{2}$	$\frac{2\pi}{3}$	$\frac{3\pi}{4}$	$\frac{5\pi}{6}$	π
cosine	1	$\frac{\sqrt{3}}{2}$	$\frac{1}{\sqrt{2}}$	$\frac{1}{2}$	0	$-\frac{1}{2}$	$-\frac{1}{\sqrt{2}}$	$-\frac{\sqrt{3}}{2}$	-1
sine	0	$\frac{1}{2}$	$\frac{1}{\sqrt{2}}$	$\frac{\sqrt{3}}{2}$	1	$\frac{\sqrt{3}}{2}$	$\frac{1}{\sqrt{2}}$	$\frac{1}{2}$	0

While decimal approximations to the values of sine and cosine can be found in mathematical tables or by using a scientific calculator, it is useful to remember the values for angles 0, $\frac{\pi}{6}$, $\frac{\pi}{4}$, $\frac{\pi}{3}$, and $\frac{\pi}{2}$ (as presented in the table above) as they occur frequently in applications. When using a scientific calculator to calculate any trigonometric functions, be sure you have selected the proper angular mode: degrees or radians. Some calculators also have a third mode, grads (100 grads = 90 degrees = $\frac{\pi}{2}$ radians). We will make no use of grads. The graphs of cosine and sine are shown in Figures 3.1.9 and 3.1.10. Both functions are continuous on the whole real line.

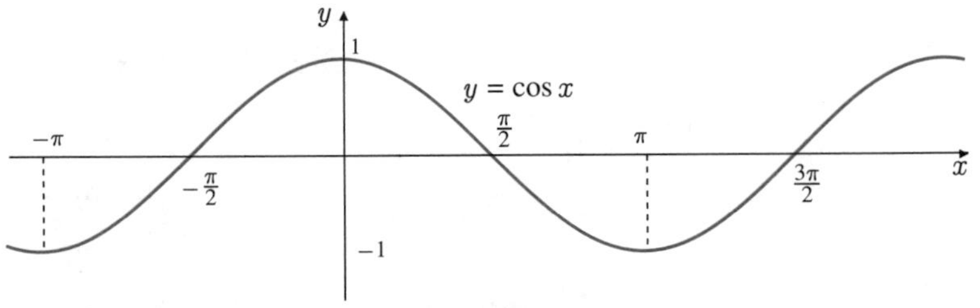

FIGURE 3.1.9

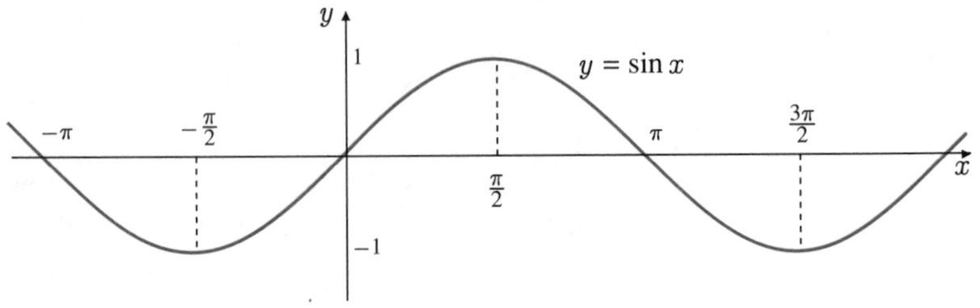

FIGURE 3.1.10

The Addition Formulas

The following formulas enable us to determine the cosine and sine of a sum or difference of two angles in terms of the cosines and sines of those angles.

3.1.7 The Addition Formulas

$$\cos(s+t) = \cos s \cos t - \sin s \sin t$$
$$\sin(s+t) = \sin s \cos t + \cos s \sin t$$
$$\cos(s-t) = \cos s \cos t + \sin s \sin t$$
$$\sin(s-t) = \sin s \cos t - \cos s \sin t$$

We prove the third of these formulas as follows: Let s and t be real numbers and consider the points

$$P_t = (\cos t, \sin t) \qquad P_{s-t} = (\cos(s-t), \sin(s-t))$$
$$P_s = (\cos s, \sin s) \qquad A = (1, 0)$$

as shown in Fig. 3.1.11.

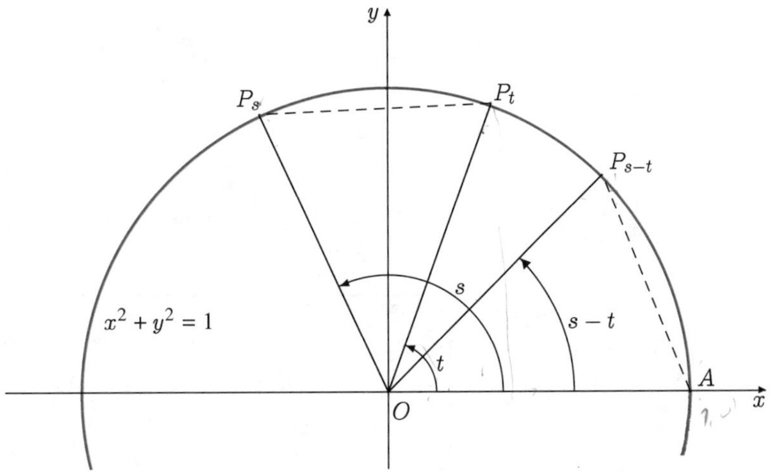

FIGURE 3.1.11

Angle $P_tOP_s = s - t$ radians = angle AOP_{s-t}; therefore,

$$(\text{distance } P_s \text{ to } P_t)^2 = (\text{distance } P_{s-t} \text{ to } A)^2.$$

Expressing these squared distances in terms of coordinates, we obtain

$$(\cos s - \cos t)^2 + (\sin s - \sin t)^2 = (\cos(s-t) - 1)^2 + \sin^2(s-t),$$

$$\cos^2 s - 2\cos s\cos t + \cos^2 t + \sin^2 s - 2\sin s\sin t + \sin^2 t$$
$$= \cos^2(s-t) - 2\cos(s-t) + 1 + \sin^2(s-t).$$

Since $\cos^2 x + \sin^2 x = 1$ for every x, this reduces to

$$\cos(s-t) = \cos s\cos t + \sin s\sin t.$$

Replacing t with $-t$ in the formula above, and recalling that $\cos(-t) = \cos t$ and $\sin(-t) = -\sin t$, we have

$$\cos(s+t) = \cos s\cos t - \sin s\sin t.$$

The complementary angle formulas can be used to obtain either of the addition formulas for sine:

$$\begin{aligned}\sin(s+t) &= \cos\left(\frac{\pi}{2} - (s+t)\right)\\ &= \cos\left(\left(\frac{\pi}{2} - s\right) - t\right)\\ &= \cos\left(\frac{\pi}{2} - s\right)\cos t + \sin\left(\frac{\pi}{2} - s\right)\sin t\\ &= \sin s\cos t + \cos s\sin t,\end{aligned}$$

and the other formula again follows if we replace t with $-t$.

EXAMPLE 3.1.8 Find the value of $\cos(\pi/12) = \cos 15°$.

SOLUTION

$$\begin{aligned}\cos\frac{\pi}{12} = \cos\left(\frac{\pi}{3} - \frac{\pi}{4}\right) &= \cos\frac{\pi}{3}\cos\frac{\pi}{4} + \sin\frac{\pi}{3}\sin\frac{\pi}{4}\\ &= \left(\frac{1}{2}\right)\left(\frac{1}{\sqrt{2}}\right) + \left(\frac{\sqrt{3}}{2}\right)\left(\frac{1}{\sqrt{2}}\right) = \frac{1+\sqrt{3}}{2\sqrt{2}}\end{aligned}$$

From the addition formulas, we obtain as special cases certain useful formulas called **half-angle formulas**. Put $s = t$ in the addition formulas for $\sin(s+t)$ and $\cos(s+t)$ to get

3.1.9 Half-Angle Formulas

$$\begin{aligned}\sin 2t &= 2\sin t\cos t \qquad \text{and}\\ \cos 2t &= \cos^2 t - \sin^2 t\\ &= 2\cos^2 t - 1 \qquad (\text{using } \sin^2 t + \cos^2 t = 1)\\ &= 1 - 2\sin^2 t\end{aligned}$$

From these latter,

$$\cos^2 t = \frac{1+\cos 2t}{2}, \qquad \text{and } \sin^2 t = \frac{1-\cos 2t}{2}.$$

Later we will find these formulas useful when we have to integrate even powers of $\cos x$ and $\sin x$.

Other Trigonometric Functions

The remaining four trigonometric functions, tangent (tan), cotangent (cot), secant (sec), and cosecant (csc), are defined in terms of cosine and sine. Their graphs are shown in Figures 3.1.12–3.1.15.

3.1.10
The Functions tan, cot, sec and csc

$$\tan t = \frac{\sin t}{\cos t} \qquad \sec t = \frac{1}{\cos t}$$
$$\cot t = \frac{\cos t}{\sin t} = \frac{1}{\tan t} \qquad \csc t = \frac{1}{\sin t}$$

Domain & Ranges of Functions

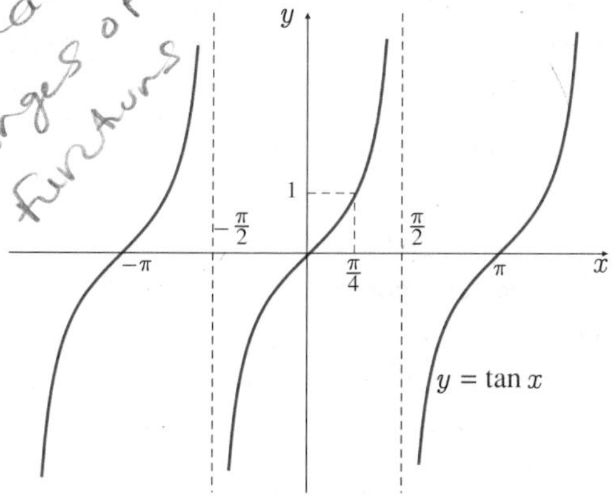

FIGURE 3.1.12

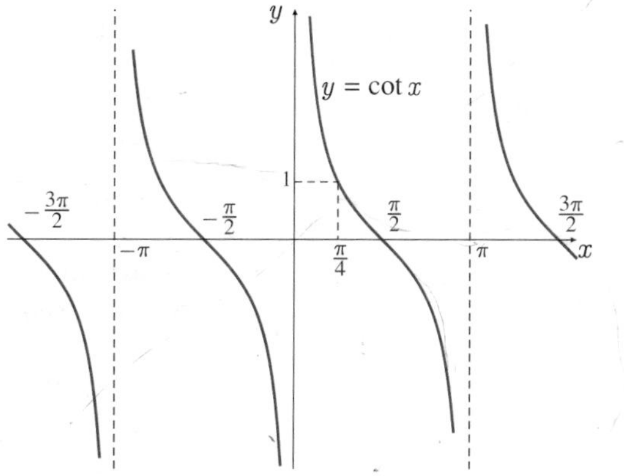

FIGURE 3.1.13

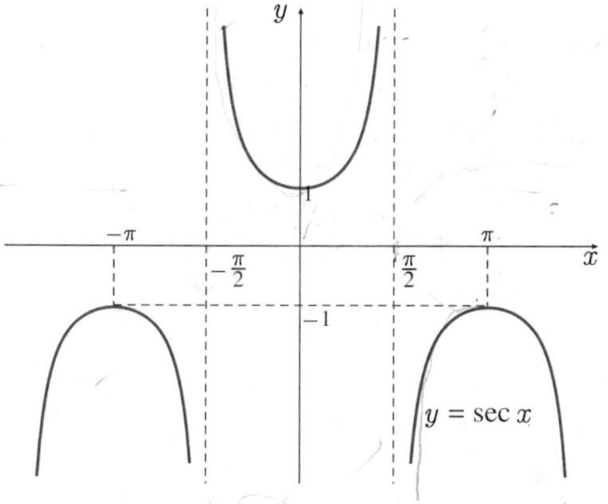

FIGURE 3.1.14

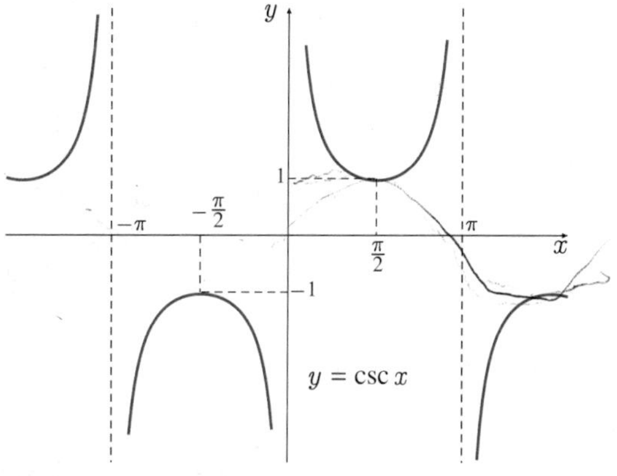

FIGURE 3.1.15

Observe that each of these functions is undefined (and its graph has vertical asymptotes) at points where the function in the denominator of its defining fraction has value 0. Observe also that tangent, cotangent, and cosecant are odd functions and the secant is an even function. Since $|\sin x| \le 1$ and $|\cos x| \le 1$ for all x, $|\csc x| \ge 1$ and $|\sec x| \ge 1$ for all x where they are defined.

The three functions sine, cosine, and tangent are called the **primary trigonometric functions** while their reciprocals, cosecant, secant and cotangent are called **secondary trigonometric functions**. Figure 3.1.16 provides a useful mnemonic called the "CAST RULE" to help you remember where the primary functions are positive. All three are positive in the first quadrant, marked A. Of the three, only sine is positive in the second quadrant S, only tangent in the third quadrant T, and only cosine in the fourth quadrant C.

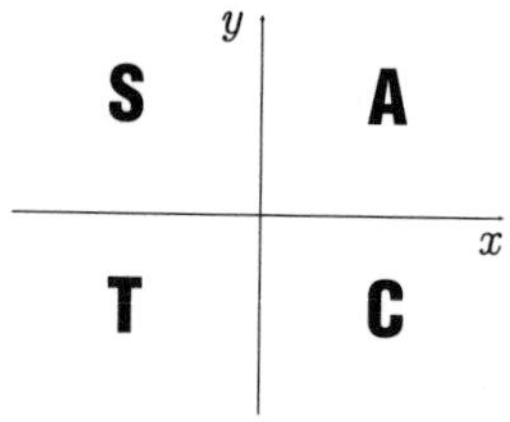

FIGURE 3.1.16

Like their reciprocals cosine and sine, the functions secant and cosecant are periodic with period 2π. So are tangent and cotangent, but these also have period π:

$$\tan(x+\pi) = \frac{\sin(x+\pi)}{\cos(x+\pi)} = \frac{\sin x \cos\pi + \cos x \sin\pi}{\cos x \cos\pi - \sin x \sin\pi} = \frac{-\sin x}{-\cos x} = \tan x.$$

Division of the Pythagorean identity $\sin^2 x + \cos^2 x = 1$ by $\sin^2 x$ and $\cos^2 x$, respectively, leads to the two alternate versions of that identity:

$$1 + \cot^2 x = \csc^2 x, \qquad 1 + \tan^2 x = \sec^2 x.$$

Addition formulas for tangent and cotangent can be obtained from those for sine and cosine. For example,

$$\tan(s+t) = \frac{\sin(s+t)}{\cos(s+t)} = \frac{\sin s \cos t + \cos s \sin t}{\cos s \cos t - \sin s \sin t} = \frac{\tan s + \tan t}{1 - \tan s \tan t}.$$

Similarly,

$$\tan(s-t) = \frac{\tan s - \tan t}{1 + \tan s \tan t}.$$

Some Trigonometry

The trigonometric functions are so called because of their usefulness in expressing the relationships between the sides and angles of a triangle.

The sides of a right-angled triangle may be designated with respect to one of the acute angles of the triangle as hyp (hypotenuse), opp (side opposite the angle), and adj (side adjacent to the angle). If the angle is t, as shown in Fig. 3.1.17, then the triangle is similar to the one with hypotenuse 1, side opposite having length $\sin t$, and side adjacent having length $\cos t$ (Fig. 3.1.18). Thus,

$$\sin t = \frac{\text{opp}}{\text{hyp}}, \qquad \cos t = \frac{\text{adj}}{\text{hyp}}, \qquad \tan t = \frac{\text{opp}}{\text{adj}}.$$

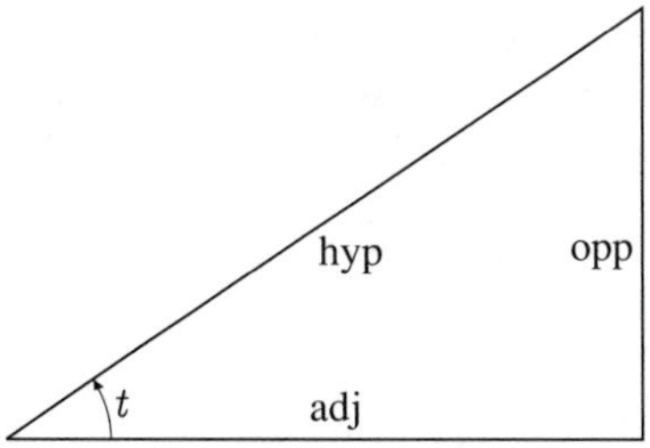

FIGURE 3.1.17

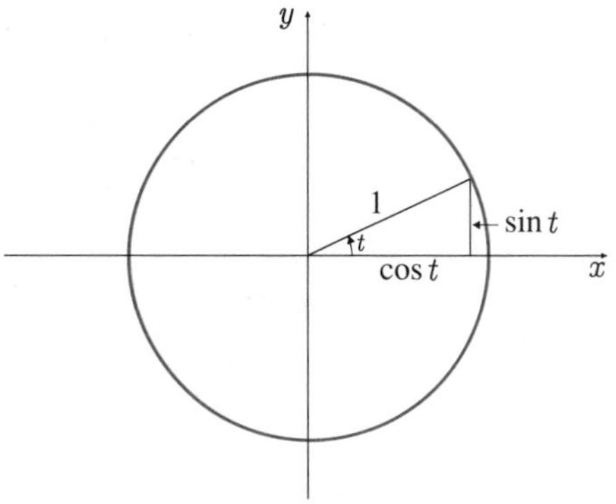

FIGURE 3.1.18

EXAMPLE 3.1.11 Find the unknown sides x and y in the triangles in Figures 3.1.19 and 3.1.20.

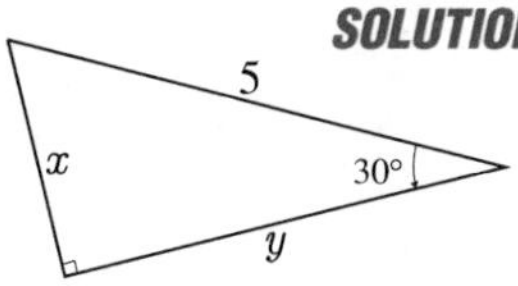

FIGURE 3.1.19

SOLUTION For the first triangle, x is the side opposite and y is the side adjacent the 30° angle. The hypotenuse of the triangle is 5 units. Thus

$$\frac{x}{5} = \sin 30° = \frac{1}{2}, \qquad \frac{y}{5} = \cos 30° = \frac{\sqrt{3}}{2},$$

so $x = \dfrac{5}{2}$ units and $y = \dfrac{5\sqrt{3}}{2}$ units.

For the second triangle, x is the side opposite the angle θ and y is the hypotenuse. The side adjacent θ is a. Thus

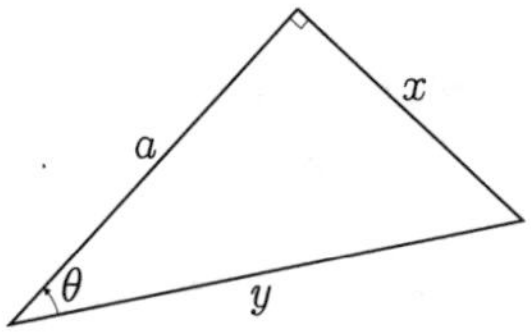

FIGURE 3.1.20

$$\frac{x}{a} = \tan\theta, \qquad \frac{a}{y} = \cos\theta.$$

Hence $x = a\ \tan\theta$ and $y = \dfrac{a}{\cos\theta} = a\ \sec\theta$.

For an arbitrary triangle ABC with sides a, b, and c and opposite angles A, B, and C, respectively, the following theorem presents two results, called the **Sine Law** and the **Cosine Law**, that describe relationships in the triangle.

THEOREM 3.1.12

$$\frac{\sin A}{a} = \frac{\sin B}{b} = \frac{\sin C}{c} \qquad \text{Sine Law}$$

$$c^2 = a^2 + b^2 - 2ab\cos C \qquad \text{Cosine Law}$$

PROOF See Figures 3.1.21 and 3.1.22. Let h be the length of the perpendicular from A to the side BC. From right-angled triangles (and using $\sin(\pi - t) = \sin t$ if required), we get $c \sin B = h = b \sin C$. Thus $(\sin B)/b = (\sin C)/c$. By symmetry of the formulas (or by dropping a perpendicular to another side), both of these fractions must be equal to $(\sin A)/a$, so the Sine Law is proved. For the cosine law observe that

$$\begin{aligned} c^2 &= \begin{cases} h^2 + (a - b\cos C)^2 & \text{if } C \le \dfrac{\pi}{2} \\ h^2 + (a + b\cos(\pi - C))^2 & \text{if } C > \dfrac{\pi}{2} \end{cases} \\ &= h^2 + (a - b\cos C)^2 \qquad (\text{since } \cos(\pi - C) = -\cos C) \\ &= b^2 \sin^2 C + a^2 - 2ab\cos C + b^2 \cos^2 C \\ &= a^2 + b^2 - 2ab\cos C. \ \square \end{aligned}$$

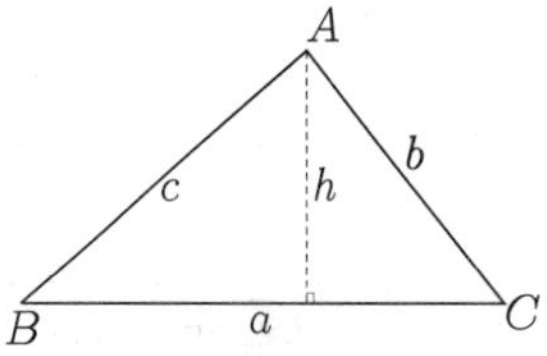

FIGURE 3.1.21

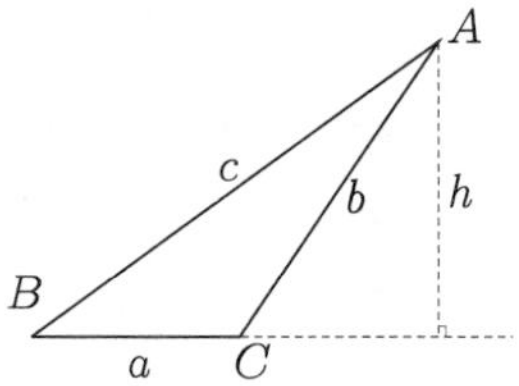

FIGURE 3.1.22

EXERCISES

Find the values of the quantities in Exercises 1–12 by using various formulas presented in this section. Do not use tables or a calculator.

1. $\cos \dfrac{3\pi}{4}$

2. $\tan \dfrac{5\pi}{4}$

3. $\sin \dfrac{2\pi}{3}$

4. $\cos \dfrac{5\pi}{6}$

5. $\sin \dfrac{7\pi}{12}$

6. $\cos \dfrac{5\pi}{12}$

7. $\cos \dfrac{7\pi}{12}$

8. $\sin \dfrac{11\pi}{12}$

9. $\sec \dfrac{11\pi}{12}$

10. $\tan \dfrac{-\pi}{3}$

11. $\sin \dfrac{4\pi}{3}$

12. $\sec \left(-\dfrac{\pi}{12}\right)$

In Exercises 13–20 express the given quantity in terms of $\sin x$ and $\cos x$.

13. $\cos(\pi + x)$

14. $\sin(2\pi - x)$

15. $\sin \left(\dfrac{3\pi}{2} - x\right)$

16. $\cos \left(\dfrac{3\pi}{2} + x\right)$

17. $\tan x + \cot x$

18. $\sec x - \tan x$

19. $\sec^2 x + \csc^2 x$

20. $\dfrac{\tan x - \cot x}{\tan x + \cot x}$

In Exercises 21–23 prove the given identities.

21. $\dfrac{1 - \cos x}{\sin x} = \dfrac{\sin x}{1 + \cos x} = \tan \dfrac{x}{2}$

22. $\dfrac{1 - \cos x}{1 + \cos x} = \tan^2 \dfrac{x}{2}$

23. $\dfrac{\cos x - \sin x}{\cos x + \sin x} = \sec 2x - \tan 2x$

24. Express $\sin 3x$ and $\cos 3x$ in terms of $\sin x$ and $\cos x$.

In Exercises 25–28 sketch the graph of the given function. What is the period of the function?

25. $\cos 2x$

26. $\sin \dfrac{x}{2}$

27. $\sin \pi x$

28. $\cos \dfrac{\pi x}{2}$

In Exercises 29–34 $\sin x$, $\cos x$, or $\tan x$ is given. Find the other two primary trigonometric functions of x if x lies in the specified interval.

29. $\sin x = \frac{3}{5}, \quad x \text{ in } \left[\frac{\pi}{2}, \pi\right]$

30. $\tan x = 2, \quad x \text{ in } \left[0, \frac{\pi}{2}\right]$

31. $\cos x = \frac{1}{3}, \quad x \text{ in } \left[-\frac{\pi}{2}, 0\right]$

32. $\cos x = -\frac{5}{13}, \quad x \text{ in } \left[\frac{\pi}{2}, \pi\right]$

33. $\sin x = \frac{-1}{2}, \quad x \text{ in } \left[\pi, \frac{3\pi}{2}\right]$

34. $\tan x = \frac{1}{2}, \quad x \text{ in } \left[\pi, \frac{3\pi}{2}\right]$

In Exercises 35–46 ABC is a triangle with right angle at C. The sides opposite angles A, B, and C are a, b, and c, respectively.

35. Find a and b if $c = 2$, $B = \frac{\pi}{3}$.

36. Find a and c if $b = 2$, $B = \frac{\pi}{3}$.

37. Find b and c if $a = 5$, $B = \frac{\pi}{6}$.

38. Express a in terms of A and c.

39. Express a in terms of A and b.

40. Express a in terms of B and c.

41. Express a in terms of B and b.

42. Express c in terms of A and a.

43. Express c in terms of A and b.

44. Express $\sin A$ in terms of a and c.

45. Express $\sin A$ in terms of b and c.

46. Express $\sin A$ in terms of a and b.

In Exercises 47–56 ABC is an arbitrary triangle with sides a, b, and c, opposite to angles A, B, and C, respectively. Find the indicated quantities. Use tables or a scientific calculator if necessary.

47. Find $\sin B$ if $a = 4$, $b = 3$, $A = \frac{\pi}{4}$.

48. Find $\cos A$ if $a = 2$, $b = 2$, $c = 3$.

49. Find $\sin B$ if $a = 2$, $b = 3$, $c = 4$.

50. Find c if $a = 2$, $b = 3$, $C = \frac{\pi}{4}$.

51. Find a if $c = 3$, $A = \frac{\pi}{4}$, $B = \frac{\pi}{3}$.

52. Find c if $a = 2$, $b = 3$, $C = 35°$.

53. Find b if $a = 4$, $B = 40°$, $C = 70°$.

54.*Find c if $a = 1$, $b = \sqrt{2}$, $A = 30°$. (There are two possible answers.)

55. Show that the area of triangle ABC is given by $(1/2)ab\sin C = (1/2)bc\sin A = (1/2)ca\sin B$.

56.*Show that the area of triangle ABC is given by $\sqrt{s(s-a)(s-b)(s-c)}$ where $s = (a+b+c)/2$ is the semi-perimeter of the triangle.

57. Two guy wires stretch from the top T of a vertical pole to points B and C on the ground, where C is 10 m closer to the base of the pole than is B. If wire BT makes angle of 35° with the horizontal, and wire CT makes an angle of 50° with the horizontal, how high is the pole?

58. Observers at position A and B 2km apart simultaneously measure the angle of elevation of a weather balloon to be 40° and 70°, respectively. If the ballon is directly above a point on the line segment between A and B, find the height of the ballon.

3.2 DERIVATIVES OF THE TRIGONOMETRIC FUNCTIONS

A careful inspection of the graphs of sine and cosine (Figures 3.1.9 and 3.1.10) might lead one to guess the following formulas:

3.2.1
Derivatives of sin and cos

$$\frac{d}{dx}\sin x = \cos x, \qquad \frac{d}{dx}\cos x = -\sin x$$

In order to prove that these formulas are indeed correct we require the following result, which amounts to the assertion that at $x = 0$, $\sin x$ has derivative 1, and $\cos x$ has derivative 0.

THEOREM 3.2.2

$$\lim_{h\to 0} \frac{\sin h}{h} = 1, \qquad \lim_{h\to 0} \frac{\cos h - 1}{h} = 0$$

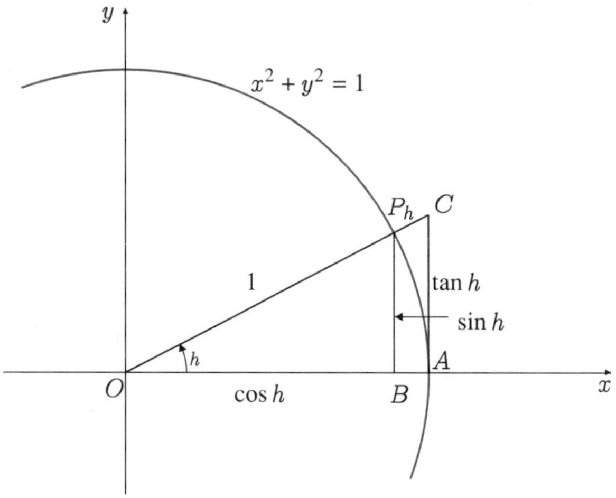

FIGURE 3.2.1

PROOF Since $\sin h$ and h are odd functions, $(\sin h)/h$ is an *even function* of h, and it is sufficient to prove the first limit for $h \to 0+$; the limit as $h \to 0-$ must necessarily be the same. (Why?) Let h be a small positive number (between 0 and $\frac{\pi}{2}$ say), and represent h as angle AOP_h in the unit circle (see Fig. 3.2.1). Thus $A = (1, 0)$ and $P_h = (\cos h, \sin h)$. If C lies on the line through O and P_h and CA is perpendicular to OA, then, by similar triangles,

$$\frac{CA}{OA} = \frac{P_hB}{OB} = \frac{\sin h}{\cos h} = \tan h.$$

Hence $C = (1, (\sin h/\cos h)) = (1, \tan h)$. Observe in the diagram that

Area of triangle P_hOB < Area of circular sector P_hOA < Area of triangle COA.

$$\frac{1}{2}\sin h \cos h < \frac{1}{2}h < \frac{1}{2}\frac{\sin h}{\cos h}.$$

Since h and $\sin h$ are greater than 0, it follows that

$$\cos h < \frac{h}{\sin h} < \frac{1}{\cos h}.$$

Since $\lim_{h\to 0} \cos h = 1$, we have, by the Squeeze Theorem (Theorem 1.5.8),

$$\lim_{h\to 0+} \frac{h}{\sin h} = 1, \quad \text{and therefore,} \quad \lim_{h\to 0+} \frac{\sin h}{h} = \frac{1}{1} = 1.$$

The second limit follows from the first via one of the half-angle formulas and Theorem 1.5.7:

$$\begin{aligned}\lim_{h\to 0}\frac{\cos h - 1}{h} &= \lim_{h\to 0} -2\frac{\sin^2(h/2)}{h}\\ &= \lim_{h/2\to 0} -\frac{\sin(h/2)}{h/2}\sin(h/2) = (-1)\times(0) = 0. \quad \square\end{aligned}$$

Calculating the Derivatives of sin and cos

We now use the two limits calculated above to determine the derivative of the sine function.

$$\begin{aligned}\frac{d}{dx}\sin x &= \lim_{h\to 0}\frac{\sin(x+h)-\sin x}{h} && \text{(the definition of derivative)}\\ &= \lim_{h\to 0}\frac{\sin x\cos h + \cos x \sin h - \sin x}{h} && \text{(addition formula for sin)}\\ &= \lim_{h\to 0}\left(\sin x\frac{\cos h - 1}{h} + \cos x\frac{\sin h}{h}\right)\\ &= (\sin x)(0) + (\cos x)(1) = \cos x && \text{(by the two limits above)}\end{aligned}$$

We could calculate the derivative of cosine in the same manner, using the same two limits. An alternative method makes use of the complementary angle formulas and the chain rule:

$$\frac{d}{dx}\cos x = \frac{d}{dx}\sin\left(\frac{\pi}{2}-x\right) = -\cos\left(\frac{\pi}{2}-x\right) = -\sin x.$$

EXAMPLE 3.2.3 Find the derivatives of $\cos(5x-2)$, $\sin(x^2+1)$, $\sin^3 \pi x$, and $\cos(3\cos(5\cos 7x))$.

SOLUTION

$$\frac{d}{dx}\cos(5x-2) = \big(-\sin(5x-2)\big)(5) = -5\sin(5x-2)$$

$$\frac{d}{dx}\sin(x^2+1) = \big(\cos(x^2+1)\big)(2x) = 2x\cos(x^2+1)$$

$$\frac{d}{dx}\sin^3 \pi x = (3\sin^2\pi x)(\cos \pi x)(\pi) = 3\pi(\sin^2\pi x)(\cos\pi x)$$

$$\begin{aligned}\frac{d}{dx}\cos\big(3\cos(5\cos 7x)\big) &= \big[-\sin(3\cos\big(5\cos 7x)\big)\big]\big[-3\sin(5\cos 7x)\big]\big[-5\sin 7x\big](7)\\ &= -105\sin 7x\sin(5\cos 7x)\sin\big(3\cos(5\cos 7x)\big)\end{aligned}$$

EXAMPLE 3.2.4 Derive *and memorize* the following formulas.

$$\frac{d}{dx}\tan x = \sec^2 x \qquad \frac{d}{dx}\cot x = -\csc^2 x$$

$$\frac{d}{dx}\sec x = \sec x\tan x \qquad \frac{d}{dx}\csc x = -\csc x\cot x$$

SOLUTION Using the Quotient rule and the Pythagorean identity we calculate

$$\frac{d}{dx}\tan x = \frac{d}{dx}\frac{\sin x}{\cos x} = \frac{\cos x\cos x - \sin x(-\sin x)}{\cos^2 x} = \frac{1}{\cos^2 x} = \sec^2 x,$$

Similarly, by three applications of the Reciprocal Rule,

$$\frac{d}{dx}\cot x = \frac{d}{dx}\frac{1}{\tan x} = \frac{-1}{\tan^2 x}\sec^2 x = \frac{-1}{\sin^2 x} = -\csc^2 x,$$
$$\frac{d}{dx}\sec x = \frac{d}{dx}\frac{1}{\cos x} = \frac{-1}{\cos^2 x}(-\sin x) = \sec x\tan x,$$
$$\frac{d}{dx}\csc x = \frac{d}{dx}\frac{1}{\sin x} = \frac{-1}{\sin^2 x}\cos x = -\csc x\cot x.$$

The fact that the functions sine, tangent, and secant do not have an explicit negative sign in their derivatives, while the "cofunctions" cosine, cosecant, and cotangent do, is a useful mnemonic for helping you to remember these derivatives. Also observe the patterns pairing sine and cosine, secant and tangent, and cosecant and cotangent.

It is very important to remember that the formulas obtained above for the derivatives of the trigonometric functions of x were obtained under the assumption that x was *measured in radians*. Specifically, we used the formula for the area of a sector of a circle in the proof of Theorem 3.2.2. This formula assumes the angle is in radians.

EXAMPLE 3.2.5 Calculate the derivatives of $\sin(x^\circ)$, $\cos(x^\circ)$ and $\tan(x^\circ)$.

SOLUTION Since $1^\circ = \frac{\pi}{180}$ radians, we have $x^\circ = \frac{\pi x}{180}$ radians, and so

$$\frac{d}{dx}\sin(x^\circ) = \frac{d}{dx}\sin\left(\frac{\pi x}{180}\right) = \frac{\pi}{180}\cos\left(\frac{\pi x}{180}\right) = \frac{\pi}{180}\cos(x^\circ).$$

Similarly,

$$\frac{d}{dx}\cos(x^\circ) = -\frac{\pi}{180}\sin(x^\circ) \qquad \text{and} \qquad \frac{d}{dx}\tan(x^\circ) = \frac{\pi}{180}\sec^2(x^\circ).$$

EXAMPLE 3.2.6 Find the equation of the straight line tangent to the curve

$$y - \tan\left(\frac{\pi x}{y}\right) = 3$$

at the point $(1, 4)$.

SOLUTION First observe that, since $\tan\frac{\pi}{4} = 1$, the point $(1, 4)$ does indeed lie on the given curve. Using implicit differentiation, we calculate

$$y' - \sec^2\left(\frac{\pi x}{y}\right)\left[\frac{\pi y - \pi x y'}{y^2}\right] = 0.$$

Now substitute $x = 1$ and $y = 4$ and solve for y':

$$y' - 2\left[\frac{\pi}{4} - \frac{\pi}{16}y'\right] = 0$$

$$y'\left(1 + \frac{\pi}{8}\right) = \frac{\pi}{2}, \qquad \text{so} \quad y'\bigg|_{(1,4)} = \frac{4\pi}{8+\pi}.$$

The required tangent line has equation

$$y = 4 + \frac{4\pi}{8+\pi}(x-1).$$

Finally, let us gather together the indefinite integrals that correspond to the differentiation formulas for the trigonometric functions.

3.2.7
Some Trigonometric Integrals

$$\int \sin x\,dx = -\cos x + C \qquad \int \cos x\,dx = \sin x + C$$

$$\int \sec^2 x\,dx = \tan x + C \qquad \int \csc^2 x\,dx = -\cot x + C$$

$$\int \sec x \tan x\,dx = \sec x + C \qquad \int \csc x \cot x\,dx = -\csc x + C$$

We do not yet know the antiderivatives of $\tan x$, $\cot x$, $\sec x$ and $\csc x$.

Simple Harmonic Motion

Many natural phenomena exhibit periodic behavior. The swinging of a clock pendulum, the vibrating of a plucked guitar string or drum membrane, the height of a rider on a rotating ferris wheel, the motion of an object floating in wavy seas, and the voltage produced by an alternating current generator are but a few examples where quantities depend on time in a periodic way. Being periodic, the circular functions sine and cosine provide a useful model for such behavior.

It often happens that a quantity displaced from an equilibrium value experiences a restoring force tending to move it back in the direction of its equilibrium. Besides the obvious examples of elastic motions in physics, one can imagine such a model applying, say, to a biological population in equilibrium with its food supply, or the price of a commodity in an elastic economy where increasing price causes increasing supply, which in turn causes decreasing demand and hence decreasing price. In the simplest models, the restoring force is proportional to the amount of displacement from equilibrium. Such a force causes the quantity to oscillate sinusoidally; we say that it executes *simple harmonic motion*.

As a specific example, suppose a mass m is suspended by an elastic spring so that it hangs unmoving in its equilibrium position. If it is displaced vertically by an amount y from this position, a force $-ky$, $(k > 0)$, is exerted by the spring (Hooke's law), tending to restore the mass to its equilibrium position (see Fig. 3.2.2). Assuming the spring to be weightless, this force imparts to the mass m an acceleration $\dfrac{d^2y}{dt^2} = y''$ which satisfies, by Newton's Law of Motion, $my'' = -ky$, (mass × acceleration = force). Thus

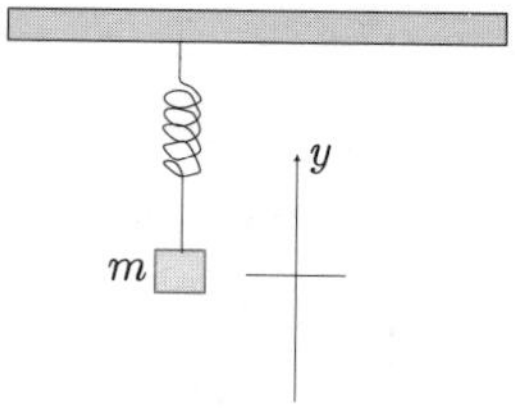

FIGURE 3.2.2

$$y'' + \omega^2 y = 0, \qquad \text{where} \quad \omega^2 = \frac{k}{m}.$$

The second-order differential equation $y'' + \omega^2 y = 0$ is called the **equation of simple harmonic motion**. One can readily check by differentiation that if $A, B, R,$ and t_o are constants, then *either of the functions*

$$y = A\cos\omega t + B\sin\omega t,$$
$$y = R\sin\big(\omega(t - t_0)\big)$$

is a *general solution* of the differential equation. This means that any function $y = f(t)$ which satisfies the differential equation $y'' + \omega^2 y = 0$ can be expressed in either of the above forms with particular choices for the constants A and B or R and t_0. (See Exercises 49–53 below.) The two forms do not represent different functions; they are just different ways of writing the same function. If we expand the second version using the addition formula, we get

$$\begin{aligned} y &= R\sin\omega t\cos\omega t_0 - R\cos\omega t\sin\omega t_0 \\ &= A\cos\omega t + B\sin\omega t. \end{aligned}$$

Thus the constants A and B are related to the constants R and t_0 by the equations

$$A = -R\sin\omega t_0 \qquad B = R\cos\omega t_0$$
$$R^2 = A^2 + B^2 \qquad \tan\omega t_0 = -A/B.$$

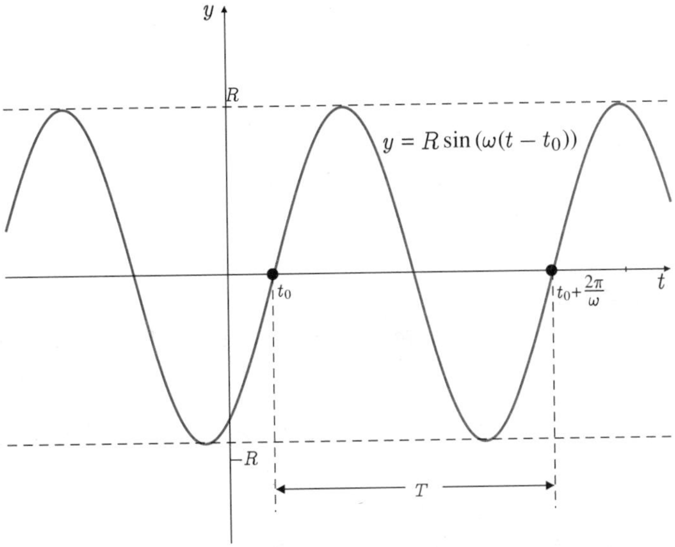

FIGURE 3.2.3

The constants A and B are related to the position y_0 and the velocity v_0 of the mass m at time $t = 0$:

$$y_0 = y(0) = A\cos 0 + B\sin 0 = A,$$
$$v_0 = y'(0) = -A\omega\sin 0 + B\omega\cos 0 = B\omega.$$

The constant $R = \sqrt{A^2 + B^2}$ is called the **amplitude** of the motion. Because $\sin x$ oscillates between -1 and 1, the displacement y varies between $-R$ and R. Note in Fig. 3.2.3 that the graph of the displacement as a function of time is that of the sine curve $y = R\sin\omega t$ displaced t_0 units to the right. Thus t_0 is called a *time-shift*. (The related quantity ωt_0 is called the *phase-shift*.) The period of this curve is $T = 2\pi/\omega$; it is the time interval between consecutive instants when the mass is at the same height moving in the same direction. The reciprocal of the period, $\dfrac{1}{T}$, is called the **frequency** of the motion. It is usually measured in Hertz (Hz), that is, cycles per second. The quantity $\omega = 2\pi/T$ is called the **circular frequency**. It is measured in radians per second since 1 cycle = 1 revolution = 2π radians.

EXAMPLE 3.2.8 Solve the initial-value problem $\begin{cases} y'' + 16y = 0 \\ y(0) = -6 \\ y'(0) = 32 \end{cases}$. (Note that we are using $y = y(t)$ to denote the solution function.) Find the amplitude, frequency and period of the solution.

SOLUTION Here $\omega^2 = 16$ so $\omega = 4$. The solution is

$$y = A\cos(4t) + B\sin(4t),$$

where $-6 = y(0) = A$ and $32 = y'(0) = 4B$. Thus the solution is

$$y = -6\cos(4t) + 8\sin(4t).$$

The amplitude is $\sqrt{36+64} = 10$, the frequency is $\omega/(2\pi) \approx 0.637$, and the period is $2\pi/\omega \approx 1.57$.

EXAMPLE 3.2.9 Suppose that a 100-g mass is suspended from a spring and that a force of 3×10^4 dynes (3×10^4 g cm/s^2) is required to produce a displacement from equilibrium of 1/3 cm. At time $t = 0$ the mass is pulled down 2 cm below equilibrium and flicked upwards with a velocity of 60 cm/s. Find its subsequent displacement at any time $t > 0$. Find the frequency, period, amplitude, and time-shift of the motion. Express the position of the mass at time t in terms of the amplitude and the time-shift.

SOLUTION The spring constant k is determined from Hooke's law, $F = -ky$. Here $F = -3 \times 10^4$ g cm/s^2 is the force of the spring on the mass displaced 1/3 cm:

$$-3 \times 10^4 = -\frac{1}{3}k,$$

so $k = 9 \times 10^4$ g/s^2. Hence the circular frequency is $\omega = \sqrt{k/m} = 30$ radians per second, the frequency is $\omega/2\pi = 15/\pi \approx 4.77$ Hz, and the period is $2\pi/\omega \approx 0.209$ seconds.

Since the displacement at time $t = 0$ is $y_0 = -2$ and the velocity at that time is $v_0 = 60$, the subsequent displacement is $y = A\cos(30t) + B\sin(30t)$ where $A = y_0 = -2$ and $B = v_0/\omega = 60/30 = 2$. Thus

$$y = -2\cos(30t) + 2\sin(30t), \qquad (y \text{ in cm, } t \text{ in sec}).$$

The amplitude of the motion is $R = \sqrt{(-2)^2 + 2^2} = 2\sqrt{2} \approx 2.83$ cm. The time-shift t_0 must satisfy

$$-2 = A = -R\sin(\omega t_0) = -2\sqrt{2}\sin(30t_0),$$
$$2 = B = R\cos(\omega t_0) = 2\sqrt{2}\cos(30t_0),$$

so $\sin(30t_0) = 1/\sqrt{2} = \cos(30t_0)$. Hence the phase-shift is $30t_0 = \pi/4$ radians, and the time-shift is $t_0 = \pi/120 \approx 0.0262$ seconds. The position of the mass at time $t > 0$ is also given by

$$y = 2\sqrt{2}\sin\big(30(t - \frac{\pi}{120})\big).$$

EXERCISES

Differentiate the functions given in Exercises 1–28.

1. $y = \cos 3x$

2. $y = \sin 2x$

3. $y = \tan \pi x$

4. $y = \sin \frac{x}{5}$

5. $y = \cos \frac{2x}{\pi}$

6. $y = \sec ax$

7. $y = \cot(4 - 3x)$

8. $y = \sin(x^2 - 4x - 1)$

9. $y = \sin(Ax + B)$

10. $y = \csc \frac{3}{x}$

11. $f(x) = \cos(s - rx)$

12. $y = \tan(1 - x^2)$

13. $F(t) = \sin at \cos at$

14. $G(\theta) = \frac{\sin a\theta}{\cos b\theta}$

15. $u = \sec(x^2 - 4x)$

16. $f(t) = \sin^3\left(\cos \frac{t}{3}\right)$

17. $y = \frac{\tan^2 x}{x}$

18. $y = \cos(\sqrt{1 + \sin^2 x})$

19. $y = x \sin \frac{1}{x}$

20. $y = (1 - x)^2 \cos \frac{1}{x}$

21. $f(\theta) = \sqrt{\cos 2\theta}$

22. $y = \cos\sqrt{1 + x^4}$

23. $y = \sin t \cos 2t \tan 3t$

24. $y = \frac{\sin x + \cos x}{\cos x - \sin x}$

25. $f(t) = \sqrt{\frac{1 - \cos t}{1 + \cos t}}$

26. $u = \frac{\sin^2 x}{x^2}$

27. $y = x^2 \cos(x^2 + 1)$

28. $G(t) = \sin\sqrt{|t|}$

29. Evaluate the integrals $\int \cos(2x)\,dx$, $\int \sin \frac{x}{3}\,dx$, $\int \cos(ax)\,dx$ and $\int \sin(ax)\,dx$.

30. Use the Half-angle Formulas from Section 3.1 to help you evaluate the integrals $\int \sin x \cos x\,dx$, $\int \cos^2 x\,dx$ and $\int \sin^2 x\,dx$.

31. Find an equation of the line tangent to the curve $y = \sin(x^\circ)$ at the point where $x = 45$.

32. Find an equation of the straight line normal to $y = \sec(x^\circ)$ at the point where $x = 60$.

33. Find the derivative of $f(x) = |\sin x|$ and sketch the graph of f.

34. Find the critical points of $f(x) = x + \sin x$. Sketch the graph of f between $x = -2\pi$ and $x = 2\pi$.

35. Repeat the previous exercise for $f(x) = x + 2\sin x$.

36. If $z = \tan \frac{x}{2}$ show that $\frac{dx}{dz} = \frac{2}{1 + z^2}$, $\sin x = \frac{2z}{1 + z^2}$, and $\cos x = \frac{1 - z^2}{1 + z^2}$.

In Exercises 37–40 find an equation of the line tangent to the given curve at the given point.

37. $2x + y - \sqrt{2}\sin(xy) = \pi/2$ at $\left(\frac{\pi}{4}, 1\right)$

38. $\tan(xy^2) = \frac{2xy}{\pi}$ at $\left(-\pi, \frac{1}{2}\right)$

39. $x\sin(xy - y^2) = x^2 - 1$ at $(1, 1)$

40. $\cos\left(\frac{\pi y}{x}\right) = \frac{x^2}{y} - \frac{17}{2}$ at $(3, 1)$

41. If $y = \tan kx$, show that $y'' = 2k^2 y(1 + y^2)$.

42. If $y = \sec kx$, show that $y'' = k^2 y(2y^2 - 1)$.

43. Given that $\sin 2x = 2\sin x \cos x$, deduce that $\cos 2x = \cos^2 x - \sin^2 x$.

44. Given that $\cos 2x = \cos^2 x - \sin^2 x$, deduce that $\sin 2x = 2\sin x \cos x$.

45. Find a formula for the nth derivative of $\sin 2x$.

46. Find a formula for the nth derivative of $\cos ax$.

47.*Find a formula for the nth derivative of $x \sin x$.

48.*Find a formula for the nth derivative of $x^2 \cos x$.

Exercises 49–53 all refer to the differential equation of Simple Harmonic Motion:

$$\frac{d^2y}{dt^2} + \omega^2 y = 0. \qquad (*)$$

Together they show that $y = A\cos\omega t + B\sin\omega t$ is a *general solution* of this equation, that is, that every solution is of this form for some choice of the constants A and B.

49.†Show that if $y = f(t)$ and $y = g(t)$ are solutions of the differential equation $(*)$, then so is $y = Af(t) + Bg(t)$ for any constants A and B.

50.†Show that $y = A\cos\omega t + B\sin\omega t$ is a solution of $(*)$.

51.†If $f(t)$ is any solution of $(*)$, show that $\omega^2(f(t))^2 + (f'(t))^2$ is constant.

52.†If $g(t)$ is a solution of $(*)$ satisfying $g(0) = g'(0) = 0$, show that $g(t)$ is identically 0.

53.†Suppose that $f(t)$ is any solution of the differential equation $(*)$. Show that $f(t) = A\cos\omega t + B\sin\omega t$, where $A = f(0)$ and $B\omega = f'(0)$. (*Hint:* Let $g(t) = f(t) - A\cos\omega t - B\sin\omega t$.)

In Exercises 54–56 solve the given initial-value problems. For each problem determine the circular frequency, the frequency, the period, and the amplitude of the solution.

54.† $\begin{cases} y'' + 4y = 0 \\ y(0) = 2 \\ y'(0) = -5 \end{cases}$

55.† $\begin{cases} y'' + 100y = 0 \\ y(0) = 0 \\ y'(0) = 3 \end{cases}$

56.† $\begin{cases} y'' + \dfrac{1}{T^2}y = 0 \\ y(0) = A \\ y'(0) = 0 \end{cases}$

57.†Show that $y = \mathcal{A}\cos(\omega(t - c)) + \mathcal{B}\sin(\omega(t - c))$ is a solution of the differential equation $y'' + \omega^2 y = 0$, and that it satisfies $y(c) = \mathcal{A}$ and $y'(c) = \mathcal{B}\omega$. Express the solution in the form $y = A\cos(\omega t) + B\sin(\omega t)$ for certain values of the constants A and B depenαing on $\mathcal{A}$, $\mathcal{B}$ and ω.

58.†Solve $\begin{cases} y'' + y = 0 \\ y(2) = 3 \\ y'(2) = -4 \end{cases}$

59.†Solve $\begin{cases} y'' + \omega^2 y = 0 \\ y(a) = A \\ y'(a) = B \end{cases}$

60.†What mass should be suspended from the spring in Example 3.2.9 to provide a system whose natural frequency of oscillation is 10 Hz? Find the displacement of such a mass from its equilibrium position t s after it is pulled down 1 cm from equilibrium and flicked upward with a speed of 2cm/s. What is the amplitude of this motion?

61.†A mass of 400 g suspended from a certain elastic spring will oscillate with a frequency of 24 Hz. What would be the frequency if the 400 g mass were replaced with a 900 g mass? A100 g mass?

3.3 THE INVERSE TRIGONOMETRIC FUNCTIONS

Recall (from Section 1.4) that a function f is *one-to-one* if $f(x_1) \neq f(x_2)$ whenever $x_1 \neq x_2$ (and x_1 and x_2 belong to the domain of f). For such a function we can define an *inverse function*, f^{-1}, whose domain is the range of f and whose range is the domain of f, such that

$$y = f^{-1}(x) \iff x = f(y),$$
$$f^{-1}(f(x)) = x \quad \text{for all } x \text{ in the domain of } f,$$
$$f(f^{-1}(x)) = x \quad \text{for all } x \text{ in the range of } f.$$

The graph of f^{-1} is the mirror image of the graph of f in the line $y = x$. (Review Section 1.4 if you are unfamiliar with these concepts.)

All six trigonometric functions are periodic, and hence not one-to-one. However, as we did with the function x^2 in Section 1.4, we can restrict their domains so that they become one-to-one and invertible.

The Inverse Sine Function

Let us define a function $\operatorname{Sin} x$ (note the capital letter) to be $\sin x$, restricted so that its domain is the interval $-\frac{\pi}{2} \leq x \leq \frac{\pi}{2}$. By the definition of sin in Section 3.1, Sin is one-to-one:

3.3.1
The Function Sin

$$\operatorname{Sin} x = \sin x \qquad \text{if } -\frac{\pi}{2} \leq x \leq \frac{\pi}{2}.$$

Sin has domain $\left[-\frac{\pi}{2}, \frac{\pi}{2}\right]$ and range $[-1, 1]$. Its graph is shown in Fig. 3.3.1. Being one-to-one, Sin has an inverse which is denoted by $\sin^{-1}$ (or in some books by arcsin or Arcsin) and which is called the **inverse sine function**.

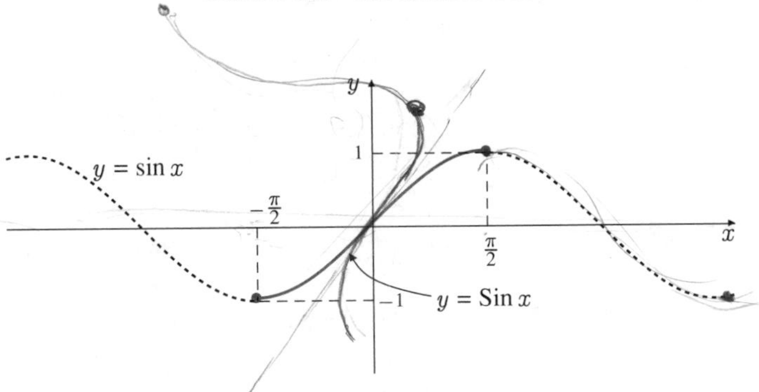

FIGURE 3.3.1

3.3.2
The inverse sine function $\sin^{-1}$

a) $y = \sin^{-1} x \Longleftrightarrow x = \operatorname{Sin} y$

$\Longleftrightarrow x = \sin y \quad \text{and} \quad -\dfrac{\pi}{2} \leq y \leq \dfrac{\pi}{2}$

b) $\sin^{-1}(\sin x) = x \quad \text{for } -\dfrac{\pi}{2} \leq x \leq \dfrac{\pi}{2}$

c) $\sin(\sin^{-1} x) = x \quad \text{for } -1 \leq x \leq 1$

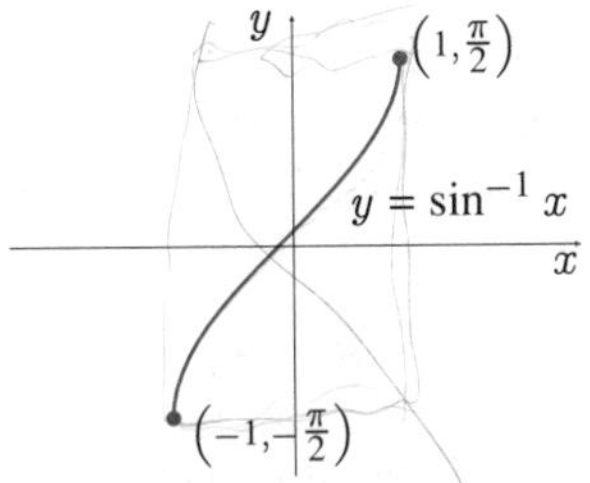

FIGURE 3.3.2

The graph of $\sin^{-1}$ is shown in Fig. 3.3.2; it is the mirror image of the graph of Sin in the line $y = x$. The domain of $\sin^{-1}$ is $[-1, 1]$ (the range of Sin), and the range of $\sin^{-1}$ is $\left[-\frac{\pi}{2}, \frac{\pi}{2}\right]$ (the domain of Sin).

REMARK: $\sin^{-1} x$ does *not* represent the *reciprocal* $\dfrac{1}{\sin x}$ of $\sin x$. (We already have a perfectly good name for the reciprocal of $\sin x$; we call it $\csc x$.) We should think of $\sin^{-1} x$ as "the angle between $-\frac{\pi}{2}$ and $\frac{\pi}{2}$ whose sine is x."

EXAMPLE 3.3.3

a) $\sin^{-1} \frac{1}{2} = \frac{\pi}{6}$ (because $\sin \frac{\pi}{6} = \frac{1}{2}$ and $-\frac{\pi}{2} < \frac{\pi}{6} < \frac{\pi}{2}$).

b) $\sin^{-1}\left(-\frac{1}{\sqrt{2}}\right) = -\frac{\pi}{4}$ (because $\sin\left(-\frac{\pi}{4}\right) = -\frac{1}{\sqrt{2}}$ and $-\frac{\pi}{2} < -\frac{\pi}{4} < \frac{\pi}{2}$).

c) $\sin^{-1}(-1) = -\frac{\pi}{2}$ (because $\sin\left(-\frac{\pi}{2}\right) = -1$.)

d) $\sin^{-1} 2$ is not defined. (2 is not in the range of sine.)

EXAMPLE 3.3.4 Find a) $\sin\left(\sin^{-1} 0.7\right)$, b) $\sin^{-1}(\sin 0.3)$, c) $\sin^{-1}\left(\sin \dfrac{4\pi}{5}\right)$, d) $\cos\left(\sin^{-1} 0.6\right)$.

SOLUTION

a) $\sin\left(\sin^{-1} 0.7\right) = 0.7$ (by formula 3.3.2(c)).

b) $\sin^{-1}(\sin 0.3) = 0.3$ (by formula 3.3.2(b)).

c) $\frac{4\pi}{5}$ does not lie in $\left[-\frac{\pi}{2}, \frac{\pi}{2}\right]$ so we can't apply formula 3.3.2(b) directly. However, by the supplementary angle identity, $\sin \frac{4\pi}{5} = \sin\left(\pi - \frac{\pi}{5}\right) = \sin \frac{\pi}{5}$. Therefore, $\sin^{-1}\left(\sin \frac{4\pi}{5}\right) = \sin^{-1}\left(\sin \frac{\pi}{5}\right) = \frac{\pi}{5}$ (by formula 3.3.2(b)).

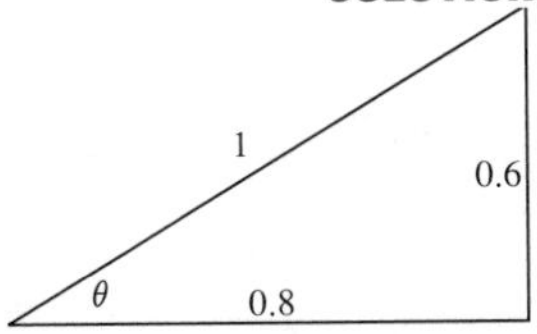

FIGURE 3.3.3

d) Let $\theta = \sin^{-1} 0.6$, as shown in the right triangle in Fig. 3.3.3, which has hypotenuse 1 and side opposite θ equal to 0.6. By the Pythagorean Theorem, the side adjacent θ is $\sqrt{1-(0.6)^2} = 0.8$. Thus $\cos\left(\sin^{-1} 0.6\right) = \cos\theta = 0.8$.

EXAMPLE 3.3.5 Simplify the expression $\tan(\sin^{-1} x)$.

SOLUTION Suppose first that $0 \le x < 1$. As in Example 3.3.4(d) we draw a right triangle (Fig. 3.3.4) with one angle $\theta = \sin^{-1} x$, opposite side x, and hypotenuse 1. The remaining side is $\sqrt{1-x^2}$ and we have

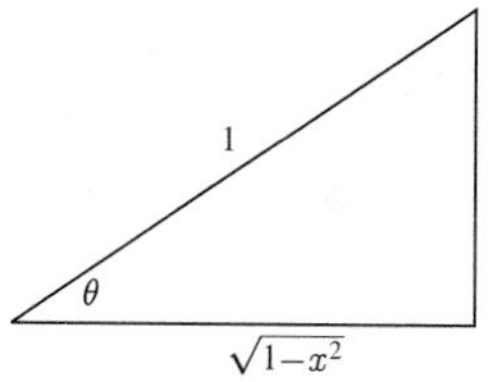

FIGURE 3.3.4

$$\tan(\sin^{-1} x) = \tan\theta = \frac{x}{\sqrt{1-x^2}}.$$

Because both sides of the above equation are odd functions of x, the same result holds for $-1 < x < 0$.

Now let us use implicit differentiation to find the derivative of the inverse sine function. If $y = \sin^{-1} x$ then $x = \sin y$ and $-\frac{\pi}{2} \le y \le \frac{\pi}{2}$. Differentiating with respect to x, we obtain

$$1 = (\cos y)\frac{dy}{dx}.$$

Since $-\frac{\pi}{2} \le y \le \frac{\pi}{2}$ we know that $\cos y \ge 0$. Therefore $\cos y = \sqrt{1-\sin^2 y} = \sqrt{1-x^2}$, and $dy/dx = 1/\cos y = 1/\sqrt{1-x^2}$.

3.3.6
The Derivative of $\sin^{-1}$

$$\frac{d}{dx}\sin^{-1} x = \frac{1}{\sqrt{1-x^2}}.$$

Note that inverse sine is differentiable only on the *open* interval $(-1, 1)$; the slope of its graph approaches infinity as $x \to -1+$ or $x \to 1-$.

EXAMPLE 3.3.7 Find the derivative of $\sin^{-1}\left(\frac{x}{a}\right)$ and hence evaluate $\displaystyle\int \frac{dx}{\sqrt{a^2-x^2}}$ where $a > 0$.

SOLUTION

$$\frac{d}{dx}\sin^{-1}\frac{x}{a} = \frac{1}{\sqrt{1-\dfrac{x^2}{a^2}}}\,\frac{1}{a} = \frac{1}{\sqrt{a^2-x^2}} \qquad \text{if } a > 0.$$

Hence,

$$\int \frac{dx}{\sqrt{a^2-x^2}} = \sin^{-1}\frac{x}{a} + C \qquad (a > 0).$$

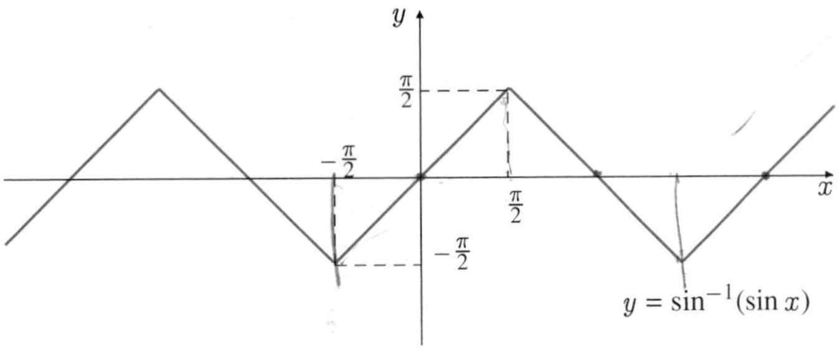

FIGURE 3.3.5

EXAMPLE 3.3.8 Find the solution y of the following initial-value problem:

$$\begin{cases} y' = \dfrac{4}{\sqrt{2 - x^2}} & (-\sqrt{2} < x < \sqrt{2}) \\ y(1) = 2\pi \end{cases}.$$

SOLUTION Using the integral from the previous example, we have

$$y = 4\int \frac{dx}{\sqrt{2 - x^2}} = 4\sin^{-1}\left(\frac{x}{\sqrt{2}}\right) + C$$

for some constant C. Also $2\pi = y(1) = 4\sin^{-1}(1/\sqrt{2}) + C = 4\left(\frac{\pi}{4}\right) + C = \pi + C$. Thus $C = \pi$ and $y = 4\sin^{-1}(x/\sqrt{2}) + \pi$.

EXAMPLE 3.3.9 Simplify and sketch the graph of the function $f(x) = \sin^{-1}(\sin x)$. Where is f continuous? Where is f differentiable? Find $f'(x)$.

SOLUTION The function $f(x)$ is defined for all real x and is periodic with period 2π because $\sin x$ is. By formula 3.3.2(b), $f(x) = x$ if $-\frac{\pi}{2} \le x \le \frac{\pi}{2}$. If $\frac{\pi}{2} < x < \frac{3\pi}{2}$ then $-\frac{\pi}{2} < \pi - x < \frac{\pi}{2}$, and, by the supplementary angle identity, $f(x) = \sin^{-1}(\sin(\pi - x)) = \pi - x$. Thus, on the interval $\left[-\frac{\pi}{2}, \frac{3\pi}{2}\right)$ we have

$$f(x) = \begin{cases} x & \text{if } -\dfrac{\pi}{2} \le x \le \dfrac{\pi}{2} \\ \pi - x & \text{if } \dfrac{\pi}{2} < x < \dfrac{3\pi}{2} \end{cases} = \frac{\pi}{2} - \left|x - \frac{\pi}{2}\right|.$$

!!DANGER!!

This is quite a difficult example. Go slowly!

Outside the interval $\left[-\frac{\pi}{2}, \frac{3\pi}{2}\right)$, (which has length 2π), $f(x)$ is defined by periodicity:

$$\text{If } -\frac{\pi}{2} + 2n\pi \le x < \frac{3\pi}{2} + 2n\pi, \quad \text{then} \quad f(x) = f(x - 2n\pi) = \frac{\pi}{2} - \left|x - \frac{\pi}{2} - 2n\pi\right|.$$

The graph of f is the saw-tooth line in Fig. 3.3.5. f is continuous on the real line, but fails to be differentiable at the odd multiples of $\pi/2$, that is, at points x where $\cos x = 0$. In fact

$$f'(x) = \frac{\cos x}{\sqrt{1 - \sin^2 x}} = \frac{\cos x}{\sqrt{\cos^2 x}} = \frac{\cos x}{|\cos x|} = \begin{cases} 1 & \text{if } \cos x > 0 \\ -1 & \text{if } \cos x < 0 \end{cases}.$$

The Inverse Tangent Function

The inverse tangent function is defined in a manner similar to the inverse sine. We begin by restricting tangent to an interval where it is one-to-one; in this case we use the open interval $\left(-\frac{\pi}{2}, \frac{\pi}{2}\right)$.

3.3.10
The Function Tan

$$\operatorname{Tan} x = \tan x \qquad \text{if } -\frac{\pi}{2} < x < \frac{\pi}{2}.$$

See Fig. 3.3.6.

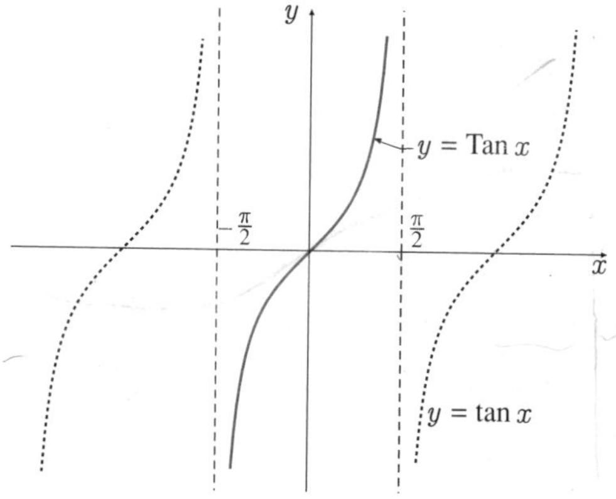

FIGURE 3.3.6

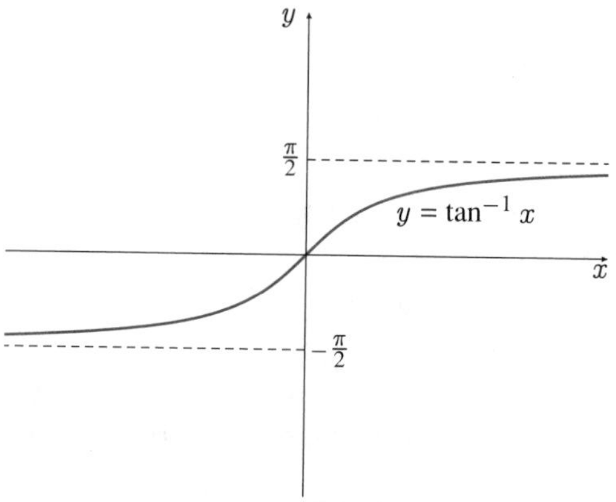

FIGURE 3.3.7

The inverse of the function Tan is called the **inverse tangent** function and is denoted $\tan^{-1}$ (or arctan or Arctan). The function $\tan^{-1}$ has domain the whole real line (the range of Tan). Its range is the open interval $\left(-\frac{\pi}{2}, \frac{\pi}{2}\right)$.

3.3.11
The inverse tangent function $\tan^{-1}$

a) $y = \tan^{-1} x \iff x = \operatorname{Tan} y$

$\iff x = \tan y$ and $-\frac{\pi}{2} < y < \frac{\pi}{2}$

b) $\tan^{-1}(\tan x) = x$ for $-\frac{\pi}{2} < x < \frac{\pi}{2}$

c) $\tan(\tan^{-1} x) = x$ for $-\infty < x < \infty$

The graph of $\tan^{-1}$ is shown in Fig. 3.3.7; it is the mirror image of the graph of Tan in the line $y = x$.

EXAMPLE 3.3.12 Evaluate a) $\tan(\tan^{-1} 3)$, b) $\tan^{-1}\left(\tan \frac{3\pi}{4}\right)$, c) $\cos(\tan^{-1} 2)$.

SOLUTION a) $\tan(\tan^{-1} 3) = 3$ by formula 3.3.11(c).

b) $\tan^{-1}\left(\tan \frac{3\pi}{4}\right) = \tan^{-1}(-1) = -\frac{\pi}{4}$.

c) $\cos(\tan^{-1} 2) = \cos\theta = \frac{1}{\sqrt{5}}$ via the triangle in Fig. 3.3.8. Alternatively, we have $\tan(\tan^{-1} 2) = 2$, so $\sec^2(\tan^{-1} 2) = 1 + 2^2 = 5$ and $\cos(\tan^{-1} 2) = 1/\sqrt{5}$ since cosine is positive on the range of $\tan^{-1}$.

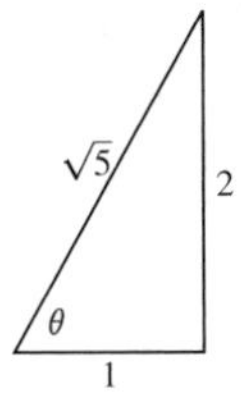

FIGURE 3.3.8

The derivative of the inverse tangent function is also found by implicit differentiation: if $y = \tan^{-1} x$ then $x = \tan y$ and

$$1 = (\sec^2 y)\frac{dy}{dx} = (1 + \tan^2 y)\frac{dy}{dx} = (1 + x^2)\frac{dy}{dx}.$$

Thus

3.3.13 The Derivative of $\tan^{-1}$

$$\frac{d}{dx}\tan^{-1} x = \frac{1}{1 + x^2}.$$

Note that $\frac{d}{dx}\tan^{-1}\left(\frac{x}{a}\right) = \frac{1}{1 + \frac{x^2}{a^2}}\frac{1}{a} = \frac{a}{a^2 + x^2}$, and hence

$$\int \frac{dx}{a^2 + x^2} = \frac{1}{a}\tan^{-1}\left(\frac{x}{a}\right) + C.$$

EXAMPLE 3.3.14 Prove that

$$\tan^{-1}\left(\frac{x - 1}{x + 1}\right) = \tan^{-1} x - \frac{\pi}{4} \qquad \text{for } x > -1.$$

SOLUTION Let $f(x) = \tan^{-1}\left(\frac{x - 1}{x + 1}\right) - \tan^{-1} x$. On the interval $(-1, \infty)$ we have

$$\begin{aligned} f'(x) &= \frac{1}{1 + \left(\frac{x - 1}{x + 1}\right)^2}\frac{(x + 1) - (x - 1)}{(x + 1)^2} - \frac{1}{1 + x^2} \\ &= \frac{2}{2 + 2x^2} - \frac{1}{1 + x^2} = 0. \end{aligned}$$

Hence $f(x) = C$ (constant) on that interval. We can find C by finding $f(0)$:

$$C = f(0) = \tan^{-1}(-1) - \arctan 0 = -\frac{\pi}{4}.$$

Hence the given identity holds on $(-1, \infty)$.

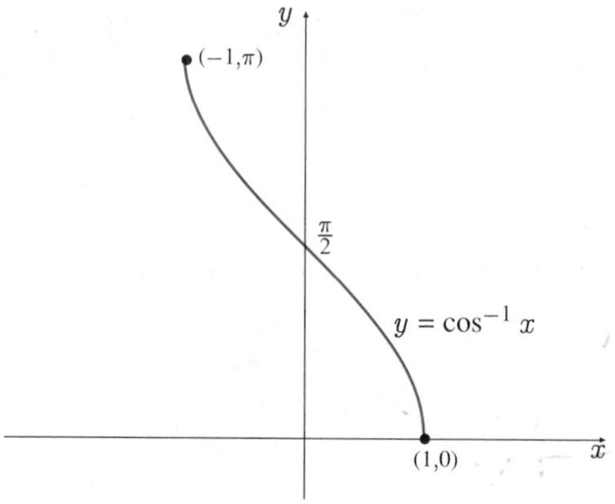

FIGURE 3.3.9

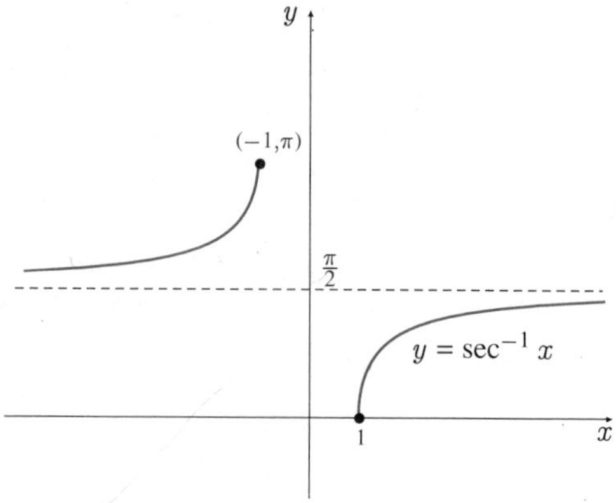

FIGURE 3.3.10

Other Inverse Trigonometric Functions

The function $\cos x$ is one-to-one on the interval $[0, \pi]$ so we could define the **inverse cosine function**, $\cos^{-1} x$, so that

$$y = \cos^{-1} x \Longleftrightarrow x = \cos y \quad \text{and} \quad 0 \leq y \leq \pi.$$

However, the complementary angle identities enable us to give a simpler equivalent definition.

3.3.15
The Function $\cos^{-1}$

$$\cos^{-1} x = \frac{\pi}{2} - \sin^{-1} x \qquad \text{for} \quad -1 \leq x \leq 1.$$

If $y = \cos^{-1} x$ then $0 \leq y \leq \pi$ (because $-\frac{\pi}{2} \leq \sin^{-1} x \leq \frac{\pi}{2}$), so

$$\cos(\cos^{-1} x) = \cos\left(\frac{\pi}{2} - \sin^{-1} x\right) = \sin(\sin^{-1} x) = x \qquad \text{for} \quad -1 \leq x \leq 1.$$

Similarly, if $z = \dfrac{\pi}{2} - y$ then $\sin z = \cos y$ so

$$\cos^{-1}(\cos y) = \frac{\pi}{2} - \sin^{-1}(\cos y) = \frac{\pi}{2} - \sin^{-1}(\sin z) = \frac{\pi}{2} - z = y \qquad \text{for} \quad 0 \leq y \leq \pi.$$

From 3.1.15, the derivative of $\cos^{-1} x$ is the negative of that of $\sin^{-1} x$:

3.3.16
The Derivative of $\cos^{-1}$

$$\frac{d}{dx} \cos^{-1} x = -\frac{1}{\sqrt{1 - x^2}}.$$

The graph of $\cos^{-1}$ is shown in Fig. 3.3.9.

The inverses of the secondary trigonometric functions are most easily expressed in terms of those of their reciprocal functions. For example, we define:

3.3.17
The Function $\sec^{-1}$

$$\sec^{-1} x = \cos^{-1}\left(\frac{1}{x}\right) \qquad \text{for } |x| \geq 1.$$

Observe that

$$\sec(\sec^{-1} x) = \sec\left(\cos^{-1}\left(\frac{1}{x}\right)\right) = \frac{1}{\cos\left(\cos^{-1}\left(\frac{1}{x}\right)\right)} = \frac{1}{\frac{1}{x}} = x \qquad \text{for } |x| \geq 1,$$

$$\sec^{-1}(\sec x) = \cos^{-1}\left(\frac{1}{\sec x}\right) = \cos^{-1}(\cos x) = x \qquad \text{for } x \text{ in } [0, \pi],\ x \neq \frac{\pi}{2}.$$

We calculate the derivative of $\sec^{-1} x$:

$$\frac{d}{dx}\sec^{-1} x = \frac{d}{dx}\cos^{-1}\left(\frac{1}{x}\right) = \frac{-1}{\sqrt{1-\frac{1}{x^2}}}\left(-\frac{1}{x^2}\right) = \frac{1}{x^2}\sqrt{\frac{x^2}{x^2-1}} = \frac{1}{|x|\sqrt{x^2-1}}.$$

Note that we had to use $\sqrt{x^2} = |x|$, there being negative values of x in the domain of $\sec^{-1}$. Observe in Fig. 3.3.10 that the slope of $y = \sec^{-1}(x)$ is always positive.

Finally, we note that $\csc^{-1}$ and $\cot^{-1}$ are defined similarly to $\sec^{-1}$:

3.3.18
The functions $\csc^{-1}$ and $\cot^{-1}$

$$\csc^{-1} x = \sin^{-1}\left(\frac{1}{x}\right), \quad (|x| \geq 1); \qquad \cot^{-1} x = \tan^{-1}\left(\frac{1}{x}\right), \quad (x \neq 0).$$

EXERCISES

In Exercises 1–12 evaluate the given expression.

1. $\sin^{-1}\frac{\sqrt{3}}{2}$.
2. $\cos^{-1}\left(\frac{-1}{2}\right)$
3. $\tan^{-1}(-1)$
4. $\sec^{-1}\sqrt{2}$
5. $\sin(\sin^{-1} 0.7)$
6. $\cos(\sin^{-1} 0.7)$
7. $\tan^{-1}\left(\tan\frac{2\pi}{3}\right)$
8. $\sin^{-1}(\cos 40°)$
9. $\cos^{-1}(\sin(-0.2))$
10. $\sin\left(\cos^{-1}\left(\frac{-1}{3}\right)\right)$
11. $\cos\left(\tan^{-1}\frac{1}{2}\right)$
12. $\tan(\tan^{-1} 200)$

In Exercises 13–20 simplify the given expression.

13. $\sin(\cos^{-1} x)$
14. $\cos(\sin^{-1} x)$
15. $\cos(\tan^{-1} x)$
16. $\sin(\tan^{-1} x)$
17. $\tan(\cos^{-1} x)$
18. $\tan(\sec^{-1} x)$

19.* $\cos^{-1}(\cos x)$

20.* $\sin^{-1}(\cos x)$

In Exercises 21–34 differentiate the given function and simplify the answer whenever possible.

21. $y = \sin^{-1}\dfrac{2x-1}{3}$

22. $y = \tan^{-1}(ax+b)$

23. $y = \cos^{-1}\dfrac{x-b}{a}$

24. $f(x) = x\sin^{-1}x$

25. $f(t) = t\tan^{-1}t$

26. $u = z^2\sec^{-1}(1+z^2)$

27. $F(x) = (1+x^2)\tan^{-1}x$

28. $y = \sin^{-1}\dfrac{a}{x}$

29. $G(x) = \dfrac{\sin^{-1}x}{\sin^{-1}2x}$

30. $H(t) = \dfrac{\sin^{-1}t}{\sin t}$

31. $f(x) = (\sin^{-1}x^2)^{1/2}$

32. $y = \cos^{-1}\dfrac{a}{\sqrt{a^2+x^2}}$

33. $y = \sqrt{a^2-x^2} + a\sin^{-1}\dfrac{x}{a} \quad (a>0)$

34. $y = a\cos^{-1}\left(1-\dfrac{x}{a}\right) - \sqrt{2ax-x^2} \quad (a>0)$

35. Find the slope of the curve $\tan^{-1}\left(\dfrac{2x}{y}\right) = \dfrac{\pi x}{y^2}$ at the point $(1,2)$.

36. Find equations of two straight lines tangent to the graph of $y = \sin^{-1}x$ and having slope 2.

37. Show that $\sin^{-1}$ and $\tan^{-1}$ are increasing functions and $\cos^{-1}$ is a decreasing function.

38. The derivative of $\sec^{-1}x$ is positive for every x in the domain of $\sec^{-1}$. Does this imply that $\sec^{-1}$ is increasing on its domain? Why?

39. Sketch the graph of $\csc^{-1}x$ and find its derivative.

40. Sketch the graph of $\cot^{-1}x$ and find its derivative.

41. Show that $\tan^{-1}x + \cot^{-1}x = \frac{\pi}{2}$ for $x>0$. What is the sum if $x<0$?

42.* Find the derivative of $g(x) = \tan(\tan^{-1}x)$ and $h(x) = \tan^{-1}(\tan x)$ and sketch the graph of each function.

43.* Show that the function $f(x)$ of Example 3.3.14 is also constant on the interval $(-\infty,-1)$. Find the value of the constant. (*Hint:* Find $\lim_{x\to-\infty} f(x)$.)

44.* Find the derivative of $f(x) = x - \tan^{-1}(\tan x)$. What does your answer imply about $f(x)$? Calculate $f(0)$ and $f(\pi)$. Is there a contradiction here?

45.* Find the derivative of $f(x) = x - \sin^{-1}(\sin x)$ for $-\pi \le x \le \pi$ and sketch the graph of f on that interval.

In Exercises 46–49 solve the initial-value problems.

46.† $\begin{cases} y' = \dfrac{1}{1+x^2} \\ y(0) = 1 \end{cases}$

47.† $\begin{cases} y' = \dfrac{1}{9+x^2} \\ y(3) = 2 \end{cases}$

48.† $\begin{cases} y' = \dfrac{1}{\sqrt{1-x^2}} \\ y(1/2) = 1 \end{cases}$

49.† $\begin{cases} y' = \dfrac{4}{\sqrt{25-x^2}} \\ y(0) = 0 \end{cases}$

3.4 EXPONENTIAL AND LOGARITHMIC FUNCTIONS

An **exponential function** is a function of the form $f(x) = a^x$, where the **base** a is a positive constant and the **exponent** x is the variable. Do not confuse such functions with **power** functions like $f(x) = x^a$ where the base is variable and the exponent is constant. The exponential function a^x can be defined for integer and rational exponents x as follows:

3.4.1
Definition of Exponentials

$$a^0 = 1$$
$$a^n = \underbrace{a\cdot a\cdot a\cdots a}_{n\text{ factors}} \qquad \text{if } n = 1,2,3,\ldots$$
$$a^{-n} = \frac{1}{a^n} \qquad \text{if } n = 1,2,3,\ldots$$
$$a^{m/n} = \sqrt[n]{a^m} \qquad \text{if } n = 1,2,3,\ldots \text{ and } m = \pm1,\pm2,\pm3,\ldots,$$

where, for $a > 0$, $\sqrt[n]{a} = b$ means that $b > 0$ and $b^n = a$. How should we define a^x if x is not rational? For example, what does 2^π mean? In order to calculate a derivative of a^x we will want the function to be defined for all real numbers x, not just rational ones.

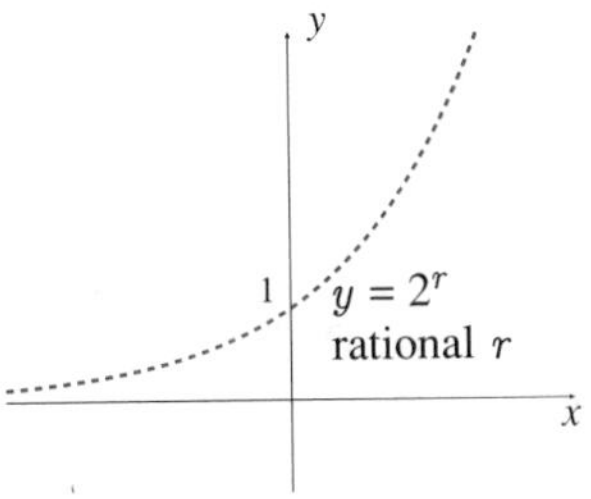

FIGURE 3.4.1

In Fig. 3.4.1 we plot points with coordinates $(x, 2^x)$ for many, closely spaced rational values of x. They appear to lie on a smooth curve. Let us *assume* for now that a^x can be defined for irrational x in such a way that the graph of $y = a^x$ is a smooth curve; that is, assume that a^x is a continuous and even differentiable function of x on the whole real line. It would be difficult to prove this now; in Section 3.6 we will approach the definition of a^x in a different way so as to avoid this difficulty. For the moment, if x is irrational we can regard a^x as being the limit of values a^r for rational numbers r approaching x:

$$a^x = \lim_{\substack{r \to x \\ r \text{ rational}}} a^r.$$

For example,

$$2^\pi = \lim_{n\to\infty} 2^{r_n}, \qquad \text{where } r_1 = \frac{31}{10},\ r_2 = \frac{314}{100},\ r_3 = \frac{3141}{1000}, \ldots,$$

where r_n is the decimal expansion of $\pi = 3.141592654\cdots$ truncated (i.e. cut off) to n digits after the decimal point.

Exponential functions satisfy several **laws of exponents**; if $a, b > 0$, then

3.4.2
Laws of Exponents

$$a^0 = 1 \qquad a^{x+y} = a^x a^y$$
$$a^{-x} = \frac{1}{a^x} \qquad a^{x-y} = \frac{a^x}{a^y}$$
$$(a^x)^y = a^{xy} \qquad (ab)^x = a^x b^x$$

These identities can be proved for rational exponents using the definitions above, and for irrational exponents by taking limits.

If $a = 1$ then $a^x = 1^x = 1$ for every x. If $a > 1$ then a^x is an increasing function of x; if $0 < a < 1$, then a^x is decreasing. The graphs of some typical exponential functions are shown in Fig. 3.4.2. They all pass through the point (0,1) since $a^0 = 1$ for every $a > 0$. Observe that $a^x > 0$ for all $a > 0$ and all real x, and that

$$\text{If } a > 1 \text{ then } \lim_{x\to-\infty} a^x = 0 \text{ and } \lim_{x\to\infty} a^x = \infty.$$
$$\text{If } 0 < a < 1 \text{ then } \lim_{x\to-\infty} a^x = \infty \text{ and } \lim_{x\to\infty} a^x = 0.$$

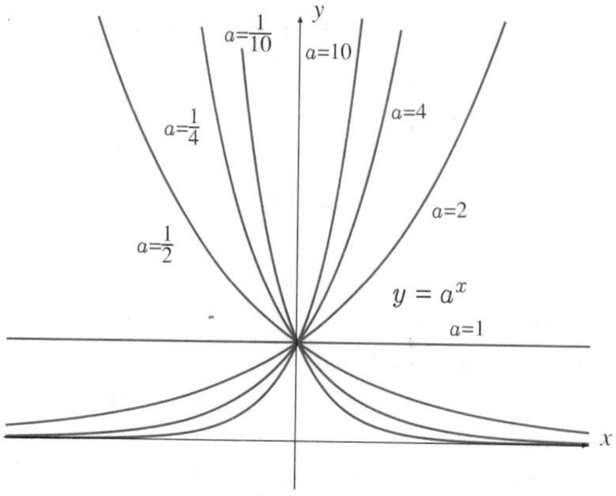

FIGURE 3.4.2

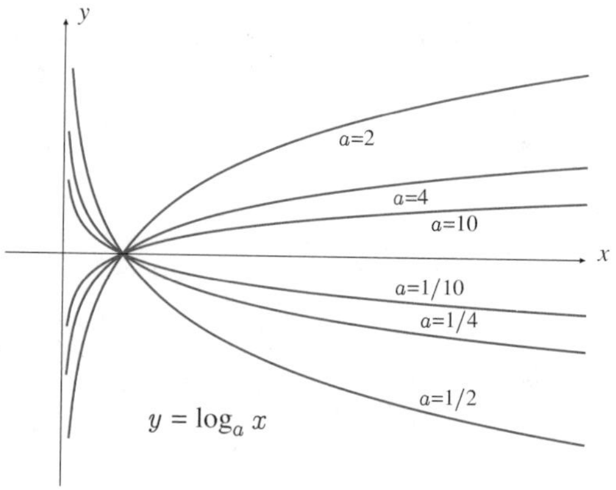

FIGURE 3.4.3

If $a > 0$ and $a \neq 1$, then $f(x) = a^x$ is a one-to-one function and so has an inverse function which we denote by $f^{-1}(x) = \log_a x$ and call **the logarithm of x to the base a.** Thus

3.4.3
Definition of Logarithms

$$y = \log_a x \quad \text{means the same as} \quad x = a^y, \quad (a > 0, \quad a \neq 1).$$

Since $x = a^y$ takes only positive values, therefore the domain of $\log_a x$ is the interval $(0, \infty)$. Since a^y is defined for all real y, therefore $y = \log_a x$ takes on all real values. We have

$$\log_a (a^x) = x \quad \text{for all real } x$$
$$a^{\log_a x} = x \quad \text{for all } \quad x > 0.$$

The graphs of some typical logarithmic functions are shown in Fig. 3.4.3.

Corresponding to the laws of exponents we have several **laws of logarithms**; if $a,\ b > 0$ then

3.4.4
Laws of Logarithms

$$\log_a 1 = 0 \qquad \log_a (xy) = \log_a x + \log_a y$$
$$\log_a \left(\frac{1}{x}\right) = -\log_a x \qquad \log_a \left(\frac{x}{y}\right) = \log_a x - \log_a y$$
$$\log_a (x^y) = y \log_a x \qquad \log_a x = \frac{\log_b x}{\log_b a}$$

These identities can be derived from the laws of exponents.

EXAMPLE 3.4.5 If $a > 0$, $x > 0$ and $y > 0$ show that $\log_a(xy) = \log_a x + \log_a y$.

SOLUTION Let $u = \log_a x$ and $v = \log_a y$. Then $x = a^u$, $y = a^v$ and $xy = a^u a^v = a^{u+v}$. Therefore $\log_a(xy) = u + v = \log_a x + \log_a y$.

EXAMPLE 3.4.6 Simplify (a) $\log_2 10 + \log_2 12 - \log_2 15$, (b) $\log_{a^2} a^3$, and (c) $3^{\log_9 4}$.

SOLUTION (a) $\log_2 10 + \log_2 12 - \log_2 15 = \log_2 \dfrac{10 \times 12}{15} = \log_2 8 = \log_2 2^3 = 3$.

(b) $\log_{a^2} a^3 = 3 \log_{a^2} a = \dfrac{3}{2} \log_{a^2} a^2 = \dfrac{3}{2}$.

(c) $3^{\log_9 4} = 3^{(\log_3 4)/(\log_3 9)} = 4^{1/\log_3 3^2} = 4^{1/2} = 2$.

EXERCISES

Simplify the expressions in Exercises 1–18.

1. $\dfrac{3^3}{\sqrt{3^5}}$
2. $2^{1/2}8^{1/2}$
3. $\left(x^{-3}\right)^{-2}$
4. $\left(\dfrac{1}{2}\right)^x 4^{x/2}$
5. $\log_5 125$
6. $\log_4 \left(\dfrac{1}{8}\right)$
7. $\log_{1/3} 3^{2x}$
8. $2^{\log_4 8}$
9. $10^{-\log_{10}(1/x)}$
10. $x^{1/(\log_a x)}$
11. $(\log_a b)(\log_b a)$
12. $\log_x \left(x(\log_y y^2)\right)$
13. $(\log_4 16)(\log_4 2)$
14. $\log_{15} 75 + \log_{15} 3$
15. $\log_6 9 + \log_6 4$
16. $2 \log_3 12 - 4 \log_3 6$
17. $\log_a(x^4 + 3x^2 + 2) + \log_a(x^4 + 5x^2 + 6)$
$-4 \log_a \sqrt{x^2 + 2}$
18. $\log_\pi(1 - \cos x) + \log_\pi(1 + \cos x) - 2 \log_\pi \sin x$
19. Assume that you have a calculator which gives values of exponentials and logarithms to base 10 only. Explain how you would use it to find a) the value of $3^{\sqrt{2}}$, b) $\log_3 5$, and c) the solution of the exponential equation $2^{2x} = 5^{x+1}$.

Use the Laws of Exponents to prove the Laws of Logarithms in Exercises 20–23.

20. $\log_a \left(\dfrac{1}{x}\right) = -\log_a x$
21. $\log_a \left(\dfrac{x}{y}\right) = \log_a x - \log_a y$
22. $\log_a(x^y) = y \log_a x$
23. $\log_a x = (\log_b x)/(\log_b a)$

3.5 DERIVATIVES OF EXPONENTIALS AND LOGARITHMS

Using the definition of derivative and the properties of the exponential function given above, we can calculate the derivative of $f(x) = a^x$ as follows:

$$f'(x) = \lim_{h \to 0} \frac{a^{x+h} - a^x}{h} = \lim_{h \to 0} a^x \frac{a^h - 1}{h} = f'(0)\, a^x.$$

Thus the derivative of an exponential function is that function multiplied by a constant, the value of the derivative at 0. Of course, the constant $f'(0)$ depends on the base a; let us define

$$L(a) = f'(0) = \lim_{h \to 0} \frac{a^h - 1}{h}.$$

Clearly $L(1) = 0$. Examining the slopes at $x = 0$ of the various exponential graphs in Fig. 3.4.2, we surmise that $L(a)$ increases from $-\infty$ to ∞ as a increases from 0 to ∞. We can show, for example, that $\frac{1}{2} \le L(2) \le 1$ and $1 \le L(4) \le 2$. (See Exercises 80–81 at the end of this section.) We are going to show now that $L(x)$ is the logarithm of x to a certain base.

THEOREM 3.5.1 There exists a unique number $e > 1$ such that $L(x) = \log_e x$ for $x > 0$.

PROOF If $a > 0$ and $r \neq 0$ observe that

$$L(a^r) = \lim_{h\to 0} \frac{a^{rh} - 1}{h} = r \lim_{h\to 0} \frac{a^{rh} - 1}{rh} = r \lim_{rh\to 0} \frac{a^{rh} - 1}{rh} = r\, L(a).$$

(The same identity holds if $r = 0$ since $L(1) = 0$.) Let c be *any* number* such that $L(c) \neq 0$, (for example, let $c = 2$), and let $e = c^{1/L(c)}$. Since $L(2) > 0$, we have $e = 2^{1/L(2)} > 1$. Now

$$L(e) = L\left(c^{1/L(c)}\right) = \frac{1}{L(c)} L(c) = 1.$$

If $x > 0$ is arbitrary then $x = e^{\log_e x}$, so

$$L(x) = L\left(e^{\log_e x}\right) = (\log_e x) L(e) = \log_e x.$$

It might appear that e depends on the particular choice made for c. This is not so. If a is such that $L(a) \neq 0$ and $b = a^{1/L(a)}$ then the same proof shows that $L(b) = 1$ and $L(x) = \log_b x$. Therefore $\log_b e = L(e) = 1$ and $e = b^1 = b$. Thus e is a uniquely determined. □

The function $\log_e x$ is called the **natural logarithm** of x, and is usually denoted $\ln x$ (pronounced "lawn x") rather than $\log_e x$. We have shown that

3.5.2
Derivatives of Exponential Functions

$$\frac{d}{dx} a^x = a^x \ln a, \quad \text{and therefore} \quad \int a^x\, dx = \frac{a^x}{\ln a} + C.$$

In particular, since $\ln e = \log_e e = 1$, the exponential function with base e is its own derivative:

3.5.3
Derivative of e^x

$$\frac{d}{dx} e^x = e^x, \quad \text{and} \quad \int e^x\, dx = e^x + C.$$

---

* This idea is due to R. Christian, Two Year College Math. Journ., 1983.

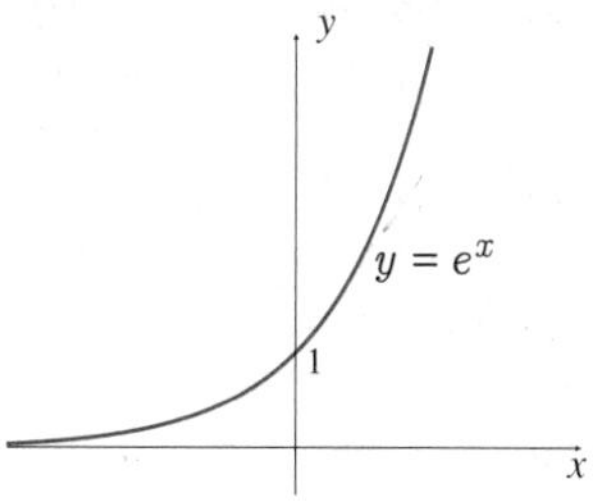

FIGURE 3.5.1

The graph of $y = e^x$ is shown in Fig. 3.5.1. Its slope at any point is equal to the height of that point above the x-axis.

Like π, the number e is irrational. Its decimal expansion is

$$e = 2.718281828459045\cdots.$$

Later on we will discover some formulas for calculating e to any desired degree of accuracy.

Because it is so important, the function e^x is usually called *the* exponential function. It is sometimes denoted $\exp x$, especially if x is a complicated expression: thus

$$\exp x = e^x, \qquad \frac{d}{dx}\exp x = \exp x.$$

The function $\ln x$ is the inverse of the function e^x; its graph is shown in Fig. 3-5-2.

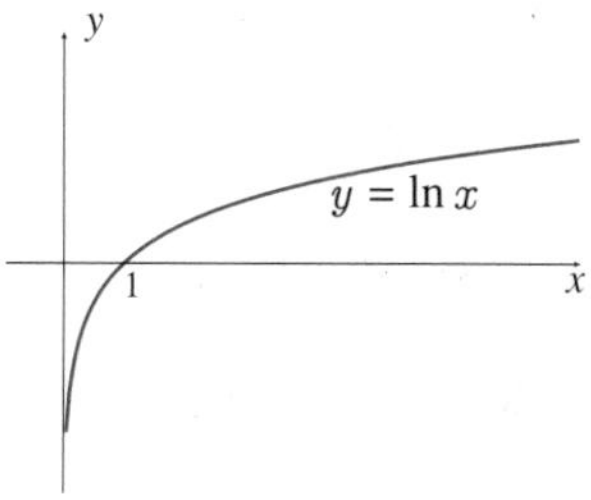

FIGURE 3.5.2

$$y = \ln x = \log_e x \qquad \text{means the same as} \qquad x = e^y = \exp y.$$

In particular,

$$\ln(e^x) = x \qquad \text{for all real } x,$$
$$e^{\ln x} = x \qquad \text{for all } x > 0.$$

Let us calculate the derivative of $y = \ln x$ by rewriting that equation in the form $x = e^y$ and differentiating implicitly with respect to x; $1 = e^y y'$ so $y' = 1/e^y = 1/x$. Thus

3.5.4
Derivative of ln

$$\frac{d}{dx}\ln x = \frac{1}{x}, \qquad (x > 0).$$

One of the laws of logarithms shows that any logarithm can be expressed in terms of the natural logarithm:

$$\log_a x = \frac{\ln x}{\ln a}, \qquad \text{and so} \qquad \frac{d}{dx}\log_a x = \frac{1}{x\ln a}.$$

Most scientific calculators provide two exponential functions, e^x and 10^x, and their corresponding inverses, $\ln x = \log_e x$ and $\log x = \log_{10} x$. Although $10^x = e^{x \ln 10}$ and $\log_{10} x = (\ln x)/(\ln 10)$, the base 10 functions are included because many quantities of interest in other scientific disciplines are defined in terms of base 10. Examples include the pH of solutions in Chemistry, and the decibel measure of sound energy in accoustics. Many calculators also have the function x^y, which is, however, implemented as $e^{y \ln x}$. Electronic computers represent numbers internally using a binary (base 2) format; therefore in Computer Science, log usually refers to $\log_2$.

EXAMPLE 3.5.5 Solve the equation $3^{4x} = 2^{x+5}$.

SOLUTION Take the natural logarithm of both sides to obtain

$$4x \ln 3 = (x+5) \ln 2,$$

so that

$$x = \frac{5 \ln 2}{4 \ln 3 - \ln 2} = \frac{5 \ln 2}{\ln(81/2)} \approx 0.936356.$$

(We used a calculator in the last step.) Logarithms to any other base could also have been used to solve the equation.

EXAMPLE 3.5.6 Find the derivative and the indefinite integral of e^{ax}.

SOLUTION By the Chain Rule we have

$$\frac{d}{dx} e^{ax} = a\, e^{ax}. \qquad \text{Thus} \qquad \int e^{ax}\, dx = \frac{e^{ax}}{a} + C.$$

EXAMPLE 3.5.7 Find the nth derivatives of the functions $f(x) = e^x$ and $g(x) = e^{3-2x}$.

SOLUTION We have

$$\begin{aligned}
f'(x) &= e^x & g'(x) &= -2\, e^{3-2x} \\
f''(x) &= e^x & g''(x) &= (-2)^2\, e^{3-2x} = 4\, e^{3-2x} \\
f^{(3)}(x) &= e^x & g^{(3)}(x) &= (-2)^3\, e^{3-2x} = -8\, e^{3-2x} \\
&\vdots & &\vdots \\
f^{(n)}(x) &= e^x & g^{(n)}(x) &= (-2)^n\, e^{3-2x} = (-1)^n 2^n\, e^{3-2x}
\end{aligned}$$

EXAMPLE 3.5.8 Find the derivatives of a) $P(t) = t^5 - 5^t$, and b) $Q(s) = \log_5 s - \log_s 5$, $(s > 0)$.

SOLUTION a) $P'(t) = 5t^4 - 5^t \ln 5$.

b) $Q'(s) = \dfrac{d}{ds}\left(\dfrac{\ln s}{\ln 5} - \dfrac{\ln 5}{\ln s}\right) = \dfrac{1}{s \ln 5} + \dfrac{\ln 5}{s(\ln s)^2}$.

Note that an arbitrary real power of a postive base can be expressed in terms of the exponential and natural logarithm:

3.5.9 General Exponentials

$$x^a = e^{a \ln x}.$$

Thus, for example, $2^\pi = e^{\pi \ln 2}$. (Compare this way of determining 2^π with that given just preceding Box 3.4.2.)

EXAMPLE 3.5.10 Show that the General Power Rule, $\dfrac{d}{dx} x^a = a\, x^{a-1}$, holds for all real numbers a if $x > 0$.

SOLUTION If $x > 0$ then $x = e^{\ln x}$, so $x^a = (e^{\ln x})^a = e^{a \ln x}$. Thus

$$\frac{d}{dx} x^a = e^{a \ln x}\left(\frac{a}{x}\right) = x^a\left(\frac{a}{x}\right) = a\, x^{a-1}.$$

EXAMPLE 3.5.11 Show that $\dfrac{d}{dx} \ln|x| = \dfrac{1}{x}$ for any $x \neq 0$. Find $\displaystyle\int \frac{dx}{x}$.

SOLUTION We already know that the derivative of $\ln x$ is $1/x$ for $x > 0$. If $x < 0$ then $|x| = -x$, so, by the Chain Rule,

$$\frac{d}{dx} \ln|x| = \frac{d}{dx} \ln(-x) = \frac{1}{-x}(-1) = \frac{1}{x}.$$

Thus, if $x \neq 0$,

$$\frac{d}{dx} \ln|x| = \frac{1}{x}, \qquad \text{and} \qquad \int \frac{dx}{x} = \ln|x| + C.$$

EXAMPLE 3.5.12 Evaluate the derivative of a) $\ln|\sec x|$, and b) $\ln\left(x + \sqrt{x^2+1}\right)$. Use the results to evaluate appropriate indefinite integrals.

SOLUTION a) $\dfrac{d}{dx} \ln|\sec x| = \dfrac{1}{\sec x}(\sec x \tan x) = \tan x$. Thus $\displaystyle\int \tan x\, dx = \ln|\sec x| + C$.

b) $$\frac{d}{dx}\ln(x+\sqrt{x^2+1}) = \frac{1+\dfrac{2x}{2\sqrt{x^2+1}}}{x+\sqrt{x^2+1}} = \left(\frac{1}{x+\sqrt{x^2+1}}\right)\left(\frac{\sqrt{x^2+1}+x}{\sqrt{x^2+1}}\right) = \frac{1}{\sqrt{x^2+1}}.$$

Thus $\displaystyle\int \frac{dx}{\sqrt{x^2+1}} = \ln\left(x+\sqrt{x^2+1}\right) + C.$

Logarithmic Differentiation

Suppose we wish to differentiate a function of the form

$$y = (f(x))^{g(x)}, \qquad (\text{for } f(x) > 0),$$

where the variable appears in both the base and the exponent so that neither the general power rule, $\dfrac{d}{dx}x^a = ax^{a-1}$, nor the exponential rule, $\dfrac{d}{dx}a^x = a^x \ln a$, can be directly applied. One method for finding the derivative of such a function is to express it in the form

$$y = e^{g(x)\ln f(x)}.$$

and then differentiate, using the Product Rule to handle the exponent.

EXAMPLE 3.5.13 $\displaystyle \frac{d}{dx}x^x = \frac{d}{dx}e^{x\ln x} = e^{x\ln x}\left(\ln x + x\left(\frac{1}{x}\right)\right) = x^x(1+\ln x),\ (x>0).$

The same result can be obtained by taking natural logarithms of both sides of the equation $y = x^x$ and differentiating implicitly:

$$\begin{aligned} \ln y &= x\ln x \\ \frac{1}{y}\frac{dy}{dx} &= \ln x + \frac{x}{x} = 1 + \ln x \\ \frac{dy}{dx} &= y(1+\ln x) = x^x(1+\ln x). \end{aligned}$$

This latter technique is called **logarithmic differentiation**.

EXAMPLE 3.5.14 Find $f'(t)$ if $f(t) = \left(\sin t\right)^{\ln t}$ where $0 < t < \pi$.

SOLUTION We have $\ln y = \ln t \, \ln\sin t$. Thus

$$\begin{aligned} \frac{1}{y}\frac{dy}{dt} &= \frac{1}{t}\ln\sin t + \ln t\frac{\cos t}{\sin t} \\ \frac{dy}{dt} &= y\left(\frac{\ln\sin t}{t} + \ln t\,\cot t\right) \\ &= (\sin t)^{\ln t}\left(\frac{\ln\sin t}{t} + \ln t\,\cot t\right). \end{aligned}$$

Logarithmic differentiation is also useful for finding the derivatives of functions expressed as products and quotients of many factors. Taking logarithms reduces these products and quotients to sums and differences. This usually makes the calculation easier than it would be using the Product and Quotient Rules, especially if the derivative is to be evaluated at a specific point.

EXAMPLE 3.5.15 Differentiate $y = [(x+1)(x+2)(x+3)]/(x+4)$.

SOLUTION $\ln y = \ln(x+1) + \ln(x+2) + \ln(x+3) - \ln(x+4)$. thus

$$\begin{aligned}\frac{1}{y}y' &= \frac{1}{x+1} + \frac{1}{x+2} + \frac{1}{x+3} - \frac{1}{x+4}\\ y' &= \frac{(x+1)(x+2)(x+3)}{x+4}\left(\frac{1}{x+1} + \frac{1}{x+2} + \frac{1}{x+3} - \frac{1}{x+4}\right)\\ &= \frac{(x+2)(x+3)}{x+4} + \frac{(x+1)(x+3)}{x+4} + \frac{(x+1)(x+2)}{x+4} - \frac{(x+1)(x+2)(x+3)}{(x+4)^2}.\end{aligned}$$

EXAMPLE 3.5.16 Find $\left.\dfrac{du}{dx}\right|_{x=1}$ if $u = \sqrt{(x+1)(x^2+1)(x^3+1)}$.

SOLUTION

$$\begin{aligned}\ln u &= \frac{1}{2}\Big(\ln(x+1) + \ln(x^2+1) + \ln(x^3+1)\Big)\\ \frac{1}{u}\frac{du}{dx} &= \frac{1}{2}\left(\frac{1}{x+1} + \frac{2x}{x^2+1} + \frac{3x^2}{x^3+1}\right)\end{aligned}$$

At $x = 1$ we have $u = \sqrt{8} = 2\sqrt{2}$. Hence

$$\left.\frac{du}{dx}\right|_{x=1} = \sqrt{2}\left(\frac{1}{2} + 1 + \frac{3}{2}\right) = 3\sqrt{2}.$$

The Growth of Exponentials and Logarithms

The straight line $y = x - 1$ is tangent to the curve $y = \ln x$ at the point $(1, 0)$. The following theorem asserts that the curve never lies above that line. (See Fig. 3.5.3.)

THEOREM 3.5.17 If $x > 0$ then $\ln x \le x - 1$.

PROOF Let $g(x) = \ln x - (x - 1)$ for $x > 0$. Then $g(1) = 0$ and

$$g'(x) = \frac{1}{x} - 1 \quad \begin{cases} > 0 & \text{if } 0 < x < 1\\ < 0 & \text{if } x > 1.\end{cases}$$

As observed in Section 2.6, these inequalities imply that g is increasing on $(0, 1)$ and decreasing on $(1, \infty)$. Thus $g(x) \le g(1) = 0$ for all $x > 0$, and so $\ln x \le x - 1$ for all such x. □

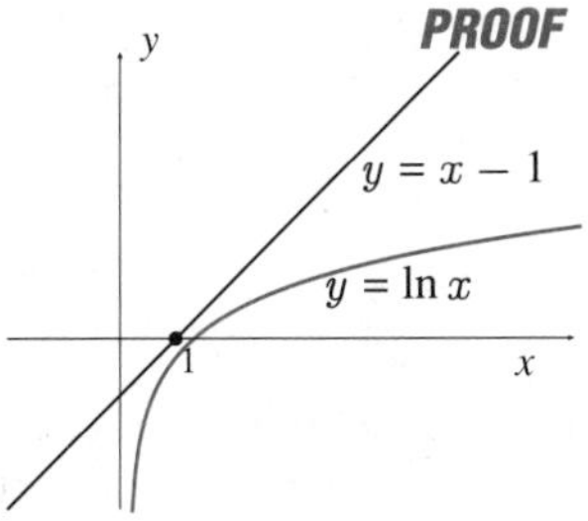

FIGURE 3.5.3

Both e^x and $\ln x$ grow large (approach infinity) as x grows large. However, e^x increases very rapidly as x increases, and $\ln x$ increases very slowly. In fact, e^x increases, for large x, faster than any positive power of x (no matter how large the power), while $\ln x$ increases more slowly than any positive power of x (no matter how small the power).

THEOREM 3.5.18 If $a > 0$ then

$$\text{a)}\ \lim_{x\to\infty} \frac{e^x}{x^a} = \infty, \qquad \text{b)}\ \lim_{x\to\infty} \frac{\ln x}{x^a} = 0,$$
$$\text{c)}\ \lim_{x\to-\infty} |x|^a e^x = 0, \qquad \text{d)}\ \lim_{x\to 0+} x^a \ln x = 0.$$

We can paraphrase Theorem 3.5.18 as follows: "in a struggle between a power and an exponential, the exponential always wins." (For example, in (a) the power x^a in the denominator of the fraction is trying to make the fraction become smaller while the exponential e^x in the numerator is trying to make it grow larger. The exponential wins; the fraction approaches infinity.) Similarly, "in a struggle between a power and a logarithm the power always wins."

Theorem 3.5.17 can be used to prove Theorem 3.5.18. (See Exercise 79 below.) We will see another proof in Section 5.6.

EXERCISES

Simplify the expressions given in Exercises 1–12.

1. $e^3/\sqrt{e^5}$

2. $\ln(e^{1/2}e^{2/3})$

3. $e^{5\ln x}$

4. $e^{(3\ln 9)/2}$

5. $\ln \dfrac{1}{e^{3x}}$

6. $\ln(4e)^{1/2}$

7. $e^{3\ln 5}e^{-\ln 25}$

8. $e^{2\ln\cos x} + \left(\ln e^{\sin x}\right)^2$

9. $3\ln 4 - 4\ln 3$

10. $4\ln\sqrt{x} + 6\ln(x^{1/3})$

11. $2\ln x + 5\ln(x-2)$

12. $\ln(x^2 + 6x + 9)$

Solve the equations in Exercises 13–16 for x.

13. $2^{x+1} = 3^x$

14. $3^x = 9^{1-x}$

15. $\dfrac{1}{2^x} = \dfrac{5}{8^{x+3}}$

16. $2^{x^2-3} = 4^x$

In Exercises 17–54 differentiate the given functions. If possible, simplify your answers.

17. $y = e^{5x}$

18. $y = xe^x - x$

19. $y = \dfrac{x}{e^{2x}}$

20. $y = x^2e^{x/2}$

21. $y = \ln(3x-2)$

22. $y = \ln|3x-2|$

23. $y = \ln(1+e^x)$

24. $y = 2\ln\sqrt{x^2+2}$

25. $y = \dfrac{e^x + e^{-x}}{2}$

26. $f(x) = e^{(x^2)}$

27. $y = e^{(e^x)}$

28. $y = \exp\exp\exp x$

29. $g(t) = t^r e^{at}$

30. $x = e^{3t}\ln t$

31. $y = \dfrac{e^x}{1+e^x}$

32. $f(x) = \dfrac{e^x - e^{-x}}{e^x + e^{-x}}$

33. $y = e^x \sin x$

34. $y = e^{-x}\cos x$

35. $F(u) = \sin^2 e^u$

36. $z = ye^y \sin y$

37. $y = \ln\ln x$

38. $y = x\ln x - x$

39. $y = x^2\ln x - \dfrac{x^2}{2}$

40. $y = \ln|\sin x|$

41. $y = 5^{2x+1}$

42. $y = 2^{(x^2-3x+8)}$

43. $g(x) = t^x x^t$

44. $h(t) = t^x - x^t$

45. $f(s) = \log_a(bs+c)$

46. $g(x) = \log_x(2x+3)$

47. $y = x^{\sqrt{x}}$

48. $y = (1/x)^{\ln x}$

49. $y = \ln|\sec x + \tan x|$

50. $y = \ln|x + \sqrt{x^2 - a^2}|$

51. $y = \ln(\sqrt{x^2+a^2} - x)$

52. $y = (x + \sqrt{\ln x})^3$

53. $y = \ln\left(\dfrac{a + \sqrt{a^2+x^2}}{x}\right)$

54. $y = (\cos x)^x - x^{\cos x}$

55. Find the nth derivative of $f(x) = xe^{ax}$.

56. Show that the nth derivative of $(ax^2 + bx + c)e^x$ is a function of the same form but with different constants.

57. Find the first four derivatives of e^{x^2}.

58. Find the nth derivative of $\ln(2x+1)$.

59. Differentiate a) $f(x) = (x^x)^x$, and b) $g(x) = x^{(x^x)}$. Which function grows most rapidly as x grows large?

60. *Solve the equation $x^{x^{x^{\cdot^{\cdot^{\cdot}}}}} = a$, where $a > 0$. The exponent tower goes on forever.

Use logarithmic differentiation to find the required derivatives in Exercises 61–63.

61. $f(x) = (x-1)(x-2)(x-3)(x-4)$. Find $f'(x)$.

62. $F(x) = \dfrac{\sqrt{1+x}(1-x)^{1/3}}{(1+5x)^{4/5}}$. Find $F'(0)$.

63. $f(x) = \dfrac{(x^2-1)(x^2-2)(x^2-3)}{(x^2+1)(x^2+2)(x^2+3)}$. Find $f'(2)$. Also find $f'(1)$.

64. On the same set of coordinate axes, sketch the graphs of $y = e^x, y = e^{2x}$, and $y = e^{-x}$.

65. Let $f(x) = xe^{-x}$. Determine where f is increasing and where it is decreasing. Sketch the graph of f.

66. At what points does the graph $y = x^2e^{-x^2}$ have a horizontal tangent line?

67. Find an equation of the straight line tangent to the curve $y = e^{2x}$ at the point $(2, e^4)$.

68. Find the equation of a straight line of slope 4 that is tangent to the graph of $y = \ln x$.

69. Find an equation of the straight line tangent to the curve $y = e^x$ and passing through the origin. Sketch the curve and tangent.

70. Find an equation of the straight line tangent to the curve $y = \ln x$ and passing through the origin. Sketch the curve and tangent.

71. Find an equation of the straight line that is tangent to $y = 2^x$ and that passes through the point $(1, 0)$.

72. For what values of $a > 0$ does the curve $y = a^x$ intersect the straight line $y = x$?

73. Find the slope of the curve $e^{xy}\ln\dfrac{x}{y} = x + \dfrac{1}{y}$ at the point $\left(e, \dfrac{1}{e}\right)$.

74. Find an equation of the straight line tangent to the curve $xe^y + y - 2x = \ln 2$ at the point $(1, \ln 2)$.

75. Find the derivative of $f(x) = Ax\cos\ln x + Bx\sin\ln x$. Use the result to help you find the indefinite integrals $\int \cos\ln x\,dx$ and $\int \sin\ln x\,dx$.

76. *Let $F_{A,B}(x) = Ae^x\cos x + Be^x\sin x$. Show that

$$\frac{d}{dx}F_{A,B}(x) = F_{A+B,B-A}(x).$$

77. *Using the results of Exercise 76, find a) $(d^2/dx^2)F_{A,B}(x)$, and b) $(d^3/dx^3)e^x\cos x$.

78. *Find $\dfrac{d}{dx}(Ae^{ax}\cos bx + Be^{ax}\sin bx)$ and use the answer to help you evaluate a) $\int e^{ax}\cos bx\,dx$ and b) $\int e^{ax}\sin bx\,dx$.

79. *Prove Theorem 3.5.18 of this section as follows. Let $a > 0$ and let $s = a/2$. Use Theorem 3.5.17 to show that $\ln x < \dfrac{1}{s}x^s$ for $x > 0$. Hence verify part b) of Theorem 3.5.18. Replace x with $\dfrac{1}{x}$ in part b) and thereby prove part d). Also deduce from part b) that $\lim_{t\to\infty}(\ln t)^a/t = 0$. Now replace t with e^x to obtain part a). Finally, replace x with $-x$ in part a) to prove part c).

80. *Let n be a positive integer. Substitute $a = 2^{1/n}$ in the identity

$$a - 1 = \frac{a^n - 1}{a^{n-1} + a^{n-2} + a^{n-3} + \cdots + a + 1}$$

and hence show that $\dfrac{1}{2} < n\left(2^{1/n} - 1\right) \le 1$. Assuming that $L(2) = \lim_{h\to0}(2^h - 1)/h$ exists, show that $\dfrac{1}{2} \le L(2) \le 1$.

81. *If $L(a) = \lim_{h\to0}(a^h - 1)/h$ for $a > 0$, show that $L(a^2) = 2L(a)$. The previous exercise therefore implies that $1 \le L(4) \le 2$. Thus $2 \le e \le 4$.

3.6 AN ALTERNATE APPROACH TO EXP AND LN

This is an *optional* section. In order to find the derivatives of exponential functions in Section 3.5, we had to *assume*, without proof, that these functions were defined and continuous for *all real* x, and had derivatives, at least at $x = 0$. In this section we outline an alternate way to define these functions. The approach taken here avoids the difficulties mentioned above, but instead requires that we have an intuitive idea of what "area" means. (We will discuss area in Chapter 6.)

$f(x)$	$f'(x)$
$\vdots$	$\vdots$
x^3	$3x^2$
x^2	$2x^1$
x^1	x^0
x^0	0
x^{-1}	$-x^{-2}$
x^{-2}	$-2x^{-3}$
x^{-3}	$-3x^{-4}$
$\vdots$	$\vdots$

Consider the table of derivatives appearing at the left. Every integer power of x appears in the left column, but $x^{-1} = 1/x$ is conspicuously absent from the right column. We are going to redefine the function $f(x) = \ln x$ for $x > 0$ so that it will satisfy the *initial value problem*

$$\begin{cases} f'(x) = \dfrac{1}{x} \\ f(0) = 1 \end{cases}$$

and then prove that it is indeed a logarithm. Our new definition of $\ln x$ is expressed in terms of area under the graph of its derivative. This definition suggests a relationship between the problem of finding derivatives and that of finding areas. We will explore this relationship in Section 6.3. Theorem 3.6.2 below is a special case of the *Fundamental Theorem of Calculus* presented in that section.

3.6.1 Alternate Definition of ln

For $x > 0$, let A_x be the area of the plane region bounded by the curve $y = 1/t$, the t-axis, and the vertical lines $t = 1$ and $t = x$. The function $\ln x$ is defined by

$$\ln x = \begin{cases} A_x & \text{if } x \geq 1, \\ -A_x & \text{if } 0 < x < 1, \end{cases}$$

as shown in Figures 3.6.1 and 3.6.2.

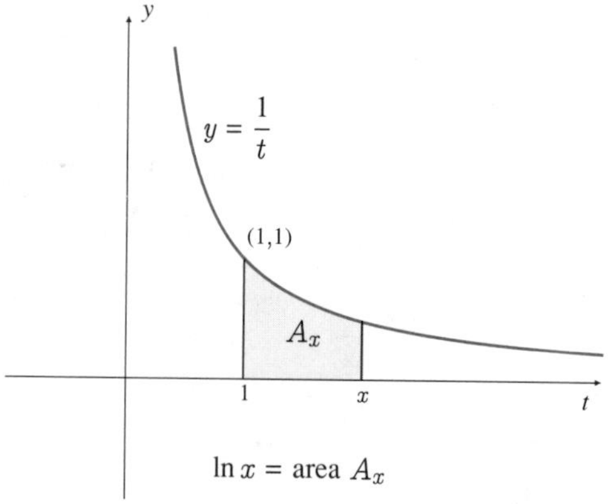

FIGURE 3.6.1

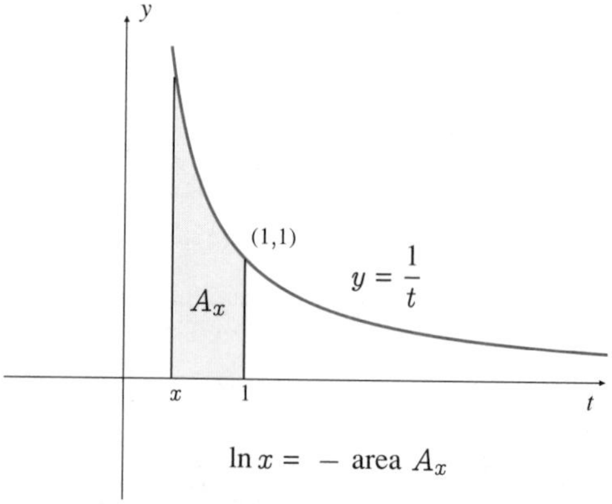

FIGURE 3.6.2

It is obvious from the definition that $\ln 1 = 0$, that $\ln x > 0$ if $x > 1$, that $\ln x < 0$ if $0 < x < 1$, and that ln is a one-to-one function. We now show that $y = \ln x$ satisfies the differential equation $y' = 1/x$.

THEOREM 3.6.2

$$\frac{d}{dx}\ln x = \frac{1}{x} \qquad \text{if } x > 0$$

PROOF For $x > 0$ and $h > 0$, $\ln(x+h) - \ln x$ is the area of the plane region bounded by $y = 1/t$, $y = 0$, and the vertical lines $t = x$ and $t = x + h$; it is the shaded area in Fig. 3.6.3. Comparing this area with that of two rectangles, we see that

$$\frac{h}{x+h} < \text{shaded area} < \frac{h}{x}.$$

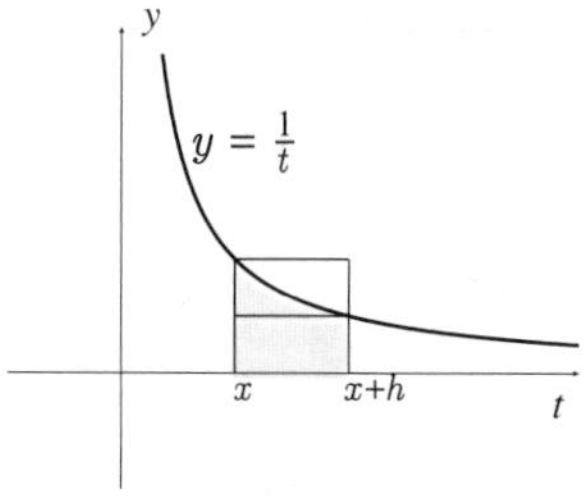

FIGURE 3.6.3

Hence the Newton quotient for $\ln x$ satisfies

$$\frac{1}{x+h} < \frac{\ln(x+h) - \ln x}{h} < \frac{1}{x}.$$

Letting h approach 0 from the right, we obtain (by the Squeeze Theorem applied to one-sided limits)

$$\lim_{h \to 0+} \frac{\ln(x+h) - \ln x}{h} = \frac{1}{x}.$$

A similar argument shows that if $0 < x + h < x$, then

$$\frac{1}{x} < \frac{\ln(x+h) - \ln x}{h} < \frac{1}{x+h},$$

so that

$$\lim_{h \to 0-} \frac{\ln(x+h) - \ln x}{h} = \frac{1}{x}.$$

Combining these two one-sided limits we get the desired result:

$$\frac{d}{dx}\ln x = \lim_{h \to 0} \frac{\ln(x+h) - \ln x}{h} = \frac{1}{x}. \square$$

The two properties $(d/dx)\ln x = 1/x$ and $\ln 1 = 0$ are sufficient to characterize the function $\ln x$ completely. We can deduce from these two properties that $\ln x$ satisfies the appropriate laws of logarithms.

EXAMPLE 3.6.3 Show that $\ln ax = \ln a + \ln x$ if $a > 0$ and $x > 0$.

SOLUTION $\frac{d}{dx}(\ln(ax) - \ln x) = \frac{a}{ax} - \frac{1}{x} = 0$ for all $x > 0$. Therefore $(\ln(ax) - \ln x) = C$ (a constant) for $x > 0$. Puting $x = 1$ we get $C = \ln a$ and the identity follows. $\square$

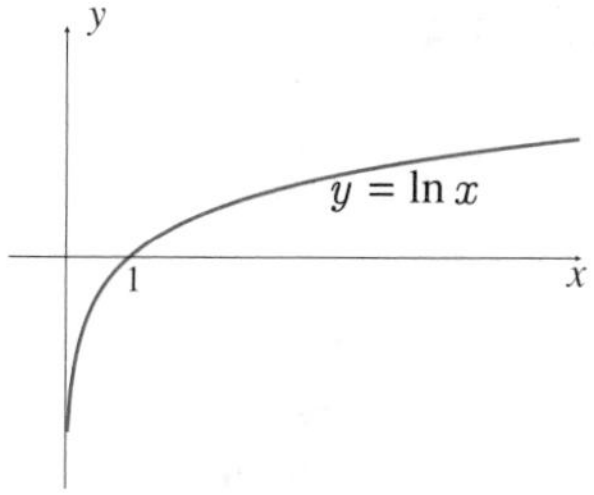

FIGURE 3.6.4

The laws $\ln(1/x) = -\ln x$ and $\ln(x/y) = \ln x - \ln y$ can be proved similarly. So can the law $\ln(x^r) = r\ln x$, but this latter law should be regarded as holding only for rational r for the moment, since we don't want to assume we know what irrational powers mean yet in this section. We now have $\ln(2^n) = n\ln 2 \to \infty$ and $\ln(1/2)^n = -n\ln 2 \to -\infty$ as $n \to \infty$. Since $(d/dx)\ln x = 1/x > 0$ for $x > 0$, it follows that $\ln x$ is increasing, so we must have (see Fig. 3.6.4)

$$\lim_{x \to \infty} \ln x = \infty, \qquad \lim_{x \to 0+} \ln x = -\infty.$$

Also, $\ln x$ is one-to-one and hence has an inverse, which we will call $\exp x$:

$$y = \exp x \qquad \text{means the same as} \qquad x = \ln y.$$

Since $\ln 1 = 0$, therefore $\exp 0 = 1$. The domain of exp is $(-\infty, \infty)$ which is the range of ln. The range of exp is $(0, \infty)$ which is the domain of ln. We have

$$\ln(\exp x) = x \quad \text{for all real } x, \qquad \text{and} \qquad \exp(\ln x) = x \quad \text{for } x > 0.$$

We can deduce various properties of exp from corresponding properties of ln. Not surprisingly, they are properties we would expect an exponential function to have.

EXAMPLE 3.6.4 Show that $(\exp x)^r = \exp(rx)$ for any rational r.

SOLUTION Let $y = (\exp x)^r$. Thus $\ln y = r\ln(\exp x) = rx$ and so $y = \exp(rx)$. □

Now we make an important definition:

3.6.5
Definition of e

Let $e = \exp(1)$.

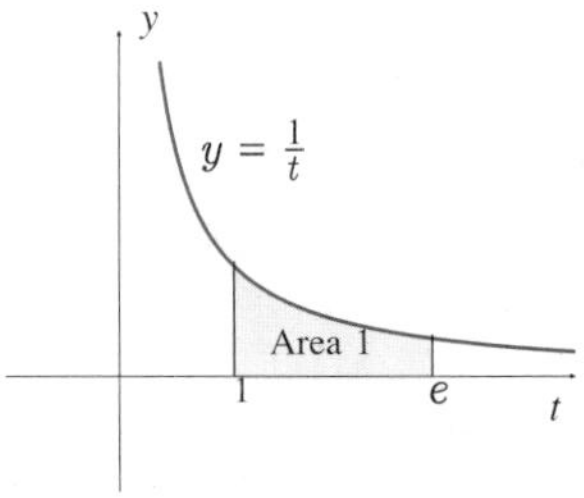

FIGURE 3.6.5

The number e satisfies $\ln e = 1$ so the area bounded by the curve $y = 1/t$, the t-axis, and the vertical lines $t = 1$ and $t = e$ must be equal to 1 square unit.

The example above now shows that $\exp r = \exp(1r) = (\exp 1)^r = e^r$ for any rational number r. Since $\exp x$ is defined for all *real* x we can use it as a *definition* of what is meant by e^x and there will be no contradiction if x is rational and we already know what e^x means.

3.6.6
Definition of e^x

$e^x = \exp x$ for all real x.

Since $\exp x = e^x$ is actually an exponential function, its inverse is actually a logarithm: $\ln x = \log_e x$.

We can verify that e^x satisfies the other laws of exponents, namely $e^{x+y} = e^x e^y$, $e^{-x} = 1/e^x$ and $e^{x-y} = e^x/e^y$, by rephrasing these laws in terms of the corresponding laws for $\ln x$. For example, if $u = e^x$ and $v = e^y$, then $x = \ln u$ and $y = \ln v$. Therefore $x + y = \ln u + \ln v = \ln(uv)$ and $e^{x+y} = uv = e^x e^y$.

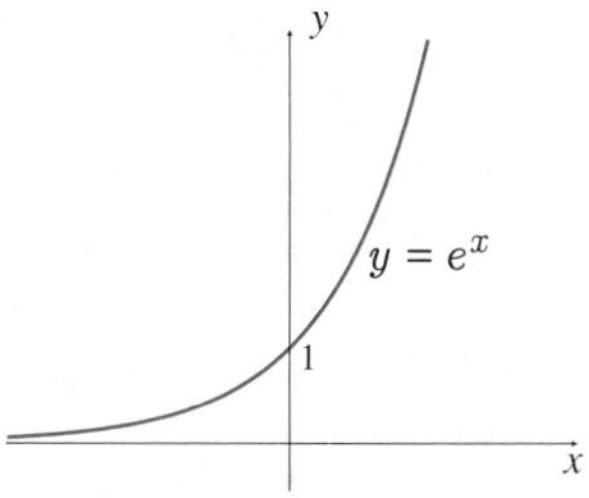

FIGURE 3.6.6

The derivative of $y = e^x$ is calculated by implicitly differentiating the equivalent equation $x = \ln y$ with respect to x. We get

$$1 = \frac{1}{y}\frac{dy}{dx}, \qquad \text{so} \qquad \frac{dy}{dx} = y = e^x.$$

We can use the fact that e^x is now defined for all *real* x to define the arbitrary exponential a^x (where $a > 0$) for all real x. If r is rational we know that

$$\ln(a^r) = r \ln a, \qquad \text{so} \qquad a^r = e^{r \ln a}.$$

Since $e^{x \ln a}$ is defined for all real x, we can use it as a definition of a^x with no possibility of contradiction arising if x is rational.

3.6.7
Definition of a^x

$$a^x = e^{x \ln a} \text{ for } a > 0 \text{ and } x \text{ real.}$$

The laws of exponents for a^x as presented in Section 3.4 can now be obtained from those for e^x, as can the derivative

$$\frac{d}{dx} a^x = \frac{d}{dx} e^{x \ln a} = e^{x \ln a} \ln a = a^x \ln a.$$

Finally, observe that $\frac{d}{dx} a^x$ is negative for all x if $0 < a < 1$ and is positive for all x if $a > 1$. Thus a^x has an inverse function, $\log_a x$, provided $a > 0$ and $a \neq 1$. Its properties follow in the same way as in the previous sections.

EXERCISES

1. Use the facts $\frac{d}{dx} \ln x = \frac{1}{x}$ and $\ln 1 = 0$ to verify that a) $\ln(1/x) = -\ln x$, b) $\ln(x/a) = \ln x - \ln a$, and c) $\ln(x^a) = a \ln x$.

2.*Let $x > 0$ and let $F(x)$ be the area bounded by the curve $y = t^2$, the t-axis, and the vertical lines $t = 0$ and $t = x$. Using the method of the proof of Theorem 3.6.2, show that $F'(x) = x^2$. Hence find an explicit formula for $F(x)$. What is the area of the region bounded by $y = t^2$, $y = 0$, $t = 0$ and $t = 2$?

3.*At this stage in your study of calculus you are not likely able to guess a function whose derivative is e^{x^2}. However, explain how you could use areas to define a function $F(x)$ such that $F'(x) = e^{x^2}$ and $F(-2) = 0$.

3.7 GROWTH AND DECAY PROBLEMS

Many natural processes involve quantities that increase or decrease at a rate proportional to their size. For example, the mass of a culture of bacteria growing in a medium supplying adequate nourishment will increase at a rate proportional to that mass. The value of an investment bearing interest that is continuously compounding increases at a rate proportional to that value. The mass of undecayed radioactive material in a sample decreases at a rate proportional to that mass.

All of these phenomena, and others exhibiting similar behavior, can be modeled mathematically in the same way. If $y = y(t)$ denotes the value of a quantity y at time t, and if y changes at a rate proportional to its size, then

$$\frac{dy}{dt} = ky,$$

where k is the constant of proportionality. The above equation is called the **differential equation of exponential growth or decay**, because, for any value of the constant C, it is obvious that $y = Ce^{kt}$ satisfies the equation. In fact, if $y(t)$ is any solution of the differential equation $y' = ky$ then

$$\frac{d}{dt}\left(\frac{y(t)}{e^{kt}}\right) = \frac{e^{kt}y'(t) - ke^{kt}y(t)}{e^{2kt}} = \frac{y'(t) - ky(t)}{e^{kt}} = 0 \quad \text{for all } t.$$

Thus $y(t)/e^{kt} = C$, a constant, and $y(t) = Ce^{kt}$. Since $y(0) = Ce^0 = C$, we have shown that

> The Initial Value Problem $\begin{cases} \dfrac{dy}{dt} = ky \\ y(0) = y_0 \end{cases}$ has unique solution $y = y_0e^{kt}$.

If $y_0 > 0$, then $y(t)$ is an increasing function of t if $k > 0$ and a decreasing function of t if $k < 0$. We say that the quantity y exhibits **exponential growth** if $k > 0$, and **exponential decay** if $k < 0$.

EXAMPLE 3.7.1 A certain cell culture grows at a rate proportional to the number of cells present. If the culture contains 500 cells initially, and 800 after 24 hours, how many cells will be present after a further 12 hours?

SOLUTION Let $y(t)$ be the number of cells present t hours after there were 500 cells. Thus $y(0) = 500$ and $y(24) = 800$. Because $dy/dt = ky$, we have

$$y(t) = y(0)e^{kt} = 500e^{kt}.$$

Thus $800 = y(24) = 500e^{24k}$, and so $24k = \ln \frac{800}{500} = \ln(1.6)$. Thus

$$y(t) = 500e^{(t/24)\ln(1.6)} = 500(1.6)^{t/24}.$$

We want to know y when $t = 36$: $y(36) = 500e^{(36/24)\ln(1.6)} \approx 1012$. (We used a scientific calculator to evaluate the answer.) The cell count grew to about 1012 in the 12 hours after it was 800.

Exponential growth is characterized by **fixed doubling time**. If T is the time at which y has doubled from its size at $t = 0$, then $2y(0) = y(T) = y(0)e^{kT}$, and so $e^{kT} = 2$. Therefore, since $y(t) = y(0)e^{kt}$, we have

$$y(t+T) = y(0)e^{k(t+T)} = e^{kT}y(0)e^{kt} = 2y(t),$$

that is, T units of time are required for y to double from any value. Similarly, exponential decay involves fixed halving time (usually called **half-life**). If $y(T) = \frac{1}{2}y(0)$, then $e^{kT} = \dfrac{1}{2}$ and

$$y(t+T) = y(0)e^{k(t+T)} = \tfrac{1}{2}y(t).$$

EXAMPLE 3.7.2 A radioactive material has a half-life of 1200 years. What percentage of the original radioactivity of a sample is left after 10 years? How many years are required to reduce the radioactivity by 10 percent?

SOLUTION Let $p(t)$ be the percentage of the original radioactivity left after t years. Thus $p(0) = 100$ and $p(1200) = 50$. Since the radioactivity decreases at a rate proportional to itself, $dp/dt = kp$ and

$$p(t) = 100e^{kt}. \qquad \text{Therefore } 50 = p(1200) = 100e^{1200k},$$

and $k = -(\ln 2)/1200$. The percentage left after 10 years is

$$p(10) = 100e^{10k} = 100e^{-(\ln 2)/120} \approx 99.424.$$

If after t years 90 percent of the radioactivity is left, then

$$\begin{aligned} 90 &= 100e^{kt}, \\ kt &= \ln \frac{90}{100}, \\ t &= -\frac{1200}{\ln 2} \ln \frac{9}{10} \approx 182.4, \end{aligned}$$

so it will take a little over 182 years to reduce the radioactivity by 10 percent.

Sometimes an exponential growth or decay problem will involve a quantity that changes at a rate proportional to the difference between itself and a fixed value:

$$\frac{dy}{dt} = k(y - a).$$

In this case the change of dependent variable $u = y - a$ should be used to convert the differential equation to the standard form. Observe that u has the same rate of change as the original quantity y (that is, $du/dt = dy/dt$) and so it satisfies

$$\frac{du}{dt} = ku.$$

EXAMPLE 3.7.3 According to Newton's law of cooling, a hot object introduced into a cooler environment will cool at a rate proportional to the excess of its temperature above that of its environment. If a cup of coffee sitting in a room maintained at a temperature of 20°C cools from 80°C to 50°C in five minutes, how much longer will it take to cool to 40°C?

SOLUTION Let $y(t)$ be the temperature of the coffee t minutes after it was 80°C. Thus $y(0) = 80$ and $y(5) = 50$. Newton's law says that $dy/dt = k(y - 20)$ in this case, so let $u(t) = y(t) - 20$. Thus $u(0) = 60$ and $u(5) = 30$. We have

$$\frac{du}{dt} = \frac{dy}{dt} = k(y - 20) = ku.$$

Thus

$$\begin{aligned} u(t) &= 60e^{kt}, \\ 30 = u(5) &= 60e^{5k}, \\ 5k &= \log \tfrac{1}{2} = -\ln 2. \end{aligned}$$

We wish to know t such that $y(t) = 40$, that is, $u(t) = 20$.

$$\begin{aligned} 20 = u(t) &= 60e^{-(t/5)\ln 2} \\ -\frac{t}{5}\ln 2 = \ln \frac{20}{60} &= -\ln 3, \\ t = 5\,\frac{\ln 3}{\ln 2} &\approx 7.92. \end{aligned}$$

The coffee will take about $7.92 - 5 = 2.92$ minutes to cool from 50°C to 40°C.

Interest on Investments

Suppose that \$10,000.00 is invested at an annual rate of interest of 8%. Thus the value of the investment at the end of 1 year will be \$10,000.00(1.08)= \$10,800.00. If this amount remains invested for a second year at the same rate it will grow to \$10,000.00$(1.08)^2$=\$11,664.00, and in general, n years after the original investment was made it will be worth \$10,000.00$(1.08)^n$.

Now suppose that the 8% rate is *compounded semiannually* so that the interest is actually paid at a rate of 4% per six-month period. After 1 year the \$10,000.00 will grow to \$10,000.00$(1.04)^2$=\$10,816.00. This is \$16.00 more than was obtained when the 8% was compounded only once per year. The extra \$16.00 is the interest paid in the second six month period on the \$400 interest earned in the first six month period. Continuing in this way, we see that if the 8% interest is compounded *monthly* (12 periods per year and $\frac{8}{12}$% paid per period) or *daily* (365 periods per year and $\frac{8}{365}$% paid per period) then the original \$10,000.00 would grow in 1 year to $\$10{,}000.00\left(1 + \frac{8}{1200}\right)^{12} = \$10,830.00$ or $\$10{,}000.00\left(1 + \frac{8}{36500}\right)^{365} = \$10,832.78$ respectively.

Clearly for any given *nominal* interest rate, the investment grows more if the compounding period is shorter. In general, an original investment of $\$A$ invested at $r\%$ per annum compounded n times per year grows in one year to

$$\$A\left(1 + \frac{r}{100n}\right)^n.$$

It is natural to ask how well we can do with our investment if we let the number of periods in a year approach infinity, that is, we compound the interest *continuously*. The answer is that in 1 year the $\$A$ will grow to

$$\$A \lim_{n\to\infty}\left(1+\frac{r}{100n}\right)^n = \$Ae^{r/100}.$$

For example, at 8% per annum compounded continuously, our \$10,000.00 will grow in 1 year to $\$10,000.00e^{0.08} \approx \$10,832.87$. (Note that this is just a few cents more than we get compounding daily.) To justify this result we need the following theorem.

THEOREM 3.7.4

$$e^x = \lim_{n\to\infty}\left(1+\frac{x}{n}\right)^n \qquad \text{for every real } x$$

PROOF If $x = 0$, there is nothing to prove; both sides of the identity are 1. If $x \neq 0$ let $h = x/n$. As n tends to infinity h approaches 0. Thus

$$\begin{aligned}
\lim_{n\to\infty} \ln\left(1+\frac{x}{n}\right)^n &= \lim_{n\to\infty} n\ln\left(1+\frac{x}{n}\right) \\
&= \lim_{n\to\infty} x\,\frac{\ln\left(1+\frac{x}{n}\right)}{\frac{x}{n}} \\
&= x\lim_{h\to 0}\frac{\ln(1+h)}{h} && \left(\text{where } h=\frac{x}{n}\right) \\
&= x\lim_{h\to 0}\frac{\ln(1+h)-\ln 1}{h} && (\text{since } \ln 1 = 0) \\
&= x\left(\frac{d}{dt}\ln t\right)\bigg|_{t=1} && (\text{by the definition of derivative}) \\
&= x\,\frac{1}{t}\bigg|_{t=1} = x.
\end{aligned}$$

Since ln is differentiable, it is continuous by Theorem 2.3.1. Hence, by Theorem 1.7.5(b),

$$\ln\left(\lim_{n\to\infty}\left(1+\frac{x}{n}\right)^n\right) = \lim_{n\to\infty}\ln\left(1+\frac{x}{n}\right)^n = x.$$

Since the exponential function is the inverse of the natural logarithm, we conclude

$$\lim_{n\to\infty}\left(1+\frac{x}{n}\right)^n = e^x. \ \square$$

In case $x = 1$ the formula given in Theorem 3.7.4 takes the following form:

$$e = \lim_{n\to\infty}\left(1+\frac{1}{n}\right)^n.$$

n	$\left(1+\frac{1}{n}\right)^n$
1	2
10	2.59374 ⋯
100	2.70481 ⋯
1000	2.71692 ⋯
10000	2.71815 ⋯
100000	2.71827 ⋯

We can use this formula to compute approximations to e, as shown in the table at the left. In a sense we have cheated in obtaining the numbers in this table; they were produced using the y^x function on a scientific calculator. However, this function is actually computed as $e^{x \ln y}$. In any event, the formula in this table is not a very efficient way to calculate e to any great accuracy. A much better way is to use the series

$$e = 1 + \frac{1}{1!} + \frac{1}{2!} + \frac{1}{3!} + \frac{1}{4!} + \cdots = 1 + 1 + \frac{1}{2} + \frac{1}{6} + \frac{1}{24} + \cdots,$$

which we will establish in Chapter 11. (See Example 5.4.6 also.)

A final word about interest rates. Financial institutions sometimes quote *effective* rates of interest rather than *nominal* rates. The effective rate tells you what the actual effect of the interest rate will be after one year. Thus \$10,000.00 invested at an effective rate of 8% will grow to \$10,800.00 in one year regardless of the compounding period. Thus a nominal rate of 8% per annum compounded daily is equivalent to an effective rate of about 8.3278%.

Logistic Growth

Few quantities in nature can sustain exponential growth over extended periods of time; the growth is usually limited by external constraints. For example, suppose a small number of rabbits (of both sexes) is introduced to a small island where there were no rabbits previously, and where there are no predators who eat rabbits. By virtue of natural fertility, the number of rabbits might be expected to grow exponentially, but this growth will eventually be limited by the food supply available to the rabbits. Suppose the island can grow enough food to supply a population of L rabbits indefinitely. If there are $y(t)$ rabbits in the population at time t, we would expect $y(t)$ to grow at a rate proportional to $y(t)$ provided $y(t)$ is quite small (much less than L). But as the numbers increase, it will be harder for the rabbits to find enough food, and we would expect the rate of increase to approach 0 as $y(t)$ gets closer and closer to L. One possible model for such behavior is the differential equation

3.7.5
The Logistic Equation

$$\frac{dy}{dt} = ky\left(1 - \frac{y}{L}\right),$$

which is called the **logistic equation** since it models growth that is limited by the supply of necessary resourses. Observe that $dy/dt > 0$ if $0 < y < L$ and that this rate becomes small if y is small (there are few rabbits to reproduce) or if y is close to L (there are almost as many rabbits as the available resourses can feed). Observe also that $dy/dt < 0$ if $y > L$; there being more animals than the resourses can feed, the rabbits die at a greater rate than they are born. Of course the steady-state populations $y = 0$ and $y = L$ are solutions of the logistic equation; for both of these $dy/dt = 0$. We will examine techniques for solving differential equations like the logistic equation in Section 8.8. For now, we invite the reader to verify by differentiation that the solution satisfying $y(0) = y_0$, is

$$y = \frac{Ly_0}{y_0 + (L - y_0)e^{-kt}}.$$

Observe that, as expected, if $0 < y_0 < L$ then

$$\lim_{t\to\infty} y(t) = L, \qquad \lim_{t\to-\infty} y(t) = 0.$$

The solution given in the box above also holds for $y_0 > L$. However, the solution does not approach 0 as t approaches $-\infty$ in this case. It has a vertical asymptote at a certain negative value of t. (See Exercise 17 below.) The graphs of solutions of the logistic equation for various positive values of y_0 are given Fig. 3.7.1.

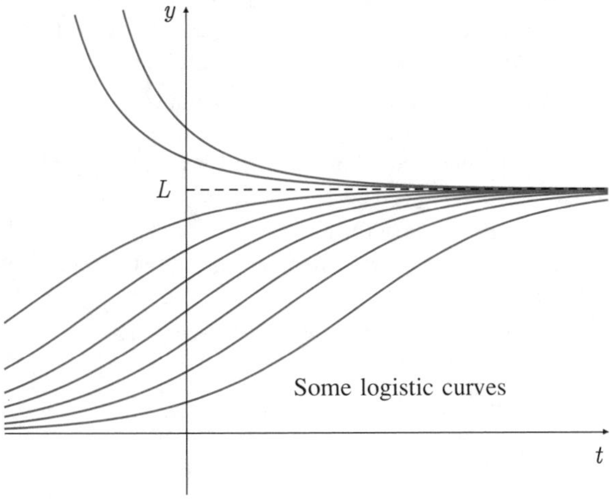

FIGURE 3.7.1

EXERCISES

1. Bacteria grow in a certain culture at a rate proportional to the amount present. If there are 100 bacteria present initially and the amount doubles in 1 hour, how many will there be after a further $1\frac{1}{2}$ hours?

2. Sugar decomposes in water at a rate proportional to the amount still undecomposed. If there were 50 kg of sugar present initially, and at the end of 5 hours only 20 kg is left, how much longer will it take until 90 percent of the sugar is decomposed?

3. A radioactive substance decays at a rate proportional to the amount present. If 30 percent of such a substance decays in 15 years, what is the half-life of the substance?

4. If the half-life of radium is 1690 years, what percentage of the amount present now will be remaining after (a) 100 years, (b) 1000 years?

5. Find the half-life of a radioactive substance if after 1 year 99.57 percent of an initial amount still remains.

6. In a certain culture where the rate of growth of bacteria is proportional to the number present, the number triples in 3 days. If at the end of 7 days there are 10 million bacteria present in the culture, how many were present initially?

7. In the first few weeks after birth, a baby gains weight at a rate proportional to its weight. A baby weighing 4 kg at birth weighs 4.4 kg after two weeks. How much did it weigh 5 days after birth?

8. When a simple electrical circuit containing inductance and resistance but no capacitance has the electromotive force removed, the rate of decrease of the current is proportional to the current. If the current is $I(t)$ amperes t seconds after cutoff, and if $I = 40$ when $t = 0$, and $I = 15$ when $t = 0.01$, find a formula for $I(t)$.

9. Use Newton's law of cooling to determine the reading on a thermometer 5 minutes after it is taken from an oven at 72°C to the outdoors where the temperature is 20°C, if the reading dropped to 48°C after one minute.

10. An object is brought into a freezer maintained at a temperature of −5°C. If the object cools from 45°C to 20°C in 40 minutes, how many more minutes will it take to cool to 0°C?

11. If a body in a room warms up from 5°C to 10°C in 4 minutes, and if the room is being maintained at 20°C, how much longer will the body take to warm up to 15°C? Assume the body warms at a rate proportional to the difference between its temperature and room temperature.

12. Money invested at compound interest (with instantaneous compounding) accumulates at a rate proportional to the amount present. If an initial investment of \$1,000 grows to \$1,500 in exactly 5 years, find (a) the doubling time for the investment, and (b) the effective annual rate of interest being paid.

13. If the purchasing power of the dollar is decreasing at an effective rate of 9 percent annually, how long will it take for the purchasing power to be reduced to 25 cents?

14.*A bank claims to pay interest at an effective rate of 9.5% on an investment account. If the interest is actually being compounded monthly, what is the nominal rate of interest being paid on the account?

15.*Suppose the quantity $y(t)$ exhibits logistic growth. If the values of $y(t)$ at times $t = 0$, $t = 1$, and $t = 2$ are y_0, y_1, and y_2, respectively, find an equation satisfied by the limiting value L of $y(t)$, ans solve it for L. If $y_0 = 3, y_1 = 5$, and $y_2 = 6$, find L.

16.*Show that a solution $y(t)$ of the logistic equation having $0 < y(0) < L$ is increasing most rapidly when its value is $L/2$. (*Hint:* You do not need to use the formula for the solution to see this.)

17.*If $y_0 > L$, find the interval on which the given solution of the logistic equation is valid. What happens to the solution as t approaches the left endpoint of this interval?

18.*If $y_0 < 0$, find the interval on which the given solution of the logistic equation is valid. What happens to the solution as t approaches the right endpoint of this interval?

3.8 THE HYPERBOLIC FUNCTIONS AND THEIR INVERSES

This Section is *optional.* Any function defined on the real line can be expressed (in a unique way) as the sum of an even function and an odd function. (See Exercise 30 of Section 1.3.) The **hyperbolic functions** $\cosh x$ and $\sinh x$ are respectively the even and odd functions whose sum is the exponential function e^x.

3.8.1
Hyperbolic Sine and Cosine

For any real x the **hyperbolic cosine**, $\cosh x$, and the **hyperbolic sine**, $\sinh x$ (usually pronounced "shine x" or "sinch x"), are defined by

$$\cosh x = \frac{e^x + e^{-x}}{2}, \qquad \sinh x = \frac{e^x - e^{-x}}{2}.$$

Recall that cosine and sine are called *circular functions* because, for any t, the point $(\cos t, \sin t)$ lies on the circle with equation $x^2 + y^2 = 1$. Similarly, cosh and sinh are called *hyperbolic functions* because the point $(\cosh t, \sinh t)$ lies on the rectangular hyperbola with equation $x^2 - y^2 = 1$. To see this, observe that

$$\begin{aligned}\cosh^2 t - \sinh^2 t &= \left(\frac{e^t + e^{-t}}{2}\right)^2 - \left(\frac{e^t - e^{-t}}{2}\right)^2 \\ &= \frac{1}{4}(e^{2t} + 2 + e^{-2t} - (e^{2t} - 2 + e^{-2t})) \\ &= \frac{1}{4}(2 + 2) = 1.\end{aligned}$$

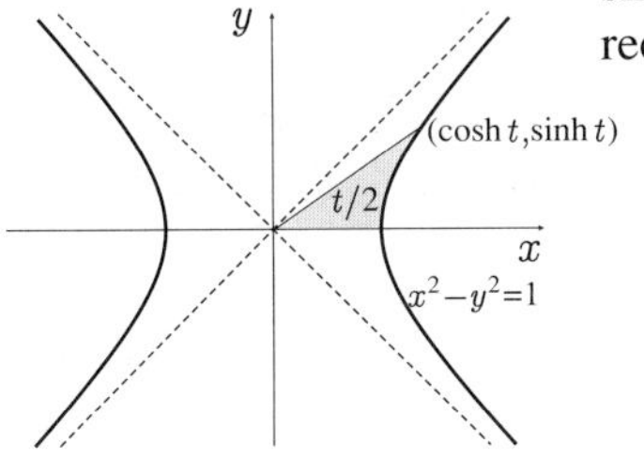

FIGURE 3.8.1

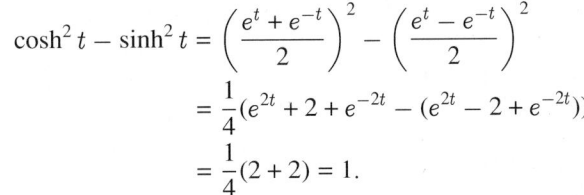

There is no interpretation of t as an arc length or angle as there was in the circular case; however, the *area* of the *hyperbolic sector* bounded by $y = 0$, the hyperbola $x^2 - y^2 = 1$ and the ray from the origin to $(\cosh t, \sinh t)$ is $t/2$ square units (see Exercise 21 of Section 9.4), just as is the area of the circular sector bounded by $y = 0$, the circle $x^2 + y^2 = 1$ and the ray from the origin to $(\cos t, \sin t)$. (See Figures 3.8.1 and 3.8.2.)

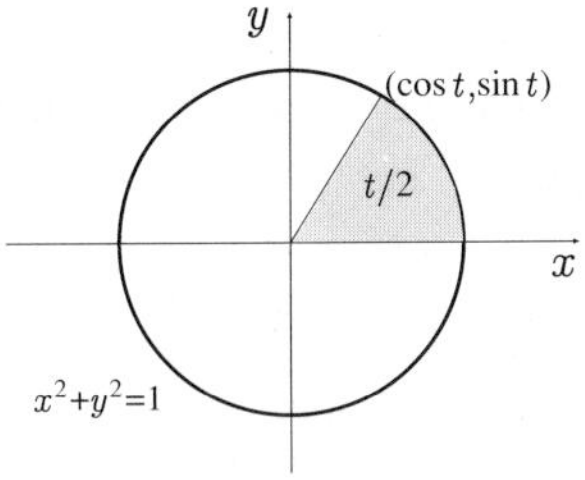

FIGURE 3.8.2

The hyperbolic functions have many properties which resemble properties of the corresponding circular functions:

3.8.2
Properties of cosh and sinh

$$\cosh 0 = 1 \qquad \sinh 0 = 0$$

$$\cosh(-x) = \cosh x \qquad \sinh(-x) = -\sinh x$$

$$\frac{d}{dx}\cosh x = \sinh x \qquad \frac{d}{dx}\sinh x = \cosh x$$

$$\cosh(x + y) = \cosh x \cosh y + \sinh x \sinh y$$

$$\sinh(x + y) = \sinh x \cosh y + \cosh x \sinh y$$

$$\cosh(2x) = \cosh^2 x + \sinh^2 x = 1 + 2\sinh^2 x = 2\cosh^2 x - 1$$

$$\sinh(2x) = 2\sinh x \cosh x$$

All of these formulas are easily verified by direct calculation using the definitions of $\cosh x$ and $\sinh x$ and the laws of exponents.

The graphs of cosh and sinh are shown in Fig. 3.8.3. The graph $y = \cosh x$ is called a **catenary**. A chain suspended at its ends will assume the shape of a catenary.

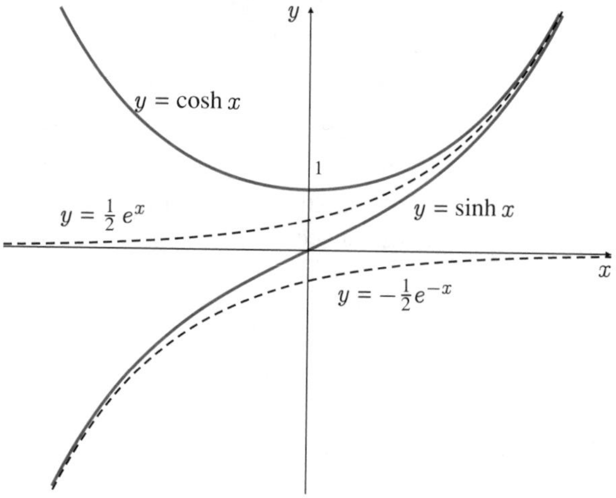

FIGURE 3.8.3

By analogy with the trigonometric functions, four other hyperbolic functions can be defined in terms of cosh and sinh.

3.8.3 Other Hyperbolic Functions

$$\tanh x = \frac{\sinh x}{\cosh x} = \frac{e^x - e^{-x}}{e^x + e^{-x}} \qquad \operatorname{sech} x = \frac{1}{\cosh x} = \frac{2}{e^x + e^{-x}}$$

$$\coth x = \frac{\cosh x}{\sinh x} = \frac{e^x + e^{-x}}{e^x - e^{-x}} \qquad \operatorname{csch} x = \frac{1}{\sinh x} = \frac{2}{e^x - e^{-x}}$$

Observe that

$$\lim_{x\to\infty} \tanh x = \lim_{x\to\infty} \frac{1 - e^{-2x}}{1 + e^{-2x}} = 1$$

$$\lim_{x\to-\infty} \tanh x = \lim_{x\to-\infty} \frac{e^{2x} - 1}{e^{2x} + 1} = -1,$$

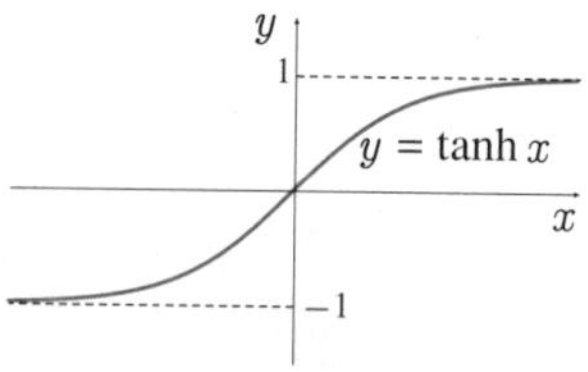

FIGURE 3.8.4

so that the graph of $y = \tanh x$ has two horizontal asymptotes. The graph resembles that of the inverse tangent function in shape, as you can see in Fig. 3.8.4.

Since the hyperbolic functions are expressible in terms of exponentials, it is not surprising that their inverses are expressible in terms of logarithms. The functions sinh and tanh are increasing and therefore one-to-one and invertible on the whole real line. Their inverses are denoted $\sinh^{-1}$ and $\tanh^{-1}$ respectively:

$$y = \sinh^{-1} x \quad \text{means the same as} \quad x = \sinh y,$$
$$y = \tanh^{-1} x \quad \text{means the same as} \quad x = \tanh y.$$

To express $y = \sinh^{-1} x$ in terms of logarithms, observe that

$$x = \sinh y = \frac{e^y - e^{-y}}{2} = \frac{e^{2y} - 1}{2e^y},$$
$$\text{so} \quad e^{2y} - 2xe^y - 1 = 0.$$

This is a quadratic equation in e^y and can be solved by the quadratic formula:

$$e^y = \frac{2x \pm \sqrt{4x^2 + 4}}{2} = x + \sqrt{x^2 + 1}.$$

We need the positive square root since e^y cannot be negative. Hence

$$\sinh^{-1} x = \ln\left(x + \sqrt{x^2 + 1}\right).$$

Similarly, we can express $y = \tanh^{-1} x$ using a logarithm:

$$x = \tanh y = \frac{e^y - e^{-y}}{e^y + e^{-y}} = \frac{e^{2y} - 1}{e^{2y} + 1} \qquad (-1 < x < 1),$$
$$xe^{2y} + x = e^{2y} - 1,$$
$$e^{2y} = \frac{1 + x}{1 - x}, \qquad y = \frac{1}{2} \ln\left(\frac{1 + x}{1 - x}\right).$$

$$\tanh^{-1} x = \frac{1}{2} \ln\left(\frac{1 + x}{1 - x}\right), \qquad (-1 < x < 1)$$

Since cosh is not one-to-one its domain must be restricted before an inverse can be defined. Let us define the principal value of cosh to be

$$\operatorname{Cosh} x = \cosh x \qquad (x \ge 0).$$

The inverse, Cosh^{-1}, is then defined by

$$y = \operatorname{Cosh}^{-1} x \qquad \text{means the same as} \qquad x = \cosh y \qquad (y \geq 0).$$

As we did for $\sinh^{-1}$, we can obtain the formula

$$\operatorname{Cosh}^{-1} x = \ln(x + \sqrt{x^2 - 1}), \qquad (x \geq 1).$$

EXERCISES

1. Calculate the derivatives of all six hyperbolic functions using the definitions of those functions as given in this section. Express each derivative in terms of hyperbolic functions.

2. Verify the addition formulas for $\cosh(x+y)$ and $\sinh(x+y)$ given in box 3.8.2. Proceed by expanding the right-hand side of each in terms of exponentials. Find similar formulas for $\cosh(x-y)$ and $\sinh(x-y)$.

3. Obtain addition formulas for $\tanh(x+y)$ and $\tanh(x-y)$ from those for sinh and cosh.

4. Sketch the graphs of $y = \coth x$, $y = \operatorname{sech} x$, and $y = \operatorname{csch} x$, showing any asymptotes.

5. Calculate the derivatives of $\sinh^{-1} x$, $\operatorname{Cosh}^{-1} x$, and $\tanh^{-1} x$. Hence express each of the indefinite integrals

$$\int \frac{dx}{\sqrt{x^2+1}}, \qquad \int \frac{dx}{\sqrt{x^2-1}}, \qquad \int \frac{dx}{1-x^2}$$

in terms of inverse hyperbolic functions.

6. Calculate the derivatives of the functions $\sinh^{-1}(x/a)$, $\operatorname{Cosh}^{-1}(x/a)$, and $\tanh^{-1}(x/a)$ (where $a > 0$) and use your answers to provide formulas for certain indefinite integrals.

7. Simplify the following expressions: a) $\sinh \ln x$, b) $\cosh \ln x$, c) $\tanh \ln x$, d) $\dfrac{\cosh \ln x + \sinh \ln x}{\cosh \ln x - \sinh \ln x}$

8.*Define $\operatorname{csch}^{-1} x$, find its domain, range, and derivative, and sketch its graph. Show that $\operatorname{csch}^{-1} x = \sinh^{-1}(1/x)$.

9.*Do an analogous version of Exercise 8 for the function $\coth^{-1} x$.

10.*Define $\operatorname{Sech} x$ at a suitably restricted version of $\operatorname{sech} x$ and repeat Exercise 8 for the function $\operatorname{Sech}^{-1} x$.

11.†Show that the functions $f_{A,B}(x) = Ae^{kx} + Be^{-kx}$ and $g_{C,D}(x) = C \cosh kx + D \sinh kx$ are both solutions of the differential equation $y'' - k^2 y = 0$. (They are both general solutions.) Express $f_{A,B}$ in terms of $g_{C,D}$, and express $g_{C,D}$ in terms of $f_{A,B}$.

12.†Show that the function $h_{L,M}(x) = L \cosh k(x-a) + M \sinh k(x-a)$ is also a solution of the differential equation in the previous exercise. Express $h_{L,M}$ in terms of the function $f_{A,B}$ above.

13.†Solve the initial-value problem

$$\begin{cases} y'' - k^2 y = 0 \\ y(a) = y_0 \\ y'(a) = v_0 \end{cases}$$

Express the solution in terms of the function $h_{L,M}$ of Exercise 12.

3.9 SECOND ORDER CONSTANT COEFFICIENT DIFFERENTIAL EQUATIONS

This is an *optional* section. In this section we will solve differential equations of the form

3.9.1
The Differential Equation

$$a\,y'' + b\,y' + cy = 0,$$

where a, b, and c are constants and $a \neq 0$. This is a *second order, linear, homogeneous* differential equation. The "second order" refers to the presence of a second derivative, the terms "linear, homogeneous" refer to the fact that if $y_1(t)$ and $y_2(t)$ are two solutions of the equation then so is $y(t) = Ay_1(t)+By_2(t)$ for any constants A and B. (Throughout this section we will assume that the independent variable in our functions is t rather than x, so the "$'$" refers to the derivative d/dt. This is because in most applications of such equations the independent variable is time.)

Equations of type 3.9.1 arise in many applications of mathematics. In Section 3.2 we encountered the equation of Simple Harmonic Motion,

$$y'' + \omega^2 y = 0,$$

(which is of type 3.9.1 with $a = 1$, $b = 0$ and $c = \omega^2$) governing the vibration of a mass suspended by an elastic spring. If the motion of the mass is impeded by a resistance proportional to its velocity (supplied, say, friction) then the equation of motion would have looked like 3.9.1 with positive constant b. The mass may still vibrate if the resistance is not too large ($b^2 < 4ac$) but the amplitude of vibration will decay exponentially as we shall see shortly.

Equation 3.9.1 also arises in the study of current flowing in an electric circuit having resistors, capacitors and inductors connected in series. In this case a measures the inductance, b the resistance and c the reciprocal of the capacitance of the circuit. In both the mechanical (vibrating mass) and electrical applications the constants a, b, and c are positive.

Let us try to find a solution of equation 3.9.1 having the form $y = e^{rt}$. Substituting this expression into the equation, we obtain

$$ar^2e^{rt} + bre^{rt} + ce^{rt} = 0.$$

Since e^{rt} does not vanish, $y = e^{rt}$ will be a solution of the differential equation 3.9.1 if and only if r satisfies the quadratic **auxiliary equation**

3.9.2
The Auxiliary Equation

$$ar^2 + br + c = 0,$$

which has roots given by the *quadratic formula:*

$$r = \frac{-b \pm \sqrt{b^2 - 4ac}}{2a} = -\frac{b}{2a} \pm \frac{\sqrt{D}}{2a},$$

where $D = b^2 - 4ac$ is the **discriminant** of the auxiliary equation 3.9.2.

There are three cases to consider, depending on whether D is positive, zero or negative.

CASE I. Suppose $D = b^2 - 4ac > 0$. Then the auxiliary equation has two different real roots, r_1 and r_2 given by

$$r_1 = \frac{-b - \sqrt{D}}{2a}, \qquad r_2 = \frac{-b + \sqrt{D}}{2a}.$$

(Often these roots can be found easily by factoring the left side of the auxiliary equation.) In this case both $y = y_1(t) = e^{r_1 t}$ and $y = y_2(t) = e^{r_2 t}$ are solutions of the differential equation, and so is

$$y = A\,e^{r_1 t} + B\,e^{r_2 t},$$

for any choice of the constants A and B. Since the differential equation is of second order and this solution involves two arbitrary constants we suspect it is the **general solution**, that is, that every solution of the differential equation can be written in this form. Exercise 17 at the end of this section outlines a way to prove this.

EXAMPLE 3.9.3 Find the general solution of $y'' + y' - 2y = 0$.

SOLUTION The auxiliary equation is $r^2 + r - 2 = 0$, or $(r + 2)(r - 1) = 0$. The auxiliary roots are $r_1 = -2$ and $r_2 = 1$. Hence the general solution of the differential equation is

$$y = A\,e^{-2t} + Be^{t}.$$

In order to facilitate discussion of the other two cases, $D = 0$, and $D < 0$, let us make a change of dependent variable in the differential equation 3.9.1; let $k = -\dfrac{b}{2a}$ and let $y(t) = e^{kt}u(t)$. Hence

$$y'(t) = e^{kt}[u'(t) + ku(t)]$$
$$y''(t) = e^{kt}[u''(t) + 2ku'(t) + k^2u(t)].$$

Substituting these expressions into the differential equation 3.9.1, we obtain

$$e^{kt}\left(au'' + (2ak + b)u' + (ak^2 + bk + c)u\right) = 0.$$

Since $2ak = -b$, the coefficient of u' in this equation is 0 and that of u is

$$ak^2 + bk + c = \frac{ab^2}{4a^2} - \frac{b^2}{2a} + c = \frac{4ac - b^2}{4a} = \frac{-D}{4a}.$$

Dividing by a and e^{kt}, we are left with the **reduced equation**

3.9.4 The Reduced Equation

$$u'' - \left(\frac{D}{4a^2}\right) u = 0.$$

CASE II. Suppose $D = b^2 - 4ac = 0$. Then the auxiliary equation has two equal roots $r_1 = r_2 = -b/(2a) = k$. Moreover, the reduced equation is $u'' = 0$, which has general solution $u = A + Bt$. Here A and B are any constants. Thus, the general solution to the differential equation 3.9.1 for the case $D = 0$ is

$$y = A\,e^{-(b/2a)t} + Bt\,e^{-(b/2a)t}.$$

EXAMPLE 3.9.5 Find the general solution of $y'' + 6y' + 9y = 0$.

SOLUTION The auxiliary equation is $r^2 + 6r + 9 = 0$, or $(r + 3)^2 = 0$, which has identical roots $r = -3$. According to the discussion above, the general solution of the differential equation is

$$y = A\,e^{-3t} + Bt\,e^{-3t}.$$

CASE III. Suppose $D = b^2 - 4ac < 0$. Then the auxiliary equation has complex roots, and the reduced equation for u is

$$u'' + \omega^2 u = 0, \qquad \text{where} \quad \omega = \frac{\sqrt{-D}}{2a}.$$

We recognize this as the equation of Simple Harmonic Motion, solved in Section 3.2. It has general solution

$$u = A\,\cos(\omega t) + B\,\sin(\omega t),$$

where A and B are arbitrary constants. Therefore the given differential equation has general solution

$$y = A\,e^{-(b/2a)t}\cos(\omega t) + B\,e^{-(b/2a)t}\sin(\omega t), \qquad \text{where} \quad \omega = \frac{\sqrt{4ac - b^2}}{2a}.$$

EXAMPLE 3.9.6 Find the general solution of $y'' + 4y' + 13y = 0$.

SOLUTION The auxiliary equation is $r^2 + 4r + 13 = 0$. Since $a = 1$, $b = 4$, and $c = 13$, the discriminant is $D = b^2 - 4ac = 16 - 52 = -36 < 0$ and Case III applies. We have $-(b/2a) = -2$ and $\omega = \sqrt{36}/2 = 3$ so the general solution of the given differential equation is

$$y = A\,e^{-2t}\cos(3t) + B\,e^{-2t}\sin(3t).$$

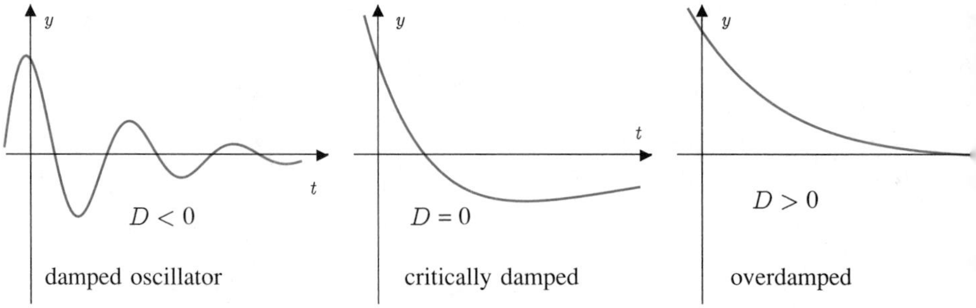

FIGURE 3.9.1

REMARK: If a and c are positive, but $b = 0$, then 3.9.1 is the differential equation of Simple Harmonic Motion and has oscillatory solutions of fixed amplitude as shown in Section 3.2. If $b > 0$ but $b^2 < 4ac$ (Case III) the solutions still oscillate, but the amplitude diminishes exponentially as $t \to \infty$ because of the factor $e^{-(b/2a)t}$. A system whose behavior is modelled by such an equation is said to exhibit **damped harmonic motion.** If $b > 0$ and $b^2 = 4ac$ (Case II) the system is **critically damped**, and if $b^2 > 4ac$ (Case I) it is **overdamped**. In these latter cases the behavior is no longer oscillatory. (See Fig. 3.9.1. Imagine a spring-suspended mass in a jar of molasses.)

We conclude with a typical initial-value problem for a differential equation of type 3.9.1.

EXAMPLE 3.9.7 Solve the initial-value problem $\begin{cases} y'' + 2y' + 2y = 0 \\ y(0) = 2 \\ y'(0) = -3. \end{cases}$

SOLUTION We have $a = 1$, $b = 2$, and $c = 2$, so $b^2 - 4ac = 4 - 8 = -4 < 0$ and Case III applies. We have $-(b/2a) = -1$ and $\omega = \sqrt{4}/2 = 1$. Thus the differential equation has general solution

$$y = A\,e^{-t}\cos t + B\,e^{-t}\sin t.$$

Also

$$y' = e^{-t}\big(-A\cos t - B\sin t - A\sin t + B\cos t\big) = (B-A)\,e^{-t}\cos t - (A+B)\,e^{-t}\sin t.$$

Applying the initial conditions $y(0) = 2$ and $y'(0) = -3$, we obtain $A = 2$ and $B - A = -3$. Hence $B = -1$ and the initial-value problem has solution

$$y = 2\,e^{-t}\cos t - e^{-t}\sin t.$$

EXERCISES

In 1–12 find the general solutions for the given equations.

1. $y'' + 7y' + 10y = 0$ **2.** $y'' - 2y' - 3y = 0$

3. $y'' + 2y' = 0$ **4.** $4y'' - 4y' - 3y = 0$

5. $y'' + 8y' + 16y = 0$ **6.** $y'' - 2y' + y = 0$

7. $y'' - 6y' + 10y = 0$ **8.** $9y'' + 6y' + y = 0$

9. $y'' + 2y' + 5y = 0$ **10.** $y'' - 4y' + 5y = 0$

11. $y'' + 2y' + 3y = 0$ **12.** $y'' + y' + y = 0$

In 13–15 solve the given initial-value problems.

13. $\begin{cases} 2y'' + 5y' - 3y = 0 \\ y(0) = 1 \\ y'(0) = 0. \end{cases}$ **14.** $\begin{cases} y'' + 10y' + 25y = 0 \\ y(1) = 0 \\ y'(1) = 2. \end{cases}$

15. $\begin{cases} y'' + 4y' + 5y = 0 \\ y(0) = 2 \\ y'(0) = 2. \end{cases}$

16. If $a > 0$, $b > 0$, and $c > 0$, prove that all solutions of the differential equation $ay'' + by' + cy = 0$ satisfy $\lim_{t\to\infty} y(t) = 0$.

17.*Prove that the solution given in the discussion of Case I, namely $y = A\,e^{r_1 t} + B\,e^{r_2 t}$, is the general solution for that case as follows: first let $y = e^{r_1 t}u$ and show that u satisfies the equation

$$u'' - (r_2 - r_1)u' = 0.$$

Then let $v = u'$, so that v must satisfy $v' = (r_2 - r_1)v$. The general solution of this equation is $v = C\,e^{(r_2 - r_1)t}$, as shown in the discussion of the equation $y' = ky$ at the beginning of Section 3.7. Hence find u and y.

Exercises 18–23 deal with the second-order, linear, homogeneous differential equation

$$at^2y'' + bty' + cy = 0 \qquad (t > 0) \qquad (*)$$

and its associated auxiliary equation (a quadratic equation in r)

$$ar(r - 1) + br + c = 0. \qquad (**)$$

The differential equation $(*)$ is called an **Euler equation**.

18. Show that if $y = y_1(t)$ and $y = y_2(t)$ are two solutions of $(*)$, then so is $y = Ay_1(t) + By_2(t)$ for any choice of constants A and B.

19. Show that $y = t^r$ is a solution of $(*)$ if r is a real solution of $(**)$.

20. If $(b - a)^2 > 4ac$, show that $(*)$ has a solution of the form $y = A\,t^{r_1} + B\,t^{r_2}$ where r_1 and r_2 are distinct real solutions of $(**)$. This is a general solution.

21. If $(b - a)^2 = 4ac$ so that $(**)$ has a double root $r = (a - b)/2a$, show that $(*)$ has a solution of the form $y = A\,t^r + B\,t^r \ln t$.

22. If $(b - a)^2 < 4ac$ and $k = (a - b)/2a$ show that $y = t^k u$ satisfies $(*)$ if and only if u satisfies the reduced equation

$$t^2u'' + tu' + \omega^2 u = 0,$$

where $\omega^2 = (4ac - (a - b)^2)/4a^2$.

23. Show that the reduced equation in the previous exercise has a solution of the form $y = A\cos(\omega \ln t) + B\sin(\omega \ln t)$, and hence that if $(b - a)^2 < 4ac$, then $(*)$ has solution

$$y = A\,t^k \cos(\omega \ln t) + B\,t^k \sin(\omega \ln t),$$

where $k = (a - b)/2a$ and $\omega^2 = (4ac - (a - b)^2)/4a^2$.

The solutions to $(*)$ given in the above exercises are general solutions, that is, every solution is of the specified form for some choice of the constants. Find general solutions to the differential equations in Exercises 24–29.

24. $t^2y'' - ty' - 3y = 0$ **25.** $t^2y'' - ty' + y = 0$

26. $t^2y'' - ty' + 5y = 0$ **27.** $t^2y'' + ty' - y = 0$

28. $t^2y'' + ty' + y = 0$ **29.** $t^2y'' + ty' = 0$

CHAPTER 4

The Mean-Value Theorem and Curve Sketching

4.1 THE MEAN-VALUE THEOREM

Suppose that A and B are two points on a smooth curve. It seems obvious that there ought to be at least one point C on the curve between A and B such that the tangent line to the curve at C is parallel to the chord line AB. See Fig. 4.1.1

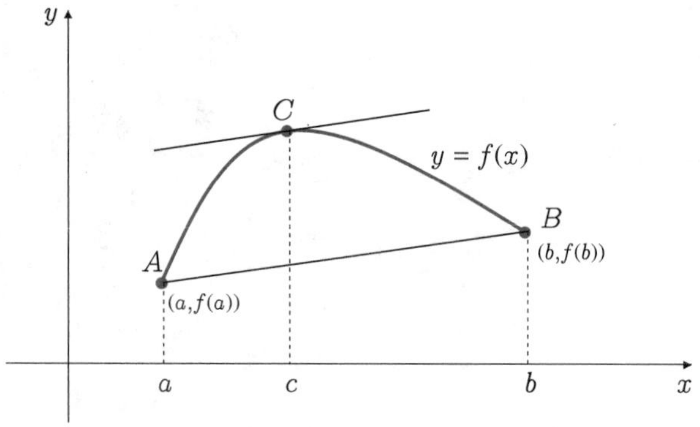

FIGURE 4.1.1

This intuitive principle is stated more precisely in the following theorem.

THEOREM 4.1.1 **(The Mean-Value Theorem)** Suppose that the function f is continuous on the closed, finite interval $[a, b]$ and that it is differentiable on the open interval (a, b). Then there exists a point c in the open interval (a, b) such that

$$\frac{f(b) - f(a)}{b - a} = f'(c).$$

This says that the slope of the chord line joining the points $(a, f(a))$ and $(b, f(b))$ is equal to the slope of the tangent line to the curve $y = f(x)$ at the point $(c, f(c))$, so the two lines are parallel. □

We will prove the Mean-Value Theorem at the end of this section. For now we make several observations.

The hypotheses of the Mean-Value Theorem are all necessary for the conclusion; if f fails to be continuous at even one point of $[a, b]$, or fails to be differentiable at even one point of (a, b), then there may be no point where the tangent line is parallel to the secant line AB. See Fig. 4.1.2.

The Mean-Value Theorem gives no indication of how many points C there may be on the curve between A and B where the tangent is parallel to AB. If the curve is itself the straight line AB, then every point on the line between A and B has the required property. In general there may be more than one point (see Fig. 4.1.3); the Mean-Value Theorem asserts only that there must be at least one.

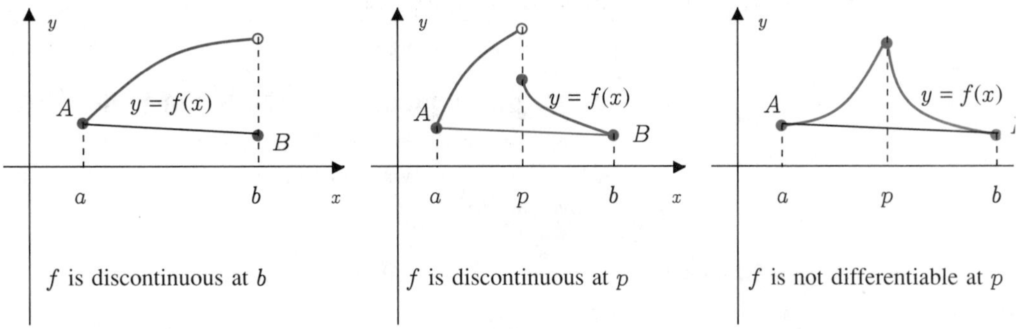

FIGURE 4.1.2

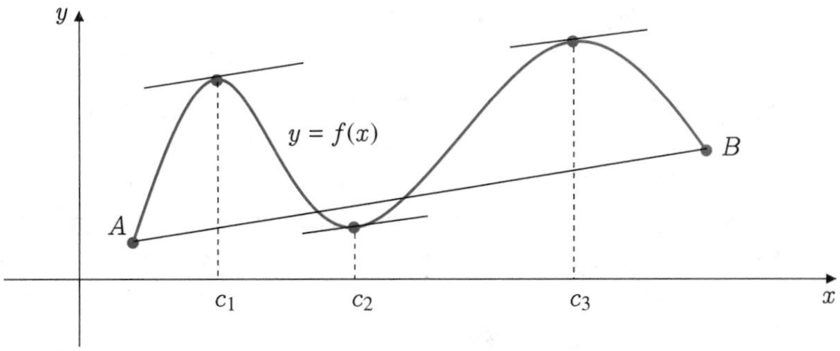

FIGURE 4.1.3

The Mean-Value Theorem gives us no information on how to find the point c, which it asserts must exist. For some simple functions it is possible to calculate c (see the following example), but doing so is usually of no practical value. As we shall see, the importance of the Mean-Value Theorem lies in its use as a theoretical tool. It belongs to a class of theorems called *existence theorems*, as does the Intermediate-Value Theorem (Theorem 1.7.10).

EXAMPLE 4.1.2 Verify the conclusion of the Mean-Value Theorem for $f(x) = \sqrt{x}$ on the interval $[a, b]$, where $0 \leq a < b$.

SOLUTION The theorem asserts that there must be a number c satisfying $a < c < b$ and

$$\frac{1}{2\sqrt{c}} = f'(c) = \frac{f(b) - f(a)}{b - a} = \frac{\sqrt{b} - \sqrt{a}}{b - a} = \frac{1}{\sqrt{b} + \sqrt{a}}.$$

Clearly, $c = \left(\frac{\sqrt{b} + \sqrt{a}}{2}\right)^2$ satisfies this equation. Since $a < b$ we have

$$a = \left(\frac{\sqrt{a} + \sqrt{a}}{2}\right)^2 < c < \left(\frac{\sqrt{b} + \sqrt{b}}{2}\right)^2 = b,$$

so c lies in the interval (a, b). (We remark again, however, that in most significant applications of the Mean-Value Theorem the point c cannot be readily calculated.)

Some Consequences of the Mean-Value Theorem

When we introduced antiderivatives in Section 2.8, we asserted that a function whose derivative was zero on an interval had to be constant there. We can now prove this formally.

THEOREM 4.1.3 If $f'(x) = 0$ everywhere on an interval I, then $f(x) = C$, a constant, on I.

PROOF Pick any point x_0 in I and let $C = f(x_0)$. If x is any other point of I, then by the Mean-Value Theorem there exists a point c in I between x_0 and x such that

$$\frac{f(x) - f(x_0)}{x - x_0} = f'(c).$$

But $f'(c) = 0$ because f' vanishes identically on I. Thus $f(x) = f(x_0) = C$ for all x in I as claimed. □

When considering the interpretation of derivatives in Section 2.6 we made the observation that functions were increasing (or decreasing) where their derivatives were positive (or negative). We can now make this observation more precise.

4.1.4 Increasing and Decreasing Functions

Suppose that the function f is defined on an interval I, and that x_1 and x_2 are two points of I.
a) If $f(x_1) < f(x_2)$ whenever $x_1 < x_2$ we say f is **increasing** on I.
b) If $f(x_1) > f(x_2)$ whenever $x_1 < x_2$ we say f is **decreasing** on I.
c) If $f(x_1) \leq f(x_2)$ whenever $x_1 < x_2$ we say f is **nondecreasing** on I.
d) If $f(x_1) \geq f(x_2)$ whenever $x_1 < x_2$ we say f is **nonincreasing** on I.

examples explain

Figure 4.1.4 demonstrates such functions. Note the distinction between increasing and nondecreasing. If a function is increasing (or decreasing) on an interval it is one-to-one there. A nondecreasing function (or a nonincreasing function) may be constant on all or part of an interval, and may therefore not be one-to-one.

THEOREM 4.1.5 Suppose f is differentiable on an interval I.
a) If $f'(x) > 0$ for all x in I then f is increasing on I.
b) If $f'(x) < 0$ for all x in I then f is decreasing on I.
c) If $f'(x) \geq 0$ for all x in I then f is nondecreasing on I.
d) If $f'(x) \leq 0$ for all x in I then f is nonincreasing on I.

PROOF Let x_1 and x_2 be points of I with $x_1 < x_2$. By the Mean-Value Theorem there exists a point c between x_1 and x_2 (and therefore in I) such that

$$\frac{f(x_2) - f(x_1)}{x_2 - x_1} = f'(c),$$

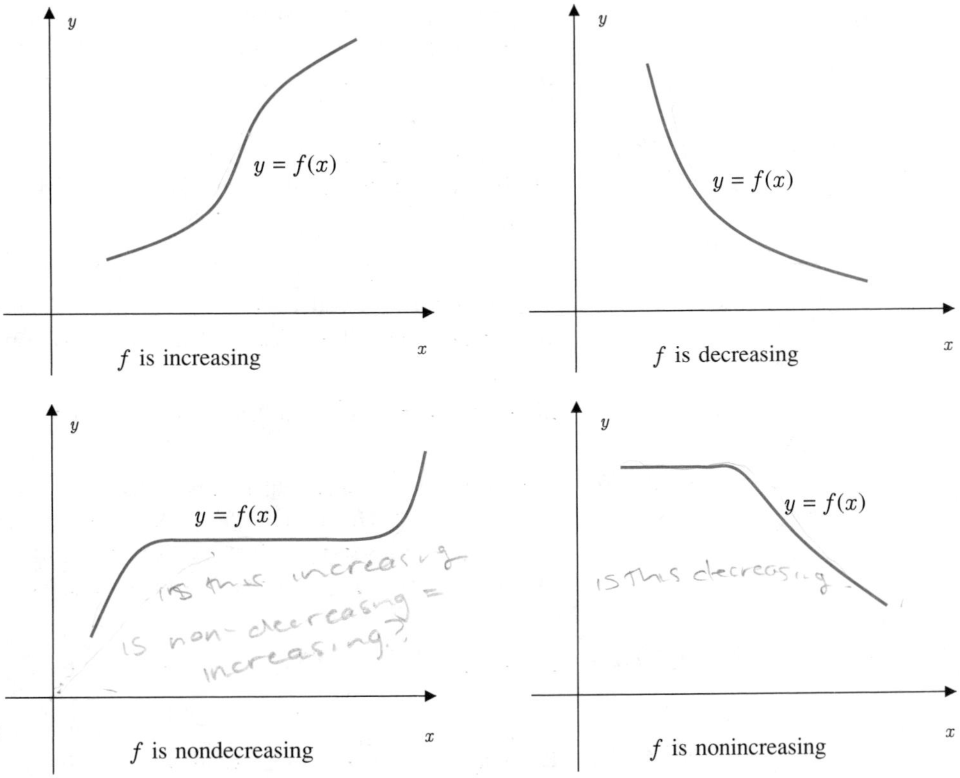

FIGURE 4.1.4

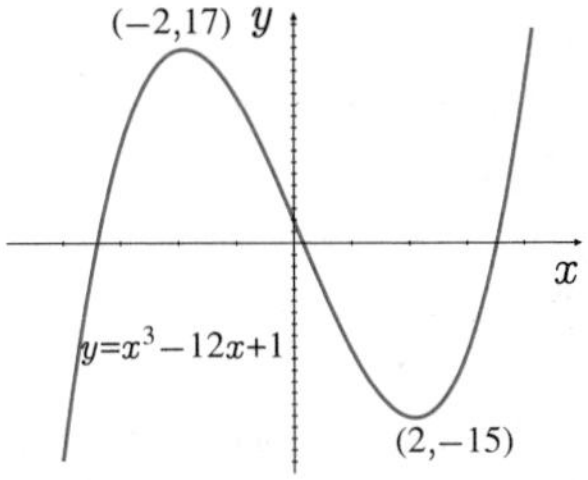

FIGURE 4.1.5

and hence $f(x_2) - f(x_1) = (x_2 - x_1) f'(c)$. Since $x_2 - x_1 > 0$, the difference $f(x_2) - f(x_1)$ has the same sign as $f'(c)$, and may be zero if $f'(c)$ is zero. Thus all four conclusions follow from the corresponding parts of Definition 4.1.4. □

EXAMPLE 4.1.6 On what intervals is the function $f(x) = x^3 - 12x + 1$ increasing? On what intervals is it decreasing?

SOLUTION We have $f'(x) = 3x^2 - 12 = 3(x-2)(x+2)$. Evidently $f'(x) > 0$ if $x < -2$ or $x > 2$ and $f'(x) < 0$ if $-2 < x < 2$. Therefore f is increasing on the intervals $(-\infty, -2)$ and $(2, \infty)$, and is decreasing on the interval $(-2, 2)$. See Fig. 4.1.5.

The following theorem is a refinement of Theorem 4.1.5 above. It is useful for obtaining strict inequalities for functions like those in Examples 4.1.8–10.

THEOREM 4.1.7 Suppose f is continuous on the closed interval $[a, b]$ and differentiable on the open interval (a, b).

a) If $f'(x) > 0$ on (a, b) and $f(a) \geq 0$ then $f(x) > 0$ on $(a, b]$.

b) If $f'(x) \geq 0$ on (a, b) and $f(a) > 0$ then $f(x) > 0$ on $[a, b]$.
c) If $f'(x) > 0$ on (a, b) and $f(b) \leq 0$ then $f(x) < 0$ on $[a, b)$.
d) If $f'(x) \geq 0$ on (a, b) and $f(b) < 0$ then $f(x) < 0$ on $[a, b]$.

PROOF We will only prove part (a); the other parts are similar. If $a < x \leq b$ then the Mean-Value Theorem assures us that there exists a number c satisfying $a < c < x$ such that

$$f(x) - f(a) = (x - a)f'(c).$$

The right side is positive since f' is assumed positive everywhere in (a, b), and since $x > a$. Therefore $f(x) > f(a)$. But $f(a) \geq 0$, so $f(x) > 0$. □

EXAMPLE 4.1.8 Show that $f(x) = x^3$ is increasing on any interval.

SOLUTION Let x_1 and x_2 be any real numbers such that $x_1 < x_2$. Since $f'(x) = 3x^2 > 0$ except at $x = 0$, Theorem 4.1.5(a) tells us that $f(x_1) < f(x_2)$ if either $x_1 < x_2 < 0$ or $0 < x_1 < x_2$. If $x_1 < 0 < x_2$ then parts (c) and (a) of Theorem 4.1.7 tell us that $f(x_1) < 0 < f(x_2)$. Thus f is increasing on every interval.

EXAMPLE 4.1.9 Show that $\sqrt{1+x} < 1 + \dfrac{x}{2}$ for $x > 0$ and for $-1 \leq x < 0$.

SOLUTION Let $f(x) = \sqrt{1+x} - 1 - \dfrac{x}{2}$. Evidently $f(0) = 0$ and $f(-1) = -1/2 < 0$. We have

$$f'(x) = \frac{1}{2\sqrt{1+x}} - \frac{1}{2} \quad \begin{cases} < 0 & \text{if } x > 0 \\ > 0 & \text{if } -1 < x < 0. \end{cases}$$

Again we apply Theorem 4.1.7(c) to get $f(x) < 0$ on the interval $(-1, 0)$, and Theorem 4.1.7(a) (to the function $-f$) to get $f(x) < 0$ on the interval $(0, \infty)$.

EXAMPLE 4.1.10 Show that for all $x > 0$,

a) $\sin x < x$ b) $\cos x > 1 - \dfrac{x^2}{2}$

c) $\sin x > x - \dfrac{x^3}{6}$ d) $\cos x < 1 - \dfrac{x^2}{2} + \dfrac{x^4}{24}$

SOLUTION Define four functions as follows:

$$f(x) = x - \sin x, \qquad g(x) = \cos x - 1 + \frac{x^2}{2},$$

$$h(x) = \sin x - x + \frac{x^3}{6}, \qquad k(x) = 1 - \frac{x^2}{2} + \frac{x^4}{24} - \cos x.$$

Evidently $f(0) = g(0) = h(0) = k(0) = 0$, and $g'(x) = f(x)$, $h'(x) = g(x)$ and $k'(x) = h(x)$. Parts (b), (c), and (d) will therefore follow by Theorem 4.1.7(a) if we can prove that $f(x) > 0$ for $x > 0$, that is, part (a).

Now $f'(x) = 1 - \cos x > 0$ if $0 < x < 2\pi$. Since $f(0) = 0$, Theorem 4.1.7(a) implies that $f(x) > 0$ for $0 < x \leq 2\pi$. But $f'(x) \geq 0$ for all $x \geq 2\pi$ and $f(2\pi) > 0$, so Theorem 4.1.7(b) implies that $f(x) > 0$ for all $x > 2\pi$. Therefore $f(x) > 0$ for all $x > 0$.

Proof of the Mean-Value Theorem

The mean-value theorem is one of those deeper results that is based on the completeness of the real number system via the fact that a continuous function on a closed, finite interval takes on a maximum and minimum value (Theorem 1.7.7). Before giving the proof we establish the following preliminary result.

THEOREM 4.1.11 If f is defined on an open interval (a, b) and achieves a maximum (or minimum) value at the point c in (a, b), and if $f'(c)$ exists, then $f'(c) = 0$, that is, c is a **critical point** of f.

PROOF Suppose f has a maximum value at c, where $a < c < b$. Then $f(x) \leq f(c)$ for all x in (a, b). Therefore $f(c + h) - f(c) \leq 0$ for h near 0. If $h > 0$, then $[f(c + h) - f(x)]/h \leq 0$ and so by Theorem 1.5.7(vi),

$$f'(c) = \lim_{h \to 0+} \frac{f(c + h) - f(c)}{h} \leq 0.$$

Similarly, if $h < 0$, then $[f(c + h) - f(c)]/h \geq 0$ so

$$f'(c) = \lim_{h \to 0-} \frac{f(c + h) - f(c)}{h} \geq 0.$$

Thus $f'(c) = 0$, as claimed. A similar proof works for a minimum. □

Proof of the Mean-Value Theorem Suppose f satisfies the conditions of the mean-value theorem. Let

$$g(x) = f(x) - f(a) - \frac{f(b) - f(a)}{b - a}(x - a).$$

(For $a \leq x \leq b$, $g(x)$ is the vertical displacement between the curve $y = f(x)$ and the chord line joining $(a, f(a))$ and $(b, f(b))$.) Clearly g is also continuous on $[a, b]$ and differentiable on (a, b). In addition, $g(a) = g(b) = 0$. We will show that there is some point c in (a, b) where $g'(c) = 0$. Since

$$g'(x) = f'(x) - \frac{f(b) - f(a)}{b - a},$$

it will follow that

$$f'(c) = \frac{f(b) - f(a)}{b - a}.$$

If $g(x) = 0$ for every x in $[a, b]$ then $g'(x)$ vanishes identically, so any c in (a, b) will do. Otherwise there is a point x_0 in (a, b) such that $g(x_0) \neq 0$. Either $g(x_0) > 0$ or $g(x_0) < 0$. To be specific, assume that $g(x_0) > 0$. By Theorem 1.7.7(ii), there is some point c in $[a, b]$ such that $g(x) \leq g(c)$ for every x in $[a, b]$ that is, g has a maximum value at c. Since $g(c) \geq g(x_0) > 0$, c cannot be a or b so c belongs to (a, b). Hence $g'(c) = 0$ by Theorem 4.1.11 above. □

REMARK: The special case of the Mean-Value Theorem for a function g continuous on $[a, b]$, differentiable on (a, b), and satisfying $g(a) = g(b) = 0$ is called **Rolle's theorem**.

EXERCISES

In Exercises 1–3 illustrate the mean-value theorem by finding any points in the open interval (a, b) where the tangent line to $y = f(x)$ is parallel to the chord line joining $(a, f(a))$ and $(b, f(b))$.

1. $f(x) = x^2$ on $[a, b]$ **2.** $f(x) = \dfrac{1}{x}$ on $[1, 2]$

3. $f(x) = x^3 - 3x + 1$ on $[-2, 2]$

Find the intervals of increase and decrease of the functions in Exercises 4–9.

4. $f(x) = x^2 + 2x + 2$ **5.** $f(x) = x^3 - 4x + 1$

6. $f(x) = x^3 + 4x + 1$ **7.** $f(x) = (x^2 - 4)^2$

8. $f(x) = \dfrac{1}{x^2 + 1}$ **9.** $f(x) = x^3(5 - x)^2$

10. If $0 < r < 1$, and either $-1 \leq x < 0$ or $x > 0$, show that $(1 + x)^r < 1 + rx$.

11. If $r > 1$, and either $-1 \leq x < 0$ or $x > 0$, show that $(1 + x)^r > 1 + rx$.

12. If $r < 0$, and either $-1 \leq x < 0$ or $x > 0$, show that $(1 + x)^r > 1 + rx$.

13. Show that $e^x > 1 + x$ if $x \neq 0$.

14. Show that $\ln(2 + x) < \ln 2 + \dfrac{x}{2}$ if $x > -2$ and $x \neq 0$.

15. Show that $e^x > 1 + x + \dfrac{x^2}{2}$ if $x > 0$.

16. Show that $e^x < 1 + x + \dfrac{x^2}{2}$ if $x < 0$.

17.*Continue Example 4.1.10 by proving that, for $x > 0$,

$$\sin x < x - \frac{x^3}{3!} + \frac{x^5}{5!}, \qquad \cos x > 1 - \frac{x^2}{2!} + \frac{x^4}{4!} - \frac{x^6}{6!}.$$

Try to generalize these results. What would the next pair of inequalities in the series be?

18. If $0 < x < \pi/2$, show that (a) $\tan x > x$, and (b) $\tan x > x + \dfrac{x^3}{3}$.

19.*If $f(x)$ is differentiable on an interval I and vanishes at $n \geq 2$ distinct points of I, prove that $f'(x)$ must vanish at at least $n - 1$ points in I.

20.*If $f''(x)$ exists on an interval I and if f vanishes at at least three distinct points of I, prove that f'' must vanish at some point in I.

21.*Generalize Exercise 20 to a function for which $f^{(n)}$ exists on I and for which f vanishes at at least $n + 1$ distinct points in I.

22.*Suppose f is twice differentiable on an interval I (that is, f'' exists on I). Suppose that the points 0 and 2 belong to I and that $f(0) = f(1) = 0$ and $f(2) = 1$. Prove that:

a) $f'(a) = \dfrac{1}{2}$ for some point a in I.

b) $f''(b) > \dfrac{1}{2}$ for some point b in I.

c) $f'(c) = \dfrac{1}{7}$ for some point c in I.

4.2 CRITICAL POINTS AND EXTREME VALUES

The first derivative of a function is a source of much useful information about the behavior of the function. As we have already seen, the sign of f' tells us whether f is increasing or decreasing. This section examines carefully what we can learn about a function and the shape of its graph by analyzing its first derivative.

We will consider functions defined on intervals, open or closed, or on the union of finitely many such intervals. The domains of such functions do not contain isolated points.

Maximum and Minimum Values

Recall (from Section 1.7) that a function has a maximum value at x_0 if $f(x) \leq f(x_0)$ for all x in the domain of f. The maximum value is $f(x_0)$. To be more precise, we should call such a maximum value an *absolute* or *global* maximum because it is largest value that f attains anywhere on its entire domain.

4.2.1
Absolute Extreme Values

> Function f has an **absolute maximum value** $f(x_0)$ at the point x_0 in its domain if $f(x) \leq f(x_0)$ holds for every x in the domain of f.
> Similarly, f has an **absolute minimum value** $f(x_1)$ at the point x_1 in its domain if $f(x) \geq f(x_1)$ holds for every x in the domain of f.

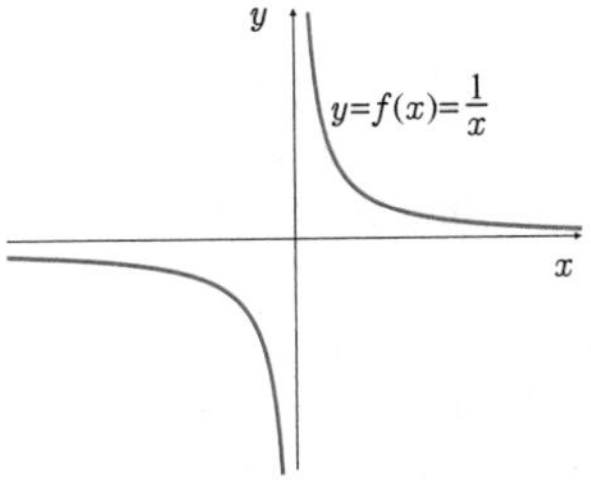

FIGURE 4.2.1

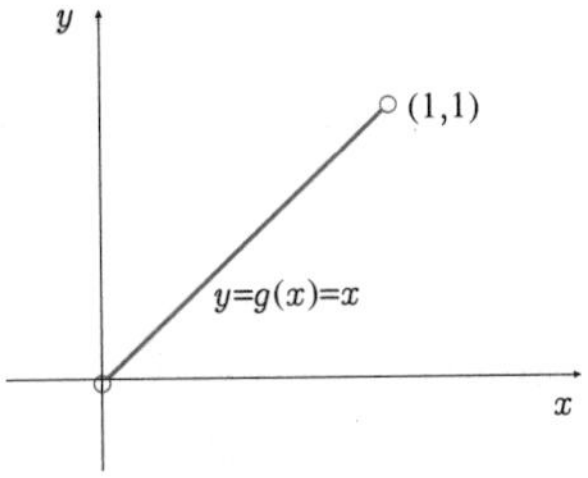

FIGURE 4.2.2

A function can have at most one absolute maximum or minimum value, though this value can be assumed at many points. For example, $f(x) = \sin x$ has absolute maximum value 1 occurring at every point of the form $x = \dfrac{\pi}{2} + 2n\pi$ where n is any integer. Of course a function need not have any absolute extreme values. The function $f(x) = 1/x$ becomes arbitrarily large as x approaches 0 from the right, so has no finite absolute maximum. (See Fig. 4.2.1. Remember, ∞ is not a number, and so is not a value of f.) Even a bounded function may not have an absolute maximum or minimum value. The function $g(x) = x$ with domain specified to be the *open* interval $(0, 1)$ has neither; the range of g is also the interval $(0, 1)$ and there is no largest or smallest number in this interval. (See Fig. 4.2.2.) Of course, if the domain of g were extended to be the *closed* interval $[0, 1]$ then g would have both a maximum value (1) and a minimum value (0).

Maximum and minimum values of a function are collectively referred to as *extreme values*. The following theorem is a restatement (and slight generalization) of Theorem 1.7.7. It will prove very useful in some circumstances when we want to find extreme values.

THEOREM 4.2.2 If the domain of the function f is a *closed, finite interval*, or a union of finitely many such intervals, and if f is *continuous* on that domain then f must have an absolute maximum value and an absolute minimum value. □

Consider the graph $y = f(x)$ shown in Fig. 4.2.3. Evidently the absolute maximum value of f is $f(x_2)$ and the absolute minimum value is $f(x_3)$. In addition to these extreme values, f has several other "local" maximum and minimum values corresponding to points on the graph which are higher or lower than neighbouring points. We say that f has *local maximum values* at a, x_2, x_4 and x_6, and local minimum values at x_1, x_3, x_5 and b. The absolute maximum is the highest of the local maxima; the absolute minimum is the lowest of the local minima.

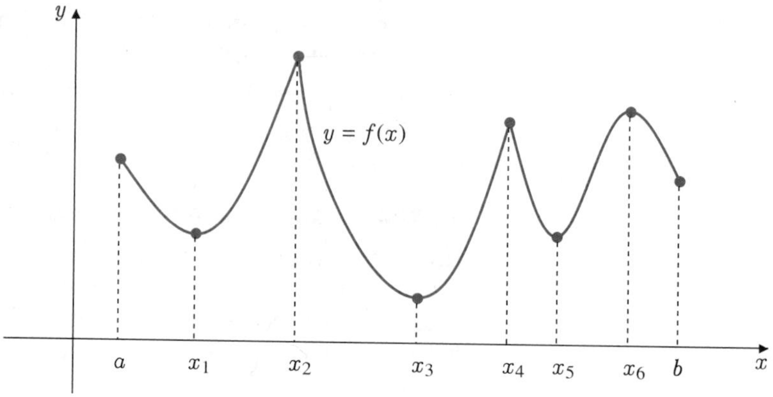

FIGURE 4.2.3

4.2.3
Local
Extreme Values

Function f has a **local maximum value** (loc max) $f(x_0)$ at the point x_0 in its domain provided there exists a number $h > 0$ such that $f(x) \leq f(x_0)$ whenever x is in the domain of f and $|x - x_0| < h$.
Similarly, f has a **local minimum value** (loc min) $f(x_1)$ at the point x_1 in its domain provided there exists a number $h > 0$ such that $f(x) \geq f(x_1)$ whenever x is in the domain of f and $|x - x_1| < h$.

Thus f has a local maximum (or minimum) value at x if it has an absolute maximum (or minimum) value at x when its domain is restricted to points sufficiently near x. Geometrically, the graph of f is at least as high (or low) at x as it is at nearby points.

Endpoints, Critical Points and Singular Points

Reexamine Fig. 4.2.3. Each point where the function f has a local extreme value belongs to one of three classes:

a) **endpoints** of the domain of f, (points in the domain of f which do not belong to any open interval contained in the domain of f),

b) **critical points** of f, (points x where $f'(x) = 0$), and

c) **singular points** of f, (points x where $f'(x)$ is not defined).

In Fig. 4.2.3, a and b are endpoints, x_1, x_3, x_5 and x_6 are critical points, and x_2 and x_4 are singular points. Strictly speaking, according to the definition of the derivative given in Section 2.2, f is not differentiable at the endpoints a and b, so these might be called singular points as well. It is possible, however, to extend the concept of derivative to allow one-sided derivatives at endpoints of domains (see Exercise 40 at the end of Section 2.2), so we will continue to make a distinction between endpoints and other singular points.

The following theorem shows that if we want to find the local extreme values of a function we need only look at points belonging to one of the three classes listed above.

THEOREM 4.2.4 If the function f is defined on an interval I and has a local maximum (or local minimum) value at point $x = x_0$ in I, then x_0 must be either an endpoint of I or a critical point of f or a singular point of f.

PROOF Suppose that f has a local maximum or local minimum value at x_0, and that x_0 is neither an endpoint of the domain of f nor a singular point of f. Then $f'(x_0)$ exists. By Theorem 4.1.11, x_0 must be a critical point of f. □

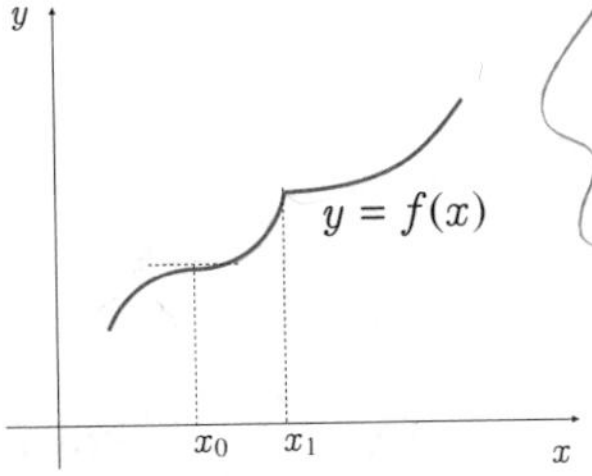

FIGURE 4.2.4

Although a function cannot have extreme values anywhere other than at endpoints, critical points and singular points, it need not have extreme values at such points. Fig. 4.2.4 shows the graph of a function with a critical point x_0 and a singular point x_1 at neither of which it has an extreme value. It is more difficult to draw the graph of a function whose domain has an endpoint at which the function fails to have an extreme value. See Exercise 47 at the end of this section for an example of such a function.

The First Derivative Test

Most functions you will encounter in elementary calculus have nonzero derivatives everywhere on their domains except possibly at a finite number of endpoints, critical points, and singular points. On intervals between these points the derivative exists and is not zero, so the function is either increasing or decreasing there. This information can be used to determine where maximum and minimum values occur.

THEOREM 4.2.5 (**The First Derivative Test**) Suppose that the function f is continuous at x_0 and that x_0 is either a critical point or a singular point of f.

i) If there exists an open interval (a, b) containing x_0 such that $f'(x) > 0$ on (a, x_0) and $f'(x) < 0$ on (x_0, b), then f has a local maximum value at x_0.

ii) If there exists an open interval (a, b) containing x_0 such that $f'(x) < 0$ on (a, x_0) and $f'(x) > 0$ (x_0, b) then f has a local minimum value at x_0.

Graphically, if the curve $y = f(x)$ rises to the left of x_0 and falls to the right, then f has a local maximum at x_0; if it falls to the left and rises to the right, then f has a local minimum at x_0.

PROOF We prove only Part (i); Part (ii) is similar. Let $a < x < x_0$. Since f is differentiable on (a, x_0) and continuous at x_0, it is continuous on $[x, x_0]$. By the Mean-Value Theorem, there exists c in (x, x_0) such that

$$\frac{f(x) - f(x_0)}{x - x_0} = f'(c).$$

By assumption, $f'(c) > 0$. Since $x - x_0 < 0$, we have $f(x) - f(x_0) < 0$, that is, $f(x) < f(x_0)$. A similar argument shows that $f(x) < f(x_0)$ for x in (x_0, b). Therefore, $f(x) \le f(x_0)$ on an interval extending some distance to either side of x_0, so f must have a local maximum value at x_0. □

REMARK: If f' is positive (or negative) on *both* sides of a critical point then f has neither a maximum nor a minimum value at that point. In this case the critical point will usually (but not always) be an *inflection point*, a concept defined in the next section.

EXAMPLE 4.2.6 Locate and classify the extreme values of $f(x) = x^4 - 2x^2 - 3$ on the interval $[-2, 2]$. Sketch the graph of f.

SOLUTION We begin by calculating and factoring the derivative $f'(x)$:

$$f'(x) = 4x^3 - 4x = 4x(x^2 - 1) = 4x(x - 1)(x + 1).$$

The critical points are 0, -1, and 1. The corresponding values are $f(0) = -3$, $f(-1) = f(1) = -4$. There are no singular points. The values of f at the endpoints -2 and 2 are $f(-2) = f(2) = 5$. The factored form of $f'(x)$ is also convenient for determining the sign of $f'(x)$ on intervals between these endpoints and critical points. On $(1, 2)$ all three factors are positive; on $(0, 1)$ the factor $(x - 1)$ is negative while the other factors are still positive, so $f'(x) < 0$ there. Similarly, $f'(x) > 0$ on $(-1, 0)$ because exactly two of its factors are negative, and $f'(x) < 0$ on $(-2, -1)$ because three of its factors are negative on that interval. This information about the sign of $f'(x)$ on intervals between points where $f'(x)$ is 0 can also be obtained (provided $f'(x)$ is continuous, as it is here) by using the Intermediate-Value Theorem (Theorem 1.7.10). We summarize the positive/negative properties of $f'(x)$ and the implied increasing/decreasing behavior of $f(x)$ in chart form:

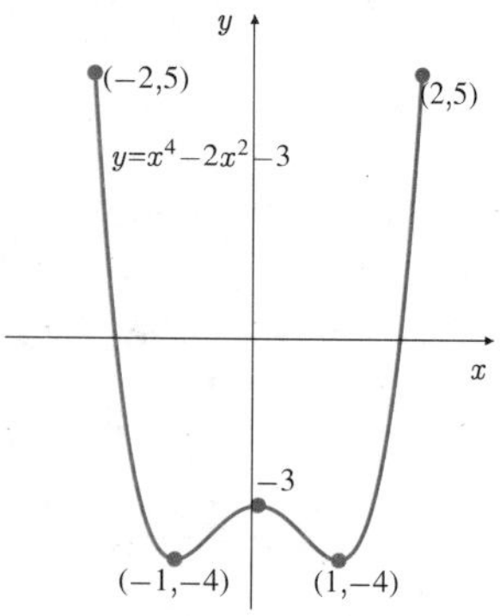

FIGURE 4.2.5

	EP		CP		CP		CP		EP	
f'	-2	$-$	-1	$+$	0	$-$	1	$+$	2	$\to x$
f	loc max	↘	loc min	↗	loc max	↘	loc min	↗	loc max	

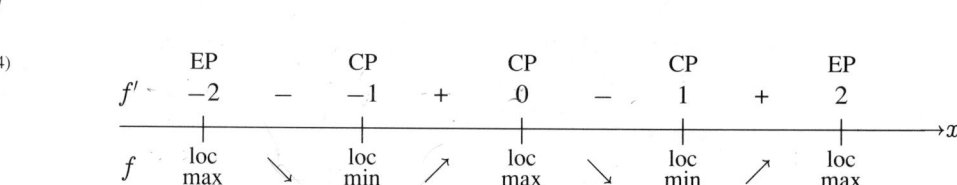

Note how the sloping arrows indicate visually the appropriate classification of the endpoints (EP) and critical points (CP) as determined by the First Derivative Test. We will make extensive use of such charts in future sections. The graph of f is shown in Fig. 4.2.5. Since the domain is a closed, finite interval, f must have absolute maximum and minimum values. Evidently these are 5 (at ± 2) and -4 (at ± 1).

EXAMPLE 4.2.7 Let $f(x) = x\, e^{-x^2}$. Find and classify the critical points of f, evaluate $\lim_{x\to\pm\infty} f(x)$, and use these results to help you sketch the graph of f.

SOLUTION $f'(x) = e^{-x^2}(1 - 2x^2) = 0$ only if $1 - 2x^2 = 0$ since the exponential is always positive. Thus the critical points are $\pm\frac{1}{\sqrt{2}}$. We have $f\left(\pm\frac{1}{\sqrt{2}}\right) = \pm\frac{1}{\sqrt{2e}}$. f' is positive (or negative) when $1 - 2x^2$ is positive (or negative). We summarize the intervals where f is increasing and decreasing in chart form:

		CP		CP	
$f'(x)$	$-$	$-\frac{1}{\sqrt{2}}$	$+$	$\frac{1}{\sqrt{2}}$	$-$
$f(x)$	↘	loc min	↗	loc max	↘

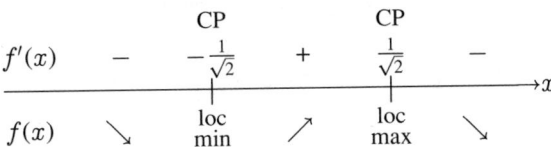

Note that $f(0) = 0$ and that f is an odd function ($f(-x) = -f(x)$) so the graph is symmetric about the origin. By Theorem 3.5.18,

$$\lim_{x\to\pm\infty} x\,e^{-x^2} = \lim_{x\to\pm\infty} \frac{1}{x}\frac{x^2}{e^{x^2}} = 0.$$

The graph is shown in Fig. 4.2.6. The x-axis is an asymptote as $x \to \pm\infty$.

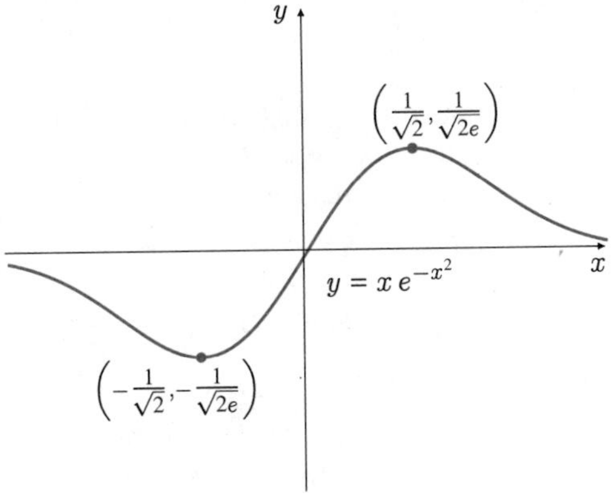

FIGURE 4.2.6

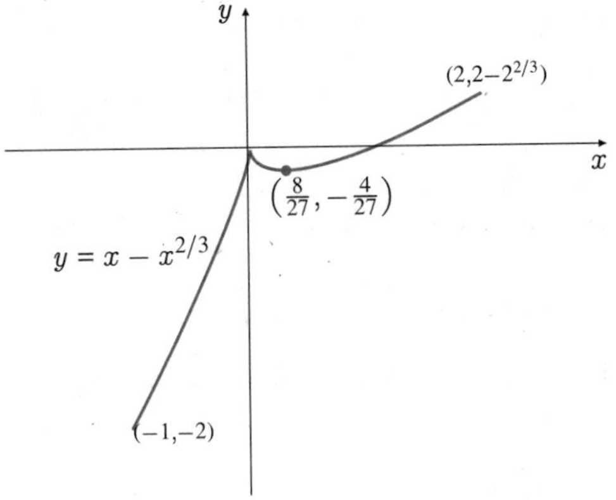

FIGURE 4.2.7

EXAMPLE 4.2.8 Find and classify the local extreme values of $f(x) = x - x^{2/3}$ on $[-1, 2]$. Sketch the graph of f.

SOLUTION $f'(x) = 1 - \frac{2}{3}x^{-1/3} = \left(x^{1/3} - \frac{2}{3}\right)/x^{1/3}$. There is a singular point, $x = 0$, and a critical point, $x = 8/27$. The endpoints are $x = -1$ and $x = 2$. The values of f at these points are $f(-1) = -2, f(0) = 0, f(8/27) = -4/27$, and $f(2) = 2 - 2^{2/3} \approx 0.4126$ (see Fig. 4.2.7).

	EP		SP		CP		EP
f'	-1	$+$	0	$-$	$\frac{8}{27}$	$+$	2
f	loc min	↗	loc max	↘	loc min	↗	loc max

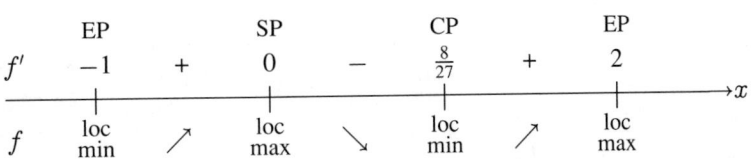

There are two local minima and two local maxima. The absolute maximum of f is $2 - 2^{2/3}$ at $x = 2$; the absolute minimum is -2 at $x = -1$.

EXERCISES

In Exercises 1–15 determine whether the given function has any local or absolute extreme values, and find those values if possible.

1. $f(x) = x + 2$ on $[-1, 1]$ **2.** $f(x) = x + 2$ on $(-\infty, 0]$

3. $f(x) = x + 2$ on $[-1, 1)$ **4.** $f(x) = x^2 - 1$

5. $f(x) = x^2 - 1$ on $[-2, 3]$ **6.** $f(x) = x^2 - 1$ on $(2, 3)$

7. $f(x) = x^3 + x - 4$ on $[a, b]$

8. $f(x) = x^3 + x - 4$ on (a, b)

9. $f(x) = x^5 + x^3 + 2x$ on $(a, b]$

10. $f(x) = \dfrac{1}{x-1}$ **11.** $f(x) = \dfrac{1}{x-1}$ on $(0, 1)$

12. $f(x) = \dfrac{1}{x-1}$ on $[2, 3]$ **13.** $f(x) = \dfrac{1}{x^2+1}$

14. $f(x) = (x + 2)^{2/3}$ **15.** $f(x) = (x - 2)^{1/3}$

In Exercises 16–43 locate and classify all local extreme values of the given function. Determine whether any of these extreme values are absolute. Sketch the graph of the function.

16. $f(x) = x^2 + 2x$ **17.** $f(x) = x^3 - 3x - 2$

18. $f(x) = (x^2 - 4)^2$ **19.** $f(x) = x(x - 1)^2$

20. $f(x) = x^4 + 4x$ **21.** $f(x) = x^3(x - 1)^2$

22. $f(x) = x^2(x - 1)^2$ **23.** $f(x) = x(x^2 - 1)^2$

24. $f(x) = \dfrac{x}{x^2+1}$ **25.** $f(x) = \dfrac{x^2}{x^2+1}$

26. $f(x) = \dfrac{x}{\sqrt{x^4+1}}$ **27.** $f(x) = x\sqrt{2 - x^2}$

28. $f(x) = x + \sin x$ **29.** $f(x) = x - 2\sin x$

30. $f(x) = x - 2\tan^{-1} x$ **31.** $f(x) = 2x - \sin^{-1} x$

32. $f(x) = e^{-x^2/2}$ **33.** $f(x) = x\,2^{-x}$

34. $f(x) = x^2 e^{-x^2}$ **35.** $f(x) = \dfrac{\ln x}{x}$

36. $f(x) = |x + 1|$ **37.** $f(x) = |x^2 - 1|$

38. $f(x) = \sin |x|$ **39.** $f(x) = |\sin x|$

40.* $f(x) = (x - 1)^{2/3} - (x + 1)^{2/3}$

41.* $f(x) = (x - 1)^{1/3} + (x + 1)^{1/3}$

42.* $f(x) = x - x^{1/3}$

43.* $f(x) = (x - 1)^{1/3} + (x + 1)^{2/3}$

44. If a function has an absolute maximum value must it have any local maximum values? If a function has a local maximum value must it have an absolute maximum value? Give reasons for your answers.

45. If the function f has an absolute maximum value and $g(x) = |f(x)|$, must g have an absolute maximum value? Justify your answer.

46.*Prove Theorem 4.2.2. (*Hint:* Apply part (ii) of Theorem 1.7.7 to each separate closed interval of the finite union of such intervals that make up the domain of f. Use the maximum of the finitely many maximum values obtained this way.)

47.*Let

$$f(x) = \begin{cases} x \sin \dfrac{1}{x} & \text{if } x > 0, \\ 0 & \text{if } x = 0. \end{cases}$$

Show that f is continuous on $[0, \infty)$ and differentiable on $(0, \infty)$ but that it has neither a local maximum nor a local minimum value at the endpoint $x = 0$.

48.*Suppose that f is continuous on (a, b) where $-\infty \le a < b \le \infty$ and that

$$\lim_{x\to a+} f(x) = \lim_{x\to b-} f(x) = \infty.$$

Show that f has an absolute minimum value in (a, b).

49.*If the function f of Exercise 48 satisfies instead

$$\lim_{x\to a+} f(x) = \lim_{x\to b-} f(x) = -\infty.$$

show that f must have an absolute maximum value in (a, b).

50.*Suppose f is continuous on the interval (a, b) where $-\infty \le a < b \le \infty$, and suppose that

$$\lim_{x\to a+} f(x) = \lim_{x\to b-} f(x) = L$$

exists. Prove the following.

a) If $f(x) > L$ at any point of (a, b), then f has an absolute maximum value on (a, b).

b) If $f(x) < L$ at any point of (a, b), then f has an absolute minimum value on (a, b).

c) f must have at least one extreme value on (a, b).

4.3 CONCAVITY AND INFLECTIONS

Like the first derivative, the second derivative of a function also provides useful information about the behavior of the function and the shape of its graph; it determines whether the graph is *bending upward* (that is, has increasing slope) or *bending downward* (has decreasing slope) as we move along the graph toward the right.

4.3.1
Definition of Concavity

> We say that the differentiable function f is **concave up** on an open interval I if the derivative f' is an increasing function on I. Similarly, f is **concave down** on I if f' is a decreasing function on I.

The terms "concave up" and "concave down" are used to describe the graph of the function as well as the function itself.

Note that concavity is defined only for differentiable functions, and even for those, only on intervals on which their derivatives are not constant. According to the above definition, a function is neither concave up nor concave down on an interval where its graph is a straight line segment. We say the function has no concavity on such an interval. We also say a function has opposite concavity on two intervals if it is concave up on one interval and concave down on the other.

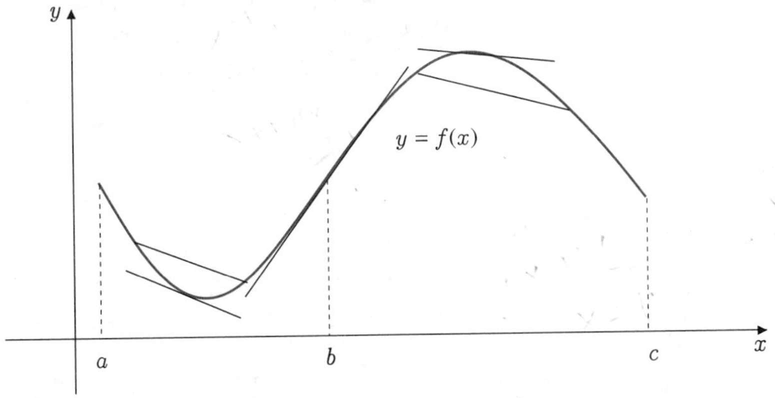

FIGURE 4.3.1

EXAMPLE 4.3.2 The function f whose graph is shown in Fig. 4.3.1 is concave up on the interval (a, b) and concave down on the interval (b, c).

Some geometric observations can be made about concavity:

i) If f is concave up on an interval, then its graph lies above any tangent lines drawn at points in that interval, and any chord lines between points in that interval lie above the graph.

ii) If f is concave down on an interval, then its graph lies below any tangent line drawn at points in that interval, and any chord lines between points in that interval lie below the graph.

iii) If the graph of f has a tangent line at a point, and if the concavity of f is opposite on opposite sides of that point, then the graph crosses its tangent line at that point. (This occurs at point b in Fig. 4.3.1. Such a point is called an *inflection point* of the function f.)

4.3.3
Inflection Points

A point x_0 is called an **inflection point** (or simply an **inflection**) of the function f (and of its graph, the curve $y = f(x)$), if
a) the graph $y = f(x)$ has a tangent line at x_0, and
b) the concavity of f is opposite on opposite sides of x_0.

Note that (a) implies that either f is differentiable at x_0 or its graph has a vertical tangent line there and that (b) implies that the graph crosses its tangent line at x_0.

EXAMPLE 4.3.4 In Fig. 4.3.2, the function $f(x) = x^3$ has an inflection point at $x = 0$; the function is concave down on $(-\infty, 0)$ and concave up on $(0, \infty)$. The function g in Fig. 4.3.3 is concave down to the left of $x = a$ and concave up to the right of $x = a$, but $x = a$ is not an inflection point of g because the graph of g has no tangent line there. ($x = a$ is a singular point of g.)

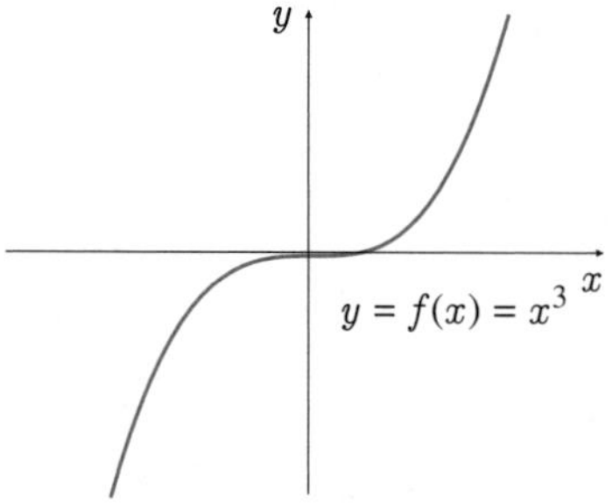

FIGURE 4.3.2

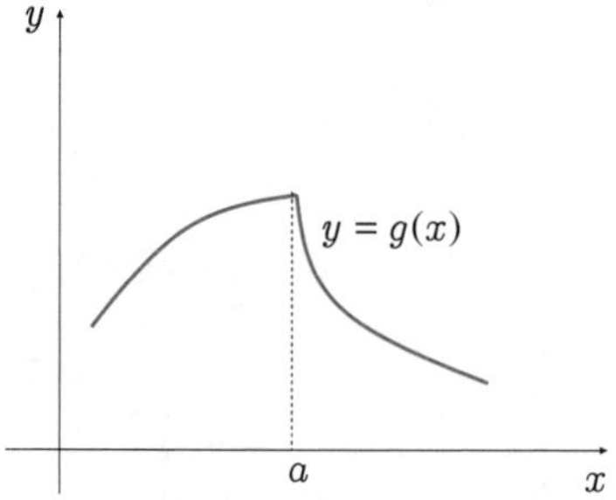

FIGURE 4.3.3

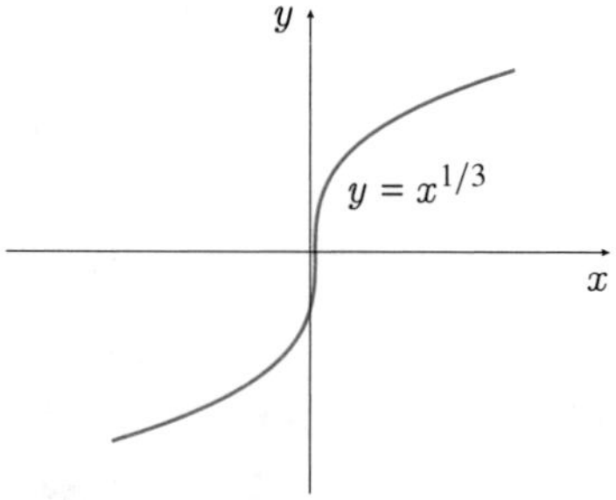

FIGURE 4.3.4

An inflection point of a function f may or may not coincide with a critical point or singular point. The point $x = 0$ is both a critical point and an inflection point for the function x^3 in Fig. 4.3.2. A function can have an inflection at a singular point only if its graph has a vertical tangent line there. For example, the graph of $y = x^{1/3}$, shown in Fig. 4.3.4, has a vertical tangent line at the origin, so $x = 0$ is a singular point of $f(x) = x^{1/3}$. Since the graph is concave up to the left and concave down to the right of the origin, $x = 0$ is also an inflection point of f.

THEOREM 4.3.5
a) If $f''(x) > 0$ on interval I, then f is concave up on I.
b) If $f''(x) < 0$ on interval I, then f is concave down on I.
c) If f has an inflection point at x_0 and $f''(x_0)$ exists, then $f''(x_0) = 0$.

PROOF Parts (a) and (b) follow from applying Theorem 4.1.5 to the derivative f' of f. If x_0 is an inflection point of f and $f''(x_0)$ exists, then f must be differentiable in an open interval containing x_0. Since f' is increasing on one side of x_0 and decreasing on the other side, it must have a local maximum or minimum value at x_0. By Theorem 4.1.11, $f''(x_0) = 0$. □

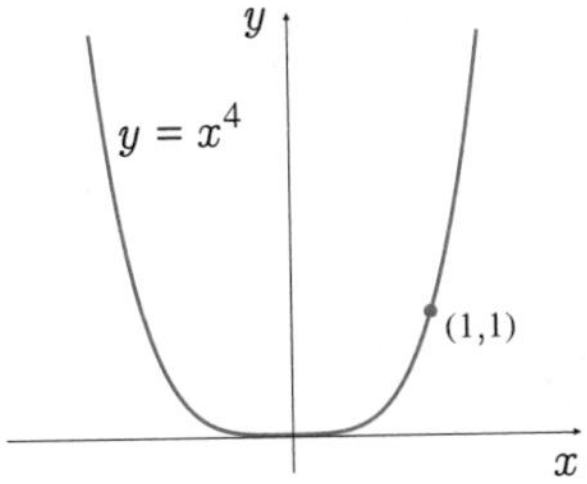

FIGURE 4.3.5

Theorem 4.3.5 tells us that to find inflection points of a twice differentiable function f we need only look at points where $f''(x) = 0$. Of course, not every such point has to be an inflection point. For example, $f(x) = x^4$, whose graph is shown in Fig. 4.3.5, does not have an inflection point at $x = 0$ even though $f''(0) = 12x^2|_{x=0} = 0$. In fact, x^4 is concave up on every interval.

EXAMPLE 4.3.6 Determine the intervals of concavity and the inflections of $f(x) = x^6 - 10x^4$.

SOLUTION We have

$$f'(x) = 6x^5 - 40x^3,$$
$$f''(x) = 30x^4 - 120x^2 = 30x^2(x-2)(x+2).$$

Having factored $f''(x)$ in this manner, we can see that it vanishes only at $x = -2$, $x = 0$ and $x = 2$. On the intervals $(-\infty, -2)$ and $(2, \infty)$, $f''(x) > 0$ so f is concave up. On $(-2, 0)$ and $(0, 2)$, $f''(x) < 0$ so f is concave down. $f''(x)$ changes sign as we pass through -2 and 2 so these are inflection points of f. No sign change occurs at $x = 0$ since $x^2 > 0$ for both positive and negative x. Thus 0 is not an inflection point of f. As was the case for the first derivative, information about the sign of $f''(x)$ and the consequent concavity of f can be conveniently conveyed in a chart:

									x
f''	+	-2	−	0	−	2	+		
f	⌣	infl	⌢		⌢	infl	⌣		

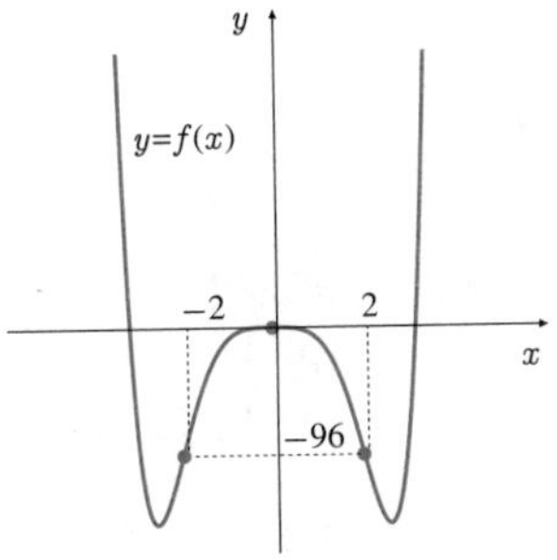

FIGURE 4.3.6

The graph of f is sketched in Fig 4.3.6.

EXAMPLE 4.3.7 Determine the intervals of increase and decrease, the local extreme values, and the concavity and inflections of $f(x) = x^4 - 2x^3 + 1$. Use the information to sketch the graph of f.

SOLUTION

$$f'(x) = 4x^3 - 6x^2 = 2x^2(2x-3) = 0 \quad \text{at } x = 0 \text{ and } x = 3/2,$$
$$f''(x) = 12x^2 - 12x = 12x(x-1) = 0 \quad \text{at } x = 0 \text{ and } x = 1.$$

The behavior of f is summarized in the following charts:

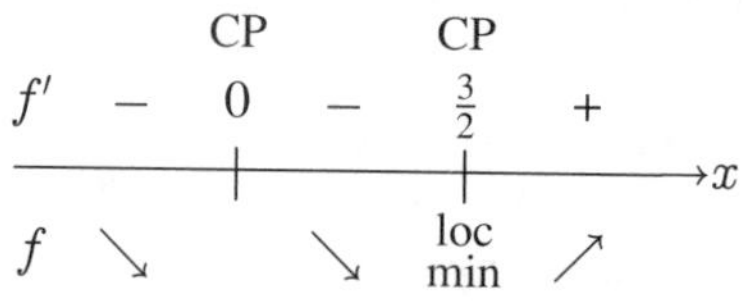

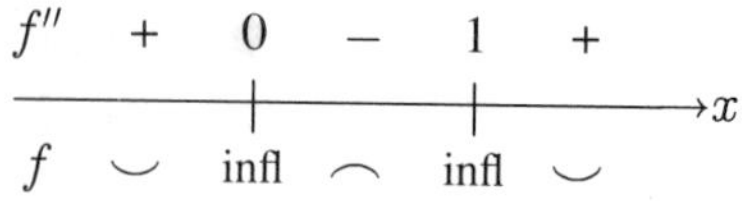

Note that $x = 0$ is both a critical point and an inflection point. We calculate the values of f at the critical and inflection points:

$$f(0) = 1, \qquad f(1) = 0, \qquad f\left(\tfrac{3}{2}\right) = -\tfrac{11}{16}.$$

The graph of f is sketched in Fig. 4.3.7.

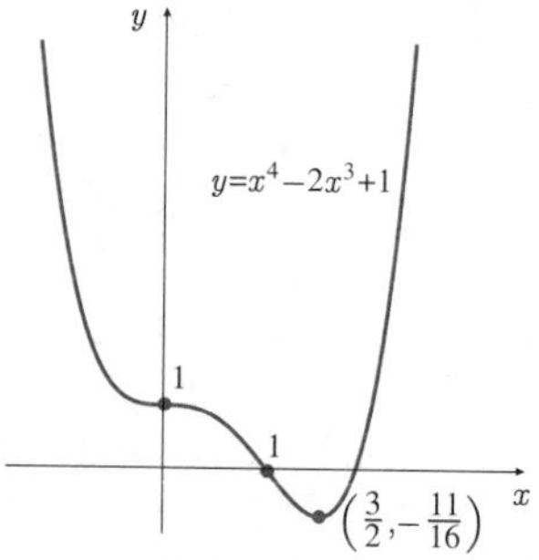

FIGURE 4.3.7

The Second Derivative Test

A function f will have a local maximum (or minimum) value at a critical point if its graph is concave downward (or upward) in an interval containing that point. In fact, we can often use the value of the second derivative at the critical point to determine whether the function has a local maximum or a local minimum value there.

THEOREM 4.3.8 **(The Second Derivative Test)**

a) If $f'(x_0) = 0$ and $f''(x_0) < 0$, then f has a local maximum value at x_0.
b) If $f'(x_0) = 0$ and $f''(x_0) > 0$, then f has a local minimum value at x_0.

PROOF Suppose that $f'(x_0) = 0$ and $f''(x_0) < 0$. Since

$$\lim_{h\to 0} \frac{f'(x_0 + h)}{h} = \lim_{h\to 0} \frac{f'(x_0 + h) - f'(x_0)}{h} = f''(x_0) < 0,$$

it follows that $f'(x_0+h) < 0$ for all sufficiently small positive h, and $f'(x_0+h) > 0$ for all sufficiently small negative h. By the First Derivative Test (Theorem 4.2.5), f must have a local maximum value at x_0. The proof of the local minimum case is similar. □

EXAMPLE 4.3.9 Find and classify the critical points of $f(x) = x^2 e^{-x}$.

SOLUTION

$$\begin{aligned} f'(x) &= (2x - x^2)e^{-x} = x(2 - x)e^{-x} = 0 \quad \text{at } x = 0 \text{ and } x = 2, \\ f''(x) &= (2 - 4x + x^2)e^{-x} \\ f''(0) &= 2 > 0, \qquad f''(2) = -2e^{-2} < 0. \end{aligned}$$

Thus f has a local minimum value at $x = 0$ and a local maximum value at $x = 2$. See Fig. 4.3.8.

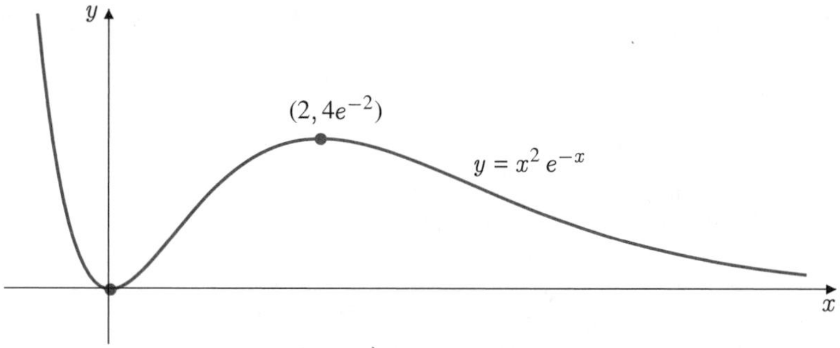

FIGURE 4.3.8

For many functions the second derivative may be more complicated to calculate than the first derivative, so the First Derivative Test is likely to be of more use in classifying critical points than is the Second Derivative Test. Also note that the First Derivative Test can classify local extreme values that occur at singular points as well as at critical points.

The Second Derivative Test makes no assertion about what can happen at x_0 if $f'(x_0) = 0$ and $f''(x_0) = 0$. In fact, at such a point, f may have a local maximum or a local minimum value, or it may have neither, but an inflection instead. The functions $-x^4$, x^4, and x^3 are examples of these three possibilities. It is possible to generalize the Second Derivative Test to obtain a higher derivative test to deal with some such situations. (See Exercise 48 at the end of this section.)

EXERCISES

In Exercises 1–30 determine the intervals of constant concavity of the given function and locate any inflection points.

1. $f(x) = \sqrt{x}$
2. $f(x) = 2x - x^2$
3. $f(x) = x^2 + 2x + 3$
4. $f(x) = x^2 + x^4$
5. $f(x) = x + x^3$
6. $f(x) = x - x^3$
7. $f(x) = 10x^3 - 3x^5$
8. $f(x) = 10x^3 + 3x^5$
9. $f(x) = 1 - x^2 + x^4$
10. $f(x) = x(2 - x)^2$
11. $f(x) = (3 - x^2)^2$
12. $f(x) = (2 + 2x - x^2)^2$
13. $f(x) = 8x^3 - 6x + 1$
14. $f(x) = 8x^3 - 6x^2 + 1$
15. $f(x) = (x^2 - 4)^3$
16. $f(x) = \dfrac{x}{x^2 + 3}$
17. $f(x) = \sin x$
18. $f(x) = \cos 3x$
19. $f(x) = x + \sin 2x$
20. $f(x) = x - 2 \sin x$
21. $f(x) = \tan^{-1} x$
22. $f(x) = x\, e^x$
23. $f(x) = e^{-x^2}$
24. $f(x) = x^2 e^{-x^2}$
25. $f(x) = x \ln(x^2)$
26. $f(x) = \dfrac{\ln(x^2)}{x}$
27. $f(x) = \ln(1 + x^2)$
28. $f(x) = (\ln x)^2$
29. $f(x) = \dfrac{x^3}{3} - 4x^2 + 12x - \dfrac{25}{3}$
30. $f(x) = (x - 1)^{1/3} + (x + 1)^{1/3}$
31. Discuss the concavity of the linear function $f(x) = ax + b$. Does it have any inflections?

Classify the critical points of the functions in Exercises 32–43 using the Second Derivative Test whenever possible.

32. $f(x) = 3x^3 - 36x - 3$
33. $f(x) = x(x - 2)^2 + 1$
34. $f(x) = x + \dfrac{4}{x}$
35. $f(x) = x^3 + \dfrac{1}{x}$
36. $f(x) = \dfrac{x}{2^x}$
37. $f(x) = \dfrac{x}{1 + x^2}$
38. $f(x) = xe^x$
39. $f(x) = x \ln x$
40. $f(x) = (x^2 - 4)^2$
41. $f(x) = (x^2 - 4)^3$
42. $f(x) = (x^2 - 3)e^x$
43. $f(x) = x^2 e^{-2x^2}$

44. Let $f(x) = x^2$ if $x \geq 0$ and $f(x) = -x^2$ if $x < 0$. Is 0 a critical point of f? Is it an inflection point? Is $f''(0) = 0$? If a function has a nonvertical tangent line at an inflection point, does the second derivative of the function necessarily vanish at that point?

45.* Verify that if f is concave up on an interval, then its graph lies above its tangent lines on that interval. (*Hint:* Suppose f is concave up on an open interval containing x_0. Let $h(x) = f(x) - f(x_0) - f'(x_0)(x - x_0)$. Show that h has a local minimum value at x_0 and hence that $h(x) \geq 0$ on the interval. Show that $h(x) > 0$ if $x \neq x_0$.)

46.* Verify that the graph $y = f(x)$ crosses its tangent line at an inflection point. (*Hint:* Consider separately the cases where the tangent line is vertical and nonvertical.)

47. For $f_n(x) = x^n$ and $g_n(x) = -x^n$, $(n = 2, 3, 4, \ldots)$, determine whether $x = 0$ is a local maximum, a local minimum, or an inflection point.

48.* Use your conclusions from the previous Exercise to suggest a generalization of the second-derivative test that applies when

$$f'(x_0) = f''(x_0) = \ldots = f^{(k-1)}(x_0) = 0, \quad f^{(k)}(x_0) \neq 0,$$

for some $k \geq 2$.

49.* This problem shows that no test based solely on the signs of derivatives at x_0 can determine whether every function with a critical point at x_0 has a local maximum or minimum, or an inflection point there. Let

$$f(x) = \begin{cases} e^{-1/x^2} & \text{if } x \neq 0, \\ 0 & \text{if } x = 0. \end{cases}$$

Prove the following

a) $\lim_{x \to 0} x^{-n} f(x) = 0$ for $n = 0, 1, 2, 3, \ldots$

b) $\lim_{x \to 0} P(1/x) f(x) = 0$ for every polynomial P.

c) For $x \neq 0, f^{(k)}(x) = P_k(1/x) f(x) (k = 1, 2, 3, \ldots)$ where P_k is a polynomial.

d) $f^{(k)}(0)$ exists and equals 0 for $k = 1, 2, 3, \ldots$

e) f has a local minimum at $x = 0$; $-f$ has a local maximum at $x = 0$.

f) If $g(x) = x f(x)$, then $g^{(k)}(0) = 0$ for every positive integer k and g has an inflection point at $x = 0$.

50.* A critical point of a function may be neither a local maximum nor a local minimum nor an inflection point. Show this by considering the following function.

$$f(x) = \begin{cases} x^2 \sin \dfrac{1}{x} & \text{if } x \neq 0 \\ 0 & \text{if } x = 0 \end{cases}$$

Show that $f'(0) = f(0) = 0$, so the x-axis is tangent to the graph of f at $x = 0$; but $f'(x)$ is not continuous at $x = 0$, so $f''(0)$ does not exist. Show that the concavity of f is not constant on any interval with endpoint 0.

4.4 SKETCHING THE GRAPH OF A FUNCTION

When sketching the graph $y = f(x)$ of a function f we have three sources of useful information:

i) **the function f itself**, from which we determine the coordinates of some points on the graph, the symmetry of the graph, and any asymptotes,

ii) **the first derivative**, f', from which we determine the intervals of increase and decrease and any local extreme values, and

iii) **the second derivative**, f'', from which we determine the concavity of the graph and any inflection points.

Items (ii) and (iii) have been explored in the previous two sections. In this section we consider what we can learn from the function itself about the shape of its graph, and then we illustrate the entire sketching procedure with several examples using all three sources of information.

A graph may be sketched by plotting the coordinates of many points on it and joining them by a suitably smooth curve. This simplistic approach is at best tedious, and at worst (see Example 1.3.10) can fail to reveal the most interesting aspects of the graph (singular points, extreme values, and so on). We could also compute the slope at each of the plotted points and, by drawing short line segments through these points with the appropriate slopes, ensure that the sketched graph passes through each plotted point with the correct slope. A better procedure is to obtain the coordinates of only a few points and use qualitative information from the function and its first and second derivatives to determine the shape of the graph between these points.

Besides critical and singular points and inflections, a graph may have other "interesting" points. The **intercepts** (points at which the graph intersects the coordinate axes) are usually among these. When sketching any graph it is wise to try to find all such intercepts, that is, all points with coordinates $(x, 0)$ and $(0, y)$ that lie on the graph. Of course, not every graph will have such points, and even when they do exist it may not always be possible to compute them exactly. Whenever a graph is made up of several disconnected pieces (called **components**), the coordinates of *at least one point on each component* must be obtained. Vertical asymptotes (discussed below) usually break the graph of a function into such components. It can sometimes be useful to determine the slopes at those points too.

Realizing that a given function possesses some symmetry can aid greatly in obtaining a good sketch of its graph. In Section 1.3 we discussed odd and even functions and observed that odd functions have graphs that are symmetric about the origin, while even functions have graphs that are symmetric about the y-axis, as shown in Figures 4.4.1 and 4.4.2. These are the symmetries you are most likely to notice, but functions can have other symmetries. For example, the graph of $f(x) = 2 + (x - 1)^2$ will certainly be symmetric about the line $x = 1$, and the graph of $\sin x$ is symmetric about the line $x = \pi/2$.

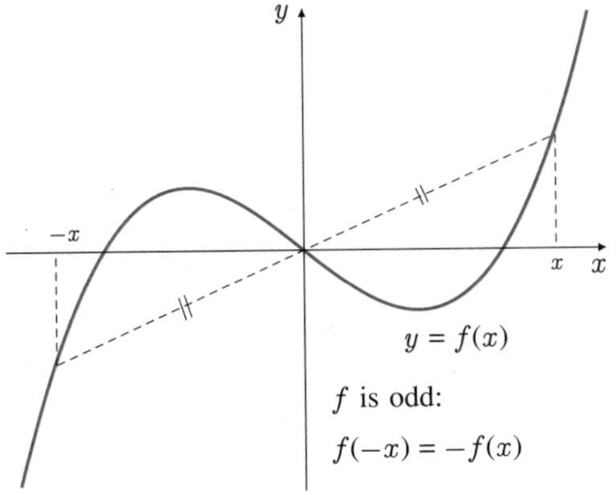

FIGURE 4.4.1

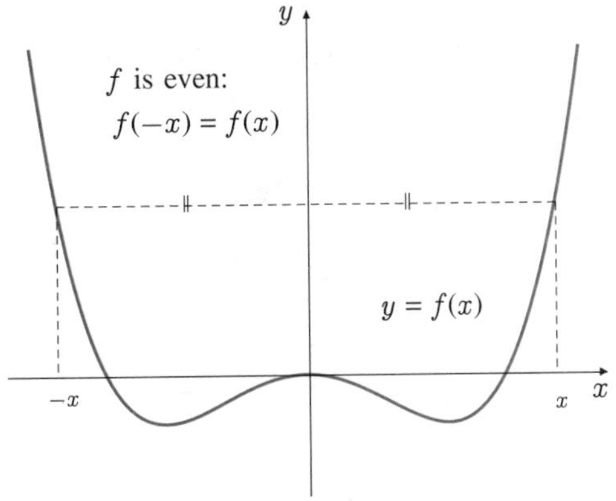

FIGURE 4.4.2

Asymptotes

Some of the curves we have sketched in previous sections have had **asymptotes**, that is, straight lines to which the curve draws arbitrarily near as it recedes to infinite distance from the origin. Asymptotes are of three types: vertical, horizontal, and oblique.

4.4.1
Vertical Asymptotes

> The graph of $y = f(x)$ has a **vertical asymptote** at $x = a$ if
>
> $$\lim_{x \to a-} f(x) = \pm\infty, \qquad \text{or} \quad \lim_{x \to a+} f(x) = \pm\infty, \qquad \text{or both.}$$

This situation tends to arise when $f(x)$ is a quotient of two expressions and the denominator is zero at $x = a$.

EXAMPLE 4.4.2 Let $f(x) = \dfrac{1}{x^2 - x} = \dfrac{1}{x(x-1)}$ has vertical asymptotes at $x = 0$ and $x = 1$. (See Fig. 4.4.3.) Note that

$$\lim_{x \to 0-} \frac{1}{x^2 - x} = \infty, \quad \lim_{x \to 0+} \frac{1}{x^2 - x} = -\infty, \quad \lim_{x \to 1-} \frac{1}{x^2 - x} = -\infty, \quad \lim_{x \to 1+} \frac{1}{x^2 - x} = \infty.$$

4.4.3
Horizontal Asymptotes

> The graph of $y = f(x)$ has a **horizontal asymptote** $y = L$ if
>
> $$\lim_{x \to \infty} f(x) = L, \qquad \text{or} \quad \lim_{x \to -\infty} f(x) = L, \qquad \text{or both.}$$

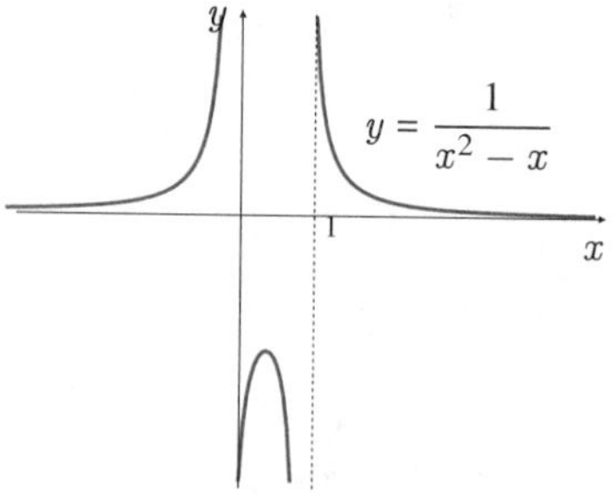

FIGURE 4.4.3

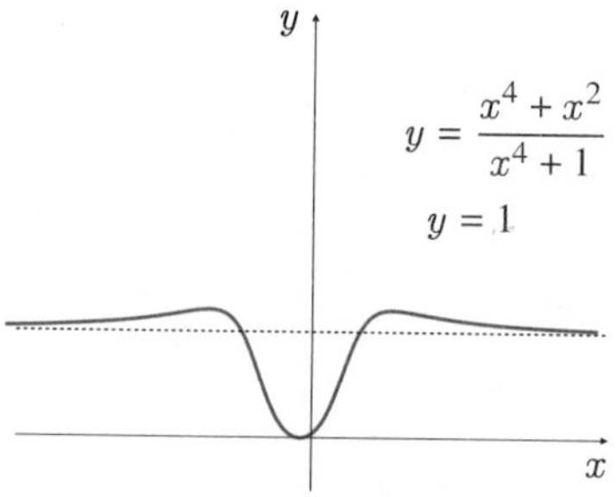

FIGURE 4.4.4

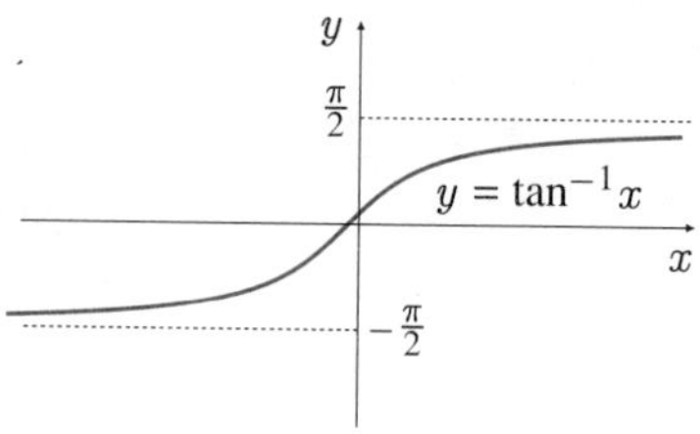

FIGURE 4.4.5

EXAMPLE 4.4.4 a) $f(x) = \dfrac{1}{x^2 - x}$ has horizontal asymptote $y = 0$; $\lim_{x\to\pm\infty} \dfrac{1}{x^2 - x} = 0$. See Fig. 4.4.3.

b) $g(x) = \dfrac{x^4 + x^2}{x^4 + 1}$ has horizontal asymptote $y = 1$; $\lim_{x\to\pm\infty} \dfrac{x^4 + x^2}{x^4 + 1} = 1$. See Fig. 4.4.4. Observe that the graph of g crosses its asymptote twice. (There is a popular misconception among students that curves cannot cross their asymptotes.) Exercise 39 below gives an example of a curve which crosses its asymptote infinitely often.

In both (a) and (b) the asymptotes are **two-sided**, which means that the graphs approach the asymptotes as x approaches both infinity and negative infinity. The function $tan^{-1}x$ has two **one-sided** asymptotes, $y = \pi/2$ (as $x \to \infty$) and $y = -(\pi/2)$ (as $x \to -\infty$). See Fig. 4.4.5.

It can also happen that the graph of a function $f(x)$ approaches a nonhorizontal straight line as x approaches ∞ or $-\infty$ (or both). Such a line is called an *oblique asymptote* of the graph.

4.4.5 Oblique Asymptotes

The straight line $y = ax + b$, (where $a \neq 0$), is an **oblique asymptote** of the graph of $y = f(x)$ if

$$\lim_{x\to-\infty} \big(f(x) - (ax + b)\big) = 0, \qquad \text{or} \quad \lim_{x\to\infty} \big(f(x) - (ax + b)\big) = 0, \qquad \text{or both.}$$

EXAMPLE 4.4.6 a) Consider the function $f(x) = \dfrac{x^2 + 1}{x} = x + \dfrac{1}{x}$, whose graph is given in Fig. 4.4.6. The straight line $y = x$ is a *two-sided* oblique asymptote of the graph of f because

$$\lim_{x\to\pm\infty} \big(f(x) - x\big) = \lim_{x\to\pm\infty} \frac{1}{x} = 0.$$

b) The graph of $y = \dfrac{x\,e^x}{1 + e^x}$ is shown in Fig. 4.4.7. It has a horizontal asymptote $y = 0$ at the left and an oblique asymptote $y = x$ at the right.

Recall that a **rational function** is a function of the form

$$f(x) = P(x)/Q(x),$$

where P and Q are polynomials. (A **polynomial** is a function of the form

$$P(x) = p_0 + p_1x + p_2x^2 + \cdots + p_nx^n,$$

where $p_0, p_1, p_2, \ldots p_n$ are constants.) If $p_n \neq 0$ we say the polynomial P has **degree** n. It is possible to be quite specific about the asymptotes of a rational function.

4.4.7
Asymptotes of a Rational Function

Suppose that

$$f(x) = \frac{P_m(x)}{Q_n(x)},$$

where P_m and Q_n are polynomials of degree m amd n, respectively. Suppose also that P_m and Q_n have no common linear factors. Then

a) The graph of f has a vertical asymptote at every position x such that $Q_n(x) = 0$.

b) The graph of f has a two-sided horizontal asymptote $y = 0$ if $m < n$.

c) The graph of f has a two-sided horizontal asymptote $y = L$, $(L \neq 0)$ if $m = n$. L can be found by evaluating the limit of $f(x)$ as x approaches $\pm\infty$.

d) The graph of f has a two-sided oblique asymptote if $m = n + 1$. This asymptote can be found by dividing Q_n into P_m to obtain a linear quotient, $ax + b$, and remainder, R, a polynomial of degree at most $n - 1$. That is,

$$f(x) = ax + b + \frac{R(x)}{Q_n(x)}.$$

The asymptote is $y = ax + b$.

e) The graph of f has no horizontal or oblique asymptotes if $n > m + 1$.

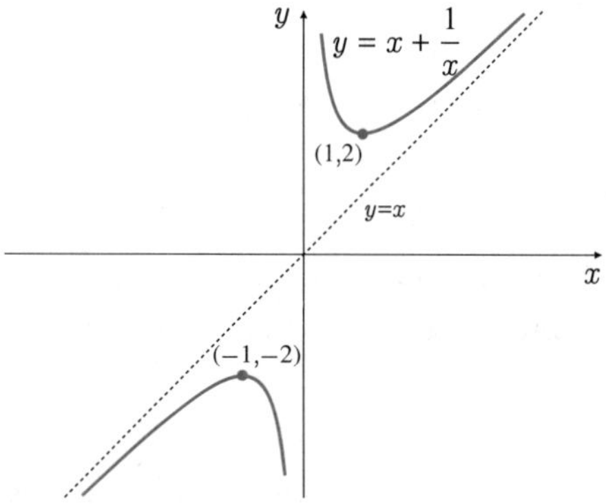

FIGURE 4.4.6

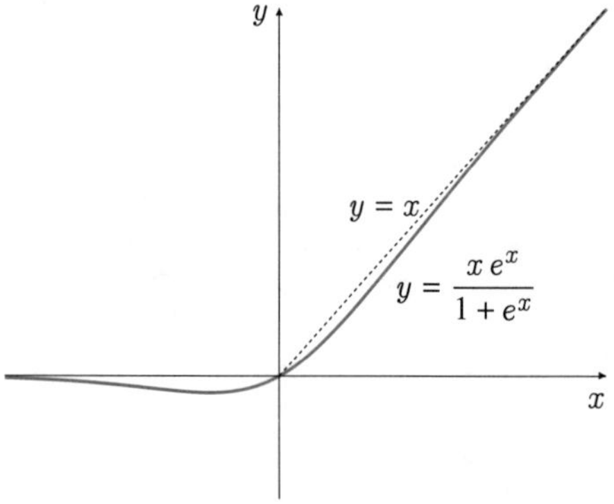

FIGURE 4.4.7

EXAMPLE 4.4.8 Find the oblique asymptote of $y = \dfrac{x^3}{x^2 + x + 1}$.

SOLUTION We can either obtain the quotient by long division:

$$\begin{array}{r|l} & x \;-\; 1 \\ \hline x^2 + x + 1 & x^3 \\ & x^3 + x^2 + x \\ \hline & \quad - x^2 - x \\ & \quad - x^2 - x - 1 \\ \hline & \qquad\qquad 1 \end{array} \qquad \frac{x^3}{x^2+x+1} = x - 1 + \frac{1}{x^2+x+1},$$

or we can obtain the same result by "short division:"

$$\frac{x^3}{x^2+x+1} = \frac{x^3 + x^2 + x - x^2 - x - 1 + 1}{x^2+x+1} = x - 1 + \frac{1}{x^2+x+1}.$$

In any event we see that the oblique asymptote has equation $y = x - 1$.

Examples of Formal Curve Sketching

Here is a check-list of things to consider when you are asked to make a careful sketch of the graph $y = f(x)$. It will, of course, not always be possible to obtain every item of information mentioned in the list.

1. Calculate $f'(x)$ and $f''(x)$, and express the results in factored form.
2. Examine $f(x)$ to determine its domain and the following items:
   a) any vertical asymptotes. (Look for zeros of denominators.)
   b) any horizontal or oblique asymptotes. (Consider $\lim_{x\to\pm\infty} f(x)$.)
   c) any obvious symmetry. (Is f even or odd?)
   d) any easily calculated intercepts (points with coordinates $(x, 0)$ or $(0, y)$), or endpoints or other "obvious" points. You will add to this list when you know any critical points, singular points and inflection points. Eventually you should make sure you know the coordinates of at least one point on every component of the graph.
3. Examine $f'(x)$ for the following:
   a) any critical points.
   b) any points where f' is not defined. (This will include singular points, endpoints of the domain of f, and vertical asymptotes.)
   c) intervals on which f' is positive or negative. It's a good idea to convey this information in the form of a chart such as those used in the examples. Conclusions about where f is increasing and decreasing and classification of some critical and singular points as local maxima and minima can also be indicated on the chart.

4. Examine $f''(x)$ for the following:
   a) points where $f''(x) = 0$.
   b) points where $f''(x)$ is undefined. (This will include singular points, end-points, vertical asymptotes, and possibly other points as well, where f' is defined but f'' isn't.)
   c) intervals where f'' is positive and negative, and where f is therefore concave up or down. Use a chart.
   d) any inflection points.

When you have obtained as much of this information as possible, make a careful sketch that reflects *everything* you have learned about the function. Consider where best to place the axes and what scale to use on each so the "interesting features" of the graph show up most clearly. Be alert for seeming inconsistencies in the information — that is a strong suggestion you may have made an error somewhere. For example, if you have determined that $f(x) \to \infty$ as x approaches the vertical asymptote $x = a$ from the right, and also that f is decreasing and concave down on the interval (a, b) then you have very likely made an error. (Try to sketch such a situation to see why.)

EXAMPLE 4.4.9 Sketch the graph of $y = \dfrac{x^2 + 2x + 4}{2x} = \dfrac{x}{2} + 1 + \dfrac{2}{x}$.

SOLUTION First calculate and factor the derivatives y' and y'':

$$y' = \frac{1}{2} - \frac{2}{x^2} = \frac{x^2 - 4}{2x^2}, \qquad y'' = \frac{4}{x^3}.$$

From y: Domain: all x except 0. Vertical asymptote: $x = 0$,
Oblique asymptote: $y = \dfrac{x}{2} + 1$, $\quad y - \left(\dfrac{x}{2} + 1\right) = \dfrac{2}{x} \to 0$ as $x \to \pm\infty$.
Symmetry: none (y is neither odd nor even).
Intercepts: none. y is not defined at $x = 0$ and $x^2+2x+4 = (x+1)^2+3 \ge 3$ for all x.

From y': Critical points: $x = \pm 2$; points $(-2, -1)$ and $(2, 3)$.
y' not defined at $x = 0$ (vertical asymptote).

		CP		ASY		CP		
y'	+	-2	−	0	−	2	+	$\to x$
y	↗	loc max	↘		↘	loc min	↗	

From y'': $y'' = 0$ nowhere; y'' undefined at $x = 0$.

		ASY		
y''	−	0	+	$\to x$
y	⌢		⌣	

The graph is shown in Fig. 4.4.8.

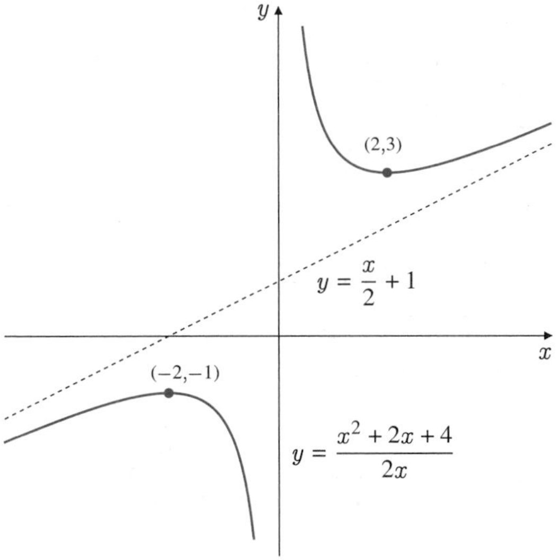

FIGURE 4.4.8

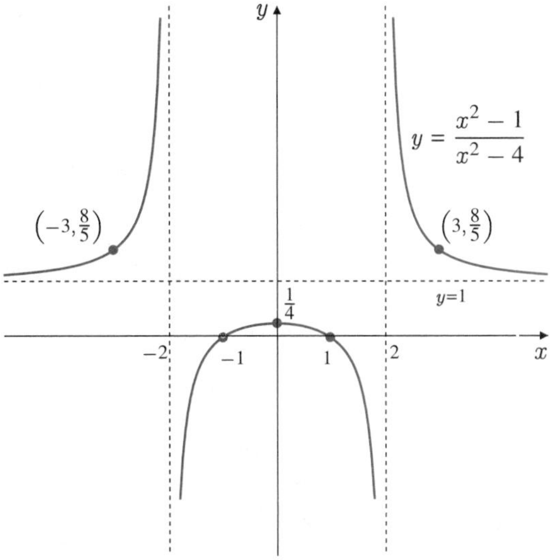

FIGURE 4.4.9

EXAMPLE 4.4.10 Sketch the graph of $f(x) = \dfrac{x^2 - 1}{x^2 - 4}$.

SOLUTION We have

$$f'(x) = \frac{-6x}{(x^2 - 4)^2}, \qquad f''(x) = \frac{6(3x^2 + 4)}{(x^2 - 4)^3}.$$

From f: Domain: all x except ± 2. Vertical asymptotes: $x = -2$ and $x = 2$.
Horizontal asymptote: $y = 1$ (as $x \to \pm\infty$).
Symmetry: about the y-axis (y is even).
Intercepts: $(0, 1/4)$, $(-1, 0)$ and $(1, 0)$.
Other points: $(-3, 8/5)$, $(3, 8/5)$. (The two vertical asymptotes divide the graph into three components; we need points on each. The outer components require points with $|x| > 2$.)

From f': Critical point: $x = 0$; f' not defined at $x = 2$ or $x = -2$.

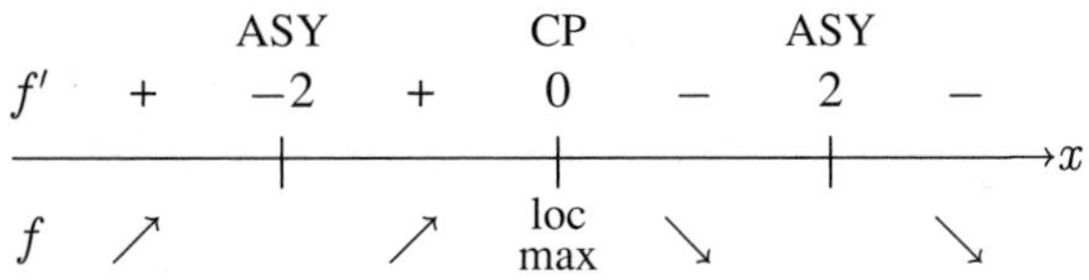

From f'': $f''(x) = 0$ nowhere; f'' not defined at $x = 2$ or $x = -2$.

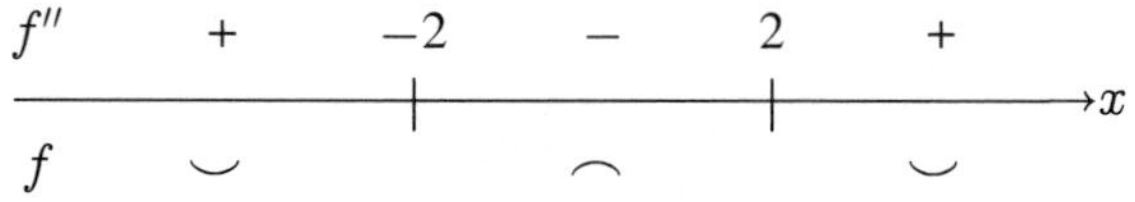

The graph is shown in Fig. 4.4.9.

EXAMPLE 4.4.11 Sketch the graph of the function $f(x) = \dfrac{3x^2 - 1}{x^3} = \dfrac{3}{x} - \dfrac{1}{x^3}$.

SOLUTION It is easier to differentiate f as a sum of two fractions and then recombine the fractions after differentiating:

$$f'(x) = -\frac{3}{x^2} + \frac{3}{x^4} = \frac{-3(x^2-1)}{x^4} = \frac{-3(x-1)(x+1)}{x^4}$$
$$f''(x) = \frac{6}{x^3} - \frac{12}{x^5} = \frac{6(x^2-2)}{x^5} = \frac{6(x-\sqrt{2})(x+\sqrt{2})}{x^5}$$

From f: Domain: all x except 0. Vertical asymptote: $x = 0$.
Horizontal asymptote: $y = 0$ (as $x \to \pm\infty$).
Symmetry: about the origin (f is odd).
Intercepts: $\left(\pm\dfrac{1}{\sqrt{3}}, 0\right) \approx (\pm 0.58, 0)$.

From f': Critical points: $x = -1$ and 1; points $(-1, -2)$ and $(1,2)$.
f' not defined at $x = 0$ (vertical asymptote).

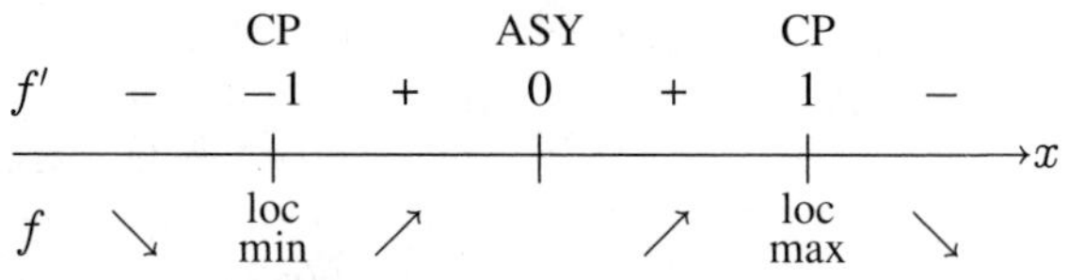

From f'': $f''(x) = 0$ at $x = -\sqrt{2}$ and $x = \sqrt{2}$; not defined at $x = 0$.
Inflection points: $\left(\pm\sqrt{2}, \pm\dfrac{5}{2\sqrt{2}}\right) \approx (\pm 1.41, \pm 1.77)$.

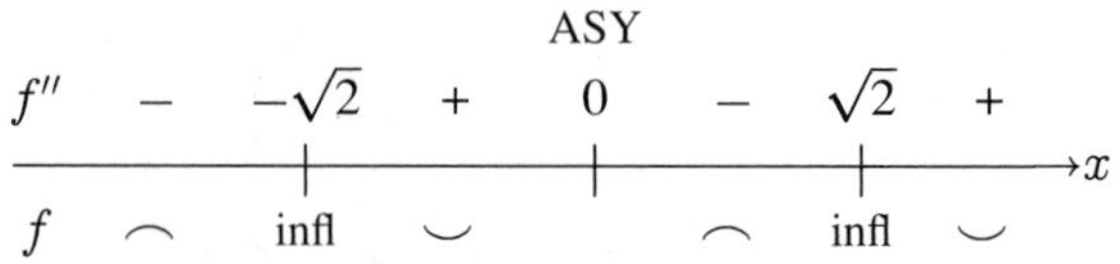

The graph is shown in Fig. 4.4.10.

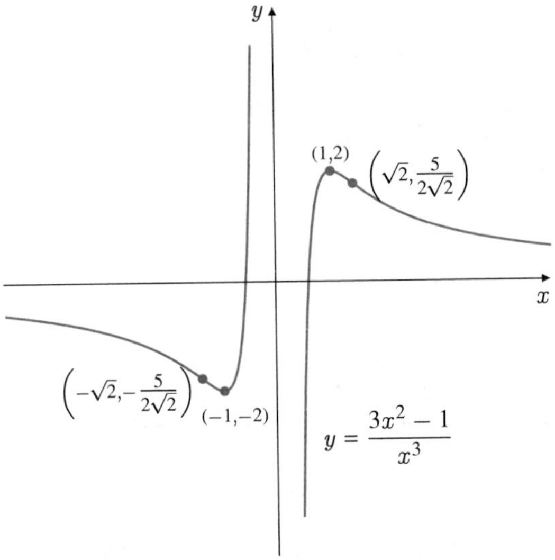

FIGURE 4.4.10

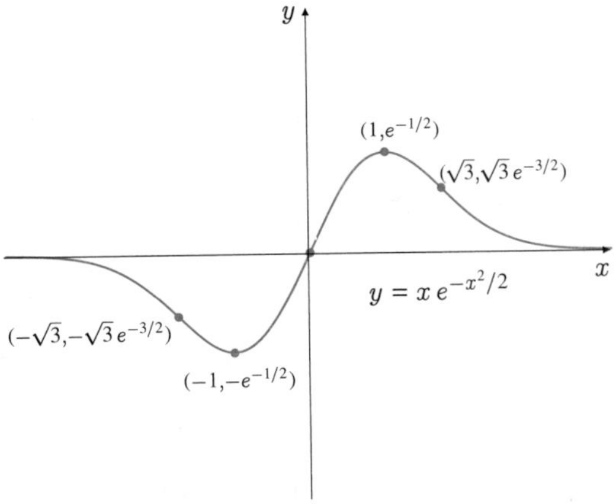

FIGURE 4.4.11

EXAMPLE 4.4.12 Sketch the graph of $y = xe^{-x^2/2}$.

SOLUTION We have $y' = (1 - x^2)e^{-x^2/2}$, $y'' = x(x^2 - 3)e^{-x^2/2}$.

From y: Domain: all x.
Horizontal asymptote: $y = 0$. Note that if $t = x^2/2$, then $|xe^{-x^2/2}| = \sqrt{2t}\, e^{-t} \to 0$ as $t \to \infty$ (hence as $x \to \pm\infty$). See Theorem 3.5.18.
Symmetry: about the origin (y is odd). Intercepts:$(0, 0)$.

From y': Critical points: $x = \pm 1$; points $(\pm 1, \pm 1/\sqrt{e}) \approx (\pm 1, \pm 0.61)$.

		CP		CP		
y'	−	−1	+	1	−	→ x
y	↘	loc min	↗	loc max	↘	

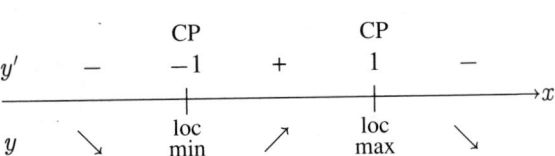

From y'': $y'' = 0$ at $x = 0$ and $x = \pm\sqrt{3}$; points $(0, 0)$, $(\pm\sqrt{3}, \pm\sqrt{3}e^{-3/2}) \approx (\pm 1.73, \pm 0.39)$.

y''	−	$-\sqrt{3}$	+	0	−	$\sqrt{3}$	+	→ x
y	⌢	infl	⌣	infl	⌢	infl	⌣	

The graph is shown in Fig. 4.4.11.

EXAMPLE 4.4.13 Sketch the graph of $f(x) = (x^2 - 1)^{2/3}$.

SOLUTION $f'(x) = \dfrac{4}{3}\dfrac{x}{(x^2-1)^{1/3}}, \qquad f''(x) = \dfrac{4}{9}\dfrac{x^2-3}{(x^2-1)^{4/3}}.$

From f: Domain: all x.
Asymptotes: none. ($f(x)$ grows like $x^{4/3}$ as $x \to \pm\infty$).
Symmetry: about the y-axis (f is an even function).
Intercepts: $(\pm 1, 0)$, $(0, 1)$.

From f': Critical points: $x = 0$; singular points: $x = \pm 1$.

		SP		CP		SP	
f'	$-$	-1	$+$	0	$-$	1	$+$
f	↘	loc min	↗	loc max	↘	loc min	↗

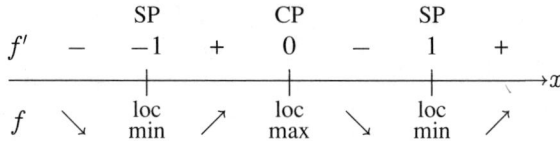

From f'': $f''(x) = 0$ at $x = \pm\sqrt{3}$; points $(\pm\sqrt{3}, 2^{2/3}) \approx (\pm 1.73, 1.59)$;
$f''(x)$ not defined at $x = \pm 1$.

f''	$+$	$-\sqrt{3}$	$-$	-1	$-$	1	$-$	$\sqrt{3}$	$+$
f	⌣	infl	⌢		⌢		⌢	infl	⌣

The graph is sketched in Fig. 4.4.12.

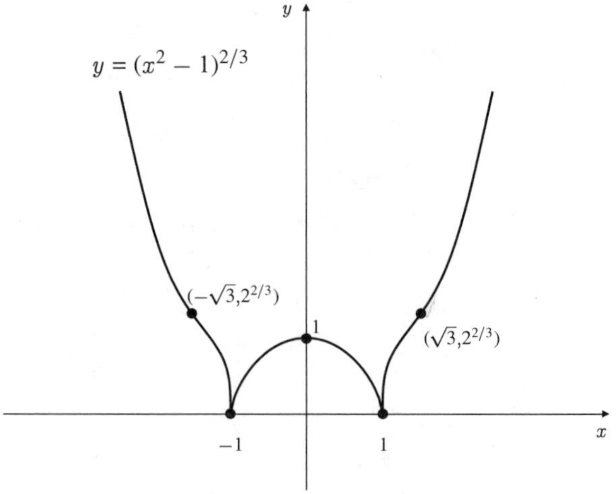

FIGURE 4.4.12

EXERCISES

In Exercises 1–2 sketch the graph of a function which has the given properties. Identify any critical points, singular points, local maxima and minima and inflection points. Assume derivatives exist everywhere unless the contrary is implied or explicitly stated.

1. $f(0) = 1$, $f(\pm 1) = 0$, $f(2) = 1$, $\lim_{x\to\infty} f(x) = 2$, $\lim_{x\to-\infty} f(x) = -1$, $f'(x) > 0$ on $(-\infty, 0)$ and on $(1, \infty)$, $f'(x) < 0$ on $(0, 1)$, $f''(x) > 0$ on $(-\infty, 0)$ and on $(0, 2)$, and $f''(x) < 0$ on $(2, \infty)$.
2. $f(-1) = 0$, $f(0) = 2$, $f(1) = 1$, $f(2) = 0$, $f(3) = 1$, $\lim_{x\to\pm\infty}(f(x) + 1 - x) = 0$, $f'(x) > 0$ on $(-\infty, -1)$, $(-1, 0)$ and $(2, \infty)$, $f'(x) < 0$ on $(0, 2)$, $\lim_{x\to-1} f'(x) = \infty$, $f''(x) > 0$ on $(-\infty, -1)$ and on $(1, 3)$, and $f''(x) < 0$ on $(-1, 1)$ and on $(3, \infty)$.

In Exercises 3–37 sketch the graphs of the given functions, making use of any suitable information you can obtain from the function and its first and second derivatives.

3. $y = (x^2 - 1)^3$
4. $y = x(x^2 - 1)^2$
5. $y = \dfrac{2 - x}{x}$
6. $y = \dfrac{x - 1}{x + 1}$
7. $y = \dfrac{x^3}{1 + x}$
8. $y = \dfrac{1}{4 + x^2}$
9. $y = \dfrac{1}{2 - x^2}$
10. $y = \dfrac{x}{x^2 - 1}$
11. $y = \dfrac{x^2}{x^2 - 1}$
12. $y = \dfrac{x^3}{x^2 - 1}$
13. $y = \dfrac{x^3}{x^2 + 1}$
14. $y = \dfrac{x^2}{x^2 + 1}$
15. $y = \dfrac{x^2 - 4}{x + 1}$
16. $y = \dfrac{x^2 - 2}{x^2 - 1}$
17. $y = \dfrac{x^3 - 4x}{x^2 - 1}$
18. $y = \dfrac{x^2 - 1}{x^2}$
19. $y = \dfrac{x^5}{(x^2 - 1)^2}$
20. $y = \dfrac{(2 - x)^2}{x^3}$
21. $y = \dfrac{1}{x^3 - 4x}$
22. $y = \dfrac{x}{x^2 + x - 2}$
23. $y = \dfrac{x^4 - 4x^2 + 4}{x^3}$
24. $y = x + \sin x$
25. $y = x + 2\sin x$
26. $y = e^{-x^2}$
27. $y = xe^x$
28. $y = e^{-x}\sin x$, $(x \geq 0)$
29. $y = x^2 e^{-x^2}$
30. $y = x^2 e^x$
31. $y = \dfrac{\ln x}{x}$, $(x > 0)$
32. $y = \dfrac{\ln x}{x^2}$, $(x > 0)$
33. $y = \dfrac{1}{\sqrt{4 - x^2}}$
34. $y = \dfrac{x}{\sqrt{x^2 + 1}}$
35. $y = (x^2 - 1)^{1/3}$
36.* $y = x^{1/3} + (x - 1)^{2/3}$
37.* $y = (x - 1)^{2/3} - (x + 1)^{2/3}$
38.* What is $\lim_{x\to 0+} x \ln x$? $\lim_{x\to 0} x \ln |x|$? If $f(x) = x \ln |x|$ for $x \neq 0$ is it possible to define $f(0)$ in such a way that f is continuous on the whole real line? Sketch the graph of f.
39. What straight line is an asymptote of the curve $y = \dfrac{\sin x}{1 + x^2}$? At what points does the curve cross this asymptote?

CHAPTER 5

Applications of Differentiation

Differential calculus can be used to analyze many kinds of problems and situations that arise in applied disciplines. Calculus can and has made significant contributions to every field of human endeavour that uses quantitative measurement to further its aims. From economics to physics and from biology to sociology, problems can be found whose solution can be aided by the use of some calculus.

In this chapter we will examine several kinds of problems to which the techniques we have already learned can be applied. These problems arise both outside and within mathematics. We will deal with the following kinds of problems:

i) optimization problems – where a quantity is to be maximized or minimized,

ii) related rates problems – where the rates of change of related quantities are analyzed,

iii) approximation problems – where complicated functions are approximated by simpler ones,

iv) root finding methods – where we try to find numerical solutions of equations,

v) indeterminate forms – where we evaluate certain non-obvious limits.

It would be naïve to assume that the problems we present here are "real world" problems; most of the latter are much too complex to be treated in a general calculus course. However, the problems we do consider, though somewhat artificial, do suggest how calculus can be applied in concrete situations.

5.1 OPTIMIZATION PROBLEMS

In this section we solve various word problems that, when translated into mathematical terms, require the finding of a maximum or minimum value of a function of one variable. Such problems can range from simple to very complex and difficult; they can be phrased in terminology appropriate to some other discipline or can be already partially translated into a more mathematical context. We have already encountered a few such problems in earlier chapters. See, for instance, Examples 1.7.8 and 2.6.10.

Let us consider a few more examples before attempting to abstract any general principles for dealing with such problems.

EXAMPLE 5.1.1 A rectangular enclosure is to be constructed having one side along an existing long wall and the other three sides fenced. If 100 metres of fence are available, what is the largest possible area for the enclosure?

SOLUTION This problem, like many others, is essentially a geometric one. A sketch should be made at the outset, as we have done in Fig. 5.1.1. Let the length and width of the enclosure be x and y m, respectively, and let its area be A m^2. Thus $A = xy$.

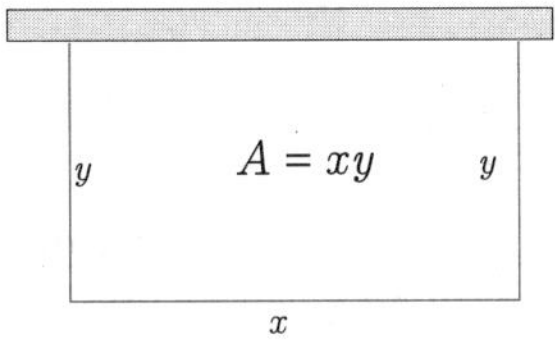

FIGURE 5.1.1

Since the total length of the fence is 100 m, we must have $x + 2y = 100$. A appears to be a function of two variables, x and y, but these variables are not independent; they are related by the *constraint* $x + 2y = 100$. This constraint equation can be solved for one variable in terms of the other, and A can therefore be written as a function of only one variable:

$$x = 100 - 2y,$$
$$A = A(y) = (100 - 2y)y = 100y - 2y^2.$$

Evidently we require $y \geq 0$ and $y \leq 50$, (that is, $x \geq 0$), in order that the area make sense. (It would otherwise be negative.) Thus we must maximize the function $A(y)$ on the interval $[0, 50]$. Being continuous on this closed, finite interval, A must have a maximum value, by Theorem 1.7.7 (or Theorem 4.2.2). Clearly $A(0) = A(50) = 0$ and $A(y) > 0$ for $0 < y < 50$. Hence the maximum cannot occur at an endpoint. Since A has no singular points, the maximum occurs at a critical point. To find any critical points, we set

$$0 = A'(y) = 100 - 4y. \qquad \text{Therefore } y = 25.$$

Since A must have a maximum value and there is only one possible point where it can exist, the maximum must occur at $y = 25$. The greatest possible area for the enclosure is therefore $A(25) = 1250$ m^2.

EXAMPLE 5.1.2 A lighthouse L is located on a small island 5 km north of a point A on a straight east-west shoreline. A cable is to be laid from L to point B on the shoreline 10 km east of A. The cable will be laid through the water in a straight line from L to a point C on the shoreline between A and B and from there to B along the shoreline. If the part of the cable lying in the water costs \$5000/km and the part along the shoreline costs \$3000/km, where should C be chosen to minimize the total cost of the cable? Where should C be chosen if B is only 3 km from A?

SOLUTION Let C be x km from A toward B. Thus $0 \leq x \leq 10$. The length of LC is $\sqrt{25 + x^2}$ km and that of CB is $10 - x$ km, as illustrated in Fig. 5.1.2. Hence the total cost of the cable is \$$T$ where

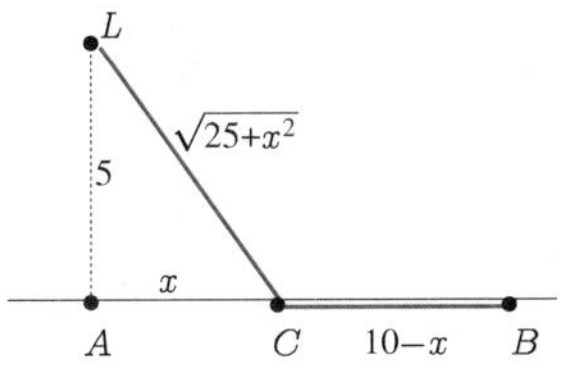

FIGURE 5.1.2

$$T = T(x) = 5000\sqrt{25 + x^2} + 3000(10 - x), \qquad (0 \leq x \leq 10).$$

T is continuous on the closed, finite interval $[0, 10]$, and so has a minimum value that may occur at one of the endpoints $x = 0$ or $x = 10$, or at a critical point in the interval $(0, 10)$. (T has no singular points.) To find any critical points, we set

$$0 = \frac{dT}{dx} = \frac{5000x}{\sqrt{25 + x^2}} - 3000.$$

$$\text{Thus} \quad \frac{5x}{3} = \sqrt{25 + x^2}$$

$$\frac{25x^2}{9} = 25 + x^2$$

$$x^2 = \frac{225}{16}.$$

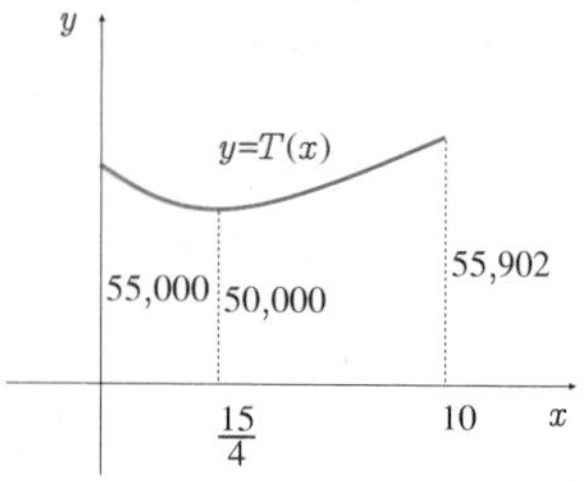

FIGURE 5.1.3

Only one critical point, $x = 15/4 = 3.75$, lies in the interval $(0, 10)$. Since $T(0) = 55,000, T(15/4) = 50,000$, and $T(10) \approx 55,902$, the critical point evidently provides the minimum value for $T(x)$. Alternatively, the second derivative of T is

$$T''(x) = 125,000/(25 + x^2)^{3/2} > 0,$$

so the graph of T is concave upward, and the critical point must yield a minimum. (See Fig. 5.1.3.) For minimal cost, C should be 3.75 km from A.

If B is 3 km from A, the corresponding total cost function is

$$T(x) = 5000\sqrt{25 + x^2} + 3000(3 - x), \quad (0 \le x \le 3),$$

which differs from the total cost function $T(x)$ used above only by the added constant (9000 rather than 30,000). It therefore has the same critical points, which do not lie in the interval $(0, 3)$. Since $T(0) > T(3)$, in this case C should be chosen at B; to minimize the total cost the cable should go straight from L to B.

EXAMPLE 5.1.3 Find the length of the shortest ladder that can extend from a vertical wall, over a fence 2 m high located 1 m away from the wall, to a point on the ground outside the fence.

SOLUTION Let θ be the angle of inclination of the ladder, as shown in Fig. 5.1.4. Using the two right-angled triangles in the figure, we obtain the length L of the ladder as a function of θ:

$$L = L(\theta) = \frac{1}{\cos\theta} + \frac{2}{\sin\theta},$$

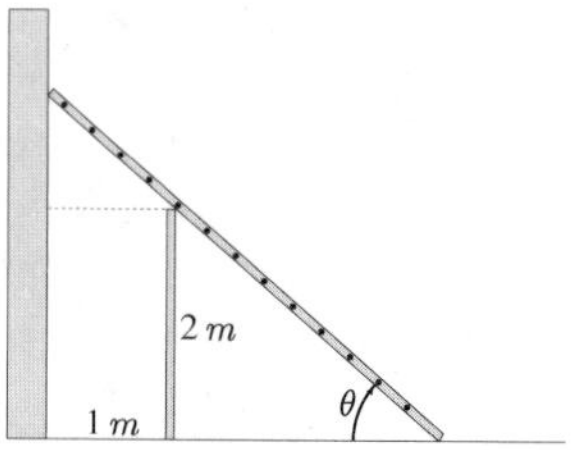

FIGURE 5.1.4

where $0 < \theta < \pi/2$. Since

$$\lim_{\theta\to(\pi/2)-} L(\theta) = \infty \quad \text{and} \quad \lim_{\theta\to 0+} L(\theta) = \infty,$$

(see Fig. 5.1.5) any minimum value for $L(\theta)$ can occur only at a critical point; there are no singular points in $(0, \pi/2)$. To find critical any points, we set

$$0 = L'(\theta) = \frac{\sin\theta}{\cos^2\theta} - \frac{2\cos\theta}{\sin^2\theta}.$$

It follows that $\tan^3\theta = 2$. We could solve this equation for $\theta = \tan^{-1}(2^{1/3})$, but it is really the corresponding value of $L(\theta)$ that we want. Observe that

$$\sec^2\theta = 1 + \tan^2\theta = 1 + 2^{2/3}.$$

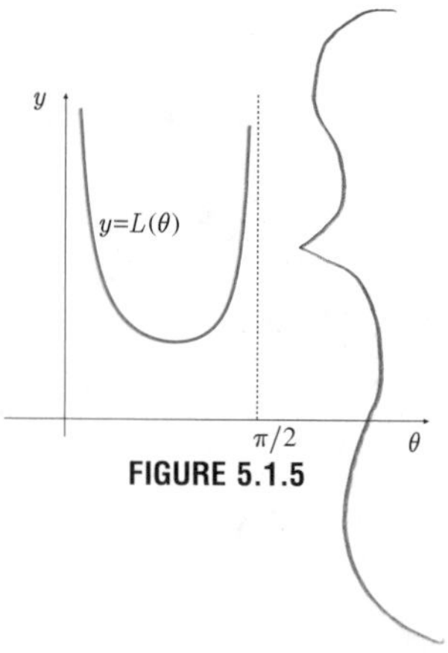

FIGURE 5.1.5

It follows that

$$\cos\theta = \frac{1}{(1 + 2^{2/3})^{1/2}}, \quad \text{and} \quad \sin\theta = \tan\theta\cos\theta = \frac{2^{1/3}}{(1 + 2^{2/3})^{1/2}}.$$

Therefore the minimal value of $L(\theta)$ is

$$\frac{1}{\cos\theta} + \frac{2}{\sin\theta} = (1 + 2^{2/3})^{1/2} + 2\,\frac{(1 + 2^{2/3})^{1/2}}{2^{1/3}} = \left(1 + 2^{2/3}\right)^{3/2} \approx 4.16.$$

The shortest ladder which can extend from the wall over the fence to the ground outside is about 4.16 m long.

Procedure for Solving Optimization Problems

Based on our experience with the examples above we can formulate a check-list of steps involved in solving optimization problems.

5.1.4 Solving Optimization Problems

1. Read the problem very carefully, perhaps more than once. You must understand clearly what is given and what must be found.
2. Define any symbols you wish to use that are not already specified in the statement of the problem.
3. Make a diagram if appropriate. Many problems have a geometric component and a good diagram can often be an essential part of the solution process.
4. Express the quantity Q to be maximized or minimized as a function of one or more variables.
5. If Q depends on more than one variable (say n) find $n - 1$ equations (constraints) linking these variables. (If this cannot be done, the problem cannot be solved by single-variable techniques.)
6. Use the constraints to eliminate variables and hence express Q as a function of only one variable. Determine the interval(s) in which this variable must lie for the problem to make sense. Alternatively, regard the constraints as defining $n-1$ of the variables, and hence Q, as functions of the remaining variable implicitly. (It is usually better to avoid this implicit method in an optimization problem if you can.)
7. Find all the local extreme values of Q, considering any critical points, singular points and endpoints.
8. Give some justification that one particular value is the desired (absolute) extreme value. (The first- or second-derivative test may be useful in this context, as may Theorem 1.7.7, Theorem 4.2.2, or Exercises 48–50 of Section 4.2.) Perhaps the easiest way is to draw a rough sketch of the graph of Q.
9. Make a concluding statement answering the question asked. Is your answer for the question "reasonable?" If not, check back through the solution to see what went wrong.

EXAMPLE 5.1.5 Find the largest possible volume of a right circular cone that can be inscribed in a sphere of radius R.

SOLUTION Let r, h and V denote the radius, height, and volume of the cone respectively. The volume of a cone is one-third the base area times the height, so

$$V = \frac{1}{3}\pi r^2 h.$$

From the small right-angled triangle in Fig. 5.1.6,

$$(h-R)^2 + r^2 = R^2.$$

Thus $r^2 = R^2 - (h-R)^2$ and

$$V = V(h) = \frac{\pi}{3} h\big(R^2 - (h-R)^2\big) = \frac{\pi}{3}\big(2Rh^2 - h^3\big).$$

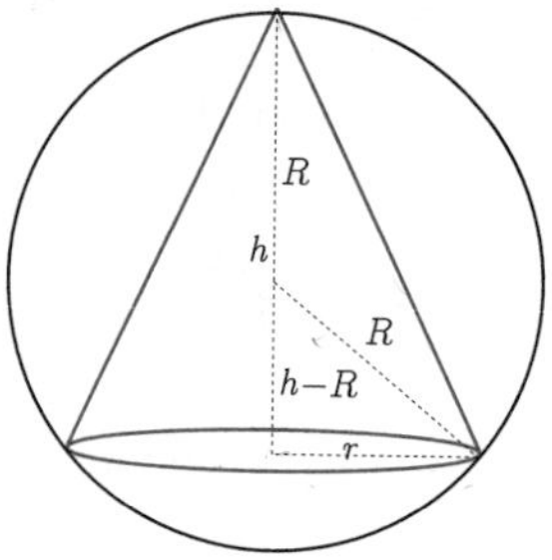

FIGURE 5.1.6

The height of any inscribed cone cannot exceed the diameter of the sphere, so $0 \le h \le 2R$. Being continuous, $V(h)$ must have a maximum value on this interval. Since $V = 0$ when $h = 0$ or $h = 2R$, and $V > 0$ if $0 < h < 2R$, the maximum value of V must occur at a critical point. (V has no singular points.) For a critical point,

$$0 = V'(h) = \frac{\pi}{3}(4Rh - 3h^2) = \frac{\pi}{3} h(4R - 3h),$$

$$h = 0 \quad \text{or} \quad h = \frac{4R}{3}.$$

$V'(h) > 0$ if $0 < h < 4R/3$ and $V'(h) < 0$ if $4R/3 < h < 2R$. Hence $h = 4R/3$ does indeed give the maximum value for V. The volume of the largest cone can be inscribed in a sphere of radius R is

$$V\left(\frac{4R}{3}\right) = \frac{\pi}{3}\left(2R\left(\frac{4R}{3}\right)^2 - \left(\frac{4R}{3}\right)^3\right) = \frac{32}{81}\pi R^3 \text{ cubic units.}$$

EXAMPLE 5.1.6 Find the most economical shape of a cylindrical tin can.

SOLUTION This example is stated in a rather vague way. We must consider what is meant by "most economical" and even "shape." Wanting further information, we can take one of two points of view:

i) the volume of the tin can is to be regarded as given and we must choose the dimensions to minimize the total surface area, or

ii) the total surface area is given (we can use just so much metal) and we must choose the dimensions to maximize the volume.

We will discuss other possible interpretations later.

Since a cylinder is determined by its radius and height, its shape is determined by the ratio radius/height (see Fig. 5.1.7). Let r, h, S and V denote, respectively, the radius, height, total surface area and volume of the can. The volume of a cylinder is the base area times the height, so

$$V = \pi r^2 h.$$

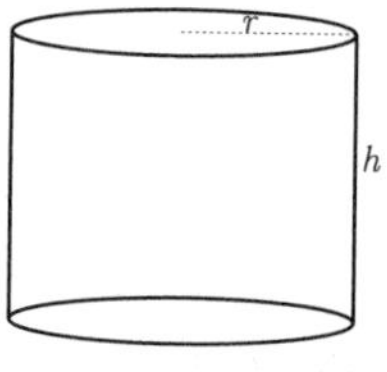

FIGURE 5.1.7

The surface of the can is made up of the cylindrical wall and circular discs for the top and bottom. The discs each have area πr^2 and the cylindrical wall is really just a rolled-up rectangle with base $2\pi r$ (the circumference of the can) and height h. Therefore the total surface area of the can is

$$S = 2\pi r^2 + 2\pi r h.$$

Suppose we use interpretation (i); V is a given constant, and S is to be minimized. We can use the equation for V to eliminate one of the two variables r and h on which S depends. Say we solve for $h = V/(\pi r^2)$ and substitute into the equation for S to obtain S as a function of r alone:

$$S = S(r) = 2\pi r \frac{V}{\pi r^2} + 2\pi r^2 = \frac{2V}{r} + 2\pi r^2 \qquad (0 < r < \infty).$$

Evidently, $\lim_{r\to 0+} S(r) = \infty$ and $\lim_{r\to\infty} S(r) = \infty$. Thus any minimum value of $S(r)$ must occur at a critical point. To find any critical points,

$$\begin{aligned} 0 = S'(r) &= -\frac{2V}{r^2} + 4\pi r, \\ r^3 = \frac{V}{2\pi} &= \frac{1}{2\pi}\pi r^2 h = \frac{1}{2} r^2 h. \end{aligned}$$

Thus $h = 2r$ at the critical point of S. Since

$$S''(r) = \frac{4V}{r^3} + 4\pi > 0 \quad \text{if} \quad r > 0,$$

the graph of $S(r)$ is concave up on $(0, \infty)$, and the critical point $r = h/2$ gives a local and absolute minimum value.

Thus, under interpretation (i), the most economical shape of a cylindrical tin can has base diameter equal to height. We encourage you to solve the problem using interpretation (ii) and see that the same solution results.

There is another way to obtain the solution that shows directly that interpretations (i) and (ii) must give the same solution. Again we start from the two equations

$$V = \pi r^2 h, \qquad S = 2\pi r h + 2\pi r^2.$$

If we regard h as a function of r and differentiate implicitly, we obtain

$$\begin{aligned} \frac{dV}{dr} &= 2\pi r h + \pi r^2 \frac{dh}{dr}, \\ \frac{dS}{dr} &= 2\pi h + 2\pi r \frac{dh}{dr} + 4\pi r. \end{aligned}$$

Under interpretation (i) V is constant and we want a critical point of S; under interpretation (ii), S is constant and we want a critical point of V. In either case, $dV/dr = 0$ and $dS/dr = 0$. Hence both interpretations yield

$$\begin{aligned} 2\pi r h + \pi r^2 \frac{dh}{dr} &= 0, \\ 2\pi h + 4\pi r + 2\pi r \frac{dh}{dr} &= 0. \end{aligned}$$

If we divide the first equation by πr^2 and the second equation by $2\pi r$ and subtract to eliminate dh/dr we again get $h = 2r$.

Given the sparse information provided in the statement of the problem, interpretations (i) and (ii) are the best we can do. The problem could be made more meaningful economically (from the point of view, say, of a tin can manufacturer) if more elements were brought into it. For example:

a) Most cans use thicker material for the cylindrical wall than for the top and bottom discs. If the cylindrical wall material costs \$$A$ per unit area and that for the top and bottom costs \$$B$ per unit area, we might prefer to minimize the total cost for materials for a can of given volume.

b) Large numbers of cans are to be manufactured. The material is probably being cut out of sheets of metal. The cylindrical walls are made by bending up rectangles, and rectangles can be cut from the sheet with little or no waste. There will, however, always be a proportion of material wasted when the discs are cut out. The exact proportion will depend on how the discs are arranged; two possible arrangements are shown in Fig. 5.1.8.

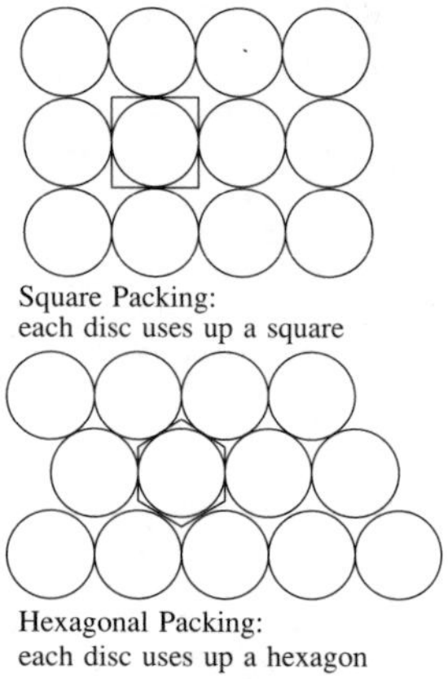

FIGURE 5.1.8

Any such modification of the original problem will alter the optimal shape to some extent. Another complication arises if the manufacturer has a can-making machine that can only make cans of one particular radius, or one of a small set of possible radii. In this case the problem might more properly be interpreted to choose h so that the ratio V/S is as large as possible.

In all of the examples above the maximum or minimum value being sought always occurred at a critical point. Our final example shows that this is not always the case.

EXAMPLE 5.1.7 A man can run twice as fast as he can swim. He is standing at point A on the edge of a circular swimming pool 40 m in diameter, and he wishes to get to the diametrically opposite point B as quickly as possible. He can run around the edge to point C, then swim directly from C to B. Where should C be chosen to minimize the total time taken to get from A to B?

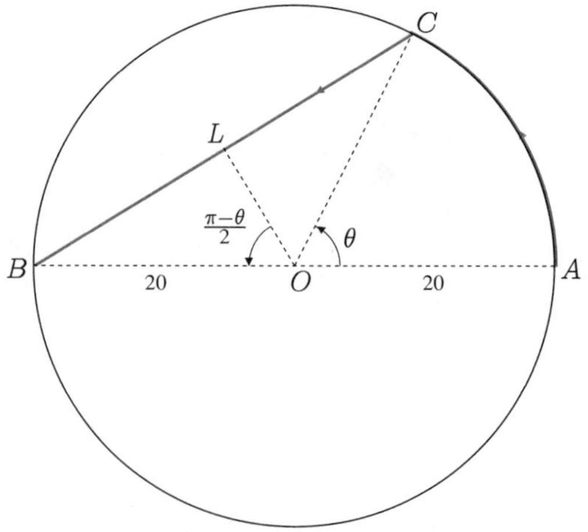

FIGURE 5.1.9

SOLUTION It is convenient to describe the position of C in terms of the angle AOC where O is the centre of the pool (see Fig. 5.1.9). Let θ denote this angle. Clearly $0 \le \theta \le \pi$. (If $\theta = 0$ the man swims the whole way; if $\theta = \pi$ he runs the whole way.) Thus arc $AC = 20\theta$, and since angle $BOC = \pi - \theta$, chord $BC = 2BL = 40 \sin \dfrac{\pi - \theta}{2}$. Suppose the man swims at a rate k m/s and therefore runs at a rate $2k$ m/s. If t is the total time he takes to get from A to B, then

$$t = t(\theta) = \text{time running} \quad + \quad \text{time swimming}$$
$$= \frac{20\theta}{2k} + \frac{40}{k} \sin \frac{\pi - \theta}{2}.$$

(We are assuming that no time is wasted in jumping into the water at C.) The domain of t is $[0, \pi]$ and t has no singular points. Since t is continuous on a closed, finite interval it must have a minimum value, and that value must occur at a critical point or an endpoint. For critical points,

$$0 = t'(\theta) = \frac{10}{k} - \frac{20}{k} \cos \frac{\pi - \theta}{2}$$

$$\text{Thus} \quad \cos \frac{\pi - \theta}{2} = \frac{1}{2}, \qquad \frac{\pi - \theta}{2} = \frac{\pi}{3}, \qquad \theta = \frac{\pi}{3}.$$

This is the only value of θ lying in the interval $[0, \pi]$. Observe, however, that

$$t''(\theta) = -\frac{10}{k} \sin \frac{\pi - \theta}{2} < 0 \quad \text{on} \quad (0, \pi),$$

so $\theta = \pi/3$ gives, in fact, a local *maximum* rather than a local *minimum* value of t. Thus the minimum time must occur at one of the endpoints $\theta = 0$ or $\theta = \pi$. Since $t(0) = 40/k$ and $t(\pi) = 10\pi/k < 40/k$, the man should run the entire distance.

This problem shows how important it is to check every candidate point to see whether it gives a loc max or loc min. Here the critical point yielded the worst possible strategy; running one-third of the way around and then swimming the remainder would take the greatest time, not the least. In the above solution we used the second-derivative test to test the critical point, but the first-derivative test could have been used as well. Both endpoints give local minima to t and only comparison of the actual values showed which was in fact the absolute minimum.

EXERCISES

1. Two nonnegative numbers have sum 7. What is the largest possible value for their product?

2. Two positive numbers have product 8. What is the smallest possible value for their sum?

3. Two nonnegative numbers have sum 60. What are the numbers if the product of one of them and the square of the other is maximal?

4. Two numbers have sum 16. What are the numbers if the product of the cube of one and the fifth power of the other is as large as possible?

5. Among all rectangles of given area show that the square has the least perimeter.

6. Among all rectangles of given perimeter show that the square has the greatest area.

7. Among all isosceles triangles of given perimeter, show that the equilateral triangle has the greatest area.

8. Find the largest possible area for an isosceles triangle if the sum of the lengths of its two equal sides is 20 m.

9. Find the area of the largest rectangle that can be inscribed in a semicircle of radius R if one side of the rectangle lies along the diameter of the semicircle.

10. Find the largest possible perimeter of a rectangle inscribed in a semicircle of radius R if one side of the rectangle lies along the diameter of the semicircle. (It is interesting to note that the rectangle with the largest perimeter has a different shape than the one with the largest area, obtained in the previous Exercise.)

11. A rectangle with sides parallel to the coordinate axes is inscribed in the ellipse

$$\frac{x^2}{a^2} + \frac{y^2}{b^2} = 1.$$

Find the largest possible area for this rectangle.

12. Let ABC be a triangle right-angled at C and having area S. Find the maximum area of a rectangle inscribed in the triangle if (a) one corner of the rectangle lies at C, or (b) one side of the rectangle lies along the hypotenuse, AB.

13. A billboard is to be made with 100 m^2 of printed area, and with margins of 2 m at the top and bottom and 4 m on each side. Find the outside dimensions of the billboard if its total area is to be a minimum.

14. A box is to be made from a rectangular sheet of cardboard 70 cm by 150 cm by cutting equal squares out of the four corners and bending up the resulting four flaps to make the sides of the box. (The box has no top.) What is the largest possible volume of the box?

15. An automobile manufacturer sells 2,000 cars per month at an average profit of \$1,000 per car. Market research indicates that for each \$50 of "factory rebate" the manufacturer offers to buyers it can expect to sell 200 more cars each month. How much of a rebate should it offer to maximize its monthly profit?

16. All 80 rooms in a motel will be rented each night if the manager charges \$40 or less per room. If he charges \$$(40 + x)$ per room, then $2x$ rooms will remain vacant. If each rented room costs the manager \$10 per day and each unrented room \$2 per day in overhead, how much should the manager charge per room to maximize his daily profit?

17. You are in a dune buggy in the desert 12 km due south of the nearest point A on a straight east-west road. You wish to get to point B on the road 10 km east of A. If your dune buggy can average 15 km/h traveling over the desert, and 39 km/h traveling on the road, toward what point on the road should you head in order to minimize your travel time to B?

18. Repeat the previous Exercise, but assume that B is only 4 km from A.

19. A 1-metre length of stiff wire is cut into two pieces. One piece is bent into a circle, the other piece into a square. Find the length of the part used for the square if the sum of the areas of the circle and the square is (a) maximum, (b) minimum.

20. Find the area of the largest rectangle which can be drawn so that each of its sides passes through one vertex of a rectangle having sides a and b.

21. What is the length of the shortest line segment having one end on the x-axis, the other end on the y-axis, and passing through the point $(9, \sqrt{3})$?

22. Find the length of the longest beam that can be carried horizontally around the corner from a hallway of width a m to a hallway of width b m, as shown in Fig. 5.1.10. (Assume the beam has no width.)

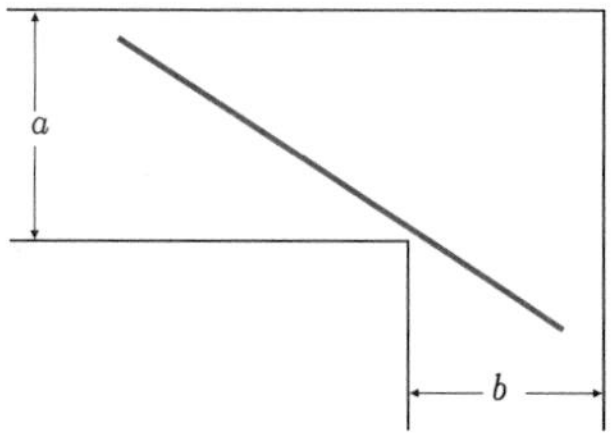

FIGURE 5.1.10

23. If the height of both hallways in the previous exercise is c m, and if the beam need not be carried horizontally, how long can it be and still get around the corner? (Hint: you can use the result of the previous exercise to do this one easily.)

24. The fence in Example 5.1.3 is demolished and a new fence is built 2 m away from the wall. How high can the fence be if a 6 metre ladder must be able to extend from the wall, over the fence, to the ground outside?

25. Find the shortest distance from the origin to the curve $x^2y^4 = 1$.

26. Find the shortest distance from the point $(8, 1)$ to the curve $y = 1 + x^{3/2}$.

27. Find the dimensions of the largest right-circular cylinder that can be inscribed in a sphere of radius R.

28. Find the dimensions of the circular cylinder of greatest volume that can be inscribed in a cone of base radius R and height H if the base of the cylinder lies in the base of the cone.

29. A box with square base and no top is to have a volume of 4 m^3. Find the dimensions of the most economical box.

30. Refer to Example 5.1.6. Find the most economical shape of the tin can if the material used for the cylindrical walls is twice as expensive per unit area as that used for the top and bottom discs.

31. Suppose the discs to make the tops and bottoms of cans are cut from sheet metal in a square packing arrangement (see Fig. 5.1.8), and the waste is discarded. Find the most economical shape for the can in Example 5.1.6 under these circumstances.

32. Repeat the previous exercise, but use the hexagonal packing of discs.

33. A window has perimeter 10 m and is in the shape of a rectangle with top edge replaced by a semicircle. Find the dimensions of the rectangle if the window admits the greatest amount of light.

34. A fuel tank is made of a cylindrical part capped by hemispheres at each end. If the hemispheres are twice as expensive per unit area as the cylindrical wall, and if the volume of the tank is V, find the radius and height of the cylindrical part to minimize the total cost. The surface area of a sphere of radius r is $4\pi r^2$; its volume is $\frac{4}{3}\pi r^3$.

35. Light travels in such a way that it requires the minimum possible time to get from one point to another. A ray of light from C reflects off a plane mirror AB at X and then passes through D (see Fig. 5.1.11). Show that the rays CX and XD make equal angles with the normal to AB at X. (*Remark:* You may wish to give a proof based on elementary geometry without using any calculus, or you can minimize the travel time on CXD.)

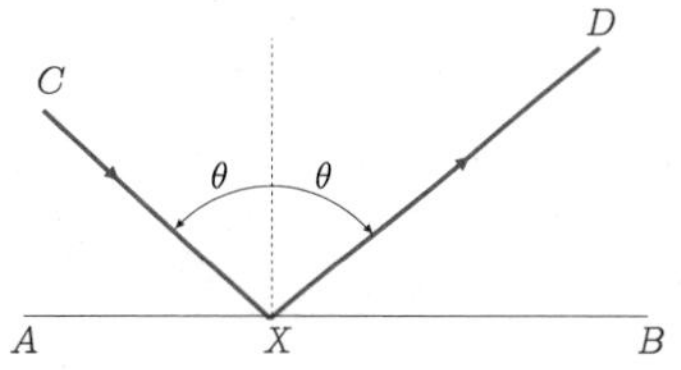

FIGURE 5.1.11

36. *If light travels with speed v_1 in one medium and speed v_2 in a second medium, and if the two media are separated by a plane interface, show that a ray of light passing from point A in one medium to point B in the other is bent at the interface in such a way that

$$\frac{\sin i}{\sin r} = \frac{v_1}{v_2},$$

where i and r are the angles of incidence and refraction, as is shown in Fig. 5.1.12. This is known as **Snell's law**. Deduce it from the least time principle stated in the previous exercise.

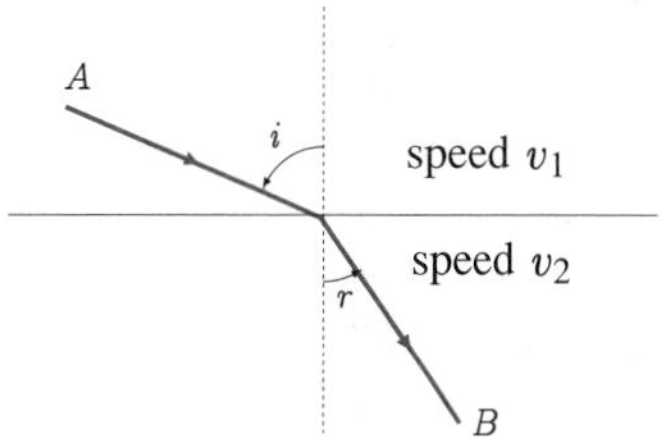

FIGURE 5.1.12

37. The stiffness of a wooden beam of rectangular cross section is proportional to the product of the width and the cube of the depth of the cross section. Find the width and depth of the stiffest beam that can be cut out of a circular log of radius R.

38. Find the equation of the straight line of maximum slope tangent to the curve $y = 1 + 2x - x^3$.

39. A quantity Q grows according to the differential equation

$$\frac{dQ}{dt} = kQ^3(L - Q)^5$$

where k and L are positive constants. How large is Q when it is growing most rapidly?

40.*Find the smallest possible volume of a right circular cone which can contain a sphere of radius R. (The volume of a cone of base radius r and height h is $\frac{1}{3}\pi r^2 h$.)

41.*How far back from a mural should one stand to view it best if the mural is 10 ft high and the bottom of it is 2 ft above eye level? (See Fig. 5.1.13.)

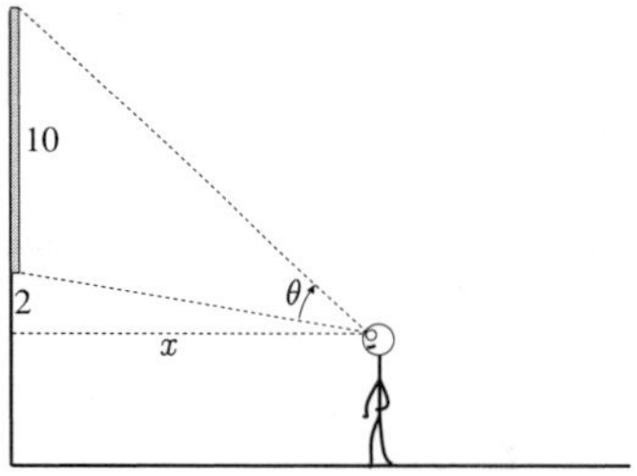

FIGURE 5.1.13

42.*An enclosure is to be constructed having part of its boundary along an existing straight wall. The other part of the boundary is to be fenced in the shape of an arc of a circle. If 100 m of fencing are available, what is the area of the largest possible enclosure? Into what fraction of a circle is the fence bent?

43.*A sector is cut out of a circular disc of radius R as shown in Fig. 5-1-14 and the remaining part of the disc is bent up so that the two edges join and a cone is formed. What is the largest possible volume for the cone?

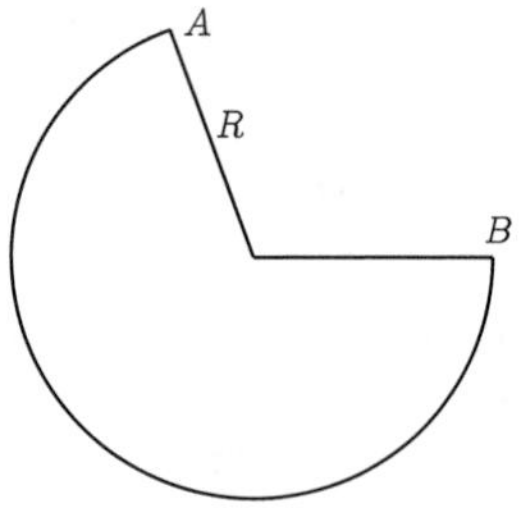

FIGURE 5.1.14

44.*One corner of a strip of paper a cm wide is folded up so that it lies along the opposite edge, as shown in Fig. 5.1.15. Find the least possible length for the fold line.

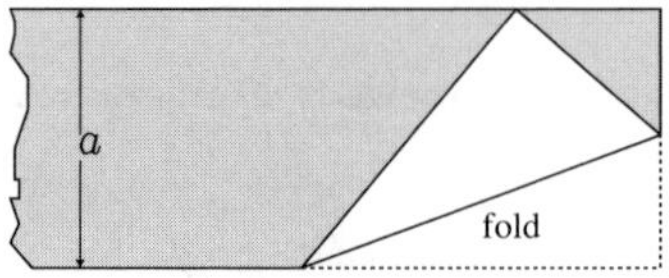

FIGURE 5.1.15

5.2 RELATED RATES

When two or more quantities that change with time are linked by an equation, that equation can be differentiated with respect to time to produce an equation linking the rates of change of the quantities. Any one of these rates may then be determined when the others, and the values of the quantities themselves are known. Again we will consider a couple of examples before formulating a list of procedures for dealing with such problems.

EXAMPLE 5.2.1 A policeman is standing beside a highway using a radar gun to catch speeders. He aims the gun at a car that has just passed his position and, when the gun is pointing at an angle of 45° to the direction of the highway, notes that the distance between the car and the gun is increasing at a rate of 100 km/h. How fast is the car travelling?

SOLUTION We begin by making a sketch of the situation. See Fig. 5-2-1. Let P be the position of the policeman, and let A be the point on the highway nearest him. We do not know the distance AP; let it be k. Let x and s denote the distance from the car C to A and P respectively. From the right-angled triangle,

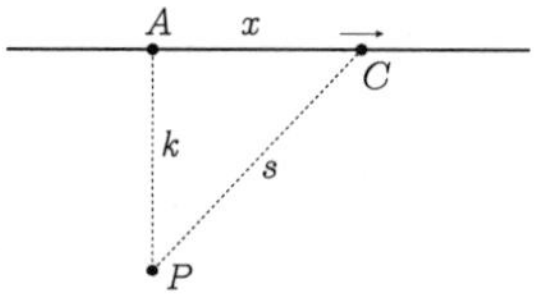

FIGURE 5.2.1

$$s^2 = x^2 + k^2.$$

Both x and s depend on time t, but k does not. If we differentiate the above equation implicitly with respect to t we obtain

$$2s\,\frac{ds}{dt} = 2x\,\frac{dx}{dt}.$$

We are told that $\frac{ds}{dt} = 100$ at the instant when the angle $\angle PCA = 45°$. At this instant $x = k$ and $s = \sqrt{2}\,k$ so

$$2\sqrt{2}\,k \times 100 = 2\,k\,\frac{dx}{dt}, \qquad \text{and} \qquad \frac{dx}{dt} = 100\sqrt{2} \approx 141.$$

The car is travelling at about 141 km/h.

EXAMPLE 5.2.2 How fast is the area of a rectangle changing if one side is 10 cm long and is increasing at a rate of 2 cm/s and the other side is 8 cm long and is decreasing at a rate of 3 cm/s?

SOLUTION Let the lengths of the sides of the rectangle at time t be x cm and y cm, respectively. Thus the area at time t is $A = xy$ cm^2. (See Fig. 5.2.2.) We want to know the value of dA/dt when $x = 10$ and $y = 8$, given that $dx/dt = 2$ and $dy/dt = -3$. (Note the negative sign to indicate that y is decreasing.) Since all the quantities in the equation $A = xy$ are functions of time, we can differentiate that equation implicitly with respect to time and obtain

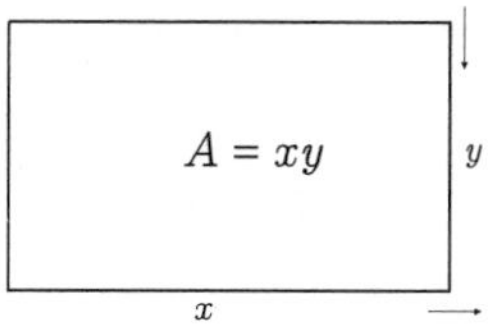

FIGURE 5.2.2

$$\frac{dA}{dt} = \frac{dx}{dt}\,y + x\,\frac{dy}{dt}.$$

$$\text{Thus} \quad \left.\frac{dA}{dt}\right|_{\substack{x=10,\\ y=8}} = 2(8) + 10(-3) = -14.$$

At the time in question, the area of the rectangle is decreasing at a rate of 14 cm^2/s.

Procedures for Related Rates Problems

In view of these examples we can formulate a few general procedures for dealing with related-rates problems similar to those listed for optimization problems in Section 5.1.

5.2.3 Solving Related Rates Problems

1. Read the problem very carefully. Try to understand the relationships among the variable quantities.
2. Define any symbols you want to use that are not defined in the statement of the problem, and express given and required quantities and rates in terms of these symbols.
3. Make a sketch if appropriate.
4. Discover from a careful reading of the problem or consideration of the sketch one or more equations linking the variable quantities. (You will need as many equations as quantities or rates to be found in the problem.)
5. Differentiate the equation or equations implicitly with respect to time, regarding all variable quantities as functions of time. You can manipulate the equations algebraically before the differentiation is performed (for instance, they could be solved for the quantities whose rates are to be found), but it is usually better to differentiate the equations as they are originally obtained and solve for the desired items later.
6. Substitute any given values for the quantities and their rates and solve the resulting equation or equations for the unknown quantities and rates.
7. Make a concluding statement answering the question asked. Is your answer "reasonable?" If not, check back through your solution to see what went wrong.

EXAMPLE 5.2.4 Air is being pumped into a spherical balloon. The volume of the balloon is increasing at a rate of 20 cm^3/s when the radius is 30 cm. How fast is the radius increasing at that time?

SOLUTION Let r and V denote the radius and volume of the balloon, respectively, at time t. A diagram is not of much help in this problem, but we do have to know the formula for the volume of a sphere:

$$V = \frac{4}{3}\pi r^3.$$

Differentiating this equation implicitly with respect to time t, we obtain

$$\frac{dV}{dt} = 4\pi r^2 \frac{dr}{dt}.$$

We are given that $dV/dt = 20$ when $r = 30$. At this time,

$$20 = 4\pi \times 900 \frac{dr}{dt}, \qquad \text{so} \qquad \frac{dr}{dt} = \frac{1}{180\pi}.$$

The radius is increasing at $1/180\pi \approx 0.00177$ cm/s when it is 30 cm.

Observe that while we wanted dr/dt ultimately, we did not solve the initial equation $V = \frac{4}{3}\pi r^3$ for r before differentiating. It is usually easier to differentiate the relationship between the variables in its original form and then solve for the desired rate.

EXAMPLE 5.2.5 A lighthouse L is located on a small island 2 km from the nearest point A on a long, straight shoreline. If the lighthouse lamp rotates at 3 revolutions per minute, how fast is the illuminated spot P on the shoreline moving along the shoreline when it is 4 km from A?

SOLUTION Referring to Fig. 5.2.3, let x be the distance AP and let θ be the angle $\angle PLA$. Then $x = 2\tan\theta$ and

$$\frac{dx}{dt} = 2\sec^2\theta\,\frac{d\theta}{dt}.$$

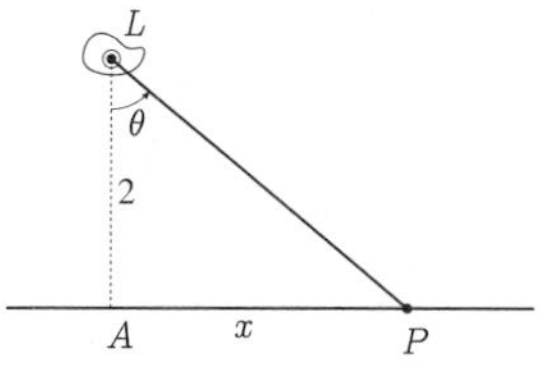

FIGURE 5.2.3

Now

$$\frac{d\theta}{dt} = 3 \text{ rev/min} \times 2\pi \text{ radians/rev} = 6\pi \text{ radians/min}.$$

When $x = 4$ we have $\tan\theta = 2$ and $\sec^2\theta = 1 + \tan^2\theta = 5$. Thus

$$\frac{dx}{dt} = 2 \times 5 \times 6\pi = 60\pi \approx 188.5.$$

The spot of light is moving along the shoreline at a rate of about 188.5 km/min when it is 4 km from A.

(Note that it was essential to convert the rate of change of θ from revolutions per minute to radians per minute. If θ were not measured in radians we could not assert that $(d/d\theta)\tan\theta = \sec^2\theta$.)

EXAMPLE 5.2.6 A water tank is in the shape of an inverted right circular cone with depth 5 m and top radius 2 m. Water leaks out of the tank at a rate proportional to the depth of the water in the tank. When the water in the tank is 4 m deep it is leaking out at a rate $\frac{1}{12}$ m^3/min; how fast is the water level in the tank dropping at that time?

SOLUTION Let r and h denote the surface radius and depth of water in the tank at time t (both measured in metres). Thus the volume V (in m^3) of water in the tank at time t is

$$V = \frac{1}{3}\pi r^2 h.$$

Using similar triangles in Fig. 5.2.4 we can find a relationship between r and h:

$$\frac{r}{h} = \frac{2}{5}, \qquad \text{so} \qquad r = \frac{2h}{5}.$$

Thus

$$V = \frac{1}{3}\pi\left(\frac{2h}{5}\right)^2 h = \frac{4\pi}{75}h^3.$$

Differentiating this equation with respect to t we obtain

$$\frac{dV}{dt} = \frac{4\pi}{25}h^2\frac{dh}{dt}.$$

We are told that the volume of water in the tank is decreasing at a rate proportional to h, that is, $dV/dt = kh$, and that the rate is $1/12$ m^3/min when $h = 4$ m. Thus $-(1/12) = 4k$ and $k = -(1/48)$. Hence

$$-\frac{1}{48}h = \frac{dV}{dt} = \frac{4\pi}{25}h^2\frac{dh}{dt}.$$

It follows that

$$\frac{dh}{dt} = -\frac{25}{192\pi}\frac{1}{h},$$

and

$$\left.\frac{dh}{dt}\right|_{h=4} = -\frac{25}{768\pi}.$$

When the water in the tank is 4 m deep its level is dropping at a rate of $25/768\pi$ m/min, or about 1.036 cm/min.

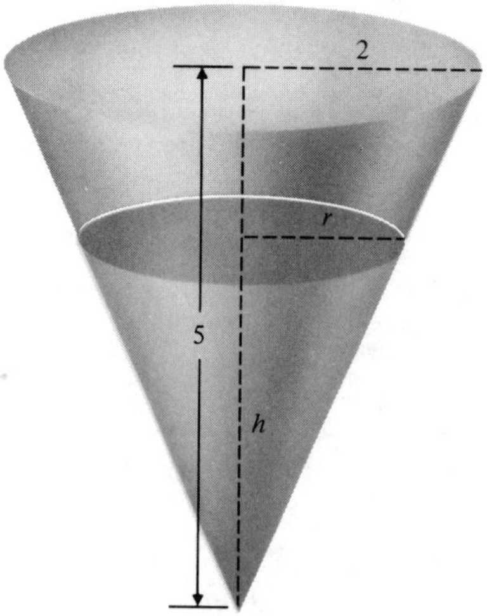

FIGURE 5.2.4

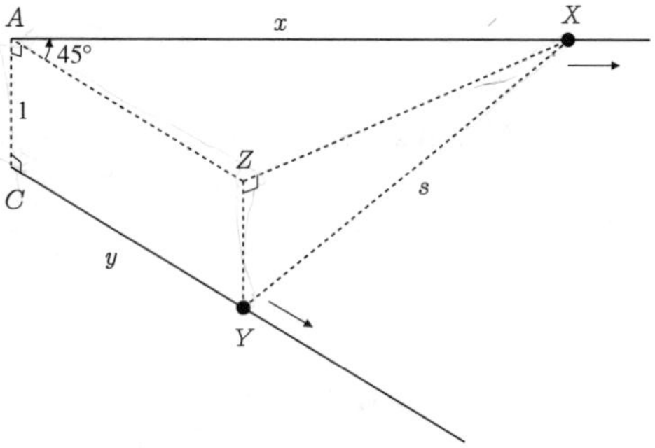

FIGURE 5.2.5

EXAMPLE 5.2.7 At a certain instant an aircraft flying due east at 400 km/h passes directly over a car traveling due southeast at 100 km/h on a straight, level road. If the aircraft is flying at an altitude of 1 km, how fast is the distance between the aircraft and the car increasing 36 s after the aircraft passes directly over the car?

SOLUTION A good diagram is essential here. See Fig. 5-2-5. Let time t be measured in hours from the time the aircraft was at position A directly above the car at position C. Let X and Y be the positions of the aircraft and the car respectively, at time t. Let x be the distance AX, y be the distance CY, and let s be the distance XY, all measured in kilometres. Let Z be the point 1 km above Y. Since angle $XAZ = 45°$, we have using the Pythagorean theorem and Cosine Law (Theorem 3.1.12),

$$\begin{aligned} s^2 &= 1 + (ZX)^2 = 1 + x^2 + y^2 - 2xy\cos 45° \\ &= 1 + x^2 + y^2 - \sqrt{2}xy. \end{aligned}$$

$$\text{Thus} \qquad \begin{aligned} 2s\frac{ds}{dt} &= 2x\frac{dx}{dt} + 2y\frac{dy}{dt} - \sqrt{2}\frac{dx}{dt}y - \sqrt{2}x\frac{dy}{dt} \\ &= 400(2x - \sqrt{2}y) + 100(2y - \sqrt{2}x), \end{aligned}$$

since $dx/dt = 400$ and $dy/dt = 100$. When $t = 1/100$ (that is, 36 s after $t = 0$) we have $x = 4$ and $y = 1$. Hence

$$\begin{aligned} s^2 &= 1 + 16 + 1 - 4\sqrt{2} = 18 - 4\sqrt{2} \\ s &\approx 3.5133. \end{aligned}$$

$$\text{Thus} \qquad \frac{ds}{dt} = \frac{1}{2s}\left(400(8 - \sqrt{2}) + 100(2 - 4\sqrt{2})\right) \approx 322.86.$$

The aircraft and the car are separating at a rate of approximately 323 km/h after 36 s.

(Note that it was necessary to convert 36 s to hours in the solution. In general all measurements should be in compatible units.)

EXERCISES

1. A pebble dropped into a pond causes a circular ripple to expand outward from the point of impact. How fast is the area enclosed by the ripple increasing when the radius is 20 cm and is increasing at a rate of 4 cm/s?

2. The area of a circle is decreasing at a rate of 2 cm^2/min. How fast is the radius of the circle changing when the area is 100 cm^2?

3. At a certain instant the length of a rectangle is 16 m and the width is 12 m. The width is increasing at 3 m/s. How fast is the length changing if the area of the rectangle is not changing?

4. The volume of a right circular cylinder is 60 cm^3 and is increasing at 2 cm^3/min at a time when the radius is 5 cm and is increasing at 1 cm/min. How fast is the height of the cylinder changing at that time?

5. A lump of modelling clay is being rolled out so that it maintains the shape of a circular cylinder. If the length is increasing at a rate proportional to itself, show that the radius is decreasing at a rate proportional to itself.

6. A radar gun measures the rate of change of distance between itself and the object at which it is aimed. If it is aimed at a car travelling at 90 km/h along a straight road, what will its reading be at an instant when it is aimed in a direction making an angle of 30° with the road?

7. How fast is the volume of a rectangular box changing when the length is 6 cm, the width is 5 cm, and the depth is 4 cm, if the length and depth are both increasing at a rate of 1 cm/s and the width is decreasing at a rate of 2 cm/s?

8. The area of a rectangle is increasing at a rate of 5 m^2/s while the length is increasing at a rate of 10 m/s. If the length is 20 m and the width is 16 m, how fast is the width changing?

9. A point moves on the curve $y = x^2$. How fast is y changing when $x = -2$ and x is decreasing at a rate 3?

10. A point is moving to the right along the first-quadrant portion of the curve $x^2y^3 = 72$. When the point has coordinates $(3, 2)$ its horizontal velocity is 2 units/s. What is its vertical velocity?

11. The point P moves so that at time t it is at the intersection of the curves $xy = t$ and $y = tx^2$. How fast is the distance of P from the origin changing at time $t = 2$?

12. A man 2 m tall walks toward a lamppost on level ground at a rate of 0.5 m/s. If the lamp is 5 m high on the post, how fast is the length of the man's shadow decreasing when he is 3 m from the post? How fast is the shadow of his head moving at that time?

13. The top of a ladder 5 m long rests against a vertical wall. If the base of the ladder is being pulled away from the base of the wall at a rate of 1/3 m/s, how fast is the top of the ladder slipping down the wall when it is 3 m above the base of the wall?

14. An aircraft is flying horizontally at a rate of 13 km/min at an altitude of 5 km. The aircraft passes directly over a radio beacon at 3:00 p.m. How fast is the distance between the aircraft and the beacon increasing 1 min later?

15. At 1:00 p.m. ship A is 25 km due north of ship B. If ship A is sailing west at a rate of 16 km/h and ship B is sailing south at 20 km/h, find the rate at which the distance between the two ships is changing at 1:30 p.m.

16. What is the first time after 3 o'clock that the hands of the clock are together?

17. A balloon released at point A rises vertically with a constant speed of 5 m/s. Point B is level with and 100 m distant from point A. How fast is the angle of elevation of the balloon at B changing when the balloon is 200 m above A?

18. Sawdust is falling onto a pile at a rate of 1/2 m^3/min. If the pile maintains the shape of a right circular cone with height equal to half the diameter of its base, how fast is the height of the pile increasing when it is 3 m?

19. A water tank is in the shape of an inverted right circular cone with top radius 10 m and depth 8 m. Water is flowing in at a rate of 1/10 m^3/min. How fast is the depth of water in the tank increasing when the water is 4 m deep?

20. Repeat Exercise 19 with the added assumption that water is leaking out of the bottom of the tank at a rate of $h^3/1000$ m^3/min when the depth of water in the tank is h m. How full can the tank get in this case?

21. Water is pouring into a leaky tank at a rate of 10 m^3/h. The tank is a cone with vertex down, 9 m in depth and 6 m in diameter at the top. The surface of water in the tank is rising at a rate of 20 cm/h when the depth is 6 m. How fast is the water leaking out at that time?

22. How fast must you let out line if the kite you are flying is 30 m high, 40 m horizontally away from you, and moving horizontally away from you at a rate of 10 m/min?

23. You are riding on a Ferris wheel of diameter 20 m. The wheel is rotating at 1 revolution per minute. How fast are you rising or falling when you are 6 m horizontally away from the vertical line passing through the centre of the wheel?

24. An aircraft is 144 km east of an airport and is traveling west at 200 km/h. Simultaneously a second aircraft at the same altitude is 60 km north of the airport and travelling north at 150 km/h. How fast is the distance between the aircraft changing?

25. A boat is being pulled toward a pier by a rope attached to its bow. A person on the pier is pulling in the rope at a rate of 6 m/min. If the person's hands are 5 m higher than the bow of the boat, how fast is the boat moving toward the pier when there are still 13 m of rope out?

26. A lamp is located at point $(3, 0)$ in the xy-plane. An ant is crawling in the first quadrant of the plane and the lamp casts its shadow onto the y-axis. How fast is the ant's shadow moving along the y-axis when the ant is at position $(1, 2)$ and moving so that its x-coordinate is increasing at rate 1/3 units/s and its y-coordinate is decreasing at 1/4 units/s?

27. A straight highway and a straight canal intersect at right angles, the highway crossing over the canal on a bridge 20 m above the water. A boat traveling at 20 km/h passes under the bridge just as a car travelling at 80 km/h passes over it. How fast are the boat and car separating after one minute?

28. The cross section of a water trough is an equilateral triangle with top edge horizontal. If the trough is 10 m long and 30 cm deep, and if water is flowing in at a rate of 1/4 m^3/min, how fast is the water level rising when the water is 20 cm deep at the deepest?

29. A rectangular swimming pool is 8 m wide and 20 m long. Its bottom is a sloping plane, the depth increasing from 1 m at the shallow end to 3 m at the deep end. (See Fig. 5.2.6.) Water is draining out of the pool at a rate of 1 m^3/min. How fast is the surface of the water falling when the depth of water at the deep end is (a) 2.5 m? (b) 1 m?

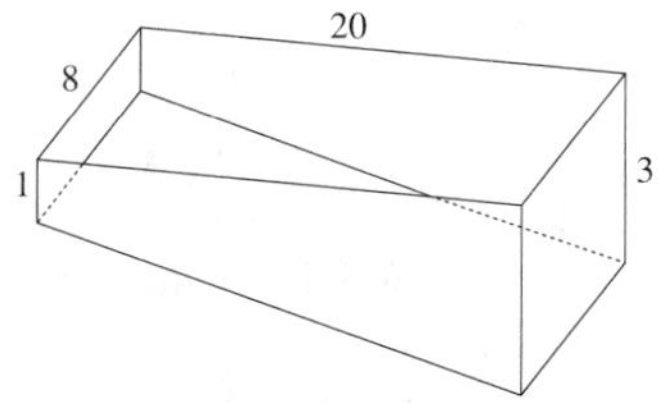

FIGURE 5.2.6

30.*One end of a 10 m long ladder is on the ground and the ladder is supported part way along its length by resting on top of a 3 m high fence. (See Fig. 5.2.7.) If the bottom of the ladder is 4 m from the base of the fence and is being dragged along the ground away from the fence at a rate of 1/5 m/s, how fast is the free top end of the ladder moving (a) vertically, and (b) horizontally?

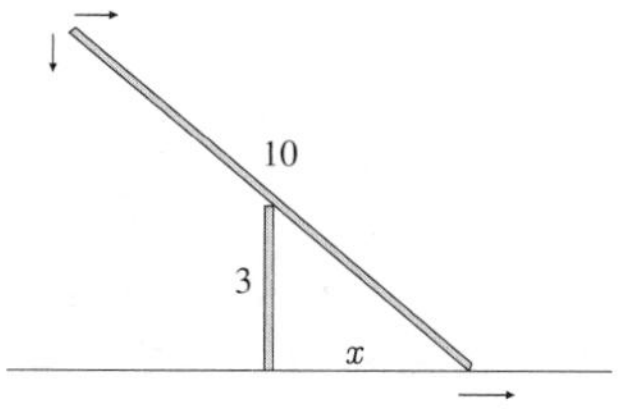

FIGURE 5.2.7

31.*Two crates, A and B are on the floor of a warehouse. The crates are joined by a rope 15 m long, each crate being hooked at floor level to an end of the rope. The rope is stretched tight and pulled over a pulley P that is attached to a rafter 4 m above a point Q on the floor directly between the two crates. (See Fig. 5.2.8.) If crate A is 3 m from Q and is being pulled directly away from Q at a rate of 1/2 m/s, how fast is crate B moving toward Q?

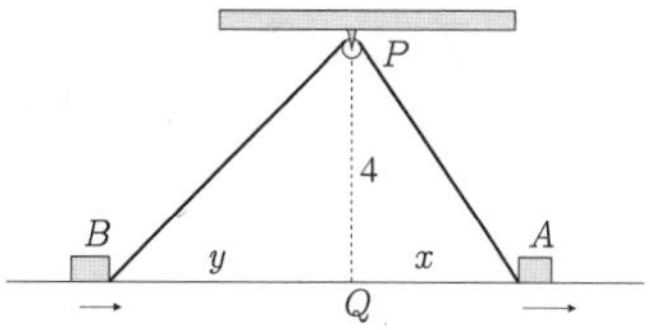

FIGURE 5.2.8

32. Shortly after launch, a rocket is 100 km high and 50 km downrange. If it is traveling at 4 km/s at an angle of 30° above the horizontal, how fast is its angle of elevation, as measured at the launch site, changing?

33. A lamp is 20 m high on a pole. At time $t = 0$ a ball is dropped from a point level with the lamp and 10 m away from it. The ball falls under gravity (acceleration 9.8 m/s^2) until it hits the ground. How fast is the shadow of the ball moving along the ground (a) one second after it is dropped? (b) just as the ball hits the ground?

34. A rocket blasts off at time $t = 0$ and climbs vertically with acceleration 10 m/s^2. The progress of the rocket is monitored by a tracking station located 2 km horizontally away from the launch pad. How fast is the tracking station antenna rotating upward 10 seconds after launch?

35. What is the maximum rate at which the antenna in the previous exercise must be able to turn in order to track the rocket during its entire vertical ascent?

36.*(*A Review of Calculus!*) You are in a tank (the military variety) moving down the y-axis toward the origin. At time $t = 0$ you are 4 km from the origin, and 10 min later you are 2 km from the origin. Your speed is decreasing; it is proportional to your distance from the origin. You know that an enemy tank is waiting somewhere on the positive x-axis, but there is a high wall along the curve $xy = 1$ (all distances in km) preventing you from seeing just where it is. How fast must your gun turret be capable of turning to maximize your chances of surviving the encounter?

5.3 THE TANGENT-LINE APPROXIMATION

Many problems in applied mathematics are too difficult to be solved exactly — all we can hope to do is find approximate solutions which are correct to within some acceptably small tolerance. In this section and the next we will examine how knowledge of the value of a function and its derivatives at one point can help us find approximate values for the function at other points.

The tangent to the graph $y = f(x)$ at $x = a$ (see Fig. 5.3.1) describes the behavior of that graph near the point $P = (a, f(a))$ better than does any other straight line through P, because it goes through P in the same direction as does the curve $y = f(x)$. We exploit this fact by using the height to the tangent line to calculate approximate values of $f(x)$ for values of x near a. The tangent line has equation $y = f(a) + f'(a)(x - a)$, so the approximation is

5.3.1
The Tangent-Line Approximation

$$f(x) \approx f(a) + f'(a)(x - a).$$

Observe that at points x "near" a, the vertical distance between the curve and the tangent line is small *compared to the horizontal distance between x and a.*

EXAMPLE 5.3.2 Find an approximate value for $\sqrt{26}$ using a suitable tangent line.

SOLUTION If $f(x) = \sqrt{x}$, then $f'(x) = 1/(2\sqrt{x})$. Take $a = 25$ (the nearest point to 26 that is a perfect square) and obtain $f(a) = 5$ and $f'(a) = 1/10$. Thus

$$\sqrt{x} = f(x) \approx 5 + \frac{1}{10}(x - 25)$$

for x near 25. Putting $x = 26$, we get $\sqrt{26} \approx 5.1$. We can say a little more. Since f is increasing, $\sqrt{26} > 5$, and since $f''(x) = -\frac{1}{4}x^{-3/2} < 0$ if $x > 0$, the graph of f is concave down and lies below its tangent lines. Hence $\sqrt{26} < 5.1$, so

$$5 < \sqrt{26} < 5.1.$$

(The actual value of $\sqrt{26}$ is $5.0990\cdots$, but if we knew that we would have had no need of an approximation!)

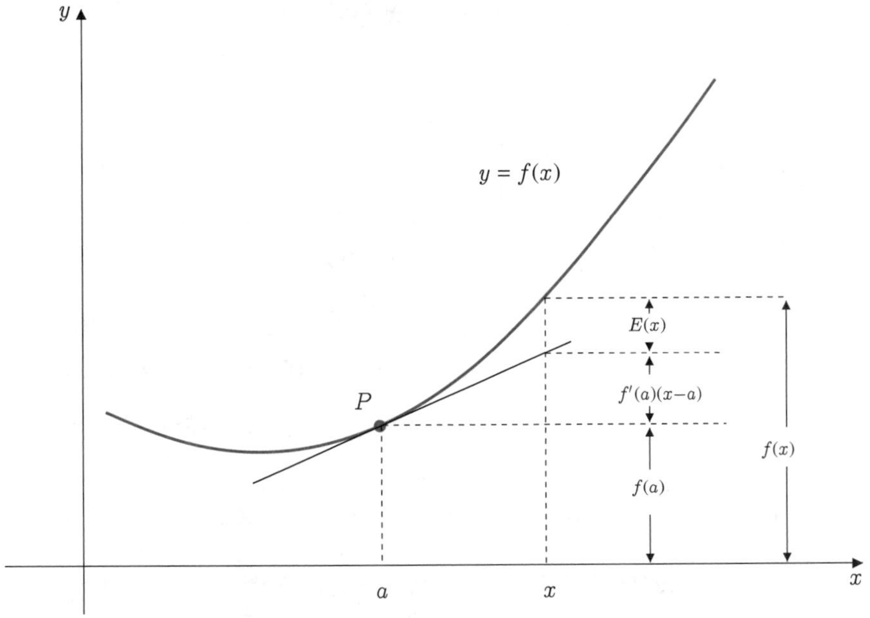

FIGURE 5.3.1

EXAMPLE 5.3.3 Find an approximate value for $\cos 62°$.

SOLUTION If $f(x) = \cos x$ and $a = 60° = \frac{\pi}{3}$, then $f(a) = \cos\frac{\pi}{3} = \frac{1}{2}$ and $f'(a) = -\sin\frac{\pi}{3} = -\frac{\sqrt{3}}{2}$. Hence

$$\cos x = f(x) \approx \frac{1}{2} - \frac{\sqrt{3}}{2}\left(x - \frac{\pi}{3}\right),$$

for x near $\frac{\pi}{3}$. Since $62° = \frac{\pi}{3} + 2\frac{\pi}{180} = \frac{\pi}{3} + \frac{\pi}{90}$, we have

$$\cos 62° = f\left(\frac{\pi}{3} + \frac{\pi}{90}\right) \approx \frac{1}{2} - \frac{\sqrt{3}}{2}\frac{\pi}{90} \approx 0.469770\cdots.$$

(We have used a calculator with square-root function to find the decimal value.) Again $f''(x) = -\cos x < 0$ for x near $\pi/3$, so the true value is less than the height to the tangent line: $\cos 62° < 0.46978$.

If we use the cosine funtion on a scientific calculator we can obtain the "true value" (actually, just another approximation, though presumably a rather better one): $\cos 62° = 0.46947\cdots$, but if we have such a calculator we don't need the approximation in the first place. Approximations are useful when there is no easy way to obtain the true value. However, if we don't know the true value, we would like to have some way of determining how good the approximation must be; that is, we want an *estimate for the error*. After all, *any number* is an approximation to $\cos 62°$; but the error may be unacceptably large. For instance, the size of the error in the approximation $\cos 62° \approx 1,000,000$ is greater than 999,999.

The Error Estimate

In any approximation, the **error** is defined by

$$\text{error} = \text{true value} - \text{approximate value}.$$

If the tangent line to $y = f(x)$ at $x = a$ is used to approximate $f(x)$ near a, that is,

$$f(x) \approx f(a) + f'(a)(x-a),$$

then the error $E(x)$ in this approximation is

$$E(x) = f(x) - f(a) - f'(a)(x-a).$$

It is the vertical distance between the graph of f and the tangent line, as shown in Fig. 5.3.1. The following theorem gives us a bound on the size of this error.

THEOREM 5.3.4 If $f''(t)$ exists for all t in an interval containing a and x, then there exists some point X between a and x such that

$$E(x) = \frac{f''(X)}{2}(x-a)^2.$$

In particular, if $f''(t)$ has constant sign on an interval containing a and x then $E(x)$ has the same sign, and if $|f''(t)| \le K$ on the interval (where K is some constant) then

$$|E(x)| \le \frac{K}{2}(x-a)^2.$$

We will prove this theorem a little later in this section. First, we illustrate its application to estimating errors in tangent-line approximations.

EXAMPLE 5.3.5 Obtain an estimate for the error in the approximation $\sqrt{26} \approx 5.1$ obtained in Example 5.3.2. Find the smallest interval you can that contains $\sqrt{26}$.

SOLUTION $f(x) = x^{1/2}$, $f'(x) = \frac{1}{2}x^{-1/2}$, $f''(x) = -\frac{1}{4}x^{-3/2}$. Clearly

$$|f''(t)| \le \frac{1}{4}25^{-3/2} = \frac{1}{500} \quad \text{if} \quad t \ge 25,$$

(In particular this inequality holds if $25 \le t \le 26$.) Hence

$$|E(26)| \le \frac{1}{2} \times \frac{1}{500} \times (26-25)^2 = 0.001.$$

Since concavity ($f''(x) < 0$) implies $\sqrt{26} \le 5.1$, we can assert that $5.1 - 0.001 \le \sqrt{26} \le 5.1$, that is,

$$5.099 \le \sqrt{26} \le 5.1.$$

EXAMPLE 5.3.6 Obtain an estimate for the error in the approximation to $\cos 62°$ obtained in Example 5.3.3. Give the smallest interval you can which contains $\cos 62°$.

SOLUTION $f(x) = \cos x$, $f'(x) = -\sin x$, $f''(x) = -\cos x$. We have $|f''(t)| = |\cos t| \leq 1/2$ if $\pi/3 \leq t \leq \pi/2$ (and in particular if $60° \leq t \leq 62°$). Hence,

$$|E(62°)| \leq \frac{1}{4}\left(\frac{\pi}{90}\right)^2 < 0.00031.$$

Again concavity ($f''(x) < 0$) indicates that $\cos 62°$ is less than the approximate value calculated $(0.46977\cdots)$, so

$$0.46977 - 0.00031 \leq \cos 62° \leq 0.46978,$$

that is,

$$0.46946 \leq \cos 62° \leq 0.46978.$$

EXAMPLE 5.3.7 Find an approximate value for $(61)^{1/3}$ and a suitable bound for the error.

SOLUTION Let $f(x) = x^{1/3}$; then $f'(x) = \frac{1}{3}x^{-2/3}$ and $f''(x) = -\frac{2}{9}x^{-5/3}$. We take $a = 64$ (the closest perfect cube) and so obtain

$$x^{1/3} = f(x) \approx f(64) + f'(64)(x - 64) = 4 + \frac{1}{48}(x - 64).$$

Thus $61^{1/3} \approx 4 - \dfrac{3}{48} = 3.9375$. Now

$$|f''(t)| = \frac{2}{9t^{5/3}} \leq \frac{2}{9 \times 27^{5/3}} = \frac{2}{9 \times 243} < 0.0009145$$

if $t \geq 27$, and in particular if $61 \leq t \leq 64$. (Note that we could not use the value $t = 64$ to get an estimate this time because the values of $|f''(t)|$ for t to the left of 64 are greater than the value at 64. We went all the way down to 27 so as to hit another perfect cube. Clearly a better estimate could be found. See Exercise 18 at the end of this section.) We have

$$|E(61)| \leq \frac{0.0009145}{2}(61 - 64)^2 < 0.0042.$$

Since $f''(x)$ is negative, f is concave down and the tangent line lies above the curve. Hence

$$3.9375 - 0.0042 < 61^{1/3} < 3.9375$$
$$3.9333 < 61^{1/3} < 3.9375.$$

(The actual value of $(61)^{1/3}$ is $3.9364\cdots$.)

REMARK: The error in the tangent line approximation can be interpreted in terms of differentials (review the last three paragraphs of Section 2.2) as follows. If $x - a = \Delta x = dx$, then the change in height to the graph of $y = f(x)$ as we pass from $x = a$ to $x = a + \Delta x$ is $f(a+\Delta x) - f(a) = \Delta y$, and the corresponding change in height to the tangent line is $f'(a)(x-a) = f'(a)\,dx$, which is just the value at $x = a$ of the differential $dy = f'(x)\,dx$. (See Fig. 5.3.1.) Thus

$$E(x) = \Delta y - dy.$$

The fact that the approximating line is tangent to the graph of f at $x = a$ implies that the error is small compared with Δx as Δx approaches 0. In fact,

$$\lim_{\Delta x \to 0} \frac{\Delta y - dy}{\Delta x} = \lim_{\Delta x \to 0} \left(\frac{\Delta y}{\Delta x} - \frac{dy}{dx} \right) = \frac{dy}{dx} - \frac{dy}{dx} = 0.$$

If $|f''(t)| \le K$ near $t = a$, we can assert a stronger result than this, namely,

$$\left| \frac{\Delta y - dy}{(\Delta x)^2} \right| = \left| \frac{E(x)}{(\Delta x)^2} \right| \le \frac{K}{2} = \text{constant},$$

so $|\Delta y - dy| \le \dfrac{K}{2}(\Delta x)^2$.

Errors in Measurement

Suppose that a quantity x is obtained by measurement and that a second quantity y is determined as a function of x; $y = f(x)$. Any error involved in the measurement of x will result in an error in the calculated value of y as well. It is convenient to use the differential $dx = \Delta x$ to denote the error involved in the measurement of x. The corresponding error in y is $\Delta y = f(x + \Delta x) - f(x)$. The tangent-line approximation gives us a handy first-order approximation for Δy:

$$\Delta y \approx dy = f'(x)\,dx.$$

The errors $dx = \Delta x$ and $dy \approx \Delta y$ are absolute errors. Errors frequently represented in relative terms, that is, as a fraction or percentage of the size of the quantity being expressed. The relative error in a measurement of x which has absolute error dx is $\dfrac{dx}{x}$.

EXAMPLE 5.3.8 The length x of the side of a square is measured with less than 1 percent error. By approximately what percentage will the calculated area $A = x^2$ of the square be in error?

SOLUTION We have $A = x^2$, so $\Delta A \approx dA = 2x\,dx$. Thus

$$\frac{\Delta A}{A} \approx \frac{2x\,dx}{x^2} = 2\frac{dx}{x}.$$

Since we are given that dx is no bigger than about 1 percent of x we have

$$\left|\frac{\Delta A}{A}\right| \approx \left|\frac{2x\,dx}{x^2}\right| = 2\left|\frac{dx}{x}\right| < \frac{2}{100}.$$

Thus the area is in error by less than approximately 2 percent.

It must be stressed that $dA = 2x\,dx$ does not give the exact error in A. The exact error is $\Delta A = (x + dx)^2 - x^2 = 2x\,dx + (dx)^2$. Evidently if dx is small, then the term $(dx)^2$ is much smaller than the term $2x\,dx$, so the approximation of ΔA by dA is a good one. In Fig. 5.3.2, dA is the sum of the areas of two small rectangles of length x and width dx, while ΔA is this sum plus the area of a small square of side dx.

EXAMPLE 5.3.9 The fraction of a radioactive sample remaining undecayed after one year is measured to be 0.998. If this measurement involves a possible error of up to 0.0001, find the approximate half-life of the sample and give an approximate maximum size for the error in this half-life.

SOLUTION If $y(t)$ is the fraction of the original sample remaining undecayed after t years, then $y(t) = e^{kt}$. (See Section 3.7.) The constant k is determined from the fraction p remaining undecayed after one year: $p = e^k$, so $k = \ln p$.

For $p = 0.998$ and $dp = 0.0001$, we have $k = \ln 0.998$ and $dk = dp/p = 0.0001/0.998$. The half-life T of the sample is determined by $\frac{1}{2} = e^{kT}$. Thus

$$T = \frac{1}{k}\ln\frac{1}{2} = -\frac{\ln 2}{k} \approx 346.2 \text{ years.}$$

Now

$$\Delta T \approx dT = \frac{dT}{dk}\,dk = \frac{\ln 2}{k^2}\,dk = -\frac{T}{k}\frac{dp}{p} \approx 17.3.$$

Thus the half-life of the sample is approximately 364.2 years with error no larger than about 17.3 years.

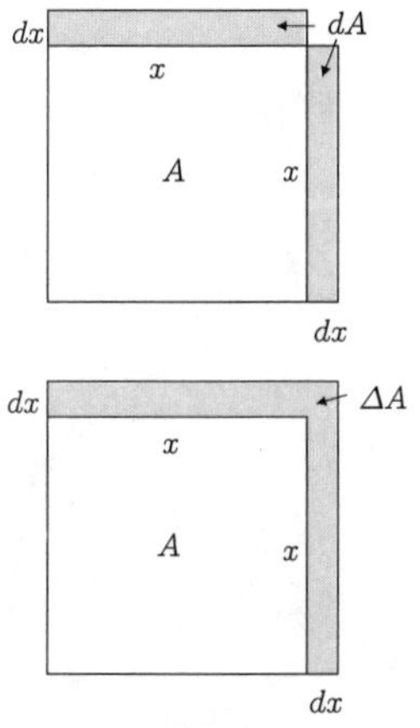

FIGURE 5.3.2

The Generalized Mean-Value Theorem

The following generalization of the Mean-Value Theorem (Theorem 4.1.1) will be used several times in the rest of this chapter.

THEOREM 5.3.10 (**The Generalized Mean-Value Theorem**) Suppose the functions f and g are both continuous on the closed interval $[a, b]$ and differentiable on the open interval (a, b). Suppose also that $g'(x)$ is never 0 for any x in (a, b). Then there exists a number c in (a, b) such that

$$\frac{f(b) - f(a)}{g(b) - g(a)} = \frac{f'(c)}{g'(c)}.$$

PROOF First note that g must be one-to-one on $[a,b]$, for if there were two points where it had the same value then the Mean-Value Theorem (Theorem 4.1.1) would imply that $g'(x) = 0$ at some point between those two points. Therefore neither denominator in the equation displayed above can be zero. Let

$$h(x) = \big(f(b) - f(a)\big)\big(g(x) - g(a)\big) - \big(g(b) - g(a)\big)\big(f(x) - f(a)\big).$$

The function h is of the form $Af(x) + Bg(x) + C$ for certain constants A, B, and C, and so is continuous on $[a,b]$ and differentiable on (a,b) because f and g are. Moreover, $h(a) = h(b) = 0$. By the Mean-Value Theorem there exists a number c in (a,b) such that $h'(c) = 0$. Thus

$$\big(f(b) - f(a)\big)\, g'(c) - (g(b) - g(a))\, f'(c) = 0,$$

and the required result follows on division by the g factors. □

Proof of the Error Estimate (Theorem 5.3.4)

Suppose $x > a$; a similar proof can be given for the case $x < a$. For $a \le t \le x$ we have

$$E(t) = f(t) - f(a) - f'(a)(t-a). \qquad \text{Thus} \quad E(a) = 0.$$

We apply the Generalized Mean-Value Theorem to the functions $E(t)$ and $(t-a)^2$ on $[a,x]$ and obtain a number c for which

$$\frac{E(x)}{(x-a)^2} = \frac{E(x) - E(a)}{(x-a)^2 - (a-a)^2} = \frac{E'(c)}{2(c-a)} = \frac{f'(c) - f'(a)}{2(c-a)}.$$

Now apply the Generalized Mean-Value Theorem again, this time to $f'(t)$ and $2t$ on the interval $[a,c]$ and obtain a number X in (a,c) (and therefore in (a,x)) such that

$$\frac{f'(c) - f'(a)}{2(c-a)} = \frac{f''(X)}{2}.$$

Thus $E(x) = \dfrac{f''(X)}{2}(x-a)^2$, as required. □

EXERCISES

In Exercises 1–12 find a suitable tangent-line approximation for the indicated value. Determine the sign of the error and estimate its size. What is the smallest interval you can be sure contains the value?

1. $\sqrt{50}$

2. $\sqrt{47}$

3. $\sqrt[4]{85}$

4. $\dfrac{1}{2.003}$

5. $e^{-1/10}$

6. $\sin 33°$

7. $\cos 46°$

8. $\sin \dfrac{\pi}{5}$

9. $\tan^{-1}(1.05)$

10. $\sin^{-1}(0.48)$

11. $\ln(0.94)$

12. $\cosh(0.02)$

13. If $f(2) = 4$, $f'(2) = -1$ and $0 \le f''(x) \le 1/x$ for all $x > 0$, find the smallest interval you can that contains $f(3)$.

14. Approximately what percentage error can result in the calculation of the volume of a cube if the edge is measured to within 2 percent tolerance?

15. By approximately how much can the area of a circle be in error if the radius is 10 cm, with possible error of 0.005 cm? About how large is the relative error in this case?

16. If the radius and height of a circular cylinder are both measured with relative error of less than 1 percent in absolute value, by about what percentage can the calculated volume of the cylinder be in error?

17. A spherical ball of ice melts so that the radius decreases from 20.00 cm to 19.80 cm in 1 hour. Approximately what volume of ice has melted in that hour?

18. Use the result $3.9333 < 61^{1/3} < 3.9375$ obtained in Example 5.3.7 to find a number K such that $|f''(t)| \leq K$ for $61 \leq t \leq 64$, smaller than the number $K = (2/9)27^{-5/3}$ used in that Example. Hence find a smaller interval which must contain $61^{1/3}$.

19. What is wrong with the following "proof" of the Generalized Mean-Value Theorem 5.3.10? "By the Mean-Value Theorem 4.1.1, $f(b)-f(a) = (b-a)f'(c)$ for some c between a and b, and similarly $g(b)-g(a) = (b-a)g'(c)$ for some such c. Hence $(f(b) - f(a))/(g(b) - g(a)) = f'(c)/g'(c)$, as required."

5.4 HIGHER ORDER APPROXIMATIONS -- TAYLOR'S FORMULA

Among all straight lines passing through the point $(a, f(a))$, the tangent line $y = f(a) + f'(a)(x - a)$ best approximates the behavior of the curve $y = f(x)$ near $x = a$. If we want a better approximation than is provided by the tangent line we try to use other "simple" curves to approximate the graph of f. A straight line is the graph of a *linear* function, that is, a function of the form $y = A + Bx$, where A and B are constants. The next simplest kind of function is a quadratic function $y = A + Bx + Cx^2$ whose graph is a parabola with vertical axis.

Suppose we want to find the quadratic function g which best approximates a given function f near a point a in the domain of f. It is convenient to express $g(x)$ as a quadratic function of $x - a$ rather than x; this can always be done by a change of variable $t = x - a$. Therefore we take $g(x)$ of the form

$$g(x) = A + B(x - a) + C(x - a)^2.$$

How should the constants A, B, and C be chosen to make g best approximate f near the point a? Clearly we want the graph of g to pass through $(a, f(a))$, so

$$f(a) = g(a) = A.$$

We also want the two graphs to have the same slope at a. Since $g'(x) = B+2C(x-a)$, this implies

$$f'(a) = g'(a) = B.$$

Finally, it seems reasonable to require that the slopes of both curves change at the same rate at a. Thus f and g must have the same second derivative at a. Since $g''(x) = 2C$, this implies

$$f''(a) = g''(a) = 2C.$$

Assuming that $f''(a)$ exists, the three conditions above have determined the three constants and we have

$$g(x) = f(a) + f'(a)(x - a) + \frac{f''(a)}{2}(x - a)^2.$$

We can use $g(x)$ as an approximation to $f(x)$ in the same way we used the tangent line in Section 5.3.

5.4.1
The Quadratic Approximation

The **quadratic** or **second order** approximation to $f(x)$ near a is

$$f(x) \approx f(a) + f'(a)(x-a) + \frac{f''(a)}{2}(x-a)^2.$$

EXAMPLE 5.4.2 Find the quadratic approximation to $\sqrt{26}$ based on values of $f(x) = \sqrt{x}$ and its derivatives at 25.

SOLUTION Since $f'(x) = \frac{1}{2}x^{-1/2}$ and $f''(x) = -\frac{1}{4}x^{-3/2}$, the approximation is

$$\begin{aligned}\sqrt{26} = f(26) &\approx f(25) + f'(25)(26-25) + \frac{f''(25)}{2}(26-25)^2\\ &= 5 + \frac{1}{10} - \frac{1}{2\times 4\times 125} = 5.09900.\end{aligned}$$

This may be compared with the true value of $5.09902\cdots$.

We can obtain an error estimate for the quadratic approximation by the same method used to prove the error estimate for the tangent-line approximation in the previous section. Let us rewrite the assertion of Theorem 5.3.4 in the following form:

5.4.3
Case $n = 1$ of Taylor's Formula

If $f''(t)$ exists for all t in an interval containing a and x, then

$$f(x) = f(a)+f'(a)(x-a)+R_1(f;a,x) \quad \text{where} \quad R_1(f;a,x) = \frac{f''(X)}{2}(x-a)^2,$$

for some X between a and x.

We are now using a more complicated expression $R_1(f;a,x)$ for the error we denoted $E(x)$ in Section 5.3. This is because we need to be able to express the error for different functions and different intervals in what follows. The formula on the left in the box above is the special case $n = 1$ of **Taylor's Formula**; we will present the general case below. The term $R_1(f;a,x)$ is called the **remainder term** or the **error term**. The real significance of Theorem 5.3.4 is that it gave the formula on the right in the box above for that error.

Now let us do the same thing for the quadratic approximation. Assume $f'''(t)$ exists on an interval containing a and x. For the quadratic approximation the error function is

$$R_2(f;a,x) = f(x) - f(a) - f'(a)(x-a) - \frac{f''(a)}{2}(x-a)^2.$$

Applying the Generalized Mean-Value Theorem to $R_2(f; a, t)$ and $(t-a)^3$ on the interval $a \le t \le x$ (assuming $x > a$) we obtain c in (a, x) such that

$$\frac{R_2(f; a, x) - R_2(f; a, a)}{(x-a)^3} = \frac{R_2'(f; a, c)}{3(c-a)^2} = \frac{1}{3}\,\frac{f'(c) - f'(a) - f''(a)(c-a)}{(c-a)^2}.$$

The numerator in the last fraction on the right is just the error function $R_1(f'; a, c)$ for the function f' instead of f, and is therefore equal to $\frac{1}{2} f'''(X)(c-a)^2$ for some X between a and c. Since $R_2(f; a, a) = 0$, we have proved

5.4.4
Case $n = 2$ of Taylor's Formula

> If $f'''(t)$ exists for all t in an interval containing a and x then
>
> $$f(x) = f(a) + f'(a)(x-a) + \frac{f''(a)}{2}(x-a)^2 + R_2(f; a, x),$$
> $$\text{where} \quad R_2(f; a, x) = \frac{f'''(X)}{6}(x-a)^3 = \frac{f'''(X)}{3!}(x-a)^3,$$
>
> for some X between a and x.

In particular, if $|f'''(t)| \le K$ for t between a and x, then

$$|R_2(f; a, x)| \le \frac{K}{3!}|x-a|^3.$$

Taylor's Formula

We can continue trying to find better and better approximations to $f(x)$ near a by using polynomials of higher and higher degree. For example, the cubic (degree 3) polynomial which best matches the behavior of a function f near a is

$$h(x) = f(a) + f'(a)(x-a) + \frac{f''(a)}{2}(x-a)^2 + \frac{f'''(a)}{3!}(x-a)^3,$$

and the error in the approximation $f(x) \approx h(x)$ is

$$R_3(f; a, x) = f(x) - h(x) = \frac{f^{(4)}(X)}{4!}(x-a)^4.$$

where X is some number between a and x.

The pattern in these approximations should be obvious by now. We state the general result in the following theorem, which is one version of **Taylor's Theorem**. The formula for the error term $R_n(f; a, x)$ is known as the **Lagrange Remainder**. We will encounter another version of Taylor's Theorem with a different form of remainder term Chapter 11.

THEOREM 5.4.5 (**Taylor's Theorem**) If the $(n+1)$-st order derivative, $f^{(n+1)}(t)$, exists for all t in an interval containing a and x, then

$$f(x) = f(a) + f'(a)(x-a) + \frac{f''(a)}{2!}(x-a)^2 + \cdots + \frac{f^{(n)}(a)}{n!}(x-a)^n + R_n(f;a,x),$$

Taylor's Formula,

where the error term $R_n(f;a,x)$ is given by

$$R_n(f;a,x) = \frac{f^{(n+1)}(X)}{(n+1)!}(x-a)^{n+1}, \qquad \textbf{The Lagrange Remainder},$$

where X is some number between a and x. □

PROOF We are going to use *mathematical induction.* (See Example 2.7.7 or Appendix I for a discussion of induction.) We have already proved the cases $n = 1$ and $n = 2$ of Taylor's theorem. (In fact we have proved the case $n = 0$ also; it is just the Mean-Value Theorem.) Suppose, therefore, that we have proved the case $n = k-1$ where $k \geq 2$ is an integer. Thus we are assuming that if f is any function whose kth derivative exists on an interval containing a and x, then

$$R_{k-1}(f;a,x) = \frac{f^{(k)}(X)}{k!}(x-a)^k,$$

where X is some number between a and x. Let us consider the next higher case: $n = k$. We have, by the Generalized Mean-Value Theorem (Theorem 5.3.10),

$$\frac{R_k(f;a,x)}{(x-a)^{k+1}} = \frac{R_k(f;a,x) - R_k(f;a,a)}{(x-a)^{k+1} - (a-a)^{k+1}} = \frac{R_k'(f;a,c)}{(k+1)(c-a)^k},$$

for some c between a and x. Now

$$\begin{aligned} &R_k'(f;a,c) \\ &= \frac{d}{dt}\left(f(t) - f(a) - f'(a)(t-a) - \frac{f''(a)}{2!}(t-a)^2 - \cdots - \frac{f^{(k)}(a)}{k!}(t-a)^k \right)\Bigg|_{t=c} \\ &= f'(c) - f'(a) - f''(a)(c-a) - \cdots - \frac{f^{(k)}(a)}{(k-1)!}(c-a)^{k-1} \\ &= R_{k-1}(f';a,c) = \frac{f^{(k+1)}(X)}{k!}(c-a)^k, \end{aligned}$$

for some X between a and c. Therefore

$$R_k(f;a,x) = \frac{f^{(k+1)}(X)}{(k+1)!}(x-a)^{k+1}.$$

We have shown that the case $n = k$ of Taylor's theorem is true if the case $n = k-1$ is true, and the inductive proof is complete. □

We will delve more deeply into the implications and uses of Taylor's Formula in Chapter 11. For the moment we will provide one simple example.

EXAMPLE 5.4.6 Write Taylor's Formula for $f(x) = e^x$ with $a = 0$. Assuming that $e < 3$, how many terms do you need to be sure you can calculate e correct to 3 decimal places?

SOLUTION We have $f'(x) = e^x$, $f''(x) = e^x$, ..., $f^{(n)}(x) = e^x$ for every integer $n \geq 0$. Hence $f^{(n)}(0) = 1$. Taylor's Formula gives

$$e^x = 1 + x + \frac{x^2}{2!} + \frac{x^3}{3!} + \cdots + \frac{x^n}{n!} + R_n(f; 0, x),$$

where

$$R_n(f; 0, x) = \frac{e^X}{(n+1)!}x^{n+1}, \quad \text{for some } X \text{ between 0 and } x.$$

If $x = 1$ then $0 < X < 1$ so $e^X < e < 3$. In this case

$$0 < R_n(f; 0, 1) < \frac{3}{(n+1)!}.$$

To get an approximation for $e = e^1$ correct to three decimal places, we need to have $R_n(f; 0, 1) < 0.0005$. Since $3/(8!) = 3/40320 \approx 0.000074$, but $3/(7!) = 3/5040 \approx 0.00059$, we can be sure $n = 7$ will do, but not $n = 6$:

$$e \approx 1 + 1 + \frac{1}{2!} + \frac{1}{3!} + \frac{1}{4!} + \frac{1}{5!} + \frac{1}{6!} + \frac{1}{7!} \approx 2.7182 \approx 2.718$$

to three decimal places.

EXERCISES

Write quadratic approximations for the given function near the point specified, and use it to approximate the indicated value. Estimate the error and write the smallest interval you can be sure contains the value.

1. $f(x) = x^{1/3}$ near 8; approximate $9^{1/3}$.
2. $f(x) = \sqrt{x}$ near 64; approximate $\sqrt{61}$.
3. $f(x) = \dfrac{1}{x}$ near 1; approximate $\dfrac{1}{1.02}$.
4. $f(x) = \tan^{-1} x$ near 1; approximate $\tan^{-1}(0.97)$.
5. $f(x) = e^x$ near 0; approximate $e^{-0.5}$.
6. $f(x) = \sin x$ near $\pi/4$; approximate $\sin(47°)$.

In Exercises 7–12 write the indicated case of Taylor's Formula for the given function. What is the Lagrange remainder in each case?

7. $f(x) = \sin x, \ a = 0, \ n = 7$
8. $f(x) = \cos x, \ a = 0, \ n = 6$
9. $f(x) = \sin x, \ a = \pi/4, \ n = 4$
10. $f(x) = \dfrac{1}{1-x}, \ a = 0, \ n = 6$
11. $f(x) = \ln x, \ a = 1, n = 6$
12. $f(x) = \tan x, \ a = 0, \ n = 3$
13. Write Taylor's Formula for $f(x) = e^{-x}$ with $a = 0$ and use it to calculate $1/e$ to 5 decimal places. (You may use a calculator, but not the e^x function on it.)
14. *Write the general form of Taylor's Formula for $f(x) = \sin x$ with $a = 0$. How large need n be taken to ensure that the corresponding approximation (Taylor's Formula without the remainder term) will give the sine of 1 radian correct to 5 decimal places?
15. What is the best quadratic approximation to the function $f(x) = (x-1)^2$ near $x = 0$? What is the error in this approximation? Now answer the same questions for $g(x) = x^3 + 2x^2 + 3x + 4$. Can the constant $1/6 = 1/3!$, in the error estimate for the quadratic approximation, be improved (i.e., made smaller)?

5.5 FINDING ROOTS

In this section we are concerned with finding approximate numerical solutions of equations. Specifically, we will be concerned with equations of the form

$$f(x) = 0,$$

where f is a given function. A solution of this equation is called a **root** of the equation or a **zero** of the function f. We will examine three methods for finding roots:

a) the Bisection Method, which requires only that f be continuous,

b) Newton's Method, which requires f be differentiable, and is usually the most efficient of the three methods, and

c) the Fixed Point Iteration Method, which is concerned with equations of a more special form: $f(x) = x$.

All three methods require that we have rough idea beforehand of approximately where a root can be found, and they generate sequences of approximations which get closer and closer to the root.

The Bisection Method

If the function f is continuous on an interval $[a, b]$ and if $f(a)$ and $f(b)$ have opposite signs, then the Intermediate-Value Theorem (Theorem 1.7.10) assures us that f must have at least one zero r in the interval (a, b). Let $c = (a + b)/2$, the mid-point of $[a, b]$. Calculate $f(c)$. If $f(c) = 0$ we have found a root; if $f(c) \neq 0$ then either $f(c)$ has the same sign as $f(a)$ in which case there is a root in $[c, b]$ or $f(c)$ has the same sign as $f(b)$ in which case there is a root in $[a, c]$. In either case we have obtained a new interval half as long as the original interval which must contain a zero of f. We can repeat this bisection process over and over, each time obtaining either an exact zero at the midpoint, or an interval containing one which is half as long as the interval at the previous step. When we have reduced the length of the interval to less than twice the acceptable error in the root, we can stop and use the midpoint of the last interval as an approximation to the root r.

The bisection method can be carried out with a calculator, but is best performed by a computer. Here is a description of the algorithm in a form which can be adapted for inclusion in a computer program. The algorithm assumes $f(a)$ and $f(b)$ have opposite sign, and it repeatedly assigns values to x and y such that $f(x) < 0$ and $f(y) > 0$ in such a way that the distance d between x and y halves at each repetition. The process stops when $f(m) = 0$, where $m = (x + y)/2$, or when d is less than some prescribed tolerance. If the tolerance is 10^{-n}, the algorithm will produce a value r for a root of f in $[a, b]$ correct to n decimal places.

5.5.1 Algorithm for the Bisection Method

```
LET tolerance=0.01
IF f(a) < 0 THEN LET x = a, y = b ELSE LET x = b, y = a
REPEAT
    LET m = (x + y)/2
    LET d = |x − y|
    LET f = f(m)
    IF f < 0 THEN LET x = m
    IF f > 0 THEN LET y = m
UNTIL f = 0 OR d <tolerance
LET r = (x + y)/2
STOP
```

One obvious drawback of the Bisection Method is that it takes many steps to get much improvement in accuracy. Specifically, since $\log_2 10 \approx 3.32$, it takes, on average, about 3.3 bisections to gain one extra decimal place of accuracy.

EXAMPLE 5.5.2 Show that the equation $x^3 + x - 1 = 0$ has exactly one real root, and find it correct to 2 decimal places using the Bisection Method.

SOLUTION The function $f(x) = x^3 + x - 1$ is continuous everywhere. Since $f(0) = -1 < 0$ and $f(1) = 1 > 0$, f must have a zero r in the interval $(0, 1)$. Since $f'(x) = 3x^2 + 1 > 0$ for all x, f is an increasing function on the whole real line and its graph can cross the x-axis only once. We implement the algorithm box above with tolerance set at 0.01. The results are recorded in the following table; each line gives the values of the variables at the end of an iteration of the REPEAT...UNTIL loop:

x	y	$m = \dfrac{x+y}{2}$	$d = \|x - y\|$	$f = f(m)$
0.0	1.0	0.5	1.0	−0.375
0.5	1.0	0.75	0.5	0.172
0.5	0.75	0.625	0.25	−0.131
0.625	0.75	0.6875	0.125	0.0125
0.625	0.6875	0.65625	0.0625	−0.0611
0.65625	0.6875	0.671875	0.03125	−0.0248
0.671875	0.6875	0.679688	0.015625	−0.0063
0.679688	0.6875	0.683594	0.007813	0.0030

In the last line $d < 0.01$, so the process stops. The required root is

$$r \approx 0.683594 \approx 0.68$$

rounded to two decimal places.

Newton's Method

Again we want to find a root r of the equation

$$f(x) = 0.$$

If f is differentiable near the root, then tangent lines can be used to produce a sequence of approximations to the root that approaches the root quite quickly. The idea is as follows. (See Fig. 5.5.1.) Make an initial guess at the root, say $x = x_0$. Let x_1 be the x-intercept of the tangent line to $y = f(x)$ at $(x_0, f(x_0))$. Under certain circumstances we can expect x_1 to be closer to the root than x_0 was. The process can be repeated over and over to get closer and closer; x_{n+1} is the x-intercept of the tangent line at $(x_n, f(x_n))$.

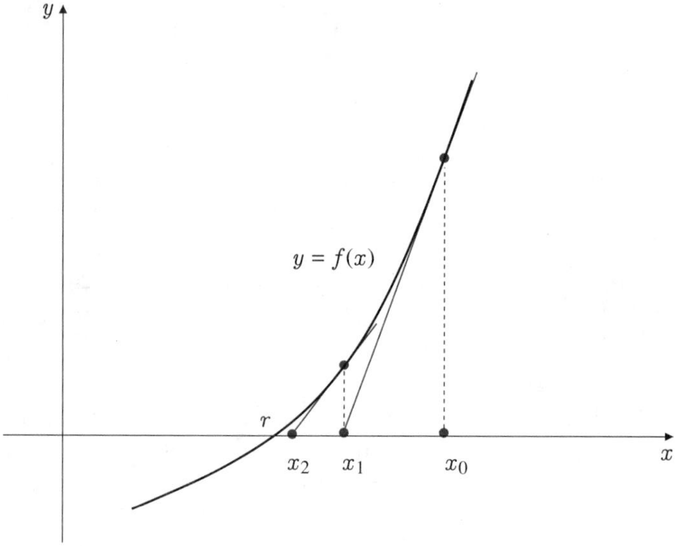

FIGURE 5.5.1

The tangent line at $x = x_0$ has equation

$$y = f(x_0) + f'(x_0)(x - x_0).$$

Since the point $(x_1, 0)$ lies on this line, we have $0 = f(x_0) + f'(x_0)(x_1 - x_0)$. Hence

$$x_1 = x_0 - \frac{f(x_0)}{f'(x_0)},$$

and, in general

5.5.3
Newtons' Method Formula

$$x_{n+1} = x_n - \frac{f(x_n)}{f'(x_n)}.$$

One generally uses a calculator or computer to calculate the successive approximations $x_1, x_2, x_3, \ldots$, and observes whether they appear to converge to a limit. If $\lim_{n\to\infty} x_n = r$ exists, and if f/f' is continuous near r, then r must be a root of f because

$$\lim_{n\to\infty} x_{n+1} = \lim_{n\to\infty} x_n - \lim_{n\to\infty} \frac{f(x_n)}{f'(x_n)}$$

$$r = r - \frac{f(r)}{f'(r)},$$

from which it follows that $f(r) = 0$. This method is known as **Newton's Method**.

EXAMPLE 5.5.4 Use Newton's Method to find the real root of the equation $x^3 + x - 1 = 0$ correct to six decimal places.

SOLUTION We have $f(x) = x^3 + x - 1$ and $f'(x) = 3x^2 + 1$. As noted in Example 5.5.2, the equation has a root in the interval $[0, 1]$. Let us make the initial guess $x_0 = 0.5$. Newton's formula here is

$$x_{n+1} = x_n - \frac{x_n^3 + x_n - 1}{3x_n^2 + 1} = \frac{2x_n^3 + 1}{3x_n^2 + 1}.$$

Thus

$$\begin{aligned} x_0 &= 0.5 \\ x_1 &= \frac{2(0.5)^3 + 1}{3(0.5)^2 + 1} = 0.7142857\cdots \\ x_2 &= 0.6831797\cdots \\ x_3 &= 0.6823284\cdots \\ x_4 &= 0.6823278\cdots \\ x_5 &= 0.6823278\cdots \end{aligned}$$

Evidently $r = 0.682328$ correctly rounded to six decimal places.

Observe the behavior of the numbers x_n. By x_2 we have apparently achieved an accuracy of two decimal places, and by x_3 about five decimal places. It is characteristic of Newton's Method that when one begins to get close to the root the convergence is very rapid. Compare these results with those obtained by the Bisection Method in Example 5.5.2; there we achieved only 2 decimal place accuracy after 8 iterations.

Newton's Method does not always work as well as it does in the preceding example. If the first derivative f' is very small near the root, or if the second derivative f'' is very large near the root, a single iteration of the formula can take us from quite close to the root to quite far away. Figure 5.5.2 illustrates this possibility.

The following theorem gives sufficient conditions for the Newton approximations to converge to a root r of the equation $f(x) = 0$ if the initial guess x_0 is sufficiently close to that root.

THEOREM 5.5.5 Suppose that f, f', and f'' are continuous on an interval containing x_n, x_{n+1}, and a root r of $f(x) = 0$. Suppose also that there exist constants K and $L > 0$ such that for all x in I we have

i) $|f''(x)| \leq K$, and

ii) $|f'(x)| \geq L$.

Then

a) $|x_{n+1} - r| \leq \dfrac{K}{2L}|x_{n+1} - x_n|^2$, and

b) $|x_{n+1} - r| \leq \dfrac{K}{2L}|x_n - r|^2$. □

Conditions i) and ii) assert that near r the slope of $y = f(x)$ is not to small in size and does not change to rapidly. We leave the proof of the theorem to you; see Exercises 19 and 20 at the end of this section. If $K/2L$ is not large (say $K/2L < 1$), the theorem states that x_n converges quickly to r once n becomes large enough that $|x_n - r| < 1$.

Theorem 5.5.5 is of considerable theoretical significance but little practical use. In practice, we compute successive approximations using Newton's formula and observe whether they seem to converge to a limit. If they do, and if the values of f at these approximations approach 0, we can be confident that we have located a root.

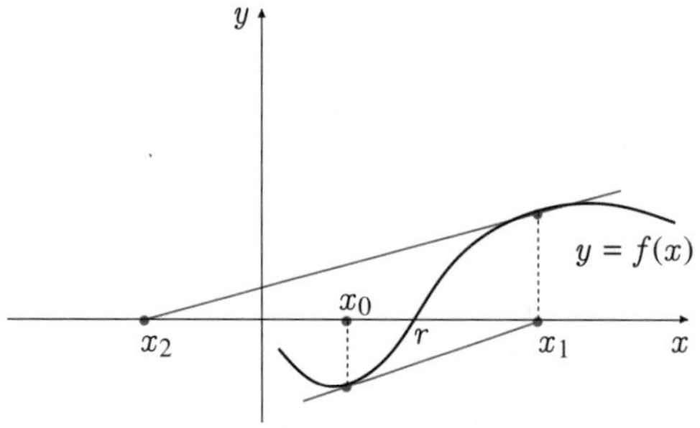

FIGURE 5.5.2

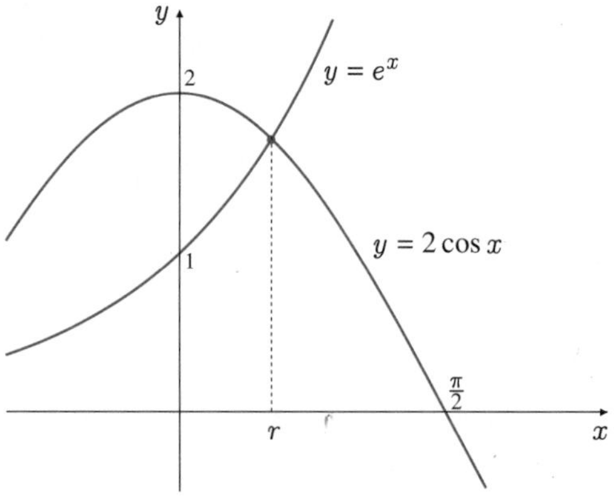

FIGURE 5.5.3

EXAMPLE 5.5.6 Find the positive solution of the equation $e^x = 2\cos x$.

SOLUTION The graphs of e^x and $2\cos x$ evidently cross at a point $x = r$ between 0 and $\pi/2$ (see Fig. 5.5.3). Since $\pi/2 \approx 1.5$ and r appears to be less than half way between 0 and $\pi/2$, let us try $x_0 = 0.7$ as an initial guess. Let

$$f(x) = e^x - 2\cos x.$$

Then

$$f'(x) = e^x + 2\sin x.$$

Newton's formula becomes

$$x_{n+1} = x_n - \frac{e^{x_n} - 2\cos x_n}{e^{x_n} + 2\sin x_n}.$$

We calculate successive approximations using a scientific calculator (remembering that the angular mode must be set in radians).

$$\begin{aligned} x_0 &= 0.7 \\ x_1 &= 0.5534098\cdots \\ x_2 &= 0.5398995\cdots \\ x_3 &= 0.5397851\cdots \\ x_4 &= 0.5397851\cdots \end{aligned}$$

The root is evidently 0.539785 rounded to six decimal places.

Fixed Point Iteration

A number r satisfying the equation $f(r) = r$ is called a **fixed point** of the function f. For certain kinds of functions, fixed points can be found by starting with an initial "guess" x_0 and calculating successive approximations $x_1 = f(x_0)$, $x_2 = f(x_1)$, ...; in general,

$$x_{n+1} = f(x_n), \qquad (n = 0, 1, 2, \ldots).$$

Let us begin by investigating a particular example:

EXAMPLE 5.5.7 Find a root of the equation $e^{-x} = x$.

SOLUTION Let $f(x) = e^{-x}$. The graphs $y = f(x)$ and $y = x$ are shown in Fig. 5.5.4. The two graphs appear to cross at a point in the vicinity of $x = 0.5$. Letting $x_0 = 0.5$ we calculate

$$x_1 = f(x_0) = 0.6065\cdots, \qquad x_2 = f(x_1) = 0.5452\cdots, \qquad x_3 = f(x_2) = 0.5797\cdots.$$

n	x_n
0	0.5000
1	0.6065
2	0.5452
3	0.5797
4	0.5601
5	0.5712
6	0.5649
7	0.5684
8	0.5664
9	0.5676
10	0.5669
11	0.5673
12	0.5671
13	0.5672
14	0.5671
15	0.5672

Subsequent values of x_n are shown in the table at the left. These values were made with a pocket calculator and recorded to four decimal places. As n increases, the numbers x_n appear to be getting closer and closer to $r = 0.567\cdots$. We infer that this is the desired root (correct to three decimal places).

Why did the method used in the above example work? Will it work for any function f? In order to answer these questions, examine the polygonal line in Fig. 5.5.4. Starting at x_0 it goes vertically to the curve $y = f(x)$, the height there being x_1. Then it goes horizontally to the line $y = x$, meeting that line at a point whose x-coordinate must therefore be x_1. Then the process repeats; the line goes vertically to the curve $y = f(x)$ and horizontally to $y = x$, arriving at $x = x_2$. The line continues in this way "spiraling" closer and closer to the desired root. Each value of x_n is closer to the root than the previous value.

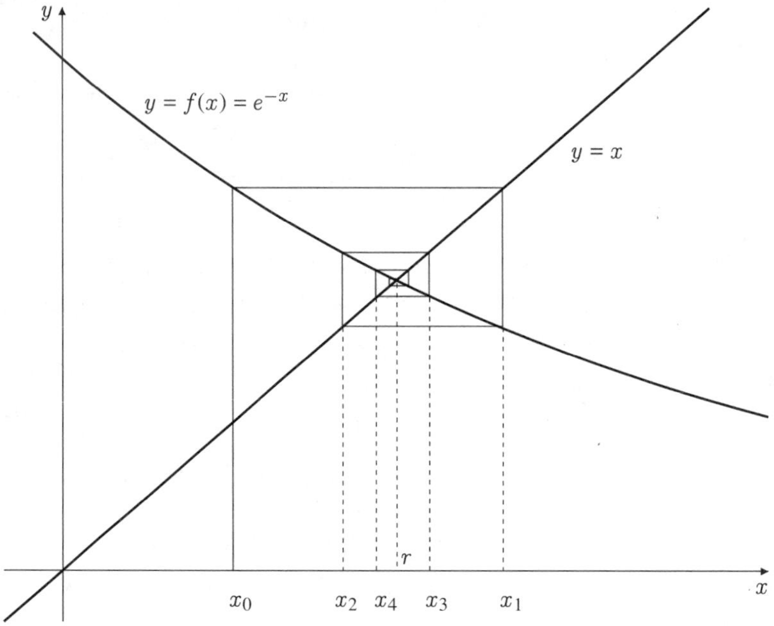

FIGURE 5.5.4

Now consider the function f whose graph appears in Fig. 5.5.5. If we try the same method there, starting with x_0, the polygonal line spirals outward, away from the root, and the resulting values x_n will not "converge" to the root as they did in the previous example. To see why the method works in for the function in Fig. 5.5.4 but not for the function in Fig. 5.5.5, observe the slopes of the two graphs $y = f(x)$, near the root r. Both slopes are negative, but in Fig. 5.5.4 the absolute value of the slope is less than 1 ($|f'(x)| = |-e^{-x}| = e^{-x} < 1$ if $x > 0$) while the absolute value of the slope of f in Fig. 5.5.5 is greater than 1. Close consideration of the graphs should convince you that it is this fact that caused the points x_n to get closer to r in Fig. 5.5.4 and farther from r in Fig. 5.5.5.

A third example, Fig. 5.5.6, shows that the method can be expected to work for functions whose graphs have positive slope near the fixed point r, provided that the slope is less than 1. In this case the polygonal line forms a "staircase" rather than a "spiral" and the successive approximations x_n increase towards the root if $x_0 < r$ and decrease towards it if $x_0 > r$.

The following theorem states a general result along these lines.

THEOREM 5.5.8 Suppose that f is a function defined on an interval $I = [a, b]$ and x_0 is a point in I. Suppose also that

i) $f(x)$ belongs to I whenever x belongs to I, and

ii) there exists a number $K < 1$ such that $|f'(x)| \leq K$ for all x in I.

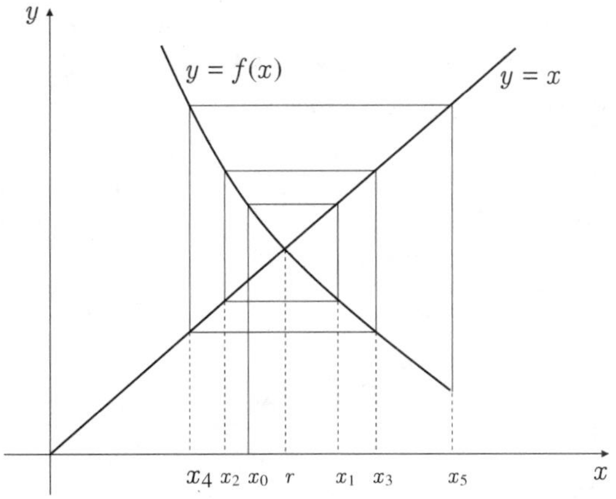

FIGURE 5.5.5

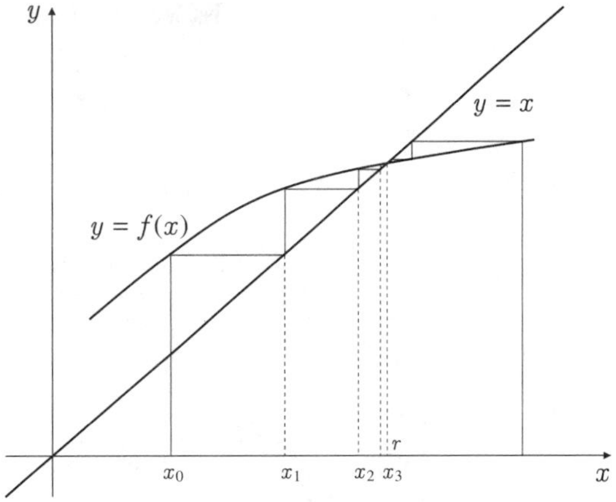

FIGURE 5.5.6

Then f has a fixed point r in I, and the *sequence* of numbers x_0, x_1, x_2, ... defined by

$$x_{n+1} = f(x_n), \quad \text{for } n \geq 0,$$

has limit r as $n \to \infty$.

PROOF Since f is differentiable on I, it is continuous there, and therefore so is the function $g(x) = f(x) - x$. Since $f(x)$ belongs to $I = [a, b]$ whenever x belongs to I, we have $f(a) \geq a$ and $f(b) \leq b$. Therefore $g(a) \geq 0$ and $g(b) \leq 0$. By the Intermediate-Value Theorem (Theorem 1.7.10) there exists a number r in $[a, b]$ such that $g(r) = 0$, that is, $f(r) = r$. Therefore f has a fixed point in I.

By the Mean-Value Theorem (Theorem 4.1.1) we have, since $x_{n+1} = f(x_n)$,

$$x_{n+1} - r = f(x_n) - f(r) = (x_n - r)f'(c),$$

for some c between x_n and r. Therefore c belongs to I and

$$|x_{n+1} - r| = |f'(c)||x_n - r| \leq K\,|x_n - r|, \qquad (n = 0, 1, 2, \ldots).$$

This inequality implies

$$\begin{aligned}
|x_1 - r| &\leq K\,|x_0 - r|, \\
|x_2 - r| &\leq K\,|x_1 - r| \leq K^2\,|x_0 - r|, \\
|x_3 - r| &\leq K\,|x_2 - r| \leq K^3\,|x_0 - r|, \\
&\vdots \\
|x_n - r| &\leq K^n\,|x_0 - r|.
\end{aligned}$$

Since $0 \leq K < 1$, $\lim_{n\to\infty} K^n = 0$ so $x_n \to r$ as $n \to \infty$. □

REMARK: It is evident from the proof that we could replace condition (ii) in the statement of the theorem with the weaker **Lipschitz condition**:

ii′) there exists a constant $K < 1$ such that $|f(x) - f(t)| \leq K|x - t|$ holds for all x and t in I,

which does not require that f be differentiable on I.

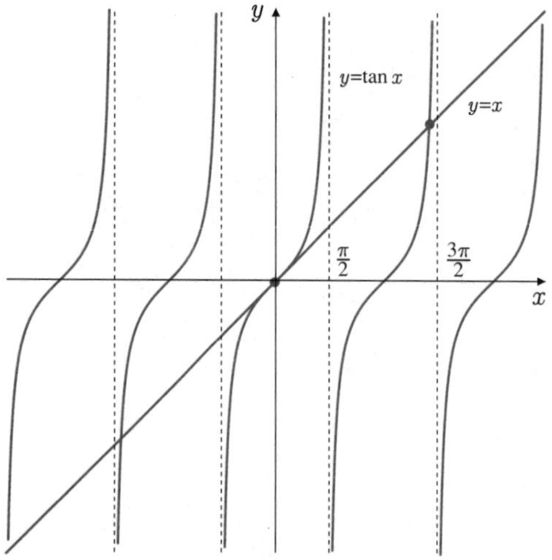

FIGURE 5.5.7

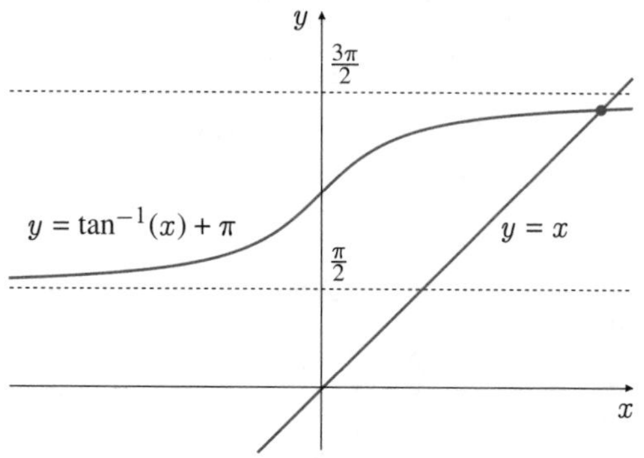

FIGURE 5.5.8

EXAMPLE 5.5.9 Find the first positive solution of the equation $\tan x = x$.

SOLUTION Consider the graphs of $y = \tan x$ and $y = x$ shown in Fig. 5.5.7. The first *positive* value of x where these graphs intersect is between $\frac{1}{2}\pi$ and $\frac{3}{2}\pi$, closer to the latter point. (They cross at $x = 0$, but 0 is not positive.) We can't apply the method of fixed point iteration directly, because $\tan x$ has derivative greater than 1 near the root. We can however transform the given equation to the form

$$x = \tan^{-1} x + \pi.$$

n	x_n
0	4.500000
1	4.493720
2	4.493424
3	4.493410
4	4.493410

This is because the one-to-one function $g(x) = \tan x$, $\frac{1}{2}\pi < x < \frac{3}{2}\pi$ has graph lying π units to the right of the graph $y = \operatorname{Tan} x$, the function whose inverse is $\tan^{-1}$; it follows that $g^{-1}(x)$ must have graph lying π units above the graph of $y = \tan^{-1} x$. (See Fig. 5.5.8.) Accordingly, we iterate

$$x_{n+1} = \tan^{-1}(x_n) + \pi,$$

starting with, say, $x_0 = 4.5$, a number slightly smaller than $\frac{3}{2}\pi$. The results are shown in the table at the left. The required solution is $x = 4.493410$ to six decimal places. The convergence to the fixed point is much faster than was the case in Example 5.5.7 because the graph of $\tan^{-1}(x) + \pi$ has very small slope near the fixed point.

EXERCISES

Use the Bisection Method to find roots of the equations in Exercises 1–2. Show that a root lies in the given interval and begin the search with that interval. Find the root to 3 decimal places.

1. $x^2 - 2 = 0, \quad [1.4, 1.5]$ **2.** $e^x = 2 - x, \quad [0.3, 0.5]$

3. Use Newton's Method to solve the equation in Exercise 1. Start with $x_0 = 1.45$.

4. Use Newton's Method to solve the equation in Exercise 2. Start with $x_0 = 0.4$.

In Exercises 5–13 find the indicated roots, by Newton's Method, to six decimal place precision, using a scientific calculator. If you have a programmable calculator you can program the Newton's Method formula and obtain the successive approximations easily. Make as good an initial guess x_0 as you can; a sketch is often useful in this regard.

5. $x^3 - 2 = 0$. (Show there is only one real root.)

6. $x^3 + 3x - 8 = 0$. (Show there is only one real root.)

7. $x^3 + 3x^2 - 2 = 0$. (Find all three real roots).

8. $x^5 + 5x - 3 = 0$. (Show there is only one real root.)

9. $\sin x = 1 - x^2$. (Find the root between 0 and $\pi/2$.)

10. $x \ln x = 1$ **11.** $x\,e^x = 1$

12. $x = e^{-x^2}$ **13.** $\tan^{-1} x = 1 - x$

14. Use the method of Fixed Point Iteration to find the second and third positive roots of $\tan x = x$. See Example 5.5.9.

Use Fixed Point Iteration to solve the equations in Exercises 15–18. Obtain 5 decimal place precision.

15. $\cos \dfrac{x}{3} = x$ **16.** $\ln x = 3 - x$

17. $\dfrac{1}{2 + x^2} = x$ **18.** $e^{-6x} = x + 2$

19.*Prove conclusion (a) of Theorem 5.5.5 as follows. Since x_{n+1} is the x-intercept of the tangent line to $y = f(x)$ at $x = x_n$, we have $f(x_n) + f'(x_n)(x_{n+1} - x_n) = 0$. Since $f(r) = 0$ also, we conclude that

$$f(x_{n+1}) - f(r) = f(x_{n+1}) - f(x_n) - f'(x_n)(x_{n+1} - x_n).$$

Apply the Mean-Value Theorem to the left-hand side of this equation and Theorem 5.3.4 to the right-hand side, and thus complete the proof.

20.*Prove conclusion (b) of Theorem 5.5.5 as follows. Show that

$$x_{n+1} - r = \frac{f(r) - f(x_n) - f'(x_n)(r - x_n)}{f'(x_n)},$$

and then apply Theorem 5.3.4 to the right-hand side.

5.6 INDETERMINATE FORMS AND L'HÔPITAL'S RULES

In Section 3.2 we showed that

$$\lim_{x \to 0} \frac{\sin x}{x} = 1.$$

We could not readily see this by substituting $x = 0$ into the function $(\sin x)/x$ because, although both $\sin x$ and x are continuous at $x = 0$, nevertheless both vanish there. We call $(\sin x)/x$ an **indeterminate form** of type [0/0]. The limit of such an indeterminate form can be any number. For instance,

$$\lim_{x \to 0} \frac{kx}{x} = k, \qquad \lim_{x \to 0} \frac{|x|}{x^2} = \infty, \qquad \lim_{x \to 0} \frac{x^3}{x^2} = 0.$$

There are other types of indeterminate forms, specifically $[\infty/\infty]$, $[0\cdot\infty]$, $[\infty-\infty]$, $[0^0]$, $[\infty^0]$, and $[1^\infty]$. Following are some examples:

$$\lim_{x\to 0}\frac{\ln x^2}{\cot x^2}, \qquad \text{type } [-\infty/\infty]$$

$$\lim_{x\to 0+} x\ln\frac{1}{x}, \qquad \text{type } [0\cdot\infty]$$

$$\lim_{x\to(\pi/2)-}\left(\tan x - \frac{1}{\pi - 2x}\right), \qquad \text{type } [\infty-\infty]$$

$$\lim_{x\to 0+} x^x, \qquad \text{type } [0^0]$$

$$\lim_{x\to(\pi/2)-}(\tan x)^{\cos x}, \qquad \text{type } [\infty^0]$$

$$\lim_{x\to\infty}\left(1+\frac{1}{x}\right)^x, \qquad \text{type } [1^\infty].$$

In this section we develop techniques called l'Hôpital's Rules* for evaluating some limits of indeterminate forms of the types $[0/0]$ and $[\infty/\infty]$. The other types can usually be reduced to one of these two by algebraic manipulation and the taking of logarithms. Many indeterminate forms of type $[0/0]$ or $[\infty/\infty]$ can be evaluated using simple algebra, typically by cancelling common factors. Examples can be found in Sections 1.5 and 1.6; l'Hôpital's rules should only be used when such simple methods cannot be used.

l'Hôpital's Rules

THEOREM 5.6.1 (**The First l'Hôpital Rule**) Suppose the functions f and g are differentiable on the interval (a, b), and $g'(x) \neq 0$ there. Suppose also that

i) $\lim_{x\to a+} f(x) = \lim_{x\to a+} g(x) = 0$, and

ii) $\lim_{x\to a+} \dfrac{f'(x)}{g'(x)} = L$ (where L is finite or ∞ or $-\infty$).

Then

$$\lim_{x\to a+}\frac{f(x)}{g(x)} = L.$$

Similar results hold if every occurrence of $\lim_{x\to a+}$ is replaced by $\lim_{x\to b-}$ or even $\lim_{x\to c}$ where $a < c < b$. The cases $a = -\infty$ and $b = \infty$ are also allowed.

PROOF We prove the case involving $\lim_{x\to a+}$ for finite a. Define

$$F(x) = \begin{cases} f(x) & \text{if } a < x < b \\ 0 & \text{if } x = a \end{cases} \qquad \text{and} \qquad G(x) = \begin{cases} g(x) & \text{if } a < x < b \\ 0 & \text{if } x = a \end{cases}$$

* The Marquis de l'Hôpital, for whom these rules are named, lived before the French Revolution, and published the first calculus book. The circumflex (ˆ) came into use in the French language after the Revolution. Thus the Marquis would have written his name l'Hospital.

Then F and G are continuous on the interval $[a, x]$ and differentiable on the interval (a, x) for every x in (a, b). By the Generalized Mean-Value Theorem (Theorem 5.3.10) there exists a number c in (a, x) such that

$$\frac{f(x)}{g(x)} = \frac{F(x)}{G(x)} = \frac{F(x) - F(a)}{G(x) - G(a)} = \frac{F'(c)}{G'(c)} = \frac{f'(c)}{g'(c)}.$$

Since $a < c < x$, if $x \to a+$, then necessarily $c \to a+$, so we have

$$\lim_{x \to a+} \frac{f(x)}{g(x)} = \lim_{c \to a+} \frac{f'(c)}{g'(c)} = L.$$

The case involving $\lim_{x \to b-}$ for finite b is proved similarly. The cases where $a = -\infty$ or $b = \infty$ follow from the cases already considered via the change of variable $x = 1/t$:

$$\lim_{x \to \infty} \frac{f(x)}{g(x)} = \lim_{t \to 0+} \frac{f\left(\frac{1}{t}\right)}{g\left(\frac{1}{t}\right)} = \lim_{t \to 0+} \frac{f'\left(\frac{1}{t}\right)\left(\frac{-1}{t^2}\right)}{g'\left(\frac{1}{t}\right)\left(\frac{-1}{t^2}\right)} = \lim_{x \to \infty} \frac{f'(x)}{g'(x)} = L. \ \square$$

Note that in applying l'Hôpital's Rule we calculate the quotient of the derivatives, *not* the derivative of the quotient.

EXAMPLE 5.6.2

$$\begin{aligned} &\lim_{x \to 1} \frac{\ln x}{x^2 - 1} \qquad \left[\frac{0}{0}\right] \\ &= \lim_{x \to 1} \frac{\frac{1}{x}}{2x} = \lim_{x \to 1} \frac{1}{2x^2} = \frac{1}{2}. \end{aligned}$$

This example illustrates how calculations based on l'Hôpital's Rule are carried out. Having identified the limit as that of a $[0/0]$ indeterminate form, we replace it by the limit of the quotient of derivatives; the existence of this latter limit justifies the equality. It is possible that the limit of the quotient of derivatives may still be indeterminate, in which case a second application of l'Hôpital's Rule can be made. Such applications may be strung out until a limit can finally be extracted, which then justifies all the previous applications of the rule.

EXAMPLE 5.6.3

$$\begin{aligned} &\lim_{x \to 0} \frac{1 - e^{3x^2}}{x \sin 2x} \qquad \left[\frac{0}{0}\right] \\ &= \lim_{x \to 0} \frac{-6xe^{3x^2}}{\sin 2x + 2x \cos 2x} \qquad \left[\frac{0}{0}\right] \\ &= \lim_{x \to 0} \frac{-6e^{3x^2} - 36x^2 e^{3x^2}}{4 \cos 2x - 4x \sin 2x} = \frac{-6 - 0}{4 - 0} = -\frac{3}{2} \end{aligned}$$

Note that l'Hôpital's Rule was used twice here.

EXAMPLE 5.6.4

$$\begin{aligned} &\lim_{x\to(\pi/2)-} \frac{2x-\pi}{\cos^2 x} \qquad \left[\frac{0}{0}\right] \\ =& \lim_{x\to(\pi/2)-} \frac{2}{-2\sin x\cos x} = -\infty \end{aligned}$$

EXAMPLE 5.6.5

$$\lim_{x\to 1+} \frac{x}{\ln x} = \infty$$

Note that l'Hôpital's Rule was not used here at all; $x/(\ln x)$ is not an indeterminant form. (Had we tried to apply l'Hôpital's Rule, we would have been led to $\lim_{x\to 1+}(1/(1/x)) = 1$, an erroneous answer.)

EXAMPLE 5.6.6

$$\begin{aligned} &\lim_{x\to 0+}\left(\frac{1}{x}-\frac{1}{\sin x}\right) \qquad [\infty-\infty] \\ =& \lim_{x\to 0+}\frac{\sin x - x}{x\sin x} \qquad \left[\frac{0}{0}\right] \\ =& \lim_{x\to 0+}\frac{\cos x - 1}{\sin x + x\cos x} \qquad \left[\frac{0}{0}\right] \\ =& \lim_{x\to 0+}\frac{-\sin x}{2\cos x - x\sin x} = \frac{-0}{2} = 0. \end{aligned}$$

THEOREM 5.6.7 (**The Second l'Hôpital Rule**) Suppose that f and g are differentiable on the interval (a,b) and that $g'(x) \neq 0$ there. Suppose also that

i) $\lim_{x\to a+} g(x) = \pm\infty$, and

ii) $\lim_{x\to a+} \dfrac{f'(x)}{g'(x)} = L$ (where L is finite, or ∞ or $-\infty$).

Then

$$\lim_{x\to a+}\frac{f(x)}{g(x)} = L.$$

Again, similar results hold for $\lim_{x\to b-}$ and for $\lim_{x\to c}$, and the cases $a=-\infty$ and $b=\infty$ are allowed. □

The proof of the Second l'Hôpital Rule is technically rather more difficult than that of the First Rule and we will not give it here. A sketch of the proof is outlined in Exercise 34 at the end of this section.

REMARK: Do **not** try to use l'Hôpital's Rules to evaluate limits which are not indeterminate of type $[0/0]$ or $[\infty/\infty]$; such attempts will almost always lead to false conclusions. (See Example 5.6.5.)

(Note that, strictly speaking, the second l'Hôpital Rule can be applied to the form $[a/\infty]$ whether or not a is infinite. However, there is no reason to apply it if a is not infinite, since the limit is obviously 0 in that case.)

EXAMPLE 5.6.8

$$\begin{aligned} &\lim_{x\to\infty} \frac{x^2}{e^x} \qquad \left[\frac{\infty}{\infty}\right] \\ &= \lim_{x\to\infty} \frac{2x}{e^x} \qquad \text{still } \left[\frac{\infty}{\infty}\right] \\ &= \lim_{x\to\infty} \frac{2}{e^x} = 0 \end{aligned}$$

Similarly, one can show that $\lim_{x\to\infty} x^n/e^x = 0$ for any positive integer n by repeated applications of l'Hôpital's Rule and hence effect a proof of part of Theorem 3.5.18. The following Example proves another part of that theorem.

EXAMPLE 5.6.9

$$\begin{aligned} &\lim_{x\to 0+} x^a \ln x \qquad (a > 0) \qquad [0 \cdot (-\infty)] \\ &= \lim_{x\to 0+} \frac{\ln x}{x^{-a}} \qquad \left[\frac{-\infty}{\infty}\right] \\ &= \lim_{x\to 0+} \frac{1/x}{-ax^{-a-1}} = \lim_{x\to 0+} \frac{x^a}{-a} = 0 \end{aligned}$$

EXAMPLE 5.6.10 Evaluate $\lim_{x\to 0+} x^x$. $[0^0]$

SOLUTION Let $y = x^x$. Then $\lim_{x\to 0+} \ln y = \lim_{x\to 0+} x \ln x = 0$, by Example 5.6.9. Hence $\lim_{x\to 0} x^x = \lim_{x\to 0+} y = e^0 = 1$.

EXAMPLE 5.6.11 Evaluate $\lim_{x\to(\pi/2)-} (\tan x)^{\cos x}$. $[\infty^0]$

SOLUTION Let $y = (\tan x)^{\cos x}$. Then

$$\begin{aligned} \lim_{x\to(\pi/2)-} \ln y &= \lim_{x\to(\pi/2)-} \cos x \ln \tan x \qquad [0 \cdot \infty] \\ &= \lim_{x\to(\pi/2)-} \frac{\ln \tan x}{\sec x} \qquad \left[\frac{\infty}{\infty}\right] \\ &= \lim_{x\to(\pi/2)-} \frac{\dfrac{1}{\tan x} \sec^2 x}{\sec x \tan x} \\ &= \lim_{x\to(\pi/2)-} \frac{\sec x}{\tan^2 x} = \lim_{x\to(\pi/2)-} \frac{\cos x}{\sin^2 x} = 0 \end{aligned}$$

Thus $\lim_{x\to(\pi/2)-}(\tan x)^{\cos x} = e^0 = 1.$

EXAMPLE 5.6.12 Evaluate $\displaystyle\lim_{x\to\infty}\left(1+\sin\frac{3}{x}\right)^x$. $[1^\infty]$

SOLUTION Let $y = \left(1+\sin\frac{3}{x}\right)^x$. Then

$$\begin{aligned}
\lim_{x\to\infty}\ln y &= \lim_{x\to\infty} x\ln\left(1+\sin\frac{3}{x}\right) && [\infty\cdot 0]\\
&= \lim_{x\to\infty}\frac{\ln\left(1+\sin\frac{3}{x}\right)}{\frac{1}{x}} && \left[\frac{0}{0}\right]\\
&= \lim_{x\to\infty}\frac{\frac{1}{1+\sin\frac{3}{x}}\left(\cos\frac{3}{x}\right)\left(-\frac{3}{x^2}\right)}{-\frac{1}{x^2}}\\
&= \lim_{x\to\infty}\frac{3\cos\frac{3}{x}}{1+\sin\frac{3}{x}} = 3
\end{aligned}$$

Hence $\displaystyle\lim_{x\to\infty}\left(1+\sin\frac{3}{x}\right)^x = e^3.$

EXERCISES

Evaluate the limits in Exercises 1–32.

1. $\displaystyle\lim_{x\to0}\frac{3x}{\tan 4x}$
2. $\displaystyle\lim_{x\to2}\frac{\ln(2x-3)}{x^2-4}$
3. $\displaystyle\lim_{x\to0}\frac{\sin ax}{\sin bx}$
4. $\displaystyle\lim_{x\to0}\frac{1-\cos ax}{1-\cos bx}$
5. $\displaystyle\lim_{x\to0}\frac{\sin^{-1}x}{\tan^{-1}x}$
6. $\displaystyle\lim_{x\to1}\frac{x^{1/3}-1}{x^{2/3}-1}$
7. $\displaystyle\lim_{x\to0} x\cot x$
8. $\displaystyle\lim_{x\to0}\frac{1-\cos x}{\ln(1+x^2)}$
9. $\displaystyle\lim_{t\to\pi}\frac{\sin^2 t}{t-\pi}$
10. $\displaystyle\lim_{x\to0}\frac{10^x-e^x}{x}$
11. $\displaystyle\lim_{x\to\pi/2}\frac{\cos 3x}{\pi-2x}$
12. $\displaystyle\lim_{x\to1}\frac{\ln(ex)-1}{\sin\pi x}$
13. $\displaystyle\lim_{x\to\infty} x\sin\frac{1}{x}$
14. $\displaystyle\lim_{x\to0}\frac{x-\sin x}{x^3}$
15. $\displaystyle\lim_{x\to0}\frac{x-\sin x}{x-\tan x}$
16. $\displaystyle\lim_{x\to0}\frac{2-x^2-2\cos x}{x^4}$
17. $\displaystyle\lim_{x\to0+}\frac{\sin^2 x}{\tan x - x}$
18. $\displaystyle\lim_{r\to\pi/2}\frac{\ln\sin r}{\cos r}$
19. $\displaystyle\lim_{t\to\pi/2}\frac{\sin t}{t}$
20. $\displaystyle\lim_{x\to1-}\frac{\cos^{-1}x}{x-1}$
21. $\displaystyle\lim_{x\to\infty} x(2\tan^{-1}x-\pi)$
22. $\displaystyle\lim_{t\to(\pi/2)-}(\sec t-\tan t)$
23. $\displaystyle\lim_{t\to0}\left(\frac{1}{t}-\frac{1}{te^{at}}\right)$
24. $\displaystyle\lim_{x\to0+} x^{\sqrt{x}}$
25.* $\displaystyle\lim_{x\to0+}(\csc x)^{\sin^2 x}$
26.* $\displaystyle\lim_{x\to1+}\left(\frac{x}{x-1}-\frac{1}{\ln x}\right)$
27.* $\displaystyle\lim_{t\to0}\frac{3\sin t-\sin 3t}{3\tan t-\tan 3t}$
28.* $\displaystyle\lim_{x\to0}\left(\frac{\sin x}{x}\right)^{1/x^2}$

29.* $\lim_{t\to 0}(\cos 2t)^{1/t^2}$

30.* $\lim_{x\to 0+} \dfrac{\csc x}{\ln x}$

31.* $\lim_{x\to 1-} \dfrac{\ln \sin \pi x}{\csc \pi x}$

32.* $\lim_{x\to 0}(1+\tan x)^{1/x}$

33. Evaluate $\lim_{h\to 0} \dfrac{f(x+h)-2f(x)+f(x-h)}{h^2}$ if f is twice differentiable.

34.*Fill in the details of the following outline of a proof of the second l'Hôpital Rule (Theorem 5.6.7) for the case where a and L are both finite. Let $a < x < t < b$ and show that there exists c in (x, t) such that

$$\frac{f(x)-f(t)}{g(x)-g(t)} = \frac{f'(c)}{g'(c)}.$$

Now juggle the above equation algebraically into the form

$$\frac{f(x)}{g(x)} - L = \frac{f'(c)}{g'(c)} - L + \frac{1}{g(x)}\left(f(t) - g(t)\frac{f'(c)}{g'(c)}\right).$$

It follows that

$$\left|\frac{f(x)}{g(x)} - L\right| \le \left|\frac{f'(c)}{g'(c)} - L\right| + \frac{1}{|g(x)|}\left(|f(t)| + |g(t)|\left|\frac{f'(c)}{g'(c)}\right|\right).$$

Now show that the right side of the above inequality can be made as small as you wish (say less than a positive number ϵ) by choosing first t then x close enough to a. Remember, you are given that $\lim_{c\to a+}\big(f'(c)/g'(c)\big) = L$ and $\lim_{x\to a+}|g(x)| = \infty$.

CHAPTER 6

Integration

The second fundemental problem addressed by calculus is the problem of areas, that is, the problem of determining the area of a region of the plane bounded by various curves. Like the problem of tangents considered in Chapter 2, the solution of the problem of areas necessarily involves the notion of limits. On the surface the problem of areas appears unrelated to the problem of tangents. However, we will see that the two problems are very closely related; one is the inverse of the other. Finding an area is equivalent to finding an antiderivative or, as we prefer to say, finding an integral.

The relationship between areas and antiderivatives is called the Fundamental Theorem of Calculus. When we have proved it we will be able to find areas at will, provided only that we can integrate (that is, antidifferentiate) the various functions we encounter.

We would clearly like to have at our disposal a "calculus" of integrals similar to the calculus of derivatives developed in Chapter 2. We can find the derivative of any differentiable function using the differentiation rules established there. Unfortunately, integration is generally more difficult; indeed, some fairly simple functions are not themselves derivatives of simple functions. For example, e^{x^2} is not the derivative of any finite combination of elementary functions. Nevertheless, we expend considerable effort in this chapter to develop techniques for integrating as many functions as possible. Other functions such as exp e^{x^2} can be integrated using infinite series techniques, which we develop in Chapter 11.

6.1 AREAS UNDER CURVES

We began the study of derivatives in Chapter 2 by defining what is meant by a tangent line to a curve at a particular point. It would seem appropriate to begin the study of integrals by defining what is meant by the **area** of a plane region. However, a definition of area is harder to give than a definition of tangency. Let us assume, therefore, that we know intuitively what area means and list some of its properties. (See Fig. 6.1.1.)

i) The area of a plane region is a nonnegative real number of *square units*.
ii) The area of a rectangle with width w and height h is $A = wh$.
iii) The areas of congruent plane regions are equal.
iv) If region S is contained in region R, then the area of S is less than or equal to that of R.
v) If region R is a union of (finitely many) non-overlapping regions, then the area of R is the sum of the areas of those regions.

Using these five properties we can calculate the area of any **polygon** (a region bounded by straight line segments). First we note that properties (iii) and (v) show that the area of a parallelogram is the same of that of a rectangle having the same base width and height. Any triangle can be butted against a congruent copy of itself to form a parallelogram, so a triangle has area half the base width times the height. Finally, any polygon can be subdivided into finitely many non-overlapping triangles so its area is the sum of the areas of those triangles.

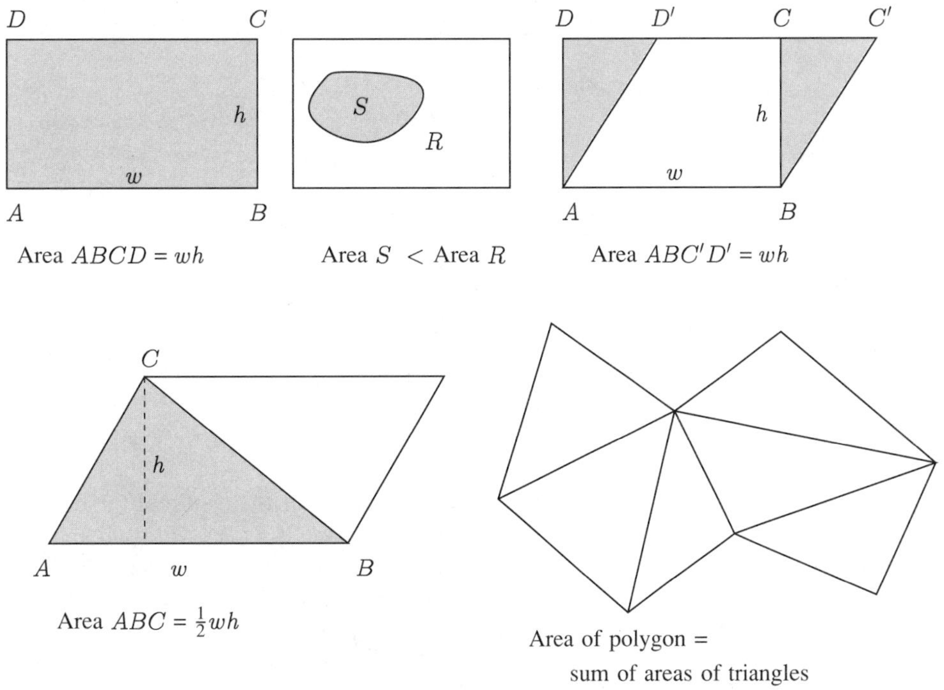

FIGURE 6.1.1

We can't go beyond polygons without taking limits. If a region has a curved boundary its area can only be approximated by using rectangles or triangles; calculating the exact area requires the evaluation of a limit.

As our first example of such a limiting process let us derive the formula for the area of a circle of radius r:

$$A = \pi r^2.$$

You may have "known" this formula for a long time, but it is likely that you have not seen any proof of it. The number π is *defined* to be the ratio of the circumference of a circle to its diameter. (All circles are "similar" geometric objects so that ratio is a constant.) It is not obvious that the ratio of the area of a circle to the square of its radius should be the same number.

EXAMPLE 6.1.1 Find the area of a circle of radius r.

SOLUTION Inscribe a regular polygon having n equal sides in the circle as shown in Fig. 6.1.2. Evidently the area A_n of the polygon is less than the area A of the circle, but $A_n \to A$ as $n \to \infty$. Also the perimeter P_n of the polygon approaches the circumference $C = 2\pi r$ of the circle as $n \to \infty$. If X and Y are adjacent vertices

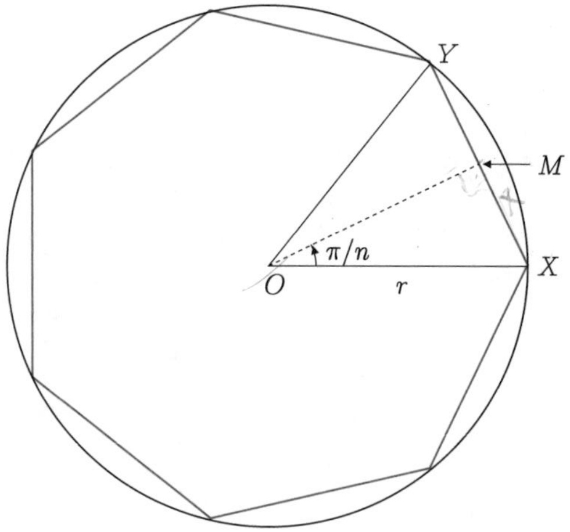

FIGURE 6.1.2

of the polygon, and O is the centre of the circle, then the area of the polygon is n times the area of triangle XOY. Now angle $XOY = 2\pi/n$ radians, and is bisected by the line OM drawn perpendicular to XY from O, so angle $XOM = \pi/n$ radians, $XM = r\sin(\pi/n)$ and $OM = r\cos(\pi/n)$. Thus triangle XOY has area $r^2 \sin(\pi/n)\cos(\pi/n)$ and

$$A = \lim_{n\to\infty} A_n = \lim_{n\to\infty} nr^2 \sin\frac{\pi}{n}\cos\frac{\pi}{n}.$$

Similarly, the circumference of the circle is n times the length of XY:

$$2\pi r = C = \lim_{n\to\infty} n\left(2r\sin\frac{\pi}{n}\right),$$

so $\lim_{n\to\infty} n\sin(\pi/n) = \pi$. Since $\lim_{n\to\infty}\cos(\pi/n) = \cos 0 = 1$ we have, at last,

$$A = r^2 \times \pi \times 1 = \pi r^2 \text{ square units.}$$

The basic problem we consider in this section is how to find the area of a region R lying under the graph $y = f(x)$ of a nonnegative-valued, continuous function f, above the x-axis and between the vertical lines $x = a$ and $x = b$, where $a < b$. (See Fig. 6.1.3.) To accomplish this we proceed as follows. Divide the interval $[a, b]$ into n subintervals by using division points

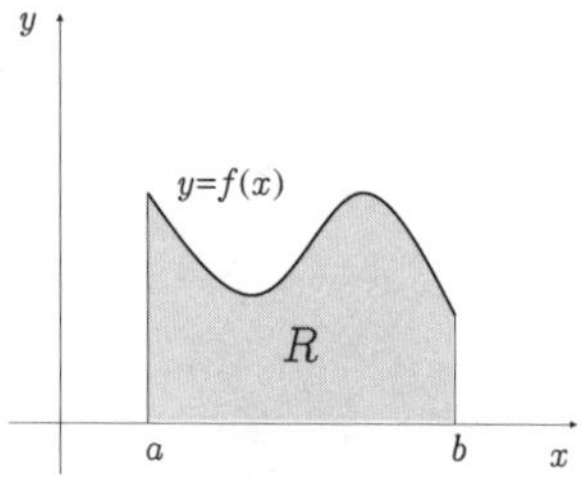

FIGURE 6.1.3

$$a = x_0 < x_1 < x_2 < x_3 < \cdots < x_{n-1} < x_n = b.$$

Denote by Δx_i the length of the ith subinterval $[x_{i-1}, x_i]$:

$$\Delta x_i = x_i - x_{i-1}, \qquad (i = 1, 2, 3, \ldots, n).$$

Vertically above each subinterval $[x_{i-1}, x_i]$, build a rectangle whose base has length Δx_i and whose height is $f(x_i)$. The area of this rectangle is $f(x_i)\,\Delta x_i$. Form the sum of these areas:

$$S_n = f(x_1)\,\Delta x_1 + f(x_2)\,\Delta x_2 + f(x_3)\,\Delta x_3 + \cdots + f(x_n)\,\Delta x_n.$$

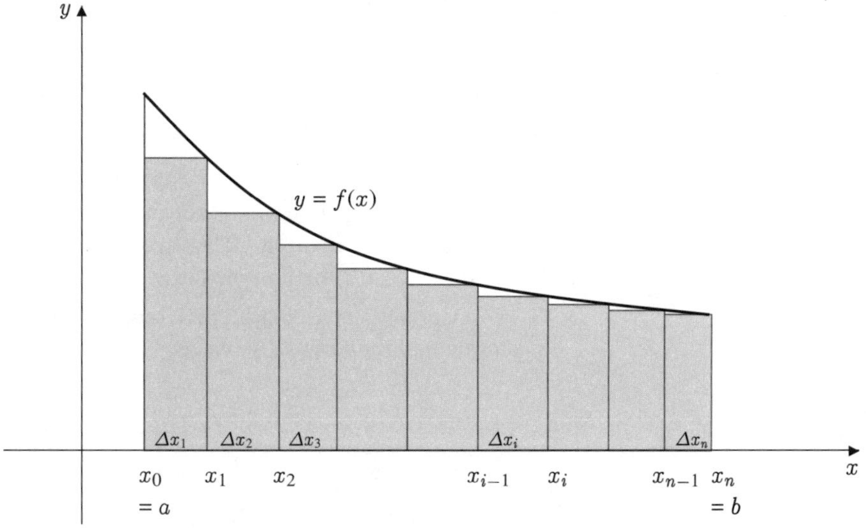

FIGURE 6.1.4

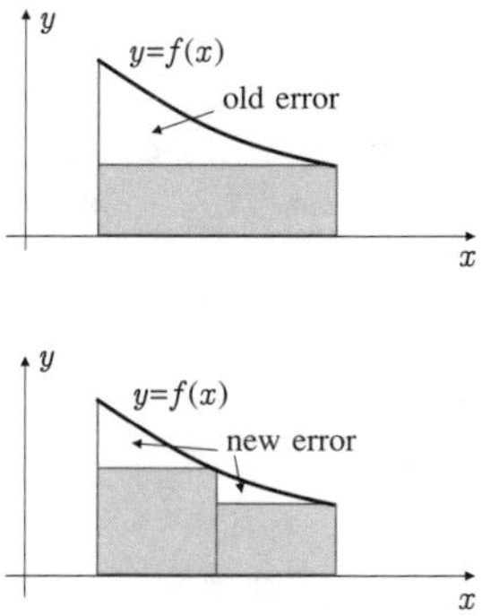

FIGURE 6.1.5

The rectangles are shown shaded in Fig. 6.1.4 for a decreasing function f. For an increasing function, the tops of the rectangles would lie above the graph of f rather than below it. Evidently S_n is an approximation to the area of the region R, and the approximation gets better and better as n increases, provided we choose the points $a = x_0 < x_1 < \cdots < x_n = b$ in such a way that all of the rectangles have widths Δx_i approaching zero. (Observe in Fig. 6.1.5, for example, that subdividing a subinterval into two smaller subintervals reduces the "error" in the approximation by reducing that part of the area under the curve which is not contained in the rectangles.) It is reasonable, therefore, to calculate the area of R by finding the limit of S_n as $n \to \infty$ with the restriction that the largest of the subinterval widths Δx_i must approach zero:

$$\text{Area of } R = \lim_{\substack{n\to\infty \\ \max \Delta x_i \to 0}} S_n.$$

Summation Notation

We frequently encounter sums where the terms are all of the same form except for an integer variable which takes on consecutive values. The definition of S_n above is an example; all the terms are of the form $f(x_i)\,\Delta x_i$ with i ranging from 1 up to n. There is a very useful notation for representing such a sum:

$$S_n = \sum_{i=1}^{n} f(x_i)\,\Delta x_i.$$

The symbol $\sum$ is called a **summation sign**; it is the capital Greek letter "sigma." In general, the symbol $\sum_{j=k}^{m} c(j)$ stands for the sum of terms $c(j)$ from $j = k$ up to $j = m$ inclusive:

$$\sum_{j=k}^{m} c(j) = c(k) + c(k+1) + c(k+2) + \cdots + c(m).$$

(It is assumed that $m \geq k$.) Note that the sum depends on k and m, and on the form of the function c, but not on j, which is called the **index of summation** and is a **dummy variable**. Thus the sum $\sum_{i=k}^{m} c(i)$ represents for exactly the same sum; it doesn't matter whether we call the index j or i.

Certain rules of arithmetic apply to summations. Since the order of adding terms is unimportant, we have

$$\sum_{i=k}^{m} \big(f(i) + g(i)\big) = \sum_{i=k}^{m} f(i) + \sum_{i=k}^{m} g(i).$$

Also, constant common factors can be removed outside the sum as in $ab+ac+ad = a(b+c+d)$; for summations this becomes

$$\sum_{i=k}^{m} af(i) = a \sum_{i=k}^{m} f(i).$$

The following theorem evaluates certain sums which we will use later in this section.

THEOREM 6.1.2

(a) $$\sum_{i=1}^{n} 1 = \overbrace{1 + 1 + 1 + \cdots + 1}^{n \text{ terms}} = n$$

(b) $$\sum_{i=1}^{n} i = 1 + 2 + 3 + \cdots + n = \frac{n(n+1)}{2}$$

(c) $$\sum_{i=1}^{n} i^2 = 1^2 + 2^2 + 3^2 + \cdots + n^2 = \frac{n(n+1)(2n+1)}{6}$$

(d) $$\sum_{i=1}^{n} r^{i-1} = 1 + r + r^2 + r^3 + \cdots + r^{n-1} = \frac{r^n - 1}{r - 1} \quad \text{provided } r \neq 1.$$

PROOF Formula (a) is trivial; the sum of n ones is n.

Let $S = \sum_{i=1}^{n} i$. We write the terms in S forwards and backwards and then add:

$$\begin{array}{rcccccccccccc}
S = & 1 & + & 2 & + & 3 & + \cdots + & (n-2) & + & (n-1) & + & n \\
S = & n & + & (n-1) & + & (n-2) & + \cdots + & 3 & + & 2 & + & 1 \\
\hline
2S = & (n+1) & + & (n+1) & + & (n+1) & + \cdots + & (n+1) & + & (n+1) & + & (n+1) = n(n+1)
\end{array}$$

Formula (b) follows on division by 2.

To prove (c) we write n copies of the identity

$$(k+1)^3 - k^3 = 3k^2 + 3k + 1,$$

one for each value of k from 1 to n, and add them up:

$$\begin{array}{rcrcrcrcl}
2^3 & - & 1^3 & = & 3\times 1^2 & + & 3\times 1 & + & 1\\
3^3 & - & 2^3 & = & 3\times 2^2 & + & 3\times 2 & + & 1\\
4^3 & - & 3^3 & = & 3\times 3^2 & + & 3\times 3 & + & 1\\
\vdots & & \vdots & & \vdots & & \vdots & & \vdots\\
n^3 & - & (n-1)^3 & = & 3(n-1)^2 & + & 3(n-1) & + & 1\\
(n+1)^3 & - & n^3 & = & 3\,n^2 & + & 3n & + & 1\\
\hline
(n+1)^3 & - & 1^3 & = & 3\left(\sum_{i=1}^n i^2\right) & + & 3\left(\sum_{i=1}^n i\right) & + & n\\
 & & & = & 3\left(\sum_{i=1}^n i^2\right) & + & \dfrac{3n(n+1)}{2} & + & n.
\end{array}$$

We used formula (b) in the last line. The final equation can be solved for the desired sum to give formula (c).

To prove formula (d), let $s = \sum_{i=1}^n r^{i-1}$ and subtract s from rs:

$$\begin{aligned}(r-1)s = rs - s &= (r + r^2 + r^3 + \cdots + r^n) - (1 + r + r^2 + \cdots + r^{n-1})\\ &= r^n - 1.\end{aligned}$$

The result follows on division by $r - 1$. □

Some Area Calculations

We devote the rest of this section to some examples in which we apply the technique described above for finding areas under graphs of functions by approximating with rectangles. Let us begin with a region for which we already know the area so we can satisfy ourselves that the method does give the correct value.

EXAMPLE 6.1.3 Find the area of the region R lying under the straight line $y = x + 1$, above the x-axis and between the lines $x = 0$ and $x = 2$.

SOLUTION The region R is shown in Fig. 6.1.6. It is a *trapezoid* (a four-sided polygon with one pair of parallel sides) and has area 4 square units. (It can be divided into a rectangle and a triangle each of area 2 square units.) Let us calculate the area by the method discussed earlier in this section. Divide the interval $[0, 2]$ into n subintervals *of equal length* by points

$$x_0 = 0,\ x_1 = \frac{2}{n},\ x_2 = \frac{4}{n},\ x_3 = \frac{6}{n},\ \ldots\ x_n = \frac{2n}{n} = 2.$$

The value of $y = x+1$ at $x = x_i$ is $x_i+1 = \frac{2i}{n}+1$ and ith subinterval, $\left[\frac{2(i-1)}{n}, \frac{2i}{n}\right]$ has length $\Delta x_i = \frac{2}{n}$. Clearly $\Delta x_i \to 0$ and $n \to \infty$. The sum of the areas of the approximating rectangles shown in Fig. 6.1.6 is

$$\begin{aligned} S_n &= \sum_{i=1}^{n} \left(\frac{2i}{n} + 1\right) \frac{2}{n} \\ &= \left(\frac{2}{n}\right) \left[\frac{2}{n} \sum_{i=1}^{n} i + \sum_{i=1}^{n} 1\right] \\ &= \left(\frac{2}{n}\right) \left[\frac{2}{n} \frac{n(n+1)}{2} + n\right] \\ &= 2\,\frac{n+1}{n} + 2. \end{aligned}$$

Therefore the area A of R is given by

$$\begin{aligned} A = \lim_{n\to\infty} S_n &= \lim_{n\to\infty} \left(2\,\frac{n+1}{n} + 2\right) \\ &= 2 + 2 = 4 \text{ sq. units.} \end{aligned}$$

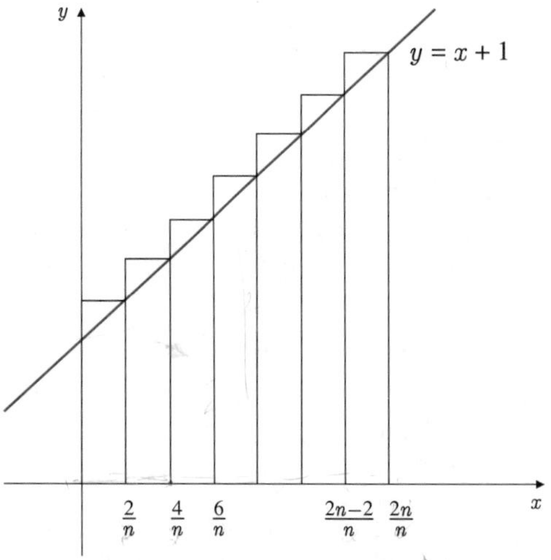

FIGURE 6.1.6

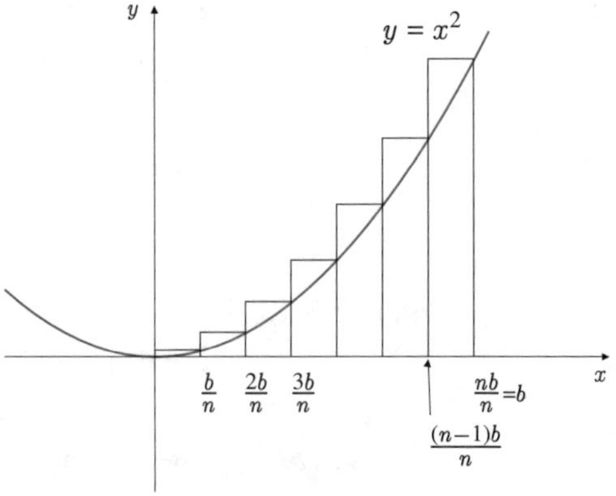

FIGURE 6.1.7

EXAMPLE 6.1.4 Find the area of the region bounded by the parabola $y = x^2$ and the straight lines $y = 0$, $x = 0$, and $x = b > 0$.

SOLUTION The area A of the region is the limit of the sum S_n of areas of the rectangles shown in Fig. 6.1.7. Again we have used equal subintervals, each of length b/n. The height of the ith rectangle is $(ib/n)^2$. Thus

$$S_n = \sum_{i=1}^{n} \left(\frac{ib}{n}\right)^2 \frac{b}{n} = \frac{b^3}{n^3} \sum_{i=1}^{n} i^2 = \frac{b^3}{n^3} \frac{n(n+1)(2n+1)}{6},$$

by formula (c) of Theorem 6.1.2. Hence the required area is

$$A = \lim_{n\to\infty} S_n = \lim_{n\to\infty} b^3 \frac{(n+1)(2n+1)}{6n^2} = \frac{b^3}{3} \text{ sq. units.}$$

Finding an area under the graph of $y = x^k$ over an interval I becomes more and more difficult as k increases if we continue to try to subdivide I into subintervals of equal length. (See Exercise 9 at the end of this section for the case $k = 3$.) It is, however, possible to find the area for arbitrary k if we subdivide the interval I into subintervals whose lengths increase in geometric progression. Our final example illustrates this.

EXAMPLE 6.1.5 Let $b > a > 0$ and let k be any real number except -1. Show that the area A of the region bounded by $y = x^k$, $y = 0$, $x = a$ and $x = b$ is

$$A = \frac{b^{k+1} - a^{k+1}}{k+1} \text{ square units.}$$

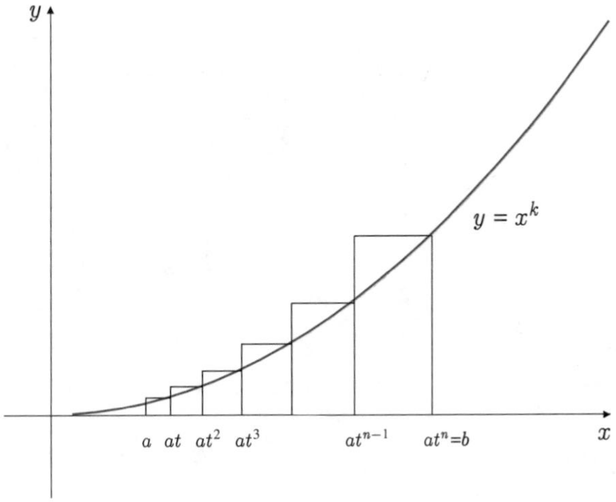

FIGURE 6.1.8

SOLUTION Let $t = (b/a)^{1/n}$ and let

$$x_0 = a,\ x_1 = at,\ x_2 = at^2,\ x_3 = at^3,\ \ldots\ x_n = at^n = b.$$

These points subdivide the interval $[a, b]$ into n subintervals of which the ith, $[x_{i-1}, x_i]$, has length $\Delta x_i = at^{i-1}(t-1)$. If $f(x) = x^k$, then $f(x_i) = a^k t^{ki}$. The sum of the areas of the rectangles shown in Fig. 6.1.8 is

!!DANGER!!

This is a long and difficult example. Either skip over it or take your time and check each step carefully.

$$\begin{aligned}
S_n &= \sum_{i=1}^{n} f(x_i)\,\Delta x_i \\
&= \sum_{i=1}^{n} a^k\, t^{ki}\, at^{i-1}(t-1) \\
&= a^{k+1}(t-1)\,t^k \sum_{i=1}^{n} t^{(k+1)(i-1)} \\
&= a^{k+1}(t-1)\,t^k \sum_{i=1}^{n} r^{(i-1)} \qquad \text{where } r = t^{k+1} \\
&= a^{k+1}(t-1)\,t^k\,\frac{r^n - 1}{r - 1} \qquad \text{(by Theorem 6.1.2(d))} \\
&= a^{k+1}(t-1)\,t^k\,\frac{t^{(k+1)n} - 1}{t^{k+1} - 1}.
\end{aligned}$$

Now replace t with its value $(b/a)^{1/n}$ and rearrange factors to obtain

$$\begin{aligned}
S_n &= a^{k+1}\left(\left(\frac{b}{a}\right)^{1/n} - 1\right)\left(\frac{b}{a}\right)^{k/n} \frac{\left(\frac{b}{a}\right)^{k+1} - 1}{\left(\frac{b}{a}\right)^{(k+1)/n} - 1} \\
&= \left(b^{k+1} - a^{k+1}\right) c^{k/n}\,\frac{c^{1/n} - 1}{c^{(k+1)/n} - 1}, \qquad \text{where } c = \frac{b}{a}.
\end{aligned}$$

Of the three factors on the right side of the final line above, the first does not depend on n, and the second, $c^{k/n}$, approaches $c^0 = 1$ as $n \to \infty$. The third factor is an indeterminate form of type $[0/0]$ which we evaluate using l'Hôpital's Rule. First let $u = 1/n$. Then

$$\begin{aligned}
\lim_{n\to\infty} \frac{c^{1/n} - 1}{c^{(k+1)/n} - 1} &= \lim_{u\to 0+} \frac{c^u - 1}{c^{(k+1)u} - 1} \qquad \left[\frac{0}{0}\right] \\
&= \lim_{u\to 0+} \frac{c^u \ln c}{(k+1)\,c^{(k+1)u} \ln c} = \frac{1}{k+1}.
\end{aligned}$$

The required area is, therefore,

$$A = \lim_{n\to\infty} S_n = \left(b^{k+1} - a^{k+1}\right) \times 1 \times \frac{1}{k+1} = \frac{b^{k+1} - a^{k+1}}{k+1} \text{ sq. units.} \ \square$$

It is, as you can see, rather difficult to calculate areas bounded by curves by the methods developed above. Fortunately, there is an easier way, as we shall discover in Section 6.3.

REMARK: It was necessary to assume $a > 0$ in the above example for technical reasons. The result is also valid for $a = 0$ provided $k > -1$. In this case we have $\lim_{a \to 0+} a^{k+1} = 0$, so the area under $y = x^k$, above $y = 0$ between $x = 0$ and $x = b > 0$ is $A = b^{k+1}/(k+1)$ square units. For $k = 2$ this agrees with the result of Example 6.1.4.

EXERCISES

Use the techniques of Examples 6.1.3–4 (with subintervals of equal length) to find the areas of the regions specified in Exercises 1–7.

1. Below $y = 3x$, above $y = 0$, from $x = 0$ to $x = 1$.

2. Below $y = ax + b$, above $y = 0$, between $x = c$ and $x = d$. Assume $ax + b \geq 0$ on $[c, d]$.

3. Below $y = x^2$, above $y = 0$, from $x = 1$ to $x = 3$.

4. Below $y = x^2 + 1$, above $y = 0$, from $x = 0$ to $x = a > 0$.

5.*Below $y = x^2 + 2x + 3$, above $y = 0$, from $x = -1$ to $x = 2$.

6.*Below $y = e^x$, above $y = 0$, from $x = 0$ to $x = b > 0$.

7.*Below $y = 2^x$, above $y = 0$, from $x = -1$ to $x = 1$.

8.*Adapt the method used to prove Part (c) of Theorem 6.1.2 to show that

$$\sum_{j=1}^{n} j^3 = 1^3 + 2^3 + 3^3 + \ldots + n^3 = \frac{n^2(n+1)^2}{4}.$$

(Hint: start with the expansion of $(k+1)^4 - k^4$.)

9. Use the result of the previous exercise and the method of Example 6.1.4 to find the area of the region lying under $y = x^3$, above the x-axis, and between the vertical lines at $x = 0$ and $x = b > 0$.

10.*Use the subdivision of $[a, b]$ given in Example 6.1.5 to find the area under $y = 1/x$, above $y = 0$ from $x = a > 0$ to $x = b > a$.

11.*Identify the expression $S_n = \sum_{j=1}^{n} \frac{1}{n}\sqrt{1 - (j/n)^2}$ as a sum of areas of rectangles approximating the area of a certain region in the plane. Hence evaluate $\lim_{n \to \infty} S_n$.

6.2 THE DEFINITE INTEGRAL

In this section we generalize and make more precise the procedure used for finding areas developed in the previous section, and we use it to define the *definite integral* of a function f on an interval I. Let us assume, for the time being, that $f(x)$ is defined and continuous on the closed, finite interval $[a, b]$. We no longer assume f is nonnegative-valued.

Let P be a finite set of points arranged in order between a and b, say

$$P = \{a = x_0 < x_1 < x_2 < x_3 < \ldots < x_{n-1} < x_n = b\}.$$

P is called a **partition** of $[a, b]$; it divides $[a, b]$ into n subintervals of which the ith is $[x_{i-1}, x_i]$. The number n depends on the particular partition. We denote by

$$\Delta x_i = x_i - x_{i-1},$$

the length of the ith subinterval of P.

Since f is continuous on each subinterval $[x_{i-1}, x_i]$, it takes on maximum and minimum values at points of that interval (Theorem 1.7.7). Thus there are numbers u_i and l_i in $[x_{i-1}, x_i]$ such that

$$f(l_i) \leq f(x) \leq f(u_i) \qquad \text{whenever } x_{i-1} \leq x \leq x_i.$$

If $f(x) \geq 0$ on $[a, b]$, then $f(l_i)\,\Delta x_i$ and $f(u_i)\,\Delta x_i$ represent the areas of rectangles having the interval $[x_{i-1}, x_i]$ on the x-axis as base, and having tops passing through the lowest and highest points, respectively, on the graph of f on that interval. If A_i is that part of the area under $y = f(x)$ and above the x-axis which lies in the vertical strip between $x = x_{i-1}$ and $x = x_i$, then

$$f(l_i)\,\Delta x_i \leq A_i \leq f(u_i)\,\Delta x_i.$$

If f can take on negative values, then one or both of $f(l_i)\,\Delta x_i$ and $f(u_i)\,\Delta x_i$ can take on negative values; representing the negatives of areas of rectangles lying below the x-axis. In any event we always have $f(l_i)\,\Delta x_i \leq f(u_i)\,\Delta x_i$.

6.2.1
Upper and Lower Riemann Sums

The **lower (Riemann) sum** $L(f, P)$ and the **upper (Riemann) sum** $U(f, P)$ are defined for the function f and the partition P by:

$$L(f, P) = f(l_1)\,\Delta x_1 + f(l_2)\,\Delta x_2 + \cdots + f(l_n)\,\Delta x_n = \sum_{i=1}^{n} f(l_i)\,\Delta x_i,$$

$$U(f, P) = f(u_1)\,\Delta x_1 + f(u_2)\,\Delta x_2 + \cdots + f(u_n)\,\Delta x_n = \sum_{i=1}^{n} f(u_i)\Delta x_i.$$

Fig. 6.2.1 illustrates these Riemann sums as sums of *signed* areas of rectangles; any such areas which lie below the x-axis are counted as negative.

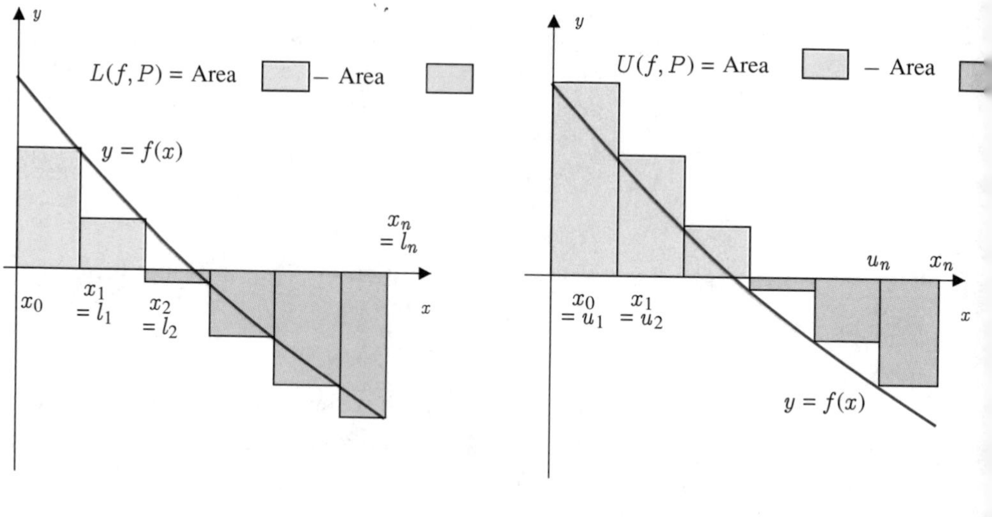

FIGURE 6.2.1

If we calculate $L(f,P)$ and $U(f,P)$ for partitions P having more and more points spaced closer and closer together, we expect that, in the limit, these Riemann sums will "converge" to a common value (which will be the area bounded by $y = f(x)$, $y = 0$, $x = a$, and $x = b$ if $f(x) \geq 0$ on $[a,b]$). This is indeed the case, but we cannot fully prove it yet. We can, however, describe a condition which will guarantee it.

If P_1 and P_2 are two partitions of $[a,b]$ such that every point of P_1 also belongs to P_2, then we say that P_2 is a **refinement** of P_1. It is not difficult to show that in this case

$$L(f,P_1) \leq L(f,P_2) \leq U(f,P_2) \leq U(f,P_1);$$

adding more points to a partition increases the lower sum and decreases the upper sum. (See Exercise 25 at the end of this section.) Given any two partitions, P_1 and P_2, we can form their **common refinement** P which consists of all of the points of P_1 and P_2. Thus

$$L(f,P_1) \leq L(f,P) \leq U(f,P) \leq U(f,P_2).$$

Hence every lower sum is less than or equal to every upper sum. Since the real numbers are complete, there must exist *at least one* real number I such that

$$L(f,P) \leq I \leq U(f,P), \qquad \text{for every partition } P.$$

If there is *only one* such number, we will call it the definite integral of f on $[a,b]$.

6.2.2 The Definite Integral

Suppose there is exactly one number I such that for every partition P of $[a,b]$ we have

$$L(f,P) \leq I \leq U(f,P).$$

Then we say that the function f is **integrable** on $[a,b]$, and we call I the **definite integral** of f on $[a,b]$. The definite integral is denoted by the symbol

$$I = \int_a^b f(x)\,dx.$$

We stress at once that the definite integral of $f(x)$ over $[a,b]$ is a *number*; it is not a function of x. It depends on the numbers a and b and on the particular function f, but not on the variable x (which is a **dummy variable** like the variable i in the symbol $\sum_{i=1}^{n} c(i)$), and can be replaced with any other variable without changing the value of the integral:

$$\int_a^b f(x)\,dx = \int_a^b f(t)\,dt.$$

For all partitions P of $[a, b]$ we have

$$L(f, P) \leq \int_a^b f(x)\, dx \leq U(f, P).$$

If $f(x) \geq 0$ on $[a, b]$, then the area of the region R bounded by the graph of $y = f(x)$, the x-axis, and the lines $x = a$ and $x = b$ is A square units, where $A = \int_a^b f(x)\, dx$. If $f(x) \leq 0$ on $[a, b]$, the area of R is $-\int_a^b f(x)\, dx$ square units. For general f, $\int_a^b f(x)\, dx$ is the area of that part of R lying above the x-axis less the area of that part lying below the x-axis (see Fig. 6.2.2). You can think of $\int_a^b f(x)\, dx$ as a "sum" of "areas" of infinitely many rectangles with heights $f(x)$ and "infinitesimally small widths" dx; in this sense it is a limit of the upper and lower Riemann sums.

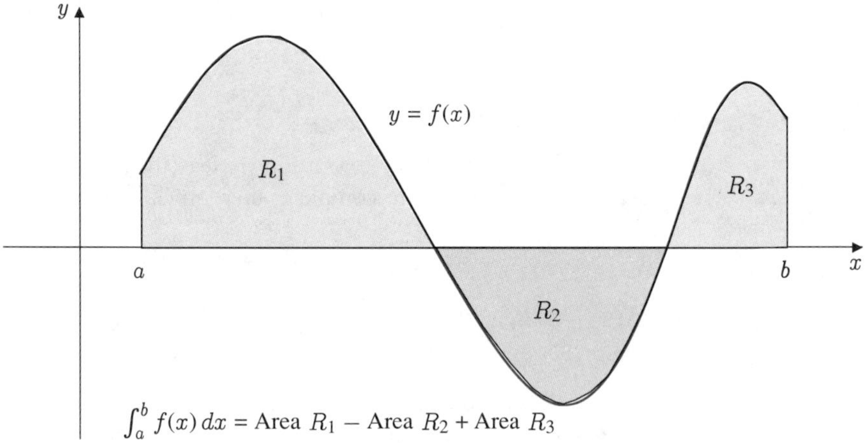

FIGURE 6.2.2

The various parts of the symbol $\int_a^b f(x)\, dx$ have their own names:

i) $\int$ is called the **integral sign**; it resembles the letter S since it represents the limit of a sum.

ii) a and b are called the **limits of integration**; a is the **lower limit**, b is the **upper limit**.

iii) The function f is the **integrand**; x is the **variable of integration**.

iv) dx is the **differential of** x. It replaces Δx in the Riemann sums. If an integrand depends on more than one variable, the differential tells you which one is the variable of integration.

If $P = \{a = x_0 < x_1 < x_2 < \cdots < x_n = b\}$ is a partition of $[a, b]$ and c_i is any point in the ith subinterval $[x_{i-1}, x_i]$ then the sum

$$R(f, P) = f(c_1)\, \Delta x_1 + f(c_2)\, \Delta x_2 + f(c_3)\, \Delta x_3 + \cdots + f(c_n)\, \Delta x_n = \sum_{i=1}^{n} f(c_i)\, \Delta x_i$$

is called a **Riemann sum** of f on $[a, b]$. Note in Fig. 6.2.3 that $R(f, P)$ is a sum of *signed* areas of rectangles between the x-axis and the curve $y = f(x)$. Evidently

$$L(f, P) \leq R(f, P) \leq U(f, P)$$

for any choice of the points c_i, $(1 \leq i \leq n)$. Therefore, if f is integrable on $[a, b]$ then

$$\lim R(f, P) = \int_a^b f(x)\, dx,$$

where the limit is taken as the number of subintervals of P increases to infinity in such a way the lengths of the subintervals approach zero.

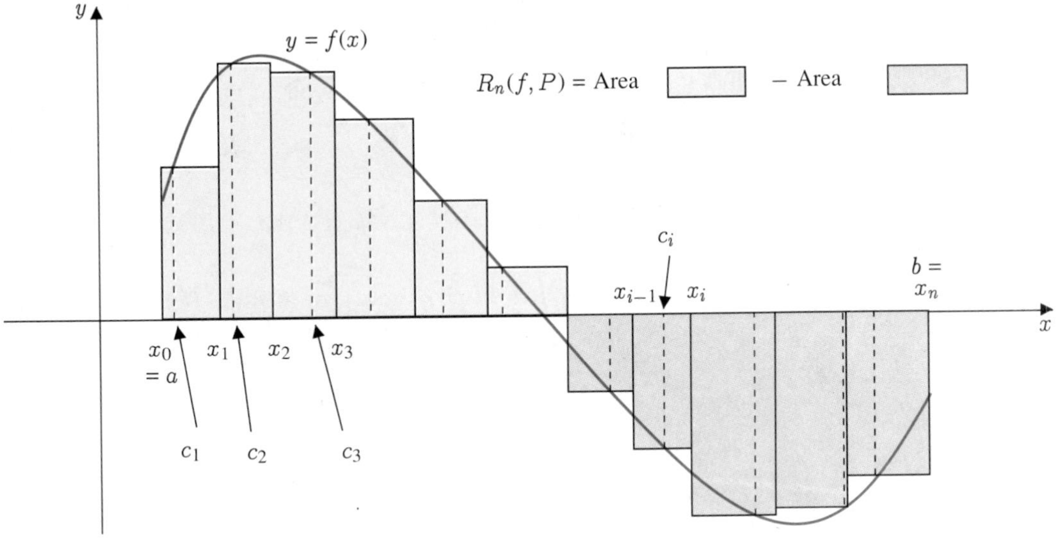

FIGURE 6.2.3

THEOREM 6.2.3 If f is continuous on $[a, b]$, then f is integrable on $[a, b]$. □

As noted above, we cannot prove this theorem in its full generality now. The proof makes use of the completeness property of the real numbers, and is given in Appendix 3. We can, however, make the following observation. In order to prove that f is integrable on $[a, b]$, it is sufficient that, for any given positive number ϵ, we should be able to find a partition P of $[a, b]$ for which $U(f, P) - L(f, P) < \epsilon$. This condition prevents there being more than one number I which is both greater than every lower sum and less than every upper sum. It is not difficult to find such a partition if the function f is nondecreasing (or if it is nonincreasing) on $[a, b]$. (See Exercise 24 at the end of this section.) Therefore nondecreasing and nonincreasing continuous functions are integrable; so, therefore, is any continuous function which is the sum of a nondecreasing and a nonincreasing function. This class of functions includes any continuous functions we are likely to encounter in concrete applications of calculus but, unfortunately, does not include all continuous functions.

Later in this section we will extend the definition of the definite integral to certain kinds of functions which are not continuous.

Properties of the Definite Integral

It is convenient to extend the definition of the definite integral $\int_a^b f(x)\,dx$ to allow $a = b$ and $a > b$ as well as $a < b$. The extension still involves partitions P having $x_0 = a$ and $x_n = b$ with intermediate points in order between these, so if $a = b$ we must have $\Delta x_i = 0$ for every i, and hence the integral is zero. If $a > b$ we have $\Delta x_i < 0$ for each i, so the integral will be negative for positive functions f and vice versa.

Some of the most important properties of the definite integral are summarized in the following theorem.

THEOREM 6.2.4 Let f and g be integrable on an interval containing the points a, b, and c. Then

a) $\displaystyle\int_a^a f(x)\,dx = 0.$

b) $\displaystyle\int_b^a f(x)\,dx = -\int_a^b f(x)\,dx.$

c) $\displaystyle\int_a^b \big(Af(x) + Bg(x)\big)\,dx = A\int_a^b f(x)\,dx + B\int_a^b g(x)\,dx$
(A and B are constants.)

d) $\displaystyle\int_a^b f(x)\,dx + \int_b^c f(x)\,dx = \int_a^c f(x)\,dx.$

e) If $f(x) \le g(x)$ and $a \le b$, then $\displaystyle\int_a^b f(x)\,dx \le \int_a^b g(x)\,dx.$

f) If $a \le b$, then $\displaystyle\left|\int_a^b f(x)\,dx\right| \le \int_a^b |f(x)|\,dx.$

g) If f is an odd function ($f(-x) = -f(x)$), then $\displaystyle\int_{-a}^a f(x)\,dx = 0.$

h) If f is an even function ($f(-x) = f(x)$), then $\displaystyle\int_{-a}^a f(x)\,dx = 2\int_0^a f(x)\,dx.$ □

All these properties can be deduced from the definition of definite integral. Most of them should appear intuitively obvious if you regard the integrals as representing (signed) areas. For instance, properties (d) and (e) are, respectively, properties (v) and (iv) of areas mentioned in the first paragraph of Section 6.1 (see Fig. 6.2.4). Property (f) is a generalization of the triangle inequality for numbers:

$$|x+y| \le |x| + |y|, \quad \text{or more generally,} \quad \left|\sum_{i=1}^n x_i\right| \le \sum_{i=1}^n |x_i|.$$

It follows from property (e), (assuming $|f|$ is integrable on $[a, b]$), since $-|f(x)| \le f(x) \le |f(x)|$. The symmetry properties (g) and (h), which are illustrated in Fig. 6.2.5, are particularly useful and should always be kept in mind when you are evaluating definite integrals as they can save you much unnecessary work.

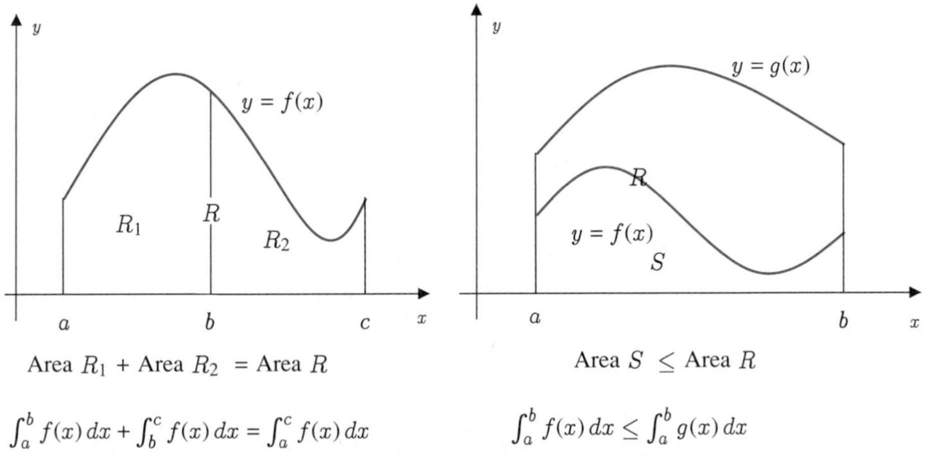

FIGURE 6.2.4

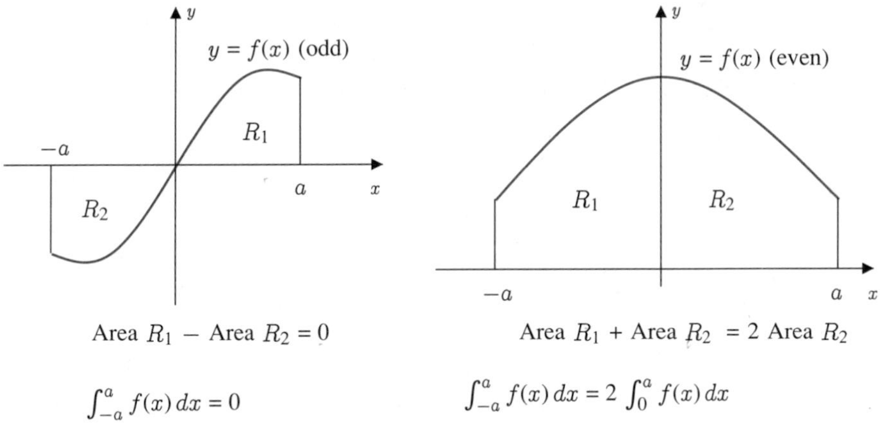

FIGURE 6.2.5

As yet we have no easy method for evaluating definite integrals. However, by using the above properties and interpreting the integrals as areas, we can sometimes find values of definite integrals by inspection.

EXAMPLE 6.2.5 Figs. 6.2.6–7, 6.2.8 and 6.2.9 illustrate parts (a)–(c), respectively.

a) $\int_{-2}^{2}(1+x)\,dx = \int_{-2}^{2} 1\,dx + \int_{-2}^{2} x\,dx = 4+0 = 4$. (area of rectangle plus difference of areas of two congruent triangles)

b) $\int_{-3}^{3}\sqrt{9-x^2}\,dx = 2\int_{0}^{3}\sqrt{9-x^2}\,dx = \dfrac{9\pi}{2}$. (area of a semicircle of radius 3)

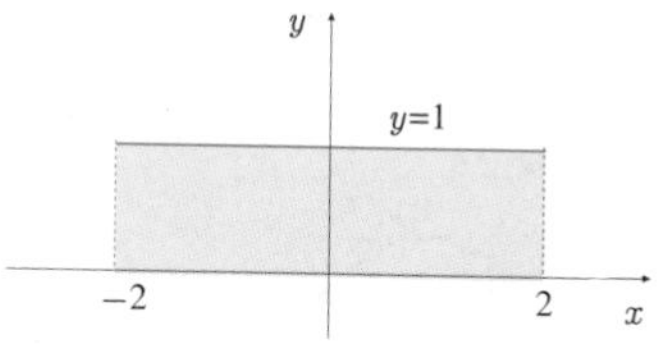

FIGURE 6.2.6

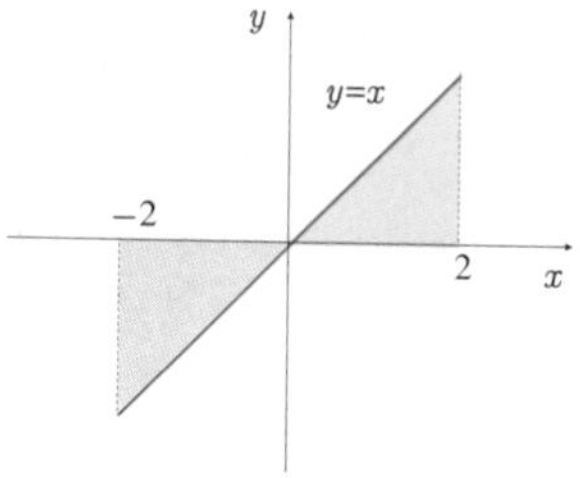

FIGURE 6.2.7

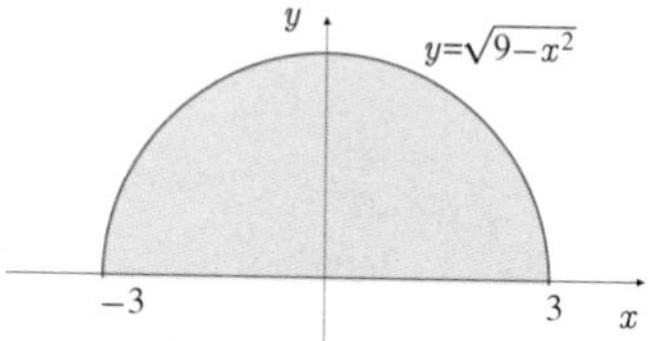

FIGURE 6.2.8

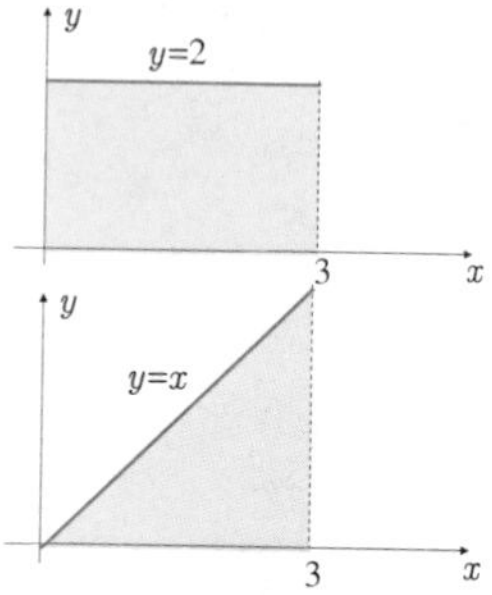

FIGURE 6.2.9

c) $\int_0^3 (2+x)\,dx = \int_0^3 2\,dx + \int_0^3 x\,dx = 6 + \frac{1}{2}(3)(3) = \frac{21}{2}$. (sum of areas of rectangle and triangle)

The Mean-Value Theorem for Integrals

Let f be a function continuous on the interval $[a, b]$. Then f assumes maximum and minimum values on the interval, say

$$m = f(l) \le f(x) \le f(u) = M \qquad \text{for all } x \text{ in } [a, b].$$

For the 2-point partition P of $[a, b]$ having $x_0 = a$ and $x_1 = b$, we have

$$m(b-a) = L(f, P) \le \int_a^b f(x)\,dx \le U(f, P) = M(b-a).$$

Therefore

$$f(l) = m \le \frac{1}{b-a}\int_a^b f(x)\,dx \le M = f(u).$$

By the Intermediate-Value Theorem 1.7.10, $f(x)$ must take on every value between the two values $f(l)$ and $f(u)$ at some point between l and u (Fig. 6.2.10). Hence there is a number c between l and u such that

$$f(c) = \frac{1}{b-a}\int_a^b f(x)\,dx,$$

that is, $\int_a^b f(x)\,dx$ is equal to the area $(b-a)f(c)$ of a rectangle with base width $b-a$ and height $f(c)$ for some c between a and b. This is the Mean-Value Theorem for integrals.

THEOREM 6.2.6 (**The Mean-Value Theorem for Integrals**) If f is continuous on $[a, b]$, then there exists a point c in $[a, b]$ such that

$$\int_a^b f(x)\,dx = (b-a)f(c). \ \square$$

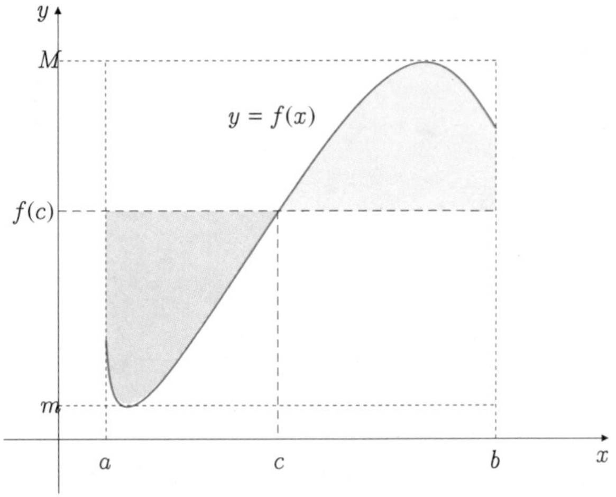

FIGURE 6.2.10

Observe in Fig. 6.2.10 that the area below the curve $y = f(x)$ and above the line $y = f(c)$ is equal to the area above $y = f(x)$ and below $y = f(c)$. In this sense, $f(c)$ is the average value of the function $f(x)$ on the interval $[a, b]$.

6.2.7
Average Value of a Function

If f is integrable on $[a, b]$, then the **average value** or **mean value** of f on $[a, b]$ is

$$\frac{1}{b-a}\int_a^b f(x)\,dx.$$

EXAMPLE 6.2.8 The average value of $f(x) = 2x$ on $[1, 5]$ is (see Fig. 6.2.11)

$$\frac{1}{5-1}\int_1^5 2x\,dx = \frac{1}{4}\left(4 \times 2 + \tfrac{1}{2}(4 \times 8)\right) = 6.$$

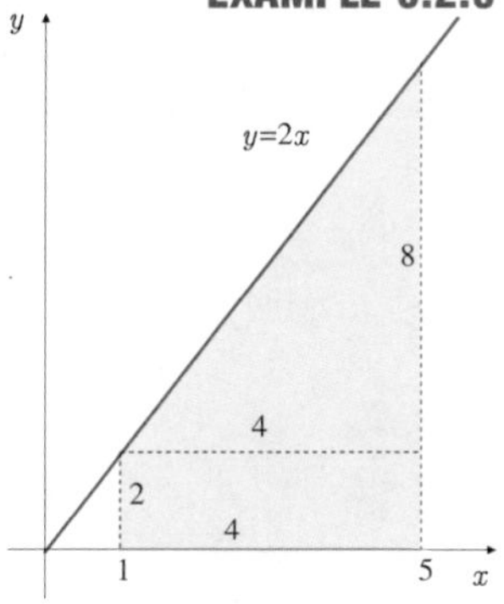

FIGURE 6.2.11

Definite Integrals of Piecewise Continuous Functions.

The definition of integrability and the definite integral given above can be extended to a wider class than the continuous functions. One simple but very important extension is to the class of *piecewise continuous functions*.

Consider the graph $y = f(x)$ shown in Fig. 6.2.12. Although f is not continuous at all points of $[a, b]$, (it is discontinuous at c_1 and c_2), it is clear that the region lying under the graph and above the x-axis between $x = a$ and $x = b$ does have an area. We would like to represent this area as

$$\int_a^{c_1} f(x)\,dx + \int_{c_1}^{c_2} f(x)\,dx + \int_{c_2}^{b} f(x)\,dx.$$

This is reasonable because there are continuous functions on $[a, c_1]$, $[c_1, c_2]$ and $[c_2, b]$ equal to $f(x)$ on the corresponding open intervals, (a, c_1), (c_1, c_2) and (c_2, b).

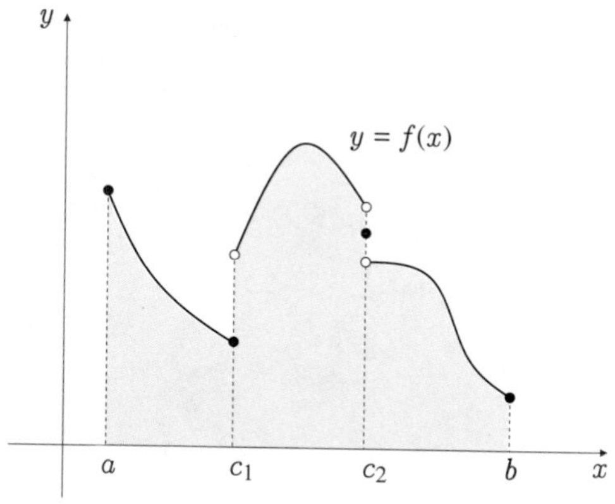

FIGURE 6.2.12

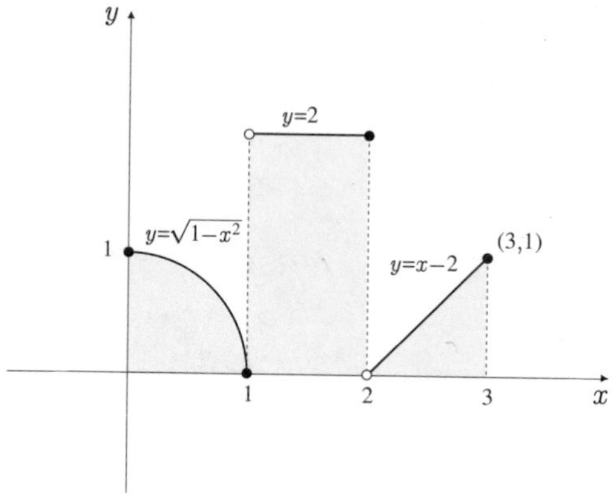

FIGURE 6.2.13

6.2.9
Piecewise Continuous Functions

Let $c_0 < c_1 < c_2 < \cdots < c_n$ be a finite set of points on the real line. A function f defined on $[c_0, c_n]$ except possibly at some of the points c_i, $(0 \leq i \leq n)$, is called **piecewise continuous** on that interval if for each i, $(1 \leq i \leq n)$, there exists a function F_i continuous on $[c_{i-1}, c_i]$ such that

$$f(x) = F_i(x) \qquad \text{on the interval} \quad (c_{i-1}, c_i).$$

In this case, we define the definite integral of f on from c_0 to c_n to be

$$\int_{c_0}^{c_n} f(x)\,dx = \sum_{i=1}^{n} \int_{c_{i-1}}^{c_i} F_i(x)\,dx.$$

This definition looks more complicated than it is. Here is an example.

EXAMPLE 6.2.10 Find $\displaystyle\int_0^3 f(x)\,dx$ where $f(x) = \begin{cases} \sqrt{1-x^2} & \text{if } 0 \leq x \leq 1 \\ 2 & \text{if } 1 < x \leq 2 \\ x-2 & \text{if } 2 < x \leq 3. \end{cases}$

SOLUTION See Fig. 6.2.13. The value of the integral is the sum of the shaded areas:

$$\int_0^3 f(x)\,dx = \int_0^1 \sqrt{1-x^2}\,dx + \int_1^2 2\,dx + \int_2^3 (x-2)\,dx$$

$$= \left(\tfrac{1}{4} \times \pi \times 1^2\right) + (2 \times 1) + \left(\tfrac{1}{2} \times 1 \times 1\right) = \frac{\pi + 10}{4} \text{ square units.}$$

EXERCISES

In Exercises 1–6 let P_n denote the partition of the given interval $[a, b]$ into n subintervals of equal length $\Delta x_i = (b-a)/n$. Evaluate $L(f, P_n)$ and $U(f, P_n)$ for the given functions f and the given values of n.

1. $f(x) = x$ on $[0, 2]$, with $n = 8$
2. $f(x) = x^2$ on $[0, 4]$, with $n = 4$
3. $f(x) = e^x$ on $[-2, 2]$, with $n = 4$
4. $f(x) = \ln x$ on $[1, 2]$, with $n = 5$
5. $f(x) = \sin x$ on $[0, \pi]$, with $n = 6$
6. $f(x) = \cos x$ on $[0, 2\pi]$, $n = 4$
7. Calculate $L(f, P_n)$ and $U(f, P_n)$ for the function $f(x) = x$ on the interval $[0, 1]$, where P_n is the partition of the interval into equal subintervals each of length $1/n$. Show that $\lim_{n\to\infty} L(f, P_n) = \lim_{n\to\infty} U(f, P_n)$. Hence f is integrable on $[0, 1]$. (Why?) What is $\int_0^1 f(x)\,dx$?
8. Repeat the previous exercise for $f(x) = x^2$.

Evaluate the integrals in Exercises 9–16 by using symmetry and/or interpreting them as areas.

9. $\int_{-2}^{2} (x+2)\,dx$
10. $\int_{0}^{2} (3x+1)\,dx$
11. $\int_{a}^{b} x\,dx$
12. $\int_{-\sqrt{2}}^{0} \sqrt{2-x^2}\,dx$
13. $\int_{-\pi}^{\pi} \sin(x^3)\,dx$
14. $\int_{-a}^{a} (a-|x|)\,dx$
15. $\int_{-1}^{1} (x^5 - 3x^3 + \pi)\,dx$
16. $\int_{0}^{2} \sqrt{2x-x^2}\,dx$
17. Find the average value of $f(x) = x + 2$ over $[a, b]$.
18. Find the average value of $f(x) = (4-x^2)^{1/2}$ over the interval $[0, 2]$.
19. Evaluate $\int_{-1}^{2} \operatorname{sgn} x\,dx$. Recall that $\operatorname{sgn} x$ is 1 if $x > 0$ and -1 if $x < 0$.
20. Find $\int_{-3}^{2} f(x)\,dx$ where $f(x) = \begin{cases} 1+x & \text{if } x < 0 \\ 2 & \text{if } x \ge 0 \end{cases}$.
21.*Evaluate $\int_{0}^{2} \sqrt{4-x^2}\,\operatorname{sgn}(x-1)\,dx$.
22. Evaluate $\int_{0}^{3.5} [x]\,dx$ where $[x]$ is the greatest integer less than or equal to x. (See Example 1.6.2(b).)
23. Find the average value of the function $f(x) = |x+1|\operatorname{sgn} x$ on the interval $[-2, 2]$.
24.*If f is continuous and nondecreasing on $[a, b]$, and P_n is the partition of $[a, b]$ into n subintervals of equal length ($\Delta x_i = (b-a)/n$ for $1 \le i \le n$), show that

$$U(f, P_n) - L(f, P_n) = \frac{(b-a)\big(f(b) - f(a)\big)}{n}.$$

Since we can make the right side as small as we please by choosing n large enough, f must be integrable on $[a, b]$.

25.*Let $P = \{a = x_0 < x_1 < x_2 < \cdots < x_n = b\}$ be a partition of $[a, b]$ and let P' be a refinement of P having one more point, x', satisfying, say, $x_{j-1} < x' < x_j$ for some j between 1 and n. Show that $L(f, P) \le L(f, P') \le U(f, P') \le U(f, P)$ for any continuous function f. (Hint: consider the maximum and minimum values of f on the intervals $[x_{i-1}, x_i]$, $[x_{i-1}, x']$ and $[x', x_i]$.) Hence deduce that $L(f, P) \le L(f, P") \le U(f, P") \le U(f, P)$ if $P"$ is *any* refinement of P.

6.3 THE FUNDAMENTAL THEOREM OF CALCULUS

In this section we demonstrate the relationship between the definite integral defined in Section 6.2 and the indefinite integral (or general antiderivative) introduced in Section 2.8. A consequence of this relationship is that we will be able to calculate the areas under graphs of functions whose antiderivatives we can find.

In Section 3.6 we wanted to find a function whose derivative was $1/x$. We solved this problem by defining a function ($\ln x$) in terms of the area under the graph of $y = 1/x$. This idea motivates, and is a special case of, the following theorem.

THEOREM 6.3.1 (**The Fundamental Theorem of Calculus**) Suppose that the function $f(x)$ is continuous on an interval I containing the point a. Let a function $F(x)$ be defined on I by

$$F(x) = \int_a^x f(t)\,dt.$$

Then F is differentiable on I and $F'(x) = f(x)$; symbolically,

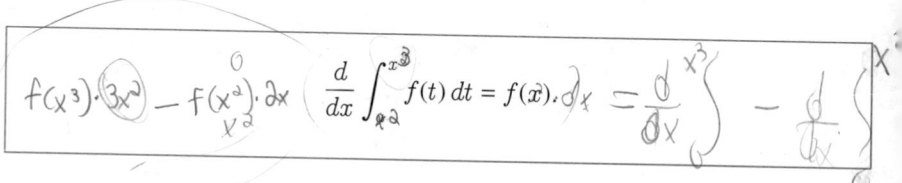

$$\frac{d}{dx}\int_a^x f(t)\,dt = f(x).$$

If $G(x)$ is *any* antiderivative of $f(x)$ on I (that is, if $G'(x) = f(x)$ on I), then for any b in I we have

$$\int_a^b f(x)\,dx = G(b) - G(a).$$

PROOF Using the definition of the derivative, we calculate

$$\begin{aligned} F'(x) = \lim_{h\to 0}\frac{F(x+h)-F(x)}{h} &= \lim_{h\to 0}\frac{1}{h}\left(\int_a^{x+h} f(t)\,dt - \int_a^x f(t)\,dt\right) \\ &= \lim_{h\to 0}\frac{1}{h}\int_x^{x+h} f(t)\,dt \qquad \text{by Theorem 6.2.4(d)} \\ &= \lim_{h\to 0}\frac{1}{h}\,hf(c) \qquad \text{for some } c \text{ between } x \text{ and } x+h \text{ (Theorem 6.2.6)} \\ &= \lim_{c\to x} f(c) = f(x) \qquad \text{since } c\to x \text{ as } h\to 0 \text{ and } f \text{ is continuous.} \end{aligned}$$

Also, if $G'(x) = f(x)$, then $F(x) = G(x) + C$ on I for some constant C (by Theorem 2.8.2). Hence

$$\int_a^x f(t)\,dt = F(x) = G(x) + C.$$

Let $x = a$ and obtain $0 = G(a) + C$ via Theorem 6.2.4(a), so $C = -G(a)$. Now let $x = b$ to get

$$\int_a^b f(t)\,dt = G(b) - G(a).$$

Of course, we may now replace t with x (or any other variable) as the variable of integration on the left-hand side. □

REMARK: You should remember *both* conclusions of the Fundamental Theorem; they are both useful. The first conclusion tells you how to differentiate a definite integral with respect to its upper limit. The second conclusion tells you how to evaluate a definite integral if you can find an antiderivative of the integrand.

6.3.2 Evaluation Symbol

To facilitate the evaluation of definite integrals using the Fundamental Theorem, we define the **evaluation symbol**

$$F(x)\Big|_a^b = F(b) - F(a).$$

Thus

$$\int_a^b f(x)\,dx = \left(\int f(x)\,dx\right)\Big|_a^b$$

where $\int f(x)\,dx$ denotes the indefinite integral or general antiderivative of f (see Section 2.8). When evaluating a definite integral this way, we will omit the constant of integration ($+C$) from the indefinite integral because it cancels out in the subtraction:

$$(F(x)+C)\Big|_a^b = F(b)+C-(F(a)+C) = F(b)-F(a) = F(x)\Big|_a^b.$$

Thus *any* antiderivative of f can be used to calculate the definite integral.

EXAMPLE 6.3.3

$$\begin{aligned}\int_{-1}^{2}(x^2-3x+2)\,dx &= \left(\frac{1}{3}x^3-\frac{3}{2}x^2+2x\right)\Big|_{-1}^{2}\\ &= \frac{1}{3}(8)-\frac{3}{2}(4)+4-\left(\frac{1}{3}(-1)-\frac{3}{2}(1)+(-2)\right)=\frac{9}{2}.\end{aligned}$$

EXAMPLE 6.3.4 Find the area of the plane region lying above the x-axis and under the curve $y = 3x - x^2$.

SOLUTION We need to find the points where the curve $y = 3x - x^2$ meets the x-axis. These are solutions of the equation.

$$0 = 3x - x^2 = x(3-x).$$

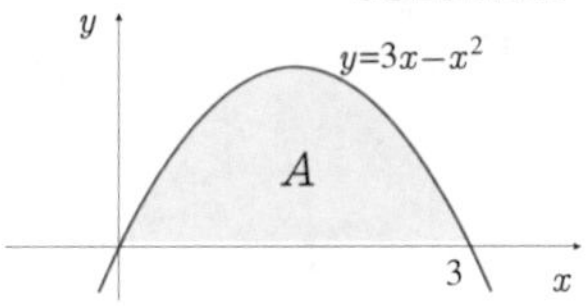

FIGURE 6.3.1

The only roots are $x = 0$ and $x = 3$ (see Fig. 6.3.1). Hence the area of the region is given by

$$\begin{aligned}A = \int_0^3 (3x-x^2)\,dx &= \left(\frac{3}{2}x^2-\frac{1}{3}x^3\right)\Big|_0^3\\ &= \frac{27}{2}-\frac{27}{3}=\frac{27}{6}=\frac{9}{2}\text{ square units.}\end{aligned}$$

EXAMPLE 6.3.5 Find the area under the curve $y = \sin x$, above $y = 0$ from $x = 0$ to $x = \pi$.

SOLUTION The required area, illustrated in Fig. 6.3.2, is

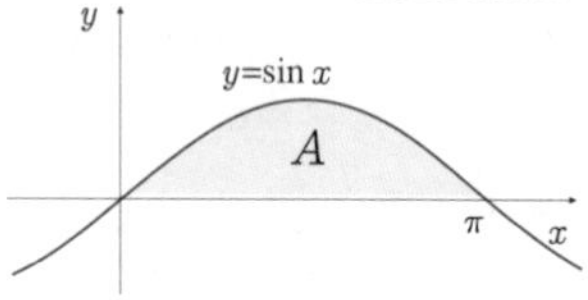

FIGURE 6.3.2

$$A = \int_0^\pi \sin x\,dx = -\cos x\Big|_0^\pi$$
$$= -(-1 - (1)) = 2 \text{ square units.}$$

Note that while the definite integral is a pure number, an area is a geometric quantity implicitly involving units. Where the units of length along the x-axis and y-axis are not specified, areas should be quoted in square units.

EXAMPLE 6.3.6 Find the area of the region R lying above the line $y = 1$ and below the curve $y = 5/(x^2 + 1)$.

SOLUTION The region is shaded in Fig. 6.3.3. To find the intersections of $y = 1$ and $y = 5/(x^2 + 1)$, we must solve these equations simultaneously:

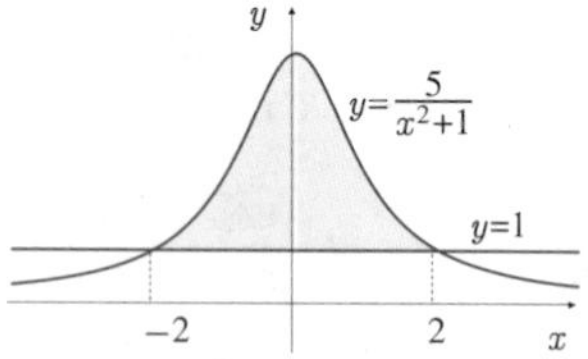

FIGURE 6.3.3

$$1 = \frac{5}{x^2 + 1}$$

so $x^2 + 1 = 5$, $x^2 = 4$, and $x = \pm 2$.

The area of the shaded region is

$$A = \int_{-2}^{2} \frac{5}{x^2 + 1}\,dx - 4 = 2\int_0^2 \frac{5}{x^2 + 1}\,dx - 4$$
$$= 10\tan^{-1} x\Big|_0^2 - 4 = 10\tan^{-1} 2 - 4 \text{ square units.}$$

Observe the use of symmetry to replace the lower limit of integration by 0. It is easier to substitute 0 into the antiderivative than -2.

EXAMPLE 6.3.7 Find the average value of $f(x) = e^{-x} + \cos x$ on the interval $[-\pi/2, 0]$.

SOLUTION The average value is

$$\frac{1}{0 - \left(-\frac{\pi}{2}\right)} \int_{-(\pi/2)}^{0} (e^{-x} + \cos x)\,dx = \frac{2}{\pi}\left(-e^{-x} + \sin x\right)\Big|_{-(\pi/2)}^{0}$$
$$= \frac{2}{\pi}\left(-1 + 0 + e^{\pi/2} - (-1)1\right) = \frac{2}{\pi}e^{\pi/2}.$$

Beware of integrals of the form $\int_a^b f(x)\,dx$ where f is not continuous at *all* points of the interval $[a, b]$. The fundamental theorem does not apply in such cases.

EXAMPLE 6.3.8 We know that $\frac{d}{dx} \ln|x| = \frac{1}{x}$ if $x \neq 0$. It is *incorrect*, however, to state that

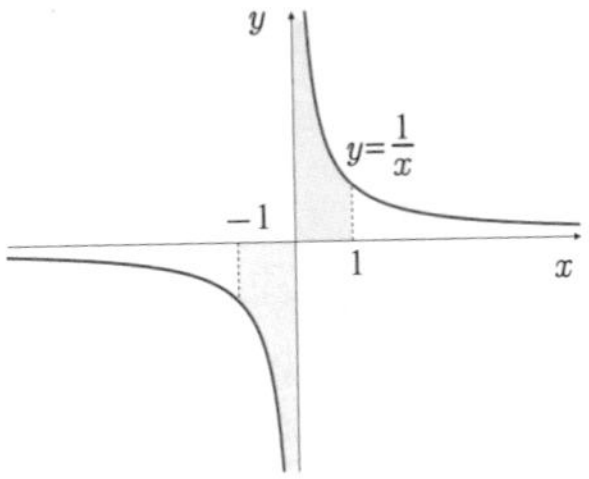

FIGURE 6.3.4

$$\int_{-1}^{1} \frac{dx}{x} = \ln|x| \bigg|_{-1}^{1} = 0 - 0 = 0,$$

even though $1/x$ is an odd function. In fact, $1/x$ is discontinuous and unbounded on $[-1, 1]$, and it is not integrable on $[-1, 0)$ or $(0, 1]$ (Fig. 6.3.4). Observe that

$$\lim_{c \to 0+} \int_{c}^{1} \frac{1}{x}\, dx = \lim_{c \to 0+} -\ln c = \infty,$$

so both shaded regions in Fig. 6.3.4 have infinite area. Integrals of this type are called **improper integrals**. We deal with them in Section 6.8.

EXAMPLE 6.3.9 Find the derivatives of the following functions.

a) $F(x) = \int_{x}^{3} e^{-t^2}\, dt,$ b) $G(x) = \sin x \int_{-4}^{5x} e^{-t^2}\, dt,$ c) $H(x) = \int_{x^2}^{x^3} e^{-t^2}\, dt.$

SOLUTION These are direct applications of the first conclusion of the Fundamental Theorem together with other differentiation rules as appropriate.

a) Observe that $F(x) = -\int_3^x e^{-t^2}\, dt$ (by Theorem 6.2.4(b)). Therefore, by the Fundamental Theorem, $F'(x) = -e^{-x^2}$.

b) In this example we use the Product Rule and the Chain Rule as well as the Fundamental Theorem:

$$G'(x) = \cos x \int_{-4}^{5x} e^{-t^2}\, dt + 5 \sin x\, e^{-25x^2}.$$

c) First we split the integral into a difference of integrals in each of which the variable x appears only in the upper limit. Then we differentiate each using the Fundamental Theorem and the Chain Rule:

$$H(x) = \int_{0}^{x^3} e^{-t^2}\, dt - \int_{0}^{x^2} e^{-t^2}\, dt$$
$$H'(x) = 3x^2 e^{-x^6} - 2x\, e^{-x^4}.$$

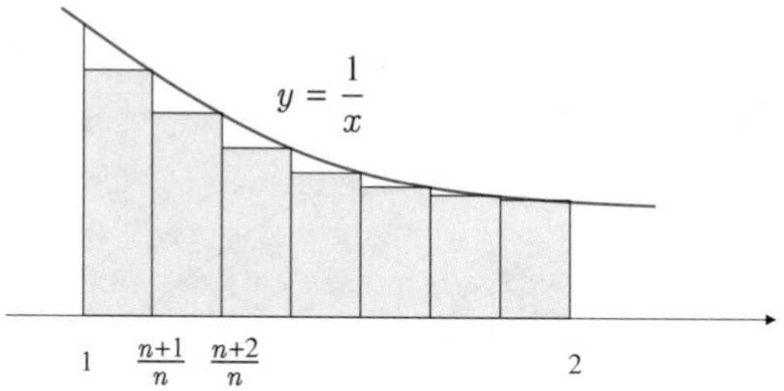

FIGURE 6.3.5

EXAMPLE 6.3.10 Evaluate $\lim_{n\to\infty} \frac{1}{n}\left(\frac{n}{n+1} + \frac{n}{n+2} + \frac{n}{n+3} + \cdots + \frac{n}{n+n}\right)$.

SOLUTION The sum looks as though it might be a Riemann sum for some function. Indeed, it is just

$$\frac{1}{n}\left(\frac{1}{1+\frac{1}{n}} + \frac{1}{1+\frac{2}{n}} + \frac{1}{1+\frac{3}{n}} + \cdots + \frac{1}{1+\frac{n}{n}}\right),$$

which is the lower sum $L(f, P)$ for the function $f(x) = \frac{1}{x}$ and the partition P of the interval $[1, 2]$ into n equal subintervals. See Fig. 6.3.5. Hence the limit of the sum is

$$\int_1^2 \frac{dx}{x} = \ln x \Big|_1^2 = \ln 2 - \ln 1 = \ln 2.$$

EXERCISES

Evaluate the definite integrals in Exercises 1–12.

1. $\int_0^1 (x^5 + x^3)\, dx$
2. $\int_0^4 \sqrt{x}\, dx$
3. $\int_{-\pi/4}^{-\pi/6} \cos x\, dx$
4. $\int_{-2}^{-1} \left(\frac{1}{x^2} - \frac{1}{x^3}\right) dx$
5. $\int_{-1}^{1} \frac{dx}{\sqrt{4 - x^2}}$
6. $\int_0^{\pi/3} \sec^2 \theta\, d\theta$
7. $\int_{-2}^{0} \frac{dx}{4 + x^2}$
8. $\int_4^9 \left(\sqrt{x} - \frac{1}{\sqrt{x}}\right) dx$
9. $\int_{-2}^{2} (x^2 + 3)^2\, dx$
10. $\int_{-2}^{-1} \frac{dt}{t}$
11. $\int_0^3 \frac{2x}{x^2 + 1}\, dx$
12. $\int_{-1}^{1} 2^x\, dx$

Find the area of the region R specified in Exercises 13–24. Sketch each region.

13. Bounded by $y = x^4$, $y = 0$, $x = 0$, and $x = 1$
14. Bounded by $y = 1/x$, $y = 0$, $x = e$, and $x = e^2$
15. Above $y = x^2 - 4x$ and below the x-axis
16. Bounded by $y = 5 - 2x - 3x^2$, $y = 0$, $x = -1$, and $x = 1$
17. Bounded by $y = x^2 - 3x + 3$ and $y = 1$
18. Below $y = \sqrt{x}$ and above $y = \frac{x}{2}$
19. Above $y = x^2$ and to the right of $x = y^2$
20. Above $y = |x|$ and below $y = 12 - x^2$
21. Bounded by $y = x^{1/3} - x^{1/2}$, $y = 0$, $x = 0$, and $x = 1$
22. Under $y = e^{-x}$ and above $y = 0$ from $x = -a$ to $x = 0$
23. Below $y = 1 - \cos x$ and above $y = 0$ between two consecutive intersections of these graphs

24. Below $y = x^{-1/3}$ and above $y = 0$ from $x = 1$ to $x = 27$

25. Bounded by $y = x^{-3}$, $y = 0$, $x = -4$, and $x = -2$

In Exercises 26–30 find the average values of the given functions over the intervals specified.

26. $f(x) = e^{3x}$ over $[-2, 2]$

27. $f(x) = 1 + x + x^2 + x^3$ over $[0, 2]$

28. $f(x) = 1 + \sin x$ over $[0, \pi]$ and over $[-\pi, \pi]$

29. $f(x) = \sqrt{1 - x^2}$ over $[-1, 1]$

30. $g(t) = \begin{cases} 0 & \text{if } 0 \le t \le 1 \\ 1 & \text{if } 1 < t \le 3 \end{cases}$ over $[0, 3]$

Find the indicated derivatives in Exercises 31–38.

31. $\dfrac{d}{dx}\displaystyle\int_2^x \frac{\sin t}{t}\,dt$

32. $\dfrac{d}{dt}\displaystyle\int_t^3 \frac{\sin x}{x}\,dx$

33. $\dfrac{d}{dx}\displaystyle\int_{x^2}^0 \frac{\sin t}{t}\,dt$

34. $\dfrac{d}{dx}x^2\displaystyle\int_0^{x^2} \frac{\sin u}{u}\,du$

35. $\dfrac{d}{dt}\displaystyle\int_{-\pi}^t \frac{\cos y}{1 + y^2}\,dy$

36. $\dfrac{d}{dt}\dfrac{\int_{2t}^{10}\sqrt{1+s}\,ds}{\sqrt{1+2t}}$

37. $\dfrac{d}{dx}\displaystyle\int_{ax}^{bx} \frac{dt}{5 + t^4}$

38. $\dfrac{d}{d\theta}\displaystyle\int_{\sin\theta}^{\cos\theta} \frac{1}{1 - x^2}\,dx$

39. $\dfrac{d}{dx}F(\sqrt{x})$, if $F(t) = \displaystyle\int_0^t \cos(x^2)\,dx$

40. $H'(2)$, if $H(x) = 3x\displaystyle\int_4^{x^2} e^{-\sqrt{t}}\,dt$

41.*Criticize the following erroneous calculation:

$$\int_{-1}^{1} \frac{dx}{x^2} = -\frac{1}{x}\Bigg|_{-1}^{1} = -1 + \frac{1}{-1} = -2.$$

Exactly where did the error occur? Why is -2 an unreasonable value for the integral?

42.*Use a definite integral to define a function $F(x)$ having derivative $\dfrac{\sin x}{1 + x^2}$ for all x and satisfying $F(17) = 0$.

43.*Does the function $F(x) = \displaystyle\int_0^{2x - x^2} \cos\left(\frac{1}{1 + t^2}\right) dt$ have a maximum or a minimum value? Justify your answer.

Evaluate the limits in Exercises 44–46.

44.* $\displaystyle\lim_{n\to\infty} \frac{1}{n}\left(\left(1 + \frac{1}{n}\right)^5 + \left(1 + \frac{2}{n}\right)^5 + \cdots + \left(1 + \frac{n}{n}\right)^5\right).$

45.* $\displaystyle\lim_{n\to\infty} \frac{\pi}{n}\left(\sin\frac{\pi}{n} + \sin\frac{2\pi}{n} + \sin\frac{3\pi}{n} + \cdots + \sin\frac{n\pi}{n}\right).$

46.* $\displaystyle\lim_{n\to\infty} \left(\frac{n}{n^2 + 1} + \frac{n}{n^2 + 4} + \frac{n}{n^2 + 9} + \cdots + \frac{n}{2n^2}\right).$

6.4 THE METHOD OF SUBSTITUTION

As we have seen, the evaluation of definite integrals is most easily carried out if we can antidifferentiate the integrand. In the next four sections we develop some *techniques of integration*, that is, methods for finding antiderivatives of functions. Although the techniques we develop can be used for a large class of functions, they will not work for all functions we might want to integrate. If a definite integral involves an integrand whose antiderivative is either impossible or very difficult to find, we may wish, instead, to approximate the definite integral by numerical means. Techniques for doing that will be presented in Chapter 7.

Let us begin by assembling some known indefinite integrals. These results have all emerged during our development of differentiation formulas for elementary functions. You should *memorize* them.

6.4.1 Some Elementary Integrals

1. $\int 1\,dx = x + C$	2. $\int x\,dx = \frac{1}{2}x^2 + C$
3. $\int x^2\,dx = \frac{1}{3}x^3 + C$	4. $\int \frac{1}{x^2}\,dx = -\frac{1}{x} + C$
5. $\int \frac{1}{\sqrt{x}}\,dx = 2\sqrt{x} + C$	6. $\int \frac{1}{x}\,dx = \ln\|x\| + C$
7. $\int x^r\,dx = \frac{1}{r+1}x^{r+1} + C \qquad (r \neq -1)$	
8. $\int \sin ax\,dx = -\frac{1}{a}\cos ax + C$	9. $\int \cos ax\,dx = \frac{1}{a}\sin ax + C$
10. $\int \sec^2 ax\,dx = \frac{1}{a}\tan ax + C$	11. $\int \csc^2 ax\,dx = -\frac{1}{a}\cot ax + C$
12. $\int \sec ax\,\tan ax\,dx = \frac{1}{a}\sec ax + C$	
13. $\int \csc ax\,\cot ax\,dx = -\frac{1}{a}\csc ax + C$	
14. $\int \frac{1}{a^2 + x^2}\,dx = \frac{1}{a}\tan^{-1}\frac{x}{a} + C$	
15. $\int \frac{1}{\sqrt{a^2 - x^2}}\,dx = \sin^{-1}\frac{x}{a} + C \qquad (a > 0)$	
16. $\int e^{ax}\,dx = \frac{1}{a}e^{ax} + C$	17. $\int b^{ax}\,dx = \frac{1}{a\ln b}b^{ax} + C$
18. $\int \cosh ax\,dx = \frac{1}{a}\sinh ax + C$	19. $\int \sinh ax\,dx = \frac{1}{a}\cosh ax + C$

Note that formulas 1–5 are special cases of formula 7, which holds on any interval where x^r makes sense.

The linearity formula

$$\int (A\,f(x) + B\,g(x))\,dx = A\int f(x)\,dx + B\int g(x)\,dx$$

makes it possible to integrate sums and constant multiples of other functions.

EXAMPLE 6.4.2

a) $\int (x^4 - 3x^3 + 8x^2 - 6x - 7)\,dx = \frac{x^5}{5} - \frac{3x^4}{4} + \frac{8x^3}{3} - 3x^2 - 7x + C$

b) $\int \left(5x^{3/5} - \frac{3}{2 + x^2}\right)dx = \frac{25}{8}x^{8/5} - \frac{3}{\sqrt{2}}\tan^{-1}\frac{x}{\sqrt{2}} + C$

c) $\int (4\cos 5x - 5\sin 3x)\,dx = \frac{4}{5}\sin 5x + \frac{5}{3}\cos 3x + C$

d) $\displaystyle\int \left(\frac{1}{\pi x} + a^{\pi x}\right) dx = \frac{1}{\pi} \ln|x| + \frac{1}{\pi \ln a} a^{\pi x} + C, \quad (a > 0).$

When an integral cannot be evaluated by inspection, as those in Example 6.4.2 can, we require one or more special techniques. The principal of these techniques is the **Method of Substitution**, the integral version of the Chain Rule. If we rewrite the Chain Rule,

$$\frac{d}{dx} f(g(x)) = f'\left(g(x)\right) g'(x),$$

in integral form, we obtain

$$\int f'\left(g(x)\right) g'(x)\, dx = f\left(g(x)\right) + C.$$

Observe that the following formalism would produce this latter formula even if we did not already know it was true:

Let $u = g(x)$. Then $du/dx = g'(x)$. Rewrite this in differential form: $du = g'(x)\, dx$. Thus

$$\int f'\left(g(x)\right) g'(x)\, dx = \int f'(u)\, du = f(u) + C = f\left(g(x)\right) + C.$$

EXAMPLE 6.4.3 a) $\displaystyle\int \frac{x}{x^2+1}\, dx$ Let $u = x^2 + 1$.
Then $du = 2x\, dx$.
Thus $x\, dx = \frac{1}{2}\, du$.

$$= \frac{1}{2}\int \frac{du}{u} = \frac{1}{2} \ln|u| + C = \frac{1}{2} \ln(x^2+1) + C = \ln\sqrt{x^2+1} + C.$$

(Both versions of the final answer are equally acceptable.)

b) $\displaystyle\int \frac{\sin(3\ln x)}{x}\, dx$ Let $u = 3\ln x$.
Then $du = \dfrac{3}{x}\, dx$.

$$= \frac{1}{3}\int \sin u\, du = -\frac{1}{3}\cos u + C = -\frac{1}{3}\cos(3\ln x) + C.$$

c) $\displaystyle\int e^x\sqrt{1+e^x}\, dx$ Let $v = 1 + e^x$.
Then $dv = e^x\, dx$.

$$= \int v^{1/2}\, dv = \frac{2}{3} v^{3/2} + C = \frac{2}{3}(1+e^x)^{3/2} + C.$$

Sometimes the appropriate substitution is not as obvious as it was in (a)–(c), and it may be necessary to play with the integrand a bit to put it into a better form for substitution.

d)
$$\int \frac{dx}{x^2+4x+5} = \int \frac{dx}{(x+2)^2+1} \qquad \text{Let } t = x+2,\ dt = dx.$$
$$= \int \frac{dt}{t^2+1}$$
$$= \tan^{-1} t + C = \tan^{-1}(x+2) + C.$$

e)
$$\int \frac{dx}{\sqrt{e^{2x}-1}} = \int \frac{dx}{e^x\sqrt{1-e^{-2x}}}$$
$$= \int \frac{e^{-x}\,dx}{\sqrt{1-(e^{-x})^2}} \qquad \text{Let } u = e^{-x},\ du = -e^{-x}\,dx.$$
$$= -\int \frac{du}{\sqrt{1-u^2}}$$
$$= -\sin^{-1} u + C = -\sin^{-1}\left(e^{-x}\right) + C.$$

The method of substitution cannot be "forced" to work. There is no substitution that will do much good with the integral $\int x(2+x^7)^{1/5}\,dx$, for instance. However, the integral $\int x^6(2+x^7)^{1/5}\,dx$ is quite amenable to the subtitution $u = 2+x^7$. The substitution $u = g(x)$ is more likely to work if $g'(x)$ is a factor of the integrand.

The following theorem simplifies the use of the method of substitution in definite integrals.

THEOREM 6.4.4 Suppose that g is a differentiable function on $[a,b]$, and satisfies $g(a) = A$ and $g(b) = B$. Suppose that f is continuous on the range of g. Then

$$\int_a^b f(g(x))\,g'(x)\,dx = \int_A^B f(u)\,du.$$

PROOF Let F be an antiderivative of f; $F'(u) = f(u)$. Then

$$\frac{d}{dx} F\big(g(x)\big) = F'\big(g(x)\big)\,g'(x) = f\big(g(x)\big)\,g'(x).$$

Thus

$$\int_a^b f\big(g(x)\big)\,g'(x)\,dx = F\big(g(x)\big)\Big|_a^b = F\big(g(b)\big) - F\big(g(a)\big)$$
$$= F(B) - F(A) = F(u)\Big|_A^B = \int_A^B f(u)\,du. \ \square$$

EXAMPLE 6.4.5 Evaluate the integral

$$I = \int_0^8 \frac{\cos\sqrt{x+1}}{\sqrt{x+1}}\, dx.$$

SOLUTION **Method I.** Let $u = \sqrt{x+1}$. Then $du = \dfrac{dx}{2\sqrt{x+1}}$. If $x = 0$, then $u = 1$; if $x = 8$, then $u = 3$. Thus

$$I = 2\int_1^3 \cos u\, du = 2\sin u\Big|_1^3 = 2\sin 3 - 2\sin 1.$$

Method II. We use the same substitution as in the Method I, but we do not transform the limits of integration from x values to u values. Hence we must return to the variable x before substituting in the limits:

$$I = 2\int_{x=0}^{x=8} \cos u\, du = 2\sin u\Big|_{x=0}^{x=8} = 2\sin\sqrt{x+1}\Big|_0^8 = 2\sin 3 - 2\sin 1.$$

Note that the limits *must* be written $x = 0$ and $x = 8$ at any stage where the variable is not x. It would be wrong to write

$$I = 2\int_0^8 \cos u\, du$$

because this would imply that u goes from 0 to 8 rather than that x goes from 0 to 8. Method I gives the shorter solution, and is therefore preferable. However, in cases where the transformed limits (the u-limits) are very complicated to write, one might prefer to use Method II.

EXAMPLE 6.4.6 Find the area of the region bounded by $y = \left(2 + \sin\dfrac{x}{2}\right)^2 \cos\dfrac{x}{2}$, $y = 0$, $x = 0$, and $x = \pi$.

SOLUTION Because $y \geq 0$ when $0 \leq x \leq \pi$, the area is

$$A = \int_0^{\pi} \left(2 + \sin\frac{x}{2}\right)^2 \cos\frac{x}{2}\, dx \qquad \text{Let } v = 2 + \sin\tfrac{x}{2},\ dv = \tfrac{1}{2}\cos\tfrac{x}{2}\, dx.$$

$$= 2\int_2^3 v^2\, dv = \frac{2}{3}v^3\Big|_2^3 = \frac{2}{3}(27 - 8) = \frac{38}{3} \text{ square units.}$$

REMARK: The condition that f be continuous on the range of the function $u = g(x)$ (for $a \le x \le b$) is essential in Theorem 6.4.4. Using the substitution $u = x^2$ in the integral $\int_{-1}^{1} x\csc(x^2)\,dx$ leads to the erroneous conclusion

$$\int_{-1}^{1} x\ \csc(x^2)\,dx = \frac{1}{2}\int_{1}^{1} \csc u\, du = 0.$$

Although $x\ \csc(x^2)$ is an odd function, it is not continuous at 0, and it happens that the given integral represents the difference of *infinite* areas. If we assume that f is continuous on an interval containing A and B, then it suffices to know that $u = g(x)$ is one-to-one as well as differentiable. In this case the range of g will lie between A and B so the condition of Theorem 6.4.4 will be satisfied.

Trigonometric Integrals

The method of substitution is often useful for evaluating trigonometric integrals. We begin by integrating the four trigonometric functions whose integrals we have not yet seen.

$$\begin{aligned}
\int \tan x\, dx &= \int \frac{\sin x}{\cos x}\, dx \qquad \text{Let } u = \cos x, \\
&\qquad\qquad\qquad\qquad\quad du = -\sin x\, dx \\
&= -\int \frac{du}{u} = -\ln|u| + C \\
&= -\ln|\cos x| + C = \ln\left|\frac{1}{\cos x}\right| + C = \ln|\sec x| + C.
\end{aligned}$$

Similarly, the substitution $u = \sin x$ leads to

$$\int \cot x\, dx = \int \frac{\cos x}{\sin x}\, dx = \int \frac{du}{u} = \ln|u| + C = \ln|\sin x| + C.$$

Observe that

$$\frac{d}{dx}\ln|\sec x + \tan x| = \frac{\sec x\tan x + \sec^2 x}{\sec x + \tan x} = \sec x.$$

Similarly,

$$\frac{d}{dx}\ln|\csc x + \cot x| = -\frac{\csc x\cot x + \csc^2 x}{\csc x + \cot x} = -\csc x.$$

Thus we have the four formulas

6.4.7
Integrals of tan, cot, sec and csc

$$\int \tan x\,dx = \ln|\sec x| + C$$

$$\int \cot x\,dx = \ln|\sin x| + C$$

$$\int \sec x\,dx = \ln|\sec x + \tan x| + C$$

$$\int \csc x\,dx = -\ln|\csc x + \cot x| + C$$
$$= \ln|\csc x - \cot x| + C$$

(Show that the two versions of the integral of $\csc x$ are equivalent!)

All four integrals given above are frequently encountered and should be committed to memory.

We now consider integrals of the form

$$\int \sin^m x \cos^n x\,dx.$$

If either m or n is an odd, positive integer, the integral can be done easily by substitution. If, say, $n = 2k+1$ where k is an integer, then we can use the identity $\sin^2 x + \cos^2 x = 1$ to rewrite the integral in the form

$$\int \sin^m x\,(1-\sin^2 x)^k \cos x\,dx,$$

which can be integrated using the substitution $u = \sin x$. Similarly, $u = \cos x$ can be used if m is an odd integer.

EXAMPLE 6.4.8 a)

$$\int \sin^3 x \cos^8 x\,dx = \int (1-\cos^2 x)\cos^8 x \sin x\,dx \qquad \text{Let } u = \cos x,\ du = -\sin x\,dx.$$
$$= -\int (1-u^2)\,u^8\,du = \int (u^{10}-u^8)\,du$$
$$= \frac{u^{11}}{11} - \frac{u^9}{9} + C = \frac{1}{11}\cos^{11} x - \frac{1}{9}\cos^9 x + C$$

b)

$$\int \cos^5 ax\,dx = \int (1-\sin^2 ax)^2 \cos ax\,dx \qquad \text{Let } u = \sin ax,\ du = a\cos ax\,dx.$$
$$= \frac{1}{a}\int (1-u^2)^2\,du = \frac{1}{a}\int (1-2u^2+u^4)\,du$$
$$= \frac{1}{a}\left(u - \frac{2}{3}u^3 + \frac{1}{5}u^5\right) + C$$
$$= \frac{1}{a}\left(\sin ax - \frac{2}{3}\sin^3 ax + \frac{1}{5}\sin^5 ax\right) + C$$

If the powers of $\sin x$ and $\cos x$ are both even, then we can make use of the *half-angle formulas*, (see Section 3.1):

$$\cos^2 x = \frac{1}{2}(1 + \cos 2x), \qquad \sin^2 x = \frac{1}{2}(1 - \cos 2x).$$

EXAMPLE 6.4.9 a)

$$\int \cos^2 x\,dx = \frac{1}{2}\int (1 + \cos 2x)\,dx = \frac{x}{2} + \frac{1}{4}\sin 2x + C$$
$$= \frac{x}{2} + \frac{1}{2}\sin x \cos x + C$$

b)

$$\int \sin^2 x\,dx = \frac{1}{2}\int (1 - \cos 2x)\,dx = \frac{x}{2} - \frac{1}{2}\sin x \cos x + C$$

c)

$$\int \sin^4 x\,dx = \frac{1}{4}\int (1 - \cos 2x)^2\,dx = \frac{1}{4}\int (1 - 2\cos 2x + \cos^2 2x)\,dx$$
$$= \frac{x}{4} - \frac{1}{4}\sin 2x + \frac{1}{8}\int (1 + \cos 4x)\,dx$$
$$= \frac{x}{4} - \frac{1}{4}\sin 2x + \frac{x}{8} + \frac{1}{32}\sin 4x + C$$
$$= \frac{3}{8}x - \frac{1}{4}\sin 2x + \frac{1}{32}\sin 4x + C$$

(Note that there is no point in inserting the constant of integration C until the last integral has been evaluated.)

Using the identities $\sec^2 x = 1 + \tan^2 x$ and $\csc^2 x = 1 + \cot^2 x$ and one of the substitutions $u = \sec x$, $u = \tan x$, $u = \csc x$, or $u = \cot x$, we can evaluate integrals of the form

$$\int \sec^m x \tan^n x\,dx \qquad \text{or} \qquad \int \csc^m x \cot^n x\,dx,$$

unless m is odd and n is even. (If this is the case, these integrals can be handled by Integration by Parts; see Section 6.6.)

EXAMPLE 6.4.10 a)

$$\int \sec^2 x \tan^2 x\,dx \qquad \text{Let } u = \tan x,\quad du = \sec^2 x\,dx.$$
$$= \int u^2\,du = \frac{u^3}{3} + C = \frac{1}{3}\tan^3 x + C$$

b)

$$\int \sec^3 x \tan^3 x\,dx$$
$$= \int \sec^2 x\,(\sec^2 x - 1)\sec x \tan x\,dx \qquad \text{Let } u = \sec x,\quad du = \sec x \tan x\,dx.$$
$$= \int (u^4 - u^2)\,du = \frac{u^5}{5} - \frac{u^3}{3} + C = \frac{1}{5}\sec^5 x - \frac{1}{3}\sec^3 x + C.$$

EXERCISES

Evaluate the integrals in Exercises 1–52. Remember to include a constant of integration with the indefinite integrals. Your answers may appear different from those provided but still be correct. For example, evaluating $I = \int \sin x \cos x\,dx$ using the substitution $u = \sin x$ leads to the answer $I = \frac{1}{2}\sin^2 x + C$; using $u = \cos x$ leads to $I = -\frac{1}{2}\cos^2 x + C$; and rewriting $I = \frac{1}{2}\int \sin(2x)\,dx$ leads to $I = -\frac{1}{4}\cos(2x) + C$. These answers are all actually equivalent up to different choices for the constant of integration: $\frac{1}{2}\sin^2 x = -\frac{1}{2}\cos^2 + \frac{1}{2} = -\frac{1}{4}\cos(2x) + \frac{1}{4}$.

You can always check your own answer to an indefinite integral by differentiating it to get back to the integrand. This is often easier than comparing your answer with the answer in the back of the book. You may find integrals that you can't do, but you should not make mistakes in those you can do, because the answer is so easily checked. (This is a good thing to remember in exams.)

1. $\displaystyle\int e^{5-2x}\,dx$

2. $\displaystyle\int \cos(ax+b)\,dx$

3. $\displaystyle\int \sqrt{3x+4}\,dx$

4. $\displaystyle\int e^{2x}\sin(e^{2x})\,dx$

5. $\displaystyle\int \frac{x^2}{(x^3+2)^{5/2}}\,dx$

6. $\displaystyle\int (x+2)(x^2+4x+9)^{1/3}\,dx$

7. $\displaystyle\int \frac{x\,dx}{(4x^2+1)^5}$

8. $\displaystyle\int \frac{dx}{2x^2+3}$

9. $\displaystyle\int \sin x \sin(\cos x)\,dx$

10. $\displaystyle\int \frac{\sin\sqrt{x}}{\sqrt{x}}\,dx$

11. $\displaystyle\int x\,e^{x^2}\,dx$

12. $\displaystyle\int x^2 2^{x^3+1}\,dx$

13. $\displaystyle\int \frac{\cos x}{4+\sin^2 x}\,dx$

14. $\displaystyle\int \frac{\sec^2 x}{\sqrt{1-\tan^2 x}}\,dx$

15. $\displaystyle\int \frac{e^x+1}{e^x-1}\,dx$

16. $\displaystyle\int \frac{\ln t}{t}\,dt$

17. $\displaystyle\int \frac{ds}{\sqrt{4-5s}}$

18. $\displaystyle\int \frac{x+1}{\sqrt{x^2+2x+3}}\,dx$

19. $\displaystyle\int \frac{t\,dt}{\sqrt{4-t^4}}$

20. $\displaystyle\int \frac{x^2\,dx}{2+x^6}$

21. $\displaystyle\int \frac{dx}{e^x+1}$

22. $\displaystyle\int \frac{dx}{e^x+e^{-x}}$

23. $\displaystyle\int \tan x \ln\cos x\,dx$

24. $\displaystyle\int \frac{x+1}{\sqrt{1-x^2}}\,dx$

25. $\displaystyle\int \frac{2t+3}{t^2+9}\,dt$

26. $\displaystyle\int \frac{ax+b}{\sqrt{A^2-B^2x^2}}\,dx$

27. $\displaystyle\int \frac{dx}{x^2+6x+13}$

28. $\displaystyle\int \frac{dx}{\sqrt{4+2x-x^2}}$

29. $\displaystyle\int \sin^3 x\cos^5 x\,dx$

30. $\displaystyle\int \sin^4 t\cos^5 t\,dt$

31. $\displaystyle\int \sin ax\cos^2 ax\,dx$

32. $\displaystyle\int x\sin^2 x^2\cos^3 x^2\,dx$

33. $\displaystyle\int \sin^2 x\cos^2 x\,dx$

34. $\displaystyle\int \sin^4 x\cos^2 x\,dx$

35. $\displaystyle\int \sin^6 x\,dx$

36. $\displaystyle\int \cos^4 x\,dx$

37. $\displaystyle\int \sec^5 x\tan x\,dx$

38. $\displaystyle\int \sec^6 x\tan^2 x\,dx$

39. $\displaystyle\int \sqrt{\tan x}\sec^4 x\,dx$

40. $\displaystyle\int \sin^{-2/3} x\cos^3 x\,dx$

41. $\displaystyle\int \cos x\sin^4(\sin x)\,dx$

42. $\displaystyle\int \frac{\sin^3 \ln x\cos^3 \ln x}{x}\,dx$

43. $\displaystyle\int \frac{\sin^2 x}{\cos^4 x}\,dx$

44. $\displaystyle\int \frac{\sin^3 x}{\cos^4 x}\,dx$

45. $\displaystyle\int \csc^5 x\cot^5 x\,dx$

46. $\displaystyle\int \frac{\cos^4 x}{\sin^8 x}\,dx$

47. $\displaystyle\int_0^4 x^3(x^2+1)^{-\frac{1}{2}}\,dx$

48. $\displaystyle\int_1^{\sqrt{e}} \frac{\sin(\pi\ln x)}{x}\,dx$

49. $\displaystyle\int_0^{\pi/2} \sin^4 x\,dx$

50. $\displaystyle\int_{\pi/4}^{\pi} \sin^5 x\,dx$

51. $\displaystyle\int_e^{e^2} \frac{dt}{t\ln t}$

52. $\displaystyle\int_{\frac{\pi^2}{16}}^{\frac{\pi^2}{9}} \frac{2^{\sin\sqrt{x}}\cos\sqrt{x}}{\sqrt{x}}\,dx$

53. Use the identities $\cos 2\theta = 2\cos^2\theta - 1 = 1 - 2\sin^2\theta$ and $\sin\theta = \cos\left(\frac{\pi}{2}-\theta\right)$ to help you evaluate the following:

$$\int_0^{\pi/2}\sqrt{1+\cos x}\,dx, \quad \text{and} \quad \int_0^{\pi/2}\sqrt{1-\sin x}\,dx$$

54. Find the area of the region bounded by $y = x/(x^2+16)$, $y = 0$, $x = 0$, and $x = 2$.

55. Find the area of the region bounded by $y = x/(x^4 + 16)$, $y = 0$, $x = 0$, and $x = 2$.

56. Express the area bounded by the ellipse $(x^2/a^2)+(y^2/b^2) = 1$ as a definite integral. Make a substitution that converts this integral into one representing the area of a circle, and hence evaluate it.

57.*Use the addition formulas for $\sin(x \pm y)$ and $\cos(x \pm y)$ (Formulas 3.1.7) to establish the following identities:

$$\cos x \cos y = \frac{1}{2}\big(\cos(x-y) + \cos(x+y)\big),$$
$$\sin x \sin y = \frac{1}{2}\big(\cos(x-y) - \cos(x+y)\big),$$
$$\sin x \cos y = \frac{1}{2}\big(\sin(x+y) + \sin(x-y)\big).$$

58.*Use the identities established in the previous exercise to calculate the following integrals:

$$\int \cos ax \cos bx \, dx, \quad \int \sin ax \sin bx \, dx,$$

and $\displaystyle\int \sin ax \cos bx \, dx$.

59.*If m and n are integers, show that:

i) $\displaystyle\int_{-\pi}^{\pi} \cos mx \cos nx \, dx = 0$ if $m \neq n$,

ii) $\displaystyle\int_{-\pi}^{\pi} \sin mx \sin nx \, dx = 0$ if $m \neq n$,

iii) $\displaystyle\int_{-\pi}^{\pi} \sin mx \cos nx \, dx = 0$.

6.5 INVERSE SUBSTITUTIONS

The substitutions considered in the previous section were direct substitutions in the sense that one simplified an integrand by replacing an expression appearing in it with a single variable. In this section we consider the reverse approach; we replace the variable of integration with a function of a new variable. Such substitutions, called *inverse substitutions*, would appear on the surface to make the integral more complicated. That is, substituting $x = g(u)$ in the integral

$$\int_a^b f(x)\, dx$$

leads to the more "complicated" integral

$$\int_{x=a}^{x=b} f\big(g(u)\big)\, g'(u)\, du$$

As we will see, however, sometimes such substitutions can actually simplify an integrand, transforming the integral into one that can be evaluated by inspection or to which other techniques can readily be applied.

The Inverse Trigonometric Substitutions

Three very useful inverse substitutions are

$$x = a \sin\theta, \qquad x = a \tan\theta, \qquad x = a \sec\theta.$$

These correspond to the direct substitutions

$$\theta = \sin^{-1}\frac{x}{a}, \qquad \theta = \tan^{-1}\frac{x}{a}, \qquad \theta = \sec^{-1}\frac{x}{a} = \cos^{-1}\frac{a}{x}.$$

6.5.1 The Inverse Sine Substitution

> Integrals involving $\sqrt{a^2 - x^2}$ (where $a > 0$) can frequently be reduced to a simpler form by means of the substitution
>
> $$\theta = \sin^{-1} \frac{x}{a}, \quad \text{or equivalently,} \quad x = a \sin \theta.$$

Observe that $\sqrt{a^2 - x^2}$ makes sense only if $-a \le x \le a$, in which case we have $-\pi/2 \le \theta \le \pi/2$. Since $\cos\theta \ge 0$ for such θ, we have

$$\sqrt{a^2 - x^2} = \sqrt{a^2(1 - \sin^2\theta)} = \sqrt{a^2 \cos^2\theta} = a\cos\theta.$$

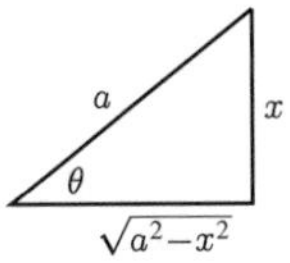

FIGURE 6.5.1

(If $\cos\theta$ were not nonnegative, we would have obtained $a|\cos\theta|$ instead.) If needed, the other trigonometric functions of θ can be recovered in terms of x by examining a right-angled triangle labeled to correspond to the substitution (see Fig. 6.5.1):

$$\cos\theta = \frac{\sqrt{a^2 - x^2}}{a}, \qquad \tan\theta = \frac{x}{\sqrt{a^2 - x^2}}.$$

EXAMPLE 6.5.2 Refer to Fig. 6.5.2.

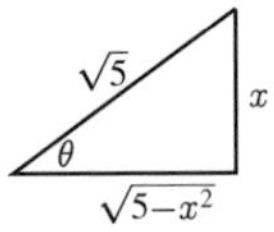

FIGURE 6.5.2

$$\int \frac{dx}{(5 - x^2)^{3/2}} \qquad \begin{aligned} &\text{Let } x = \sqrt{5}\sin\theta, \\ &dx = \sqrt{5}\cos\theta\, d\theta \end{aligned}$$

$$= \int \frac{\sqrt{5}\cos\theta\, d\theta}{5^{3/2}\cos^3\theta}$$

$$= \frac{1}{5}\int \sec^2\theta\, d\theta = \frac{1}{5}\tan\theta + C = \frac{1}{5}\frac{x}{\sqrt{5 - x^2}} + C$$

EXAMPLE 6.5.3 Find the area of the circular sector shaded in Fig. 6.5.3.

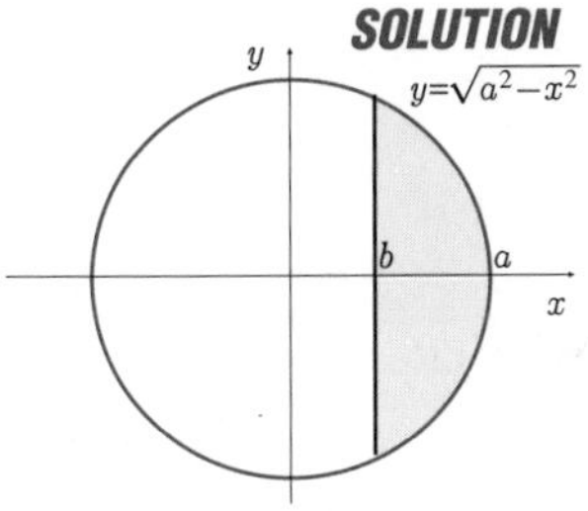

FIGURE 6.5.3

SOLUTION The area is

$$A = 2\int_b^a \sqrt{a^2 - x^2}\, dx \qquad \begin{aligned} &\text{Let } x = a\sin\theta, \\ &dx = a\cos\theta\, d\theta \end{aligned}$$

$$= 2\int_{x=b}^{x=a} a^2 \cos^2\theta\, d\theta$$

$$= a^2 \left(\theta + \sin\theta\cos\theta\right)\Big|_{x=b}^{x=a} \qquad \text{(as in Example 6.4.9(a))}$$

$$= a^2 \left(\sin^{-1}\frac{x}{a} + \frac{x\sqrt{a^2 - x^2}}{a^2}\right)\Bigg|_b^a \qquad \text{(See Fig. 6.5.1.)}$$

$$= \frac{\pi}{2}a^2 - a^2 \sin^{-1}\frac{b}{a} - b\sqrt{a^2 - b^2} \text{ square units.}$$

6.5.4
The Inverse Tangent Substitution

> Integrals involving $\sqrt{a^2+x^2}$ or $\dfrac{1}{x^2+a^2}$ (where $a > 0$) are often simplified by the substitution
>
> $$\theta = \tan^{-1}\frac{x}{a}, \qquad \text{or equivalently,} \qquad x = a\tan\theta.$$

Since x can take any real value, we have $-\pi/2 < \theta < \pi/2$, so $\sec\theta > 0$ and

$$\sqrt{a^2+x^2} = a\sqrt{1+\tan^2\theta} = a\sec\theta.$$

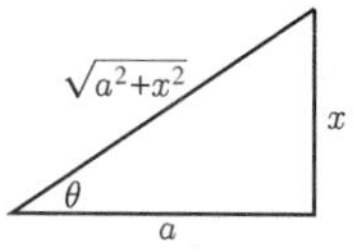

FIGURE 6.5.4

Other trigonometric functions of θ can be expressed in terms of x by referring to Fig. 6.5.4:

$$\sin\theta = \frac{x}{\sqrt{a^2+x^2}}, \qquad \cos\theta = \frac{a}{\sqrt{a^2+x^2}}.$$

EXAMPLE 6.5.5 Figures 6.5.5 and 6.5.6 illustrate parts (a) and (b), respectively.

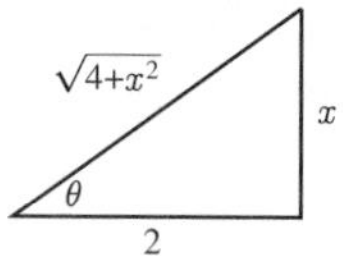

FIGURE 6.5.5

a) $\displaystyle\int \frac{dx}{\sqrt{4+x^2}}$ Let $x = 2\tan\theta$, $dx = 2\sec^2\theta\, d\theta$

$$= \int \frac{2\sec^2\theta\, d\theta}{2\sec\theta} = \int \sec\theta\, d\theta$$

$$= \ln|\sec\theta + \tan\theta| + C = \ln\left|\frac{\sqrt{4+x^2}}{2} + \frac{x}{2}\right| + C = \ln\left(\sqrt{4+x^2}+x\right) + C_1,$$

where $C_1 = C - \ln 2$. (Note that $\sqrt{4+x^2}+x > 0$ for all x, so we do not need an absolute value on it.)

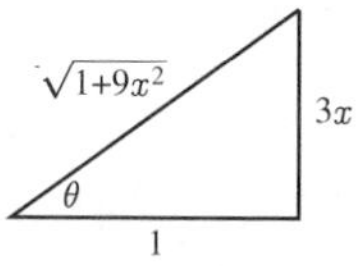

FIGURE 6.5.6

b) $\displaystyle\int \frac{dx}{(1+9x^2)^2}$ Let $3x = \tan\theta$, $3dx = \sec^2\theta\, d\theta$, $1+9x^2 = \sec^2\theta$

$$= \frac{1}{3}\int \frac{\sec^2\theta\, d\theta}{\sec^4\theta}$$

$$= \frac{1}{3}\int \cos^2\theta\, d\theta = \frac{1}{6}\int (1+\cos 2\theta)\, d\theta$$

$$= \frac{\theta}{6} + \frac{\sin 2\theta}{12} + C = \frac{\theta}{6} + \frac{1}{6}\sin\theta\cos\theta + C$$

$$= \frac{1}{6}\tan^{-1}(3x) + \frac{1}{6}\frac{3x}{\sqrt{1+9x^2}}\frac{1}{\sqrt{1+9x^2}} + C$$

$$= \frac{1}{6}\tan^{-1}(3x) + \frac{1}{2}\frac{x}{1+9x^2} + C$$

6.5.6 The Inverse Secant Substitution

> Integrals involving $\sqrt{x^2 - a^2}$ (where $a > 0$) can frequently be simplified by using the substitution
>
> $$\theta = \sec^{-1}\frac{x}{a}, \quad \text{or equivalently,} \quad x = a\sec\theta.$$

One must be more careful with this substitution. Although

$$\sqrt{x^2 - a^2} = a\sqrt{\sec^2\theta - 1} = a\sqrt{\tan^2\theta} = a|\tan\theta|,$$

we cannot always drop the absolute value from the tangent. Observe that $\sqrt{x^2 - a^2}$ makes sense for $x \geq a$ and for $x \leq -a$.

$$\text{If } x \geq a \quad \text{then } 0 \leq \theta = \sec^{-1}\frac{x}{a} = \cos^{-1}\frac{a}{x} < \frac{\pi}{2}, \text{ and } \tan\theta \geq 0.$$
$$\text{If } x \leq -a \text{ then } \frac{\pi}{2} < \theta = \sec^{-1}\frac{x}{a} = \cos^{-1}\frac{a}{x} \leq \pi, \text{ and } \tan\theta \leq 0.$$

In the first case $\sqrt{x^2 - a^2} = a\tan\theta$; in the second case $\sqrt{x^2 - a^2} = -a\tan\theta$.

EXAMPLE 6.5.7 Find $I = \displaystyle\int \frac{dx}{\sqrt{x^2 - a^2}}$ where $a > 0$.

SOLUTION For the moment assume that $x \geq a$. If $x = a\sec\theta$, then $dx = a\sec\theta\tan\theta\,d\theta$ and $\sqrt{x^2 - a^2} = a\tan\theta$ (Fig. 6.5.7). Thus

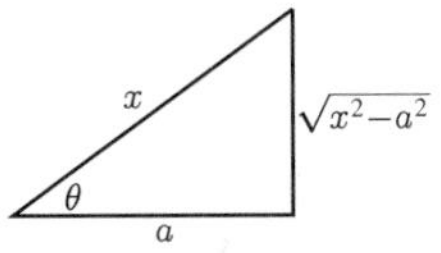

FIGURE 6.5.7

$$\begin{aligned} I &= \int \sec\theta\,d\theta = \ln|\sec\theta + \tan\theta| + C \\ &= \ln\left|\frac{x}{a} + \frac{\sqrt{x^2 - a^2}}{a}\right| + C = \ln|x + \sqrt{x^2 - a^2}| + C_1, \end{aligned}$$

where $C_1 = C - \ln a$. If $x \leq -a$, let $u = -x$ so that $u \geq a$. We have

$$\begin{aligned} I &= -\int \frac{du}{\sqrt{u^2 - a^2}} = -\ln|u + \sqrt{u^2 - a^2}| + C_1 \\ &= \ln\left|\frac{1}{-x + \sqrt{x^2 - a^2}}\,\frac{x + \sqrt{x^2 - a^2}}{x + \sqrt{x^2 - a^2}}\right| + C_1 \\ &= \ln\left|\frac{x + \sqrt{x^2 - a^2}}{-a^2}\right| + C_1 = \ln|x + \sqrt{x^2 - a^2}| + C_2, \end{aligned}$$

where $C_2 = C_1 - 2\ln a$. Thus, in either case, we have

$$I = \ln|x + \sqrt{x^2 - a^2}| + C.$$

Completing the Square

Quadratic expressions of the form $Ax^2 + Bx + C$ are often found in integrands. These can be written as sums or differences of squares using the procedure of completing the square. First factor out A so that the remaining expression begins with $x^2 + 2bx$, where $2b = B/A$. These are the first two terms of $(x + b)^2$.

$$\begin{aligned}
Ax^2 + Bx + C &= A\left(x^2 + \frac{B}{A}x + \frac{C}{A}\right) \\
&= A\left(x^2 + \frac{B}{A}x + \frac{B^2}{4A^2} + \frac{C}{A} - \frac{B^2}{4A^2}\right) \\
&= A\left(x + \frac{B}{2A}\right)^2 + \frac{4AC - B^2}{4A}
\end{aligned}$$

The substitution $u = x + \dfrac{B}{2A}$ should then be made.

EXAMPLE 6.5.8 a)

$$\begin{aligned}
\int \frac{dx}{\sqrt{2x - x^2}} &= \int \frac{dx}{\sqrt{1 - (1 - 2x + x^2)}} \\
&= \int \frac{dx}{\sqrt{1 - (x - 1)^2}} \qquad \text{Let } u = x - 1,\ du = dx \\
&= \int \frac{du}{\sqrt{1 - u^2}} = \sin^{-1} u + C = \sin^{-1}(x - 1) + C
\end{aligned}$$

b)

$$\begin{aligned}
\int \frac{x\,dx}{4x^2 + 12x + 13} &= \int \frac{x\,dx}{4\left(x^2 + 3x + \frac{9}{4} + 1\right)} \\
&= \frac{1}{4}\int \frac{x\,dx}{\left(x + \frac{3}{2}\right)^2 + 1} \qquad \text{Let } u = x + (3/2),\ du = dx,\ x = (2u - 3)/2 \\
&= \frac{1}{4}\int \frac{u\,du}{u^2 + 1} - \frac{3}{8}\int \frac{du}{u^2 + 1} \qquad \text{In the first integral let } v = u^2 + 1,\ dv = 2u\,du \\
&= \frac{1}{8}\int \frac{dv}{v} - \frac{3}{8}\tan^{-1} u \\
&= \frac{1}{8}\ln|v| - \frac{3}{8}\tan^{-1} u + C \\
&= \frac{1}{8}\ln(4x^2 + 12x + 13) - \frac{3}{8}\tan^{-1}\left(x + \frac{3}{2}\right) + C_1,
\end{aligned}$$

where $C_1 = C - (\ln 4)/8$.

Other Inverse Substitutions

Integrals involving $\sqrt{ax+b}$ can sometimes be made simpler with the substitution $ax + b = u^2$.

EXAMPLE 6.5.9

$$\int \frac{dx}{1+\sqrt{2x}} \qquad \begin{aligned} &\text{Let } 2x = u^2,\\ &2\,dx = 2u\,du \end{aligned}$$

$$= \int \frac{u\,du}{1+u} = \int \frac{1+u-1}{1+u}\,du = \int \left(1 - \frac{1}{1+u}\right) du \qquad \begin{aligned} &\text{Let } v = 1+u,\\ &dv = du \end{aligned}$$

$$= u - \int \frac{dv}{v} = u - \ln|v| + C$$

$$= \sqrt{2x} - \ln\left(1+\sqrt{2x}\right) + C$$

If more than one fractional power is present, it may be possible to eliminate all of them at once.

EXAMPLE 6.5.10

$$\int \frac{dx}{x^{1/2}(1+x^{1/3})} \qquad \begin{aligned} &\text{Let } x = u^6,\\ &dx = 6u^5\,du \end{aligned}$$

$$= 6\int \frac{u^5\,du}{u^3(1+u^2)} = 6\int \frac{u^2}{1+u^2}\,du = 6\int \left(1 - \frac{1}{1+u^2}\right) du$$

$$= 6\,(u - \tan^{-1} u) + C = 6\,(x^{1/6} - \tan^{-1} x^{1/6}) + C$$

The tan($\theta/2$) Substitution

There is a certain special substitution that can transform an integral whose integrand is a rational function of $\sin\theta$ and $\cos\theta$ (that is, a quotient of polynomials in $\sin\theta$ and $\cos\theta$) into a rational function of x. The substitution is

$$x = \tan\frac{\theta}{2}, \qquad \text{or equivalently,} \qquad \theta = 2\tan^{-1} x.$$

Observe that

$$\cos^2\frac{\theta}{2} = \frac{1}{\sec^2\frac{\theta}{2}} = \frac{1}{1+\tan^2\frac{\theta}{2}} = \frac{1}{1+x^2},$$

so

$$\cos\theta = 2\cos^2\frac{\theta}{2} - 1 = \frac{2}{1+x^2} - 1 = \frac{1-x^2}{1+x^2}$$

$$\sin\theta = 2\sin\frac{\theta}{2}\cos\frac{\theta}{2} = 2\tan\frac{\theta}{2}\cos^2\frac{\theta}{2} = \frac{2x}{1+x^2}.$$

Also, $dx = \frac{1}{2}\sec^2\frac{\theta}{2}\,d\theta$, so

$$d\theta = 2\cos^2\frac{\theta}{2}\,dx = \frac{2\,dx}{1+x^2}.$$

In summary:

6.5.11
The $\tan(\theta/2)$ Substitution

> If $x = \tan(\theta/2)$ then
>
> $$\cos\theta = \frac{1-x^2}{1+x^2}, \qquad \sin\theta = \frac{2x}{1+x^2}, \qquad d\theta = \frac{2\,dx}{1+x^2}.$$

Note that $\cos\theta$, $\sin\theta$, and $d\theta$ all involve only rational functions of x. We will be investigating general techniques for integrating rational functions of x in Section 6.7.

EXAMPLE 6.5.12

$$\int \frac{d\theta}{2+\cos\theta}$$

Let $x = \tan(\theta/2)$, so
$\cos\theta = \frac{1-x^2}{1+x^2}$,
$d\theta = (2\,dx)/(1+x^2)$

$$= \int \frac{\frac{2\,dx}{1+x^2}}{2+\frac{1-x^2}{1+x^2}} = 2\int \frac{dx}{3+x^2} = \frac{2}{\sqrt{3}}\tan^{-1}\frac{x}{\sqrt{3}} + C$$

$$= \frac{2}{\sqrt{3}}\tan^{-1}\left(\frac{1}{\sqrt{3}}\tan\frac{\theta}{2}\right) + C.$$

EXERCISES

Evaluate the integrals in Exercises 1–44

1. $\displaystyle\int \frac{dx}{\sqrt{1-4x^2}}$
2. $\displaystyle\int \frac{x\,dx}{\sqrt{1-4x^2}}$
3. $\displaystyle\int \frac{x^2\,dx}{\sqrt{1-4x^2}}$
4. $\displaystyle\int \frac{dx}{x\sqrt{1-4x^2}}$
5. $\displaystyle\int x^3\sqrt{9-x^2}\,dx$
6. $\displaystyle\int x^2\sqrt{9-x^2}\,dx$
7. $\displaystyle\int \frac{x^2\,dx}{\sqrt{9-x^2}}$
8. $\displaystyle\int \frac{x^3\,dx}{\sqrt{9-x^2}}$
9. $\displaystyle\int \frac{dx}{x^2\sqrt{9-x^2}}$
10. $\displaystyle\int \frac{dx}{x\sqrt{9-x^2}}$
11. $\displaystyle\int \frac{x+1}{\sqrt{9-x^2}}\,dx$
12. $\displaystyle\int \frac{dx}{9-x^2}$
13. $\displaystyle\int \frac{dx}{\sqrt{9+x^2}}$
14. $\displaystyle\int \frac{x\,dx}{\sqrt{9+x^2}}$
15. $\displaystyle\int \frac{x^3\,dx}{\sqrt{9+x^2}}$
16. $\displaystyle\int \frac{\sqrt{9+x^2}}{x^4}\,dx$
17. $\displaystyle\int \frac{dx}{(a^2-x^2)^{3/2}}$
18. $\displaystyle\int \frac{dx}{(a^2+x^2)^{3/2}}$

19. $\displaystyle\int \frac{x^2\,dx}{(a^2-x^2)^{3/2}}$

20. $\displaystyle\int \frac{dx}{(1+2x^2)^{5/2}}$

21. $\displaystyle\int \frac{\sqrt{1-x^2}}{x^2}\,dx$

22. $\displaystyle\int \frac{dx}{1-2x^2}$

23. $\displaystyle\int \frac{dx}{x\sqrt{x^2-a^2}}$

24. $\displaystyle\int \frac{dx}{x^2\sqrt{x^2-a^2}}$

25. $\displaystyle\int \frac{dx}{x^2+2x+10}$

26. $\displaystyle\int \frac{dx}{x^2+x+1}$

27. $\displaystyle\int \frac{dx}{(4x^2+4x+5)^2}$

28. $\displaystyle\int \frac{x\,dx}{x^2-2x+3}$

29. $\displaystyle\int \frac{x\,dx}{\sqrt{2ax-x^2}}$

30. $\displaystyle\int \frac{dx}{(4x-x^2)^{3/2}}$

31. $\displaystyle\int \frac{x\,dx}{(3-2x-x^2)^{3/2}}$

32.* $\displaystyle\int \frac{x\,dx}{(4x^2+12x+13)^{5/2}}$

33. $\displaystyle\int \frac{dx}{(4x^2+1)^2}$

34. $\displaystyle\int \frac{dx}{(x^2+2x+2)^2}$

35. $\displaystyle\int \frac{dx}{(1+x^2)^3}$

36. $\displaystyle\int \frac{x^2\,dx}{(1+x^2)^2}$

37. $\displaystyle\int \frac{dx}{2+\sqrt{x}}$

38. $\displaystyle\int \frac{dx}{1+x^{1/3}}$

39.* $\displaystyle\int \frac{1+x^{1/2}}{1+x^{1/3}}\,dx$

40.* $\displaystyle\int \frac{x\sqrt{2-x^2}}{\sqrt{x^2+1}}\,dx$

41. $\displaystyle\int_{-\ln 2}^{0} e^x\sqrt{1-e^{2x}}\,dx$

42. $\displaystyle\int_0^{\pi/2} \frac{\cos x}{\sqrt{1+\sin^2 x}}\,dx$

43. $\displaystyle\int_{-1}^{\sqrt{3}-1} \frac{dx}{x^2+2x+2}$

44. $\displaystyle\int_1^2 \frac{dx}{x^2\sqrt{9-x^2}}$

In Exercises 45–47 evaluate the integral using the special substitution $x = \tan\dfrac{\theta}{2}$ as in Example 6.5.12.

45.* $\displaystyle\int \frac{d\theta}{2+\sin\theta}$

46.* $\displaystyle\int_0^{\pi/2} \frac{d\theta}{1+\cos\theta+\sin\theta}$

47.* $\displaystyle\int \frac{d\theta}{3+2\cos\theta}$

48. Find the area of the region bounded by $y = (2x-x^2)^{-1/2}$, $y = 0$, $x = 1/2$, and $x = 1$.

49. Find the area of the region lying below $y = 9/(x^4+4x^2+4)$ and above $y = 1$.

50. Find the average value of the function $f(x) = (x^2-4x+8)^{-3/2}$ over the interval $[0, 4]$.

51.*Evaluate the integrals

$$\int \frac{dx}{\sqrt{x^2-a^2}} \quad \text{and} \quad \int \frac{dx}{x^2\sqrt{x^2-a^2}},$$

using the subtitution $x = a\cosh u$. (*Hint:* Review the properties of the hyperbolic functions in Section 3.8.) This substitution is an alternative to $x = a\sec\theta$ when dealing with $\sqrt{x^2-a^2}$.

6.6 INTEGRATION BY PARTS

Our next general method for antidifferentiation is called **integration by parts** or **partial integration**. Just as the method of substitution can be regarded as inverse to the Chain Rule for differentiation, so the method for integration by parts is inverse to the Product Rule for differentiation.

Suppose that $U(x)$ and $V(x)$ are two differentiable functions. According to the product rule,

$$\frac{d}{dx}\big(U(x)V(x)\big) = U(x)\frac{dV}{dx} + V(x)\frac{dU}{dx}.$$

Integrating both sides of this equation and transposing terms, we obtain

$$\int U(x)\frac{dV}{dx}\,dx = U(x)V(x) - \int V(x)\frac{dU}{dx}\,dx$$

or, more simply,

6.6.1 Integration by Parts

$$\int U\,dV = UV - \int V\,dU.$$

The above formula serves as a *pattern* for carrying out integration by parts, as we will see in the examples below. In each application of the method, we break up the given integrand into a product of two pieces, U and V', where V' is readily integrated and where $\int VU'\,dx$ is usually (but not always) a "simpler" integral than $\int UV'\,dx$. The technique is called integration by parts because it replaces one integral with the sum of an integrated term and another integral that remains to be evaluated. That is, it accomplishes only "part" of the original integration.

EXAMPLE 6.6.2

$$\int xe^x\,dx \qquad \text{Let} \quad U = x \quad dV = e^x\,dx$$
$$\qquad\qquad \text{Then} \quad dU = dx \quad V = e^x$$
$$= xe^x - \int e^x\,dx \qquad \text{(this is, } UV - \textstyle\int V\,dU)$$
$$= xe^x - e^x + C.$$

Note the form in which integration by parts is carried out. We indicate at the side what choices we are making for U and dV and then calculate dU and V from these. However, we do not actually substitute U and V into the integral; instead we use the formula $\int U\,dV = UV - \int V\,dU$ as a pattern or mnemonic device to replace the given integral by the equivalent partially integrated form on the second line.

Note also that had we included a constant of integration with V, (say, $V = e^x + K$), that constant would cancel out in the next step:

$$\int xe^x\,dx = x(e^x + K) - \int (e^x + K)\,dx$$
$$= xe^x + Kx - e^x - Kx + C = xe^x - e^x + C.$$

In general, do not include a constant of integration on the right-hand side until the last integral has been evaluated.

EXAMPLE 6.6.3 a)

$$\int \ln x\,dx \qquad \text{Let} \quad U = \ln x \quad dV = dx$$
$$\qquad\qquad \text{Then} \quad dU = dx/x \quad V = x$$
$$= x\ln x - \int x\,\frac{1}{x}\,dx$$
$$= x\ln x - x + C.$$

b) $\displaystyle\int x^2 \sin x\,dx$ Let $U = x^2$ $dV = \sin x\,dx$
Then $dU = 2x\,dx$ $V = -\cos x$

$\displaystyle = -x^2 \cos x + 2\int x \cos x\,dx$ Let $U = x$ $dV = \cos x\,dx$
Then $dU = dx$ $V = \sin x$

$$= -x^2 \cos x + 2\left(x \sin x - \int \sin x\,dx\right)$$

$$= -x^2 \cos x + 2x \sin x + 2 \cos x + C$$

c) $\displaystyle\int x \tan^{-1} x\,dx$ Let $U = \tan^{-1} x$ $dV = x\,dx$
Then $dU = dx/(1+x^2)$ $V = \frac{1}{2}x^2$

$$= \frac{1}{2}x^2 \tan^{-1} x - \frac{1}{2}\int \frac{x^2}{1+x^2}\,dx$$

$$= \frac{1}{2}x^2 \tan^{-1} x - \frac{1}{2}\int \left(1 - \frac{1}{1+x^2}\right)dx$$

$$= \frac{1}{2}x^2 \tan^{-1} x - \frac{1}{2}x + \frac{1}{2}\tan^{-1} x + C$$

d) $\displaystyle\int \sin^{-1} x\,dx$ Let $U = \sin^{-1} x$ $dV = dx$
Then $dU = dx/\sqrt{1-x^2}$ $V = x$

$\displaystyle = x \sin^{-1} x - \int \frac{x}{\sqrt{1-x^2}}\,dx$ Let $u = 1 - x^2$,
$du = -2x\,dx$

$$= x \sin^{-1} x + \frac{1}{2}\int u^{-1/2}\,du$$

$$= x \sin^{-1} x + u^{1/2} + C = x \sin^{-1} x + \sqrt{1-x^2} + C$$

e) $\displaystyle I = \int \sec^3 x\,dx$ Let $U = \sec x$ $dV = \sec^2 x\,dx$
Then $dU = \sec x \tan x\,dx$ $V = \tan x$

$$= \sec x \tan x - \int \sec x \tan^2 x\,dx$$

$$= \sec x \tan x - \int \sec x(\sec^2 x - 1)\,dx$$

$$= \sec x \tan x - \int \sec^3 x\,dx + \int \sec x\,dx$$

$$= \sec x \tan x - I + \ln|\sec x + \tan x|$$

This is an equation that can be solved for the desired integral I:

$$\int \sec^3 x\,dx = I = \frac{1}{2}\sec x \tan x + \frac{1}{2}\ln|\sec x + \tan x| + C.$$

Study the preceding examples carefully; they show the various ways in which integration by parts is used, and they give some insights into what choices should be made for U and dV in various situations. An improper choice can result in making an integral harder rather than easier. Look for a factor of the integrand that is easily integrated, and include dx with that factor to make up dV. Then U is the remaining factor of the integrand. Sometimes it is necessary to take $dV = dx$ only. When breaking up an integrand using integration by parts, choose U and dV so that, if possible, $V\,dU$ is "simpler" (easier to integrate) than $U\,dV$. The following are two useful rules of thumb:

i) If the integrand involves a polynomial multiplied by an exponential, a sine or a cosine, or some other readily integrable function, try U = the polynomial, dV = the rest.

ii) If the integrand involves a logarithm, an inverse trigonometric function, or some other function that is not readily integrable but whose derivative is readily calculated, try that function for U and let dV equal the rest.

(Of course, these "rules" come with no guarantee. They may fail to be helpful if "the rest" is not of suitable form. There remain many integrals that cannot be evaluated by any of the standard techniques presented in this chapter.)

Example 6.6.3(e) illustrates a frequently occuring and very useful phenomenon. It may happen after one or two integrations by parts, with the possible application of some known identity, that the original integral reappears on the right-hand side. Unless its coefficient there is 1, we have an equation that can be solved for that integral.

EXAMPLE 6.6.4 Find $I = \displaystyle\int e^{ax} \cos bx\,dx$.

SOLUTION If either $a = 0$ or $b = 0$ the integral is easy to do, so let us assume $a \neq 0$ and $b \neq 0$. We have

$$I = \int e^{ax} \cos bx\,dx, \qquad \begin{array}{lll} \text{Let} & U = e^{ax} & dV = \cos bx\,dx \\ \text{Then} & dU = a\,e^{ax}\,dx & V = (1/b)\sin bx \end{array}$$

$$= \frac{1}{b} e^{ax} \sin bx - \frac{a}{b}\int e^{ax} \sin bx\,dx \qquad \begin{array}{lll} \text{Let} & U = e^{ax} & dV = \sin bx\,dx \\ \text{Then} & dU = ae^{ax}dx & V = -(\cos bx)/b \end{array}$$

$$= \frac{1}{b} e^{ax} \sin bx - \frac{a}{b}\left(-\frac{1}{b} e^{ax} \cos bx + \frac{a}{b}\int e^{ax} \cos bx\,dx\right)$$

$$= \frac{1}{b} e^{ax} \sin bx + \frac{a}{b^2} e^{ax} \cos bx - \frac{a^2}{b^2} I$$

Thus

$$\left(1 + \frac{a^2}{b^2}\right) I = \frac{1}{b} e^{ax} \sin bx + \frac{a}{b^2} e^{ax} \cos bx + C_1$$

and

$$\int e^{ax} \cos bx\,dx = I = \frac{b\,e^{ax} \sin bx + a\,e^{ax} \cos bx}{b^2 + a^2} + C.$$

Observe that after the first integration by parts we had an integral that was different from, but no simpler than the original. At this point we might have become discouraged and given up on this method. However, perseverance proved worthwhile; a second integration by parts returned the original integral I in an equation that could be solved for I. Having chosen to let U be the exponential in the first integration by parts (we could have let it be the cosine), we made the same choice for U in the second integration by parts. Had we "switched horses in midstream" and decided to let U be the trigonometric function the second time, we would have obtained

$$I = \frac{1}{b} e^{ax} \sin bx - \frac{1}{b} e^{ax} \sin bx + I;$$

we would have *undone* what we accomplished in the first step.

If we wish to evaluate a definite integral by the method of integration by parts, we must remember to include the appropriate evaluation symbol with the integrated term.

EXAMPLE 6.6.5 $\int_1^e x^3 (\ln x)^2\,dx$ Let $U = (\ln x)^2$ $dV = x^3\,dx$; Then $dU = (2\ln x\,dx)/x$ $V = x^4/4$

$$= \frac{x^4}{4}(\ln x)^2\Big|_1^e - \frac{1}{2}\int_1^e x^3 \ln x\,dx$$

Let $U = \ln x$ $dV = x^3\,dx$; Then $dU = dx/x$ $V = x^4/4$

$$\frac{e^4}{4}(1^2) - 0 - \frac{1}{2}\left(\frac{x^4}{4}\ln x\Big|_1^e - \frac{1}{4}\int_1^e x^3\,dx\right)$$

$$= \frac{e^4}{4} - \frac{e^4}{8} + \frac{1}{8}\frac{x^4}{4}\Big|_1^e = \frac{e^4}{8} + \frac{e^4}{32} - \frac{1}{32} = \frac{5}{32}e^4 - \frac{1}{32}.$$

Reduction Formulas

Consider the problem of finding $\int x^{10} e^{-x}\,dx$. We can, of course, proceed by using integration by parts 10 times. Each time will reduce the power of x by 1. Since this is repetitive and tedious, we prefer the following approach. For $n \geq 0$ let

$$I_n = \int x^n e^{-x}\,dx.$$

We want to find I_{10}. If we integrate by parts, we obtain a formula for I_n in terms of I_{n-1}:

$$I_n = \int x^n e^{-x}\,dx$$

Let $U = x^n$ $dV = e^{-x}\,dx$; Then $dU = nx^{n-1}\,dx$ $V = -e^{-x}$

$$= -x^n e^{-x} + n\int x^{n-1} e^{-x}\,dx = -x^n e^{-x} + nI_{n-1}$$

The formula

$$I_n = -x^n e^{-x} + nI_{n-1}$$

is called a **reduction formula** because it gives the value of the integral I_n in terms of I_{n-1}, an integral corresponding to a reduced value of the exponent n. We can apply the formula over and over (10 times) to get

$$\begin{aligned} I_{10} &= -x^{10}e^{-x} + 10I_9 \\ &= -x^{10}e^{-x} + 10(-x^9e^{-x} + 9I_8) \\ &= -e^{-x}(x^{10} + 10x^9) + 10\cdot 9(-x^8e^{-x} + 8I_7) \\ &= -e^{-x}(x^{10} + 10x^9 + 10\cdot 9x^8 + 10\cdot 9\cdot 8x^7) + 10\cdot 9\cdot 8\cdot 7I_6 \\ &= \cdots \\ &= -e^{-x}(x^{10} + 10x^9 + 10\cdot 9x^8 + 10\cdot 9\cdot 8x^7 + \cdots + 10\cdot 9\cdots 3\cdot 2x) + 10!I_0 \end{aligned}$$

Since $I_0 = \int e^{-x}\,dx = -e^{-x} + C$, we have

$$\int x^{10}e^{-x}\,dx = -e^{-x}\left(x^{10} + \frac{10!}{9!}x^9 + \frac{10!}{8!}x^8 + \cdots + \frac{10!}{1!}x + 10!\right) + C.$$

EXAMPLE 6.6.6 Obtain and use a reduction formula to evaluate

$$I_n = \int_0^{\pi/2} \cos^n x\,dx \qquad (n \ge 0).$$

SOLUTION Observe first that

$$I_0 = \int_0^{\pi/2} dx = \frac{\pi}{2}, \qquad I_1 = \int_0^{\pi/2} \cos x\,dx = \sin x\Big|_0^{\pi/2} = 1.$$

Now let $n \ge 2$:

$$\begin{aligned} I_n &= \int_0^{\pi/2} \cos^{n-1} x\cos x\,dx \qquad \begin{array}{ll} U = \cos^{n-1} x, & dV = \cos x\,dx \\ dU = -(n-1)\cos^{n-2} x\,\sin x\,dx, & V = \sin x \end{array} \\ &= \sin x\,\cos^{n-1} x\Big|_0^{\pi/2} + (n-1)\int_0^{\pi/2} \cos^{n-2} x\,\sin^2 x\,dx \\ &= 0 - 0 + (n-1)\int_0^{\pi/2} \cos^{n-2} x\,(1 - \cos^2 x)\,dx \\ &= (n-1)I_{n-2} - (n-1)I_n \end{aligned}$$

Transposing the term $-(n-1)I_n$, we obtain $nI_n = (n-1)I_{n-2}$, or

$$I_n = \frac{n-1}{n}I_{n-2},$$

which is the required reduction formula. It is valid for $n \geq 2$. (Where have we used this assumption in deriving the formula?) If $n \geq 2$ is an *even integer*, we have

$$\begin{aligned} I_n &= \frac{n-1}{n} I_{n-2} = \frac{n-1}{n} \cdot \frac{n-3}{n-2} I_{n-4} = \cdots \\ &= \frac{n-1}{n} \cdot \frac{n-3}{n-2} \cdot \frac{n-5}{n-4} \cdots \frac{5}{6} \cdot \frac{3}{4} \cdot \frac{1}{2} \cdot I_0 \\ &= \frac{n-1}{n} \cdot \frac{n-3}{n-2} \cdot \frac{n-5}{n-4} \cdots \frac{5}{6} \cdot \frac{3}{4} \cdot \frac{1}{2} \cdot \frac{\pi}{2}. \end{aligned}$$

If $n \geq 3$ is an *odd* integer, we have

$$\begin{aligned} I_n &= \frac{n-1}{n} \cdot \frac{n-3}{n-2} \cdot \frac{n-5}{n-4} \cdots \frac{6}{7} \cdot \frac{4}{5} \cdot \frac{2}{3} \cdot I_1 \\ &= \frac{n-1}{n} \cdot \frac{n-3}{n-2} \cdot \frac{n-5}{n-4} \cdots \frac{6}{7} \cdot \frac{4}{5} \cdot \frac{2}{3}. \end{aligned}$$

EXERCISES

Evaluate the integrals in Exercises 1–32.

1. $\displaystyle\int x \cos x\, dx$
2. $\displaystyle\int (x+3)e^{2x}\, dx$
3. $\displaystyle\int x^2 \cos \pi x\, dx$
4. $\displaystyle\int (x^2 - 2x)e^{kx}\, dx$
5. $\displaystyle\int x^3 \ln x\, dx$
6. $\displaystyle\int x(\ln x)^3\, dx$
7. $\displaystyle\int \tan^{-1} x\, dx$
8. $\displaystyle\int x^2 \tan^{-1} x\, dx$
9. $\displaystyle\int x \sin^{-1} x\, dx$
10. $\displaystyle\int x^2 \sin^{-1} x\, dx$
11. $\displaystyle\int \frac{\tan^{-1} x}{1+x^2}\, dx$
12. $\displaystyle\int x^5 e^{-x^2}\, dx$
13. $\displaystyle\int_0^{\pi/4} \sec^5 x\, dx$
14. $\displaystyle\int \tan^2 x \sec x\, dx$
15. $\displaystyle\int e^{2x} \sin 3x\, dx$
16. $\displaystyle\int x e^{\sqrt{x}}\, dx$
17. $\displaystyle\int_{1/2}^{1} \frac{\sin^{-1} x}{x^2}\, dx$
18. $\displaystyle\int_0^1 \sqrt{x} \sin(\pi\sqrt{x})\, dx$
19. $\displaystyle\int x \sec^2 x\, dx$
20. $\displaystyle\int x \sin^2 x\, dx$
21. $\displaystyle\int \cos(\ln x)\, dx$
22. $\displaystyle\int_1^e \sin(\ln x)\, dx$
23. $\displaystyle\int \frac{\ln(\ln x)}{x}\, dx$
24. $\displaystyle\int_0^4 \sqrt{x} e^{\sqrt{x}}\, dx$
25. $\displaystyle\int \cos^{-1} x\, dx$
26. $\displaystyle\int x \sec^{-1} x\, dx$
27. $\displaystyle\int_1^2 \sec^{-1} x\, dx$
28. $\displaystyle\int \sqrt{9+x^2}\, dx$
29.* $\displaystyle\int \frac{\sqrt{1-x^2}}{x^3}\, dx$
30.* $\displaystyle\int (\sin^{-1} x)^2\, dx$
31.* $\displaystyle\int x(\tan^{-1} x)^2\, dx$
32.* $\displaystyle\int x e^x \cos x\, dx$
33. Obtain a reduction formula for $I_n = \int (\ln x)^n\, dx$ and use it to evaluate I_4.
34. Obtain a reduction formula for $I_n = \int_0^{\pi/2} x^n \sin x\, dx$ and use it to evaluate I_6.
35. Obtain a reduction formula for $I_n = \int \sin^n x\, dx$, (where $n \geq 2$), and use it to find I_6 and I_7.
36. Obtain a reduction formula for $I_n = \int \sec^n x\, dx$, where $n \geq 3$), and use it to find I_6 and I_7.
37.*By writing

$$\begin{aligned} I_n &= \int \frac{dx}{(x^2+a^2)^n} \\ &= \frac{1}{a^2} \int \frac{dx}{(x^2+a^2)^{n-1}} - \frac{1}{a^2} \int x \frac{x}{(x^2+a^2)^n}\, dx \end{aligned}$$

and integrating the last integral by parts, using $U = x$, obtain a reduction formula for I_n. Use this formula to find I_3.

38. *If f is twice differentiable on $[a, b]$ and $f(a) = f(b) = 0$, show that

$$\int_a^b (x-a)(b-x)f''(x)\,dx = -2\int_a^b f(x)\,dx.$$

(*Hint:* Use integration by parts on the left-hand side twice.) This formula will be used in the next chapter to construct an error estimate for the Trapezoid Rule approximation formula.

39. *If f and g are two functions having continuous second derivatives on the interval $[a, b]$, and if $f(a) = g(a) = f(b) = g(b) = 0$, show that

$$\int_a^b f(x)\,g''(x)\,dx = \int_a^b f''(x)\,g(x)\,dx.$$

What other assumptions about the values of f and g at a and b would give the same result?

6.7 THE METHOD OF PARTIAL FRACTIONS

In this final section on techniques of integration we are concerned with integrals of the form

$$\int \frac{P(x)}{Q(x)}\,dx,$$

where P and Q are polynomials. Recall that a **polynomial** is a function P of the form

$$P(x) = a_n x^n + a_{n-1}x^{n-1} + \cdots + a_2x^2 + a_1x + a_0$$

where n is a nonnegative integer, $a_0, a_1, a_2, \ldots, a_n$ are constants, and $a_n \neq 0$. We call n the **degree** of P. A quotient $P(x)/Q(x)$ of two polynomials is called a **rational function**. We need normally concern ourselves only with rational functions $P(x)/Q(x)$ where the degree of P is less than that of Q. If the degree of P equals or exceeds the degree of Q, then we can use long division or some equivalent procedure to express the fraction $P(x)/Q(x)$ as a polynomial quotient plus another fraction $R(x)/Q(x)$ where R, the remainder in the division, has degree less than that of Q.

EXAMPLE 6.7.1 Evaluate $I = \displaystyle\int \frac{x^3+3x^2}{x^2+1}\,dx$.

SOLUTION The numerator has degree 3 and the denominator has degree 2 so we need to divide. We use long division:

$$\begin{array}{r|llll}
 & x & +\,3x^2 \\
\hline
x^2+1 & x^3 & +\,3x^2 & & \\
 & x^3 & & +\,x & \\
\hline
 & & 3x^2 & -\,x & \\
 & & 3x^2 & & +\,3 \\
\hline
 & & & -\,x & -\,3
\end{array}$$

$$\frac{x^3+3x^2}{x^2+1} = x + 3 - \frac{x+3}{x^2+1}.$$

Thus

$$\begin{aligned} I &= \int (x+3)\,dx - \int \frac{x}{x^2+1}\,dx - 3\int \frac{dx}{x^2+1} \\ &= \frac{1}{2}x^2 + 3x - \frac{1}{2}\ln(x^2+1) - 3\tan^{-1}x + C. \end{aligned}$$

EXAMPLE 6.7.2 Evaluate $I = \displaystyle\int \frac{x}{2x-1}\,dx$.

SOLUTION The numerator and denominator have the same degree, 1, so division is again required. In this case the "long division" can be carried out by manipulation of the integrand:

$$\begin{aligned} I = \frac{1}{2}\int \frac{2x-1+1}{2x-1}\,dx &= \frac{1}{2}\int \left(1 + \frac{1}{2x-1}\right)dx \\ &= \frac{x}{2} + \frac{1}{4}\ln|2x-1| + C. \end{aligned}$$

In the discussion that follows, we always assume that any necessary division has been performed and the quotient polynomial has been integrated. The remaining basic problem with which we will deal in this section is the following:

6.7.3 The Basic Problem

> We want to integrate a rational function $f(x) = \dfrac{P(x)}{Q(x)}$ where
>
> the degree of $P <$ the degree of Q.

The complexity of this problem depends on the degree of Q.

Suppose that $Q(x)$ has degree 1; $Q(x) = ax + b$ where $a \neq 0$. Then $P(x)$ must have degree 0, and be a constant c. We have $f(x) = c/(ax+b)$. The substitution $u = ax + b$ leads to

$$\int \frac{dx}{ax+b} = \frac{1}{a}\int \frac{du}{u} = \frac{1}{a}\ln|u| + C.$$

Thus

6.7.4 The Case of Linear Q

> $$\int \frac{dx}{ax+b} = \frac{1}{a}\ln|ax+b| + C.$$

Now suppose that $Q(x)$ is quadratic, that is, has degree 2. We can assume for purposes of this discussion that $Q(x)$ is either of the form $x^2 + a^2$ or of the form $x^2 - a^2$, since completing the square and making the appropriate change of variable can always reduce a quadratic denominator to this form, as shown in Section 6.5. Since $P(x)$ can be at most a linear function, $P(x) = Ax + B$, we are led to consider the following four integrals:

$$\int \frac{x\,dx}{x^2 + a^2}, \qquad \int \frac{x\,dx}{x^2 - a^2}, \qquad \int \frac{dx}{x^2 + a^2}, \qquad \text{and} \qquad \int \frac{dx}{x^2 - a^2}.$$

(If $a = 0$, there are only two integrals, each easily evaluated.) The first two integrals yield to the substitution $u = x^2 \pm a^2$; the third is a known integral. The fourth integral can be done with the substitution $x = a \sin\theta$ if $|x| < |a|$, and with the substitution $x = a \sec\theta$ if $|x| > |a|$, but we will evaluate it by a different method below. The values of all four integrals are given in the following box:

6.7.5 The Case of Quadratic Q

$$\int \frac{x\,dx}{x^2 + a^2} = \frac{1}{2} \ln(x^2 + a^2) + C$$
$$\int \frac{x\,dx}{x^2 - a^2} = \frac{1}{2} \ln|x^2 - a^2| + C,$$
$$\int \frac{dx}{x^2 + a^2} = \frac{1}{a} \tan^{-1} \frac{x}{a} + C,$$
$$\int \frac{dx}{x^2 - a^2} = \frac{1}{2a} \ln\left|\frac{x - a}{x + a}\right| + C.$$

To obtain the last formula in the box let us try to write the integrand as a sum of two fractions with linear denominators:

$$\frac{1}{x^2 - a^2} = \frac{1}{(x - a)(x + a)} = \frac{A}{x - a} + \frac{B}{x + a} = \frac{Ax + Aa + Bx - Ba}{x^2 - a^2},$$

where we have added the two fractions together again in the last step. If this equation is to hold identically for all x (except $x = \pm a$), then the numerators on the left and right sides must be identical as polynomials in x. The equation $(A + B)x + (Aa - Ba) = 1 = 0x + 1$ can hold for all x only if

$$\begin{aligned} A + B &= 0 \qquad \text{(the coefficient of } x\text{)} \\ Aa - Ba &= 1 \qquad \text{(the constant term)} \end{aligned}$$

Solving this pair of linear equations for the unknowns A and B, we get $A = 1/(2a)$ and $B = -1/(2a)$. Therefore,

$$\begin{aligned} \int \frac{dx}{x^2 - a^2} &= \frac{1}{2a} \int \frac{dx}{x - a} - \frac{1}{2a} \int \frac{dx}{x + a} \\ &= \frac{1}{2a} \ln|x - a| - \frac{1}{2a} \ln|x + a| + C \\ &= \frac{1}{2a} \ln\left|\frac{x - a}{x + a}\right| + C. \end{aligned}$$

The technique used above, involving the writing of a complicated fraction as a sum of simpler fractions, is called the **method of partial fractions**. Suppose that a polynomial $Q(x)$ is of degree n and that its highest degree term is x^n (with coefficient 1). Suppose also that Q factors into a product of n *distinct* linear (degree 1) factors, say

$$Q(x) = (x - a_1)(x - a_2)\cdots(x - a_n),$$

where $a_i \neq a_j$ if $i \neq j$, $1 \leq i, j \leq n$. If $P(x)$ is a polynomial of degree smaller than n, then $P(x)/Q(x)$ has a **partial fraction decomposition** of the form

$$\frac{P(x)}{Q(x)} = \frac{A_1}{x - a_1} + \frac{A_2}{x - a_2} + \cdots + \frac{A_n}{x - a_n}$$

for certain values of the constants $A_1, A_2, \ldots, A_n$. We do not attempt to give any formal proof of this assertion here; such a proof belongs in an algebra course. (See Theorem 6.7.12 for the statement of a more general result.)

Given that $P(x)/Q(x)$ has a partial fraction decomposition as claimed above, there are two methods for determining the constants $A_1, A_2, \ldots, A_n$. The first of these methods, and one that generalizes most easily to the more complicated decompositions considered below, is to add up the fractions in the decomposition, obtaining a new fraction $S(x)/Q(x)$ with numerator $S(x)$, a polynomial of degree one less than that of $Q(x)$. This new fraction will be identical to the original fraction $P(x)/Q(x)$ if S and P are identical polynomials. The constants $A_1, A_2, \ldots, A_n$ are determined by solving the n linear equations resulting from equating the coefficients of like powers of x in these two polynomials.

The second method depends on the following observation: If we multiply the partial fraction decomposition by $x - a_j$, we get

$$(x-a_j)\frac{P(x)}{Q(x)} = A_1\frac{x - a_j}{x - a_1} + \cdots + A_{j-1}\frac{x - a_j}{x - a_{j-1}} + A_j + A_{j+1}\frac{x - a_j}{x - a_{j+1}} + \cdots + A_n\frac{x - a_j}{x - a_n}.$$

All terms on the right side are 0 at $x = a_j$ except the jth term, A_j. Hence

$$A_j = \lim_{x \to a_j}(x - a_j)\frac{P(x)}{Q(x)} = \frac{P(a_j)}{(a_j - a_1)\cdots(a_j - a_{j-1})(a_j - a_{j+1})\cdots(a_j - a_n)},$$

for $1 \leq j \leq n$. In practice, this method is carried out by adding the partial fractions together, equating the numerator of the sum to $P(x)$, and substituting $x = a_1$, $x = a_2$, ..., $x = a_n$ to find $A_1, A_2, \ldots, A_n$.

EXAMPLE 6.7.6 Evaluate $I = \displaystyle\int \frac{(x + 4)\,dx}{x^2 - 5x + 6}$.

SOLUTION

$$\frac{x + 4}{x^2 - 5x + 6} = \frac{x + 4}{(x - 2)(x - 3)} = \frac{A}{x - 2} + \frac{B}{x - 3}$$

How?

We calculate A and B by both methods suggested above. Add the fractions:

Explain

$$\frac{x+4}{x^2-5x+6} = \frac{Ax-3A+Bx-2B}{(x-2)(x-3)}.$$

Method I

Equate the coefficients of x and the constant terms in the numerators on both sides to obtain

$$A+B=1, \quad -3A-2B=4.$$

Solve these equations to get $A=-6$ and $B=7$.

Method II

Equate the numerators on both sides to get

$$x+4 = Ax-3A+Bx-2B.$$

Substitute $x=2$ and $x=3$ into this last equation and obtain, respectively, $A = -6$ and $B=7$.

In either case we have

$$I = -6\int \frac{dx}{x-2} + 7\int \frac{dx}{x-3} = -6\ln|x-2| + 7\ln|x-3| + C.$$

EXAMPLE 6.7.7 Evaluate $I = \displaystyle\int \frac{x^3+2}{x^3-x}\,dx$.

SOLUTION Since the numerator does not have degree smaller than the denominator, we must write

How do you manipulate?

$$I = \int \frac{x^3-x+x+2}{x^3-x}\,dx = \int\left(1+\frac{x+2}{x^3-x}\right)dx = x + \int \frac{x+2}{x^3-x}\,dx.$$

Now we can use the method of partial fractions.

$$\frac{x+2}{x^3-x} = \frac{x+2}{x(x-1)(x+1)} = \frac{A}{x} + \frac{B}{x-1} + \frac{C}{x+1}$$
$$= \frac{A(x^2-1)+B(x^2+x)+C(x^2-x)}{x(x-1)(x+1)}$$

We have

$$\begin{aligned} A + B + C &= 0 \qquad &&\text{(coefficient of } x^2\text{)}\\ B - C &= 1 &&\text{(coefficient of } x\text{)}\\ -A &= 2 &&\text{(constant term).}\end{aligned}$$

It follows that $A=-2$, $B=3/2$, and $C=1/2$. (We could also have found these by using the second method of Example 6.7.6: substitute $x=0$, 1 and -1, respectively, into the equation $x+2 = A(x^2-1)+B(x^2+x)+C(x^2-x)$.) Finally, we have

$$I = x - 2\int\frac{dx}{x} + \frac{3}{2}\int\frac{dx}{x-1} + \frac{1}{2}\int\frac{dx}{x+1}$$
$$= x - 2\ln|x| + \frac{3}{2}\ln|x-1| + \frac{1}{2}\ln|x+1| + C.$$

Next we consider a rational function whose denominator has a quadratic factor that is equivalent to a sum of squares and that cannot, therefore, be further factored into a product of real linear factors.

EXAMPLE 6.7.8 Evaluate $I = \displaystyle\int \frac{2+3x+x^2}{x(x^2+1)}\,dx$.

SOLUTION Note that the numerator has degree 2 and the denominator degree 3, so no division is necessary. If we decompose the integrand as a sum of two simpler fractions, we want one with denominator x and one with denominator x^2+1. The appropriate form of the decomposition turns out to be

$$\frac{2+3x+x^2}{x(x^2+1)} = \frac{A}{x} + \frac{Bx+C}{x^2+1} = \frac{A(x^2+1)+Bx^2+Cx}{x(x^2+1)}.$$

Note that corresponding to the quadratic (degree 2) denominator we use a linear (degree 1) numerator. Equating coefficients in the two numerators, we obtain

$$\begin{aligned} A + B \quad &= 1 \qquad \text{(coefficient of } x^2\text{)}\\ C &= 3 \qquad \text{(coefficient of } x\text{)}\\ A \quad &= 2 \qquad \text{(constant term).}\end{aligned}$$

Hence $A = 2$, $B = -1$, and $C = 3$. We have, therefore,

$$\begin{aligned} I &= 2\int \frac{dx}{x} - \int \frac{x\,dx}{x^2+1} + 3\int \frac{dx}{x^2+1}\\ &= 2\ln|x| - \frac{1}{2}\ln(x^2+1) + 3\tan^{-1}x + C.\end{aligned}$$

We remark that addition of the fractions is the only reasonable real-variable method for determining the constants A, B, and C here. We could determine A by a limit procedure such as in Example 6.7.6, but there is no simple equivalent way of finding B or C without using complex numbers.

EXAMPLE 6.7.9 Evaluate $I = \displaystyle\int \frac{dx}{x^3+1}$.

SOLUTION Here $Q(x) = x^3 + 1 = (x+1)(x^2-x+1)$, and the latter factor has no real roots so no real linear subfactors. We have

$$\begin{aligned}\frac{1}{x^3+1} &= \frac{1}{(x+1)(x^2-x+1)} = \frac{A}{x+1} + \frac{Bx+C}{x^2-x+1}\\ &= \frac{A(x^2-x+1)+B(x^2+x)+C(x+1)}{(x+1)(x^2-x+1)}\end{aligned}$$

$$\begin{aligned} A + B \quad &= 0 \qquad \text{(coefficient of } x^2\text{)}\\ -A + B + C &= 0 \qquad \text{(coefficient of } x\text{)}\\ A \quad + C &= 1 \qquad \text{(constant term).}\end{aligned}$$

Hence $A = 1/3$, $B = -(1/3)$, and $C = 2/3$. We have

$$\begin{aligned}
I &= \frac{1}{3}\int \frac{dx}{x+1} - \frac{1}{3}\int \frac{x-2}{x^2-x+1}\,dx \\
&= \frac{1}{3}\ln|x+1| - \frac{1}{3}\int \frac{x - \frac{1}{2} - \frac{3}{2}}{\left(x-\frac{1}{2}\right)^2 + \frac{3}{4}}\,dx \qquad \text{Let } u = x - 1/2,\ du = dx \\
&= \frac{1}{3}\ln|x+1| - \frac{1}{3}\int \frac{u\,du}{u^2+\frac{3}{4}} + \frac{1}{2}\int \frac{du}{u^2+\frac{3}{4}} \\
&= \frac{1}{3}\ln|x+1| - \frac{1}{6}\ln\left(u^2+\frac{3}{4}\right) + \frac{1}{2}\frac{2}{\sqrt{3}}\tan^{-1}\left(\frac{2u}{\sqrt{3}}\right) + C \\
&= \frac{1}{3}\ln|x+1| - \frac{1}{6}\ln(x^2-x+1) + \frac{1}{\sqrt{3}}\tan^{-1}\left(\frac{2x-1}{\sqrt{3}}\right) + C.
\end{aligned}$$

We require one final refinement of the method of partial fractions. If any of the linear or quadratic factors of $Q(x)$ is *repeated* (say m times), then the partial fraction decomposition of $P(x)/Q(x)$ requires m distinct fractions corresponding to that factor. The denominators of these fractions have exponents increasing from 1 to m, and the numerators are all constants where the repeated factor is linear, or linear where the repeated factor is quadratic. (See Theorem 6.7.12 below.)

EXAMPLE 6.7.10 Evaluate $I = \displaystyle\int \frac{dx}{x(x-1)^2}$.

SOLUTION The appropriate partial fraction decomposition here is

$$\frac{1}{x(x-1)^2} = \frac{A}{x} + \frac{B}{x-1} + \frac{C}{(x-1)^2} = \frac{A(x^2-2x+1) + B(x^2-x) + Cx}{x(x-1)^2}.$$

Equating coefficients of x^2, x, and 1 in the numerators of both sides, we get

$$\begin{aligned}
A + B &= 0 \qquad \text{(coefficient of } x^2\text{)} \\
-2A - B + C &= 0 \qquad \text{(coefficient of } x\text{)} \\
A &= 1 \qquad \text{(constant term).}
\end{aligned}$$

Hence $A = 1$, $B = -1$, $C = 1$, and

$$\begin{aligned}
I = \int \frac{dx}{x} - \int \frac{dx}{x-1} + \int \frac{dx}{(x-1)^2} &= \ln|x| - \ln|x-1| - \frac{1}{x-1} + C \\
&= \ln\left|\frac{x}{x-1}\right| - \frac{1}{x-1} + C.
\end{aligned}$$

EXAMPLE 6.7.11 Evaluate $I = \displaystyle\int \frac{x^2+2}{4x^5+4x^3+x}\,dx$.

SOLUTION The denominator factors to $x(2x^2+1)^2$ so the appropriate partial fraction decomposition is

$$\begin{aligned}\frac{x^2+2}{x(2x^2+1)^2} &= \frac{A}{x} + \frac{Bx+C}{2x^2+1} + \frac{Dx+E}{(2x^2+1)^2}\\ &= \frac{A(4x^4+4x^2+1)+B(2x^4+x^2)+C(2x^3+x)+Dx^2+Ex}{x(2x^2+1)^2}.\end{aligned}$$

Thus

$$\begin{array}{lllll} 4A + 2B & & & = 0 & \text{(coefficient of } x^4)\\ & 2C & & = 0 & \text{(coefficient of } x^3)\\ 4A + B & & + D & = 1 & \text{(coefficient of } x^2)\\ & C & + E & = 0 & \text{(coefficient of } x)\\ A & & & = 2 & \text{(constant term).}\end{array}$$

Solving these equations, we get $A = 2$, $B = -4$, $C = 0$, $D = -3$, $E = 0$.

$$\begin{aligned} I &= 2\int\frac{dx}{x} - 4\int\frac{x\,dx}{2x^2+1} - 3\int\frac{x\,dx}{(2x^2+1)^2} \qquad \text{Let } u = 2x^2+1,\ du = 4x\,dx\\ &= 2\ln|x| - \int\frac{du}{u} - \frac{3}{4}\int\frac{du}{u^2}\\ &= 2\ln|x| - \ln|u| + \frac{3}{4u} + C\\ &= \ln\left(\frac{x^2}{2x^2+1}\right) + \frac{3}{4}\frac{1}{2x^2+1} + C\end{aligned}$$

The following theorem summarizes the various aspects of the method of partial fractions.

THEOREM 6.7.12 Let P and Q be real polynomials with real coefficients, and suppose that the degree of P is less than the degree of Q. Then

a) $Q(x)$ can be factored into the product of a constant k, real linear factors of the form $x - a_i$, and real quadratic factors of the form $x^2 + b_i x + c_i$ having no real roots. The linear and quadratic factors may be repeated.

$$Q(x) = k(x-a_1)^{m_1}(x-a_2)^{m_2}\cdots(x-a_j)^{m_j}(x^2+b_1x+c_1)^{n_1}\cdots(x^2+b_kx+c_k)^{n_k}.$$

The degree of Q is $m_1 + m_2 + \cdots + m_j + 2n_1 + 2n_2 + \cdots + 2n_k$.

b) The rational function $P(x)/Q(x)$ can be expressed as a sum of partial fractions as follows:

i) corresponding to each factor $(x-a)^m$ of $Q(x)$ the decomposition contains a sum of fractions of the form

$$\frac{A_1}{x-a}+\frac{A_2}{(x-a)^2}+\cdots+\frac{A_m}{(x-a)^m};$$

ii) corresponding to each factor $(x^2+bx+c)^n$ of $Q(x)$ the decomposition contains a sum of fractions of the form

$$\frac{B_1x+C_1}{x^2+bx+c}+\frac{B_2x+C_2}{(x^2+bx+c)^2}+\cdots+\frac{B_nx+C_n}{(x^2+bx+c)^n}.$$

The constants $A_1, A_2, \ldots, A_m, B_1, B_2, \ldots, B_n, C_1, C_2, \ldots, C_n$ can be determined by adding up the fractions in the decomposition and equating the coefficients of like powers of x in the numerator of the sum with those in $P(x)$. □

We will not attempt to prove this theorem here; the proof belongs in a study of polynomial algebra. The proof of part (a) can often be found in textbooks on complex analysis (calculus involving functions of a variable that is complex rather than real).

Also, note that part (a) does not tell us how to find the factors of $Q(x)$; it tells us only what form they have. We must know the factors of Q before we can make use of partial fractions to integrate the rational function $P(x)/Q(x)$. Partial fraction decompositions are also used in other mathematical situations, in particular to solve certain problems involving differential equations

EXERCISES

Evaluate the integrals in Exercises 1–35.

1. $\displaystyle\int \frac{2\,dx}{2x-3}$

2. $\displaystyle\int \frac{dx}{5-4x}$

3. $\displaystyle\int \frac{x\,dx}{\pi x+2}$

4. $\displaystyle\int \frac{x^2}{x-4}\,dx$

5. $\displaystyle\int \frac{1}{x^2-9}\,dx$

6. $\displaystyle\int \frac{dx}{5-x^2}$

7. $\displaystyle\int \frac{dx}{a^2-x^2}$

8. $\displaystyle\int \frac{dx}{b^2-a^2x^2}$

9. $\displaystyle\int \frac{x^2\,dx}{x^2+x-2}$

10. $\displaystyle\int \frac{x\,dx}{3x^2+8x-3}$

11. $\displaystyle\int \frac{x-2}{x^2+x}\,dx$

12. $\displaystyle\int \frac{dx}{x^3+9x}$

13. $\displaystyle\int \frac{dx}{1-6x+9x^2}$

14. $\displaystyle\int \frac{x\,dx}{2+6x+9x^2}$

15. $\displaystyle\int \frac{x^2+1}{6x-9x^2}\,dx$

16. $\displaystyle\int \frac{x^3+1}{12+7x+x^2}\,dx$

17. $\displaystyle\int \frac{dx}{x(x^2-a^2)}$

18. $\displaystyle\int \frac{dx}{x^4-a^4}$

19.* $\displaystyle\int \frac{x^3\,dx}{x^3-a^3}$

20. $\displaystyle\int \frac{dx}{x^3+2x^2+2x}$

21. $\displaystyle\int \frac{dx}{x^3-4x^2+3x}$

22. $\displaystyle\int \frac{x^2+1}{x^3+8}\,dx$

23. $\displaystyle\int \frac{dx}{(x^2-1)^2}$

24. $\displaystyle\int \frac{x^2\,dx}{(x^2-1)(x^2-4)}$

25. $\displaystyle\int \frac{dx}{x^4-3x^3}$

26. $\displaystyle\int \frac{x\,dx}{(x^2-x+1)^2}$

27.* $\displaystyle\int \frac{t\,dt}{(t+1)(t^2+1)^2}$

28.* $\displaystyle\int \frac{dt}{(t-1)(t^2-1)^2}$

29.* $\displaystyle\int \frac{dx}{x(3+x^2)\sqrt{1-x^2}}$

30.* $\displaystyle\int \frac{dx}{e^{2x}-4e^x+4}$

31.* $\displaystyle\int \frac{dx}{x(1+x^2)^{3/2}}$

32.* $\displaystyle\int \frac{dx}{x(1-x^2)^{3/2}}$

33.* $\displaystyle\int \frac{dx}{x^2(x^2-1)^{3/2}}$

34.* $\displaystyle\int \frac{d\theta}{\cos\theta(1+\sin\theta)}$

35.* $\displaystyle\int \frac{d\theta}{\sin\theta(1+\sin\theta)}$

36.*Suppose that P and Q are polynomials such that the degree of P is smaller than that of Q. If

$$Q(x) = (x-a_1)(x-a_2)\cdots(x-a_n),$$

where $a_i \neq a_j$ if $i \neq j (1 \leq i, j \leq n)$, so that $P(x)/Q(x)$ has partial fraction decomposition

$$\frac{P(x)}{Q(x)} = \frac{A_1}{x-a_1} + \frac{A_2}{x-a_2} + \cdots + \frac{A_n}{x-a_n},$$

show that

$$A_j = \frac{P(a_j)}{Q'(a_j)} \qquad (1 \leq j \leq n).$$

This gives yet another method for computing the constants in a partial fraction decomposition if the denominator factors completely into distinct linear factors.

Summary of Techniques of Integration

Students sometimes have difficulty deciding which method to use to evaluate a given integral. Often no one method will suffice to produce the whole solution, but one method may lead to a different, possibly simpler integral that can then be dealt with on its own merits. Here are a few guidelines:

1. First, and at all stages, be alert for simplifying substitutions. Even when these don't accomplish the whole integration, they can lead to integrals to which some other method can be applied.
2. If the integral involves a quadratic expression $Ax^2 + Bx + C$ with $A \neq 0$ and $B \neq 0$, complete the square. A simple substitution then reduces the quadratic expression to a sum or difference of squares.
3. Integrals of products of trigonometric functions can sometimes be evaluated or rendered simpler by the use of appropriate trigonometric identities such as $\sin^2 x + \cos^2 x = 1$, $\sec^2 x = 1 + \tan^2 x$, $\csc^2 x = 1 + \cot^2 x$, $\sin x \cos x = \frac{1}{2} \sin 2x$, $\sin^2 x = \frac{1}{2}(1 - \cos 2x)$, and $\cos^2 x = \frac{1}{2}(1 + \cos 2x)$.
4. Integrals involving $(a^2 - x^2)^{1/2}$ can be transformed using $x = a\sin\theta$. Integrals involving $(a^2 + x^2)^{1/2}$ or $1/(a^2 + x^2)$ may yield to $x = a\tan\theta$. Integrals involving $(x^2 - a^2)^{1/2}$ can be transformed using $x = a\sec\theta$.
5. Use integration by parts for integrals of functions such as products of polynomials and transcendental functions, and for inverse trigonometric functions and logarithms. Be alert for ways of using integration by parts to get formulas representing complicated integrals in terms of simpler ones.
6. Use partial fractions to integrate rational functions whose denominators can be factored into real linear and quadratic factors. Remember to divide the polynomials first, if necessary, to reduce the fraction to one whose numerator has degree smaller than that of its denominator.
7. There is a table of indefinite integrals at the back of this book. If you can't do an integral directly, try to use the methods above to convert it to the form of one of the integrals in the table.

8. If you can't find any way to evaluate a definite integral for which you neec a numerical value, consider using a computer or calculator and one of the numerical methods presented in Chapter 7.

Review Exercises on Techniques of Integration

Here is an opportunity to get more practice evaluating integrals. Unlike the ex ercises in Sections 6.4–6.7, which concentrated on the technique of the particula section, these exercises are grouped randomly so you will have to decide whicl techniques to use.

EXERCISES

Evaluate these integrals, which review the techniqes of Sections 6.4–6.7.

1. $\int \frac{x\,dx}{2x^2+5x+2}$
2. $\int \frac{x\,dx}{(x-1)^3}$
3. $\int \sin^3 x \cos^3 x\,dx$
4. $\int \frac{(1+\sqrt{x})^{1/3}}{\sqrt{x}}\,dx$
5. $\int \frac{3\,dx}{4x^2-1}$
6. $\int (x^2+x-2)\sin 3x\,dx$
7. $\int \frac{\sqrt{1-x^2}}{x^4}\,dx$
8. $\int x^3\cos(x^2)\,dx$
9. $\int \frac{x^2\,dx}{(5x^3-2)^{2/3}}$
10. $\int \frac{dx}{x^2+2x-15}$
11. $\int \frac{dx}{(4+x^2)^2}$
12. $\int (\sin x+\cos x)^2\,dx$
13. $\int 2^x\sqrt{1+4^x}\,dx$
14. $\int \frac{\cos x}{1+\sin^2 x}\,dx$
15. $\int \frac{\sin^3 x}{\cos^7 x}\,dx$
16. $\int \frac{x^2\,dx}{(3+5x^2)^{3/2}}$
17. $\int e^{-x}\sin(2x)\,dx$
18. $\int \frac{2x^2+4x-3}{x^2+5x}\,dx$
19. $\int \cos(3\ln x)\,dx$
20. $\int \frac{dx}{4x^3+x}$
21. $\int \frac{x\ln(1+x^2)}{1+x^2}\,dx$
22. $\int \sin^2 x\cos^4 x\,dx$
23. $\int \frac{x^2}{\sqrt{2-x^2}}\,dx$
24. $\int \tan^4 x\sec x\,dx$
25. $\int \frac{x^2\,dx}{(4x+1)^{10}}$
26. $\int x\sin^{-1}\frac{x}{2}\,dx$
27. $\int \sin^5(4x)\,dx$
28. $\int \frac{dx}{x^5-2x^3+x}$
29. $\int \frac{dx}{2+e^x}$
30. $\int x^3 3^x\,dx$
31. $\int \frac{\sin^2 x\cos x}{2-\sin x}\,dx$
32. $\int \frac{x^2+1}{x^2+2x+2}\,dx$
33. $\int \frac{dx}{x^2\sqrt{1-x^2}}$
34. $\int x^3(\ln x)^2\,dx$
35. $\int \frac{x^3}{\sqrt{1-4x^2}}\,dx$
36. $\int \frac{e^{1/x}\,dx}{x^2}$
37. $\int \frac{x+1}{\sqrt{x^2+1}}\,dx$
38. $\int e^{(x^{1/3})}\,dx$
39. $\int \frac{x^3-3}{x^3-9x}\,dx$
40. $\int \frac{10^{\sqrt{x+2}}}{\sqrt{x+2}}\,dx$
41. $\int \sin^5 x\cos^9 x\,dx$
42. $\int \frac{x^2\,dx}{\sqrt{x^2-1}}$
43. $\int \frac{x\,dx}{x^2+2x-1}$
44. $\int \frac{2x-3}{\sqrt{4-3x+x^2}}\,dx$
45. $\int x^2\sin^{-1}(2x)\,dx$
46. $\int \frac{\sqrt{3x^2-1}}{x}\,dx$
47. $\int \cos^4 x\sin^4 x\,dx$
48. $\int \sqrt{x-x^2}\,dx$
49. $\int \frac{dx}{(4+x)\sqrt{x}}$
50. $\int x\tan^{-1}\frac{x}{3}\,dx$
51. $\int \frac{x^4-1}{x^3+2x^2}\,dx$
52. $\int \frac{dx}{x(x^2+4)^2}$
53. $\int \frac{\sin(2\ln x)}{x}\,dx$
54. $\int \frac{\sin(\ln x)}{x^2}\,dx$

55. $\displaystyle\int \frac{e^{2\tan^{-1}x}}{1+x^2}\,dx$

56. $\displaystyle\int \frac{x^3+x-2}{x^2-7}\,dx$

57. $\displaystyle\int \frac{\ln(3+x^2)}{3+x^2}\,x\,dx$

58. $\displaystyle\int \cos^7 x\,dx$

59. $\displaystyle\int \frac{\sin^{-1}(x/2)}{(4-x^2)^{1/2}}\,dx$

60. $\displaystyle\int \tan^4(\pi x)\,dx$

61. $\displaystyle\int \frac{(x+1)\,dx}{\sqrt{x^2+6x+10}}$

62. $\displaystyle\int e^x(1-e^{2x})^{5/2}\,dx$

63. $\displaystyle\int \frac{x^3\,dx}{(x^2+2)^{7/2}}$

64. $\displaystyle\int \frac{x^2}{2x^2-3}\,dx$

65. $\displaystyle\int \frac{x^{1/2}}{1+x^{1/3}}\,dx$

66. $\displaystyle\int \frac{dx}{x(x^2+x+1)^{1/2}}$

67. $\displaystyle\int \frac{1+x}{1+\sqrt{x}}\,dx$

68. $\displaystyle\int \frac{x\,dx}{4x^4+4x^2+5}$

69. $\displaystyle\int \frac{x\,dx}{(x^2-4)^2}$

70. $\displaystyle\int \frac{dx}{x^3+x^2+x}$

71. $\displaystyle\int x^2\tan^{-1}x\,dx$

72. $\displaystyle\int e^x\sec(e^x)\,dx$

73. $\displaystyle\int \frac{dx}{4\sin x-3\cos x}$

74. $\displaystyle\int \frac{dx}{x^{1/3}-1}$

75. $\displaystyle\int \frac{dx}{\tan x+\sin x}$

76. $\displaystyle\int \frac{x\,dx}{\sqrt{3-4x-4x^2}}$

77. $\displaystyle\int \frac{\sqrt{x}}{1+x}\,dx$

78. $\displaystyle\int \sqrt{1+e^x}\,dx$

79. $\displaystyle\int \frac{x^4\,dx}{x^3-8}$

80. $\displaystyle\int xe^x\cos x\,dx$

6.8 IMPROPER INTEGRALS

Up to this point we have considered definite integrals of the form

$$I = \int_a^b f(x)\,dx$$

where the integrand f is *continuous* on the *closed, finite* interval $[a, b]$. Since such a function is necessarily *bounded*, the integral I is necessarily a finite number; for positive f it corresponds to the area of a **bounded region** of the plane, a region contained inside some disc of finite radius with centre at the origin. Such integrals are also called **proper integrals**. We are now going to generalize the definite integral to allow for two possibilities excluded in the situation described above:

i) We may have $a = -\infty$ or $b = \infty$ or both.

ii) f may become unbounded as x approaches a or b or both.

Integrals satisfying (i) are called **improper integrals of type I**; integrals satisfying (ii) are called **improper integrals of type II**. Either type of improper integral corresponds (for positive f) to the area of a region in the plane that "extends to infinity" in some direction and therefore is *unbounded*. As we will see, such integrals may or may not have finite values. The ideas involved are best introduced by examples.

EXAMPLE 6.8.1 Suppose we want to find the area of A the region lying under the curve $y = 1/x^2$ and above the x-axis to the right of $x = 1$. (See Fig. 6.8.1.) We would like to calculate the area via an integral

$$A = \int_1^\infty \frac{dx}{x^2}.$$

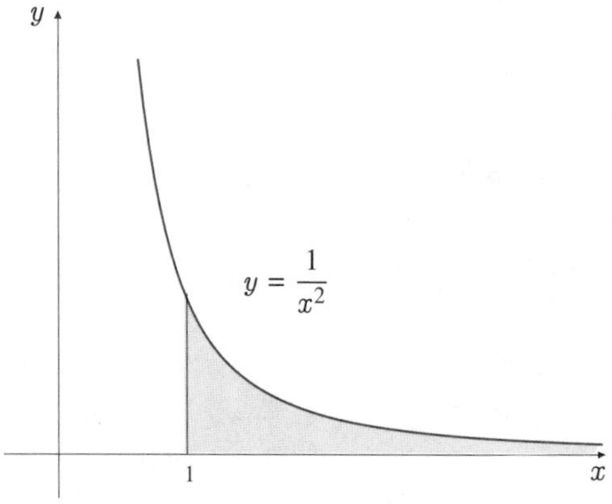

FIGURE 6.8.1

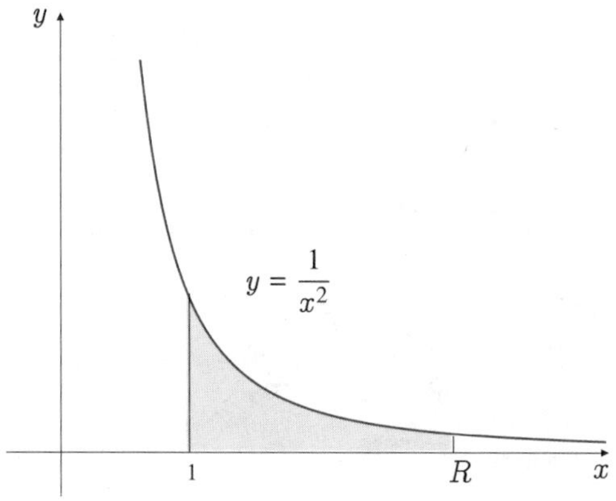

FIGURE 6.8.2

This integral is improper of type I since the interval of integration is infinite. It is not immediately obvious whether the area is finite or not; the region has an infinitely long "spike" along the x-axis, but this spike becomes infinitely thin as x approaches ∞. In order to evaluate this improper integral, we interpret it as a limit of proper integrals over intervals $[1, R]$ as $R \to \infty$ (see Fig. 6.8.2).

$$\begin{aligned} A = \int_1^\infty \frac{dx}{x^2} &= \lim_{R\to\infty} \int_1^R \frac{dx}{x^2} = \lim_{R\to\infty} \left(-\frac{1}{x}\right)\Bigg|_1^R \\ &= \lim_{R\to\infty} \left(-\frac{1}{R} + 1\right) = 1 \end{aligned}$$

Since the limit exists (is finite), we say that the improper integral *converges*. The region has finite area, $A = 1$ square unit.

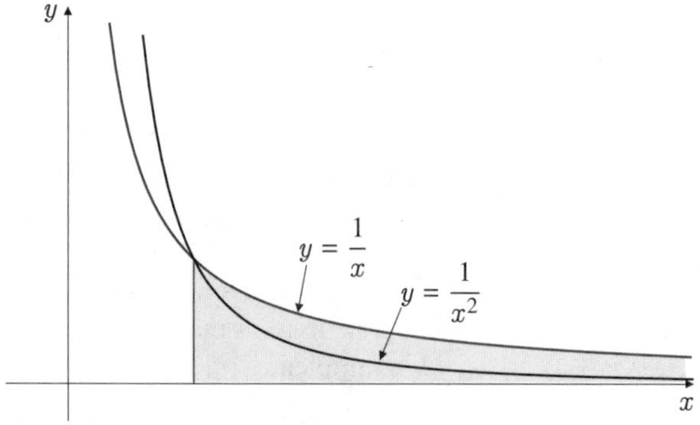

FIGURE 6.8.3

EXAMPLE 6.8.2 We carry out a similar calculation to find the area of the region under $y = 1/x$ above $y = 0$ to the right of $x = 1$. (See Fig. 6.8.3.) This area is given by the improper integral

$$A = \int_1^\infty \frac{dx}{x} = \lim_{R\to\infty} \int_1^R \frac{dx}{x} = \lim_{R\to\infty} \ln x \Big|_1^R = \lim_{R\to\infty} \ln R = \infty.$$

We say that this improper integral *diverges to infinity*. Observe that the region has a similar shape to the region under $y = 1/x^2$ considered in the above example, but its "spike" is somewhat thicker at each value of $x > 1$. Evidently this extra thickness makes a difference; this region has infinite area.

6.8.3
Improper Integrals of Type I

If f is continuous on $[a, \infty)$, we define the improper integral as the limit of proper integrals thus:

$$\int_a^\infty f(x)\,dx = \lim_{R\to\infty} \int_a^R f(x)\,dx.$$

Similarly, if f is continuous on $(-\infty, b]$ then we define

$$\int_{-\infty}^b f(x)\,dx = \lim_{R\to-\infty} \int_R^b f(x)\,dx.$$

In either case, if the limit is a finite number, we say that the improper integral **converges**; if the limit does not exist, or if it is ∞ or $-\infty$, we say that the improper integral **diverges** or **diverges to infinity** or **negative infinity**.

The integral $\int_{-\infty}^\infty f(x)\,dx$ is, for f continuous on the real line, improper of type I at both endpoints. We break it into two separate integrals:

$$\int_{-\infty}^\infty f(x)\,dx = \int_{-\infty}^0 f(x)\,dx + \int_0^\infty f(x)\,dx.$$

The integral on the left converges if and only if both integrals on the right converge.

EXAMPLE 6.8.4 a)

$$\begin{aligned}\int_{-\infty}^\infty \frac{dx}{1+x^2} &= \int_{-\infty}^0 \frac{dx}{1+x^2} + \int_0^\infty \frac{dx}{1+x^2}\\ &= 2\lim_{R\to\infty}\int_0^R \frac{dx}{1+x^2} \qquad \text{(by symmetry)}\\ &= 2\lim_{R\to\infty} \tan^{-1} R = 2\left(\frac{\pi}{2}\right) = \pi\end{aligned}$$

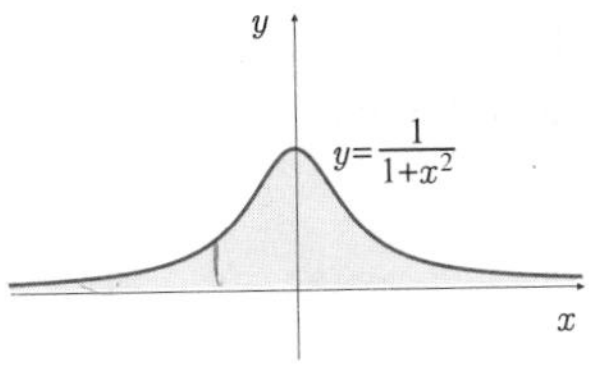

FIGURE 6.8.4

The use of symmetry here requires some justification. (See Fig. 6.8.4.) At the time we used it we did not know whether each of the half-line integrals was finite or infinite. However, since both are positive, even if they are infinite, their sum would still be twice one of them. If one had been positive and the other negative, we would not have been justified in cancelling them to get 0 until we knew that they were finite. ($\infty + \infty = \infty$, but $\infty - \infty$ is not defined.) In any event, the given integral converges to π.

EXAMPLE 6.8.5 b)

$$\int_0^\infty \cos x\, dx = \lim_{R\to\infty} \int_0^R \cos x\, dx = \lim_{R\to\infty} \sin R$$

This limit does not exist (and is not ∞ or $-\infty$), so all we can say is that the given integral diverges. See Fig. 6.8.5. As R increases the integral alternately adds and subtracts the areas of the hills and valleys but does not approach any unique limit.

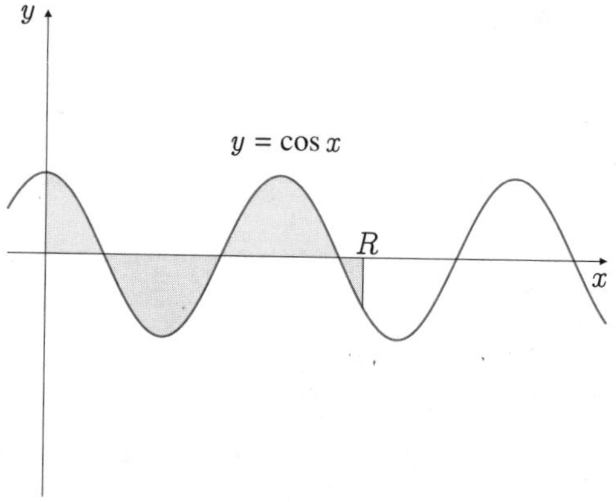

FIGURE 6.8.5

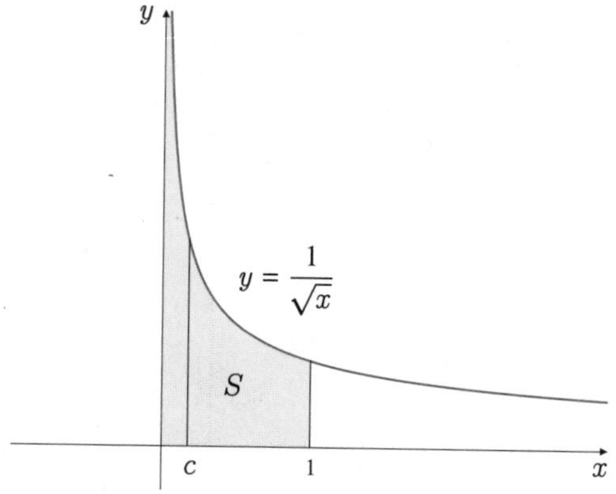

FIGURE 6.8.6

EXAMPLE 6.8.6 Consider the problem of finding the area of the region S lying under $y = 1/\sqrt{x}$, above the x-axis, between $x = 0$ and $x = 1$. The area A is given by

$$A = \int_0^1 dx/\sqrt{x},$$

which is an improper integral of type II since the integrand becomes unbounded near $x = 0$. The region S has a "spike" extending to infinity along the y-axis, as shown in Fig. 6.8.6. As for improper integrals of type I, we express such integrals as limits of proper integrals.

$$A = \lim_{c\to 0+} \int_c^1 x^{-1/2}\, dx = \lim_{c\to 0+} 2x^{1/2}\Big|_c^1 = \lim_{c\to 0+} (2 - 2\sqrt{c}) = 2.$$

This integral converges, and S has a finite area of 2 square units.

6.8.7
Improper Integrals of Type II

> If f is continuous on the interval $(a, b]$ and possibly unbounded near a, we define the improper integral
>
> $$\int_a^b f(x)\,dx = \lim_{c\to a+} \int_c^b f(x)\,dx.$$
>
> Similarly, if f is continuous on $[a, b)$ and possibly unbounded near b, we define
>
> $$\int_a^b f(x)\,dx = \lim_{c\to b-} \int_a^c f(x)\,dx.$$
>
> Again the possibilities are convergence, divergence, divergence to infinity, and divergence to negative infinity.

While improper integrals of type I are always easily recognized because of the infinite limits of integration, improper integrals of type II can be somewhat harder to spot. You should be alert for singularities of integrands and especially points where they have vertical asymptotes. It may be necessary to break an improper integral into several improper integrals if it is improper at both endpoints or at points inside the interval of integration. For example,

$$\int_0^2 \frac{dx}{\sqrt{x|x-1|}} = \int_0^{1/2} \frac{dx}{\sqrt{x(1-x)}} + \int_{1/2}^1 \frac{dx}{\sqrt{x(1-x)}} + \int_1^2 \frac{dx}{\sqrt{x(x-1)}}.$$

Each integral on the right is improper because of a singularity of its integrand at one endpoint.

EXAMPLE 6.8.8 a)

$$\int_0^1 \frac{dx}{x} = \lim_{c\to 0+} \int_c^1 \frac{dx}{x} = \lim_{c\to 0+} (-\ln c) = \infty$$

This integral diverges to infinity.

b)

$$\int_0^1 \frac{dx}{(1-x)^{1/3}} = \lim_{c\to 1-} \int_0^c \frac{dx}{(1-x)^{1/3}} \qquad \text{Let } u = 1-x,\ du = -dx$$

$$= \lim_{c\to 1-} -\int_1^{1-c} u^{-1/3}\,du = \lim_{c\to 1-} -\frac{3}{2}u^{2/3}\Big|_1^{1-c} = \frac{3}{2}$$

This integral converges to $3/2$.

c)

$$\int_0^2 \frac{dx}{\sqrt{2x-x^2}} = \int_0^2 \frac{dx}{\sqrt{1-(x-1)^2}} \qquad \text{Let } u = x-1,\ du = dx$$

$$= \int_{-1}^1 \frac{du}{\sqrt{1-u^2}} = 2\int_0^1 \frac{du}{\sqrt{1-u^2}} \qquad \text{(by symmetry)}$$

$$= 2 \lim_{c\to 1-} \sin^{-1} u \Big|_0^c = 2 \lim_{c\to 1-} \sin^{-1} c = \pi$$

This integral converges to π.

Parts (b) and (c) show how a change of variable can be made after or before an improper integral is expressed as a limit of proper integrals.

d) $\displaystyle\int_0^1 \ln x\,dx = \lim_{c\to0+}\int_c^1 \ln x\,dx$ (See Example 6.6.3(a) for the evaluation of the indefinite integral.)

$$\begin{aligned}
&= \lim_{c\to0+}(x\ln x - x)\Big|_c^1\\
&= \lim_{c\to0+}(-c\ln c - 1 + c)\\
&= -1 - \lim_{c\to0+}\frac{\ln c}{1/c} \qquad \left[\frac{-\infty}{\infty}\right]\\
&= -1 - \lim_{c\to0+}\frac{1/c}{-(1/c^2)} \qquad \text{(by l'Hôpital's Rule)}\\
&= -1 - \lim_{c\to0+}(-c) = -1.
\end{aligned}$$

The integral converges to -1.

The following theorem summarizes the behavior of improper integrals of types I and II for powers of x.

THEOREM 6.8.9 If $0 < a < \infty$, then

$$\int_a^\infty x^{-p}\,dx \quad \begin{cases} \text{converges to } \dfrac{a^{1-p}}{p-1} & \text{if } p > 1\\ \text{diverges to } \infty & \text{if } p \le 1\end{cases}$$

$$\int_0^a x^{-p}\,dx \quad \begin{cases} \text{converges to } \dfrac{a^{1-p}}{1-p} & \text{if } p < 1\\ \text{diverges to } \infty & \text{if } p \ge 1. \ \square\end{cases}$$

The proof just involves calculating the two integrals and is left to the student. These integrals are called p- **integrals**. Knowledge of when they converge and diverge is very useful when you have to decide whether certain other improper integrals converge or not and you can't find the appropriate antiderivatives. (See the discussion of estimating convergence below.) Note that $\int_0^\infty x^{-p}\,dx$ does not converge for any value of p.

REMARK: If f is continuous on the interval $[a, b]$ so that $\int_a^b f(x)\,dx$ is a proper definite integral, then treating the integral as improper will lead to the same value:

$$\lim_{c \to a+} \int_c^b f(x)\,dx = \int_a^b f(x)\,dx = \lim_{c \to b-} \int_a^c f(x)\,dx.$$

This justifies the definition of the definite integral of a piecewise continuous function (Definition 6.2.9) given in Section 6.2. To integrate a function defined to be different continuous functions on different intervals, we merely add the integrals of the various component functions over their respective intervals. Any of these integrals may be proper or improper; if any are improper all must converge or the given integral diverges.

EXAMPLE 6.8.10 Evaluate $\displaystyle\int_0^2 f(x)\,dx$ where $f(x) = \begin{cases} 1/\sqrt{x} & \text{if } 0 < x \le 1 \\ x - 1 & \text{if } 1 < x \le 2 \end{cases}$

SOLUTION The graph of f is shown in Fig. 6.8.7. We have

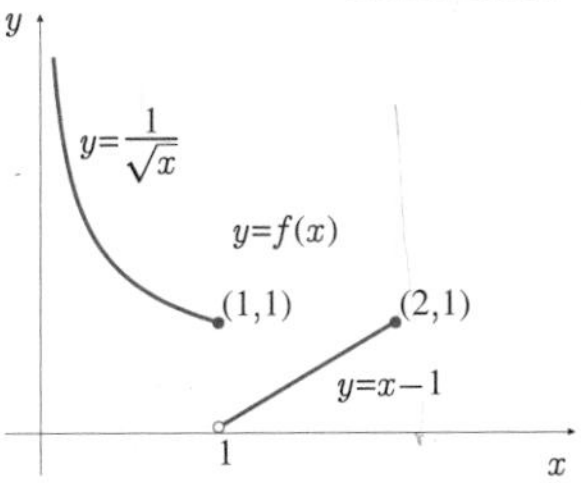

FIGURE 6.8.7

$$\begin{aligned}\int_0^2 f(x)\,dx &= \int_0^1 \frac{dx}{\sqrt{x}} + \int_1^2 (x-1)\,dx \\ &= \lim_{c \to 0+} \int_c^1 \frac{dx}{\sqrt{x}} + \left(\frac{x^2}{2} - x\right)\Bigg|_1^2 = 2 + (2 - 2 - \tfrac{1}{2} + 1) = \tfrac{5}{2};\end{aligned}$$

the first integral in the right is improper but convergent (see Example 6.8.6 above) and the second is proper.

Estimating Convergence and Divergence

When an improper integral cannot be evaluated by the Fundamental Theorem of Calculus because an antiderivative can't be found, we may still be able to determine whether the integral converges by comparing it with simpler integrals. The following theorem is central to this approach.

THEOREM 6.8.11 (**A Comparison Theorem for Integrals**) Let $-\infty \le a < b \le \infty$ and suppose that functions f and g are continuous on the interval (a, b) and satisfy $0 \le f(x) \le Kg(x)$ there for some positive constant K. If $\int_a^b g(x)\,dx$ converges, then so does $\int_a^b f(x)\,dx$, and

$$\int_a^b f(x)\,dx \le K \int_a^b g(x)\,dx.$$

Equivalently, if $\int_a^b f(x)\,dx$ diverges to ∞, then so does $\int_a^b g(x)\,dx$.

PROOF Since both integrands are nonnegative, there are only two possibilities for each integral: It can converge to a nonnegative number or diverge to ∞. Since $f(x) \le Kg(x)$ on (a, b), it follows by Theorem 6.2.4(e) that if $a < r < s < b$, then

$$\int_r^s f(x)\,dx \le K \int_r^s g(x)\,dx.$$

This theorem now follows by taking limits as $r \to a+$ and $s \to b-$. □

EXAMPLE 6.8.12 Show that $\int_0^\infty e^{-x^2}\,dx$ converges and find an upper bound for its value.

SOLUTION On $[0,1]$ we have $0 < e^{-x^2} \le 1$, so

$$0 < \int_0^1 e^{-x^2}\,dx \le \int_0^1 dx = 1.$$

On $[1,\infty)$ we have $x^2 \ge x$ so that $0 < e^{-x^2} \le e^{-x}$. Thus

$$0 < \int_1^\infty e^{-x^2}\,dx \le \int_1^\infty e^{-x}\,dx = \lim_{R\to\infty} \left.\frac{e^{-x}}{-1}\right|_1^R$$
$$= \lim_{R\to\infty}\left(\frac{1}{e} - \frac{1}{e^R}\right) = \frac{1}{e}.$$

Hence $\int_0^\infty e^{-x^2}\,dx$ converges and its value is not greater then $1 + (1/e)$.

We remark that the above integral is in fact equal to $\frac{1}{2}\sqrt{\pi}$ although we cannot prove this now. It is normally proved in advanced calculus courses using multivariable techniques.

EXAMPLE 6.8.13 Determine whether

$$\int_0^\infty \frac{dx}{\sqrt{x + x^3}}$$

converges.

SOLUTION The integral is improper of both types, so we write

$$\int_0^\infty \frac{dx}{\sqrt{x+x^3}} = \int_0^1 \frac{dx}{\sqrt{x+x^3}} + \int_1^\infty \frac{dx}{\sqrt{x+x^3}} = I_1 + I_2.$$

On $(0,1]$ we have $\sqrt{x+x^3} > \sqrt{x}$, so

$$I_1 < \int_0^1 \frac{dx}{\sqrt{x}} = 2 \qquad \text{(by Example 6.8.6).}$$

On $[1,\infty)$ we have $\sqrt{x+x^3} > \sqrt{x^3}$, so

$$I_2 < \int_1^\infty x^{-3/2}\,dx = 2 \qquad \text{(by Theorem 6.8.9).}$$

Hence the given integral converges and its value is less than 4.

EXAMPLE 6.8.14 Determine whether

$$\int_2^\infty \frac{dx}{\ln x}$$

converges.

SOLUTION For $x > 2$ we have $\ln x < x$, so

$$\int_2^\infty \frac{dx}{\ln x} > \int_2^\infty \frac{dx}{x},$$

which diverges to infinity by Theorem 6.8.9. Hence the given integral diverges to infinity by Theorem 6.8.11.

EXERCISES

why isn't this a proper integral? because $\sin\frac{\pi}{2}=1$ ∴ denominator is $(1-1)=0$ ∴ $\frac{1}{0}=\infty$ ∴ improper

In Exercises 1–16 evaluate the given integral or show that it diverges.

1. $\displaystyle\int_0^\infty e^{-2x}\,dx$

2. $\displaystyle\int_{-\infty}^{-1} \frac{dx}{x^2+1}$

3. $\displaystyle\int_{-1}^{1} \frac{dx}{(x+1)^{2/3}}$

4. $\displaystyle\int_0^a \frac{dx}{a^2-x^2}$

5. $\displaystyle\int_0^{\pi/2} \frac{\cos x\,dx}{(1-\sin x)^{2/3}}$

6. $\displaystyle\int_1^\infty \frac{dx}{(x-1)^2}$

7. $\displaystyle\int_0^\infty x\,e^{-x}\,dx$

8. $\displaystyle\int_0^1 x\ln x\,dx$

9. $\displaystyle\int_0^1 \frac{dx}{\sqrt{x(1-x)}}$

10. $\displaystyle\int_0^\infty \frac{x}{1+2x^2}\,dx$

11. $\displaystyle\int_0^\infty \frac{x\,dx}{(1+2x^2)^{3/2}}$

12. $\displaystyle\int_0^{\pi/2} \sec x\,dx$

13. $\displaystyle\int_0^{\pi/2} \tan x\,dx$

14. $\displaystyle\int_e^\infty \frac{dx}{x\ln x}$

15. $\displaystyle\int_1^e \frac{dx}{x\sqrt{\ln x}}$

16. $\displaystyle\int_e^\infty \frac{dx}{x(\ln x)^2}$

17. Find the area of a region that lies above $y=0$, to the right of $x=1$, and under the curve $y=\dfrac{4}{2x+1}-\dfrac{2}{x+2}$.

18. Find the area of the plane region that lies under the graph of $y=x^{-2}e^{-1/x}$, above the x-axis, and to the right of the y-axis.

19. Prove Theorem 6.8.9 by directly evaluating the integrals involved. Note: the case $p=1$ was covered in Examples 6.8.2 and 6.8.8(a).

20. Evaluate $\int_{-1}^{1}(x\ \mathrm{sgn}\ x)/(x+2)\,dx$. Recall that $\mathrm{sgn}\ x = x/|x|$.

21. Evaluate $\int_0^2 x^2\ \mathrm{sgn}\,(x-1)\,dx$.

22. Evaluate $\int_{-2}^{3} \mathrm{sgn}\,(1-|x|)\,dx$.

23. Evaluate $\int_0^3 x^2[x]\,dx$ where $[t]$ denotes the greatest integer $\le t$.

In Exercises 24–35 state whether the given integral converges or diverges and justify your claim.

24. $\displaystyle\int_0^\infty \frac{x^2}{x^5+1}\,dx$

25. $\displaystyle\int_0^\infty \frac{dx}{1+\sqrt{x}}$

26. $\displaystyle\int_2^\infty \frac{x\sqrt{x}\,dx}{x^2-1}$

27. $\displaystyle\int_0^\infty e^{-x^3}\,dx$

28. $\displaystyle\int_0^\infty \frac{dx}{\sqrt{x}+x^2}$

29. $\displaystyle\int_{-1}^{1} \frac{e^x}{x+1}\,dx$

30. $\displaystyle\int_0^\pi \frac{\sin x}{x}\,dx$

31.* $\displaystyle\int_0^\infty \frac{|\sin x|}{x^2}\,dx$

32.* $\displaystyle\int_0^{\pi^2} \frac{dx}{1-\cos\sqrt{x}}$

33.* $\displaystyle\int_{-\pi/2}^{\pi/2} \csc x\,dx$

34.* $\displaystyle\int_2^\infty \frac{dx}{\sqrt{x}\ln x}$

35.* $\displaystyle\int_0^\infty \frac{dx}{xe^x}$

36.* Given that $\int_0^\infty e^{-x^2}\,dx=\frac{1}{2}\sqrt{\pi}$, evaluate

(a) $\displaystyle\int_0^\infty x^2e^{-x^2}\,dx$, and (b) $\displaystyle\int_0^\infty x^4e^{-x^2}\,dx$

37.* If f is continuous on $[a,b]$, show that

$$\lim_{c\to a+}\int_c^b f(x)\,dx=\int_a^b f(x)\,dx.$$

Hint: use Theorems 1.7.7(i), 6.2.4(d) and 6.2.4(f) and show that

$$\lim_{c\to a+}\int_a^c f(x)\,dx=0.$$

Similarly, show that

$$\lim_{c\to b-}\int_a^c f(x)\,dx=\int_a^b f(x)\,dx.$$

CHAPTER 7

Numerical Integration

Most of the applications of integration, within and outside of mathematics, involve the definite integral

$$I = \int_a^b f(x)\,dx.$$

Thanks to the Fundamental Theorem of Calculus, we can evaluate such definite integrals by first finding an antiderivative of f. This is why we have spent considerable time on developing techniques of integration in Chapter 6. There are, however, two obstacles that can prevent our calculating I in this way:

i) Finding an antiderivative of f may be impossible, or at least very difficult.

ii) We may not be given $f(x)$ explicitly as a function of x; for instance, $f(x)$ may be an unknown function whose values at certain points of the interval $[a, b]$ have been determined by experimental measurement.

In this section we shall investigate the problem of approximating the value of the definite integral I using only the values of $f(x)$ at finitely many points of $[a, b]$. Obtaining such an approximation is called **numerical integration**, or **numerical quadrature**.

There are many techniques for performing numerical integration. One simple method is to calculate a Riemann sum for the integral. We consider this and three other techniques in this chapter, the Trapezoid Rule, Simpson's Rule, and the Romberg Method. Both the Trapezoid Rule and Simpson's Rule are special cases of the more sophisticated Romberg Method. All the methods are easily implemented on a small computer or by using a scientific calculator. The ubiquitousness of these devices makes numerical integration a steadily more important tool for the user of mathematics.

All the techniques require calculation of the values of $f(x)$ at a set of equally spaced points in $[a, b]$. The computational "expense" involved in determining an approximate value for the integral I will be roughly proportional to the number of function values required, so that the fewer function evaluations needed to achieve a desired degree of accuracy for the integral the better we will regard the technique. Even in the world of computers, "time is money."

7.1 THE TRAPEZOID AND MIDPOINT RULES

We assume that $f(x)$ is continuous on $[a, b]$ and subdivide $[a, b]$ into n subintervals of equal length $h = (b - a)/n$ using the $n + 1$ points

$$x_0 = a, \quad x_1 = a + h, \quad x_2 = a + 2h, \quad \ldots, \quad x_n = a + nh = b.$$

We assume that the value of $f(x)$ at each of these points is known:

$$y_0 = f(x_0), \quad y_1 = f(x_1), \quad y_2 = f(x_2), \quad \ldots, \quad y_n = f(x_n).$$

The Trapezoid Rule approximates $\int_a^b f(x)\,dx$ by using straight line segments between the points (x_{j-1}, y_{j-1}) and (x_j, y_j) $(1 \le j \le n)$, to approximate the graph of f, as shown in Fig. 7.1.1, and summing the areas of the resulting n *trapezoids*. A **trapezoid** is a four-sided polygon with one pair of parallel sides. (For our discussion we assume f is positive so we can talk about "areas," but the resulting formulas apply to any continuous function f.

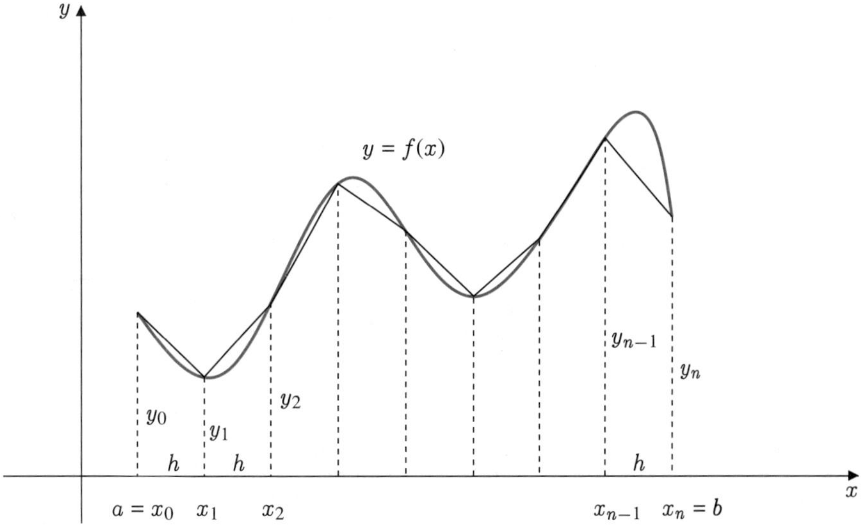

FIGURE 7.1.1

The first trapezoid has vertices $(x_0, 0)$, (x_0, y_0), (x_1, y_1), and $(x_1, 0)$. The two parallel sides are vertical, and have lengths y_0 and y_1 with a perpendicular distance $h = x_1 - x_0$ between them. (See Fig. 7.1.2.) The area of this trapezoid is h times the average of the parallel sides:

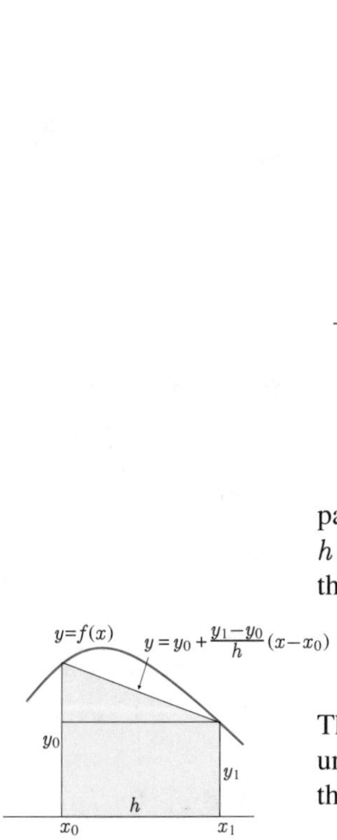

FIGURE 7.1.2

$$h\,\frac{y_0 + y_1}{2} \text{ square units.}$$

This can be seen geometrically by considering the trapezoid as the non-overlapping union of a rectangle and a triangle, or by observing that the top (slanted) side of the trapezoid is a straight line with equation

$$y = y_0 + \frac{y_1 - y_0}{h}\,(x - x_0),$$

and calculating

$$\begin{aligned}\int_{x_0}^{x_1} \left(y_0 + \frac{y_1 - y_0}{h}\,(x - x_0) \right) dx &= y_0(x_1 - x_0) + \frac{y_1 - y_0}{2h}\,(x_1 - x_0)^2 \\ &= y_0 h + \frac{y_1 - y_0}{2h}\,h^2 = h\,\frac{y_0 + y_1}{2}.\end{aligned}$$

We use this trapezoidal area to approximate the integral of f over the first subinterval $[x_0, x_1]$:

$$\int_{x_0}^{x_1} f(x)\,dx \approx h\,\frac{y_0 + y_1}{2}.$$

We can approximate the integral of f over any subinterval in the same way:

$$\int_{x_{j-1}}^{x_j} f(x)\,dx \approx h\,\frac{y_{j-1} + y_j}{2}, \qquad (1 \le j \le n).$$

It follows that the original integral I can be approximated by the sum of these trapezoidal areas:

$$\int_a^b f(x)\,dx \approx h\left(\frac{y_0+y_1}{2}+\frac{y_1+y_2}{2}+\frac{y_2+y_3}{2}+\cdots+\frac{y_{n-1}+y_n}{2}\right)$$
$$= h\left(\frac{1}{2}\,y_0+y_1+y_2+y_3+\cdots+y_{n-1}+\frac{1}{2}\,y_n\right).$$

7.1.1
The Trapezoid Rule

The n-subinterval **Trapezoid Rule** approximation to $\int_a^b f(x)\,dx$, denoted T_n, is given by

$$T_n = h\left(\frac{1}{2}\,y_0+y_1+y_2+y_3+\cdots+y_{n-1}+\frac{1}{2}\,y_n\right) = \frac{h}{2}\left(y_0+2\sum_{j=1}^{n-1} y_j + y_n\right).$$

Let us illustrate the Trapezoid Rule by using it to approximate an integral whose value we already know:

$$I = \int_1^2 \frac{dx}{x} = \ln 2 = 0.69314718\cdots.$$

(This value, and those of all the approximations quoted in this chapter, were calculated using a scientific calculator.) We will use the same integral to illustrate other methods for approximating definite integrals later.

EXAMPLE 7.1.2 Calculate the Trapezoid Rule approximations, T_4, T_8 and T_{16} for $I = \displaystyle\int_1^2 \frac{dx}{x}$.

SOLUTION For $n = 4$ we have $h = (2-1)/4 = 1/4$; for $n = 8$ we have $h = 1/8$; for $n = 16$ we have $h = 1/16$. We have

$$T_4 = \frac{1}{4}\left[\frac{1}{2}(1)+\frac{4}{5}+\frac{2}{3}+\frac{4}{7}+\frac{1}{2}\left(\frac{1}{2}\right)\right] = 0.69702381\cdots$$
$$T_8 = \frac{1}{8}\left[\frac{1}{2}(1)+\frac{8}{9}+\frac{4}{5}+\frac{8}{11}+\frac{2}{3}+\frac{8}{13}+\frac{4}{7}+\frac{8}{15}+\frac{1}{2}\left(\frac{1}{2}\right)\right]$$
$$= \frac{1}{8}\left[4T_4+\frac{8}{9}+\frac{8}{11}+\frac{8}{13}+\frac{8}{15}\right] = 0.69412185\cdots$$
$$T_{16} = \frac{1}{16}\left[8T_8+\frac{16}{17}+\frac{16}{19}+\frac{16}{21}+\frac{16}{23}+\frac{16}{25}+\frac{16}{27}+\frac{16}{29}+\frac{16}{31}\right] = 0.69339120$$

Note how the function values used to calculate T_4 were reused in the calculation of T_8, and similarly how those in T_8 were reused for T_{16}. When several approximations are needed it is very useful to double the number of subintervals for each new calculation, so that previously calculated values of f can be reused.

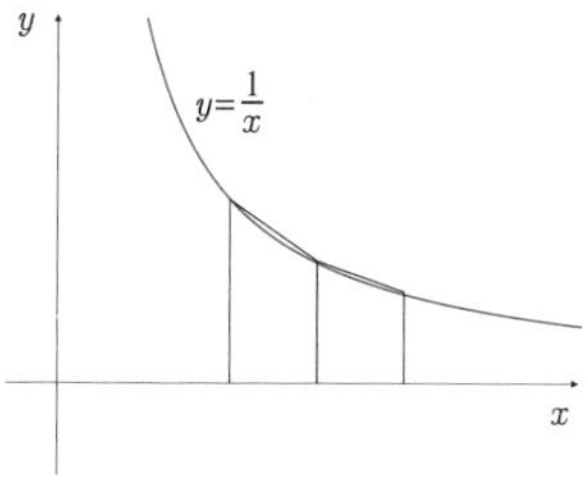

FIGURE 7.1.3

It is clear that all Trapezoid Rule approximations to I are greater than the true value of I. This is because the graph of $y = 1/x$ is concave up on $[1, 2]$, and therefore the tops of the approximating trapezoids lie above the curve. (See Fig. 7.1.3.)

We can calculate the exact errors in the three approximations since we know that $I = \ln 2 = 0.69314718\cdots$

$$I - T_4 = 0.69314718\cdots - 0.69702381\cdots = -0.00387663$$
$$I - T_8 = 0.69314718\cdots - 0.69412185\cdots = -0.00097467$$
$$I - T_{16} = 0.69314718\cdots - 0.69339120\cdots = -0.00024402$$

Observe that the size of the error decreases to about a quarter of its previous value each time we double n. We will show below that this is to be expected for a "well-behaved" function like $1/x$.

Example 7.1.2 is somewhat artificial in the sense that we know the actual value of the integral so we really don't need an approximation. In practical applications of numerical integration we do not know the actual value. It is tempting to calculate several approximations for increasing values of n until the two most recent ones agree to within a prescribed error tolerance. For example, we would have some inclination to claim that $\ln 2 \approx 0.69\cdots$ from examining T_4 and T_8, and comparison of T_{16} with T_8 suggests that the third decimal place is probably 3: $I \approx 0.693\cdots$. Though this approach cannot be justified in general, it is frequently used in practice.

The Midpoint Rule

A somewhat simpler approximation to $\int_a^b f(x)\,dx$ based on the partition of $[a, b]$ into n equal subintervals involves forming a Riemann sum of the areas of rectangles whose heights are taken at the midpoints of the n subintervals.

7.1.3
The Midpoint Rule

If $h = (b - a)/n$ let $m_j = a + \left(j - \frac{1}{2}\right) h$ for $1 \le j \le n$. The **Midpoint Rule** approximation to $\int_a^b f(x)\,dx$ is denoted M_n and is given by

$$M_n = h\big(f(m_1) + f(m_2) + \cdots + f(m_n)\big) = h \sum_{j=1}^{n} f(m_j).$$

M_n is the sum of the areas of the rectangles shown in Fig. 7.1.4.

EXAMPLE 7.1.4 The Midpoint Rule approximations M_4 and M_8 for $I = \int_1^2 \frac{dx}{x}$ are

$$M_4 = \frac{1}{4}\left[\frac{8}{9} + \frac{8}{11} + \frac{8}{13} + \frac{8}{15}\right] = 0.69121989\cdots$$
$$M_8 = \frac{1}{8}\left[\frac{16}{17} + \frac{16}{19} + \frac{16}{21} + \frac{16}{23} + \frac{16}{25} + \frac{16}{27} + \frac{16}{29} + \frac{16}{31}\right] = 0.69266055\cdots$$

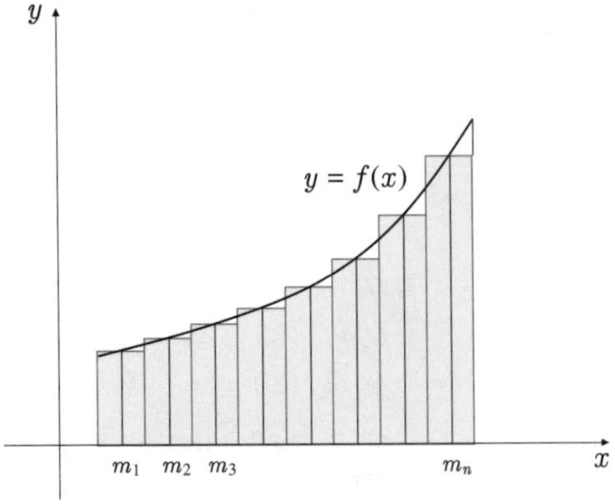

FIGURE 7.1.4

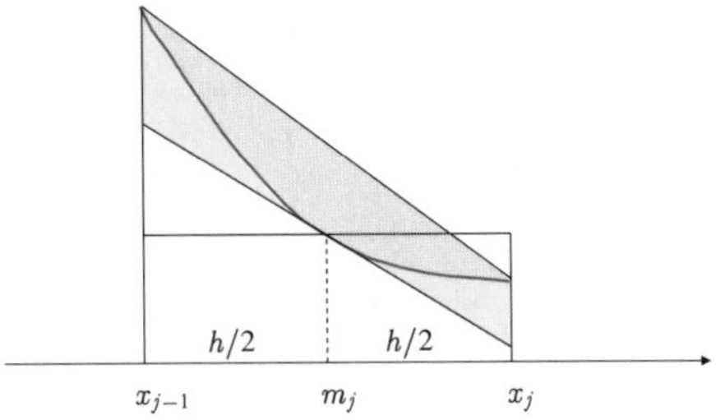

FIGURE 7.1.5

The errors in these approximations are

$$I - M_4 = 0.69314718\cdots - 0.69121989\cdots = 0.00192729\cdots$$
$$I - M_8 = 0.69314718\cdots - 0.69266055\cdots = 0.00048663\cdots$$

These errors are of opposite sign and about *half the size* of the corresponding Trapezoid Rule errors $I - T_4$ and $I - T_8$. Fig. 7.1.5 suggests the reason for this. The rectangular area $hf(m_j)$ is equal to the area of the trapezoid formed by the tangent line to $y = f(x)$ at $(m_j, f(m_j))$. The shaded region above the curve is the part of the Trapezoid Rule error due to the jth subinterval. The shaded area below the curve is the corresponding Midpoint Rule error. One drawback of the Midpoint Rule is that we cannot reuse values of f calculated for M_n when we calculate M_{2n}. However, to calculate T_{2n} we can use all the data values already calculated for M_n, specifically, $T_{2n} = \frac{1}{2}(T_n + M_n)$. A good strategy for using these methods to obtain a value for and integral I to a desired degree of accuracy is to calculate successively

$$T_n, \quad M_n, \quad T_{2n} = \frac{T_n + M_n}{2}, \quad M_{2n}, \quad T_{4n} = \frac{T_{2n} + M_{2n}}{2}, \quad M_{4n}, \quad \cdots$$

until two consecutive terms agree sufficiently closely. If a single quick approximation is needed, M_n is a better choice than T_n.

Error Estimates

The following theorem provides a bound for the error in the Trapezoid and Midpoint Rule approximations in terms of the second derivative of the integrand.

THEOREM 7.1.5 (**Error Estimates for the Trapezoid and Midpoint Rules**) If f has a continuous second derivative on $[a,b]$ and satisfies $|f''(x)| \le K$ there, then

$$\left|\int_a^b f(x)\,dx - T_n\right| \le \frac{K(b-a)}{12}\,h^2 = \frac{K(b-a)^3}{12n^2},$$
$$\left|\int_a^b f(x)\,dx - M_n\right| \le \frac{K(b-a)}{24}\,h^2 = \frac{K(b-a)^3}{24n^2},$$

where $h = (b-a)/n$. Note that these error bounds decrease like the square of the subinterval length as n increases.

!!DANGER!!
Hard Proof

PROOF We will prove here only the Trapezoid Rule error estimate. The one for the Midpoint Rule is a little easier to prove; the method is suggested in Exercise 10 below. Let $y = A + Bx$ be the straight line approximating $y = f(x)$ in the first subinterval, $[x_0, x_1]$, as considered above. ($A = y_0$, $B = (y_1 - y_0)/h$.) Let

$$g(x) = f(x) - (A + Bx).$$

Then g is twice differentiable, $g''(x) = f''(x)$, and $g(x_0) = g(x_1) = 0$. Two integrations by parts (see Exercise 38 of Section 6.6) show that

$$\begin{aligned}\int_{x_0}^{x_1} (x-x_0)(x_1-x)\,f''(x)\,dx &= \int_{x_0}^{x_1} (x-x_0)(x_1-x)\,g''(x)\,dx\\ &= -2\int_{x_0}^{x_1} g(x)\,dx\\ &= -2\left(\int_{x_0}^{x_1} f(x)\,dx - h\,\frac{y_0+y_1}{2}\right).\end{aligned}$$

Therefore, by Theorem 6.2.4(f),

$$\begin{aligned}\left|\int_{x_0}^{x_1} f(x)\,dx - h\,\frac{y_0+y_1}{2}\right| &\le \frac{1}{2}\int_{x_0}^{x_1} (x-x_0)(x_1-x)\,|f''(x)|\,dx.\\ &\le \frac{K}{2}\int_{x_0}^{x_1}\left(-x^2 + (x_0+x_1)x - x_0x_1\right)dx\\ &= \frac{K}{12}(x_1-x_0)^3 = \frac{K}{12}\,h^3.\end{aligned}$$

(We have omitted the details in the evaluation of the last integral above.) A similar estimate holds on each subinterval $[x_{j-1}, x_j]$ $(1 \le j \le n)$. Therefore,

$$\begin{aligned}\left|\int_a^b f(x)\,dx - T_n\right| &= \left|\sum_{j=1}^{n}\left(\int_{x_{j-1}}^{x_j} f(x)\,dx - h\,\frac{y_{j-1}+y_j}{2}\right)\right|\\ &\le \sum_{j=1}^{n}\left|\int_{x_{j-1}}^{x_j} f(x)\,dx - h\,\frac{y_{j-1}+y_j}{2}\right|\\ &= \sum_{j=1}^{n}\frac{K}{12}\,h^3 = \frac{K}{12}\,nh^3 = \frac{K(b-a)}{12}\,h^2,\end{aligned}$$

since $nh = b - a$. □

We illustrate this error estimate for the approximations of Example 7.1.2 above.

EXAMPLE 7.1.6 Obtain bounds for the errors for T_4, T_8, T_{16}, M_4, and M_8 for $I = \int_1^2 \frac{dx}{x}$.

SOLUTION If $f(x) = 1/x$, then $f'(x) = -1/x^2$ and $f''(x) = 2/x^3$. On $[1, 2]$ we have $|f''(x)| \le 2$, so we may take $K = 2$ in the estimate. Thus

$$|I - T_4| \le \frac{2(2-1)}{12}\left(\frac{1}{4}\right)^2 = 0.0104\cdots, \qquad |I - M_4| \le \frac{2(2-1)}{24}\left(\frac{1}{4}\right)^2 = 0.0052\cdots,$$
$$|I - T_8| \le \frac{2(2-1)}{12}\left(\frac{1}{8}\right)^2 = 0.0026\cdots, \qquad |I - M_8| \le \frac{2(2-1)}{24}\left(\frac{1}{8}\right)^2 | = 0.0013\cdots,$$
$$|I - T_{16}| \le \frac{2(2-1)}{12}\left(\frac{1}{16}\right)^2 = 0.00065\cdots.$$

The actual errors calculated earlier are considerably smaller than these bounds, because $|f''(x)|$ isn't usually as large as K.

REMARK: For any function f with a bounded second derivative on $[a, b]$, the error bounds for the Trapezoid and Midpoint Rule approximations T_n and M_n decrease like $1/n^2$ as n increases; in particular the error bound corresponding to using $2n$ subintervals is one quarter that corresponding to using n subintervals. The actual errors calculated for the approximations of $\int_1^2 \frac{dx}{x}$ behaved somewhat similarly. Of course, actual errors are not equal to the error bounds so they won't always be cut to exactly a quarter of their size when we double n.

There is a notation called "big-O notation" frequently used to describe error estimates qualitatively.

7.1.7
Big-O Notation

We say that $F(t) = G(t) + O(|H(t)|)$ if

$$|F(t) - G(t)| \le C\,|H(t)|$$

for some constant C.

Theorem 7.1.5 says that $I = T_n + O\left(\frac{1}{n^2}\right)$ and $I = M_n + O\left(\frac{1}{n^2}\right)$, that is,

$$|I - T_n| \le \frac{C_1}{n^2}, \qquad \text{and} \qquad |I - M_n| \le \frac{C_2}{n^2}.$$

EXERCISES

A scientific calculator should be used in these exercises. Students with programming experience and access to a computer may wish write short programs to carry out the calculations. In Exercises 1–4, calculate the approximations T_4, M_4, T_8, M_8 and T_{16} for the given integrals. Also calculate the exact value of each integral and so determine the exact error in each approximation. Compare these exact errors with the bounds for the size of the error supplied by Theorems 7.1.5.

1. $I = \int_0^2 (1 + x^2)\, dx$

2. $I = \int_0^1 e^{-x}\, dx$

3. $I = \int_0^{\pi/2} \sin x\, dx$

4. $I = \int_0^1 \frac{dx}{1 + x^2}$

5. Find T_4, M_4, T_8, M_8 and T_{16} for $\int_0^{1.6} f(x)\, dx$ for the function f whose values are tabulated:

x	$f(x)$	x	$f(x)$
0.0	1.4142	0.1	1.4124
0.2	1.4071	0.3	1.3983
0.4	1.3860	0.5	1.3702
0.6	1.3510	0.7	1.3285
0.8	1.3026	0.9	1.2734
1.0	1.2411	1.1	1.2057
1.2	1.1772	1.3	1.1258
1.4	1.0817	1.5	1.0348
1.6	1.9853		

6. Find the approximations M_8 and T_{16} for $\int_0^1 e^{-x^2}\, dx$. Quote a value for the integral to as many decimal places as you feel is justified.

7. Repeat the previous exercise for $\int_0^{\pi/2} \frac{\sin x}{x}\, dx$. (The integrand is 1 at $x = 0$.)

8. Compute the actual error in the approximation $\int_0^1 x^2\, dx \approx T_1$ and use it to show that the constant 12 in the estimate of Theorem 7.1.5 cannot be improved. That is, show that the absolute value of the actual error is as large as allowed by that estimate.

9. Repeat the previous exercise for M_1.

10.*Prove the error estimate for the Midpoint Rule in Theorem 7.1.5 as follows: If $x_1 - x_0 = h$ and m_1 is the midpoint of $[x_0, x_1]$ use the error estimate for the tangent line approximation (Theorem 5.3.4) to show that

$$|f(x) - f(m_1) - f'(m_1)(x - m_1)| \le \frac{K}{2}(x - m_1)^2.$$

Use this inequality to show that

$$\begin{aligned} &\left| \int_{x_0}^{x_1} f(x)\, dx - f(m_1)h \right| \\ &= \left| \int_{x_0}^{x_1} \big(f(x) - f(m_1) - f'(m_1)(x - m_1)\big)\, dx \right| \\ &\le \frac{K}{24} h^3. \end{aligned}$$

Complete the proof the same way we did for the Trapezoid Rule estimate in Theorem 7.1.5.

7.2 SIMPSON'S RULE

The Trapezoid Rule approximation to $\int_a^b f(x)\, dx$ results from approximating the graph of f by straight line segments through adjacent pairs of data points on that graph. Intuitively we would expect to do better if we approximate the graph by more general curves. Since straight lines are the graphs of linear functions, the simplest obvious generalization is to use the class of quadratic functions, that is, to approximate the graph of f by segments of parabolas. This is the basis of Simpson's Rule.

Suppose that we are given three points in the plane, one on each of three equally spaced vertical lines, spaced, say, h units apart. If we choose the middle of these lines as the y-axis, then the coordinates of the three points will be, say $(-h, y_L)$, $(0, y_M)$, and (h, y_R), as illustrated in Fig. 7.2.1.

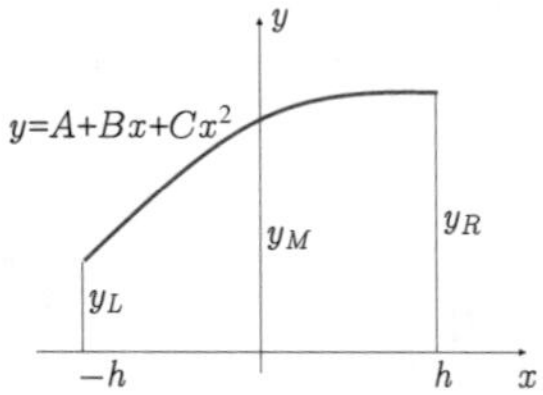

FIGURE 7.2.1

Constants A, B, and C can be chosen so that the parabola $y = A + Bx + Cx^2$ passes through these points; evidently

$$\left.\begin{aligned} y_L &= A - Bh + Ch^2 \\ y_M &= A \\ y_R &= A + Bh + Ch^2 \end{aligned}\right\} \quad \Rightarrow \quad A = y_M, \qquad 2Ch^2 = y_L - 2y_M + y_R.$$

Now we have

$$\begin{aligned} \int_{-h}^{h} (A + Bx + Cx^2)\,dx &= \left(Ax + \frac{B}{2}x^2 + \frac{C}{3}x^3\right)\Big|_{-h}^{h} = 2Ah + \frac{2}{3}Ch^3 \\ &= h\left(2y_M + \frac{1}{3}(y_L - 2y_M + y_R)\right) = \frac{h}{3}(y_L + 4y_M + y_R). \end{aligned}$$

Thus the area of the plane region bounded by the parabolic arc, the interval of length $2h$ on the x-axis, and the left and right vertical lines is equal to $(h/3)$ times the sum of the heights of the region at the left and right edges and four times the height at the middle. (It is independent of the position of the y-axis.)

Now suppose that we are given the same data for f as we were given for the Trapezoid Rule, that is, we know the values $y_j = f(x_j)$ $(0 \le j \le n)$ at $n+1$ equally spaced points

$$x_0 = a, \quad x_1 = a + h, \quad x_2 = a + 2h, \quad \ldots, \quad x_n = a + nh = b,$$

where $h = (b-a)/n$. We can approximate the graph of f over *pairs* of the subintervals $[x_{j-1}, x_j]$ using parabolic segments, and use the integrals of the corresponding quadratic functions to approximate the integrals of f over these subintervals. Since we need to use the subintervals two at a time we must assume that n is *even*. Using the integral computed for the parabolic segment above, we have

$$\begin{aligned} \int_{x_0}^{x_2} f(x)\,dx &\approx \frac{h}{3}(y_0 + 4y_1 + y_2) \\ \int_{x_2}^{x_4} f(x)\,dx &\approx \frac{h}{3}(y_2 + 4y_3 + y_4) \\ &\vdots \\ \int_{x_{n-2}}^{x_n} f(x)\,dx &\approx \frac{h}{3}(y_{n-2} + 4y_{n-1} + y_n). \end{aligned}$$

Adding these $n/2$ individual approximations we get the Simpson's Rule approximation to the integral $\int_a^b f(x)\,dx$.

7.2.1
Simpson's Rule

The **Simpson's Rule** approximation to $\int_a^b f(x)\,dx$ based on a subdivision of $[a, b]$ into an even number n of subintervals of equal length is denoted S_n:

$$\begin{aligned}\int_a^b f(x)\,dx &\approx S_n \\ &= \frac{h}{3}\left(y_0 + 4y_1 + 2y_2 + 4y_3 + 2y_4 + \cdots + 2y_{n-2} + 4y_{n-1} + y_n\right) \\ &= \frac{h}{3}\left(y_{\text{“ends”}} + 4y_{\text{“odds”}} + 2y_{\text{“evens”}}\right)\end{aligned}$$

Note that the Simpson's Rule approximation S_n requires no more data than does the Trapezoid Rule approximation T_n; both require the values of $f(x)$ at $n + 1$ equally spaced points. However, Simpson's Rule treats the data differently, weighting successive values either $1/3$, $2/3$, or $4/3$. As we will see, this can produce a much better approximations to the integral of f.

EXAMPLE 7.2.2 Calculate the approximations S_4, S_8 and S_{16} for $I = \int_1^2 dx/x$ and compare them with the actual value $I = \ln 2 = 0.69314718\cdots$, and with the values of T_4, T_8 and T_{16} obtained in Example 7.1.2

SOLUTION We calculate

$$S_4 = \frac{1}{12}\left[1 + 4\left(\frac{4}{5}\right) + 2\left(\frac{2}{3}\right) + 4\left(\frac{4}{7}\right) + \frac{1}{2}\right] = 0.69325397\cdots$$

$$\begin{aligned}S_8 = \frac{1}{24}\Bigg[1 + \frac{1}{2} &+ 4\left(\frac{8}{9} + \frac{8}{11} + \frac{8}{13} + \frac{8}{15}\right) \\ &+ 2\left(\frac{4}{5} + \frac{2}{3} + \frac{4}{7}\right)\Bigg] = 0.69315453\cdots.\end{aligned}$$

$$\begin{aligned}S_{16} = \frac{1}{48}\Bigg[1 + \frac{1}{2} \\ &+4\left(\frac{16}{17} + \frac{16}{19} + \frac{16}{21} + \frac{16}{23} + \frac{16}{25} + \frac{16}{27} + \frac{16}{29} + \frac{16}{31}\right) \\ &+2\left(\frac{8}{9} + \frac{4}{5} + \frac{8}{11} + \frac{2}{3} + \frac{8}{13} + \frac{4}{7} + \frac{8}{15}\right)\Bigg] = 0.69314765\cdots.\end{aligned}$$

The errors are

$$\begin{aligned}I - S_4 &= 0.69314718\cdots - 0.69325397\cdots = -0.00010679 \\ I - S_8 &= 0.69314718\cdots - 0.69315453\cdots = -0.00000735 \\ I - S_{16} &= 0.69314718\cdots - 0.69314765\cdots = -0.00000047\end{aligned}$$

These errors are evidently much smaller than the corresponding errors for the corresponding Trapezoid or Midpoint Rule approximations.

Obtaining an error estimate for Simpson's Rule is more difficult than it was for the Trapezoid Rule. We state the appropriate estimate in the following theorem, but we do not attempt any proof. Proofs can be found in textbooks on numerical analysis.

THEOREM 7.2.3 (*Error Estimate for Simpson's Rule*) If f has a continuous fourth derivative on the interval $[a, b]$, and satisfies $|f^{(4)}(x)| \le K$ there, then

$$\left| \int_a^b f(x)\,dx - S_n \right| \le \frac{K(b-a)}{180} h^4 = \frac{K(b-a)^5}{180 n^4},$$

where $h = (b-a)/n$. □

Observe that, as n increases, the error decreases as the fourth power of h, and hence as $1/n^4$. Using the big-O notation we have

$$\left| \int_a^b f(x)\,dx - S_n \right| = O\left(\frac{1}{n^4}\right).$$

This accounts for the fact that S_n is a much better approximation than is T_n, provided h is small and $|f^{(4)}(x)|$ is not unduly large compared to $|f''(x)|$. Note also that for any (even) n, S_n gives the exact value of the integral of any *cubic* function $f(x) = A + Bx + Cx^2 + Dx^3$; $f^{(4)}(x) = 0$ identically for such f, so we can take $K = 0$ in the error estimate.

EXAMPLE 7.2.4 Obtain bounds for the absolute values of the errors in the approximations of Example 7.2.2.

SOLUTION If $f(x) = 1/x$ then

$$f'(x) = -\frac{1}{x^2}, \qquad f''(x) = \frac{2}{x^3}, \qquad f^{(3)}(x) = -\frac{6}{x^4}, \qquad f^{(4)}(x) = \frac{24}{x^5}.$$

Clearly, $|f^{(4)}(x)| \le 24$ on $[1, 2]$, so we can take $K = 24$ in the estimate of Theorem 7.2.3. We have

$$|I - S_4| \le \frac{24(2-1)}{180}\left(\frac{1}{4}\right)^4 \approx 0.00052083,$$

$$|I - S_8| \le \frac{24(2-1)}{180}\left(\frac{1}{8}\right)^4 \approx 0.00003255,$$

$$|I - S_{16}| \le \frac{24(2-1)}{180}\left(\frac{1}{16}\right)^4 \approx 0.00000203.$$

Again we observe that the actual errors are well within these bounds.

EXAMPLE 7.2.5 A function f is known to satisfy $|f^{(4)}(x)| \leq 7$ on the interval $[1, 3]$. Given that f has values $f(1.0) = 0.1860$, $f(1.5) = 0.9411$, $f(2.0) = 1.1550$, $f(2.5) = 1.4511$, $f(3.0) = 1.2144$, find the best Simpson's rule approximation you can for $I = \int_1^3 f(x)\,dx$, and give a bound for the size of the error. Specify the smallest interval you can which must contain the value of I.

SOLUTION We take $n = 4$, $h = 0.5$ and obtain

$$\begin{aligned} I &= \int_1^3 f(x)\,dx \approx S_4 \\ &= \frac{0.5}{3}\big(0.1860 + 4(0.9411 + 1.4511) + 2(1.1550) + 1.2144\big) \\ &= 2.2132. \end{aligned}$$

Since $|f^{(4)}(x)| \leq 7$ on $[1, 3]$ we have

$$|I - S_4| \leq \frac{7(3-1)}{180}(0.5)^4 < 0.0049.$$

I must therefore satisfy

$$2.2132 - 0.0049 < I < 2.2132 + 0.0049, \quad \text{or} \quad 2.2083 < I < 2.2181.$$

EXERCISES

In 1–4 find Simpson's Rule approximations S_4, and S_8 for the given functions, and compare your results with the actual values of the integrals and with the corresponding Trapezoid Rule approximations obtained in Exercises 1–4 of Section 7.1.

1. $I = \displaystyle\int_0^2 (1 + x^2)\,dx$ **2.** $I = \displaystyle\int_0^1 e^{-x}\,dx$

3. $I = \displaystyle\int_0^{\pi/2} \sin x\,dx$ **4.** $I = \displaystyle\int_0^1 \frac{dx}{1 + x^2}$

5. Use Theorem 7.2.3 to obtain bounds for the errors in the approximations obtained in Exercises 2 and 3 above.

6. Verify that $S_{2n} = \dfrac{T_n + 2M_n}{3} = \dfrac{2T_{2n} + M_n}{3}$, where T_n and M_n refer to the appropriate Trapezoid and Midpoint Rule approximations.

7. Find S_4, S_8 and S_{16} for $\int_0^{1.6} f(x)\,dx$ for the function f whose values are tabulated in Exercise 5 of Section 7.1.

8. Find the Simpson's Rule approximations S_8 and S_{16} for $\displaystyle\int_0^1 e^{-x^2}\,dx$. Quote a value for the integral to the number of decimal places you feel is justified based on comparing the two approximations.

9.* Compute the actual error in the approximation $\int_0^1 x^4\,dx \approx S_2$ and use it to show that the constant 180 in the estimate of Theorem 7.2.3 cannot be improved.

10.* Since Simpson's Rule is based on quadratic approximation, it is not surprising that it should give an exact value for an integral of $A + Bx + Cx^2$. It is more surprising that it is exact for a cubic function as well. Verify by direct calculation that $\int_0^1 x^3\,dx = S_2$.

7.3 ROMBERG INTEGRATION

The increased accuracy in the approximations of $\int_a^b f(x)\,dx$ using Simpson's Rule over those using the Trapezoid Rule suggests that we could do even better by approximating f with polynomials of degree greater than 2. This is certainly the case, but we will approach the task from a different direction. We will use a certain formula for the error in the Trapezoid Rule approximation to generate a sequence of successively better and better approximations. This method is known as **Romberg integration**.

Let $I = \int_a^b f(x)\,dx$ and let T_n be the Trapezoid Rule approximation to I based on n equal subintervals:

$$T_n = h\left(\frac{y_0}{2} + y_1 + y_2 + \cdots + y_{n-1} + \frac{y_n}{2}\right),$$

where $y_j = f(x_j)$, $x_j = a + jh$, $(0 \le j \le n)$, and $h = (b-a)/n$. Our starting point is the following theorem which gives a formula for the error $E_n = I - T_n$ in the Trapezoid Rule approximation.

THEOREM 7.3.1 If f has continuous derivatives up to order $2m+2$ on the interval $[a, b]$, then

$$E_n = I - T_n = \sum_{j=1}^{m} \frac{C_j}{n^{2j}} + O\left(\frac{1}{n^{2m+2}}\right),$$

where $C_j = K_j \int_a^b f^{(2j)}(x)\,dx$, and the numbers K_j are constants not depending on n or f. (The big-O notation is explained after Box 7.1.7.) □

The proof of this theorem is quite difficult and we postpone it to the end of this section. You don't have to read it to learn how to use the method.

We are going to construct Trapezoid Rule approximations for values of n which are powers of 2: $n = 1, 2, 4, 8, \ldots$. Accordingly, let us define

$$T_k^0 = T_{2^k}. \qquad \text{Thus} \quad T_0^0 = T_1, \quad T_1^0 = T_2, \quad T_2^0 = T_4, \quad \ldots.$$

Using the formula for $T_{2^k} = I - E_{2^k}$ supplied by Theorem 7.3.1, we write

> **!!DANGER!!**
> This is tricky. If you can't follow it skip ahead to "Using the Romberg Method."

$$T_k^0 = I - \frac{C_1}{4^k} - \frac{C_2}{4^{2k}} - \cdots - \frac{C_m}{4^{mk}} + O\left(\frac{1}{4^{(m+1)k}}\right).$$

Similarly, replacing k by $k+1$, we get

$$T_{k+1}^0 = I - \frac{C_1}{4^{k+1}} - \frac{C_2}{4^{2(k+1)}} - \cdots - \frac{C_m}{4^{m(k+1)}} + O\left(\frac{1}{4^{(m+1)(k+1)}}\right).$$

If we multiply the formula for T_{k+1}^0 by 4 and subtract the formula for T_k^0, the term involving C_1 will cancel out. The first term on the right will be $4I - I = 3I$ so let us also divide by 3 and define T_{k+1}^1 to be the result:

$$T_{k+1}^1 = \frac{4T_{k+1}^0 - T_k^0}{3} = I - \frac{C_2'}{4^{2k}} - \frac{C_3'}{4^{3k}} - \cdots - \cdots - \frac{C_m'}{4^{mk}} + O\left(\frac{1}{4^{(m+1)k}}\right).$$

(The C_j' are new constants.) T_{k+1}^1 ought to be a better approximation to I than T_{k+1}^0 since we have eliminated the lowest order (and therefore the largest) of the error terms, $C_1/4^{k+1}$.

Let $h = (b-a)/2^{k+1}$ and $y_j = f(a+jh)$ for $0 \le j \le 2^{k+1}$. Then

$$\begin{aligned} T_{k+1}^1 &= \frac{1}{3}\left[4h\left(\frac{y_0}{2} + y_1 + y_2 + \cdots + \frac{y_{2^{k+1}}}{2}\right)\right. \\ &\qquad \left. -2h\left(\frac{y_0}{2} + y_2 + y_4 + \cdots + \frac{y_{2^{k+1}}}{2}\right)\right] \\ &= \frac{h}{3}\left[y_0 + 4y_1 + 2y_2 + 4y_3 + \cdots + 4y_{2^{k+1}-1} + y_{2^{k+1}}\right] \\ &= S_{2^{k+1}}. \end{aligned}$$

Thus T_{k+1}^1 is just the Simpson's Rule approximation based on 2^{k+1} subintervals.

We can continue the process of eliminating error terms begun above:

$$\begin{aligned} T_{k+1}^1 &= I - \frac{C_2'}{4^{2k}} - \frac{C_3'}{4^{3k}} - \cdots - \frac{C_m'}{4^{mk}} + O\left(\frac{1}{4^{(m+1)k}}\right) \\ T_{k+2}^1 &= I - \frac{C_2'}{4^{2(k+1)}} - \frac{C_3'}{4^{3(k+1)}} - \cdots - \frac{C_m'}{4^{m(k+1)}} + O\left(\frac{1}{4^{(m+1)(k+1)}}\right) \end{aligned}$$

To eliminate C_2' we can multiply the bottom formula by 16 and subtract the top formula: define

$$T_{k+2}^2 = \frac{16T_{k+2}^1 - T_{k+1}^1}{15} = I - \frac{C_3''}{4^{3k}} - \cdots - \cdots - \frac{C_m''}{4^{mk}} + O\left(\frac{1}{4^{(m+1)k}}\right).$$

We can proceed in this way, eliminating one error term after another. In general, for $j < m$,

$$T_{k+j}^j = \frac{4^j T_{k+j}^{j-1} - T_{k+j-1}^{j-1}}{4^j - 1} = I - \frac{C_{j+1}^j}{4^{(j+1)k}} - \cdots - \frac{C_m^j}{4^{mk}} + O\left(\frac{1}{4^{(m+1)k}}\right).$$

Using the Romberg Method

All this looks very complicated, but it is not difficult to carry out in practice. Let $R_j = T_j^j$ and calculate the entries in the following table in order from left to right and down each column when you come to it:

7.3.2
Scheme for Calculating Romberg Approximations

$$
\begin{array}{ccccccc}
T_0^0 = T_1 = R_0 & \longrightarrow & T_1^0 = T_2 & \longrightarrow & T_2^0 = T_4 & \longrightarrow & T_3^0 = T_8 & \longrightarrow \\
 & & \downarrow & & \downarrow & & \downarrow & \\
 & & T_1^1 = S_2 = R_1 & & T_2^1 = S_4 & & T_3^1 = S_8 & \\
 & & & & \downarrow & & \downarrow & \\
 & & & & T_2^2 = R_2 & & T_3^2 & \\
 & & & & & & \downarrow & \\
 & & & & & & T_3^3 = R_3 &
\end{array}
$$

Stop when T_j^{j-1} and R_j differ by less than the acceptable error, and quote R_j as the Romberg approximation to $\int_a^b f(x)\,dx$.

The top line in the table is made up of the Trapezoid Rule approximations T_1, T_2, T_4, T_8, ... Elements in subsequent rows are calculated by the formulas

7.3.3
Formulas for Calculating Romberg Approximations

$$
T_1^1 = \frac{4T_1^0 - T_0^0}{3} \qquad T_2^1 = \frac{4T_2^0 - T_1^0}{3} \qquad T_3^1 = \frac{4T_3^0 - T_2^0}{3} \qquad \ldots
$$

$$
T_2^2 = \frac{16T_2^1 - T_1^1}{15} \qquad T_3^2 = \frac{16T_3^1 - T_2^1}{15} \qquad \ldots
$$

$$
T_3^3 = \frac{64T_3^2 - T_2^2}{63} \qquad \ldots
$$

In general, if $k < j$, $T_j^k = \dfrac{4^k T_j^{k-1} - T_{j-1}^{k-1}}{4^k - 1}$.

Each new entry is calculated from the one above and the one to the left of that one.

EXAMPLE 7.3.4 Calculate the Romberg approximations R_0, R_1, R_2, R_3 and R_4 for the integral $I = \int_1^2 \frac{dx}{x}$.

SOLUTION We will carry all calculations to 8 decimal places. Since we must obtain R_4, we will need to find all the entries in the first five columns of the scheme. First we calculate the first two Trapezoid Rule approximations:

$$
R_0 = T_0^0 = T_1 = \frac{1}{2} + \frac{1}{4} = 0.75000000
$$

$$
T_1^0 = T_2 = \frac{1}{2}\left[\frac{1}{2}(1) + \frac{2}{3} + \frac{1}{2}\left(\frac{1}{2}\right)\right] = 0.70833333
$$

The remaining required Trapezoid Rule approximations were calculated in Example 7.1.2, so we may as well record them now:

$$T_2^0 = T_4 = 0.69702381$$
$$T_3^0 = T_8 = 0.69412185$$
$$T_4^0 = T_{16} = 0.69339120$$

Now we calculate down the columns from left to right. For the second column:

$$R_1 = S_2 = T_1^1 = \frac{4T_1^0 - T_0^0}{3} = 0.69444444$$

the third column

$$S_4 = T_2^1 = \frac{4T_2^0 - T_1^0}{3} = 0.69325397$$
$$R_2 = T_2^2 = \frac{16T_2^1 - T_1^1}{15} = 0.69317460$$

the fourth column

$$S_8 = T_3^1 = \frac{4T_3^0 - T_2^0}{3} = 0.69315453$$
$$T_3^2 = \frac{16T_3^1 - T_2^1}{15} = 0.69314790$$
$$R_3 = T_3^3 = \frac{64T_3^2 - T_2^2}{63} = 0.69314748$$

and the fifth column

$$S_{16} = T_4^1 = \frac{4T_4^0 - T_3^0}{3} = 0.69314765$$
$$T_4^2 = \frac{16T_4^1 - T_3^1}{15} = 0.69314719$$
$$T_4^3 = \frac{64T_4^2 - T_3^2}{63} = 0.69314718$$
$$R_4 = T_4^4 = \frac{256T_4^3 - T_3^3}{255} = 0.69314718$$

Since T_4^3 and R_4 agree to the 8 decimal places we are calculating, we conclude that

$$I = \int_1^2 \frac{dx}{x} = \ln 2 \approx 0.69314718\cdots,$$

correct to eight decimal places.

The various approximations calculated above suggest that for any given value of $n = 2^k$, the Romberg approximation R_n should give the best value obtainable for the integral based on the $n+1$ data values $y_0, y_1, \ldots, y_n$. This is so only if the derivatives $f^{(n)}(x)$ do not grow rapidly as n increases.

Proof of Theorem 7.3.1

!!DANGER!!

Very Hard Proof. You should probably skip it.

The following proof is *very difficult*. You are advised to skip it unless you really want to know how the theorem is proved. In the proof we are going to make extensive use of Taylor's Formula with Lagrange Remainder (see Theorem 5.4.5) to express functions f on the interval $[-a, a]$:

$$f(x) = f(0) + f'(0)\,x + \frac{f''(0)}{2!}\,x^2 + \cdots + \frac{f^{(n)}(0)}{n!}\,x^n + R_n(f; 0, x),$$

where $R_n(f; 0, x) = \dfrac{f^{(n+1)}(X)}{(n+1)!}\,x^{n+1}$ for some X between 0 and x. Using summation and big-O notation we can write this formula in more compact form:

$$f(x) = \sum_{j=0}^{n} \frac{f^{(j)}(0)}{j!}\,x^j + O(|x|^{n+1}).$$

We are given in the statement of the theorem that f has continuous derivatives up to order $2m+2$ on the interval $[a, b]$. We can make a change of variable $u = x - \dfrac{a+b}{2}$ to move the origin to the midpoint of the interval $[a, b]$. For the moment, therefore, we assume that the interval is, in fact $[-a, a]$, an interval symmetric about the origin. Accordingly, the integral we want to approximate is

$$I = \int_{-a}^{a} f(x)\,dx.$$

If T_1 is the Trapezoid Rule approximation to I based on the whole interval $[-a, a]$, then $T_1 = a\big(f(-a)+f(a)\big)$. We expand this using Taylor's Formula (with n replaced by $2m+1$)

$$T_1 = a\left(\sum_{j=0}^{2m+1} \frac{f^{(j)}(0)}{j!}\left[(-a)^j + a^j\right] + O(|a|^{2m+2})\right).$$

The term in the square brackets is 0 if j is an odd integer, so we can restrict the sum to even integers, $j = 2k$, and obtain

$$T_1 = 2\sum_{k=0}^{m} \frac{f^{(2k)}(0)}{(2k)!}\,a^{2k+1} + O(|a|^{2m+3}).$$

Also, by Taylor's Formula,

$$\begin{aligned} I = \int_{-a}^{a} f(x)\,dx &= \sum_{j=0}^{2m+1} \frac{f^{(j)}(0)}{j!} \int_{-a}^{a} x^j\,dx + O(|a|^{2m+3}) \\ &= 2\sum_{k=0}^{m} \frac{f^{(2k)}(0)}{(2k+1)!}\,a^{2k+1} + O(|a|^{2m+3}); \qquad (*) \end{aligned}$$

again the terms for odd j cancelled out. Subtracting the formulas above for T_1 and I, we obtain an expression for the error E_1:

$$E_1 = I - T_1 = 2\sum_{k=0}^{m} f^{(2k)}(0)\left(\frac{1}{(2k+1)!} - \frac{1}{(2k)!}\right) a^{2k+1} + O(|a|^{2m+3})$$

$$= \sum_{k=1}^{m} A_k\, f^{(2k)}(0)\, a^{2k+1} + O(|a|^{2m+3}).$$

Observe that the last sum starts at $k = 1$ because the term for $k = 0$ cancelled out. ($0! = 1 = 1!$.) The numbers A_k are independent of f.

Now we apply formula (*) to the second derivative $f^{(2)}$, with $m - 1$ in place of m:

$$\int_{-a}^{a} f^{(2)}(x)\,dx = 2\sum_{j=0}^{m-1} \frac{f^{(2+2j)}(0)}{(2j+1)!}\, a^{2j+1} + O(|a|^{2m+1})$$

$$= 2af^{(2)}(0) + 2\sum_{j=1}^{m-1} \frac{f^{(2+2j)}(0)}{(2j+1)!}\, a^{2j+1} + O(|a|^{2m+1}).$$

Solve for $2af^{(2)}(0)$ and replace $j + 1$ with k.

$$2af^{(2)}(0) = \int_{-a}^{a} f^{(2)}(x)\,dx - 2\sum_{k=2}^{m} \frac{f^{(2k)}(0)}{(2k-1)!}\, a^{2k-1} + O(|a|^{2m+1}).$$

Now substitute this into the formula for E_1 obtained above, and get

$$E_1 = K_1(2a)^2 \int_{-a}^{a} f^{(2)}(x)\,dx + 2\sum_{k=2}^{m} B_k\, f^{(2k)}(0)\, a^{2k+1} + O(|a|^{2m+3}),$$

where $K_1 = A_1/4$, and the B_k are other numbers independent of f.

We can continue in the same way, applying (*) to $f^{(4)}$, and then $f^{(6)}$, and so on, until all the terms $f^{(2k)}(0)$ have been replaced by integrals $\int_{-a}^{a} f^{(2k)}(x)\,dx$. Thus we obtain the formula

$$E_1 = \sum_{j=1}^{m} K_j(2a)^{2j} \int_{-a}^{a} f^{(2j)}(x)\,dx + O(|a|^{2m+3}).$$

For functions on the more general interval $[a, b]$, this formula translates to

$$E_1 = \sum_{j=1}^{m} K_j(b-a)^{2j} \int_{a}^{b} f^{(2j)}(x)\,dx + O(|b-a|^{2m+3}).$$

Finally, if $x_0 = a$, $x_1 = a + h$, $x_2 = a + 2h$, ..., $x_n = a + nh = b$, where $h = (b - a)/n$, then the error $E_n = I - T_n = \int_a^b f(x)\,dx - T_n$ is obtained by summing the errors E_1 as obtained above for the Trapezoid Rule approximations for $\displaystyle\int_{x_{i-1}}^{x_i} f(x)\,dx$. We have

$$\begin{aligned}
E_n &= \sum_{i=1}^{n}\left(\sum_{j=1}^{m} K_j h^{2j} \int_{x_{i-1}}^{x_i} f^{(2j)}(x)\,dx + O(h^{2m+3})\right)\\
&= \sum_{j=1}^{m} K_j \frac{(b-a)^{2j}}{n^{2j}} \int_a^b f^{(2j)}(x)\,dx + n\,O\left(\frac{(b-a)^{2m+3}}{n^{2m+3}}\right)\\
&= \sum_{j=1}^{m} \frac{C_j}{n^{2j}} + O\left(\frac{1}{n^{2m+2}}\right).
\end{aligned}$$

This completes the proof. □

EXERCISES

1. Calculate sufficient Romberg approximations R_1, R_2, R_3, ... for $\displaystyle\int_0^1 e^{-x^2}\,dx$ to be confident you have evaluated the integral correctly to six decimal places.

2. Use the values of $f(x)$ given in the table accompanying Exercise 5 in Section 7.1 to calculate Romberg approximations R_1, R_2 and R_3 for the integral $\displaystyle\int_0^{1.6} f(x)\,dx$ in that exercise.

3.*The Romberg approximation R_2 for $\displaystyle\int_a^b f(x)\,dx$ requires 5 values of f, $y_0 = f(a)$, $y_1 = f(a+h)$, ..., $y_4 = f(x + 4h) = f(b)$, where $h = (b - a)/4$. Write the formula for R_2 explicitly in terms of these five values.

7.4 OTHER ASPECTS OF APPROXIMATE INTEGRATION

The numerical methods described in Sections 7.1–7.3 are suitable for finding approximate values for integrals of the form

$$I = \int_a^b f(x)\,dx,$$

where $[a, b]$ is a finite interval and the integrand f is "well-behaved" on $[a, b]$. In particular, I must be a *proper* integral. There are many other methods for dealing with such integrals, some of which we mention later in this section. First, however, we consider what can be done if the function f isn't "well-behaved" on $[a, b]$. We mean by this that either the integral is improper, or f doesn't have sufficiently many continuous derivatives on $[a, b]$ to justify whatever numerical methods we want to use.

The ideas of this section are best presented by means of concrete examples.

EXAMPLE 7.4.1 Consider the integral $I = \int_0^1 \sqrt{x}\, e^x\, dx$. Although I is a proper integral, with integrand $f(x) = \sqrt{x}\, e^x$ satisfying $f(x) \to 0$ as $x \to 0+$, nevertheless the standard numerical methods can be expected to perform poorly for I because the derivatives of f are not bounded near 0. This problem is easily remedied; just make a change of variable $x = t^2$ and rewrite I in the form

$$I = 2\int_0^1 t^2\, e^{t^2}\, dt,$$

whose integrand $g(t) = t^2\, e^{t^2}$ has bounded derivatives near 0.

Approximating Improper Integrals

EXAMPLE 7.4.2 Describe how to evaluate $I = \int_0^1 \frac{\cos x}{\sqrt{x}}\, dx$ numerically.

SOLUTION The integral is improper, but convergent because on $[0, 1]$,

$$0 < \frac{\cos x}{\sqrt{x}} \le \frac{1}{\sqrt{x}}, \qquad \text{and} \qquad \int_0^1 \frac{dx}{\sqrt{x}} = 2.$$

However, since $\lim_{x\to 0+} \dfrac{\cos x}{\sqrt{x}} = \infty$, we cannot directly apply any of the techniques developed in Sections 7.1–7.3. (y_0 is infinite.)

The substitution $x = t^2$ removes this difficulty:

$$I = \int_0^1 \frac{\cos t^2}{t}\, 2t\, dt = 2\int_0^1 \cos t^2\, dt.$$

The latter integral is not improper, and is well-behaved. Numerical techniques can be applied to evaluate it.

EXAMPLE 7.4.3 Describe how to evaluate $I = \int_0^2 f(x)\, dx$ numerically, where $f(x) = \dfrac{1}{\sqrt{x(4-x^2)}}$.

SOLUTION This integral is improper because the integrand f approaches infinity at both ends of the interval $(0, 2)$. First break the integral into two integrals each with only one singularity:

$$I = \int_0^1 f(x)\, dx + \int_1^2 f(x)\, dx = I_1 + I_2.$$

In I_1 make the substitution $x = t^2$:

$$I_1 = \int_0^1 \frac{dx}{\sqrt{x(4-x^2)}} = \int_0^1 \frac{2t\, dt}{\sqrt{t^2(4-t^4)}} = 2\int_0^1 \frac{dt}{\sqrt{4-t^4}}.$$

I_2 is improper at $x = 2$ since the denominator involves the factor $\sqrt{2-x}$. In this integral the appropriate substitution is $2 - x = t^2$, or $x = 2 - t^2$. In this case

$$x(4-x^2) = (2-t^2)\big(4-(2-t^2)^2\big) = t^2(2-t^2)(4-t^2).$$

Thus we have

$$I_2 = \int_1^2 \frac{dx}{\sqrt{x(4-x^2)}} = \int_0^1 \frac{2t\,dt}{t\sqrt{(2-t^2)(4-t^2)}} = 2\int_0^1 \frac{dt}{\sqrt{(2-t^2)(4-t^2)}}.$$

Recombining these two integrals, we obtain

$$I = 2\int_0^1 \left(\frac{1}{\sqrt{4-t^4}} + \frac{1}{\sqrt{(2-t^2)(4-t^2)}}\right) dt.$$

The integrand is now well-behaved on $[0, 1]$, and the integral can be evaluated by numerical methods.

EXAMPLE 7.4.4 Show how to evaluate $I = \displaystyle\int_0^\infty \frac{dx}{\sqrt{2+x^2+x^4}}$ by numerical means.

SOLUTION Here the integral is improper of type I; the interval of integration is infinite. Although there is no singularity at $x = 0$, it is still useful to break the integral into two parts:

$$I = \int_0^1 \frac{dx}{\sqrt{2+x^2+x^4}} + \int_1^\infty \frac{dx}{\sqrt{2+x^2+x^4}} = I_1 + I_2.$$

I_1 is proper. In I_2 make the change of variable $x = 1/t$:

$$I_2 = \int_0^1 \frac{dt}{t^2\sqrt{2+\dfrac{1}{t^2}+\dfrac{1}{t^4}}} = \int_0^1 \frac{dt}{\sqrt{2t^4+t^2+1}}.$$

This is also a proper integral. If desired, I_1 and I_2 can be recombined into a single integral before application of numerical methods:

$$I = \int_0^1 \left(\frac{1}{\sqrt{2+x^2+x^4}} + \frac{1}{\sqrt{2x^4+x^2+1}}\right) dx.$$

Example 7.4.4 suggests that when an integral is taken over an infinite interval, a change of variable should be made to convert the integral to a finite interval.

Other Methods

Taylor's Formula (Theorem 5.4.5) can sometimes be useful for evaluating integrals. We will be looking into this method in some detail when we study series representations of functions in Chapter 11, but will give one example now to illustrate the idea.

EXAMPLE 7.4.5 Use Taylor's Formula for $f(x) = e^x$ obtained in Example 5.4.6 to evaluate the integral

$$\int_0^1 e^{x^2}\,dx$$

with error less than 10^{-4}.

SOLUTION In Example 5.4.6 we showed that

$$f(x) = e^x = 1 + x + \frac{x^2}{2!} + \frac{x^3}{3!} + \cdots + \frac{x^n}{n!} + R_n(f;0,x),$$

where

$$R_n(f;0,x) = \frac{e^X}{(n+1)!}\,x^{n+1}$$

for some X between 0 and x. If $0 \le x \le 1$, then $0 \le X \le 1$ so $e^X \le e < 3$. Therefore

$$|R_n(f;0,x)| \le \frac{3}{(n+1)!}\,x^{n+1}.$$

Now replace x by x^2 in the formula for e^x above and integrate from 0 to 1:

$$\begin{aligned}\int_0^1 e^{x^2}\,dx &= \int_0^1 \left(1 + x^2 + \frac{x^4}{2!} + \cdots + \frac{x^{2n}}{n!}\right) dx + \int_0^1 R_n(f;0,x^2)\,dx \\ &= 1 + \frac{1}{3} + \frac{1}{5\times 2!} + \cdots + \frac{1}{(2n+1)n!} + \int_0^1 R_n(f;0,x^2)\,dx\end{aligned}$$

We want the error less than 10^{-4}, so we estimate the remainder term:

$$\left|\int_0^1 R_n(f;0,x^2)\,dx\right| \le \frac{3}{(n+1)!}\int_0^1 x^{2(n+1)}\,dx = \frac{3}{(n+1)!(2n+3)} < 10^{-4}$$

provided $(2n+3)(n+1)! > 30{,}000$. Since $13 \times 6! = 9{,}360$ and $15 \times 7! = 75{,}600$ we need $n = 6$. Thus

$$\int_0^1 e^{x^2}\,dx = 1 + \frac{1}{3} + \frac{1}{5\times 2!} + \frac{1}{7\times 3!} + \frac{1}{9\times 4!} + \frac{1}{11\times 5!} + \frac{1}{13\times 6!} \approx 1.46264$$

with error less than 10^{-4}.

The numerical methods developed in Sections 7.1–7.3 all involved using equal subdivisions of the interval $[a,b]$. There are other methods which avoid this restriction. In particular, one method, known as **Gaussian quadrature** selects evaluation points and weights in an optimal way so as to give the most accurate results for "well-behaved" functions. The interested reader should consult a text on numerical analysis to learn more about this method.

Finally, we note that even when one of the methods of Sections 7.1–7.3 is applied, it may be advisable to break up the integral into two or more integrals over smaller intervals and then use different subinterval lengths h for each of the different integrals. You want to evaluate the integrand more often in an interval where its graph is quite irregular than in an interval where the graph is smoother.

EXERCISES

Rewrite the integrals in Exercises 1–6 in a form to which numerical methods can be readily applied.

1. $\displaystyle\int_0^1 \frac{dx}{x^{1/3}(1+x)}$

2. $\displaystyle\int_0^1 \frac{e^x}{\sqrt{1-x}}\,dx$

3. $\displaystyle\int_{-1}^1 \frac{e^x}{\sqrt{1-x^2}}\,dx$

4. $\displaystyle\int_1^\infty \frac{dx}{x^2+\sqrt{x}+1}$

5.* $\displaystyle\int_0^{\pi/2} \frac{dx}{\sqrt{\sin x}}$

6. $\displaystyle\int_0^\infty \frac{dx}{x^4+1}$

7. Find T_2, T_4, T_8, and T_{16} for $\displaystyle\int_0^1 \sqrt{x}\,dx$ and find the actual errors in these approximations. Do the errors decrease like $1/n^2$ as n increases? Why?

8. Transform the integral $I = \displaystyle\int_1^\infty e^{-x^2}\,dx$ using the substitution $x = 1/t$ and calculate the Simpson's Rule approximations S_2, S_4 and S_8 for the resulting integral (whose integrand has limit 0 as $t \to 0+$). Quote the value of I to the accuracy you feel is justified. Do the approximations converge as quickly as you might expect? Can you think of a reason why they might not?

9. Evaluate $I = \displaystyle\int_0^1 e^{-x^2}\,dx$, by the Taylor's Formula method of Example 7.4.5, to within an error of 10^{-4}.

10. Recall that $\displaystyle\int_0^\infty e^{-x^2}\,dx = \frac{1}{2}\sqrt{\pi}$. Combine this fact with the result of the previous exercise to evaluate $I = \displaystyle\int_1^\infty e^{-x^2}\,dx$ to 3 decimal places.

11.*If $f(x) = \dfrac{\sin x}{x}$ for $x \neq 0$ and $f(0) = 1$ show that $f''(x)$ has a finite limit as $x \to 0$. Hence f'' is bounded on finite intervals $[0, a]$ and Trapezoid approximations T_n to $\displaystyle\int_0^a \frac{\sin x}{x}\,dx$ converge suitably quickly as n increases. Higher derivatives are also bounded (Taylor's formula is useful for showing this) so Simpson's Rule and higher order approximations can also be used effectively.

12.*Explain why the change of variable $x = 1/t$ is not suitable for transforming the integral $\displaystyle\int_\pi^\infty \frac{\sin x}{1+x^2}\,dx$ into a form to which numerical methods can be applied. Try to devise a method whereby this integral could be approximated to any desired degree of accuracy.

CHAPTER 8

Applications of Integration

Numerous mathematical and physical quantities can be conveniently represented by definite integrals, both proper and improper. In addition to measuring plane areas, the problem that motivated the definition of the definite integral, we can use these integrals to express volumes of solids, lengths of curves, areas of surfaces, probabilities, forces, work, energy, pressure, and variety of other quantities that are in one sense or another equivalent to areas under graphs.

In this chapter we examine some of these applications.

8.1 AREAS OF PLANE REGIONS

In this section we review and extend the use of definite integrals to represent plane areas. Recall that the integral $\int_a^b f(x)\,dx$ measures the area between the graph of f and the x-axis from $x = a$ to $x = b$, but it counts as *negative* any part of this area that lies below the x-axis. (We are assuming that $a < b$.) In order to express the total area bounded by $y = f(x)$, $y = 0$, $x = a$, and $x = b$, counting all of the area positively, we should integrate the *absolute value* of f (see Fig. 8.1.1):

$$\int_a^b f(x)\,dx = A_1 - A_2,$$

$$\int_a^b |f(x)|\,dx = A_1 + A_2.$$

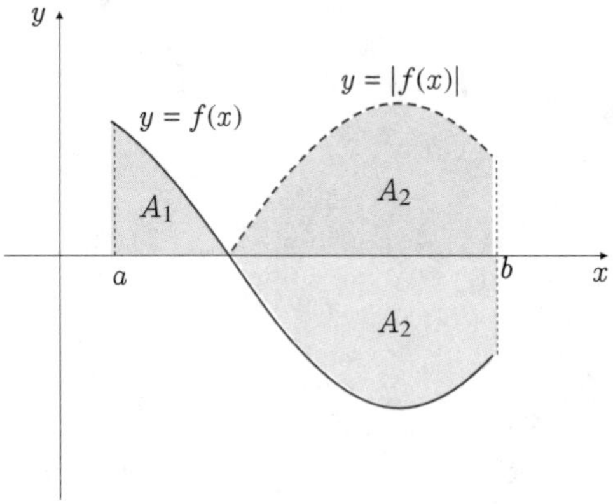

FIGURE 8.1.1

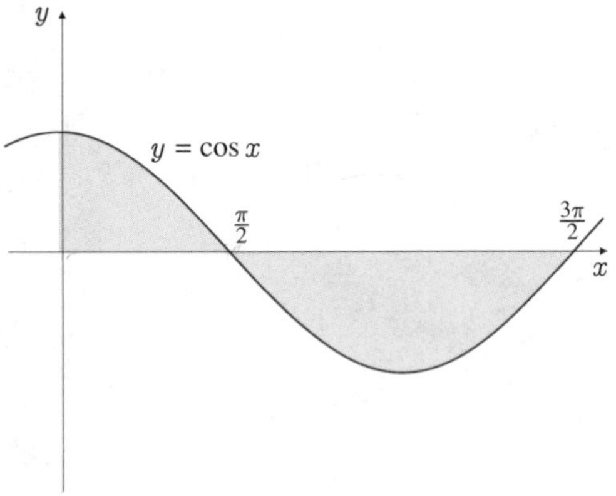

FIGURE 8.1.2

EXAMPLE 8.1.1 The area bounded by $y = \cos x$, $y = 0$, $x = 0$, and $x = 3\pi/2$ (see Fig. 8.1.2) is

$$\begin{aligned} A &= \int_0^{3\pi/2} |\cos x|\,dx \\ &= \int_0^{\pi/2} \cos x\,dx + \int_{\pi/2}^{3\pi/2} (-\cos x)\,dx \end{aligned}$$

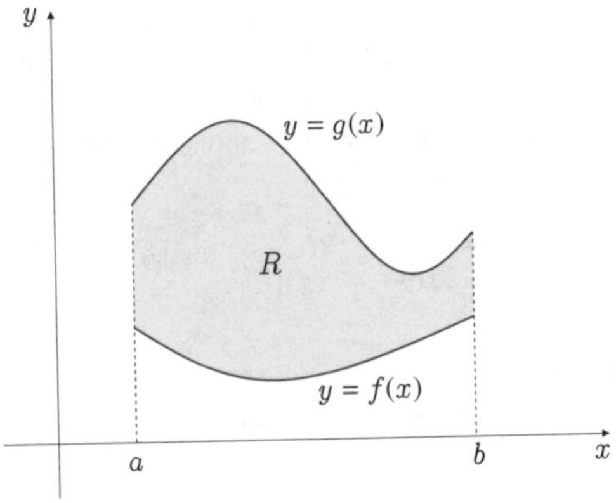

FIGURE 8.1.3

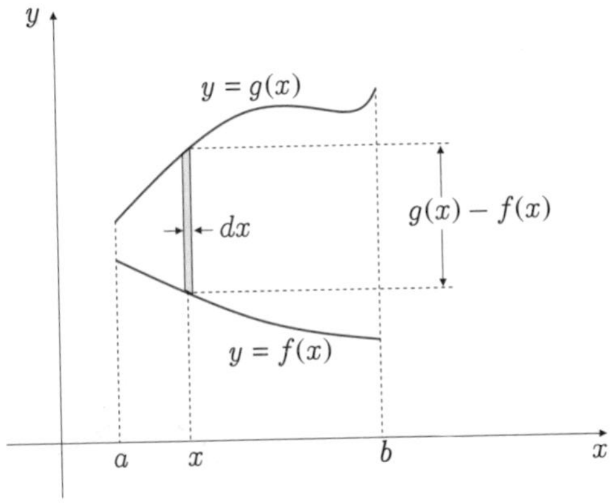

FIGURE 8.1.4

$$= \sin x\Big|_0^{\pi/2} - \sin x\Big|_{\pi/2}^{3\pi/2}$$
$$= (1 - 0) - (-1 - 1) = 3 \text{ sq units.}$$

Note that there is no "rule" for integrating $\int_a^b |f(x)|\,dx$; one must break the integral into a sum of integrals over intervals where $f(x) > 0$ (so $|f(x)| = f(x)$), and intervals where $f(x) < 0$ (so $|f(x)| = -f(x)$).

Areas Between Two Curves

Suppose that a plane region R is bounded by the graphs of two continuous functions, $y = f(x)$ and $y = g(x)$, and the vertical straight lines $x = a$ and $x = b$, as shown in Fig. 8.1.3. Assume that $a < b$ and that $f(x) \le g(x)$ on $[a, b]$, so the graph of f lies below that of g. If f is nonnegative-valued on $[a, b]$, the area A of R is evidently given by

$$A = \int_a^b g(x)\,dx - \int_a^b f(x)\,dx = \int_a^b (g(x) - f(x))\,dx.$$

It is useful to regard this formula as expressing A as the "sum" (that is, the integral) of *infinitely many* **area elements**

$$dA = (g(x) - f(x))\,dx,$$

corresponding to values of x between a and b. Each such area element is the area of an infinitely thin vertical rectangle of width dx and height $g(x) - f(x)$ located at position x (see Fig. 8.1.4). Even if f and g can take on negative values on $[a, b]$, this interpretation, and the resulting area formula

$$A = \int_a^b (g(x) - f(x))\,dx$$

remain valid, provided that $f(x) \le g(x)$ on $[a, b]$ so that all the area elements dA have positive area. Using integrals to represent a quantity as a "sum" of "differential elements" (i.e., a sum of little bits of the quantity) is a very helpful approach. We will do this often in this chapter. Of course, what we are really doing is identifying the integral as a *limit* of a suitable Riemann sum.

More generally, if the restriction $f(x) \leq g(x)$ is removed, then the vertical rectangle of width dx at position x extending between the graphs of f and g has height $|f(x) - g(x)|$ and hence area $dA = |f(x) - g(x)|\, dx$ (see Fig. 8.1.5). Hence the total area lying between the graphs $y = f(x)$ and $y = g(x)$ and between the vertical lines $x = a$ and $x = b > a$ is given by

$$A = \int_a^b |f(x) - g(x)|\, dx.$$

In order to evaluate this integral, we have to determine the intervals on which $f(x) > g(x)$ or $f(x) < g(x)$, and break the integral into a sum of integrals over each of these pieces.

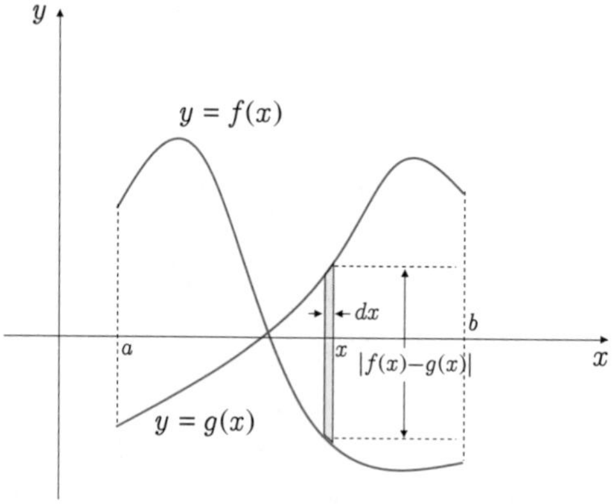

FIGURE 8.1.5

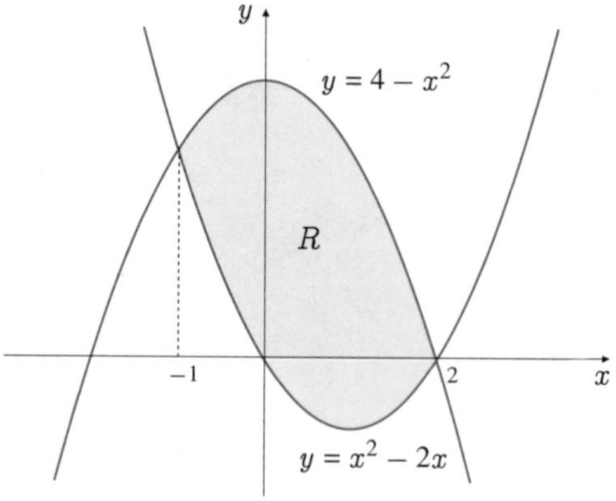

FIGURE 8.1.6

EXAMPLE 8.1.2 Find the area of the bounded, plane region R lying between the curves $y = x^2 - 2x$ and $y = 4 - x^2$.

SOLUTION We must find the intersections of the curves, so we solve the equations simultaneously:

$$\begin{aligned} x^2 - 2x &= y = 4 - x^2 \\ 2x^2 - 2x - 4 &= 0 \\ 2(x-2)(x+1) &= 0 \quad \text{so } x = 2 \text{ or } x = -1. \end{aligned}$$

The curves are sketched in Fig. 8.1.6, and the bounded (finite) region between them is shaded. (A sketch should always be made in problems of this sort.) Since $4 - x^2 \geq x^2 - 2x$ for $-1 \leq x \leq 2$, the area A of R is given by

$$\begin{aligned} A &= \int_{-1}^{2} \big((4 - x^2) - (x^2 - 2x)\big)\, dx \\ &= \int_{-1}^{2} (4 - 2x^2 + 2x)\, dx \end{aligned}$$

$$= \left(4x - \frac{2}{3}x^3 + x^2\right)\Bigg|_{-1}^{2}$$

$$= 4(2) - \frac{2}{3}(8) + 4 - \left(-4 + \frac{2}{3} + 1\right) = 9 \text{ sq units.}$$

Note that in representing the area as an integral we *must subtract the height y to the lower curve from the height y to the upper curve* to get a positive area element dA. Subtracting the wrong way would have produced a negative value for the area.

EXAMPLE 8.1.3 Find the total area A lying between the curves $y = \sin x$ and $y = \cos x$ from $x = 0$ to $x = 2\pi$.

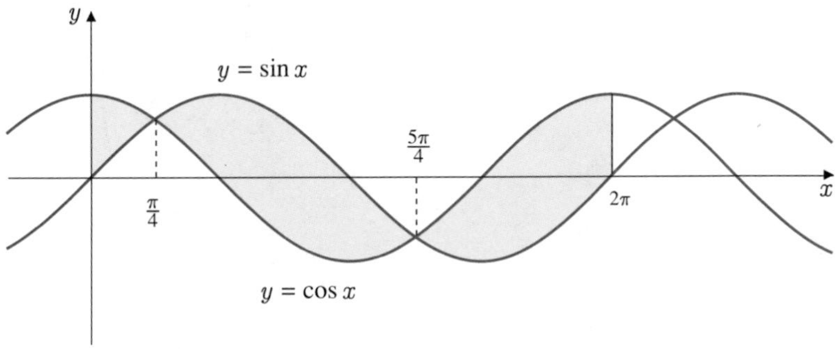

FIGURE 8.1.7

SOLUTION The region is shaded in Fig. 8.1.7. Evidently the graphs of sine and cosine cross at $x = \pi/4$ and $x = 5\pi/4$. The required area is

$$A = \int_0^{\pi/4} (\cos x - \sin x)\,dx + \int_{\pi/4}^{5\pi/4} (\sin x - \cos x)\,dx + \int_{5\pi/4}^{2\pi} (\cos x - \sin x)\,dx$$

$$= (\sin x + \cos x)\Bigg|_0^{\pi/4} - (\cos x + \sin x)\Bigg|_{\pi/4}^{5\pi/4} + (\sin x + \cos x)\Bigg|_{5\pi/4}^{2\pi}$$

$$= (\sqrt{2} - 1) + (\sqrt{2} + \sqrt{2}) + (1 + \sqrt{2}) = 4\sqrt{2} \text{ sq units.}$$

The formula for area between curves also holds for unbounded regions; we can regard the area of such a region as the limit of areas of bounded regions and hence express the area as an improper integral.

EXAMPLE 8.1.4 Find the area of the plane region R bounded by the curve $y = e^x/(1 + e^x)$ and the line $y = 1$ and lying to the right of the y-axis.

SOLUTION The region R is shown in Fig. 8.1.8. Its area is

$$A = \int_0^{\infty} \left(1 - \frac{e^x}{1 + e^x}\right) dx = \int_0^{\infty} \frac{dx}{1 + e^x}$$

$$= \lim_{R\to\infty} \int_0^R \frac{e^{-x}}{e^{-x} + 1}\,dx.$$

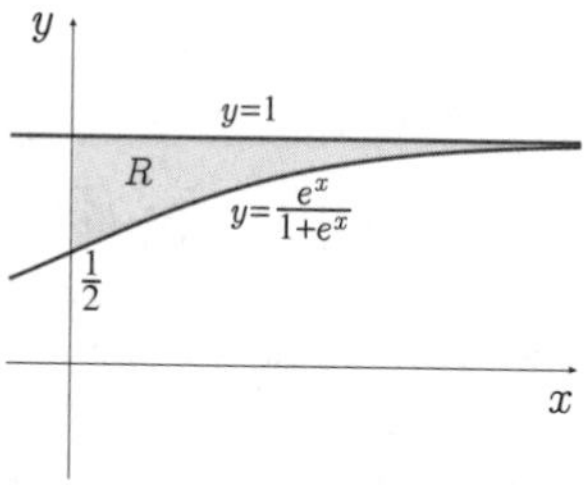

FIGURE 8.1.8

Written in this last form, the integral can be evaluated via the substitution $u = e^{-x} + 1$, $du = -e^{-x}\,dx$. We have

$$A = -\lim_{R\to\infty}\int_2^{e^{-R}+1}\frac{du}{u} = -\lim_{R\to\infty}\left(\ln(e^{-R}+1) - \ln 2\right) = \ln 2 \text{ sq units.}$$

(As $R \to \infty$ we have $\ln(e^{-R} + 1) \to \ln(1) = 0$.)

It is sometimes more convenient to use horizontal area elements instead of vertical ones, and integrate over an interval of the y-axis instead of the x-axis. This is usually the case if the region whose area we want to find is bounded by curves whose equations are written in terms of functions of y. The region R lying to the right of $x = f(y)$ and to the left of $x = g(y)$, and between the horizontal lines $y = c$ and $y = d > c$ (see Fig. 8.1.9) has area element $dA = \big(g(y) - f(y)\big)\,dy$ and its area is

$$A = \int_c^d \big(g(y) - f(y)\big)\,dy.$$

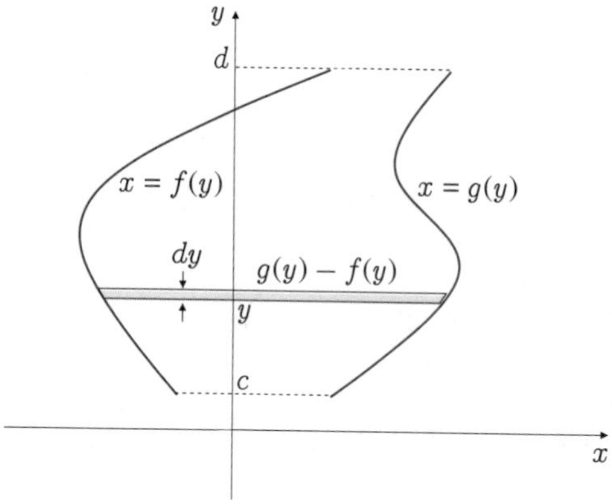

FIGURE 8.1.9

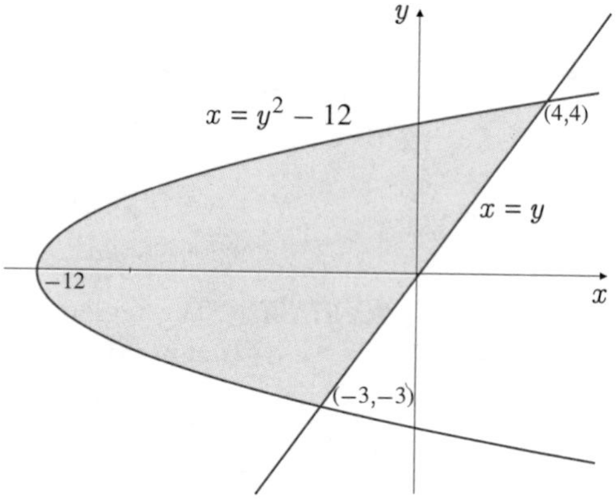

FIGURE 8.1.10

EXAMPLE 8.1.5 Find the area of the plane region lying to the right of the parabola $x = y^2 - 12$ and to the left of the straight line $y = x$, as illustrated in Fig. 8.1.10.

SOLUTION For the intersections of the curves:

$$\begin{aligned} &y^2 - 12 = x = y \\ &y^2 - y - 12 = 0 \\ &(y-4)(y+3) = 0 \quad \text{so } y = 4 \text{ or } y = -3. \end{aligned}$$

Observe that $y^2 - 12 \le y$ for $-3 \le y \le 4$. Thus the area is

$$A = \int_{-3}^4 \left(y - (y^2 - 12)\right) dy = \left(\frac{y^2}{2} - \frac{y^3}{3} + 12y\right)\Bigg|_{-3}^4 = \frac{343}{6} \text{ sq units.}$$

Of course, the same result could have been obtained by integrating in the x direction, but the integral would have been more complicated:

$$A = \int_{-12}^{-3} (\sqrt{12+x} - (-\sqrt{12+x}))\, dx + \int_{-3}^{4} (\sqrt{12+x} - x)\, dx.$$

EXERCISES

In Exercises 1–15 find the area of the plane region bounded by the given curves. In each case you should make a sketch of the region.

1. $y = x, \quad y = x^2$

2. $y = \sqrt{x}, \quad y = x^2$

3. $y = x^2 - 5, \quad y = 3 - x^2$

4. $y = x^2 - 2x, \quad y = 6x - x^2$

5. $2y = 4x - x^2, \quad 2y + 3x = 6$

6. $x - y = 7, \quad x = 2y^2 - y + 3$

7. $y = x^3, \quad y = x$

8. $y = x^3, \quad y = x^2$

9. $y = x^3, \quad x = y^2$

10. $x = y^2, \quad x = 2y^2 - y - 2$

11. $y = \dfrac{1}{x}, \quad 2x + 2y = 5$

12. $y = (x^2 - 1)^2, \quad y = 1 - x^2$

13. $y = \dfrac{1}{2}x^2, \quad y = \dfrac{1}{x^2+1}$

14. $y = \dfrac{4x}{3+x^2}, \quad y = 1$

15. $y = \dfrac{4}{x^2}, \quad y = 5 - x^2$

Sketch and find the areas of the regions described in Exercises 16–24.

16. In the first quadrant, above the hyperbola $xy = 12$ and inside the circle $x^2 + y^2 = 25$.

17. Bounded by $y = \sin x$ and $y = \cos x$, and between two consecutive intersections of these curves.

18. Under $y = 4x/\pi$ and above $y = \tan x$, between $x = 0$ and the first intersection of the curves to the right of $x = 0$.

19. Inside the circle $x^2 + y^2 = a^2$ and above the line $y = b$, $(-a \le b \le a)$.

20. To the left of $\dfrac{x^2}{a^2} + \dfrac{y^2}{b^2} = 1$ and to the right of the line $x = c$, where $-a \le c \le a$

21. *Below $y = e^{-x}\sin x$ and above $y = 0$ from $x = 0$ to $x = \pi$

22. Below $y = e^{-x}$ and above $y = e^{-2x}$ to the right of $x = 0$

23. Below $y = 0$ and above $y = \ln x$, to the right of $x = 0$

24. Inside both of the circles $x^2+y^2 = 1$ and $(x-2)^2+y^2 = 4$

25. Find the total area enclosed by the curve $y^2 = x^2 - x^4$.

26. Find the area of the closed loop of the curve $y^2 = x^4(2+x)$ that lies to the left of the origin.

27. Find the area of the finite plane region that is bounded by the curve $y = e^x$, the line $x = 0$, and the tangent line to $y = e^x$ at $x = 1$.

28. Find the area of the finite plane region bounded by the curve $y = \ln x$, the line $y = 1$, and the tangent line to $y = \ln x$ at $x = 1$.

29. *Find the area of the finite plane region bounded by the curve $y = x^3$ and the tangent line to that curve at the point $(1, 1)$. (*Hint:* Find the other point at which that tangent line meets the curve.)

30. Find the total area bounded by the curve $y = \sin x$ and the straight line $y = x/2$. (You will have to use numerical means to find one of the three intersections.)

31. *Find the area of the region bounded by the x-axis, the hyperbola $x^2 - y^2 = 1$ and the straight line from the origin to the point $(\sqrt{1+Y^2}, Y)$ on that hyperbola. (Assume $Y > 0$.) In particular, show that the area is $t/2$ square units if $Y = \sinh t$.

8.2 VOLUMES

In this section we show how volumes of certain three-dimensional regions (or "solids") can be expressed as definite integrals and thereby determined. We will not attempt to give a definition of "volume," but will rely on our intuition and experience with solid objects to provide enough insight for us to specify the volumes of certain simple solids. For example, if the base of a rectangular box is a rectangle of length l and width w (and therefore area $A = lw$), and if the box has height h, then its volume is $V = Ah = lwh$. (If l, w and h are measured in "units" then the volume is expressed in "cubic units.")

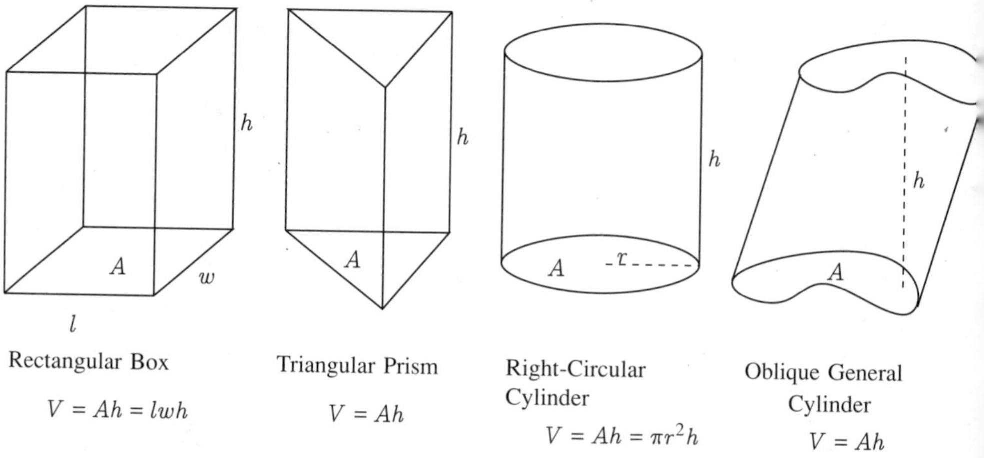

FIGURE 8.2.1

A rectangular box is a special case of a solid called a prism or cylinder. (See Fig. 8.2.1.) Such a solid S has a flat base, occupying a plane region R having area A. Every cross-section of S in a plane parallel to the base is congruent to R so has the same area. If S has height h (so that its top is in a plane parallel to the base and h units above it), then the volume of S is $V = Ah$. Such solids are usually called *prisms* if the base region R is bounded by straight lines, and are called *cylinders* if R is bounded by curves. In particular, if R is a circular disc with base radius r and the top is directly above the base then the solid is a *right-circular cylinder*. If it has height h then its volume is $V = \pi r^2 h$ cubic units. Cylinders (and prisms) are said to be *right* if their side walls are perpendicular to their bases; otherwise they are *oblique*. This has no effect on the volume $V = Ah$, for which h is always measured in a direction perpendicular to the base.

Volumes by Slicing

Knowing the volume of a cylinder enables us to determine the volumes of some more general solids. We can divide solids into thin "slices" by parallel planes. (Think of a loaf of sliced bread.) Each slice is approximately a cylinder of very small "height;" the height is the thickness of the slice. If we know the cross-sectional area of each slice we can determine its volume, and sum these volumes to find the volume of the solid.

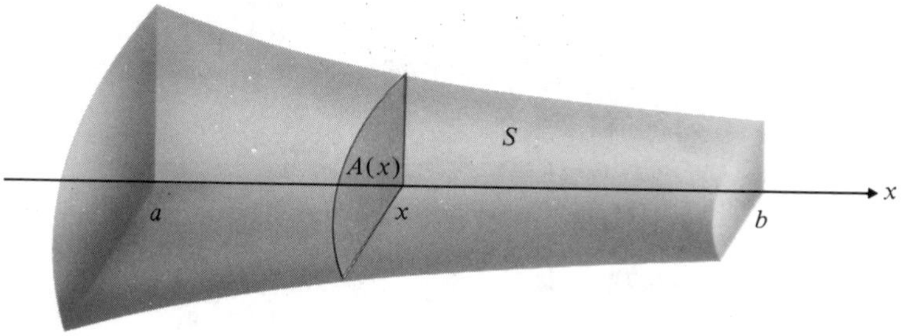

FIGURE 8.2.2

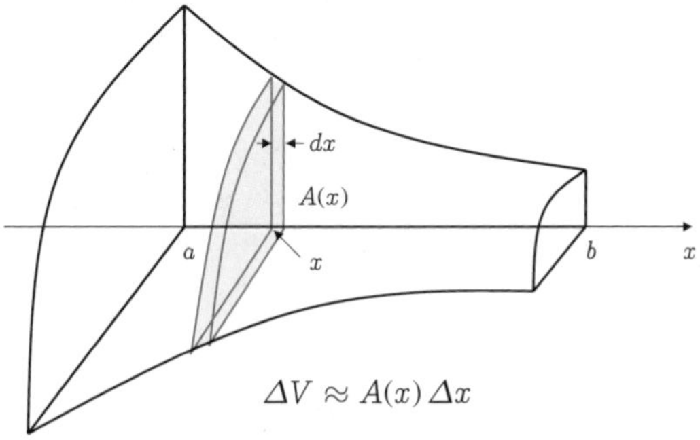

FIGURE 8.2.3

To be specific, suppose that the solid S lies between planes perpendicular to the x-axis at positions $x = a$ and $x = b$ and that the cross-sectional area of S in the plane perpendicular to the x-axis at x is a known function $A(x)$, for $a \le x \le b$. See Fig. 8.2.2. We assume that $A(x)$ is continuous on $[a, b]$. A slice of the solid, perpendicular to the x-axis between x and $x + \Delta x$ has thickness Δx and base area $A(x)$. Since $A(x)$ is continuous it doesn't change much in a short interval, so if Δx is small then the cross-sectional area of the slice is almost constant, and if ΔV is the volume of the slice, then the error in the approximation

$$\Delta V \approx A(x)\,\Delta x$$

is small compared to the size of ΔV. (See Fig. 8.2.3.) This suggests, correctly, that the **volume element**, that is, the volume of an infinitely thin slice of thickness dx is $dV = A(x)\,dx$, and that the volume of the solid is the "sum" (that is, the integral) of these volume elements between the two ends of the solid, $x = a$ and $x = b$.

8.2.1
Volume by Slicing

> Suppose a solid lies between the planes $x = a$ and $x = b$ and has cross-sectional area given by the continuous function $A(x)$ at every x in $[a, b]$. Then the volume of the solid is
>
> $$V = \int_a^b A(x)\,dx.$$

The argument given above does not really constitute a "proof" of the formula in the box, but you are strongly encouraged to think of the formula that way; the volume is the integral of the volume elements. If we wanted to give a proof, we would consider a partition of $[a, b]$ into subintervals $[x_{j-1}, x_j]$ of lengths Δx_j, $(1 \le j \le n)$, and use the Intermediate-Value Theorem (1.7.10) to show that the slice of thickness Δx_j between the planes $x = x_{j-1}$ and $x = x_j$ has volume $\Delta V_j = A(c_j)\,\Delta x_j$ for some c_j in $[x_{j-1}, x_j]$. The volume of S is the sum of the volumes of these slices,

$$V = \sum_{j=1}^{n} A(c_j)\,\Delta x_j,$$

which is a Riemann sum for $\int_a^b A(x)\,dx$.

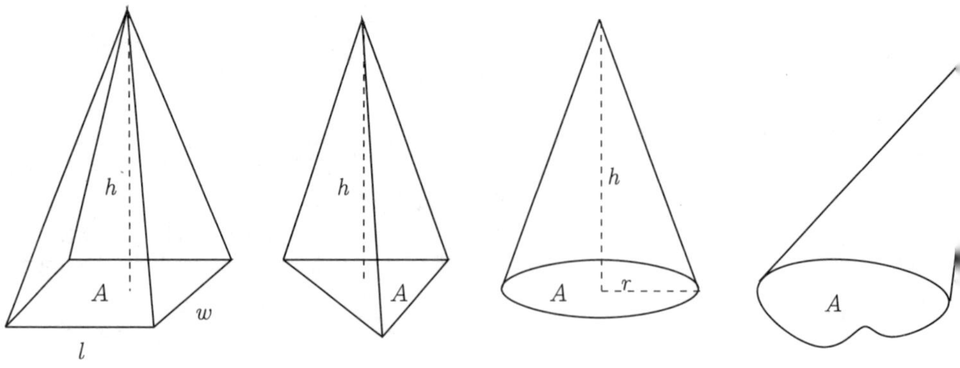

Rectangular Pyramid $V = \frac{1}{3}Ah = \frac{1}{3}lwh$ | Triangular Pyramid $V = \frac{1}{3}Ah$ | Right-Circular Cone $V = \frac{1}{3}Ah = \frac{1}{3}\pi r^2 h$ | Oblique General Cone $V = \frac{1}{3}Ah$

FIGURE 8.2.4

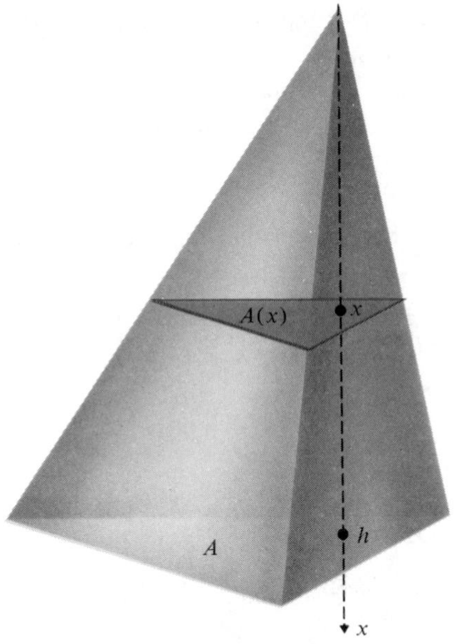

FIGURE 8.2.5

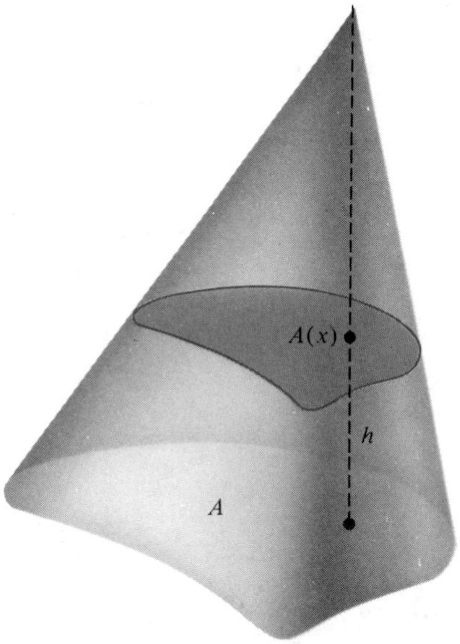

FIGURE 8.2.6

Pyramids and cones are solids consisting of all points on line segments which join a fixed point, the *vertex*, to all the points in a region R lying in a plane not containing the vertex. R is called the *base* of the pyramid or cone. Some pyramids and cones are shown in Fig. 8.2.4. If R is bounded by straight lines, the solid is a pyramid; if R has a curved boundary the solid is called a cone. We show that all pyramids and cones have volume

$$V = \frac{1}{3} Ah,$$

where A is the area of the base and h is the height measured perpendicularly from the vertex to the plane of the base.

EXAMPLE 8.2.2 If a pyramid has a polygonal base of area A and has height h, then cross sections of the pyramid in planes parallel to the base are similar polygons. If the origin is at the vertex of the pyramid and the x-axis is perpendicular to the base, then the cross section at position x is similar in shape to the base and has linear dimensions x/h times those of the base (see Fig. 8.2.5). Thus the area of the cross section at x is

$$A(x) = \left(\frac{x}{h}\right)^2 A.$$

The volume of the pyramid is therefore

$$V = \int_0^h \left(\frac{x}{h}\right)^2 A\,dx = \frac{A}{h^2}\frac{x^3}{3}\bigg|_0^h = \frac{1}{3}Ah \text{ cu units.}$$

A similar argument, using the same formula, holds for a cone, that is, a pyramid with a more general (curved) shape to its base, such as that in Fig. 8.2.6.

EXAMPLE 8.2.3 A tent has a circular base of radius a metres, and is supported by a horizontal ridge bar held at height b metres above a diameter of the base by vertical supports at each end of the diameter. The material of the tent is stretched tight so that each cross section perpendicular to the ridge bar is an isoceles triangle. (See Fig. 8.2.7.) Find the volume of the tent.

SOLUTION Let the x-axis be the diameter of the base under the ridge bar. The cross-section at position x has base length $2\sqrt{a^2 - x^2}$, and so has area

$$A(x) = \frac{1}{2}\left(2\sqrt{a^2 - x^2}\right)b.$$

Thus the volume of the solid is

$$V = \int_{-a}^{a} b\sqrt{a^2 - x^2}\,dx = 2b\int_{0}^{a} \sqrt{a^2 - x^2}\,dx = 2b\frac{\pi a^2}{4} = \frac{\pi}{2}a^2 b \text{ m}^3.$$

Note that we evaluated the last integral by inspection. It is the area of a quarter of a circle of radius a.

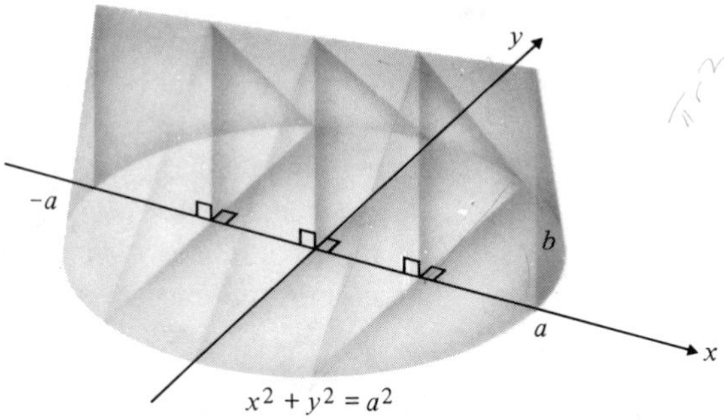

FIGURE 8.2.7

EXAMPLE 8.2.4 Two circular cylinders, each having radius a, intersect so that their axes meet at right angles. Find the volume of the region lying inside both cylinders.

SOLUTION We represent the cylinders in a three-dimensional Cartesian coordinate system where the plane containing the x- and y-axes is horizontal and the z-axis is vertical. One eighth of the solid is shown in Fig. 8.2.8, that part corresponding to all three coordinates being positive. The two cylinders have axes along the x-axis and y-axis, respectively. The cylinder with axis along the x-axis intersects the plane of the y- and z-axes in a circle of radius a. Similarly, the other cylinder meets the plane of the x- and z-axes in a circle of radius a. It follows that if the region lying inside both cylinders (and having $x \geq 0$, $y \geq 0$, and $z \geq 0$) is sliced by a horizontal plane at height z above the xy-plane, then the cross section in that plane

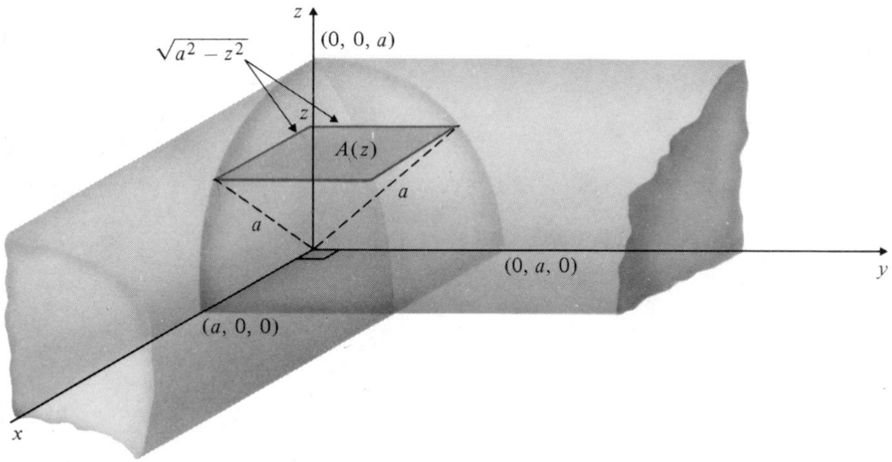

FIGURE 8.2.8

is a square of side $\sqrt{a^2 - z^2}$ and has area $A(z) = a^2 - z^2$. The volume V of the whole region, being eight times that of the part shown, is

$$V = 8\int_0^a (a^2 - z^2)\,dz = 8\left(a^2 z - \frac{z^3}{3}\right)\Bigg|_0^a = \frac{16}{3}a^3 \text{ cu units.}$$

Solids of Revolution

Many common solids have circular cross sections in planes perpendicular to some axis. Such solids are called **solids of revolution** because they can be generated by rotating a plane region about an axis in that plane so that it sweeps out the solid. (See Fig. 8.2.9.) For example, a solid ball is generated by rotating a half-disc about the diameter of that half-disc (Fig. 8.2.10). Similarly, a solid right-circular cone is generated by rotating a right-angled triangle about one of its legs (Fig. 8.2.11).

If the region R bounded by $y = f(x)$, $y = 0$, $x = a$ and $x = b$ is rotated about the x-axis, then the cross section of the solid generated in the plane perpendicular to the x-axis at x is a circle of radius $|f(x)|$. Thus the volume of such a solid of revolution is

$$V = \pi \int_a^b (f(x))^2\,dx.$$

EXAMPLE 8.2.5 If the half-disc, $0 \le y \le \sqrt{a^2 - x^2}$, is rotated about the x-axis, it generates a ball of radius a (Fig. 8.2.10) having volume

$$V = \pi \int_{-a}^{a} (\sqrt{a^2 - x^2})^2\,dx = 2\pi \int_0^a (a^2 - x^2)\,dx$$
$$= 2\pi \left(a^2 x - \frac{x^3}{3}\right)\Bigg|_0^a = 2\pi\left(a^3 - \frac{1}{3}a^3\right) = \frac{4}{3}\pi a^3 \text{ cu units.}$$

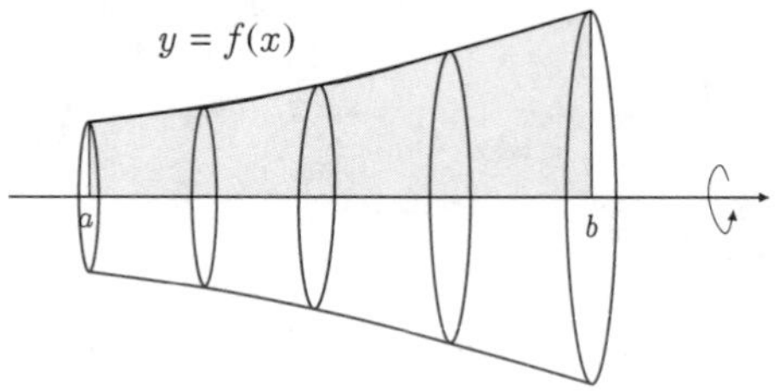

FIGURE 8.2.9

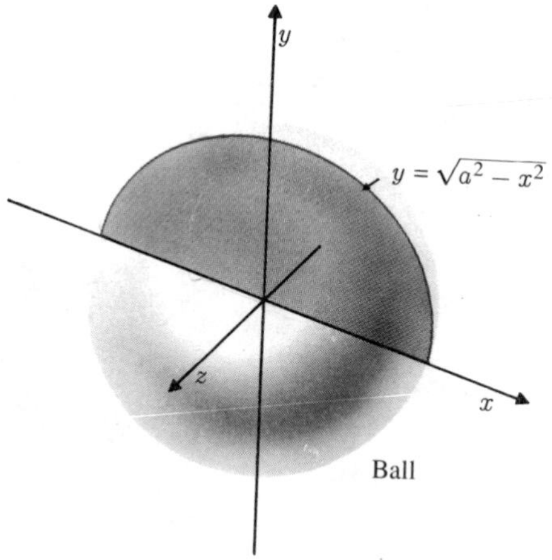

FIGURE 8.2.10

EXAMPLE 8.2.6 If the triangular region bounded by $y = 0$, $y = rx/h$, and $x = h$ is rotated about the x-axis, it sweeps out a right-circular cone of base radius r and height h. (See Fig. 8.2.11.) This cone has volume

$$V = \pi \int_0^h \left(\frac{rx}{h}\right)^2 dx = \pi \left(\frac{r}{h}\right)^2 \frac{x^3}{3}\bigg|_0^h = \frac{1}{3}\pi r^2 h \text{ cu units.}$$

(Note that this is just a special case of Example 8.2.2.)

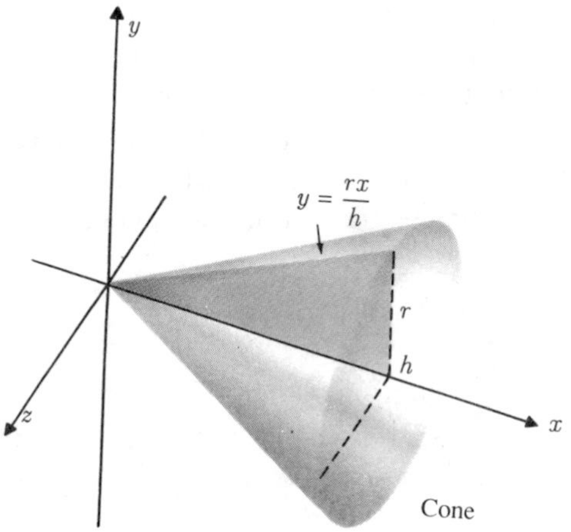

FIGURE 8.2.11

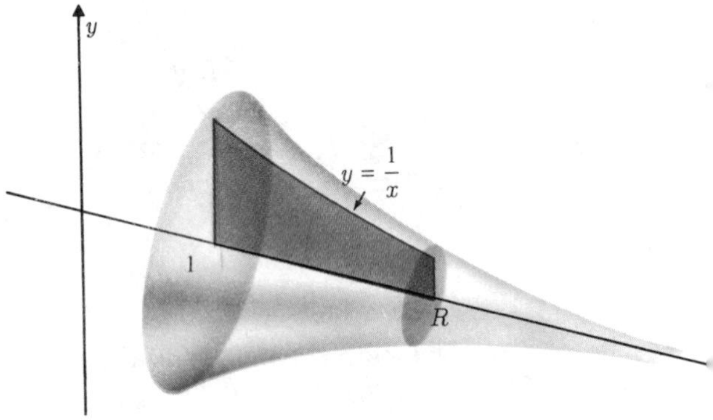

FIGURE 8.2.12

EXAMPLE 8.2.7 Find the volume of the infinitely long horn generated by rotating about the x-axis the region bounded by $y = 1/x$ and $y = 0$ and lying to the right of $x = 1$. The horn is illustrated in Fig. 8.2.12.

SOLUTION The volume is

$$V = \pi \int_1^\infty \left(\frac{1}{x}\right)^2 dx = \pi \lim_{R\to\infty} \int_1^R \frac{1}{x^2}\, dx$$
$$= -\pi \lim_{R\to\infty} \frac{1}{x}\bigg|_1^R = \pi \text{ cu units.}$$

It is interesting to note that this finite volume arises from rotating a region that itself has infinite area: $\int_1^\infty dx/x = \infty$. We have a paradox: It takes infinitely much paint to paint the region, but only finitely much to fill the horn obtained by rotating the region. (How can you resolve this paradox?)

The following example shows how to deal with a problem where the axis of rotation is not the x-axis. Just rotate a suitable area element about the axis to form a volume element.

EXAMPLE 8.2.8 The region R, lying above the parabola $y = x^2$ and below the line $y = 1$, is rotated about the line $y = 2$ to generate a solid of revolution having the shape of a rope float. Find its volume.

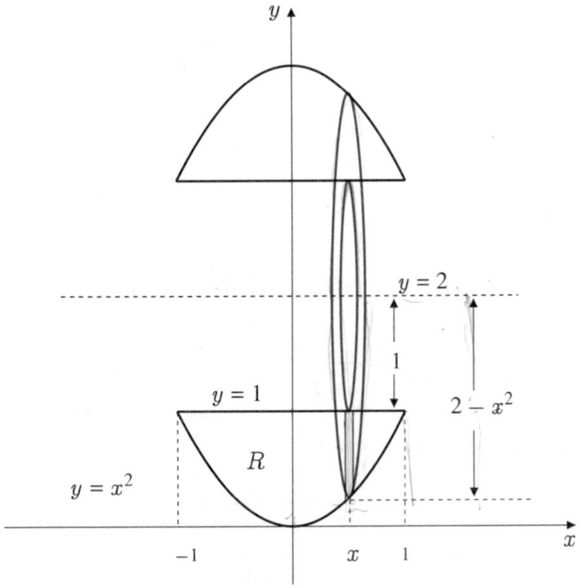

FIGURE 8.2.13

SOLUTION The area element of R at position x is a vertical strip of width dx extending upward from $y = x^2$ to $y = 1$ as shown in Fig. 8.2.13. When R is rotated about the line $y = 2$, this area element sweeps out a thin, washer-shaped volume element of thickness dx and radius $2 - x^2$, having a hole of radius 1 through the middle. This volume element has volume

$$dV = \left(\pi(2 - x^2)^2 - \pi(1)^2\right)dx = \pi(3 - 4x^2 + x^4)\,dx.$$

Since the solid extends from $x = -1$ to $x = 1$, its volume is

$$\begin{aligned} V &= \pi \int_{-1}^{1} (3 - 4x^2 + x^4)\,dx = 2\pi \int_0^1 (3 - 4x^2 + x^4)\,dx \\ &= 2\pi \left(3x - \frac{4x^3}{3} + \frac{x^5}{5}\right)\Bigg|_0^1 = 2\pi\left(3 - \frac{4}{3} + \frac{1}{5}\right) = \frac{56\pi}{15} \text{ cu units.} \end{aligned}$$

Cylindrical Shells

Suppose that the region R bounded by $y = f(x)$, $y = 0$, $x = a \geq 0$, and $x = b > a$ is rotated about the y-axis to generate a solid of revolution S. In order to evaluate the volume of S using (plane) slices, we would need to know the cross-sectional area $A(y)$ in each plane of height y, and this would entail solving the equation $y = f(x)$ for one or more solutions of the form $x = g(y)$. In practice this can be inconvenient or impossible.

Consider the standard area element of R at position x. It is a vertical strip of width dx, height $f(x)$, and area $dA = f(x)\,dx$. When R is rotated about the y-axis this strip sweeps out a volume element in the shape of a circular **cylindrical shell** having radius x, height $f(x)$ and thickness dx. (See Fig. 8.2.14.) Regard this shell as a rolled up rectangular slab with dimensions $2\pi x$, $f(x)$ and dx; evidently it has volume

$$dV = 2\pi x\, f(x)\,dx.$$

The volume of the solid of revolution is the "sum" of the volumes of such shells with radii ranging from a to b:

8.2.9
Volumes by Cylindrical Shells

$$V = 2\pi \int_a^b x\, f(x)\,dx.$$

EXAMPLE 8.2.10 A disc of radius a has centre at the point $(b, 0)$ where $b > a$. The disc is rotated about the y-axis to generate a **torus**, (a doughnut-shaped solid), illustrated in Fig. 8.2.15. Find its volume.

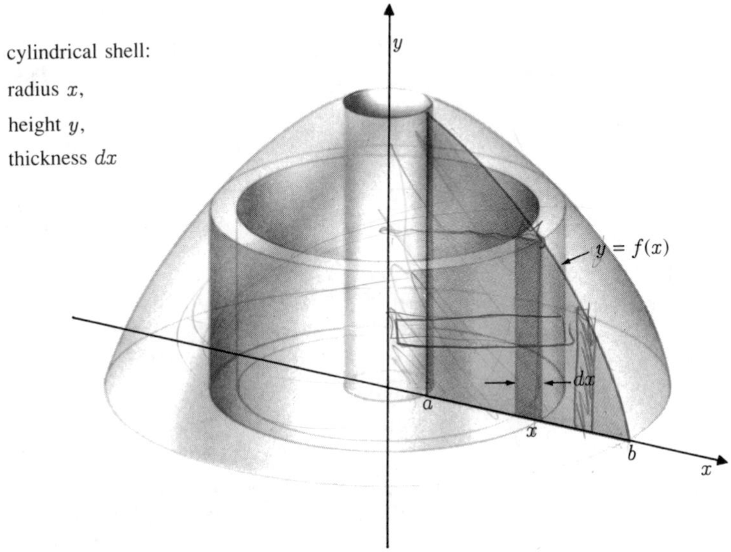

FIGURE 8.2.14

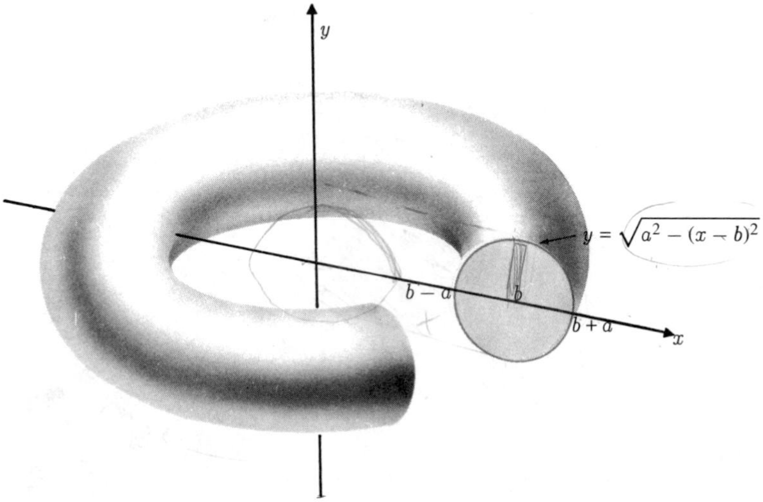

FIGURE 8.2.15

SOLUTION We will double the volume of the upper half of the torus, which is generated by rotating the half-disc $0 \le y \le \sqrt{a^2 - (x-b)^2}$, $b - a \le x \le b + a$ about the y-axis. The volume of the torus is

$$V = 2 \times 2\pi \int_{b-a}^{b+a} x\sqrt{a^2 - (x-b)^2}\, dx \qquad \text{Let } u = x - b,\ du = dx.$$

$$= 4\pi \int_{-a}^{a} (u + b)\sqrt{a^2 - u^2}\, du$$

$$= 4\pi \int_{-a}^{a} u\sqrt{a^2 - u^2}\, du + 4\pi b \int_{-a}^{a} \sqrt{a^2 - u^2}\, du$$

$$= 0 + 4\pi b \frac{\pi a^2}{2} = 2\pi^2 a^2 b \text{ cu units.}$$

(The first of the final two integrals is 0 because the integrand is odd and the interval symmetric about 0; the second is the area of a semicircle of radius a.)

We have described two methods for determining the volume of a solid of revolution. The choice of method for a particular solid is usually dictated by the form of the equations defining the region which is to be rotated, and by the axis of rotation. The volume element dV can always be determined by rotating a suitable area element dA about the axis of rotation. If the region is bounded by vertical lines and one or more graphs of the form $y = f(x)$, the appropriate area element is a vertical strip of width dx. If the rotation is about the x-axis or any other horizontal line, this strip generates a disc- or washer-shaped slice of thickness dx. If the rotation is about the y-axis or any other vertical line, the strip generates a cylindrical shell of thickness dx. (See Fig. 8.2.16.) On the other hand, if the region being rotated is bounded by horizontal lines and one or more graphs of the form $x = g(y)$, it is easier to use a horizontal strip of width dy as area element, which generates a slice if the rotation is about a vertical line and a cylindrical shell if the rotation is about a horizontal line. For very simple regions either method can be made to work easily.

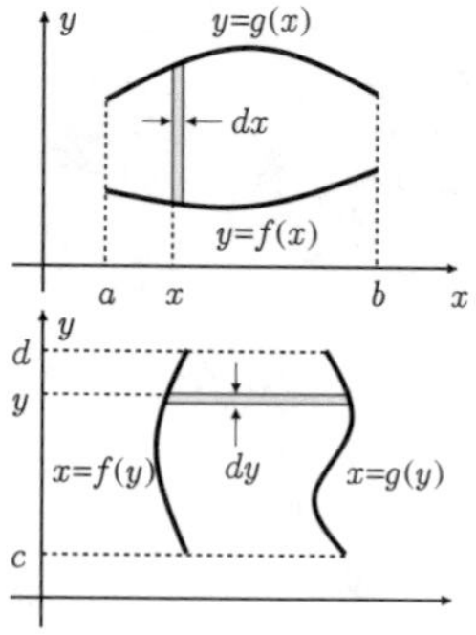

FIGURE 8.2.16

EXAMPLE 8.2.11 The triangular region bounded by $y = x$, $y = 0$, and $x = a > 0$ is rotated about the line $x = b > a$. (See Fig. 8.2.17.) Find the volume of the solid so generated.

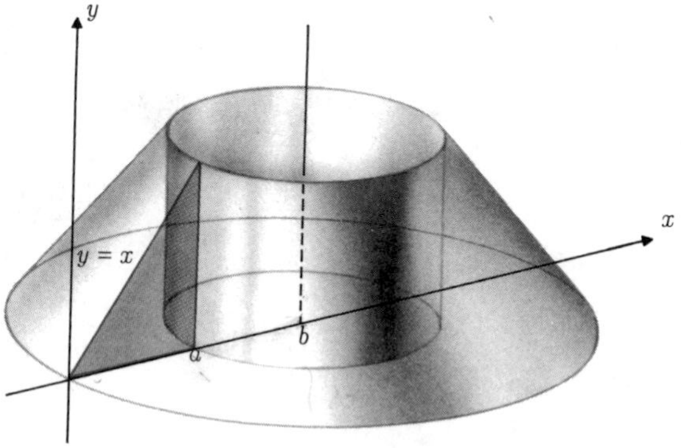

FIGURE 8.2.17

SOLUTION Here the vertical area element at x generates a cylindrical shell of radius $b - x$, height x and thickness dx. Its volume is $dV = 2\pi(b - x)\, x\, dx$, and the volume of the solid is

$$V = 2\pi \int_0^a (b - x)\, x\, dx = 2\pi \left(\frac{bx^2}{2} - \frac{x^3}{3} \right) \Bigg|_0^a = \pi \left(a^2 b - \frac{2a^3}{3} \right) \text{ cu units.}$$

EXERCISES

Find the volume of each solid S in Exercises 1–4 in two ways, using the method of slicing and also the method of cylindrical shells.

1. S is generated by rotating about the x-axis the region bounded by $y = x^2$, $y = 0$, and $x = 1$.

2. S is generated by rotating the region of Exercise 1 about the y-axis.

3. S is generated by rotating about the x-axis the region bounded by $y = x^2$ and $y = \sqrt{x}$ between $x = 0$ and $x = 1$.

4. S is generated by rotating the region of Exercise 3 about the y-axis.

Find the volumes of the solids obtained if the plane regions R described in Exercises 5–10 are rotated about (a) the x-axis, and (b) the y-axis.

5. R is bounded by $y = x(2 - x)$ and $y = 0$ between $x = 0$ and $x = 2$.

6. R is the finite region bounded by $y = x$ and $y = x^2$.

7. R is the finite region bounded by $y = x$ and $x = 4y - y^2$.

8. R is bounded by $y = 1 + \sin x$ and $y = 1$ from $x = 0$ to $x = \pi$.

9. R is bounded by $y = 1/(1+x^2)$, $y = 2$, $x = 0$, and $x = 1$.

10. R is the finite region bounded by $y = 1/x$ and $3x+3y = 10$.

11. The triangular region with vertices $(0, -1)$, $(1, 0)$ and $(0, 1)$ is rotated about the line $x = 2$. Find the volume of the solid so generated.

12. Find the volume of the solid generated by rotating the region $0 \le y \le 1 - x^2$ about the line $y = 1$.

13. A solid has a circular base of radius r. All sections of the solid perpendicular to a particular diameter of the base are squares. Find the volume of the solid.

14. Repeat the previous exercise but with sections that are equilateral triangles instead of squares.

15. What percentage of the volume of a ball of radius 2 is removed if a hole of radius 1 is drilled through the centre of the ball?

16. A cylindrical hole is bored through the centre of a ball of radius R. If the length of the hole is L, show that the volume of the remaining part of the ball depends only on L and not on R.

17. A cylindrical hole of radius a is bored through a solid right-circular cone of height h and base radius $b > a$. If the axis of the hole lies along that of the cone, find the volume of the remaining part of the cone.

18. Find the volume of the solid obtained by rotating a circular disk about one of its tangent lines.

19. A plane slices a ball of radius a into two pieces. If the plane passes b units away from the centre of the ball (where $b < a$), find the volume of the smaller piece.

20. Water partially fills a hemispherical bowl of radius 30 cm so that the maximum depth of the water is 20 cm. What volume of water is in the bowl?

21. Find the volume of the ellipsoid of revolution obtained by rotating the ellipse $(x^2/a^2) + (y^2/b^2) = 1$ about the x-axis.

22. The elliptical disc bounded by the ellipse of the previous exercise is rotated about the line $y = c > b$. Find the volume of the solid generated.

23.*A sphere of radius R and a right-circular cone of radius r and height $h > R$ intersect so that the vertex of the cone is at the centre of the sphere. Find the volume of the region lying inside both the sphere and the cone.

24. A solid has a circular base of radius r and a vertical wall. Its top is a plane inclined at an angle to the horizontal. If the lowest and highest points on the top are at heights a and b, respectively, above the base, find the volume of the solid.

25. The region R bounded by $y = e^{-x}$ and $y = 0$ and lying to the right of $x = 0$ is rotated (a) about the x-axis, (b) about the y-axis. Find the volume of the solid of revolution generated in each case.

26. The region R bounded by $y = x^{-k}$ and $y = 0$ and lying to the right of $x = 1$ is rotated about the x-axis. Find all real values of k for which the solid so generated has finite volume.

27. Repeat the previous exercise with rotation about the y-axis.

28.*Find the volume enclosed by the ellipsoid

$$\frac{x^2}{a^2} + \frac{y^2}{b^2} + \frac{z^2}{c^2} = 1.$$

(*Hint:* this is not a solid of revolution. As in Example 8.2.4, the z-axis is perpendicular to the plane of the x- and y-axes. Each plane $z = k(-c \le k \le c)$ intersects the ellipsoid in an ellipse $(x/a)^2 + (y/b)^2 = 1 - (k/c)^2$. Thus $dV = dz\times$ the area of this ellipse. The area of the ellipse $(x/a)^2 + (y/b)^2 = 1$ is πab.)

29.*A 45° notch is cut to the centre of a cylindrical log having radius 20 cm as shown in Fig. 8.2.18. One plane face of the notch is perpendicular to the axis of the log. What volume of wood was removed from the log by cutting the notch?

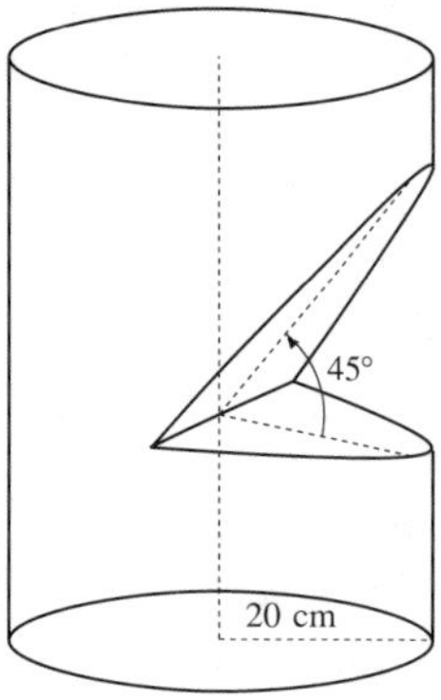

FIGURE 8.2.18

30.*The axes of two circular cylinders intersect at right angles. If the radii of the cylinders are a and b $(a > b > 0)$, show that the region lying inside both cylinders has volume

$$V = 8\int_0^b \sqrt{b^2 - z^2}\sqrt{a^2 - z^2}\,dz$$

(*Hint:* Review Example 8.2.4. Try to make a similar diagram, showing only one eighth of the region. The integral is not easily evaluated.)

31.*Given that the surface area of a sphere of radius r is kr^2 for some constant k, express the volume of a ball of radius R as a "sum" of volume elements that are the volumes of spherical shells of varying radii and infinitesimal thickness dr. Hence find k.

The following problems are *very difficult*. You will need some ingenuity and a lot of hard work to solve them by the techniques at our disposal.

32.*The finite plane region bounded by the curve $xy = 1$ and the straight line $2x + 2y = 5$ is rotated about that line to generate a solid of revolution. Find the volume of that solid.

33.*A wine glass in the shape of a right-circular cone of height h and semi-vertical angle α (see Fig. 8.2.19) is filled with wine. Slowly a ball is lowered into the glass, displacing wine and causing it to overflow. Find the radius R of the ball which causes the greatest volume of wine to overflow out of the glass.

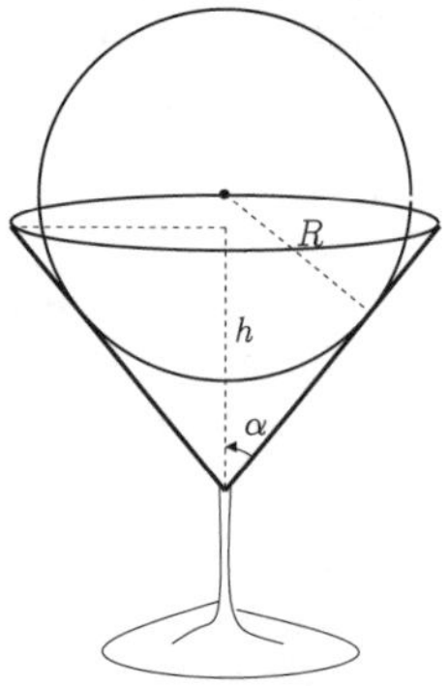

FIGURE 8.2.19

8.3 ARC LENGTH AND SURFACE AREA

In this section we consider how integrals can be used to express the lengths of curves and the areas of the surfaces of solids of revolution.

Arc Length

If A and B are two points in the plane, let $|AB|$ denote the distance between A and B, that is, the length of the straight line segment AB.

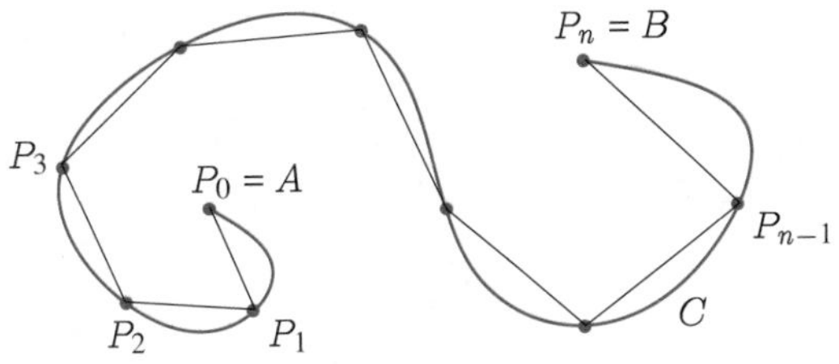

FIGURE 8.3.1

Given a curve C joining two points A and B, we can form a polygonal line $P_0P_1P_2 \cdots P_{n-1}P_n$ by choosing points $A = P_0, P_1, P_2, \ldots, P_{n-1}$ and $P_n = B$ in order along the curve, as shown in Fig. 8.3.1. The lengths

$$L_n = |P_0P_1| + |P_1P_2| + \cdots + |P_{n-1}P_n| = \sum_{k=1}^{n} |P_{k-1}P_k|$$

of such polygonal lines approximate the length of the curve and generally increase as n increases provided the lengths of all the individual segments $P_{k-1}P_k$ decrease toward 0. If L_n has a limit under these circumstances, we call that limit the arc length of the curve C. Arc length is frequently denoted s.

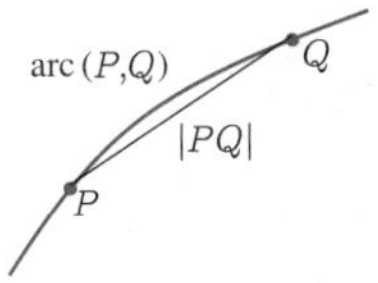

FIGURE 8.3.2

It is possible to construct continuous curves that are bounded (they do not go off to infinity anywhere) but do not have finite arc lengths. To avoid such pathological examples, we assume that our curves are **smooth**; we mean by this that they have continuously turning tangent lines. "Very short" pieces of a smooth curve are "almost straight." This condition suffices to guarantee finite arc length between any two points on the curve. It is intuitively clear that if P and Q are points on a smooth curve, and arc (P, Q) denotes the arc length along the curve from P to Q (Fig. 8.3.2), then

8.3.1
Condition for a Smooth Curve

$$\lim_{Q \to P} \frac{\text{arc}\,(P, Q)}{|PQ|} = 1.$$

Now suppose that C is the graph of a function f that has a continuous derivative f' on $[a,b]$. Then C certainly has a continuously turning tangent line there. Let P_x denote the point on C having coordinates $(x, f(x))$. See Fig. 8.3.3. If $s(x)$ denotes the arc length along C from $A = P_a$ to P_x, then

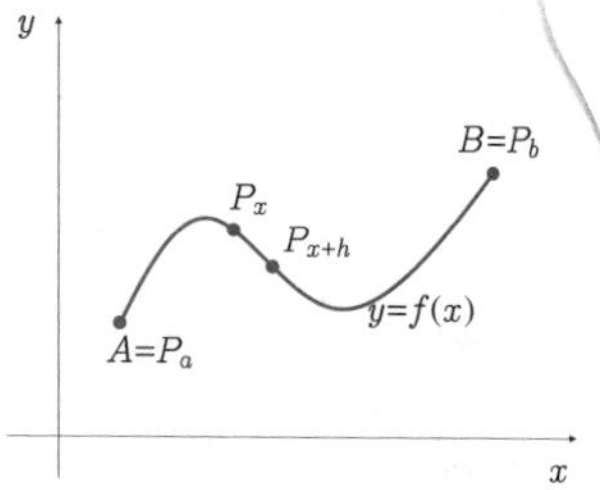

FIGURE 8.3.3

$$\begin{aligned}\frac{s(x+h)-s(x)}{h} &= \frac{\operatorname{arc}(P_{x+h},P_x)}{h} = \frac{\operatorname{arc}(P_{x+h},P_x)}{|P_{x+h}P_x|}\,\frac{|P_{x+h}P_x|}{h}\\ &= \frac{\operatorname{arc}(P_{x+h},P_x)}{|P_{x+h}P_x|}\,\frac{\sqrt{h^2+(f(x+h)-f(x))^2}}{h}\\ &= \frac{\operatorname{arc}(P_{x+h},P_x)}{|P_{x+h}P_x|}\sqrt{1+\left(\frac{f(x+h)-f(x)}{h}\right)^2}\\ &\to (1)\sqrt{1+(f'(x))^2} \quad \text{as } h\to 0.\end{aligned}$$

Thus $ds/dx = \sqrt{1+(f'(x))^2}$. Since $s(a)=0$, the arc length s of C is $s(b)$:

8.3.2
Arc Length of $y = f(x)$

$$s = \int_a^b \sqrt{1+(f'(x))^2}\,dx.$$

The formula for ds/dx can be written in the form $(ds/dx)^2 = 1 + (dy/dx)^2$ or, in terms of differentials,

$$(ds)^2 = (dx)^2 + (dy)^2.$$

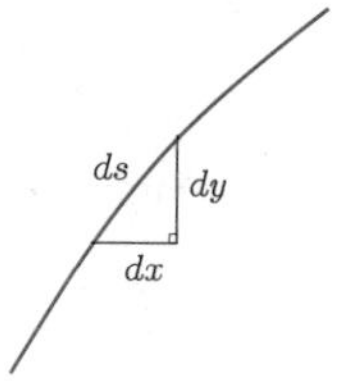

FIGURE 8.3.4

This formula is easily remembered in association with the "differential triangle" in Fig. 8.3.4, which provides a useful mnemonic device for obtaining the formula for s:

$$s = \int_{x=a}^{x=b} ds = \int_a^b \frac{ds}{dx}\,dx = \int_a^b \sqrt{1+\left(\frac{dy}{dx}\right)^2}\,dx.$$

EXAMPLE 8.3.3 Find the length of the curve $y = x^{2/3}$ from $x = 1$ to $x = 8$.

SOLUTION Since $dy/dx = \frac{2}{3}x^{-1/3}$, the length of the curve is given by

$$\begin{aligned}s &= \int_1^8 \sqrt{1+\frac{4}{9}x^{-2/3}}\,dx\\ &= \int_1^8 \frac{\sqrt{9x^{2/3}+4}}{3x^{1/3}}\,dx \qquad \text{Let } u = 9x^{2/3}+4,\\ &\qquad\qquad\qquad\qquad\qquad du = 6x^{-1/3}\,dx\\ &= \frac{1}{18}\int_{13}^{40} u^{1/2}\,du = \frac{1}{27}u^{3/2}\Big|_{13}^{40} = \frac{40\sqrt{40}-13\sqrt{13}}{27} \text{ units.}\end{aligned}$$

EXAMPLE 8.3.4 Find the length of the curve $y = x^4 + \dfrac{1}{32x^2}$ from $x = 1$ to $x = 2$.

SOLUTION Here $\dfrac{dy}{dx} = 4x^3 - \dfrac{1}{16x^3}$ and

$$1 + \left(\frac{dy}{dx}\right)^2 = 1 + (4x^3)^2 - \frac{1}{2} + \left(\frac{1}{16x^3}\right)^2$$
$$= (4x^3)^2 + \frac{1}{2} + \left(\frac{1}{16x^3}\right)^2 = \left(4x^3 + \frac{1}{16x^3}\right)^2.$$

Hence the length of the curve is

$$s = \int_1^2 \left(4x^3 + \frac{1}{16x^3}\right) dx = \left(x^4 - \frac{1}{32x^2}\right)\Bigg|_1^2 = 16 - \frac{1}{128} - \left(1 - \frac{1}{32}\right)$$
$$= 15 + \frac{3}{128} \text{ units.}$$

The examples above are deceptively simple; the curves were chosen in such a way that the arc length integrals were easily evaluated. For instance, the number 32 in the curve in Example 8.3.4 was chosen just so the expression $1 + (dy/dx)^2$ would turn out to be a perfect square and its square root would cause no problems. Because of the square root in the formula, arc length problems for most curves lead to integrals that can only be evaluated by numerical techniques.

EXAMPLE 8.3.5 Flat rectangular sheets of metal 2 m wide are to be formed into corrugated roofing panels 2 m wide by bending them into the sinusoidal shape shown in Fig. 8.3.5. The period of the cross-sectional sine curve is 20 cm. Its amplitude is 5 cm, so the panel is 10 cm thick. How long should the flat sheets be cut if the resulting panels must be 5 m long?

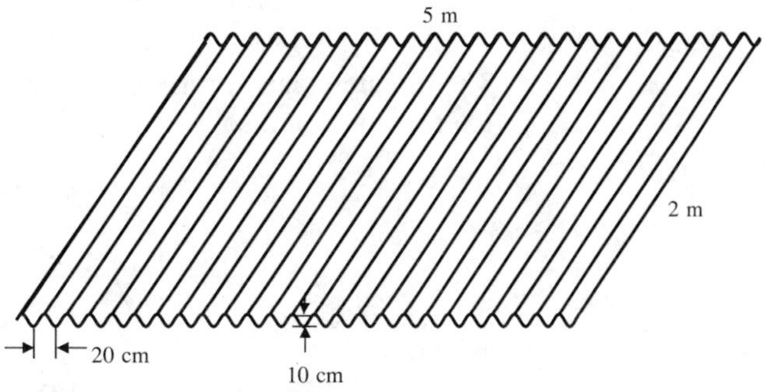

FIGURE 8.3.5

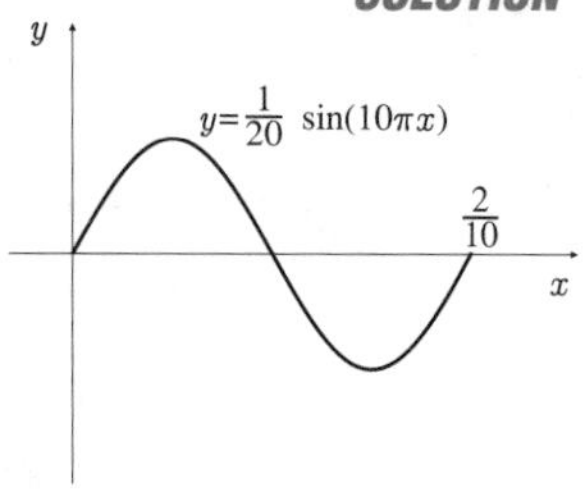

FIGURE 8.3.6

SOLUTION One period of the sinusoidal cross-section is shown in Fig. 8.3.6. The distances are in metres; the curve has equation

$$y = \frac{1}{20}\sin(10\pi x).$$

Evidently 25 periods are required to produce 5 m long panels. The length of the flat sheets required is 25 times the length of one period of the sine curve:

$$\begin{aligned} s &= 25\int_0^{2/10}\sqrt{1+\left(\frac{\pi}{2}\cos(10\pi x)\right)^2}\,dx \qquad \begin{aligned}&\text{Let } t = 10\pi x,\\ &dt = 10\pi\,dx\end{aligned}\\ &= \frac{5}{2\pi}\int_0^{2\pi}\sqrt{1+\frac{\pi^2}{4}\cos^2 t}\,dt = \frac{10}{\pi}\int_0^{\pi/2}\sqrt{1+\frac{\pi^2}{4}\cos^2 t}\,dt. \end{aligned}$$

The integral can be evaluated numerically by any of the standard methods of Chapter 7; we obtain $s \approx 7.32$. The flat metal sheets should be about 7.32 m long to yield 5 m long finished panels.

EXAMPLE 8.3.6 Find the circumference of the ellipse $\frac{x^2}{a^2}+\frac{y^2}{b^2} = 1$, where $a \ge b > 0$. See Fig. 8.3.7.

SOLUTION The upper half of the ellipse has equation $y = b\sqrt{1-\frac{x^2}{a^2}} = \frac{b}{a}\sqrt{a^2-x^2}$. Hence

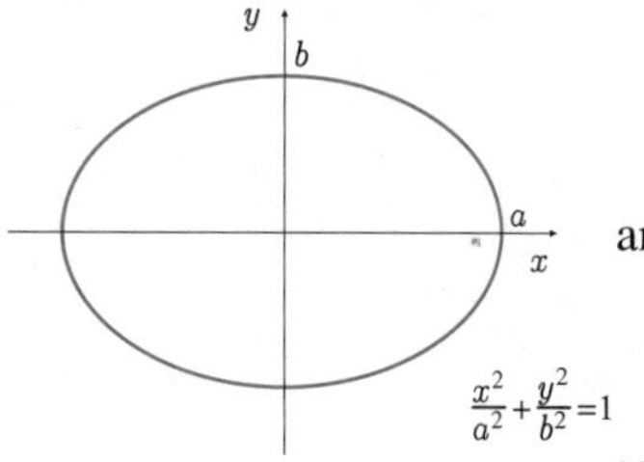

FIGURE 8.3.7

$$\frac{dy}{dx} = -\frac{b}{a}\frac{x}{\sqrt{a^2-x^2}}$$

and so

$$\begin{aligned} 1+\left(\frac{dy}{dx}\right)^2 &= 1+\frac{b^2}{a^2}\frac{x^2}{a^2-x^2}\\ &= \frac{a^2-\dfrac{a^2-b^2}{a^2}x^2}{a^2-x^2}. \end{aligned}$$

The circumference of the ellipse is four times the arc length of that part lying in the first quadrant, so

$$\begin{aligned} s &= 4\int_0^a \frac{\sqrt{a^2-\dfrac{a^2-b^2}{a^2}x^2}}{\sqrt{a^2-x^2}}\,dx \qquad \begin{aligned}&\text{Let } x = a\sin t,\\ &dx = a\cos t\,dt\end{aligned}\\ &= 4\int_0^{\pi/2}\sqrt{a^2-(a^2-b^2)\sin^2 t}\,dt\\ &= 4a\int_0^{\pi/2}\sqrt{1-\epsilon^2\sin^2 t}\,dt \text{ units,} \end{aligned}$$

where $\epsilon = (\sqrt{(a^2 - b^2)})/a$ is the *eccentricity* of the ellipse. (Refer to Section 9.1 for a discussion of ellipses.) Note that $0 < \epsilon < 1$. The function $E(\epsilon)$ defined by

$$E(\epsilon) = \int_0^{\pi/2} \sqrt{1 - \epsilon^2 \sin^2 t}\, dt$$

is called the **complete elliptic integral of the second kind**. The integral cannot be evaluated by elementry techniques for general ϵ, though numerical methods can be applied to find approximate values for any given value of ϵ. Tables of values of $E(\epsilon)$ for various values of ϵ can be found in collections of mathematical tables. As shown above, the circumference of the ellipse is given by $4aE(\epsilon)$. Note that for $a = b$ we have $\epsilon = 0$, and the formula returns the circumference of a circle; $s = 4a(\pi/2) = 2\pi a$ units.

Areas of Surfaces of Revolution

When a plane curve is rotated (in three dimensions) about a line in the plane of the curve, it sweeps out a **surface of revolution**. For instance, a sphere of radius a is generated by rotating a semicircle of radius a about the diameter of that semicircle. The area of a surface of revolution can be found by integrating differential area elements dS constructed by rotating about the given line differential elements ds of arc length along the curve. If the radius of rotation of an arc length element ds is r, then

$$dS = 2\pi r\, ds,$$

as illustrated in Fig. 8.3.8.

8.3.7
Surface Area of Revolution

If the smooth graph $y = f(x)$, $(a \le x \le b)$, is rotated about the x-axis, the surface area generated is given by

$$S = 2\pi \int_{x=a}^{x=b} |y|\, ds = 2\pi \int_a^b |f(x)|\sqrt{1 + (f'(x))^2}\, dx.$$

If the rotation is about the y-axis, the surface area is

$$S = 2\pi \int_{x=a}^{x=b} |x|\, ds = 2\pi \int_a^b |x|\sqrt{1 + (f'(x))^2}\, dx.$$

EXAMPLE 8.3.8 Find the area of the surface of a sphere of radius a.

SOLUTION Such a sphere can be generated by rotating the semicircle with equation $y = \sqrt{a^2 - x^2}$, $(-a \le x \le a)$, about the x-axis. See Fig. 8.3.9. Since

$$\frac{dy}{dx} = -\frac{x}{\sqrt{a^2 - x^2}} = -\frac{x}{y},$$

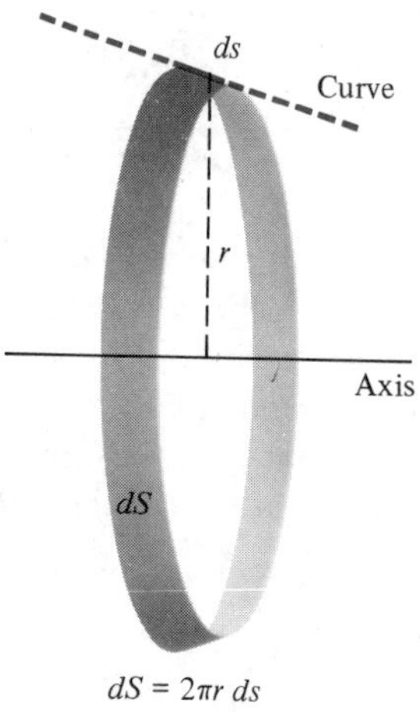

FIGURE 8.3.8

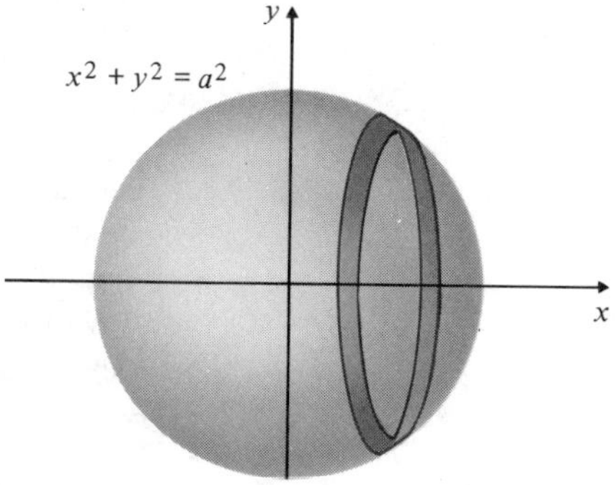

FIGURE 8.3.9

the area of the sphere is given by

$$S = 2\pi \int_{-a}^{a} y\sqrt{1 + \left(\frac{x}{y}\right)^2}\, dx = 4\pi \int_0^a \sqrt{y^2 + x^2}\, dx$$
$$= 4\pi a \int_0^a dx = 4\pi a x\Big|_0^a = 4\pi a^2 \text{ sq units.}$$

EXAMPLE 8.3.9 Find the surface area of a parabolic reflector whose shape is obtained by rotating the parabolic arc $y = x^2$, $(0 \le x \le 1)$, about the y-axis, as illustrated in Fig. 8.3.10.

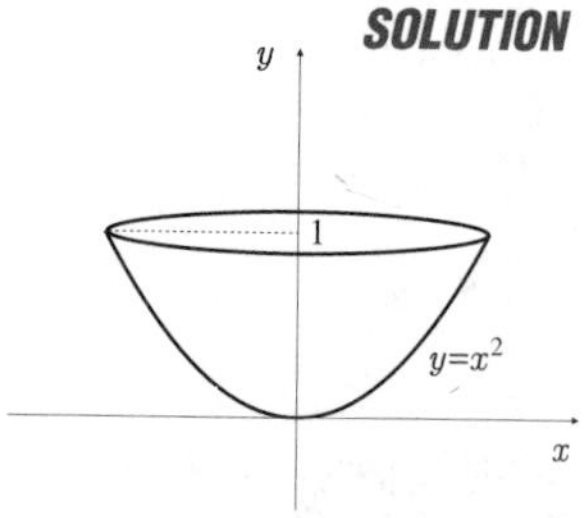

FIGURE 8.3.10

SOLUTION The area is

$$S = 2\pi \int_0^1 x\sqrt{1 + 4x^2}\, dx \qquad \text{Let } u = 1 + 4x^2,\ du = 8x\, dx$$
$$= \frac{\pi}{4} \int_1^5 u^{1/2}\, du$$
$$= \frac{\pi}{6} u^{3/2}\Big|_1^5 = \frac{\pi}{6}(5\sqrt{5} - 1) \text{ sq units.}$$

EXERCISES

In Exercises 1–10 find the lengths of the given curves.

1. $y = ax + b$ from $x = A$ to $x = B$

2. $y = x^{3/2}$ from $x = 1$ to $x = 9$

3. $y = \frac{x^3}{12} + \frac{1}{x}$ from $x = 1$ to $x = 4$

4. $y = \frac{e^x + e^{-x}}{2}$ $(= \cosh x)$ from $x = 0$ to $x = a$

5. $y = \ln(1 - x^2)$ from $x = -(1/2)$ to $x = 1/2$

6. $y = \ln \cos x$ from $x = \pi/6$ to $x = \pi/4$

7. $y = x^2$ from $x = 0$ to $x = 2$

8. $y = x^2 - \frac{\ln x}{8}$ from $x = 1$ to $x = 2$

9.* $y = \ln \frac{e^x - 1}{e^x + 1}$ from $x = 2$ ro $x = 4$

10.* $y = \ln x$ from $x = 1$ to $x = e$

11. Find the circumference of the closed curve $x^{2/3} + y^{2/3} = a^{2/3}$.

Use numerical methods to find the lengths of the curves in Exercises 12–13 to four decimal places.

12. $y = x^4$ from $x = 0$ to $x = 1$

13. $y = x^{1/3}$ from $x = 1$ to $x = 2$

In Exercises 14–19 find the areas of the surfaces obtained by rotating the given curve about the indicated lines.

14. $y = x^2$, $(0 \le x \le 2)$, about the y-axis

15. $y = x^{3/2}$, $(0 \le x \le 1)$, about the y-axis

16. $y = e^x$, $(0 \le x \le 1)$, about the x-axis

17. $y = \sin x$, $(0 \le x \le \pi)$, about the x-axis

18. $y = \frac{x^3}{12} + \frac{1}{x}$, $(1 \le x \le 4)$, about the x-axis

19. $y = \frac{x^3}{12} + \frac{1}{x}$, $(1 \le x \le 4)$, about the y-axis

20. By rotating the line segment $y = hx/r$, $(0 \le x \le r)$, about the y-axis, find the area of the curved surface of a right-circular cone of base radius r and height h.

21. Find the surface area of the torus (doughnut) obtained by rotating the circle $x^2 + y^2 = a^2$ about the line $x = b > a$.

22. Find the area of the surface obtained by rotating the ellipse $x^2 + 4y^2 = 4$ about the x-axis.

23. Find the area of the surface obtained by rotating the ellipse $x^2 + 4y^2 = 4$ about the y-axis.

24. The ellipse of Example 8.3.6 is rotated about the line $y = c > b$ to generate a doughnut with elliptical cross-sections. Express the surface area of this doughnut in terms of the complete elliptic integral function $E(\epsilon)$ introduced in that example.

25. A hollow container in the shape of an infinitely long horn is generated by rotating the curve $y = 1/x$, $(1 \le x < \infty)$, about the x-axis.

    a) Find the volume of the container.

    b) Show that the container has infinite surface area.

    c) How do you explain the "paradox" that the container can be filled with a finite volume of paint but requires infinitely much paint to cover its surface?

26. The curve $y = \ln x$, $(0 < x \le 1)$, is rotated about the y-axis. Find the area of the area of the horn-shaped surface so generated.

27. For what real values of k does the surface generated by rotating the curve $y = x^k$, $(0 < x \le 1)$, about the y-axis have finite surface area?

28. If two parallel planes intersect a sphere, show that the surface area of that part of the sphere lying between the two planes depends only on the radius of the sphere and the distance between the planes and not on the position of the planes.

8.4 MASS, MOMENTS, AND CENTRE OF MASS

If a solid object is made of a homogeneous material, we would expect different parts of the solid that have the same volume to have the same mass as well. We express this homogeneity by saying that the object has constant density, that density being the mass divided by the volume for the whole object or for any part of it. Thus a rectangular brick with dimensions 20 cm, 10 cm, and 8 cm would have volume $V = 20 \times 10 \times 8 = 1600$ cm^3, and if it was made of material having constant density $\rho = 3$ g/cm^3, it would have mass $m = \rho V = 3 \times 1600 = 4800$ g. (Here ρ is the Greek letter "rho" – pronounced "roh.")

If the density of the material constituting a solid object is not constant but varies from point to point in the object, no such simple relationship exists between mass and volume. If the density $\rho = \rho(P)$ is a *continuous* function of position P we could subdivide the solid into many small volume elements and, by regarding ρ as approximately constant over each such element, determine the masses of all the elements and add them up to get the mass of the solid. The mass Δm of a volume element ΔV containing the point P would satisfy

$$\Delta m \approx \rho(P)\,\Delta V,$$

so the mass m of the solid can be approximated:

$$m = \sum \Delta m \approx \sum \rho(P)\,\Delta V.$$

Such approximations become exact as we pass to the limit of differential elements $dm = \rho(P)\,dV$, so we expect to be able to calculate masses as integrals, that is, as the limits of such sums:

$$m = \int dm = \int \rho(P)\,dV.$$

EXAMPLE 8.4.1 A solid vertical cylinder of height H cm and base area A cm^2 has density given by $\rho = \rho_0(1+h)$ g/cm^3 where h is the height in centimeters above the base and ρ_0 is a constant. Find the mass of the cylinder.

SOLUTION See Fig. 8.4.1. A slice of the solid at height h above the base and having thickness dh is a cylindrical disc of volume $dV = A\,dh$. The mass of this volume element is

$$dm = \rho\,dV = \rho_0(1+h)\,A\,dh.$$

Therefore the mass of the whole cylinder is

$$V = \int_0^H \rho_0 A(1+h)\,dh = \rho_0 A\left(H + \frac{H^2}{2}\right) \text{ g}.$$

EXAMPLE 8.4.2 A spherical planet of radius R km has density varying with distance r from the centre according to the formula

$$\rho = \frac{\rho_0}{1+r^2} \text{ kg/km}^3.$$

Find the mass of the planet.

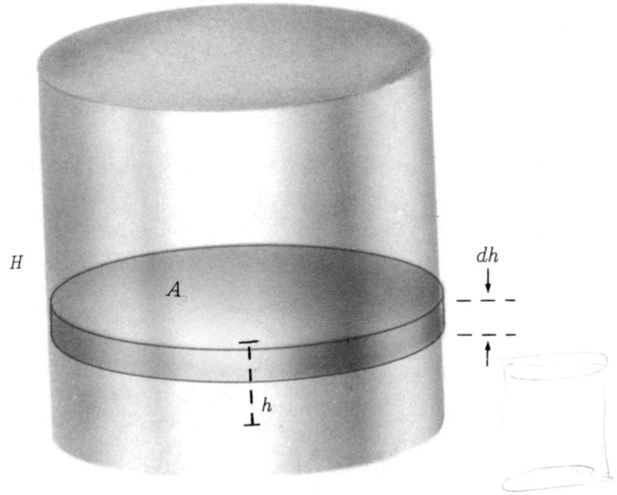

FIGURE 8.4.1

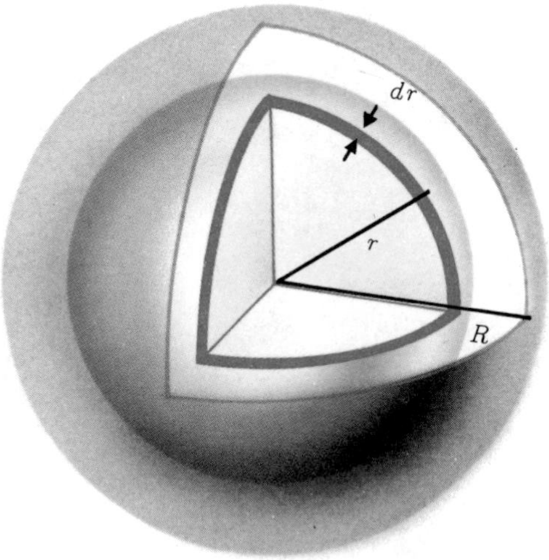

FIGURE 8.4.2

SOLUTION A spherical shell of thickness dr and radius r (see Fig. 8.4.2) has volume

$$dV = 4\pi r^2\, dr$$

and so has mass

$$dm = \rho\, dV = 4\pi\rho_0 \frac{r^2}{1+r^2}\, dr.$$

Hence the mass of the planet is

$$\begin{aligned} m = 4\pi\rho_0 \int_0^R \frac{r^2}{1+r^2}\, dr &= 4\pi\rho_0 \int_0^R \left(1 - \frac{1}{1+r^2}\right) dr \\ &= 4\pi\rho_0 (r - \tan^{-1} r)\Big|_0^R \\ &= 4\pi\rho_0 (R - \tan^{-1} R) \text{ kg}. \end{aligned}$$

Similar techniques can be applied to find masses of one- and two-dimensional objects such as wires and thin plates that have variable densities.

EXAMPLE 8.4.3 A wire of variable composition is stretched along the x-axis from $x = 0$ to $x = L$ cm. Find the mass of the wire if the density at position x is $\rho(x) = x(L - x)$ g/cm.

SOLUTION An element of the wire of length dx at position x has mass $dm = \rho(x)\, dx = (Lx - x^2)\, dx$. Thus the mass of the wire is

$$m = \int_0^L (Lx - x^2)\, dx = \left(L\frac{x^2}{2} - \frac{x^3}{3}\right)\Big|_0^L = \frac{L^3}{6} \text{ g}.$$

EXAMPLE 8.4.4 Find the mass of a disc of radius a cm having centre at the origin of the xy-plane if the density at position (x, y) is $\sigma = k(2a + x)$ g/cm^2.

SOLUTION A vertical strip of thickness dx at x has area $dA = 2\sqrt{a^2 - x^2}\,dx$ (see Fig. 8.4.3), and so has mass

$$dm = \sigma\,dA = 2k(2a + x)\sqrt{a^2 - x^2}\,dx.$$

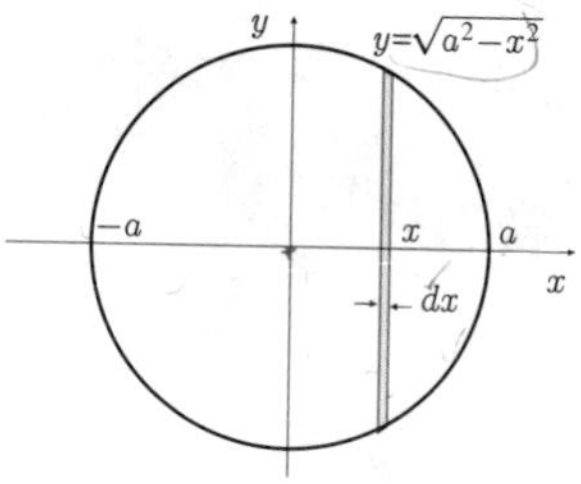

FIGURE 8.4.3

Hence the mass of the disc is

$$\begin{aligned} m &= \int_{x=-a}^{x=a} dm = 2k\int_{-a}^{a} (2a + x)\sqrt{a^2 - x^2}\,dx \\ &= 4ak\int_{-a}^{a} \sqrt{a^2 - x^2}\,dx + 2k\int_{-a}^{a} x\sqrt{a^2 - x^2}\,dx \\ &= 4ak\frac{\pi a^2}{2} + 0 = 2\pi k a^3 \text{ g.} \end{aligned}$$

(We used the area of a semicircle to evaluate the first integral. The second integral is zero because the integrand is odd and the interval is symmetric about $x = 0$.)

Observe that in Example 8.4.3 the density ρ was specified in units of mass per unit length, that is, it was a **line density**, and the mass was determined from $dm = \rho\,dx$. In Example 8.4.4 the given density was an **areal density** (mass per unit area), so the mass was determined from $dm = \rho\,dA$. Distributions of mass along one-dimensional structures (lines or curves) necessarily lead to integrals of functions of one variable, but distributions of mass on a surface or in space can lead to integrals involving functions of more than one variable. Such integrals are studied in courses on advanced (multivariable) calculus. In Examples 8.4.1, 8.4.2, and 8.4.4 above, the given densities were functions of only one variable, so these problems, though higher dimensional in nature, led to integrals of functions of only one variable and could be solved by the methods at hand.

Moments and Centres of Mass

A mass m located at position x on the x-axis is said to have **moment** xm about the point $x = 0$ or, more generally, moment $(x - x_0)m$ about the point $x = x_0$. If several masses $m_1, m_2, m_3, \ldots, m_n$ are located at the points $x_1, x_2, x_3, \ldots, x_n$, respectively, then the total moment of the system of masses about the point $x = x_0$ is the sum of the individual moments (see Fig. 8.4.4):

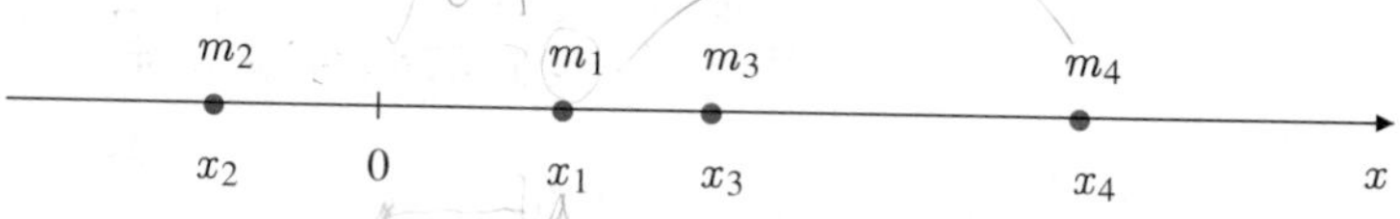

FIGURE 8.4.4

$$M_{x=x_0} = (x_1 - x_0)m_1 + (x_2 - x_0)m_2 + \cdots + (x_n - x_0)m_n = \sum_{j=1}^{n}(x_j - x_0)m_j.$$

The **centre of mass** of the system of masses is the point $\bar{x}$ about which the system has **moment zero**. Thus

$$0 = \sum_{j=1}^{n}(x_j - \bar{x})m_j = \sum_{j=1}^{n} x_j m_j - \bar{x}\sum_{j=1}^{n} m_j,$$

and the centre of mass is therefore given by

$$\bar{x} = \frac{\sum_{j=1}^{n} x_j m_j}{\sum_{j=1}^{n} m_j} = \frac{M_{x=0}}{m},$$

where m is the total mass of the system and $M_{x=0}$ is the total moment about $x = 0$. If you think of the x-axis as being a weightless wire supporting the masses, $\bar{x}$ is the point at which the wire could be suspended and remain in perfect balance (equilibrium), not tipping either way. Even if the axis represents a nonweightless support, say a seesaw, supported at $x = \bar{x}$, it will remain balanced after the masses are added, provided it was balanced before they were placed on.

Now suppose that a one-dimensional distribution of mass with continuously variable line density $\rho(x)$ lies along the interval $[0, a]$ of the x-axis. An element of length dx at position x contains mass $dm = \rho(x)\,dx$ and so has moment $dM_{x=0} = x\,dm = x\rho(x)\,dx$ about the origin. The total moment about the origin is

$$M_{x=0} = \int_0^a x\rho(x)\,dx.$$

Since the total mass is

$$m = \int_0^a \rho(x)\,dx,$$

the centre of mass is given by

8.4.5
Centre of Mass for a Line Distribution

$$\bar{x} = \frac{M_{x=0}}{m} = \frac{\int_0^a x\rho(x)\,dx}{\int_0^a \rho(x)\,dx}$$

m₁x₁ + m₂x₂ + ...

mass of wire

EXAMPLE 8.4.6 At what point can the wire of Example 8.4.3 be suspended so that it will balance?

SOLUTION In Example 8.4.3 we evaluated the mass of the wire to be $L^3/6$ g. Its moment about $x = 0$ is

$$M_{x=0} = \int_0^L x\rho(x)\,dx = \int_0^L x^2(L-x)\,dx = \left(\frac{x^3L}{3} - \frac{x^4}{4}\right)\Bigg|_0^L = \frac{L^4}{12}\ \text{g}\cdot\text{cm}.$$

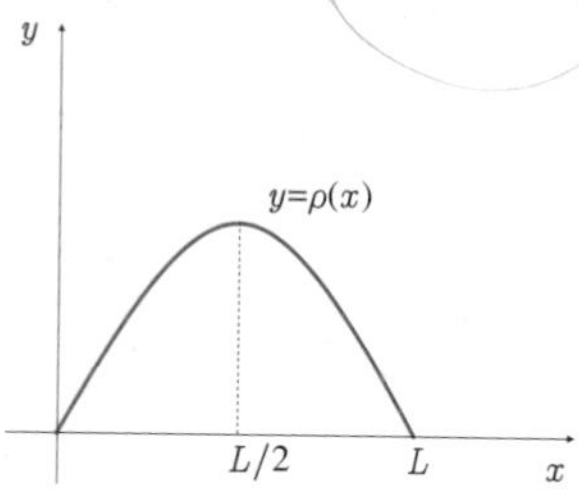

FIGURE 8.4.5

Thus the centre of mass is $\bar{x} = (L^4/12)/(L^3/6) = L/2$. This wire will balance at its midpoint, at distance $L/2$ cm from the end $x = 0$, as shown in Fig. 8.4.5. The centre of mass turned out to be the midpoint because the mass, though varying in density from point to point, is neverless distributed symmetrically about the midpoint $x = L/2$.

Such symmetry can frequently be useful in determining centres of mass without doing unnecessary integrals.

EXAMPLE 8.4.7 Determine the centre of mass of a wire of length L lying on the interval $[0, L]$ if the density at point x is $\rho(x) = x$.

SOLUTION Here the mass is not distributed symmetrically; the density increases as we move to the right along the wire. Thus we expect to centre of mass to be closer to the right end than the left. The mass of the wire is

$$m = \int_0^L x\,dx = \frac{L^2}{2},$$

and the moment about the origin is

$$M_{x=0} = \int_0^L x^2\,dx = \frac{L^3}{3}.$$

Hence the centre of mass is located at $\bar{x} = M_{x=0}/m = 2L/3$, two-thirds of the way along the wire from the left end.

Two- and Three-Dimensional Examples

The systems of mass considered in Examples 8.4.6 and 8.4.7 are all one-dimensional and lie along straight lines. If mass is distributed in a plane or in space, similar considerations prevail. For a system of masses m_1 at (x_1, y_1), m_2 at (x_2, y_2), ..., m_n at (x_n, y_n), we define the **moment about** $x = 0$, (that is, about the y-axis), to be

$$M_{x=0} = x_1m_1 + x_2m_2 + \cdots + x_nm_n = \sum_{j=1}^{n} x_jm_j,$$

and **the moment about** $y = 0$ (that is, about the x-axis), to be

$$M_{y=0} = y_1m_1 + y_2m_2 + \cdots + y_nm_n = \sum_{j=1}^{n} y_jm_j.$$

The centre of mass is the point $(\bar{x}, \bar{y})$ where

$$\bar{x} = \frac{M_{x=0}}{m} = \frac{\sum_{j=1}^n x_jm_j}{\sum_{j=1}^n m_j}, \qquad \bar{y} = \frac{M_{y=0}}{m} = \frac{\sum_{j=1}^n y_jm_j}{\sum_{j=1}^n m_j}.$$

For continuous distributions of mass, the sums become appropriate integrals.

EXAMPLE 8.4.8 Find the centre of mass of a rectangular plate occupying the region $0 \le x \le a$, $0 \le y \le b$ if the areal density of the material in the plate at position (x, y) is kx.

SOLUTION Since the density is independent of y and the rectangle is symmetric about the line $y = b/2$, the y-coordinate of the centre of mass must be $\bar{y} = b/2$.

A thin vertical strip of width dx at position x (see Fig. 8.4.6) has mass $dm = bkx\,dx$. The moment of this strip about $x = 0$ is $dM_{x=0} = x\,dm = kbx^2\,dx$. Hence the mass and moment about $x = 0$ of the whole plate are

$$m = kb\int_0^a x\,dx = \frac{kba^2}{2}$$

$$M_{x=0} = kb\int_0^a x^2\,dx = \frac{kba^3}{3}.$$

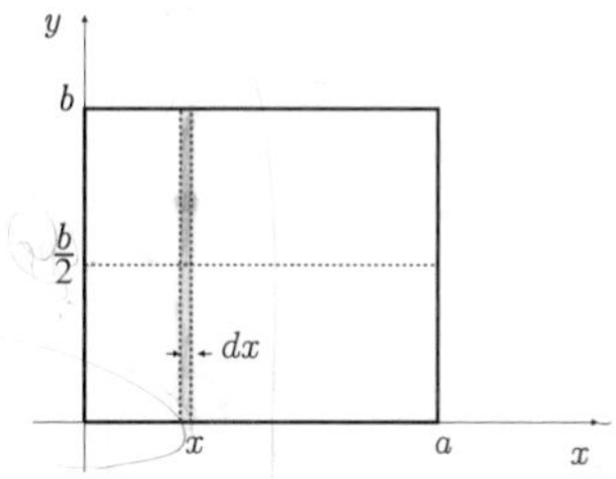

FIGURE 8.4.6

Hence $\bar{x} = M_{x=0}/m = 2a/3$, and the centre of mass of the plate is $(2a/3, b/2)$. The plate would be balanced if supported at this point.

For distributions of mass in three-dimentional space one defines, analogously, the moments $M_{x=0}$, $M_{y=0}$, and $M_{z=0}$ of the system of mass about the planes $x = 0$, $y = 0$, and $z = 0$, respectively. For a discrete system of masses m_1 at (x_1, y_1, z_1), ..., m_n at (x_n, y_n, z_n), we have

$$M_{x=0} = x_1m_1 + x_2m_2 + \cdots + x_nm_n,$$

with similar formulas for $M_{y=0}$ and $M_{z=0}$. The centre of mass is $(\bar{x}, \bar{y}, \bar{z})$ where

$$\bar{x} = \frac{M_{x=0}}{m}, \qquad \bar{y} = \frac{M_{y=0}}{m}, \qquad \bar{z} = \frac{M_{z=0}}{m},$$

m being the total mass: $m = m_1 + m_2 + \cdots + m_n$.

EXAMPLE 8.4.9 Find the centre of mass of a solid hemisphere of radius R if the density at height h above the base plane of the hemisphere is $\rho_0 h$.

SOLUTION The solid is symmetric about the vertical axis (let us call it the z-axis), and the density is constant in planes perpendicular to this axis. Therefore the centre of mass must lie somewhere on this axis. A slice of the solid at height z above the base, and having thickness dz, is a disc of radius $\sqrt{R^2 - z^2}$ (see Fig. 8.4.7). Its volume is $dV = \pi(R^2 - z^2)\,dz$, and its mass is $dm = \rho_0 z\,dV = \rho_0\pi(R^2z - z^3)\,dz$. Its moment about the base plane $z = 0$ is $dM_{z=0} = z\,dm = \rho_0\pi(R^2z^2 - z^4)\,dz$. The mass of the solid hemisphere is

$$m = \rho_0\pi\int_0^R (R^2z - z^3)\,dz = \rho_0\pi\left(\frac{R^2z^2}{2} - \frac{z^4}{4}\right)\Bigg|_0^R = \frac{\pi}{4}\rho_0R^4.$$

The moment of the hemisphere about the plane $z = 0$ is

$$M_{z=0} = \rho_0\pi\int_0^R (R^2z^2 - z^4)\,dz = \rho_0\pi\left(\frac{R^2z^3}{3} - \frac{z^5}{5}\right)\Bigg|_0^R = \frac{2\pi}{15}\rho_0R^5.$$

The centre of mass therefore lies along the axis of symmetry of the hemisphere at height $\bar{z} = M_{z=0}/m = 8R/15$ above the base of the hemisphere.

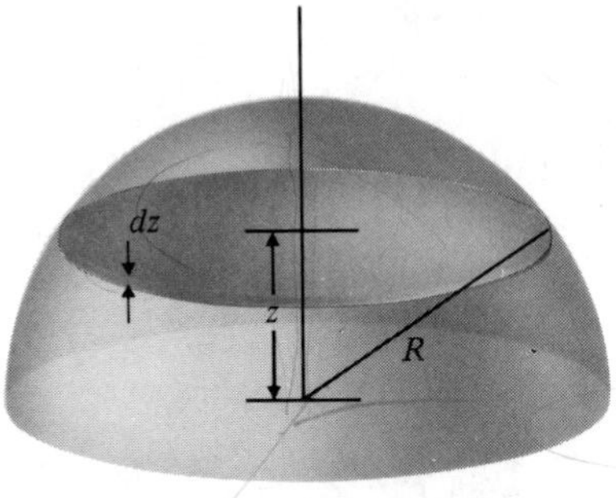

FIGURE 8.4.7

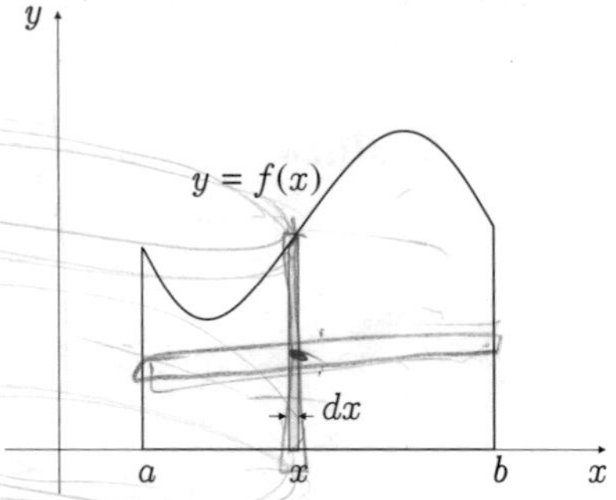

FIGURE 8.4.8

EXAMPLE 8.4.10 Find the centre of mass of a plate occupying the region $a \le x \le b$, $0 \le y \le f(x)$, if the density at any point (x, y) is $\rho(x)$.

SOLUTION The appropriate area element is shown in Fig. 8.4.8. It has area $f(x)\,dx$, mass

$$dm = \rho(x)f(x)\,dx,$$

and moment about $x = 0$

$$dM_{x=0} = x\rho(x)f(x)\,dx.$$

Since the density is constant in this mass element dm, the y-coordinate of *its* centre of mass is at its midpoint: $\bar{y}_{dm} = \frac{1}{2}f(x)$. Therefore the moment of the mass element dm about $y = 0$ is

$$dM_{y=0} = \bar{y}_{dm}\,dm = \frac{1}{2}\,\rho(x)\big(f(x)\big)^2\,dx.$$

Thus the coordinates of the centre of mass of the plate are $\bar{x} = \dfrac{M_{x=0}}{m}$ and $\bar{y} = \dfrac{M_{y=0}}{m}$ where

$$m = \int_a^b \rho(x)f(x)\,dx,$$

$$M_{x=0} = \int_a^b x\rho(x)f(x)\,dx,$$

$$M_{y=0} = \frac{1}{2}\int_a^b \rho(x)\big(f(x)\big)^2\,dx.$$

REMARK: Similar formulas can be obtained if the density depends on y instead of x provided that the region admits a suitable horizontal area element. (For example, the region might be specified by $c \le y \le d$, $0 \le x \le g(y)$.) Finding centres of mass for plates occupying regions specified by functions of x, but where the density depends on y, generally requires "double integrals." Such problems are therefore studied in multi-variable calculus.

EXERCISES

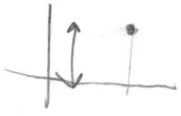

Find the mass and centre of mass for the systems in Exercises 1–16. Be alert for symmetries.

1. A straight wire of length L cm, where the density at distance s cm from one end is $\rho(s) = \sin \pi s/L$ g/cm

2. A straight wire along the x-axis from $x = 0$ to $x = L$ if the density is constant ρ_0, but the cross-sectional radius of the wire varies so that its value at x is $a + bx$.

3. A quarter-circular plate of radius a occupying the region $x^2 + y^2 \le a^2$, $x \ge 0$, $y \ge 0$, having constant areal density ρ_0.

4. A quarter-circular plate of radius a occupying the region $x^2 + y^2 \le a^2$, $x \ge 0$, $y \ge 0$, having areal density $\rho(x) = \rho_0 x$.

5. A plate occupying the region $0 \le y \le 4 - x^2$ if the areal density at (x, y) is ky.

6. A right-triangular plate with legs 2 m and 3 m if the areal density at any point P is $5h$ kg/m^2, h being the distance of P from the shorter leg.

7. A square plate of edge a cm if the areal density at P is kx g/cm^2 where x is the distance from P to one edge of the square.

8. The plate in the previous exercise, but with areal density kr g/cm^2 where r is the distance (in cm) from P to one of the diagonals of the square.

9. A plate of density $\rho(x)$ occupying the region $a \le x \le b$, $f(x) \le y \le g(x)$.

10. A rectangular brick of dimensions 20 cm, 10 cm, and 5 cm if the density at P is kx g/cm^3 where x is the distance from P to one of the 10×5 faces

11. A solid ball of radius R m if the density at P is z kg/m^3 where z is the distance from P to a plane at distance $2R$ m from the centre of the ball

12. A right-circular cone of base radius a cm and height b cm if the density at point P is kz g/cm^3 where z is the distance of P from the base of the cone.

13.*The solid occupying the quarter of a ball of radius a centred at the origin having as base the region $x^2+y^2 \le a^2$, $x \ge 0$ in the xy-plane, if the density at height z above the base is $\rho_0 z$.

14.*The cone of Exercise 12, but with density at P equal to ks g/cm^3 where s is the distance of P from the central axis of the cone. (*Hint:* First find the mass of a disc-shaped slice of the cone perpendicular to the axis of the cone. You will still need an integral to do this.)

15.*A semicircular plate occupying the region $x^2+y^2 \le a^2$, $y \ge 0$, if the density at distance s from the origin is ks g/cm^2.

16.*The wire in Exercise 1 if it is bent in a semicircle.

8.5 CENTROIDS

If matter is distributed uniformly in a system so that the density ρ is constant, then that density cancels out of the numerator and denominator in sum or integral expressions for coordinates of the centre of mass. In such cases the centre of mass depends only on the *shape* of the object, that is, on geometric properties of the region occupied by the object, and we call it the **centroid** of the region.

Centroids are calculated using the same formulas as those used for centres of mass, except that the density (being constant) is taken to be unity, so the mass is just the length, area, or volume of the region, and the moments are referred to as moments of the region rather than of any mass occupying the region. If we set $\rho(x) = 1$ in the formulas obtained in Example 8.4.10 we obtain the following result:

8.5.1
Centroid of a Standard Plane Region

The centroid of the plane region $a \le x \le b$, $0 \le y \le f(x)$, is $(\bar{x}, \bar{y})$, where $\bar{x} = \dfrac{M_{x=0}}{A}$, $\bar{y} = \dfrac{M_{y=0}}{A}$, and

$$A = \int_a^b f(x)\,dx, \qquad M_{x=0} = \int_a^b x f(x)\,dx, \qquad M_{y=0} = \frac{1}{2}\int_a^b \left(f(x)\right)^2 dx.$$

EXAMPLE 8.5.2 Find the centroid of the half-disc $-a \le x \le a$, $0 \le y \le \sqrt{a^2 - x^2}$.

SOLUTION By symmetry the x-coordinate of the centroid is $\bar{x} = 0$. Since $A = \frac{1}{2}\pi a^2$, we hav

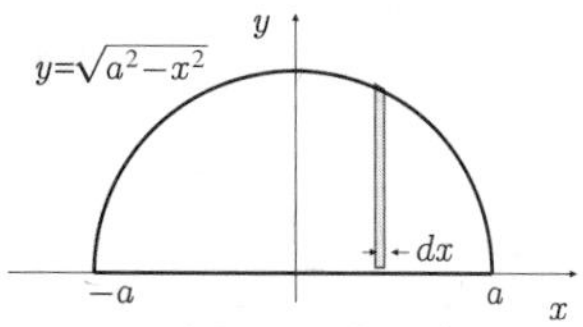

FIGURE 8.5.1

$$\bar{y} = \frac{M_{y=0}}{A} = \frac{2}{\pi a^2}\,\frac{1}{2}\int_{-a}^{a} (a^2 - x^2)\,dx = \frac{2}{\pi a^2}\,\frac{2a^3}{3} = \frac{4a}{3\pi}.$$

The centroid of the half-disc is $\left(0, \dfrac{4a}{3\pi}\right)$.

EXAMPLE 8.5.3 Find the centroid of the semicircle $y = \sqrt{a^2 - x^2}$.

SOLUTION Again $\bar{x} = 0$ by symmetry. A short arc of length ds at height y on the semicirc has moment $dM_{y=0} = y\,ds$ about $y = 0$. Since

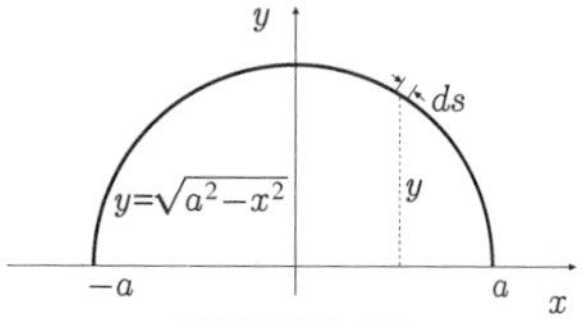

FIGURE 8.5.2

$$ds = \sqrt{1 + \left(\frac{dy}{dx}\right)^2}\,dx = \sqrt{1 + \frac{x^2}{a^2 - x^2}}\,dx = \frac{a\,dx}{\sqrt{a^2 - x^2}},$$

and since $y = \sqrt{a^2 - x^2}$ on the semicircle, we have

$$M_{y=0} = \int_{-a}^{a} \sqrt{a^2 - x^2}\,\frac{a\,dx}{\sqrt{a^2 - x^2}} = a\int_{-a}^{a} dx = 2a^2.$$

Since the length of the semicircle is πa, we have $\bar{y} = \dfrac{M_{y=0}}{\pi a} = \dfrac{2a}{\pi}$, and the centroi of the semicircle is $\left(0, \dfrac{2a}{\pi}\right)$. Note that the centroid of a semicircle of radius is not the same as that of half-disc of radius a. Note also that the centroid of th semicircle does not lie on the semicircle itself.

THEOREM 8.5.4 All three medians of a triangle pass through the centroid of the triangle.

PROOF Recall that a median of a triangle is a straight line joining one vertex of the triang to the midpoint of the opposite side. Given any median of a triangle we will sho that the centroid lies on that median. Thus the centroid must lie on all thre medians.

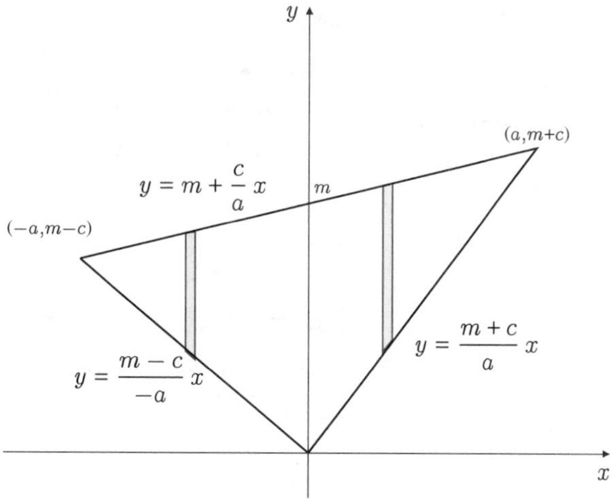

FIGURE 8.5.3

Adopt a coordinate system where the median in question lies along the y-axis, and such that a vertex of the triangle is at the origin. (See Fig. 8.5.3.) Let the midpoint of the opposite side be $(0, m)$. Then the other two vertices of the triangle must have coordinates of the form $(-a, m-c)$ and $(a, m+c)$ so that $(0, m)$ will be the midpoint between them. The three sides of the triangle have equations

$$y = m + \frac{c}{a}\,x, \qquad y = \frac{m-c}{-a}\,x, \qquad y = \frac{m+c}{a}\,x,$$

as shown in the figure. We calculate the moment of the traingle about $x = 0$:

$$\begin{aligned} M_{x=0} &= \int_{-a}^{0} x\left(m + \frac{c}{a}x - \frac{m-c}{-a}x\right)dx + \int_0^a x\left(m + \frac{c}{a}x - \frac{m+c}{a}x\right)dx \\ &= m\int_{-a}^{0}\left(x + \frac{x^2}{a}\right)dx + m\int_0^a\left(x - \frac{x^2}{a}\right)dx \\ &= -m\int_0^a\left(x - \frac{x^2}{a}\right)dx + m\int_0^a\left(x - \frac{x^2}{a}\right)dx = 0. \end{aligned}$$

(We substituted $-x$ for x in the first integral on second last line above.) Since $M_{x=0}$ therefore $\bar{x} = 0$ and the centroid lies on the median (the y-axis). □

REMARK: By solving simultaneously the equations of any two medians of a triangle, we can verify the following formula:

8.5.5
The Centroid of a Triangle

The coordinates of the centroid of a triangle are the averages of the corresponding coordinates of the three vertices of the triangle. The triangle with vertices (x_1, y_1), (x_2, y_2) and (x_3, y_3) has centroid

$$(\bar{x}, \bar{y}) = \left(\frac{x_1 + x_2 + x_3}{3}, \frac{y_1 + y_2 + y_3}{3}\right).$$

If a region is a union of non-overlapping subregions, then any moment of the region is the sum of the corresponding moments of the subregions. This fact enables us to calculate the centroid of the region if we know the centroids and areas of all the subregions.

EXAMPLE 8.5.6 Find the centroid of the trapezoid with vertices (0,0), (1,0), (1,2) and (0,1).

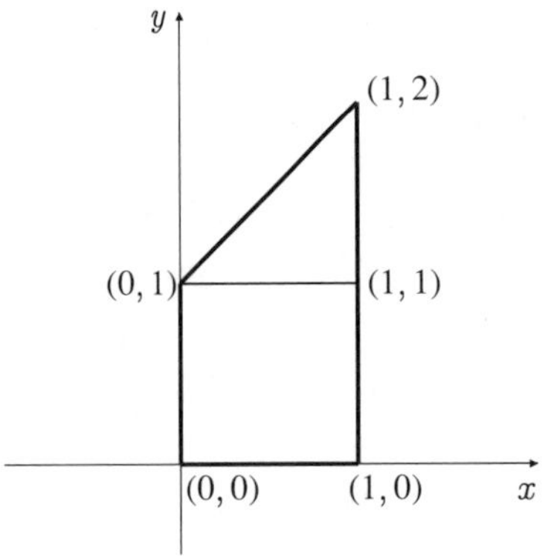

FIGURE 8.5.4

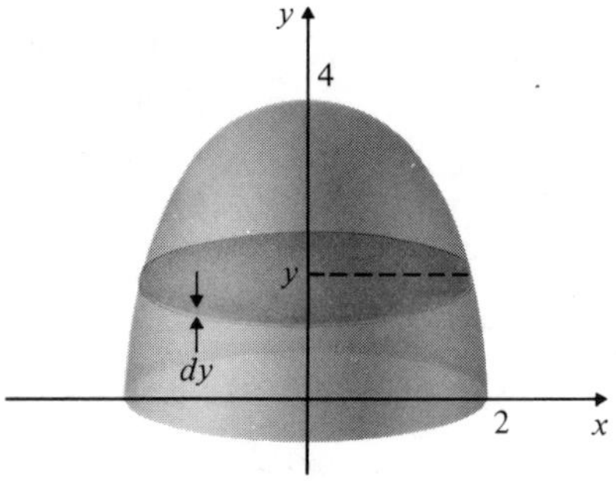

FIGURE 8.5.5

SOLUTION The trapezoid is the union of a square and a (non-overlapping) triangle as shown in Fig. 8.5.4. By symmetry, the square has centroid $(\bar{x}_S, \bar{y}_S) = \left(\frac{1}{2}, \frac{1}{2}\right)$, and its area is $A_S = 1$. The triangle has centroid $(\bar{x}_T, \bar{y}_T) = \left(\frac{2}{3}, \frac{4}{3}\right)$, and area $A_T = \frac{1}{2}$. Continuing to use subscripts S and T to denote the square and triangle respectively, we calculate

$$M_{x=0} = M_{S;x=0} + M_{T;x=0} = A_s\bar{x}_S + A_T\bar{x}_T = \frac{1}{2} + \frac{1}{2} \times \frac{2}{3} = \frac{5}{6}$$

$$M_{y=0} = M_{S;y=0} + M_{T;y=0} = A_s\bar{y}_S + A_T\bar{y}_T = \frac{1}{2} + \frac{1}{2} \times \frac{4}{3} = \frac{7}{6}$$

Since the area of the trapezoid is $A = A_S + A_T = \frac{3}{2}$, its centroid is

$$(\bar{x}, \bar{y}) = \left(\frac{5}{6} \times \frac{2}{3}, \frac{7}{6} \times \frac{2}{3}\right) = \left(\frac{5}{9}, \frac{7}{9}\right).$$

EXAMPLE 8.5.7 Find the centroid of the solid region obtained by rotating about the y-axis the first quadrant region lying between the x-axis and the parabola $y = 4 - x^2$.

SOLUTION By symmetry the centroid of the parabolic solid will lie on the axis of symmetry, the y-axis. A thin disc-shaped slice of the solid at height y and having thickness dy (see Fig. 8.5.5) has volume

$$dV = \pi x^2\, dy = \pi(4 - y)\, dy,$$

and has moment about the base plane

$$dM_{y=0} = y\, dV = \pi(4y - y^2)\, dy.$$

Hence the volume of the solid is

$$V = \pi \int_0^4 (4 - y)\, dy = \pi \left(4y - \frac{y^2}{2}\right)\Bigg|_0^4 = \pi(16 - 8) = 8\pi \text{ cu units.}$$

The moment of the solid about $y = 0$ is

$$M_{y=0} = \pi \int_0^4 (4y - y^2)\, dy = \pi \left(2y^2 - \frac{y^3}{3}\right)\Bigg|_0^4 = \pi \left(32 - \frac{64}{3}\right) = \frac{32}{3}\pi.$$

Hence the centroid is located at $\bar{y} = \dfrac{32\pi}{3} \times \dfrac{1}{8\pi} = \dfrac{4}{3}$.

The Pappus Theorem

The following theorem relates volumes or surface areas of revolution to the centroid of the region or curve being rotated.

THEOREM 8.5.8 **(Pappus's Theorem)**

a) If a plane region R lies on one side of a line L in that plane, and if R is rotated about L to generate a solid of revolution, the volume V of that solid is the product of the area of R and the distance traveled by the centroid of R under the rotation; that is,

$$V = 2\pi\bar{r}A,$$

where A is the area of R and $\bar{r}$ is the perpendicular distance from the centroid of R to L.

b) If a plane curve C, lying on one side of a line L in the plane, is rotated about that line to generate a surface of revolution, then the area S of that surface is the length of C times the distance traveled by the centroid of C;

$$S = 2\pi\bar{r}s,$$

where s is the length of the curve C and $\bar{r}$ is the perpendicular distance from the centroid of C to the line L.

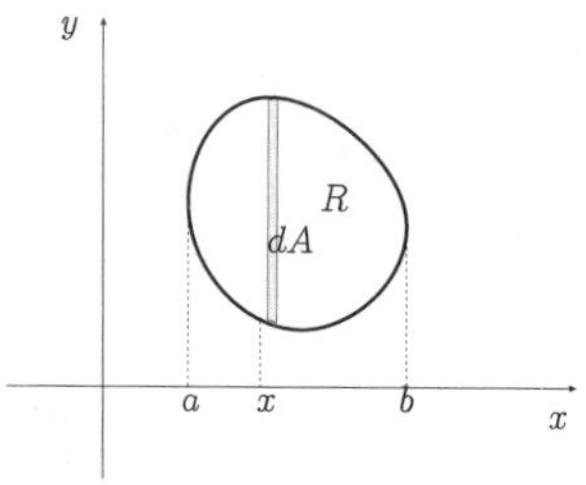

FIGURE 8.5.6

PROOF a) Let us take L to be the y-axis and suppose that R lies between $x = a$ and $x = b$ where $0 \le a < b$. Thus $\bar{r} = \bar{x}$, the x-coordinate of the centroid of R. Let dA denote the area of a thin strip of R at position x and having width dx. This strip generates, on rotation about L, a cylindrical shell of volume $dV = 2\pi x\, dA$, so the volume of the solid of revolution is

$$V = 2\pi \int_{x=a}^{x=b} x\, dA = 2\pi M_{x=0} = 2\pi \bar{x} A = 2\pi \bar{r} A.$$

The proof of part (b) is similar and is left to the student as an exercise. □

As the following examples illustrate, Pappus's theorem can be used in two ways; either the centroid can be determined when the appropriate volume or surface area is known, or the volume or surface area can be determined if the centroid of the rotating region or curve is known.

EXAMPLE 8.5.9 The centroid of the semicircle $y = \sqrt{a^2 - x^2}$ lies on the axis of symmetry of that semicircle, so it is located at a point $(0, \bar{y})$. Since the semicircle has length πa units and generates, on rotation about the x-axis, a sphere having area $4\pi a^2$ sq units, we obtain, using part (b) of Pappus's theorem,

$$4\pi a^2 = 2\pi(\pi a)\bar{y},$$

so $\bar{y} = 2a/\pi$, as shown previously in Example 8.5.3.

EXAMPLE 8.5.10 Rotating a disc of radius a about a line at distance $b > a$ from the centre of the disc generates a solid torus (doughnut) of volume

$$V = 2\pi(\pi a^2)b = 2\pi^2 a^2 b \text{ cubic units,}$$

since the centroid of the disc is clearly at the centre of the disc. The surface area of the torus (in case you want to have icing on the doughnut) is

$$S = 2\pi(2\pi a)b = 4\pi^2 ab \text{ sq units,}$$

the surface being obtained by rotating a circle of radius a.

EXERCISES

Find the centroids of the geometric structures in Exercises 1–17. Be alert for symmetries.

1. The quarter-disc $x^2 + y^2 \le r^2, x \ge 0, y \ge 0$

2. The region $0 \le y \le 9 - x^2$

3. The region $0 \le x \le 1, 0 \le y \le \dfrac{1}{\sqrt{1 + x^2}}$

4. The circular disc sector $x^2 + y^2 \le r^2, 0 \le y \le x$

5. The circular disc segment $0 \le y \le \sqrt{4 - x^2} - 1$

6. The semi-elliptic disc $0 \le y \le b\sqrt{1 - (x/a)^2}$

7. The quadrilateral with vertices (in clockwise order) $(0, 0)$, $(3, 1)$, $(4, 0)$, and $(2, -2)$

8. The region bounded by the semicircle $y = \sqrt{1 - (x - 1)^2}$, the y-axis, and the line $y = x - 2$.

9. A hemispherical surface of radius r

10. A solid hemisphere of radius r

11. A solid cone of base radius r and height h

12. A conical surface of base radius r and height h

13. The plane region $0 \le y \le \sin x,\ 0 \le x \le \pi$

14. The plane region $0 \le y \le \cos x,\ 0 \le x \le \pi/2$

15. The quarter-circle arc $x^2 + y^2 = r^2, x \ge 0,\ y \ge 0$

16. The arc $y = e^{|x|},\ -1 \le x \le 1$

17. The solid obtained by rotating the plane region $0 \le y \le 2x - x^2$ about the line $y = -2$.

18. The line segment from $(1, 0)$ to $(0, 1)$ is rotated about the line $x = 2$ to generate part of a conical surface. Find the area of that surface.

19. The triangle with vertices $(0, 0)$, $(1, 0)$, and $(0, 1)$ is rotated about the line $x = 2$ to generate a certain solid. Find the volume of that solid.

20. An equilateral triangle of edge s cm is rotated about one of its edges to generate a solid. Find the volume and surface area of that solid.

21. Find the centroid of the infinitely long spike-shaped region lying between the x-axis and the curve $y = (x + 1)^{-3}$ and to the right of the y-axis.

22. Show that the curve $y = e^{-x^2}$ $(-\infty < x < \infty)$ generates a surface of finite area when rotated about the x-axis. What does this imply about the location of the centroid of this infinitely long curve?

23. Obtain formulas for the coordinates of the centroid of the plane region $c \le y \le d,\ 0 < f(y) \le x \le g(y)$.

24. Prove part (b) of Theorem 8.5.8.

8.6 OTHER PHYSICAL APPLICATIONS

Hydrostatic Pressure

The **pressure** (that is, force per unit area) that a liquid exerts on a surface immersed in it increases with depth and is proportional to depth and to the density of the liquid. At a depth h the pressure is given by

8.6.1 Pressure at Depth h

$$p = \rho g h,$$

where ρ is the density of the liquid, and g is the acceleration produced by gravity where the fluid is located. For water at the surface of the earth we have, approximately, $\rho = 1000$ kg/m^3 and $g = 9.8$ m/s^2, so the pressure at depth h metres is

$$p = 9800h \text{ N/m}^2,$$

the unit of force used here is the Newton (N); 1 N = 1 kg·m/s^2, the force which imparts an acceleration of 1 m/s^2 to a mass of 1 kg.

The molecules in a liquid interact in such a way that the pressure at any depth acts equally in all directions; the pressure against a vertical surface is the same as that against a horizontal surface at the same depth. This is known as Pascal's principle.

The total force exerted by a liquid on a horizontal surface (say the bottom of a tank holding the liquid) is found by multiplying the area of that surface by the depth of the surface below the top of the liquid. For nonhorizontal surfaces, however the pressure is not constant over the whole surface, and the total force cannot be determined so easily. In this case we divide the surface into area elements dA, each at some particular depth h, and we then sum (that is, integrate) the corresponding force elements $dF = \rho g h\, dA$ to find the total force.

EXAMPLE 8.6.2 One vertical wall of a water trough is a semicircular plate of radius R metres with curved edge downward. If the trough is full, so that the water comes up to the top of the plate, find the total force of the water on that plate.

SOLUTION A horizontal strip of the surface of the plate at depth h m and having width dh m (see Fig. 8.6.1), has length $2\sqrt{R^2 - h^2}$ m and hence area $dA = 2\sqrt{R^2 - h^2}\, dh$ m^2. The force of the water on this strip is

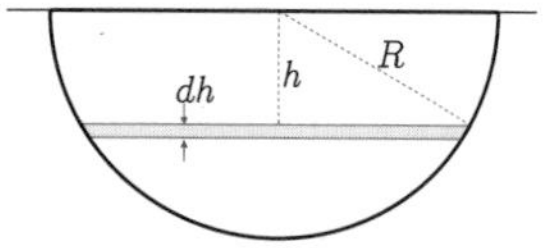

FIGURE 8.6.1

$$dF = \rho g h\, dA = 2\rho g h\sqrt{R^2 - h^2}\, dh.$$

Thus the total force on the plate is

$$\begin{aligned} F &= 2\rho g \int_0^R h\sqrt{R^2 - h^2}\, dh \qquad \text{Let } u = R^2 - h^2, \\ &\qquad\qquad\qquad\qquad\qquad\quad du = -2h\, dh \\ &= \rho g \int_0^{R^2} u^{1/2}\, du = \rho g\, \frac{2}{3} u^{3/2}\bigg|_0^{R^2} \approx \frac{2}{3} \times 9800R^3 \approx 6533R^3 \text{ N}. \end{aligned}$$

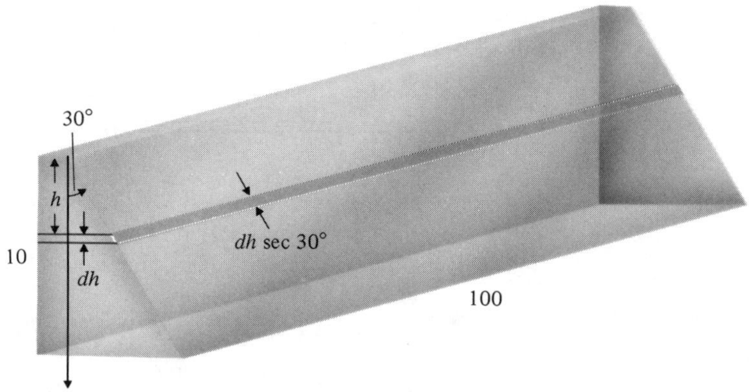

FIGURE 8.6.2

EXAMPLE 8.6.3 Find the total force on a section of a dike 100 m long and having a vertical height of 10 m, if the surface holding back the water is inclined at an angle of 30° to the vertical and the water comes up to the top of the dike.

SOLUTION The water in a horizontal layer of thickness dh at depth h actually makes contact with the section of dike along a slanted strip of width $dh \sec 30° = (2/\sqrt{3})\, dh$. See Fig. 8.6.2. The area of this strip is $dA = (200/\sqrt{3})\, dh$ m^2, and the force of water against the strip is $dF = (200/\sqrt{3})\, \rho g h\, dh$. The total force on the dike section is therefore

$$F = \frac{200}{\sqrt{3}} \rho g \int_0^{10} h\, dh = \frac{200}{\sqrt{3}} \rho g \frac{h^2}{2}\bigg|_0^{10} \approx \frac{10,000}{\sqrt{3}} \times 9800 \approx 5.66 \times 10^7 \text{ N}.$$

Work

When a force acts on an object to move that object, it is said to have done **work** on the object. The amount of work done is measured by the product of the force and the distance through which it moves the object, assuming that the force is constant, and that it is in the direction of the motion.

$$\text{Work} = \text{Force} \times \text{Distance}.$$

Work is always related to a particular force. If other forces acting on an object cause it to move in a direction opposite to the force F, then work is said to have been done *against* the force F.

Suppose that a force in the direction of the x-axis moves an object from $x = a$ to $x = b$ on that axis, and that the force varies continuously with the position x of the object; that is, $F = F(x)$ is a continuous function. The element of work done by the force in moving the object through a very short distance from x to $x + dx$ is $dW = F(x)\, dx$ so the total work done by the force is

8.6.4 Work

$$W = \int_a^b F(x)\, dx.$$

EXAMPLE 8.6.5 The force required to extend (or compress) an elastic spring to x units longer (or shorter) than its natural length is proportional to x (at least for sufficiently small values of x). This is known as Hooke's law. If we denote the force by $F(x)$, then

$$F(x) = kx,$$

where k is the **spring constant** for the particular spring. If a force of 2000 N is required to extend a certain spring to 4 cm longer than its natural length, how much work must be done to extend that far?

SOLUTION Here $F(x) = kx$ where, since $4k = F(4) = 2000$, we have $k = 500$ N/cm. The work done in extending the spring 4 cm is

$$W = \int_0^4 kx\, dx = k\frac{x^2}{2}\bigg|_0^4 = 8k = 4000 \text{ N·cm} = 40 \text{ N·m}.$$

40 Newton-metres (joules) of work must be done to stretch the spring 4 cm.

EXAMPLE 8.6.6 Water fills a tank in the shape of a right-circular cone with top radius 3 m and depth 4 m. How much work must be done (against gravity) to pump all the water out of the tank over the top edge of the tank?

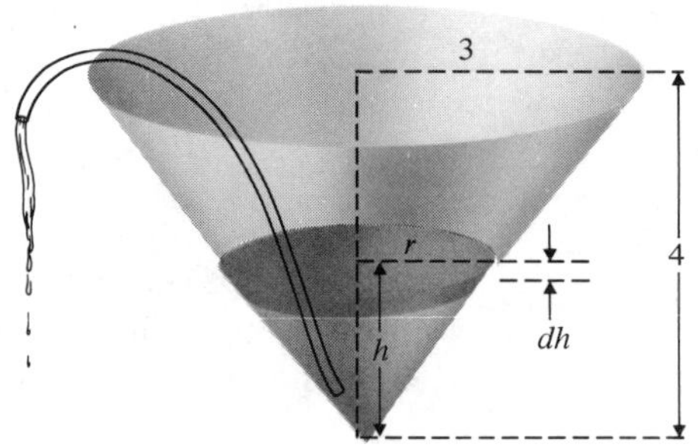

FIGURE 8.6.3

SOLUTION A thin disc of water at height h above the vertex of the tank has radius r (see Fig. 8.6.3), where $r = \frac{3}{4}h$ (by similar triangles). The volume of this disc is

$$dV = \pi r^2\,dh = \frac{9}{16}\pi h^2\,dh$$

and its *weight* (the force of gravity on the mass of water in the disc) is

$$dF = \rho g\,dV = \frac{9}{16}\rho g\,\pi h^2\,dh.$$

The water in this disc must be raised (against gravity) a distance $(4 - h)$ m by the pump. The work required to do this is

$$dW = \frac{9}{16}\rho g\,\pi(4-h)h^2\,dh.$$

The total work that must be done to empty the tank is the sum (integral) of all these elements of work for discs at depths between 0 and 4 metres:

$$\begin{aligned} W = \int_0^4 \frac{9}{16}\rho g\,\pi(4h^2 - h^3)\,dh &= \frac{9}{16}\rho g\,\pi\left(\frac{4h^3}{3} - \frac{h^4}{4}\right)\Bigg|_0^4 \\ &= \frac{9\pi}{16} \times 1000 \times 9.8 \times \frac{64}{3} \approx 3.69 \times 10^5 \text{ N·m}. \end{aligned}$$

EXAMPLE 8.6.7 The gravitaional force of the earth on a mass m located at height h above the surface of the earth is given by

$$F(h) = \frac{Km}{(R+h)^2},$$

where R is the radius of the earth and K is a constant independent of m and h. Determine, in terms of K and R the work that must be done against gravity to raise an object from the surface of the earth to

a) a height H above the surface of the earth, and

b) an infinite height above the surface of the earth.

SOLUTION The work done to raise the mass m from height h to height $h + dh$ is

$$dW = \frac{Km}{(R+h)^2}\,dh.$$

The total work to raise it from height $h = 0$ to height $h = H$ is

$$W = \int_0^H \frac{Km}{(R+h)^2}\,dh = \left.\frac{-Km}{R+h}\right|_0^H = Km\left(\frac{1}{R} - \frac{1}{R+H}\right).$$

(If R and H are measured in metres and F is measured in Newtons, then W is measured in Newton-metres (N·m).) The total work necessary to raise the mass m to infinite height is

$$W = \int_0^\infty \frac{Km}{(R+h)^2}\,dh = \lim_{H\to\infty} Km\left(\frac{1}{R} - \frac{1}{R+H}\right) = \frac{Km}{R}.$$

Potential and Kinetic Energy

Note that the units of work (force × distance) are the same as those of energy. Work done against a force may be regarded as storing up energy for future use or for conversion to other forms. Such stored energy is called **potential energy** (P.E.). For instance, in extending or compressing an elastic spring, we are doing work against the tension in the spring and hence storing energy in the spring. When work is done against a (variable) force $F(x)$ to move an object from $x = a$ to $x = b$, the potential energy stored is

8.6.8
Potential Energy

$$\text{P.E.} = -\int_a^b F(x)\,dx.$$

Since the work is being done against F, the signs of $F(x)$ and $b-a$ are opposite, so the integral is negative; the explicit negative sign is included so that the calculated potential energy will be positive.

One of the forms of energy into which potential energy can be converted is **kinetic energy** (K.E.), the energy of motion. If an object of mass m is moving with velocity v, it has kinetic energy

8.6.9
Kinetic Energy

$$\text{K.E.} = \frac{1}{2} m v^2.$$

For example, if an object is raised and then dropped, it accelerates downward under gravity as more and more of the potential energy stored in it when it was raised is converted to kinetic energy.

Consider the change in potential energy stored in a mass m as it moves along the x-axis from a to b under the influence of a force $F(x)$ depending only on x:

$$\text{P.E.}(b) - \text{P.E.}(a) = -\int_a^b F(x)\,dx.$$

(The change in P.E. is negative if m is moving in the direction of F.) According to Newton's second law of motion, the force $F(x)$ causes the mass m to accelerate, with acceleration dv/dt given by

$$F(x) = m\frac{dv}{dt} \qquad \text{(force = mass} \times \text{acceleration).}$$

By the chain rule we can rewrite dv/dt in the form

$$\frac{dv}{dt} = \frac{dv}{dx}\frac{dx}{dt} = v\frac{dv}{dx},$$

so $F(x) = mv\dfrac{dv}{dx}$. Hence

$$\begin{aligned}
\text{P.E.}(b) - \text{P.E.}(a) &= -\int_a^b mv\frac{dv}{dx}\,dx \\
&= -m\int_{x=a}^{x=b} v\,dv \\
&= -\frac{1}{2}mv^2\Big|_{x=a}^{x=b} \\
&= \text{K.E.}(a) - \text{K.E.}(b).
\end{aligned}$$

Thus

$$\text{P.E.}(b) + \text{K.E.}(b) = \text{P.E.}(a) + \text{K.E.}(a).$$

This shows that the total energy (potential + kinetic) remains constant as the mass m moves under the influence of a force F, *depending only on position.* Such a force is said to be **conservative** and the above result is called the **law of conservation of energy**.

EXAMPLE 8.6.10 We will use the result of Example 8.6.7 together with the following known values,

a) the radius R of the earth is approximately 6400 km, or 6.4×10^6 m,

b) the acceleration of gravity g at the surface of the earth is approximately 9.8 $\mathrm{m/s^2}$,

to determine the constant K in the gravitational force formula of Example 8.6.7, and hence to determine the escape velocity for a projectile fired vertically from the surface of the earth. The escape velocity is the (minimum) speed that such a projectile must have at firing to ensure that it will continue to recede farther and farther from the surface of the earth and not fall back.

According to the formula of Example 8.6.7, the force of gravity on a mass m kg at the surface of the earth ($h = 0$) is

$$F = \frac{Km}{(R+0)^2} = \frac{Km}{R^2}.$$

According to Newton's second law of motion, this force is related to the acceleration of gravity there (g) by the equation $F = mg$. Thus

$$\frac{Km}{R^2} = mg$$

and $K = gR^2$.

According to the law of conservation of energy, the projectile must have sufficient kinetic energy at firing to do the work necessary to raise the mass m to infinite height (that is, to supply sufficient potential energy to raise it to that height). By the result of Example 8.6.7, this required energy is Km/R. If the initial velocity of the projectile is v, we want

$$\frac{1}{2}mv^2 \geq \frac{Km}{R}.$$

Thus v must satisfy

$$v \geq \sqrt{\frac{2K}{R}} = \sqrt{2gR} \approx \sqrt{2 \times 9.8 \times 6.4 \times 10^6} \approx 1.12 \times 10^4 \text{ m/s}.$$

Thus the escape velocity is approximately 11.2 km/s, and is independent of the mass m. In this calculation we have neglected any air resistance near the surface of the earth. Such resistance depends on velocity rather than on position, so it is not a conservative force. The effect of such resistance would be to use up (convert to heat) some of the initial kinetic energy and so raise the escape velocity.

EXERCISES

1. A tank has a square base 2 m on each side and vertical sides 6 m high. If the tank is filled with water, find the total force exerted by the water (a) on the bottom of the tank, and (b) on one of the four vertical walls of the tank.

2. A swimmimg pool 20 m long and 8 m wide has a sloping plane bottom so that the depth of the pool at one end is 1 m and at the other end is 3 m. Find the total force exerted on the bottom of the pool if the pool is full of water.

3. A dam 200 m long and 24 m high presents a sloping face of 26 m slant height to the water in a reservoir behind the dam (see Fig. 8.6.4). If the surface of the water is level with the top of the dam, what is the total force of the water on the dam?

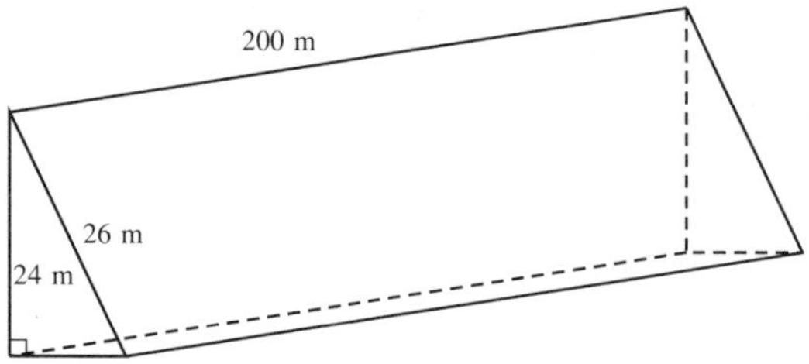

FIGURE 8.6.4

4. A pyramid with a square base 4 metres on each side and four equilateral triangular faces sits on the level bottom of a lake at a place where the lake is 10 m deep. Find the total force of the water on each of the triangular faces.

5. A lock on a canal has a gate in the shape of a vertical rectangle 5 m wide and 20 m high. If the water on one side of the gate comes up to the top of the gate and the water on the other side comes only 6 m up the gate, find the total force that must be exerted to hold the gate in place.

6. If 100 N·cm of work must be done to compress an elastic spring to 3 cm shorter than its natural length, how much work must be done to compress it 1 cm further?

7. Find the total work that must be done to pump all the water in the tank of Exercise 1 out over the top of the tank.

8. Find the total work that must be done to pump all the water in the swimming pool of Exercise 2 out over the top edge of the pool.

9. Find the work that must be done to pump all the water in a full hemispherical bowl of radius a m to a height h m above the top of the bowl.

10.*A bucket is raised vertically from ground level at a constant speed of 2 m/min by a winch. If the bucket weighs 1 kg and contains 15 kg of water when it starts up, but loses water by leakage at a rate of 1 kg/min thereafter, how much work must be done by the winch to raise the bucket to a height of 10 m?

8.7 PROBABILITY

Probability theory is a very important field of application of the definite integral. This subject cannot, of course, be developed thoroughly here—an adequate presentation requires one or more whole courses—but we can give a brief introduction that suggests some of the ways integrals are used in probability theory.

The **probability** of an event occuring is a real number between 0 and 1 that measures the proportion of times the event can be expected to occur in a large number of trials. If the occurance of an event is certain, its probability is 1; if the event cannot possibly occur, its probability is 0. For example, the probability that a tossed coin will land heads is 1/2 because we would expect it to land heads about half the time if it were tossed a great many times. In such a tossing of a coin there are only two possible outcomes, heads or tails, each equally likely, that is, each having probability 1/2. (We are assuming the coin won't ever land standing on its edge.) For any toss, let $X = 0$ if the outcome is heads, and let $X = 1$ if the outcome is tails. X is called a **discrete random variable**. The probability that $X = 0$ is 1/2 and the probability that $X = 1$ is 1/2, so we write

$$\Pr(X = 0) = \frac{1}{2}, \qquad \Pr(X = 1) = \frac{1}{2}.$$

Note that $\Pr(X = 0) + \Pr(X = 1) = 1$ since it is certain that the coin will land either heads or tails.

EXAMPLE 8.7.1 A single die is rolled so that it will show one of the numbers 1 to 6 on top when it stops. If X denotes the number showing on any roll, then X is a discrete random variable. Assuming no one value of X is any more likely than any other, the probability that the number showing is x must be 1/6 for each possible value of x; that is,

$$\Pr(X = x) = \frac{1}{6} \qquad \text{for each } x \text{ in } \{1,2,3,4,5,6\}.$$

The discrete random variable X is therefore said to be distributed **uniformly**. Again we note that

$$\sum_{n=1}^{6} \Pr(X = n) = 1,$$

reflecting the fact that the rolled die must certainly give one of the six possible outcomes. The probability that a roll will produce a value from 1 to 4 is

$$\Pr(1 \le X \le 4) = \sum_{n=1}^{4} \Pr(X = n) = \frac{1}{6} + \frac{1}{6} + \frac{1}{6} + \frac{1}{6} = \frac{2}{3}.$$

Now we consider an example with a continuous range of possible outcomes.

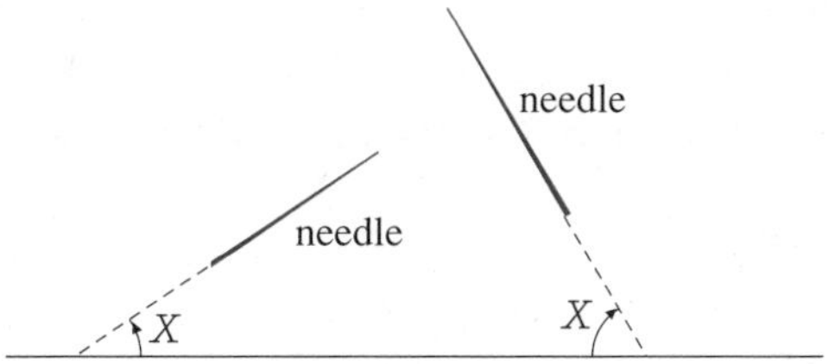

FIGURE 8.7.1

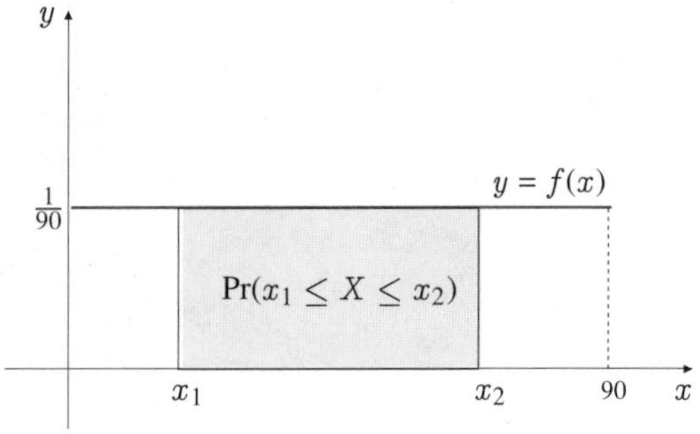

FIGURE 8.7.2

EXAMPLE 8.7.2 Suppose that a needle is dropped at random on a flat table with a straight line drawn on it. For each drop, let X be the number of degrees in the (acute) angle that the needle makes with the line (see Fig. 8.7.1). Evidently X can take any real value in the interval $[0, 90]$; therefore X is called a **continuous random variable**. The probability that X takes on any particular real value is 0. (There are infinitely many real numbers in $[0, 90]$ and none is more likely than any other.) However, the probability that X lies in some interval, say $[10, 20]$, is the same as the probability that it lies in any other interval of the same length. Since the interval has length 10 and the interval of all possible values of X has length 90, this probability is

$$\Pr(10 \le X \le 20) = \frac{10}{90} = \frac{1}{9}.$$

More generally, if $0 \le x_1 \le x_2 \le 90$, then

$$\Pr(x_1 \le X \le x_2) = \frac{1}{90}(x_2 - x_1).$$

This situation can be conveniently represented as follows: Let $f(x)$ be defined on the interval $[0, 90]$, taking at each point the constant value 1/90:

$$f(x) = \frac{1}{90}, \qquad 0 \le x \le 90.$$

The area under the graph of f is 1, and $\Pr(x_1 \le X \le x_2)$ is equal to the area under that part of the graph lying over the interval $[x_1, x_2]$ (see Fig. 8.7.2). The function $f(x)$ is called the **probability density function** for the random variable X. Since $f(x)$ is constant on its domain, X is said to be **uniformly distributed**.

8.7.3 Probability Density Functions

> A function defined on an interval $[a, b]$ is a probability density function for a continuous random variable X distributed on $[a, b]$ if, whenever $a \le x_1 \le x_2 \le b$ we have
>
> $$\Pr(x_1 \le X \le x_2) = \int_{x_1}^{x_2} f(x)\,dx.$$

In order to be such a probability density function, f must satisfy two conditions:

a) $f(x) \ge 0$ on $[a, b]$, (probability cannot be negative), and

b) $\int_a^b f(x)\,dx = 1$, $(\Pr(a \le X \le b) = 1)$.

These ideas extend to random variables distributed on semi-infinite or infinite intervals, but the integrals appearing will be improper in those cases.

In the example of the dropping needle, the probability density function has a horizontal straight line graph, and we termed such a probability distribution uniform. The uniform probability density function on the interval $[a, b]$ is

$$f(x) = \begin{cases} \dfrac{1}{b-a} & \text{if } a \le x \le b \\ 0 & \text{otherwise.} \end{cases}$$

Many other functions are commonly encountered as density functions for continuous random variables.

EXAMPLE 8.7.4 The length of time T that any particular atom in a radioactive sample survives before decaying is a random variable taking values in $[0, \infty)$. It has been observed that the proportion of atoms that survive to time t becomes small exponentially at t increases; thus

$$\Pr(T \geq t) = Ce^{-kt}.$$

Let f be the probability density function for the random variable T. Then

$$\int_t^\infty f(x)\,dx = \Pr(T \geq t) = Ce^{-kt}.$$

Differentiating this equation with respect to t (using the Fundamental Theorem of Calculus), we obtain $-f(t) = -Cke^{-kt}$, so $f(t) = Cke^{-kt}$. C is determined by the requirement that $\int_0^\infty f(t)\,dt = 1$. We have

$$1 = Ck\int_0^\infty e^{-kt}\,dt = \lim_{R\to\infty} Ck\int_0^R e^{-kt}\,dt = -C\lim_{R\to\infty}(e^{-kR} - 1) = C.$$

Thus $C = 1$ and $f(t) = ke^{-kt}$. Note that $\Pr(T \geq (\ln 2)/k) = e^{-k(\ln 2)/k} = 1/2$, reflecting the fact that the half-life of such a radioactive sample is $(\ln 2)/k$.

EXAMPLE 8.7.5 For what value of C is $f(x) = C(1 - x^2)$ a probability density function on $[-1, 1]$? If X is a random variable with this density what is the probability that $X \leq 1/2$?

SOLUTION Evidently $f(x) \geq 0$ on $[-1, 1]$ if $C \geq 0$. Since

$$\int_{-1}^1 f(x)\,dx = C\int_{-1}^1 (1 - x^2)\,dx = 2C\left(x - \frac{x^3}{3}\right)\bigg|_0^1 = \frac{4C}{3},$$

$f(x)$ will be a probability density function if $C = 3/4$. In this case

$$\begin{aligned}\Pr\left(X \leq \frac{1}{2}\right) &= \frac{3}{4}\int_{-1}^{1/2}(1 - x^2)\,dx = \frac{3}{4}\left(x - \frac{x^3}{3}\right)\bigg|_{-1}^{1/2}\\ &= \frac{3}{4}\left(\frac{1}{2} - \frac{1}{24} - (-1) + \frac{-1}{3}\right)\\ &= \frac{27}{32}.\end{aligned}$$

Expectation, Mean, Variance, and Standard Deviation

Consider a simple gambling game in which the player pays the house C dollars for the privilege of rolling a single die and in which he wins X dollars where X is the number showing on top of the rolled die. In each game the possible winnings are 1, 2, 3, 4, 5 or 6 dollars, each with probability 1/6. In n games the player can expect to win about $n/6 + 2n/6 + 3n/6 + 4n/6 + 5n/6 + 6n/6 = 21n/6 = 7n/2$ dollars, so that his expected *average winnings per game* are 7/2 dollars, \$3.50. If $C > 3.5$, the player can expect, on average, to lose money. The amount 3.5 is called the **expectation** or **mean** of the discrete random variable X.

In general, if a random variable can take on values x_1 with probability p_1, x_2 with probability p_2, ..., and x_n with probability p_n (where $p_1 + p_2 + \cdots + p_n = 1$), the mean μ or expectation $E(X)$ of that random variable X is given by

$$\mu = E(X) = \sum_{i=1}^{n} x_i p_i.$$

We formulate an analogous definition for the mean or expectation of a continuous random variable as follows:

8.7.6
Mean or Expectation

If X is a continuous random variable on $[a, b]$ with probability density function $f(x)$, the **mean** (denoted μ), or **expectation** (denoted $E(X)$) of X, is

$$\mu = E(X) = \int_a^b x f(x)\, dx.$$

(Note that in this usage $E(X)$ does not define a function of X but a constant (parameter) associated with the probability distribution of X. Note also that if $f(x)$ were a mass density such as that studied in Section 8.4, then μ would be the moment of the mass about 0 and, since the total mass would be $\int_a^b f(x)\, dx = 1$, μ would in fact be the centre of mass.)

More generally, the **expectation** of any function $g(X)$ of the random variable X is defined to be

$$E(g(X)) = \int_a^b g(x) f(x)\, dx.$$

8.7.7
Variance

The **variance** of a random variable X with density $f(x)$ on $[a, b]$ is the expectation of the square of the distance from X to its mean μ. The variance is denoted σ^2 or $\mathrm{Var}(X)$.

$$\sigma^2 = \mathrm{Var}(X) = E((X - \mu)^2) = \int_a^b (x - \mu)^2 f(x)\, dx.$$

Since $\int_a^b f(x)\, dx = 1$, the expression above for the variance can be rewritten as follows:

$$\begin{aligned}\sigma^2 = \mathrm{Var}(X) &= \int_a^b x^2 f(x)\, dx - 2\mu \int_a^b x f(x)\, dx + \mu^2 \int_a^b f(x)\, dx \\ &= \int_a^b x^2 f(x)\, dx - 2\mu^2 + \mu^2 = E(X^2) - \mu^2,\end{aligned}$$

that is,

$$\sigma^2 = \text{Var}(X) = E(X^2) - \mu^2 = E(X^2) - (E(X))^2.$$

8.7.8 Standard Deviation

> The **standard deviation** of the random variable X is σ, the square root of the variance. Thus it is the square root of the mean square deviation of X from its mean:
>
> $$\sigma = \left(\int_a^b (x - \mu)^2 f(x)\, dx \right)^{1/2} = \sqrt{E(X^2) - \mu^2}.$$

The standard deviation gives a measure of how spread out the probability distribution of X is. The smaller the standard deviation, the more concentrated is the area under the density curve around the mean, and so the smaller is the probability that a value of X will be far away from the mean. See Fig. 8.7.3.

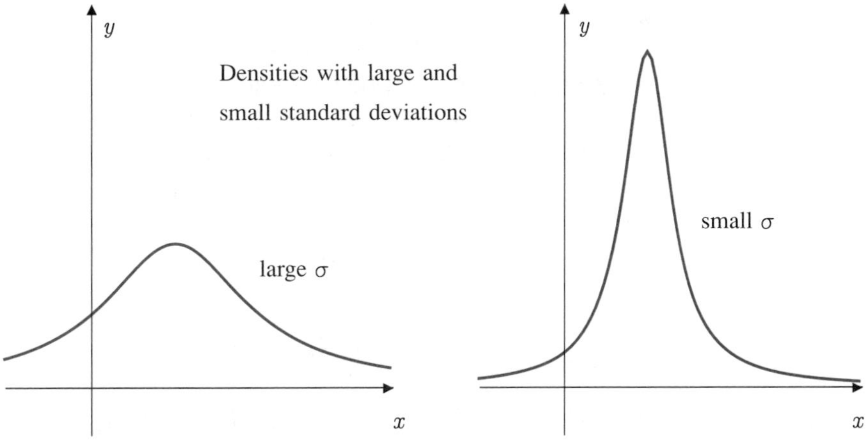

FIGURE 8.7.3

EXAMPLE 8.7.9 Find the mean μ and the standard deviation σ of a random variable X distributed uniformly on the interval $[a, b]$. Find $\Pr(\mu - \sigma \le X \le \mu + \sigma)$.

SOLUTION The probability density function is $f(x) = 1/(b - a)$ on $[a, b]$, so the mean is given by

$$\mu = E(X) = \int_a^b \frac{x}{b-a}\, dx = \frac{1}{b-a} \frac{x^2}{2} \bigg|_a^b = \frac{1}{2} \frac{b^2 - a^2}{b - a} = \frac{b + a}{2}.$$

Hence the mean is, as might have been anticipated, the midpoint of $[a, b]$. The expectation of X^2 is given by

$$E(X^2) = \int_a^b \frac{x^2}{b-a}\,dx = \frac{1}{b-a}\left.\frac{x^3}{3}\right|_a^b = \frac{1}{3}\frac{b^3-a^3}{b-a} = \frac{b^2+ab+a^2}{3}.$$

Hence the variance is

$$\sigma^2 = E(X^2) - \mu^2 = \frac{b^2+ab+a^2}{3} - \frac{b^2+2ab+a^2}{4} = \frac{(b-a)^2}{12},$$

and the standard deviation is

$$\sigma = \frac{b-a}{2\sqrt{3}} \approx 0.29(b-a).$$

Finally, $\Pr(\mu - \sigma \le X \le \mu + \sigma) = \int_{\mu-\sigma}^{\mu+\sigma} \frac{dx}{b-a} = \frac{1}{b-a}\frac{2(b-a)}{2\sqrt{3}} = \frac{1}{\sqrt{3}} \approx 0.577.$

EXAMPLE 8.7.10 Find the mean μ and the standard deviation σ of a random variable X distributed exponentially with density function $f(x) = ke^{-kx}$ on the interval $[0, \infty)$. Find $\Pr(\mu - \sigma \le X \le \mu + \sigma)$.

SOLUTION We use integration by parts to find the mean:

$$\begin{aligned}
\mu = E(X) &= k\int_0^\infty xe^{-kx}\,dx \\
&= \lim_{R\to\infty} k\int_0^R xe^{-kx}\,dx \qquad \begin{matrix}\text{Let} & U = x & dV = e^{-kx}\,dx \\ \text{Then} & dU = dx & V = -e^{-kx}/k\end{matrix} \\
&= \lim_{R\to\infty}\left(-xe^{-kx}\Big|_0^R + \int_0^R e^{-kx}\,dx\right) \\
&= \lim_{R\to\infty}\left(-Re^{-kR} - \frac{1}{k}\left(e^{-kR}-1\right)\right) = \frac{1}{k} \qquad \text{(since } k > 0\text{)}.
\end{aligned}$$

Thus the mean of the exponential distribution is $1/k$. This fact can be quite useful in determining the value of k for an exponentially distributed random variable. A similar integration by parts enables us to evaluate

$$E(X^2) = k\int_0^\infty x^2 e^{-kx}\,dx = 2\int_0^\infty xe^{-kx}\,dx = \frac{2}{k^2},$$

so that the variance of the exponential distribution is

$$\sigma^2 = E(X^2) - \mu^2 = \frac{1}{k^2}$$

and the standard deviation is equal to the mean

$$\sigma = \mu = \frac{1}{k}.$$

Now we have

$$\begin{aligned}\Pr(\mu - \sigma \le X \le \mu + \sigma) &= \Pr(0 \le X \le 2/k) \\ &= k \int_0^{2/k} e^{-kx}\, dx \\ &= -e^{-kx} \Big|_0^{2/k} \\ &= 1 - e^{-2} \approx 0.86,\end{aligned}$$

which is independent of the value of k. Exponential densities for small and large values of k are sketched in Fig. 8.7.4.

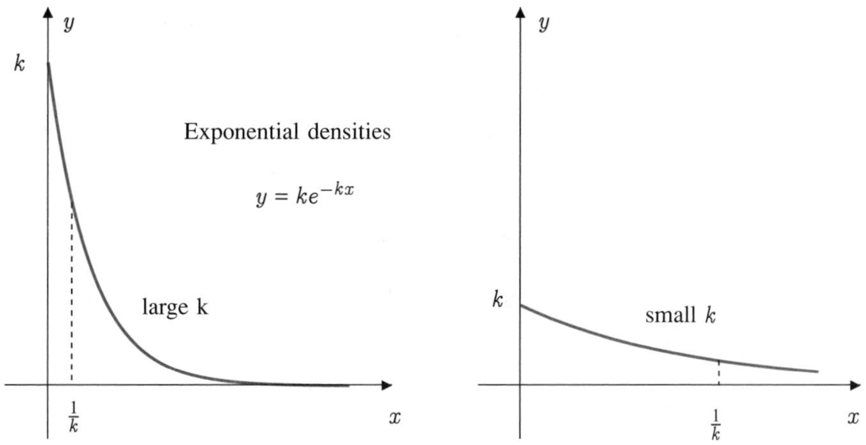

FIGURE 8.7.4

The Normal distribution

The most important probability distributions are the so-called **normal** or **Gaussian** distributions. Such distributions govern the behavior of many interesting random variables, in particular those associated with random errors in measurements. There is a family of normal distributions, all related to the particular normal distribution called the **standard normal distribution**, which has probability density function

8.7.11
The Standard
Normal Distribution

$$f(z) = \frac{1}{\sqrt{2\pi}} e^{-z^2/2}, \qquad -\infty < z < \infty.$$

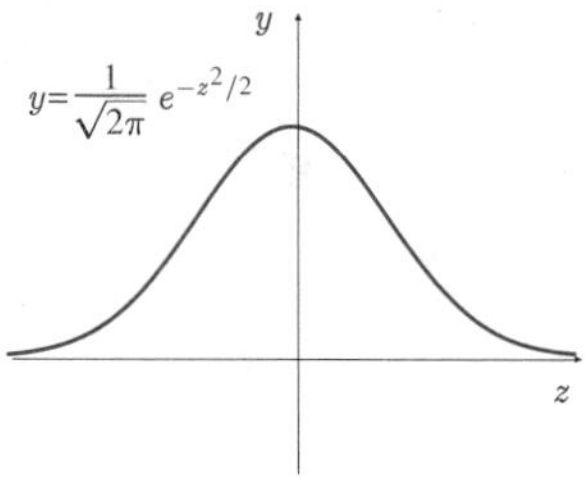

FIGURE 8.7.5

It is common to use z to denote the random variable in the standard normal distribution; the other normal distributions are obtained from this one by a change of variable. The graph of the standard normal density has a pleasant bell shape, as shown in Fig. 8.7.5.

As we have noted previously, the function e^{-z^2} has no elementary antiderivative, so the improper integral

$$I = \int_{-\infty}^{\infty} e^{-z^2/2}\,dz$$

cannot be easily evaluated using the Fundamental Theorem of Calculus, though it is easily shown to be a convergent improper integral. The integral can be evaluated by using techniques of advanced calculus involving integrals of functions of two variables. The value is $I = \sqrt{2\pi}$, which ensures that the above-defined standard normal density $f(z)$ is indeed a probability density function:

$$\int_{-\infty}^{\infty} f(z)\,dz = \frac{1}{\sqrt{2\pi}}\int_{-\infty}^{\infty} e^{-z^2/2}\,dz = 1.$$

Since $ze^{-z^2/2}$ is an odd function of z, the mean of the standard normal distribution is 0:

$$\mu = E(Z) = \frac{1}{\sqrt{2\pi}}\int_{-\infty}^{\infty} ze^{-z^2/2}\,dz = \frac{1}{\sqrt{2\pi}}\lim_{R\to\infty}\int_{-R}^{R} ze^{-z^2/2}\,dz = 0.$$

We calculate the variance of the standard normal distribution using integration by parts as follows:

$$\begin{aligned}
\sigma^2 = E(Z^2) &= \frac{1}{\sqrt{2\pi}}\int_{-\infty}^{\infty} z^2e^{-z^2/2}\,dz \\
&= \frac{1}{\sqrt{2\pi}}\lim_{R\to\infty}\int_{-R}^{R} z^2e^{-z^2/2}\,dz \qquad \text{Let } U = z \quad dV = ze^{-z^2/2}\,dz \\
& \qquad\qquad\qquad\qquad\qquad\qquad\quad \text{Then } dU = dz \quad V = -e^{-z^2/2} \\
&= \frac{1}{\sqrt{2\pi}}\lim_{R\to\infty}\left(-ze^{-z^2/2}\bigg|_{-R}^{R} + \int_{-R}^{R} e^{-z^2/2}\,dz\right) \\
&= \frac{1}{\sqrt{2\pi}}\lim_{R\to\infty}(-2Re^{-R^2/2}) + \frac{1}{\sqrt{2\pi}}\int_{-\infty}^{\infty} e^{-z^2/2}\,dz \\
&= 0 + 1 = 1.
\end{aligned}$$

Hence the standard deviation of the standard normal distribution is 1.

Other normal distributions are obtained from the standard normal distribution by a change of variable. A random variable X on $(-\infty, \infty)$ is said to be *normally distributed with mean* μ and *standard deviation* σ (where μ is any real number and $\sigma > 0$) if its probability density function is

8.7.12
General Normal Distribution

$$f_{\mu,\sigma}(x) = \frac{1}{\sigma} f\left(\frac{x-\mu}{\sigma}\right) = \frac{1}{\sigma\sqrt{2\pi}} e^{-(x-\mu)^2/(2\sigma^2)}.$$

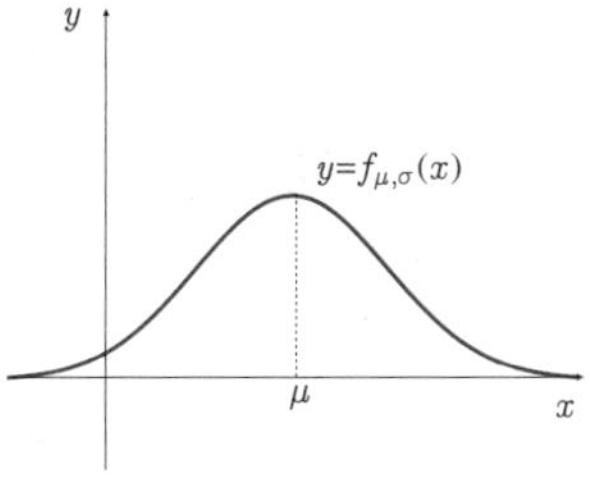

FIGURE 8.7.6

See Fig. 8.7.6. Using the change of variable $z = (x - \mu)/\sigma$, $dz = dx/\sigma$, you can readily verify that

$$\int_{-\infty}^{\infty} f_{\mu,\sigma}(x)\,dx = \int_{-\infty}^{\infty} f(z)\,dz = 1,$$

so $f_{\mu,\sigma}(x)$ is indeed a probability density function. Using the same change of variable, we can readily show that

$$E(X) = \mu, \qquad \text{and} \qquad E((X - \mu)^2) = \sigma^2.$$

Hence $f_{\mu,\sigma}(x)$ does indeed have a mean μ and a standard deviation σ.

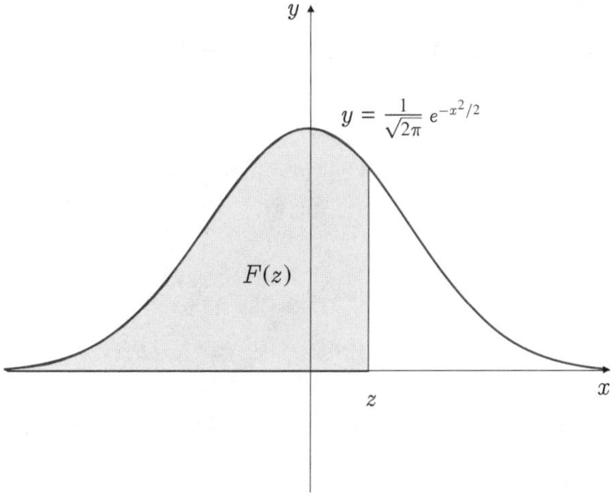

FIGURE 8.7.7

Because $e^{-z^2/2}$ cannot be readily antidifferentiated, we cannot determine probabilities (that is, areas) for normal curves by using the Fundamental Theorem of Calculus. Numerical integrations can be performed, or one can consult a book of statistical tables for computed areas under the standard normal curve. Specifically, these tables usually provide values for what is called the **cumulative distribution function** of a random variable with standard normal distribution. This is the function

8.7.13
Distribution Function for the Standard Normal Distribution

$$F(z) = \frac{1}{\sqrt{2\pi}} \int_{-\infty}^{z} e^{-x^2/2}\,dx = \Pr(Z \leq z),$$

which represents the area under the standard normal density function from $-\infty$ up to z, as shown in Fig. 8.7.7.

For use in the following examples and exercises, we include here an abbreviated version of such a table.

Values of F(z) (rounded to 3 decimal places)

z	0.0	0.1	0.2	0.3	0.4	0.5	0.6	0.7	0.8	0.9
−3.0	0.001	0.001	0.001	0.001	0.000	0.000	0.000	0.000	0.000	0.000
−2.0	0.023	0.018	0.014	0.011	0.008	0.006	0.005	0.004	0.003	0.002
−1.0	0.159	0.136	0.115	0.097	0.081	0.067	0.055	0.045	0.036	0.029
−0.0	0.500	0.460	0.421	0.382	0.345	0.309	0.274	0.242	0.212	0.184
0.0	0.500	0.540	0.579	0.618	0.655	0.692	0.726	0.758	0.788	0.816
1.0	0.841	0.864	0.885	0.903	0.919	0.933	0.945	0.955	0.964	0.971
2.0	0.977	0.982	0.986	0.989	0.992	0.994	0.995	0.997	0.997	0.998
3.0	0.999	0.999	0.999	1.000	1.000	1.000	1.000	1.000	1.000	1.000

EXAMPLE 8.7.14 If Z is a standard normal random variable, find

a) $\Pr(-1.2 \le Z \le 2.0)$, and (b) $\Pr(Z \ge 1.5)$.

SOLUTION a) Using values from the table

$$\begin{aligned}\Pr(-1.2 \le Z \le 2.0) &= \Pr(Z \le 2.0) - \Pr(Z < -1.2)\\ &= F(2.0) - F(-1.2) \approx 0.977 - 0.115 = 0.862\end{aligned}$$

b) $\Pr(Z \ge 1.5) = 1 - \Pr(Z < 1.5) = 1 - F(1.5) \approx 1 - 0.933 = 0.067.$

EXAMPLE 8.7.15 A random variable X is distributed normally with mean 2 and standard deviation 0.4. Find

a) $\Pr(1.8 \le X \le 2.4)$, and b) $\Pr(X > 2.4)$.

SOLUTION Since X is distributed normally with mean 2 and standard deviation 0.4, $Z = (X-2)/0.4$ is distributed according to the standard normal distribution (with mean 0 and standard deviation 1). Accordingly,

a) $$\begin{aligned}\Pr(1.8 \le X \le 2.4) &= \Pr(-0.5 \le Z \le 1) = F(1) - F(-0.5)\\ &\approx 0.841 - 0.309 = 0.532,\end{aligned}$$

b) $$\begin{aligned}\Pr(X > 2.4) &= \Pr(Z > 1) = 1 - \Pr(Z \le 1) = 1 - F(1)\\ &\approx 1 - 0.841 = 0.159.\end{aligned}$$

EXERCISES

For each function $f(x)$ in Exercises 1–7 find the following:

a) the value of C for which f is a probability density on the given interval,

b) the mean μ, variance σ^2, and standard deviation σ of the probability density f,

c) $\Pr(\mu - \sigma \le X \le \mu + \sigma)$, that is, the probability that the random variable X is no further than one standard deviation away from its mean.

1. $f(x) = Cx$ on $[0, 3]$

2. $f(x) = Cx$ on $[1, 2]$

3. $f(x) = Cx^2$ on $[0, 1]$

4. $f(x) = C \sin x$ on $[0, \pi]$

5. $f(x) = C(x - x^2)$ on $[0, 1]$

6. $f(x) = C\,xe^{-kx}$ on $[0, \infty)$, $(k > 0)$

7. $f(x) = C\,e^{-x^2}$ on $[0, \infty)$. (*Hint:* Use properties of the standard normal density to show that $\int_0^\infty e^{-x^2}\,dx = \sqrt{\pi}/2$.)

8. Is it possible for a random variable to be uniformly distributed on the whole real line? Explain why.

9. Carry out the calculations to show that the normal density $f_{\mu,\sigma}(x)$ defined in the text is a probability density function and has mean μ and standard deviation σ.

10.*Show that $f(x) = \dfrac{2}{\pi(1+x^2)}$ is a probability density on $[0, \infty)$. Find the expectation of X for this density. If a machine generates values of a random variable X distributed with density $f(x)$, how much would you be willing to pay, per game, to play a game in which you operate the machine to produce a value of X and win X dollars? Explain.

11. Calculate $\Pr(|X - \mu| \ge 2\sigma)$ for

a) the uniform distribution on $[a, b]$,

b) the exponential distribution with density $f(x) = ke^{-kx}$ on $[0, \infty)$,

c) the normal distribution with density $f_{\mu,\sigma}(x)$.

12. The length of time T (in hours) between malfunctions of a computer system is an exponentially distributed random variable. If the average length of time between successive malfunctions is 20 hours, find the probability that the system, having just had a malfunction corrected, will operate without malfunction for at least 12 hours.

13. The number X of metres of cable produced any day by a cable-making company is a normally distributed random variable with mean 5000 and standard deviation 200. On what fraction of the days the company operates will the number of metres of cable produced exceed 5500?

8.8 FIRST-ORDER SEPARABLE AND LINEAR DIFFERENTIAL EQUATIONS

This final section on applications of integration concentrates on application of the indefinite integral rather than of the definite integral. We can use the techniques of integration developed in Chapter 6 to solve certain kinds of first-order differential equations that arise in a variety of modeling situations. We have already seen some examples of applications of differential equations to modeling growth and decay phenomena in Section 3.7.

Separable Equations

Consider the logistic equation introduced in Section 3.7 to model the growth of an animal population with limited food supply:

$$\frac{dy}{dt} = ky\left(1 - \frac{y}{L}\right),$$

where $y(t)$ is the size of the population at time t, k is a positive constant related to the fertility of the population, and L is the steady-state population size that can be sustained by the available food supply. This equation is an example of a class of first-order differential equations called **separable equations** because when they are written in terms of differentials they can be separated with only the dependent variable on one side of the equation and only the independent variable on the other. The logistic equation can be written in the form

$$\frac{L\,dy}{y(L-y)} = k\,dt,$$

and solved by integrating both sides. Expanding the left side in partial fractions and integrating, we get

$$\int \left(\frac{1}{y} + \frac{1}{L-y}\right) dy = kt + C.$$

Assuming that $0 < y < L$, we therefore obtain

$$\ln\left(\frac{y}{L-y}\right) = \ln y - \ln(L-y) = kt + C,$$

and this equation can be solved for y to give

$$y = \frac{C_1 L e^{kt}}{1 + C_1 e^{kt}}, \qquad \text{where } C_1 = e^C.$$

Generally, separable equations are of the form

$$\frac{dy}{dx} = f(x)g(y).$$

We solve them by rewriting them in the form

$$\frac{dy}{g(y)} = f(x)\,dx$$

and integrating both sides.

EXAMPLE 8.8.1 Solve the equation $\dfrac{dy}{dx} = \dfrac{x}{y}$.

SOLUTION We have $y\,dy = x\,dx$, so, on integrating both sides, we get $\frac{1}{2}y^2 = \frac{1}{2}x^2 + C$ or $y^2 - x^2 = C_1$ where $C_1 = 2C$ is an arbitrary constant. The solution curves are rectangular hyperbolas with asymptotes $y = x$ and $y = -x$.

EXAMPLE 8.8.2 Find the function $y(x)$ that satisfies $\dfrac{dy}{dx} = x^2y^3$ and the initial condition $y(1) = 3$.

SOLUTION We have $\dfrac{dy}{y^3} = x^2\,dx$, so $\displaystyle\int \frac{dy}{y^3} = \int x^2\,dx$ and

$$\frac{-1}{2y^2} = \frac{x^3}{3} + C.$$

Since $y = 3$ when $x = 1$, we have $-\frac{1}{18} = \frac{1}{3} + C$ and $C = -\frac{7}{18}$. Substituting this value into the above solution and solving for y, we obtain

$$y(x) = \frac{3}{\sqrt{7 - 6x^3}}.$$

This solution is valid for $x \leq \left(\frac{7}{6}\right)^{1/3}$.

EXAMPLE 8.8.3 (**A Solution Concentration Problem**) Initially a tank contains 1000 L of brine with 50 kg of dissolved salt. If brine containing 10 g of salt per litre is flowing into the tank at a constant rate of 10 L/min, if the contents of the tank are kept thoroughly mixed at all times, and if the solution also flows out at 10 L/min, how much salt remains in the tank at the end of 40 min?

SOLUTION Let $x(t)$ be the number of kg of salt in solution in the tank after t minutes. Thus $x(0) = 50$. Salt is coming into the tank at a rate of 10 g/L $\times$ 10 L/min = 100 g/min = 1/10 kg/min. At all times the tank contains 1000 L of liquid, so the concentration of salt in the tank at time t is $x/1000$ kg/L. Since the contents flow out at 10 L/min, salt is being removed at a rate of $10x/1000 = x/100$ kg/min. Therefore,

$$\frac{dx}{dt} = \text{rate in} - \text{rate out} = \frac{1}{10} - \frac{x}{100} = \frac{10 - x}{100}$$

or

$$\frac{dx}{10 - x} = \frac{dt}{100}.$$

Integrating both sides of this equation, we obtain

$$-\ln|10 - x| = \frac{t}{100} + C.$$

Observe that $x(t) \neq 10$ for any finite time t (since $\ln 0$ is not defined). Since $x(0) = 50 > 10$, it follows that $x(t) > 10$ for all $t > 0$. ($x(t)$ is necessarily continuous so it cannot take any value less than 10 without somewhere taking the value 10 by the Intermediate-Value Theorem.) Hence

$$\ln(x - 10) = -\frac{t}{100} - C.$$

Since $x(0) = 50$, we have $-C = \ln 40$ and

$$x = x(t) = 10 + 40e^{-t/100}.$$

After 40 min there will be $10 + 40e^{-0.4} \approx 36.8$ kg of salt in the tank.

EXAMPLE 8.8.4 (**A Rate of Reaction Problem**) In a chemical reaction that goes to completion in solution, one molecule of each of two reactants, A and B, combines to form each molecule of the product C. According to the law of mass action, the reaction proceeds at a rate proportional to the product of the concentrations of A and B in the solution. Thus if there were initially present a molecules/cm^3 of A and b molecules/cm^3 of B, then the number $x(t)$ of molecules/cm^3 of C present at time t thereafter is determined by the differential equation

$$\frac{dx}{dt} = k(a - x)(b - x).$$

We solve this equation by the technique of partial fraction decomposition under the assumption that $b \neq a$.

$$\int \frac{dx}{(a-x)(b-x)} = k\int dt = kt + C.$$

Since $\dfrac{1}{(a-x)(b-x)} = \dfrac{1}{b-a}\left(\dfrac{1}{a-x} - \dfrac{1}{b-x}\right)$, and since necessarily $x \leq a$ and $x \leq b$, we have

$$\frac{1}{b-a}\big(-\ln(a-x) + \ln(b-x)\big) = kt + C,$$

or

$$\ln\left(\frac{b-x}{a-x}\right) = (b-a)\,kt + C_1 \qquad (C_1 = (b-a)C).$$

By assumption, $x(0) = 0$, so $C_1 = \ln(b/a)$ and

$$\ln\frac{a}{b}\left(\frac{b-x}{a-x}\right) = (b-a)\,kt.$$

This equation can be solved for x to yield

$$x = x(t) = \frac{ab(e^{(b-a)kt} - 1)}{be^{(b-a)kt} - a}.$$

EXAMPLE 8.8.5 Find a family of curves, each of which intersects each parabola with equation of the form $y = Cx^2$ at right angles.

SOLUTION The family of parabolas $y = Cx^2$ satisfies the differential equation

$$\frac{d}{dx}\left(\frac{y}{x^2}\right) = \frac{d}{dx}C = 0,$$

that is,

$$x^2\frac{dy}{dx} - 2xy = 0 \qquad \text{or} \qquad \frac{dy}{dx} = \frac{2y}{x}.$$

Any curve that meets the parabolas $y = Cx^2$ at right angles must, at any point (x, y) on it, have slope equal to the negative reciprocal of the slope of the particular parabola passing through that point. Thus such a curve must satisfy

$$\frac{dy}{dx} = -\frac{x}{2y}.$$

Separation of the variables leads to $2y\,dy = -x\,dx$, and integration of both sides then yields $y^2 = -\frac{1}{2}x^2 + C_1$ or $x^2 + 2y^2 = C_2$, $(C_2 = 2C_1)$. This equation represents a family of ellipses centred at the origin. Each ellipse meets each parabola at right angles, as shown in the Fig. 8.8.1.

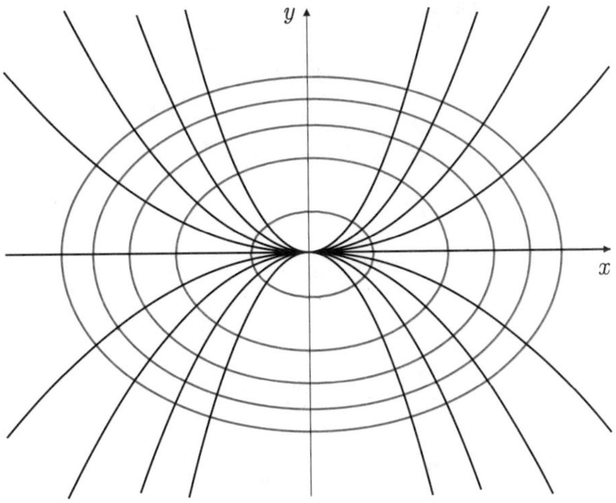

FIGURE 8.8.1

First-Order Linear Equations

A first-order **linear** differential equation is one of the type

$$\frac{dy}{dx} + p(x)y = q(x),$$

where $p(x)$ and $q(x)$ are given functions, which we assume to be continuous. We can solve such equations (that is, find y as a function of x) by the following procedure.

Let $\mu(x)$ be any antiderivative of $p(x)$; thus $d\mu/dx = p(x)$. If $y = y(x)$ satisfies the given equation, then we calculate, using the product rule,

$$\begin{aligned}\frac{d}{dx}\left(e^{\mu(x)}y(x)\right) &= e^{\mu(x)}\frac{dy}{dx} + e^{\mu(x)}\frac{d\mu}{dx}\,y(x)\\ &= e^{\mu(x)}\left(\frac{dy}{dx} + p(x)y\right) = e^{\mu(x)}\,q(x).\end{aligned}$$

Therefore,

$$e^{\mu(x)}\,y(x) = \int e^{\mu(x)}\,q(x)\,dx$$

or

$$y(x) = e^{-\mu(x)}\int e^{\mu(x)}\,q(x)\,dx.$$

We reuse this method, rather than the final formula, in the examples below. $\mu(x)$ is called an **integrating factor**.

EXAMPLE 8.8.6 Solve $\dfrac{dy}{dx} + \dfrac{y}{x} = 1$ for $x > 0$.

SOLUTION Here we want $d\mu/dx = 1/x$, so $\mu(x) = \ln x$ (for $x > 0$). Thus $e^{\mu(x)} = x$ and we calculate

$$\frac{d}{dx}(xy) = x\frac{dy}{dx} + y = x\left(\frac{dy}{dx} + \frac{y}{x}\right) = x,$$

and

$$xy = \int x\,dx = \frac{1}{2}x^2 + C.$$

Finally,

$$y = \frac{1}{x}\left(\frac{1}{2}x^2 + C\right) = \frac{x}{2} + \frac{C}{x}.$$

This function is a solution of the given equation for any value of the constant C.

EXAMPLE 8.8.7 Solve $\dfrac{dy}{dx} + xy = x^3$.

SOLUTION Here $p(x) = x$ so $\mu(x) = x^2/2$. We calculate

$$\frac{d}{dx}\left(e^{x^2/2}y\right) = e^{x^2/2}\frac{dy}{dx} + e^{x^2/2}xy = x^3 e^{x^2/2}.$$

Thus,

$$\begin{aligned} e^{x^2/2}y &= \int x^3 e^{x^2/2}\,dx \qquad \text{Let} \quad U = x^2 \qquad dV = x\,e^{x^2/2}\,dx \\ & \qquad\qquad\qquad\qquad \text{Then} \quad dU = 2x\,dx \qquad V = e^{x^2/2} \\ &= x^2 e^{x^2/2} - 2\int x\,e^{x^2/2}\,dx \\ &= x^2 e^{x^2/2} - 2e^{x^2/2} + C, \end{aligned}$$

and, finally,

$$y = x^2 - 2 + Ce^{-x^2/2}.$$

EXAMPLE 8.8.8 A savings account is opened with a deposit of A dollars. At any time t years thereafter, money is being continually deposited into the account at a rate of $(C+Dt)$ dollars per year. If interest is also being paid into the account at a nominal rate of $100R\%$ per year, compounded continuously, find the balance $B(t)$ dollars in the account after t years. Illustrate the solution for the data $A = 5000$, $C = 1000$, $D = 200$, $R = 0.13$, and $t = 5$.

SOLUTION As noted in Section 3.7, continuous compounding of interest at a nominal rate of $100R\%$ causes \$1.00 to grow to $\$e^{Rt}$ in t years. Without subsequent deposits, the balance in the account would grow according to the differential equation of exponential growth:

$$\frac{dB}{dt} = RB.$$

Allowing for additional growth due to the continual deposits, we obtain that B must satisfy the differential equation

$$\frac{dB}{dt} = RB + (C + Dt)$$

or, equivalently, $dB/dt - RB = C + Dt$. This is a linear equation for B having $p(t) = -R$. Hence we may take $\mu(t) = -Rt$ and $e^{\mu(t)} = e^{-Rt}$. We now calculate

$$\frac{d}{dt}\left(e^{-Rt}B(t)\right) = e^{-Rt}\frac{dB}{dt} - Re^{-Rt}B(t) = (C + Dt)\,e^{-Rt},$$

and

$$\begin{aligned} e^{-Rt}B(t) &= \int (C + Dt)e^{-Rt}\,dt \qquad \text{Let } U = C + Dt \quad dV = e^{-Rt}\,dt \\ &\qquad\qquad\qquad\qquad\quad \text{Then } dU = D\,dt \quad V = -e^{-Rt}/R \\ &= -\frac{C + Dt}{R}e^{-Rt} + \frac{D}{R}\int e^{-Rt}\,dt \\ &= -\frac{C + Dt}{R}e^{-Rt} - \frac{D}{R^2}e^{-Rt} + K \qquad (K = \text{constant}). \end{aligned}$$

Hence

$$B(t) = -\frac{C + Dt}{R} - \frac{D}{R^2} + Ke^{Rt}.$$

Since $A = B(0) = -\frac{C}{R} - \frac{D}{R^2} + K$, we have $K = A + \frac{C}{R} + \frac{D}{R^2}$ and

$$B(t) = \left(A + \frac{C}{R} + \frac{D}{R^2}\right)e^{Rt} - \frac{C + Dt}{R} - \frac{D}{R^2}.$$

For the illustration $A = 5000$, $C = 1000$, $D = 200$, $R = 0.13$, and $t = 5$, we calculate, using a scientific calculator, $B(5) = 19,762.82$. Thus the account will contain \$19,762.82, after 5 years, under these circumstances.

EXERCISES

Solve the differential equations in Exercises 1–16

1. $\frac{dy}{dx} = \frac{y}{2x}$
2. $\frac{dy}{dx} = \frac{3y - 1}{x}$
3. $\frac{dy}{dx} = \frac{x^2}{y^2}$
4. $\frac{dy}{dx} = x^2y^2$
5. $\frac{dY}{dt} = tY$
6. $\frac{dx}{dt} = e^x \sin t$
7. $\frac{dy}{dx} = 1 - y^2$
8. $\frac{dy}{dx} = 1 + y^2$
9. $\frac{dy}{dt} = 2 + e^y$
10. $\frac{dy}{dx} = y^2(1 - y)$
11. $\frac{dy}{dx} - \frac{2y}{x} = x^2$
12. $\frac{dy}{dx} + \frac{2y}{x} = \frac{1}{x^2}$
13. $\frac{dy}{dx} + 2y = 3$
14. $\frac{dy}{dx} + y = e^x$
15. $\frac{dy}{dx} + y = x$
16. $\frac{dy}{dx} + 2e^x y = e^x$
17. Why is the solution given for the chemical reaction rate problem in Example 8.8.4 not valid for $a = b$? Find the solution for the case $a = b$.

18. An object of mass m falling near the surface of the earth is retarded by air resistance proportional to its velocity so that, according to Newton's second law of motion,

$$m\frac{dv}{dt} = mg - kv,$$

where $v = v(t)$ is the velocity of the object at time t, and g is the acceleration of gravity near the surface of the earth. Assuming that the object falls from rest at time $t = 0$, that is, $v(0) = 0$, find the velocity $v(t)$ for any $t > 0$ (up until the object strikes the ground). Show $v(t)$ approaches a limit as $t \to \infty$. Do you need the explicit formula for $v(t)$ to determine this limiting velocity?

19. Repeat Exercise 18 except assuming that the air resistance is proportional to the square of the velocity so that the equation of motion is

$$m\frac{dv}{dt} = mg - kv^2.$$

20. Find the amount in a savings account after 1 year if th initial balance in the account was $1000, if the interes is paid continuously into the account at a nominal rat of 10% per annum, compounded continually, and if th account is being continuously depleted (by taxes, say) a a rate of $y^2/1,000,000$ dollars per year, where $y = y(t$ is the balance in the account after t years. How larg can the account grow? How long will it take the accoun to grow to half this balance?

21. Find the family of curves each of which intersects al of the hyperbolas $xy = C$ at right angles.

22. Repeat the solution concentration problem in Exampl 8.8.3, changing the rate of inflow of brine into the tan to 12 L/min but leaving all the other data as they wer in that example. Note that the volume of liquid in th tank is no longer constant as time increases.

CHAPTER 9

Plane Curves

9.1 CONICS

Circles, ellipses, parabolas and hyperbolas are called **conic sections**, (or, more simply, just **conics**), because they are curves in which planes intersect right-circular cones.

To be specific, suppose that a line A is fixed in space, and V is a point fixed on A. The right-circular cone having axis A, vertex V, and **semi-vertical angle** α is the surface consisting of all points on straight lines through V which make angle α with the line A. (See Fig. 9.1.1.1.) The cone has two halves (called **nappes**) lying on opposite sides of the vertex V. Any plane P which does not pass through V will intersect the cone (one or both nappes) in a curve $\mathcal{C}$. (See Fig. 9.1.1.2.) If a line normal (that is, perpendicular) to P makes angle θ with the axis A of the cone, where $0 \le \theta \le \pi/2$, then

$$\begin{array}{ll} \mathcal{C} \text{ is a circle if} & \theta = 0 \\ \mathcal{C} \text{ is an ellipse if} & 0 < \theta < \dfrac{\pi}{2} - \alpha \\ \mathcal{C} \text{ is a parabola if} & \theta = \dfrac{\pi}{2} - \alpha \\ \mathcal{C} \text{ is a hyperbola if} & \theta > \dfrac{\pi}{2} - \alpha. \end{array}$$

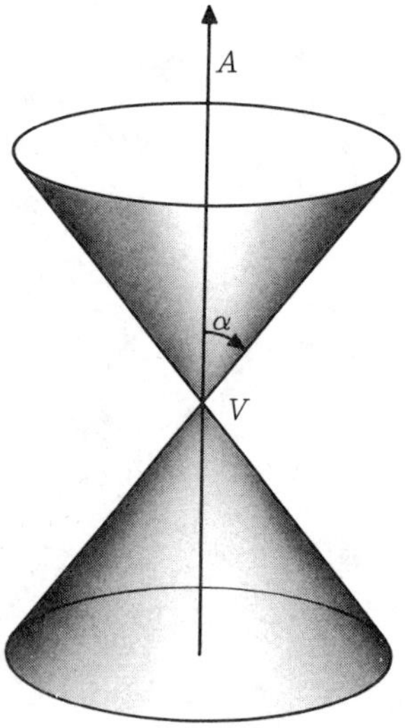

FIGURE 9.1.1

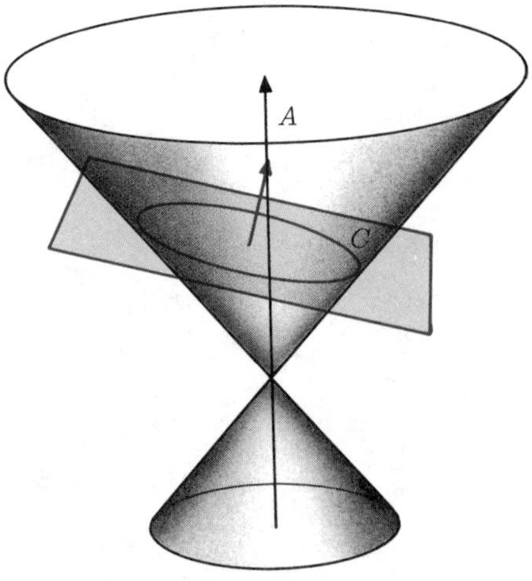

FIGURE 9.1.2

Since planes are represented by first degree equations and cones by second degree equations, all conics can be represented analytically (in terms of Cartesian coordinates x and y in the plane of the conic) by a second degree equation of the general form

$$Ax^2 + Bxy + Cy^2 + Dx + Ey + F = 0,$$

where A, B, ..., F are constants. However, any such an equation can represent a conic, the empty set, a single point, or, if the left-hand side factors into linear factors,

$$(A_1x + B_1y + C_1)(A_2x + B_2y + C_2) = 0,$$

one or two straight lines.

After straight lines the conic sections are the simplest of plane curves. They have many important properties, most of which were discovered by the Greek geometer, Appolonius of Perga, about 200 BC. It is remarkable that he was able to obtain these properties using only the techniques of classical Euclidean geometry; today most of these properties are expressed more conveniently using analytic geometry and specific coordinate systems.

The rest of this section is devoted to giving alternative definitions of the conics and developing some of their more useful properties. We will not attempt to prove every assertion, but, in the spirit of Appolonius, we will present proofs of the focal properties of the conics by elementary geometry.

Circles

9.1.1 Circles

> A **circle** consists of all points in a plane which are at constant distance (the **radius**) from a fixed point (the **centre**).

EXAMPLE 9.1.2 The circle with centre at the point (a, b) and radius r has equation

$$(x - a)^2 + (y - b)^2 = r^2,$$

or, equivalently,

$$x^2 + y^2 - 2ax - 2by + c = 0,$$

where $c = a^2 + b^2 - r^2$. (See Fig. 9.1.1.3.) Note that in the equation of a circle, the coefficients of x^2 and y^2 are equal and there is no xy term.

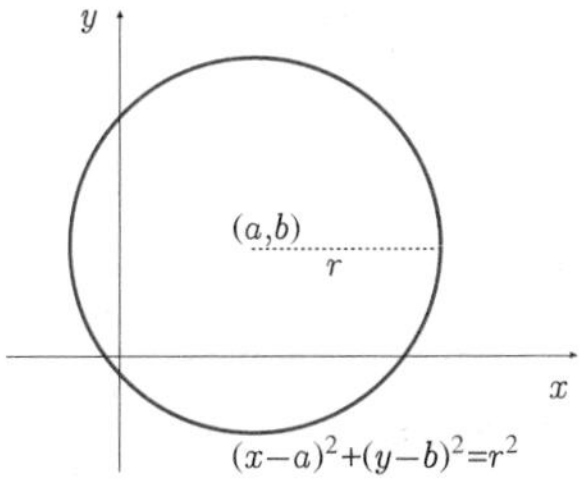

FIGURE 9.1.3

Circles have many useful geometrical properties. Of particular importance is the fact that a *radial line* drawn from the centre to a point on the circle is perpendicular to the line tangent to the circle at that point.

Parabolas

9.1.3 Parabolas

> A **parabola** consists of points in the plane which are equidistant from a given point (the **focus**) and a given straight line (the **directrix**). The line through the focus perpendicular to the directrix is called the **principal axis** (or simply the **axis**) of the parabola. The **vertex** of the parabola is the point where the parabola crosses its principal axis. It is on the axis halfway between the focus and the directrix.

EXAMPLE 9.1.4 The parabola with focus at the point $F = (a, 0)$ and directrix the line L with equation $x = -a$ has axis along the x-axis and vertex at the origin (see Fig. 9.1.4). If $P = (x, y)$ is any point on the parabola, then the distance from P to F is equal to the perpendicular distance from P to L. Thus

$$\sqrt{(x-a)^2 + y^2} = x + a$$
$$\text{or } x^2 - 2ax + a^2 + y^2 = x^2 + 2ax + a^2,$$

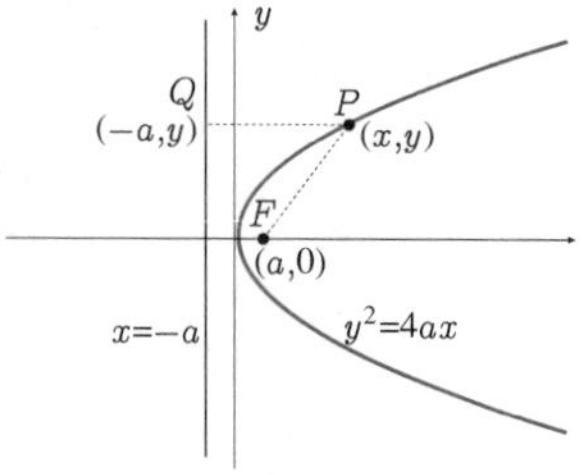

FIGURE 9.1.4

or, upon simplification,

$$y^2 = 4ax.$$

Similarly we can obtain standard equations for parabolas with vertices at the origin and foci at $(-a, 0)$, $(0, a)$ and $(0, -a)$:

Focus	Directrix	Equation
$(a, 0)$	$x = -a$	$y^2 = 4ax$
$(-a, 0)$	$x = a$	$y^2 = -4ax$
$(0, a)$	$y = -a$	$x^2 = 4ay$
$(0, -a)$	$y = a$	$x^2 = -4ay$

Parabolas (and other conics), whose axes of symmetry are not parallel to one of the coordinate axes, may have a *cross term* xy in their equations.

EXAMPLE 9.1.5 Show that the equation $x^2 + y^2 + 2xy - x + y = 0$ represents a parabola with vertex at the origin and axis along the line $x + y = 0$. Find its focus and directrix.

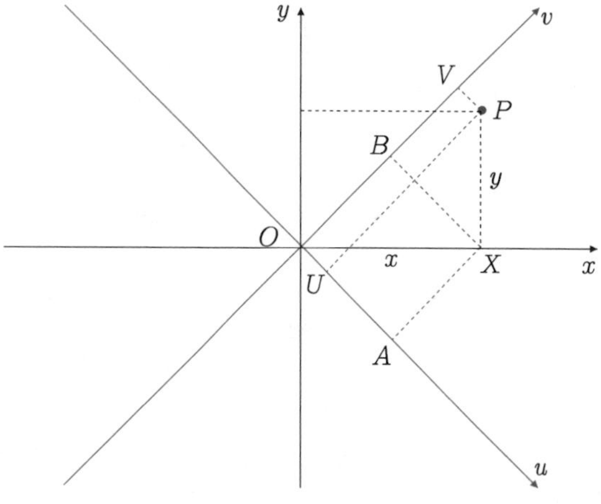

FIGURE 9.1.5

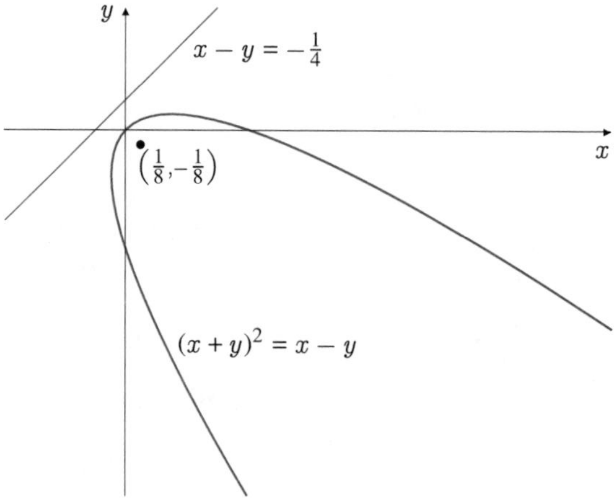

FIGURE 9.1.6

SOLUTION The equation can be rewritten with completed square in the form

$$(x + y)^2 = x - y.$$

Let (u, v) be the Cartesian coordinates of a point P with respect to axes u and v rotated $-45°$ from the x- and y-axes. (See Fig. 9.1.1.5.) (The u- and v-axes lie along the perpendicular lines $x + y = 0$ and $x - y = 0$ respectively.) As can be observed in the figure,

$$u = OU = OA - UA = \frac{x}{\sqrt{2}} - \frac{y}{\sqrt{2}}, \qquad v = OV = OB + BV = \frac{x}{\sqrt{2}} + \frac{y}{\sqrt{2}}.$$

With respect to these new coordinates, the given curve has equation

$$v^2 = \frac{u}{\sqrt{2}},$$

which is evidently a parabola with vertex at the origin, focus at the point $(u, v) = (\frac{1}{4\sqrt{2}}, 0)$ and directrix $u = -\frac{1}{4\sqrt{2}}$. In terms of the original coordinates (x, y) the focus is $(\frac{1}{8}, -\frac{1}{8})$ and the directrix is $x - y = -\frac{1}{4}$. (See Fig. 9.1.1.6.)

The Focal Property of a Parabola

All of the conic sections have interesting and useful focal properties relating to the way they would reflect light if they were used to generate surfaces of revolution which were mirrors. For instance, a circle will clearly reflect back along the same path any ray of light incident along a line passing through its centre. The focal properties of parabolas, ellipses and hyperbolas can be derived from the reflecting property of a straight line (that is, a plane mirror) by elementary geometrical arguments.

It is well-known that in a medium of constant optical density, (one where light travels with constant speed), light travels in straight lines. This is a consequence of the physical Principle of Least Action which asserts that in travelling between two points, light takes the path requiring the minimum travel time. Given a straight line L in a plane and two points A and B in the plane on the same side of L, the point P on L for which the sum of the distances $AP + PB$ is minimum is such that AP and PB make equal angles with L, or equivalently, with the normal to L at P. (See Fig. 9.1.1.7.) If B' is the point such that L is the right bisector of the line segment BB', then P is the intersection of L and AB'. Since one side of a triangle cannot exceed the sum of the other two sides,

$$AP + PB = AP + PB' \leq AQ + QB' = AQ + QB.$$

9.1.6
Reflection in a Straight Line

> The point P on L at which a ray from A would reflect so as to pass through B is just the point which minimizes the sum of the distances $AP + PB$.

Now consider a parabola with focus F and directrix D. Let P be on the parabola and let T be the line tangent to the parabola at P. (See Fig. 9.1.1.8.) Let Q be any point on T. Then FQ meets the parabola at a point X between F and Q. Let M and N be points on D such that MX and NP are perpendicular to D, and let A be a point on the line through N and P which lies on the same side of the parabola as F. We have

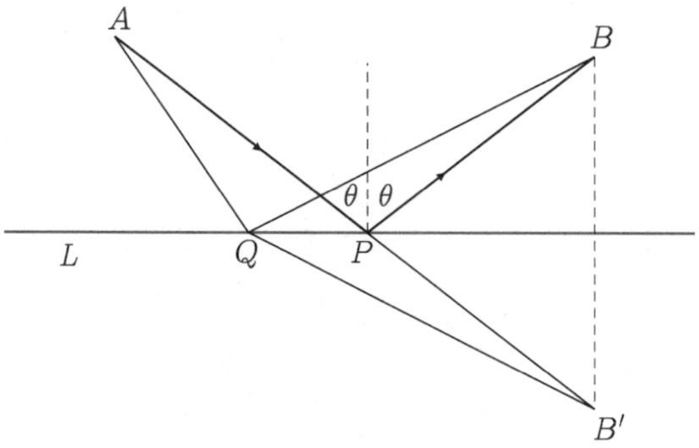

FIGURE 9.1.7

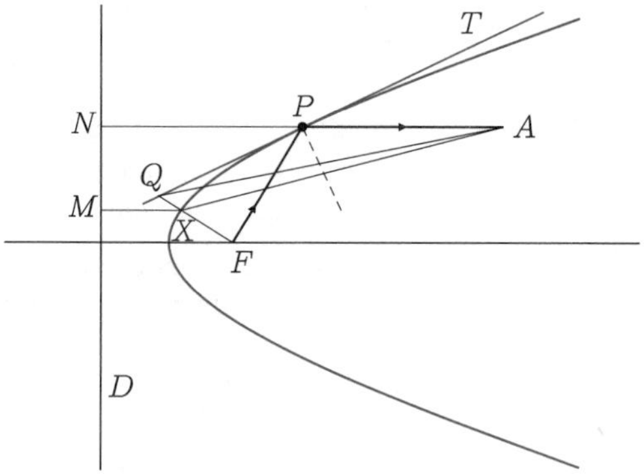

FIGURE 9.1.8

$$\begin{aligned} FP + PA = NP + PA = NA &\leq MX + XA = FX + XA \\ &\leq FX + XQ + QA = FQ + QA. \end{aligned}$$

Thus, among all points Q on the line T, $Q = P$ is the one which minimizes the sum of distances $FQ + QA$. By the observation made for straight lines above, FP and PA make equal angles with T and so also with the normal to the parabola at P. (The parabola and the tangent line have the same normal at P.)

9.1.7
Reflection by a Parabola

> Any ray from the focus will be reflected parallel to the axis of the parabola. Equivalently, any ray incident parallel to the axis of the parabola will be reflected through the focus.

Ellipses

9.1.8
Ellipses

> An **ellipse** consists of all points in the plane, the sum of whose distances from two fixed points (the **foci**) is constant.

EXAMPLE 9.1.9 If the foci of an ellipse are at the points $(-c, 0)$ and $(c, 0)$ and the sum of the distances from any point $P = (x, y)$ on the ellipse to these two foci is the constant $2a$ (where $0 < c < a$) then the ellipse clearly passes through the four points $(a, 0)$, $(-a, 0)$, $(0, b)$ and $(0, -b)$, where $b^2 = a^2 - c^2$. (See Fig. 9.1.1.9.) Also,

$$\sqrt{(x-c)^2 + y^2} + \sqrt{(x+c)^2 + y^2} = 2a.$$

Transposing one term from the left side to the right side, squaring, cancelling terms, transposing and squaring again leads to (the reader should verify this!)

$$\frac{x^2}{a^2} + \frac{y^2}{b^2} = 1.$$

The following quantities describe this ellipse:

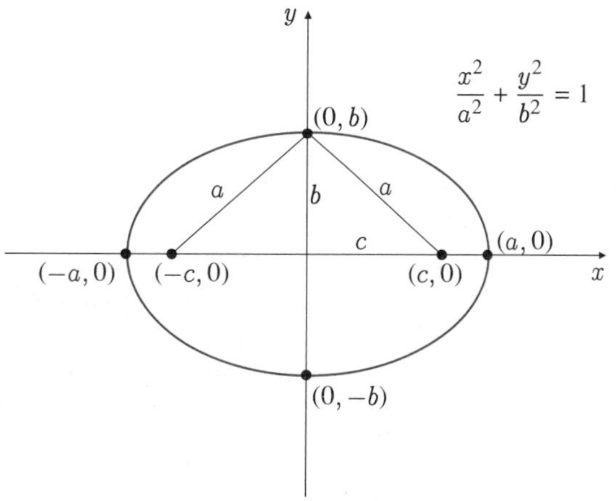

FIGURE 9.1.9

a	is the **semi-major axis**
b	is the **semi-minor axis**
$c = \sqrt{a^2 - b^2}$	is the **semi-focal separation**.

The point halfway between the foci is called the **centre** of the ellipse. In the Example above it is the origin. Note that $a > b$ in this Example. If $a < b$ then the ellipse has semi-major axis along the y-axis instead of the x-axis and its foci are at $(0, c)$ and $(0, -c)$, where $c = \sqrt{b^2 - a^2}$. In general, the line containing the foci (the **major axis**), and the line through the centre perpendicular to that line (the **minor axis**) are called the **principal axes** of the ellipse.

The **eccentricity** of an ellipse is the ratio of the semi-focal separation to the semi-major axis. We denote the eccentricity ε. For the ellipse $\dfrac{x^2}{a^2} + \dfrac{y^2}{b^2} = 1$ with $a > b$

$$\varepsilon = \frac{c}{a} = \frac{\sqrt{a^2 - b^2}}{a}.$$

Evidently $\varepsilon < 1$ for any ellipse; the greater the value of ε the more elongated (less circular) is the ellipse. If $\varepsilon = 0$ so that $a = b$ and $c = 0$, the two foci coincide and the ellipse is a circle.

EXAMPLE 9.1.10 The equation

$$\frac{(x-2)^2}{4} + \frac{(y+3)^2}{9} = 1$$

represents an ellipse with centre at the point $(2, -3)$, semi-major axis 3, semi-minor axis 2, and semi-focal separation $\sqrt{9 - 4} = \sqrt{5}$. The foci are at the points $(2, -3 \pm \sqrt{5})$. The eccentricity is $\varepsilon = \sqrt{5}/3$.

The Focal Property of an Ellipse

Let P be any point on an ellipse having foci F_1 and F_2. The normal to the ellipse at P bisects the angle between the lines F_1P and F_2P.

9.1.11 Reflection by an Ellipse

> Any ray coming from one focus of an ellipse will be reflected through the other focus.

To see this, observe that if Q is any point on the line T tangent to the ellipse at P then F_1Q meets the ellipse at a point X between F_1 and Q (see Fig. 9.1.1.10) and so

$$F_1P + PF_2 = F_1X + XF_2 \leq F_1X + XQ + QF_2 = F_1Q + QF_2.$$

Among all points on T, P is the one which minimizes the sum of the distances to F_1 and F_2. This implies that the normal to the ellipse at P bisects the angle F_1PF_2.

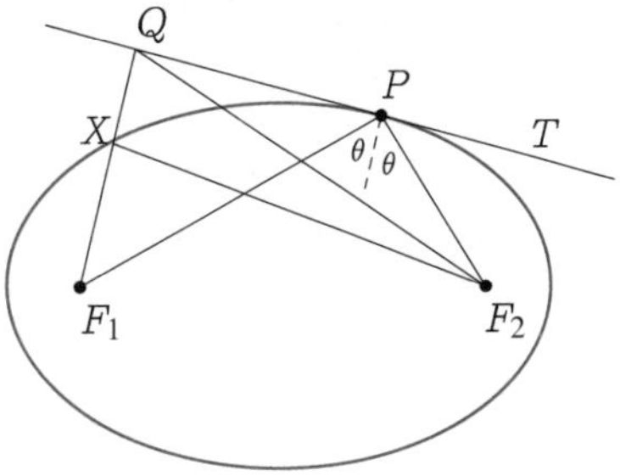

FIGURE 9.1.10

Directrices

If $a > b > 0$, each of the lines $x = a/\varepsilon$ and $x = -a/\varepsilon$ is called a **directrix** of the ellipse $\dfrac{x^2}{a^2} + \dfrac{y^2}{b^2} = 1$. If P is on the ellipse then the ratio of the distance from P to a focus to its distance from the corresponding directrix is equal to the eccentricity ε. If $P = (x, y)$, F is the focus $(c, 0)$, Q is on the corresponding directrix $x = a/\varepsilon$ and PQ is perpendicular to the directrix, then (see Fig. 9.1.1.11)

FIGURE 9.1.11

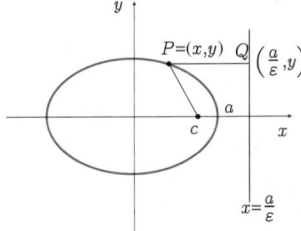

$$
\begin{aligned}
PF^2 &= (x - c)^2 + y^2 \\
&= x^2 - 2cx + c^2 + b^2\left(1 - \frac{x^2}{a^2}\right) \\
&= x^2\left(\frac{a^2 - b^2}{a^2}\right) - 2cx + a^2 - b^2 + b^2 \\
&= \varepsilon^2 x^2 - 2\varepsilon a x + a^2 && \text{(because } c = \varepsilon a\text{)} \\
&= (a - \varepsilon x)^2.
\end{aligned}
$$

Thus $PF = a - \varepsilon x$. Also, $QP = (a/\varepsilon) - x = (a - \varepsilon x)/\varepsilon$. Therefore $PF/QP = \varepsilon$ as asserted.

A parabola may be considered as the limiting case of an ellipse whose eccentricity has increased to 1. The distance between the foci is infinite, so the centre, one focus and its corresponding directrix have moved off to infinity leaving only one focus and its directrix in the finite plane.

Hyperbolas

9.1.12
Hyperbolas

A **hyperbola** consists of all points in the plane, the difference of whose distances from two fixed points (the **foci**) is constant.

EXAMPLE 9.1.13 If the foci are $F_1 = (c, 0)$ and $F_2 = (-c, 0)$, and the difference of the distances from $P = (x, y)$ to these foci is $2a$ (where $a < c$) then

$$PF_2 - PF_1 = \sqrt{(x+c)^2 + y^2} - \sqrt{(x-c)^2 + y^2} = \begin{cases} 2a & \text{(right branch)} \\ -2a & \text{(left branch).} \end{cases}$$

Simplifying this equation by squaring and transposing as was suggested for the ellipse (do it!), we are led to the standard equation for the hyperbola:

$$\frac{x^2}{a^2} - \frac{y^2}{b^2} = 1,$$

where $b^2 = c^2 - a^2$. The points $(a, 0)$ and $(-a, 0)$ (called **vertices**) lie on the hyperbola, one on each branch. (The two branches correspond to the intersections of the plane of the hyperbola with the two nappes of a cone.) Parameters used to describe the hyperbola are

a	the **semi-transverse axis**
b	the **semi-conjugate axis**
$c = \sqrt{a^2 + b^2}$	the **semi-focal separation**.

The midpoint of the line segment F_1F_2 (in this case the origin) is again called the centre of the hyperbola. If a rectangle with sides $2a$ and $2b$ is drawn centred at the centre of the hyperbola and with two sides parallel to the transverse axis (the line containing the foci and the vertices), the two straight lines which are diagonals of the box are called **asymptotes** of the hyperbola. They have equations $(x/a) \pm (y/b) = 0$; that is, they are solutions of the degenerate equation

$$\frac{x^2}{a^2} - \frac{y^2}{b^2} = 0.$$

The hyperbola approaches arbitrarily close to these lines as it recedes from the origin. (See Fig. 9.1.1.12.) A **rectangular** hyperbola is one whose asymptotes are perpendicular lines.

The equation

$$\frac{x^2}{a^2} - \frac{y^2}{b^2} = -1$$

represents a hyperbola with the same asymptotes as the hyperbola in the above example, but with transverse axis along the y-axis, vertices at $(0, b)$ and $(0, -b)$ and foci at $(0, c)$ and $(0, -c)$. The two hyperbolas are said to be **conjugate** to one another. (See Fig. 9.1.12.) The **conjugate axis** of a hyperbola is the transverse axis of the conjugate hyperbola. Together, the transverse and conjugate axes of a hyperbola are called its **principal axes**.

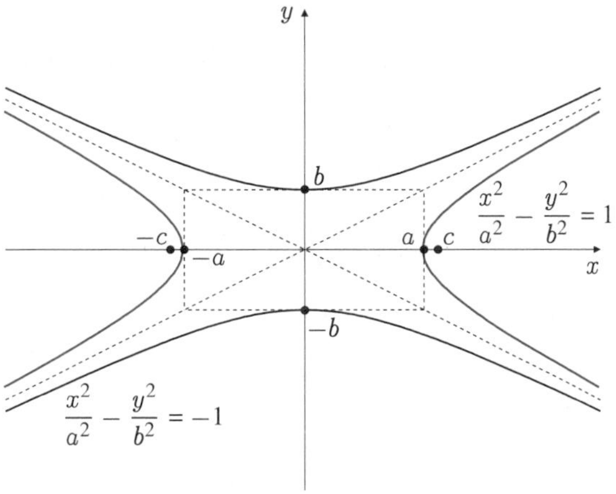

FIGURE 9.1.12

The eccentricity of the hyperbola is

$$\varepsilon = \frac{c}{a} = \frac{\sqrt{a^2 + b^2}}{a}.$$

Evidently $\varepsilon > 1$. The lines $x = \pm(a/\varepsilon)$ are called **directrices** of the hyperbola $(x^2/a^2) - (y^2/b^2) = 1$. In a manner similar to that used for the ellipse one can show that if P is on the hyperbola then

$$\frac{\text{distance from } P \text{ to a focus}}{\text{distance from } P \text{ to the corresponding directrix}} = \varepsilon.$$

The eccentricity of a rectangular hyperbola is $\sqrt{2}$.

The Focal Property of the Hyperbola

Let P be any point on a hyperbola with foci F_1 and F_2. Then the tangent line to the hyperbola at P bisects the angle between the lines F_1P and F_2P.

9.1.14 Reflection by a Hyperbola

> A ray from one focus of a hyperbola is reflected by the hyperbola so that it appears to have come from the other focus.

To see this, let P be on the right branch, let T be the line tangent to the hyperbola at P, and let C be a circle of large radius centred at F_2 (see Fig. 9.1.13). Let F_2P intersect this circle at D. Let Q be any point on T. Then QF_1 meets the hyperbola at X between Q and F_1, and F_2X meets C at E. Since X is on the radial line F_2E, it is closer to E than it is to other points on C. That is, $XE \leq XD$. Thus

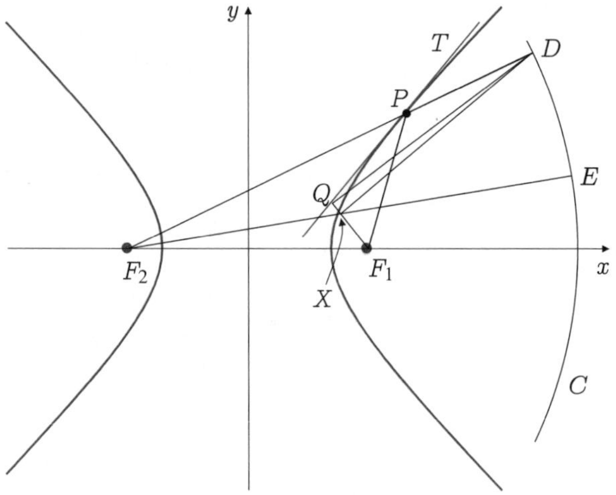

FIGURE 9.1.13

$$\begin{aligned} F_1P + PD = F_1P + F_2D - F_2P &= F_2D - (F_2P - F_1P) \\ &= F_2E - (F_2X - F_1X) \\ &= F_1X + F_2E - F_2X \\ &= F_1X + XE \\ &\le F_1X + XD \\ &\le F_1X + XQ + QD = F_1Q + QD. \end{aligned}$$

P is the point on T which minimizes the sum of distances to F_1 and D, and therefore the normal to the hyperbola at P bisects the angle F_1PD. Therefore T bisects the angle F_1PF_2.

Classifying General Conics

A second degree equation in two variables,

$$Ax^2 + Bxy + Cy^2 + Dx + Ey + F = 0, \qquad (A^2 + B^2 + C^2 > 0)$$

generally represents a conic curve, but in certain degenerate cases may represent two straight lines (e.g. $x^2 - y^2 = 0$ represents the lines $x = y$ and $x = -y$), one straight line (e.g. $x^2 = 0$ represents the line $x = 0$), a single point (e.g. $x^2 + y^2 = 0$ represents the origin), or no points at all (e.g. $x^2 + y^2 = -1$ is not satisfied by any points in the plane).

The nature of the set of points represented by a given second degree equation can be determined by rewriting the equation in a standard form which can be recognized as one of the standard types. If $B = 0$ this rewriting can be accomplished by completing the squares in x and y.

EXAMPLE 9.1.15 Describe the curve with equation $x^2 + 2y^2 + 6x - 4y + 7 = 0$.

SOLUTION The equation can be written in the form

$$x^2 + 6x + 9 + 2(y^2 - 2y + 1) = 9 + 2 - 7 = 4,$$

and hence in the form

$$\frac{(x+3)^2}{4} + \frac{(y-1)^2}{2} = 1.$$

Therefore it represents an ellipse with centre at $(-3, 1)$, semi-major axis 2, semi-minor axis $\sqrt{2}$, and foci at $(-3 \pm \sqrt{2}, 1)$.

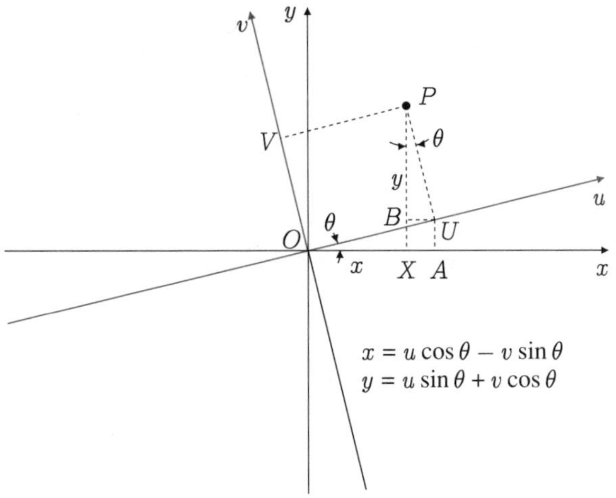

FIGURE 9.1.14

If $B \neq 0$, the equation has an xy term, and it cannot represent a circle. To see what it does represent, we can rotate the coordinate axes to produce an equation with no xy term. Let new coordinate axes (a u-axis and an v-axis) have the same origin but be rotated an angle θ from the x- and y-axes respectively. (See Fig. 9.1.1.14.) If point P has coordinates (x, y) with respect to the old axes and coordinates (u, v) with respect to the new axes, then an analysis of triangles in the figure shows that

$$\begin{aligned} x &= OA - XA = OU\cos\theta - OV\sin\theta = u\cos\theta - v\sin\theta \\ y &= XB + BP = OU\sin\theta + OV\cos\theta = u\sin\theta + v\cos\theta. \end{aligned}$$

Substituting these expressions into the equation

!!DANGER!!

Long calculations are needed here.

$$Ax^2 + Bxy + Cy^2 + Dx + Ey + F = 0, \qquad (A^2 + B^2 + C^2 > 0)$$

leads to a new equation

$$A'u^2 + B'uv + C'v^2 + D'u + E'v + F = 0$$

where

$$\begin{aligned}
A' &= A\cos^2\theta + B\cos\theta\sin\theta + C\sin^2\theta \\
B' &= (2C - 2A)\cos\theta\sin\theta + B(\cos^2\theta - \sin^2\theta) \\
C' &= A\sin^2\theta - B\cos\theta\sin\theta + C\cos^2\theta \\
D' &= D\cos\theta + E\sin\theta \\
E' &= -D\sin\theta + E\cos\theta.
\end{aligned}$$

Note that F remains unchanged. If we choose θ so that

$$\tan 2\theta = \frac{B}{A - C}, \qquad (\theta = \frac{\pi}{4} \text{ if } A = C,\ B \neq 0),$$

then $B' = 0$, and the new equation can then be analysed as described previously.

EXAMPLE 9.1.16 What curve is represented by the equation $xy = 1$?

SOLUTION The reader is likely well aware that the given equation represents a rectangular hyperbola with the coordinate axes as asymptotes. Since the given equation involves $A = C = D = E = 0$ and $B = 1$, it is appropriate to rotate the axes through angle $\pi/4$ so that

$$\begin{aligned}
x &= \frac{1}{\sqrt{2}}(u - v) \\
y &= \frac{1}{\sqrt{2}}(u + v).
\end{aligned}$$

The transformed equation is

$$\frac{u^2}{2} - \frac{v^2}{2} = 1$$

which is, as suspected, a rectangular hyperbola with vertices at $u = \pm\sqrt{2}$, $v = 0$, foci at $u = \pm 2$, $v = 0$, and asymptotes $u = \pm v$. Hence $xy = 1$ represents a rectangular hyperbola with coordinate axes as asymptotes, vertices at $(1, 1)$ and $(-1, -1)$, and foci at $(\sqrt{2}, \sqrt{2})$ and $(-\sqrt{2}, -\sqrt{2})$.

EXAMPLE 9.1.17 Show that $x^2 + 2\sqrt{3}xy - y^2 + 2\sqrt{3}x + 2y = 6$ represents a hyperbola. What are its principal axes? its vertices? its foci? its asymptotes?

SOLUTION Here $A = 1$, $B = 2\sqrt{3}$, $C = -1$, $D = 2\sqrt{3}$, $E = 2$ and $F = -6$. We rotate the axes through angle θ satisfying $\tan 2\theta = B/(A - C) = \sqrt{3}$, that is, $\theta = \pi/6$, and obtain $A' = 2$, $B' = 0$, $C' = -2$, $D' = 4$, and $E' = 0$. The transformed equation is

$$2u^2 - 2v^2 + 4u = 6,$$

or, with completed square,

$$(u + 1)^2 - v^2 = 4.$$

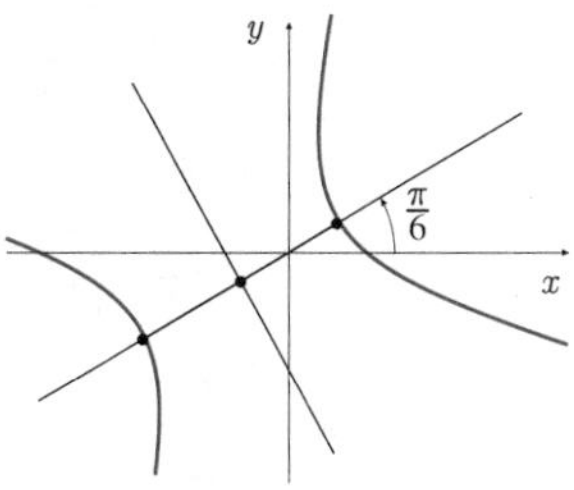

FIGURE 9.1.15

This is a rectangular hyperbola with centre at $u = -1$, $v = 0$, vertices at $u = -1 \pm 2$, $v = 0$, foci at $u = -1 \pm 2\sqrt{2}$, $v = 0$, and asymptotes $u + 1 \pm v = 0$. Since the rotation of axes is specified by

$$x = \frac{\sqrt{3}}{2}u - \frac{1}{2}v, \qquad y = \frac{u}{2} + \frac{\sqrt{3}}{2}v,$$

the given equation represents a rectangular hyperbola with centre $\left(-\frac{\sqrt{3}}{2}, -\frac{1}{2}\right)$ and transverse and conjugate axes through that point inclined at angle $\frac{\pi}{6}$ to the x- and y-axes respectively. (See Fig. 9.1.1.15.) Its vertices are at the points $\left(\frac{\sqrt{3}}{2}, \frac{1}{2}\right)$ and $\left(-\frac{3\sqrt{3}}{2}, -\frac{3}{2}\right)$, its foci at the points $\left(-\frac{\sqrt{3}(1+2\sqrt{2})}{2}, -\frac{1+2\sqrt{2}}{2}\right)$ and $\left(\frac{\sqrt{3}(-1+2\sqrt{2})}{2}, \frac{-1+2\sqrt{2}}{2}\right)$, and its asymptotes are $\sqrt{3}x + y + 2 \pm (\sqrt{3}y - x) = 0$.

EXERCISES

In Exercises 1–9 identify and sketch the set of points in the plane satisfying the given equation. Specify the asymptotes of any hyperbolas.

1. $x^2 + y^2 + 2x = -1$

2. $x^2 + 4y^2 - 4y = 0$

3. $4x^2 + y^2 - 4y = 0$

4. $4x^2 - y^2 - 4y = 0$

5. $x^2 + 2x - y = 3$

6. $x + 2y + 2y^2 = 1$

7. $x^2 - 2y^2 + 3x + 4y = 2$

8. $9x^2 + 4y^2 - 18x + 8y = -13$

9. $9x^2 + 4y^2 - 18x + 8y = 23$

10. Identify and sketch the curve which is the graph of the equation $(x - y)^2 - (x + y)^2 = 1$.

In Exercises 11–16 identify the conic and find its centre, principal axes, foci and eccentricity. Specify the asymptotes of any hyperbolas.

11. $xy + x - y = 2$

12.* $x^2 + 4xy + y^2 - 2\sqrt{2}(x - y) = 2$

13.* $x^2 + xy = \dfrac{1}{2\sqrt{2}}$

14.* $x^2 + 2xy + y^2 = 4x - 4y + 4$

15.* $8x^2 + 12xy + 17y^2 = 20$

16.* $x^2 - 4xy + 4y^2 + 2x + y = 0$

17. The *focus-directrix definition of a conic* defines a conic as a set of points P in the plane which satisfy the condition

$$\frac{\text{distance from } P \text{ to } F}{\text{distance from } P \text{ to } D} = \varepsilon$$

where F is a fixed point, D a fixed straight line, and ε a fixed positive number. The conic is an ellipse, a parabola or a hyperbola according to whether $\varepsilon < 1$, $\varepsilon = 1$ or $\varepsilon > 1$. Find the equation of the conic if F is the origin and D is the line $x = -p$.

Another parameter associated with conics is the **semi-latus rectum**, usually denoted ℓ. For a circle it is equal to the radius. For other conics it is half the length of the chord through a focus and perpendicular to the axis (for a parabola), the major axis (for an ellipse), or the transverse axis (for a hyperbola). That chord is called the **latus rectum** of the conic.

18. Show that the semi-latus rectum of the parabola is twice the distance from the vertex to the focus.

19. Show that the semi-latus rectum for an ellipse with semi-major axis a and semi-minor axis b is $\ell = b^2/a$.

20. Show that the formula in the above exercise also gives the semi-latus rectum of a hyperbola with semi-transverse axis a and semi-conjugate axis b.

21.*Suppose a plane intersects a right-circular cone in an ellipse, and that two spheres (one on each side of the plane) are inscribed between the cone and the plane so that each is tangent to the cone around a circle and is also tangent to the plane at a point. Show that the points where these two spheres touch the plane are the foci of the ellipse. Hints: All tangent lines drawn to a sphere from a given point outside the sphere are equal in length. The distance between the two circles in which the spheres intersect the cone, measured along generators of the cone (i.e. straight lines lying on the cone), is the same for all generators.

22.*State and prove a result analogous to that in the above exercise but pertaining to a hyperbola.

23.*Suppose a plane intersects a right-circular cone in a parabola with vertex at V. Suppose that a sphere is inscribed between the cone and the plane as in the previous exercises, and is tangent to the plane of the parabola at point F. Show that the chord to the parabola through F which is perpendicular to FV has length equal to that of the latus rectum of the parabola. Therefore F is the focus of the parabola.

9.2 POLAR COORDINATES AND POLAR CURVES

The **polar coordinate system** is an alternative to the rectangular (Cartesian) coordinate system for describing the location of points in a plane. In the polar coordinate system there is an origin (or **pole**), O, and a **polar axis**, a ray (that is, a half-line) extending from O horizontally to the right. The position of any point P in the plane is then determined by its polar coordinates (r, θ), where

i) r is the distance from P to O, and

ii) θ is the angle that the ray OP makes with the polar axis (counterclockwise angles being considered positive).

Figure 9.2.1 shows some points with their polar coordinates. The rectangular axes x and y are usually shown on a polar graph. The polar axis coincides with the positive x-axis.

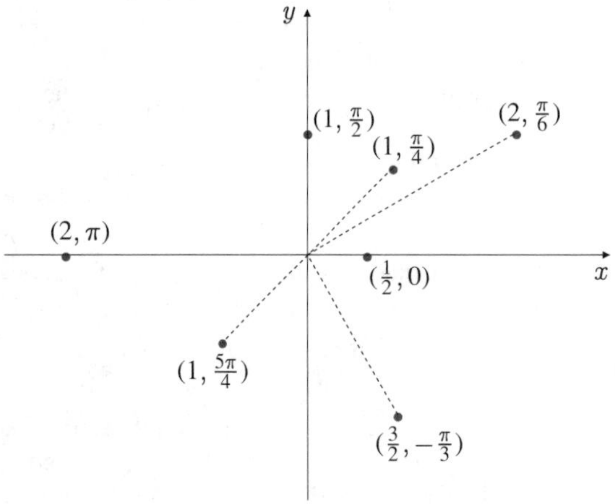

FIGURE 9.2.1

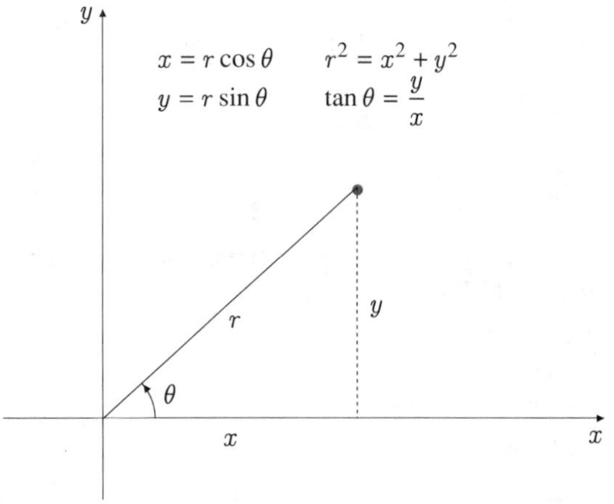

FIGURE 9.2.2

Unlike rectangular coordinates, the polar coordinates of a point are not unique. The polar coordinates (r, θ_1) and (r, θ_2) represent the same point provided θ_1 and θ_2 differ by an integer multiple of 2π:

$$\theta_2 = \theta_1 + 2n\pi, \qquad \text{where } n = 0, \pm 1, \pm 2, \ldots.$$

For instance, the polar coordinates

$$\left(1, \frac{\pi}{4}\right) = \left(1, \frac{9\pi}{4}\right) = \left(1, -\frac{7\pi}{4}\right)$$

all represent the point with Cartesian coordinates $\left(\frac{1}{\sqrt{2}}, \frac{1}{\sqrt{2}}\right)$. Similarly, $(2, \pi) = (2, -\pi)$ and $(4, 0) = (4, 2\pi)$. In addition, the origin O has polar coordinates $(0, \theta)$ for any value of θ. (If we go zero distance from O, it doesn't matter in what direction we go.)

Sometimes we need to interpret polar coordinates (r, θ) where $r < 0$. The appropriate interpretation for this "negative distance" r is that it represents a positive distance $-r$ measured in the *opposite direction* (that is, in the direction $\theta + \pi$):

$$(-r, \theta) = (r, \theta + \pi).$$

For example, $(-1, \pi/4) = (1, 5\pi/4)$.

If we wish to consider both rectangular and polar coordinate systems in the same plane, and we choose the positive x-axis as the polar axis, then the relationships between the rectangular coordinates of a point and its polar coordinates are as shown in Fig. 9.2.2.

9.2.1
Polar-Rectangular Conversion

$$x = r\cos\theta, \qquad x^2 + y^2 = r^2$$
$$y = r\sin\theta, \qquad \tan\theta = \frac{y}{x}$$

A single equation in x and y generally represents some curve in the plane with respect to the rectangular coordinate system. Similarly, a single equation in r and θ generally represents a curve with respect to the polar coordinate system. The relationships given in the Box 9.2.1 can be used to convert one representation of a curve into the other.

EXAMPLE 9.2.2 The straight line $x - 2y = 5$ has polar equation $r(\cos\theta - 2\sin\theta) = 5$.

EXAMPLE 9.2.3 The polar equation $r = 2a\cos\theta$ can be transformed to rectangular coordinates if we first multiply it by r:

$$r^2 = 2ar\cos\theta$$
$$x^2 + y^2 = 2ax$$
$$(x - a)^2 + y^2 = a^2$$

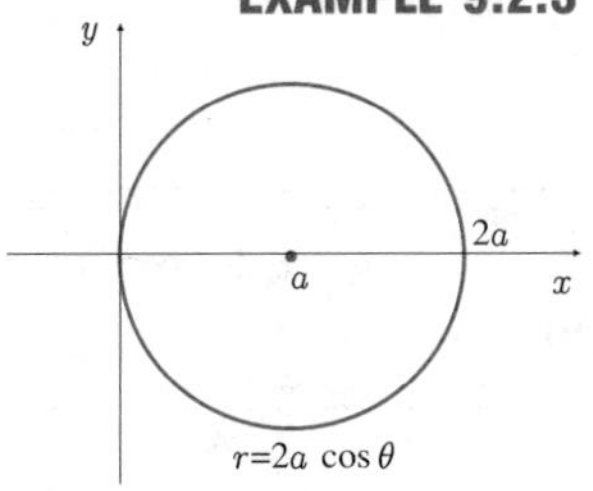

FIGURE 9.2.3

The given polar equation $r = 2a\cos\theta$ thus represents a circle with centre $(a, 0)$ and radius a as shown in Fig. 9.2.3. Observe from the equation that $r \to 0$ as $\theta \to \pm\pi/2$. This shows up in the figure; the circle approaches the origin in the vertical direction.

Some Polar Curves

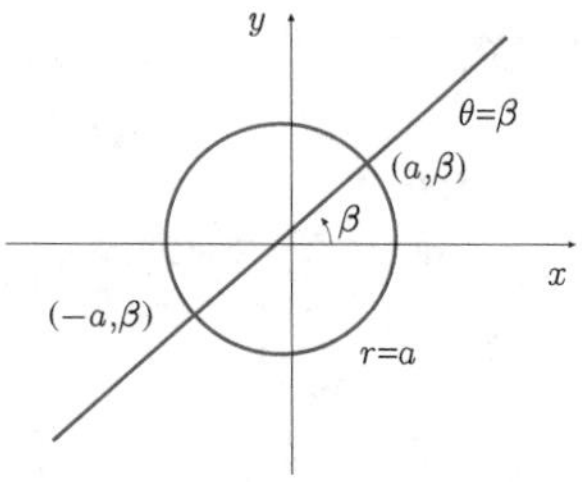

FIGURE 9.2.4

Fig. 9.2.4 shows the graphs of the polar equations $r = a$ and $\theta = \beta$ (where a and β are constants). These are, respectively, the circle with radius $|a|$ centred at the origin and a line through the origin making angle β with the polar axis. Note that the line and the circle meet in two points, with polar coordinates (a, β) and $(-a, \beta)$. The "coordinate curves" for polar coordinates, that is, the curves $r =$ constant and $\theta =$ constant, are circles and lines through the origin respectively. The "coordinate curves" for Cartesian coordinates, $x =$ constant and $y =$ constant, are vertical and horizontal straight lines. Cartesian graph paper is ruled by vertical and horizontal lines; polar graph paper is ruled in concentric circles and radial lines emanating from the origin.

The graph of an equation of the form $r = f(\theta)$ is called the **polar graph** of the function f. Some polar graphs can be recognized easily if the polar equation is transformed to rectangular form. For others, this transformation does not help; the rectangular equation may be too complicated to be recognizable. In these cases one must resort to constructing a table of values and plotting points.

EXAMPLE 9.2.4 Sketch and identify the curve $r = 2a\cos(\theta - \alpha)$.

SOLUTION We proceed as in Example 9.2.3.

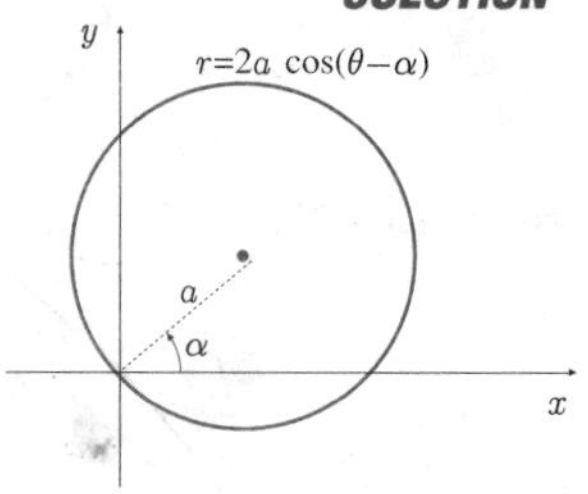

FIGURE 9.2.5

$$
\begin{aligned}
r^2 &= 2ar\cos(\theta - \alpha) = 2ar\cos\alpha\cos\theta + 2ar\sin\alpha\sin\theta \\
x^2 + y^2 &= 2a\cos\alpha x + 2a\sin\alpha y \\
x^2 - 2a\cos\alpha x + a^2\cos^2\alpha + y^2 - 2a\sin\alpha y + a^2\sin^2\alpha &= a^2 \\
(x - a\cos\alpha)^2 + (y - a\sin\alpha)^2 &= a^2
\end{aligned}
$$

This is a circle of radius a which passes through the origin in those directions $\theta = \alpha \pm \frac{\pi}{2}$, which make $r = 0$ (see Fig. 9.2.5). Its centre has Cartesian coordinates $(a\cos\alpha, a\sin\alpha)$ and hence polar coordinates (a, α). For $\alpha = \pi/2$ we have $r = 2a\sin\theta$ as the equation of a circle of radius a centred on the y-axis.

9.2.5 Rotating a Polar Graph

The polar graph with equation $r = f(\theta - \alpha)$ is the polar graph with equation $r = f(\theta)$ rotated through angle α about the origin.

EXAMPLE 9.2.6 Sketch the *cardioid* $r = a(1 - \cos\theta)$, where $a > 0$.

SOLUTION Transformation to rectangular coordinates is not much help here; the resulting equation is $(x^2 + y^2 + ax)^2 = a^2(x^2 + y^2)$ (verify this) which we do not recognize. Instead we make a table of values and plot some points.

θ	0	$\pm\frac{\pi}{6}$	$\pm\frac{\pi}{4}$	$\pm\frac{\pi}{3}$	$\pm\frac{\pi}{2}$	$\pm\frac{2\pi}{3}$	$\pm\frac{3\pi}{4}$	$\pm\frac{5\pi}{6}$	π
r	0	$0.13a$	$0.29a$	$0.5a$	a	$1.5a$	$1.71a$	$1.87a$	$2a$

Observe the cusp at the origin in Fig. 9.2.6. As in the previous example, the curve enters the origin in the direction θ which makes $r = f(\theta) = 0$. It is important, when sketching polar graphs, to show clearly any directions of approach to the origin.

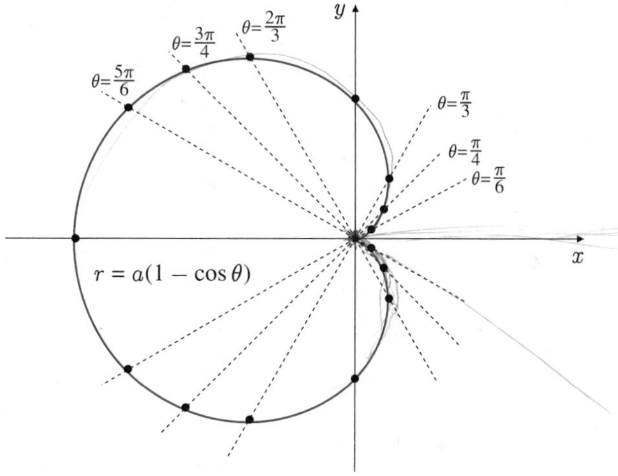

FIGURE 9.2.6

9.2.7
Polar Graphs Near the Origin

> A polar graph $r = f(\theta)$ approaches the origin from the direction θ for which $f(\theta) = 0$.

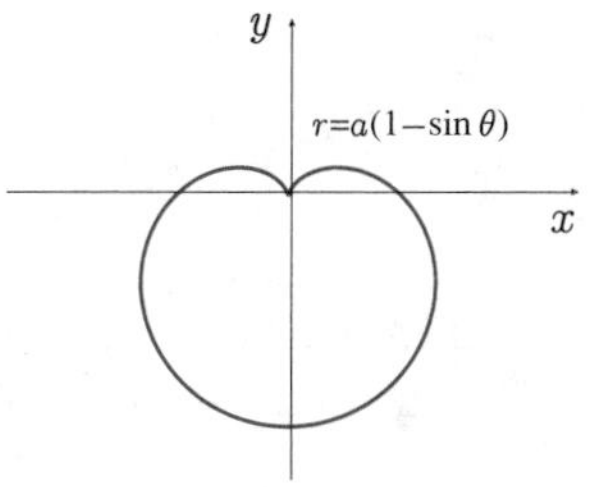

FIGURE 9.2.7

As indicated in Box 9.2.5, the equation $r = a(1 - \cos(\theta - \alpha))$ represents a cardioid of the same size and shape as that in Fig. 9.2.6, but rotated through an angle α counterclockwise about the origin. Its cusp is in the direction $\theta = \alpha$. In particular, $r = a(1 - \sin\theta)$ has vertical cusp as shown in Fig. 9.2.7

It is not usually necessary to make a detailed table of values to sketch a polar curve with a simple equation of the form $r = f(\theta)$. It is essential to determine those values of θ for which $r = 0$ and indicate them on the graph with rays. It is also useful to determine points where the curve is farthest from the origin. (Where is $f(\theta)$ maximum or minimum?) Except possibly at the origin, polar curves will be smooth wherever $f(\theta)$ is a smooth (i.e. differentiable) function of θ.

EXAMPLE 9.2.8 Sketch the polar graphs (a) $r = \cos(2\theta)$, (b) $r = \sin(3\theta)$, and (c) $r^2 = \cos(2\theta)$.

SOLUTION The graphs are shown in Figs. 9.2.8–10. Observe how the curves (a) and (c) approach the origin in the directions $\theta = \pm\frac{\pi}{4}$ and $\theta = \pm\frac{3\pi}{4}$, and curve (b) approaches in the directions $\theta = 0, \pi, \pm\frac{\pi}{3}$ and $\pm\frac{2\pi}{3}$. This curve is traced out twice as θ increases from $-\pi$ to π. So is curve (c) if we allow both square roots $r = \pm\sqrt{\cos(2\theta)}$. Note that there are no points on curve (c) between $\theta = \pm\frac{\pi}{4}$ and $\theta = \pm\frac{3\pi}{4}$ because r^2 cannot be negative.

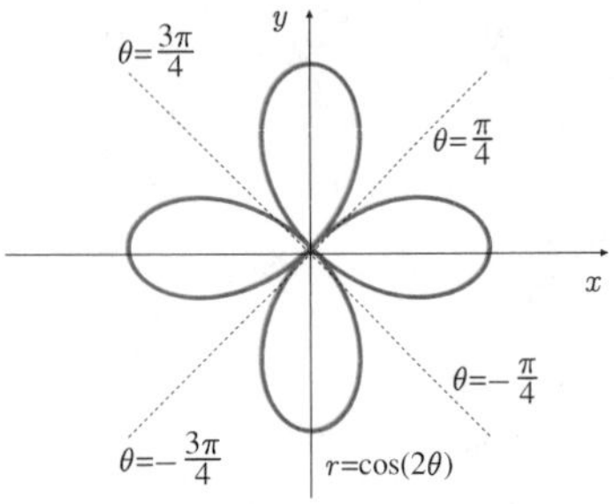

FIGURE 9.2.8

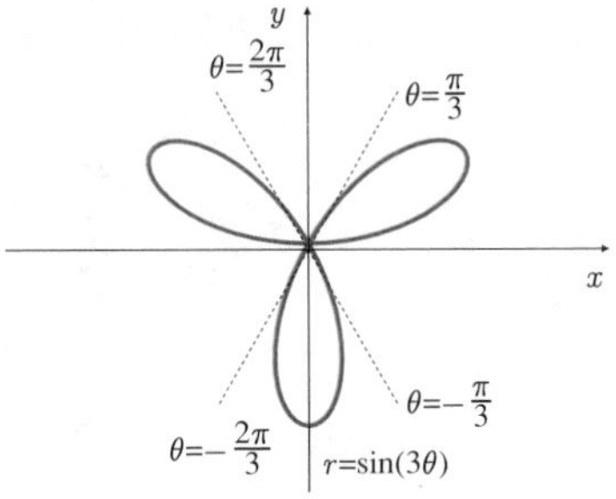

FIGURE 9.2.9

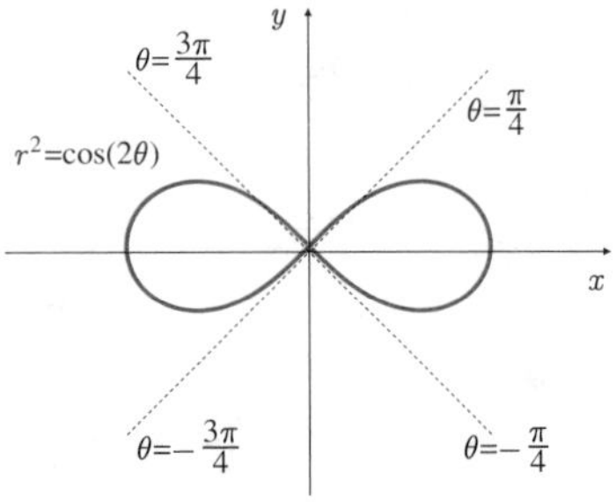

FIGURE 9.2.10

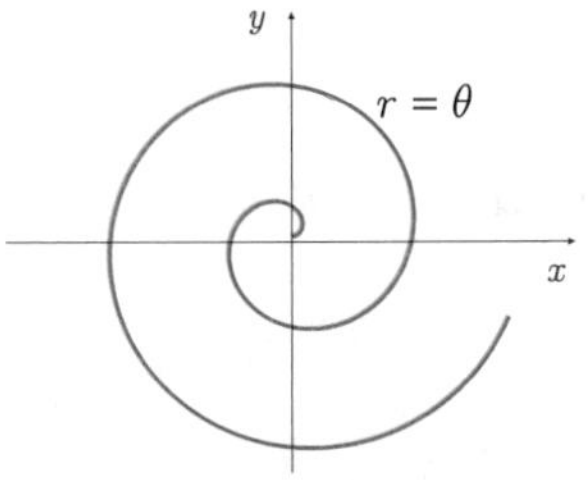

FIGURE 9.2.11

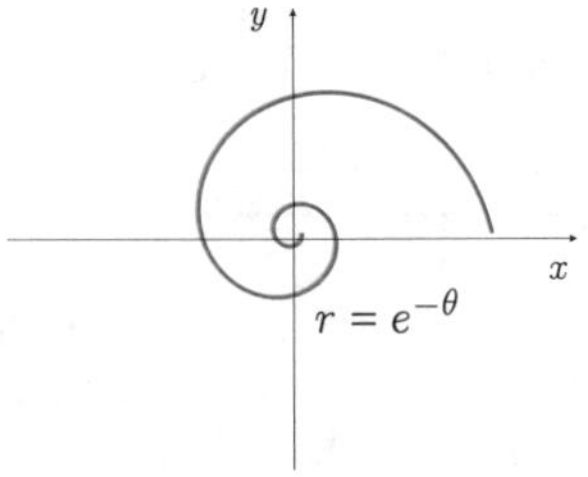

FIGURE 9.2.12

Curve (c) is called a **lemniscate**. Lemniscates are curves consisting of points P such that the product of the distances from P to certain fixed points is constant. For the curve (c) these fixed points are $\left(\pm\frac{1}{\sqrt{2}}, 0\right)$.

In all of the examples above the functions $f(\theta)$ are periodic and 2π is a period of each, so each line through the origin could meet the polar graph at most twice. (θ and $\theta + \pi$ determine the same line.) If $f(\theta)$ does not have period 2π, then the curve can wind around the origin many times. Two such "spirals" are shown in Figs. 9.2.11–12, the equiangular spiral $r = \theta$ sketched for $\theta \geq 0$, and the exponential spiral $r = e^{-\theta}$ sketched for $0 \leq \theta \leq 4\pi$.

Polar Conics

Let D be the vertical straight line $x = -p$ and let ε be a positive real number. The set of points P in the plane which satisfy the condition

$$\frac{\text{distance of } P \text{ from the origin}}{\text{perpendicular distance from } P \text{ to } D} = \varepsilon$$

is a conic section with eccentricity ε, focus at the origin, and corresponding directrix D, as observed in Section 9.1. (It is an ellipse if $\varepsilon < 1$, a parabola if $\varepsilon = 1$ and a hyperbola if $\varepsilon > 1$.) If P has polar coordinates (r, θ), then the condition above becomes (see Fig. 9.2.2.13)

$$\frac{r}{p + r\cos\theta} = \varepsilon,$$

or, solving for r,

9.2.9
Polar Equation of a Conic

$$r = \frac{\varepsilon p}{1 - \varepsilon \cos\theta}.$$

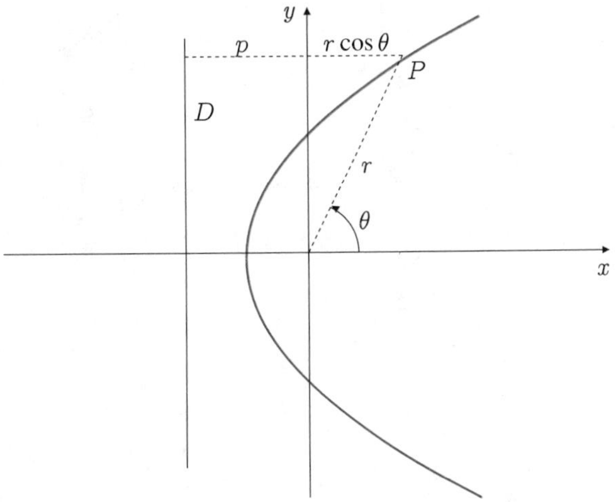

FIGURE 9.2.13

Examples of the three possibilities (ellipse, parabola and hyperbola) are shown in Figs. 9.2.14–16. Note that for the hyperbola, the directions of the asymptotes are the angles which make the denominator $1 - \varepsilon \cos\theta = 0$.

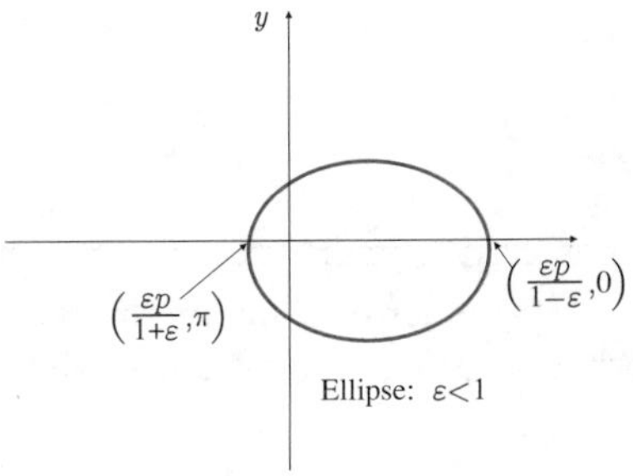

FIGURE 9.2.14

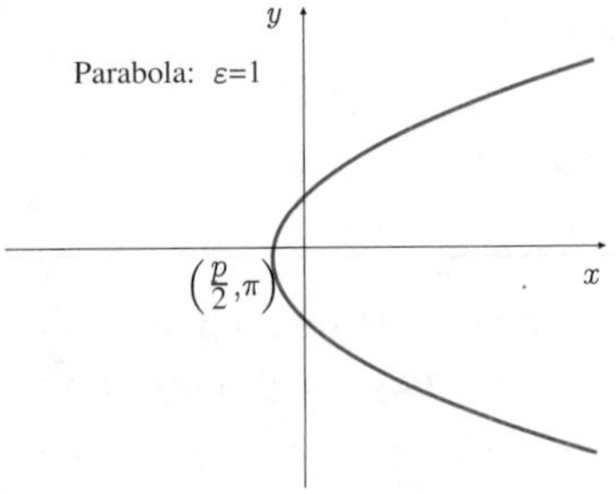

FIGURE 9.2.15

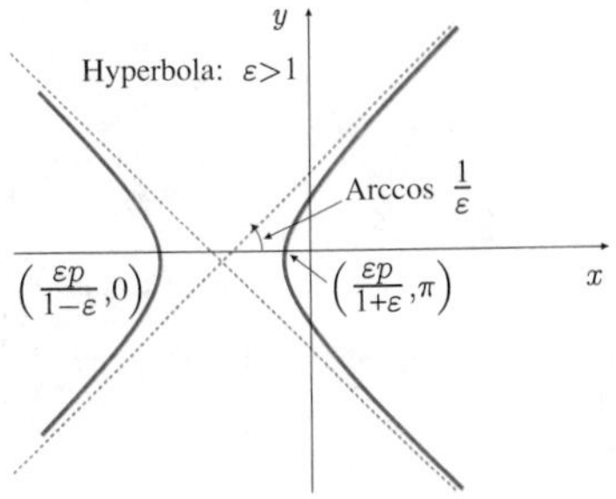

FIGURE 9.2.16

The Slope of a Polar Curve

There is a simple formula that can be used to determine the direction of the tangent line to a polar curve $r = f(\theta)$ at a point $P = (r, \theta)$ other than the origin. Let Q be a point on the curve nearby P corresponding to polar angle $\theta + h$. Draw PS perpendicular to OQ. Observe that $PS = f(\theta)\sin h$ and $SQ = OQ - OS = f(\theta + h) - f(\theta)\cos h$. If the tangent line to $r = f(\theta)$ at P makes angle ψ with the radial line OP as shown in Fig. 9.2.17, then ψ is the limit of the angle SQP as $h \to 0$. Thus

$$\begin{aligned}\tan\psi &= \lim_{h\to 0}\frac{PS}{SQ} = \lim_{h\to 0}\frac{f(\theta)\sin h}{f(\theta+h) - f(\theta)\cos h} && \left[\frac{0}{0}\right]\\ &= \lim_{h\to 0}\frac{f(\theta)\cos h}{f'(\theta+h) + f(\theta)\sin h} && \text{(by l'Hôpital's rule)}\\ &= \frac{f(\theta)}{f'(\theta)} = \frac{r}{dr/d\theta}.\end{aligned}$$

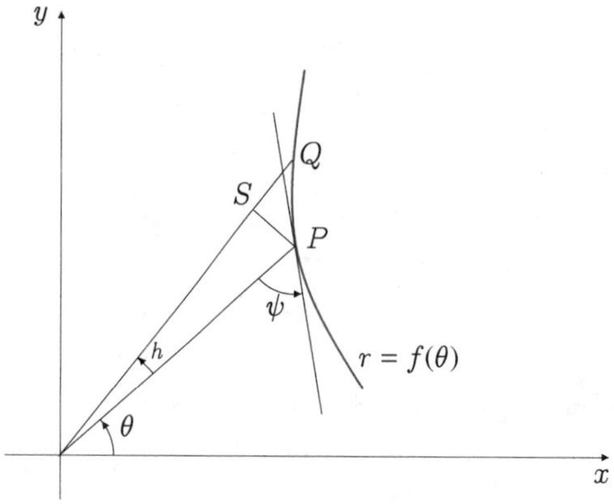

FIGURE 9.2.17

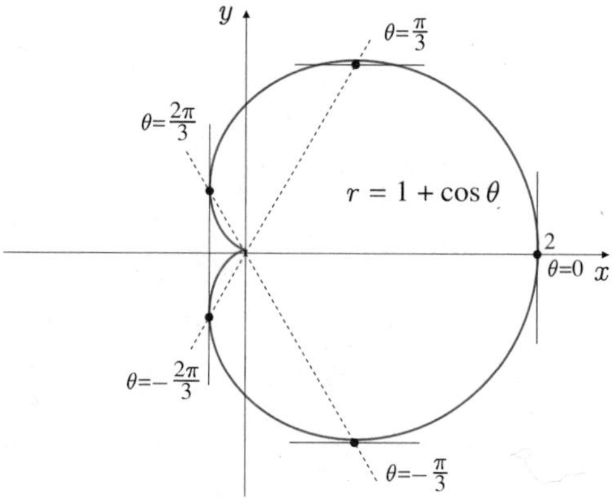

FIGURE 9.2.18

9.2.10
Tangent Direction for a Polar Curve

At any point P on the polar curve $r = f(\theta)$ other than the origin, the angle ψ between the radial line from the origin to P and the tangent to the curve is given by

$$\tan\psi = \frac{f(\theta)}{f'(\theta)}.$$

If $f(\theta_0) = 0$ and the curve has a tangent line at θ_0 then that tangent line has equation $\theta = \theta_0$.

The formula in the box above can be used to find points where a polar graph has horizontal or vertical tangents:

$$\psi + \theta = \pi, \quad \text{so } \tan\psi = -\tan\theta \quad \text{for a horizontal tangent,}$$
$$\psi + \theta = \frac{\pi}{2}, \quad \text{so } \tan\psi = \cot\theta \quad \text{for a vertical tangent.}$$

However, it is usually easier to find the critical points of $y = f(\theta)\sin\theta$ for horizontal tangents, and of $x = f(\theta)\cos\theta$ for vertical tangents.

EXAMPLE 9.2.11 Find the points on the cardioid $r = 1 + \cos\theta$ where the tangent line is vertical or horizontal.

SOLUTION We have $y = (1 + \cos\theta)\sin\theta$ and $x = (1 + \cos\theta)\cos\theta$. For horizontal tangents

$$\begin{aligned} 0 = \frac{dy}{d\theta} &= -\sin^2\theta + \cos^2\theta + \cos\theta \\ &= 2\cos^2\theta + \cos\theta - 1 \\ &= (2\cos\theta - 1)(\cos\theta + 1). \end{aligned}$$

The solutions are $\cos\theta = \frac{1}{2}$ and $\cos\theta = -1$. There are horizontal tangents at $\left(\frac{3}{2}, \pm\frac{\pi}{3}\right)$. For $\cos\theta = -1$, we have $r = 0$. The curve does not have a tangent line at the origin (it has a cusp). (See Fig. 9.2.2.18.)

For vertical tangents

$$0 = \frac{dx}{d\theta} = -\sin\theta - 2\cos\theta\sin\theta = -\sin\theta(1 + 2\cos\theta).$$

The solutions are $\sin\theta = 0$ and $\cos\theta = -\frac{1}{2}$. There are vertical tangent lines at $(2, 0)$ and $\left(\frac{1}{2}, \pm\frac{2\pi}{3}\right)$.

Areas Bounded by Polar Curves

The basic area problem in polar coordinates is that of finding the area of the region R bounded by the polar graph $r = f(\theta)$ and the two rays $\theta = \alpha$ and $\theta = \beta$. We assume that $\beta > \alpha$ and that f is continuous on $[\alpha, \beta]$. (See Fig. 9.2.2.19.)

A suitable area element in this case is a sector of angular width $d\theta$ as shown in Fig. 9.2.19. For infinitesimal $d\theta$ this is just a sector of a circle of radius $r = f(\theta)$:

$$dA = \frac{d\theta}{2\pi}\pi r^2 = \frac{1}{2}r^2\,d\theta = \frac{1}{2}\left(f(\theta)\right)^2 d\theta.$$

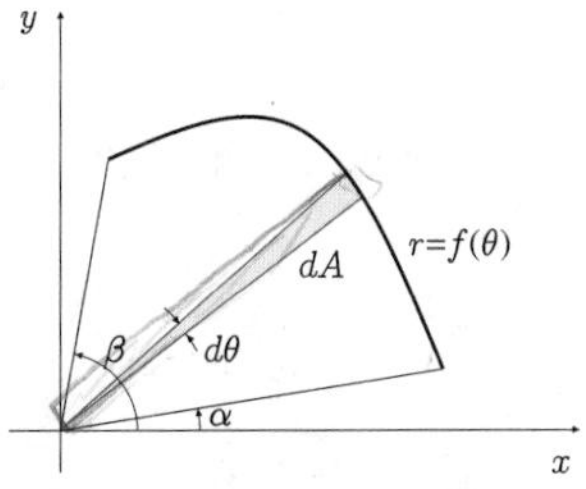

FIGURE 9.2.19

9.2.12 Area in Polar Coordinates

The region bounded by $r = f(\theta)$ and the rays $\theta = \alpha$ and $\theta = \beta$, $(\alpha < \beta)$, has area $$A = \frac{1}{2}\int_{\alpha}^{\beta}\left(f(\theta)\right)^2 d\theta.$$

EXAMPLE 9.2.13 Find the area bounded by the cardioid $r = a(1+\cos\theta)$, as illustrated in Fig. 9.2.20.

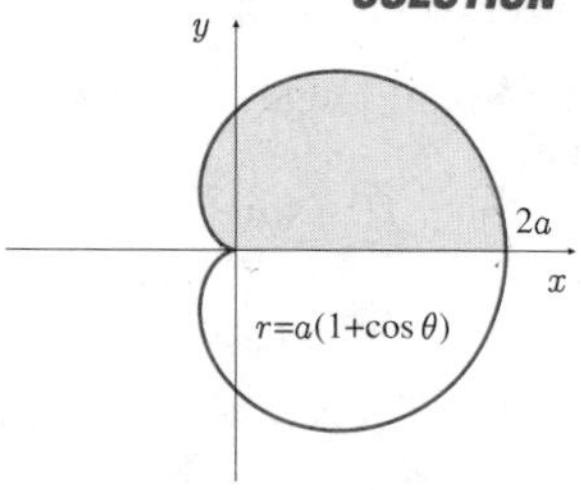

FIGURE 9.2.20

SOLUTION By symmetry, the area is twice that of the top half:

$$\begin{aligned} A &= 2 \times \frac{1}{2} \int_0^{\pi} a^2(1+\cos\theta)^2\, d\theta \\ &= a^2 \int_0^{\pi} (1 + 2\cos\theta + \cos^2\theta)\, d\theta \\ &= a^2 \int_0^{\pi} \left(1 + 2\cos\theta + \frac{1+\cos 2\theta}{2}\right) d\theta \\ &= a^2 \left(\frac{3}{2}\theta + 2\sin\theta + \frac{1}{4}\sin 2\theta\right)\Bigg|_0^{\pi} \\ &= \frac{3}{2}\pi a^2 \text{ sq units.} \end{aligned}$$

EXAMPLE 9.2.14 Find the area of the region which lies inside the circle $r = \sqrt{2}\sin\theta$ and inside the lemniscate $r^2 = \sin 2\theta$.

SOLUTION The region is shaded in Fig. 9.2.21. Besides intersecting at the origin, the curves intersect at the first quadrant point satisfying

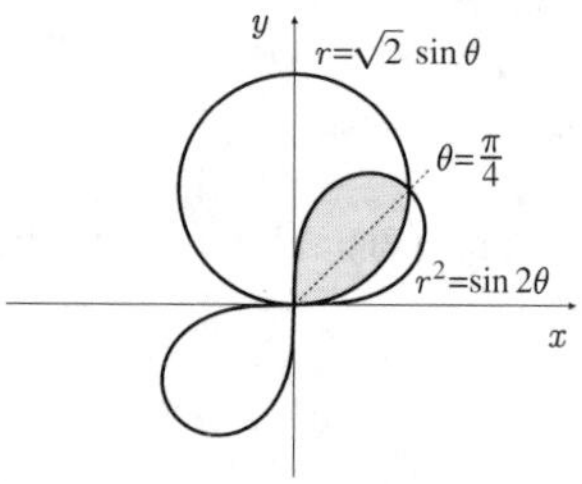

FIGURE 9.2.21

$$2\sin^2\theta = \sin 2\theta = 2\sin\theta\cos\theta.$$

Thus $\sin\theta = \cos\theta$ and $\theta = \frac{\pi}{4}$. The required area is

$$\begin{aligned} A &= \frac{1}{2}\int_0^{\pi/4} 2\sin^2\theta\, d\theta + \frac{1}{2}\int_{\pi/4}^{\pi/2} \sin 2\theta\, d\theta \\ &= \int_0^{\pi/4} \frac{1-\cos 2\theta}{2}\, d\theta - \frac{1}{4}\cos 2\theta\Bigg|_{\pi/4}^{\pi/2} \\ &= \frac{\pi}{8} - \frac{1}{4}\sin 2\theta\Bigg|_0^{\pi/4} + \frac{1}{4} = \frac{\pi}{8} - \frac{1}{4} + \frac{1}{4} = \frac{\pi}{8} \quad \text{sq units.} \end{aligned}$$

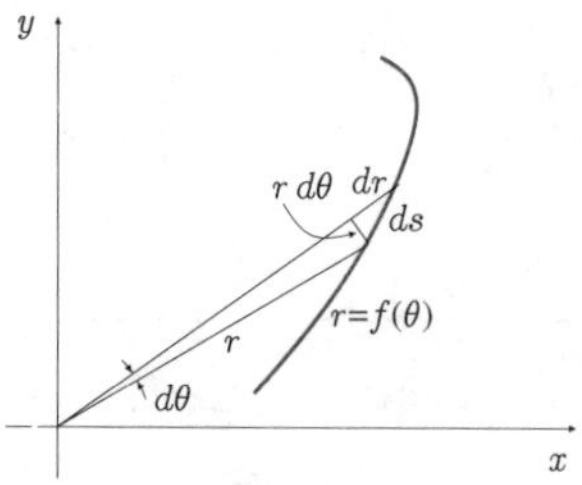

FIGURE 9.2.22

Arc Length of Polar Curves

The arc length element for the polar curve $r = f(\theta)$ can be determined from the differential triangle shown in Fig. 9.2.22. The leg $r\,d\theta$ of the triangle is obtained as the arc length of a circular arc of radius r subtending angle $d\theta$ at the origin. We have

$$(ds)^2 = (dr)^2 + r^2(d\theta)^2 = \left[\left(\frac{dr}{d\theta}\right)^2 + r^2\right](d\theta)^2,$$

and so we obtain the following formula:

9.2.15
Arc Length Element for Polar Curves

The arc length element for the polar curve $r = f(\theta)$ is

$$ds = \sqrt{\left(\frac{dr}{d\theta}\right)^2 + r^2}\, d\theta = \sqrt{\left(f'(\theta)\right)^2 + \left(f(\theta)\right)^2}\, d\theta.$$

EXAMPLE 9.2.16 Find the total length of the cardioid $r = a(1 + \cos\theta)$.

SOLUTION The total length is twice the length from $\theta = 0$ to $\theta = \pi$ (review Fig. 9.2.20):

$$\begin{aligned} s &= 2\int_0^{\pi} \sqrt{a^2 \sin^2\theta + a^2(1+\cos\theta)^2}\, d\theta \\ &= 2\int_0^{\pi} \sqrt{2a^2 + 2a^2\cos\theta}\, d\theta \qquad (\text{but } 1+\cos\theta = 2\cos^2(\theta/2)) \\ &= 2\sqrt{2}a \int_0^{\pi} \sqrt{2\cos^2\frac{\theta}{2}}\, d\theta = 4a\int_0^{\pi} \cos\frac{\theta}{2}\, d\theta = 8a \sin\frac{\theta}{2}\Big|_0^{\pi} = 8a \text{ units.} \end{aligned}$$

EXERCISES

In Exercises 1–12, transform the given polar equation to rectangular coordinates and hence identify the curve represented.

1. $r = 3\sec\theta$
2. $r = -2\csc\theta$
3. $r = \dfrac{5}{3\sin\theta - 4\cos\theta}$
4. $r = \sin\theta + \cos\theta$
5. $r^2 = \csc 2\theta$
6. $r = \sec\theta\tan\theta$
7. $r = \sec\theta(1 + \tan\theta)$
8. $r = \dfrac{2}{\sqrt{\cos^2\theta + 4\sin^2\theta}}$
9. $r = \dfrac{1}{1 - \cos\theta}$
10. $r = \dfrac{2}{2 - \cos\theta}$
11. $r = \dfrac{2}{1 - 2\sin\theta}$
12. $r = \dfrac{2}{1 + \sin\theta}$

In Exercises 13–24 sketch the polar graphs of the given equations.

13. $r = 1 + \sin\theta$
14. $r = 1 - \cos(\theta + \frac{\pi}{4})$
15. $r = 1 + 2\cos\theta$
16. $r = 1 - 2\sin\theta$
17. $r = 2 + \cos\theta$
18. $r = 2\sin 2\theta$
19. $r = \cos 3\theta$
20. $r = 2\cos 4\theta$
21. $r^2 = 4\sin 2\theta$
22. $r^2 = 4\cos 3\theta$
23. $r^2 = \sin 3\theta$
24. $r = \ln\theta$

25.*Sketch the graph of the equation $r = 1/\theta$, $\theta > 0$. Show that this curve has a horizontal asymptote. Does $r = 1/(\theta - \alpha)$ have an asymptote?

26. How many leaves does the curve $r = \cos n\theta$ have? The curve $r^2 = \cos n\theta$? Distinguish the cases where n is odd and even.

In Exercises 27–37 sketch and find the areas of the given polar regions R.

27. R lies between the origin and the spiral $r = \sqrt{\theta}$, $0 \le \theta \le 2\pi$.

28. R lies between the origin and the spiral $r = \theta$, $0 \le \theta \le 2\pi$.

29. R is bounded by the curve $r^2 = a^2\cos 2\theta$.

30. R is one leaf of the curve $r = \sin 3\theta$.

31. R is bounded by the curve $r = \cos 4\theta$.

32. R lies inside both of the circles $r = a$ and $r = 2a\cos\theta$.

33. R lies inside the cardioid $r = 1 - \cos\theta$ and outside the circle $r = 1$.

34. R lies inside the cardioid $r = a(1 - \sin\theta)$ and inside the circle $r = a$.

35. R lies inside the cardioid $r = 1 + \cos\theta$ and outside the circle $r = 3\cos\theta$.

36. R is bounded by the lemniscate $r^2 = 2\cos 2\theta$ and is outside the circle $r = 1$.

37. R is bounded by the smaller loop of the curve $r = 1 + 2\cos\theta$.

Find the lengths of the polar curves in Exercises 38–40

38. $r = \theta^2,\ 0 \le \theta \le \pi$

39. $r = e^{a\theta},\ -\pi \le \theta \le \pi$

40. $r = a\theta,\ 0 \le \theta \le 2\pi$

41. Show that the total arc length of the lemniscate $r^2 = \cos 2\theta$ is $4\int_0^{\pi/4} \sqrt{\sec 2\theta}\, d\theta$.

42. One leaf of the lemniscate $r^2 = \cos 2\theta$ is rotated about (a) the x-axis, (b) the y-axis. Find the area of the surface generated in each case.

43.*Determine the angles at which the straight line $\theta = \pi/4$ intersects the cardioid $r = 1 + \sin\theta$.

44.*At what points do the curves $r^2 = 2\sin 2\theta$ and $r = 2\cos\theta$ intersect? At what angle do the curves intersect at each of these points?

45.*At what points do the curves $r = 1 - \cos\theta$ and $r = 1 - \sin\theta$ intersect? At what angle do the curves intersect at each of these points?

In Exercises 46–51 find all points on the given curve where the tangent line is horizontal, vertical, or does not exist.

46.*$r = \cos\theta + \sin\theta$

47.*$r = 2\cos\theta$

48.*$r^2 = \cos 2\theta$

49.*$r = \sin 2\theta$

50.*$r = e^{\theta}$

51.*$r = 2(1 - \sin\theta)$

9.3 PARAMETRIC CURVES

Suppose that an object moves around in the xy-plane so that the coordinates of its position at any time t are functions of the variable t:

$$x = f(t), \qquad y = g(t).$$

The path followed by the moving object is a curve C in the plane which is specified by the two equations above. We call these equations, parametric equations of C. A curve specified by a particular pair of parametric equations is called a parametric curve.

9.3.1 Parametric Curve

A **parametric curve** C in the plane consists of an interval I together with an ordered pair (f, g) of continuous functions defined on I. The equations

$$x = f(t), \qquad y = g(t), \qquad \text{for } t \text{ in } I$$

are called **the parametric equations** of C. The independent variable t is called the **parameter**.

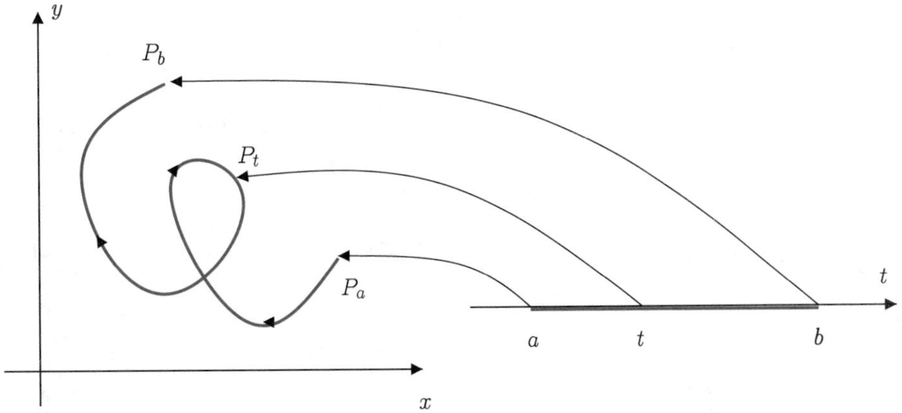

FIGURE 9.3.1

Note that the parametric curve C was *not* defined as a set of points in the plane, but rather as the ordered pair of functions whose range is that set of points. Nevertheless, we will often refer to the set of points (the path traced out by (x, y) as t traverses I) as the curve C. The axis (real line) of the parameter t is considered to be distinct from the coordinate axes of the plane of the curve. (See Fig. 9.3.1.) We will usually denote the parameter by t; in many applications the parameter represents time. Because f and g are assumed continuous, the curve $x = f(t)$, $y = g(t)$ has no breaks in it. A parametric curve has a *direction*, (indicated, say, by an arrowhead), namely the direction corresponding to increasing values of the parameter t.

EXAMPLE 9.3.2 Sketch and identify the parametric curve

$$x = t^2 - 1, \qquad y = t + 1 \qquad (-\infty < t < \infty).$$

SOLUTION We could construct a table of values of x and y for various values of t, thus getting the coordinates of a number of points on a curve. However, for this example it is easier to *eliminate the parameter* from the pair of parametric equations, thus producing a single equation in x and y whose graph is the desired curve:

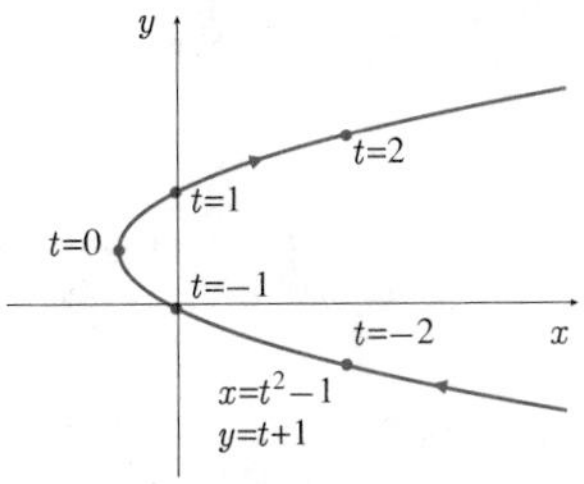

FIGURE 9.3.2

$$t = y - 1, \qquad x = t^2 - 1 = (y - 1)^2 - 1 = y^2 - 2y.$$

All points on the curve lie on the parabola $x = y^2 - 2y$. Since $y \to \pm\infty$ as $t \to \pm\infty$, the parametric curve is the whole parabola. (See Fig. 9.3.2.)

Although the curve in this example is more easily identified when the parameter is eliminated, there is a loss of information in going to the nonparametric form. Specifically, we lose the sense of the curve as the path of a moving point and hence also the direction of the curve. If the t in the parametric form denotes the time at which an object is at the point (x, y), the nonparametric equation $x = y^2 - 2y$ no longer tells us where the object is at any particular time t.

EXAMPLE 9.3.3 The straight line passing through the two points $P_0 = (x_0, y_0)$ and $P_1 = (x_1, y_1)$ (see Fig. 9.3.3) has parametric equations

$$\begin{cases} x = x_0 + t(x_1 - x_0) \\ y = y_0 + t(y_1 - y_0) \end{cases} \qquad -\infty < t < \infty.$$

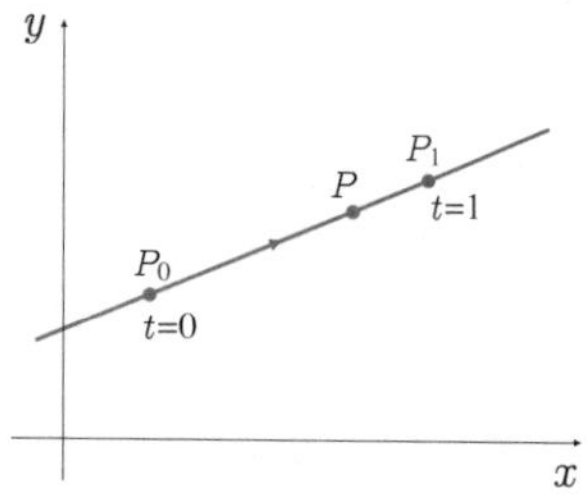

FIGURE 9.3.3

To see that these equations represent a straight line, note that

$$\frac{y - y_0}{x - x_0} = \frac{y_1 - y_0}{x_1 - x_0} = \text{constant} \qquad (\text{assuming } x_1 \neq x_0).$$

The point $P = (x, y)$ is at position P_0 when $t = 0$ and at P_1 when $t = 1$. If $t = 1/2$, then P is the midpoint between P_0 and P_1. Note that the line segment from P_0 to P_1 corresponds to values of t between 0 and 1.

EXAMPLE 9.3.4 Sketch and identify the curve $x = 3\cos t$, $y = 3\sin t$, $(0 \le t \le 3\pi/2)$.

SOLUTION Since $x^2 + y^2 = 9\cos^2 t + 9\sin^2 t = 9$, all points on the curve lie on the circle $x^2 + y^2 = 9$. As t increases from 0 through $\pi/2$ and π to $3\pi/2$, the point (x, y) moves from $(3, 0)$ through $(0, 3)$ and $(-3, 0)$ to $(0, -3)$. See Fig. 9.3.4. Thus the parametric curve is three-quarters of a circle.

EXAMPLE 9.3.5 Sketch and identify the curve $x = a\cos s$, $y = b\sin s$, $(0 \le s \le 2\pi)$, where $a > b > 0$.

SOLUTION Since $\frac{x^2}{a^2} + \frac{y^2}{b^2} = \cos^2 s + \sin^2 s = 1$, the curve is all or part of an ellipse with major axis from $(-a, 0)$ to $(a, 0)$ and minor axis from $(0, -b)$ to $(0, b)$. As s increases from 0 to 2π, the point (x, y) moves counterclockwise around the ellipse starting from $(a, 0)$ and returning to the same point. Thus the curve is the whole ellipse.

Figure 9.3.5 shows how the parameter s can be interpreted as an angle and how the points on the ellipse can be obtained using circles of radii a and b. Since the curve starts and ends at the same point, it is called a **closed curve**.

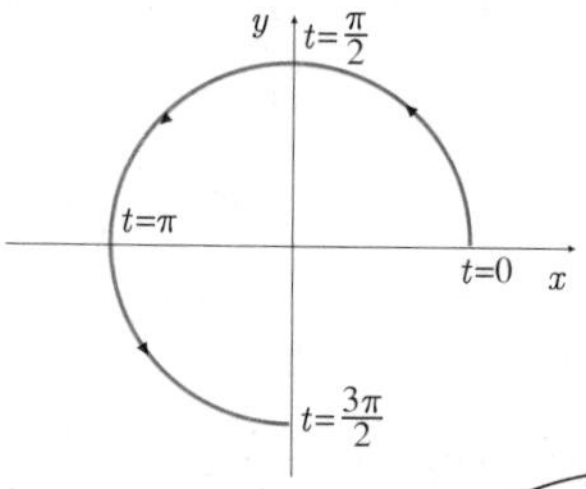

FIGURE 9.3.4

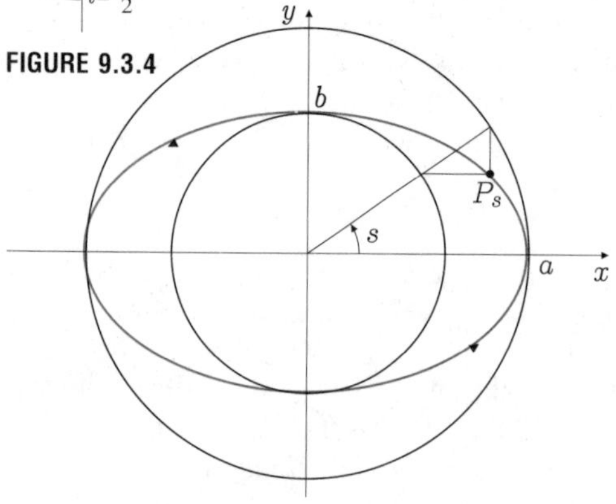

FIGURE 9.3.5

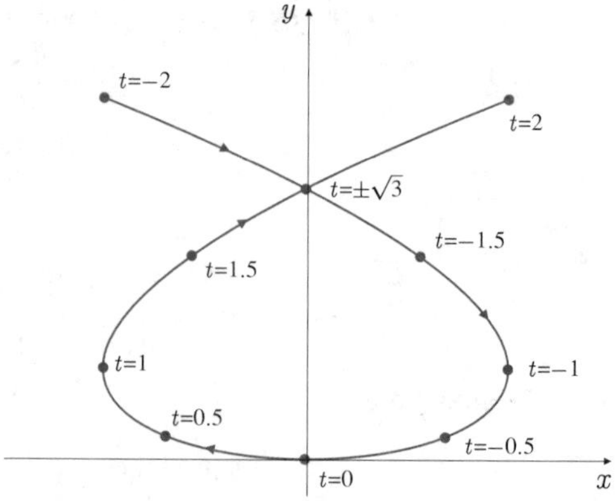

FIGURE 9.3.6

EXAMPLE 9.3.6 Sketch the parametric curve

$$x = t^3 - 3t, \qquad y = t^2 \qquad (-2 \le t \le 2).$$

SOLUTION We could eliminate the parameter and obtain

$$x^2 = t^2(t^2 - 3)^2 = y(y - 3)^2,$$

but this doesn't help much since we do not recognize this curve from its Cartesian equation. Instead let us obtain some points:

t	-2	$-\frac{3}{2}$	-1	$-\frac{1}{2}$	0	$\frac{1}{2}$	1	$\frac{3}{2}$	2
x	-2	$\frac{9}{8}$	2	$\frac{11}{8}$	0	$-\frac{11}{8}$	-2	$-\frac{9}{8}$	2
y	4	$\frac{9}{4}$	1	$\frac{1}{4}$	0	$\frac{1}{4}$	1	$\frac{9}{4}$	4

Note that the curve is symmetric about the y-axis because x is an odd function of t and y is an even function of t. (At t and $-t$, x has opposite values but y has the same value.)

The curve intersects itself on the y-axis (see Fig. 9.3.6). To find this self-intersection set $x = 0$:

$$0 = x = t^3 - 3t = t(t - \sqrt{3})(t + \sqrt{3}).$$

For $t = 0$ the curve is at $(0, 0)$, but for $t = \pm\sqrt{3}$ the curve is at $(0, 3)$. The self-intersection occurs because the curve passes through the same point for two different values of the parameter.

General Plane Curves: Parametrizations

According to the definition 9.3.1, a parametric curve always involves a particular set of parametric equations; it is not just a set of points in the plane. When we are interested in considering a curve solely as a set of points, (a *geometric object*), we need not be concerned with any particular pair of parametric equations representing that curve. In this case we call the curve simply a *plane curve*.

9.3.7
Plane Curves

> A **plane curve** is a set of points (x, y) in the plane such that $x = f(t)$ and $y = g(t)$ for some t in an interval I, where f and g are continuous functions defined on I. Any such interval I and function pair (f, g) which so generates the points of C is called a **parametrization** of C.

Since a plane curve does not involve any specific parametrization, it has no specific direction.

EXAMPLE 9.3.8 The circle $x^2 + y^2 = 1$ is a plane curve. Each of the following is a possible parametrization of C:

i) $x = \cos t,\ y = \sin t, \qquad (0 \le t \le 2\pi)$

ii) $x = \sin s^2,\ y = \cos s^2, \qquad (0 \le s \le \sqrt{2\pi})$

iii) $x = \cos(\pi u + 1), y = \sin(\pi u + 1), \qquad (-1 \le u \le 1)$

iv) $x = 1 - t^2, y = t\sqrt{2 - t^2}, \qquad (-\sqrt{2} \le t \le \sqrt{2})$

There are, of course, infinitely many other possible parametrizations of this curve.

EXAMPLE 9.3.9 If f is a function continuous on an interval I, then the graph of f is a plane curve. One obvious parametrization of this curve is

$$x = t, \qquad y = f(t), \qquad (t \text{ in } I).$$

Some Interesting Plane Curves

EXAMPLE 9.3.10 (**The Involute of a Circle**) A string is wound around a fixed circle. One end is unwound in such a way that part of the string not lying on the circle is extended in a straight line. The curve followed by this free end of the string is called an **involute** of the circle. Suppose the circle has equation $x^2+y^2 = a^2$ and suppose the end of the string being unwound starts at the point $A = (a, 0)$. At some subsequent time let P be the position of the end of the string and let T be the point where the string leaves the circle. Clearly PT is tangent to the circle at T. We shall parametrize the path of P in terms of the angle AOT, which we denote by t.

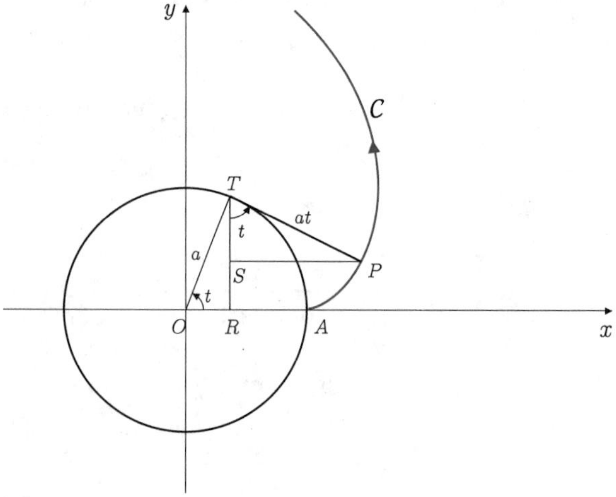

FIGURE 9.3.7

Let points R on OA and S on TR be as shown in Fig. 9.3.7. TR is perpendicular to OA and to PS. Clearly

$OR = OT\cos t = a\cos t,$
$RT = a\sin t,$
$TP = \text{ arc } TA = at$ (the string does not stretch or slip on the circle),
angle $OTP = \dfrac{\pi}{2}$ (a tangent to a circle is perpendicular to the radial line to the point of contact),
angle $STP =$ angle $ROT = t$ (similar triangles),
$SP = TP\sin t = at\sin t,$
$ST = at\cos t.$

If P has coordinates (x, y), then $x = OR + SP$, and $y = RT - ST$:

$$x = a\cos t + at\sin t, \qquad y = a\sin t - at\cos t, \qquad (t \geq 0).$$

These are parametric equations of the involute.

EXAMPLE 9.3.11 (**The Cycloid**) If a circle rolls without slipping along a straight line, the path followed by a point fixed on the circle is called a **cycloid**. Suppose that the line is the x-axis, that the circle has radius a and lies above the line, and that the point whose motion we follow is originally at the origin O. See Fig. 9.3.8.

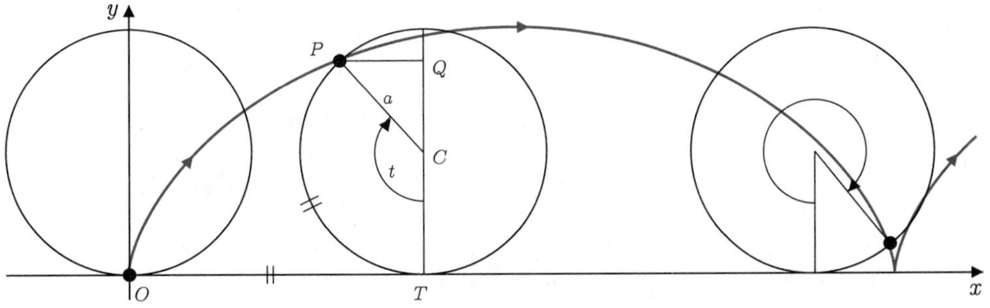

FIGURE 9.3.8

The point has moved to position P when the circle has rolled through an angle t and is tangent to the line at T. Since no slipping occurs,

$$\text{segment } OT = \text{ arc } PT = at.$$

Let PQ be perpendicular to TC, as shown in the figure. If P has coordinates (x, y), then

$$x = OT - PQ = at - a\sin(\pi - t) = at - a\sin t,$$
$$y = TC + CQ = a + a\cos(\pi - t) = a - a\cos t.$$

The parametric equations of the cycloid are therefore

$$x = a(t - \sin t), \qquad y = a(1 - \cos t).$$

Observe that the cycloid has a cusp at the points where it returns to the x-axis, i.e., at points corresponding to $t = 2n\pi$ where n is an integer. Even though the functions x and y are everywhere differentiable functions of t, the curve is not smooth everywhere. We shall look at such matters in the next section.

EXERCISES

In Exercises 1–12 sketch the given parametric curve, showing its direction with an arrow. Eliminate the parameter to give an ordinary equation in x and y whose graph contains the parametric curve.

1. $x = t,\ y = 1 - t,\ (0 \le t \le 1)$

2. $x = 2 - t,\ y = t + 1,\ (0 \le t < \infty)$

3. $x = 1 + 2t,\ y = t^2,\ (-\infty < t < \infty)$

4. $x = t^2,\ y = 3t^2 - 2,\ (-a \le t \le a)$

5. $x = \dfrac{1}{t},\ y = t - 1,\ (0 < t < 4)$

6. $x = \dfrac{1}{1+t^2},\ y = \dfrac{t}{1+t^2},\ (-\infty < t < \infty)$

7. $x = 3\sin 2t,\ y = 3\cos 2t,\ \left(0 \le t \le \dfrac{\pi}{3}\right)$

8. $x = a\sec t,\ y = b\tan t,\ \left(-\dfrac{\pi}{2} < t < \dfrac{\pi}{2}\right)$

9. $x = 3\sin \pi t,\ y = 4\cos \pi t,\ (-1 \le t \le 1)$

10. $x = \cos\sin s,\ y = \sin\sin s,\ (-\infty < s < \infty)$

11. $x = \cos^3 t,\ y = \sin^3 t,\ (0 \le t \le 2\pi)$

12. $x = t\cos t,\ y = t\sin t,\ (0 \le t \le 4\pi)$

13. Show that the polar graph $r = f(\theta)$ (where f is continuous) can be written as a parametric curve with parameter θ.

14. Show that each of the following sets of parametric equations represents a different arc of the parabola $2(x+y) = 1 + (x-y)^2$.

i) $x = \cos^4 t,\ y = \sin^4 t,$

ii) $x = \sec^4 t,\ y = \tan^4 t,$

iii) $x = \tan^4 t,\ y = \sec^4 t$

15. Find a parametrization of the parabola $y = x^2$ using as parameter the slope of the tangent line at the general point.

16. Find a parametrization of the circle $x^2 + y^2 = R^2$ using as parameter the slope m of the line joining the general point to the point $(R, 0)$. Does the parametrization fail to give any point on the circle?

17.*Eliminate the parameter from the parametric equations

$$x = \frac{3t}{1+t^3}, \qquad y = \frac{3t^2}{1+t^3} \qquad (t \ne -1),$$

and hence find an ordinary equation in x and y for this curve (called a **folium of Descartes**). The parameter t can be interpreted as the slope of the line joining the general point (x, y) to the origin. Sketch the curve and show that the line $x + y = -1$ is an asymptote.

18.*A railroad wheel has a flange extending below the level of the track on which the wheel rolls. If the radius of the wheel is a, and that of the flange is $b > a$, find parametric equations of the path of a point P at the circumference of the flange as the wheel rolls along the track. This curve is called a **prolate cycloid**. Note that for a portion of each revolution of the wheel, P is moving backward. Try to sketch the graph of a prolate cycloid.

19.*(**Hypocycloids**) If a circle of radius b rolls, without slipping, around the inside of a fixed circle of radius $a > b$, a point on the circumference of the rolling circle traces a curve called a hypocycloid. If the fixed circle is centred at the origin and the point tracing the curve starts at $(a, 0)$, show that the hypocycloid has parametric equations

$$x = (a-b)\cos t + b\cos\left(\frac{a-b}{b}t\right),$$

$$y = (a-b)\sin t - b\sin\left(\frac{a-b}{b}t\right),$$

where t is the angle between the positive x-axis and the line from the origin to the point at which the rolling circle touches the fixed circle.

If $a = 2$ and $b = 1$, show that the hypocycloid becomes a straight line segment.

If $a = 4$ and $b = 1$, show that the parametric equations of the hypocycloid simplify to $x = 4\cos^3 t$, $y = 4\sin^3 t$. This curve is called a hypocycloid of four cusps or an *astroid*. It has Cartesian equation $x^{2/3} + +y^{2/3} = 4^{2/3}$. (See Fig. 9.3.9)

Hypocycloids resemble the curves produced by a popular children's toy called Spirograph, but Spirograph curves result from following a point inside the disc of the rolling circle rather than on its circumference, and they therefore do not have sharp cusps.

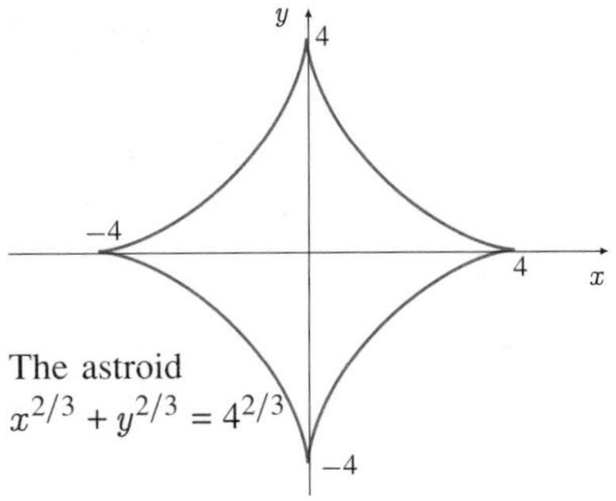

FIGURE 9.3.9

9.4 SMOOTH CURVES

We say that a plane curve is *smooth* if it has a tangent line at each point P and this tangent turns in a continuous way as P moves along the curve. (That is, the angle between the tangent line at P and some fixed line, the x-axis say, is a continuous function of the position of P.)

If the curve C is the graph of function f, then C is certainly smooth on any interval where the derivative $f'(x)$ exists and is a continuous function of x. It may also be smooth on intervals containing isolated singular points; for example, the curve $y = x^{1/3}$ is smooth everywhere even though dy/dx does not exist at $x = 0$.

For parametric curves $x = f(t)$, $y = g(t)$, the situation is more complicated. Even if f and g have continuous derivatives everywhere, such curves may fail to be smooth at certain points, specifically points where $f'(t) = g'(t) = 0$.

EXAMPLE 9.4.1 Consider the parametric curve $x = f(t) = t^2$, $y = g(t) = t^3$. Even though $f'(t) = 2t$ and $g'(t) = 3t^2$ are continuous for all t, the curve is not smooth at $t = 0$. (See Fig. 9.4.1.) Observe that f' and g' both vanish at $t = 0$: $f'(0) = g'(0) = 0$. If we regard the parametric equations as specifying the position at time t of a moving point P, then the horizontal velocity is $f'(t)$ and the vertical velocity is $g'(t)$. Both velocities are 0 at $t = 0$, so P has come to a stop at that instant. When it starts moving again it need not move in the direction it was going before it stopped. The cycloid of Example 9.3.11 is another example where a paramteric curve is not smooth at points where dx/dt and dy/dt both vanish.

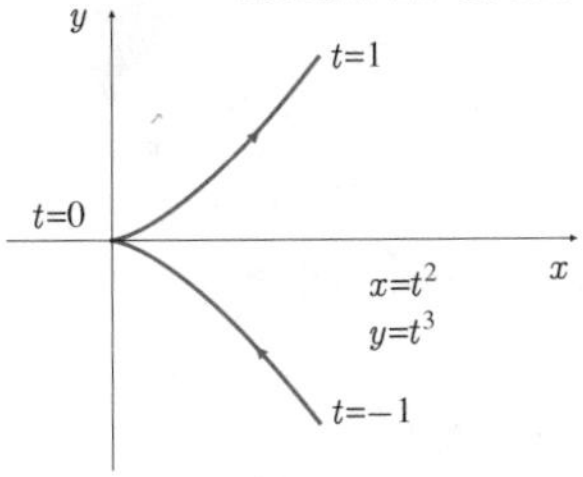

FIGURE 9.4.1

The Slope of a Parametric Curve

THEOREM 9.4.2 Suppose that f' and g' are continuous on an interval I. If $f'(t) > 0$ on I (or $f'(t) < 0$ on I), then the curve C with parametric equations $x = f(t)$, $y = g(t)$ is smooth for t in I and has, at the point with parameter value t, a tangent line with slope

$$\frac{dy}{dx} = \frac{g'(t)}{f'(t)}.$$

If $g'(t) > 0$ on I (or $g'(t) < 0$ on I), then C is smooth for t in I and has, at the point with parameter value t, a normal line with slope

$$-\frac{dx}{dy} = -\frac{f'(t)}{g'(t)}.$$

Thus C is smooth except possibly at points where $f'(t)$ and $g'(t)$ are both 0.

PROOF If $f'(t) > 0$ on I (or $f'(t) < 0$ on$I$), then f is increasing (or decreasing) on I and so is one-to-one and invertible. The part of C corresponding to values of t in I has ordinary equation $y = g\big(f^{-1}(x)\big)$ and hence slope

$$\frac{dy}{dx} = g'\big(f^{-1}(x)\big)\,\frac{d}{dx}f^{-1}(x) = \frac{g'\big(f^{-1}(x)\big)}{f'\big(f^{-1}(x)\big)} = \frac{g'(t)}{f'(t)},$$

(where we have used the derivative of an inverse function,

$$\frac{d}{dx}f^{-1}(x) = \frac{1}{f'\big(f^{-1}(x)\big)},$$

obtained in Section 2.5.) This slope is a continuous function of t, so the tangent to C turns continuously for t in I. The proof for $g'(t) > 0$ (or < 0) is similar. In this case the slope of the normal is a continuous function of t, so the normal turns continuously. Therefore, so does the tangent. □

If f' and g' are continuous, and both vanish at some point t_0, then the curve $x = f(t)$, $y = g(t)$ *may or may not* be smooth around t_0. Example 1 was an example of a curve that was not smooth at such a point.

EXAMPLE 9.4.3 The curve with parametrization $x = t^3$, $y = t^6$ is just the parabola $y = x^2$ so it is smooth everywhere although $dx/dt = 3t^2$ and $dy/dt = 6t^5$ both vanish at $t = 0$.

9.4.4 Tangents and Normals

If f' and g' are continuous and not both 0 at t_0, then the parametric equations

$$x = f(t_0) + sf'(t_0), \qquad y = g(t_0) + sg'(t_0), \qquad (-\infty < s < \infty),$$

represent the tangent line to the parametric curve $x = f(t)$, $y = g(t)$ at the point $\big(f(t_0), g(t_0)\big)$. The normal line there has parametric equations

$$x = f(t_0) + sg'(t_0), \qquad y = g(t_0) - sf'(t_0), \qquad (-\infty < s < \infty).$$

EXAMPLE 9.4.5 Find the tangent and normal lines to the parametric curve $x = t^2 - t,\ y = t^2 + t$ at the point where $t = 2$.

SOLUTION At $t = 2$ we have $x = 2$, $y = 6$ and

$$\frac{dx}{dt} = 2t - 1 = 3, \qquad \frac{dy}{dx} = 2t + 1 = 5.$$

Hence the tangent and the normal lines have parametric equations

$$\text{Tangent: } \begin{cases} x = 2 + 3s \\ y = 6 + 5s \end{cases} \qquad \text{Normal: } \begin{cases} x = 2 + 5s \\ y = 6 - 3s \end{cases}$$

The concavity of a parametric curve can be determined using the second derivatives of the parametric equations. The procedure is just to calculate d^2y/dx^2 using the chain rule:

$$\frac{d^2y}{dx^2} = \frac{d}{dx}\frac{dy}{dx} = \frac{d}{dx}\frac{g'(t)}{f'(t)} = \frac{dt}{dx}\frac{d}{dt}\frac{g'(t)}{f'(t)} = \frac{1}{f'(t)}\frac{f'(t)g''(t) - g'(t)f''(t)}{(f'(t))^2}.$$

9.4.6
Concavity of a Parametric Curve

> On an interval where $f'(t) \neq 0$, the parametric curve $x = f(t)$, $y = g(t)$ has concavity determined by
>
> $$\frac{d^2y}{dx^2} = \frac{f'(t)g''(t) - g'(t)f''(t)}{(f'(t))^3}.$$

Sketching Parametric Curves

As in the case of graphs of functions, derivatives provide useful information about the shape of a parametric curve. At points where $dy/dt = 0$ but $dx/dt \neq 0$, the tangent is horizontal; at points where $dx/dt = 0$ but $dy/dt \neq 0$, the tangent is vertical. For points where $dx/dt = dy/dt = 0$, anything can happen; it is wise to calculate left- and right-hand limits of the slope dy/dx as the parameter t approaches one of these points. Concavity can be determined using Formula 9.4.6 above. We illustrate these ideas by reconsidering a parametric curve encountered in the previous section.

EXAMPLE 9.4.7 Reconsider Example 9.3.6. The curve has parametric equations

$$x = f(t) = t^3 - 3t, \qquad y = g(t) = t^2. \qquad (-2 \leq t \leq 2).$$

Therefore $f'(t) = 3(t^2 - 1) = 3(t - 1)(t + 1)$ and $g'(t) = 2t$. The curve has a horizontal tangent at $t = 0$, that is, at $(0, 0)$, and vertical tangents at $t = \pm 1$, that is, at $(2, 1)$ and $(-2, 1)$. Directional information for the curve between these points is summarized in the following chart.

t	-2 to -1	-1 to 0	0 to 1	1 to 2
$f'(t)$	$+$	$-$	$-$	$+$
$g'(t)$	$-$	$-$	$+$	$+$
x	$\rightarrow$	$\leftarrow$	$\leftarrow$	$\rightarrow$
y	$\downarrow$	$\downarrow$	$\uparrow$	$\uparrow$
curve	$\searrow$	$\swarrow$	$\nwarrow$	$\nearrow$

For concavity we calculate d^2y/dx^2 by formula 9.4.6. Since $f''(t) = 6t$ and $g''(t) = 2$, we have

$$\frac{d^2y}{dx^2} = \frac{3(t^2 - 1)(2) - 2t(6t)}{[3(t^2 - 1)]^3} = -\frac{2}{9}\frac{t^2 + 1}{(t^2 - 1)^3},$$

which is never zero but which fails to be defined at $t = \pm 1$. The concavity is shown in the following chart:

t	$t < -1$	$-1 < t < 1$	$t > 1$
d^2y/dx^2	$-$	$+$	$-$
curve	$\frown$	$\smile$	$\frown$

The curve was sketched in the previous section: see Fig. 9.3.6.

Arc Lengths of Parametric Curves

Let C be a smooth parametric curve with equations

$$x = f(t), \qquad y = g(t), \qquad a \le t \le b.$$

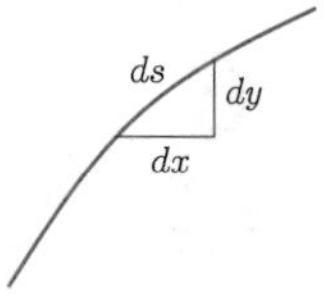

FIGURE 9.4.2

(We assume that $f'(t)$ and $g'(t)$ are continuous on the interval $[a, b]$ and never both zero.) From the differential triangle with legs dx and dy and hypotenuse ds (see Fig. 9.4.2), we have $(ds)^2 = (dx)^2 + (dy)^2$, so

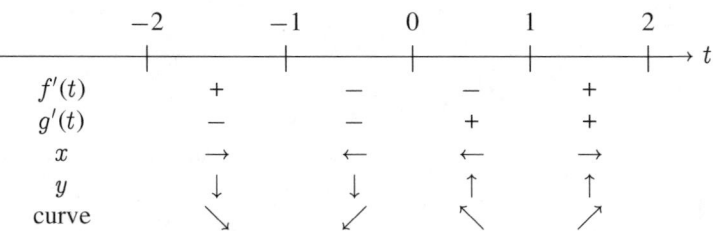

9.4.8 Arc Length Element for Parametric Curves

$$ds = \frac{ds}{dt}\,dt = \sqrt{\left(\frac{ds}{dt}\right)^2}\,dt = \sqrt{\left(\frac{dx}{dt}\right)^2 + \left(\frac{dy}{dt}\right)^2}\,dt$$

The length of the curve C is given by

$$s = \int_{t=a}^{t=b} ds = \int_a^b \sqrt{\left(\frac{dx}{dt}\right)^2 + \left(\frac{dy}{dt}\right)^2}\,dt.$$

EXAMPLE 9.4.9 Find the length of the parametric curve

$$x = e^t \cos t, \qquad y = e^t \sin t, \qquad (0 \le t \le 2).$$

SOLUTION We have

$$\frac{dx}{dt} = e^t(\cos t - \sin t), \qquad \frac{dy}{dt} = e^t(\sin t + \cos t).$$

Squaring these formulas, adding and simplifying, we get

$$\left(\frac{ds}{dt}\right)^2 = e^{2t}(\cos t - \sin t)^2 + e^{2t}(\sin t + \cos t)^2 = 2e^{2t}.$$

The length of the curve is therefore

$$s = \int_0^2 \sqrt{2e^{2t}}\, dt = \sqrt{2}\int_0^2 e^t\, dt = \sqrt{2}\,(e^2 - 1) \text{ units.}$$

EXAMPLE 9.4.10 The arc length formula 9.2.15 for a polar curve can be derived directly from that for parametric curves. The polar curve $r = f(\theta)$, $(\alpha \le \theta \le \beta)$ can be parametrized

$$x = r\cos\theta = f(\theta)\cos\theta, \qquad y = r\sin\theta = f(\theta)\sin\theta.$$

Thus

$$\frac{dx}{d\theta} = f'(\theta)\cos\theta - f(\theta)\sin\theta$$
$$\frac{dy}{d\theta} = f'(\theta)\sin\theta + f(\theta)\cos\theta.$$

Squaring and adding these, we obtain $\left(\frac{ds}{d\theta}\right)^2 = \big(f'(\theta)\big)^2 + \big(f(\theta)\big)^2$, so

$$ds = \sqrt{\big(f'(\theta)\big)^2 + \big(f(\theta)\big)^2}\, d\theta.$$

Parametric curves can be rotated around various axes to generate surfaces of revolution. The areas of these surfaces can be found by the same procedure used for graphs of functions, with the appropriate version of ds.

If the smooth parametric curve with equations

$$x = f(t), \qquad y = g(t), \qquad a \le t \le b,$$

is rotated about the x-axis, the area of S of the surface so generated is given by

$$S = 2\pi\int_{t=a}^{t=b} |y|\, ds = 2\pi\int_a^b |g(t)|\sqrt{(f'(t))^2 + (g'(t))^2}\, dt.$$

If the rotation is about the y-axis, then the area is

$$S = 2\pi\int_{t=a}^{t=b} |x|\, ds = 2\pi\int_a^b |f(t)|\sqrt{(f'(t))^2 + (g'(t))^2}\, dt$$

Areas Bounded by Parametric Curves

Consider the parametric curve C with equations $x = f(t)$, $y = g(t)$, $(a \le t \le b)$, where f is differentiable and g is continuous on $[a, b]$. For the moment let us also assume that $f'(t) \ge 0$ and $g(t) \ge 0$ on $[a, b]$, so C has no points below the x-axis and is traversed from left to right as t increases from a to b.

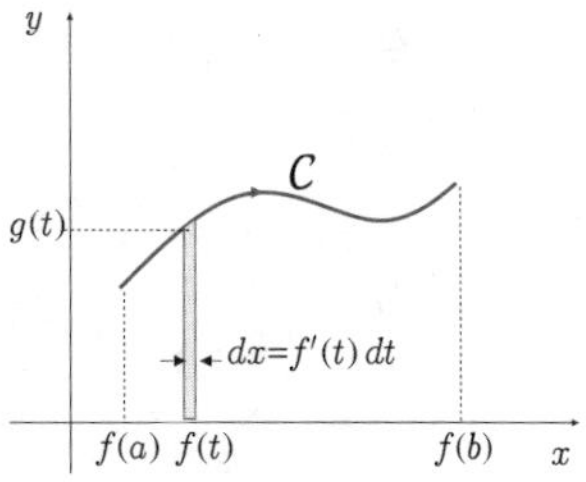

FIGURE 9.4.3

The region under C and above the x-axis has area element $dA = y\,dx = g(t)f'(t)\,dt$ and so its area (see Fig. 9.4.3) is

$$A = \int_a^b g(t)f'(t)\,dt.$$

Similar arguments can be given for three other cases:

If $f'(t) \ge 0$ and $g(t) \le 0$ on $[a, b]$, then $A = -\int_a^b g(t)f'(t)\,dt$,

If $f'(t) \le 0$ and $g(t) \ge 0$ on $[a, b]$, then $A = -\int_a^b g(t)f'(t)\,dt$,

If $f'(t) \le 0$ and $g(t) \le 0$ on $[a, b]$, then $A = \int_a^b g(t)f'(t)\,dt$,

where A is the area bounded by C, the x-axis, and the vertical lines $x = f(a)$ and $x = f(b)$. Combining these results we can see that

$$\int_a^b g(t)f'(t)\,dt = A_1 - A_2,$$

where A_1 is the area lying vertically between C and that part of the x-axis consisting of points $x = f(t)$ such that $g(t)f'(t) \ge 0$ and A_2 is a similar area corresponding to points where $g(t)f'(t) < 0$. See Fig. 9.4.4. This formula is valid for arbitrary

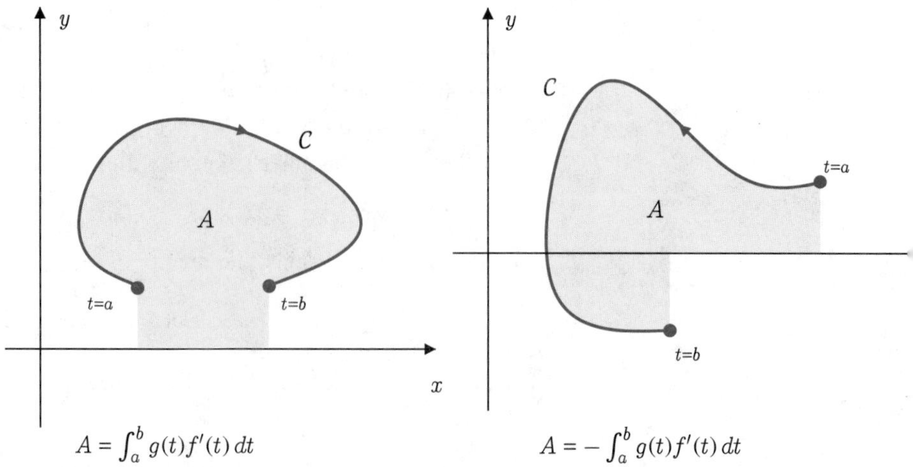

FIGURE 9.4.4

continuous g and differentiable f. In particular, if C is a non-self-intersecting closed curve, then the area of the region bounded by C is given by

$$A = \int_a^b g(t)f'(t)\,dt \qquad \text{if } C \text{ is traversed clockwise as } t \text{ increases,}$$

$$A = -\int_a^b g(t)f'(t)\,dt \qquad \text{if } C \text{ is traversed counterclockwise,}$$

both of which are illustrated in Fig. 9.4.5.

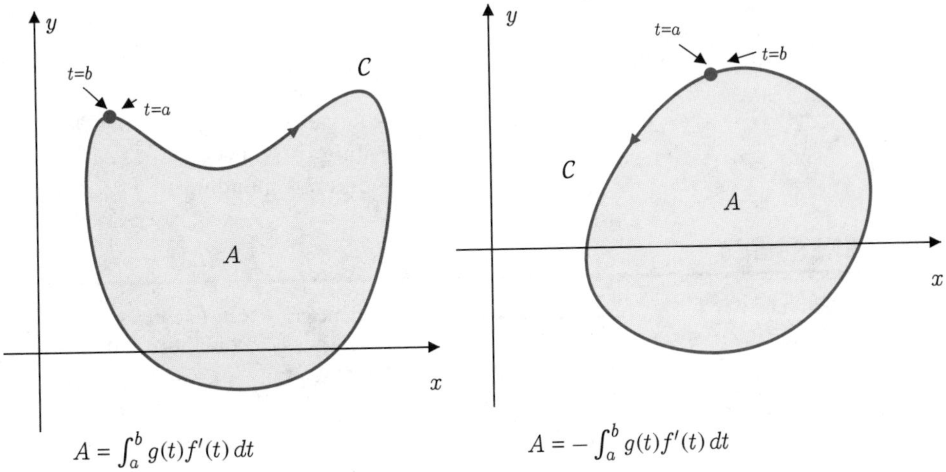

FIGURE 9.4.5

EXAMPLE 9.4.11 Find the area bounded by the ellipse $x = a\cos s$, $y = b\sin s$, $(0 \le s \le 2\pi)$.

SOLUTION This ellipse is traversed counterclockwise. (See Example 9.3.5.) The area enclosed is

$$A = -\int_0^{2\pi} b\sin s(-a\sin s)\,ds = \frac{ab}{2}\int_0^{2\pi}(1-\cos 2s)\,ds$$
$$= \frac{ab}{2}s\Big|_0^{2\pi} - \frac{ab}{4}\sin 2s\Big|_0^{2\pi} = \pi ab \text{ sq units.}$$

EXAMPLE 9.4.12 Find the area above the x-axis and under one arch of the cycloid $x = at - a\sin t$, $y = a - a\cos t$.

SOLUTION Part of the cycloid is shown in Figure 9.3.8 in the previous section. One arch corresponds to $0 \le t \le 2\pi$. Since $y = a(1-\cos t) \ge 0$ and $dx/dt = a(1-\cos t) \ge 0$, the area under one arch is

$$A = \int_0^{2\pi} a^2(1-\cos t)^2\,dt = a^2\int_0^{2\pi}\left(1 - 2\cos t + \frac{1+\cos 2t}{2}\right)dt$$
$$= a^2\left(t - 2\sin t + \frac{t}{2} + \frac{\sin 2t}{4}\right)\Bigg|_0^{2\pi} = 3\pi a^2 \text{ sq units.}$$

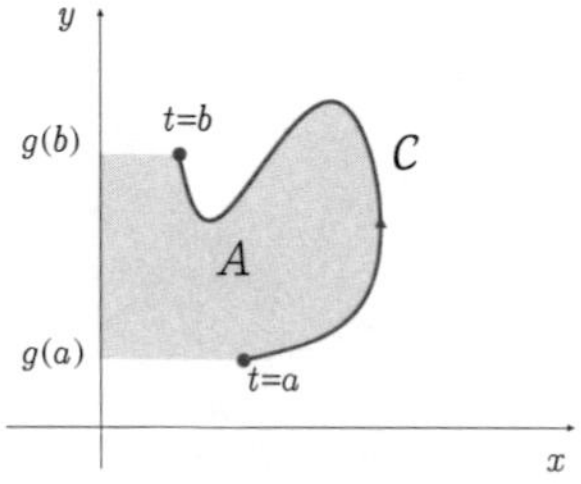

FIGURE 9.4.6

Similar arguments to those used above show that if f is continuous and g is differentiable, then we can also interpret

$$\int_a^b f(t)g'(t)\,dt = \int_{t=a}^{t=b} x\,dy = A_1 - A_2,$$

where A_1 is the area of the region lying *horizontally* between C and that part of the y-axis consisting of points $y = g(t)$ such that $f(t)g'(t) \ge 0$, and A_2 is the area of a similar region corresponding to $f(t)g'(t) < 0$. See Fig. 9.4.6.

EXERCISES

In Exercises 1–7 sketch the graphs of the given parametric curves, making use of information from the first two derivatives. Unless otherwise stated, the parameter interval for each curve is $\mathbb{R}$.

1. $x = t^2 - 2t,\ y = t^2 - 4t$

2. $x = t^3,\ y = 3t^2 - 1$

3. $x = t^3 + 3t,\ y = t^2$

4. $x = t^3 - 3t - 2,\ y = t^2 - t - 2$

5. $x = t^3 - 3t,\ y = \dfrac{2}{1+t^2}$

6. $x = 1 - \cos t - \sin t, y = 1 - \cos t + \sin t.$

7. $x = \cos t + t\sin t,\ y = \sin t - t\cos t,\ (t \ge 0)$. (See Example 9.3.10.)

Find the lengths of the curves in Exercises 8–11.

8. $x = 1 + t^3,\ y = 1 - t^2,\ (-1 \le t \le 2)$

9. $x = a\cos^3 t,\ y = a\sin^3 t,\ (0 \le t \le 2\pi)$

10. $x = \ln(1+t^2),\ y = 2\tan^{-1} t,\ (0 \le t \le 1)$

11. $x = t^2 \sin t,\ y = t^2\cos t,\ (0 \le t \le 2\pi)$

12. Find the area of the surfaces obtained by rotating one arch of the cycloid $x = at - a\sin t$, $y = a - a\cos t$ about (a) the x-axis, (b) the y-axis. (One arch corresponds to $0 \le t \le 2\pi$.)

In Exercises 13–18 sketch and find the area of the region R described in terms of given parametric curves.

13. R is the closed loop bounded by $x = t^3 - 4t$, $y = t^2$, $(-2 \le t \le 2)$.

14. R is bounded by the astroid $x = a\cos^3 t$, $y = a\sin^3 t$, $(0 \le t \le 2\pi)$.

15. R is bounded by the coordinate axes and the parabolic arc $x = \sin^4 t$, $y = \cos^4 t$.

16. R is bounded by $x = \cos s \sin s$, $y = \sin^2 s$, $(0 \le s \le \pi/2)$, and the y-axis.

17. R is bounded by the oval $x = (2 + \sin t)\cos t$, $y = (2 + \sin t)\sin t$.

18.* R is bounded by the x-axis, the hyperbola $x = \sec t$, $y = \tan t$ and the ray joining the origin to the point $(\sec t_0, \tan t_0)$.

19. Show that the region bounded by the x-axis and the hyperbola $x = \cosh t$, $y = \sinh t$, $(t > 0)$ and the ray from the origin to the point $(\cosh t_0, \sinh t_0)$ has area $t_0/2$ square units. This proves a claim made at the beginning of Section 3.8.

20. Find the volume of the solid obtained by rotating about the x-axis the region bounded by that axis and one arch of the cycloid (see Example 9.3.11) $x = at - a\sin t$, $y = a - a\cos t$.

9.5 VELOCITY, ACCELERATION AND PLANE VECTORS

In Section 2.6 we examined the role of derivatives in describing motion in one dimension; velocity and acceleration were seen to be derivatives of position. Parametric equations provide a framework for extending these ideas to the study of motion in more than one dimension. In this section we consider motion in the xy-plane.

Suppose that $P = (x, y)$ is the position at time t of a point moving in the plane. The parametric equations of the path of the point,

$$x = f(t), \qquad y = g(t),$$

represent independent motions of the point in directions parallel to the x-axis and y-axis respectively; each equation determines a *component* of the overall motion. Similarly, the velocity and acceleration of the moving point have, at time t, two components. The x and y components of velocity are

$$v_x = \frac{dx}{dt} = f'(t), \qquad v_y = \frac{dy}{dt} = g'(t).$$

The x and y components of acceleration are

$$a_x = \frac{d^2x}{dt^2} = f''(t), \qquad a_y = \frac{d^2y}{dt^2} = g''(t).$$

In spite of the fact that we have used two quantities (v_x and v_y) to represent the velocity, we clearly think of this velocity as being a single quantity possessing two attributes, size and direction. (At a specific time the point is moving in a specific direction with a specific speed.) In order to deal effectively with velocity and acceleration as single quantities, we need to develop the notion of a vector.

Plane Vectors

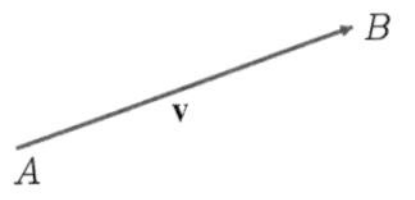

FIGURE 9.5.1

A **vector** is a quantity that involves both **magnitude** (size or length) and **direction**. Such quantities are conveniently represented geometrically by arrows (directed line segments) and are often actually identified with these arrows. For instance, the vector $\overrightarrow{AB}$ is an arrow with tail at the point A and head at the point B. We often denote such a vector by a single boldface letter,

$$\mathbf{v} = \overrightarrow{AB},$$

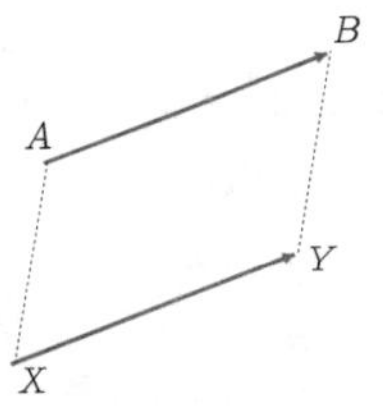

FIGURE 9.5.2

(see Fig. 9.5.1), though you may prefer to use an arrow over a letter ($\overrightarrow{v} = \overrightarrow{AB}$) when you are writing vectors. The magnitude of $\mathbf{v}$ is the length of the arrow and is denoted $|\mathbf{v}|$ or $|\overrightarrow{AB}|$.

While vectors have magnitude and direction, they do not generally have *position*, that is, they are not regarded as being in a particular place. We consider as equal two vectors $\mathbf{u}$ and $\mathbf{v}$ that have *the same length and the same direction*, even if their representative arrows do not coincide. The arrows must be parallel, have the same length, and point in the same direction. In Fig. 9.5.2, for example, if $ABYX$ is a parallelogram, then $\overrightarrow{AB} = \overrightarrow{XY}$.

In this section we consider only plane vectors, that is, vectors whose repre sentative arrows lie in a plane. If we introduce a Cartesian coordinate system int the plane, we can talk about the x and y components of any vector. If $A = (a, b$ and $P = (p, q)$, as shown in Fig. 9.5.3, then the x and y components of $\overrightarrow{AP}$ are respectively, $p - a$ and $q - b$. Note that if O is the origin and X is the poin $(p - a, q - b)$, then

$$|\overrightarrow{AP}| = \sqrt{(p-a)^2 + (q-b)^2} = |\overrightarrow{OX}|$$

$$\text{slope of } \overrightarrow{AP} = \frac{q-b}{p-a} = \text{slope of } \overrightarrow{OX}.$$

Hence $\overrightarrow{AP} = \overrightarrow{OX}$. In general, two vectors are equal if and only if they have the same x components and y components.

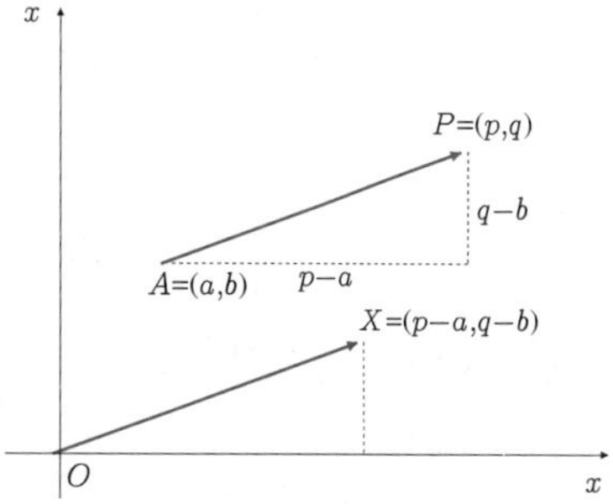

FIGURE 9.5.3

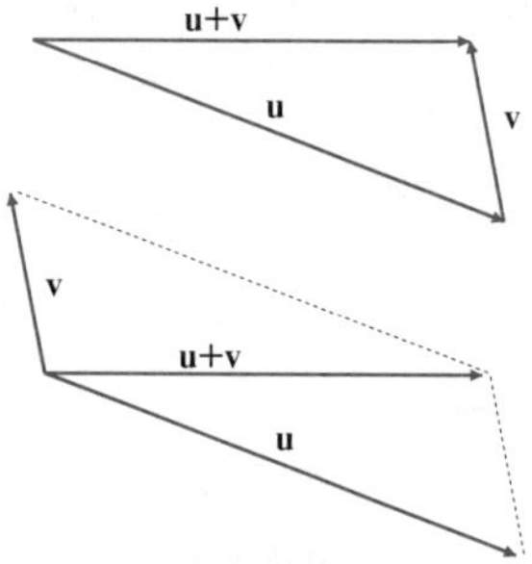

FIGURE 9.5.4

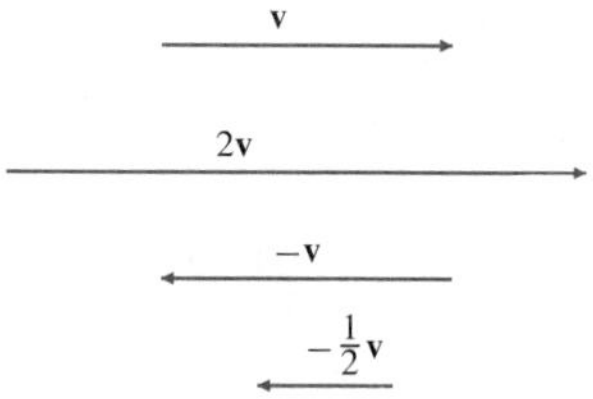

FIGURE 9.5.5

There are two important algebraic operations defined for vectors: addition and scalar multiplication.

9.5.1
Vector Addition

> Given two vectors **u** and **v**, their **sum u + v** is defined as follows. If an arrow representing **v** is placed with its tail at the head of an arrow representing **u**, then an arrow from the tail of **u** to the head of **v** represents **u**+**v**. Equivalently, if **u** and **v** have tails at the same point, then **u** + **v** is represented by an arrow with its tail at that point and its head at the opposite vertex of the parallelogram spanned by **u** and **v**. This is shown in Fig. 9.5.4.

9.5.2
Scalar Multiplication

> If **v** is a vector and t is a real number (also called a **scalar**), then the **scalar multiple** t**v** is a vector with magnitude $|t|$ times that of **v** and direction the same as **v** if $t > 0$, or opposite to that of **v** if $t < 0$. See Fig. 9.5.5. If $t = 0$, then t**v** has zero length and therefore no particular direction. It is the **zero vector**, denoted **0**.

Suppose that **u** has components a and b and that **v** has components x and y. Then the components of $\mathbf{u}+\mathbf{v}$ are $a+x$ and $b+y$, and those of $t\mathbf{v}$ are tx and ty. See Fig. 9.5.6.

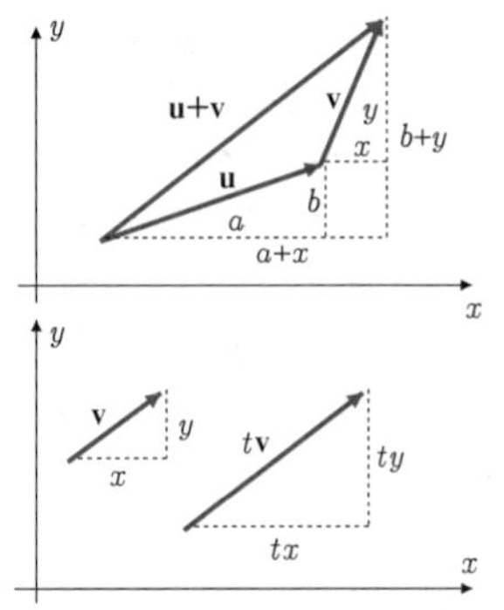

FIGURE 9.5.6

In the Cartesian plane we single out two particular vectors for special attention. They are

i) the vector **i** from the origin to the point $(1, 0)$, and

ii) the vector **j** from the origin to the point $(0, 1)$.

Thus **i** has components 1 and 0, and **j** has components 0 and 1. These vectors are called the **standard basis vectors** in the plane. If **v** is a vector with components x and y, then **v** can be expressed in the form

$$\mathbf{v} = x\mathbf{i} + y\mathbf{j}.$$

We say that we have written **v** as a **linear combination of the standard basis vectors**. (See Fig. 9.5.7.) The length of **v** is $|\mathbf{v}| = \sqrt{x^2+y^2}$.

EXAMPLE 9.5.3 If $A = (2, -1)$, $B = (-1, 4)$, and $C = (0, 2)$, express each of the following vectors as a linear combination of the standard basis vectors.

a) $\overrightarrow{AB}$ b) $\overrightarrow{BC}$ c) $\overrightarrow{AC}$ d) $\overrightarrow{AB} + \overrightarrow{BC}$ e) $2\overrightarrow{AC} - 3\overrightarrow{CB}$

SOLUTION

a) $\overrightarrow{AB} = (-1-2)\mathbf{i} + (4-(-1))\mathbf{j} = -3\mathbf{i} + 5\mathbf{j}$

b) $\overrightarrow{BC} = (0-(-1))\mathbf{i} + (2-4)\mathbf{j} = \mathbf{i} - 2\mathbf{j}$

c) $\overrightarrow{AC} = (0-2)\mathbf{i} + (2-(-1))\mathbf{j} = -2\mathbf{i} + 3\mathbf{j}$

d) $\overrightarrow{AB} + \overrightarrow{BC} = \overrightarrow{AC} = -2\mathbf{i} + 3\mathbf{j}$

e) $2\overrightarrow{AC} - 3\overrightarrow{CB} = 2(-2\mathbf{i} + 3\mathbf{j}) - 3(-\mathbf{i} + 2\mathbf{j}) = -\mathbf{i}$

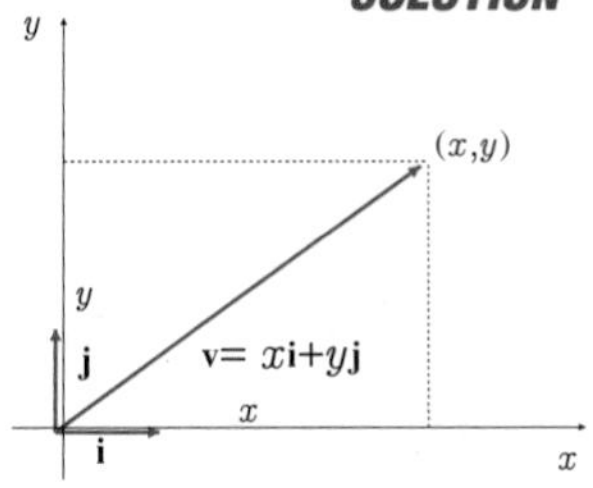

FIGURE 9.5.7

Implicit in the above example is the fact that the operations of addition and scalar multiplication obey appropriate algebraic rules, such as

$$\begin{aligned} \mathbf{u}+\mathbf{v} &= \mathbf{v}+\mathbf{u} \\ (\mathbf{u}+\mathbf{v})+\mathbf{w} &= \mathbf{u}+(\mathbf{v}+\mathbf{w}) \\ \mathbf{u}-\mathbf{v} &= \mathbf{u}+(-1)\mathbf{v} \\ t(\mathbf{u}+\mathbf{v}) &= t\mathbf{u}+t\mathbf{v} \end{aligned}$$

We require one additional operation on vectors, the dot product.

9.5.4 The Dot Product of Two Vectors

Given two vectors, $\mathbf{u} = a\mathbf{i} + b\mathbf{j}$ and $\mathbf{v} = x\mathbf{i} + y\mathbf{j}$, we define their **dot product** as the *number*

$$\mathbf{u} \bullet \mathbf{v} = ax + by.$$

Note that the dot product of two vectors is not a vector but a number (a scalar). Moreover, the arrows representing the vectors **u** and **v** are perpendicular if and only if $\mathbf{u} \bullet \mathbf{v} = 0$. To see this, observe that the slopes of the arrows **u** and **v** are b/a and y/x, respectively (assuming a and x are not 0), and the product of these slopes is -1 if and only if $ax + by = 0$.

The dot product has the following algebraic properties

$$\begin{aligned}\mathbf{u} \bullet \mathbf{v} &= \mathbf{v} \bullet \mathbf{u}\\ \mathbf{u} \bullet (\mathbf{v} + \mathbf{w}) &= \mathbf{u} \bullet \mathbf{v} + \mathbf{u} \bullet \mathbf{w}\\ \mathbf{v} \bullet \mathbf{v} &= |\mathbf{v}|^2\\ (t\mathbf{u}) \bullet \mathbf{v} &= \mathbf{u} \bullet (t\mathbf{v}) = t(\mathbf{u} \bullet \mathbf{v}),\end{aligned}$$

all of which are easily verified using the definition of dot product.

Finally we show that

$$\mathbf{u} \bullet \mathbf{v} = |\mathbf{u}||\mathbf{v}| \cos\theta,$$

where θ is the angle between the directions of **u** and **v** ($0 \le \theta \le \pi$). To see this refer to Fig. 9.5.8 and apply the Cosine Law (Theorem 3.1.12) to the triangle with the arrows **u**, **v**, and **u** − **v** as sides.

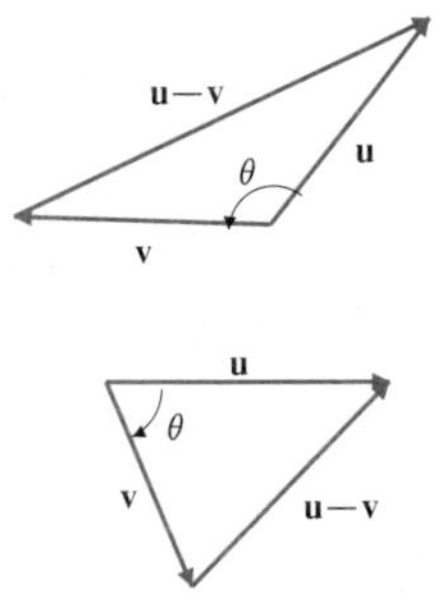

FIGURE 9.5.8

$$\begin{aligned}|\mathbf{u}|^2 + |\mathbf{v}|^2 - 2|\mathbf{u}|\,|\mathbf{v}| \cos\theta &= |\mathbf{u} - \mathbf{v}|^2 = (\mathbf{u} - \mathbf{v}) \bullet (\mathbf{u} - \mathbf{v})\\ &= \mathbf{u} \bullet (\mathbf{u} - \mathbf{v}) - \mathbf{v} \bullet (\mathbf{u} - \mathbf{v})\\ &= \mathbf{u} \bullet \mathbf{u} - \mathbf{u} \bullet \mathbf{v} - \mathbf{v} \bullet \mathbf{u} + \mathbf{v} \bullet \mathbf{v}\\ &= |\mathbf{u}|^2 + |\mathbf{v}|^2 - 2\mathbf{u} \bullet \mathbf{v}\end{aligned}$$

Hence $|\mathbf{u}||\mathbf{v}| \cos\theta = \mathbf{u} \bullet \mathbf{v}$ as claimed.

EXAMPLE 9.5.5 The angle θ between the vectors $\mathbf{i} + \mathbf{j}$ and $\mathbf{i} - 3\mathbf{j}$ satisfies

$$\cos\theta = \frac{(\mathbf{i} + \mathbf{j}) \bullet (\mathbf{i} - 3\mathbf{j})}{|\mathbf{i} + \mathbf{j}|\,|\mathbf{i} - 3\mathbf{j}|} = \frac{1 - 3}{\sqrt{2}\sqrt{10}} = -\frac{2}{\sqrt{2}\sqrt{10}} = -\frac{1}{\sqrt{5}}.$$

Thus $\theta = \cos^{-1}(-1/\sqrt{5}) \approx 116.565°$.

Position, Velocity, and Acceleration as Vectors

A **position vector** is an arrow that has its tail at the origin; the head of the arrow indicates the position of some point in the plane, and the components of the vector are the coordinates of that point.

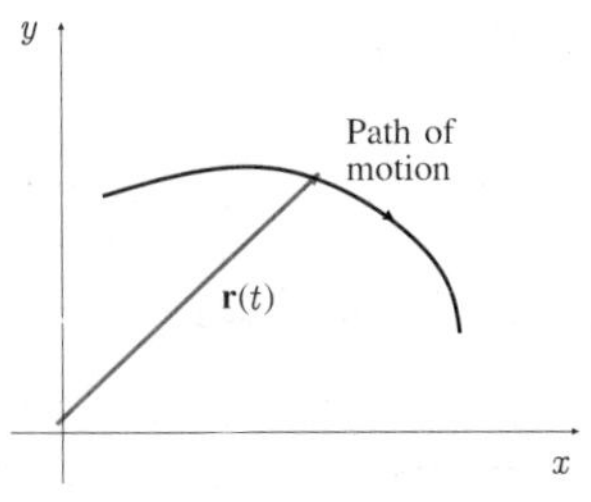

FIGURE 9.5.9

If a particle moves in the plane so that its position at time t is given by the parametric equations $x = f(t)$, $y = g(t)$, we can use these functions as components of a **position vector function r** = **r**(t) defined by (see Fig. 9.5.9)

$$\mathbf{r} = x\mathbf{i} + y\mathbf{j} = f(t)\,\mathbf{i} + g(t)\,\mathbf{j}.$$

The velocity of the particle at time t is also a vector since it embodies the direction as well as the speed of motion. The velocity is the rate of change of the position vector with respect to time:

$$\begin{aligned}\mathbf{v}(t) = \frac{d\mathbf{r}}{dt} &= \lim_{h\to 0} \frac{\mathbf{r}(t+h) - \mathbf{r}(t)}{h} \\ &= \lim_{h\to 0} \left(\frac{f(t+h) - f(t)}{h}\,\mathbf{i} + \frac{g(t+h) - g(t)}{h}\,\mathbf{j} \right) \\ &= f'(t)\mathbf{i} + g'(t)\mathbf{j}.\end{aligned}$$

As we might have anticipated, the components of the velocity vector are just the horizontal and verical components of velocity, as considered earlier. The *speed* $s(t)$ of the particle at time t is the length of the velocity vector:

$$s(t) = |\mathbf{v}(t)| = \left| \frac{d\mathbf{r}}{dt} \right| = \sqrt{(f'(t))^2 + (g'(t))^2}.$$

Note that the length of the path traced by the moving point in the time interval $[a, b]$ is just the integral of the speed over that interval.

The Newton quotient $\dfrac{\mathbf{r}(t+h) - \mathbf{r}(t)}{h}$ is a vector along a secant line to the path of motion. Hence its limit, the velocity vector $\mathbf{v}(t)$, is tangent to the path of motion at the position $\mathbf{r}(t)$, as shown in Fig. 9.5.10. Evidently $\mathbf{v}(t)$ points in the direction in which the particle is moving at time t.

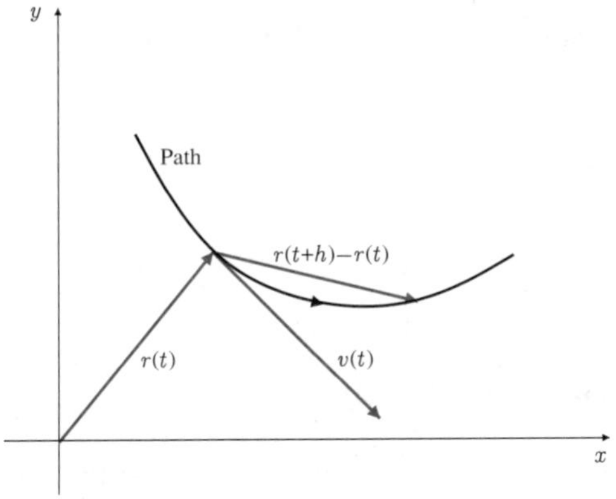

FIGURE 9.5.10

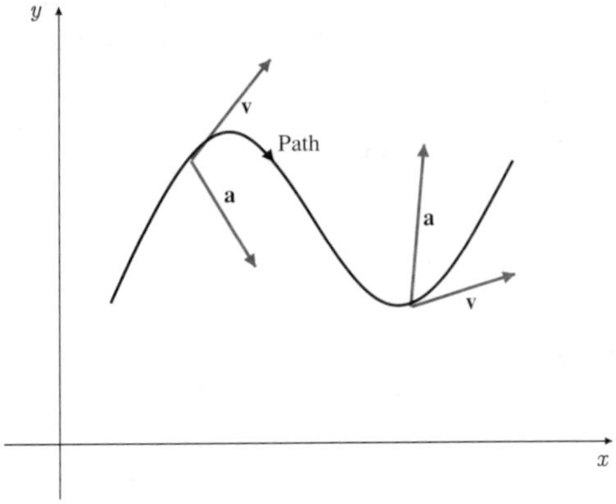

FIGURE 9.5.11

Similarly, the acceleration vector $\mathbf{a}(t)$ is the derivative of the velocity vector:

$$\mathbf{a}(t) = \frac{d\mathbf{v}}{dt} = \frac{d^2\mathbf{r}}{dt^2} = f''(t)\,\mathbf{i} + g''(t)\,\mathbf{j}.$$

The acceleration is also represented by an arrow with tail at $\mathbf{r}(t)$. The position of this arrow with respect to the velocity arrow indicates whether the speed is increasing or decreasing and which way the curve of motion is bending. If the angle between $\mathbf{v}$ and $\mathbf{a}$ at a point is less than $\pi/2$, then the speed is increasing; if it is greater than $\pi/2$, the speed is decreasing. (See Fig. 9.5.11 and Exercise 22 at the end of this section.)

The algebraic rules governing dot product ensure that the product rule for differentiation also holds for a dot product of vector functions: If $\mathbf{u}(t) = a(t)\mathbf{i} + b(t)\mathbf{j}$ and $\mathbf{w}(t) = x(t)\mathbf{i} + y(t)\mathbf{j}$, then

$$\begin{aligned}\frac{d}{dt}\big[\mathbf{u}(t) \bullet \mathbf{w}(t)\big] &= \frac{d}{dt}\big(a(t)x(t) + b(t)y(t)\big)\\ &= a'(t)x(t) + a(t)x'(t) + b'(t)y(t) + b(t)y'(t)\\ &= \frac{d\mathbf{u}}{dt} \bullet \mathbf{w}(t) + \mathbf{u}(t) \bullet \frac{d\mathbf{w}}{dt}.\end{aligned}$$

If the speed of a moving particle is constant over an interval of time, then

$$0 = \frac{d}{dt}|\mathbf{v}(t)|^2 = \frac{d}{dt}\big(\mathbf{v}(t) \bullet \mathbf{v}(t)\big) = \frac{d\mathbf{v}}{dt} \bullet \mathbf{v}(t) + \mathbf{v}(t) \bullet \frac{d\mathbf{v}}{dt} = 2\mathbf{a}(t) \bullet \mathbf{v}(t).$$

Thus $\mathbf{a}(t)$ is perpendicular to $\mathbf{v}(t)$. Conversely, if $\mathbf{a}$ is perpendicular to $\mathbf{v}$ over a time interval, then the speed is constant over that interval. Note that constant speed does *not* imply constant velocity; $\mathbf{a}$ need not be $\mathbf{0}$.

EXAMPLE 9.5.6 An object is moving in the plane so that its position at time t is given by

$$\mathbf{r} = t^3\,\mathbf{i} - (4t^2 - 8t + 1)\,\mathbf{j}.$$

Find the velocity, speed, and acceleration of the object at any time t. Where is the object at time $t = 1$, in what direction and how fast is it moving at that time? Is it speeding up or slowing down?

SOLUTION The velocity, speed, and acceleration at time t are given by

$$\begin{aligned}\mathbf{v} &= \frac{d\mathbf{r}}{dt} = 3t^2\,\mathbf{i} - 8(t-1)\,\mathbf{j}\\ s &= |\mathbf{v}| = \sqrt{9t^4 + 64(t-1)^2}\\ \mathbf{a} &= \frac{d\mathbf{v}}{dt} = 6t\,\mathbf{i} - 8\mathbf{j}\end{aligned}$$

At time $t = 1$ we have $\mathbf{r} = \mathbf{i} + 3\mathbf{j}$, $\mathbf{v} = 3\mathbf{i}$, and $\mathbf{a} = 6\mathbf{i} - 8\mathbf{j}$. At that time the object is at the point $(1, 3)$, and is moving in the direction of the positive x-axis with speed 3. Since $\mathbf{v} \bullet \mathbf{a} = 18 > 0$ at $t = 1$, the angle between the velocity and the acceleration is less than $90°$, so the object is speeding up.

EXAMPLE 9.5.7 Find the position at time t of an object moving in the plane if its acceleration is proportional in magnitude to its position vector and is directed in the opposite direction. Assume that the position $\mathbf{r}_0$ and the velocity $\mathbf{v}_0$ at time $t = 0$ are given.

SOLUTION The relation between acceleration and position can be expressed by the differential equation

$$\frac{d^2\mathbf{r}}{dt^2} = -k^2\mathbf{r}$$

where k^2 is a positive constant. This is simply the vector form of the equation of Simple Harmonic Motion. As for the scalar case considered in Section 3.2, we can verify by differentiation that

$$\mathbf{r} = \mathbf{A}\cos kt + \mathbf{B}\sin kt$$

is a solution for any constant vectors $\mathbf{A} = a_1\mathbf{i} + a_2\mathbf{j}$ and $\mathbf{B} = b_1\mathbf{i} + b_2\mathbf{j}$. Since

$$\mathbf{r}_0 = \mathbf{r}(0) = \mathbf{A} \quad \text{and} \quad \mathbf{v}_0 = \mathbf{v}(0) = (-k\mathbf{A}\sin kt + k\mathbf{B}\cos kt)\Big|_{t=0} = k\mathbf{B},$$

we have

$$\mathbf{r} = \mathbf{r}(t) = \mathbf{r}_0 \cos kt + \frac{1}{k}\,\mathbf{v}_0 \sin kt.$$

For typical constant vectors $\mathbf{r}_0$ and $\mathbf{v}_0$, the path of the object is an ellipse. (See Exercise 19.)

The Projectile Problem

Suppose that an object is thrown or fired at time $t = 0$ with an initial speed s_0 in a direction making an angle α above the horizontal. The object will follow an arched path until it strikes the ground some distance away from its firing point (which we assume also to be at ground level). (See Fig. 9.5.12.)

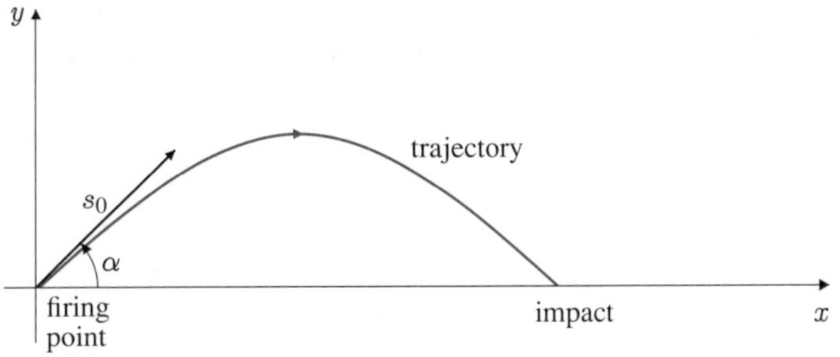

FIGURE 9.5.12

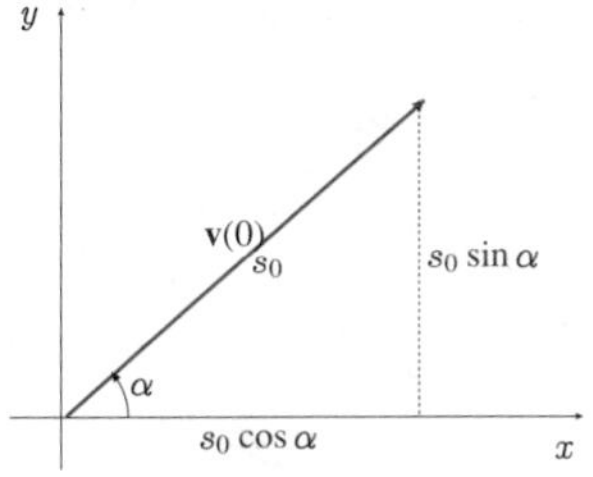

FIGURE 9.5.13

Assuming level ground, no air resistance, and s_0 small enough that during its flight the object may be regarded as subject only to constant downward acceleration due to gravity, we can find the equation of the trajectory (that is, path) of the object. Choosing the firing point as origin, the y-axis vertical, and the x-axis horizontal in the plane of motion, let $\mathbf{r}(t)$ denote the position of the object at time $t > 0$. We have

$$\mathbf{r}(0) = \mathbf{0} = 0\mathbf{i} + 0\mathbf{j}.$$

The initial velocity has horizontal component $s_0 \cos\alpha$ and vertical component $s_0 \sin\alpha$ as shown in Fig. 9.5.13. Thus

$$\mathbf{v}(0) = s_0 \cos\alpha\,\mathbf{i} + s_0 \sin\alpha\,\mathbf{j}.$$

Since the acceleration of object has constant magnitude (g say), and is in the downward direction, $\mathbf{r}(t)$ must satisfy the differential equation

$$\frac{d^2\mathbf{r}}{dt^2} = \mathbf{a}(t) = -g\,\mathbf{j}.$$

Integrating this vector equation once, we obtain

$$\frac{d\mathbf{r}}{dt} = \mathbf{v}(t) = -gt\,\mathbf{j} + \mathbf{C}_1.$$

Note that the constant of integration is a constant vector. Using the value for $\mathbf{v}(0)$ given above, we determine that

$$\mathbf{C}_1 = \mathbf{v}(0) = s_0 \cos\alpha\,\mathbf{i} + s_0 \sin\alpha\,\mathbf{j},$$

and hence that

$$\frac{d\mathbf{r}}{dt} = \mathbf{v}(t) = s_0 \cos\alpha\,\mathbf{i} + (s_0 \sin\alpha - gt)\,\mathbf{j}.$$

Now integrate again to get

$$\mathbf{r}(t) = (s_0 \cos\alpha)t\,\mathbf{i} + \left((s_0 \sin\alpha)t - \frac{gt^2}{2}\right)\mathbf{j} + \mathbf{C}_2.$$

Since $\mathbf{r}(0) = \mathbf{0}$ we have $\mathbf{C}_2 = \mathbf{0}$. Thus the position of the object at any time during its flight is given by

$$\mathbf{r}(t) = (s_0 \cos\alpha)t\,\mathbf{i} + \left((s_0 \sin\alpha)t - \frac{gt^2}{2}\right)\mathbf{j}.$$

The parametric equations of the trajectory are

$$x = (s_0 \cos\alpha)t, \qquad \text{and} \qquad y = -\frac{1}{2}gt^2 + (s_0 \sin\alpha)t.$$

The ordinary Cartesian equation of the trajectory can be found by eliminating the parameter between this pair of equations. Solving the first equation for $t = x/(s_0 \cos\alpha)$ and substituting into the second equation, we obtain

$$y = -\left(\frac{g}{2s_0^2 \cos^2\alpha}\right)x^2 + (\tan\alpha)x.$$

Evidently the trajectory is a parabola.

We can use the parametric equations to calculate the *maximum height*, the *time of flight*, and the *range* of the projectile. (The range is the horizontal distance traveled by the projectile before impact.)

Clearly the maximum height is attained when the vertical component of the velocity is 0, that is, at time $t = (s_0 \sin\alpha)/g$. The maximum height is

$$y_{\max} = -\frac{g}{2}\left(\frac{s_0 \sin\alpha}{g}\right)^2 + (s_0 \sin\alpha)\frac{s_0 \sin\alpha}{g} = \frac{s_0^2}{2g}\sin^2\alpha.$$

Since the projectile leaves ground level at time $t = 0$, the time of flight (the elapsed time until impact with the ground) is the positive time t for which $y = 0$:

$$-\frac{1}{2}gt^2 + (s_0 \sin\alpha)t = 0$$
$$t\left(s_0 \sin\alpha - \frac{gt}{2}\right) = 0.$$

Thus the time of flight is $t_{\max} = (2s_0 \sin\alpha)/g$. (Why do we ignore the root $t = 0$?) The range is the value of x at $t = t_{\max}$. Thus the range is

$$x_{\max} = (s_0 \cos\alpha)\frac{2s_0 \sin\alpha}{g} = \frac{s_0^2}{g}\sin 2\alpha.$$

For given initial speed s_0, the range $x_{\max}$ will be maximum if $\sin 2\alpha = 1$, that is, if $\alpha = \frac{\pi}{4} = 45°$. If the projectile is a shell being fired from a gun, this maximum range, s_0^2/g, increases as the square of the muzzle speed s_0.

EXERCISES

1. Let $A = (-1,2)$, $B = (2,0)$, $C = (1,-3)$, $D = (0,4)$. Express each of the following vectors as a linear combination of the standard basis vectors.

a) $\overrightarrow{AB}$, b) $\overrightarrow{BA}$, c) $\overrightarrow{AC}$, d) $\overrightarrow{BD}$, e) $\overrightarrow{DA}$,

f) $\overrightarrow{AB} - \overrightarrow{BC}$, g) $\overrightarrow{AC} - 2\overrightarrow{AB} + 3\overrightarrow{CD}$,

h) $\dfrac{\overrightarrow{AB} + \overrightarrow{AC} + \overrightarrow{AD}}{3}$.

In Exercises 2–7, find $\mathbf{u} + \mathbf{v}$, $\mathbf{u} - \mathbf{v}$, $|\mathbf{u}|$, $|\mathbf{v}|$, $\mathbf{u} \bullet \mathbf{v}$, and the angle between $\mathbf{u}$ and $\mathbf{v}$. Sketch arrows for $\mathbf{u}$, $\mathbf{v}$, $\mathbf{u} + \mathbf{v}$, and $\mathbf{u} - \mathbf{v}$.

2. $\mathbf{u} = \mathbf{i}$, $\mathbf{v} = \mathbf{j}$

3. $\mathbf{u} = \mathbf{i} + \mathbf{j}$, $\mathbf{v} = \mathbf{i} - \mathbf{j}$

4. $\mathbf{u} = 3\mathbf{i}$, $\mathbf{v} = 4\mathbf{i} - 3\mathbf{j}$

5. $\mathbf{u} = \mathbf{i} - 2\mathbf{j}$, $\mathbf{v} = -\mathbf{i} + 3\mathbf{j}$

6. $\mathbf{u} = \mathbf{j}$, $\mathbf{v} = -\mathbf{i} - \mathbf{j}$

7. $\mathbf{u} = \mathbf{i} + 2\mathbf{j}$, $\mathbf{v} = 2\mathbf{i} - \mathbf{j}$

8. Use vectors to show that the triangle with vertices $(-1, 1)$, $(2, 5)$, and $(10, -1)$ is right-angled.

In Exercises 9–12 $\mathbf{r}$ is the position of a point moving in the plane. Sketch the path of motion, and show on the sketch the velocity and acceleration vectors at time $t = 1$ and $t = 2$.

9. $\mathbf{r} = t\mathbf{i} - \frac{1}{2}t^2\mathbf{j}$

10. $\mathbf{r} = \sin\frac{\pi t}{2}\mathbf{i} + \cos\frac{\pi t}{2}\mathbf{j}$

11. $\mathbf{r} = t\mathbf{i} + (4t - t^3)\mathbf{j}$

12. $\mathbf{r} = \frac{1}{t}\mathbf{i} - t\mathbf{j}$

13. Describe the motion of an object whose position vector is $\mathbf{r}(t) = \mathbf{r}_0 + t\mathbf{u}$, where $\mathbf{u}$ is a nonzero constant vector. What are the parametric equations of the path?

In Exercises 14–17 find the position vector at time t of a point that moves with given acceleration $\mathbf{a}(t)$ and has given position and velocity at the time indicated.

14. $\mathbf{a}(t) = 2\mathbf{i} - \mathbf{j}$, $\mathbf{v}(0) = \mathbf{i} + \mathbf{j}$, $\mathbf{r}(0) = -3\mathbf{j}$

15. $\mathbf{a}(t) = t\mathbf{i}$, $\mathbf{v}(1) = \mathbf{0}$, $\mathbf{r}(1) = -\mathbf{i} + \mathbf{j}$

16. $\mathbf{a}(t) = 6t\mathbf{i} - 6t^2\mathbf{j}$, $\mathbf{v}(0) = 2\mathbf{i}$, $\mathbf{r}(0) = \mathbf{0}$

17. $\mathbf{a}(t) = \sin t\,\mathbf{i} + e^{-t}\mathbf{j}$, $\mathbf{v}(0) = \mathbf{i}$, $\mathbf{r}(0) = \mathbf{0}$

18. The acceleration due to gravity, g, is approximately 9.8 m/s^2. Find the muzzle speed s_0 of a shell fired at an angle 30° above the horizontal if the shell strikes the ground at a point 2 km horizontally away from its firing point.

19.*Verify that the path of the object in Example 9.5.7 is an ellipse if $\mathbf{r}_0$ is not parallel to $\mathbf{v}_0$. Hint: eliminate t from the parametric equations of the path. Under what conditions is the ellipse a circle?

20. Find the acceleration at time t of a point whose position is given by $\mathbf{r} = a(t - \sin t)\mathbf{i} + a(1 - \cos t)\mathbf{j}$. (The point is moving on a cycloid. See Example 9.3.11.) Show that the acceleration has constant magnitude. What is the direction of motion at the cusps of the cycloid? At the peaks?

21.*A point moves on the curve $y = e^x$ in the direction of increasing x. If the speed is constant (say k), find the velocity and acceleration vectors as functions of x.

22.*Let $\mathbf{v}(t)$ and $\mathbf{a}(t)$ be the velocity and acceleration, respectively, of an object moving in the plane. Show that the following three conditions are equivalent:

i) the speed of the object is increasing at time t,

ii) $\mathbf{v}(t) \bullet \mathbf{a}(t) > 0$,

iii) the angle between $\mathbf{v}(t)$ and $\mathbf{a}(t)$ is less than $\pi/2$.

State an analogous result for decreasing speed.

In Exercises 23–26 prove the stated geometric result using vectors.

23. The line segment joining the midpoints of two sides of a triangle is parallel to and half as long as the third side.

24. If P, Q, R, S are midpoints of sides AB, BC, CD, and DA, respectively, of quadrilateral $ABCD$, then $PQRS$ is a parallelogram.

25.*The diagonals of any parallelogram bisect each other.

26.*The medians of any triangle meet in a common point. (A median is a line joining one vertex to the midpoint of the opposite side. The common point is the *centroid* of the triangle.)

CHAPTER 10

Sequences and Series

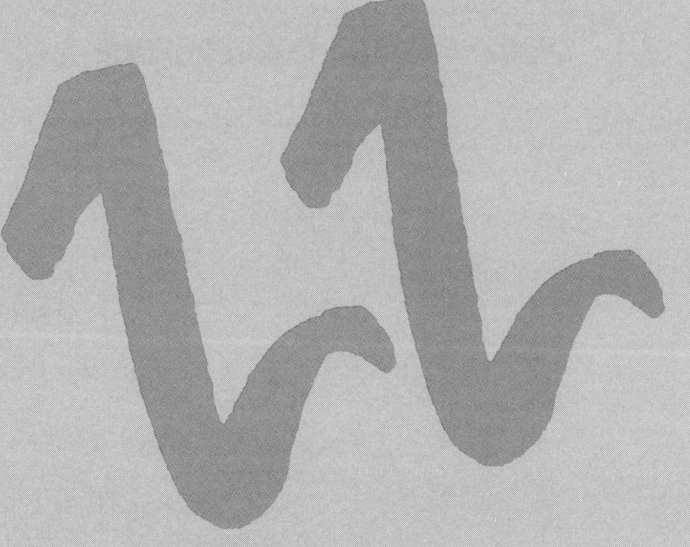

An infinite series is a sum involving infinitely many terms. Since addition is carried out on two numbers at a time, the evaluation of the sum of an infinite series necessarily involves finding a limit. Complicated functions $f(x)$ can frequently be expressed as series of powers of x, that is, as polynomials of infinite degree. Since such series can be differentiated and integrated term by term, they play a very important role in the study of calculus. This role is developed in Chapter 11.

Much of the material in this chapter can be omitted without hindering the understanding of power series in Chapter 11. Section 10.2 should, however, not be omitted. Our presentation of power series will make some reference to the ratio test covered in Section 10.3 and the concepts of conditional and absolute convergence in Section 10.4, but these topics are not essential to understanding or using power series. The discussion of infinite sequences and series of real numbers presented in this chapter involves limit and convergence notions similar to, but somewhat simpler than, those for functions and integrals. Accordingly, we proceed with slightly more rigor in developing them.

10.1 SEQUENCES AND CONVERGENCE

By a **sequence** (or **infinite sequence**) we mean an ordered list having a first element but no last element. For our purposes the elements (called **terms**) of a sequence will always be real numbers, though much of our discussion could be applied to complex numbers as well. Examples of sequences are:

$\{1, 2, 3, 4, 5, \ldots\}$ the sequence of positive integers,

$\left\{-\frac{1}{2}, \frac{1}{4}, -\frac{1}{8}, \frac{1}{16}, \ldots\right\}$ the sequence of positive integer powers of $-\frac{1}{2}$.

It is convenient to list the terms of a sequence in braces { } as shown. The ellipsis (...) should be read "and so on."

The concept of infinite sequence is a special case of the concept of function. The sequence $\{a_1, a_2, a_3, a_4, \ldots\}$ can be regarded as a function a with domain the set of positive integers that takes the value $a(n) = a_n$ at each integer n in its domain. A sequence can be specified in three ways:

i) We can list the first few terms followed by ... *if the pattern is obvious.*

ii) We can provide a formula for the **general term** a_n as a function of n. Unless the contrary is stated, it is assumed that the first term is a_1, the second term a_2, and so on.

iii) We can provide a formula for calculating the term a_n as a function of earlier terms $a_1, a_2, \ldots, a_{n-1}$ and specify a_1 so the process of computing higher terms can begin.

In each case it must be possible to determine any term of the sequence.

EXAMPLE 10.1.1 a) $\{n\} = \{1, 2, 3, 4, 5, \ldots\}$

b) $\left\{\left(-\frac{1}{2}\right)^n\right\} = \left\{-\frac{1}{2}, \frac{1}{4}, -\frac{1}{8}, \frac{1}{16}, \ldots\right\}$

c) $\left\{\dfrac{n-1}{n}\right\} = \left\{0, \dfrac{1}{2}, \dfrac{2}{3}, \dfrac{3}{4}, \dfrac{4}{5}, \ldots\right\}$

d) $\{(-1)^{n-1}\} = \{1, -1, 1, -1, 1, \ldots\}$

e) $\left\{\dfrac{n^2}{2^n}\right\} = \left\{\dfrac{1}{2}, 1, \dfrac{9}{8}, 1, \dfrac{25}{32}, \dfrac{36}{64}, \dfrac{49}{128}, \ldots\right\}$

f) $\left\{\left(1+\dfrac{1}{n}\right)^n\right\} = \left\{2, \left(\dfrac{3}{2}\right)^2, \left(\dfrac{4}{3}\right)^3, \left(\dfrac{5}{4}\right)^4, \ldots\right\}$

g) $\left\{\dfrac{\cos(n\pi/2)}{n}\right\} = \left\{0, -\dfrac{1}{2}, 0, \dfrac{1}{4}, 0, -\dfrac{1}{6}, 0, \dfrac{1}{8}, 0, \ldots\right\}$

h) $a_1 = 1,\ a_{n+1} = \sqrt{6+a_n},\ (n = 1, 2, 3, \ldots)$, so $\{a_n\} = \{1, \sqrt{7}, \sqrt{6+\sqrt{7}}, \ldots\}$. Note that there is no "obvious" formula for a_n as an explicit function of n in this case, but we can still calculate a_n for any desired value of n provided we first calculate all the earlier values $a_2, a_3, \ldots, a_{n-1}$.

In Examples 1(a)–1(g) the formulas on the left sides define the general term of each sequence $\{a_n\}$ as an explicit function of n. In 1(h) we say the sequence $\{a_n\}$ is defined **recursively** or **inductively**; each term must be calculated from the previous one rather than directly as a function of n.

The following definitions introduce terminology used to describe various properties of sequences.

10.1.2 Describing Sequences

a) The sequence $\{a_n\}$ is **bounded below** by L, and L is a **lower bound** for $\{a_n\}$, if $a_n \geq L$ for every $n = 1, 2, 3, \ldots$. Similarly, the sequence is **bounded above** by M, and M is an **upper bound**, if $a_n \leq M$ for every n.

   The sequence $\{a_n\}$ is **bounded** if it is both bounded above and bounded below. In this case there is a constant K such that $|a_n| \leq K$ for every $n = 1, 2, 3, \ldots$. (We may take K to be the larger of $-L$ and M.)

b) The sequence $\{a_n\}$ is **positive** if it is bounded below by zero, that is, if $a_n \geq 0$ for every $n = 1, 2, 3, \ldots$; it is **negative** if $a_n \leq 0$ for every n.

c) The sequence $\{a_n\}$ is **increasing** if $a_{n+1} \geq a_n$ for every $n = 1, 2, 3, \ldots$; it is **decreasing** if $a_{n+1} \leq a_n$ for every such n. The sequence is said to be **monotonic** if it is either increasing or decreasing.

d) The sequence $\{a_n\}$ is **alternating** if $a_n a_{n+1} < 0$ for every $n = 1, 2, \ldots$, that is, if any two consecutive terms have opposite sign.

EXAMPLE 10.1.3 a) The sequence $\{n\} = \{1, 2, 3, \ldots\}$ is increasing and bounded below; 1 is a lower bound, as is any smaller number. The sequence is not bounded above.

b) $\left\{\frac{n-1}{n}\right\} = \left\{0, \frac{1}{2}, \frac{2}{3}, \frac{3}{4}, \ldots\right\}$ is bounded. Evidently 0 is a lower bound and 1 is an upper bound. The sequence is positive and increasing.

c) $\left\{\left(-\frac{1}{2}\right)^n\right\} = \left\{-\frac{1}{2}, \frac{1}{4}, -\frac{1}{8}, \frac{1}{16}, \ldots\right\}$ is alternating and bounded. ($-\frac{1}{2}$ is a lower bound and $\frac{1}{4}$ is an upper bound.)

d) $\{(-1)^n n\} = \{-1, 2, -3, 4, -5, \ldots\}$ is alternating but not bounded either above or below.

The sequence $\left\{\frac{n^2}{2^n}\right\} = \left\{\frac{1}{2}, 1, \frac{9}{8}, 1, \frac{25}{32}, \frac{36}{64}, \frac{49}{128}, \ldots\right\}$ is obviously positive therefore bounded below. It seems clear that from the fourth term on, all the terms are getting smaller. However, $a_2 > a_1$ and $a_3 > a_2$. Since $a_{n+1} \le a_n$ only if $n \ge 3$, we say that this sequence is **ultimately decreasing**. The adverb *ultimately* is used to describe any termwise property of a sequence that the terms have from some point on, but not necessarily at the beginning of the sequence. Thus the sequence

$$\{n-100\} = \{-99, -98, \ldots, -2, -1, 0, 1, 2, 3, \ldots\}$$

is *ultimately positive*, and the sequence

$$\left\{(-1)^n + \frac{4}{n}\right\} = \left\{3, 3, \frac{1}{3}, 2, -\frac{1}{5}, \frac{5}{3}, -\frac{3}{7}, \frac{3}{2}, \ldots\right\}$$

is *ultimately alternating* even though the first few terms do not alternate.

Central to the study of sequences is the notion of convergence. The concept of limit of a sequence is a special case of the concept of limit of a function $f(x)$ as $x \to \infty$. We say that the sequence $\{a_n\}$ **converges to the limit** L, and we write $\lim a_n = L$, provided the distance from a_n to L on the real line approaches 0 as n increases toward ∞. We state this definition somewhat more formally as follows:

10.1.4
Limit of a Sequence

> We say that the sequence $\{a_n\}$ converges to the limit L, and write $\lim a_n = L$, if for every positive real number ϵ there exists an integer N (which may depend on ϵ) such that if $n \ge N$, then $|a_n - L| < \epsilon$.

This definition is illustrated in Fig. 10.1.1.

Every sequence $\{a_n\}$ must either **converge** to a finite limit L or **diverge**. That is, either $\lim a_n = L$ exists, or $\lim a_n$ does not exist. If $\lim a_n = \infty$, we can say that the sequence diverges to ∞; if $\lim a_n = -\infty$, we can say that it diverges to $-\infty$. If $\lim a_n$ simply does not exist (but is not ∞ or $-\infty$), we can only say that the sequence diverges.

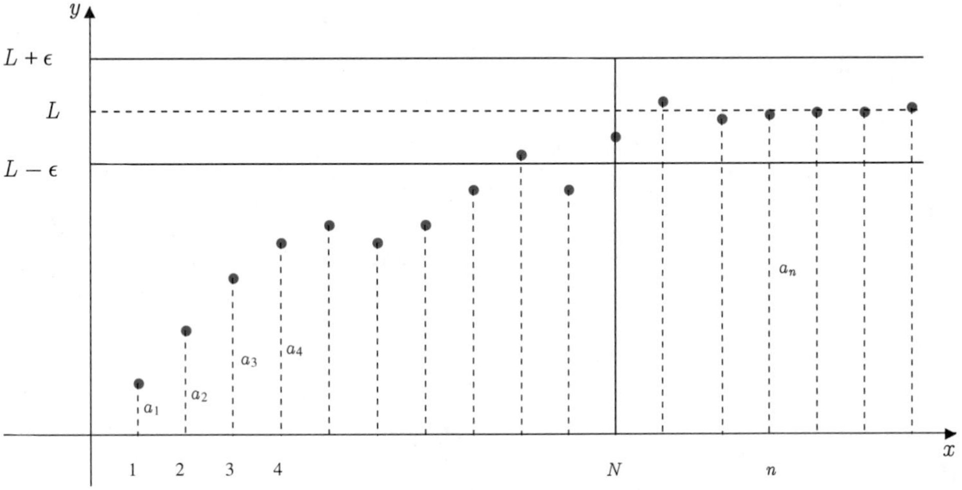

FIGURE 10.1.1

EXAMPLE 10.1.5 a) $\left\{\dfrac{n-1}{n}\right\}$ converges to 1; $\lim \dfrac{n-1}{n} = \lim\left(1-\dfrac{1}{n}\right) = 1$.

b) $\{n\} = \{1, 2, 3, 4, \ldots\}$ diverges to ∞.

c) $\{-n\} = \{-1, -2, -3, -4, \ldots\}$ diverges to $-\infty$.

d) $\{(-1)^n\} = \{-1, 1, -1, 1, -1, \ldots\}$ simply diverges.

e) $\{(-1)^n n\} = \{-1, 2, -3, 4, -5, \ldots\}$ diverges (but not to ∞ or $-\infty$ even though $\lim |a_n| = \infty$).

All the standard properties of limits (Theorems 1.5.6–8 of Section 1.5) apply to the limits of sequences, with the appropriate changes of notation. Thus, if $\{a_n\}$ and $\{b_n\}$ converge, then

$$\begin{aligned}
\lim(a_n \pm b_n) &= \lim a_n \pm \lim b_n, \\
\lim ca_n &= c \lim a_n, \\
\lim a_n b_n &= (\lim a_n)(\lim b_n), \\
\lim \frac{a_n}{b_n} &= \frac{\lim a_n}{\lim b_n} \qquad \text{assuming } \lim b_n \neq 0.
\end{aligned}$$

If $a_n \le b_n$ ultimately, then $\lim a_n \le \lim b_n$.

The limits of many explicitly defined sequences can be evaluated using these properties in a manner analogous to the methods used for limits of the form $\lim_{x\to\infty} f(x)$. See, for example, Examples 1.6.5 and 1.6.6.

EXAMPLE 10.1.6 a) $\left\{\dfrac{2n^2-n-1}{5n^2+n-3}\right\}$ converges to $\dfrac{2}{5}$ since

$$\lim \frac{2n^2-n-1}{5n^2+n-3} = \lim \frac{2-(1/n)-(1/n^2)}{5+(1/n)-(3/n^2)} = \frac{2}{5}.$$

b) $\{\sqrt{n^2+2n}-n\}$ converges to 1 since

$$\begin{aligned}\lim(\sqrt{n^2+2n}-n) &= \lim \frac{(\sqrt{n^2+2n}-n)(\sqrt{n^2+2n}+n)}{\sqrt{n^2+2n}+n} \\ &= \lim \frac{2n}{\sqrt{n^2+2n}+n} = \lim \frac{2}{\sqrt{1+(2/n)}+1} = 1.\end{aligned}$$

It often happens that $a_n = f(n)$ where f is a continuous function on $[1, \infty)$ and $\lim_{x\to\infty} f(x)$ exists or is ∞ or $-\infty$. In this case,

$$\lim a_n = \lim_{x\to\infty} f(x),$$

and the latter can be evaluated by standard techniques for functions, for instance, l'Hôpital's rules.

EXAMPLE 10.1.7

$$\begin{aligned}\lim n \tan^{-1}\left(\frac{1}{n}\right) &= \lim_{x\to\infty} \frac{\tan^{-1}\left(\dfrac{1}{x}\right)}{\dfrac{1}{x}} = \lim_{y\to 0+} \frac{\tan^{-1} y}{y} \qquad \left[\frac{0}{0}\right] \\ &= \lim_{y\to 0+} \frac{\dfrac{1}{y^2+1}}{1} = 1\end{aligned}$$

THEOREM 10.1.8 If $\{a_n\}$ converges, then $\{a_n\}$ is bounded.

PROOF If $\lim a_n = L$, then according to Definition 10.1.4, for $\epsilon = 1$ there exists a number N such that if $n \geq N$ then $|a_n - L| < 1$; therefore $|a_n| < 1 + |L|$ for such n. (Why is this true?) If K denotes the largest of the numbers $|a_1|$, $|a_2|$, $\ldots$, $|a_{N-1}|$, $1+|L|$, then $|a_n| \leq K$ for every $n = 1, 2, 3, \ldots$. Hence $\{a_n\}$ is bounded. □

The converse of Theorem 10.1.8 is false; the sequence $\{(-1)^n\}$ is bounded but does not converge.

The *completeness property* of the real number system (see Section 1.1) can be reformulated in terms of sequences to read as follows:

10.1.9
Bounded Monotonic Sequences Converge

> If the sequence $\{a_n\}$ is bounded above and is (ultimately) increasing, then it converges. The same conclusion holds if $\{a_n\}$ is bounded below and is (ultimately) decreasing.

Thus, a bounded, ultimately monotonic sequence is convergent. See Fig. 10.1.2.

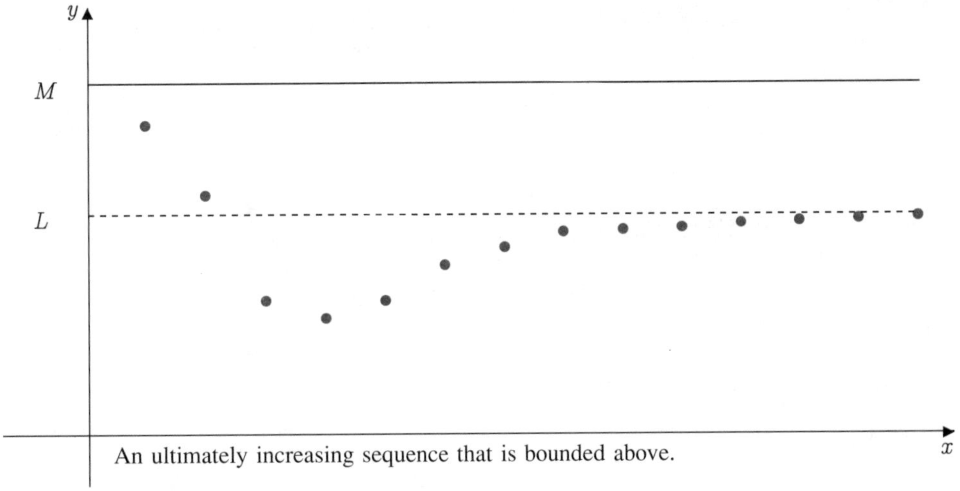
y
M
L
x
An ultimately increasing sequence that is bounded above.

FIGURE 10.1.2

EXAMPLE 10.1.10 Let a_n be defined recursively by

$$a_1 = 1, \qquad a_{n+1} = \sqrt{6 + a_n} \qquad (n = 1, 2, 3, \ldots).$$

Show that $\lim a_n$ exists and find its value.

SOLUTION Observe that $a_2 = \sqrt{6+1} = \sqrt{7} > a_1$. If $a_{k+1} > a_k$, then $a_{k+2} = \sqrt{6 + a_{k+1}} > \sqrt{6 + a_k} = a_{k+1}$, so $\{a_n\}$ is increasing, by induction. (See Appendix I.) Now observe that $a_1 = 1 < 3$. If $a_k < 3$, then $a_{k+1} = \sqrt{6 + a_k} < \sqrt{6+3} = 3$, so $a_n < 3$ for every n by induction. Since $\{a_n\}$ is increasing and bounded above, $\lim a_n = a$ exists, by completeness. Since $\sqrt{6+x}$ is a continuous function of x, we have

$$a = \lim a_{n+1} = \lim \sqrt{6 + a_n} = \sqrt{6 + \lim a_n} = \sqrt{6 + a}.$$

Thus $a^2 = 6 + a$, or $a^2 - a - 6 = 0$, or $(a - 3)(a + 2) = 0$. Since $a_n \geq 1$ for every n, we must have $a \geq 1$. Therefore, $a = 3$, and $\lim a_n = 3$.

EXAMPLE 10.1.11 Does $\left\{\left(1 + \dfrac{1}{n}\right)^n\right\}$ converge or not?

SOLUTION We could expend considerable effort to show that the given sequence is, in fact, increasing and bounded above. (See Exercise 32 at the end of this section.) However, we already know the answer. By Theorem 3.7.4 of Section 3.7,

$$\lim \left(1 + \frac{1}{n}\right)^n = e^1 = e.$$

THEOREM 10.1.12 If $\{a_n\}$ is (ultimately) increasing, then either it is bounded above, and therefore convergent, or else it is not bounded above and diverges to infinity. □

The proof of this theorem is left as an exercise.

The following two examples find frequent application in the study of series.

EXAMPLE 10.1.13

$$\lim x^n = 0 \qquad \text{if } |x| < 1.$$

To see this, observe that

$$\lim \ln |x|^n = \lim n \ln |x| = -\infty,$$

since $\ln |x| < 0$ when $|x| < 1$. Accordingly, since e^x is continuous,

$$\lim |x|^n = \lim e^{\ln |x|^n} = e^{\lim \ln |x|^n} = 0.$$

EXAMPLE 10.1.14

$$\lim \frac{x^n}{n!} = 0 \qquad \text{for every real } x.$$

To see this, pick any x and let N be an integer such that $N > |x|$. If $n > N$ we have

$$\begin{aligned}\left|\frac{x^n}{n!}\right| &= \frac{|x|}{1}\frac{|x|}{2}\frac{|x|}{3}\cdots\frac{|x|}{N-1}\frac{|x|}{N}\frac{|x|}{N+1}\cdots\frac{|x|}{n} \\ &< \frac{|x|^{N-1}}{(N-1)!}\frac{|x|}{N}\frac{|x|}{N}\frac{|x|}{N}\cdots\frac{|x|}{N} \\ &= \frac{|x|^{N-1}}{(N-1)!}\left(\frac{|x|}{N}\right)^{n-N+1} = K\left(\frac{|x|}{N}\right)^n,\end{aligned}$$

where $K = \dfrac{|x|^{N-1}}{(N-1)!}\left(\dfrac{|x|}{N}\right)^{1-N}$ is a constant independent of n. Since $|x|/N < 1$, we have $\lim(|x|/N)^n = 0$ by Example 10.1.13 above. Thus $\lim |x^n/n!| = 0$, and so $\lim x^n/n! = 0$.

EXERCISES

In Exercises 1–13 determine whether the given sequence is (a) bounded (above or below), (b) positive or negative (ultimately), (c) increasing, decreasing, or alternating, (d) convergent, divergent, divergent to ∞ or $-\infty$.

1. $\left\{\dfrac{2n^2}{n^2+1}\right\}$

2. $\left\{\dfrac{2n}{n^2+1}\right\}$

3. $\left\{4 - \dfrac{(-1)^n}{n}\right\}$

4. $\left\{\sin\dfrac{1}{n}\right\}$

5. $\left\{\dfrac{n^2-1}{n}\right\}$

6. $\left\{\dfrac{e^n}{\pi^n}\right\}$

7. $\left\{\dfrac{e^n}{\pi^{n/2}}\right\}$

8. $\left\{\dfrac{(-1)^n n}{e^n}\right\}$

9. $\left\{\dfrac{2^n}{n^n}\right\}$

10. $\left\{\dfrac{(n!)^2}{(2n)!}\right\}$

11. $\left\{ n \cos\left(\frac{n\pi}{2}\right) \right\}$

12. $\left\{ \frac{\sin n}{n} \right\}$

13. $\{1, 1, -2, 3, 3, -4, 5, 5, -6, \ldots\}$

In Exercises 14–29 evaluate, wherever possible, the limit of the sequence $\{a_n\}$.

14. $a_n = \dfrac{5 - 2n}{3n - 7}$

15. $a_n = \dfrac{n^2 - 4}{n + 5}$

16. $a_n = \dfrac{n^2}{n^3 + 1}$

17. $a_n = (-1)^n \dfrac{n}{n^3 + 1}$

18. $a_n = \dfrac{n^2 - 2\sqrt{n} + 1}{1 - n - 3n^2}$

19. $a_n = \dfrac{e^n - e^{-n}}{e^n + e^{-n}}$

20. $a_n = n \sin \dfrac{1}{n}$

21. $a_n = \left(\dfrac{n - 3}{n}\right)^n$

22. $a_n = \dfrac{n}{\ln(n + 1)}$

23. $a_n = \sqrt{n + 1} - \sqrt{n}$

24. $a_n = n - \sqrt{n^2 - 4n}$

25. $a_n = \sqrt{n^2 + n} - \sqrt{n^2 - 1}$

26. $a_n = \left(\dfrac{n - 1}{n + 1}\right)^n$

27. $a_n = \dfrac{(n!)^2}{(2n)!}$

28. $a_n = \dfrac{n^2 2^n}{n!}$

29. $a_n = \dfrac{\pi^n}{1 + 2^{2n}}$

review evaluation of limits

30. Let $a_1 = 1$ and $a_{n+1} = \sqrt{1 + 2a_n}$ $(n = 1, 2, 3, \ldots)$. Show that $\{a_n\}$ is increasing and bounded above. (*Hint:* Show that 3 is an upper bound.) Hence conclude that the sequence converges, and find its limit.

31.*Repeat the previous exercise for the sequence defined by $a_1 = 3$, $a_{n+1} = \sqrt{15 + 2a_n}$, $n = 1, 2, 3, \ldots$. This time you will have to guess an upper bound.

32.*Let $a_n = \left(1 + \dfrac{1}{n}\right)^n$ so that $\ln a_n = n \ln\left(1 + \dfrac{1}{n}\right)$. Use properties of the logarithm function to show that (a) $\{a_n\}$ is increasing, and (b) e is an upper bound for $\{a_n\}$.

33.*Prove Theorem 10.1.12. Also, state an analogous theorem pertaining to ultimately decreasing sequences.

34.*Which of the following statements are true and which are false? Justify your answers.

a) If $\lim a_n = \infty$ and $\lim b_n = L > 0$, then $\lim a_n b_n = \infty$.

b) If $\lim a_n = \infty$ and $\lim b_n = -\infty$, then $\lim(a_n + b_n) = 0$.

c) If $\lim a_n = \infty$ and $\lim b_n = -\infty$, then $\lim a_n b_n = -\infty$.

d) If neither $\{a_n\}$ nor $\{b_n\}$ converges, then $\{a_n b_n\}$ does not converge.

10.2 INFINITE SERIES

An **infinite series**, usually just called a **series**, is a formal sum of infinitely many terms; for instance

$$a_1 + a_2 + a_3 + a_4 + \cdots$$

is a series formed by adding the terms of the sequence $\{a_n\}$. This series is also denoted $\sum_{n=1}^{\infty} a_n$ or, when confusion is not likely to occur, simply $\sum a_n$.

$$\sum a_n = \sum_{n=1}^{\infty} a_n = a_1 + a_2 + a_3 + a_4 + \cdots$$

For example,

$$\sum \frac{1}{n} = 1 + \frac{1}{2} + \frac{1}{3} + \frac{1}{4} + \cdots$$
$$\sum \frac{(-1)^{n-1}}{2^{n-1}} = 1 - \frac{1}{2} + \frac{1}{4} - \frac{1}{8} + \frac{1}{16} - \cdots.$$

It is sometimes necessary or convenient to start the sum from some index other than 1; when we do we will always show the index limits specifically.

$$\sum_{n=0}^{\infty} a^n = 1 + a + a^2 + a^3 + \cdots$$
$$\sum_{n=2}^{\infty} \frac{1}{\ln n} = \frac{1}{\ln 2} + \frac{1}{\ln 3} + \frac{1}{\ln 4} + \cdots.$$

Note that the latter series would make no sense if we had started the sum from $n = 1$; the first term would have been undefined.

Addition is an operation that is carried out on two numbers at a time. If we want to calculate the finite sum

$$a_1 + a_2 + a_3,$$

we could proceed by adding $a_1 + a_2$ and then adding a_3 to this sum, or else we might first add $a_2 + a_3$ and then add a_1 to the sum. Of course the associative law for addition assures us we will get the same answer both ways. This is the reason the symbol $a_1+a_2+a_3$ makes sense; we would otherwise have to write $(a_1+a_2)+a_3$ or $a_1 + (a_2 + a_3)$. This reasoning extends to any sum $a_1 + a_2 + \cdots + a_n$ of finitely many terms, but it is not obvious what should be meant by a sum with infinitely many terms:

$$a_1 + a_2 + a_3 + a_4 + \cdots.$$

We no longer have any assurance that the terms can be added up in any order to yield the same sum. The interpretation we place on the infinite sum is that of adding from left to right as suggested by the grouping

$$\cdots((((a_1 + a_2) + a_3) + a_4) + a_5) + \cdots.$$

We formalize this in the following definition.

10.2.1
Partial Sums and Convergence of Series

Corresponding to the **infinite series** $\sum_{n=1}^{\infty} a_n$ we form the **sequence** $\{s_n\}$ of **partial sums** of the series:

$$\begin{aligned} s_1 &= a_1 \\ s_2 &= a_1 + a_2 \\ s_3 &= a_1 + a_2 + a_3 \\ &\vdots \\ s_n &= a_1 + a_2 + a_3 + \cdots + a_n = \sum_{j=1}^{n} a_j . \\ &\vdots \end{aligned}$$

We say that the series $\sum_{n=1}^{\infty} a_n$ **converges to the sum** s, and we write

$$\sum_{n=1}^{\infty} a_n = s,$$

if $\lim s_n = s$. The *series* $\sum a_n$ converges if and only if the *sequence* $\{s_n\}$ of partial sums converges.

Similarly, a series is said to diverge to infinity, diverge to negative infinity, or simply diverge if its sequence of partial sums does so. It must be stressed that the convergence of the series $\sum_{n=1}^{\infty} a_n$ depends on the convergence of the sequence $\{s_n\} = \{\sum_{j=1}^{n} a_j\}$, *not* the sequence $\{a_n\}$.

Geometric Series

10.2.2
Geometric Series

A series whose nth term is $a_n = a\,r^{n-1}$ has the form

$$\sum_{n=1}^{\infty} a\,r^{n-1} = a + ar + ar^2 + ar^3 + \cdots .$$

It is called a **geometric series**. The number a is the first term. The number r is called the common ratio of the series since it is the value of the ratio of the $(n+1)$st term to the nth term:

$$\frac{a_{n+1}}{a_n} = \frac{ar^n}{ar^{n-1}} = r \qquad n = 1, 2, 3, \ldots$$

The nth partial sum s_n of a geometric series is calculated as follows:

$$s_n = a + ar + ar^2 + ar^3 + \cdots + ar^{n-1}$$
$$rs_n = \qquad ar + ar^2 + ar^3 + \cdots + a^{n-1} + ar^n$$

The second equation is obtained by multiplying the first by r. Subtracting these two equations (note the cancellations), we get $(1-r)s_n = a - ar^n$. If $r \neq 1$, we can divide by $1-r$ and get a formula for s_n.

10.2.3
Partial Sums of Geometric Series

> If $r \neq 1$, then the nth partial sum of a geometric series is
>
> $$s_n = a + ar + ar^2 + \cdots + ar^{n-1} = \frac{a(1-r^n)}{1-r}.$$
>
> If $r = 1$, we have, evidently, $s_n = na$.

If $|r| < 1$, then $\lim r^n = 0$, so $\lim s_n = a/(1-r)$. If $r > 1$, then $\lim r^n = \infty$, and $\lim s_n = \infty$ if $a > 0$, or $\lim s_n = -\infty$ if $a < 0$. The same conclusion holds if $r = 1$ since $s_n = na$ in this case. If $r \leq -1$, $\lim r^n$ does not exist and neither does $\lim s_n$. Hence we conclude that

10.2.4
Convergence of Geometric Series

$$\sum_{n=1}^{\infty} ar^{n-1} \begin{cases} \text{converges to } \dfrac{a}{1-r} & \text{if } -1 < r < 1 \\ \text{diverges to } \infty & \text{if } r \geq 1 \text{ and } a > 0 \\ \text{diverges to } -\infty & \text{if } r \geq 1 \text{ and } a < 0 \\ \text{diverges} & \text{if } r \leq -1. \end{cases}$$

EXAMPLE 10.2.5

a) $1 + \frac{1}{2} + \frac{1}{4} + \frac{1}{8} + \cdots = \sum_{n=1}^{\infty} \left(\frac{1}{2}\right)^{n-1} = \frac{1}{1-\frac{1}{2}} = 2.$

b) $\pi - e + \frac{e^2}{\pi} - \frac{e^3}{\pi^2} + \cdots = \sum_{n=1}^{\infty} \pi\left(-\frac{e}{\pi}\right)^{n-1} = \frac{\pi}{1-\left(-\frac{e}{\pi}\right)} = \frac{\pi^2}{\pi + e}$ since $\left|-\frac{e}{\pi}\right| < 1.$

c) $1 + 2^{1/2} + 2 + 2^{3/2} + \cdots = \sum_{n=1}^{\infty} (\sqrt{2})^{n-1}$ diverges to ∞ since $\sqrt{2} > 1$.

d) $1 - 1 + 1 - 1 + 1 - \cdots = \sum_{n=1}^{\infty} (-1)^{n-1}$ diverges.

e) Let $x = 0.323232\cdots = 0.\overline{32}$; then

$$x = \frac{32}{100} + \frac{32}{100^2} + \frac{32}{100^3} + \cdots = \frac{32}{100}\,\frac{1}{1-\frac{1}{100}} = \frac{32}{99}.$$

This is an alternative to the method of Section 1.2 (Example 1.2.2) for representing repeating decimals as quotients of integers.

EXAMPLE 10.2.6 If money earns interest at a constant effective rate of 8% per year, how much should you pay today for an annuity which will pay you (a) \$1,000 at the end of each of the next 10 years, (b) \$1,000 at the end of every year forever?

SOLUTION A payment of \$1,000 to be received n years from now has present value $\$1,000 \times \left(\frac{1}{1.08}\right)^n$ (since $\$A$ would grow to $\$A(1.08)^n$ in n years). Thus \$1,000 payments at the end of each of the next n years are worth $\$s_n$ at the present time, where

$$s_n = \frac{1,000}{1.08}\left[1 + \frac{1}{1.08} + \left(\frac{1}{1.08}\right)^2 + \cdots + \left(\frac{1}{1.08}\right)^{n-1}\right]$$
$$= \frac{1,000}{1.08}\,\frac{1-\left(\frac{1}{1.08}\right)^n}{1-\frac{1}{1.08}} = \frac{1,000}{0.08}\left[1-\left(\frac{1}{1.08}\right)^n\right]$$

The present value of 10 future payments is $\$s_{10} = \$6,710.08$. The present value of future payments continuing forever is $\$\lim_{n\to\infty} s_n = \frac{\$1,000}{0.08} = \$12,500$.

Telescoping Series and Harmonic Series

EXAMPLE 10.2.7 Consider the series

$$\sum_{n=1}^{\infty} \frac{1}{n(n+1)} = \frac{1}{1\times 2} + \frac{1}{2\times 3} + \frac{1}{3\times 4} + \frac{1}{4\times 5} + \cdots$$

Since $\frac{1}{n(n+1)} = \frac{1}{n} - \frac{1}{n+1}$ we can write the partial sum s_n in the form

$$s_n = \frac{1}{1\times 2} + \frac{1}{2\times 3} + \frac{1}{3\times 4} + \cdots + \frac{1}{(n-1)n} + \frac{1}{n(n+1)}$$
$$= \left(1-\frac{1}{2}\right) + \left(\frac{1}{2}-\frac{1}{3}\right) + \left(\frac{1}{3}-\frac{1}{4}\right) + \cdots + \left(\frac{1}{n-1}-\frac{1}{n}\right) + \left(\frac{1}{n}-\frac{1}{n+1}\right)$$
$$= 1 - \left(\frac{1}{2}-\frac{1}{2}\right) - \left(\frac{1}{3}-\frac{1}{3}\right) - \cdots - \left(\frac{1}{n}-\frac{1}{n}\right) - \frac{1}{n+1}$$
$$= 1 - \frac{1}{n+1}.$$

Therefore $\lim s_n = 1$ and the series converges to 1:

$$\sum_{n=1}^{\infty} \frac{1}{n(n+1)} = 1.$$

This is an example of a **telescoping series**, so called because the partial sums "fold up" into a simple form when the terms are expanded in partial fractions. Other examples can be found in the exercises at the end of this section. As these examples show, the method of partial fractions can be a useful tool for series as well as for integrals.

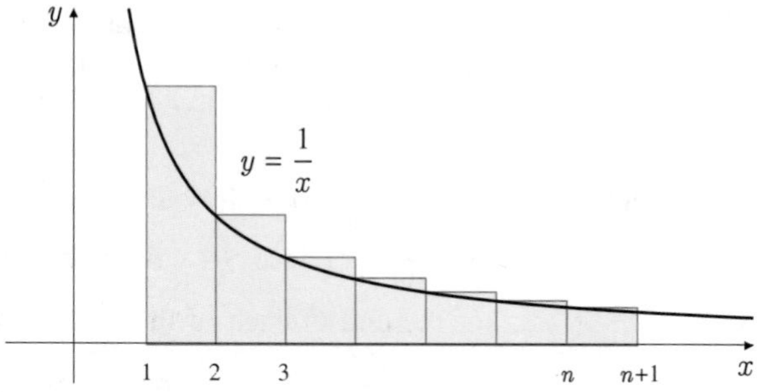

FIGURE 10.2.1

EXAMPLE 10.2.8 The series

$$\sum_{n=1}^{\infty} \frac{1}{n} = 1 + \frac{1}{2} + \frac{1}{3} + \frac{1}{4} + \cdots$$

is called a **harmonic series**. If s_n is its nth partial sum then

$$\begin{aligned} s_n &= 1 + \frac{1}{2} + \frac{1}{3} + \cdots + \frac{1}{n} \\ &= \text{sum of areas of rectangles shaded in Fig. 10.2.1.} \\ &> \text{area under } y = \frac{1}{x} \text{ from } 1 \text{ to } n+1 \\ &= \int_1^{n+1} \frac{dx}{x} = \ln(n+1). \end{aligned}$$

Therefore $\lim s_n = \infty$ and $\sum_{n=1}^{\infty} \frac{1}{n} = 1 + \frac{1}{2} + \frac{1}{3} + \cdots$ diverges to infinity.

Some Theorems About Series

THEOREM 10.2.9 If $\sum a_n$ converges, then $\lim a_n = 0$.

PROOF If $s_n = a_1 + a_2 + \cdots + a_n$, then $s_n - s_{n-1} = a_n$. If $\sum a_n$ converges, then $\lim s_n = s$ exists, and $\lim s_{n-1} = s$. Hence $\lim a_n = s - s = 0$. □

REMARK: Theorem 10.2.9 is *very important* for the understanding of infinite series. Students often err either in forgetting that *a series cannot converge if its terms do not approach zero* or in confusing this result with its *converse*, which is false. The converse would say that if $\lim a_n = 0$, then $\sum a_n$ must converge. The harmonic series is a counterexample showing the falsehood of this assertion:

$$\lim \frac{1}{n} = 0, \qquad \text{but} \qquad \sum \frac{1}{n} \text{ diverges to infinity.}$$

When considering whether a given series converges, the first question you should ask yourself is "Does the nth term approach 0 as n approaches ∞?" If the answer is "no" the series cannot converge. If the answer is "yes" the series *may or may not* converge. If the sequence of terms $\{a_n\}$ tends to a nonzero limit L, then $\sum a_n$ diverges to infinity if $L > 0$ and diverges to negative infinity if $L < 0$.

EXAMPLE 10.2.10 a) $\sum \frac{n}{2n-1}$ diverges to infinity since $\lim \frac{n}{2n+1} = 1/2 > 0$.

b) $\sum (-1)^n n \sin(1/n)$ diverges since

$$\lim \left|(-1)^n n \sin \frac{1}{n}\right| = \lim_{n\to\infty} \frac{\sin(1/n)}{1/n} = \lim_{x\to 0+} \frac{\sin x}{x} = 1 \neq 0.$$

The following theorem asserts that it is only the *ultimate* behavior of $\{a_n\}$ that determines whether $\sum a_n$ converges. Any finite number of terms can be dropped from the beginning of a series without affecting the convergence; the convergence depends only on the "tail" of the series. Of course, the sum of the series depends on *all* the terms.

THEOREM 10.2.11 $\sum_{n=1}^{\infty} a_n$ converges if and only if $\sum_{n=N}^{\infty} a_n$ converges for any integer $N \geq 1$. □

THEOREM 10.2.12 If $\{a_n\}$ is ultimately positive, then the series $\sum a_n$ must either converge (if its partial sums are bounded above) or diverge to infinity (if its partial sums are not bounded above). □

The proofs of these two theorems are left as exercises at the end of this section. The following theorem just a reformulation of standard laws of limits.

THEOREM 10.2.13 If $\sum_{n=1}^{\infty} a_n$ and $\sum_{n=1}^{\infty} b_n$ converge to A and B, respectively, then

a) $\sum_{n=1}^{\infty} ca_n$ converges to cA (where c is any constant);

b) $\sum_{n=1}^{\infty}(a_n \pm b_n)$ converges to $A \pm B$;

c) If $a_n \leq b_n$ for all $n = 1, 2, 3, \ldots$ then $A \leq B$. □

EXERCISES

In Exercises 1–14 find the sum of the given series, or show that the series diverges (possibly to infinity or negative infinity). Exercises 7–10 are telescoping series and should be done by partial fractions as suggested in Example 10.2.7 in the text.

1. $\frac{1}{3} + \frac{1}{9} + \frac{1}{27} + \cdots = \sum_{n=1}^{\infty} \frac{1}{3^n}$

2. $3 - \frac{3}{4} + \frac{3}{16} - \frac{3}{64} + \cdots = \sum_{n=1}^{\infty} 3\left(-\frac{1}{4}\right)^{n-1}$

3. $\sum_{n=5}^{\infty} \frac{1}{(2+\pi)^{2n}}$

4. $\sum_{n=0}^{\infty} \frac{5}{10^{3n}}$

5. $\sum_{n=2}^{\infty} \frac{(-5)^n}{8^{2n}}$

6. $\sum_{n=0}^{\infty} \frac{1}{e^n}$

7. $\sum_{n=1}^{\infty} \frac{1}{n(n+2)} = \frac{1}{1\times 3} + \frac{1}{2\times 4} + \frac{1}{3\times 5} + \cdots$

8. $\sum_{n=1}^{\infty} \frac{1}{(2n-1)(2n+1)} = \frac{1}{1\times 3} + \frac{1}{3\times 5} + \frac{1}{5\times 7} + \cdots$

9. $\sum_{n=1}^{\infty} \frac{1}{(3n-2)(3n+1)} = \frac{1}{1\times 4} + \frac{1}{4\times 7} + \frac{1}{7\times 10} + \cdots$

10.* $\sum_{n=1}^{\infty} \frac{1}{n(n+1)(n+2)}$
$= \frac{1}{1\times 2\times 3} + \frac{1}{2\times 3\times 4} + \frac{1}{3\times 4\times 5} + \cdots$

11. $\sum_{n=1}^{\infty} \frac{1}{2n-1}$

12. $\sum_{n=1}^{\infty} \frac{n}{n+2}$

13. $\displaystyle\sum_{n=1}^{\infty} n^{-1/2}$

14. $\displaystyle\sum_{n=1}^{\infty} \frac{2}{n+1}$

15. Obtain a simple expression for the partial sum s_n of the series $\sum_{n=1}^{\infty}(-1)^n$ and use it to show that this series diverges.

16.*Find a simple expression for the partial sum s_n of the series $\displaystyle\sum_{n=2}^{\infty}\left(3e^{-n} - \frac{2}{n^2-1}\right)$. Also find the sum of the series.

17. When dropped, an elastic ball bounces back up to a height three-quarters of that from which it fell. If the ball is dropped from a height of 2 m and allowed to bounce up and down indefinitely, what is the total distance it travels before coming to rest?

18. If a bank account pays 10% simple interest into an account once a year, what is the balance in the account at the end of 8 years if $1000 is deposited into the account at the beginning of each of the eight years? (Assum there was no balance in the account initially.)

19.*Prove Theorem 10.2.11 **20.***Prove Theorem 10.2.12.

21.*State a theorem analogous to Theorem 10.2.12 but fo a negative sequence.

In Exercises 22–27 decide whether the given statement true or false. If it is true, prove it. If it is false, give counterexample showing the falsehood.

22.*If $a_n = 0$ for every n, then $\sum a_n$ converges.

23.*If $\sum a_n$ converges, then $\sum(1/a_n)$ diverges to infinit

24.*If $\sum a_n$ and $\sum b_n$ both diverge, then so does $\sum(a_n$ $b_n)$.

25.*If $a_n \geq c > 0$ for every n, then $\sum a_n$ diverges infinity.

26.*If $\sum a_n$ diverges and $\{b_n\}$ is bounded, then $\sum a_n b$ diverges.

27.*If $a_n > 0$ and $\sum a_n$ converges, then $\sum(a_n)^2$ co verges.

10.3 CONVERGENCE TESTS FOR POSITIVE SERIES

In the previous section we saw a few examples of convergent series (geometri and telescoping series) whose sums could be determined exactly because the parti sums s_n could be algebraically manipulated into explicit functions of n whose limi as $n \to \infty$ could be easily evaluated. It is not usually possible to do this wi a given series, and therefore it is not usually possible to determine the sum the series exactly. However, there are many techniques for determining wheth a given series converges and, if it does, for approximating the sum to any desire degree of accuracy. We shall investigate some of these techniques in the remainin sections of this chapter.

In this section we deal exclusively with *positive series*, that is, series of th form

$$\sum_{n=1}^{\infty} a_n = a_1 + a_2 + a_3 + \cdots$$

where $a_n \geq 0$ for all $n \geq 1$. As noted in Theorem 10.2.12, such a series wi converge if its partial sums are bounded above and will diverge to infinity otherwis All our results apply equally well to *ultimately* positive series, but for simplici we ignore this in our discussion.

Comparison Tests

The first test we consider is analogous to the Comparison Test for improper inte grals, (see Theorem 6.8.11.) It enables us to determine convergence or divergenc of one series by comparing it with another series that is known to converge o diverge.

THEOREM 10.3.1 (**A Comparison Test**) Let $\{a_n\}$ and $\{b_n\}$ be positive sequences, and suppose there exists a positive constant K such that

$$0 \le a_n \le K b_n \qquad \text{for all } n \ge 1.$$

a) If the series $\sum_{n=1}^{\infty} b_n$ converges, then so does the series $\sum_{n=1}^{\infty} a_n$.
b) If $\sum_{n=1}^{\infty} a_n$ diverges to infinity, then so does $\sum_{n=1}^{\infty} b_n$.

PROOF Let $s_n = a_1 + a_2 + \cdots + a_n$ and $S_n = b_1 + b_2 + \cdots + b_n$. Then $s_n \le K S_n$. If $\sum b_n$ converges, then $\{S_n\}$ is convergent and hence bounded by Theorem 10.1.8. Hence $\{s_n\}$ is bounded above. By Theorem 10.2.12, $\sum a_n$ converges. Since the convergence of $\sum b_n$ guarentees that of $\sum a_n$, if the latter series diverges to infinity, then the former cannot converge either, so it must diverge to infinity too. □

REMARK: Theorem 10.3.1 does *not* say that if $\sum a_n$ converges, then $\sum b_n$ converges. It is possible that the "smaller" sum may be finite while the "larger" one is infinite. (Do not confuse a theorem with its converse.)

EXAMPLE 10.3.2 Do the following series converge or not? Give reasons. Obtain an upper bound for the sums of the convergent series.

$$\text{a) } \sum_{n=1}^{\infty} \frac{1}{2^n + 1}, \qquad \text{b) } \sum_{n=1}^{\infty} \frac{1}{n^2}, \qquad \text{c) } \sum_{n=2}^{\infty} \frac{1}{\ln n}.$$

SOLUTION a) Since $0 < \dfrac{1}{2^n + 1} < \dfrac{1}{2^n}$ for $n = 1, 2, 3, \ldots$, and since $\sum_{n=1}^{\infty}(1/2^n)$ is a convergent geometric series, the series $\sum_{n=1}^{\infty}(1/(2^n+1))$ also converges by comparison. Evidently in this case,

$$0 < \sum_{n=1}^{\infty} \frac{1}{2^n + 1} < \sum_{n=1}^{\infty} \frac{1}{2^n} = \frac{\frac{1}{2}}{1 - \frac{1}{2}} = 1.$$

b) Since $0 < \dfrac{1}{n^2} < \dfrac{1}{(n-1)n}$ for $n = 2, 3, 4, \ldots$, and since

$$\sum_{n=2}^{\infty} \frac{1}{(n-1)n} = \frac{1}{1 \times 2} + \frac{1}{2 \times 3} + \frac{1}{3 \times 4} + \cdots = 1$$

(by Example 10.2.7), therefore $\sum_{n=1}^{\infty}(1/n^2)$ converges by comparison. In this case we have the bound

$$\sum_{n=1}^{\infty} \frac{1}{n^2} = 1 + \sum_{n=2}^{\infty} \frac{1}{n^2} < 1 + \sum_{n=2}^{\infty} \frac{1}{(n-1)n} = 1 + 1 = 2.$$

c) For $n = 2, 3, 4, \ldots$ we have $0 < \ln n < n$. Thus $\dfrac{1}{\ln n} > \dfrac{1}{n}$. Since $\sum_{n=2}^{\infty} \dfrac{1}{n}$ diverges to infinity (it is a harmonic series), so does $\sum_{n=2}^{\infty} \dfrac{1}{\ln n}$ by comparison.

The following theorem provides a version of the Comparison Test that is not quite as general as Theorem 10.3.1 but is often easier to apply in specific cases.

THEOREM 10.3.3 (**A Limit Comparison Test**) Suppose that $\{a_n\}$ and $\{b_n\}$ are positive sequences and that

$$\lim \frac{a_n}{b_n} = L,$$

where L is either a nonnegative finite number or $+\infty$.

a) If $L < \infty$ and $\sum b_n$ converges, then $\sum a_n$ also converges.

b) If $L > 0$ and $\sum b_n$ diverges to infinity, then so does $\sum a_n$.

PROOF If $L < \infty$, then for n sufficiently large, we have $b_n > 0$ and

$$0 \le \frac{a_n}{b_n} \le L + 1,$$

so $0 \le a_n \le (L + 1)b_n$. Hence $\sum a_n$ converges if $\sum b_n$ converges, by Theorem 10.3.1(a). If $L > 0$, then for n sufficiently large

$$\frac{a_n}{b_n} \ge \frac{L}{2}.$$

Therefore $0 < b_n \le (2/L)a_n$, and $\sum a_n$ diverges to infinity if $\sum b_n$ does, by theorem 10.3.1(b). □

EXAMPLE 10.3.4 Do the following series converge or not? Give reasons.

$$\text{a)} \sum_{n=1}^{\infty} \frac{1}{1+\sqrt{n}}, \qquad \text{b)} \sum_{n=1}^{\infty} \frac{n+5}{n^3 - 2n + 3}, \qquad \text{c)} \sum_{n=1}^{\infty} \frac{2^n + 1}{3^n - 1}.$$

SOLUTION a) Since the sequence $\{\sqrt{n}\}$ does not grow as fast as $\{n\}$, we suspect that the series diverges to infinity by comparison with $\sum \dfrac{1}{n}$. Observe that

$$L = \lim \frac{\dfrac{1}{1+\sqrt{n}}}{\dfrac{1}{n}} = \lim \frac{n}{1+\sqrt{n}} = \infty > 0.$$

Therefore $\sum \dfrac{1}{1+\sqrt{n}}$ does diverge to infinity by comparison with $\sum \dfrac{1}{n}$.

b) For large n, the terms behave like $\dfrac{n}{n^3}$ so let us compare with the series $\sum 1/n^2$, which we know converges by Example 10.3.2(b). ($\sum 1/n(n+1)$ would also do.)

$$L = \lim \frac{\dfrac{n+5}{n^3-2n+3}}{\dfrac{1}{n^2}} = \lim \frac{n^3+5n^2}{n^3-2n+3} = 1.$$

Since $L < \infty$, the series $\sum_{n=1}^{\infty} \dfrac{n+5}{n^3-2n+3}$ converges by comparison with $\sum \dfrac{1}{n^2}$.

c) In this case the terms behave like $(2/3)^n$ for large n. We have

$$L = \lim \frac{\dfrac{2^n+1}{3^n-1}}{\left(\dfrac{2}{3}\right)^n} = \lim \frac{2^n+1}{2^n}\,\frac{3^n}{3^n-1} = \lim \frac{1+\dfrac{1}{2^n}}{1-\dfrac{1}{3^n}} = 1.$$

Again $L < \infty$ and the series $\sum_{n=1}^{\infty} \dfrac{2^n+1}{3^n-1}$ converges by comparison with the convergent geometric series $\sum_{n=1}^{\infty} \left(\dfrac{2}{3}\right)^n$.

In order to apply the original version of the Comparison Test (Theorem 10.3.1) successfully, it is important to have an intuitive feeling for whether the given series converges or diverges. The form of the comparison will depend on whether you are trying to prove convergence or divergence. For instance, if you did not know intuitively that $\sum_{n=1}^{\infty} \dfrac{1}{100n+20,000}$ would have to diverge to infinity, you might try to argue

$$\frac{1}{100n+20,000} < \frac{1}{n} \qquad \text{for } n = 1, 2, 3, \ldots.$$

While true, this doesn't help at all. $\sum 1/n$ diverges to infinity, and therefore Theorem 10.3.1 yields no information from this comparison. We could, of course, argue instead that

$$\frac{1}{100n+20,000} \geq \frac{1}{20,100n} \qquad \text{if } n \geq 1,$$

and conclude by Theorem 10.3.1 that $\sum_{n=1}^{\infty}(1/(100n+20,000))$ diverges to infinity by a comparison with the divergent series $\sum \dfrac{1}{n}$. An easier way is to use Theorem 10.3.3 and the fact that

$$L = \lim \frac{\dfrac{1}{100n+20,000}}{\dfrac{1}{n}} = \lim \frac{n}{100n+20,000} = \frac{1}{100} > 0.$$

The limit Comparison Test has two disadvantages, however:

i) Its use does not lead to any obvious bounds for the sum of a convergent series such as those bounds given in Examples 10.3.2(a) and (b).

ii) The Limit Comparison Test can fail in certain cases because the limit L does not exist. In such cases it is possible that the ordinary Comparison Test may still work.

EXAMPLE 10.3.5 Test the series $\sum_{n=1}^{\infty} \dfrac{1+\sin n}{n^2}$ for convergence.

SOLUTION Since $\lim \dfrac{\frac{1+\sin n}{n^2}}{\frac{1}{n^2}} = \lim(1+\sin n)$ does not exist, Theorem 10.3.3 gives no information. However, it is evident that

$$0 \le \frac{1+\sin n}{n^2} \le \frac{2}{n^2} \qquad \text{for } n = 1, 2, 3, \ldots,$$

so that the given series does, in fact, converge by comparison with $\sum \dfrac{1}{n^2}$, using Theorem 10.3.1.

The Integral Test

The Integral Test provides a means for determining whether an ultimately positive series converges or diverges by comparison with an improper integral that behaves similarly. Example 10.2.8 was an example of the use of this technique. We formalize the method in the following theorem.

THEOREM 10.3.6 (**The Integral Test**) Suppose that for $n \ge N$, we have $a_n = f(n)$ where $f(x)$ is positive, continuous, and nonincreasing on $[N, \infty)$. Then

$$\sum_{n=1}^{\infty} a_n \qquad \text{and} \qquad \int_N^{\infty} f(t)\,dt$$

either both converge or both diverge to infinity.

PROOF Let $s_n = a_1 + a_2 + \cdots + a_n$. Then if $n > N$ we have

$$\begin{aligned} s_n &= s_N + a_{N+1} + a_{N+2} + \cdots a_n \\ &= s_N + f(N+1) + f(N+2) + \cdots + f(n) \\ &= s_N + \text{ sum of areas of rectangles shaded in Fig. 10.3.1} \\ &\le s_N + \int_N^{\infty} f(t)\,dt. \end{aligned}$$

Thus if the improper integral $\int_N^{\infty} f(t)\,dt$ converges, then $\{s_n\}$ is bounded above and $\sum a_n$ converges.

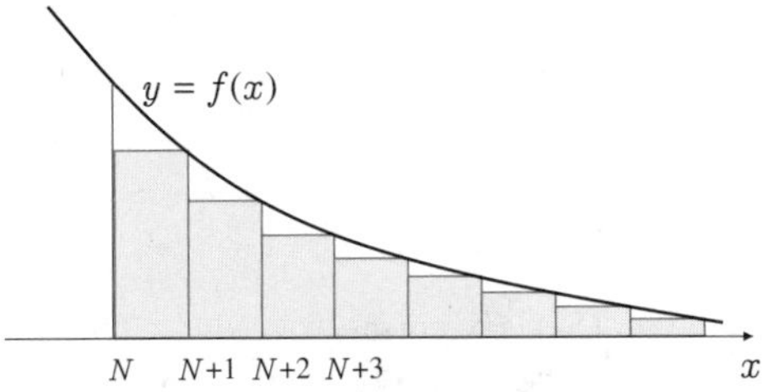

FIGURE 10.3.1

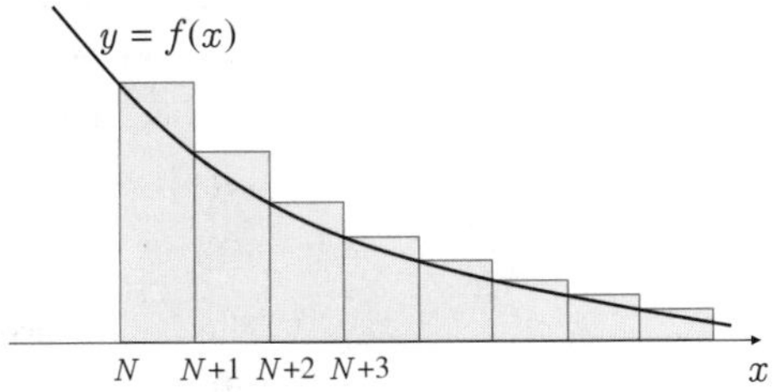

FIGURE 10.3.2

Conversely, suppose that $\sum_{n=1}^{\infty} a_n$ converges to the sum s. Then

$$\begin{aligned}\int_N^\infty f(t)\,dt &= \text{ area under } y = f(t) \text{ above } y = 0 \text{ from } t = N \text{ to } t = \infty \\ &\le \text{ sum of areas of shaded rectangles in Fig. 10.3.2} \\ &= a_N + a_{N+1} + a_{N+2} + \cdots \\ &= s - s_{N-1} < \infty,\end{aligned}$$

so the improper integral represents a finite area and is thus convergent. (We omit the remaining details showing that $\lim_{R\to\infty} \int_N^R f(t)\,dt$ exists; like the series case, the argument depends on completeness of the real numbers.) □

The principal use of the Integral Test is to establish the following result concerning the series $\sum_{n=0}^{\infty} n^{-p}$, which is called a p-**series**. This result should be memorized, as p-series are very convenient for use with the Comparison Test.

10.3.7
p-Series

$$\sum_{n=1}^{\infty} n^{-p} = \sum_{n=1}^{\infty} \frac{1}{n^p} \begin{cases} \text{converges if } p > 1 \\ \text{diverges to infinity if } p \le 1 \end{cases}$$

Observe that if $p > 0$, then $f(x) = x^{-p}$ is positive, continuous, and decreasing on $[1, \infty)$. Hence the given series converges for $p > 1$ and diverges for $0 < p \le 1$ by comparison with $\int_1^\infty x^{-p}\,dx$. (See Theorem 6.8.9.) If $p \le 0$, then $\lim(1/n^p) \ne 0$, so the series cannot converge in this case. Being a positive series, it must diverge to infinity.

The harmonic series $\sum \dfrac{1}{n}$ (the case $p = 1$ of the p-series) is on the borderline between convergence and divergence, although it diverges. The series

$$\sum_{n=2}^{\infty} \frac{1}{n(\ln n)^p} \quad \text{converges if } p > 1 \text{ and diverges to infinity if } p \le 1.$$

Note that the sum starts from $n = 2$ because $\ln 1 = 0$. Once again this result follows by comparison with an improper integral,

$$\int_2^\infty \frac{dx}{x(\ln x)^p}.$$

The details are left to the student. (See Exercise 29 below.) This result is a "fine tuning" of the result in Box 10.3.7. Since $\ln n$ grows more slowly than any positive power of n, if $r > 1$ and $p > 0$, we have, ultimately,

$$\frac{1}{n^r} < \frac{1}{n(\ln n)^p} < \frac{1}{n}.$$

This fine-tuning can be successfully refined even further (see Exercise 30 below).

EXAMPLE 10.3.8 Test for convergence

$$\sum_{n=1}^\infty \frac{n^3 - n^2 + 5n + 22}{n^4\sqrt{n} + 7n^2 - \sqrt{n} + 3}.$$

SOLUTION Inspection of the highest powers of n occuring in the numerator and denominator leads us to compare the series with $\sum_{n=1}^\infty \frac{1}{n^{3/2}}$, a convergent p-series. Since

$$\lim \frac{n^3 - n^2 + 5n + 22}{n^4\sqrt{n} + 7n^2 - \sqrt{n} + 3} \bigg/ \frac{1}{n^{3/2}} = 1,$$

the given series is also convergent by the limit comparison test.

The method of comparing the tail of a series with an improper integral can be used to obtain estimates for the error if a positive series is approximated using its partial sums. We look into this further in Section 10.5.

The Ratio Test

THEOREM 10.3.9 (**The Ratio Test**) Suppose that $a_n > 0$ (ultimately) and that

$$\rho = \lim_{n\to\infty} \frac{a_{n+1}}{a_n} \qquad \text{exists or is } +\infty.$$

a) If $0 \le \rho < 1$, then $\sum_{n=1}^\infty a_n$ converges.

b) If $1 < \rho \le \infty$, then $\lim a_n = \infty$ and $\sum_{n=1}^\infty a_n$ diverges to infinity.

c) If $\rho = 1$, this test gives no information; the series may either converge or diverge to infinity.

PROOF a) Suppose $\rho < 1$. Pick a number r such that $\rho < r < 1$. Since $\lim(a_{n+1}/a_n) = \rho$, therefore $a_{n+1}/a_n \le r$ for n sufficiently large; that is, $a_{n+1} \le ra_n$ for $n \ge N$, say. In particular,

$$\begin{aligned} a_{N+1} &\le ra_N \\ a_{N+2} &\le ra_{N+1} \le r^2 a_N \\ a_{N+3} &\le ra_{N+2} \le r^3 a_N \\ &\vdots \\ a_{N+k} &\le r^k a_N \qquad (k = 0, 1, 2, 3, \ldots) \end{aligned}$$

Hence $\sum_{n=N}^{\infty} a_n$ converges by comparison with the convergent geometric series $\sum_{k=0}^{\infty} r^k$. It follows that $\sum_{n=1}^{\infty} a_n = \sum_{n=1}^{N-1} a_n + \sum_{n=N}^{\infty} a_n$ must converge.

b) Now suppose that $\rho > 1$. Pick a number r such that $1 < r < \rho$. Since $\lim a_{n+1}/a_n = \rho$, therefore $a_{n+1}/a_n \ge r$ for n sufficiently large, say for $n \ge N$. We assume N is chosen large enough that $a_N > 0$. It follows by an argument similar to that used in part (a) that $a_{N+k} \ge r^k a_N$ for $k = 0, 1, 2, \ldots$, and since $r > 1$, $\lim a_n = \infty$.

c) If ρ is computed for each of the series $\sum(1/n)$ and $\sum(1/n^2)$, we get $\rho = 1$ in each case. Since the first series diverges to infinity and the second converges, it is apparent that the Ratio Test cannot distinguish between convergence and divergence if $\rho = 1$. □

All p-series fall into the indecisive category where $\rho = 1$. The Ratio Test is most useful for series whose terms decrease at least exponentially fast. The presence of factorials in a term also suggests the use of the Ratio Test.

EXAMPLE 10.3.10 Test for convergence

a) $\sum_{n=1}^{\infty} \dfrac{99^n}{n!}$, b) $\displaystyle\sum_{n=1}^{\infty} \frac{n^5}{2^n}$, c) $\displaystyle\sum_{n=1}^{\infty} \frac{n!}{n^n}$, d) $\displaystyle\sum_{n=1}^{\infty} \frac{(2n)!}{(n!)^2}$

SOLUTION a) $\rho = \lim \dfrac{99^{n+1}}{(n+1)!} \Big/ \dfrac{99^n}{n!} = \lim \dfrac{99}{n+1} = 0 < 1$. Thus $\sum_{n=1}^{\infty}(99^n/n!)$ converges by the Ratio Test.

b) $\rho = \lim \dfrac{(n+1)^5}{2^{n+1}} \Big/ \dfrac{n^5}{2^n} = \lim \dfrac{1}{2}\left(\dfrac{n+1}{n}\right)^5 = \dfrac{1}{2} < 1$. Hence $\sum_{n=1}^{\infty}(n^5/2^n)$ converges by the Ratio Test.

c) $$\rho = \lim \frac{(n+1)!}{(n+1)^{n+1}} \Big/ \frac{n!}{n^n} = \lim \frac{(n+1)!\,n^n}{(n+1)^{n+1} n!} = \lim \left(\frac{n}{n+1}\right)^n$$

$$= \lim \frac{1}{\left(1 + \dfrac{1}{n}\right)^n} = \frac{1}{e} < 1.$$

Thus $\sum_{n=1}^{\infty}(n!/n^n)$ converges by the Ratio Test.

d) $\rho = \lim \dfrac{(2(n+1))!}{((n+1)!)^2} \bigg/ \dfrac{(2n)!}{(n!)^2} = \lim \dfrac{(2n+2)(2n+1)}{(n+1)^2} = 4 > 1$. Thus $\sum_{n=1}^{\infty} \dfrac{(2n)!}{(n!)^2}$ diverges to infinity by the Ratio Test.

EXERCISES

In Exercises 1–28 determine whether the given series converges or diverges by using any appropriate test. The p-series (Box 10.3.7) can be used for comparison, as can geometric series. Be alert for series whose terms do not approach 0.

1. $\displaystyle\sum_{n=1}^{\infty} \frac{1}{n^2+1}$

2. $\displaystyle\sum_{n=1}^{\infty} \frac{n}{n^4-2}$

3. $\displaystyle\sum_{n=1}^{\infty} \frac{n^2+1}{n^3+1}$

4. $\displaystyle\sum_{n=1}^{\infty} \frac{\sqrt{n}}{n^2+n+1}$

5. $\displaystyle\sum_{n=1}^{\infty} \left|\sin \frac{1}{n^2}\right|$

6. $\displaystyle\sum_{n=8}^{\infty} \frac{1}{\pi^n+5}$

7. $\displaystyle\sum_{n=2}^{\infty} \frac{1}{(\ln n)^3}$

8. $\displaystyle\sum_{n=1}^{\infty} \frac{1}{\ln(3n)}$

9. $\displaystyle\sum_{n=1}^{\infty} \frac{1}{\pi^n - n^\pi}$

10. $\displaystyle\sum_{n=0}^{\infty} \frac{1+n}{2+n}$

11. $\displaystyle\sum_{n=1}^{\infty} \frac{1+n^{4/3}}{2+n^{5/3}}$

12. $\displaystyle\sum_{n=1}^{\infty} \frac{n^2}{1+n\sqrt{n}}$

13. $\displaystyle\sum_{n=3}^{\infty} \frac{1}{n \ln n \sqrt{\ln \ln n}}$

14. $\displaystyle\sum_{n=2}^{\infty} \frac{1}{n \ln n (\ln \ln n)^2}$

15. $\displaystyle\sum_{n=1}^{\infty} \frac{1-(-1)^n}{n^4}$

16. $\displaystyle\sum_{n=1}^{\infty} \frac{1+(-1)^n}{\sqrt{n}}$

17. $\displaystyle\sum_{n=1}^{\infty} \frac{2+\cos n}{n+\ln n}$

18. $\displaystyle\sum_{n=1}^{\infty} \frac{e^n \cos^2 n}{1+\pi^n}$

19. $\displaystyle\sum_{n=1}^{\infty} \frac{1}{2^n(n+1)}$

20. $\displaystyle\sum_{n=1}^{\infty} \frac{n^4}{n!}$

21. $\displaystyle\sum_{n=1}^{\infty} \frac{n!}{n^2 e^n}$

22. $\displaystyle\sum_{n=1}^{\infty} \frac{(2n)!6^n}{(3n)!}$

23. $\displaystyle\sum_{n=2}^{\infty} \frac{\sqrt{n}}{3^n \ln n}$

24. $\displaystyle\sum_{n=0}^{\infty} \frac{n^{100}2^n}{\sqrt{n!}}$

25. $\displaystyle\sum_{n=1}^{\infty} \frac{(2n)!}{(n!)^3}$

26. $\displaystyle\sum_{n=1}^{\infty} \frac{1+n!}{(1+n)!}$

27. $\displaystyle\sum_{n=4}^{\infty} \frac{2^n}{3^n - n^3}$

28. $\displaystyle\sum_{n=1}^{\infty} \frac{n^n}{\pi^n n!}$

29. Complete Example 10.3.7 by showing that

$$\int_2^{\infty} \frac{dx}{x(\ln x)^p} \begin{cases} \text{converges if } p > 1 \\ \text{diverges to infinity if } p \le 1. \end{cases}$$

30.*Show that $\sum_{n=3}^{\infty}(1/(n \ln n (\ln \ln n)^p))$ converges if and only if $p > 1$. Generalize this result to series of the form

$$\sum_{n=N}^{\infty} \frac{1}{n(\ln n)(\ln \ln n)\cdots(\ln_j n)(\ln_{j+1} n)^p}$$

where $\ln_j n = \underbrace{\ln \ln \ln \ln \cdots \ln}_{j \ln' s} n$.

31.*(**The Root Test**) Suppose that $a_n > 0$ (ultimately) and that $\sigma = \lim_{n\to\infty}(a_n)^{1/n}$ exists or is $+\infty$. Show that $\sum_{n=1}^{\infty} a_n$ converges if $\sigma < 1$ and diverges to infinity if $\sigma > 1$. Show that the series may either converge or diverge if $\sigma = 1$. (*Hint:* Mimic the proof of the Ratio Test.)

32.*Test for convergence $\displaystyle\sum_{n=1}^{\infty} \left(\frac{n}{n+1}\right)^{n^2}$.

33.*Try to use the Ratio Test to determine whether $\displaystyle\sum_{n=1}^{\infty} \frac{2^{2n}(n!)^2}{(2n)!}$ converges. What happens? Now observe that

$$\frac{2^{2n}(n!)^2}{(2n)!} = \frac{[2n(2n-2)(2n-4)\cdots 6\times 4\times 2]^2}{2n(2n-1)(2n-2)\cdots 4\times 3\times 2\times 1}$$
$$= \frac{2n}{2n-1}\times\frac{2n-2}{2n-3}\times\cdots\times\frac{4}{3}\times\frac{2}{1}.$$

Does the given series converge or not? Why?

34.*Try to decide whether the series $\sum_{n=1}^{\infty}\frac{(2n)!}{2^{2n}(n!)^2}$ converges. (*Hint:* Proceed as in the previous exercise. Show that $a_n \geq \frac{1}{2n}$.)

10.4 ABSOLUTE AND CONDITIONAL CONVERGENCE

All of the series $\sum_{n=1}^{\infty} a_n$ considered in the previous section were ultimately positive; that is, $a_n \geq 0$ for n sufficiently large. We now drop this restriction and allow arbitrary real terms a_n. We can, however, always obtain a positive series from any given series by replacing all the terms with their absolute values.

10.4.1 Absolute Convergence

> The series $\sum_{n=1}^{\infty} a_n$ is said to be **absolutely convergent** if $\sum_{n=1}^{\infty} |a_n|$ converges.

The series

$$s = \sum_{n=1}^{\infty}\frac{(-1)^n}{n^2} = -1 + \frac{1}{4} - \frac{1}{9} + \frac{1}{16} - \cdots$$

converges absolutely since

$$S = \sum_{n=1}^{\infty}\left|\frac{(-1)^n}{n^2}\right| = \sum_{n=1}^{\infty}\frac{1}{n^2} = 1 + \frac{1}{4} + \frac{1}{9} + \frac{1}{16} + \cdots$$

converges. It seems intuitively clear that the first series must converge, and its sum s must satisfy $-S \leq s \leq S$. In general, the cancellation that occurs because some terms are negative and others positive makes it "easier" for a series to converge than if all the terms are of one sign. We verify this intuition in the following theorem.

THEOREM 10.4.2 If a series converges absolutely, then it converges.

PROOF Let $\sum_{n=1}^{\infty} a_n$ be absolutely convergent. Let

$$s_n = a_1 + a_2 + a_3 + \cdots + a_n$$
$$S_n = |a_1| + |a_2| + |a_3| + \cdots + |a_n|.$$

Then $\lim S_n = S$ exists. Since $\{S_n\}$ is increasing, we have $S_n \leq S$ for every n. We want to show that $\lim s_n$ exists.

For $n = 1, 2, 3, \ldots$ let

$$p_n = \begin{cases} a_n & \text{if } a_n \geq 0 \\ 0 & \text{if } a_n < 0, \end{cases} \qquad q_n = \begin{cases} 0 & \text{if } a_n \geq 0 \\ -a_n & \text{if } a_n < 0. \end{cases}$$

Evidently $a_n = p_n - q_n$ and $|a_n| = p_n + q_n$ for each n. Let

$$P_n = p_1 + p_2 + p_3 + \cdots + p_n \qquad \text{and} \qquad Q_n = q_1 + q_2 + q_3 + \cdots + q_n.$$

Then $s_n = P_n - Q_n$ and $S_n = P_n + Q_n$ for each n.

Since $p_n \geq 0$ and $q_n \geq 0$ for every n, $\{P_n\}$ and $\{Q_n\}$ are increasing sequences. Both of these sequences are bounded above since $P_n \leq S_n \leq S$ and $Q_n \leq S_n \leq S$. Hence $P = \lim P_n$ and $Q = \lim Q_n$ exist by completeness (Box 10.1.9). It follows that $\lim s_n = \lim P_n - \lim Q_n = P - Q$ exists. Therefore $\sum_{n=1}^{\infty} a_n$ converges. □

REMARK: Again you are cautioned not to confuse the statement of Theorem 10.4.2 with the converse statement which is false. We will show later in this section that the **alternating harmonic series**

$$\sum_{n=1}^{\infty} \frac{(-1)^{n-1}}{n} = 1 - \frac{1}{2} + \frac{1}{3} - \frac{1}{4} + \frac{1}{5} - \cdots$$

converges, although it does not converge absolutely. If we replace all the terms by their absolute values we get the divergent harmonic series:

$$\sum_{n=1}^{\infty} \frac{1}{n} = 1 + \frac{1}{2} + \frac{1}{3} + \frac{1}{4} + \cdots = \infty.$$

10.4.3 Conditional Convergence

> If $\sum_{n=1}^{\infty} a_n$ is convergent, but not absolutely convergent, then we say that it is **conditionally convergent** or that it **converges conditionally**.

The alternating harmonic series is an example of a conditionally convergent series.

The Comparison Tests, the Integral Test and the Ratio Test, can each be used to test for absolute convergence. They should be applied to the series $\sum_{n=1}^{\infty} |a_n|$. For the Ratio Test we calculate $\rho = \lim |a_{n+1}/a_n|$. If $\rho < 1$, then $\sum a_n$ converges absolutely. If $\rho > 1$, then $\lim |a_n| = \infty$, so both $\sum |a_n|$ and $\sum a_n$ must diverge. If $\rho = 1$ we get no information.

EXAMPLE 10.4.4 Test for absolute convergence a) $\sum_{n=1}^{\infty} \dfrac{(-1)^{n-1}}{2n-1}$, b) $\sum_{n=1}^{\infty} \dfrac{n\cos(n\pi)}{2^n}$

SOLUTION a) $\lim \dfrac{\left|\dfrac{(-1)^{n-1}}{2n-1}\right|}{\dfrac{1}{n}} = \lim \dfrac{n}{2n-1} = \dfrac{1}{2} > 0$. Since the harmonic series $\sum_{n=1}^{\infty}(1/n)$ diverges to infinity, therefore $\sum_{n=1}^{\infty}((-1)^{n-1}/(2n-1))$ does not converge absolutely.

b) $\rho = \lim \left|\dfrac{\dfrac{(n+1)\cos((n+1)\pi)}{2^{n+1}}}{\dfrac{n\cos(n\pi)}{2^n}}\right| = \lim \dfrac{n+1}{2n} = \dfrac{1}{2} < 1$. (Note that $\cos(n\pi)$ is just a fancy way of writing $(-1)^n$.) Therefore $\sum_{n=1}^{\infty}((n\cos(n\pi))/2^n)$ converges absolutely by the Ratio Test.

The Alternating Series Test

We cannot use any of the previously developed tests to show that the alternating harmonic series converges; all of those tests apply only to (ultimately) positive series, and so they can test only for absolute convergence. Demonstrating convergence that is not absolute is generally harder to do. We present only one test that can establish such convergence; this test can only be used on a very special kind of series.

THEOREM 10.4.5 (**The Alternating Series Test**) Suppose that the sequence $\{a_n\}$ is positive, decreasing, and tends to 0, that is, suppose

i) $a_n \geq 0$ for $n = 1, 2, 3, \ldots$;

ii) $a_{n+1} \leq a_n$ for $n = 1, 2, 3, \ldots$;

iii) $\lim a_n = 0$.

Then the alternating series

$$\sum_{n=1}^{\infty}(-1)^{n-1}a_n = a_1 - a_2 + a_3 - a_4 + a_5 - \cdots$$

converges. Moreover, if the sum of the series is s and if s_n denotes the nth partial sum,

$$s_n = \sum_{k=1}^{n}(-1)^{k-1}a_k = a_1 - a_2 + a_3 - \cdots + (-1)^{n-1}a_n,$$

then for every n we have

$$s_{2n} \leq s \leq s_{2n-1}, \qquad \text{and} \qquad s_{2n} \leq s \leq s_{2n+1}.$$

In particular,

$$|s - s_n| \leq a_{n+1}.$$

Thus the error incurred in using a partial sum to approximate the sum of the series does not exceed the first omitted term in absolute value.

PROOF Since the sequence $\{a_n\}$ is decreasing, we have $a_{2n+1} \geq a_{2n+2}$. Therefore $s_{2n+2} = s_{2n} + a_{2n+1} - a_{2n+2} \geq s_{2n}$ for $n = 1, 2, 3, \ldots$; the even partial sums $\{s_n\}$ form an increasing sequence. Similarly $s_{2n+1} = s_{2n-1} - a_{2n} + a_{2n+1} \leq s_{2n-1}$, so the old partial sums $\{s_{2n-1}\}$ form a decreasing sequence. Since $s_{2n} = s_{2n-1} - a_{2n} \leq s_{2n-1}$, we can say, for any n, that

!!DANGER!!

If you read this proof do it slowly and think about why each statement is true.

$$s_2 \leq s_4 \leq s_6 \leq \cdots \leq s_{2n} \leq s_{2n-1} \leq s_{2n-3} \leq \cdots s_5 \leq s_3 \leq s_1.$$

Hence s_2 is a lower bound for the decreasing sequence $\{s_{2n-1}\}$ and s_1 is an upper bound for the increasing sequence $\{s_{2n}\}$. Both of these sequences therefore converge by completeness of the real numbers (Box 10.1.9):

$$\lim_{n\to\infty} s_{2n-1} = s_{\text{odd}}, \qquad \lim_{n\to\infty} s_{2n} = s_{\text{even}}.$$

Since $0 = \lim a_{2n} = \lim(s_{2n-1} - s_{2n}) = s_{\text{odd}} - s_{\text{even}}$, we have $s_{\text{odd}} = s_{\text{even}} = s$, say. Thus $\lim s_n = s$ exists and the series $\sum(-1)^{n-1}a_n$ converges to this sum s.

It is evident that every even partial sum is less than or equal to s and every odd partial sum is greater than or equal to s. That is, s lies between s_{2n} and either s_{2n-1} or s_{2n+1}. It follows that

$$|s - s_n| \leq |s_{n+1} - s_n| = a_{n+1}. \;\square$$

Note that the series $\sum_{n=1}^{\infty}(-1)^{n-1}a_n$ begins with a positive term, a_1. The conclusions of Theorem 10.4.5 also hold for the series $\sum_{n=1}^{\infty}(-1)^n a_n$, which starts with a negative term, $-a_1$. (It is just the negative of the first series.) In this case, however, we have

$$s_{2n-1} \leq s \leq s_{2n} \qquad \text{and} \qquad s_{2n+1} \leq s \leq s_{2n}.$$

The theorem also remains valid if conditions (i) and (ii) in its statement are replaced by the corresponding "ultimate" versions:

$$\text{i)}\;\; a_n \geq 0 \quad \text{and} \quad \text{ii)}\; a_{n+1} \leq a_n \qquad \text{for } n = N, N+1, N+2, \ldots.$$

The inequalities bounding s in terms of partial sums s_n are then only valid for $n > N$.

EXAMPLE 10.4.6 Each of the series a) $\sum_{n=1}^{\infty} \dfrac{(-1)^{n-1}}{n}$, b) $\sum_{n=2}^{\infty} \dfrac{\cos(n\pi)}{\ln n}$, and c) $\sum_{n=1}^{\infty} \dfrac{(-1)^{n-1}}{n^4}$ satisfies the conditions of the Alternating Series Test, and so converges. Evidently (a) and (b) do not converge absolutely (for (b) note that $1/\ln n > 1/n$), so these are both conditionally convergent series. Series (c), however, is absolutely convergent so we really do not need the Alternating Series Test to show that it converges; Theorem 10.4.2 would do as well. When determining the convergence of a given series, it is best to consider first whether the series converges absolutely. If it does not, then there remains the possibility of conditional convergence.

EXAMPLE 10.4.7 For what values of x does the series $\sum_{n=1}^{\infty} \dfrac{(x-5)^n}{n\,2^n}$ converge absolutely? Converge conditionally? Diverge?

SOLUTION For such series whose terms involve functions of a variable x, it is usually wisest to begin testing for absolute convergence with the ratio test. We have

$$\rho = \lim \left| \frac{(x-5)^{n+1}}{(n+1)2^{n+1}} \Big/ \frac{(x-5)^n}{n\,2^n} \right| = \lim \frac{n}{n+1} \left| \frac{x-5}{2} \right| = \left| \frac{x-5}{2} \right|.$$

The series converges absolutely if $|(x-5)/2| < 1$. This inequality is equivalent to $|x-5| < 2$, (the distance from x to 5 is less than 2), that is, $3 < x < 7$. If $x < 3$ or $x > 7$, then $|(x-5)/2| > 1$. The series diverges; its terms do not approach zero.

If $x = 3$, the series is $\sum_{n=1}^{\infty}((-1)^n/n)$, which converges conditionally (it is an alternating harmonic series); if $x = 7$, the series is the harmonic series $\sum_{n=1}^{\infty}(1/n)$, which diverges to infinity. Hence the given series converges absolutely on the interval $(3, 7)$, converges conditionally at $x = 3$, and diverges everywhere else.

EXAMPLE 10.4.8 For what values of x does the series $\sum_{n=0}^{\infty}(n+1)^2\left(\dfrac{x}{x+2}\right)^n$ converge absolutely? Converge conditionally? Diverge?

SOLUTION Again we begin with the ratio test.

$$\rho = \lim \left| (n+2)^2 \left(\frac{x}{x+2}\right)^{n+1} \Big/ (n+1)^2 \left(\frac{x}{x+2}\right)^n \right| = \lim \left(\frac{n+2}{n+1}\right)^2 \left|\frac{x}{x+2}\right| = \left|\frac{x}{x+2}\right|$$

The series converges absolutely if $\left|\dfrac{x}{x+2}\right| < 1$. This condition says that the distance from x to 0 is less than the distance from x to -2. Hence $x > -1$. The series diverges if $\left|\dfrac{x}{x+2}\right| > 1$, that is, if $x < -1$. If $x = -1$, the series is $\sum_{n=0}^{\infty}(-1)^n(n+1)^2$, which diverges. We conclude that the series converges absolutely for $x > -1$, converges conditionally nowhere, and diverges for $x \le -1$.

When using the Alternating Series Test, it is important to verify (at least mentally) that *all three conditions* (i)–(iii) are satisfied. (As mentioned above, (i) and (ii) need only be satisfied ultimately.)

EXAMPLE 10.4.9 a) Consider the series $\displaystyle\sum_{n=1}^{\infty}(-1)^{n-1}\frac{n+1}{n}$. Here $a_n = (n+1)/n$ is positive and decreases as n increases. However, $\lim a_n = 1 \ne 0$. The test fails. In fact the given series diverges, since its terms do not approach 0.

b) Consider the series

$$1 - \frac{1}{4} + \frac{1}{3} - \frac{1}{16} + \frac{1}{5} - \cdots = \sum_{n=1}^{\infty}(-1)^{n-1}a_n,$$

where

$$a_n = \begin{cases} \dfrac{1}{n} & \text{if } n \text{ is odd,} \\ \dfrac{1}{n^2} & \text{if } n \text{ is even.} \end{cases}$$

This series alternates, a_n is positive, and $\lim a_n = 0$. However, $\{a_n\}$ is not decreasing (even ultimately). Once again the Alternating Series Test cannot be applied. In fact, since

$$-\frac{1}{4} - \frac{1}{16} - \cdots - \frac{1}{(2n)^2} - \cdots \qquad \text{converges,}$$

and

$$1 + \frac{1}{3} + \frac{1}{5} + \cdots + \frac{1}{2n-1} + \cdots \qquad \text{diverges to infinity,}$$

it is readily seen that the given series diverges to infinity.

The basic difference between absolute and conditional convergence is that when a series $\sum a_n$ converges absolutely it does so because its terms $\{a_n\}$ decrease in size fast enough that their sum can be finite even if no cancellation occurs due to terms of opposite sign. If cancellation is required to make the series converge (because the terms decrease slowly), then the series can only converge conditionally.

Consider the alternating harmonic series

$$1 - \frac{1}{2} + \frac{1}{3} - \frac{1}{4} + \frac{1}{5} - \frac{1}{6} + \cdots.$$

This series converges, but only conditionally. If we take the subseries containing only the positive terms, we get the series

$$1 + \frac{1}{3} + \frac{1}{5} + \frac{1}{7} + \cdots,$$

which diverges to infinity. Similarly, the subseries of negative terms

$$-\frac{1}{2} - \frac{1}{4} - \frac{1}{6} - \frac{1}{8} - \cdots$$

diverges to negative infinity.

If a series converges absolutely, the subseries consisting of positive terms and the subseries consisting of negative terms must each converge to a finite sum. If a series converges conditionally the positive and negative subseries will both diverge, to ∞ and $-\infty$ respectively. The following theorem makes use of this fact.

THEOREM 10.4.10 a) If the terms of an absolutely convergent series are rearranged so that the additions occur in a different order, the rearranged series still converges to the same sum as the original series.

b) If a series is conditionally convergent, and L is any real number, then the terms of the series can be rearranged so as to make the series converge (conditionally) to the sum L. It can also be rearranged so as to diverge to ∞ or to $-\infty$, or just to diverge. □

We will not present a proof. (See, however, Exercise 30 below.) Part (b) shows that conditional convergence is a rather suspect kind of convergence, being dependent on the order in which the terms are added.

EXERCISES

Determine whether the series in Exercises 1–12 converge absolutely, converge conditionally, or diverge.

1. $\sum_{n=1}^{\infty} \frac{(-1)^{n-1}}{\sqrt{n}}$

2. $\sum_{n=1}^{\infty} \frac{(-1)^n}{n^2 + \ln n}$

3. $\sum_{n=1}^{\infty} \frac{\cos(n\pi)}{(n+1)\ln(n+1)}$

4. $\sum_{n=1}^{\infty} \frac{(-1)^{2n}}{2^n}$

5. $\sum_{n=0}^{\infty} \frac{(-1)^n(n^2-1)}{n^2+1}$

6. $\sum_{n=1}^{\infty} \frac{(-2)^n}{n!}$

7. $\sum_{n=1}^{\infty} \frac{(-1)^n}{n\pi^n}$

8. $\sum_{n=0}^{\infty} \frac{-n}{n^2+1}$

9. $\sum_{n=1}^{\infty} (-1)^n \frac{20n^2 - n - 1}{n^3 + n^2 + 33}$

10. $\sum_{n=1}^{\infty} \frac{100\cos(n\pi)}{2n+3}$

11. $\sum_{n=1}^{\infty} \frac{n!}{(-100)^n}$

12. $\sum_{n=10}^{\infty} \frac{\sin(n+1/2)\pi}{\ln\ln n}$

Determine the values of x for which the series in Examples 13–24 converge absolutely, converge conditionally, or diverge.

13. $\sum_{n=0}^{\infty} \frac{x^n}{\sqrt{n+1}}$

14. $\sum_{n=1}^{\infty} \frac{(x-2)^n}{n^2 2^{2n}}$

15. $\sum_{n=0}^{\infty} (-1)^n \frac{(x-1)^n}{2n+3}$

16. $\sum_{n=1}^{\infty} \frac{1}{2n-1}\left(\frac{3x+2}{-5}\right)^n$

17. $\sum_{n=2}^{\infty} \frac{x^n}{2^n \ln n}$

18. $\sum_{n=1}^{\infty} \frac{(4x+1)^n}{n^3}$

19. $\sum_{n=1}^{\infty} \frac{(2x+3)^n}{n^{1/3}4^n}$

20. $\sum_{n=1}^{\infty} \frac{1}{n}\left(1+\frac{1}{x}\right)^n$

21. $\sum_{n=1}^{\infty} \frac{1}{n^2}\left(1-\frac{1}{x}\right)^n$

22. $\sum_{n=1}^{\infty} \frac{(x^2-1)^n}{\sqrt{n}}$

23. $\sum_{n=0}^{\infty} n(x^2-x-1)^n$

24. $\sum_{n=1}^{\infty} \frac{\sin^n x}{n}$

25.*Does the Alternating Series Test apply directly to the series $\sum_{n=1}^{\infty} \frac{\sin(n\pi/2)}{n}$? Determine whether the series converges.

26.*Show that the series $\sum_{n=1}^{\infty} a_n$ converges absolutely if $a_n = 10/n^2$ for even n and $a_n = -1/10n^3$ for odd n.

27.*Which of the following statements are true and which are false? Justify your assertion of truth, or give a counterexample to show falsehood.

a) If $\sum_{n=1}^{\infty} a_n$ converges, then $\sum_{n=1}^{\infty}(-1)^n a_n$ converges.

b) If $\sum_{n=1}^{\infty} a_n$ converges and $\sum_{n=1}^{\infty}(-1)^n a_n$ converges, then $\sum_{n=1}^{\infty} a_n$ converges absolutely.

c) If $\sum_{n=1}^{\infty} a_n$ converges absolutely, then $\sum_{n=1}^{\infty}(-1)^n a_n$ converges absolutely.

28.*a) Use a Riemann sum argument to show that

$$\ln n! \geq \int_1^n \ln t\,dt = n\ln n - n + 1.$$

b) For what values of x does the series $\sum_{n=1}^{\infty} \frac{n!x^n}{n^n}$ converge absolutely? Converge conditionally? Diverge? (*Hint:* First use the Ratio Test. To test the cases where $\rho = 1$, you may find the inequality in part (a) useful.)

29. *For what values of x does the series $\sum_{n=1}^{\infty} \frac{(2n)!x^n}{2^{2n}(n!)^2}$ converge absolutely? Converge conditionally? Diverge? (*Hint:* See Exercise 34 of Section 10.3.)

30. *Devise a procedure for rearranging the terms of the alternating harmonic series so that the rearranged series converges to 8.

10.5 ESTIMATING THE SUM OF A SERIES

So far we have been concerned mainly with determining whether a given series converges or diverges, that is, whether or not it has a finite sum. Except for very special series (geometric series, telescoping series) we have not actually calculated the sums of the series we have shown to converge. Indeed, there are no elementary techniques for calculating the sums of most series, though we will learn how to find the sums of some other series in the next chapter.

When it is not practical to find the exact sum of a convergent series, we can still approximate that sum by using a partial sum s_n of the series:

$$s = \sum_{k=1}^{\infty} a_k = a_1 + a_2 + a_3 + \cdots$$

$$s \approx s_n = \sum_{k=1}^{n} a_k = a_1 + a_2 + a_3 + \cdots + a_n.$$

Of course, such an approximation is of little use unless we know how good an approximation it is, that is, unless we have a bound for the size of the error, $|s - s_n|$.

Observe that the error $s - s_n$ is just the *tail* of the given series; it is the sum of the rest of the terms of the series beyond those included in s_n:

$$s - s_n = \sum_{k=n+1}^{\infty} a_k = a_{n+1} + a_{n+2} + a_{n+3} + \cdots$$

If the given series converges, then the error $s - s_n$ approaches 0 as n increases toward infinity (since $\lim s_n = s$), but we would still like to know how fast that approach is and how large n need be to ensure that the error is within tolerable bounds. Various convergence tests involve, implicitly or explicitly, techniques for finding bounds for the error. We shall investigate error-bounding techniques arising from the Integral Test, the Ratio Test, and the Alternating Series Test.

Integral Bounds

Suppose that $a_k = f(k)$ for $k = n+1, n+2, n+3, \ldots$ where f is a positive, continuous function, decreasing on the interval $[n, \infty)$. Evidently,

$$\begin{aligned} s - s_n &= \sum_{k=n+1}^{\infty} f(k) \\ &= \text{sum of areas of rectangles shaded in Fig. 10.5.1} \\ &\le \int_n^{\infty} f(x)\,dx. \end{aligned}$$

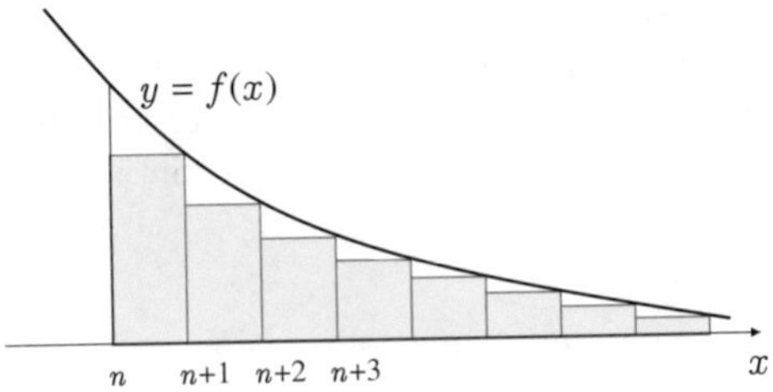

FIGURE 10.5.1

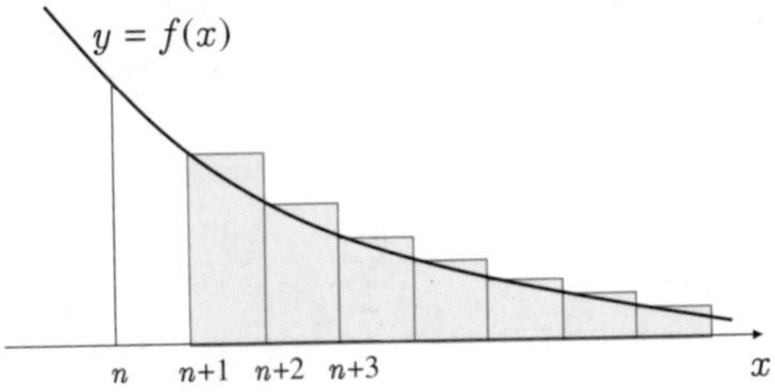

FIGURE 10.5.2

Similarly,

$$\begin{aligned} s - s_n &= \text{ sum of areas of rectangles in Fig. 10.5.2} \\ &\geq \int_{n+1}^{\infty} f(x)\,dx. \end{aligned}$$

If we define

$$A_n = \int_n^{\infty} f(x)\,dx$$

then we can combine the above inequalities to obtain

$$A_{n+1} \leq s - s_n \leq A_n,$$

or, equivalently,

$$s_n + A_{n+1} \leq s \leq s_n + A_n.$$

The error in the approximation $s \approx s_n$ satisfies $0 \leq s - s_n \leq A_n$. However, since s must lie in the interval $[s_n + A_{n+1}, s_n + A_n]$, we can do better by using the midpoint of this interval as an approximation for s:

10.5.1 Integral Approximation

$$s \approx s_n^* = s_n + \frac{A_{n+1} + A_n}{2}.$$

Evidently the size of the error in this approximation, $s - s_n^*$, cannot exceed half the length of the interval:

10.5.2 Error Estimate

$$|s - s_n^*| \leq \frac{A_n - A_{n+1}}{2}.$$

(Whenever an unknown quantity is known to lie in a certain interval, the midpoint of that interval can be used to approximate the quantity, and the absolute value of the error in that approximation does not exceed half the length of the interval.)

EXAMPLE 10.5.3 Find the best approximation s_n^* to the sum s of the series $\sum_{n=1}^{\infty}(1/n^2)$ making use of the partial sum s_n of the first n terms. How large should n be taken to ensure that this approximation $s \approx s_n^*$ has error less than 0.001 in absolute value? How large would n be have to be to ensure that the approximation $s \approx s_n$ has error less than 0.001 in absolute value?

SOLUTION Since $f(x) = 1/x^2$ is positive, continuous, and decreasing on $[1, \infty)$ for any $n = 1, 2, 3, \ldots$, we have

$$s_n + A_{n+1} \le s \le s_n + A_n$$

where

$$A_n = \int_n^{\infty} \frac{dx}{x^2} = \lim_{R\to\infty} \left(-\frac{1}{x}\right)\Bigg|_n^{\infty} = \frac{1}{n}.$$

The best approximation to s using s_n is

$$\begin{aligned} s_n^* &= s_n + \frac{1}{2}\left(\frac{1}{n+1} + \frac{1}{n}\right) = s_n + \frac{2n+1}{2n(n+1)} \\ &= 1 + \frac{1}{4} + \frac{1}{9} + \cdots + \frac{1}{n^2} + \frac{2n+1}{2n(n+1)}. \end{aligned}$$

The error in this approximation satisfies

$$|s - s_n^*| \le \frac{1}{2}\left(\frac{1}{n} - \frac{1}{n+1}\right) = \frac{1}{2n(n+1)} \le 0.001$$

provided $2n(n+1) \ge 1/0.001 = 1000$. It is easily checked that this condition is satisfied if $n \ge 22$; the approximation

$$s \approx s_{22}^* = 1 + \frac{1}{4} + \frac{1}{9} + \cdots + \frac{1}{22^2} + \frac{45}{44 \times 23}$$

will have error with absolute value not exceeding 0.001.

Had we used the approximation $s \approx s_n$ we could only have concluded that

$$0 \le s - s_n \le A_n = \frac{1}{n} < 0.001$$

provided $n > 1000$; we would need 1000 terms of the series to get the desired accuracy.

EXAMPLE 10.5.4 Find the sum of the series $\sum_{n=1}^{\infty}(1/n^4)$ with error less than 0.001.

SOLUTION Let $f(x) = 1/x^4$, which is positive, continuous, and decreasing on $[1, \infty)$. Let

$$A_n = \int_n^{\infty} \frac{dx}{x^4} = \lim_{R\to\infty} \left(-\frac{1}{3x^3}\right)\Bigg|_n^{R} = \frac{1}{3n^3}.$$

We use the approximation

$$s \approx s_n^* = s_n + \frac{1}{2}\left(\frac{1}{3(n+1)^3} + \frac{1}{3n^3}\right).$$

The error satisfies

$$|s - s_n^*| \le \frac{1}{2}\left(\frac{1}{3n^3} - \frac{1}{3(n+1)^3}\right) = \frac{1}{6}\,\frac{(n+1)^3 - n^3}{n^3(n+1)^3} = \frac{1}{6}\,\frac{3n^2+3n+1}{n^3(n+1)^3} < \frac{7}{6n^4}.$$

We have used $3n^2+3n+1 \le 7n^2$ and $n^3(n+1)^3 > n^6$ to obtain the last inequality. We will have $|s - s_n^*| < 0.001$ provided

$$\frac{7}{6n^4} < 0.001,$$

that is, if $n^4 > 7000/6$. Since $6^4 = 1296 > 7000/6$, $n = 6$ will do. Thus

$$\begin{aligned}\sum_{n=1}^{\infty} \frac{1}{n^4} &\approx s_6^* = 1 + \frac{1}{2^4} + \frac{1}{3^4} + \frac{1}{4^4} + \frac{1}{5^4} + \frac{1}{6^4} + \frac{1}{6}\left(\frac{1}{7^3} + \frac{1}{6^3}\right)\\ &\approx 1.082 \qquad \text{with error less than 0.001 in absolute value.}\end{aligned}$$

Observe that only six terms are required to enable us to estimate the sum of the series $\sum_{n=1}^{\infty}(1/n^4)$ in Example 10.5.4 within the same tolerance for which we needed 22 terms of the series $\sum_{n=1}^{\infty}(1/n^2)$ in Example 10.5.3. Evidently the series $\sum_{n=1}^{\infty}(1/n^2)$ *converges more slowly* than does $\sum_{n=1}^{\infty}(1/n^4)$. The use of integrals to bound the tails of series is appropriate for those series for which the Integral Test would be appropriate to demonstrate convergence. In particular, this class includes series whose terms are rational functions of n.

Geometric Bounds

Suppose that an inequality of the form

$$0 \le a_k \le K r^k$$

holds for $k = n+1, n+2, n+3, \ldots$, where K and r are constants and $r < 1$. We can then use a geometric series to bound the tail of $\sum_{n=1}^{\infty} a_n$.

$$\begin{aligned}0 \le s - s_n = \sum_{k=n+1}^{\infty} a_k &\le \sum_{k=n+1}^{\infty} K r^k\\ &= K r^{n+1}(1 + r + r^2 + \cdots)\\ &= \frac{K r^{n+1}}{1-r}.\end{aligned}$$

Since $r < 1$ the series converges and the error approaches 0 at an exponential rate as n increases.

EXAMPLE 10.5.5 In Chapter 11 we will show that

$$e = \frac{1}{0!} + \frac{1}{1!} + \frac{1}{2!} + \frac{1}{3!} + \cdots = \sum_{n=0}^{\infty} \frac{1}{n!}$$

(Recall that $0! = 1$.) Estimate the error if the sum s_n of the first n terms of the series is used to approximate e. Find e to three decimal place accuracy using the series.

SOLUTION We have

$$\begin{aligned} s_n &= \frac{1}{0!} + \frac{1}{1!} + \frac{1}{2!} + \frac{1}{3!} + \cdots + \frac{1}{(n-1)!} \\ &= 1 + 1 + \frac{1}{2} + \frac{1}{6} + \frac{1}{24} + \cdots + \frac{1}{(n-1)!}. \end{aligned}$$

(Since the series starts with $n = 0$, then the nth term is $1/(n-1)!$.) We can estimate the error in the approximation $s \approx s_n$ as follows:

$$\begin{aligned} 0 < s - s_n &= \frac{1}{n!} + \frac{1}{(n+1)!} + \frac{1}{(n+2)!} + \frac{1}{(n+3)!} + \cdots \\ &= \frac{1}{n!}\left(1 + \frac{1}{n+1} + \frac{1}{(n+1)(n+2)} + \frac{1}{(n+1)(n+2)(n+3)} + \cdots\right) \\ &< \frac{1}{n!}\left(1 + \frac{1}{n+1} + \frac{1}{(n+1)^2} + \frac{1}{(n+1)^3} + \cdots\right) \end{aligned}$$

since $n + 2 > n + 1$, $n + 3 > n + 1$, and so on. Thus

$$0 < s - s_n < \frac{1}{n!} \frac{1}{1 - \dfrac{1}{n+1}} = \frac{n+1}{n!n}.$$

If we wish to evaluate e accurately to three decimal places, then we would like to ensure that the error is less than 5 in the fourth decimal place, that is, that the error is less than 0.0005. Hence we want

$$\frac{n+1}{n} \frac{1}{n!} < 0.0005 = \frac{1}{2000}.$$

Since $7! = 5040$, but $6! = 720$, we can use $n = 7$ but no smaller. We have

$$\begin{aligned} e &\approx 1 + 1 + \frac{1}{2!} + \frac{1}{3!} + \frac{1}{4!} + \frac{1}{5!} + \frac{1}{6!} \\ &= 2 + \frac{1}{2} + \frac{1}{6} + \frac{1}{24} + \frac{1}{120} + \frac{1}{720} \approx 2.718 \qquad \text{(rounded to three decimal places).} \end{aligned}$$

It is appropriate to use geometric series to bound the tails of positive series whose convergence would be demonstrated by the Ratio Test. Such series converge ultimately faster than any p-series $\sum_{n=1}^{\infty} n^{-p}$, for which the limit ratio is $\rho = 1$.

Alternating Series Bounds

The Alternating Series Test (Theorem 10.4.5) provides a direct estimate for the tail of any series that satisfies its conditions. Suppose that

i) $a_k \geq 0$ for $k = n,\ n+1,\ n+2,\ \ldots,$

ii) $a_{k+1} \leq a_k$ for $k = n,\ n+1,\ n+2,\ \ldots,$ and

iii) $\lim a_k = 0.$

The error $s - s_n$, encountered if we use the sum of the first n terms to approximate the sum of the alternating series $s = \sum_{k=1}^{\infty}(-1)^{k-1}a_k$, does not exceed the $(n+1)$st term in absolute value and is, in fact, of the same sign as that term.

EXAMPLE 10.5.6 How many terms are sufficient to approximate the sum of the alternating harmonic series to within 0.001?

SOLUTION

$$s = 1 - \frac{1}{2} + \frac{1}{3} - \frac{1}{4} + \cdots = s_n + \frac{(-1)^n}{n+1} + \frac{(-1)^{n+1}}{n+2} + \cdots$$

We have $|s - s_n| \leq 1/(n+1) < 0.001$, provided $n + 1 > 1000$. To be sure that the approximation error is less than 0.001, we need at least 1000 terms. (Evidently there is not much future in computing the sum of this series to any great accuracy; it converges very slowly!)

EXAMPLE 10.5.7 Find an approximate value for $\sin 10^\circ = \sin(\pi/18)$ accurate to four decimal places, using the series

$$\sin x = x - \frac{x^3}{3!} + \frac{x^5}{5!} - \cdots = \sum_{k=0}^{\infty}(-1)^k \frac{x^{2k+1}}{(2k+1)!}$$

which will be established in Chapter 11.

SOLUTION

$$\sin 10^\circ = \sin\frac{\pi}{18} = \frac{\pi}{18} - \frac{1}{3!}\left(\frac{\pi}{18}\right)^3 + \frac{1}{5!}\left(\frac{\pi}{18}\right)^5 - \cdots$$

The terms of this alternating series certainly satisfy the three conditions listed above. If we stop with the term $(-1)^{n-1}\dfrac{1}{(2n-1)!}\left(\dfrac{\pi}{18}\right)^{2n-1}$, the error E will satisfy

$$|E| \leq \frac{1}{(2n+1)!}\left(\frac{\pi}{18}\right)^{2n+1}.$$

Now $\dfrac{\pi}{18} < \dfrac{4}{16} = \dfrac{1}{4}$, so

$$|E| < \frac{1}{4^{2n+1}(2n+1)!} < 0.00005 = 1/20,000$$

provided $4^{2n+1}(2n+1)! > 20,000$. This is satisfied even for $n = 2$ ($4^5 5! = 122,880$). Hence

$$\sin 10^\circ \approx \frac{\pi}{18} - \frac{1}{6}\left(\frac{\pi}{18}\right)^3 \approx 0.1736 \qquad \text{(to four decimal places).}$$

EXERCISES

In Exercises 1–6 let s_n denote the sum of the first n terms of the given series. Using s_n, find the smallest interval that you can be sure contains the sum s of the series. If the midpoint s_n^* of this interval is used to approximate s, how large should n be chosen to ensure that the error is less than 0.001?

1. $\displaystyle\sum_{k=1}^{\infty} \frac{1}{k^{10}}$

2. $\displaystyle\sum_{k=1}^{\infty} \frac{1}{k^3}$

3. $\displaystyle\sum_{k=1}^{\infty} \frac{1}{k^{3/2}}$

4. $\displaystyle\sum_{k=2}^{\infty} \frac{1}{k(\ln k)^2}$

5. $\displaystyle\sum_{k=1}^{\infty} \frac{1}{e^k}$

6. $\displaystyle\sum_{k=1}^{\infty} \frac{1}{k^2+4}$

For each positive series in Exercises 7–10 find the best upper bound you can for the error $s - s_n$ encountered if the partial sum s_n is used to approximate the sum s of the series. How many terms of each series do you need to be sure that the approximation has error less than 0.001?

7. $\displaystyle\sum_{k=1}^{\infty} \frac{1}{2^k k!}$

8. $\displaystyle\sum_{n=1}^{\infty} \frac{1}{(2n-1)!}$

9. $\displaystyle\sum_{n=0}^{\infty} \frac{2^n}{(2n)!}$

10. $\displaystyle\sum_{n=1}^{\infty} \frac{1}{n^n}$

For the series in Exercises 11–14 determine how many terms are sufficient to ensure that the partial sum s_n approximates the sum s with error less than 0.001 in absolute value.

11. $\displaystyle\sum_{n=1}^{\infty}(-1)^{n-1}\frac{n}{n^2+1}$

12. $\displaystyle\sum_{n=0}^{\infty} \frac{(-1)^n}{(2n)!}$

13. $\displaystyle\sum_{n=1}^{\infty}(-1)^{n-1}\frac{n}{2^n}$

14. $\displaystyle\sum_{n=0}^{\infty}(-1)^n\frac{3^n}{n!}$

15.*a) Show that if $k > 0$ and n is a positive integer, then $n < \frac{1}{k}(1+k)^n$.
b) Use the estimate in (a) with $0 < k < 1$ to obtain an upper bound for the sum of the series $\sum_{n=0}^{\infty}(n/2^n)$. For what value of k is this bound least?
c) If we use the sum s_n of the first n terms to approximate the sum s of the series in (b), obtain an upper bound for the error $s - s_n$ using the inequality from (a). For given n, find k to minimize this upper bound.

16.* **(Improving the Convergence of a Series)** We know (from Example 10.2.7) that $\sum_{n=0}^{\infty}\frac{1}{n(n+1)} = 1$. Since $\frac{1}{n^2} = \frac{1}{n(n+1)} + c_n$, where $c_n = \frac{1}{n^2(n+1)}$, we have $\sum_{n=1}^{\infty}\frac{1}{n^2} = 1 + \displaystyle\sum_{n=1}^{\infty} c_n$.
Evidently, $\sum c_n$ converges more rapidly than does $\sum(1/n^2)$. Hence fewer terms of that series will be needed to compute $\sum(1/n^2)$ to any desired degree of accuracy than would be needed if we calculated with $\sum(1/n^2)$ directly. Using integral upper and lower bounds, determine a value of n for which the modified partial sum s_n^* for the series $\sum_{n=1}^{\infty} c_n$ approximates the sum of that series with error less than 0.001 in absolute value. Hence determine $\sum_{n=1}^{\infty}(1/n^2)$ to within 0.001 of its true value.
(The technique exibited in this exercise is known as *improving the convergence* of a series. It can be applied to estimating the sum $\sum a_n$ if we know the sum $\sum b_n$ and if $a_n - b_n = c_n$ where $|c_n|$ decreases faster than $|a_n|$ as n tends to infinity.)

CHAPTER 11

Representing Functions by Power Series

The main reason for studying infinite series in an elementary calculus course is that many of the most important elementary functions $f(x)$ have useful representations as series whose terms are constant multiples of nonnegative integer powers of x. Such series are called power series; they generalize polynomials that are sums of finitely many such terms. Among the functions studied in calculus, polynomials are the easiest to manipulate, to calculate with, to differentiate, and to integrate. Power series, behaving somewhat like "polynomials of infinite degree," inherit many of these benign qualities. Therefore, the representation of such transcendental functions as e^x, $\sin x$, $\tan^{-1} x$, and others as power series renders these functions even more useful in the mathematical modeling of concrete problems. Power series can be used to solve easily many kinds of problems that are intractable by other methods.

In this chapter we develop the machinery of power series and establish several power series representations of elementary functions. We begin by reexamining polynomial approximations to functions, a study begun in Sections 5.3 and 5.4. Except for Section 11.2, the material of Sections 10.3–10.5 is not required for understanding the material in this chapter. (Section 10.2 is required.) If you have not studied the material in Sections 10.3 and 10.4 skip Section 11.2.

In addition to power series, there are many other useful series representations of functions, in particular trigonometric series (Fourier series), that are beyond the scope of this book. They are encountered in higher level courses in mathematical analysis and differential equations and, like power series, constitute a basic tool of the mathematician, the engineer, and other mathematically oriented scientists.

11.1 TAYLOR POLYNOMIALS AND TAYLOR'S FORMULA

If the function $f(x)$ has derivatives up to and including order n at the point $x = c$ we can form the polynomial

11.1.1
The Taylor Polynomial

$$P_n(x) = f(c) + f'(c)(x-c) + \frac{f''(c)}{2!}(x-c)^2 + \cdots + \frac{f^{(n)}(c)}{n!}(x-c)^n$$
$$= \sum_{k=0}^{n} \frac{f^{(k)}(c)}{k!}(x-c)^k.$$

$P_n(x)$ is called the **Taylor polynomial of degree** n for $f(x)$ "about $x = c$," or "in powers of $(x - c)$." (If $c = 0$, P_n is called a **Maclaurin polynomial**.) Note that in using the compact summation formula for P_n we are using the definitions $0! =$ and $f^{(0)}(x) = f(x)$.

Among all polynomials of degree at most n, the Taylor polynomial $P_n(x)$ best describes the behavior of $f(x)$ near $x = c$ in the sense that

$$P_n(c) = f(c), \quad P_n'(c) = f'(c), \quad P_n''(c) = f''(c), \quad \ldots, \quad P_n^{(n)}(c) = f^{(n)}(c).$$

You can verify that $P^{(k)}(c) = f^{(k)}(c)$ for $0 \leq k \leq n$ by differentiating the formula for $P_n(x)$ given in Box 11.1.1 k times and then substituting $x = c$.

EXAMPLE 11.1.2 Find the Maclaurin polynomial of degree n for $f(x) = \dfrac{1}{1-x}$.

SOLUTION We have

$$f'(x) = \frac{1}{(1-x)^2}, \quad f''(x) = \frac{2!}{(1-x)^3}, \quad f'''(x) = \frac{3!}{(1-x)^4}, \quad \ldots$$
$$f^{(k)}(x) = \frac{k!}{(1-x)^{k+1}}.$$

Therefore $f^{(k)}(0) = k!$ for $k = 0, 1, 2, \ldots$. The Maclaurin polynomial of degree n for f is

$$P_n(x) = \sum_{k=0}^{n} \frac{k!}{k!} x^k = \sum_{k=0}^{n} x^k = 1 + x + x^2 + x^3 + \cdots + x^n.$$

EXAMPLE 11.1.3 Find the Taylor polynomial of degree n for $f(x) = e^x$ about the point $x = c$.

SOLUTION Since $f^{(k)}(x) = e^x$ for $k = 0, 1, 2, \ldots$, therefore $f^{(k)}(c) = e^c$ and the required Taylor polynomial is

$$P_n(x) = \sum_{k=0}^{n} \frac{e^c}{k!}(x-c)^k = e^c + e^c(x-c) + \frac{e^c}{2!}(x-c)^2 + \cdots + \frac{e^c}{n!}(x-c)^n.$$

EXAMPLE 11.1.4 Find the Maclaurin polynomials of degree 5 and 6 for $f(x) = \sin x$.

SOLUTION We have

$$f'(x) = \cos x, \qquad f''(x) = -\sin x, \qquad f'''(x) = -\cos x,$$
$$f^{(4)}(x) = \sin x, \qquad f^{(5)}(x) = \cos x, \qquad f^{(6)}(x) = -\sin x.$$

Hence $f(0) = f''(0) = f^{(4)}(0) = f^{(6)}(0) = 0$, and $f'(0) = 1$, $f'''(0) = -1$, $f^{(5)}(0) = 1$. Both the Maclaurin polynomials P_5 and P_6 therefore have nonzero terms up to degree 5 only:

$$P_5(x) = P_6(x) = x - \frac{x^3}{3!} + \frac{x^5}{5!} = x - \frac{x^3}{6} + \frac{x^5}{120}.$$

Note that Maclaurin polynomials for the *odd function* $\sin x$ involve only *odd* powers of x. Graphs of the Maclaurin polynomials $P_1(x)$, $P_3(x)$ and $P_5(x)$ for $\sin x$ are shown in Fig. 11.1.1.

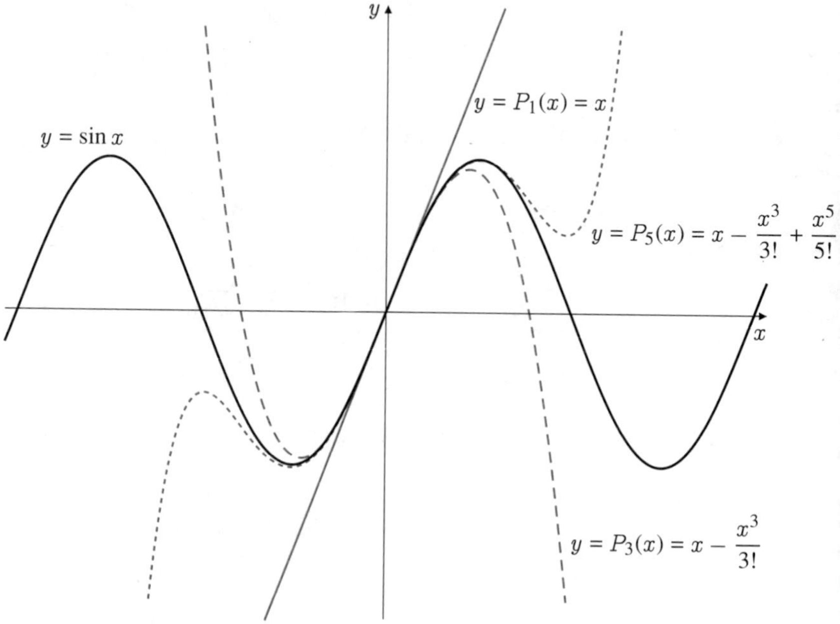

FIGURE 11.1.1

Taylor and Maclaurin polynomials are not always as easy to calculate as those in the above examples. Calculating higher derivatives can become very difficult.

EXAMPLE 11.1.5 Find the fifth degree Maclaurin polynomial for $f(x) = \tan x$.

SOLUTION Since $\tan x$ is an odd function, its Maclaurin polynomials involve only odd powers of x. If $f(x) = \tan x$, we have $f(0) = 0$ and

$$\begin{aligned}
f'(x) &= \sec^2 x & f'(0) &= 1\\
f''(x) &= 2\sec^2 x \tan x & f''(0) &= 0\\
f^{(3)}(x) &= 2\sec^4 x + 4\sec^2 x \tan^2 x & f^{(3)}(0) &= 2\\
f^{(4)}(x) &= 16\sec^4 x \tan x + 8\sec^2 x \tan^3 x & f^{(4)}(0) &= 0\\
f^{(5)}(x) &= 88\sec^4 x \tan^2 x + 16\sec^6 x + 16\sec^2 x \tan^4 x. & f^{(5)}(0) &= 16
\end{aligned}$$

Thus the Maclaurin polynomial of degree 5 is

$$P_5(x) = x + \frac{2}{3!}x^3 + \frac{16}{5!}x^5 = x + \frac{x^3}{3} + \frac{2x^5}{15}.$$

Because the derivatives of the Taylor polynomial $P_n(x)$ up to order n match those of $f(x)$ at $x = c$, $P_n(x)$ is the polynomial of degree n which "best approximates" $f(x)$ near $x = c$. Let

$$R_n(x) = f(x) - P_n(x)$$

denote the *error* in this approximation. The formula

11.1.6
Taylor's Formula

$$f(x) = P_n(x) + R_n(x) = \sum_{k=0}^{n} \frac{f^{(k)}(c)}{k!}(x-c)^k + R_n(x)$$

is known as **Taylor's Formula with Remainder**, or simply just Taylor's Formula. Various versions of **Taylor's Theorem** provide specific expressions for the remainder term, $R_n(x)$, in Taylor's Formula. We have already seen one version of Taylor's Theorem in Section 5.4, namely Theorem 5.4.5 which we restate in our current context as follows.

THEOREM 11.1.7 (**Taylor's Theorem with Lagrange Remainder**) If the $(n+1)$st derivative of f exists on an interval containing c and x, and if $P_n(x)$ is the Taylor polynomial of degree n for f about the point $x = c$, then the remainder $R_n(x) = f(x) - P_n(x)$ in Taylor's Formula is given by

$$R_n(x) = \frac{f^{(n+1)}(X)}{(n+1)!}(x-c)^{n+1}$$

for some X between c and x. □

Observe that the Lagrange form of the remainder, $R_n(x)$, looks just like the term of $(n+1)$st degree term in $P_{n+1}(x)$, except that c in $f^{(n+1)}(c)$ has been replaced by an unknown number X between c and x. As was noted in Section 5.4, the cases $n = 0$ and $n = 1$ of Taylor's Formula with Lagrange Remainder are just the Mean-Value Theorem (Theorem 4.1.1) and the error formula for the tangent line approximation (Theorem 5.3.4) respectively.

EXAMPLE 11.1.8 Estimate the error in the approximation $\tan(0.5) \approx 0.5 + \dfrac{(0.5)^3}{3}$.

SOLUTION As observed in Example 11.1.5 above, the Maclaurin series for $\tan x$ involves only odd powers of x and the Maclaurin polynomials $P_3(x)$ and $P_4(x)$ coincide:

$$P_3(x) = P_4(x) = x + \frac{x^3}{3}.$$

If we use the approximation $\tan(0.5) \approx P_4(0.5) = 0.5 + (0.5)^3/3$, then the error is

$$R_4(0.5) = \frac{f^{(5)}(X)}{5!}(0.5)^5 \qquad \text{for some } X \text{ between 0 and 0.5.}$$

Since $0.5 < \pi/6$, if $0 \le X \le 0.5$ we have

$$0 \le \sec X \le \sec(0.5) < \sec\frac{\pi}{6} = \frac{2}{\sqrt{3}},$$
$$0 \le \tan X \le \tan(0.5) < \tan\frac{\pi}{6} = \frac{1}{\sqrt{3}}.$$

Using the formula for $f^{(5)}(x)$ obtained in Example 11.1.5 we estimate

$$\begin{aligned}|f^{(5)}(X)| &= |88\sec^4 X\ \tan^2 X + 16\sec^6 X + 16\sec^2 X\ \tan^4 X| \\ &\le 88\left(\frac{2}{\sqrt{3}}\right)^4\left(\frac{1}{\sqrt{3}}\right)^2 + 16\left(\frac{2}{\sqrt{3}}\right)^6 + 16\left(\frac{2}{\sqrt{3}}\right)^2\left(\frac{1}{\sqrt{3}}\right)^4 \\ &= \frac{1}{27}(88\times 16 + 16\times 64 + 16\times 4) \approx 92.44 < 93.\end{aligned}$$

Thus

$$|R_4(0.5)| < \frac{93}{5!}(0.5)^5 \approx 0.024$$

and the error in the approximation $\tan(0.5) \approx 0.5 + (0.5)^3/3 \approx 0.542$ is less than 0.025 in absolute value.

EXAMPLE 11.1.9 Use Taylor's Formula to prove the Binomial Theorem: if n is a positive integer then

$$\begin{aligned}(a+x)^n &= a^n + n\,a^{n-1}x + \frac{n(n-1)}{2!}a^{n-2}x^2 + \cdots + n\,ax^{n-1} + x^n \\ &= \sum_{k=0}^{n}\binom{n}{k}a^{n-k}x^k,\end{aligned}$$

where $\displaystyle\binom{n}{k} = \frac{n!}{(n-k)!k!}$.

SOLUTION Let $f(x) = (a+x)^n$. Then

$$\begin{aligned}f'(x) &= n(a+x)^{n-1} = \frac{n!}{(n-1)!}(a+x)^{n-1} \\ f''(x) &= \frac{n!}{(n-1)!}(n-1)(a+x)^{n-2} = \frac{n!}{(n-2)!}(a+x)^{n-2} \\ &\ \vdots \\ f^{(k)}(x) &= \frac{n!}{(n-k)!}(a+x)^{n-k} \qquad (\text{if } 0 \le k \le n)\end{aligned}$$

In particular, $f^{(n)}(x) = \dfrac{n!}{0!}(a+x)^{n-n} = n!$, a constant, and

$$f^{(k)}(x) = 0 \qquad \text{for all } x, \text{ if } k > n.$$

For $0 \le k \le n$ we have $f^{(k)}(0) = \dfrac{n!}{(n-k)!}a^{n-k}$, so Taylor's Formula with Lagrange Remainder gives

$$\begin{aligned}(a+x)^n = f(x) &= \sum_{k=0}^{n}\frac{f^{(k)}(0)}{k!}x^k + \frac{f^{(n+1)}(X)}{(n+1)!}x^{n+1} \\ &= \sum_{k=0}^{n}\frac{n!}{(n-k)!k!}a^{n-k}x^k + 0 = \sum_{k=0}^{n}\binom{n}{k}a^{n-k}x^k.\end{aligned}$$

REMARK: If $f(x) = (a + x)^r$ where $a > 0$ and r is any real number, then calculations similar to those above show that the Maclaurin polynomial of degree n for f is

$$P_n(x) = a^r + \sum_{k=1}^{n} \frac{r(r-1)(r-2)\cdots(r-k+1)}{k!} a^{r-k} x^k.$$

If r is not a positive integer, there will be no n for which the remainder $R_n(x) = f(x) - P_n(x)$ vanishes identically.

Taylor and Maclaurin Series

If a function f has derivatives of all orders, we can write Taylor's formula for any n:

$$f(x) = P_n(x) + R_n(x).$$

If we can show that $\lim_{n\to\infty} R_n(x) = 0$ for all x in some interval I, then we are entitled to conclude, for x in I, that

$$f(x) = \lim_{n\to\infty} P_n(x) = \sum_{k=0}^{\infty} \frac{f^{(k)}(c)}{k!}(x-c)^k, .$$

that is, we will have expressed $f(x)$ as the sum of an infinite series of terms which are multiples of positive integer powers of $(x - c)$, and the series converges for all x in I. Such "infinite degree polynomials" are called **power series** or **Taylor series** (or **Maclaurin series** if $c = 0$.)

EXAMPLE 11.1.10 Find the Maclaurin series for $f(x) = e^x$. Where does it converge to $f(x)$?

SOLUTION Since e^x is positive and increasing, $e^X \le e^{|x|}$ for any $X \le |x|$. Since $f^{(k)}(x) = e^x$ for any k we have, taking $c = 0$ in the Lagrange Remainder in Taylor's Formula,

$$\left|\frac{f^{(n+1)}(X)}{(n+1)!}x^{n+1}\right| \le \frac{e^X}{(n+1)!}|x|^{n+1} \le e^{|x|}\frac{|x|^{n+1}}{(n+1)!} \to 0 \text{ as } n \to \infty$$

for any real x, as shown in Example 10.1.14. Therefore

$$e^x = \sum_{k=0}^{\infty} \frac{x^k}{k!} = 1 + x + \frac{x^2}{2!} + \frac{x^3}{3!} + \cdots,$$

and the series converges to e^x for all real numbers x.

We will investigate Taylor and Maclaurin series further in the remaining sections of this chapter.

Taylor's Theorem with Integral Remainder

The following theorem is another version of Taylor's Theorem where the Remainder in Taylor's Formula is expressed as an integral. This form of the Remainder will be useful when we later establish Binomial series.

THEOREM 11.1.11 (**Taylor's Theorem with Integral Remainder**) If the $n+1$st derivative of f is continuous on an interval containing c and x, and if $P_n(x)$ is the Taylor polynomial of degree n for f about the point $x = c$, then the remainder $R_n(x) = f(x) - P_n(x)$ in Taylor's Formula is given by

$$R_n(x) = \frac{1}{n!} \int_c^x (x-t)^n f^{(n+1)}(t)\,dt.$$

PROOF We start with the Fundamental Theorem of Calculus written in the form

$$f(x) = f(c) + \int_c^x f'(t)\,dt = P_0(x) + R_0(x).$$

Note that the Fundamental Theorem is just the special case $n = 0$ of Taylor's Formula with Integral Remainder. We now apply integration by parts to the integral, setting

$$\begin{aligned} U &= f'(t), & dV &= dt \\ dU &= f''(t)\,dt, & V &= -(x-t). \end{aligned}$$

(Note that we have broken our usual rule about not including a constant of integration with V. In this case we have included the constant $-x$ in V in order to have V vanish when $t = x$.) We have

$$\begin{aligned} f(x) &= f(c) - f'(t)(x-t)\Big|_{t=c}^{t=x} + \int_c^x (x-t) f''(t)\,dt \\ &= f(c) + f'(c)(x-c) + \int_c^x (x-t) f''(t)\,dt \\ &= P_1(x) + R_1(x). \end{aligned}$$

We have thus proved the case $n = 1$ of Taylor's Formula with Integral Remainder.

Let us complete the proof for general n by induction (see Appendix I). Suppose that Taylor's Formula holds with Integral Remainder for some $n = k$:

$$f(x) = P_k(x) + R_k(x) = P_k(x) + \frac{1}{k!} \int_c^x (x-t)^k f^{(k+1)}(t)\,dt.$$

Again we integrate by parts. Let

$$\begin{aligned} U &= f^{(k+1)}(t), & dV &= (x-t)^k\,dt, \\ dU &= f^{(k+2)}(t)\,dt, & V &= \frac{-1}{k+1}(x-t)^{k+1}. \end{aligned}$$

We have

$$\begin{aligned} f(x) &= P_k(x) + \frac{1}{k!}\left(-\frac{f^{(k+1)}(t)(x-t)^{k+1}}{k+1}\Bigg|_{t=c}^{t=x} + \int_c^x \frac{(x-t)^{k+1} f^{(k+2)}(t)}{k+1}\,dt \right) \\ &= P_k(x) + \frac{f^{(k+1)}(c)}{(k+1)!}(x-c)^{k+1} + \frac{1}{(k+1)!}\int_c^x (x-t)^{k+1} f^{(k+2)}(t)\,dt \\ &= P_{k+1}(x) + R_{k+1}(x). \end{aligned}$$

Thus Taylor's formula with Integral Remainder is valid for $n = k+1$ if it is valid for $n = k$. Having been shown to be valid for $n = 0$ (and $n = 1$), it must therefore be valid for every positive integer n. □

EXERCISES

Find the Taylor or Maclaurin polynomials of the specified degrees for the functions in Exercises 1–10.

1. $f(x) = 1/x$ in powers of $x - 2$, degree 4

2. $f(x) = \sqrt{x}$ about $x = 1$, degree 4

3. $f(x) = \cos x$ in powers of x, degree 4

4. $f(x) = \sin x$ in powers of x, degree 5

5. $f(x) = \sin x$ in powers of $x - (\pi/4)$, degree 6

6. $f(x) = e^{-x}$ in powers of x, degree n

7. $f(x) = e^{-x}$ in powers of $x + 2$, degree n

8. $f(x) = \ln x$ about $x = 2$, degree n

9. $f(x) = \sec x$ about $x = 0$, degree 4

10. $f(x) = \tan^{-1} x$ in powers of $x - 1$, degree 3

11. Let $f(x) = x^2 + 2x + 3$. Find

a) the Maclaurin polynomial of degree 2 for $f(x)$,

b) the Maclaurin polynomial of degree 99 for $f(x)$,

c) the Taylor polynomial of degree 2 for $f(x)$ in powers of $x + 2$,

d) the Taylor polynomial of degree 99 for $f(x)$ in powers of $x - 3$.

12. Find the Taylor polynomial of degree 6 for $f(x) = x^6$ about the point of $x = 1$.

13. For what functions $f(x)$ is it true that $f(x)$ is identically equal to its Maclaurin polynomial of degree n?

14. For what functions $f(x)$ is it true that the Taylor polynomial of degree n in powers of $x - a$ vanishes identically?

15. Estimate the error if the Maclaurin polynomial of degree 5 for $\sin x$ is used to approximate $\sin(0.2)$.

16. Estimate the error if the Maclaurin polynomial of degree 6 for $\cos x$ is used to approximate $\cos(1)$.

17. Estimate the error if the Maclaurin polynomial of degree 4 for e^{-x} is used to approximate $e^{-0.5}$.

18. Estimate the error if the Maclaurin polynomial of degree 2 for $\sec x$ is used to approximate $\sec(0.2)$.

19. Estimate the error if the Maclaurin polynomial of degree 3 for $\ln(\cos x)$ is used to approximate $\ln(\cos 0.1)$.

20. Estimate the error if the Taylor polynomial of degree 3 for $\tan^{-1} x$ in powers of $x - 1$ is used to approximate $\tan^{-1} 0.99$.

21. Estimate the error if the Taylor polynomial of degree 4 for $\ln x$ in powers of $x - 2$ is used to approximate $\ln(1.95)$.

Use Taylor's formula to establish the Maclaurin series for the functions in Exercises 22–29.

22. e^{-x}

23. 2^x

24. $\cos x$

25. $\sin x$

26. $\sin^2 x$

27. $\dfrac{1}{1-x}$

28.*$\ln(1 + x)$ (Use Theorem 11.1.11.)

29.*$\dfrac{x}{2+3x}$ (Use Exercise 27.)

Use Taylor's formula to obtain the Taylor series indicated in Exercises 30–35.

30. for e^x in powers of $x - a$

31. for $\sin x$ in powers of $x - (\pi/6)$

32. for $\cos x$ in powers of $x - (\pi/4)$

33.*for $\ln x$ in powers of $x - 1$ (Use Theorem 11.1.11.)

34.*for $\ln x$ in powers of $x - 2$

35. for $1/x$ in powers of $x + 2$ (Use Exercise 27.)

11.2 POWER SERIES

This section is concerned with a special kind of infinite series called a *power series* which may be thought of as polynomials of infinite degree.

The material of Sections 10.3 and 10.4 is necessary for understanding this section. If you have not read Sections 10.3 and 10.4, skip this section and go on to Section 11.3.

11.2.1
Power Series

A series of the form

$$\sum_{n=0}^{\infty} a_n(x-c)^n = a_0 + a_1(x-c) + a_2(x-c)^2 + a_3(x-c)^3 + \cdots$$

is called a **power series in powers of** $x - c$ or a **power series about the point** $x = c$. The constants $a_0, a_1, a_2, \ldots$ are called the **coefficients** of the power series.

The Taylor and Maclaurin series introduced in Section 11.1 are power series. Since the terms of a power series are functions of a variable x, the series may or may not converge for each value of x. For those values of x for which the series does converge, the sum defines a function of x. For example, if $-1 < x < 1$, then

$$1 + x + x^2 + x^3 + \cdots = \frac{1}{1-x}.$$

The geometric series on the left hand side is a power series *representation* of the function $1/(1-x)$ in powers of x (or about the point $x = 0$). Note that the representation is valid only in the open interval $(-1, 1)$ even though $1/(1-x)$ is defined for all real x except $x = 1$. For $x = -1$ and for $|x| > 1$ the series does not converge, so it cannot represent $1/(1-x)$ at these points.

The point c is the **centre of convergence** of the power series $\sum_{n=0}^{\infty} a_n(x-c)^n$. Evidently the series converges (to a_0) at $x = c$. (All the terms except possibly the first are 0.) Theorem 11.2.2 below shows that if the series converges anywhere else, then it converges on an interval (possibly infinite) centred at $x = c$, and it converges absolutely everywhere in that interval except possibly at one or both of the endpoints if the interval is finite. An example of this situation is the series

$$\sum_{n=1}^{\infty} \frac{1}{n\,2^n}(x-5)^n,$$

which we discussed in Example 10.4.7. We showed that the series converges in the interval $[3, 7)$, an interval with centre $x = 5$, and that the convergence was absolute on the interval $(3, 7)$ but only conditional at the endpoint $x = 3$.

THEOREM 11.2.2 Suppose the series $\sum_{n=0}^{\infty} a_n(x_0-c)^n$ converges, where $x_0 \neq c$. If x satisfies $|x-c| < |x_0-c|$, that is, if x is closer to c than x_0 is, then $\sum_{n=0}^{\infty} a_n(x-c)^n$ converges absolutely.

PROOF Since $\sum_{n=0}^{\infty} a_n(x_0-c)^n$ converges, $\lim a_n(x_0-c)^n = 0$, and so $|a_n(x_0-c)^n| \leq K$ for all n, where K is some constant (Theorem 10.1.8). If $r = \dfrac{|x-c|}{|x_0-c|} < 1$, then

$$\sum_{n=0}^{\infty} |a_n(x-c)^n| = \sum_{n=0}^{\infty} |a_n(x_0-c)^n| \left|\frac{x-c}{x_0-c}\right|^n \le K \sum_{n=0}^{\infty} r^n = \frac{K}{1-r} < \infty.$$

Thus $\sum_{n=0}^{\infty} a_n(x-c)^n$ converges absolutely. □

Theorem 11.2.2 shows that the set of values x for which the power series $\sum_{n=0}^{\infty} a_n(x-c)^n$ converges is an interval centred at $x = c$. We call this interval the **interval of convergence** of the power series. It must be of one of the following forms:

i) the isolated point $x = c$,

ii) $[c-R, c+R]$, or $[c-R, c+R)$, or $(c-R, c+R]$, or $(c-R, c+R)$,

iii) the entire line.

The number R in (ii) is called the **radius of convergence** of the power series. In case (i) we say the radius of convergence is 0; in case (iii) it is infinite.

The radius of convergence, R, can often be found by using the ratio test on the power series: the series $\sum a_n(x-c)^n$ converges absolutely if

$$1 > \rho = \lim \left|\frac{a_{n+1}(x-c)^{n+1}}{a_n(x-c)^n}\right| = \lim \left|\frac{a_{n+1}}{a_n}\right| |x-c|,$$

that is, if $|x-c| < \lim \left|\dfrac{a_n}{a_{n+1}}\right| = R.$

11.2.3
Radius of Convergence

> If $R = \lim_{n\to\infty} \left|\dfrac{a_n}{a_{n+1}}\right|$ exists or is ∞, then the power series $\sum_{n=0}^{\infty} a_n(x-c)^n$ has radius of convergence R.

EXAMPLE 11.2.4 Determine the centre, radius, and interval of convergence of

$$\sum_{n=0}^{\infty} \frac{(2x+5)^n}{(n^2+1)3^n}.$$

SOLUTION The series can be rewritten

$$\sum_{n=0}^{\infty} \left(\frac{2}{3}\right)^n \frac{1}{n^2+1} \left(x + \frac{5}{2}\right)^n.$$

The centre of convergence is $x = -5/2$. The radius of convergence, R, is given by

$$R = \lim \left| \frac{\left(\frac{2}{3}\right)^n \frac{1}{n^2+1}}{\left(\frac{2}{3}\right)^{n+1} \frac{1}{(n+1)^2+1}} \right| = \lim \frac{3}{2} \frac{(n+1)^2+1}{n^2+1} = \frac{3}{2}.$$

The series converges absolutely on $(-5/2 - 3/2, -5/2 + 3/2) = (-4, -1)$, and it diverges on $(-\infty, -4)$ and on $(-1, \infty)$. At $x = -1$ the series is $\sum_{n=0}^{\infty} 1/(n^2+1)$; at $x = -4$ it is $\sum_{n=0}^{\infty}(-1)^n/(n^2+1)$. Both series converge (absolutely). The interval of convergence of the given power series is therefore $[-4, -1]$.

EXAMPLE 11.2.5 Determine the radii of convergence of the series (a) $\sum_{n=0}^{\infty} \dfrac{x^n}{n!}$, (b) $\sum_{n=0}^{\infty} n!x^n$.

SOLUTION a) $R = \left| \lim \dfrac{1}{n!} \bigg/ \dfrac{1}{(n+1)!} \right| = \lim \dfrac{(n+1)!}{n!} = \lim n + 1 = \infty.$

This series converges (absolutely) for all x. The sum is e^x as was shown in Example 11.1.10.

b) $R = \lim \dfrac{n!}{(n+1)!} = \lim \dfrac{1}{n+1} = 0.$

This series converges only at $x = 0$.

Algebraic Operations on Power Series

To simplify the following discussion, we will consider only power series with $x = 0$ as centre of convergence, that is, series of the form

$$\sum_{n=0}^{\infty} a_n x^n = a_0 + a_1 x + a_2 x^2 + a_3 x^3 + \cdots.$$

Any properties we demonstrate for such series extend automatically to more general series of the form $\sum_{n=0}^{\infty} a_n(y-c)^n$ via the change of variable $x = y - c$.

First we observe that series having the same centre of convergence can be added or subtracted on whatever interval is common to their intervals of convergence. The following theorems are simple consequences of Theorem 10.2.13 and do not require proofs.

THEOREM 11.2.6 $\sum_{n=0}^{\infty}(a_n \pm b_n)x^n = \sum_{n=0}^{\infty} a_n x^n \pm \sum_{n=0}^{\infty} b_n x^n$. The series on the left side converges whenever both series on the right side converge. If the two series on the right have radii of convergence R_a and R_b, respectively, then the series on the left has radius of convergence $R \geq \min\{R_a, R_b\}$ (it is, at least as large as the smaller of the numbers R_a and R_b). □

THEOREM 11.2.7 If $\sum_{n=0}^{\infty} a_n x^n$ has radius of convergence R, then so does $\sum_{n=0}^{\infty}(ca_n)x^n$ for any nonzero constant c. Moreover,

$$\sum_{n=0}^{\infty}(ca_n)x^n = c\sum_{n=0}^{\infty} a_n x^n. \ \square$$

The situation regarding multiplication and division of power series is more complicated. We will mention only the results and not attempt any proofs of our assertions. You can refer to a textbook in mathematical analysis for more details.

Long multiplication of the form

$$\begin{aligned}(a_0 + a_1x + a_2x^2 + \cdots)&(b_0 + b_1x + b_2x^2 + \cdots)\\ &= a_0b_0 + (a_0b_1 + a_1b_0)x + (a_0b_2 + a_1b_1 + a_2b_0)x^2 + \cdots\end{aligned}$$

leads one to conjecture the formula

$$\left(\sum_{n=0}^{\infty} a_n x^n\right)\left(\sum_{n=0}^{\infty} b_n x^n\right) = \sum_{n=0}^{\infty} c_n x^n,$$

where

$$c_n = a_0b_n + a_1b_{n-1} + \cdots + a_nb_0 = \sum_{j=0}^{n} a_j b_{n-j}.$$

The series $\sum_{n=0}^{\infty} c_n x^n$ is called the **Cauchy product** of the series $\sum_{n=0}^{\infty} a_n x^n$ and $\sum_{n=0}^{\infty} b_n x^n$. The Cauchy product also has radius of convergence at least equal to the lesser of those of the factor series.

EXAMPLE 11.2.8 Since

$$\frac{1}{1-x} = 1 + x + x^2 + x^3 + \cdots = \sum_{n=0}^{\infty} x^n$$

holds for $-1 < x < 1$, we can determine a power series representation for $1/(1-x)^2$ by taking the Cauchy product of this series with itself. Since $a_n = b_n = 1$ for $n = 0, 1, 2, \ldots$ we have

$$c_n = \sum_{j=0}^{n} 1 = n + 1$$

and

$$\frac{1}{(1-x)^2} = 1 + 2x + 3x^2 + 4x^3 + \cdots \sum_{n=0}^{\infty} (n+1)x^n,$$

which must also hold for $-1 < x < 1$. The same series can be obtained by direct long multiplication of the series:

$$\begin{array}{rccccccccc}
 & 1 & + & x & + & x^2 & + & x^3 & + & \cdots \\
\times & 1 & + & x & + & x^2 & + & x^3 & + & \cdots \\
\hline
 & 1 & + & x & + & x^2 & + & x^3 & + & \cdots \\
 & & & x & + & x^2 & + & x^3 & + & \cdots \\
 & & & & & x^2 & + & x^3 & + & \cdots \\
 & & & & & & & x^3 & + & \cdots \\
 & & & & & & & & & \cdots \\
\hline
 & 1 & + & 2x & + & 3x^2 & + & 4x^3 & + & \cdots
\end{array}$$

Long division can also be performed on power series, but there is no simple rule for determining the coefficients of the quotient series. The radius of convergence of the quotient series is not less than the least of the three numbers R_1, R_2, and R_3 where R_1 and R_2 are the radii of convergence of the divisor and dividend series and R_3 is the distance from the centre of convergence to the nearest complex number where the divisor series has sum equal to 0.

Differentiation and Integration of Power Series

If a power series has a positive radius of convergence, it can be differentiated or integrated term by term. The resulting series will converge to the appropriate derivative or integral of the sum of the original series everywhere except possibly at the endpoints of the interval of convergence of the original series. This very important fact ensures that, for purposes of calculation, power series behave just like polynomials, the easiest functions to differentiate and integrate. We formalize the differentiation and integration properties of power series in the following theorem.

THEOREM 11.2.9 If the series $\sum_{n=0}^{\infty} a_n x^n$ converges to the sum $f(x)$ on an interval $(-R, R)$, where $R > 0$; that is,

$$f(x) = \sum_{n=0}^{\infty} a_n x^n = a_0 + a_1 x + a_2 x^2 + a_3 x^3 + \cdots, \qquad (-R < x < R),$$

then f is differentiable on $(-R, R)$ and

$$f'(x) = \sum_{n=1}^{\infty} n a_n x^{n-1} = a_1 + 2a_2 x + 3a_3 x^2 + \cdots, \qquad (-R < x < R).$$

Also, f is integrable over any closed subinterval of $(-R, R)$, and if $|x| \le R$, then

$$\int_0^x f(t)\,dt = \sum_{n=0}^{\infty} \frac{a_n}{n+1} x^{n+1} = a_0 x + \frac{a_1}{2} x^2 + \frac{a_2}{3} x^3 + \cdots.$$

PROOF Let x satisfy $-R < x < R$ and choose $H > 0$ such that $|x| + H < R$. By Theorem 11.2.22, we then have*

$$\sum_{n=1}^{\infty} |a_n|(|x| + H)^n = K < \infty.$$

The Binomial Theorem (Example 11.1.9) shows that if $n \ge 1$ then

$$(x+h)^n = x^n + n x^{n-1} h + \sum_{k=2}^{n} \binom{n}{k} x^{n-k} h^k.$$

Therefore, if $|h| \le H$ we have

$$\begin{aligned}
|(x+h)^n - x^n - n x^{n-1} h| &= \left| \sum_{k=2}^{n} \binom{n}{k} x^{n-k} h^k \right| \\
&\le \sum_{k=2}^{n} \binom{n}{k} |x|^{n-k} \frac{|h|^k}{H^k} H^k \\
&\le \frac{|h|^2}{H^2} \sum_{k=0}^{n} \binom{n}{k} |x|^{n-k} H^k \\
&= \frac{|h|^2}{H^2} (|x| + H)^n.
\end{aligned}$$

* This proof is due to R. Výborný, American Mathematical Monthly, April 1987.

Also

$$|nx^{n-1}| = \frac{n|x|^{n-1}H}{H} \le \frac{1}{H}\big(|x| + H\big)^n.$$

Thus

$$\sum_{n=1}^{\infty} |na_n x^{n-1}| \le \frac{1}{H}\sum_{n=1}^{\infty} |a_n|(|x| + H)^n = \frac{K}{H} < \infty$$

so the series $\sum_{n=1}^{\infty} na_n x^{n-1}$ converges (absolutely) to $g(x)$, say. Now

$$\begin{aligned} \left|\frac{f(x+h) - f(x)}{h} - g(x)\right| &= \left|\sum_{n=1}^{\infty} \frac{a_n(x+h)^n - a_n x^n - na_n x^{n-1}h}{h}\right| \\ &\le \frac{1}{|h|}\sum_{n=1}^{\infty} |a_n||(x+h)^n - x^n - nx^{n-1}h| \\ &\le \frac{|h|}{H^2}\sum_{n=1}^{\infty} |a_n|\big(|x| + H\big)^n \le \frac{K|h|}{H^2}. \end{aligned}$$

Letting h approach zero we obtain $|f'(x) - g(x)| \le 0$, so $f'(x) = g(x)$ as required.

Now observe that since $|a_n/(n+1)| \le |a_n|$, the series

$$h(x) = \sum_{n=0}^{\infty} \frac{a_n}{n+1} x^{n+1}$$

converges (absolutely) at least on the interval $(-R, R)$. Using the differentiation result proved above, we obtain

$$h'(x) = \sum_{n=0}^{\infty} a_n x^n = f(x).$$

Since $h(0) = 0$, we have

$$\int_0^x f(t)\,dt = \int_0^x h'(t)\,dt = h(t)\Big|_0^x = h(x),$$

as required. □

Together these results imply that the termwise differentiated or integrated series have the same radius of convergence as the given series. In fact, as the following examples illustrate, the interval of convergence of the differentiated series is the same as that of the original series except for the *possible* loss of an endpoint if the original series converges at an endpoint of its interval of convergence. Similarly, the integrated series will converge everywhere on the interval of convergence of the original series and possibly at one or both endpoints of that interval even if the original series does not converge at the endpoints.

Being differentiable on $(-R, R)$, where R is the radius of convergence, the sum $f(x)$ of a power series is necessarily continuous on that open interval. If the series happens to converge at either or both of the endpoints $-R$ and R, then f is also continuous (on one side) up to these endpoints. This result is stated formally in the following theorem. We will not prove it here; the interested reader is referred to textbooks on mathematical analysis for a proof.

THEOREM 11.2.10 (**Abel's Theorem**) The sum of a power series is a continuous function everywhere on the interval of convergence of that series. In particular, if $\sum_{n=0}^{\infty} a_n R^n$ converges for some $R > 0$, then

$$\lim_{x \to R-} \sum_{n=0}^{\infty} a_n x^n = \sum_{n=0}^{\infty} a_n R^n,$$

and if $\sum_{n=0}^{\infty} a_n (-R)^n$ converges, then

$$\lim_{x \to -R+} \sum_{n=0}^{\infty} a_n x^n = \sum_{n=0}^{\infty} a_n (-R)^n. \ \square$$

The following Examples show how the above theorems are applied to obtain power series representations for functions. The series obtained in Examples 11.2.11(a) and (c) and 11.2.12 are very important and should be memorized. We remark that these series are just Maclaurin series as defined in the previous section. We will consider this matter further in Section 11.3

EXAMPLE 11.2.11 Use the geometric series

$$\frac{1}{1-x} = \sum_{n=0}^{\infty} x^n = 1 + x + x^2 + x^3 + \cdots \qquad \text{if } -1 < x < 1$$

to obtain power series representations for (a) $\dfrac{1}{(1-x)^2}$, (b) $\dfrac{1}{(1-x)^3}$, (c) $\ln(1+x)$.
Where is each series valid?

SOLUTION a) Differentiating the geometric series term by term according to Theorem 11.2.9, we obtain

$$\frac{1}{(1-x)^2} = \sum_{n=1}^{\infty} n x^{n-1} = 1 + 2x + 3x^2 + 4x^3 + \cdots \qquad \text{if } -1 < x < 1.$$

This is the same result obtained by multiplication of series in Example 11.2.8 above.

b) Differentiating again we get

$$\frac{2}{(1-x)^3} = \sum_{n=2}^{\infty} n(n-1)\, x^{n-2} = (1\times 2)+(2\times 3)x+(3\times 4)x^2+\cdots \quad \text{if } -1 < x < 1.$$

Now divide by 2:

$$\frac{1}{(1-x)^3} = \sum_{n=2}^{\infty} \frac{n(n-1)}{2} x^{n-2} = 1 + 3x + 6x^2 + 10x^3 + \cdots \quad \text{if } -1 < x < 1.$$

c) Substitute $-t$ in place of x in the original geometric series:

$$\frac{1}{1+t} = \sum_{n=0}^{\infty}(-1)^n t^n = 1 - t + t^2 - t^3 + t^4 - \cdots \qquad \text{if } -1 < t < 1.$$

Integrating from 0 to x, where $|x| < 1$, we get

$$\begin{aligned}\ln(1+x) &= \int_0^x \frac{dt}{1+t} = \sum_{n=0}^{\infty}(-1)^n \int_0^x t^n\,dt \\ &= \sum_{n=0}^{\infty}(-1)^n \frac{x^{n+1}}{n+1} = x - \frac{x^2}{2} + \frac{x^3}{3} - \frac{x^4}{4} + \cdots \qquad \text{if } -1 < x \le 1.\end{aligned}$$

Note that the latter series converges (conditionally) at the endpoint $x = 1$ as well as on the interval $-1 < x < 1$. Since $\ln(1+x)$ is continuous at $x = 1$, Theorem 11.2.10 assures us the series must converge to that function at $x = 1$ also. In particular, therefore, the alternating harmonic series converges to $\ln 2$:

$$\ln 2 = 1 - \frac{1}{2} + \frac{1}{3} - \frac{1}{4} + \frac{1}{5} - \cdots = \sum_{n=0}^{\infty} \frac{(-1)^n}{n+1}.$$

This would not, however, be a very useful formula for calculating the value of $\ln 2$. (Why not?)

EXAMPLE 11.2.12 Use the geometric series of the previous example to find a power series representation for $\tan^{-1} x$.

SOLUTION Substitute $-t^2$ for x in the geometric series. Since $0 \le t^2 < 1$ whenever $-1 < t < 1$, we obtain

$$\frac{1}{1+t^2} = 1 - t^2 + t^4 - t^6 + t^8 - \cdots \qquad \text{for } -1 < t < 1.$$

Now integrate from 0 to x, where $|x| < 1$:

$$\begin{aligned}\tan^{-1} x &= \int_0^x \frac{dt}{1+t^2} = \int_0^x (1 - t^2 + t^4 - t^6 + t^8 - \cdots)\,dt \\ &= x - \frac{x^3}{3} + \frac{x^5}{5} - \frac{x^7}{7} + \frac{x^9}{9} - \cdots \\ &= \sum_{n=0}^{\infty}(-1)^n \frac{x^{2n+1}}{2n+1} \qquad \text{if } -1 < x < 1.\end{aligned}$$

However, note that the series also converges (conditionally) at $x = -1$ and 1. Since $\tan^{-1}$ is continuous at ± 1, the above series representation for $\tan^{-1} x$ also holds for these values, by Theorem 11.2.10. Letting $x = 1$ we get an interesting series:

$$\frac{\pi}{4} = 1 - \frac{1}{3} + \frac{1}{5} - \frac{1}{7} + \frac{1}{9} - \cdots.$$

Again, however, this would not be a good formula with which to calculate a numerical value of π. (Why not?)

EXAMPLE 11.2.13 Find the sum of the series $\sum_{n=1}^{\infty} \frac{n^2}{2^n}$ by first finding the sum of the power series

$$\sum_{n=1}^{\infty} n^2 x^n = x + 4x^2 + 9x^3 + 16x^4 + \cdots .$$

SOLUTION Observing (in Example 11.2.11(a)) how the process of differentiating the geometric series produces a series with coefficients 1, 2, 3, ..., we start with the series obtained for $1/(1-x)^2$ and multiply it by x to obtain

$$\sum_{n=1}^{\infty} nx^n = x + 2x^2 + 3x^3 + 4x^4 + \cdots = \frac{x}{(1-x)^2}.$$

Now differentiate again to get a series with coefficients 1^2, 2^2, 3^2, ...

$$\sum_{n=1}^{\infty} n^2 x^{n-1} = 1 + 4x + 9x^2 + 16x^3 + \cdots = \frac{d}{dx}\frac{x}{(x-1)^2} = \frac{1+x}{(1-x)^3}.$$

Multiplication by x again gives the desired power series

$$\sum_{n=1}^{\infty} n^2 x^n = x + 4x^2 + 9x^3 + 16x^4 + \cdots = \frac{x(1+x)}{(1-x)^3}.$$

Differentiation and multiplication by x do not change the radius of convergence, so this series converges to the indicated function for $-1 < x < 1$. Putting $x = 1/2$ we get

$$\sum_{n=1}^{\infty} \frac{n^2}{2^n} = \frac{\frac{1}{2} \times \frac{3}{2}}{\frac{1}{8}} = 6.$$

EXAMPLE 11.2.14 Find a series representation of $f(x) = 1/(2+x)$ in powers of $x - 1$. What is the interval of convergence of this series?

SOLUTION Let $t = x - 1$ so that $x = t + 1$. We have

$$\begin{aligned}
\frac{1}{2+x} = \frac{1}{3+t} &= \frac{1}{3}\,\frac{1}{1+\frac{t}{3}} \\
&= \frac{1}{3}\left(1 - \frac{t}{3} + \frac{t^2}{3^2} - \frac{t^3}{3^3} + \cdots\right) && \text{for } -1 < \frac{t}{3} < 1 \\
&= \sum_{n=0}^{\infty} (-1)^n \frac{t^n}{3^{n+1}} && \text{for } -3 < t < 3 \\
&= \sum_{n=0}^{\infty} (-1)^n \frac{(x-1)^n}{3^{n+1}} && \text{for } -2 < x < 4.
\end{aligned}$$

EXERCISES

Determine the centre, radius, and interval of convergence of the power series in Exercises 1–8.

1. $\displaystyle\sum_{n=0}^{\infty} \frac{x^{2n}}{\sqrt{n+1}}$

2. $\displaystyle\sum_{n=0}^{\infty} 3n\,(x+1)^n$

3. $\displaystyle\sum_{n=1}^{\infty} \frac{1}{n}\left(\frac{x+2}{2}\right)^n$

4. $\displaystyle\sum_{n=1}^{\infty} \frac{(-1)^n}{n^4 2^{2n}}\, x^n$

5. $\displaystyle\sum_{n=0}^{\infty} n^3(2x-3)^n$

6. $\displaystyle\sum_{n=1}^{\infty} \frac{e^n}{n^3}(4-x)^n$

7. $\displaystyle\sum_{n=0}^{\infty} \frac{(1+5^n)}{n!}\, x^n$

8. $\displaystyle\sum_{n=1}^{\infty} \frac{(4x-1)^n}{n^n}$

9. Use multiplication of series to find a power series representation of $1/(1-x)^3$ valid in the interval $(-1, 1)$.

10. Determine the Cauchy product of the series $1 + x + x^2 + x^3 + \cdots$ and $1 - x + x^2 - x^3 + \cdots$. On what interval and to what function does the product series converge?

11. Determine the power series expansion of $1/(1-x)^2$ by formally dividing $1 - 2x + x^2$ into 1.

Starting with the power series representation

$$\frac{1}{1-x} = 1 + x + x^2 + x^3 + \cdots \qquad (-1 < x < 1)$$

determine power series representations for the functions indicated in Exercises 12–20. On what interval is each representation valid?

12. $\dfrac{1}{2-x}$ in powers of x

13. $\dfrac{1}{(2-x)^2}$ in powers of x

14. $\dfrac{1}{1+2x}$ in powers of x

15. $\ln(2-x)$ in powers of x

16. $\dfrac{1}{x}$ in powers of $x-1$

17. $\dfrac{1}{x^2}$ in powers of $x+2$

18. $\dfrac{1-x}{1+x}$ in powers of x

19. $\dfrac{x^3}{1-2x^2}$ in powers of x

20. $\ln x$ in powers of $x-4$

Determine the interval of convergence and the sum of the series in Exercises 21–25

21. $1 - 4x + 16x^2 - 64x^3 + \cdots = \displaystyle\sum_{n=0}^{\infty} (-1)^n(4x)^n$

22.* $3 + 4x + 5x^2 + 6x^3 + \cdots = \displaystyle\sum_{n=0}^{\infty} (n+3)x^n$

23.* $\dfrac{1}{3} + \dfrac{x}{4} + \dfrac{x^2}{5} + \dfrac{x^3}{6} + \cdots = \displaystyle\sum_{n=0}^{\infty} \frac{x^n}{n+3}$

24.* $1 \times 3 - 2 \times 4x + 3 \times 5x^2 - 4 \times 6x^3 + \cdots$
$= \displaystyle\sum_{n=0}^{\infty} (-1)^n(n+1)(n+3)\, x^n$

25.* $2 + 4x^2 + 6x^4 + 8x^6 + 10x^8 + \cdots = \displaystyle\sum_{n=0}^{\infty} 2(n+1)\, x^{2n}$

11.3 TAYLOR AND MACLAURIN SERIES

11.3.1 Taylor and Maclaurin Series

If $f(x)$ has derivatives of all orders at $x = c$, (that is, if $f^{(k)}(c)$ exists for $k = 0, 1, 2, 3, \ldots$), then the series

$$\sum_{k=0}^{\infty} \frac{f^{(k)}(c)}{k!}(x-c)^k$$

$$= f(c) + f'(c)(x-c) + \frac{f''(c)}{2!}(x-c)^2 + \frac{f^{(3)}(c)}{3!}(x-c)^3 + \cdots$$

is called the **Taylor series of f about $x = c$** (or the **Taylor series of f in powers of $x - c$**). If $c = 0$, the term **Maclaurin series** is usually used in place of Taylor series.

Note that the definition of Taylor series makes no requirement that the series should converge anywhere except at the point $x = c$ where the series is just $f(0) + 0 + 0 + \cdots$. The series exists provided all the derivatives of f exist $x = c$; in practice this means that each derivative must exist in an open interval containing $x = c$. (Why?) However, the series may converge nowhere except at $x = c$, and if it does converge elsewhere, it may converge to something other than $f(x)$.

11.3.2 Analytic Functions

If the Taylor series for $f(x)$ in powers of $x - c$ converges to $f(x)$ in an open interval containing $x = c$, then we say that f is **analytic at $x = c$**. If f is analytic at each point of an open interval I, then we say it is analytic on the interval I

The Taylor series is a power series as defined in Section 11.2. Theorem 11.2.2 implies that c must be the centre of any interval on which such a series converges. Most of the elementary functions encountered in calculus are analytic wherever they have derivatives of all orders. However, an example of a function that does not have this property is given in Exercise 38 at the end of the section. Thus not every function having infinitely many derivatives is analytic. On the other hand, whenever a power series converges on an open interval, its sum $f(x)$ is analytic at the centre of the interval, and the given series is the Taylor series of $f(x)$ about that point.

THEOREM 11.3.3 Suppose the series

$$f(x) = \sum_{n=0}^{\infty} a_n(x-c)^n = a_0 + a_1(x-c) + a_2(x-c)^2 + a_3(x-c)^3 + \cdots$$

converges for $c - R < x < c + R$, where $R > 0$. Then

$$a_k = \frac{f^{(k)}(c)}{k!} \qquad \text{for } k = 0, 1, 2, 3, \ldots,$$

so f is analytic at $x = c$. In particular, a given function can have at most one power series representation about any particular point. If

$$f(x) = \sum_{k=0}^{\infty} a_k (x - c)^k = \sum_{k=0}^{\infty} b_k (x - c)^k,$$

then $a_k = b_k$ for $k = 0, 1, 2, \ldots$.

PROOF This proof requires that we differentiate the series for $f(x)$ term by term several times, a process justified by Theorem 11.2.9 (suitably reformulated for powers of $x - c$):

$$f'(x) = \sum_{n=1}^{\infty} n a_n (x - c)^{n-1} = a_1 + 2a_2(x - c) + 3a_3(x - c)^2 + \cdots$$

$$f''(x) = \sum_{n=2}^{\infty} n(n-1) a_n (x - c)^{n-2} = 2a_2 + 6a_3(x - c) + 12a_4(x - c)^2 + \cdots$$

$$\vdots$$

$$\begin{aligned} f^{(k)}(x) &= \sum_{n=k}^{\infty} n(n-1)(n-2)\cdots(n-k+1) a_n (x - c)^{n-k} \\ &= k!a_k + \frac{(k+1)!}{1!} a_{k+1}(x - c) + \frac{(k+2)!}{2!} a_{k+2}(x - c)^2 + \cdots. \end{aligned}$$

Each series converges for $c - R < x < c + R$. Setting $x = c$ we obtain $f^{(k)}(c) = k!a_k$, as required. □

Maclaurin Series for Some Elementary Functions

Calculating Taylor and Maclaurin series for a function f directly from Definition 11.3.1 is practical only when we can find a formula for the nth derivative of f. Examples of such functions include $(ax + b)^r$, e^{ax+b}, $\ln(ax + b)$, $\sin(ax + b)$, $\cos(ax + b)$ and sums of such functions. (See Exercises 22–35 in Section 11.1.) Taylor's Theorem can be used to show that the series converge to the appropriate functions on appropriate intervals, and therefore that the functions are analytic. Theorem 11.3.3 shows that we can use any available means to find a power series converging to a given function on an interval, and the series obtained will turn out to be the Taylor series. In Section 11.2 several series were constructed by manipulating a geometric series. These include

11.3.4
Some Maclaurin Series

$$\frac{1}{1-x} = \sum_{n=0}^{\infty} x^n = 1 + x + x^2 + x^3 + \cdots \qquad (-1 < x < 1)$$

$$\frac{1}{(1-x)^2} = \sum_{n=1}^{\infty} n x^{n-1} = 1 + 2x + 3x^2 + 4x^3 + \cdots \qquad (-1 < x < 1)$$

$$\ln(1+x) = \sum_{n=1}^{\infty} \frac{(-1)^{n-1}}{n} x^n = x - \frac{x^2}{2} + \frac{x^3}{3} - \frac{x^4}{4} + \cdots \qquad (-1 < x \le 1)$$

$$\tan^{-1} x = \sum_{n=0}^{\infty} \frac{(-1)^n}{2n+1} x^{2n+1} = x - \frac{x^3}{3} + \frac{x^5}{5} - \frac{x^7}{7} + \cdots \qquad (-1 \le x \le 1)$$

These series, together with the intervals on which they converge, should be memorized as they are frequently used hereafter.

In Example 11.1.10 we established the Maclaurin series for e^x. Series for e^{-x}, and the hyperbolic functions $\cosh x$ and $\sinh x$ follow from it by substitution ($-x$ in place of x), addition and subtraction of series. We collect these four series in a box; they should also be memorized.

11.3.5
Maclaurin Series for Exponentials

$$e^x = \sum_{n=0}^{\infty} \frac{x^n}{n!} = 1 + x + \frac{x^2}{2!} + \frac{x^3}{3!} + \cdots \qquad \text{(for all real } x\text{)},$$

$$e^{-x} = \sum_{n=0}^{\infty} \frac{(-1)^n}{n!} x^n = 1 - x + \frac{x^2}{2!} - \frac{x^3}{3!} + \frac{x^4}{4!} - \cdots \qquad \text{(for all real } x\text{)},$$

$$\cosh x = \frac{e^x + e^{-x}}{2} = \sum_{n=0}^{\infty} \frac{x^{2n}}{(2n)!} = 1 + \frac{x^2}{2!} + \frac{x^4}{4!} + \cdots \qquad \text{(for all real } x\text{)},$$

$$\sinh x = \frac{e^x - e^{-x}}{2} = \sum_{n=0}^{\infty} \frac{x^{2n+1}}{(2n+1)!} = x + \frac{x^3}{3!} + \frac{x^5}{5!} + \cdots \qquad \text{(for all real } x\text{)}.$$

Taylor's Theorem with Lagrange Remainder can be applied to the functions $\cos x$ and $\sin x$ in a manner similar to that used for e^x in Example 11.1.10 to obtain the series

11.3.6
Maclaurin Series for cos and sin

$$\cos x = \sum_{n=0}^{\infty} \frac{(-1)^n}{(2n)!} x^{2n} = 1 - \frac{x^2}{2!} + \frac{x^4}{4!} - \frac{x^6}{6!} + \cdots \quad \text{(for all real } x\text{)},$$

$$\sin x = \sum_{n=0}^{\infty} \frac{(-1)^n}{(2n+1)!} x^{2n+1} = x - \frac{x^3}{3!} + \frac{x^5}{5!} - \frac{x^7}{7!} + \cdots \quad \text{(for all real } x\text{)}.$$

Instead, however, we will use differentiation of series to obtain them. (Both series converge absolutely for all real x by the Ratio Test.) Let their sums be $f(x)$ and $g(x)$, respectively:

$$f(x) = \sum_{n=0}^{\infty} \frac{(-1)^n}{(2n)!} x^{2n}, \qquad g(x) = \sum_{n=0}^{\infty} \frac{(-1)^n}{(2n+1)!} x^{2n+1}.$$

Differentiating these series we obtain

$$g'(x) = \frac{d}{dx}\left(x - \frac{x^3}{3!} + \frac{x^5}{5!} - \frac{x^7}{7!} + \cdots\right) = 1 - \frac{x^2}{2!} + \frac{x^4}{4!} - \frac{x^6}{6!} + \cdots = f(x)$$

$$g''(x) = f'(x) = \frac{d}{dx}\left(1 - \frac{x^2}{2!} + \frac{x^4}{4!} - \frac{x^6}{6!} + \cdots\right) = -x + \frac{x^3}{3!} - \frac{x^5}{5!} + \cdots = -g(x).$$

Since $g''(x) + g(x) = 0$ for all real x, we conclude that

$$g(x) = A\cos x + B\sin x.$$

(See the discussion of Simple Harmonic Motion in Section 3.2 and also Exercises 49–53 of that section.) But $0 = g(0) = A$ and $1 = f(0) = g'(0) = B$, so $g(x) = \sin x$, as desired. Hence also $f(x) = g'(x) = \cos x$.

REMARK: Observe the similarity between the series for $\sin x$ and $\sinh x$ and between those for $\cos x$ and $\cosh x$. If we were to allow complex numbers (numbers of the form $z = x + iy$ where $i^2 = -1$ and x and y are real) as arguments for our functions, and if we were to demonstrate that our operations on series could be extended to series of complex numbers, we would see that $\cos x = \cosh(ix)$ and $\sin x = -i\sinh(ix)$. In fact,

$$e^{ix} = \cos x + i\sin x, \qquad e^{-ix} = \cos x - i\sin x,$$

so

$$\cos x = \frac{e^{ix} + e^{-ix}}{2}, \qquad \sin x = \frac{e^{ix} - e^{-ix}}{2i}.$$

Such formulas are established in courses (and textbooks) on functions of a complex variable; from the complex point of view the trigonometric and exponential functions are just different manifestations of the same basic function, a complex exponential $e^z = e^{x+iy}$. We content ourselves here with having mentioned the interesting relationships above and invite the reader to verify them formally by calculating with series. (Such formal calculations do not, of course, constitute a proof since we have not established the various rules covering series of complex numbers.)

Other Maclaurin and Taylor Series

We now show how other Maclaurin and Taylor series can be obtained from the ones we already know by various operations. We have already seen some examples above.

EXAMPLE 11.3.7 Obtain Maclaurin series for the following: (a) $\dfrac{\sin(x^2)}{x}$, (b) $\sin^2 x$, (c) $e^{-x^2/3}$.

SOLUTION a) For all real $x \neq 0$ we have

$$\begin{aligned}\frac{\sin x^2}{x} &= \frac{1}{x}\left(x^2 - \frac{(x^2)^3}{3!} + \frac{(x^2)^5}{5!} - \cdots\right)\\ &= x - \frac{x^5}{3!} + \frac{x^9}{5!} - \cdots = \sum_{n=0}^{\infty}(-1)^n \frac{x^{4n+1}}{(2n+1)!}.\end{aligned}$$

Note that $(\sin(x^2))/x$ is not defined as $x = 0$ but does have a limit (0) as x approaches 0. The series converges to $(\sin(x^2))/x$ for all $x \neq 0$, and to that limit at $x = 0$.

b)
$$\begin{aligned}\sin^2 x = \frac{1-\cos 2x}{2} &= \frac{1}{2} - \frac{1}{2}\left(1 - \frac{(2x)^2}{2!} + \frac{(2x)^4}{4!} - \cdots\right)\\ &= \frac{1}{2}\left(\frac{(2x)^2}{2!} - \frac{(2x)^4}{4!} + \frac{(2x)^6}{6!} - \cdots\right)\\ &= \sum_{n=0}^{\infty}(-1)^n \frac{2^{2n+1}}{(2n+2)!}\, x^{2n+2} \qquad \text{(for all real } x\text{)}\end{aligned}$$

c)
$$\begin{aligned}e^{-x^2/3} &= 1 - \frac{x^2}{3} + \frac{1}{2!}\left(\frac{x^2}{3}\right)^2 - \frac{1}{3!}\left(\frac{x^2}{3}\right)^3 + \cdots\\ &= \sum_{n=0}^{\infty}(-1)^n \frac{1}{3^n n!}\, x^{2n} \qquad \text{(for all real } x\text{)}\end{aligned}$$

Sometimes it is quite difficult, if not impossible, to obtain all the terms (that is, the general term) of a Maclaurin or Taylor series. In such cases it is usually possible to obtain the first few terms before the calculations get too cumbersome. Had we attempted to do Example 11.3.7(b) by multiplying the series for $\sin x$ by itself we might have found ourselves in this bind. Other examples occur when it is necessary to substitute one series into another or divide one by another.

EXAMPLE 11.3.8 Obtain the first three nonzero terms of the Maclaurin series for (a) $\ln \cos x$, (b) $\tan x$.

SOLUTION a)

$$
\begin{aligned}
\ln \cos x &= \ln\left(1 + \left(-\frac{x^2}{2!} + \frac{x^4}{4!} - \frac{x^6}{6!} + \cdots\right)\right) \\
&= \left(-\frac{x^2}{2!} + \frac{x^4}{4!} - \frac{x^6}{6!} + \cdots\right) - \frac{1}{2}\left(-\frac{x^2}{2!} + \frac{x^4}{4!} - \frac{x^6}{6!} + \cdots\right)^2 \\
&\quad + \frac{1}{3}\left(-\frac{x^2}{2!} + \frac{x^4}{4!} - \frac{x^6}{6!} + \cdots\right)^3 - \cdots \\
&= -\frac{x^2}{2} + \frac{x^4}{24} - \frac{x^6}{720} + \cdots - \frac{1}{2}\left(\frac{x^4}{4} - \frac{x^6}{24} + \cdots\right) \\
&\quad + \frac{1}{3}\left(-\frac{x^6}{8} + \cdots\right) - \cdots \\
&= -\frac{x^2}{2} - \frac{x^4}{12} - \frac{x^6}{45} - \cdots
\end{aligned}
$$

Note that at each stage of the calculation we kept only enough terms to ensure that we could get all the terms with powers up to x^6. Being an even function, $\ln \cos x$ has only even powers in its Maclaurin series. We cannot find the general term of this series, and only with considerable computational effort can we find many more terms than we have already found. One can also try to calculate terms by using the formula $a_k = f^{(k)}(0)/k!$ but even this becomes difficult after the first few values of k.

b) $\tan x = (\sin x)/(\cos x)$. We can obtain the first three terms of the Maclaurin series for $\tan x$ by long division of the series for $\cos x$ into that for $\sin x$:

$$
\begin{array}{r|l}
 & x \;+\; \frac{x^3}{3} \;+\; \frac{2}{15}x^5 \;+\; \cdots \\
\hline
1 - \frac{x^2}{2} + \frac{x^4}{24} & x \;-\; \frac{x^3}{6} \;+\; \frac{x^5}{120} \;-\; \cdots \\
 & x \;-\; \frac{x^3}{2} \;+\; \frac{x^5}{24} \;-\; \cdots \\
\cline{2-2}
 & \qquad \frac{x^3}{3} \;-\; \frac{x^5}{30} \;+\; \cdots \\
 & \qquad \frac{x^3}{3} \;-\; \frac{x^5}{6} \;+\; \cdots \\
\cline{2-2}
 & \qquad\qquad \frac{2x^5}{15} \;-\; \cdots \\
 & \qquad\qquad \frac{2x^5}{15} \;-\; \cdots \\
\cline{2-2}
\end{array}
$$

Thus $\tan x = x + \frac{1}{3}x^3 + \frac{2}{15}x^5 + \cdots$.

Again we cannot easily find all the terms of the series. This Maclaurin series for $\tan x$ converges for $|x| < \pi/2$, but we cannot demonstrate this fact by the techniques we have at our disposal now. Note that the series for $\tan x$ could also have been derived from that of $\ln\cos x$ obtained in part (a) because we have $\tan x = -\dfrac{d}{dx}\ln\cos x$.

Elementary Maclaurin series can also be used to obtain Taylor series about points other than 0.

EXAMPLE 11.3.9 Find the Taylor series for (a) e^x in powers of $x - c$, (b) $\cos x$ about the point $x = \pi/3$.

SOLUTION a) We have

$$\begin{aligned} e^x = e^{x-c+c} &= e^c\, e^{x-c} \\ &= e^c\left(1 + (x-c) + \frac{(x-c)^2}{2!} + \frac{(x-c)^3}{3!} + \cdots\right) \\ &= \sum_{n=0}^{\infty} \frac{e^c}{n!}(x-c)^n \end{aligned}$$

This representation is valid for all x, and e^x is analytic at every point of the real line.

b) We use the addition formula for cosine:

$$\begin{aligned} \cos x = \cos\left(x - \frac{\pi}{3} + \frac{\pi}{3}\right) &= \cos\left(x - \frac{\pi}{3}\right)\cos\frac{\pi}{3} - \sin\left(x - \frac{\pi}{3}\right)\sin\frac{\pi}{3} \\ &= \frac{1}{2}\left[1 - \frac{1}{2!}\left(x - \frac{\pi}{3}\right)^2 + \frac{1}{4!}\left(x - \frac{\pi}{3}\right)^4 - \cdots\right] \\ &\quad - \frac{\sqrt{3}}{2}\left[\left(x - \frac{\pi}{3}\right) - \frac{1}{3!}\left(x - \frac{\pi}{3}\right)^3 + \cdots\right] \\ &= \frac{1}{2} - \frac{\sqrt{3}}{2}\left(x - \frac{\pi}{3}\right) - \frac{1}{2}\frac{1}{2!}\left(x - \frac{\pi}{3}\right)^2 + \frac{\sqrt{3}}{2}\frac{1}{3!}\left(x - \frac{\pi}{3}\right)^3 \\ &\quad + \frac{1}{2}\frac{1}{4!}\left(x - \frac{\pi}{3}\right)^4 - \cdots. \end{aligned}$$

This series representation is valid for all x. A similar calculation would enable us to expand $\cos x$ or $\sin x$ in powers of $x - c$ for any real c; both functions are analytic at every point of the real line.

The Binomial Series

If $f(x) = (1 + x)^r$, where r is any real number and $x > -1$, then the kth derivative of f is

$$f^{(k)}(x) = r(r-1)(r-2)\cdots(r-k+1)(1+x)^{r-k}, \qquad k = 1, 2, \ldots.$$

We write Taylor's Formula with Integral Remainder for f with $c = 0$ (see Theorem 11.1.11):

$$(1+x)^r = 1 + \sum_{k=1}^{n} \frac{r(r-1)(r-2)\cdots(r-k+1)}{k!} x^k + R_n(x)$$

where

$$R_n(x) = \frac{1}{n!}\int_0^x (x-t)^n f^{(n+1)}(t)\,dt = \frac{r(r-1)(r-2)\cdots(r-n)}{n!}\int_0^x \frac{(x-t)^n}{(1+t)^{n+1-r}}\,dt.$$

We are going to show that $\lim_{n\to\infty} R_n(x) = 0$ for each x in the interval $(-1, 1)$. Thus we will have verified the Binomial Series

11.3.10
Binomial Series

$$(1+x)^r = 1 + rx + \frac{r(r-1)}{2!}x^2 + \frac{r(r-1)(r-2)}{3!}x^3 + \cdots$$
$$= 1 + \sum_{n=1}^{\infty} \frac{r(r-1)(r-2)\cdots(r-n+1)}{n!} x^n \quad (-1 < x < 1).$$

The proof is rather complicated, and involves concepts from Section 10.3, so proceed with caution!

!!DANGER!!
Difficult proof

Select an x such that $-1 < x < 1$. Since the variable of integration, t, in the formula for $R_n(x)$ lies between 0 and x, we have

$$\begin{cases} 1 \le 1+t \le 1+x < 2 & \text{if } 0 \le x < 1 \\ 0 < 1+x \le 1+t \le 1 & \text{if } -1 < x \le 0. \end{cases}$$

In either case we can find a constant K, depending on x and r, but not on n, such that

$$|1+t|^{r-1} \le K \qquad \text{for all } t \text{ between 0 and } x.$$

Also, if $0 \le x < 1$, then

$$\left|\frac{x-t}{1+t}\right| = \frac{x-t}{1+t} \le x - t \le x = |x|.$$

If $-1 < x \le 0$, then

$$\left|\frac{x-t}{1+t}\right| = \frac{t-x}{1+t} \le |x|$$

because $(t - x)/(1 + t)$ increases from 0 to $-x = |x|$ as t increases from x to 0. Thus we have

$$\begin{aligned}|R_n(x)| &= \left|\frac{r(r-1)(r-2)\cdots(r-n)}{n!}\right| \left|\int_0^x (1+t)^{r-1}\left(\frac{x-t}{1+t}\right)^n dt\right| \\ &\leq K\left|\frac{r(r-1)(r-2)\cdots(r-n)}{n!}\right| |x|^n \left|\int_0^x dt\right| \\ &= K\left|\frac{r(r-1)(r-2)\cdots(r-n)}{n!}\right| |x|^{n+1}.\end{aligned}$$

Now apply the ratio test to the series $\sum_{n=1}^{\infty} \dfrac{r(r-1)(r-2)\cdots(r-n)}{n!} x^{n+1}$:

$$\begin{aligned}\rho &= \lim_{n\to\infty} \left|\frac{\dfrac{r(r-1)(r-2)\cdots(r-n)(r-n-1)}{(n+1)!}x^{n+2}}{\dfrac{r(r-1)(r-2)\cdots(r-n)}{n!}x^{n+1}}\right| \\ &= \lim_{n\to\infty}\left|\frac{r-n-1}{n+1}\right| |x| = |x| < 1.\end{aligned}$$

Thus the series converges and so its nth term must approach 0:

$$\lim_{n\to\infty}\left|\frac{r(r-1)(r-2)\cdots(r-n)}{n!}\right| |x|^{n+1} = 0.$$

Therefore $\lim_{n\to\infty} R_n(x) = 0$, and the Binomial Series representation holds at x as claimed.

EXAMPLE 11.3.11 Find the Maclaurin series for $\dfrac{1}{\sqrt{1+x}}$.

SOLUTION Here $r = -(1/2)$:

$$\begin{aligned}\frac{1}{\sqrt{1+x}} &= (1+x)^{-1/2} \\ &= 1 - \frac{1}{2}x + \frac{1}{2!}\left(-\frac{1}{2}\right)\left(-\frac{3}{2}\right)x^2 + \frac{1}{3!}\left(-\frac{1}{2}\right)\left(-\frac{3}{2}\right)\left(-\frac{5}{2}\right)x^3 + \cdots \\ &= 1 - \frac{1}{2}x + \frac{1\times 3}{2^2 2!}x^2 - \frac{1\times 3\times 5}{2^3 3!}x^3 + \cdots \\ &= 1 + \sum_{n=1}^{\infty}(-1)^n \frac{1\times 3\times 5\times\cdots\times(2n-1)}{2^n n!}x^n.\end{aligned}$$

This series converges for $-1 < x \leq 1$. (Use the Alternating Series Test to get the endpoint $x = 1$.)

EXAMPLE 11.3.12 Find the Maclaurin series for $\sin^{-1} x$.

SOLUTION Replace x with $-x^2$ in the series obtained in the above example to get

$$\frac{1}{\sqrt{1-x^2}} = 1 + \sum_{n=1}^{\infty} \frac{1 \times 3 \times 5 \times \cdots \times (2n-1)}{2^n n!} x^{2n} \qquad (-1 < x < 1).$$

$$\begin{aligned}
\sin^{-1} x &= \int_0^x \frac{dt}{\sqrt{1-t^2}} = \int_0^x \left(1 + \sum_{n=1}^{\infty} \frac{1 \times 3 \times 5 \times \cdots \times (2n-1)}{2^n n!} t^{2n}\right) dt \\
&= x + \sum_{n=1}^{\infty} \frac{1 \times 3 \times 5 \times \cdots \times (2n-1)}{2^n n!(2n+1)} x^{2n+1} \\
&= x + \frac{x^3}{6} + \frac{3}{40} x^5 + \cdots \qquad (-1 < x < 1).
\end{aligned}$$

EXERCISES

Find Maclaurin series representations for the functions in Exercises 1–16. For what values of x is each representation valid?

1. e^{3x+1}

2. $\cos(2x^3)$

3. $\sin(x - \pi/4)$

4. $\cos(2x - \pi)$

5. $x^2 \sin(x/3)$

6. $\cos^2(x/2)$

7. $\sin x \cos x$

8. $\tan^{-1}(5x^2)$

9. $\dfrac{1+x^3}{1+x^2}$

10. $\ln(2 + x^2)$

11. $\ln \dfrac{1-x}{1+x}$

12. $(e^{2x^2} - 1)/x^2$

13. $\cosh x - \cos x$

14. $\sqrt{1-x}$

15. $\sqrt{4+x}$

16. $\dfrac{1}{\sqrt{4+x^2}}$

Find the required Taylor series in Exercises 17–24. Where is each series representation valid?

17. for $f(x) = e^{-2x}$ about the point $x = -1$

18. for $f(x) = \sin x$ about the point $x = \pi/2$

19. for $f(x) = \cos x$ in powers of $x - \pi$

20. for $f(x) = \ln x$ in powers of $x - 3$

21. for $f(x) = \sin x - \cos x$ about $x = \dfrac{\pi}{4}$

22. for $f(x) = \cos^2 x$ about $x = \dfrac{\pi}{8}$

23. for $f(x) = 1/x^2$ in powers of $x + 2$

24. for $f(x) = \dfrac{x}{1+x}$ in powers of $x - 1$

Find the first three nonzero terms in the Maclaurin series for the functions in Exercises 25–28.

25. $\sec x$

26. $\sec x \tan x$

27. $\tan^{-1}(e^x - 1)$

28. $e^{\tan^{-1} x} - 1$

29.*Use the fact that $(\sqrt{1+x})^2 = 1 + x$ to find the first three nonzero terms of the Maclaurin series for $\sqrt{1+x}$.

30. Does $\csc x$ have a Maclaurin series? Why? Find the first three nonzero terms of the Taylor series for $\csc x$ about the point $x = \pi/2$.

Find the sums of the series in Exercises 31–34.

31. $1 + x^2 + \dfrac{x^4}{2!} + \dfrac{x^6}{3!} + \dfrac{x^8}{4!} + \cdots$

32.*$x^3 - \dfrac{x^9}{3! \times 4} + \dfrac{x^{15}}{5! \times 16} - \dfrac{x^{21}}{7! \times 64} + \dfrac{x^{27}}{9! \times 256} - \cdots$

33. $1 + \dfrac{x^2}{3!} + \dfrac{x^4}{5!} + \dfrac{x^6}{7!} + \dfrac{x^8}{9!} + \cdots$

34.*$1 + \dfrac{1}{2 \times 2!} + \dfrac{1}{4 \times 3!} + \dfrac{1}{8 \times 4!} + \cdots$

35. Let $P(x) = 1 + x + x^2$. Find (a) the Maclaurin series for $P(x)$, (b) the Taylor series for $P(x)$ about $x = 1$.

36.*Verify for direct calculation that $f(x) = 1/x$ is analytic at $x = a$ for every $a \neq 0$.

37.*Verify by direct calculation that $\ln x$ is analytic at $x = a$ for every $a > 0$.

38. *Review Exercise 49 of Section 4.3. It shows that the function

$$f(x) = \begin{cases} e^{-1/x^2} & \text{if } x \neq 0 \\ 0 & \text{if } x = 0 \end{cases}$$

has derivatives of all orders at every point of the real line, and $f^{(k)}(0) = 0$ for every positive integer k. What is the Maclaurin series for $f(x)$? What is the interval of convergence of this Maclaurin series? On what interval does the series converge to $f(x)$? Is f analytic at $x = 0$?

39. *By direct multiplication of the Maclaurin series for e^x and e^y show that $e^x e^y = e^{x+y}$.

11.4 APPLICATIONS OF TAYLOR AND MACLAURIN SERIES

Approximating the Values of Functions

Partial sums of Taylor and Maclaurin series can be used as polynomial approximations to more complicated functions. Example 5.4.6 was an example of this. In that example the Lagrange Remainder in Taylor's Formula was used to determine how many terms of the Maclaurin series for e^x were needed to calculate $e^1 = e$ correct to 3 decimal places. In the following examples we use other methods (originally described in Section 10.5) to estimate the error when a partial sum is used to approximate an infinite series.

EXAMPLE 11.4.1 Use Maclaurin series to find $\sqrt{e}$ correct to four decimal places, that is, with error less than 0.00005 in absolute value.

SOLUTION

$$\begin{aligned}\sqrt{e} = e^{1/2} &= 1 + \frac{1}{2} + \frac{1}{2!}\frac{1}{2^2} + \frac{1}{3!}\frac{1}{2^3} + \cdots \\ &\approx 1 + \frac{1}{2} + \frac{1}{2!}\frac{1}{2^2} + \frac{1}{3!}\frac{1}{2^3} + \cdots + \frac{1}{(n-1)!}\frac{1}{2^{n-1}} = s_n\end{aligned}$$

The error in this approximation can be bounded by a geometric series if we use $n+1 < n+2 < n+3 < \cdots$.

$$\begin{aligned}0 &< \sqrt{e} - s_n \\ &= \frac{1}{n!}\frac{1}{2^n} + \frac{1}{(n+1)!}\frac{1}{2^{n+1}} + \frac{1}{(n+2)!}\frac{1}{2^{n+2}} + \cdots \\ &= \frac{1}{n!}\frac{1}{2^n}\left[1 + \frac{1}{2(n+1)} + \frac{1}{2^2(n+1)(n+2)} + \frac{1}{2^3(n+1)(n+2)(n+3)} + \cdots\right] \\ &< \frac{1}{n!}\frac{1}{2^n}\left[1 + \frac{1}{2(n+1)} + \left(\frac{1}{2(n+1)}\right)^2 + \left(\frac{1}{2(n+1)}\right)^3 + \cdots\right] \\ &= \frac{1}{n!}\frac{1}{2^n}\frac{1}{1 - \dfrac{1}{2(n+1)}} = \frac{2(n+1)}{(2n+1)n!\,2^n}.\end{aligned}$$

We want this error less than $0.00005 = 1/20,000$. Thus we want

$$\frac{2^n n!(2n+1)}{2n+2} > 20,000.$$

We can get by with $n = 6$; $2^6 6!(13)/14 \approx 42,789$. Thus

$$\sqrt{e} \approx 1 + \frac{1}{2} + \frac{1}{8} + \frac{1}{48} + \frac{1}{384} + \frac{1}{3840} \approx 1.6487,$$

rounded to four decimal places.

When the terms a_n of a series (i) alternate in sign, (ii) decrease steadily in size, and (iii) approach zero as $n \to \infty$ then the error involved in using a partial sum of the series as an approximation to the sum of the series has the same sign as, and is smaller in absolute value than the first omitted term. This was proved in connection with the Alternating Series Test (Theorem 10.4.5) and provides an easy estimate for the error when the three conditions (i)–(iii) are satisfied.

EXAMPLE 11.4.2 Find $\cos 43°$ with error less than $1/10,000$.

SOLUTION We give two alternative solutions:

i) We can use the Maclaurin series:

$$\cos 43° = \cos\frac{43\pi}{180} = 1 - \frac{1}{2!}\left(\frac{43\pi}{180}\right)^2 + \frac{1}{4!}\left(\frac{43\pi}{180}\right)^4 - \cdots.$$

Now $43\pi/180 \approx 0.75049\cdots < 1$, so the series above evidently satisfies the conditions (i)–(iii) mentioned above. If we truncate the series after the term

$$(-1)^{n-1}\frac{1}{(2n-2)!}\left(\frac{43\pi}{180}\right)^{2n-2},$$

then the error E will satisfy

$$|E| \le \frac{1}{(2n)!}\left(\frac{43\pi}{180}\right)^{2n} < \frac{1}{(2n)!}.$$

The error will not exceed $1/10,000$ if $(2n)! > 10,000$, so $n = 4$ will do; $(8! = 40,320)$.

$$\cos 43° \approx 1 - \frac{1}{2!}\left(\frac{43\pi}{180}\right)^2 + \frac{1}{4!}\left(\frac{43\pi}{180}\right)^4 - \frac{1}{6!}\left(\frac{43\pi}{180}\right)^6 \approx 0.73135\cdots$$

ii) Since $43°$ is close to $45°$ we can do a bit better by using the Taylor series about $x = \pi/4$ instead of the Maclaurin series.

$$\begin{aligned}&\cos 43°\\ &= \cos\left(\frac{\pi}{4} - \frac{\pi}{90}\right) = \cos\frac{\pi}{4}\cos\frac{\pi}{90} + \sin\frac{\pi}{4}\sin\frac{\pi}{90}\\ &= \frac{1}{\sqrt{2}}\left(\left(1 - \frac{1}{2!}\left(\frac{\pi}{90}\right)^2 + \frac{1}{4!}\left(\frac{\pi}{90}\right)^4 - \cdots\right) + \left(\frac{\pi}{90} - \frac{1}{3!}\left(\frac{\pi}{90}\right)^3 + \cdots\right)\right)\end{aligned}$$

Since

$$\frac{1}{4!}\left(\frac{\pi}{90}\right)^4 < \frac{1}{3!}\left(\frac{\pi}{90}\right)^3 < \frac{1}{20000},$$

we need only the first two terms of the first series and the first term of the second series:

$$\cos 43^\circ \approx \frac{1}{\sqrt{2}}\left(1 + \frac{\pi}{90} - \frac{1}{2}\left(\frac{\pi}{90}\right)^2\right) \approx 0.731358\cdots.$$

(In fact, $\cos 43^\circ = 0.7313537\cdots$.)

When finding approximate values of functions, it is best, whenever possible, to use a power series about a point as close as possible to the point where the approximation is desired.

Functions Defined by Integrals

As we saw in Chapter 6, many functions expressible as simple combinations of elementary functions cannot be antidifferentiated by elementary techniques; their antiderivatives are not simple combinations of elementary functions. We can, however, often find the Taylor series for the antiderivatives of such functions and hence approximate their definite integrals.

EXAMPLE 11.4.3 Find the Taylor series for

$$E(x) = \int_0^x e^{-t^2}\,dt$$

and use it to evaluate $E(1)$ correct to three decimal places.

SOLUTION We have, for all real x,

$$\begin{aligned} E(x) &= \int_0^x \left(1 - t^2 + \frac{t^4}{2!} - \frac{t^6}{3!} + \frac{t^8}{4!} - \cdots\right) dt \\ &= \left(t - \frac{t^3}{3} + \frac{t^5}{5\times 2!} - \frac{t^7}{7\times 3!} + \frac{t^9}{9\times 4!} - \cdots\right)\Bigg|_0^x \\ &= x - \frac{x^3}{3} + \frac{x^5}{5\times 2!} - \frac{x^7}{7\times 3!} + \frac{x^9}{9\times 4!} - \cdots = \sum_{n=0}^{n}(-1)^n\frac{x^{2n+1}}{(2n+1)n!}. \end{aligned}$$

Now we have

$$\begin{aligned} E(1) &= 1 - \frac{1}{3} + \frac{1}{5\times 2!} - \frac{1}{7\times 3!} + \cdots \\ &\approx 1 - \frac{1}{3} + \frac{1}{5\times 2!} - \frac{1}{7\times 3!} + \cdots + \frac{(-1)^n}{(2n+1)n!}. \end{aligned}$$

The error in this approximation does not exceed the first omitted term, so it will be less than 0.0005 provided $(2n+3)(n+1)! > 2000$. Since $13\times 6! = 9360$, $n = 5$ will do. Thus

$$E(1) \approx 1 - \frac{1}{3} + \frac{1}{10} - \frac{1}{42} + \frac{1}{216} - \frac{1}{1320} \approx 0.747$$

rounded to three decimal places.

Indeterminate Forms

Maclaurin and Taylor series can provide a useful alternative to l'Hôpital's rules for evaluating limits of indeterminate forms.

EXAMPLE 11.4.4 Evaluate (a) $\lim_{x\to 0} \frac{x - \sin x}{x^3}$, and (b) $\lim_{x\to 0} \frac{(e^{2x}-1)\ln(1+x^3)}{(1-\cos 3x)^2}$

SOLUTION a)

$$\begin{aligned}
&\lim_{x\to 0} \frac{x - \sin x}{x^3} \qquad \left[\frac{0}{0}\right] \\
&= \lim_{x\to 0} \frac{x - \left(x - \frac{x^3}{3!} + \frac{x^5}{5!} - \cdots\right)}{x^3} \\
&= \lim_{x\to 0} \frac{\frac{x^3}{3!} - \frac{x^5}{5!} + \cdots}{x^3} \\
&= \lim_{x\to 0} \left(\frac{1}{3!} - \frac{x^2}{5!} + \cdots\right) = \frac{1}{3!} = \frac{1}{6}.
\end{aligned}$$

b)

$$\begin{aligned}
&\lim_{x\to 0} \frac{(e^{2x}-1)\ln(1+x^3)}{(1-\cos 3x)^2} \qquad \left[\frac{0}{0}\right] \\
&= \lim_{x\to 0} \frac{\left(1 + (2x) + \frac{(2x)^2}{2!} + \frac{(2x)^3}{3!} + \cdots - 1\right)\left(x^3 - \frac{x^6}{2} + \cdots\right)}{\left(1 - \left(1 - \frac{(3x)^2}{2!} + \frac{(3x)^4}{4!} - \cdots\right)\right)^2} \\
&= \lim_{x\to 0} \frac{2x^4 + 2x^5 + \cdots}{\left(\frac{9}{2}x^2 - \frac{3^4}{4!}x^4 + \cdots\right)^2} \\
&= \lim_{x\to 0} \frac{2 + 2x + \cdots}{\left(\frac{9}{2} - \frac{3^4}{4!}x^2 + \cdots\right)^2} = \frac{2}{\left(\frac{9}{2}\right)^2} = \frac{8}{81}.
\end{aligned}$$

The student can check that the latter of these examples will require much more laborious calculation if attempted using l'Hôpital's rule.

EXERCISES

Use Maclaurin or Taylor series to calculate the function values indicated in Exercises 1–12 with error less than 5×10^{-5} in absolute value.

1. $e^{0.2}$

2. $1/e$

3. $e^{1.2}$

4. $\sin(0.1)$

5. $\cos 5°$

6. $\ln(6/5)$

7. $\ln(0.9)$

8. $\sin 80°$

9. $\cos 65°$

10. $\tan^{-1} 0.2$

11. $\cosh(1)$

12. $\ln(3/2)$

Find Maclaurin series for the functions in Exercises 13–17.

13. $I(x) = \displaystyle\int_0^x \frac{\sin t}{t}\, dt$

14. $J(x) = \displaystyle\int_0^x \frac{e^t - 1}{t}\, dt$

15. $K(x) = \displaystyle\int_1^{1+x} \frac{\ln t}{t-1}\, dt$

16. $L(x) = \displaystyle\int_0^x \cos(t^2)\, dt$

17. $M(x) = \displaystyle\int_0^x \frac{\tan^{-1} t^2}{t^2}\, dt$

18. Find $L(0.5)$ correct to three decimal places, with L defined as in Exercise 16.

19. Find $I(1)$ correct to three decimal places, with I defined as in Exercise 13.

Evaluate the limits in Exercises 20–25

20. $\displaystyle\lim_{x\to 0} \frac{\sin(x^2)}{\sinh x}$

21. $\displaystyle\lim_{x\to 0} \frac{1 - \cos(x^2)}{(1 - \cos x)^2}$

22. $\displaystyle\lim_{x\to 0} \frac{(e^x - 1 - x)^2}{x^2 - \ln(1 + x^2)}$

23. $\displaystyle\lim_{x\to 0} \frac{2\sin 3x - 3\sin 2x}{5x - \tan^{-1} 5x}$

24. $\displaystyle\lim_{x\to 0} \frac{\sin(\sin x) - x}{x(\cos(\sin x) - 1)}$

25. $\displaystyle\lim_{x\to 0} \frac{\sinh x - \sin x}{\cosh x - \cos x}$

26.†Series can provide a useful tool for solving differential equations. Substitute the series $y = \sum_{n=0}^{\infty} a_n x^n$ into the differential equation

$$y'' + xy' + y = 0$$

and thereby determine a relationship between a_{n+2} and a_n that enables all the coefficients in the series to be determined in terms of a_0 and a_1. Solve the initial-value problem

$$\begin{cases} y'' + xy' + y = 0 \\ y(0) = 1 \\ y'(0) = 0 \end{cases}$$

27.†Solve the initial-value problem

$$\begin{cases} y'' + xy' + 2y = 0 \\ y(0) = 1 \\ y'(0) = 2 \end{cases}$$

by the method suggested in the previous exercise.

CHAPTER 12

Partial Differentiation

It may seem odd to have a chapter on the differentiation of functions of several variables in a book called "Single-Variable Calculus." A complete course in real-variable calculus is usually spread over three or four semesters with the first two concerned mainly with functions of a single variable and the remainder devoted to differentiation and integration of functions of two or more variables. Nevertheless, it sometimes happens that a short introduction to the differentiation of functions of several variables is included towards the end of the second semester of a single-variable course. This chapter is intended for such situations.

12.1 FUNCTIONS OF SEVERAL VARIABLES

We have been using the notation $y = f(x)$ to denote that the variable y is a function of the single real variable x. The domain of such a function f is a set of real numbers. Many quantities can be regarded as depending on more than one real variable, and thus to be functions of more than one variable. For example, the volume of a circular cylinder of radius r and height h is given by $V = \pi r^2 h$; that is, V is a function of the two variables r and h. If we choose to denote this function by f we could write $V = f(r, h)$ where

$$f(r, h) = \pi r^2 h, \qquad (r > 0,\ h > 0).$$

Thus f is a function of two variables having domain the set of points in the rh-plane with coordinates (r, h) satisfying $r > 0$ and $h > 0$. Similarly, the relationship $w = f(x, y, z) = x + 2y - 3z$ defines w as a function of the three variables x, y, and z, with domain the three dimensional space of coordinates (x, y, z), or, if we state explicitly, some particular subset of that space.

By analogy with Definition 1.3.1 we define a function of n variables as follows:

12.1.1
Functions of n Variables

> A **function** f of n real variables is a rule which assigns to each n-tuple of real numbers $(x_1, x_2, \ldots, x_n)$ in some set $\mathcal{D}(f)$ of the n-dimensional space of coordinates $(x_1, x_2, \ldots, x_n)$, a **unique** real number $f(x_1, x_2, \ldots, x_n)$. The set $\mathcal{D}(f)$ is called the **domain** of f; the set of all real numbers $f(x_1, x_2, \ldots, x_n)$ obtained from points in the domain is called the **range** of f.

As for functions of one variable, it is conventional to take the domain of a function of n variables to be the largest set of points with coordinates $(x_1, x_2, \ldots, x_n)$ for which $f(x_1, x_2, \ldots, x_n)$ makes sense as a real number, unless the domain is explicitly stated to be a smaller set.

Most of the examples we consider in this chapter will be functions of **two** independent variables. We shall usually denote these variables x and y, and shall denote by z the dependent variable giving the value of the function. Thus we speak of the function $z = f(x, y)$. Theorems will also be stated in the two-variable case; the extensions to three or more variables will usually be obvious.

Graphical Representations

The graph of a function of one variable requires a two-dimensional coordinate system, one dimension each for the independent and dependent variables. The graph of a function of two variables, that is, the graph of the equation $z = f(x, y)$, requires a three-dimensional coordinate system with mutually perpendicular x-, y- and z-axes. Most commonly these axes are drawn as in Fig. 12.1.1, with the x- and y-axes horizontal and the z-axis vertical. Such a coordinate system is called right-handed because the thumb, forefinger and middle finger of the right hand can be extended so as to point in the direction of the x-axis, y-axis and z-axis respectively. The graph of $z = f(x, y)$ is, in general, a surface lying above (or below) the region $\mathcal{D}(f)$ in the xy-plane.

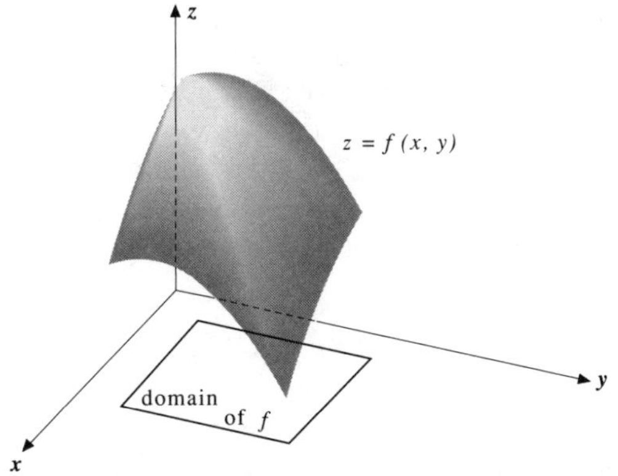

FIGURE 12.1.1

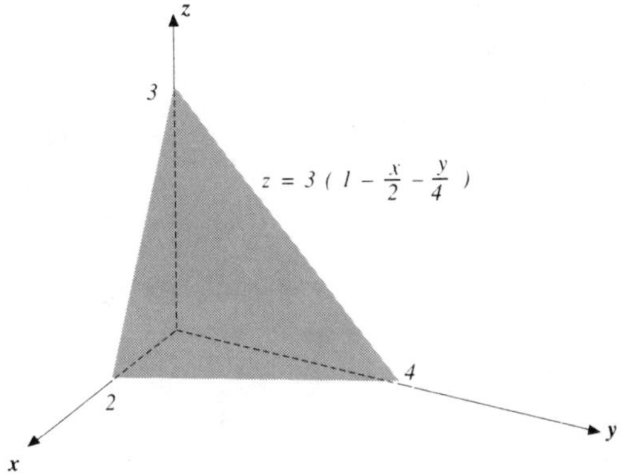

FIGURE 12.1.2

EXAMPLE 12.1.2 The graph of the function

$$f(x, y) = 3\left(1 - \frac{x}{2} - \frac{y}{4}\right), \qquad (0 \le x \le 2,\ 0 \le y \le 4 - 2x)$$

is the plane triangular surface with vertices at $(2, 0, 0)$, $(0, 4, 0)$ and $(0, 0, 3)$. See Fig. 12.1.2. If the domain of f had not been specified explicitly as a subset of the xy-plane, the graph would have been the whole plane through these three points.

EXAMPLE 12.1.3 The function $f(x, y) = \sqrt{9 - x^2 - y^2}$ has domain the disc $x^2 + y^2 \le 9$. The graph of f is a hemisphere of radius 3 centred at the origin and lying above the xy-plane. See Fig. 12.1.3.

Since it is necessary to project the surface $z = f(x, y)$ onto a two-dimensional page, most such graphs are difficult to sketch without considerable artistic talent and training. Another way of representing the function $f(x, y)$ graphically is to produce a two-dimensional "topographic" map of the surface $z = f(x, y)$. Thus we sketch, in the xy-plane, the curves $f(x, y) = C$ for various choices of the constant C. These curves are called **level curves** of f because their vertical projections

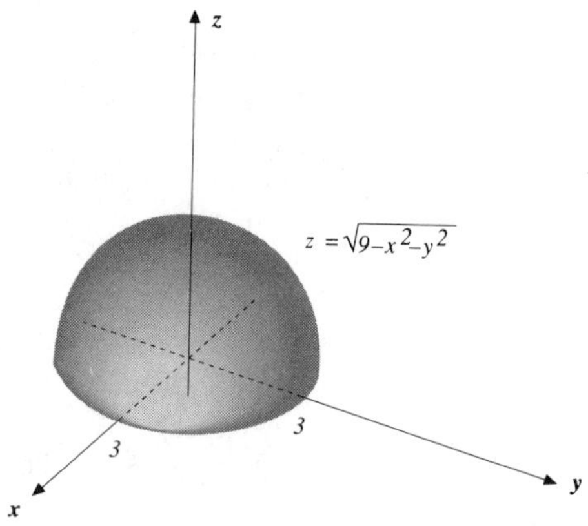

FIGURE 12.1.3

onto the surface $z = f(x, y)$ are curves which lie in horizontal planes $z = C$ and so remain at the same distance above (or below) the xy-plane. The graph and some level curves of the function $f(x, y) = x^2 + y^2$ are shown in Fig. 12.1.4. The contour curves in the topographic map in Fig. 12.1.5 shows elevations, in 200 ft increments above sea level, on part of Nelson Island on the British Columbia coast.

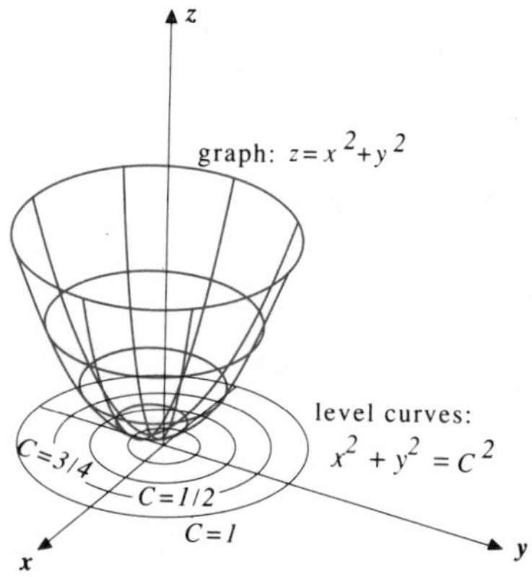

FIGURE 12.1.4

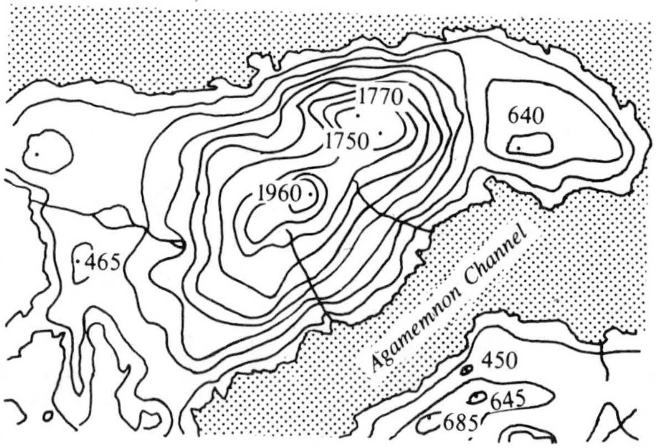

FIGURE 12.1.5

EXAMPLE 12.1.4 The level curves of the function $f(x, y) = 3\left(1 - \frac{x}{2} - \frac{y}{4}\right)$ of Example 12.1.2 are the segments of the straight lines

$$3\left(1 - \frac{x}{2} - \frac{y}{4}\right) = C, \qquad \text{or} \qquad \frac{x}{2} + \frac{y}{4} = 1 - \frac{C}{3}, \qquad (0 \le C \le 3)$$

which lie in the first quadrant: $x \ge 0$, $y \ge 0$. See Fig. 12.1.6.

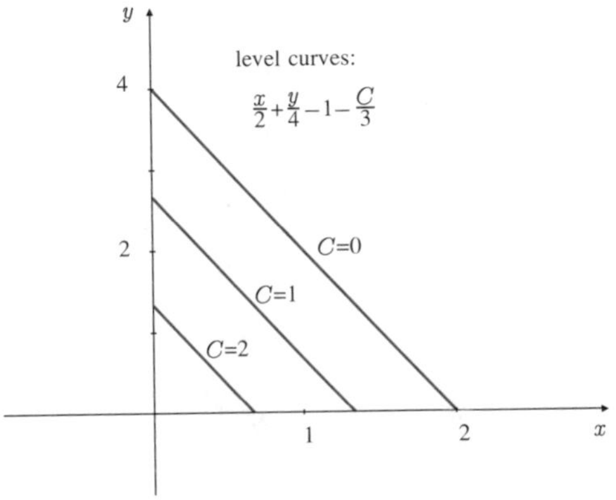

FIGURE 12.1.6

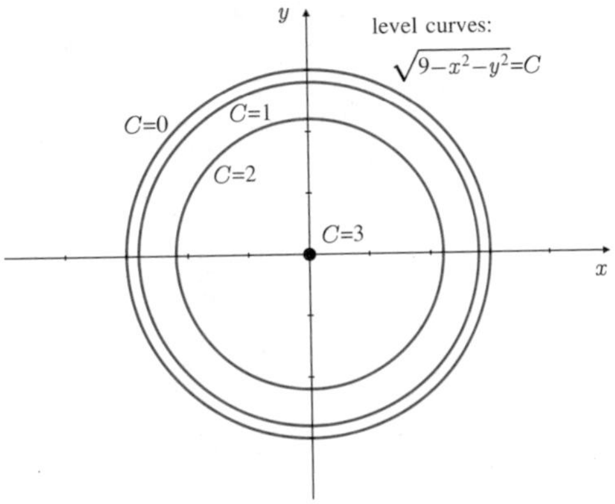

FIGURE 12.1.7

EXAMPLE 12.1.5 The level curves of the function $f(x,y) = \sqrt{9-x^2-y^2}$ of Example 12.1.3 are the concentric circles (see Fig. 12.1.7):

$$\sqrt{9-x^2-y^2} = C, \qquad \text{or} \qquad x^2+y^2 = 9-C^2, \qquad (0 \le C \le 3).$$

EXAMPLE 12.1.6 The level curves of the function $f(x,y) = x^2 - y^2$ are the curves $x^2 - y^2 = C$. For $C = 0$ the level "curve" is the pair of straight lines $x = y$ and $x = -y$. For other values of C they are rectangular hyperbolas with these lines as asymptotes. See Fig. 12.1.8. The graph of $z = x^2 - y^2$ is somewhat more difficult to sketch. It is the saddle-like surface of Fig. 12.1.9.

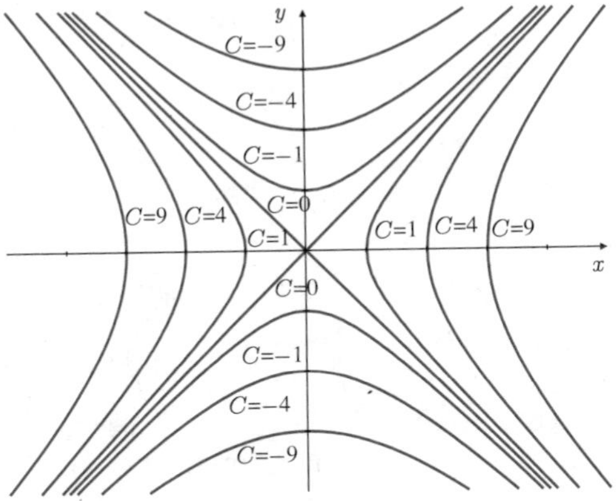

FIGURE 12.1.8

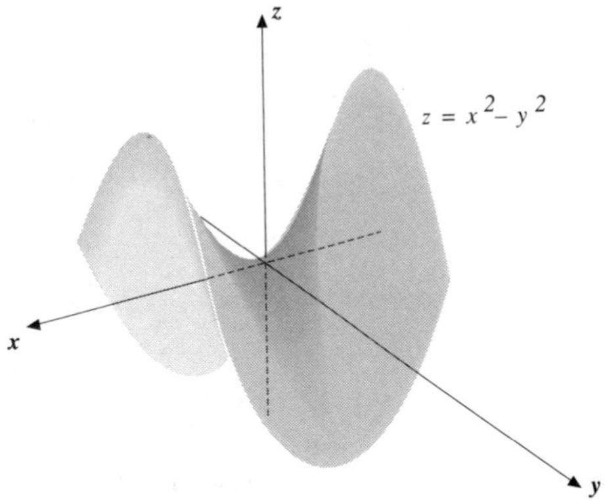

FIGURE 12.1.9

Limits and Continuity

The concept of limit of a function of several variables is similar to that for functions of one variable discussed in Section 1.5. For clarity we present the definition for functions of two variables only; the general case is similar.

12.1.7
Limits

> We say that $f(x, y)$ approaches the limit L as the point (x, y) approaches the point (a, b), and we write
>
> $$\lim_{(x,y)\to(a,b)} f(x, y) = L,$$
>
> if all points of a disc of positive radius centred at (a, b) except possibly the point (a, b) itself belong to the domain of f, and if $f(x, y)$ approaches arbitrarily close to L as (x, y) approaches (a, b) along any curve.

More formally, $\lim_{(x,y)\to(a,b)} f(x, y) = L$ if and only if for every positive number ϵ there exists a positive number δ (depending on ϵ) such that

$$0 < \sqrt{(x-a)^2 + (y-b)^2} < \delta \quad \text{implies} \quad |f(x, y) - L| < \epsilon.$$

Just as the existence of $\lim_{x\to a} f(x)$ implies that $f(x)$ approaches the same limit as x approaches a from either the right or the left, so also here the limit cannot exist unless $f(x, y)$ approaches L no matter how (x, y) approaches (a, b). The examples below illustrate this.

All the usual laws of limits (see Theorems 1.5.6, 1.5.7 and 1.5.8) extend to limits of functions of several variables in the obvious way.

EXAMPLE 12.1.8 a) $\lim_{(x,y)\to(2,3)} (2x - y^2) = 4 - 9 = -5$.

b) $\lim_{(x,y)\to(a,b)} x^2y = a^2b$.

c) $\lim_{(x,y)\to((\pi/3),2)} y\sin(x/y) = 2\sin(\pi/6) = 1$.

EXAMPLE 12.1.9 Let $f(x, y) = \dfrac{2xy}{x^2 + y^2}$. Note that $f(x, y)$ is defined at all points of the xy-plane except the origin $(0, 0)$. We can still ask whether $\lim_{(x,y)\to(0,0)} f(x, y)$ exists or not. If we let (x, y) approach $(0, 0)$ along the x-axis we get $f(x, y) = f(x, 0) = 0$ so the limit must be 0 if it exists at all. Similarly, at all the points of the y-axis we have $f(x, y) = f(0, y) = 0$. However, at points of the line $x = y$, f has a different constant value: $f(x, x) = 1$. Hence the limit of $f(x, y)$ is 1 as the point (x, y) approaches the origin along this line. It follows that $f(x, y)$ cannot have a (unique) limit at the origin; $\lim_{(x,y)\to(0,0)} f(x, y)$ does not exist. Observe that $f(x, y)$ has a constant values on any ray through the origin (on the ray $y = kx$ the value is $2k/(1 + k^2)$), but these values differ on different rays. The level curves of $f(x, y)$ are the rays from the origin, with the origin itself removed. It is rather difficult to sketch the graph of f near the origin. It is the "hood-shaped" surface in Fig. 12.1.10.

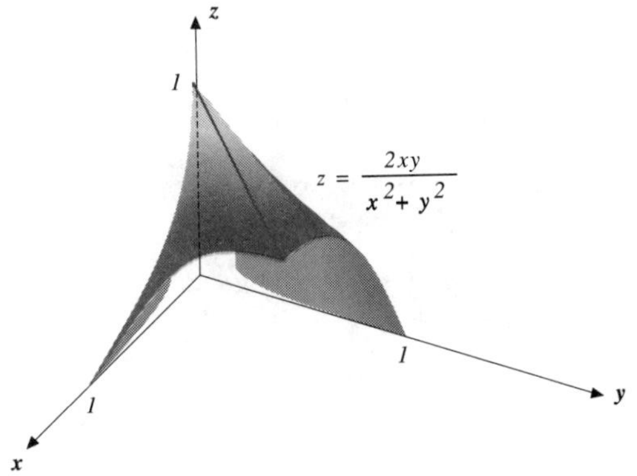

FIGURE 12.1.10

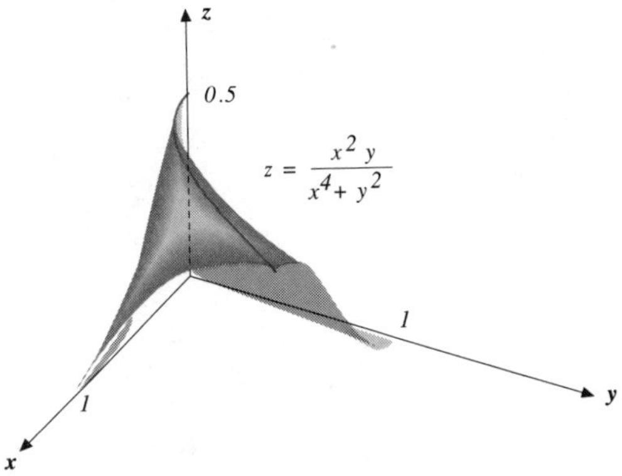

FIGURE 12.1.11

EXAMPLE 12.1.10 Let $f(x, y) = \dfrac{x^2 y}{x^4 + y^2}$. As in the previous example, $f(x, y)$ vanishes identically on the coordinate axes so $\lim_{(x,y)\to(0,0)} f(x, y)$ must be 0 if the limit exists at all. If we examine $f(x, y)$ at points of the ray $y = kx$, we obtain

$$f(x, kx) = \frac{kx^3}{x^4 + k^2x^2} = \frac{kx}{x^2 + k^2} \to 0 \quad \text{as} \quad x \to 0 \quad (k \neq 0).$$

Thus $f(x, y) \to 0$ as $(x, y) \to (0, 0)$ along *any* straight line through the origin. We might jump to the conclusion that $\lim_{(x,y)\to(0,0)} f(x, y) = 0$, but this would be incorrect. Observe the behavior of $f(x, y)$ along the parabola $y = x^2$:

$$f(x, x^2) = \frac{x^4}{x^4 + x^4} = 1/2,$$

so that $f(x, y)$ does not approach 0 as $(x, y) \to (0, 0)$ along this curve. Thus $\lim_{(x,y)\to(0,0)} f(x, y)$ does not exist. The graph of f is shown in Fig. 12.1.11. The level curves of f are the coordinate axes and pairs of parabolas, $y = kx^2$ and $y = x^2/k$, all with the origin removed.

EXAMPLE 12.1.11 Let $f(x, y) = \dfrac{x^2 y}{x^2 + y^2}$. In this case we really do have a limit: $\lim_{(x,y)\to(0,0)} f(x, y) = 0$. To see this, observe that since $|x| \le \sqrt{x^2 + y^2}$ and $|y| \le \sqrt{x^2 + y^2}$, therefore $|f(x, y)| \le \sqrt{x^2 + y^2}$ which approaches zero as $(x, y) \to (0, 0)$. See Fig. 12.1.12.

12.1.12
Continuity

> The function $f(x, y)$ is continuous at the point (a, b) if $\lim_{(x,y)\to(a,b)} f(x, y)$ exists and equals $f(a, b)$.

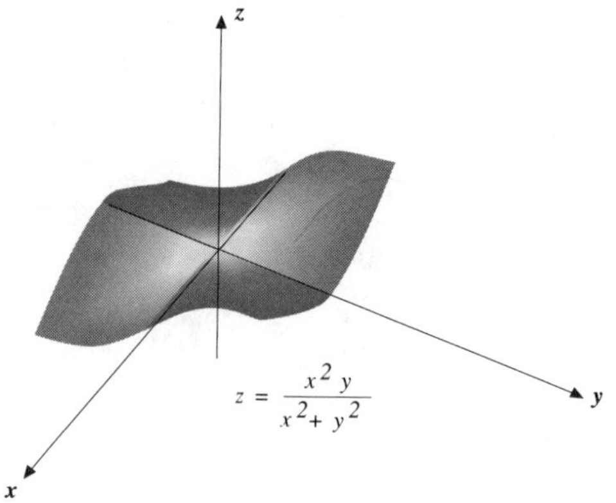

FIGURE 12.1.12

The graph of a continuous function is an unbroken surface. As for functions of one variable, suitable combinations of continuous functions are continuous. (See Theorem 1.7.5.) The functions of Examples 12.1.9–11 above are continuous wherever they are defined, that is, at all points except the origin. There is no way to define $f(0,0)$ so that the functions $f(x,y)$ of Examples 12.1.9 and 10 become continuous at the origin, but the function of Example 12.1.11 will be continuous at $(0,0)$ if we define $f(0,0) = 0$.

EXERCISES

Specify the domains of the functions in Exercises 1–8.

1. $f(x,y) = \dfrac{x+y}{x-y}$

2. $f(x,y) = \sqrt{xy}$

3. $f(x,y) = \dfrac{x}{x^2+y^2}$

4. $f(x,y) = \dfrac{xy}{x^2-y^2}$

5. $f(x,y) = \dfrac{1}{\sqrt{x^2-y^2}}$

6. $f(x,y) = \ln(1+xy)$

7. $f(x,y) = \sin^{-1}(x+y)$

8. $f(x,y,z) = \dfrac{xyz}{x^2+y^2+z^2}$

Sketch the graphs of the functions in Exercises 9–18.

9. $f(x,y) = x, \quad (0 \le x \le 2,\ 0 \le y \le 3)$

10. $f(x,y) = 4 - x^2 - y^2, \ (x^2+y^2 \le 4,\ x \ge 0,\ y \ge 0)$

11. $f(x,y) = y^2$

12. $f(x,y) = 4x^2 + y^2$

13. $f(x,y) = \sqrt{x^2+y^2}$

14. $f(x,y) = 6 - x - 2y$

15. $f(x,y) = |x| + |y|$

16. $f(x,y) = \sin x$

17. $f(x,y) = \dfrac{1}{x^2+y^2}$

18. $f(x,y) = xy$

Sketch some level curves of the functions in Exercises 19–26.

19. $f(x,y) = x - y$

20. $f(x,y) = x^2 + 2y^2$

21. $f(x,y) = xy$

22. $f(x,y) = \dfrac{x^2}{y}$

23. $f(x,y) = \dfrac{x-y}{x+y}$

24. $f(x,y) = \dfrac{y}{x^2+y^2}$

25. $f(x,y) = \sin(x+y)$

26. $f(x,y) = x\,e^{-y}$

27.*Find $f(x,y)$ if each level curve $f(x,y) = C$ is a circle with centre the origin and radius
(a) C, (b) C^2, (c) $\sqrt{C}$, (d) $\ln(C)$.

In Exercises 28–35 evaluate the indicated limit or explain why it does not exist.

28. $\displaystyle\lim_{(x,y)\to(2,-1)} (xy + x^2)$

29. $\displaystyle\lim_{(x,y)\to(0,0)} \frac{x}{x^2+y^2}$

30. $\displaystyle\lim_{(x,y)\to(0,0)} \frac{y^3}{x^2+y^2}$

31. $\displaystyle\lim_{(x,y)\to(0,0)} \frac{\sin(x-y)}{\cos(x+y)}$

32. $\displaystyle\lim_{(x,y)\to(1,2)} \frac{2x^2-xy}{4x^2-y^2}$

33. $\displaystyle\lim_{(x,y)\to(0,0)} \frac{x^2y^2}{x^2+y^4}$

34. $\displaystyle\lim_{(x,y)\to(0,0)} \frac{x^2y^2}{2x^4+y^4}$

35. $\displaystyle\lim_{(x,y)\to(0,0)} \frac{\sin(xy)}{x^2+y^2}$

36. How can the function $f(x,y) = \dfrac{x^2+y^2-x^3y^3}{x^2+y^2}$ be defined at the origin so that the resulting function becomes continuous at all points in the xy-plane?

37.*What is the domain of $f(x,y) = \dfrac{x-y}{x^2-y^2}$? Does $f(x,y)$ have a limit as $(x,y) \to (1,1)$? How can the domain of f be extended so that the resulting function is continuous at $(1,1)$?

38.*The definition of limit (Definition 12.1.7) can be modified to the following analogue of the definition of **one-sided limits** given in Section 1.6: we say that $f(x,y)$ tends to the limit L as (x,y) approaches (a,b) if

i) every disc of positive radius about (a,b) contains points of the domain of f distinct from (a,b), and

ii) for every positive number ϵ there exists a positive number δ such that $|f(x,y)-L| < \epsilon$ whenever (x,y) lies in the domain of f and $0 < \sqrt{(x-a)^2+(y-b)^2} < \delta$.

Show that the function $f(x,y)$ of Exercise 37 does have a limit as $(x,y) \to (1,1)$ according to this definition.

39.*Given a function $f(x,y)$ and a point (a,b) in its domain we define single-variable functions g and h as follows: $g(x) = f(x,b)$, $h(y) = f(a,y)$. If g is continuous at $x = a$ and h is continuous at $y = b$ does it follow that f is continuous at (a,b)? Justify your answer.

12.2 PARTIAL DERIVATIVES

In attempting to extend the concepts and techniques of single-variable calculus to functions of more than one variable, it is convenient to begin by isolating the dependence of such functions on one variable at a time. Thus we define two **partial derivatives** of a functions of two variables—one with respect to each independent variable.

12.2.1
Partial Derivatives

The **first partial derivative** of the function $f(x,y)$ **with respect to the variable** x is the function $f_1(x,y)$ given by

$$f_1(x,y) = \lim_{h\to 0} \frac{f(x+h,y)-f(x,y)}{h}$$

provided the limit exists. Similarly, the **first partial derivative of** f **with respect to** y is the function $f_2(x,y)$ given by

$$f_2(x,y) = \lim_{k\to 0} \frac{f(x,y+k)-f(x,y)}{k}$$

provided the limit exists.

We evaluate $f_1(x,y)$ by differentiating $f(x,y)$ with respect to x, using all the usual differentiation rules, and treating y as a constant. At any particular point (a,b), $f_1(a,b)$ measures the rate of change of $f(x,y)$ with respect to x at $x = a$ while y is held fixed at b. In geometric terms, the surface $z = f(x,y)$ intersects the vertical plane $y = b$ in a curve. If we take horizontal and vertical lines through the point $(0,b,0)$ as x- and z-coordinate axes in that vertical plane, the curve has equation $z = f(x,b)$, and its slope at $x = a$ is $f_1(a,b)$. See Fig. 12.2.1. Similarly, we evaluate $f_2(x, y)$ by differentiating $f(x, y)$ with respect to y, treating x as a constant. The surface $z = f(x, y)$ intersects the vertical plane $x = a$ in a curve $z = f(a, y)$ whose slope at $y = b$ is $f_2(a, b)$. See Fig. 12.2.2.

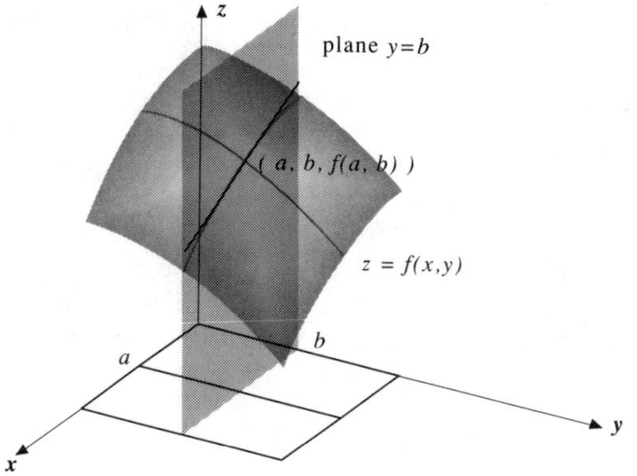

FIGURE 12.2.1

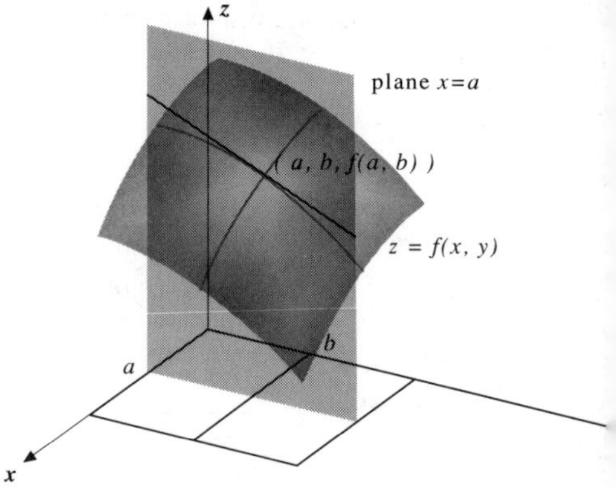

FIGURE 12.2.2

Various other notations are used to denote the partial derivatives of $z = f(x, y)$ considered as functions of x and y:

$$\frac{\partial z}{\partial x} = \frac{\partial}{\partial x} f(x, y) = f_1(x, y) = D_1 f(x, y),$$
$$\frac{\partial z}{\partial y} = \frac{\partial}{\partial y} f(x, y) = f_2(x, y) = D_2 f(x, y).$$

The subscripts "1" and "2" in these notations refer to the first and second variables of f. We prefer the notation f_1 to an alternative notation f_x found in some books. This avoids problems when we consider derivatives of compositions of functions. For example, $f_1(x^2, x^3)$ refers to the derivative of $f(u, v)$ with respect to u evaluated at $u = x^2$, $v = x^3$, while $f_x(x^2, x^3)$ might refer to the same thing *or* might refer to the derivative with respect to x of the function $f(x^2, x^3)$, a function of one variable. For similar reasons we avoid using the notation $\frac{\partial f}{\partial x}$ or $\frac{\partial f}{\partial y}$, though these are often encountered elsewhere.

Values at a particular point (a, b) are denoted similarly:

$$\left.\frac{\partial z}{\partial x}\right|_{(a,b)} = \left.\frac{\partial}{\partial x} f(x, y)\right|_{(a,b)} = f_1(a, b) = D_1 f(a, b),$$
$$\left.\frac{\partial z}{\partial y}\right|_{(a,b)} = \left.\frac{\partial}{\partial y} f(x, y)\right|_{(a,b)} = f_2(a, b) = D_2 f(a, b).$$

EXAMPLE 12.2.2 Find $\dfrac{\partial z}{\partial x}$ and $\dfrac{\partial z}{\partial y}$ if $z = x^3y^2 + x^4y + y^4$.

SOLUTION $\dfrac{\partial z}{\partial x} = 3x^2y^2 + 4x^3y, \quad \dfrac{\partial z}{\partial y} = 2x^3y + x^4 + 4y^3.$

EXAMPLE 12.2.3 Find $f_1(1, \pi)$ and $f_2(1, \pi)$ if $f(x, y) = e^{\pi x - y}\cos(xy)$.

SOLUTION $f_1(x, y) = e^{\pi x - y}\big[\pi\cos(xy) - y\sin(xy)\big]$, so $f_1(1, \pi) = 1(-\pi - 0) = -\pi$.
$f_2(x, y) = e^{\pi x - y}\big[-\cos(xy) - x\sin(xy)\big]$, so $f_2(1, \pi) = 1(1 - 0) = 1$.

EXAMPLE 12.2.4 If f is an everywhere differentiable function of one variable, show that (wherever $y \neq 0$), $z = f\left(\dfrac{x}{y}\right)$ satisfies the **partial differential equation**

$$x\frac{\partial z}{\partial x} + y\frac{\partial z}{\partial y} = 0.$$

SOLUTION By the Chain Rule $\dfrac{\partial z}{\partial x} = f'\left(\dfrac{x}{y}\right)\left(\dfrac{1}{y}\right)$ and $\dfrac{\partial z}{\partial y} = f'\left(\dfrac{x}{y}\right)\left(-\dfrac{x}{y^2}\right)$, so

$$x\frac{\partial z}{\partial x} + y\frac{\partial z}{\partial y} = f'\left(\frac{x}{y}\right)\left(\frac{x}{y} - \frac{x}{y}\right) = 0.$$

Definition 12.2.1 can be extended in the obvious way to cover functions of more than two variables. If f is a function of the n variables, $x_1, x_2, \ldots, x_n$ then f has n first partial derivatives, $f_1(x_1, x_2, \ldots, x_n)$, $f_2(x_1, x_2, \ldots, x_n)$, ..., $f_n(x_1, x_2, \ldots, x_n)$, one with respect to each variable. Of course, all the standard rules for derivatives of sums, products and quotients apply to partial derivatives.

EXAMPLE 12.2.5 $\dfrac{\partial}{\partial z}\dfrac{2xy}{1 + xz + yz} = -\dfrac{2xy}{(1 + xz + yz)^2}(x + y).$

Tangent Planes

If the graph $z = f(x, y)$ is a smooth surface near the point P with coordinates $(a, b, f(a, b))$, then that graph will have a **tangent plane** at P. (We are using the word "smooth" informally here, and won't attempt to define it precisely. Smooth surfaces have no kinks, sharp bends or cusps, and do have tangent planes.) This tangent plane will contain the tangent lines to all smooth curves through P which lie in the surface $z = f(x, y)$. In particular, the two straight lines which are tangent at P to the two curves in which the surface $z = f(x, y)$ intersects the vertical planes $y = b$ and $x = a$ will lie in that tangent plane. See Figs. 12.2.1, 12.2.2 and 12.2.3.

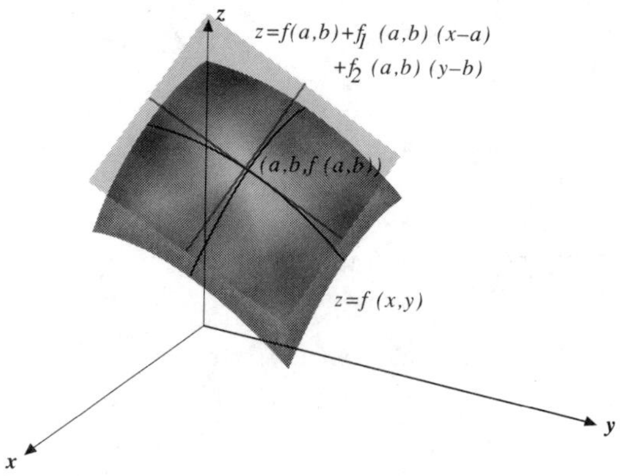

FIGURE 12.2.3

In general, planes in 3-dimensional space have *linear equations* of the form

$$Ax + By + Cz = D$$

where A, B, C and D are constants. A non-vertical tangent plane to $z = f(x, y)$ at $(a, b, f(a, b))$ will have a linear equation of the more specialized form

$$z = f(a, b) + A(x - a) + B(y - b)$$

with certain constants A and B. The intersection of this plane with the vertical plane $y = b$ is the straight line with equations

$$z = f(a, b) + A(x - a), \qquad y = b.$$

As observed above, this line is tangent to the curve

$$z = f(x, b), \qquad y = b,$$

so we must have $A = f_1(a, b)$, the slope of that curve at $x = a$. Similarly, the tangent plane meets the vertical plane $x = a$ in the straight line

$$z = f(a, b) + B(y - b), \qquad x = a,$$

which is tangent to the curve

$$z = f(a, y), \qquad x = a.$$

Hence $B = f_2(a, b)$. Therefore the tangent plane has equation

12.2.6
Equation of Tangent Plane

$$z = f(a, b) + f_1(a, b)(x - a) + f_2(a, b)(y - b).$$

EXAMPLE 12.2.7 Find an equation of the tangent plane to the graph $z = \sin(xy)$ at the point $\left(\frac{\pi}{3}, -1\right)$.

SOLUTION We have $\dfrac{\partial z}{\partial x} = y\cos(xy)$, $\dfrac{\partial z}{\partial y} = x\cos(xy)$.

At $(\pi/3, -1)$ we have $z = -\sqrt{3}/2$ and $\partial z/\partial x = -1/2$, and $\partial z/\partial y = \pi/6$. Thus the tangent plane has equation

$$z = \frac{-\sqrt{3}}{2} - \frac{1}{2}\left(x - \frac{\pi}{3}\right) + \frac{\pi}{6}(y+1),$$

or,

$$z = -\frac{x}{2} + \frac{\pi y}{6} - \frac{\sqrt{3}}{2} + \frac{\pi}{3}.$$

Higher Order Derivatives

Second and higher order partial derivatives are calculated by taking partial derivatives of already calculated partial derivatives. The order in which the differentiations are performed is indicated by the notations in an obvious way. If $z = f(x, y)$, then we can calculate four partial derivatives of second order:

Two **pure** second partial derivatives with respect to x or y,

12.2.8
Pure Second Partial Derivatives

$$\frac{\partial^2 z}{\partial x^2} = \frac{\partial}{\partial x}\frac{\partial z}{\partial x} = f_{11}(x, y), \qquad \frac{\partial^2}{\partial y^2} = \frac{\partial}{\partial y}\frac{\partial z}{\partial y} = f_{22}(x, y),$$

and two **mixed** second partial derivatives with respect to x and y,

12.2.9
Mixed Second Partial Derivatives

$$\frac{\partial^2 z}{\partial x \partial y} = \frac{\partial}{\partial x}\frac{\partial z}{\partial y} = f_{21}(x, y), \qquad \frac{\partial^2 z}{\partial y \partial x} = \frac{\partial}{\partial y}\frac{\partial z}{\partial x} = f_{12}(x, y).$$

Similarly, if $w = f(x, y, z)$ then

$$\frac{\partial^5 w}{\partial y \partial x \partial y^2 \partial z} = f_{32212}(x, y, z) = \frac{\partial}{\partial y}\frac{\partial}{\partial x}\frac{\partial}{\partial y}\frac{\partial}{\partial y}\frac{\partial w}{\partial z}.$$

EXAMPLE 12.2.10 Find the four second partial derivatives of $f(x,y) = x^3y^4$.

SOLUTION

$$f_1(x,y) = 3x^2y^4, \qquad f_2(x,y) = 4x^3y^3,$$
$$f_{11}(x,y) = \frac{\partial}{\partial x}(3x^2y^4) = 6xy^4, \qquad f_{21}(x,y) = \frac{\partial}{\partial x}(4x^3y^3) = 12x^2y^3,$$
$$f_{12}(x,y) = \frac{\partial}{\partial y}(3x^2y^4) = 12x^2y^3, \qquad f_{22}(x,y) = \frac{\partial}{\partial y}(4x^3y^3) = 12x^3y^2.$$

EXAMPLE 12.2.11 Calculate $f_{322}(x,y,z)$ and $f_{223}(x,y,z)$ for the function $f(x,y,z) = e^{x-2y+3z}$.

SOLUTION

$$f_{322}(x,y,z) = \frac{\partial}{\partial y}\frac{\partial}{\partial y}\frac{\partial}{\partial z}[e^{x-2y+3z}] = \frac{\partial}{\partial y}\frac{\partial}{\partial y}[3e^{x-2y+3z}] = \frac{\partial}{\partial y}[-6e^{x-2y+3z}] = 12e^{x-2y+3z}$$
$$f_{223}(x,y,z) = \frac{\partial}{\partial z}\frac{\partial}{\partial y}\frac{\partial}{\partial y}[e^{x-2y+3z}] = \frac{\partial}{\partial z}\frac{\partial}{\partial y}[-2e^{x-2y+3z}]\frac{\partial}{\partial z}[4e^{x-2y+3z}] = 12e^{x-2y+3z}.$$

In both of the examples above, observe that the two mixed partial derivatives taken with respect to the same variables but in different orders turned out to be equal. This is not a coincidence. It will occur whenever the mixed partial derivatives involved are **continuous**. The following theorem states formally the simplest case of this important phenomenon.

THEOREM 12.2.12 (**Equality of mixed partials**) If f, f_1 and f_2 are continuous throughout a disc of positive radius centred at the point (a,b), and if the mixed second partials f_{12} and f_{21} exist throughout that disc and are continuous at (a,b), then $f_{12}(a,b) = f_{21}(a,b)$.

PROOF Let h and k have absolute values sufficiently small that the point $(a+h, b+k)$ lies in the disc referred to in the statement. Then so do all points of the rectangle with these two points as diagonally opposite corners. See Fig. 12.2.4.

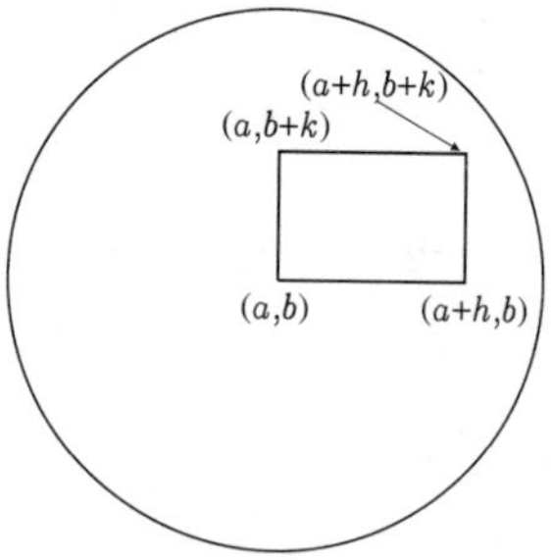

FIGURE 12.2.4

Let $Q = f(a+h, b+k) - f(a+h, b) - f(a, b+k) + f(a,b)$, and define single-variable functions $u(x)$ and $v(y)$ by

$$u(x) = f(x, b+k) - f(x,b), \qquad v(y) = f(a+h, y) - f(a,y).$$

Evidently $Q = u(a+h) - u(a)$ and also $Q = v(b+k) - v(b)$. By the Mean-Value Theorem (Theorem 4.1.1) there is a point $a + \theta_1 h$ between a and $a+h$ (that is, satisfying $0 < \theta_1 < 1$) such that

$$Q = u(a+h) - u(a) = hu'(a+\theta_1 h)$$
$$= h\big[f_1(a+\theta_1 h, b+k) - f_1(a+\theta_1 h, b)\big].$$

Now apply the Mean-Value Theorem again to f_1 considered as a function of its second variable, and obtain another number θ_2 satisfying $0 < \theta_2 < 1$ such that

$$f_1(a+\theta_1, b+k) - f_1(a+\theta_1, b) = kf_{12}(a+\theta_1 h, b+\theta_2 k).$$

Thus $Q = hkf_{12}(a + \theta_1 h, b + \theta_2 k)$. Two similar applications of the Mean-Value Theorem starting from $Q = v(b + k) - v(b)$ yield two numbers θ_3 and θ_4, each between 0 and 1 such that $Q = khf_{21}(a + \theta_4 h, b + \theta_3 k)$. Equating these two expressions for Q and dividing by hk, we obtain

$$f_{12}(a + \theta_1 h, b + \theta_2 k) = f_{21}(a + \theta_4 h, b + \theta_3 k).$$

Since f_{12} and f_{21} are continuous at (a, b), we can let $h \to 0$ and $k \to 0$ to obtain $f_{12}(a, b) = f_{21}(a, b)$, as required. □

EXAMPLE 12.2.13 Show that the function $z = e^{kx}\cos(ky)$ satisfies the partial differential equation

$$\frac{\partial^2 z}{\partial x^2} + \frac{\partial^2 z}{\partial y^2} = 0$$

at every point (x, y).

SOLUTION We have

$$\frac{\partial z}{\partial x} = k\,e^{kx}\cos(ky), \qquad \frac{\partial z}{\partial y} = -k\,e^{kx}\sin(ky),$$
$$\frac{\partial^2 z}{\partial x^2} = k^2\,e^{kx}\cos(ky), \qquad \frac{\partial^2 z}{\partial y^2} = -k^2\,e^{kx}\cos(ky).$$

Thus $\dfrac{\partial^2 z}{\partial x^2} + \dfrac{\partial^2 z}{\partial y^2} = k^2\,e^{kx}\cos(ky) - k^2\,e^{kx}\cos(ky) = 0.$

REMARK: The partial differential equation in Example 12.2.13 is called the (two-dimensional) **Laplace equation**. A function having continuous second partial derivatives in a region of the plane is said to be **harmonic** there if it satisfies the Laplace equation. Harmonic functions play an important role in many areas of mathematics and its applications. Besides being central to the study of differentiable functions of a **complex variable**, such functions are used to model various physical quantities such as steady-state temperature distributions, fluid flows, and electric and magnetic potential fields. Laplace's equation, and therefore harmonic functions, can be considered in any number of dimensions. (See Exercises 37 and 38.)

EXAMPLE 12.2.14 If f and g are any twice differentiable functions of one variable, show that the function $w = f(x - ct) + g(x + ct)$ satisfies the differential equation

$$\frac{\partial^2 w}{\partial t^2} = c^2 \frac{\partial^2 w}{\partial x^2}.$$

SOLUTION Using the (single-variable) Chain Rule we calculate

$$\frac{\partial w}{\partial t} = -cf'(x - ct) + cg'(x + ct), \qquad \frac{\partial w}{\partial x} = f'(x - ct) + g'(x + ct),$$
$$\frac{\partial^2 w}{\partial t^2} = c^2 f''(x - ct) + c^2 g''(x + ct), \qquad \frac{\partial^2 w}{\partial x^2} = f''(x - ct) + g''(x + ct).$$

Evidently w satisfies the given differential equation.

REMARK: The differential equation of Example 12.2.14 is called the (one-dimensional) **wave equation**. If t measures time then $f(x - ct)$ represents a waveform travelling to the right along the x-axis with speed c. See Fig. 12.2.5. Similarly, $g(x + ct)$ represents a waveform travelling to the left with speed c.

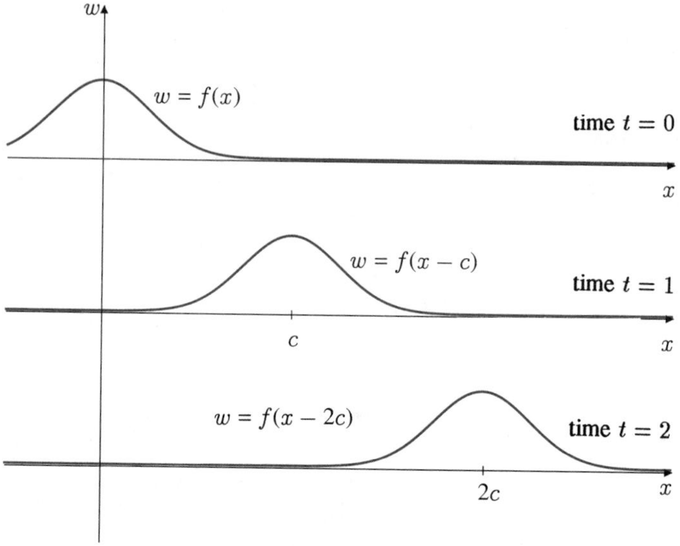

FIGURE 12.2.5

EXERCISES

In Exercises 1–9 find all the first partial derivatives of the functions specified. Also evaluate these at the given point.

1. $f(x, y) = x - y + 2$; $(3, 2)$

2. $f(x, y) = xy + x^2$; $(2, 0)$

3. $f(x, y, z) = x^3 y^4 z^5$; $(0, -1, -1)$

4. $g(x, y, z) = \dfrac{yz}{x + z}$; $(1, 1, 1)$

5. $z = \tan^{-1} \dfrac{y}{x}$; $(-1, 1)$

6. $w = \ln(1 + e^{xyz})$; $(2, 0, -1)$

7. $f(x, y) = \sin(x\sqrt{y})$; $\left(\dfrac{\pi}{3}, 4\right)$

8. $f(x, y) = \dfrac{1}{\sqrt{x^2 + y^2}}$; $(-3, 4)$

9. $w = x^{y \ln z}$; $(e, 2, e)$

10. Find $f_1(0, 0)$ and $f_2(0, 0)$ if

$$f(x, y) = \begin{cases} \dfrac{x^2 - 2y^2}{x - y} & \text{if } x \neq y \\ 0 & \text{if } x = y \end{cases}$$

In Exercises 11–19 find the equation of the tangent plane to the graph of the given function at the point specified.

11. $f(x, y) = x^2 - y^2$ at $(-2, 1)$

12. $f(x, y) = \dfrac{x - y}{x + y}$ at $(1, 1)$

13. $f(x, y) = \cos \dfrac{x}{y}$ at $(\pi, 4)$

14. $f(x, y) = e^{xy}$ at $(2, 0)$

15. $f(x, y) = \dfrac{x}{x^2 + y^2}$ at $(1, 2)$

16. $f(x, y) = y\, e^{-x^2}$ at $(-1, 1)$

17. $f(x, y) = \ln(x^2 + y^2)$ at $(1, -2)$

18. $f(x, y) = \tan^{-1}(x/y)$ at $(1, -1)$

19. $f(x,y)=\sqrt{1+x^3y^2}$ at $(2,1)$

In Exercises 20–23 show that the given function satisfies the given partial differential equation.

20. $z=\dfrac{x+y}{x-y}$, $\quad x\dfrac{\partial z}{\partial x}+y\dfrac{\partial z}{\partial y}=0$

21. $z=\sqrt{x^2+y^2}$, $\quad x\dfrac{\partial z}{\partial x}+y\dfrac{\partial z}{\partial y}=z$

22. $w=x^2+yz$, $\quad x\dfrac{\partial w}{\partial x}+y\dfrac{\partial w}{\partial y}+z\dfrac{\partial w}{\partial z}=2w$

23. $w=\dfrac{1}{x^2+y^2+z^2}$, $\quad x\dfrac{\partial w}{\partial x}+y\dfrac{\partial w}{\partial y}+z\dfrac{\partial w}{\partial z}=-2w$

24. If $z=f(x^2+y^2)$, where f is a differentiable function of 1 variable, show that $y\dfrac{\partial z}{\partial x}-x\dfrac{\partial z}{\partial y}=0$.

In Exercises 25–30 find all the second partial derivatives of the given function.

25. $f(x,y)=x^2+y^2$ **26.** $z=x^2(1+y^2)$

27. $z=\sqrt{3x^2+y^2}$ **28.** $w=x^3y^3z^3$

29. $f(x,y)=\ln(1+\sin(xy))$ **30.** $z=x\,e^y-y\,e^x$

31. How many partial derivatives of order 3 can a function of three variables have? If they are all continuous, how many different values can they have at one point? Find the three mixed partial derivatives of order 3 for $f(x,y,z)=x\,e^{xy}\cos(xz)$ which involve two differentiations with respect to z and one with respect to x.

Show that the functions in Exercises 32–36 are harmonic in the plane regions indicated. (See Example 12.2.13 and the Remark following it.)

32. $f(x,y)=e^{kx}\sin(ky)$ in the whole plane.

33. $f(x,y)=A(x^2-y^2)+Bxy$ in the whole plane. (A, B are constants.)

34. $f(x,y)=3x^2y-y^3$ in the whole plane. (Can you think of another polynomial of degree 3 in x and y which is also harmonic?)

35. $f(x,y)=\ln(x^2+y^2)$ everywhere except at the origin

36. $f(x,y)=\tan^{-1}(y/x)$ except at points on the y-axis

37. Show that $w=e^{3x+4y}\sin(5z)$ is a harmonic function in the whole 3-dimensional space. That is, show that it satisfies the 3-dimensional Laplace equation

$$\frac{\partial^2 w}{\partial x^2}+\frac{\partial^2 w}{\partial y^2}+\frac{\partial^2 w}{\partial z^2}=0.$$

38. Show that if $f(x,y)$ is harmonic in the plane, then each of the following functions is harmonic in three dimensional space: (a) $z\,f(x,y)$, (b) $x\,f(y,z)$, (c) $y\,f(z,x)$.
What condition should the constants a, b and c satisfy to ensure that $f(ax+by,cz)$ is harmonic?

39. Suppose the functions $u(x,y)$ and $v(x,y)$ have continuous second partial derivatives and satisfy the **Cauchy-Riemann equations**

$$\frac{\partial u}{\partial x}=\frac{\partial v}{\partial y},\qquad \frac{\partial v}{\partial x}=-\frac{\partial u}{\partial y}.$$

Show that u and v are both harmonic.

40.†Show that $u(x,t)=t^{-1/2}e^{-x^2/4t}$ satisfies the partial differential equation

$$\frac{\partial u}{\partial t}=\frac{\partial^2 u}{\partial x^2}.$$

This equation is called the (one-dimensional) **heat equation** because it models heat diffusion in an insulated rod (with u representing the temperature at position x at time t) and other similar phenomena.

41.†Show that $u(x,y,t)=t^{-1}e^{-(x^2+y^2)/4t}$ satisfies the two-dimensional heat equation

$$\frac{\partial u}{\partial t}=\frac{\partial^2 u}{\partial x^2}+\frac{\partial^2 u}{\partial y^2}.$$

Comparing this result with that of the previous exercise, guess a solution of the three-dimensional heat equation

$$\frac{\partial u}{\partial t}=\frac{\partial^2 u}{\partial x^2}+\frac{\partial^2 u}{\partial y^2}+\frac{\partial^2 u}{\partial z^2}$$

and verify your guess.

42.*Let $f(x,y) = \begin{cases} \dfrac{2xy}{x^2+y^2} & \text{if } (x,y) \neq (0,0) \\ 0 & \text{if } (x,y) = (0,0) \end{cases}$. As noted in Example 12.1.9, f is not continuous at $(0,0)$. Therefore its graph is not smooth. Show, however, that $f_1(0,0)$ and $f_2(0,0)$ both exist. (Use Definition 12.2.1.) Hence the existence of partial derivatives does not imply continuity for functions of several variables. This is in contrast to the single-variable case—see Theorem 2.3.1.

43.*Let $g(x,y) = (x^2 - y^2)f(x,y)$ where f is the function defined in the previous exercise. Calculate $g_1(x,y)$ and $g_2(x,y)$ for $(x,y) \neq (0,0)$. Also calculate $g_1(0,0)$ and $g_2(0,0)$. Finally, show that $g_{21}(0,0) = 2$ but $g_{12}(0,0) = -2$. Does this result contradict Theorem 12.2.12? Explain why.

12.3 THE CHAIN RULE

The Chain Rule for functions of one variable is a formula giving the derivative of a composition $f(g(x))$ of two functions f and g:

$$\frac{d}{dx} f(g(x)) = f'(g(x))\, g'(x).$$

The situation for several variables is more complicated. If f depends on more than one variable, and each of those variables can depend on other variables, we cannot expect a single formula to cover all the possibilities. Rather we must come to think of the Chain Rule as a method for differentiating compositions. We begin by quoting Chain Rules in several different contexts without any proof. Later we give the proof of a simple, but representative case.

Let us start with a function of two variables each of which depends on two other variables:

$$z = f(x,y) \qquad \text{where} \qquad x = u(s,t), \;\; y = v(s,t).$$

We may then form the composite function

$$z = g(s,t) = f\big(u(s,t), v(s,t)\big).$$

Assume that f, u, and v have continuous first partial derivatives with respect to their respective variables. Then so does g, and the first partial derivatives of g are given by

$$g_1(s,t) = f_1\big(u(s,t), v(s,t)\big)u_1(s,t) + f_2\big(u(s,t), v(s,t)\big)v_1(s,t),$$
$$g_2(s,t) = f_1\big(u(s,t), v(s,t)\big)u_2(s,t) + f_2\big(u(s,t), v(s,t)\big)v_2(s,t),$$

or, expressed more simply using Leibniz notation,

12.3.1
A Chain Rule

$$\frac{\partial z}{\partial s} = \frac{\partial z}{\partial x}\frac{\partial x}{\partial s} + \frac{\partial z}{\partial y}\frac{\partial y}{\partial s}, \qquad \frac{\partial z}{\partial t} = \frac{\partial z}{\partial x}\frac{\partial x}{\partial t} + \frac{\partial z}{\partial y}\frac{\partial y}{\partial t}.$$

The first term in the expression for $g_1(s,t)$ (or in that for $\partial z/\partial s$) represents the contribution to the rate of change of z with respect to s which comes from the dependence of z on s through the first variable of f; the second term is the contribution coming from the second variable of f. Note the significance of the subscripts on the various partial derivatives: the "1" in $g_1(s,t)$ refers to differentiation with respect to s, the first variable on which g depends; by contrast, the "1" in $f_1(u(s,t),v(s,t))$ refers to differentiation with respect to x, the first variable on which f depends.

EXAMPLE 12.3.2 If $z = \sin(x^2y)$ where $x = st^2$ and $y = s^2 + \dfrac{1}{t}$, find $\dfrac{\partial z}{\partial s}$ and $\dfrac{\partial z}{\partial t}$

a) by direct substitution and the single-variable Chain Rule, and

b) by using the two-variable Chain Rule.

SOLUTION a) By direct substitution and the single-variable Chain Rule,

$$z = \sin\left((st^2)^2\left(s^2+\frac{1}{t}\right)\right) = \sin(s^4t^4+s^2t^3)$$

$$\frac{\partial z}{\partial s} = (4s^3t^4+2st^3)\cos(s^4t^4+s^2t^3),$$

$$\frac{\partial z}{\partial t} = (4s^4t^3+3s^2t^2)\cos(s^4t^4+s^2t^3).$$

b) Using the two-variable Chain Rule,

$$\begin{aligned}\frac{\partial z}{\partial s} = \frac{\partial z}{\partial x}\frac{\partial x}{\partial s}+\frac{\partial z}{\partial y}\frac{\partial y}{\partial s} &= \left[2xy\cos(x^2y)\right]t^2+\left[x^2\cos(x^2y)\right]2s\\ &= \left[2st^2\left(s^2+\frac{1}{t}\right)t^2+2s^3t^4\right]\cos(s^4t^4+s^2t^3)\\ &= (4s^3t^4+2st^3)\cos(s^4t^4+s^2t^3),\\ \frac{\partial z}{\partial t} = \frac{\partial z}{\partial x}\frac{\partial x}{\partial t}+\frac{\partial z}{\partial y}\frac{\partial y}{\partial t} &= [2xy\cos(x^2y)]2st+[x^2\cos(x^2y)]\left(\frac{-1}{t^2}\right)\\ &= \left[2st^2\left(s^2+\frac{1}{t}\right)2st+s^2t^4\left(\frac{-1}{t^2}\right)\right]\cos(s^4t^4+s^2t^3)\\ &= (4s^4t^3+3s^2t^2)\cos(s^4t^4+s^2t^3).\end{aligned}$$

Note that we still had to use direct substitution on the derivatives obtained using the Chain Rule in order to show they were in fact the same as those obtained without using the two-variable form of the Chain Rule. (Of course, the one-variable form of the Chain Rule was used even in (b).)

Now suppose that $z = f(x,y)$ where $x = g(t)$ and $y = h(t)$. Then $z = Q(t) = f(g(t),h(t))$, and z may be regarded as a function of the single variable t. Since z depends on t through both of the variables of f, the Chain Rule for dz/dt still has two terms:

$$\frac{dz}{dt} = \frac{\partial z}{\partial x}\frac{dx}{dt}+\frac{\partial z}{\partial y}\frac{dy}{dt}.$$

Note the use of the straight "d" for derivatives of functions of only one variable: dz/dt means $Q'(t)$ while $\partial z/\partial x$ means $f_1(x, y)$.

Similarly, if $z = f(x)$ where $x = g(s, t)$, then we can regard z as being a function of the two variables s and t:

$$z = h(s, t) = f(g(s, t)).$$

Using the Chain Rule here gives

$$\begin{aligned}\frac{\partial z}{\partial s} &= h_1(s, t) = \frac{dz}{dx}\frac{\partial x}{\partial s} = f'(g(s, t))\, g_1(s, t),\\ \frac{\partial z}{\partial t} &= h_2(s, t) = \frac{dz}{dx}\frac{\partial x}{\partial t} = f'(g(s, t))\, g_2(s, t).\end{aligned}$$

In this case we are really only using the single-variable form of the Chain Rule.

The following example involves a "hybrid" application of the Chain Rule to a function depending both directly and indirectly on the variable of differentiation.

EXAMPLE 12.3.3 Find dz/dt where $z = f(x, y, t)$ and $x = g(t)$, $y = h(t)$. (Assume that f, g, and h all have continuous derivatives.)

SOLUTION z depends on t through each of the three variables of f so there will be three terms in the appropriate Chain Rule:

$$\begin{aligned}\frac{dz}{dt} &= \frac{\partial z}{\partial x}\frac{dx}{dt} + \frac{\partial z}{\partial y}\frac{dy}{dt} + \frac{\partial z}{\partial t}\\ &= f_1(x, y, t)g'(t) + f_2(x, y, t)h'(t) + f_3(x, y, t).\end{aligned}$$

REMARK: In Example 12.3.3 we can easily distinguish between the meaning of the symbols dz/dt and $\partial z/\partial t$. If, however, we had been dealing with a function of the form $z = f(x, y, s, t)$ where $x = g(s, t)$ and $y = h(s, t)$ then the meaning of the symbol $\partial z/\partial t$ would be unclear; it could refer to either the simple partial derivative of f with respect to its fourth variable, ignoring x, y and s, or to the derivative of the composite function $f\big(g(s, t), h(s, t), s, t\big)$ with only s ignored. We distinguish these notationally by showing which variables are ignored as follows:

$$\begin{aligned}\left.\frac{\partial z}{\partial t}\right|_{x,y,s} &= \frac{\partial}{\partial t} f(x, y, s, t) = f_4(x, y, s, t),\\ \left.\frac{\partial z}{\partial t}\right|_{s} &= \frac{\partial}{\partial t} f\big(g(s, t), h(s, t), s, t\big) = f_1 g_2 + f_2 h_2 + f_4.\end{aligned}$$

In applications, the variables to be ignored in a given differentiation will usually be clear from the context.

EXAMPLE 12.3.4 Atmospheric temperature depends on position and time. If we denote position by means of three spatial coordinates (x, y, z) and time by t, then the temperature T can be represented by a function of four variables, $T(x, y, z, t)$. If a thermometer is attached to a weather balloon and moves through the atmosphere on a path with parametric equations $x = f(t)$, $y = g(t)$, $z = h(t)$, what is the rate of change of the temperature recorded by the thermometer at time t? Find the rate of change of the temperature at time $t = 1$ if

$$T(x, y, z, t) = \frac{e^{-z} \sin t}{5 + x^2 + y^2},$$

and if the balloon moves along the curve

$$x = f(t) = t, \quad y = g(t) = 2t, \quad z = h(t) = t - t^4.$$

SOLUTION Here the rate must take into account the change of position of the thermometer as well as increasing time. It is given by

$$\frac{dT}{dt} = \frac{\partial T}{\partial x}\frac{dx}{dt} + \frac{\partial T}{\partial y}\frac{dy}{dt} + \frac{\partial T}{\partial z}\frac{dz}{dt} + \frac{\partial T}{\partial t}.$$

Note that the partial derivative $\partial T/\partial t$ refers only to the rate of change of temperature with respect to time at a fixed position in the atmosphere. For the special case we have, at $t = 1$, $x = 1$, $y = 2$, $z = 0$, $\partial x/\partial t = 1$, $\partial y/\partial t = 2$, $\partial z/\partial t = -3$ and

$$\begin{aligned}
\frac{\partial T}{\partial x} &= -2x\frac{e^{-z}\sin t}{(5 + x^2 + y^2)^2} = -2\frac{\sin(1)}{100} = -\frac{\sin(1)}{50} \\
\frac{\partial T}{\partial y} &= -2y\frac{e^{-z}\sin t}{(5 + x^2 + y^2)^2} = -4\frac{\sin(1)}{100} = -\frac{\sin(1)}{25} \\
\frac{\partial T}{\partial z} &= -\frac{e^{-z}\sin t}{5 + x^2 + y^2} = -\frac{\sin(1)}{10} \\
\frac{\partial T}{\partial t} &= \frac{e^{-z}\cos t}{5 + x^2 + y^2} = \frac{\cos(1)}{10}.
\end{aligned}$$

Thus

$$\left.\frac{dT}{dt}\right|_{t=1} = -\sin(1)\left(\frac{1}{50} + \frac{2}{25} - \frac{3}{10}\right) + \frac{\cos(1)}{10} \approx 0.2223$$

The discussion and examples above show that the Chain Rule for functions of several variables can take many different forms depending on the nature of the various functions being composed and the numbers of their variables. As a mnemonic device for determining the correct form of the Chain Rule in a given situation one can construct a simple chart showing the dependence of the various variables on one another. Fig. 12.3.1 shows such a chart for the temperature function of Example 12.3.4.

The Chain Rule for dT/dt involves a term for every route from T to t in the chart; the route from T through x to t produces the term $\dfrac{\partial T}{\partial x}\dfrac{dx}{dt}$ and so on.

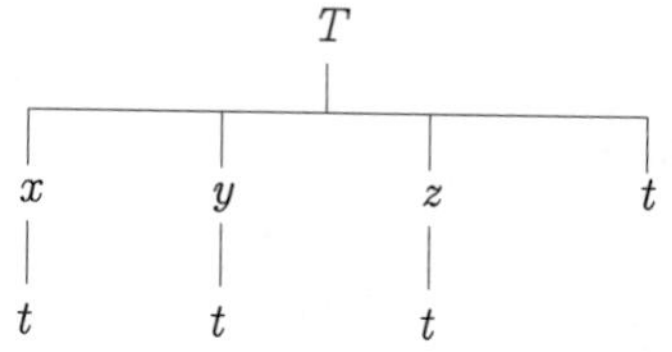

FIGURE 12.3.1

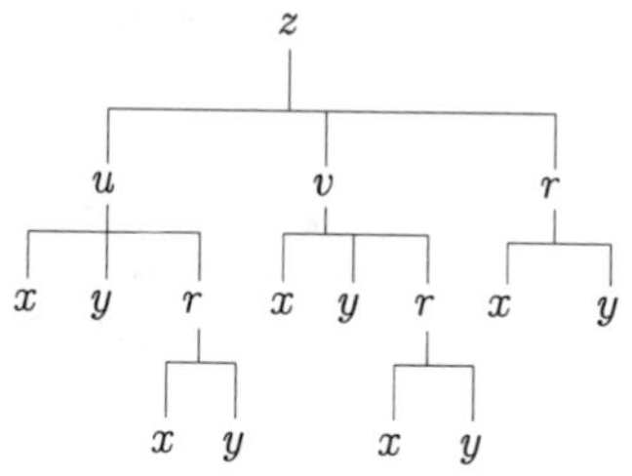

FIGURE 12.3.2

EXAMPLE 12.3.5 Write the appropriate Chain Rule for $\left.\dfrac{\partial z}{\partial x}\right|_y$, where $z = f(u, v, r)$, $u = g(x, y, r)$, $v = h(x, y, r)$ and $r = k(x, y)$.

SOLUTION With only y held fixed, the appropriate chart is shown in Fig. 12.3.2. There are five routes from z to x:

$$\begin{aligned}\left.\frac{\partial z}{\partial x}\right|_y &= \frac{\partial z}{\partial u}\frac{\partial u}{\partial x} + \frac{\partial z}{\partial u}\frac{\partial u}{\partial r}\frac{\partial r}{\partial x} + \frac{\partial z}{\partial v}\frac{\partial v}{\partial x} + \frac{\partial z}{\partial v}\frac{\partial v}{\partial r}\frac{\partial r}{\partial x} + \frac{\partial z}{\partial r}\frac{\partial r}{\partial x}\\ &= f_1 g_1 + f_1 g_3 k_1 + f_2 h_1 + f_2 h_3 k_1 + f_3 k_1.\end{aligned}$$

Homogeneous Functions

A function $f(x_1, x_2, \ldots, x_n)$ is said to be **positively homogeneous of degree** k if, for all values of its variables, and any positive real number t, we have

$$f(tx_1, tx_2, \ldots tx_n) = t^k f(x_1, x_2, \ldots, x_n).$$

For example,

$$\begin{aligned} f(x, y) &= x^2 + xy - y^2 \text{ is positively homogeneous of degree 2,}\\ f(x, y) &= \frac{2xy}{x^2 + y^2} \text{ is positively homogeneous of degree 0,}\\ f(x, y, z) &= \frac{x - y + z}{xy + z^2} \text{ is positively homogeneous of degree } -1.\end{aligned}$$

Observe that a positively homogeneous function of degree 0 remains constant along rays from the origin. More generally, along such rays a positively homogeneous function of degree k grows or decays proportionally to the kth power of distance from the origin. If f is sufficiently smooth, this observation is equivalent to the following theorem.

THEOREM 12.3.6 (**Euler's Theorem**) If $f(x_1, x_2, \dots, x_n)$ has continuous first partial derivatives, and is positively homogeneous of degree k then

$$x_1 f_1(x_1, \dots, x_n) + x_2 f_2(x_1, \dots, x_n) + \cdots + x_n f_n(x_1, \dots, x_n) = k f(x_1, \dots, x_n).$$

PROOF Differentiate the equation $f(tx_1, \dots, tx_n) = t^k f(x_1, \dots, x_n)$ with respect to t to get

$$x_1 f_1(tx_1, \dots, tx_n) + x_2 f_2(tx_1, \dots, tx_n) + \cdots + x_n f_n(tx_1, \dots, tx_n) = k\, t^{k-1} f(x_1, \dots, x_n).$$

Now substitute $t = 1$ to get the desired result. □

Note that Exercises 20–23 of Section 12.2 are illustrations of this theorem.

Higher Order Derivatives

Applications of the Chain Rule can become quite complicated when higher order derivatives are involved. It is important to keep in mind at every stage which variables are independent of one another. The following example shows that a smooth function $v(r, \theta)$ expressed in terms of polar coordinates $x = r\cos\theta$, $y = r\sin\theta$, is harmonic (i.e., satisfies the two dimensional Laplace equation—see Example 12.2.13) provided

$$\frac{\partial^2 v}{\partial r^2} + \frac{1}{r}\frac{\partial v}{\partial r} + \frac{1}{r^2}\frac{\partial^2 v}{\partial \theta^2} = 0.$$

EXAMPLE 12.3.7 If $u(x, y)$ has continuous partial derivatives of second order, and if $v(r, \theta) = u(x, y)$ where $x = r\cos\theta$ and $y = r\sin\theta$, show that

$$\frac{\partial^2 v}{\partial r^2} + \frac{1}{r}\frac{\partial v}{\partial r} + \frac{1}{r^2}\frac{\partial^2 v}{\partial \theta^2} = \frac{\partial^2 u}{\partial x^2} + \frac{\partial^2 u}{\partial y^2}.$$

SOLUTION First note that since $x = r\cos\theta$ and $y = r\sin\theta$,

$$\frac{\partial x}{\partial r} = \cos\theta, \qquad \frac{\partial x}{\partial \theta} = -r\sin\theta, \qquad \frac{\partial y}{\partial r} = \sin\theta, \qquad \frac{\partial y}{\partial \theta} = r\cos\theta.$$

Thus

$$\frac{\partial v}{\partial r} = \frac{\partial u}{\partial x}\frac{\partial x}{\partial r} + \frac{\partial u}{\partial y}\frac{\partial y}{\partial r} = \cos\theta\frac{\partial u}{\partial x} + \sin\theta\frac{\partial u}{\partial y}.$$

Now differentiate with respect to r again. Remember that r and θ are independent variables, so the factors $\cos\theta$ and $\sin\theta$ can be regarded as constants. However $\partial u/\partial x$ and $\partial u/\partial y$ depend on x and y and therefore on r and θ.

$$\begin{aligned}
\frac{\partial^2 v}{\partial r^2} &= \cos\theta\,\frac{\partial}{\partial r}\frac{\partial u}{\partial x} + \sin\theta\,\frac{\partial}{\partial r}\frac{\partial u}{\partial y} \\
&= \cos\theta\left[\cos\theta\,\frac{\partial^2 u}{\partial x^2} + \sin\theta\,\frac{\partial^2 u}{\partial y\partial x}\right] + \sin\theta\left[\cos\theta\,\frac{\partial^2 u}{\partial x\partial y} + \sin\theta\,\frac{\partial^2 u}{\partial y^2}\right] \\
&= \cos^2\theta\,\frac{\partial^2 u}{\partial x^2} + 2\cos\theta\sin\theta\,\frac{\partial^2 u}{\partial x\partial y} + \sin^2\theta\,\frac{\partial^2 u}{\partial y^2}.
\end{aligned}$$

We have used the equality of mixed partials to obtain the last line. Similarly,

$$\frac{\partial v}{\partial \theta} = -r\sin\theta\,\frac{\partial u}{\partial x} + r\cos\theta\,\frac{\partial u}{\partial y}.$$

When we differentiate with respect to θ a second time, we can regard r as constant, but each term is still a product of two functions which depend on θ:

$$\begin{aligned}
\frac{\partial^2 v}{\partial \theta^2} &= -r\left[\cos\theta\,\frac{\partial u}{\partial x} + \sin\theta\,\frac{\partial}{\partial \theta}\frac{\partial u}{\partial x}\right] + r\left[-\sin\theta\,\frac{\partial u}{\partial y} + \cos\theta\,\frac{\partial}{\partial \theta}\frac{\partial u}{\partial y}\right] \\
&= -r\frac{\partial v}{\partial r} - r\sin\theta\left[-r\sin\theta\,\frac{\partial^2 u}{\partial x^2} + r\cos\theta\,\frac{\partial^2 u}{\partial y \partial x}\right] \\
&\qquad + r\cos\theta\left[-r\sin\theta\,\frac{\partial^2 u}{\partial x \partial y} + r\cos\theta\,\frac{\partial^2 u}{\partial y^2}\right] \\
&= -r\frac{\partial v}{\partial r} + r^2\sin^2\theta\,\frac{\partial^2 u}{\partial x^2} - 2r^2\sin\theta\cos\theta\,\frac{\partial^2 u}{\partial x \partial y} + r^2\cos^2\theta\,\frac{\partial^2 u}{\partial y^2}.
\end{aligned}$$

Combining these results we obtain,

$$\frac{\partial^2 v}{\partial r^2} + \frac{1}{r}\frac{\partial v}{\partial r} + \frac{1}{r^2}\frac{\partial^2 v}{\partial \theta^2} = \frac{\partial^2 u}{\partial x^2} + \frac{\partial^2 u}{\partial y^2}.$$

Approximations and Differentiability

The tangent line to the graph $y = f(x)$ at $x = a$ provides a convenient way of approximating the value of $f(x)$ for x near a:

$$f(a+h) \approx f(a) + f'(a)h.$$

The mere existence of $f'(a)$ is sufficient to guarantee that the error in this approximation is small compared to the distance h between a and $a + h$, that is to say,

$$\lim_{h\to 0} \frac{f(a+h) - f(a) - f'(a)h}{h} = f'(a) - f'(a) = 0.$$

Similarly, we can use the height to the tangent plane

$$z = f(a,b) + f_1(a,b)(x-a) + f_2(a,b)(y-b)$$

to the graph $z = f(x,y)$ at (a,b) to approximate the value of $f(x,y)$ for (x,y) near (a,b): putting $x = a + h$ and $y = b + k$, we have

12.3.8
Tangent Plane Approximation

$$f(a+h, b+k) \approx f(a,b) + f_1(a,b)h + f_2(a,b)k.$$

EXAMPLE 12.3.9 Find an approximate value for $f(x, y) = \sqrt{2x^2 + e^{2y}}$ at $(2.2, -0.2)$.

SOLUTION It is convenient to use the tangent plane at $(2, 0)$:

$$\begin{aligned} & & f(2,0) &= 3, \\ f_1(x,y) &= \frac{2x}{\sqrt{2x^2+e^{2y}}}, & f_1(2,0) &= \frac{4}{3}, \\ f_2(x,y) &= \frac{e^{2y}}{\sqrt{2x^2+e^{2y}}}, & f_2(2,0) &= \frac{1}{3}. \end{aligned}$$

Taking $h = 0.2$ and $k = -0.2$, we obtain

$$\begin{aligned} f(2.2, -0.2) = f(2+h, 0+k) &\approx f(2,0) + f_1(2,0)h + f_2(2,0)k \\ &= 3 + \frac{0.8}{3} - \frac{0.2}{3} = 3.2\,. \end{aligned}$$

Unlike the single-variable case, the mere existence of the partial derivatives $f_1(a, b)$ and $f_2(a, b)$ does not even imply that f is continuous at (a, b), let alone that the error in the approximation 12.3.8 is small compared to the distance $\sqrt{h^2 + k^2}$ between (a, b) and $(a + h, b + k)$. We adopt this latter condition as our definition of what it means for a function to be differentiable at a point.

12.3.10
Definition of Differentiability

We say that the function $f(x, y)$ is **differentiable** at the point (a, b) if

$$\lim_{h,k\to 0} \frac{f(a+h, b+k) - f(a,b) - h\, f_1(a,b) - k f_2(a,b)}{\sqrt{h^2+k^2}} = 0.$$

This definition and the following theorems can be generalized to functions any number of variables in the obvious way. We state them for the two-variable case only for the sake of simplicity.

The function $f(x, y)$ is differentiable if and only if the surface $z = f(x, y)$ has a nonvertical tangent plane at (a, b). (Compare this with the single-variable situation.) In particular, the function is *continuous* wherever it is differentiable. We shall now establish a certain two-variable version of the Mean-Value Theorem, and use it to show that functions with *continuous* first partial derivatives are differentiable.

THEOREM 12.3.11 (*A Mean-Value Theorem*) If $f_1(x, y)$ and $f_2(x, y)$ are continuous in a disc of positive radius centred at (a, b) and if h and k have sufficiently small absolute value, then there exist numbers θ_1 and θ_2 between 0 and 1 such that

$$f(a+h, b+k) - f(a,b) = h f_1(a + \theta_1 h, b+k) + k f_2(a, b + \theta_2 k).$$

PROOF The proof is very similar to that of Theorem 12.2.12 so we give only a sketch of it here. Write

$$f(a+h,b+k)-f(a,b)=\big[f(a+h,b+k)-f(a,b+k)\big]+\big[f(a,b+k)-f(a,b)\big]$$

and apply the one-variable Mean-Value Theorem separately to $f(x,b+k)$ on the interval $[a,a+h]$ and to $f(a,y)$ on the interval $[b,b+k]$ to get the desired result. □

THEOREM 12.3.12 If f_1 and f_2 are continuous in a disc of positive radius centred at the point (a,b), then f is differentiable at (a,b).

PROOF Using Theorem 12.3.11 and the fact that $\left|\dfrac{h}{\sqrt{h^2+k^2}}\right|\le 1$ and $\left|\dfrac{k}{\sqrt{h^2+k^2}}\right|\le 1$, we estimate

$$\begin{aligned}&\left|\frac{f(a+h,b+k)-f(a,b)-hf_1(a,b)-kf_2(a,b)}{\sqrt{h^2+k^2}}\right|\\ &=\left|\frac{h}{\sqrt{h^2+k^2}}\big[f_1(a+\theta_1h,b+k)-f_1(a,b)\big]+\frac{k}{\sqrt{h^2+k^2}}\big[f_2(a,b+\theta_2k)-f_2(a,b)\big]\right|\\ &\le\big|f_1(a+\theta_1h,b+k)-f_1(a,b)\big|+\big|f_2(a,b+\theta_2k)-f_2(a,b)\big|.\end{aligned}$$

Since f_1 and f_2 are continuous at (a,b), each of these latter terms approaches zero as $(h,k)\to(0,0)$. This is what we wanted to prove. □

Let us illustrate differentiability with an example where we can calculate directly the error in the tangent plane approximation.

EXAMPLE 12.3.13 Calculate $f(x+h,y+k)-f(x,y)-f_1(x,y)h-f_2(x,y)k$ if $f(x,y)=x^3+xy^2$.

SOLUTION Since $f_1(x,y)=3x^2+y^2$ and $f_2(x,y)=2xy$, we have

$$\begin{aligned}&f(x+h,y+k)-f(x,y)-f_1(x,y)h-f_2(x,y)k\\ &\qquad=(x+h)^3+(x+h)(y+k)^2-x^3-xy^2-(3x^2+y^2)h-2xyk\\ &\qquad=3xh^2+h^3+2yhk+hk^2+xk^2.\end{aligned}$$

Observe that the result is a polynomial in h and k with no term of less than second degree in these variables. Evidently this difference approaches zero like the square of the distance from (h,k) to $(0,0)$ as $(h,k)\to(0,0)$, so f is certainly differentiable. Yhis quadratic behavior holds for any function f with continuous second partial derivatives. (See Exercise 31.)

Proof of the Chain Rule

We conclude this section by stating formally, and proving a simple but representative case of the Chain Rule for multivariate functions.

THEOREM 12.3.14 Let $z = f(x, y)$, where $x = u(s, t)$ and $y = v(s, t)$. Suppose that

i) $u(a, b) = p$ and $v(a, b) = q$,

ii) The first partial derivatives of u and v exist at (a, b).

iii) $f_1(x, y)$ and $f_2(x, y)$ are continuous in a disc of positive radius centred at (p, q).

Then $z = w(s, t) = f\big(u(s, t), v(s, t)\big)$ has first partial derivatives with respect to s and t at (a, b), and

$$\begin{aligned} w_1(a, b) &= f_1(p, q)u_1(a, b) + f_2(p, q)v_1(a, b) \\ w_2(a, b) &= f_1(p, q)u_2(a, b) + f_2(p, q)v_2(a, b), \end{aligned}$$

that is,

$$\frac{\partial z}{\partial s} = \frac{\partial z}{\partial x}\frac{\partial x}{\partial s} + \frac{\partial z}{\partial y}\frac{\partial y}{\partial s}, \qquad \frac{\partial z}{\partial t} = \frac{\partial z}{\partial x}\frac{\partial x}{\partial t} + \frac{\partial z}{\partial y}\frac{\partial y}{\partial t}.$$

PROOF We mimic the proof of the one-variable Chain Rule in Section 2.4. Let

$$\begin{aligned} E(0, 0) &= 0, \\ E(h, k) &= \frac{f(p + h, q + k) - f(p, q) - hf_1(p, q) - kf_2(p, q)}{\sqrt{h^2 + k^2}} \quad \text{if } (h, k) \neq (0, 0). \end{aligned}$$

Since f is differentiable at (p, q) by Theorem 12.3.12, $E(h, k)$ is continuous at $(0, 0)$. Now

$$f(p + h, q + k) - f(p, q) = hf_1(p, q) + kf_2(p, q) + \sqrt{h^2 + k^2}\, E(h, k).$$

In this formula put $h = u(a + \alpha, b) - u(a, b)$, $k = v(a + \alpha, b) - v(a, b)$ and divide by α to get

$$\begin{aligned} \frac{w(a + \alpha, b) - w(a, b)}{\alpha} &= \frac{f\big(u(a + \alpha, b), v(a + \alpha, b)\big) - f\big(u(a, b), v(a, b)\big)}{\alpha} \\ &= f_1(p, q)\frac{h}{\alpha} + f_2(p, q)\frac{k}{\alpha} + \sqrt{\left(\frac{h}{\alpha}\right)^2 + \left(\frac{k}{\alpha}\right)^2}\, E(h, k). \end{aligned}$$

Since $\lim_{\alpha \to 0} h/\alpha = u_1(a, b)$ and $\lim_{\alpha \to 0} k/\alpha = v_1(a, b)$ we have, letting $\alpha \to 0$,

$$w_1(a, b) = f_1(p, q)u_1(a, b) + f_2(p, q)v_1(a, b).$$

The proof for w_2 is similar. □

EXERCISES

In Exercises 1–6 write appropriate versions of the Chain Rule for the indicated derivatives.

1. $\dfrac{\partial w}{\partial t}$ if $w = f(x, y, z)$, $x = g(s, t)$, $y = h(s, t)$, $z = k(s, t)$.

2. $\dfrac{\partial w}{\partial t}$ if $w = f(x, y, z)$, $x = g(s)$, $y = h(s, t)$, $z = k(t)$.

3. $\dfrac{\partial z}{\partial u}$ if $z = g(x, y)$, $y = f(x)$, $x = h(u, v)$.

4. $\dfrac{dw}{dx}$ if $w = f(x, y, z)$, $y = g(x, z)$, $z = h(x)$.

5. $\left.\dfrac{\partial w}{\partial x}\right|_z$ if $w = f(x, y, z)$, $y = g(x, z)$.

6. $\dfrac{dw}{dt}$ if $w = f(x, y)$, $x = g(r, s)$, $y = h(r, t)$, $r = k(s, t)$, $s = m(t)$.

7. If $w = f(x, y, z)$, $x = g(r, s)$, $y = h(x, r, s)$, and $z = k(x, r, s)$, state appropriate versions of the Chain Rule for

(a) $\left.\dfrac{\partial w}{\partial r}\right|_{x,s}$, (b) $\left.\dfrac{\partial w}{\partial r}\right|_{y,s}$, (c) $\left.\dfrac{\partial w}{\partial r}\right|_{x,y,s}$, (d) $\left.\dfrac{\partial w}{\partial r}\right|_{s}$.

8. Find $\dfrac{\partial u}{\partial t}$ if $u = \sqrt{x^2 + y^2}$, $x = e^{st}$, $y = 1 + s^2 \cos t$.

9. Find $\dfrac{\partial z}{\partial x}$ if $z = \tan^{-1} \dfrac{u}{v}$, $u = 2x + y$, $v = 3x - y$.

10. Find $\dfrac{dz}{dt}$ if $z = txy^2$, $x = t + \ln(y + t^2)$, $y = e^t$.

In Exercises 11–14 find the indicated derivatives, assuming that the function $f(x, y)$ has continuous first partial derivatives.

11. $\dfrac{\partial}{\partial v} f(u^2, v^2)$

12. $\dfrac{\partial}{\partial t} f(st^2, s^2 + t)$

13. $\dfrac{\partial}{\partial x} f\big(f(x, y), f(x, y)\big)$

14. $\dfrac{\partial}{\partial y} f\big(yf(x, t), f(y, t)\big)$

15. Suppose that the temperature T in a certain liquid varies with depth z and time t according to the formula $T = e^{-t}z$. Find the rate of change of temperature with respect to time at a point which is moving through the liquid of a path whose depth at time t is $f(t)$. What is the rate if $f(t) = e^t$? What is happening in this case?

16. Suppose the strength S of an electric field in space depends on position (x, y, z) and time t according to the formula $S = f(x, y, z, t)$, where f has continuous first partial derivatives. Find the rate of change of the electric field strength with respect to time measured at a point moving through space along the spiral path $x = \sin t$, $y = \cos t$, $z = t$.

17. If $z = f(x, y)$ where f has continuous partial derivatives of all orders and $x = 2s + 3t$ and $y = 3s - 2t$, find
(a) $\dfrac{\partial^2 z}{\partial s^2}$, (b) $\dfrac{\partial^2 z}{\partial s \partial t}$, (c) $\dfrac{\partial^2 z}{\partial t^2}$, (d) $\dfrac{\partial^3 z}{\partial s^3}$.
Guess, and verify by induction, a formula for $\partial^n z/\partial s^n$, $n = 1, 2, 3, \ldots$

In Exercises 18–23 suppose f has continuous partial derivatives of all orders. Express the indicated derivatives in terms of the partial derivatives f_1, f_2, f_{11}, ... of f.

18. $\dfrac{\partial^2}{\partial s^2} f(st, t - s)$

19. $\dfrac{\partial^2}{\partial s\, \partial t} f(s^2, s + t^2)$

20. $\dfrac{\partial^2}{\partial x^2} x\, f(y, x)$

21. $\dfrac{\partial^2}{\partial s\, \partial t} f(t \sin s, t \cos s)$

22. If $f(x, y)$ is harmonic, show that $f(ax + by, bx - ay)$ is also harmonic.

23. If $f(x, y)$ is harmonic, show that $f(x^2 - y^2, 2xy)$ is also harmonic.

24.*If $x = e^s \cos t$ and $y = e^s \sin t$, show that

$$\frac{\partial^2 z}{\partial s^2} + \frac{\partial^2 z}{\partial t^2} = (x^2 + y^2)\left(\frac{\partial^2 z}{\partial x^2} + \frac{\partial^2 z}{\partial y^2}\right).$$

25. If $f(x, y)$ is positively homogeneous of degree k and has continuous partial derivatives of second order, show that

$$x^2 f_{11}(x, y) + 2xy f_{12}(x, y) + y^2 f_{22}(x, y) = k(k-1) f(x, y).$$

In Exercises 26–28 find approximate values of the given functions at the points indicated.

26. $f(x, y) = \sin(\pi xy + \ln(y))$ at $(0.01, 1.05)$

27. $f(x, y) = \dfrac{24}{x^2 + xy + y^2}$ at $(2.1, 1.8)$

28. $f(x, y, z) = \sqrt{x + 2y + 3z}$ at $(1.9, 1.8, 1.1)$

29.*By approximately what percentage will $f(x, y) = \dfrac{xy^2}{x^2 + y^2}$ increase or decrease if $x = 3$ and decreases by 2%, and $y = 4$ and increases by 1%?

30.*Prove the following version of the Mean-Value Theorem: if $f(x, y)$ has continuous first partial derivatives continuous in a disc centred at (a, b) and if h and h are sufficiently small in absolute value, then there exists a number θ satisfying $0 < \theta < 1$ such that

$$f(a + h, b + k) = f(a, b) + hf_1(a + \theta h, b + \theta k) + kf_2(a + \theta h, b + \theta k).$$

Hint: apply the one-variable Mean-Value Theorem to $g(t) = f(a+th, b+tk)$. Why could we not have used this result in place of Theorem 12.3.11 to prove Theorem 12.3.12 and hence the Chain Rule?

31. *Generalize the previous exercise as follows: show that if $f(x, y)$ has continuous partial derivatives of second order near the point (a, b), then there exists a number θ satisfying $0 < \theta < 1$ such that for h and k sufficiently small in absolute value,

$$f(a+h, b+k) = f(a, b) + hf_1(a, b) + kf_2(a, b) + \frac{1}{2}\Big(h^2 f_{11}(a+\theta h, b+\theta k) + 2hkf_{12}(a+\theta h, b+\theta k) + k^2 f_{22}(a+\theta h, b+\theta k)\Big).$$

Hence show that there is a constant K such that for sufficiently small h and k,

$$|f(a+h, b+k) - f(a, b) - hf_1(a, b) - kf_2(a, b)| \le K(h^2 + k^2).$$

12.4 GRADIENTS AND DIRECTIONAL DERIVATIVES

It is often useful to combine the first partial derivatives of a function into a single vector called the **gradient**. We will examine carefully the development and interpretation of the gradient vector in two dimensions, and will discuss extensions to three and higher dimensions at the end of the section. The reader may wish to review the material on plane vectors in Section 9.5 before reading this section.

12.4.1 The Gradient Vector

At any point where the partial derivatives $f_1(x, y)$ and $f_2(x, y)$ of the function $f(x, y)$ exist, we define the **gradient vector** ∇f by

$$\nabla f(x, y) = f_1(x, y)\,\mathbf{i} + f_2(x, y)\,\mathbf{j}.$$

Recall that $\mathbf{i}$ and $\mathbf{j}$ denote the unit basis vectors joining the origin $(0, 0)$ to the points $(1, 0)$ and $(0, 1)$ respectively.

EXAMPLE 12.4.2 If $f(x, y) = x^2 + y^2$, then $\nabla f(x, y) = 2x\,\mathbf{i} + 2y\,\mathbf{j}$. In particular, $\nabla f(1, 2) = 2\mathbf{i} + 4\mathbf{j}$. Observe that this vector is perpendicular to the straight line $x + 2y = 5$ which is tangent at $(1, 2)$ to the circle $x^2 + y^2 = 5$, the level curve of $f(x, y)$ which passes through the point $(1, 2)$. See Fig. 12.4.1. This is no coincidence, as the following theorem shows.

THEOREM 12.4.3 If $f(x, y)$ is differentiable at (a, b) and $\nabla f(a, b) \neq \mathbf{0}$, then $\nabla f(a, b)$ is a **normal vector** to the level curve of f which passes through (a, b). (That is, $\nabla f(a, b)$ is perpendicular to the level curve at that point.)

PROOF The angle θ between the vector $\nabla f(a, b)$ and the vector $h\mathbf{i}+k\mathbf{j}$ from the point (a, b) to the point $(a+h, b+k)$ (see Fig. 12.4.2) satisfies

$$\cos\theta = \frac{\nabla f(a, b) \bullet (h\mathbf{i} + k\mathbf{j})}{|\nabla f(a, b)|\sqrt{h^2 + k^2}} = \frac{1}{|\nabla f(a, b)|}\,\frac{hf_1(a, b) + kf_2(a, b)}{\sqrt{h^2 + k^2}}.$$

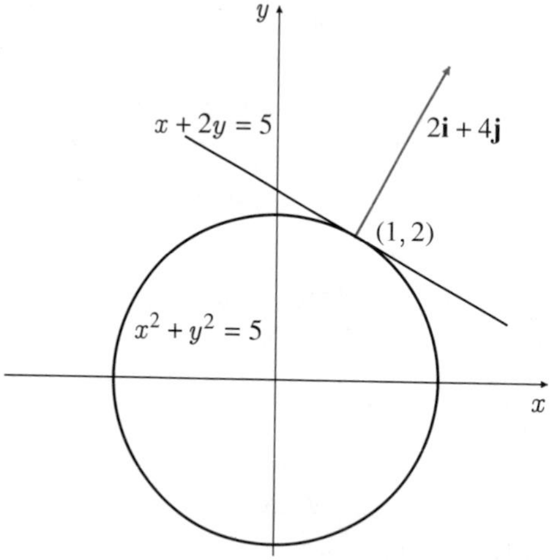

FIGURE 12.4.1

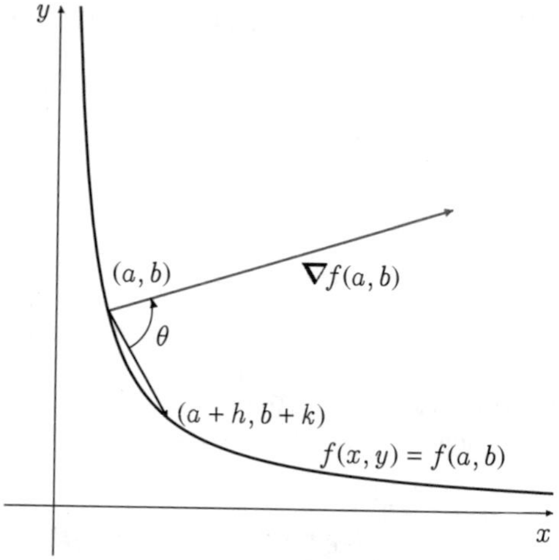

FIGURE 12.4.2

If $(a+h, b+k)$ lies on the level curve of f which passes through (a, b), then we have $f(a+h, b+k) = f(a, b)$, and, since f is differentiable at (a, b) (see Definition 12.3.10), it follows that

$$\lim_{(h,k)\to(0,0)} \frac{hf_1(a,b) + kf_2(a,b)}{\sqrt{h^2+k^2}} = -\lim_{(h,k)\to(0,0)} \frac{f(a+h,b+k) - f(a,b) - hf_1(a,b) - kf_2(a,b)}{\sqrt{h^2+k^2}} = 0.$$

Hence $\cos\theta \to 0$ and $\theta \to \pi/2$ as $(h, k) \to (0, 0)$; that is, $\nabla f(a, b)$ is perpendicular to the limit of short secant vectors to the level curve $f(x, y) = f(a, b)$. □

Directional Derivatives

The first partial derivatives $f_1(a, b)$ and $f_2(a, b)$ give the rates of change of $f(x, y)$ at (a, b) measured in the directions of the positive x- and y-axes respectively. If we want to know how fast $f(x, y)$ changes value as we move through (a, b) in some other direction in the plane, we require a more general **directional derivative**. If $\mathbf{v}$ is any nonzero vector, the directional derivative of f at (a, b) in the direction of $\mathbf{v}$ is the rate of change of f with respect to distance measured in that direction. The directional derivative depends only on the direction of $\mathbf{v}$, not on its length; for calculation it is useful to replace $\mathbf{v}$ by a unit vector $\mathbf{u}$ (a vector of length 1) in the same direction:

$$\mathbf{u} = \frac{\mathbf{v}}{|\mathbf{v}|} = u\mathbf{i} + v\mathbf{j}, \quad \text{where} \quad u^2 + v^2 = 1.$$

12.4.4
The Directional Derivative

If f has first partial derivatives at (a, b) and $\mathbf{u} = u\mathbf{i} + v\mathbf{j}$ is a unit vector (i.e., $|\mathbf{u}| = 1$), then the directional derivative of f in the direction of $\mathbf{u}$ is defined by

$$D_{\mathbf{u}} f(a,b) = \lim_{h\to 0} \frac{f(a+hu, b+hv) - f(a,b)}{h}.$$

The following theorem shows how the gradient can be used to calculate any directional derivative.

THEOREM 12.4.5 If $f(x, y)$ has continuous first partial derivatives in a disc of positive radius centred at (a, b) and if $\mathbf{u}$ is a unit vector, then the directional derivative of f in the direction of $\mathbf{u}$ is given by

$$D_{\mathbf{u}} f(a, b) = \mathbf{u} \bullet \nabla f(a, b).$$

PROOF For sufficiently small h, the first partial derivatives of f are continuous near all points of the straight line segment from (a, b) to $(a + hu, b + hv)$. If we apply the (one-dimensional) Mean-Value Theorem to the function $g(t) = f(a + thu, b + thv)$ on the interval $[0, 1]$, we obtain the formula $g(1) - g(0) = g'(\theta)$ for some θ between 0 and 1, that is,

$$f(a + hu, b + hv) - f(a, b) = huf_1(a + \theta hu, b + \theta hv) + hvf_2(a + \theta hu, b + \theta hv).$$

Hence

$$\begin{aligned} D_{\mathbf{u}} f(a, b) &= \lim_{h \to 0} \left[uf_1(a + \theta hu, b + \theta hv) + vf_2(a + \theta hu, b + \theta hv) \right] \\ &= uf_1(a, b) + vf_2(a, b) = \mathbf{u} \bullet \nabla f(a, b). \ \square \end{aligned}$$

REMARK: The first part of the above proof is essentially the proof of the Mean-Value Theorem referred to in Exercise 30 of Section 12.3.

EXAMPLE 12.4.6 Find the rate of change of $f(x, y) = y^4 + 2xy^3 + x^2y^2$ at $(0, 1)$ measured in the directions: (a) $\mathbf{i} + 2\mathbf{j}$, (b) $\mathbf{j} - 2\mathbf{i}$, (c) $\mathbf{i}$, (d) $\mathbf{i} + \mathbf{j}$.

SOLUTION $\nabla f(x, y) = (2y^3 + 2xy^2)\mathbf{i} + (4y^3 + 6xy^2 + 2x^2y)\mathbf{j}$, so $\nabla f(0, 1) = 2\mathbf{i} + 4\mathbf{j}$.

a) A unit vector in the direction of $\mathbf{i} + 2\mathbf{j}$ is $\mathbf{u} = \dfrac{1}{\sqrt{5}}(\mathbf{i} + 2\mathbf{j})$. The rate of change of f in this direction is the directional derivative

$$D_{\mathbf{u}} f(0, 1) = \frac{1}{\sqrt{5}}(\mathbf{i} + 2\mathbf{j}) \bullet (2\mathbf{i} + 4\mathbf{j}) = \frac{2 + 8}{\sqrt{5}} = 2\sqrt{5}.$$

Observe that $\mathbf{i} + 2\mathbf{j}$ points in the same direction as $\nabla f(0, 1)$ so the rate of change is positive and equal to the length of $\nabla f(0, 1)$.

b) Similarly, the directional derivative of f at $(0, 1)$ in the direction of $\mathbf{j} - 2\mathbf{i}$ is

$$D_{(\mathbf{j} - 2\mathbf{i})/\sqrt{5}} f(0, 1) = \frac{(-2\mathbf{i} + \mathbf{j}) \bullet (2\mathbf{i} + 4\mathbf{j})}{\sqrt{5}} = \frac{-4 + 4}{\sqrt{5}} = 0.$$

Since $\mathbf{j} - 2\mathbf{i}$ is perpendicular to $\nabla f(0,1)$, it is tangent to the level curve of f through $(0,1)$ and so the rate of change of f in that direction is zero.

c) $D_{\mathbf{i}} f(0,1) = \mathbf{i} \bullet (2\mathbf{i} + 4\mathbf{j}) = 2$. As noted previously, the directional derivative of f in the direction of the positive x-axis is just the partial derivative f_1.

d) A unit vector in the direction of $\mathbf{i} + \mathbf{j}$ is $\mathbf{u} = \dfrac{\mathbf{i}+\mathbf{j}}{|\mathbf{i}+\mathbf{j}|} = \dfrac{\mathbf{i}+\mathbf{j}}{\sqrt{2}}$. The directional derivative of f at $(0,1)$ in this direction is

$$D_{\mathbf{u}} f(0,1) = \frac{(\mathbf{i}+\mathbf{j}) \bullet (2\mathbf{i}+4\mathbf{j})}{\sqrt{2}} = \frac{2+4}{\sqrt{2}} = 3\sqrt{2}.$$

If we move along the surface $z = f(x,y)$ through the point $(0,1,1)$ in a direction making horizontal angles of 45 degrees with the positive directions of the x- and y-axes, we would be rising at a rate of $3\sqrt{2}$ vertical units per horizontal unit moved.

As observed in the previous example, Theorem 12.4.5 provides a useful interpretation for the gradient vector. For any unit vector $\mathbf{u}$ we have

$$D_{\mathbf{u}} f(a,b) = \mathbf{u} \bullet \nabla f(a,b) = |\nabla f(a,b)| \cos\theta,$$

Where θ is the angle between the vectors $\mathbf{u}$ and $\nabla f(a,b)$. Since $-1 \le \cos\theta \le 1$ for any θ, $D_{\mathbf{u}} f(a,b)$ only takes values between $-|\nabla f(a,b)|$ and $|\nabla f(a,b)|$. We will have $D_{\mathbf{u}} f(a,b) = |\nabla f(a,b)|$ if and only if $\cos\theta = 1$, that is, $\mathbf{u}$ points in the same direction as $\nabla f(a,b)$. Thus $\nabla f(a,b)$ points in the direction in which $f(x,y)$ increases most rapidly at (a,b), and the length of $\nabla f(a,b)$ is equal to this maximum rate of increase:

$$|\nabla f(a,b)| = \max_{|\mathbf{u}|=1} D_{\mathbf{u}} f(a,b).$$

Similarly, the minimum value of a directional derivative of f at the point (a,b) is $-|\nabla f(a,b)|$, and this value occurs in the direction opposite to the gradient. For example, the streams shown on the topographic map in Fig. 12.1.5 flow in the direction of steepest descent at all points, that is, in the direction of $-\nabla f$ where f is the function giving the elevation of the land. The streams therefore cross the contours (the level curves of f) at right angles.

The directional derivative $D_{\mathbf{u}} f(a,b)$ is zero when $\theta = \pi/2$, that is, in the direction of the tangent line to the level curve of f at (a,b).

EXAMPLE 12.4.7 In what direction at the point $(2,1)$ does the function $f(x,y) = x^2 e^{-y}$ increase most rapidly? What is the rate of increase of f in this direction?

SOLUTION $\nabla f(x,y) = 2xe^{-y}\,\mathbf{i} - x^2 e^{-y}\,\mathbf{j}$, so $\nabla f(2,1) = \dfrac{4}{e}\mathbf{i} - \dfrac{4}{e}\mathbf{j} = \dfrac{4}{e}(\mathbf{i} - \mathbf{j})$.

At $(2,1)$, $f(x,y)$ increases most rapidly in the direction of the vector $\mathbf{i} - \mathbf{j}$. The rate of increase in this direction in $|\nabla f(2,1)| = 4\sqrt{2}/e$.

EXAMPLE 12.4.8 The temperature T (in °C) at position (x, y) in the plane is given by

$$T(x, y) = 50e^{(9-x^2-2y^2)/10}.$$

Here x and y are measured in metres. A bug, crawling in the plane, is at position $(-1, 2)$.

a) If the bug wants to cool off as quickly as possible in what direction should she crawl?

b) If she crawls in that direction at a rate of 2 metres per minute, at what rate will she experience the temperature decreasing as she leaves $(-1, 2)$?

SOLUTION We have

$$\nabla T(x, y) = 50e^{(9-x^2-2y^2)/10}\left(\frac{-2x\mathbf{i} - 4y\mathbf{j}}{10}\right),$$

so $\nabla T(-1, 2) = 10(\mathbf{i} - 4\mathbf{j})$.

a) In order to cool off as quickly as possible the bug should crawl in the direction of $-\nabla T(-1, 2)$, that is, in the direction $-\mathbf{i} + 4\mathbf{j}$.

b) In the direction $-\mathbf{i} + 4\mathbf{j}$, T decreases at a rate $|\nabla T(-1, 2)| = 10\sqrt{17}$ degrees/metre. Since the bug crawls at a rate of 2 metres/minute, she will experience temperature decreasing at a rate $20\sqrt{17} \approx 82.5$°/min.

Tangent Lines to Level Curves

If $f(x, y)$ is differentiable at (a, b) and $\nabla f(a, b) \neq \mathbf{0}$, then the level curve of f passing through (a, b), that is, the curve with equation $f(x, y) = f(a, b)$, is **smooth** at (a, b); since it has a nonzero normal vector, namely the vector $\nabla f(a, b)$, it must also have a tangent line. If (x, y) is any point on the tangent line to the level curve at (a, b), then the vector $(x - a)\mathbf{i} + (y - b)\mathbf{j}$ is perpendicular to $\nabla f(a, b)$. Thus the equation of the tangent line can be written in the form

$$\nabla f(a, b) \bullet \left[(x - a)\mathbf{i} + (y - b)\mathbf{j}\right] = 0$$

which simplifies to $Ax + By = C$ where $A = f_1(a, b)$, $B = f_2(a, b)$ and $C = Aa + Bb$.

This equation for the tangent line can also be found by using implicit differentiation as in Section 2.5. Since $\nabla f(a, b)$ is assumed not to vanish, one or other of the first partial derivatives of f is nonzero at (a, b). To be specific, suppose $f_2(a, b) \neq 0$. It can be shown that this implies that the equation $f(x, y) = f(a, b)$ defines y implicitly as a function of x near (a, b). Differentiating the equation of the level curve implicitly with respect to x (using the Chain Rule), we obtain

$$f_1(x, y) + f_2(x, y)\frac{dy}{dx} = 0.$$

Thus the slope of the tangent line at (a, b) is

$$\left.\frac{dy}{dx}\right|_{x=a,\, y=b} = -\frac{f_1(a, b)}{f_2(a, b)} = -\frac{A}{B},$$

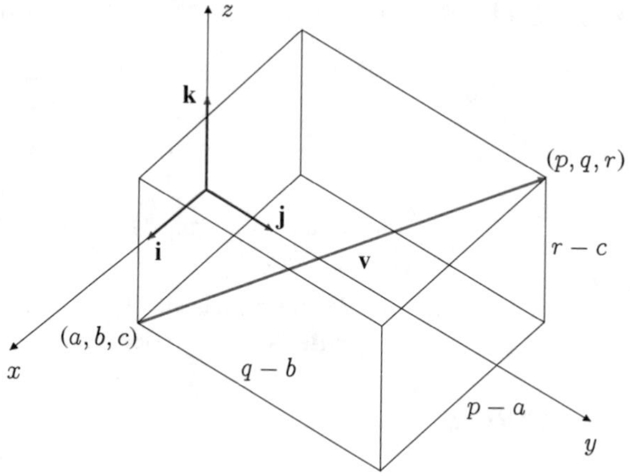

FIGURE 12.4.3

and once again we obtain $Ax + By = C$ as the equation of the tangent line. The same result would have occurred if we had assumed instead that $f_1(a, b) \neq 0$. In this case we would have regarded $f(x, y) = f(a, b)$ as defining x as a function of y and would have differentiated with respect to y to obtain dx/dy whose reciprocal is the slope of the tangent.

EXAMPLE 12.4.9 Find $\nabla f(1, 2)$ if $f(x, y) = x^2y^3$. Find an equation of the tangent line to the curve $x^2y^3 = 8$ at $(1, 2)$.

SOLUTION $\nabla f(x, y) = 2xy^3\mathbf{i} + 3x^2y^2\mathbf{j}$, so $\nabla f(1, 2) = 16\mathbf{i} + 12\mathbf{j}$. The equation of the tangent line must be

$$0 = \big[(x - 1)\mathbf{i} + (y - 2)\mathbf{j}\big] \bullet \big[16\mathbf{i} + 12\mathbf{j}\big] = 16(x - 1) + 12(y - 2),$$

or, more simply, $4x + 3y = 10$.

Higher Dimensional Vectors

Section 9.5 contains an introduction to vectors in the plane. Most of the concepts developed there extend in a natural way to vectors in three or more dimensions. For example, an arrow from the point $A = (a, b, c)$ to the point $P = (p, q, r)$ in 3-dimensional space is a vector $\mathbf{v} = \overrightarrow{AP}$ having length

$$|\mathbf{v}| = \sqrt{(p - a)^2 + (q - b)^2 + (r - c)^2}$$

and which can be written in terms of standard basis vectors,

$$\mathbf{v} = (p - a)\mathbf{i} + (q - b)\mathbf{j} + (r - c)\mathbf{k},$$

where $\mathbf{i}$, $\mathbf{j}$ and $\mathbf{k}$ are the mutually perpendicular unit vectors joining the origin $(0, 0, 0)$ to the points $(1, 0, 0)$, $(0, 1, 0)$ and $(0, 0, 1)$ respectively. See Fig. 12.4.3. $\mathbf{v}$ has components $(p - a, q - b, r - c)$.

Addition and scalar multiplication can be defined using components in the same way as for plane vectors: if $\mathbf{a} = a\mathbf{i} + b\mathbf{j} + c\mathbf{k}$ and $\mathbf{u} = u\mathbf{i} + v\mathbf{j} + w\mathbf{k}$, then

$$\mathbf{a} + \mathbf{u} = (a+u)\mathbf{i} + (b+v)\mathbf{j} + (c+w)\mathbf{k},$$
$$t\mathbf{A} = ta\mathbf{i} + tb\mathbf{j} + tc\mathbf{k}.$$

The dot product is defined similarly:

$$\mathbf{a} \bullet \mathbf{u} = au + bv + cw,$$

and has all the properties mentioned in Section 9.5. In particular, if θ is the angle between $\mathbf{a}$ and $\mathbf{u}$ then

$$\mathbf{a} \bullet \mathbf{u} = |\mathbf{a}||\mathbf{u}| \cos\theta.$$

Finally, we remark that all these concepts extend to spaces of dimension greater than three in the obvious way, although it is more difficult to visualize vectors of dimension higher than three.

The Gradient in Higher Dimensions

All of the machinery developed earlier for functions of two variables extends in an obvious way to functions of n variables. We shall summarize here the major results as they apply to a function $f(x, y, z)$ of three variables. The reader is invited to formulate the results for functions $f(x_1, x_2, \ldots, x_n)$ of n variables.

The function $f(x, y, z)$ is said to be differentiable at the point (a, b, c) if

$$\lim_{(h,k,m)\to(0,0,0)} \frac{f(a+h, b+k, c+m) - f(a,b,c) - hf_1(a,b,c) - kf_2(a,b,c) - mf_3(a,b,c)}{\sqrt{h^2+k^2+m^2}} = 0,$$

a condition which will be satisfied if all three first partial derivatives of f are continuous near the point (a, b, c). The gradient of f is the three-dimensional vector

$$\nabla f = f_1\,\mathbf{i} + f_2\,\mathbf{j} + f_3\,\mathbf{k}.$$

If f is differentiable at (a, b, c) and $\nabla f(a, b, c)$ is not the zero vector, then it is normal to the level surface of f passing through (a, b, c), that is, to the surface with equation $f(x, y, z) = f(a, b, c)$.

If $\mathbf{u} = u\mathbf{i}+v\mathbf{j}+w\mathbf{k}$ is a unit vector (so that $u^2+v^2+w^2 = 1$), then the directional derivative of f in the direction of $\mathbf{u}$ is

$$D_{\mathbf{u}}(a, b, c) = \lim_{h\to 0} \frac{f(a+hu, b+hv, c+hw) - f(a,b,c)}{h}.$$

If f has continuous first partial derivatives near (a, b, c), then the directional derivative is also given by

$$D_{\mathbf{u}}f(a, b, c) = \mathbf{u} \bullet \nabla f(a, b, c).$$

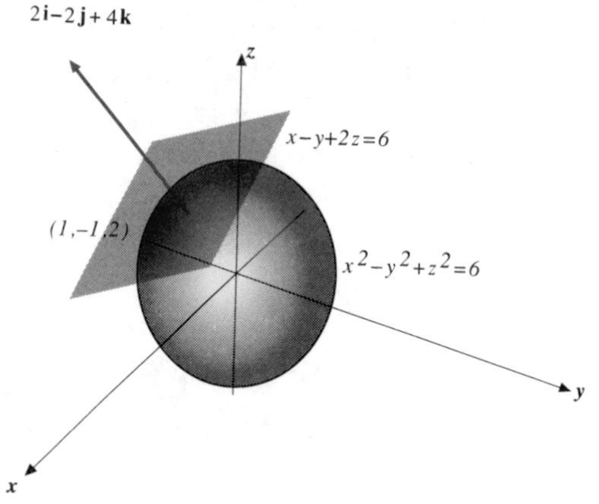

FIGURE 12.4.4

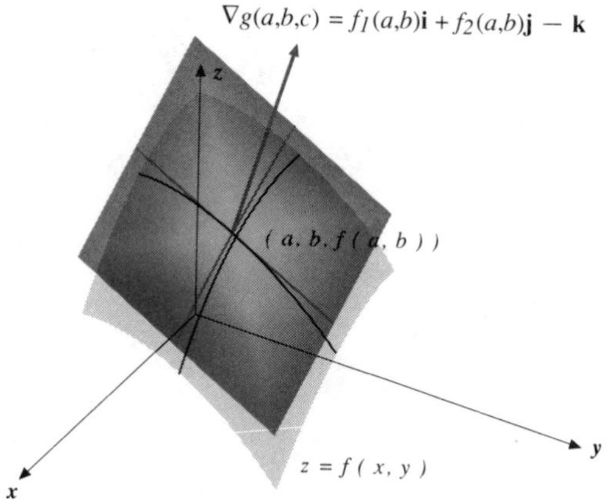

FIGURE 12.4.5

The level surface $f(x, y, z) = f(a, b, c)$ has a nonzero normal vector at (a, b, c) provided f is differentiable there and the gradient vector $\nabla f(a, b, c)$ is nonzero. Therefore the surface must be smooth and must have a tangent plane at (a, b, c), namely the unique plane passing through the point (a, b, c) which is perpendicular to $\nabla f(a, b, c)$. If (x, y, z) is any point on that tangent plane, then

$$\nabla f(a, b, c) \bullet \big[(x - a)\mathbf{i} + (y - b)\mathbf{j} + (z - c)\mathbf{k}\big] = 0,$$

which is therefore an equation of the tangent plane. This equation can be rewritten in the form $Ax+By+Cz = D$ where $A_1 = f_1(a, b, c)$, $B = f_2(a, b, c)$, $C = f_3(a, b, c)$ and $D = Aa + Bb + Cc$.

EXAMPLE 12.4.10 Let $f(x, y, z) = x^2 + y^2 + z^2$.

a) Find $\nabla f(1, -1, 2)$ and use it to find an equation of the tangent plane to the sphere $x^2 + y^2 + z^2 = 6$ at the point $(1, -1, 2)$.

b) What is the maximum rate of increase of f at $(1, -1, 2)$ and in what direction does it occur?

c) What is the rate of change of $f(x, y, z)$ at $(1, -1, 2)$ in the direction towards the point $(3, 1, 1)$?

SOLUTION a) $\nabla f = 2x\mathbf{i} + 2y\mathbf{j} + 2z\mathbf{k}$ so $\nabla f(1, -1, 2) = 2\mathbf{i} - 2\mathbf{j} + 4\mathbf{k}$. The tangent plane has equation $2(x - 1) - 2(y + 1) + 4(z - 2) = 0$, or $x - y + 2z = 6$. See Fig. 12.4.4.

b) The maximum rate of increase of f at $(1, -1, 2)$ is $|\nabla f(1, -1, 2)| = \sqrt{24} = 2\sqrt{6}$, and it occurs in the direction of $\nabla f(1, -1, 2)$, that is, in the direction of the vector $\mathbf{i} - \mathbf{j} + 2\mathbf{k}$.

c) The rate of change of f in the direction from $(1, -1, 2)$ towards $(3, 1, 1)$, that is, in the direction of the vector $(3 - 1)\mathbf{i} + (1 - (-1))\mathbf{j} + (1 - 2)\mathbf{k} = 2\mathbf{i} + 2\mathbf{j} - \mathbf{k}$, is

$$\frac{2\mathbf{i} + 2\mathbf{j} - \mathbf{k}}{\sqrt{(4 + 4 + 1)}} \bullet \big(2\mathbf{i} - 2\mathbf{j} + 4\mathbf{k}\big) = \frac{4 - 4 - 4}{3} = -\frac{4}{3},$$

so f is decreasing at a rate $4/3$ in this direction.

Tangents to the Graphs of Functions

In Section 12.2 we derived an equation of the tangent plane to the graph of a function $f(x, y)$ by an argument based on direct interpretation of the partial derivatives. This equation can also be obtained by considering gradients.

The graph of the function $f(x, y)$ is the surface $z = f(x, y)$ in three-dimensional space. This surface can be thought of as the level surface $g(x, y, z) = 0$ of the 3-variable function

$$g(x, y, z) = f(x, y) - z.$$

Since $g_1(x, y, z) = f_1(x, y)$, $g_2(x, y, z) = f_2(x, y)$ and $g_3(x, y, z) = -1$, g will be differentiable at any point (a, b, c) if f is differentiable at (a, b), and will have a nonzero gradient vector in this case:

$$\nabla g(a, b, c) = f_1(a, b)\mathbf{i} + f_2(a, b)\mathbf{j} - \mathbf{k}.$$

Therefore, at any point (a, b) where $f(x, y)$ is differentiable, the graph $z = f(x, y)$ of this function is smooth and has a tangent plane (see Fig. 12.4.5) whose equation is given by

$$\big[f_1(a, b)\mathbf{i} + f_2(a, b)\mathbf{j} - \mathbf{k}\big] \bullet \big[(x - a)\mathbf{i} + (y - b)\mathbf{j} + (z - f(a, b))\mathbf{k}\big] = 0,$$

or, more simply,

$$z = f(a, b) + f_1(a, b)(x - a) + f_2(a, b)(y - b).$$

This is the same equation derived in Section 12.2.

EXERCISES

In Exercises 1–10 find the gradients of the given functions at a general point ((x, y) or (x, y, z)) and at the specific point indicated.

1. $f(x, y) = x^2 - y^2$ at $(-2, 1)$
2. $f(x, y) = (x - y)/(x + y)$ at $(1, 1)$
3. $f(x, y) = \cos(x/y)$ at $(\pi, 4)$
4. $f(x, y) = e^{xy}$ at $(2, 0)$
5. $f(x, y) = \dfrac{x}{x^2 + y^2}$ at $(1, 2)$
6. $f(x, y) = \ln(x^2 + y^2)$ at $(1, -2)$
7. $f(x, y) = \dfrac{2xy}{x^2 + y^2}$ at $(0, 2)$
8. $f(x, y, z) = y\,e^{-x^2} \sin z$ at $\left(0, 1, \dfrac{\pi}{3}\right)$
9. $f(x, y, z) = x^2y + y^2z + z^2x$ at $(1, -1, 1)$
10. $f(x, y, z) = \cos(x + 2y + 3z)$ at (π, π, π)

In Exercises 11–16 find the rate of change of the given function at the given point in the given direction.

11. $f(x, y) = 3x - 4y$ at $(0, 2)$ in the direction of the vector $-\mathbf{i}$
12. $f(x, y) = x^2y$ at $(-1, -1)$ in the direction of the vector $\mathbf{i} + 2\mathbf{j}$
13. $f(x, y) = \dfrac{x}{1 + y}$ at $(0, 0)$ in the direction of the vector $\mathbf{i} - \mathbf{j}$
14. $f(x, y) = x^2 + y^2$ at $(1, 2)$ in a direction making a positive angle of $60°$ with the positive x-axis.
15. $f(x, y) = e^{x+2y}$ at the origin in the direction towards the point $(4, -3)$.
16. $f(x, y) = x^2 - y^2$ at $(1, 0)$ in the direction towards the point $(2, -1)$.

17. In what directions at the point $(1,1)$ does the function $f(x,y) = xy^2$ (a) increase most rapidly? (b) decrease most rapidly? (c) instantaneously neither increase nor decrease? (d) increase at rate 2?

18.*The height of land in the vicinity of a hill is given in terms of horizontal coordinates x and y by $h(x,y) = 80/(3+x^2+2y^2)$. where h is measured in the same units as x and y. In what (horizontal) direction is a stream flowing as it passes through the point where $x = 3$ and $y = 2$? At what (horizontal) angle to the path of the stream should a hiker at (3,2) set out if she wants to ascend the hill at a 45° angle to the horizontal?

19. Find a vector normal to the surface with equation $x^2 + 2y^2 + 3z^2 = 6$ at the point $(1,1,1)$.

20. Find the rate of change of the function $f(x,y,z) = xy^2z^3$ at the point $(0,-1,1)$ in the direction from this point towards the point $(2,-2,-1)$.

In Exercises 21–26 find an equation of the straight line tangent (at the point indicated) to the level curve of the given function passing through that point. (Use the results of Exercises 1–6.)

21. $f(x,y) = x^2 - y^2$ at $(-2,1)$

22. $f(x,y) = (x-y)/(x+y)$ at $(1,1)$

23. $f(x,y) = \cos(x/y)$ at $(\pi,4)$

24. $f(x,y) = e^{xy}$ at $(2,0)$ **25.** $f(x,y) = \dfrac{x}{x^2+y^2}$ at $(1,2)$

26. $f(x,y) = \ln(x^2+y^2)$ at $(1,-2)$

27. Find an equation of the tangent plane at $(1,-1,1)$ to the level surface of $f(x,y,z) = x^2y + y^2z + z^2x$ passing through that point. (Use Exercise 9.)

28. Find an equation of the tangent plane to the ellipsoid $x^2 + 2y^2 + 3z^2 = 20$ at the point $(3,-2,-1)$.

29.*Why do we require $\nabla f(a,b) \neq \mathbf{0}$ in Theorem 12.4.3? (Consider the function $f(x,y) = y^3 - x^2$ at $(0,0)$.)

30.*If $\nabla f(x,y) = 0$ throughout the disc $x^2+y^2 < r^2$, prove that $f(x,y) = C$ (constant) throughout that disc.

31.*Find $\nabla f(a,b)$ for a differentiable function $f(x,y)$ given that the directional derivatives of f at (a,b) in the directions $\mathbf{i}+\mathbf{j}$ and $3\mathbf{i} - 4\mathbf{j}$ are $3\sqrt{2}$ and 5 respectively.

32.*Let $f(x,y)$ have continuous first partial derivatives in a disc of positive radius centred at (a,b). If the directional derivatives of f at (a,b) is zero in each of two non-parallel directions, show that $\nabla f(a,b) = 0$.

12.5 EXTREME VALUES

The subject of maximum and minimum values of functions of several variables is interesting and important in applications of mathematics. Unfortunately, it is also rather complicated. We cannot give an extensive treatment in the context of this introductory chapter, so we shall limit our consideration mainly to functions of two variables; the discussion can be extended to functions of three or more variables but the form of the extension will not always be obvious.

Recall that a single-variable function $f(x)$ has a *local maximum value* (or a *local minimum value*) at a point a in its domain if $f(x) \le f(a)$ (or $f(x) \ge f(a)$) for all x in the domain of f which are sufficiently close to a. (If the appropriate inequality holds for all x in the domain of f, we can call the maximum or minimum value *absolute* instead of local.) Moreover, local (or absolute) extreme values can occur only at points which are

critical points — where the derivative $f'(x) = 0$, or
singular points — where $f'(x)$ does not exist, or
endpoints of the domain of f.

A similar situation obtains for functions of several variables.

12.5.1
Maximum and Minimum Values

> We say that $f(x, y)$ has a **local maximum** (or **minimum**) value $f(a, b)$ at the point (a, b) if $f(x, y) \leq f(a, b)$ (or $f(x, y) \geq f(a, b)$) for all points (x, y) in the domain of f which are sufficiently close to the point (a, b). (If the condition holds for all (x, y) in the domain of f, we can call the maximum or minimum value **absolute**.)

12.5.2
Location of Extreme Values

> A function $f(x, y)$ can have local or absolute extreme values only at points which are
>
> critical points — where $\nabla f(x, y) = \mathbf{0}$, i.e. $f_1(x, y) = f_2(x, y) = 0$, or
> singular points — where $\nabla f(x, y)$ fails to exist, or
> boundary points of the domain of f.

A point P is a **boundary point** of a set S in the plane if every disc of positive radius centred at P contains points in S and points not in S. A boundary point may or may not belong to the set. If all boundary points of S belong to S then we say that S is a **closed set**. Of course, if a function $f(x, y)$ is to have a maximum value at a boundary point of its domain then that boundary point must belong to the domain. Points of a set which are not boundary points are called **interior points** of the set. For example, if S is the set of all points in the plane which satisfy $x^2 + y^2 \leq 1$ (S is called a **closed disc**), then the boundary of S is the circle $x^2 + y^2 = 1$, and the interior of S consists of points satisfying $x^2 + y^2 < 1$.

We can prove the assertion in Box 12.5.2 as follows. If (a, b) is not a boundary point of f, and if $\nabla f(a, b)$ exists and is not the zero vector, then $f(x, y)$ is increasing as we move away from (a, b) in the direction of $\nabla f(a, b)$, and is decreasing if we move in the opposite direction. Thus f cannot have either a maximum or a minimum value at (a, b).

EXAMPLE 12.5.3 The function $f(x, y) = 1 + x^2 + y^2$ has a critical point at $(0, 0)$ since $f_1(x, y) = 2x$ and $f_2(x, y) = 2y$ both vanish at that point. Evidently $f(x, y) > 1 = f(0, 0)$ if $(x, y) \neq (0, 0)$, so f has an absolute minimum value at $(0, 0)$. See Fig. 12.5.1. Similarly, $g(x, y) = 1 - x^2 - y^2$ has an absolute maximum value at its critical point $(0, 0)$. See Fig. 12.5.2.

EXAMPLE 12.5.4 The function $f(x, y) = 1 + x^2 - y^2$ has a critical point at $(0, 0)$ but has neither a local maximum nor a local minimum value at this point: $f(0, 0) = 1$ but $f(x, 0) > 1$ and $f(0, y) < 1$ for all nonzero values of x and y. In view of the shape of the graph of f (see Fig. 12.5.3) we say that f has a **saddle point** at $(0, 0)$.

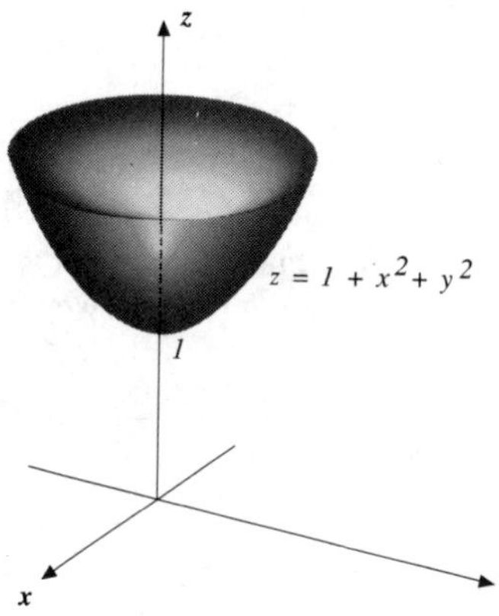

FIGURE 12.5.1

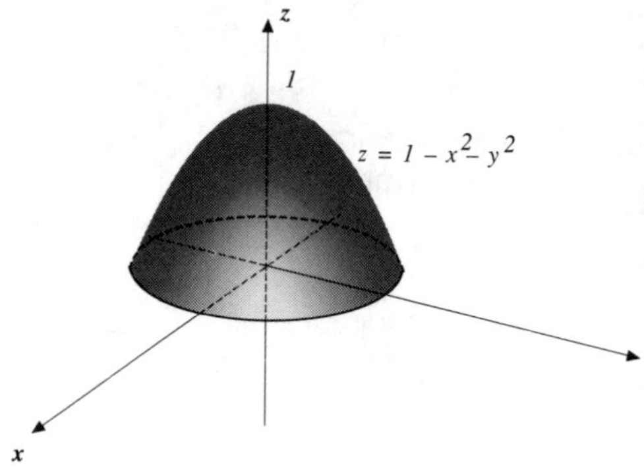

FIGURE 12.5.2

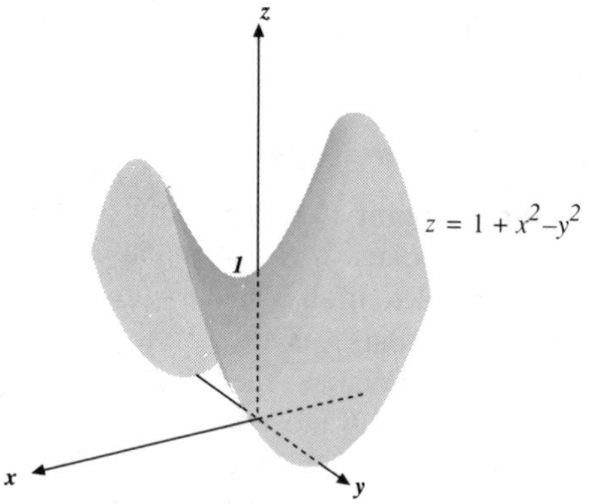

FIGURE 12.5.3

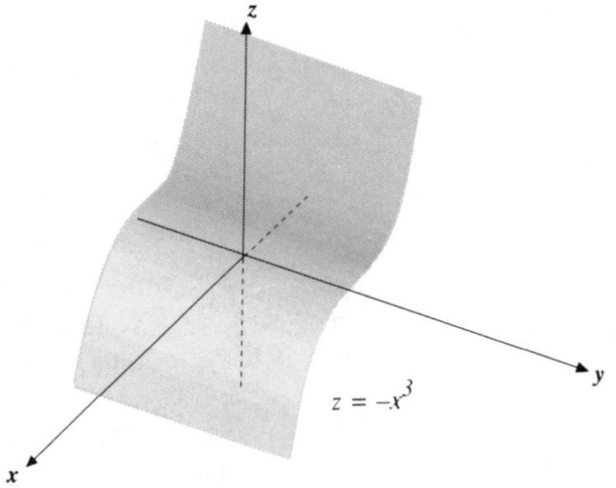

FIGURE 12.5.4

In general we shall call any interior critical point of $f(x,y)$ a **saddle point** if f does not have a local extreme value there. For example, $f(x,y) = -x^3$ has a whole line of saddle points along the y-axis though its graph (see Fig. 12.5.4) does not resemble a saddle near them. Saddle points generalize the horizontal inflection points of functions of one variable.

EXAMPLE 12.5.5 The function $f(x,y) = \sqrt{x^2+y^2}$ has no critical points, but it has a singular point at $(0,0)$ where it has a local (and absolute) minimum value. The graph of f is a circular cone. See Fig. 12.5.5.

EXAMPLE 12.5.6 The function $f(x,y) = 1 - x$ has no critical or singular points ($\nabla(x,y) = -\mathbf{i}$ at every point (x,y)), and so has no local extreme values. However, since the values of $f(x,y)$ decrease as x increases, if we restrict the domain of f to consist only of points in the disc $x^2+y^2 \le 1$ then f has local (and absolute) maximum value

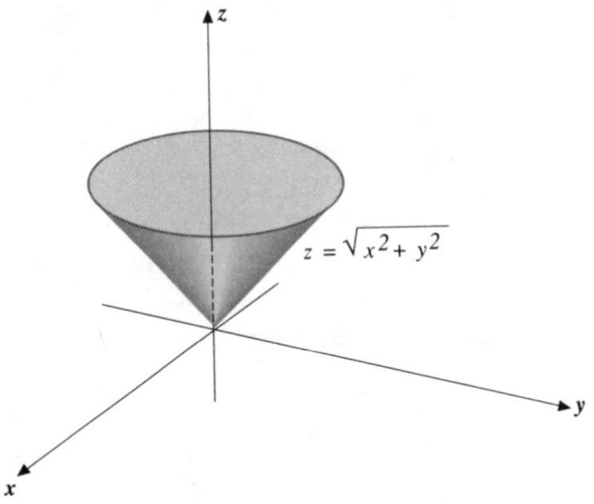

FIGURE 12.5.5

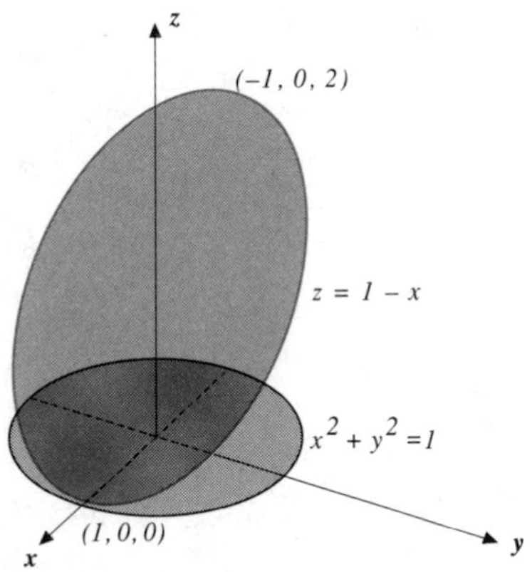

FIGURE 12.5.6

2 at the disc's leftmost boundary point $(-1, 0)$ and minimum value 0 at the disc's rightmost boundary point $(1, 0)$. See Fig. 12.5.6.

Classifying Critical Points

The following theorem is a generalization of the Second Derivative Test (Theorem 4.3.8) for extreme values of functions of one variable. In most cases it enables us to determine whether an interior critical point is a local maximum or local minimum or a saddle point.

THEOREM 12.5.7 (**A Second Derivative Test**) Suppose that (a, b) is a critical point of $f(x, y)$ and is an interior point of the domain of f. Suppose also that all the second partial derivatives of f exist and are continuous near (a, b). Let

$$A = f_{11}(a, b), \qquad B = f_{12}(a, b) = f_{21}(a, b), \qquad C = f_{22}(a, b).$$

a) If $B^2 - AC < 0$ and $A > 0$, then f has local minimum value at (a, b).

b) If $B^2 - AC < 0$ and $A < 0$, then f has local maximum value at (a, b).

c) If $B^2 - AC > 0$, then f has a saddle point at (a, b).

The test provides no information if $B^2 - AC = 0$.

PROOF We will apply the single-variable Second Derivative Test to the function

$$g(t) = f(a + th, b + tk)$$

where $(h, k) \neq (0, 0)$. Observe that $g'(0) = hf_1(a, b) + kf_2(a, b) = 0$. If $g''(0) > 0$ (or $g''(0) < 0$) for all such (h, k), then f will have a local maximum (or local minimum) value at (a, b). If there exist points (h, k) for which $g''(0) > 0$ and other such points for which $g''(0) < 0$, then f must have a saddle point at (a, b). It is easily calculated that

$$g''(0) = Ah^2 + 2Bhk + Ck^2.$$

Assume, for the moment, that $A \neq 0$. Completing the square in the above expression we obtain

$$g''(0) = A\left[\left(h + \frac{B}{A}k\right)^2 + \left(\frac{AC - B^2}{A^2}\right)k^2\right].$$

If $B^2 - AC < 0$ the expression in the large brackets is a sum of squares and so it is positive for all $(h, k) \neq (0, 0)$. Thus we will have $g''(0) > 0$ if $A > 0$ and $g''(0) < 0$ if $A < 0$. If $B^2 - AC > 0$ then the expression in the large brackets is a difference of squares and can be positive or negative for various values of h and k as close to 0 as we like. This completes the proof of the theorem provided $A \neq 0$. If $A = 0$ but $C \neq 0$ a similar proof works. If $A = C = 0$ but $B \neq 0$, then $g''(0) = 2Bhk$ which can clearly take on positive or negative values and so corresponds to a saddle point of f. □

EXAMPLE 12.5.8 Find and classify the critical points of the functions

a) $f(x, y) = x^3 + y^3 - 3xy$, (b) $g(x, y) = xye^{-(x^2+y^2)/2}$.

SOLUTION a) $f_1(x, y) = 3x^2 - 3y$, $f_2(x, y) = 3y^2 - 3x$.

For critical points $x^2 = y$ and $y^2 = x$. Thus $x^4 = x$ so $x = 0$ (and $y = 0$) or $x = 1$ (and $y = 1$). The critical points are $(0, 0)$ and $(1, 1)$.

Now $f_{11}(x, y) = 6x$, $f_{12}(x, y) = -3$, $f_{22}(x, y) = 6y$.

At $(0, 0)$: $A = 0 = C$, $B = -3$. Thus $B^2 - AC = 9 > 0$ and f has a saddle point at $(0, 0)$.

At $(1, 1)$: $A = 6 = C$, $B = -3$. Thus $B^2 - AC = -27 < 0$ and f has a local minimum at $(1, 1)$.

b) $g_1(x, y) = y(1 - x^2)e^{-(x^2+y^2)/2}$, $g_2(x, y) = x(1 - y^2)e^{-(x^2+y^2)/2}$.

The exponential can never vanish, so the critical points must satisfy $y(1-x^2) = 0$ and $x(1 - y^2) = 0$. They are $(0, 0)$, $(1, 1)$, $(1, -1)$, $(-1, 1)$ and $(-1, -1)$. Now

$$g_{11}(x, y) = xy(x^2 - 3)e^{-(x^2+y^2)/2},$$
$$g_{12}(x, y) = (1 - x^2)(1 - y^2)e^{-(x^2+y^2)/2},$$
$$g_{22}(x, y) = xy(y^2 - 3)e^{-(x^2+y^2)/2}.$$

At $(0, 0)$: $A = C = 0$, $B = 1$. Thus $B^2 - AC > 0$ and g has a saddle point at $(0, 0)$.

At $(1, 1)$ and $(-1, -1)$: $A = C = -2e^{-1} < 0$, $B = 0$. Thus $B^2 - AC < 0$ and g has local maximum values at these points.

At $(-1, 1)$ and $(1, -1)$: $A = C = 2e^{-1} > 0$, $B = 0$. Thus $B^2 - AC < 0$ and g has local minimum values at these points.

Extreme Values of Functions Defined on Closed, Bounded Sets

Theorem 1.7.7 states that if a function $f(x)$ is continuous on a closed, finite interval then it assumes absolute maximum and minimum values at points of that interval. (This is proved in Appendix II.) A similar result holds for functions of several variables. We state the two-dimensional version; we say that a set S in the plane is **bounded** if S is contained inside some disc which has finite radius.

THEOREM 12.5.9 If $f(x, y)$ is continuous on a closed, bounded set in the plane then f takes on absolute maximum and minimum values on S. That is, there exist points (x_0, y_0) and (x_1, y_1) in S such that for every point (x, y) in S we have

$$f(x_0, y_0) \leq f(x, y) \leq f(x_1, y_1). \ \square$$

We will not attempt to prove this, but note that the points (x_0, y_0) and (x_1, y_1) must be critical points or singular points of f, or boundary points of the domain of f.

EXAMPLE 12.5.10 Find the maximum and minimum values of $f(x, y) = xy - y^2$ on the disc $x^2 + y^2 \leq 1$.

SOLUTION Since f is continuous and the disc is a closed, bounded set, f must have absolute maximum and minimum values at some points of the disc. The partial derivatives of f are

$$f_1(x, y) = y, \qquad \text{and} \qquad f_2(x, y) = x - 2y.$$

Both of these derivatives are continuous everywhere so f has no singular points. Any critical point must satisfy $y = 0$, $x - 2y = 0$, so the origin, an interior point of the domain of f, is its only critical point. We have $f(0, 0) = 0$.

We must also look at boundary points of the disc, that is, points of the circle $x^2 + y^2 = 1$. We can express $f(x, y)$ as a function of one variable on this circle by using a convenient parametrization of the circle, say

$$x = \cos t, \ y = \sin t, \ 0 \leq t \leq 2\pi.$$

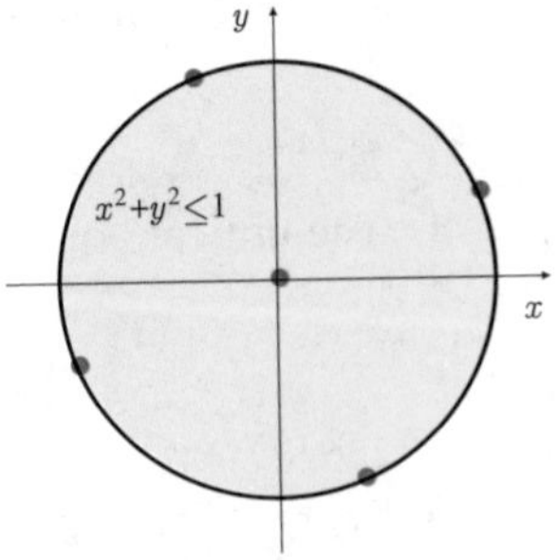

FIGURE 12.5.7

We have

$$\begin{aligned} f(\cos t, \sin t) &= \cos t \sin t - \sin^2 t \\ &= \frac{\sin 2t}{2} - \frac{1 - \cos 2t}{2} = g(t). \end{aligned}$$

In order to find any extreme values of $g(t)$, we look at its critical points:

$$\begin{aligned} &0 = g'(t) = \cos 2t - \sin 2t \quad \Leftrightarrow \quad \tan 2t = 1, \\ &2t = \frac{\pi}{4}, \ \frac{5\pi}{4}, \ \frac{9\pi}{4}, \ \frac{13\pi}{4}, \qquad \text{so} \quad t = \frac{\pi}{8}, \ \frac{5\pi}{8}, \ \frac{9\pi}{8}, \ \frac{13\pi}{8}. \end{aligned}$$

Evidently g is periodic with period π so

$$g\left(\frac{\pi}{8}\right) = g\left(\frac{9\pi}{8}\right) = \frac{2 - \sqrt{2}}{2\sqrt{2}} > 0, \qquad g\left(\frac{5\pi}{8}\right) = g\left(\frac{13\pi}{8}\right) = -\frac{2 + \sqrt{2}}{2\sqrt{2}} < 0.$$

Because g is periodic, the endpoints $t = 0$ and $t = 2\pi$ will not contribute any local extreme values of f unless they are critical points of g. (Why?) Thus the (absolute) maximum value of f over the disc must be $\dfrac{2 - \sqrt{2}}{2\sqrt{2}}$, and the (absolute) minimum value must be $-\dfrac{2 + \sqrt{2}}{2\sqrt{2}}$, each value occurring at two boundary points. (See Fig. 12.5.7.) The second derivative test shows that, in this example, the interior critical point, $(0, 0)$, is a saddle point of f.

EXAMPLE 12.5.11 Find the extreme values of the function $f(x,y) = x^2ye^{-(x+y)}$ on the triangular region $x \geq 0,\ y \geq 0,\ x + y \leq 4$.

SOLUTION First we look for critical points:

$$0 = f_1(x,y) = (2xy - x^2y)e^{-(x+y)} \Leftrightarrow xy(2-x) = 0 \Leftrightarrow x = 0,\ y = 0,\ \text{or } x = 2.$$
$$0 = f_2(x,y) = (x^2 - x^2y)e^{-(x+y)} \Leftrightarrow x^2(1-y) = 0 \Leftrightarrow x = 0 \text{ or } y = 1.$$

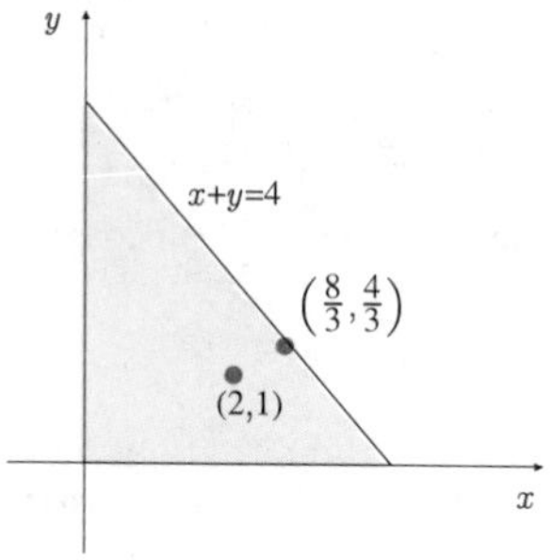

FIGURE 12.5.8

These equations are satisfied for any y if $x = 0$, or, if $x = 2$ and $y = 1$. The critical points are $(0, y)$ for any y, and $(2, 1)$. Note that $(2, 1)$ is interior to the triangular region. See Fig. 12.5.8. There are no singular points. We have $f(0, y) = 0$, $f(2, 1) = 4/e^3 \approx 0.199$. The boundary of the domain of f consists of three straight line segments:

i) $x = 0,\ 0 \leq y \leq 4$,

ii) $y = 0,\ 0 \leq x \leq 4$, and

iii) $x + y = 4,\ 0 \leq x \leq 4$. Note that $f(x, y) = 0$ at all points of the first two segments, and $f(x, y) > 0$ at all other points in the triangular region. On the third boundary segment we can express $f(x, y)$ as a function $g(x)$ of the single variable x:

$$g(x) = f(x, 4-x) = x^2(4-x)e^{-4}.$$

Clearly $g(0) = g(4) = 0$, and $g(x) > 0$ if $0 < x < 4$. The critical points of g are given by

$$0 = g'(x) = (8x - 3x^2)e^{-4},$$

so they are $x = 0$ and $x = 8/3$. We have $g(8/3) = (256/27)e^{-4} \approx 0.174 < f(2, 1)$. We conclude that the maximum value of f over the triangular region is $4/e^3$ and it occurs at the interior point $(2, 1)$. The minimum value of f over the region is 0 and this value occurs at all points of the two perpendicular sides of the triangular boundary. Note that f has neither a local maximum nor a local minimum at the boundary point $(8/3, 4/3)$ although g has a local maximum there.

Extreme Value Problems with Constraints

A **constrained extreme value problem** is one in which the variables of the function to be maximized or minimized are not independent of one another, but must satisfy one or more constraint equations or inequalities. For example, the problem

$$\textbf{maximize } f(x,y) \textbf{ subject to } g(x,y) = C,$$

has an equation constraint and the problem

$$\textbf{maximize } f(x,y) \textbf{ subject to } g(x,y) \leq C$$

has an inequality constraint.

Generally, inequality constraints restrict the domain of the function to be extremized to a set which still has interior points. Examples 12.5.10 and 12.5.11 deal with problems of this sort. In the solutions we looked for "free" critical points in the interior of the domain, and then we looked for other extreme values subject to "constraint equations" restricting our attention to points on the boundary of the domain.

For the rest of this section we shall be dealing with problems with equation constraints. Such constraints usually force us to look for extreme values of a function on some curve in space, a set with no interior points. If the constraint equations can be solved for some variables in terms of others, then the problem can be reduced by substitution to a free (that is, unconstrained) problem for a function with fewer variables. (This is what we did in Examples 12.5.10 and 12.5.11 to deal with the boundary cases.) If the constraint equation $g(x, y) = C$ can be solved for $y = h(x)$ then the problem

$$\textbf{maximize } f(x, y) \textbf{ subject to } g(x, y) = C$$

can be reduced to the unconstrained, single-variable problem

$$\textbf{maximize } F(x) = f(x, h(x)).$$

However, this approach to equation constraints is not very satisfactory because it is often difficult, if not impossible, to solve the constraint equations.

In order to discover a somewhat better approach to dealing with equation constraints, we look at a specific, concrete example.

EXAMPLE 12.5.12 Find the shortest distance from the origin to the curve $x^2y = 16$.

SOLUTION It is sufficient to minimize the square of the distance from the point (x, y) on the curve $x^2y = 16$ to the origin, that is, to solve the problem

$$\textbf{minimize } f(x, y) = x^2 + y^2 \textbf{ subject to } g(x, y) = x^2y = 16.$$

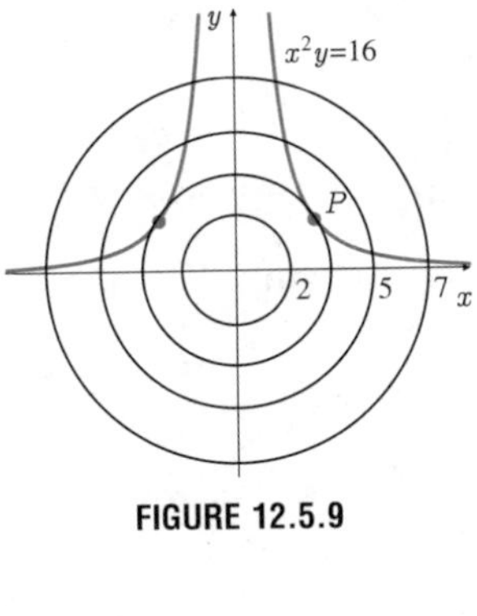

FIGURE 12.5.9

As noted above, we could solve the constraint equation for $y = 16/x^2$ and substitute into f, thus reducing the problem to one of finding the unconstrained minimum value of $F(x) = f(x, 16/x^2) = x^2 + (256/x^4)$. Instead let us examine the problem geometrically. Fig. 12.5.9 shows the graph of the curve C with equation $x^2y = 16$, as well as some level curves of f. (The latter are circles centred at the origin.) Let P be a point on C which is closest to the origin. (By symmetry, there are two such points.) Evidently the level curve of f passing through P is tangent to C at P. Thus the gradient vectors of $f(x, y)$ and $g(x, y)$ at P, which are normal to these two curves at P, must be parallel to each other.

Now $\nabla f(x, y) = 2x\mathbf{i}+2y\mathbf{j}$ and $\nabla g(x, y) = 2xy\mathbf{i}+x^2\mathbf{j}$. These vectors are parallel if one is a scalar multiple of the other, say

$$\nabla f(x, y) = -\lambda \nabla g(x, y) \text{ for some real number } \lambda,$$

that is,

$$2x = -\lambda 2xy, \qquad \text{and} \qquad 2y = -\lambda x^2.$$

Eliminating λ from these two equations we get $\dfrac{x}{y} = \dfrac{2xy}{x^2}$, or $x^2 = 2y^2$. The coordinates of P must satisfy this equation, and also the constraint equation $x^2y = 16$. Thus $2y^3 = 16$ and $y = 2$. Finally, $x = \pm 2\sqrt{2}$ and the two points on C closest to the origin are $(\pm 2\sqrt{2}, 2)$. The shortest distance from C to the origin is $\sqrt{8+4} = 2\sqrt{3}$ units.

The crux of the above solution was the observation that at the point where the extreme value occurred, the constraint curve and the level curve of f were tangent and so had parallel gradients; $\nabla f = -\lambda \nabla g$, or $\nabla(f + \lambda g) = 0$. Thus the extreme value occurs at critical points of the function $L = f(x, y) + \lambda g(x, y)$. The function L is called the **Lagrangian** for the constrained extreme value problem, the number λ is called a **Lagrange multiplier**, and the method used to solve the problem is called the **method of Lagrange multipliers**.

The method of Lagrange multipliers can be applied to find extreme values of functions of n variables subject to at most $n - 1$ equation constraints in those variables. The general procedure involves defining a Lagrangian function by adding constant multiples of the constraint functions to the function to be extremized (a different bultiplier is used for each constraint), and then looking for critical or singular points of the Lagrangian as well as endpoints or edgepoints of the constraint set.

EXAMPLE 12.5.13 Find the maximum value of $f(x, y) = x + y$ on the ellipse $x^2 + 2y^2 = 6$.

SOLUTION The method of Lagrange multipliers requires that we look for critical points of

$$L = x + y + \lambda(x^2 + 2y^2 - 6).$$

We have

$$\begin{aligned} 0 &= \frac{\partial L}{\partial x} = 1 + 2\lambda x, \\ 0 &= \frac{\partial L}{\partial y} = 1 + 4\lambda y, \\ 0 &= \frac{\partial L}{\partial \lambda} = x^2 + 2y^2 - 6, \quad \text{the constraint equation.} \end{aligned}$$

Now eliminate λ between the first two equations:

$$2x = -\frac{1}{\lambda} = 4y, \qquad \text{so} \qquad x = 2y.$$

Substitute this relation into the constraint equation to get $6y^2 = 6$, or $y = \pm 1$, whence $x = \pm 2$. There are two candidates for points where extreme values can occur, $(2, 1)$ and $(-2, -1)$. Since $f(2, 1) = 2+1 = 3$ and $f(-2, -1) = -3$, therefore the maximum value of $f(x, y) = x + y$ on the ellipse is 3. (The minimum value is -3.)

EXERCISES

In Exercises 1–10 find and classify the critical points of the given functions.

1. $f(x,y) = x^2 + 2y^2 - 4x + 4y$
2. $f(x,y) = xy - x + y$
3. $f(x,y) = 2x^3 - 6xy + 3y^2$
4. $f(x,y) = x^4 + y^4 - 4xy$
5. $f(x,y) = \dfrac{x}{y} + \dfrac{8}{x} - y$
6. $f(x,y) = \cos(x+y)$
7. $f(x,y) = x \sin y$
8. $f(x,y) = \cos x + \cos y$
9. $f(x,y) = xye^{-(x^2+y^2)/2}$
10. $f(x,y) = \dfrac{xy}{2 + x^4 + y^4}$
11. Find the maximum and minimum values of $f(x,y) = x + 2y$ on the disc $x^2 + y^2 \le 1$.
12. Find the maximum value of $f(x,y) = xy - x^3y^2$ on the square $0 \le x \le 1$, $0 \le y \le 1$.
13. Find the maximum and minimum values of $f(x,y) = \sin x \cos y$ on the closed triangular region bounded by the coordinate axes and the line $x + y = 2\pi$.
14. Find the maximum and minimum values of the function $f(x,y) = x^2 y$ on the disc $x^2 + y^2 \le 1$.
15. Find the shortest distance from the point $(3,0)$ to the curve $y = x^2$
    a) by reducing to an unconstrained problem in one variable.
    b) by using the method of Lagrange multipliers.
16. Find the shortest distance from the origin $(0,0,0)$ to the plane with equation $x + 2y + 2z = 3$
    a) by using geometric arguments (vectors, but no calculus).
    b) by reducing the problem to an unconstrained problem in two variables.
    c) by using the method of Lagrange multipliers.

It is suggested that Lagrange multipliers be used in Exercises 17–25.

17. Find the maximum and minimum values of $2x - y$ on the circle $x^2 + y^2 = 5$.
18. Find the maximum and minimum values of xy on the ellipse $x^2 + 2y^2 = 4$.
19. Find the points on the ellipse $3x^2 + 2xy + 3y^2 = 16$ which are (a) nearest the origin, and (b) farthest from the origin. The centre of the ellipse is at the origin. Where are its major and minor axes?
20. Find the maximum and minimum values of $f(x,y,z) = x + y - z$ over the sphere $x^2 + y^2 + z^2 = 1$.
21. Find the shortest distance from the origin to the surface $xyz^2 = 2$.
22.*Find the maximum and minimum values of $f(x,y,z) = x + y^2 z$ subject to the two constraints $y^2 + z^2 = 2$ and $z = x$.
23. Find the volume of the largest rectangular box with sides parallel to the coordinate planes which can be inscribed inside the ellipsoid $4x^2 + y^2 + z^2 = 12$.
24. Show that among all rectangular boxes with specified volume V, the one which is a cube has the least total surface area.
25.*If x, y, z are the angles of a triangle, show that

$$\sin\frac{x}{2}\sin\frac{y}{2}\sin\frac{z}{2} \le \frac{1}{8}.$$

For what triangles does equality occur?

26. What is meant by "boundary point" of a set in 3-dimensional space? closed set? interior of a set? If S is the set of points in 3-space which form the plane with equation $x + y + z = 0$, what is the boundary of S? the interior of S? Is S a closed set?
27.*Find the critical points of the function $f(x,y,z) = xyz - x^2 - y^2 - z^2$. Determine whether each critical point is a local maximum, a local minimum, or neither. Hint: Let $\mathbf{u} = u\mathbf{i} + v\mathbf{j} + w\mathbf{k}$ be a unit vector. Calculate the **second directional derivative** of f in the direction of $\mathbf{u}$, that is, $D_{\mathbf{u}}(D_{\mathbf{u}}f(x,y,z))$, and evaluate it at each critical point of f.
28.*Suppose that f and g are differentiable functions in the plane, and suppose that $g_2(a,b) \ne 0$. This implies that the equation $g(x,y) = C$ defines y implicitly as a (differentiable) function of x near $x = a$. Use the chain rule to show that if $f(x,y)$ has a local extreme value at (a,b) subject to the constraint $g(x,y) = C$ then (a,b) is a critical point of the function $L = f(x,y) + \lambda g(x,y)$. This constitutes a more formal justification for the method of Lagrange multipliers in this case.

Appendices

APPENDIX I MATHEMATICAL INDUCTION

Mathematical induction is a technique for proving the validity of statements about the integer n that are suspected to be true for all integers greater than or equal to some starting integer n_0.

EXAMPLE A1.1 Show that for $n = 1, 2, 3, \ldots$, the sum of integers from 1 up to n is $n(n+1)/2$, that is,

$$\sum_{k=1}^{n} k = 1 + 2 + 3 + \cdots + n = \frac{n(n+1)}{2} \qquad (*)$$

PROOF A direct proof of this fact was given in Section 6.1. An alternate inductive proof can be given as follows.

First, observe that the formula $(*)$ is true for $n = 1$; in this case the left side is 1 and the right side is $(1+1)/2 = 1$. Now *assume* that formula $(*)$ is true for *some* value of $n \geq 1$, say for $n = k$; that is, we are assuming that

$$1 + 2 + 3 + \cdots + k = \frac{k(k+1)}{2}.$$

Consider the sum of the numbers from 1 up to $k+1$. Using the assumption above we calculate

$$\begin{aligned}1+2+3+\cdots+k+(k+1) &= (1+2+3+\cdots+k)+(k+1)\\ &= \frac{k(k+1)}{2}+(k+1)=\frac{(k+1)(k+2)}{2}.\end{aligned}$$

This is just what formula $(*)$ says if $n=k+1$. Therefore, if formula $(*)$ is true for $n=k$ then it must also be true for $n=k+1$.

Since $(*)$ is true for $n=1$, therefore it is true for $n=2$; since it is true for $n=2$, therefore it is true for $n=3$, and so on. We summarize this last sentence by saying formula $(*)$ holds for all $n\geq 1$ *by induction*. □

As suggested in Example A1.1, mathematical induction is carried out in the following way:

A1.2 Technique for Mathematical Induction

If S is a statement about the integer n that is to be proved for $n=n_0, n_0+1, n_0+2, \ldots$ (that is, for all integers $n\geq n_0$), then we should

i) prove by direct means that S is true for $n=n_0$;

ii) assume that S is true for $n=k$ for some $k\geq n_0$ (or even for all n between n_0 and some such k); this is called the **induction hypothesis**;

iii) using the induction hypothesis, show that S must also be true for $n=k+1$; this is called the **induction step**;

iv) conclude that S is true for all integers $n\geq n_0$ **by induction**.

The justification for mathematical induction rests on the fact that every set of positive integers which is not empty contains a smallest integer. If the statement S to be proved is false for some $n\geq n_0$, then there must be a smallest n, say n_1, for which it is false. $n_1>n_0$ since S is true for n_0. S is true for n_1-1 since n_1 is the smallest n for which it is false. But then the induction step shows that S is true for n_1 and we have a contradiction. Since there cannot be any smallest false value of n, there cannot be any $n\geq n_0$ for which S is false.

Students are sometimes bothered by the fact that in an inductive proof we seem to be assuming what we want to prove. This is not so; we merely show that *if* S holds for $n=k$, *then* it also holds for $n=k+1$.

EXAMPLE A1.3 Show that $\frac{d^n}{dx^n}x^n=n!$ (for $n\geq 1$).

PROOF The assertion holds for $n=1$ since $\frac{d}{dx}x=1=1!$. Assume that for some $k\geq 1$ we have $\frac{d^k}{dx^k}x^k=k!$. Observe that

$$\frac{d^{k+1}}{dx^{k+1}}x^{k+1}=\frac{d^k}{dx^k}\frac{d}{dx}x^{k+1}=\frac{d^k}{dx^k}(k+1)x^k=(k+1)\frac{d^k}{dx^k}x^k=(k+1)k!=(k+1)!.$$

Since the assertion holds for $n = k+1$ if it holds for $n = k$, and it holds for $n = 1$, it must hold for all $n \geq 1$ by induction. □

We remark that the inductive technique can also be used in definitions. For instance, $n!$ is defined inductively:

$$0! = 1$$
$$(n+1)! = (n+1)n! \qquad \text{if } n = 0,\ 1,\ 2,\ 3,\ \ldots.$$

There is no ready formula for computing $n!$ directly for any n. We must actually multiply $1 \times 2 \times 3 \times \cdots \times n$, that is, we must calculate $1!, 2!, 3!, \ldots$ up to $(n-1)!$ and finally $n!$ to find the value of $n!$. See Example 10.1.10 for another example of an inductive definition.

EXERCISES

1. Use mathematical induction to prove that $\dfrac{d}{dx}\, x^n = nx^{n-1}$ for $n = 1, 2, 3, \ldots.$
2. Use mathematical induction and the product rule to prove that $\dfrac{d}{dx}\, x^{-n} = -nx^{-n-1}$ for $n = 1, 2, 3, \ldots.$
3. Use induction to prove the General Product Rule

$$\big(f_1 f_2 \cdots f_n\big)' = f_1' f_2 \cdots f_n + f_1 f_2' \cdots f_n + \cdots + f_1 f_2 \cdots f_n'.$$

4. Use induction to prove that

$$\sum_{k+1}^{n} k^2 = 1^2 + 2^2 + 3^2 + \cdots + n^2 = \frac{n(n+1)(2n+1)}{6}.$$

5. Prove that

$$\sum_{k=1}^{n} (2k-1) = 1 + 3 + 5 + \cdots + (2n-1) = n^2.$$

6. Use induction to prove the formula for a finite geometric sum:

$$\sum_{k=0}^{n} r^k = 1 + r + r^2 + \cdots + r^n = \frac{1 - r^{n+1}}{1 - r} \qquad (r \neq 1).$$

7. Prove the formula

$$\sum_{k=1}^{n} k^3 = 1^3 + 2^3 + 3^3 + \cdots + n^3 = \left(\frac{n(n+1)}{2}\right)^2.$$

It is interesting that $\sum_{k=1}^{n} k^3 = (\sum_{k=1}^{n} k)^2$.

8. Prove that $\displaystyle\sum_{k=1}^{n} (-1)^{k-1} k^2 = (-1)^{n-1} \sum_{k=1}^{n} k.$
9. If $r \geq -1$ and $n = 1, 2, 3, \ldots$ show that $(1+r)^n \geq 1 + nr$.
10. Prove that $\displaystyle\sum_{k=1}^{n} \frac{k}{2^k} = \frac{1}{2} + \frac{2}{2^2} + \cdots + \frac{n}{2^n} = 2 - \frac{n+2}{2^n}.$
11. Guess a formula for $f^{(n)}(x)$ where $f(x) = \sqrt{1+x}$ and prove it by induction.
12. Guess a formula for $\dfrac{d^n}{dx^n}\, xe^{-x}$ and prove it by induction.
13. Do not use induction in this problem; direct proofs are easier. The results will be needed in Exercises 14 and 15. For $0 \leq k \leq n$ (k and n are nonnegative integers) we define the **binomial coefficients** $\dbinom{n}{k}$

$$\binom{n}{k} = \frac{n!}{k!\,(n-k)!}.$$

For example, $\dbinom{5}{2} = \dfrac{5!}{2!\,3!} = \dfrac{120}{12} = 10$. Show that

i) $\dbinom{n}{0} = \dbinom{n}{n} = 1$ for every n, and

ii) if $0 \leq k \leq n$, then $\dbinom{n}{k-1} + \dbinom{n}{k} = \dbinom{n+1}{k}.$

It follows that, for fixed $n \geq 1$, the binomial coefficients $\binom{n}{0}, \binom{n}{1}, \binom{n}{2}, \ldots, \binom{n}{n}$ are the elements of the nth row of **Pascal's triangle**

$$
\begin{array}{ccccccccccc}
 & & & & 1 & & 1 & & & & \\
 & & & 1 & & 2 & & 1 & & & \\
 & & 1 & & 3 & & 3 & & 1 & & \\
 & 1 & & 4 & & 6 & & 4 & & 1 & \\
1 & & 5 & & 10 & & 10 & & 5 & & 1 \\
\cdot & & & & & & & & & & \cdot \\
\end{array}
$$

where each element with value > 1 is the sum of the two diagonally above it.

14. Use mathematical induction and the results of Exercise 13 to prove the Binomial Theorem:

$$
\begin{aligned}
(a+b)^n &= \sum_{k=0}^{n} \binom{n}{k} a^{n-k} b^k \\
&= a^n + na^{n-1}b + \binom{n}{2} a^{n-2}b^2 + \binom{n}{3} a^{n-3}b^3 + \cdots + b^n.
\end{aligned}
$$

15. Use mathematical induction, the product rule, and Exercise 13 to verify the **Leibniz rule** for the nth derivative of a product of two functions:

$$
\begin{aligned}
(fg)^{(n)} &= \sum_{k=0}^{n} \binom{n}{k} f^{(n-k)} g^{(k)} \\
&= f^{(n)}g + nf^{(n-1)}g' + \binom{n}{2} f^{(n-2)}g'' \\
&\quad + \binom{n}{3} f^{(n-3)}g^{(3)} + \cdots + fg^{(n)}.
\end{aligned}
$$

16. Show that every positive integer is interesting.

APPENDIX II THE THEORETICAL FOUNDATIONS OF CALCULUS

The development of calculus depends in an essential way on the concept of limit of a function and thereby on properties of the real number system. In Chapter 1 we presented these notions in an intuitive way and did not attempt to prove any of the quoted results on limits or continuity of functions. Indeed, these results seem quite obvious and most students and users of calculus are not bothered by applying them without proof.

Nevertheless, mathematics is a highly logical and rigorous discipline, and any statement, however obvious, that cannot be proved by strictly logical arguments from acceptable assumptions must be considered suspect. In this appendix we present formal proofs of the properties of limits and continuous functions given in Chapter 1. The branch of mathematics that deals with such proofs is called Mathematical Analysis. This subject is not usually pursued by students in introductory calculus courses, but is postponed to higher years and studied by students in majors or honors programs in mathematics. We include here only the basics of Mathematical Analysis, enough to get us as far as the proof that a function continuous on a closed, finite interval assumes a maximum and a minimum value and has the Intermediate-Value Property. It is hoped that some of this material will be of value to honors-level calculus courses and individual students with a deeper interest in understanding calculus. The reader should still refer back to Sections 1.5 and 1.7 for more discussion of the topics covered in this appendix.

Limits of Functions

At the heart of Mathematical Analysis is the formal definition of limit, Definition 1.5.3, which we restate as follows:

A2.1
Definition of Limit

> We say that $\lim_{x\to a} f(x) = L$ if for every positive number ϵ there exists a positive number δ, depending on ϵ, (that is, $\delta = \delta(\epsilon)$), such that $|f(x)-L| < \epsilon$ whenever $0 < |x-a| < \delta$.

(It is traditional to use the Greek letters ϵ (epsilon) and δ (delta) in this context.) Note in Fig. A2.1 that $\lim_{x\to a} f(x) = L$ can exist even if $f(x)$ is not defined at $x = a$, and even if $f(a)$ is defined we need not have $f(a) = L$.

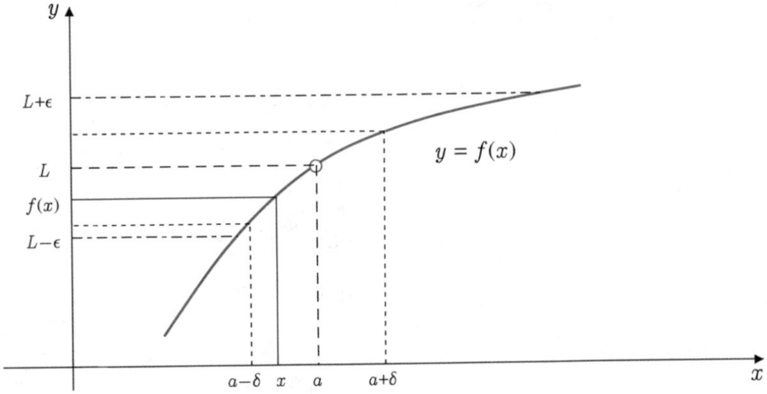

Fig. A2.1

Let us begin by illustrating this definition with a few examples.

EXAMPLE A2.2 Prove that $\lim_{x\to 2}(5-2x) = 1$.

PROOF Let ϵ be any given positive number. Referring to Fig. A2.2 we have

$$|(5-2x)-1| = |4-2x| = 2|2-x| = 2|x-2| < \epsilon,$$

provided $|x-2| < \epsilon/2$. Thus we may take $\delta = \epsilon/2$ and ensure that $|(5-2x)-1| < \epsilon$ by taking x to satisfy $0 < |x-2| < \delta$. (Indeed, in this case we can even allow $|x-2| = 0$.) Thus $\lim_{x\to 2}(5-2x) = 1$. □

EXAMPLE A2.3 Show that $\lim_{x\to 2}(x^2-4x+4) = 0$.

PROOF Let ϵ be any positive number. We have

$$|(x^2-4x+4)-0| = |(x-2)^2| = |x-2|^2 < \epsilon,$$

provided $0 < |x-2| < \delta = \sqrt{\epsilon}$. Thus $\lim_{x\to 2}(x^2-4x+4) = 0$. See Fig. A2.3. □

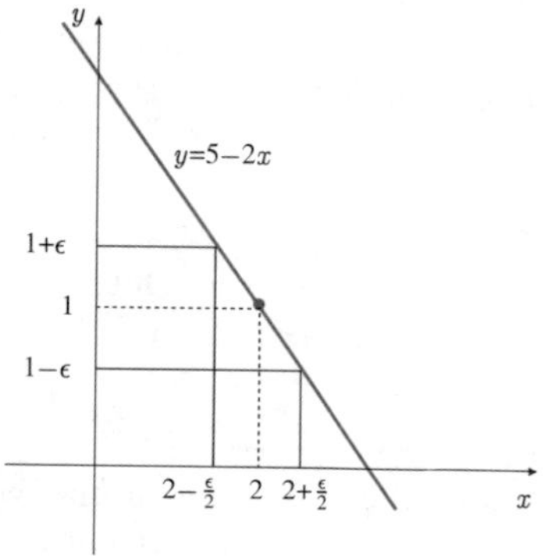

Fig. A2.2

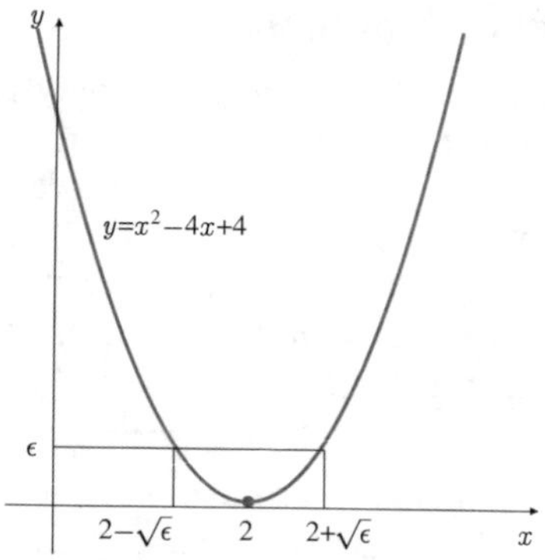

Fig. A2.3

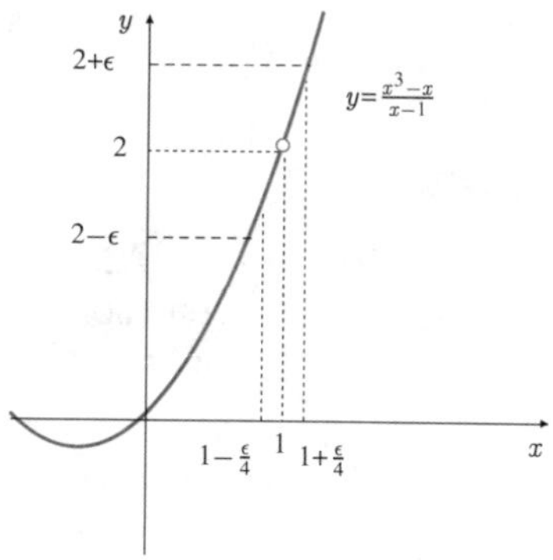

Fig. A2.4

EXAMPLE A2.4 Show that $\lim_{x\to 4}\sqrt{x} = 2$.

PROOF Let ϵ be any positive number. If $|x-4| < 4$, then $x > 0$ and

$$|\sqrt{x}-2| = \left|\frac{(\sqrt{x}-2)(\sqrt{x}+2)}{\sqrt{x}+2}\right| = \frac{1}{\sqrt{x}+2}\,|x-4| < \frac{1}{2}\,|x-4|.$$

Let $\delta = \min\{4, 2\epsilon\}$, that is, whichever of the numbers 4 and 2ϵ is smaller. If $0 < |x-4| < \delta$, then $|\sqrt{x}-2| < \frac{1}{2}2\epsilon = \epsilon$. Thus $\lim_{x\to 4}\sqrt{x} = 2$. □

EXAMPLE A2.5 Show that $\lim_{x\to 1}\dfrac{x^3-x}{x-1} = 2$.

PROOF Let ϵ be any positive number. If $x \neq 1$ we have

$$\left|\frac{x^3-x}{x-1} - 2\right| = |x(x+1)-2| = |x^2+x-2| = |x-1||x+2|.$$

If $|x-1| < 1$, then $0 < x < 2$ and $2 < x+2 < 4$, so $|x+2| < 4$. Hence, if $0 < |x-1| < \delta = \min\left\{1, \dfrac{\epsilon}{4}\right\}$, we have $|x+2| < 4$ and $|x-1| < \dfrac{\epsilon}{4}$, so $\left|\dfrac{x^3-x}{x-1} - 2\right| < \epsilon$. Hence $\lim_{x\to 1}\dfrac{x^3-x}{x-1} = 2$. See Fig. A2.4. Observe that in this example we really must restrict $|x-1| > 0$; the fraction is not defined at $x = 1$. □

While the definition of limit provides a means of checking whether a given number is the limit of a particular function at a particular point (as in the four examples above), it provides no means for finding an unknown limit. The limit in each of the examples was intuitively obvious. The main reason for having a formal definition of limit is to prove general theorems about limits such as Theorems 1.5.6, 1.5.7, and 1.5.8 to which we now turn our attention.

THEOREM A2.6 **(Uniqueness of Limits)** If $\lim_{x\to a} f(x) = L$ and $\lim_{x\to a} f(x) = M$, then, necessarily, $L = M$.

PROOF It may seem to the student that nothing needs to be proved here; things that are equal to the same thing are necessarily equal to each other. There is, however, the possibility that the definition might have been satisfied for more than one number L. If this were the case we should really have used $\lim_{x\to a} f(x)$ to denote a *set* of numbers rather than just *one number*.

It suffices for us to prove that for any positive number ϵ we have $|L-M| < \epsilon$. It is then impossible that $L \neq M$. (Why?) Let $\epsilon > 0$ be given. Since $\lim_{x\to a} f(x) = L$, there exists a number δ_1 such that if $0 < |x-a| < \delta_1$ then $|f(x)-L| < \epsilon/2$. Similarly, since $\lim_{x\to a} f(x) = M$, there exists a number δ_2 such that if $0 < |x-a| < \delta_2$ then $|f(x)-M| < \epsilon/2$. Now if $0 < |x-a| < \delta = \min\{\delta_1, \delta_2\}$, then we obtain, using the triangle inequality for real numbers,

$$|L-M| = |(L-f(x)) + (f(x)-M)| \le |f(x)-L| + |f(x)-M| < \frac{\epsilon}{2} + \frac{\epsilon}{2} = \epsilon.$$

Therefore $L = M$. □

THEOREM A2.7 If $\lim_{x\to a} f(x) = L$ and $\lim_{x\to a} g(x) = M$, then the following conclusions hold.

i) $\lim_{x\to a}(f(x)+g(x)) = L+M$ — The limit of a sum is the sum of limits.

ii) $\lim_{x\to a}(f(x)-g(x)) = L-M$ — The limit of a difference is the difference of the limits.

iii) $\lim_{x\to a}(f(x)g(x)) = LM$ — The limit of a product is the product of limits.

iv) $\displaystyle\lim_{x\to a}\frac{f(x)}{g(x)} = \frac{L}{M}$, provided $M \neq 0$ — The limit of a quotient is the quotient of the limits provided the limit of the denominator is not 0.

v) $\lim_{x\to a} cf(x) = cL$ for any constant c.

vi) If $f(x) \le g(x)$ near a, then $L \le M$.

PROOF We will prove only parts (i) and (iii) to demonstrate the techniques. The remaining parts are left as exercises. (See Exercises 7–9 below.)

i) Let $\epsilon > 0$ be given. Since $\lim_{x\to a} f(x) = L$, there exists a positive number δ_1 such that if $0 < |x-a| < \delta_1$ then $|f(x)-L| < \epsilon/2$. Similarly, since $\lim_{x\to a} g(x) = M$, there exists a positive number δ_2 such that if $0 < |x-a| < \delta_2$ then $|g(x)-M| < \epsilon/2$. Let $\delta = \min\{\delta_1, \delta_2\}$. If $0 < |x-a| < \delta$, then

$$\begin{aligned} |(f(x)+g(x)) - (L+M)| &= |(f(x)-L) + (g(x)-M)| \\ &\le |f(x)-L| + |g(x)-M| < \frac{\epsilon}{2} + \frac{\epsilon}{2} = \epsilon. \end{aligned}$$

Thus $\lim_{x\to a}(f(x)+g(x)) = L+M$. □

iii) Let $\epsilon > 0$ be given. Since $\lim_{x\to a} f(x) = L$, there exists a positive number δ_1 such that if $0 < |x - a| < \delta_1$ then $|f(x) - L| < 1$, and hence $|f(x)| = |f(x) - L + L| < 1 + |L|$. (This says that f is bounded near the point a except possibly at a.) Also, there exists a positive number δ_2 such that if $0 < |x - a| < \delta_2$, then $|f(x) - L| < \epsilon/(2(|M| + 1))$. Since $\lim_{x\to a} g(x) = M$, there exists a positive number δ_3 such that if $0 < |x - a| < \delta_3$ then $|g(x) - M| < \epsilon/(2(|L| + 1))$. Let $\delta = \min\{\delta_1, \delta_2, \delta_3\}$. If $0 < |x - a| < \delta$, then

$$\begin{aligned} |f(x)g(x) - LM| &= |f(x)g(x) - f(x)M + f(x)M - LM| \\ &\le |f(x)(g(x) - M)| + |M(f(x) - L)| \\ &= |f(x)||g(x) - M| + |M||f(x) - L| \\ &< (|L| + 1)\frac{\epsilon}{2(|L| + 1)} + |M|\frac{\epsilon}{2(|M| + 1)} < \epsilon. \end{aligned}$$

Hence $\lim_{x\to a} f(x)g(x) = LM$. □

It should be remarked that one does not construct a proof such as this by starting at the top and pulling numbers like $\epsilon/2$ and $\epsilon/(2(|M| + 1))$ out of a hat. One really starts by considering the expressions $|(f(x) + g(x)) - (L + M)|$ and $|f(x)g(x) - LM|$, which must be made less than ϵ, and rewriting them to show their explicit dependence on $|f(x) - L|$ and $|g(x) - M|$. Then one arranges to take x sufficiently close to a to ensure that each of these quantities is sufficiently small that the desired combination is in fact less than ϵ.

THEOREM A2.8 **(The Squeeze Theorem)** If $f(x) \le g(x) \le h(x)$ for $0 < |x - a| < k$ (for some $k > 0$) and if

$$\lim_{x\to a} f(x) = L \qquad \text{and} \qquad \lim_{x\to a} h(x) = L,$$

then $\lim_{x\to a} g(x)$ exists and equals L also. (See Fig. A2.5.)

PROOF First estimate

$$\begin{aligned} |g(x) - L| &= |g(x) - f(x) + f(x) - L| \\ &\le \big(g(x) - f(x)\big) + |f(x) - L| \qquad \text{if } 0 < |x - a| < k \\ &\le \big(h(x) - f(x)\big) + |f(x) - L| \\ &\le |h(x) - L| + |f(x) - L| + |f(x) - L| \\ &= |h(x) - L| + 2|f(x) - L|. \end{aligned}$$

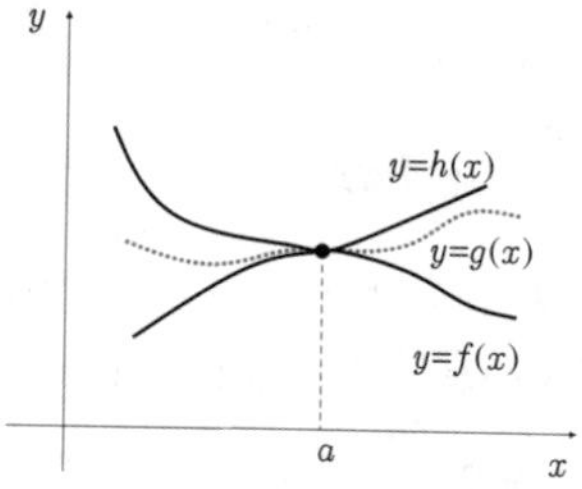

Fig. A2.5

Let $\epsilon > 0$ be given. Since $\lim_{x\to a} f(x) = \lim_{x\to a} h(x) = L$, there exist positive numbers δ_1 and δ_2 such that if $0 < |x - a| < \delta_1$ then $|f(x) - L| < \epsilon/3$ and if $0 < |x - a| < \delta_2$ then $|h(x) - L| < \epsilon/3$. Hence if $0 < |x - a| < \delta = \min\{k, \delta_1, \delta_2\}$, then $|g(x) - L| < \epsilon/3 + 2\epsilon/3 = \epsilon$. Therefore $\lim_{x\to a} g(x) = L$. □

Minor modifications can be made to the definition of limit to allow for one-sided limits, infinite limits, and limits at infinity.

A2.9
One-Sided Limits

> We say that $\lim_{x\to a+} f(x) = L$ if for every positive number ϵ there exists a positive number $\delta = \delta(\epsilon)$ such that $|f(x) - L| < \epsilon$ holds whenever $a < x < a + \delta$. See Fig. A2.6.

The definition for $\lim_{x\to a-} f(x) = L$ is similar except that $a < x < a + \delta$ is replaced by $a - \delta < x < a$.

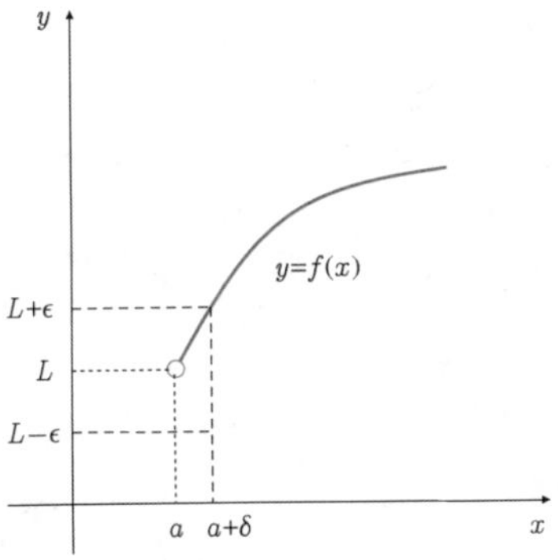

Fig. A2.6

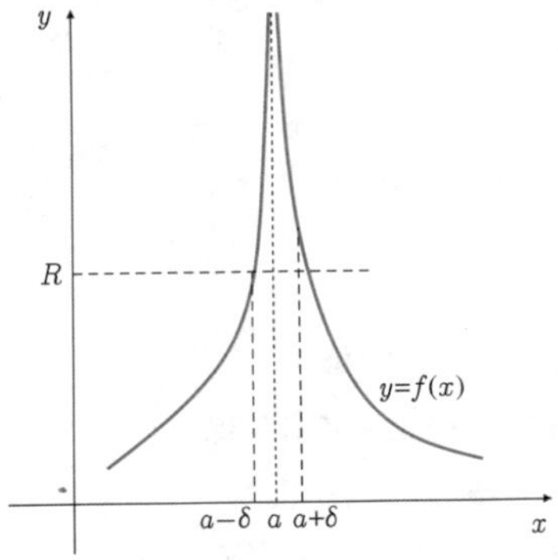

Fig. A2.7

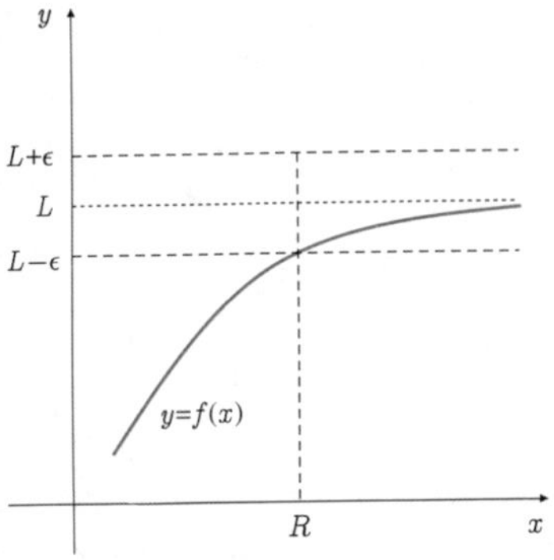

Fig. A2.8

A2.10
Infinite Limits

> We say that $\lim_{x\to a} f(x) = \infty$ if for every positive number R (no matter how large) there exists a positive number $\delta = \delta(R)$ such that $f(x) > R$ holds whenever $0 < |x - a| < \delta$. See Fig. A2.7.

The definition for $\lim_{x\to a} f(x) = -\infty$ is similar except that we replace the condition $f(x) > R$ with $f(x) < -R$.

A2.11
Limits at Infinity

> We say $\lim_{x\to\infty} f(x) = L$ if for every positive number ϵ there exists a positive number $R = R(\epsilon)$ such that $|f(x) - L| < \epsilon$ holds whenever $x > R$. See Fig. A2.8

The definition for $\lim_{x\to-\infty} f(x) = L$ is similar with $x < -R$ replacing $x > R$.

Appropriate versions of Theorems A2.7–8 hold for such extensions of the limit concept.

EXAMPLE A2.12 Show that $\lim_{x \to 0+} \sqrt{x} = 0$.

PROOF Let $\epsilon > 0$ be given. We have $|\sqrt{x} - 0| = \sqrt{x} < \epsilon$ if $0 < x < \epsilon^2 = \delta$. Hence $\lim_{x \to 0+} \sqrt{x} = 0$. □

EXAMPLE A2.13 Show that $\lim_{x \to 2+} \dfrac{x}{x-2} = \infty$.

PROOF Let $R > 0$ be given. If $0 < x - 2 < \delta = 2/R$, then $x > 2$ and $\dfrac{x}{x-2} > \dfrac{2}{x-2} > 2\dfrac{R}{2} = R$. Hence $\lim_{x \to 2+} \dfrac{x}{x-2} = \infty$. □

EXAMPLE A2.14 Show that $\displaystyle\lim_{x \to \infty} \frac{x}{x-1} = 1$.

PROOF Let $\epsilon > 0$ be given. If $x > 1 + \dfrac{1}{\epsilon}$, then

$$\left|\frac{x}{x-1} - 1\right| = \left|\frac{1}{x-1}\right| = \frac{1}{x-1} < \epsilon. \ \square$$

Continuous Functions

A2.15
Continuity at a Point

A function f defined on an open interval containing the point a is said to be continuous at the point a if

$$\lim_{x \to a} f(x) = f(a),$$

that is, if for every $\epsilon > 0$ there exists $\delta > 0$ such that if $|x - a| < \delta$, then $|f(x) - f(a)| < \epsilon$.

A2.16
Continuity on an Interval

A function f is continuous on an interval if it is continuous at every point of that interval. In the case of an endpoint of a closed interval, f need only be continuous on one side. Thus, f is continuous on the interval $[a, b]$ if

$$\lim_{t \to x} f(t) = f(x)$$

for each x satisfying $a < x < b$, and

$$\lim_{t \to a+} f(t) = f(a) \quad \text{and} \quad \lim_{t \to b-} f(t) = f(b).$$

These concepts are illustrated in Fig. A2.9

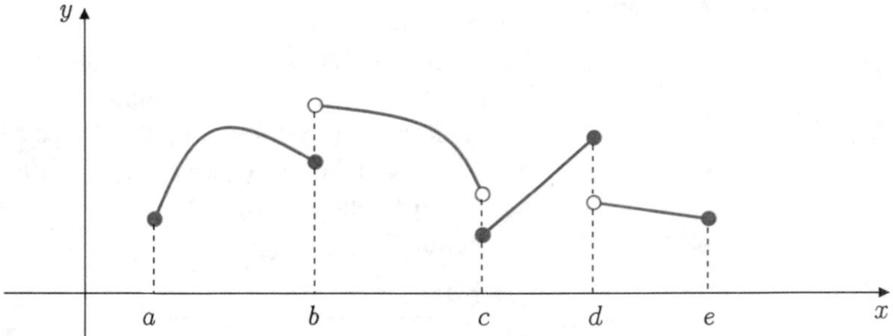

f is continuous on the intervals $[a, b]$, (b, c), $[c, d]$ and $(d, e]$

Fig. A2.9

Some important results about continuous functions are collected in Theorem 1.7.5, which we restate here

THEOREM A2.17

a) If f and g are continuous at the point a, then so are $f + g$, $f - g$, fg, and, if $g(a) \neq 0$, f/g.

b) If f is continuous at the point L and if $\lim_{x\to a} g(x) = L$, then $\lim_{x\to a} f\big(g(x)\big) = f(L) = f\big(\lim_{x\to a} g(x)\big)$. In particular, if g is continuous at the point a (so that $L = g(a)$), then $\lim_{x\to a} f\big(g(x)\big) = f\big(g(a)\big)$, that is, $f \circ g(x) = f\big(g(x)\big)$ is continuous at $x = a$.

c) The function $f(x) = C$ (constant) is continuous on the whole real line.

d) For any rational number r the function $f(x) = x^r$ is continuous at every real number where it is defined.

PROOF Part (a) is merely a restatement of various parts of Theorem A2.7 above; for example,

$$\lim_{x\to a} f(x)g(x) = (\lim_{x\to a} f(x))(\lim_{x\to a} g(x)) = f(a)g(a).$$

Part (b) can be proved as follows. Let $\epsilon > 0$ be given. Since f is continuous at L, there exists $k > 0$ such that $|f\big(g(x)\big) - f(L)| < \epsilon$ whenever $|g(x) - L| < k$. Since $\lim_{x\to a} g(x) = L$, there exists $\delta > 0$ such that if $0 < |x - a| < \delta$, then $|g(x) - L| < k$. Hence if $0 < |x - a| < \delta$, then $|f\big(g(x)\big) - f(L)| < \epsilon$, and $\lim_{x\to a} f\big(g(x)\big) = f(L)$.

The proofs of (c) and (d) are left to the student in Exercises 11–13 at the end of this Appendix. □

Completeness and Sequential Limits

A real number u is said to be an **upper bound** for a nonempty set S of real numbers if $x \le u$ for every x in S. The number u^* is called the **least upper bound** of S if u^* is an upper bound for S and $u^* \le u$ for every upper bound u of S.

Similarly, ℓ is a **lower bound** for S if $\ell \le x$ for every x in S and ℓ^* is the **greatest lower bound** of S if ℓ^* is a lower bound and $\ell \le \ell^*$ for every lower bound ℓ of S.

EXAMPLE A2.18 Set $S_1 = [2, 3]$ and $S_2 = (2, \infty)$. Any number $u \geq 3$ is an upper bound for S_1. S_2 has no upper bound; we say that it is not bounded above. The least upper bound of S_1 is 3. Any real number $\ell \leq 2$ is a lower bound for both S_1 and S_2. $\ell^* = 2$ is the greatest lower bound for each set. Note that the least upper bound and greatest lower bound of a set may or may not belong to that set. We now recall the **Completeness Axiom** for the real number system, which we discussed briefly in Section 1.2.

A2.19 Completeness Axiom

> A nonempty set of real numbers that has an upper bound must have a least upper bound.

Equivalently, a nonempty set of real numbers having a lower bound must have a greatest lower bound.

We stress that this is an *axiom* to be assumed without proof. It cannot be deduced from the more elementary algebraic and order properties of the real numbers. These other properties are shared by the rational numbers, a set that is not complete. The completeness axiom is essential for the proof of the most important results about continuous functions, in particular those in Theorems 1.7.7 and 1.7.10. Before attempting these proofs, however, we must develop a little more machinery.

In Section 10.1 there is stated a version of the completeness axiom that pertains to *sequences* of real numbers; specifically, an increasing sequence that is bounded above converges to a limit. We begin by verifying that this follows from Axiom A2.19. (Both statements are, in fact, equivalent.) As noted in Section 10.1, the sequence

$$\{x_n\} = \{x_1, x_2, x_3, \ldots\}$$

is a function on the positive integers, that is, $x_n = x(n)$. We say that the sequence converges to the limit L, and we write $\lim x_n = L$, if the corresponding function $x(t)$ satisfies $\lim_{t\to\infty} x(t) = L$ as defined above. More formally,

A2.20 Limit of a Sequence

> We say that $\lim x_n = L$ if for every positive number ϵ there exists a positive number $N = N(\epsilon)$ such that $|x_n - L| < \epsilon$ holds whenever $n \geq N$.

THEOREM A2.21 If $\{x_n\}$ is an increasing sequence that is bounded above, that is,

$$x_{n+1} \geq x_n \qquad \text{and} \qquad x_n \leq K \qquad \text{for } n = 1, 2, 3, \ldots,$$

then $\lim x_n = L$ exists. (Equivalently, if $\{x_n\}$ is decreasing and bounded below, then $\lim x_n$ exists.)

PROOF Let $\{x_n\}$ be increasing and bounded above. The set S of real numbers x_n has an upper bound, K, and so has a least upper bound, say L. Thus $x_n \leq L$ for every n, and if $\epsilon > 0$ then there exists a positive integer N such that $x_N > L - \epsilon$. (Otherwise $L - \epsilon$ would be an upper bound for S that is lower than the least upper bound.) If $n \geq N$, then we have $L - \epsilon < x_N \leq x_n \leq L$, so $|x_n - L| < \epsilon$. Thus $\lim x_n = L$. The proof for a decreasing sequence that is bounded below is similar. □

THEOREM A2.22 If $a \leq x_n \leq b$ for each n, and if $\lim x_n = L$, then $a \leq L \leq b$.

PROOF Suppose that $L > b$. Let $\epsilon = L - b$. Since $\lim x_n = L$, there exists n such that $|x_n - L| < \epsilon$. Thus $x_n > L - \epsilon = L - (L - b) = b$, which is a contradiction since we are given that $x_n \leq b$. Thus $L \leq b$. A similar argument shows that $L \geq a$. □

THEOREM A2.23 If f is continuous on $[a, b]$, if $a \leq x_n \leq b$ for each n, and if $\lim x_n = L$, then $\lim f(x_n) = f(L)$. □

The proof is similar to that of Theorem A2.17(b) and is left as Exercise 16 at the end of this Appendix.

Continuous Functions on a Closed, Finite Interval

We are now in a position to prove Theorems 1.7.7 and 1.7.10, restated as follows.

THEOREM A2.24 If f is continuous on $[a, b]$, then f is bounded there; that is, there exists a constant K such that $|f(x)| \leq K$ if $a \leq x \leq b$.

PROOF We show that f is bounded above; a similar proof shows that f is bounded below. For each positive integer n let S_n be the set of points x in $[a, b]$ such that $f(x) > n$:

$$S_n = \{x : a \leq x \leq b \quad \text{and} \quad f(x) > n\}.$$

We would like to show that S_n is empty for some n. It would then follow that $f(x) \leq n$ for all x in $[a, b]$; that is, n would be an upper bound for f on $[a, b]$.

Suppose to the contrary, that S_n is nonempty for every n. We will show that this leads to a contradiction. Since S_n is bounded below, (a is a lower bound), by completeness S_n has a greatest lower bound; call it x_n. See Fig. A2.10. Evidently $a \leq x_n$. Since $f(x) > n$ at some point of $[a, b]$ and f is continuous at that point, $f(x) > n$ on some interval contained in $[a, b]$. Hence $x_n < b$. It follows that $f(x_n) \geq n$. (If $f(x_n) < n$ then by continuity $f(x) < n$ for some distance to the right of x_n, and x_n could not be the greatest lower bound of S_n.)

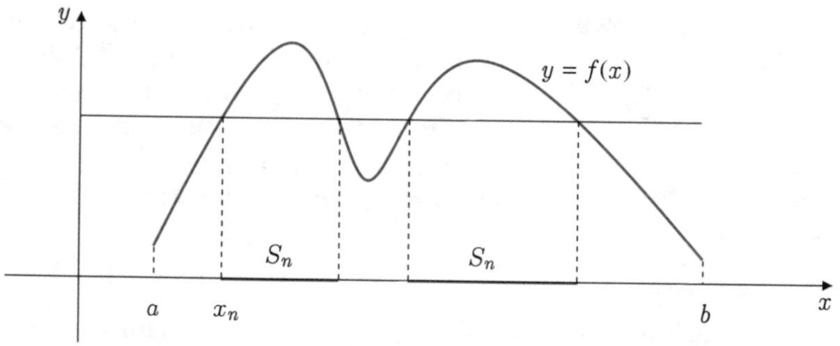

Fig. A2.10

For each n we have $S_{n+1} \subset S_n$. Therefore $x_{n+1} \geq x_n$ and $\{x_n\}$ is an increasing sequence. Being bounded above (b is an upper bound) this sequence converges, by Theorem A2.21. Let $\lim x_n = L$. By Theorem A2.22, $a \leq L \leq b$. Since f is continuous at L, $\lim f(x_n) = f(L)$ exists by Theorem A2.23. But since $f(x_n) \geq n$, $\lim f(x_n)$ cannot exist. This contradiction completes the proof. □

THEOREM A2.25 If f is continuous on $[a, b]$, then there are points v and u in $[a, b]$ such that for any x in $[a, b]$ we have

$$f(v) \leq f(x) \leq f(u);$$

that is, f assumes maximum and minimum values on $[a, b]$.

PROOF By Theorem A2.24 we know that the set $S = \{f(x) : a \leq x \leq b\}$ has an upper bound and therefore, by the Completeness Axiom, a least upper bound. Call this least upper bound M. Suppose there exists no point u in $[a, b]$ such that $f(u) = M$. Then by Theorem A2.17(a), $1/(M - f(x))$ is continuous on $[a, b]$. By Theorem A2.24, there exists a constant K such that $1/(M - f(x)) \leq K$ for all x in $[a, b]$. Thus $f(x) \leq M - 1/K$, which contradicts the fact that M is the *least* upper bound for the values of f. Hence there must exist some point u in $[a, b]$ such that $f(u) = M$. Since M is an upper bound for the values of f on $[a, b]$, we have $f(x) \leq f(u) = M$ for all x in $[a, b]$.

The proof that there must exist a point v in $[a, b]$ such that $f(x) \geq f(v)$ for all x in $[a, b]$ is similar. □

THEOREM A2.26 (**The Intermediate-Value Theorem**) If f is continuous on $[a, b]$ and s is a real number lying between the numbers $f(a)$ and $f(b)$, then there exists a point c in $[a, b]$ such that $f(c) = s$.

PROOF To be specific, we assume that $f(a) < s < f(b)$. (The proof for $f(a) > s > f(b)$ is similar.) Let $S = \{x : a \leq x \leq b \text{ and } f(x) \leq s\}$. S is nonempty (a belongs to S) and bounded above (b is an upper bound) so by completeness S has a least upper bound; call it c.

Suppose that $f(c) > s$. Then $c \neq a$ and, by continuity, $f(x) > s$ on some interval $(c-\delta, c]$ where $\delta > 0$. But this says $c - \delta$ is an upper bound for S lower than the least upper bound, which is impossible. Thus $f(c) \leq s$.

Suppose $f(c) < s$. Then $c \neq b$ and, by continuity, $f(x) < s$ on some interval of the form $[c, c+\delta)$ for some $\delta > 0$. But this says that $[c, c+\delta) \subset S$, which contradicts the fact that c is an upper bound for S. Hence we cannot have $f(c) < s$. Therefore, $f(c) = s$. □

For more discussion of Theorems A2.24–26 and some applications, see Section 1.7.

EXERCISES

Use the formal definition of limit to verify the limits asserted in Exercises 1–6.

1. $\lim_{x\to 2}(3x-1) = 5$

2. $\lim_{x\to -2}(x^2+3x) = -2$

3. $\lim_{x\to 0}\dfrac{x-2}{x+1} = -2$

4. $\lim_{x\to -1}\dfrac{(x+1)^2}{x^2-1} = 0$

5. $\lim_{x\to 1}\dfrac{x^2+x-2}{x^2+2x-3} = \dfrac{3}{4}$

6. $\lim_{x\to 2}\sqrt{x^2+4} = 2\sqrt{2}$

7. Prove parts (ii) and (v) of Theorem A2.7.

8. If $\lim_{x\to a} g(x) = M$ where $M \neq 0$, prove that there exists $\delta_1 > 0$ such that $|g(x)| > |M|/2$ whenever $0 < |x-a| < \delta_1$. Then prove part (iv) of Theorem A2.7.

9. Prove part (vi) of Theorem A2.7 by contradiction; assume that $L > M$ and deduce that $f(x) > g(x)$ for all x sufficiently near a. This contradicts the condition that $f(x) \leq g(x)$ for $0 < |x-a| < \delta$.

10. If $f(x) \leq K$ on the intervals $[a, b)$ and $(b, c]$, and if $\lim_{x\to b} f(x) = L$, prove that $L \leq K$.

11. Prove that $\lim_{x\to 0+} x^r = 0$ for any positive, rational number r.

12. Prove the following:

a) $f(x) = C$ (constant) and $g(x) = x$ are both continuous on the whole real line.

b) Every polynomial is continuous on the whole real line.

c) A rational function (quotient of polynomials) is continuous everywhere except where the denominator is 0.

13. Prove the following:

a) If n is a positive integer and $a > 0$, then $f(x) = x^{1/n}$ is continuous at $x = a$.

b) If $r = m/n$ is a rational number, then $g(x) = x^r$ is continuous at every point $a > 0$.

c) If $r = m/n$ where m and n are integers and n is odd, show that $g(x) = x^r$ in continuous at every point $a < 0$. If $r \geq 0$ show that g is continuous at 0 also.

14. Prove that $f(x) = |x|$ is continuous on the real line.

15. Use the definitions of $\cos x$, $\sin x$, e^x, and $\ln x$ given in Chapter 3 to prove that these functions are continuous on their respective domains.

16. Prove Theorem A2.23.

17. Suppose that every function that is continuous and bounded on $[a, b]$ must assume a maximum value and a minimum value on that interval. Without using Theorem A2.24, prove that every function f that is continuous on $[a, b]$ must be bounded on that interval. (*Hint:* Show that $g(t) = t/(1+|t|)$ is continuous and increasing on the real line. Then consider $g\big(f(x)\big)$.)

APPENDIX III THE RIEMANN INTEGRAL

In Section 6.2 we defined the definite integral $\int_a^b f(x)\,dx$ of a function f which is continuous on the finite, closed interval $[a, b]$. The integral was defined as a kind of "limit" of Riemann sums formed by partitioning the interval $[a, b]$ into small subintervals. In this appendix we will reformulate the definition of the integral so that it can be used for some functions which are not necessarily continuous; in the following discussion we assume only that f is **bounded** on $[a, b]$. Later we will prove Theorem 6.2.3 which asserts that any continuous function is integrable.

Recall that a **partition** P of $[a, b]$ is a finite, ordered set of points $P = \{x_0, x_1, x_2, \ldots, x_n\}$ where $a = x_0 < x_1 < x_2 < \cdots < x_{n-1} < x_n = b$. Such a partition subdivides $[a, b]$ into n subintervals $[x_0, x_1]$, $[x_1, x_2]$, $\ldots$, $[x_{n-1}, x_n]$ where $n = n(P)$ depends on the partition. The length of the jth subinterval $[x_{j-1}, x_j]$ is $\Delta x_j = x_j - x_{j-1}$.

Suppose that the function f is bounded on $[a, b]$. Given any partition P, the n sets $S_j = \{f(x) : x_{j-1} \le x \le x_j\}$ have least upper bounds M_j and greatest lower bounds m_j, $(1 \le j \le n)$, so that

$$m_j \le f(x) \le M_j \qquad \text{on } [x_{j-1}, x_j].$$

We define upper and lower Riemann sums for f corresponding to the partition P to be

$$U(f, P) = \sum_{j=1}^{n(P)} M_j \Delta x_j, \qquad \text{and}$$

$$L(f, P) = \sum_{j=1}^{n(P)} m_j \Delta x_j.$$

(See Fig. A3.1) Note that if f is continuous on $[a, b]$, then m_j and M_j are in fact the minimum and maximum values of f over $[x_{j-1}, x_j]$ (by Theorem A2.25); that is,

$$m_j = f(l_j) \text{ and } M_j = f(u_j), \quad \text{where} \quad f(l_j) \le f(x) \le f(u_j) \text{ for } x_{j-1} \le x \le x_j.$$

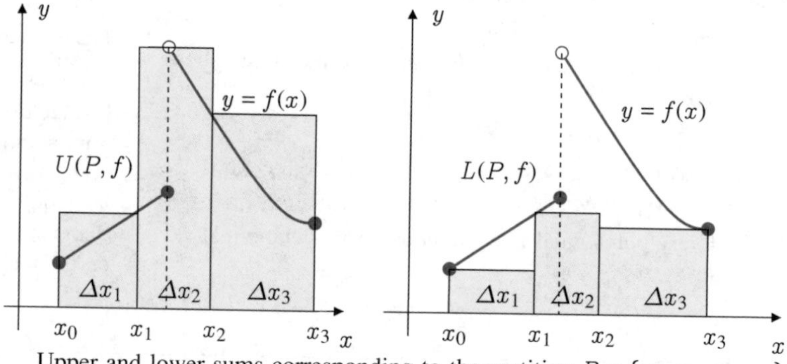

Upper and lower sums corresponding to the partition $P = \{x_0, x_1, x_2, x_3\}$

Fig. A3.1

If P is any partition of $[a, b]$ and we create a new partition P^* by adding new subdivision points to those of P, thus subdividing the subintervals of P into smaller ones, then we call P^* a **refinement** of P.

THEOREM A3.1 If P^* is a refinement of P, then $L(f, P^*) \geq L(f, P)$ and $U(f, P^*) \leq U(f, P)$.

PROOF If S and T are sets of real numbers, and $S \subset T$, then any lower bound (or upper bound) of T is also a lower bound (or upper bound) of S. Hence the greatest lower bound of S is at least as large as that of T, and the least upper bound of S is no greater than that of T.

Let P be a given partition of $[a, b]$ and form a new partition P' by adding one subdivision point to those of P, say the point k dividing the jth subinterval $[x_{j-1}, x_j]$ of P into two subintervals $[x_{j-1}, k]$ and $[k, x_j]$. See Fig. A3.2. Let m_j, m_j', and m_j'' be the greatest lower bounds of the sets of values of $f(x)$ on the intervals $[x_{j-1}, x_j]$, $[x_{j-1}, k]$, and $[k, x_j]$, respectively. Then $m_j \leq m_j'$ and $m_j \leq m_j''$. Thus $m_j(x_j - x_{j-1}) \leq m_j'(k - x_{j-1}) + m_j''(x_j - k)$, and so $L(f, P) \leq L(f, P')$.

If P^* is a refinement of P, it can be obtained by adding one point at a time to those of P and thus $L(f, P) \leq L(f, P^*)$. We can prove that $U(f, P) \geq U(f, P^*)$ in a similar manner. □

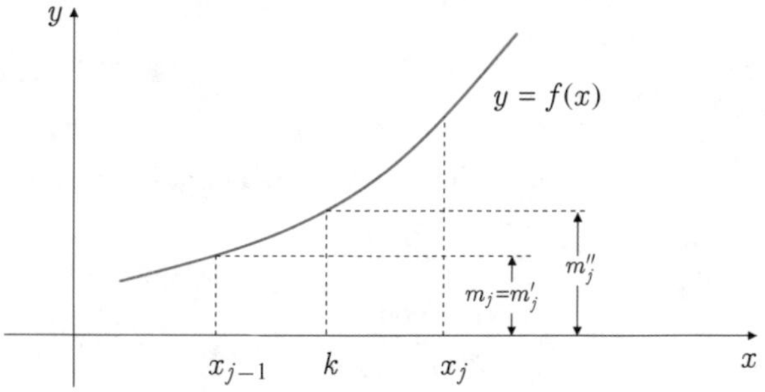

Fig. A3.2

THEOREM A3.2 If P and P' are any two partitions of $[a, b]$, then $L(f, P) \leq U(f, P')$.

PROOF Combine the subdivision points of P and P' to form a new partition P^*, which is a refinement of both P and P'. Then by Theorem A3.1,

$$L(f, P) \leq L(f, P^*) \leq U(f, P^*) \leq U(f, P'). \; \square$$

Theorem A3.2 shows that the set of values of $L(f, P)$ for fixed f and various partitions P of $[a, b]$ is a bounded set; any upper sum is an upper bound for this set. By completeness, the set has a least upper bound, which we shall denote I_*. Thus $L(f, P) \leq I_*$ for any partition P. Similarly, there exists a greatest lower bound I^* for the set of values of $U(f, P)$ corresponding to different partitions P. It follows that $I_* \leq I^*$. (Exercise 4.)

A3.3
The Riemann Integral

> If f is bounded on $[a, b]$ and $I_* = I^*$, then we say that f is **Riemann integrable**, or simply **integrable** on $[a, b]$ and denote by
>
> $$\int_a^b f(x)\,dx = I_* = I^*$$
>
> the **(Riemann) integral** of f on $[a, b]$.

The following theorem provides a convenient test for determining whether a given bounded function is integrable.

THEOREM A3.4 The bounded function f is integrable on $[a, b]$ if and only if for every positive number ϵ there exists a partition P of $[a, b]$ such that $U(f, P) - L(f, P) < \epsilon$.

PROOF If for every $\epsilon > 0$ there exists a partition P such that $U(f, P) - L(f, P) < \epsilon$, then

$$I^* \le U(f, P) < L(f, P) + \epsilon \le I_* + \epsilon.$$

Since $I^* < I_* + \epsilon$ must hold for every $\epsilon > 0$, it follows that $I^* \le I_*$. Since we already know that $I^* \ge I_*$, we have $I^* = I_*$ and f is integrable on $[a, b]$.

Conversely, if $I^* = I_*$ and $\epsilon > 0$ are given, we can find a partition P' such that $L(f, P') > I_* - \epsilon/2$, and another partition P'', such that $U(f, P'') < I^* + \epsilon/2$. If P is a common refinement of P' and P'', then by Theorem A3.1, $U(f, P) - L(f, P) \le U(f, P'') - L(f, P') < \dfrac{\epsilon}{2} + \dfrac{\epsilon}{2} = \epsilon$, as required. □

EXAMPLE A3.5 Let $f(x) = \begin{cases} 0 & \text{if } 0 \le x < 1 \text{ or } 1 < x \le 2 \\ 1 & \text{if } x = 1. \end{cases}$
Show that f is integrable on $[0, 2]$ and find $\int_0^2 f(x)\,dx$.

SOLUTION Let $\epsilon > 0$ be given. Let $P = \{0,\ 1 - \epsilon/3,\ 1 + \epsilon/3,\ 2\}$. Then $L(f, P) = 0$ since $f(x) = 0$ at points of each of these subintervals into which P subdivides $[0, 2]$. See Fig. A3.3. Since $f(1) = 1$, we have

$$U(f, P) = 0\left(1 - \frac{\epsilon}{3}\right) + 1\left(\frac{2\epsilon}{3}\right) + 0\left(2 - \left(1 - \frac{\epsilon}{3}\right)\right) = \frac{2\epsilon}{3}.$$

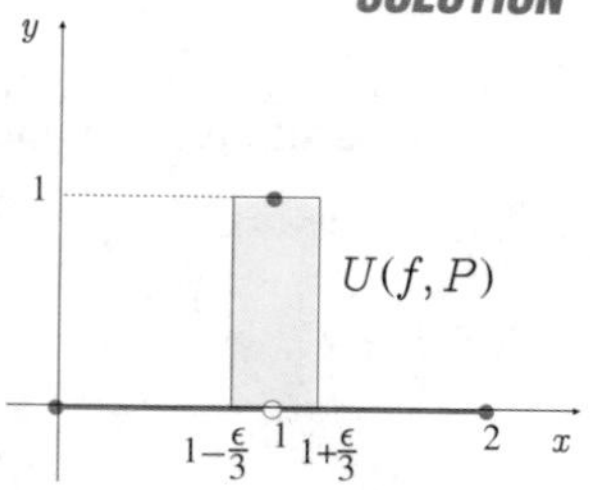

Fig. A3.3

Hence $U(f, P) - L(f, P) < \epsilon$ and f is integrable on $[0, 2]$. Evidently $\int_0^2 f(x)\,dx = I_* = 0$.

EXAMPLE A3.6 Let $f(x)$ be defined on $[0, 1]$ by

$$f(x) = \begin{cases} 1 & \text{if } x \text{ is rational} \\ 0 & \text{if } x \text{ is irrational.} \end{cases}$$

Show that f is not integrable on $[0, 1]$.

PROOF Every subinterval of $[0, 1]$ having positive length contains both rational and irrational numbers. Hence, for any partition P of $[0, 1]$ we have $L(f, P) = 0$ and $U(f, P) = 1$. Thus $I_* = 0$ and $I^* = 1$, so f is not integrable on $[0, 1]$. □

Uniform Continuity

When we assert that a function f is continuous on the interval $[a, b]$, we imply that for every x in that interval, and every $\epsilon > 0$, we can find a positive number δ (depending on *both* x and ϵ) such that $|f(y) - f(x)| < \epsilon$ whenever $|y - x| < \delta$ and y lies in $[a, b]$. In fact, however, it is possible to find a number δ *depending only on* ϵ such that $|f(y) - f(x)| < \epsilon$ holds whenever x and y belong to $[a, b]$ and satisfy $|y - x| < \delta$. We describe this phenomenon by saying that f is **uniformly continuous** on the interval $[a, b]$.

THEOREM A3.7 If f is continuous on the closed, finite interval $[a, b]$, then f is uniformly continuous on that interval.

PROOF Let $\epsilon > 0$ be given. Define numbers x_n in $[a, b]$ and subsets S_n of $[a, b]$ as follows:

$$x_1 = a$$
$$S_1 = \left\{x : x_1 < x \le b \text{ and } |f(x) - f(x_1)| \ge \frac{\epsilon}{3}\right\}.$$

If S_1 is empty, stop; otherwise let

$$x_2 = \text{ the greatest lower bound of } S_1$$
$$S_2 = \left\{x : x_2 < x \le b \text{ and } |f(x) - f(x_2)| \ge \frac{\epsilon}{3}\right\}.$$

If S_2 is empty, stop; otherwise proceed to define x_3 and S_3 analogously. We proceed in this way as long as we can; if x_n and S_n have been defined and S_n is not empty, we define

$$x_{n+1} = \text{ the greatest lower bound of } S_n$$
$$S_{n+1} = \left\{x : x_{n+1} < x \le b \text{ and } |f(x) - f(x_{n+1})| \ge \frac{\epsilon}{3}\right\}.$$

At any stage where S_n is not empty, the continuity of f at x_n assures us that $x_{n+1} > x_n$ and $|f(x_{n+1}) - f(x_n)| = \epsilon/3$.

We must consider two possibilities for the above procedure: Either S_n is empty for some n, or S_n is nonempty for every n.

Suppose S_n is nonempty for every n. Then we have constructed an infinite, increasing sequence $\{x_n\}$ in $[a, b]$ that, being bounded above (by b), must have a limit by completeness (Theorem A2.21). Let $\lim x_n = x^*$. We have $a \le x^* \le b$. Since f is continuous at x^* there exists $\delta > 0$ such that $|f(x) - f(x^*)| < \epsilon/8$ whenever $|x - x^*| < \delta$ and x lies in $[a, b]$. Since $\lim x_n = x^*$, there exists a positive integer N such that $|x_n - x^*| < \delta$ whenever $n \ge N$. For such n we have

$$\begin{aligned}\frac{\epsilon}{3} = |f(x_{n+1}) - f(x_n)| &= |f(x_{n+1}) - f(x^*) + f(x^*) - f(x_n)| \\ &\le |f(x_{n+1}) - f(x^*)| + |f(x_n) - f(x^*)| \\ &< \frac{\epsilon}{8} + \frac{\epsilon}{8} = \frac{\epsilon}{4},\end{aligned}$$

which is clearly impossible. Thus S_n must, in fact, be empty for some n.

Suppose that S_N is empty. Thus S_n is nonempty for $n < N$, and the procedure for defining x_n stops with x_N. Since S_{N-1} is not empty, $x_N < b$. In this case define $x_{N+1} = b$ and let

$$\delta = \min\{x_2 - x_1,\ x_3 - x_2,\ \ldots,\ x_{N+1} - x_N\}.$$

The minimum of a finite set of positive numbers is a positive number so $\delta > 0$. If x lies in $[a, b]$, then x lies in one of the intervals $[x_1, x_2]$, $[x_2, x_3]$, ..., $[x_N, x_{N+1}]$. Suppose x lies in $[x_k, x_{k+1}]$. If y is in $[a, b]$ and $|y - x| < \delta$, then y lies in either the same subinterval as x or in an adjacent one; that is, y lies in $[x_j, x_{j+1}]$ where $j = k - 1$, k, or $k + 1$. Thus

$$\begin{aligned}|f(y) - f(x)| &= |f(y) - f(x_j) + f(x_j) - f(x_k) + f(x_k) - f(x)|\\ &\le |f(y) - f(x_j)| + |f(x_j) - f(x_k)| + |f(x_k) - f(x)|\\ &< \frac{\epsilon}{3} + \frac{\epsilon}{3} + \frac{\epsilon}{3} = \epsilon,\end{aligned}$$

which was to be proved. □

We are now in a position to prove that a continuous function is integrable. This is Theorem 6.2.3.

THEOREM A3.8 If f is continuous on $[a, b]$, then f is integrable on $[a, b]$.

PROOF By Theorem A3.7, f is uniformly continuous on $[a, b]$. Let $\epsilon > 0$ be given. Let $\delta > 0$ be such that $|f(x) - f(y)| < \epsilon/(b - a + 1)$ whenever $|x - y| < \delta$ and x and y belong to $[a, b]$. Choose a partition $P = \{x_0, x_1, \ldots, x_n\}$ of $[a, b]$ for which each subinterval $[x_{j-1}, x_j]$ has length $\Delta x_j < \delta$. Then the greatest lower bound, m_j, and the least upper bound, M_j, of the set of values of $f(x)$ on $[x_{j-1}, x_j]$ satisfy $M_j - m_j \le \epsilon/(b - a + 1) < \epsilon/(b - a)$. Accordingly,

$$U(f, P) - L(f, P) < \frac{\epsilon}{b - a} \sum_{j=1}^{n(P)} \Delta x_j = \frac{\epsilon}{b - a}(b - a) = \epsilon.$$

Thus f is integrable on $[a, b]$ as asserted. □

EXERCISES

1. Let $f(x) = \begin{cases} 1 & \text{if } 0 \le x \le 1 \\ 0 & \text{if } 1 < x \le 2 \end{cases}$. Prove that f is integrable on $[0, 2]$ and find the value of $\int_0^2 f(x)\,dx$.

2. Let $f(x) = \begin{cases} 1 & \text{if } x = 1/n, \quad n = 1, 2, 3, \ldots \\ 0 & \text{for all other values of } x \end{cases}$.
Show that f is integrable over $[0, 1]$ and find the value of the integral $\int_0^1 f(x)\,dx$.

3.*Let $f(x) = 1/n$ if $x = m/n$ where m, n are integers having no common factors, and let $f(x) = 0$ if x is an irrational number. Thus $f(1/2) = 1/2$, $f(1/3) = f(2/3) = 1/3$, $f(1/4) = f(3/4) = 1/4$, etc. Show that

f is integrable on $[0, 1]$ and find $\displaystyle\int_0^1 f(x)\,dx$. (*Hint:* Show that for any $\epsilon > 0$, only finitely many points of the graph of f over $[0, 1]$ lie above the line $y = \epsilon$.)

4. Prove that I_* and I^* defined in the paragraph following Theorem A3.2 satisfy $I_* \leq I^*$ as claimed there.

5. Prove parts (c), (d), (e), (f), (g), and (h) of Theorem 6.2.4 (Section 6.2) for the Riemann Integral.

6. Use the definition of uniform continuity given in the paragraph preceding Theorem A3.7 to prove that $f(x) = \sqrt{x}$ is uniformly continuous on $[0, 1]$. Do not use Theorem A3.7.

7. Show directly from the definition of uniform continuity (without using Theorem A2.24) that a function f uniformly continuous on a closed, finite interval is necessarily bounded there.

8. If f is bounded and integrable on $[a, b]$, prove that $F(x) = \int_a^x f(t)\,dt$ is uniformly continuous on $[a, b]$. (If f were continuous we would have a stronger result; F would be differentiable on (a, b) and $F'(x) = f(x)$ (the Fundemental Theorem of Calculus.))

ANSWERS TO ODD-NUMBERED EXERCISES

CHAPTER 1. FUNCTIONS, LIMITS AND CONTINUITY

Section 1.2 (page 13)

1. (a) $\frac{100}{11}$, (b) $\frac{1589}{495}$, (c) $\frac{11}{24975}$, (d) $\frac{2}{7}$

3. $-3 \le x \le 3$

5. $x > 0$

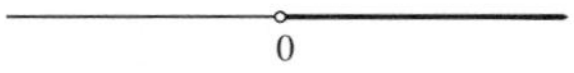

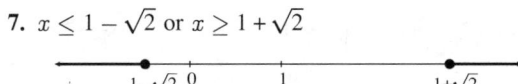

7. $x \le 1 - \sqrt{2}$ or $x \ge 1 + \sqrt{2}$

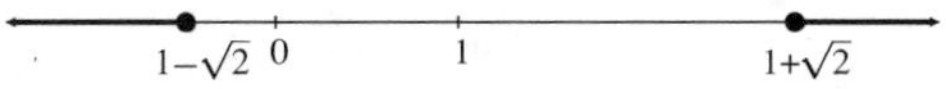

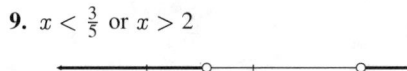

9. $x < \frac{3}{5}$ or $x > 2$

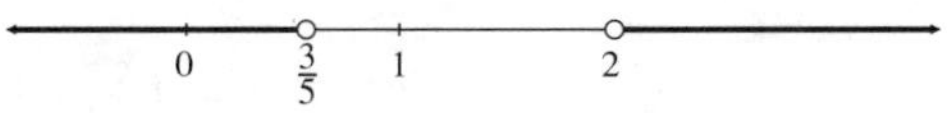

11. $0 < x < 1$

13. $x < -5$ or $-1 < x < 1$

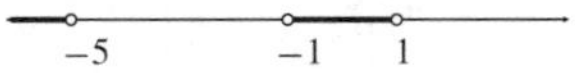

15. $x \le -\frac{5}{2}$

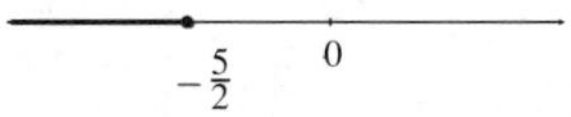

17. $x \le -3$ or $x \ge 1$

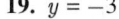

19. $y = -3$

21. $y = 2x - 3$

23. $x + 3y = 0$

25. $x + 3y = 5$

27. line, slope -1

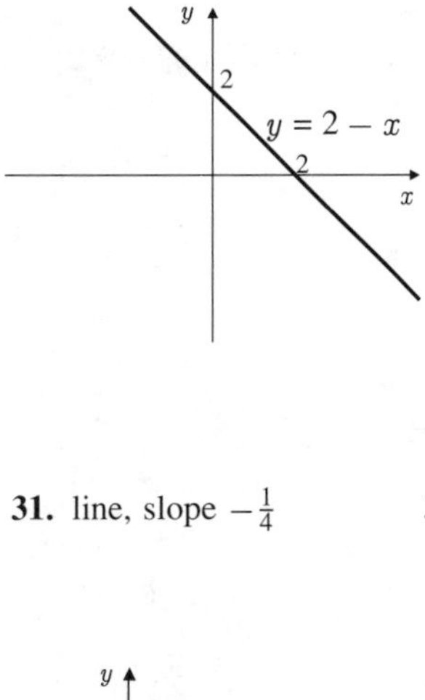

29. vertical line

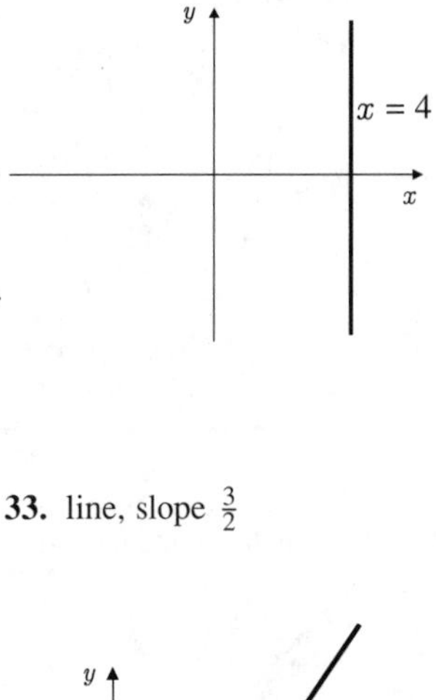

31. line, slope $-\frac{1}{4}$

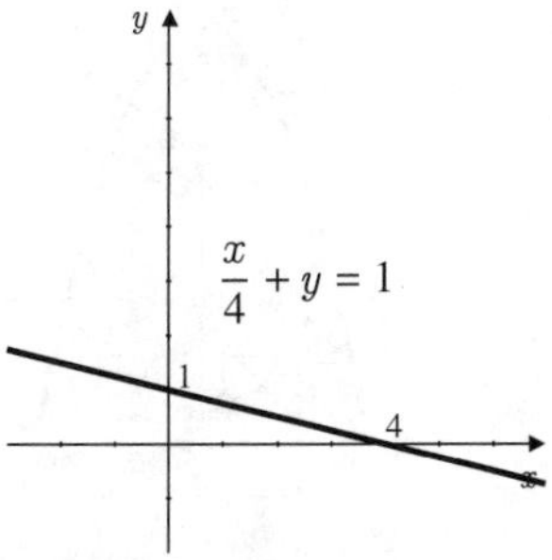

33. line, slope $\frac{3}{2}$

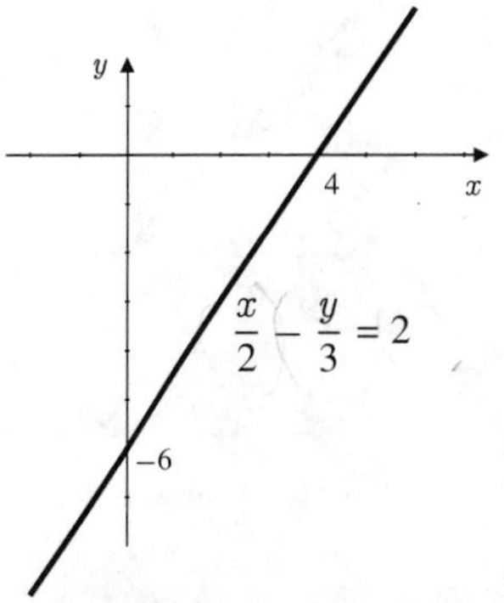

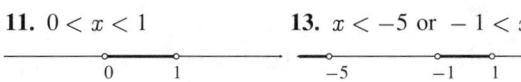

35. circle, centre $(0,0)$, radius $\sqrt{5}$

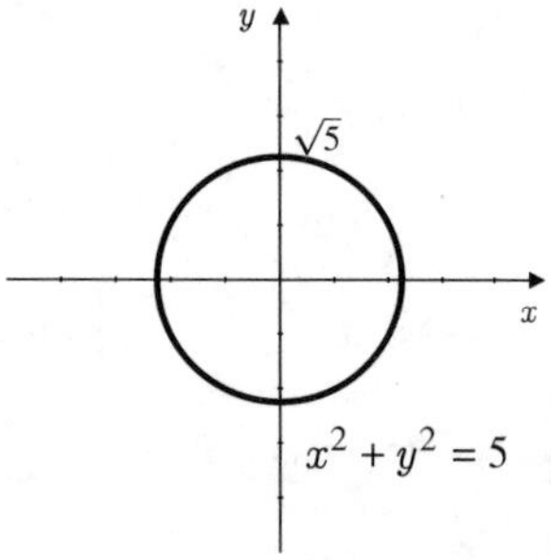

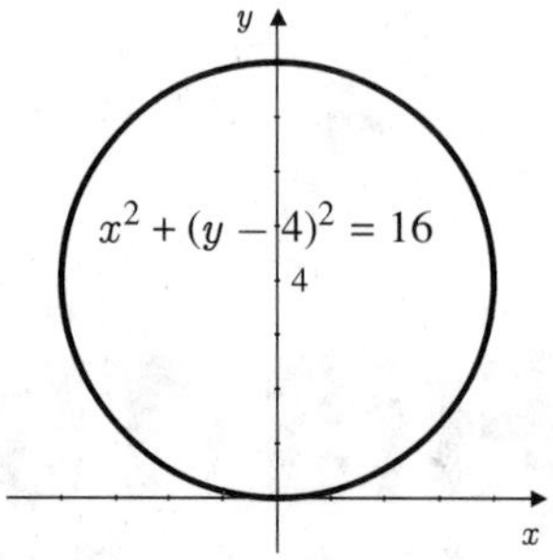

37. circle, centre $(0,4)$, radius 4

39. circle, centre $(-1,1)$, radius 2

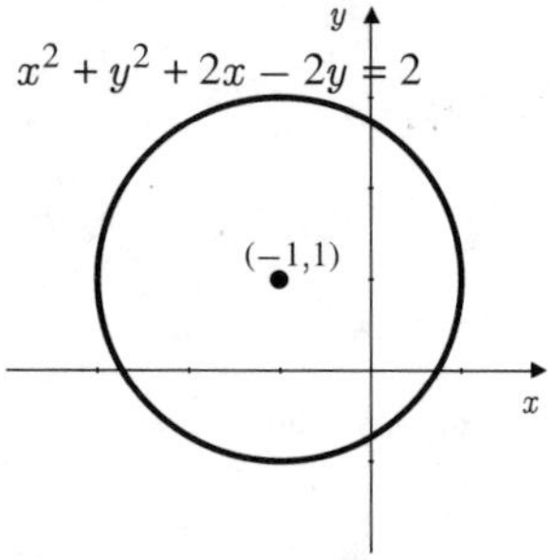

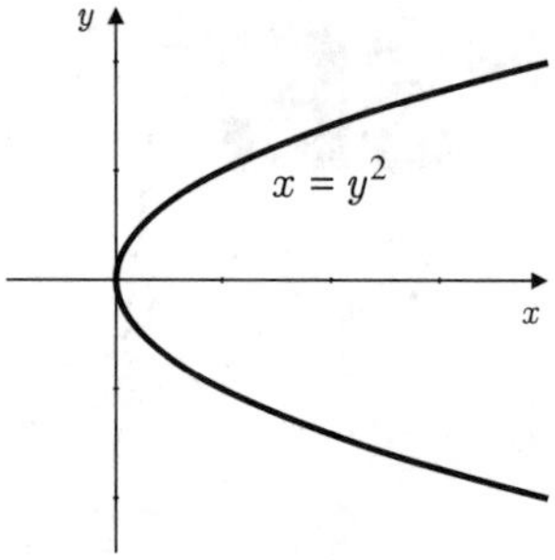

41. parabola, vertex $(0,0)$

43. parabola, vertex $(-1,1)$ **45.** single point $(0,0)$

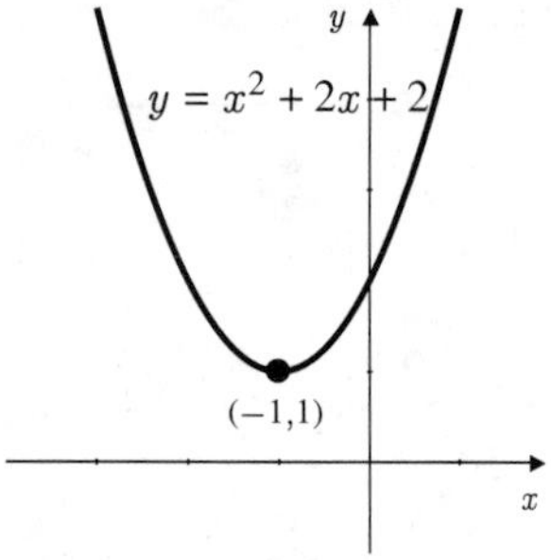

47. half-plane under line $x + y = 2$

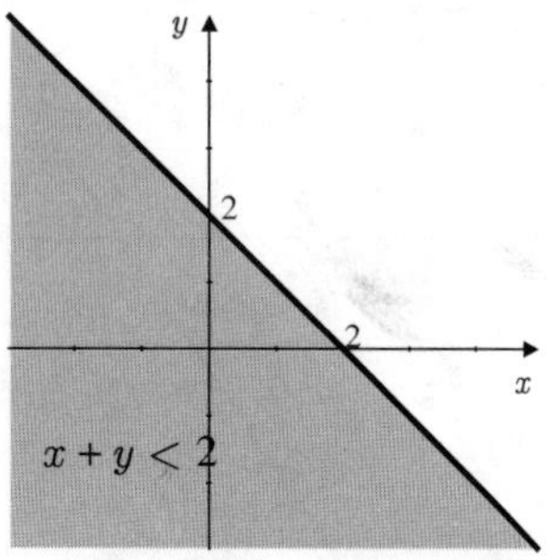

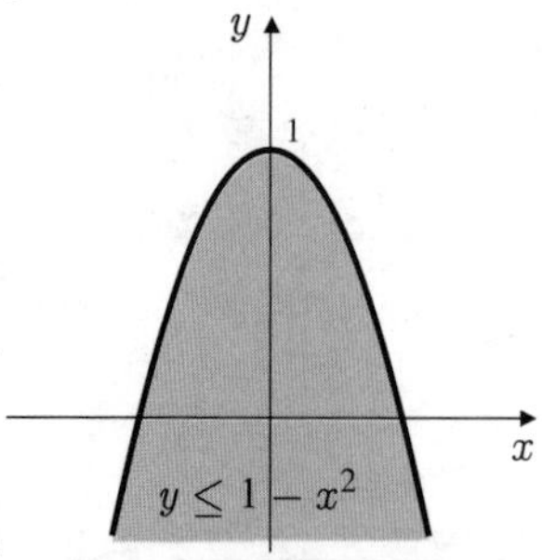

49. region on and under parabola $y = 1 - x^2$

Section 1.3 (page 21)

1. $\mathcal{D}(f) = \mathbb{R}$, $\mathcal{R}(f) = \{5\}$ **3.** $\mathcal{D}(F) = \mathbb{R}$, $\mathcal{R}(F) = (-\infty, 2$

5. $\mathcal{D}(f) = (-\infty) \cup (0,\infty)$, $\mathcal{R}(f) = (-\infty, 1) \cup (1, \infty)$

7. $\mathcal{D}(k) = (-\infty, 2]$, $\mathcal{R}(k) = [0,\infty)$

9. $\mathcal{D}(g) = \mathbb{R}$, $\mathcal{R}(g) = (0,1]$

11. $\mathcal{D}(H) = (-\infty, 0] \cup [1,\infty)$, $\mathcal{R}(H) = [0,\infty)$

13. The functions in Example 1.3.7(a) and (c) are even; those in (b) and (d) are neither even nor odd.

15. f is odd

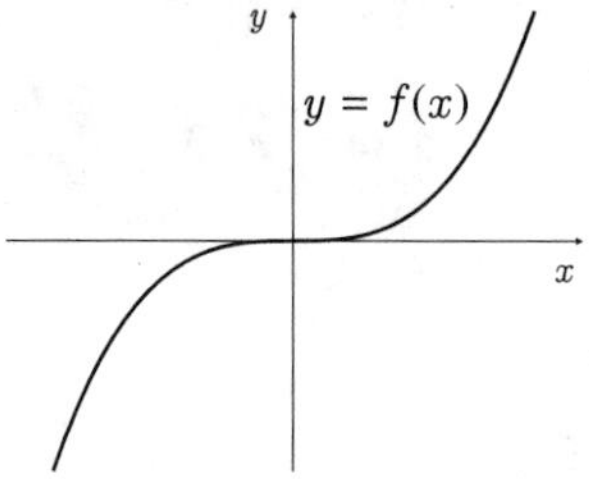

17. g is neither even nor odd

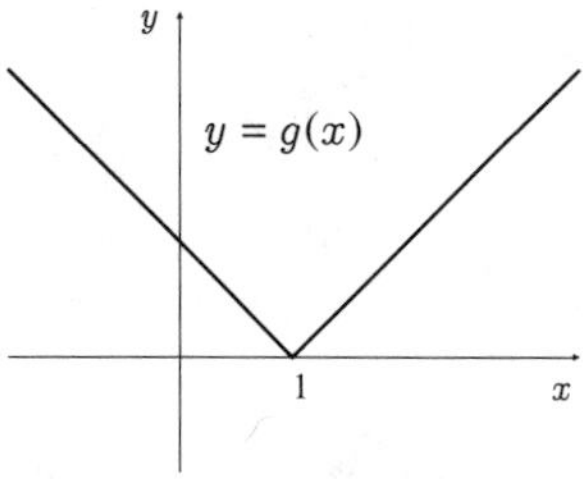

19. G is even

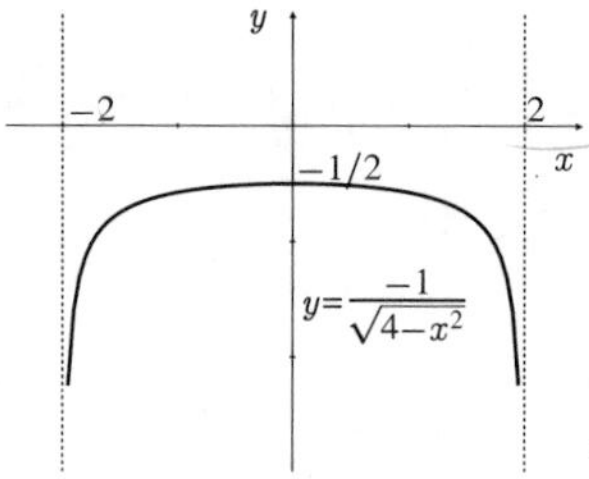

21. f is neither even nor odd

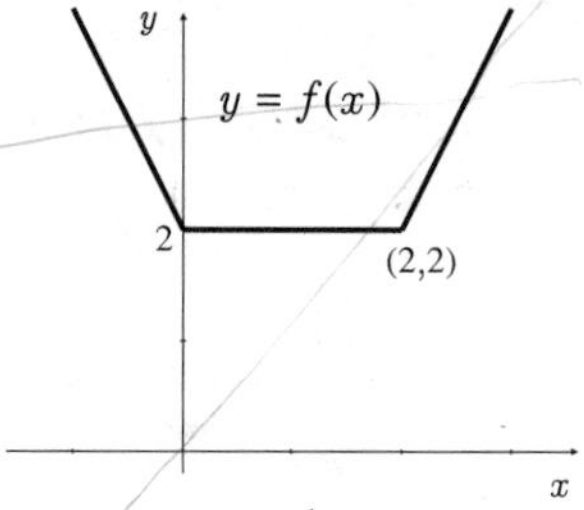

23. $(f+g)(x) = \dfrac{1}{x-1} + \dfrac{1}{x}$, $(fg)(x) = \dfrac{1}{(x-1)x}$,
$\left(\dfrac{g}{f}\right)(x) = \dfrac{x-1}{x}$
$f \circ f(x) = \dfrac{x-1}{2-x}$, $f \circ g(x) = \dfrac{x}{1-x}$,
$g \circ f(x) = x - 1$, $g \circ g(x) = x$,
$\mathcal{D}(f+g) = \mathcal{D}(fg) = \mathcal{D}(g/f) = \mathcal{D}(f \circ g) = \{x : x \neq 0, 1\}$,
$\mathcal{D}(f \circ f) = \{x : x \neq 1, 2\}$,
$\mathcal{D}(g \circ f) = \{x : x \neq 1\}$, $\mathcal{D}(g \circ g) = \{x : x \neq 0\}$

25. $(f+g)(x) = \sqrt{1-x^2} + 2 + x$,
$(fg)(x) = \sqrt{1-x^2}(2+x)$, $\left(\dfrac{g}{f}\right)(x) = \dfrac{2+x}{\sqrt{1-x^2}}$,
$f \circ f(x) = |x|$, $f \circ g(x) = \sqrt{1-(2+x)^2}$,
$g \circ f(x) = 2 + \sqrt{1-x^2}$, $g \circ g(x) = 4 + x$,
$\mathcal{D}(f+g) = \mathcal{D}(fg) = \mathcal{D}(f \circ f) = \mathcal{D}(g \circ f) = [-1,1]$,
$\mathcal{D}(g/f) = (-1,1)$, $\mathcal{D}(f \circ g) = [-3,-1]$, $\mathcal{D}(g \circ g) = \mathbb{R}$

27.

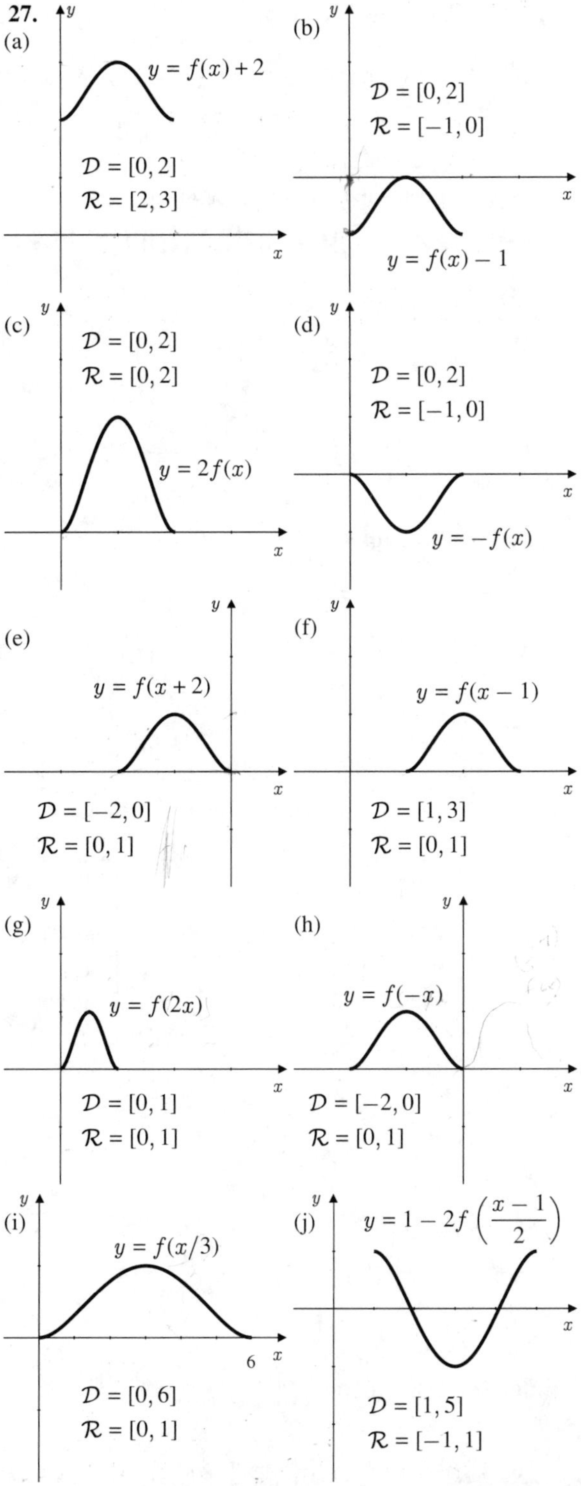

Section 1.4 (page 26)

1. $f^{-1}(x) = x + 1$
$\mathcal{D}(f^{-1}) = \mathcal{R}(f) = \mathcal{R}(f^{-1}) = \mathcal{D}(f) = \mathbb{R}$

3. $f^{-1}(x) = \dfrac{1+x}{2}$
$\mathcal{D}(f^{-1}) = \mathcal{R}(f) = \mathcal{R}(f^{-1}) = \mathcal{D}(f) = \mathbb{R}$

5. $f^{-1}(x) = x^{1/3}$
$\mathcal{D}(f^{-1}) = \mathcal{R}(f) = \mathcal{R}(f^{-1}) = \mathcal{D}(f) = \mathbb{R}$

7. $f^{-1}(x) = -\sqrt{x}$, $\mathcal{D}(f^{-1}) = \mathcal{R}(f) = [0, \infty)$,
$\mathcal{R}(f^{-1}) = \mathcal{D}(f) = (-\infty, 0]$

9. $f^{-1}(x) = \dfrac{1}{x} - 1$, $\mathcal{D}(f^{-1}) = \mathcal{R}(f) = \{x : x \neq 0\}$,
$\mathcal{R}(f^{-1}) = \mathcal{D}(f) = \{x : x \neq -1\}$

11. $f^{-1}(x) = \dfrac{1-x}{2+x}$, $\mathcal{D}(f^{-1}) = \mathcal{R}(f) = \{x : x \neq -2\}$,
$\mathcal{R}(f^{-1}) = \mathcal{D}(f) = \{x : x \neq -1\}$

13. $g^{-1}(x) = f^{-1}(x+2)$ **15.** $k^{-1}(x) = f^{-1}\left(-\dfrac{x}{3}\right)$

17. $p^{-1}(x) = f^{-1}\left(\dfrac{1}{x} - 1\right)$

19. $r^{-1}(x) = \dfrac{1}{4}\left(3 - f^{-1}\left(\dfrac{1-x}{2}\right)\right)$

21. $f^{-1}(x) = \begin{cases} \sqrt{x-1} & \text{if } x >= 1 \\ x - 1 & \text{if } x < 1 \end{cases}$

23. $c = 1$. a and b may have any values.

Section 1.5 (page 32)

1. 1 **3.** 0 **5.** 0
7. -3 **9.** 0 **11.** -1
13. does not exist **15.** $-1/2$
17. $3/8$ **19.** $8/3$ **21.** $3/16$
23. 4 **25.** $1/4$ **27.** $1/\sqrt{2}$
29. $\lim_{x\to 1} f(x) = 1/3$
31. $\lim_{x\to -1} f(x) = \lim_{x\to 1} f(x) = 1$, $\lim_{x\to 0} f(x) = 0$
33. $\lim_{x\to a} f(x) = f(a)$

Section 1.6 (page 39)

1. 0 **3.** 2
5. does not exist **7.** $+\infty$
9. does not exist **11.** does not exist
13. $-2/21$ **15.** $-\infty$
17. $1/(2a)$ **19.** does not exist
21. ∞ **23.** $-\infty$ **25.** ∞
27. 0 **29.** ∞
31. does not exist **33.** $-\infty$

35. $-3/5$ **37.** 0 **39.** ∞

41. -3 **43.** $-2/\sqrt{3}$ **45.** $-\sqrt{2}/4$

47. -2 **49.** -1

51. $C(t)$ has a limit at every real t except at the integers. $\lim_{t\to t_0-} C(t) = C(t_0)$ everywhere, but
$$\lim_{t\to t_0+} C(t) = \begin{cases} C(t_0) & \text{if } t_0 \text{ is not an integer} \\ C(t_0+1) & \text{if } t_0 \text{ is an integer} \end{cases}$$

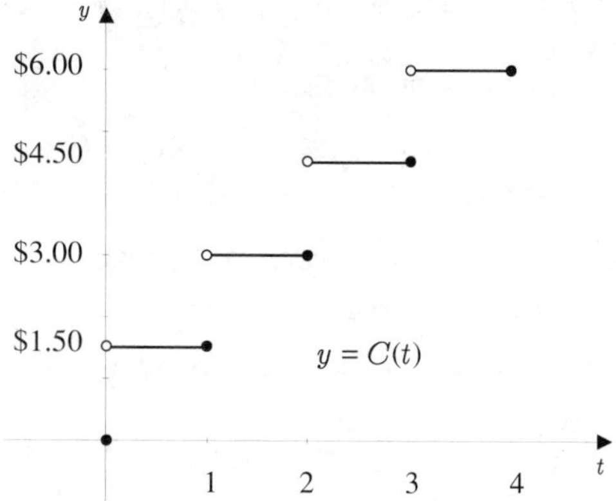

53. (a) B, (b) A, (c) A, (d) A

Section 1.7 (page 46)

1. continuous everywhere

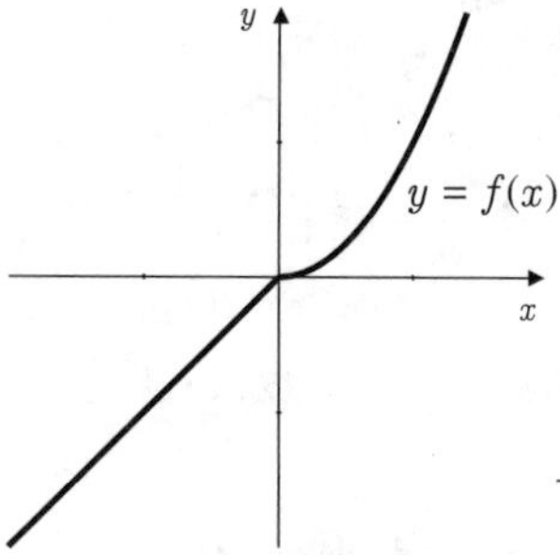

3. continuous except at $x = 2$

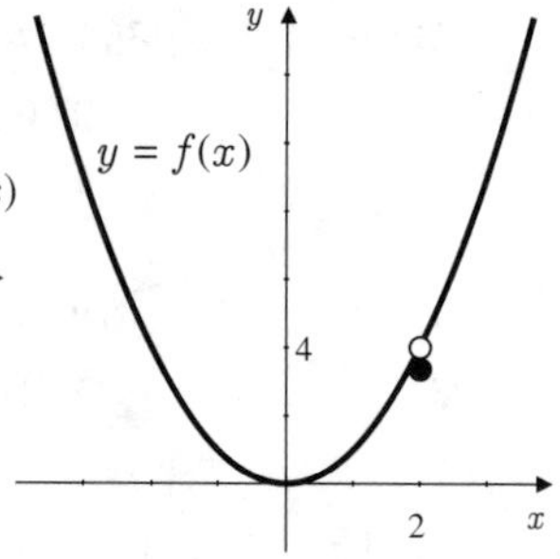

5. continuous except at $x = 0$; right continuous at $x = 0$

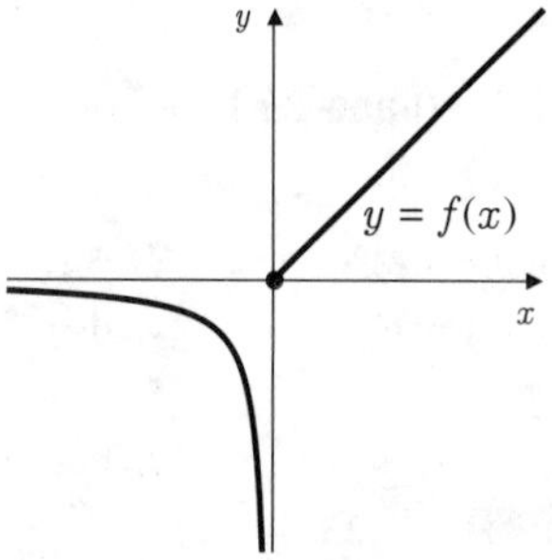

7. continuous except at the integers; left continuous at the integers

9. $\left(\frac{1}{2}, \frac{1}{2\sqrt{3}}\right)$

11. f positive on $(-1,0)$ and $(1,\infty)$; f negative on $(-\infty,-1)$ and $(0,1)$

13. f positive on $(-\infty,-2)$, $(-1,1)$ and $(2,\infty)$; f negative on $(-2,-1)$ and $(1,2)$

15. f has zeros in $(-4,-3)$, $(-3,1)$ and $(1,4)$.

CHAPTER 2. DIFFERENTIATION

Section 2.1 (page 53)

1. $y = 8x - 13$ **3.** $y = 1$ **5.** $y = 12x - 24$

7. $2x + 27y = 9$ **9.** $x - 4y = -5$ **11.** $x - 4y = -2$

13. $y = 2x_0x - x_0^2$ **15.** $y = (2ax_0 + b)x - ax_0^2 + c$

17. (a) $3a^2$; (b) $y = 3x - 2$ and $y = 3x + 2$

19. No. See Fig. 2.1.7.

Section 2.2 (page 61)

1. $2x - 3$ **3.** $3x^2$ **5.** $-\dfrac{4}{(2+x)^2}$

7. $\dfrac{1}{\sqrt{2t+1}}$ **9.** $1 - \dfrac{1}{x^2}$ **11.** $-\dfrac{1}{t^2} - \dfrac{1}{2t^{3/2}}$

13. $-\dfrac{1}{2(1+x)^{3/2}}$ **15.** $-\dfrac{x}{(1+x^2)^{3/2}}$

17. All three are differentiable everywhere except at $x = 0$. $\dfrac{d}{dx}\operatorname{sgn} x = 0$, $\dfrac{d}{dx}x\operatorname{sgn} x = \operatorname{sgn} x$, $\dfrac{d}{dx}x^2\operatorname{sgn} x = 2x\operatorname{sgn} x$

19.

x	$\dfrac{f(x)-f(2)}{x-2}$
1.9	−0.26316
1.99	−0.25126
1.999	−0.25013
1.9999	−0.25001

x	$\dfrac{f(x)-f(2)}{x-2}$
2.1	−0.23810
2.01	−0.24876
2.001	−0.24988
2.0001	−0.24999

$$\frac{d}{dx}\left(\frac{1}{x}\right)\bigg|_{x=2} = -\frac{1}{4}$$

21. $x - 6y = -15$

23. $y = \dfrac{2}{a^2+a} - \dfrac{2(2a+1)}{(a^2+a)^2}(t-a)$

25. $22t^{21}$ **27.** -16 **29.** $y = a^2x - a^3 + \dfrac{1}{a}$

31. $y = 6x - 9$ and $y = -2x - 1$

33. $\dfrac{1}{2\sqrt{2}}$ **37.** $f'(x) = \frac{1}{3}x^{-2/3}$

Section 2.3 (page 67)

1. $6x - 5$ **3.** $x^2 - x + 1$

5. $2Ax + B$ **7.** $\frac{1}{3}s^4 - \frac{1}{5}s^2$

9. $\frac{1}{3}t^{-2/3} + \frac{1}{2}t^{-3/4} + \frac{3}{5}t^{-4/5}$

11. $\dfrac{5}{2\sqrt{x}} - \frac{3}{2}\sqrt{x} - \frac{5}{6}x^{3/2}$

13. $-\dfrac{2x+5}{(x^2+5x)^2}$ **15.** $\dfrac{\pi^2}{(2-\pi t)^2}$

17. $-\dfrac{3(2\sqrt{x}+1)}{2\sqrt{x}(x+\sqrt{x})^2}$ **19.** $-\dfrac{24}{(3+4x)^2}$

21. $\dfrac{1}{\sqrt{t}(1-\sqrt{t})^2}$ **23.** $\dfrac{ad-bc}{(cx+d)^2}$

25. $10 + 70x + 150x^2 + 96x^3$

27. $2x(\sqrt{x}+1)(5x^{2/3}-2) + \dfrac{1}{2\sqrt{x}}(x^2+4)(5x^{2/3}-2)$
$+\frac{10}{3}x^{-1/3}(x^2+4)(\sqrt{x}+1)$

29. $\dfrac{6x+1}{(6x^2+2x+1)^2}$ **31.** $-\dfrac{1}{2}$

33. $-\dfrac{1}{18\sqrt{2}}$ **35.** $y = 4x - 6$

37. $(1,2)$ and $(-1,-2)$ **39.** $\left(-\frac{1}{2}, \frac{4}{7}\right)$

41. $y = b - \dfrac{b^2x}{4}$

43. $\dfrac{d}{dx}\sqrt{f(x)} = \dfrac{f'(x)}{2\sqrt{f(x)}}$; $\dfrac{d}{dx}\sqrt{x^2+1} = \dfrac{x}{\sqrt{x^2+1}}$

Section 2.4 (page 72)

1. $12(2x+3)^5$ **3.** $-20x(4-x^2)^9$

5. $Ar(Ax+B)^{r-1}$ **7.** $\dfrac{12}{(5-4x)^2}$

9. $\dfrac{5}{2\sqrt{5x+3}}$ **11.** $-\dfrac{30}{t^2}\left(2+\dfrac{3}{t}\right)^9$

13. $-2x^3(x^4+1)^{-3/2}$ **15.** $\dfrac{4x^4+6x^2}{\sqrt{x^2+2}}$

17. $-\dfrac{1}{3(t+1)^2}$ **19.** $4 + 4\,\mathrm{sgn}\,(4x-1) = \begin{cases} 8 & \text{if } x > 1/4 \\ 0 & \text{if } x < 1/4 \end{cases}$

21. $\dfrac{-3}{2\sqrt{3x+4}(2+\sqrt{3x+4})^2}$ **23.** $\sqrt{3-t}\left(6t - \frac{7}{2}t^2 - 3\right)$

25. $-\dfrac{5}{3}\left(1 - \dfrac{1}{(u-1)^2}\right)\left(u + \dfrac{1}{u-1}\right)^{-8/3}$

27.

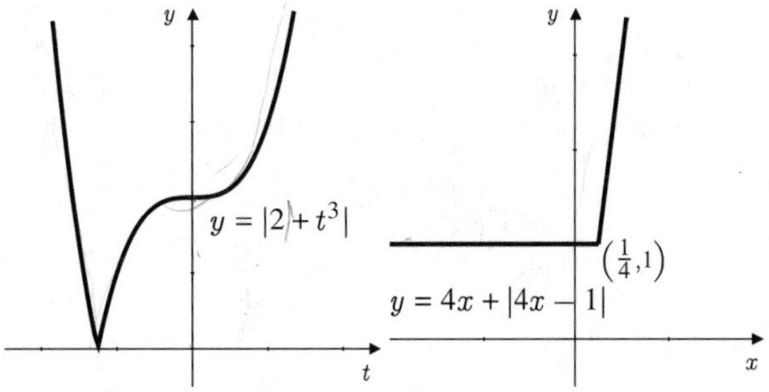

29. $(5-2x)f'(5x-x^2)$ **31.** $\dfrac{f'(x)}{\sqrt{3+2f(x)}}$

33. $\dfrac{1}{\sqrt{x}}f'(3+2\sqrt{x})$ **35.** $15f'(4-5t)f'(2-3f(4-5t))$

37. $\dfrac{3}{2\sqrt{2}}$ **39.** 102

41. $-6\left(1 - \frac{15}{2}(3x)^4\left((3x)^5-2\right)^{-3/2}\right)$
$\times\left(x + \left((3x)^5-2\right)^{-1/2}\right)^{-7}$

43. $y = 2^{3/2} + \sqrt{2}(x+1)$ **45.** $y = \frac{1}{27} + \frac{5}{162}(x+2)$

47. It may happen that $k = g(x+h) - g(x) = 0$ for values of h arbitrarily close to 0 so that division by k in the "proof" is not justified.

Section 2.5 (page 77)

1. $\dfrac{1-y}{2+x}$ **3.** $\dfrac{2x+y}{3y^2-x}$ **5.** $\dfrac{2-2xy^3}{3x^2y^2+1}$

7. $-\dfrac{3x^2+2xy}{x^2+4y}$ **9.** $2x+3y=5$ **11.** $y=x$

15. $y' = -\dfrac{2x+y}{4y+x}$ Points on the given curve must satisfy $(x,y) \neq (0,0)$ and $x^2+2y^2+xy=0$, and hence also $\left(x+\frac{1}{2}y\right)^2 + \frac{7}{4}y^2 = 0$. There are no such points.

17. $1/[6f^{-1}(x)]^2$

Section 2.6 (page 82)

1. $\dfrac{dA}{ds} = 2s$ **3.** $\dfrac{dV}{dr} = 4\pi r^2$ **5.** $\dfrac{dC}{dA} = \sqrt{\dfrac{\pi}{A}}$

7. (a) $t > 2$, (b) $t < 0$, (c) all t, (d) no t, (e) $t > 2$, (f) $t < 2$, (g) 2, (h) 0

9. (a) $t < -2/\sqrt{3}$ or $t > 2/\sqrt{3}$, (b) $-2/\sqrt{3} < t < 2/\sqrt{3}$, (c) $t > 0$, (d) $t < 0$, (e) $t > 2/\sqrt{3}$ or $-2/\sqrt{3} < t < 0$, (f) $t < -2/\sqrt{3}$ or $0 < t < 2/\sqrt{3}$, (g) $\pm 12/\sqrt{3}$ at $t = \pm 2/\sqrt{3}$, (h) 12

11. acc=9.8 m/s^2 downward at all times; max height=4.9 m; ball strikes ground at 9.8 m/s

13. time 27.8 s; distance 771.6 m

15. $v = \begin{cases} 2t & \text{if } 0 < t \le 2 \\ 4 & \text{if } 2 < t < 8 \\ 20 - 2t & \text{if } 8 \le t < 10 \end{cases}$

v is continuous for $0 < t < 10$.

$a = \begin{cases} 2 & \text{if } 0 < t < 2 \\ 0 & \text{if } 2 < t < 8 \\ -2 & \text{if } 8 < t < 10 \end{cases}$

a is continuous except at $t = 2$ and $t = 8$.
Maximum velocity 4 is attained for $2 \le t \le 8$.

17. 1500 tonnes

19. The "extra cost of producing one more item" is $\dfrac{\Delta C}{\Delta x}$ where $\Delta x = 1$. The marginal cost of production is $\dfrac{dC}{dx} = \lim\limits_{\Delta x \to 0} \dfrac{\Delta C}{\Delta x}$. In the sense that $\Delta x = 1$ is "small," the extra cost and marginal cost are approximately the same.

Section 2.7 (page 86)

1. $\begin{cases} y' = -14(3-2x)^6, \\ y'' = 168(3-2x)^5, \\ y''' = -1680(3-2x)^4 \end{cases}$

3. $\begin{cases} y' = -12(x-1)^{-3}, \\ y'' = 36(x-1)^{-4}, \\ y''' = -144(x-1)^{-5} \end{cases}$

5. $\begin{cases} y' = \frac{1}{3}x^{-2/3} + \frac{1}{3}x^{-4/3}, \\ y'' = -\frac{2}{9}x^{-5/3} - \frac{4}{9}x^{-7/3} \\ y''' = \frac{10}{27}x^{-8/3} + \frac{28}{27}x^{-10/3} \end{cases}$

7. $\begin{cases} y' = \frac{5}{2}x^{3/2} + \frac{3}{2}x^{-1/2} \\ y'' = \frac{15}{4}x^{1/2} - \frac{3}{4}x^{-3/2} \\ y''' = \frac{15}{8}x^{-1/2} + \frac{9}{8}x^{-5/2} \end{cases}$

9. $(-1)^n n! x^{-(n+1)}$ **11.** $n!(2-x)^{-(n+1)}$

13. $(-1)^n n! b^n (a+bx)^{-(n+1)}$

15. $-\dfrac{1 \times 3 \times 5 \times \cdots \times (2n-3)}{2^n} 3^n (1-3x)^{-(2n-1)/2}$, $(n = 2, 3, \ldots)$

17. $\dfrac{2(y-1)}{(1-x)^2}$

19. $\dfrac{(2-6y)(1-3x^2)^2}{(3y^2-2y)^3} - \dfrac{6x}{3y^2-2y}$

21. $-a^2/y^3$

Section 2.8 (page 92)

1. $5x + C$ **3.** $\frac{2}{3}x^{3/2} + C$ **5.** $\frac{1}{4}x^4 + C$

7. $a^2x - \frac{1}{3}x^3 + C$ **9.** $\frac{4}{3}x^{3/2} + \frac{9}{4}x^{4/3} + C$

11. $6x^5 + 5x^6 + C$ **13.** $\frac{1}{12}x^4 - \frac{1}{6}x^3 + \frac{1}{2}x^2 - x + C$

15. $\dfrac{1}{a(r+1)}(ax+b)^{r+1} + C$ **17.** $\frac{1}{3}(2x+3)^{3/2} + C$

19. $y = \frac{1}{2}x^2 - 2x + 3$, all x **21.** $y = 2x^{3/2} - 15$, $(x > 0)$

23. $y = \dfrac{A}{3}(x^3 - 1) + \dfrac{B}{2}(x^2 - 1) + C(x-1) + 1$, (all x)

25. $y = x^2 + 5x - 3$, (all x) **27.** $y = \dfrac{x^5}{20} - \dfrac{x^2}{2} + 8$, (all x)

29. $y = 3x - \dfrac{1}{x}$, (all x) **33.** $y = -\dfrac{7\sqrt{x}}{2} + \dfrac{18}{\sqrt{x}}$, $(x > 0)$

CHAPTER 3. THE ELEMENTARY TRANSCENDENTAL FUNCTIONS

Section 3.1 (page 104)

1. $-\dfrac{1}{\sqrt{2}}$ **3.** $\dfrac{\sqrt{3}}{2}$ **5.** $\dfrac{1+\sqrt{3}}{2\sqrt{2}}$

7. $\dfrac{1-\sqrt{3}}{2\sqrt{2}}$ **9.** $-\dfrac{2\sqrt{2}}{1+\sqrt{3}}$ **11.** $-\dfrac{\sqrt{3}}{2}$

13. $-\cos x$ **15.** $-\cos x$ **17.** $\dfrac{1}{\cos x \sin x}$

19. $\dfrac{1}{(\cos x \sin x)^2}$

25. $\cos 2x$; period π **27.** $\sin \pi x$; period 2

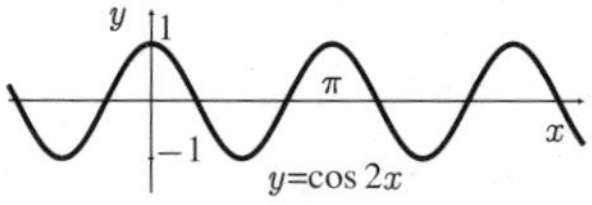

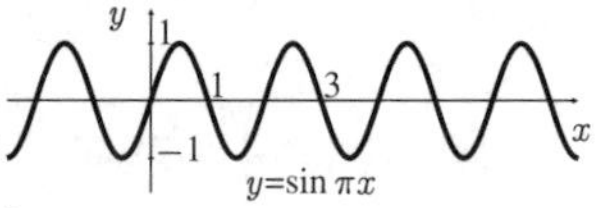

29. $\cos x = -\dfrac{4}{5}$; $\tan x = -\dfrac{3}{4}$

31. $\sin x = -\dfrac{2\sqrt{2}}{3}$; $\tan x = -2\sqrt{2}$

33. $\cos x = -\dfrac{\sqrt{3}}{2}$; $\tan x = \dfrac{1}{\sqrt{3}}$

35. $a = 1$; $b = \sqrt{3}$ **37.** $b = \dfrac{5}{\sqrt{3}}$; $c = \dfrac{10}{\sqrt{3}}$

39. $a = b \tan A$ **41.** $a = b \cot B$

43. $c = b \sec A$ **45.** $\sin A = \dfrac{\sqrt{c^2 - b^2}}{c}$

47. $\sin B = \dfrac{3}{4\sqrt{2}}$ **49.** $\sin B = \dfrac{\sqrt{135}}{16}$

51. $a = \dfrac{6}{1+\sqrt{3}}$ **53.** $b = 4\dfrac{\sin 40°}{\sin 70°} \approx 2.736$

57. $\dfrac{10 \tan 50° \tan 35°}{\tan 50° + \tan 35°} \approx 4.41m$

Section 3.2 (page 113)

1. $-3\sin 3x$ **3.** $\pi\sec^2\pi x$ **5.** $-\dfrac{2}{\pi}\sin\dfrac{2x}{\pi}$

7. $3\csc^2(4-3x)$ **9.** $A\cos(Ax+B)$

11. $r\sin(s-rx)$ **13.** $a\cos 2at$

15. $(2x-4)\sec(x^2-4x)\tan(x^2-4x)$

17. $\dfrac{x(2\tan x\sec^2 x)-\tan^2 x}{x^2}$

19. $\sin\dfrac{1}{x}-\dfrac{1}{x}\cos\dfrac{1}{x}$ **21.** $-\dfrac{\sin 2\theta}{\sqrt{\cos 2\theta}}$

23. $\cos t\cos 2t\tan 3t-2\sin t\sin 2t\tan 3t$
$+3\sin t\cos 2t\sec^2 3t$

25. $\dfrac{\operatorname{sgn}(\sin t)}{1+\cos t}$

27. $2x\cos(x^2+1)-2x^3\sin(x^2+1)$

29. $\dfrac{\sin 2x}{2}+C;\ -3\cos\dfrac{x}{3}+C;\ \dfrac{\sin ax}{a}+C;\ -\dfrac{\cos ax}{a}+C$

31. $y=\dfrac{1}{\sqrt{2}}+\dfrac{\pi}{180\sqrt{2}}(x-45)$

33. $\cos x\operatorname{sgn}(\sin x)$

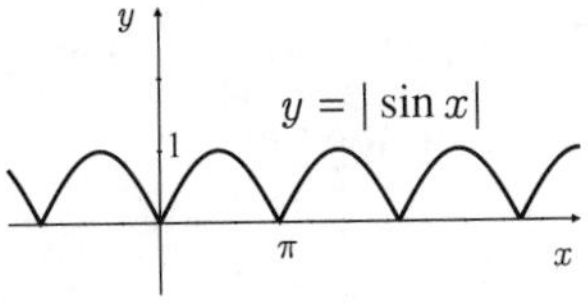

35. CP: $x=\pm\dfrac{2\pi}{3}+2n\pi,\ (n=0,\pm1,\pm2,\ldots)$

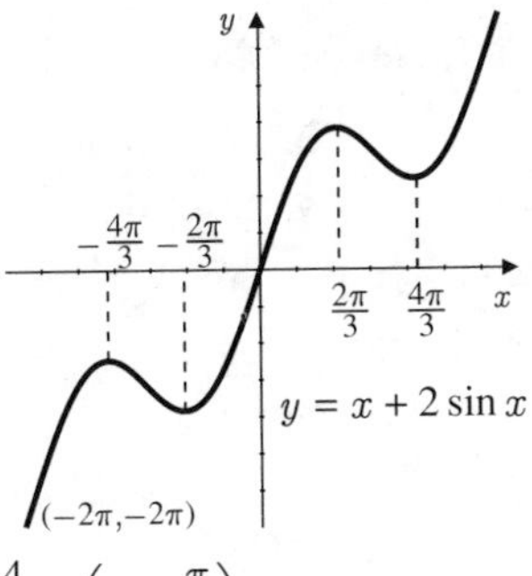

37. $y=1-\dfrac{4}{4-\pi}\left(x-\dfrac{\pi}{4}\right)$ **39.** $y=2-x$

45. $f^{(n)}(x)=\begin{cases}2^n\sin 2x & \text{if } n=4k\\ 2^n\cos 2x & \text{if } n=4k+1\\ -2^n\sin 2x & \text{if } n=4k+2\\ -2^n\cos 2x & \text{if } n=4k+3\end{cases}$
where $k=0,1,2,\ldots$

47. $f^{(n)}(x)=\begin{cases}-n\cos x+x\sin x & \text{if } n=4k\\ n\sin x+x\cos x & \text{if } n=4k+1\\ n\cos x-x\sin x & \text{if } n=4k+2\\ -n\sin x-x\cos x & \text{if } n=4k+3\end{cases}$
where $k=0,1,2,\ldots$

55. $y=\frac{3}{10}\sin(10t)$

57. $A=\mathcal{A}\cos(\omega c)-\mathcal{B}\sin(\omega c)$
$B=\mathcal{A}\sin(\omega c)+\mathcal{B}\cos(\omega c)$

59. $y=A\cos\big(\omega(t-a)\big)+\dfrac{B}{\omega}\sin\big(\omega(t-a)\big)$

61. 16 Hz for 900gm; 48 Hz for 100 gm

Section 3.3 (page 121)

1. $\pi/3$ **3.** $-\pi/4$ **5.** 0.7

7. $-\pi/3$ **9.** $\dfrac{\pi}{2}+0.2$ **11.** $2/\sqrt{5}$

13. $\sqrt{1-x^2}$ **15.** $\dfrac{1}{\sqrt{1+x^2}}$ **17.** $\dfrac{\sqrt{1-x^2}}{x}$

19. $|x-2n\pi|$, where n is the unique integer such that $(2n-1)\pi<x\le(2n+1)\pi$

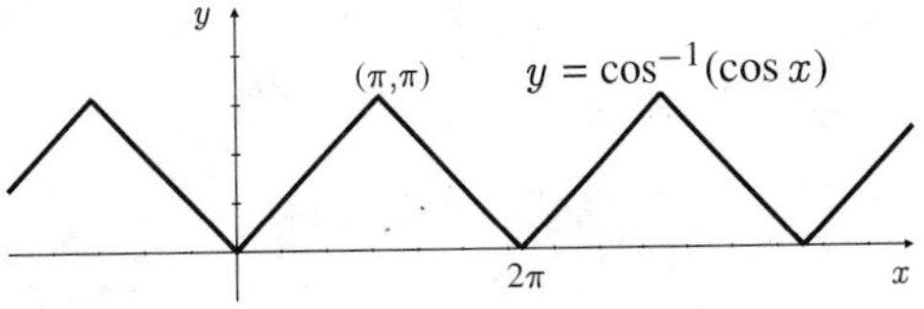

21. $\dfrac{1}{\sqrt{2+x-x^2}}$ **23.** $\dfrac{-1}{\sqrt{a^2-(x-b)^2}},\ (a>0)$

25. $\tan^{-1}t+\dfrac{t}{1+t^2}$ **27.** $2x\tan^{-1}x+1$

29. $\dfrac{\sqrt{1-4x^2}\sin^{-1}2x-2\sqrt{1-x^2}\sin^{-1}x}{\sqrt{1-x^2}\sqrt{1-4x^2}\left(\sin^{-1}2x\right)^2}$

31. $\dfrac{x}{\sqrt{(1-x^4)\sin^{-1}x^2}}$ **33.** $\sqrt{\dfrac{a-x}{a+x}}$

35. $\dfrac{\pi-2}{\pi-1}$

39. $\dfrac{d}{dx}\csc^{-1}x=-\dfrac{1}{|x|\sqrt{x^2-1}}$

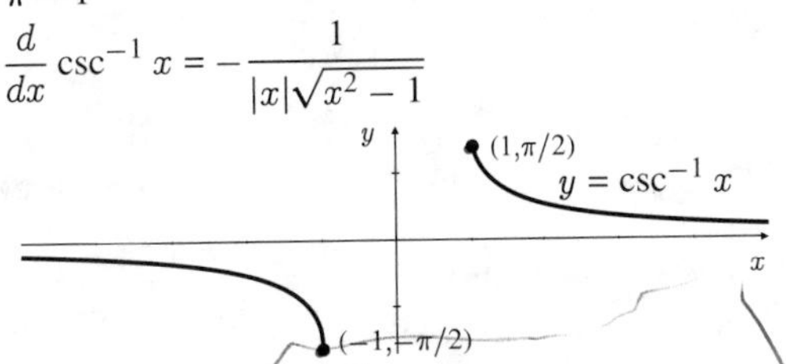

41. $\tan^{-1}x+\cot^{-1}x=-\dfrac{\pi}{2}$ for $x<0$

43. $\tan^{-1}\left(\dfrac{x-1}{x+1}\right)-\tan^{-1}x=\dfrac{3\pi}{4}$ on $(-\infty,-1)$

45. $f'(x) = 1 - \text{sgn}(\cos x)$

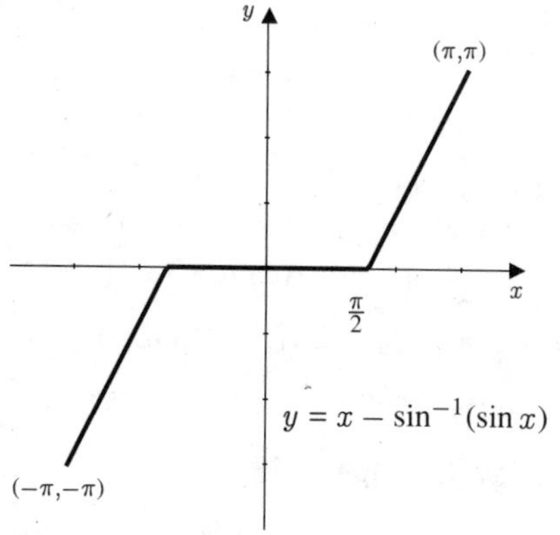

47. $y = \frac{1}{3}\tan^{-1}\frac{x}{3} + 2 - \frac{\pi}{12}$ **49.** $y = 4\sin^{-1}\frac{x}{5}$

Section 3.4 (page 125)

1. $\sqrt{3}$ **3.** x^6 **5.** 3
7. $-2x$ **9.** x **11.** 1
13. 1 **15.** 2 **17.** $\log_a(x^4 + 4x^2 + 3)$

19. $s^{\sqrt{2}} = 10^{\sqrt{2}\log_{10} 3}$; $\log_3 5 = \frac{\log_{10} 5}{\log_{10} 3}$;
$x = \frac{\log_{10} 5}{2\log_{10} 2 - \log_{10} 5}$

Section 3.5 (page 132)

1. $\sqrt{e}$ **3.** x^5 **5.** $-3x$
7. 5 **9.** $\ln\frac{64}{81}$ **11.** $\ln\left(x^2(x-2)^5\right)$

13. $x = \frac{\ln 2}{\ln(3/2)}$ **15.** $x = \frac{\ln 5 - 9\ln 2}{2\ln 2}$

17. $5e^{5x}$ **19.** $(1-2x)e^{-2x}$ **21.** $\frac{3}{3x-2}$

23. $\frac{e^x}{1+e^x}$ **25.** $\frac{e^x - e^{-x}}{2}$ **27.** e^{x+e^x}

29. $t^{r-1}e^{at}(r - at)$ **31.** $\frac{e^x}{(1+e^x)^2}$

33. $e^x(\sin x - \cos x)$ **35.** $e^u \sin(2e^u)$

37. $\frac{1}{x\ln x}$ **39.** $2x\ln x$ **41.** $(2\ln 5)5^{2x+1}$

43. $t^x x^t \ln t + t^{x+1}x^{t-1}$ **45.** $\frac{b}{(bs+c)\ln a}$

47. $x^{\sqrt{x}}\left(\frac{1}{\sqrt{x}}\left(\frac{1}{2}\ln x + 1\right)\right)$ **49.** $\sec x$

51. $-\frac{1}{\sqrt{x^2+a^2}}$ **53.** $-\frac{a}{x\sqrt{a^2+x^2}}$

55. $f^{(n)}(x) = e^{ax}(na^{n-1} + a^n x)$, $n = 1, 2, 3, \ldots$

57. $y' = 2xe^{x^2}$, $y'' = 2(1+2x^2)e^{x^2}$,
$y''' = 4(3x + 2x^3)e^{x^2}$, $y^{(4)} = 4(3 + 12x^2 + 4x^4)e^{x^2}$

59. $f'(x) = x^{x^2+1}(2\ln x + 1)$,
$g'(x) = x^x\left(1 + \ln x + \frac{1}{x\ln x}\right)$;
g grows more rapidly than does f.

61. $f'(x) = f(x)\left(\frac{1}{x-1} + \frac{1}{x-2} + \frac{1}{x-3} + \frac{1}{x-4}\right)$

63. $f'(2) = \frac{556}{3675}$, $f'(1) = \frac{1}{6}$

65. f inc. for $x < 1$, dec. for $x > 1$

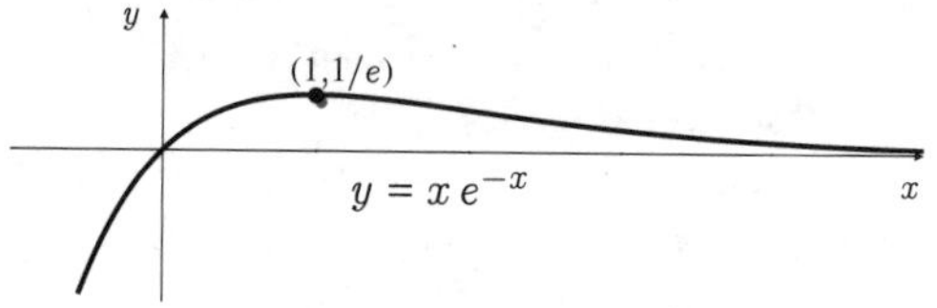

67. $y = 2e^4x - 3e^4$ **69.** $y = ex$
71. $y = 2e\ln 2(x-1)$ **73.** $-1/e^2$

75. $f'(x) = (A+B)\cos\ln x + (B-A)\sin\ln x$,
$\int \cos\ln x\,dx = \frac{x}{2}(\cos\ln x + \sin\ln x)$,
$\int \sin\ln x\,dx = \frac{x}{2}(\sin\ln x - \cos\ln x)$

77. (a) $F_{2B,-2A}(x)$; (b) $-2e^x(\cos x + \sin x)$

Section 3.6 (page 137)

3. Define $F(x) = \begin{cases} A_x & \text{if } x \ge -2 \\ -A_x & \text{if } x < -2 \end{cases}$ where A_x is the area under the graph $y = e^{x^2}$, above $y = 0$ between vertical lines at -2 and x.

Section 3.7 (page 143)

1. 566 **3.** 29.15 years **5.** 160.85 years
7. 4139 gm **9.** 22.35°C **11.** 6.84 min
13. about 14.7 years

15. $L = \frac{y_1^2(y_0 + y_2) - 2y_0y_1y_2}{y_1^2 - y_0y_2}$; $L = \frac{45}{7}$

17. (t_0, ∞) where $t_0 = -\frac{1}{k}\ln\frac{y_0}{y_0 - L}$; $\lim_{t\to t_0+} y(t) = \infty$

Section 3.8 (page 148)

1. $\frac{d}{dx}\cosh x = \sinh x$, $\frac{d}{dx}\sinh x = \cosh x$,
$\frac{d}{dx}\tanh x = \text{sech}^2 x$, $\frac{d}{dx}\text{sech}\, x = -\text{sech}\, x\tanh x$,
$\frac{d}{dx}\text{csch}\, x = -\text{csch}\, x\coth x$, $\frac{d}{dx}\coth x = -\text{csch}^2 x$

3. $\tanh(x+y) = \dfrac{\tanh x + \tanh y}{1 + \tanh x \tanh y}$

$\tanh(x-y) = \dfrac{\tanh x - \tanh y}{1 - \tanh x \tanh y}$

5. $\dfrac{d}{dx}\sinh^{-1}(x) = \dfrac{1}{\sqrt{x^2+1}}$,

$\dfrac{d}{dx}\operatorname{Cosh}^{-1}(x) = \dfrac{1}{\sqrt{x^2-1}}$,

$\dfrac{d}{dx}\tanh^{-1}(x) = \dfrac{1}{1-x^2}$,

$\displaystyle\int \frac{dx}{\sqrt{x^2+1}} = \sinh^{-1}(x) + C$,

$\displaystyle\int \frac{dx}{\sqrt{x^2-1}} = \operatorname{Cosh}^{-1}(x) + C \quad (x > 1)$,

$\displaystyle\int \frac{dx}{1-x^2} = \tanh^{-1}(x) + C \quad (-1 < x < 1)$

7. (a) $\dfrac{x^2-1}{2x}$; (b) $\dfrac{x^2+1}{2x}$; (c) $\dfrac{x^2-1}{x^2+1}$; (d) x^2

9. $\coth^{-1} x = \tanh^{-1}\dfrac{1}{x} = \dfrac{1}{2}\ln\left(\dfrac{x+1}{x-1}\right)$, $(|x| > 1)$

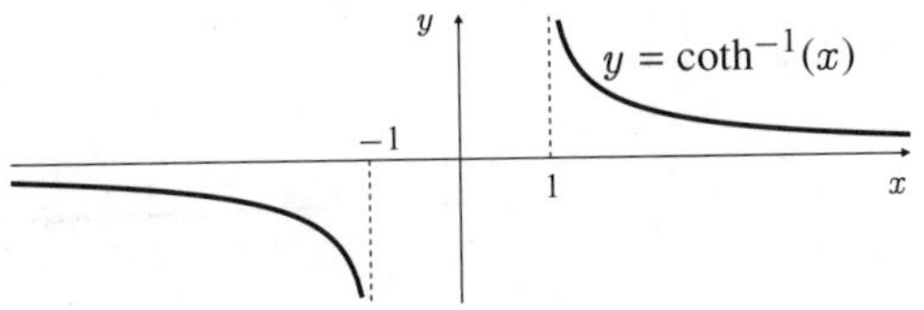

11. $f_{A,B} = g_{A+B,A-B}$; $g_{C,D} = f_{(C+D)/2,(C-D)/2}$

13. $y = y_0 \cosh k(x-a) + \dfrac{v_0}{k}\sinh k(x-a)$

Section 3.9 (page 153)

1. $y = Ae^{-5t} + Be^{-2t}$ **3.** $y = A + Be^{-2t}$

5. $y = (A + Bt)e^{-4t}$ **7.** $y = (A\cos t + B\sin t)e^{3t}$

9. $y = (A\cos 2t + B\sin 2t)e^{t}$ **13.** $y = \frac{6}{7}e^{t/2} + \frac{1}{7}e^{-3t}$

15. $y = e^{-2t}(2\cos t + 6\sin t)$ **25.** $y = At + Bt\ln t$

27. $y = At + \dfrac{B}{t}$ **29.** $y = A + B\ln t$

CHAPTER 4. THE MEAN-VALUE THEOREM AND CURVE SKETCHING

Section 4.1 (page 162)

1. $c = \dfrac{a+b}{2}$ **3.** $c = \pm\dfrac{2}{\sqrt{3}}$

5. inc. on $\left(-\infty, -\dfrac{2}{\sqrt{3}}\right)$ and $\left(\dfrac{2}{\sqrt{3}}, \infty\right)$, dec. on $\left(-\dfrac{2}{\sqrt{3}}, \dfrac{2}{\sqrt{3}}\right)$

7. inc. on $(-2, 0)$ and $(2, \infty)$; dec. on $(-\infty, -2)$ and $(0, 2)$

9. inc. on $(-\infty, 3)$ and $(5, \infty)$; dec. on $(3, 5)$

17. Next pair:

$\sin x > x - \dfrac{x^3}{3!} + \dfrac{x^5}{5!} - \dfrac{x^7}{7!}$

$\cos x < 1 - \dfrac{x^2}{2!} + \dfrac{x^4}{4!} - \dfrac{x^6}{6!} + \dfrac{x^8}{8!}$

Generalization: if n is even then

$\sin x < x - \dfrac{x^3}{3!} + \cdots + \dfrac{x^{2n+1}}{(2n+1)!}$

$\cos x < 1 - \dfrac{x^2}{2!} + \dfrac{x^4}{4!} - \cdots + \dfrac{x^{2n}}{(2n)!}$

If n is odd then

$\sin x > x - \dfrac{x^3}{3!} + \cdots - \dfrac{x^{2n+1}}{(2n+1)!}$

$\cos x > 1 - \dfrac{x^2}{2!} + \dfrac{x^4}{4!} - \cdots - \dfrac{x^{2n}}{(2n)!}$

21. If $f^{(n)}$ exists on an interval I and f vanishes at $n+1$ distinct points of I then $f^{(n)}$ vanishes at at least one point of I.

Section 4.2 (page 168)

1. abs min 1 at $x = -1$; abs max 3 at $x = 1$

3. abs min 1 at $x = -1$; no max

5. abs min -1 at $x = 0$; abs max 8 at $x = 3$; loc max 3 at $x = -2$

7. abs min $a^3 + a - 4$ at $x = a$; abs max $b^3 + b - 4$ at $x = b$

9. abs max $b^5 + b^3 + 2b$ at $x = b$; no min value

11. no max or min values

13. abs max 1 at $x = 0$; no min value

15. no max or min value

17. loc max at $x = -1$; loc min at $x = 1$

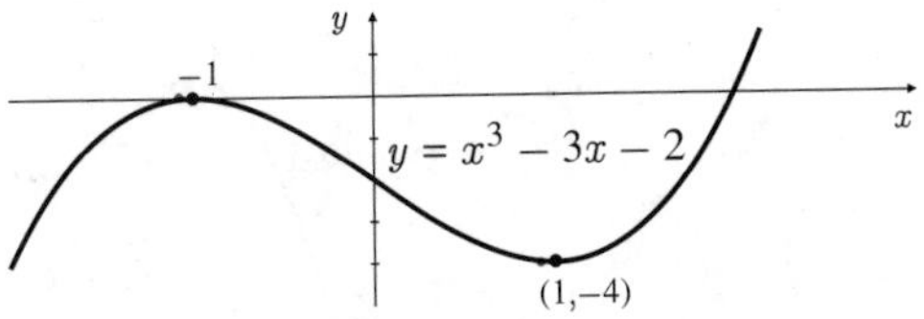

19. loc max at $x = \frac{1}{3}$; loc min at $x = 1$

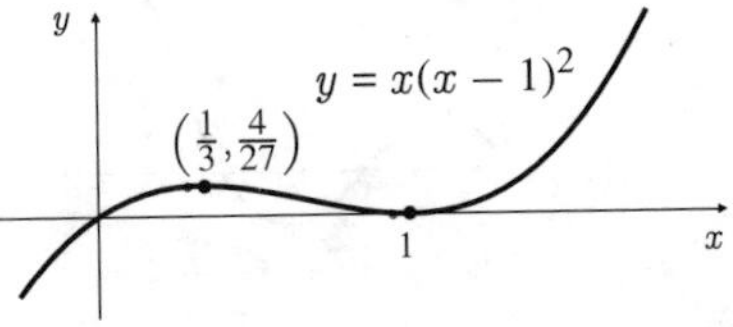

21. loc max at $x = \frac{3}{5}$; loc min at $x = 1$; critical point $x = 0$ is neither max nor min

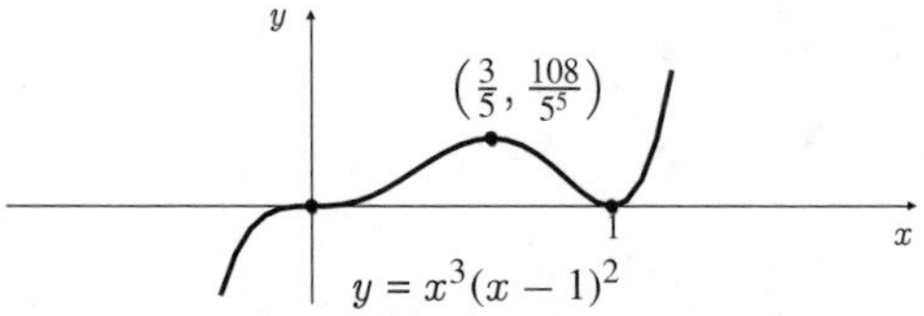

23. loc max at $x = -1$ and $x = 1/\sqrt{5}$; loc min at $x = 1$ and $x = -1/\sqrt{5}$

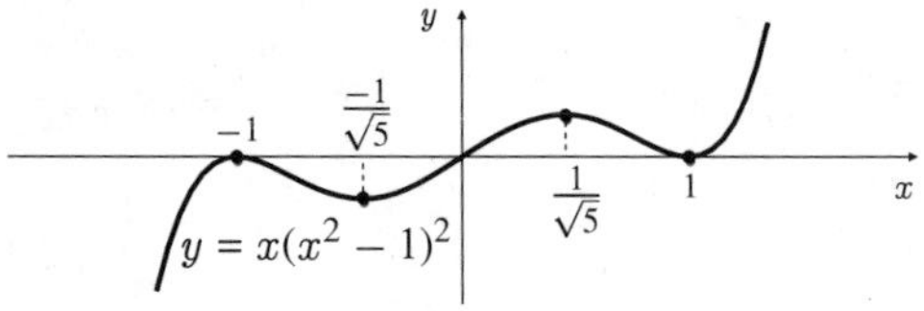

25. abs min at $x = 0$

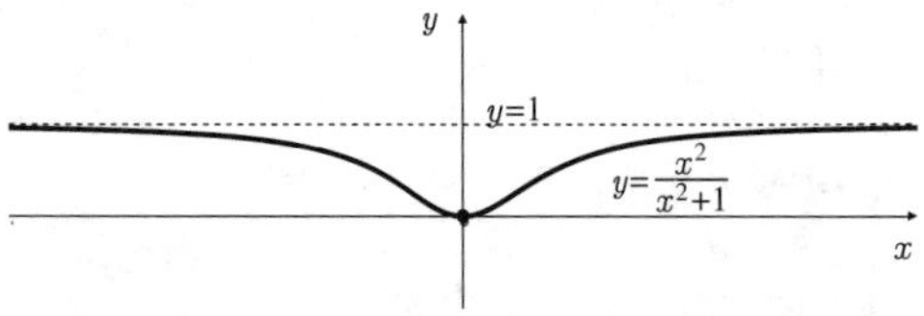

27. loc min at CP $x = -1$ and endpoint SP $x = \sqrt{2}$; loc max at CP $x = 1$ and endpoint SP $x = -\sqrt{2}$

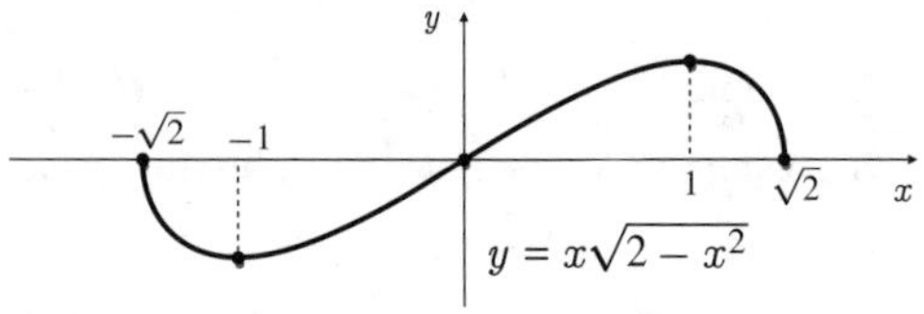

29. loc max at $x = 2n\pi - \frac{\pi}{3}$; loc min at $x = 2n\pi + \frac{\pi}{3}$ $(n = 0, \pm 1, \pm 2, \ldots)$

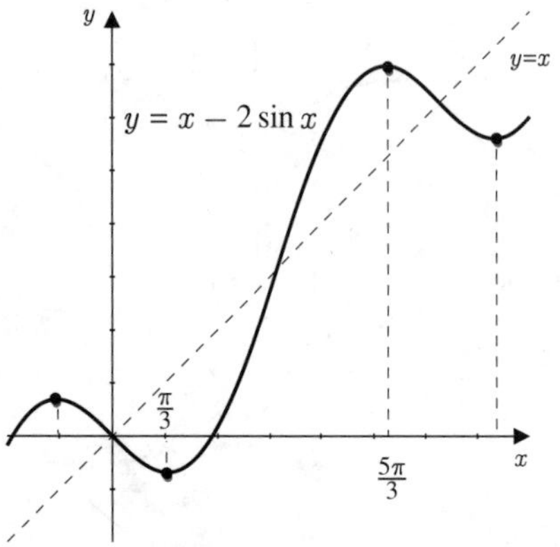

31. loc max at CP $x = \sqrt{3}/2$ and endpoint SP $x = -1$; loc min at CP $x = -\sqrt{3}/2$ and endpoint SP $x = 1$

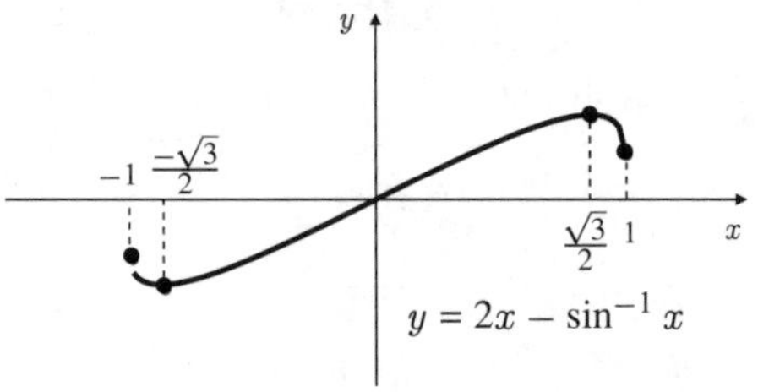

33. abs max at $x = 1/\ln 2$

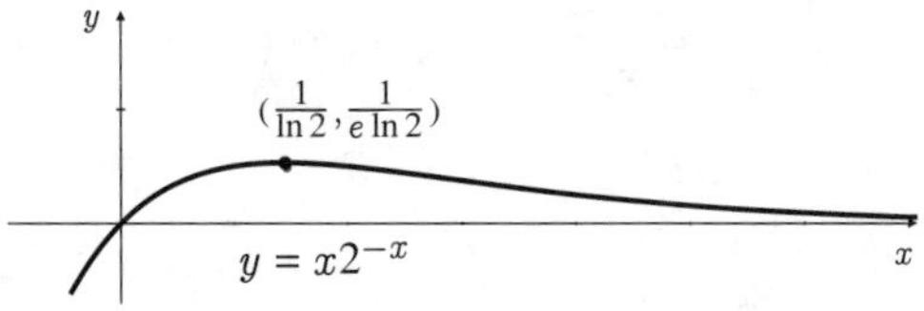

35. abs max at $x = e$

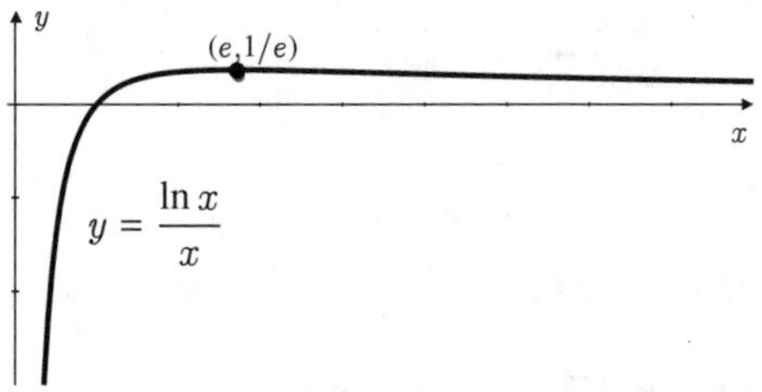

37. loc max at CP $x = 0$; abs min at SPs $x = \pm 1$

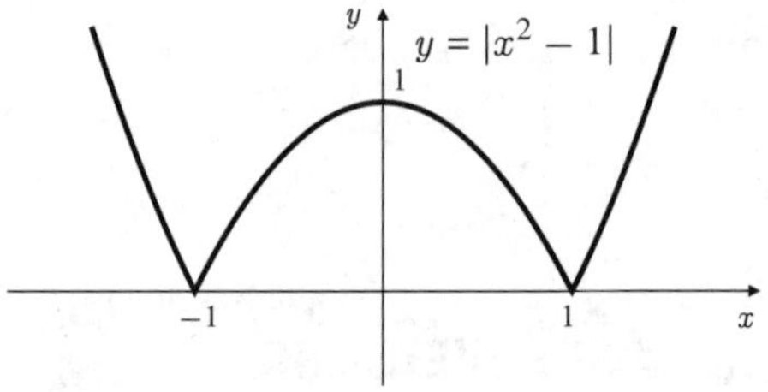

39. abs max at CPs $x = (2n+1)\pi/2$; abs min at SPs $x = n\pi$ $(n = 0, \pm 1, \pm 2, \ldots)$

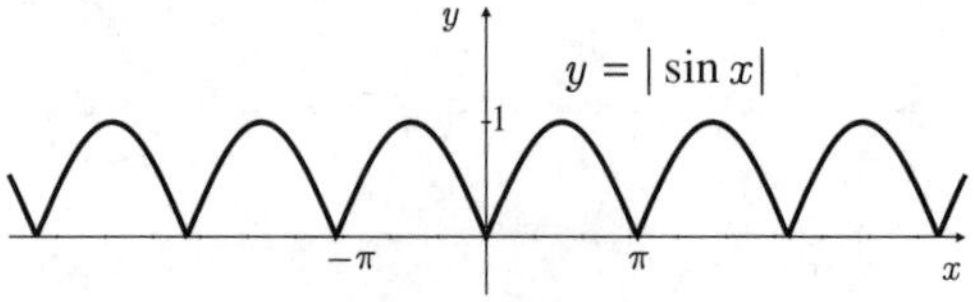

41. no max or min; SPs at $x = \pm 1$

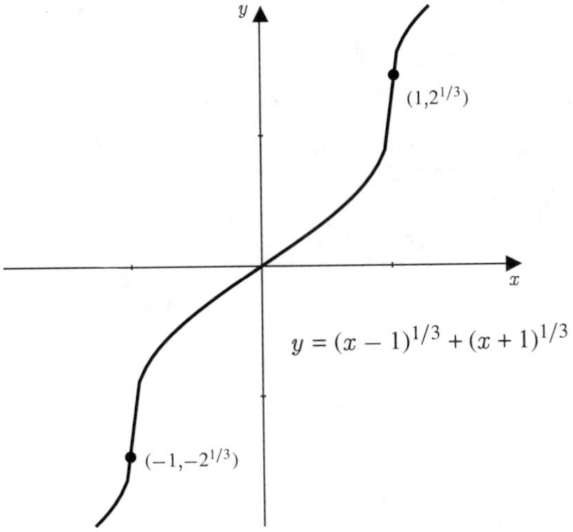

43. abs min at SP $x = -1$; no max or min at SP $x = 1$

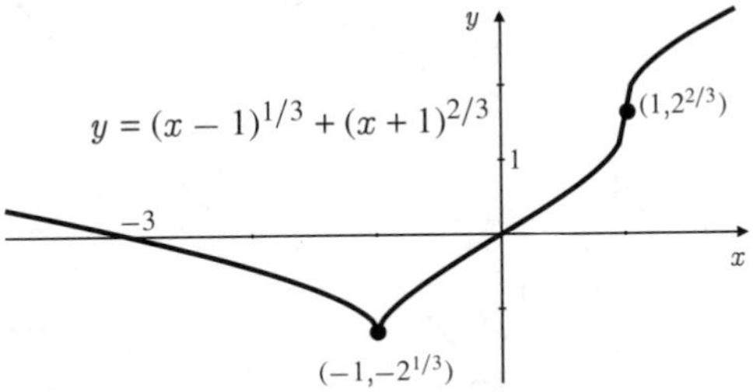

45. No. Consider $f(x) = -x^2$.

Section 4.3 (page 173)

1. conc down on $(0, \infty)$

3. conc up on $\mathbb{R}$

5. conc down on $(-\infty, 0)$, up on $(0, \infty)$; infl $x = 0$

7. conc down on $(-1, 0)$ and $(1, \infty)$; conc up on $(-\infty, -1)$ and $(0, 1)$; infl $x = -1, 0, 1$

9. conc down on $(-1/\sqrt{6}, 1/\sqrt{6})$; conc up on $(-\infty, -1/\sqrt{6})$ and $(1/\sqrt{6}, \infty)$; infl $x = \pm 1/\sqrt{6}$

11. conc down on $(-1, 1)$; conc up on $(-\infty, -1)$ and $(1, \infty)$; infl $x = \pm 1$

13. conc down on $(-\infty, 0)$, up on $(0, \infty)$; infl $x = 0$

15. conc down on $(-2, -2/\sqrt{5})$ and $(2/\sqrt{5}, 2)$; conc up on $(-\infty, -2)$, $(-2/\sqrt{5}, 2/\sqrt{5})$ and $(2, \infty)$; infl $x = \pm 2$, $\pm 2/\sqrt{5}$

17. conc down on $(2n\pi, (2n+1)\pi)$; conc up on $((2n-1)\pi, 2n\pi)$, $(n = 0, \pm 1, \pm 2, \ldots)$; infl $x = n\pi$

19. conc down on $\left(n\pi, (n + \frac{1}{2})\pi\right)$; conc up on $\left((n - \frac{1}{2})\pi, n\pi\right)$; infl $x = n\pi/2$, $(n = 0, \pm 1, \pm 2, \ldots)$

21. conc down on $(0, \infty)$, up on $(-\infty, 0)$; infl $x = 0$

23. conc down on $(-1/\sqrt{2}, 1/\sqrt{2})$, up on $(-\infty, -1/\sqrt{2})$ and $(1/\sqrt{2}, \infty)$; infl $x = \pm 1/\sqrt{2}$

25. conc down on $(-\infty, 0)$, up on $(0, \infty)$; $x = 0$ will be a singular inflection point if we define $f(0) = 0$.

27. conc down on $(-\infty, -1)$ and $(1, \infty)$; conc up on $(-1, 1)$; infl $x = \pm 1$

29. conc down on $(-\infty, 4)$, up on $(4, \infty)$; infl $x = 4$

31. no concavity, no inflections

33. loc min at $x = 2$; loc max at $x = \frac{2}{3}$

35. loc min at $x = 1/\sqrt[4]{3}$; loc max at $-1/\sqrt[4]{3}$

37. loc max at $x = 1$; loc min at $x = -1$ (both abs)

39. loc (and abs) min at $x = 1/e$

41. loc min at $x = 0$; inflections at $x = \pm 2$ (not discernible by Second Derivative Test)

43. loc min at $x = 0$; loc (and abs) max at $x = \pm 1/\sqrt{2}$

47. If n is even, f_n has a min and g_n has a max at $x = 0$. If n is odd both have inflections at $x = 0$.

Section 4.4 (page p 185)

1.

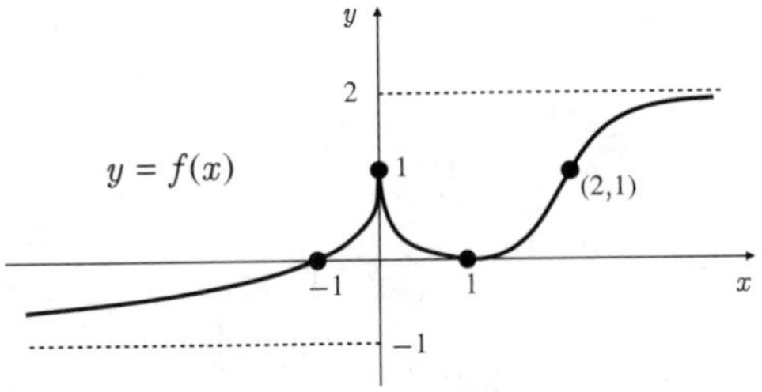

3.

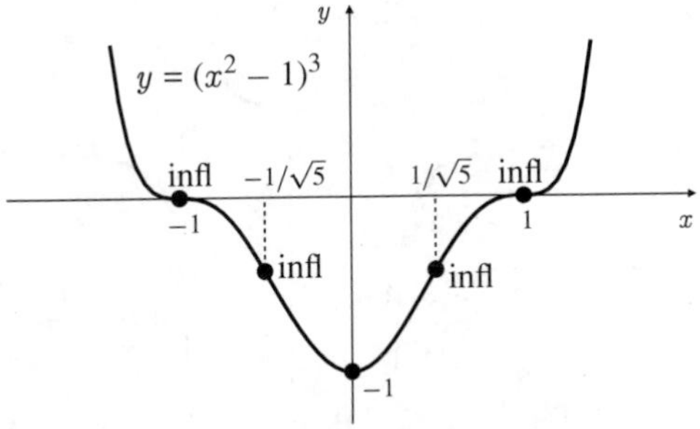

5.

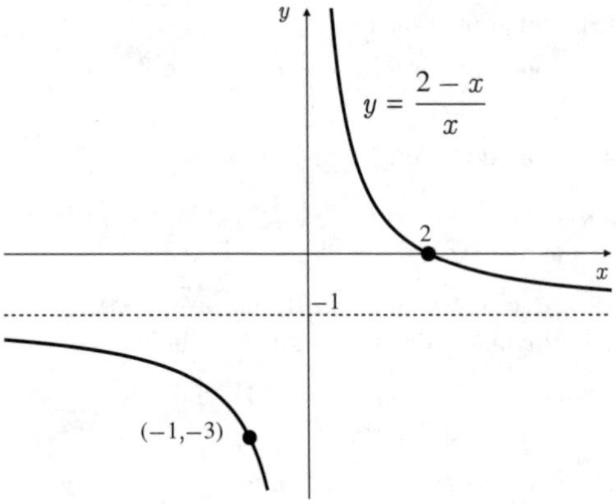
y
x
$y = \frac{2-x}{x}$
2
−1
(−1,−3)

11.

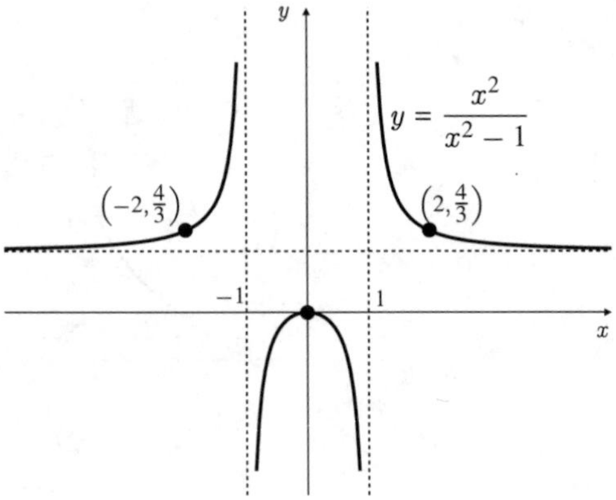
y
x
$y = \frac{x^2}{x^2-1}$
$\left(-2, \frac{4}{3}\right)$
$\left(2, \frac{4}{3}\right)$
−1
1

7.

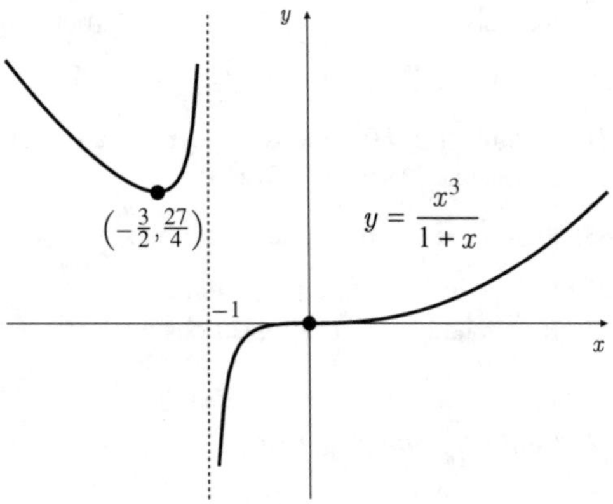
y
x
$y = \frac{x^3}{1+x}$
$\left(-\frac{3}{2}, \frac{27}{4}\right)$
−1

13.

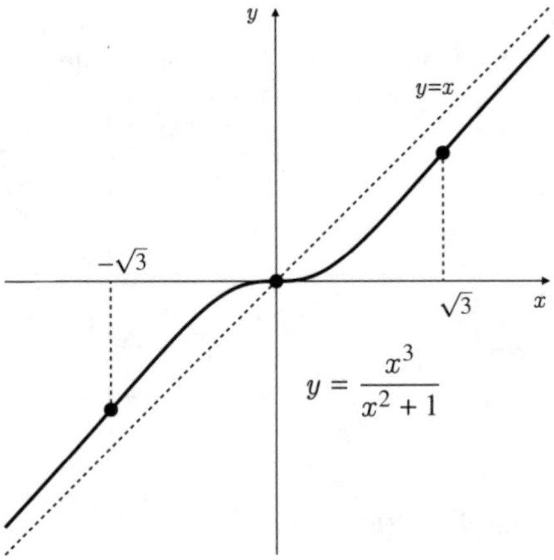
y
x
y=x
$-\sqrt{3}$
$\sqrt{3}$
$y = \frac{x^3}{x^2+1}$

9.

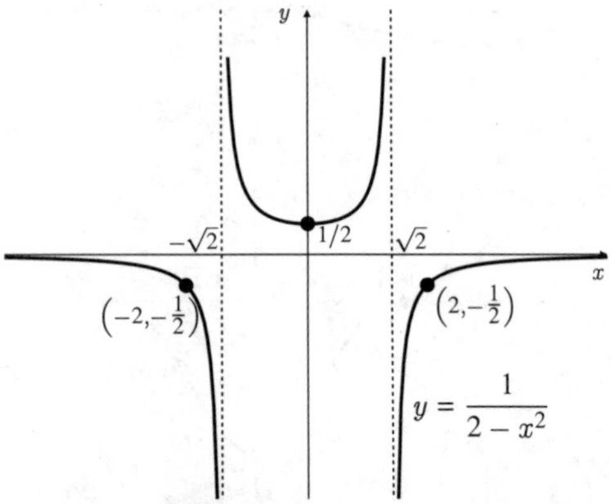
y
x
$-\sqrt{2}$
1/2
$\sqrt{2}$
$\left(-2, -\frac{1}{2}\right)$
$\left(2, -\frac{1}{2}\right)$
$y = \frac{1}{2-x^2}$

15.

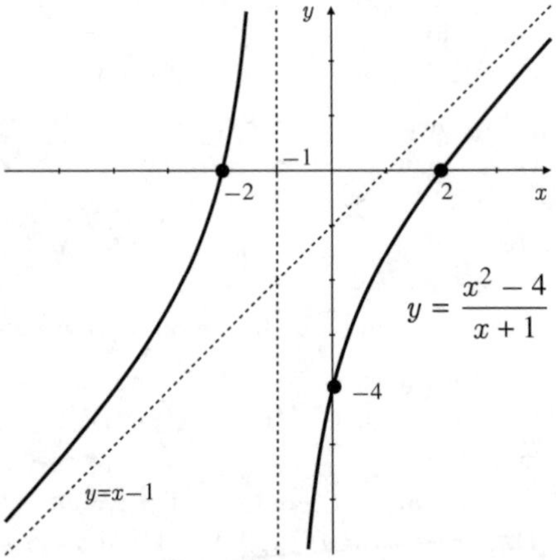
y
x
−1
−2
2
$y = \frac{x^2-4}{x+1}$
−4
y=x−1

17.

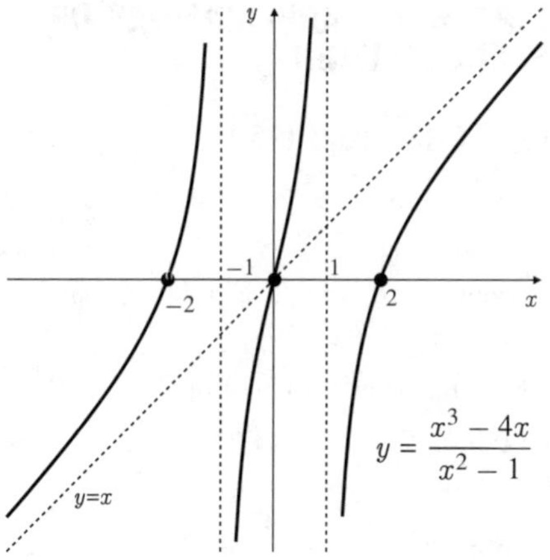
y
−2
−1
1
2
x
y=x
$y = \frac{x^3 - 4x}{x^2 - 1}$

19.

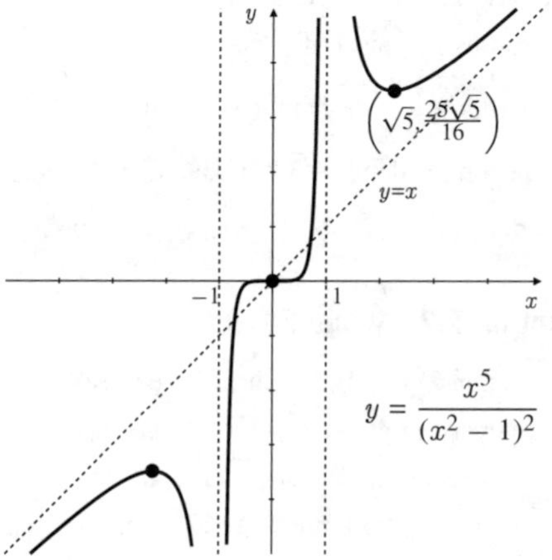
y
$\left(\sqrt{5}, \frac{25\sqrt{5}}{16}\right)$
y=x
−1
1
x
$y = \frac{x^5}{(x^2 - 1)^2}$

21.

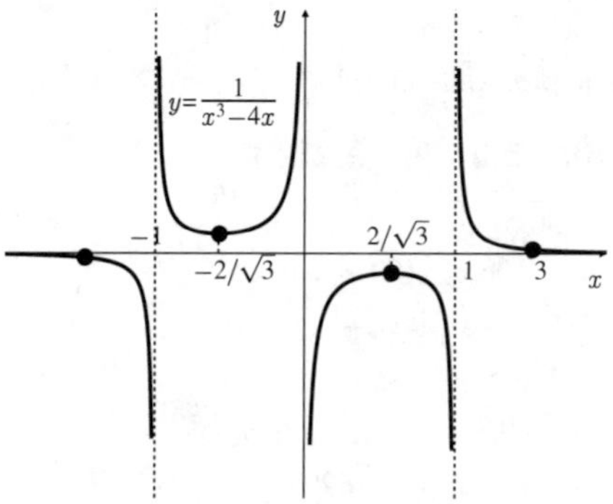
y
$y=\frac{1}{x^3-4x}$
−1
$-2/\sqrt{3}$
$2/\sqrt{3}$
1
3
x

23.

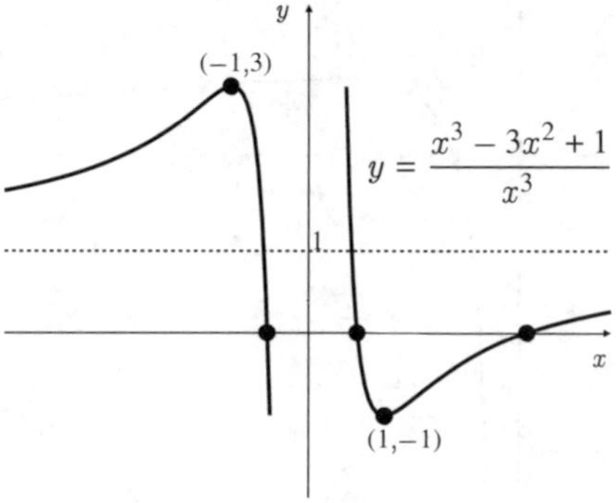
y
(−1,3)
$y = \frac{x^3 - 3x^2 + 1}{x^3}$
1
x
(1,−1)

25.

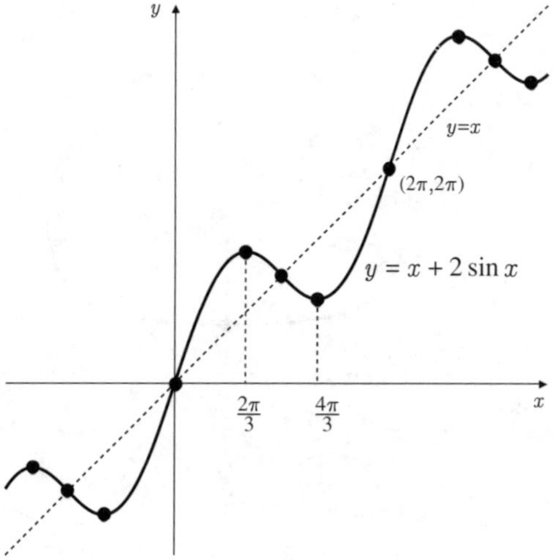
y
y=x
(2π,2π)
$y = x + 2\sin x$
$\frac{2\pi}{3}$
$\frac{4\pi}{3}$
x

27.

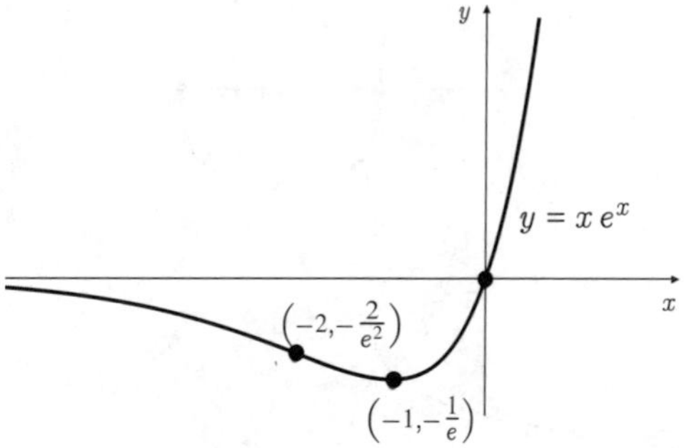
y
$y = x\,e^x$
x
$\left(-2, -\frac{2}{e^2}\right)$
$\left(-1, -\frac{1}{e}\right)$

29.

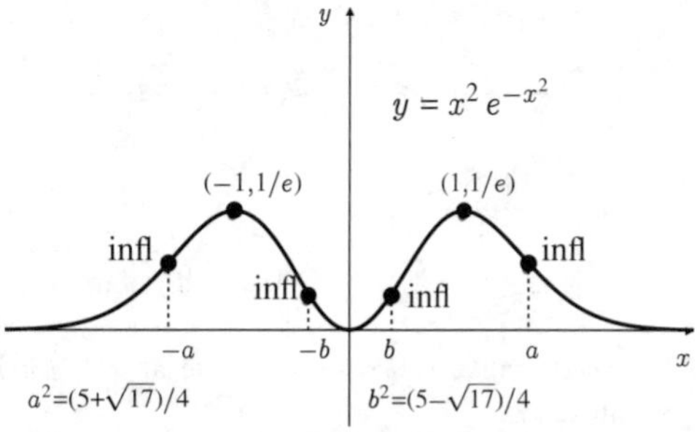
y
$y = x^2\,e^{-x^2}$
(−1,1/e)
(1,1/e)
infl
infl
infl
infl
−a
−b
b
a
x
$a^2=(5+\sqrt{17})/4$
$b^2=(5-\sqrt{17})/4$

31.

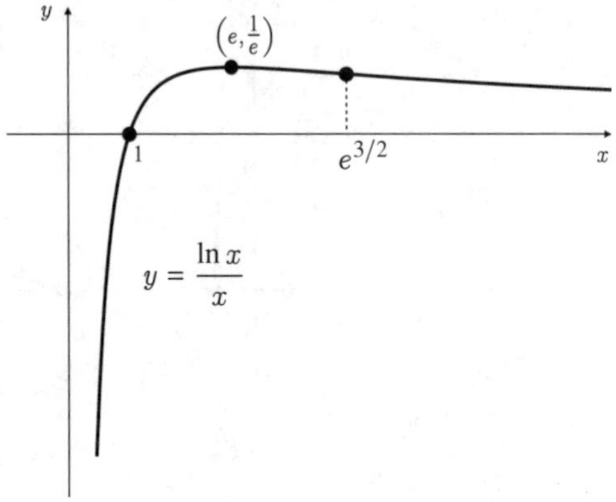

33.

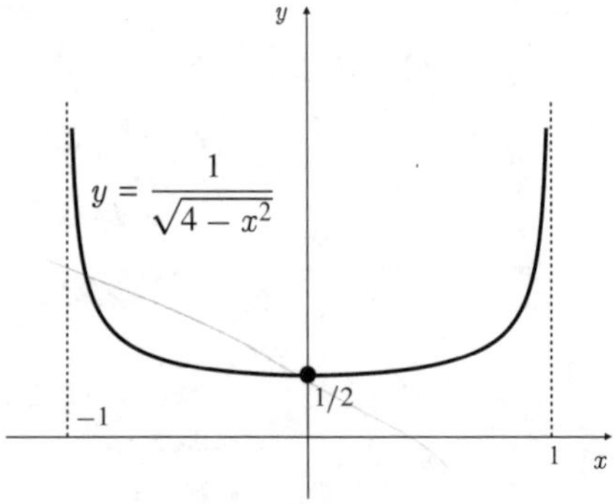

35.

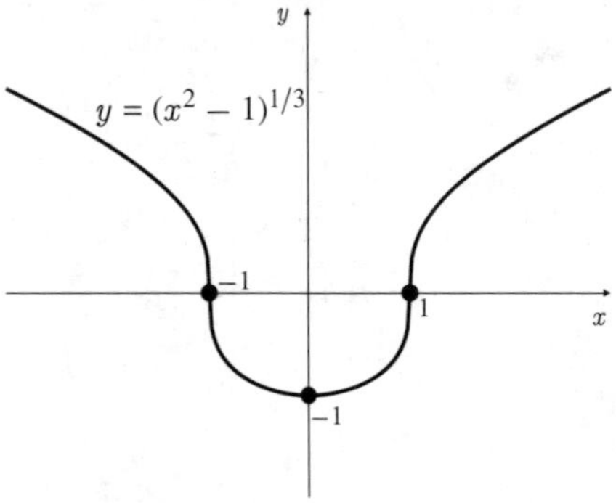

37.

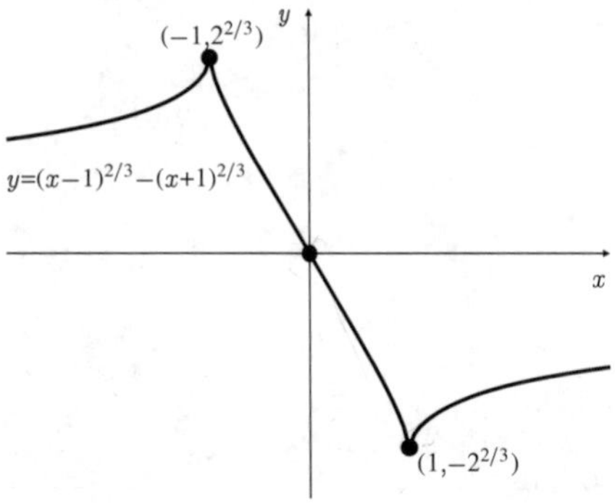

39. $y = 0$. curve crosses asymptote at $x = n\pi$ for every integer n.

CHAPTER 5. APPLICATIONS OF DIFFERENTIATION

Section 5.1 (page 195)

1. 49/4 **3.** 20 and 40

9. R^2 sq. units **11.** $2ab$ square units

13. width $8 + 10\sqrt{2}$ m, height $4 + 5\sqrt{2}$ m

15. rebate \$250

17. head for point 5 km east of A

19. (a) 0 m, (b) $\pi/(4 + \pi)$ m

21. $8\sqrt{3}$ units

23. $\left[(a^{2/3} + b^{2/3})^3 + c^2\right]^{1/2}$ units

25. $3^{1/2}/2^{1/3}$ units

27. height $\dfrac{2R}{\sqrt{3}}$, radius $\sqrt{\dfrac{2}{3}}\,R$ units

29. base 2m×2m, height 1m

31. radius/height=$\pi/8$

33. width $\dfrac{20}{4+\pi}$ m, height $\dfrac{10}{4+\pi}$ m

37. width R, depth $\sqrt{3}R$ **39.** $Q = 3L/8$

41. $2\sqrt{6}$ ft **43.** $\dfrac{2\pi}{9\sqrt{3}} R^3$ cubic units

Section 5.2 (page 203)

1. increasing at 160π cm²/s **3.** decreasing at 4 m/s

7. increasing at 2 cm³/s **9.** increasing at rate 12

11. increasing at rate $2/\sqrt{5}$ **13.** 4/9 m/s

15. approx 23.06 km/h **17.** increasing at 1/100 rad/s

19. $1/(250\pi)$ m/min **21.** 7.49 m³/h

23. 12π m/min **25.** 13/2 m/min **27.** 82.45 km/h

29. (a) $\frac{1}{160}$ m/min, (b) $\frac{1}{80}$ m/min

31. approx 0.237 m/s

33. (a) 13.245 m/s, (b) 4.95 m/s

35. approx 0.0255 rad/s

Section 5.3 (page 212)

1. $\sqrt{50} \approx \frac{99}{14} \approx 7.071429$, $-\frac{1}{2744} \le \text{error} < 0$,
$7.07106 \le \sqrt{50} < 7.071439$

3. $\sqrt[4]{85} \approx \frac{82}{27}$, $-\frac{1}{2\times 3^6} \le \text{error} < 0$,
$3.03635 \le \sqrt[4]{85} < 3.03704$

5. $e^{-1/10} \approx 0.9$, $0 < \text{error} \le \frac{1}{200}$,
$0.9 < e^{-1/10} \le 0.905$

7. $\cos 46° \approx \frac{1}{\sqrt{2}}\left(1 - \frac{\pi}{180}\right) \approx 0.694765$,
$-\frac{1}{2\sqrt{2}}\left(\frac{\pi}{180}\right)^2 \leq \text{error} < 0$,
$0.694658 \leq \cos 46° < 0.694765$

9. $\arctan 1.05 \approx \frac{\pi}{4} + \frac{0.05}{2}$,
$-\frac{1}{4}(0.05)^2 < \text{error} < 0$,
$0.80977 < \arctan 1.05 < 0.81040$

11. $\ln 0.94 \approx -0.06$, $\quad -\frac{(0.06)^2}{2(0.94)^2} \leq \text{error} < 0$,
$0.06204 \leq \ln 0.94 < -0.06$

13. $3 \leq f(3) \leq 13/4$

15. about 0.1π cm^2, about $\frac{1}{10}$ of 1 percent

17. about 1005 cm^3

19. The two separate applications of MVT cannot be expected to give the same value of c.

Section 5.4 (page 217)

1. $x^{1/3} \approx 2 + \frac{1}{12}(x-8) - \frac{1}{288}(x-8)^2$,
$9^{1/3} \approx 2.07986$, $\quad 0 < \text{error} \leq 5/(81 \times 256)$,
$2.07986 < 9^{1/3} < 2.08010$

3. $\frac{1}{x} \approx 1 - (x-1) + (x-1)^2$, $\quad \frac{1}{1.02} \approx 0.9804$,
$-(0.02)^3 \leq \text{error} < 0$, $\quad 0.980392 \leq \frac{1}{1.02} < 0.9804$

5. $e^x \approx 1 + x + \frac{1}{2}x^2$, $\quad e^{-0.5} \approx 0.625$,
$-\frac{1}{6}(0.5)^3 \leq \text{error} < 0$, $\quad 0.596 \leq e^{-0.5} < 0.625$

7. $\sin x = x - \frac{x^3}{3!} + \frac{x^5}{5!} - \frac{x^7}{7!} + R_7$;
$R_7 = \frac{\sin X}{8!}x^8$ for some X between 0 and x

9. $\sin x = \frac{1}{\sqrt{2}}\left[1 + \left(x - \frac{\pi}{4}\right) - \frac{1}{2!}\left(x - \frac{\pi}{4}\right)^2 - \frac{1}{3!}\left(x - \frac{\pi}{4}\right)^3 + \frac{1}{4!}\left(x - \frac{\pi}{4}\right)^4\right] + R_4$;
where $R_4 = \frac{\cos X}{5!}\left(x - \frac{\pi}{4}\right)^5$ for some X between x and $\pi/4$

11. $\ln x = (x-1) - \frac{(x-1)^2}{2} + \frac{(x-1)^3}{3} - \frac{(x-11)^4}{4} + \frac{(x-1)^5}{5} - \frac{(x-1)^6}{6} + R_6$;
where $R_6 = \frac{(x-1)^7}{7X^7}$ for some X between 1 and x

13. $e^{-x} = 1 - x + \frac{x^2}{2!} - \frac{x^3}{3!} + \cdots + (-1)^n\frac{x^n}{n!} + R_n$;
where $R_n = (-1)^{n+1}\frac{X^{n+1}}{(n+1)!}$ for some X between 0 and x;
$\frac{1}{e} \approx \frac{1}{2!} - \frac{1}{3!} + \cdots + \frac{1}{8!} \approx 0.36788$

15. $1 - 2x + x^2$ (f is its own best quadratic approximation; $g(x) \approx 4 + 3x + 2x^2$; error $= x^3$;
since $g'''(x) = 6 = 3!$, therefore error $= \frac{g'''(X)}{3!}x^3$; no improvement possible.

Section 5.5 (page 227)

1. 1.414 **3.** 1.414 **5.** 1.259921
7. -1.000000, -2.732051, 0.732051
9. 0.636733 **11.** 0.567143
13. 0.520269 **15.** 0.95025
17. 0.45340

Section 5.6 (page 232)

1. $3/4$ **3.** a/b **5.** 1
7. 1 **9.** 0 **11.** $-3/2$
13. 1 **15.** $-1/2$ **17.** ∞
19. $2/\pi$ **21.** -2 **23.** a
25. 0 **27.** $-1/2$ **29.** e^{-2}
31. 0 **33.** $f''(x)$

CHAPTER 6. INTEGRATION

Section 6.1 (page 245)

1. 3/2 sq. un. **3.** 26/3 sq. un. **5.** 15 sq. un.
7. $\frac{3}{2\ln 2}$ sq. un. **9.** $\frac{1}{4}b^4$ sq. un. **11.** $\pi/4$

Section 6.2 (page 255)

1. $L(f, P_8) = \frac{7}{4}$, $\quad U(f, P_8) = \frac{9}{4}$

3. $L(f, P_4) = \frac{e^4 - 1}{e^2(e-1)} \approx 4.22$,
$U(f, P_4) = \frac{e^4 - 1}{e(e-1)} \approx 11.48$

5. $L(f, P_6) = \frac{\pi}{6}(1 + \sqrt{3}) \approx 1.43$,
$U(f, P_6) = \frac{\pi}{6}(3 + \sqrt{3}) \approx 2.48$

7. $L(f, P_n = \frac{n-1}{2n}$, $U(f, P_n) = \frac{n+1}{2n}$, $\int = \frac{1}{2}$

9. 8 **11.** $(b^2-a^2)/2$ **13.** 0

15. 2π **17.** $(4+a+b)/2$ **19.** 1

21. $\dfrac{\pi}{3}-2\sqrt{3}$ **23.** 3/4

Section 6.3 (page 260)

1. $\dfrac{5}{12}$ **3.** $\dfrac{2-\sqrt{2}}{2\sqrt{2}}$ **5.** $\dfrac{\pi}{3}$

7. $\dfrac{\pi}{8}$ **9.** $80\frac{4}{5}$ **11.** $\ln 10$

13. $\frac{1}{5}$ sq. un. **15.** $\frac{32}{3}$ sq. un. **17.** $\frac{1}{6}$ sq. un.

19. $\frac{1}{3}$ sq. un. **21.** $\frac{1}{12}$ sq. un. **23.** 2π sq. un.

25. $\frac{3}{32}$ sq. un. **27.** $\frac{16}{3}$ **29.** $\pi/4$

31. $\dfrac{\sin x}{x}$ **33.** $-2\dfrac{\sin x^2}{x}$

35. $\dfrac{\cos t}{1+t^2}$ **37.** $\dfrac{b}{5+b^4x^4}-\dfrac{a}{5+a^4x^4}$

39. $\cos x$

41. $1/x^2$ is not continuous (or even defined) at $x=0$ so the Fundamental Theorem cannot be applied over $[-1,1]$; Since $1/x^2>0$ on its domain, we would expect the integral to be positive if it exists at all. (It doesn't.)

43. $F(x)$ has a maximum value at $x=1$ but no minimum value.

45. 2

Section 6.4 (page 269)

1. $-\frac{1}{2}e^{5-2x}+C$ **3.** $\frac{2}{9}(3x+4)^{3/2}+C$

5. $-\frac{2}{9}(x^3+2)^{-3/2}+C$ **7.** $-\frac{1}{32}(4x^2+1)^{-4}+C$

9. $\cos(\cos x)+C$ **11.** $\frac{1}{2}e^{x^2}+C$

13. $\frac{1}{2}\tan^{-1}\left(\frac{1}{2}\sin x\right)+C$

15. $2\ln\left|e^{x/2}-e^{-x/2}\right|+C=\ln\left|e^x-2+e^{-x}\right|+C$

17. $-\frac{2}{5}\sqrt{4-5s}+C$ **19.** $\frac{1}{2}\sin^{-1}\left(\dfrac{t^2}{2}\right)+C$

21. $-\ln\left(1+e^{-x}\right)+C$ **23.** $-\frac{1}{2}(\ln\cos x)^2+C$

25. $\tan^{-1}\dfrac{t}{3}+\ln(t^2+9)+C$ **27.** $\frac{1}{2}\tan^{-1}\dfrac{x+3}{2}+C$

29. $\frac{1}{8}\cos^8 x-\frac{1}{6}\cos^6 x+C$ **31.** $-\dfrac{1}{3a}\cos^3 ax+C$

33. $\frac{1}{8}x-\frac{1}{32}\sin 4x+C$

35. $\frac{5}{16}x-\frac{1}{4}\sin 2x+\frac{3}{64}\sin 4x+\frac{1}{48}\sin^3 2x+C$

37. $\frac{1}{5}\sec^5 x+C$

39. $\frac{2}{3}(\tan x)^{3/2}+\frac{2}{7}(\tan x)^{7/2}+C$

41. $\frac{3}{8}\sin x-\frac{1}{4}\sin(2\sin x)+\frac{1}{32}\sin(4\sin x)+C$

43. $\frac{1}{3}\tan^3 x+C$

45. $-\frac{1}{9}\csc^9 x+\frac{2}{7}\csc^7 x-\frac{1}{5}\csc^5 x+C$

47. $\frac{14}{3}\sqrt{17}+\frac{2}{3}$ **49.** $3\pi/16$

51. $\ln 2$ **53.** $2(\sqrt{2}-1)$ **55.** $\pi/32$ sq. un.

Section 6.5 (page 276)

1. $\frac{1}{2}\sin^{-1}(2x)+C$

3. $\frac{1}{16}\sin^{-1}(2x)-\frac{1}{8}x\sqrt{1-4x^2}+C$

5. $\frac{1}{5}(9-x^2)^{5/2}-3(9-x^2)^{3/2}+C$

7. $\frac{9}{2}\sin^{-1}\dfrac{x}{3}-\frac{1}{2}x\sqrt{9-x^2}+C$

9. $-\dfrac{\sqrt{9-x^2}}{9x}+C$

11. $-\sqrt{9-x^2}+\sin^{-1}\dfrac{x}{3}+C$

13. $\ln\left(x+\sqrt{9+x^2}\right)+C$

15. $\frac{1}{3}(9+x^2)^{3/2}-9\sqrt{9+x^2}+C$

17. $\dfrac{1}{a^2}\dfrac{x}{\sqrt{a^2-x^2}}+C$

19. $\dfrac{x}{\sqrt{a^2-x^2}}-\sin^{-1}\dfrac{x}{a}+C$

21. $-\sin^{-1}x-\dfrac{\sqrt{1-x^2}}{x}+C$

23. $\dfrac{1}{a}\sec^{-1}\dfrac{x}{a}+C$ **25.** $\frac{1}{3}\tan^{-1}\dfrac{x+1}{3}+C$

27. $\frac{1}{32}\tan^{-1}\dfrac{2x+1}{2}+\dfrac{1}{16}\dfrac{2x+1}{4x^2+4x+5}+C$

29. $a\sin^{-1}\dfrac{x-a}{a}-\sqrt{2ax-x^2}+C$

31. $\dfrac{3-x}{4\sqrt{3-2x-x^2}}+C$

33. $\frac{1}{4}\tan^{-1}(2x)+\dfrac{x}{2(1+4x^2)}+C$

35. $\frac{3}{8}\tan^{-1}x+\dfrac{3x^3+5x}{8(1+x^2)^2}+C$

37. $2\sqrt{x}-4\ln(2+\sqrt{x})+C$

39. $\frac{6}{7}x^{7/6}-\frac{6}{5}x^{5/6}+\frac{3}{2}x^{2/3}+2x^{1/2}$
$-3x^{1/3}-6x^{1/6}+\frac{1}{2}\ln(1+x^{1/3})+6\tan^{-1}x^{1/6}+C$

41. $\dfrac{\pi}{6}-\dfrac{\sqrt{3}}{8}$ **43.** $\pi/3$

45. $\dfrac{2}{\sqrt{3}}\tan^{-1}\left(\dfrac{2\tan(\theta/2)+1}{\sqrt{3}}\right)+C$

47. $\dfrac{2}{\sqrt{5}}\tan^{-1}\left(\dfrac{\tan(\theta/2)}{\sqrt{5}}\right)+C$

49. $\dfrac{9}{2\sqrt{2}}\tan^{-1}\dfrac{1}{\sqrt{2}} - \dfrac{1}{2}$ square units

51. $\ln(x+\sqrt{x^2-a^2})+C, \quad \dfrac{\sqrt{x^2-a^2}}{a^2x}+C$

Section 6.6 (page 283)

1. $x\sin x + \cos x + C$

3. $\dfrac{1}{\pi}x^2\sin\pi x + \dfrac{2}{\pi^2}x\cos\pi x - \dfrac{2}{\pi^3}\sin\pi x + C$

5. $\frac{1}{4}x^4\ln x - \frac{1}{16}x^4 + C$

7. $x\tan^{-1}x - \frac{1}{2}\ln(1+x^2)+C$

9. $\left(\frac{1}{2}x^2-\frac{1}{4}\right)\sin^{-1}x + \frac{1}{4}x\sqrt{1-x^2}+C$

11. $\frac{1}{2}\left(\tan^{-1}x\right)^2+C$ **13.** $\frac{7}{8}\sqrt{2}+\frac{3}{8}\ln(1+\sqrt{2})$

15. $\frac{1}{13}e^{2x}(2\sin 3x - 3\cos 3x)+C$

17. $\ln(2+\sqrt{3}) - \dfrac{\pi}{6}$

19. $x\tan x - \ln|\sec x| + C$

21. $\dfrac{x}{2}\left[\cos(\ln x)+\sin(\ln x)\right]+C$

23. $\ln x\left(\ln(\ln x)-1\right)+C$

25. $x\cos^{-1}x - \sqrt{1-x^2}+C$

27. $\dfrac{2\pi}{3} - \ln(2+\sqrt{3})$

29. $\ln\left(1+\sqrt{1-x^2}\right) - \ln|x| - \dfrac{\sqrt{1-x^2}}{2x^2}+C$

31. $\frac{1}{2}(x^2+1)\left(\tan^{-1}x\right)^2 - x\tan^{-1}x + \frac{1}{2}\ln(1+x^2)+C$

33. $I_n = x(\ln x)^n - nI_{n-1}$,
$I_4 = x\left[(\ln x)^4 - 4(\ln x)^3 + 12(\ln x)^2 - 24(\ln x) + 24\right]+C$

35. $I_n = -\dfrac{1}{n}\sin^{n-1}x\cos x + \dfrac{n-1}{n}I_{n-2}$,
$I_6 = \dfrac{5x}{16} - \cos x\left[\frac{1}{6}\sin^5 x + \frac{5}{24}\sin^3 x + \frac{5}{16}\sin x\right]+c$,
$I_7 = -\cos x\left[\frac{1}{7}\sin^6 x + \frac{6}{35}\sin^4 x + \frac{8}{35}\sin^2 x + \frac{16}{35}\right]+C$

37. $I_n = \dfrac{x}{2a^2(n-1)(x^2+a^2)^{n-1}} + \dfrac{2n-3}{2a^2(n-1)}I_{n-1}$,
$I_3 = \dfrac{x}{4a^2(x^2+a^2)^2} + \dfrac{3x}{8a^4(x^2+a^2)} + \dfrac{3}{8a^5}\tan^{-1}\dfrac{x}{a}+C$

39. Any conditions which guarantee that $f(b)g'(b) - f'(b)g(b) = f(a)g'(a) - f'(a)g(a)$ will suffice.

Section 6.7 (page 292)

1. $\ln|2x-3|+C$ **3.** $\dfrac{x}{\pi} - \dfrac{2}{\pi^2}\ln|\pi x+2|+C$

5. $\dfrac{1}{6}\ln\left|\dfrac{x-3}{x+3}\right|+C$ **7.** $\dfrac{1}{2a}\ln\left|\dfrac{a+x}{a-x}\right|+C$

9. $x - \frac{4}{3}\ln|x+2| + \frac{1}{3}\ln|x-1|+C$

11. $3\ln|x+1| - 2\ln|x|+C$

13. $\dfrac{1}{3(1-3x)}+C$

15. $-\frac{1}{9}x - \dfrac{13}{54}\ln|2-3x| + \frac{1}{6}\ln|x|+C$

17. $\dfrac{1}{2a^2}\ln\dfrac{|x^2-a^2|}{x^2}+C$

19. $x+\dfrac{a}{3}\ln|x-a| - \dfrac{a}{6}\ln(x^2+ax+a^2) - \dfrac{a}{\sqrt{3}}\tan^{-1}\dfrac{2x+a}{\sqrt{3}a}+C$

21. $\frac{1}{3}\ln|x| - \frac{1}{2}\ln|x-1| + \frac{1}{6}\ln|x-3|+C$

23. $\dfrac{1}{4}\ln\left|\dfrac{x+1}{x-1}\right| - \dfrac{x}{2(x^2-1)}+C$

25. $\dfrac{1}{27}\ln\left|\dfrac{x-3}{x}\right| + \dfrac{1}{9x} + \dfrac{1}{6x^2}+C$

27. $\dfrac{t-1}{4(t^2+1)} - \frac{1}{4}\ln|t+1| + \frac{1}{8}\ln(t^2+1)+C$

29. $\dfrac{1}{3}\ln\left|\dfrac{1-\sqrt{1-x^2}}{x}\right| + \dfrac{1}{12}\ln\left(\dfrac{\left(2+\sqrt{1-x^2}\right)^2}{3+x^2}\right)+C$

31. $\dfrac{1}{\sqrt{1+x^2}} + \dfrac{1}{2}\ln\left|\dfrac{1-\sqrt{1+x^2}}{1+\sqrt{1+x^2}}\right|+C$

33. $\dfrac{1-2x^2}{x\sqrt{x^2-1}}+C$

35. $\ln|\tan(\theta/2)| + \dfrac{2}{1+\tan(\theta/2)}+C$

Review Exercises on Techniques of Integration (page 294)

1. $\frac{2}{3}\ln|x+2| - \frac{1}{6}\ln|2x+1|+C$

3. $\frac{1}{4}\sin^4 x - \frac{1}{6}\sin^6 x+C$

5. $\dfrac{3}{4}\ln\left|\dfrac{2x-1}{2x+1}\right|+C$ **7.** $-\dfrac{1}{3}\left(\dfrac{\sqrt{1-x^2}}{x}\right)^3+C$

9. $\frac{1}{5}\left(5x^3-s\right)^{1/3}+C$

11. $\frac{1}{16}\tan^{-1}\dfrac{x}{2} + \dfrac{x}{8(4+x^2)}+C$

13. $\dfrac{1}{2\ln 2}\left(2^x\sqrt{1+4^x} + \ln(2^x+\sqrt{1+4^x})\right)+C$

15. $\frac{1}{4}\tan^4 x + \frac{1}{6}\tan^6 x+C$

17. $-e^{-x}\left(\frac{2}{5}\cos 2x + \frac{1}{5}\sin 2x\right)+C$

19. $\dfrac{x}{10}\left(cos(3\ln x) + 3\sin(3\ln x)\right)+C$

21. $\frac{1}{4}\left(\ln(1+x^2)\right)^2+C$

23. $\sin^{-1}\dfrac{x}{\sqrt{2}} - \dfrac{x\sqrt{2-x^2}}{2}+C$

25. $\frac{1}{64}\left(-\frac{1}{7(4x+1)^7}+\frac{1}{4(4x+1)^8}-\frac{1}{9(4x+1)^9}\right)+C$

27. $-\frac{1}{4}\cos 4x+\frac{1}{6}\cos^3 4x-\frac{1}{20}\cos^5 4x+C$

29. $-\frac{1}{2}\ln(2e^{-x}+1)+C$

31. $-\frac{1}{2}\sin^2 x-2\sin x-4\ln(2-\sin x)+C$

33. $-\frac{\sqrt{1-x^2}}{x}+C$

35. $\frac{1}{48}(1-4x^2)^{3/2}-\frac{1}{16}\sqrt{1-4x^2}+C$

37. $\sqrt{x^2+1}+\ln(x+\sqrt{x^2+1})+C$

39. $x-\frac{1}{3}\ln|x|+\frac{4}{3}\ln|x-3|-\frac{5}{3}\ln|x+3|+C$

41. $-\frac{1}{10}\cos^{10}x+\frac{1}{6}\cos^{12}x-\frac{1}{14}\cos^{14}x+C$

43. $\frac{1}{2}\ln|x^2+2x-1|-\frac{1}{2\sqrt{2}}\ln\left|\frac{x+1-\sqrt{2}}{x+1+\sqrt{2}}\right|+C$

45. $\frac{1}{3}x^3\sin^{-1}2x+\frac{1}{24}\sqrt{1-4x^2}-\frac{1}{72}(1-4x^2)^{3/2}+C$

49. $\tan^{-1}(\sqrt{x}/2)+C$

51. $\frac{x^2}{2}-2x+\frac{1}{4}\ln|x|+\frac{1}{2x}+\frac{15}{4}\ln|x+2|+C$

53. $-\frac{1}{2}\cos(2\ln x)+C$ **55.** $\frac{1}{2}\exp\big(2\tan^{-1}x\big)+C$

57. $\frac{1}{4}\big(\ln(3+x^2)\big)^2+C$ **59.** $\frac{1}{2}\big(\sin^{-1}(x/2)\big)^2+C$

61. $\sqrt{x^2+6x+10}-2\ln(x+3+\sqrt{x^2+6x+10})+C$

63. $\frac{2}{5(2+x^2)^{5/2}}-\frac{1}{3(2+x^2)^{3/2}}+C$

65. $\frac{6}{7}x^{7/6}-\frac{6}{5}x^{5/6}+2\sqrt{x}-6x^{1/6}+6\tan^{-1}x^{1/6}+C$

67. $\frac{2}{3}x^{3/2}-x+4\sqrt{x}-4\ln(1+\sqrt{x})+C$

69. $\frac{1}{2(4-x^2)}+C$

71. $\frac{1}{3}x^3\tan^{-1}x-\frac{1}{6}x^2+\frac{1}{6}\ln(1+x^2)+C$

73. $\frac{1}{5}\ln\left|\frac{3\tan^{-1}(x/2)-1}{\tan^{-1}(x/2)+3}\right|+C$

75. $\frac{1}{2}\ln|\tan(x/2)|-\frac{1}{4}\big(\tan^{-1}(x/2)\big)^2+C$
$=\frac{1}{4}\left(\ln\left|\frac{1-\cos x}{1+\cos x}\right|-\frac{1-\cos x}{1+\cos x}\right)+C$

77. $2\sqrt{x}-2\tan^{-1}\sqrt{x}+C$

79. $\frac{1}{2}x^2+\frac{4}{3}\ln|x-2|-\frac{2}{3}\ln(x^2+2x+4)+\frac{4}{\sqrt{3}}\tan^{-1}\frac{x+1}{\sqrt{3}}+C$

Section 6.8 (page 303)

1. 1/2 **3.** $3\times 2^{1/3}$ **5.** 3

7. 1 **9.** π **11.** 1/2

13. diverges to ∞ **15.** 2

17. $2\ln 2$ square units **21.** 2

23. 15 **25.** diverges to ∞

27. converges **29.** diverges to ∞

31. diverges to ∞ **33.** diverges

35. diverges to ∞

CHAPTER 7. NUMERICAL INTEGRATION

Section 7.1 (page 313)

1. $T_4=4.75$,
$M_4=4.625$,
$T_8=4.6875$,
$M_8=4.65625$,
$T_{16}=4.671875$,
Actual errors:
$I-T_4\approx-0.0833333$,
$I-M_4\approx 0.0416667$,
$I-T_8\approx-0.0208333$,
$I-M_8\approx 0.0104167$,
$I-T_{16}\approx-0.0052083$
Error estimates:
$|I-T_4|\le 0.0833334$,
$|I-M_4|\le 0.0416667$,
$|I-T_8|\le 0.0208334$,
$|I-M_8|\le 0.0104167$,
$|I-T_{16}|\le 0.0052084$

3. $T_4=0.9871158$,
$M_4=1.0064545$,
$T_8=0.9967852$,
$M_8=1.0016082$,
$T_{16}=0.9991967$,
Actual errors:
$I-T_4\approx 0.0128842$,
$I-M_4\approx-0.0064545$,
$I-T_8\approx 0.0032148$,
$I-M_8\approx-0.0016082$,
$I-T_{16}\approx 0.0008033$
Error estimates:
$|I-T_4|\le 0.020186$,
$|I-M_4|\le 0.010093$,
$|I-T_8|\le 0.005045$,
$|I-M_8|\le 0.002523$,
$|I-T_{16}|\le 0.001262$

5. $T_4\approx 2.22622$, $M_4\approx 2.03236$,
$T_8\approx 2.12929$, $M_8\approx 2.02982$,
$T_{16}\approx 2.07956$

7. $M_8\approx 1.3714136$, $T_{16}\approx 1.3704366$,
$I\approx 1.371$

Section 7.2 (page 317)

1. $S_4=S_8=I$, Errors = 0

3. $S_4\approx 1.0001346$, $S_8\approx 1.0000083$,
$I-S_4\approx-0.0001346$, $I-S_8\approx-0.0000083$

5. For $f(x)=e^{-x}$:
$|I-S_4|\le 0.000023$, $|I-S_8|\le 0.0000014$;
for $f(x)=\sin x$,
$|I-S_4|\le 0.00021$,
$|I-S_8|\le 0.000013$

7. $S_4\approx 2.167667$, $S_8\approx 2.08898$,
$S_{16}\approx 2.058977$

Section 7.3 (page 324)

1. $R_1\approx 0.7471805$, $R_2\approx 0.7468337$,
$R_3\approx 0.7468241$, $I\approx 0.746824$

3. $R_2=\frac{2h}{45}\big(7y_0+32y_2+12y_2+32y_3+7y_4\big)$

Section 7.4 (page 328)

1. $3\int_0^1 \frac{u\,du}{1+u^3}$

3. $\int_{-\pi/2}^{\pi/2} e^{\sin\theta}\,d\theta$, or $2\int_0^1 \frac{e^{1-u^2}+e^{u^2-1}}{\sqrt{2-u^2}}\,du$

5. $4\int_0^1 \frac{dv}{\sqrt{(2-v^2)(2-2v^2+v^4)}}$

7. $T_2 \approx 0.603553$ $T_4 \approx 0.643283$, $T_8 \approx 0.658130$, $T_{16} \approx 0.663581$; Errors: $I - T_2 \approx 0.0631$, $I - T_4 \approx 0.0239$, $I - T_8 \approx 0.0085$, $I - T_{16} \approx 0.0031$; Errors do not decrease lile $1/n^2$ because the second derivative of $f(x) = \sqrt{x}$ is not bounded on $[0, 1]$.

9. $I \approx 0.74684$ with error less than 10^{-4}; seven terms of the series are needed.

CHAPTER 8. APPLICATIONS OF INTEGRATION

Section 8.1 (page 335)

1. $\frac{1}{6}$ sq. un. **3.** $\frac{64}{3}$ sq. un. **5.** $\frac{125}{12}$ sq. un.

7. $\frac{1}{2}$ sq. un. **9.** $\frac{5}{12}$ sq. un. **11.** $\frac{15}{8} - 2\ln 2$ sq. un.

13. $\frac{\pi}{2} - \frac{1}{3}$ sq. un. **15.** $\frac{4}{3}$ sq. un. **17.** $2\sqrt{2}$ sq. un.

19. $a^2\cos^{-1}\left(\frac{b}{a}\right) - b\sqrt{a^2-b^2}$ sq. un.

21. $\frac{1+e^{-\pi}}{2}$ sq. un. **23.** 1 sq. un.

25. $\frac{4}{3}$ sq. un. **27.** $\frac{e}{2}$ sq. un.

29. $\frac{27}{4}$ sq. un. **31.** $\frac{\ln(Y+\sqrt{1+Y^2})}{2}$ sq. un.

Section 8.2 (page 347)

1. $\frac{\pi}{5}$ cu. un. **3.** $\frac{3\pi}{10}$ cu. un.

5. (a) $\frac{16\pi}{15}$ cu. un., (b) $\frac{8\pi}{3}$ cu. un.

7. (a) $\frac{27}{2}$ cu. un., (b) $\frac{108\pi}{5}$ cu. un.

9. (a) $\frac{15\pi}{4} - \frac{\pi^2}{8}$ cu. un., (b) $\pi(2 - \ln 2)$ cu. un.

11. $\frac{34\pi}{3}$ cu. un. **13.** $\frac{16r^3}{3}$ cu. un. **15.** about 35%

17. $\frac{\pi h}{3}\left(b^2 - 3a^2 + \frac{2a^3}{b}\right)$ cu. un.

19. $\frac{\pi}{3}(a-b)^2(2a+b)$ cu. un.

21. $\frac{4\pi ab^2}{3}$ cu. un.

23. $\frac{1\pi R^3}{3}\left(1 - \frac{h}{\sqrt{r^2+h^2}}\right)$ cu. un.

25. (a) $\pi/2$ cu. un., (b) 2π cu. un.

27. $k > 2$ **29.** $\frac{16,000}{3}$ cu. un.

31. Vol. of ball $= \int_0^R kr^2\,dr = \frac{kR^3}{3}$; $k = 4\pi$

33. $R = \frac{h\sin\alpha}{\sin\alpha + \cos 2\alpha}$

Section 8.3 (page 355)

1. $\sqrt{1+a^2}(B-A)$ units **3.** 6 units

5. $2\ln 3 - 1$ units **7.** $\sqrt{7} + \frac{1}{4}\ln(4+\sqrt{17})$ units

9. $\ln(e^2 + e^{-2})$ units **11.** $6a$ units

13. 1.0338 units

15. $\frac{64\pi}{81}\left[\frac{(13/9)^{5/2}-1}{5} - \frac{(13/9)^{3/2}-1}{3}\right]$ sq. un.

17. $2\pi\left(\sqrt{2} + \ln(1+\sqrt{2})\right)$ sq. un.

19. $2\pi\left(\frac{255}{16} + \ln 4\right)$ sq. un.

21. $4\pi^2 ab$ sq. un.

23. $8\pi\left(1 + \frac{\ln(2+\sqrt{3})}{2\sqrt{3}}\right)$ sq. un.

25. (a) π cu. un.; (c) "Covering" a surface with paint requires putting on a layer of constant thickness. Far enough to the right, the horn is thinner than any prescirbed constant, so it can contain less paint than would be necessary to paint its surface.

27. $k > -1$

Section 8.4 (page 363)

1. mass $\frac{2L}{\pi}$; centre of mass at $\bar{s} = \frac{L}{2}$

3. $m = \frac{1}{4}\pi\rho_0 a^2$; $\bar{x} = \bar{y} = \frac{4a}{3\pi}$

5. $m = \frac{256k}{15}$; $\bar{x} = 0$, $\bar{y} = \frac{16}{7}$

7. $m = \frac{ka^3}{2}$; $\bar{x} = \frac{2a}{3}$, $\bar{y} = \frac{a}{2}$

9. $m = \int_a^b \rho(x)\big(g(x) - f(x)\big)\,dx;$

$M_{x=0} = \int_a^b x\rho(x)\big(g(x) - f(x)\big)\,dx, \quad \bar{x} = M_{x=0}/m,$

$M_{y=0} = \frac{1}{2}\int_a^b \rho(x)\big((g(x))^2 - (f(x))^2\big)\,dx,$

$\bar{y} = M_{y=0}/m$

11. Mass is $\frac{8}{3}\pi R^4$ kg. The centre of mass is along the line through the centre of the ball perpendicular to the plane, at a distance $R/10$ m from the centre of the ball on the side opposite the plane.

13. $m = \frac{1}{16}\pi\rho_0 a^4; \quad \bar{x} = \bar{y} = \bar{z} = 8a/15$

15. $m = \frac{1}{3}k\pi a^3; \quad \bar{x} = 0, \qquad y = \dfrac{3a}{2\pi}$

Section 8.5 (page 368)

1. $\left(\dfrac{4r}{3\pi}, \dfrac{4r}{3\pi}\right)$

3. $\left(\dfrac{\sqrt{2}-1}{\ln(1+\sqrt{2})}, \dfrac{\pi}{8\ln(1+\sqrt{2})}\right)$

5. $\left(0, \dfrac{9\sqrt{3}-4\pi}{4\pi-\sqrt{3}}\right)$ **7.** $\left(\dfrac{16}{9}, -\dfrac{1}{3}\right)$

9. The centroid is on the axis of symmetry of the hemisphere half way between the base plane and the vertex.

11. The centroid is on the axis of the cone, one quarter of the cone's height above the base plane.

13. $\left(\dfrac{\pi}{2}, \dfrac{\pi}{8}\right)$ **15.** $\left(\dfrac{2r}{\pi}, \dfrac{2r}{\pi}\right)$

17. $(1, -2)$ **19.** $\dfrac{5\pi}{3}$ cu. un. **21.** $\left(1, \frac{1}{5}\right)$

23. $\bar{x} = \dfrac{M_{x=0}}{A}, \ \bar{y} = \dfrac{M_{y=0}}{A},$

where $A = \int_c^d \big(g(y) - f(y)\big)\,dy,$

$M_{x=0} = \frac{1}{2}\int_c^d \big((g(y))^2 - (f(y))^2\big)\,dy,$

$M_{y=0} = \int_c^d y\big(g(y) - f(y)\big)\,dy$

Section 8.6 (page 376)

1. (a) 235,200 N, (b) 352,800 N

3. 6.12×10^8 N **5.** 8.92×10^6 N

7. 7.056×10^5 N-m **9.** $2450\pi a^3\left(a + \dfrac{8h}{3}\right)$ N-m

Section 8.7 (page 386)

1. (a) $\dfrac{2}{9}$, (b) $\mu = 2, \sigma^2 = \dfrac{1}{2}, \sigma = \dfrac{1}{\sqrt{2}}$, (c) $\dfrac{8}{9\sqrt{2}} \approx 0.63$

3. (a) 3, (b) $\mu = \dfrac{3}{4}, \sigma^2 = \dfrac{3}{80}, \sigma = \sqrt{\dfrac{3}{80}}$,

(c) $\dfrac{69}{20}\sqrt{\dfrac{3}{80}} \approx 0.668$

5. (a) 6, (b) $\mu = \dfrac{1}{2}, \sigma^2 = \dfrac{1}{20}, \sigma = \sqrt{\dfrac{1}{20}}$,

(c) $\dfrac{7}{5\sqrt{5}} \approx 0.626$

7. (a) $\dfrac{2}{\sqrt{\pi}}$, (b) $\mu = \dfrac{1}{\sqrt{\pi}} \approx 0.0.564, \sigma^2 = \dfrac{\pi-2}{2\pi}$,

$\sigma = \sqrt{\dfrac{\pi-2}{2\pi}} \approx 0.426$, (c) Pr$\approx$ 0.52

11. (a) 0, (b) $e^{-3} \approx 0.05$, (c) ≈ 0.046

13. approximately 0.006

Section 8.9 (page 393)

1. $y^2 = Cx$ **3.** $x^3 - y^3 = C$

5. $y = Ce^{t^2/2}$ **7.** $y = \dfrac{Ce^{2x}-1}{Ce^{2x}+1}$

9. $y = -\ln\left(Ce^{-2t} - \frac{1}{2}\right)$ **11.** $y = x^3 + Cx^2$

13. $y = \frac{3}{2} + Ce^{-2x}$ **15.** $y = x - 1 + Ce^{-x}$

17. If $a = b$ the given solution is indeterminate $0/0$; in this case the solution is $x = a^2kt/(1+akt)$.

19. $v = \sqrt{\dfrac{mg}{k}}\,\dfrac{e^{2\sqrt{kg/m}t}-1}{e^{2\sqrt{kg/m}t}+1}, \quad v \to \sqrt{\dfrac{mg}{k}}$

CHAPTER 9. PLANE CURVES

Section 9.1 (page 409)

1. single point $(-1, 0)$

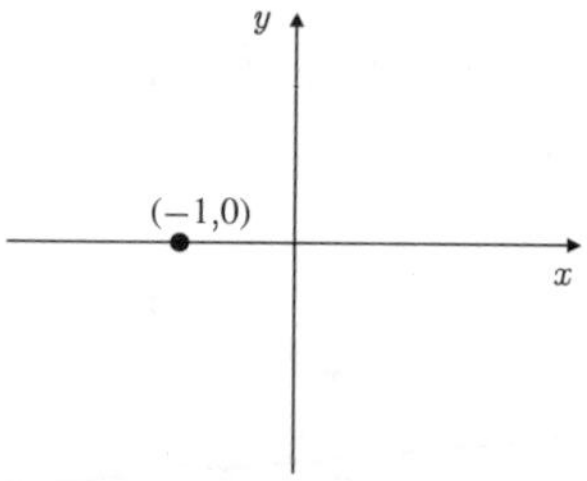

3. ellipse, centre (0,2)

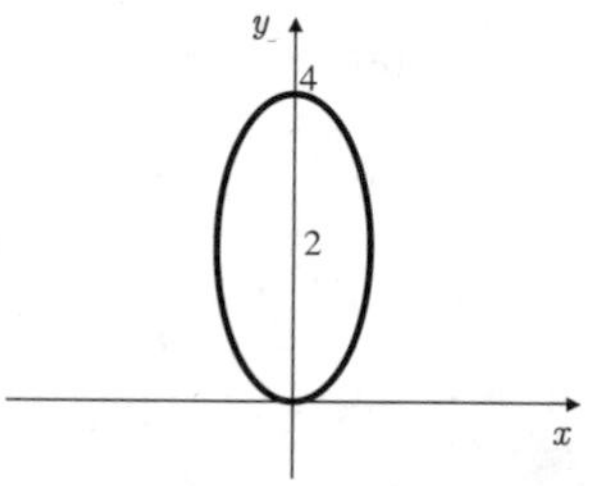

5. parabola, vertex $(-1,-4)$

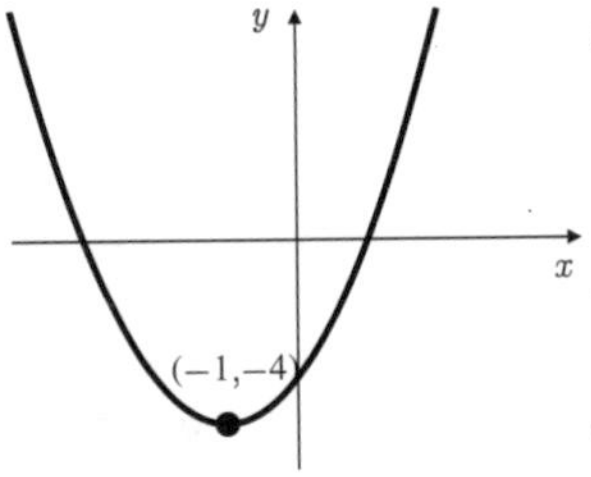

7. hyperbola, centre $\left(-\frac{3}{2},1\right)$
asymptotes
$2x+3=\pm 2^{3/2}(y-1)$

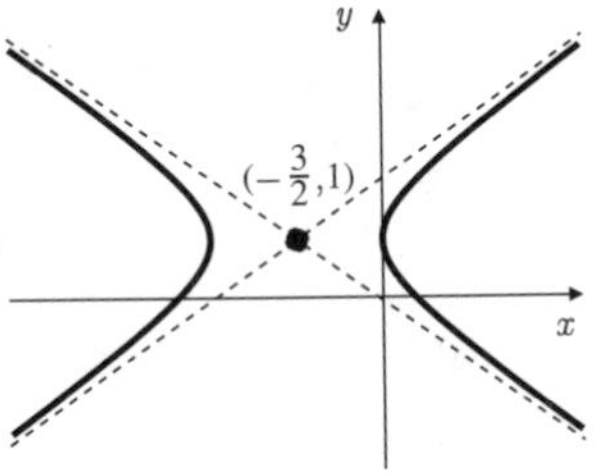

9. ellipse, centre $(1,-1)$

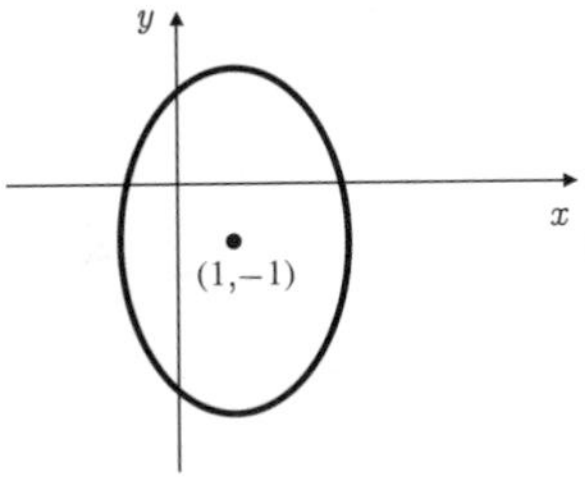

11. rectangular hyperbola, centre $(1,-1)$,
semiaxes $a=b=\sqrt{2}$,
eccentricity $\sqrt{2}$,
foci $(\sqrt{2}+1,\sqrt{2}-1)$,
$(-\sqrt{2}+1,-\sqrt{2}-1)$,
asymptotes $x=1$, $y=-1$

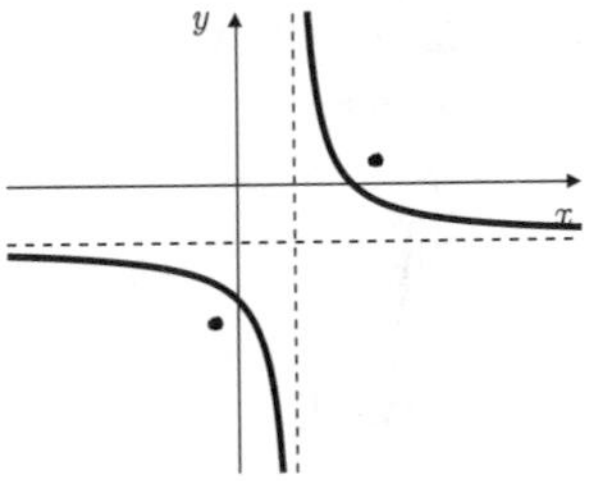

13. hyperbola, centre (0,0),
semiaxes
$a=1/\sqrt{2+\sqrt{2}}$,
$b=1/\sqrt{2-\sqrt{2}}$,
foci
$\pm\big(\cos(\pi/8),\sin(\pi/8)\big)$,
asymptotes $x=0$, $y=-x$

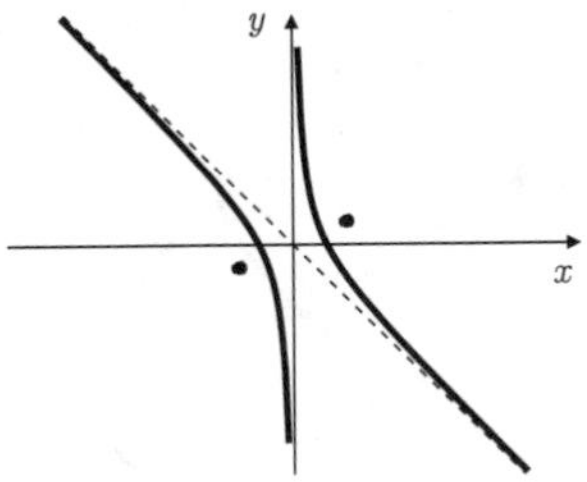

15. ellipse, centre (0,0),
semi-axes $a=2$, $b=10/7$,
foci $\left(\frac{8\sqrt{6}}{7\sqrt{5}},-\frac{4\sqrt{6}}{7\sqrt{5}}\right)$,
$\left(-\frac{8\sqrt{6}}{7\sqrt{5}},\frac{4\sqrt{6}}{7\sqrt{5}}\right)$

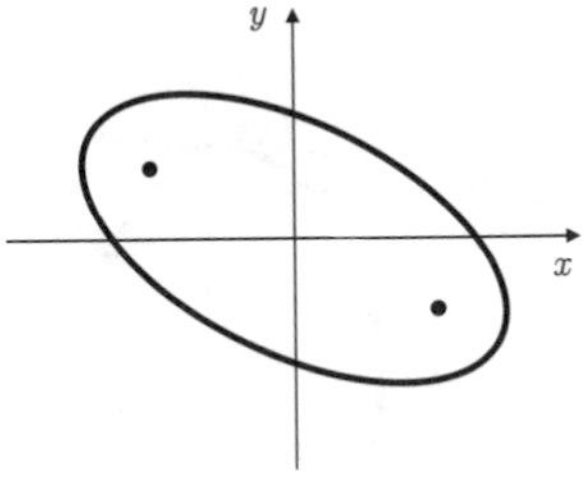

17. $(1-\epsilon^2)x^2+y^2-2p\epsilon^2x=\epsilon^2p^2$

Section 9.2 (page 419)

1. $x=3$,
vertical st. line

3. $3y-4x=5$,
straight line

5. $2xy=1$,
rectangular hyperbola

7. $y=x^2-x$,
a parabola

9. $y^2=1+2x$,
a parabola

11. $x^2-3y^2-8y=4$,
a hyperbola

13.

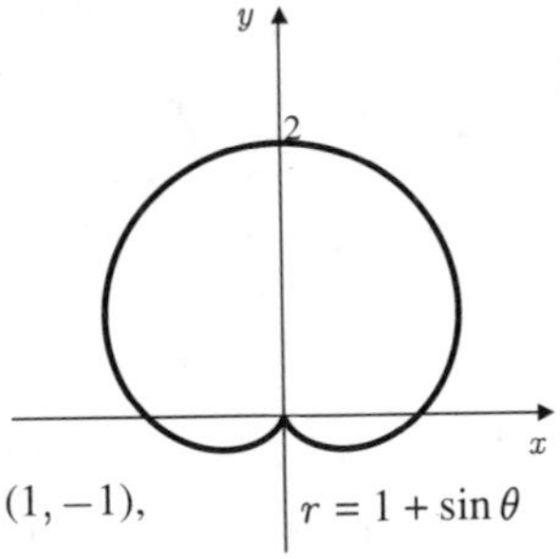

15.

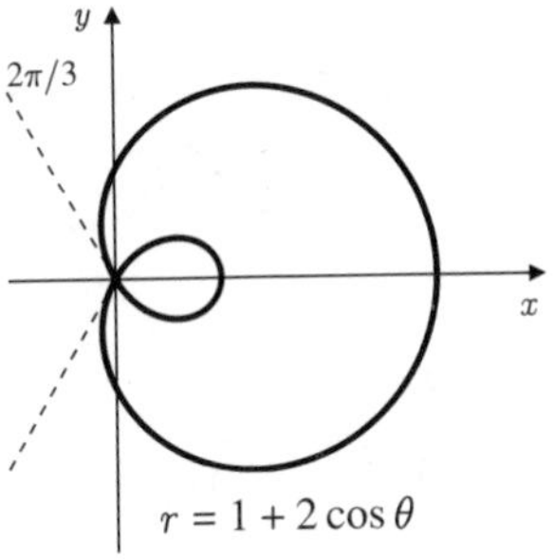

17.

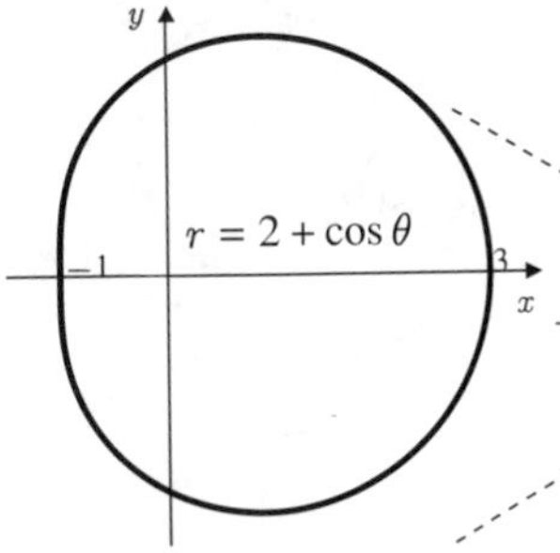

19.

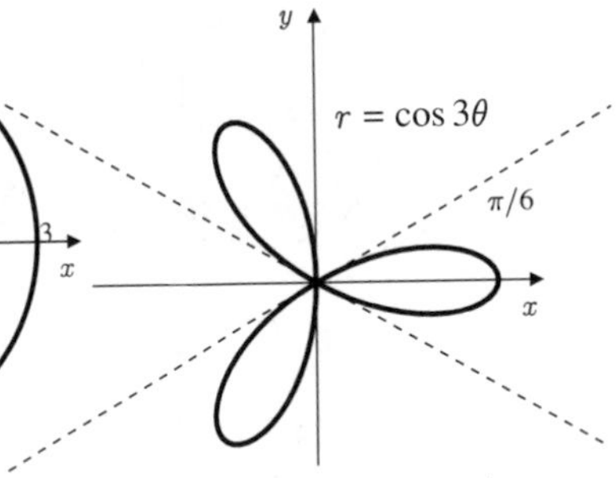

21.

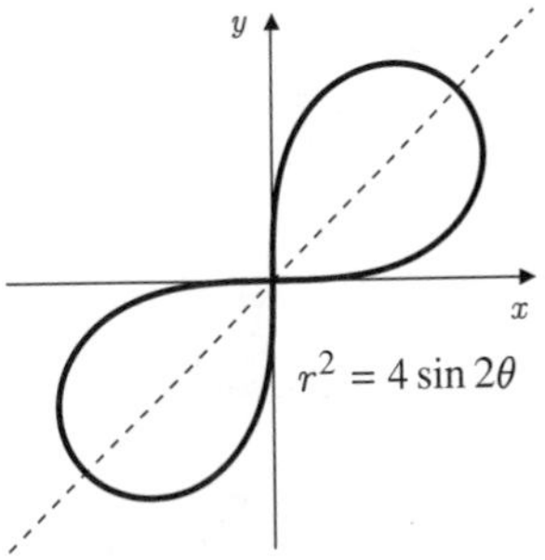

23. $r=\pm\sqrt{\sin 3\theta}$

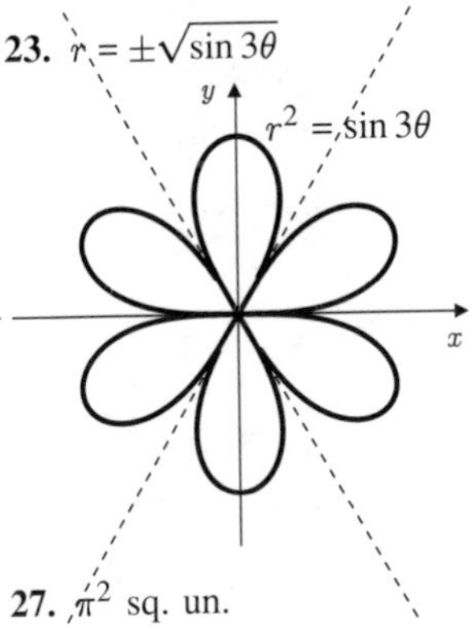

25. asymptote $y=1$,
$r=1/(\theta-\alpha)$ has
asymptote $(\cos\alpha)y-(\sin\alpha)x=1$

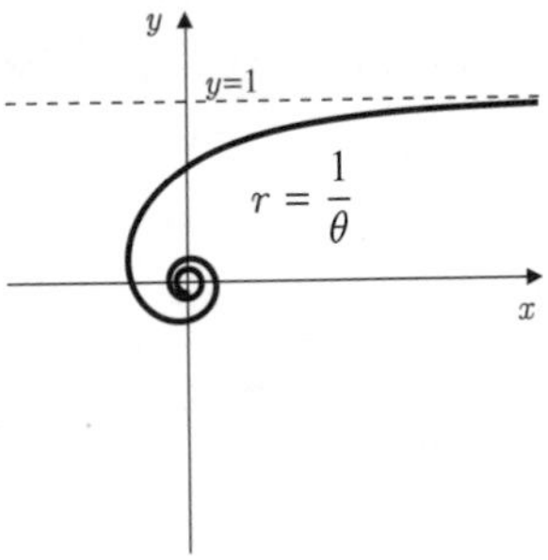

27. π^2 sq. un.

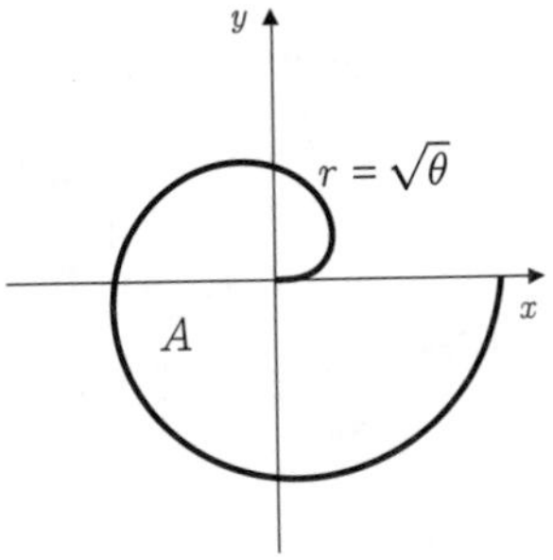

29. a^2 sq. un.

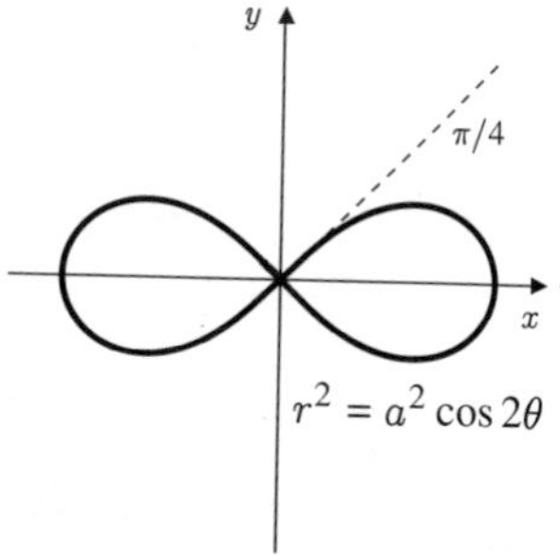

31. $\pi/2$ sq. un.

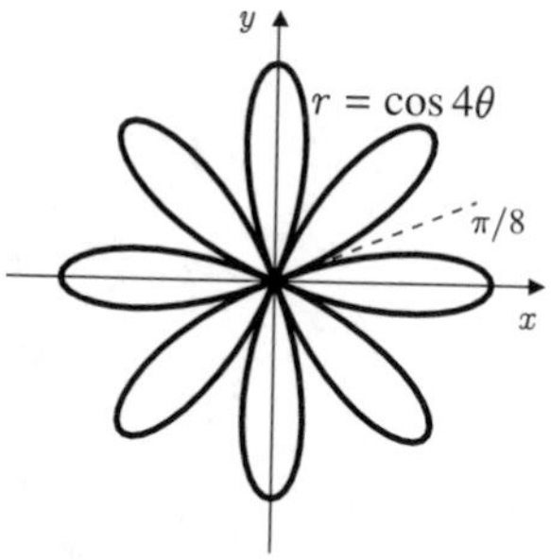

33. $2 + (\pi/4)$ sq. un.

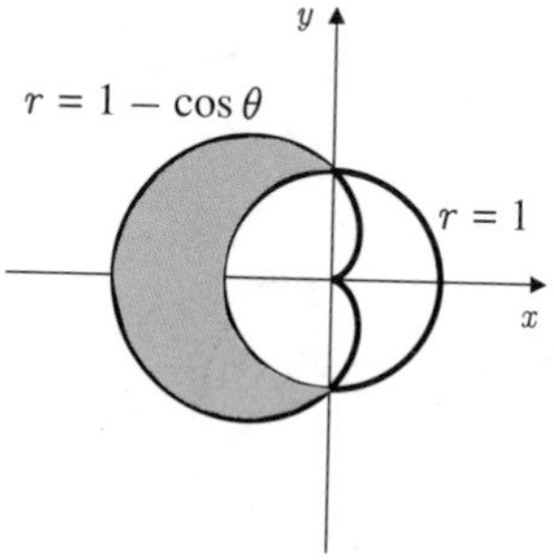

35. $\pi/4$ sq. un.

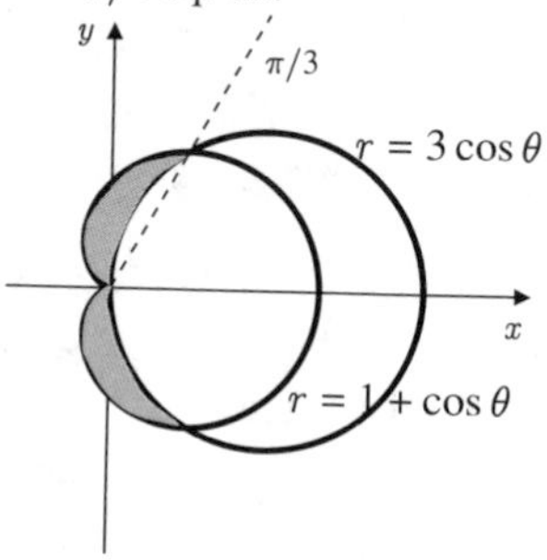

37. $\pi - \frac{3}{2}\sqrt{3}$ sq. un.

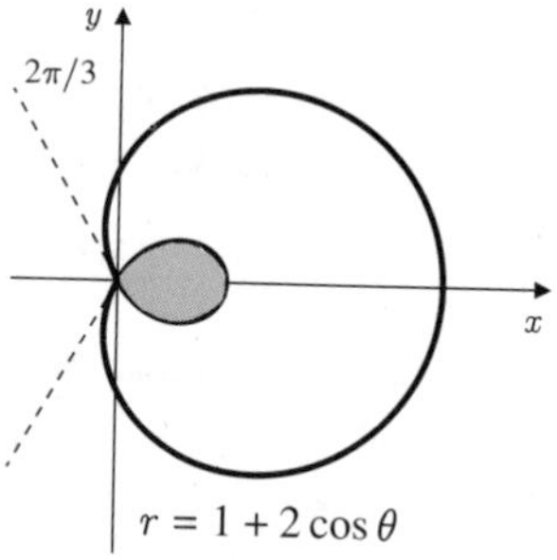

39. $\dfrac{\sqrt{1+a^2}}{a}\left(e^{a\pi} - e^{-a\pi}\right)$ units

43. 67.5°, −22.5°

45. 90° at (0,0),
$\pm 45°$ at $\left(1 - \dfrac{1}{\sqrt{2}}, \dfrac{\pi}{4}\right)$,
$\pm 135°$ at $\left(1 + \dfrac{1}{\sqrt{2}}, \dfrac{5\pi}{4}\right)$

47. horizontal at $\left(\pm\frac{\pi}{4}, \sqrt{2}\right)$, vertical at (2, 0) and the origin

49. horizontal at (0,0), $\left(\frac{2}{3}\sqrt{2}, \pm\tan^{-1}\sqrt{2}\right)$, $\left(\frac{2}{3}\sqrt{2}, \pi \pm \tan^{-1}\sqrt{2}\right)$, vertical at $\left(0, \frac{\pi}{2}\right)$, $\left(\frac{2}{3}\sqrt{2}, \pm\tan^{-1}(1/\sqrt{2})\right)$, $\left(\frac{2}{3}\sqrt{2}, \pi \pm \tan^{-1}(1/\sqrt{2})\right)$

51. horizontal at $\left(2, -\frac{\pi}{2}\right)$, $\left(1, \frac{\pi}{6}\right)$, $\left(1, \frac{5\pi}{6}\right)$, vertical at $\left(1, -\frac{\pi}{6}\right)$, $\left(1, -\frac{5\pi}{6}\right)$, no tangent at $\left(0, \frac{\pi}{2}\right)$

Section 9.3 (page 426)

1. $x + y = 1$

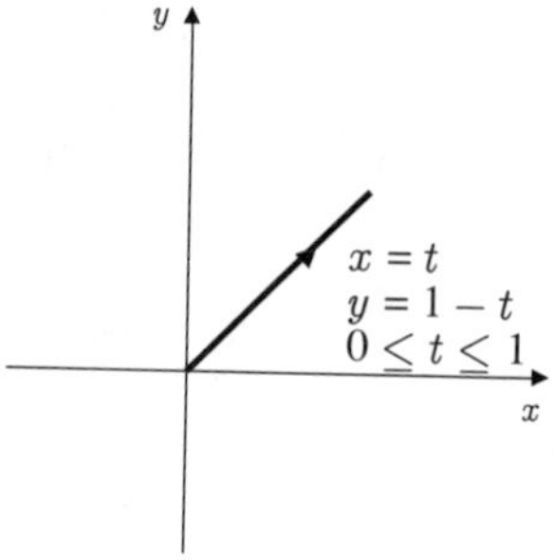

3. $y = (x-1)^2/4$

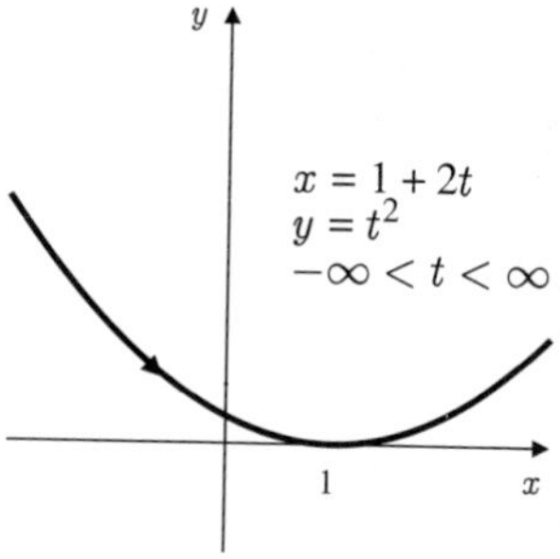

5. $y = (1/x) - 1$

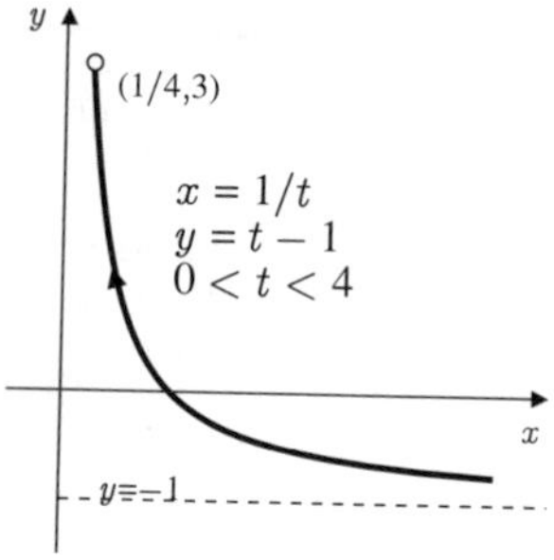

7. $x^2 + y^2 = 9$

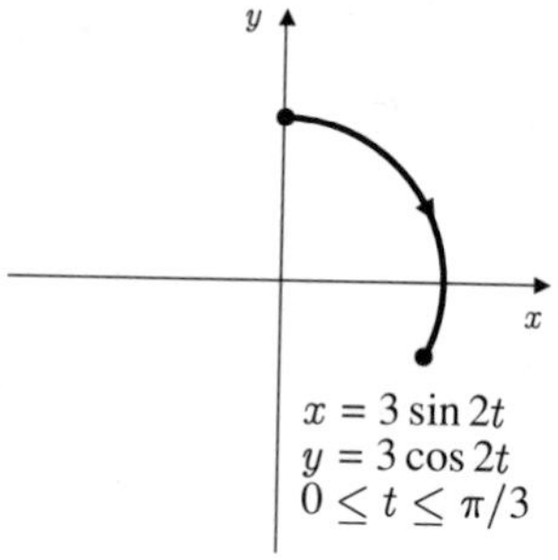

9. $\dfrac{x^2}{9} + \dfrac{y^2}{16} = 1$

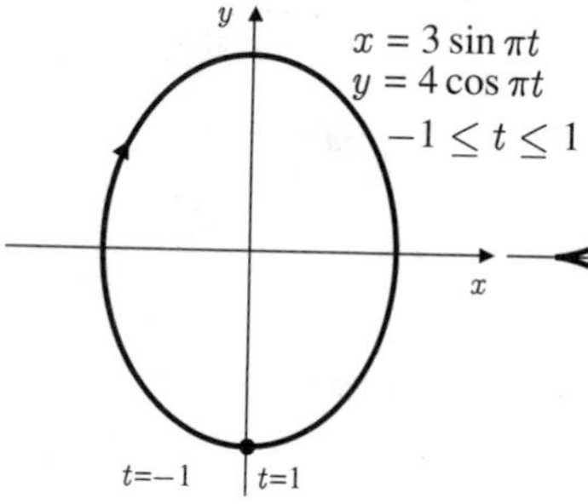

11. $x^{2/3} + y^{2/3} = 1$

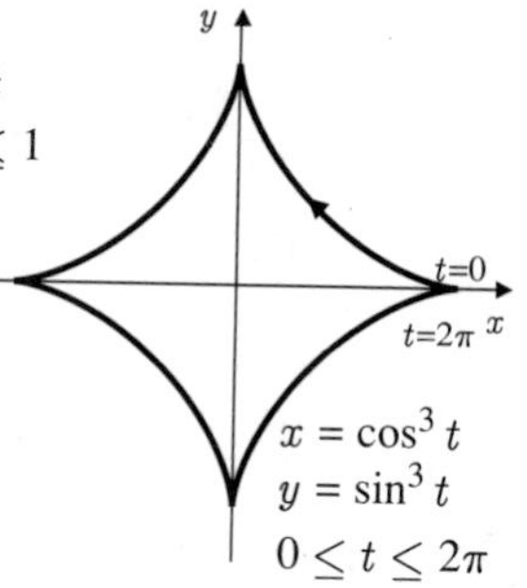

13. $x = f(\theta)\cos\theta, \quad y = f(\theta)\sin\theta$

15. $x = m/2, \quad y = m^2/4, \quad (-\infty < m < \infty)$

17. $x^3 + y^3 = 3xy$

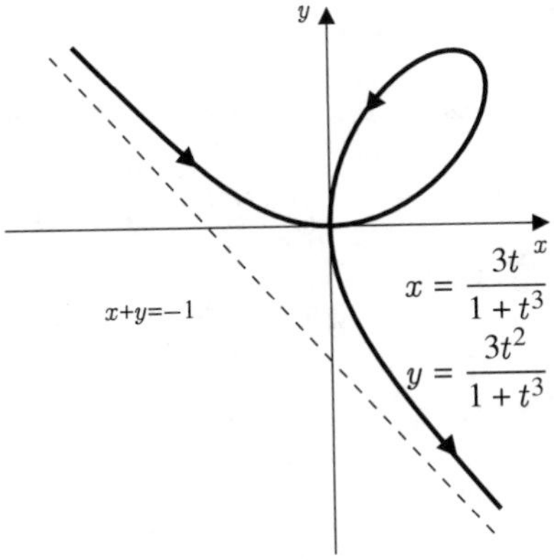

13. 256/15 sq. un.

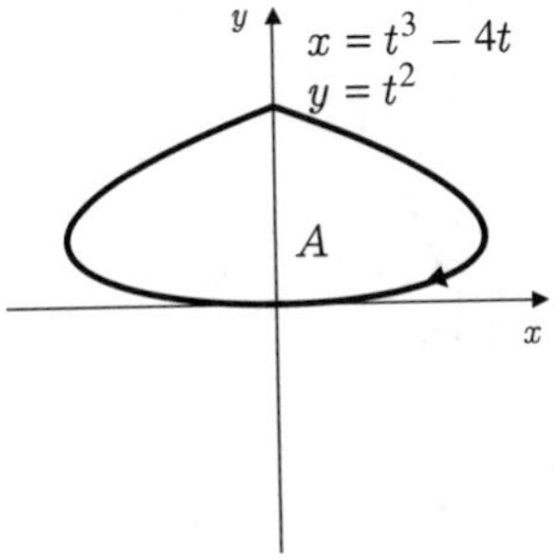

15. 1/6 sq. un.

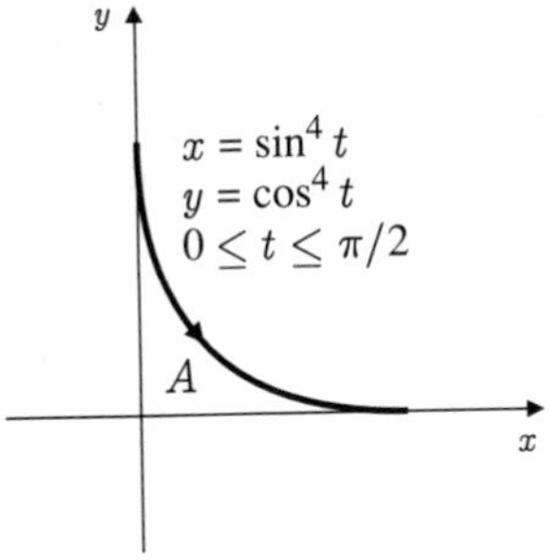

17. $9\pi/2$ sq. un.

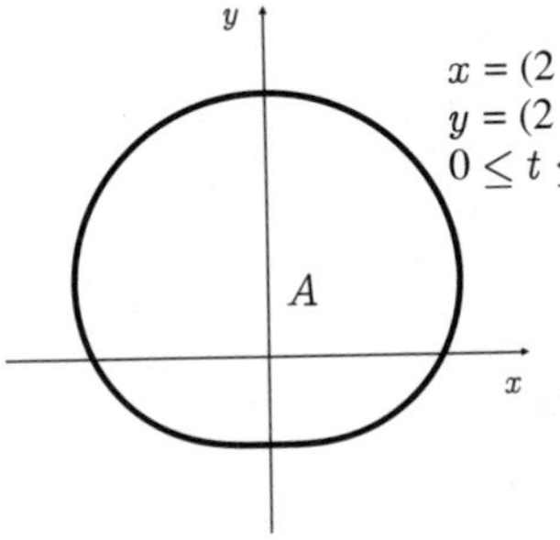

Section 9.4 (page 434)

1.

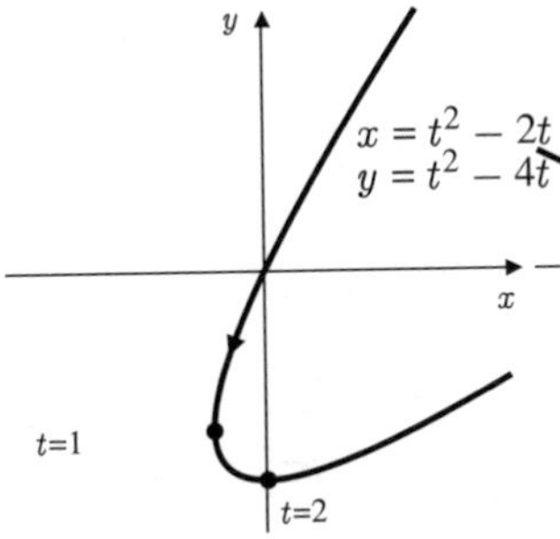

3.

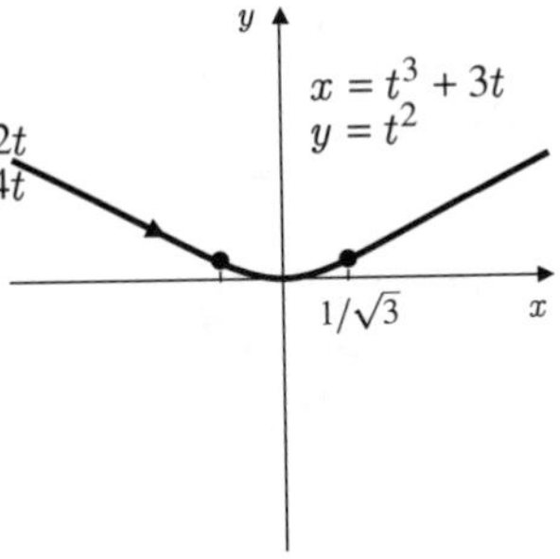

5.

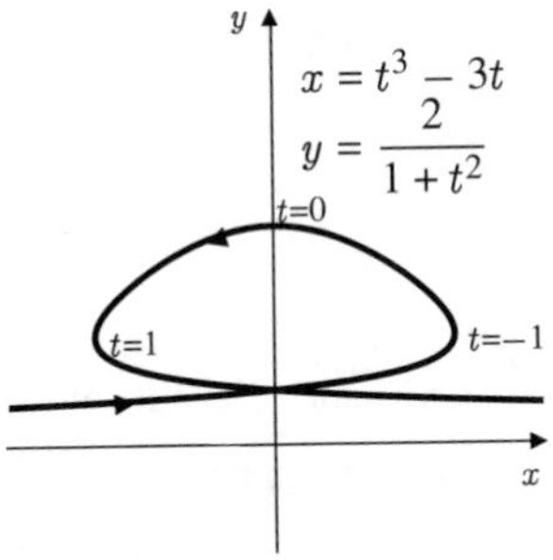

7.

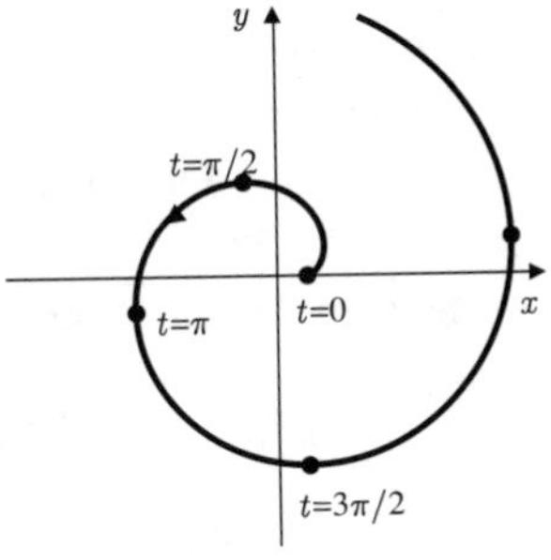

9. $6a$ units

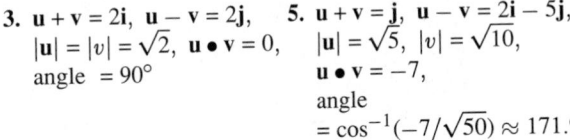

11. $\frac{8}{3}\left((1+\pi^2)^{3/2} - 1\right)$ units

Section 9.5 (page 443)

1. (a) $3\mathbf{i} - 2\mathbf{j}$, (b) $-3\mathbf{i} + 2\mathbf{j}$, (c) $2\mathbf{i} - 5\mathbf{j}$, (d) $-2\mathbf{i} + 4\mathbf{j}$, (e) $-\mathbf{i} - 2\mathbf{j}$, (f) $4\mathbf{i} + \mathbf{j}$, (g) $-7\mathbf{i} + 20\mathbf{j}$, (h) $2\mathbf{i} - (5/3)\mathbf{j}$

3. $\mathbf{u} + \mathbf{v} = 2\mathbf{i}$, $\mathbf{u} - \mathbf{v} = 2\mathbf{j}$, $|\mathbf{u}| = |v| = \sqrt{2}$, $\mathbf{u} \bullet \mathbf{v} = 0$, angle $= 90^\circ$

5. $\mathbf{u} + \mathbf{v} = \mathbf{j}$, $\mathbf{u} - \mathbf{v} = 2\mathbf{i} - 5\mathbf{j}$, $|\mathbf{u}| = \sqrt{5}$, $|v| = \sqrt{10}$, $\mathbf{u} \bullet \mathbf{v} = -7$, angle $= \cos^{-1}(-7/\sqrt{50}) \approx 171.9^\circ$

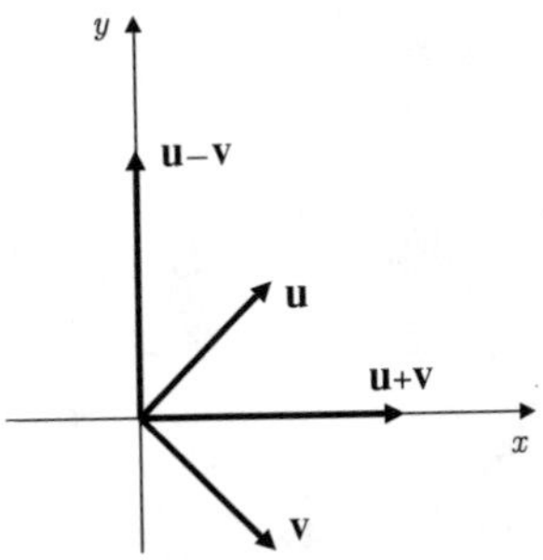

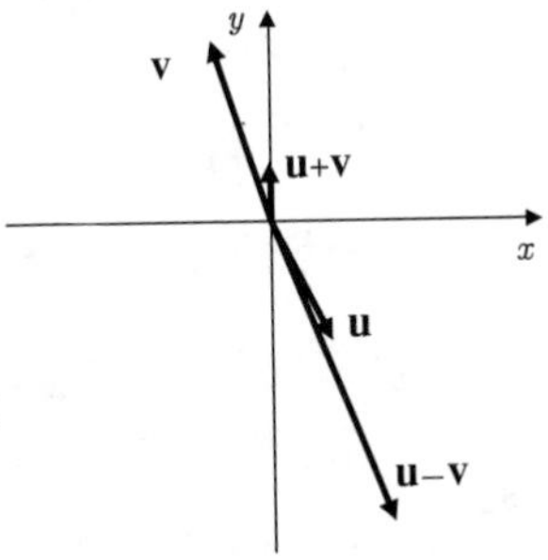

7. $\mathbf{u} + \mathbf{v} = 3\mathbf{i} + \mathbf{j}$, $\mathbf{u} - \mathbf{v} = -\mathbf{i} + 3\mathbf{j}$, $|\mathbf{u}| = |v| = \sqrt{5}$, $\mathbf{u} \bullet \mathbf{v} = 0$, angle $= 90°$

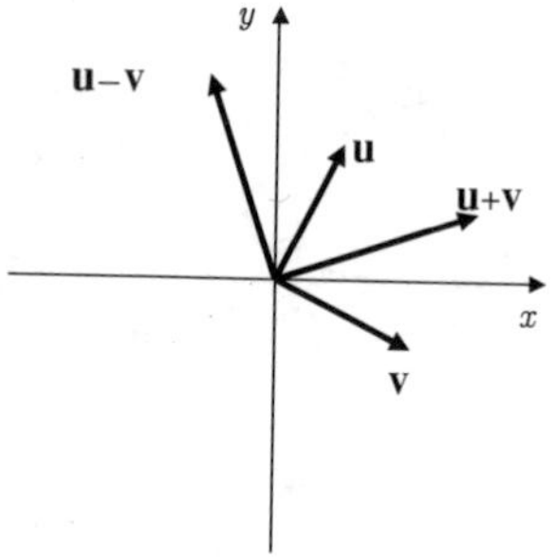

9. $\mathbf{v}(1) = \mathbf{i} - \mathbf{j}$, $\mathbf{a}(1) = -\mathbf{j}$,
$\mathbf{v}(2) = \mathbf{i} - 2\mathbf{j}$, $\mathbf{a}(2) = -\mathbf{j}$
11$\mathbf{v}(1)$=$\mathbf{i}$+$\mathbf{j}$, $\mathbf{a}(1)$=-6$\mathbf{j}$,
$\mathbf{v}(2)$=$\mathbf{i}$-8$\mathbf{j}$, $\mathbf{a}(2)$=-12$\mathbf{j}$

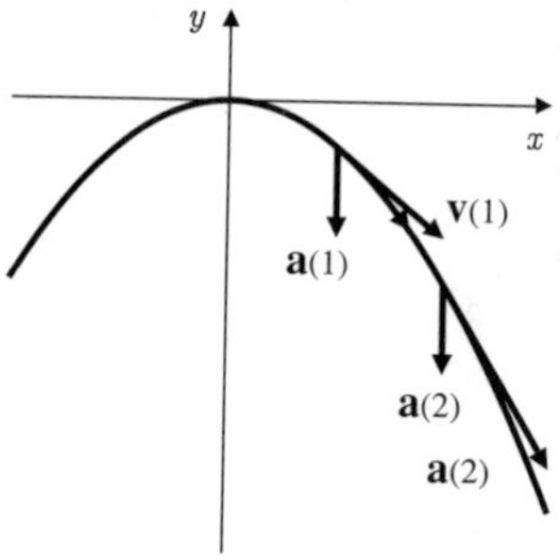

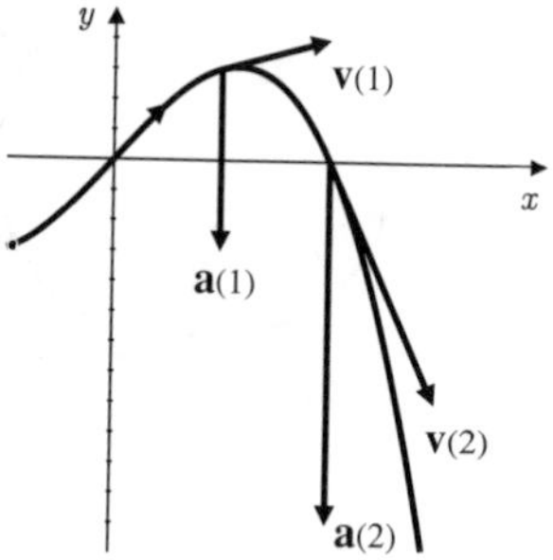

13. path is a straight line through (x_0, y_0) parallel to $\mathbf{u}$; parametric equations $x = x_0 + at$, $y = y_0 + bt$, where $\mathbf{r}_0 = x_0\mathbf{i} + y_0\mathbf{j}$ and $\mathbf{u} = a\mathbf{i} + b\mathbf{j}$

15. $\mathbf{r} = \left(\dfrac{t^3}{6} - \dfrac{t}{2} - \dfrac{2}{3}\right)\mathbf{i} + \mathbf{j}$

17. $\mathbf{r} = (2t - \sin t)\mathbf{i} + (t + e^{-t} - 1)\mathbf{j}$

19. if $\mathbf{r}_0 = x_0\mathbf{i} + y_0\mathbf{j}$ and $\mathbf{v}_0 = ka\mathbf{i} + kb\mathbf{j}$ then the solution simplifies to
$(y_0x - x_0y)^2 + (bx - ay)^2 = (ay_0 - bx_0)^2$,
which is a bounded quadratic curve, so is an ellipse or circle; it is a circle if $x_0y_0 + ab = 0$ and $x_0^2 + a^2 = y_0^2 + b^2$.

21. $\mathbf{v} = \dfrac{k}{\sqrt{1 + e^{2x}}}(\mathbf{i} + e^x\mathbf{j})$,
$\mathbf{a} = \dfrac{k^2}{(1 + e^{2x})^2}\left(-e^{2x}\mathbf{i} + e^x\mathbf{j}\right)$

CHAPTER 10. SEQUENCES AND SERIES

Section 10.1 (page 452)

1. bounded, positive, increasing, convergent to 2

3. bounded, positive, convergent to 4

5. bounded below, ultimately positive, increasing, divergent to infinity

7. bounded below, positive, increasing, divergent to infinity

9. bounded, positive, decreasing, convergent to 0

11. divergent **13.** divergent

15. ∞ **17.** 0 **19.** 1

21. e^{-3} **23.** 0 **25.** 1/2

27. 0 **29.** 0 **31.** $\lim a_n = 5$

33. If $\{a_n\}$ is (ultimately) decreasing, then either it is bounded below, and therefore convergent, or else it is unbounded below and therefore divergent to negative infinity.

Section 10.2 (page 459)

1. $\dfrac{1}{2}$ **3.** $\dfrac{1}{(2+\pi)^8\left((2+\pi)^2 - 1\right)}$

5. $\dfrac{25}{4416}$ **7.** $\dfrac{3}{4}$ **9.** $\dfrac{1}{3}$

11. div. to ∞ **13.** div. to ∞

15. diverges **17.** 14 m

21. If $\{a_n\}$ is ultimately negative, then the series $\sum a_n$ must either converge (if its partial sums are bounded below), or diverge to $-\infty$ (if its partial sums are not bounded below).

23. false, e.g. $\sum \dfrac{(-1)^n}{2^n}$ **25.** true

27. true

Section 10.3 (page 468)

1. converges **3.** diverges to ∞

5. converges **7.** diverges to ∞

9. converges **11.** diverges to ∞

13. diverges to ∞ **15.** converges

17. diverges to ∞ **19.** converges

21. diverges to ∞ **23.** converges

25. converges **27.** converges

33. no info from Ratio test, but series diverges to infinity since all terms exceed 1.

Section 10.4 (page 475)

1. conv. conditionally **3.** conv. conditionally

5. diverges **7.** conv. absolutely

9. conv. conditionally **11.** diverges

13. converges absolutely if $-1 < x < 1$, conditionally if $x = -1$, diverges elsewhere

15. converges absolutely if $0 < x < 2$, conditionally if $x = 2$, diverges elsewhere

17. converges absolutely if $-2 < x < 2$, conditionally if $x = -2$, diverges elsewhere

19. converges absolutely if $-\frac{7}{2} < x < \frac{1}{2}$, conditionally if $x = -\frac{7}{2}$, diverges elsewhere

21. converges absolutely if $x \geq \frac{1}{2}$, diverges if $x < \frac{1}{2}$, $x \neq 0$, undefined at $x = 0$

23. converges absolutely if $-1 < x < 0$ or $1 < x < 2$, diverges elsewhere

25. AST does not apply directly, but does if we remove all the 0 terms; series converges conditionally

27. (a) false, e.g. $a_n = \dfrac{(-1)^n}{n}$,
(b) false, e.g. $a_n = \dfrac{\sin(n\pi/2)}{n}$, (see Exercise 25),
(c)true

29. converges absolutely for $-1 < x < 1$, conditionally if $x = -1$, diverges elsewhere

Section 10.5 (page 482)

1. $s_n + \dfrac{1}{9(n+1)^9} \leq s \leq s_n + \dfrac{1}{9n^9}$; $n = 2$

3. $s_n + \dfrac{2}{\sqrt{n+1}} \leq s \leq s_n + \dfrac{2}{\sqrt{n}}$; $n = 63$

5. $s_n + \dfrac{1}{e^{n+1}} \leq s \leq s_n + \dfrac{1}{e^n}$; $n = 6$

7. $0 < s - s_n \leq \dfrac{n+2}{2^n(n+1)!(2n+3)}$; $n = 4$

9. $0 < s - s_n \leq \dfrac{2^n(4n^2+6n+2)}{(2n)!(4n^2+6n)}$; $n = 4$

11. $|s - s_n| < \dfrac{n+1}{n^2+2n+2}$; $n = 999$

13. $|s - s_n| < \dfrac{n+1}{2^{n+1}}$; $n = 13$

15. (b) $s \leq \dfrac{2}{k(1-k)}$, $k = \frac{1}{2}$,
(c) $0 < s - s_n < \dfrac{(1+k)^{n+1}}{2^n k(1-k)}$, $k = \dfrac{n+2-\sqrt{n^2+8}}{2(n-1)}$
for $n \geq 2$

CHAPTER 11. REPRESENTING FUNCTIONS BY POWER SERIES

Section 11.1 (page 491)

1. $\dfrac{1}{2} - \dfrac{x-2}{4} + \dfrac{(x-2)^2}{8} - \dfrac{(x-2)^3}{16} + \dfrac{(x-2)^4}{32}$

3. $1 + \dfrac{x^2}{2} + \dfrac{x^4}{24}$

5. $\dfrac{1}{\sqrt{2}}\left[1 + \left(x - \dfrac{\pi}{4}\right) - \dfrac{1}{2!}\left(x - \dfrac{\pi}{4}\right)^2 - \dfrac{1}{3!}\left(x - \dfrac{\pi}{4}\right)^3 + \dfrac{1}{4!}\left(x - \dfrac{\pi}{4}\right)^4 + \dfrac{1}{5!}\left(x - \dfrac{\pi}{4}\right)^5 - \dfrac{1}{6!}\left(x - \dfrac{\pi}{4}\right)^6\right]$

7. $e^2\left[1 - (x+2) + \dfrac{x+2)^2}{2!} - \cdots + (-1)^n\dfrac{(x+2)^n}{n!}\right]$

9. $1 + \dfrac{x^2}{2} + \dfrac{5x^4}{24}$

11. (a) $3+2x+x^2$, (b) $3+2x+x^2$, (c) $3-2(x+2)+(x+2)^2$, (d) $18+8(x-3)+(x-3)^2$

13. polynomials of degree at most n

15. $\dfrac{1}{720}(0.2)^7$ **17.** $\dfrac{1}{120}(0.5)^5$

19. $\dfrac{4\sec^2(0.1)\tan^2(0.1) + 2\sec^4(0.1)}{4!\,10^4}$

21. $\dfrac{12}{120(1.95)^5(20)^5}$

23. $2^x = \sum_{n=0}^{\infty} \dfrac{x \ln 2)^n}{n!}$, all x

25. $\sin x = \sum_{n=0}^{\infty}(-1)^n \dfrac{x^{2n+1}}{(2n+1)!}$, all x

27. $\dfrac{1}{1-x} = \displaystyle\sum_{n=0}^{\infty} x^n$, $-1 < x < 1$

29. $\dfrac{x}{2+3x} = \displaystyle\sum_{n=1}^{\infty}(-1)^{n-1}3^{n-1}\left(\dfrac{x}{2}\right)^n$, $-\dfrac{2}{3} < x < \dfrac{2}{3}$

31. $\sin x = \dfrac{1}{2}\displaystyle\sum_{n=0}^{\infty}\dfrac{c_n}{n!}\left(x - \dfrac{\pi}{6}\right)^n$, (for all x), where $c_n = (-1)^{n/2}$ if n is even, and $c_n = (-1)^{(n-1)/2}\sqrt{3}$ if n is odd

33. $\ln x = \sum_{n=1}^{\infty}(-1)^{n-1}\dfrac{(x-1)^n}{n}$, $0 < x \leq 2$

35. $\dfrac{1}{x} = -\dfrac{1}{2}\displaystyle\sum_{n=0}^{\infty}\left(\dfrac{x+2}{2}\right)^n$, $-4 < x < 0$

Section 11.2 (page 501)

1. centre 0, radius 1, interval $(-1, 1)$

3. centre -2, radius 2, interval $[-4, 0)$

5. centre $\frac{3}{2}$, radius $\frac{1}{2}$, interval $(1, 2)$

7. centre 0, radius ∞, interval $(-\infty, \infty)$

9. $\dfrac{1}{(1-x)^3} = \displaystyle\sum_{n=0}^{\infty}\dfrac{(n+1)(n+2)}{2}x^n$, $(-1 < x < 1)$

11. $\dfrac{1}{(1-x)^2} = \displaystyle\sum_{n=0}^{\infty}(n+1)x^n$, $(-1 < x < 1)$

13. $\dfrac{1}{(2-x)^2} = \sum_{n=0}^{\infty} \dfrac{n+1}{2^{n+2}} x^n, \quad (-2 < x < 2)$

15. $\ln(2-x) = \ln 2 - \sum_{n=1}^{\infty} \dfrac{x^n}{2^n n}, \quad (-2 \le x < 2)$

17. $\dfrac{1}{x^2} = \sum_{n=0}^{\infty} \dfrac{n+1}{2^{n+2}} (x+2)^n, \quad (-4 < x < 0)$

19. $\dfrac{x^3}{1-2x^2} = \sum_{n=0}^{\infty} 2^n x^{2n+3}, \quad \left(-\dfrac{1}{\sqrt{2}} < x < \dfrac{1}{\sqrt{2}}\right)$

21. $\left(-\frac{1}{4}, \frac{1}{4}\right); \quad \dfrac{1}{1+4x}$

23. $[-1, 1); \quad \frac{1}{3}$ if $x = 0$,
$-\dfrac{1}{x^3}\ln(1-x) - \dfrac{1}{x^2} - \dfrac{1}{2x}$ otherwise

25. $(-1, 1); \quad \dfrac{2}{(1-x^2)^2}$

Section 11.3 (page 511)

1. $e^{3x+1} = \sum_{n=0}^{\infty} \dfrac{3^n e}{n!} x^n, \quad$ all x

3. $\sin\left(x - \dfrac{\pi}{4}\right) = \dfrac{1}{\sqrt{2}} \sum_{n=0}^{\infty} (-1)^n \left[-\dfrac{x^{2n}}{(2n)!} + \dfrac{x^{2n+1}}{(2n+1)!}\right],$
all x

5. $x^2 \sin\left(\dfrac{x}{3}\right) = \sum_{n=0}^{\infty} \dfrac{(-1)^n}{3^{2n+1}(2n+1)!} x^{2n+3}, \quad$ all x

7. $\sin x \cos x = \sum_{n=0}^{\infty} \dfrac{(-1)^n 2^{2n}}{(2n+1)!} x^{2n+1}, \quad$ all x

9. $\dfrac{1+x^3}{1+x^2} = 1 - x^2 + \sum_{n=2}^{\infty} (-1)^n \left(x^{2n-1} + x^{2n}\right),$
$(-1 < x < 1)$

11. $\ln \dfrac{1-x}{1+x} = -2 \sum_{n=1}^{\infty} \dfrac{x^{2n-1}}{2n-1}, \quad (-1 < x < 1)$

13. $\cosh x - \cos x = 2 \sum_{n=0}^{\infty} \dfrac{x^{4n}}{(4n)!}, \quad$ all x

15. $\sqrt{4+x}$
$= 2 + \dfrac{x}{4} + 2 \sum_{n=2}^{\infty} (-1)^{n-1} \dfrac{1 \times 3 \times 5 \times \cdots \times (2n-3)}{2^{3n} n!} x^n,$
$(-4 < x \le 4)$

17. $e^{-2x} = e^2 \sum_{n=0}^{\infty} \dfrac{(-1)^n 2^n}{n!} (x+1)^n, \quad$ all x

19. $\cos x = \sum_{n=0}^{\infty} \dfrac{(-1)^{n+1}}{(2n)!} (x-\pi)^{2n}, \quad$ all x

21. $\sin x - \cos x = \sqrt{2} \sum_{n=0}^{\infty} \dfrac{(-1)^n}{(2n+1)!} \left(x - \dfrac{\pi}{4}\right)^{2n+1}, \quad$ all x

23. $\dfrac{1}{x^2} = \dfrac{1}{4} \sum_{n=0}^{\infty} \dfrac{n+1}{2^n} (x+2)^n, \quad (-4 \le x < 0)$

25. $1 + \dfrac{x^2}{2} + \dfrac{5x^4}{24}$

27. $x + \dfrac{x^2}{2} - \dfrac{x^3}{6}$

29. $1 + \dfrac{x}{2} - \dfrac{x^2}{8}$

31. e^{x^2} all x

33. $\dfrac{e^x - e^{-x}}{2x} = \dfrac{\sinh x}{x}$ if $x \ne 0$, 1 if $x = 0$

35. (a) $1 + x + x^2$, (b) $3 + 3(x-1) + (x-1)^2$

Section 11.4 (page 516)

1. 1.22140 **3.** 3.32011 **5.** 0.99619
7. −0.10533 **9.** 0.42262 **11.** 1.54306

13. $I(x) = \sum_{n=0}^{\infty} \dfrac{(-1)^n}{(2n+1)(2n+1)!} x^{2n+1}, \quad$ all x

15. $K(x) = \sum_{n=0}^{\infty} \dfrac{(-1)^n}{(n+1)^2} x^{n+1}, \quad (-1 \le x \le 1)$

17. $M(x) = \sum_{n=0}^{\infty} \dfrac{(-1)^n}{(2n+1)(4n+1)} x^{4n+1}, \quad (-1 \le x \le 1)$

19. 0.946

21. 2 **23.** −3/25 **25.** 0

27. $y = \sum_{n=0}^{\infty} (-1)^n \left[\dfrac{2^n n!}{(2n)!} x^{2n} + \dfrac{1}{2^{n-1} n!} x^{2n+1}\right]$

CHAPTER 12. PARTIAL DIFFERENTIATION

Section 12.1 (page 524)

1. $\{(x,y) : x \ne y\}$ **3.** $\{(x,y) : (x,y) \ne (0,0)\}$
5. $\{(x,y) : x \ne \pm y\}$ **7.** $\{(x,y) : -1 < x + y < 1\}$
9. $z = f(x,y) = x$ **11.** $z = f(x,y) = y^2$

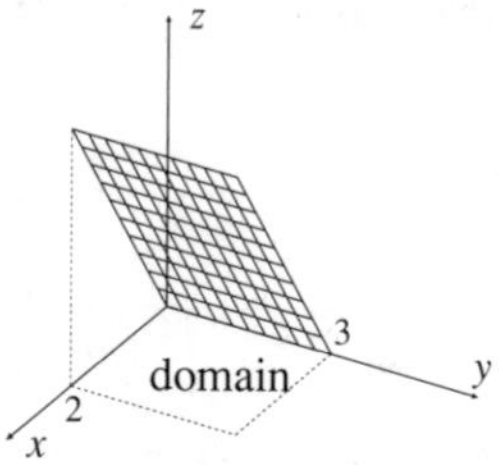

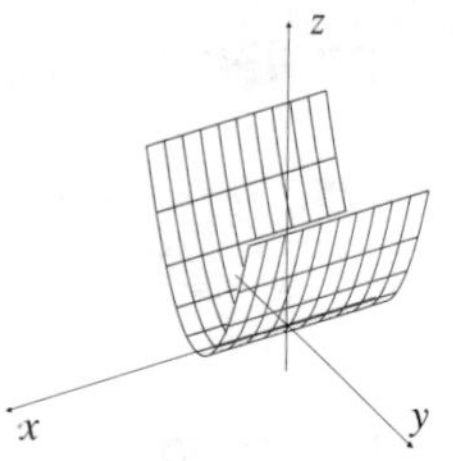

13. $z = f(x, y) = \sqrt{x^2 + y^2}$

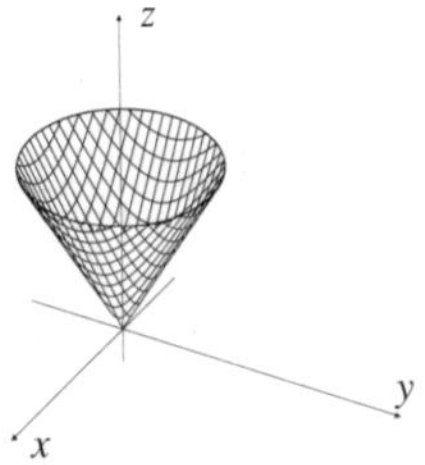

15. $z = f(x, y) = |x| + |y|$

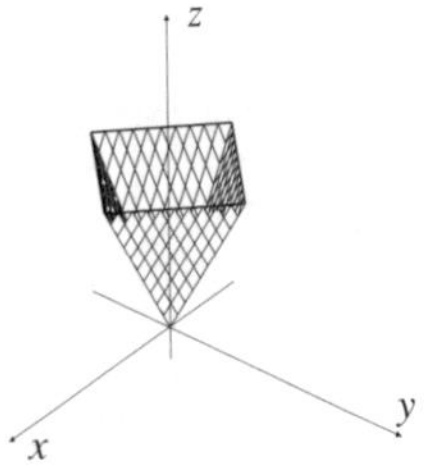

17. $z = f(x, y) = \dfrac{1}{x^2 + y^2}$

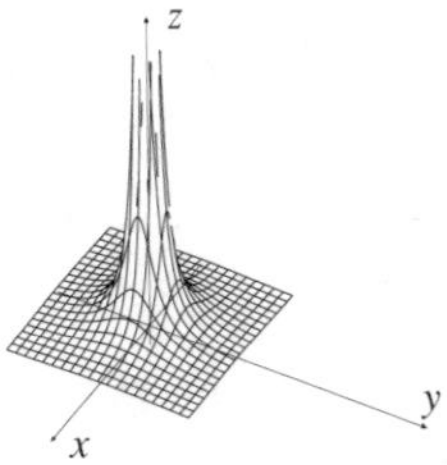

19. $f(x, y) = x - y = C$

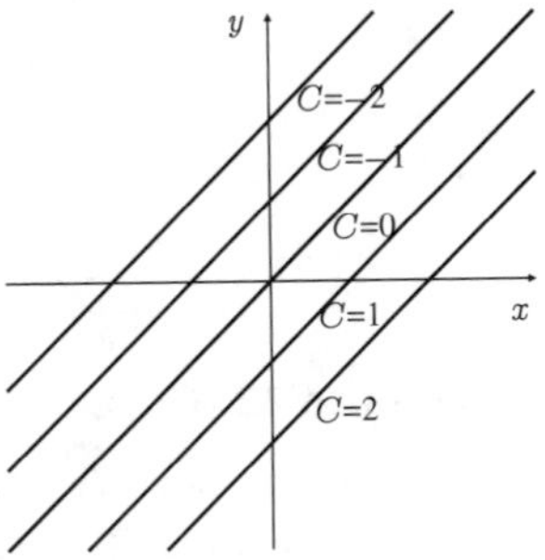

21. $f(x, y) = xy = C$

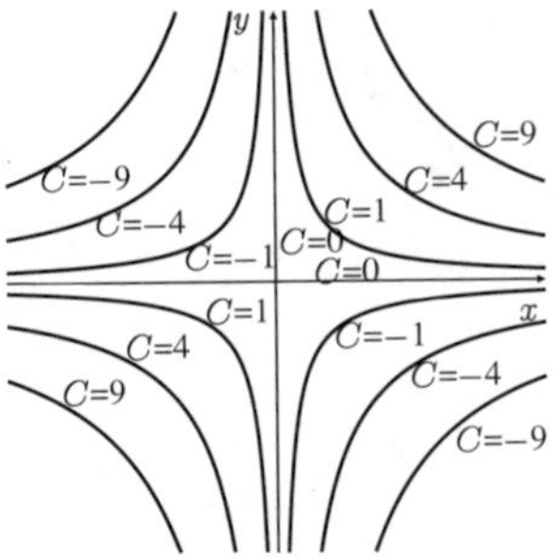

23. $f(x, y) = \dfrac{x - y}{x + y} = C$

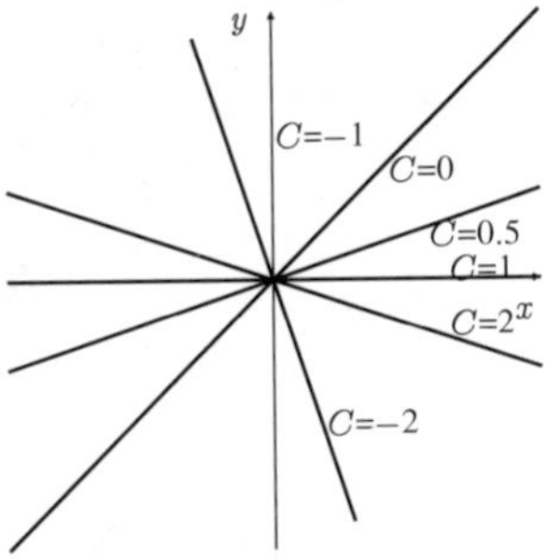

25. $f(x, y) = \sin(x + y) = C$

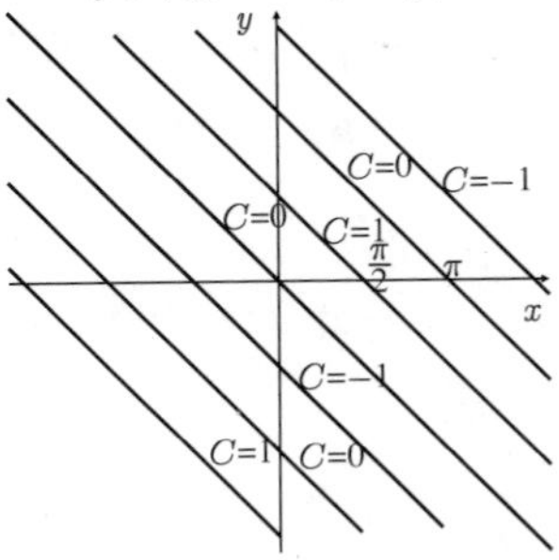

27. (a) $f(x, y) = \sqrt{x^2 + y^2}$, (b) $f(x, y) = (x^2 + y^2)^{1/4}$, (c) $f(x, y) = x^2 + y^2$, (d) $f(x, y) = e^{\sqrt{x^2+y^2}}$

29. does not exist

31. 0

33. 0

35. does not exist

37. domain $\{(x, y) : x \neq \pm y\}$; limit does not exist; define $f(x, x) = \dfrac{1}{2x}$ for $x \neq 0$

39. no; counterexample $f(x, y) = \begin{cases} \dfrac{2xy}{x^2 + y^2} & \text{if } (x, y) \neq (0, 0), \\ 0 & \text{if } (x, y) = (0, 0) \end{cases}$

Section 12.2 (page 532)

1. $f_1(x, y) = 1 = f_1(3, 2)$, $f_2(x, y) = -1 = f_2(3, 2)$

3. $f_1(x, y, z) = 3x^2y^4z^5$
$f_2(x, y, z) = 4x^3y^3z^5$
$f_3(x, y, z) = 5x^3y^4z^4$
all 0 at $(0, -1, -1)$

5. $\dfrac{\partial z}{\partial x} = -\dfrac{y}{x^2 + y^2}$
$\dfrac{\partial z}{\partial y} = \dfrac{x}{x^2 + y^2}$
$\dfrac{\partial z}{\partial x} = \dfrac{\partial z}{\partial y} = -\dfrac{1}{2}$ at $(-1, 1)$

7. $f_1(x, y) = \sqrt{y}\cos(x\sqrt{y})$
$f_2(x, y) = \dfrac{x}{2\sqrt{y}}\cos(x\sqrt{y})$
$f_1\left(\dfrac{\pi}{3}, 4\right) = -1$
$f_2\left(\dfrac{\pi}{3}, 4\right) = -\dfrac{\pi}{24}$

9. $\dfrac{\partial w}{\partial x} = y \ln z\, x^{y \ln z - 1}$
$\dfrac{\partial w}{\partial y} = \ln x \ln z\, x^{y \ln z}$
$\dfrac{\partial w}{\partial z} = \dfrac{y \ln x}{z} x^{y \ln z}$
at $(e, 2, e)$, $\dfrac{\partial w}{\partial x} = \dfrac{\partial w}{\partial z} = 2e$
$\dfrac{\partial w}{\partial y} = e^2$

11. $4x + 2y + z = -3$

13. $4x - \pi y + 16\sqrt{2}z = 16$

15. $3x - 4y - 25z = -10$

17. $2x - 4y - 5z = 10 - 5\ln 5$

19. $6x + 8y - 3z = 11$

25. $f_{11}(x, y) = f_{22}(x, y) = 2$,
$f_{12}(x, y) = f_{2,1}(x, y) = 0$

27. $\dfrac{\partial^2 z}{\partial x^2} = \dfrac{3y^2}{(3x^2 + y^2)^{3/2}}$
$\dfrac{\partial^2 z}{\partial y^2} = \dfrac{3x^2}{(3x^2 + y^2)^{3/2}}$
$\dfrac{\partial^2 z}{\partial x \partial y} = \dfrac{\partial^2 z}{\partial y \partial x}$
$= -\dfrac{3xy}{(3x^2 + y^2)^{3/2}}$

29. $f_{11}(x, y) = -\dfrac{y^2}{1 + \sin xy}$
$f_{22}(x, y) = -\dfrac{x^2}{1 + \sin xy}$
$f_{12}(x, y) = f_{21}(x, y)$
$= \dfrac{\cos(xy) - xy}{1 + \sin xy}$

31. 27; 10; $f_{133} = f_{313} = f_{331}$
$= -3x^2e^{xy}\cos xz - x^3ye^{xy}\cos xz + x^3ze^{xy}\sin xz$

41. three dimensional solution
$u(x, y, z, t) = t^{-3/2}\, e^{-(x^2+y^2+z^2)/4t}$

Section 12.3 (page 544)

1. $\dfrac{\partial w}{\partial t} = f_1(x, y, z)g_2(s, t) + f_2(x, y, z)h_2(s, t)$
$+ f_3(x, y, z)k_2(s, t)$

3. $\dfrac{\partial z}{\partial u} = g_1(x, y)h_1(u, v) + g_2(x, y)f'(x)h_1(u, v)$

5. $\left.\dfrac{\partial w}{\partial x}\right|_z = f_1(x, y, z) + f_2(x, y, z)g_1(x, z)$

7. (a) $\left.\dfrac{\partial w}{\partial r}\right|_{x,s} = f_2(x,y,z)h_2(x,r,s)+f_3(x,y,z)k_2(x,r,s)$

(b) $\left.\dfrac{\partial w}{\partial r}\right|_{y,s} = f_1(x,y,z)g_1(r,s)$
$+ f_3(x,y,z)\big(k_2(x,r,s)+k_1(x,r,s)g_1(r,s)\big)$

(c) $\left.\dfrac{\partial w}{\partial r}\right|_{x,y,s} = f_3(x,y,z)\big(k_2(x,r,s)$

(d) $\left.\dfrac{\partial w}{\partial r}\right|_{s} = f_1(x,y,z)g_1(r,s)$
$+ f_2(x,y,z)\big(h_2(x,r,s)+h_1(x,r,s)g_1(r,s)\big)$
$+ f_3(x,y,z)\big(k_2(x,r,s)+k_1(x,r,s)g_1(r,s)\big)$

9. $\dfrac{2v-3u}{u^2+v^2} = -\dfrac{5y}{13x^2-2xy+y^2}$

11. $2vf_2(u^2,v^2)$

13. $f_1\big(f(x,y),f(x,y)\big)f_1(x,y)+f_2\big(f(x,y),f(x,y)\big)f_1(x,y)$

15. $e^{-t}(f'(t)-f(t))$; 0, because rate of increase of temperature as depth increases is balanced by rate of decrease of temperature as time increases

17. (a) $4f_{11}+12f_{12}+9f_{22}$, (b) $6f_{11}+5f_{12}-6f_{22}$,
(c) $9f_{11}-12f_{12}+4f_{22}$,
(d) $8f_{111}+36f_{112}+54f_{122}+27f_{222}$,
where all derivatives of f are evaluated at (x,y);
$\dfrac{\partial^n z}{\partial s^n} = \displaystyle\sum_{k=0}^{n} \frac{n!2^{n-k}3^k}{k!(n-k)!}\frac{\partial^n}{\partial x^{n-k}\partial y^k}f(x,y).$

19. $4stf_{21}(s^2,s+t^2)+2tf_{22}(s^2,s+t^2)$

21. $(\cos s)f_1-(\sin s)f_2+(\cos s\sin s)(f_{11}-f_{22})$
$+(\cos^2 x-\sin^2 s)f_{12}$,
where the partials of f are all evaluated at the point $(t\sin s, t\cos s)$

27. 2.1 **29.** increases by $\approx 0.16\%$

Section 12.4 (page 553)

1. $2x\mathbf{i}-2y\mathbf{j}$; $-4\mathbf{i}-2\mathbf{j}$

3. $-\sin\left(\dfrac{x}{y}\right)\left[\dfrac{1}{y}\mathbf{i}-\dfrac{x}{y^2}\mathbf{j}\right]$; $\dfrac{1}{16\sqrt{2}}(-4\mathbf{i}+\pi\mathbf{j})$

5. $\dfrac{1}{(x^2+y^2)^2}\big((y^2-x^2)\mathbf{i}-2xy\mathbf{j}\big)$; $\dfrac{3}{25}\mathbf{i}-\dfrac{4}{25}\mathbf{j}$

7. $\dfrac{2(y^2-x^2)}{(x^2+y^2)^2}(y\mathbf{i}-x\mathbf{j})$; $\mathbf{i}$

9. $(2xy+z^2)\mathbf{i}+(2yz+x^2)\mathbf{j}+(2zx+y^2)\mathbf{k}$,
$-\mathbf{i}-\mathbf{j}+3\mathbf{k}$

11. -3 **13.** $1/\sqrt{2}$ **15.** $-2/5$

17. (a) $\mathbf{i}+2\mathbf{j}$, (b) $-\mathbf{i}-2\mathbf{j}$,
(c) $\pm(2\mathbf{i}-\mathbf{j})$, (d) $\mathbf{j}$ or $4\mathbf{i}+3\mathbf{j}$

19. $\mathbf{i}+2\mathbf{j}+3\mathbf{k}$ **21.** $2x+y=-3$

23. $4x-\pi y=0$ **25.** $3x-4y=-5$

27. $x+y-3z=-3$

29. if $\nabla f(a,b)=\mathbf{0}$ the level curve may not be smooth at (a,b)

31. $7\mathbf{i}-\mathbf{j}$

Section 12.5 (page 563)

1. $(2,-1)$ local (and absolute) minimum

3. $(0,0)$ saddle point, $(1,1)$ loc min

5. $(-4,2)$ loc max

7. $(0,n\pi)$ (n an integer) all saddle points

9. $(0,0)$ saddle point, $(1,1)$ and $(-1,-1)$ loc (abs) max, $(1,-1)$ and $(-1,1)$ loc (abs) min

11. max $\sqrt{5}$ at $\left(\dfrac{1}{\sqrt{5}},\dfrac{2}{\sqrt{5}}\right)$, min $-\sqrt{5}$ at $\left(-\dfrac{1}{\sqrt{5}},-\dfrac{2}{\sqrt{5}}\right)$

13. max $\dfrac{2}{3\sqrt{3}}$ at $\left(\dfrac{1}{\sqrt{3}},1\right)$, min 0 on $x=0$, $y=0$ and at $(1,1)$

15. $\sqrt{5}$ units

17. max 5 at $(2,-1)$, min -5 at $(-2,1)$

19. nearest $\pm(\sqrt{2},\sqrt{2})$, farthest $\pm(2,-2)$, major axis $y=-x$, minor axis $y=x$

21. 2 units **23.** 32 cubic units

25. equality for equilateral triangles

27. $(0,0,0)$ loc max, $(2,2,2)$, $(-2,-2,2)$, $(-2,2,-2)$ and $(2,-2,-2)$ all saddle points

INDEX

INTEGRALS INVOLVING $\sqrt{x^2 \pm a^2}$

$$\int \sqrt{x^2 \pm a^2}\, dx = \frac{x}{2}\sqrt{x^2 \pm a^2} \pm \frac{a^2}{2} \ln |x + \sqrt{x^2 \pm a^2}| + C$$

$$\int \frac{dx}{\sqrt{x^2 \pm a^2}} = \ln |x + \sqrt{x^2 \pm a^2}| + C$$

$$\int \frac{\sqrt{x^2 + a^2}}{x}\, dx = \sqrt{x^2 + a^2} - a \ln \left(\frac{a + \sqrt{x^2 + a^2}}{x} \right) + C$$

$$\int \frac{\sqrt{x^2 - a^2}}{x}\, dx = \sqrt{x^2 - a^2} - a \sec^{-1} \frac{x}{a} + C$$

$$\int x^2 \sqrt{x^2 \pm a^2}\, dx = \frac{x}{8}(2x^2 \pm a^2)\sqrt{x^2 \pm a^2} - \frac{a^4}{8} \ln |x + \sqrt{x^2 \pm a^2}| + C$$

$$\int \frac{x^2}{\sqrt{x^2 \pm a^2}}\, dx = \frac{x}{2}\sqrt{x^2 \pm a^2} \mp \frac{a^2}{2} \ln |x + \sqrt{x^2 \pm a^2}| + C$$

$$\int \frac{\sqrt{x^2 \pm a^2}}{x^2}\, dx = -\frac{\sqrt{x^2 \pm a^2}}{x} + \ln |x + \sqrt{x^2 \pm a^2}| + C$$

$$\int \frac{dx}{x^2\sqrt{x^2 \pm a^2}} = \mp \frac{\sqrt{x^2 \pm a^2}}{a^2 x} + C$$

$$\int \frac{dx}{(x^2 \pm a^2)^{3/2}} = \frac{\pm x}{a^2\sqrt{x^2 \pm a^2}} + C$$

$$\int (x^2 \pm a^2)^{3/2}\, dx = \frac{x}{8}(2x^2 \pm 5a^2)\sqrt{x^2 \pm a^2} + \frac{3a^4}{8} \ln |x + \sqrt{x^2 \pm a^2}| + C$$

INTEGRALS INVOLVING $\sqrt{a^2 - x^2}$

$$\int \sqrt{a^2 - x^2}\, dx = \frac{x}{2}\sqrt{a^2 - x^2} + \frac{a^2}{2} \sin^{-1} \frac{x}{a} + C$$

$$\int \frac{\sqrt{a^2 - x^2}}{x}\, dx = \sqrt{a^2 - x^2} - a \ln \left| \frac{a + \sqrt{a^2 - x^2}}{x} \right| + C$$

$$\int \frac{x^2}{\sqrt{a^2 - x^2}}\, dx = -\frac{x}{2}\sqrt{a^2 - x^2} + \frac{a^2}{2} \sin^{-1} \frac{x}{a} + C$$

$$\int x^2\sqrt{a^2 - x^2}\, dx = \frac{x}{8}(2x^2 - a^2)\sqrt{a^2 - x^2} + \frac{a^4}{8} \sin^{-1} \frac{x}{a} + C$$

$$\int \frac{dx}{x^2\sqrt{a^2 - x^2}} = -\frac{\sqrt{a^2 - x^2}}{a^2 x} + C$$

$$\int \frac{\sqrt{a^2 - x^2}}{x^2}\, dx = -\frac{\sqrt{a^2 - x^2}}{x} - \sin^{-1} \frac{x}{a} + C$$

$$\int \frac{dx}{x\sqrt{a^2 - x^2}} = -\frac{1}{a} \ln \left| \frac{a + \sqrt{a^2 - x^2}}{x} \right| + C$$

$$\int \frac{dx}{(a^2 - x^2)^{3/2}} = \frac{x}{a^2\sqrt{a^2 - x^2}} + C$$

$$\int (a^2 - x^2)^{3/2}\, dx = \frac{x}{8}(5a^2 - 2x^2)\sqrt{a^2 - x^2} + \frac{3a^4}{8} \sin^{-1} \frac{x}{a} + C$$

INTEGRALS OF INVERSE TRIGONOMETRIC FUNCTIONS

$$\int \sin^{-1} x\, dx = x \sin^{-1} x + \sqrt{1 - x^2} + C$$

$$\int \tan^{-1} x\, dx = x \tan^{-1} x - \frac{1}{2} \ln(1 + x^2) + C$$

$$\int \sec^{-1} x\, dx = x \sec^{-1} x - \ln |x + \sqrt{x^2 - 1}| + C$$

$$\int x \sin^{-1} x\, dx = \frac{1}{4}(2x^2 - 1)\sin^{-1} x + \frac{x}{4}\sqrt{1 - x^2} + C$$

$$\int x \tan^{-1} x\, dx = \frac{1}{2}(x^2 + 1)\tan^{-1} x - \frac{x}{2} + C$$

$$\int x \sec^{-1} x\, dx = \frac{x^2}{2} \sec^{-1} x - \frac{1}{2}\sqrt{x^2 - 1} + C$$

$$\int x^n \sin^{-1} x\, dx = \frac{x^{n+1}}{n+1} \sin^{-1} x - \frac{1}{n+1} \int \frac{x^{n+1}}{\sqrt{1 - x^2}}\, dx + C \text{ if } n \neq -1$$

$$\int x^n \tan^{-1} x\, dx = \frac{x^{n+1}}{n+1} \tan^{-1} x - \frac{1}{n+1} \int \frac{x^{n+1}}{1 + x^2}\, dx + C \text{ if } n \neq -1$$

$$\int x^n \sec^{-1} x\, dx = \frac{x^{n+1}}{n+1} \sec^{-1} x - \frac{1}{n+1} \int \frac{x^n}{\sqrt{x^2 - 1}}\, dx + C \text{ if } n \neq -1$$

EXPONENTIAL AND LOGARITHMIC INTEGRALS

$$\int xe^x\, dx = (x - 1)e^x + C$$

$$\int x^n e^x\, dx = x^n e^x - n \int x^{n-1} e^x\, dx$$

$$\int \ln x\, dx = x \ln x - x + C$$

$$\int x^n \ln x\, dx = \frac{x^{n+1}}{n+1} \ln x - \frac{x^{n+1}}{(n+1)^2} + C$$

$$\int x^n (\ln x)^m\, dx = \frac{x^{n+1}}{n+1} (\ln x)^m - \frac{m}{n+1} \int x^n (\ln x)^{m-1}\, dx$$

$$\int e^{ax} \sin bx\, dx = \frac{e^{ax}}{a^2 + b^2}(a \sin bx - b \cos bx) + C$$

$$\int e^{ax} \cos bx\, dx = \frac{e^{ax}}{a^2 + b^2}(a \cos bx + b \sin bx) + C$$

INTEGRALS OF HYPERBOLIC FUNCTIONS

$$\int \sinh x\, dx = \cosh x + C$$

$$\int \cosh x\, dx = \sinh x + C$$

$$\int \tanh x\, dx = \ln(\cosh x) + C$$

$$\int \coth x\, dx = \ln |\sinh x| + C$$

$$\int \operatorname{sech} x\, dx = \tan^{-1} |\sinh x| + C$$

$$\int \operatorname{csch} x\, dx = \ln \left| \tanh \frac{x}{2} \right| + C$$

$$\int \sinh^2 x\, dx = \frac{1}{4} \sinh 2x - \frac{x}{2} + C$$

$$\int \cosh^2 x\, dx = \frac{1}{4} \sinh 2x + \frac{x}{2} + C$$

$$\int \tanh^2 x\, dx = x - \tanh x + C$$

$$\int \coth^2 x\, dx = x - \coth x + C$$

$$\int \operatorname{sech}^2 x\, dx = \tanh x + C$$

$$\int \operatorname{csch}^2 x\, dx = -\coth x + C$$

$$\int \operatorname{sech} x \tanh x\, dx = -\operatorname{sech} x + C$$

$$\int \operatorname{csch} x \coth x\, dx = -\operatorname{csch} x + C$$

P9-CSB-352

F	future value, future worth
f	inflation rate per year
FW	future worth
g	growth rate for geometric gradient
i	actual interest rate
I	interest amount
i'	real interest rate
I_c	compound interest amount
i_e	effective interest rate
i_s	interest rate per sub-period
I_s	simple interest amount
$i°$	growth adjusted interest rate
IRR	internal rate of return
$\mathbf{IRR_A}$	actual dollar IRR
$\mathbf{IRR_R}$	real dollar IRR
i^*	internal rate of return
i^*_e	external rate of return
i^*_{ea}	approximate external rate of return
$I_{0,N}$	the value of a global price index at year N, relative to year 0
m	number of sub-periods in a period
MARR	minimum acceptable rate of return
$\mathbf{MARR_A}$	actual dollar MARR
$\mathbf{MARR_R}$	real dollar MARR
MAUT	multi-attribute utility theory
MCDM	multi-criterion decision making
N	number of periods, useful life of an asset
P	present value, present worth, purchase price, principal amount
PCM	pairwise comparison matrix
PW	present worth
$p(x)$	probability distribution
$\Pr\{X = x_i\}$	alternative expression of probability distribution
r	nominal interest rate, rating for a decision matrix
$R_{0,N}$	real dollar equivalent to A_N relative to year 0, the base year
RI	random index
S	salvage value
t	tax rate
UCC	undepreciated capital cost
X	random variable
$\mathbf{w}$	an eigenvector
λ_{max}	the maximun eigenvalue
λ	an eigenvalue
π_{01}	Laspeyres price index

Engineering Economics in Canada

Engineering Economics in Canada

Second Edition

Niall M. Fraser
University of Waterloo

Irwin Bernhardt
University of Waterloo

Elizabeth M. Jewkes
University of Waterloo

May Tajima
University of Waterloo

Prentice Hall Canada Inc.
Scarborough, Ontario

Canadian Cataloguing in Publication Data

Main entry under title:

Engineering economics in Canada

Previous edition by Niall M. Fraser, Irwin Bernhardt, Elizabeth M. Jewkes.
Includes index.
ISBN 0-13-013844-4

1. Engineering economy — Canada. I. Fraser, Niall M. (Niall Morris), 1952– .

TA177.4.F725 2000 658.15 C99-930501-8

Prentice-Hall, Inc., Upper Saddle River, New Jersey
Prentice-Hall International (UK) Limited, London
Prentice-Hall of Australia, Pty. Limited, Sydney
Prentice-Hall Hispanoamericana, S.A., Mexico City
Prentice-Hall of India Private Limited, New Delhi
Prentice-Hall of Japan, Inc., Tokyo
Simon & Schuster Southeast Asia Private Limited, Singapore
Editora Prentice-Hall do Brasil, Ltda., Rio de Janeiro

ISBN 0-13-013844-4

Vice President and Editorial Director: Patrick Ferrier
Acquisitions Editor: Dave Ward
Developmental Editor: Maurice Esses
Marketing Manager: Don Thompson
Production Editor: Nicole Mellow
Copy Editor: Rosemary Tanner
Production Coordinator: Kathrine Pummell
Art Director: Mary Opper
Cover Image: Courtesy of the Ontario Power Generation Corp.
Cover and Interior Design: Lisa LaPointe
Page Layout and Illustrator: Nelson Gonzalez

1 2 3 4 5 04 03 02 01 00

Printed and bound in Canada.

Contents

CHAPTER 3: Cash Flow Analysis *53*

CHAPTER 4: Comparison Methods Part 1 *103*

Preface

Courses on engineering economics are traditionally found in engineering curricula in Canada and throughout the world. The courses generally deal with deciding among alternatives with respect to expected costs and benefits. In Canada, the Canadian Engineering Accreditation Board requires that all accredited Professional Engineering programs provide at least one course in engineering economics. Many engineers have found that a course in engineering economics can be as useful in their practice as any of their more technical courses.

There are several stages to making a good decision. One stage is being able to determine whether a solution to a problem is technically feasible. This is one of the roles of the engineer, who has specialized training to make such technical judgments. Another stage is deciding which of several technically feasible alternatives is best. Deciding among alternatives often does not require the technical competence needed to determine which alternatives are feasible, but it is equally important in making the final choice. Some engineers have found that choosing among alternatives can be more difficult than deciding what alternatives exist.

The role of engineers in Canadian society is changing. In the past, engineers tended to have a fairly narrow focus, concentrating on the technical aspects of a problem and on strictly computational aspects of engineering economics. As a result, many engineering economics texts focused on the mathematics of the subject. Today, engineers are more likely to be the decision makers, and they need to be able to take into account strategic and policy issues.

This book is designed for teaching a course on engineering economics to match engineering practice in Canada today. It recognizes the role of the engineer as a decision maker who has to make and defend sensible decisions. Such decisions must not only take into account a correct assessment of costs and benefits. They must also reflect an understanding of the environment in which the decisions are made.

Canadian engineers have a unique set of circumstances that warrant a text with a specific Canadian focus. Canadian firms make decisions according to norms and standards that reflect Canadian views on social responsibility, environmental concerns, and cultural diversity. This perspective is reflected in the content and tone of much of the material in this book. Furthermore, Canadian tax regulations are complicated and directly affect engineering economic analysis. These regulations and their effect on decision making are covered in detail in Chapter 8.

This book also relates to students' everyday lives. In addition to examples and problems with an engineering focus, there are a number that involve decisions that many students might face, such as renting an apartment, getting a job, or buying a car. Other examples in the text are adapted from familiar sources such as Canadian newspapers and well-known Canadian companies.

Content and Organization

Because the mathematics of finance has not changed dramatically over the past number of years, there is a natural order to the course material. Nevertheless, a modern view of the role of the engineer flavours this entire book and provides a new, balanced exposure to the subject.

Chapter 1 frames the problem of engineering decision making as one involving many issues. Manipulating the cash flows associated with an engineering project is an important process for which useful mathematical tools exist. These tools form the bulk of the remaining chapters. However, throughout the text, students are kept aware of the fact that the eventual decision depends not only on the cash flows, but also on less easily quantifiable considerations of business policy, social responsibility, and ethics.

Chapters 2 and 3 present tools for manipulating monetary values over time. Chapters 4 and 5 show how the students can use their knowledge of manipulating cash flows to make comparisons among alternative engineering projects. Chapter 6 provides an understanding of the environment in which the decisions are made by examining depreciation and the role it plays in the financial functioning of a company and in financial accounting.

Chapter 7 deals with the analysis of replacement decisions. Chapters 8 and 9 are concerned with taxes and inflation, which affect decisions based on cash flows. Chapter 10 provides an introduction to public sector decision making.

Most engineering projects involve estimates of future cash flows. Since the future is unknown, it is important for engineers to take uncertainty and risk into account as completely as possible. Chapter 11 deals with uncertainty, with a focus on sensitivity analysis. Chapter 12 deals with risk, using some of the tools of probability analysis.

Chapter 13 picks up an important thread running throughout the book: a good engineering decision cannot be based only on selecting the least-cost alternative. The increasing influence on decision making of health and safety issues, environmental responsibility, and human relations, among others, makes it necessary for the engineer to understand some of the basic principles of multicriteria decision making.

New to this Edition

In addition to clarifying explanations, improving readability, updating all the material, and correcting errors, we have made the following important changes for the second edition:

- An entirely new Chapter 12 has been added to introduce students to some of the tools of probability analysis that can be applied to decision making under risk. Chapter 12 (Dealing with Risk: Probability Analysis) provides enough basic probability theory to introduce the practical tools of decision trees and Monte Carlo simulation into an engineering economic analysis where risk is a component of the decision process. Thus, students in the first or second year of an engineering degree who have had little or no exposure to probability theory will be able follow the discussion. At the same time, the material on probability theory is concise enough to serve as a helpful review for students in the third or fourth year.
- The old chapter on sensitivity analysis has been shifted and revised to form the new Chapter 11 (Dealing with Uncertainty: Sensitivity Analysis).
- A second Extended Case has been added directly following Chapter 11. Like the first Extended Case (directly following Chapter 6), this new case provides a comprehensive problem-solving experience for students and requires them to bring together material covered over several chapters.
- The number of Problems at the end of each chapter has been dramatically increased. They provide important practice at various levels of difficulty.
- A Canadian Mini-Case has been added at the end of each chapter. Each of these cases includes open-ended discussion questions concerning the application of engineering economics in real Canadian scenarios. They also provide an opportunity to consider engineering economics as more than simply solving numerical questions.
- Finally, a new design has been created for this edition, making it more accessible for today's student.

Features

We have created special features for this book in order to facilitate the learning of the material and an understanding of its applications:

- Each chapter begins with a list of the major sections to provide an overview of the material that follows.
- Engineering in Action Boxes near the beginning and end of each chapter recount the experiences of a young engineer at a Canadian company. These fictional vignettes reflect and support the chapter material. The first Box in each chapter usually portrays one of the characters trying to deal with a practical problem. The second Box, near the end of the chapter, demonstrates how the character has solved the problem by applying material discussed in the chapter. All these vignettes are linked to form a narrative that runs throughout the book. The main character is Naomi, a recent engineering graduate. In the first chapter, she starts her job in the engineering department at Canadian Widget Industries and is given a decision problem by her supervisor. Over the course of the book, Naomi learns about engineering economics on the job as the students learn from the book. There

are several characters that relate to one another in various ways, exposing the students to practical, ethical, and social issues as well as mathematical problems.

- Key terms are boldfaced where they are defined in the body of the text. For easy reference, all the key terms with their definitions are collated in a Glossary near the back of the book.
- Close-Up Boxes in the chapters present additional material about concepts that are important but that are not essential to the chapter.
- Other additional material is presented in Appendices at the ends of Chapters 3, 4, 5, 8, 9, and 13.
- Numerous worked-out Examples are given throughout the chapters. Although the decisions have often been simplified for clarity, most of the examples are based on real situations encountered in the authors' consulting experiences.
- Worked-out Review Problems near the end of each chapter provide more complex examples that integrate the chapter material.
- A concise prose Summary is given for each chapter.
- Each chapter has 30 to 50 Problems of various levels of difficulty covering all of the material presented. Like the worked-out Examples, many of the problems have been adapted from real situations.
- At the end of each chapter, a Canadian Mini-Case, complete with Discussion Questions, relates interesting stories about how familiar Canadian companies have used engineering economic principles in practice.
- Two Extended Cases are provided—one directly following Chapter 6 and the other directly following Chapter 11. They concern complex situations that incorporate much of the material in the preceding chapters. Unlike chapter examples, which are usually directed at a particular concept being taught, the Extended Cases require the students to integrate what they have learned over a number of chapters. The Extended Cases can be used for assignments, class discussions, or independent study.
- A special spreadsheet icon indicates where Examples or Problems involve spreadsheets, which are available on disks packaged with the *Instructor's Solutions Manual*. The use of computers by engineers is now as commonplace as the use of slide rules was thirty years ago. Students using this book will likely be very familiar with spreadsheet software. Consequently, such knowledge is assumed rather than taught in this book. The spreadsheet Examples and Problems are presented in such a manner that they can be done using any popular spreadsheet program, such as Excel, Lotus 1-2-3, or Quattro Pro.
- Tables of Interest Factors are provided in Appendix A, Appendix B, and Appendix C.
- Answers to Selected Problems are provided in Appendix D.
- For convenience, a List of Symbols used in the book is given on the inside of the front cover, and a List of Formulas is given on the inside of the back cover.

Course Designs

This book is ideal for a one-term course, but it can also be used for a two-term course with supplemental material. It is intended to meet the needs of students in all engineering programs, including, but not limited to, Aeronautical, Chemical, Computer, Electrical, Industrial, Mechanical, Mining, and Systems Engineering. Certain programs emphasizing public projects may wish to supplement Chapter 10 on Public Projects with additional material.

A course based on this book can be taught in the first, second, third, or fourth year of an engineering program. The book is also suitable for college technology programs. No more than high-school mathematics is required as a prerequisite for a course based on this text. The probability theory required to understand and apply the tools of risk analysis are provided in Chapter 12. Prior knowledge of calculus or linear algebra is not needed, except for working through the appendix to Chapter 13.

This book is also suitable for self-study by a practitioner or anybody interested in the economic aspects of decision making. It is easy to read and self contained, with many clear examples. It can serve as a permanent resource for practising engineers or anyone involved in decision making.

Supplement

An *Instructor's Solution Manual with Transparency Masters and Two Spreadsheet Disks* is available to facilitate instruction. It includes complete solutions for all the problems in the text, as well as for the Extended Cases following Chapters 6 and 11. It also provides printouts of any spreadsheets involved in the solutions. The Transparency Masters comprise all the figures and many of the tables that occur in the text. The two Spreadsheet Disks have been prepared in Excel format, but they can be read and used by most popular spreadsheet programs (such as Lotus 1-2-3 and Quattro Pro). One disk contains spreadsheets from the body of the text; the other disk has those that can be used to solve some of the end-of-chapter problems. Each spreadsheet is identified in the textbook by a special spreadsheet icon.

Acknowledgments

The authors wish to acknowledge the contributions of a number of individuals who assisted in the development of this text. First and foremost are the hundreds of engineering students at the University of Waterloo who have provided us with feedback on passages they found hard to understand, typographical errors, and examples that they thought could be improved. There are too many individuals to name in person, but we are very thankful to each of them for their patience and diligence.

We are very grateful for the constructive suggestions provided by David Fuller and John Moore as they taught Engineering Economics to engineering students in the first, second, and third years at the University of Waterloo. We would also like to thank Tim Nye for his suggestions on the drafts leading up to the first edition and for his continued feedback on the changes for the second edition, especially his contributions to the second extended case.

During the development process for the new edition, Prentice-Hall Canada arranged for the anonymous review of parts of the manuscript by instructors of engineering economics courses at other Canadian universities. These reviews were extremely beneficial to us, and many of the best ideas incorporated in the final text originated with these reviewers. We can now thank them by name:

Francis Hartman (University of Calgary)
Isobel Heathcote (University of Guelph)
Quinn Hoang, CGA
J. Charles Lemmon (The University of Western Ontario)
John Morrall (University of Calgary)
Leo Michelis (Ryerson Polytechnic University)
John Whittaker (University of Alberta)

Finally, we want to express our appreciation to the various editors at Prentice-Hall Canada for their professionalism and support during the writing of this book. Maurice Esses, our Developmental Editor, had a particularly strong role in bringing the projects of both the first and second editions to completion.

To all of the above, thank you again for your help. To those of you we forgot to thank: our appreciation is just as great even if our memories fail us. And to the reader, any errors that remain cannot be blamed on those who helped us. The remaining errors are perhaps the only things for which we can claim sole credit.

Niall M. Fraser
Irwin Bernhardt
Elizabeth M. Jewkes
May Tajima

CHAPTER 1

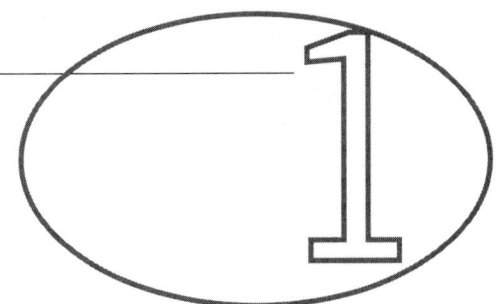

Engineering Decision Making

Engineering Economics in Action, Part 1A: Naomi Arrives

Naomi's first day on the job wasn't really her first day on the job. Ever since she had received the acceptance letter three weeks earlier, she had been reading and rereading all her notes about the company. Somehow she had arranged to walk past the plant entrance going on errands that never would have taken her that exact route in the past. So today wasn't the first time she had walked through that tidy brick entrance to the main offices of Canadian Widget Industries—she had done it the same way in her imagination a hundred times before.

Clement Sheng, the Engineering Manager who had interviewed Naomi for the job, was waiting for her at the reception desk. His warm smile and easy manner did a lot to break the ice. He suggested that they could go through the plant on the way to her desk. She agreed enthusiastically. "I hope you remember the engineering economics you learned in school," he said.

Naomi did, but rather than sound like a know-it-all, she replied, "I think so, and I still have my old textbook. I suppose you're telling me I'm going to use it."

"Yes. That's where we'll start you out, anyhow. It's a good way for you to learn how things work around here. We've got some projects lined up for you already, and they involve some pretty big decisions for Canadian Widgets. We'll keep you busy."

1.1 Engineering Decision Making

Engineering is a noble profession with a long history. The first engineers supported the military, using practical know-how to build bridges, fortifications, and assault equipment. In fact, the term "civil" engineer was coined to make the distinction between engineers who worked on civilian projects and engineers who worked on military problems.

In the beginning, all that engineers had to know was the technical aspects of their jobs. Military commanders, for example, would have wanted a strong bridge built quickly. The engineer would be challenged to find a solution to the technical problem, and would not have been particularly concerned about the costs, safety, or environmental impacts of the project. As years went by, however, the engineer's job became far more complicated.

All engineering projects use resources, such as raw materials, money, labour, and time. Any particular project can be undertaken in a variety of ways, each one calling for a different mix of resources. For example, a standard light bulb requires inexpensive raw materials and little labour, but is inefficient in its use of electricity and does not last very long. On the other hand, a high-efficiency light bulb uses more expensive raw materials and is more expensive to manufacture, but consumes less electricity and lasts longer. Both products provide light, but choosing which is better in a particular situation depends on how the costs and benefits are compared.

Historically, as the kinds of projects engineers worked on evolved and technology provided more than one way of solving technical problems, engineers were faced more often with having to choose among alternative solutions to a problem. If two solutions both dealt with a problem effectively, clearly the cheaper of the two was preferred. The practical science of engineering economics was originally developed specifically to deal with determining which of several alternatives was, in fact, the cheapest.

Choosing the cheapest alternative, though, is not the entire story. Though a project might be technically feasible and the cheapest solution to a problem, if the money isn't available to do it, it can't be done. The engineer has to become aware of the *financial* constraints on the problem, particularly if resources are very limited. In addition, an engineering project can meet all other criteria, but may cause detrimental *environmental* effects. Finally, any project can be affected by *social* and *political* constraints. For example, a large irrigation project called the Garrison Diversion Unit in North Dakota was effectively cancelled because of political action by Canadians and environmental groups, even though over $2 000 000 000 had been spent.

Engineers today must make decisions in an extremely complex environment. The heart of an engineer's skill set is still technical competence in a particular field. This permits the determination of possible solutions to a problem. However, necessary to all engineering is the ability to choose among several technically feasible solutions and to defend that choice credibly. The skills permitting the selection of a good choice are common to all engineers and, for the most part, are independent of which engineering field is involved. These skills form the discipline of engineering economics.

1.2 What Is Engineering Economics?

Just as the role of the engineer in society has changed over the years, so has the nature of engineering economics. Originally, engineering economics was the body of knowledge that allowed the engineer to determine which of several alternatives was economically best—the cheapest, or perhaps the most profitable. In order to make this determination properly, the engineer needed to understand the mathematics governing the relationship between time and money. Most of this book deals with teaching and using this knowledge. Also, for many kinds of decisions the costs and benefits are the most important factors affecting the decision, so concentrating on determining the economically "best" alternative is appropriate.

In earlier times, an engineer would be responsible for making a recommendation based on technical and analytic knowledge, including the knowledge of engineering economics, and then a manager would decide what should be done. A manager's decision would often be different from the engineer's recommendation because the manager would take into account issues outside the engineer's range of expertise. Recently, however, the trend has been for managers to become more reliant on the technical skills of the engineers, or the engineers are themselves the managers. Products are often very complex; manufacturing processes are fine-tuned to optimize productivity;

and even understanding the market sometimes requires the analytic skills of an engineer. As a result, it is often only the engineer who has sufficient depth of knowledge to make a competent decision.

Consequently, understanding how to compare costs, although still of vital importance, is not the only skill needed to make suitable engineering decisions. One must also be able to take into account all the other considerations that affect a decision, and to do so in a reasonable and defensible manner.

Engineering economics then can be defined as that science which deals with techniques of quantitative analysis useful for selecting a preferable alternative from several technically viable ones.

The evaluation of costs and benefits is very important and has formed the primary content of engineering economics in the past. The mathematics for doing this evaluation is well developed, and it still forms the bulk of studies of engineering economics. However, the modern engineer must be able to recognize the limits and applicability of these economic calculations, and must be able to take into account the inherent complexity of the real world.

1.3 Making Decisions

All decisions, except perhaps the most routine and automatic ones, or those that are institutionalized in large organizations, are made, in the end, on the basis of belief as opposed to logic. People, even highly trained engineers, do what *feels* like the right thing to do. This is not to suggest that one should trust only one's intuition and not one's intellect, but rather to point out something true about human nature and the function of engineering economics studies.

Figure 1.1 is a useful illustration of how decisions are made. At the top of the pyramid are preferences, which directly control the choices made. Preferences are the beliefs about what is best, and are often hard to explain coherently. They sometimes have an emotional basis and include criteria and issues that are difficult to verbalize.

The next tier is composed of people and politics. Politics in this context means the use of power (intentional or not) in organizations. For example, if the owner of a factory has a strong opinion that automation is important, this has a great effect on engineering decisions on the plant floor. Similarly, an influential personality can affect decision making. It's difficult to make a decision without the support, either real or imagined, of other people. This support can be manipulated, for example, by a persuasive salesperson or a persistent lobbyist. Support might just be a general understanding communicated through subtle messages.

The next tier is a collection of "facts." The facts, which may or may not be valid or verifiable, contribute to the politics and the people, and indirectly to the preferences. At the bottom of the pyramid are the activities that contribute to the facts. These include the history of previous similar decisions, statistics of various sorts, and, among other things, a determination of costs.

Figure 1.1 Decision Pyramid

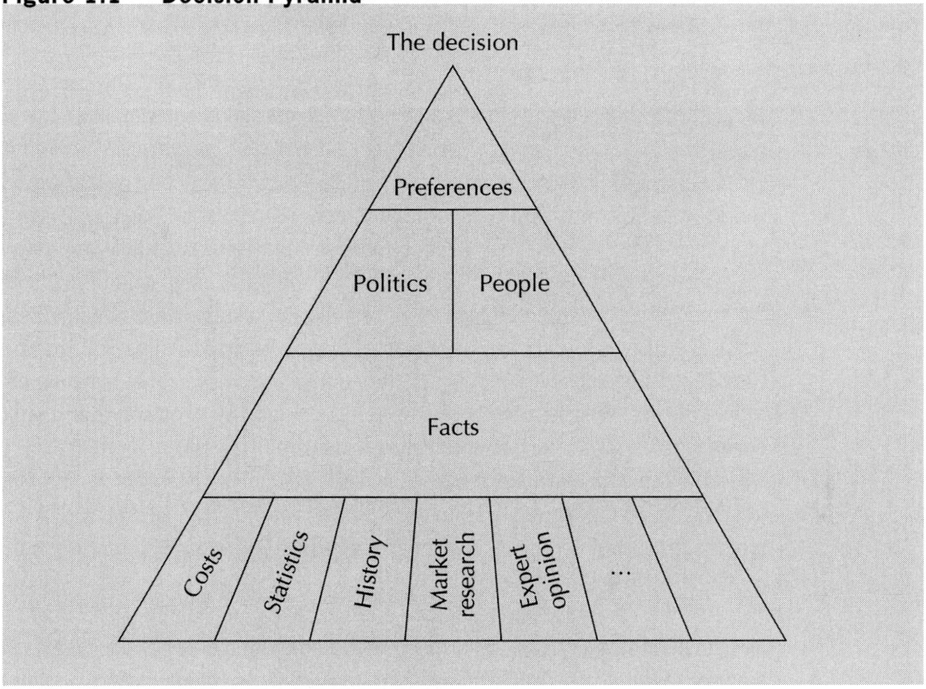

In this view of decisions, engineering economics is not very important. It deals essentially with facts and, in particular, with determining costs. There are many other facts that affect the final decision, and even then the decision may be made on the basis of politics, personality, or unstated preferences. However, this is an extreme view.

Although preferences, politics, and people can outweigh facts, usually the relationship is the other way around. The facts tend to control the politics, the people, and the preferences. It is facts that allow an individual to develop a strong opinion, which then may be used to influence others. Facts accumulated over time create intuition and experience that control our "gut feeling" about a decision. Facts, and particularly the activities that develop the facts, form the foundation for the pyramid in Figure 1.1. Without the foundation, the pyramid would collapse.

Engineering economics is important because it facilitates the establishment of verifiable facts about a decision. The facts are important and necessary for the decision to be made. However, the decision eventually made may be contrary to that suggested by analysis. For example, a study of several methods of treating effluent might determine that Method A is most efficient and cheapest, but Method B might in fact be chosen because it requires a visible change to the plant which, it is felt, will contribute to the company's image in environmental issues. Such a final decision is appropriate because it takes into account facts beyond those dealt with in the economic analysis.

Engineering Economics in Action, Part 1B: Naomi Settles In

As Naomi and Clement were walking, they passed the loading docks. A honk from behind told them to move over so that a forklift could get through. The operator waved in passing and continued on with the task of moving coils of sheet metal into the warehouse. Naomi noticed shelves and shelves of packaging material, dies, spare parts, and other items that she didn't recognize. She would find out more soon enough. They continued to walk. As they passed a welding area, Clem pointed out the newest recycling project in Canadian Widgets: the water used to degrease the metal was now being cleaned and recycled rather than being used only once.

Naomi became aware of a pervasive, pulsating noise emanating from the distance. Suddenly the corridor opened up to the main part of the plant, and the noise became a bedlam of clanging metal and thumping machinery. Her senses were assaulted. The ceiling was very high, and there were rows of humpbacked metal monsters unlike any presses she had seen before. The tang of mill oil overwhelmed her sense of smell, and she felt the throbbing from the floor knocking her bones together. Clem handed her hearing and eye protectors.

"These are our main press lines." Clem was yelling right into Naomi's ear, but she had to strain to hear. "We go right from sheet metal to finished widgets in 12 operations." A passing forklift blew propane exhaust at her, momentarily replacing the mill-oil odour with hot-engine odour. "Engineering is off to the left there."

As they went through the double doors into the Engineering Department, the din subsided and the ceiling came down to normal height. Removing the safety equipment, they stopped for a moment to get some juice at the vending machines. As Naomi looked around, she saw computers on desks more or less sectioned off by acoustic room dividers. As Clem led her farther, they stopped long enough for him to introduce Naomi to Carole Brown, the receptionist and secretary. Just past Carole's desk and around the corner was Naomi's desk. It was a nondescript metal desk with a long row of empty shelving above. Clem said that her computer would arrive within the week. Naomi noticed that the desk next to hers was empty, too.

"Am I sharing with someone?" she asked.

"Well, you will be. That's for your co-op student."

"My co-op student?"

"Yep. Don't worry, we have enough to do to keep you both busy. Why don't you take a few minutes to settle in, while I take care of a couple of things. I'll be back in, say, fifteen minutes. I'll take you over to Human Resources. You'll need a security pass, and I'm sure they have lots of paperwork for you to fill out."

Clem left. Naomi sat down and opened the briefcase she had carefully packed that morning. Alongside the brown-bag lunch was an engineering economics textbook. She took it out and placed it on the empty shelf above the desk. "I thought I might need you," she said to herself. "Now, let's get this place organized!" ◎

1.4 Dealing with Abstractions

The world is far more complicated than what can ever be described in words, or even thought about. Whenever one deals with reality, it is done through models or abstractions. For example, consider the following description:

Naomi watched the roll of sheet metal pass through the first press. The die descended and punched six oval shapes from the sheet. These "blanks" dropped through a chute into a large metal bin. The strip of sheet metal jerked forward into the die and the press came down again. Pounding like a massive heart 30 times a minute, the machine kept the operator busy full time just providing the giant coils of metal, removing the waste skeleton scrap, and stacking blanks in racks for transport to the next operation.

This gives a description of a manufacturing process that is reasonably complete, in that it permits one to visualize the process. But it is not absolutely complete. For example, how large and thick were the blanks? How big was the metal bin? How heavy was the press? How long did it take to change a die? These questions could be answered, but no matter how many questions are asked, it is impossible to express all of the complexity of the real world. It is also undesirable to do so.

When one describes something, one does so for a purpose. In the description, one selects those aspects of the real world that are relevant to that purpose. This is appropriate since it would be very confusing if a great deal of unnecessary information were given every time something was talked or written about. For example, if the purpose of the above description were to explain the exact nature of the blanks, there would be considerably less emphasis on the process, and many more details about the blanks themselves.

This process of simplifying the complexities of the real world is necessary for any engineering analysis. For example, in designing a truss for a building, it is usually assumed that the members exhibit uniform characteristics. However, in the real world these members would be lumber with individual variations: some would be stronger than average and some would be weaker. Since it is impractical to measure the characteristics of each piece of wood, a simplification is made. As another example, the various components of an electric circuit, such as resistors and capacitors, have values that differ from their nominal specifications because of manufacturing tolerances, but such differences are often ignored and the nominal values are the ones that are used in calculations.

Figure 1.2 illustrates the basic process of modelling that applies in so much of what humans do, and applies especially to engineering. The world is too complicated to express completely, as represented by the amorphous shape at the top of the figure. People extract from the real world a simplification (in other words, a model) which captures information that is useful and appropriate for a given purpose. Once the model is developed, it is used to analyze a situation, and perhaps make some predictions about the real world. The analysis and the predictions are then related back to the real world to make sure that the model is valid. As a result, the model might need some modification, so that it more accurately reflects the relevant features of the real world.

The process illustrated in Figure 1.2 is exactly what is done in engineering economics. The model is often a mathematical one that simplifies a more complicated situation, but does so in a reasonable way. The analysis of the model provides some information, such as which solution to a problem is

Figure 1.2 The Modelling Process

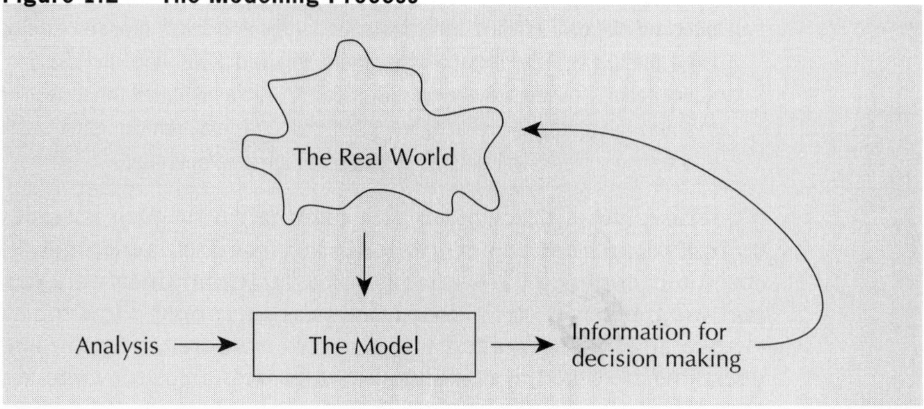

cheapest. This information must always be related back to the real problem, however, to take into account the aspects of the real world that may have been ignored in the original modelling effort. For example, the economic model might not have included taxes or inflation, and an examination of the result might suggest that taxes and inflation should not be ignored. Or, as already pointed out, environmental, political, or other considerations might modify any conclusions drawn from the mathematical model.

Example 1.1

Naomi's brother Ben has been given a one-year assignment in the Yukon, and he wants to buy a car just for the time he is there. He has three choices, as illustrated in Table 1.1. For each alternative, there is a purchase price, an operating cost (including gas, insurance, and repairs), and an estimated resale value at the end of the year. Which should Ben buy?

Table 1.1 Buying a Car

	57 Chevy	97 Neon	93 Mercedes
Purchase	$12 000	$7 000	$20 000
Operation	200/month	50/month	150/month
Resale	13 000	5 000	20 000

The next few chapters of this book will show how to take the information from Table 1.1 and determine which alternative is *economically* best. As it turns out, under most circumstances, the Chevy is best. However, in constructing a model of the decision, we must make a number of important assumptions.

For example, how can one be sure of the resale value of something until one actually tries to sell it? Along the same lines, who can tell what the actual maintenance costs will be? There is a lot of uncertainty about future events that is generally ignored in these kinds of calculations. Despite this uncertainty, estimates can provide insights into the appropriate decision.

Another problem for Ben is getting the money to buy a car. Ben is fairly young, and would find it very difficult to raise even $7000, perhaps impossible to raise $20 000. The Chevy might be the best value, but, if the money isn't available to take advantage of the opportunity, it doesn't matter. In order to do an economic analysis, we may assume that he has the money available.

If an economic model is judged to be appropriate, does that mean that Ben should buy the Chevy? Maybe not.

A person who has to drive to work every morning would probably not want to drive an antique car. It is too important that the car be reliable (especially in the Yukon in the winter). The operating costs for the Chevy are high, reflecting the need for more maintenance than with the other cars, and there are indirect effects of low reliability that are hard to capture in dollars.

If Ben were very tall, he would be extremely uncomfortable in the compact Neon car, so that, even if it were economically best, he would hesitate to resign himself to driving with his knees on either side of the steering wheel.

Ben might have strong feelings about the environmental record of one of the car manufacturers, and might want to avoid driving that car as a way of making a statement.

Clearly, there are so many *intangibles* involved in a decision like this that it is impossible for anyone but Ben himself to make such a personal choice. An outsider can point out to Ben the results of a quantitative analysis, given certain assumptions, but cannot authoritatively determine the best choice for Ben. ■

1.5 The Moral Question: Three True Stories

Complex decisions often have an ethical component to them. Recognizing this ethical component is important for engineers, since society relies on them for so many things. The following three anecdotes concern real Canadian companies—although names and details have been altered for anonymity—and illustrate some extreme examples of the forces acting on engineering decision making.

Example 1.2

The process of making sandpaper is similar to that of making a photocopy. A two-metre-wide roll of paper is coated with glue and given a negative electric charge. It is then passed over sand (of a particular type) which has a positive charge. The sand is attracted to the paper and sticks on the glue. The fact that all of the bits of sand have the same type of charge makes sure that the grains are evenly spaced. The paper then passes through a long heated chamber to cure the glue. Although the process sounds fairly simple, the machine that does this, called a *maker*, is very complicated and expensive. One such machine, costing several million dollars, can support a factory employing hundreds of workers.

Preston Sandpapers was the Canadian subsidiary of a large United States firm, and was located in a small town. Its maker was almost 30 years old and desperately needed replacement. However, rather than replace it, the parent company could choose to close down the Canadian plant and transfer production to one of the American sister plants.

The chief engineer had a problem. The costs for installing a new maker were extremely high, and it was difficult to justify a new maker economically. However, if he could not do so, the plant would close and hundreds of workers would be out of a job, including perhaps himself. What he chose to do was lie. He fabricated figures, ignored important costs, and exaggerated benefits to justify the expenditures. The investment was made, and the Canadian plant is still operating. ■

Example 1.3

Hespeler Meats is a medium-sized meat processor specializing in deli-style cold cuts and European process meats. Hoping to expand their product offerings, they decided to add a line of canned patés. They were eligible for a government grant to cover some of the purchase price of the necessary production equipment.

Government support for manufacturing is generally fairly sensible. Support is also usually not given for projects that are clearly very profitable, since the company should be able to justify such an expense itself. On the other hand, support is also usually not given for projects that are clearly not very profitable, because taxpayers' money should not be wasted. Support is directed at projects that the company would not otherwise undertake, but which have good potential to create jobs and expand the economy.

Hespeler Meats had to provide a detailed justification for their canned paté project in order to qualify for the government grant. Their problem was that they had to predict both the expenditures and the receipts for the following five years. This was a product line with which they had no experience, and which, in fact, had not been offered in North America by any meat processor. They had absolutely no idea what their sales would be. Any numbers they picked would be guesses, but to get the grant they had to give numbers.

What they did was select an estimate of sales that, given the equipment expenditures expected, fell exactly within that range of profitability that made the project suitable for government support. They got the money. As it turned out, the product line was a flop, and the canning equipment was sold as scrap five years later. ■

Example 1.4

When a large metal casting is made, as for the engine block of a car, it has only a rough exterior, and often has *flash*, which is ragged edges of metal formed where molten metal seeped between the two halves of the mould. The first step in finishing the casting is to grind off the flash, and to grind flat surfaces so that the casting can be held properly for subsequent machining.

Galt Casting Grinders (GCG) made the complex specialized equipment for this operation. It had once commanded the world market for this product, but lost market share to Japanese competitors. The competitors did not have a better product than GCG. However, they were able to increase market share by adding fancy display panels with coloured lights, dials, and switches that looked very sophisticated.

GCG's problem was that their idea of sensible design was to omit the features the competitors included (or the customers wanted). GCG reasoned that these features added nothing to the capability of the equipment, but did add a lot to the manufacturing cost and to the maintenance costs that would be borne by the purchaser. They had no doubt that it was unwise, and poor engineering design, to make such unnecessarily complicated displays, so they made no changes.

GCG went bankrupt several years later. ■

In each of these three examples, the technical issues are overwhelmed by the non-technical ones. For Preston Sandpapers, the chief engineer was pressured by his social responsibility and self-interest to lie and recommend a decision that was not justified by the facts. In the Hespeler Meats case, the engineer had to choose between stating the truth—that future sales were unknown—which would deny the company a very useful grant, and selecting a convenient number that would encourage government support. For Galt Casting Grinders, the issue was marketing. They did not recognize that a product must be more than technically good; it must also be salable.

Beyond these principles, however, there is a moral component to each of these anecdotes. As guardians of knowledge, engineers have a vital responsibility to society to behave ethically and responsibly in all ways. When so many different issues must be taken into account in engineering decision making, it is often difficult to determine what course of action is ethical.

For Preston Sandpapers, most people would probably say that what the chief engineer did was unethical. However, he did not exploit his position simply for personal gain. He was, in his mind, saving a town. Is the principle of honesty more important than several hundred jobs? Perhaps it is, but when the job holders are friends and family it is understandable that unethical choices can be made.

For Hespeler Meats, the issue is more subtle. Is it ethical to choose figures that match the ideal ones to gain a government grant? It is, strictly speaking, a lie, or at least misleading, since there is no estimate of sales. On the other hand, the bureaucracy demands that some numbers be given, so why not pick ones that suit your case?

In the Galt Casting Grinders case, the engineers apparently did no wrong. The ethical question concerns the competitors' actions. Is it ethical to put features on equipment that do no good, add cost, and decrease reliability? In this case and for many other products, this is often done, whether ethical or not. If it is unethical, then the ethical suppliers will sometimes go out of business.

There are no general answers to difficult moral questions. Practising engineers often have to make choices with an ethical component, and can sometimes rely on no stronger foundation than their own sense of right and wrong. More information about ethical issues for engineers can be obtained from provincial professional engineering associations.

1.6 Uncertainty and Sensitivity Analysis

Whenever people predict the future, errors occur. Sometimes predictions are correct, whether the predictions are about the weather, a ball game, or company cash flow. On the other hand, it would be unrealistic to expect anyone to be always right about something that hasn't happened yet.

Although one cannot expect an engineer to precisely predict the future, approximations are very useful. A Canadian weather forecaster can dependably say that it will not snow in July, for example, even though it may be more difficult to forecast the exact temperature. Similarly, an engineer may not be able to precisely predict the scrap rate of a testing process, but may be able to determine a range of likely rates to help in a decision making process.

Engineering economics analyses are quantitative in nature, and most of the time the quantities used in economic evaluations are estimates. The fact that we don't have precise values for some quantities may be very important since decisions may have expensive consequences and significant health and environmental effects. How can the impact of this uncertainty be minimized?

One way to control this uncertainty is to make sure that the information being used is valid and as accurate as possible. The GIGO rule—Garbage In, Garbage Out—applies here. There is little else as useless or potentially dangerous as a precise calculation made from inaccurate data. However, even accurate data from the past is only of limited value when predicting the future. Even with sure knowledge of past events, the future is still uncertain.

Sensitivity analysis involves assessing the effect of uncertainty on a decision. It is very useful in engineering economics. The idea is that, although a particular value for a parameter can be known with only a limited degree of certainty, a range of values can be assessed with reasonable certainty. In sensitivity analysis, the calculations are done several times, varying each important parameter over its range of possible values. Usually only one parameter at a time is changed, so that the effect of each change on the conclusion can be assessed independently of the effect of other changes.

In Example 1.1, Naomi's brother Ben had to choose a car. He made an estimate of the resale value of each of the alternative cars, but the *actual* resale amount is unknown until the car is sold. Similarly, the operating costs are not known with certainty until the cars are driven for a while. Before concluding that the Chevy is the right car to buy (on economic grounds at least), Ben should assess the sensitivity of this decision by varying the resale values and operating costs within a range from the minimum likely amount to the maximum likely amount. Since these calculations are often done on spreadsheets, this assessment is not hard to do, even with many different parameters to vary.

Sensitivity analysis is an integral part of all engineering economics decisions because data regarding future activities are always uncertain. In this text, emphasis is usually given to the structure and formulation of problems rather than to verifying whether the result is robust. In this context, robust means that the same decision will be made over a wide range of parameter values. It should be remembered that no decision is properly made unless the sensitivity of that decision to variation in the underlying data is assessed.

A related issue is the number of significant digits in a calculation. Modern calculators and computers can carry out calculations to a large number of decimal places of precision. For most purposes, such precision is meaningless. For example, a cost calculated as $1.0014613076 is of no more use than $1.00 in most applications. It is useful, though, to carry as many decimal places as convenient to reduce the magnitude of accumulated round-off errors.

In this book, all calculations have been done with as many significant digits as could conveniently be carried, even though the intermediate values are shown with three to six digits. As a rule, only three significant digits are assumed in the final value. For decision-making purposes, this is plenty.

1.7 How This Book Is Organized

There are twelve chapters remaining in this book. The first block of chapters, Chapters 2 to 5, forms the core material of the book. Chapters 2 and 3 provide the mathematics needed to manipulate monetary values over time. Chapters 4 and 5 deal with comparing alternative projects. Chapter 4 illustrates present worth, annual worth, and payback period comparisons, and Chapter 5 covers the internal rate of return (IRR) method of comparison.

The second block of chapters, Chapters 6 to 8, broadens the core material. It covers depreciation and analysis of a company's financial statements, when to replace equipment (replacement analysis), and taxation.

The third block includes Chapters 9 to 13 which provide supporting material for the previous chapters. Chapter 9 concerns the effect of inflation on engineering decisions, and Chapter 10 explores how decision making is done for projects owned by or affecting the public, rather than an individual or firm. Chapter 11 deals with handling uncertainty about important information through sensitivity analysis, while Chapter 12 deals with situations where exact parameter values are not known, but probability distributions for them are known. Finally, Chapter 13 provides some formal methods for taking into account the intangible components of an engineering decision.

Each chapter begins with a story about Naomi and her experiences at Canadian Widgets. There are several purposes to these stories. They provide an understanding of engineering practice that is impossible to convey with short examples. In each chapter, the story has been chosen to make clear why the ideas being discussed are important. It is also hoped that the stories make the material taught a little more interesting.

There are two extended cases in the text. The first, located between Chapter 6 and Chapter 7, is a problem that is too complicated to include in a particular chapter, but it reflects a realistic situation that would likely be encountered in engineering practice. The second case, located between Chapter 11 and Chapter 12, builds on the first case to use some of the more sophisticated ideas presented in the later chapters of the book.

Throughout the text are boxes which contain information that is associated with, and complements, the text material. One set of boxes contains *Close-Ups*, which focus on topics of relevance to the chapter material. These appear in each chapter in the appropriate section. Another set of boxes presents *Canadian Mini-Cases*, which appear at the end of each chapter, following the problem set. These cases report how engineering economics is used in familiar Canadian companies, and include questions designed for classroom discussion or individual reflection.

End-of-chapter appendices contain relevant but more advanced material. Appendices at the back of the book provide tables of important and useful values and answers to selected chapter-end problems.

Engineering Economics in Action, Part 1C: A Taste of What Is to Come

Naomi was just putting on her newly laminated security pass when Clem came rushing in. "Sorry to be late," he puffed. "I got caught up in a discussion with someone in Marketing. Are you ready for lunch?" She certainly was. She had spent the better part of the morning going through the benefits package offered by Canadian Widgets and was a bit overwhelmed by the paperwork. Dental plan options, pension plan beneficiaries, and tax forms swam in front of her eyes. The thought of food sounded awfully good.

As they walked to the lunchroom, Clem continued to talk. "Maybe you will be able to help out once you get settled in, Naomi."

"What's the problem?" asked Naomi. Obviously Clem was still thinking about his discussion with this person from Marketing.

"Well," said Clem, "currently we buy small aluminum parts from a subcontractor. The cost is quite reasonable, but we should consider making the parts ourselves, as our volumes are increasing and the fabrication process would not be difficult for us to bring in-house. We might be able to make the parts at a lower cost. Of course, we'd have to buy some new equipment. That's why I was up in the Marketing Department talking to Prabha."

"What do you mean?" asked Naomi, still a little unsure. "What does this have to do with Marketing?"

Clem realized that he was making a lot of assumptions about Naomi's knowledge of Canadian Widgets. "Sorry," he said, "I need to explain. I was up in Marketing to ask for some demand forecasts so that we would have a better handle on the volumes of these aluminum parts that we might need in the next few years. That, combined with some digging on possible equipment costs, would allow us to do an analysis of whether we should make the parts in-house or continue to buy them."

Things made much more sense to Naomi now. Her engineering economics text was certainly going to come in handy.

PROBLEMS

1.1 In which of the following situations would engineering economics analysis play a strong role and why?

(a) Buying new equipment

(b) Changing design specifications for a product

(c) Deciding the paint colour for the factory floor

(d) Hiring a new engineer

(e) Deciding when to replace old equipment with new equipment of the same type

(f) Extending the cafeteria business hours

(g) Deciding which invoice forms to use

(h) Changing the 8-hour work shift to a 12-hour shift

(i) Deciding how much to budget for Research and Development programs

(j) Deciding how much to donate for the town's new library

(k) Building a new factory

(l) Downsizing the company

1.2 Starting a new business requires many decisions. List five examples of decisions that can be assisted by engineering economics analysis.

1.3 For each of the following items, describe how the design might differ if the costs of manufacturing, use, and maintenance were not important. On the basis of these descriptions, is it important to consider costs in engineering design?

(a) a car

(b) a television set

(c) a light bulb

(d) a book

1.4 Leslie and Sandy, recently married students, are going to rent their first apartment. Leslie has carefully researched the market and has decided that, all things considered, there is only one reasonable choice. The two-bedroom apartment in the building at the corner of University and Erb Streets is the best value for the money and is also close to school. Sandy, on the other hand, has just fallen in love with the top half of a duplexed house on Dunbar Road. Which apartment should they move into? Why? Which do you think they will move into? Why?

1.5 Describe the process of using the telephone as you might describe it to a six-year-old child using it for the first time to call a friend from school. Describe using the telephone to an electrical engineer who just happens never to have seen one before. Which is the correct way to describe a telephone?

1.6 **(a)** Karen has to decide which of several computers to buy for school use. Should she buy the cheapest one? Can she make the best choice on price alone?

(b) Several computers offer essentially the same features, reliability, service, etc. Among these, can she decide the best choice on price alone?

1.7 For each of the following situations, describe what you think you *should* do. In each case *would* you do this?

(a) A fellow student, who is a friend, is copying assignments and submitting them as his own work.

(b) A fellow student, who is *not* a friend, is copying assignments and submitting them as her own work.

(c) A fellow student, who is your only competitor for an important academic award, is copying assignments and submitting them as his own work.

(d) A friend wants to hire you to write an essay for school for her. You are dead broke and the pay is excellent.

(e) A friend wants to hire you to write an essay for school for him. You have lots of money, but the pay is excellent.

(f) A friend wants to hire you to write an essay for school for her. You have lots of money, and the pay is poor.

(g) Your car was in an accident. The insurance adjuster says that the car was totalled and they will give you only the "blue book" value for it as scrap. They will pick up the car in a week. A friend points out that in the meantime you could sell the almost-new tires and replace them with bald ones from the scrap yard, and perhaps sell some other parts, too.

(h) The CD player from your car has been stolen. The insurance adjuster asks you how much it was worth. It was a very cheap one, of poor quality.

(i) The engineer you work for has told you that the meter measuring effluent discharged from a production process exaggerates, and the measured value must be halved for record keeping.

(j) The engineer you work for has told you that part of your job is to make up realistic-looking figures reporting effluent discharged from a production process.

(k) You observe unmetered and apparently unreported effluent discharged from a production process.

(l) An engineer where you work is copying directly from a manufacturer's brochure machine-tool specifications to be included in a purchase request. These specifications limit the possible purchase to the particular one specified.

(m) An engineer where you work is copying directly from a manufacturer's brochure machine-tool specifications to be included in a purchase request. These specifications limit the possible purchase to the particular one specified. You know that the engineer's best friend is the salesman for that manufacturer.

1.8 Ciel is trying to decide whether now is a good time to expand her manufacturing plant. The viability of expansion depends on the Canadian economy (an expanding economy means more sales), the relative value of the Canadian dollar (a lower dollar means more exports), and changes in international trade agreements (lower tariffs also mean more exports). These factors, however, may be highly unpredictable. What two things can she do to help make sure she makes a good decision?

1.9 Trevor started a high-tech business two years ago, and now wants to sell out to one of his larger competitors. Two different buyers have made firm offers. They are similar in all but two respects. They differ in price: the Investco offer would result in Trevor's walking away with $2 000 000, while the Venture Corporation offer would give him $3 000 000. The other way they differ is that Investco say they will recapitalize Trevor's company to increase growth, while Trevor thinks that Venture Corporation will close down the business so that it doesn't compete with several of Venture Corporation's other divisions. What would you do if you were Trevor, and why?

1.10 Telekom Company is considering the development of a new type of cell phone based on a brand new, emerging technology. If successful, Telekom will be able to offer a cell phone which works over long distances and even in hilly areas. Before proceeding with the project, however, what uncertainties associated with the new technology should they be aware of? Can sensitivity analysis help address these uncertainties?

Bata Industries Ltd. and Trioxide Inc.

CANADIAN 1.1 MINI-CASE

Bata Industries Ltd. of Toronto is one of the largest manufacturers of footwear in the world. Although one might think shoe production would be a relatively benign business, on February 2, 1992, the Ontario Court of Justice found the firm liable for significant environmental damage. Metal drums filled with toxic waste were improperly managed and the contents leaked. Bata had to pay \$500 000 in clean-up costs, the corporation was fined \$90 000, and both the company president and vice-president were fined \$6000.

This was the first time in Canada that directors of a large company were convicted of an environmental offence. The court held that both the president and vice-president had sufficient knowledge and control over the company's daily activities that they should be held responsible for the damage.

The largest penalty to date for environmental damage in Canada was to Trioxide Inc. The corporation and five of its directors pleaded guilty to charges of permitting the discharge of waste water into the St. Lawrence River. The company was fined \$1 000 000 and was ordered to donate \$3 000 000 to environmental protection programs.

Source: Canadian Consulting Engineer, July/August 1994.

Discussion

There is often strong motivation for companies to commit environmental offenses. Companies primarily focus on profits, and preventing environmental damage is always a cost. It benefits society to have a clean environment, but almost never benefits the company directly. Sometimes, in spite of the efforts of upper management to be good citizens, the search for profit at a lower level may result in environmental damage anyway. In a large company it can be difficult for one person to know what is happening everywhere in the firm.

Older companies have an additional problem. An older company may have been producing goods in a certain way for years, and have established ways to dispose of waste. Changes in society now makes those traditional methods unacceptable. Even when a source of pollution has been identified, it may not be easy to fix. There may be decades of accumulated damage to correct. A production process may not be easily changed in a way that would allow the company to stay in business. Loss of jobs and the effect on the local economy may create strong political pressure to keep the company running.

The government has an important role to offset the profit motive for large companies, for example, by taking action through the courts. Unfortunately, that alone will not be enough to prevent some firms from continuing to cause environmental damage. Economics and politics will occasionally win out.

Questions

1) There are probably several companies in your city or province that are known to pollute. Name some of these. For each:
- **(a)** What sort of damage do they do?
- **(b)** How long have they been doing it?
- **(c)** Why is this company still permitted to pollute?
- **(d)** What would happen if this company was forced to shut down? Is it ethically correct to allow the company to continue to pollute?

2) Does it make more sense to fine a company for environmental damage or to fine management personally for environmental damage caused by a company? Why?

3) Should the fines for environmental damage be raised high enough so that no company is tempted to pollute? Why or why not?

4) Governments can impose fines, give tax breaks, and take other actions that use economics to control the behaviour of companies. Is it necessary to do this whenever a company that pursues profits might do some harm to society as a whole? Why might a company do the socially correct thing even if profits are lost?

C H A P T E R 2

Time Value of Money

Engineering Economics in Action, Part 2A: A Steal For Steel

"Naomi, can you check this for me?" Terry's request broke the relative silence as Naomi and Terry worked together one Tuesday afternoon. "I was just reviewing our J-class line for Clem, and it seems to me that we could save a lot of money there."

"O.K., tell me about it." Since Naomi and Terry had met two weeks earlier, just after Naomi started her job, things had being going very well. Terry, an engineering student at the local university, was on a four-month co-op work term at Canadian Widgets.

"Well, mostly we use the heavy rolled stock on that line. According to the pricing memo we have for that kind of steel, there is a big price break at a volume that could supply our needs for six months. We've been buying this stuff on a week-by-week basis. It just makes sense to me to take advantage of that price break."

"Interesting idea, Terry. Have you got data about how we have ordered before?"

"Yep, right here."

"Let's take a closer look."

"Well," Terry said, as he and Naomi looked over his figures, "the way we have been paying doesn't make too much sense. We order about a week's supply. The cost of this is added to our account. Every six months we pay off our account. Meanwhile, the supplier is charging us 2% of our outstanding amount at the end of each month!"

"Well, at least it looks as if it might make more sense for us to pay off our bills more often," Naomi replied.

"Now look at this. In the six months ending last December, we ordered steel for a total cost of \$1 600 000. If we had bought this steel at the beginning of July, it would have only cost \$1 400 000. That's a saving of \$200 000!"

"Good observation, Terry, but I don't think buying in advance is the right thing to do. If you think about it . . ." ◎

2.1 Introduction

Engineering decisions frequently involve evaluating trade-offs among costs and benefits that occur at different times. A typical situation is when we invest in a project today in order to obtain benefits from the project in the future. This chapter discusses the economic methods used to compare benefits and costs that occur at different times. The key to making these comparisons is the use of an interest rate. In Sections 2.2 to 2.5, we illustrate the comparison process with examples and introduce some interest and interest rate terminology. Section 2.6 deals with cash flow diagrams, which are graphical representations of the magnitude and timing of cash flows over time. Section 2.7 explains the equivalence of benefits and costs that occur at different times.

2.2 Interest and Interest Rates

Everyone is familiar with the idea of interest from their everyday activities:

From the furniture store ad: *Pay no interest until next year!*
From the bank: *Now 2.6% daily interest on passbook accounts!*

Why are there interest rates? If people are given the choice between having money today and the same amount of money one year from now, most would prefer the money today. If they had the money today, they could do something productive with it in hopes of benefit in the future. For example, they could buy an asset like a machine today, and could use it to make money from their initial investment. Or they may want to buy a consumer good like a new stereo system and start enjoying it immediately. What this means is that one dollar today is worth more than one dollar in the future. This is because a dollar today can be invested for productive use while that opportunity is lost or diminished if the dollar is not available until some time in the future.

The observation that a dollar today is worth more than a dollar in the future means that people must be compensated for lending money. They are giving up the opportunity to invest their money for productive purposes now on the promise of getting more money in the future. The compensation for loaning money is in the form of an interest payment, say I. More formally, **interest** is the difference between the amount of money lent and the amount of money later repaid. It is the compensation for giving up the use of the money for the duration of the loan.

An amount of money today, P (also called the *principal amount*), can be related to a *future amount* F by the interest amount I or interest rate i. This relationship is illustrated graphically in Figure 2.1 and can be expressed as $F = P + I$. The interest I can also be expressed as an *interest rate* i with respect to the principal amount so that $I = Pi$. Thus

$$\begin{aligned} F &= P + Pi \\ &= P(1 + i) \end{aligned}$$

Figure 2.1 Present and Future Worth

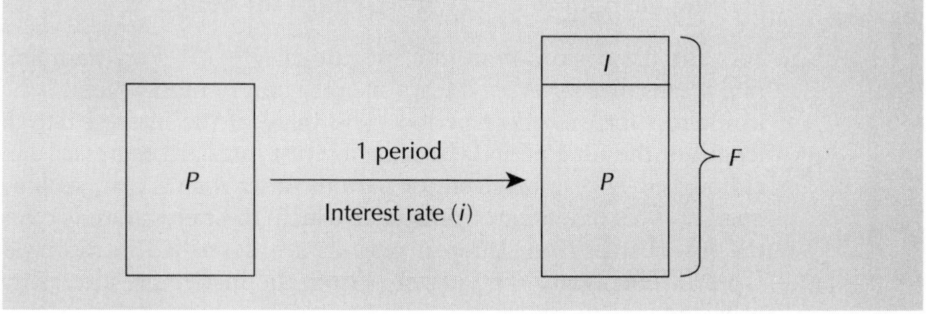

Example 2.1

Samuel bought a one-year guaranteed investment certificate for \$5000 from a bank on May 15, 1999. The bank was paying 10% on one-year guaranteed investment certificates at the time. One year later, on May 15, 2000, Samuel cashed in his certificate for \$5500.

We may think of the interest payment that Samuel got from the bank as compensation for giving up the use of money. When Samuel bought the guaranteed investment certificate for \$5000, he gave up the opportunity to use the money in some other way during the following year. On the other hand, the bank got use of the money for the year. In effect, Samuel lent \$5000 to the bank for a year. The \$500 interest was payment by the bank to Samuel for the loan. The bank wanted the loan so that it could use the money for the year. (It may have lent the money to someone else at a higher interest rate.) ■

This leads to a formal definition of interest rates. Divide time into periods like days, months, or years. If the right to P at the beginning of a time period exchanges for the right to F at the end of the period, where $F = P(1 + i)$, i is the **interest rate** per time period. In this definition, P is called the **present worth** of F, and F is called the **future worth** of P.

Example 2.1 Restated

Samuel bought a one-year guaranteed investment certificate for \$5000 from a bank on May 15, 1999. The bank was paying 10% on one-year guaranteed investment certificates at the time. The certificate gave Samuel the right to claim \$5500 from the bank on May 15, 2000.

Notice in this example that there was a transaction between Samuel and the bank on May 15, 1999. There was an exchange of \$5000 on May 15, 1999, for the right to collect \$5500 on May 15, 2000. On May 15, 1999, the bank got the \$5000 and Samuel got the right to collect \$5500 one year later. Evidently, having a dollar on May 15, 1999, was worth more than the right to collect a dollar a year later. Each dollar on May 15, 1999, was worth the right to collect $\frac{5500}{5000} = 1.1$ dollars a year later. This 1.1 may be written as $1 + 0.1$ where 0.1 is the interest rate. The interest rate, then, gives the rate of exchange between money at the beginning of a period (one year in this example) and the right to money at the end of the period. ■

The dimension of an interest rate is $\frac{\text{dollars/dollars}}{\text{time period}}$. For example, a 9% interest rate means that for every dollar (or other unit of money) lent, 0.09 dollars is paid in interest for each time period. The value of the interest rate depends on the length of the time period. Usually, interest rates are expressed on a yearly basis, although they may be given for periods other than a year, such as a month or a quarter. This base unit of time over which an interest rate is calculated is called the **interest period**. Interest periods are described in more detail in Close-Up 2.1. The longer the interest period, the higher the interest rate must be to provide the same return.

CLOSE-UP 2.1 Interest Periods

The most commonly used interest period is one year. If we say, for example, "6% interest" without specifying an interest period, the assumption is that 6% interest is paid for a one-year period. However, interest periods can be of any duration. Here are some other common interest periods:

Interest period	*Interest is calculated*
Semi-annual	twice per year, or once each six months
Quarterly	four times a year, or once each three months
Monthly	12 times per year
Weekly	52 times per year
Daily	365 times per year
Continuous	for infinitesimally small periods

Interest concerns the lending and borrowing of money. It is a parameter that allows an exchange of a larger amount of money in the future for a smaller amount of money in the present, and vice versa. As we will see in Chapter 3, it also allows us to evaluate very complicated exchanges of money over time.

Interest also has a physical basis. Money can be invested in financial instruments that pay interest, such as a bond or a savings account, and money can also be invested directly in industrial processes or services that generate wealth. In fact, the money invested in financial instruments is also, indirectly, invested in productive activities by the organization providing the instrument. Consequently, the root source of interest is the productive use of money, as this is what makes the money actually increase in value. The actual return generated by a specific productive investment varies enormously, as will be seen in Chapter 4.

2.3 Compound and Simple Interest

We have seen that if an amount, P, is lent for one interest period at the interest rate, i, the amount that must be repaid at the end of the period is $F = P(1 + i)$. But loans may be for several periods. How is the quantity of money that must be repaid computed when the loan is for N interest periods? The usual way is "one period at a time." Suppose that the amount P is borrowed for N periods at the interest rate i. The amount that must be repaid at the end of the N periods is $P(1 + i)^N$, that is

$$F = P(1 + i)^N \quad (2.1)$$

This is derived as shown in Table 2.1.

Table 2.1 Compound Interest Computations

Beginning of Period	Amount Lent	Interest Amount	Amount Owed at Period End
1	P	$+\ Pi$	$= P + Pi = P(1 + i)$
2	$P(1 + i)$	$+\ P(1 + i)i$	$= P(1 + i) + P(1 + i)i$
3	$P(1 + i)^2$	$+\ P(1 + i)^2 i$	$= P(1 + i)^2 + P(1 + i)^2 i$
⋮	⋮		
N	$P(1 + i)^{N-1}$	$+\ [P(1 + i)^{N-1}]i$	$= P(1 + i)^N$

This method of computing interest is called *compounding*. Compounding assumes that there are N sequential one-period loans. At the end of the first interest period, the borrower owes $P(1 + i)$. This is the amount borrowed for the second period. Interest is required on this larger amount. At the end of the second period $[P(1 + i)](1 + i)$ is owed. This is the amount borrowed for the third period. This continues so that at the end of the $(N-1)^{th}$ period, $P(1 + i)^{N-1}$ is owed. The interest on this is $[P(1 + i)^{N-1}]i$. The total interest on the loan over the N periods is

$$I_c = P(1 + i)^N - P \tag{2.2}$$

I_c is called **compound interest**. It is the standard method of computing interest where interest accumulated in one interest period is added to the principal amount used to calculate interest in the next period. The interest period when compounding is used to compute interest is called the **compounding period**.

Example 2.2

If you were to lend \$100 for three years at 10% per year compound interest, how much interest would you get at the end of the three years?

If you lend \$100 for three years at 10% compound interest per year, you will earn \$10 in interest in the first year. That \$10 will be lent, along with the original \$100, for the second year. Thus, in the second year, the interest earned will be \$11 = \$110(0.10). The \$11 is lent for the third year. This makes the loan for the third year \$121; \$12.10 = \$121(0.10) in interest will be earned in the third year. At the end of the three years, the amount you are owed will be \$133.10. The interest received is then \$33.10. This can also be calculated from Equation (2.2):

$$I_c = \$100(1 + 0.1)^3 - \$100 = \$33.10$$

Table 2.2 summarizes the compounding process. ■

Table 2.2 Compound Interest Computations for Example 2.2

Beginning of Year	Amount Lent	Interest Amount	Amount Owed at Year End
1	100	+ $100 × 0.1	= $110
2	110	+ $110 × 0.1	= $121
3	121	+ $121 × 0.1	= $133.10

If the interest payment for an N-period loan at the interest rate i per period is computed without compounding, the interest amount, I_s, is called *simple interest*. It is computed as

$$I_s = PiN$$

Simple interest is a method of computing interest where interest earned during an interest period is not added to the principal amount used to calculate interest in the next period. Simple interest is rarely used in practice, except as a method of calculating approximate interest.

Example 2.3

If you were to lend \$100 for three years at 10% per year simple interest, how much interest would you get at the end of the three years?

The total amount of interest earned on the \$100 over the three years would be \$30. This can be calculated by using $I_s = PiN$:

$$I_s = PiN = \$100(0.10)(3) = \$30 \blacksquare$$

Interest amounts computed with simple interest and compound interest will yield the same results only when the number of interest periods is one. As the number of periods increases, the difference between the accumulated interest amounts for the two methods increases exponentially.

When the number of interest periods is significantly greater than one, the difference between simple interest and compound interest can be very great. In April, 1993, a couple in Nevada presented the state government with a \$1000 bond issued by the state in 1865. The bond carried an annual interest rate of 24%. The couple claimed the bond was now worth several trillion dollars (*Source: Newsweek*, August 9, 1993, p. 8). If one takes the length of time from 1865 to the time the couple presented the bond to the state as 127 years, the value of the bond could have been \$732 trillion = $\$1000(1 + 0.24)^{127}$.

If, instead of compound interest, a simple interest rate given by $iN = (24\%)127 = 3048\%$ were used, the bond would be worth only \$31 480 = \$1000(1 + 30.48). Thus, the difference between compound

Figure 2.2 Compound and Simple Interest at 24% per Year for 20 Years

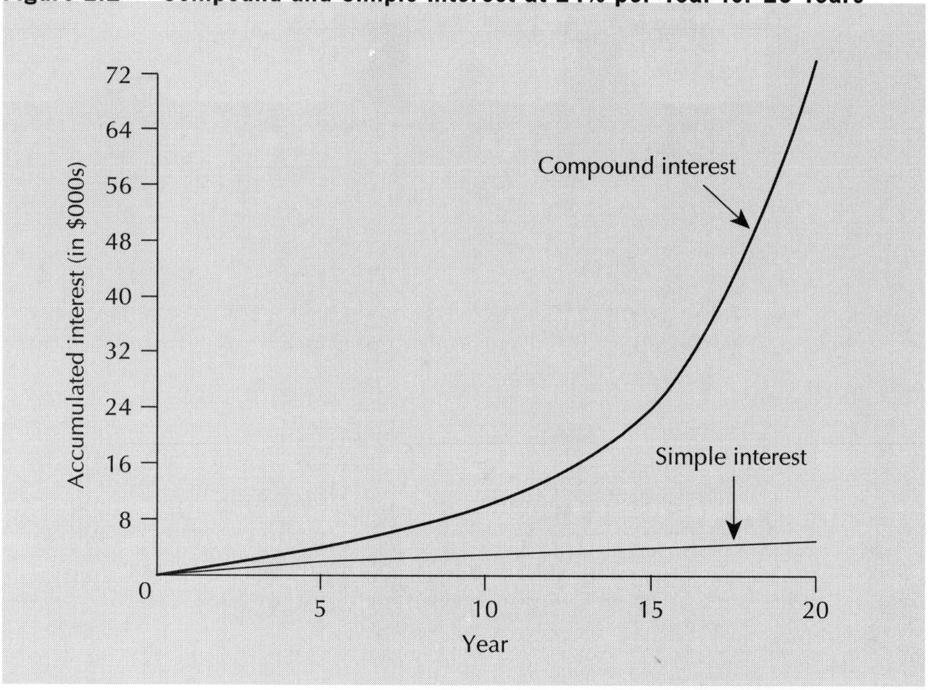

and simple interest can be dramatic, especially when the interest rate is high and the number of periods is large. The graph in Figure 2.2 shows the difference between compound interest and simple interest for the first twenty years of the bond example. As for the couple in Nevada, the $1000 bond was worthless after all—a state judge ruled that the bond had to have been cashed by 1872.

The conventional approach for computing interest is the compound interest method rather than simple interest. Simple interest is rarely used, except perhaps as an intuitive (yet incorrect!) way of thinking of compound interest. We mention simple interest primarily to contrast it with compound interest and to indicate that the difference between the two methods can be large.

2.4 Effective and Nominal Interest Rates

Interest rates may be stated for some period, like a year, while the computation of interest is based on shorter compounding sub-periods like months. In this section we consider the relation between the *nominal* interest rate that is stated for the full period and the *effective* interest rate that results from the compounding based on the sub-periods. This relation between nominal and effective interest rates must be understood to answer questions like this one: How would you choose between two investments, one bearing 12% per year interest, compounded yearly, with another with 1% per month interest, compounded monthly? Are they the same?

Nominal interest rate is the conventional method of stating the annual interest rate. It is calculated by multiplying the interest rate per compounding period by the number of compounding periods per year. Suppose that a time period is divided into m equal sub-periods. Let there be stated a nominal interest rate, r, for the full period. By convention, for nominal interest, the interest rate for each sub-period is calculated as $i_s = \frac{r}{m}$. For example, a nominal interest rate of 18% per year, compounded monthly, is the same as $\frac{0.18}{12} = 0.015$ or 1.5% per month.

Effective interest rate is the actual but not usually stated interest rate, found by converting a given interest rate with an arbitrary compounding period (normally less than a year) to an equivalent interest rate with a one-year compounding period. What is the effective interest rate, i_e, for the full period that will yield the same amount as compounding at the end of each sub-period, i_s? If we compound interest every sub-period, we have

$$F = P(1 + i_s)^m$$

We want to find the effective interest rate, i_e, that yields the same future amount F at the end of the full period from the present amount P. Set

$$P(1 + i_s)^m = P(1 + i_e)$$

Then

$$(1 + i_s)^m = 1 + i_e$$

$$i_e = (1 + i_s)^m - 1 \qquad (2.3)$$

Note that Equation (2.3) allows the conversion between the interest rate over a compounding sub-period, i_s, and the effective interest rate over a longer period, i_e, by using the number of sub-periods, m, in the longer period.

Example 2.4

What interest rate per year, compounded yearly, is equivalent to 1% interest per month, compounded monthly?

Since the month is the shorter compounding period, we let $i_s = 0.01$ and $m = 12$. Then i_e refers to the effective interest rate per year. Substitution into Equation 2.3 then gives

$$\begin{aligned} i_e &= (1 + i_s)^m - 1 \\ &= (1 + 0.01)^{12} - 1 \\ &= 0.126825 \\ &\approx 0.127 \text{ or } 12.7\% \end{aligned}$$

An interest rate of 1% per month, compounded monthly, is equivalent to an effective rate of approximately 12.7% per year, compounded yearly. The answer to our previously posed question is that an investment bearing 12% per year interest, compounded yearly, pays less than an investment bearing 1% per month interest, compounded monthly.

Interest rates are normally given as nominal rates. We may get the effective (yearly) rate by substituting $i_s = \frac{r}{m}$ into Equation (2.3). We then obtain a direct means of computing an effective interest rate, given a nominal rate and the number of compounding periods per year:

$$i_e = \left(1 + \frac{r}{m}\right)^m - 1 \qquad (2.4)$$

This formula is suitable only for converting from a nominal rate r to an annual effective rate. If the effective rate desired is for a period longer than a year, then Equation (2.3) must be used. ■

Example 2.5

Leona the loan shark lends money to clients at the rate of 5% interest per week! What is the nominal interest rate for these loans? What is the effective annual interest rate?

The nominal interest rate is 5% × 52 = 260%. Recall that nominal interest rates are usually expressed on a yearly basis. The effective yearly interest rate can be found by substitution into Equation (2.3):

$$i_e = (1 + 0.05)^{52} - 1 = 11.6$$

Leona charges an effective annual interest rate of about 1160% on her loans. ■

Example 2.6

The Cardex Credit Card Company charges a nominal 24% interest on overdue accounts, compounded daily. What is the effective interest rate?

Assuming that there are 365 days per year, we can calculate the interest rate per day using either Equation (2.3) with $i_s = \frac{r}{m} = \frac{0.24}{365} = 0.0006575$ or by the use of Equation (2.4) directly. The effective interest rate (per year) is

$$\begin{aligned} i_e &= (1 + 0.0006575)^{365} - 1 \\ &= 0.271 \text{ or } 27.1\% \end{aligned}$$

With a nominal rate of 24% compounded daily, the Cardex Credit Card Company is actually charging an effective rate of about 27.1% per year. ■

Although there are laws which may require that the effective interest rate be disclosed for loans and investments, it is still very common for nominal interest rates to be quoted. Since the nominal rate will be less than the effective rate whenever the number of compounding periods per year exceeds one, there is an advantage to quoting loans using the nominal rates, since it makes the loan look more attractive. This is particularly true when interest rates are high and compounding occurs frequently.

2.5 Continuous Compounding

As has been seen, compounding can be done yearly, quarterly, monthly, or daily. The periods can be made even smaller, as small as desired; the main disadvantage in having very small periods is having to do more calculations. If the period is made infinitesimally small, we say that interest is compounded *continuously*. There are situations where very frequent compounding makes sense. For instance, an improvement in material handling may reduce downtime on machinery. There will be benefits in the form of increased output that may be used immediately. If there are several additional runs a day, there will be benefits several times a day. Another example is trading on the stock market. Personal and corporate investments are often in the form of mutual funds. Mutual funds represent a changing set of stocks and bonds, where transactions occur very frequently, often many times a day.

A formula for **continuous compounding** can be developed from Equation (2.3) by allowing the number of compounding periods per year to become infinitely large:

$$i_e = \lim_{m \to \infty} \left(1 + \frac{r}{m}\right)^m - 1$$

By noting from a definition of the natural exponential function, e, that

$$\lim_{m \to \infty} \left(1 + \frac{r}{m}\right)^m = e^r$$

we get

$$i_e = e^r - 1 \tag{2.5}$$

Example 2.7

Cash flow at the Arctic Oil Company is continuously reinvested. An investment in a new data logging system is expected to return a nominal interest of 40%, compounded continuously. What is the effective interest rate earned by this investment?

The nominal interest rate is given as $r = 0.40$. From Equation (2.5),

$$i_e = e^{0.4} - 1$$
$$= 1.492 - 1 = 0.492 \text{ or } 49.2\%$$

The effective interest rate earned on this investment is about 49.2%. ■

Although continuous compounding makes sense in some circumstances, it is rarely used. As with effective interest and nominal interest, in the days before calculators and computers, calculations involving continuous compounding were difficult to do. Consequently, discrete compounding is, by convention, the norm. As illustrated in Figure 2.3, the difference between continuous compounding and discrete compounding is relatively insignificant, even at a fairly high interest rate.

2.6 Cash Flow Diagrams

Sometimes a set of cash flows can be sufficiently complicated that it is useful to have a graphical representation. A **cash flow diagram** is a chart that summarizes the timing and magnitude of cash flows as they occur over time.

A cash flow diagram is actually a graph, although the vertical axis is not shown explicitly. The horizontal (X) axis represents time, measured in periods, and the vertical (Y) axis represents the size and direction of the cash flows. Individual cash flows are indicated by arrows pointing up or down from the horizontal axis, as indicated in Figure 2.4. The arrows that point up represent positive cash flows, or receipts. The downward pointing arrows represent neg-

Figure 2.3 Growth in Value of $1 at 30% Interest, at Various Compounding Periods

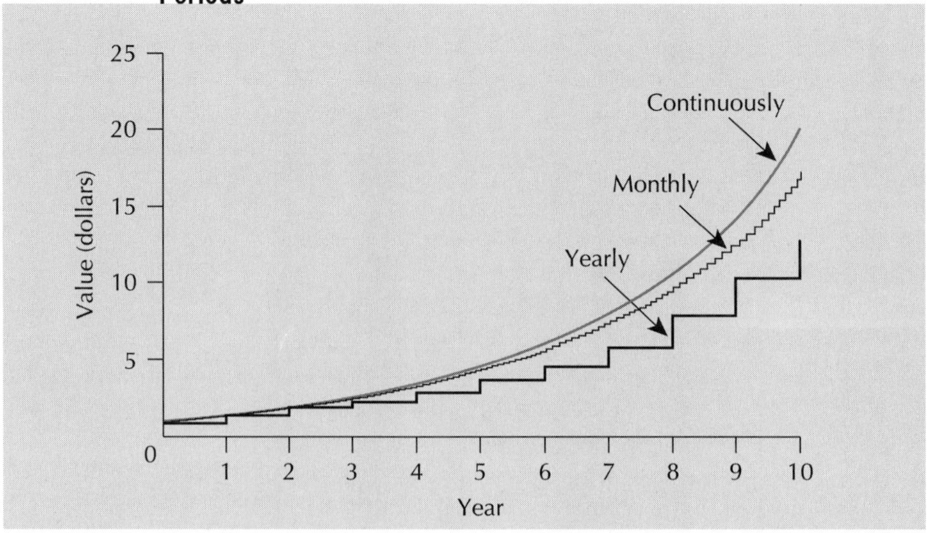

Figure 2.4 Cash Flow Diagram

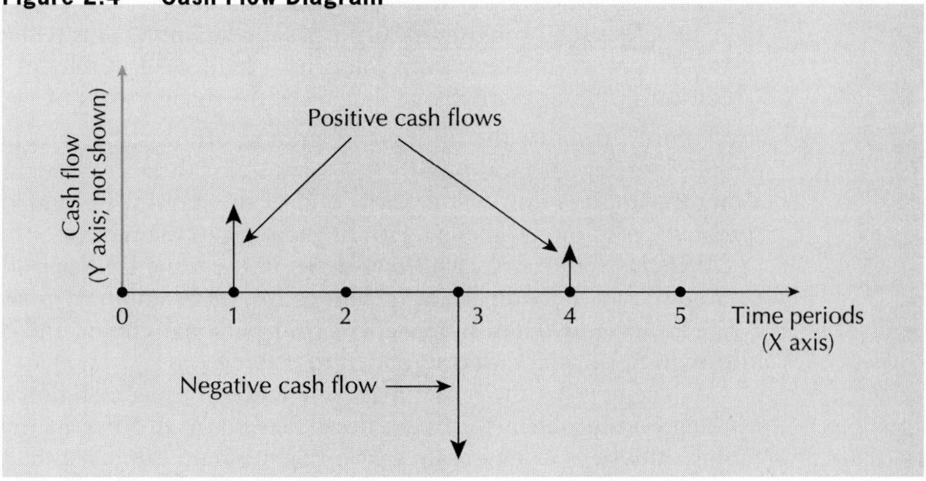

ative cash flows, or disbursements. See Close-Up 2.2 for some conventions pertaining to the beginning and ending of periods.

Example 2.8

Consider Ashok, a recent university graduate who is trying to summarize typical cash flows for each month. His monthly income is \$2200, received at the end of each month. Out of this he pays for rent, food, entertainment, telephone charges, and a credit card bill for all other purchases. Rent is \$700 per month

CLOSE-UP 2.2 Beginning and Ending of Periods

As illustrated in a cash flow diagram (see Figure 2.5), the end of one period is exactly the same point in time as the beginning of the next period. Now is time 0, which is the end of period −1 and also the beginning of period 1. The end of period 1 is the same as the beginning of period 2, and so on. N years from now is the end of period N and the beginning of period $(N + 1)$.

Figure 2.5 Beginning and Ending of Periods

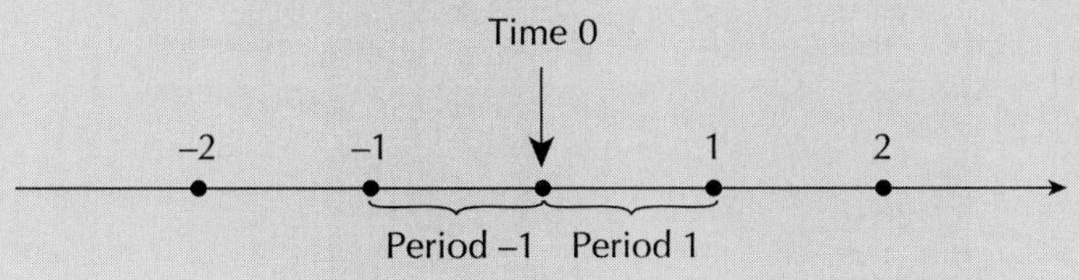

(including utilities), due at the end of each month. Weekly food and entertainment expenses total roughly $120, a typical telephone bill is $40 (due at the end of the first week in the month), and his credit card purchases average $300. Credit card payments are due at the end of the second week of each month.

Figure 2.6 shows the timing and amount of the disbursements and the single receipt over a typical month. It assumes that there are exactly four weeks in a month, and it is now just past the end of the month. Each arrow, which represents a cash flow, is labelled with the amount of the receipt or disbursement.

When two or more cash flows occur in the same time period, the amounts may be shown individually, as in Figure 2.6, or in summary form, as in Figure 2.7. The level of detail used depends on personal choice and the amount of information the diagram is intended to convey.

We suggest that the reader make a practice of using cash flow diagrams when working on a problem with cash flows that occur at different times. Just going through the steps in setting up a cash flow diagram can make the problem structure clearer. Seeing the pattern of cash flows in the completed diagram gives a "feel" for the problem. ■

Figure 2.6 Cash Flow Diagram for Example 2.8

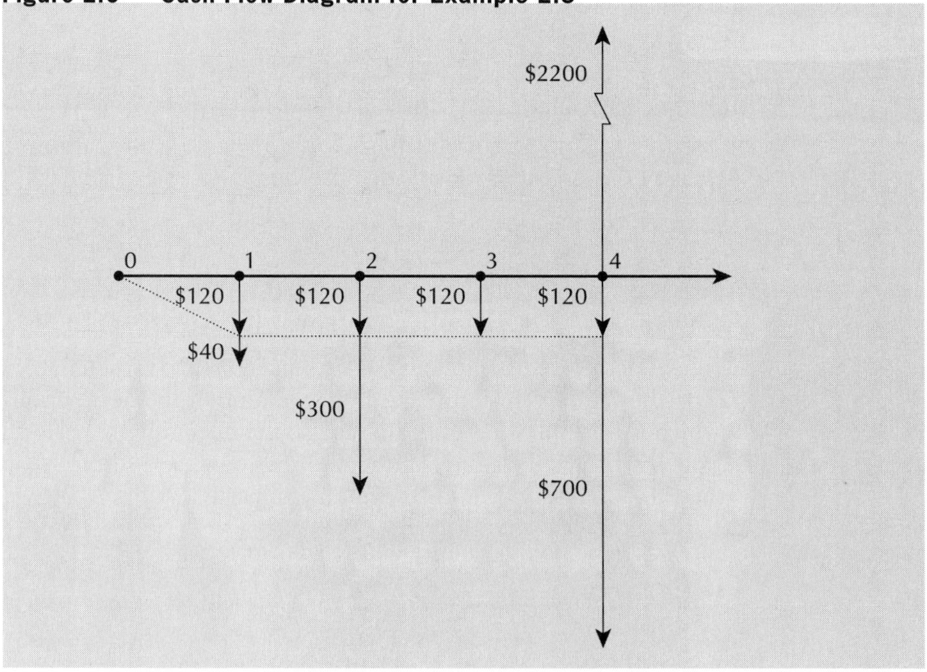

Figure 2.7 Cash Flow Diagram for Example 2.8 in Summary Form

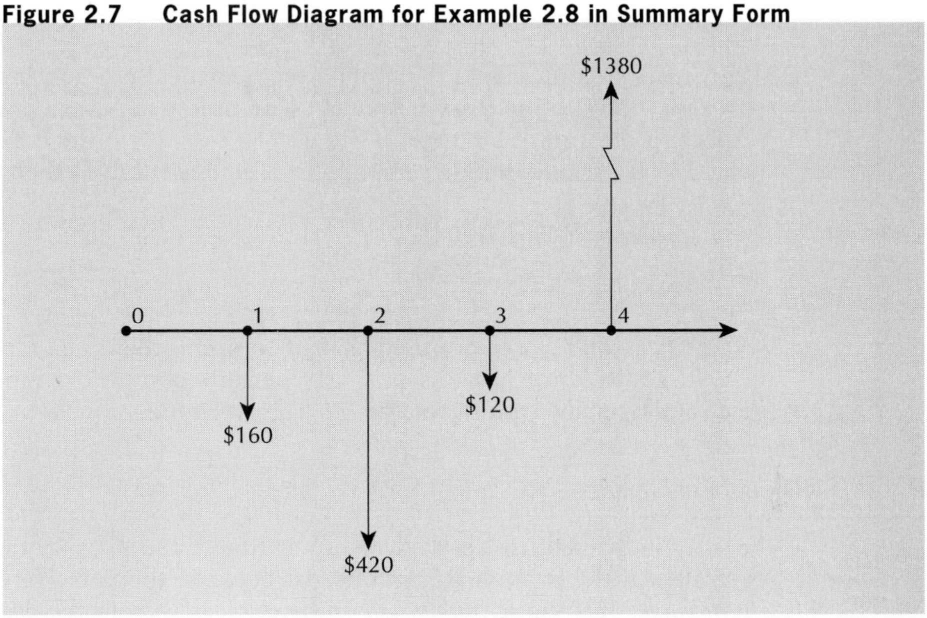

2.7 Equivalence

We started this chapter by pointing out that many engineering decisions involve costs and benefits that occur at different times. Making these decisions requires that the costs and benefits at different times be compared. To make these comparisons, we must be able to say that certain values at different times are *equivalent*. **Equivalence** is a condition that exists when the value of a cost at one time is equivalent to the value of the related benefit received at a different time. In this section we distinguish three concepts of equivalence that may underlie comparisons of costs and benefits at different times.

With **mathematical equivalence**, equivalence is a consequence of the mathematical relationship between time and money. This is the form of equivalence used in $F = P(1 + i)^N$.

With **decisional equivalence**, equivalence is a consequence of indifference on the part of a decision maker among available choices.

With **market equivalence**, equivalence is a consequence of the ability to exchange one cash flow for another at zero cost.

Although the mathematics governing money is the same regardless of which form of equivalence is most appropriate for a given situation, it can be important to be aware of what assumptions must be made for the mathematical operations to be meaningful.

2.7.1 Mathematical Equivalence

Mathematical equivalence is simply a mathematical relationship. It says that two cash flows, P_t at time t, and F_{t+N} at time $t + N$, are equivalent with respect to the interest rate, i, if $F_{t+N} = P_t(1 + i)^N$. Notice that if F_{t+N+M} (where M is a second number of periods) is equivalent to P_t then

$$F_{t+N+M} = P_t(1 + i)^{N+M}$$
$$F_{t+N+M} = F_{t+N}(1 + i)^M$$

so that F_{t+N} and F_{t+N+M} are equivalent to each other. The fact that mathematical equivalence has this property permits complex comparisons to be made among many cash flows which occur over time.

2.7.2 Decisional Equivalence

For any individual, two cash flows, P_t at time t and F_{t+N} at time $t + N$, are equivalent if the individual is indifferent between the two. Here, the implied interest rate relating P_t and F_{t+N} can be calculated from the decision that the cash flows are equivalent, as opposed to mathematical equivalence in which the interest rate determines whether the cash flows are equivalent. This can be illustrated best through an example.

Example 2.9

Bildmet is an extruder of aluminum shapes used in construction. The company buys aluminum from Alpure, an outfit that recovers aluminum from scrap. When Bildmet's purchasing manager, Greta Kehl, called in an order for 1000 kilograms of metal on August 15, she was told that Alpure was having production difficulties and was running behind schedule. Alpure's manager, Masaaki Sawada, said that he could ship the order immediately if Bildmet required it. If Alpure shipped Bildmet's order, they would not be able to fill an order for another user whom Mr. Sawada was anxious to impress with Alpure's reliability. Mr. Sawada suggested that, if Ms. Kehl would wait a week until August 22, he would show his appreciation by shipping 1100 kilograms then at the same cost to Bildmet as 1000 kilograms now. In either case, payment would be due at the end of the month. Should Ms. Kehl accept Alpure's offer?

The rate of exchange, 1100 kg to 1000 kg, may be written as (1 + 0.1) to 1 where the 0.1 = 10% is an interest rate for the one-week period. (This is equivalent to an effective interest rate of more than 14 000% per year!) Whether or not Ms. Kehl accepts the offer from Alpure depends on her situation. There is some chance of Bildmet's running out of metal if they don't get supplied for a week. This would require Ms. Kehl to do some scrambling to find other sources of metal in order to ship to her own customers on time. Ms. Kehl would prefer the 1000 kilograms on the fifteenth to 1000 kilograms on the twenty-second. But there is some minimum amount, larger than 1000 kilograms, that she would accept on the twenty-second in exchange for 1000 kilograms on the fifteenth. This amount would take into account both the measurable costs and also unmeasurable costs, such as inconvenience and anxiety.

Let the minimum rate at which Ms. Kehl would be willing to make the exchange be one kilogram on the fifteenth for $(1+x)$ kilograms on the twenty-second. In this case, if $x < 10\%$, Ms. Kehl should accept Alpure's offer of 1100 kilograms on the twenty-second.

In Example 2.9, the aluminum is a capital good that can be used productively by Bildmet. There is value in that use, and that value can be measured by Greta's willingness to postpone receiving the aluminum. It can be seen that interest is not necessarily a function of exchanges of money at different points in time. However, money is a convenient measure of the worth of a variety of goods, and so interest is usually expressed in terms of money. ■

2.7.3 Market Equivalence

Market equivalence is based on the idea that there is a market for money that permits cash flows in the future to be exchanged for cash flows in the present, and vice versa. Converting a future cash flow, F, to a present cash flow, P, is called borrowing money, while converting P to F is called lending or investing money. The market equivalence of two cash flows P and F means that they can be exchanged, one for the other, at zero cost.

The interest rate associated with an individual's borrowing money is usually a lot higher than the interest rate applied when that individual lends money. For example, the interest rate a bank pays on deposits is lower than what it charges to lend money to clients. The difference between these interest rates provides the bank with income. This means that, for an individual, market equivalence does not exist. An individual can exchange a present worth for a future worth by investing money but, if he or she were to try to borrow against that future worth to obtain money now, the resulting present worth would be less than the original amount invested. Moreover, each time either borrowing or lending occurred, transaction costs (the fees charged or cost incurred) would further diminish the capital.

Example 2.10

This morning, Averill bought a $5000 one-year guaranteed investment certificate (GIC) at his local bank. It has an effective interest rate of 7% per year. At the end of a year, the GIC will be worth $5350. On the way home from the bank, Averill unexpectedly discovered a valuable piece of art he had been seeking for some time. He wanted to buy it, but all his spare capital was "tied up" in the GIC. So he went back to the bank again, this time to negotiate a one-year loan for $5000, the cost of the piece of art. He figured that, if the loan came due at the same time as the GIC, he would simply pay off the loan with the proceeds of the GIC.

Unfortunately, Averill found out that the bank charges 10% effective interest per year on loans. Considering the proceeds from the GIC of $5350 one year from now, the amount the bank would give him today is only $5350/1.1 = $4864 (roughly), less any fees applicable to the loan. Averill has discovered that, for him, market equivalence does not hold. He cannot exchange $5000 today for $5350 one year from now, and vice versa, at zero cost. ■

Large companies with good records have opportunities that differ from those of individuals. Large companies borrow and invest money in so many ways, both internally and externally, that the interest rates for borrowing and for lending are very close to being the same, and also the transaction costs are negligible. They can shift funds from the future to the present by raising new money or by avoiding investment in a marginal project that would earn only the rate that they pay on new money. They can shift funds from the present to the future by undertaking an additional project or investing externally.

But how large is a "large company"? Established businesses of almost any size, and even individuals with some wealth and good credit, can acquire cash and invest at about the same interest rate, provided that the amounts are small relative to their total assets. For these companies and individuals, market equivalence is a reasonable model. Assuming market equivalence makes calculations easier and still generally results in good decisions.

For most of the remainder of the book, we will be making two broad assumptions with respect to equivalence: first, that market equivalence holds, and, second, that decisional equivalence can be expressed entirely in monetary terms. If these two assumptions are reasonably valid, mathematical equivalence can be used as an accurate model of how costs and benefits relate to one another over time. In several sections of the book, when we cover how firms raise capital and how to incorporate non-monetary aspects of a situation into a decision, we will discuss the validity of these two assumptions. In the meantime, however, mathematical equivalence is used to relate cash flows which occur at different points in time.

REVIEW PROBLEMS

REVIEW PROBLEM 2.1

Atsushi has had \$800 stashed under his mattress for 30 years. How much money has he lost by not putting it in a bank account at 8% annual compound interest all these years?

ANSWER

Since Atsushi has kept the \$800 under his mattress, he has not earned any interest over the 30 years. Had he put the money into an interest-bearing account, he would have far more today. We can think of the \$800 as a present amount and the amount in 30 years as the future amount.

Given: $P = \$800$

$i = 0.08$ per year

$N = 30$ years

Formula: $F = P(1 + i)^N$

$= \$800(1 + 0.08)^{30}$

$= \$8050.13$

Atsushi would have \$8050.13 in the bank account today had he deposited his \$800 at 8% annual compound interest. Instead, he has only \$800. He has suffered an opportunity cost of \$8050.13 − \$800 = \$7250.13 by not investing the money. ■

REVIEW PROBLEM 2.2

You want to buy a new computer, but you are \$1000 short of the amount you need. Your aunt has agreed to lend you the \$1000 you need now, provided you pay her \$1200 two years from now. She compounds interest monthly. Another place from which you can borrow \$1000 is the bank. There is, however, a loan processing fee of \$20, which will be included in the loan amount. The bank is expecting to receive \$1220 two years from now based on monthly compounding of interest.

(a) What monthly rate is your aunt charging you for the loan? What is the bank charging?

(b) What effective annual rate is your aunt charging? What is the bank charging?

(c) Would you prefer to borrow from your aunt or from the bank?

ANSWER

(a) *Your aunt*

Given: $P = \$1000$

$F = \$1200$

$N = 24$ months (since compounding is done monthly)

Formula: $F = P(1 + i)^N$

The formula $F = P(1 + i)^N$ must be solved in terms of i to answer the question.

$$\begin{aligned} i &= \sqrt[N]{F/P} - 1 \\ &= \sqrt[24]{\$1200/\$1000} - 1 \\ &= 0.007626 \end{aligned}$$

Your aunt is charging interest at a rate of approximately 0.76% per month.

The bank

Given: $P = \$1020$ (since the fee is included in the loan amount)

$F = \$1220$

$N = 24$ months (since compounding is done monthly)

$$\begin{aligned} i &= \sqrt[N]{F/P} - 1 \\ &= \sqrt[24]{\$1220/\$1020} - 1 \\ &= 0.007488 \end{aligned}$$

The bank is charging interest at a rate of approximately 0.75% per month.

(b) The effective annual rate can be found with the formula $i_e = (1 + r/m)^m - 1$, where r is the nominal rate per year and m is the number of compounding periods per year. Since the number of compounding periods per year is 12, notice that r/m is simply the interest rate charged per month.

Your aunt

$i = 0.007626$ per month

Then

$$\begin{aligned} i_e &= (1 + r/m)^m - 1 \\ &= (1 + 0.007626)^{12} - 1 \\ &= 0.095445 \end{aligned}$$

The effective annual rate your aunt is charging is approximately 9.54%.

The bank

$$i = 0.007488 \text{ per month}$$

Then

$$\begin{aligned} i_e &= (1 + r/m)^m - 1 \\ &= (1 + 0.007488)^{12} - 1 \\ &= 0.09365 \end{aligned}$$

The effective annual rate for the bank is approximately 9.37%.

(c) The bank appears to be charging a lower interest rate than does your aunt. This can be concluded by comparing the two monthly rates or the effective annual rates the two charge. If you were to base your decision only upon who charged the lower interest rate, you would pick the bank, despite the fact they have a fee. However, although you are borrowing \$1020 from the bank, you are getting only \$1000 since the bank immediately gets its \$20 fee. The cost of money for you from the bank is better calculated as:

Given: $P = \$1000$

$F = \$1220$

$N = 24$ months (since compounding is done monthly)

$$\begin{aligned} i &= \sqrt[N]{F/P} - 1 \\ &= \sqrt[24]{\$1220/\$1000} - 1 \\ &= 0.00832 \end{aligned}$$

From this point of view, the bank is charging interest at a rate of approximately 0.83% per month, and you would be better off borrowing from your aunt. ■

REVIEW PROBLEM 2.3

At the end of four years, you would like to have \$5000 in a bank account to purchase a used car. What you need to know is how much to deposit in the bank account now. The account pays daily interest. Create a spreadsheet and plot the necessary deposit today as a function of interest rate. Consider nominal interest rates ranging from 5% to 15% per year, and assume that there are 365 days per year.

ANSWER

From the formula $F = P(1 + i)^N$, we have $\$5000 = P(1 + i)^{365\times4}$. This gives

$$P = \$5000 \times \frac{1}{(1 + i)^{365 \times 4}}$$

Table 2.3 is an excerpt from a sample spreadsheet. It shows the necessary deposit to accumulate $5000 over four years at a variety of interest rates. The following is the calculation for cell B2 (i.e., the second row, second column):

$$\$5000 \times \frac{1}{\left[1 + \left(\frac{\text{A2}}{365}\right)\right]^{365 \times 4}}$$

The specific implementation of this formula will vary, depending on the particular spreadsheet program used. Figure 2.8 is a diagram of the necessary deposits plotted against interest rates. ■

Table 2.3 Necessary Deposits for a Range of Interest Rates

Interest Rate (%)	Necessary Deposit ($)
0.05	4114
0.06	3957
0.07	3805
0.08	3660
0.09	3520
0.10	3385
0.11	3256
0.12	3131
0.13	3011
0.14	2896
0.15	2785

Figure 2.8 Graph for Review Problem 2.3

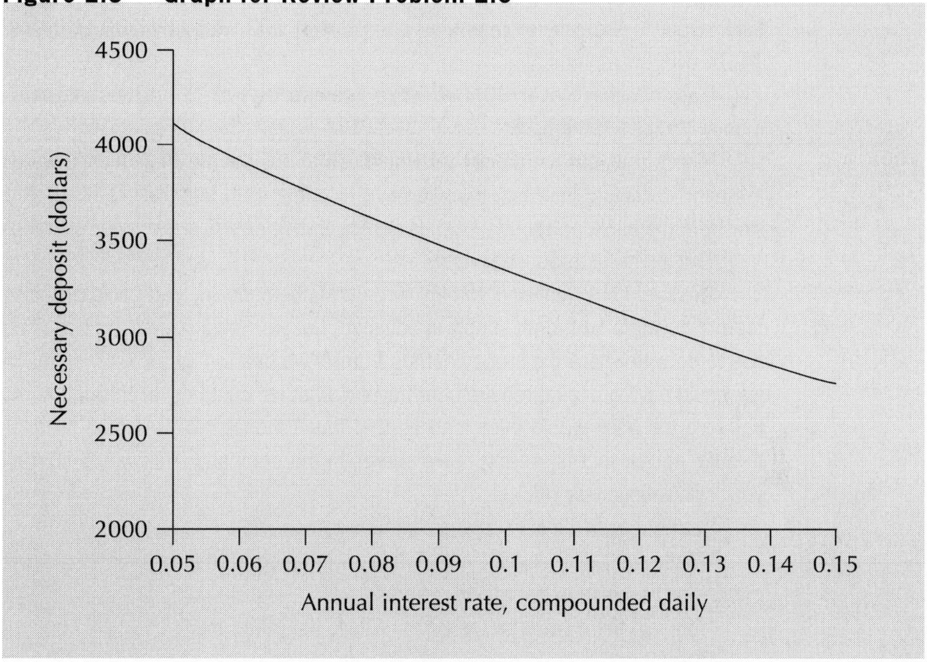

SUMMARY

This chapter has provided an introduction to interest, interest rate terminology, and interest rate conventions. Through a series of examples, the mechanics of working with simple and compound interest, nominal and effective interest rates, and continuous compounding were illustrated. Cash flow diagrams were introduced in order to represent graphically monetary transactions at various points in time. The final part of the chapter contained a discussion of various forms of cash flow equivalence: mathematical, decisional, and market equivalence. With the assumption that mathematical equivalence can be used as an accurate model of how costs and benefits relate to one another over time, we now move on to Chapter 3, in which equivalence formulas for a variety of cash flow patterns are presented.

Engineering Economics in Action, Part 2B: You Just Have to Know When

Naomi and Terry were looking at the steel orders for the J-class line. Terry thought money could be saved by ordering in advance. "Now look at this," Terry said. "In the three months ending last December, we ordered steel for a total cost of $1 600 000. If we had bought this steel at the beginning of July, it would have cost only $1 400 000. That's a savings of $200 000!"

"Good observation, Terry, but I don't think buying in advance is the right thing to do. If you think about it, the rate of return on our $1 400 000 would be 200 000/1 400 000 or about 14.3% over six months."

"Yes, but that's over 30% effective interest, isn't it? I'll bet we only make 8% or 10% for money we keep in the bank."

"That's true, but the money we would use to buy the steel in advance we don't have sitting in the bank collecting interest. It would have to come from somewhere else, either money we borrow from the bank, at about 14% plus administrative costs, or from our shareholders."

"But it's still a good idea, right?"

"Well, you are right and you are wrong. Mathematically, you could probably show the advantage of buying a six-month supply in advance. But we wouldn't do it for two reasons. The first one has to do with where the money comes from. If we had to pay for six months of steel in advance, we would have a capital requirement beyond what we could cover through normal cash flows. I'm not sure the bank would even lend us that much money, so we would probably have to raise it through equity, that is, selling more shares in the company. This would cost a lot, and throw all your calculations way off."

"Just because it's such a large amount of money?"

"That's right. Our regular calculations are based on the assumption that the capital requirements don't take an extraordinary effort."

"You said there were two reasons. What's the other one?"

"The other reason is that we just wouldn't do it."

"Huh?"

"We just wouldn't do it. Right now the steel company's taking the risk — if we can't pay, they are in trouble. If we buy in advance, it's the other way around — if our widget orders dropped, we would be stuck with some pretty expensive raw materials. We would also have the problem of where to store the steel, and other practical difficulties. It makes sense mathematically, but I'm pretty sure we just wouldn't do it."

Terry looked a little dejected. Naomi continued, "But your figures make sense. The first thing to do is find out why we are carrying that account so long before we pay it off. The second thing to do is see if we can't get that price break, retroactively. We are good customers, and I'll bet we can convince them to give us the price break anyhow, without changing our ordering pattern. Let's talk to Clem about it."

"But, Naomi, why use the mathematical calculations at all, if they don't work?"

"But they do work, Terry. You just have to know when."

PROBLEMS

2.1 Using 12% simple interest per year, how much interest will be owed on a loan of $500 at the end of two years?

2.2 If a sum of $3000 is borrowed for six months at 9% simple interest per year, what is the total amount due (principal and interest) at the end of six months?

2.3 What principal amount will yield \$150 in interest at the end of three months when the interest rate is 1% simple interest per month?

2.4 If \$2400 interest is paid on a two-year simple-interest loan of \$12 000, what is the interest rate per year?

2.5 Simple interest of \$190.67 is owed on a loan of \$550 after four years and four months. What is the annual interest rate?

2.6 How much will be in a bank account at the end of five years if \$2000 is invested today at 12% interest per annum, compounded yearly?

2.7 How much is accumulated in each of these savings plans over two years?

(a) Deposit \$1000 today at 10% compounded annually.

(b) Deposit \$900 today at 12% compounded monthly.

2.8 Greg wants to have \$50 000 in five years. The bank is offering five-year investment certificates that pay 8% nominal interest, compounded quarterly. How much money should he invest in the certificates to reach his goal?

2.9 Greg wants to have \$50 000 in five years. He has \$20 000 today to invest. The bank is offering five-year investment certificates that pay interest compounded quarterly. What is the minimum nominal interest rate he would have to receive to reach his goal?

2.10 Greg wants to have \$50 000. He will invest \$20 000 today in investment certificates that pay 8% nominal interest, compounded quarterly. How long will it take him to reach his goal?

2.11 Greg will invest \$20 000 today in five-year investment certificates that pay 8% nominal interest, compounded quarterly. How much money will this be in five years?

2.12 You bought an antique car three years ago for \$50 000. Today it is worth \$65 000.

(a) What annual interest rate did you earn if interest is compounded yearly?

(b) What monthly interest rate did you earn if interest is compounded monthly?

2.13 You have a bank deposit now worth \$5000. How long will it take for your deposit to be worth more than \$8000 if

(a) the account pays 5% actual interest every half year, and is compounded?

(b) the account pays 5% nominal interest, compounded semi-annually?

2.14 Some time ago, you put \$500 into a bank account for a "rainy day." Since then, the bank has been paying you 1% per month, compounded monthly. Today, you checked the balance, and found it to be \$708.31. How long ago did you deposit the \$500?

2.15 **(a)** If you put \$1000 in a bank account today that pays 10% interest per year, how much money could be withdrawn 20 years from now?

(b) If you put \$1000 in a bank account today that pays 10% *simple* interest per year, how much money could be withdrawn 20 years from now?

2.16 How long will it take any sum to double itself,

(a) with an 11% simple interest rate?

(b) with an 11% interest rate, compounded annually?

(c) with an 11% interest rate, compounded continuously?

2.17 Compute the effective annual interest rate on each of these investments:

(a) 25% nominal interest, compounded semi-annually

(b) 25% nominal interest, compounded quarterly

(c) 25% nominal interest, compounded continuously

2.18 For a 15% effective annual interest rate, what is the nominal interest rate if

(a) interest is compounded monthly?

(b) interest is compounded daily (assume 365 days per year)?

(c) interest is compounded continuously?

2.19 A Studebaker automobile that cost \$665 in 1934 was sold as an antique car at \$14 800 in 1998. What was the rate of return on this "investment"?

2.20 Clifford has \$X right now. In 5 years, X will be \$3500 if it is invested at 7.5%, compounded annually. Determine the present value of X. If Clifford invested \$X at 7.5%, compounded daily, how much would the value of X be in 10 years?

2.21 You have just won a lottery prize of \$1 000 000 collectable in ten yearly instalments of \$100 000, starting today. Why is this prize not really \$1 000 000? What is it really worth today if money can be invested at 10% annual interest, compounded monthly? Use a spreadsheet to construct a table showing the present worth of each instalment, and the total present worth of the prize.

2.22 Suppose in Problem 2.21 that you have a large mortgage you want to pay off now. You propose an alternate, but equivalent, payment scheme. You would like \$300 000 today, and the balance of the prize in 5 years when you intend to purchase a large piece of waterfront property. How much will the payment be in 5 years? Assume that annual interest is 10%, compounded monthly.

2.23 You are looking at purchasing a new computer for your four-year undergraduate program. Brand 1 costs \$4000 now, and you expect it will last throughout your program without any upgrades. Brand 2 costs \$2500 now and will need an upgrade at the end of 2 years, which you expect to be \$1700. With 8% annual interest, compounded monthly, which is the less expensive alternative, if they both provide the same level of service and will be worthless at the end of the four years?

2.24 The Bank of Edmonton advertises savings account interest as 6% compounded daily. What is the effective interest rate?

2.25 The Bank of Kitchener is offering a new savings account that pays a nominal 7.99% interest, compounded continuously. Will your money earn more in this account than in a daily interest account that pays 8%?

2.26 You are comparing two investments. The first pays 1% interest per month, compounded monthly, and the second pays 6% interest per six months, compounded every six months.

(a) What is the effective semi-annual interest rate for each investment?

(b) What is the effective annual interest rate for each investment?

(c) Based on interest rate, which investment do you prefer? Does your decision depend on whether you make the comparison based on an effective six-month rate or an effective one-year rate?

2.27 The Bank of Calgary advertises savings account interest as 5.5% compounded weekly and chequing account interest at 7% compounded monthly. What are the effective interest rates for the two types of account?

2.28 Victory Visa, Magnificent Master Card, and Amazing Express are credit card companies which charge different interest on overdue accounts. Victory Visa charges 26% compounded daily, Magnificent Master Card charges 28% compounded weekly, and Amazing Express charges 30% compounded monthly. Based on interest rate, which credit card has the best deal?

2.29 April has a bank deposit now worth \$796.25. A year ago, it was \$750. What was the nominal monthly interest rate on her account?

2.30 You have $50 000 to invest in the stock market and have sought the advice of Adam, an experienced colleague who is willing to advise you, for a fee. Adam has told you that he has found a one-year investment for you that provides 15% interest, compounded monthly.

(a) What is the effective annual interest rate, based on a 15% nominal annual rate and monthly compounding?

(b) Adam says that he will make the investment for you for a modest fee of 2% of the investment's value one year from now. If you invest the $50 000 today, how much will you have at the end of one year (before Adam's fee)?

(c) What is the effective annual interest rate of this investment including Adam's fee?

2.31 May has $2000 in her bank account right now. She wanted to know how much it would be in one year, so she calculated and came up with $2140.73. Then she realized that she made a mistake. She wanted to use the formula for monthly compounding, but instead, she used the continuous compounding formula. Redo the calculation for May and find out how much will actually be in her account a year from now.

2.32 Hans now has $6000. In three months, he will receive a cheque for $2000. He must pay $900 at the end of each month (starting exactly one month from now). Draw a single cash flow diagram illustrating all of these payments for a total of six monthly periods. Include his cash on hand as a payment at time 0.

2.33 Margaret is considering an investment that will cost her $500 today. It will pay her $100 at the end of each of the next 12 months, and cost her another $300 one year from today. Illustrate these cash flows in two cash flow diagrams. The first should show each cash flow element separately, and the second should show only the net cash flows in each period.

2.34 Heddy is considering working on a project that will cost her $20 000 today. It will pay her $10 000 at the end of each of the next 12 months, and cost her another $15 000 at the end of each quarter. An extra $10 000 will be received at the end of the project, one year from now. Illustrate these cash flows in two cash flow diagrams. The first should show each cash flow element separately, and the second should show only the net cash flow in each period.

2.35 Illustrate the following cash flows over 12 months in a cash flow diagram. Show only the net cash flow in each period.

Cash Payments	\$20 every 3 months, starting now
Cash Receipts	Receive \$30 at the end of the first month, and from that point on, receive 10% more than the previous month at the end of each month

2.36 There are two possible investments, A and B. Their cash flows are shown in the table below. Illustrate these cash flows over 12 months in two cash flow diagrams. Show only the net cash flow in each period. Just by looking at the diagrams, would you prefer one investment over the other? Comment on this.

	Investment A	**Investment B**
Payments	\$2400 now and a closing fee of \$200 at the end of month 12	\$500 every 2 months, starting two months from now
Receipts	\$250 monthly payment at the end of each month	Receive \$50 at the end of the first month, and from that point on, receive \$50 more than the previous month at the end of each month

2.37 You are indifferent between receiving \$100 today and \$110 one year from now. The bank pays you 6% interest on deposits and charges you 8% for loans. Name the three types of equivalence and comment (one sentence for each) on whether or not each exists for this situation and why.

2.38 June has a small house on a small street in a small town. If she sells the house now, she will likely get \$110 000 for it. If she waits for one year, she will likely get more, say, \$120 000. If she sells the house now, she can invest the money in one-year guaranteed investment certificate (GIC) that pays 8% interest, compounded monthly. If she keeps the house, then the interest on the mortgage payments is 8% compounded daily. June is indifferent between the two options: selling the house now and keeping the house for another year. Discuss whether each of the three types of equivalence exists in this case.

2.39 Using a spreadsheet, construct graphs for the loan described in (a) below.

(a) Plot the amount owed (principal plus interest) on a simple interest loan of \$100 for N years for $N = 1, 2, \ldots 10$. On the same graph, plot the amount owed on a compound interest loan of \$100 for N years for $N = 1, 2, \ldots 10$. The interest rate is 6% per year for each loan.

(b) Repeat part (a), but use an interest rate of 18%. Observe the dramatic effect compounding has on the amount owed at the higher interest rate.

2.40 **(a)** At 12% interest per annum, how long will it take for a penny to become a million dollars? How long will it take at 18%?

(b) Show the growth in values on a spreadsheet using ten-year time intervals.

2.41 Use a spreadsheet to determine how long it will take for a $100 deposit to double in value for each of the following interest rates and compounding periods. For each, plot the size of the deposit over time, for as many periods as necessary for the original sum to double.

(a) 8% per year, compounded monthly

(b) 11% per year, compounded semi-annually

(c) 12% per year, compounded continuously

2.42 Construct a graph showing how the effective interest rate for the following nominal rates increases as the compounding period becomes shorter and shorter. Consider a range of compounding periods of your choice from daily compounding to annual compounding.

(a) 6% per year

(b) 10% per year

(c) 20% per year

2.43 Today, an investment you made three years ago has matured and is now worth $3000. It was a three-year deposit which bore an interest rate of 10% per year, compounded monthly. You knew at the time that you took a risk in making such an investment because interest rates vary over time and you "locked in" at 10% for three years.

(a) How much was your initial deposit? Plot the value of your investment over the three-year period.

(b) Looking back over the past three years, interest rates for similar one-year investments did indeed vary. The interest rates were 8% the first year, 10% the second, and 14% the third. Plot the value of your initial deposit over time as if you had invested at this set of rates, rather than for a constant 10% rate. Did you lose out by having locked into the 10% investment? If so, by how much?

Canada Trust's Powerline Account

CANADIAN MINI-CASE 2.1

Canada Trust is a consumer financial services company headquartered in London, Ontario. It offers a "Powerline" line of credit which allows customers to borrow money with the ease of using a chequing account. One form is referred to as a "secured" line of credit. In this case, a valuable asset, usually the customer's home, is used to guarantee the loan. It is thus a form of mortgage where the amount borrowed can freely vary up to some maximum amount.

Each monthly statement displays the interest rate used to calculate that month's interest. The January 1999 statement contained the following information:

Interest is calculated at an annual rate of 6.75% (0.555% per 30-day month, or 0.0185 per day).

Although the printed information does not use the terms *nominal* or *effective*, or define the compounding period, these points can easily be deduced. Noting that $6.75/365 = 0.0185$ and that $0.0185 \times 30 = 0.555$, it is clear that the quoted annual rate is nominal, and the compounding period is daily. The actual effective interest rate is $(1 + 0.000185)^{365} - 1 = 6.985\%$.

Discussion

Interest information must be disclosed by law, but lenders and borrowers have some latitude as to how and where they disclose it. Moreover, there is a natural desire to make the interest rate look lower than it really is for borrowers, and higher than it really is for lenders.

In the Canada Trust example, the effective interest rate is 6.985%. However, the uninformed bank client would certainly get the impression that the actual cost of borrowing was roughly one-quarter of one percent less. If a consumer compared interest rates between institutions, and one quoted the nominal rate and the other the effective rate, the one quoting the nominal rate could seem to be better even if it was worse in some cases.

Questions

1) Go to your local bank branch and find out the interest rate paid for various kinds of savings accounts, chequing accounts, and loans. For each interest rate quoted, determine if it is a nominal or effective rate. If it is nominal, determine the compounding period and calculate the effective interest rate.
2) Have a contest with your classmates to see who can find the organization that will lend money to a student like you at the cheapest effective interest rate, or that will take investments which provide a guaranteed return at the highest effective interest rate. The valid rates must be generally available, not tied to particular behaviour by the client, and not secured to an asset (like a mortgage).
3) If you borrowed $1000 at the best rate you could find and invested it at the best rate you could find, how much money would you make or lose in a year? Explain why the result of your calculation could not have the opposite sign.

C H A P T E R 3

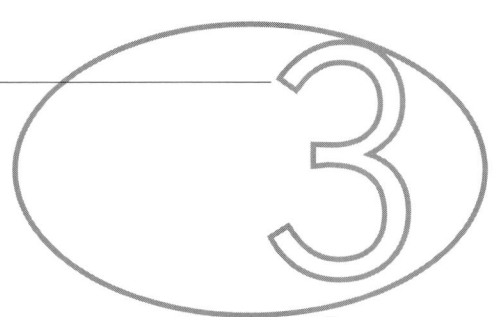

Cash Flow Analysis

Engineering Economics in Action, Part 3A: Apples and Oranges

The flyer was slick, all right. The information was laid out so anybody could see that leasing palletizing equipment through the Provincial Finance Company (PFC) made much more sense than buying it. It was something Naomi could copy right into her report to Clem.

Naomi had been asked to check out options for automating part of the Shipping Department. Parts were to be stacked and bound on plastic pallets, then loaded onto trucks and sent to one of Canadian Widget's sister companies. The saleswoman for the company whose equipment seemed most suitable for Canadian Widget's needs included the leasing flyer with her quote.

Naomi looked at the figures again. They seemed to make sense, but there was something that didn't seem right to her. For one thing, if it was cheaper to lease, why didn't everybody lease everything? She knew that some things, like automobiles and airplanes, are often leased instead of bought, but generally companies buy assets. Second, where was the money coming from to give the finance company a profit? If the seller was getting the same amount and the buyer was paying less, how could PFC make money?

"Got a recommendation on that palletizer yet, Naomi?" Clem's voice was cheery as he suddenly appeared at her doorway. Naomi knew that the Shipping Department was the focus of Clem's attention right now and he wanted to get improvements in place as soon as possible.

"Yes, I do. There's really only one that will do the job, and it does it well at a good price. There is something I'm trying to figure out, though. Christine sent me some information about leasing it instead of buying it, and I'm trying to figure out where the catch is. There has got to be one, but I can't see it right now."

"O.K., let me give you a hint: apples and oranges. You can't add them. Now, let's get the paperwork started for that palletizer. The shipping department is just too much of a bottleneck." Clem disappeared from her door as quickly as he had arrived, leaving Naomi musing to herself.

"Apples and ORANGES? APPLES and oranges? Ahh... apples and oranges, of course!"

3.1 Introduction

Chapter 2 showed that interest is the basis for determining whether or not different patterns of cash flows are equivalent. Rather than comparing patterns of cash flows from first principles, it is usually easier to use functions that define *mathematical* equivalence among certain common cash flow patterns. These functions are called *compound interest factors*. We discuss a number of these common cash flow patterns along with their associated compound interest factors in this chapter. These compound interest factors are used throughout the remainder of the book. It is, therefore, particularly important to understand their use before proceeding to subsequent chapters.

This chapter opens with an explanation of how cash flow patterns that engineers commonly use are simplified approximations of complex reality. Next, we discuss four simple, discrete cash flow patterns and the compound interest factors that relate them to each other. There is then a brief discussion of the case in which the number of time periods considered is so large that it is treated as though the relevant cash flows continue indefinitely. Appendix 3A discusses modelling cash flow patterns when the interval between disbursements or receipts is short enough that we may view the flows as being continuous. Appendix 3B presents mathematical derivations of the compound interest factors.

3.2 Timing of Cash Flows and Modelling

The actual timing of cash flows can be very complicated and irregular. Unless some simple approximation is used, comparisons of different cash flow sequences will be very difficult and impractical. Consider, for example, the cash flows generated by a relatively simple operation like a service station that sells gasoline and supplies, and also services cars. Some cash flows, like sales of gasoline and minor supplies, will be almost continuous during the time the station is open. Other flows, like receipts for the servicing of cars, will be on a daily basis. Disbursements for wages may be on a weekly basis. Some disbursements, like those for a manager's salary and for purchases of gasoline and supplies, may be monthly. Disbursements for insurance and taxes may be quarterly or semi-annual. Other receipts and disbursements, like receipts for major repairs or disbursements for used parts, may be irregular.

An analyst trying to make a comparison of two projects with different, irregular timings of cash flows might have to record each of the flows of the projects, then, on a one-by-one basis, find summary equivalent values like present worth that would be used in the comparison. This activity would be very time consuming and tedious if it could be done, but it probably could not be done because the necessary data would not exist. If the projects were potential rather than actual, the cash flows would have to be predicted. This could not be done with great precision for either size or timing of the flows. Even if the analysis were of the past performances of ongoing operations, it is unlikely that it would be worthwhile to maintain a data bank that contained the exact timing of all cash flows.

Because of the difficulties of making precise calculations of complex and irregular cash flows, engineers usually work with fairly simple models of cash flow patterns. The most common type of model assumes that all cash flows and all compounding of cash flows occur at the ends of conventionally defined periods like months or years. Models that make this assumption are called **discrete models**. In some cases, analysts use models that assume cash flows and their compounding occur continuously over time; such models are called **continuous models**. Whether the analyst uses discrete modelling or continuous modelling, the model is usually an approximation. Cash flows do not occur only at the ends of conventionally defined periods, nor are they actually continuous. We shall emphasize discrete models throughout the book because they are more common and more readily understood by persons of varied backgrounds. Discrete cash flow models are discussed in the main body of this chapter, and continuous models are presented in Appendix 3A, at the end of this chapter.

3.3 Compound Interest Factors for Discrete Compounding

Compound interest factors are formulas that define mathematical equivalence for specific common cash flow patterns. The compound interest factors permit cash flow analysis to be done more conveniently because tables or spreadsheet functions can be used instead of complicated formulas. This section presents compound interest factors for four discrete cash flow patterns that are commonly used to model the timing of receipts and disbursements in engineering economic analysis. The four patterns are:

1. A single disbursement or receipt
2. A set of equal disbursements or receipts over a sequence of periods, referred to as an **annuity**
3. A set of disbursements or receipts that change by a constant *amount* from one period to the next in a sequence of periods, referred to as an **arithmetic gradient series**
4. A set of disbursements or receipts that change by a constant *proportion* from one period to the next in a sequence of periods, referred to as a **geometric gradient series**

The principle of discrete compounding requires several assumptions:

1. Compounding periods are of equal length.
2. Each disbursement and receipt occurs at the end of a period. A payment at time zero can be considered to occur at the end of period -1.
3. Annuities and gradients coincide with the ends of sequential periods. (Section 3.8 suggests several methods for dealing with annuities and gradients that do not coincide with the ends of sequential periods.)

Mathematical derivations of six of the compound interest factors are given in Appendix 3B at the end of this chapter.

3.4 Compound Interest Factors for Single Disbursements or Receipts

In many situations, a single disbursement or receipt is an appropriate model of cash flows. For example, the salvage value of production equipment with a limited service life will be a single receipt at some future date. An investment today to be redeemed at some future date is another example.

Figure 3.1 illustrates the general form of a single disbursement or receipt. Two commonly used factors relate a single cash flow in one period to another single cash flow in a different period. They are the *compound amount factor* and the *present worth factor*.

Figure 3.1 Single Receipt at End of Period N

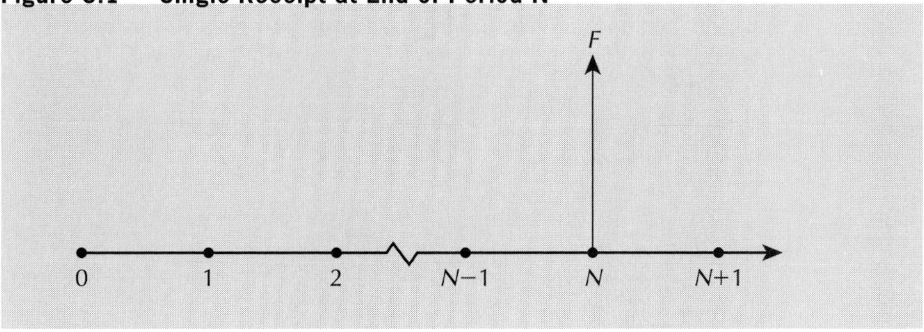

The **compound amount factor**, denoted by $(F/P,i,N)$, gives the future amount, F, that is equivalent to a present amount, P, when the interest rate is i and the number of periods is N. The value of the compound amount factor is easily seen as coming from Equation (2.1), the compound interest equation, which relates present and future values,

$$F = P(1 + i)^N$$

In the symbolic convention used for compound interest factors, this is written

$$F = P(1 + i)^N = P(F/P,i,N)$$

so that the compound amount factor is

$$(F/P,i,N) = (1 + i)^N$$

A handy way of thinking of the notation is (reading from left to right): "What is F, given P, i, and N?"

The compound amount factor is useful in determining the future value of an investment made today if the number of periods and the interest rate are known.

The **present worth factor**, denoted by $(P/F,i,N)$, gives the present amount, P, that is equivalent to a future amount, F, when the interest rate is i and the number of periods is N. The present worth factor is the inverse of the compound amount factor, $(F/P,i,N)$. That is, while the compound amount factor gives the future amount, F, that is equivalent to a present amount, P, the present worth factor goes in the other direction. It gives the present worth, P, of a future amount, F. Since $(F/P,i,N) = (1 + i)^N$,

$$(P/F,i,N) = \frac{1}{(1 + i)^N}$$

The compound amount factor and the present worth factor are fundamental to engineering economic analysis. Their most basic use is to convert a single cash flow which occurs at one point in time to an equivalent cash flow at another point in time. When comparing several individual cash flows which occur at different points in time, an analyst would apply the compound amount factor or the present worth factor, as necessary, to determine the equivalent cash flows at a common reference point in time. In this way, each of the cash flows is stated as an amount at one particular time. Example 3.1 illustrates this process.

Although the compound amount factor and the present worth factor are relatively easy to calculate, some of the other factors discussed in this chapter are more complicated, and it is therefore desirable to have an easier way to determine their values. The compound interest factors are sometimes available as functions in calculators and spreadsheets, but often these functions are provided in an awkward format that makes them relatively difficult to use. They can, however, be fairly easily programmed in a calculator or spreadsheet.

A traditional and still useful method for determining the value of a compound interest factor is to use tables. Appendix A at the back of this book lists values for all the compound interest factors for a selection of interest rates for discrete compounding periods. The desired compound interest factor can be determined by looking in the appropriate table.

Example 3.1

How much money will be in a bank account at the end of 15 years if \$100 is invested today and the nominal interest rate is 8% compounded semi-annually?

Since a present amount is given and a future amount is to be calculated, the appropriate factor to use is the compound amount factor, $(F/P,i,N)$. There are several ways of choosing i and N to solve this problem. The first method is to observe that, since interest is compounded semi-annually, the number of compounding periods, N, is 30. The interest rate per six-month period is 4%. Then

$$\begin{aligned} F &= \$100(F/P, 4\%, 30) \\ &= \$100(1 + 0.04)^{30} \\ &= \$324.34 \end{aligned}$$

The bank account will hold \$324.34 at the end of 15 years.

Alternatively, we can obtain the same results by using the interest factor tables.

$$\begin{aligned} F &= \$100(3.2434) \quad \text{(from Appendix A)} \\ &= \$324.34 \end{aligned}$$

A second solution to the problem is to calculate the *effective* yearly interest rate and then compound over 15 years at this rate. Recall from Equation (2.4) that the effective interest rate per year is

$$i_e = \left(1 + \frac{r}{m}\right)^m - 1$$

where i_e = the effective annual interest rate
r = the nominal rate per year
m = the number of periods in a year

$$i_e = (1 + 0.08/2)^2 - 1 = 0.0816$$

where $r = 0.08$
$m = 2$

When the effective yearly rate for each of 15 years is applied to the future worth computation, the future worth is

$$\begin{aligned} F &= P(F/P,i,N) \\ &= P(1 + i)^N \\ &= \$100(1 + 0.0816)^{15} \\ &= \$324.34 \end{aligned}$$

Once again, we conclude that the balance will be $324.34. ■

3.5 Compound Interest Factors for Annuities

The next four factors involve a series of uniform receipts or disbursements that start at the end of the first period and continue over N periods, as illustrated in Figure 3.2. This pattern of cash flows is called an annuity. Mortgage or lease payments and maintenance contract fees are examples of the annuity cash flow pattern. Annuities may also be used to model series of cash flows that fluctuate over time around some average value. Here the average value would be the constant uniform cash flow. This would be done if the fluctuations were unknown or deemed to be unimportant for the problem.

The **sinking fund factor**, denoted by $(A/F,i,N)$, gives the size, A, of a repeated receipt or disbursement that is equivalent to a future amount, F, if the interest rate is i and the number of periods is N. The name of the factor comes from the term **sinking fund**. A sinking fund is an interest-bearing account into which regular deposits are made in order to accumulate some amount.

The equation for the sinking fund factor can be found by decomposing the series of disbursements or receipts made at times 1, 2, . . . , N, and summing to produce a total future value. The formula for the sinking fund factor is

$$(A/F,i,N) = \frac{i}{(1 + i)^N - 1}$$

Figure 3.2 Annuity over N Periods

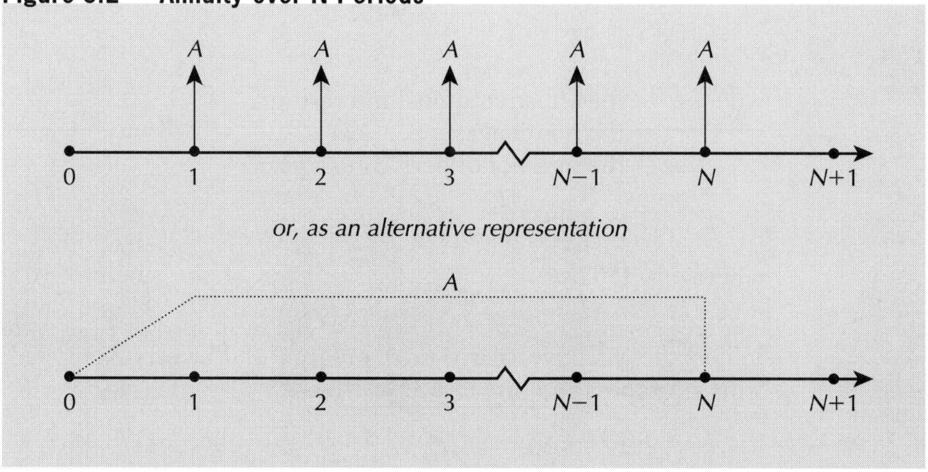

The sinking fund factor is commonly used to determine how much has to be set aside or saved per period to accumulate an amount F at the end of N periods at an interest rate i. The amount F might be used, for example, to purchase new or replacement equipment, to pay for renovations, or to cover capacity expansion costs. In more general terms, the sinking fund factor allows us to convert a single future amount into a series of equal-sized payments, made over N equally spaced intervals, with the use of a given interest rate i.

The **uniform series compound amount factor**, denoted by $(F/A,i,N)$, gives the future value, F, that is equivalent to a series of equal-sized receipts or disbursements, A, when the interest rate is i and the number of periods is N. Since the uniform series compound amount factor is the inverse of the sinking fund factor,

$$(F/A,i,N) = \frac{(1+i)^N - 1}{i}$$

The **capital recovery factor**, denoted by $(A/P,i,N)$, gives the value, A, of the equal periodic payments or receipts that is equivalent to a present amount, P, when the interest rate is i and the number of periods is N. The capital recovery factor is easily derived from the sinking fund factor and the compound amount factor:

$$\begin{aligned}(A/P,i,N) &= (A/F,i,N)(F/P,i,N)\\ &= \frac{i}{(1+i)^N-1}(1+i)^N\\ &= \frac{i(1+i)^N}{(1+i)^N-1}\end{aligned}$$

The capital recovery factor can be used to find out, for example, how much money must be saved over N future periods to "recover" a capital investment of P today. The capital recovery factor for the purchase cost of something is sometimes combined with the sinking fund factor for its salvage value after N years to compose the *capital recovery formula*. See Close-Up 3.1.

CLOSE-UP 3.1 Capital Recovery Formula

Industrial equipment and other assets are often purchased at a cost of P on the basis that they will incur savings of A per period for the firm. At the end of their useful life, they will be sold for some salvage value S. The expression to determine A for a given P and S combines the capital recovery factor (for P) with the sinking fund factor (for S):

$$A = P(A/P,i,N) - S(A/F,i,N)$$

Since

$$(A/F,i,N) = \frac{i}{(1+i)^N - 1} = \frac{i}{(1+i)^N - 1} + i - i$$

$$= \frac{i}{(1+i)^N - 1} + \frac{i[(1+i)^N - 1]}{(1+i)^N - 1} - i$$

$$= \frac{i + i(1+i)^N - i}{(1+i)^N - 1} - i = \frac{i(1+i)^N}{(1+i)^N - 1} - i$$

$$= (A/P,i,N) - i$$

then

$$A = P(A/P,i,N) - S[(A/P,i,N) - i]$$

$$= (P - S)(A/P,i,N) + Si$$

This is the capital recovery formula, which can be used to calculate the savings necessary to justify a capital purchase of cost P and salvage value S after N periods at interest rate i.

The capital recovery formula is also used to determine an annual amount which captures the loss in value of an asset over the time it is owned. Chapter 7 treats this use of the capital recovery formula more fully.

The **series present worth factor**, denoted by $(P/A,i,N)$, gives the present amount, P, that is equivalent to an annuity with disbursements or receipts in the amount, A, where the interest rate is i and the number of periods is N. It is the reciprocal of the capital recovery factor:

$$(P/A,i,N) = \frac{(1+i)^N - 1}{i(1+i)^N}$$

Example 3.2

The Hanover Go-Kart Club has decided to build a clubhouse and track five years from now. It must accumulate \$50 000 by the end of five years by setting aside a uniform amount from its dues at the end of each year. If the interest rate is 10%, how much must be set aside each year?

Since the problem requires that we calculate an annuity amount given a future value, the solution can be obtained using the sinking fund factor where $i = 10\%$, $F = \$50\ 000$, $N = 5$, and A is unknown.

$$\begin{aligned} A &= \$50\ 000(A/F, 10\%, 5) \\ &= \$50\ 000(0.1638) \\ &= \$8190.00 \end{aligned}$$

The Go-Kart Club must set aside \$8190 at the end of each year to accumulate \$50 000 in five years. ■

Example 3.3

A car loan requires 30 monthly payments of \$199.00, starting *today*. At an annual rate of 12% compounded monthly, how much money is being lent?

This cash flow pattern is referred to as an **annuity due**. It differs from a standard annuity in that the first of the N payments occurs at time 0 (now) rather than at the end of the first time period. Annuity dues are uncommon—not often will one make the first payment on a loan on the date the loan is received! Unless otherwise stated, it is reasonable to assume that any annuity starts at the end of the first period.

Two simple methods of analyzing an annuity due will be used for this example.

Method 1. Count the first payment as a present worth and the next 29 payments as an annuity:

$$P = \$199 + A(P/A,i,N)$$

where $A = \$199$, $i = 12\%/12 = 1\%$, and $N = 29$

$$P = \$199 + \$199(P/A, 1\%, 29)$$
$$= \$199 + \$199(25.066)$$
$$= \$199 + \$4988.13$$
$$= \$5187.13$$

The present worth of the loan is the current payment, \$199.00, plus the present worth of the subsequent 29 payments, \$4988.13, a total of about \$5187.

Method 2. Determine the present worth of a standard annuity at time −1, and then find its worth at time 0 (now). The worth at time −1 is

$$P_{-1} = A(P/A,i,N)$$
$$= \$199(P/A, 1\%, 30)$$
$$= \$199(25.807)$$
$$= \$5135.79$$

Then the present worth now (time 0) is

$$P_0 = P_{-1}(F/P,i,N)$$
$$= \$5135.79(F/P,1\%,1)$$
$$= \$5135.79(1.01)$$
$$= \$5187.15$$

The second method gives the same result as the first, allowing a small margin for the effects of rounding. ■

It is worth noting here that although it is natural to think about the symbol P as meaning a cash flow at time 0, the present, and F as meaning a cash flow in the future, in fact these symbols can be more general in meaning. As illustrated in the last example, we can consider any point in time to be the "present" for calculation purposes, and similarly any point in time to be the "future," provided P is some point in time earlier than F. This observation gives us substantial flexibility in analyzing cash flows.

Example 3.4

Clarence bought a house for \$94 000 in 1982. He made a \$14 000 down payment and negotiated a mortgage from the previous owner for the balance. Clarence agreed to pay the previous owner \$2000 per month at 12% nominal interest, compounded monthly. How long did it take him to pay back the mortgage?

Clarence borrowed only \$80 000, since he made a \$14 000 down payment.

The \$2000 payments form an annuity over N months where N is unknown. The interest rate per month is 1%. We must find the value of N such that

$$P = A(P/A,i,N) = A\left(\frac{(1+i)^N - 1}{i(1+i)^N}\right)$$

or, alternatively, the value of N such that

$$A = P(A/P,i,N) = P\left(\frac{i(1+i)^N}{(1+i)^N - 1}\right)$$

where $P = \$80\,000$, $A = \$2000$, and $i = 0.01$

By substituting the known quantities into either expression, some manipulation is required to find N. For illustration, the capital recovery factor has been used.

$$A = P\left(\frac{i(1+i)^N}{(1+i)^N - 1}\right)$$

$$\$2000 = \$80\,000\left(\frac{0.01(1.01)^N}{1.01^N - 1}\right)$$

$$2.5 = \frac{(1.01)^N}{(1.01)^N - 1}$$

$$2.5/1.5 = (1.01)^N$$

$$N[ln(1.01)] = ln(2.5/1.5)$$

$$N = 51.34 \text{ months}$$

It will take Clarence four years and four months to pay off the mortgage. He will make 51 full payments of \$2000 and will be left with only a fraction of a full payment for his fifty-second and last monthly instalment. Problem 3.34 asks what his final payment will be. Note also that mortgages can be confusing because of the different terms used. See Close-Up 3.2. ■

In Example 3.4, it was possible to use the formula for the compound interest factor to solve for the unknown quantity directly. It is not always possible to do this when the number of periods or the interest rate is unknown. We can proceed in several ways. One possibility is to determine the unknown value by trial and error with a spreadsheet. Another approach is to find the nearest values using tables, and then to interpolate linearly to determine an approximate value. Some calculators will perform the interpolation automatically. See Close-Up 3.3 and Figure 3.3 for a reminder of how linear interpolation works.

CLOSE-UP 3.2 Canadian Mortgages

Canadian mortgages can be a little confusing because of the terms used. The interest rate is a nominal rate, usually compounded monthly. The **amortization period** is the duration over which the original loan is calculated to be repaid. The **term** is the duration over which the loan agreement is valid.

For example, Salim has just bought a house for \$135 000. He paid \$25 000 down, and the rest of the cost has been obtained from a mortgage. The mortgage has a nominal interest rate of 9.5% compounded monthly with a 20-year amortization period. The term of the mortgage is three years. What are Salim's monthly payments? How much does he owe after three years?

Salim's monthly payments can be calculated as

$$\begin{aligned} A &= (\$135\,000 - \$25\,000)(A/P, 9.5/12\%, [20 \times 12]) \\ &= \$110\,000(A/P, 0.7917\%, 240) \\ &= \$110\,000(0.00932) \\ &= \$1025.20 \end{aligned}$$

Salim's monthly payments would be about \$1025.20. After three years he would have to renegotiate his mortgage at whatever was the current interest rate at that time. The amount owed would be

$$\begin{aligned} F &= \$110\,000(F/P, 9.5/12\%, 36) - \$1025.20(F/A, 9.5/12\%, 36) \\ &= \$110\,000(1.3283) - \$1025.20(41.47) \\ &\cong \$103\,598 \end{aligned}$$

After three years, Salim still owes \$103 598.

Example 3.5

Clarence paid off an \$80 000 mortgage completely in 48 months. He paid \$2000 per month, and at the end of the first year made an extra payment of \$7000. What interest rate was he charged on the mortgage?

Using the series present worth factor and the present worth factor, this can be formulated for an unknown interest rate:

$$\$80\,000 = \$2000(P/A, i, 48) + \$7000(P/F, i, 12)$$

$$2(P/A, i, 48) + 7(P/F, i, 12) = 80$$

$$2\left[\frac{(1+i)^{48}-1}{i(1+i)^{48}}\right] + 7\left[\frac{1}{(1+i)^{12}}\right] = 80 \qquad (3.1)$$

CLOSE-UP 3.3 Linear Interpolation

Linear interpolation is the process of approximating a complicated function by a straight line in order to estimate a value for the independent variable based on two sample pairs of independent and dependent variables and an instance of the dependent variable. For example, the function f in Figure 3.3 relates the dependent variable y to the independent variable x. Two sample points, (x_1, y_1) and (x_2, y_2), and an instance of y, y^* are known, but the actual shape of f is not. An estimate of the value for x^* can be made by drawing a straight line between (x_1, y_1) and (x_2, y_2).

Because the line between (x_1, y_1) and (x_2, y_2) is assumed to be straight, the following ratios must be equal:

$$\frac{x^* - x_1}{x_2 - x_1} = \frac{y^* - y_1}{y_2 - y_1}$$

Isolating the x^* gives the linear interpolation formula:

$$x^* = x_1 + (x_2 - x_1)\left[\frac{y^* - y_1}{y_2 - y_1}\right]$$

Figure 3.3 Linear Interpolation

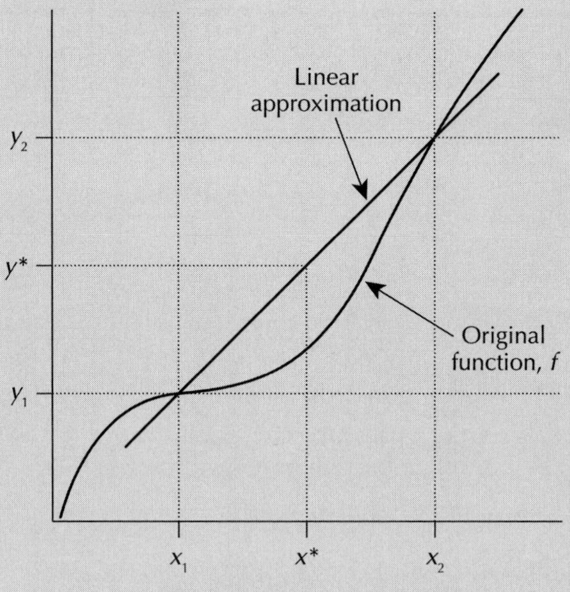

Solving such an equation for i directly is generally not possible. However, using a spreadsheet as illustrated in Table 3.1 can establish some close values for the left-hand side of Equation (3.1), and a similar process can be done using either tables or a calculator. Using a spreadsheet program or calculator, trials can establish a value for the unknown interest rate to the desired number of significant digits.

Table 3.1 Trials to Determine an Unknown Interest Rate

Interest rate i	$2(P/A, i, 48) + 7(P/F, i, 12)$
0.5%	91.7540
0.6%	89.7128
0.7%	87.7350
0.8%	85.8185
0.9%	83.9608
1.0%	82.1601
1.1%	80.4141
1.2%	78.7209
1.3%	77.0787
1.4%	75.4855
1.5%	73.9398

Once the approximate values for the interest rate are found, linear interpolation can be used to find a more precise answer. For instance, working from the values of the interest rate which give the LHS (left-hand side) value closest to the RHS (right-hand side) value of 80, which are 1.1% and 1.2%,

$$i = 1.1 + (1.2 - 1.1)\left[\frac{80 - 80.4141}{78.7209 - 80.4141}\right]$$

$$= 1.1 + 0.02 = 1.12\% \text{ per month}$$

The nominal interest rate was $1.12 \times 12 = 13.44\%$.
The effective interest rate was $(1.0112)^{12} - 1 = 14.30\%$. ■

Another interesting application of compound interest factors is calculating the value of a bond. See Close-Up 3.4.

CLOSE-UP 3.4 Bonds

Bonds are investments that provide an annuity and a future value in return for a cost today. They have a *par* or *face* value, which is the amount for which they can be redeemed after a certain period of time. They also have a *coupon rate*, meaning that they pay the bearer an annuity, usually semi-annually, calculated as a percentage of the face value. For example, a coupon rate of 10% on a bond with an \$8000 face value would pay an annuity of \$400 each six months. Bonds can sell at more or less than the face value, depending on how buyers perceive them as investments.

To calculate the worth of a bond today, sum together the present worth of the face value (a future amount) and the coupons (an annuity) at an appropriate interest rate. For example, if money can earn 12% compounded semi-annually, a bond maturing in 15 years with a face value of \$5000 and a coupon rate of 7% is today worth

$$\begin{aligned} P &= \$5000(P/F, 6\%, 30) + (\$5000 \times 0.07/2)(P/A, 6\%, 30) \\ &= \$5000(0.17411) + \$175(13.765) \\ &= \$3279.43 \end{aligned}$$

The bond is worth about \$3279 today.

3.6 Conversion Factor for Arithmetic Gradient Series

An **arithmetic gradient series** is a series of receipts or disbursements that start at zero at the end of the first period and then increase by a constant *amount* from period to period. Figure 3.4 illustrates an arithmetic gradient series of receipts. Figure 3.5 shows an arithmetic gradient series of disburse-

Figure 3.4 Arithmetic Gradient Series of Receipts

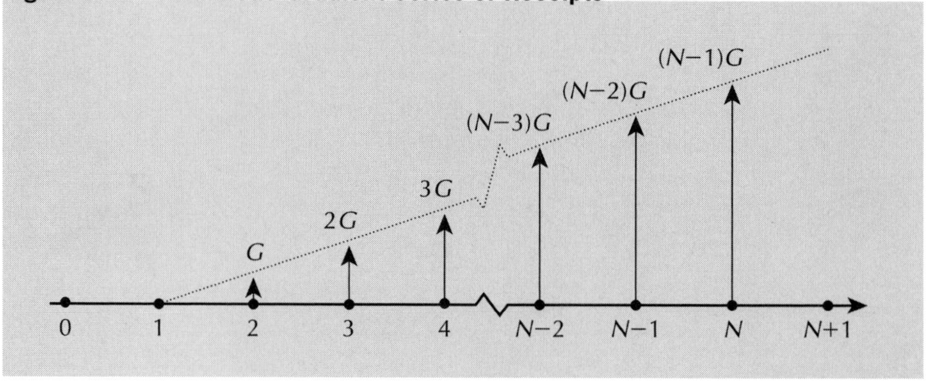

Figure 3.5 Arithmetic Gradient Series of Disbursements

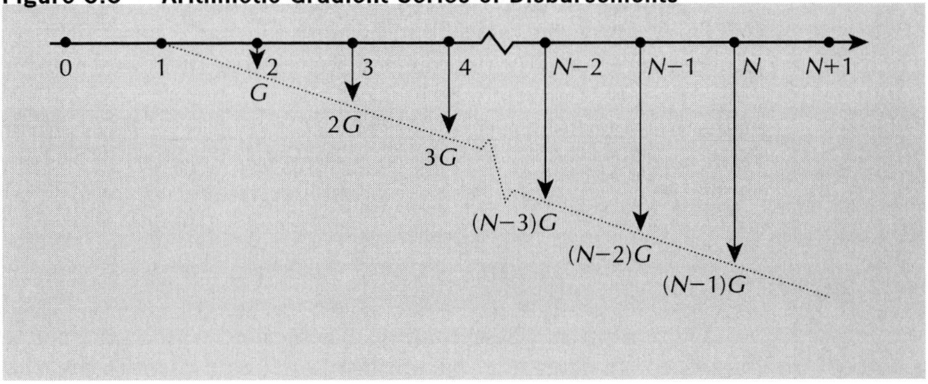

ments. As an example, we may model a pattern of increasing operating costs for an aging machine as an arithmetic gradient series if the costs are increasing by (approximately) the same amount each period. Note carefully that the first non-zero cash flow of a gradient occurs at the end of the *second* compounding period, not the first.

The sum of an annuity plus an arithmetic gradient series is a common pattern. The annuity is a base to which the arithmetic gradient series is added. This is shown in Figure 3.6. A constant-amount increase to a base level of receipts may occur where the increase in receipts is due to adding capacity and where the ability to add capacity is limited. For example, a company that specializes in outfitting warehouses for grocery chains can expand by adding work crews. But the crews must be trained by managers who have time to train only one crew member every six months. Hence, we would have a base amount and a constant amount of growth in cash flows each period.

Figure 3.6 Arithmetic Gradient Series with Base Annuity

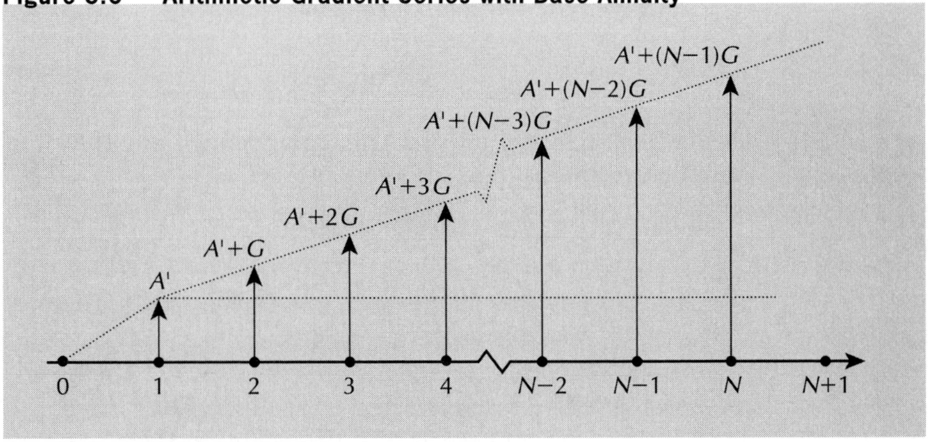

The **arithmetic gradient to annuity conversion factor**, denoted by $(A/G,i,N)$, gives the value of an annuity, A, that is equivalent to an arithmetic gradient series where the constant increase in receipts or disbursements is G per period, the interest rate is i, and the number of periods is N. That is, the arithmetic gradient series, $0G, 1G, 2G, \ldots, (N-1)G$ is given and the uniform cash flow, A, over N periods is found. Problem 3.29 asks the reader to show that the equation for the arithmetic gradient to annuity factor is

$$(A/G,i,N) = \frac{1}{i} - \frac{N}{(1+i)^N - 1}$$

There is often a base annuity A' associated with a gradient, as illustrated in Figure 3.6. To determine the uniform series equivalent to the *total* cash flow, the base annuity A' must be included to give the overall annuity:

$$A_{tot} = A' + G(A/G,i,N)$$

Example 3.6

Susan Ng owns an eight-year-old Jetta automobile. She wants to find the present worth of repair bills over the four years that she expects to keep the car. Susan has the car in for repairs every six months. Repair costs are expected to increase by \$50 every six months over the next four years, starting with \$500 six months from now, \$550 six months later, and so on. What is the present worth of the repair costs over the next four years if the interest rate is 12% compounded monthly?

First, observe that there will be $N = 8$ repair bills over four years and that the base annuity payment, A', is \$500. The arithmetic gradient component of the bills, G, is \$50, and hence the arithmetic gradient series is \$0, \$50, \$100, and so on. The present worth of the repair bills can be obtained in a two-step process:

Step 1. Find the total uniform annuity, A_{tot}, equivalent to the sum of the base annuity, $A' = \$500$, and the arithmetic gradient series with $G = \$50$ over $N = 8$ periods.

Step 2. Find the present worth of A_{tot}, using the series present worth factor.

The 12% nominal interest rate, compounded monthly, is 1% per month. The effective interest rate per six month period is

$$i_{6month} = (1 + 0.12/12)^6 - 1 = 0.06152 \text{ or } 6.152\%$$

Step 1

$$A_{tot} = A' + G(A/G,i,N)$$

$$= \$500 + \$50\left(\frac{1}{i} - \frac{N}{(1+i)^N - 1}\right)$$

$$= \$500 + \$50\left(\frac{1}{0.06152} - \frac{8}{(1.06152)^8 - 1}\right)$$

$$= \$659.39$$

Step 2

$$P = A_{tot}(P/A,i,N) = A_{tot}\left(\frac{(1+i)^N - 1}{i(1+i)^N}\right)$$

$$= \$659.39\left(\frac{(1.06152)^8 - 1}{0.06152(1.06152)^8}\right)$$

$$= \$4070.09$$

The present worth of the repair costs is about \$4070. ■

3.7 Conversion Factor for Geometric Gradient Series

A **geometric gradient series** is a series of cash flows that increase or decrease by a constant *percentage* each period. The geometric gradient series may be used to model inflation or deflation, productivity improvement or degradation, and growth or shrinkage of market size, as well as many other phenomena.

In a geometric series, the base value of the series is A and the "growth" rate in the series (the rate of increase or decrease) is referred to as g. The terms in such a series are given by A, $A(1 + g)$, $A(1 + g)^2$, . . ., $A(1 + g)^{N-1}$ at the ends of periods 1, 2, 3, . . ., N, respectively. If the rate of growth, g, is positive, the terms are increasing in value. If the rate of growth, g, is negative, the terms are decreasing. Figure 3.7 shows a series of receipts where g is positive. Figure 3.8 shows a series of receipts where g is negative.

The **geometric gradient to present worth conversion factor**, denoted by $(P/A,g,i,N)$, gives the present worth, P, that is equivalent to a geometric gradient series where the base receipt or disbursement is A, and where the rate of growth is g, the interest rate is i, and the number of periods is N.

The present worth of a geometric series is

$$P = \frac{A}{1+i} + \frac{A(1+g)}{(1+i)^2} + \ldots + \frac{A(1+g)^{N-1}}{(1+i)^N}$$

where A = the base amount
g = the rate of growth
i = the interest rate
N = the number of periods
P = the present worth

We can define a **growth adjusted interest rate**, i°, as

Figure 3.7 Geometric Gradient Series for Receipts with Positive Growth

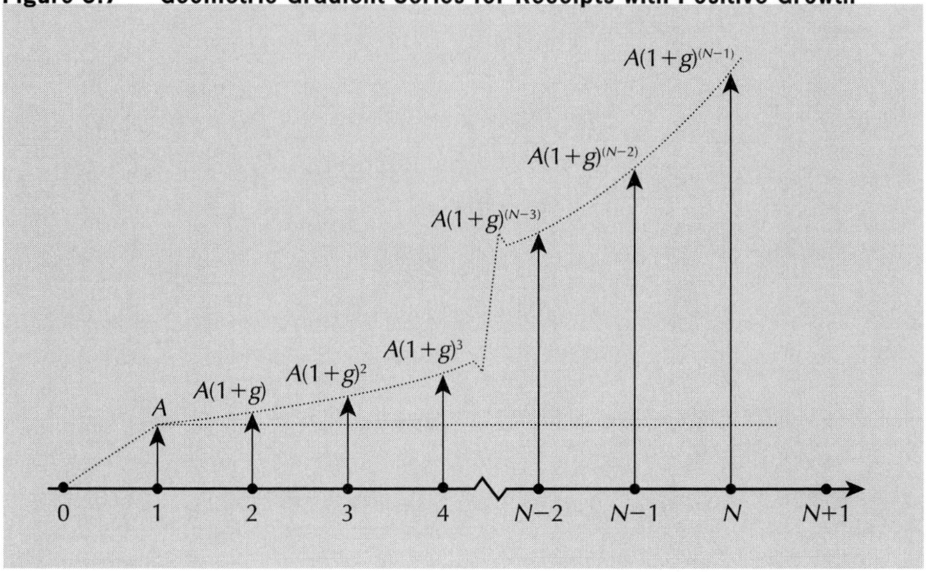

$$i^\circ = \frac{1+i}{1+g} - 1$$

so that

$$\frac{1}{1+i^\circ} = \frac{1+g}{1+i}$$

Then the geometric gradient series to present worth conversion factor is given by

$$(P/A,g,i,N) = \frac{(P/A,i^\circ,N)}{1+g} \text{ or}$$

$$(P/A,g,i,N) = \left(\frac{(1+i^\circ)^N - 1}{i^\circ(1+i^\circ)^N}\right)\frac{1}{1+g}$$

Care must be taken in using the geometric gradient to present worth conversion factor. Four cases may be distinguished.

1. $i > g > 0$. *Growth is positive, but less than the rate of interest.* The growth adjusted interest rate, i°, is positive. Tables or functions built into software may be used to find the conversion factor.

2. $g > i > 0$. *Growth is positive and greater than the interest rate.* The growth adjusted interest rate, i°, is negative. It is necessary to compute the conversion factor directly from the formula.

Figure 3.8 Geometric Gradient Series for Receipts with Negative Growth

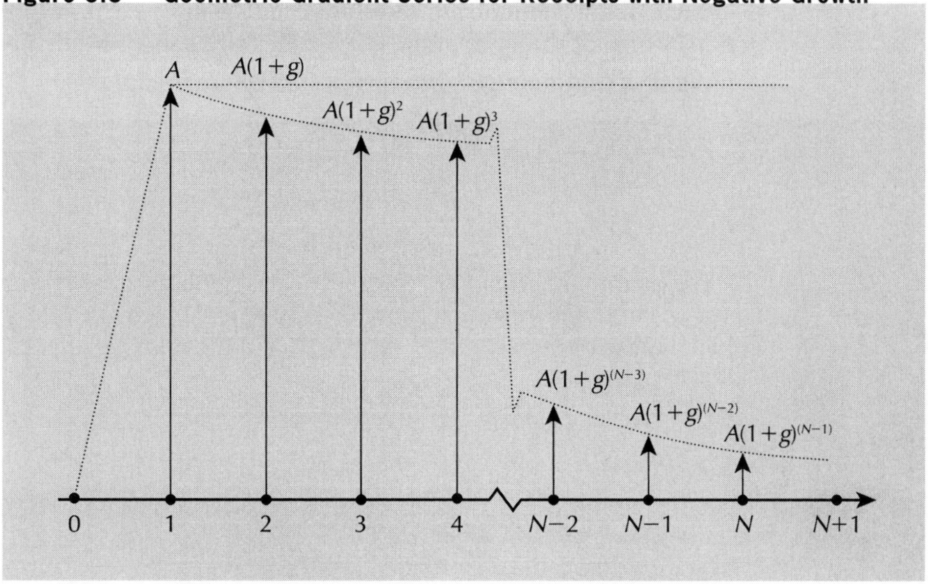

3. $g = i > 0$. *Growth is positive and exactly equal to the interest rate.* The growth adjusted interest rate $i^\circ = 0$. As with any case where the interest rate is zero, the present worth of the series with constant terms, $A/(1 + g)$, is simply the sum of all the N terms

$$P = N\left(\frac{A}{1+g}\right)$$

4. $g < 0$. *Growth is negative.* In other words, the series is decreasing. The growth adjusted interest rate, i°, is positive. Tables or functions built into software may be used to find the conversion factor.

Example 3.7

Tru-Test is in the business of assembling and packaging automotive and marine testing equipment to be sold through retailers to "do-it-yourselfers" and small repair shops. One of their products is tire pressure gauges. This operation has some excess capacity. Tru-Test is considering using this excess capacity to add engine compression gauges to their line. They can sell engine pressure gauges to retailers for \$8 per gauge. They expect to be able to produce about 1000 gauges in the first month of production. They also expect that, as the workers learn how to do the work more efficiently, productivity will rise by 0.25% per month for the first two years. In other words, each month's output of gauges will be 0.25% more than the previous month's.The interest rate is 1.5% per month. All gauges are sold in the month in which they are produced, and receipts from sales are at the end of each month. What is the present worth of the sales of the engine pressure gauges in the first two years?

We first compute the growth adjusted interest rate, i°.

$$i^\circ = \frac{1+i}{1+g} - 1 = \frac{1.015}{1.0025} - 1 = 0.01247$$

$$i^\circ \cong 1.25\%$$

We then make use of the geometric gradient to present worth conversion factor with the uniform cash flow $A = \$8000$, the growth rate $g = 0.0025$, the growth adjusted interest rate $i^\circ = 0.0125$, and the number of periods $N = 24$.

$$P = A(P/A,g,i,N) = A\left(\frac{(P/A,i^\circ,N)}{1+g}\right)$$

$$= \$8000\left(\frac{(P/A,\ 1.25\%,\ 24)}{1.0025}\right)$$

From the interest rate tables we get

$$P = \$8000\left(\frac{20.624}{1.0025}\right)$$

$$= \$164\,580$$

The present worth of sales of engine compression gauges over the two-year period would be about \$165 000. Recall that we worked with an *approximate* growth-adjusted interest rate of 1.25% when the correct rate was a bit less than 1.25%. This means that \$164 580 is a slight understatement of the present worth. ■

Example 3.8

Emery's company, Dry-All, produces control systems for drying grain. Proprietary technology has allowed Dry-All to maintain steady growth in the US market in spite of numerous competitors. Company dividends, all paid to Emery, are expected to rise at a rate of 10% per year over the next 10 years. Dividends at the end of this year are expected to total \$110 000. If all dividends are invested at 10% interest, how much will Emery accumulate in 10 years?

If we calculate the growth adjusted interest rate, we get:

$$i^\circ = \frac{1.1}{1.1} - 1 = 0$$

and it is natural to think that the present worth is simply the first year's dividends multiplied by 10. However, recall that in the case where $g = i$ the present worth is given by

$$P = N\left(\frac{A}{1+g}\right) = 10\left(\frac{\$110\ 000}{1.1}\right) = \$1\ 000\ 000$$

Intuitively, dividing by $(1 + g)$ compensates for the fact that growth is considered to start after the end of the first period, but the interest rate applies to all periods.

We want the future worth of this amount after 10 years:

$$F = \$1\ 000\ 000\ (F/P, 10\%, 10) = \$1\ 000\ 000\ (2.5937) = \$2\ 593\ 700$$

Emery will accumulate \$2 593 700 in dividends and interest. ■

3.8 Non-Standard Annuities and Gradients

As discussed in Section 3.3, the standard assumption for annuities and gradients is that the payment period and compounding period are the same. If they are not, the formulas given in this chapter cannot be applied directly. There are three methods for dealing with this situation:

1. Treat each cash flow in the annuity or gradient individually. This is most useful when the annuity or gradient series is not large.
2. Convert the non-standard annuity or gradient to standard form by changing the compounding period.
3. Convert the non-standard annuity to standard form by finding an equivalent standard annuity for the compounding period. This method cannot be used for gradients.

Example 3.9

How much is accumulated over 20 years in a fund that pays 4% interest, compounded yearly, if \$1000 is deposited at the end of every fourth year?

The cash flow diagram for this set of payments is shown in Figure 3.9.

Figure 3.9 Non-standard Annuity for Example 3.9

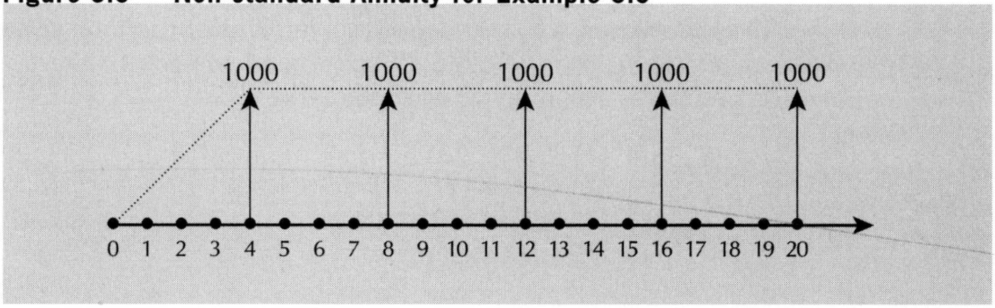

Method 1: Consider the annuities as separate future payments.

Formula: $F = P(F/P,i,N)$
Known values: $P = \$1000$, $i = 0.04$, $N = 16, 12, 8, 4$, and 0

Year	***Future Value***		
4	$1000(*F*/*P*, 4%, 16)	= $1000(1.8729)	= $1873
8	$1000(*F*/*P*, 4%, 12)	= $1000(1.6010)	= $1601
12	$1000(*F*/*P*, 4%, 8)	= $1000(1.3685)	= $1369
16	$1000(*F*/*P*, 4%, 4)	= $1000(1.1698)	= $1170
20	$1000		= $1000
	Total future value		= $7013

About $7013 is accumulated over the 20 years.

Method 2: Convert the compounding period from yearly to every four years. This can be done with the effective interest rate formula.

$$i_e = (1 + 0.04)^4 - 1$$
$$= 16.99\%$$

The future value is then

$$F = \$1000(F/A, 16.99\%, 5) = \$1000\ (7.013)$$
$$= \$7013$$

Method 3: Convert the annuity to an equivalent yearly annuity. This can be done by considering the first payment as a future value over the first four-year period, and finding the equivalent annuity over that period, using the sinking fund factor:

$$A = \$1000(A/F, 4\%, 4)$$
$$= \$1000(0.23549)$$
$$= \$235.49$$

In other words, a $1000 deposit at the end of the four years is equivalent to four equal deposits of $235.49 at the end of each of the four years. This yearly annuity is accumulated over the 20 years.

$$F = \$235.49(F/A, 4\%, 20)$$
$$= \$235.49(29.777)$$
$$= \$7012$$

Note that each method produces the same amount, allowing for rounding. When you have a choice in methods as in this example, your choice will depend on what you find convenient, or what is the most efficient computationally. ■

Example 3.10

This year's electrical engineering class has decided to save up for a class party. Each of the 90 people in the class is to contribute \$0.25 per day which will be placed in a daily interest (7 days a week, 365 days a year) savings account that pays a nominal 8% interest. Contributions will be made *five* days a week, Monday through Friday, beginning on Monday. The money is put into the account at the beginning of each day, and thus earns interest for the day. The class party is in 14 weeks (a full 14 weeks of payments will be made), and the money will be withdrawn on the Monday morning of the fifteenth week. How much will be saved, assuming everybody makes payments on time?

There are several good ways to solve this problem. One way is to convert each days' contribution to a weekly amount on Sunday evening/Monday morning, and then accumulate the weekly amounts over the 14 weeks:

Total contribution per day is $\$0.25 \times 90 = \$22.50$

The interest rate per day is $\dfrac{0.08}{365} = 0.000219$

The effective interest rate for a 1-week period is

$$i = (1 + 0.08/365)^7 - 1 = 0.00154$$

Value of 1 week's contribution on Friday evening (*annuity due* formula):

$$\$22.50 \times (F/P, 0.08/365, 1) \times (F/A, 0.08/365, 5)$$

On Sunday evening this is worth

$$[\$22.50\ (F/P, 0.08/365, 1)\ (F/A, 0.08/365, 5)] \times (F/P, 0.08/365, 2)$$
$$= \$22.50\ (F/P, 0.08/365, 3)\ (F/A, 0.08/365, 5)$$

Then the total amount accumulated by Monday morning of the 15th week is given by:

$$[\$22.50\ (F/P, 0.08/365, 3)\ (F/A, 0.08/365, 5)]\ (F/A, (1 + 0.08/365)^7 - 1, 14)$$
$$= [\$22.50\ (1.000\ 658)\ (5.002\ 19)]\ (14.1406)$$
$$= \$1592.56$$

The total amount saved would be \$1592.56. ■

3.9 Present Worth Computations when $N \to \infty$

We have until now assumed that the cash flows of a project occur over some fixed, finite number of periods. For long-lived projects, it may be reasonable to model the cash flows as though they continued indefinitely. The present worth of an infinitely long uniform series of cash flows is called the **capitalized value** of the series. We can get the capitalized value of a series by allowing the number of periods, N, in the series present worth factor to go to infinity:

$$\begin{aligned} P &= \lim_{N\to\infty} A(P/A,i,N) \\ &= A \lim_{N\to\infty} \left[\frac{(1+i)^N - 1}{i(1+i)^N}\right] \\ &= A \lim_{N\to\infty} \left[\frac{1 - \dfrac{1}{(1+i)^N}}{i}\right] \\ &= \frac{A}{i} \end{aligned}$$

Example 3.11

The town of South Battleford is considering building a by-pass for truck traffic around the downtown commercial area. The by-pass will provide merchants and shoppers with benefits that have an estimated value of \$500 000 per year. Maintenance costs will be \$125 000 per year. If the by-pass is properly maintained, it will provide benefits for a very long time. The actual life of the by-pass will depend on factors like future economic conditions that cannot be forecast at the time the by-pass is being considered. It is, therefore, reasonable to model the flow of benefits as though they continued indefinitely. If the interest rate is 10%, what is the present worth of benefits minus maintenance costs?

$$P = \frac{A}{i} = \frac{\$500\,000 - \$125\,000}{0.1} = \$3\,750\,000$$

The present worth of benefits net of maintenance costs is \$3 750 000. ■

REVIEW PROBLEMS

REVIEW PROBLEM 3.1

The benefits of a revised production schedule for a seasonal manufacturer will not be realized until the peak summer months. Net savings will be \$1100, \$1200, \$1300, \$1400, and \$1500 at the ends of months 5, 6, 7, 8, and 9, respectively. It is now the beginning of month 1. Assume 365 days per year, 30 days per month. What is the present worth (PW) of the savings if nominal interest is

(a) 12% per year, compounded monthly?

(b) 12% per year, compounded daily?

ANSWER

(a) $A = \$1100$

$G = \$100$

$i = 0.12/12 = 0.01$ per month $= 1\%$

$$\begin{aligned}\text{PW(end of period 4)} &= (P/A, 1\%, 5)[\$1100 + \$100(A/G, 1\%, 5)] \\ &= 4.8528[\$1100 + \$100(1.9801)] \\ &= \$6298.98 \\ \text{PW(at time 0)} &= \text{PW(end of period 4)}(P/F, 1\%, 4) \\ &= \$6298.98/(1.01)^4 = \$6053.20\end{aligned}$$

The present worth is about \$6053.

(b) Effective interest rate $i = (1 + 0.12/365)^{30} - 1 = 0.0099102$

$$\begin{aligned}\text{PW(at time 0)} &= \text{PW(end of period 4)}(P/F, i, 4) \\ &= (P/A, i, 5)[\$1100 + \$100(A/G, i, 5)](P/F, i, 4) \\ &= \$4.8547[\$1100 + \$100(1.98023)](0.9613) \\ &= \$6057.80\end{aligned}$$

The present worth is about \$6058. ■

REVIEW PROBLEM 3.2

It is January 1 of this year. You are starting your new job tomorrow, having just finished your engineering degree at the end of last term. Your take-home pay for this year will be \$36 000. It will be paid to you in equal amounts at the end of each month, starting at the end of January. There is a cost-of-living clause in your contract that says that each subsequent January you will get an increase of 3% in your yearly salary (i.e., your take-home pay for next year will be 1.03 × \$36 000). In addition to your salary, a wealthy relative regularly sends you a \$2000 birthday present at the end of each June.

Recognizing that you are not likely to have any government pension, you have decided to start saving 10% of your monthly salary and 50% of your birthday present for your retirement. Interest is 1% per month, compounded monthly. How much will you have saved at the end of five years?

ANSWER

Yearly pay is a geometric gradient; convert your monthly salary into a yearly amount by the use of an effective yearly rate. The birthday present can be dealt with separately.

Salary: The future worth (FW) of the salary at the end of the first year is

$$\text{FW(salary, year 1)} = \$3000(F/A, 1\%, 12) = \$38\,040.00$$

This forms the base of the geometric gradient; all subsequent years increase by 3% per year. Savings are 10% of salary, which implies that $A = \$3804.00$.

$$A = \$3804.00 \qquad g = 0.03$$

Effective yearly interest rate $i_e = (1 + 0.01)^{12} - 1 = 0.1268$ per year

$$i^{\circ} = \frac{1 + i_e}{1 + g} - 1 = \frac{1 + 0.1268}{1 + 0.03} - 1 = 0.093981$$

$$\begin{aligned}\text{PW(gradient)} &= A\,(P/A, i^{\circ}, 5)/(1 + g) = \$3804(3.8498)/1.03\\ &= \$14\,218\end{aligned}$$

$$\begin{aligned}\text{FW(gradient, end of five years)} &= \text{PW(gradient)}(F/P, i_e, 5)\\ &= \$14\,218(1.1268)^5 = \$25\,827\end{aligned}$$

Birthday present: The present arrives in the middle of each year. To get the total value of the five gifts, we can find the present worth of an annuity of five payments of \$2000(0.5) as of six months prior to employment:

$$\text{PW}(-6 \text{ months}) = \$2000(0.5)(P/A, i_e, 5) = \$3544.90$$

The future worth $= 5 \times 12 + 6 = 66$ months later is

$$\text{FW(end of five years)} = \$3544.9(1.01)^{66} = \$6836$$

Total amount saved $= \$6836 + \$25\,827 = \$32\,663$ ■

REVIEW PROBLEM 3.3

The Easy Loan Company advertises a "10%" loan. You need to borrow \$1000, and the deal you are offered is the following: You pay \$1100 (\$1000 plus \$100 interest) in 11 equal \$100 amounts, starting one month from today. In addition, there is a \$25 administration fee for the loan, payable immediately, and a processing fee of \$10 per payment. Furthermore, there is a

$20 non-optional closing fee to be included in the last payment. Recognizing fees as a form of interest payment, what is the actual effective interest rate?

ANSWER

Since the $25 administration fee is paid immediately, you are only getting $975. The remaining payments amount to an annuity of $110 per month, plus a $20 future payment 11 months from now.

Formulas: $P = A(P/A,i,N)$, $P = F(P/F,i,N)$

Known values: $P = \$975$, $A = \$110$, $F = \$20$, $N = 11$

$$\$975 = \$110(P/A, i, 11) + \$20(P/F, i, 11)$$

At $i = 4\%$

$$\$110(P/A, 4\%, 11) + \$20(P/F, 4\%, 11)$$
$$= \$110(8.7603) + \$20(0.64958)$$
$$= \$976.62$$

At $i = 5\%$

$$\$110(P/A, 5\%, 11) + \$20(P/F, 5\%, 11)$$
$$= \$110(8.3062) + \$20(0.58469)$$
$$= \$925.37$$

Linearly interpolating gives

$$i = 4 + (5 - 4)\,(975 - 976.62)/(925.37 - 976.62)$$
$$= 4.03$$

The effective interest rate is then

$$i = (1 + 0.0403)^{12} - 1$$
$$= 60.69\% \text{ per annum (!)}$$

Although the loan is advertised as a "10%" loan, the actual effective rate is over 60%. ■

REVIEW PROBLEM 3.4

Ming wants to retire as soon as she has enough money invested in a special bank account (paying 14% interest, compounded annually) to provide her with an annual income of $25 000. She is able to save $10 000 per year, and the account now holds $5000. If she just turned 20, and expects to die in 50 years, how old will she be when she retires? There should be no money left when she turns 70.

ANSWER

Let Ming's retirement age be $20 + x$ so that

$$\$5000(F/P, 14\%, x) + \$10\,000(F/A, 14\%, x) = \$25\,000(P/A, 14\%, 50 - x)$$

Dividing both sides by \$5000,

$$(F/P, 14\%, x) + 2(F/A, 14\%, x) - 5(P/A, 14\%, 50 - x) = 0$$

At $x = 5$

$$(F/P, 14\%, 5) + 2(F/A, 14\%, 5) - 5(P/A, 14\%, 45) = (1.9254) + 2(6.6101) - 5(7.1232) = -20.4704$$

At $x = 10$

$$(F/P, 14\%, 10) + 2(F/A, 14\%, 10) - 5(P/A, 14\%, 40) = (3.7072) + 2(19.337) - 5(7.1050) = 6.8562$$

Linearly interpolating,

$$x = 5 + 5 \times (20.4704)/(6.8562 + 20.4704) = 8.7$$

Ming can retire at age 20 + 8.7 = 28.7 years old. ■

SUMMARY

In Chapter 3 we considered ways of modelling patterns of cash flows that enable easy comparisons of the worths of projects. The emphasis was on discrete models. Four basic patterns of discrete cash flows were considered:

1. Flows at a single point
2. Flows that are constant over time
3. Flows that grow or decrease at a constant arithmetic rate
4. Flows that grow or decrease at a constant geometric rate

Compound interest factors were presented that defined mathematical equivalence among the basic patterns of cash flows. A list of these factors with their names, symbols, and formulas appears in Table 3.2. The chapter also addressed the issue of how to analyze non-standard annuities and gradients as well as the ideas of capital recovery and capitalized value.

For those who are interested in continuous compounding and continuous cash flows, Appendix 3A contains a summary of relevant notation and interest factors.

Table 3.2 Summary of Useful Formulas for Discrete Models

Name	Symbol and Formula
Compound amount factor	$(F/P,i,N) = (1+i)^N$
Present worth factor	$(P/F,i,N) = \dfrac{1}{(1+i)^N}$
Sinking fund factor	$(A/F,i,N) = \dfrac{i}{(1+i)^N - 1}$
Uniform series compound amount factor	$(F/A,i,N) = \dfrac{(1+i)^N - 1}{i}$
Capital recovery factor	$(A/P,i,N) = \dfrac{i(1+i)^N}{(1+i)^N - 1}$
Series present worth factor	$(P/A,i,N) = \dfrac{(1+i)^N - 1}{i(1+i)^N}$
Arithmetic gradient to annuity conversion factor	$(A/G,i,N) = \dfrac{1}{i} - \dfrac{N}{(1+i)^N - 1}$
Geometric gradient to present worth conversion factor	$(P/A,g,i,N) = \dfrac{(P/A,i^\circ,N)}{1+g}$ $(P/A,g,i,N) = \left(\dfrac{(1+i^\circ)^N - 1}{i^\circ(1+i^\circ)^N}\right)\dfrac{1}{1+g}$ $i^\circ = \dfrac{1+i}{1+g} - 1$
Capitalized value formula	$P = \dfrac{A}{i}$
Capitalized recovery formula	$A = (P - S)(A/P,i,N) + Si$

Engineering Economics in Action, Part 3B: No Free Lunch

This time it was Naomi who stuck her head in Clem's doorway. "Here's the recommendation on the shipping palletizer. Oh, and thanks for the hint on the leasing figures. It cleared up my confusion right away."

"No problem. What did you figure out?" Clem had his "mentor" expression on his face, so Naomi knew he was expecting a clear explanation of the trick used by the leasing company.

"Well, as you hinted, they were adding apples and oranges. They listed the various costs for each choice over time, including interest charges, taxes, and so on. But then, for the final comparison, they added up these costs. When they added, leasing was cheaper."

"So what's wrong with that?" Clem prompted.

"They're adding apples and oranges. We're used to thinking of money as being just money, without remembering that money always has a 'when' associated with it. If you add money at different points in time, you might as well be adding apples and oranges; you have a number but it doesn't mean anything. In order to compare leasing with buying, you first have to change the cash flows into the same money, that is, at the same point in time. That's a little harder to do, especially when there's a complicated set of cash flows."

"So were you able to do it?"

"Yes. I identified various components of the cash flows as annuities, gradients, and present and future worths. Then I converted all of these to a present worth for each alternative and summed them. This is the correct way to compare them. If you do that, buying is cheaper, even when borrowing money to do so. And of course it has to be—that leasing company has to pay for those slick brochures somehow. There's no free lunch."

Clem nodded. "I think you've covered it. Mind you, there are some circumstances where leasing is worthwhile. For example, we lease our company cars to save us the time and trouble of reselling them when we're finished with them. Leasing can be good when it's hard to raise the capital for very large purchases, too. But almost always, buying is better. And you know, it amazes me how easy it is to fall for simplistic cash flow calculations that fail to take into account the time value of money. I've even seen articles in the newspaper quoting accountants who make the same mistake, and you'd think they would know better."

"Engineers can make that mistake, too, Clem. I almost did." ◎

PROBLEMS

3.1 St. Agatha Kennels provides dog breeding and boarding services for the nearby city of Kitchener, Ontario. Most of the income is derived from boarding, with typical boarding stays being one or two weeks. Customers pay at the end of the dog's stay. Boarding is offered only during the months of May to September. Other income is received from breeding golden retrievers, with two litters of about 8 dogs each being produced per year, spring and fall. Expenses include heating, water, and sewage, which are paid monthly, and food, bought in bulk each spring. The business has been neither growing nor shrinking over the past few years.

Joan, the owner of St. Agatha Kennels, wants to model the cash flows for the business over the next 10 years. What cash flow elements (e.g., single payments, annuities, gradients) would she likely consider, and how would she estimate their value? Consider the present to be the first of May. For example, one cash flow element is food. It would be modelled as an *annuity due* over 10 years, and estimated by the amount paid for food over the last few years.

3.2 It is September, and Marco has to watch his expenses while he is going to school. Over the next eight months, he wants to estimate his cash flows. He pays rent once a month. He takes the bus to and from school. A couple of times a week he goes to the grocery store for food, and eats lunch in the cafeteria at school every school day. At the end of each four-month term, he will have printing and copying expenses because of reports that will be due. After the first term, over the Christmas holidays, he will have extra expenses for buying presents, but will also get some extra cash from his parents. What cash flow elements (e.g., single payments, annuities, gradients) would Marco likely consider in his estimates? How would he estimate them?

3.3 How much money will be in a bank account at the end of 15 years if \$100 is deposited today and the interest rate is 8% compounded annually?

3.4 How much should you invest today at 12% interest to accumulate \$1 000 000 in 30 years?

3.5 Martin and Marcy McCormack have just become proud parents of Canada's first septuplets. They have savings of \$5000. They want to invest their savings so that they can partially support the children's university education. Martin and Marcy hope to provide \$20 000 for each child by the time the children turn 18. What does the annual rate of return have to be on the investment for Martin and Marcy to meet their goal?

3.6 You have \$1725 to invest. You know that a particular investment will double your money in 5 years. How much will you have in 10 years if you invest in this investment, assuming that the annual rate of return is guaranteed for the time period?

3.7 Morris paid \$500 a month for 20 years to pay off the mortgage on his house. If his down payment was \$5000 and the interest rate was 6% compounded monthly, how much did the house cost?

3.8 An investment pays \$10 000 every five years, starting in seven years, for a total of four payments. If interest is 9%, how much is this investment worth today?

3.9 An industrial juicer costs \$45 000. It will be used for five years and then sold to a remarketer for \$25 000. If interest is 15%, what net yearly savings are needed to justify its purchase?

3.10 Fred wants to save up for a car. How much must he put in his bank account each month to save \$10 000 in two years if the bank pays 6% interest compounded monthly?

3.11 It is May 1. You have just bought \$2000 worth of furniture. You will pay for it in 24 equal monthly payments, starting at the end of May next year. Interest is 6% nominal per year, compounded monthly. How much will your payments be?

3.12 What is the present worth of the total of 20 payments, occurring at the end of every four months (the first payment is in four months), which are \$400, \$500, \$600, increasing arithmetically? Interest is 12% nominal per year, compounded continuously.

3.13 What is the total value of the sum of the present worths of all the payments and receipts mentioned in Problem 2.32, at an interest rate of 0.5% per month?

3.14 How much is accumulated in each of the following savings plans over two years?

(a) \$40 at the end of each month for 24 months at 12% compounded monthly

(b) \$30 at the end of the first month, \$31 at the end of the second month, and so forth, increasing by \$1 per month, at 12% compounded monthly

3.15 What interest rate will result in \$5000 seven years from now, starting with \$2300 today?

3.16 Refer back to the Hanover Go-Kart problem of Example 3.2. The club members determined that it is possible to set aside only \$7000 each year, and that they will have to put off building the clubhouse until they have saved the \$50 000 necessary. How long will it take to save a total of \$50 000, assuming that the interest rate is 10%? (*Hint:* Use logarithms to simplify the sinking fund factor.)

3.17 Gwen just bought a satellite dish, which provides her with exactly the same service as cable TV. The dish cost \$2000, and the cable service she has now cancelled cost her \$40 per month. How long will it take her to recoup her investment in the dish, if she can earn 12% interest, compounded monthly, on her money?

3.18 Yoko has just bought a new computer ($2000), a printer ($350), and a scanner ($210). She wants to take the monthly payment option. There is a monthly interest of 3% on her purchase.

(a) If Yoko pays $100 per month, how long does it take to complete her payment?

(b) If Yoko wants to finish paying in 24 months, how much will her monthly payment be?

3.19 Rinku has just finished her first year of university. She wants to go to Europe when she graduates in 3 years. By having a part-time job through the school year and a summer job during the summer, she plans to make regular weekly deposits into a savings account, which bears 18% interest, compounded monthly.

(a) If Rinku deposits $15 per week, how much will she save in 3 years? How about $20 per week?

(b) Find out exactly how much Rinku needs to deposit every week if she wants to save $5000 in 3 years.

3.20 Seema is looking at an investment in upgrading an inspection line at her plant. The initial cost would be $140 000 with a salvage value of $37 000 after 5 years. Use the *Capital Recovery Formula* to determine how much money must be saved each year to justify the investment, at an interest rate of 14%.

3.21 Trenny has asked her assistant to prepare estimates of cost of two different sizes of power plants. The assistant reports that the cost of the 100 MW plant is $20 000 000, while the cost of the 200 MW plant is $36 000 000. If Trenny has a budget of only $30 000 000, estimate how large a power plant she could afford using linear interpolation.

3.22 Enrique has determined that investing $500 per month will enable him to accumulate $11 350 in 12 years, and that investing $800 per month will enable him to accumulate $18 950 over the same period. Estimate using linear interpolation how much he would have to invest each month to accumulate exactly $15 000.

3.23 A lottery prize pays $1000 at the end of the first year, $2000 the second, $3000 the third, and so on for 20 years. If there is only one prize in the lottery, 10 000 tickets are sold, and you could invest your money elsewhere at 15% interest, how much is each ticket worth, on average?

3.24 Joseph and three other friends bought a $110 000 house close to the university at the end of August last year. At that time they put down a deposit of $10 000 and took out a mortgage for the balance. Their mortgage payments are due at the end of each month (September 30, last year, was the date of the first payment) and are based on the assumption that Joseph and friends will take 20 years to pay off the debt. Annual nominal interest is 12%, compounded monthly. It is now February. Joseph and friends have made all their fall-term payments and have just made the January 31 payment for this year. How much do they still owe?

3.25 A new software package is expected to improve productivity at Saskatoon Insurance. However, because of training and implementation costs, savings are not expected to occur until the third year of operation. At that time, savings of \$10 000 are expected, increasing by \$1000 per year for the following five years. After this time (eight years from implementation), the software will be abandoned with no scrap value. How much is the software worth today, at 15% interest?

3.26 Clem is saving for a car, in a bank account that pays 12% interest, compounded monthly. The balance is now \$2400. Clem will be saving \$120 per month from his salary, and once every four months (starting in four months) he adds \$200 in dividends from an investment. Bank fees, currently \$10 per month, are expected to increase by \$1 per month henceforth. How much will Clem have saved in two years?

3.27 Yogajothi is thinking of investing in a rental house. The total cost to purchase the house, including legal fees and taxes, is \$115 000. All but \$15 000 of this amount will be mortgaged. He will pay \$800 per month in mortgage payments. At the end of two years, he will sell the house, and at that time expects to clear \$20 000 after paying off the remaining mortgage principal (in other words, he will pay off all his debts for the house, and still have \$20 000 left). Rents will earn him \$1000 per month for the first year, and \$1200 per month for the second year. The house is in fairly good condition now so that he doesn't expect to have any maintenance costs for the first six months. For the seventh month, Yogajothi has budgeted \$200. This figure will be increased by \$20 per month thereafter (e.g., the expected month 7 expense will be \$200, month 8, \$220, month 9, \$240, etc.). If interest is 6% compounded monthly, what is the present worth of this investment? Given that Yogajothi's estimates of revenue and expenses are correct, should Yogajothi buy the house?

3.28 You have been paying off a mortgage in quarterly payments at a 24% nominal annual rate, compounded quarterly. Your bank is now offering an alternative payment plan, so you have a choice of two methods—continuing to pay as before, or switching to the new plan. Under the new plan, you would make monthly payments, 30% of the size of your current payments. The interest rate would be 24% nominal, compounded monthly. The time until the end of the mortgage would not change, regardless of the method chosen.

(a) Which plan would you choose, given that you naturally wish to minimize the level of your payment costs? (*Hint:* Look at the costs over a three-month period.)

(b) Under which plan would you be paying a higher effective yearly interest rate?

3.29 Derive the arithmetic gradient conversion to a uniform series formula. (*Hint:* Convert each period's gradient amount to its future value, and then look for a substitution from the other compound amount factors.)

3.30 Derive the geometric gradient to present worth conversion factor. (*Hint:* Divide and multiply the present worth of a geometric series by $[1 + g]$ and then substitute in the growth adjusted interest rate.)

3.31 Reginald is expecting steady growth of 10% per year in profits from his new company. All profits are going to be invested at 20% interest. If profits for this year (at the end of the year) total $100 000, how much will be saved at the end of 10 years?

3.32 Reginald is expecting steady growth in profits from his new company of 20% per year. All profits are going to be invested at 10% interest. If profits for this year (at the end of the year) total $100 000, how much will be saved at the end of 10 years?

3.33 Ruby's business has been growing quickly over the past few years, with sales increasing at about 50% per year. She has been approached by a buyer for the business. She has decided she will sell it for 1/2 of the value of the estimated sales for the next five years. This year she will sell products worth $1 456 988. Use the geometric gradient factor to calculate her selling price for an interest rate of 5%.

3.34 In Example 3.4, Clarence bought a $94 000 house with a $14 000 down payment and took out a mortgage for the remaining $80 000 at 12% nominal interest, compounded monthly. We determined that he would make 51 $2000 payments and then a final payment. What is his final payment?

3.35 A new wave-soldering machine is expected to save Yukon Circuit Boards $15 000 per year through reduced labour costs and increased quality. The device will have a life of eight years, and have no salvage value after this time. If the company can generally expect to get 12% return on its capital, how much could it afford to pay for the wave-soldering machine?

3.36 Gail has won a lottery that pays her $100 000 at the end of this year, $110 000 at the end of next year, $120 000 the following year, and so on, for 30 years.

Leon has offered Gail \$2 500 000 today in exchange for all the money she will receive. If Gail can get 8% interest on her savings, is this a good deal?

3.37 Gail has won a lottery that pays her \$100 000 at the end of this year, and increases by 10% per year thereafter for 30 years. Leon has offered Gail \$2 500 000 today in exchange for all the money she will receive. If Gail can get 8% interest on her savings, is this a good deal?

3.38 Tina has saved up \$20 000 from her summer jobs. Rather than work for a living, she plans to buy an annuity from a trust company and become a beachcomber in Fiji. An annuity will pay her a certain amount each month for the rest of her life, and is calculated at 7% interest, compounded monthly, over Tina's 55 remaining years. Tina calculates that she needs at least \$5 per day to live in Fiji, and she needs \$1200 for air fare. Can she retire now? How much would she have available to spend each day?

3.39 The Regional Municipality of Kitchener is studying a water supply plan for the tri-city and surrounding area to the end of year 2040. To satisfy the water demand, one suggestion is to construct a pipeline from one of the Great Lakes. Construction would start in 2000 and take five years at a cost of \$20 million per year. The cost of maintenance and repairs starts after completion of construction and for the first year is \$2 million, increasing by 1% per year thereafter. At an interest rate of 6%, what is the present worth of this project?

Assume that all cash flows take place at year-end. Consider the present to be the end of 1995 / beginning of 1996. Assume that there is no salvage value at the end of year 2040.

3.40 Clem has a \$50 000 loan. The interest rate offered is 8% compounded annually, and the repayment period is 15 years. Payments are to be received in equal instalments at the end of each year. Construct a spreadsheet (you must use a spreadsheet program) similar to the sample on page 91 that shows the amount received each year, the portion that is interest, the portion that is unrecovered capital, and the amount that is outstanding (i.e., unrecovered). Also, compute the total recovered capital which must equal the original capital amount; this can serve as a check on your solution. Design the spreadsheet so that the capital amount and the interest rate can be changed by updating only one cell for each. Construct:

(a) the completed spreadsheet for the amount, interest rate, and repayment period indicated

(b) the same spreadsheet, but for \$75 000 at 10% interest (same repayment period)

(c) a listing showing the formulas used

Sample Capital Recovery Calculations				
Capital amount		$50 000.00		
Annual interest rate		8.00%		
Number of years to repay		15		
Payment Periods	**Annual Payment**	**Interest Received**	**Recovered Capital**	**Unrecovered Capital**
0				$ 50 000.00
1	$ 5 841.48	$ 4 000.00	$ 1 841.48	$ 48 158.52
2				
⋮				
15				0.00
Total			50 000.00	

3.41 A software genius has been offered $10 000 per year for the next five years and then $20 000 per year for the following 10 years for the rights to his new video game. At 9% interest, how much is this worth today?

3.42 A bank offers a personal loan called "The Eight Per Cent Plan." The bank adds 8% to the amount borrowed; the borrower pays back one-twelfth of this total at the end of each month for a year. On a loan of $500, the monthly payment is 540/12 = $45. There is also an administrative fee of $45, payable now. What is the actual effective interest rate on a $500 loan?

3.43 Coastal Shipping is setting aside capital to fund an expansion project. Funds earmarked for the project will accumulate at the rate of $50 000 per month until the project is completed. The project will take two years to complete. Once the project starts, costs will be incurred monthly at the rate of $150 000 per month over 24 months. Coastal currently has $250 000 saved. What is the minimum number of months they will have to wait before they can start if money is worth 18% nominal, compounded monthly? Assume that:

1. Cash flows are all at the ends of months.
2. The first $50 000 savings occurs one month from today.
3. The first $150 000 payment occurs one month after the start of the project.
4. The project must start at the beginning of a month.

3.44 A company is about to invest in a joint venture research and development project with another company. The project is expected to last eight years, but yearly payments the company makes will begin immediately (i.e., a payment is made today, and the last payment is eight years from today). Salaries will account for $40 000 of each payment. The remainder of each payment will cover equipment costs and facility overhead. The initial (immediate) equipment and facility cost is $26 000. Each subsequent year the figure will drop by $3000 until a cost of $14 000 is reached, after which the costs will remain constant until the end of the project.

(a) Draw a cash flow diagram to illustrate the cash flows for this situation.

(b) At an interest rate of 7%, what is the total future worth of all project payments at the end of the eight years?

3.45 Shamsir's business has been growing slowly. He has noticed that his monthly profit increases by 1% every two months. Suppose that the profit at the end of this month is $1000. What is the present value of all his profit over the next 2 years? Annual nominal interest is 18%, compounded monthly.

3.46 Xiaohang is conducting a biochemical experiment for the next 12 months. In the first month, the expenses are estimated to be $1500. As the experiment progresses, the expenses are expected to increase by 5% each month. Xiaohang plans to pay for the experiment by a government grant, which are received in six monthly instalments, starting a month after the experiment completion date. Determine the amount of the monthly instalment so that the total of the six instalments pay for all expenses occurred during the experiment. Annual nominal interest is 12%, compounded monthly.

3.47 City engineers are considering several plans for building municipal aqueduct tunnels. They use an interest rate of 8%. One plan calls for a full capacity tunnel that will meet the needs of the city forever. The cost is $3 000 000 now, and $100 000 every 10 years thereafter for repairs. What is the total present worth of the costs of building and maintaining the aqueduct?

3.48 The City of Surrey is installing a new swimming pool in the downtown recreation centre. One design being considered is a reinforced concrete pool which will cost $1 500 000 to install. Thereafter, the inner surface of the pool will need to be refinished and painted every 10 years at a cost of $200 000 per refinishing. Assuming that the pool will have essentially an infinite life, what is the present worth of the costs associated with this pool design? The city uses a 5% interest rate.

3.49 Goderich Automotive (GA) wants to donate a vacant lot next door to their plant to the city for use as a public park and ball field. The city will accept only if GA will also donate enough cash to maintain the park forever. The estimated maintenance costs are \$18 000 per year and interest is 7%. How much cash must GA donate?

3.50 A 7%, 20-year municipal bond has a \$10 000 face value. I want to receive at least 10% compounded semi-annually on this investment. How much should I pay for the bond?

3.51 A Paradorian bond pays \$500 (Paradorian dollars) twice each year and \$5000 five years from now. I want to earn at least 300% *annual* (effective) interest on this investment (to compensate for the very high inflation in Parador). How much should I pay for this bond now?

3.52 If money is worth 8% compounded semi-annually, how much is a bond maturing in nine years, with a face value of \$10 000 and a coupon rate of 9%, worth today?

3.53 A bond with a face value of \$5000 pays quarterly interest of 1.5% each period. Twenty-six interest payments remain before the bond matures. How much would you be willing to pay for this bond today if the next interest payment is due now and you want to earn 8% compounded quarterly on your money?

A New Distribution Station for Kitchener-Wilmot Hydro

CANADIAN 3.1 MINI-CASE

Kitchener-Wilmot Hydro provides electricity to the City of Kitchener, Ontario, and nearby Wilmot Township as a regulated monopoly. It purchases power from the provincial utility Ontario Hydro and distributes it to homes and industries in these areas.

Recently, the utility undertook an investigation of the locations and capacities of the distribution stations in Wilmot Township. These distribution stations are essentially transformers which reduce voltage from 27.6 kV (thousand volts), which is efficient for long-distance distribution, to 8.3 kV, which is better for local distribution. Near each user, it is further transformed to 120 V or 600 V.

A particular problem was distribution station (DS) #5, located northwest of the town of St. Agatha. This station had old technology that required upgrading to match the other distribution stations. Three alternatives were identified. In short, they were:

1. Upgrade DS#5.
2. Build a new distribution station at the town of Philipsburg, using the salvaged transformer from former DS#5.
3. Build a new distribution station at the town of New Prussia, using the salvaged transformer from former DS#5.

Data were gathered concerning current and projected usage and costs for the three alternatives. A present worth analysis was done, using a discount factor

continued

(interest rate) of 10%. The least expensive choice turned out to be alternative 1, which was recommended for implementation.

Discussion

In order to compare alternatives, it is necessary to determine the future cash flows associated with each. Of course nobody knows for sure what the future holds, so future cash flows are always estimates. In some cases the estimates will turn out to be significantly different from the actual cash flows, and therefore result in an incorrect decision.

The electricity demands in the distribution area of DS#5 were fast changing at the time Kitchener-Wilmot Hydro had to make this decision. Former villages were expanding to become suburban communities for the booming twin cities of Kitchener-Waterloo, and affluent country estates were replacing farms. The changing population in turn results in predictable changes in the location and amount of electricity demand. On the other hand, predicting electricity demand itself is tricky—for example, an economic downturn would likely change the growth rate of the Kitchener-Waterloo area.

There are sophisticated ways to deal with uncertainty about future cash flows, some of which are discussed in Chapters 11 and 12. In many cases it makes sense to assume that the future cash flows are treated as if they were certain because there is no particular reason to think they are not. On the other hand, in some cases even estimating future cash flows can be very difficult or impossible.

Questions

1) For each of the following, comment on how sensible it is to estimate the precise value of future cash flows:
   (a) Your rent for the next six months
   (b) Your food bill for the next six months
   (c) Your medical bills for the next six months
   (d) A company's payroll for the next six months
   (e) A company's raw material costs for the next six months
   (f) A company's legal costs for liability lawsuits for the next six months
   (g) Canada's costs for funding university research for the next six months
   (h) Canada's costs for Employment Insurance for the next six months
   (i) Canada's costs for Emergency Management for the next six months

2) Your company is looking at the possibility of buying a new widget grinder for the widget line. The future cash flows associated with the purchase of the grinder are fairly predictable, except for one factor. A significant benefit is achieved with the higher production volume of widgets, which depends on a contract to be signed with a particular important customer. This won't happen for several months, but you must make the decision about the widget-grinder now. Discuss some sensible ways of dealing with this issue.

Appendix 3A: Continuous Compounding and Continuous Cash Flows

We now consider compound interest factors for continuous models. Two forms of continuous modelling are of interest. The first form has discrete cash flows with continuous compounding. In some cases, projects generate discrete, end-of-period cash flows within an organization in which other cash flows and compounding occur many times per period. In this case, a reasonable model is to assume that the project's cash flows are discrete, but compounding is continuous. The second form has continuous cash flows with continuous compounding. Where a project generates many cash flows per period, we could model this as continuous cash flows with continuous compounding.

We first consider models with discrete cash flows with continuous compounding. We can obtain formulas for compound interest factors for these projects from formulas for discrete compounding simply by substituting the effective continuous interest rate with continuous compounding for the effective rate with discrete compounding.

Recall that for a given nominal interest rate, r, when the number of compounding periods per year becomes infinitely large, the effective interest rate per period is $i_e = e^r - 1$. This implies $1 + i_e = e^r$ and $(1 + i_e)^N = e^{rN}$. The various compound interest factors for continuous compounding can be obtained by substituting $e^r - 1$ for i in the formulas for the factors.

For example, the series present worth factor with discrete compounding is

$$(P/A,i,N) = \frac{(1 + i)^N - 1}{i(1 + i)^N}$$

If we substitute $e^r - 1$ for i and e^{rN} for $(1 + i)^N$, we get the series present worth factor for continuous compounding

$$(P/A,r,N) = \frac{e^{rN} - 1}{(e^r - 1)e^{rN}}$$

Similar substitutions can be made in each of the other compound interest factor formulas to get the compound interest factor for continuous rather than discrete compounding. The formulas are shown in Table 3A.1. Tables of values for these formulas are available in Appendix B at the end of the book.

Table 3A.1 Compound Interest Formulas for Discrete Cash Flow with Continuous Compounding

Name	Symbol and Formula
Compound amount factor	$(F/P,r,N) = e^{rN}$
Present worth factor	$(P/F,r,N) = \dfrac{1}{e^{rN}}$
Sinking fund factor	$(A/F,r,N) = \dfrac{e^{r} - 1}{e^{rN} - 1}$
Uniform series compound amount factor	$(F/A,r,N) = \dfrac{e^{rN} - 1}{e^{r} - 1}$
Capital recovery factor	$(A/P,r,N) = \dfrac{(e^{r} - 1)e^{rN}}{e^{rN} - 1}$
Series present worth factor	$(P/A,r,N) = \dfrac{e^{rN} - 1}{(e^{r} - 1)e^{rN}}$
Arithmetic gradient to annuity conversion factor	$(A/G,r,N) = \dfrac{1}{e^{r} - 1} - \dfrac{N}{e^{rN} - 1}$

Example 3A.1

Yoram Gershon is saving to buy a new sound system. He plans to deposit \$100 each month for the next 24 months in the Bank of Montrose. The nominal interest rate at the Bank of Montrose is 0.5% per month, compounded continuously. How much will Yoram have at the end of the 24 months?

We start by computing the uniform series compound amount factor for continuous compounding. Recall that the factor for discrete compounding is

$$(F/A,i,N) = \frac{(1 + i)^{N} - 1}{i}$$

Substituting $e^{r} - 1$ for i and e^{rN} for $(1 + i)^{N}$ gives the series compound amount, when compounding is continuous, as

$$(F/A,r,N) = \frac{e^{rN} - 1}{e^{r} - 1}$$

The amount Yoram will have at the end of 24 months, F, is given by

$$F = \$100\left(\frac{e^{(0.005)24} - 1}{e^{0.005} - 1}\right)$$

$$F = \$100\left(\frac{1.127497 - 1}{1.00501 - 1}\right)$$

$$= \$2544.85$$

Yoram will have about \$2545 saved at the end of the 24 months. ■

The formulas for *continuous cash flows with continuous compounding* are derived using integral calculus. The continuous *series present worth* factor, denoted by $(P/\bar{A},r,T)$, for a continuous flow, $\bar{A}$, over a period length, T, where the nominal interest rate is r, is given by

$$P = \bar{A}\left(\frac{e^{rT} - 1}{re^{rT}}\right)$$

so that

$$(P/\bar{A},r,T) = \frac{e^{rT} - 1}{re^{rT}}$$

It is then easy to derive the formula for the continuous *uniform series compound amount factor*, denoted by $(F/\bar{A},r,T)$, by multiplying the series present worth factor by e^{rT} to get the future worth of a present value, P.

$$(F/\bar{A},r,T) = \frac{e^{rT} - 1}{r}$$

We can get the continuous *capital recovery factor*, denoted by $(\bar{A}/P,r,T)$, as the inverse of the continuous series present worth factor. The *continuous sinking fund factor* $(F/\bar{A},r,T)$ is the inverse of the continuous uniform series compound amount factor. A summary of the formulas for continuous cash flow and continuous compounding is shown in Table 3A.2. Tables of values for these formulas are available in Appendix C at the back of the book.

Table 3A.2 Compound Interest Formulas for Continuous Cash Flow with Continuous Compounding

Name	Symbol and Formula
Sinking fund factor	$(\bar{A},F,r,T) = \frac{r}{e^{rT} - 1}$
Uniform series compound amount factor	$(F/\bar{A},r,T) = \frac{e^{rT} - 1}{r}$
Capital recovery factor	$(\bar{A},/P,r,T) = \frac{re^{rT}}{e^{rT} - 1}$
Series present worth factor	$(P/\bar{A},r,T) = \frac{e^{rT} - 1}{re^{rT}}$

Example 3A.2

Savings from a new widget grinder are estimated to be $10 000 per year. The grinder will last 20 years and will have no scrap value at the end of that time. Assume that the savings are generated as a continuous flow. The *effective* interest rate is 15% compounded continuously. What is the present worth of the savings?

From the problem statement, we know that $\bar{A} = \$10\ 000$, $i = 0.15$, and $T = 20$. From the relation $i = e^r - 1$, for $i = 0.15$ the interest rate to apply for continuously compounding is $r = 0.13976$. The present worth computations are

$$P = \bar{A}(P/\bar{A},r,T)$$
$$= \$10\ 000\left(\frac{e^{(0.13976)20} - 1}{(0.13976)e^{(0.13976)20}}\right)$$
$$= \$67\ 180$$

The present worth of the savings is $67 180. Note that if we had used discrete compounding factors for the present worth computations we would have obtained a lower value.

$$P = A(P/A,i,N)$$
$$= \$10\ 000(6.2593)$$
$$= \$62\ 593 \ \blacksquare$$

REVIEW PROBLEM 3A.1 FOR APPENDIX 3A

Mr. Big is thinking of buying the MQM Grand Hotel in Las Vegas. The hotel has continuous net receipts totalling $120 000 000 per year (Vegas hotels run 24 hours per day). This money could be immediately reinvested in Mr. Big's many other ventures, all of which earn a nominal 10% interest. The hotel will likely be out of style in about eight years, and could then be sold for about $200 000 000. What is the maximum Mr. Big should pay for the hotel today?

ANSWER

$$P = \$120\ 000\ 000(P/\bar{A}, 10\%, 8) + \$200\ 000\ 000e^{-(0.1)(8)}$$
$$= \$120\ 000\ 000\frac{e^{(0.1)(8)} - 1}{(0.1)e^{(0.1)(8)}} + \$200\ 000\ 000e^{-(0.1)(8)}$$
$$= \$701\ 184\ 547$$

Mr. Big should not pay more than about $700 000 000. ■

PROBLEMS

3A.1 An investment in new data logging technology is expected to generate extra revenue continuously for Calgary Petroleum Services. The initial cost is \$300 000, but extra revenues total \$75 000 per year. If the effective interest rate is 10% compounded continuously, does the present worth of the savings over five years exceed the original purchase cost? By how much does one exceed the other?

3A.2 Desmond earns \$25 000 continuously over a year from an investment that pays 8% nominal interest, compounded continuously. How much money does he have at the end of the year?

3A.3 Gina intently plays the stock market, so that any capital she has can be considered to be compounding continuously. At the end of 1999, Gina had \$10 000. How much did she have at the beginning of 1999, if she earned a nominal interest rate of 18%?

3A.4 Gina (from Problem 3A.3) has earned a nominal interest rate of 18% on the stock market every year since she started with an initial investment of \$100. What year did she start investing?

Appendix 3B: Derivation of Discrete Compound Interest Factors

This appendix derives six discrete compound interest factors presented in this chapter. All of them can be derived from the compound interest equation

$$F = P(1 + i)^N$$

3B.1 Compound Amount Factor

In the symbolic convention used for compound interest factors, the compound interest equation can be written

$$F = P(1 + i)^N = P(F/P,i,N)$$

so that the compound amount factor is

$$(F/P,i,N) = (1 + i)^N \qquad (3B.1)$$

3B.2 Present Worth Factor

The present worth factor, $(P/F,i,N)$, converts a future amount F to a present amount P:

$$P = F(P/F,i,N)$$

$$\Rightarrow F = P\left(\frac{1}{(P/F,i,N)}\right)$$

Thus the present worth factor is the reciprocal of the compound amount factor. From Equation (3B.1),

$$(P/F,i,N) = \frac{1}{(1 + i)^N}$$

3B.3 Sinking Fund Factor

If a series of payments A follows the pattern of a standard annuity of N payments in length, then the future value of the payment in the ith period, from Equation (3B.1), is:

$$F = A(1 + i)^{N-j}$$

The future value of all of the annuity payments is then

$$F = A(1 + i)^{N-1} + A(1 + i)^{N-2} + \ldots + A(1 + i)^1 + A$$

Factoring out the annuity amount gives

$$F = A[(1 + i)^{N-1} + (1 + i)^{N-2} + \ldots + (1 + i)^1 + 1] \tag{3B.2}$$

Multiplying Equation (3B.2) by $(1 + i)$ gives

$$F(1+i) = A[(1 + i)^{N-1} + (1 + i)^{N-2} + \ldots + (1 + i)^1 + 1]\,(1 + i)$$

$$F(1+i) = A[(1 + i)^N + (1 + i)^{N-1} + \ldots + (1 + i)^2 + (1 + i)] \tag{3B.3}$$

Subtracting Equation (3B.2) from Equation (3B.3) gives

$$F(1+i) - F = A[(1 + i)^N - 1]$$

$$Fi = A[(1 + i)^N - 1]$$

$$A = F\left[\frac{i}{(1 + i)^N - 1}\right]$$

Thus the sinking fund factor is given by

$$(A/F,i,N) = \frac{i}{(1 + i)^N - 1} \tag{3B.4}$$

3B.4 Uniform Series Compound Amount Factor

The uniform series compound amount factor, $(F/A,i,N)$, converts an annuity A into a future amount F:

$$F = A(F/A,i,N)$$

$$\Rightarrow A = F\left(\frac{1}{(F/A,i,N)}\right)$$

Thus the uniform series compound amount factor is the reciprocal of the sinking fund factor. From Equation (3B.4),

$$(F/A,i,N) = \frac{(1+i)^N - 1}{i}$$

3B.5 Capital Recovery Factor

If a series of payments A follows the pattern of a standard annuity of N payments in length, then the present value of the payment in the jth period is

$$P = A\frac{1}{(1+i)^j}$$

The present value of the total of all the annuity payments is

$$P = A\left(\frac{1}{(1+i)}\right) + A\left(\frac{1}{(1+i)^2}\right) + \ldots + A\left(\frac{1}{(1+i)^{N-1}}\right) + A\left(\frac{1}{(1+i)^N}\right)$$

Factoring out the annuity amount gives

$$P = A\left[\left(\frac{1}{(1+i)}\right) + \left(\frac{1}{(1+i)^2}\right) + \ldots + \left(\frac{1}{(1+i)^{N-1}}\right) + \left(\frac{1}{(1+i)^N}\right)\right] \tag{3B.5}$$

Multiplying both sides of Equation (3B.5) by $(1 + i)$ gives

$$P(1+i) = A\left[1 + \left(\frac{1}{(1+i)}\right) + \ldots + \left(\frac{1}{(1+i)^{N-2}}\right) + \left(\frac{1}{(1+i)^{N-1}}\right)\right] \tag{3B.6}$$

Subtracting Equation (3B.5) from Equation (3B.6) gives

$$Pi = A\left[1 - \left(\frac{1}{(1+i)^N}\right)\right]$$

$$P = A\left[\frac{(1+i)^N - 1}{i(1+i)^N}\right]$$

$$A = P\left[\frac{i(1+i)^N}{(1+i)^N - 1}\right]$$

Thus the capital recovery factor is given by

$$(A/P,i,N) = \frac{i(1+i)^N}{(1+i)^N - 1} \tag{3B.7}$$

3B.6 Series Present Worth Factor

The series present worth factor, $(P/A,i,N)$, converts an annuity A into a present amount P:

$$P = A(P/A,i,N)$$

$$\Rightarrow A = P\left(\frac{1}{(P/A,i,N)}\right)$$

Thus the uniform series compound amount factor is the reciprocal of the sinking fund factor. From Equation (3B.7),

$$(P/A,i,N) = \frac{(1+i)^N - 1}{i(1+i)^N}$$

3B.7 Arithmetic and Geometric Gradients

The derivation of the arithmetic gradient to annuity conversion factor and the geometric gradient to present worth conversion factor are left as problems for the student. See Problems 3.29 and 3.30.

C H A P T E R 4

Comparison Methods Part 1

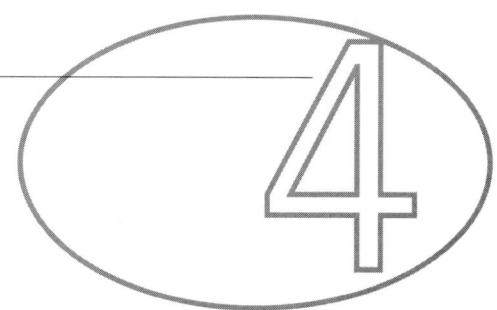

Engineering Economics in Action, Part 4A: What's Best?

Naomi waved hello as she breezed by Carole Brown, the receptionist, on her way in from the parking lot one Monday morning. She stopped as Carole caught her eye. "Clem wants to see you right away. Good morning."

After a moment of socializing, Clem got right to the point. "I have a job for you. Put aside the vehicle-life project for a couple of days."

"O.K., but you wanted a report by Friday."

"This is more important. You know that drop forging hammer in the South Shop? The beast is about 50 years old. I don't remember the exact age. We got it used four years ago. We were having quality control problems with the parts we were buying on contract and decided to bring production in-house. Stinson Brothers sold it to us cheap when they upgraded their forging operation. Fundamentally the machine is still sound, but the guides are worn out. The production people are spending too much time fiddling with it instead of turning out parts. Something has to be done. I have to make a recommendation to Ed Burns and Anna Kulkowski, who are going to be making decisions on investments for the next quarter. I'd like you to handle it." Ed Burns was the manager of manufacturing, and Anna Kulkowski was, among other things, the president of Canadian Widgets.

"What's the time frame?" Naomi asked. She was shifting job priorities in her mind and deciding what she would need to postpone.

"I want a report by tomorrow morning. I'd like to have a chance to review what you've done and submit a recommendation to Burns and Kulkowski for their Wednesday meeting." Clem sat back and gave Naomi his best big smile.

Naomi's return smile was a bit weak, as she was preoccupied with trying to sort out where to begin.

Clem laughed and continued with, "It's really not so bad. Dave Sullivan has done most of the work. But he's away and can't finish. His father-in-law had a heart attack on Friday, and he and Helena have gone to Florida to see him."

"What's involved?" asked Naomi.

"Not much, really. Dave has estimated all the cash flows. He's put everything on a spreadsheet. Essentially, there are three major possibilities. We can refurbish and upgrade the existing machine. We can get a manually operated mechanical press that will use less energy and be a lot quieter. Or we can go for an automated mechanical press.

"Since there is going to be down time while we are changing the unit, we might also want to replace the material-handling equipment at the same time. If we get the automated press, there is the possibility of going whole hog and integrating material handling with the press. But even if we automate, we could stay with a separate material-handling setup.

"Basically, you're looking at a fairly small first cost to upgrade the current beast, versus a large first cost for the automated equipment. But, if you take the high first cost route, you will get big savings down the road. All you have to do is decide what's best."

4.1 Introduction

The essential idea of investing is to give up something valuable now for the expectation of receiving something of greater value later. An investment may be thought of as an exchange of resources now for an expected flow of benefits in the future. Business firms, other organizations, and individuals all have opportunities to make such exchanges. A company may be able to use funds to install equipment that will reduce labour costs in the future. These funds might otherwise have been used on another project or returned to the shareholders or owners. An individual may be able to study to become an engineer. Studying requires that time be given up that could have been used to earn money or to travel. The benefit of study, though, is the expectation of a good income from an interesting job in the future.

Not all investment opportunities *should* be taken. The company considering a labour-saving investment may find that the value of the savings is less than the cost of installing the equipment. Not all investment opportunities *can* be taken. The person spending the next four years studying engineering cannot also spend that time getting a degree in business.

Engineers play a major role in making decisions about investment opportunities. In many cases, they are the ones who estimate the expected costs of and returns from an investment. They then must decide whether the expected returns outweigh the costs to see if the opportunity is potentially acceptable. They may also have to examine competing investment opportunities to see which is best. Engineers frequently refer to investment opportunities as **projects**. Throughout the rest of this text, the term "project" will be used to mean "investment opportunity."

In this chapter and in Chapter 5, we deal with methods of evaluating and comparing projects, sometimes called **comparison methods**. We start in this chapter with a scheme for classifying groups of projects. This classification system permits the appropriate use of any of the comparison methods. We then turn to a consideration of several widely used methods for evaluating opportunities. The **present worth method** compares projects by looking at the present worth of all cash flows associated with the projects. The **annual worth method** is similar, but converts all cash flows to a uniform series, i.e., an annuity. The **payback period method** estimates how long it takes to "pay back" investments. The study of comparison methods is continued in Chapter 5, which deals with the internal rate of return.

We have made six assumptions about all the situations presented in this chapter and in Chapter 5:

1. We have assumed that costs and benefits are always measurable in terms of money. In reality, costs and benefits need not be measurable in terms of money. For example, providing safe working conditions has many benefits, including improvement of worker morale. However, it would be difficult to express the value of improved worker morale objectively in dollars and cents. Such other benefits as the pleasure

gained from appreciating beautiful design may not be measurable quantitatively. We shall consider qualitative criteria and multiple objectives in Chapter 13.

2. We have assumed that future cash flows are known with certainty. In reality, future cash flows can only be estimated. Usually the further into the future we try to forecast, the less certain our estimates become. We look at methods of assessing the impact of uncertainty and risks in Chapters 11 and 12.
3. We have assumed that cash flows are unaffected by inflation or deflation. In reality, the purchasing power of money typically declines over time. We shall consider how inflation affects decision making in Chapter 9.
4. Unless otherwise stated, we have assumed that sufficient funds are available to implement all projects. In reality, cash constraints on investments may be very important, especially for new enterprises with limited ability to raise capital. We look at methods of raising capital in Appendix 4A.
5. We have assumed that taxes are not applicable. In reality, taxes are pervasive. We shall show how to include taxes in the decision-making process in Chapter 8.
6. Unless otherwise stated, we shall assume that all investments have a cash outflow at the start. These outflows are called *first costs*. We also assume that projects with first costs have cash inflows after the first costs that are at least as great in total as the first costs. In reality, some projects have cash inflows at the start, but involve a commitment of cash outflows at a later period. For example, a consulting engineer may receive an advance payment from a client, a cash inflow, to cover some of the costs of a project, but to complete the project the engineer will have to make disbursements over the project's life. We shall consider evaluation of such projects in Chapter 5.

4.2 Relations Among Projects

Companies and individuals are often faced with a large number of investment opportunities at the same time. Relations among these opportunities can range from the simple to the complex. We can distinguish three types of connections among projects that cover all the possibilities. Projects may be

(1) independent,

(2) mutually exclusive, or

(3) related but not mutually exclusive.

The simplest relation between projects occurs when they are **independent**. Two projects are independent if the expected costs and the expected benefits of each project do not depend on whether or not the other one is chosen. A student considering the purchase of a vacuum cleaner and the purchase of a personal computer would probably find that the expected costs and benefits of the computer did not depend on whether he or she bought the vacuum cleaner. Similarly, the benefits and costs of the vacuum cleaner would be the same, whether or not the computer was purchased. If there are more than two projects under consideration, they are said to be independent if all possible pairs of projects in the set are independent. When two or more projects are independent, evaluation is simple. Consider each opportunity one at a time, and accept or reject it on its own merits.

Projects are **mutually exclusive** if, in the process of choosing one, all the other alternatives are excluded. In other words, two projects are mutually exclusive if it is impossible to do both or it clearly would not make sense to do both. For example, suppose Bismuth Realty Company wants to develop downtown office space on a specific piece of land. They are considering two potential projects. The first is a low-rise poured-concrete building. The second is a high-rise steel-frame structure with the same capacity as the low-rise building, but it has a small park at the entrance. It is impossible for Bismuth to have both buildings on the same site.

As another example, consider a student about to invest in a computer printer. She can get an ink-jet printer or a laser printer, but it would not make sense to get both. She would consider the options to be mutually exclusive.

The third class of projects consists of those that are **related but not mutually exclusive**. For pairs of projects in this category, the expected costs and benefits of one project depend on whether the other one is chosen. For example, Klamath Petroleum may be considering a service station at Fourth Avenue and Main Street as well as one at Twelfth and Main. The costs and benefits from either station will clearly depend on whether or not the other is built, but it may be possible, and may make sense, to have both stations.

Evaluation of related but not mutually exclusive projects can be simplified by combining them into exhaustive, mutually exclusive sets. For example, the two projects being considered by Klamath can be put into four mutually exclusive sets:

1. Neither station — the "do nothing" option
2. Just the station at Fourth and Main
3. Just the station at Twelfth and Main
4. Both stations

In general, n related projects can be put into 2^n sets including the "do nothing" option. Once the related projects are put into mutually exclusive sets, the analyst treats these sets as the alternatives. We can make 2^n mutually exclusive sets with n related projects by noting that for any single set there are exactly two possibilities for each project. The project may be *in* or *out* of that set. To get the total number of sets, we multiply the n twos to get 2^n. In the Klamath example, there were two possibilities for the station at Fourth and

Main — accept or reject. These are combined with the two possibilities for the station at Twelfth and Main, to give the four sets that we listed.

A special case of related projects is where one project is *contingent* on another. Consider the case where project A could be done alone or A and B could be done together, but B could not be done by itself. Project B is then contingent on Project A because it cannot be taken unless A is taken first. For example, the Athens and Manchester Development Company is considering building a shopping mall on the outskirts of Moncton. They are also considering building a parking garage to avoid long outdoor walks by patrons. Clearly, they would not build the parking garage unless they were also building the mall.

Sometimes two or more projects are mutually exclusive, not because of physical restrictions, but because of resource constraints. Usually the constraint is financial. For example, Bismuth may be considering two office buildings at different sites, where the expected costs and benefits of the two are unrelated, but Bismuth may be able to finance only one building. The two office-building projects would then be mutually exclusive because of financial constraints. If there are more than two projects, then all of the sets of projects that meet the budget form a mutually exclusive set of alternatives.

When there are several related projects, the number of logically possible combinations becomes quite large. If there are four related projects, there are $2^4 = 16$ mutually exclusive sets, including the "do-nothing" alternative. If there are five related projects, the number of alternatives doubles to 32. A good way to keep track of these alternatives is to construct a table with all possible combinations of projects. The potential projects are in rows. The alternatives, which are sets of projects, are in columns. An x in a cell indicates that a project is in the alternative represented by that column. Not all logical combinations of projects represent feasible alternatives, as seen in the special cases of contingent alternatives or budget constraints. A last row, below the potential-project rows, indicates whether or not the sets are feasible alternatives.

Example 4.1

The Small Street residential association wants to improve their district. Four ideas for renovation projects have been proposed: (1) converting part of the roadway to gardens, (2) adding old-fashioned light standards, (3) replacing the pavement with cobblestones, and (4) making the street one-way. However, there are a number of restrictions. The residential association can afford to do only two of the first three projects together. Also, gardens are possible only if the street is one-way. Finally, old-fashioned light standards would look out of place unless the pavement was replaced with cobblestones. The residential association feels it must do something. They do not want simply to leave things the way they are. What mutually exclusive alternatives are possible?

Since the association does not want to "do nothing," only $15 = 2^4 - 1$ alternatives will be considered. These are shown in Table 4.1.

Table 4.1 Potential Alternatives for the Small Street Renovation

Potential Alternative	1	2	3	4	5	6	7	8	9	10	11	12	13	14	15
Gardens	x				x	x	x				x	x	x		x
Lights		x			x			x	x		x	x		x	x
Cobblestones			x			x		x		x	x		x	x	x
One-way				x			x		x	x		x	x	x	x
Feasible?	No	No	Yes	Yes	No	No	Yes	Yes	No	Yes	No	No	Yes	Yes	No

The result is that there are seven feasible mutually exclusive alternatives:

1. Cobblestones (alternative 3)
2. One-way street (alternative 4)
3. One-way street with gardens (alternative 7)
4. Cobblestones with lights (alternative 8)
5. One-way street with cobblestones (alternative 10)
6. One-way street with cobblestones and gardens (alternative 13)
7. One-way street with cobblestones and lights (alternative 14) ■

To summarize our investigation of possible relations among projects, we have a threefold classification system: (1) independent projects, (2) mutually exclusive projects, and (3) related but not mutually exclusive projects. We can, however, arrange related projects into mutually exclusive sets and treat the sets as mutually exclusive alternatives. This reduces the system to two categories, independent and mutually exclusive. (See Figure 4.1.) Therefore, in the remainder of this chapter we consider only independent and mutually exclusive projects.

Figure 4.1 Possible Relations Among Projects and How to Treat Them

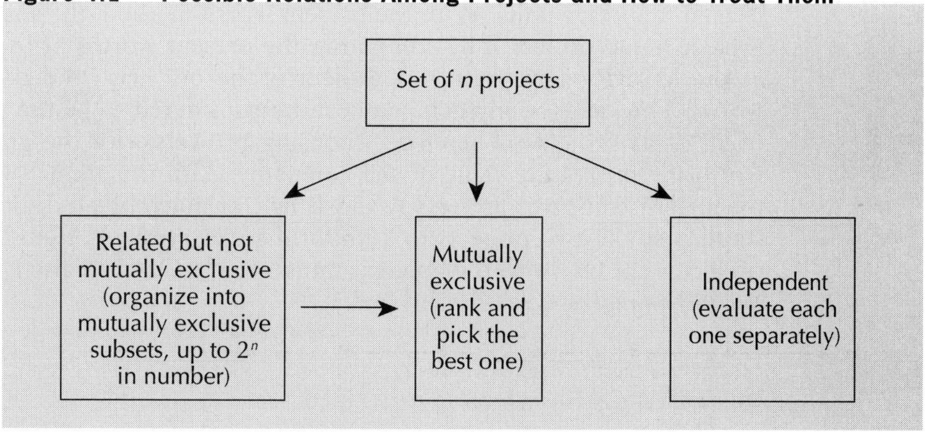

4.3 Minimum Acceptable Rate of Return (MARR)

A company evaluating projects will set for itself a lower limit for investment acceptability known as the **minimum acceptable rate of return (MARR)**. The MARR is an interest rate that must be earned for any project to be accepted. Projects that earn at least the MARR are desirable, since this means that the money is earning at least as much as can be earned elsewhere. Projects that earn less than the MARR are not desirable, since investing money in these projects denies the opportunity to use the money more profitably elsewhere.

The MARR can also be viewed as the rate of return required to get investors to invest in a business. If a company accepts projects that earn less than the MARR, investors will not be willing to put money into the company. This minimum return required to induce investors to invest in the company is the company's **cost of capital**. Methods for determining the cost of capital are presented in Appendix 4A.

The MARR is thus an opportunity cost in two senses. First, investors have investment opportunities outside any given company. Investing in a given company implies foregoing the opportunity of investing elsewhere. Second, once a company sets a MARR, investing in a given project implies giving up the opportunity of using company funds to invest in other projects that pay at least the MARR.

We shall show in this chapter and in Chapter 5 how the MARR is used in calculations involving the present worth, annual worth, or internal rate of return to evaluate projects. Henceforth, it is assumed that a value for the MARR has been supplied.

4.4 Present Worth (PW) and Annual Worth (AW) Comparisons

The present worth (PW) comparison method and the annual worth (AW) comparison method are based on finding a comparable basis to evaluate projects in monetary units. With the present worth method, the analyst compares project A and project B by computing the present worths of the two projects at the MARR. The preferred project is the one with the greater present worth. The value of any company can be considered to be the present worth of all of its projects. Therefore, choosing projects with the greatest present worth maximizes the value of the company. With the annual worth method, the analyst compares projects A and B by transforming all disbursements and receipts of the two projects to a uniform series at the MARR. The preferred project is the one with the greater annual worth. One can also speak of *present cost* and *annual cost*. See Close-Up 4.1.

CLOSE-UP 4.1 Present Cost and Annual Cost

Sometimes mutually exclusive projects are compared in terms of present cost or annual cost. That is, the best project is the one with minimum cost as opposed to maximum worth. Two conditions should hold for this to be valid: (1) all projects have the same major benefit, and (2) the estimated value of the major benefit clearly outweighs the projects' costs, even if that estimate is imprecise. Therefore, the "do nothing" option is rejected. The value of the major benefit is ignored in further calculations since it is the same for all projects. We choose the project with the lowest cost, considering secondary benefits as offsets to costs.

4.4.1 Present Worth for Independent Projects

The alternative to investing money in an independent project is to "do nothing." Doing nothing doesn't mean that the money is not used productively. In fact, it would be used for some other project, earning interest at a rate at least equal to the MARR. However, the present worth of any money invested at the MARR is zero, since the present worth of future receipts would exactly offset the current disbursement. Consequently, if an independent project has a present worth greater than zero, it is acceptable. If an independent project has a present worth less than zero, it is unacceptable. If an independent project has a present worth of exactly zero, it is considered *marginally* acceptable.

Example 4.2

Steve Chen, a third-year electrical engineering student at Seaforth University, has noticed that the networked personal computers provided by the university for its students are frequently fully utilized, so that students often have to wait to get on a machine. The university has a building plan that will create more space for network computers, but the new facilities won't be available for five years. In the meantime, Steve sees the opportunity to create an alternative network in a mall near the campus. The first cost for equipment, furniture, and software is expected to be $70 000. Students would be able to rent time on computers by the hour and to use the printers at a charge per page. Annual net cash flow from computer rentals and other charges, after paying for labour, supplies, and other costs, is expected to be $30 000 a year for five years. When the University opens new facilities at the end of five years, business at Steve's network would fall off and net cash flow would turn negative. Therefore, the plan is to dismantle the network after five years. The five-year-old equipment and furniture are expected to have zero value. If investors in this type of service enterprise demand a return of 20% per year, is this a good investment?

The present worth of the project is

$$\begin{aligned} PW &= -\$70\,000 + \$30\,000(P/A, 20\%, 5) \\ &= -\$70\,000 + \$30\,000(2.9906) = \$19\,718 \\ &\cong \$20\,000 \end{aligned}$$

The project is acceptable since the present worth is greater than zero.

Another way to look at the project is to suppose that, once Steve has set up the network off campus, he tries to sell it. If he can convince potential investors, who demand a return of 20% a year, that the expectation of a $30 000 per year cash flow for five years is accurate, how much would they be willing to pay for the network? Investors would calculate the present worth of a 20% annuity paying $30 000 for five years. This is given by

$$\begin{aligned} PW &= \$30\,000(P/A, 20\%, 5) \\ &= \$30\,000(2.9906) \\ &= \$89\,718 \\ &\cong \$90\,000 \end{aligned}$$

Investors would be willing to pay about $90 000. Steve will have taken $70 000, the first cost, and used it to create an asset worth $90 000. As illustrated in Figure 4.2, the $20 000 difference may be viewed as profit. ■

Let us now consider an example in which the benefit of an investment is a reduction in cost.

Figure 4.2 Present Worth as a Measure of Profit

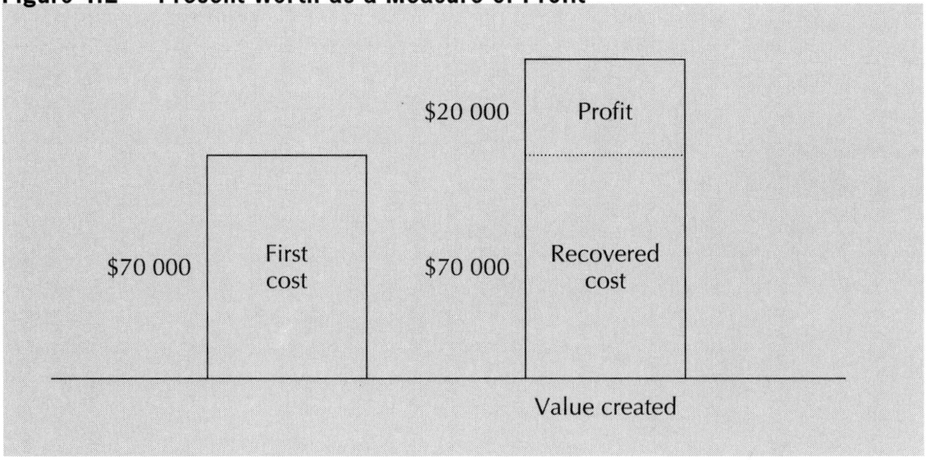

Example 4.3

A mechanical engineer is considering building automated material-handling equipment for a production line. On one hand, the equipment would substantially reduce the manual labour currently required to move items from one part of the production process to the next. On the other hand, the equipment would consume energy, require insurance, and need periodic maintenance.

Alternative 1: Continue to use the current method. Yearly labour costs are $9200.

Alternative 2: Build automated material-handling equipment with an expected service life of 10 years.

First cost	$15 000
Labour	$3 300 per year
Power	$400 per year
Maintenance	$2 400 per year
Taxes and insurance	$300 per year

If the MARR is 9%, which alternative is better? Use a present worth comparison.

The investment of $15 000 can be viewed as yielding a positive cash flow of $2800 = $9200 − ($3300 + $400 + $2400 + $300) per year in the form of a reduction in cost.

$$\begin{aligned} PW &= -\$15\,000 \\ &\quad + [\$9200 - (\$3300 + \$400 + \$2400 + \$300)](P/A, 9\%, 10) \\ &= -\$15\,000 + \$2800(P/A, 9\%, 10) \\ &= -\$15\,000 + \$2800(6.4176) \\ &= \$2969.44 \\ &\cong \$3000 \end{aligned}$$

The present worth of the cost savings is about $3000 greater than the $15 000 first cost. Therefore, the project is worth implementing. ■

4.4.2 Present Worth for Mutually Exclusive Projects

It is very easy to use the present worth method to choose the best project among a set of mutually exclusive projects *when the service lives are the same.* One just computes the present worth of each project using the MARR. The project with the greatest present worth is the preferred project because it is the one with the greatest profit.

Example 4.4

Fly-by-Night Aircraft must purchase a new lathe. It is considering four lathes, each of which has a life of 10 years with no scrap value.

Lathe	1	2	3	4
First cost	\$100 000	\$150 000	\$200 000	\$255 000
Annual savings	25 000	34 000	46 000	55 000

Given a MARR of 15%, which alternative should be taken?

The present worths are:

Lathe 1: $\text{PW} = -\$100\,000 + \$25\,000(P/A, 15\%, 10)$

$= -\$100\,000 + \$25\,000(5.0187) = \$25\,468$

Lathe 2: $\text{PW} = -\$150\,000 + \$34\,000(P/A, 15\%, 10)$

$= -\$150\,000 + \$34\,000(5.0187) = \$20\,636$

Lathe 3: $\text{PW} = -\$200\,000 + \$46\,000(P/A, 15\%, 10)$

$= -\$200\,000 + \$46\,000(5.0187) = \$30\,860$

Lathe 4: $\text{PW} = -\$255\,000 + \$55\,000(P/A, 15\%, 10)$

$= -\$155\,000 + \$55\,000(5.0187) = \$21\,029$

Lathe 3 has the greatest present worth, and is therefore the preferred alternative. ■

4.4.3 Annual Worth Comparisons

Annual worth comparisons are essentially the same as present worth comparisons, except that all disbursements and receipts are transformed to a uniform series at the MARR, rather than to the present worth. Any present worth P can be converted to an annuity A by the capital recovery factor $(A/P,i,N)$. Therefore, a comparison of two projects *that have the same life* by the present worth and annual worth methods will always indicate the same preferred alternative. Note that, although the method is called annual worth, the uniform series is not necessarily on a yearly basis.

Present worth comparisons make sense because they compare the worth today of each alternative, but annual worth comparisons can sometimes be more easily grasped mentally. For example, to say that operating an automobile over five years has a present cost of \$20 000 is less meaningful than saying that it will cost about \$5300 per year for each of the following five years.

Sometimes there is no clear justification for preferring either the present worth method or the annual worth method. Then it is reasonable to use the one that requires less conversion. For example, if most receipts or costs are given as annuities or gradients, one can more easily perform an annual worth comparison. Sometimes it can be useful to compare projects on the basis of future worths. See Close-Up 4.2.

CLOSE-UP 4.2 Future Worth

Sometimes it may be desirable to compare projects with the **future worth method**, on the basis of the future worth of each project. This is most likely to be true for cases where money is being saved for some future expense.

For example, two investment plans are being compared to see which accumulates more money for retirement. Plan A consists of a payment of \$10 000 today and then \$2000 per year over 20 years. Plan B is \$3000 per year over 20 years. Interest for both plans is 10%. Rather than convert these cash flows to either present worth or annual worth, it is sensible to compare the future worths of the plans, since the actual dollar value in 20 years has particular meaning.

$$\begin{aligned} FW_A &= \$10\,000(F/P, 10\%, 20) + \$2000(F/A, 10\%, 20) \\ &= \$10\,000(6.7275) + \$2000(57.275) \\ &= \$181\,825 \end{aligned}$$

$$\begin{aligned} FW_B &= \$3000(F/A, 10\%, 20) \\ &= \$3000(57.275) \\ &= \$171\,825 \end{aligned}$$

Plan A is the better choice. It will accumulate to \$181 825 over the next 20 years.

Example 4.5

Sweat University is considering two alternative types of bleachers for a new athletic stadium:

Alternative 1: Concrete bleachers. The first cost is \$350 000. The expected life of the concrete bleachers is 90 years and the annual upkeep costs are \$2500.

Alternative 2: Wooden bleachers on earth fill. The first cost of \$200 000 consists of \$100 000 for earth fill and \$100 000 for the wooden bleachers. The annual painting costs are \$5000. The wooden bleachers must be replaced every 30 years at a cost of \$100 000. The earth fill will last the entire 90 years.

One of the two alternatives will be chosen. It is assumed that the receipts and other benefits of the stadium are the same for both construction methods. Therefore, the greatest net benefit is obtained by choosing the alternative with the lower cost. The University uses a MARR of 7%. Which of the two alternatives is better?

For this example, let us base the analysis on annual worth. Since both alternatives have a life of 90 years, we shall get the equivalent annual costs over 90 years for both at an interest rate of 7%.

Alternative 1: Concrete bleachers

The equivalent annual cost over the 90-year life span of the concrete bleachers is

$$\begin{aligned} AW &= \$350\,000(A/P, 7\%, 90) + \$2500 \\ &= \$350\,000(0.07016) + \$2500 \\ &= \$27\,056 \text{ per year} \\ &\cong \$27\,000 \text{ per year} \end{aligned}$$

Alternative 2: Wooden bleachers on earth fill

The total annual costs can be broken into three components: A_1 (for the earth fill), A_2 (for the bleachers), and A_3 (for the painting). The equivalent annual cost of the earth fill is

$$AW_1 = \$100\,000(A/P, 7\%, 90)$$

The equivalent annual cost of the bleachers is easy to determine. The first set of bleachers is put in at the start of the project, the second set at the end of 30 years, and the third set at the end of 60 years, but the cost of the bleachers is the same at each installation. Therefore, we need to get only the cost of the first installation.

$$AW_2 = \$100\,000(A/P, 7\%, 30)$$

The last expense is for annual painting:

$$AW_3 = \$5000$$

The total equivalent annual cost for alternative 2, wooden bleachers on earth fill, is the sum of AW_1, AW_2, and AW_3:

$$\begin{aligned} AW &= AW_1 + AW_2 + AW_3 \\ &= \$100\,000[(A/P, 7\%, 90) + (A/P, 7\%, 30)] + \$5000 \\ &= \$100\,000(0.07016 + 0.08059) + \$5000 \\ &= \$20\,075 \\ &\cong \$20\,000 \end{aligned}$$

The concrete bleachers have an equivalent annual cost of about \$7000 more than the wooden ones. Therefore, the wooden bleachers are the better choice. ■

4.4.4 Comparison of Alternatives with Unequal Lives

When making present worth comparisons, we must always use the same time period in order to take into account the full benefits and costs of each alternative. If the lives of the alternatives are not the same, we can transform them to equal lives with one of the following two methods:

1. Repeat the *service life* of each alternative to arrive at a common time period for all alternatives. Here we assume that each alternative can be repeated with the same costs and benefits in the future — an assumption known as **repeated lives**. Usually we use the *least common multiple* of the lives of the various alternatives. Sometimes it is convenient to assume that the lives of the various alternatives are repeated indefinitely. Note that the assumption of repeated lives may not be valid where it is reasonable to expect technological improvements.
2. Adopt a specified **study period** — a time period that is given for the analysis. To set an appropriate study period, a company will usually take into account the time of required service, or the length of time they can be relatively certain of their forecasts. The study period method necessitates an additional assumption about *salvage value* whenever the life of one of the alternatives exceeds that of the given study period. Arriving at a reliable estimate of salvage value may be difficult sometimes.

Because they rest on different assumptions, the repeated lives and the study period methods can lead to different conclusions when applied to a particular project choice.

Example 4.6 (Modification of Example 4.3)

A mechanical engineer has decided to introduce automated material-handling equipment for a production line. She must choose between two alternatives, building the equipment or buying the equipment off the shelf. Each alternative has a different service life and a different set of costs.

Alternative 1: Build custom automated material-handling equipment.

First cost	\$15 000	
Labour	\$ 3 300	per year
Power	\$ 400	per year
Maintenance	\$ 2 400	per year
Taxes and insurance	\$ 300	per year
Service life	10	years

Alternative 2: Buy off-the-shelf standard automated material-handling equipment.

First cost	\$25 000	
Labour	\$ 1 450	per year
Power	\$ 600	per year
Maintenance	\$ 3 075	per year
Taxes and insurance	\$ 500	per year
Service life	15	years

If the MARR is 9%, which alternative is better?

The present worth of the custom system over its 10-year life is

$$\begin{aligned} PW(1) &= -\$15\,000 - (\$3300 + \$400 + \$2400 + \$300)(P/A, 9\%, 10) \\ &= -\$15\,000 - \$6400(6.4176) \\ &= -\$56\,073 \\ &\cong -\$56\,100 \end{aligned}$$

The present worth of the off-the-shelf system over its 15-year life is:

$$\begin{aligned} PW(2) &= -\$25\,000 - (\$1450 + \$600 + \$3075 + \$500)(P/A, 9\%, 15) \\ &= -\$25\,000 - \$5625(8.0606) \\ &= -\$70\,341 \\ &\cong -\$70\,300 \end{aligned}$$

The custom system has a lower cost for its 10-year life than the off-the-shelf system for its 15-year life, but it would be *wrong* to conclude from these calculations that the custom system should be preferred. The custom system yields benefits for only 10 years, whereas the off-the-shelf system lasts 15 years. It would be surprising if the cost of 15 years of benefits were not higher than the cost of 10 years of benefits. A fair comparison of the costs can be made only if equal lives are compared.

Let us apply the *repeated lives method*. If each alternative is repeated enough times, there will be a point in time where their service lives are simultaneously completed. This will happen first at the time equal to the *least common multiple* of the service lives. The least common multiple of 10 years and 15 years is 30 years. Alternative 1 will be repeated twice (after 10 years and after 20 years), while alternative 2 will be repeated once (after 15 years) during the 30-year period. At the end of 30 years, both alternatives will be completed simultaneously. See Figure 4.3.

Figure 4.3 Least Common Multiple of the Service Lives

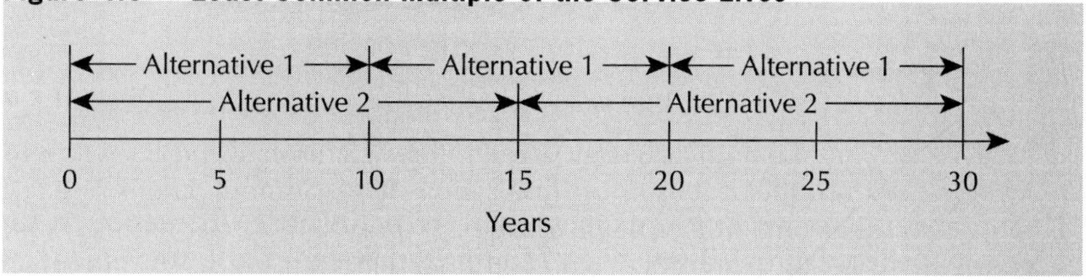

With the same time period of 30 years for both alternatives, we can now compare present worths.

Alternative 1: Build custom automated material-handling equipment and repeat twice.

$$\begin{aligned} \text{PW}(1) &= -\$15\,000 - \$15\,000(P/F, 9\%, 10) - \$15\,000(P/F, 9\%, 20) \\ &\quad - (\$3300 + \$400 + \$2400 + \$300)(P/A, 9\%, 30) \\ &= -\$15\,000 - \$15\,000(0.42241) - \$15\,000(0.17843) \\ &\quad - \$6400(10.273) \\ &= -\$89\,760 \\ &\cong -\$89\,800 \end{aligned}$$

Alternative 2: Buy off-the-shelf standard automated material-handling equipment and repeat once.

$$\begin{aligned} \text{PW}(2) &= -\$25\,000 - \$25\,000(P/F, 9\%, 15) \\ &\quad - (\$1450 + \$600 + \$3075 + \$500)(P/A, 9\%, 30) \\ &= -\$25\,000 - \$25\,000(0.27454) - \$5625(10.273) \\ &= -\$89\,649 \\ &\cong -\$89\,600 \end{aligned}$$

Using the repeated lives method, we find little difference between the alternatives. An annual worth comparison can also be done over a period of time equal to the least common multiple of the service lives by multiplying each of these present worths by the capital recovery factor for 30 years.

$$\begin{aligned} \text{AW}(1) &= -\$89\,760(A/P, 9\%, 30) \\ &= -\$89\,760(0.09734) \\ &= -\$8737 \\ &\cong -\$8700 \end{aligned}$$

$$\begin{aligned} AW(2) &= -\$89\,649(A/P, 9\%, 30) \\ &= -\$89\,649(0.09734) \\ &= -\$8726 \\ &\cong -\$8700 \end{aligned}$$

As we would expect, there is again little difference in the annual cost between the alternatives. However, there is a more convenient approach for an annual worth comparison if it can be assumed that the alternatives are repeated indefinitely. Since the annual costs of an alternative remain the same no matter how many times it is repeated, it is not necessary to determine the least common multiple of the service lives. The annual worth of each alternative can be assessed for whatever time period is most convenient for each alternative.

Alternative 1: Build custom automated material-handling equipment.

$$\begin{aligned} AW(1) &= -\$15\,000(A/P, 9\%, 10) - \$6400 \\ &= -\$15\,000(0.15582) - \$6400 \\ &= -\$8737 \\ &\cong -\$8700 \end{aligned}$$

Alternative 2: Buy off-the-shelf standard automated material-handling equipment.

$$\begin{aligned} AW(2) &= -\$25\,000(A/P, 9\%, 15) - \$5625 \\ &= -\$25\,000(0.12406) - \$5625 \\ &= -\$8726 \\ &\cong -\$8700 \end{aligned}$$

If it cannot be assumed that the alternatives can be repeated to permit a calculation over the least common multiple of their service lives, then it is necessary to use the *study period method.*

Suppose that the given study period is 10 years, because the engineer is uncertain about costs past that time. The service life of the off-the-shelf system (15 years) is greater than the study period (10 years). Therefore, we have to make an assumption about the salvage value of the off-the-shelf system after 10 years. Suppose the engineer judges that its salvage value will be \$5000. We can now proceed with the comparison.

Alternative 1: Build custom automated material-handling equipment (10-year study period).

$$\begin{aligned} PW(1) &= -\$15\,000 - (\$3300 + \$400 + \$2400 + \$300)(P/A, 9\%, 10) \\ &= -\$15\,000 - \$6400(6.4176) \\ &= -\$56\,073 \\ &\cong -\$56\,100 \end{aligned}$$

Alternative 2: Buy off-the-shelf standard automated material-handling equipment (10-year study period).

$$\begin{aligned}\text{PW}(2) &= -\$25\,000 - (\$1450 + \$600 + \$3075 + \$500)(P/A, 9\%, 10) \\ &\quad + \$5000(P/F, 9\%, 10) \\ &= -\$25\,000 - \$5625(6.4176) + \$5000(0.42241) \\ &= -\$58\,987 \\ &\cong -\$59\,000\end{aligned}$$

Using the study period method of comparison, alternative 1 has the smaller present cost and is, therefore, preferred.

Note that here the study period method gives a different answer than the repeated lives method gives. The study period method is often sensitive to the chosen salvage value. A larger salvage value tends to make an alternative with a life longer than the study period more attractive, and a smaller value tends to make it less attractive.

In some instances, it may be difficult to arrive at a reliable estimate of salvage value. Given the sensitivity of the study period method to the salvage value estimate, the analyst may be uncertain about the validity of the results. One way of circumventing this problem is to avoid estimating the salvage value at the outset. Instead we calculate what salvage value would make the alternatives equal in value. Then we decide whether the actual salvage value will be above or below the break-even value found. Applying this approach to our example, we set PW(2) = PW(1) so that

$$\begin{aligned}\text{PW}(2) &= -\$25\,000 - \$5625(6.4176) + S(0.42241) \\ &= -\$56\,100\end{aligned}$$

where S is the salvage value.

Solving for S, we find $S \cong \$12\,100$. Is a reasonable estimate of the salvage value above or below \$12 100? If it is above \$12 100, then we conclude that the off-the-shelf system is the preferred choice. If it is below \$12 100, then we conclude that the custom system is preferable. ■

The study period can also be used for the annual worth method if the assumption of being able to indefinitely repeat the choice of alternatives is not justified.

Example 4.7

Joan is renting an apartment while on a one-year assignment in a distant city. The apartment does not have a refrigerator. She can rent one for a \$100 deposit (returned in a year) and \$15 per month (paid at the end of each month). Alternatively, she can buy a used refrigerator for \$300, which she would sell in a year when she leaves. For how much would Joan have to be able to sell the used refrigerator in one year when she leaves, in order to be better off buying the used refrigerator than renting one? Interest is at 6% nominal, compounded monthly.

Let S stand for the unknown salvage value (i.e., the amount Joan will be able to sell the refrigerator for in a year). We then equate the present worth of the rental alternative with the present worth of the purchase alternative for the one-year study period:

$$-\$100 - \$15(P/A, 0.5\%, 12) + \$100(P/F, 0.5\%, 12)$$
$$= -\$300 + S(P/F, 0.5\%, 12)$$
$$-\$100 - \$15(11.616) + \$100(0.94192)$$
$$= -\$300 + S(0.94192)$$
$$S = \$127.35$$

If Joan can sell the used refrigerator for more than about \$127 after one year's use, she is better off buying it rather than renting one. ■

4.5 Payback Period

The simplest method for judging the economic viability of projects is the payback period method. It is a rough measure of the time it takes for an investment to pay for itself. More precisely, the **payback period** is the number of years it takes for an investment to be recouped when the interest rate is assumed to be zero. It is usually calculated as follows:

$$\text{Payback period} = \frac{\text{First cost}}{\text{Annual savings}}$$

For example, if a first cost of \$20 000 yielded a return of \$8000 per year, then the payback period would be 2.5 years (i.e., \$20 000/\$8000).

If the annual savings are not constant, we can calculate the payback period by deducting each year of savings from the first cost until the first cost is recovered. The number of years of savings required to do this is the payback period. For example, suppose the saving from a \$20 000 first cost is \$5000 the first year, increasing by \$1000 each year thereafter. By adding the annual savings one year at a time, we see that it would take a just over three years to pay back the first cost (\$5000 + \$6000 + \$7000 + \$8000 = \$26 000). The payback period would then be stated as either four years (if we assume that the \$8000 is received at the end of the fourth year) or 3.25 years (if we assume that the \$8000 is received throughout the fourth year).

According to the payback period method of comparison, the project with the shorter payback period is the preferred investment. A company may have a policy of rejecting projects for which the payback period exceeds some preset number of years. The length of the maximum payback period depends on the type of project and the company's financial situation. If the company expects a cash constraint in the near future, or if a project's returns are highly uncertain after more than a few periods, the company will set a maximum payback period that is relatively short. As a rule of thumb, a payback period of two years is often considered acceptable, while one of more than four years is unacceptable. Accordingly, government grant programs often target projects with payback periods of between two and four years on the rationale that in this range the grant can justify economically feasible projects that a company with limited cash flow would otherwise be unwilling to undertake.

The payback period need not, and perhaps should not, be used as the sole criterion for evaluating projects. It is a rough method of comparison and possesses some glaring weaknesses (as we shall discuss after the examples). Nevertheless, the payback period method can be used effectively as a preliminary filter. All projects with paybacks within the minimum would then be evaluated, using either rate of return methods (Chapter 5) or present/annual worth methods.

Example 4.8

Elyse runs a second-hand book business out of her home where she advertises and sells the books over the Internet. Her small business is becoming quite successful and she is considering purchasing an upgrade to her computer system which will give her more reliable up-time. The cost is \$5000. She expects that the investment will bring about an annual savings of \$2000, due to the fact that her system will no longer suffer long failures and thus she will be able to sell more books. What is the payback period on her investment, assuming that the savings accrue over the whole year?

$$\text{Payback period} = \frac{\text{First Cost}}{\text{Annual Savings}} = \frac{\$5000}{\$2000} = 2.5 \text{ years.} \blacksquare$$

Example 4.9

Pizza-in-a-Hurry operates a pizza delivery service to its customers with two eight-year-old vehicles, both of which are large, consume a great deal of gas and are starting to cost a lot to repair. The owner, Ray, is thinking of replacing one of the cars with a smaller, three-year-old car that his sister-in-law is selling for \$8000. Ray figures he can save \$3000, \$2000, and \$1500 per year for the next 3 years and \$1000 per year for the following two years by purchasing the smaller car. What is the payback period for this decision?

The payback period is the number of years of savings required to pay back the initial cost. After three years, \$3000 + \$2000 + \$1500 = \$6500 has been paid back, and this amount is \$7500 after four years and \$8500 after five years. The payback period would be stated as five years if the savings are assumed to occur at the end of each year, or 4.5 years if the savings accrue continuously throughout the year. ■

The payback period method has four main advantages:

1. It is very easy to understand. One of the goals of engineering decision making is to communicate the reasons for a decision to managers or clients with a variety of backgrounds. The reasons behind the payback period and its conclusions are very easy to explain.
2. The payback period is very easy to calculate. It can usually be done without even using a calculator, so projects can be very quickly assessed.
3. It accounts for the need to recover capital quickly. Cash flow is almost always a problem for small- to medium-sized companies. Even large companies sometimes can't tie up their money in long-term projects.
4. The future is unknown. The future benefits from an investment may be estimated imprecisely. It may not make much sense to use precise methods like present worth on numbers that are imprecise to start with. A rule of thumb like the payback period may be good enough for most purposes.

But the payback period method has three important disadvantages:

1. It discriminates against long-term projects. No houses or highways would ever be built if they had to pay themselves off in two years.
2. It ignores the effect of the timing of cash flows within the payback period. It disregards interest rates and takes no account of the time value of money. (Occasionally, a discounted payback period is used to overcome this disadvantage. See Close-Up 4.3.)
3. It ignores the expected service life. It disregards the benefits that accrue after the end of the payback period.

CLOSE-UP 4.3 Discounted Payback Period

In a discounted payback period calculation, the present worth of each year's savings is subtracted from the first cost until the first cost is diminished to zero. The number of years of savings required to do this is the discounted payback period. The main disadvantages of using a discounted payback period include the more complicated calculations and the need for an interest rate.

For instance, in Example 4.8, Elyse had an investment of $5000 recouped by annual savings of $2000. If interest were at 10%, the present worth of savings would be:

Year	*Present Worth*	*Cumulative*
Year 1	$\$2000(P/F, 10\%, 1) = \$2000(0.90909) = \$1818$	$1 818
Year 2	$\$2000(P/F, 10\%, 2) = \$2000(0.82645) = \$1653$	$3 471
Year 3	$\$2000(P/F, 10\%, 3) = \$2000(0.75131) = \$1503$	$4 974
Year 4	$\$2000(P/F, 10\%, 4) = \$2000(0.68301) = \$1366$	$6 340

Thus the discounted payback period is over 3 years, compared with 2.5 years calculated for the standard payback period.

Example 4.10 illustrates how the payback period method can ignore future cash flows.

Example 4.10

Self Defence Systems is going to upgrade their paper-shredding facility. They have a choice between two models. Model 007, with a first cost of \$50 000 and a service life of seven years, would save them \$10 000 per year. Model MX, with a first cost of \$10 000 and an expected service life of 20 years, would save them \$1500 per year. If the company's MARR is 8%, which model is the better buy?

Using payback period as the sole criterion

Model 007: Payback period = \$50 000/\$10 000 = 5 years

Model MX: Payback period = \$10 000/\$1500 = 6.6 years

It appears that the 007 model is better.

Using annual worth

Model 007: AW = − \$50 000(A/P, 8%, 7) + \$10 000 = \$396.50

Model MX: AW = − \$10 000(A/P, 8%, 20) + \$1500= \$481.50

Here, the model MX is substantially better.

The difference in the results from the two comparison methods is that the payback period method has ignored the benefits of the models that occur after the models have paid themselves off. This is illustrated in Figure 4.4. For model MX, about

Figure 4.4 Flows Ignored by the Payback Period

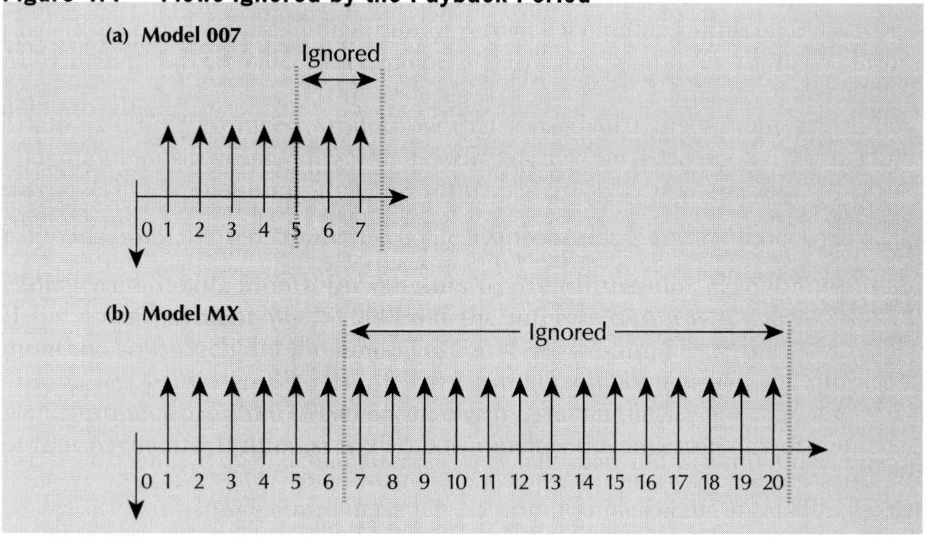

14 years of benefits have been omitted, whereas for model 007, only two years of benefits have been left out. ■

REVIEW PROBLEMS

REVIEW PROBLEM 4.1

Tilson Dairies operates several cheese plants. The plants are all old and in need of renovation. Tilson's engineers have developed plans to renovate all the plants. Each project would have a positive present worth at the company's MARR. Tilson has $3.5 million available to invest in these projects. The following facts about the potential renovation projects are available:

Plant	First Cost	Present Worth
A	$0.8 million	$1.1 million
B	$1.2 million	$1.7 million
C	$1.4 million	$1.8 million
D	$2.0 million	$2.7 million

Which projects should Tilson accept?

ANSWER

Table 4.2 shows the possible mutually exclusive projects that Tilson can consider.

Tilson should accept projects A, B, and C. They have a combined present worth of $4.6 million. Other feasible combinations that come close to using all available funds are B and D with a total present worth of $4.4 million, and C and D with a total present worth of $4.5 million.

Note that it is not necessary to consider explicitly the "leftovers" of the $3.5 million budget when comparing the present worths. The assumption is that any leftover part of the budget will be invested and provide interest at the MARR, resulting in a zero present worth for that part. Therefore, it is best to choose the combination of projects that has the largest total present worth and stays within the budget constraint.

Table 4.2 Mutually Exclusive Projects for Tilson Dairies

Project	Total First Cost	Total Present Worth	Feasibility
Do nothing	\$0.0 million	\$0.0 million	Feasible
A	\$0.8 million	\$1.1 million	Feasible
B	\$1.2 million	\$1.7 million	Feasible
C	\$1.4 million	\$1.8 million	Feasible
D	\$2.0 million	\$2.7 million	Feasible
A and B	\$2.0 million	\$2.8 million	Feasible
A and C	\$2.2 million	\$2.9 million	Feasible
A and D	\$2.8 million	\$3.8 million	Feasible
B and C	\$2.6 million	\$3.5 million	Feasible
B and D	\$3.2 million	\$4.4 million	Feasible
C and D	\$3.4 million	\$4.5 million	Feasible
A, B, and C	\$3.4 million	\$4.6 million	Feasible
A, B, and D	\$4.0 million	\$5.5 million	Not feasible
A, C, and D	\$4.2 million	\$5.6 million	Not feasible
B, C, and D	\$4.6 million	\$6.2 million	Not feasible
A, B, C and D	\$5.4 million	\$7.3 million	Not feasible

■

REVIEW PROBLEM 4.2

City engineers are considering two plans for municipal aqueduct tunnels. They are to decide between the two, using an interest rate of 8%.

Plan A is a full-capacity tunnel that will meet the needs of the city forever. Its cost is \$3 000 000 now, and \$100 000 every 10 years for lining repairs.

Plan B involves building a half-capacity tunnel now and a second half-capacity tunnel in 20 years, when the extra capacity will be needed. Each of the half-capacity tunnels costs \$2 000 000. Maintenance costs for each tunnel are \$80 000 every 10 years. There is also an additional \$15 000 per tunnel per year required to pay for extra pumping costs caused by greater friction in the smaller tunnels.

(a) Which alternative is preferred? Use a present worth comparison.

(b) Which alternative is preferred? Use an annual worth comparison.

ANSWER

(a) *Plan A: Full-Capacity Tunnel*

First, the \$100 000 paid at the end of 10 years can be thought of as a future amount which has an equivalent annuity.

$$AW = \$100\ 000(A/F, 8\%, 10) = \$100\ 000(0.06903) = \$6903$$

Thus, at 8% interest, \$100 000 every 10 years is equivalent to \$6903 every year.

Since the tunnel will have (approximately) an infinite life, the present cost of the lining repairs can be found using the capitalized cost formula, giving a total cost of

$$PW(\text{Plan A}) = \$3\ 000\ 000 + \$6903/0.08 = \$3\ 086\ 288$$

Plan B: Half-Capacity Tunnels

For the first tunnel, the equivalent annuity for the maintenance and pumping costs is

$$AW = \$15\ 000 + \$80\ 000(0.06903) = \$20\ 522$$

The present cost is then found with the capitalized cost formula, giving a total cost of

$$PW1 = \$2\ 000\ 000 + \$20\ 522/0.08 = \$2\ 256\ 525$$

Now, for the second tunnel, basically the same calculation is used, except that the present worth calculated must be discounted by 20 years at 8%, since the second tunnel will be built 20 years in the future.

$$PW2 = \{\$2\ 000\ 000 + [\$15\ 000 + \$80\ 000(0.06903)]/0.08\}(P/F, 8\%, 20)$$
$$= \$2\ 256\ 525(0.21455) = \$484\ 137$$

$$PW(\text{Plan B}) = PW1 + PW2 = \$2\ 740\ 662$$

Consequently, the two half-capacity aqueducts are economically preferable.

(b) *Plan A: Full-Capacity Tunnel*

First, the \$100 000 paid at the end of 10 years can be thought of as a future amount which has an equivalent annuity of

$$AW = \$100\ 000(A/F, 8\%, 10) = \$100\ 000(0.06903) = \$6903$$

Thus, at 8% interest, \$100 000 every 10 years is equivalent to \$6903 every year.

Since the tunnel will have (approximately) an infinite life, an annuity equivalent to the initial cost can be found using the capitalized cost formula, giving a total annual cost of

$$AW(\text{Plan A}) = \$3\ 000\ 000(0.08) + \$6903 = \$246\ 903$$

Plan B: Half Capacity Tunnels

For the first tunnel, the equivalent annuity for the maintenance and pumping costs is

$$AW = \$15\,000 + \$80\,000(0.06903) = \$20\,522$$

The annual equivalent of the initial cost is then found with the capitalized cost formula, giving a total cost of

$$AW1 = \$2\,000\,000(0.08) + \$20\,522 = \$180\,522$$

Now, for the second tunnel, basically the same calculation is used, except that the annuity must be discounted by 20 years at 8%, since the second tunnel will be built 20 years in the future.

$$AW2 = AW1(P/F, 8\%, 20)$$
$$= \$180\,522(0.21455) = \$38\,731$$

$$AW(\text{Plan B}) = AW1 + AW2 = \$180\,522 + \$38\,731 = \$219\,253$$

Consequently, the two half-capacity aqueducts are economically preferable.

■

REVIEW PROBLEM 4.3

Constantine Fernando, an engineer at Corner Brook Manufacturing, has a $100 000 budget for plant improvements. He has identified four mutually exclusive investments, all of five years' duration, which have the cash flows shown in Table 4.3. For each alternative, he wants to determine the payback period and the present worth. For his recommendation report, he will order the alternatives from most preferred to least preferred in each case. Corner Brook Manufacturing uses an 8% MARR for such decisions.

Table 4.3 Cash Flows for Review Problem 4.3

	Cash Flow at the End of Each Year					
Alternative	0	1	2	3	4	5
A	$100 000	$25 000	$25 000	$25 000	$25 000	$25 000
B	100 000	5 000	10 000	20 000	40 000	80 000
C	100 000	50 000	50 000	10 000	0	0
D	100 000	0	0	0	0	1 000 000

ANSWER

The payback period can be found by decrementing yearly. The payback periods for the alternatives are then

A: 4 years
B: 4.3125 or 5 years
C: 2 years
D: 4.1 or 5 years

The order of the alternatives from most preferred to least preferred using the payback period method with yearly decrementing is: C, A, D, B. The present worth computations for each alternative are:

$$\begin{aligned}
\text{A:}\ \ \text{PW} &= -\ \$100\,000 + \$25\,000(P/A, 8\%, 5)\\
&= -\ \$100\,000 + \$25\,000(3.9926)\\
&= -\ \$185
\end{aligned}$$

$$\begin{aligned}
\text{B:}\ \ \text{PW} &= -\ \$100\,000 + \$5000(P/F, 8\%, 1) + \$10\,000(P/F, 8\%, 2)\\
&\quad + \$20\,000(P/F, 8\%, 3) + \$40\,000(P/F, 8\%, 4)\\
&\quad + \$80\,000(P/F, 8\%, 5)\\
&= -\ \$100\,000 + \$5000(0.92593) + \$10\,000(0.85734)\\
&\quad + \$20\,000(0.79383) + \$40\,000(0.73503) + \$80\,000(0.68059)\\
&= \$12\,982
\end{aligned}$$

$$\begin{aligned}
\text{C:}\ \ \text{PW} &= -\ \$100\,000 + \$50\,000(P/F, 8\%, 1) + \$50\,000(P/F, 8\%, 2)\\
&\quad + \$10\,000(P/F, 8\%, 3)\\
&= -\ \$100\,000 + \$50\,000(0.92593) + \$50\,000(0.85734)\\
&\quad + \$10\,000(0.79283)\\
&= -\ \$2908
\end{aligned}$$

$$\begin{aligned}
\text{D:}\ \ \text{PW} &= -\ \$100\,000 + \$1\,000\,000(P/F, 8\%, 5)\\
&= -\ \$100\,000 + \$1\,000\,000(0.68059)\\
&= \$580\,590
\end{aligned}$$

The order of the alternatives from most preferred to least preferred using the present worth method is: D, B, A, C. ■

SUMMARY

This chapter discussed relations among projects, and the present worth, annual worth, and payback period methods for evaluating projects. There are three classes of relations among projects, (1) independent, (2) mutually exclusive, and (3) related but not mutually exclusive. We then showed how the third class of projects, those that are related but not mutually exclusive, could be combined into sets of mutually exclusive projects. This enabled us to limit the discussion to the first two classes, independent and mutually exclusive. Independent projects are considered one at a time and are either accepted or rejected. Only the best of a set of mutually exclusive projects is chosen.

The present worth method compares projects on the basis of converting all cash flows for the project to a present worth. An independent project is acceptable if its present worth is greater than zero. The mutually exclusive project with the highest present worth should be taken. Projects with unequal lives must be compared by assuming that the projects are repeated or by specifying a study period. Annual worth is similar to present worth, except that the cash flows are converted to a uniform series. The annual worth method may be more meaningful, and also does not require more complicated calculations when the projects have different service lives.

The payback period is a rule-of-thumb method that calculates the length of time it takes to pay back an initial investment. It is inaccurate but very easy to calculate.

Engineering Economics in Action, Part 4B: Doing It Right

Naomi stopped for coffee on her way back from Clem's office. She needed time to think about how to decide which potential forge shop investments were best. She wasn't sure that she knew what "best" meant. She got down her engineering economics text and looked at the table of contents. There were a couple of chapters on comparison methods that seemed to be what she wanted. She sat down with the coffee in her right hand and the text on her lap, and hoped for an uninterrupted hour.

One read through the chapters was enough to remind Naomi of the main relevant ideas that she had learned in school. The first thing she had to do was decide whether the investments were independent or not. They clearly were not independent. It would not make sense to refurbish the current forging hammer and replace it with a mechanical press. Where potential investments were not independent, it was easiest to form mutually exclusive combinations as investment options. Naomi came up with seven options. She ranked the options by first cost, starting with the one with the lowest cost:

1. Refurbish the current machine.
2. Refurbish the current machine plus replace the material-handling equipment.
3. Buy a manually operated mechanical press.
4. Buy a manual mechanical press plus replace the material-handling equipment.
5. Buy an automated mechanical press.
6. Buy an automated mechanical press plus replace the material-handling equipment.
7. Buy an automated mechanical press plus integrate it with the material-handling equipment.

At this point, Naomi wasn't sure what to do next. There were different ways of comparing the options.

Naomi wanted a break from thinking about theory. She decided to take a look at Dave Sullivan's diskette. She started up her computer and put in the diskette. One of the files was called "Naomi." In it Dave apologized for dumping the work on her and invited Naomi to call him in Florida if she needed help. Naomi decided to call him. The phone was answered by Dave's wife, Helena. After telling Naomi that her father was out of intensive care and was in good spirits, Helena turned the phone over to Dave.

"Hi, Naomi. How's it going?"

"Well, I'm trying to finish off the forge project that you started. And I'm taking you up on your offer to consult."

"You have my attention. What's the problem?"

"Well, I've gotten started. I have formed seven mutually exclusive combinations of potential investments." Naomi went on to explain her selection of alternatives.

"That sounds right, Naomi. I like the way you've organized that. Now, how are you going to make the choice?"

"I've just reread the present worth, annual worth, and payback period stuff, and of those three, present worth makes the most sense to me. I can just compare the present worths of the cash flows for each alternative, and the one whose present worth is highest is the best one. Annual worth is the same, but I don't see any good reason in this case to look at the costs on an annual basis."

"What about internal rate of return?"

"Well, actually, Dave, I haven't reviewed IRR yet. I'll need it, will I?"

"You will. Have a look at it, and also remember that your recommendation is for Burns and Kulkowski. Think about how they will be looking at your information."

"Thanks, Dave. I appreciate your help."

"No problem. This first one is important for you; let's make sure we do it right." ◎

PROBLEMS

4.1 IQ Computer assembles Unix workstations at its plant just outside St. Catharines, Ontario. Their current product line is nearing the end of its marketing life, and it is time to start production of one or more new products. The data for several candidates are shown below.

	Potential Product			
	A	B	C	D
Research and development costs	$120 000	$60 000	$150 000	$75 000
Lead time	1 year	2 years	1 year	2 years
Resource draw	60%	50%	40%	30%

The maximum budget for research and development is $300 000. A minimum of $200 000 should be spent on these projects. It is desirable to spread out the introduction of new products, so if two products are to be developed together, they should have different lead times. Resource draw refers to the labour and space that are available to the new products; it cannot exceed 100%.

Based on the above information, determine the set of feasible mutually exclusive alternative projects that IQ Computers should consider.

4.2 The Alabama Alabaster Marble Company (AAM) is considering opening three new quarries. One, designated T, is in Tusksarelooser County; a second, L, is in Lefant County; the third, M, is in Marxbro County. Marble is shipped mainly within a 500-kilometre range of its quarry because of its weight. The market within this range is limited. The returns that AAM can expect from any of the quarries depends on how many quarries AAM opens. Therefore, these potential projects are related.

(a) Construct a set of mutually exclusive alternatives from these three potential projects.

(b) The Lefant County quarry has very rich deposits of marble. This makes the purchase of mechanized cutter-loaders a reasonable investment at this quarry. Such loaders would not be considered at the other quarries. Construct a set of mutually exclusive alternatives from the set of quarry projects augmented by the potential mechanized cutter-loader project.

(c) AAM has decided to invest no more than $2.5 million in the potential quarries. The first costs are as follows:

Project	First Cost
T quarry	$0.9 million
L quarry	$1.4 million
M quarry	$1.0 million
Cutter-loader	$0.4 million

Construct a set of mutually exclusive alternatives that are feasible, given the investment limitation.

4.3 Chatham Automotive has $100 000 to invest in internal projects. The choices are:

Project	Cost
1. Line improvements	$20 000
2. New manual tester	30 000
3. New automatic tester	60 000
4. Overhauling press	50 000

Only one tester may be bought and the press will not need overhauling if the line improvements are not made. What mutually exclusive project combinations are available if Chatham will invest in at least one project?

4.4 The intersection of King and Main Streets needs widening and improvement. The possibilities include:

1. Widening King
2. Widening Main
3. Left turn lane on King
4. Left turn lane on Main
5. Stoplights
6. Stoplights with advanced green for Main
7. Stoplights with advanced green for King

A left turn lane can be installed only if the street in question is widened. Left turn lanes are necessary on each street that has stoplights installed with an advanced green. The city cannot afford to widen both streets. How many mutually exclusive projects are there?

4.5 Yun is deciding among a number of business opportunities. She can

(a) Sell the X division of her company, Yunco

(b) Buy Barzoo's company, Barco.

(c) Get new financing

(d) Expand into the United States

There is no sense in getting new financing unless she is either buying Barco or expanding into the States. She can only buy Barco if she gets financing or sells the X division. She can only expand into the States if she has purchased Barco. The X division is necessary to compete in the US. What are the feasible projects she should consider?

4.6 Nottawasaga Printing has four printing lines, each of which consists of three printing stations, A, B, and C. They have allocated $20 000 for upgrading the printing stations. Station A costs $7000 and takes 10 days to upgrade. Station B costs $5000 and takes 5 days, and station C costs $3000 and takes 3 days. Due to the limited number of technicians, Nottawasaga can only upgrade one printing station at a time. That is, if they decide to upgrade two Bs, the total down time will be 10 days. During the upgrading period, the down time should not exceed 14 days in total. Also, at least two printing lines must be available at all times to satisfy the current customer demand. The entire line will not be available if any of the printing stations is turned off for upgrading. Nottawasaga Printing wants to know which line and which printing station to upgrade. Determine the feasible mutually exclusive combinations of lines and stations for Nottawasaga Printing.

4.7 Margaret has a project with a \$28 000 first cost that returns \$5000 per year over its 10-year life. It has a salvage value of \$3000 at the end of 10 years. If the MARR is 15%:

(a) What is the present worth of this project?

(b) What is the annual worth of this project?

(c) What is the future worth of this project after 10 years?

(d) What is the payback period of this project?

(e) What is the discounted payback period for this project?

4.8 Appledale Dairy is considering upgrading an old ice-cream maker. Upgrading is available at two levels: moderate and extensive. Moderate upgrading costs \$6500 now and yields annual savings of \$3300 in the first year, \$3000 in the second year, \$2700 in the third year, and so on. Extensive upgrading costs \$10 550 and saves \$7600 in the first year. The savings then decrease by 20% each year thereafter. If the upgraded ice-cream maker will last for 7 years, which upgrading option is better? Use a present worth comparison. Appledale's MARR is 8%.

4.9 Kiwidale Dairy is considering purchasing a new ice-cream maker. Two models, Smoothie and Creamy, are available and their information is given below.

	Smoothie	**Creamy**
First Cost	\$15 000	\$36 000
Service Life	12 years	12 years
Annual Profit	\$4200	\$10 800
Annual Operating Cost	\$1200	\$3520
Salvage Value	\$2250	\$5000

(a) What is Kiwidale's MARR that makes the two alternatives equivalent? Use a present worth comparison.

(b) It turned out that the service life of Smoothie was 14 years. Which alternative is better based on the MARR computed in (a)? Assume that each alternative can be repeated indefinitely.

4.10 Nabil is considering buying a house while he is at university. The house costs \$100 000 today. Renting out part of the house and living in the rest over his five years at school will net, after expenses, \$1000 per month. He estimates that he will sell the house after five years for \$105 000. If Nabil's MARR is 18%, compounded monthly, should he buy the house?

4.11 A software genius is selling the rights to a new video game he has developed. Two companies have offered him contracts. The first contract offers \$10 000 at the end of each year for the next five years, and then \$20 000 per year for the following 10 years. The second offers ten payments, starting with \$10 000 at the end of the first year, \$13 000 at the end of the second, and so forth, increasing by \$3000 each year (i.e., the tenth payment will be \$10 000 + 9 × \$3000). Assume the genius uses a MARR of 9%. Which contract should the genius choose? Use a present worth comparison.

4.12 Sam is considering buying a new lawnmower. He has a choice between a "Lawn Guy" mower or a Bargain Joe's "Clip Job" mower. Sam has a MARR of 5%. The salvage value of each mower at the end of its service life is zero.

	Lawn Guy	**Clip Job**
First cost	\$350	\$120
Life	10 years	4 years
Annual gas	\$60	\$40
Annual maintenance	\$30	\$60

(a) Determine which alternative is preferable. Use a present worth comparison and the least common multiple of the service lives.

(b) For a four-year study period, what salvage value for the Lawn Guy mower would result in its being the preferred choice? What salvage value for the Lawn Guy would result in the Clip Job being the preferred choice?

4.13 Water supply for an irrigation system can be obtained from a stream in some nearby mountains. Two alternatives are being considered, both of which have essentially infinite lives, provided proper maintenance is performed. The first is a concrete reservoir with a steel pipe system and the second is an earthen dam with a wooden aqueduct. Below are the costs associated with each.

	Concrete reservoir	**Earthen dam**
First cost	\$500 000	\$200 000
Annual maintenance costs	2 000	12 000
Replacing the wood portion of the aqueduct each 15 years	N/A	100 000

Compare the present worths of the two alternatives, using an interest rate of 8%. Which alternative should be chosen?

4.14 CB Electronix needs to expand its capacity. It has two feasible alternatives under consideration. Both alternatives will have essentially infinite lives.

Alternative 1: Construct a new building of 200 000 square feet now. The first cost will be $2 000 000. Annual maintenance costs will be $10 000. In addition, the building will need to be painted every 15 years (starting in 15 years) at a cost of $15 000.

Alternative 2: Construct a new building of 125 000 square feet now and an additional 75 000 square feet in 10 years. The first cost of the 125 000-square-foot building will be $1 250 000. The annual maintenance costs will be $5000 for the first 10 years (that is, until the addition is built). The 75 000-square-foot addition will have a first cost of $1 000 000. Annual maintenance costs of the renovated building (the original building and the addition) will be $11 000. The renovated building will cost $15 000 to repaint every 15 years (starting 15 years after the addition is done).

Carry out an annual worth comparison of the two alternatives. Which is preferred if the MARR is 15%?

4.15 Katie's project had a five-year term, a first cost, no salvage value and annual savings of $20 000 per year. After doing present worth and annual worth calculations with a 15% interest rate, Katie noticed that the calculated annual worth for the project was exactly three times the present worth. What was the project's present worth and annual worth? Should Katie undertake the project?

4.16 Newmarket Supermarket wants to replace their cash registers and are currently evaluating two models that seem reasonable to them. The information on the two alternatives, CR1000 and CRX, is shown below.

	CR1000	CRX
First Cost	$680	$1100
Annual Savings	$245	$440
Annual Maintenance Cost	$35 in year 1, increasing by $10 each year thereafter	$60
Service Life	4 years	6 years
Scrap Value	$100	$250

(a) If Newmarket Supermarket's MARR is 10%, which type of cash register should they choose? Use the present worth method.

(b) For the less preferred type of cash register found in (a), what scrap value would make it the preferred choice?

4.17 Midland Metalworking, Inc., is examining a 750-tonne hydraulic press and a 600-tonne molding press for purchase. Midland has only enough budget for one of them. If Midland's MARR is 12% and the relevant information is as given below, which press should they purchase? Use an annual worth comparison.

	Hydraulic Press	Molding Press
Initial Cost	\$275 000	\$185 000
Annual Savings	\$33 000	\$24 500
Annual Maintenance Cost	\$2000, increasing by 15% each year thereafter	\$1000, increasing by \$350 each year thereafter
Life	15 years	10 years
Salvage Value	\$19 250	\$14 800

4.18 Westmount Waxworks is considering buying a new wax melter for their line of replicas of statues of government leaders. They have two choices of supplier, Finedetail and Simplicity. Their proposals are as follows:

	Finedetail	Simplicity
Expected life	7 years	10 years
First cost	\$200 000	\$350 000
Maintenance	\$10 000/year + \$0.05/unit	\$20 000/year + \$0.01/unit
Labour	\$1.25/unit	\$0.50/unit
Other costs	\$6500/year + \$0.95/unit	\$15 500/year + \$0.55/unit
Salvage value	\$5000	\$20 000

Management thinks they will sell about 30 000 replicas per year if there is stability in world governments. If the world becomes very unsettled so that there are frequent overturns of governments, sales may be as high as 200 000 units a year. Westmount Waxworks uses a MARR of 15% for equipment projects.

(a) Who is the preferred supplier if sales are 30 000 units per year? Use an annual worth comparison.

(b) Who is the preferred supplier if sales are 200 000 units per year? Use an annual worth comparison.

(c) How sensitive is the choice of supplier to sales level? Experiment with sales levels between 30 000 and 200 000 units per year. At what sales level will the costs of the two melters be equal?

4.19 The City of Hanover is installing a new swimming pool in the downtown recreation centre. There are two designs under consideration, both of which are to be permanent (i.e., lasting forever). The first design is for a reinforced concrete pool which has a first cost of \$1 500 000. Every 10 years the inner surface of the pool would have to be refinished and painted at a cost of \$200 000.

The second design consists of a metal frame and a plastic liner, which would have an initial cost of \$500 000. For this alternative, the plastic liner must be replaced every five years at a cost of \$100 000, and every 15 years the metal frame would need replacement at a cost of \$150 000. Extra insurance of \$5000 per year is required for the plastic liner (to cover repair costs if the liner leaks). The city's cost of long-term funds is 5%.

Determine which swimming pool design has the lower present cost.

4.20 Sam is buying a refrigerator. He has two choices. A used one, at \$475, should last him about three years. A new one, at \$1250, would likely last eight years. Both have a scrap value of \$0. The interest rate is 8%.

(a) Which refrigerator has a lower cost? (Use a present worth analysis with repeated lives. Assume operating costs are the same.)

(b) If Sam knew that he could resell the new refrigerator after three years for \$1000, would this change the answer in part (a)? (Use a present worth analysis with a three-year study period. Assume operating costs are the same.)

4.21 Val is considering purchasing a display panel to allow her to project her notebook computer through an overhead projector. One model, the XJ3, costs \$4500 new, while another, the Y19, sells for \$3200. Val figures that the XJ3 will last about three years, at which point it could be sold for \$1000, while the Y19 will last for only two years and will also sell for \$1000. Both panels give similar service, except that the Y19 is not suitable for client presentations. If she buys the Y19, about four times a year she will have to rent one similar to the XJ3, at a total year-end cost of about \$300. Using present worth and the least common multiple of the service lives, determine which display panel Val should buy. Val's MARR is 10%.

4.22 For Problem 4.21, Val has determined that the salvage value of the XJ3 after two years of service is \$1900. Which display panel is the better choice, based on present worth with a two-year study period?

4.23 Tom is considering buying a \$24 000 car. After five years, he will be able to sell the car for \$8000. Gas costs will be \$2000 per year, insurance \$600 per year, and parking \$600 per year. Maintenance costs for the first year are \$1000, rising by \$400 per year thereafter.

The alternative is for Tom to take taxis everywhere. This will cost an estimated \$6000 per year. Tom will rent a car each year at a total cost (to year-end) of $600 for the family vacation, if he has no car. If Tom values money at 11% annual interest, should he buy the car? Use an annual worth comparison method.

4.24 A new gizmo costs $10 000. Maintenance costs $2000 per year, and labour savings are $6567 per year. What is its payback period?

4.25 Building a bridge will cost $65 million. A round-trip toll of $12 will be charged to all vehicles. Traffic projections are estimated to be 5000 per day. The operating and maintenance costs will be 20% of the toll revenue. Find the payback period (in years) for this project.

4.26 A new packaging machine will save the Greene Cheese Company $3000 per year in reduced spoilage, $2500 per year in labour, and $1000 per year in packaging material. The new machine will have additional expenses of $700 per year in maintenance and $200 per year in energy. If it costs $20 000 to purchase, what is its payback period? Assume that the savings are earned throughout the year, not just at year end.

4.27 Diana usually uses a three-year payback period to determine if a project is acceptable. A recent project with uniform yearly savings over a five-year life had a payback period of almost exactly three years, so Diana decided to find the project's present worth to help determine if the project was truly justifiable. However, that calculation didn't help either since the present worth was exactly 0. What interest rate was Diana using to calculate the present worth? The project has no salvage value at the end of its five-year life.

4.28 The Biltmore Garage has lights in places that are difficult to reach. Management estimates that it costs about $2 to change a bulb. Standard 100-watt bulbs with an expected life of 1000 hours are now used. Standard bulbs cost $1. A long-life bulb that requires 90 watts for the same effective level of light is available. Long-life bulbs cost $3. The bulbs that are difficult to reach are in use for about 500 hours a month. Electricity costs $0.08/kilowatt-hour payable at the end of each month. Biltmore uses a 12% MARR (1% per month) for projects involving supplies.

(a) What minimum life for the long-life bulb would make its cost lower?

(b) If the cost of changing bulbs is ignored, what is the minimum life for the long-life bulb for them to have a lower cost?

(c) If the solutions are obtained by linear interpolation of the capital recovery factor, will the approximations understate or overstate the required life?

4.29 A chemical recovery system costs $30 000 and saves $5280 each year of its seven-year life. The salvage value is estimated at $7500. The MARR is 9%. What is the net annual benefit or cost of purchasing the chemical recovery system? Use the capital recovery formula.

4.30 Savings of $5600 per year can be achieved through either a $14 000 machine (A) with a seven-year service life and a $2000 salvage value, or a $25 000 machine (B) with a ten-year service life and a $10 000 salvage value. If the MARR is 9%, which machine is a better choice, and for what annual benefit or cost? Use annual worth and the capital recovery formula.

4.31 **(a)** Ridgely Custom Metal Products (RCMP) must purchase a new tube bender. There are two models:

Model	First Cost	Economic Life	Yearly Net Savings	Salvage Value
T	$100 000	5 years	$50 000	$20 000
A	150 000	5 years	60 000	30 000

RCMP's MARR is 11%. Using the *present worth* method, which tube bender should they buy?

(b) RCMP has discovered a third alternative, which has been added to the table below. Now which tube bender should they buy?

Model	First Cost	Economic Life	Yearly Net Savings	Salvage Value
T	$100 000	5 years	$50 000	$ 20 000
A	150 000	5 years	60 000	30 000
X	200 000	3 years	75 000	100 000

4.32 RCMP [see Problem 4.31 (b)] can forecast demand for its products for only three years in advance. The salvage value after three years is $40 000 for model T and $80 000 for model A. Using the *study period* method, which of the three alternatives is best?

4.33 Using the *annual worth* method, which of the three tube benders should RCMP buy? The MARR is 11%. Use the data from Problem 4.31 (b).

4.34 What is the payback period for each of the three alternatives from the RCMP problem? Use the data from Problem 4.31 (b).

4.35 Data for two independent investment opportunities are shown below.

	Machine A	Machine B
Initial cost	$15 000	$20 000
Revenues (annual)	$ 9 000	$11 000
Costs (annual)	$ 6 000	$ 8 000
Scrap value	$ 1 000	$ 2 000
Service life	5 years	10 years

(a) For a MARR of 8%, should either, both, or neither machine be purchased? Use the annual worth method.

(b) For a MARR of 8%, should either, both, or neither machine be purchased? Use the present worth method.

(c) What are the payback periods for these machines? Should either, both, or neither machine be purchased, based on the payback periods? The required payback period for investments of this type is three years.

4.36 Xaviera is comparing two mutually exclusive projects, A and B, that have the same initial investment and the same present worth over their service lives. Wolfgang points out that, using the annual worth method, A is clearly better than B. What can be said about the service lives for the two projects?

4.37 Xaviera noticed that two mutually exclusive projects, A and B, have the same payback period and the same economic life, but A has a larger present worth than B does. What can be said about the size of the annual savings for the two projects?

4.38 Two plans have been proposed for accumulating money for capital projects at Thunder Bay Lighting. One idea is to put aside $10 000 per year, independent of growth. The second is to start with a smaller amount, $8000 per year, but to increase this in proportion to the expected company growth. The money will accumulate interest at 10%, and the company is expected to grow about 5% per year. Which plan will accumulate more money in 10 years?

4.39 Crystal City Environmental Services is evaluating two alternative methods of disposing of municipal waste. The first involves developing a landfill site near the city. Costs of the site include $1 000 000 start-up costs, $100 000 close-down costs 30 years from now, and operating costs of $20 000 per year. Starting in 10 years, it is expected that there will be revenues from user fees of $30 000 per year. The alternative is to ship the waste out of the province. A United States firm will agree to a long-term contract to dispose of the waste for $130 000 per year. Using the *annual worth* method, which alternative is economically preferred for a MARR of 11%? Would this likely be the actual preferred choice?

4.40 Peterborough Auto Parts is considering investing in a new forming line for their grille assemblies. For a five-year study period, the cash flows for two separate designs are shown below. Create a spreadsheet which will calculate the present worths for each project for a variable MARR. Through trial and error, establish the MARR at which the present worths of the two projects are exactly the same.

Cash Flows for Grille Assembly Project

	Automated Line			Manual Line		
Year	**Disburse-ments**	**Receipts**	**Net Cash Flow**	**Disburse-ments**	**Receipts**	**Net Cash Flow**
0	$1 500 000	$ 0	– $1 500 000	$1 000 000	$ 0	– $1 000 000
1	50 000	300 000	250 000	20 000	200 000	180 000
2	60 000	300 000	240 000	25 000	200 000	175 000
3	70 000	300 000	230 000	30 000	200 000	170 000
4	80 000	300 000	220 000	35 000	200 000	165 000
5	90 000	800 000	710 000	40 000	200 000	160 000

4.41 Stayner Catering is considering setting up a temporary division to handle demand created by a special tourist promotion being made by City of Toronto during the year 2000. They will invest in tables, serving equipment and trucks for a one-year period. Labour is employed on a monthly basis. Warehouse space is rented monthly, and revenue is generated monthly. The items purchased will be sold at the end of the year, but the salvage values are somewhat uncertain. Given below are the known or expected cash flows for the project.

Month	**Purchase**	**Labour Expenses**	**Warehouse Expenses**	**Revenue**
January (beginning)	$200 000			
January (end)		$ 2 000	$3 000	$ 2 000
February		2 000	3 000	2 000
March		2 000	3 000	2 000
April		2 000	3 000	2 000
May		4 000	3 000	10 000
June		10 000	6 000	40 000
July		10 000	6 000	110 000
August		10 000	6 000	60 000
September		4 000	3 000	30 000
October		2 000	3 000	10 000
November		2 000	3 000	5 000
December	Salvage?	2 000	3 000	2 000

For an interest rate of 12% compounded monthly, create a spreadsheet that calculates and graphs the present worth of the project for a range of salvage values of the purchased items from 0% to 100% of the purchase price. Should Stayner Catering go ahead with this project?

4.42 Peterborough Auto Parts has two options for increasing efficiency. They can expand the current building or keep the same building but remodel the inside layout. For a five-year study period, the cash flows for the two options are shown below. Construct a spreadsheet which will calculate the present worth for each option for a variable MARR. By trial and error, determine the MARR at which the present worths of the two options are equivalent.

	Expansion Option			Remodelling Option		
Year	**Disburse-ments**	**Receipts**	**Net Cash Flow**	**Disburse-ments**	**Receipts**	**Net Cash Flow**
0	$850 000	$ 0	$–850 000	$230 000	$ 0	$–230 000
1	25 000	200 000	175 000	9000	80 000	71 000
2	30 000	225 000	195 000	11 700	80 000	68 300
3	35 000	250 000	215 000	15 210	80 000	64 790
4	40 000	275 000	235 000	19 773	80 000	60 227
5	45 000	300 000	255 000	25 705	80 000	54 295

4.43 Derek has two choices for a heat-loss prevention system for the shipping doors at Kirkland Manufacturing. He can isolate the shipping department from the rest of the plant, or he can curtain off each shipping door separately. Isolation consists of building a permanent wall around the shipping area. It will cost $60 000 and will save $10 000 in heating costs per year. Plastic curtains around each shipping door will have a total cost of about $5000, but will have to be replaced about once every two years. Savings in heating costs for installing the curtains will be about $3000 per year. Use the payback period method to determine which alternative is better. Comment on the use of the payback period for making this decision.

4.44 Assuming that the wall built to isolate the shipping department in Problem 4.43 will last forever, and that the curtains have zero salvage value, compare the annual worths of the two alternatives. The MARR for Kirkland Manufacturing is 11%. Which alternative is better?

4.45 Crystal City Environmental Services is considering investing in a new water treatment system. Based on the information given below for two alternatives, a fully automated and a partially automated system, construct a spreadsheet for computing the annual worths for each alternative with a variable MARR. Through trial and error, determine the MARR at which the annual worths of the two alternatives are equivalent.

	Fully Automated System			Partially Automated System		
Year	**Disbursements**	**Receipts**	**Net Cash Flow**	**Disbursements**	**Receipts**	**Net Cash Flow**
0	$1 000 000	$ 0	$–1 000 000	$650 000	$ 0	$–650 000
1	30 000	300 000	270 000	30 000	220 000	190 000
2	30 000	300 000	270 000	30 000	220 000	190 000
3	80 000	300 000	220 000	35 000	220 000	185 000
4	30 000	300 000	270 000	35 000	220 000	185 000
5	30 000	300 000	270 000	40 000	220 000	180 000
6	80 000	300 000	220 000	40 000	220 000	180 000
7	30 000	300 000	270 000	45 000	220 000	175 000
8	30 000	300 000	270 000	45 000	220 000	175 000
9	80 000	300 000	220 000	50 000	220 000	170 000
10	30 000	300 000	270 000	50 000	220 000	170 000

Rockwell International of Canada

CANADIAN 4.1 MINI-CASE

The Light Vehicle Division of Rockwell International of Canada makes seat-slide assemblies for the automotive industry. They have two major classifications for investment opportunities: developing new products to be manufactured and sold and developing new machines to improve production. The overall approach to assessing whether or not an investment should be made depends on the nature of the project.

In evaluating a new product, they consider the following:

1. *Marketing strategy:* Does it fit the business plan for the company?
2. *Work force:* How will it affect human resources?
3. *Margins:* The product should generate appropriate profits.
4. *Cash flow:* Positive cash flow is expected within two years.

continued

In evaluating a new machine, they consider the following:

1. *Cash flow:* Positive cash flow is expected within a limited time period.
2. *Quality issues:* For issues of quality, justification is based on cost avoidance rather than positive cash flow.
3. *Cost avoidance:* Savings should pay back an investment within one year.

Discussion

All companies consider more than just the economics of a decision. Most take into account the other issues—often called *intangibles*—by using managerial judgement in an informal process. Others, like Rockwell International of Canada, explicitly consider a selection of intangible issues.

The trend today is to carefully consider several intangible issues, either implicitly or explicitly. Human resource issues are particularly important since employee enthusiasm and commitment have significant repercussions. Environmental impacts of a decision can affect the image of the company. Health and safety is another intangible with significant effects.

However, the economics of the decision is usually (but not always) the single most important factor in a decision. Also, economics is the factor that is usually the easiest to measure.

Questions

1) Why do you think Rockwell International of Canada has different issues to consider depending on whether an investment is a new product or a new machine?
2) For each of the issues mentioned, describe how it would be measured. How would you determine if it was worth investing in a new product or new machine with respect to that issue?
3) There are two kinds of errors that can be made. The first is that an investment is made when it should not, and the second is that an investment is not made when it should. Describe examples of both kinds of errors for both products and machines (four examples in total) if the issues listed for Rockwell International of Canada are strictly followed. What are some sensible ways to prevent such errors?

Appendix 4A: The MARR and the Cost of Capital

For a business to survive, it must be able to earn a high enough return to induce investors to put money into the company. The minimum rate of return required to get investors to invest in a business is that business's **cost of capital**. A company's cost of capital is also its minimum acceptable rate of return for projects, its MARR. This appendix reviews how the cost of capital is determined. We first look at the relation between risk and the cost of capital. Next, we discuss sources of capital for large businesses and small businesses.

4A.1 Risk and the Cost of Capital

There are two main forms of investment in a company, *debt* and *equity*. Investors in a company's debt are lending money to the company. The loans are contracts that give lenders rights to repayment of their loans, and to interest at predetermined interest rates. Investors in a company's equity are the owners of the company. They hold rights to the residual after all contractual payments, including those to lenders, are made.

Investing in equity is more risky than investing in debt. Equity owners are paid only if the company first meets its contractual obligations to lenders. This higher risk means that equity owners require an expectation of a greater return on average than the interest rate paid to debt holders. Consider a simple case in which a company has three possible performance levels — weak results, normal results, and strong results. Investors do not know which level will actually occur. Each level is equally probable. To keep the example simple, we assume that all after-tax income is paid to equity holders as dividends so that there is no growth. The data are shown in Table 4A.1.

Table 4A.1 Cost of Capital Example

	Possible Performance Levels		
	Weak Results	Normal Results	Strong Results
Net operating income ($/year)[1]	40 000	100 000	160 000
Interest payment ($/year)	10 000	10 000	10 000
Net income before tax ($/year)	30 000	90 000	150 000
Tax at 40% ($/year)	12 000	36 000	60 000
After-tax income = Dividends ($/year)	18 000	54 000	90 000
Debt ($)	100 000	100 000	100 000
Value of shares ($)	327 273	327 273	327 273

[1] Net operating income per year is revenue per year minus cost per year.

We see that, no matter what happens, lenders will get a return of 10%:

$$0.1 = \frac{10\,000}{100\,000}$$

Owners get one of three possible returns:

$$5.5\% \left(0.055 = \frac{18\,000}{327\,273}\right),$$

$$16.5\% \left(0.165 = \frac{54\,000}{327\,273}\right), \text{ or}$$

$$27.5\% \left(0.275 = \frac{90\,000}{327\,273}\right)$$

These three possibilities average out to 16.5%. If things are good, owners do better than lenders. If things are bad, owners do worse. But their average return is greater than returns to lenders.

The lower rate of return to lenders means that companies would like to get their capital with debt. However, reliance on debt is limited for two reasons.

1. If a company increases the share of capital from debt, it increases the chance that it will not be able to meet the contractual obligations to lenders. This means the company will be bankrupt. Bankruptcy may lead to reorganizing the company or possibly closing the company. In either case, bankruptcy costs may be high.
2. Lenders are aware of the dangers of high reliance on debt and will, therefore, limit the amount they lend to the company.

4A.2 Company Size and Sources of Capital

Large, well-known companies, like Dofasco or Molson Breweries, can secure capital both by borrowing and by selling ownership shares. Large companies will seek ratios of debt to equity that balance the marginal advantages and disadvantages of debt financing. Because the companies are well known, there will be ready markets for their shares as well as any debt instruments, like bonds, they may issue. Well-known companies can raise additional capital in their desired proportions of debt and equity. Or they can reduce their capital by repaying loans and buying back shares while keeping debt and equity at their desired proportions.

The cost of capital for large well-known companies is a weighted average of the costs of borrowing and of selling shares. The weights are the fractions of total capital that come from the different sources. This cost of capital is called the **weighted average cost of capital**. If market conditions do not change, a large company that seeks to raise a moderate amount of additional capital can do so at a stable cost of capital. This cost of capital is that company's MARR. We can compute the after-tax cost of capital for the example shown in Table 4A.1.

Weighted average cost of capital

$$= 0.1\left(\frac{100\,000}{427\,273}\right) + 0.165\left(\frac{327\,273}{427\,273}\right) = 0.150$$

This company has a cost of capital of about 15%.

The cost of capital for small companies is greater than for large well-known companies. As well, the cost depends on the amounts small companies seek to raise. Most investors in large companies are not willing to invest in unknown small companies. At start-up, a small company may rely entirely on the capital of the owners and their friends and relatives. Here the cost of capital is the opportunity cost for the investors. The only alternative available to the owners may be bank deposits or publicly traded securities. This cost will be moderate.

If a new company seeks to grow more rapidly than the owners' investment plus cash flow permits, more funds will be needed. The next source of capital with the lowest cost is usually a bank loan. Bank loans are limited because banks are usually not willing to lend more than some fraction of the amount an owner puts into a business. If a small company needs more capital than is available from the owners or from bank loans, the company will have to sell new shares of the company.

There are two major sources of new equity investments for small companies. One of these sources is venture capitalists. Venture capitalists are investors who specialize in investing in new companies. The cost of evaluating new companies is usually high. The risk of investing in them is also high. Together, these factors usually lead venture capitalists to want to put enough money into a small company to enable the venture capitalist to control the company.

The second source of additional equity capital for small companies is selling shares through stock exchanges, like the Vancouver Stock Exchange or the Alberta Stock Exchange. These exchanges specialize in small, speculative companies. In either case, new equity investment is usually very expensive. Studies have shown that venture capitalists typically require the expectation of at least a 35% rate of return after tax. Raising funds on a stock exchange is usually even more expensive than getting funding from a venture capitalist.

C H A P T E R 5

Comparison Methods Part 2

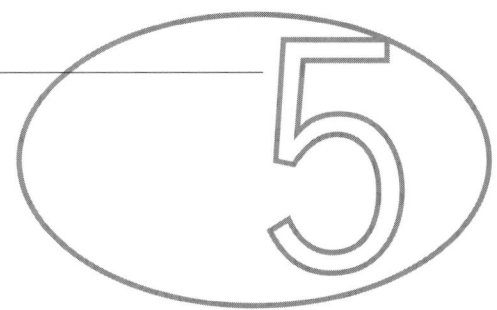

Engineering Economics in Action, Part 5A: What's Best? Revisited

Clem had said, "I have to make a recommendation to Ed Burns and Anna Kulkowski for their Wednesday meeting on this forging hammer in the South Shop. I'd like you to handle it." Dave Sullivan, who had started the project, had gone to Florida to see his sick father-in-law. Naomi welcomed the opportunity, but she still had to figure out exactly what to recommend.

Naomi looked carefully at the list of seven mutually exclusive alternatives for replacing or refurbishing the machine. Present worth could tell her which of the seven was "best," but present worth was just one of several comparison methods. Which of the comparison methods should she use?

Dave would help more, if she asked him. In fact, he could no doubt tell her exactly what to do, if she wanted. But this one she knew she could handle, and it was a matter of pride to do it herself. Opening her engineering economics textbook, she read on.

5.1 Introduction

In Chapter 4, we showed how to structure projects so that they were either independent or mutually exclusive. The present worth, annual worth, and payback period methods for evaluating projects were also introduced. This chapter continues on the theme of comparing projects by presenting a commonly used but somewhat more complicated comparison method called the *internal rate of return*, or IRR.

Although the IRR method is widely used, all of the comparison methods have value in particular circumstances. Selecting the method to use is also covered in this chapter. It is also shown that the present worth, annual worth, and IRR methods all result in the same recommendations for the same problem. We close this chapter with a chart summarizing the strengths and weaknesses of the four comparison methods presented in Chapters 4 and 5.

5.2 The Internal Rate of Return

Investments are undertaken with the expectation of a return in the form of future earnings. One way to measure the return from an investment is as a rate of return per dollar invested, or in other words as an interest rate. The rate of return usually calculated for a project is known as the *internal rate of return (IRR)*. The adjective "internal" refers to the fact that the internal rate of return depends only on the cash flows due to the investment. The internal rate of return is that interest rate at which a project just breaks even. The meaning of the IRR is most easily seen with a simple example.

Example 5.1

Suppose \$100 is invested today in a project that returns \$110 in one year. We can calculate the IRR by finding the interest rate at which \$100 now is equivalent to \$110 at the end of one year:

$$P = F(P/F, i^*, 1)$$

$$\$100 = \$110(P/F, i^*, 1)$$

$$\$100 = \frac{\$110}{1 + i^*}$$

where i^* is the internal rate of return.

Solving this equation gives a rate of return of 10%. In a simple example like this, the process of finding an internal rate of return is finding the interest rate that makes the present worth of benefits equal to the first cost. This interest rate is the IRR. ■

Of course, cash flows associated with a project will usually be more complicated than in the example above. A more formal definition of the IRR is: The **internal rate of return (IRR)** on an investment is that interest rate, i^*, such that, when all cash flows associated with the project are discounted at i^*, the present worth of the cash inflows equals the present worth of the cash outflows. That is, the project just breaks even. An equation that expresses this is

$$\sum_{t=0}^{T} \frac{(R_t - D_t)}{(1 + i^*)^t} = 0 \qquad (5.1)$$

where

R_t = the cash inflow (receipts) in period t
D_t = the cash outflow (disbursements) in period t
T = the number of time periods
i^* = the internal rate of return

Since Equation (5.1) can also be expressed as

$$\sum_{t=0}^{T} R_t(1 + i^*)^{-t} = \sum_{t=0}^{T} D_t(1 + i^*)^{-t}$$

it can be seen that, in order to calculate the IRR, one sets the disbursements equal to the receipts and solves for the unknown interest rate. For this to be done, the disbursements and receipts must be comparable, as a present worth, a uniform series, or a future worth. That is, use

PW(disbursements) = PW(receipts) and solve for the unknown i^*,

AW(disbursements) = AW(receipts) and solve for the unknown i^*, or

FW(disbursements) = FW(receipts) and solve for the unknown i^*.

The IRR is usually positive, but can be negative as well. A negative IRR means that the project is losing money rather than earning it.

We usually solve the equations for the IRR by trial and error as there is no explicit means of solving Equation (5.1) for projects where the number of periods is large. A spreadsheet provides a quick way to perform trial and error calculations; most spreadsheet programs also include a built-in IRR function.

Example 5.2

Clem is considering buying a tuxedo. It would cost \$500, but would save him \$160 per year in rental charges over its five-year life. What is the IRR for this investment?

As illustrated in Figure 5.1, Clem's initial cash outflow for the purchase would be \$500. This is an up-front outlay relative to continuing to rent tuxedos. The investment would create a saving of \$160 per year over the five-year life of the tuxedo. These savings can be viewed as a series of receipts relative to rentals. The IRR of Clem's investment can be found by determining what interest rate makes the present worth of the disbursements equal to the present worth of the cash inflows.

Present worth of disbursements = \$500
Present worth of receipts = \$160(P/A, i^*, 5)

Setting the two equal,

$$\$500 = \$160(P/A, i^*, 5)$$
$$(P/A, i^*, 5) = \$500/\$160$$
$$= 3.125$$

Figure 5.1 Clem's Tuxedo

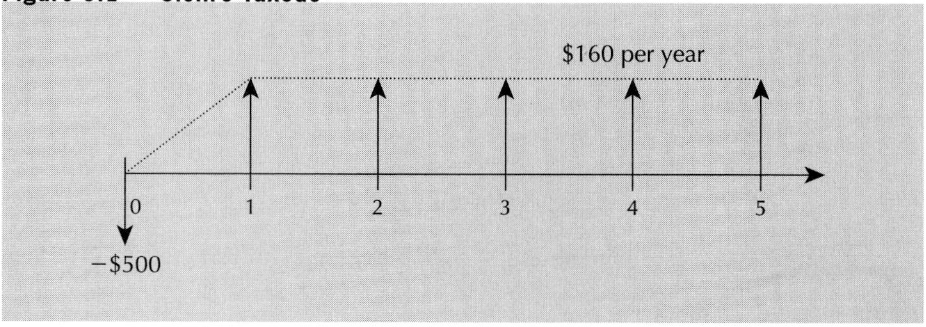

From the interest rate tables, we find that

$$(P/A, 15\%, 5) = 3.3521$$
$$(P/A, 20\%, 5) = 2.9906$$

Interpolating between $(P/A, 15\%, 5)$ and $(P/A, 20\%, 5)$ gives

$$i^* = 15\% + (5\%)[(3.125 - 3.3521)/(2.9906 - 3.3521)]$$
$$= 18.14\%$$

An alternative way to get the IRR for this problem is to convert all cash outflows and inflows to equivalent annuities over the five-year period. This will yield the same result as when present worth was used.

Annuity equivalent to the disbursements = $\$500(A/P, i^*, 5)$

Annuity equivalent to the receipts = $160

Again, setting the two equal,

$$\$500(A/P, i^*, 5) = \$160$$
$$(A/P, i^*, 5) = \$160/\$500$$
$$= 0.32$$

From the interest rate tables,

$$(A/P, 15\%, 5) = 0.29832$$
$$(A/P, 20\%, 5) = 0.33438$$

An interpolation gives

$$i^* = 15\% + 5\%[(0.32 - 0.29832)/(0.33438 - 0.29832)]$$
$$= 18.01\%$$
$$\cong 18.0\%$$

Note that there is a slight difference in the answers, depending on whether the disbursements and receipts were compared as present worths or as annuities. This difference is due to the small error induced by the linear interpolation. ■

5.3 Internal Rate of Return Comparisons

In this section, we show how the internal rate of return can be used to decide whether or not a project should be accepted. We first show how to use the IRR to evaluate independent projects. Then we show how to use the IRR to decide which of a group of mutually exclusive alternatives to accept. We then show that it is possible for a project to have more than one IRR. Finally, we show how to handle this difficulty by using an *external rate of return.*

5.3.1 IRR for Independent Projects

Recall from Chapter 4 that projects under consideration are evaluated using the MARR and that any independent project, which has a present or annual worth equal to or exceeding zero, should be accepted. The basic principle for the IRR method is analogous. We will invest in any project that has an IRR equal to or exceeding the MARR. Just as projects with a zero present or annual worth are marginally acceptable, projects with IRR = MARR have a marginally acceptable rate of return (by definition of the MARR).

Also analogous to Chapter 4, when we do a rate of return comparison on several independent projects, the projects must have equal lives. If this is not the case, then the approaches covered in Section 4.4.4 (Comparison of Alternatives with Unequal Lives) must be used.

Example 5.3

The High Society Baked Bean Co. is considering a new canner. The canner costs \$120 000, and will have a scrap value of \$5000 after its 10-year life. Given the expected increases in sales, the total savings due to the new canner, compared with continuing with the current operation, will be \$15 000 the first year, increasing by \$5000 each year thereafter. Total extra costs due to the more complex equipment will be \$10 000 per year. The MARR for High Society is 12%. Should they invest in the new canner?

The cash inflows and outflows for this problem are summarized in Figure 5.2. We need to compute the internal rate of return in order to decide if High Society should buy the canner. There are several ways we can do this. In this problem, equating annual outflows and receipts appears to be the easiest approach, because most of the cash flows are already stated on a yearly basis.

$$\$5000(A/F, i^*, 10) + \$15\,000 + \$5000(A/G, i^*, 10) - \$120\,000(A/P, i^*, 10) - \$10\,000 = 0$$

Dividing by 5000,

$$(A/F, i^*, 10) + 1 + (A/G, i^*, 10) - 24(A/P, i^*, 10) = 0$$

The IRR can be found by trial and error alone, by trial and error and linear interpolation, or by a spreadsheet IRR function. A trial and error process is particularly easy using a spreadsheet, so this is often the best approach. A good starting point for the trial and error process is at zero interest. A graph (Figure 5.3) derived from the spreadsheet indicates that the IRR is between

Figure 5.2 High Society Baked Bean Canner

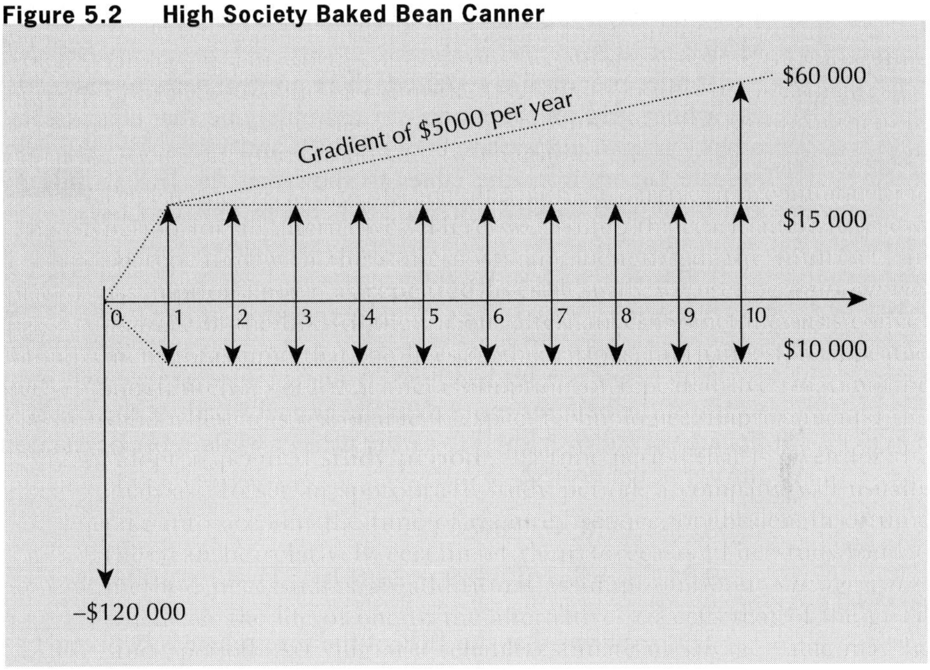

Figure 5.3 Estimating the IRR for Example 5.3

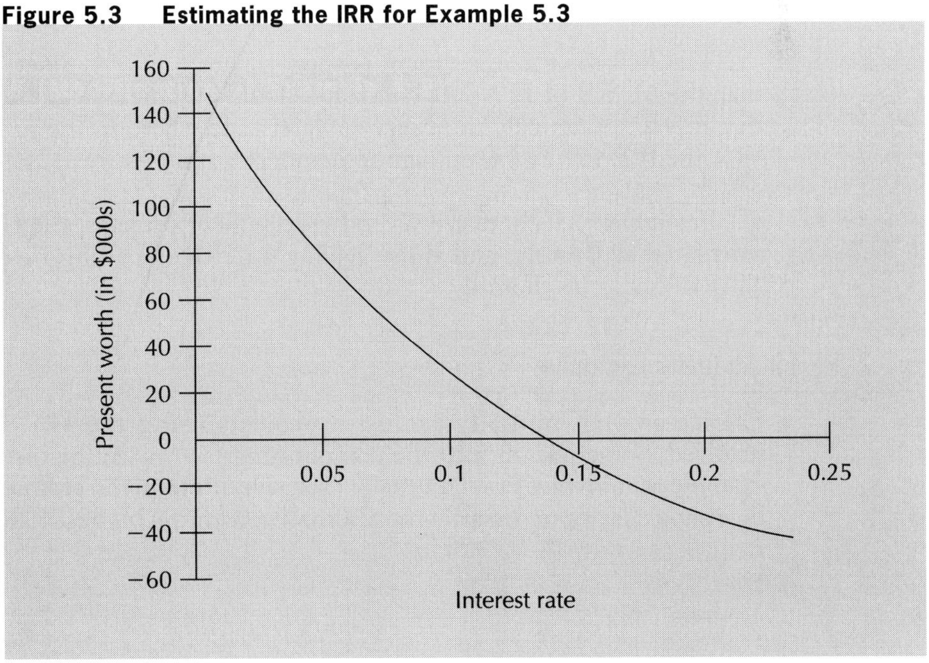

13% and 14%. This may be good enough for a decision, since it exceeds the MARR of 12%.

If finer precision is required, there are two ways to proceed. One way is to use a finer grid on the graph, for example, one that covers 13% to 14%. The other way is to interpolate between 13% and 14%. We shall first use the interest rate factors from the tables to show that the IRR is indeed between 13% and 14%. Next we shall interpolate between 13% and 14%.

First, at $i = 13\%$, we have

$$0.05429 + 1 + 3.5161 - 24(0.18429) = 0.1474$$

The result is a bit too high. A higher interest rate will reduce the annual worth of the benefits more than the annual worth of the costs, since the benefits are spread over the life of the project while most of the costs are early in the life of the project.

At $i = 14\%$, we have

$$0.05171 + 1 + 3.4489 - 24(0.19171) = -0.1004$$

This confirms that the IRR of the investment is between 13% and 14%. A good approximation to the IRR can be found by linearly interpolating:

$$i^* = 13\% + (0 - 0.1474)/(0.1004 - 0.1474)$$
$$\cong 13.6\%$$

The IRR for the investment is approximately 13.6%. Since this is greater than the MARR of 12%, the company should buy the new canner. Note again that it was not actually necessary to determine where in the range of 13% to 14% the IRR fell. It was enough to demonstrate that it was 12% or more. ■

In summary, if there are several independent projects, the IRR for each is calculated separately, and those having an IRR equalling or exceeding the MARR should be chosen.

5.3.2 IRR for Mutually Exclusive Projects

Choice among mutually exclusive projects using the IRR is a bit more involved. Some insight into the complicating factors can be obtained from an example that involves two mutually exclusive alternatives. It illustrates that the best choice is not necessarily the alternative with the highest IRR.

Example 5.4

Consider two investments. The first costs \$1 today and returns \$2 in one year. The second costs \$1000 and returns \$1900 in one year. Which is the preferred investment? Your MARR is 70%.

The first project has an IRR of 100%:

$$-\$1 + \$2(P/F, i^*, 1) = 0$$

$$(P/F, i^*, 1) = \$1/\$2 = 0.5$$

$$i^* = 100\%$$

The second has an IRR of 90%:

$$-\$1000 + \$1900(P/F, i^*, 1) = 0$$

$$(P/F, i^*, 1) = \$1000/\$1900 = 0.52631$$

$$i^* = 90\%$$

It might be tempting to choose the first project, the alternative with the larger rate of return. However, this approach is incorrect because it can overlook projects that have a rate of return equal to or greater than the MARR, but don't have the maximum IRR. In the example, the correct approach is to observe that the least cost investment provides a rate of return that exceeds MARR. The next step is to find the rate of return on the more expensive investment to see if the *incremental* investment has a rate of return equal to or exceeding the MARR. The incremental investment is the additional \$999 that would be invested if the second investment was taken instead of the first:

$$-(\$1000 - \$1) + (\$1900 - \$2)(P/F, i^*, 1) = 0$$

$$(P/F, i^*, 1) = \$999/\$1898 = 0.52634$$

$$i^* = 89.98\%$$

Indeed, the incremental investment has an IRR exceeding 70% and thus the second investment should be chosen. ■

The next example illustrates the process again, showing this time how the incremental investment is not justified.

Example 5.5

Monster Meats can buy a new meat slicer system for \$50 000. They estimate it will save them \$11 000 per year in labour and operating costs. The same system with an automatic loader is \$68 000, and will save approximately \$14 000 per year. The life of either system is thought to be eight years. Monster Meats has three feasible alternatives:

Alternative	First Cost	Annual Savings
"Do nothing"	$ 0	$ 0
Meat slicer alone	50 000	11 000
Meat slicer with automatic loader	68 000	14 000

Monster Meats uses a MARR of 12% for this type of project. Which alternative is better?

We first consider the system without the loader. Its IRR is 14.5%, which exceeds the MARR of 12%. This can be seen by solving for i^* in

$$-\$50\,000 + \$11\,000(P/A, i^*, 8) = 0$$

$$(P/A, i^*, 8) = \$50\,000/\$11\,000$$

$$= 4.545$$

From the interest rate tables, or by trial and error with a spreadsheet,

$$(P/A, 14\%, 8) = 4.6388$$

$$(P/A, 15\%, 8) = 4.4873$$

By interpolation or further trial and error,

$$i^* \cong 14.5\%$$

The slicer alone is thus economically justified and is better than the "do nothing" alternative.

We now consider the system with the slicer and loader. Its IRR is 12.5%, which may be seen by solving for i^* in

$$-\$68\,000 + \$14\,000(P/A, i^*, 8) = 0$$

$$(P/A, i^*, 8) = \$68\,000/\$14\,000$$

$$(P/A, i^*, 8) = 4.857$$

$$(P/A, 12\%, 8) = 4.9676$$

$$(P/A, 13\%, 8) = 4.7987$$

$$i^* \cong 12.5\%$$

The IRR of the meat slicer and automatic loader is about 12.5%, which on the surface appears to meet the 12% MARR requirement. But, on the incremental investment, Monster Meats would be earning only 7%. This may be seen by looking at the IRR on the *extra,* or *incremental,* $18 000 spent on the loader.

Figure 5.4 Monster Meats

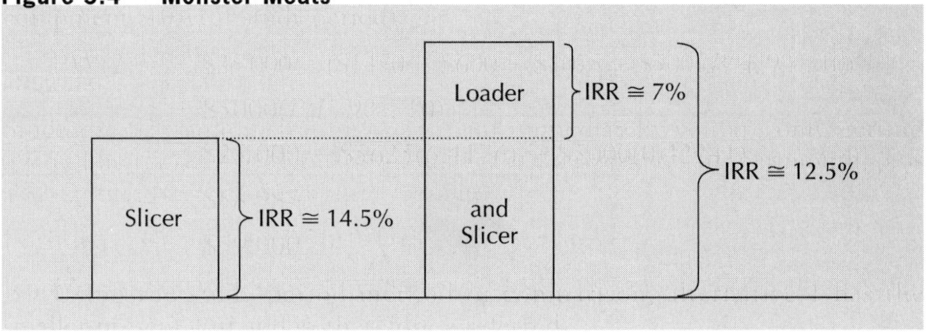

$$-(\$68\,000 - \$50\,000) + (\$14\,000 - \$11\,000)(P/A, i^*, 8) = 0$$
$$-\$18\,000 + \$3000(P/A, i^*, 8) = 0$$
$$(P/A, i^*, 8) = \$18\,000/\$3000$$
$$(P/A, i^*, 8) = 6$$
$$i^* \cong 7\%$$

This is less than the MARR; therefore, Monster Meats should not buy the automated loader.

When the IRR was calculated for the system including the loader, the surplus return on investment earned by the slicer alone essentially subsidized the loader. The slicer investment made enough of a return so that, even when it was coupled with the money-losing loader, the whole machine still seemed to be a good buy. In fact, the extra \$18 000 would be better spent on some other project at the MARR or higher. The relation between the potential projects is shown in Figure 5.4 above. ■

The fundamental principle illustrated by the two examples is that, to use the IRR to compare two mutually exclusive alternatives properly, not only must both of the alternatives exceed the MARR, but also the *incremental* investment is relevant. When there are more than two alternatives, it is necessary to have a systematic means of determining which pairs of comparisons to make. Note that before undertaking an analysis of mutually exclusive alternatives with the IRR method, you should ensure that the alternatives have equal lives. If they do not have equal lives, then the methods of Section 4.4.4 (study period or repeated lives methods) must be applied first to set up comparable cash flows.

The first step in the process of comparing several mutually exclusive alternatives using the IRR is to order the alternatives from the smallest first cost to the largest first cost. Start with the alternative with the smallest first cost (which may be the "do nothing" alternative with \$0 first cost). The alternative with the smallest first cost becomes what is referred to as the *current best alternative* (denoted by an A in Figure 5.5). Note that the current best alternative

Figure 5.5 Flowchart for Comparing Mutually Exclusive Alternatives

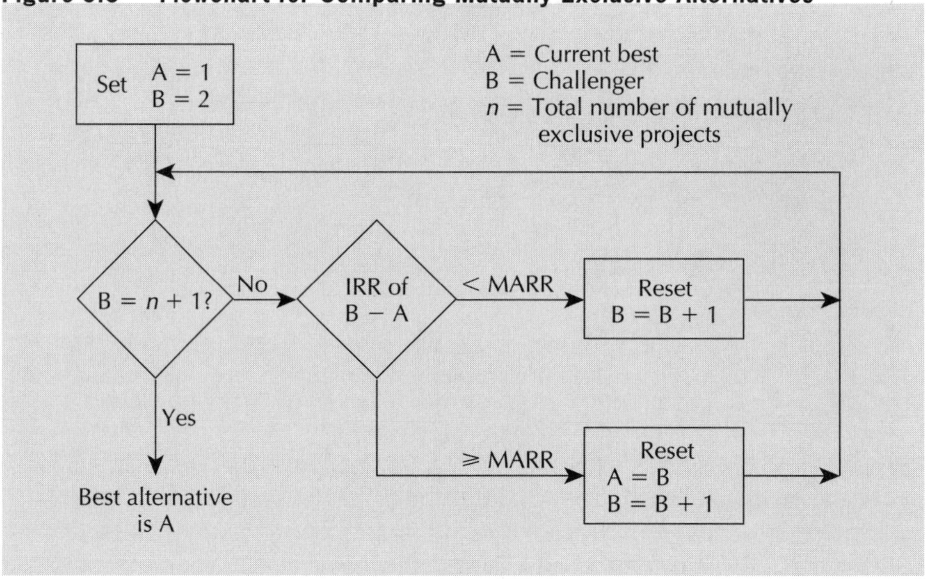

may have an IRR *less* than the MARR if the "do nothing" alternative is not possible. For this reason, we don't have to check the IRR of any of the individual alternatives. Instead, we rely on the step-by-step procedure described in Figure 5.5.

Assume that there are n projects and they are ranked from 1 (the current best) to n, in increasing order of first costs. The analysis consists of looking at the *incremental* investments of alternatives that have a higher first cost than A. The current best is "challenged" by the project ranked second. One of two things occurs:

1. The incremental investment to implement the challenger does not have an IRR at least equal to the MARR. In this case, the challenger is excluded from further consideration and the current best is challenged by the investment that is ranked third.
2. The incremental investment to implement the challenger has an IRR at least as high as the MARR. In this case, the challenger replaces the current best. It then is challenged by the alternative ranked third.

The process then continues with the next alternative challenging the current best until all alternatives have been compared. In Figure 5.5, this is indicated by showing $B = n + 1$, implying that the process is complete because there are then no more challengers to the current best alternative. The current best alternative remaining at the end of the process is then selected.

In the next section, the issue of multiple IRRs is discussed, and methods for identifying and eliminating them are given. Note that the process described in Figure 5.5 requires that a single IRR (or ERR, as discussed later) be determined for each incremental investment. If there are multiple IRRs, they must be dealt with for *each* increment of investment.

Example 5.6 (Reprise of Example 4.4)

Fly-by-Night Aircraft must purchase a new lathe. It is considering one of four new lathes, each of which has a life of 10 years with no scrap value. Given a MARR of 15%, which alternative should be chosen?

Lathe	1	2	3	4
First cost	\$100 000	\$150 000	\$200 000	\$255 000
Annual savings	25 000	34 000	46 000	55 000

The alternatives have already been ordered from lathe 1, which has the smallest first cost, to lathe 4, which has the greatest first cost. The current best alternative is lathe 1. Calculating the IRR for alternative 1, although not necessary, is shown as follows:

$$\$100\,000 = \$25\,000(P/A, i^*, 10)$$

$$(P/A, i^*, 10) = 4$$

An approximate IRR is obtained by trial and error with a spreadsheet.

$$i^* \cong 21.4\%$$

The current best alternative and the remaining alternatives with higher first costs are then retained for the "challenger" analysis. The first challenger, lathe 2, has the next highest first cost compared with the current best. The next step is to see if the incremental investment from lathe 1 to lathe 2 is justified. This is done by finding the IRR on the incremental investment:

$$(\$150\,000 - \$100\,000) - (\$34\,000 - \$25\,000)(P/A, i^*, 10) = 0$$

or

$$[\$150\,000 - \$34\,000(P/A, i^*, 10)]$$

$$- [\$100\,000 - \$25\,000(P/A, i^*, 10)] = 0$$

$$(P/A, i^*, 10) = \$50\,000/\$9000 = 5.556$$

An approximate IRR is obtained by trial and error.

$$i^* \cong 12.4\%$$

Since the IRR of the incremental investment falls below the MARR, lathe 2 fails the challenge to become the current best alternative. The reader can verify that lathe 2 alone has an IRR of approximately 18.7%. Even so, lathe 2 is not considered a viable alternative. In other words, the incremental investment of \$50 000 could be put to better use elsewhere. Lathe 1 remains the current best and the next challenger is lathe 3.

As before, the incremental IRR is the interest rate at which the present worth of lathe 3 less the present worth of lathe 1 is 0:

$$[\$200\,000 - \$46\,000(P/A, i^*, 10)]$$
$$- [\$100\,000 - \$25\,000(P/A, i^*, 10)] = 0$$
$$(P/A, i^*, 10) = \$100\,000/\$21\,000 = 4.762$$

An approximate IRR is obtained by trial and error.

$$i^* \cong 16.4\%$$

The IRR on the incremental investment exceeds the MARR, and therefore lathe 3 is preferred to lathe 1. Lathe 3 now becomes the current best. The new challenger is lathe 4. The IRR on the incremental investment is

$$[\$255\,000 - \$55\,000(P/A, i^*, 10)]$$
$$- [\$200\,000 - \$46\,000(P/A, i^*, 10)] = 0$$
$$(P/A, i^*, 10) = \$55\,000/\$9000 = 6.11$$
$$i^* \cong 10.1\%$$

The additional investment from lathe 3 to lathe 4 is not justified. The reader can verify that the IRR of lathe 4 alone is about 17%. Once again, we have a challenger with an IRR greater than the MARR, but it fails as a challenger because the incremental investment from the current best does not have an IRR at least equal to the MARR. The current best remains lathe 3. There are no new challengers, and so the best overall alternative is lathe 3. ■

5.3.3 Multiple IRRs

A problem with implementing the internal rate of return method is that there may be more than one internal rate of return. Consider the following example.

Example 5.7

A project pays \$1000 today, costs \$5000 a year from now, and pays \$6000 in two years. (See Figure 5.6.) What is its IRR?

Equating the present worths of disbursements and receipts and solving for the IRR gives the following:

$$\$1000 - \$5000(P/F, i^*, 1) + \$6000(P/F, i^*, 2) = 0$$

Figure 5.6 Multiple IRR Example

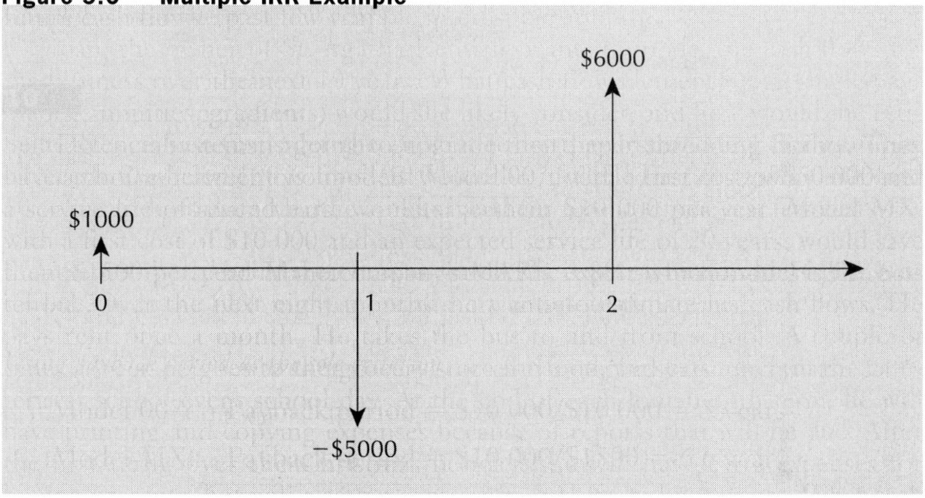

Recalling that $(P/F,i^*,N)$ stands for $\frac{1}{(1 + i^*)^N}$, we have

$$1 - \frac{5}{1 + i^*} + \frac{6}{(1 + i^*)^2} = 0$$

$$(1 + i^*)^2 - 5(1 + i^*) + 6 = 0$$

$$(1 + 2i^* + i^*) - 5i^* + 1 = 0$$

$$i^* - 3i^* + 2 = 0$$

$$(i^* - 1)(i^* - 2) = 0$$

The roots of this equation are $i^*_1 = 1$ and $i^*_2 = 2$. In other words, this project has two IRRs: 100% and 200%! ■

The multiple internal rates of return problem may be stated more generally. Consider a project that has cash flows over T periods. The **net cash flow**, A_t, associated with period t is the difference between cash inflows and outflows for the period (i.e., $A_t = R_t - D_t$ where R_t is cash inflow in period t and D_t is cash outflow in period t). We set the present worth of the net cash flows over the entire life of the project equal to zero to find the IRR(s). We have

$$A_0 + A_1(1 + i)^{-1} + A_2(1 + i)^{-2} + \ldots + A_T(1 + i)^{-T} = 0 \qquad (5.2)$$

Any value of i that solves Equation (5.2) is an internal rate of return for that project. That there may be multiple solutions to Equation (5.2) can be seen if we rewrite the equation as

$$A_0 + A_1x + A_2x^2 + \ldots + A_Tx^T = 0 \tag{5.3}$$

where $x = (1 + i)^{-1}$

Solving the Tth degree polynomial of Equation (5.3) is the same as solving for the internal rates of return in Equation (5.2). In general, when finding the roots of Equation (5.3), there may be as many positive real solutions for x as there are sign changes in the coefficients, the As. Thus, there may be as many IRRs as there are sign changes in the As.

We can see the meaning of multiple roots most easily with the concept of **project balance**. If a project has a sequence of net cash flows $A_0, A_1, A_2, \ldots, A_T$, and the interest rate is i', there are $T + 1$ project balances, $B_0, B_1, B_2, \ldots, B_T$, one at the end of each period t, $t = 0, 1, \ldots, T$. A project balance, B_t, is the accumulated future value of all cash flows, up to the end of period t, compounded at the rate, i'. That is,

$$B_0 = A_0$$

$$B_1 = A_0(1 + i') + A_1$$

$$B_2 = A_0(1 + i')^2 + A_1(1 + i') + A_2$$

$$\vdots$$

$$B_T = A_0(1 + i')^T + A_1(1 + i')^{T-1} + \ldots + A_T$$

Table 5.1 shows the project balances at the end of each year for both 100% and 200% interest rates for the project in Example 5.7. The project starts with a cash inflow of $1000. At a 100% interest rate, the $1000 increases to $2000 over the first year. At the end of the first year, there is a $5000 disbursement, leaving a negative project balance of $3000. At 100% interest, this negative balance increases to $6000 over the second year. This negative $6000 is offset exactly by the $6000 inflow. This makes the project balance zero at the end of the second year. The project balance at the end of the project is the future worth of all the cash flows in the project. When the future worth at the end of the project life is zero, the present worth is also zero. This verifies that the 100% is an IRR.

Now consider the 200% interest rate. Over the first year, the $1000 inflow increases to $3000. At the end of the first year, $5000 is paid out, leaving a negative project balance of $2000. This negative balance grows at 200% to $6000 over the second year. This is offset exactly by the $6000 inflow so that the project balance is zero at the end of the second year. This verifies that the 200% is also an IRR!

Table 5.1 Project Balances for Example 5.7

End of Year	At $i' = 100\%$	At $i' = 200\%$
0	\$1000	\$1000
1	\$1000(1 + 1) − \$5000 = −\$3000	\$1000(1 + 2) − \$5000 = −\$2000
2	−\$3000(1 + 1) + \$6000 = 0	−\$2000(1 + 2) + \$6000 = 0

Looking at Table 5.1, it's actually fairly obvious that an important assumption is being made about the initial \$1000 received. *The IRR computation implicitly assumes that the \$1000 is invested during the first period at either 100% or 200%, one of the two IRRs*. During the first period, the project is not an investment. The project balance is positive. The project is *providing* money, not using it. This money is not reinvested immediately in the project. It is simply cash on hand. The \$1000 must be invested elsewhere for one year if it is to earn any return. It is unlikely, however, that the \$1000 provided by the project in this example would be invested in something else at 100% or 200%. More likely, it would be invested at a rate at or near the company's MARR.

5.3.4 External Rate of Return Methods

To resolve the multiple IRR difficulty, we need to consider what return is earned by money associated with a project that is not invested in the project. The usual assumption is that the funds are invested elsewhere and earn an *explicit rate of return* equal to the MARR. This makes sense, because when there is cash on hand that is not invested in the project under study, it will be used elsewhere. These funds would, by definition, gain interest at a rate at least equal to the MARR. The **external rate of return (ERR)**, denoted by i^*_e, is the rate of return on a project where any cash flows that are not invested in the project are assumed to earn interest at a predetermined explicit rate (usually the MARR). For a given explicit rate of return, a project can have only one value for its ERR.

It is possible to calculate a precise ERR that is comparable to the IRRs of other projects using an explicit interest rate exactly when necessary. Because the cash flows of Example 5.7 are fairly simple, let us use them to illustrate how to calculate the ERR precisely.

Example 5.8 (Example 5.7 Revisited (ERR))

A project pays \$1000 today, costs \$5000 a year from now, and pays \$6000 in two years. What is its rate of return? Assume that the MARR is 25%.

The first \$1000 is not invested immediately in the project. Therefore, we assume that it is invested outside the project for one year at the MARR. Thus, the cumulative cash flow for year 1 is

$$\$1000(F/P, 25\%, 1) - \$5000 = \$1250 - \$5000 = -\$3750$$

With this calculation, we transform the cash-flow diagram representing this problem from that in Figure 5.7(a) to that in Figure 5.7(b). These cash flows provide a single (precise) ERR, as follows:

$$-\$3750 + \$6000(P/F, i^*_e, 1) = 0$$

$$(P/F, i^*_e, 1) = \$3750/\$6000 = 0.625$$

$$\frac{1}{1+i^*_e} = 0.625$$

$$1+i^*_e = \frac{1}{0.625} = 1.6$$

$$i^*_e = 0.6$$

$$ERR = 60\%$$ ■

In general, computing a precise ERR can be a complex procedure because of the difficulty in determining exactly when the explicit interest rate should be applied. In order to do such a calculation, project balances have to be computed for trial ERRs. In periods in which project balances are positive for the trial ERR, the project is a source of funds. These funds would have to be invested outside the project at the MARR. During periods when the project balance is negative for the trial ERR, any receipts would be invested in the

Figure 5.7 Multiple IRR Solved

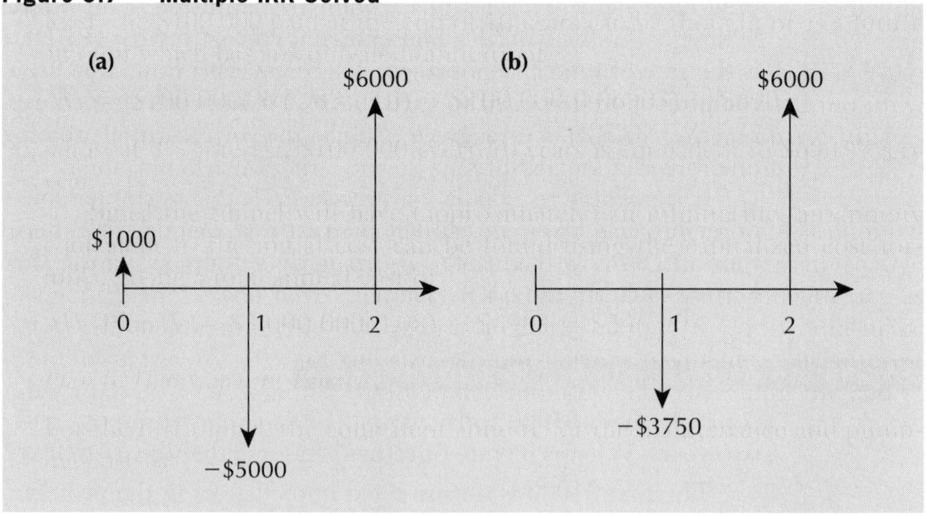

project and will typically yield more than the MARR. Whether the project balances are negative or positive will depend on the trial ERRs. This implies that the calculation process requires much experimenting with trial ERRs before an ERR is found that makes the future worth zero. A more convenient, but approximate, method is to use the following procedure:

1. Take all *net* receipts forward at the MARR to the time of the last cash flow.
2. Take all *net* disbursements forward at an unknown interest rate, i^*_{ea}, also to the time of the last cash flow.
3. Equate the future value of the receipts from Step 1 to the future value of the disbursements from Step 2 and solve for i^*_{ea}.
4. The value for i^*_{ea} is the *approximate ERR* for the project.

Example 5.9 (Example 5.7 Revisited Again (An Approximate ERR))

To approximate the ERR, we compute the interest rate that gives a zero future worth at the end of the project when all receipts are brought forward at the MARR. In Example 5.7, the \$1000 is thus assumed to be reinvested at the MARR for two years, the life of the project. The disbursements are taken forward to the end of the two years at an unknown interest rate, i^*_{ea}. With a MARR of 25%, the revised calculation is

$$\$1000(F/P, 25\%, 2) + \$6000 = \$5000(F/P, i^*_{ea}, 1)$$

$$(F/P, i^*_{ea}, 1) = [\$1000(1.5625) + \$6000]/\$5000$$

$$(F/P, i^*_{ea}, 1) = 1.5125$$

$$1 + i^*_{ea} = 1.5125$$

$$i^*_{ea} = 0.5125 \textit{ or}$$

$$i^*_{ea} = 51.25\%$$

$$ERR \cong 51\% \blacksquare$$

The ERR calculated using this method is an approximation, since all receipts, not just those that occur when the project balance is positive, are assumed to be invested at the MARR. Note that the precise ERR of 60% is different from the approximate ERR of 51%. Fortunately, it can be shown that the approximate ERR will always be between the precise ERR and the MARR. This means that whenever the precise ERR is above the MARR, the approximate ERR will also be above the MARR and whenever the precise ERR is below the MARR, the approximation will be below the MARR as well. This implies that using the approximate ERR will always lead to the correct decision. It should also be noted that an acceptable project will earn *at least* the rate given by the approximate ERR. Therefore, even though an approximate

ERR is inaccurate, it is often used in practice because it provides the correct decision as well as a lower bound on the return on an investment while being easy to calculate.

5.3.5 When to Use the ERR

The ERR (approximate or precise) must be used whenever there are multiple IRRs possible. Unfortunately, it can be difficult to know in advance whether there will be multiple IRRs. On the other hand, it is fortunate that most ordinary projects have a structure which precludes multiple IRRs.

Most projects consist of one or more periods of outflows at the start, followed only by one or more periods of inflows. Such projects are called **simple investments**. The cash flow diagram for a simple investment takes the general form shown in Figure 5.8. In terms of Equations (5.2) and (5.3), there is only one change of sign, from negative to positive in the *A*s, the sequence of coefficients. Hence, a simple investment always has a unique IRR.

If a project is not a simple investment, there may or may not be multiple IRRs — there is no way of knowing for sure without further analysis. In practice, it may be reasonable to use an approximate ERR whenever the project is not a simple investment, since it can be used to evaluate any project. Recall from Section 5.3.4 that the approximate ERR will always provide a correct decision, whether its use is required or not, since it will understate the true rate of return.

However, it is generally desirable to compute an IRR whenever it is possible to do so, and to use an approximate ERR only when there may be multiple IRRs. In this way, the computations will be as accurate as possible. When it is desirable to know for sure whether there will be only one IRR, there are several steps of analysis that can be undertaken. These are covered in detail in Appendix 5A.

Figure 5.8 The General Form of Simple Investments

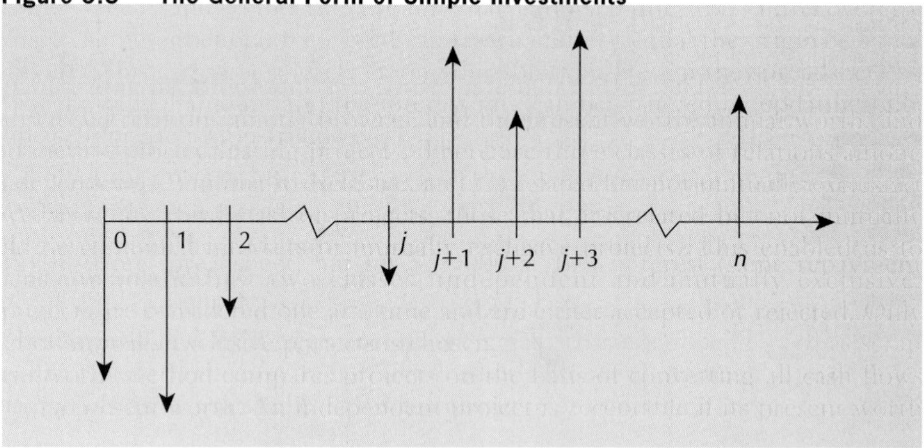

To reiterate, the approximate ERR can be used to evaluate any project, whether it is a simple investment or not. However, the approximate ERR will tend to be a less accurate rate than the IRR. The inaccuracy will tend to be similar for projects with cash flows of a similar structure, and either method will result in the same decision in the end.

5.4 Rate of Return and Present/Annual Worth Methods Compared

A comparison of rate of return and present/annual worth methods leads to two important conclusions:

1. The two sets of methods, when properly used, give the same decisions.
2. Each set of methods has its own advantages and disadvantages.

Let us consider each of these conclusions in more detail.

5.4.1 Equivalence of Rate of Return and Present/Annual Worth Methods

If an independent project has a unique IRR, the IRR method and the present worth method give the same decision. Consider Figure 5.9. It shows the present worth as a function of the interest rate for a project with a unique IRR. The maximum of the curve lies at the vertical axis (where the interest rate = 0) at the

Figure 5.9 Present Worth (PW) as a Function of Interest Rate (*i*) for a Simple Investment

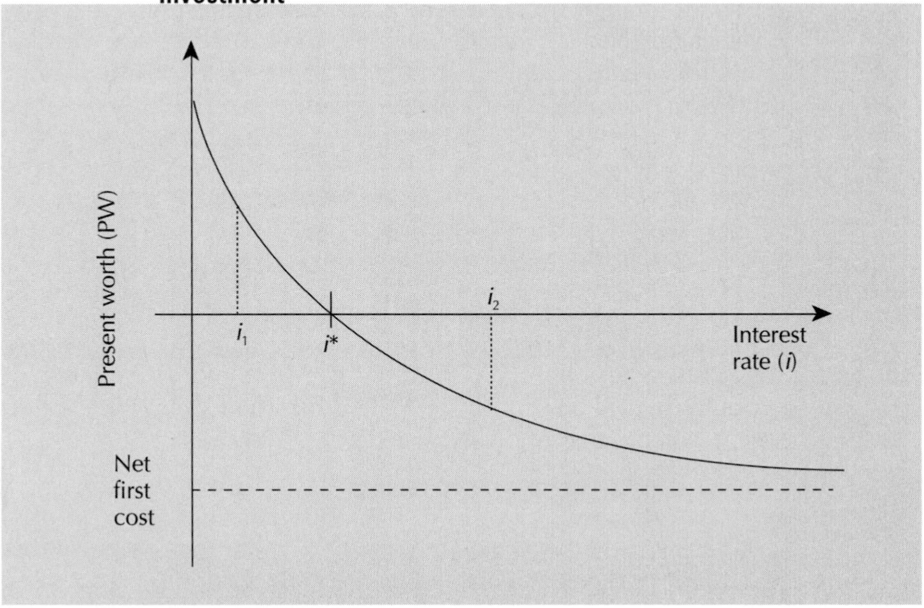

point given by the sum of all undiscounted net cash flows. (We assume that the sum of all the undiscounted net cash flows is positive.) As the interest rate increases, the present worth of all cash flows after the first cost decreases. Therefore, the present worth curve slopes down to the right. To determine what happens as the interest rate increases indefinitely, let us recall the general equation for present worth

$$\mathrm{PW} = \sum_{t=0}^{T} A_t(1+i)^{-t} \tag{5.1}$$

where

i = the interest rate
A_t = the net cash flow in period t
T = the number of periods

Letting $i \to \infty$, we have

$$\lim_{i \to \infty} \frac{1}{(1+i)^t} = 0$$

Therefore, as the interest rate becomes indefinitely large, all terms in Equation (5.1) approach zero except the first term (where $t = 0$), which remains at A_0. In Figure 5.9, this is shown by the asymptotic approach of the curve to the first cost, which, being negative, is below the horizontal axis.

The interest rate at which the curve crosses the horizontal axis (i^* in Figure 5.9), where the present worth is zero, is, by definition, the IRR.

To demonstrate the equivalence of the rate of return and the present/annual worth methods for decision making, let us consider possible values for the MARR. First, suppose the MARR = i_1, where $i_1 < i^*$. In Figure 5.9, this MARR would lie to the left of the IRR. From the graph we see that the present worth is positive at i_1. In other words, we have

$$\mathrm{IRR} > \mathrm{MARR}$$

and

$$\mathrm{PW} > 0$$

Thus, in this case, both the IRR and PW methods lead to the same conclusion: Accept the project.

Second, suppose the MARR = i_2, where $i_2 > i^*$. In Figure 5.9, this MARR would lie to the right of the IRR. From the graph we see that, at i_2, the present worth is negative. Thus we have

$$\mathrm{IRR} < \mathrm{MARR}$$

and

$$\mathrm{PW} < 0$$

Here, too, both the IRR and PW methods lead to the same conclusion: Reject the project.

Now consider two simple, mutually exclusive projects, A and B, where the first cost of B is greater than the first cost of A. If the increment from A to B has a unique IRR, then we can readily demonstrate that the IRR and PW methods lead to the same decision. See Figure 5.10(a), which shows the present worths of projects A and B as a function of the interest rate. Since the first cost of B is greater than that of A, the curve for project B asymptotically approaches a lower present worth than does the curve for project A as the interest rate becomes indefinitely large, and thus the two curves must cross at some point.

Figure 5.10 Present Worth (PW) as a Function of Interest Rate (*i*) for Two Simple, Mutually Exclusive Projects

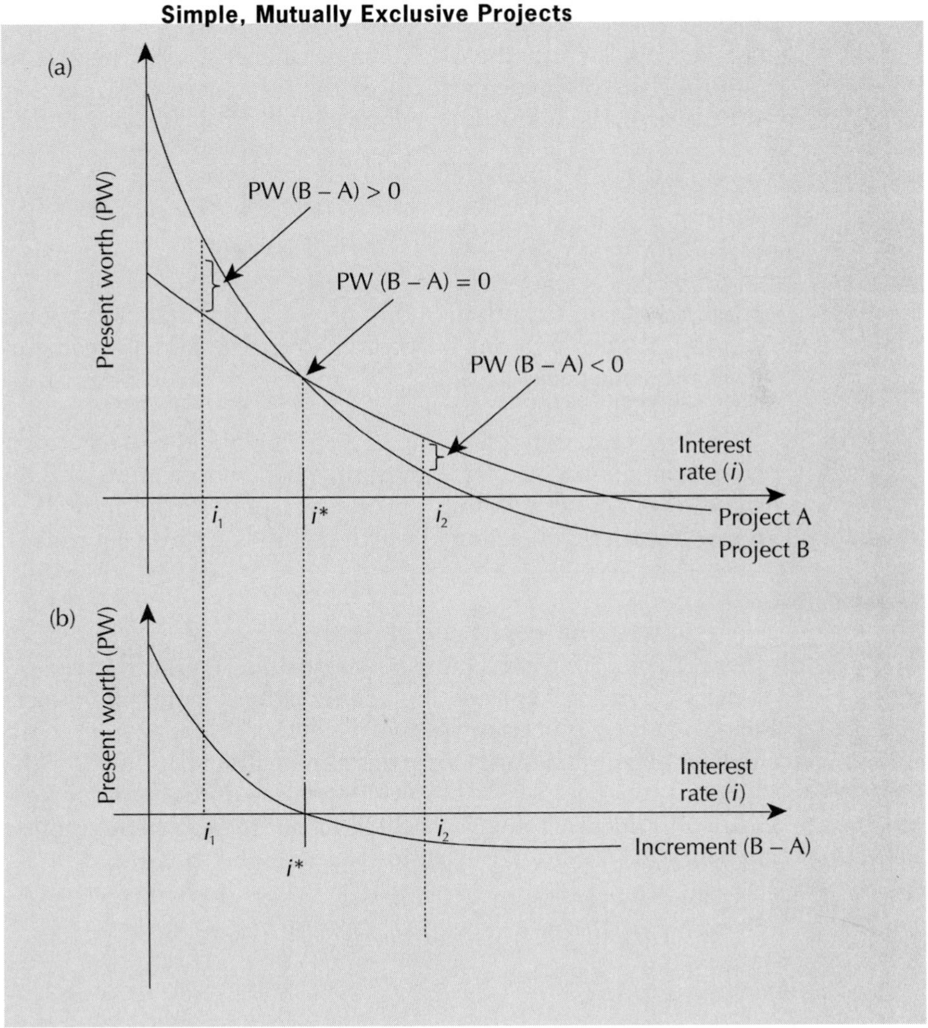

To apply the IRR method, we must consider the increment (denoted by B–A). The present worth of the increment (B–A) will be zero where the two curves cross. This point of intersection is marked by the interest rate, i^*. We have plotted the curve for the increment (B–A) in Figure 5.10(b) to clarify the relationships.

Let us again deal with possible values of the MARR. First, suppose the MARR (i_1) is less than i^*. Then, as we see in Figure 5.10(b), the present worth of (B–A) is positive at i_1. That is, the following conditions hold:

$$\text{IRR(B–A)} > \text{MARR}$$

and

$$\text{PW(B–A)} > 0$$

Thus, according to both the IRR method and the PW method, project B is better than project A.

Second, suppose the MARR = i_2, where $i_2 > i^*$. Then we see from Figure 5.10(b) that the present worth of the increment (B–A) is negative at i_2. In other words, the following conditions hold:

$$\text{IRR(B–A)} < \text{MARR}$$

and

$$\text{PW(B–A)} < 0$$

Thus, according to both methods, project A is better than project B.

In a similar fashion, we could show that the approximate ERR method gives the same decisions as the PW method in those cases where there may be multiple IRRs.

We already noted that the annual worth and present worth methods are equivalent. Therefore, by extension, our demonstration of the equivalence of the rate of return methods and the present worth methods means that the rate of return and the annual worth methods are also equivalent.

Example 5.10

Bracebridge Enterprises operates a resort in the tourist area north of Toronto, Ontario. They are considering adding either a parasailing operation or canoe rentals to their other activities. Available space limits them to one of these two choices. The initial costs for parasailing will be \$100 000, with net returns of \$15 000 for the 15-year life of the project. Initial costs for canoeing will be \$10 000, with net returns of \$2000 for its 15-year life. Assume that both projects have a \$0 salvage value after 15 years, and the MARR is 10%.

(a) Using present worth analysis, which alternative is better?

(b) Using IRR, which alternative is better?

(a) The present worths of the two projects are calculated as follows:

$$\begin{aligned} PW_{para} &= -\$100\ 000 + \$15\ 000(P/A, 10\%, 15) \\ &= -\$100\ 000 + \$15\ 000(7.6061) \\ &= \$14\ 091.50 \end{aligned}$$

$$\begin{aligned} PW_{can} &= -\$10\ 000 + \$2000(P/A, 10\%, 15) \\ &= -\$10\ 000 + \$2000(7.6061) \\ &= \$5212.20 \end{aligned}$$

The parasailing venture has a higher present worth and is thus preferred.

(b) The IRRs of the two projects are calculated as follows:

Parasailing

$$\$100\ 000 = \$15\ 000(P/A, i^*, 15)$$

$$(P/A, i^*, 15) = \$100\ 000/\$15\ 000 = 6.67 \Rightarrow i^*_{para} = 12.4\%$$

Canoeing

$$\$10\ 000 = \$2000(P/A, i^*, 15)$$

$$(P/A, i^*, 15) = 5 \Rightarrow i^*_{can} = 18.4\%$$

One might conclude that, because IRR_{can} is larger, Bracebridge Enterprises should invest in the canoeing project, but this is *wrong*. When done correctly, a present worth analysis and an IRR analysis will always agree. The error here is that the parasailing project was assessed without consideration of the increment from the canoeing project. Checking the IRR of the increment:

$$(\$100\ 000 - \$10\ 000) = (\$15\ 000 - \$2000)(P/A, i^*, 15)$$

$$(P/A, i^*, 15) = \$90\ 000/\$13\ 000 = 6.923 \Rightarrow i^*_{can-para} = 11.7\%$$

Since the increment from the canoeing project also exceeds the MARR, the larger parasailing project should be taken. ■

5.4.2 Why Choose One Method over the Other?

Although rate of return methods and present worth/annual worth methods give the same decisions, each set of methods has its own advantages and disadvantages. The choice of method may depend on the way the results are to be used and the sort of data the decision makers prefer to consider. In fact, many companies, by policy, require that several methods be applied so that a more complete picture of the situation is presented. A summary of the advantages and disadvantages of each method is given in Table 5.2.

Rate of return methods state results in terms of *rates*, while present/annual worth methods state results in absolute figures. Many managers prefer rates to absolute figures because rates facilitate direct comparisons of projects whose sizes are quite different. For example, a petroleum company comparing performances of a refining division and a distribution division would not look at the typical values of present or annual worth for projects in the two divisions. A refining project may have first costs in the range of hundreds of *millions* of dollars, while distribution projects may have first costs in the range of *thousands* of dollars. It would not be meaningful to compare the absolute profits between a refining project and a distribution project. The absolute profits of refining projects will almost certainly be larger than those of distribution projects. Expressing project performance in terms of rates of return permits understandable comparisons. A disadvantage of rate of return methods, however, is the possible complication that there may be more than one rate of return. Under such circumstances, it is necessary to calculate an ERR.

In contrast to a rate of return, a present or annual worth computation gives a direct measure of the profit provided by a project. A company's main goal is likely to earn profits for its owners. The present and annual worth methods state the contribution of a project toward that goal. Another reason that managers prefer these methods is that present worth and annual worth methods are typically easier to apply than rate of return methods.

For completeness of coverage, we note that the Payback Period method may not give results consistent with rate of return or present/annual worth methods as it ignores the time value of money and the service life of projects. It is, however, a method commonly used in practice due to its ease of use and intuitive appeal.

Table 5.2 Advantages and Disadvantages of Comparison Methods

Method	Advantages	Disadvantages
IRR	Facilitates comparisons of projects of different sizes Commonly used	Relatively difficult to calculate Multiple IRRs may exist
Present worth	Gives explicit measure of profit contribution	Difficult to compare projects of different sizes
Annual worth	Annual cash flows may have familiar meanings to decision makers	Difficult to compare projects of different sizes
Payback period	Very easy to calculate Commonly used Takes into account the need to have capital recovered quickly	Discriminates against long-term projects Ignores time value of money Ignores the expected service life

Example 5.11

Each of the following scenarios suggests a best choice of comparison method:

1. Edward has his own small firm that will lease injection-moulding equipment to make polyethylene containers. He must decide on the specific model to lease. He has estimates of future monthly sales.

   The annual worth method makes sense here, because Edward's cash flows, including sales receipts and leasing expenses, will probably all be on a monthly basis. As a sole proprietor, Edward need not report his conclusions to others.

2. Ramesh works for a large power company and must assess the viability of locating a transformer station at various sites in the city. He is looking at the cost of the building lot, power lines, and power losses for the various locations. He has fairly accurate data about costs and future demand for electricity.

   As part of a large firm, Ramesh will probably be obliged to use a specific comparison method. This would probably be IRR. A power company makes many large and small investments, and the IRR method allows them to be compared fairly. Ramesh has the data necessary for the IRR calculations.

3. Sehdev must buy a relatively inexpensive log splitter for his agricultural firm. There are several different types that require a higher or lower degree of manual assistance. He has only rough estimates of how this machine will affect future cash flows.

   This relatively inexpensive purchase is a good candidate for the payback period method. The fact that it is inexpensive means that extensive data gathering and analysis are probably not warranted. Also, since future cash flows are relatively uncertain, there is no justification for using a particularly precise comparison method.

4. Ziva will be living in Inuvik for six months, testing her company's equipment under hostile weather conditions. She needs a field office and must determine which of the following choices is economically best: (1) renting space in an industrial building; (2) buying and outfitting a trailer; (3) renting a hotel room for the purpose.

   For this decision, a present worth analysis would be appropriate. The cash flows for each of the alternatives are of different types, and bringing them to present worth would be a fair way to compare them. It would also provide an accurate estimate to Ziva's firm of the expected cost of the remote office for planning purposes. ■

REVIEW PROBLEMS

REVIEW PROBLEM 5.1

Wei-Ping's consulting firm needs new quarters. A downtown office building is ideal. The company can either buy or lease it. To buy the office building will cost \$6 000 000. If the building is leased, the lease fee is \$400 000 payable at the beginning of each year. In either case, the company must pay city taxes, maintenance, and utilities.

Wei-Ping figures that the company needs the office space for only 15 years. Therefore, they will either sign a 15-year lease or buy the building. If they buy the building, they will then sell it after 15 years. The value of the building at that time is estimated to be \$15 000 000.

What rate of return will Wei-Ping's firm receive by buying the office building instead of leasing it?

ANSWER

The rate of return can be calculated as the IRR on the incremental investment necessary to buy the building rather than lease it.

The IRR on the incremental investment is found by solving for i^* in

$$(\$6\,000\,000 - \$400\,000) - \$15\,000\,000(P/F, i^*, 15)$$
$$= \$400\,000(P/A, i^*, 14)$$

$$4(P/A, i^*, 14) + 150(P/F, i^*, 15) = 56$$

For $i^* = 11\%$, the result is

$$4(P/A, 11\%, 14) + 150(P/F, 11\%, 15)$$
$$= 4(6.9819) + 150(0.20900)$$
$$= 59.2781$$

For $i^* = 12\%$,

$$4(P/A, 12\%, 14) + 150(P/F, 12\%, 15)$$
$$= 4(6.6282) + 150(0.1827)$$
$$= 53.9171$$

A linear interpolation between 11% and 12% gives the IRR

$$i^* = 11\% + (59.2781-56)/(59.2781-53.9171) = 11.6115\%$$

By investing their money in buying the building rather than leasing, Wei-Ping's firm is earning an IRR of about 11.6%. ■

REVIEW PROBLEM 5.2

The Real S. Tate Company is considering investing in one of four rental properties. Real S. Tate will rent out whatever property they buy for four years and then sell it at the end of that period. The following data concerning the properties are available:

Rental Property	Purchase Price	Net Annual Rental Income	Sale Price at the End of Four Years
1	\$100 000	\$ 7 200	\$100 000
2	120 000	9 600	130 000
3	150 000	10 800	160 000
4	200 000	12 000	230 000

Based on the purchase prices, rental incomes, and sale prices at the end of the four years, answer the following questions:

(a) Which property, if any, should Tate invest in? Real S. Tate uses a MARR of 8% for projects of this type.

(b) Construct a graph which depicts the present worth of each alternative as a function of interest rates ranging from 0% to 20%. (A spreadsheet would be helpful in answering this part of the problem.)

(c) From your graph, determine the range of interest rates for which your choice in part (a) is the best investment. If the MARR were 9%, which rental property would be the best investment? Comment on the sensitivity of your choice to the MARR used by the Real S. Tate Company.

ANSWER

(a) Since the "do nothing" alternative is feasible and it has the least first-cost, it become the current best alternative. The IRR on the incremental investment for property 1 is given by:

$$-\$100\,000 + \$100\,000(P/F, i^*, 4) + \$7200(P/A, i^*, 4) = 0$$

The IRR on the incremental investment is 7.2%. Because this is less than the MARR of 8%, property 1 is discarded from further consideration. The IRR for the incremental investment for property 2, the alternative with the next highest first cost, is found by solving for i^* in

$$-\$120\,000 + \$130\,000(P/F, i^*, 4) + \$9600(P/A, i^*, 4) = 0$$

The interest rate that solves the above equation is 9.8%. Since an IRR of 9.8% exceeds the MARR, property 2 becomes the current best alternative. Now the incremental investments over and above the first cost of property 2 are analyzed. First, property 3 challenges the current best. The IRR in the incremental investment to property 3 is

$$(-\$150\,000 + \$120\,000) + (\$160\,000 - \$130\,000)(P/F, i^*, 4)$$
$$+ (\$10\,800 - \$9600)(P/A, i^*, 4) = 0$$

$$-\$30\,000 + \$30\,000(P/F, i^*, 4) + \$1200(P/A, i^*, 4) = 0$$

This gives an IRR of only 4%, which is below the MARR. Property 2 remains the current best alternative and property 3 is discarded. Next, property 4 challenges the current best. The IRR on the incremental investment from property 2 to property 4 is

$$(-\$200\,000 + \$120\,000) + (\$230\,000 - \$130\,000)(P/F, i^*, 4) + (\$12\,000 - \$9600)(P/A, i^*, 4) = 0$$

$$-\$80\,000 + \$100\,000(P/F, i^*, 4) + \$2400(P/A, i^*, 4) = 0$$

The IRR on the incremental investment is 8.5%, which is above the MARR. Property 4 becomes the current best choice. Since there are no further challengers, the choice based on IRR is the current best, property 4.

(b) The graph for part (b) is shown in Figure 5.11.

(c) From the graph, one can see that, provided the MARR is between 0% and 8.5%, property 4 is the best alternative. This is the range of interest rates over which property 4 has the largest present worth.

If the MARR is 9%, the best alternative is property 2. This can be seen by going back to the original IRR computations and observing that the results of the analysis are essentially the same, except that the incremental investment from property 2 to property 4 no longer has a return exceeding the MARR. This can be confirmed from the diagram (Figure 5.11) as well, since the property with the largest present worth at 9% is property 2.

With respect to sensitivity analysis, the graph shows that, for a MARR between 0% and 8.5%, property 4 is the best choice and, for a MARR between 8.5% and 9.8%, property 2 is the best choice. If the MARR is

Figure 5.11 Present Worths for Review Problem 5.2

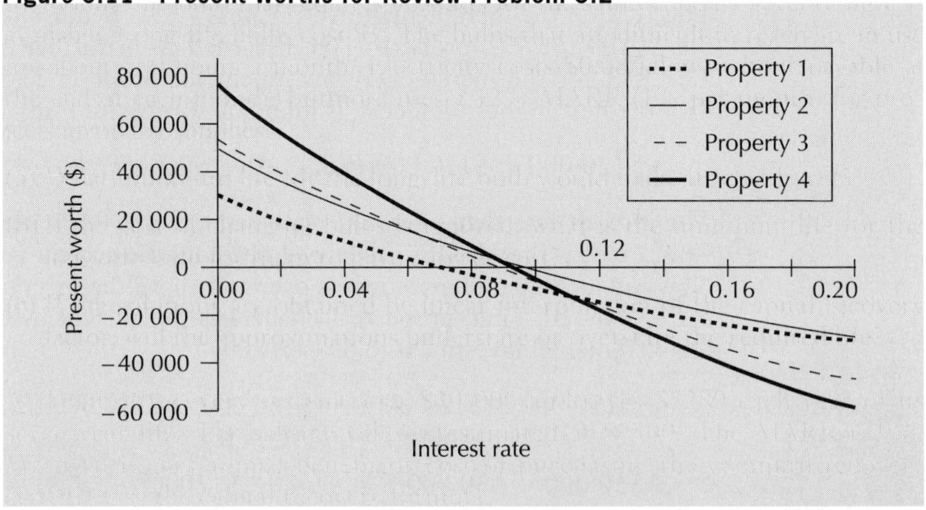

above 9.8%, no property has an acceptable return on investment, and the "do nothing" alternative would be chosen. ■

REVIEW PROBLEM 5.3

You are in the process of arranging a marketing contract for a new browser that will allow easy access to the "web." It still needs more development, so your contract will pay you \$5000 today to finish the prototype. You will then get royalties of \$10 000 at the end of each of the second and third years. At the end of each of the first and fourth years, you will be required to spend \$20 000 and \$10 000 in upgrades, respectively. What is the (approximate) ERR on this project, assuming a MARR of 20%? Should you accept the contract ?

ANSWER

To calculate the approximate ERR, set

$$\text{FW(receipts @ MARR)} = \text{FW(disbursements @ ERR)}$$

$$\$5000(F/P, 20\%, 4) + \$10\,000(F/P, 20\%, 2) + \$10\,000(F/P, 20\%, 1) = \$20\,000(F/P, i^*_{ea}, 3) + \$10\,000$$

$$\$5000(2.0736) + \$10\,000(1.44) + \$10\,000(1.2) = \$20\,000(F/P, i^*_{ea}, 3) + \$10\,000$$

$$(F/P, i^*_{ea}, 3) = 1.3384$$

At ERR = 10%, $(F/P, i, 3) = 1.3310$

At ERR = 11%, $(F/P, i, 3) = 1.3676$

Interpolating:

$$\begin{aligned} i^*_{ea} &= 10\% + (1.3384 - 1.3310)(11 - 10)/(1.3676 - 1.3310) \\ &= 10\% + 0.0074/0.0366 \cong 10.2\% \end{aligned}$$

The (approximate) ERR is 10.2%. Since this is below the MARR of 20%, the contract should not be accepted. ■

SUMMARY

This chapter presented the IRR method for evaluating projects and also discussed the relationship among the present worth, annual worth, payback period, and IRR methods.

The IRR method consists of determining the rate of return for a sequence of cash flows. For an independent project, the calculated IRR is compared with a MARR, and if it is equal to or exceeds the MARR it is an acceptable project. To determine the best project of several mutually exclusive ones, it is necessary to determine the IRR of each increment of investment.

The IRR selection procedure is complicated by the possibility of having more than one rate of return because of a cash flow structure that, over the course of a project, requires that capital, eventually invested in the project at some point, be invested externally. Under such circumstances, it is necessary to calculate an ERR.

The present worth and annual worth methods are closely related, and both give results identical to those of the IRR method. Rate of return measures are readily understandable, especially when comparing projects of unequal sizes, whereas present/annual worth measures give an explicit expression of the profit contribution of a project. The main advantage of the payback period method is that it is easy to implement and understand, and takes into account the need to have capital recovered quickly.

Engineering Economics in Action, Part 5B: The Invisible Hand

"Hello." Dave's voice was clear enough over the phone that he could have been in his office down the hall, but Naomi could tell from his relaxed tone that the office was not on his mind.

"Hi, Dave, it's Naomi. Can I bend your ear about that drop forge project again?"

"Oh, hi, Naomi. Sure, what have you got?"

"Well, as I see it, IRR has got to be the way to go. Of course, present worth or annual worth will give the same answer, but I'm sure Ed Burns and Anna Kulkowski would prefer IRR. They have to compare potential investments across different parts of the organization. It's kind of hard to compare net present worths of investments in information systems, where you rarely get above a first cost of $100 000, with forge investments where you can easily get up to a few hundred thousand first cost. And, as I said before, the drop forge operation isn't one in which the annual cost has any particular significance."

There was a short pause. Naomi suddenly regretted speaking as if she was so sure of herself — but, darn it, she was sure on this one.

"Exactly right," Dave replied. Naomi could feel an invisible hand pat her on the back. "So how exactly would you proceed?"

"Well, I have the options ranked by first cost. The first one is just refurbishing the existing machine. There is no test on that one unless we are willing to stop making our own parts, and Clem told me that was out . . ."

Dave interjected with, "You don't mean that you're automatically going to refurbish the existing machine, do you?"

"No, no. The simple refurbishing option is the base. I then go to the next option which is to refurbish the drop forging hammer and replace the material-handling system. I compare this with the just-refurbish option by looking at the incremental first cost. I will check to see if the additional first cost has an IRR of at least 15% after tax, which, Clem tells me, is the minimum acceptable rate of return. If the incremental first cost has an IRR of at least 15%, the combination of refurbishing and replacing the material-handling system is better than just refurbishing. I then consider the next option, which is to buy the manually operated mechanical press with no change in material handling. I look at the incremental investment here and see if its IRR is at least 15%. To go back a step, if the IRR of replacing material handling plus refurbishing the old machine did not pay off at 15%, I would have rejected that and compared the manually operated mechanical press with the first option, just refurbishing the old machine. I then work my way, option by option, up to the seventh. How does that sound?"

"Well, that sounds great, as far as it goes. Have you checked for problems with multiple IRRs?"

"Well, so far each set of cash flows has been a simple investment, but I will be careful."

"I would also compute payback periods for them in case we are having cash flow problems. They may not necessarily take an option, even if its incremental IRR is above their 15% MARR, if the payback is too long."

Naomi considered this for a second. "One other question, Dave. What should I do about intangibles?"

"You mean the noise from the forging hammer?"

"Yes. It's important, but you can't evaluate it in dollars and cents."

"Just remind them of it in your report. If they want a more formal analysis, they'll come back to you."

"Thanks, Dave. You've been a big help."

As Naomi hung up the phone, she couldn't help smiling ruefully to herself. She had ignored the payback period altogether — after all, it didn't take either interest or service life into account. "I guess that's what they call practical experience," she said to herself as she got out her laptop.

PROBLEMS

5.1 What is the IRR for a $1000 investment that returns $200 at the end of each of the next

(a) 7 years?

(b) 6 years?

(c) 100 years?

(d) 2 years?

5.2 New windows are expected to save $400 per year in energy costs over their 30-year life for Nottawasaga Fabricating. At an initial cost of $8000 and zero salvage value, are they a good investment? NF's MARR is 8%.

5.3 An advertising campaign will cost $2 000 000 for planning and $400 000 in each of the next six years. It is expected to increase revenues permanently by $400 000 per year. Additional revenues will be gained in the pattern of an arithmetic gradient with $200 000 in the first year, declining by $50 000 per year to zero in the fifth year. What is the IRR of this investment? If the company's MARR is 12%, is this a good investment?

5.4 Aline has three contracts from which to choose. The first contract will require an outlay of $100 000 but will return $150 000 one year from now. The second contract requires an outlay of $200 000 and will return $300 000 one year from now. The third contract requires an outlay of $250 000 and will return $355 000 one year from now. Only one contract can be accepted. If her MARR is 20%, which one should she choose?

5.5 Refer to Review Problem 4.3. Assuming the four investments are independent, use the IRR method to select which, if any, should be chosen. Use a MARR of 8%.

5.6 Antigonish Footwear can invest in one of two different automated clicker cutters. The first, A, has a $10 000 first cost. A similar one with many extra features, B, has a $40 000 first cost. A will save $5000 per year over the cutter now in use. B will save $15 000 per year. Each clicker cutter will last five years. If the MARR is 10%, which alternative is better? Use an IRR comparison.

5.7 CB Electronix must buy a piece of equipment to place electronic components on the printed circuit boards it assembles. The proposed equipment has a 10-year life with no scrap value.

The supplier has given CB several purchase alternatives. The first is to purchase the equipment for $850 000. The second is to pay for the equipment in 10 equal instalments of $135 000 each, starting one year from now. The third is to pay $200 000 now and $95 000 at the end of each year for the next 10 years.

(a) Which alternative should CB choose if their MARR is 11% per year? Use an IRR comparison approach.

(b) Below what MARR does it make sense for CB to buy the equipment now for $850 000?

5.8 The following table summarizes information for four projects:

Project	First Cost	IRR on Overall Investment	IRR on Increments of Investment Compared with Project 1	2	3
1	$100 000	19%			
2	175 000	15%	9%		
3	200 000	18%	17%	23%	
4	250 000	16%	12%	17%	13%

The data can be interpreted in the following way: The IRR on the incremental investment between project 4 and project 3 is 13%.

(a) If the projects are independent, which projects should be undertaken if the MARR is 16%?

(b) If the projects are mutually exclusive, which project should be undertaken if the MARR is 15%? Indicate what logic you have used.

(c) If the projects are mutually exclusive, which project should be undertaken if the MARR is 17%? Indicate what logic you have used.

5.9 There are several mutually exclusive ways Rimouski Dairy can meet a requirement for a filling machine for their creamer line. One choice is to buy a machine. This would cost \$65 000 and last for six years with a salvage value of \$10 000. Alternatively, they could contract with a packaging supplier to get a machine free. In this case, the extra costs for packaging supplies would amount to \$15 000 per year over the six-year life (after which the supplier gets the machine back with no salvage value for Rimouski). The third alternative is to buy a used machine for \$30 000 with zero salvage value after six years. The used machine has extra maintenance costs of \$3000 in the first year, increasing by \$2500 per year. In all cases, there are installation costs of \$6000 and revenues of \$20 000 per year. Using the IRR method, determine which is the best alternative. The MARR is 10%.

5.10 Project X has an IRR of 16% and a first cost of \$20 000. Project Y has an IRR of 17% and a first cost of \$18 000. The MARR is 15%. What can be said about which (if either) of the two projects should be undertaken if (a) the projects are independent, and (b) the projects are mutually exclusive?

5.11 Charlie has a project for which he had determined a present worth of \$56 740. He now has to calculate the IRR for the project, but unfortunately he has lost complete information about the cash flows. He knows only that the project has a five-year service life and a first cost of \$180 000, that a set of equal cash flows occurred at the end of each year, and that the MARR used was 10%. What is the IRR for this project?

5.12 Lucy's project has a first cost P, annual savings A, and a salvage value of \$1000 at the end of the 10-year service life. She has calculated the present worth as \$20 000, the annual worth as \$4000, and the payback period as three years. What is the IRR for this project?

5.13 Patti's project has an IRR of 15%, first cost P, and annual savings A. She observed that the salvage value S at the end of the five-year life of the project was exactly half of the purchase price, and that the present worth of the project was exactly double the annual savings. What was Patti's MARR?

5.14 Jerry has an opportunity to buy a bond with a face value of \$10 000 and a coupon rate of 14%, payable semi-annually.

(a) If the bond matures in five years and Jerry can buy one now for \$3500, what is his IRR for this investment?

(b) If his MARR for this type of investment is 20%, should he buy the bond?

5.15 The following cash flows result from a potential construction contract for Estevan Engineering:

1. Receipts of $500 000 at the start of the contract and $1 200 000 at the end of the fourth year
2. Expenditures at the end of the first year of $400 000 and at the end of the second year of $900 000
3. A net cash flow of $0 at the end of the third year

Using an appropriate rate of return method, for a MARR of 25%, should Estevan Engineering accept this project?

5.16 Samiran has entered into an agreement to develop and maintain a computer program for symbolic mathematics. Under the terms of the agreement, he will pay $90 000 in royalties to the investor at the end of the fifth, tenth, and fifteenth years, with the investor paying Samiran $45 000 now, and then $65 000 at the end of the twelfth year.

Samiran's MARR for this type of investment is 20%. Calculate the ERR of this project. Should he accept this agreement, based on these disbursements and receipts alone? Are you sure that the ERR you calculated is the only ERR? Why? Are you sure that your recommendation to Samiran is correct? Justify your answer.

5.17 Refer to Problem 4.12. Find which alternative is preferable using the IRR method and a MARR of 5%. Assume that one of the alternatives must be chosen. Answer the following questions by using present worth computations to find the IRRs. Use the least common multiple of service lives.

(a) What are the cash flows for each year of the comparison period (i.e., the least common multiple of service lives)?

(b) Are you able to conclude that there is a single IRR on the incremental investment? Why or why not?

(c) Which of the two alternatives should be chosen? Use the ERR method if necessary.

5.18 Refer to Example 4.6 in which a mechanical engineer has decided to introduce automated material-handling equipment to a production line. Use a present worth approach with an IRR analysis to determine which of the two alternatives is best. The MARR is 9%. Use the repeated lives method to deal with the fact that the service lives of the two alternatives are not equal.

5.19 Refer to Problem 4.20. Use an IRR analysis to determine which of the two alternatives is best. The MARR is 8%. Use the repeated lives method to deal with the unequal service lives of the two alternatives.

5.20 Refer to Problem 4.21. Val has determined that the salvage value of the XJ3 after two years of service is $1900. Using the IRR method, which display panel is the better choice? Use a two-year study period. She must choose one of the alternatives.

5.21 Yee Swian has received an advance of $2000 on a software program she is writing. She will spend $12 000 this year writing it (consider the money to have been spent at the end of year 1), and then receive $10 000 at the end of the second year. The MARR is 12%.

(a) What is the IRR for this project? Does the result make sense?

(b) What is the precise ERR?

(c) What is the approximate ERR?

5.22 Zhe develops truss analysis software for civil engineers. He has the opportunity to contract with at most one of two clients who have approached him with development proposals. One contract pays him $15 000 immediately, and then $22 000 at the end of the project three years from now. The other possibility pays $20 000 now and $5000 at the end of each of the three years. In either case, his expenses will be $10 000 per year. For a MARR of 10%, which project should Zhe accept? Use an appropriate rate of return method.

5.23 The following table summarizes cash flows for a project:

Year	Cash Flow at End of Year
0	−$5000
1	3000
2	4000
3	−1000

(a) Write out the expression you need to solve to find the IRR(s) for this set of cash flows. Do not solve.

(b) What is the maximum number of solutions for the IRR that could be found in part (a)? Explain your answer in one sentence.

(c) You have found that an IRR of 14.58% solves the expression in part (a). Compute the project balances for each year.

(d) Can you tell (without further computations) if there is a unique IRR from this set of cash flows? Explain in one sentence.

5.24 Orillia Properties screens various projects using the payback period method. For renovation decisions, the minimum acceptable payback period is five years. Renovation projects are characterized by an immediate investment of

P dollars which is recouped as an annuity of A dollars per year over 20 years. They are considering changing to the IRR method for such decisions. If they changed to the IRR method, what MARR would result in exactly the same decisions as their current policy using payback period?

5.25 Six mutually exclusive projects, A, B, C, D, E, and F, are being considered. They have been ordered by first costs so that project A has the smallest first cost, F the largest. The data in the table below applies to these projects. The data can be interpreted as follows: The IRR on the incremental investment between project D and project C is 6%. Which project should be chosen using a MARR of 15%?

Project	IRR on Overall Investment	IRR on Increments of Investment Compared with Project A	B	C	D	E
A	20%					
B	15%	12%				
C	24%	30%	35%			
D	16%	18%	22%	6%		
E	17%	16%	19%	15%	16%	
F	21%	20%	21%	19%	18%	11%

5.26 Three mutually exclusive designs for a by-pass are under consideration. The by-pass has a 10-year life. The first design incurs a cost of $1.2 million for a net savings of $300 000 per annum. The second design would cost $1.5 million for a net savings of $400 000 per annum. The third has a cost of $2.1 million for a net savings of $500 000 per annum. For each of the alternatives, what range of values for the MARR results in its being chosen? It is not necessary that any be chosen.

5.27 Linus's project has cash flows at times 0, 1 and 2. He notices that for a MARR of 12%, the ERR falls exactly half-way between the MARR and the IRR, while for a MARR of 18%, the ERR falls exactly 1/4 of the way between the MARR and the IRR. If the cash flow is $2000 at time 2 and negative at time 0, what are the possible values of the cash flow at time 1?

5.28 Three construction jobs are being considered by Crystal City Construction (see the following table). Each is characterized by an initial deposit paid by the client to CCC, a yearly cost incurred by CCC at the end of each of three years, and a final payment to CCC by the client at the end of three years. CCC only has the capacity to do one of these contracts. Use IRR to determine which they should do. Their MARR is 10%.

Job	Deposit ($)	Cost per Year ($)	Final Payment ($)
1	100 000	75 000	200 000
2	150 000	100 000	230 000
3	175 000	150 000	300 000

5.29 Kenora Karavans is considering three investment proposals. Each of them is characterized by an initial cost, annual savings over four years, and no salvage value, as illustrated in the following table. They can only invest in two of these proposals. If their MARR is 12%, which two should they choose?

Proposal	First Cost ($)	Annual Savings ($)
A	40 000	20 000
B	110 000	30 000
C	130 000	45 000

5.30 Development projects done by Produits Trois Rivières are subsidized by a government grant program. The program pays 30% of the total cost of the project (costs summed without discounting, i.e., the interest rate is zero), half at the beginning of the project and half at the end, up to a maximum of $100 000. There are two projects being considered. One is a customized checkweigher for cheese products, and the other is an automated production scheduling system. Each project has a service life of five years. Costs and benefits for both projects, not including grant income, are shown below. Only one can be done, and the grant money is certain. PTR has a MARR of 15% for projects of this type. Using an appropriate rate of return method, which project should be chosen?

Checkweigher

First cost	$30 000
Yearly costs	5 000
Yearly benefits	14 000
Salvage value	8 000

Scheduler

First cost	$10 000
Yearly costs	12 000
Yearly benefits	17 000
Salvage value	0

5.31 Jacob is considering the replacement of the heating system for his building. There are three alternatives. All are natural gas-fired furnaces, but they vary in energy efficiency. Model A is leased at a cost of $500 per year over a 10-year study period. There are installation charges of \$500 and no salvage value. It is expected to provide energy savings of \$200 per year. Model B is purchased for a total cost of \$3600, including installation. It has a salvage value of \$1000 after 10 years of service, and is expected to provide energy savings of \$500 per year. Model C is also purchased, for a total cost of \$8000, including installation. However, half of this cost is paid now, and the other half is paid at the end of two years. It has a salvage value of \$1000 after 10 years, and is expected to provide energy savings of \$1000 per year. For a MARR of 12% and using a rate of return method, which heating system should be installed? One model must be chosen.

5.32 Calgary Cartage leases trucks to service its shipping contracts. Larger trucks have cheaper operating costs if there is sufficient business, but are more expensive if they are not full. CC has estimates of monthly shipping demand. What comparison method(s) would be appropriate for choosing which trucks to lease?

5.33 The bottom flaps of shipping cartons for Yonge Auto Parts are fastened with industrial staples. Yonge needs to buy a new stapler. What comparison method(s) would be appropriate for choosing which stapler to buy?

5.34 Joan runs a dog kennel. She is considering installing a heating system for the interior runs which will allow her to operate all year. What comparison method(s) would be appropriate for choosing which heating system to buy?

5.35 A large Canadian food company is considering replacing a scale on its packaging line with a more accurate one. What comparison method(s) would be appropriate for choosing which scale to buy?

5.36 Mona runs a one-person company producing custom paints for hobbyists. She is considering buying printing equipment to produce her own labels. What comparison method(s) would be appropriate for choosing which equipment to buy?

5.37 Peter is the president of a rapidly growing company. There are dozens of important things to do, and cash flow is tight. What comparison method(s) would be appropriate for Peter to make acquisition decisions?

5.38 Lemuel is an engineer working for Ontario Hydro. He must compare several routes for transmission lines from the Darlington nuclear plant to new industrial parks north of Toronto. What comparison method(s) is he likely to use?

5.39 Vicky runs a music store that has been suffering from thefts. She is considering installing a magnetic tag system. What comparison method(s) would be best for her to use to choose among competing leased systems?

5.40 Thanh's company is growing very fast and has a hard time meeting its orders. An opportunity to purchase additional production equipment has arisen. What comparison method(s) would Thanh use to justify to her manager that the equipment purchase was prudent?

Air Canada Supply and Stores

CANADIAN 5.1 MINI-CASE

Air Canada's Supply and Stores facility in Dorval, Quebec, is the main maintenance base for the airline. The facility houses over 140 000 parts worth \$200 million, and supplies service and replacement parts for Air Canada's operations. To improve the use of the existing warehouse space and to increase the efficiency of the parts-picking process, two automated storage and retrieval systems (AS/RS) were installed. Each AS/RS system consists of many aisles of stacked parts storage bins, several cranes used for parts retrieval, and a conveyor loop to transport the parts to the requesting workstation. Both AS/RS systems were configured so that the parts needed most often were stored in positions that would facilitate fast retrieval.

The warehousing software that controls the storage systems communicates with the Air Canada inventory control system and has boosted efficiency by prioritizing the parts-picking sequence. The priorities range from "aircraft-on-ground with passengers aboard" to requests for forms or stationery. Other improvements with the use of the AS/RS systems have been increased accuracy of parts counting and reduced costs of inventory purchasing and holding. Air Canada reported that the increase in productivity due to the AS/RS system provided a 3.5-year payback period.

Source: Industrial Engineering, June 1994

Discussion

Companies have a choice of how to calculate the benefits of a project in order to determine if it is worth doing. They also have a choice of how to report the benefits of a project to others. For Air Canada Supply and Stores, a 3.5-year payback period was reported to indicate the value of the AS/RS system to their operations. One would expect that more detailed calculations were done to actually determine whether the project was viable, even if they weren't reported to others.

The 3.5-year payback period that was reported leaves many questions unanswered. For example, the project could, in theory, have had a negative present worth or unsatisfactory IRR, even with a 3.5-year payback period. It also raises the question of whether the AS/RS system has a realistic life of 3.5 years, given the rate of change of computer software and the uncertainties of the aviation industry.

continued

Questions

1) Why would Air Canada report the payback period for the AS/RS system as opposed to the present worth, annual worth, or IRR?
2) Construct a set of hypothetical cash flows that indicate a 3.5-year payback period for a project with a negative present worth for a positive MARR. What is that project's IRR? Is it likely that the AS/RS system had a cash flow structure similar to the one you developed? Why or why not?
3) How would you judge the desirability of a project that depends significantly on a technology that is changing rapidly, such as computer software? Similarly, how would you take into account the effects of a rapidly changing environment in which a project decision must be made?

Appendix 5A: Tests for Multiple IRRs

When the IRR method is used to evaluate projects, we have to test for multiple IRRs. If there are undetected multiple IRRs, an IRR might be calculated that seems correct, but is in error. We consider three tests for multiple IRRs, forming essentially a three-step procedure. In the first test, the signs of the cash flows are examined to see if the project is a *simple investment.* In the second test, the present worth of the project is plotted against the interest rate to search for interest rates at which the present value is zero. In the third test, project balances are calculated. Each test *can* have a definite outcome: (1) There is definitely a unique IRR or (2) There definitely are two or more IRRs. But each test *may* have an outcome that is consistent with either a single, unique IRR, or with multiple IRRs. The tests are applied sequentially. The second test is applied only if the outcome of the first test is not clear. The third test is applied only if the outcomes of the first two are not clear. Even after all three tests have been applied, there are still three possible outcomes:

1. There is definitely a unique IRR.
2. There are definitely multiple IRRs because two or more IRRs have been found.
3. The test outcomes are ambiguous. A unique IRR or multiple IRRs are both possible.

Recall that most projects consist of one or more periods of outflows at the start, followed only by one or more periods of inflows; these are called simple investments. Although simple investments guarantee a single IRR, a project that is not simple may have a single IRR or multiple IRRs. Some investment projects have large cash outflows during their lives or at the ends of their lives that cause net cash flows to be negative after years of positive net cash flows. For example, a project that involves the construction of a manufacturing plant may involve a planned expansion of the plant requiring a large expenditure some years after its initial operation. As another example, a nuclear electricity plant may have planned large cash outflows for disposal of spent fuel at the end of its life. Such a project may have a unique IRR, but it may also have multiple IRRs. Consequently we must examine such projects further.

Where a project is not simple, we go to a second test. The second test consists of making a graph plotting present worth against interest rate. Points at which the present worth crosses or just touches the interest-rate axis (i.e., where present worth = 0) are IRRs. (We assume that there is at least one IRR.) If there are more than one such point, we know that there are more than one IRR. A convenient way to produce such a graph is using a spreadsheet. Table 5A.1 was obtained by computing the present worth of the cash flows in Example 5.7 for a variety of interest rates. Figure 5.12 shows the graph of the values in Table 5A.1.

Example 5A.1 (Example 5.7 Restated)

A project pays \$1000 today, costs \$5000 a year from now, and pays \$6000 in two years. What is its IRR?

Table 5A.1 The Spreadsheet Cells Used to Construct Figure 5A.1

Interest Rate, i	Present Worth (\$)
0.6	218.8
0.8	74.1
1.0	0.0
1.2	−33.1
1.4	−41.7
1.6	−35.5
1.8	−20.4
2.0	0.0
2.2	23.4
2.4	48.4

While finding multiple IRRs in a plot ensures that the project does indeed have multiple IRRs, failure to find multiple IRRs does not necessarily mean that multiple IRRs do not exist. *Any* plot will cover only a finite set of points. There may be values of the interest rate for which the present worth of the project is zero that are not in the range of interest rates used.

Where the project is not simple and a plot does not show multiple IRRs, we apply the third test. The third test entails calculating the project balances.

Figure 5A.1 Illustration of Two IRRs for Example 5A.1

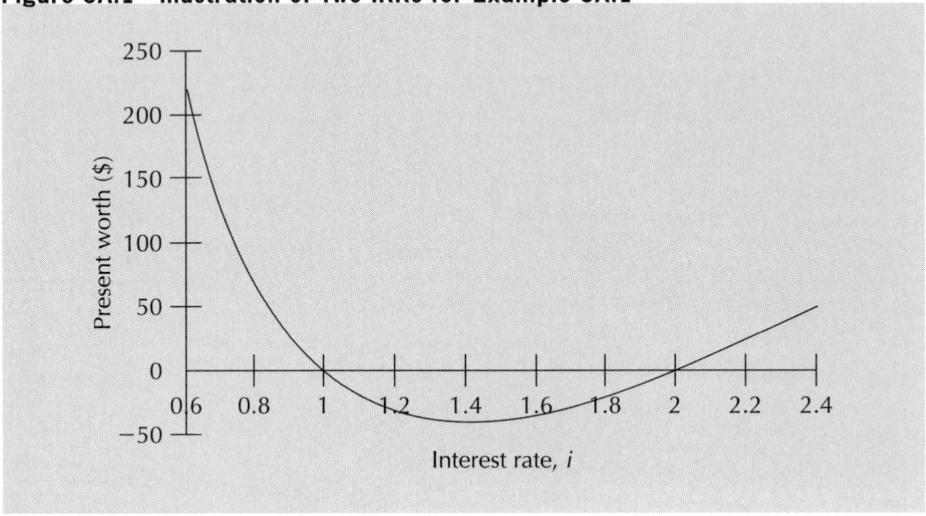

As we mentioned earlier, project balances refer to the cumulative net cash flows at the end of each time period. For an IRR to be unique, there should be no time when the project balances, computed using that IRR, are positive. This means that there is no extra cash not reinvested in the project. This is a sufficient condition for there to be a unique IRR. (Recall that it is the cash generated by a project, but not reinvested in the project, that creates the possibility of multiple IRRs.)

We now present three examples. All three examples involve projects that are not simple investments. In the first, a plot shows multiple IRRs. In the second, the plot shows only a single IRR. This is inconclusive, so project balances are computed. None of the project balances is positive, so we know that there is a single IRR. In the third example, the plot shows only one IRR, so the project balances are computed. One of these is positive, so the results of all tests are inconclusive. ■

Example 5A.2

Wellington Woods is considering buying land that they will log for three years. In the second year, they expect to develop the area that they clear as a residential subdivision that will entail considerable costs. Thus, in the second year, the net cash flow will be negative. In the third year, they expect to sell the developed land at a profit. The net cash flows that are expected for the project are:

End of Year	Cash Flow
0	−$100 000
1	440 000
2	− 639 000
3	306 000

The negative net cash flow in the second period implies that this is not a simple project. Therefore, we apply the second test. We plot the present worth against interest rates to search for IRRs. (See Figure 5A.2.) At 0% interest, the present worth is a small positive amount, $7000. The present worth is then 0 at 20%, 50%, and 70%. Each of these values is an IRR. The spreadsheet cells that were used for the plot are shown in Table 5A.2.

In this example, a moderately fine grid of two percentage points was used. Depending on the problem, the analyst may wish to use a finer or coarser grid.

The correct decision in this case can be obtained by the approximate ERR method. Suppose the MARR is 15%; then the approximate ERR is the interest rate that makes the future worth of outlays equal to the future worth of receipts when the receipts earn 15%. In other words, the approximate ERR is the value of i that solves

Figure 5A.2 Wellington Woods Present Worth

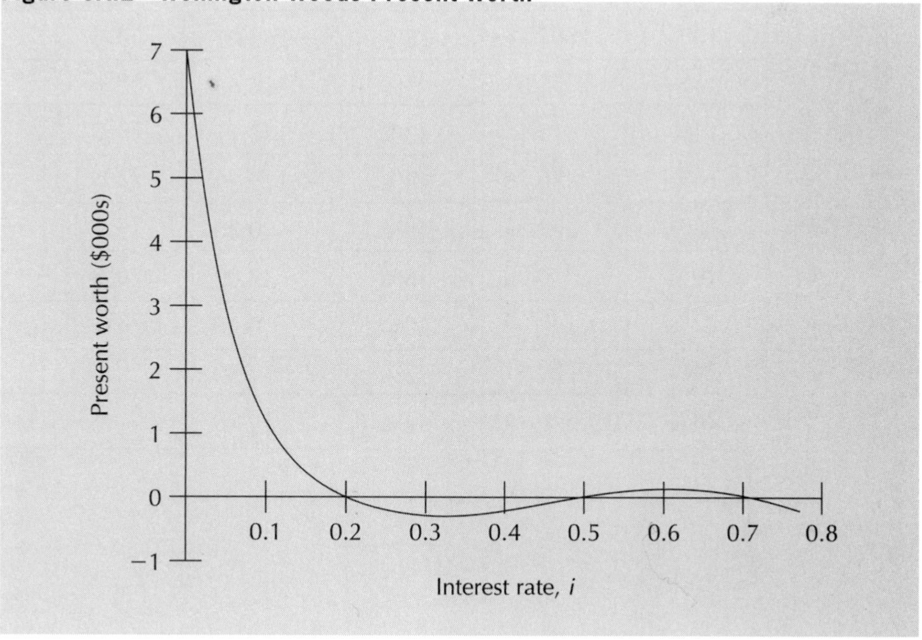

$$\$100\,000(1 + i)^3 + \$639\,000(1 + i) = \$440\,000(1.15)^2 + \$306\,000$$

$$\$100\,000(1 + i)^3 + \$639\,000(1 + i) = \$887\,900$$

Try 15% for i^*_{ea}. Using the tables for the left-hand side of the above equation, we have

$$\$100\,000(F/P, 15\%, 3) + \$639\,000(F/P, 15\%, 1)$$

$$= \$100\,000(1.5208) + \$639\,000(1.15)$$

$$= \$887\,000 < \$887\,900$$

Thus, the approximate ERR is slightly above 15%. The project is (marginally) acceptable by this calculation, because the approximate ERR, which is a conservative estimate of the correct ERR, is above the MARR. ■

Table 5A.2 Spreadsheet Cells Used to Generate Figure 5A.2

Interest Rate	Present Worth	Interest Rate	Present Worth	Interest Rate	Present Worth
0%	$7000.00	28%	– 352.48	56%	$ 79.65
2%	5536.33	30%	– 364.13	58%	92.49
4%	4318.39	32%	– 356.87	60%	97.66
6%	3310.12	34%	– 335.15	62%	94.84
8%	2480.57	36%	– 302.77	64%	83.79
10%	1803.16	38%	– 263.01	66%	64.36
12%	1255.01	40%	– 218.66	68%	36.44
14%	816.45	42%	– 172.11	70%	0.00
16%	470.50	44%	– 125.39	72%	– 44.96
18%	202.55	46%	– 80.20	74%	– 98.41
20%	0.00	48%	– 38.00	76%	– 160.24
22%	– 148.03	50%	0.00	78%	– 230.36
24%	– 250.91	52%	32.80		
26%	– 316.74	54%	59.58		

Example 5A.3

Investment in a new office coffeemaker has the following effects:

1. There is a three-month rental fee of $40 for the equipment, payable immediately and in three months.
2. A rebate of $30 from the supplier is given immediately for an exclusive six-month contract.
3. Supplies will cost $20 per month, payable at the beginning of each month.
4. Income from sales will be $30 per month, received at the end of each month.

Will there be more than one IRR for this problem?

We apply the first test by calculating net cash flows for each time period. The net cash flows for this project are as follows:

End of Month	Receipts	Disbursements	Net Cash Flow
0	+ $30	− $40 + (− $20)	− $30
1	+ 30	− 20	+ 10
2	+ 30	− 20	+ 10
3	+ 30	− 40 + (− 20)	− 30
4	+ 30	− 20	+ 10
5	+ 30	− 20	+ 10
6	+ 30	0	+ 30

As illustrated in Figure 5A.3, the net cash flows for this problem do not follow the pattern of a simple investment. Therefore, there may be more than one IRR for this problem. Accordingly, we apply the second test.

Figure 5A.3 Net Cash Flows for the Coffeemaker

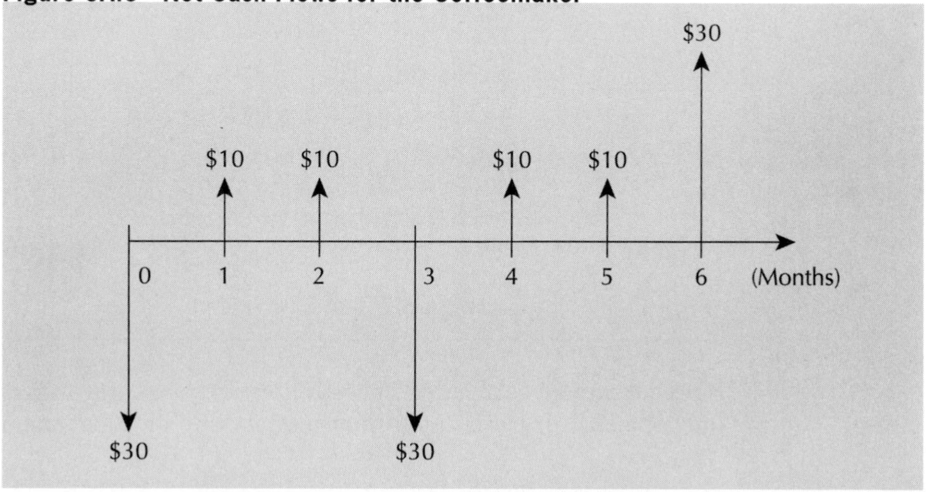

Figure 5A.4 IRR for New Coffeemaker

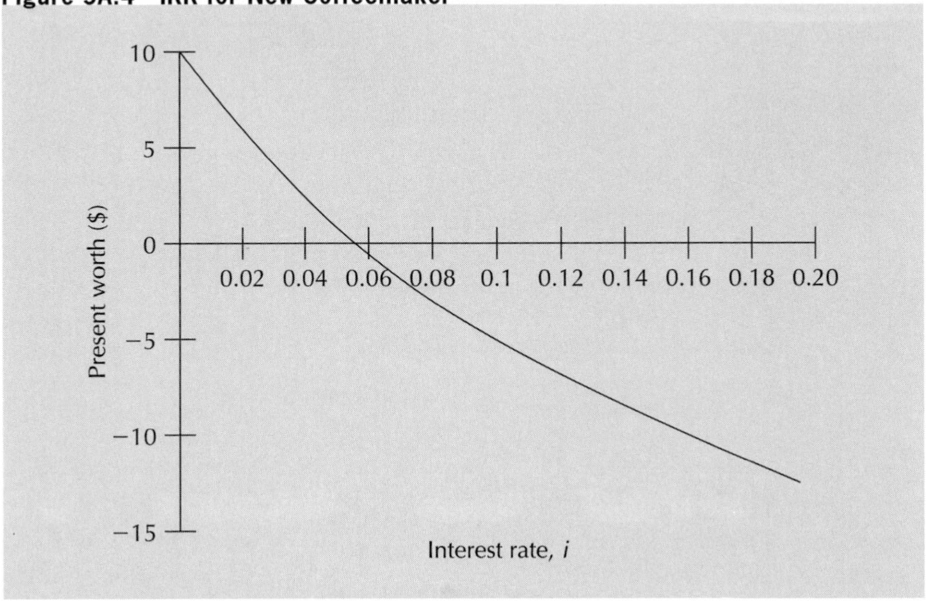

A plot of present worth against the interest rate is shown in Figure 5A.4. The plot starts with a zero interest rate where the present worth is just the arithmetic sum of all the net cash flows over the project's life. This is a positive \$10. The present worth as a function of the interest rate decreases as the interest rate increases from zero. The plot continues down, and passes through the interest-rate axis at about 5.8%. There is only one IRR in the range plotted. We need to apply the third test by computing project balances at the 5.8% interest rate.

The project balances at the 5.8% interest rate are as follows:

Month	Project Balance
0	$B_0 = -\$30$
1	$B_1 = -\$30.0(1.058) + \$10 = -\$21.7$
2	$B_2 = -\$21.7(1.058) + \$10 = -\$13.0$
3	$B_3 = -\$13.0(1.058) - \$30 = -\$43.7$
4	$B_4 = -\$43.7(1.058) + \$10 = -\$36.3$
5	$B_5 = -\$36.3(1.058) + \$10 = -\$28.4$
6	$B_6 = -\$28.4(1.058) + \$30 = \$0$

Since all project balances are negative or zero, we know that this investment has only one IRR. It is 5.8% per month or about 69.3% per year. ■

Example 5A.4

Green Woods, like Wellington Woods, is considering buying land that they will log for three years. In the second year, they also expect to develop the area that they have logged as a residential subdivision, which again will entail considerable costs. Thus, in the second year, the net cash flow will be negative. In the third year, they expect to sell the developed land at a profit. But Green Woods expect to do much better than Wellington Woods in the sale of the land. The net cash flows that are expected for the project are:

Year	Cash Flow
0	−\$100 000
1	455 000
2	− 667 500
3	650 000

The negative net cash flow in the second period implies that this is not a simple project. We now plot the present worth against interest rate. See Figure 5A.4. At zero interest rate, the present worth is a positive \$337 500.

Figure 5A.4 Present Worths for Example 5A.4

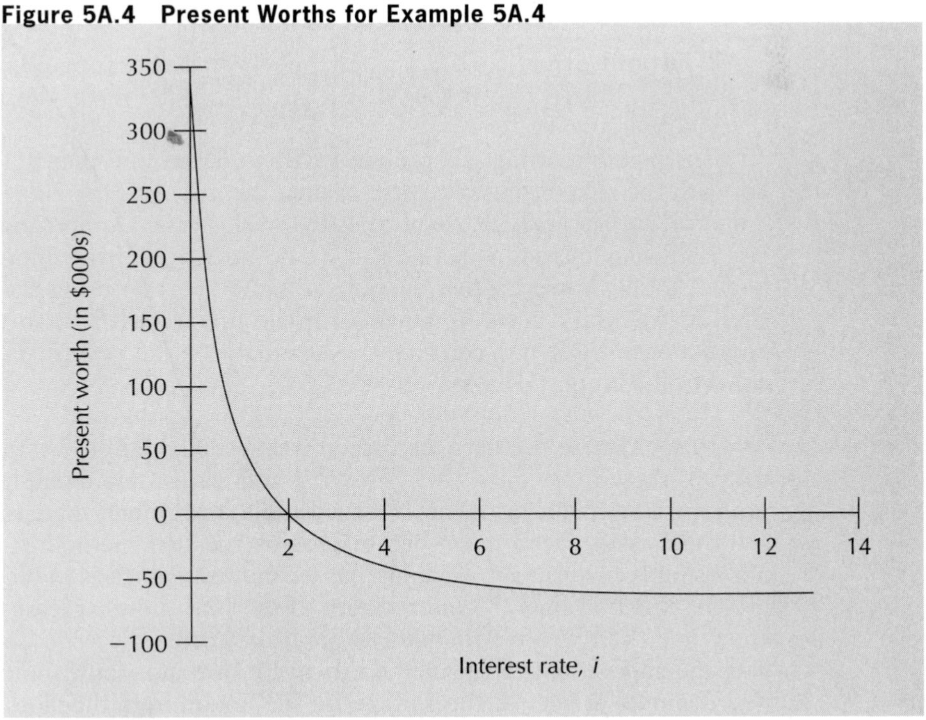

The present worth falls as the interest rate rises. It crosses the interest-rate axis at about 206.4%. There are no further crossings of the interest-rate axis in the range plotted, but since this is not conclusive we compute project balances.

Year	Project Balance
0	$B_0 = -\ \$100\ 000$
1	$B_1 = -\ \$100\ 000(3.064) + \$455\ 000 = \$148\ 600$
2	$B_2 = -\ \$148\ 600(3.064) - \$667\ 500 = -\ \$2\ 121\ 900$
3	$B_3 = -\ \$2\ 121\ 900(3.064) + \$650\ 000 = -\ \$1500$

We note that the project balance is positive at the end of the first period. This means that a unique IRR is *not* guaranteed. We have gone as far as the three tests can take us. On the basis of the three tests, there may be only the single IRR that we have found, 206.4%, or there may be multiple IRRs.

In this case, we use the approximate ERR to get a decision. Suppose the MARR is 30%. The approximate ERR, then, is the interest rate that makes the future worth of outlays equal to the future worth of receipts when the receipts earn 30%. That is, the approximate ERR is the value of i that solves the following:

$$\$100\ 000(1 + i)^3 + \$667\ 500(1 + i) = \$455\ 000(1.3)^2 + \$650\ 000$$

Trial and error with a spreadsheet gives the approximate ERR as $i^*_{ea} \cong 57\%$. This is above the MARR of 30%. Therefore, the investment is acceptable.

It is possible, using the precise ERR, to determine that the IRR that we got with the plot of present worth against the interest rate, 206.4%, is, in fact, unique. The precise ERR equals the IRR in this case. Computation of the precise ERR may be cumbersome, and we do not cover this computation in this book. Notice, however, that we got the same decision using the approximate ERR as we would have obtained with the precise ERR. Also note that the approximate ERR is a conservative estimate of the precise ERR, which is equal to the unique IRR. ■

To summarize, we have discussed three tests that are to be applied sequentially, as shown in Figure 5A.5. The first and easiest test to apply is to see if a project is a simple investment. If it is a simple investment, there is a single IRR, and the correct decision can be obtained by the IRR method. If the project is not a simple investment, we apply the second test, which is to plot the project's present worth against the interest rate. If the plot shows at least two IRRs, we know that there is not a unique IRR, and the correct decision can be obtained with the approximate ERR method. If a plot does not show multiple IRRs, we next compute project balances using the IRR found from the plot. If none of the

Figure 5A.5 Tests for Multiple IRR

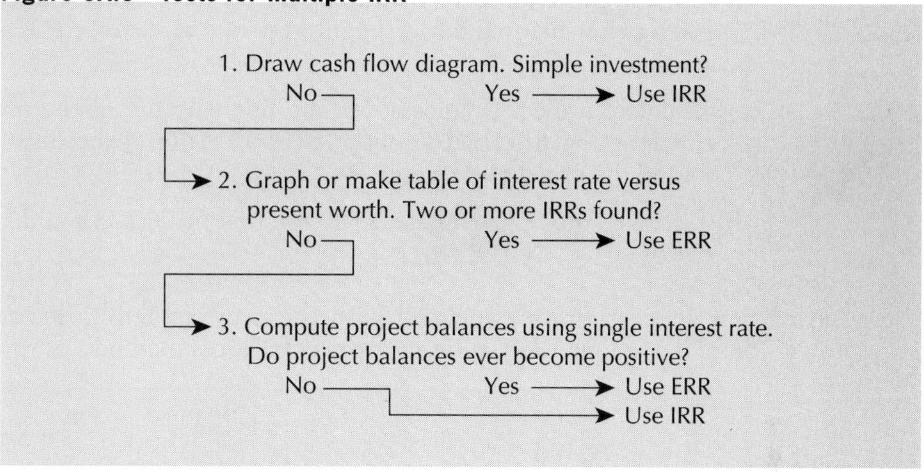
1. Draw cash flow diagram. Simple investment?
No
Yes
Use IRR
2. Graph or make table of interest rate versus present worth. Two or more IRRs found?
No
Yes
Use ERR
3. Compute project balances using single interest rate. Do project balances ever become positive?
No
Yes
Use ERR
Use IRR

project balances is positive, the IRR is unique, and the correct decision can be obtained with the IRR method. If one or more of the project balances are positive, we don't know whether or not there is a unique IRR. Accordingly, we use the approximate ERR method which always will yield a correct decision.

PROBLEMS

5A.1 A five-year construction project for Wawa Engineering receives staged payments in years 2 and 5. The resulting net cash flows are as follows:

Year	Cash Flow
0	−$300 000
1	− 500 000
2	700 000
3	− 400 000
4	− 100 000
5	900 000

The MARR for Wawa Engineering is 15%.

(a) Is this a simple project?

(b) Plot the present worth of the project against interest rates from 0% to 100%. How many times is the interest-rate axis crossed? How many IRRs are there?

(c) Calculate project balances over the five-year life of the project. Can we conclude that the IRR(s) observed in (b) is (are) the only IRR(s)? If so, should the project be accepted?

(d) Calculate the approximate ERR for this project. Should the project be accepted?

5A.2 For the cash flows associated with the projects below, determine whether there is a unique IRR, using the project balances method.

		Project	
End of Period	**1**	**2**	**3**
0	−$3000	−$1500	$ 600
1	900	7000	− 2000
2	900	− 9000	500
3	900	2900	500
4	900	500	1000

5A.3 A mining opportunity in a third-world country has the following cash flows:

1. $10 000 000 is received at time 0 as an advance against expenses.
2. Costs in the first year are $20 000 000, and in the second year $10 000 000.
3. Over years 3 to 10, annual revenues of $5 000 000 are expected.

After 10 years, the site reverts to government ownership. MARR is 30%.

(a) Is this a simple project?

(b) Plot the present worth of the project against interest rates from 0% to 100%. How many times is the interest-rate axis crossed? How many IRRs are there?

(c) Calculate project balances over the 10-year life of the project. Can we conclude that the IRR(s) observed in (b) is (are) the only IRR(s)? If so, should the project be accepted?

(d) Calculate the approximate ERR for this project. Should the project be accepted?

(e) Calculate the exact ERR for this project. Should the project be accepted?

C H A P T E R 6

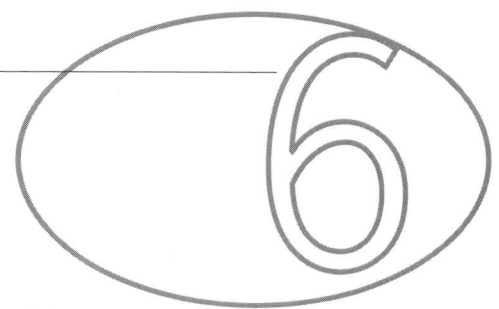

Depreciation and Financial Accounting

Engineering Economics in Action, Part 6A: The Pit Bull

Naomi liked to think of Terry as a pit bull. Terry had this endearing habit of finding some detail that irked him, and not letting go of it until he was satisfied that things were done properly. Naomi had seen this several times in the months they had worked together. Terry would sink his teeth into some quirk of Canadian Widgets' operating procedures and, just like a fighting dog, not let go until the fight was over.

This time, it was about the disposal of some computing equipment. Papers in hand, he quietly approached Naomi and earnestly started to explain his concern. "Naomi, I don't know what Bill Fisher is doing, but something's definitely not right here. Look at this."

Terry displayed two documents to Naomi. One was an accounting statement showing the book value of various equipment, including some CAD/CAM computers that had been sold for scrap the previous week. The other was a copy of a sales receipt from a local salvage firm for that same equipment.

"I don't like criticizing my fellow workers, but I really am afraid that Bill might be doing something wrong." Bill Fisher was the buyer responsible for capital equipment at Canadian Widgets, and he also disposed of surplus assets. "You know the CAD/CAM workstations they had in Engineering Design? Well, they were replaced recently and sold. Here is the problem. They were only three years old, and our own accounting department estimated their value as about $5000 each." Terry's finger pointed to the evidence on the accounting statement. "But here," his finger moving to the guilty figure on the sales receipt, "they were actually sold for $300 each!" Terry sat back in his chair. "How about that!"

Naomi smiled. Unfortunately, she would have to pry his teeth out of this one. "Interesting observation, Terry. But you know, I think it's probably O.K. Let me explain."

6.1 Introduction

Engineering projects often involve an investment in equipment, buildings, or other assets which are put to productive use. As time passes, these assets lose value, or depreciate. The first part of this chapter is concerned with the concept of depreciation and several methods that are commonly used to model depreciation. Depreciation is taken into account when a firm states the value of its assets in its financial statements, as seen in the second half of this chapter. It also forms an important part of the decision of when to replace an aging asset and when to make cyclic replacements, as will be seen in Chapter 7, and has an important impact on taxation, as we will see in Chapter 8.

With the growth in importance of small technology-based enterprises, many engineers have taken on broad managerial responsibilities that include financial accounting. Financial accounting is concerned with recording and organizing the financial data of businesses. The data cover both *flows over time*, like revenues and expenses, and *levels*, like an enterprise's resources and the claims on those resources at a given date. Even engineers who do not have broad managerial responsibilities need to know the elements of financial accounting to understand the enterprises with which they work.

In the second part of this chapter, we explain three basic financial statements used to summarize the financial dimensions of a business. We then explain how these statements can be used to make inferences about the financial health of the firm.

6.2 Depreciation and Depreciation Accounting

6.2.1 Reasons for Depreciation

An asset starts to lose value as soon as it is purchased. For example, a car bought for $20 000 today may be worth $18 000 next week, $15 000 next year, and $1000 in 10 years. This loss in value, called **depreciation**, occurs for several reasons.

Use-related physical loss: As something is used, parts wear out. An automobile engine has a limited life span because the metal parts within it wear out. This is one reason why a car diminishes in value over time. Often, use-related physical loss is measured with respect to *units of production*, such as thousands of kilometres for a car, hours of use for a light bulb, or thousands of cycles for a punch press.

Time-related physical loss: Even if something is not used, there can be a physical loss over time. This can be due to environmental factors affecting the asset or to endogenous physical factors. For example, an unused car can rust and thus lose value over time. Time-related physical loss is expressed in units of time, such as a 10-year-old car or a 40-year-old sewage treatment plant.

Functional loss: Losses can occur without any physical changes. For example, a car can lose value over time because styles change so that it is no longer fashionable. Other examples of causes of loss of value include legislative changes, such as for pollution control or safety devices, and technical changes. Functional loss is usually expressed simply in terms of the particular unsatisfied function.

6.2.2 Value of an Asset

Models of depreciation can be used to estimate the loss in value of an asset over time, and also to determine the remaining value of the asset at any point in time. This remaining value has several names, depending on the circumstances.

Market value is usually taken as the actual value an asset can be sold for in an open market. Of course, the only way to determine the actual market value for an asset is to sell it. Consequently, the term *market value* usually means an *estimate* of the market value. One way to make such an estimation is by using a depreciation model that reasonably captures the true loss in value of an asset.

Book value is the depreciated value of an asset for accounting purposes, as calculated with a depreciation model. The book value may be more or less than market value. The depreciation model used to arrive at a book value might be controlled by regulation for some purposes, such as taxation, or simply by the desirability of an easy calculation scheme. There might be several different book values for the same asset, depending on the purpose and depreciation model applied. We shall see how book values are reported in financial statements later in this chapter.

Scrap value can be either the actual value of an asset at the end of its physical life (when it is broken up for the material value of its parts) or an estimate of the scrap value calculated using a depreciation model.

Salvage value can be either the actual value of an asset at the end of its useful life (when it is sold) or an estimate of the salvage value calculated using a depreciation model.

It is desirable to be able to construct a good model of depreciation in order to state a book value of an asset for a variety of reasons:

1. In order to make many managerial decisions, it is necessary to know the value of owned assets. For example, money may be borrowed using the firm's assets as collateral. In order to demonstrate to the lender that there is security for the loan, a credible estimate of the assets' value must be made. A depreciation model permits this to be done. The use of depreciation for this purpose is explored more thoroughly in the second part of this chapter.
2. One needs an estimate of the value of owned assets for planning purposes. In order to decide whether to keep an asset or replace it, you have to be able to judge how much it is worth. More than that, you have to be able to assess how much it will be worth at some time in the future. The impact of depreciation in replacement studies is covered in Chapter 7.
3. Government tax legislation requires that taxes be paid on company profits. Because there can be many ways of calculating profits, strict rules are made concerning how to calculate income and expenses. These rules include a particular scheme for determining depreciation expenses. This use of depreciation is discussed more thoroughly in Chapter 8.

To match the way in which certain assets depreciate and to meet regulatory or accuracy requirements, many different depreciation models have been developed over time. Of the large number of depreciation schemes available (see Close-Up 6.1), straight-line and declining-balance are certainly the most commonly used in Canada. Straight-line depreciation is popular primarily because it is particularly easy to calculate. The declining-balance method is required by Canadian tax law for determining corporate taxes, as is discussed in Chapter 8. In the United States, tax laws prior to 1954 required the use of straight-line depreciation, and between 1954 and 1981, several other methods were permitted. Under current United States law, things are more complicated, but the main depreciation methods used are straight-line and declining-balance. Consequently, these are the only depreciation methods presented in detail in this book.

CLOSE-UP 6.1 Depreciation Methods

Method	*Description*
Straight-line	The book value of an asset diminishes by an equal *amount* each year.
Declining-balance	The book value of an asset diminishes by an equal *proportion* each year.
Sum-of-the-year's-digits	An accelerated method, like declining-balance, in which the depreciation rate is calculated as the ratio of the remaining years of life to the sum of the digits corresponding to the years of life
Double-declining-balance	A declining-balance method in which the depreciation rate is calculated as 2/N for an asset with service life N years
150%-declining-balance	A declining-balance method in which the depreciation rate is calculated as 1.5/N for an asset with service life N years
Units-of-production	Depreciation rate is determined per unit of production by distributing the initial cost over the estimated lifetime of the production capacity.

6.2.3 Straight-Line Depreciation

The **straight-line method of depreciation** assumes that the rate of loss in value of an asset is constant over its useful life. This is illustrated in Figure 6.1 for an asset worth $1000 at the time of purchase and $200 eight years later. Graphically, straight-line depreciation is determined by drawing a straight line between the first cost of the asset and its salvage or scrap value.

Algebraically, the assumption is that the rate of loss in asset value is constant and is based on its original cost and salvage value. This gives rise to a simple expression for the depreciation amount per period. We estimate the depreciation per period from the asset's current value and its estimated salvage value at the end of its useful life, N periods from now, by

$$D_{sl}(n) = \frac{P - S}{N} \tag{6.1}$$

where

$D_{sl}(n)$ = the depreciation amount for period n using the straight-line method

Figure 6.1 Book Value under Straight-Line Depreciation ($1000 Purchase and $200 Salvage Value after Eight Years)

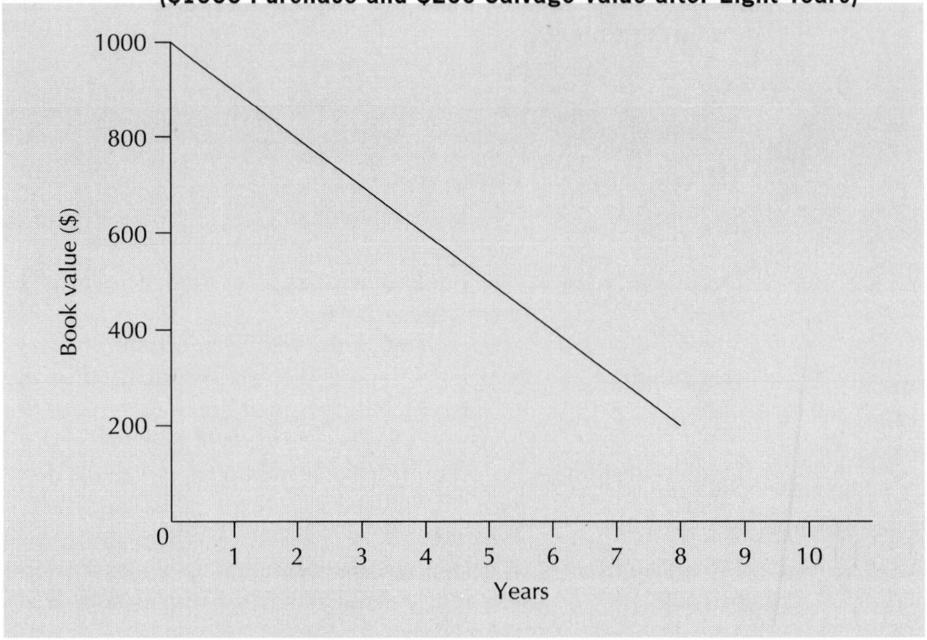

P = the purchase price or current market value

S = the salvage value after N periods

N = the useful life of the asset, in periods

Similarly, the book value at the end of any particular period is easy to calculate:

$$BV_{sl}(n) = P - n\left[\frac{P - S}{N}\right] \tag{6.2}$$

where

$BV_{sl}(n)$ = the book value at the end of period n using straight-line depreciation

Example 6.1

A laser cutting machine was purchased four years ago for \$380 000. It will have a salvage value of \$30 000 two years from now. If we believe a constant rate of depreciation is a reasonable means of approximating book value, what is its current book value?

From Equation (6.2), with $P = \$380\,000$, $S = \$30\,000$, $N = 6$, and $n = 4$,

$$BV_{sl}(4) = \$380\,000 - 4\left[\frac{\$380\,000 - \$30\,000}{6}\right]$$

$$BV_{sl}(4) = \$146\,667$$

The current book value for the cutting machine is \$146 667. ■

The straight-line depreciation method has the great advantage of being easy to calculate. It also is easy to understand and is in common use. The main problem with the method is that its assumption of a constant rate of loss in asset value is often not valid. Thus, book values calculated using straight-line depreciation will frequently be different from market values. For example, the loss in value of a car over its first year (say from \$20 000 to \$15 000) is clearly more than its loss in value over its fifth year (say from \$6000 to \$5000). The declining-balance method of depreciation covered in the next section allows for "faster" depreciation in earlier years of an asset's life.

6.2.4 Declining-Balance Depreciation

Declining-balance depreciation models the loss in value of an asset over a period as a constant proportion of the asset's current value. In other words, the depreciation amount in a particular period is a constant percentage (called the depreciation rate) of its closing book value from the previous period. The effect of various depreciation rates on estimated book values is illustrated in Figure 6.2.

Algebraically, the depreciation charge for period n is simply the depreciation rate multiplied by the book value from the end of period $(n - 1)$. Noting that $BV_{db}(0) = P$,

$$D_{db}(n) = BV_{db}(n - 1) \times d \qquad (6.3)$$

where

$D_{db}(n)$ = the depreciation amount in period n using the declining balance method

$BV_{db}(n)$ = the book value at the end of period n using the declining balance method

P = the purchase price or current market value

d = the depreciation rate

Similarly, the book value at the end of any particular period is easy to calculate, by noting that the remaining value after each period is $(1 - d)$ times the value at the end of the previous period.

Figure 6.2 Book Value under Declining-Balance Depreciation ($1000 Purchase with Various Depreciation Rates)

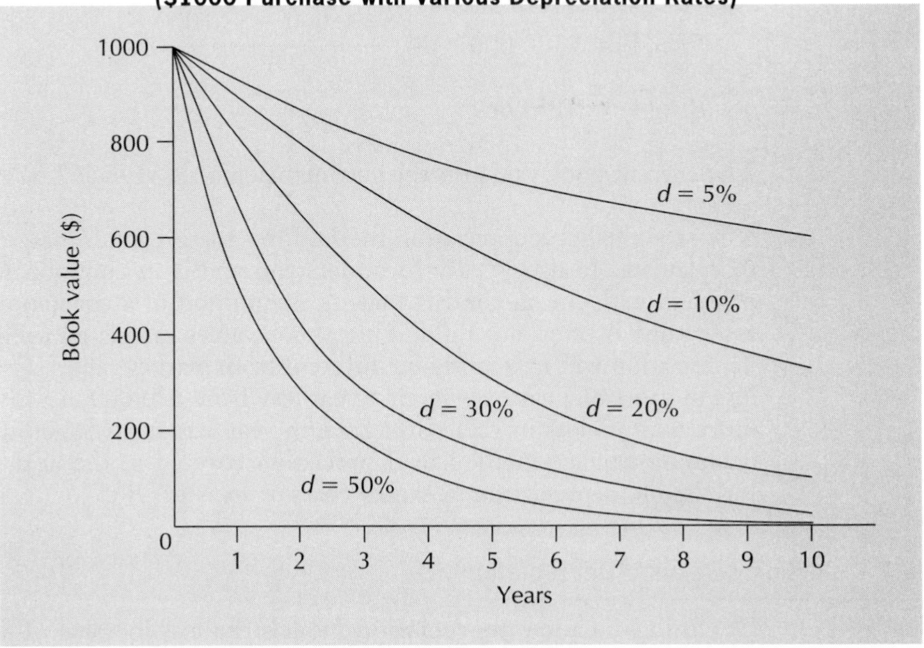

$$BV_{db}(n) = P(1 - d)^n \tag{6.4}$$

In order to use the declining-balance method of depreciation, we must determine a reasonable depreciation rate. By using an asset's current value, P, and a salvage value, S, n periods from now, we can use Equation (6.4) to find the declining balance rate that relates P and S.

$$BV_{db}(n) = S = P(1 - d)^n$$

$$(1 - d) = \sqrt[n]{\frac{S}{P}}$$

$$d = 1 - \sqrt[n]{\frac{S}{P}} \tag{6.5}$$

Example 6.2

Paquita wants to estimate the scrap value of a smokehouse twenty years after purchase. She feels that the depreciation is best represented using the declining-balance method, but she doesn't know what depreciation rate to use. She observes that the purchase price of the smokehouse was $245 000 three years ago, and an estimate of its current salvage value is $180 000. What is a good estimate of the value of the smokehouse after 20 years?

From Equation (6.5),

$$d = 1 - \sqrt[n]{\frac{S}{P}}$$

$$d = 1 - \sqrt[3]{\frac{\$180\ 000}{\$245\ 000}}$$

$$d = 0.097663$$

Then, by using Equation (6.4), we have

$$BV_{db}(20) = \$245\ 000(1 - 0.097663)^{20}$$
$$= \$31\ 372$$

An estimate of the salvage value of the smokehouse after 20 years using the declining-balance method of depreciation is \$31 372. ■

The declining-balance method has a number of useful features. For one thing, it matches the observable loss in value that many assets have over time. The rate of loss is expressed in one parameter, the depreciation rate. It is relatively easy to calculate, although perhaps not quite as easy as the straight-line method. In particular, it is required to be used in Canada for taxation purposes, as discussed in detail in Chapter 8.

Example 6.3

Sherbrooke Data Services has purchased a new mass storage system for \$250 000. It is expected to last six years, with a \$10 000 salvage value. Using both the straight-line and declining-balance methods, determine the following:

(a) The depreciation amount during year 1

(b) The depreciation amount during year 6

(c) The book value at the end of year 4

(d) The accumulated depreciation at the end of year 4

This is an ideal application for a spreadsheet solution. Table 6.1 illustrates a spreadsheet that calculates the book value, depreciation charge, and accumulated depreciation for both depreciation methods over the six-year life of the system.

The depreciation amount for each year with the *straight-line* method is \$40 000:

$$D_{sl}(n) = (\$250\ 000 - \$10\ 000)/6 = \$40\ 000$$

The depreciation rate for the *declining-balance* method is

$$d = 1 - \sqrt[n]{\frac{S}{P}} = 1 - \sqrt[6]{\frac{\$10\ 000}{\$250\ 000}} = 0.4152$$

The detailed calculation for each of the given questions is as follows:

(a) The depreciation amount during year 1

$D_{sl}(1) = (\$250\ 000 - \$10\ 000)/6 = \$40\ 000$

$D_{db}(1) = BV_{db}(0)d = \$250\ 000(0.4152) = \$103\ 799.11$

Table 6.1 Spreadsheet for Example 6.3

	Straight-Line Depreciation		
Year	Depreciation Charge	Accumulated Depreciation	Book Value
0			$250 000
1	$40 000	$ 40 000	210 000
2	40 000	80 000	170 000
3	40 000	120 000	130 000
4	40 000	160 000	90 000
5	40 000	200 000	50 000
6	40 000	240 000	10 000

	Declining-Balance Depreciation		
Year	Depreciation Charge	Accumulated Depreciation	Book Value
0			$250 000
1	$103 799	$103 799	146 201
2	60 702	164 501	85 499
3	35 499	200 000	50 000
4	20 760	220 760	29 240
5	12 140	232 900	17 100
6	7 100	240 000	10 000

(b) The depreciation amount during year 6

$D_{sl}(6) = D_{sl}(1) = \$40\ 000$

$D_{db}(6) = BV_{db}(5)d = \$250\ 000(0.5848)^5(0.4152) = \$7099.82$

(c) The book value at the end of year 4

$BV_{sl}(4) = \$250\ 000 - 4(\$250\ 000 - \$10\ 000)/6 = \$90\ 000$

$BV_{db}(4) = \$250\ 000(1 - 0.4152)^4 = \$29\ 240.17$

(d) The accumulated depreciation at the end of year 4

Straight-line accumulated depreciation at the end of year 4 is

$P - BV_{sl}(4) = \$160\ 000$

Declining-balance accumulated depreciation at the end of year 4

$P - BV_{db}(4) = \$220\ 759.83$ ■

In summary, depreciation affects economic analyses in several ways. First, it allows us to estimate the value of an owned asset, as illustrated in the above examples. We shall see in the next part of this chapter how these values are reported in a firm's financial statements. Next, the capability of estimating the value of an asset is particularly useful in replacement studies, which is the topic of Chapter 7. Finally, in Chapter 8, we cover aspects of the Canadian tax system which affect decision making; in particular we look at the effect of depreciation expenses.

6.3 Elements of Financial Accounting

How well is a business doing? Can it survive an unforeseen temporary drop in cash flows? How does a business compare with others of its size in the industry? Answering these questions and others like them is part of the accounting function. The accounting function has two parts, financial accounting and management accounting. **Financial accounting** is concerned with recording and organizing the financial data of a business, which include revenues and expenses, and an enterprise's resources and the claims on those resources. **Management accounting** is concerned with the costs and benefits of the various activities of an enterprise. The goal of management accounting is to provide managers with information to help in decision making.

Engineers have always played a major role in management accounting, especially in a part of management accounting called *cost* accounting. They have not, for the most part, had significant responsibility for financial accounting until recently. With the growth in importance of small technology-based enterprises, many engineers have taken on broad managerial responsibilities that include financial accounting. Even engineers who do not have broad managerial responsibilities need to know the elements of financial accounting to understand the enterprises in which they work. Management accounting is not covered in this text because it is difficult to provide useful information without taking more than a single chapter. Instead, we focus on financial accounting.

The object of financial accounting is to provide information to internal management and interested external parties. Internally, management uses financial accounting information for processes such as budgeting, cash management, and management of long-term debt. External users include actual and potential investors and creditors who wish to make rational decisions about an enterprise. External users also include government agencies concerned with taxes and regulation.

Areas of interest to all these groups include an enterprise's revenues and expenses, and assets (resources held by the enterprise) and liabilities (claims on those resources).

In the next few sections, we discuss two basic summary financial statements that give information about these matters. These are

(a) the balance sheet, and

(b) the income statement.

These financial statements form the basis of a financial report, which is usually produced on a monthly, quarterly, or yearly basis. We also briefly discuss one additional financial statement, the statement of changes in financial position. It is useful for understanding the cash status of the firm.

Following the discussion of financial statements, we shall consider the use of information in these statements when making inferences about an enterprise's performance compared with industry standards and with its own performance over time.

6.3.1 Measuring the Performance of a Firm

The flow of money in a company is much like the flow of water in a network of pipes or the flow of electricity in an electrical circuit, as illustrated in Figure 6.3. In order to measure the performance of a water system, we need to determine the flow through the system and the pressure in the system. For an electrical circuit, the analogous parameters are current and voltage. Flow and current are referred to as *through variables*, and are measured with respect to time (flow is litres per second and current is amperes, which are coulombs per second). Pressure and voltage are referred to as *across variables*, and are measured at a point in time.

The flow of money in an organization is measured in a similar way with the income statement and balance sheet. The income statement represents a *through variable* because it summarizes revenues and expenses over a period of time. It is prepared by listing the revenues earned during a period and the expenses incurred during the same period, and by subtracting total expenses

Figure 6.3 Through and Across Variables

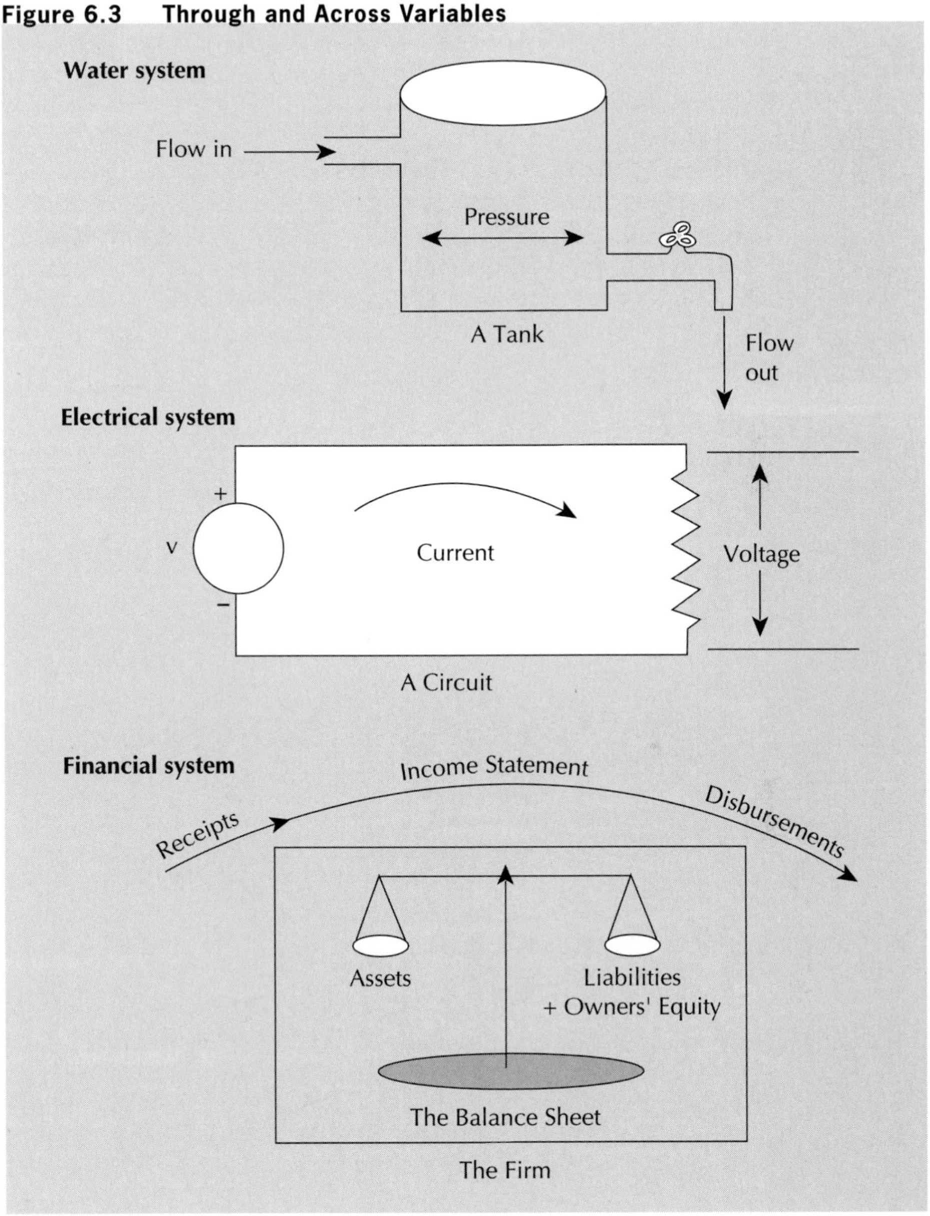

from total revenues, arriving at a net income. An income statement is always associated with a particular period of time, be it a month, quarter, or year.

The balance sheet, in contrast to the income statement, is a snapshot of the financial position of a firm at a particular point in time, and so represents an *across variable*. The financial position is summarized by listing the assets of the firm, its liabilities (debts), and the equity of the owner or owners.

6.3.2 The Balance Sheet

A **balance sheet** (often called a position statement) is a snapshot of an enterprise's financial position at a particular point in time, normally the last day of an accounting period. A firm's financial position is summarized in a balance sheet by listing its assets, liabilities, and owners' equity. The heading of the balance sheet gives the name of the enterprise and the date.

Example 6.4

Table 6.2 shows a balance sheet for the Major Electric Company, a manufacturer of small electrical appliances.

The first category of financial information in a balance sheet reports the **assets** of the enterprise; these are the economic resources owned by the enterprise. Assets are classified on a balance sheet as current assets and long-term assets. **Current assets** are cash and other assets that could be converted to cash within a relatively short period of time, usually a year or less. Inventory and accounts receivable are examples of non-cash current assets. **Long-term assets** are assets that are not expected to be converted to cash in the short term, usually taken to be one year. Indeed, it may be difficult to convert long-term assets into cash without selling the business as a going concern. Equipment, land, and buildings are examples of long-term assets.

An enterprise's **liabilities** are claims, other than those of the owners, on a business's assets. Debts are usually the most important liabilities on a balance sheet. There may be other forms of liabilities as well. A commitment to employees' pensions is an example of a non-debt liability. As with assets, liabilities may be classified as current or long-term. **Current liabilities** are liabilities that are due within some short period of time, usually a year or less. Examples of current liabilities are debts that are close to maturity, accounts payable to suppliers, and taxes due. **Long-term liabilities** are liabilities that are not expected to draw on the business's current assets. Long-term loans and bonds issued by the business are examples of long-term liabilities.

The difference between a business's assets and its liabilities is the amount due to the owners, i.e., their equity in the business. That is, owners' equity is what is left over from assets after claims of others are deducted. We have, therefore,

Owners' Equity = Assets − Liabilities *or*
Assets = Liabilities + Owners' Equity

Owners' equity is the interest of the owner or owners of a firm in its assets. Owners' equity usually appears as two components on a balance sheet of a corporation. The first is the par value of the owners' shares. When an

Table 6.2 A Balance Sheet for the Major Electric Company

Major Electric Company Balance Sheet as of November 30, 2000		
Assets		
Current Assets		
Cash		$ 39 000
Accounts receivable		27 000
Raw materials inventory		52 000
Finished goods inventory		683 000
Total Current Assets		$ 801 000
Long-Term Assets		
Equipment	$6 500 000	
Less accumulated depreciation	4 000 000	2 500 000
Buildings	1 750 000	
Less accumulated depreciation	150 000	1 600 000
Land		500 000
Total Long-Term Assets		$ 4 600 000
Total Assets		**$5 401 000**
Liabilities and Owners' Equity		
Current Liabilities		
Accounts payable		$ 15 000
Loan due December 31, 2000		75 000
Total Current Liabilities		90 000
Long-Term Liabilities		
Loan due December 31, 2003		1 000 000
Total Liabilities		**$1 090 000**
Owners' Equity		
Common stock: 1 000 000 shares at $3 par value per share		$ 3 000 000
Retained earnings		1 311 000
Total Owners' Equity		**$4 311 000**
Total Liabilities and Owners' Equity		**$5 401 000**

enterprise is first organized, it is authorized to issue a certain number of shares. **Par value** is the price per share set by the corporation at the time the shares are originally issued. At any time after the first sale, the shares may be traded at prices that are greater than or less than the par value, depending on investors' expectations of the return that will be earned by the business in the future. There is no reason to expect the market price to equal the par value for very long after the shares were first sold. Nonetheless, the amount recorded in the balance sheet is the original par value. New shares sold anytime after the first issue may have a par value of their own, distinct from the original issue.

The second part of owners' equity usually shown on the balance sheet is retained earnings. **Retained earnings** includes the cumulative sum of earnings from normal operations, in addition to gains (or losses) from transactions like the sale of plant assets or investments, the proceeds of which have been reinvested in the business (i.e., not paid out as dividends). Firms retain earnings mainly to expand operations through the purchase of additional assets. Contrary to what one may think, retained earnings do not represent cash. They may be invested in assets such as equipment and inventory.

The balance sheet gets its name from the idea that the total assets are equal in value to or *balanced by* the sum of the total liabilities and the owners' equity. A simple way of thinking about it is that the money used to buy each asset has to come from debt (Liabilities) and/or equity (Owners' Equity). ■

Example 6.5

Ian Claymore is the accountant at Major Electric. He has just realized that he forgot to include in the balance sheet for November 30, 2000, a government loan of $10 000 to help in the purchase of a $25 000 test stand (which he also forgot to include). The loan is due to be repaid in two years. When he revises the statement, what changes should he make?

The government loan is a long-term liability because it is due in more than one year. Consequently, an extra $10 000 would appear in long-term liabilities. This extra $10 000 must also appear as an asset for the balance to be maintained. The $25 000 value of the test stand would appear as equipment, increasing the equipment amount to $6 525 000. The $15 000 extra must come from a diminution in cash from $39 000 to $24 000.

Depending on the timing of the purchase, depreciation for the test stand might also be recognized in the balance sheet. Depreciation would reduce the net value of the equipment by the depreciation amount. The same amount would be balanced in the liabilities section by a reduction in retained earnings. ■

6.3.3 The Income Statement

An **income statement** summarizes an enterprise's revenues and expenses over a specified accounting period. Months, quarters, and years are commonly used as reporting periods. As with the balance sheet, the heading of an income statement gives the name of the enterprise and the reporting period. The income statement first lists revenues, by type, followed by expenses. Expenses are then subtracted from revenues to give profit (or income) before taxes. Income taxes are then deducted to obtain net profit.

Example 6.6

The income statement for the Major Electric Company for the year ended November 30, 2000, is shown in Table 6.3.

Table 6.3 Income Statement for the Major Electric Company

Major Electric Company Income Statement for the Year Ended November 30, 2000		
Revenues		
Sales	$7 536 000	
Management fees earned	106 000	
Total Revenues		**$7 642 000**
Expenses		
Cost of goods sold	$6 007 000	
Selling costs	285 000	
Administrative expenses	757 000	
Interest paid	86 000	
Total Expenses		**$7 135 000**
Profit before taxes		$507 000
Taxes (at 40%)		202 800
Net Profit		**$304 200**

We see that Major Electric's largest source of revenue was the sale of goods. They also earned revenue from the sale of management services to other companies. The largest expense was the cost of the goods sold. This includes the cost of raw materials, production costs, and other costs incurred to produce the items sold. Sometimes firms choose to list cost of goods sold as a *negative revenue.* The cost of goods sold will be subtracted from the sales to give a net sales figure. The net sales amount is the listed revenue, and the cost of goods sold does not appear as an expense.

The particular entries listed as either revenues or expenses will vary, depending on the nature of the business and the activities carried out. All revenues and expenses appear in one of the listed categories. For Major Electric, for example, the next item on the list of expenses, selling cost, includes delivery cost and other expenses such as salespersons' pay. Administrative expenses include those costs incurred in running the company that are not directly associated with manufacturing. Payroll administration is an example of an administrative expense.

Subtracting the expenses from the revenues gives a measure of profit for the company over the accounting period, one year in the example. However, this profit is taxed at a rate that depends on the company's particular circumstances.

For Major Electric, the tax rate is 40%, so the company's profit is reduced by that amount. ■

Example 6.7

Refer back to Example 6.5. Ian Claymore also forgot to include the effects of the loan and test stand purchase in the income statement shown in Example 6.6. When he revises the statement, what changes should he make?

Neither the loan nor the purchase of the asset appears directly in the income statement. The loan itself is neither income nor expense; if interest is paid on it, this is an expense. The test stand is a depreciable asset, which means only the depreciation for the test stand appears as an expense. ■

6.3.4 The Statement of Changes in Financial Position

The income statement summarizes the revenues and expenses of a business over a period of time. However, it does not directly give information about the generation of cash. It can be useful to augment information contained in the income statement. The **statement of changes in financial position** (sometimes called a statement of sources and uses of funds or a funds statement) is an accounting report that shows how much cash was generated by a company's operation and by other sources during an accounting period. The statement also shows the amounts of cash used for investments as well as for other non-operating disbursements.

Example 6.8

The statement of changes in financial position for Major Electric for year 2000 is shown in Table 6.4.

We usually expect cash provided by operations to be substantially greater than after-tax profits. This is because of expenses, like depreciation, that do not require cash outlays. These expenses are deducted from revenue to arrive at profits. However, since they do not require cash outlays, they are not deducted from revenues to arrive at cash flow. In Major Electric's case, we find that there is only a small difference between after-tax profit and cash flow provided by operations. The small difference between Major Electric's after-tax profit of $304 200 and the amount of cash provided by operations, $305 200, is due to two opposing forces that nearly cancel each other.

First, two expenses that did not require cash, depreciation and drawing down the materials inventory, made cash generated by operations larger than after-tax profits. This is normally expected. Second, the use of resources to produce finished goods that were not sold reduced the amount of cash generated by operations. However, this did not reduce reported profits because only the cost of goods actually sold was deducted from revenue. Such buildups of inventory and the resulting near equality between after-tax profits and cash generated by opera-

Table 6.4 Statement of Changes in Financial Position for Major Electric

Major Electric Company Statement of Changes in Financial Position for the Year Ended November 30, 2000		
Sources of Cash		
Cash from operating activities		
Net profit after taxes	$304 200	
Add depreciation	620 000	
Add decrease in materials inventory	2 000	
Less increase in finished goods inventory	(621 000)	
Total cash from operating activities		$305 200
Cash from financing activities		
Loan		1 000 000
Total Sources of Cash		$1 305 200
Uses of Cash		
Purchases of equipment	$1 300 000	
Dividend at $0.10 per share	100 000	
Total Uses of Cash		$1 400 000
Increase (Decrease) in Cash		$ (94 800)

tions are not typical. Inventories are expensive and businesses try to limit them. Perhaps Major Electric should look into the reasons why they have had a buildup of inventory. ■

Example 6.9

Refer back to Example 6.5. Ian Claymore wants to include the effects of the loan and test stand purchase in the statement of changes in financial position. When he revises the statement, what changes should he make?

The government loan is a source of cash, so an extra $10 000 would appear in that section of the statement. Similarly, the purchase of the test stand would increase the amount of purchases of equipment by $25 000 to $1 325 000. ■

6.3.5 Estimated Values in Financial Statements

The values in financial statements appear to be authoritative. However, many of the values in financial statements are estimates that are based on the **cost principle of accounting**. The cost principle of accounting states that assets are to be valued on the basis of their cost as opposed to market or other values. For example, the $500 000 given as the value of the land held by Major Electric was what Major Electric paid for the land. The market value of the land may now be greater or less than $500 000.

The value of plant and equipment is also based on cost. The value reported in the balance sheet is given by the initial cost minus accumulated depreciation. If Major Electric tried to sell the equipment, they might get more or less than this because depreciation models only approximate market value. For example, if there were a significant improvement in new equipment offered now by equipment suppliers compared with when Major Electric bought their equipment, Major Electric might get less than the $2 500 000 shown on the balance sheet.

Consider the finished goods inventory as another example of an estimated value. The number reported is Major Electric's manufacturing cost for producing the items. The implicit assumption being made is that Major Electric will be able to sell these goods at approximately their normal prices. Their estimated value may be reduced in later balance sheets if it appears that Major Electric cannot sell the goods easily.

The value of accounts receivable is clearly an estimate. Some fraction of accounts receivable may never be collected by Major. The figure in the balance sheet reflects what the accountant believes to be a conservative estimate based on experience.

In summary, when examining financial statement data, it is important to remember that many reported figures are estimates. Most firms include their accounting methods and assumptions within their periodic reports to assist in the interpretation of the figures.

6.3.6 Financial Ratio Analysis

Performance measures are calculated values that allow conclusions to be drawn from data. Performance measures drawn from financial statements can be used to answer such questions as:

1. Is the firm able to meet its short-term financial obligations?
2. Are sufficient profits being generated from the firm's assets?
3. How dependent is the firm on its creditors?

Financial ratios are one kind of performance measure that can answer these questions. They give an analyst a framework for asking questions about the firm's liquidity, asset management, leverage, and profitability. Financial ratios are ratios between key amounts taken from the financial statements of the firm. While financial ratios are simple to compute, they do require some skill to interpret, and they may be used for different purposes. For example, internal management may be concerned with the firm's ability to pay its current liabilities or the effect of long-term borrowing for a plant expansion. An external investor may be interested in past and current earnings to judge the wisdom of investing in the firm's stock. A bank will assess the riskiness of lending money to a firm before extending credit.

To properly interpret financial ratios, analysts commonly make comparisons with ratios computed for the same company from previous financial

Table 6.5 Industry-Total Balance Sheet Financial Data for Year-End 1996 (in Millions of $)

Balance Sheet (as of Year-End 1996)	Telecommunications Carriers	Electronic Equipment and Computer Services
Assets		
Assests		
Cash and deposits	$ 483	$ 2 547
Accounts receivable and accrued revenue	3 844	9 169
Inventories	411	4 344
Investments and accounts with affiliates	3 729	10 066
Portfolio investments	872	1 299
Loans	62	839
Capital assets, net	32 506	4 966
Other assets	3 108	2 898
Total assets	$45 015	$36 127
Liabilities and Owners' Equity		
Liabilities		
Accounts payable and accrued liabilities	$ 4 293	$ 8 487
Loans and accounts with affiliates	5 111	3 645
Borrowings:		
Loans and overdrafts from banks	509	1 408
Loans and overdrafts from others	250	1 345
Bankers' acceptances and paper	680	232
Bonds and debentures	12 174	2 487
Mortgages	88	137
Deferred income tax	3 075	(247)
Other liabilities	2 217	1 931
Total liabilities	$28 398	$19 426
Owner's Equity		
Share capital	10 665	9 177
Contributed surplus and other	1 576	415
Retained earnings	4 376	7 110
Total owners' equity	$16 617	$16 702
Current assets	$ 5 334	$17 663
Current liabilities	9 555	11 692

statements (a **trend analysis**) and with industry standard ratios. This is referred to as **financial ratio analysis.**

Industry standards can be obtained from catalogues routinely published by Dun & Bradstreet Canada and by Statistics Canada. The *Financial and Taxation Statistics for Enterprises* is a Statistics Canada publication that lists financial data from the balance sheets, income statements, and statements of changes in financial position for numerous industries. Also reported are selected financial ratios. Examples of industry total balance sheet and income statement figures (in millions of dollars) from 1996 are shown in Tables 6.5 on page 223 and 6.6. For individual firms operating within the telecommunications and electronic equipment indus-

Table 6.6 Industry-Total Income Statement Financial Statistics for the Year 1996 (Millions of $)

Income Statement (for the Year Ending in 1996)		
	Telecommunications Carriers	**Electronic Equipment and Computer Services**
Operating revenues	**$22 509**	**$50 681**
Sales of goods and services	22 091	49 357
Other operating revenues	419	1 324
Operating expenses	**18 514**	**48 091**
Depreciation, depletion, and amortization	4 443	1 501
Other operating expenses	14 071	46 590
Operating profit	**3 995**	**2 590**
Other revenue	**229**	**309**
Interest and dividends	229	309
Other expenses	**1 613**	**646**
Interest on short-term debt	192	168
Interest on long-term debt	1 421	478
Gains (Losses)	**3**	**(122)**
On sale of assets	90	25
Others	(87)	(148)
Profit before income tax	**2 615**	**2 130**
Income tax	1 138	715
Equity in affiliates' earnings	(32)	123
Profit before extraordinary gains	**1 445**	**1 538**
Extraordinary gains/losses	—	—
Net profit	**1 445**	**1 538**

tries, these figures allow an analyst to compare the firm's financial statements with those of the appropriate industry.

We shall see in the next section how the industry total financial data can be used to assess the health of a firm.

6.3.7 Financial Ratios

There are numerous financial ratios used in a financial analysis of a firm. We introduce five commonly used ratios to illustrate their use in a comparison with industry standards and in trend analysis. To facilitate the discussion, we shall use the balance sheets and income statements for Electco Electronics, a small electronics equipment manufacturer.

Example 6.10

Tables 6.7 and 6.8 give the balance sheet and income statement figures for Electco Electronics for the years 1996, 1997, and 1998.

The first two financial ratios we address are referred to as **liquidity ratios**. Liquidity ratios evaluate the ability of a business to meet its current liability obligations. In other words, they help us evaluate its ability to weather unforeseen fluctuations in cash flows. A company that does not have a reserve of cash, or other assets that can be converted to cash easily, may not be able to fulfil its short-term obligations.

A company's reserve of cash and assets easily converted to cash is called its working capital. **Working capital** is simply the difference between total current assets and total current liabilities:

$$\text{Working capital} = \text{Current assets} - \text{Current liabilities}$$

The adequacy of working capital is commonly measured with two ratios. The first, the **current ratio**, is the ratio of all current assets relative to all current liabilities. The current ratio may also be referred to as the **working capital ratio**.

$$\text{Current ratio} = \frac{\text{Current assets}}{\text{Current liabilities}}$$

Electco Electronics had a current ratio of \$4314/\$2489 = 1.73 in 1996. Ordinarily, a current ratio of 2 is considered adequate, although this determination may depend a great deal on the composition of current assets. It also may depend on industry standards. In the case of Electco, the industry standard can be obtained from Table 6.5. It is \$17 663/\$11 692 = 1.51. It would appear that Electco had a reasonable amount of liquidity in 1996, certainly more than the industry average.

A second ratio, the acid-test ratio, is more conservative than the current ratio. The **acid-test ratio** (also known as the **quick ratio**) is the ratio of quick

Table 6.7 Balance Sheets for Electco Electronics at Year-Ends 1996, 1997, 1998

Electco Electronics Year-End Balance Sheets *(in thousands of dollars)*	1996	1997	1998
Assets			
Current assets			
Cash	431	340	320
Accounts receivable	2489	2723	2756
Inventories	1244	2034	2965
Prepaid services	150	145	149
Total current assets	4314	5242	6190
Long-term assets			
Buildings and equipment (net of depreciation)	3461	2907	2464
Land	521	521	521
Total long-term assets	3982	3428	2985
Total Assets	8296	8670	9175
Liabilities			
Current liabilities			
Accounts payable	1493	1780	2245
Bank overdraft	971	984	992
Accrued taxes	25	27	27
Total current liabilities	2489	2791	3264
Long-term liabilities			
Mortgage	2489	2455	2417
Total Liabilities	4978	5246	5681
Owners' Equity			
Share capital	1825	1825	1825
Retained earnings	1493	1599	1669
Total Owners' Equity	3318	3424	3497
Total Liabilities and Owners' Equity	8296	8670	9175

Table 6.8 Income Statements for Electco Electronics for Years Ended 1996, 1997, 1998

Electco Electronics Income Statements *(in thousands of dollars)*	1996	1997	1998
Revenues			
Sales	12 440	11 934	12 100
Total Revenues	12 440	11 934	12 100
Expenses			
Cost of goods sold (excluding depreciation)	10 100	10 879	11 200
Depreciation	692	554	443
Interest paid	346	344	341
Total Expenses	11 138	11 777	11 984
Profit before taxes	1 302	157	116
Taxes (at 40%)	521	63	46
Profit before extraordinary items	781	94	70
Extraordinary gains/losses	70		
Profit after taxes	**851**	**94**	**70**

assets to current liabilities:

$$\text{Acid-test ratio} = \frac{\text{Quick assets}}{\text{Current liabilities}}$$

The acid-test ratio recognizes that some current assets, for example, inventory and prepaid expenses, may be more difficult to turn into cash than others. *Quick assets* are cash, accounts receivable, notes receivable, and temporary investments in marketable securities—those current assets considered to be highly *liquid*.

The acid-test ratio for Electco for 1996 was (\$431 + \$2489)/\$2489 = 1.17. Normally, an acid-test ratio of 1 is considered adequate, as this indicates that a firm could meet all its current liabilities with the use of its *quick* current assets if it were necessary. Electco appears to meet this requirement.

The current ratio and the acid-test ratio provide important information about how liquid a firm is, or how well it is able to meet its current financial obligations. The extent to which a firm relies on debt for its operations can be captured by what are called **leverage** or **debt-management ratios**. An example of such a ratio is the **equity ratio**. It is the ratio of total owners' equity to total assets. The smaller this ratio, the more dependent the firm is on debt for its operations and the higher are the risks the company faces.

$$\text{Equity ratio} = \frac{\text{Total owners' equity}}{\text{Total liabilities + Total equity}}$$
$$= \frac{\text{Total owners' equity}}{\text{Total assets}}$$

The equity ratio for Electco in 1996 was \$3318/\$8296 = 0.40 and the industry average was \$16 702/\$36 127 = 0.46. Electco has paid for roughly 60% of its assets with debt; the remaining 40% represents equity. This is close to the industry average and would appear acceptable.

Another group of ratios is called **asset-management** or **efficiency ratios**. They assess how efficiently a firm is using its assets. Inventory-turnover ratio is an example. **Inventory-turnover ratio** specifically looks at how efficiently a firm is using its resources to manage its inventories. This is reflected in the number of times that its inventories are replaced (or turned over) per year. The inventory-turnover ratio provides a measure of whether the firm has more or less inventory than normal.

$$\text{Inventory-turnover ratio} = \frac{\text{Sales}}{\text{Inventories}}$$

Electco's turnover ratio for 1996 was \$12 440/\$1244 = 10 turns per year. This is reasonably close to the industry average of \$40 357/\$4344 = 11.36 turns per year. In 1996, Electco invested roughly the same amount in inventory per dollar of sales as the industry did, on average.

Two points should be observed about the inventory-turnover ratio. First, the sales amount in the numerator has been generated over a period of time, while the inventory amount in the denominator is for one point in time. A more accurate measure of inventory turns would be to approximate the average inventory over the period in which sales were generated.

A second point is that sales refer to market prices, while inventories are listed at cost. The result is that the inventory-turnover ratio as computed above will be an overstatement of the true turnover. It may be more reasonable to compute inventory turnover based on the ratio of cost of goods sold to inventories. Despite this observation, traditional financial analysis uses the sales to inventories ratio.

The next ratio gives evidence of how productively assets have been employed in producing a profit. The **return-on-total-assets** or **net-profit ratio** is an example of a **profitability ratio**:

$$\text{Return-on-total-assets ratio} = \frac{\text{Profits after taxes (but before extraordinary items)}}{\text{Total assets}}$$

Electco had a return on total assets of \$781/\$8296 = 0.0941 or 9.41%, roughly twice the industry average. The industry average return on total assets for 1996 was $1538/$36 127 = 0.0426, or 4.26%. Note the comments on extraordinary items in Close-Up 6.2.

CLOSE-UP 6.2 Extraordinary Items

Extraordinary items are gains and losses that do not typically result from a company's normal business activities, are not expected to occur regularly, and are not recurring factors in any evaluations of the ordinary operations of the business. For example, cost or loss of income caused by natural disasters (floods, tornadoes, ice storms, etc.) would be extraordinay loss. Revenue created by the sale of a division of a firm is an example of extraordinary gain. Extraordinary items are reported separately from regular items and are listed net of applicable taxes.

Overall, Electco's performance in 1996 is very similar to that of the electronics equipment industry as a whole. The one exception is that Electco generated higher profits than the norm; it may have been extremely efficient in its operations that year.

The rosy profit picture painted for Electco in 1996 does not appear to extend into 1997 and 1998, as a trend analysis shows. Table 6.9 shows the financial ratios computed for 1997 and 1998 with those of 1996 and the industry standard of 1996.

For more convenient reference, we have summarized the five financial ratios we have dealt with and their definitions in Table 6.10.

Electco's return on total assets has dropped significantly over the three-year period. Though the current and quick ratios indicate that Electco should be able to meet its short-term liabilities, there has been a significant buildup of inventories over the period. Electco is not selling what it is manufacturing. This would explain the drop in Electco's inventory turns.

Table 6.9 1996 Industry Standard Ratios and Financial Ratios for Electco Electronics for 1996, 1997, and 1998.

Financial Ratio	Industry Standard	Electco Electronics		
	1996	**1996**	**1997**	**1998**
Current ratio	1.51	1.73	1.88	1.90
Quick ratio	—	1.17	1.10	0.94
Equity ratio	0.46	0.40	0.39	0.38
Inventory-turnover ratio	11.36	10.00	5.87	4.08
Return-on-total-assets ratio	4.26%	9.41%	1.09%	0.76%

Table 6.10 A Summary of Financial Ratios and Definitions

Ratio	Definition	Comments
Current ratio (Working capital ratio)	$\frac{\text{Current assets}}{\text{Current liabilities}}$	A liquidity ratio
Acid-test ratio (Quick ratio)	$\frac{\text{Quick assets}}{\text{Current liabilities}}$	A liquidity ratio (quick assets = current assets − inventories − prepaid items)
Equity ratio	$\frac{\text{Total equity}}{\text{Total assets}}$	A leverage or debt-management ratio
Inventory-turnover ratio	$\frac{\text{Sales}}{\text{Inventories}}$	An asset management or efficiency ratio
Return-on-total-assets ratio (Net-profit ratio)	$\frac{\text{Profits after taxes}}{\text{Total assets}}$	A profitability ratio (excludes extraordinary items)

Coupled with rising inventory levels is a slight increase in the cost of goods sold over the three years. From the building and equipment figures in the balance sheet, we know that no major capital expenditures on new equipment have occurred during the period. Electco's equipment may be aging and need replacement, though further analysis of what is happening is necessary before any conclusions on this issue can be drawn.

A final observation is that Electco's accounts receivable seems to be growing over the three-year period. Since there may be risks associated with the possibility of bad debt, Electco should probably investigate the matter.

In summary, Electco's main problem appears to be a mismatch between production levels and sales levels. Other areas deserving attention are the increasing cost of goods sold and possible problems with accounts receivable collection. These matters need investigation if Electco is to recover from its current slump in profitability. ■

We close the section on financial ratios with some cautionary notes on their use. First, since financial statement figures are often approximations, we need to interpret the financial ratios accordingly. In addition, accounting practices vary from firm to firm and may lead to differences in ratio values. Wherever possible, look for explanations of how figures are derived in financial reports.

A second problem encountered in comparing a firm's financial ratios with an industry standard is that it may be difficult to determine what industry the firm best fits into. Furthermore, within each industry, large variations exist. In some cases, an analyst may construct a relevant "average" by searching out a small number of similar firms (in size and business type) that may be used to form a customized industry average.

Finally, it is important to recognize the effect of seasonality on the financial ratios calculated. Many firms have highly seasonal operations with natural high and low periods of activity. An analyst needs to judge these fluctuations in context. One solution to this problem is to adjust the data seasonally through the use of averages. Another is to collect financial data from several seasons so that any deviations from the normal pattern of activity can be picked up.

Despite our cautionary words on the use of financial ratios, they do provide a useful framework for analyzing the financial health of a firm and for answering many questions about its operations.

REVIEW PROBLEMS

REVIEW PROBLEM 6.1

Joan is the sole proprietor of a small lawn care service. Last year, she purchased an eight-horsepower chipper-shredder to make mulch out of small tree branches and leaves. At the time it cost $760. She expects that the value of the chipper-shredder will decline by a constant amount each year over the next six years. At the end of six years, she thinks that it will have a salvage value of $100.

Construct a table that gives the book value of the chipper-shredder at the end of each year, for six years. Also indicate the accumulated depreciation at the end of each year. A spreadsheet may be helpful.

ANSWER

The depreciation amount for each year is

$$D_{sl}(n) = \frac{(P - S)}{N} = \frac{\$760 - \$100}{6} = \$110 \qquad n = 1, \ldots, 6$$

This is the requested table.

Year	Depreciation Amount	Book Value	Accumulated Depreciation
0		$760	
1	$110	650	$110
2	110	540	220
3	110	430	330
4	110	320	440
5	110	210	550
6	110	100	660

■

REVIEW PROBLEM 6.2

A three-year-old extruder used in making plastic yogurt cups has a current book value of \$168 750. The declining-balance method of depreciation with a rate $d = 0.25$ is used to determine depreciation amounts. What was its original price? What will be its book value two years from now?

ANSWER

Let the original price of the extruder be P. The book value three years after purchase is \$168 750. This means that the original price was

$$BV_{db}(3) = P(1 - d)^3$$

$$\$168\ 750 = P(1 - 0.25)^3$$

$$P = \$400\ 000$$

The original price was \$400 000.

Two years from now, the book value will be the book value five years after purchase.

$$BV_{db}(5) = \$400\ 000(1 - 0.25)^5$$
$$= \$94\ 921.88$$

or

$$BV_{db}(3) = \$168\ 750(1 - 0.25)^2$$
$$= \$94\ 921.88$$

The book value two years from now will be \$94 921.88. ■

REVIEW PROBLEM 6.3

You have been given the following data from the Fine Fishing Factory for the year ending December 31, 1999. Construct an income statement and a balance sheet from the data.

Accounts payable	\$ 27 500
Accounts receivable	32 000
Advertising expense	2 500
Bad debts expense	1 100
Buildings, net	14 000
Cash	45 250
Common stock	125 000
Cost of goods sold	311 250
Depreciation expense, buildings	900
Government bonds	25 000
Income taxes	9 350
Insurance expense	600

Interest expense	500
Inventory, December 31, 1999	42 000
Land	25 000
Machinery, net	3 400
Mortgage due May 30, 2002	5 000
Office equipment, net	5 250
Office supplies expense	2 025
Other expenses	7 000
Prepaid expenses	3 000
Retained earnings	?
Salaries expense	69 025
Sales	421 400
Taxes payable	2 500
Wages payable	600

ANSWER

Solving this problem consists of sorting through the listed amounts and identifying which are balance sheet entries and which are income statement entries. Then, assets can be separated from liabilities and owners' equity, and revenues from expenses.

Fine Fishing Factory
Balance Sheet
as of December 31, 1999

Assets	
Current Assets	
Cash	$ 45 250
Accounts receivable	32 000
Inventory, December 31, 1999	42 000
Prepaid expenses	3 000
Total Current Assets	$122 250
Long-Term Assets	
Land	25 000
Government bonds	25 000
Machinery, net	3 400
Office equipment, net	5 250
Buildings, net	14 000
Total Long-Term Assets	$ 72 650
Total Assets	**$194 900**

Liabilities	
Current Liabilities	
Accounts payable	$ 27 500
Taxes payable	2 500
Wages payable	600
Total Current Liabilities	$ 30 600
Long-Term Liabilities	
Mortgage due May 30, 2002	5 000
Total Long-Term Liabilities	$ 5 000
Total liabilities	**$ 35 600**
Owners' Equity	
Common stock	$125 000
Retained earnings	34 300
Total Owners' Equity	**$159 300**
Total Liabilities and Owners' Equity	**$194 900**

Fine Fishing Factory
Income Statement
for the Year Ending December 31, 1999

Revenues	
Sales	$421 400
Cost of goods sold	311 250
Net revenue from sales	$110 150
Expenses	
Salaries expense	69 025
Bad debts expense	1 100
Advertising expense	2 500
Interest expense	500
Insurance expense	600
Office supplies expense	2 025
Other expenses	7 000
Depreciation expense, buildings	900
Depreciation expense, office equipment	850
Total Expenses	$ 84 500
Profit before taxes	$ 25 650
Income taxes	9 350
Profit after taxes	**$ 16 300**

■

REVIEW PROBLEM 6.4

Perform a ratio analysis for the Major Electric Company using the balance sheet and income statement from Sections 6.3.2 and 6.3.3. Industry standards for the ratios are as follows:

Ratio	Industry Standard
Current ratio	1.80
Acid-test ratio	0.92
Equity ratio	0.71
Inventory-turnover ratio	14.21
Return-on-total-assets ratio	7.91%

ANSWER

The ratio computations for Major Electric are:

$$\text{Current ratio} = \frac{\text{Current assets}}{\text{Current liabilities}} = \frac{\$801\ 000}{\$90\ 000} = 8.9$$

$$\text{Acid-test ratio} = \frac{\text{Quick assets}}{\text{Current liabilities}} = \frac{\$66\ 000}{\$90\ 000} = 0.73$$

$$\text{Equity ratio} = \frac{\text{Total equity}}{\text{Total assets}} = \frac{\$4\ 311\ 000}{\$5\ 401\ 000} = 0.7982 \cong 0.80$$

$$\text{Inventory-turnover ratio} = \frac{\text{Sales}}{\text{Inventories}} = \frac{\$7\ 536\ 000}{\$683\ 000}$$
$$= 11.03 \text{ turns per year}$$

$$\text{Return-on-total-assets ratio} = \frac{\text{profits after taxes}}{\text{total assets}} = \frac{\$304\ 200}{\$5\ 401\ 000}$$
$$= 0.0563 \text{ or } 5.63\% \text{ per year}$$

A summary of the ratio analysis results follows:

Ratio	Industry Standard	Major Electric
Current ratio	1.80	8.90
Acid-test ratio	0.92	0.73
Equity ratio	0.71	0.80
Inventory-turnover ratio	14.21	11.03
Return-on-total-assets ratio	7.91%	5.63%

Major Electric's current ratio is well above the industry average and well above the general guideline of 2. They appear to be quite liquid. However, the acid-test ratio, with a value of 0.73, gives a slightly different view of Major Electric's liquidity. Most of Major Electric's current assets are inventory; thus, the current ratio is somewhat misleading. If we look at the acid test, Major Electric's quick assets are only 73% of their current liabilities. They may have a difficult time meeting their current debt obligations if they have unforeseen difficulties with their cash flows.

Major Electric's equity ratio of 0.80 indicates that it is not heavily reliant on debt and therefore is not at high risk of going bankrupt. Major Electric's inventory turns are lower than the industry norm, as is their return on total assets.

Taken together, Major Electric appears to be in reasonable financial shape. One matter they should probably investigate is why their inventories are so high. With lower inventories, they could improve their inventory turns as well as their return on total assets. ■

SUMMARY

This chapter opened with a discussion of the concept of depreciation and various reasons why assets lose value. Two popular depreciation models, straight-line and declining-balance, were then presented as methods commonly used to approximate book value of capital assets and depreciation amounts.

The second part of the chapter dealt with the elements of financial accounting. We first presented the two main financial statements: the balance sheet and the income statement. The statement of changes in financial position was also introduced as a means of identifying changes in cash available to the firm. Next, we showed how these statements can be used to assess the financial health of a firm through the use of ratios. Comparisons with industry norms and trend analysis are normally part of financial analysis. We closed with cautionary notes on the interpretation of the ratios.

The significance of the material in this chapter is twofold. First, it sets the groundwork for material in Chapters 7 and 8, replacement analysis and taxation. Second, and perhaps more importantly, it is increasingly necessary for all engineers to have an understanding of depreciation and financial accounting as they become more and more involved in decisions that affect the financial positions of the organizations in which they work.

Engineering Economics in Action, Part 6B: Usually the Truth

Terry had shown Naomi what he thought was evidence of wrongdoing by a fellow employee. Naomi said, "Interesting observation, Terry. But you know, I think it's probably O.K. Let me explain. The main problem is that you are looking at two kinds of evaluation here, book value and market value. The book value is an estimate of what an asset is worth, while the market value is what it sells for."

Terry nodded."Yes, I know that. That's true about anything you sell. But this is different. We've got a $5000 estimate against a $300 sale. You can't tell me that our guess about the sales price can be that far out!"

"Yes, it can, and I'll tell you why. That book value is an estimate that has been calculated according to very particular rules. The Canadian tax rules require us to use a declining-balance depreciation method for almost all of our assets. They also specify which declining-balance rate to use, which they call the CCA, which stands for capital cost allowance. For most equipment it's 20%. Now, the reality is that things decline in value at different rates, and computers lose value really quickly. We could, for our own purposes, determine a book value for any asset that is a better estimate of its market value, but sometimes it's too much trouble to keep one set of figures for tax reasons and another for other purposes. So often everything is given a book value according to the tax rules, and consequently sometimes the difference between the book value and the market value can be a lot."

"But surely the government can see that computers in particular don't match that 20% rate. Or are they just ripping us off?"

"Well, they can see that. Until a few years ago, the CCA rate for computers was 20%. But because it was painfully obvious that this rate was too low, it was revised to 30%. This is still too low, as you can see from the sale of our own computers, but it is better than it was."

Terry smiled ruefully, "So our accounting statements don't really show the truth?"

Naomi smiled back, "I guess not, if by 'truth' you mean market value. But usually they're close. Usually." ◎

PROBLEMS

6.1 For each of the following, state whether the loss in value is due to use-related physical loss, time-related physical loss, or functional loss:

(a) Albert sold his two-year old computer for $500, but he paid $4000 for it new. It wasn't fast enough for the new software he wanted.

(b) Beatrice threw out her old tennis shoes because the soles had worn thin.

(c) Claudia threw out her old tennis shoes because she is jogging now instead.

(d) Day-old bread is sold at half price at the neighbourhood bakery.

(e) Egbert sold his old lawnmower to his neighbour for $20.

(f) Fred picked up a used overcoat at the thrift store for less than 10% of the new price.

(g) Greg notices that newspapers cost $0.50 on the day of purchase, but are worth less than $0.01 each as recyclable newsprint.

(h) Harold couldn't get the price he wanted for his house because the exterior paint was faded and flaking.

6.2 For each of the following, state whether the value is a market value, book value, scrap value, or salvage value:

(a) Inta can buy a new stove for $800 at Joe's Appliances.

(b) Jack can sell his used stove to Inta for $200.

(c) Kitty can sell her used stove to the recyclers for $20.

(d) Liam can buy Jack's used stove for $200.

(e) Mick is adding up the value of the things he owns. He figures his stove is worth at least $200.

6.3 A new industrial sewing machine costs in the range of $5000 to $10 000. Technological change in sewing machines does not occur very quickly, nor is there much change in the functional requirements of a sewing machine. A machine can operate well for many years with proper care and maintenance. Discuss the different reasons for depreciation and which you think would be most appropriate for a sewing machine.

6.4 Communications network switches are changing dramatically in price and functionality as changes in technology occur in the communications industry. Prices drop frequently as more functionality and capacity are achieved. A switch only six months old will have depreciated since it was installed. Discuss the different reasons for depreciation and which you think would be most appropriate for this switch.

6.5 Ryan owns a five hectare plot of land in the countryside. He has been planning to build a cottage on the site for some time, but has not been able to afford it yet. However, five years ago, he dug a pond to collect rainwater as a water supply for the cottage. It has never been used, and is beginning to fill in with plant life and garbage that has been dumped there. Ryan realizes that his investment in the pond has depreciated in value since he dug it. Discuss the different reasons for depreciation and which you think would be most appropriate for the pond.

6.6 A company that sells a particular type of web-indexing software has had two larger firms approach it for a possible buyout. The current value of the company, based on recent financial statements, is $4.5 million. The two bids were for $4 million and $7 million, respectively. Both bids were bona fide, in that they were real offers. What is the market value of the company? Its book value?

6.7 An asset costs $14 000 and has a scrap value of $3000 after seven years. Calculate its book value using straight-line depreciation:

(a) after one year

(b) after four years

(c) after seven years

6.8 An asset costs \$14 000. At a depreciation rate of 20%, calculate its book value using declining-balance depreciation:

(a) after one year

(b) after four years

(c) after seven years

6.9 An asset costs \$14 000 and has a scrap value of \$3000 after seven years.

(a) What depreciation rate could be used to estimate the book value of the asset using the declining-balance method?

(b) Using this depreciation rate, what is the book value of the asset after four years?

6.10 Using a spreadsheet program, chart the book value of a \$14 000 asset over a seven-year life using declining-balance depreciation (d = 0.2). On the same chart, show the book value of the \$14 000 asset, using straight-line depreciation, with a scrap value of \$3000 after seven years.

6.11 Using a spreadsheet program, chart the book value of a \$150 000 asset for the first 10 years of its life at depreciation rates of 5%, 20%, and 30%.

6.12 A machine has a life of 30 years, costs \$245 000, and has a salvage value of \$10 000 using straight-line depreciation. What depreciation rate will result in the machine's having the same book value for both the declining-balance and straight-line methods in year 20?

6.13 A new press brake costs Medicine Hat Steel \$780 000. It is expected to last 20 years, with a \$60 000 salvage value. What rate of depreciation for the declining-balance method will produce a book value after 20 years that equals the salvage value of the press?

6.14 **(a)** Using straight-line depreciation, what is the book value after four years for an asset costing \$150 000 that has a salvage value of \$25 000 after 10 years? What is the depreciation charge during the fifth year?

(b) Using declining-balance depreciation with d = 20%, what is the book value after four years for an asset costing \$150 000? What is the depreciation charge during the fifth year?

(c) What is the depreciation rate using declining-balance for an asset costing \$150 000 that has a salvage value of \$25 000 after 10 years?

6.15 Julia must choose between two different designs for a safety enclosure, which will be in use indefinitely. Model A has a life of three years, a first cost of \$8000, and maintenance of \$1000 per year. Model B will last four years, has a first cost of \$10 000, and has maintenance of \$800 per year. A salvage value can be estimated for Model A using a depreciation rate of 40% and declining-balance depreciation, while a salvage value for Model B can be estimated using straight-line depreciation and the knowledge that after one year its salvage value will be \$7500. Interest is at 14%. Using a present worth analysis, which design is better?

6.16 A company had net sales of \$20 000 last month. From the balance sheet for the end of last month, and an income statement for the month, you have determined that the current ratio was 2.0, the acid-test ratio was 1.2, and the inventory turnover was 2 per month. What was the value of the company's current assets?

6.17 Adventure Airline's new baggage handling conveyor costs \$250 000 and has a service life of 10 years. For the first six years, depreciation of the conveyor is calculated using the declining-balance method at the rate of 30%. During the last four years, the straight-line method is used for accounting purposes in order to have a book value of 0 at the end of the service life. What is the book value of the conveyor after 7, 8, 9, and 10 years?

6.18 Molly inherited \$5000 and decided to start a lawn-mowing service. With her inheritance and a bank loan of \$5000, she bought a used ride-on lawn mower and a used truck. For five years, Molly had a gross income of \$30 000, which covered the annual operating costs and the loan payment, both of which totalled \$4500. She spent the rest of her income personally. At the end of five years, Molly found that her loan was paid off but the equipment was wearing out.

(a) If the equipment (lawn mower and truck) was depreciating at the rate of 50% according to the declining-balance method, what is its book value after five years?

(b) If Molly wanted to avoid being left with a worthless lawn mower and truck and with no money for renewing them at the end of five years, how much should she have saved annually toward the second set of used lawn mower and used truck of the same initial value as the first set? Assume that Molly's first lawn mower and truck could be sold at their book value. Interest rate is 7%.

6.19 Enrique is planning for a trip around the world in three years. He will sell all of his possessions at that time to fund the trip. Two years ago, he bought a used car for \$12 500. He observes that the market value for the car now is about \$8300. He needs to know how much money his car will add to his stash for his trip when he sells it three years from now. Use the declining-balance depreciation method to tell him.

6.20 Ben is choosing between two different industrial fryers using an annual worth calculation. Fryer 1 has a five-year service life and a first cost of $400 000. It will generate net year-end savings of $128 000 per year. Fryer 2 has an eight-year service life and a first cost of $600 000. It will generate net year-end savings of $135 000 per year. If the salvage value is estimated using declining-balance depreciation with a 20% depreciation rate, and the MARR is 12%, which fryer should Ben buy?

6.21 Dick noticed that the book value for an asset he owned was exactly $500 higher if the value was calculated by straight-line depreciation over declining-balance depreciation, exactly half-way through the asset's service life, and the scrap value at the end of the service life was the same by either method. If the scrap value was $100, what was the purchase price of the asset?

6.22 In the last quarter, the financial-analysis report for XYZ Company revealed that the current, quick, and equity ratios were 1.9, 0.8, and 0.37, respectively. In order to improve their financial health, based on these financial ratios, the following strategies are considered by XYZ for the current quarter:

(i) Reduce inventory

(ii) Pay back short-term loans

(iii) Increase retained earnings

(a) Which strategy (or strategies) is effective for improving each of the three financial ratios?

(b) If only one strategy is considered by XYZ, which one seems to be most effective? Assume no other information is available for analysis.

6.23 The end-of-quarter balance sheets for XYZ Company indicated that the current ratio was 1.8 and the equity ratio was 0.45. They also indicated that the long-term assets were twice as much as the current assets, and half of the current assets were highly liquid. The total equity was $68 000. Since the current ratio was close to 2, XYZ feels that the company had a reasonable amount of liquidity in the last quarter. However, if XYZ wants more assurance, which financial ratio would provide further information? Using the information provided, compute the appropriate ratio and comment on XYZ's concern.

6.24 A potentially very large customer for Chicoutimi Metals wants to fully assess the financial health of the company in order to decide whether to commit to a long-term, high-volume relationship. You have been asked by the company president, Roch, to review the company's financial performance over the last three years and make a complete report to him. He will then select from your report information to present to the customer. Consequently, your report should be as thorough and honest as possible.

Research has revealed that in your industry (sheet metal products), the average value of the current ratio is 2.79, the equity ratio is 0.54, the inventory turnover is 4.9, and the net-profit ratio is 3.87. Beyond that information, you have access to only the balance sheet shown on this page and the income statement shown on the next page, and should make no further assumptions. Your report should be limited to the equivalent of about 300 words.

Chicoutimi Metals
Consolidated Balance Sheets
December 31, 1996, 1997, and 1998
(in thousands of dollars)

Assets

	1998	**1997**	**1996**
Current Assets			
Cash	19	19	24
Accounts receivable	779	884	1 176
Inventories	3 563	3 155	2 722
	4 361	4 058	3 922
Fixed Assets			
Land	1 136	1 064	243
Buildings and equipment	2 386	4 682	2 801
	3 552	5 746	3 044
Other Assets	413	....	3
Total Assets	8 296	9 804	6 969

Liabilities and Owners' Equity

	1998	**1997**	**1996**
Current Liabilities			
Due to bank	1 431	1 929	2 040
Accounts payable	1 644	1 349	455
Wages payable	341	312	333
Income tax payable	562	362	147
Long-Term Debt	2 338	4 743	2 528
Total Liabilities	6 316	8 695	5 503
Owners' Equity			
Capital stock	1 194	1 191	1 091
Retained earnings	786	(82)	375
Total Owner's Equity	1 980	1 109	1 466
Total Liability and Owners' Equity	8 296	9 804	6 969

Chicoutimi Metals
Income Statement
for the Years Ending December 31, 1996, 1997, and 1998
(in thousands of dollars)

	1998	1997	1996
Total revenue	9 355	9 961	8 470
Less: Costs	8 281	9 632	7 654
Net revenue	1 074	329	816
Less: Depreciation	447	431	398
Interest	412	334	426
Income taxes	117	21	156
Net income from operations	98	(457)	(164)
Add: Extraordinary item	770		(1 832)
Net income	868	(457)	(1 996)

6.25 The Chicoutimi Metals income statement and balance sheets shown in Problem 6.24 were in error. A piece of production equipment was sold for \$100 000 cash in 1998 and was not accounted for. Which items on these statements must be changed, and (if known) by how much?

6.26 The Chicoutimi Metals income statement and balance sheet shown in Problem 6.24 were in error. An extra \$100 000 in sales was made in 1998 and not accounted for. Only half of the payments for these sales have been received. Which items on these statements must be changed, and (if known) by how much?

6.27 At the end of last month, Estevan Manufacturing had \$45 954 in the bank. They owed the bank, because of their mortgage, \$224 000. They also had a working capital loan of \$30 000. Their customers owed them \$22 943, and they owed their suppliers \$12 992. The company owned property worth \$250 000. They had \$123 000 in finished goods, \$102 000 in raw materials, and \$40 000 in work in progress. Their production equipment was worth \$450 000 when new (partially paid for by a large government loan due to be paid back in three years) but had accumulated a total of \$240 000 in depreciation, \$34 000 worth last month.

The company has investors who put up \$100 000 for their ownership. It has been reasonably profitable; this month the gross income from sales was \$220 000, and the cost of the sales was only \$40 000. Expenses were also relatively low; salaries were \$45 000 last month, while the other expenses were depreciation, maintenance at \$1500, advertising at \$3400, and insurance at \$300. In spite of \$32 909 in accrued taxes (they pay taxes at 55%), the company had retained earnings of \$135 000.

Construct a balance sheet (at the end of this month) and income statement (for this month) for Estevan Manufacturing. Should the company release some of its retained earnings through dividends at this time?

6.28 Brandon Industries bought land and built its plant 20 years ago. The depreciation on the building is calculated using the straight-line method, with a life of 30 years and a salvage value of $50 000. The land is not depreciating. The depreciation for the equipment, all of which was purchased at the same time the plant was constructed, is calculated using declining-balance at 20%. Brandon currently has two outstanding loans, one for $50 000 due December 31, this year, and another one for which the next payment is due in four years.

Brandon Industries Balance Sheet as of June 30, 1998		
Assets		
[ ]		
Cash		$ 350 000
Accounts receivable		2 820 000
Inventories		2 003 000
Prepaid services		160 000
Total Current Assets		[ ]
Long-Term Assets		
Building	$200 000	
Less accumulated depreciation	[ ]	[ ]
Equipment	$480 000	
Less accumulated depreciation	[ ]	[ ]
Land		540 000
Total Long-Term Assets		[ ]
Total Assets		[ ]
Liabilities and Owners' Equity		
Current Liabilities		
Accounts payable		$ 921 534
[ ]		[ ]
Accrued taxes		29 000
Total [ ]		[ ]
Long-Term Liabilities		
Mortgage		$1 200 000
[ ]		318 000
Total Long-Term Liabilities		[ ]
Total [ ]		[ ]
Owners' Equity		
Capital stock		$1 920 000
[ ]		[ ]
Total Owners' Equity		[ ]
Total Liabilities and Owners' Equity		[ ]

During April 1998, there was a flood in the building because a nearby river overflowed its banks after unusually heavy rain. Pumping out the water and cleaning up the basement and the first floor of the building took a week. Manufacturing was suspended during this period, and some inventory was damaged. Because of lack of adequate insurance, this unusual and unexpected event cost the company $100 000 net.

(a) Fill in the blanks and complete a copy of the balance sheet on page 244 and the income statement below, using any of the above information you feel necessary.

(b) Show how information from financial ratios can indicate whether Brandon Industries can manage an unusual and unexpected event such as the flood without threatening its existence as a viable business.

Brandon Industries
Income Statement for the Year Ended June 30, 1998

Income		
Gross income from sales	$8 635 000	
Less [ ]	7 490 000	[ ]
Total income		[ ]
[ ]		
Depreciation		70 000
Interest paid		240 000
Other expenses		100 000
Total expenses		[ ]
Profit before taxes		[ ]
Taxes at 40%		[ ]
[ ]		[ ]
[ ]		[ ]
Profit after taxes		[ ]

6.29 Movit Manufacturing has the following alphabetized income statement and balance sheet entries from the year 1998. Construct an income statement and a balance sheet from the information given.

Accounts payable	$ 7 500
Accounts receivable	15 000
Accrued wages	2 850
Cash	2 100
Common shares	150
Contributed capital	3 000
Cost of goods sold	57 000
Current assets	

Current liabilities	
Deferred income taxes	2 250
Depreciation expense	750
General expense	8 100
GIC's	450
Income taxes	1 800
Interest expense	1 500
Inventories	18 000
Land	3 000
Less: Accumulated depreciation	10 950
Long-term assets	
Long-term bonds	4 350
Long-term liabilities	
Mortgage	9 450
Net income after taxes	2 700
Net income before taxes	4 500
Net plant and equipment	7 500
Net sales	76 500
Operating expenses	
Owners' equity	
Prepaid expenses	450
Selling expenses	4 650
Total assets	46 500
Total current assets	36 000
Total current liabilities	15 000
Total expenses	15 000
Total liabilities and owners' equity	46 500
Total long-term assets	10 500
Total long-term liabilities	16 050
Total owners' equity	15 450
Working capital loan	4 650

6.30 Calculate for Movit Manufacturing in Problem 6.29 the financial ratios listed in the table below. Using these ratios and those provided for 1996 and 1997, conduct a short analysis of Movit's financial health.

Movit Manufacturing Financial Ratios			
Ratio	**1998**	**1997**	**1996**
Current ratio		1.90	1.60
Acid-test ratio		0.90	0.75
Equity ratio		0.40	0.55
Inventory-turnover ratio		7.00	12.00
Return-on-total-assets ratio		0.08	0.10

6.31 Léger Lites' balance sheets for year-ends 1998 and 1999 and their 1999 income statement follow. Construct a statement of changes in financial position for the year ended in 1999. Use the major headings given on page 248 to summarize the changes in financial position.

Léger Lites Company
Comparative Balance Sheets
for the Years Ending December 31, 1998 and 1999
(in thousands of dollars)

	1998	1999
Assets		
Current Assets		
Cash	3 750	3 500
GIC's	750	750
Accounts receivable	21 250	25 000
Inventories	28 250	30 000
Prepaid expenses	500	750
Total Current Assets	54 500	60 000
Long-Term Assets		
Land	5 000	5 000
Plant and equipment	28 000	30 750
Less: Accumulated depreciation	17 000	18 250
Net Plant and Equipment	11 000	12 500
Total Long-Term Assets	16 000	17 500
Total Assets	70 500	77 500
Liabilities and Owner's Equity		
Current Liabilities		
Accounts payable	9 250	12 500
Accrued wages	2 250	4 750
Working capital loan	9 500	7 750
Total Current Liabilities	21 000	25 000
Long-Term Liabilities		
Deferred income taxes	3 500	3 750
Mortgage	16 500	15 750
Long term bonds	7 500	7 250
Total Long-Term Liabilities	27 500	26 750
Owners' Equity		
Common shares	250	250
Contributed capital	5 000	5 000
Retained earnings	16 750	20 500
Total Owners' Equity	22 000	25 750
Total Liabilities and Owners' Equity	70 500	77 500

Léger Lites Income Statement for the Year Ended December 31, 1999 *(in thousands of dollars)*	
Net sales	127 500
Cost of goods sold	(95 000)
Gross profit	32 500
Operating expenses	
Selling expenses	7 750
Depreciation expense	1 250
General expense	13 500
Interest expense	2 500
Total expenses	25 000
Net income before taxes	7 500
Income taxes	3 000
Net income after taxes	4 500
Common dividends paid	750
Increase in retained earnings	3 750

1. Cash from operating activities (Include all inflows and outflows.)
2. Cash from financial activities (This includes new debt or debt repayment, and dividend payments.)
3. Cash from investing activities (Sale or acquisition of assets)
4. Total cash increase (decrease)

Verify that the total cash increase or decrease is consistent with the cash amounts which appear on Léger Lites' balance sheets for year-ends 1998 and 1999.

6.32 Refer back to the year-end balance sheets and income statements for Electco Electronics (Tables 6.7 and 6.8). Construct a statement of changes in financial position for the 1998 year-end. Include the following major sections in your statement:

1. Cash from operating activities (Include all inflows and outflows.)
2. Cash from financial activities (This includes new debt or debt repayment, and dividend payments.)
3. Cash from investing activities (Sale or acquisition of assets)
4. Total cash increase (decrease)

Verify that the total cash increase or decrease is consistent with the cash amounts which appear in Electco Electronics' balance sheets for year ends 1997 and 1998.

6.33 Fraser Phraser operates a small publishing company. He is interested in getting a loan for expanding his computer systems. The bank has asked Phraser to supply them with his financial statements from the past two years. His statements appear below. Comment on Phraser's financial position with regard to the loan, using a financial ratio analysis.

Fraser Phraser Company
Comparative Balance Sheets
for the Years Ending in 1998 and 1999
(in thousands of dollars)

	1998	**1999**
Assets		
Current Assets		
Cash	22 500	1 250
Accounts receivable	31 250	40 000
Inventories	72 500	113 750
Total Current Assets	126 250	155 000
Long-Term Assets		
Land	50 000	65 000
Plant and equipment	175 000	250 000
Less: Accumulated depreciation	70 000	95 000
Net Plant and equipment	105 000	155 000
Total Long-Term Assets	155 000	220 000
Total Assets	281 250	375 000
Liabilities and Owners' Equity		
Current Liabilities		
Accounts payable	26 250	55 000
Working capital loan	42 500	117 500
Total Current Liabilities	68 750	172 500
Long-Term Liabilities		
Mortgage	71 875	57 375
Total Long-Term Liabilities	71 875	57 375
Owners' Equity		
Common shares	78 750	78 750
Retained earnings	61 875	66 375
Total Owners' Equity	140 625	145 125
Total Liabilities and Owners' Equity	281 250	375 000

Fraser Phraser Income Statements for Years Ending in 1998 and 1999 *(in thousands of dollars)*		
	1998	**1999**
Revenues		
Sales	156 250	200 000
Cost of goods sold	93 750	120 000
Net revenue from sales	62 500	80 000
Expenses		
Operating expenses	41 875	46 250
Depreciation expense	5 625	12 500
Interest expense	3 750	7 625
Total expenses	51 250	66 375
Profit before Taxes	11 250	13 625
Income taxes	5 625	6 813
Profit after Taxes	5 625	6 813

6.34 Complete a statement of changes in financial position for 1999 from the financial statements for Fraser Phraser from Problem 6.33. Does this provide you with additional information for the bank with respect to its decision to give Phraser the loan?

6.35 A friend of yours is thinking of purchasing shares in Petit Ourson Ltée in the near future, and decided that it would be prudent to examine its financial statements for the past two years before making a phone call to his stockbroker. The statements are shown on pages 251 and 252.

Your friend has asked you to help him conduct a financial ratio analysis. Fill out the financial ratio information on a copy of the table below. After comparison with industry standards, what advice would you give your friend?

Petit Ourson Ltée Financial Ratios			
	Industry Norm	**1997**	**1998**
Current ratio	4.50		
Acid-test ratio	2.75		
Equity ratio	0.60		
Inventory-turnover ratio	2.20		
Return-on-total-assets ratio	0.09		

Petit Ourson Ltée Comparative Balance Sheets for the Years Ending 1997 and 1998 *(in thousands of dollars)*		
	1997	**1998**
Assets		
Current Assets		
Cash	500	375
Accounts receivable	1 125	1 063
Inventories	1 375	1 563
Total Current Assets	3 000	3 000
Long-Term Assets		
Plant and equipment	5 500	6 500
Less: Accumulated depreciation	2 500	3 000
Net Plant and Equipment	3 000	3 500
Total Long-Term Assets	3 000	3 500
Total Assets	6 000	6 500
Liabilities and Owners' Equity		
Current Liabilities		
Accounts payable	500	375
Working capital loan	000	375
Total Current Liabilities	500	750
Long-Term Liabilities		
Bonds	1 500	1 500
Total Long-Term Liabilities	1 500	1 500
Owners' Equity		
Common shares	750	750
Contributed capital	1 500	1 500
Retained earnings	1 750	2 000
Total Owners' Equity	4 000	4 250
Total Liabilities and Owners' Equity	6 000	6 500

6.36 Construct an income statement and a balance sheet from the scrambled entries for Paradise Pond Company from the years 1997 and 1998, shown in the table at the bottom of page 252.

Petit Ourson Ltée
Income Statements
for the Years Ending in 1997 and 1998
(in thousands of dollars)

	1997	1998
Revenues		
Sales	3 000	3 625
Cost of goods sold	1 750	2 125
Net revenue from sales	1 250	1 500
Expenses		
Operating expenses	75	100
Depreciation expense	550	500
Interest expense	125	150
Total expenses	750	750
Profit before Taxes	500	750
Income taxes	200	300
Profit after Taxes	300	450

Paradise Pond Company	**1997**	**1998**
Accounts receivable	$ 675	$ 638
Less: Accumulated depreciation	1 500	1 800
Accounts payable	300	225
Bonds	900	900
Cash	300	225
Common shares	450	450
Contributed capital	900	900
Cost of goods sold	1 750	2 125
Depreciation expense	550	500
Income taxes	200	300
Interest expense	125	150
Inventories	825	938
Net plant and equipment	1 800	2 100
Net revenue from sales	1 250	1 500
Operating expenses	075	100
Plant and equipment	3 300	3 900
Profit after taxes	300	450
Profit before taxes	500	750
Retained earnings	1 050	1 200
Sales	3 000	3 625
Total assets	3 600	3 900
Total current assets	1 800	1 800
Total current liabilities	300	450
Total expenses	750	750
Total liabilities and owners' equity	3 600	3 900
Total long-term assets	1 800	2 100
Total long-term liabilities	900	900
Total owners' equity	2 400	2 550
Working capital loan	000	225

Capital Expenditure or Business Expense?

CANADIAN MINI-CASE 6.1

When a Calgary shopping centre began to experience considerable congestion on the roadways which provided access to the centre, it approached the local municipality to see if the roadways close to the mall could be improved. An agreement was struck with the municipality, and Oxford Shopping Centres Ltd. paid the municipality \$450 050 to make the improvements in lieu of any increase in local tax rates which might have otherwise been charged by the municipality.

Was the \$450 050 a business expense or a capital expenditure? The issue at hand was whether Oxford Shopping Centres Ltd. could claim the \$450 050 as a current expense in the year in which it was paid, or whether the amount had to be amortized over a period of years, with only a fraction of the total amount claimed as an expense each year. The owners of the shopping centre argued that the aim was to increase the popularity of the mall and that the outlay related to the business as a whole rather than an improvement to the physical premises. With this logic, Oxford Shopping Centres Ltd. was allowed to claim the entire amount as an expense in the year in which it was paid.

Discussion

Calculating depreciation is made difficult by many factors. First, the value of an asset can change over time in many complicated ways. Age, wear and tear, and functional changes all have their effects, and often unpredictably. A 30-year old VW Beetle, for example, can suddenly increase in value because a new Beetle is introduced by the manufacturer.

A second complication is created by tax laws that make it desirable for companies to depreciate things quickly, while at the same time restricting the way they calculate depreciation. As will be seen in Chapter 8, tax laws require that companies, at least for tax purposes, use declining-balance depreciation with a specified depreciation rate.

A third complication is that, in real life, it is sometimes hard to determine what is an asset and what is not an asset. In the case of the Calgary shopping centre, it was unclear whether the road improvements were part of the asset — the physical shopping centre — or simply the equivalent of maintenance expenses.

In the end, depreciation calculations are simply estimates that are useful only with a clear understanding of the assumptions underlying them.

Questions

1) For each of the following, indicate whether the straight-line method or the declining-balance method would likely give the most accurate estimate of how the asset's value changes over time, or explain why neither method is suitable.

(a) A \$20 bill

(b) A \$2 bill

continued

(c) A pair of shoes
(d) A haircut
(e) An engineering degree
(f) A Van Gogh painting

2) What differences would have occurred to the balance sheet and income statement for Oxford Shopping Centres Ltd. if the $450 050 expense had been considered a capital expenditure (depreciating at, for example, 5% per year) as opposed to an expense? Would the company's profit be greater or less in the first year? How does the total profit over the life of the asset compare?

Extended Case: **Part 1**

Welcome to the Real World

A.1 Introduction

Clem looked up from his computer as Naomi walked into his office. "Hi, Naomi. Sit down. Just let me save this stuff."

After a few seconds Clem turned around showing a grin. "I'm working on our report for the last quarter's operations. Things went pretty well. We exceeded our targets on defect reductions and on reducing overtime. And we shipped everything required—over 90% on time."

Naomi caught a bit of Clem's exuberance. "Sounds like a report you don't mind writing."

"Yeah, well, it was a team job. Everyone did good work. Talking about doing good work, I should have told you this before, but I didn't think about it at the right time. Ed Burns and Anna Kulkowski were really impressed with the report you did on the forge project."

Naomi leaned forward. "But they didn't follow my recommendation to get a new manual forging press. I assumed there was something wrong with what I did."

"I read your report carefully before I sent it over to them. If there had been something wrong with it, you would have heard right away. Trust me. I'm not shy. It's just that we were a little short of cash at the time. We could stay in business with just fixing up the guides on the old forging hammer. And Burns and Kulkowski decided there were more important things to do with our money.

"If I didn't have confidence in you, you wouldn't be here this morning. I'm going to ask you and Dave Sullivan to look into an important strategic issue concerning whether or not we continue to buy small aluminum parts or whether we make them ourselves. We're just waiting for Dave to show up."

"O.K. Thanks, Clem. But please tell me next time if what I do is all right. I'm still finding my way around here."

"You're right. I guess that I'm still more of an engineer than a manager."

Voices carried into Clem's office from the corridor. "That sounds like Dave in the hall saying hello to Carole," Naomi observed. "It looks like we can get started."

Dave Sullivan came in with long strides and dropped into a chair. "Good morning, everybody. It is still morning, barely. Sorry to be late. What's up?"

Clem looked at Dave and started talking. "What's up is this. I want you and Naomi to look into our policy about buying or making small aluminum parts. We now use about 200 000 pieces a month. Most of these, like bolts and sleeves, are cold formed.

"Prabha Vaidyanathan has just done a market projection for us. If she's right, our demand for these parts will continue to grow. Unfortunately, she wasn't very precise about the *rate* of growth. Her estimate was for anything between 5% and 15% a year. We now contract this work out. But even if growth is only 5%, we may be at the level where it pays for us to start doing this work ourselves.

"You remember we had a couple of engineers from Hamilton Tools looking over our processes last week? Well, they've come back to us with an offer to sell us a cold former. They have two possibilities. One is a high-volume job that is a version of their Model E2. The other is a low-volume machine based on their Model E1.

"The E2 will do about 2000 pieces an hour, depending on the sizes of the parts and the number of change-overs we make. The E1 will do about 1000 pieces an hour."

Naomi asked, "About how many hours per year will these formers run?"

"Well, with our two shifts, I think we're talking about 3600 hours a year for either model."

Dave came in with, "Hold it. If my third grade arithmetic still works, that sounds like either 3.6 million or 7.2 million pieces a year. You say that we are using only 2.4 million pieces a year."

Clem answered with, "That's right. Ms. Vaidyanathan has an answer to that one. She says we can sell excess capacity until we need it ourselves. Again, unfortunately, she isn't very precise about what this means. We now pay about five cents a piece. Metal cost is in addition to that. We pay for that by weight. She says that we won't get as much as five cents because we don't have the market connections. But she says we should be able to find a broker so that we net somewhere between three cents and four cents a piece, again plus metal."

Naomi spoke. "That's a pretty wide range, Clem."

"I know. Prabha says that she couldn't do any better with the budget Burns and Kulkowski gave her. For another $5000, she says that she can narrow the range on *either* the growth rate or the potential prices for selling pieces from any excess capacity. Or, for about $7500, she could do both. I spoke to Anna Kulkowski about this. Anna says that they won't approve anything over $5000. One of the things I want you two to look at is whether or not it's necessary to get more information. If you do recommend spending on market research, it has to be for just one of either the selling price range or the growth rate.

"I have the proposal from Hamilton Tools here. It has information on the two formers. This is Wednesday. I'd like a report from you by Friday afternoon so that I can look at it over the weekend.

"Did I leave anything out?"

Naomi asked, "Are we still working with a 15% after-tax MARR?"

Clem hesitated. "This is just a first cut. Don't worry about details on taxes. We can do a more precise calculation before we actually make a decision. Just bump up the MARR to 25% before tax. That will about cover our 40% marginal tax rate."[1]

Dave asked, "What time frame should we use in our calculations?"

"Right. Use 10 years. Either of these models should last at least that long. But I wouldn't want to stretch Prabha's market projections beyond 10 years."

Dave stood up and announced, "It's about a quarter to one." He turned to Naomi. "Do you want to start on this over at the Grand China Restaurant? It's past the lunch rush, and we'll be able to talk while we eat. I think Clem will buy us lunch out of his budget."

Clem interjected. "All right, Dave. Just don't order the most expensive thing on the menu."

Naomi laughed. "I'm glad we have one big spender around here."

[1] Businesses in Canada and the United States are required to pay tax on their earnings. Determining the effect of taxes on before-tax earnings may involve extensive computation. Managers frequently approximate the effect of taxes on decisions by increasing the MARR. However, it is good practice to do precise calculations before actually making a decision.

A.2 Problem Definition

About forty minutes later, Dave and Naomi were most of the way through their main courses. Dave suggested that they get started. He took a pad from his brief case and said that he would take notes. Naomi agreed to let him do that.

Dave started with, "O.K. What are our options?"

"Well, I did a bit of arithmetic on my calculator while you were on the phone before lunch. It looks as though, even if the demand growth rate is only 5%, a single small former will not have enough capacity to see us through 10 years. This means that there are four options. The first is a sequence of two low-capacity formers. The second is just a high-capacity former. The third, which would kick in only if the growth rate is high, would be a sequence of three low-capacity formers. The fourth is a low-capacity and a high-capacity. I'm not sure of the sequence for that."

Dave thought for a bit. "I don't think so. I assume that, even if we put in our own former, we could contract out requirements that our own shop couldn't handle. That might be the way to go. That is, you wouldn't want to add to capacity if there was only one year left in the 10-year horizon. There probably would not be enough unsatisfied output requirement to amortize the first cost of a second small former."

"That sounds as though there are a whole bunch of options. It looks like we're in for a couple of long nights." Naomi sounded dejected.

"Well, maybe not."

"Maybe not what?"

"Maybe we won't have to spend those long nights. I think we can look at just three options at the start. We could look at a sequence of two small formers. We could look at a small former followed by outsourcing when capacity is exhausted. And we could look at a large former, possibly followed by outsourcing if capacity is exhausted. If we can rule out the two-small-formers option, we can certainly rule out three small formers or a big former combined with a small one."

"Smart, Dave. How should we proceed?"

"Well, there are a couple of ways of doing this. But, given that we have only 10 years to look at, it's probably easiest to use a spreadsheet to develop cash flow sequences for the three options. The two options in which only a single machine is bought at time zero are pretty straightforward. The one with a sequence of two small formers is a bit more complicated. One of us can do the two easy ones. The other can do the sequence. For each option we need to look at, say, nine outcomes: three possible demand growth rates—5%, 10%, and 15%—times three possible prices—3¢, 3.5¢, and 4¢—for selling excess capacity. That should be enough to show us what's happening. What part do you want to do?"

"I'll take the hard one, the sequence of two. I'd like the practice. I'll let you check my analysis when I'm finished."

"O.K. That's fine. But notice that the two-low-capacity-formers sequence is not a simple investment, so let's stick with present worths at this stage. Also, we are going to have to make some decision about how we record the timing of the purchase of the second former if we run out of capacity during a year. I suggest that we assume the second former is bought at the end of the year before we run out of capacity."

"O.K."

Dave continued with, "We need to put together a simple table on the specifications for these two machines. I'll do that, since you are doing the hard job with cash flows. Why don't we go back to the plant. I'll make up the table. I'll get you a copy later this afternoon."

A.3 Crunch Time

Naomi was sitting in her office thinking about structuring the cash flows for the two-small-machine sequence. Dave knocked and came in.

Table A.1 Specifications for the Two Cold Formers

Characteristics	Model E1 (Small)	Model E2 (Large)
First cost ($)	125 000	225 000
Hourly running cost ($)	35.00	61.25
Average number of pieces/hour	1000	2000
Hours/year	3600	3600
Depreciation	20%/year declining balance, for both machines	
Market facts	Buying price: $0.05/piece plus cost of metal Selling price: $0.03 to $0.04 per piece plus metal Demand growth: 5% to 15%	

He handed Naomi a sheet of paper.

"Here's the table, as promised. Shall we meet tomorrow morning to compare results?" (See Table A.1.)

Naomi glanced at the sheet of paper, and motioned Dave to stay. "This is a bit new to me. Would it be okay if we get all of our assumptions down before we go too far?"

"Sure, we can do that," replied Dave as he grabbed a seat beside Naomi's desk. Naomi already had a pad and pencil out.

Dave went on, "O.K., first, what's our goal?"

"Well, right now we are faced with a 'make-or-buy' decision," suggested Naomi. "We need to find out if it is cheaper to make these parts or to continue to buy them, and if making them is cheaper we need to choose which machine or machines we should buy."

"Pretty close, Naomi, but I think that may be more than we need at this point. Remember, Clem is primarily interested in whether this project is worth pursuing as a full-blown proposal to top management. So we can use a greatly simplified model now to answer that question, and get into the details if Clem decides it's worth pursuing."

"Right. If we look at the present worths of the three options we came up with at lunch, we should see quickly enough if investing in our own machines is feasible. Positive present worths indicate buying machines is the best course; negative values would suggest it's probably not worth pursuing further."

Dave nodded. "O.K.," continued Naomi, "so we want to do a 'first-approximation' calculation of present worths of each of our three options. Now, what assumptions should we use?"

"Let's see. We want to consider only a 10-year study period for the moment. I think we can assume the machines will have no salvage value at the end of the study period, so we don't have to worry about estimating depreciation and salvage values. This will be a conservative estimate in any case."

"Next," Dave continued, "we will ignore tax implications, and use a before-tax MARR of 25%. The 'sequence of machines' option has a complex cash flow, but by only considering present worths we can avoid dealing with possible multiple IRRs."

"You know," interjected Naomi, "it just occurred to me that the small machine makes 1000 pieces per hour at an operating cost of $35.00."

"And this is news to you?" said a bemused Dave.

"No, Dave. What I mean is that the operating cost is $0.035 per part on this machine, while one scenario calls for us selling excess capacity at only $0.03. We'll be losing half a cent on every part made for sale."

"Hmmm. You know, you're right. And look at this. Even the bigger machine's operating cost is $0.030625 per piece. We'd be losing money on that one too."

"When we run the different cases, it looks like we'll have to consider whether excess capacity will be sold or whether the machine will just be shut down. Just when I thought this was going to be easy..." trailed Naomi.

"Maybe, but maybe not. Look at this another way. If the small machine sells its excess capacity at a loss of half a cent per piece, that comes to five bucks per hour. On the other hand, if the machine is shut down, we'll have an operator standing around at a cost of a lot more than that."

"But can't an idle operator be given other work to do?"

"Again, maybe, but maybe not. With the job security clause the union has now, if we lay the operator off for any period of time, the company has to pay most of the wages anyway, so there isn't much savings that way. On the other hand, the operator would probably be idled at unpredictable times, so other departments would have a difficult job of scheduling work for him or her to do. I'm not saying it's impossible, but it's a job in itself to figure out if we can place the operator in productive work when trying to idle the machine."

"Boy, they didn't talk about these problems in engineering school!"

"Welcome to the real world, Naomi, where nothing is simple."

"Then how about if we do this?" Naomi continued. "We assume that since an idle operator is probably more costly to the company than $5.00 per hour, the machines will run a full 3600 hours per year each regardless of whether the excess capacity is being sold at a price above operating costs or not."

"Sounds good, Naomi. In fact, we could claim an indirect benefit of this otherwise unprofitable operation since our operators will gain more experience with doing quick set-ups and Statistical Process Control. That should make the Brass happy," Dave said with a quick wink.

"I guess we can also assume that a machine is purchased and paid for at the beginning of a year while savings and operating costs accrue at the end of the year," continued Naomi. "And that the demand for parts is uniform for each month within a year but grows by a fixed percentage from one year to the next."

Naomi finished scribbling down the assumptions, including the ones they had discussed with Clem earlier that day. Turning to Dave, she said "so how does this look?"

"Looks fine to me. Should we get together tomorrow and compare results?"

"O.K. What time?"

"Why don't we exchange results first thing in the morning and then meet about nine-thirty in my office?"

"That's fine. See you then."

As Dave left, Naomi looked more closely at the table he had given her (Table A.1). "Time to crunch those numbers," she said to herself.

QUESTIONS

1. Construct spreadsheets for calculating present worths of the three proposals. For each proposal, you need to calculate PWs for each of 5%, 10% and 15% demand growth and $0.03, $0.035 and $0.04 selling price (nine combinations in total). Present the results in tabular and/or graphical format to support your analysis. A portion of a sample spreadsheet layout is given in Table A.2.

Table A.2 Portion of a Present Worth Spreadsheet

Sequence: E1 Machine purchased at time 0					
Projected Demand Growth	5%	/year			
Projected Selling Price	$0.03	/piece			
Year	0	1	2	3	4
Projected Demand (units)		2 400 000	2 520 000	2 646 000	2 778 300
Capacity (units)		3 600 000	3 600 000	3 600 000	3 600 000
Purchases (units)		0	0	0	0
Sales (units)		1 200 000	1 080 000	954 000	821 700
Operating Cost	$0	$126 000	–$126 000	–$126 000	–$126 000
Purchasing Savings	$0	$120 000	$126 000	$132 300	$138 915
Sales Revenue	$0	$36 000	$32 400	$28 620	$24 651
Capital Expenditure	–$125 000	$0	$0	$0	$0
Net Cash Flow	–$125 000	$30 000	$32 400	$34 920	$37 566
Present Worth	$8 490				

2. Write a memo to Clem containing your findings. The goal of the analysis is to determine if bringing production in-house appears feasible, and if so, which machine purchase sequence(s) should be studied in further detail. The memo should contain a tentative recommendation about which option looks best and what additional research, if any, should be done. Keep the memo as brief and concise as possible. ◎

C H A P T E R 7

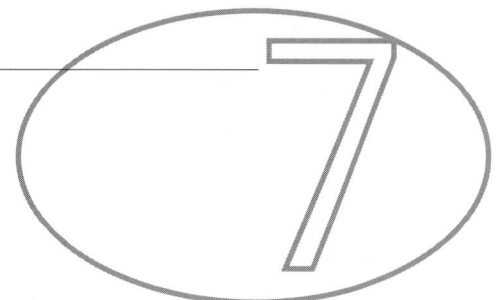

Replacement Decisions

Engineering Economics in Action, Part 7A: You Need the Facts

You know the five-stage progressive die that we use for the Admiral Motors rocker arm contract?" Naomi was speaking to Terry, her co-op student, one Tuesday afternoon. "Clem asked me to look into replacing it with a 10-stage progressive die that would reduce the hand finishing substantially. It's mostly a matter of labour cost saving, but there is likely to be some quality improvement with the 10-stage die as well. I would like you to come up with a ballpark estimate of the cost of switching to the 10-stage progressive die."

Terry asked, "Don't you have the cost from the supplier?"

"Yes, but not really," said Naomi. "The supplier is Hamilton Tools. They've given us a price for the machine, but there are a lot of other costs involved in replacing one production process with another."

"You mean things like putting the machine in place?" Terry asked.

"Well, there's that," responded Naomi. "But there is also a lot more. For example, we will lose production during the changeover. That's going to cost us something."

"Is that part of the cost of the 10-stage die?"

"It's part of the first cost of switching to the 10-stage die," Naomi said. "If we decide to go ahead with the 10-stage die and incur these costs, we'll never recover them — they are sunk. We have already incurred those costs for the five-stage die and it's only two years old. It still has lots of life in it. If the first costs of the 10-stage die are large, it's going to be hard to make a cost justification for switching to the 10-stage die at this time."

"O.K. How do I go about this?" Terry asked.

Naomi sat back and chewed on her yellow pencil for about 15 seconds. She leaned forward and began. "Let's start with order-of-magnitude estimates of what it's going to cost to get the 10-stage die in place. If it looks as if there is no way that the 10-stage die is going to be cost effective now, we can just stop there."

"It sounds like a lot of fuzzy work," said Terry.

"Terry, I know you like to be working with mathematical models. I'm also sure that you can read the appropriate sections on replacement models in an engineering economics book. But none of those models is worth anything unless you have data to put in it. You need the models to know what information to look for. And once you have the information, you will make better decisions using the models. But you do need the facts." ◎

7.1 Introduction

Survival of businesses in a competitive environment requires regular evaluation of the plant and equipment used in production. As these assets age, they may not provide adequate quality, or their costs may become excessive. When a plant or equipment is evaluated, one of four mutually exclusive choices will be made:

1. An existing asset may be kept in its current use without major change.
2. An existing asset may be overhauled so as to improve its performance.
3. An existing asset may be removed from use without replacement by another asset.

4. An existing asset may be replaced with another asset.

This chapter is concerned with methods of making choices about possible replacement of long-lived assets. While the comparison methods developed in Chapters 4 and 5 for choosing between alternatives are often used for making these choices, the issues of replacement deserve a separate chapter for several reasons. First, the relevant costs for making replacement decisions are not always obvious, since there are costs associated with taking the replaced assets out of service that should be considered. This was ignored in the studies in Chapters 4 and 5. Second, the service lives of assets were provided to the reader in Chapters 4 and 5. As seen in this chapter, the principles of replacement allow the calculation of these service lives. Third, assumptions about how an asset might be replaced in the future can affect a decision now. Some of these assumptions are implicit in the methods analysts use to make the choices. It is therefore important to be aware of these assumptions when making replacement decisions.

The chapter starts with a consideration of the reasons why a long-lived asset may be replaced. In the following section we consider the built-in cost advantage of existing assets. Then, the idea of the *economic life* of an asset is developed. This is the service life that minimizes the average cost of using the asset. We then consider replacement of an asset with a new asset that is identical to the current asset. This is followed by a discussion of replacement with an asset that differs from the current asset.

We shall not consider the implications of taxes for replacement decisions in this chapter. This is postponed until Chapter 8. We shall assume in this chapter that no future price changes are expected. The effect of expected price changes on replacement decisions will be considered in Chapter 9.

7.2 Reasons for Replacement or Retirement

An existing asset is **retired** if it is removed from use without being replaced. This can happen if the service that the asset provides is no longer needed. Changes in customer demand, changes in production methods, or changes in technology may result in an asset's no longer being necessary. For example, the growth in the use of compact discs for audio recordings has led manufacturers of cassette tapes to retire production equipment since the service it provided is no longer needed.

If there is an ongoing need for the service an asset provides, at some point it will need *replacement*. Replacement becomes necessary if there is a cheaper way to get the service the asset provides, or if the service provided by the existing asset is no longer adequate.

There may be a cheaper way to get the service provided by the existing asset for several reasons. First, productive assets often deteriorate over time because of wearing out in use, or simply because of the effect of time. As a familiar example, an automobile becomes less valuable with age (older cars, unless they are collectors' cars, are worth less than newer cars with the same mileage) or if it has a high mileage (the kilometres driven reflect the wear on the vehicle). Similarly, production equipment may become less productive or more costly to operate over time. It is usually more expensive to maintain older assets. They need fixing more often, and parts may be harder to find and may cost more.

Technological or organizational change can also bring about cheaper methods of providing service than the method used by an existing asset. For example, the technological changes associated with the use of computers have improved productivity. Organizational changes, both within a company and in markets outside the company, can lead to lower-cost methods of production. A form of organizational change that has become very popular is the specialist company. These companies take on parts of the production activities of other companies. Their specialization may enable them to have lower costs than the companies can attain themselves. See Close-Up 7.1.

CLOSE-UP 7.1 Specialist Companies

Specialist companies concentrate on a limited range of very specialized products. They develop the expertise to produce these products at minimal cost. Larger firms often find it more economical to contract out production of low volume components instead of manufacturing the components themselves.

In some industries, the use of specialist companies is so pervasive that the companies apparently producing a product actually are simply assembling it; the manufacturing takes place at dozens or sometimes hundreds of supplier firms.

A good example of this is the automotive industry. In North America, auto makers support an extremely large network of specialist companies, linked by computer. A single specialist company might supply brake pads, for example, to all three major auto manufacturers. Producing brake pads in huge quantities, the specialist firm can refine its production process to extremes of efficiency and profitability.

The second major reason why a current asset may be replaced is inadequacy. An asset used in production can become inadequate because it has insufficient capacity to meet growing demand or because it no longer produces items of high-enough quality. A company may have a choice between adding new capacity parallel to the existing capacity, or replacing the existing asset with a higher capacity asset, perhaps one with more advanced technology. If higher quality is required, there may be a choice between upgrading an existing piece of equipment or replacing it with equipment that will yield the higher quality. In either case, contracting out the work to a specialist is one possibility.

In summary, there are two main reasons for replacing an existing asset. First, an existing asset will be replaced if there is a cheaper way to get the service that the asset provides. This can occur because the asset ages or because of technological or organizational changes. Second, an existing asset will be replaced if the service it provides is inadequate in either quantity or quality.

7.3 The Relevance of Capacity Costs

When a decision is made to acquire a new asset, it is essentially a decision to purchase the capacity to perform tasks or produce output. **Capacity** is the ability to produce, often measured in units of production per time period. Although production requires capacity, it is also important to understand that just acquiring the capacity entails costs that are incurred whether or not there is actual production. Furthermore, a large portion of the capacity cost is incurred early in the life of the capacity. There are two main reasons for this:

1. Part of the cost of acquiring capacity is the depreciation expense incurred over time because the assets required for that capacity gradually lose their value. This depreciation expense is often called the **capital cost** of the asset. It is incurred by the difference between what is paid for the assets required for a particular capacity and what the assets could be resold for some time after purchase. The largest portion of the capital costs typically occurs early in the lives of the assets.

2. Installing a new piece of equipment or new plant involves substantial up-front costs, called **installation costs**. These are the costs of aquiring capacity, excluding the purchase cost, which may include disruption of production, training of workers, and perhaps a reorganization of other processes. Installation costs are not reversible once the capacity has been put in place.

It is worth noting that the total cost of a new asset includes the installation costs and the cost of purchasing the asset. When we compute the capital costs of an asset over a period of time, the first cost (usually denoted by P) includes the installation costs. However, when we compute a salvage value for the asset as it ages, we do *not* include the installation costs as part of the depreciable value of the asset, since these costs are expended upon installation and cannot be recovered at a later time.

The large influence of capital costs associated with acquiring new capacity means that, once the capacity has been installed, the *incremental* cost of continuing to use that capacity during its planned life is relatively low. This gives an existing asset, the **defender**, a cost advantage during its planned life over a potential replacement, a **challenger**. This up-front weighting of capital costs also gives a defender a cost advantage over the alternative of contracting out the service performed by the asset, as illustrated by Example 7.1.

Example 7.1

The Jiffy Printer Company produces printers for home use. Currently, they pay a custom moulder \$0.25 per piece (excluding material costs) to produce parts for their printers. Demand is forecast to be 200 000 parts per year. Jiffy is considering installing an automated plastic moulding system to produce the parts themselves. The moulder itself costs \$20 000 and the installation costs are estimated to be \$5000. The expected life of the system is six years. Operating and maintenance costs are expected to be \$30 000 in the first year and to rise at the rate of 5% per year. Jiffy estimates its capital costs with a declining-balance depreciation model with a rate of 40%, and uses a MARR of 15% for such investments.

In Jiffy's situation, the *defender* is the current technology: a subcontractor. The *challenger* is the automated plastic moulding system. In order to decide whether Jiffy is better off with the defender or the challenger, we need to compute the cost per piece of production with the moulder. We first find the present worth of the total cost associated with acquiring and using the moulder over its six-year life. We then convert the present cost into an equivalent annual amount, and then into a cost per piece.

The present worth of acquiring the moulder and using it over its six-year life is the sum of its installation costs, capital costs, and operating and maintenance costs:

$$\text{PW(moulder)} = \text{PW(capital)} + \text{PW(operating and maintenance)}$$

The present worth of the capital cost of the moulding system is the loss in value of the asset over the six years over which it will be operated:

$$\begin{aligned}\text{PW(capital)} &= \$20\,000 + \$5000 - BV_{db}(6)(P/F, 15\%, 6)\\ &= \$25\,000 - \$20\,000(1-0.4)^6(0.4323)\\ &= \$25\,000 - \$933.12(0.4323)\\ &\cong \$24\,597\end{aligned}$$

Finally, the present worth of operating and maintenance costs over the six-year life can be obtained using the geometric gradient to present worth conversion factor (see Section 3.7):

$$\text{PW(operating and maintenance)} = \frac{\$30\,000(P/A, i^\circ, 6)}{1+g}$$

The growth-adjusted interest rate, i°, is given by

$$i^\circ = \frac{1+i}{1+g} - 1 = \frac{1.15}{1.05} - 1 = 0.09524$$

With $g = 0.05$ and the series present worth formula, we get

$$\text{PW(operating and maintenance)} = \$30\,000(4.4167)/1.05$$
$$\cong \$126\,191$$

This gives a present worth of cost for the six-year planned life of the moulding system of \$24 597 + \$126 191 = \$150 788.

To determine the costs on a per piece basis, we convert the present worth to an **equivalent annual cost (EAC)** and divide by the number of pieces per year. Notice that by converting to an equivalent annual cost we are assuming that all yearly payments occur at the end of each year. This is an approximation to the periodic costs associated with the moulder that are actually spread out over the year.

$$\text{EAC(moulder)} = \$150\,788(A/P, 15\%, 6)$$
$$= \$150\,788(0.2642)$$
$$\cong \$39\,838$$

Finally, the cost per piece is \$39 838/200 000 = \$0.1992

When the cost per piece of in-house production is compared with the \$0.25 cost per piece of contracting out the work, it appears that Jiffy should replace the *defender* (contracting out) with in-house production (the challenger). ■

This example has illustrated the basic idea behind a replacement analysis when we are considering the purchase of a *new* asset as a replacement to current technology. The cost of the replacement must take into account the capital costs (including installation) and the operating and maintenance costs over the life of the new asset.

In the next section, we see how some costs are no longer relevant in the decision to replace an *existing* asset.

7.4 The Irrelevance of Sunk Costs

Once an asset has been installed and has been operating for some time, the costs of installation and all other costs incurred up to that time are no longer relevant to any decision to replace the current asset. These costs are called **sunk costs**. Only those costs that will be incurred in keeping and operating the asset from this time on are relevant. This is best illustrated with an example.

Example 7.2

Two years have passed since the Jiffy Printer Company from Example 7.1 installed an automated moulding system to produce parts for their printers. At the time of installation, they expected to be producing about 200 000 pieces per year, which justified the investment. However, their actual production has turned out to be only about 150 000 pieces per year. Their cost per piece is \$39 838/150 000 = \$0.2656 rather than the \$0.1992 they had expected. They estimate the market value of the moulder now at \$7200. Maintenance costs do not depend, in this case, on the actual production rate. Should Jiffy sell the moulding system and go back to buying from the custom moulder at \$0.25 per piece?

In the context of a replacement problem, Jiffy is looking at replacing the existing system (the defender) with a different technology (the challenger). Since Jiffy already has the moulder and has already expended considerable capital on putting it into place, it may be better for Jiffy to keep the current moulder for some time longer. Let us calculate the cost to Jiffy of keeping the moulding system one more year. This may not be the optimal length of time to continue to use the system, but if the cost is less than \$0.25 per piece it is cheaper than retiring or replacing it now.

The reason that the cost of keeping the moulder an additional year may be low is that the capital costs for the two-year-old system are now low compared with the costs of putting the capacity in place. The capital cost for the third year is simply the loss in value of the moulder over the third year. This is the difference between what Jiffy can get for the system now, at the end of the second year, and what they can get a year from now when the system will be three years old. Jiffy can get \$7200 for the system now. Using the declining-balance depreciation rate of 40% to calculate a salvage value, we can determine the annual capital cost for keeping the moulder to the end of the third year. Applying the *capital recovery formula* from Chapter 3, the EAC for capital costs is

$$
\begin{aligned}
&\text{EAC(capital costs, third year)}\\
&\qquad = (P - S)(A/P, 15\%, 1) + Si\\
&\qquad = [\$7200 - 0.6(\$7200)](1.15) + 0.6(\$7200)(0.15)\\
&\qquad = \$3960
\end{aligned}
$$

Recall that the operating and maintenance costs started at \$30 000 and rose at 5% each year. The operating and maintenance costs for the third year are

$$
\begin{aligned}
\text{EAC(operating and maintenance, third year)} &= \$30\ 000(1.05)^2\\
&= \$33\ 075
\end{aligned}
$$

The total annual cost of keeping the moulder for the third year is the sum of the capital costs and the operating and maintenance costs:

$$
\begin{aligned}
\text{EAC(third year)} &= \text{EAC(capital costs, third year)}\\
&\quad + \text{EAC(operating and maintenance, third year)}\\
&= \$3960 + \$33\ 075\\
&= \$37\ 035
\end{aligned}
$$

Dividing the annual costs for the third year by 150 000 gives us a cost per piece of $0.247 for moulding in-house during the third year. Not only is this lower than the average cost over a six-year life of the system, it is also lower than what the custom moulder charges. Similar computations would show that Jiffy could go two more years after the third year with in-house moulding. Only then would the increase in operating and maintenance costs cause total cost per piece to rise above $0.25. The computations for this are shown in Section 7.5.

We see that installing the automated moulding system was a mistake for Jiffy. The average lifetime costs for in-house moulding were greater than the cost of contracting out, but, once the system was installed, it was not optimal to go back to contracting out immediately. This is because the sunk costs for the initial period of use of an asset are disproportionately large as compared with the costs of using the asset once it is in place. ■

That a defender has a cost advantage over a challenger, or over contracting out during the planned life of the defender, is important. It means that if a defender is to be removed from service during its life for cost reasons, the average lifetime costs for the challenger or the costs of contracting out must be considerably lower than the average lifetime costs of the defender.

Just because well-functioning defenders are not often retired for cost reasons does not mean that they will not be retired at all. Changes in markets or technology may make the quantity or quality of output produced by a defender inadequate. This can lead to their retirement or replacement even when they are functioning well.

7.5 Economic Life and Cyclic Replacement

All long-lived assets eventually require replacement. Consequently, the issue in replacement studies is not *whether* to replace an asset, but *when* to replace it. In this section we consider the case where there is an ongoing need for a service provided by an asset and where the asset technology is not changing rapidly. There are several assumptions made:

1. The defender and challenger are assumed to be technologically identical. It is also assumed that this constancy of technology remains for the company's entire planning horizon.
2. The lives of these identical assets are assumed to be short relative to the time horizon over which the assets are required.
3. Relative prices and interest rates are assumed to be constant over the company's time horizon.

These assumptions are quite restrictive. In fact, there are only a few cases where the assumptions strictly hold (cable used for electric power delivery is an example). Nonetheless, the idea of economic life of an asset is still useful to our understanding of replacement analysis.

Assumptions 1 and 2 imply that we may model the replacement decision as being repeated an indefinitely large number of times. The objective is then to determine a minimum-cost lifetime for the assets, a lifetime that will be the same for all the assets in the sequence of replacements over the company's time horizon.

We have seen that the relevant costs associated with acquiring a new asset are the capital costs, installation costs (which are often pooled with the capital costs), and operating and maintenance costs. It is usually true that operating and maintenance costs of assets — plant or equipment — rise with the age of the asset. Offsetting increases in operating and maintenance costs is the fall in capital costs per year that usually occurs as the asset life is extended and the capital costs are spread over a greater number of years. The rise in operating and maintenance costs per year and the fall in capital costs per year as the life of an asset is extended work in opposite directions. In the early years of an asset's life, the capital costs per year (although decreasing) usually, but not always, dominate total yearly costs. As the asset ages, increasing operating and maintenance costs usually overtake the declining annual capital costs. This means that there is a lifetime that will minimize the *average* cost (adjusting for the time value of money) per year of owning and using long-lived assets. This is referred to as the **economic life** of the asset. These ideas are illustrated in Figure 7.1 where we see that the economic life of an asset is found at the point where the rate of increase in operating and maintenance costs per year equals the rate of decrease in capital costs per year.

Example 7.3

Refer back to Example 7.1 where the Jiffy Printer Company is considering the purchase of an automated moulding system based on an expected production level of 200 000 pieces per year. It is economically justified for them to purchase the system, as was shown in Example 7.1. Assuming that there will be an ongoing need for the moulder, and assuming that the technology does not change (i.e., no cheaper or better method will arise), how long should Jiffy keep a moulder before replacing it with a new model? In other words, what is the economic life of the automated moulding system?

Determining the economic life of an asset is most easily done with a spreadsheet. Table 7.1 shows the development of the equivalent annual costs for the automated plastic moulding system of Example 7.1.

In the first column is the life of the asset, in years. The second column shows the salvage value of the moulding system as it ages. The equipment costs \$20 000 originally, and as the system ages the value declines by 40% of current value each year, giving the estimated salvage values listed in Table 7.1. For example, the salvage value at the end of year 4 is

$$\begin{aligned} BV_{db}(4) &= \$20\,000(1 - 0.4)^4 \\ &= \$2592 \end{aligned}$$

Figure 7.1 Cost Components for Replacement Studies

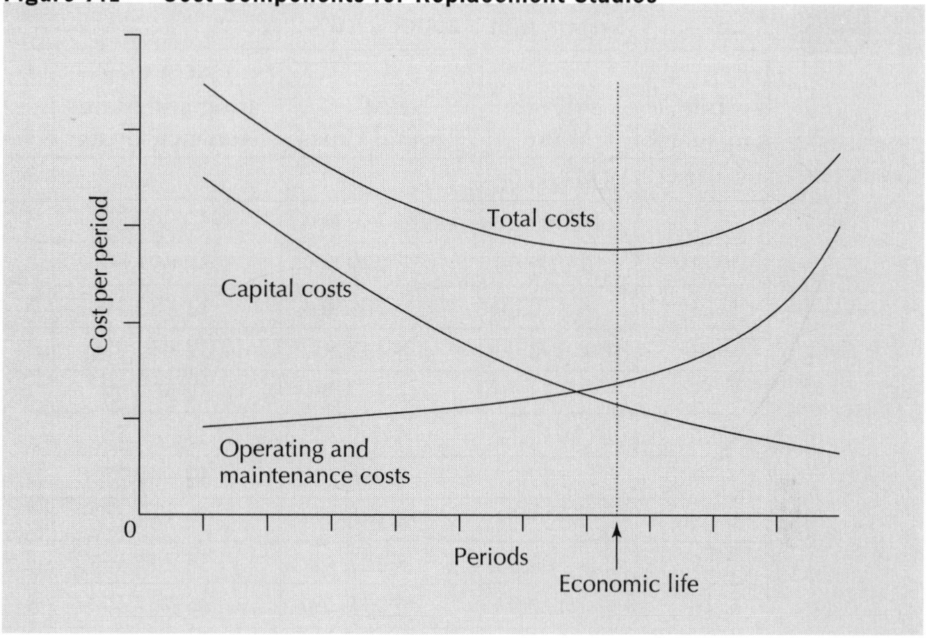

The next column gives the equivalent annual capital costs if the asset is kept for n years, $n = 1, \ldots, 10$. This captures the loss in value of the asset over the time it is kept in service. As an example of the computations, the equivalent annual capital costs of keeping the moulding system for four years is

$$\begin{aligned}&\text{EAC(capital costs over four years)}\\&= (P - S)(A/P, 15\%, 4) + Si\\&= (\$20\,000 + \$5000 - \$2592)(0.35027) + \$2592(0.15)\\&\cong \$8238\end{aligned}$$

Note that the installation costs have been included in the capital costs, as these are expenses incurred at the time the asset is originally put into service. Table 7.1 illustrates that the equivalent annual capital costs decline as the asset's life is extended.

Next, the equivalent annual operating and maintenance costs are found by converting the stream of operating and maintenance costs (which are increasing by 5% per year) into a stream of equal-sized annual amounts. Continuing with our sample calculations, the EAC of operating and maintenance costs when the moulding system is kept for four years is

$$\begin{aligned}&\text{EAC(operating and maintenance costs over four years)}\\&= \$30\,000\,[(P/F, 15\%, 1) + (1.05)(P/F, 15\%, 2)\\&\quad + (1.05)^2(P/F, 15\%, 3) + (1.05)^3(P/F, 15\%, 4)]\,(A/P, 15\%, 4)\\&\cong \$32\,052\end{aligned}$$

Table 7.1 Computation of Total Equivalent Annual Costs of the Moulding System with a MARR = 15%

Life in Years	Salvage Value	EAC Capital Costs	EAC Operating and Maintenance Costs	EAC Total
0	$20 000.00			
1	12 000.00	$16 750.00	$30 000.00	$46 750.00
2	7 200.00	12 029.07	30 697.67	42 726.74
3	4 320.00	9 705.36	31 382.29	41 087.65
4	2 592.00	8 237.55	32 052.47	40 290.02
5	1 555.20	7 227.23	32 706.94	39 934.17
6	933.12	6 499.33	33 344.56	39 843.88
7	559.87	5 958.42	33 964.28	39 922.70
8	335.92	5 546.78	34 565.20	40 111.98
9	201.55	5 227.34	35 146.55	40 373.89
10	120.93	4 975.35	35 707.69	40 683.04

Notice that the equivalent annual operating and maintenance costs increase as the asset life of the moulding system increases.

Finally, we obtain the equivalent annual total cost of the moulding system by adding the equivalent annual capital costs and the equivalent annual operating and maintenance costs. This is shown in the last column of Table 7.1. We see that at a six-year life the declining equivalent annual installation and capital costs offset the increasing operating and maintenance costs. In other words, the economic life of the moulder is six years. ■

While it is *usually* true that capital cost per year falls with increasing life, it is not always true. Capital costs per year can rise at some point in the life of an asset if the decline in value of the asset is not smooth or if the asset requires a major overhaul.

If there is a large drop in the value of the asset in some year during the asset's life, the cost of holding the asset over that year will be high. Consider the following example.

Example 7.4

An asset costs $50 000 to buy and install. The asset has a resale value of $40 000 after installation. It then declines in value by 20% per year until the fourth year when its value drops from over $20 000 to $5000 because of a predictable wearing out of a major component. Determine the equivalent annual capital cost of this asset for lives ranging from one to four years. The MARR is 15%.

The computations are summarized in Table 7.2. The first column gives the life of the asset in years. The second gives the salvage value of the asset as it ages. The asset loses 20% of its previous year's value each year except the fourth, when its value drops to \$5000. The last column summarizes the equivalent annual capital cost of the asset. Sample computations for the third and fourth years are

$$\begin{aligned}&\text{EAC(capital costs, three-year life)}\\&\quad= (P - S)(A/P,15\%,3) + Si\\&\quad= (\$40\,000 + \$10\,000 - \$20\,480)(0.43798) + \$20\,480(0.15)\\&\quad\cong \$16\,001\end{aligned}$$

$$\begin{aligned}&\text{EAC(capital costs, four-year life)}\\&\quad= (P - S)(A/P, 15\%, 4) + Si\\&\quad= (\$40\,000 + \$10\,000 - \$5000)(0.35027) + \$5000(0.15)\\&\quad\cong \$16\,512\end{aligned}$$

Table 7.2 EAC of Capital Costs for Example 7.4

Life in Years	Salvage Value	EAC Capital Costs
0	\$40 000	
1	32 000	\$25 500
2	25 600	18 849
3	20 480	16 001
4	5 000	16 512

The large drop in value in the fourth year means that there is a high cost of holding the asset in the fourth year. This is enough to raise the average capital cost per year. ■

In summary, when we replace one asset with another asset with an identical technology, it makes sense to speak of its economic life. This is the lifetime of an individual asset that will minimize the average cost per year of owning and using it. In the next section, we deal with the case where the challenger is different from the defender.

7.6 Challenger Is Different from Defender

In this section we explore a more common situation in which there is a challenger that is different from the defender. We distinguish between two such cases. In the first case we assume that, although the challenger is different from the defender, all succeeding challengers are identical to the first challenger. The next case is more general. We recognize that there will be non-identical challengers after the current challenger. We expect them to be better than the current challenger. We shall see in Section 7.6.2 that the solution to this latter problem is quite complex.

7.6.1 Sequence of Identical Challengers

The decision rule that minimizes cost in the case where a defender is faced by a challenger that is expected to be followed by a sequence of identical challengers is as follows:

1. Determine the economic life of the challenger and its associated equivalent annual cost.
2. Determine the cost of using the defender one more year. If the cost of using the defender one more year is less than the equivalent annual cost of installing and using the challenger over its economic life, then keep the defender at least one more year.
3. If the cost of using the defender one more year is greater than the cost of installing and using the challenger over its economic life, see if there is a life longer than one year over which the cost of using the defender is less than the cost of installing and using the challenger. If there is such a life, keep the defender for that remaining economic life. If there is no incremental life for the defender for which its cost per year is less than the cost of installing and using the challenger for its economic life, replace the defender immediately.

Steps 2 and 3 are considered separately because operating costs are typically assumed to increase smoothly and monotonically over time, while capital costs decrease smoothly and monotonically. When this is true, it is only necessary to check whether the defender needs replacing due to costs over the next year, because in subsequent years the case for the defender will only get worse, not better. Consequently, Step 3 needs to be taken only when the operating or capital costs are not smooth and monotonic. This might occur, for example, when periodic overhauls are required. Consider this example concerning the potential replacement of a generator.

Example 7.5

The Colossal Construction Company uses a generator to produce power at remote sites. The existing generator is now three years old. It cost $11 000 when purchased. Its current salvage value is $2400, expected to fall to $1400 next year and $980 the year after, and to continue declining at 30% of current value per year. Its operating and maintenance costs are now $2000 per year and are expected to rise by $500 per year.

New fuel-efficient generators have been developed, and Colossal is thinking of replacing its existing generator. It is expected that the new generator technology will be the best available for the foreseeable future. The new generator sells for $9500. Installation costs are negligible. Other data for the new generator are summarized in Table 7.3.

Should Colossal replace the existing generator with the new type? The MARR is 12%.

Table 7.3 Salvage Values and Operating Costs for New Generators

End of Year	Salvage Value	Operating Cost
0	$9 500	
1	8 000	$1 000
2	7 000	1 000
3	6 000	1 200
4	5 000	1 500
5	4 000	2 000
6	3 000	2 000
7	2 000	2 000
8	1 000	3 000

We first determine the economic life for the challenger. The calculations are shown in Table 7.4.

Table 7.4 Economic Life of the Generator

End of year	Salvage Value	Operating Costs	EAC
1	$8 000	$1 000	$3 640.00
2	7 000	1 000	3 319.25
3	6 000	1 200	3 236.50
4	5 000	1 500	**3 233.07***
5	4 000	2 000	3 290.81
6	3 000	2 000	3 314.16
7	2 000	2 000	3 318.68
8	1 000	3 000	3 393.52

*Lowest equivalent annual cost

Sample calculations for the EAC of keeping the challenger for one, two, and three years are as follows:

$$\begin{aligned}\text{EAC(1 year)} &= (P - S)(A/P,12\%,1) + Si + \$1000\\ &= (\$9500 - \$8000)(1.12) + \$8000(0.12) + \$1000\\ &= \$3640\end{aligned}$$

$$\begin{aligned}\text{EAC(2 years)} &= (P - S)(A/P, 12\%, 2) + Si + \$1000\\ &= (\$9500 - \$7000)(0.5917) + \$7000(0.12) + \$1000\\ &\cong \$3319\end{aligned}$$

$$\begin{aligned}\text{EAC(3 years)} &= (P - S)(A/P, 12\%, 3) + Si + \$1000\\ &\quad + \$200(A/F, 12\%, 3)\\ &= (\$9500 - \$6000)(0.41635) + \$6000(0.12)\\ &\quad + \$1000 + \$200(0.29635)\\ &\cong \$3237\end{aligned}$$

As the number of years increases, this approach for calculating the EAC becomes more difficult, especially since in this case the operating costs are neither a standard annuity nor an arithmetic gradient. An alternative is to calculate the present worths of the operating costs for each year. The EAC of the operating costs can found by applying the capital recovery factor to the sum of the present worths. This approach is particularly handy when using spreadsheets.

By either calculation, we see in Table 7.4 that the economic life of the generator is four years.

Next, to see if and when the defender should be replaced, we get the costs of keeping the defender for one more year. Using the capital recovery formula:

$$\begin{aligned}&\text{EAC(keep defender 1 more year)}\\ &\quad = \text{EAC(capital costs)} + \text{EAC(operating costs)}\\ &\quad = (\$2400 - \$1400)(A/P, 12\%, 1) + \$1400(0.12) + \$2000\\ &\quad = \$3288\end{aligned}$$

The equivalent annual cost of using the defender one more year is \$3288. This is more than the yearly cost of installing and using the challenger over its economic life. Since the operating costs are not smoothly increasing, we therefore see if there is a longer life for the defender for which its costs are lower than for the challenger. This can be done with a spreadsheet, as shown in Table 7.5.

We see that, for an additional life of two years, the defender has a lower cost per year than the challenger, when the challenger is kept over its economic life. Therefore, the defender should be kept for at least two more years. At this time, a new evaluation should be performed. ■

Table 7.5 Equivalent Annual Cost of Additional Life for the Defender

Additional Life in Years	Salvage Value	Operating Costs	EAC
0	$2400		
1	1400	$2000	$3288
2	980	2500	**3194**
3	686	3000	3258
4	480	3500	3369
5	336	4000	3500
6	235	4500	3641

7.6.2 Challenger Is Not Repeated

In this section, we no longer assume that challengers are alike. We recognize that future challengers will be available and we expect them to be better than the current challenger. We must then decide if the defender should be replaced by the current challenger. Furthermore, if it is to be replaced by the current challenger, *when* should the replacement occur? This problem is quite complex. The reason for the complexity is that, if we believe that challengers will be improving, we may be better off skipping the current challenger and waiting until the next improved challenger arrives. The difficulties are outlined in Example 7.6.

Example 7.6

Rita is examining the possibility of replacing the kiln controllers at the Burnaby Insulators plant. She has information about the existing controllers and the best replacement on the market. She also has information about a new controller design that will be available in three years. Rita has a five-year time horizon for the problem. What replacement alternatives should Rita consider?

One way to determine the minimum cost over the five-year horizon is to determine the costs of *all* possible combinations of the defender and the two challengers. This is impossible, since the defender and challengers could replace one another at any time. However, it is reasonable to consider only the combinations of the period length of one year. Any period length could be used, but a year is a natural choice because investment decisions tend, by convention, to follow a yearly cycle. These combinations form a mutually exclusive set of investment opportunities (see Section 4.2). If no time horizon was given in the problem, we would have had to assume one, to limit the number of possible alternatives.

The possible decisions that need to be evaluated in this case are shown in Table 7.6.

Table 7.6 Possible Decisions for Burnaby Insulators

Decision Alternative	Defender Life in Years	First Challenger Life in Years	Second Challenger Life in Years
1	5	0	0
2	4	1	0
3	4	0	1
4	3	2	0
5	3	1	1
6	3	0	2
7	2	3	0
8	2	2	1
9	2	1	2
10	1	4	0
11	1	3	1
12	1	2	2
13	0	5	0
14	0	4	1
15	0	3	2

To choose among these possible decisions, we need information about the following for the defender and both challengers:

1. Costs of installing the challengers
2. Salvage values for different possible lives for all three kiln controllers
3. Operating and maintenance costs for all possible ages for all three

With this information, the minimum cost solution is obtained by computing the costs for all possible decision alternatives. Since these are mutually exclusive projects, any of the comparison methods of Chapters 4 and 5 are appropriate, including present worth, annual worth, or IRR. The effects of sunk costs are already included in the enumeration of the various replacement possibilities, so looking at the benefits of keeping the defender is already automatically taken into account. ■

The difficulty with this approach is that the computational burden becomes great if the number of years in the time horizon is large. On the other hand, it is unlikely that information about a future challenger will be available under normal circumstances. In Example 7.6, Rita knew about a controller that wouldn't be available for three years. In real life, even if somehow Rita had inside information on the supplier research and marketing plans, it is unlikely that she would be confident enough of events three years away to use the information with complete assurance. Normally, if the information were available, the challenger itself would be available, too. Consequently, in many cases it is reasonable to assume that challengers in the planning future will be identical to the current challenger, and the decision procedure to use is the simpler one presented in the previous section.

REVIEW PROBLEMS

REVIEW PROBLEM 7.1

Kenwood Limousines runs a fleet of vans which ferry people from several outlying cities to a major international airport. New vans cost $45 000 each and depreciate at a declining-balance rate of 30% per year. Maintenance for each van is quite expensive, because they are in use 24 hours a day, seven days a week. Maintenance costs, which are about $3000 the first year, double each year the vehicle is in use. Given a MARR of 8%, what is the economic life of a van?

ANSWER

Table 7.7 shows the various components of this problem for replacement periods from one to five years. It can be seen that the replacement period with the minimum equivalent annual cost is two years. Therefore, the economic life is two years.

Table 7.7 Summary Computations for Review Problem 7.1

			Equivalent Annual Costs		
Year	Salvage Value	Maintenance Costs	Capital	Maintenance	Total
0	$45 000				
1	31 500	$ 3 000	$17 100	$ 3 000	$20 100
2	22 050	6 000	14 634	4 442	**19 076**
3	15 435	12 000	12 707	6 770	19 477
4	10 805	24 000	11 189	10 594	21 783
5	7 563	48 000	9 981	16 970	26 951

As an example, the calculation for a three-year period is:

EAC(capital costs)

$= (\$45\ 000 - \$15\ 435)\,(A/P, 8\%, 3) + \$15\ 435(0.08)$

$= \$29\ 565(0.38803) + \$15\ 435(0.08)$

$\cong \$12\ 707$

EAC(maintenance costs)

$= [\$3000(F/P, 8\%, 2) + \$6000(F/P, 8\%, 1) + \$12\ 000]\,(A/F, 8\%, 3)$

$= [\$3000(1.1664) + \$6000(1.08) + \$12\ 000)](0.30804)$

$\cong \$6770$

EAC(total) = EAC(capital costs) + EAC(maintenance costs)

$= \$12\ 707 + \$6770 = \$19\ 477$ ■

REVIEW PROBLEM 7.2

Canadian Widgets makes rocker arms for car engines. The manufacturing process consists of punching blanks from raw stock, forming the rocker arm in a five-stage progressive die, and finishing in a sequence of operations using hand tools. A recently developed 10-stage die can eliminate many of the finishing operations for high-volume production.

The existing five-stage die could be used for a different product, and in this case would have a salvage value of \$20 000. Maintenance costs of the five-stage die will total \$3500 this year, and are expected to increase by \$3500 per year. The 10-stage die will cost \$89 000, and will incur maintenance costs of \$4000 this year, increasing by \$2700 per year thereafter. Both dies depreciate at a declining-balance rate of 20% per year. The net yearly benefit of the automation of the finishing operations is expected to be \$16 000 per year. The MARR is 10%. Should the five-stage die be replaced?

ANSWER

Since there is no information about subsequent challengers, it is reasonable to assume that the 10-stage die would be repeated. The EAC of using the 10-stage die for various periods is shown in Table 7.8.

A sample EAC computation for keeping the 10-stage die for two years is as follows:

EAC(capital costs, two-year life)

$= (P - S)(A/P, 10\%, 2) + Si$

$= (\$89\ 000 - \$56\ 960)(0.57619) + \$56\ 960(0.10)$

$\cong \$24\ 157$

EAC(maintenance costs, two-year life)

$= [\$4000(F/P, 10\%, 1) + \$6700](A/F, 10\%, 2)$

$= [\$4000(1.1) + \$6700](0.47619)$

$\cong \$5286$

Table 7.8 EAC Computations for the Challenger in Review Problem 7.2

Life in Years	Salvage Value	Maintenance Costs	Equivalent Annual Costs		
			Capital	Maintenance	Total
0	$89 000				
1	71 200	$ 4 000	$26 700	$ 4 000	$30 700
2	56 960	6 700	24 157	5 286	29 443
3	45 568	9 400	22 021	6 529	28 550
4	36 454	12 100	20 222	7 729	27 951
5	29 164	14 800	18 701	8 887	27 589
6	23 331	17 500	17 411	10 004	27 415
7	18 665	20 200	16 314	11 079	**27 393**
8	14 932	22 900	15 377	12 113	27 490

$$\text{EAC(total, two-year life)} = \$24\ 157 + \$5286$$
$$= \$29\ 443$$

Completing similar computations for other lifetimes shows that the economic life of the 10-stage die is seven years and the associated equivalent annual costs are $27 393.

The next step in the replacement analysis is to consider the annual cost of the five-stage die (the defender) over the next year. This cost is to be compared with the economic life EAC of the 10-stage die, i.e., $27 393. Note that the cost analysis of the defender should include the benefits generated by the 10-stage die as an operating cost for the five-stage die as this $16 000 is a cost of *not* changing to the 10-stage die. The EAC of the capital and operating costs of keeping the defender one additional year are found as follows:

The salvage value of the five-stage die after one year is

$$\$20\ 000(1 - 0.2) = \$16\ 000$$

EAC(capital costs, one additional year)

$$= (P - S)(A/P, 10\%, 1) + Si$$
$$= (\$20\ 000 - \$16\ 000)(1.10) + \$16\ 000(0.10)$$
$$= \$6000$$

EAC(maintenance and operating costs, one additional year)

$$= \$3500 + \$16\ 000$$
$$= \$19\ 500$$

EAC(total, one additional year)

$= \$19\,500 + \$6000$

$= \$25\,500$

The five-stage die should not be replaced this year because the EAC of keeping it one additional year ($25 500) is less than the optimal EAC of the 10-stage die ($27 393). The knowledge that the five-stage die should not be replaced this year is usually sufficient for the immediate replacement decision. However, if a different challenger appears in the future, we would want to reassess the replacement decision.

It may also be desirable to estimate when in the future the defender might be replaced, even if it is not being replaced now. This can be done by calculating the equivalent annual cost of keeping the defender additional years until the time we can determine when it should be replaced. Table 7.9 summarizes those calculations for additional years of operating the five-stage die.

Table 7.9 EAC Computations for Keeping the Defender Additional Years

Additional Life in Years	Salvage Value	Maintenance and Operating Costs	Equivalent Annual Costs		
			Capital	Operating	Total
0	$20 000				
1	16 000	$19 500	$6 000	$19 500	$25 500
2	12 800	23 000	5 429	21 167	26 595
3	10 240	26 500	4 949	22 778	27 727
4	8 192	30 000	4 544	24 334	28 878
5	6 554	33 500	4 202	25 836	30 038
6	5 243	37 000	3 913	27 283	31 196
7	4 194	40 500	3 666	28 677	32 343
8	3 355	44 000	3 455	30 018	33 473

As an example of the computations, the EAC of keeping the defender for two additional years is calculated as

Salvage value of five-stage die after two years:

$\$16\,000(1 - 0.2) = \$12\,800$

EAC(capital costs, two additional years)

$= (P - S)(A/P, 10\%, 2) + Si$

$= (\$20\,000 - \$12\,800)(0.57619) + \$12\,800(0.10)$

$\cong \$5429$

$$\text{EAC(maintenance and operating costs, two additional years)}$$
$$= [\$19\,500(F/P, 10\%, 1) + (\$16\,000 + \$7000)](A/F, 10\%, 2)$$
$$= [\$19\,500(1.1) + \$23\,000](0.47619)$$
$$\cong \$21\,167$$

$$\text{EAC(total, two additional years)}$$
$$= \$5429 + \$21\,167 = \$26\,595$$

Continuing calculations in this manner will predict that the defender should be replaced at the end of the second year, given that the challenger remains the same during this time. This is because the EAC of keeping the defender for two years is less than the optimal EAC of the 10-stage die, but keeping the defender three years or more is more costly. ■

REVIEW PROBLEM 7.3

Avril bought a computer three years ago for \$3000, which she can now sell on the open market for \$300. The local Mr. Computer store will sell her a new HAL computer for \$4000, including the new accounting package she wants. Her own computer will probably last another two years, and then would be worthless. The new computer would have a salvage value of \$300 at the end of its economic life of five years. The net benefits to Avril of the new accounting package and other features of the new computer amount to \$800 per year. An additional feature is that Mr. Computer will give Avril a \$500 trade-in on her current computer. Interest is 15%. What should Avril do?

ANSWER

There are a couple of things to note about this problem. First, the cost of the new computer should be taken as \$3800 rather than \$4000. This is because, although the price was quoted as \$4000, the dealer was willing to give Avril a \$500 trade-in on a used computer that had a market value of only \$300. This amounts to discounting the price of the new computer to \$3800. Similarly, the used computer should be taken to be worth \$300, and not \$500. The \$500 figure does not represent the market value of the used computer, but rather the value of the used computer combined with the discount on the new computer. One must sometimes be careful to extract from the available information the best estimates of the values and costs for the various components of a replacement study.

First, we need to determine the EAC of the challenger over its economic life. We are told that the economic life is five years and hence the EAC computations are as follows:

$$\text{EAC(capital costs)} = (\$3800 - \$300)(A/P, 15\%, 5) + \$300(0.15)$$
$$= \$3500(0.29832) + \$45$$
$$= \$1089$$

$$\text{EAC(operating costs)} = \$0$$

$$\text{EAC(challenger, total)} = \$1089$$

Now we need to check the equivalent annual cost of keeping the existing computer one additional year. A salvage value for the computer for one year was not given. However, we can check to see if the EAC for the defender over two years is less than for the challenger. If it is, this is sufficient to retain the old computer.

$$\begin{aligned} \text{EAC(capital costs)} &= (\$300 - 0)(A/P, 15\%, 2) + \$0(0.15) \\ &= \$300(0.61512) + \$0 \\ &\cong \$185 \end{aligned}$$

$$\text{EAC(operating costs)} = \$800$$

$$\text{EAC(defender, total over 2 years)} = \$985$$

Avril should hang on to her current computer because its EAC over two years is less than the EAC of the challenger over its five-year economic life. ■

SUMMARY

This chapter is concerned with replacement and retirement decisions. Replacement can be required because there may be a cheaper way to provide the same service, or the nature of the service may have changed. Retirement can be required if there is no longer a need for the asset.

Defenders that are still functioning well have a significant cost advantage over challengers or over obtaining the service performed by the defender from another source. This is because there are installation costs and because the capital cost per year of an asset diminishes over time.

If an asset is replaced by a stream of identical assets, it is useful to determine the economic life of the asset, which is the replacement interval that provides the minimum annual cost. The asset should then be replaced at the end of its economic life.

If there is a challenger which is different from the defender, and future changes in technology are not known, one can determine the minimum EAC of the challenger and compare this with the cost of keeping the defender. If keeping the defender for any period of time is cheaper than the minimum EAC of the challenger, the defender should be kept. Normally, it is sufficient to assess the cost of keeping the defender for one more year.

Where future changes in technology are expected, decisions about when and whether to replace defenders are more complex. In this case, possible replacement decisions must be enumerated as a set of mutually exclusive alternatives and subjected to any of the standard comparison methods.

Figure 7.2 provides a summary of the overall procedure for assessing a replacement decision.

Figure 7.2 The Replacement Process

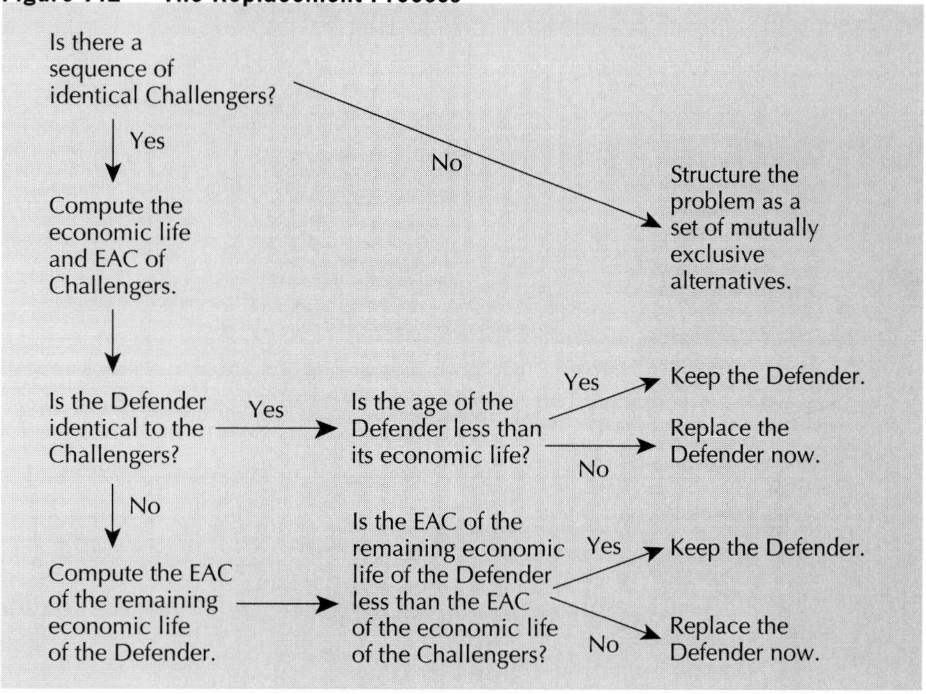

Engineering Economics in Action, Part 7B: Decision Time

Naomi, Dave, and Clem were meeting in Clem's office. They had just finished a discussion of their steel-ordering policy. Clem turned to Naomi and said, "O.K. Let's look at the 10-stage progressive die. Where does that stand?"

Naomi said, "It looks possible. Did you get a chance to read Terry's report?"

"Yes, I did," Clem answered. "Was it his idea to use the five-stage die for small runs so that we don't have to take a big hit from scrapping it?"

"Actually, it was," Naomi said.

"The kid may be a little intense," Clem said, "but he does good work. So where does that leave us?"

"Well, as I said, it looks possible that the 10-stage die will pay off," Naomi responded. "We have to decide what the correct time horizon is for making the analysis. Then we need more precise estimates of the costs and salvage value for the 10-stage die."

Clem turned in his chair and asked, "What do you think, Dave?"

Dave straightened himself in his chair and said, "I really don't know. How much experience has Hamilton Tools had at making dies this complicated?"

Naomi answered. "Not much. If we took them up on their proposal, we'd be their second or third customer."

"What do you have in mind, Dave?" Clem asked.

Dave said, "Well, if it's only the second or third time they've done something like this, I think we can expect some improvements over the next couple of years. So maybe we ought to wait."

"That makes sense," Clem responded. "I'd like you two to work on this. Give Tan Wang at Hamilton Tools a call. He'll know if anything is in the works. Get him to give you an estimate of what to expect. Then I want you to consider some possibilities. You know: 'Replace now.' 'Wait one year.' 'Wait two years.' And so on. Don't make it too complicated. Then, evaluate the different possibilities. I want a recommendation for next week's meeting. It's getting to be decision time."

PROBLEMS

7.1 Freeport Brothers have recently purchased a network computer system. Cabling installed in the office walls connects workstations in each employee's office to a central server. The costs of this system included the following:

5 workstations	\$4500 each
New server	\$6000
60 m cable	\$11.40 per metre
Cabling hardware	\$188
Workstation software	\$1190 per workstation
Server software	\$1950
10 hours of hardware installer time	\$20 per hour
25 hours of software installer time	\$60 per hour

If Freeport Brothers wanted to calculate a replacement interval for such a computer system, what would be the capital cost for the first year at a depreciation rate of 30%?

7.2 Last year, Clairbrook Canning Co. bought a fancy colour printer that cost \$20 000, for special printing jobs. Fast changes in colour printing technology have resulted in almost identical printers being available today for about 1/4 of the cost. Should CCC consider selling their printer and buying one of the new ones?

7.3 Maryhill Mines has a pelletizer that they are considering for replacement. Every three years it is overhauled at considerable cost. It is due for an overhaul this year. Evelyn has calculated that the sum of the operating and capital costs for this year for the pelletizer are significantly more than the average EAC for a new pelletizer over the new pelletizer's service life. Should the existing pelletizer be replaced?

7.4 Determine the economic life for each of the items listed below. Salvage values can be estimated by a declining-balance rate of 20%. The MARR is 8%.

	Purchase	Installation	Operating
Item 1	\$10 000	\$2000	\$300 first year, increasing by \$300 per year
Item 2	\$20 000	\$2000	\$200 first year, increasing by \$200 per year
Item 3	\$30 000	\$3000	\$2000 first year, increasing by \$2000 per year

7.5 A new bottle-capping machine costs \$45 000, including \$5000 for installation. The machine is expected to have a useful life of eight years with no salvage value at that time (assume straight-line depreciation). Operating and maintenance costs are expected to be \$3000 for the first year, increasing by \$1000 each year thereafter. Interest is 12%.

(a) Construct a spreadsheet which has the following headings: Year, Salvage Value, Maintenance Costs, EAC(Capital Costs), EAC(Operating Costs), and EAC(Total Costs). Compute the EAC(Total Costs) if the bottle capper is kept for n years, $n = 1, \ldots, 8$.

(b) Construct a chart showing the EAC(Capital Costs), EAC(Operating Costs), and EAC(Total Costs) if the bottle capper were to be kept for n years, $n = 1, \ldots, 8$.

(c) What is the economic life of the bottle capper?

7.6 Gerry likes driving small cars and buys nearly identical ones whenever the old one needs replacing. Typically, he trades in his old car for a new one costing about \$15 000. A new car warranty covers all repair costs above standard maintenance (standard maintenance costs are constant over the life of the car) for the first two years. After that, his records show an average repair expense (over standard maintenance) of \$2500 in the third year (at the end of the year), increasing by 50% per year thereafter. If a 30% declining balance depreciation rate is used, and interest is 8%, how often should Gerry get a new car?

7.7 Gerry (see Problem 7.6) has observed that the cars he buys are somewhat more reliable now than in the past. A better estimate of the repair costs is \$1500 in the third year, increasing by 50% per year thereafter, with all other information in Problem 7.6 being the same. Now how often should Gerry get a new car?

7.8 If the operating costs for an asset are $\$500 \times 2^n$ and the capital costs are $\$10\,000 \times (0.8)^n$, where n is the life in years, what is the economic life of the asset?

7.9 A roller conveyor system used to transport cardboard boxes along an order-filling line costs $100 000 plus $20 000 to install. It is estimated to depreciate at a declining-balance rate of 25% per year over its 15-year useful life. Annual maintenance costs are estimated to be $6000 for the first year, increasing by 20% each year thereafter. In addition, every third year, the rollers must be replaced at a cost of $7000. Interest is at 10%.

(a) Construct a spreadsheet which has the following headings: Year, Salvage Value, Maintenance Costs, EAC(Capital Costs), EAC(Maintenance Costs), and EAC(Total Costs). Compute the EAC(Total Costs) if the conveyor were to be kept for n years, $n = 1, \ldots, 15$.

(b) Construct a chart showing the EAC(Capital Costs), EAC(Maintenance Costs), and EAC(Total Costs) if the conveyor were to be kept for n years, $n = 1, \ldots, 15$.

(c) What is the economic life of the roller conveyor system?

7.10 Brockville Brackets (BB) has a three-year-old robot which welds small brackets onto car-frame assemblies. At the time the robot was purchased, it cost $300 000 and an additional $50 000 was spent on installation. BB acquired the robot as part of an eight-year contract to produce the car-frame assemblies. The useful life of the robot is 12 years, and its value is estimated to decline by 20% of current value per year, as shown in the table below. Operating and maintenance costs estimated when the robot was purchased are also in the table.

Defender, When New		
Life (Years)	**Salvage Value**	**Operating and Maintenance Costs**
0	$300 000	
1	240 000	$40 000
2	192 000	40 000
3	153 600	40 000
4	122 880	40 000
5	98 304	44 000
6	78 643	48 400
7	62 915	53 240
8	50 332	58 564
9	40 265	64 420
10	32 212	70 862
11	25 770	77 949
12	20 616	85 744

BB has found that the operating and maintenance costs for the robot have been higher than anticipated. At the end of the third year, new estimates of the operating and maintenance costs are as follows.

Costs for 3-Year-Old Defender		
Additional Life (Years)	**Salvage Value**	**Operating and Maintenance Costs**
0	$153 600	
1	122 880	$50 000
2	98 304	55 000
3	78 643	60 500
4	62 915	66 550
5	50 332	73 205

BB has determined that the reason their operating and maintenance costs were in error was that they positioned the robot too close to existing equipment so that the mechanics could not easily and quickly repair it. BB is considering moving the robot farther away from some adjacent equipment so that mechanics can get easier access for repairs. To move the robot will cause BB to lose valuable production time, which they estimate to have a cost of $25 000. However, once complete, the move will lower maintenance costs to what they originally had expected for the remainder of the contract (e.g., $40 000 for the fourth year, increasing by 10% per year thereafter). Moving the robot will not affect its salvage value.

If BB uses a MARR of 15%, should they move the robot? If so, when? Remember that the contract exists only for a further five years.

7.11 Consider Brockville Brackets from Problem 7.10 but assume that they have a contract to produce the car assemblies for an indefinite period. If they do not move the robot, their operating and maintenance costs will be higher than expected. If they do move the robot (at a cost of $25 000), their operating and maintenance costs are expected to be what they originally expected for the robot. Furthermore, BB expects to be able to obtain new versions of the existing robot for an indefinite period in the future; each is expected to have an installation cost of $50 000.

(a) Construct a spreadsheet table showing the EAC(total costs) if BB keeps the current robot in its current position for n more years, $n = 1, \ldots, 9$.

(b) Construct a spreadsheet table showing the EAC(total costs) if BB moves the current robot and then keeps it for n more years, $n = 1, \ldots, 9$

(c) Construct a spreadsheet table showing the EAC(total costs) if BB is to buy a new robot and keep it for n years, $n = 1, \ldots, 9$.

(d) Based on your answers for parts (a) through (c), what do you advise BB to do?

7.12 Nico has a 20-year-old oil-fired hot air furnace in his house. He is considering replacing it with a new high-efficiency natural gas furnace. The oil-fired furnace has a scrap value of $500, which it will retain indefinitely. A maintenance contract costs $300 per year, plus parts. Nico estimates that parts will cost $200 this year, increasing by $100 per year in subsequent years. The new gas furnace will cost $4500 to buy and $500 to install. It will save $500 per year in energy costs. The maintenance costs for the gas furnace are covered under guarantee for the first five years. The market value of the gas furnace can be estimated from straight-line depreciation with a salvage value of $500 after 10 years. Using a MARR of 10%, should the oil furnace be replaced?

7.13 A certain machine cost $25 000 to purchase and install. It has salvage values and operating costs as shown in the table below. The salvage value of $20 000 listed at the end of year 0 reflects the loss of the installation costs at the time of installation. The MARR is 12%.

Costs and Salvage Values for Various Lives		
Life in Years	**Salvage Value**	**Operating Cost**
0	$20 000.00	
1	16 000.00	$ 3 000.00
2	12 800.00	3 225.00
3	10 240.00	3 466.88
4	8 192.00	3 726.89
5	6 553.60	4 006.41
6	5 242.88	4 306.89
7	4 194.30	4 629.90
8	3 355.44	4 977.15
9	2 684.35	5 350.43
10	2 147.48	5 751.72
11	1 717.99	6 183.09
12	1 374.39	6 646.83
13	1 099.51	7 145.34
14	879.61	7 681.24
15	703.69	8 257.33
16	562.95	8 876.63
17	450.36	9 542.38
18	360.29	10 258.06

(a) What is the economic life of the machine?

(b) What is the equivalent annual cost over that life?

Now assume that the MARR is 5%.

(c) What is the economic life of the machine?

(d) What is the equivalent annual cost over that life?

(e) Explain the effect of decreasing the MARR.

7.14 Jack and Jill live on a hill in Deep Cove. Jack is a self-employed house painter who works out of their house. Jill works in Burnaby, to which she regularly commutes by car. The car is a four-year-old Japanese import. Jill could commute by bus. They are considering selling the car and getting by with the van Jack uses for work.

The car cost $12 000 new. It dropped about 20% in value in the first year. After that it fell by about 15% per year. The car is now worth about $5900. They expect it to continue to decline in value by about 15% of current value each year. Operating and other costs are about $2670 per year. They expect this to rise by about 7.5% per year. A commuter pass costs $112 per month, and is not expected to increase in cost.

Jack and Jill have a MARR of 10%, which is what Jack earns on his business investment. Their time horizon is two years because Jill expects to quit work at that time.

(a) Will commuting by bus save money?

(b) Can you advise Jack and Jill about retiring the car?

7.15 Ener-G purchases new turbines at a cost of $100 000. Each has a 15-year useful life and must be overhauled periodically at a cost of $10 000. The salvage value of a turbine declines 15% of current value each year, and operating and maintenance costs (including the cost of the overhauls) of a typical turbine are as shown in the table on page 292 (the costs for the fifth and tenth years include a $10 000 overhaul, but an overhaul is not done in the fifteenth year since this is the end of the turbine's useful life).

(a) Construct a spreadsheet which gives, for each year, the EAC(Operating and Maintenance Costs), EAC(Capital Costs), and EAC(Total Costs) for the turbines. Interest is 15%. How long should Ener-G keep each turbine before replacing it, given a five-year overhaul schedule? What are the associated equivalent annual costs?

(b) If Ener-G were to overhaul its turbines every six years (at the same cost), the salvage value and operating and maintenance costs would be as shown in the spreadsheet on page 293. Should Ener-G switch to a six-year overhaul cycle?

Defender, When New, Overhaul Every Five Years		
Life (Years)	**Salvage Value**	**Operating and Maintenance Costs**
0	$100 000	
1	85 000	$15 000
2	72 250	20 000
3	61 413	25 000
4	52 201	30 000
5	44 371	45 000
6	37 715	20 000
7	32 058	25 000
8	27 249	30 000
9	23 162	35 000
10	19 687	50 000
11	16 734	25 000
12	14 224	30 000
13	12 091	35 000
14	10 277	40 000
15	8 735	45 000

7.16 The Burnaby Machine Company makes small parts under contract for manufacturers in the Vancouver area. The company makes a group of metal parts on a turret lathe for a local ski manufacturer. The current lathe is now six years old. It has a planned further life of three years. The contract with the ski manufacturer has three more years to run as well. A new, improved lathe has become available. The challenger will have lower operating costs than the defender.

The defender can now be sold for $1200 in the used-equipment market. The challenger will cost $25 000 including installation. Its salvage value after installation, but before use, will be $20 000. Further data for the defender and the challenger are shown in the tables at the bottom of page 293 and at the top of page 294.

Burnaby Machine is not sure if the contract it has with the ski company will be renewed. Therefore, Burnaby wants to make the decision about replacing the defender with the challenger using a three-year study period. Burnaby Machine uses a 12% MARR for this type of investment.

Defender, When New, Overhaul Every Six Years		
Life (Years)	**Salvage Value**	**Maintenance Costs**
0	$100 000	
1	85 000	$15 000
2	72 250	20 000
3	61 413	25 000
4	52 201	30 000
5	44 371	35 000
6	37 715	50 000
7	32 058	20 000
8	27 249	25 000
9	23 162	30 000
10	19 687	35 000
11	16 734	40 000
12	14 224	55 000
13	12 091	25 000
14	10 277	30 000
15	8 735	35 000

(a) What is the present worth of costs over the next three years for the defender?

(b) What is the present worth of costs over the next three years for the challenger?

(c) Now suppose that Burnaby did not have a good estimate of the salvage value of the challenger at the end of three years. What minimum salvage value for the challenger at the end of three years would make the present worth of costs for the challenger equal to that for the defender?

Defender		
Additional Life in Years	**Salvage Value**	**Operating Cost**
0	$1 200	
1	600	$20 000
2	300	20 500
3	150	21 012.50

Challenger		
Life in Years	Salvage Value	Operating Cost
0	20 000	
1	14 000	13 875
2	9 800	14 360.63
3	6 860	14 863.25

7.17 Suppose, in the situation described in Problem 7.16, Burnaby Machine believed that the contract with the ski company would be renewed. Burnaby also believed that all challengers after the current challenger would be identical to the current challenger. Further data concerning these challengers are given below. Recall that a new challenger costs $25 000 installed.

Challenger		
Life in Years	Salvage Value	Operating Cost
0	$20 000.00	
1	14 000.00	$13 875.00
2	9 800.00	14 369.63
3	6 860.00	14 863.25
4	4 802.00	15 383.46
5	3 361.40	15 921.88
6	2 352.98	16 479.15
7	1 647.09	17 055.92
8	1 152.96	17 652.87
9	807.07	18 270.73
10	564.95	18 910.20
11	395.47	19 572.06
12	276.83	20 257.08

Burnaby were also advised that machines identical to the defender would be available indefinitely. New copies of the defender would cost $17 500, including installation. Further data concerning new defenders are shown in the table at the top of page 295. The MARR is 12%.

(a) Find the economic life of the challenger. What is the equivalent annual cost over that life?

Defender When New		
Life in Years	**Salvage Value**	**Operating Cost**
0	$15 000.00	
1	9 846.45	$17 250.00
2	6 463.51	17 681.25
3	4 242.84	18 123.28
4	2 785.13	18 576.36
5	1 828.24	19 040.77
6	1 200.11	19 516.79
7	600.00	20 004.71
8	300.00	20 504.83
9	150.00	21 017.45
10	150.00	21 542.89
11	150.00	22 081.46
12	150.00	22 633.49
13	150.00	23 199.33

(b) Should the defender be replaced with the challenger or with a new defender?

(c) When should this be done?

7.18 You own several copiers which are currently valued at $10 000 all together. Annual operating and maintenance costs for all copiers are estimated at $9000 next year, increasing by 10% each year thereafter. Declining-balance depreciation on the existing equipment is about 20% per year.

You are considering replacing your existing copiers with new ones which have a suggested retail price of $25 000. Operating and maintenance costs for the new equipment will be $6000 per year, increasing by 10% each year. The salvage value of the new equipment is well approximated by a 20% drop from the suggested retail price per year. Furthermore, you can get a trade-in allowance of $12 000 for your equipment if you purchase the new equipment at its suggested retail price. Your MARR is 8%. Should you replace your existing equipment now?

7.19 An existing piece of equipment has the following pattern of salvage values and operating and maintenance costs:

Defender					
Additional Life (Years)	Salvage Value	Maintenance Costs	EAC Capital Costs	EAC Operating and Maintenance Costs	EAC Total
0	$10 000				
1	8 000	$2 000	$3 500	$2 000	$5 500
2	6 400	2 500	3 174	2 233	5 407
3	5 120	3 000	2 905	2 454	5 359
4	4 096	3 500	2 682	2 663	5 345
5	3 277	4 000	2 497	2 861	5 359
6	2 621	4 500	2 343	3 049	5 391
7	2 097	5 000	2 214	3 225	5 439
8	1 678	5 500	2 106	3 391	5 497
9	1 342	6 000	2 016	3 546	5 562

A replacement asset is being considered. Its relevant costs over the next nine years are:

Challenger					
Additional Life (Years)	Salvage Value	Maintenance Costs	EAC Capital Costs	EAC Operating and Maintenance Costs	EAC Total
0	$12 000				
1	9 600	$1 500	$4 200	$1 500	$5 700
2	7 680	1 900	3 809	1 686	5 495
3	6 144	2 300	3 486	1 863	5 349
4	4 915	2 700	3 219	2 031	5 249
5	3 932	3 100	2 997	2 189	5 186
6	3 146	3 500	2 811	2 339	5 150
7	2 517	3 900	2 657	2 480	5 137
8	2 013	4 300	2 528	2 613	5 140
9	1 611	4 700	2 419	2 737	5 156

There is need for the asset (either the defender or the challenger) for the next nine years.

(a) What replacement alternatives are there?

(b) What replacement timing do you recommend?

7.20 The Brunswick Table Top Company makes tops for tables and desks. The company now owns a seven-year-old planer that is experiencing increasing operating costs. The defender has a maximum additional life of five years. They are considering replacing the defender with a new planer.

Defender		
Additional Life in Years	**Salvage Value**	**Operating Cost**
0	$4 000	
1	3 000	$20 000
2	2 000	25 000
3	1 000	30 000
4	500	35 000
5	500	40 000

First Challenger		
Life in Years	**Salvage Value**	**Operating Cost**
0	$25 000	
1	20 000	$16 800
2	16 000	17 640
3	12 800	18 522
4	10 240	19 448
5	8 192	20 421
6	6 554	21 442
7	5 243	22 514
8	4 194	23 639
9	3 355	24 821
10	2 684	26 062

The new planer would cost $30 000 installed. Its value after installation, but before use, would be about $25 000. The company has been told that there will be a new model planer available in two years. The new model is expected to have the same first costs as the current challenger. However, it is expected to have lower operating costs. Data concerning the defender and the two challengers are shown in the tables on page 297 and below. Brunswick Table has a 10-year planning period and uses a MARR of 10%.

Second Challenger		
Life in Years	**Salvage Value**	**Operating Cost**
0	$25 000	
1	20 000	$12 000
2	16 000	12 600
3	12 800	13 230
4	10 240	13 892
5	8 192	14 586
6	6 554	15 315
7	5 243	16 081
8	4 194	16 885
9	3 355	17 729
10	2 684	18 616

(a) What are the combinations of planers that Brunswick can use to cover requirements for the next 10 years? For example, Brunswick may keep the defender one more year, then install the first challenger and keep it for nine years. Notice that the first challenger will not be installed after the second year when the second challenger becomes available. You may ignore combinations that involve installing the first challenger after the second becomes available. Recall also that the maximum additional life for the defender is five years.

(b) What is the best combination?

7.21 You estimate that your two-year-old car is now worth $12 000 and that it will decline in value by 25% of its current value each year of its eight-year remaining useful life. You also estimate that its operating and maintenance costs will be $2100, increasing by 20% per year thereafter. Your MARR is 12%.

(a) Construct a spreadsheet showing (1) additional life in years, (2) salvage value, (3) operating and maintenance costs, (4) EAC(operating and maintenance costs), (5) EAC(capital costs), and (6) EAC(total costs). What additional life minimizes the EAC(total costs)?

(b) Now you are considering the possibility of painting the car in three years' time for $2000. Painting the car will increase its salvage value. By how much will the salvage value have to increase before painting the car is economically justified? Modify the spreadsheet you developed for part (a) to show this salvage value and the EAC(total costs) for each additional year of life. Will painting the car extend its economic life?

7.22 A long-standing principle of computer innovations is that computers double in power for the same price, or, equivalently, halve in cost for the same power, every 18 months. Barrie Data Services (BDS) owns a single computer which is at the end of its third year of service. BDS will continue to buy computers of the same power as its current one. Its current computer would cost $80 000 to buy today, excluding installation. Given that a new model is released every 18 months, what replacement policy should BDS adopt for computers over the next three years? Other facts to be considered are:

1. Installation cost is 15% of purchase price.
2. Salvage values are computed at a declining-balance depreciation rate of 50%.
3. Maintenance is estimated as 10% of accumulated depreciation per year or as 15% of accumulated depreciation per 18-month period.
4. BDS uses a MARR of 12%.

7.23 A water pump to be used by the city's maintenance department costs $10 000 new. A running-in period, costing $1000 immediately, is required for a new pump. Operating and maintenance costs average $500 the first year, increasing by $300 per year thereafter. The salvage value of the pump at any time can be estimated by the declining-balance rate of 20%. Interest is at 10%. Using a spreadsheet, calculate the EAC for replacing the pump after one year, two years, etc. How often should the pump be replaced?

7.24 The water pump from Problem 7.23 is being considered to replace an existing one. The current one has a salvage value of $1000 and will retain this salvage value indefinitely.

(a) Operating costs are currently $2500 per year and are rising by $400 per year. Should the current pump be replaced? When?

(b) Operating costs are currently $3500 per year and are rising by $200 per year. Should the current pump be replaced? When?

7.25 Chatham Automotive purchased new electric forklifts to move steel automobile parts two years ago. These cost $75 000 each, including the charging stand. In practice, it was found that they do not hold a charge as long as claimed by the manufacturer, so operating costs are very high. This also results in their currently having a salvage value of about $10 000. Chatham

Automotive is considering replacing them with propane models. The new ones cost $58 000. After one year, they have a salvage value of $40 000, and thereafter decline in value at a declining-balance depreciation rate of 20%, as does the electric model from this time on. The MARR is 8%. Operating costs for the electric model are currently $20 000 per year, rising by 12% per year. Operating costs for the propane model initially will be $10 000 per year, rising by 12% per year. Should Chatham Automotive replace the forklifts now?

7.26 Suppose that Chatham Automotive (Problem 7.25) can get a $14 000 trade-in value for their current electric model when they purchase a new propane model. Should they replace the electric forklifts now?

7.27 A joint former cost $60 000 to purchase and $10 000 to install seven years ago. The market value now is $33 000, and this will decline by 12% of current value each year for the next three years. Operating and maintenance costs are estimated to be $3400 this year, and are expected to increase by $500 per year.

(a) How much should the EAC of a new joint former be over its economic life to justify replacing the old one sometime in the next three years? The MARR is 10%.

(b) The EAC for a new joint former turns out to be $10 300 for a 10-year life. Should the old joint former be replaced within the next three years? If so, when?

(c) Is it necessary to consider replacing the old joint former more than three years from now, given that a new one has an EAC of $10 300?

7.28 Northwest Aerocomposite manufactures fibreglass and carbon fibre fairings. Their largest water-jet cutter will have to be replaced some time before the end of four years. The old cutter is currently worth $49 000. Other cost data for the current and replacement cutters can be found in the tables on page 301. The MARR is 15%. What is the economic life of the new cutter, and what is the equivalent annual cost for that life? When should the new cutter replace the old?

7.29 The water pump from Problem 7.23 has an option to be overhauled once. It costs $1000 to overhaul a three-year-old pump and $2000 to overhaul a five-year-old pump. The major advantage of an overhaul is that it reduces the operating and maintenance costs to $500, which will increase again by $300 per year thereafter. Should the pump be overhauled? If so, should it be overhauled in three years or five years?

7.30 Northwest Aerocomposite in Problem 7.28 found out that their old water-jet cutter may be overhauled at a cost of $14 000 now. The cost information for the old cutter after an overhaul is as shown in the table at the bottom of page 301.

Problem 7.28: Challenger		
Life in Years	Salvage Value	Operating and Maintenance Costs
0	$90 000	
1	72 000	$12 000
2	57 600	14 400
3	46 080	17 280
4	36 864	20 736
5	29 491	24 883
6	23 593	29 860
7	18 874	35 832
8	15 099	42 998
9	12 080	51 598

Problem 7.28: Defender		
Life in Years	Salvage Value	Operating and Maintenance Costs
0	$49 000	
1	36 500	$17 000
2	19 875	21 320
3	15 656	26 806
4	6 742	33 774

Problem 7.30: Defender with an Overhaul		
Life (Years)	Salvage Value	Operating and Maintenance Costs
0	$55 000	
1	40 970	$16 500
2	22 310	20 690
3	17 574	26 013
4	7 568	32 775

Should Northwest overhaul the old cutter? If an overhaul takes place, when should the new cutter replace the old? Assume that the cost information for the replacement cutter is as given in Problem 7.28.

7.31 Tiny Bay Freight Company wants to begin business with one delivery truck. After two years of operation, the company plans to increase the number of trucks to two, and after four years, plans to increase the number to three. TBFC currently has no trucks. The company is considering purchasing one type of truck which costs $30 000. The operating and maintenance costs are estimated to be $7200 per year. The resale value of the truck will decline each year by 40% of the current value. The company will consider replacing a truck every two years. That is, the company may keep a truck for two years, four years, six years, and so on. TBFC's MARR is 12%.

(a) What are the possible combinations for purchasing and replacing trucks over the next five years so that Tiny Bay Freight Company will meet their expansion and will have three trucks in hand at the end of five years?

(b) Which purchase/replacement combination is the best?

Problems 7.32 through 7.35 are concerned with the economic life of assets where there is a sequence of identical assets. The problems explore the sensitivity of the economic life to four parameters: the MARR, the level of operating cost, the rate of increase in operating cost, and the level of first cost. In each problem there is a pair of assets. The assets differ in only a single parameter. The problem asks you to determine the effect of this difference on the economic life and to explain the result. All assets decline in value by 20% of current value each year. Installation costs are zero for all assets. Further data concerning the four pairs of assets are given in the table which follows.

Asset Number	First Cost	Initial Operating Cost	Rate of Operating Cost Increase	MARR
A1	$125 000	$30 000	12.5%/year	5%
B1	125 000	30 000	12.5%/year	25%
A2	100 000	30 000	$2000/year	15%
B2	100 000	40 000	$2000/year	15%
A3	100 000	30 000	5%/year	15%
B3	100 000	30 000	12.5%/year	15%
A4	75 000	30 000	5%/year	15%
B4	150 000	30 000	5%/year	15%

7.32 Consider Assets A1 and B1. They differ only in the MARR.

(a) Determine the economic lives for Assets A1 and B1.

(b) Create a diagram showing the EAC(capital), the EAC(operating), and the EAC(total) for Assets A1 and B1.

(c) Explain the difference in economic life between A1 and B1.

7.33 Consider Assets A2 and B2. They differ only in the level of operating cost.

(a) Determine the economic lives for Assets A2 and B2.

(b) Create a diagram showing the EAC(capital), the EAC(operating), and the EAC(total) for Assets A2 and B2.

(c) Explain the difference in economic life between A2 and B2.

7.34 Consider Assets A3 and B3. They differ only in the rate of increase of operating cost.

(a) Determine the economic lives for Assets A3 and B3.

(b) Create a diagram showing the EAC(capital), the EAC(operating), and the EAC(total) for Assets A3 and B3.

(c) Explain the difference in economic life between A3 and B3.

7.35 Consider Assets A4 and B4. They differ only in the level of first cost.

(a) Determine the economic lives for Assets A4 and B4.

(b) Create a diagram showing the EAC(capital), the EAC(operating), and the EAC(total) for Assets A4 and B4.

(c) Explain the difference in economic life between A4 and B4.

7.36 This problem concerns the economic life of assets where there is a sequence of identical assets. In this case there is an opportunity to overhaul equipment. Two issues are explored. The first issue concerns the optimal life of equipment. The second issue concerns the decision as to whether or not to replace equipment that is past its economic life.

Consider a piece of equipment that costs \$40 000 to buy and install. The equipment has a maximum life of 15 years. Overhaul is required in the fourth, eighth, and twelfth years. The company uses a MARR of 20%. Further information is given in the table on page 304.

(a) Show that the economic life for this equipment is seven years.

(b) Suppose that the equipment is overhauled in the eighth year rather than replaced. Show that keeping the equipment for three more years (after the eighth year), until it next comes up for overhaul, has lower cost than replacing the equipment immediately.

Year	Salvage Value	Operating Cost	Overhaul Cost
0	$15 000		
1	12 000	$2 000	
2	9 600	2 200	
3	7 680	2 420	
4	7 500	2 662	$2 500
5	6 000	2 000	
6	4 800	2 200	
7	3 840	2 420	
8	4 500	2 662	32 500
9	3 600	2 000	
10	2 880	2 800	
11	2 304	3 920	
12	2 000	5 488	17 500
13	1 200	4 000	
14	720	8 000	
15	432	16 000	

Hint for (b): The comparison must be done fairly and carefully. Assume that under either plan the replacement is kept for its optimal life of seven years. It is easier to compare the plans if they cover the same number of years. One way to do this is to consider an 11-year period as shown on page 305.

First, show that the present worth of costs over the 11 years is lower under Plan A than under Plan B. Second, point out that the equipment that is in place at the end of the eleventh year is newer under Plan A than under Plan B.

(c) Why is it necessary to take into account the age of the equipment at the end of the 11-year period?

Year	Plan A	Plan B
0		
1	Defender	Replacement #1
2	Defender	Replacement #1
3	Defender	Replacement #1
4	Replacement #1	Replacement #1
5	Replacement #1	Replacement #1
6	Replacement #1	Replacement #1
7	Replacement #1	Replacement #1
8	Replacement #1	Replacement #2
9	Replacement #1	Replacement #2
10	Replacement #1	Replacement #2
11	Replacement #2	Replacement #2

7.37 Northfield Metal Works is a Manitoba household appliance parts manufacturer which has just won a contract with a major appliance company to supply replacement parts to service shops. The contract is for five years. Northfield is considering using three existing manual punch presses or a new automatic press for part of the work. The new press would cost \$225 000 installed. Northfield is using a five-year time horizon for the project. The MARR is 25% for projects of this type. Further data concerning the two options are shown below and on page 306.

Note that the hand-fed press values are for each of the three presses. Costs must be multiplied by three to get the costs for three presses. Northfield is not sure of the salvage values for the new press. What salvage value at the end of five years would make the two options equal?

Automatic Punch Press		
Life in Years	**Salvage Value**	**Operating Cost**
0	\$125 000	
1	100 000	\$25 000
2	80 000	23 750
3	64 000	22 563
4	51 200	21 434
5	40 960	20 363

Hand-Fed Press		
Additional Life in Years	Salvage Value	Operating Cost
0	$10 000	
1	9 000	$25 000
2	8 000	25 000
3	7000	25 000
4	6 000	25 000
5	5 000	25 000

Canadians Holding Onto their Cars Longer

CANADIAN 7.1 MINI-CASE

A typical Canadian motorist now drives a car at least five years old with close to 100 000 kilometres on it. Based on a survey done by the Canadian Automobile Association in 1995, people are keeping their cars longer because they're built better and last longer, and fewer people can afford the cost of driving a new car.

One reason people are driving older cars is that the cost of repairs for old cars is not that much higher than for newer ones. The survey found that owners of 1992-model-year vehicles spent an average of $514 on repairs and maintenance in 1995, while owners of cars built in 1986 or earlier spent an average of $870.

Another reason why people are driving older cars is that the cost of car ownership goes down with the length of time you keep the vehicle. This is because you lose less on depreciation each year on an older car than on a newer one. For example, the $7715 average cost of driving a new 1995 Chevrolet Cavalier for the year is largely due to its large depreciation expense of $3500. The balance includes the cost of repairs, insurance, and gas. Older cars have lower depreciation expenses and thus cost less to keep on the road.

Source: Article by Michael Prentice, adapted from the *Ottawa Citizen*, in the *Kitchener-Waterloo Record*, February 7, 1996.

Discussion

People buy and sell things for many reasons. A car might be desirable because it exhibits a sporty image, is known to be reliable, or is big enough for a large family. Purchase price, resale value, and operating costs are also very important, but people often make purchase decisions that are not economically wise.

In particular, the cost of a new car is hard to justify in many cases. A new car depreciates very quickly in terms of resale value over the first year of use. However, the functional difference between a new car and a year-old car is often very small. On the other hand, if nobody bought new cars, there wouldn't be any old cars available to buy!

Questions

1) Draw a graph of the purchase or resale value of a particular make and model of car over time. Estimate the values you use from personal experience, newspaper want-ads, internet sites, etc.
2) For the same make and model of car, estimate the operating cost per year, including insurance, repairs, fuel, etc., over the same number of year as done in Question 1.
3) If you were going to buy this make and model, brand new or used, and keep it for four years, what would be the most economical four-year period to do so?
4) Are there reasons why the answer in Question 3 would not be done in practice?

CHAPTER 8

Taxes

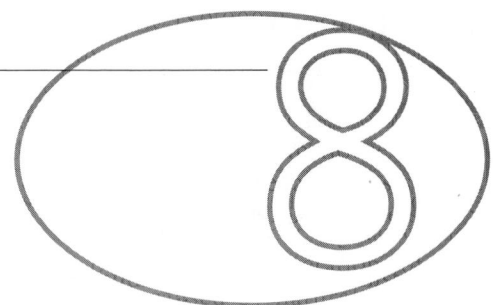

Engineering Economics in Action, Part 8A: It's in the Details

"Details, Terry. Sometimes it's all in the details." Naomi pursed her lips and nodded sagely. Terry and Naomi were sitting in the coffee room together. The main break periods for the line workers had passed, so they were alone except for a maintenance person on the other side of the room who was enjoying either a late breakfast or an early lunch.

"Uh, O.K., Naomi. What is?"

"Well," Naomi replied, "you know that rocker arm die deal? The one where we're upgrading to a 10-stage die? The rough replacement study you did seems to have worked out O.K. We're going to do something, sometime, but now we have to be a little more precise."

"O.K. What do we do?" Terry was interested now. Terry liked things precise and detailed.

"The main thing is to make sure we are working with the best numbers we can get. I'm getting good cost figures from Tan Wang at Hamilton Tools for this die and future possibilities, and we'll also work out our own costs for the changeover. Your cost calculations are going to have to be more accurate, too."

"You mean to more significant digits?" Naomi couldn't tell whether Terry was making a joke or the idea of more significant digits really did thrill him. She decided it was the former.

"Ha. Ha. No, I mean that we had better explicitly look at the tax implications of the purchase. Our rough calculations ignored taxes, and taxes can have a significant effect on the choice of best alternative. And when lots of money is at stake, the details matter."

8.1 Introduction

In Canada, the federal and provincial governments levy taxes on both individuals and corporations. These taxes can have a significant impact on the economic viability of a project. A vital component of a thorough economic analysis will therefore include the tax implications of an investment decision. This chapter provides an introduction to the tax environment in Canada and shows how this environment can affect engineering economics decisions.

When a firm makes an investment, the income from the project will affect the company's cash flows. If the investment yields a profit, the profits will be taxed. Since the taxes result as a direct consequence of the investment, they reduce the net profits associated with that investment. In this sense, taxes associated with a project are a disbursement. If the investment yields a loss, the company may be able to offset the loss from this project against the profits from another, and end up paying less tax overall. As a result, when evaluating a loss-generating project, the net savings in tax can be viewed as a negative disbursement.

The most significant kind of tax for economic comparisons is *income tax*, which is charged on *net* income. In this context, net income is the difference between expenses and receipts for the company as a whole. Corporate tax rates in Canada are generally between 35% and 60% of net income. The actual tax rate applied is fairly complicated and can depend on the size of the firm, whether or not it is a manufacturer, its location, and a variety of other factors. For example, to encourage the development of new technology, special rules were designed to reduce the tax burden on research and development projects. Our concern here is with the basic approach used in determining the impact of taxes on a project. For special tax rules, it is best to check with Revenue Canada or a tax specialist.

It is also worth noting that tax rules can change suddenly. As seen below, the current tax rules have applied only since 1981. Also, the Canadian tax rules are unique to Canada, so that the methods presented in this chapter may not apply elsewhere. For comparison, Close-Up 8.1 reviews the tax procedures used in the United States.

CLOSE-UP 8.1 United States Tax Rules

In the U.S. tax system, all depreciable assets are designated as belonging to a "Modified Accelerated Cost Recovery System (MACRS) Class." The MACRS Class determines the declining-balance rate (usually double-declining-balance or 150% declining-balance) and the **recovery period**, which is the designated service life for depreciation calculation purposes.

The declining-balance method is not used for the entire recovery period. All assets are required to attain a book value of \$0 at the end of the recovery period. Since, under a declining-balance depreciation method, a book value of \$0 is never reached, at an appropriate point the depreciation method switches from declining-balance to straight-line.

8.2 Before- and After-tax MARR

Taxes have a significant effect on engineering decision making, so much that they cannot be ignored. In this text so far, it seems that they have been ignored, since no specific tax calculations have been done. In fact, they have been implicitly incorporated into the computations through the use of a *before-tax* MARR, though we have not called it such.

The basic logic is as follows: since taxes have the effect of reducing profits associated with a project, we need to make sure that we set the MARR for project acceptability accordingly. If we do not explicitly account for the impact of taxes in the project cash flows, then we need to set a MARR which is high enough to recognize that taxes will need to be paid. This is the *before-tax* MARR. If, on the other hand, we explicitly account for the impact of taxes in the cash flows of a project (i.e., reduce the cash flows by the tax rate), then the MARR used for the project should be lower, since the cash flows already take into account the payment of taxes. This is the *after-tax* MARR.

In fact, we can express an approximate relationship

$$\text{MARR}_{\text{after-tax}} \cong \text{MARR}_{\text{before-tax}} \times (1 - t) \tag{8.1}$$

where t is the corporate tax rate. The *before-tax* MARR means that the MARR has been chosen high enough to provide an acceptable rate of return without explicitly considering taxes. In other words, since all profits are taxed at the rate *t*, the *before-tax* MARR has to include enough returns to meet the *after-tax* MARR and, in addition, provide the amount to be paid in taxes. As we can see from Equation (8.1), the after-tax MARR will generally be lower than the before-tax MARR. We will see later in this chapter how the relationship given in Equation (8.1) is a simplification but a reasonable approximation of the effect of taxes. In practice, the before- and after-tax MARRs are often chosen independently and are not directly related by Equation (8.1). Generally speaking, if a MARR is given without specifying whether it is a before- or after-tax MARR, it can be assumed to be a before-tax MARR.

Example 8.1

Prince Rupert Gold Mines has been selecting projects for investments based on a before-tax MARR of 12%. Sherri feels that some good projects have been missed because the effects of taxation on the projects has not been examined in enough detail, so she proposed reviewing the projects on an after-tax basis. What would be a good choice of after-tax MARR for her review? PRGM pays 45% corporate taxes.

Although the issue of selecting an after-tax MARR is likely to be more complicated, a reasonable choice for Sherri would be to use Equation (8.1) as a way of calculating an after-tax MARR for her review. This gives:

$$\text{MARR}_{\text{after-tax}} \cong 0.12 \times (1 - 0.45) = 0.066 = 6.6\%$$

A reasonable choice for after-tax MARR would be 6.6%. ■

8.3 The Capital Cost Allowance (CCA) System

When a firm buys a depreciable asset for use in its business, a **capital expense** is incurred. It is the expenditure associated with the purchase of a long-term depreciable asset. (Almost all tangible assets are depreciable. The primary exception to this is land.) Since the asset deteriorates through the passage of time, the firm must deduct the capital expense over a period of years. This is done by claiming a depreciation expense each year of the asset's useful life, as its value declines. The depreciation is recorded by accountants in the balance sheet as a reduction in the book value of the asset. It is also recorded as an expense on the income statement. Depreciation thus reduces the before-tax income even though, in reality, there has been no cash expense.

In general, a firm will want to "write off" (i.e., depreciate) an investment as fast as possible. This is because depreciation is considered an expense and offsets revenue so as to reduce net income. Since net income is taxed, taxes can be deferred or reduced by depreciating assets quickly. The effect of deferring the taxes can be considerable. To counter this effect, the Canadian tax system defines a specific amount of depreciation that companies may claim in any year for any one depreciable asset. This amount is called the *capital cost allowance (CCA)*. In this section, we examine how to apply CCA rules to investment decisions and compare the CCA to depreciation claimed for accounting records.

Example 8.2

In the imaginary country of Monovia, companies can depreciate their capital asset purchases as fast as they want. Clive Cutler, owner of Monovia Manufacturing, has just bought $200 000 worth of equipment. He has made up two spreadsheets, shown in Tables 8.1 and 8.2, to illustrate the effect of different depreciation strategies over the next five years under the following assumptions:

1. Income is $300 000 per year.
2. Expenses excluding depreciation are $100 000 per year.
3. The tax rate is 50%.
4. Available cash is invested at 10% interest.
5. The salvage value of the equipment after five years is $0.

Table 8.1 Full Depreciation in One Year

Year	1	2	3	4	5
Income	$300 000	$300 000	$300 000	$300 000	$300 000
Expenses excluding depreciation	100 000	100 000	100 000	100 000	100 000
Depreciation expense	200 000	0	0	0	0
Net income	0	200 000	200 000	200 000	200 000
Taxes	0	100 000	100 000	100 000	100 000
Profit	0	100 000	100 000	100 000	100 000
Cash	200 000	100 000	100 000	100 000	100 000
Accumulated cash	200 000	320 000	452 000	597 200	756 920

Table 8.2 Straight-Line Depreciation over Five Years

Year	1	2	3	4	5
Income	\$300 000	\$300 000	\$300 000	\$300 000	\$300 000
Expenses excluding depreciation	100 000	100 000	100 000	100 000	100 000
Depreciation expense	40 000	40 000	40 000	40 000	40 000
Net income	160 000	160 000	160 000	160 000	160 000
Taxes	80 000	80 000	80 000	80 000	80 000
Profit	80 000	80 000	80 000	80 000	80 000
Cash	120 000	120 000	120 000	120 000	120 000
Accumulated cash	120 000	252 000	397 200	556 920	732 612

Table 8.1 illustrates the case where the equipment is fully depreciated in the first year, although it generates revenue over its five-year life. In Table 8.2, straight-line depreciation is used over the five-year life.

When we look at the effects of depreciation on economic analyses, it is important to distinguish between expenses that represent a cash outflow and expenses that do not. Purchasing an asset, like a piece of equipment, will produce a cash outflow at the time the purchase is made. In particular, the balance sheet will reflect a transfer out of current assets (cash) and a transfer into fixed assets (equipment) and perhaps to current liabilities (bank loan).

Depreciation, on the other hand, does not actually represent a cash outflow, although it is recorded as an expense in the income statement. For example, in Table 8.1, writing off (depreciating) the entire cost of the equipment in its first year produced a depreciation expense of \$200 000 in that year. There was no actual cash outflow due to the depreciation (although there was for the actual purchase of the asset), but depreciation caused the net income to be reduced to zero for that year, even though \$200 000 in cash was actually available. Investing the \$200 000 for the second year at 10% interest produces accumulated cash of \$220 000. Adding this to the profit of \$100 000 for the second year gives accumulated cash of \$320 000 at the end of the second year. Continuing in this fashion produces accumulated cash of \$756 920 at the end of the five-year period.

In contrast, if the equipment is depreciated on a straight-line basis over five years, only \$732 612 in cash is accumulated. This can be seen by working through the expenses, net income, taxes, and profit for each year. For example, in year 1, a (straight-line) depreciation expense of \$200 000/5 = \$40 000 is claimed. This reduces net income by \$40 000, to \$160 000, and leaves after-tax profits of \$80 000. Now, as before, the depreciation expense of \$40 000 is not a cash outflow, so the cash actually available to invest at the end of the first year is the $80 000 profit *plus* $40 000. Since the depreciation expense is constant with the straight-line method, cash of $120 000 will be available to invest at the end of each of the five years. Thus, at the end of five years, the accumulated cash will be $732 612. This is $24 308 less than when the equipment was fully depreciated in the first year because taxes were delayed by depreciating more of the asset's value earlier. The extra income that was available for investment early allowed more interest to accumulate over the five-year period. The $24 308 is significant, and illustrates why faster depreciation is preferred to slower depreciation. ■

As seen in Chapter 6, there are several generally accepted depreciation methods. The most prevalent methods in Canada are straight line and declining balance. For the purposes of preparing financial statements for investors, a firm may use any one (or all) of the generally accepted methods for calculating depreciation expenses, provided that the method used is the same from period to period. However, if companies had the freedom to depreciate as they wanted to for tax purposes, they would depreciate their assets immediately, since in that way they would get the largest benefit because of tax savings.

Governments have a different perspective, since they would prefer to receive the tax money as quickly as possible. They would want companies to depreciate assets as slowly as possible to keep taxable income as high as possible and produce the most taxes. In order to limit the depreciation amount which companies use for tax purposes, the Canadian government established a maximum level of capital cost expense (i.e., depreciation) which a company can claim each year. This maximum amount is referred to as the firm's **capital cost allowance (CCA)**. The system set up to allow firms to compute their capital cost allowance is called the **capital cost allowance (CCA) system**. This system, established by the Canadian government, specifies the amount and timing of depreciation expenses on capital assets. According to the CCA system, the declining-balance method of depreciation must be used for claiming capital costs associated with most tangible assets. Straight-line depreciation is used for intangible assets. We are mainly concerned with tangible assets, so our discussion will focus on declining-balance depreciation.

The CCA system specifies the maximum rate a company can use to depreciate its assets for tax calculations; this is referred to as the **capital cost allowance (CCA) rate**. To implement the CCA system, a firm's assets are grouped by **capital cost allowance (CCA) asset class** for which a specified CCA rate is used to compute CCA. For example, all assets classified as office equipment (desks, chairs, filing cabinets, copiers, and the like) are grouped together, and depreciation expenses are based on the total remaining undepreciated cost of all assets in that class. Some examples of CCA rates and CCA asset classes are given in Table 8.3. Note that a 100% CCA rate means that the assets are **expensed**, that is, treated as an operating cost rather than a capital cost.

Table 8.3 Sample CCA Rates and Classes

CCA Rate (%)	Class	Description
5 to 10	1, 3, 6	Buildings and additions
20	8	Machinery, office furniture, and equipment which cost $200 or more
25	9	Aircraft, aircraft furniture, and equipment
30	10	Passenger vehicles, vans, trucks, computers, and systems software
40	16	Taxis, rental cars, and freight trucks
50	22	Most power-operated, movable equipment bought before 1988
100	12	Dies, tools, and instruments which cost less than $200

CCA vs. depreciation: CCA is kinda like a fake dep. rate which allows corporations to maximize their expenses, ∴ minimizing net taxable income.

In addition to these standard rates, sometimes, as part of government policy, special rates are set to encourage certain kinds of investments. This is a form of *incentive*. Close-up 8.2 discusses the general idea of government incentives.

Figure 8.1 illustrates how the remaining value of an asset subject to taxes diminishes within the standard range of CCA rates.

CLOSE-UP 8.2 Incentives

Federal and provincial governments sometimes try to influence corporate behaviour through the use of *incentives*. These incentives include grants to certain types of projects, to projects done in particular geographic areas, or to projects providing employment to certain categories of people.

Other incentives take the form of tax relief. For example, in recent years the CCA rate for pollution control equipment has been 100%. The ability to depreciate the full cost of pollution equipment in the year of purchase makes it a desirable investment for a company, and a beneficial investment for society as a whole.

The exact form of incentives changes from year to year as governments change and as the political interests of society change. In most companies there is an individual or department that keeps track of possible programs affecting company projects.

Incentives must be considered when assessing the viability of a project. Grant incentives provide an additional cash flow to the project that can be taken into account like any other cash flow element. Tax incentives may be more difficult to assess since sometimes, for example, they use other forms of depreciation calculation, or may give different tax rates for different parts of the project.

Figure 8.1 Effect of Different CCA Rates

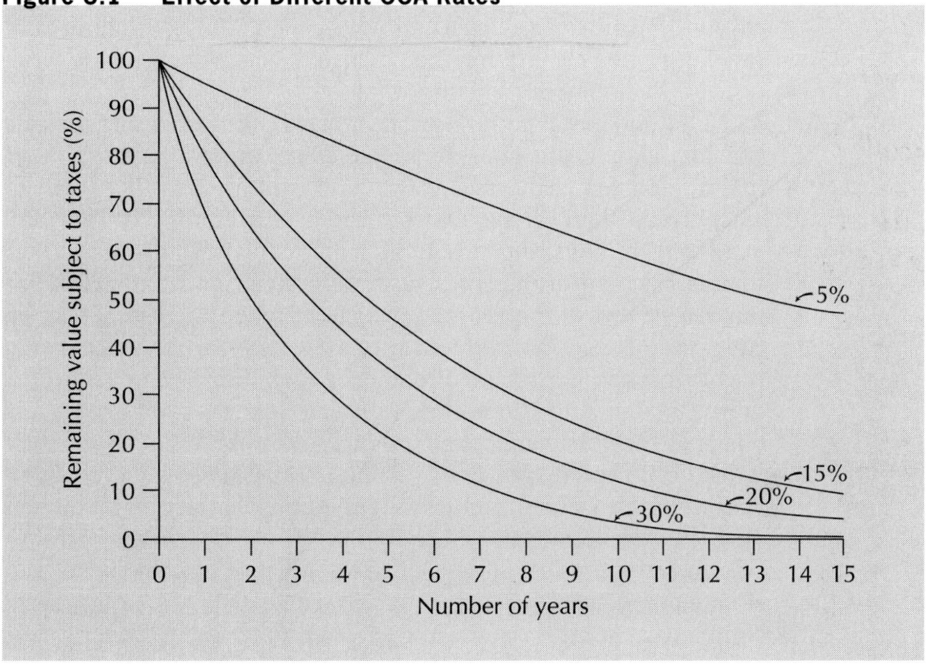

While capital cost allowance and depreciation are conceptually similar, it is important to distinguish between the two terms. Recall from Chapter 6 that, in determining net income, we deduct depreciation expenses from the revenues to arrive at net income. This is the net income for accounting purposes. For tax purposes, we need to determine taxable net income. To establish taxable net income, we start with the accounting net income, add back the depreciation expense for accounting purposes, and then deduct the CCA. Such accounting adjustments are common in determining the amount of tax that a company needs to pay. Given the complexities of the tax system, it is possible that net income for accounting purposes differs from net income for tax purposes by a large amount. For our purposes, we need only to distinguish between the depreciation for accounting purposes and the capital cost allowance, which is depreciation for tax purposes.

8.4 Undepreciated Capital Cost and the Half-Year Rule

The basis for calculating the capital cost allowance for assets in a particular asset class is the total *undepreciated capital cost (UCC)* of the assets included in that class. The capital cost of an asset when it is purchased is the total cost of acquiring the asset. This includes the purchase price, legal fees, accounting costs, and possibly other costs over and above the purchase price. As the asset is depreciated, companies keep track of the undepreciated portion of the original capital cost through an **undepreciated capital cost (UCC)** account. The UCC is the remaining book value for the assets subject to depreciation for taxation purposes, which may or may not differ from the market or salvage value.

The undepreciated capital cost for each asset is not recorded individually within a class; instead, assets in each class are pooled and only one account is maintained for each asset class. The capital cost allowance for a particular asset class is then calculated from the CCA rate for that class and its UCC.

Prior to November 13, 1981, a company was allowed to include in its base for the calculation of capital cost allowance the full purchase price of an asset purchased within the year, regardless of when the asset was purchased during the taxation year. Consequently, there was considerable motivation for companies to purchase assets at the end of their fiscal year. The Canadian government, recognizing the considerable tax losses brought about in this manner, changed the rules, effective Friday, November 13, 1981. Since that date, only half of the capital cost of acquiring an asset is considered in the CCA in the year of purchase of an item, while the other half is then included in the second year. This is commonly referred to as "the half-year rule" in the CCA system.

To see the effect of this change and to illustrate the UCC account, consider a company that has just purchased a \$1 000 000 piece of equipment. For simplicity, we will assume that this equipment is the only equipment in its class. The CCA rate for the equipment is 20%, and the company's tax rate is 50%. Table 8.4 shows the company's UCC amounts for the first four years of the asset's life, assuming that the purchase occurred before the 1981 tax rule change. Table 8.5 shows equivalent figures, assuming that the purchase occurred after the 1981 tax change.

To explain some of the amounts, we will start with Table 8.4, showing the pre-1981 rules. The UCC at the beginning of the year in which the asset was bought was the purchase cost of the asset. The CCA rate for the equipment was 20%. Thus the company could claim a capital cost allowance of 20% of the \$1 000 000 for the first year, leaving a UCC of \$800 000 at the end of the first year. In the second year, the CCA amount was 20% of the UCC from the end of the previous year: 20% of \$800 000 = \$160 000. The UCC of the asset thus declined by 20% of the *current* book value each year as the CCA rate was applied to the undepreciated capital cost from the previous year.

Table 8.5 shows what happens to the UCC account if it is assumed that the purchase occurred *after* the 1981 CCA tax regulation change. In the first year of the asset's life, the full first cost of \$1 000 000 can be added to the UCC account, but only half of that amount is subject to a CCA claim. Thus the CCA amount in the first year is 20% of \$500 000, leaving a balance of \$900 000 of undepreciated capital cost. The CCA amount for the second year is then 20% of \$900 000, or \$180 000. The remainder of the CCA calculations are computed as usual.

Notice that the CCA expenses generate tax savings by reducing taxable income. At a 10% interest rate, the present worth of the tax savings using the

Table 8.4 UCC Amounts Using Pre-1981 CCA Rules

Year	Adjustments to UCC from Purchases and Dispositions	Base UCC Amount for Capital Cost Allowance	Capital Cost Allowance	Remaining UCC	Tax Savings Due to the CCA
1	$1 000 000	$1 000 000	$200 000	$800 000	$100 000
2	0	800 000	160 000	640 000	80 000
3	0	640 000	128 000	512 000	64 000
4	0	512 000	102 400	409 600	51 200

Table 8.5 UCC Amounts Using Post-1981 CCA Rules

Year	Adjustments to UCC from Purchases and Dispositions	Base UCC Amount for Capital Cost Allowance	Capital Cost Allowance	Remaining UCC	Tax Savings Due to the CCA
1	$1 000 000	$500 000	$100 000	$900 000	$50 000
2	0	900 000	180 000	720 000	90 000
3	0	720 000	144 000	576 000	72 000
4	0	576 000	115 200	460 800	57 600

pre-1981 rules is $240 079, while that for the post-1981 rules is only $213 271.

Since the change in the tax law pertaining to "the half-year rule" in 1981, there still remains an incentive to purchase equipment at the end of the (fiscal) year. However, the incentive has been reduced as the tax effects have been diminished.

The previous example illustrated a simple case where only one asset was purchased. In fact, a company typically purchases assets over time and disposes of them when they are no longer required. It is important to note that, if an asset is disposed of in the same year as another one in the same CCA class is purchased, the disposal amount (for the class) is subtracted from the purchase amount (for the class) *before* applying the half-year rule. For any given year, the UCC balance can be calculated as follows:

$$UCC_{opening} + additions - disposals - CCA = UCC_{ending}$$

Table 8.6 Half-Year Rule Summary

Component	Treatment
Purchase	Add only 1/2 of the purchase cost of an asset to the Base UCC Amount for its CCA class in the year of purchase. After the CCA calculation, add the other 1/2 to the remaining UCC (note that the second 1/2 is not considered an acquisition in the following year).
Disposition	Subtract the full amount received for a disposition of an asset from the Base UCC Amount for its CCA class.
Purchases and Dispositions in Same Year	Subtract total dispositions from total purchases for a CCA class. If the remainder is positive, treat it as a Purchase. If the remainder is negative, treat it as a Disposition.

Table 8.6 summarizes the effect of the half-year rule under various circumstances.

To illustrate the use of UCC accounts when several assets of the same class are acquired and then disposed of, consider Example 8.3.

Example 8.3

Egonomical Corporation, an injection-moulding firm, is planning to set up business. It will purchase two used injection moulders for $5000 each in 1998, a new, full-featured moulder for $20 000 in 1999, and a computer controller for the new moulder for $5000 in 2003. One used moulder will be salvaged for $2000 in 2003, and the other for the same amount in 2004. If the CCA rate for all these assets is 20%, determine the balance of the UCC account for years 1998 to 2005.

Table 8.7 illustrates the calculations for the UCC balance for Example 8.2. It can be assumed that the original balance is zero, since the company is just starting up. In 1998, purchases totalling $10 000 were made. However, only half of that amount, $5000, is used for the CCA calculations because of the half-year rule. At 20%, the CCA amount is then $1000. The UCC account for that class is increased by the full amount of the purchase, so subtracting the $1000 from the UCC results in a balance of $9000. In 1999, a purchase of $20 000 increases the amount subject to the UCC to $19 000 (since only half of the cost of the new purchase can be included in 1999), resulting in a CCA amount for that year of $3800. The UCC balance is calculated as $9000 + $20 000 − $3800 = $25 200. For the years 2000 to 2002, the CCA amount is simply 20% of the UCC balance for the previous year, since no acquisitions or disposals are made.

In 2003, two things happen at the same time. A computer controller is purchased for $5000, and a moulder is salvaged for $2000. This results in a net positive adjustment to the UCC of $3000. It is this $3000 which is subject to the

Table 8.7 UCC Computations with Several Changes in Asset Holdings

Year	Adjustments to UCC from Purchases and Dispositions	Base UCC Amount for Capital Cost Allowance	Capital Cost Allowance	Remaining UCC
1998	$10 000	$ 5 000	$1 000	$ 9 000
1999	20 000	19 000	3 800	25 200
2000	0	25 200	5 040	20 160
2001	0	20 160	4 032	16 128
2002	0	16 128	3 226	12 902
2003	3 000	14 402	2 880	13 022
2004	(2 000)	11 022	2 204	8 817
2005	0	8 817	1 763	7 054

half-year rule. Thus the UCC amount for CCA calculations is half of this amount plus the UCC balance from the previous year: $1500 + $12 902 = $14 402. The UCC balance at the end of 2003 includes the total $3000 amount: $12 902 + $3000 − $2880 = $13 022. The negative adjustment in 2004 is not subject to the half-year rule since the half-year rule only applies to net purchases over the year, so the UCC amount for CCA calculations is $2000 less than the previous year's balance. In 2005, there are no adjustments to the UCC other than the CCA amount, leaving a final balance in 2005 of $7054.

8.5 The Capital Cost Tax Factor

From the example illustrated in Tables 8.4 and 8.5, it is clear that the CCA creates tax savings. For example, if an asset with a CCA rate of 20% is purchased for $100 000, in the first year this provides a CCA amount of $10 000. With a tax rate of 50%, this deduction from income saves $5000 in taxes. This saving would not have occurred if the $100 000 had not been spent for the asset in the first place. Therefore, the present worth of the first cost of the asset is actually less than $100 000; it is reduced by the present worth of all of the tax savings that result from its depreciation in all future years. In this example, the tax saving for each year of the asset's life is shown in Table 8.8.

Table 8.8 Tax Savings Due to the Capital Cost Allowance (50% Tax Rate)

Year	Base UCC Amount for Capital Cost Allowance	Capital Cost Allowance	Remaining UCC	Tax Savings Due to the CCA
1	\$50 000	\$10 000	\$90 000	\$5 000
2	90 000	18 000	72 000	9 000
3	72 000	14 400	57 600	7 200
4	57 600	11 520	46 080	5 760
5	46 080	9 216	36 864	4 608

The present worth of the savings is

$$\begin{aligned}\text{Present worth of tax savings} &= \$5000(P/F, i, 1) + \$9000(P/F, i, 2)\\ &\quad + \$7200(P/F, i, 3) + \$5760(P/F, i, 4)\\ &\quad + \$4608(P/F, i, 5) + \cdots\end{aligned}$$

The present worth of the tax savings essentially reduces the first cost of the investment because making the investment and depreciating it over time brings about tax benefits. The **capital cost tax factor (CCTF)** is a value that summarizes the effect of the future benefit of tax savings due to CCA and allows analysts to take these benefits into account when calculating the present worth of an asset. The CCTF remains constant for a given CCA rate, interest rate, and tax rate, and allows the determination of the present worth independently of the actual first cost of the asset. This makes it a very useful number.

Because of the change in tax laws on November 13, 1981, there are two CCTFs: the old CCTF (CCTF_{old}) and the new CCTF (CCTF_{new}). The CCTF_{old} is valid for purchases made before November 13, 1981, and, as seen later in this chapter, is also used for the salvage of assets regardless of the time at which the salvage is made. The CCTF_{new} is used for capital purchases on or after November 13, 1981.

As shown in detail in Appendix 8A, the CCTF_{old} is given by

$$\text{CCTF}_{\text{old}} = 1 - \frac{td}{(i + d)}$$

where

t = taxation rate
d = CCA rate
i = after-tax interest rate

The $CCTF_{new}$ is also derived in Appendix 8A as

$$CCTF_{new} = 1 - \frac{td\left(1 + \frac{i}{2}\right)}{(i + d)(1 + i)}$$

For an asset purchased after 1981, the present worth of the first cost (at the time of purchase) is found by multiplying the first cost by the $CCTF_{new}$. This takes into account the tax benefits forever. When an asset is salvaged or scrapped, we need to terminate the remaining stream of tax savings. This is done by applying the $CCTF_{old}$. Examples in Section 8.6 will clarify the process.

Example 8.4

An automobile purchased this year by Lestev Corporation for \$25 000 has a CCA rate of 30%. Lestev is subject to 43% corporate taxes and the corporate (after-tax) MARR is 12%. What is the present worth of the first cost of this automobile, taking into account the future tax benefits of depreciation?

The car is purchased this year, so the $CCTF_{new}$ applies. The $CCTF_{new}$ is calculated as:

$$\begin{aligned} CCTF_{new} &= 1 - (0.43)(0.3)(1 + 0.06)/[(0.12 + 0.3)(1 + 0.12)] \\ &= 0.709311 \end{aligned}$$

The present worth of the first cost of the car is then calculated as:

$$\begin{aligned} PW &= 0.709311(\$25\ 000) \\ &\cong \$17\ 733 \end{aligned}$$

The present worth of the first cost of the car, taking into account all future tax benefits due to depreciation, is about \$17 733. The tax benefit due to claiming CCA has effectively reduced the cost of the car from \$25 000 to \$17 733 in terms of present worth. ■

It may seem strange that the effective cost of purchasing an asset is less than its first cost. However, bear in mind that the first cost is not the only effect that the purchase of an asset has on cash flows. The purchase will also likely generate savings. These savings are income, which is also taxed. Taking taxes into account when determining the present worth or annual worth of an asset will affect the present or annual worth *positively* because of the tax benefits resulting from future CCA, but also *negatively* because of the taxation of future savings.

8.6 Components of a Complete Tax Calculation

As discussed in Section 8.2, in evaluating projects with the explicit consideration of taxes, it is important to recognize that there is a difference between a before-tax MARR and an after-tax MARR. A before-tax MARR is chosen to reflect the fact that taxes are not explicitly taken into account in the economic calculations, and conversely the after-tax MARR is used where taxes are explicitly taken into account.

Evaluating the economic impact of purchasing a depreciable asset goes beyond the impact of taxes on the first cost. There are two other components to a complete economic analysis. First, we need to assess the tax implications of the savings or additional expenses brought about by the asset over its useful life. Second, when the asset is disposed of, we no longer can take advantage of its capital cost allowance and thus must terminate the stream of tax savings resulting from depreciating the asset.

Each of these components has a tax effect that has to be taken into account when doing a cash flow analysis such as determining present worth or annual worth. A summary of the procedure for a present worth computation is shown in Table 8.9.

Table 8.9 Components of a Complete Present-Worth Tax Calculation

Component	Treatment
First cost	Multiply by the $CCTF_{new}$.
Savings or expenses	Multiply by $(1 - t)$. Convert to present worth.
Salvage value	Multiply by the $CCTF_{old}$. Convert to present worth.

First cost: As presented in Section 8.5, the first cost of an asset purchased after 1981 is reduced by the tax savings due to CCA. Multiply the first cost by the $CCTF_{new}$ to find the after-tax first cost.

Savings or expenses: Reduce savings or expenses by the tax rate by multiplying by $(1 - t)$. There is an assumption that the company is making a profit, so that taxes are paid on all the savings at the rate t, and expenses will reduce taxes at the rate t.

Salvage value: Apply $CCTF_{old}$. When an asset is disposed of, the salvage value reduces the UCC for the *full* amount in the year of disposal (at least in the absence of a corresponding purchase in the same year). The effect of reducing the UCC is the same in magnitude but opposite in sign as increasing the UCC. The $CCTF_{new}$ has built into it a delay in depreciating half of the value of the asset, whereas the full effect occurs immediately when disposing of an asset. Consequently, the $CCTF_{old}$ is the one to use at the time an asset is sold.

Note that technically, when assets are disposed of in a given year, they are netted against any additions for the year before the half-year rule is applied.

However, in project analysis, we generally want to evaluate the project independently, at least in preliminary evaluation. Thus, when we determine the salvage value we do not consider the effects of other additions or disposals that the company may also be planning for the same year. Nevertheless, it is worth noting that significant tax advantages can be made by properly planning the timing of investment additions and disposals. Our goal is to decide on the merits of the project on a more basic level at this time.

Example 8.5

The owner of a spring-water bottling company in Erbsville has just purchased an automated bottle capper. What is the after-tax present worth of the new automated bottle capper if it costs \$10 000 and saves \$4000 per year over its five-year life? Assume a \$2000 salvage value and a 50% tax rate. The after-tax MARR is 12%.

A CCA rate is not given in this question. As production equipment, the new bottle capper can be assumed to be in CCA Class 8, with a rate of 20%.

The present worth of the first cost (assuming that the purchase took place after November 13, 1981) must take into account the tax benefits of CCA. The after-tax first cost is

$$PW(\text{first cost}) = -\$10\,000(CCTF_{new})$$

where the $CCTF_{new}$ is calculated as

$$CCTF_{new} = 1 - \frac{(0.5)(0.2)(1 + 0.06)}{(0.12 + 0.2)(1 + 0.12)} = 0.70424$$

Therefore, the present worth of the first cost is

$$PW(\text{first cost}) = -\$10\,000(0.70424) \cong -\$7042$$

The annual savings are taxed at 50%, so the present worth of the savings is

$$\begin{aligned} PW(\text{annual savings}) &= \$4000(P/A, 12\%, 5)(1 - t) \\ &= \$4000(3.6047)(0.5) \\ &\cong \$7\,209 \end{aligned}$$

The salvage value is not simply \$2000 five years from now. It reduces the UCC and thus diminishes the tax benefits resulting from the CCA on the original purchase. The after-tax benefits can be determined by applying the pre-November 13, 1981, CCTF:

$$\text{PW(salvage value)} = \$2000(P/F, 12\%, 5)\text{CCTF}_{\text{old}}$$

$$\text{CCTF}_{\text{old}} = 1 - \frac{(0.5)(0.2)}{(0.12 + 0.2)}$$

$$= 0.6875$$

$$\text{PW(salvage value)} = \$2000(0.56743)(0.6875)$$

$$\cong \$780$$

Summing the present worths,

$$\text{PW} = \text{PW(first cost)} + \text{PW(annual savings)} + \text{PW(salvage value)}$$

$$= -\ \$7042 + \$7209 + \$780$$

$$= \$947$$

The present worth, after taxes, for the new bottle capper is \$947. ■

Example 8.5 illustrated the complete effect of taxes for a present worth analysis. Similar adjustments are made to an annual worth computation, as illustrated in Table 8.10.

Table 8.10 Components of a Complete Annual-Worth Tax Calculation

Component	Treatment
First cost	Multiply by the CCTF_{new}. Convert to annual worth.
Savings or expenses	Multiply by $(1 - t)$.
Salvage value	Multiply by the CCTF_{old}. Convert to annual worth.

Example 8.6

A small device used to test printed circuit boards has a first cost of \$45 000. The tester is expected to reduce labour costs and improve the defect detection rate so as to bring about a saving of \$23 000 per year. Additional operating costs are expected to be \$7300 per year. The salvage value of the tester will be \$5000 in five years. With an after-tax MARR of 12%, a CCA rate of 20%, and a tax rate of 42%, what is the annual worth of the tester, taking into account the effect of taxes?

The basic process for adjusting for tax effects in an annual worth comparison is similar to a present worth analysis. First, we apply the CCTF_{new} to the first cost, and convert it into an annual amount over five years. Next, the annual savings and expenses are multiplied by $(1 - t)$. Finally, the salvage value at the end of five years is multiplied by CCTF_{old} and then converted into an annual amount:

$$\begin{aligned}\text{AW(tester)} &= -\$45\,000(A/P, 12\%, 5)\text{CCTF}_{\text{new}} \\ &\quad + (\$23\,000 - \$7300)(1 - t) \\ &\quad + \$5000(A/F, 12\%, 5)\text{CCTF}_{\text{old}}\end{aligned}$$

Using $d = 0.20$, $t = 0.42$, and $i = 0.12$, we have

$$\text{CCTF}_{\text{new}} = 0.7516$$
$$\text{CCTF}_{\text{old}} = 0.7375$$

Therefore

$$\begin{aligned}\text{AW(tester)} &= -\$45\,000(0.27741)(0.7516) \\ &\quad + (\$23\,000 - \$7300)(0.58) + \$5000(0.15741)(0.7375) \\ &\cong \$304\end{aligned}$$

The annual worth, taking into account taxes, is \$304. ■

As the previous two examples show, taking taxes into account for present worth and annual worth analyses is relatively straightforward. IRR computations are a bit more involved, however, as the next example illustrates.

Example 8.7

Find the after-tax IRR for the testing equipment described in Example 8.6.

First, observe that, if we solve for i in AW(receipts) − AW(disbursements) = 0, or PW(receipts) − PW(disbursements) = 0, the resulting rate will be an after-tax IRR as the amounts will have been adjusted for taxes. Since Example 8.6 was expressed in terms of annual amounts, we will find the after-tax IRR by solving for i in AW(receipts) − AW(disbursements) = 0, using the operations listed in Table 8.10:

$$\begin{aligned}&(\$23\,000 - \$7300)(1 - t) + \$5000(A/F, i, 5)\text{CCTF}_{\text{old}} \\ &- \$45\,000(A/P, i, 5)\text{CCTF}_{\text{new}} = 0\end{aligned}$$

In order to solve the above equation, a trial and error approach is necessary because the interest rate i appears in the capital cost tax factors as well as in the compound interest factors. Table 8.11 shows the result of this process obtained through the use of a spreadsheet.

Table 8.11 Trial and Error Process for Finding the After-Tax IRR

i	AW(receipts) – AW(disbursements)
0.10	$997.5485
0.11	651.3784
0.12	304.3646
0.13	– 43.6201
0.14	– 392.682

From the spreadsheet computations, we can see that the after-tax IRR on the testing equipment is between 12% and 13%. Additional trial and error iterations with the spreadsheet program give an after-tax IRR of 12.87%. ■

8.7 Approximate After-Tax Rate-of-Return Calculations

The IRR is probably one of the most popular ways to assess the desirability of an investment. Unfortunately, as we saw in Section 8.6, a detailed analysis can be somewhat involved. However, an approximate IRR analysis when taxes are explicitly considered can be very easy. The formula to use is:

$$IRR_{\text{after-tax}} \cong IRR_{\text{before-tax}} \times (1 - t)$$

The reasons for this are exactly the same as described in Section 8.2 for the before- and after-tax MARR. It is an approximation that works because the IRR represents the percentage of the total investment that is net income. Since the tax rate is applied to net income, it correspondingly reduces the IRR by the same proportion. It is in error because it assumes that expenses offset receipts in the year that they occur, rather than being spread out over time (as CCA deductions) as required by the Canadian tax laws.

It should therefore be noted that this after-tax IRR will tend to be somewhat *higher* than it would be if calculated in a perfectly accurate manner. Thus, if the after-tax IRR is close to the after-tax MARR, a more precise calculation is required.

Example 8.8

What is the approximate IRR on the testing equipment described in Example 8.6?

First, we find the $IRR_{\text{before-tax}}$ by solving for i in

$$\text{AW(receipts)} - \text{AW(disbursements)} = 0$$

$$(\$23\ 000 - \$7300) + \$5000(A/F, i, 5) - \$45\ 000(A/P, i, 5) = 0$$

Through trial and error, we find that the $IRR_{before\text{-}tax}$ is 23.8%. The $IRR_{after\text{-}tax}$ is then approximately 13.8%:

$$\begin{aligned} IRR_{after\text{-}tax} &\cong IRR_{before\text{-}tax}(1 - t) \\ &= 0.238(1 - 0.42) \\ &= 0.13804 \end{aligned}$$

Notice that it is a little higher than the precise IRR of 12.87%. ■

When doing an after-tax IRR computation in practice, the approximate after-tax IRR can be used as a first pass on the IRR computation. If the approximate after-tax IRR turns out to be close to the after-tax MARR, a precise after-tax IRR computation may be required to make a fully informed decision about the project.

Example 8.9

Waterloo Industries pays 40% corporate income taxes. Their after-tax MARR is 18%. A project has a before-tax IRR of 24%. Should the project be approved? What would your decision be if the after-tax MARR were 14%?

$$\begin{aligned} IRR_{after\text{-}tax} &\cong IRR_{before\text{-}tax} \times (1 - t) \\ &= 0.24 \times (1 - 0.40) \\ &= 0.144 \end{aligned}$$

The after-tax IRR is approximately 14.4%. For an after-tax MARR of 18%, the project should not be approved. However, for an after-tax MARR of 14%, since the after-tax IRR is an approximation, a more detailed examination would be advisable. ■

In summary, we can simplify after-tax IRR computations by using an easy approximation. The approximate after-tax IRR may be adequate for decision making in many cases, but in others a detailed after-tax analysis may be necessary.

REVIEW PROBLEMS

REVIEW PROBLEM 8.1: UCC COMPUTATIONS

Angus and his sister Oona operate a small charter flight service that takes tourists on sightseeing tours over the beautiful Margaree River on Cape Breton Island. At the end of 1993, they had one four-seater plane in the aircraft asset class with a UCC of $30 000. In 1994, they purchased a second plane for $50 000. Business was going well in 1995, so they sold the old plane they had in 1993 for $15 000 and bought a newer version for $64 000. What was the

UCC balance in the aircraft asset class at the end of 1996? The CCA rate for aircraft is 25%.

ANSWER

Table 8.12 shows the fluctuation in the UCC balance over time. At the end of 1993, the UCC for the aircraft asset class was $30 000. In 1994, half of the capital cost of the airplane purchased in 1994 (= $25 000) contributed to the CCA calculation. The CCA rate of 25% gave a CCA amount of $13 750 and resulted in a UCC balance of $66 250 at the end of 1994. In 1995, the net positive adjustment to the UCC due to the capital cost of $64 000 for the new plane and $15 000 benefit from the sale of the old plane was $49 000. Half of this amount, $24 500, contributed to the CCA calculation. After subtracting the CCA amount of $22 688 for 1994, the remaining UCC was $92 563 (= $66 250 + $49 000 − $22 688). Finally, in 1996, there were no further adjustments to the UCC, and after the CCA was deducted the closing UCC account balance was $69 422. ■

Table 8.12 Summary of UCC Computations for Review Problem 8.1

Year	Adjustments to UCC from Purchases or Dispositions	Base UCC Amount for Capital Cost Allowance	CCA Allowance	Remaining UCC
1993	$30 000			
1994	50 000	$55 000	$13 750	$66 250
1995	49 000	90 750	22 688	92 563
1996	0	92 563	23 141	69 422

REVIEW PROBLEM 8.2: A BUY OR LEASE DECISION AND TAXES

David Cosgrove has just started a management consulting firm that he operates out of his home at Paradise Lake. As part of his new business, David is considering buying a new $30 000 van, which will be used 90% or more of the time for earning business income (if this were not the case, special limits on the allowable CCA would apply). He estimates that the expenses associated with operating the van will be $3000 per year in gas, $1200 per year for insurance, $600 annually for parking, and maintenance costs of $1000 for the first year, rising $400 per year thereafter. He expects to keep the van for five years. At the end of this time, he estimates a salvage value of $6000. The CCA rate for vans is 30%.

The alternative for David is to lease the van. With a lease arrangement, he will have to pay for parking, gas, and insurance, but the leasing company will pay for the repairs. The lease costs are $10 500 per year.

David estimates his after-tax cost of capital to be 12% per year and his tax rate is 40%. Based on an annual worth analysis over the five years, should David buy the van or lease it?

ANSWER

The approach will be to find the after-tax annual worth of each alternative. Since the parking, insurance, and gas costs are the same for both alternatives, we can exclude them from the analysis.

The after-tax annual costs of purchasing the van are

$$\begin{aligned}\text{AW(van)} = {} & \$30\,000(A/P, 12\%, 5)\,\text{CCTF}_{\text{new}} \\ & - \$6000(A/F, 12\%, 5)\text{CCTF}_{\text{old}} \\ & + [\$1000 + \$400(A/G, 12\%, 5)](1 - t)\end{aligned}$$

We can calculate that

$$\begin{aligned}\text{CCTF}_{\text{new}} &= 1 - \frac{td\left(1 + \frac{i}{2}\right)}{(i + d)(1 + i)} \\ &= 1 - \frac{0.4(0.3)(1 + 0.12/2)}{(0.12 + 0.30)(1 + 0.12)} \\ &= 0.7296\end{aligned}$$

$$\begin{aligned}\text{CCTF}_{\text{old}} &= 1 - \frac{td}{(i + d)} \\ &= 1 - \frac{0.4(0.3)}{(0.12 + 0.30)} \\ &= 0.71429\end{aligned}$$

Thus

$$\begin{aligned}\text{AW(van)} &= \$30\,000(0.27741)(0.7296) - \$6000(0.15741)(0.71429) \\ &\quad + [\$1000 + \$400(1.7745)](1 - 0.4) \\ &\cong \$6423\end{aligned}$$

The annual cost of purchasing and operating the van over a five-year period is a little over $6400.

There is a large difference between buying and leasing. When we lease, we do not have a depreciable asset on which to claim depreciation expenses. We only have lease payment expenses. Therefore, the impact of taxes on the lease expense is simply to multiply the leasing costs by $(1 - t)$:

$$\begin{aligned}\text{AW(lease)} &= \$10\,500(1 - 0.4) \\ &= \$10\,500(0.6) \\ &= \$6300\end{aligned}$$

The after-tax annual cost of leasing is \$6300. It is less expensive to lease the van than it is to buy it. Therefore, David should lease the van. It is, however, worth noting that this example is a simplified version of real life, since numerous tax rules relating to the eligibility of expenses for automobiles have been ignored for illustration purposes. ■

REVIEW PROBLEM 8.3: AN IRR COMPARISON AND TAXES

Putco does subcontracting for an electronics firm that assembles printed circuit boards. Business has been good lately, and Putco is thinking of purchasing a new IC chip-placement machine. It has a first cost of \$450 000 and is expected to save them \$125 000 per year in labour and operating costs compared with the manual system they have now. A similar system that also automates the circuit board loading and unloading process costs \$550 000 and will save about \$155 000 per year. The life of either system is expected to be four years. The salvage value of the \$450 000 machine will be \$180 000, and that of the \$550 000 machine will be \$200 000. Putco uses an after-tax MARR of 9% to make decisions about such projects. On the basis of an IRR comparison, which alternative (if either) should they choose? Putco pays taxes at a rate of 40%, and the CCA rate for the equipment is 20%.

ANSWER

Putco has three mutually exclusive alternatives:

(1) do nothing;

(2) buy the chip-placement machine; or

(3) buy a similar chip-placement machine with an automated loading and unloading process.

Following the procedure from Chapter 5, the projects are already ordered, based on first cost. We therefore begin with the first alternative: the before-tax (and thus the after-tax) IRR of the do-nothing alternative is 0%. Next, the before-tax IRR on the incremental investment to the second alternative can be found by solving for i in

$$-\$450\,000 + \$125\,000(P/A, i, 4) + \$180\,000(P/F, i, 4) = 0$$

By trial and error, we obtain an $\text{IRR}_{\text{before-tax}}$ of 15.92%. This gives an approximate $\text{IRR}_{\text{after-tax}}$ of $0.1592(1 - 0.40) = 0.0944$ or 9.44%. With an after-tax MARR of 9%, it would appear that this alternative is acceptable, though a detailed after-tax computation may be in order. We need to solve for i in

$$(-\$450\,000)\text{CCTF}_{\text{new}} + \$125\,000(P/A, i, 4)(1 - t)$$
$$+ \$180\,000(P/F, i, 4)\text{CCTF}_{\text{old}} = 0$$

Doing so gives an $\text{IRR}_{\text{after-tax}}$ of 9.5%. Since this exceeds the required after-tax MARR of 9%, this alternative becomes the current best. We next find the $\text{IRR}_{\text{after-tax}}$ on the incremental investment required for the third alternative. The $\text{IRR}_{\text{before-tax}}$ is first found by solving for i in

$$-(\$550\,000 - \$450\,000) + (\$155\,000 - \$125\,000)(P/A, i, 4)$$
$$+ (\$200\,000 - \$180\,000)(P/F, i, 4) = 0$$

This gives an $IRR_{\text{before-tax}}$ of 7.13%, or an approximate $IRR_{\text{after-tax}}$ of 4.28%. This is sufficiently below the required after-tax MARR of 9% to warrant rejection of the third alternative without a detailed incremental $IRR_{\text{after-tax}}$ computation. Putco should therefore select the second alternative. ■

REVIEW PROBLEM 8.4: CYCLIC REPLACEMENT AND TAXES

David Cosgrove (from Review Problem 8.2) is still thinking over whether or not to buy a van. Assuming he remains in business for the foreseeable future, he will need a vehicle for transportation indefinitely, whether he owns or leases it. In his original analysis, he assumed that the van would be replaced at the end of five years. Because appearances are important to David, he would not consider keeping a vehicle for longer than five years, but he now recognizes that the economic life of the van may be *shorter* than five years. Assuming that the van depreciates in value by a constant proportion each year, determine how frequently David should replace it. The CCA rate is 30% and his tax rate is 40%.

ANSWER

The first step in the solution is to recognize that David is facing a cyclic replacement problem, since it is reasonable to assume that he will replace each van with one similar to the previous, indefinitely. We now need to assess the annual cost of replacing a van each year, every two years, and so on, up to replacement every five years. Before proceeding, however, we need to determine the depreciation rate to use so that we can determine the approximate value of the van when it is n years old for $n = 1, 2, 3, 4$, and 5. Referring back to Chapter 6, we have for the declining-balance method of depreciation

$$d = 1 - \sqrt[n]{\frac{S}{P}} = 1 - \sqrt[5]{\frac{\$6000}{\$30\,000}} = 0.27522$$

Using the formula $BV_{db}(n) = P(1 - d)^n$, we find that the book value of the van at the end of each year is:

End of Year	Book Value
0	\$30 000
1	21 743
2	15 759
3	11 422
4	8 278
5	6 000

Note that these are book values, not the UCC balances. The book values are estimates of the market value, which is needed to judge when the asset should be replaced. A UCC balance is similar to a book value, but is used for calculating the CCA only.

Now the annual worth computations can be done using the CCTF values calculated in Review Problem 8.2:

AW(replace every year) = AW(capital recovery) + AW(operating)

$$= \$30\,000(A/P, 12\%, 1)\text{CCTF}_{\text{new}} - \$21\,743(A/F, 12\%, 1)\text{CCTF}_{\text{old}} + (\$5800)(1 - t)$$

$$= \$12\,463$$

AW(replace every two years)

$$= \$30\,000(A/P, 12\%, 2)\text{CCTF}_{\text{new}} - \$15\,759(A/F, 12\%, 2)\text{CCTF}_{\text{old}} + \$5800(1 - t) + \$400(A/F, 12\%, 2)(1 - t)$$

$$= \$11\,235$$

AW(replace every three years)

$$= \$30\,000(A/P, 12\%, 3)\text{CCTF}_{\text{new}} - \$11\,422(A/F, 12\%, 3)\text{CCTF}_{\text{old}} + \$5800(1 - t) + [\$400(F/P, 12\%, 1) + \$800](A/F, 12\%, 3)(1 - t)$$

$$= \$10\,397$$

AW(replace every four years)

$$= \$30\,000(A/P, 12\%, 4)\text{CCTF}_{\text{new}} - \$8278(A/F, 12\%, 4)\text{CCTF}_{\text{old}} + \$5800(1 - t) + [\$400(F/P,12\%,2) + \$800(F/P, 12\%, 1) + \$1200](A/F, 12\%, 4)(1 - t)$$

$$= \$9775$$

AW(replace every five years)

$$= \$30\,000(A/P, 12\%, 5)\text{CCTF}_{\text{new}} - \$6000(A/F, 12\%, 5)\text{CCTF}_{\text{old}} + \$5800(1 - t) + [\$400(F/P, 12\%, 3) + \$800(F/P, 12\%, 2) + \$1200(F/P, 12\%, 1) + \$1600](A/F, 12\%, 5)(1 - t)$$

$$= \$9303$$

Based on these calculations, it is best for David to replace the van at the end of every five years. Its economic life may be longer than five years, but as far as David is concerned, a five-year-old van has reached the end of its useful life and must be replaced. ■

SUMMARY

Income taxes can have a significant effect on engineering economics decisions. In particular, taxes reduce the effective cost of an asset, the savings generated, and the value of the sale of an asset.

In this chapter, we provided a basic introduction to the Canadian capital cost allowance (CCA) system and the use of undepreciated capital cost (UCC) accounts. The CCA rate is a declining-balance rate that is mandated for use in calculating the depreciation expenses for capital assets. These depreciation expenses are then used in determining the amount of taxes owing for the year. Assets are designated as belonging to a particular CCA class. The book values for taxation purposes calculated for all the assets in each class are accumulated into a UCC account.

The future CCA claims that arise from the purchase of an asset are benefits that reduce the after-tax first cost. The $CCTF_{new}$ permits the quick calculation of the net effect of these benefits, while similarly the $CCTF_{old}$ permits the calculation of the net effect of future loss of CCA claims for assets that are sold or scrapped.

It was noted that, for after-tax calculations, an after-tax MARR must be used. After-tax calculations were illustrated for present worth, annual worth, and IRR evaluations. An approximate IRR comparison method was also given.

The review problems at the end of the chapter illustrated how taxes affect present worth and annual worth comparisons, replacement analysis, and internal rate of return computations.

Engineering Economics in Action, Part 8B: The Work Report

"So what is this, anyhow?" Clem was looking at the report that Naomi had handed him, "A consulting report?"

"Sorry, chief, it is a bit thick." Naomi looked a little embarrassed. "You see, Terry has to do a work report for his university. It's part of the co-op program that they have. He got interested in the 10-stage die problem and asked me if he could make that study his work report. I said O.K., subject to its perhaps being confidential. I didn't expect it to be so thick, either. But he's got a good executive summary at the front."

"Hmm..." The room was quiet for a few minutes while Clem read the summary. He then leafed through the remaining parts of the report. "Have you read this through? It looks really quite good."

"I have. He has done a very professional job, er, at least what seems to me to be very professional." Naomi suddenly remembered that she hadn't yet gained her professional engineer's designation. She also hadn't been working at Canadian Widgets much longer than Terry. "I gathered most of the data for him, but he did an excellent job of analyzing it. As you can see from the summary, he set up the replacement problem as a set of mutually exclusive alternatives, involving upgrading the die now or later and even more than once. He did a nice job on the taxes, too."

"How did he handle the taxes?"

"Quite well. I had to hold his hand a bit to make sure he understood how the UCC accounts work, but once he had that everything else seemed to fall into place. He reduced the purchase price by the benefits of future CCA claims. The installation cost and future savings were reduced by the taxation rate, and the salvage values were reduced for loss of future CCA claims."

"Did he understand about the old and new CCTFs?"

"Yes, he did."

"Not bad. I think we've got a winner there, Naomi. Let's make sure we get him back for his next work term."

Naomi nodded. "What about his work report, Clem? Should we ask him to keep it confidential?"

Clem laughed, "Well, I think we should, and not just because there are trade secrets in the report. I don't want anyone else knowing what a gem we have in Terry!"

PROBLEMS

8.1 Go to the Revenue Canada web site and search for the T2 Corporation Income Tax Guide. In the guide, find the section dealing with CCA rates. Identify each of the following according to their CCA class(es) and CCA rate(s).

(a) New software for inventory control purposes costing $70 000

(b) Communications network switching equipment which costs $56 000

(c) A heated, rigid-frame greenhouse worth $57 000

(d) A small tractor with attachments for earthmoving and snow removal worth $24 000

(e) Cash register and bar-code scanning device used for point-of-sales data collection. Cost is $12 000 for the cash register and $200 for the scanning device.

8.2 The MARR generally used by Collingwood Caskets is a before-tax MARR of 14%. Vincent wants to do a detailed calculation of the cash flows associated with a new planer for the assembly line. What would be an appropriate after-tax MARR for him to use if Collingwood Caskets pays

(a) 40% corporate taxes?

(b) 50% corporate taxes?

(c) 60% corporate taxes?

8.3 Enrique has just completed a detailed analysis of the IRR of a wastewater treatment plant for Gimli Meat Products. The 8.7% after-tax IRR he calculated compared favourably with a 5.2% after-tax MARR. For reporting to upper management, he wants to present this information as a before-tax IRR.

If Gimli Meat Products pays 53% corporate taxes, what figures will Enrique report to upper management?

8.4 A company's first year's operations (in 1976) can be summarized as follows:

Revenues: $110 000

Expenses (except CCA): $65 000

Their capital asset purchases in the first year totalled $100 000. With a CCA rate of 20% and a tax rate of 55%, how much income tax did they pay?

8.5 A company's first year's operations (in 1996) can be summarized as follows:

Revenues: $110 000

Expenses (except CCA): $65 000

Their capital asset purchases in the first year totalled $100 000. With a CCA rate of 20% and a tax rate of 55%, how much income tax did they pay?

8.6 What is the after-tax present worth of a chip placer if it costs $55 000 and saves $17 000 per year? After-tax interest is at 10%. Assume the device will be sold for its $1000 salvage value at the end of its six-year life. The CCA rate is 20%, and the corporate income tax rate is 54%.

8.7 Quebec Widgets is looking at a $400 000 digital midget rigid widget gadget (CCA class 8). It is expected to save $85 000 per year over its 10-year life, with no scrap value. Their tax rate is 45%, and their after-tax MARR is 15%. Using an approximate IRR, should they invest in this gadget?

8.8 Canada Widgets is looking at a $400 000 digital midget rigid widget gadget (CCA class 8). It is expected to save $85 000 per year over its 10-year life, with no scrap value. Their tax rate is 45%, and their after-tax MARR is 15%. Using an exact IRR, should they invest in this gadget?

8.9 The UCC for a firm's automobile fleet at the end of 1996 was $10 000. There was one truck in service at this time. At the beginning of 1997, they purchased two trucks for a total of $50 000. At the beginning of 1999, they purchased another truck for $20 000. At the beginning of 2000, the truck owned in 1996 was sold for $3000. The CCA rate for automobiles is 30%. What was the firm's UCC at the end of 2000?

8.10 Go to the Revenue Canada web site and find the form T2S(8), a worksheet for calculating UCC balances. Use the sheet to make the following calculations (separately). In all cases, there are no adjustments (e.g., for GST rebates or for investment tax credits).

(a) The UCC balance at the end of the previous year is $10 000. Assets purchased in class 10 for the current year amount to $30 000 for the year. Find the UCC at the end of the year.

(b) The UCC balance at the end of the previous year is \$10 000. Assets purchased in class 10 for the current year amount to \$20 000. Dispositions were \$5 000. Find the UCC at the end of the year.

(c) The UCC balance at the end of the previous year for class 8 was \$20 000. This year, an asset worth \$20 000 was added to class 8 and another worth \$15 000 was disposed of. What is the year-end UCC for this year?

8.11 Churchill Metal Products opened for business in 1984. Over the following years, their transactions for CCA Class 8 assets consisted of the following:

Date	Item	Activity	Amount
March 11, 1984	Machine 1	Purchase	\$ 50 000
April 24, 1984	Machine 2	Purchase	150 000
November 3, 1987	Machine 3	Purchase	250 000
November 22, 1987	Machine 1	Sale	10 000
May 20, 1991	Machine 4	Purchase	60 000
August 3, 1996	Machine 5	Purchase	345 000
September 12, 1997	Machine 3	Sale	45 000

What CCA amount did Churchill Metal Products claim for the 20% UCC account in 1998?

8.12 Calculate the $CCTF_{old}$ and $CCTF_{new}$ for each of the following:

(a) Tax rate of 50%, CCA of 20%, and an after-tax MARR of 9%

(b) Tax rate of 35%, CCA of 30%, and an after-tax MARR of 12%

(c) Tax rate of 55%, CCA of 5%, and an after-tax MARR of 6%

8.13 Use a spreadsheet program to create a chart showing how the values of the $CCTF_{old}$ and the $CCTF_{new}$ change for after-tax MARRs of 0% to 30%. Assume a fixed tax rate of 50% and a CCA rate of 20%.

8.14 What is the approximate after-tax IRR on a project for which the first cost is \$12 000, savings are \$5000 in the first year and \$10 000 in the second year, and taxes are at 40%.

8.15 What is the exact after-tax IRR on a project for which the first cost is \$12 000, savings are \$5000 in the first year and \$10 000 in the second year, taxes are at 40%, and the CCA rate is 30%? What would the exact after-tax IRR have been if this analysis had been done in 1975?

8.16 What is the total after-tax annual cost of a machine with a first cost of $45 000 and operating and maintenance costs of $0.22 per unit? It will be sold for $4500 at the end of five years. Production volumes are 750 units per day, 250 days per year. The CCA rate is 30%, the after-tax MARR is 20%, and the corporate income tax rate is 40%.

8.17 Refer to Problem 5.9. Rimouski Dairies has a corporate tax rate of 40% and the filling machine for the dairy line has a CCA rate of 30%. They have an after-tax MARR of 10%. Using an exact IRR method, determine which alternative Rimouski Dairies should choose.

8.18 In 1965, the Sackville Furniture Company bought a new band saw for $360 000. Aside from depreciation expenses, their yearly expenses totalled $1 300 000 versus $1 600 000 in income. How much tax (at 50%) would they have paid for 1965 if they had been permitted to use each of the following depreciation schemes?

(a) Straight line, with a life of 10 years and a 0 salvage value

(b) Straight line, with a life of five years and a 0 salvage value

(c) Declining balance, at 20%

(d) Declining balance, at 40%

(e) Fully expensed that year

8.19 Mulroney Brothers Salvage had several equipment purchases in the 70s. Their first asset was a tow truck bought in 1972 for $25 000. In 1974, a van was purchased for $14 000. A second tow truck was bought in 1977 for $28 000, and the first one was sold the following year for $5000. What was the value of their UCC at the end of 1979, with a 30% CCA rate (automobiles, trucks, and vans)?

8.20 Chrétien Brothers Salvage had several equipment purchases in the '80s. Their first asset was a tow truck bought in 1982 for $25 000. In 1984, a van was purchased for $14 000. A second tow truck was bought in 1987 for $28 000, and the first one was sold the following year for $5000. What was the value of their UCC at the end of 1989, with a 30% CCA rate (automobiles, trucks, and vans)?

8.21 Whitehorse Construction has just bought a crane for $380 000. At a CCA rate of 20%, what is the present worth of the crane, taking into account the future benefits of CCA? Whitehorse has a tax rate of 35% and an after-tax MARR of 6%.

8.22 Hull Hulls is considering the purchase of a 30-tonne hoist. The first cost is expected to be $230 000. Net savings will be $35 000 per year over a 12-year life. It will be salvaged for $30 000. If their after-tax MARR is 8% and they are taxed at 45%, what is the present worth of this project?

8.23 Kanata Konstruction is considering the purchase of a truck. Over its five-year life, it will provide net revenues of $15 000 per year, at an initial cost of $65 000 and a salvage value of $20 000. KK pays 35% in taxes, the CCA rate for trucks is 30%, and their after-tax MARR is 12%. What is the annual cost or worth of this purchase?

8.24 A new binder will cost Revelstoke Printing $17 000, incur net savings of $3000 per year over a seven-year life, and be salvaged for $1000. Revelstoke's before-tax MARR is 10%, they are taxed at 40%, and the binder has a 20% CCA rate. What is their approximate after-tax IRR on this investment? Should the investment be made?

8.25 A new binder will cost Revelstoke Printing $17 000, incur net savings of $3000 per year over a seven-year life, and be salvaged for $1000. Revelstoke's before-tax MARR is 10%, they are taxed at 40%, and the binder has a 20% CCA rate. What is their exact after-tax IRR on this investment? Should the investment be made?

8.26 A slitter for sheet sandpaper owned by Abbotsford Abrasives (AA) requires regular maintenance costing $7500 per year. Every five years it is overhauled at a cost of $25 000. The original capital cost was $200 000, with an additional $25 000 in non-capital expenses that occurred at the time of installation. The machine has an expected life of 20 years and a $15 000 salvage value. The machine is not overhauled at the end of its life. AA pays taxes at a rate of 45% and expects an after-tax rate of return of 10% on investments. Recalling that the CCA rate for production equipment is 20%, what is the after-tax annual cost of the slitter?

8.27 Rodney has discovered that, for the last three years, his company has been classifying as Class 8 items costing between $100 and $200 that should be in CCA Class 12. If an estimated $10 000 of assets per year were misclassified, what is the present worth today of the cost of this mistake? Assume that the mistake can only be corrected for assets bought in the future. Rodney's company pays taxes at 50% and their after-tax MARR is 9%.

8.28 Identify each of the following according to their CCA class(es) and CCA rate(s):

(a) A soldering gun costing $75

(b) A garage used to store spare parts

(c) A new computer

(d) A 100-tonne punch press

(e) A crop dusting attachment for a small airplane

(f) An oscilloscope worth exactly $200

8.29 Roch bought a $100 000 machine (Machine A) on November 12, 1981. As a CCA Class 8 asset, what was its book value, measured as its contribution to the UCC for that class, at the end of 1991? Roch purchased an identical $100 000 machine (Machine B) on November 14, 1981. What was its book value at the end of 1991?

8.30 A chemical recovery system costs $30 000 and saves $5280 each year of its seven-year life. The salvage value is estimated at $7500. The after-tax MARR is 9% and taxes are at 45%. What is the net after-tax annual benefit or cost of purchasing the chemical recovery system?

8.31 CB Electronix needs to expand its capacity. It has two feasible alternatives under consideration. Both alternatives will have essentially infinite lives.

Alternative 1: Construct a new building of 200 000 square feet now. The first cost will be $2 000 000. Annual maintenance costs will be $10 000. In addition, the building will need to be painted every 15 years (starting in 15 years) at a cost of $15 000.

Alternative 2: Construct a new building of 125 000 square feet now and an addition of 75 000 square feet in 10 years. The first cost of the 125 000-square-foot building will be $1 250 000. The annual maintenance costs will be $5000 for the first 10 years (i.e., until the addition is built). The 75 000-square-foot addition will have a first cost of $1 000 000. Annual maintenance costs of the renovated building (the original building and the addition) will be $11 000. The renovated building will cost $15 000 to repaint every 15 years (starting 15 years after the addition is done).

Given a CCA rate of 5% for the buildings, a corporate tax rate of 45%, and an after-tax MARR of 15%, carry out an annual-worth comparison of the two alternatives. Which is preferred?

The following Ridgely Custom Metal Products (RCMP) case is used for Problems 8.32 to 8.37. RCMP must purchase a new tube bender. There are three models:

Model	First Cost	Economic Life	Yearly Net Savings	Salvage Value
T	$100 000	5 years	$50 000	$20 000
A	150 000	5 years	60 000	30 000
X	200 000	3 years	75 000	100 000

RCMP's after-tax MARR is 11% and the corporate tax rate is 52%. A tube bender is a CCA Class 8 asset.

8.32 Using the present-worth method and the least-cost multiple of the service lives, which tube bender should they buy?

8.33 RCMP realizes that it can forecast demand for its products for only three years in advance. The salvage value for model T after three years is $40 000 and for model A, $80 000. Using the present-worth method and a three-year study period, which of the three alternatives is now best?

8.34 Using the annual-worth method, which tube bender should Ridgely buy?

8.35 What is the approximate after-tax IRR for *each* of the tube benders?

8.36 What is the exact after-tax IRR for *each* of the tube benders?

8.37 Using the approximate after-tax IRR comparison method, which of the tube benders should Ridgely buy? (*Reminder:* You must look at the increment of investment.)

8.38 Refer to Problem 5.9. Rimouski Dairies has a corporate tax rate of 40% and the filling machine for the dairy line has a CCA rate of 30%. They have an after-tax MARR of 10%. Using an approximate IRR approach, determine which alternative Rimouski Dairies should choose.

8.39 Salim is considering the purchase of a backhoe for his pipeline contracting firm. The machine will cost $110 000, last six years with a salvage value of $20 000, and reduce annual maintenance, insurance, and labour costs by $30 000 per year. The after-tax MARR is 9%, and Salim's corporate tax rate is 55%. What is the exact after-tax IRR for this investment? What is the approximate after-tax IRR for this investment? Should Salim buy the backhoe?

MT&T Tries to Reduce Its Taxes

CANADIAN 8.1 MINI-CASE

Maritime Telephone and Telegraph Company Limited (MT&T) is the telephone utility that supplies telecommunications services to the province of Nova Scotia. It is headquartered in Halifax.

MT&T bills its customers on a monthly basis for local and long distance charges. Prior to 1984, MT&T reported its income using the *earned method*, for both taxation purposes and as part of its obligations to the Canadian Radio-Television and Telecommunications Commission (CRTC). In the earned method, the billable income earned during a period is reported, even if it has not yet been billed. In the 1984 taxation year, MT&T changed the reporting method used for taxation purposes. Although it continued to use the earned method for the CRTC, it used the *billed method* for taxation purposes. In the billed method, income is recognized only when billed, not when earned. Using the billed method was an advantage for MT&T, since earned income that would normally be reported in one period could be delayed until the next. The delay provided a consequent temporary saving in taxes, or, more accurately, the use of the money for a period during which it would otherwise have been paid in taxes.

Unfortunately for MT&T, its use of the billed method was not permitted. Several court cases concluded that the earned method produced a truer picture of the company's income. In the case of telecommunications services, there was an exact record of the service performed, so there was no uncertainty about the income. MT&T had to return to the use of the earned method, and retroactively pay the extra taxes due for the time it had incorrectly used the billed method.

Discussion

The tax system is constantly changing. Companies and individuals always have a motivation to search for ways to reduce the taxes they pay, since it is perfectly legal to *avoid* taxes, as opposed to the illegal act of *evading* taxes. Similarly, governments always have a motivation to clarify acceptable behaviour and close up "loopholes" that permit taxes to be avoided. Taxpayers and the government are thus in an ongoing confrontation that generally creates increasing complications.

Another cause of changes in the tax system is that society itself changes. For example, in recent years we have seen the growth of computing technology and the aging of the baby boomers as important social trends that have had significant effects on the tax system.

Changes in the tax system can effect all decision making fundamentally, or may just affect certain individual decisions. For example, the introduction of the half-year rule in 1981 was a fundamental change that affected all subsequent decisions. On the other hand, accelerated cost recovery incentives usually apply only to specific types of assets.

Questions

Revenue Canada maintains a database containing recent tax changes and incentives. Find the location of the database on the world wide web and review the information there.

continued

1) What recent changes have been made to the tax rules? Would you classify these as fundamental changes or changes that might affect only selected projects?
2) For each change noted, can you suggest a reason? Is the change a result of a confrontation with a taxpayer? Has there been a change in society that led to this change in the tax rules?
3) For each change noted, construct a hypothetical project that would be affected by the change. Calculate the loss or gain in present worth of the project due to the tax-rule change.

Appendix 8A: Deriving the Capital Cost Tax Factors

The change in tax laws of November 13, 1981, has made the formula for the CCTF a little complicated. To derive the CCTF formula, it is easiest to look at the situation before the laws were changed.

Before November 13, 1981, the tax *benefit* that could be obtained for a depreciable asset with a CCA rate d and a first cost P, when the company was paying tax at rate t is

Ptd	for the first year
$Ptd(1-d)$	for the second year
$Ptd(1-d)^{N-1}$	for the Nth year

Taking the present worth of each of these benefits and summing gives

$$\text{PW(benefits)} = Ptd\left(\frac{1}{(1+i)} + \frac{(1-d)}{(1+i)^2} + \ldots \frac{(1-d)^{N-1}}{(1+i)^N} + \ldots\right)$$

$$= \frac{Ptd}{(1+i)}\left(1 + \frac{(1-d)}{(1+i)} + \frac{(1-d)^2}{(1+i)^2} + \ldots + \frac{(1-d)^N}{(1+i)^N} + \ldots\right)$$

Noting that for $q < 1$

$$\lim_{n \to \infty} (1 + q + q^2 + \ldots + q^n) = \frac{1}{1-q}$$

and

$$\frac{1-d}{1+i} < 1$$

then

$$\text{PW(benefits)} = \frac{Ptd}{1+i}\left[\frac{1}{1-\dfrac{(1-d)}{(1+i)}}\right]$$

$$= \frac{Ptd}{(1+i)}\left[\frac{1}{\dfrac{(1+i)}{(1+i)}-\dfrac{(1-d)}{(1+i)}}\right]$$

$$= \frac{Ptd}{(1+i)}\left(\frac{(1+i)}{(i+d)}\right)$$

$$= \frac{Ptd}{(i+d)}$$

If we subtract the present worth of the tax benefits from the first cost, it will give us the present worth of the asset, taking into account all tax benefits from depreciation forever.

$$\text{PW(asset)} = P - \frac{Ptd}{i+d}$$

$$= P\left(1-\frac{td}{i+d}\right)$$

The factor $1 - \dfrac{td}{(i+d)}$ is called the *old* capital cost tax factor ($CCTF_{old}$), and was the formula in use before November 13, 1981.

The new tax rules mean that since November 13, 1981, only half of the first cost of an asset can be used in the calculations for the first year. By recognizing that the net effect of the new law is to delay the tax benefits of half of the first cost by one year, the present worth of the benefits is then

$$\text{PW(benefits)} = 0.5\,\frac{Ptd}{i+d} + 0.5\left(\frac{Ptd}{1+d}\right)\left(\frac{1}{1+i}\right)$$

$$= 0.5\,\frac{Ptd}{i+d}\left(1+\frac{1}{1+i}\right)$$

$$= 0.5\,\frac{Ptd}{i+d}\left(\frac{1+i}{1+i}+\frac{1}{1+i}\right)$$

$$= 0.5\,\frac{Ptd}{i+d}\left(\frac{2+i}{1+i}\right)$$

$$= P \frac{td\left(1 + \frac{i}{2}\right)}{(i + d)(1 + i)}$$

And the present worth of the asset itself is

$$\text{PW}(asset) = P - P \frac{td\left(\frac{1 + i}{2}\right)}{(i + d)(1 + i)}$$

$$= P\left[1 - \frac{td\left(1 + \frac{i}{2}\right)}{(i + d)(1 + i)}\right]$$

Thus the CCTF_{new} is

$$\text{CCTF}_{\text{new}} = 1 - \frac{td\left(1 + \frac{i}{2}\right)}{(i + d)(1 + i)}$$

C H A P T E R 9

Inflation

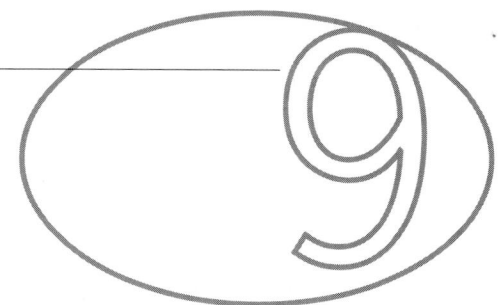

Engineering Economics in Action, Part 9A: The Inflated Expert

Terry left Canadian Widgets to go back for his last term of school. Naomi and Terry had cleaned up a lot of backed-up projects in the last few months, and Naomi had been increasingly taking part in projects involving sister companies of Canadian Widgets; all were owned by Canadian Conglomerate Inc., often referred to as "head office."

"There's a guy from head office to see you, Naomi." It was Carole announcing the expected visitor, Bill Astad. Bill was one of the company trouble shooters. His current interest concerned a sister company, Mexifab, a maquiladora on the Mexican border with Texas. (A maquiladora is an assembly plant that manufactures finished goods in Northern Mexico under special tariff and tax rules.) After a few minutes of socializing, Bill explained the concern.

"It's the variability in the Mexican inflation rate that causes the problems. Mexico gets a new president every six years, and usually, about the time the president changes, the economy goes out of whack. And we can't price everything to U.S. or Canadian dollars. We do some of that, but we are located in Mexico and so we have to use Mexican money for a lot of our transactions.

"I understand from Anna Kulkowski that you know something about how to treat problems like that," Bill continued.

Naomi smiled to herself. She had written a memo a few weeks earlier pointing out how Canadian Widgets had been missing some good projects by failing to take advantage of the current very low inflation rates, and suddenly she was the expert!

"Well," she said, "I might be able to help. What you can do is this."

9.1 Introduction

Prices of goods and services bought and sold by individuals and firms change over time. Some prices, like those of agricultural commodities, may change several times a day. Other prices, like those for electric power, change infrequently. While prices for some consumer goods and services occasionally decrease (as with high-tech products), on average it is more typical for prices to increase over time. In fact, on a yearly basis, average prices of consumer goods and services in Canada have risen in every year but one since 1940.

Inflation is the increase, over time, in average prices of goods and services. It can also be described as a decrease in the purchasing power of money over time. While Canada has experienced inflation in most years since World War II, there have been short periods when average prices in Canada have fallen. A decrease, over time, in average prices is called **deflation**. It can also be viewed as an increase in the purchasing power of money over time.

Because of inflation or deflation, prices are likely to change over the lives of most engineering projects. These changes will affect cash flows associated with the projects. Engineers may have to take predicted price changes into account during project evaluation to prevent the changes from distorting decisions.

In this chapter, we shall discuss how to incorporate an expectation of inflation into project evaluation. We focus on inflation because it has been the dominant pattern of price changes since the beginning of the twentieth century. The chapter begins with a discussion of how inflation is measured. We then show how to convert cash flows which occur at different points in time into dollars with the same purchasing power. We then consider how inflation affects the MARR, the internal rate of return, and the present worth of a project.

9.2 Measuring the Inflation Rate

The **inflation rate** is the rate of increase in average prices of goods and services over a specified time period, usually a year. If prices of all goods and services moved up and down together, determining the inflation rate would be trivial. If all prices increased by 2% over a year, it would be clear that the average inflation rate would also be 2%. But prices do not move in perfect synchronization. In any period, some prices will increase, others will fall, and some will remain about the same. For example, candy bars are about ten times as costly now as they were in the 1960's, but television sets are about the same price or cheaper.

Because prices do not move in perfect synchronization, a variety of methods have been developed to measure the inflation rate. Statistics Canada tracks movement of average prices for a number of different collections of goods and services and calculates inflation rates from the changes in prices in these collections over time.

One set of prices tracked by Statistics Canada consists of goods and services bought by Canadian consumers. This set forms the basis of the **consumer price index (CPI)**. The CPI for a given period relates the average price of a fixed "basket" of these goods in the given period to the average price of the same basket in a *base period.* The current CPI has 1992 as the base year. The base year index is set at 100. The index for any other year indicates the number of dollars needed in that year to buy the fixed basket of goods that cost $100 in 1992. Figure 9.1 shows the CPI for the period 1961 to 1997. It shows that the basket of goods, which cost $100 in 1992 (the base year), would have cost $18.70 in 1961 and approximately $108 in 1997. More information about the CPI can be found at the Statistics Canada web site at http://www.statcan.ca.

A national inflation rate can be estimated directly from the CPI by expressing the changes in the CPI as a year-by-year percentage change. This is probably the most commonly used estimate of a national inflation rate. Figure 9.2 shows the national inflation rate for the period from 1961 to 1997 as derived from the CPI quantities in Figure 9.1.

It is important to note that, although the CPI is a commonly accepted inflation index, many different indices are used to measure inflation. The value of an index depends on the method used to compute the index and the set of goods and services for which the index measures price changes. To

Figure 9.1 Canadian CPI 1961–1997

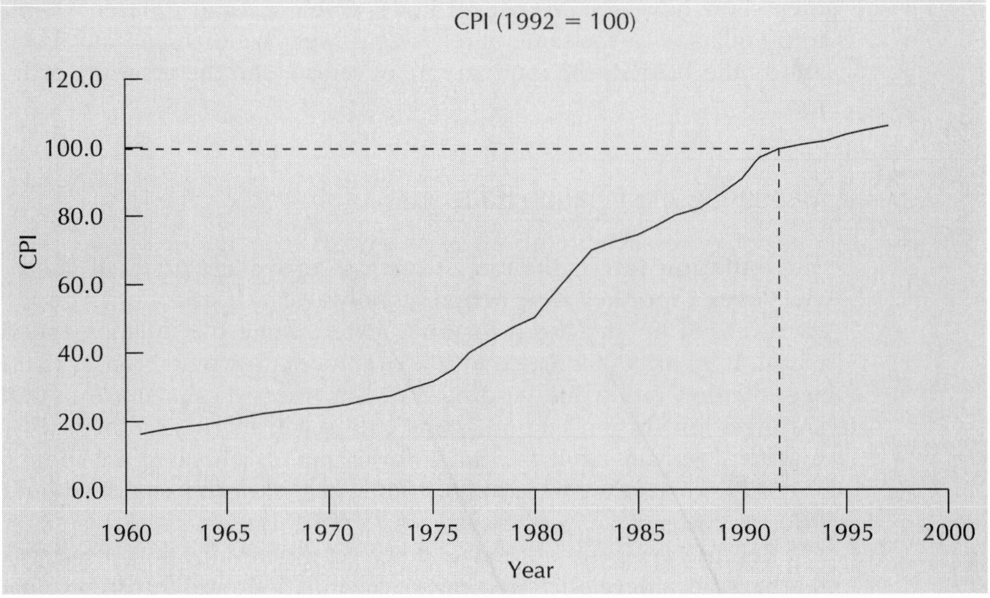

judge whether an index is appropriate for a particular purpose, the analyst should know how the goods and services for which he or she is estimating inflation compare with the set of goods and services used to compute the index. For this reason, we provide Appendix 9A, in which we illustrate the computation of one popularly used index.

Inflation rates in Canada have varied considerably over the last 40 years from a high of 12.4% in 1981 to a low of 0.2% in 1994 (see Figure 9.2). Low expected rates of inflation may be safely ignored, given the typical imprecision of predicted future cash flows. However, when expected inflation is high, it is necessary to include inflation in detailed economic calculations to avoid rejecting good projects.

Throughout the rest of this chapter, we assume that an analyst is able to obtain estimates for expected inflation rates over the life of a project and that project cash flows will change at the same rate as average prices. Consequently, the cash flows for a project can be assumed to increase at approximately the rate of inflation per year.

Figure 9.2 Canadian Inflation Rate 1961–1997

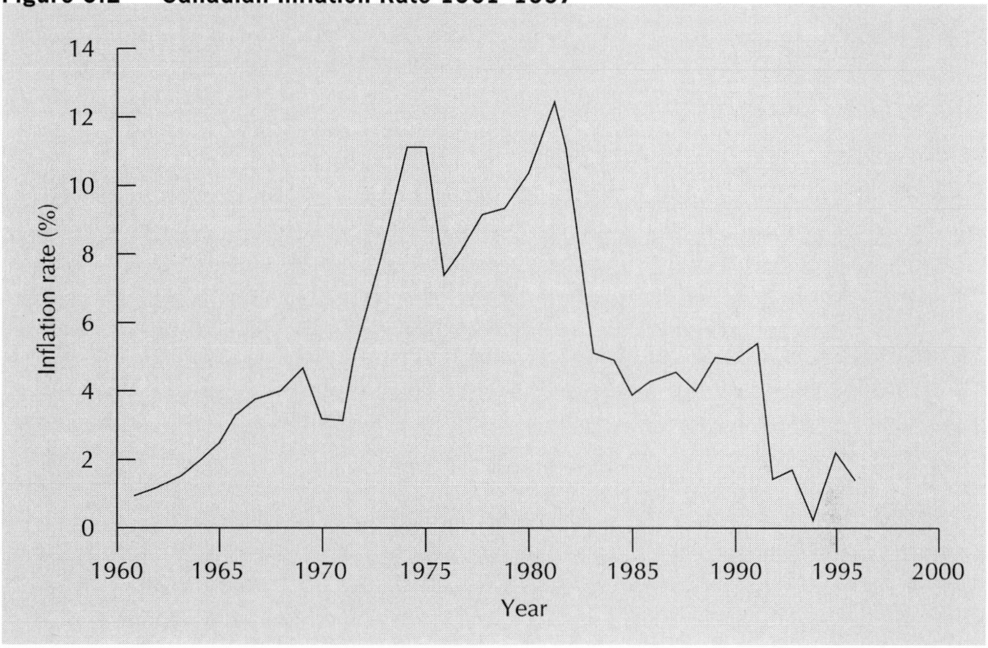

9.3 Economic Evaluation with Inflation

When prices change, the amount of goods a dollar will buy changes too. If prices fall, more goods may be bought for a given number of dollars, and the value of a dollar has risen. If prices rise, fewer goods may be bought for a given number of dollars, and the value of a dollar has fallen.

In project evaluation, we cannot make comparisons of dollar values across time without taking the price changes into account. We want dollars, not for themselves, but for what we can get for them. Workers are not directly interested in the money wages they will earn in a job. They are interested in how many hours of work it will take to cover expenses for their families, or how long it will take them to accumulate enough to make down payments on houses. Similarly, investors want to know if making an investment now will enable them to buy more real goods in the future, and by how much the amount they can buy in the future will increase. To know if an investment will lead to an increase in the amount they can buy in the future, they must take into account expected price changes.

We can take price changes into account in an approximate way by measuring the cash flows associated with a project in monetary units of constant purchasing power called **real dollars** (sometimes called **constant dollars**). This is in contrast to the **actual dollars** (sometimes called **current** or **nominal dollars**), which are expressed in the monetary units at the time the cash flows occur.

For example, if a photocopier will cost $2200 one year from now, the $2200 represents *actual* dollars since that is the amount that would be paid at that time. If inflation is expected to be 10% over the year, the $2200 is equivalent to $2000 real (today) dollars. Although the term "dollar" is used by convention when speaking of inflation, the principles apply to any monetary unit.

Real dollars always need to be associated with a particular date (usually a year), called the **base year**. The base year need not be the present; it could be any time. People speak of "1990 dollars" or "1985 dollars" to indicate that real dollars are being used as well as indicating the base year associated with them. Provided that cash flows which occur at different times are converted into real dollars with the same base year, they can be compared fairly in terms of buying power.

9.3.1 Converting between Real and Actual Dollars

Converting actual dollars in year N into real dollars in year N relative to a base year 0 is straightforward, provided that the value of a global price index like the CPI at year N relative to the base year is available. Let

A_N = actual dollars in year N

$R_{0,N}$ = real dollars equivalent to A_N relative to year 0, the base year

$I_{0,N}$ = the value of a global price index (like the CPI) at year N, relative to year 0

Then the conversion from actual to real dollars is

$$R_{0,N} = \frac{A_N}{I_{0,N}/100} \tag{9.1}$$

Note that in Equation (9.1), $I_{0,N}$ is divided by 100 to convert it into percentage terms; recall that the CPI for the base year is set at 100.

Transforming actual dollar values into real dollars gives only an approximate offset to the effect of inflation. The reason is that there may be no readily available price index that accurately matches the "basket" of goods and services being evaluated. Despite the fact that available price indices are approximate, they do provide a reasonable means of converting actual cash flows to real cash flows relative to a base year.

An alternative means of converting actual dollars to real dollars is available if we have an estimate for the average yearly inflation between now (year 0) and year N. Let

A_N = actual dollars in year N

$R_{0,N}$ = real dollars equivalent to A_N relative to year 0, the base year

f = the inflation rate per year, assumed to be constant from year 0 to year N

Then the conversion from actual dollars in year N to real dollars in year N relative to the base year 0 is

$$R_{0,N} = \frac{A_N}{(1+f)^N}$$

When the base year is omitted from the notation for real dollars, it is understood that the current year (year 0) is the base year, as in

$$R_N = \frac{A_N}{(1+f)^N} \qquad (9.2)$$

Equation (9.2) can also conveniently be written in terms of the present worth compound interest factor

$$R_N = A_N(P/F,f,N) \qquad (9.3)$$

Note that here A_N is the actual dollar amount in year N, i.e., a future value. It should not be confused with an annuity A.

Example 9.1

Elliot Weisgerber's income rose from \$40 000 per year in 1990 to \$42 000 in 1993. At the same time the CPI rose from 93.3 in 1990 to 101.8 in 1993. Was Elliot worse off or better off in 1993 compared with 1990?

We can convert Elliot's actual dollar income in 1990 and 1993 into real dollars. This will tell us if his total purchasing power increased or decreased over the period from 1990 to 1993. Since the current base year for the CPI is 1992, we shall compare his 1990 and 1993 incomes in terms of real 1992 dollars.

His real incomes in 1990 and 1993 in terms of 1992 dollars, using Equation (9.1), were

$$R_{92,90} = \frac{\$40\ 000}{0.933} = \$42\ 872$$

$$R_{92,93} = \frac{\$42\ 000}{1.018} = \$41\ 257$$

Even though Elliot's actual dollar income rose between 1990 and 1993, his purchasing power fell since the real dollar value of his income, based on the CPI, fell about 4%. ■

Example 9.2

The cost of replacing a storage tank one year from now is expected to be \$2 000 000. If inflation is assumed to be 5% per year, what is the cost of replacing the storage tank in real (today) dollars?

First, note that the \$2 000 000 is expressed in actual dollars one year from today. The cost of replacing the tank in real (today) dollars can be found by letting

A_1 = \$2 000 000 = the actual cost 1 year from the base year (today)

R_1 = the real dollar cost of the storage tank in 1 year

f = the inflation rate per year

Then, with Equation (9.2)

$$R_1 = \frac{A_1}{1+f} = \frac{\$2\ 000\ 000}{1.05} = \$1\ 904\ 762$$

Alternatively, Equation (9.3) gives

$$R_1 = A_1(P/F, 5\%, 1) = \$2\ 000\ 000\ (0.9524) = \$1\ 904\ 762$$

The \$2 000 000 actual cost is equivalent to \$1 904 762 real (today) dollars at the end of one year. ■

Example 9.3

The cost of replacing a storage tank in 15 years is expected to be \$2 000 000. If inflation is assumed to be 5% per year, what is the cost of replacing the storage tank 15 years from now in real (today) dollars?

The cost of the tank 15 years from now in real dollars can be found by letting

A_{15} = \$2 000 000 = the actual cost 15 years from the base year (today)

R_{15} = the real dollar cost of the storage tank in 15 years

f = the inflation rate per year

Then, with the use of Equation (9.2), we have

$$R_{15} = \frac{A_{15}}{(1+f)^{15}} = \frac{\$2\ 000\ 000}{(1.05)^{15}} = \$962\ 040$$

Alternatively, Equation (9.3) gives

$$R_{15} = A_{15}(P/F, 5\%, 15) = \$2\ 000\ 000\ (0.48102) = \$962\ 040$$

In 15 years, the storage tank will cost \$962 040 in real (today) dollars. Note that this \$962 040 is money to be paid 15 years from now. What this means is that the new storage tank can be replaced at a cost that would have the same purchasing power as about \$962 040 today. ■

Now that we have the ability to convert from actual to real dollars using an index or an inflation rate, we turn to the question of how inflation affects project evaluation.

9.4 The Effect of Correctly Anticipated Inflation

The main observation made in this section is that engineers must be aware of potential changes in price levels over the life of a project. We shall see that when future inflation is expected over the life of a project, the MARR needs to be increased. Engineers need to recognize this effect of inflation on the MARR to avoid rejecting good projects.

9.4.1 The Effect of Inflation on the MARR

If we expect inflation, the number of actual dollars that will be returned in the future does not tell us the value, in terms of purchasing power, of the future cash flow. The purchasing power of the earnings from an investment depends on the *real* dollar value of those earnings.

The **actual interest rate** is the stated or observed interest rate based on actual dollars. If we wish to earn interest at the actual interest rate, i, on a one-year investment, and we invest \$$M$, the investment will yield \$$M(1 + i)$ at the end of the year. If the inflation rate over the next year is f, the real value of our cash flow is \$$M\left(\frac{1+i}{1+f}\right)$. We can use this to define the *real interest rate*, i'. The **real interest rate**, i', is the interest rate that would yield the same number of real dollars in the absence of inflation as the actual interest rate yields in the presence of inflation.

$$M(1 + i') = M\left(\frac{1+i}{1+f}\right)$$

$$i' = \frac{1+i}{1+f} - 1 \qquad (9.4)$$

We may see terms like "real rate of return" or "real discount rate." These are just special cases of the real interest rate.

The definition of the real interest rate can be turned around by asking the following question: If an investor wants a real rate of return, i', over the next year, and the inflation rate is expected to be f, what actual interest rate, i, must be realized to get a real rate of return of i'?

The answer can be obtained with some manipulation of the definition of the real interest rate in Equation (9.4):

$$i = (1 + i')(1 + f) - 1 \text{ or, equivalently, } i = i' + f + i'f \qquad (9.5)$$

Therefore, an investor who desires a real rate of return i' and who expects inflation at a rate of f will require an actual interest rate $i = i' + f + i'f$. This has implications for the actual MARR used in economic analyses. The **actual MARR** is the minimum acceptable rate of return when cash flows are expressed in actual dollars. If investors expect inflation, they require higher actual rates of return on their investments than if inflation were not expected. The actual MARR then will be the real MARR plus an upwards adjustment which reflects the effect of inflation. The **real MARR** is the minimum acceptable rate of return when cash flows are expressed in real, or constant, dollars.

If we denote the actual MARR by $MARR_A$ and the real MARR by $MARR_R$, we have from Equation (9.5)

$$MARR_A = MARR_R + f + MARR_R \times f \qquad (9.6)$$

Note that if $MARR_R$ and f are small, the term $MARR_R \times f$ may be ignored and $MARR_A = MARR_R + f$ can be used as a "back of the envelope" approximation.

The real MARR can also be expressed as a function of the actual MARR and the expected inflation rate:

$$MARR_R = \frac{1 + MARR_A}{1 + f} - 1 \qquad (9.7)$$

Figure 9.3 shows the Canadian experience with inflation, the actual prime interest rate and the real interest rate for the 1961–1996 period. From 1961 to 1971, when inflation was moderate and stable, the real interest rate was also stable, except for one blip in 1967. In the 1970s, conditions were very different when inflation exploded. This was due partly to large jumps in energy prices. Real interest rates were negative for the period 1972 to 1975 and 1977 to 1978. In the 1980s and early 1990s, real interest rates were quite high.

The high inflation rates of the 1970s were very unusual. Inflation in the range of 2% to 4% per year is more typical of Canadian experience. Averages of real interest rates and inflation rates over sub-periods are shown in Table 9.1.

Table 9.1 Canadian Averages of Real Interest Rate (%) and Inflation Rate (%)

Period	Real Interest Rate (%) Average	Inflation Rate (%) Average
1961–1971	1.82	2.75
1972–1981	0.08	8.47
1982–1991	4.84	5.29
1992–1996	3.89	1.43

Figure 9.3 Canadian Inflation Rates and Actual and Real Interest Rates 1961–1996

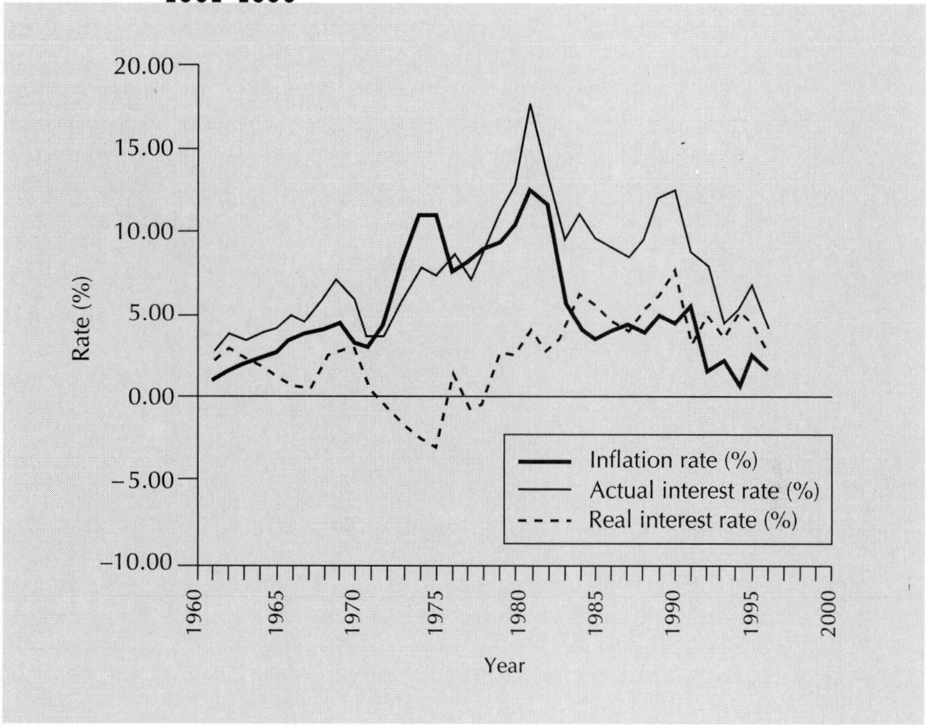

Example 9.4

Security Trust is paying 12% on one-year guaranteed investment certificates (GICs). The inflation rate is expected to be 5% over the next year. What is the real rate of interest? For a \$5000 GIC, what will be the real dollar value of the amount received at the end of the year?

The real interest rate is

$$i' = \frac{1 + i}{1 + f} - 1 = \frac{1.12}{1.05} - 1 = 0.067, \text{ or } 6.7\%$$

A \$5000 GIC will return \$5600 at the end of the year. The real value of the \$5600 *in today's dollars* is \$5600/1.05 = \$5333. This is the same as if there were no inflation and the investment earned 6.7% interest. ■

Example 9.5

Susan got a \$1000 present from her aunt on her sixteenth birthday. She has noticed that Security Trust offers 6.5% on one-year guaranteed investment certificates (GICs). Her mother's *National Post* indicates that analysts are predicting an inflation rate of about 3.5% for the coming year. Susan's real MARR for such investments is 4%. If the analysts are correct, what is Susan's actual MARR? Should she invest?

If the analysts are correct, Susan's actual MARR is

$$\begin{aligned} MARR_A &= MARR_R + f + MARR_R \times f \\ &= 0.04 + 0.035 + (0.04)(0.035) \\ &= 0.0764 \end{aligned}$$

Susan's actual MARR is about 7.64%. Since the actual interest rate on the GIC is only 6.5%, she should not invest in a GIC. ■

9.4.2 The Effect of Inflation on the IRR

The effect of expected inflation on the actual internal rate of return of a project is similar to the effect of inflation on the actual MARR. Suppose that we are considering a T-year investment. The actual rate of return on the project, IRR_A, is the rate of return of the project based on actual dollar cash flows. It can be found by solving for i^* in

$$\sum_{t=0}^{T} \frac{A_t}{(1+i^*)^t} = 0$$

where:

A_t = the actual cash flow in period t (receipts – disbursements)
T = the number of time periods
i^* = the actual internal rate of return

Suppose further that a yearly inflation rate of f is expected over the T-year life of the project. In terms of real dollars (with a base year of the time of the first cost), the actual cash flow in period t can be written as $A_t = R_t(1+f)^t$ where R_t refers to the *real* dollar amount equivalent to the cash flow A_t. The expression which gives the actual internal rate of return can be rewritten as:

$$\sum_{t=0}^{T} \frac{R_t(1+f)^t}{(1+i^*)^t} = 0 \qquad (9.8)$$

In contrast, the *real* internal rate of return for the project, IRR_R, is the rate of return obtained on the real dollar cash flows associated with the project. It is the solution for i' in:

$$ {}_{t=0} \frac{R_t}{(1+i')^t} = 0 \qquad (9.9)$$

What is the relationship between IRR_R and IRR_A? We have from Equations (9.4) and (9.5)

$$\frac{1}{1+i'} = \frac{1+f}{1+i} \text{ or } i = i' + f + i'f$$

and thus, analogous to Equation (9.5)

$$IRR_A = IRR_R + f + IRR_R \times f \qquad (9.10)$$

Or, analogous to Equation (9.4), the real IRR can be expressed in terms of the actual IRR and the inflation rate:

$$IRR_R = \frac{1 + IRR_A}{1+f} - 1 \qquad (9.11)$$

In summary, the effect of inflation on the IRR is that the actual IRR will be the real IRR plus an upward adjustment which reflects the effect of inflation. ■

Example 9.6

Consider a two-year project which has a \$10 000 first cost and which is expected to bring about a saving of \$15 000 at the end of the two years. If inflation is expected to be 5% per year and the real MARR is 13%, should the project be undertaken? Base your answer on an IRR analysis.

From the information given, $A_0 = -\$10\ 000$, $A_2 = \$15\ 000$, and $f = 0.05$. The actual IRR can be found by solving for i in

$$A_0 + \frac{A_2}{(1+i)^2} = 0$$

$$\$10\ 000 = \$15\ 000/(1+i)^2$$

which leads to an actual IRR of 22.475%.

The real IRR is then

$$IRR_R = \frac{1 + IRR_A}{1+f} - 1$$

$$= \frac{1 + 0.22475}{1 + 0.05} - 1$$

$$= 0.1664 \text{ or } 16.64\%$$

Since the real IRR exceeds the real MARR, the project should be done. ■

In conclusion, the impact of inflation on the actual MARR and the actual IRR is that both have an adjustment for expected inflation implicitly included in them. The main implication of this observation is that, since both the actual MARR and the actual IRR increase in the same fashion, any project that was acceptable without inflation remains acceptable when inflation is expected. Any project that was unacceptable remains unacceptable.

9.5 Project Evaluation Methods with Inflation

The engineer typically starts a project evaluation with an observed (actual) MARR and projections of cash flows. As we have seen, the actual MARR has two parts, the real rate of return on investment that investors require to put money into the company, plus an adjustment for the expected rate of inflation. The engineer usually observes only the sum and not the individual parts.

As for the projected cash flows, these are typically based on current prices. Because the projected cash flows are based on prices of the period in which evaluations are being carried out, they are in *real* dollars. They do not incorporate the effect of inflation. In this case, the challenge for the engineer is to correctly analyze the project when he or she has an *actual* MARR (which incorporates inflation implicitly) and *real* cash flows (which do not take inflation into account).

Though common, it is not always the case that the engineer or analyst starts out with an *actual* MARR and *real* cash flows. The cash flows may already have inflation implicitly factored in (in which case the cash flows are said to be actual amounts). To carry out a project evaluation properly, the analyst must know whether inflation has been accounted for already in the MARR and the cash flows or whether it needs to be dealt with explicitly.

As a brief example, consider a one-year project which requires an investment of \$1000 today and which yields \$1200 in one year. The actual MARR is 25%. Whether the project is considered acceptable will depend on whether the \$1200 in one year is understood to be the actual cash flow in one year or if it is the real value of the cash flow received in one year.

If the \$1200 is taken to be the actual cash flow, the actual internal rate of return, IRR_A, is found by solving for i^* in:

$$-\$1000 + \frac{\$1200}{(1 + i^*)} = 0$$

$$i^* = \text{IRR}_A = 20\%$$

Hence, the project would not be considered economical. However, if the \$1200 is taken to be the real value of the cash flow in one year, and inflation is expected to be 5% over the year, then the actual internal rate of return is found by solving for i^* in:

$$-\$1000 + \frac{\$1200\,(1 + 0.05)}{(1 + i^*)} = 0$$

$$i^* = \text{IRR}_\text{A} = 26\%$$

Hence, the project *would* be considered acceptable. As seen in this example, the economic viability of the project may depend on whether or not the \$1200 in one year has inflation implicitly factored in (i.e., is taken to be the actual amount). This is why it is important to know what type of cash flows you are dealing with.

If the engineer has an estimate of inflation, there are two equivalent ways to carry out a project evaluation properly. The first is to work with actual values for cash flows and actual interest rates. The second is to work with real values for cash flows and real interest rates. *The two methods should not be mixed.*

These two methods of dealing with expected inflation, as well as two incorrect methods, are shown in Table 9.2.

Table 9.2 Methods of Incorporating Inflation into Project Evaluation

1. Real MARR and Real Cash Flows
The real MARR does not include the effect of expected inflation. Cash flows are determined by today's prices. **Correct**
2. Actual MARR and Actual Cash Flows
The actual MARR includes the effect of anticipated inflation. Cash flows include increases due to inflation. **Correct**
3. Actual MARR and Real Cash Flows
The actual MARR includes the effect of anticipated inflation. Cash flows are determined by today's prices. **Incorrect:** Biased against investments
4. Real MARR and Actual Cash Flows
The real MARR does not include the effect of expected inflation. Cash flows include increases due to inflation. **Incorrect:** Biased in favour of investments

The engineer must have a forecast of the inflation rate over the life of the project in order to adjust the MARR or cash flows for inflation. The best source of such forecasts may be the estimates of experts. Financial publications like the *Report on Business* of the *Globe and Mail* regularly report such predictions for relatively short periods of up to one year. Because there is evidence that even the short-term estimates are not totally reliable, and estimates for longer periods will necessarily be imprecise, it is good practice to determine a range of possible inflation values for both long- and short-term projects. The engineer should test for sensitivity of the decision to values in the range. The subject of sensitivity analysis is addressed more fully in Chapter 11. Close-Up 9.1 discusses the problem of price changes that are specific to an industry.

CLOSE-UP 9.1 Relative Price Changes

Engineers usually expect prices associated with a project to move together with the general inflation rate. However, there are situations where it makes sense to expect prices associated with a project to move differently from the average. This can happen when there are atypical forces affecting either the supply or the demand for the goods. Here are some examples:

1. Reductions in the availability of logs in North America have caused a decrease in the supply of wood for construction, furniture, and pulp and paper. Average wood prices in Canada more than doubled between 1986 and 1995. This is about twice the increase in the CPI over that period.
2. Product development and increases in productivity have led to increases in the supply of computers. This, in turn, has led to reductions in the relative price of computing power.
3. Reductions in family size and in the proportion of the population in the 30- to 40-year-old age group have caused a drop in the demand for housing in Canada and the United States. This has made the relative price of housing fall in most cities.

Changes in the relative prices of the goods sold by a specific industry will generally not have a noticeable effect on a MARR because investors are concerned with the overall purchasing power of the dollars they receive from an investment. Changes in the relative prices of the goods of one industry will not have much effect on investors' abilities to buy what they want.

Because the relative price changes will not affect the MARR, the analyst must incorporate expected relative price changes directly into the expected cash flows associated with a project. If the rate of relative price change is expected to be constant over the life of the project, this can be done using a geometric gradient to present worth conversion factor.

Example 9.7

Jagdeep can put his money into an investment that will pay him \$1000 a year for the next four years and \$10 000 at the end of the fifth year. Inflation is expected to be 5% over the next five years. Jagdeep's real MARR is 8%. What is the present worth of this investment?

The present worth may be obtained with real dollar cash flows and a real MARR or with actual dollar cash flows and an actual MARR.

The first solution approach will be to use real dollars and MARR_R. The real dollar cash flows *in terms of today's dollars* are

$$R_1, R_2, R_3, R_4, R_5 = \frac{A_1}{(1+f)}, \frac{A_2}{(1+f)^2}, \frac{A_3}{(1+f)^3}, \frac{A_4}{(1+f)^4}, \frac{A_5}{(1+f)^5}$$

$$= \frac{\$1000}{(1.05)}, \frac{\$1000}{(1.05)^2}, \frac{\$1000}{(1.05)^3}, \frac{\$1000}{(1.05)^4}, \frac{\$10\,000}{(1.05)^5}$$

The present worth of the real cash flows, discounted by $\text{MARR}_\text{R} = 8\%$, is

$$\text{PW} = \frac{\$1000}{(1.05)(1.08)} + \frac{\$1000}{(1.05)^2(1.08)^2} + \frac{\$1000}{(1.05)^3(1.08)^3}$$

$$+ \frac{\$1000}{(1.05)^4(1.08)^4} + \frac{\$10\,000}{(1.05)^5(1.08)^5}$$

$$\cong \$8282$$

The present worth of Jagdeep's investment is about \$8282.

Alternatively, the present worth can be found in terms of actual dollars and MARR_A:

$$\text{PW} = \frac{\$1000}{(1+\text{MARR}_\text{A})} + \frac{\$1000}{(1+\text{MARR}_\text{A})^2} + \frac{\$1000}{(1+\text{MARR}_\text{A})^3}$$

$$+ \frac{\$1000}{(1+\text{MARR}_\text{A})^4} + \frac{\$10\,000}{(1+\text{MARR}_\text{A})^5}$$

where

$$\text{MARR}_\text{A} = \text{MARR}_\text{R} + f + \text{MARR}_\text{R} \times f$$

Note that this is the sum of a four-period annuity with equal payments of \$1000 for four years and a single payment of \$10 000 in period 5. With this observation, the present worth computation can be simplified by the use of compound interest formulas:

$$\text{PW} = \$1000(P/A, \text{MARR}_\text{A}, 4) + \$10\,000(P/F, \text{MARR}_\text{A}, 5)$$

With a real MARR of 8% and an inflation rate of 5%, the actual MARR is then

$$\begin{aligned}\text{MARR}_A &= \text{MARR}_R + f + \text{MARR}_R \times f\\ &= 0.08 + 0.05 + (0.08)(0.05)\\ &= 0.134\end{aligned}$$

and the present worth of Jagdeep's investment is

$$\begin{aligned}\text{PW} &= \$1000(P/A, 13.4\%, 4) + \$10\,000(P/F, 13.4\%, 5)\\ &\cong \$8282\end{aligned}$$

The present worth of Jagdeep's investment is about \$8282, as was obtained through the use of the real MARR and a conversion from actual to real dollars. ■

Though there are two distinct means of correctly adjusting for inflation in project analysis, the norm for engineering analysis in Canada is to make comparisons with the actual MARR. One reason this is done has to do with how a MARR is chosen. As discussed in Chapter 3, the MARR is based on, among other things, the cost of capital. Since lenders and investors recognize the need to have a return on their investments higher than the expected inflation rate, they will lend to or invest in companies only at a rate that exceeds the inflation rate. In other words, the cost of capital already has inflation included. A MARR based on this cost of capital already includes, to some extent, inflation.

Consequently, if inflation is fairly static (even if it is high), an *actual* dollar MARR is sensible and will arise naturally. On the other hand, if changes in inflation are foreseen, or if sensitivity analysis specifically for inflation is desired, it may be wise to set a *real* dollar MARR and recognize an inflation rate explicitly in the analysis.

Example 9.8

Lethbridge Communications is considering an investment in plastic moulding equipment for its product casings. The project involves \$150 000 in first costs and is expected to generate net savings (in actual dollars) of \$65 000 per year over its three-year life. They forecast an inflation rate of 15% over the next year, and then inflation of 10% thereafter. Their real dollar MARR is 5%. Should this project be accepted on the basis of an IRR analysis?

In this problem, the inflation rate is not constant over the life of the project, so it is easiest to consider the cash flows for each year separately and to work in real dollars. First, as shown in Table 9.3, the actual cash flows are converted into real cash flows.

Table 9.3 Converting from Actual to Real Dollars for Lethbridge Communications

Year	Actual Dollars	Real Dollars
0	−\$150 000	−\$150 000
1	65 000	56 522 $= \$65\,000(P/F, 15\%, 1)$ $= \$65\,000(0.86957)$
2	65 000	51 384 $= \$65\,000(P/F, 15\%, 1)(P/F, 10\%, 1)$ $= \$65\,000(0.86957)(0.90909)$
3	65 000	46 713 $= \$65\,000(P/F, 15\%, 1)(P/F, 10\%, 2)$ $= \$65\,000(0.86957)(0.82645)$

Then, the real IRR can be found by solving for i' in

$$\$56\,522(P/F, i', 1) + \$51\,384(P/F, i', 2) + \$46\,713(P/F, i', 3) = \$150\,000$$

At $i' = 1\%$, LHS (left-hand side) = \$151 673

At $i' = 2\%$, LHS = \$148 821

The real IRR is between 1% and 2%. This is less than the real dollar MARR of 5%, so the project should not be done. ■

Example 9.9

New Glasgow Resources (NGR) has been offered a contract to sell land to the government at the end of 20 years. The contract states that NGR will get \$500 000 20 years from today, with no costs or benefits in the intervening years. A financial analyst for the firm believes that the inflation rate will be 4% for the next two years, rise to 15% for the succeeding 10 years, and then go down to 10%, where it will stay forever. NGR's real dollar MARR is 10%. What is the present worth of the contract?

In this case, it is easiest to proceed by calculating the actual dollar MARR for each of the different inflation periods.

$$\text{MARR}_A\text{, years 13 to 20} = 0.10 + 0.10 + (0.10)(0.10) = 0.21 \text{ or } 21\%$$

$$\text{MARR}_A\text{, years 3 to 12} = 0.10 + 0.15 + (0.10)(0.15) = 0.265 \text{ or } 26.5\%$$

$$\text{MARR}_A\text{, years 0 to 2} = 0.10 + 0.04 + (0.10)(0.04) = 0.144 \text{ or } 14.4\%$$

With the individual MARRs, the present worth of the \$500 000 for each of years 12, 2, and 0 can be found.

$$\text{PW(year 12)} = \$500\,000(P/F, 21\%, 8) = \$500\,000 \times 1/(1.21)^8$$
$$= \$108\,815$$
$$\text{PW(year 2)} = \$108\,815\ (P/F, 26.5\%, 10) = \$108\,815 \times 1/(1.265)^{10}$$
$$= \$10\,370$$
$$\text{PW(year 0)} = \$10\,370\ (P/F, 14.4\%, 2) = \$10\,370 \times 1/(1.144)^2$$
$$= \$7924$$

The present worth of the contract is approximately \$7924. ■

Example 9.10

Bildmet is an extruder of aluminum shapes used in construction. They are experiencing a high scrap rate of 5%. The manager, Greta Kehl, estimates that reprocessing scrap costs about \$0.30 per kilogram. The high scrap rate is due partly to operator error. Ms. Kehl believes that a short training course for the operator would reduce the scrap rate to about 4%. The course would cost about \$1100. Bildmet is now working with a before-tax actual MARR of 22%. Past experience suggests that operators quit their jobs after about five years; the correct time horizon for the retraining project is, therefore, five years. The data pertaining to the training course are summarized in Table 9.4. Should Bildmet retrain its operator?

Table 9.4 Training Course Data

Output (kilograms/year)	125 000
Scrap (kilograms/year)	6 250
Reprocessing cost (\$/kilogram)	0.30
Scrap cost (\$/year)	1875
Savings due to training (\$/year)	375
First cost of training (\$)	1 100
Inflation rate (%/year)	5
Actual MARR (%/year)	22

First, note that the actual MARR $i = 22\%$ incorporates an estimate by investors of inflation of $f = 5\%$ per year over the next five years. If this estimate of future inflation is correct, Ms. Kehl needs to make an adjustment to take inflation into account. Either the projected annual saving from reduced scrap needs to be increased by the 5% rate of inflation, or she needs to reduce the MARR to its real value. We shall illustrate the first approach with actual cash flows and the actual MARR.

Table 9.5 Savings Due to the Training Course

Year	Actual Savings	Present Worth
1	\$393.75	\$322.75
2	413.44	277.77
3	434.11	239.07
4	455.81	205.75
5	478.61	177.08

Increasing savings to take inflation into account leads to projected (actual) savings as shown in the Table 9.5.

For example, using Equation (9.2), the expected saving in year 3 is $\$375(1 + f)^3 = \$434.11$. The present worth of the savings in year 3 is $\$434.11/(1 + i)^3 = \$239.07$.

The present worth of the savings over the five-year time frame, when discounted at the actual MARR of 22%, is \$1222. This makes the project viable since its cost is \$1100.

We note that the same result could have been reached by working with the real MARR and the constant cost savings of \$375 per year. MARR_R is given by

$$\text{MARR}_\text{R} = 1 + \frac{\text{MARR}_\text{A}}{1 + f} - 1 = \frac{1.22}{1.05} - 1 = 0.1619$$

The present worth of the real stream of returns, when these are discounted by the real MARR, is given by

$$\text{PW} = \$375(P/A, 0.1619, 5) \cong \$1222$$

which is the same result obtained with the actual MARR and actual cash flows. ■

REVIEW PROBLEMS

REVIEW PROBLEM 9.1

Athabaska Engineering was paid \$100 000 to manage a construction project in 1970. How much would the same job have cost in 1990 if the average annual inflation rate between 1970 and 1990 were 5%?

ANSWER

The compound amount factor can be used to calculate the value of 100 000 1970 dollars in 1990 dollars, using the inflation rate as an interest rate:

$$\begin{aligned}\text{1990 dollars} &= \$100\ 000(F/P, 5\%, 20)\\ &= \$100\ 000\ (2.6533)\\ &= \$265\ 330\end{aligned}$$

The same job would have cost about $265 330 in 1990 dollars. ■

REVIEW PROBLEM 9.2

A computerized course drop-and-add program is being developed for a local community college. It will cost $300 000 to develop and is expected to save $50 000 per year in administrative costs over its 10-year life. If inflation is expected to be 4% per year for the next 10 years and a real MARR of 5% is required, should the project be done?

ANSWER

First, assuming that $50 000 in administrative costs are actual dollars, we can calculate the actual IRR for the project. The actual IRR is the solution for i in

$$\$300\ 000 = \$50\ 000(P/A, i, 10)$$

$$(P/A, i, 10) = 6$$

$$\text{For } i = 11\%, (P/A, i, 10) = 5.8892$$

$$\text{For } i = 10\%, (P/A, i, 10) = 6.1445$$

The actual IRR of 10.55% is found by interpolating between these two points. We then convert the actual IRR into a real IRR to determine if the project is viable:

$$IRR_R = \frac{1 + IRR_A}{1 + f} - 1 = \frac{1.1055}{1.04} - 1 = 0.06298 \text{ or } 6.3\%$$

Since the real IRR of 6.3% exceeds the MARR of 5%, the project should be done. ■

REVIEW PROBLEM 9.3

Robert is considering purchasing a bond with a face value of $5000 and a coupon rate of 8%, due in 10 years. Inflation is expected to be 5% over the next 10 years. Robert's real MARR is 10%, compounded semi-annually. What is the present worth of this bond to Robert?

ANSWER

This problem can be done with either real interest and real cash flows or actual interest and actual cash flows. It is somewhat easier to work with actual cash flows, so we must first convert the real interest rate given to an actual interest rate.

Robert's annual real MARR is $(1 + 0.10/2)^2 - 1 = 0.1025$. (Recall that the 10% is a nominal rate, compounded semi-annually.)

If annual inflation is 5%, Robert's actual *annual* MARR is

$$\begin{aligned} \text{MARR}_\text{A} &= \text{MARR}_\text{R} + f + \text{MARR}_\text{R} \times f \\ &= 0.1025 + 0.05 + (0.1025)(0.05) \\ &= 0.15763 \text{ or } 15.763\% \end{aligned}$$

The present worth of the $5000 Robert will get in 10 years is then

$$\begin{aligned} \text{PW} &= \$5000(P/F, \text{MARR}_\text{A}, 10) \\ &= \$5000(0.23138) \cong \$1157 \end{aligned}$$

Next, the bond pays an annuity of $\$5000 \times 0.08/2 = \$200$ every six months. To convert the annuity payments to their present worth, we need an actual six-month MARR. This can be obtained with a six-month inflation rate and Robert's six-month real MARR of 10%/2 = 5%. With $f = 5\%$ per annum, the inflation rate per six-month period can be calculated with

$$\begin{aligned} f_{12} &= (1 + f_6)^2 - 1 \\ f_6 &= (1 + f_{12})^{1/2} - 1 \\ &= (1 + 0.05)^{1/2} - 1 = 0.0247 = 2.47\% \end{aligned}$$

The actual MARR per six-month period is then given by

$$\begin{aligned} \text{MARR}_\text{A} &= \text{MARR}_\text{R} + f + \text{MARR}_\text{R} \times f \\ &= 0.05 + 0.0247 + (0.05)(0.0247) \\ &= 0.07593 \text{ or } 7.593\% \end{aligned}$$

The present worth of the dividend payments is

$$\begin{aligned} \text{PW(dividends)} &= \$200(P/A, 7.59\%, 20) \\ &= \$200(10.125) \\ &= \$2025 \end{aligned}$$

Finally,

$$\begin{aligned} \text{PW(bond)} &= \$1157 + \$2025 \\ &= \$3182 \end{aligned}$$

The present worth of the bond is $3182. ■

REVIEW PROBLEM 9.4

Trimfit, a Southern Ontario manufacturer of automobile interior trim, is considering the addition of a new product to their line. Data concerning the project are given below. Should Trimfit accept the project?

New Product Line Information	
First cost ($)	11 500 000
Planned output (units/year)	275 000
Actual MARR	20%
Range of possible inflation rates	0% to 4%
Study period	10 years

Current Year 2000 Prices ($/unit)	
Raw materials	16.00
Labour	6.25
Product sales price	32.00

ANSWER

First, we note that the expected net revenue per unit (not counting amortization of first costs) is $9.75 = $32 – $16 – $6.25. The project is potentially viable.

In doing the project evaluation, we can proceed with either actual dollars or real dollars. Since we do not know what the inflation rate will be, the easiest way to account for inflation is to keep all prices in real 2000 dollars and adjust the actual MARR to a real MARR by using values for the inflation rate within the potential range given. The project can then be evaluated with one of the standard methods. Since many of the figures are given in terms of annual amounts, an annual worth analysis will be carried out. Inflation rates of 0%, 1%, and 4% will be used. The results are shown below in Table 9.6.

In Table 9.6, the annual worth of the project depends on the inflation rate assumed. Since the actual MARR of 20% implicitly includes anticipated inflation, different trial inflation rates imply different values for the real MARR. For example, at 1% inflation, the real MARR implied is

$$\text{MARR}_\text{R} = \frac{1 + \text{MARR}_\text{A}}{1 + f} - 1 = \frac{1.20}{1.01} - 1 = 0.1881 \text{ or } 18.81\%$$

The fixed cost per year is obtained by finding the annual amount over 10 years equivalent to the first cost when the appropriate real MARR is used. For example, with 1% inflation, the fixed cost per year is

Table 9.6 Annual Worth Computations for Trimfit

Annual Worth Comparisons for Various Inflation Rates					
Inflation Rate per Year	Real MARR	Fixed Cost per Year ($)	Variable Cost per Year ($)	Revenue per Year ($)	Annual Worth (Profit) per Year ($)
0%	20.00%	2 743 012	6 118 750	8 800 000	− 61 762
1%	18.81%	2 633 122	6 118 750	8 800 000	48 128
4%	15.38%	2 325 083	6 118 750	8 800 000	356 167

$$A = P(A/P, \text{MARR}_\text{R}, 10) = \$11\ 500\ 000\left(\frac{0.1881\ (1.1881)^{10}}{(1.1881)^{10}} - 1\right)$$
$$= \$2\ 633\ 122$$

Next, the variable cost per year is the sum of the raw material cost and the labour cost per unit multiplied by the total expected output per year, i.e., $22.25 × 275 000 = $6 118 750. Revenue per year is the sales price multiplied by the expected output: $32 × 275 000 = $8 800 000. Notice that the variable cost and the revenue are the same for all three values of the inflation rate. This is because they are given in constant 2000 dollars.

Finally, the annual worth of the project is determined by the revenue per year less the fixed and variable costs per year. The annual worth is negative for zero inflation, but is positive for both 1% and 4% inflation rates. Since periods of at least 10 years in which there has been zero inflation have been rare in the twentieth century, it is probably safe to assume that there will be some inflation over the life of the project. Therefore, the project appears to be acceptable since its annual worth will be positive if inflation is at least 1%. ■

SUMMARY

In this chapter, the concept of inflation was introduced, and we considered the impact that inflation has on project evaluation. We began by discussing methods of measuring inflation. The main result here was that there are many possible measures, all of which are only approximate.

The concept of actual cash flows and interest rates, and real cash flows and interest rates was introduced. Actual dollars are in currency at the time of payment or receipt, while real dollars are constant over time and are expressed with respect to a base year. Compound amount factors can be used to convert single payments between real and actual dollars.

Most of the chapter was concerned with the effect of correctly anticipated inflation on project evaluation and on how to incorporate inflation into project evaluation correctly. We showed that, where engineers have no reason to believe project prices will behave differently from average prices, project decisions are the same with or without correctly anticipated inflation. Finally, we pointed out that predicting inflation is very difficult. This implies that engineers should work with ranges of values for possible future inflation rates. The engineer should test for sensitivity of decisions to possible inflation rates.

Engineering Economics in Action, Part 9B: Exploiting Volatility

Bill Astad of head office had been asking Naomi about how to deal with the variable inflation rates experienced by a sister company in Mexico. "O.K., Naomi, let's see if I have this straight. For long-term projects, of say six years or more, it makes sense to use a single inflation figure — the average rate. I can just add that to the real MARR to get an actual MARR. Boy, it's easy to get confused between the real and the actual. But I do understand the principle."

"And the short-term projects?" Naomi prompted.

"For the short-term ones, it makes more sense to break them up into time periods. For each period, select a 'best guess' inflation rate, and do a stepwise calculation from period to period. So the inflation rate in the middle of the presidential cycle would be relatively low, while near the changeover time it would be a higher estimate. Of course, the actual values used would depend on the political and economic situation at the time the decision is made. I understand that one, too, but it is complicated."

"I agree," said Naomi. "I guess we're lucky things are more predictable here."

"We are," Bill replied. "On the other hand, if we can make good decisions in spite of a volatile economy in Mexico, Mexifab may have an advantage over its competitors. Thanks for your help, Naomi."

PROBLEMS

9.1 Which of the following are real dollars and which are actual dollars?

(a) Allyson has been promised a \$10 000 inheritance when her Uncle Bill dies.

(b) Bette's auto insurance will pay the cost of a new windshield if her current one breaks.

(c) Cory's meal allowance while he is in university is \$2000 per term.

(d) Dieter's company promises that its prices will always be the same as they were in 1975.

(e) Engworth will construct a house for Zolda, and Zolda will pay Engworth \$150 000 when the house is finished.

(f) Fran's current salary is \$3000 per month.

(g) Greta's retirement plan will pay her \$1500 per month, adjusted for the cost of living.

9.2 Find the real dollars (with today as the base year) corresponding to the actual dollars shown below, for a 4% inflation rate.

(a) \$400 three year from now

(b) \$400 three years ago

(c) \$10 next year

(d) \$350 983 ten years from now

(e) \$1 one thousand years ago

(f) \$1 000 000 000 three hundred years from now

9.3 Find the present worth today in real dollars corresponding to the actual dollars shown below, for a 4% inflation rate and a 4% interest rate.

(a) \$400 three year from now

(b) \$400 three years ago

(c) \$10 next year

(d) \$350 983 ten years from now

(e) \$1 one thousand years ago

(f) \$1 000 000 000 three hundred years from now

9.4 An investment pays \$10 000 in five years.

(a) If inflation is 10% per year, what is the real value of the \$10 000 in today's dollars?

(b) If inflation is 10% and the real MARR is 10%, what is the present worth?

(c) What actual dollar MARR is equivalent to a 10% real MARR when inflation is 10%?

(d) Compute the present worth using the actual dollar MARR from part (c).

9.5 An annuity pays \$1000 per year for 10 years. Inflation is 6% per year.

(a) If the real MARR is 8%, what is the actual dollar MARR?

(b) Using the actual dollar MARR from part (a), calculate the present worth of the annuity.

9.6 An annuity pays \$1000 per year for 12 years. Inflation is 6% per year. The annuity costs \$7500 now.

(a) What is the actual dollar internal rate of return?

(b) What is the real internal rate of return?

9.7 A bond pays $10 000 per year for the next ten years. The bond costs $90 000 now. Inflation is expected to be 5% over the next 10 years.

(a) What is the actual dollar internal rate of return?

(b) What is the real internal rate of return?

9.8 Inflation is expected to average about 4% over the next 50 years. How much would we expect to pay 50 years from now for each of the following?

(a) $1.59 hamburger

(b) $15 000 automobile

(c) $180 000 house

9.9 The average Canadian now has assets totaling $38 000. If the average real wealth per Canadian remains the same, and if inflation averages 5% in the future, when will the average Canadian become a millionaire?

9.10 How much is the present worth of $10 000 ten years from now under each of the following patterns of inflation, if interest is at 5%? Based on your answers, is it generally reasonable to use an average inflation rate in economic calculations?

(a) Inflation is 4%.

(b) Inflation is 0% for five years, and then 8% for five years.

(c) Inflation is 8% for five years, and then 0% for five years.

(d) Inflation is 6% for five years and then 2% for five years.

(e) Inflation is 0% for nine years and then 40% for one year.

9.11 The actual dollar MARR for Jungle Products Ltd. of Parador is 300%. The inflation rate in Parador is 250%. What is the company's real MARR?

9.12 Krystyna has a long-term consulting contract with an insurance company that guarantees her $25 000 per year for five years. Krystyna believes inflation will be 3% this year and 5% next year, and then will stay at 10% indefinitely. Krystyna's real dollar MARR is 12%. What is the present worth of this contract?

9.13 I have a bond that will pay me $2000 every year for the next 30 years. My first payment will be a year from today. I expect inflation to average 3% over the next 30 years. My real MARR is 10%. What is the present worth of this bond?

9.14 Ken will receive a \$15 000 annual payment from a family trust. This will continue until Ken is 30; he is now 20. Inflation averages 4%, and Ken's real MARR is 8%. If the first payment is a year from now and a total of 10 payments are to be made, what is the present worth of his remaining income from the trust?

9.15 Inflation in Russistan currently averages 40% per month. It is expected to diminish to 20% per month following the presidential elections 12 months from now. The Russistan Oil Company (ROC) has just signed an agreement with the Canadian Petroleum Group for the sale of future shipments. The ROC will receive 500 million rubles per month over the next two years, and also 500 million rubles per month indexed to inflation (i.e., real rubles). If the ROC has a real MARR of 1.5% per month, what is the total present worth of this contract?

9.16 The widget industry maintains a price index for a standard collection of widgets. The base year was 1992 until 1999, when the index was recomputed with 1999 as the base year. The following data concerning prices for the years 1997 to 2000 are available:

Year	Price Index 1992 Base	Price Index 1999 Base
1997	125	n.a.
1998	127	n.a.
1999	130	100
2000	n.a.	110

What was the percentage increase in prices of widgets between 1997 and 2000?

9.17 A group of farmers in Inverness is considering building an irrigation system from a water supply in some nearby mountains. They want to build a concrete reservoir with a steel pipe system. The first cost would be \$200 000 with (actual) annual maintenance costs of \$2000. They expect the irrigation system will bring them \$22 000 per year in additional (actual) revenues due to better crop production. Their real dollar MARR is 4% and they anticipate inflation to be 3% per year. Assume that the reservoir will have a 20-year life.

(a) Using the actual cash flows, find the actual IRR on this project.

(b) What is the actual MARR?

(c) Should they invest?

9.18 Refer to Problem 9.17.

(a) Convert the actual cash flows into real cash flows.

(b) Find the present worth of the project using the real MARR.

(c) Should they invest?

9.19 Go to the Statistics Canada web site (http://www.statcan.ca) and find the document "Your Guide to the Consumer Price Index."

(a) Summarize, in several paragraphs, what the CPI is, misconceptions about the CPI, and four of the price indexes used other than the CPI.

(b) Summarize, in several paragraphs, what the commodities in the CPI basket are, the relative importance of the commodities in the CPI basket, how the CPI basket is updated, and how prices are collected for the CPI.

9.20 Go to the Statistics Canada web site (http://www.statcan.ca) and find the CPI Indexes for Canada and the Provinces.

(a) For the most recent time period reported, find the province in Canada with the highest "all items" CPI.

(b) For the province identified in (a), which of the subcategories (e.g., food, transportation, etc.) appears to be contributing the most to the overall CPI index?

(c) Discuss why you think this province has the highest CPI.

9.21 Bosco Consulting of Calgary is considering a potential contract with the Upper Sobonian government to advise them on exploration for oil in Upper Sobonia. Bosco would make an investment of 1 500 000 Sobonian zerts to set up a Sobonian office in 2000. The Upper Sobonian government would pay Bosco 300 000 zerts in 2001. In the years 2002 to 2007, the actual zerts value of the payments would increase at the rate of inflation in Upper Sobonia. The following data are available concerning the project:

Investment in Upper Sobonia	
Expected Sobonian inflation rate (2000–2007)	15%/year
Expected Canadian inflation rate (2000–2007)	3%/year
Value of Sobonian zerts in 2000	$0.25
Expected decline in value of zerts (2000–2007)	10%/year
First cost in 2000 (zerts)	1 500 000
Cash flows in 2001–2007 (real 2001 zerts)	300 000
Bosco's actual dollar MARR	22%

(a) Construct a table with the following items:

Real (2000) zerts cash flows
Actual zerts cash flows
Actual dollar cash flows
Real dollar cash flows

(b) What is the present worth in 2000 dollars of this project?

9.22 Bildkit, an Alberta building products company, is considering an agreement with a distributor in Maloria to supply kits for constructing houses in Maloria. Sales would start next year. The expected receipts from the sale of the kits next year is 30 000 000 Malorian yen. The number of units sold is expected to grow by 10% per year over the life of the contract. The actual yen price is expected to grow at the rate of Malorian inflation.

There will be a first cost for Bildkit. As well, there will be operating costs over the life of the contract. Operating cost per unit will be constant in real dollars over the life of the contract. Since the number of units sold will rise by 10% per year, real operating costs will rise by 10% per year. Actual operating costs per unit will rise at the rate of inflation in Canada.

The value of the Malorian yen is expected to increase over the life of the contract. Data concerning the proposed contract are shown in the table below.

Bildkit in Maloria

Receipts in first year (actual yen)	30 000 000	First cost now (actual $)	200 000
Growth of receipts (real yen)	10%/year	Operating cost in first year of operation (actual $)	350 000
Malorian inflation rate	1%/year	Canadian inflation rate	3%
Value of yen year 0 ($)	0.015	Actual dollar MARR	22%
Rate of increase in value of yen	2%	Study period	8 years

(a) What is the present worth of receipts in dollars?

(b) What is the present worth of the cash outflows in dollars?

9.23 Leftway Information Systems of Saint John, New Brunswick, is considering a contract with the Ibernian government to supply consulting services over a five-year period. The following real Ibernian pounds cash flows are expected:

Cash Flows in Year 2000 Ibernian Pounds	
First cost	1 800 000
Net revenue 2001 to 2005	550 000

Further information is in the table below:

Expected Ibernian inflation rate	10%
Value of Ibernian pound in 2000 (Canadian $)	1.25
Expected annual rate of decline in the value of the Ibernian pound	5%
Expected Canadian inflation rate	2.50%
Leftway's real MARR	15%

(a) What is the real Ibernian pound internal rate of return on this project? (*Hint:* Canada can be ignored in answering this question.)

(b) What is the actual pound internal rate of return? (*Hint:* Canada can be ignored in answering this question.)

(c) Use the internal rate of return in Canadian dollars to decide if Leftway should accept the proposed contract.

9.24 Sonar warning devices are being purchased by the St. James Bay department store chain to help trucks back up at store loading docks. The total cost of purchase and installation is $220 000. There are two types of saving from the system. Faster turn-around time at the congested loading docks will save $50 000 per year in today's dollars. Reduced damage to the loading docks will save $30 000 per year in today's dollars. St. James Bay has an observed actual dollar MARR of 18%. The sonar system has a life of four years. Its scrap value in today's dollars is $20 000. The inflation rate is expected to be 6% per year over the next four years.

(a) What is St. James Bay's real MARR?

(b) What is the real internal rate of return? (This is most easily done with a spreadsheet.)

(c) Compute the actual internal rate of return using Equation (9.10).

(d) Compute the actual internal rate of return from the actual dollar cash flows. (This is most easily done with a spreadsheet.)

(e) What is the present worth of the system?

9.25 Johnson Products, a manufacturer in Wolfville, Nova Scotia, now buys a certain part for its chain saws. They are considering the production of the part in-house. They can install a production system that would have a life of five years with no salvage value. They believe that over the next five years the real price of purchased parts will remain fixed. They expect the real price of labour and other inputs to production in Wolfville to rise over the next five years. Further information about the situation is in the table below.

Annual cost of purchase ($/year)	750 000
Expected real change in cost of purchase	0%
Expected real change in labour cost	4%
Expected real change in other operating costs	2%
Labour cost/unit (first year of operation) ($)	10.5
Other operating cost/unit (first year of operation) ($)	9
In-house first cost ($)	200 000
Use rate (units/year)	25 000
Actual dollar MARR	20%
Study period (years)	5

(a) Assume inflation is 2% per year in the first year of operation. What will be the actual dollar cost of labour for in-house production in the second year?

(b) Assume inflation is 2% per year in the first two years of operation. What will be the actual dollar cost of other operating inputs for in-house production in the third year?

(c) Assume that inflation averages 2% per year over the five-year life of the project. What is the present worth of costs for purchase and for in-house production?

9.26 Lifewear, a Winnipeg manufacturer of women's sports clothes, is considering adding a line of skirts and jackets. The production would take place in a part of their factory that is now not being used. The first output would be available in time for the following year, for the 2001 fall season. The following information is available:

New Product Line Information	
First cost in 2000 ($)	15 500 000
Planned output (units/year)	325 000
Observed, actual dollar MARR before tax	0.25
Study period	6 years

Year 2000 Prices ($/unit)	
Materials	12
Labour	7.75
Output	35

(a) What is the real internal rate of return?

(b) What inflation rate will make the real MARR equal to the real internal rate of return?

(c) Calculate the present worths of the project under three possible future inflation rates. Assume that the inflation rate will be 1%, 2%, or 3% per year.

(d) Decide if Lifewear should add this new line of skirts and jackets. Explain your answer.

9.27 Century Foods, a Saskatoon producer of frozen meat products, is considering a new plant near Calgary for its sausage rolls and frozen meat pies. The company has estimates of production cost and selling prices in the first year. It expects the real value of operating costs per unit to fall because of improved operating methods. It also expects competitive pressures to cause the real value of product prices to fall. The following data are available:

Century Food Plant Data	
Output price in 2000 ($/box)	22
Operating cost in 2000 ($/box)	15.5
Planned output rate (boxes/year)	275 000
Fall in real output price	1.5% per year
Fall in real operating cost per box	1.0% per year
First cost in 1999 ($)	7 500 000
Study period	10 years
Actual dollar MARR before tax	20%

(a) Assume that there is zero inflation. What is the present worth, in 1999, of the project?

(b) Assume that there is zero inflation. What is the internal rate of return? (This is most easily done with a spreadsheet.)

(c) At what inflation rate would the actual dollar internal rate of return equal 20%?

(d) Should Century Foods build the new plant? Explain your answer.

9.28 Metcan Ltd.'s Newfoundland smelter produces its own electric power. The plant's power capacity exceeds its current requirements. Metcan has been offered a contract to sell excess power to a nearby utility company. Metcan would supply the utility company with 17 500 MWh/year for 10 years. The contract would specify a price of \$22.75/MWh for the first year of supply. The price would rise by 1% per year after this. This is independent of the actual rate of inflation over the 10 years.

Metcan would incur a first cost to connect its plant to the utility system. There would also be operating costs attributable to the contract. Metcan believes these costs would track the actual inflation rate. The terms of the contract and Metcan's costs are shown in the tables below.

Metcan Sale of Power	
Output price in 2000 (\$/MWh)	22.75
Price adjustment (2001–2009)	1% per year
Power to be supplied (MWh/year)	17 500
Contract length	10 years

Metcan's Costs	
First cost in 1999 (\$)	175 000
Operating cost in 2000 (\$)	332 500
Actual dollar MARR before tax	20%

(a) Find the present worth of the contract under the assumption that there is no inflation over the life of the contract.

(b) Find the present worths of the contract under four assumptions:

Inflation is 1% per year.
Inflation is 2% per year.
Inflation is 3% per year.
Inflation is 4% per year.

(c) Should Metcan accept the contract?

9.29 Clarkwood is a British Columbia wood products manufacturer. They are considering a modification to their production line that would enable an increase in their output. One of Clarkwood's concerns is that the price of wood is rising more rapidly than inflation. They expect that because of this their operating cost per unit will rise at a rate 4% higher than the rate of inflation. That is, if the rate of inflation is f, Clarkwood's operating cost will rise at the rate $f_c = 1.04(1 + f) - 1$. However, competitive pressures from plastics will prevent the prices of their products from rising more than 1% above the inflation rate. The particulars of the project are shown in the table below.

Clarkwood's Project	
Output price in 2000 ($/unit)	30
Price increase	2% above inflation
Operating cost in 2000 ($/unit)	24
Operating cost increase	4% above inflation
Expected output due to project (units/year)	50 000
First cost in 1999 ($)	900 000
Observed actual dollar MARR	0.25
Time horizon (years)	10

(a) Find the present worth of the project under the assumption of zero inflation.

(b) Find the present worth of the project under these assumptions:

The expected inflation rate is 1%.

The expected inflation rate is 2%.

(c) Should Clarkwood accept the project?

9.30 Smooth-Top is a British Columbia manufacturer of desktops. They are considering an increase of their capacity. Consulting engineers have submitted two routes to do this. (1) Install a new production line that would produce wood desktops finished with hardwood veneer. (2) Install a new production line that would produce wood desktops finished with simulated wood made from hard plastic.

Smooth-Top is concerned about the price of hardwood veneer. They believe that the price of veneer will rise over the next ten years. However, they believe the price of veneer-finished desktops will rise by less than the rate at which the price of veneer rises. Information about the two potential projects is in the following table.

Smooth-Top Desktop Project	
Plastic-finish real price and real cost change	0%
Veneer-finish expected real price change	1%
Veneer-finish expected real cost change	5%
Wood cost/unit ($)	12.5
Plastic cost/unit ($)	9
Wood price/unit ($)	32
Plastic price/unit ($)	26
Wood first cost ($)	2 050 000
Plastic first cost ($)	2 700 000
Wood output rate (units/year)	30 000
Plastic output rate (units/year)	45 000
Study period	10 years
Actual dollar MARR	25%

(a) Compute the present worth of each option under the assumption that the real price of hardwood-finished desktops and real cost of hardwood veneer do not change (rather than as stated in the table). Assume zero inflation.

(b) Compute the present worth of each option under the assumption that the real price of hardwood-finished desktops and the real value of hardwood veneer desktop operating costs increase as indicated in the table. Assume that inflation is expected to be 2% over the study period.

9.31 Belmont Grocers has a distribution centre in Fredericton. The manual materials-handling system at the centre has deteriorated to the point at which it must be either replaced or substantially refurbished. Replacement with an automated system would cost about $240 000. Refurbishing the manual system would cost about $50 000. In either case, capital expenditures would take place this year. Operating either the new system or the refurbished system would begin next year. It is expected that either the new system or the refurbished system will operate for ten years with no further capital expenditures. Belmont is concerned that labour costs in Fredericton may rise in real terms over the next ten years. The range of increases in real terms that appears possible is from 4% to 7% per year. Inflation rates between 2% and 4% are expected over the next ten years. Complete data on the two alternatives are in the table below.

Materials Handling Data	
Automated expected real operating cost change	0%
Manual expected real operating cost change	4% to 7%
Manual operating cost/unit (first year of operation) ($)	10.5
Automated operating cost/unit (first year of operation) ($)	9
Manual first cost ($)	50 000
Automated first cost ($)	240 000
Output rate (units/year)	15 000
Actual dollar MARR	20%
Study period	10 years
Possible inflation rates	2% to 4%

(a) Find the total costs per unit for each of the two alternatives under the assumption of zero inflation and no increase in costs for the manual system.

(b) Make a recommendation as to which alternative to adopt. Base the recommendation on the present worth of costs for the two systems under various assumptions concerning inflation and the rate of change in the real operating cost of the manual system. Explain your recommendation.

9.32 The United Gum Workers have a cost-of-living clause in their contract with Mont-Gum-Ery Foods. The contract is for two years. The contract states that, if the inflation rate in the first year exceeds 1%, wages in the second year will increase by the inflation rate of the first year. Does this clause increase or decrease risk? Explain.

9.33 Free Wheels Manitoba has a plant that assembles bicycles in Louisbourg, Manitoba. The plant now has a small cafeteria for the workers. The kitchen equipment is in need of substantial overhaul. Free Wheels has been offered a contract by Besteats to supply food to their workers. The particulars of the situation are shown in the table. Should Free Wheels continue with the in-house food service or contract the service to Besteats?

Food Service: In-House Versus Contract	
Food service labour (hours/year)	6 000
Wage rate (real, time 1, $/hour)	7.5
Overhead cost (real, time 1, $/year)	18 000
Kitchen equipment first cost (actual, time 0, $)	25 000
Contract cost, years 1 to 3 (actual $)	55 000
Contract cost, years 4 to 6 (actual $)	63 700
Actual dollar MARR	22%
Expected annual inflation rate	5%
Study period (years)	6

Economic Comparision of High Pressure and Conventional Pipelines: Associated Engineering

CANADIAN 9.1 MINI-CASE

Associated Engineering of Toronto conducted an evaluation of sources of water supply for an Ontario municipality. One of the considerations was the choice of high-pressure or conventional pipelines for transmitting treated water from one of the Great Lakes to the municipality.

Conventional pipelines, most often made of concrete, have a limited maximum tensile strength, which for analysis purposes was taken to be 200 psi (pounds per square inch). High-pressure pipe, made of steel, can withstand up to 60 000 psi, although the pipe examined by Associated Engineering had a strength of 42 000 psi.

The major advantage of the steel pipe is that fewer pumping stations are needed than with the concrete pipe. The distance to be pumped is 85 kilometres; this requires either one pumping station for high-pressure pipe or six pumping stations for concrete pipe.

Each pipeline type was analyzed over a range of pipeline diameters ranging from 24″ to 72″. Construction costs included the pipe, pumping stations, and a reception reservoir, with the time of the cost taken to be the commissioning date of 2025. Operating and maintenance costs starting in 2026 were included, and administration, engineering fees, contingencies, and taxes were also accounted for.

Real 1993 dollars were used; an inflation rate of 2% was assumed for the period of study. The best alternative was chosen on the basis of a present worth comparison with a 4% discount rate. The result was that a 36″-diameter high-pressure pipeline was economically best, at a present cost $7.5 million lower than for the best conventional pipeline.

continued

Discussion

Estimating future inflation is difficult. The average inflation in Canada over the last 50 years has been about 4%, but over the last five years it has been only 1 or 2%. For some five-year periods, inflation has averaged over 10%. How can we estimate future inflation?

One way is to simply assume that inflation will remain at the current value. This is probably wrong; as has been seen, inflation typically changes over time. However, there are factors that are controlling the inflation rate. Lacking knowledge of any reason why these controlling factors might change, the current rate seems to be a reasonable choice.

A second approach is to use the long-term average. Knowing that inflation will change over time suggests that the long-term average is a good choice even if inflation is lower or higher than the average right now. After all, those controlling factors have changed in the past and are likely to change again.

A third way is to take into account the controlling factors for inflation. These include government policy: a government committed to social welfare is likely to induce more inflation than one committed to fiscal responsibility. Trends in business and consumer behaviour affect inflation: large labour contract increases presage inflation, as does high consumer borrowing. Social trends like the aging of the baby boomers also has an effect on inflation.

Understanding the effect of the controlling factors for inflation in detail is very difficult. So usually decision makers make a broad judgement based on both the current inflation rate and the historical average, perhaps informed by a general understanding of the contributing factors.

Questions

1) How significant would the difference have been to the savings of the high-pressure pipeline if an inflation rate of 4% had been used instead? Assume the only difference between the concrete- and steel-pipe systems was the capital cost, expended in 2025. Would the decision be any different? Could it be any different for any assumed inflation rate?
2) Design two cash flow structures for projects that start in 2025 so that the present worth (in 2000) at a discount rate of 4% is higher for one project than the other at an inflation rate of 2%, but lower at an inflation rate of 4%. Is there a significant opportunity to control the best choice in a decision situation by selecting the appropriate inflation rate?
3) Why would the analysts have chosen to separate the inflation rate from the discount rate for this problem, rather than combining them to an actual dollar discount rate? Do you think the analysts estimated the actual dollar cost of the alternatives in 2025, or would they have used the real costs?

Appendix 9A: Computing a Price Index

We can represent changes in average prices over time with a **price index**. A price index relates the average price of a given set of goods in some time period to the average price of the same set of goods in another period. Commonly used price indexes work with weighted averages, because simple averages do not reflect the differences in importance of the various goods and services in which we are interested.

Many different ways of weighting changes in prices may be used, and each method leads to a different price index. We shall discuss only the most commonly used index, the **Laspeyres price index**. It can be explained as follows:

Suppose there are n goods in which we are interested. We want to represent their prices at a time, t_1, relative to a **base period**, t_0, the period from which the expenditure shares are calculated.

The prices of the n goods at times t_0 and t_1 are denoted by $p_{01}, p_{02}, \ldots, p_{0n}$ and $p_{11}, p_{12}, \ldots, p_{1n}$. The quantities of the n goods purchased at t_0 are denoted by $q_{01}, q_{02}, \ldots, q_{0n}$. The share, s_{0j}, of good j in the total expenditure for the period, t_0, is defined as

$$s_{0j} = \frac{p_{0j}q_{0j}}{p_{01}q_{01} + p_{02}q_{02} + \ldots + p_{0n}q_{0n}}$$

Note that

$$\sum_{j=1}^{n} s_{0j} = 1$$

A Laspeyres price index, π_{01}, is defined as a weighted average of relative prices.

$$\pi_{01} = \left(\frac{p_{11}}{p_{01}} s_{01} + \frac{p_{12}}{p_{02}} s_{02} + \ldots + \frac{p_{1n}}{p_{0n}} s_{0n} \right) \times 100$$

The term in the brackets is a weighted average because the weights (the expenditure shares in the base period) sum to one. The relative prices are the prices of the individual goods in period t_1 relative to the base period, t_0. The weighted average is multiplied by 100 to put the index in percentage terms.

Example 9A.1

A student uses four foods for hamburgers: (1) ground beef, (2) hamburger buns, (3) onions, and (4) breath mints. Suppose that, in one year, the price of ground beef fell by 10%, the price of buns fell by 1%, the price of onions increased by 5%, and breath mints rose in price by 50%.

The price and quantity data for the student's hamburger are shown in Table 9A.1.

Table 9A.1

	Quantity at t_0	Price at t_0 (\$)	Price at t_1 (\$)
Ground beef (kg)	0.25	3.5/kg	3.15/kg
Buns	1	0.40	0.396
Onions	1	0.20	0.21
Breath mints	1	0.10	0.15

The Laspeyres price index is calculated in four steps:

1. Compute the base period expenditure for each ingredient.
2. Compute the share of each ingredient in the total base period expenditure.
3. Then compute the relative price for each ingredient.
4. Use the shares to form a weighted average of the relative prices.

These computations are shown in Table 9A.2.

Table 9A.2

	Price at t_0 (\$)	Share at t_0	Relative Price	Weighted Relative Price
Ground beef (kg)	0.875	0.556	0.900	0.500
Buns	0.400	0.254	0.990	0.251
Onions	0.200	0.127	1.050	0.133
Breath mints	0.100	0.063	1.500	0.095
Sums	1.575	1.000		0.980

As an example of the computations, the price of the ground beef per hamburger at t_0 is found by multiplying the price per kilogram by the weight of the hamburger used: \$3.50/kg $\times$ 0.25 kg = \$0.875. Similar computations for each of the other ingredients lead to a total cost of \$1.575 per hamburger. The ground beef then represents a share of 0.875/1.575 = 0.556 of the total cost. The relative price for the hamburger is 3.15/3.5 = 0.9 and thus the weighted relative price is 0.556 $\times$ 0.9 = 0.50. Similar computations for the other ingredients lead to a total weighted average of 0.98. After multiplying by 100, the Laspeyres price index is 98 (it is understood that this is a percentage). Therefore, the cost of the hamburger ingredients at t_1 was 2% lower than in the base period. ■

Statistics Canada compiles many Laspeyres price indexes. The consumer price index (CPI) is a Laspeyres price index in which the weights are the shares of urban consumers' budgets in the base year, currently 1992. Another well-known Laspeyres price index is the GNP deflator. For the GNP deflator, the weights are the shares of total output in the base year (also 1992).

The CPI and the GNP deflator are global indexes in that they represent an economy-wide set of prices. As well, Statistics Canada produces sector Laspeyres price indexes. For example, there are price indexes for durable consumer goods, for exports, and for investment by businesses. It is up to the analyst to know the composition of the different indexes and to decide which is best for his or her purpose.

Example 9A.2

We can classify consumer goods and services into four classes: durable goods, semi-durable goods, non-durable goods, and services. Groups of these goods and services were formed. They had the following prices in 1986 and 1993:

Category	1986	1993
Durable	2.421	2.818
Semi-durable	2.849	3.715
Non-durable	4.926	6.404
Services	4.608	6.263

Quantities in 1986 were:

Quantity in 1986 (units)	
Durable	21.304
Semi-durable	11.315
Non-durable	19.159
Services	31.422

Find the Laspeyres price index for 1993 with 1986 as a base.

We first calculate the relative prices.

	Price		
Category	1986	1993	Relative Price
Durable	2.421	2.818	1.164
Semi-durable	2.849	3.715	1.304
Non-durable	4.926	6.404	1.300
Services	4.608	6.263	1.359

We next determine expenditure shares in 1986.

Category	Expenditure	Share
Durable	51.583	0.1597
Semi-durable	32.235	0.0998
Non-durable	94.381	0.2922
Services	144.801	0.4483
Total	323.000	

We then multiply the relative prices by the shares and sum. We get the index by multiplying the sum by 100. For example, the term for durable goods is given by 1.164(0.1597) = 0.186

Index	
Durable	0.186
Semi-durable	0.130
Non-durable	0.380
Services	0.609
Total	1.305

This gives a Laspeyres price index of 130.5. ■

C H A P T E R 10

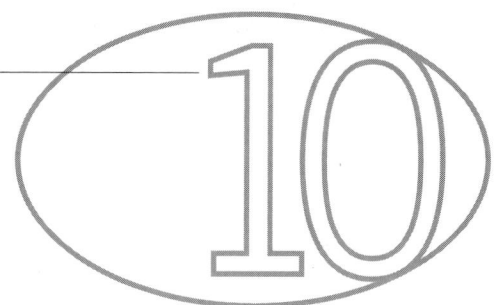

Public Sector Decision Making

Engineering Economics in Action, Part 10A: Trouble in Lotus Land

"Hi, Naomi. How's it going in Lotus Land?" Naomi could easily imagine Bill's feet up on his desk, leaning back in his chair, telephone wedged against his ear. Naomi was in Vancouver, checking out how British Columbia's regulations on absorbable organic halides (AOX) in effluent will affect the Edgemont Pulp Mill, a sister company of Canadian Widgets.

Naomi answered, "Lotus Land is great. I can look out my window and see green grass and rhododendrons with new buds. Not bad for late February. But the situation at the Edgemont Pulp Mill is not so bright. Basically, what I found is that the regulations are really tough. It's not clear that Edgemont can meet them and still be competitive in the bleached pulp export market."

"What's going on?" Bill asked. "We've spent over 40 million in the last few years making sure that mill is up-to-date. You're not telling me that all that's going down the drain, are you?"

"Well," Naomi began, "by the end of 2002 pulp mills must have AOX down to zero. That's the tough part. They will have to go to ozone and hydrogen peroxide bleaching rather than the chlorine bleaching method they use now. You can get the AOX level very low, say a half kilogram per tonne, by improving conventional chlorine bleaching. But you can't get it to zero without going to ozone and hydrogen peroxide. The cost of converting will be very high for Edgemont. Worse still, most of their competitors in other countries will not have such stringent regulations. Only Sweden is moving as fast as B.C. Of course, that could change because of pressure from environmental groups. But I don't think we should hold our collective breaths."

"Can we get the government to soften up on the regulations?" asked Bill.

"Maybe if we did a broadly focussed benefit-cost analysis. Can we do that?" asked Naomi.

"I'm not sure," Bill responded. "Let's explore the idea. Talk to the people at the mill today. I'll try to get some ideas from the people here. See if you can get a flight back this afternoon. I'll try to set up a meeting for us with Anna Kulkowski in the next couple of days. I suspect she won't want to quit on the Edgemont mill just yet."

10.1 Introduction

All companies, and the engineers who work for them, must take into account the effects of what they do on society as a whole. Consider these two examples.

1. MacMillan Bloedel (MB) is a British Columbia-based forest products company. In the spring of 1995, MB was faced with a campaign that included boycotts of their customers, demonstrations, and even a bomb threat against MB's headquarters. The campaign was a protest against MB's timber cutting practices in the Clayoquot area of Vancouver Island. The protesters were concerned about the impact of MB's clear-cutting on the ecological balance in Clayoquot. The object of the boycott was to pressure MB into changing its cutting practices. By the spring of 1995, the boycott had already cost MB a \$5-million contract to supply a British subsidiary of the Scott Paper Company, and other customers were being pressured to stop buying from MB.

In May 1995, MB announced that it would adhere to the recommendations of a panel set up by the British Columbia government. In June 1998, MB stated that it would phase out clear cutting. Whether or not MB's new cutting policy will be socially beneficial depends on how it is implemented.

2. In August 1982, the Manville Corporation had over 25 000 employees in the United States and Canada. It had over $2 billion in assets and stood 181st on the Fortune 500 list. Nevertheless, on August 26, 1982, the Manville Corporation filed for reorganization under the United States Bankruptcy Code. They did this to enable the company to continue operating in the face of claims that far exceeded the value of their assets. Manville was the defendant in over 16 000 lawsuits related to the health effects of asbestos, and they expected over 50 000 more lawsuits. Most of these suits came from men, or families of men, who had worked as installers of asbestos insulation made by Manville. These men had contracted asbestosis and other lung diseases as a result of long-term exposure to asbestos dust.

The MB and Manville examples illustrate a recent phenomenon that has important implications for engineers. It is not enough to produce goods and services at a cost that customers are willing to pay. Engineers must also pay attention to broader social values. This is because the market prices that guide most production decisions may not reflect all the social benefits and costs of engineering decisions adequately. Where markets fail to reflect all social benefits and costs, society uses other means of attaining social values.

In this chapter, we shall look at the social aspects of engineering decision making. First, we shall consider the reasons markets may fail in such areas as the environment and health. We shall also consider different methods that society may use to correct these failures. Second, we shall consider decision making in the public sector. Here we shall be concerned mainly with government projects or government-supported projects.

10.2 Market Failure

A **market** is a group of buyers and sellers linked by trade in a particular product or service. The prices in a market that guide most production decisions usually reflect all the social benefits and costs of engineering decisions adequately. This is not always true, however. Where prices do not reflect all social benefits and costs of a decision, we say that there has been market failure. When this occurs, society will seek a means of correcting for the failure. In this section, we define market failure and give examples of the effects of market failure. Then we discuss a number of ways in which society seeks remedies for market failure.

10.2.1 Market Failure Defined

Most decisions that lead to market behaviour have desirable outcomes. This is because these decisions affect mainly those people who are party to those decisions. Since people can generally freely choose to participate in markets, it is reasonable to assume that the individuals who participate must somehow benefit by their actions. In most cases, this is the end of the story.

In some cases, however, decisions have important effects on people who are not party to the decisions. In these cases, it is possible that the gains to the decision makers, and any others who might benefit from the decisions, are less than the losses imposed on those who are affected by the decisions. Such situations are clearly undesirable. These decisions are instances of market failure. **Market failure** occurs where decisions are made in which aggregate benefits to all persons who benefit from the decision are less than aggregate costs imposed on persons who bear costs that result from the decision. Market failure can occur if the decision maker does not correctly take into account the gains and losses imposed on others by the consequences of a decision.

There are several reasons for market failure. First, there may be no market through which those affected by the decision can induce the decision maker to take their situations into account. Or losses may exceed gains, even where there is a market, if the market gives the wrong signals about the gains and losses resulting from decisions. Market failure can even occur where someone decides *not* to do something that would create benefits to others. The market would fail if the cost of creating the benefits was less than the value of the benefits.

Acid rain is an example of the effects of market failure. The burning of high-sulphur coal by thermal-electric power plants is believed to be one of the causes of acid rain. These plants could burn low-sulphur fuels, but they do not, partly because low-sulphur fuels are more expensive than high-sulphur fuels. If a market existed through which those affected by acid rain could buy a reduction in power plant sulphur emissions, they could try to make a deal with the power plants. They would be able to offer the power plants enough to offset the higher costs of low-sulphur fuel and still come out ahead. But there is no such market. The reason for this is that there is no single private individual or group whose loss from acid rain is large enough to make it worthwhile to offer the power plants payment to reduce sulphur emissions. It would require a large number of those affected by acid rain to form a coalition to make the offer. There are markets for electric power and for coal. However, they do not lead to socially desirable decisions about the use of power or coal. This is because the market prices for power and coal do not reflect the costs related to acid rain. If the prices for power and high-sulphur coal reflected these costs, less power would be used, and less of it would be made with high-sulphur coal. Both would reduce acid rain.

The health damage to Manville's asbestos-insulation workers is another example of the effects of market failure. There was market failure in this case in that the health costs to the installers probably far outweighed all the benefits of asbestos insulation to all parties. The problem was not that the work was dangerous. Some jobs are inherently dangerous, but their objectively estimated expected benefits are greater than their expected costs. The market failed in this case because the installers of asbestos insulation whose health was damaged did not have the information necessary to evaluate the risks of this type of work. The insulation producers did not actually have the relevant information either. They could have obtained the information at moderate cost. The main gap in knowledge was the actual level of exposure of installers to asbestos dust under various conditions. The companies made no effort to get the information. Had they done so, they would have found that installers' exposure exceeded industry standards. If the insulation producers had warned the workers of the dangers involved, the market would not have failed. The workers probably would not have agreed to work as installers without higher pay and better protection against exposure to dust. In this case, if the companies provided higher wages and better protection against dust, and if the workers agreed to work, we can assume that the objectively estimated expected benefits of the deal would have been greater than expected costs for both sides.

We can see how market failure has caused socially undesirable outcomes such as acid rain and health problems for asbestos workers. When markets fail, as in cases such as these, society will seek to remedy these problems through a variety of mechanisms. These remedies are the subject of the next section.

10.2.2 Remedies for Market Failure

There are four main formal methods of eliminating or reducing the impact of market failure. They are:

1. Economic regulation by government
2. Monetary incentives or monetary deterrents
3. Permitting persons or companies who are adversely affected by the actions of others to seek compensation in courts
4. Government provision of goods and services

We shall discuss the first three methods in this section. Government provision of goods and services is discussed in the next section under decision making in the public sector.

The first and most common means of trying to deter or reduce the effects of market failure is *economic regulation*. Economic regulation is the imposition of rules by government that are intended to modify behaviour. The rules are backed by the use of penalties for failure to obey the rules. We have regulations concerning such widely differing areas as product labelling, automobile emissions, and the use of bodies of water to dispose of waste.

A challenge associated with developing economic regulations is that they may be inefficient. For example, suppose that we wish to improve the quality of a lake by reducing the amount of effluents dumped into it. These effluents may contain material with excessive biological oxygen demand (BOD). A typical regulation to control dumping would require all those who dump to meet the same BOD standards. But the costs of meeting these standards are likely to differ among the producers of the effluent. To attain a regulated reduction in BOD in their effluent, some producers will have to make expensive changes to their procedures, while others can respond with a lower cost. The most efficient way to obtain the reduction in BOD in the lake is to have those with low effluent-cleaning costs make the greatest reduction in BOD. Uniform regulation is not likely to do this.

A second means of overcoming market failure is to offer *monetary incentives* or *deterrents* to induce desired behaviour. Monetary incentives or deterrents, often more efficient than economic regulation, may be subsidies or special tax treatments. For example, referring back to effluent dumping, there could be subsidies for the installation of equipment that would reduce dumping, or a charge for dumping effluent. This way, producers for whom the cost of reducing BOD is low will do so since this would be cheaper than paying fees. By setting an appropriate price, the desired reduction can be attained.

A third means of reducing the effects of market failure is *litigation*. The use of courts as a means of reducing the health and safety effects of market failure has grown in both Canada and the United States. Since the 1970s in Canada (the exact year depending on the province), courts have held that regular sellers of a product implicitly guarantee that the product is fit for reasonable use. Where the cost of reducing a risk in the use of their product is less than the objectively estimated expected loss, sellers are supposed to reduce the risk. Moreover, these sellers are held legally responsible for having expertise in the production and use of the products. It is not enough for sellers to say they did not know that use of the product was risky. Sellers are supposed to make reasonable efforts to determine potential risks in the use of products they sell. This is the basis of Manville's loss in the suits against them.

The fourth formal method of reducing the effects of market failure is *government provision of goods and services*. This provision may be direct, as in the case of police services, or indirect, as in the case of health care given by physicians. Market failure is remedied when public sector analysts take into account all parties affected by a decision through a comprehensive assessment of total costs and benefits of a decision. Health care provision, transportation, municipal services and electric and gas utilities are some examples of goods and services provided by the public sector. Each service requires numerous economic decisions to be made in the best interests of the public. This is of sufficient importance for us to devote a separate section of this chapter to the topic.

In addition to formal methods for dealing with market failure, *informal methods* can be used. For example, the boycotts against MB mentioned in the introduction of this chapter may have been an effective means of getting MB to alter its clear-cutting behaviour.

10.3 Decision Making in the Public Sector

This section is devoted to the decision-making process for public provision of goods and services. Public (or government) production in Canada has occurred mainly in two classes of goods and services. The first class includes those services for which there is no market because it is not practical to require people to pay for the service. Police and fire protection, defence, and the maintenance of city streets are examples of government services where it is not practical for users to pay for use.

The second class includes those services for which scale economies make it inefficient to have more than one provider. Where there is only a single provider of a service, there is no market competition to enforce efficiency and low prices. There is a danger that the single provider, called a monopolist, will charge excessive prices and/or be inefficient. To ameliorate this potential problem, governments are often the provider. It is also possible for the government to monitor and regulate the performance of a private monopolist.

For example, local deliveries of natural gas and electric power are situations where economies of scale are important enough that having more than one provider is not efficient. These services may be provided by publicly owned or privately owned monopolies. For example, Toronto Hydro is a publicly owned company that distributes electric power in the Toronto area. Centra Gas Manitoba is a privately owned gas distribution company whose prices and quality of service are regulated by the Manitoba Public Utilities Commission.

In this section, we shall consider three major issues concerning engineering projects where there is government provision of services. These issues are

(1) the valuation of costs and benefits where government provides services,

(2) the use of benefit-cost ratios to evaluate government projects, and

(3) the choice of the MARR for government projects.

10.3.1 Measuring the Costs and Benefits of Government Services

The most important and frequently the most difficult part of evaluating government projects is measuring costs and benefits in the government sector. There are two reasons: First, it may be difficult to identify all the costs and benefits. Second, actual measurement of the costs and benefits that have been identified is frequently difficult.

Identifying all the costs and benefits associated with a project may be difficult because some of the costs and benefits of public projects may not be reflected in the monetary flows of the project. We are concerned with the real effects of a project, but the cash flows may or may not reflect all these real effects. For example, in building a road, there are cash flows for the wages of the workers who construct the road. This is a real cost, and the wages reflect these costs well. In contrast, consider the cost of disruption of traffic during road construction. These costs are not reflected in the cash flows of a project, but are, nonetheless, an important cost of putting the road in place.

There may be cash flows associated with a project that reflect, from a social point of view, neither costs nor benefits. Consider the case of a toll for the use of a new road. The revenue from the tolls goes to the government, which it uses to reduce taxes. This reduction in taxes can be counted as a benefit to taxpayers as a whole, but there is an equal offsetting cost to those who pay the tolls. From a social point of view, the tolls represent neither a gain nor a loss.

Close-Up 10.1 describes BOOT projects, in which the private sector companies invest in the public sector projects and recoup their investments through user-based tolls, fees, or tariffs. BOOT projects seem to represent the latest trend in major Canadian public sector projects.

Another reason that measuring costs and benefits of government services may be difficult is that there may be no prices to reflect their values. With goods and services that are distributed through markets, we have prices to measure values. This may not be the case with public services. Consider the following three examples.

Example 10.1

A major improvement in a highway near an urban area, like Highway 401 which goes through Toronto, is being considered. What are the costs and benefits? How are they measured?

Obvious costs are the labour, materials, and equipment used for the project. But one major cost is not explicitly measured. This is the cost of traffic disruption during the work. Measuring this cost requires an evaluation of the time delays incurred by car passengers and trucks. There are approximations to these delay costs based on earnings per hour of passengers and the hourly cost of running trucks. The approximation for the value of travellers' time is based on the idea that a person who can earn, for example, \$35 an hour working should be willing to pay \$35 an hour for time saved travelling to work. The disruption costs may be large enough that they make it more efficient to have the work done at night, despite the fact that this will raise the explicit construction cost.

There is a long list of benefits from a highway improvement. These include reduced travel time, lower vehicle operating and maintenance costs, and improved safety. Similar to the calculation of the cost of traffic disruption, the value of reduced travel time can be estimated as earnings regained by travellers who otherwise could be working. Lower vehicle costs are based on crude approximations to the average cost under different road conditions. The value of improved safety may be difficult to estimate and is often qualitatively valued as an extra benefit in such projects. ■

CLOSE-UP 10.1 BOOT Projects

Traditionally, large-scale public infrastructure projects, such as roads, bridges, power plants, and public utilities, have been financed, built, and maintained by government. In a Build-Own-Operate-Transfer (BOOT) scheme, a private company or consortium builds such large-scale infrastructure, and owns and operates it for a certain concessionary period (e.g., 20 years). During this period, the investment made by the private company is recovered through user-based tolls, fees, sales, or tariffs. After the concession period is over, the ownership and operational responsibilities are transferred to the government. BOOT projects are designed to reduce the taxpayers' financial burden, improve efficiency, and stimulate innovation by letting the private sector directly invest in large-scale public projects. The benefits of BOOT projects, however, still remain in theory since, to date, no major BOOT projects have successfully completed all stages of the cycle.

BOOT projects have been gaining global recognition in the past decade or so. In Canada, the Confederation Bridge in Prince Edward Island and Highway 407 in Ontario are two examples of recent BOOT projects.

Confederation Bridge: The Confederation Bridge, officially opened in May 1997, links Prince Edward Island and New Brunswick over the Northumberland Strait. This 13-kilometre-long, high-level, $840-million bridge serves as an all-weather roadway between the island and the mainland, providing cost savings over the subsidised ferry operation. The bridge was built and is currently owned by a consortium formed by Strait Crossing Inc. of Calgary. State-of-the-art high-performance concrete was employed in the construction. The service life of the bridge is 100 years, and Public Works Canada will take over ownership in 35 years. Currently, a round-trip toll is charged to all vehicles.

407 ETR: 407 Express Toll Route (ETR) is a multi-lane highway that runs 69 kilometres across southern Ontario's busiest traffic region between Highway 403 in Oakville and Highway 48 in Markham. The estimated traffic congestion-related costs in Ontario were $2 billion per year prior to the opening of the 407 ETR in June 1997. The Canadian Highways International Corporation (CHIC), a consortium of Ontario-based firms, designed and built the highway in four years. 407 ETR employs the world's first all-electronic toll-collection system. The traffic projections in the first year of operation are 110 000 trips per day with an average revenue of approximately $6 million per month. The highway is expected to pay for itself in 25 years.

Sources: **http://138.25.138.94/signposts/index.html**
http://www.stantech.com/transportation/index.htm
http://www.407etr.com

Example 10.2

What are the costs and benefits of a university education? There are obvious costs for such items as textbooks, tuition fees, and so on. The major benefit is the increased productivity of students. Are there other costs and benefits? How are these costs and benefits measured?

A major cost that is not explicitly measured is the value of students' time. This is usually measured by foregone earnings. That is, study is viewed as a form of work. This work could have earned a certain amount that is foregone when the time is spent at university. When student time is measured by foregone earnings, it accounts for over half of the costs of university education in Canada. Measuring foregone earnings requires estimating what the students could have earned during the time they were in university had they not gone to university.

Increased productivity of university graduates is difficult to measure. The measurement is based on the idea that students would have worked with or without education. If they have a university education, their productivity is, presumably, greater and this will result in greater output, which, in turn, will yield higher incomes for students and their employers. The actual measurement of increased productivity is usually based on the assumption that students capture the entire benefit of increased productivity in higher incomes. Benefit measurement then consists of estimating what students could have earned over their whole lifetimes without the education versus what they are expected to earn over their lifetimes with a university education. The benefit stream is then estimated as the difference between these two earnings streams.

Another benefit of publicly supported university education is the social benefit of having a society in which there are opportunities for self advancement for large parts of the population. It is not possible to put a reasonable monetary value on this benefit, but it may be the main justification for subsidizing university education in Canada. There is no obvious reason for public subsidy to increase the productivity of students. The students, and the companies for which they will work, should be willing to pay for the increased productivity. ■

Example 10.3

Consider the construction of a bridge across a narrow part of a lake that gives access to a provincial park. The major benefit of the bridge will be reduced travel time to get to the park from a nearby urban centre. This will lower the cost of camping trips at the park. As well, more people are expected to use the park because of the lower cost per visit. How can these benefits be measured?

Data concerning the number of week-long visits and their costs are shown in Table 10.1.

Table 10.1 Average Cost per Visit and Number of Visits per Year

	Without Bridge	With Bridge
Travel cost per visit ($)	140.00	87.50
Use of equipment ($)	50.00	50.00
Food cost per visit ($)	100.00	100.00
Total cost per visit ($)	290.00	237.50
Number of visits/year	8000	11 000

First, the reduction in cost for the 8000 visits that would have been made even without the bridge creates a straightforward benefit:

$$\text{Travel cost saving on 8000 visits} = (\$140 - \$87.50) \times 8000 = \$420\,000$$

There is a benefit of $420 000 per year from reduced travel cost on the 8000 visits that would have been taken even without the bridge.

Next, we see that the number of visits to the park is expected to rise from 8000 per year to 11 000 per year. But how much is this worth? We do not have prices for park visits, but we do have data that enable estimates of actual costs to campers. These costs may be used to infer the value of visits to campers.

We see that before the bridge, the cost of a week-long park visit, including travel and other costs, averaged $290. It is reasonable to assume that a week spent camping was worth at least $290 to anyone who incurred that cost. The average cost of a week-long visit to a park would fall from $290 per visit to $237.50 per visit if the bridge were built. We are concerned with the value of the incremental 3000 visits per year. Clearly, none of these visits would be made if the cost were $290 per trip. And each of them is worth at least $237.50 or else the trip would not have been taken. The standard approximation in cases like this is halfway between the highest and lowest possible values. This gives an aggregate benefit of the increased use of the park of

$$\frac{(\$290.00 + \$237.50)}{2} \times 3000 = \$791\,250$$

Therefore, the value of the incremental 3000 visits per year is estimated as approximately $791 000 per year. However, there is also a cost of $237.50 per visit. The net benefit of the incremental 3000 visits is therefore

$$\$791\,250 - \$237.50(3000) = \$78\,750$$

The total value of benefits of the bridge is the sum of the reduced travel cost plus increased use:

$$\$420\,000 + \$78\,750 = \$498\,750$$

Figure 10.1 Benefits from Bridge

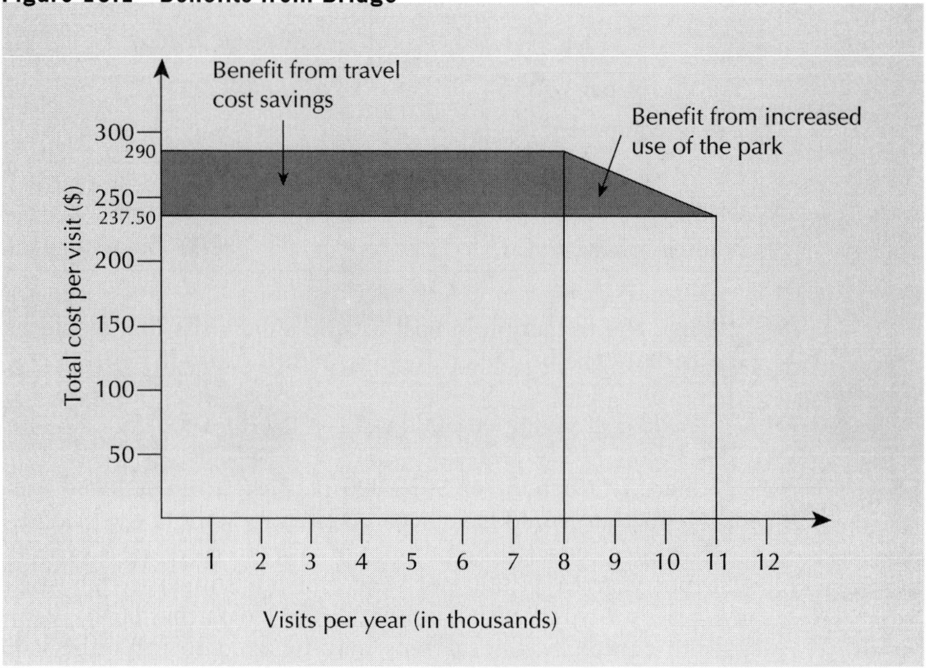

The total value of the benefits yielded by the bridge is almost $500 000 per year. These benefits are the shaded area shown in Figure 10.1. These benefits need to be weighed against the costs of the bridge. ■

10.3.2 Benefit-Cost Ratios

The same comparison methods that are used for private sector projects are appropriate for government sector projects. That is, we can use the present worth, annual worth, and internal-rate-of-return performance measures in the same ways in both the private and government sectors. It is important to emphasize this, because other methods based on ratios of benefits to costs have been used frequently in government project evaluations, almost to the exclusion of present worth, annual worth, and internal-rate-of-return methods. Because of the prevalent use of these ratios, this section is devoted to a discussion of several benefit-cost ratios that are commonly used in public sector decision making. We then point out several problems associated with the use of benefit-cost ratios so that the reader is aware of them and understands the correct and incorrect ways of using them.

Benefit-cost ratios can be based on either the present worths or the annual worths of benefits and costs of projects. The resulting ratios are equivalent in that they lead to the same decisions. We shall discuss the ratios in terms of present worths, but the reader should be aware that everything we say about ratios based on present worths applies to ratios based on annual worths.

The conventional **benefit-cost ratio**, denoted BCR, is given by the ratio of the present worth of benefits to the present worth of costs for a project, that is,

$$\text{BCR} = \frac{\text{PW(benefits)}}{\text{PW(costs)}}$$

A **modified benefit-cost ratio**, also in common use, denoted BCRM, is given by the ratio of the present worth of benefits minus the present worth of operating costs to the present worth of capital costs, that is,

$$\text{BCRM} = \frac{\text{PW(benefits)} - \text{PW(operating costs)}}{\text{PW(capital costs)}}$$

In general, a project is considered desirable if its benefit-cost ratio exceeds one, which is to say, its benefits exceed its costs. If we hope to go a step further and compare two projects with the use of benefit-cost ratios, the comparison becomes tricky because the ratios are not unique. This is a significant problem with the use of benefit-cost ratios. Benefit-cost ratios (either conventional or modified) are not unique because it is not clear whether certain positive effects of projects are benefits or reductions in cost, or whether certain negative effects are reductions in benefits or increases in costs. For this reason the Treasury Board of Canada's *Benefit-Cost Analysis Guide*, which was written for internal federal government use, recommends that benefit-cost ratios not be used for project comparisons. Despite this recommendation, benefit-cost ratios are still used for many public projects. Since they are still in use, we present readers with enough material in this section so that they may properly construct and interpret benefit-cost ratios and may understand the problems associated with their use.

To illustrate the ambiguity associated with the use of a benefit-cost ratio, consider this example adapted from the Treasury Board's *Guide*. The example is concerned with a *negative effect* of a project that may be classified as a reduction in benefit or increase in cost.

Example 10.4

The town of Helen Lake, Manitoba, has limited parking for cars near its main shopping street. The town plans to pave a lot for parking near Main Street. The main beneficiaries will be the merchants and their customers. The present worth of expected benefits is \$3 000 000. The cost of buying the lot, clearing, paving, and painting is expected to be \$500 000. The present worth of expected maintenance costs over the lifetime of the project is \$50 000. In the period in which the lot is being cleared and paved, there will be some disruption of traffic in the Main Street area. The value of the disruption is expected to be about $75 000. What is the benefit-cost ratio?

The answer depends on whether the effect of the disruption is counted as a reduction in benefits or an increase in costs. If we consider the disruption to be a reduction in benefits, the benefit-cost ratio is

$$BCR_1 = \frac{\$3\ 000\ 000 - \$75\ 000}{(\$500\ 000 + \$50\ 000)} = 5.3$$

If we consider the disruption to be a cost, the benefit-cost ratio is

$$BCR_2 = \frac{\$3\ 000\ 000}{(\$500\ 000 + \$50\ 000 + \$75\ 000)} = 4.8 \ ■$$

We see from this example that some ambiguity arises in the use of benefit-cost ratios when a negative effect exists. The ambiguity can give rise to different values for the benefit-cost ratios, depending on how the negative effect is classified.

A similar problem arises with the classification of *positive effects* of a project. Example 10.5 concerns a positive effect that may be classified as a benefit or a reduction in capital cost.

Example 10.5

A fire department in a medium-sized city is considering a new dispatch system for its fire-fighting equipment responding to calls. The system would select routes, taking into account recently updated traffic conditions. This would reduce response times. A side effect would be a reduction in required fire-fighting equipment. Ignore the value of reducing equipment requirements for the moment. The present worth of benefits and costs are

PW(benefits) = $37 500 000
PW(operating costs) = $3 750 000
PW(capital costs) = $13 500 000

The present worth of reducing fire-fighting equipment requirements is $2 250 000. What is the modified benefit-cost ratio?

The answer depends on how the reduction in equipment requirements is treated. If this is seen as a benefit, we get a modified benefit-cost ratio of 2.67 as follows.

$$BCRM_1 = \frac{(\$37\ 500\ 000 + \$2\ 250\ 000 - \$3\ 750\ 000)}{\$13\ 500\ 000} = 2.67$$

If the effect of the reduction in equipment requirements is treated as reduction in capital cost, we get a modified benefit-cost ratio of 3.

$$BCRM_2 = \frac{(\$37\ 500\ 000 - \$3\ 750\ 000)}{(\$13\ 500\ 000 - \$2\ 250\ 000)} = 3.00 \ ■$$

This lack of uniqueness of benefit-cost ratios means that *a comparison of the benefit-cost ratios of two projects is meaningless.* That is, if we compare the benefit-cost ratios of two projects, A and B, we may find that project A has a higher benefit-cost ratio. However, there may be another way of classifying some of the negative effects of the two projects as costs or reductions in benefits. As well, there may be another way of classifying the positive effects as benefits or as reductions in costs. The different classifications can give project B a higher benefit-cost ratio.

The lack of uniqueness of benefit-cost ratios does not mean that the ratios cannot be used. Comparison methods based on benefit-cost ratios remain valid because the comparison methods depend only on whether the benefit-cost ratio is less than or greater than one. Whether the benefit-cost ratio is greater than one or less than one does not depend upon how positive and negative effects are classified. This is clearly illustrated in the following example.

A certain project has present worth of benefits, B, and present worth of costs, C. As well, there is a positive effect with a present worth of d; the analyst is unsure of whether this positive effect is a benefit or a reduction in cost. There are two possible benefit-cost ratios,

$$\text{BCR}_1 = \frac{B + d}{C}$$

and

$$\text{BCR}_2 = \frac{B}{C - d}$$

For a ratio to exceed one, the numerator must be greater than the denominator. This means that, for BCR_1,

$$\text{BCR}_1 = \frac{B + d}{C} > 1 \quad \Leftrightarrow \quad B + d > C \quad \Leftrightarrow \quad B > C - d$$

But this is the same as

$$\frac{B}{C - d} = \text{BCR}_2 > 1$$

Consequently, any BCR that is greater than one or less than one will be so regardless of whether any positive effects are treated as positive benefits or as negative costs. A similar analysis would show that the choice of classification of a negative effect as a cost or as a reduction in benefits does not affect whether the benefit-cost ratio is greater or less than one.

Notice also that if the conventional benefit-cost ratio is greater (less) than one, it follows that the modified benefit-cost ratio will be greater (less) than one. This means that the two types of benefit-cost ratio will lead to the same decision. We present both because the reader may encounter either ratio.

For *independent* projects then, the following decision rule may be used: *Accept all projects with a benefit-cost ratio greater than one.* In other words, accept a project if

$$\text{BCR} = \frac{\text{PW(benefits)}}{\text{PW(costs)}} > 1, \text{ or}$$

$$\text{BCRM} = \frac{\text{PW(benefits)} - \text{PW(operating costs)}}{\text{PW(capital costs)}} > 1$$

This rule, using either the benefit-cost ratio or the modified benefit-cost ratio, is equivalent to the rule that all projects with a present worth of benefits greater than the present worth of costs should be accepted. This is, then, the same rule that was presented in Chapter 4, which accepts a project if its present worth is positive. Also recall that, as shown in Chapter 5, the present/annual worth method and the internal rate of return method give the same conclusion if an independent project has a unique IRR.

To use benefit-cost ratios to choose among *mutually exclusive* projects, we must evaluate the increment between projects. Suppose we have two mutually exclusive projects, X and Y, with present worths of benefits and costs given by B_X, B_Y, C_X, and C_Y. We first check to see if the separate benefit-cost ratios are greater than one. We discard a project with a benefit-cost ratio of less than one. If both projects have benefit-cost ratios greater than one, we rank the projects in ascending order by the present worths of costs. Suppose $C_X \geq C_Y$. We then form the ratio of the differences in benefits and costs,

$$\text{BCR(X} - \text{Y)} = \frac{B_X - B_Y}{C_X - C_Y}$$

If this ratio is greater than one, we choose project X. If it is less than one, we choose project Y. If $C_X = C_Y$ (in which case the ratio is undefined), we choose the project with the greater present worth of benefits. If $B_X = B_Y$, the ratio is zero, and we choose the project with the lower present worth of costs, project Y.

This rule is the same as comparing two mutually exclusive projects using the internal rate of return method, which was presented in Chapter 5. In order to choose a project, we saw in Chapter 5 that not only the IRR of the individual project, but also the IRR of the incremental investment, must exceed the MARR. This rule is also the same as choosing the project for which the value of the present worth is the largest, as presented in Chapter 4, provided the present worth is positive.

As was pointed out before, the classification of positive effects as benefits or cost reductions and negative effects as reductions in benefits or increases in costs does not affect the decision under the increment ratio rule.

The following example, concerning two mutually exclusive projects, summarizes our discussion of benefit-cost ratios and illustrates their correct use.

Example 10.6

A medium-sized city is considering increasing its airport capacity. At the current airport, flights are frequently delayed and congestion at the terminal has limited the number of flights. Two mutually exclusive alternatives are being considered. Alternative A is to construct a new airport 65 kilometres from the city. Alternative B is to enlarge and otherwise upgrade the current airport that is only 15 kilometres from the city. The advantage of a new airport is that there are essentially no limits on its size. The disadvantage is that it will require travellers to spend additional travel time to and from the airport.

There are two disadvantages for upgrading and enlarging the current airport. One disadvantage is that the increase in size is limited by existing development. The second disadvantage is that the noise of airplanes in a new flight path to the current airport will reduce the value of homes near that flight path. A thousand homes will be affected, with the average loss in value about \$25 000. Such losses are large enough so that it is not reasonable to expect the owners' losses will be offset by gains elsewhere. If the city wishes to ensure that their losses are offset, it must compensate these owners. Note that such compensation would not be an additional social cost. It would merely be a transfer from taxpayers to the owners. Benefit and cost data are shown in Table 10.2.

Table 10.2 Airport Benefits and Costs

Effect	Alternative A New Airport (millions of \$)	Alternative B Current Airport (millions of \$)
Improved service/year	55	28.5
Increased travel cost/year	15	0
Cost of highway improvements	50	10
Construction costs	150	115
Reduced value of houses	0	25

The city uses a MARR of 10%. It is using a 10-year time horizon for this project. What are the benefit-cost ratios for the two alternatives? Which alternative should be accepted?

Before we start the computation of benefit-cost ratios, we need to convert all values to present worths. Two effects are shown on an annual basis: improved service, which occurs under both alternatives, and increased travel cost, which appears only with the new airport. We get the present worths (in millions of \$) by multiplying the relevant terms by the series present worth factor:

$$\begin{aligned}\text{PW(improved service of A)} &= \$55(P/A, 10\%, 10)\\ &= \$55(6.1446) = \$337.95\end{aligned}$$

$$\text{PW(improved service of B)} = \$28.5(6.1446) = \$175.10$$

$$\text{PW(increased travel cost of A)} = \$15(6.1446) = \$92.17$$

The remainder of the costs are already in terms of present worth.

There is some ambiguity as to the correct benefit-cost ratios. For both airports, some effects could be counted as costs or as reduced benefits. For the new airport, it is not clear if we should count the increased travel time as a reduction of benefits or as a cost. For the current airport, it is not clear if we should count the reduction in the value of residential properties as a cost or as a reduction of benefits. We get two reasonable benefit-cost ratios for each alternative. They are shown in Table 10.3.

Table 10.3 Airport Benefit-Cost Ratios

Method of Counting Effect	Alternative A New Airport	Alternative B Current Airport
Count as cost	1.16	1.17
Count as benefit reduction	1.23	1.20

For example, the BCR for alternative A, counting the increased travel time as a cost, is

$$\text{BCR(alternative A)} = \frac{\$337.95}{(\$150 + \$50 + \$92.17)} \cong 1.16$$

and the BCR of alternative A, counting the increased travel time as a benefit reduction, is

$$\text{BCR(alternative A)} = \frac{(\$337.95 - \$92.17)}{(\$150 + \$50)} \cong 1.23$$

First, in Table 10.3, note that all benefit-cost ratios exceed one. Both alternatives are viable. We must, therefore, choose between the two alternatives. If we try to choose by ranking the alternatives by benefit-cost ratios, we are unable to choose. If we count the two effects as costs, improving the current airport has a slightly higher benefit-cost ratio. If we count the effects as reductions in benefits, the new airport has a slightly higher benefit-cost ratio. Neither of these comparisons gives much insight.

To decide which alternative is better, we must compute the benefit-cost ratio of the incremental investment between the alternatives.

$$\text{BCR (A} - \text{B)} = \frac{B_A - B_B}{C_A - C_B}$$

$$= \frac{\$337.95 - \$175.10}{(\$92.17 + \$50 + \$150) - (\$10 + \$115 + \$25)} = 1.15$$

The ratio of the benefits of the new airport minus the benefits of the current airport modification over the difference in their costs is greater than one. We interpret this to mean that the benefit-cost ratio ranks the new airport ahead of the current airport.

We also note that the same ranking would result if we were to compare the present worths of the two alternatives. The present worth of the new airport is \$45.78 = \$337.95 − \$92.17 − \$150 − \$50 (in millions of \$). The present worth of modifying the current airport is \$25.10 = \$175.10 − \$10 − \$115 − \$25 (in millions of \$).

In addition to the quantifiable aspects of each of the two airports, recall that the new airport is preferred in terms of the unmeasured value of space for future growth. As well, the new airport does not require compensations to home owners for loss in value of their homes. Together with the benefit-cost ratios (or equivalently, the present worths), this means that the city should build the new airport. ■

In summary of this section, we can say several things about the evaluation of public sector projects. First, the comparison methods developed in Chapters 4 and 5 are fully applicable to public sector projects. Second, the reader may encounter the use of benefit-cost ratios for public projects, despite the fact that benefit-cost ratios can be ambiguous. Next, the reader should be wary of decisions based on the relative sizes of benefit-cost ratios, as these ratios may be changed by reclassification of some of the positive effects of projects as benefits or as reductions in cost, or by reclassification of some of the negative effects as reductions in benefits or increases in costs. However, since the classifications do not affect whether a ratio is greater or less than one, it is possible to use benefit-cost ratios to give the same results that would come from use of the methods of Chapters 4 and 5.

10.3.3 The MARR in the Public Sector

There are significant differences between the private sector and public (government) sector of society with respect to investment. As observed in the previous sections, public sector organizations provide a mechanism for resources to be allocated to projects which are believed beneficial to society in general. These include projects for which scale economies make it inefficient to have more than one supplier, and projects in markets which would otherwise suffer market failure. The government also regulates markets and collects and redistributes taxes toward the goal of maximizing social benefits. Typical projects include health, safety, and education programs; cultural development; and infrastructure development. Profits generated by public projects are not taxed.

Private institutions and individuals, in contrast to public institutions, are more concerned with generating wealth (profits) and are taxed by the government on this wealth. We therefore would expect that the MARR for a public institution should be lower than for a private institution because the latter has a substantial extra expense acting to reduce its net profit.

In evaluating public sector projects, the MARR is used in the same way as in evaluating private sector projects — it captures the time value of money. In the private sector, the MARR expresses the minimum rate of return required on projects, taking into account that those profits will ultimately be taxed. The MARR used for public projects, often called the "social discount rate," reflects the more general investment goal of maximizing social benefits. Of course, as with private projects, public projects will sometimes be chosen on the basis of lowest present cost (or the equivalent) rather than highest present worth.

There is substantial debate as to what an appropriate social discount rate should be. Several viewpoints can help shed some light on this issue.

The first viewpoint is simply that the MARR should be the interest rate on capital borrowed to fund a project. In the public sector, funds are typically raised by issuing government bonds. Hence, the current bond rate might be an appropriate MARR to use. The second viewpoint is that the MARR used to evaluate public projects should take into account that government spending on public projects consumes capital that might otherwise be used by taxpayers for private purposes. This viewpoint says that funds taken away from individuals and private organizations in the form of taxes could be otherwise used to fund private projects. This line of thinking would lead to a social discount rate which is the same as the MARR for the taxpayers. The two viewpoints are lower and upper bounds, respectively, to what could be seen as reasonable MARRs to apply when evaluating public projects.

In practice, the two viewpoints are captured by the *Benefit-Cost Analysis Guide* (Federal Treasury Board of Canada) recommendation that public projects be evaluated at a 10% social discount rate and that the evaluation should include a sensitivity analysis at lower and upper bounds of 5% and 15% percent, respectively. (See Close-Up 10.2.)

CLOSE-UP 10.2 MARR Used in the Canadian Public Sector

Public projects in Canada are evaluated using various MARRs. Some examples of the MARR applied in the public sector are shown in the table below. Due to uncertainty about what the *correct* MARR is, sensitivity analysis is recommended for the public projects. For example, in the *Benefit-Cost Analysis Guide*, the Treasury Board suggests that one should use 10% as the base case and vary it in the range of 5% to 15%. Similarly, the Government of British Columbia suggests using 6% and 10% in addition to 8% in Land and Resource Management Planning projects. Finally, the Canadian Coordinating Office for Health Technology Assessment (CCOHTA) proposes the economic analysis using 0% and 3% as the MARR in addition to the base case of 5%.

Project Type	Government Level	MARR	Suggested Sensitivity Values
Benefit-Cost Analysis Guide	Treasury Board of Canada	10%	5–15%
Pharmaceutical	Canadian Coordinating Office for Health Technology Assessment (CCOHTA)	5%	0%, 3%
Land and Resource Management Planning	Government of British Columbia	8%	6%, 10%
Agricultural	Government of Alberta	13%	—
Assessment of Damages in Personal Injury and Fatal Accident Litigation	Provincial Governments (British Columbia, Saskachewan, Manitoba, Ontario, New Brunswick, Nova Scotia, Prince Edward Island)	2.5–3.5%	—

REVIEW PROBLEMS

REVIEW PROBLEM 10.1

This review problem is adapted from an example in the Treasury Board's *Benefit-Cost Analysis Guide*.

There are periodic floods in the spring and drought conditions in the summer that cause losses in a 15 000-square-kilometre Prairie river basin which has a population of 50 000 people. The area is mostly farmland, but there are several towns. Several flood control and irrigation alternatives are being considered:

1. Dam the river to provide flood control, irrigation, and recreation.
2. Dam the river to provide flood control and irrigation without recreation.
3. Control flooding with a joint Canada-United States water control project on the river.
4. Develop alternative land uses that would not be affected by flooding.

The constraints faced by the government are the following:

1. The project must not reduce arable land.
2. Joint Canada-United States projects are subject to delays caused by legal and political obstacles.
3. Damming of the river in the United States will cause damage to wildlife refuges.
4. The target date for completion is three years.

Taking into account the constraints, alternatives 3 and 4 above were eliminated, leaving two:

1. Construct a dam for flood control, irrigation, and recreation.
2. Construct a dam for flood control and irrigation only.

A number of assumptions were made with respect to the dam and the recreational facilities:

1. An earthen dam will have a 50-year useful life.
2. Population and demand for recreational facilities will grow by 3.25% per year.
3. A three-year planning and construction period is reasonable for the dam.
4. Operating and maintenance costs for the dam will be constant in real dollars.
5. Recreational facilities will be constructed in year 2.
6. It will be necessary to replace the recreational facilities every 10 years. This will occur in years 12, 22, 32, and 42. Replacement costs will be constant in real dollars.
7. Operating and maintenance costs for the recreational facilities will be constant in real dollars.
8. The real dollar opportunity cost of funds used for this project is estimated to be in the range of 5% to 15%.

The benefits and costs of the two projects are shown in Tables 10.4 and 10.5.

Notice that the benefits and costs are estimated averages. For example, the value of reduced flood damages will vary from year to year, depending on such factors as rainfall and snowmelt. It is not possible to predict actual values for a 50-year period.

(a) What is the present worth of building the dam only? What is the benefit-cost ratio? What is the modified benefit-cost ratio? Use 10% as the MARR.

(b) What is the present worth of building the dam plus the recreational facilities? Use 10% as the MARR.

(c) What is the benefit-cost ratio for building the dam and recreation facilities together? What is the modified benefit-cost ratio?

(d) Which project, 1 or 2, is preferred, based on your benefit-cost analysis? Use 10% as the MARR.

Table 10.4 Estimated Average Benefits of the Two Projects

Year	Flood Damage Reduction	Irrigation Benefits	Recreation Benefits
0	$ 0	$ 0	$ 0
1	0	0	0
2	0	0	0
3	182 510	200 000	27 600
4	182 510	200 000	28 497
⋮	⋮	⋮	⋮
52	182 510	200 000	132 288

Table 10.5 Estimated Average Costs of the Two Projects

Year	Dam Construction	Operating and Maintenance Dam	Recreation Construction	Operating and Maintenance Recreation
0	$ 300 000	$ 0	$ 0	$ 0
1	750 000	0	0	0
2	1 500 000	0	50 000	0
3	0	30 000	0	15 000
4	0	30 000	0	15 000
⋮	⋮	⋮	⋮	⋮
11	0	30 000	0	15 000
12	0	30 000	20 000	15 000
13	0	30 000	0	15 000
⋮	⋮	⋮	⋮	⋮
21	0	30 000	0	15 000
22	0	30 000	20 000	15 000
23	0	30 000	0	15 000
⋮	⋮	⋮	⋮	⋮
31	0	30 000	0	15 000
32	0	30 000	20 000	15 000
33	0	30 000	0	15 000
⋮	⋮	⋮	⋮	⋮
41	0	30 000	0	15 000
42	0	30 000	20 000	15 000
43	0	30 000	0	15 000
⋮	⋮	⋮	⋮	⋮
52	0	30 000	0	15 000

ANSWER

(a) We need to get the present worth of benefits and costs of the dam alone. There are two benefits from the dam alone. They are reduced flood damage and the benefits of irrigation. Both are approximated as annuities that start in year 3. We get the present worths of these benefits by multiplying the annual benefits by the series present worth factor and the present worth factor.

$$\begin{aligned} \text{PW(flood control)} &= \$182\,510(P/A, 10\%, 50)(P/F, 10\%, 2) \\ &= \frac{\$182\,510(9.99148)}{(1.1)^2} \\ &= \$1\,495\,498 \end{aligned}$$

Similar computations give the present worth of irrigation as

$$\text{PW(irrigation)} = \$1\,638\,812$$

There are two costs for the dam: capital costs that are incurred at the ends of years 0, 1, and 2, and operating and maintenance costs that are approximated as an annuity that begins in year 3. Capital costs are given by

$$\begin{aligned} \text{PW(dam, capital cost)} &= \$300\,000 + \$750\,000(P/F, 10\%, 1) \\ &\quad + \$1\,500\,000(P/F, 10\%, 2) \\ &= \$2\,221\,487 \end{aligned}$$

The present worth of operating and maintenance costs is obtained in the same way as the present worths of flood control and irrigation benefits. The result is

$$\text{PW(operating and maintenance)} = \$245\,822$$

We get the present worth of the dam alone as

$$\begin{aligned} \text{PW(dam)} &= \$1\,495\,498 + \$1\,638\,812 - (\$2\,221\,488 + \$245\,822) \\ &= \$667\,001 \end{aligned}$$

The benefit-cost ratio for the dam is

$$\text{BCR(dam)} = \frac{\$1\,495\,498 + \$1\,638\,812}{\$2\,221\,488 + \$245\,822} = 1.27$$

The modified benefit-cost ratio is given by

$$\text{BCRM(dam)} = \frac{\$1\,495\,498 + \$1\,638\,812 - \$245\,822}{\$2\,221\,488} = 1.30$$

The present worth of the dam alone is positive, and both benefit-cost ratios are greater than 1. The dam alone appears to be economically viable.

(b) We already have the present worths of benefits and costs for the dam alone. Therefore, we need only compute the present worths of benefits and costs for the recreation facilities. The capital costs for the recreation facilities consist of five outlays, in years 2, 12, 22, 32, and 42. The present worth of capital costs for the recreational facilities is given by

$$\begin{aligned}
&\text{PW(recreation facilities, capital cost)}\\
&\quad = \$50\,000(P/F, 10\%, 2)\\
&\qquad + \$20\,000[(P/F, 10\%, 12) + (P/F, 10\%, 22)\\
&\qquad + (P/F, 10\%, 32) + (P/F, 10\%, 42)]\\
&\quad = \$51\,464
\end{aligned}$$

Operating and maintenance costs are approximated by an annuity that starts in year 3. The computation is the same as the computation of similar annuities that were shown for the benefits and operating and maintenance costs of the dam. We get them as the present worth of recreation operating and maintenance costs.

$$\text{PW(recreation facilities, operating and maintenance)} = \$122\,911$$

To obtain the present worth of the benefits from recreation, we need to define a growth adjusted interest rate with $i = 10\%$ and $g = 3.25\%$ per year.

$$i^\circ = \frac{1+i}{1+g} - 1 = \frac{1+0.10}{1+0.0325} - 1 = 0.0653$$

We then use this to get the present worth geometric gradient series factor,

$$(P/A,g,i,N) = \left(\frac{(1+i^\circ)^N - 1}{i^\circ(1+i^\circ)^N}\right)\left(\frac{1}{1+g}\right)$$

$$= \left(\frac{(1.0653)^{50} - 1}{0.0653(1.0653)^{50}}\right)\left(\frac{1}{1.0325}\right) = 14.19$$

To bring this to the end of year 0, we multiply by $(P/F, 10\%, 2)$. We then multiply by the initial value to get the present worth of recreation benefits.

$$\text{PW(recreation benefits)} = [(P/A, 3.25\%, 10\%, N)(\$27\ 600)\ (P/F, 10\%, 2)]$$
$$= 14.19(\$27\ 600)(1.1)^{-2}$$
$$= \$323\ 679$$

Another way to get this result is to use a spreadsheet. First, a column that contains the benefits in each year is obtained. The benefits start at $27 600 in the third year and grow at 3.25% each year. These benefits are then multiplied by the appropriate present worth factor in another column to obtain the present worth of the benefits. The individual present worths are then summed to obtain the overall total of $323 679. Some of the spreadsheet computations are shown in Table 10.6.

Table 10.6 Spreadsheet Computations

Year	Recreation Benefits	PW of Recreation Benefits
0	$ 0.00	$ 0.00
1	0.00	0.00
2	0.00	0.00
3	28 600.00	20 736.29
4	28 497.00	19 463.83
5	29 423.15	18 269.46
6	30 379.40	17 148.38
⋮	⋮	⋮
52	132 288.48	931.33
Total PW		323 678 .84

The total present worth of the recreation facilities is

$$\text{PW(total recreation facilities)} = \$323\ 679 - \$51\ 464 - \$122\ 911$$
$$= \$149\ 304$$

We get the present worth of the dam plus recreation facilities simply by adding the present worth of the dam alone to the present worth of the recreation facility. The final result is given by

$$\text{PW(dam and recreation facility)} = \$667\ 001 + \$149\ 304$$
$$= \$816\ 305$$

In present worth terms, the present worth of the dam *and* recreation facility exceeds that of the dam alone, so the dam and the recreation facility should be chosen.

(c) The benefit-cost ratio for the dam and recreation facilities together is

$$\text{BCR(dam and recreation)} = \frac{(\$1\,495\,498 + \$1\,638\,812 + \$323\,679)}{(\$2\,221\,488 + \$245\,822 + \$51\,464 + \$122\,911)} = 1.31$$

The modified benefit-cost ratio is

$$\text{BCRM(dam and recreation)} = \frac{(\$1\,495\,498 + \$1\,638\,812 + \$323\,679 - \$245\,822 - \$122\,911)}{(\$2\,221\,488 + \$51\,464)} = 1.36$$

The dam and recreation facilities project appears to be viable, since the benefit-cost ratios are greater than one.

(d) Based on benefit-cost ratios, which of the two projects, project 1 or project 2, should be chosen? The dam and recreation facility is more costly, so the correct benefit-cost ratio for comparing the two is

$$\text{BCR} = \frac{\text{Benefits(dam and recreation facility)} - \text{Benefits(dam)}}{\text{Costs(dam and recreation facility)} - \text{Costs(dam)}}$$

$$= \frac{\$323\,679}{(\$51\,464 + \$122\,911)} = 1.86$$

The ratio exceeds one, and hence the dam and recreation facilities project should be chosen. This is consistent with the original present worth computations. ■

SUMMARY

Chapter 10 concerns decision making in the public sector. We started by considering why markets may fail to lead to efficient decisions. We presented four formal methods by which society seeks to remedy market failure. One of these methods is to have production by government. Next, we laid out three issues in decision making about government production. First, we saw that identification and measurement of costs and benefits in the public sector are more difficult than in the private sector. Identification may be difficult because there may not be cash flows that reflect the costs or benefits. Measurement may be difficult because there are no prices to indicate values. Second, we discussed the use of benefit-cost ratios in public sector project evaluation. While it is possible to use benefit-cost ratios so as to give the same results as using the comparison methods discussed in Chapters 4 and 5, the ratios may also be misused. Therefore, the use of benefit-cost ratios is not recommended by the Treasury Board of the Canadian federal government. Third and last, we considered the

MARR for government sector investments. The result of this discussion is that we can only state a range in which the opportunity cost of funds used in the public sector may lie. Therefore, the analyst should test to see if the choice of the MARR from within this range will affect the decision choice.

Engineering Economics in Action, Part 10B: Look at It from Their Side

"How was your trip, Naomi?" Anna Kulkowski asked as she walked into the conference room. Bill and Naomi were already there waiting.

"Well, I'm here," responded Naomi, "but my jets are still lagging."

"You'll get used to it." Anna looked at Bill. "So, what's it going to cost to upgrade the mill?"

"Naomi got some back-of-the-envelope figures from the people at Edgemont," Bill answered. "They're talking over a hundred million between now and 2002. If we amortize that over twenty years it's going to add about 20% to total costs per tonne of pulp. It looks as though the only way we can keep pulp bleaching viable long-term is to get the B.C. government to change the regulation."

"Well," Naomi continued, "I did talk to the people at the mill yesterday before coming home. Things may not be quite as bleak as I first told Bill. First, even if costs do go up by 20%, we may still be able to compete by finding niches that demand 'environmentally friendly' pulp. Unfortunately, if 'environmentally friendly' means only very low AOX, there will be offshore mills that will be able to underprice us, even in the niche markets."

"That doesn't sound like a big help," Anna said. "What's the other reason for hope?"

"The other reason is that we may be able to make a reasonable argument for modifying the regulation," Naomi said. "The form of the regulation, in terms of AOX per tonne of output, doesn't make sense from an environmental point of view."

"Why not?" Anna asked.

"Well," said Naomi, "we're really interested in water quality, not AOX measured at the end of the pipe. Water quality around the mill can vary greatly for a given effluent concentration. It will depend on the degree of dilution, and water flow patterns which vary over time and across mills, depending on season and location. The main implication is that it is possible that a reasonable rule in terms of water quality would not require zero AOX. There is background AOX, in any case. This is stuff that would be there even if the mills shut down."

"Where does this leave us, Naomi?" Anna asked.

"Well, I'm not sure," Naomi said. "But it certainly makes more sense, from an environmental point of view, to state the regulations in terms of water quality at some distance from mills. This would mean different effluent concentration limits for different mills, depending on location. Administratively, it would be more difficult for the government. But we're talking about fewer than twenty bleached pulp mills in the province. If a revised form of regulation enabled some of the mills to avoid zero AOX, it would save everybody a lot of money."

"That sounds interesting, Naomi," Anna said. "Why don't you write this up. I'm going to B.C. in a few weeks. I'll see if I can get a meeting with some people in the environment ministry. I'll try the idea out on them. It's worth a shot."

"By the way, Naomi," Anna continued, "good work!"

PROBLEMS

10.1 The following data are available for a project:

Present worth of benefits: $17 000 000
Present worth of operating and maintenance costs: $5 000 000
Present worth of capital costs: $6 000 000

(a) Find the benefit-cost ratio.

(b) Find the modified benefit-cost ratio.

10.2 The following data are available for two mutually exclusive projects:

	Project A	Project B
PW (benefits)	$19 000 000	$15 000 000
PW (operating and maintenance costs)	5 000 000	8 000 000
PW (capital cost)	5 000 000	1 000 000

(a) Compute the benefit-cost ratios for both projects.

(b) Compute the modified benefit-cost ratios for both projects.

(c) Compute the benefit-cost ratio for the increment between the projects.

(d) Compute the present worths of the two projects.

(e) Which is the preferred project? Explain.

10.3 The following data are available for two mutually exclusive projects:

	Project A	Project B
PW (benefits)	$17 000 000	$17 000 000
PW (operating and maintenance costs)	5 000 000	11 000 000
PW (capital cost)	6 000 000	1 000 000

(a) Compute the benefit-cost ratios for both projects.

(b) Compute the modified benefit-cost ratios for both projects.

(c) Compute the benefit-cost ratio for the increment between the projects.

(d) Compute the present worths of the two projects.

(e) Which is the preferred project? Explain.

10.4 The following data are available for two mutually exclusive projects:

	Project A	Project B
PW (benefits)	$17 000 000	$15 000 000
PW (operating and maintenance costs)	5 000 000	8 000 000
PW (capital cost)	6 000 000	3 000 000

(a) Compute the benefit-cost ratios for both projects.

(b) Compute the modified benefit-cost ratios for both projects.

(c) Compute the benefit-cost ratio for the increment between the projects.

(d) Compute the present worths of the two projects.

(e) Which is the preferred project? Explain.

10.5 There are two beef packing plants, A and B, in the town of Reybourne, Saskatchewan. Both plants dump partially treated liquid waste into Lake Jeannette. The two plants together dump over 33 000 kilograms of BOD5 per day. (BOD5 is the amount of oxygen used by micro-organisms over five days to decompose the waste.) This is more than half the total BOD5 dumped into Lake Jeannette. Reybourne town council wants to reduce the BOD5 dumped by the two plants by 10 000 kilograms per day.

The following data are available concerning the two plants:

	Outputs of the Two Plants	
	Steers/day	BOD5/steer (kg)
Plant A	20 000	1.0
Plant B	9 000	1.5

The costs of making reductions in BOD5 per steer are shown below:

	Incremental Cost of Reducing BOD ($/kg/steer)						
Reduction (kg/steer)	**0.1**	**0.2**	**0.3**	**0.4**	**0.5**	**0.6**	**0.7**
Plant A	0.05	0.08	0.12	0.25	0.45	0.65	0.95
Plant B	0.15	0.15	0.15	0.15	0.15	0.35	0.45

For example, to reduce the BOD5 of Plant A by 0.25 kg/steer, the cost is calculated as

$$(0.1 \times 0.05) + (0.1 \times 0.08) + (0.05 \times 0.12) = 0.019/\text{steer}$$

The council is considering three methods of inducing the plants to reduce their BOD5 dumping: (1) a regulation that limits BOD5 dumping to 0.81 kg/steer, (2) a tax of $0.16/kg of BOD5 dumped, and (3) a subsidy paid by the town to the plants of $0.16/kg reduction from their current levels in BOD5 dumped.

(a) Verify that, if both plants reduce their BOD5 dumping to 0.81 kg/steer, there will be a 10 000 kg/day reduction in BOD5 dumped. What will this cost?

(b) Under a tax of $0.16/kg, how much BOD5 will Plant A dump? How much will Plant B dump? (Assume that outputs of steers would not be affected by the tax.) Verify that this will lead to more than a 10 000 kg/day reduction in BOD5. What will this cost?

(c) Under a subsidy of $0.16/kg reduction in BOD5, how much will Plant A dump? How much will Plant B dump? Verify that this will lead to more than a 10 000 kg/day reduction in BOD5. What will this cost?

(d) Explain why the tax and subsidy schemes lead to the same behaviour by the meat packing plants.

(e) Explain why the tax and subsidy schemes have lower costs for the company than the regulation.

10.6 There are three petrochemical plants, A, B, and C, in Port Jayne, Ontario. The three plants produce Good Stuff. Unfortunately, they also dump Bad Stuff into the air. Data concerning their outputs of Stuff are shown below:

	Outputs	
	Good (kg/day)	**Bad/Good (cL/kg)**
Plant A	17 000	10
Plant B	11 000	15
Plant C	8 000	18

The town council wants to reduce the dumping of Bad Stuff by 150 000 cL/day. Costs for reducing the concentration of Bad Stuff in output are shown below:

	Incremental Cost of Reducing Bad Stuff/Good Stuff ($/cL/kg)							
Reduction (cL/kg)	**1**	**2**	**3**	**4**	**5**	**6**	**7**	**8**
Plant A	0.02	0.032	0.048	0.1	0.18	0.26	0.38	0.57
Plant B	0.06	0.06	0.063	0.068	0.075	0.193	0.27	0.405
Plant C	0.25	0.25	0.25	0.25	0.25	0.25	0.25	0.375

The council is considering two methods: (1) Require all plants to meet the performance level of the best-practice plant, Plant A, which is 10 centilitres of Bad Stuff per kilogram of Good Stuff. (2) Impose a tax of $0.20/cL of Bad Stuff dumped.

(a) What will be the reduction in dumping of Bad Stuff under the best-practice regulation? What will be the cost of this reduction?

(b) Under the tax, how much Bad Stuff will be dumped from the three plants combined? What will be each plant's reduction in dumping per kilogram of Good Stuff?

(c) How much will the reduction of dumping cost for each company under the tax?

10.7 In the summer of 1998, the Kitchener-Waterloo area often experienced the worst air pollution in Canada, affecting the health of hundreds of thousands of people. Explain how air pollution is an example of market failure. Give an example of how each of the four remedies for market failure listed in section 10.2.2 could be applied to the case of air pollution.

10.8 The Canadian fishing industry has been devastated in recent years because of overfishing. On the east coast, cod fishing has almost disappeared, while on the west coast, salmon fishing has been considerably reduced. How is overfishing an example of market failure? Give an example of how each of the four remedies for market failure listed in section 10.2.2 could be applied to the case of overfishing.

10.9 Consider these situations in which cutting of trees is relevant:

1) The Brown family owns a view house in West Vancouver. Their neighbours across the street, the Smith family, have trees on their lot that are obstructing the Browns' view of the Lions Gate Bridge. The Browns are the only ones affected by the Smiths' trees. The Browns asked the Smiths to top their trees, but the Smiths refused to do so. The Browns then offered to pay the Smiths for the topping and an additional $500 to cover any loss they might feel because their tall trees were topped. The Smiths still refused.

2) The Brown family, the Green family, the White family and the Blue family own view houses in West Vancouver. Their neighbours across the street, the Smith family, have trees on their lot that are obstructing everyone's view of the Lions Gate Bridge. The Browns, Greens, Whites and Blues asked the Smiths to top their trees, but the Smiths refused to do so.

3) Timber companies on Vancouver Island sometimes use clear cutting on old growth forests. Environmentalists have asked the companies to change this practice because it leads to reduced biodiversity.

Why is there no market failure in the first situation involving the Smiths and the Browns? Why may there be market failure in the situation with several families? Why is there market failure in the third situation involving the timber companies?

10.10 Technical changes in electricity supply and information transmission have made it efficient for consumers of both services to be served by suppliers using different technologies and operating in different locations. Does this increase or decrease the need for government regulation in these industries?

10.11 An electric utility company is considering a re-engineering of a major hydroelectric facility. The project would yield greater capacity and lower cost per kilowatt-hour of power. As a result of the project, the price of power would be dropped. This is expected to increase the quantity of power demanded. The following data are available:

Effect of Reduced Price of Power	
Current price ($/kWh)	0.07
Current consumption (kWh/year)	9 000 000
New price ($/kWh)	0.05
Expected consumption (kWh/year)	12 250 000

What is the annual benefit to consumers of power from this project?

10.12 Brisbane and Johnsonburg are two Prairie towns. They are separated by the Wind River. Traffic between the two towns crosses the river by a ferry run by the Johnsonburg Ferry Company. They charge a toll for crossing. The province is considering building a bridge somewhat upstream from the ferry crossing. There would be no toll on the bridge. Travel time between the towns would be about the same with the bridge as with the ferry because of the bridge's upstream location. The following information is available concerning the crossing:

Ferry/Bridge Information	
Ferry crossings (number/year)	60 000
Average cost of ferry trip ($/crossing)	1
Ferry fare ($/crossing)	1.5
Bridge toll ($/crossing)	0
Expected bridge crossings (number/year)	90 000
EAC of bridge ($/year)	85 000

Note that all data are on an annual basis. The cost of the bridge is given as the equivalent annual cost of capital and operating costs. We assume that all bridge costs are independent of use, that is, there are no costs which are due to use of the bridge. The average cost per crossing of the ferry includes capital cost and operating cost.

(a) If the bridge were built, what would be the annual benefits to travellers?

(b) How much would the owners of the Johnsonburg Ferry Company lose if the bridge were built?

(c) What would be the effect on taxpayers if the bridge were built? (Assume that Johnsonburg Ferry pays no taxes.)

(d) What would be the net social gain or loss if the bridge were built? Take into account the effects on travellers, Johnsonburg Ferry owners, and taxpayers.

(e) Would the net social gains or losses be improved if there were a toll for crossing the bridge?

10.13 It is common for municipalities to provide snowplow service for public roads. The major benefit of such services is to allow the convenient movement of vehicles over public roads following snow accumulation, at a cost of the snowplow (capital and operating) and driver. Are there other costs and benefits? List all you can think of, along with how they could be measured.

10.14 Most provinces and states provide travel information kiosks alongside major highways just across the border from neighbouring provinces and states. These kiosks provide maps and brochures on attractions, and some will make hotel and campground reservations. There are obvious costs, such as staffing costs and building capital and maintenance costs. What other costs and benefits are associated with this government service? How can these costs and benefits be measured?

10.15 The Ontario government is considering putting a car-pool parking lot at a new interchange near Cambridge, Ontario, on the main east-west highway through the province, the 401. The car-pool parking lot allows commuters to meet in separate cars, park all but one, and proceed in one car to a joint destination. Studies estimate that an average of 200 cars will be parked at the lot on weekdays, with 1/4 of that number on weekends. The average commuting distance from that intersection is 75 km, and the marginal cost of driving an average car is \$0.28 per km.

(a) If it is assumed that all the cars that are parked in the lot would otherwise have been driven to work, how much will be saved by this commuter parking lot per year? Assume an interest rate of 0.

(b) How could you find out how many would have been parked somewhere else?

(c) How could you calculate the benefit to all drivers of having fewer cars on the road?

10.16 A medium-sized city in Alberta (population 250 000, 45 000 families) is considering introducing a recycling program. The program would require them to separate newspaper, cardboard, and cans from their regular waste so that it could be collected weekly, sorted, and sold, rather than put into the local landfill site. The program would also require households to separate "wet," or compostable waste, which they would then be responsible for composting. The city would provide free composting units to the households. What kinds of potential costs and benefits can you identify for this project? How might this information be gathered?

10.17 Consider Problem 10.5. We saw that a tax on effluent BOD5 enabled the same reduction in BOD5 as a regulation, but with a lower cost. We did not consider the distribution of this cost.

(a) How much tax does Plant A pay under the tax of t = \$666.67/1000 PE's dumped? How much does Plant B pay?

(b) What are the two total effluent costs (tax plus cleaning cost) for Plant A? For Plant B?

(c) Compare these costs with the costs under regulation.

(d) Suppose that the province used the proceeds of the tax to provide benefits that had equal value to each plant. How great would these benefits have to be to ensure that both plants would be better off with the combination of tax and benefits?

10.18 A four-day school week has been advocated as a means of reducing costs. School days would be longer so as to maintain the same number of hours per week as under the current five-day-a-week system. The main cost savings that are expected are in school cleaning and maintenance and in school bus operation. The main effects that are anticipated are

1. Reduced school cleaning and maintenance.
2. Reduced use of school buses on the off-day.
3. Reduced driving to school by parents, students, and staff on the off-day.
4. Reduction in public transportation use on the off-day.
5. Some high-school students and school staff will seek part-time work for the off-day.
6. Reduced absences by students and staff. This is mainly because some required personal activities could be scheduled for the off-day.
7. Reduced subsidized school lunch requirements.

8. Greater need for daycare on the off-day for working parents. About a third of elementary schools could be opened for day care. The costs would be covered by fees.
9. Learning by elementary students may be reduced because of their limited attention spans.
10. Lower school taxes.

Which of these effects are benefits? Who receives the benefits? Which are costs? Who bears the costs? Which are neither costs nor benefits?

10.19 A new suburban development is being planned near Petroville, Alberta. There is now a two-lane road from the site of the development, along the river, to Petroville. The new development will require additional road capacity. Two alternatives are being considered. The first is to upgrade the existing road to four wide modern lanes. The second is to build a new four-lane highway through Beaver Hill tunnel. The following data are available concerning the two routes:

Route	Distance	First Cost	Operating & Maintenance Costs per Year
River	20 km	\$21 million	\$90 000
Tunnel	10 km	\$45 million	\$130 000

The planning period is 40 years for both routes. The MARR is 10%. Cars will travel at 100 km/hour along either route. Operating cost for the cars is expected to be about \$0.25/km along either route. About 400 000 trips per year are expected on either route.

(a) Which route should be built? Use only the data given. Use annual worth to make your decision.

(b) What important benefit of the Beaver Hill tunnel route has been left out of the analysis?

(c) Do you need more information about travellers to determine if the benefit that has been left out of the analysis from part (b) would change the recommendation? Explain.

(d) How would the possibility of collecting a toll on the tunnel route affect your recommendation?

10.20 The Principality of Upper Pigovia has just one export, pig crackling. The crackling is produced in two plants, Old Gloria and New Gloria. Both plants give off a delightful, mouth-watering odour while in operation. This odour has created a health problem. The citizens' appetites are huge, and, consequently, so are the citizens. Princess Juliana has decreed that the daily emission of odour from the two plants must be reduced by 7000 Odour Units (OU) per day. The Princess prides herself on her even-handedness. The decree specifies that each plant is to reduce its emission by 3500 OU/day. The following table shows the incremental cost per 1000 OU of attaining various levels of odour reduction in the two plants. To help in interpreting the table, note that the cost for Old Gloria to remove 3000 OU/day would be 2(U$25) + U$30 = U$80.

Quantity of Odour Removed (1000 OU/day)	Incremental Cost of Removing Odour in $U/1000 OU 0 to 2	Over 2	Over 4	Over 5
Old	25	30	40	50
New	20	20	25	30

(a) What is the cost per day of implementing Princess Juliana's decree?

(b) Can you suggest a tax scheme that will yield the same reduction in odour emission as the decree at a lower cost?

(c) Can you suggest a subsidy scheme that will yield the same reduction in odour emission as the decree at a lower cost?

10.21 A provincial ministry of transportation is considering the construction of a bridge over a narrow point in a lake. Traffic now goes around the lake. The bridge will save 30 km in travel distance. Three alternatives are being considered: (1) do nothing, (2) build the bridge, and (3) build the bridge and charge a toll. If the bridge is built, the present road will be maintained, but its use will drop. One effect of the reduced use of the present road is a loss of revenue by businesses along the present road. Available data is given on the following page.

(a) Identify social benefits and costs of the bridge.

(b) Which of the costs and revenues in the table are neither social benefits nor social costs?

(c) The table makes an implicit assumption about the effect of the toll on bridge traffic. This assumption is probably incorrect. What is the assumption?

(d) How would you expect the toll to affect benefits and costs? Explain.

Costs and Benefits	Do Nothing ($)	Bridge with Toll ($)	Bridge without Toll ($)
First cost	0	46 000 000	46 000 000
Annual road and/or bridge operating and maintenance costs	160 000	80 000	10 000
Annual vehicle operating cost	3 300 000	100 000	100 000
Annual driver and passenger time cost	2 500 000	500 000	500 000
Annual accident cost	500 000	10 000	10 000
Annual revenue lost by roadside businesses	0	1 000 000	1 000 000
Annual toll revenues	N/A	1 200 000	N/A

10.22 An example concerning the effect of a flood control project is found in the benefit-cost analysis chapter of an imaginary engineering economics text. The benefits of the project are stated as:

Prevented losses due to floods in the Conestogo River Basin	$480 000/year
Annual worth of increased land value in the Conestogo River Basin	$48 000/year

Comment on these two items.

10.23 A province is considering the construction of a bridge. The bridge would cross a narrow part of a lake near a provincial park. The major benefit of the bridge would be reduced travel time to go to a camp site from a nearby urban centre. This lowers the cost of camping trips at the park. As well, they expect an increase in the number of visits resulting from the lower cost per visit.

Data concerning the number of week-long visits and their costs are shown below:

	Number of Visits and Average Cost per Visit to Park	
Inputs	Without Bridge	With Bridge
Travel cost ($)	140	87.5
Use of equipment ($)	50	50
Food ($)	100	100
Total ($)	290	237.5
Number of visits/year	8000	11 000

The following data are available as well:

1. The bridge will take one year to build.
2. The bridge will have a 25-year life once it is completed. This means that the time horizon for computations is 26 years.
3. Construction cost for the bridge is \$3 750 000. Assume that this cost is incurred at the beginning of year 1.
4. Annual operating and maintenance costs for the bridge are given by

$$\text{Operating and Maintenance Cost per year} = 7500 + 0.25q$$

   where q is the number of crossings.

5. Operating and maintenance costs are incurred at the end of each year the bridge is in operation. This is at the ends of years 2, 3, . . ., 26.
6. MARR is 10%.

(Notice that the annual benefits for this project were computed as part of the discussion of Example 10.3.)

(a) Compute the net present worth of the project.

(b) Compute the benefit-cost ratio.

(c) Compute the modified benefit-cost ratio.

10.24 Continue with the bridge project of Problem 10.23. In that problem, we assumed that there would be no toll for crossing the bridge. Now suppose the province is considering a toll of \$7 per round trip over the bridge. They estimate that, if the toll is charged, the number of park visits will rise to only 10 600 per year instead of 11 000 visits per year.

(a) Compute the net present worth of the project if the toll is charged.

(b) Why is the net present worth of the project reduced by the toll?

10.25 The town of Migli Lake, Manitoba, has a new subdivision, Paradise Mountain, at the outskirts of the town. The town wants to encourage the growth of Paradise Mountain by improving transportation between Paradise Mountain and the centre of Migli Lake. Two alternatives are being considered: (1) new buses on the route between Paradise Mountain and Migli Lake centre and (2) improvement of the road between Paradise Mountain and Migli Lake centre.

Both projects will have as their main benefit improved transportation between Paradise Mountain and Migli Lake centre. Rather than measure the value of this directly to the city, engineers have estimated the benefit in terms of an increase in the value of land in Paradise Mountain. That is, potential residents are expected to show their evaluations of the present worth of improved access to the town centre by their willingness to pay for homes in Paradise Mountain.

The road improvement will entail construction cost and increased operating and maintenance costs. As well, the improved road will require construction of a parking garage in the centre of Migli Lake. The new buses will have a first cost as well as operating and maintenance costs. Information about the two alternatives is shown below.

	Road Improvement	New Buses
First cost	\$15 000 000	\$ 4 500 000
PW (operating and maintenance cost)	5 000 000	12 000 000
Parking garage cost	4 000 000	
Estimated increased land value	26 000 000	18 000 000

(a) Compute the benefit-cost ratio of the road improvement under each of two assumptions: (1) that the parking garage is a cost, and (2) that the parking garage is a reduction in benefits.

(b) Compute the incremental benefit-cost ratio under each of the two assumptions in part (a). Is the choice between projects affected by the assumption?

(c) Compute the present worths of the two alternatives. Compare the decision based on present worths with the decisions based on benefit-cost ratios.

10.26 A provincial government is considering a new two-lane road through a mountainous area. The new road would improve access to a city from farms on the other side of the mountains. The improved access would permit farmers to switch from grains to perishable soft fruits that would be either frozen at an existing plant near the city or sold in the city. Two routes are being considered. Route A is more roundabout. Even though the speed on route B would be less than on Route A, the trip on Route B would take less time. Almost all vehicles using either road would go over the full length of the road. A Department of Transport engineer has produced the table on page 431.

The province uses a 10% MARR for road projects. The road will take one year to build. The province is using a 21-year time horizon for this project, since it is not known what the market for perishable crops will be in the distant future. Comment on the engineer's list of benefits; there may be a couple of errors.

10.27 Continue with the road project in Problem 10.26. (*Note:* Correct for the errors mentioned.)

(a) Compute a benefit-cost ratio for Route A with road use costs counted as a cost.

(b) Compute a benefit-cost ratio for Route A with road use costs counted as a reduction in benefits.

Costs and Benefits of the New Road	Route A	Route B
Properties		
Distance (km)	24	16
Construction cost ($)	53 400	75 000
Operating and maintenance cost per year ($)	60	45
Resurfacing after 10 years of use ($)	3 100	2 350
Road Use		
Number of vehicles per year	1 000 000	1 200 000
Vehicle cost per km ($)	0.3	0.3
Speed (kph)	100	80
Value of time per vehicle hour ($)	15	15
Benefits		
Increased crop value per year ($)	13 500	18 000
Increased land value ($)	104 484.6	139 312.8
Increased tax collections per year ($)	811.21	1 081.61

(c) In what way are the two benefit-cost ratios consistent, even though the numerical values differ?

(d) Make a recommendation as to what the province should do regarding these two roads. Explain your answer briefly.

10.28 Find the net present worths for the dam and the dam plus recreation facilities considered in Review Problem 10.1. Use a MARR of 15%. Make a recommendation as to which option should be adopted.

10.29 The recreation department for Port Elgin is trying to decide how to develop a piece of land. They have narrowed the choices down to tennis courts or a swimming pool. The swimming pool will cost $2.5 million to construct, and will cost $300 000 per year to operate, but will bring benefits of $475 000 per year over its 25-year expected life. Tennis courts would cost $200 000 to build, cost $20 000 per year to operate, and bring $60 000 per year in benefits over its 8-year life. Both projects are assumed to have a salvage value of zero. The appropriate MARR is 5%.

(a) Which project is preferable? Use a BCR and an annual worth approach.

(b) Which project is preferable? Use a BCRM and an annual worth approach.

10.30 The environmental protection agency of a county would like to preserve a piece of land as a wilderness area. The owner of the land will lease the land to the county for a 20-year period for the sum of $1 750 000, payable immediately. The protection agency estimates that the land will generate benefits of

\$150 000 per year, but they will forgo \$20 000 per year in taxes. Assume that the MARR for the county is 5%.

(a) Calculate a BCR using annual worth and classify the foregone taxes as a cost.

(b) Repeat part (a), but consider the foregone taxes as a reduction in benefits.

(c) Using your results from part (b), determine the most the county would be willing to pay for the land (within \$10 000) if they accept projects with a BCM of 1 or more.

10.31 The data processing centre at a local government tax centre has been plagued recently by the increasing incidence of repetitive strain injuries in the workplace. Health and Safety consultants have recommended to management that they invest in upgrading computer desks and chairs at a cost of \$500 000. They advise that this would reduce the number and severity of medical costs by \$70 000 per year and that productivity losses and sick leaves could be reduced by a further \$80 000/year. The furniture has a life of eight years with zero scrap value. The city uses a MARR of 9%. Should the centre purchase the furniture? Use a BCR method.

10.32 Several new big box shopping stores have created additional congestion at an intersection in north Winnipeg. City engineers have recommended the addition of a turn lane, a computer-controlled signal, and sidewalks at an estimated cost of \$1.5 million. The annual maintenance costs at the new intersection will be \$8000, but users will save \$50 000 per year due to reduced waiting. In addition, accidents are expected to decline, representing a property and medical savings of \$175 000 per year. The renovation is expected to handle traffic adequately over a 10-year period. The city uses a MARR of 5%.

(a) What is the BCR of this project?

(b) What is the BCRM of the project?

(c) Comment on whether the project should be done.

10.33 What determines the MARR that is used on government funded projects?

10.34 How will a decrease in the tax rate on investment income affect the MARR that is used for evaluating government funded projects?

10.35 How does an expectation of inflation affect the MARR for public sector projects?

10.36 A provincial department of transportation has \$16 500 000 that it can commit to highway safety projects. The goal is to reduce loss of life due to highway accidents. The potential projects are: (1) flashing lights at 10 railroad crossings; (2) flashing lights and gates at the same 10 railroad crossings; (3) widening the roadway on an existing rural bridge from 3 meters to 3.5 meters; (4) widening the roadway on a second rural bridge from 3 meters to 3.5 meters; (5) reduction of the density of utility poles on rural roads from 30 poles to 15 poles per kilometre; and (6) building runaway lanes for trucks on steep downhills.

The data for the flashing-lights and the flashing-lights-with-gates projects reflect the costs and benefits for the entire set of 10 crossings. Portions of the projects may also be completed for individual crossings at proportional reductions in costs and savings. At any single site, the lights and lights-with-gates projects are mutually exclusive. Any fraction of the reduction of utility pole density project can be carried out. Data concerning costs and safety effects of the projects are shown below. (A life-year saved is one year of additional life for one person.)

Highway Safety Projects

		Total Cost ($)	Life-Years Saved per Year
1	Flashing lights	450 000	14
2	Flashing lights and gates	750 000	20
3	Widening bridge #1	1 200 000	14
4	Widening bridge #2	700 000	10
5	Widening bridge #3	1 100 000	18
6	Pole density reduction	3 000 000	96
7	Runaway lane #1	6 000 000	206
8	Runaway lane #2	6 000 000	156

Advise the department of transportation how best to commit the $16 500 000. Assume that the money must be used to increase highway safety.

10.37 A provincial department of transportation is considering widening lanes on major highways from 6 m to 7.5 m. The objective is to reduce the accident rate. Accidents have both material and human costs. The following data are available for highway section XYZ:

Lane Widening on Section XYZ

Accidents per 100 000 000 vehicle-km in 6 m lanes	150
Accidents per 100 000 000 vehicle-km in 7.5 m lanes	90
Serious personal injuries per accident	10%
Average non-human cost per accident ($)	2500
Annual road use (vehicles)	7 500 000
First cost per kilometre ($)	175 000
Operating and maintenance costs per km/year ($)	7500
Project life (years)	25
MARR	10%

(a) Compute the present worth of costs of lane widening.

(b) Compute the present worth of savings of non-human accident costs.

(c) What is the minimum value for a serious personal injury that would justify the project?

10.38 The federal government is considering three flood control projects in Manitoba. Projects A and B consist of permanent dikes along the Rat River near Winnipeg. Project C is a dam. The dam will have recreation and irrigation benefits as well as the flood control benefits. Some facts about the three projects are in the following table. Each project has a life of 25 years. The MARR is 10%.

	Project A	Project B	Project C
First Cost (in $million)	25	32	52
Annual benefits (in $million)	3.3	4.2	7.1
Annual operating and maintenance costs (in $million)	0.3	0.3	0.5

(a) Use present worth to choose the best project.

(b) Compute the benefit-cost ratios for the 3 projects.

(c) Use incremental benefit-cost to choose the best project.

(d) It is possible that the dam (Project C) will cause some wildlife damage. This damage could be prevented by an additional expenditure of $3 000 000 when the dam is constructed. Does this change the choice of best project? Compute the benefit-cost ratio of Project C assuming the additional $3 000 000 to be an addition to the first cost.

10.39 A provincial highway department needs to upgrade a 20 km rural road. The upgrade is needed to accommodate increased traffic. There will be 5000 cars/day in the first year after upgrading. The number of cars will increase each year by 200 cars/day for the next 20 years. A modern two-lane road will be adequate for current traffic levels, but it will be inadequate after about 10 years. At that time, a four-lane road will be required. A four-lane road will permit greater speed even at the current traffic level.

Two alternatives are being considered. One proposal is for a modern two-lane road now, followed by adding lanes to make a four-lane road after 10 years. The second alternative is for a four-lane road now. The planning horizon for both alternatives is 20 years. Costs for the alternatives are shown in the table on page 435. All values are in real dollars.

	Two Lanes	Four Lanes
First cost (in $million)	20	32
Adding lanes after 10 years (in $million)	18	0
Annual operating and maintenance costs (years 1–10) (in $000s)	30	40
Annual operating and maintenance costs (years 11–20) (in $000s)	40	40

The average value of travel time for each car is estimated at $40/hour. The distance is 20 km. Speeds will be as follows:

Years	Two Lanes	Four Lanes
1 to 5	80 kmh	100 kmh
6 to 10	70 kmh	100 kmh
11 to 20	N/A	100 kmh

(a) Which alternative has a greater present worth if the MARR is 5%?

(b) Which alternative has a greater present worth if the MARR is 20%?

(c) Suppose the real after-tax rate of return on government savings bonds is 5% and the real before-tax rate of return on investment in the private sector is 20%. Use your answers to parts (a) and (b) to decide which alternative is better. Explain your answer.

10.40 A small municipality needs to upgrade its town dump to meet provincial environment standards. Two alternatives are being considered. Alternative A has a first cost of $420 000 and annual operating and maintenance costs of $52 500. Alternative B has a first cost of $315 000 and annual operating and maintenance costs of $74 000. Both alternatives have 15-year lives with no salvage. An increase in dumping fees for households and business in the town is expected to yield $50 000 per year.

(a) Which alternative has lower cost if the MARR is 5%? Use annual worth.

(b) Which alternative has lower cost if the MARR is 20%? Use annual worth.

(c) The after-tax return on government savings bonds is 5%. The average rate of return before taxes in private sector investment is 20%. Which alternative should be chosen?

Wood Pulp Mill Effluent Treatment in Alberta

CANADIAN 10.1 MINI-CASE

The Alberta government's policy concerning wood-pulp mills has two major objectives: to encourage the development of pulp mills to provide income for Albertans, and to avoid harm to humans, plants, or wildlife that could result from the release of liquid effluent into nearby bodies of fresh water.

The most common way for a government to limit damage to the environment by industrial effluents is to establish regulations that limit the release of the effluents. The regulations usually apply uniformly to all companies and plants that fall in certain classes defined by the regulations. Uniform regulation usually is inefficient for two reasons. First, different companies have different costs of reducing effluent. This means that efficiency could be increased by having plants that have low costs of reducing effluent reduce more than plants with high costs. Second, the same amounts of effluent released have different impacts in different locations. This means that efficiency could be increased by requiring higher standards where the impact of effluent release is greatest.

Alberta has adopted an unusual approach to promoting development while preventing damage from pulp mill effluent release. The Alberta government has negotiated individual licences with the six pulp mills that were started in the province between 1957 and 1993. Infrastructure grants and effluent limitations depend on each mill's particular situation.

This approach has the potential of being more efficient than uniform regulation. There are two drawbacks, however. First, individual licensing is feasible only where the number of sites being controlled is relatively small. Second, the cost of the subsidies to the government may be high. This is because there is always some uncertainty about the cost of attaining a given level of effluent control. Engineers representing the companies will want to avoid underestimating these costs. In the process, they may overestimate them. Since their knowledge of the situation is likely to be better than the knowledge of engineers representing the government, the company engineers' overestimates are likely to prevail.

Source: K. M. Lindsay and D. W. Smith, "Factors Influencing Pulp Mill Effluent Treatment in Alberta," *Journal of Environmental Management*, Vol. 44 (1995), pp. 11-27.

Discussion

Government subsidies always have the potential to be misused. In some cases, they provide an opportunity for certain industry or geographical areas to be rewarded for political reasons. In other cases, individuals or companies can exploit poorly designed programs for their own gain. At the best of times, such as with wood pulp mill effluent treatment in Alberta, the inherent process of establishing the subsidies incurs inefficiencies. For all these reasons, the idea that government subsidies have a somewhat tarnished reputation persists in the public mind.

On the other hand, the benefits to society of correcting situations of market failure can be enormous, in spite of the inefficiencies.

Questions

List some of the services provided by your municipal, provincial, or federal government. For each of these services:

1) Describe the result if the service were provided instead by private individuals or firms. Does the government service correct a potential market failure?
2) Estimate the direct cost per served individual for the government service. Would this cost go up or down if the service were provided privately?
3) Are there other costs that would be incurred by society as a whole if the service were provided privately? Estimate these costs if you can.
4) Is society generally better off with or without government provision of this service? Can you estimate the quantitative value of the benefit?

C H A P T E R 11

Dealing with Uncertainty: Sensitivity Analysis

Engineering Economics in Action, Part 11A: Filling a Vacuum

"I have something new for you, Naomi. It's going to require some imagination and disciplined thinking at the same time." Anna's tone indicated to Naomi that something interesting was coming.

"As you may know, Canadian Widgets has been working toward getting into consumer products for some time." Anna continued, "An opportunity has come up to co-operate on a new vacuum cleaner with Powerluxe. They have some potential designs for a vacuum with electronic sensing capabilities. It will sense carpet height and density. It will then adjust the power and the angle of the head to optimize cleaning on a continuous basis. Our role in this would be to design the manufacturing system and to do the actual manufacturing for North America. Sound interesting so far?"

"Yes, Ms. Kulkowski," Naomi answered. Naomi couldn't help being respectful in front of Anna, who was Canadian Widget's president, among other things. "What would my role be?"

"For one thing, you will be working with Bill Astad from head office. I think you know him." Naomi did; she had given Bill some advice on handling variable rates of inflation a few weeks earlier. "First, we want to establish some idea of the demand for the product. We have to see how this might affect our manufacturing capacity and capital costs to determine if the whole idea is even feasible. Later on, we will have to make some design decisions, but the general feasibility comes first."

"Do we have any market studies from Powerluxe?" Naomi asked.

"Yes, but they seem to be guesswork and magic, not hard figures. After all, no one has sold a product like this before."

Naomi looked pensive. "Sounds as though we'll need to do some sensitivity analysis on this one." She muttered more to herself than Anna. Then to Anna, "Right. We'll do what we can. Thanks for the nice opportunity, Anna. This is interesting!"

11.1 Introduction

To this point in our coverage of engineering economics, we have assumed that the cash flows used in an economic analysis are known with certainty, both in timing and size. In fact, the timing and size of the cash flows are only estimates, and are subject to error. Future events may be uncertain or they may be risky. If they are uncertain, it means that we don't know what will happen, and that we have no further information about the probability of what will happen. This chapter deals with making decisions under uncertainty. The problem of making decisions under risk, where we do have further information about the probability of something happening, is discussed in the following chapter.

Making decisions under uncertainty is difficult because there are relatively few mathematical tools available. The Maximin and Minimax principles, the Minimum Regret principle and the Hurwicz rule are tools that can be used but are outside the scope of this text. Instead, we focus on sensitivity analysis to deal with uncertainty.

There are several reasons why estimates of cash flows may not match the actual outcome. Technological change can unexpectedly shorten the life of a product or piece of equipment. A change in the number of competing firms may affect sales volume or market share or the life of a product. In addition, the general economic environment may affect inflation and interest rates and general activity levels within an industry. All of these factors may result in cash flows different from what was expected in both timing and size.

Economic analyses are not complete unless we try to assess the potential effects of these uncertainties on the outcomes of the analyses. Because cash flows can be so hard to estimate, analysts usually consider a range of possible values for uncertain components of a project. There is then naturally a range of values for present worth, annual worth, or whatever the relevant performance measure is. In this way, the analyst gets a better feel for the range of possible outcomes and can make better decisions.

In this chapter, we will consider three basic methods commonly used by analysts in order to understand better the effect that uncertainties or errors in parameters such as estimated cash flows have on economic decisions. The first method is the use of *sensitivity graphs*. Sensitivity graphs illustrate the sensitivity of a particular measure (e.g., present worth or annual worth) to changes in one or more of the parameters of a project. Sensitivity graphs will reveal key parameters that have a significant impact on the performance measure, and hence we should be particularly careful to get good estimates for these key parameters.

The second method is the use of *break-even analysis*. Break-even analysis can answer such questions as: "What production level is necessary in order for the present worth of the project to be greater than zero?" or "Below what interest rate will the project have a positive annual worth?" Break-even analysis can also give insights into comparisons between projects. With break-even analysis, we can answer questions like, "What scrap value for the proposed forklift will cause us to be indifferent between replacing the old forklift and not replacing it?"

The third method we will introduce is called *scenario analysis*. Both sensitivity graphs and break-even analysis have the drawback that we can look at parameter changes only one at a time. Scenario analysis allows us to look at the overall impact of a variety of outcomes, usually "optimistic," "expected," and "pessimistic." In this way, the analyst comes to understand the range of possible economic outcomes.

Each of these three methods—sensitivity graphs, break-even analysis, and scenario analysis—is an example of **sensitivity analysis**. Each method tries to assess the sensitivity of an economic measure to uncertainties in estimates in the various parameters of the problem. A thorough economic evaluation should include aspects of all three types of analysis.

11.2 Sensitivity Graphs

The first sensitivity analysis tool we will look at is the sensitivity graph. Sensitivity graphs are used to assess the effect of changes in key parameters of a project on an economic performance measure. We begin with a "base case" where all the estimated parameters are used to evaluate the present worth, annual worth, or IRR of a project, whatever the appropriate measure is. We then vary parameters above and below the base case one at a time, *holding all other variables fixed*. A graph of the changes in a performance measure brought about by these one-at-a-time parameter changes is called a **sensitivity graph**. From the graph, the analyst can see which parameters have a significant impact on the performance measure and which do not.

Example 11.1

Cogenesis Corporation is replacing their current steam plant with a 6MW cogeneration plant that will produce both steam and electric power for their operations. The new plant will use wood as a source of fuel, and it will eliminate the need for Cogenesis to purchase a large amount of electric power from a public utility. To move to the new system, Cogenesis will have to integrate a new turbogenerator and cooling tower with their current system. The estimated first cost of the equipment and installation is \$3 000 000, though there is some uncertainty surrounding this estimate. The plant is expected to have a 20-year life and no scrap value at the end of this life. In addition to the first cost, the turbogenerator will require an overhaul with an estimated cost of \$35 000 at the end of years 4, 8, 12, and 16. The cooling tower will need an overhaul at the end of 10 years. This is expected to cost \$17 000.

The cogeneration system is expected to have higher annual operating and maintenance costs than the current system, and will require the use of chemicals to treat the water used in the new plant. These incremental costs are estimated to be \$65 000 per year. The incremental annual costs of wood fuel are estimated to be \$375 000. The cogeneration plant will save Cogenesis from having to purchase 40 000 000 kWh of electricity per year at \$0.025 /kWh, an annual savings of \$1 000 000. Cogenesis uses a MARR of 12%. What is the present worth of the incremental investment in the cogeneration plant? What is the impact of a 5% and 10% increase and decrease in each of the parameters of the problem?

$$
\begin{aligned}
&\text{PW(cogeneration plant)}\\
&\quad = -\$3\,000\,000 - (\$65\,000 + \$375\,000 - \$1\,000\,000)\\
&\qquad \times (P/A, 12\%, 20) - \$17\,000(P/F, 12\%, 10)\\
&\qquad - \$35\,000[(P/F, 12\%, 4) + (P/F, 12\%, 8) + (P/F, 12\%, 12)\\
&\qquad + (P/F, 12\%, 16)]\\
&\quad = \$1\,126\,343
\end{aligned}
$$

The present worth of the incremental investment is \$1 126 343. Based on this assessment, the project appears to be economically viable.

In order to better understand the situation, analysts for Cogenesis have also completed some sensitivity graphs which indicate how sensitive the present worth is to changes in some of the parameters. In particular, they feel that some of the cash flows may turn out to be different from their estimates, and they would like to get a feel for what impact these errors may have on the evaluation of the cogeneration plant. To investigate, they have labelled their current estimates the "base case" and have generated other cash flow estimates that are 5% and 10% above and below the base case for each major cash flow category. These are summarized in Table 11.1.

Table 11.1 Summary Data for Example 11.1

Cost Category	−10%	−5%	Base Case	+5%	+10%
Initial investment	$2 700 000	$2 850 000	$3 000 000	$3 150 000	$3 300 000
Annual chemical, operations, and maintenance costs	58 500	61 750	65 000	68 250	71 500
Cooling tower overhaul (after 10 years)	15 300	16 150	17 000	17 850	18 700
Turbogenerator overhauls (after 4, 8, 12, and 16 years)	31 500	33 250	35 000	36 750	38 500
Annual wood costs	337 500	356 250	375 000	393 750	412 500
Annual savings in electricity costs	900 000	950 000	1 000 000	1 050 000	1 100 000
MARR	0.108	0.114	0.12	0.126	0.132

For example, the initial investment may be more than the estimate of $3 000 000 if they run into unforeseen difficulties in the installation. Or the savings in electricity costs may be overestimated if the cost per kWh drops in the future. The analysts would like to get a better understanding of which of these changes would have the greatest impact on the evaluation of the plant.

To keep the illustration simple, we will consider changes to the initial investment; annual chemical, operations, and maintenance costs; the MARR; and the savings in electrical costs. Each of these is varied one at a time, leaving all other cash flow estimates at the base case values. For example, if the initial investment is 10% below the initial estimate of $3 000 000, and all other estimates are as in the base case, the present worth of the project will be $1 426 343 (see the first row of Table 11.2, under −10%). Similarly, if the first cost is 10% more than the original estimate, the present worth drops to $826 343.

Interest rate uncertainty will almost always be present in an economic analysis. If Cogenesis' MARR increases by 10% (with all other parameters at their base case values), the present worth of the project drops to $835 115, about the same impact as if the first cost ended up being 10% more than expected. Other variations are shown in Table 11.2. A sensitivity graph, shown in Figure 11.1, illustrates the impact of one-at-a-time parameter variations on the present worth.

Small changes in the annual chemical, operations, and maintenance costs do not have much of an impact on the present worth of the project, as can be seen from Table 11.2 and Figure 11.1. What appears to have the greatest impact on the viability of the project is the savings in electricity costs. A 10% drop in the savings causes the present worth of the project to drop to about one-third of the base case estimate. This change could occur because of a drop in electricity rates or a drop in demand. Alternatively, the present worth of the project increases to almost $1 900 000 if the savings are higher than anticipated. This could, once again, occur because of a change in either rates or demand for power. Clearly, if Cogenesis is to expend effort in getting better forecasts, it should be for energy consumption and power rates.

One final point about this example should be noted. If management feels that, individually, the cash flow estimates will fall within the ± 10% range, the investment looks economically viable (i.e., yields a positive present worth) and they should go ahead with it.

Table 11.2 Present Worth of Variations from Base Case in Example 11.1

Cost Category	−10%	−5%	Base Case	+5%	+10%
Initial investment	$1 426 343	$1 276 343	$1 126 343	$ 976 343	$ 826 343
Annual chemical, operations, and maintenance costs	1 174 894	1 150 619	1 126 343	1 102 067	1 077 792
Cooling tower overhaul (after 10 years)	1 126 890	1 126 617	1 126 343	1 126 069	1 125 796
Turbogenerator overhauls (after 4, 8, 12, and 16 years)	1 131 450	1 128 897	1 126 343	1 123 789	1 121 236
Annual wood costs	1 406 447	1 266 395	1 126 343	986 291	846 239
Savings in electricity costs	379 399	752 871	1 126 343	1 499 815	1 873 287
MARR	1 456 693	1 286 224	1 126 343	976 224	835 115

■

As we can see from Example 11.1, the benefit of a sensitivity graph is that it can be used to select key parameters in an economic analysis. It is easy to understand and communicates a lot of information in a single diagram. There are, however, several shortcomings of sensitivity graphs. First, they are valid

Figure 11.1 Sensitivity Graph for Example 11.1

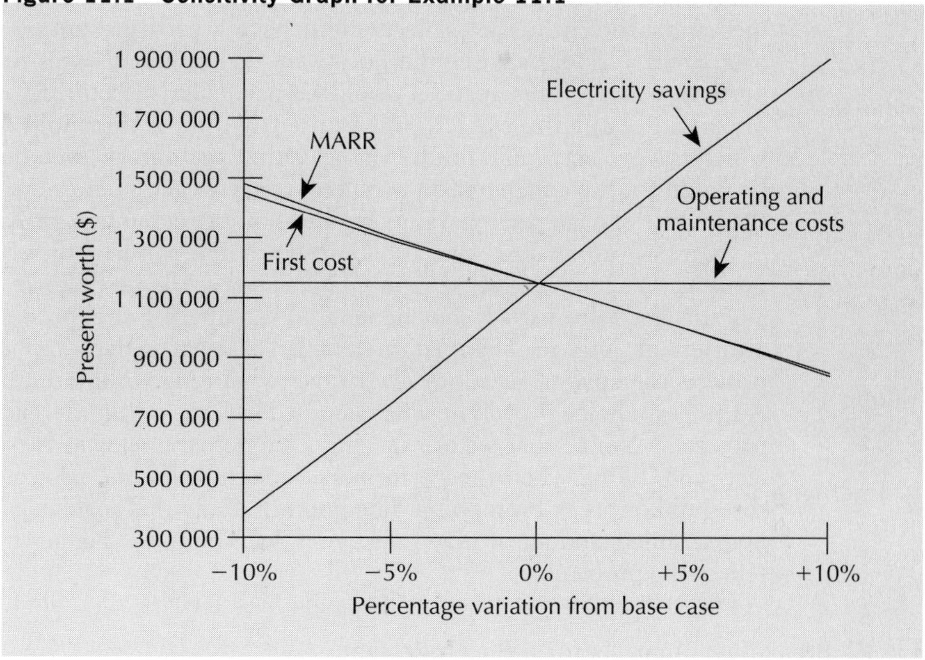

Problems w/ SA: 1. only valid over range 2. dependent parameters

only over the range of parameter values in the graph. The impact of parameter variations outside the range considered may not be simply a linear extrapolation of the lines in the graph. If you need to assess the impact of greater variations, the computations should be redone. Second, and probably the greatest drawback of sensitivity graphs, is that they do not consider the interaction between two or more parameters that may exist. You cannot simply "add up" the impact of individual changes when several parameters are varied, producing an interaction effect. We will come back to this issue in the section on scenario analysis, where we do consider entire "packages" of changes from the base case.

11.3 Break-Even Analysis

In this section, we cover a second type of sensitivity analysis called break-even analysis. Once again, we are trying to answer the question of what impact changes (or errors) in parameter estimates will have on the economic performance measures we use in our analyses, or on a decision made on the basis of an economic performance measure. In general, **break-even analysis** is the process of varying a parameter of a problem and determining what parameter value causes the performance measure to reach some threshold or "break-even" value. In Example 11.1, we saw that an increase in the MARR caused the present worth of the cogeneration plant to decrease. If the MARR were to increase sufficiently, the project might have a zero present worth. A break-even analysis could answer the question, "What MARR will result in a zero present worth?" This analysis would be particularly useful if Cogenesis were uncertain about the MARR and wanted to find a threshold MARR above which the project would not be viable. Other such break-even questions could be posed for the cogeneration problem, to try to get a better understanding of the impact of changes in parameter values on the economic analysis.

Break-even analysis can also be used in the comparison of two or more projects. We have already seen in Chapter 4 that the best choice among mutually exclusive alternatives may depend on the interest rate, production level, or a variety of other problem parameters. Break-even analysis applied to multiple projects can answer questions like, "Over what range of interest rates is project A the best choice?" or "For what output level are we indifferent between two projects?" Notice that we are varying a single parameter in two or more projects and asking when the performance measure for the projects meets some threshold or break-even point. The point of doing this analysis is to try to get a better understanding of how sensitive a decision is to changes in the parameters of the problem.

11.3.1 Break-Even Analysis for a Single Project

In this section, we show how break-even analysis can be applied to a single project to illustrate how sensitive a project evaluation is to changes in project parameters. We will continue with Example 11.1 to expand upon the information provided by the sensitivity graphs.

Example 11.2

Having completed the sensitivity graph in Example 11.1, management recognizes that the present worth of the cogeneration plant is quite sensitive to the savings in electricity costs, the MARR, and the initial costs. Since there is some uncertainty about these estimates, they want to explore further the impact of changes in these parameters on the viability of the project. You are to carry out a break-even analysis for each of these parameters to find out what range of values results in a viable project (i.e., PW > 0) and to determine the "break-even" parameter values which make the present worth of the project zero. You are also to construct a graph to illustrate the present worth of the project as a function of each parameter.

First, Figure 11.2 shows the present worth of the project as a function of the MARR. It shows that the break-even MARR is 17.73%. In other words, the project will have a positive present worth for any MARR less than 17.73% (all other parameters fixed) and a negative present worth for a MARR more than 17.73%. Notice that the break-even interest rate is, in fact, the IRR for the project.

Figure 11.2 Break-Even Chart for the MARR

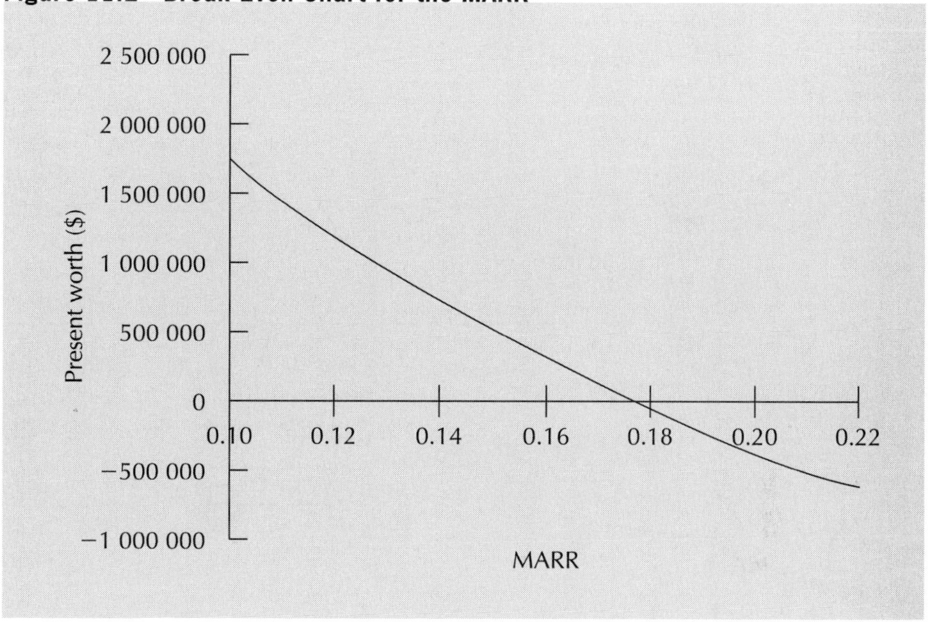

A similar break-even chart for the first cost, Figure 11.3, shows that the first cost can be as high as $4 126 350 before the present worth declines to zero. Assuming that all other cost estimates are accurate, the project will be

Figure 11.3 Break-Even Chart for First Cost

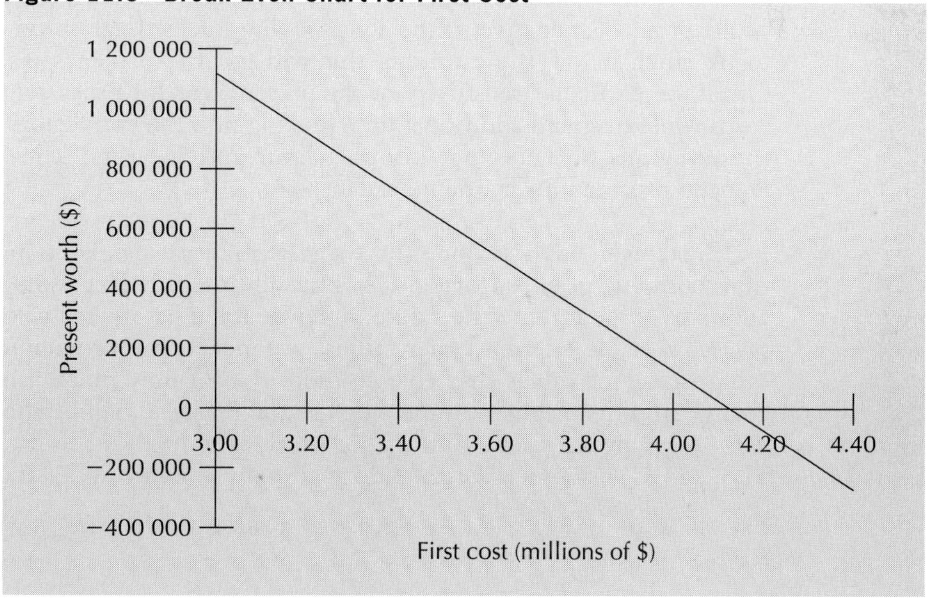

Figure 11.4 Break-Even Chart for Electricity Savings

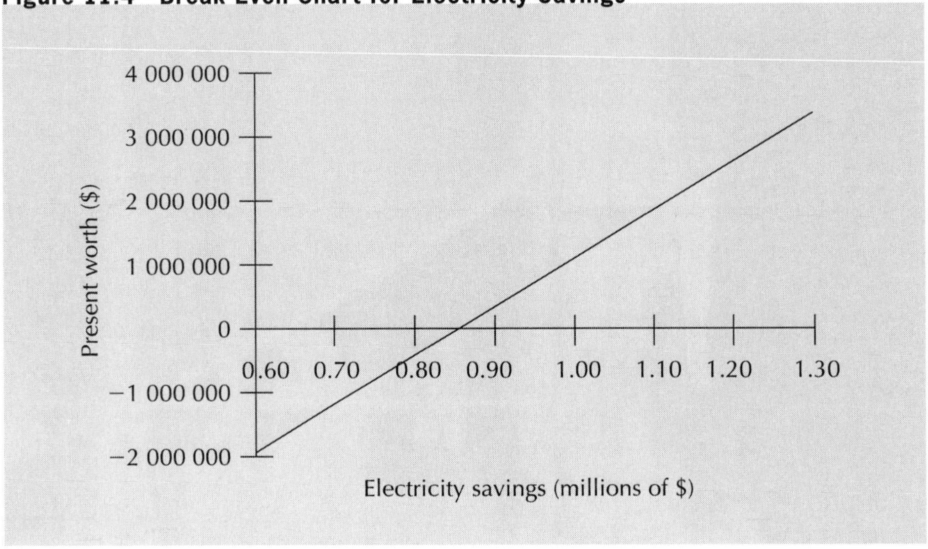

viable as long as the first cost is below this break-even amount. One issue management should assess is the likelihood that the first cost will exceed \$4 126 350.

Finally, a break-even chart for the savings in electrical power costs is shown in Figure 11.4. We have already seen from the sensitivity graph that the viability of the project is very sensitive to the savings in electricity produced by the cogeneration plant. Provided that the annual savings are above \$849 207, the project is viable. Below this break-even level, the present worth of the project is negative. If the actual saving in electrical power costs is likely to be much below the estimate, this will put the project's viability at risk. Given the particular sensitivity of the present worth to the savings, it may be worthwhile to spend additional time looking into the two factors that make up these savings: the cost per kilowatt-hour and the total kilowatt-hours of demand provided for by the new plant. ■

Break-even analysis done for a single project expands upon the information sensitivity graphs provide. It has the advantage that it is easy to apply and allows us to determine the range of values for a parameter within which the project is viable or some other criteria are met. It can provide us with break-even parameter values that give an indication of how much a parameter can change from its original estimate before the project's viability becomes a concern. Graphical presentation of the break-even analysis, as in Figures 11.2, 11.3, and 11.4, summarizes the information in an easily understood way.

11.3.2 Break-Even Analysis for Multiple Projects

In the previous section, we saw how break-even analysis can be applied to a single project in order to understand more clearly the impact of changes in parameter values on the evaluation of the project. This break-even analysis may influence a decision as to whether the project should be done or not. When there is a choice among several projects, be they independent or mutually exclusive, the basic question remains the same. We are concerned with the impact that changes in problem parameters have on the relevant economic performance measure, and, ultimately, on the decision made with respect to the projects. With one project, we are concerned with whether the project should be done and how changes in parameter values affect this decision. With multiple projects, we are concerned about how changes in parameter values affect which project or projects are chosen.

For multiple independent projects, assuming that there are sufficient funds to finance all projects, break-even analysis can be carried out on each project independently, as was done for a single project in the previous section. This will lead to insights into how robust a decision is under changes in the parameters.

For mutually exclusive projects, the best choice will seldom stand out as clearly superior from all points of view. Even if we have narrowed down the choices, it is still likely that the best choice may depend on a particular interest rate, level of output, or first cost. A break-even comparison can reveal the range over which each alternative is preferred and can show the break-even points where we are indifferent between two projects. Break-even analysis will provide a decision maker with further information about each of the projects and how they relate to one another when parameters change.

Example 11.3

Westmount Waxworks (see Problem 4.18) is considering buying a new wax melter for their line of replicas of statues of government leaders. They have two choices of suppliers, Finedetail and Simplicity. The proposals are as follows:

	Finedetail	Simplicity
Expected life	7 years	10 years
First cost	\$200 000	\$350 000
Maintenance	\$10 000/year + \$0.05/unit	\$20 000/year + \$0.01/unit
Labour	\$1.25/unit	\$0.50/unit
Other costs	\$6 500/year + \$0.95/unit	\$15 500/year + \$0.55/unit
Salvage value	\$5 000	\$20 000

The marketing manager has indicated that sales have averaged 50 000 units per year over the past five years. In addition to this information, management thinks that they will sell about 30 000 replicas per year if there is stability in world governments. If the world becomes very unsettled so that there are frequent overturns of governments, sales may be as high as 200 000 units per year. There is also some uncertainty about the "other costs" of the Simplicity wax melter. These include energy costs and an allowance for scrap. Though the costs are estimated to be \$0.55 per unit, the Simplicity model is a new technology, and the costs may be as low as \$0.45 per unit or as high as \$0.75 per unit. Westmount Waxworks would like to carry out a break-even analysis on the sales volume and on the "other costs" of the Simplicity wax melter. They want to know which the preferred supplier would be as sales vary from 30 000 per year to 200 000 per year. They also wish to know which is the preferred supplier if the "other costs" per unit for the Simplicity model are as low as \$0.45 per unit or as high as \$0.75 per unit. Westmount Waxworks uses an after-tax MARR of 15% for equipment projects. Their tax rate is 40% and the CCA rate for such equipment is 30%.

Assuming that the "other costs" of the Simplicity wax melter are \$0.55 per unit, a break-even chart which shows the present worth of the projects as a function of sales levels can give much insight into the supplier selection. Table 11.3 gives the annual cost of each of the two alternatives, and Figure 11.5 shows the break-even chart for sales level. A sample computation for the Finedetail wax melter at the 60 000 sales level is

$$\begin{aligned}\text{AW(Finedetail)} &= \text{CCTF}_{\text{new}}(\$200\,000)(A/P, 15\%, 7)\\&\quad - \text{CCTF}_{\text{old}}(\$5000)(A/F, 15\%, 7) + (1 - t)[\$10\,000\\&\quad + \$6500 + (\$0.05 + \$1.25 + \$0.95)(\text{sales level})]\\&= 0.75073(\$200\,000)(0.24036) - 0.73333(\$5000)(0.09036)\\&\quad + (1 - 0.4)[\$16\,500 + \$2.25(60\,000)]\\&\cong \$126\,658\end{aligned}$$

where

$$\text{CCTF}_{\text{old}} = 1 - \frac{td}{i+d} = 1 - \frac{0.4(0.3)}{0.15 + 0.30} \cong 0.73333$$

$$\text{CCTF}_{\text{new}} = 1 - \frac{td\left(1 + \frac{i}{2}\right)}{(i+d)(1+i)} = 1 - \frac{(0.4)(0.3)(1+0.075)}{(0.15+0.30)(1+0.15)}$$

$$\cong 0.75073$$

If sales are 30 000 units per year, the Finedetail wax melter is slightly preferred to the Simplicity melter. At a sales level of 200 000 units per year, the preference is for the Simplicity wax melter. Interpolation of the amounts in

Table 11.3 Annual Cost as a Function of Sales

Sales (Units)	Annual Costs ($)	
	Finedetail	Simplicity
20 000	72 658	85 651
60 000	126 658	111 091
100 000	180 658	136 531
140 000	234 658	161 971
180 000	288 658	187 411
220 000	342 658	212 851

Table 11.3 indicates that the break-even sales level is 38 199 units. That is to say, for sales below 38 199 per year, Finedetail is preferred, and Simplicity is preferred for sales levels of 38 199 units and above.

Since 30 000 units per year is the lowest sales will likely be, and sales have averaged 50 000 units per year over the past five years, it appears that the Simplicity wax melter would be the preferred choice, assuming that its "other costs" per unit is $0.55. The robustness of this decision may be affected by the other types of costs, such as maintenance and labour, of the Simplicity melter.

To assess the sensitivity of the choice of wax melter to the variable other costs of Simplicity, a break-even analysis similar to that for sales level can be carried out. We can vary the "other costs" from the estimate of $0.45 per unit to $0.75 per

Figure 11.5 Break-Even Chart for Sales Level

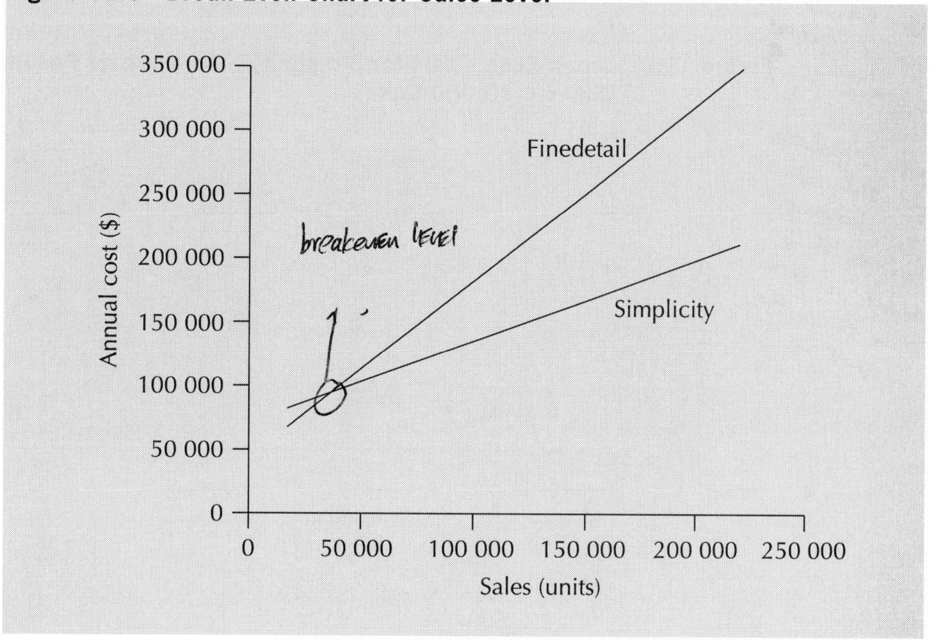

unit and observe the effect on the preferred wax melter. Table 11.4 gives the annual costs for the two wax melters as a function of the "other costs" of the Simplicity model for sales levels of 30 000, 50 000, and 200 000 units per year. In each case, we see that the best choice is not sensitive to the "other costs" of the Simplicity wax melter. In fact, for a sales level of 30 000 units per year, the break-even "other cost" is less than $0.25, as shown in Figure 11.6. This means that the other cost per unit would have to be lower than $0.25 for the best choice to change from Finedetail to Simplicity. For a sales level of 200 000 per year, the break-even "other cost" is much higher, at $1.51 per unit, and for a sales level of 50 000 units per year the break-even cost per unit is $0.83. For both of the latter sales levels, the Simplicity model is preferred.

Having done the break-even analysis for both sales level and "other costs" per unit for the Simplicity wax melter, it would appear that the Simplicity model is the better choice if sales are at all likely to exceed the break-even

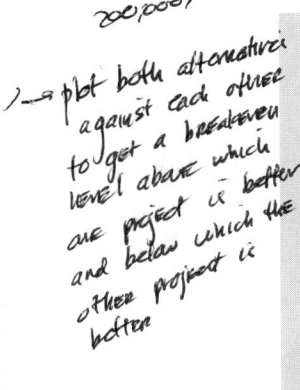

Table 11.4: Annual Cost as a Function of Simplicity's Other Costs Per Unit

Other Costs per Unit ($)	Sales = 30 000 units/year Annual Costs ($)		Sales = 50 000 units/year Annual Costs ($)		Sales = 200 000 units/year Annual Costs ($)	
	Finedetail	Simplicity	Finedetail	Simplicity	Finedetail	Simplicity
0.45	86 158	90 211	113 158	101 731	315 658	188 131
0.55	86 158	92 011	113 158	104 731	315 658	200 131
0.65	86 158	93 811	113 158	107 731	315 658	212 131
0.75	86 158	95 611	113 158	110 731	315 658	224 131

→ Finedetail better for "other costs" @ 30,000

→ Simplicity better for other costs > 30,000 (50,000 & 200,000)

→ plot both alternatives against each other to get a breakeven level above which one project is better and below which the other project is better

Figure 11.6 Break-Even Chart for Simplicity's Other Costs Per Unit (Sales = 30,000 Units)

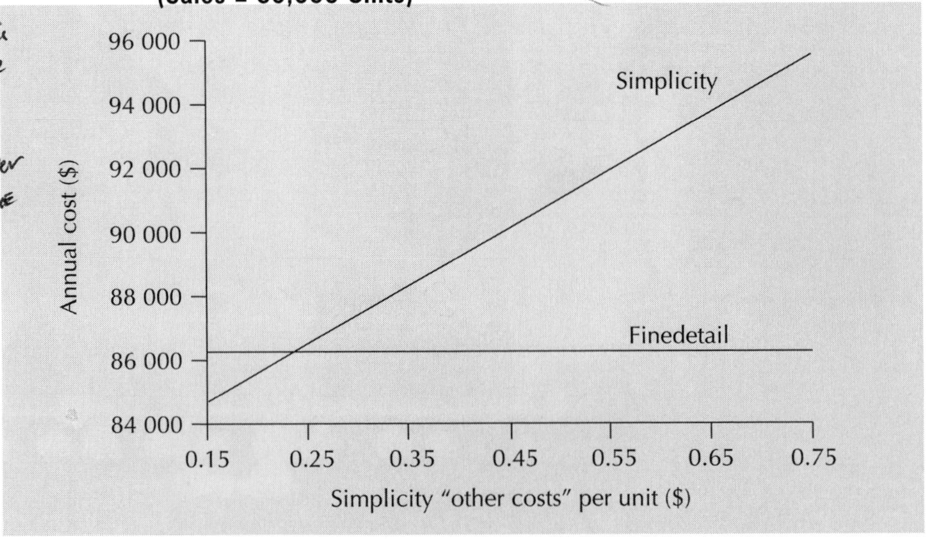

sales level of 38 199. Historically, sales have exceeded this amount. Even if sales in a particular year fall below the break-even level, the Simplicity wax melter does not have annual costs far in excess of those of the Finedetail model, so the decision would appear to be robust with respect to possible sales levels. Similarly, the decision is not sensitive to the other cost per unit of the Simplicity wax melter. ■

We have seen in this section that break-even analysis for either a single project or multiple projects is a simple tool and that it can be used to extract insights from a modest amount of data. It communicates threshold (break-even) parameter values where preference changes from one alternative to another or where a project changes from being economically justified to not justified. Break-even analysis is a popular means of assessing the impact of errors or changes in parameter values on an economic performance measure or a decision.

The main disadvantage of break-even analysis is that it cannot easily capture interdependencies among variables. Although we can vary one or two parameters at a time and graph the results, more complicated analyses are not often feasible. This disadvantage can be overcome to some degree by what is referred to as scenario analysis, the subject of the next section.

11.4 Scenario Analysis

The third type of sensitivity analysis tool that we will look at is scenario analysis. **Scenario analysis** is the process of examining the consequences of several possible sets of variables associated with a project. Scenario analysis recognizes that many estimates of cash flows or other project parameters may vary from what is projected. It is useful to look at several "what if" scenarios in order to understand the effect of changes in values of whole sets of parameters. Commonly used scenarios are the "optimistic" (or "best case") outcome, the "pessimistic" (or "worst case") outcome, and the "expected" (or "most likely") outcome. The best case and worst case outcomes can, in some sense, capture the entire range of possible outcomes for a project or a comparison among projects and provide an enriched view of the decision.

Example 11.4

Cogenesis (refer to Example 11.1) wishes to do a scenario analysis of their cogeneration problem in order to decide whether or not the project should be undertaken. They have come up with optimistic, pessimistic, and expected estimates of each of the parameters for their decision problem in order to get a better understanding of the possible range of present worth outcomes for the cogeneration plant. The three scenarios and the associated estimates are summarized in Table 11.5.

Table 11.5 Present Worth of Cogeneration Plant Scenarios

Cost Category	Pessimistic Scenario	Expected Scenario	Optimistic Scenario
Initial investment ($)	3 300 000	3 000 000	2 700 000
Annual chemical, operations, and maintenance costs ($)	75 000	65 000	60 000
Cooling tower overhaul (after 10 years) ($)	21 000	17 000	13 000
Turbogenerator overhauls (after 4, 8, 12, and 16 years) ($)	40 000	35 000	30 000
Additional annual wood costs ($)	400 000	375 000	350 000
Savings in annual electricity costs ($)	920 000	1 000 000	1 080 000
MARR	0.13	0.12	0.11
Present worth of Cogeneration plant ($)	–234 639	1 126 343	2 583 848

The scenarios capture combinations of parameter estimates which reflect the worst, best, and expected outcomes for the project. In contrast with sensitivity graphs and break-even analysis, scenario analysis allows entire groups of parameters to be changed at one time.

Evaluation of each scenario reveals that the present worth of the cogeneration plant will be negative if all parameters take on their worst case values, and hence, the project is not advisable. The major problem is that the savings in electricity costs are insufficient to make up for the high first cost of the project. In contrast, both the expected case and best case scenarios lead to positive present worths, and hence, the project would be viable. (To put the present worths into context, the expected case and best case scenarios have IRRs of 17.73% and 24.29%, respectively.) From an overall point of view, there is some risk that the cogeneration project will have a negative present worth, but this will occur only if the worst case scenario does occur. Even if the worst case outcome does occur, the loss is not huge compared with the potential gain in the other two cases. What Cogenesis needs to do if they wish to look further into the project's viability is assess the risk (or likelihood) that the worst outcome will occur. Decision making under risk is discussed further in Chapter 12. ■

As we can see from Example 11.4, scenario analysis allows us to look at the effect of multiple changes in parameter values on an individual project's viability. It can also be used to evaluate the effect of scenarios in a case where there are several alternatives.

Example 11.5

Westmount Waxworks has carried out a scenario analysis for three possible outcomes they feel represent pessimistic, optimistic, and expected outcomes for sales levels and the Simplicity wax melter's other costs per unit. The scenarios

Table 11.6 Scenario Analysis for Westmount Waxworks

	Pessimistic Scenario	Expected Scenario	Optimistic Scenario
Sales level (units)	30 000	50 000	200 000
Other costs per unit (Simplicity)	$0.75	$0.55	$0.45
Annual Cost: Finedetail	$86 158	$113 158	$315 658
Annual Cost: Simplicity	$95 611	$104 731	$188 131

and the annual costs of the two wax melters are summarized in Table 11.6. From the scenario analysis, we see that the Simplicity wax melter is the preferred choice for the expected and optimistic scenarios. The Finedetail wax melter is preferred only if the pessimistic scenario occurs. In terms of the opportunity cost of making the wrong choice, it is far larger if the optimistic outcome occurs ($315 658 − $188 301 = $127 357) than if the pessimistic outcome occurs ($95 611 − $86 158 = $9453). ■

As was seen in Examples 11.4 and 11.5, scenario analysis allows us to take into account the interrelationships among parameters when making a choice by examining likely groupings of parameter values in scenarios. The most commonly used scenarios are the pessimistic, optimistic, and expected outcomes. The use of scenarios allows an analyst to capture the range of possible outcomes for a project or group of projects. Done in combination with sensitivity graphs and break-even analysis, a great deal of information can be obtained regarding the economic viability of a project.

The one drawback common to each of the three sensitivity analysis methods covered in this chapter is that they do not capture the likelihood that a parameter will take on a certain value or the likelihood that a certain scenario will occur. This information can further guide a decision maker and is often crucial to assessing the risk of the worst case outcome. Chapter 12 will describe how these concerns are addressed.

REVIEW PROBLEMS

The following case is the basis of Review Problems 11.1 through 11.3.

Burnaby Insurance Inc. is considering two independent energy efficiency improvement projects. Each has a lifetime of 10 years and will have a scrap value of zero at the end of this time. Burnaby can afford to do both if both are economically justified. The first project involves installing high-efficiency motors in their air conditioning system. High-efficiency units use about 7% less electricity than the current motors, which represents annual savings of 70 000 kWh. They cost $28 000 to purchase and install and will require maintenance costs of $700 annually.

The second project involves installing a heat exchange unit in the current ventilation system. During the winter, the heat exchange unit transfers heat from warm room air to the cold ventilation air before the air is sent back into the building. This will save about 2 250 000 cubic feet of natural gas per year. In the summer, the heat exchange unit removes heat from the hot ventilation air before it is added to the cooler room air for recirculation. This saves about 29 000 kilowatt-hours of electricity annually. Each heat exchange unit costs \$40 000 to purchase and install and annual maintenance costs are \$3200.

Burnaby Insurance would like to evaluate the two projects, but there is some uncertainty surrounding what the electricity and natural gas prices will be over the life of the project. Current prices are \$0.07 per kilowatt-hour for electricity and \$3.50 per thousand cubic feet of natural gas, but some changes are anticipated. They use a MARR of 10%.

REVIEW PROBLEM 11.1

Construct a sensitivity graph to determine the effect that a 5% and 10% drop or increase in the cost of electricity and the cost of natural gas would have upon the present worth of each project.

ANSWER

Table 11.7 gives the costs of electricity and natural gas with 5% and 10% increases and decreases from the base case of \$0.07 per kWh for electricity and \$3.50 per 1000 cubic feet of natural gas. The table also shows the present worths of the two energy efficiency projects as the costs vary.

A sample calculation for the heat exchange unit with base case costs is

$$\begin{aligned}\text{PW(Heat exchanger)} &= -\$40\,000 + (P/A, 10\%, 10) \\ &\quad \times [\$29\,000(0.07) + \$2250(3.50) - \$3200] \\ &\cong \$1199\end{aligned}$$

Table 11.7 Costs Used as the Basis of the Sensitivity Graph for Review Problem 11.1

	−10%	−5%	0%	+5%	+10%
Cost of electricity (\$/kWh)	0.063	0.0665	0.07	0.0735	0.077
Cost of natural gas (\$/1000 cubic feet)	3.15	3.325	3.5	3.675	3.85
PW of high-efficiency motor (\$)					
With changes to electricity costs	−5204	−3698	−2193	−687	818
With changes to natural gas cost	−2193	−2193	−2193	−2193	−2193
PW of heat exchanger (\$)					
With changes to electricity costs	−48	576	1199	1823	2447
With changes to natural gas cost	−3640	−1220	1199	3619	6038

Figure 11.7 is a sensitivity graph for the high-efficiency motor. It graphically illustrates the effect of changes in the costs of electricity and natural gas on the present worth of a motor. The high-efficiency motor is not economically viable at the current prices for electricity and gas. Only if there is an increase of almost 10% in electricity costs for the life of the project will the motor produce sufficient savings for the project to have a positive present worth.

Figure 11.8 is the sensitivity graph for the heat exchange unit. The heat exchange unit has a positive present worth for the current prices, but the pre-

Figure 11.7 Sensitivity Graph for the High-Efficiency Motor

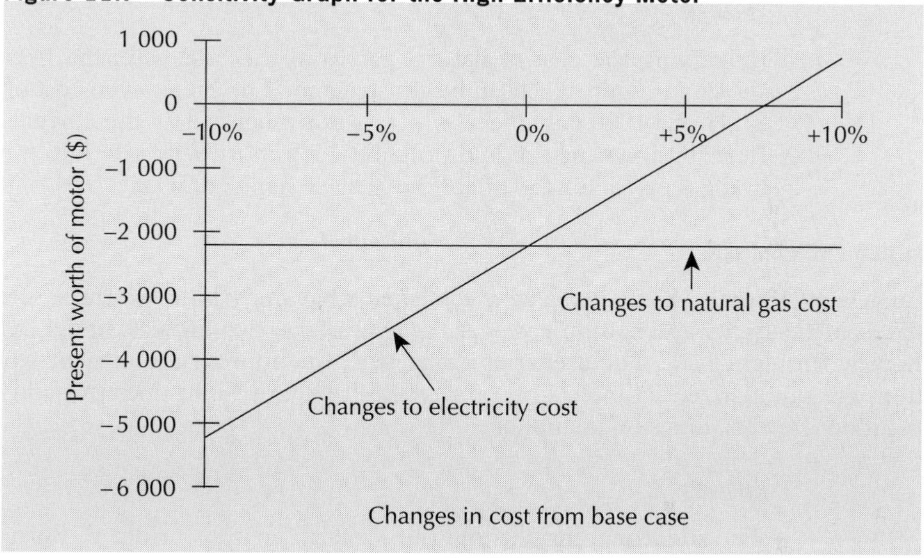

Figure 11.8 Sensitivity Graph for the Heat Exchanger

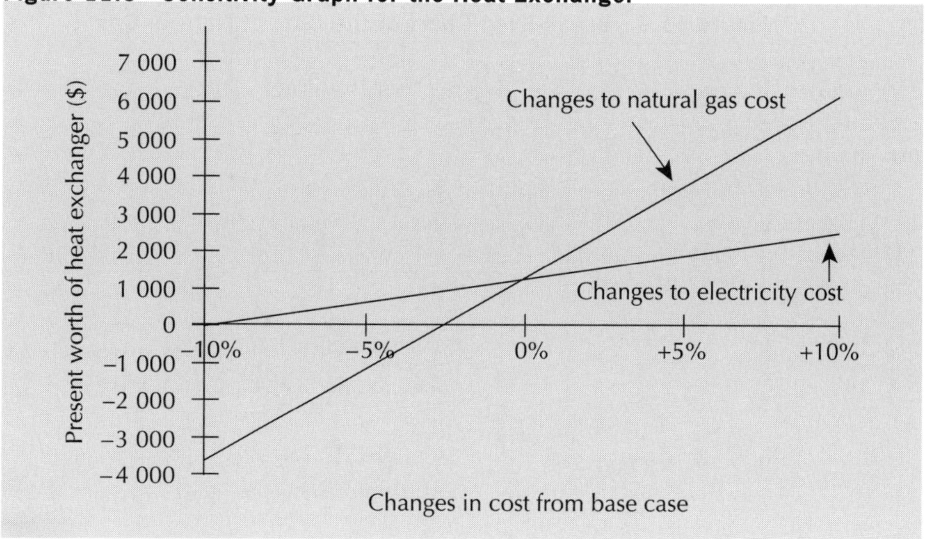

sent worth is quite sensitive to the price of natural gas. A drop in the price of natural gas in the range of only 2% to 3% (reading from the graph) will cause the project to have a negative present worth. ■

REVIEW PROBLEM 11.2

Refer to Review Problem 11.1. How much of a drop in the cost of natural gas will result in the heat exchange unit's having a present worth of zero? Construct a break-even graph to illustrate this break-even cost.

ANSWER

By varying the cost of natural gas from the base case, the break-even graph shown in Figure 11.9 can be constructed. The break-even cost of natural gas is \$3.41 per 1000 cubic feet, which is not much below the current price for gas. Burnaby Insurance should probably look more seriously into forecasts of natural gas prices for the life of the heat exchange unit. ■

REVIEW PROBLEM 11.3

Analysts at Burnaby Insurance have established what they think are three scenarios for the prices of electricity and natural gas over the lives of the two projects under consideration in Review Problem 11.1. The scenarios along with the appropriate present worth computations are summarized in Table 11.8 on page 459. What insight does this add to the investment decision for Burnaby Insurance?

ANSWER

The additional insight that the scenario analysis brings to Burnaby Insurance is the effect of changes in both electricity and natural gas costs on the two

Figure 11.9 Break-Even Chart of the Cost of Natural Gas

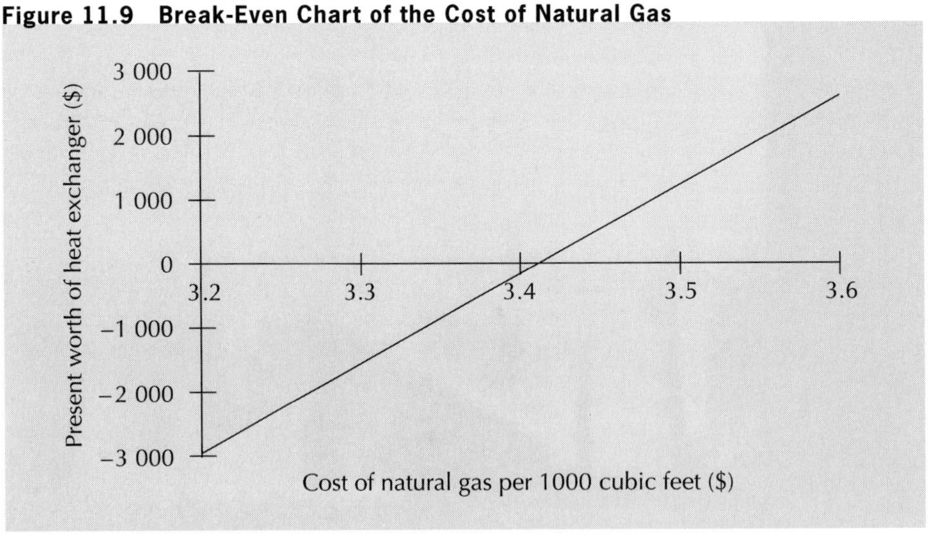

Table 11.8 Scenario Analysis for Burnaby Insurance

	Pessimistic Scenario	Expected Scenario	Optimistic Scenario
Cost of electricity ($/kWh)	0.063	0.070	0.077
Cost of natural gas ($/1000 cubit feet)	3.35	3.50	3.65
PW of high efficiency motor ($)	–5204	–2193	818
PW of heat exchange unit ($)	–2122	1199	4520

proposed projects. Sensitivity graphs and the break-even analysis can look only at the effect of one-at-a-time parameter changes on the present worth computations. It appears that the high-efficiency motor is a bad investment, as its present worth is not much above zero even if the optimistic scenario occurs. The heat exchange unit appears to be a better investment, but even that has a chance of having a negative present worth if the pessimistic outcome occurs. What Burnaby really needs to know is the likelihood of each of these scenarios occurring, or some other means of assessing the likelihood of what energy prices will be in the future. ■

SUMMARY

In this chapter, we considered three basic methods used by analysts in order to better understand the effect that uncertainties in estimated cash flows have on economic decisions. The first was the use of sensitivity graphs. Sensitivity graphs illustrate the sensitivity of a particular measure (e.g., present worth or annual worth) to changes in one or more of the parameters of a project. The second method was the use of break-even analysis for both evaluating individual projects and comparisons among projects. Finally, scenario analysis allowed us to look at the overall impact of a variety of outcomes, usually optimistic, expected, and pessimistic.

Engineering Economics in Action, Part 11B: Where the Risks Lie

Bill Astad and Naomi were working through the market demand figures provided by Powerluxe for the new self-adjusting vacuum.

"These figures are pretty ambiguous," Bill said. "We have three approaches: a set of opinions taken from focus groups and surveys of customers, the same thing from dealers and distributors, and an analysis of trends in a set of parallel products such as fuzzy-logic appliances. Like Anna said, nothing hard."

"What we really want to know," said Naomi, "is whether we have the capacity to handle the manufacturing for the product. Based on the surveys and the trend information, let's come up with three scenarios: low demand, expected demand, and high demand. If we behave according to expected demand, and the true demand is low, we will lose money because our capital investments won't be recouped as fast, and we may have passed up other opportunities. Similarly, if the demand is high, we will lose by having to pay overtime, paying for contracting out, or losing customers. But if we make money in all three cases, there really isn't much of a problem."

"And if it turns out we don't make money in all three cases?" Bill asked. "What then?"

"I know it's a lot of work, Bill, but let's do it and find out," Naomi replied. "At minimum, we will know where our risks lie."

PROBLEMS

11.1 Identify possible parameters that are involved in economic analysis for the following situations:

(a) Buying new equipment

(b) Supplying products to a foreign country with a high inflation rate

11.2 For the following examples of parameters, how would you assign a reasonable base case and a range of variation so that you can carry out sensitivity analysis? Assign specific numerical figures wherever you can.

(a) Canadian inflation rate

(b) Canadian-American exchange rate

(c) Expected annual savings from a new piece of equipment which is similar to the one you already have

(d) Expected annual revenue from an internet-based business

(e) Salvage value of a personal computer

11.3 Which sensitivity analysis method may be appropriate for analysing the following uncertain situations?

(a) Calgary Cartage leases trucks to service its shipping contracts. Larger trucks have cheaper operating costs if there is sufficient business, but are more expensive if they are not full. Calgary Cartage is not certain about their future demand.

(b) Joan runs a dog kennel. She is considering installing a heating system for the interior runs which will allow her to operate all year. Joan is not sure how much the annual heating expenses will be.

(c) Pushpa runs a one-person company producing custom paints for hobbyists. She is considering buying printing equipment to produce her own labels. However, she is not sure if she will have enough orders in the future to justify the purchase of the new equipment.

(d) Lemuel is an engineer working for Ontario Hydro. He is estimating the total cost for building transmission lines from the Darlington Nuclear Plant to new industrial parks north of Toronto. Lemuel is uncertain about the construction cost (per kilometre) of transmission lines.

(e) Thanh's company is growing very fast and has a hard time meeting its orders. An opportunity to purchase additional production equipment has arisen. She is not certain if the company will continue to grow at the same rate in the future, and she is not even certain how long the growth may last.

11.4 The Hanover Go-Kart Club has decided to build a clubhouse and track several years from now. The club needs to accumulate $50 000 by setting aside a uniform amount at the end of each year. They believe it possible to set aside $7000 each year at 10% interest. They wish to know how many years it will take to save $50 000 and how sensitive this result is to a 5% and a 10% increase or decrease in the amount saved per year and in the interest rate. Construct a sensitivity graph to illustrate the situation.

11.5 A new software package is expected to improve productivity at Saskatoon Insurance. However, because of training and implementation costs, savings are not expected to occur until the third year of operation. Annual savings of approximately $10 000 are expected, increasing by about $1000 per year for the following five years. After this time (eight years from implementation), the software will be abandoned with no scrap value. Construct a sensitivity graph showing what would happen to the present worth of the software with 7.5% and 15% increases and decreases in the interest rate, the $10 000 base savings, and the $1000 savings gradient. MARR is 15%.

11.6 The Regional Municipality of Kitchener is studying a water supply plan for the area to the end of the year 2045. To satisfy the water demand, one suggestion is to construct a pipeline from one of the Great Lakes. It is now the end of 2000. Construction would start in the year 2005 (five years from now) and take five years to complete at a cost of $20 million per year. Annual maintenance and repair costs are expected to be $2 million and will start the year following project completion (all costs are based on current estimates). From a predicted inflation rate of 3% per year, and the real MARR, city engineers have determined that a MARR of 7% per year is appropriate. Assume that all cash flows take place at the end of the year and that there is no salvage value at the end of 2040.

(a) Find the present worth of the project.

(b) Construct a sensitivity graph showing the effects of 5% and 10% increases and decreases in the construction costs, maintenance costs, and inflation rate. To which is the present worth most sensitive?

11.7 The city of Surrey is installing a new swimming pool in the downtown recreation centre. One design being considered is a reinforced concrete pool which will cost \$6 000 000 to install. Thereafter, the inner surface of the pool will need to be refinished and painted every 10 years at a cost of \$40 000 per refinishing. Assuming that the pool will have essentially an infinite life, what is the present worth of the costs associated with the pool design? The city uses a MARR of 5%. If the installation costs, refinishing costs, and MARR are subject to 5% or 10% increases or decreases, how is the present worth affected? To which parameter is the present worth most sensitive?

11.8 You and two friends are thinking about setting up a grocery delivery service for local residents to finance your last two years at university. In order to start up the business, you will need to purchase a car. You have found a used car which costs \$6000 and you expect to be able to sell it for \$3000 at the end of two years. Insurance costs are \$600 for each six months of operation, starting now. Advertising costs (e.g., flyers, newspaper advertisements) are estimated to be \$100 per month, but these could vary as much as 20% above or below the \$100, depending on the intensity of your advertising. The big questions you have now are how many customers you will have and how much of a service fee to charge per delivery. You estimate that you will have 300 deliveries each month, and are thinking of setting a \$2 per delivery fee, payable at the end of each month. The interest rate over the two-year period is expected to be 8% per year, compounded monthly, but may be 20% above or below this figure.

Using equivalent monthly worth, construct a sensitivity graph showing how sensitive the monthly worth of this project will be to the interest rate, advertising costs, and the number of deliveries you make each month. To which parameter is the equivalent monthly worth most sensitive?

11.9 Timmons Testing (TT) does subcontracting work for printed circuit board manufacturers. They perform a variety of specialized functional tests on the assembled circuit boards. TT is considering buying a new probing device which will assist the technicians in diagnosing functional defects in the printed circuit boards. Two vendors have given them quotes on first costs and expected operating costs over the life of their equipment.

	Vendor A	Vendor B
Expected life	7 years	10 years
First cost	\$200 000	\$350 000
Maintenance costs	\$10 000/year + \$0.05/unit	\$20 000/year + \$0.01/unit
Labour costs	\$1.25/unit	\$0.50/unit
Other costs	\$6 500/year + \$0.95/unit	\$15 000/year + \$0.55/unit
Salvage value	\$5 000	\$20 000

Production levels vary for TT. They may be as low as 20 000 boards per year or could be as high as 200 000 boards per year if a contract currently under negotiation comes through. They expect, however, that production quantities will be about 50 000 boards. Timmons Testing uses a MARR of 15% for equipment projects, and will be using an annual worth comparison for the two devices.

Timmons Testing is aware that the equipment vendors have given them estimates only for costs. In particular, TT would like to know how sensitive the annual worth of each device is to the first cost, annual fixed costs (maintenance + other), variable costs (maintenance + labour + other), and the salvage value.

(a) Construct a sensitivity graph for Vendor A's device, showing the effects of 5% and 10% decreases and increases in the first cost, annual fixed costs, variable costs, and the salvage value. Assume an annual production level of 50 000 units.

(b) Construct a sensitivity graph for Vendor B's device, showing the effects of 5% and 10% decreases and increases in the first cost, annual fixed costs, variable costs, and the salvage value. Assume an annual production level of 50 000 units.

11.10 Manitoba Metalworks would like to implement a local area network (LAN) for file transfer, e-mail, and database access throughout its facility. Two feasible network topologies have been identified, which they have labelled Alternative A and Alternative B. The three main components of costs for the network are (1) initial hardware and installation costs, (2) initial software development costs, and (3) software and hardware maintenance costs. The installation and hardware costs for both systems are somewhat uncertain as prices for the components are changing and Manitoba Metalworks are not sure of the installation costs for the LAN hardware. The costs for each alternative are summarized below.

Benefits from the LAN are increased productivity because of faster file transfer times, reduced data redundancy, and improved data accuracy because of the database access. The benefits were difficult to quantify and are stated below as only a range of possible values and an average.

	Alternative A	Alternative B
Initial hardware and installation costs ($)		
Optimistic estimate	70 000	86 000
Average estimate	92 500	105 500
Pessimistic estimate	115 000	125 000
Initial software cost ($)	138 750	158 250
Annual maintenance costs ($)	9 250	10 550
Annual benefits ($)		
Optimistic estimate	80 000	94 000
Average estimate	65 000	74 000
Pessimistic estimate	50 000	54 000

Manitoba Metalworks uses a 15% MARR and has established a 10-year study period for this decision. They wish to compare the projects based on annual worth.

(a) Construct a sensitivity graph for Alternative A. For the base case, use the average values for the initial hardware cost and the annual benefits. Each graph should indicate the effect of a 5% and a 10% drop or increase in the initial hardware cost and the annual benefits. Which of the two factors most affects the annual worth of Alternative A?

(b) Construct a sensitivity graph for Alternative B. For the base case, use the average values for the initial hardware cost and the annual benefits. Each graph should indicate the effect of a 5% and a 10% drop or increase in the initial hardware cost and the annual benefits. Which of the two factors most affects the annual worth of Alternative B?

11.11 **(a)** Refer back to Problem 11.8. Assuming base case figures for advertising costs and interest rates, what is the break-even number of deliveries per month? Construct a graph showing the break-even number.

(b) Assuming base case figures for advertising costs and number of deliveries per month, what is the break-even interest rate? Construct a graph illustrating the break-even interest rate.

11.12 Refer back to Problem 11.4. Members of the Go-Kart Club do not wish to wait for more than five years to build their clubhouse. They have decided to start a fund-raising campaign to increase their ability to save each year between $7000 and whatever is necessary to have $50 000 saved in five years. Construct a table and a graph which illustrate how the number of years they must wait depends on the amount they save each year. What additional funds per year will allow them to save $50,000 in five years? Use a 10% interest rate.

11.13 Refer back to Problem 11.10 in which Manitoba Metalworks is considering two LAN alternatives.

(a) For Alternative A, by how much will the installation cost have to rise before the annual worth becomes zero? In other words, what is the break-even installation cost? Is the break-even level within or above the range of likely values Manitoba Metalworks has specified?

(b) What is the break-even annual benefit for Alternative A? Use the average installation costs. Is the break-even level within or above the range of likely values Manitoba Metalworks has specified?

11.14 Repeat Problem 11.13 for Alternative B.

11.15 **(a)** Refer back to Timmons Testing, Problem 11.9. TT charges $3.25 per board tested. Assuming that costs are as in Vendor A's estimates, what production level per year would allow TT to break even if they select Vendor A's equipment? That is, for what production level would annual revenues equal annual costs? Construct a graph showing total revenues and total costs for various production levels, and indicate on it the break-even production level.

(b) Repeat (a) for Vendor B's equipment.

11.16 The Bountiful Bread Company produces home bread-making machines. Currently, they pay a custom moulder $0.19 per piece (not including material costs) for the clear plastic face on the control panel. Demand for the breadmakers is forecast to be 200 000 machines per year, but there is some uncertainty surrounding this estimate. Bountiful is considering installing a plastic moulding system to produce the parts themselves. The moulder costs $20 000 plus $7000 to install, and has an expected life of six years. Operating and maintenance costs are expected to be $30 000 in the first year and to rise at the rate of 5% per year. Bountiful estimates its capital costs using a declining-balance depreciation model with a rate of 40%, and uses a MARR of 15% for such investments.

Determine the total equivalent annual cost of the new moulder. What is the cost per unit, assuming that production is 200 000 units per year? Also, determine the break-even production quantity. That is, what is the production quantity below which it is better to continue to purchase parts and above which it is better to purchase the moulder and make the parts in-house?

11.17 Trenton Trucking (TT) is considering the purchase of a new $65 000 truck. The truck is expected to generate revenues between $12 000 and $22 000 each year, and will have a salvage value of $20 000 at the end of its five-year life. TT pays taxes at the rate of 35%. The CCA rate for trucks is 30%, and TT's after-tax MARR is 12%. Find the annual worth of the truck if the annual revenues are $12 000, and for each $1000 revenue increment up to $22 000. What is the break-even annual revenue? Provide a graph to illustrate the break-even annual revenue.

11.18 A new bottle-capping machine costs $45 000, including $5000 for installation. Operating and maintenance costs are expected to be $3000 for the first year, increasing by $1000 each year thereafter. The salvage value is calculated by straight-line depreciation where a value of 0 is assumed at the end of the service life.

(a) Construct a spreadsheet that computes the Equivalent Annual Cost for the bottle capper. What is the economic life if the expected service life is 6, 7, 8, 9, or 10 years? Interest is 12%.

(b) How sensitive is the economic life to the different length of service life? Construct a sensitivity graph to illustrate this point.

11.19 A chemical plant is considering installing a new water purification system which costs $21 500. The expected service life of the system is 10 years and the salvage value is computed using the declining-balance method with a depreciation rate of 20%. The operating and maintenance costs are estimated to be $5 per hour of operation. The expected savings is $10 per operating hour.

(a) Find the annual worth of the new water purification system if the current operating hours are 1500 per year on average. MARR is 10%.

(b) What is the break-even level of operating hours? Construct a graph showing the annual worth for various levels of operating hours.

11.20 Antigonish Footwear can invest in one of two different automated clicker cutters. The first, A, has a $10 000 first cost. A similar one, B, with many extra features, has a first cost of $40 000. A will save $5000 per year over the cutter now in use. B will save between $12 000 and $15 000 per year. Each clicker cutter will last five years and have a zero scrap value.

(a) If the MARR is 10%, and B will save $15 000 per year, which alternative is better?

(b) B will save between $12 000 and $15 000 per year. Determine the IRR for the incremental investment from A to B for this range, in increments of $500. Plot savings of B versus the IRR of the incremental investment. Over what range of savings per year is your answer from part (a) valid? What is the break-even savings for alternative B such that below this amount, A is preferred and above this amount, B is preferred?

11.21 Sam is considering buying a new lawn mower. He has a choice between a "Lawn Guy" mower or a Bargain Joe's "Clip Job" mower. Sam has a MARR of 5%. The mowers' salvage values at the end of their respective service lives is zero. Sam has collected the following information about the two mowers:

	Lawn Guy	**Clip Job**
First cost	$350	$120
Life	10 years	4 years
Annual gas	$60	$40
Annual maintenance	$30	$60

Although Sam has estimated the maintenance costs of the Clip Job at $60, he has heard that the machines have had highly variable maintenance costs. One friend claimed that her Clip Job had maintenance costs comparable to those of the Lawn Guy, but another said the maintenance costs could be as high as $80 per year. Construct a table which shows the annual worth of the Clip Job for annual maintenance costs varying from $30 per year to $80 per year. What Clip Job maintenance costs would make Sam indifferent between the two mowers, based on annual worth? Construct a graph showing the break-even maintenance costs. What mower would you recommend to Sam?

11.22 Ganesh is considering buying a \$24 000 car. After five years, he thinks he will be able to sell the car for \$8000, but this is just an estimate that he is not certain about. He is confident that gas will cost \$2000 per year, insurance \$800 per year, and parking \$600 per year, and that maintenance costs for the first year will be \$1000, rising by \$400 per year thereafter.

The alternative is for Ganesh to take taxis everywhere. This will cost an estimated \$7000 per year. If he has no car, Ganesh will rent a car for the family vacation each year at a total (year-end) cost of \$1000. Ganesh values money at 11% annual interest. If the salvage value of the car is \$8000, should he buy the car? Base your answer on annual worth. Determine the annual worth of the car for a variety of salvage values so that you can help Ganesh decide whether this uncertainty will affect his decision. For what break-even salvage value will he be indifferent between taking taxis and buying a car? Construct a break-even graph showing the annual worth of both alternatives as a function of the salvage value of the car. What advice would you give Ganesh?

11.23 Ridgely Custom Metal Products (RCMP) must purchase a new tube bender. They are considering two alternatives which have the following characteristics:

	Model T	Model A
First cost	\$100 000	\$150 000
Economic life	5 years	5 years
Yearly savings	\$50 000	\$62 000
Salvage value	\$20 000	\$30 000

Construct a break-even graph showing the present worth of each alternative as a function of interest rates between 6% and 20%. Which is the preferred choice at 8% interest? Which is the preferred choice at 16% interest? What is the break-even interest rate?

11.24 Julia must choose between two different designs for a safety enclosure. Model A has a life of three years, has a first cost of \$8000, and requires maintenance of \$1000 per year. She believes that a salvage value can be estimated for Model A using a depreciation rate of between 30% and 40% and declining-balance depreciation. Model B will last four years, has a first cost of \$10 000, and has maintenance costs of \$700 per year. A salvage value for Model B can be estimated using straight-line depreciation and the knowledge that after one year the salvage value will be \$7500. Interest is at 11%. Which of the two models would you suggest Julia choose? What break-even depreciation rate for Model A will make her indifferent between the two models? Construct a sensitivity graph showing the break-even depreciation rate.

11.25 Your neighbour, Kelly Strome, is trying to make a decision about his growing home-based copying business. He needs to acquire colour copiers able to handle maps and other large documents. He is looking at one set of copiers that will cost $15 000 to purchase. If he purchases the equipment, he will need to buy a maintenance contract that will cost $1000 for the first year, rising by $400 per year afterwards. He intends to keep the copiers for five years, and expects to salvage them for $2500. The CCA rate for office equipment is 20%.

Rather than buy the copiers, Kelly could lease them for $5500 per year with no maintenance fee. His business volume has varied over the past few years, and his tax rate has varied from a low of 20% to a high of 40%. Kelly's current cost of capital is 8%. Kelly has asked you for some help in deciding what to do. He wants to know whether he should lease or buy the copiers, and, moreover, he wants to know the impact of his tax rate on the decision. Evaluate both alternatives for him for a variety of tax rates between 20% and 40% so that you can advise him confidently. What do you advise?

11.26 Alberta Insurance wants to introduce a new accounting software package for their Human Resources department. A small-scale version is sufficient and economical if the number of employees is less than 50. A large-scale version is effective for managing 80 employees or more. All relevant information on the two packages is shown in the following table. Alberta Insurance's business is growing, and the number of employees has increased from 10 to 40 in the last 3 years. Construct a graph showing the annual worth of the two software packages as a function of the number of employees ranging from 40 to 100. Based on break-even analysis, which accounting package is a better choice for Alberta Insurance? MARR is 12%.

		Small-Scale	Large-Scale
First cost ($)		6000	10 000
Training cost at the time of installation ($)		1500	3500
Service life (years)		5	5
Salvage value		0	0
Expected annual savings ($ per employee) if the average number of employees over the next 5 years is:	less than 50	200	250
	between 50 and 80	170	300
	greater than 80	120	400

11.27 Refer back to Problem 11.10 in which Manitoba Metalworks is looking at several LAN alternatives. Conduct a scenario analysis for each alternative, using the pessimistic, expected (average), and optimistic outcomes for both installation costs and annual benefits. Which of the two alternatives would you choose? Why?

11.28 Refer back to Problem 11.4 in which the Hanover Go-Kart Club is trying to save $50 000 in order to build a clubhouse and track. They have established optimistic, expected, and pessimistic estimates for both the interest rate they will earn on their savings and the amount they will be able to save per year. Conduct a scenario analysis which shows the number of years required to save $50 000 for the three scenarios, using the data below.

Parameter	Pessimistic Scenario	Expected Scenario	Optimistic Scenario
Savings per year ($)	6000	7000	8000
Interest rate	8.00%	11.00%	12.00%
Number of years to save $50 000	6.64	5.66	4.94

11.29 Timmons Testing (refer back to Problem 11.9) has established pessimistic, expected, and optimistic figures for the first cost and other costs of the two testing devices offered by Vendors A and B. They would like you to carry out a scenario analysis to determine which of the two alternatives to choose. They charge $3.25 per board tested. What recommendations would you give Timmons Testing?

Alternative A

Parameter	Pessimistic Scenario	Expected Scenario	Optimistic Scenario
First cost ($)	220 000	200 000	190 000
Annual fixed costs ($)	18 000	16 500	13 000
Annual variable costs ($ per board)	2.35	2.25	2.20
Salvage value ($)	2 000	5 000	7 000
Annual production volume (boards)	40 000	50 000	80 000

Alternative B

Parameter	Pessimistic Scenario	Expected Scenario	Optimistic Scenario
First cost ($)	365 000	350 000	320 000
Annual fixed costs ($)	45 000	35 500	25 000
Annual variable costs ($ per board)	1.100	1.060	1.010
Salvage value ($)	17 000	20 000	23 000
Annual production volume (boards)	40 000	50 000	80 000

11.30 The Bountiful Bread Company (Problem 11.16) currently pays a custom moulder $0.19 per piece (not including material costs) for the clear plastic face on the control panel of bread-maker machines they manufacture. Demand for the bread-makers is estimated at 200 000 machines per year, but there is some uncertainty surrounding this estimate. Bountiful is considering installing a plastic moulding system to produce the parts themselves. Installation costs are $7000, and the moulder has an expected life of six years. Operating and maintenance costs are somewhat uncertain, but are expected to be $30 000 in the first year and to rise at the rate of 5% per year. Bountiful estimates its capital costs with a declining-balance depreciation model with a rate of 40%, and uses a MARR of 15% for such investments.

Parameter	Pessimistic Scenario	Expected Scenario	Optimistic Scenario
First cost ($)	25 000	20 000	18 000
Base annual operating and maintenance costs ($)	35 000	30 000	27 000
Production volume (units)	170 000	200 000	240 000

The project engineers have come up with pessimistic, expected, and optimistic figures for the first cost, operating and maintenance costs, and production levels. Determine the total equivalent annual cost of the new moulder for each scenario and then the cost per unit. What advice would you give to Bountiful regarding the purchase of the moulder?

Carlsbad Springs "Steady-Flow" Water Supply System

CANADIAN 11.1 MINI-CASE

In 1988, the rural community of Carlsbad Springs near Ottawa was experiencing widespread well-water problems both in quality and quantity. Several conventional water supply systems were initially considered by the Regional Municipality of Ottawa-Carleton. However, due to the type of soils in the area (clay, silt, and sandy soils with a high water table), the conventional systems were found to be difficult and extremely costly to install. An alternative system ("steady-flow" system) was then proposed based on a water supply system used in Alberta. The alternative system, which installs small-diameter, high-density polyethylene water mains using innovative construction techniques, would meet needs in Carlsbad Springs at significantly lower cost than traditional methods.

One of the major design challenges of the steady-flow system was finding a feasible and cost effective relationship between the flow rate to each house and the in-house storage capacity. Exhaustive sensitivity analyses were carried out to determine water usage patterns on the required flow rates, distribution pipe diameters and storage tank sizes. An optimum and affordable configuration for Carlsbad Springs was found to be a flow rate of 1.9 L/min (2700 L/day) with an in-house tank volume of 550 L.

On completion, the Carlsbad Springs water supply system was able to provide safe and potable water to 640 houses. The total project cost was $5.16 million, which translated to estimated cost savings of 70% compared to a conventional system.

Discussion

Engineering design often assumes that the world is much simpler than it really is. When an engineer designs a roof truss, for example, he or she often assumes that the lumber making up the truss behaves in a standard, predictable manner. Similarly, the engineer who designs a circuit will assume that the electrical components will behave according to their nominal values. But lumber is a natural product, and individual pieces will be weaker or stronger than expected. Electrical components, similarly, will have actual values and behaviour different, in general, from their nominal values and mathematical models.

Good engineers understand this and design accordingly. The truss-builder specifies a certain grade of lumber, or makes sure that redundant support is built into the design. The circuit designer similarly specifies the tolerances of significant components, or designs the circuit in a robust way.

The role of sensitivity analysis in economic studies is exactly the same. We don't know exactly the cash flow of a project just as we don't know exactly the behaviour of a piece of wood or a circuit component. We want to design the project to control the uncertainty of the economic elements as well as the physical ones. In the Carlsbad Springs water supply problem, sensitivity analysis was able to reconcile a novel engineering solution with small-town economics.

Questions

1) For your next job—summer, co-op, or after graduation—what salary do you expect to make? What is the lowest you would likely get? The highest? What steps could you take now to reduce the uncertainty of that salary amount?
2) Milo is considering buying a car to drive to school and back each day, and for recreational purposes. His expenses will include the car loan payment, gas, insurance, maintenance, and repairs. He is very concerned about the cost of the car, especially since his income is very limited. What considerations should he bear in mind to reduce the uncertainty of his future costs?
3) Derek has been assigned the task of designing a parking facility for an insurance company. He must keep in mind a number of different issues including land acquisition costs, building costs (if a parking building is required), expected usage, fee method (monthly fees, hourly fees, or in-and-out fees), whether the company will subsidise the facility in part or completely, etc. His boss is particularly concerned about reducing the uncertainty of the future cash flows associated with the project. How would you advise Derek?

Extended Case: Part 2

Back to the Real World

1. A Morning Meeting

Carole smiled and said a cheerful "Good morning" as Naomi and Dave came up the hallway. "Clem's not here yet, but you might as well have a seat in his office."

"Just like a manager to be late for his own meeting," observed Naomi with a smile, as she sank into her favorite green paisley chair. "Looks like we know who'll be buying the coffee this morning."

No sooner had they sat down when Clem came bounding in with a breathless "Sorry I'm late."

"No problem, Clem. Oh, don't forget the sugar in my coffee this time."

"I deserve that, Naomi," replied Clem unexpectedly. "Tell you what. You guys did such a good job on that cold former evaluation that I'll even spring for donuts today."

Dave raised his eyebrows in mock surprise. "What is this, Clem, an early Christmas?"

"Don't be so sure, Dave," Naomi interjected. "I have a feeling we're being buttered-up for something." Looking at Clem, she continued, "Do I feel some onerous task coming on?"

"'Opportunity,' Naomi. We only have 'opportunities' around here," corrected Clem, smiling. "And speaking of opportunities, it looks like we have a good one in this project." He was pointing to Naomi and Dave's cold former report. "I have a feeling that Anna Kulkowski and Ed Burns will be very happy to get this kind of cost savings."

"So we now do a full engineering economics evaluation for buying a cold former?" asked Naomi.

"Right. In fact, I'd like you to do evaluations of both the single E1 and single E2 options. No matter what you said in your report, I think that looking at these more carefully is worthwhile."

Dave thought for a minute. "Why both? If we get a better estimate of selling price from Prabha, we'll know whether the E1 or E2 is the better option."

"Yes, but," replied Clem, "I am going to recommend performing a market study on selling prices to Ed Burns right after this meeting. With the data you have here, I'm sure he'll approve the $5000 expense right away. The trouble is, it will be at least two weeks before Prabha can give us an answer. I'd like to be able to make a decision as soon as we get it."

"So...you'd like us to prepare an evaluation for each machine...leaving the selling price as a variable...so that when we get the better estimate of what the price will be, we can make an immediate decision about which machine to buy...that sounds like a 'break-even' analysis, Clem," finished Naomi.

"Can't you just see those mental wheels spinning?" said Clem to Dave with a grin. "That's exactly what I'd like. Not only would a break-even analysis show us which machine would be best at whatever selling price we end up getting, but it would also give us an idea of how sensitive the benefits of this project are to the price at which we sell excess production."

Naomi and Dave were quiet for a few seconds while they thought this over.

Dave glanced up. "Would this mean that

if we show enough extra benefit from selling our excess capacity, Marketing would be asked to put some effort into getting better prices for us?"

"That's right, Dave. A break-even analysis would give us the data to demonstrate to management how much the company would benefit and what the payoff would be as a function of the price Marketing gets for those parts."

Naomi broke in. "But for which measure should we do a break-even analysis: present worth, IRR, or payback period?"

"Anna and Ed base their decisions mostly on PW and IRR values. I know they will ask how each behaves as a function of selling price, so we should do a break-even analysis on both."

"Both?"

"Both. We'll still have to calculate payback periods as per the company capital justification procedure, but we can calculate it over a few values of selling price and demand growth if that's more convenient than a break-even analysis."

"'A few values of demand growth'?" echoed Naomi.

"That's right. We still only have a poor estimate of the demand growth rate. So, just like you've already done, we will need to do all the analyses over three possible growth rates: 5, 10, and 15%. It didn't look like the optimal decision was affected by demand growth in your analysis, but we'd better check anyway."

Naomi and Dave glanced at each other, both recognizing a few long work days coming. "Anything else?" asked Dave, with only the slightest hint of sarcasm.

"Glad you asked, Dave. Of course, these will all be after-tax calculations, and you'll have to take salvage values into account. Oh, and write up your results in a full engineering report; the President and everyone between her and us are going to read it. Am I forgetting anything?"

"Inflation, too?" Dave couldn't keep the sour tone out of his voice completely.

"No," Clem laughed. "You're off the hook on that one. Inflation looks like it will continue to stay low for the foreseeable future, so we will ignore it."

"Are we limited to just two options, or can we look for better alternatives?" Naomi asked.

"Good question, Naomi. These two are currently the most likely options to produce the best solution, depending on the selling price we get. We definitely need to do a full study of them. If you want, you can look for a better solution; if you find one it will certainly be a feather in your cap. On the other hand, remember that you still have all your other work to do."

"Don't worry, Clem, we certainly remember that we have other work to do, too," Dave said, dryly but with a smile.

"Well, I've got one last question, Clem," Dave said, a few seconds later. "Who did you say was buying the coffee today?"

"And donuts, too!" piped in Naomi.

2. Down to Details

Later that morning, Naomi met Dave in his office to divide up the work. His office was enclosed, and thankfully so, because the walls protected the rest of the offices from Dave's unique organizing style.

"Looks like we're going to be crunching numbers for some time," Dave said, while clearing a chair for Naomi to sit in.

"Maybe not, Dave. We've got most of the spreadsheets for the calculations already set up. We'll have to include salvage values and tax effects. It turns out that the spreadsheet already has PW and IRR financial functions built in."

Dave continued to clear space on the table in front of Naomi as they talked. "That will certainly save a lot of trial-and-error to come up with IRRs, but what about all the cases we have to do to make break-even graphs?"

Naomi gestured at Dave's computer, almost the only thing in the office not hidden by paper. "I've found a feature in Excel called 'Data Tables' — I suppose most spreadsheet programs have something like it. It allows us to vary the contents of several spreadsheets based on data in a table. It looks like that will allow us to calculate all the values we need in no time."

"Sounds like I know who'll be doing the spreadsheets this time," said Dave. "How about if I concentrate on writing the report?" Dave had a clump of papers in his hand that he had not yet repositioned, and was pacing about the room looking for a place to put them.

"O.K. Oh, before I forget, this equipment falls into the CCA asset class 8? CCA rate is 20%?"

"That's right. We can also get a good estimate of salvage value with a declining balance depreciation rate of 20%. By the way, what was that you said to Clem about better options?"

"Oh, I've got some ideas. I'll let you know if they pan out."

Finally Dave sat down with a clear workspace in front of both of them. "Fair enough. According to my stomach, it must be lunch time. Want to see if Clem wants to go over to the Grand China?" He jumped out of his chair again, after spending no more than three seconds sitting down.

"Sure," said Naomi. "Do we have to spread these papers out again before we leave?"

QUESTIONS

1. Using data provided in both parts of the Extended Case, update your spreadsheets to include salvage value, tax effects, PW, IRR, and Payback calculations. PW, IRR, and Payback analyses are required for each project (single E1 machine or single E2 machine purchased at time 0). Summarize PW, IRR, and Payback values in tables for each of 5, 10, and 15% demand growth and \$0.03, 0.035, and 0.04 selling price per piece.
2. Perform a break-even analysis for each of PW and IRR for each project over the range of selling prices from \$0.03 to \$0.04. Repeat for demand growth rates of 5, 10, and 15% (i.e., 2 comparison methods × 2 projects × 3 growth rates = 12 break-even calculations). Present the results graphically.
3. Based on your current analysis results, can you make a clear recommendation to the company's management which of E1 and E2 should be purchased? If so, why? If not, why not?
4. Write a full engineering economics report to the President, Anna Kulkowski, about this project. The report should follow the format of an engineering report. See the guideline given below. In the report, you should present the results of your analyses so that company's management can make a defensible decision about which machine they should purchase.

Optional

5. Currently we only consider purchasing one machine, either E1 or E2, at time 0. Find out if a better solution exists by varying the number of machines purchased at a time and/or the timing of the purchase (i.e., time 1, 2, etc.). Include your findings in the report.

Guideline for an Engineering Report

Typical contents of an engineering report may include the following:

Cover Letter ("Here is the report you ordered.")

Title Page

Table of Contents

Summary (of the contents of the report)

Introduction (e.g., description of the problem, background, and purpose)

Main Body (e.g., procedure, calculations, results, possible errors, and unanswered questions)

Conclusions

Recommendations

References

Appendices (important but too lengthy or disorderly to be included in the body)

Notes for Excel Users (similar issues exist for other spreadsheet programs)

Circular References. When calculating after-tax IRRs, you might run across a problem in Excel with Circular Reference warnings. This occurs because you use the IRR rate to find the CCTFs, which you then use to calculate the IRR. The way to get around this problem is to go to the Excel "Tools" menu, then select "Options." Select the "Calculation" tab and click on the word "Iteration," then "OK." This will set the worksheet to perform iterative calculations which will converge for the value of the IRR. You might try pressing the F9 (re-calculate) key a few times to make sure the values have converged.

IRR and NPV (Present Worth) Built-In Functions. Excel has built-in functions for calculating IRR and PW (Excel uses the name NPV: Net Present Value for Present Worth). Check the Excel help for details. Be careful with the NPV function: it calculates a PW for one period in the future. You have to adjust the value for time 0.

Data Tables. You can run analyses for many different cases very quickly by using the Data Table feature. It can be found under the "Data/Table..." menu.

CHAPTER 12

Dealing with Risk: Probability Analysis

Engineering Economics in Action, Part 12A: Trees from Another Planet

The coffee cups had been cleaned up, but the numerous brown rings that remained reported the hours of work Bill and Naomi had put in calculating the economic effects of manufacturing the new self-adjusting vacuum cleaner in partnership with the Powerluxe company.

"So, the bottom line is: if the demand is low, we lose money, but otherwise we gain. And more than that, if demand is high, we gain big-time." Bill was a bit plaintive because he knew that this did not solve their problems—he had hoped that they would make money no matter what. "So what it really comes down to is this: What is the chance that demand is low?"

Naomi looked into space for a second, and then thoughtfully started, "No, Bill, it's a little more complicated than that. It's..."

Just then, she was interrupted by Clem, who stepped in the lunchroom door. Glancing at the papers spread over the table, he remarked, "Not much of a lunch! Anyhow, I've been talking to Ms. Kulkowski, and she said that you should take into account in your Powerluxe project the potential for a competitive product."

"You mean another vacuum cleaner, just like the—what are they calling it, the 'Adaptamatic'—with the sensing capability?" asked Naomi.

"Yeah," replied Clem. "I guess the big shots are worried that those guys over at Erie Gadgets have some sort of similar thing under study. Well, that's all I had to tell you. Work hard." Clem winked and disappeared as quickly as he arrived.

Bill leaned back in his chair with his arms in the air. "For heaven's sake! We don't have any chance of coming to a clear recommendation now!"

"No, I think we're O.K., Bill. In fact, since you mention it, 'chance' is pretty important here. So are trees."

Bill looked at Naomi as if she were from another planet. "Trees! Trees?"

12.1 Introduction

In our day-to-day life, we often encounter situations where we don't know for sure what events will happen in the future. We talk about the "chance" that the weather will be rainy tomorrow, the "likelihood" that our favourite hockey team will win the Stanley Cup, or the "probability" of a railcar spill. Of course, once the event occurs, we know the outcome: either it did or did not rain, our team won or didn't, or the railcar spill happened or didn't happen.

In this section, we will define more precisely what we mean by "chance" and "likelihood" so that we are able to make useful predictions in the context of engineering economics. The branch of mathematics that formalizes this common notion of "chance" is called *probability theory*. We will start with a simple example, such as rolling a die, to illustrate some basic concepts which were the classical origins of probability theory from several hundred years ago. We then describe two tools frequently used in engineering economics decision making—*decision trees* and *Monte Carlo simulation*. Decision tree analysis addresses problems which have sequential decisions and chance events. Monte Carlo simulation is a tool for dealing with complex problems in which the likelihood of the possible outcomes is difficult to assess due to randomness of one or more input parameters.

12.2 Uncertainty and Risk

Chapter 11 dealt with sensitivity analysis tools appropriate for decision making under *uncertainty*—situations in which we can characterize a range of possible outcomes, but do not have a probability assessment associated with each outcome. Sensitivity analysis allowed us to look at the impact of this lack of probability information in a general way. Chapter 12, on the other hand, deals with decision making under *risk*, where we can characterize the range of possible outcomes and quantify the probability of each outcome. The extra probability information often permits us to draw more authoritative conclusions than we can draw using sensitivity analysis.

In this chapter we deal with two popular approaches to decision making under risk. The first, decision tree analysis, is very widely used for making the best choice among alternatives that are affected by risky future events. The second, Monte Carlo simulation, is a way to predict the outcome of many interacting risky events. Many other techniques and tools exist; for example, mean-variance analysis, aspiration level, and the most-probable future principles. These methods are beyond the scope of this book.

12.3 Basic Concepts of Probability

Suppose you are about to throw a die. You ask the likelihood, or probability, that a "two" will come up (i.e., two dots). One way to determine this probability is to roll the die many times and count the number of times a two comes up relative to the number of times you threw the die. If you compute the relative frequency of "twos" as you roll the die over and over, eventually the relative frequency will settle down to a limiting value which we refer to as a probability. In this sense, the **probability** can be defined as a long-run proportion. We often refer to probabilities as proportions because of the simplicity and intuitive appeal of this view.

By generalizing the die-rolling example, we can more formally define probability and introduce some basic properties of probability. Consider an experiment with a set of m possible outcomes, or events, referred to as e_1, e_2, …, e_m. We repeat the experiment n times, and n_i is the number of times event e_i occurs. Then we say that the probability that the outcome will be e_i is:

$$\Pr(e_i) = \lim_{n \to \infty} (n_i/n)$$

where:

e_i	is the outcome (e.g., a "two" on the roll of a die).
$\Pr(e_i)$	is the probability that the outcome will be e_i.
n_i	is the total number of times the experiment results in e_i.
n	is the total number of times the experiment is repeated.
n_i/n	is the relative frequency of e_i.

We now note some basic properties of probability. First, we know that:

$$0 \leq \Pr(e_i) \leq 1$$

since $0 \leq (n_i/n) \leq 1$ for any n, and $0 \leq \lim_{n \to \infty} (n_i/n) \leq 1$. Next, we note that:

$$\Pr(e_1) + \Pr(e_2) + \ldots + \Pr(e_m) = 1$$

since all the relative frequencies sum to 1, and therefore the limit of the relative frequencies also must sum to 1.

Going back to the die-rolling example, we can ask what is the probability that a "two" will turn up. There are six possible outcomes to this experiment, each of which is equally probable (assuming the die is fair) and their probabilities must add up to 1. The probability that a two will turn up is therefore 1/6. This example illustrates the use of classical or "symmetric" probability which was first developed for games of chance such as dice, where the outcomes are equally likely. Close-Up 12.1 summarizes different views on probability.

CLOSE-UP 12.1 Views of Probability

Classical or Symmetric Probability: This was the first view of probability and stemmed from games of chance such as dice, where the outcomes were equally likely. For example, if there are m possible outcomes, only one can and must occur, and each are equally likely, then the chance of each occurring is $1/m$. The chance of a head when flipping a coin is 1/2 because there are two sides and each is equally likely.

Relative Frequency: If an experiment is carried out M times where M is large, and a specific event occurs in n of these trials, then the probability of the event is empirically determined to be n/M. An example of this is flipping a coin 1000 times and discovering that it lands on its edge five times in 1000. The probability of it landing on its edge is then $5/1000 = 0.005$.

Subjective Probability: Subjective or personal probability is an attempt to deal with unique events which cannot be repeated and hence can't be given a frequency interpretation. In rough terms, subjective probability can be interpreted as the odds one would personally give in betting on an event, or it may be a matter of human judgment and intuition as formed by physical relationships and experimental results. An example of this is a person who judges that the chance of winning a coin toss with one of the authors of this text is very low, say 1/1000, because, based on their experience, the authors usually cheat.

Axiomatic Probability: One of the problems associated with defining probabilities is that the definition of probability requires using probability itself. To get around this circular logic, axiomatic probability makes no attempt to define probability, but simply states the rules or axioms it follows. Other properties can then be derived from the basic axioms.

Each of the above methods is the right one for the appropriate circumstances. When the physics of a process suggest a clear judgement of probability, the Classical approach makes sense. Where formal experimentation is possible, the Relative Frequency method may be justified. In many real-world cases, subjective probability supported by historical information and other data is frequently used.

12.4 Random Variables and Probability Distributions

Suppose that we are testing solder joints on a printed circuit board. We are interested in the number of open joints we discover out of (to keep it simple) three tested joints. We don't know how many "opens" there will be until after the test results become available, so the number of "opens" is a random variable which can take on the values of 0, 1, 2, or 3. A **random variable** is a variable that can take on a number of values where the probability that the random variable has a particular value is given by its probability distribution. The random variable is customarily denoted by a capital letter such as X, and a lower case letter represents its outcome. For the solder joint example, the random variable may be

X = the number of "open" joints in a sample of 3.

Since there are three joints and each joint will be either open or closed, there are four possible values of X: $x_1 = 0$, $x_2 = 1$, $x_3 = 2$, and $x_4 = 3$. Note that the "sample space," or the entire range of outcomes for this experiment, consists of eight possible outcomes or events (e_1, e_2, ..., e_8). Denoting the sequence of three joint tests by O for open and C for closed, we have e_1 = (O,O,O), e_2 = (O,O,C), e_3 = (O,C,O), e_4 = (C,O,O), e_5 = (O,C,C), e_6 = (C,C,O), e_7 = (C,O,C), and e_8 = (C,C,C). We must look through the set of outcomes to see which corresponds to $X = 0$, $X = 1$, etc. to be able to determine the probability or likelihood of each value of X.

The number of "opens" in the solder joints is an example of a *discrete* random variable. The discrete random variable can take on a finite or countable number of values. Other examples of discrete random variables are the outcome of tossing a coin (a classic!), the number of car accidents on a highway in the coming year, the number of bugs that will be found in software beta testing, and the number of spot welds a robot can successfully complete before requiring maintenance.

Symbolically, the **probability distribution** for the discrete random variable X may be denoted by $p(x)$ or $\Pr\{X = x\}$ where x is one of the values the random variable may take on. Using this notation, the probability that a "two" will turn up in the die-rolling example is $\Pr\{X = 2\} = p(2) = 1/6$. The entire probability distribution is then $p(1) = p(2) = p(3) = p(4) = p(5) = p(6) = 1/6$.

For the solder joint example, the probability distribution for X = the number of "open" joints in a sample of 3 can be calculated from the probability of an open joint for each testing. Suppose that there is 20% chance that the test discovers an open joint, and that the quality of the joint solder is the same from joint to joint (i.e., the fact that a joint is open or closed has nothing to do with another joint being open or closed—the test results are said to be *independent* of each other). Then the probability that all three tests discover an open joint is calculated by

$$\Pr(O,O,O) = \Pr(O) \times \Pr(O) \times \Pr(O) = 0.2 \times 0.2 \times 0.2 = 0.008.$$

Similar calculations yield the following probabilities for the eight possible outcomes (events) of the solder joint testing (Table 12.1).

Finally, the probability distribution of X, the number of "open" joints in the three tests, if there is a 20% chance that the test discovers an open joint, is

$$\Pr\{X = 0\} = p(0) = 0.512$$
$$\Pr\{X = 1\} = p(1) = 0.384$$
$$\Pr\{X = 2\} = p(2) = 0.096$$
$$\Pr\{X = 3\} = p(3) = 0.008$$

Table 12.1 Probability Corresponding to the Outcomes of the Solder-Joint Testing with a 20% Chance of an Open Joint

i	e_i (Outcome or Event)	Number of "Opens"	Probability
1	(O,O,O)	3	0.008
2	(O,O,C)	2	0.032
3	(O,C,O)	2	0.032
4	(C,O,O)	2	0.032
5	(O,C,C)	1	0.128
6	(C,C,O)	1	0.128
7	(C,O,C)	1	0.128
8	(C,C,C)	0	0.512

Note that the two important properties of probabilities hold: $p(x_i) \geq 0$ for all i and $\sum p(x_i) = 1$.

The probability distribution summarizes the possible values for a random variable and their associated probabilities. A histogram is often used to display the discrete probability distribution and is an excellent visual aid for describing the behaviour of the random variable. Figure 12.1 shows the probability distribution associated with the outcome of the solder joint testing. The outcome is a discrete random variable which ranges in value between 0 and 3.

In contrast to a discrete random variable, a *continuous random variable* can take on any real value within a certain interval. For example, daily demand for drinking water in a municipality might be anywhere between 10 and 200 million litres. The actual amount consumed—the outcome—is a continuous random variable with a minimum value of 10 million L and a maximum value of 200 million L.

In this chapter, we focus on applications of discrete random variables in engineering economics analysis. We do not use continuous random variables because proper treatment requires more advanced mathematical concepts such as differential and integral calculus. Also, continuous random variables can be well approximated as discrete random variables by grouping the possible output values into a number of categories or ranges. For example, rather than treating demand for drinking water as a continuous random variable, demand could be characterized as high, medium, or low, or demand that falls within a countable number of distinct ranges. Figure 12.2 shows an example of a probability distribution associated with future demand for water, approximated by a discrete random variable.

Figure 12.1 Probability Distribution for the Solder Joint Example

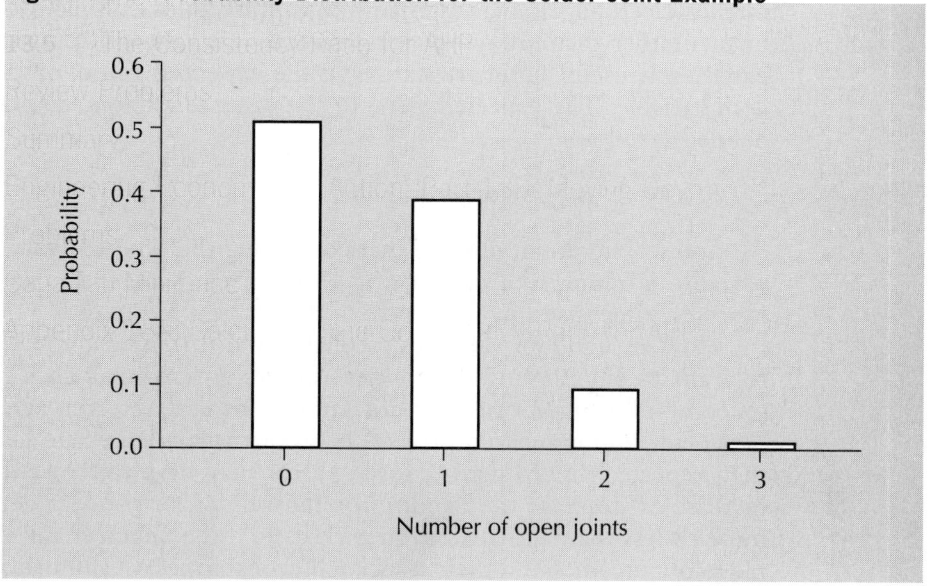

Figure 12.2 Probability Distribution of Demand for Drinking Water

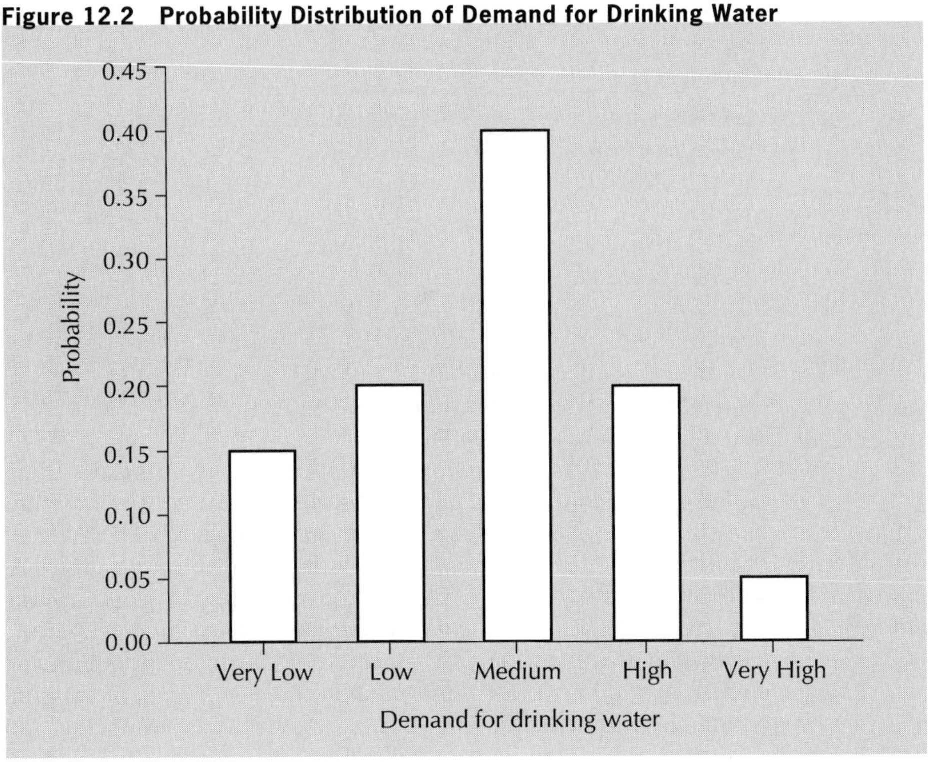

The probability distribution of a random variable contains a great deal of information that can be useful for decision-making purposes. However, we often use certain summary statistics to capture an overall picture of the distribution rather than always working directly with the entire distribution. One particularly useful summary statistic is the **expected value**, or mean, of a random variable. The expected value of the discrete random variable X, $E(X)$, is defined as follows:

$$E(X) = \sum x_i \, p(x_i) \text{ for all } i.$$

You will no doubt observe that computing the expected value of a random variable is much like computing the centre of mass for an object. The expected value is simply the centre of the probability "mass."

How are random variables, probability distributions, and expected value relevant to engineering economics? Often, an engineer does not know the outcome of a project or may not know with certainty the actual value of a parameter important to a project. A product may depreciate faster or slower than expected due to market forces, a car may have safety features that may increase or decrease the liability for the vendor, or product demand may be more or less than anticipated. In each case, if the engineer can determine the range of outcomes and their associated probabilities, this information can present a much richer view for the decision making process. Random variables and their probability distributions are, in fact, the building blocks for many tools that are useful in decision making under risk.

Examples 12.1 and 12.2 illustrate how the expected value and probability distribution information may be used in a decision making context.

Example 12.1

Recall from Example 11.5 that the management of Westmount Waxworks had some uncertainty about the future sales levels of their line of statues of government leaders. Expert opinion helped them assess the probability of the pessimistic, expected, and optimistic sales scenarios. They think that the likelihood that sales will be 50 000 per year for the next few years is roughly 50% and that the pessimistic and optimistic scenarios have likelihoods of 20% and 30%, respectively. Table 12.2 reproduces the annual cost information for the two wax melters, Finedetail and Simplicity. Based on expected annual costs, which is the best choice?

Table 12.2 Annual Cost Information for Finedetail and Simplicity

Scenario (X)	Annual Cost for Finedetail	Annual Cost for Simplicity	Probability
Pessimistic	\$ 85 314	\$ 94 381	0.2
Expected	112 314	103 501	0.5
Optimistic	314 814	186 901	0.3

The expected annual cost of the Finedetail is:

$$\begin{aligned}&\text{E(Finedetail, annual cost)}\\&\quad= (\$85\ 314)\ \Pr\{X = \text{pessimistic}\} + (\$112\ 314)\ \Pr\{X = \text{expected}\}\\&\qquad + (\$314\ 814)\ \Pr\{X = \text{optimistic}\}\\&\quad= (\$85\ 314)(0.2) + (\$112\ 314)(0.5) + (\$314\ 814)(0.3)\\&\quad= \$167\ 663\end{aligned}$$

And the expected annual cost of the Simplicity is:

$$\begin{aligned}&\text{E(Simplicity, annual cost)}\\&\quad= (\$94\ 381)(0.2) + (\$103\ 501)(0.5) + (\$186\ 901)(0.3)\\&\quad= \$126\ 697\end{aligned}$$

The expected annual cost of the Simplicity is lower than that of the Finedetail. Hence, the Simplicity is preferred on the basis of expected annual costs. ■

Example 12.2

Regional Express is a small courier service company operating in Southern Ontario. At the Toronto office, all parcels from the surrounding regions are collected, sorted, and distributed to the appropriate destinations. Regional Express is considering the purchase of a new computerized sorting device for their Toronto office. The device is so new—in fact, it is still under continuous improvement—that its maximum capacity is somewhat uncertain at the present time. They are told that the possible capacity can be 40 000, 60 000, or 80 000 parcels per month, regardless of the size of the parcels. They are also told some probabilities corresponding to the three capacity levels. Table 12.3 shows these figures. What is the expected capacity level for the new sorting device? Regional Express is growing steadily, so such a computerized sorting device will be a necessity in the future. However, if Regional Express currently deals with an average of 50 000 parcels per month, should they seriously consider purchasing the device now or should they wait?

Table 12.3 Probability Distribution for Capacity Levels of the New Sorting Device

Capacity Level (parcels/month)	Probability
40 000	0.3
60 000	0.6
80 000	0.1

The expected capacity level (X) is:

$$\begin{aligned} E(X) &= (40\,000)\,\Pr\{X = 40\,000\} + (60\,000)\,\Pr\{X = 60\,000\} \\ &\quad + (80\,000)\,\Pr\{X = 80\,000\} \\ &= (40\,000)(0.3) + (60\,000)(0.6) + (80\,000)(0.1) \\ &= 56\,000 \text{ parcels per month} \end{aligned}$$

The expected capacity level exceeds the average monthly demand of 50 000 parcels per month, so according to the expected value analysis alone, Regional Express should consider buying the sorting device now. However, by studying the probability distribution, we see that the capacity level may fall below 50 000 parcels per month with 30% chance. Perhaps Regional Express should include this information in their decision making, and ask themselves whether a 30% chance of not meeting their demand is too risky or costly for them if they decide not to purchase the sorting device. ■

12.5 Decision Tree Analysis

12.5.1 Sequential Events and Risk

Many different types of risks exist in decision making. In this section, we address problems characterized by a sequence of decisions and event outcomes. For example, a judgment about the chance of a thunderstorm tomorrow will affect your decision to plan for a picnic tomorrow afternoon. Similarly, the success or failure of a new product may largely depend on future demand for the product. As another example, a decision on the replacement interval for an asset relies on an assessment of its economic life, which can be highly uncertain if the equipment employs an emerging technology.

When a decision depends on the outcomes of a random event, the decision maker must anticipate what those outcomes might be as part of the process of analysis. This section presents a useful tool called decision tree analysis. It is particularly suited to decisions and events that have a natural sequence in time or space.

12.5.2 Decision Trees

A **decision tree** is a graphical representation of the logical structure of a decision problem in terms of the sequence of decisions and outcomes of chance events. It provides a mechanism to decompose a large and complex problem into a sequence of small and essential components. In this way, a decision tree clarifies the options a decision maker has and provides a framework in which to deal with the risk involved.

A key component to decision tree analysis is the *decision tree diagram*. Example 12.3 introduces the overall approach. A detailed explanation of the components and structure of the decision tree is included in the example.

Example 12.3

Edwin Electronics (EE) has a factory in Midland, Ontario, for assembling TVs. One of the key components is the TV screen. EE does not currently produce TV screens on-site; they are outsourced to a supplier in Barrie. Recently, EE's industrial engineering team asked if they should continue outsourcing the TV screens or produce them in-house. They realized that it was important to consider the uncertainty in demand for the company's TVs. If the future demand is low, outsourcing seems to be the reasonable option in order to save production costs. On the other hand, if the demand is high, then it may be worthwhile to produce the screens on-site, thus getting economies of scale. EE's engineers analyzed the effect of the demand uncertainty in their decision making. They represented their decision problem in a graphical manner with a decision tree. Figure 12.3 presents EE's decision tree.

Figure 12.3 Decision Tree for Edwin Electronics

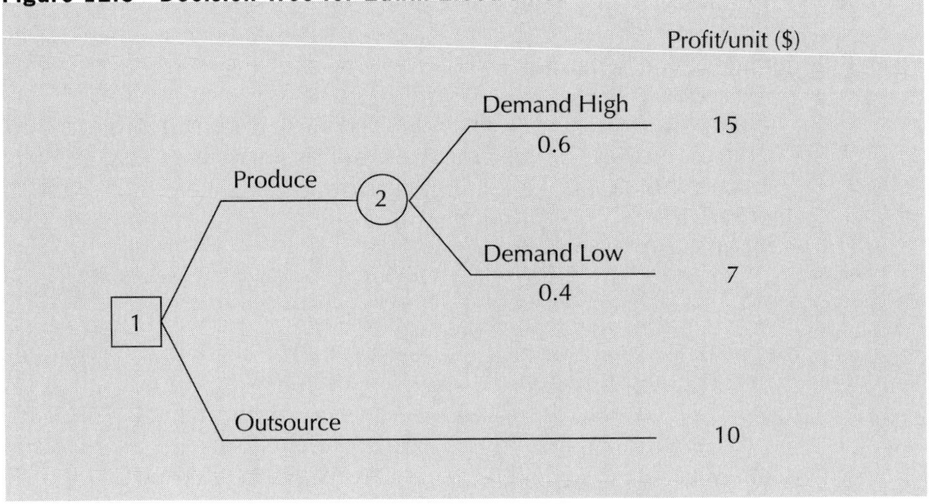

There are three main components in a decision tree: decision nodes, chance nodes, and branches. A *decision node* represents a decision to be made by the decision maker and is denoted by a square in the tree diagram. In Figure 12.3, the single square node represents the decision to produce or outsource TV screens (Node 1). A *chance node* represents an event whose outcome is uncertain, but which has to be considered during the decision making. The outcome of a chance node is a discrete random variable, as it has number of distinct outcomes and each outcome has an associated probability. A circle in the diagram denotes a chance node. The chance node in Figure 12.3 represents the uncertain demand for TV screens (Node 2). Finally, the *branches* of a tree are the lines connecting nodes from left to right, depicting the sequence of possible decisions and chance events.

A decision tree grows from left to right and usually begins with a decision node. The leftmost decision node represents an immediate decision faced by the decision maker. The branches extending from a decision node represent the decision options available for the decision maker at that node, whereas the branches extending from a chance node represent the possible outcomes of the chance event. Each branch extending from a chance node has an associated probability. In Edwin Electronics' case, the two decision options, to produce or outsource, are represented by the two branches extending from the decision node. The two branches from the chance node indicate that the future demand may be high or low. It is important in decision making that all branches out of a node, whether a decision node or chance node, constitute a set of mutually exclusive and collectively exhaustive consequences. In other words, when a decision is made, exactly one option is taken, or when uncertainty is resolved, exactly one outcome occurs as a result.

Whenever a chance node follows a decision node, as in Figure 12.3, it implies that the decision maker must anticipate the future risk in decision making. On the other hand, when a decision node follows a chance node, it implies that a decision must be made assuming a particular chance event has occurred. Finally, the rightmost branches are the terminal positions of a decision tree, indicating all possible outcomes of the overall decision situation represented by the tree. Each terminal position has an associated valuation based on a performance measure; quite typically, the performance measure is a monetary value. Edwin Electronics uses profit per TV unit as their performance measure. ■

The decision tree for Edwin Electronics in Figure 12.3 can be modified to show more complex decision situations.

Example 12.4

The EE engineering team from Example 12.3 has realized that the price per TV screen may vary in the future, especially since EE is subject to purchase conditions set by the Barrie supplier. How does this affect their decision tree?

Figure 12.4 includes the additional uncertainty in the TV screen price charged by the supplier. The price may increase, remain the same, or decrease in the future, as shown at Node 3. ■

Example 12.5

The EE engineering team then considered increasing in-house production of the TV screens if the demand is high. This raises an additional uncertainty about the ability of the market to absorb the increased production. How does this change their decision tree?

Figure 12.4 Edwin Electronics' Modified Decision Tree 1

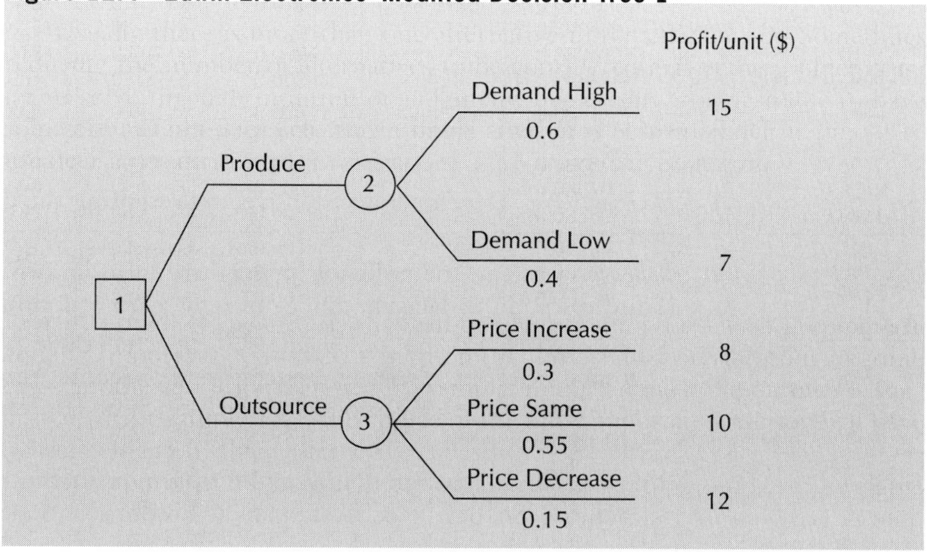

Figure 12.5 Edwin Electronics' Modified Decision Tree 2

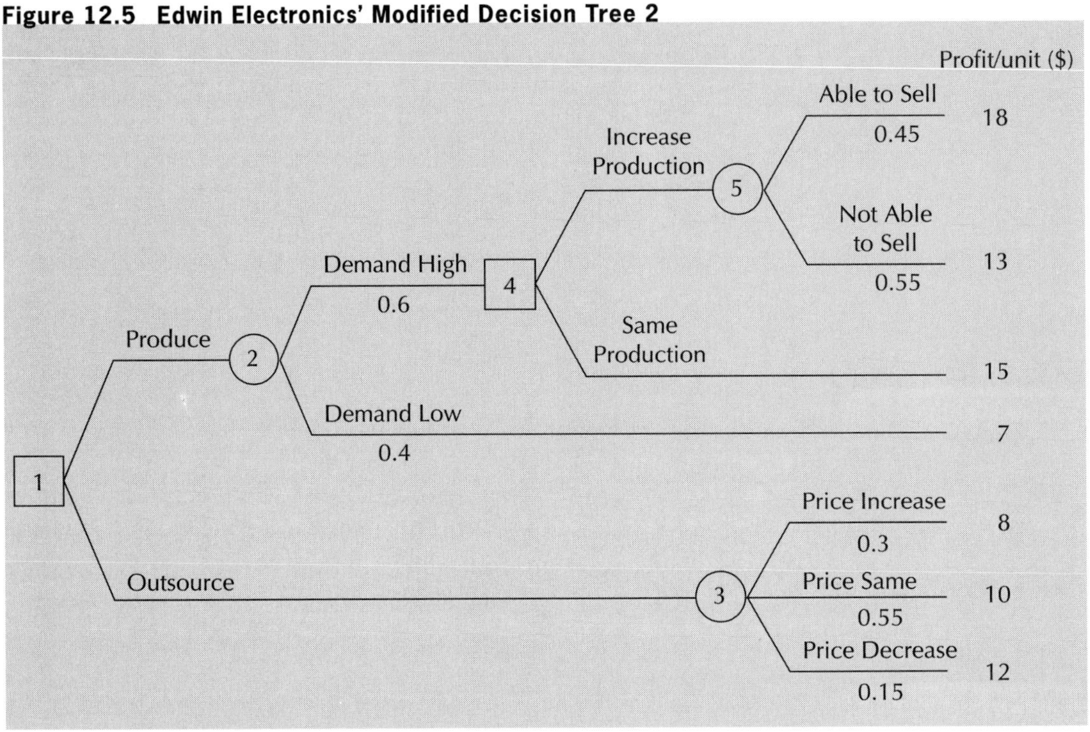

Figure 12.5 further modifies Figure 12.4 to include a new decision component and a chance event (see nodes 4 and 5). The decision component indicates the choice of whether or not to increase production. If production is increased, a new chance node captures the uncertainty about their ability to sell the excess production. ■

12.5.3 Decision Tree Analysis Procedure

Decision tree analysis is based on its structure and the idea of expected value presented in Section 12.4. It consists of three main steps, as follows:

1. *Decision Tree*. Develop a decision tree representing the decision situation in question.
2. *Rollback*. Execute the **rollback procedure** (also known as **backward induction**) on the decision tree from *right to left* as follows:
   (a) At each chance node, *compute* the expected value of the possible outcomes. The resulting expected value becomes the value associated with the chance node and the branch on the left of that node (if there is any).
   (b) At each decision node, *select* the option with the best expected value (best may be highest value or lowest cost depending on the

context). The best expected value becomes the value associated with the decision node and the branch on the left of that node (if there is any). For the option(s) not selected at this time, indicate their termination by a double-slash (//) on the corresponding branch.

(c) Continue rolling back until the leftmost node is reached.

3. *Conclusion*. The expected value associated with the final node is the expected value of the overall decision. Tracing *forward* (left to right), the non-terminated decision options indicate the set of recommended decisions at each subsequent node.

Example 12.6

Carry out a decision tree analysis on Edwin Electronics' modified tree in Figure 12.4.

Since the decision tree is already provided, Step 1 is complete. The rollback procedure described in Step 2 has two phases in this case. First, the tree is rolled back to each of the chance nodes as in Step 2(a) (phase 1). The expected values at Nodes 2 and 3 are computed as follows:

$$E(2) = 0.6(\$15) + 0.4(\$7) = \$11.80$$

$$E(3) = 0.3(\$8) + 0.55(\$10) + 0.15(\$12) = \$9.70$$

Figure 12.6 shows the rollback so far.

Figure 12.6 Phase 1 of Step 2: Rolling Back to the Chance Nodes

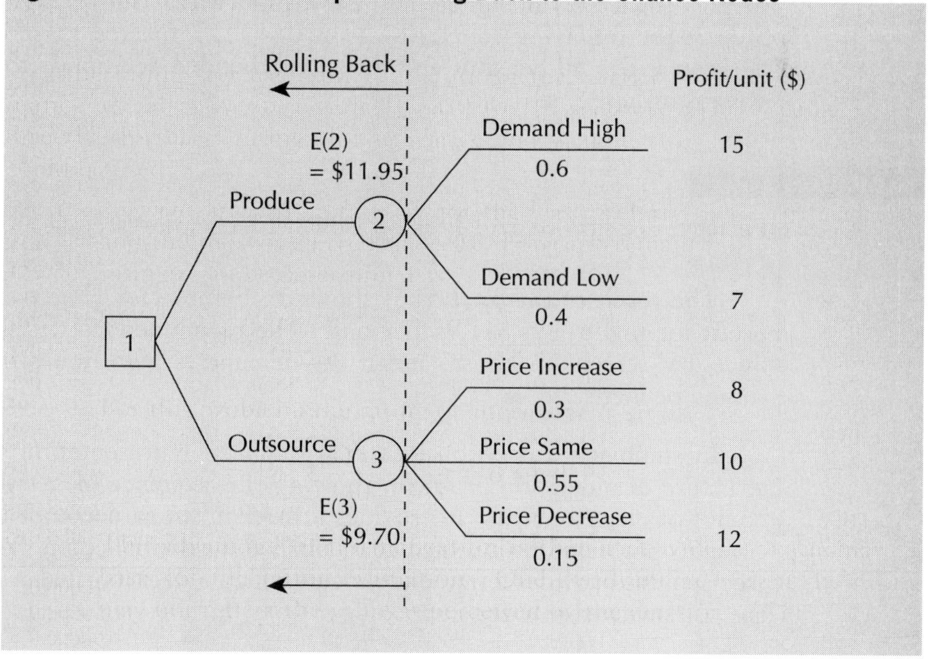

Figure 12.7 Phase 2 of Step 2: Rolling Back to the Decision Node

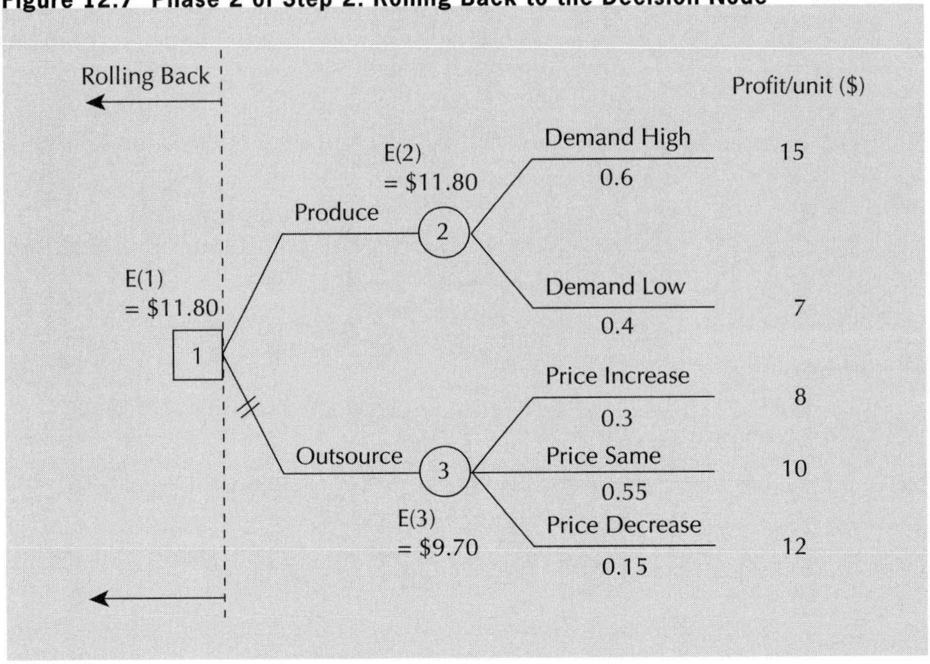

Next, the tree is further rolled back to the decision node as in Step 2(b) (phase 2). The expected value at Node 1 is then E(1) = \$11.80, which is equal to E(2) since E(2) is higher than E(3). Figure 12.7 shows this result. As for Step 3, the following conclusion is made. The expected value of the overall decision is \$11.80 per unit and the recommended decision is to produce TV screens in-house. ■

Example 12.7

Perform decision tree analysis on Edwin Electronics' second modified tree in Figure 12.5.

The result of this analysis is shown in Figure 12.8. The overall expected profit for this tree is \$11.95 per unit. The recommended decision is to produce TV screens in-house, and if the demand is high, the production level should be increased. ■

In summary, a decision tree is a graphical representation of the logical structure of a decision problem, showing the sequence of decisions and outcomes of chance events. It provides a mechanism to decompose a large and complex problem into a sequence of small and essential components. Decision tree analysis provides a framework with which a decision-maker can deal with the risk involved in a decision.

Figure 12.8 Completed Analysis for EE's Modified Decision Tree 2

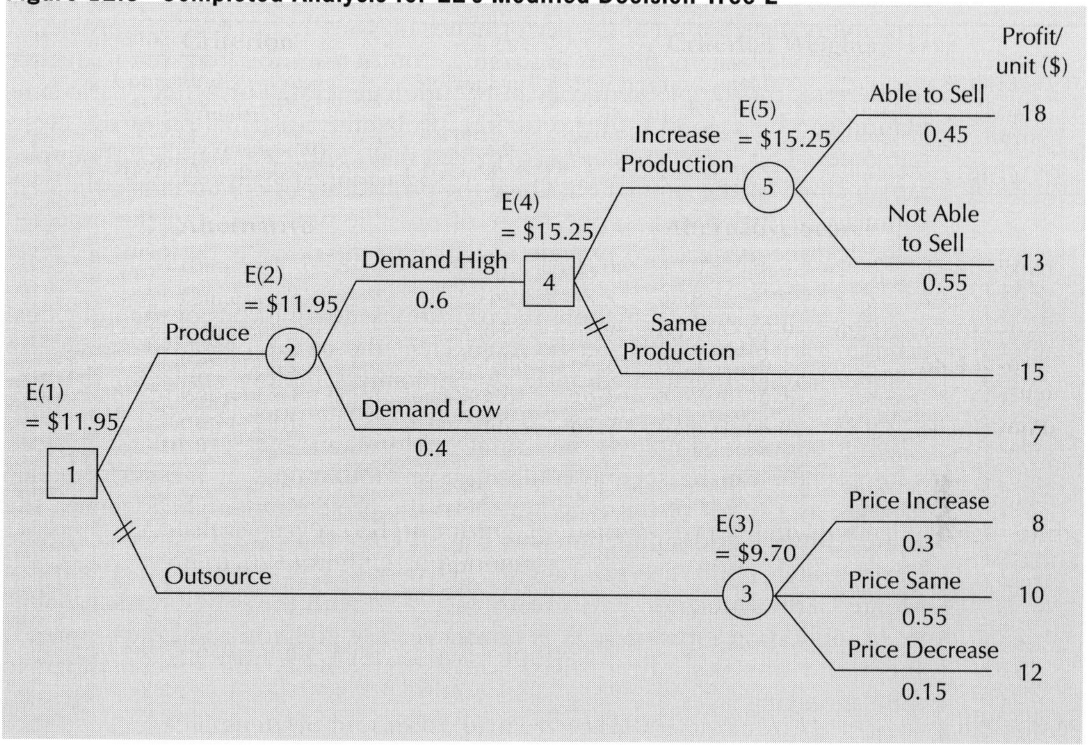

12.6 Monte Carlo Simulation

12.6.1 Dealing with Complexity

Decision tree analysis deals with risk that is present in a sequence of decisions and events. In this section, a different kind of project risk is discussed. When a project has one or more random variables or inputs, the outcome of the project itself is a random variable. It may therefore be difficult to assess the impact of the input randomness on the overall project evaluation. For instance, the present worth computation for a project may require a number of inputs, such as the initial cost, a series of revenues and savings, operating and maintenance costs, and salvage value. When one or more of these inputs are random variables, the present worth calculation inherits this randomness and may be very difficult to characterize.

12.6.2 Probability Distribution Estimation

Monte Carlo simulation attempts to construct the probability distribution of an outcome performance measure of a project (e.g., present worth) by repeatedly sampling from the input random variable probability distributions. This is a useful technique when it would otherwise be very difficult to obtain the probability distribution of the performance measure of interest for a project.

Since one basic notion of probability comes from the long-run frequency of an event, the sample frequency distribution generated by Monte Carlo simulation can be a good estimate for the probability distribution of the event outcomes. This is true, of course, provided that a sufficient number of samples are obtained by the simulation. Once the probability distribution is estimated, summary statistics such as the range of possible outcomes and the expected values can be analyzed to provide insight into the possible performance level for the project.

In Monte Carlo simulation, the probability distributions of the individual random variables, which are the input elements of the overall outcome, are assumed to be known in advance. By randomly sampling values for the random variables from the specified probability distributions, Monte Carlo simulation produces a sample of the overall performance measure for the project. This process can be seen as imitating the randomness in the performance measure as a result of the randomness of the project inputs. Incidentally, the name Monte Carlo simulation was coined after the casino games at Monte Carlo, which symbolize the random behaviour. The use and practicality of Monte Carlo simulation have greatly increased with the widespread availability of application software such as spreadsheets (e.g., Excel and Lotus), spreadsheet add-ons including @Risk and Crystal Ball, and special purpose simulation languages.

12.6.3 Monte Carlo Simulation Procedure

The following five steps are taken in Monte Carlo simulation.

1. *Analytical Model.* Identify all input random variables that affect the outcome performance measure of the project in question. Develop the equation(s) for computing the event outcome (denoted by Y) from a particular realization of the input random variables.
2. *Probability Distributions.* Establish an appropriate probability distribution for each input random variable. Note that the random variables are assumed to be independent of each other.
3. *Random Sampling.* Sample a value for each input random variable from its associated probability distribution. The following is a random sampling procedure for *discrete* random variables:
   (a) For each discrete random variable, create a table similar to Table 12.4 containing the possible outcomes, their associated probabilities, and the corresponding random number assignment ranges.
   (b) For each input random variable, generate a random number Z (see Close-Up 12.2). Find a range in which Z belongs from Table 12.4 and assign the appropriate outcome.

Table 12.4 Random Number Assignment Ranges

i	Outcome x_i	Probability $p(x_i)$	Random Number (Z) Assignment Range
1	x_1	$p(x_1)$	$0 \leq Z \leq p(x_1)$
2	x_2	$p(x_2)$	$p(x_1) \leq Z < p(x_1) + p(x_2)$
⋮	⋮	⋮	⋮
$m-1$	x_{m-1}	$p(x_{m-1})$	$p(x_1) + \ldots + p(x_{m-2}) \leq Z < p(x_1) + \ldots + p(x_{m-1})$
m	x_m	$p(x_m)$	$p(x_1) + \ldots + p(x_{m-1}) \leq Z < 1$

Table 12.5 Random Sampling Table

Step 3(b) (do for all random variables)		Step 3(c)
Random Number	Value of X	Value of Y
Generate Z	Assign value to X (one of x_i's) using Table 12.4 from Step 3(a)	Compute Y using the analytical model from Step 1

(c) Substitute the sample values of the random variables into the expression for the outcome measure, Y, and compute the value of Y. This forms one sample point in the procedure. Table 12.5 can be the basis for a spreadsheet application of random sampling.

4. *Repeat Sampling*. Continue with sampling until a sufficient sample size is obtained for the value of Y.

5. *Summary*. Summarize the frequency distribution of the sample outcomes using a histogram. Summary statistics, like the range of possible outcomes and expected values, can also be calculated from the sampled outcomes.

CLOSE-UP 12.2 Random Number Generation

In order to generate random numbers, we need a source of independent and identically distributed uniform random numbers between 0 and 1. In other words, all values between 0 and 1 need to be independent and equally likely. In the past, random number tables were used to obtain random numbers. Today, random number generation is easily achieved by using calculator functions or application software such as Excel or Lotus, which have built-in random number generators.

One point remains to be clarified. In Step 4 of Monte Carlo simulation, random sampling continues until a sufficient sample size is obtained. What is sufficient? The law of large numbers tells us that as the number of samples increases towards infinity, the frequencies will converge to the true underlying probabilities. As a practical guideline, one should aim for a sample size of at least 100 in order to obtain reasonable results. One way to ensure the validity of the end result is to monitor the expected value of the frequency distribution and continue sampling until some stability appears in it. More rigorous guidelines exist for selecting the appropriate number of samples (e.g., confidence interval methods) but are beyond the scope of this text.

Example 12.8

Pharma-Excel, a pharmaceutical company based in Halifax, is considering the worth of an R&D project that involves improvement of vitamin pills. Since this is a new research domain for Pharma, they are not certain about the related costs and benefits. As a part of the initial feasibility study, Pharma-Excel estimated the following probability distributions for the first cost and annual revenue (see Table 12.6), which are assumed to be independent quantities. Simulate the present worth of this project based on a 10-year study period using the Monte Carlo method. Does the project seem viable? Pharma-Excel's MARR is 15% for this type of project.

As the first step, the analytical expression for the present worth of the project is developed. We use the following expression for computing the present worth of the project:

$$\begin{aligned} \text{PW} &= -(\text{First Cost}) + (\text{Annual Revenue})(P/A, 15\%, 10) \\ &= -(\text{First Cost}) + (\text{Annual Revenue})(5.0188) \end{aligned}$$

Table 12.6 First Cost and Annual Revenue for Pharma-Excel's Research Project

First Cost	Probability	Annual Revenue	Probability
$1 000 000	0.2	$ 100 000	0.125
1 250 000	0.2	350 000	0.125
1 500 000	0.2	600 000	0.125
1 750 000	0.2	850 000	0.125
2 000 000	0.2	1 100 000	0.125
		1 350 000	0.125
		1 600 000	0.125
		1 850 000	0.125

Table 12.7 Random Number Assignment Ranges for the First Cost

First Cost	Probability	Random Number (Z_1) Assignment Range
\$1 000 000	0.2	$0 \leq Z_1 < 0.2$
1 250 000	0.2	$0.2 \leq Z_1 < 0.4$
1 500 000	0.2	$0.4 \leq Z_1 < 0.6$
1 750 000	0.2	$0.6 \leq Z_1 < 0.8$
2 000 000	0.2	$0.8 \leq Z_1 < 1$

Table 12.8 Random Number Assignment Ranges for the Annual Revenue

Annual Revenue	Probability	Random Number (Z_2) Assignment Range
\$100 000	0.125	$0 \leq Z_2 < 0.125$
350 000	0.125	$0.125 \leq Z_2 < 0.25$
600 000	0.125	$0.25 \leq Z_2 < 0.375$
850 000	0.125	$0.375 \leq Z_2 < 0.5$
1 100 000	0.125	$0.5 \leq Z_2 < 0.625$
1 350 000	0.125	$0.625 \leq Z_2 < 0.75$
1 600 000	0.125	$0.75 \leq Z_2 < 0.875$
1 850 000	0.125	$0.875 \leq Z_2 < 1$

The probability distributions for the first cost and annual revenue are provided in Table 12.6 (Step 2). Based on these distributions, random numbers are assigned for certain intervals of probabilities (Step 3(a)). Tables 12.7 and 12.8 summarize the random number assignment for each random variable.

Following Steps 3(b) and 3(c), the simulation results are obtained for a sample size of 200. Table 12.9 presents partial results. The histogram of the frequency distribution, shown in Figure 12.9, is also generated based on the simulation results.

Based on the simulation results, the average present worth of the research project over the 10-year study period was \$3 152 437 with a maximum value of \$8 284 780 and a minimum value of −\$1 498 120. The project exhibited a negative present worth roughly 18% of the time. From these figures, the project seems to be viable because it has a good chance of having a positive present worth which could be as high as \$8 million. ■

Table 12.9 Partial Results for the Monte Carlo Simulation

Sample Number	Random Number (Z_1)	First Cost	Random Number (Z_2)	Annual Revenue	Present Worth
1	0.076162	$1 000 000	0.605155	$1 100 000	$4 520 680
2	0.728782	1 750 000	0.293282	600 000	1 261 280
3	0.29656	1 250 000	0.747692	1 350 000	5 525 380
4	0.940748	2 000 000	0.327516	600 000	1 011 280
5	0.384964	1 250 000	0.788017	1 600 000	6 780 080
⋮	⋮	⋮	⋮	⋮	⋮

Figure 12.9 Frequency Distribution for Pharma-Excel's Research Project

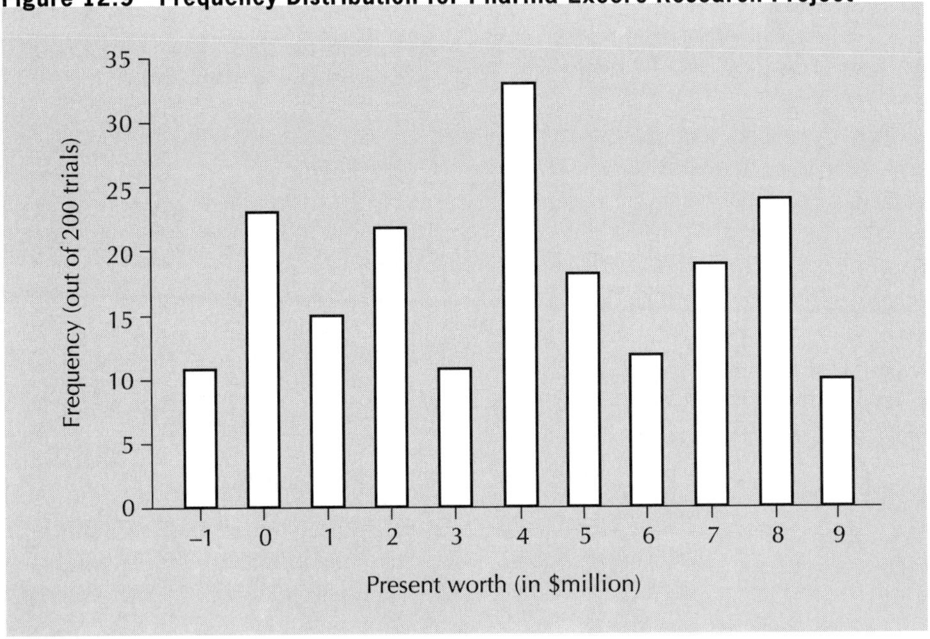

Example 12.9

While considering the cogeneration plant project outlined in Example 11.1, Cogenesis Corp. wishes to determine the probability distribution of the project's present worth to better assess the risk of a negative present worth. The three random variables they wish to investigate are the initial investment, the savings in electricity costs, and the extra wood costs. Previously, management determined that the range of possible values for the initial costs was between

$2 800 000 and $3 300 000, the range for savings in electricity costs was $920 000 to $1 080 000 per year, and the additional wood costs were $350 000 to $400 000 per year. To the best of their knowledge, management thinks that any outcome between the lower and upper bounds for the first cost, electricity savings, and additional wood costs is equally likely. Table 12.10 shows the discrete approximation of probability distributions created for the three random variables.

Table 12.10 Discrete Probability Distributions for Cogenesis' Plant Project

Initial Cost	Probability	Electricity Savings	Probability	Additional Wood Costs	Probability
$2 800 000	1/6	$ 920 000	0.2	$350 000	1/6
2 900 000	1/6	960 000	0.2	360 000	1/6
3 000 000	1/6	1 000 000	0.2	370 000	1/6
3 100 000	1/6	1 040 000	0.2	380 000	1/6
3 200 000	1/6	1 080 000	0.2	390 000	1/6
3 300 000	1/6			400 000	1/6

By randomly sampling repeatedly from each of these distributions, we can construct a probability distribution of the present worth of the project. Assuming that all other parameters are fixed at their expected scenario values, the following expression is used for computing the present worth. A portion of the Monte Carlo simulation results is shown in Table 12.11.

$$
\begin{aligned}
\text{PW} &= -(\text{First Cost}) + (\$65\,000 + \text{Wood Costs} - \text{Electricity Savings}) \\
&\quad \times (P/A, 12\%, 20) - \$17\,000(P/F, 12\%, 10) - \$35\,000[(P/F, 12\%, 4) \\
&\quad + (P/F, 12\%, 8) + (P/F, 12\%, 12) + (P/F, 12\%, 16)] \\
&= -(\text{First Cost}) + (\$65\,000 + \text{Wood Costs} - \text{Electricity Savings}) \\
&\quad \times (7.4694) - \$17\,000(0.32197) - \$35\,000(0.63552 + 0.40388 \\
&\quad + 0.25668 + 0.16312) \\
&= -(\text{First Cost}) + 7.4694(\$65\,000 + \text{Wood Costs} \\
&\quad - \text{Electricity Savings}) - \$56\,545.49
\end{aligned}
$$

By sampling a total of 200 times and computing a present worth for each sample, management arrived at the histogram shown in Figure 12.10. The average present worth was $1 007 816 with a minimum of $42 032 and a maximum of $2 110 606. There were no instances where the present worth turned out to be zero or less. The risk of this project yielding a negative present worth appears to be negligible, assuming, of course, that the probability distributions for the input parameters have been specified correctly! ■

Table 12.11 Partial Results for the Monte Carlo Simulation

Initial Cost	Electricity Savings	Additional Wood Costs	Present Worth
$3 200 000	$1 080 000	$360 000	$1 635 912
2 800 000	920 000	350 000	915 502
3 000 000	960 000	380 000	790 196
2 900 000	920 000	370 000	666 114
⋮	⋮	⋮	⋮

Figure 12.10 Frequency Distribution for Cogenesis' Plant Project

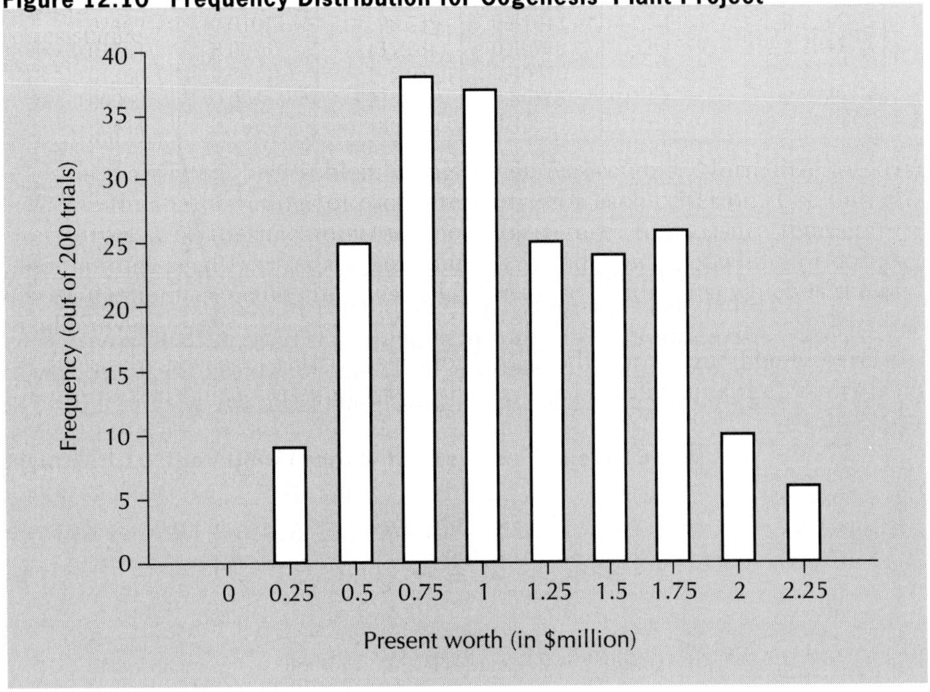

12.7 Application Issues

Decision tree analysis is designed for modelling risk in sequential events, whereas Monte Carlo simulation deals with complexity involved in risky decisions. They may have different applications, but both methods structure decision making under risk using probability theory. By explicitly modelling the randomness of components of the decision making process, engineers and analysts can structure a decision clearly and better understand the implications of the risk in decision making.

The major drawback of probability-based methods is that the probability distributions of each of the random variables must be specified. This can be a real challenge when the probabilities are highly subjective or when there is a lack of historical or experimental data upon which to base probability assessments. Therefore, we must be aware of this grey area when we interpret the output and try to put the results of the analysis in context.

Despite this drawback, as engineers we recognize that every decision has elements of uncertainty and risk. The tools we discussed in this chapter help us to put a framework around risk and to gain insights that would otherwise be ignored. In conjunction with the sensitivity analysis methods covered in Chapter 11, probability theory is a powerful tool for assessing project viability and the implications of the risk associated with every engineering economic evaluation.

REVIEW PROBLEMS

REVIEW PROBLEM 12.1

Power Tech is a company in Ottawa that specializes in building power-surge-protection devices. Power Tech has been focusing its efforts on the North American market until now. Recently, a deal with a Chinese manufacturing company has surfaced. If Power Tech decides to become partners with this manufacturing company, their market will expand to include Asia. They are, however, concerned with the uncertainty associated with possible change in North American demand and Asian demand. From studying the current economy, Power Tech feels that the chance of no change or an increase in demand in North America over the next three years is 60% and the chance of demand decrease is 40%. After discussions with their potential partners in China, Power Tech estimates that Asian demand may increase (or remain the same) with the probability of 30% and decrease with the probability of 70% in the next three years. They also estimate the revenue increase that they can expect under different scenarios if they establish the partnership. These figures are shown in Table 12.12. Conduct a decision tree analysis for Power Tech and make a recommendation regarding the partnership with China.

ANSWER

The result of the analysis is shown in Figure 12.11. The expected value calculations at each chance node are shown below.

The first phase of rolling back (in \$million):

$$E(4) = 0.3(2) + 0.7(0.75) = 1.125$$
$$E(5) = 0.3(0.5) + 0.7(-1) = -0.55$$
$$E(6) = 0.3(0.75) + 0.7(0.5) = 0.575$$
$$E(7) = 0.3(0.1) + 0.7(0.3) = 0.24$$

Table 12.12 Expected Revenue Increase for Power Tech

	North American Demand	Asian Demand	Revenue Increase (in $million)
Partnership with China	increase	increase	2
	increase	decrease	0.75
	decrease	increase	0.5
	decrease	decrease	−1
No Partnership with China	increase	increase	0.75
	increase	decrease	0.5
	decrease	increase	0.1
	decrease	decrease	0.3

Figure 12.11 Decision Tree Analysis for Power Tech

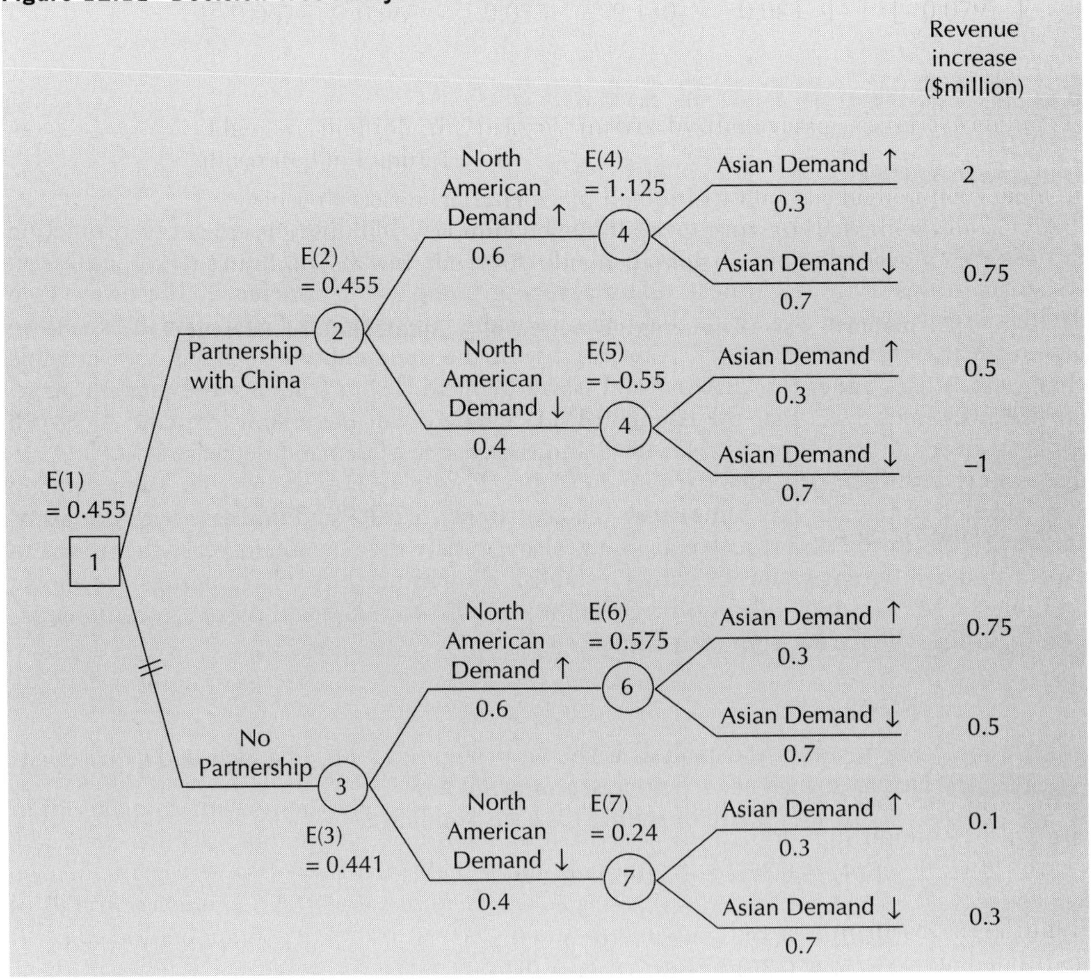

The second phase of rolling back (in $million):

$$E(2) = 0.6(1.125) + 0.4(-0.55) = 0.455$$
$$E(3) = 0.6(0.575) + 0.4(0.24) = 0.441$$

According to the expected value criterion, the partnership with China is recommended since the expected value for forming the partnership is higher than not forming. However, Power Tech should also note that these expected values have only a marginal difference. It is perhaps wise to collect more information regarding other aspects of this proposed partnership. ■

REVIEW PROBLEM 12.2

A telephone company in London, Ontario, called LOTell, thinks the introduction of an internet service package for residential customers would give it a competitive advantage over its competitors. However, a survey of the potential growth of internet home users would take at least three months. LOTell has two options at present: first, to introduce the internet package without the survey result in order to make sure that no competitors are present at the time of market entry, and second, to wait for the survey result in order to minimize the risk of failing to attract enough customers. If LOTell decides to wait for the survey result, there are three possible outcomes: the market growth is rapid (30% probability), steady (40%), or slow (30%). Depending on the survey result, LOTell may decide to introduce or not introduce the new internet service. If it decides to launch the new service after the survey, which is three months from now, then there is a 70% chance that the competitors will come up with a similar service package. What should LOTell do?

ANSWER

A decision tree for the problem is shown in Figure 12.12, which also shows the results of the analysis. The recommended decision is to introduce the internet service now because the expected value is higher, than to wait for the survey result. ■

REVIEW PROBLEM 12.3

Orla is in the rental car business. The cars she uses have a useful life of two to four years. The cars can be traded in for a new car at the dealer. The trade-in value varies anywhere from $1000 to $5000 depending on the condition of the car. Orla has estimated the probability distributions for the length of useful life and the trade-in value for a typical car. Her estimates are shown in Table 12.13. Orla wants to find out the typical annual cost of owning a car for doing this business using the Monte Carlo simulation method. Assume that the cost of a new car is $15 000 and the annual maintenance cost is $800. Her MARR is 12%.

ANSWER

Orla first comes up with the following annual cost expression for her problem:

$$\begin{aligned} AC = {} & (\text{Cost of New Car})(A/P, 12\%, \text{Useful Life}) + (\text{Maintenance Cost}) \\ & - (\text{Trade-In Value})(A/F, 12\%, \text{Useful Life}) \end{aligned}$$

Figure 12.12 Decision Tree Analysis for LOTell (Review Problem 12.2)

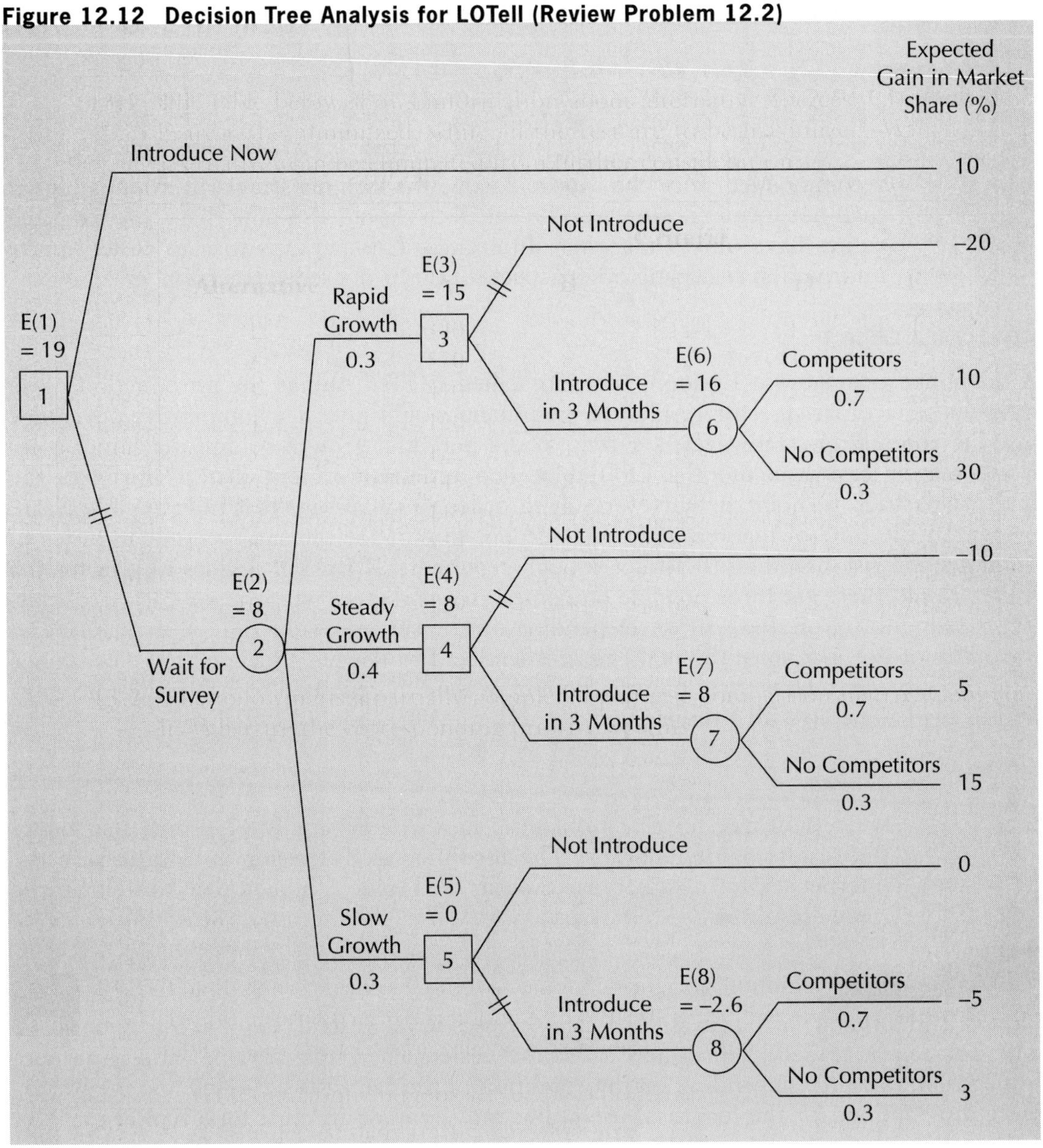

$$= \$15\,000(A/P, 12\%, \text{Useful Life}) + 800$$
$$- (\text{Trade-In Value})(A/F, 12\%, \text{Useful Life})$$

Then, she figures out the random number assignment for the useful life and the trade-in value of a car (see Table 12.14), and performs a Monte Carlo simulation and collects 200 samples. The partial result of the simulation is shown in Table 12.15 and the histogram of the frequency distribution is presented in Figure 12.13.

Table 12.13 Probability Distributions for the Useful Life and the Trade-In Value of a Rental Car

Useful Life (years)	Probability	Trade-In Value	Probability
2	0.4	\$1000	0.2
3	0.3	2000	0.2
4	0.3	3000	0.2
		4000	0.2
		5000	0.2

Table 12.14 Random Number Assignment Ranges for the Useful Life and the Trade-In Value

Useful Life	Random Number (Z_1) Assignment Range	Trade-In Value	Random Number (Z_2) Assignment Range
2	$0 \leq Z_1 < 0.4$	\$1000	$0 \leq Z_2 < 0.2$
3	$0.4 \leq Z_1 < 0.7$	2000	$0.2 \leq Z_2 < 0.4$
4	$0.7 \leq Z_1 < 1$	3000	$0.4 \leq Z_2 < 0.6$
		4000	$0.6 \leq Z_2 < 0.8$
		5000	$0.8 \leq Z_2 < 1$

Table 12.15 Partial Results for the Monte Carlo Simulation

Sample Number	Random Number (Z_1)	Useful Life	Random Number (Z_2)	Trade-In Value	Annual Cost
1	0.52275	3	0.325129	2000	5652.54
2	0.809623	4	0.423085	3000	4310.81
3	0.124285	2	0.799329	4000	6988.68
4	0.104359	2	0.207269	2000	7932.08
5	0.961704	4	0.108746	1000	4729.28
⋮	⋮	⋮	⋮	⋮	⋮

The results of the Monte Carlo simulation show that the average annual cost for each car is \$5706. In the 200 samples, the annual cost ranged from \$3892 to \$8404. ■

Figure 12.13 Frequency Distribution for the Annual Cost

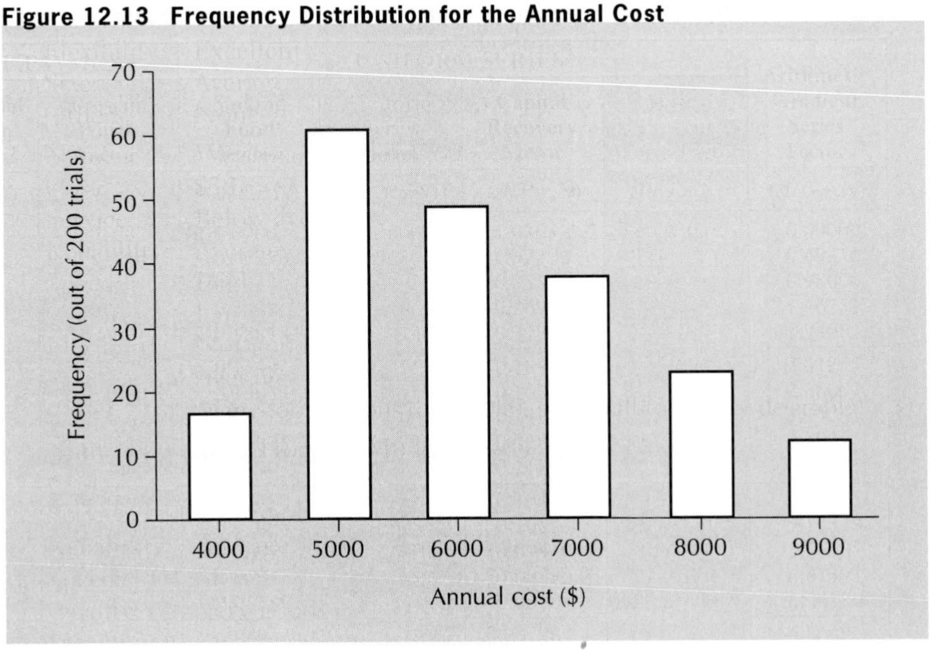

REVIEW PROBLEM 12.4

Burnaby Insurance (see Review Problems 11.1 to 11.3) has consulted several energy experts in order to further understand the implications of electricity and natural gas price changes on their two energy efficiency projects. They have estimated that the cost of electricity is a minimum of \$0.063 per kWh and a maximum of \$0.077 per kWh, and the price of natural gas has a minimum of \$3.35 and a maximum of \$3.65 per 1000 cubic feet. Both probability distributions (discrete approximation) are shown in Table 12.16. Carry out a Monte Carlo simulation to determine the probability distribution of the present worth of the two energy efficiency projects.

ANSWER

To conduct the Monte Carlo simulation, 300 samples were drawn from the electricity cost and natural gas cost distributions. The present worth of each project was computed for each simulated outcome using the expressions below, and a histogram of the results was constructed. Figure 12.14 shows the distribution of present worth of the high-efficiency motor and Figure 12.15 shows the distribution of present worth of the heat exchange unit.

PW(Efficiency motor)
$= -\$28\,000 + (P/A, 10\%, 10)[\$70\,000(\text{Electricity}) - \$700]$
$= -\$28\,000 + (6.1446)[\$70\,000(\text{Electricity}) - \$700]$

Table 12.16 Probability Distributions for the Cost of Electricity and the Price of Natural Gas

Electricity Cost (per kWh)	Probability	Natural Gas Price (per 1000 cubic feet)	Probability
\$0.063	0.125	\$3.35	1/7
0.065	0.125	3.40	1/7
0.067	0.125	3.45	1/7
0.069	0.125	3.50	1/7
0.071	0.125	3.55	1/7
0.073	0.125	3.60	1/7
0.075	0.125	3.65	1/7
0.077	0.125		

Figure 12.14 Monte Carlo Simulation Results for the High-Efficiency Motor

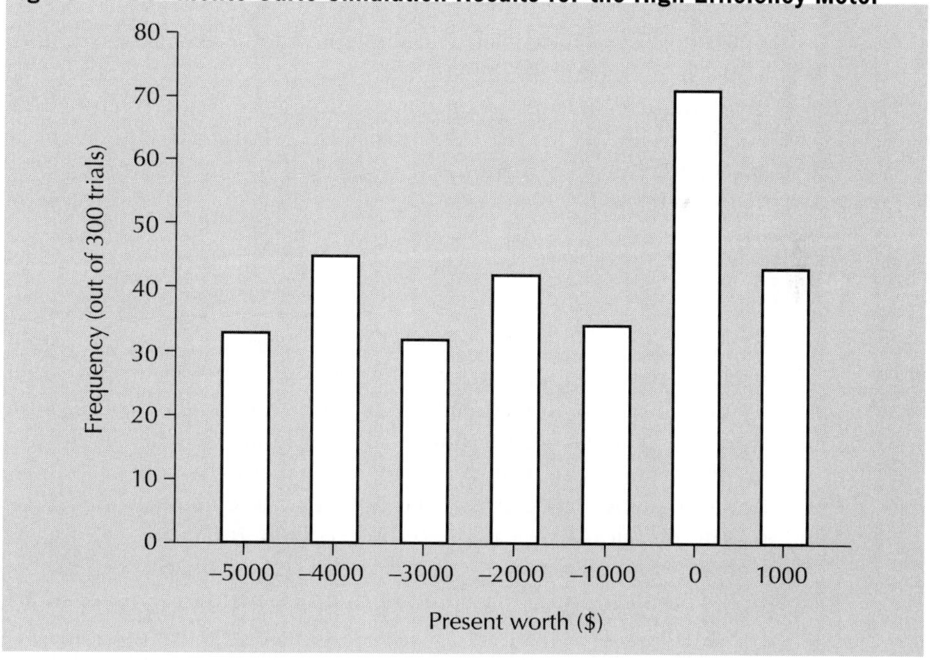

PW(Heat exchanger)

= −\$40 000 + (*P/A*, 10%, 10)[\$29 000(Electricity) +\$2250(Natural Gas) − \$3200]

= −\$40 000 + (6.1446)[\$29 000(Electricity) + \$2250(Natural Gas) − \$3200]

Figure 12.15 Monte Carlo Simulation Results for the Heat Exchange Unit

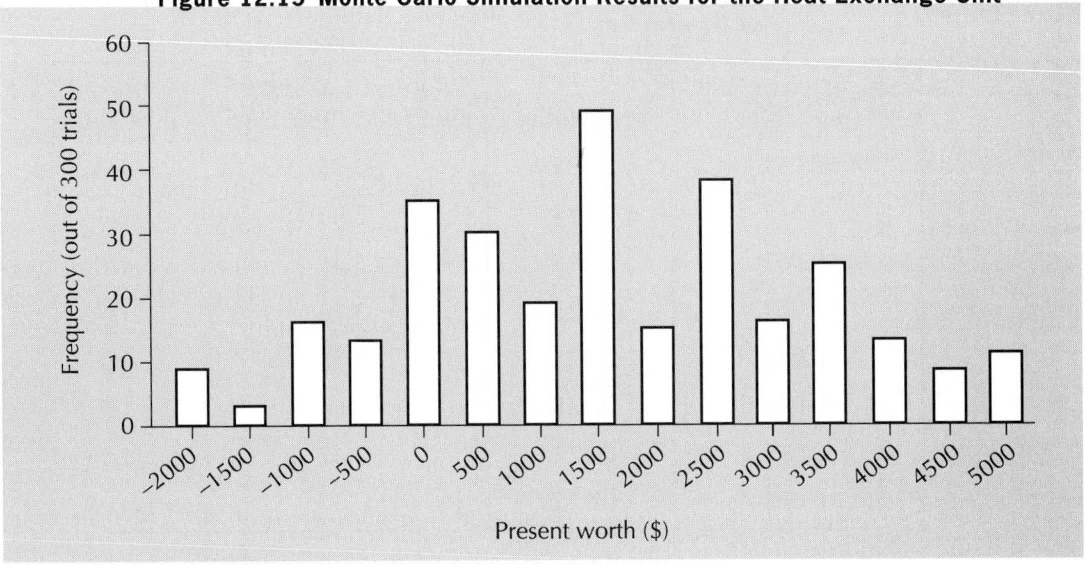

It is clear from the simulations that the high-efficiency motor has roughly a 40 percent chance of having a positive present worth and hence is not a good investment. The heat exchange unit, on the other hand, has only a small chance of having a negative present worth, and hence looks like a better choice for Burnaby Insurance. ■

SUMMARY

In this chapter, some basic probability theory was used to assist in dealing with decision making under risk. Basic concepts such as random variables, probability distributions, and expected value were introduced to help understand decision tree analysis and Monte Carlo simulation analysis. A decision tree is a graphical representation of the logical structure of a decision problem in terms of the sequence of decisions and outcomes of chance events. The analysis based on the decision tree can therefore account for the sequential nature of decisions. Monte Carlo simulation is useful for estimating the probability distribution of the economic outcome of a project when there are multiple sources of risk involved in a project. The probability distribution of the outcome performance measure can then be used to assess the economic viability of the project.

Engineering Economics in Action, Part 12B: Chances Are Good

"So let me get this straight," Clem started, a few days later. "You structured this as a tree. We—Canadian Widgets—have the first decision whether to proceed with development of the

Adaptamatic at all. Then the next node represents the probability that the Adaptamatic will have a competitor. Then there is our choice of going into production or not. Then there are the customers, which we represent as chance nodes, who can have low, medium and high demand. Finally, there are the outcomes, each of which has a dollar value associated with them. Is that right?"

"Perfect," Naomi said.

"And the dollar values came from...?"

"Well, some reasonable assumptions about the marketplace," Naomi replied. "We had enough information to quantify three levels of demand. Also, we figured we had as much consumer acceptance as any competitor because of our association with Powerluxe, so when a competitive product was in the marketplace, we assumed that we would have the same market share as anyone else."

"What about the probabilities?" Clem continued.

"That was harder." Bill answered this one. Naomi couldn't help but smile. A few days ago, she didn't know anything about marketing, and Bill didn't know about decision trees, but now she was answering the questions about marketing, and he was answering the ones about decision trees! Go figure.

"We had no hard information on probabilities," Bill continued. "We set it all up as a spreadsheet so that we could adjust the probabilities freely. First we put in our best guesses—subjective probabilities. We talked to Prabha up in Marketing and to a bunch of people at Powerluxe, and refined that down. Finally, we tested it for a whole range of possibilities. Bottom line: chances are good that we will make a killing."

Clem looked at the decision tree and at the table Bill gave him that reported the expected value of proceeding with the Adaptamatic development under various assumptions. "Are you sure this is right? How come we make more money if we have a competitor than if we don't?"

"Good observation, Clem." Bill was beaming. This was his area of expertise. "That is interesting, isn't it? That's what's called 'building the category' in marketing. The principle is that when you have a new product, you have to spend a lot of resources educating the consumer. If you have to do that yourself, it's really costly. But a competitor can work for you by taking a share of those education costs and actually making it cheaper for you. That's what happens here."

"It was the decision tree approach that revealed this dynamic in this case," Bill continued. "By ourselves, we could lose money if demand was low, even though the expected value was fairly high. However, with a competitor on the market, we can almost guarantee making money because the efforts of our competitor reduce our marketing costs and expand the market at the same time. Combine them both in the decision tree, and we have a high expected value with almost no risk. Cool, eh?"

"Really cool." Clem glanced at Naomi with an appraising look. "Speaking of chances—chances are good that Anna's going to like this. Nice job."

PROBLEMS

12.1 An investment has a possible return of 7%, 10%, and 15% in 5 years. The probabilities of attaining these rates, estimated based on the current economy, are 0.65, 0.25, and 0.1, respectively. If you have \$10 000 to invest, what is the expected return from this investment?

12.2 Rockies Adventure Wear, Inc. sells athletic and outdoor clothing through catalogue sales. They want to upgrade their order-processing centre so that they have less chance of losing customers by putting them on hold. The upgrade may result in the processing capacity of 30, 40, 50, or 60 calls per hour with the probabilities of 0.2, 0.4, 0.3, and 0.1, respectively. Market research indicates that the average number that Rockies may receive is 50 calls per hour. What is the expected loss of customers due to the lack of processing capacity?

12.3 Power Tech builds power-surge-protection devices. One of the components, a plastic moulded cover, can be produced from two automated machines, A1 and X1000. Each machine produces a number of defects with probabilities shown in the following table. Which machine is better based on the expected number of defective products?

A1		X1000	
No. of Defects (out of 100)	Probability	No. of Defects (out of 100)	Probability
0	0.3	0	0.25
1	0.28	1	0.33
2	0.15	2	0.26
3	0.15	3	0.1
4	0.1	4	0.05
5	0.02	5	0.01

12.4 Lightning City is famous for having many thunder storms during the summer months (from June to August). One of CB Electronix's factories is located in Lightning City. They have collected information, shown in the table below, regarding the number of blackouts caused by lightning.

Number of Blackouts (per month)	Probability (summer months)	Probability (non-summer months)
0	0	0.45
1	0.4	0.4
2	0.25	0.15
3	0.2	0
4	0.1	0
5	0.05	0

For the first three blackouts in a month, the cost due to suspended manufacturing is \$800 per blackout. For the fourth and fifth blackout, the cost increases to $1500 per blackout. A local insurance company offers protection against lightning-related expenses. The monthly payment is $500 for annual complete coverage. What is the expected cost related to blackouts during the summer months? During the non-summer months? Should CB consider purchasing the insurance policy? Assume each month is independent of the other months.

12.5 A new wave-soldering machine is expected to generate monthly savings of either $8000, $10 000, $12 000, or $14 000 for the next two years. The manager is not sure about the likelihood of the four savings scenarios, so she assumes that they are equally likely. What is the present worth of the expected monthly savings? Use MARR of 12%, compounded monthly, for this problem.

12.6 Regional Express is a small courier service operating in Southern Ontario. By introducing a new computerized tracking device, they anticipate some increase in revenue, currently estimated at $2.75 per parcel. The possible new revenue ranges from $2.95 to $5.00 per parcel with the following probabilities. Assuming that Regional's monthly capacity is 60 000 parcels and the monthly operating and maintenance costs are $8000, what is the present worth of the expected revenue over 12 months? Regional's MARR is 12%, compounded monthly.

Revenue per parcel	$2.95	$3.25	$3.50	$4.00	$5.00
Probability	0.1	0.35	0.3	0.15	0.1

12.7 Katrina is thinking about buying a car. She figures her monthly insurance payment is $90, gas is $30, and general maintenance $20. The car she would like to buy may last for 4, 5, or 6 years before a major repair, with the probabilities of 0.4, 0.4, and 0.2, respectively. Calculate the present worth of the monthly expenses using the expected life of the car (before a major repair). Katrina's MARR is 10%, compounded monthly.

12.8 Pharma-Excel is a pharmaceutical company based in Halifax. They are currently studying the feasibility of a research project that involves improvement of vitamin C pills. To examine the optimistic, expected, and pessimistic scenarios for this project, they gathered the data shown below. What is the expected annual cost of the vitamin C project? Assume Pharma-Excel's MARR is 15%. Note that the lead time is different for each scenario.

	Optimistic	**Expected**	**Pessimistic**
Research and development costs (at the end of research)	$75 000	$240 000	$500 000
Lead time to production (years)	1	2	3
Probability	0.15	0.5	0.35

12.9 Mega City Hospital is selling lottery tickets. All proceeds go to their cancer research program. Each ticket costs \$100, but the campaign catch-phrase promises a 1-in-1000 chance of winning the first prize. The first prize is a "dream" house, which is worth \$250 000. Based on decision tree analysis, is buying a ticket worthwhile?

12.10 Based on Problem 12.9, determine the price of a ticket so that not buying a ticket is the preferred option and determine the chance of winning so that not buying a ticket is the preferred option.

12.11 Randall at Churchill Circuits (CC) has just received an emergency order for one of CC's special purpose circuit boards. Five are in stock at the moment. However, when they were tested last week, two were defective but were mixed up with the three good ones. There is not enough time to retest the boards before shipment to the customer. Randall can either choose one of the five boards at random to ship to the customer or he can obtain a proven non-defective one from another plant. If the customer gets a bad board, the total incremental cost to CC is \$10 000. The incremental cost to CC of getting the board from another plant is \$5000.

(a) What is the chance that the customer gets a bad board if Randall sends them one of the five in stock?

(b) What is the expected value of the decision to send the customer one of the five in stock?

(c) Draw a decision tree for Randall's decision. What should he do?

12.12 St. Jacobs Cheese Factory (SJCF) is getting ready for a busy tourist season. SJCF wants to either increase production or produce the same amount as last year, depending on the demand level for the coming season. SJCF estimates the probabilities for high, medium, and low demands to be 0.4, 0.35, and 0.25, respectively, based on the number of tourists forecasted by the local recreational bureau. If SJCF increases production, the expected revenues corresponding to high, medium, and low demands are \$750 000, \$350 000, and \$100 000, respectively. If SJCF does not increase production, the expected payoffs are \$500 000, \$400 000, and \$200 000, respectively. Based on a decision tree analysis, should SJCF's production increase or not increase?

12.13 LOTell, a telephone company in London, has two options for their new internet service package. They can introduce a combined rate for the residential phone line and the internet access, or they can offer various add-on internet service rates in addition to the regular phone rate. LOTell can only afford to introduce one of the packages at this point. The expected gain in market share by introducing the internet service would likely differ for different market growth rates. LOTell has estimated that if they introduce the combined rate, they gain 30%, 15%, and 3% of the market share with rapid, steady, or slow market growth. If they introduce the add-on rates, they gain 15%, 10%, and 5% of the market share with rapid, steady, or slow market growth. Perform a decision tree analysis. Which package should LOTell introduce to the market now?

12.14 Brockville Brackets (BB) uses a robot for welding small brackets onto car-frame assemblies. BB's R&D team is proposing a new design for the welding robot. The new design should provide substantial savings to BB by increasing efficiency in the robot's mobility. However, the new design is based on the latest technology, and there is some uncertainty associated with the performance level of the robot. The R&D team estimates that the new robot may exhibit high, medium, and low performance levels with the probabilities of 0.35, 0.55, and 0.05 respectively. The annual savings corresponding to high, medium, and low performance levels are $500 000, $250 000, and $150 000 respectively. The development cost of the new robot is $550 000.

(a) Based on a five-year study period, what is the present worth of the new robot for each performance scenario? Assume BB's MARR is 12%.

(b) Perform decision tree analysis. Should BB approve the development of a new robot?

12.15 Refer to Review Problem 12.1. Power Tech is still considering the partnership with the Chinese manufacturing company. Their analysis in Review Problem 12.1 has shown that the partnership is recommended (by a marginal difference in the expected revenue increase between the two options). Power Tech now wants to further examine the possible shipping delay and quality control problems that are associated with the partnership with China. Power Tech estimates that shipping may be delayed 40% of the time due to the distance. Independent of the shipping problem, there may be a quality problem 25% of the time due to communication difficulties and lack of close supervision by Power Tech. The payoff information is estimated as shown below. Develop a decision tree for Power Tech's shipping and quality control problems and analyze it. What is the recommendation regarding the possible partnership with China now?

Shipping Problem	Quality Problem	Gain in Annual Profit
No shipping delay	Acceptable quality	$ 200 000
No shipping delay	Poor quality	25 000
Shipping delay	Acceptable quality	100 000
Shipping delay	Poor quality	−100 000

12.16 Refer to Problem 12.2. Rockies Adventure Wear, Inc. has upgraded its order-processing centre in order to improve the processing speed and customer access rate. Before completely switching to the upgraded system, Rockies has an option of testing it. The test will cost Rockies \$50 000, which includes the testing cost and loss of business due to shutting down their business for a half day. If Rockies does not test the system, there is a 55% chance of severe failure (\$150 000 repair and loss of business costs), a 35% chance of minor failure (\$35 000 repair and loss of business costs), and a 10% chance of no failure. If Rockies tests the system, the result can be favourable with the probability of 0.34, which requires no modification, and not favourable with the probability of 0.66. If the test result is not favourable, Rockies has two options: minor modification and major modification. The minor modification costs \$5000 and the major modification costs \$30 000. After the minor modification, there is still a 15% chance of severe failure (\$150 000 costs), a 45% chance of minor failure (\$35 000 costs), and a 40% chance of no failure. Finally, after the major modification, there is still a 5% chance of severe failure, a 30% chance of minor failure, and a 65% chance of no failure. What is the recommended action for Rockies to take, based on a desision tree analysis?

12.17 Refer to Problem 12.8. As a part of Pharma-Excel's feasibility study, they want to include information on the acceptance attitude of the public toward the new vitamin C product. Regardless of the optimistic, expected, and pessimistic scenarios on research and development, there is a chance that the general public may not feel comfortable with the new product because it is based on a new technology. They estimate that the likelihood of the public accepting the product (and purchasing it) is 33.3% and not accepting it is 66.7%. The expected annual profit *after* the research is \$1 000 000 if the public accepts the new product and \$200 000 if the public does not accept it.

(a) Calculate the annual worth for all possible combinations of three R&D scenarios (optimistic, expected, and pessimistic) and two scenarios on the public reaction (accept or not accept). Pharma-Excel's MARR is 15%.

(b) Using the annual worth information as the payoff information, build a decision tree for Pharma's problem. Should they proceed with the development of this new vitamin C product?

12.18 Baby Bear Beads (BBB) found themselves confronting a decision problem when a packaging line suffered a major breakdown. Ross, the manager of Maintenance; Rita, Plant Manager; and Ravi, the company President met to discuss the problem.

Ross reported that the current line could be repaired, but the cost and result were uncertain. He estimated that for \$40 000, there was an 75% chance the line would be as good as new. However, for the rest of the time, another \$100 000 might have to be spent.

Rita's studies suggested that for \$90 000, the whole line might be replaced by a new piece of equipment. However, there was a 40% chance an extra \$20 000 might be required to modify downstream operations to accept a slightly different package size.

Ravi reviewed his sales projections and revealed that there was a 30% chance that the production line wouldn't be required anymore anyhow, but that this wouldn't be known until after a replacement decision was made. Rita then pointed out that there was a 80% chance that the new equipment she proposed could easily be adapted to other purposes, so that the investment, including the modifications to downstream operations, could be completely recovered even if the line was no longer needed. On the other hand, the repaired packing line would have to be scrapped with essentially no recovery of the costs.

The present worth of the benefit of having the line running is $150 000. Use a decision tree analysis to determine what BBB should do about the packaging line.

12.19 Refer to Review Problem 12.1. Power Tech feels comfortable about their probability estimate regarding the change in North American demand. However, they would like to examine the probability estimate for Asian demand more carefully. Perform sensitivity analysis on the probability that Asian demand increases. Try the following values first, {0.1, 0.2, 0.3, 0.4, 0.5}, where 0.3 is the base case. Analyse the result and give a revised recommendation as to Power Tech's possible partnership with China.

12.20 Refer to Review Problem 12.2. LOTell is happy with the decision recommendation suggested by the previous decision tree analysis, given information on the market growth. However, LOTell feels that the uncertainty in the market growth is the most important factor in their overall decision making regarding the introduction of the internet service package. Hence, they wish to examine the sensitivity of the probability estimates for the market growth. Answer the following questions based on the decision tree that was developed for Review Problem 12.2.

(a) Let p_1 be the probability of rapid market growth, p_2 be the probability of steady growth, and p_3 be the probability of slow market growth. Express the expected value at Node 2, E(2), in terms of p_1, p_2, and p_3.

(b) If $E(2) < 10$, then the option to introduce the package now is preferred. Using the expression of E(2) that was developed in (a), graph all possible values of p_1 and p_2 which lead to the decision to introduce the package now. (You will see that p_3 is not involved in graphing from part (a).) What can you observe from the graph regarding these values of p_1 and p_2?

12.21 Kennedy Foods Company is a producer of frozen turkeys in Brampton. A new piece of freezing equipment became available in market last month. It costs $325 000. The new equipment should increase Kennedy Foods' production efficiency, and hence its annual revenue. However, the net increase in the annual revenue is somewhat uncertain because it depends on the annual operating cost of the new equipment, which is not certain at this point. Kennedy Foods estimates the possible annual revenues with the following probability distribution.

Net Increase in Annual Revenue	Probability
$25 000	0.1
30 000	0.35
35 000	0.4
40 000	0.15

(a) Express the present worth of this investment in analytical terms. Use a 10-year study period and MARR of 15%.

(b) Show the random number assignment ranges that can be used in Monte Carlo simulation.

(c) Carry out a Monte Carlo simulation of 100 samples. What is the expected present worth? What are the maximum and minimum PW in the sample frequency distribution? Construct a histogram of the present worth. Is it beneficial for Kennedy Foods to purchase the new freezing equipment?

12.22 Refer back to Problem 12.3. Power Tech has decided to use X1000 model exclusively for producing plastic moulded covers. The revenue for each non-defective unit is $0.10. For the defective units, rework costs $0.15 per unit.

(a) Show the analytical expression for the total revenue.

(b) Using the same probability distribution in Problem 12.3, create a table showing the random number assignment ranges for the number of defective units.

(c) Assume that X1000 produces moulded covers in batches of 100. Carry out a Monte Carlo simulation of 100 production runs. What is the average total revenue? What are the maximum and minimum revenue in the samples? Construct a histogram of the total revenue and comment on your observations.

12.23 Ron-Jing is starting her undergraduate study in September and needs a place to live for the next four years. Her friend, Nabil, told her about his plan to buy a house and rent out part of it. She thinks it may be a good idea to buy a house too, as long as she can get a reasonable rental income every month and can sell the house for a good price in four years. A fair-sized house, located a 15-minute walk from the university, costs $120 000. She estimates the net rental income, after expenses, will be $1050 per month, which seems reasonable. She is, however, concerned about the resale value. She figures that the resale value can be above or below $120 000 depending on the housing market in four years. She estimates the possible resale values and their likelihoods as shown below. Using the Monte Carlo method, simulate 100 samples of the present worth for buying the house. If Ron-Jing's MARR is 12%, is this a viable investment for her?

Resale Value ($)	Probability
$100 000	0.2
110 000	0.2
120 000	0.2
130 000	0.2
140 000	0.2

12.24 An oil company owns a tract of land that has good potential for containing oil. The size of the oil deposit is unknown, but from previous experience with land of similar characteristics, the geological engineers predict that an oil well there will yield an annual number of barrels between 0 (a dry well) and 100 million per year over a five-year period. The following probability distribution for the amount of yield is also estimated. The cost of drilling a well is $10 million, and the profit (after deducting production costs) is $0.50 per barrel. Interest is 10% per year. Carry out a Monte Carlo simulation of 100 wells. Construct a histogram of the present worth of drilling a well. Comment on your observations. Do you recommend drilling?

Annual Number of Barrels (in millions)	Probability
0	0.05
10	0.1
20	0.1
30	0.1
40	0.1
50	0.1
60	0.1
70	0.1
80	0.1
90	0.1
100	0.05

12.25 Refer to Problem 9.24. Before they purchase sonar warning devices to help trucks back up at store loading docks, St. James Bay Department Store is re-examining the original estimates of two types of annual savings generated by installing the devices. St. James feels that the original estimates were somewhat optimistic, and they want to include probability information in their analysis. The table below shows their revised estimates, with the probability distributions, of annual savings from faster turn-around time and reduced damage to the loading docks. The sonar system costs $220 000 and has a life of four years and a scrap value of $20 000. Carry out a Monte Carlo simulation and generate 100 random samples of the present worth for the sonar system. Make a recommendation regarding the possible purchase based on the frequency distribution. Assume that St. James' MARR is 18% and no inflation is considered.

Savings from Faster Turn-Around	Probability	Savings from Reduced Damage	Probability
$38 000	0.2	$24 000	0.25
41 000	0.2	26 000	0.25
44 000	0.2	28 000	0.25
47 000	0.2	30 000	0.25
50 000	0.2		

12.26 A fabric manufacturer has been asked to extend a line of credit to a new customer, a dress manufacturer. In the past, the mill has extended credit to customers. Although most pay back the debt, some have defaulted on the payments and the fabric manufacturer has lost money. Previous experience with similar new customers indicates that 20% of customers are bad risks, 50% are average risks, and 30% are good risks. The average length of business affiliation with bad-, average-, and good-risk customers is 2, 5, and 10 years, respectively. Previous experience also indicates that, for each risk group, the annual profit has the following probability distribution.

Bad Risk		Average Risk		Good Risk	
Annual Profit	Probability	Annual Profit	Probability	Annual Profit	Probability
$−50 000	1/7	$10 000	0.2	$20 000	1/7
−40 000	1/7	15 000	0.2	25 000	1/7
−30 000	1/7	20 000	0.2	30 000	1/7
−20 000	1/7	25 000	0.2	35 000	1/7
−10 000	1/7	30 000	0.2	40 000	1/7
0	1/7			45 000	1/7
10 000	1/7			50 000	1/7

Construct a spreadsheet with the headings used in the table at the top of page 519. Generate 100 random customers. From those 100, construct a frequency distribution for the present worth of extending a line of credit to a random customer. Use an interest rate of 10% per year.

Sample Number	Random Number 1	Risk Rating	Years of Business	Random Number 2	Annual Profit	PW
1						
2						
3						
4						
5						
⋮	⋮	⋮	⋮	⋮	⋮	⋮

12.27 Rockies Adventure Wear, Inc. has been selling athletic and outdoor clothing through catalogue sales. Most orders from customers are processed by phone and the rest by mail. Rockies is now considering expanding their market by introducing a web site-based ordering system. The first cost of setting up the web-based system is $120 000. A market expert predicts 10 000 new customers in the first year. Each new customer generates an average of $5 revenue for Rockies. There are, however, uncertainties regarding the possible market growth and annual operating and maintenance costs over the five years. The market may grow at a steady rate of 2%, 5%, 8%, 10%, or 15% from the initial estimate of 10 000, with each growth rate having a chance of 20%. The annual costs may be $10 000, $15 000, $20 000, $25 000, $30 000, and $35 000, and these estimates are equally likely. Rockies' MARR is 18% for this type of investment. Based on 100 samples generated by Monte Carlo simulation, what is the expected present worth for this web-based ordering system? Comment on the viability of the project.

12.28 Hitomi is considering buying a new lawnmower. She has a choice of a Lawn Guy or a Clip Job. Her neighbour, Sam, looked at buying a mower himself a while ago, and gave her the following information on the two types of mower.

	Lawn Guy	Clip Job
First cost	$350	$120
Life (years)	10	4
Annual gas	$60	$40
Annual maintenance	$30	$60
Salvage value	0	0

Due to the long lifespan of a Lawn Guy mower, Hitomi is reluctant to use a single estimate for its expected life and the annual cost of gas. As for the Clip Job, she is not sure about the annual maintenance cost since it has a relatively short life and it may break down easily. With help from a friend who works at a hardware store, she comes up with the following probabilistic estimates for

Lawn Guy's expected life and the cost of gas, and Clip Job's annual maintenance cost. Hitomi's MARR is 5%. Find out the expected annual cost for each mower using Monte Carlo simulation method. Collect at least 100 samples. Which mower is preferred?

Lawn Guy				Clip Job	
Life	Probability	Gas	Probability	Maintenance	Probability
7	0.25	$50	0.2	$50	0.2
8	0.25	60	0.3	60	0.5
9	0.25	70	0.4	70	0.2
10	0.25	80	0.1	80	0.1

12.29 Mountain Beer Brewery (MBB) in Mill Bay, British Columbia, currently buys 250 000 beer labels every year from a local label maker. The label maker charges MBB $0.075 apiece. A demand forecast indicates that MBB's demand may grow up to 400 000 in the near future, so MBB is considering making the labels themselves. If they did so, they would purchase a high-quality colour photocopier which costs $6000 and lasts for five years with no salvage value at the end of its life. The operating cost of the photocopier would be $4900 per year, including the cost of colour cartridges and special paper used for labels. MBB would also have to hire a label designer. The cost of labour is estimated to be $0.04 per label. Based on the following estimate on the future demand and its likelihood, simulate the present costs of production and purchase for 200 samples using the Monte Carlo method. Make a recommendation as to whether MBB should consider making their own labels. MBB's MARR is 12%.

Demand	Probability
200 000	0.2
250 000	0.2
300 000	0.2
350 000	0.2
400 000	0.2

12.30 Refer to Problem 11.29. Timmons Testing (TT) has now established probabilistic information on the annual variable costs and annual production volume for the two testing devices offered by Vendors A and B. TT would like to include Monte Carlo simulation analysis before they decide which of the two alternatives they should choose. First cost, annual fixed costs, and salvage value information are presented below in addition to the probability information. TT charge $3.25 per board tested. Their MARR is 15%. Compare the expected annual worth and its range for the two alternatives. What would you recommend to TT now?

Alternative A				Alternative B			
Annual Variable Costs ($/board)	Probability	Production Volume (boards)	Probability	Annual Variable Costs ($/board)	Probability	Production Volume (boards)	Probability
2.20	0.2	40 000	0.05	1.01	0.1	40 000	0.05
2.25	0.3	45 000	0.15	1.03	0.2	45 000	0.05
2.30	0.3	50 000	0.25	1.06	0.4	50 000	0.2
2.35	0.2	55 000	0.2	1.10	0.3	55 000	0.2
		60 000	0.15			60 000	0.2
		65 000	0.05			65 000	0.1
		70 000	0.05			70 000	0.1
		75 000	0.05			75 000	0.05
		80 000	0.05			80 000	0.05

	A	B
First cost ($)	200 000	350 000
Annual fixed costs ($)	16 500	35 500
Salvage value ($)	5000	20 000
Life	7	10

Metro Toronto's Main Treatment Plant

CANADIAN 12.1 MINI-CASE

Metro Toronto's Main SewageTreatment Plant is the largest activated-sludge plant in Canada, with a rated treatment capacity of 818 million L per day. It serves a population of 1.25 million people. Treated effluent is discharged into Lake Ontario.

A remedial action plan for this area of the lake, developed by the Ontario Ministry of Environment and Energy, would require nitrification of the effluent. A preliminary estimate of the cost of nitrification was about $220 million. In 1992, the municipality hired a consultant to assess the ability of the existing plant to meet the expected requirements and to propose any required modifications.

The year-long assessment included extensive data collection from over 50 on-line instruments. The consultants surmised that future requirements could be met by modifying existing procedures and facilities. Based on the data, a simulation model was developed to confirm the prediction made in the initial assessment and to test the response of the modified plant to different operating scenarios.

continued

The estimated capital cost of the necessary improvements to the existing facilities was less than $32 million.

Source: *Canadian Consulting Engineer*, September/October, 1994.

Discussion

A computer simulation involves making a model in software that captures the essential structure of a complex product or process. Assumptions are made about the probability distribution of uncertain parameters, and the simulation calculates the distributions of performance measures. Simulations are usually much more complex than the Monte Carlo models discussed in this chapter.

Many large projects are simulated by computer before construction or implementation. Not only can the expected design requirements be predicted, but potential problems can be discovered before irreversible decisions are made.

In the planning of Metro Toronto's Main Treatment plant, it was possible to reduce the original rough estimate of $220 million for necessary improvements to less than $32 million with validation from a detailed simulation study. Not all simulation projects will save this much money, but in any case where costs are large, the relatively small cost of a simulation can be a very good investment. Because simulations never capture all natural processes completely, the analyst must understand the sources of error and use that understanding in the larger context of analysis. Simulation experts often use sensitivity analysis to test which sources of error have the most influence on simulation results.

Questions

1) Many projects can be done without a simulation study in advance. Name a few of these, and explain why decisions about their acceptance can be made without a simulation study.
2) What are the likely characteristics of projects where simulation makes sense? Name a few projects of this nature.
3) In addition to the direct cost of a simulation study, what are other negative consequences of doing one?
4) In addition to better predicting design requirements and finding potential problems, can you think of any other benefits provided by a simulation?
5) Fredricka has a potential $20 000 000 project to build a production plant in Mexico. A consultant offers to provide her with a simulation to model the plant's performance in the North American marketplace. How would she judge how much to spend for such a simulation?

CHAPTER 13

Qualitative Considerations and Multiple Criteria

Engineering Economics in Action, Part 13A: Don't Box Them In

Naomi and Bill Astad were seated in Naomi's office. "O.K.," said Naomi. "Now that we know we can handle the demand, it's time to work on the design, right? What is the best design?" She was referring to the self-adjusting vacuum cleaner project for Powerluxe that she and Bill had been working on for several months.

"Probably the best way to find out," Bill answered, "will be to get the information from interviews with small groups of consumers."

"All right," said Naomi. "We have to know what to ask them. I guess the most important step for us is to define the relevant characteristics of vacuums."

"I agree," Bill responded. "We couldn't get meaningful responses about choices if we left out some important aspect of vacuums like suction power. One way to get the relevant characteristics will be to talk to people who have designed vacuums before, and probably to vacuum cleaner sales people, too." They both smiled at the humorous prospect of seeking out vacuum cleaner sales people, instead of trying to avoid them.

"We're going to need some technical people on the team," Naomi said. "We will have to develop a set of technically feasible possibilities."

"Exactly," Bill replied. "Moreover, we need to have working models of the feasible types. That is, we can't just ask questions about attributes in the abstract. Most people would have a hard time inferring actual performance from numbers about weight or suction power, for example. Also, consumers are not directly interested in these measurements. They don't care what the vacuum weighs. They care about what it takes to move it around and go up and down stairs. This depends on several aspects of the cleaner. It includes weight, but also the way the cleaner is balanced and the size of the wheels."

"That makes sense," Naomi said. "I assume that we would want to structure the interviews to make use of some form of MCDM approach."

"Huh?" Bill said.

13.1 Introduction

Most of this book has been concerned with making decisions based on a single economic measure such as present worth, annual worth, or internal rate of return. This is natural, since many of the decisions that are made by an individual, and most that are made by businesses, have the financial impact of a project as a primary consideration. However, rarely are costs and benefits the only consideration in evaluating a project. Sometimes other considerations are paramount.

For decisions made by and for an individual, cost may be relatively unimportant. One individual may buy vegetables based on their freshness, regardless of the cost. A dress or suit may be purchased because it is fashionable or attractive. A car may be chosen for its comfort and not its cost.

Traditionally, firms were different from individuals in this way. It was felt that all decisions for a firm *should* be made on the basis of the costs and benefits as measured in money (even if they sometimes were not, in practice), since the firm's survival depended solely on being financially competitive.

Society has changed, however. Companies now make decisions that apparently involve factors that are very difficult to measure in monetary terms. Money spent by firms on charities and good causes provides a benefit in image that is very hard to quantify. Resource companies that demonstrate a concern for the environment incur costs with no clear financial benefit. Companies that provide benefits for employees beyond statute or collective agreement norms gain something that is hard to measure.

The fact that firms are making decisions on the basis of criteria other than only money most individuals would hail as a good thing. It seems to be a good thing for the companies, too, since those that do so tend to be successful. However, it can make the process of decision making more difficult because there is no longer a single measure of value.

Money has the convenient feature that, in general, more is better. For example, of several mutually exclusive projects (of identical service lives), the one with the highest present worth is the best choice. People prefer a higher salary to a lower one. However, if there are reasons to make a choice other than just the cost, things get somewhat more difficult. For example, which is better: the project with the higher present worth but which involves clear-cutting a forest, or the one with lower present worth but which preserves the forest? Does a high salary compensate for working for a company that does business with a totalitarian government?

Although such considerations have had particular influence in recent years, the problem of including both qualitative *and* quantitative criteria in engineering decisions has always been present. This leads to the question of how a decision maker deals with multiple objectives, be they quantitative or qualitative. There are three basic approaches to the problem:

1. Model and analyze the costs alone. Leave the other considerations to be dealt with on the basis of experience and managerial judgment. In other words, consider the problem in two stages. First, treat it as if cost were the only important criterion. Subsequently, make a decision based on the refined cost information — the economic analysis — and all other considerations. The benefit of this approach is its simplicity; the methods for analyzing costs are well established and defensible. The liability is that errors can be made, since humans have only a limited ability to process information. A bad decision can be made, and, moreover, it can be hard to explain why a particular decision was made.

2. Convert other criteria to money, and then treat the problem as a cost-minimization or profit-maximization problem. Before environmental issues were recognized as being so important, the major criterion that was not easily converted to money was human health and safety. Elaborate schemes were developed to measure the cost of a lost life or injury so that good economic decisions were made. For example, one method was to estimate the money that a worker would have made if he or she had not been injured. With an estimate like this, the cost of a project in lives and injuries could be compared with the profits

obtained. A benefit of this approach is that it does take non-monetary criteria into account. A drawback is the difficult and politically sensitive task of determining the cost of a human life or the cost of cutting down a 300-year-old tree.

3. Use a **multi-criterion decision making (MCDM)** approach. There are several MCDM methods that explicitly consider multiple criteria and guide a decision maker to superior or optimal choices. The benefit of MCDM is that all important criteria can be explicitly taken into account in an appropriate manner. The main drawback is that MCDM methods take time and effort to use.

In recent years and in many circumstances, looking at only the monetary costs and benefits of projects has become inappropriate. Consequently, considerable attention has been focussed on how best to make a choice under competing criteria. The first two approaches listed above still have validity in some circumstances; in particular, when non-monetary criteria are relatively unimportant, it makes sense to look at costs alone. However, it is necessary to use an MCDM method of some sort in much of engineering decision making today.

In this chapter, we focus on three useful MCDM approaches. The first, *efficiency*, permits the identification of a subset of superior alternatives when there are multiple criteria. The second approach, *decision matrices*, is a version of multi-attribute utility theory (MAUT) which is widely practiced. The third, the *analytic hierarchy process*, is a relatively new but popular MAUT approach. It should be noted that all of these methods make assumptions about the trade-offs among criteria that may not be suitable in particular cases. They should not be applied blindly, but critically and with a strong dose of common sense.

13.2 Efficiency

When dealing with a single criterion like cost, it is usually clear which alternatives are better than others. The rule for a present worth analysis of mutually exclusive alternatives (with identical service lives) is, for example, that the highest present worth alternative is best.

All criteria can be measured in some way. The scale might be continuous, such as "weight in kilograms" or "distance from home," or discrete, such as "number of doors" or "operators needed." The measurement might be subjective, such as a rating of "excellent," "very good," "good," "fair," and "poor," or conform to an objective physical property like voltage or luminescence.

Once measured, the value of the alternative can be established with respect to that criterion. It may be that the smaller or lower measurement is better, as is often the case with cost, or the higher is better, as is the case with a criterion like "lives saved." Sometimes a target is desired, for example, a target weight or room temperature. In this case, the criterion could be adjusted to be the distance from the target, with a shorter distance being better.

Consequently, given one criterion, we can recognize which of several alternatives is best. However, once there are more than one criteria, the problem is more difficult. This is because an alternative can be highly valued with respect to one criterion and lowly valued with respect to another.

Example 13.1

Simcoe Meats will be replacing its effluent treatment system. It has evaluated several alternatives, shown in Figure 13.1. Two criteria were considered, present worth and discharge purity. Which alternatives can be eliminated from further consideration?

Consider alternatives A and E in Figure 13.1. Alternative E *dominates* alternative A because it is less costly and it provides purer discharge. If these are the only criteria to consider in making a choice, one would always choose E over A. Similarly, one can eliminate F, B, and H, all of which are dominated by D and other alternatives.

Now consider alternative G, which has the same cost as E but has poor discharge purity. One would still always choose E over G since, for the purity criterion, E is better at the same price.

Three alternatives now remain, E, D, and C. E is cheapest, but provides the least purity output of the three. C is the most expensive, but provides the

Figure 13.1 Selecting an Effluent Treatment

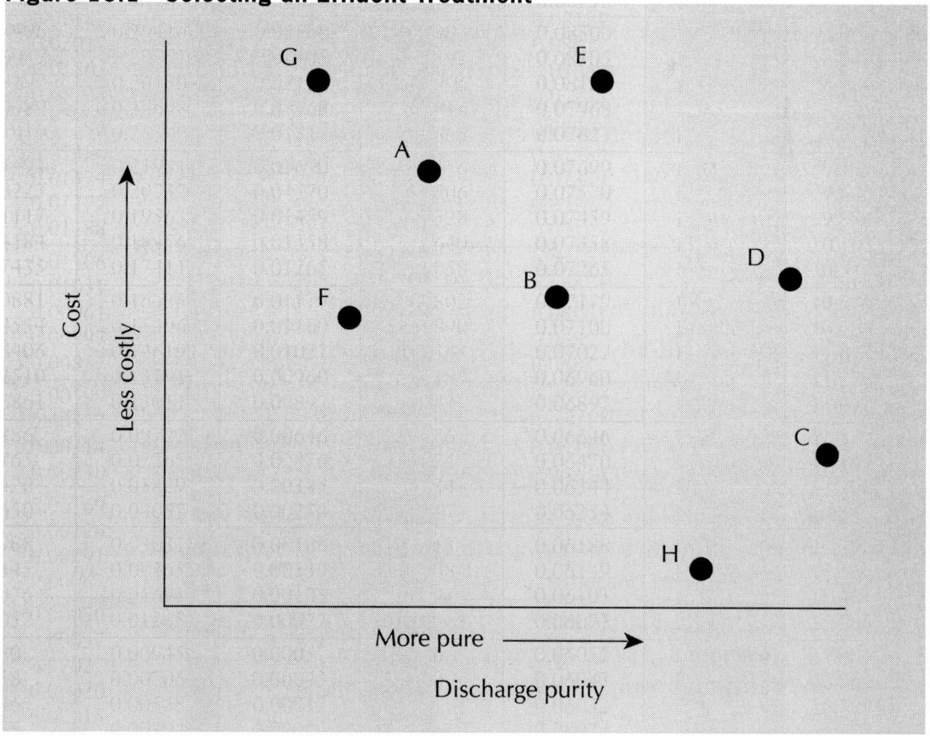

greatest purity, while D is in the middle. Certainly none of these dominates the others. There is a natural tendency to focus on D, since it seems to balance the two criteria, but it really depends on the relative importance of the criteria to the decision maker. For example, if cost were very important, E could be the best choice, since the difference in purity between E and C may be considered to be relatively small. ■

Decisions that involve only two criteria can be simplified graphically as done for Example 13.1, but when there are more than two criteria graphical methods become more difficult.

An alternative is **efficient** if no other alternative is valued as at least equal to it for all criteria and preferred to it for at least one. If an alternative is not efficient it is **inefficient**; this is the same as a **dominated** alternative.

Example 13.2

Skiven is evaluating surveillance cameras for a security system. The criteria he is taking into account are price, weight, picture clarity, and low-light performance. The details for the ten models are shown in Table 13.1. Skiven wants a camera with low cost, low weight, a high score for picture clarity, and a high score for low-light performance. Which models can be eliminated from further consideration?

To determine the efficient alternatives, the following algorithm can be used:

1. Order the alternatives according to one criterion, the *index criterion*. The cameras for Example 13.2 are already ordered by cost, so cost can be the index criterion.
2. Start with the second most preferred alternative for the index criterion. Call this the *candidate alternative*.
3. Compare the candidate alternative with each of the alternatives that are more preferred for the index criterion. (For the first candidate alternative, there is only one.)
4. If any alternative equals or exceeds the candidate for all criteria, and exceeds it for at least one, the candidate is dominated, and can be eliminated from further consideration. If no alternative equals or exceeds the candidate for all criteria and exceeds it for at least one, the candidate is efficient.
5. The next most preferred alternative for the index criterion becomes the new candidate; go to step 3. Stop if there are no more alternatives to consider.

The algorithm for Example 13.2 starts by comparing camera 2 against camera 1. It can be seen that camera 2 is better for weight and picture clarity, although worse for low-light and cost, so it is not dominated. Looking at cam-

Table 13.1 Surveillance Camera Characteristics

Camera	Price ($)	Weight (grams)	Picture Clarity (10-point scale)	Low-Light Performance (10-point scale)
1	230	900	3	6
2	243	640	5	4
3	274	910	3	5
4	313	433	5	7
5	365	450	2	4
6	415	330	6	6
7	418	552	7	5
8	565	440	3	6
9	590	630	7	4
10	765	255	9	5

era 3, it is equal to camera 1 for picture clarity, and worse than camera 1 for all other characteristics. It is dominated, since to avoid domination it would have to be better than 1 for at least one criterion. Moreover, we need not consider it for the remainder of the algorithm. We then compare camera 4 with only cameras 1 and 2. Since it is better in weight than the other two, it is not dominated, and we continue to camera 5. Camera 5 is dominated in comparison with camera 4, since it is worse in all respects.

Carrying through in this manner shows that camera 3 is dominated by camera 1, camera 5 is dominated by camera 4, camera 8 is dominated by camera 6, and camera 9 is dominated by camera 7. The set of efficient alternatives consists of cameras 1, 2, 4, 6, 7, and 10. This set clearly includes the best choice, since there is no reason to choose a dominated alternative. ■

Usually there is more than one alternative in the efficient set. Sometimes reducing the number of alternatives to be considered makes the problem easier to solve through intuition or judgment, but usually it is desirable to have some clear method for selecting a single alternative. One popular method is to use decision matrices.

13.3 Decision Matrices

Usually not all of the criteria that can be identified for a decision problem are equally important. Often cost is the most important criterion, but in some cases another criterion, safety, for example, might be most important. As suggested with Example 13.1, the choice of the most important criterion will have a direct effect on which alternative is best.

One approach to choosing the best alternative is to put numerical weights on the criteria. For example, if cost were most important, it would have a high weight, while a less important criterion could be given a low weight. If criteria are evaluated according to a scale that can be used directly as a measure of preference, then the weights and preference measures can be combined mathematically to determine a best alternative. This approach is called **multi-attribute utility theory (MAUT)**.

Many different specific techniques for making decisions are based on MAUT. This section deals with decision matrices, which are commonly used in engineering studies. The subsequent section reviews the analytic hierarchy process, a MAUT method of increasing popularity.

In a **decision matrix**, the rows of the matrix represent the criteria and the columns the alternatives. There is an extra column for the weights of the criteria. The cells of the matrix (other than the criteria weights) contain an evaluation of the alternatives on a scale from 0 to 10, where 0 is worst and 10 is best. The weights are chosen so that they sum to 10.

The following algorithm can be used:

1. Give a weight to each criterion to express its relative importance: the higher the weight, the more important the criterion. Choose the weight values so that they sum to 10.
2. For each alternative, give a rating from 0 to 10 of how well it meets each criterion. A rating of 0 is given to the worst possible fulfilment of the criterion and 10 to the best possible.
3. For each alternative, multiply each rating by the corresponding criterion weight, and sum to give an overall score.
4. The alternative with the highest score is best. The value of the score can be interpreted as the percentage of an ideal solution achieved by the alternative being evaluated.
5. Carry out some sensitivity analysis with respect to weights or rating estimates to verify the indicated decision or to determine under which conditions different choices are made.

Example 13.3

Skiven is evaluating surveillance cameras for a security system. The criteria he is taking into account, in order of importance for him, are low-light performance, picture clarity, weight, and price. The details for the six efficient models are shown in Table 13.2. Which model is best?

In order to follow the steps given above, we need to determine the criteria weights. It is usually fairly easy for a decision maker to determine which criteria are more important than others, but generally more difficult to specify particular weights. There exist many formal methods for establishing such weights in a rigorous way, but in practice, estimating weights based on careful consideration or a discussion with the decision maker is sufficient. Recall that a sensitivity analysis forms part of the overall decision process, and this compensates somewhat for the imprecision of the weights.

Table 13.2 Efficient Set of Surveillance Camera Alternatives

Camera	Price ($)	Weight (grams)	Picture Clarity (10-point scale)	Low-Light Performance (10-point scale)
1	230	900	3	6
2	243	640	5	4
4	313	433	5	7
6	415	330	6	6
7	418	552	7	5
10	765	255	9	5

Skiven suggests that weights of 1, 1.5, 3.5, and 4 for price, weight, picture clarity, and low-light performance, respectively, are appropriate weights for this problem. These weights are listed as the second column of Table 13.3.

Table 13.3 Decision Matrix for Example 13.3

Criterion	Criterion Weight	Alternatives					
		1	2	4	6	7	10
Price	1.0	10.0	9.8	8.4	6.5	6.5	0.0
Weight	1.5	0.0	4.0	7.2	8.8	5.4	10.0
Clarity	3.5	3.0	5.0	5.0	6.0	7.0	9.0
Low-light performance	4.0	6.0	4.0	7.0	6.0	5.0	5.0
Score	10.0	44.5	49.3	64.8	64.8	59.1	66.5

The ratings for each alternative for picture clarity and low-light performance are already on a scale from 0 to 10, so those ratings can be used directly. To select ratings for the price and weight, two different measures could be used:

1. *Normalization.* The rating r for the least preferred alternative (α) is 0 and the most preferred (β) is 10. For each remaining measure (γ) the rating r can be determined as:

$$r = 10 \times \frac{\gamma - \alpha}{\beta - \alpha}$$

For this problem, the rating of alternative 6 for price would be

$$r_{6,price} = 10 \times \frac{415 - 765}{230 - 765} = 6.54$$

The advantage of normalization is that it provides a mathematical basis for the rating evaluations. One disadvantage is that the rating may not reflect the value as perceived by the decision maker. A second disadvantage is that it may overrate the best alternative and underrate the worst, since these are set to the extreme values. A third disadvantage is that the addition or deletion of a single alternative (the one with the highest or lowest evaluation for a criterion) will change the entire set of ratings.

2. *Subjective evaluation.* Ask the decision maker to rate the alternatives on the 0 to 10 scale. For example, asked to rate alternative 6 for cost, Skiven might give it a 7. The advantages of subjective evaluation include that it is relatively immune to changes in the alternative set, and that it may be more accurate since it includes perceptions of worth that cannot be directly calculated from the criteria measures. Its main disadvantage is that people often make mistakes and give inconsistent evaluations.

For the ratings shown in Table 13.3, the normalization process was used. The overall score is then calculated by summing for each alternative the rating for a criterion multiplied by the weighting for that criterion. From Table 13.3, the total score for alternative 1 is calculated as

$$1 \times 10 + 1.5 \times 0 + 3.5 \times 3 + 4 \times 6 = 44.5$$

It can be seen in Table 13.3 that the highest score is for alternative 10. This means essentially that the greatest total benefit is achieved if alternative 10 is taken.

Also note that a "perfect" alternative, that is, one that rated 10 on every criterion, would have a total score of 100. Thus the 66.5 score for alternative 10 means that it is only about 66.5% of the score of a perfect alternative. The practice of making weights sum to 10 and rating the alternatives on a scale from 0 to 10 is done specifically so that the resulting score can be interpreted as a percentage of the ideal; if this is not desired, any relative weights or rating scale can be used.

Alternative 10 is the best choice for the particular weights and ratings given, but there should be some sensitivity analysis done to verify its robustness. There are several ways to do this sensitivity analysis, but the most sensible is to vary the weights of the criteria to see how the results change. This is easy to do when a spreadsheet is being used to calculate the scores.

Table 13.4 shows a range of criteria weights and the corresponding alternative scores. It can be seen that cameras 4 and 6 also can be identified as best in some of the criteria weight possibilities. For the final recommendation, it may be necessary to review these results with Skiven to let him determine which of the weight possibilities are most appropriate for him. ■

As has been seen in Example 13.3, the decision matrix approach structures information about multiple objectives of the problem. An additive utility model permits the calculation of an overall score for each alternative. A comparison of the scores permits the best one to be selected. Doing a sensitivity

Table 13.4 Sensitivity Analysis for the Surveillance Camera

Criterion	Criterion Weights						
Price	1	1	1	1	2	2.5	1
Weight	1.5	2	1	2	2	2.5	1
Picture clarity	3.5	3	3	2	2	2.5	4
Low-light performance	4	4	5	5	4	2.5	4
Alternative	**Alternative Scores**						
Camera 1	44.5	43.0	49.0	46.0	50.0	47.5	46.0
Camera 2	49.3	48.8	48.8	47.8	53.6	57.0	49.8
Camera 4	64.8	65.9	**65.7**	**67.9**	**69.4**	67.2	63.7
Camera 6	64.8	66.2	63.4	66.2	66.8	**68.4**	63.4
Camera 7	59.1	58.3	57.9	56.3	57.8	59.7	59.9
Camera 10	**66.5**	**67.7**	62.0	63.6	58.0	60.0	**66.0**

analysis may reveal promising alternatives from relatively small changes in the alternative weight assumptions.

13.4 The Analytic Hierarchy Process

The **analytic hierarchy process (AHP)** is also a MAUT approach. It offers two features beyond what is done in decision matrices. First, it provides a mechanism for structuring the problem that is particularly useful for large, complex decisions. Second, it provides a better method for establishing the criteria weights.

AHP is somewhat more complicated to carry out than decision matrices. In order to describe the procedure, we first list the basic steps. Example 13.4, which follows the list of steps, explains in more detail the operations at each step. The basic steps of AHP are as follows:

1. Identify the decision to be made, called the **goal**. Structure the goal, criteria, and alternatives into a hierarchy, as illustrated in Figure 13.2. The criteria could be more than one level (not illustrated in Figure 13.2) to provide additional structure to very complex problems.
2. Perform pairwise comparisons for alternatives. **Pairwise comparison** is an evaluation of the importance or preference of a pair of alternatives. Comparisions are made for all possible pairs of alternatives *with respect to each criterion*. This is done by giving each pair of alternatives a value according to Table 13.5 for their relationship for each criterion. These values are placed in a **pairwise comparison matrix (PCM)**.
3. **Priority weights** for the alternatives are calculated by normalizing the elements of the PCM and averaging the row entries. (The columns of priority weights together form a *priority matrix*.)

Figure 13.2 AHP Hierarchy

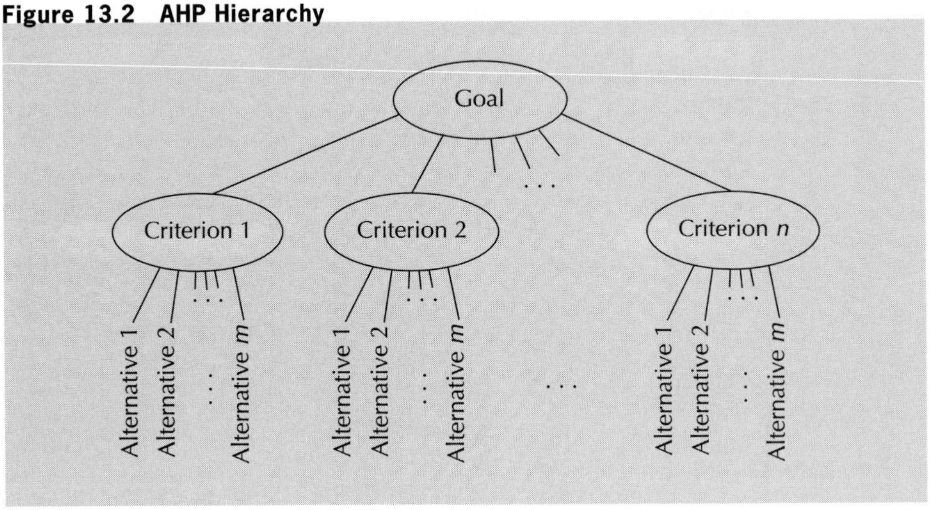

4. Perform pairwise comparisons for criteria. As in Step 2, all pairs of criteria are compared using the AHP value scale (Table 13.5). A PCM is determined, as in Step 3, and priority weights are calculated for the criteria.
5. Alternative priority weights are multiplied by the corresponding criteria priority weights and summed to give an overall alternative ranking. (That is, the priority matrix of alternatives is multiplied by the column of criteria priority weights to give a column of overall evaluations of the alternatives.)

The following example illustrates this process.

Table 13.5 The AHP Value Scale for Comparison of Two Alternatives or Two Criteria (A and B)

Value	Interpretation
1/9 = 0.111	Extreme preference/importance of B over A
1/7 = 0.1429	Very strong preference/importance of B over A
1/5 = 0.2	Strong preference/importance of B over A
1/3 = 0.333	Moderate preference/importance of B over A
1	Equal preference/importance of A and B
3	Moderate preference/importance of A over B
5	Strong preference/importance of A over B
7	Very strong preference/importance of A over B
9	Extreme preference/importance of A over B
Intermediate values	*For more detail between above values*

Example 13.4

Oksana is examining the cooling of a laboratory at Beaconsfield Pharmaceuticals. She has determined that a 12 000 BTU/hour cooling unit is suitable, but there are several models available with different features. The available quantitative data concerning the choices are shown in Table 13.6. The energy efficiency rating is a standard measure of power consumption efficiency.

Table 13.6 Cooling Unit Features

Model	Price	Energy Efficiency Rating
1	$640	9.5
2	$600	9.1
3	$959	10.0
4	$480	9.0
5	$460	9.0

Oksana also has several subjective criteria to take into account. She will use AHP to help her make this decision.

The first step for this problem is to structure the hierarchy. The goal is clear: to choose a cooling unit. The alternatives are also known, and are listed in Table 13.6. After some consideration, Oksana concludes that the following are critical to her consideration:

(1) Cost
(2) Energy consumption
(3) Loudness
(4) Perceived comfort

The resulting hierarchy is illustrated in Figure 13.3.

The second step is to construct a PCM for the alternatives with respect to each criterion. For illustration purposes, we will do this step for the criterion "perceived comfort" only.

Oksana first considers two of the alternatives only, say 1 and 2, with respect to "perceived comfort." She gives the preferred one (which is alternative 1) a rating from the scale shown in Table 13.5. In this case, Oksana judges that alternative 1 is moderately better than 2; the rating is a 3. This appears in the PCM shown in Figure 13.4 in row 1, column 2, corresponding to alternative 1 and alternative 2. Correspondingly, the reciprocal, 1/3, is put in row 2, column 1. This can be interpreted loosely as indicating that alternative 1 is three times as desirable as alternative 2 for perceived comfort, and correspondingly criterion 2 is 1/3 as desirable as alternative 1.

As another example, consider the comparison of alternatives 3 and 4. Alternative 4 is strongly preferred to 3 in "perceived comfort," so a 5 appears

Figure 13.3 The AHP Hierarchy for Example 13.4

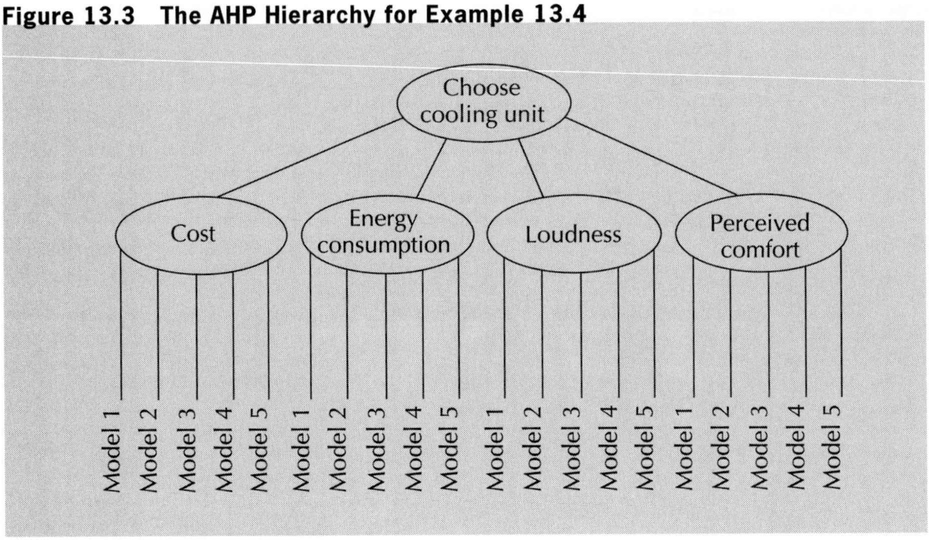

Figure 13.4 PCM for Perceived Comfort

$$\begin{bmatrix} 1 & 3 & 5 & 1 & 3 \\ \frac{1}{3} & 1 & 3 & \frac{1}{3} & 1 \\ \frac{1}{5} & \frac{1}{3} & 1 & \frac{1}{5} & \frac{1}{3} \\ 1 & 3 & 5 & 1 & 3 \\ \frac{1}{3} & 1 & 3 & \frac{1}{3} & 1 \end{bmatrix}$$

in row 4, column 3, and 1/5 appears in row 3 column 4. Similar comparisons of all pairs complete the PCM in Figure 13.4 for "perceived comfort." In summary, row 1 of Figure 13.4 shows the results of comparing alternative 1 with alternatives 1, 2, 3, 4, and 5. Row 2 shows the results of comparing alternative 2 with alternatives 1, 2, 3, 4, and 5, and so on. Note that an alternative is equally preferred to itself, so all main diagonal entries are 1. PCMs for the other three criteria are developed in exactly the same manner.

The next step is to determine priority weights for the alternatives. This is done by first normalizing the columns of the PCM and then averaging the rows. To normalize the columns, sum the column entries, and divide each entry by this sum. To average the rows, sum the rows (after normalizing) and divide by the number of entries per row.

For example, the sum of column 1 of the PCM in Figure 13.4 is 2.866.

Then the normalized entries for the column will be 1/2.866, 0.333/2.866, etc. The complete normalized PCM for "perceived comfort" is shown in Figure 13.5. The priority weights are then calculated as the average of each row of the normalized PCM, and are also illustrated in Figure 13.5.

Figure 13.5 Normalized PCM for Perceived Comfort

Normalized PCM					Average
0.349	0.360	0.294	0.349	0.360	0.342
0.116	0.120	0.176	0.116	0.120	0.130
0.070	0.040	0.059	0.070	0.040	0.056
0.349	0.360	0.294	0.349	0.360	0.342
0.116	0.120	0.176	0.116	0.120	0.130

A similar process can be carried out for the other three criteria. The four columns (one for each criterion) of priority weights form a priority matrix, shown in Figure 13.6. The first column of this matrix consists of the priority weights for cost, the second for energy efficiency, the third for noise, and the fourth for perceived comfort.

Figure 13.6 Priority Matrix for Example 13.4

0.90	0.256	0.033	0.342
0.114	0.230	0.468	0.130
0.031	0.338	0.282	0.056
0.383	0.088	0.086	0.342
0.383	0.088	0.131	0.130

The next step is to construct a PCM for the criteria themselves. This is done in the same manner as is done for the alternatives for each criterion, except that now one rates the criteria in pairwise comparisons with each other. The PCM Oksana creates is illustrated as Figure 13.7, with the rows and columns in the order: cost, energy consumption, loudness, and perceived comfort. For example, energy consumption is moderately more important than energy cost. Thus, there is a 3 in row 2, column 1 and a 1/3 in row 1, column 2. The normalized PCM and row averages are shown in Figure 13.8.

Figure 13.7 PCM for Goal

$$\begin{bmatrix} 1 & \frac{1}{3} & 5 & 1 \\ 3 & 1 & 7 & 3 \\ \frac{1}{5} & \frac{1}{7} & 1 & \frac{1}{5} \\ 1 & \frac{1}{3} & 5 & 1 \end{bmatrix}$$

Figure 13.8 Normalized PCM and Average Values for Goal

$$\begin{bmatrix} 0.192 & 0.184 & 0.278 & 0.192 \\ 0.577 & 0.553 & 0.389 & 0.577 \\ 0.038 & 0.079 & 0.056 & 0.038 \\ 0.192 & 0.184 & 0.278 & 0.192 \end{bmatrix} \qquad \overset{\textbf{Average}}{\begin{bmatrix} 0.212 \\ 0.524 \\ 0.053 \\ 0.212 \end{bmatrix}}$$

The order of the rows and columns of Figures 13.7 and 13.8 are: cost, energy efficiency, noise, and perceived comfort. Thus, the criterion with the highest priority rating is noise, at 0.524, then cost and perceived comfort identical at 0.212, and finally energy efficiency last at 0.053.

The final stage of the process consists of determining an overall score for each alternative. Note that the entire process of AHP has essentially led to the development of a decision matrix: the priority ratings for the criteria are the weights, while the priority ratings for the alternatives are the ratings of the alternatives for the criteria. Consequently, the final score is determined by multiplying each alternative priority rating by the appropriate criterion rating and then summing.

Figure 13.9 Final Alternative Scores for Example 13.4

$$\begin{bmatrix} 0.090 & 0.256 & 0.033 & 0.342 \\ 0.114 & 0.230 & 0.468 & 0.130 \\ 0.031 & 0.338 & 0.282 & 0.056 \\ 0.383 & 0.088 & 0.086 & 0.342 \\ 0.383 & 0.088 & 0.131 & 0.130 \end{bmatrix} \times \begin{bmatrix} 0.212 \\ 0.524 \\ 0.053 \\ 0.212 \end{bmatrix} = \begin{bmatrix} 0.227 \\ 0.197 \\ 0.211 \\ 0.204 \\ 0.162 \end{bmatrix}$$

This can also be viewed as matrix multiplication of the priority matrices for the alternatives by the column of priority weights of the criteria, as shown in Figure 13.9. The interpretation of the column vector on the right in Figure 13.9 is a ranking of the alternatives. The best alternative is number 1, followed by number 3, 4, and 2; number 5 is the worst.

In conclusion, the best cooling unit is model 1. Oksana should buy this one for the laboratory. ■

13.5 The Consistency Ratio for AHP

The subjective evaluation of the PCMs can be inconsistent. For example, Joe can say that alternative 1 is five times as important as alternative 2, and alternative 2 is five times as important as alternative 3, but then claim that alternative 1 is only twice as important as alternative 3. Or he might even say alternative 1 is less important than alternative 3.

The fact that the construction of PCMs includes redundant information is useful because it helps get a good estimate of the best rating for the alternative. However, there has to be a check made that the decision maker is being consistent.

A measure called the **consistency ratio** (to measure the consistency of the reported comparisons) can be calculated for any PCM. The consistency ratio ranges from 0 (perfect consistency) to 1 (no consistency). A consistency ratio of 0.1 or less is considered acceptable in practice. The calculation of the consistency ratio is briefly reviewed in Appendix 13A.

REVIEW PROBLEMS

REVIEW PROBLEM 13.1

Contrex makes thermostat controls for baseboard heaters. As part of the control manufacturing process, a 2-cm steel diaphragm is fitted to a steel cup. The diaphragm is used to open a safety switch rapidly to avoid arcing across the contacts. Currently the cup is seam welded, which is both expensive and a source of quality problems. The company wants to explore the use of adhesives to replace the welding process. Table 13.7 lists the ones examined, along with various properties for each.

High-temperature resistance is desirable, as are tensile bond strength and pressure resistance. Fast curing speeds are desirable to reduce work-in-progress inventory storage costs, and, of course, cheaper material costs are important.

Contrex wants to select a single adhesive type for comparison experiments against the current seam-welding method.

Table 13.7 Possible Adhesives for Thermostat Control

Adhesive	Maximum Temperature (°C)	Tensile Bond Strength (kPa)	Pressure Resistance (kPa)	Curing Speed	Cost
Acrylic	106	21 000	3738	Medium	Cheap
Silicone	200	3 150	560	Slow	Medium
Cyanoacrylate	250	3 500	630	Fast	Cheap
Methacrylate A	225	28 000	4984	Slow	Expensive
Methacrylate B	225	7 000	1246	Medium	Expensive

(a) Are any of the listed adhesives in Table 13.7 inefficient?

(b) Discussions with management indicate that the criteria can be weighted as follows:

Temperature resistance	1.5
Bond strength	1.5
Pressure resistance	2.5
Curing speed	3.5
Cost	1.0
Total	10.0

Create a decision matrix for this problem using the above weights. Normalize the data in Table 13.7 to estimate the ratings of each alternative for each criterion. Use only the efficient alternatives. Use the maximum temperature figures to measure temperature resistance. For curing speed, set fast as 8, medium as 5, and slow as 2, while for cost, set expensive as 10, medium as 5, and cheap as 2. Under these conditions, which is the recommended adhesive?

(c) The analytic hierarchy process (AHP) was performed for this problem. The hierarchy is shown in Figure 13.10. The priority matrix in Figure 13.11 represents the results from

Figure 13.10 The AHP Hierarchy for Review Problem 13.1(c)

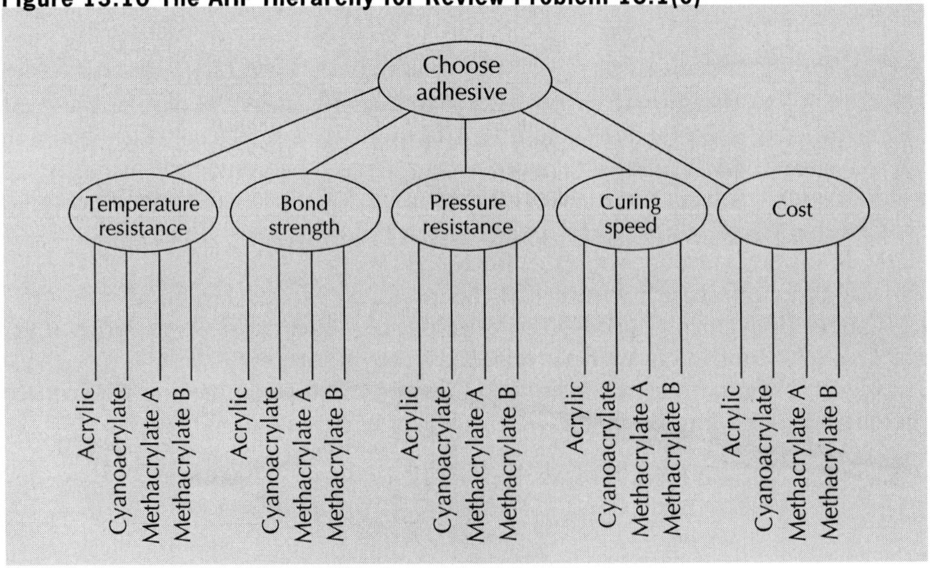

Figure 13.11 Priority Matrix for Review Problem 13.1 (c)

$$\begin{bmatrix} 0.095 & 0.368 & 0.392 & 0.213 & 0.341 \\ 0.331 & 0.077 & 0.067 & 0.502 & 0.278 \\ 0.287 & 0.445 & 0.438 & 0.061 & 0.188 \\ 0.287 & 0.110 & 0.103 & 0.224 & 0.193 \end{bmatrix}$$

Figure 13.12 PCM for Goal for Review Problem 13.1 (c)

$$\begin{bmatrix} 1 & 1 & \frac{1}{2} & \frac{1}{3} & 2 \\ 1 & 1 & \frac{1}{2} & \frac{1}{3} & 2 \\ 2 & 2 & 1 & \frac{1}{2} & 3 \\ 3 & 3 & 2 & 1 & 4 \\ \frac{1}{2} & \frac{1}{2} & \frac{1}{3} & \frac{1}{4} & 1 \end{bmatrix}$$

PCMs calculated for the different criteria. The rows of Figure 13.11 correspond to the alternatives acrylic, cyanoacrylate, methacrylate A, and methacrylate B, respectively, while the columns correspond to temperature resistance, bond strength, pressure resistance, curing speed, and cost, respectively. A PCM for the goal is shown in Figure 13.12, with the criteria in the same order as for Figure 13.11. With this information, what is the best adhesive to recommend for the experiment?

ANSWER

(a) It can be observed that silicone is dominated by cyanoacrylate in all criteria, and therefore is inefficient.

(b) Table 13.8 Decision Matrix of Adhesives

Criterion	Criterion Weight	Alternatives			
		Acrylic	Cyanoacrylate	Methacrylate A	Methacrylate B
Temperature resistance	1.5	0.0	10.0	8.3	8.3
Bond strength	1.5	7.1	0.0	10.0	1.4
Pressure resistance	2.5	7.1	0.0	10.0	1.4
Curing speed	3.5	5.0	8.0	2.0	5.0
Cost	1.0	2.0	2.0	10.0	10.0
Score	10.0	48.1	45.0	69.4	45.6

As shown in Table 13.8, under the weighting and rating conditions specified, the methacrylate A adhesive is clearly best.

(c) First we have to normalize the PCM for the goal, and then average to get the criteria weights, as shown in Figure 13.13.

Figure 13.13 Normalized PCM for Goal for Review Problem 13.1 (c)

Normalized PCM					Average
0.133	0.133	0.115	0.138	0.167	0.137
0.133	0.133	0.115	0.138	0.167	0.137
0.267	0.267	0.231	0.207	0.250	0.244
0.4	0.4	0.462	0.414	0.333	0.402
0.067	0.067	0.077	0.130	0.083	0.079

Then we multiply the priority matrix by the average criteria weights, as illustrated in Figure 13.14.

Since the second alternative in Figure 13.14 has the highest net weight, it is preferred. The recommended adhesive using AHP is the cyanoacrylate. This disagrees with the result obtained using decision matrices.

Figure 13.14 Calculating Alternative Weights for Review Problem 13.1 (c)

$$\begin{bmatrix} 0.095 & 0.368 & 0.392 & 0.213 & 0.341 \\ 0.331 & 0.077 & 0.067 & 0.502 & 0.278 \\ 0.287 & 0.445 & 0.438 & 0.061 & 0.188 \\ 0.287 & 0.110 & 0.103 & 0.224 & 0.193 \end{bmatrix} \times \begin{bmatrix} 0.137 \\ 0.137 \\ 0.244 \\ 0.402 \\ 0.079 \end{bmatrix} = \begin{bmatrix} 0.272 \\ 0.296 \\ 0.247 \\ 0.185 \end{bmatrix}$$

■

SUMMARY

In this chapter, three approaches for dealing explicitly with multiple criteria were presented. The first, *efficiency*, allows the identification of alternatives that are not dominated by others. An alternative is dominated if there is another alternative at least as good with respect to all criteria, and better in at least one.

The second approach, *decision matrices*, is in wide usage. Decision matrices are a multi-attribute utility theory (MAUT) method in which criteria are subjectively weighted and then multiplied by subjectively evaluated criteria ratings to give an overall score. The weights sum to 10 and the criteria ratings are on a scale from 0 to 10, resulting in an overall score that could range from 0 to 100. The alternative with the highest score is best, and the value of the score can be considered a percentage of an "ideal" alternative.

The third approach presented in this chapter is the *analytic hierarchy process*. AHP is also a MAUT method. Pairwise comparisons are used to extract criterion weights in a more rigorous manner than for decision matrices. Pairwise comparisons similarly are used to rate alternatives. Multiplying criterion weights and alternative ratings gives an overall evaluation for each alternative.

Engineering Economics in Action, Part 13B: Moving On

Three months later Naomi was seated in Anna Kulkowski's office. The Powerluxe project report had been submitted two days before.

"Naomi, the work you and Bill did and your report are first rate," Anna said. "We're going to start negotiations with Powerluxe to bring the Adaptamatic line of vacuum cleaners to market. I had no idea anybody could make such clear recommendations on such a complex problem. Congratulations on a good job."

Naomi thought back on all the people who had helped her in her almost two years at Canadian Widgets: how Clem had taught her practical problem solving; how Dave had shown her the ropes; how Terry had helped her realize the benefits of attention to detail. Bill had shown her how the real world mixed engineering with marketing, business, and government. Anna, too, had shown her how to manage people. "Thank you very much, Ms. Kulkowski," Naomi responded, a small break in her voice betraying her emotions.

"Do you enjoy this kind of work?" Anna asked.

"Yes, I do," Naomi replied. "It's exciting to see how the engineering relates to everything else."

"I think we have a new long-term assignment for you," Anna said. "This is just the first step for Canadian Widgets in developing new products. We have a first-rate team of engineers. We want to make better use of them. Ed Burns is going to head up a product development group. He read your report on the Adaptamatic line of vacuum cleaners and was quite impressed. He and I would like you in that development group. We need someone who understands the engineering and can relate it to markets. What do you say?"

"I'm in!" Naomi had a big grin on her face.

A few days later, Naomi answered the phone.

"Hey, Naomi, it's Terry. I hear you got promoted!"

"Hi, Terry. Nice to hear from you. Well, the money's about the same, but it sure will be interesting. How are things with you?" Naomi had fond memories of working with Terry.

"Well, I graduate next month, and I have a job. Guess where?"

Naomi knew exactly where. Clem had told her about the interviews. He wasn't really fair to the other candidates—Clem had decided to hire Terry as soon as he had applied. "Here? Really?"

PROBLEMS

13.1 The table below shows information about alternative choices. Criteria C and E are to be minimized, while all the rest are to be maximized. Which of the alternatives can be eliminated from further consideration?

	Criteria				
Alternative	A	B	C	D	E
1	340	5	11	1.2	1
2	570	8	22	3.3	1
3	410	9	22	3.2	2
4	120	4	36	0.9	3
5	122	1	46	1.3	2
6	345	8	47	0.6	3
7	119	4	57	1.1	2
8	554	2	89	2.1	3
9	317	9	117	0.9	1
10	129	5	165	1.5	3

13.2 The following is a partially completed pairwise comparison matrix. Complete it. What are the corresponding priority weights?

$$\begin{bmatrix} - & \frac{1}{2} & - & \frac{1}{9} \\ - & - & - & - \\ 4 & 2 & - & \frac{1}{2} \\ - & 4 & - & - \end{bmatrix}$$

13.3 The Toronto Transit Commission is considering building a new subway line. Twelve alternatives are being considered. All the alternatives are shown in the tables, with their criteria values. The relevant criteria are:

C1 Population and jobs served per kilometre

C2 Projected daily traffic per kilometre

C3 Capital cost per kilometre (in millions of dollars)

C4 IRR

C5 Structural effect on urbanization

It is desirable to have high population served, high traffic, low capital cost per kilometre, and a high IRR. Criterion C5 concerns the benefits for urban growth caused by the subway location, and is measured on a scale from 0 to 10, with the higher values being preferred.

	Alternative Subway Lines					
Criterion	L1	L2	L3	L4	L5	L6
C1	81 900	31 800	11 500	31 100	23 000	16 100
C2	25 500	11 600	7 100	10 500	10 200	3 500
C3	65	45	29	35	40	10
C4	8.6	6.3	4.5	14.1	13.	11.8
C5	3	7	4	6	5	4

	Alternative Subway Lines					
Criterion	L7	L8	L9	L10	L11	L12
C1	13200	28 200	36 500	24 400	18 400	13 900
C2	3 700	7 400	10 300	7 100	4 700	3 100
C3	32	30	13.2	40	43	25
C4	3.9	6	3	3.7	3.7	5.8
C5	9	10	6	9	5	4

(a) Which of these alternatives are efficient?

(b) Establish ratings for each of the efficient alternatives from part (a) for a decision matrix through normalization. The weights of the five criteria are C1: 1.5, C2: 2, C3: 2.5, C4: 3, C5: 1. Construct a decision matrix and determine the best subway route.

(c) If the criteria weights were C1: 2, C2: 2, C3: 2, C4: 2, C5: 2, would the recommended alternative be different?

13.4 Sudbury Steel is considering buying a new CNC punch press. They place high value on the reliability of this equipment, since it will be central to their production process. Speed and quality are also important to them, but not as important as reliability. As well as examining these factors, the company will make sure that the equipment will be easily adaptable to changes in production.

Construct a decision matrix for the alternatives listed and appropriate criteria. Select appropriate weightings for the criteria (state any assumptions), and determine the preferred alternative. Do a reasonable sensitivity analysis to determine the conditions under which the choice of alternative may change. A spreadsheet program should be used.

The alternatives are:

1. Name: Accumate Plasmapress
   Cost: $428 600
   Service: Average
   Reliability: Average
   Speed: High

	Quality:	Good
	Flexibility:	Excellent
	Size:	Average
	Other:	None
2.	Name:	Weissman Model 4560
	Cost:	$383 765
	Service:	Below average
	Reliability:	Average
	Speed:	High
	Quality:	Fair
	Flexibility:	Not good
	Size:	Average
	Other:	Many tool stations (desirable), but small tools (not desirable)
3.	Name:	A.D. Hockley Model 661-84
	Cost:	$533 725
	Service:	Untested
	Reliability:	Average
	Speed:	Slow
	Quality:	Very good
	Flexibility:	Very good
	Size:	Very compact
	Other:	Small turret (not desirable)
4.	Name:	Frammit Manu-Centre 1500/45
	Cost:	$393 000
	Service:	Average
	Reliability:	Below average
	Speed:	Average
	Quality:	Average
	Flexibility:	Average
	Size:	Average
	Other:	Has perforating and coining feature (very desirable) Poor torch design causes too rapid wear (undesirable)
5.	Name:	Frammit Manu-Centre Lasertool 1250/30/1500
	Cost:	$340 056
	Service:	Average
	Reliability:	Average
	Speed:	Exceptionally fast
	Quality:	Exceptionally good
	Flexibility:	Average
	Size:	Undesirably small
	Other:	Cannot handle heavy gauge metal (not desirable)
6.	Name:	Boxcab 3025/12P CNC
	Cost:	$405 232
	Service:	Untested

Reliability:	Excellent
Speed:	Average
Quality:	Very good
Flexibility:	Average
Size:	Average
Other:	None

13.5 Complete each of the following pairwise comparison matrices:

(a)
$$\begin{bmatrix} - & 1 & - & - \\ - & - & - & 2 \\ \frac{1}{5} & 3 & - & - \\ 3 & - & \frac{1}{2} & - \end{bmatrix}$$

(b)
$$\begin{bmatrix} - & 9 & - & 1 & 3 \\ - & - & 3 & - & - \\ \frac{1}{7} & - & - & \frac{1}{4} & 2 \\ - & 6 & - & - & - \\ - & 1 & - & 2 & - \end{bmatrix}$$

13.6 For each of the PCMs in Problem 13.5, compute priority weights for the alternatives.

13.7 (a) Francis has several job opportunities for his co-op work term. He would like a job with good pay that is close to home, contributes to his engineering studies, and is with a smaller company. Which of the opportunities listed below should be removed from further consideration?

	Criteria			
Job	**Pay**	**Home**	**Studies**	**Size**
1. Spinoff Consulting	1700	2	3	5
2. Nub Automotive	1600	5	3	500
3. Soutel	2200	80	4	150
4. Provincial Hydro	1800	100	3	3000
5. Fitzsimon Associates	1700	100	1	20
6. General Auto	2000	150	2	2500
7. Ring Canada	2200	250	5	300
8. Jones Mines	2700	500	3	20
9. Resources, Inc.	2700	2000	2	40

Pay: Monthly salary in dollars
Home: Distance from home in kilometres
Studies: Contribution to engineering studies, 0 = none, 5 = a lot
Size: Number of employees at that location

(b) Francis feels that the following weights can represent the importance of his four criteria:

Pay:	4.0
Home:	2.5
Studies:	2.0
Size:	1.5

Using normalization to establish the ratings, which job is best?

Problems 13.8 to 13.23 are based on the following situation.

John is considering the selection of a consultant to provide ongoing support for a computer system. John wishes to contract with one consultant only, based on the criteria of cost, reliability, familiarity with equipment, location, and quality. Cost is measured by the quoted daily rate. Reliability and quality are measured on a qualitative scale based on discussions with references and interviews with the consultants. Familiarity with equipment has three possibilities: none, some, or much. Location is measured by distance from the consultant's office to the plant site in kilometres.

The specific data for the consultants are as follows:

	Consultants				
Criterion	**A**	**B**	**C**	**D**	**E**
Cost	\$500	\$500	\$450	\$600	\$400
Reliability	Good	Good	Excellent	Good	Fair
Familiarity	None	Some	Some	Much	Some
Location	3	1	5	2	1
Quality	Excellent	Good	Fair	Excellent	Good

13.8 Is any choice of consultant inefficient?

13.9 John can assign weights to the criteria as follows:

Criterion	Weight
Cost	2
Reliability	3
Familiarity	1
Location	1
Quality	3

Using the following tables to convert qualitative evaluations to numbers and using the formula for determining ratings by normalization, construct a decision matrix for choosing a consultant. Which consultant is best?

For Reliability and Quantity	
Description	**Value**
Excellent	5
Very Good	4
Good	3
Fair	2
Poor	1

For Familiarity	
Description	**Value**
None	1
Some	2
Much	3

13.10 Draw an AHP hierarchy for this problem.

13.11 John considers a cost difference of \$100 or more to be of strong importance, and a difference of \$50 to be of moderate importance. Construct a PCM for the criterion *cost*.

13.12 John considers the difference between good reliability and fair reliability to be of moderate importance, and the difference between good and excellent to be of strong importance. He considers the difference between fair and excellent to be of very strong importance. Construct a PCM for the criterion *reliability*.

13.13 John considers the difference between no familiarity with the computing equipment and some familiarity to be of strong importance, and the difference between none and much familiarity to be of extreme importance. He considers the difference between some and much to be of strong importance. Construct a PCM for the criterion *familiarity*.

13.14 John considers each kilometre of distance worth two units on the AHP value scale. For example, the value of the location of consultant D over consultant C is $(5 - 2) \times 2 = 6$, and a 6 would be placed in the fourth row, third column, of the PCM for the criterion *location*. Construct the complete PCM for the criterion *location*.

13.15 John considers the difference between fair quality and good quality to be of strong importance, and the difference between good and excellent to be of very strong importance. He considers the difference between fair and excellent to be of extreme importance. Construct a PCM for the criterion *quality*.

13.16 Calculate the priority weights for the criterion *cost* from the PCM constructed in Problem 13.11.

13.17 Calculate the priority weights for the criterion *reliability* from the PCM constructed in Problem 13.12.

13.18 Calculate the priority weights for the criterion *familiarity* from the PCM constructed in Problem 13.13.

13.19 Calculate the priority weights for the criterion *location* from the PCM constructed in Problem 13.14.

13.20 Calculate the priority weights for the criterion *quality* from the PCM constructed in Problem 13.15.

13.21 Construct the priority matrix from the answers to Problems 13.16 to 13.20.

13.22 A partially completed PCM for the criteria is shown below. Complete the PCM and calculate the priority weights for the criteria.

$$\begin{bmatrix} - & \frac{1}{3} & - & 3 & - \\ - & - & - & - & 1 \\ \frac{1}{3} & \frac{1}{7} & - & 1 & - \\ - & \frac{1}{6} & - & - & - \\ 1 & - & 7 & 7 & - \end{bmatrix}$$

13.23 Determine the overall score for each alternative. Which consultant is best?

Problems 13.24 to 13.31 are based on the following case.

Fabian has several job opportunities for his co-op work term. The pay is in dollars per month. The distance from home is in kilometres. Relevance to studies is on a five-point scale, with a 5 meaning very relevant to studies and a 0 meaning not relevant at all. The company size refers to the number of employees at the job location.

In general, Fabian wants a job with good pay, that is close to home, contributes to his engineering studies, and is with a smaller company.

	Criteria			
Job	**Pay**	**Distance from Home**	**Relevance to Studies**	**Company Size**
1. Spinoff Consulting	1700	2	3	5
2. Soutel	2200	80	4	150
3. Ring Canada	2200	250	5	300
4. Jones Mines	2700	500	3	20

13.24 Draw an AHP hierarchy for this problem.

13.25 **(a)** Complete the PCM below for *Pay*.

$$\begin{bmatrix} - & \frac{1}{5} & \frac{1}{5} & \frac{1}{7} \\ - & - & 1 & \frac{1}{5} \\ - & - & - & \frac{1}{5} \\ - & - & - & - \end{bmatrix}$$

(b) What are the priority weights for *Pay*?

13.26 **(a)** Complete the PCM below for *Distance from Home*.

$$\begin{bmatrix} - & - & - & - \\ \frac{1}{5} & - & - & - \\ \frac{1}{7} & \frac{1}{5} & - & - \\ \frac{1}{9} & \frac{1}{7} & \frac{1}{2} & - \end{bmatrix}$$

(b) What are the priority weights for *Distance from Home*?

13.27 **(a)** Complete the PCM below for *Studies*.

$$\begin{bmatrix} - & - & \frac{1}{3} & - \\ 2 & - & - & 2 \\ - & 2 & - & - \\ 1 & - & \frac{1}{3} & - \end{bmatrix}$$

(b) What are the priority weights for *Studies*?

13.28 **(a)** Complete the PCM below for *Size*.

$$\begin{bmatrix} - & - & - & 1 \\ \frac{1}{7} & - & - & \frac{1}{7} \\ \frac{1}{9} & \frac{1}{2} & - & \frac{1}{9} \\ - & - & - & - \end{bmatrix}$$

(b) What are the priority weights for *Size*?

13.29 Form a priority matrix for the PCMs in Problems 13.25 to 13.28.

13.30 Given the following PCM for the criteria, calculate the priority weights for the criteria.

$$\begin{bmatrix} 1 & 3 & 5 & 5 \\ \frac{1}{3} & 1 & 3 & 3 \\ \frac{1}{5} & \frac{1}{3} & 1 & 2 \\ \frac{1}{5} & \frac{1}{3} & \frac{1}{2} & 1 \end{bmatrix}$$

13.31 Using the results of Problems 13.29 and 13.30, calculate the priority weights for the goal. Which job is Fabian's best choice? His second best choice?

Northwind Stoneware

CANADIAN 13.1 MINI-CASE

Northwind Stoneware of Kitchener, Ontario, makes consumer stoneware products. Stoneware is fired in a kiln, which is an enclosure made of a porous brick having heating elements designed to raise the internal temperature to over 1200°C. Clay items such as stoneware can be hardened by firing them, following a particular temperature pattern called a firing curve.

Quality and cost problems led Northwind to examine better ways to control the firing curve. Alternatives available to them included:

1. Direct human control. The temperature sensitivity of the human eye cannot be matched by any automatic control.
2. Use of a "Kilnsitter," which is a mechanical switch that shuts off the kiln at a preset temperature.
3. Use of a pyrometer, which is an electrical instrument for measuring heat, with a programmable controller.
4. Use of a pyrometer and a computer.

The criteria used to determine the best choice included installation cost, effectiveness, reliability, energy savings, maintenance costs, and other applications.

A decision matrix evaluation method was used. Under a variety of criteria weights, the use of a pyrometer and a computer was the recommended choice. The exception was when installation cost was given overwhelming weight; in this case the use of a Kilnsitter was recommended.

Discussion

Real life is always more complicated than any model used for analysis. In reading about a case where decisions are made, the complexity of the real decision process is always hidden because the very process of describing the situation is itself a model. In describing something, choices are made about what is important and worth describing and what is unimportant and not worth describing. In real life, one has to go through a process of separating from a great mass of information exactly what is important and what is not important.

The process of solving the quality and cost problems for Northwind Stoneware first involved collecting a great deal of information about the manufacturing process. Many people were interviewed, production was observed, and technical details of stoneware chemistry were researched. Market analysis was required to identify possible solutions to the process. Several person-weeks of work were required to gather the information to create the decision matrix mentioned above.

We often concentrate on the mechanics of using a decision tool instead of the mechanics of getting good data to use as input to the tool. In many cases, getting good data is 90% of the solution.

Questions

Think of an everyday decision situation that you have been involved in recently. It could be deciding which novel to buy, which apartment to rent, or which movie to go to.

1) Write down a one-page description of the decision. Include where you were, what choices you had, what you chose to do, and why.
2) Think about how you decided what to write for question 1. Was it easy to know what to include in your description, and what not to include?
3) Can you immediately identify alternatives and criteria, as formally defined in this chapter, from your one-page description? If not, why not?
4) Construct a formal decision matrix model for your problem, augmenting the alternatives and criteria if necessary. Fill in the weights and ratings. Comment on any difficulties you have coming up with the weights and ratings.
5) Calculate the overall score for your decision matrix. Did you actually choose the one with the highest score? If not, why not?

Appendix 13A: Calculating the Consistency Ratio for AHP

The **consistency ratio *(CR)*** for AHP provides a measure of the ability of a decision maker to report preferences over alternatives or criteria. Calculating the *CR* can be done without understanding the concepts underlying it, but some background can be helpful. In this appendix, a brief overview of the basis for the *CR* is given, but the main purpose is to present an algorithm for calculating the *CR*. For the background information, some understanding of linear algebra is desirable, but the algorithm for calculating the *CR* can be easily followed without this background information.

Recall that in AHP a pairwise comparison matrix (PCM) is developed by comparing, for example, the alternatives with respect to one criterion. If alternative x is compared with y and found to be twice as preferred, and y is compared with z and is three times as preferred, then it is easy to deduce that x should be six times as preferred as z. However, in AHP every pairwise comparison is made giving several judgments about the same relationship. Humans will not necessarily be perfectly consistent in reporting preferences; one of the strengths of AHP is that, by getting redundant preference information, the quality of the information is improved.

Observe that if a decision maker were perfectly consistent, the columns of a PCM would be multiples of each other. They would differ in scale because for each column n, the nth row is fixed as 1, but the relative values would be constant.

Recall that an eigenvalue λ is one of n solutions to the equation

$$\mathbf{Aw} = \lambda\mathbf{w}$$

where $\mathbf{A}$ is a square $n \times n$ matrix and $\mathbf{w}$ is an $n \times 1$ eigenvector.

There is one eigenvector corresponding to each eigenvalue λ. A PCM is unusual in that, in addition to being a square matrix, all of the entries are positive, the corresponding entries across the main diagonal are reciprocals, and there are 1s along the main diagonal. It can be shown that, in this case, if the decision maker is perfectly consistent, there will be a single non-zero eigenvalue, and $n - 1$ eigenvectors of value 0.

In practice, a PCM $\mathbf{A}$ is not perfectly consistent. However, assuming that the error is relatively small, there should be one large eigenvalue λ_{max} and $n - 1$ small ones. Further, it can be shown that, with small inconsistencies, $\lambda_{max} > n$.

The *CR* is developed from the difference between λ_{max} and n. First, this difference is divided by $n - 1$, effectively distributing it over the other, supposedly zero, eigenvalues. This gives the *consistency index* (*CI*).

$$CI = \frac{\lambda_{max} - n}{n - 1}$$

Second, the *CI* is divided by the *CI* of a random matrix of the same size, called a *random index* (*RI*), to form the consistency ratio (*CR*). The *RI* for matrices of up to 10 rows and columns is shown in Table 13A.1; these were developed by averaging the *CI*s for hundreds of randomly generated matrices.

$$CR = \frac{CI}{RI}$$

The idea is that, if the PCM were completely random, we would expect the *CI* to be equal to the *RI*. Thus, a random PCM would have a *CR* of 1. However, a consistent PCM, having a *CI* of 0, would also have a *CR* of 0. *CR*s of between 0 and 1 indicate how consistent or random the PCM is, in a manner that is independent of the size of the matrix.

Table 13A.1 Random Indexes for Various Sizes of Matrices

Size ($n \times n$)	Random Index
2	0
3	0.58
4	0.90
5	1.12
6	1.24
7	1.32
8	1.41
9	1.45
10	1.49

The *CR* is actually a well-designed statistical measure of the deviation of a particular PCM from perfect consistency. As a rule of thumb, an upper limit of 0.1 is usually used. Thus, if $CR \leq 0.1$, the PCM is *acceptably* consistent. If the $CR > 0.1$, the PCM should be re-evaluated by the decision maker.

This background can be used to construct an algorithm for calculating the *CR* for any PCM. Essentially, one determines λ_{max} and its associated eigenvector, called the **principal eigenvector**. The general algorithm for a PCM **A** of size $n \times n$ is as follows:

1. Find **w**, the eigenvector associated with λ_{max}, from normalizing the following:

$$\mathbf{w} = \lim_{k \to \infty} \frac{\mathbf{A}^k \mathbf{e}}{n}$$

   where **e** is a column vector $[111...111]^T$

   In other words, form a sequence of powers of the matrix **A**. Normalize each of the resulting matrices by dividing each element of **A** by the sum of the elements from the corresponding column. Form **w** by summing each row of normalized **A** and dividing this by n. Eventually, **w** will not noticeably change from one power of **A** to the next higher power. This value of **w** is the desired principal eigenvector.

   In the main part of this chapter, we calculated the priority weights for a PCM. That procedure was an approximate method of determining **w**; the priority weights and the eigenvector associated with λ_{max} are the same thing. The procedure mentioned here is more accurate, but more difficult to compute and time-consuming.

2. Since **w** is a vector, λ_{max} can be found by solving $\mathbf{Aw} = \lambda_{max}\mathbf{w}$ or any element of **w**, or equivalently, any row of **A**. Thus, for any i, compute:

$$\lambda_{max} = \frac{\sum_j (a_{ij} \times w_j)}{w_i}$$

3. Calculate the CI from

$$CI = \frac{\lambda_{max} - n}{n - 1}$$

4. With reference to Table 13A.1, calculate the CR from

$$CR = \frac{CI}{RI}$$

If $CR \leq 0.1$, the PCM is acceptably consistent.

Example 13A.1

Does the PCM of Figure 13.4 meet the requirement for a consistency ratio of less than or equal to 0.1?

The PCM of Figure 13.4 is reproduced as Figure 13A.1, with successive powers of the original matrix, normalized versions of the powers, and the calculated **w**. The calculations were done using a spreadsheet program; most popular spreadsheet programs can automatically find powers of matrices.

It can be seen that **w** converges very quickly to a stable set of values. Normally, the less consistent a PCM is, the longer it will take to converge. Also, note that the elements of **w** are very close to the priority weights that were calculated for this PCM in Section 13.4.

To determine λ_{max}, we must now solve $\mathbf{Aw} = \lambda_{max}\mathbf{w}$ for one row of **A**. Selecting the first row of **A** results in the following expression:

$$[1\ 3\ 5\ 1\ 3] \begin{bmatrix} 0.343 \\ 0.129 \\ 0.055 \\ 0.343 \\ 0.129 \end{bmatrix} = \lambda_{max} \times 0.343$$

Or, equivalently:

$$\lambda_{max} = \frac{(1 \times 0.343) + (3 \times 0.129) + (5 \times 0.055) + (1 \times 0.343) + (3 \times 0.129)}{0.343}$$

$$= 5.0583$$

The consistency index can then be calculated from

$$CI = \frac{\lambda_{max} - n}{n - 1}$$

$$= \frac{5.0583 - 5}{5 - 1}$$

$$= 0.0146$$

Figure 13A.1 Calculating the Principal Eigenvector w

	A					Normalized					w
	1	3	5	1	3	0.349	0.36	0.29	0.349	0.36	0.342
	0.333	1	3	0.333	1	0.116	0.12	0.18	0.116	0.12	0.130
A^1	0.2	0.333	1	0.2	0.333	0.07	0.04	0.06	0.07	0.04	0.056
	1	3	5	1	3	0.349	0.36	0.29	0.349	0.36	0.342
	0.333	1	3	0.33	1	0.116	0.12	0.18	0.116	0.12	0.130
	5	13.67	33	5	13.667	0.34	0.346	0.35	0.34	0.346	0.343
	1.933	5	13.3	1.93	5	0.132	0.126	0.13	0.132	0.126	0.129
A^2	0.822	2.2	5	0.82	2.2	0.056	0.056	0.05	0.056	0.056	0.055
	5	13.67	33	5	13.667	0.34	0.346	0.35	0.34	0.346	0.343
	1.933	5	13.3	1.93	5	0.132	0.126	0.13	0.132	0.126	0.129
	25.711	68.33	165	25.7	68.333	0.343	0.343	0.34	0.343	0.343	0.343
	9.667	25.71	61.7	9.67	25.711	0.129	0.129	0.13	0.129	0.129	01.29
A^3	4.111	1	26.4	4.11	11	0.055	0.055	0.06	0.055	0.055	0.055
	25.711	68.33	165	25.7	68.333	0.343	0.343	0.34	0.343	0.343	0.343
	9.667	25.71	61.7	9.67	25.711	0.129	0.129	0.13	0.129	0.129	01.29
	129.98	345.9	832	130	345.93	0.343	0.343	0.34	0.343	0.343	0.349
	48.807	130	313	48.8	129.98	0.129	0.129	0.13	0.129	0.129	0.129
A^4	20.84	55.47	134	20.8	55.474	0.055	0.055	0.06	0.055	0.055	0.055
	129.98	345.9	832	130	345.93	0.343	0.343	0.34	0.343	0.343	0.349
	48.807	130	313	48.8	129.98	0.129	0.129	0.13	0.129	0.129	0.129
	657	1749	4207	657	1749.1	0.343	0.343	0.34	0.343	0.343	0.343
	246.79	657	1581	247	657	0.129	0.129	0.13	0.129	0.129	0.129
A^5	105.37	280.5	675	105	280.5	0.055	0.055	0.06	0.055	0.055	0.055
	657	1749	4207	657	1749.1	0.343	0.343	0.34	0.343	0.343	0.343
	246.79	657	1581	247	657	0.129	0.129	0.13	0.129	0.129	0.129

As seen in Table 13A.1, the *RI* for a matrix of five rows and columns is 1.12. We can then calculate the consistency ratio as

$$CR = \frac{CI}{RI} = \frac{0.0146}{1.12} = 0.013$$

Clearly, the *CR* is very much less than 0.1. It can thus be concluded that the original PCM is acceptably consistent. ■

PROBLEMS FOR APPENDIX 13A

13A.1 Does the following PCM meet the requirement for a consistency ratio of less than or equal to 0.1? What are the values of the consistency index and the consistency ratio?

$$\begin{bmatrix} 1 & \frac{1}{2} & \frac{1}{4} & \frac{1}{9} \\ 2 & 1 & \frac{1}{2} & \frac{1}{4} \\ 4 & 3 & 1 & 1 \\ 9 & 7 & 1 & 1 \end{bmatrix}$$

13A.2 Does the following PCM meet the requirement for a consistency ratio of less than or equal to 0.1? What are the values of the consistency index and the consistency ratio?

$$\begin{bmatrix} 1 & \frac{1}{2} & \frac{1}{4} & \frac{1}{9} \\ 2 & 1 & \frac{1}{2} & \frac{1}{4} \\ 4 & 3 & 1 & 1 \\ 9 & 7 & 1 & 1 \end{bmatrix}$$

13A.3 Does the following PCM meet the requirement for a consistency ratio of less than or equal to 0.1? What are the values of the consistency index and the consistency ratio? How does an accurate evaluation of the principal eigenvector for this PCM compare with the priority weights calculated in Problem 13.30?

$$\begin{bmatrix} 1 & 3 & 5 & 5 \\ \frac{1}{3} & 1 & 3 & 3 \\ \frac{1}{5} & \frac{1}{3} & 1 & 2 \\ \frac{1}{5} & \frac{1}{3} & \frac{1}{2} & 1 \end{bmatrix}$$

APPENDIX A

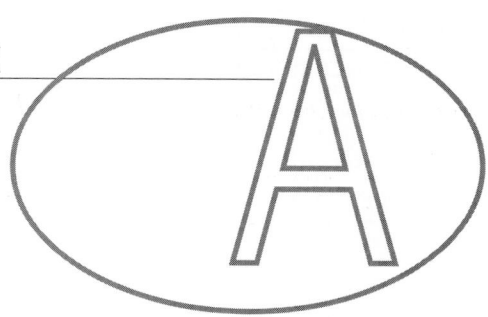

Compound Interest Factors for Discrete Compounding, Discrete Cash Flows

$i = 0.5\%$ **Discrete Compounding, Discrete Cash Flows**

	SINGLE PAYMENT		UNIFORM SERIES				Arithmetic Gradient Series Factor
	Compound Amount Factor	Present Worth Factor	Sinking Fund Factor	Uniform Series Factor	Capital Recovery Factor	Series Present Worth Factor	
N	(*F/P,i,N*)	(*P/F,i,N*)	(*A/F,i,N*)	(*F/A,i,N*)	(*A/P,i,N*)	(*P/A,i,N*)	(*A/G,i,N*)
1	1.0050	0.99502	1.0000	1.0000	1.0050	0.99502	0.00000
2	1.0100	0.99007	0.49875	2.0050	0.50375	1.9851	0.49875
3	1.0151	0.98515	0.33167	3.0150	0.33667	2.9702	0.99667
4	1.0202	0.98025	0.24813	4.0301	0.25313	3.9505	1.4938
5	1.0253	0.97537	0.19801	5.0503	0.20301	4.9259	1.9900
6	1.0304	0.97052	0.16460	6.0755	0.16960	5.8964	2.4855
7	1.0355	0.96569	0.14073	7.1059	0.14573	6.8621	2.9801
8	1.0407	0.96089	0.12283	8.1414	0.12783	7.8230	3.4738
9	1.0459	0.95610	0.10891	9.1821	0.11391	8.7791	3.9668
10	1.0511	0.95135	0.09777	10.228	0.10277	9.7304	4.4589
11	1.0564	0.94661	0.08866	11.279	0.09366	10.677	4.9501
12	1.0617	0.94191	0.08107	12.336	0.08607	11.619	5.4406
13	1.0670	0.93722	0.07464	13.397	0.07964	12.556	5.9302
14	1.0723	0.93256	0.06914	14.464	0.07414	13.489	6.4190
15	1.0777	0.92792	0.06436	15.537	0.06936	14.417	6.9069
16	1.0831	0.92330	0.06019	16.614	0.06519	15.340	7.3940
17	1.0885	0.91871	0.05651	17.697	0.06151	16.259	7.8803
18	1.0939	0.91414	0.05323	18.786	0.05823	17.173	8.3658
19	1.0994	0.90959	0.05030	19.880	0.05530	18.082	8.8504
20	1.1049	0.90506	0.04767	20.979	0.05267	18.987	9.3342
21	1.1104	0.90056	0.04528	22.084	0.05028	19.888	9.8172
22	1.1160	0.89608	0.04311	23.194	0.04811	20.784	10.299
23	1.1216	0.89162	0.04113	24.310	0.04613	21.676	10.781
24	1.1272	0.88719	0.03932	25.432	0.04432	22.563	11.261
25	1.1328	0.88277	0.03765	26.559	0.04265	23.446	11.741
26	1.1385	0.87838	0.03611	27.692	0.04111	24.324	12.220
27	1.1442	0.87401	0.03469	28.830	0.03969	25.198	12.698
28	1.1499	0.86966	0.03336	29.975	0.03836	26.068	13.175
29	1.1556	0.86533	0.03213	31.124	0.03713	26.933	13.651
30	1.1614	0.86103	0.03098	32.280	0.03598	27.794	14.126
31	1.1672	0.85675	0.02990	33.441	0.03490	28.651	14.601
32	1.1730	0.85248	0.02889	34.609	0.03389	29.503	15.075
33	1.1789	0.84824	0.02795	35.782	0.03295	30.352	15.548
34	1.1848	0.84402	0.02706	36.961	0.03206	31.196	16.020
35	1.1907	0.83982	0.02622	38.145	0.03122	32.035	16.492
40	1.2208	0.81914	0.02265	44.159	0.02765	36.172	18.836
45	1.2516	0.79896	0.01987	50.324	0.02487	40.207	21.159
50	1.2832	0.77929	0.01765	56.645	0.02265	44.143	23.462
55	1.3156	0.76009	0.01584	63.126	0.02084	47.981	25.745
60	1.3489	0.74137	0.01433	69.770	0.01933	51.726	28.006
65	1.3829	0.72311	0.01306	76.582	0.01806	55.377	30.247
70	1.4178	0.70530	0.01197	83.566	0.01697	58.939	32.468
75	1.4536	0.68793	0.01102	90.727	0.01602	62.414	34.668
80	1.4903	0.67099	0.01020	98.068	0.01520	65.802	36.847
85	1.5280	0.65446	0.00947	105.59	0.01447	69.108	39.006
90	1.5666	0.63834	0.00883	113.31	0.01383	72.331	41.145
95	1.6061	0.62262	0.00825	121.22	0.01325	75.476	43.263
100	1.6467	0.60729	0.00773	129.33	0.01273	78.543	45.361

i = 1% **Discrete Compounding, Discrete Cash Flows**

	SINGLE PAYMENT		UNIFORM SERIES				
	Compound Amount Factor	Present Worth Factor	Sinking Fund Factor	Uniform Series Factor	Capital Recovery Factor	Series Present Worth Factor	Arithmetic Gradient Series Factor
N	(*F*/*P*,*i*,*N*)	(*P*/*F*,*i*,*N*)	(*A*/*F*,*i*,*N*)	(*F*/*A*,*i*,*N*)	(*A*/*P*,*i*,*N*)	(*P*/*A*,*i*,*N*)	(*A*/*G*,*i*,*N*)
1	1.0100	0.99010	1.0000	1.0000	1.0100	0.99010	0.00000
2	1.0201	0.98030	0.49751	2.0100	0.50751	1.9704	0.49751
3	1.0303	0.97059	0.33002	3.0301	0.34002	2.9410	0.99337
4	1.0406	0.96098	0.24628	4.0604	0.25628	3.9020	1.4876
5	1.0510	0.95147	0.19604	5.1010	0.20604	4.8534	1.9801
6	1.0615	0.94205	0.16255	6.1520	0.17255	5.7955	2.4710
7	1.0721	0.93272	0.13863	7.2135	0.14863	6.7282	2.9602
8	1.0829	0.92348	0.12069	8.2857	0.13069	7.6517	3.4478
9	1.0937	0.91434	0.10674	9.3685	0.11674	8.5660	3.9337
10	1.1046	0.90529	0.09558	10.462	0.10558	9.4713	4.4179
11	1.1157	0.89632	0.08645	11.567	0.09645	10.368	4.9005
12	1.1268	0.88745	0.07885	12.683	0.08885	11.255	5.3815
13	1.1381	0.87866	0.07241	13.809	0.08241	12.134	5.8607
14	1.1495	0.86996	0.06690	14.947	0.07690	13.004	6.3384
15	1.1610	0.86135	0.06212	16.097	0.07212	13.865	6.8143
16	1.1726	0.85282	0.05794	17.258	0.06794	14.718	7.2886
17	1.1843	0.84438	0.05426	18.430	0.06426	15.562	7.7613
18	1.1961	0.83602	0.05098	19.615	0.06098	16.398	8.2323
19	1.2081	0.82774	0.04805	20.811	0.05805	17.226	8.7017
20	1.2202	0.81954	0.04542	22.019	0.05542	18.046	9.1694
21	1.2324	0.81143	0.04303	23.239	0.05303	18.857	9.6354
22	1.2447	0.80340	0.04086	24.472	0.05086	19.660	10.100
23	1.2572	0.79544	0.03889	25.716	0.04889	20.456	10.563
24	1.2697	0.78757	0.03707	26.973	0.04707	21.243	11.024
25	1.2824	0.77977	0.03541	28.243	0.04541	22.023	11.483
26	1.2953	0.77205	0.03387	29.526	0.04387	22.795	11.941
27	1.3082	0.76440	0.03245	30.821	0.04245	23.560	12.397
28	1.3213	0.75684	0.03112	32.129	0.04112	24.316	12.852
29	1.3345	0.74934	0.02990	33.450	0.03990	25.066	13.304
30	1.3478	0.74192	0.02875	34.785	0.03875	25.808	13.756
31	1.3613	0.73458	0.02768	36.133	0.03768	26.542	14.205
32	1.3749	0.72730	0.02667	37.494	0.03667	27.270	14.653
33	1.3887	0.72010	0.02573	38.869	0.03573	27.990	15.099
34	1.4026	0.71297	0.02484	40.258	0.03484	28.703	15.544
35	1.4166	0.70591	0.02400	41.660	0.03400	29.409	15.987
40	1.4889	0.67165	0.02046	48.886	0.03046	32.835	18.178
45	1.5648	0.63905	0.01771	56.481	0.02771	36.095	20.327
50	1.6446	0.60804	0.01551	64.463	0.02551	39.196	22.436
55	1.7285	0.57853	0.01373	72.852	0.02373	42.147	24.505
60	1.8167	0.55045	0.01224	81.670	0.02224	44.955	26.533
65	1.9094	0.52373	0.01100	90.937	0.02100	47.627	28.522
70	2.0068	0.49831	0.00993	100.68	0.01993	50.169	30.470
75	2.1091	0.47413	0.00902	110.91	0.01902	52.587	32.379
80	2.2167	0.45112	0.00822	121.67	0.01822	54.888	34.249
85	2.3298	0.42922	0.00752	132.98	0.01752	57.078	36.080
90	2.4486	0.40839	0.00690	144.86	0.01690	59.161	37.872
95	2.5735	0.38857	0.00636	157.35	0.01636	61.143	39.626
100	2.7048	0.36971	0.00587	170.48	0.01587	63.029	41.343

i = 1.5% **Discrete Compounding, Discrete Cash Flows**

	SINGLE PAYMENT		UNIFORM SERIES				
	Compound Amount Factor	Present Worth Factor	Sinking Fund Factor	Uniform Series Factor	Capital Recovery Factor	Series Present Worth Factor	Arithmetic Gradient Series Factor
N	(F/P,i,N)	(P/F,i,N)	(A/F,i,N)	(F/A,i,N)	(A/P,i,N)	(P/A,i,N)	(A/G,i,N)
1	1.0150	0.98522	1.0000	1.0000	1.0150	0.98522	0.00000
2	1.0302	0.97066	0.49628	2.0150	0.51128	1.9559	0.49628
3	1.0457	0.95632	0.32838	3.0452	0.34338	2.9122	0.99007
4	1.0614	0.94218	0.24444	4.0909	0.25944	3.8544	1.4814
5	1.0773	0.92826	0.19409	5.1523	0.20909	4.7826	1.9702
6	1.0934	0.91454	0.16053	6.2296	0.17553	5.6972	2.4566
7	1.1098	0.90103	0.13656	7.3230	0.15156	6.5982	2.9405
8	1.1265	0.88771	0.11858	8.4328	0.13358	7.4859	3.4219
9	1.1434	0.87459	0.10461	9.5593	0.11961	8.3605	3.9008
10	1.1605	0.86167	0.09343	10.703	0.10843	9.2222	4.3772
11	1.1779	0.84893	0.08429	11.863	0.09929	10.071	4.8512
12	1.1956	0.83639	0.07668	13.041	0.09168	10.908	5.3227
13	1.2136	0.82403	0.07024	14.237	0.08524	11.732	5.7917
14	1.2318	0.81185	0.06472	15.450	0.07972	12.543	6.2582
15	1.2502	0.79985	0.05994	16.682	0.07494	13.343	6.7223
16	1.2690	0.78803	0.05577	17.932	0.07077	14.131	7.1839
17	1.2880	0.77639	0.05208	19.201	0.06708	14.908	7.6431
18	1.3073	0.76491	0.04881	20.489	0.06381	15.673	8.0997
19	1.3270	0.75361	0.04588	21.797	0.06088	16.426	8.5539
20	1.3469	0.74247	0.04325	23.124	0.05825	17.169	9.0057
21	1.3671	0.73150	0.04087	24.471	0.05587	17.900	9.4550
22	1.3876	0.72069	0.03870	25.838	0.05370	18.621	9.9018
23	1.4084	0.71004	0.03673	27.225	0.05173	19.331	10.346
24	1.4295	0.69954	0.03492	28.634	0.04992	20.030	10.788
25	1.4509	0.68921	0.03326	30.063	0.04826	20.720	11.228
26	1.4727	0.67902	0.03173	31.514	0.04673	21.399	11.665
27	1.4948	0.66899	0.03032	32.987	0.04532	22.068	12.099
28	1.5172	0.65910	0.02900	34.481	0.04400	22.727	12.531
29	1.5400	0.64936	0.02778	35.999	0.04278	23.376	12.961
30	1.5631	0.63976	0.02664	37.539	0.04164	24.016	13.388
31	1.5865	0.63031	0.02557	39.102	0.04057	24.646	13.813
32	1.6103	0.62099	0.02458	40.688	0.03958	25.267	14.236
33	1.6345	0.61182	0.02364	42.299	0.03864	25.879	14.656
34	1.6590	0.60277	0.02276	43.933	0.03776	26.482	15.073
35	1.6839	0.59387	0.02193	45.592	0.03693	27.076	15.488
40	1.8140	0.55126	0.01843	54.268	0.03343	29.916	17.528
45	1.9542	0.51171	0.01572	63.614	0.03072	32.552	19.507
50	2.1052	0.47500	0.01357	73.683	0.02857	35.000	21.428
55	2.2679	0.44093	0.01183	84.530	0.02683	37.271	23.289
60	2.4432	0.40930	0.01039	96.215	0.02539	39.380	25.093
65	2.6320	0.37993	0.00919	108.80	0.02419	41.338	26.839
70	2.8355	0.35268	0.00817	122.36	0.02317	43.155	28.529
75	3.0546	0.32738	0.00730	136.97	0.02230	44.842	30.163
80	3.2907	0.30389	0.00655	152.71	0.02155	46.407	31.742
85	3.5450	0.28209	0.00589	169.67	0.02089	47.861	33.268
90	3.8189	0.26185	0.00532	187.93	0.02032	49.210	34.740
95	4.1141	0.24307	0.00482	207.61	0.01982	50.462	36.160
100	4.4320	0.22563	0.00437	228.80	0.01937	51.625	37.530

i = 2% **Discrete Compounding, Discrete Cash Flows**

	SINGLE PAYMENT		UNIFORM SERIES				
	Compound Amount Factor	Present Worth Factor	Sinking Fund Factor	Uniform Series Factor	Capital Recovery Factor	Series Present Worth Factor	Arithmetic Gradient Series Factor
N	(*F/P,i,N*)	(*P/F,i,N*)	(*A/F,i,N*)	(*F/A,i,N*)	(*A/P,i,N*)	(*P/A,i,N*)	(*A/G,i,N*)
1	1.0200	0.98039	1.0000	1.0000	1.0200	0.98039	0.00000
2	1.0404	0.96117	0.49505	2.0200	0.51505	1.9416	0.49505
3	1.0612	0.94232	0.32675	3.0604	0.34675	2.8839	0.98680
4	1.0824	0.92385	0.24262	4.1216	0.26262	3.8077	1.4752
5	1.1041	0.90573	0.19216	5.2040	0.21216	4.7135	1.9604
6	1.1262	0.88797	0.15853	6.3081	0.17853	5.6014	2.4423
7	1.1487	0.87056	0.13451	7.4343	0.15451	6.4720	2.9208
8	1.1717	0.85349	0.11651	8.5830	0.13651	7.3255	3.3961
9	1.1951	0.83676	0.10252	9.7546	0.12252	8.1622	3.8681
10	1.2190	0.82035	0.09133	10.950	0.11133	8.9826	4.3367
11	1.2434	0.80426	0.08218	12.169	0.10218	9.787	4.8021
12	1.2682	0.78849	0.07456	13.412	0.09456	10.575	5.2642
13	1.2936	0.77303	0.06812	14.680	0.08812	11.348	5.7231
14	1.3195	0.75788	0.06260	15.974	0.08260	12.106	6.1786
15	1.3459	0.74301	0.05783	17.293	0.07783	12.849	6.6309
16	1.3728	0.72845	0.05365	18.639	0.07365	13.578	7.0799
17	1.4002	0.71416	0.04997	20.012	0.06997	14.292	7.5256
18	1.4282	0.70016	0.04670	21.412	0.06670	14.992	7.9681
19	1.4568	0.68643	0.04378	22.841	0.06378	15.678	8.4073
20	1.4859	0.67297	0.04116	24.297	0.06116	16.351	8.8433
21	1.5157	0.65978	0.03878	25.783	0.05878	17.011	9.2760
22	1.5460	0.64684	0.03663	27.299	0.05663	17.658	9.7050
23	1.5769	0.63416	0.03467	28.845	0.05467	18.292	10.1320
24	1.6084	0.62172	0.03287	30.422	0.05287	18.914	10.5550
25	1.6406	0.60953	0.03122	32.030	0.05122	19.523	10.9740
26	1.6734	0.59758	0.02970	33.671	0.04970	20.121	11.391
27	1.7069	0.58586	0.02829	35.344	0.04829	20.707	11.804
28	1.7410	0.57437	0.02699	37.051	0.04699	21.281	12.214
29	1.7758	0.56311	0.02578	38.792	0.04578	21.844	12.621
30	1.8114	0.55207	0.02465	40.568	0.04465	22.396	13.025
31	1.8476	0.54125	0.02360	42.379	0.04360	22.938	13.426
32	1.8845	0.53063	0.02261	44.227	0.04261	23.468	13.823
33	1.9222	0.52023	0.02169	46.112	0.04169	23.989	14.217
34	1.9607	0.51003	0.02082	48.034	0.04082	24.499	14.608
35	1.9999	0.50003	0.02000	49.994	0.04000	24.999	14.996
40	2.2080	0.45289	0.01656	60.402	0.03656	27.355	16.889
45	2.4379	0.41020	0.01391	71.893	0.03391	29.490	18.703
50	2.6916	0.37153	0.01182	84.579	0.03182	31.424	20.442
55	2.9717	0.33650	0.01014	98.587	0.03014	33.175	22.106
60	3.2810	0.30478	0.00877	114.05	0.02877	34.761	23.696
65	3.6225	0.27605	0.00763	131.13	0.02763	36.197	25.215
70	3.9996	0.25003	0.00667	149.98	0.02667	37.499	26.663
75	4.4158	0.22646	0.00586	170.79	0.02586	38.677	28.043
80	4.8754	0.20511	0.00516	193.77	0.02516	39.745	29.357
85	5.3829	0.18577	0.00456	219.14	0.02456	40.711	30.606
90	5.9431	0.16826	0.00405	247.16	0.02405	41.587	31.793
95	6.5617	0.15240	0.00360	278.08	0.02360	42.380	32.919
100	7.2446	0.13803	0.00320	312.23	0.02320	43.098	33.986

$i = 3\%$ **Discrete Compounding, Discrete Cash Flows**

	SINGLE PAYMENT		UNIFORM SERIES				
	Compound Amount Factor	Present Worth Factor	Sinking Fund Factor	Uniform Series Factor	Capital Recovery Factor	Series Present Worth Factor	Arithmetic Gradient Series Factor
N	(*F/P,i,N*)	(*P/F,i,N*)	(*A/F,i,N*)	(*F/A,i,N*)	(*A/P,i,N*)	(*P/A,i,N*)	(*A/G,i,N*)
1	1.0300	0.97087	1.0000	1.0000	1.0300	0.97087	0.00000
2	1.0609	0.94260	0.49261	2.0300	0.52261	1.9135	0.49261
3	1.0927	0.91514	0.32353	3.0909	0.35353	2.8286	0.98030
4	1.1255	0.88849	0.23903	4.1836	0.26903	3.7171	1.4631
5	1.1593	0.86261	0.18835	5.3091	0.21835	4.5797	1.9409
6	1.1941	0.83748	0.15460	6.4684	0.18460	5.4172	2.4138
7	1.2299	0.81309	0.13051	7.6625	0.16051	6.2303	2.8819
8	1.2668	0.78941	0.11246	8.8923	0.14246	7.0197	3.3450
9	1.3048	0.76642	0.09843	10.159	0.12843	7.7861	3.8032
10	1.3439	0.74409	0.08723	11.464	0.11723	8.5302	4.2565
11	1.3842	0.72242	0.07808	12.808	0.10808	9.2526	4.7049
12	1.4258	0.70138	0.07046	14.192	0.10046	9.9540	5.1485
13	1.4685	0.68095	0.06403	15.618	0.09403	10.635	5.5872
14	1.5126	0.66112	0.05853	17.086	0.08853	11.296	6.0210
15	1.5580	0.64186	0.05377	18.599	0.08377	11.938	6.4500
16	1.6047	0.62317	0.04961	20.157	0.07961	12.561	6.8742
17	1.6528	0.60502	0.04595	21.762	0.07595	13.166	7.2936
18	1.7024	0.58739	0.04271	23.414	0.07271	13.754	7.7081
19	1.7535	0.57029	0.03981	25.117	0.06981	14.324	8.1179
20	1.8061	0.55368	0.03722	26.870	0.06722	14.877	8.5229
21	1.8603	0.53755	0.03487	28.676	0.06487	15.415	8.9231
22	1.9161	0.52189	0.03275	30.537	0.06275	15.937	9.3186
23	1.9736	0.50669	0.03081	32.453	0.06081	16.444	9.7093
24	2.0328	0.49193	0.02905	34.426	0.05905	16.936	10.095
25	2.0938	0.47761	0.02743	36.459	0.05743	17.413	10.477
26	2.1566	0.46369	0.02594	38.553	0.05594	17.877	10.853
27	2.2213	0.45019	0.02456	40.710	0.05456	18.327	11.226
28	2.2879	0.43708	0.02329	42.931	0.05329	18.764	11.593
29	2.3566	0.42435	0.02211	45.219	0.05211	19.188	11.956
30	2.4273	0.41199	0.02102	47.575	0.05102	19.600	12.314
31	2.5001	0.39999	0.02000	50.003	0.05000	20.000	12.668
32	2.5751	0.38834	0.01905	52.503	0.04905	20.389	13.017
33	2.6523	0.37703	0.01816	55.078	0.04816	20.766	13.362
34	2.7319	0.36604	0.01732	57.730	0.04732	21.132	13.702
35	2.8139	0.35538	0.01654	60.462	0.04654	21.487	14.037
40	3.2620	0.30656	0.01326	75.401	0.04326	23.115	15.650
45	3.7816	0.26444	0.01079	92.720	0.04079	24.519	17.156
50	4.3839	0.22811	0.00887	112.80	0.03887	25.730	18.558
55	5.0821	0.19677	0.00735	136.07	0.03735	26.774	19.860
60	5.8916	0.16973	0.00613	163.05	0.03613	27.676	21.067
65	6.8300	0.14641	0.00515	194.33	0.03515	28.453	22.184
70	7.9178	0.12630	0.00434	230.59	0.03434	29.123	23.215
75	9.1789	0.10895	0.00367	272.63	0.03367	29.702	24.163
80	10.641	0.09398	0.00311	321.36	0.03311	30.201	25.035
85	12.336	0.08107	0.00265	377.86	0.03265	30.631	25.835
90	14.300	0.06993	0.00226	443.35	0.03226	31.002	26.567
95	16.578	0.06032	0.00193	519.27	0.03193	31.323	27.235
100	19.219	0.05203	0.00165	607.29	0.03165	31.599	27.844

i = 4% **Discrete Compounding, Discrete Cash Flows**

	SINGLE PAYMENT		UNIFORM SERIES				
	Compound Amount Factor	Present Worth Factor	Sinking Fund Factor	Uniform Series Factor	Capital Recovery Factor	Series Present Worth Factor	Arithmetic Gradient Series Factor
N	(*F*/*P*,*i*,*N*)	(*P*/*F*,*i*,*N*)	(*A*/*F*,*i*,*N*)	(*F*/*A*,*i*,*N*)	(*A*/*P*,*i*,*N*)	(*P*/*A*,*i*,*N*)	(*A*/*G*,*i*,*N*)
1	1.0400	0.96154	1.0000	1.0000	1.0400	0.96154	0.00000
2	1.0816	0.92456	0.49020	2.0400	0.53020	1.8861	0.49020
3	1.1249	0.88900	0.32035	3.1216	0.36035	2.7751	0.97386
4	1.1699	0.85480	0.23549	4.2465	0.27549	3.6299	1.4510
5	1.2167	0.82193	0.18463	5.4163	0.22463	4.4518	1.9216
6	1.2653	0.79031	0.15076	6.6330	0.19076	5.2421	2.3857
7	1.3159	0.75992	0.12661	7.8983	0.16661	6.0021	2.8433
8	1.3686	0.73069	0.10853	9.2142	0.14853	6.7327	3.2944
9	1.4233	0.70259	0.09449	10.583	0.13449	7.4353	3.7391
10	1.4802	0.67556	0.08329	12.006	0.12329	8.1109	4.1773
11	1.5395	0.64958	0.07415	13.486	0.11415	8.7605	4.6090
12	1.6010	0.62460	0.06655	15.026	0.10655	9.3851	5.0343
13	1.6651	0.60057	0.06014	16.627	0.10014	9.9856	5.4533
14	1.7317	0.57748	0.05467	18.292	0.09467	10.563	5.8659
15	1.8009	0.55526	0.04994	20.024	0.08994	11.118	6.2721
16	1.8730	0.53391	0.04582	21.825	0.08582	11.652	6.6720
17	1.9479	0.51337	0.04220	23.698	0.08220	12.166	7.0656
18	2.0258	0.49363	0.03899	25.645	0.07899	12.659	7.4530
19	2.1068	0.47464	0.03614	27.671	0.07614	13.134	7.8342
20	2.1911	0.45639	0.03358	29.778	0.07358	13.590	8.2091
21	2.2788	0.43883	0.03128	31.969	0.07128	14.029	8.5779
22	2.3699	0.42196	0.02920	34.248	0.06920	14.451	8.9407
23	2.4647	0.40573	0.02731	36.618	0.06731	14.857	9.2973
24	2.5633	0.39012	0.02559	39.083	0.06559	15.247	9.6479
25	2.6658	0.37512	0.02401	41.646	0.06401	15.622	9.9925
26	2.7725	0.36069	0.02257	44.312	0.06257	15.983	10.331
27	2.8834	0.34682	0.02124	47.084	0.06124	16.330	10.664
28	2.9987	0.33348	0.02001	49.968	0.06001	16.663	10.991
29	3.1187	0.32065	0.01888	52.966	0.05888	16.984	11.312
30	3.2434	0.30832	0.01783	56.085	0.05783	17.292	11.627
31	3.3731	0.29646	0.01686	59.328	0.05686	17.588	11.937
32	3.5081	0.28506	0.01595	62.701	0.05595	17.874	12.241
33	3.6484	0.27409	0.01510	66.210	0.05510	18.148	12.540
34	3.7943	0.26355	0.01431	69.858	0.05431	18.411	12.832
35	3.9461	0.25342	0.01358	73.652	0.05358	18.665	13.120
40	4.8010	0.20829	0.01052	95.026	0.05052	19.793	14.477
45	5.8412	0.17120	0.00826	121.03	0.04826	20.720	15.705
50	7.1067	0.14071	0.00655	152.67	0.04655	21.482	16.812
55	8.6464	0.11566	0.00523	191.16	0.04523	22.109	17.807
60	10.520	0.09506	0.00420	237.99	0.04420	22.623	18.697
65	12.799	0.07813	0.00339	294.97	0.04339	23.047	19.491
70	15.572	0.06422	0.00275	364.29	0.04275	23.395	20.196
75	18.945	0.05278	0.00223	448.63	0.04223	23.680	20.821
80	23.050	0.04338	0.00181	551.24	0.04181	23.915	21.372
85	28.044	0.03566	0.00148	676.09	0.04148	24.109	21.857
90	34.119	0.02931	0.00121	827.98	0.04121	24.267	22.283
95	41.511	0.02409	0.00099	1012.8	0.04099	24.398	22.655
100	50.505	0.01980	0.00081	1237.6	0.04081	24.505	22.980

i = 5% **Discrete Compounding, Discrete Cash Flows**

	SINGLE PAYMENT		UNIFORM SERIES				
	Compound Amount Factor	Present Worth Factor	Sinking Fund Factor	Uniform Series Factor	Capital Recovery Factor	Series Present Worth Factor	Arithmetic Gradient Series Factor
N	(*F/P,i,N*)	(*P/F,i,N*)	(*A/F,i,N*)	(*F/A,i,N*)	(*A/P,i,N*)	(*P/A,i,N*)	(*A/G,i,N*)
1	1.0500	0.95238	1.0000	1.0000	1.0500	0.95238	0.00000
2	1.1025	0.90703	0.48780	2.0500	0.53780	1.8594	0.48780
3	1.1576	0.86384	0.31721	3.1525	0.36721	2.7232	0.96749
4	1.2155	0.82270	0.23201	4.3101	0.28201	3.5460	1.4391
5	1.2763	0.78353	0.18097	5.5256	0.23097	4.3295	1.9025
6	1.3401	0.74622	0.14702	6.8019	0.19702	5.0757	2.3579
7	1.4071	0.71068	0.12282	8.1420	0.17282	5.7864	2.8052
8	1.4775	0.67684	0.10472	9.5491	0.15472	6.4632	3.2445
9	1.5513	0.64461	0.09069	11.027	0.14069	7.1078	3.6758
10	1.6289	0.61391	0.07950	12.578	0.12950	7.7217	4.0991
11	1.7103	0.58468	0.07039	14.207	0.12039	8.3064	4.5144
12	1.7959	0.55684	0.06283	15.917	0.11283	8.8633	4.9219
13	1.8856	0.53032	0.05646	17.713	0.10646	9.3936	5.3215
14	1.9799	0.50507	0.05102	19.599	0.10102	9.8986	5.7133
15	2.0789	0.48102	0.04634	21.579	0.09634	10.380	6.0973
16	2.1829	0.45811	0.04227	23.657	0.09227	10.838	6.4736
17	2.2920	0.43630	0.03870	25.840	0.08870	11.274	6.8423
18	2.4066	0.41552	0.03555	28.132	0.08555	11.690	7.2034
19	2.5270	0.39573	0.03275	30.539	0.08275	12.085	7.5569
20	2.6533	0.37689	0.03024	33.066	0.08024	12.462	7.9030
21	2.7860	0.35894	0.02800	35.719	0.07800	12.821	8.2416
22	2.9253	0.34185	0.02597	38.505	0.07597	13.163	8.5730
23	3.0715	0.32557	0.02414	41.430	0.07414	13.489	8.8971
24	3.2251	0.31007	0.02247	44.502	0.07247	13.799	9.2140
25	3.3864	0.29530	0.02095	47.727	0.07095	14.094	9.5238
26	3.5557	0.28124	0.01956	51.113	0.06956	14.375	9.8266
27	3.7335	0.26785	0.01829	54.669	0.06829	14.643	10.122
28	3.9201	0.25509	0.01712	58.403	0.06712	14.898	10.411
29	4.1161	0.24295	0.01605	62.323	0.06605	15.141	10.694
30	4.3219	0.23138	0.01505	66.439	0.06505	15.372	10.969
31	4.5380	0.22036	0.01413	70.761	0.06413	15.593	11.238
32	4.7649	0.20987	0.01328	75.299	0.06328	15.803	11.501
33	5.0032	0.19987	0.01249	80.064	0.06249	16.003	11.757
34	5.2533	0.19035	0.01176	85.067	0.06176	16.193	12.006
35	5.5160	0.18129	0.01107	90.320	0.06107	16.374	12.250
40	7.0400	0.14205	0.00828	120.80	0.05828	17.159	13.377
45	8.9850	0.11130	0.00626	159.70	0.05626	17.774	14.364
50	11.467	0.08720	0.00478	209.35	0.05478	18.256	15.223
55	14.636	0.06833	0.00367	272.71	0.05367	18.633	15.966
60	18.679	0.05354	0.00283	353.58	0.05283	18.929	16.606
65	23.840	0.04195	0.00219	456.80	0.05219	19.161	17.154
70	30.426	0.03287	0.00170	588.53	0.05170	19.343	17.621
75	38.833	0.02575	0.00132	756.65	0.05132	19.485	18.018
80	49.561	0.02018	0.00103	971.23	0.05103	19.596	18.353
85	63.254	0.01581	0.00080	1245.1	0.05080	19.684	18.635
90	80.730	0.01239	0.00063	1594.6	0.05063	19.752	18.871
95	103.03	0.00971	0.00049	2040.7	0.05049	19.806	19.069
100	131.50	0.00760	0.00038	2610.0	0.05038	19.848	19.234

i = 6% **Discrete Compounding, Discrete Cash Flows**

	SINGLE PAYMENT		UNIFORM SERIES				
	Compound Amount Factor	Present Worth Factor	Sinking Fund Factor	Uniform Series Factor	Capital Recovery Factor	Series Present Worth Factor	Arithmetic Gradient Series Factor
N	(*F*/*P*,*i*,*N*)	(*P*/*F*,*i*,*N*)	(*A*/*F*,*i*,*N*)	(*F*/*A*,*i*,*N*)	(*A*/*P*,*i*,*N*)	(*P*/*A*,*i*,*N*)	(*A*/*G*,*i*,*N*)
1	1.0600	0.94340	1.0000	1.0000	1.0600	0.94340	0.00000
2	1.1236	0.89000	0.48544	2.0600	0.54544	1.8334	0.48544
3	1.1910	0.83962	0.31411	3.1836	0.37411	2.6730	0.96118
4	1.2625	0.79209	0.22859	4.3746	0.28859	3.4651	1.4272
5	1.3382	0.74726	0.17740	5.6371	0.23740	4.2124	1.8836
6	1.4185	0.70496	0.14336	6.9753	0.20336	4.9173	2.3304
7	1.5036	0.66506	0.11914	8.3938	0.17914	5.5824	2.7676
8	1.5938	0.62741	0.10104	9.8975	0.16104	6.2098	3.1952
9	1.6895	0.59190	0.08702	11.491	0.14702	6.8017	3.6133
10	1.7908	0.55839	0.07587	13.181	0.13587	7.3601	4.0220
11	1.8983	0.52679	0.06679	14.972	0.12679	7.8869	4.4213
12	2.0122	0.49697	0.05928	16.870	0.11928	8.3838	4.8113
13	2.1329	0.46884	0.05296	18.882	0.11296	8.8527	5.1920
14	2.2609	0.44230	0.04758	21.015	0.10758	9.2950	5.5635
15	2.3966	0.41727	0.04296	23.276	0.10296	9.7122	5.9260
16	2.5404	0.39365	0.03895	25.673	0.09895	10.106	6.2794
17	2.6928	0.37136	0.03544	28.213	0.09544	10.477	6.6240
18	2.8543	0.35034	0.03236	30.906	0.09236	10.828	6.9597
19	3.0256	0.33051	0.02962	33.760	0.08962	11.158	7.2867
20	3.2071	0.31180	0.02718	36.786	0.08718	11.470	7.6051
21	3.3996	0.29416	0.02500	39.993	0.08500	11.764	7.9151
22	3.6035	0.27751	0.02305	43.392	0.08305	12.042	8.2166
23	3.8197	0.26180	0.02128	46.996	0.08128	12.303	8.5099
24	4.0489	0.24698	0.01968	50.816	0.07968	12.550	8.7951
25	4.2919	0.23300	0.01823	54.865	0.07823	12.783	9.0722
26	4.5494	0.21981	0.01690	59.156	0.07690	13.003	9.3414
27	4.8223	0.20737	0.01570	63.706	0.07570	13.211	9.6029
28	5.1117	0.19563	0.01459	68.528	0.07459	13.406	9.8568
29	5.4184	0.18456	0.01358	73.640	0.07358	13.591	10.103
30	5.7435	0.17411	0.01265	79.058	0.07265	13.765	10.342
31	6.0881	0.16425	0.01179	84.802	0.07179	13.929	10.574
32	6.4534	0.15496	0.01100	90.890	0.07100	14.084	10.799
33	6.8406	0.14619	0.01027	97.343	0.07027	14.230	11.017
34	7.2510	0.13791	0.00960	104.18	0.06960	14.368	11.228
35	7.6861	0.13011	0.00897	111.43	0.06897	14.498	11.432
40	10.286	0.09722	0.00646	154.76	0.06646	15.046	12.359
45	13.765	0.07265	0.00470	212.74	0.06470	15.456	13.141
50	18.420	0.05429	0.00344	290.34	0.06344	15.762	13.796
55	24.650	0.04057	0.00254	394.17	0.06254	15.991	14.341
60	32.988	0.03031	0.00188	533.13	0.06188	16.161	14.791
65	44.145	0.02265	0.00139	719.08	0.06139	16.289	15.160
70	59.076	0.01693	0.00103	967.93	0.06103	16.385	15.461
75	79.057	0.01265	0.00077	1300.9	0.06077	16.456	15.706
80	105.80	0.00945	0.00057	1746.6	0.06057	16.509	15.903
85	141.58	0.00706	0.00043	2343.0	0.06043	16.549	16.062
90	189.46	0.00528	0.00032	3141.1	0.06032	16.579	16.189
95	253.55	0.00394	0.00024	4209.1	0.06024	16.601	16.290
100	339.30	0.00295	0.00018	5638.4	0.06018	16.618	16.371

i = 7% **Discrete Compounding, Discrete Cash Flows**

	SINGLE PAYMENT		UNIFORM SERIES				Arithmetic Gradient Series Factor
	Compound Amount Factor	Present Worth Factor	Sinking Fund Factor	Uniform Series Factor	Capital Recovery Factor	Series Present Worth Factor	
N	*(F/P,i,N)*	*(P/F,i,N)*	*(A/F,i,N)*	*(F/A,i,N)*	*(A/P,i,N)*	*(P/A,i,N)*	*(A/G,i,N)*
1	1.0700	0.93458	1.0000	1.0000	1.0700	0.93458	0.00000
2	1.1449	0.87344	0.48309	2.0700	0.55309	1.8080	0.48309
3	1.2250	0.81630	0.31105	3.2149	0.38105	2.6243	0.95493
4	1.3108	0.76290	0.22523	4.4399	0.29523	3.3872	1.4155
5	1.4026	0.71299	0.17389	5.7507	0.24389	4.1002	1.8650
6	1.5007	0.66634	0.13980	7.1533	0.20980	4.7665	2.3032
7	1.6058	0.62275	0.11555	8.6540	0.18555	5.3893	2.7304
8	1.7182	0.58201	0.09747	10.260	0.16747	5.9713	3.1465
9	1.8385	0.54393	0.08349	11.978	0.15349	6.5152	3.5517
10	1.9672	0.50835	0.07238	13.816	0.14238	7.0236	3.9461
11	2.1049	0.47509	0.06336	15.784	0.13336	7.4987	4.3296
12	2.2522	0.44401	0.05590	17.888	0.12590	7.9427	4.7025
13	2.4098	0.41496	0.04965	20.141	0.11965	8.3577	5.0648
14	2.5785	0.38782	0.04434	22.550	0.11434	8.7455	5.4167
15	2.7590	0.36245	0.03979	25.129	0.10979	9.1079	5.7583
16	2.9522	0.33873	0.03586	27.888	0.10586	9.4466	6.0897
17	3.1588	0.31657	0.03243	30.840	0.10243	9.7632	6.4110
18	3.3799	0.29586	0.02941	33.999	0.09941	10.059	6.7225
19	3.6165	0.27651	0.02675	37.379	0.09675	10.336	7.0242
20	3.8697	0.25842	0.02439	40.995	0.09439	10.594	7.3163
21	4.1406	0.24151	0.02229	44.865	0.09229	10.836	7.5990
22	4.4304	0.22571	0.02041	49.006	0.09041	11.061	7.8725
23	4.7405	0.21095	0.01871	53.436	0.08871	11.272	8.1369
24	5.0724	0.19715	0.01719	58.177	0.08719	11.469	8.3923
25	5.4274	0.18425	0.01581	63.249	0.08581	11.654	8.6391
26	5.8074	0.17220	0.01456	68.676	0.08456	11.826	8.8773
27	6.2139	0.16093	0.01343	74.484	0.08343	11.987	9.1072
28	6.6488	0.15040	0.01239	80.698	0.08239	12.137	9.3289
29	7.1143	0.14056	0.01145	87.347	0.08145	12.278	9.5427
30	7.6123	0.13137	0.01059	94.461	0.08059	12.409	9.7487
31	8.1451	0.12277	0.00980	102.07	0.07980	12.532	9.9471
32	8.7153	0.11474	0.00907	110.22	0.07907	12.647	10.138
33	9.3253	0.10723	0.00841	118.93	0.07841	12.754	10.322
34	9.9781	0.10022	0.00780	128.26	0.07780	12.854	10.499
35	10.677	0.09366	0.00723	138.24	0.07723	12.948	10.669
40	14.974	0.06678	0.00501	199.64	0.07501	13.332	11.423
45	21.002	0.04761	0.00350	285.75	0.07350	13.606	12.036
50	29.457	0.03395	0.00246	406.53	0.07246	13.801	12.529
55	41.315	0.02420	0.00174	575.93	0.07174	13.940	12.921
60	57.946	0.01726	0.00123	813.52	0.07123	14.039	13.232
65	81.273	0.01230	0.00087	1146.8	0.07087	14.110	13.476
70	113.99	0.00877	0.00062	1614.1	0.07062	14.160	13.666
75	159.88	0.00625	0.00044	2269.7	0.07044	14.196	13.814
80	224.23	0.00446	0.00031	3189.1	0.07031	14.222	13.927
85	314.50	0.00318	0.00022	4478.6	0.07022	14.240	14.015
90	441.10	0.00227	0.00016	6287.2	0.07016	14.253	14.081
95	618.67	0.00162	0.00011	8823.9	0.07011	14.263	14.132
100	867.72	0.00115	0.00008	12 382.0	0.07008	14.269	14.170

i = 8% **Discrete Compounding, Discrete Cash Flows**

	SINGLE PAYMENT		UNIFORM SERIES				Arithmetic Gradient Series Factor
	Compound Amount Factor	Present Worth Factor	Sinking Fund Factor	Uniform Series Factor	Capital Recovery Factor	Series Present Worth Factor	
N	*(F/P,i,N)*	*(P/F,i,N)*	*(A/F,i,N)*	*(F/A,i,N)*	*(A/P,i,N)*	*(P/A,i,N)*	*(A/G,i,N)*
1	1.0800	0.92593	1.0000	1.0000	1.0800	0.92593	0.00000
2	1.1664	0.85734	0.48077	2.0800	0.56077	1.7833	0.48077
3	1.2597	0.79383	0.30803	3.2464	0.38803	2.5771	0.94874
4	1.3605	0.73503	0.22192	4.5061	0.30192	3.3121	1.4040
5	1.4693	0.68058	0.17046	5.8666	0.25046	3.9927	1.8465
6	1.5869	0.63017	0.13632	7.3359	0.21632	4.6229	2.2763
7	1.7138	0.58349	0.11207	8.9228	0.19207	5.2064	2.6937
8	1.8509	0.54027	0.09401	10.637	0.17401	5.7466	3.0985
9	1.9990	0.50025	0.08008	12.488	0.16008	6.2469	3.4910
10	2.1589	0.46319	0.06903	14.487	0.14903	6.7101	3.8713
11	2.3316	0.42888	0.06008	16.645	0.14008	7.1390	4.2395
12	2.5182	0.39711	0.05270	18.977	0.13270	7.5361	4.5957
13	2.7196	0.36770	0.04652	21.495	0.12652	7.9038	4.9402
14	2.9372	0.34046	0.04130	24.215	0.12130	8.2442	5.2731
15	3.1722	0.31524	0.03683	27.152	0.11683	8.5595	5.5945
16	3.4259	0.29189	0.03298	30.324	0.11298	8.8514	5.9046
17	3.7000	0.27027	0.02963	33.750	0.10963	9.1216	6.2037
18	3.9960	0.25025	0.02670	37.450	0.10670	9.3719	6.4920
19	4.3157	0.23171	0.02413	41.446	0.10413	9.6036	6.7697
20	4.6610	0.21455	0.02185	45.762	0.10185	9.8181	7.0369
21	5.0338	0.19866	0.01983	50.423	0.09983	10.017	7.2940
22	5.4365	0.18394	0.01803	55.457	0.09803	10.201	7.5412
23	5.8715	0.17032	0.01642	60.893	0.09642	10.371	7.7786
24	6.3412	0.15770	0.01498	66.765	0.09498	10.529	8.0066
25	6.8485	0.14602	0.01368	73.106	0.09368	10.675	8.2254
26	7.3964	0.13520	0.01251	79.954	0.09251	10.810	8.4352
27	7.9881	0.12519	0.01145	87.351	0.09145	10.935	8.6363
28	8.6271	0.11591	0.01049	95.339	0.09049	11.051	8.8289
29	9.3173	0.10733	0.00962	103.97	0.08962	11.158	9.0133
30	10.063	0.09938	0.00883	113.28	0.08883	11.258	9.1897
31	10.868	0.09202	0.00811	123.35	0.08811	11.350	9.3584
32	11.737	0.08520	0.00745	134.21	0.08745	11.435	9.5197
33	12.676	0.07889	0.00685	145.95	0.08685	11.514	9.6737
34	13.690	0.07305	0.00630	158.63	0.08630	11.587	9.8208
35	14.785	0.06763	0.00580	172.32	0.08580	11.655	9.9611
40	21.725	0.04603	0.00386	259.06	0.08386	11.925	10.570
45	31.920	0.03133	0.00259	386.51	0.08259	12.108	11.045
50	46.902	0.02132	0.00174	573.77	0.08174	12.233	11.411
55	68.914	0.01451	0.00118	848.92	0.08118	12.319	11.690
60	101.26	0.00988	0.00080	1253.2	0.08080	12.377	11.902
65	148.78	0.00672	0.00054	1847.2	0.08054	12.416	12.060
70	218.61	0.00457	0.00037	2720.1	0.08037	12.443	12.178
75	321.20	0.00311	0.00025	4002.6	0.08025	12.461	12.266
80	471.95	0.00212	0.00017	5886.9	0.08017	12.474	12.330
85	693.46	0.00144	0.00012	8655.7	0.08012	12.482	12.377
90	1018.9	0.00098	0.00008	12 724.0	0.08008	12.488	12.412
95	1497.1	0.00067	0.00005	18 702.0	0.08005	12.492	12.437
100	2199.8	0.00045	0.00004	27 485.0	0.08004	12.494	12.455

i = 9% **Discrete Compounding, Discrete Cash Flows**

	SINGLE PAYMENT		UNIFORM SERIES				
	Compound Amount Factor	Present Worth Factor	Sinking Fund Factor	Uniform Series Factor	Capital Recovery Factor	Series Present Worth Factor	Arithmetic Gradient Series Factor
N	(*F*/*P*,*i*,*N*)	(*P*/*F*,*i*,*N*)	(*A*/*F*,*i*,*N*)	(*F*/*A*,*i*,*N*)	(*A*/*P*,*i*,*N*)	(*P*/*A*,*i*,*N*)	(*A*/*G*,*i*,*N*)
1	1.0900	0.91743	1.0000	1.0000	1.0900	0.91743	0.00000
2	1.1881	0.84168	0.47847	2.0900	0.56847	1.7591	0.47847
3	1.2950	0.77218	0.30505	3.2781	0.39505	2.5313	0.94262
4	1.4116	0.70843	0.21867	4.5731	0.30867	3.2397	1.3925
5	1.5386	0.64993	0.16709	5.9847	0.25709	3.8897	1.8282
6	1.6771	0.59627	0.13292	7.5233	0.22292	4.4859	2.2498
7	1.8280	0.54703	0.10869	9.2004	0.19869	5.0330	2.6574
8	1.9926	0.50187	0.09067	11.028	0.18067	5.5348	3.0512
9	2.1719	0.46043	0.07680	13.021	0.16680	5.9952	3.4312
10	2.3674	0.42241	0.06582	15.193	0.15582	6.4177	3.7978
11	2.5804	0.38753	0.05695	17.560	0.14695	6.8052	4.1510
12	2.8127	0.35553	0.04965	20.141	0.13965	7.1607	4.4910
13	3.0658	0.32618	0.04357	22.953	0.13357	7.4869	4.8182
14	3.3417	0.29925	0.03843	26.019	0.12843	7.7862	5.1326
15	3.6425	0.27454	0.03406	29.361	0.12406	8.0607	5.4346
16	3.9703	0.25187	0.03030	33.003	0.12030	8.3126	5.7245
17	4.3276	0.23107	0.02705	36.974	0.11705	8.5436	6.0024
18	4.7171	0.21199	0.02421	41.301	0.11421	8.7556	6.2687
19	5.1417	0.19449	0.02173	46.018	0.11173	8.9501	6.5236
20	5.6044	0.17843	0.01955	51.160	0.10955	9.1285	6.7674
21	6.1088	0.16370	0.01762	56.765	0.10762	9.2922	7.0006
22	6.6586	0.15018	0.01590	62.873	0.10590	9.4424	7.2232
23	7.2579	0.13778	0.01438	69.532	0.10438	9.5802	7.4357
24	7.9111	0.12640	0.01302	76.790	0.10302	9.7066	7.6384
25	8.6231	0.11597	0.01181	84.701	0.10181	9.8226	7.8316
26	9.3992	0.10639	0.01072	93.324	0.10072	9.9290	8.0156
27	10.245	0.09761	0.00973	102.72	0.09973	10.027	8.1906
28	11.167	0.08955	0.00885	112.97	0.09885	10.116	8.3571
29	12.172	0.08215	0.00806	124.14	0.09806	10.198	8.5154
30	13.268	0.07537	0.00734	136.31	0.09734	10.274	8.6657
31	14.462	0.06915	0.00669	149.58	0.09669	10.343	8.8083
32	15.763	0.06344	0.00610	164.04	0.09610	10.406	8.9436
33	17.182	0.05820	0.00556	179.80	0.09556	10.464	9.0718
34	18.728	0.05339	0.00508	196.98	0.09508	10.518	9.1933
35	20.414	0.04899	0.00464	215.71	0.09464	10.567	9.3083
40	31.409	0.03184	0.00296	337.88	0.09296	10.757	9.7957
45	48.327	0.02069	0.00190	525.86	0.09190	10.881	10.160
50	74.358	0.01345	0.00123	815.08	0.09123	10.962	10.430
55	114.41	0.00874	0.00079	1260.1	0.09079	11.014	10.626
60	176.03	0.00568	0.00051	1944.8	0.09051	11.048	10.768
65	270.85	0.00369	0.00033	2998.3	0.09033	11.070	10.870
70	416.73	0.00240	0.00022	4619.2	0.09022	11.084	10.943
75	641.19	0.00156	0.00014	7113.2	0.09014	11.094	10.994
80	986.55	0.00101	0.00009	10 951.0	0.09009	11.100	11.030
85	1517.9	0.00066	0.00006	16 855.0	0.09006	11.104	11.055
90	2335.5	0.00043	0.00004	25 939.0	0.09004	11.106	11.073
95	3593.5	0.00028	0.00003	39 917.0	0.09003	11.108	11.085
100	5529.0	0.00018	0.00002	61 423.0	0.09002	11.109	11.093

i = 10%

Discrete Compounding, Discrete Cash Flows

	SINGLE PAYMENT		UNIFORM SERIES				
	Compound Amount Factor	Present Worth Factor	Sinking Fund Factor	Uniform Series Factor	Capital Recovery Factor	Series Present Worth Factor	Arithmetic Gradient Series Factor
N	(*F/P,i,N*)	(*P/F,i,N*)	(*A/F,i,N*)	(*F/A,i,N*)	(*A/P,i,N*)	(*P/A,i,N*)	(*A/G,i,N*)
1	1.1000	0.90909	1.0000	1.0000	1.1000	0.90909	0.00000
2	1.2100	0.82645	0.47619	2.1000	0.57619	1.7355	0.47619
3	1.3310	0.75131	0.30211	3.3100	0.40211	2.4869	0.93656
4	1.4641	0.68301	0.21547	4.6410	0.31547	3.1699	1.3812
5	1.6105	0.62092	0.16380	6.1051	0.26380	3.7908	1.8101
6	1.7716	0.56447	0.12961	7.7156	0.22961	4.3553	2.2236
7	1.9487	0.51316	0.10541	9.4872	0.20541	4.8684	2.6216
8	2.1436	0.46651	0.08744	11.436	0.18744	5.3349	3.0045
9	2.3579	0.42410	0.07364	13.579	0.17364	5.7590	3.3724
10	2.5937	0.38554	0.06275	15.937	0.16275	6.1446	3.7255
11	2.8531	0.35049	0.05396	18.531	0.15396	6.4951	4.0641
12	3.1384	0.31863	0.04676	21.384	0.14676	6.8137	4.3884
13	3.4523	0.28966	0.04078	24.523	0.14078	7.1034	4.6988
14	3.7975	0.26333	0.03575	27.975	0.13575	7.3667	4.9955
15	4.1772	0.23939	0.03147	31.772	0.13147	7.6061	5.2789
16	4.5950	0.21763	0.02782	35.950	0.12782	7.8237	5.5493
17	5.0545	0.19784	0.02466	40.545	0.12466	8.0216	5.8071
18	5.5599	0.17986	0.02193	45.599	0.12193	8.2014	6.0526
19	6.1159	0.16351	0.01955	51.159	0.11955	8.3649	6.2861
20	6.7275	0.14864	0.01746	57.275	0.11746	8.5136	6.5081
21	7.4002	0.13513	0.01562	64.002	0.11562	8.6487	6.7189
22	8.1403	0.12285	0.01401	71.403	0.11401	8.7715	6.9189
23	8.9543	0.11168	0.01257	79.543	0.11257	8.8832	7.1085
24	9.8497	0.10153	0.01130	88.497	0.11130	8.9847	7.2881
25	10.835	0.09230	0.01017	98.347	0.11017	9.0770	7.4580
26	11.918	0.08391	0.00916	109.18	0.10916	9.1609	7.6186
27	13.110	0.07628	0.00826	121.10	0.10826	9.2372	7.7704
28	14.421	0.06934	0.00745	134.21	0.10745	9.3066	7.9137
29	15.863	0.06304	0.00673	148.63	0.10673	9.3696	8.0489
30	17.449	0.05731	0.00608	164.49	0.10608	9.4269	8.1762
31	19.194	0.05210	0.00550	181.94	0.10550	9.4790	8.2962
32	21.114	0.04736	0.00497	201.14	0.10497	9.5264	8.4091
33	23.225	0.04306	0.00450	222.25	0.10450	9.5694	8.5152
34	25.548	0.03914	0.00407	245.48	0.10407	9.6086	8.6149
35	28.102	0.03558	0.00369	271.02	0.10369	9.6442	8.7086
40	45.259	0.02209	0.00226	442.59	0.10226	9.7791	9.0962
45	72.890	0.01372	0.00139	718.90	0.10139	9.8628	9.3740
50	117.39	0.00852	0.00086	1163.9	0.10086	9.9148	9.5704
55	189.06	0.00529	0.00053	1880.6	0.10053	9.9471	9.7075
60	304.48	0.00328	0.00033	3034.8	0.10033	9.9672	9.8023
65	490.37	0.00204	0.00020	4893.7	0.10020	9.9796	9.8672
70	789.75	0.00127	0.00013	7887.5	0.10013	9.9873	9.9113
75	1271.9	0.00079	0.00008	12 709.0	0.10008	9.9921	9.9410

i = 11% **Discrete Compounding, Discrete Cash Flows**

	SINGLE PAYMENT		UNIFORM SERIES				
	Compound Amount Factor	Present Worth Factor	Sinking Fund Factor	Uniform Series Factor	Capital Recovery Factor	Series Present Worth Factor	Arithmetic Gradient Series Factor
N	(*F/P,i,N*)	(*P/F,i,N*)	(*A/F,i,N*)	(*F/A,i,N*)	(*A/P,i,N*)	(*P/A,i,N*)	(*A/G,i,N*)
1	1.1100	0.90090	1.0000	1.0000	1.1100	0.90090	0.00000
2	1.2321	0.81162	0.47393	2.1100	0.58393	1.7125	0.47393
3	1.3676	0.73119	0.29921	3.3421	0.40921	2.4437	0.93055
4	1.5181	0.65873	0.21233	4.7097	0.32233	3.1024	1.3700
5	1.6851	0.59345	0.16057	6.2278	0.27057	3.6959	1.7923
6	1.8704	0.53464	0.12638	7.9129	0.23638	4.2305	2.1976
7	2.0762	0.48166	0.10222	9.783	0.21222	4.7122	2.5863
8	2.3045	0.43393	0.08432	11.859	0.19432	5.1461	2.9585
9	2.5580	0.39092	0.07060	14.164	0.18060	5.5370	3.3144
10	2.8394	0.35218	0.05980	16.722	0.16980	5.8892	3.6544
11	3.1518	0.31728	0.05112	19.561	0.16112	6.2065	3.9788
12	3.4985	0.28584	0.04403	22.713	0.15403	6.4924	4.2879
13	3.8833	0.25751	0.03815	26.212	0.14815	6.7499	4.5822
14	4.3104	0.23199	0.03323	30.095	0.14323	6.9819	4.8619
15	4.7846	0.20900	0.02907	34.405	0.13907	7.1909	5.1275
16	5.3109	0.18829	0.02552	39.190	0.13552	7.3792	5.3794
17	5.8951	0.16963	0.02247	44.501	0.13247	7.5488	5.6180
18	6.5436	0.15282	0.01984	50.396	0.12984	7.7016	5.8439
19	7.2633	0.13768	0.01756	56.939	0.12756	7.8393	6.0574
20	8.0623	0.12403	0.01558	64.203	0.12558	7.9633	6.2590
21	8.949	0.11174	0.01384	72.265	0.12384	8.0751	6.4491
22	9.934	0.10067	0.01231	81.214	0.12231	8.1757	6.6283
23	11.026	0.09069	0.01097	91.15	0.12097	8.2664	6.7969
24	12.239	0.08170	0.00979	102.17	0.11979	8.3481	6.9555
25	13.585	0.07361	0.00874	114.41	0.11874	8.4217	7.1045
26	15.080	0.06631	0.00781	128.00	0.11781	8.4881	7.2443
27	16.739	0.05974	0.00699	143.08	0.11699	8.5478	7.3754
28	18.580	0.05382	0.00626	159.82	0.11626	8.6016	7.4982
29	20.624	0.04849	0.00561	178.40	0.11561	8.6501	7.6131
30	22.892	0.04368	0.00502	199.02	0.11502	8.6938	7.7206
31	25.410	0.03935	0.00451	221.91	0.11451	8.7331	7.8210
32	28.206	0.03545	0.00404	247.32	0.11404	8.7686	7.9147
33	31.308	0.03194	0.00363	275.53	0.11363	8.8005	8.0021
34	34.752	0.02878	0.00326	306.84	0.11326	8.8293	8.0836
35	38.575	0.02592	0.00293	341.59	0.11293	8.8552	8.1594
40	65.001	0.01538	0.00172	581.83	0.11172	8.9511	8.4659
45	109.53	0.00913	0.00101	986.6	0.11101	9.0079	8.6763
50	184.56	0.00542	0.00060	1668.8	0.11060	9.0417	8.8185
55	311.00	0.00322	0.00035	2818.2	0.11035	9.0617	8.9135

i = 12% **Discrete Compounding, Discrete Cash Flows**

	SINGLE PAYMENT		UNIFORM SERIES				
	Compound Amount Factor	Present Worth Factor	Sinking Fund Factor	Uniform Series Factor	Capital Recovery Factor	Series Present Worth Factor	Arithmetic Gradient Series Factor
N	*(F/P,i,N)*	*(P/F,i,N)*	*(A/F,i,N)*	*(F/A,i,N)*	*(A/P,i,N)*	*(P/A,i,N)*	*(A/G,i,N)*
1	1.1200	0.89286	1.0000	1.0000	1.1200	0.89286	0.00000
2	1.2544	0.79719	0.47170	2.1200	0.59170	1.6901	0.47170
3	1.4049	0.71178	0.29635	3.3744	0.41635	2.4018	0.92461
4	1.5735	0.63552	0.20923	4.7793	0.32923	3.0373	1.3589
5	1.7623	0.56743	0.15741	6.3528	0.27741	3.6048	1.7746
6	1.9738	0.50663	0.12323	8.1152	0.24323	4.1114	2.1720
7	2.2107	0.45235	0.09912	10.089	0.21912	4.5638	2.5515
8	2.4760	0.40388	0.08130	12.300	0.20130	4.9676	2.9131
9	2.7731	0.36061	0.06768	14.776	0.18768	5.3282	3.2574
10	3.1058	0.32197	0.05698	17.549	0.17698	5.6502	3.5847
11	3.4785	0.28748	0.04842	20.655	0.16842	5.9377	3.8953
12	3.8960	0.25668	0.04144	24.133	0.16144	6.1944	4.1897
13	4.3635	0.22917	0.03568	28.029	0.15568	6.4235	4.4683
14	4.8871	0.20462	0.03087	32.393	0.15087	6.6282	4.7317
15	5.4736	0.18270	0.02682	37.280	0.14682	6.8109	4.9803
16	6.1304	0.16312	0.02339	42.753	0.14339	6.9740	5.2147
17	6.8660	0.14564	0.02046	48.884	0.14046	7.1196	5.4353
18	7.6900	0.13004	0.01794	55.750	0.13794	7.2497	5.6427
19	8.6128	0.11611	0.01576	63.440	0.13576	7.3658	5.8375
20	9.6463	0.10367	0.01388	72.052	0.13388	7.4694	6.0202
21	10.804	0.09256	0.01224	81.699	0.13224	7.5620	6.1913
22	12.100	0.08264	0.01081	92.503	0.13081	7.6446	6.3514
23	13.552	0.07379	0.00956	104.60	0.12956	7.7184	6.5010
24	15.179	0.06588	0.00846	118.16	0.12846	7.7843	6.6406
25	17.000	0.05882	0.00750	133.33	0.12750	7.8431	6.7708
26	19.040	0.05252	0.00665	150.33	0.12665	7.8957	6.8921
27	21.325	0.04689	0.00590	169.37	0.12590	7.9426	7.0049
28	23.884	0.04187	0.00524	190.70	0.12524	7.9844	7.1098
29	26.750	0.03738	0.00466	214.58	0.12466	8.0218	7.2071
30	29.960	0.03338	0.00414	241.33	0.12414	8.0552	7.2974
31	33.555	0.02980	0.00369	271.29	0.12369	8.0850	7.3811
32	37.582	0.02661	0.00328	304.85	0.12328	8.1116	7.4586
33	42.092	0.02376	0.00292	342.43	0.12292	8.1354	7.5302
34	47.143	0.02121	0.00260	384.52	0.12260	8.1566	7.5965
35	52.800	0.01894	0.00232	431.66	0.12232	8.1755	7.6577
40	93.051	0.01075	0.00130	767.09	0.12130	8.2438	7.8988
45	163.99	0.00610	0.00074	1358.2	0.12074	8.2825	8.0572
50	289.00	0.00346	0.00042	2400.0	0.12042	8.3045	8.1597
55	509.32	0.00196	0.00024	4236.0	0.12024	8.3170	8.2251

***i* = 13%** **Discrete Compounding, Discrete Cash Flows**

	SINGLE PAYMENT		UNIFORM SERIES				Arithmetic Gradient Series Factor
	Compound Amount Factor	Present Worth Factor	Sinking Fund Factor	Uniform Series Factor	Capital Recovery Factor	Series Present Worth Factor	
N	(*F*/*P*,*i*,*N*)	(*P*/*F*,*i*,*N*)	(*A*/*F*,*i*,*N*)	(*F*/*A*,*i*,*N*)	(*A*/*P*,*i*,*N*)	(*P*/*A*,*i*,*N*)	(*A*/*G*,*i*,*N*)
1	1.1300	0.88496	1.0000	1.0000	1.1300	0.88496	0.00000
2	1.2769	0.78315	0.46948	2.1300	0.59948	1.6681	0.46948
3	1.4429	0.69305	0.29352	3.4069	0.42352	2.3612	0.91872
4	1.6305	0.61332	0.20619	4.8498	0.33619	2.9745	1.3479
5	1.8424	0.54276	0.15431	6.4803	0.28431	3.5172	1.7571
6	2.0820	0.48032	0.12015	8.3227	0.25015	3.9975	2.1468
7	2.3526	0.42506	0.09611	10.405	0.22611	4.4226	2.5171
8	2.6584	0.37616	0.07839	12.757	0.20839	4.7988	2.8685
9	3.0040	0.33288	0.06487	15.416	0.19487	5.1317	3.2014
10	3.3946	0.29459	0.05429	18.420	0.18429	5.4262	3.5162
11	3.8359	0.26070	0.04584	21.814	0.17584	5.6869	3.8134
12	4.3345	0.23071	0.03899	25.650	0.16899	5.9176	4.0936
13	4.8980	0.20416	0.03335	29.985	0.16335	6.1218	4.3573
14	5.5348	0.18068	0.02867	34.883	0.15867	6.3025	4.6050
15	6.2543	0.15989	0.02474	40.417	0.15474	6.4624	4.8375
16	7.0673	0.14150	0.02143	46.672	0.15143	6.6039	5.0552
17	7.9861	0.12522	0.01861	53.739	0.14861	6.7291	5.2589
18	9.0243	0.11081	0.01620	61.725	0.14620	6.8399	5.4491
19	10.197	0.09806	0.01413	70.749	0.14413	6.9380	5.6265
20	11.523	0.08678	0.01235	80.947	0.14235	7.0248	5.7917
21	13.021	0.07680	0.01081	92.470	0.14081	7.1016	5.9454
22	14.714	0.06796	0.00948	105.49	0.13948	7.1695	6.0881
23	16.627	0.06014	0.00832	120.20	0.13832	7.2297	6.2205
24	18.788	0.05323	0.00731	136.83	0.13731	7.2829	6.3431
25	21.231	0.04710	0.00643	155.62	0.13643	7.3300	6.4566
26	23.991	0.04168	0.00565	176.85	0.13565	7.3717	6.5614
27	27.109	0.03689	0.00498	200.84	0.13498	7.4086	6.6582
28	30.633	0.03264	0.00439	227.95	0.13439	7.4412	6.7474
29	34.616	0.02889	0.00387	258.58	0.13387	7.4701	6.8296
30	39.116	0.02557	0.00341	293.20	0.13341	7.4957	6.9052
31	44.201	0.02262	0.00301	332.32	0.13301	7.5183	6.9747
32	49.947	0.02002	0.00266	376.52	0.13266	7.5383	7.0385
33	56.440	0.01772	0.00234	426.46	0.13234	7.5560	7.0971
34	63.777	0.01568	0.00207	482.90	0.13207	7.5717	7.1507
35	72.069	0.01388	0.00183	546.68	0.13183	7.5856	7.1998
40	132.78	0.00753	0.00099	1013.7	0.13099	7.6344	7.3888
45	244.64	0.00409	0.00053	1874.2	0.13053	7.6609	7.5076
50	450.74	0.00222	0.00029	3459.5	0.13029	7.6752	7.5811
55	830.45	0.00120	0.00016	6380.4	0.13016	7.6830	7.6260

***i* = 14%** **Discrete Compounding, Discrete Cash Flows**

	SINGLE PAYMENT		UNIFORM SERIES				Arithmetic Gradient Series Factor
	Compound Amount Factor	Present Worth Factor	Sinking Fund Factor	Uniform Series Factor	Capital Recovery Factor	Series Present Worth Factor	
N	(*F/P,i,N*)	(*P/F,i,N*)	(*A/F,i,N*)	(*F/A,i,N*)	(*A/P,i,N*)	(*P/A,i,N*)	(*A/G,i,N*)
1	1.1400	0.87719	1.0000	1.0000	1.1400	0.87719	0.00000
2	1.2996	0.76947	0.46729	2.1400	0.60729	1.6467	0.46729
3	1.4815	0.67497	0.29073	3.4396	0.43073	2.3216	0.91290
4	1.6890	0.59208	0.20320	4.9211	0.34320	2.9137	1.3370
5	1.9254	0.51937	0.15128	6.6101	0.29128	3.4331	1.7399
6	2.1950	0.45559	0.11716	8.5355	0.25716	3.8887	2.1218
7	2.5023	0.39964	0.09319	10.730	0.23319	4.2883	2.4832
8	2.8526	0.35056	0.07557	13.233	0.21557	4.6389	2.8246
9	3.2519	0.30751	0.06217	16.085	0.20217	4.9464	3.1463
10	3.7072	0.26974	0.05171	19.337	0.19171	5.2161	3.4490
11	4.2262	0.23662	0.04339	23.045	0.18339	5.4527	3.7333
12	4.8179	0.20756	0.03667	27.271	0.17667	5.6603	3.9998
13	5.4924	0.18207	0.03116	32.089	0.17116	5.8424	4.2491
14	6.2613	0.15971	0.02661	37.581	0.16661	6.0021	4.4819
15	7.1379	0.14010	0.02281	43.842	0.16281	6.1422	4.6990
16	8.1372	0.12289	0.01962	50.980	0.15962	6.2651	4.9011
17	9.2765	0.10780	0.01692	59.118	0.15692	6.3729	5.0888
18	10.575	0.09456	0.01462	68.394	0.15462	6.4674	5.2630
19	12.056	0.08295	0.01266	78.969	0.15266	6.5504	5.4243
20	13.743	0.07276	0.01099	91.025	0.15099	6.6231	5.5734
21	15.668	0.06383	0.00954	104.77	0.14954	6.6870	5.7111
22	17.861	0.05599	0.00830	120.44	0.14830	6.7429	5.8381
23	20.362	0.04911	0.00723	138.30	0.14723	6.7921	5.9549
24	23.212	0.04308	0.00630	158.66	0.14630	6.8351	6.0624
25	26.462	0.03779	0.00550	181.87	0.14550	6.8729	6.1610
26	30.167	0.03315	0.00480	208.33	0.14480	6.9061	6.2514
27	34.390	0.02908	0.00419	238.50	0.14419	6.9352	6.3342
28	39.204	0.02551	0.00366	272.89	0.14366	6.9607	6.4100
29	44.693	0.02237	0.00320	312.09	0.14320	6.9830	6.4791
30	50.950	0.01963	0.00280	356.79	0.14280	7.0027	6.5423
31	58.083	0.01722	0.00245	407.74	0.14245	7.0199	6.5998
32	66.215	0.01510	0.00215	465.82	0.14215	7.0350	6.6522
33	75.485	0.01325	0.00188	532.04	0.14188	7.0482	6.6998
34	86.053	0.01162	0.00165	607.52	0.14165	7.0599	6.7431
35	98.100	0.01019	0.00144	693.57	0.14144	7.0700	6.7824
40	188.88	0.00529	0.00075	1342.0	0.14075	7.1050	6.9300
45	363.68	0.00275	0.00039	2590.6	0.14039	7.1232	7.0188
50	700.23	0.00143	0.00020	4994.5	0.14020	7.1327	7.0714
55	1348.2	0.00074	0.00010	9623.1	0.14010	7.1376	7.1020

$i = 15\%$ **Discrete Compounding, Discrete Cash Flows**

	SINGLE PAYMENT		UNIFORM SERIES				Arithmetic Gradient Series Factor
	Compound Amount Factor	Present Worth Factor	Sinking Fund Factor	Uniform Series Factor	Capital Recovery Factor	Series Present Worth Factor	
N	*(F/P,i,N)*	*(P/F,i,N)*	*(A/F,i,N)*	*(F/A,i,N)*	*(A/P,i,N)*	*(P/A,i,N)*	*(A/G,i,N)*
1	1.1500	0.86957	1.0000	1.0000	1.1500	0.86957	0.00000
2	1.3225	0.75614	0.46512	2.1500	0.61512	1.6257	0.46512
3	1.5209	0.65752	0.28798	3.4725	0.43798	2.2832	0.90713
4	1.7490	0.57175	0.20027	4.9934	0.35027	2.8550	1.3263
5	2.0114	0.49718	0.14832	6.7424	0.29832	3.3522	1.7228
6	2.3131	0.43233	0.11424	8.7537	0.26424	3.7845	2.0972
7	2.6600	0.37594	0.09036	11.067	0.24036	4.1604	2.4498
8	3.0590	0.32690	0.07285	13.727	0.22285	4.4873	2.7813
9	3.5179	0.28426	0.05957	16.786	0.20957	4.7716	3.0922
10	4.0456	0.24718	0.04925	20.304	0.19925	5.0188	3.3832
11	4.6524	0.21494	0.04107	24.349	0.19107	5.2337	3.6549
12	5.3503	0.18691	0.03448	29.002	0.18448	5.4206	3.9082
13	6.1528	0.16253	0.02911	34.352	0.17911	5.5831	4.1438
14	7.0757	0.14133	0.02469	40.505	0.17469	5.7245	4.3624
15	8.1371	0.12289	0.02102	47.580	0.17102	5.8474	4.5650
16	9.3576	0.10686	0.01795	55.717	0.16795	5.9542	4.7522
17	10.761	0.09293	0.01537	65.075	0.16537	6.0472	4.9251
18	12.375	0.08081	0.01319	75.836	0.16319	6.1280	5.0843
19	14.232	0.07027	0.01134	88.212	0.16134	6.1982	5.2307
20	16.367	0.06110	0.00976	102.44	0.15976	6.2593	5.3651
21	18.822	0.05313	0.00842	118.81	0.15842	6.3125	5.4883
22	21.645	0.04620	0.00727	137.63	0.15727	6.3587	5.6010
23	24.891	0.04017	0.00628	159.28	0.15628	6.3988	5.7040
24	28.625	0.03493	0.00543	184.17	0.15543	6.4338	5.7979
25	32.919	0.03038	0.00470	212.79	0.15470	6.4641	5.8834
26	37.857	0.02642	0.00407	245.71	0.15407	6.4906	5.9612
27	43.535	0.02297	0.00353	283.57	0.15353	6.5135	6.0319
28	50.066	0.01997	0.00306	327.10	0.15306	6.5335	6.0960
29	57.575	0.01737	0.00265	377.17	0.15265	6.5509	6.1541
30	66.212	0.01510	0.00230	434.75	0.15230	6.5660	6.2066
31	76.144	0.01313	0.00200	500.96	0.15200	6.5791	6.2541
32	87.565	0.01142	0.00173	577.10	0.15173	6.5905	6.2970
33	100.70	0.00993	0.00150	664.67	0.15150	6.6005	6.3357
34	115.80	0.00864	0.00131	765.37	0.15131	6.6091	6.3705
35	133.18	0.00751	0.00113	881.17	0.15113	6.6166	6.4019
40	267.86	0.00373	0.00056	1779.1	0.15056	6.6418	6.5168
45	538.77	0.00186	0.00028	3585.1	0.15028	6.6543	6.5830
50	1083.7	0.00092	0.00014	7217.7	0.15014	6.6605	6.6205
55	2179.6	0.00046	0.00007	14524.0	0.15007	6.6636	6.6414

i = 20% **Discrete Compounding, Discrete Cash Flows**

	SINGLE PAYMENT		UNIFORM SERIES				
	Compound Amount Factor	Present Worth Factor	Sinking Fund Factor	Uniform Series Factor	Capital Recovery Factor	Series Present Worth Factor	Arithmetic Gradient Series Factor
N	*(F/P,i,N)*	*(P/F,i,N)*	*(A/F,i,N)*	*(F/A,i,N)*	*(A/P,i,N)*	*(P/A,i,N)*	*(A/G,i,N)*
1	1.2000	0.83333	1.0000	1.0000	1.2000	0.83333	0.00000
2	1.4400	0.69444	0.45455	2.2000	0.65455	1.5278	0.45455
3	1.7280	0.57870	0.27473	3.6400	0.47473	2.1065	0.87912
4	2.0736	0.48225	0.18629	5.3680	0.38629	2.5887	1.2742
5	2.4883	0.40188	0.13438	7.4416	0.33438	2.9906	1.6405
6	2.9860	0.33490	0.10071	9.9299	0.30071	3.3255	1.9788
7	3.5832	0.27908	0.07742	12.916	0.27742	3.6046	2.2902
8	4.2998	0.23257	0.06061	16.499	0.26061	3.8372	2.5756
9	5.1598	0.19381	0.04808	20.799	0.24808	4.0310	2.8364
10	6.1917	0.16151	0.03852	25.959	0.23852	4.1925	3.0739
11	7.4301	0.13459	0.03110	32.150	0.23110	4.3271	3.2893
12	8.9161	0.11216	0.02526	39.581	0.22526	4.4392	3.4841
13	10.699	0.09346	0.02062	48.497	0.22062	4.5327	3.6597
14	12.839	0.07789	0.01689	59.196	0.21689	4.6106	3.8175
15	15.407	0.06491	0.01388	72.035	0.21388	4.6755	3.9588
16	18.488	0.05409	0.01144	87.442	0.21144	4.7296	4.0851
17	22.186	0.04507	0.00944	105.93	0.20944	4.7746	4.1976
18	26.623	0.03756	0.00781	128.12	0.20781	4.8122	4.2975
19	31.948	0.03130	0.00646	154.74	0.20646	4.8435	4.3861
20	38.338	0.02608	0.00536	186.69	0.20536	4.8696	4.4643
21	46.005	0.02174	0.00444	225.03	0.20444	4.8913	4.5334
22	55.206	0.01811	0.00369	271.03	0.20369	4.9094	4.5941
23	66.247	0.01509	0.00307	326.24	0.20307	4.9245	4.6475
24	79.497	0.01258	0.00255	392.48	0.20255	4.9371	4.6943
25	95.396	0.01048	0.00212	471.98	0.20212	4.9476	4.7352
26	114.48	0.00874	0.00176	567.38	0.20176	4.9563	4.7709
27	137.37	0.00728	0.00147	681.85	0.20147	4.9636	4.8020
28	164.84	0.00607	0.00122	819.22	0.20122	4.9697	4.8291
29	197.81	0.00506	0.00102	984.07	0.20102	4.9747	4.8527
30	237.38	0.00421	0.00085	1181.9	0.20085	4.9789	4.8731
31	284.85	0.00351	0.00070	1419.3	0.20070	4.9824	4.8908
32	341.82	0.00293	0.00059	1704.1	0.20059	4.9854	4.9061
33	410.19	0.00244	0.00049	2045.9	0.20049	4.9878	4.9194
34	492.22	0.00203	0.00041	2456.1	0.20041	4.9898	4.9308
35	590.67	0.00169	0.00034	2948.3	0.20034	4.9915	4.9406

i = 25% **Discrete Compounding, Discrete Cash Flows**

	SINGLE PAYMENT		UNIFORM SERIES				
	Compound Amount Factor	Present Worth Factor	Sinking Fund Factor	Uniform Series Factor	Capital Recovery Factor	Series Present Worth Factor	Arithmetic Gradient Series Factor
N	(*F*/*P*,*i*,*N*)	(*P*/*F*,*i*,*N*)	(*A*/*F*,*i*,*N*)	(*F*/*A*,*i*,*N*)	(*A*/*P*,*i*,*N*)	(*P*/*A*,*i*,*N*)	(*A*/*G*,*i*,*N*)
1	1.2500	0.80000	1.0000	1.0000	1.2500	0.80000	0.00000
2	1.5625	0.64000	0.44444	2.2500	0.69444	1.4400	0.44444
3	1.9531	0.51200	0.26230	3.8125	0.51230	1.9520	0.85246
4	2.4414	0.40960	0.17344	5.7656	0.42344	2.3616	1.2249
5	3.0518	0.32768	0.12185	8.2070	0.37185	2.6893	1.5631
6	3.8147	0.26214	0.08882	11.259	0.33882	2.9514	1.8683
7	4.7684	0.20972	0.06634	15.073	0.31634	3.1611	2.1424
8	5.9605	0.16777	0.05040	19.842	0.30040	3.3289	2.3872
9	7.4506	0.13422	0.03876	25.802	0.28876	3.4631	2.6048
10	9.3132	0.10737	0.03007	33.253	0.28007	3.5705	2.7971
11	11.642	0.08590	0.02349	42.566	0.27349	3.6564	2.9663
12	14.552	0.06872	0.01845	54.208	0.26845	3.7251	3.1145
13	18.190	0.05498	0.01454	68.760	0.26454	3.7801	3.2437
14	22.737	0.04398	0.01150	86.949	0.26150	3.8241	3.3559
15	28.422	0.03518	0.00912	109.69	0.25912	3.8593	3.4530
16	35.527	0.02815	0.00724	138.11	0.25724	3.8874	3.5366
17	44.409	0.02252	0.00576	173.64	0.25576	3.9099	3.6084
18	55.511	0.01801	0.00459	218.04	0.25459	3.9279	3.6698
19	69.389	0.01441	0.00366	273.56	0.25366	3.9424	3.7222
20	86.736	0.01153	0.00292	342.94	0.25292	3.9539	3.7667
21	108.42	0.00922	0.00233	429.68	0.25233	3.9631	3.8045
22	135.53	0.00738	0.00186	538.10	0.25186	3.9705	3.8365
23	169.41	0.00590	0.00148	673.63	0.25148	3.9764	3.8634
24	211.76	0.00472	0.00119	843.03	0.25119	3.9811	3.8861
25	264.70	0.00378	0.00095	1054.8	0.25095	3.9849	3.9052
26	330.87	0.00302	0.00076	1319.5	0.25076	3.9879	3.9212
27	413.59	0.00242	0.00061	1650.4	0.25061	3.9903	3.9346
28	516.99	0.00193	0.00048	2064.0	0.25048	3.9923	3.9457
29	646.23	0.00155	0.00039	2580.9	0.25039	3.9938	3.9551
30	807.79	0.00124	0.00031	3227.2	0.25031	3.9950	3.9628
31	1009.7	0.00099	0.00025	4035.0	0.25025	3.9960	3.9693
32	1262.2	0.00079	0.00020	5044.7	0.25020	3.9968	3.9746
33	1577.7	0.00063	0.00016	6306.9	0.25016	3.9975	3.9791
34	1972.2	0.00051	0.00013	7884.6	0.25013	3.9980	3.9828
35	2465.2	0.00041	0.00010	9856.8	0.25010	3.9984	3.9858

i = 30% **Discrete Compounding, Discrete Cash Flows**

	SINGLE PAYMENT		UNIFORM SERIES				
	Compound Amount Factor	Present Worth Factor	Sinking Fund Factor	Uniform Series Factor	Capital Recovery Factor	Series Present Worth Factor	Arithmetic Gradient Series Factor
N	(*F*/*P*,*i*,*N*)	(*P*/*F*,*i*,*N*)	(*A*/*F*,*i*,*N*)	(*F*/*A*,*i*,*N*)	(*A*/*P*,*i*,*N*)	(*P*/*A*,*i*,*N*)	(*A*/*G*,*i*,*N*)
1	1.3000	0.76923	1.0000	1.0000	1.3000	0.76923	0.00000
2	1.6900	0.59172	0.43478	2.3000	0.73478	1.3609	0.43478
3	2.1970	0.45517	0.25063	3.9900	0.55063	1.8161	0.82707
4	2.8561	0.35013	0.16163	6.1870	0.46163	2.1662	1.1783
5	3.7129	0.26933	0.11058	9.0431	0.41058	2.4356	1.4903
6	4.8268	0.20718	0.07839	12.756	0.37839	2.6427	1.7654
7	6.2749	0.15937	0.05687	17.583	0.35687	2.8021	2.0063
8	8.1573	0.12259	0.04192	23.858	0.34192	2.9247	2.2156
9	10.604	0.09430	0.03124	32.015	0.33124	3.0190	2.3963
10	13.786	0.07254	0.02346	42.619	0.32346	3.0915	2.5512
11	17.922	0.05580	0.01773	56.405	0.31773	3.1473	2.6833
12	23.298	0.04292	0.01345	74.327	0.31345	3.1903	2.7952
13	30.288	0.03302	0.01024	97.625	0.31024	3.2233	2.8895
14	39.374	0.02540	0.00782	127.91	0.30782	3.2487	2.9685
15	51.186	0.01954	0.00598	167.29	0.30598	3.2682	3.0344
16	66.542	0.01503	0.00458	218.47	0.30458	3.2832	3.0892
17	86.504	0.01156	0.00351	285.01	0.30351	3.2948	3.1345
18	112.46	0.00889	0.00269	371.52	0.30269	3.3037	3.1718
19	146.19	0.00684	0.00207	483.97	0.30207	3.3105	3.2025
20	190.05	0.00526	0.00159	630.17	0.30159	3.3158	3.2275
21	247.06	0.00405	0.00122	820.22	0.30122	3.3198	3.2480
22	321.18	0.00311	0.00094	1067.3	0.30094	3.3230	3.2646
23	417.54	0.00239	0.00072	1388.5	0.30072	3.3254	3.2781
24	542.80	0.00184	0.00055	1806.0	0.30055	3.3272	3.2890
25	705.64	0.00142	0.00043	2348.8	0.30043	3.3286	3.2979
26	917.33	0.00109	0.00033	3054.4	0.30033	3.3297	3.3050
27	1192.5	0.00084	0.00025	3971.8	0.30025	3.3305	3.3107
28	1550.3	0.00065	0.00019	5164.3	0.30019	3.3312	3.3153
29	2015.4	0.00050	0.00015	6714.6	0.30015	3.3317	3.3189
30	2620.0	0.00038	0.00011	8730.0	0.30011	3.3321	3.3219
31	3406.0	0.00029	0.00009	11 350.0	0.30009	3.3324	3.3242
32	4427.8	0.00023	0.00007	14 756.0	0.30007	3.3326	3.3261
33	5756.1	0.00017	0.00005	19 184.0	0.30005	3.3328	3.3276
34	7483.0	0.00013	0.00004	24 940.0	0.30004	3.3329	3.3288
35	9727.9	0.00010	0.00003	32 423.0	0.30003	3.3330	3.3297

i = 40% **Discrete Compounding, Discrete Cash Flows**

	SINGLE PAYMENT		UNIFORM SERIES				
	Compound Amount Factor	Present Worth Factor	Sinking Fund Factor	Uniform Series Factor	Capital Recovery Factor	Series Present Worth Factor	Arithmetic Gradient Series Factor
N	(*F*/*P*,*i*,*N*)	(*P*/*F*,*i*,*N*)	(*A*/*F*,*i*,*N*)	(*F*/*A*,*i*,*N*)	(*A*/*P*,*i*,*N*)	(*P*/*A*,*i*,*N*)	(*A*/*G*,*i*,*N*)
1	1.4000	0.71429	1.0000	1.0000	1.4000	0.71429	0.00000
2	1.9600	0.51020	0.41667	2.4000	0.81667	1.2245	0.41667
3	2.7440	0.36443	0.22936	4.3600	0.62936	1.5889	0.77982
4	3.8416	0.26031	0.14077	7.1040	0.54077	1.8492	1.0923
5	5.3782	0.18593	0.09136	10.946	0.49136	2.0352	1.3580
6	7.5295	0.13281	0.06126	16.324	0.46126	2.1680	1.5811
7	10.541	0.09486	0.04192	23.853	0.44192	2.2628	1.7664
8	14.758	0.06776	0.02907	34.395	0.42907	2.3306	1.9185
9	20.661	0.04840	0.02034	49.153	0.42034	2.3790	2.0422
10	28.925	0.03457	0.01432	69.814	0.41432	2.4136	2.1419
11	40.496	0.02469	0.01013	98.739	0.41013	2.4383	2.2215
12	56.694	0.01764	0.00718	139.23	0.40718	2.4559	2.2845
13	79.371	0.01260	0.00510	195.93	0.40510	2.4685	2.3341
14	111.12	0.00900	0.00363	275.30	0.40363	2.4775	2.3729
15	155.57	0.00643	0.00259	386.42	0.40259	2.4839	2.4030
16	217.80	0.00459	0.00185	541.99	0.40185	2.4885	2.4262
17	304.91	0.00328	0.00132	759.78	0.40132	2.4918	2.4441
18	426.88	0.00234	0.00094	1064.70	0.40094	2.4941	2.4577
19	597.63	0.00167	0.00067	1491.58	0.40067	2.4958	2.4682
20	836.68	0.00120	0.00048	2089.21	0.40048	2.4970	2.4761
21	1171.36	0.00085	0.00034	2925.89	0.40034	2.4979	2.4821
22	1639.90	0.00061	0.00024	4097.24	0.40024	2.4985	2.4866
23	2295.86	0.00044	0.00017	5737.14	0.40017	2.4989	2.4900
24	3214.20	0.00031	0.00012	8033.00	0.40012	2.4992	2.4925
25	4499.88	0.00022	0.00009	11 247.0	0.40009	2.4994	2.4944
26	6299.83	0.00016	0.00006	15 747.0	0.40006	2.4996	2.4959
27	8819.76	0.00011	0.00005	22 047.0	0.40005	2.4997	2.4969
28	12 348.0	0.00008	0.00003	30 867.0	0.40003	2.4998	2.4977
29	17 287.0	0.00006	0.00002	43 214.0	0.40002	2.4999	2.4983
30	24 201.0	0.00004	0.00002	60 501.0	0.40002	2.4999	2.4988
31	33 882.0	0.00003	0.00001	84 703.0	0.40001	2.4999	2.4991
32	47 435.0	0.00002	0.00001	118 585.0	0.40001	2.4999	2.4993
33	66 409.0	0.00002	0.00001	166 019.0	0.40001	2.5000	2.4995
34	92 972.0	0.00001	0.00000	232 428.0	0.40000	2.5000	2.4996
35	130 161.0	0.00001	0.00000	325 400.0	0.40000	2.5000	2.4997

$i = 50\%$ **Discrete Compounding, Discrete Cash Flows**

	SINGLE PAYMENT		UNIFORM SERIES				
	Compound Amount Factor	Present Worth Factor	Sinking Fund Factor	Uniform Series Factor	Capital Recovery Factor	Series Present Worth Factor	Arithmetic Gradient Series Factor
N	(*F*/*P*,*i*,*N*)	(*P*/*F*,*i*,*N*)	(*A*/*F*,*i*,*N*)	(*F*/*A*,*i*,*N*)	(*A*/*P*,*i*,*N*)	(*P*/*A*,*i*,*N*)	(*A*/*G*,*i*,*N*)
1	1.5000	0.66667	1.0000	1.0000	1.5000	0.66667	0.00000
2	2.2500	0.44444	0.40000	2.5000	0.90000	1.1111	0.40000
3	3.3750	0.29630	0.21053	4.7500	0.71053	1.4074	0.73684
4	5.0625	0.19753	0.12308	8.1250	0.62308	1.6049	1.0154
5	7.5938	0.13169	0.07583	13.1875	0.57583	1.7366	1.2417
6	11.3906	0.08779	0.04812	20.781	0.54812	1.8244	1.4226
7	17.0859	0.05853	0.03108	32.172	0.53108	1.8829	1.5648
8	25.6289	0.03902	0.02030	49.258	0.52030	1.9220	1.6752
9	38.443	0.02601	0.01335	74.887	0.51335	1.9480	1.7596
10	57.665	0.01734	0.00882	113.330	0.50882	1.9653	1.8235
11	86.498	0.01156	0.00585	170.995	0.50585	1.9769	1.8713
12	129.746	0.00771	0.00388	257.493	0.50388	1.9846	1.9068
13	194.620	0.00514	0.00258	387.239	0.50258	1.9897	1.9329
14	291.929	0.00343	0.00172	581.86	0.50172	1.9931	1.9519
15	437.894	0.00228	0.00114	873.79	0.50114	1.9954	1.9657
16	656.841	0.00152	0.00076	1311.68	0.50076	1.9970	1.9756
17	985.261	0.00101	0.00051	1968.52	0.50051	1.9980	1.9827
18	1477.89	0.00068	0.00034	2953.78	0.50034	1.9986	1.9878
19	2216.84	0.00045	0.00023	4431.68	0.50023	1.9991	1.9914
20	3325.26	0.00030	0.00015	6648.51	0.50015	1.9994	1.9940
21	4987.89	0.00020	0.00010	9973.77	0.50010	1.9996	1.9958
22	7481.83	0.00013	0.00007	14 962.0	0.50007	1.9997	1.9971
23	11 223.0	0.00009	0.00004	22 443.0	0.50004	1.9998	1.9980
24	16 834.0	0.00006	0.00003	33 666.0	0.50003	1.9999	1.9986
25	25 251.0	0.00004	0.00002	50 500.0	0.50002	1.9999	1.9990
26	37 877.0	0.00003	0.00001	75 752.0	0.50001	1.9999	1.9993
27	56 815.0	0.00002	0.00001	113 628.0	0.50001	2.0000	1.9995
28	85 223.0	0.00001	0.00001	170 443.0	0.50001	2.0000	1.9997
29	127 834.0	0.00001	0.00000	255 666.0	0.50000	2.0000	1.9998
30	191 751.0	0.00001	0.00000	383 500.0	0.50000	2.0000	1.9998
31	287 627.0	0.00000	0.00000	575 251.0	0.50000	2.0000	1.9999
32	431 440.0	0.00000	0.00000	862 878.0	0.50000	2.0000	1.9999
33	647 160.0	0.00000	0.00000	1 294 318.0	0.50000	2.0000	1.9999
34	970 740.0	0.00000	0.00000	1 941 477.0	0.50000	2.0000	2.0000
35	1 456 110.0	0.00000	0.00000	2 912 217.0	0.50000	2.0000	2.0000

APPENDIX B

Compound Interest Factors for Continuous Compounding, Discrete Cash Flows

r = 1% **Continuous Compounding, Discrete Cash Flows**

	SINGLE PAYMENT		UNIFORM SERIES				
	Compound Amount Factor	Present Worth Factor	Sinking Fund Factor	Uniform Series Factor	Capital Recovery Factor	Series Present Worth Factor	Arithmetic Gradient Series Factor
N	(*F*/*P*,*r*,*N*)	(*P*/*F*,*r*,*N*)	(*A*/*F*,*r*,*N*)	(*F*/*A*,*r*,*N*)	(*A*/*P*,*r*,*N*)	(*P*/*A*,*r*,*N*)	(*A*/*G*,*r*,*N*)
1	1.0101	0.99005	1.0000	1.0000	1.0101	0.99005	0.00000
2	1.0202	0.98020	0.49750	2.0101	0.50755	1.97025	0.49750
3	1.0305	0.97045	0.33001	3.0303	0.34006	2.94069	0.99333
4	1.0408	0.96079	0.24626	4.0607	0.25631	3.90148	1.48750
5	1.0513	0.95123	0.19602	5.1015	0.20607	4.85271	1.98000
6	1.0618	0.94176	0.16253	6.1528	0.17258	5.79448	2.47084
7	1.0725	0.93239	0.13861	7.2146	0.14866	6.72687	2.96000
8	1.0833	0.92312	0.12067	8.2871	0.13072	7.64999	3.44751
9	1.0942	0.91393	0.10672	9.3704	0.11677	8.56392	3.93334
10	1.1052	0.90484	0.09556	10.4646	0.10561	9.46876	4.41751
11	1.1163	0.89583	0.08643	11.5698	0.09648	10.36459	4.90002
12	1.1275	0.88692	0.07883	12.6860	0.08888	11.25151	5.38086
13	1.1388	0.87810	0.07239	13.8135	0.08244	12.12961	5.86004
14	1.1503	0.86936	0.06688	14.9524	0.07693	12.99896	6.33755
15	1.1618	0.86071	0.06210	16.1026	0.07215	13.85967	6.81340
16	1.1735	0.85214	0.05792	17.2645	0.06797	14.71182	7.28759
17	1.1853	0.84366	0.05424	18.4380	0.06429	15.55548	7.76012
18	1.1972	0.83527	0.05096	19.6233	0.06101	16.39075	8.23098
19	1.2092	0.82696	0.04803	20.8205	0.05808	17.21771	8.70018
20	1.2214	0.81873	0.04539	22.0298	0.05544	18.03644	9.16772
21	1.2337	0.81058	0.04301	23.2512	0.05306	18.84703	9.63360
22	1.2461	0.80252	0.04084	24.4848	0.05089	19.64954	10.09782
23	1.2586	0.79453	0.03886	25.7309	0.04891	20.44408	10.56039
24	1.2712	0.78663	0.03705	26.9895	0.04710	21.23071	11.02129
25	1.2840	0.77880	0.03538	28.2608	0.04543	22.00951	11.48054
26	1.2969	0.77105	0.03385	29.5448	0.04390	22.78056	11.93813
27	1.3100	0.76338	0.03242	30.8417	0.04247	23.54394	12.39407
28	1.3231	0.75578	0.03110	32.1517	0.04115	24.29972	12.84835
29	1.3364	0.74826	0.02987	33.4748	0.03992	25.04798	13.30098
30	1.3499	0.74082	0.02873	34.8112	0.03878	25.78880	13.75196
31	1.3634	0.73345	0.02765	36.1611	0.03770	26.52225	14.20128
32	1.3771	0.72615	0.02665	37.5245	0.03670	27.24840	14.64895
33	1.3910	0.71892	0.02571	38.9017	0.03576	27.96732	15.09498
34	1.4049	0.71177	0.02482	40.2926	0.03487	28.67909	15.53935
35	1.4191	0.70469	0.02398	41.6976	0.03403	29.38378	15.98208
40	1.4918	0.67032	0.02043	48.9370	0.03048	32.80343	18.17104
45	1.5683	0.63763	0.01768	56.5475	0.02773	36.05630	20.31900
50	1.6487	0.60653	0.01549	64.5483	0.02554	39.15053	22.42613
55	1.7333	0.57695	0.01371	72.9593	0.02376	42.09385	24.49262
60	1.8221	0.54881	0.01222	81.8015	0.02227	44.89362	26.51868
65	1.9155	0.52205	0.01098	91.0971	0.02103	47.55684	28.50455
70	2.0138	0.49659	0.00991	100.869	0.01996	50.09018	30.45046
75	2.1170	0.47237	0.00900	111.142	0.01905	52.49997	32.35670
80	2.2255	0.44933	0.00820	121.942	0.01825	54.79223	34.22354
85	2.3396	0.42741	0.00750	133.296	0.01755	56.97269	36.05128
90	2.4596	0.40657	0.00689	145.232	0.01694	59.04681	37.84024
95	2.5857	0.38674	0.00634	157.779	0.01639	61.01978	39.59075
100	2.7183	0.36788	0.00585	170.970	0.01582	63.21206	41.30316

r = 2% **Continuous Compounding, Discrete Cash Flows**

	SINGLE PAYMENT		UNIFORM SERIES				
	Compound Amount Factor	Present Worth Factor	Sinking Fund Factor	Uniform Series Factor	Capital Recovery Factor	Series Present Worth Factor	Arithmetic Gradient Series Factor
N	(*F*/*P*,*r*,*N*)	(*P*/*F*,*r*,*N*)	(*A*/*F*,*r*,*N*)	(*F*/*A*,*r*,*N*)	(*A*/*P*,*r*,*N*)	(*P*/*A*,*r*,*N*)	(*A*/*G*,*r*,*N*)
1	1.0202	0.98020	1.0000	1.0000	1.0202	0.98020	0.00000
2	1.0408	0.96079	0.49500	2.0202	0.51520	1.94099	0.49500
3	1.0618	0.94176	0.32669	3.0610	0.34689	2.88275	0.98667
4	1.0833	0.92312	0.24255	4.1228	0.26275	3.80587	1.47500
5	1.1052	0.90484	0.19208	5.2061	0.21228	4.71071	1.96001
6	1.1275	0.88692	0.15845	6.3113	0.17865	5.59763	2.44168
7	1.1503	0.86936	0.13443	7.4388	0.15463	6.46699	2.92003
8	1.1735	0.85214	0.11643	8.5891	0.13663	7.31913	3.39505
9	1.1972	0.83527	0.10243	9.7626	0.12263	8.15440	3.86674
10	1.2214	0.81873	0.09124	10.9598	0.11144	8.97313	4.33511
11	1.2461	0.80252	0.08209	12.1812	0.10230	9.77565	4.80016
12	1.2712	0.78663	0.07448	13.4273	0.09468	10.56228	5.26190
13	1.2969	0.77105	0.06803	14.6985	0.08824	11.33333	5.72032
14	1.3231	0.75578	0.06252	15.9955	0.08272	12.08911	6.17543
15	1.3499	0.74082	0.05774	17.3186	0.07794	12.82993	6.62723
16	1.3771	0.72615	0.05357	18.6685	0.07377	13.55608	7.07573
17	1.4049	0.71177	0.04989	20.0456	0.07009	14.26785	7.52093
18	1.4333	0.69768	0.04662	21.4505	0.06682	14.96553	7.96283
19	1.4623	0.68386	0.04370	22.8839	0.06390	15.64939	8.40144
20	1.4918	0.67032	0.04107	24.3461	0.06128	16.31971	8.83677
21	1.5220	0.65705	0.03870	25.8380	0.05890	16.97675	9.26882
22	1.5527	0.64404	0.03655	27.3599	0.05675	17.62079	9.69759
23	1.5841	0.63128	0.03459	28.9126	0.05479	18.25207	10.12309
24	1.6161	0.61878	0.03279	30.4967	0.05299	18.87086	10.54533
25	1.6487	0.60653	0.03114	32.1128	0.05134	19.47739	10.96431
26	1.6820	0.59452	0.02962	33.7615	0.04982	20.07191	11.38005
27	1.7160	0.58275	0.02821	35.4435	0.04842	20.65466	11.79253
28	1.7507	0.57121	0.02691	37.1595	0.04711	21.22587	12.20178
29	1.7860	0.55990	0.02570	38.9102	0.04590	21.78576	12.60780
30	1.8221	0.54881	0.02457	40.6963	0.04477	22.33458	13.01059
31	1.8589	0.53794	0.02352	42.5184	0.04372	22.87252	13.41017
32	1.8965	0.52729	0.02253	44.3773	0.04274	23.39981	13.80654
33	1.9348	0.51685	0.02161	46.2738	0.04181	23.91666	14.19971
34	1.9739	0.50662	0.02074	48.2086	0.04094	24.42328	14.58969
35	2.0138	0.49659	0.01993	50.1824	0.04013	24.91987	14.97648
40	2.2255	0.44933	0.01648	60.6663	0.03668	27.25913	16.86302
45	2.4596	0.40657	0.01384	72.2528	0.03404	29.37579	18.67137
50	2.7183	0.36788	0.01176	85.0578	0.03196	31.29102	20.40283
55	3.0042	0.33287	0.01008	99.2096	0.03028	33.02399	22.05883
60	3.3201	0.30119	0.00871	114.850	0.02891	34.59205	23.64090
65	3.6693	0.27253	0.00757	132.135	0.02777	36.01089	25.15068
70	4.0552	0.24660	0.00661	151.237	0.02681	37.29471	26.58991
75	4.4817	0.22313	0.00580	172.349	0.02600	38.45635	27.96040
80	4.9530	0.20190	0.00511	195.682	0.02531	39.50745	29.26404
85	5.4739	0.18268	0.00452	221.468	0.02472	40.45853	30.50278
90	6.0496	0.16530	0.00400	249.966	0.02420	41.31910	31.67864
95	6.6859	0.14957	0.00355	281.461	0.02375	42.09777	32.79365
100	7.3891	0.13534	0.00316	316.269	0.02336	42.80234	33.84990

r = 3% **Continuous Compounding, Discrete Cash Flows**

	SINGLE PAYMENT		UNIFORM SERIES				
	Compound Amount Factor	Present Worth Factor	Sinking Fund Factor	Uniform Series Factor	Capital Recovery Factor	Series Present Worth Factor	Arithmetic Gradient Series Factor
N	(*F*/*P*,*r*,*N*)	(*P*/*F*,*r*,*N*)	(*A*/*F*,*r*,*N*)	(*F*/*A*,*r*,*N*)	(*A*/*P*,*r*,*N*)	(*P*/*A*,*r*,*N*)	(*A*/*G*,*r*,*N*)
1	1.0305	0.97045	1.0000	1.0000	1.0305	0.97045	0.00000
2	1.0618	0.94176	0.49250	2.0305	0.52296	1.91221	0.49250
3	1.0942	0.91393	0.32338	3.0923	0.35384	2.82614	0.98000
4	1.1275	0.88692	0.23886	4.1865	0.26932	3.71306	1.46251
5	1.1618	0.86071	0.18818	5.3140	0.21864	4.57377	1.94002
6	1.1972	0.83527	0.15442	6.4758	0.18488	5.40904	2.41255
7	1.2337	0.81058	0.13033	7.6730	0.16078	6.21962	2.88009
8	1.2712	0.78663	0.11228	8.9067	0.14273	7.00625	3.34265
9	1.3100	0.76338	0.09825	10.1779	0.12871	7.76963	3.80025
10	1.3499	0.74082	0.08705	11.4879	0.11750	8.51045	4.25287
11	1.3910	0.71892	0.07790	12.8378	0.10835	9.22937	4.70055
12	1.4333	0.69768	0.07028	14.2287	0.10073	9.92705	5.14328
13	1.4770	0.67706	0.06385	15.6621	0.09430	10.60411	5.58107
14	1.5220	0.65705	0.05835	17.1390	0.08880	11.26115	6.01393
15	1.5683	0.63763	0.05359	18.6610	0.08404	11.89878	6.44189
16	1.6161	0.61878	0.04943	20.2293	0.07989	12.51756	6.86494
17	1.6653	0.60050	0.04578	21.8454	0.07623	13.11806	7.28311
18	1.7160	0.58275	0.04253	23.5107	0.07299	13.70081	7.69641
19	1.7683	0.56553	0.03964	25.2267	0.07010	14.26633	8.10485
20	1.8221	0.54881	0.03704	26.9950	0.06750	14.81515	8.50845
21	1.8776	0.53259	0.03470	28.8171	0.06516	15.34774	8.90722
22	1.9348	0.51685	0.03258	30.6947	0.06303	15.86459	9.30119
23	1.9937	0.50158	0.03065	32.6295	0.06110	16.36617	9.69038
24	2.0544	0.48675	0.02888	34.6232	0.05934	16.85292	10.07479
25	2.1170	0.47237	0.02726	36.6776	0.05772	17.32528	10.45445
26	2.1815	0.45841	0.02578	38.7946	0.05623	17.78369	10.82939
27	2.2479	0.44486	0.02440	40.9761	0.05486	18.22855	11.19962
28	2.3164	0.43171	0.02314	43.2240	0.05359	18.66026	11.56517
29	2.3869	0.41895	0.02196	45.5404	0.05241	19.07921	11.92605
30	2.4596	0.40657	0.02086	47.9273	0.05132	19.48578	12.28230
31	2.5345	0.39455	0.01985	50.3869	0.05030	19.88033	12.63393
32	2.6117	0.38289	0.01890	52.9214	0.04935	20.26323	12.98098
33	2.6912	0.37158	0.01801	55.5331	0.04846	20.63480	13.32346
34	2.7732	0.36059	0.01717	58.2243	0.04763	20.99540	13.66140
35	2.8577	0.34994	0.01639	60.9975	0.04685	21.34534	13.99484
40	3.3201	0.30119	0.01313	76.1830	0.04358	22.94587	15.59532
45	3.8574	0.25924	0.01066	93.8259	0.04111	24.32346	17.08739
50	4.4817	0.22313	0.00875	114.324	0.03920	25.50917	18.47499
55	5.2070	0.19205	0.00724	138.140	0.03769	26.52971	19.76232
60	6.0496	0.16530	0.00603	165.809	0.03649	27.40811	20.95382
65	7.0287	0.14227	0.00505	197.957	0.03551	28.16415	22.05405
70	8.1662	0.12246	0.00425	235.307	0.03470	28.81487	23.06771
75	9.4877	0.10540	0.00359	278.702	0.03404	29.37496	23.99955
80	11.0232	0.09072	0.00304	329.119	0.03349	29.85703	24.85433
85	12.8071	0.07808	0.00258	387.696	0.03303	30.27196	25.63678
90	14.8797	0.06721	0.00219	455.753	0.03265	30.62908	26.35156
95	17.2878	0.05784	0.00187	534.823	0.03232	30.93647	27.00324
100	20.0855	0.04979	0.00160	626.690	0.03205	31.20103	27.59626

r = 4% **Continuous Compounding, Discrete Cash Flows**

	SINGLE PAYMENT		UNIFORM SERIES				
	Compound Amount Factor	Present Worth Factor	Sinking Fund Factor	Uniform Series Factor	Capital Recovery Factor	Series Present Worth Factor	Arithmetic Gradient Series Factor
N	*(F/P,r,N)*	*(P/F,r,N)*	*(A/F,r,N)*	*(F/A,r,N)*	*(A/P,r,N)*	*(P/A,r,N)*	*(A/G,r,N)*
1	1.0408	0.96079	1.0000	1.0000	1.0408	0.96079	0.00000
2	1.0833	0.92312	0.49000	2.0408	0.53081	1.88391	0.49000
3	1.1275	0.88692	0.32009	3.1241	0.36090	2.77083	0.97334
4	1.1735	0.85214	0.23521	4.2516	0.27602	3.62297	1.45002
5	1.2214	0.81873	0.18433	5.4251	0.22514	4.44170	1.92006
6	1.2712	0.78663	0.15045	6.6465	0.19127	5.22833	2.38345
7	1.3231	0.75578	0.12630	7.9178	0.16711	5.98411	2.84021
8	1.3771	0.72615	0.10821	9.2409	0.14903	6.71026	3.29036
9	1.4333	0.69768	0.09418	10.6180	0.13499	7.40794	3.73391
10	1.4918	0.67032	0.08298	12.0513	0.12379	8.07826	4.17089
11	1.5527	0.64404	0.07384	13.5432	0.11465	8.72229	4.60130
12	1.6161	0.61878	0.06624	15.0959	0.10705	9.34108	5.02517
13	1.6820	0.59452	0.05984	16.7120	0.10065	9.93560	5.44252
14	1.7507	0.57121	0.05437	18.3940	0.09518	10.50681	5.85339
15	1.8221	0.54881	0.04964	20.1447	0.09045	11.05562	6.25780
16	1.8965	0.52729	0.04552	21.9668	0.08633	11.58291	6.65577
17	1.9739	0.50662	0.04191	23.8633	0.08272	12.08953	7.04734
18	2.0544	0.48675	0.03870	25.8371	0.07951	12.57628	7.43255
19	2.1383	0.46767	0.03585	27.8916	0.07666	13.04395	7.81143
20	2.2255	0.44933	0.03330	30.0298	0.07411	13.49328	8.18401
21	2.3164	0.43171	0.03100	32.2554	0.07181	13.92499	8.55034
22	2.4109	0.41478	0.02893	34.5717	0.06974	14.33977	8.91045
23	2.5093	0.39852	0.02704	36.9826	0.06785	14.73829	9.26438
24	2.6117	0.38289	0.02532	39.4919	0.06613	15.12118	9.61219
25	2.7183	0.36788	0.02375	42.1036	0.06456	15.48906	9.95392
26	2.8292	0.35345	0.02231	44.8219	0.06312	15.84252	10.28960
27	2.9447	0.33960	0.02099	47.6511	0.06180	16.18211	10.61930
28	3.0649	0.32628	0.01976	50.5958	0.06058	16.50839	10.94305
29	3.1899	0.31349	0.01864	53.6607	0.05945	16.82188	11.26092
30	3.3201	0.30119	0.01759	56.8506	0.05840	17.12307	11.57295
31	3.4556	0.28938	0.01662	60.1707	0.05743	17.41246	11.87920
32	3.5966	0.27804	0.01572	63.6263	0.05653	17.69049	12.17971
33	3.7434	0.26714	0.01488	67.2230	0.05569	17.95763	12.47456
34	3.8962	0.25666	0.01409	70.9664	0.05490	18.21429	12.76379
35	4.0552	0.24660	0.01336	74.8626	0.05417	18.46089	13.04745
40	4.9530	0.20190	0.01032	96.8625	0.05113	19.55620	14.38452
45	6.0496	0.16530	0.00808	123.733	0.04889	20.45296	15.59182
50	7.3891	0.13534	0.00639	156.553	0.04720	21.18717	16.67745
55	9.0250	0.11080	0.00509	196.640	0.04590	21.78829	17.64976
60	11.0232	0.09072	0.00407	245.601	0.04488	22.28044	18.51721
65	13.4637	0.07427	0.00327	305.403	0.04409	22.68338	19.28820
70	16.4446	0.06081	0.00264	378.445	0.04345	23.01328	19.97102
75	20.0855	0.04979	0.00214	467.659	0.04295	23.28338	20.57366
80	24.5325	0.04076	0.00173	576.625	0.04255	23.50452	21.10378
85	29.9641	0.03337	0.00141	709.717	0.04222	23.68558	21.56867
90	36.5982	0.02732	0.00115	872.275	0.04196	23.83381	21.97512
95	44.7012	0.02237	0.00093	1070.82	0.04174	23.95517	22.32948
100	54.5982	0.01832	0.00076	1313.33	0.04157	24.05454	22.63760

$r = 5\%$ **Continuous Compounding, Discrete Cash Flows**

	SINGLE PAYMENT		UNIFORM SERIES				
	Compound Amount Factor	Present Worth Factor	Sinking Fund Factor	Uniform Series Factor	Capital Recovery Factor	Series Present Worth Factor	Arithmetic Gradient Series Factor
N	(*F*/*P*,*r*,*N*)	(*P*/*F*,*r*,*N*)	(*A*/*F*,*r*,*N*)	(*F*/*A*,*r*,*N*)	(*A*/*P*,*r*,*N*)	(*P*/*A*,*r*,*N*)	(*A*/*G*,*r*,*N*)
1	1.0513	0.95123	1.0000	1.0000	1.0513	0.95123	0.00000
2	1.1052	0.90484	0.48750	2.0513	0.53877	1.85607	0.48750
3	1.1618	0.86071	0.31681	3.1564	0.36808	2.71677	0.96668
4	1.2214	0.81873	0.23157	4.3183	0.28284	3.53551	1.43754
5	1.2840	0.77880	0.18052	5.5397	0.23179	4.31431	1.90011
6	1.3499	0.74082	0.14655	6.8237	0.19782	5.05512	2.35439
7	1.4191	0.70469	0.12235	8.1736	0.17362	5.75981	2.80042
8	1.4918	0.67032	0.10425	9.5926	0.15552	6.43013	3.23821
9	1.5683	0.63763	0.09022	11.0845	0.14149	7.06776	3.66780
10	1.6487	0.60653	0.07903	12.6528	0.13031	7.67429	4.08923
11	1.7333	0.57695	0.06992	14.3015	0.12119	8.25124	4.50252
12	1.8221	0.54881	0.06236	16.0347	0.11364	8.80005	4.90774
13	1.9155	0.52205	0.05600	17.8569	0.10727	9.32210	5.30491
14	2.0138	0.49659	0.05058	19.7724	0.10185	9.81868	5.69409
15	2.1170	0.47237	0.04590	21.7862	0.09717	10.29105	6.07534
16	2.2255	0.44933	0.04184	23.9032	0.09311	10.74038	6.44871
17	2.3396	0.42741	0.03827	26.1287	0.08954	11.16779	6.81425
18	2.4596	0.40657	0.03513	28.4683	0.08640	11.57436	7.17205
19	2.5857	0.38674	0.03233	30.9279	0.08360	11.96111	7.52215
20	2.7183	0.36788	0.02984	33.5137	0.08111	12.32898	7.86463
21	2.8577	0.34994	0.02760	36.2319	0.07887	12.67892	8.19957
22	3.0042	0.33287	0.02558	39.0896	0.07685	13.01179	8.52703
23	3.1582	0.31664	0.02376	42.0938	0.07503	13.32843	8.84710
24	3.3201	0.30119	0.02210	45.2519	0.07337	13.62962	9.15986
25	3.4903	0.28650	0.02059	48.5721	0.07186	13.91613	9.46539
26	3.6693	0.27253	0.01921	52.0624	0.07048	14.18866	9.76377
27	3.8574	0.25924	0.01794	55.7317	0.06921	14.44790	10.05510
28	4.0552	0.24660	0.01678	59.5891	0.06805	14.69450	10.33946
29	4.2631	0.23457	0.01571	63.6443	0.06698	14.92907	10.61695
30	4.4817	0.22313	0.01473	67.9074	0.06600	15.15220	10.88766
31	4.7115	0.21225	0.01381	72.3891	0.06509	15.36445	11.15168
32	4.9530	0.20190	0.01297	77.1006	0.06424	15.56634	11.40912
33	5.2070	0.19205	0.01219	82.0536	0.06346	15.75839	11.66006
34	5.4739	0.18268	0.01146	87.2606	0.06273	15.94108	11.90461
35	5.7546	0.17377	0.01078	92.7346	0.06205	16.11485	12.14288
40	7.3891	0.13534	0.00802	124.613	0.05930	16.86456	13.24346
45	9.4877	0.10540	0.00604	165.546	0.05731	17.44844	14.20240
50	12.1825	0.08208	0.00458	218.105	0.05586	17.90317	15.03289
55	15.6426	0.06393	0.00350	285.592	0.05477	18.25731	15.74801
60	20.0855	0.04979	0.00269	372.247	0.05396	18.53311	16.36042
65	25.7903	0.03877	0.00207	483.515	0.05334	18.74791	16.88218
70	33.1155	0.03020	0.00160	626.385	0.05287	18.91519	17.32453
75	42.5211	0.02352	0.00123	809.834	0.05251	19.04547	17.69786
80	54.5982	0.01832	0.00096	1045.39	0.05223	19.14694	18.01158
85	70.1054	0.01426	0.00074	1347.84	0.05201	19.22595	18.27416
90	90.0171	0.01111	0.00058	1736.20	0.05185	19.28749	18.49313
95	115.584	0.00865	0.00045	2234.87	0.05172	19.33542	18.67508
100	148.413	0.00674	0.00035	2875.17	0.05162	19.37275	18.82580

$r = 6\%$ **Continuous Compounding, Discrete Cash Flows**

	SINGLE PAYMENT		UNIFORM SERIES				
	Compound Amount Factor	Present Worth Factor	Sinking Fund Factor	Uniform Series Factor	Capital Recovery Factor	Series Present Worth Factor	Arithmetic Gradient Series Factor
N	(*F*/*P*,*r*,*N*)	(*P*/*F*,*r*,*N*)	(*A*/*F*,*r*,*N*)	(*F*/*A*,*r*,*N*)	(*A*/*P*,*r*,*N*)	(*P*/*A*,*r*,*N*)	(*A*/*G*,*r*,*N*)
1	1.0618	0.94176	1.0000	1.0000	1.0618	0.94176	0.00000
2	1.1275	0.88692	0.48500	2.0618	0.54684	1.82868	0.48500
3	1.1972	0.83527	0.31355	3.1893	0.37538	2.66396	0.96002
4	1.2712	0.78663	0.22797	4.3866	0.28981	3.45058	1.42508
5	1.3499	0.74082	0.17675	5.6578	0.23858	4.19140	1.88019
6	1.4333	0.69768	0.14270	7.0077	0.20454	4.88908	2.32539
7	1.5220	0.65705	0.11847	8.4410	0.18031	5.54612	2.76072
8	1.6161	0.61878	0.10037	9.9629	0.16221	6.16491	3.18622
9	1.7160	0.58275	0.08636	11.5790	0.14820	6.74766	3.60195
10	1.8221	0.54881	0.07522	13.2950	0.13705	7.29647	4.00797
11	1.9348	0.51685	0.06615	15.1171	0.12799	7.81332	4.40435
12	2.0544	0.48675	0.05864	17.0519	0.12048	8.30007	4.79114
13	2.1815	0.45841	0.05234	19.1064	0.11418	8.75848	5.16845
14	2.3164	0.43171	0.04698	21.2878	0.10881	9.19019	5.53633
15	2.4596	0.40657	0.04237	23.6042	0.10420	9.59676	5.89490
16	2.6117	0.38289	0.03837	26.0638	0.10020	9.97965	6.24424
17	2.7732	0.36059	0.03487	28.6755	0.09671	10.34025	6.58445
18	2.9447	0.33960	0.03180	31.4487	0.09363	10.67984	6.91564
19	3.1268	0.31982	0.02908	34.3934	0.09091	10.99966	7.23792
20	3.3201	0.30119	0.02665	37.5202	0.08849	11.30085	7.55141
21	3.5254	0.28365	0.02449	40.8403	0.08632	11.58451	7.85622
22	3.7434	0.26714	0.02254	44.3657	0.08438	11.85164	8.15248
23	3.9749	0.25158	0.02079	48.1091	0.08262	12.10322	8.44032
24	4.2207	0.23693	0.01920	52.0840	0.08104	12.34015	8.71986
25	4.4817	0.22313	0.01776	56.3047	0.07960	12.56328	8.99124
26	4.7588	0.21014	0.01645	60.7864	0.07829	12.77342	9.25460
27	5.0531	0.19790	0.01526	65.5452	0.07709	12.97131	9.51008
28	5.3656	0.18637	0.01416	70.5983	0.07600	13.15769	9.75782
29	5.6973	0.17552	0.01316	75.9639	0.07500	13.33321	9.99796
30	6.0496	0.16530	0.01225	81.6612	0.07408	13.49851	10.23066
31	6.4237	0.15567	0.01140	87.7109	0.07324	13.65418	10.45605
32	6.8210	0.14661	0.01062	94.1346	0.07246	13.80079	10.67429
33	7.2427	0.13807	0.00991	100.956	0.07174	13.93886	10.88553
34	7.6906	0.13003	0.00924	108.198	0.07108	14.06889	11.08992
35	8.1662	0.12246	0.00863	115.889	0.07047	14.19134	11.28761
40	11.0232	0.09072	0.00617	162.091	0.06801	14.70461	12.18092
45	14.8797	0.06721	0.00446	224.458	0.06629	15.08484	12.92953
50	20.0855	0.04979	0.00324	308.645	0.06508	15.36653	13.55188
55	27.1126	0.03688	0.00237	422.285	0.06420	15.57520	14.06541
60	36.5982	0.02732	0.00174	575.683	0.06357	15.72980	14.48619
65	49.4024	0.02024	0.00128	782.748	0.06311	15.84432	14.82876
70	66.6863	0.01500	0.00094	1062.26	0.06278	15.92916	15.10600
75	90.0171	0.01111	0.00069	1439.56	0.06253	15.99202	15.32913
80	121.510	0.00823	0.00051	1948.85	0.06235	16.03858	15.50782
85	164.022	0.00610	0.00038	2636.34	0.06222	16.07307	15.65026
90	221.406	0.00452	0.00028	3564.34	0.06212	16.09863	15.76333
95	298.867	0.00335	0.00021	4817.01	0.06204	16.11756	15.85273
100	403.429	0.00248	0.00015	6507.94	0.06199	16.13158	15.92318

r = 7% **Continuous Compounding, Discrete Cash Flows**

	SINGLE PAYMENT		UNIFORM SERIES				Arithmetic Gradient Series Factor
	Compound Amount Factor	Present Worth Factor	Sinking Fund Factor	Uniform Series Factor	Capital Recovery Factor	Series Present Worth Factor	
N	(*F*/*P*,*r*,*N*)	(*P*/*F*,*r*,*N*)	(*A*/*F*,*r*,*N*)	(*F*/*A*,*r*,*N*)	(*A*/*P*,*r*,*N*)	(*P*/*A*,*r*,*N*)	(*A*/*G*,*r*,*N*)
1	1.0725	0.93239	1.0000	1.0000	1.0725	0.93239	0.00000
2	1.1503	0.86936	0.48251	2.0725	0.55502	1.80175	0.48251
3	1.2337	0.81058	0.31029	3.2228	0.38280	2.61234	0.95337
4	1.3231	0.75578	0.22439	4.4565	0.29690	3.36812	1.41262
5	1.4191	0.70469	0.17302	5.7796	0.24553	4.07281	1.86030
6	1.5220	0.65705	0.13891	7.1987	0.21142	4.72985	2.29645
7	1.6323	0.61263	0.11467	8.7206	0.18718	5.34248	2.72114
8	1.7507	0.57121	0.09659	10.3529	0.16910	5.91369	3.13444
9	1.8776	0.53259	0.08262	12.1036	0.15513	6.44628	3.53643
10	2.0138	0.49659	0.07152	13.9812	0.14403	6.94287	3.92721
11	2.1598	0.46301	0.06252	15.9950	0.13503	7.40588	4.30688
12	2.3164	0.43171	0.05508	18.1547	0.12759	7.83759	4.67555
13	2.4843	0.40252	0.04885	20.4711	0.12136	8.24012	5.03334
14	2.6645	0.37531	0.04356	22.9554	0.11607	8.61543	5.38039
15	2.8577	0.34994	0.03903	25.6199	0.11154	8.96536	5.71683
16	3.0649	0.32628	0.03512	28.4775	0.10762	9.29164	6.04282
17	3.2871	0.30422	0.03170	31.5424	0.10421	9.59587	6.35849
18	3.5254	0.28365	0.02871	34.8295	0.10122	9.87952	6.66402
19	3.7810	0.26448	0.02607	38.3549	0.09858	10.14400	6.95958
20	4.0552	0.24660	0.02373	42.1359	0.09624	10.39059	7.24533
21	4.3492	0.22993	0.02165	46.1911	0.09416	10.62052	7.52146
22	4.6646	0.21438	0.01979	50.5404	0.09229	10.83490	7.78815
23	5.0028	0.19989	0.01811	55.2050	0.09062	11.03479	8.04559
24	5.3656	0.18637	0.01661	60.2078	0.08912	11.22116	8.29397
25	5.7546	0.17377	0.01525	65.5733	0.08776	11.39494	8.53348
26	6.1719	0.16203	0.01402	71.3279	0.08653	11.55696	8.76434
27	6.6194	0.15107	0.01290	77.4998	0.08541	11.70803	8.98674
28	7.0993	0.14086	0.01189	84.1192	0.08440	11.84889	9.20088
29	7.6141	0.13134	0.01096	91.2185	0.08347	11.98023	9.40697
30	8.1662	0.12246	0.01012	98.8326	0.08263	12.10268	9.60521
31	8.7583	0.11418	0.00935	106.999	0.08185	12.21686	9.79582
32	9.3933	0.10646	0.00864	115.757	0.08115	12.32332	9.97900
33	10.0744	0.09926	0.00799	125.150	0.08050	12.42258	10.15495
34	10.8049	0.09255	0.00740	135.225	0.07990	12.51513	10.32389
35	11.5883	0.08629	0.00685	146.030	0.07936	12.60143	10.48603
40	16.4446	0.06081	0.00469	213.006	0.07720	12.95288	11.20165
45	23.3361	0.04285	0.00325	308.049	0.07575	13.20055	11.77687
50	33.1155	0.03020	0.00226	442.922	0.07477	13.37508	12.23466
55	46.9931	0.02128	0.00158	634.315	0.07408	13.49807	12.59571
60	66.6863	0.01500	0.00110	905.916	0.07361	13.58473	12.87812
65	94.6324	0.01057	0.00077	1291.34	0.07328	13.64581	13.09734
70	134.290	0.00745	0.00054	1838.27	0.07305	13.68885	13.26638
75	190.566	0.00525	0.00038	2614.41	0.07289	13.71918	13.39591
80	270.426	0.00370	0.00027	3715.81	0.07278	13.74055	13.49462
85	383.753	0.00261	0.00019	5278.76	0.07270	13.75561	13.56947
90	544.572	0.00184	0.00013	7496.70	0.07264	13.76622	13.62598
95	772.784	0.00129	0.00009	10 644.0	0.07260	13.77370	13.66846
100	1096.63	0.00091	0.00007	15 110.0	0.07257	13.77897	13.70028

r = 8% **Continuous Compounding, Discrete Cash Flows**

	SINGLE PAYMENT		UNIFORM SERIES				
	Compound Amount Factor	Present Worth Factor	Sinking Fund Factor	Uniform Series Factor	Capital Recovery Factor	Series Present Worth Factor	Arithmetic Gradient Series Factor
N	(*F/P,r,N*)	(*P/F,r,N*)	(*A/F,r,N*)	(*F/A,r,N*)	(*A/P,r,N*)	(*P/A,r,N*)	(*A/G,r,N*)
1	1.0833	0.92312	1.0000	1.0000	1.0833	0.92312	0.00000
2	1.1735	0.85214	0.48001	2.0833	0.56330	1.77526	0.48001
3	1.2712	0.78663	0.30705	3.2568	0.39034	2.56189	0.94672
4	1.3771	0.72615	0.22085	4.5280	0.30413	3.28804	1.40018
5	1.4918	0.67032	0.16934	5.9052	0.25263	3.95836	1.84044
6	1.6161	0.61878	0.13519	7.3970	0.21848	4.57714	2.26758
7	1.7507	0.57121	0.11095	9.0131	0.19424	5.14835	2.68169
8	1.8965	0.52729	0.09290	10.7637	0.17619	5.67564	3.08288
9	2.0544	0.48675	0.07899	12.6602	0.16227	6.16239	3.47127
10	2.2255	0.44933	0.06796	14.7147	0.15125	6.61172	3.84700
11	2.4109	0.41478	0.05903	16.9402	0.14232	7.02651	4.21022
12	2.6117	0.38289	0.05168	19.3511	0.13496	7.40940	4.56110
13	2.8292	0.35345	0.04553	21.9628	0.12882	7.76285	4.89980
14	3.0649	0.32628	0.04034	24.7920	0.12362	8.08913	5.22653
15	3.3201	0.30119	0.03590	27.8569	0.11918	8.39033	5.54147
16	3.5966	0.27804	0.03207	31.1770	0.11536	8.66836	5.84486
17	3.8962	0.25666	0.02876	34.7736	0.11204	8.92503	6.13689
18	4.2207	0.23693	0.02586	38.6698	0.10915	9.16195	6.41781
19	4.5722	0.21871	0.02332	42.8905	0.10660	9.38067	6.68785
20	4.9530	0.20190	0.02107	47.4627	0.10436	9.58256	6.94726
21	5.3656	0.18637	0.01908	52.4158	0.10237	9.76894	7.19628
22	5.8124	0.17204	0.01731	57.7813	0.10059	9.94098	7.43518
23	6.2965	0.15882	0.01572	63.5938	0.09901	10.09980	7.66421
24	6.8210	0.14661	0.01431	69.8903	0.09760	10.24641	7.88363
25	7.3891	0.13534	0.01304	76.7113	0.09632	10.38174	8.09372
26	8.0045	0.12493	0.01189	84.1003	0.09518	10.50667	8.29475
27	8.6711	0.11533	0.01086	92.1048	0.09414	10.62200	8.48698
28	9.3933	0.10646	0.00992	100.776	0.09321	10.72845	8.67068
29	10.1757	0.09827	0.00908	110.169	0.09236	10.82673	8.84614
30	11.0232	0.09072	0.00831	120.345	0.09160	10.91745	9.01360
31	11.9413	0.08374	0.00761	131.368	0.09090	11.00119	9.17336
32	12.9358	0.07730	0.00698	143.309	0.09026	11.07849	9.32566
33	14.0132	0.07136	0.00640	156.245	0.08969	11.14986	9.47078
34	15.1803	0.06587	0.00587	170.258	0.08916	11.21573	9.60898
35	16.4446	0.06081	0.00539	185.439	0.08868	11.27654	9.74051
40	24.5325	0.04076	0.00354	282.547	0.08683	11.51725	10.30689
45	36.5982	0.02732	0.00234	427.416	0.08563	11.67860	10.74256
50	54.5982	0.01832	0.00155	643.535	0.08484	11.78676	11.07380
55	81.4509	0.01228	0.00104	965.947	0.08432	11.85926	11.32302
60	121.510	0.00823	0.00069	1446.93	0.08398	11.90785	11.50878
65	181.272	0.00552	0.00046	2164.47	0.08375	11.94043	11.64610
70	270.426	0.00370	0.00031	3234.91	0.08360	11.96227	11.74685
75	403.429	0.00248	0.00021	4831.83	0.08349	11.97690	11.82030
80	601.845	0.00166	0.00014	7214.15	0.08343	11.98672	11.87352
85	897.847	0.00111	0.00009	10 768.0	0.08338	11.99329	11.91189
90	1339.43	0.00075	0.00006	16 070.0	0.08335	11.99770	11.93942
95	1998.20	0.00050	0.00004	23 980.0	0.08333	12.00066	11.95910
100	2980.96	0.00034	0.00003	35 779.0	0.08332	12.00264	11.97311

r = 9% **Continuous Compounding, Discrete Cash Flows**

	SINGLE PAYMENT		UNIFORM SERIES				Arithmetic Gradient Series Factor
	Compound Amount Factor	Present Worth Factor	Sinking Fund Factor	Uniform Series Factor	Capital Recovery Factor	Series Present Worth Factor	
N	(*F*/*P*,*r*,*N*)	(*P*/*F*,*r*,*N*)	(*A*/*F*,*r*,*N*)	(*F*/*A*,*r*,*N*)	(*A*/*P*,*r*,*N*)	(*P*/*A*,*r*,*N*)	(*A*/*G*,*r*,*N*)
1	1.0942	0.91393	1.0000	1.0000	1.0942	0.91393	0.00000
2	1.1972	0.83527	0.47752	2.0942	0.57169	1.74920	0.47752
3	1.3100	0.76338	0.30382	3.2914	0.39800	2.51258	0.94008
4	1.4333	0.69768	0.21733	4.6014	0.31150	3.21026	1.38776
5	1.5683	0.63763	0.16571	6.0347	0.25988	3.84789	1.82063
6	1.7160	0.58275	0.13153	7.6030	0.22570	4.43063	2.23880
7	1.8776	0.53259	0.10731	9.3190	0.20148	4.96323	2.64241
8	2.0544	0.48675	0.08931	11.1966	0.18349	5.44998	3.03160
9	2.2479	0.44486	0.07547	13.2510	0.16964	5.89484	3.40654
10	2.4596	0.40657	0.06452	15.4990	0.15869	6.30141	3.76743
11	2.6912	0.37158	0.05568	17.9586	0.14986	6.67298	4.11449
12	2.9447	0.33960	0.04843	20.6498	0.14260	7.01258	4.44793
13	3.2220	0.31037	0.04238	23.5945	0.13656	7.32294	4.76801
14	3.5254	0.28365	0.03729	26.8165	0.13146	7.60660	5.07498
15	3.8574	0.25924	0.03296	30.3419	0.12713	7.86584	5.36913
16	4.2207	0.23693	0.02924	34.1993	0.12341	8.10277	5.65074
17	4.6182	0.21654	0.02603	38.4200	0.12020	8.31930	5.92011
18	5.0531	0.19790	0.02324	43.0382	0.11741	8.51720	6.17755
19	5.5290	0.18087	0.02079	48.0913	0.11497	8.69807	6.42339
20	6.0496	0.16530	0.01865	53.6202	0.11282	8.86337	6.65794
21	6.6194	0.15107	0.01676	59.6699	0.11093	9.01444	6.88154
22	7.2427	0.13807	0.01509	66.2893	0.10926	9.15251	7.09452
23	7.9248	0.12619	0.01360	73.5320	0.10777	9.27869	7.29723
24	8.6711	0.11533	0.01228	81.4568	0.10645	9.39402	7.49000
25	9.4877	0.10540	0.01110	90.1280	0.10527	9.49942	7.67318
26	10.3812	0.09633	0.01004	99.6157	0.10421	9.59574	7.84712
27	11.3589	0.08804	0.00909	109.997	0.10327	9.68378	8.01215
28	12.4286	0.08046	0.00824	121.356	0.10241	9.76424	8.16862
29	13.5991	0.07353	0.00747	133.784	0.10165	9.83778	8.31685
30	14.8797	0.06721	0.00679	147.383	0.10096	9.90498	8.45719
31	16.2810	0.06142	0.00616	162.263	0.10034	9.96640	8.58995
32	17.8143	0.05613	0.00560	178.544	0.09978	10.02254	8.71547
33	19.4919	0.05130	0.00509	196.358	0.09927	10.07384	8.83405
34	21.3276	0.04689	0.00463	215.850	0.09881	10.12073	8.94600
35	23.3361	0.04285	0.00422	237.178	0.09839	10.16358	9.05164
40	36.5982	0.02732	0.00265	378.004	0.09682	10.32847	9.49496
45	57.3975	0.01742	0.00167	598.863	0.09584	10.43361	9.82070
50	90.0171	0.01111	0.00106	945.238	0.09523	10.50065	10.05692
55	141.175	0.00708	0.00067	1488.46	0.09485	10.54339	10.22624
60	221.406	0.00452	0.00043	2340.41	0.09460	10.57065	10.34639
65	347.234	0.00288	0.00027	3676.53	0.09445	10.58803	10.43088
70	544.572	0.00184	0.00017	5771.98	0.09435	10.59911	10.48983
75	854.059	0.00117	0.00011	9058.30	0.09428	10.60618	10.53069
80	1339.43	0.00075	0.00007	14 212.0	0.09424	10.61068	10.55884
85	2100.65	0.00048	0.00004	22 295.0	0.09422	10.61356	10.57813
90	3294.47	0.00030	0.00003	34 972.0	0.09420	10.61539	10.59128
95	5166.75	0.00019	0.00002	54 853.0	0.09419	10.61655	10.60022
100	8103.08	0.00012	0.00001	86 033.0	0.09419	10.61730	10.60627

r = 10% **Continuous Compounding, Discrete Cash Flows**

	SINGLE PAYMENT		UNIFORM SERIES				Arithmetic Gradient Series Factor
	Compound Amount Factor	Present Worth Factor	Sinking Fund Factor	Uniform Series Factor	Capital Recovery Factor	Series Present Worth Factor	
N	(*F*/*P*,*r*,*N*)	(*P*/*F*,*r*,*N*)	(*A*/*F*,*r*,*N*)	(*F*/*A*,*r*,*N*)	(*A*/*P*,*r*,*N*)	(*P*/*A*,*r*,*N*)	(*A*/*G*,*r*,*N*)
1	1.1052	0.90484	1.0000	1.0000	1.1052	0.90484	0.00000
2	1.2214	0.81873	0.47502	2.1052	0.58019	1.72357	0.47502
3	1.3499	0.74082	0.30061	3.3266	0.40578	2.46439	0.93344
4	1.4918	0.67032	0.21384	4.6764	0.31901	3.13471	1.37535
5	1.6487	0.60653	0.16212	6.1683	0.26729	3.74124	1.80086
6	1.8221	0.54881	0.12793	7.8170	0.23310	4.29005	2.21012
7	2.0138	0.49659	0.10374	9.6391	0.20892	4.78663	2.60329
8	2.2255	0.44933	0.08582	11.6528	0.19099	5.23596	2.98060
9	2.4596	0.40657	0.07205	13.8784	0.17723	5.64253	3.34227
10	2.7183	0.36788	0.06121	16.3380	0.16638	6.01041	3.68856
11	3.0042	0.33287	0.05248	19.0563	0.15765	6.34328	4.01976
12	3.3201	0.30119	0.04533	22.0604	0.15050	6.64448	4.33618
13	3.6693	0.27253	0.03940	25.3806	0.14457	6.91701	4.63814
14	4.0552	0.24660	0.03442	29.0499	0.13959	7.16361	4.92598
15	4.4817	0.22313	0.03021	33.1051	0.13538	7.38674	5.20008
16	4.9530	0.20190	0.02661	37.5867	0.13178	7.58863	5.46081
17	5.4739	0.18268	0.02351	42.5398	0.12868	7.77132	5.70856
18	6.0496	0.16530	0.02083	48.0137	0.12600	7.93662	5.94373
19	6.6859	0.14957	0.01850	54.0634	0.12367	8.08618	6.16673
20	7.3891	0.13534	0.01646	60.7493	0.12163	8.22152	6.37798
21	8.1662	0.12246	0.01468	68.1383	0.11985	8.34398	6.57790
22	9.0250	0.11080	0.01311	76.3045	0.11828	8.45478	6.76690
23	9.9742	0.10026	0.01172	85.3295	0.11689	8.55504	6.94542
24	11.0232	0.09072	0.01049	95.3037	0.11566	8.64576	7.11388
25	12.1825	0.08208	0.00940	106.327	0.11458	8.72784	7.27269
26	13.4637	0.07427	0.00844	118.509	0.11361	8.80211	7.42228
27	14.8797	0.06721	0.00758	131.973	0.11275	8.86932	7.56305
28	16.4446	0.06081	0.00681	146.853	0.11198	8.93013	7.69541
29	18.1741	0.05502	0.00612	163.297	0.11129	8.98515	7.81975
30	20.0855	0.04979	0.00551	181.472	0.11068	9.03494	7.93646
31	22.1980	0.04505	0.00496	201.557	0.11013	9.07999	8.04593
32	24.5325	0.04076	0.00447	223.755	0.10964	9.12075	8.14851
33	27.1126	0.03688	0.00403	248.288	0.10920	9.15763	8.24458
34	29.9641	0.03337	0.00363	275.400	0.10880	9.19101	8.33446
35	33.1155	0.03020	0.00327	305.364	0.10845	9.22121	8.41851
40	54.5982	0.01832	0.00196	509.629	0.10713	9.33418	8.76204
45	90.0171	0.01111	0.00118	846.404	0.10635	9.40270	9.00281
50	148.413	0.00674	0.00071	1401.65	0.10588	9.44427	9.16915
55	244.692	0.00409	0.00043	2317.10	0.10560	9.46947	9.28264
60	403.429	0.00248	0.00026	3826.43	0.10543	9.48476	9.35924
65	665.142	0.00150	0.00016	6314.88	0.10533	9.49404	9.41046
70	1096.63	0.00091	0.00010	10 418.0	0.10527	9.49966	9.44444
75	1808.04	0.00055	0.00006	17 182.0	0.10523	9.50307	9.46683

$r = 11\%$ **Continuous Compounding, Discrete Cash Flows**

	SINGLE PAYMENT		UNIFORM SERIES				
	Compound Amount Factor	Present Worth Factor	Sinking Fund Factor	Uniform Series Factor	Capital Recovery Factor	Series Present Worth Factor	Arithmetic Gradient Series Factor
N	(*F*/*P*,*r*,*N*)	(*P*/*F*,*r*,*N*)	(*A*/*F*,*r*,*N*)	(*F*/*A*,*r*,*N*)	(*A*/*P*,*r*,*N*)	(*P*/*A*,*r*,*N*)	(*A*/*G*,*r*,*N*)
1	1.1163	0.89583	1.0000	1.0000	1.1163	0.89583	0.00000
2	1.2461	0.80252	0.47253	2.1163	0.58881	1.69835	0.47253
3	1.3910	0.71892	0.29741	3.3624	0.41369	2.41728	0.92681
4	1.5527	0.64404	0.21038	4.7533	0.32666	3.06131	1.36297
5	1.7333	0.57695	0.15858	6.3060	0.27486	3.63826	1.78115
6	1.9348	0.51685	0.12439	8.0393	0.24067	4.15511	2.18154
7	2.1598	0.46301	0.10026	9.9741	0.21654	4.61813	2.56437
8	2.4109	0.41478	0.08241	12.1338	0.19869	5.03291	2.92993
9	2.6912	0.37158	0.06875	14.5447	0.18503	5.40449	3.27852
10	3.0042	0.33287	0.05802	17.2360	0.17430	5.73736	3.61047
11	3.3535	0.29820	0.04941	20.2401	0.16568	6.03556	3.92615
12	3.7434	0.26714	0.04238	23.5936	0.15866	6.30269	4.22597
13	4.1787	0.23931	0.03658	27.3370	0.15286	6.54200	4.51035
14	4.6646	0.21438	0.03173	31.5157	0.14801	6.75638	4.77973
15	5.2070	0.19205	0.02764	36.1803	0.14392	6.94843	5.03457
16	5.8124	0.17204	0.02416	41.3873	0.14044	7.12048	5.27536
17	6.4883	0.15412	0.02119	47.1998	0.13746	7.27460	5.50257
18	7.2427	0.13807	0.01863	53.6881	0.13490	7.41267	5.71673
19	8.0849	0.12369	0.01641	60.9308	0.13269	7.53636	5.91832
20	9.0250	0.11080	0.01449	69.0157	0.13077	7.64716	6.10787
21	10.0744	0.09926	0.01281	78.0407	0.12909	7.74642	6.28588
22	11.2459	0.08892	0.01135	88.1151	0.12763	7.83534	6.45287
23	12.5535	0.07966	0.01006	99.3610	0.12634	7.91500	6.60934
24	14.0132	0.07136	0.00894	111.915	0.12521	7.98636	6.75579
25	15.6426	0.06393	0.00794	125.928	0.12422	8.05029	6.89273
26	17.4615	0.05727	0.00706	141.570	0.12334	8.10756	7.02063
27	19.4919	0.05130	0.00629	159.032	0.12257	8.15886	7.13998
28	21.7584	0.04596	0.00560	178.524	0.12188	8.20482	7.25122
29	24.2884	0.04117	0.00499	200.282	0.12127	8.24599	7.35482
30	27.1126	0.03688	0.00445	224.571	0.12073	8.28288	7.45120
31	30.2652	0.03304	0.00397	251.683	0.12025	8.31592	7.54080
32	33.7844	0.02960	0.00355	281.949	0.11982	8.34552	7.62400
33	37.7128	0.02652	0.00317	315.733	0.11945	8.37203	7.70121
34	42.0980	0.02375	0.00283	353.446	0.11911	8.39579	7.77278
35	46.9931	0.02128	0.00253	395.544	0.11881	8.41707	7.83909
40	81.4509	0.01228	0.00145	691.883	0.11772	8.49449	8.10288
45	141.175	0.00708	0.00083	1205.52	0.11711	8.53916	8.27905
50	244.692	0.00409	0.00048	2095.77	0.11676	8.56493	8.39490
55	424.113	0.00236	0.00027	3638.80	0.11655	8.57980	8.47009

$r = 12\%$ **Continuous Compounding, Discrete Cash Flows**

	SINGLE PAYMENT		UNIFORM SERIES				Arithmetic Gradient Series Factor
	Compound Amount Factor	Present Worth Factor	Sinking Fund Factor	Uniform Series Factor	Capital Recovery Factor	Series Present Worth Factor	
N	*(F/P,r,N)*	*(P/F,r,N)*	*(A/F,r,N)*	*(F/A,r,N)*	*(A/P,r,N)*	*(P/A,r,N)*	*(A/G,r,N)*
1	1.1275	0.88692	1.0000	1.0000	1.1275	0.88692	0.00000
2	1.2712	0.78663	0.47004	2.1275	0.59753	1.67355	0.47004
3	1.4333	0.69768	0.29423	3.3987	0.42172	2.37122	0.92019
4	1.6161	0.61878	0.20695	4.8321	0.33445	2.99001	1.35061
5	1.8221	0.54881	0.15508	6.4481	0.28258	3.53882	1.76148
6	2.0544	0.48675	0.12092	8.2703	0.24841	4.02557	2.15307
7	2.3164	0.43171	0.09686	10.3247	0.22435	4.45728	2.52566
8	2.6117	0.38289	0.07911	12.6411	0.20660	4.84018	2.87962
9	2.9447	0.33960	0.06556	15.2528	0.19306	5.17977	3.21532
10	3.3201	0.30119	0.05495	18.1974	0.18245	5.48097	3.53320
11	3.7434	0.26714	0.04647	21.5176	0.17397	5.74810	3.83374
12	4.2207	0.23693	0.03959	25.2610	0.16708	5.98503	4.11743
13	4.7588	0.21014	0.03392	29.4817	0.16142	6.19516	4.38480
14	5.3656	0.18637	0.02921	34.2405	0.15670	6.38154	4.63641
15	6.0496	0.16530	0.02525	39.6061	0.15275	6.54684	4.87283
16	6.8210	0.14661	0.02190	45.6557	0.14940	6.69344	5.09464
17	7.6906	0.13003	0.01906	52.4767	0.14655	6.82347	5.30246
18	8.6711	0.11533	0.01662	60.1673	0.14412	6.93880	5.49687
19	9.7767	0.10228	0.01453	68.8384	0.14202	7.04108	5.67850
20	11.0232	0.09072	0.01272	78.6151	0.14022	7.13180	5.84796
21	12.4286	0.08046	0.01116	89.6383	0.13865	7.21226	6.00583
22	14.0132	0.07136	0.00980	102.067	0.13729	7.28362	6.15274
23	15.7998	0.06329	0.00861	116.080	0.13611	7.34691	6.28926
24	17.8143	0.05613	0.00758	131.880	0.13508	7.40305	6.41597
25	20.0855	0.04979	0.00668	149.694	0.13418	7.45283	6.53344
26	22.6464	0.04416	0.00589	169.780	0.13339	7.49699	6.64221
27	25.5337	0.03916	0.00520	192.426	0.13269	7.53616	6.74280
28	28.7892	0.03474	0.00459	217.960	0.13208	7.57089	6.83574
29	32.4597	0.03081	0.00405	246.749	0.13155	7.60170	6.92152
30	36.5982	0.02732	0.00358	279.209	0.13108	7.62902	7.00059
31	41.2644	0.02423	0.00317	315.807	0.13066	7.65326	7.07342
32	46.5255	0.02149	0.00280	357.071	0.13030	7.67475	7.14043
33	52.4573	0.01906	0.00248	403.597	0.12997	7.69381	7.20202
34	59.1455	0.01691	0.00219	456.054	0.12969	7.71072	7.25859
35	66.6863	0.01500	0.00194	515.200	0.12944	7.72572	7.31050
40	121.510	0.00823	0.00106	945.203	0.12855	7.77878	7.51141
45	221.406	0.00452	0.00058	1728.72	0.12808	7.80791	7.63916
50	403.429	0.00248	0.00032	3156.38	0.12781	7.82389	7.71909
55	735.095	0.00136	0.00017	5757.75	0.12767	7.83266	7.76841

r = 13% **Continuous Compounding, Discrete Cash Flows**

	SINGLE PAYMENT		UNIFORM SERIES				
	Compound Amount Factor	Present Worth Factor	Sinking Fund Factor	Uniform Series Factor	Capital Recovery Factor	Series Present Worth Factor	Arithmetic Gradient Series Factor
N	(*F*/*P*,*r*,*N*)	(*P*/*F*,*r*,*N*)	(*A*/*F*,*r*,*N*)	(*F*/*A*,*r*,*N*)	(*A*/*P*,*r*,*N*)	(*P*/*A*,*r*,*N*)	(*A*/*G*,*r*,*N*)
1	1.1388	0.87810	1.0000	1.0000	1.1388	0.87810	0.00000
2	1.2969	0.77105	0.46755	2.1388	0.60637	1.64915	0.46755
3	1.4770	0.67706	0.29106	3.4358	0.42988	2.32620	0.91358
4	1.6820	0.59452	0.20355	4.9127	0.34238	2.92072	1.33827
5	1.9155	0.52205	0.15164	6.5948	0.29046	3.44277	1.74189
6	2.1815	0.45841	0.11750	8.5103	0.25633	3.90118	2.12473
7	2.4843	0.40252	0.09353	10.6918	0.23236	4.30370	2.48718
8	2.8292	0.35345	0.07589	13.1761	0.21472	4.65716	2.82968
9	3.2220	0.31037	0.06248	16.0053	0.20131	4.96752	3.15272
10	3.6693	0.27253	0.05201	19.2273	0.19084	5.24005	3.45683
11	4.1787	0.23931	0.04367	22.8966	0.18250	5.47936	3.74260
12	4.7588	0.21014	0.03693	27.0753	0.17576	5.68950	4.01065
13	5.4195	0.18452	0.03141	31.8341	0.17024	5.87402	4.26162
14	6.1719	0.16203	0.02684	37.2536	0.16567	6.03604	4.49618
15	7.0287	0.14227	0.02303	43.4255	0.16186	6.17832	4.71503
16	8.0045	0.12493	0.01982	50.4542	0.15865	6.30325	4.91888
17	9.1157	0.10970	0.01711	58.4586	0.15593	6.41295	5.10844
18	10.3812	0.09633	0.01480	67.5743	0.15363	6.50928	5.28441
19	11.8224	0.08458	0.01283	77.9556	0.15166	6.59386	5.44753
20	13.4637	0.07427	0.01114	89.7780	0.14997	6.66814	5.59848
21	15.3329	0.06522	0.00969	103.242	0.14851	6.73335	5.73798
22	17.4615	0.05727	0.00843	118.575	0.14726	6.79062	5.86669
23	19.8857	0.05029	0.00735	136.036	0.14618	6.84091	5.98528
24	22.6464	0.04416	0.00641	155.922	0.14524	6.88507	6.09441
25	25.7903	0.03877	0.00560	178.568	0.14443	6.92384	6.19468
26	29.3708	0.03405	0.00489	204.359	0.14372	6.95789	6.28670
27	33.4483	0.02990	0.00428	233.729	0.14311	6.98779	6.37104
28	38.0918	0.02625	0.00374	267.178	0.14257	7.01404	6.44825
29	43.3801	0.02305	0.00328	305.269	0.14210	7.03709	6.51885
30	49.4024	0.02024	0.00287	348.650	0.14170	7.05733	6.58333
31	56.2609	0.01777	0.00251	398.052	0.14134	7.07511	6.64216
32	64.0715	0.01561	0.00220	454.313	0.14103	7.09071	6.69578
33	72.9665	0.01370	0.00193	518.384	0.14076	7.10442	6.74459
34	83.0963	0.01203	0.00169	591.351	0.14052	7.11645	6.78899
35	94.6324	0.01057	0.00148	674.447	0.14031	7.12702	6.82934
40	181.272	0.00552	0.00077	1298.53	0.13960	7.16340	6.98125
45	347.234	0.00288	0.00040	2493.97	0.13923	7.18239	7.07317
50	665.142	0.00150	0.00021	4783.90	0.13904	7.19231	7.12785
55	1274.11	0.00078	0.00011	9170.36	0.13894	7.19748	7.15994

$r = 14\%$ **Continuous Compounding, Discrete Cash Flows**

	SINGLE PAYMENT		UNIFORM SERIES				Arithmetic Gradient Series Factor
	Compound Amount Factor	Present Worth Factor	Sinking Fund Factor	Uniform Series Factor	Capital Recovery Factor	Series Present Worth Factor	
N	(*F*/*P*,*r*,*N*)	(*P*/*F*,*r*,*N*)	(*A*/*F*,*r*,*N*)	(*F*/*A*,*r*,*N*)	(*A*/*P*,*r*,*N*)	(*P*/*A*,*r*,*N*)	(*A*/*G*,*r*,*N*)
1	1.1503	0.86936	1.0000	1.0000	1.1503	0.86936	0.00000
2	1.3231	0.75578	0.46506	2.1503	0.61533	1.62514	0.46506
3	1.5220	0.65705	0.28790	3.4734	0.43818	2.28219	0.90697
4	1.7507	0.57121	0.20019	4.9954	0.35046	2.85340	1.32596
5	2.0138	0.49659	0.14824	6.7460	0.29851	3.34998	1.72235
6	2.3164	0.43171	0.11416	8.7598	0.26443	3.78169	2.09652
7	2.6645	0.37531	0.09028	11.0762	0.24056	4.15700	2.44894
8	3.0649	0.32628	0.07278	13.7406	0.22305	4.48328	2.78015
9	3.5254	0.28365	0.05950	16.8055	0.20978	4.76694	3.09076
10	4.0552	0.24660	0.04919	20.3309	0.19946	5.01354	3.38141
11	4.6646	0.21438	0.04101	24.3861	0.19128	5.22792	3.65282
12	5.3656	0.18637	0.03442	29.0507	0.18470	5.41429	3.90573
13	6.1719	0.16203	0.02906	34.4162	0.17933	5.57632	4.14092
14	7.0993	0.14086	0.02464	40.5881	0.17491	5.71717	4.35918
15	8.1662	0.12246	0.02097	47.6874	0.17124	5.83963	4.56135
16	9.3933	0.10646	0.01790	55.8536	0.16818	5.94609	4.74824
17	10.8049	0.09255	0.01533	65.2469	0.16560	6.03864	4.92069
18	12.4286	0.08046	0.01315	76.0518	0.16342	6.11910	5.07952
19	14.2963	0.06995	0.01130	88.4804	0.16158	6.18905	5.22555
20	16.4446	0.06081	0.00973	102.777	0.16000	6.24986	5.35957
21	18.9158	0.05287	0.00839	119.221	0.15866	6.30272	5.48237
22	21.7584	0.04596	0.00724	138.137	0.15751	6.34868	5.59471
23	25.0281	0.03996	0.00625	159.896	0.15653	6.38864	5.69731
24	28.7892	0.03474	0.00541	184.924	0.15568	6.42337	5.79087
25	33.1155	0.03020	0.00468	213.713	0.15495	6.45357	5.87608
26	38.0918	0.02625	0.00405	246.828	0.15433	6.47982	5.95356
27	43.8160	0.02282	0.00351	284.920	0.15378	6.50265	6.02392
28	50.4004	0.01984	0.00304	328.736	0.15332	6.52249	6.08772
29	57.9743	0.01725	0.00264	379.137	0.15291	6.53974	6.14552
30	66.6863	0.01500	0.00229	437.111	0.15256	6.55473	6.19780
31	76.7075	0.01304	0.00198	503.797	0.15226	6.56777	6.24505
32	88.2347	0.01133	0.00172	580.505	0.15200	6.57910	6.28769
33	101.494	0.00985	0.00150	668.740	0.15177	6.58895	6.32614
34	116.746	0.00857	0.00130	770.234	0.15157	6.59752	6.36077
35	134.290	0.00745	0.00113	886.980	0.15140	6.60497	6.39193
40	270.426	0.00370	0.00056	1792.90	0.15083	6.62991	6.50606
45	544.572	0.00184	0.00028	3617.21	0.15055	6.64230	6.57173
50	1096.63	0.00091	0.00014	7290.91	0.15041	6.64845	6.60888
55	2208.35	0.00045	0.00007	14689.0	0.15034	6.65151	6.62960

r = 15% **Continuous Compounding, Discrete Cash Flows**

	SINGLE PAYMENT		UNIFORM SERIES				
	Compound Amount Factor	Present Worth Factor	Sinking Fund Factor	Uniform Series Factor	Capital Recovery Factor	Series Present Worth Factor	Arithmetic Gradient Series Factor
N	(*F*/*P*,*r*,*N*)	(*P*/*F*,*r*,*N*)	(*A*/*F*,*r*,*N*)	(*F*/*A*,*r*,*N*)	(*A*/*P*,*r*,*N*)	(*P*/*A*,*r*,*N*)	(*A*/*G*,*r*,*N*)
1	1.1618	0.86071	1.0000	1.0000	1.1618	0.86071	0.00000
2	1.3499	0.74082	0.46257	2.1618	0.62440	1.60153	0.46257
3	1.5683	0.63763	0.28476	3.5117	0.44660	2.23915	0.90037
4	1.8221	0.54881	0.19685	5.0800	0.35868	2.78797	1.31369
5	2.1170	0.47237	0.14488	6.9021	0.30672	3.26033	1.70289
6	2.4596	0.40657	0.11088	9.0191	0.27271	3.66690	2.06846
7	2.8577	0.34994	0.08712	11.4787	0.24895	4.01684	2.41096
8	3.3201	0.30119	0.06975	14.3364	0.23159	4.31803	2.73106
9	3.8574	0.25924	0.05664	17.6565	0.21847	4.57727	3.02947
10	4.4817	0.22313	0.04648	21.5139	0.20832	4.80040	3.30699
11	5.2070	0.19205	0.03847	25.9956	0.20030	4.99245	3.56446
12	6.0496	0.16530	0.03205	31.2026	0.19388	5.15775	3.80276
13	7.0287	0.14227	0.02684	37.2522	0.18868	5.30003	4.02281
14	8.1662	0.12246	0.02258	44.2809	0.18442	5.42248	4.22554
15	9.4877	0.10540	0.01907	52.4471	0.18090	5.52788	4.41191
16	11.0232	0.09072	0.01615	61.9348	0.17798	5.61860	4.58286
17	12.8071	0.07808	0.01371	72.9580	0.17554	5.69668	4.73935
18	14.8797	0.06721	0.01166	85.7651	0.17349	5.76389	4.88231
19	17.2878	0.05784	0.00994	100.645	0.17177	5.82173	5.01264
20	20.0855	0.04979	0.00848	117.933	0.17031	5.87152	5.13125
21	23.3361	0.04285	0.00725	138.018	0.16908	5.91437	5.23898
22	27.1126	0.03688	0.00620	161.354	0.16803	5.95125	5.33666
23	31.5004	0.03175	0.00531	188.467	0.16714	5.98300	5.42507
24	36.5982	0.02732	0.00455	219.967	0.16638	6.01032	5.50497
25	42.5211	0.02352	0.00390	256.565	0.16573	6.03384	5.57706
26	49.4024	0.02024	0.00334	299.087	0.16518	6.05408	5.64200
27	57.3975	0.01742	0.00287	348.489	0.16470	6.07151	5.70042
28	66.6863	0.01500	0.00246	405.886	0.16430	6.08650	5.75289
29	77.4785	0.01291	0.00212	472.573	0.16395	6.09941	5.79997
30	90.0171	0.01111	0.00182	550.051	0.16365	6.11052	5.84215
31	104.585	0.00956	0.00156	640.068	0.16340	6.12008	5.87989
32	121.510	0.00823	0.00134	744.653	0.16318	6.12831	5.91362
33	141.175	0.00708	0.00115	866.164	0.16299	6.13539	5.94374
34	164.022	0.00610	0.00099	1007.34	0.16283	6.14149	5.97060
35	190.566	0.00525	0.00085	1171.36	0.16269	6.14674	5.99453

r = 20% **Continuous Compounding, Discrete Cash Flows**

	SINGLE PAYMENT		UNIFORM SERIES				Arithmetic Gradient Series Factor
	Compound Amount Factor	Present Worth Factor	Sinking Fund Factor	Uniform Series Factor	Capital Recovery Factor	Series Present Worth Factor	
N	(*F*/*P*,*r*,*N*)	(*P*/*F*,*r*,*N*)	(*A*/*F*,*r*,*N*)	(*F*/*A*,*r*,*N*)	(*A*/*P*,*r*,*N*)	(*P*/*A*,*r*,*N*)	(*A*/*G*,*r*,*N*)
1	1.2214	0.81873	1.0000	1.0000	1.2214	0.81873	0.00000
2	1.4918	0.67032	0.45017	2.2214	0.67157	1.48905	0.45017
3	1.8221	0.54881	0.26931	3.7132	0.49071	2.03786	0.86755
4	2.2255	0.44933	0.18066	5.5353	0.40206	2.48719	1.25279
5	2.7183	0.36788	0.12885	7.7609	0.35025	2.85507	1.60677
6	3.3201	0.30119	0.09543	10.4792	0.31683	3.15627	1.93058
7	4.0552	0.24660	0.07247	13.7993	0.29387	3.40286	2.22548
8	4.9530	0.20190	0.05601	17.8545	0.27741	3.60476	2.49289
9	6.0496	0.16530	0.04385	22.8075	0.26525	3.77006	2.73435
10	7.3891	0.13534	0.03465	28.8572	0.25606	3.90539	2.95148
11	9.0250	0.11080	0.02759	36.2462	0.24899	4.01620	3.14594
12	11.0232	0.09072	0.02209	45.2712	0.24349	4.10691	3.31943
13	13.4637	0.07427	0.01776	56.2944	0.23917	4.18119	3.47363
14	16.4446	0.06081	0.01434	69.7581	0.23574	4.24200	3.61019
15	20.0855	0.04979	0.01160	86.2028	0.23300	4.29178	3.73072
16	24.5325	0.04076	0.00941	106.288	0.23081	4.33255	3.83675
17	29.9641	0.03337	0.00764	130.821	0.22905	4.36592	3.92972
18	36.5982	0.02732	0.00622	160.785	0.22762	4.39324	4.01101
19	44.7012	0.02237	0.00507	197.383	0.22647	4.41561	4.08188
20	54.5982	0.01832	0.00413	242.084	0.22553	4.43393	4.14351
21	66.6863	0.01500	0.00337	296.683	0.22477	4.44893	4.19695
22	81.4509	0.01228	0.00275	363.369	0.22415	4.46120	4.24320
23	99.4843	0.01005	0.00225	444.820	0.22365	4.47125	4.28312
24	121.510	0.00823	0.00184	544.304	0.22324	4.47948	4.31750
25	148.413	0.00674	0.00150	665.814	0.22290	4.48622	4.34706
26	181.272	0.00552	0.00123	814.228	0.22263	4.49174	4.37243
27	221.406	0.00452	0.00100	995.500	0.22241	4.49626	4.39415
28	270.426	0.00370	0.00082	1216.91	0.22222	4.49995	4.41273
29	330.300	0.00303	0.00067	1487.33	0.22208	4.50298	4.42859
30	403.429	0.00248	0.00055	1817.63	0.22195	4.50546	4.44211
31	492.749	0.00203	0.00045	2221.06	0.22185	4.50749	4.45362
32	601.845	0.00166	0.00037	2713.81	0.22177	4.50915	4.46340
33	735.095	0.00136	0.00030	3315.66	0.22170	4.51051	4.47170
34	897.847	0.00111	0.00025	4050.75	0.22165	4.51163	4.47874
35	1096.63	0.00091	0.00020	4948.60	0.22160	4.51254	4.48471

r = 25% **Continuous Compounding, Discrete Cash Flows**

	SINGLE PAYMENT		UNIFORM SERIES				Arithmetic Gradient Series Factor
	Compound Amount Factor	Present Worth Factor	Sinking Fund Factor	Uniform Series Factor	Capital Recovery Factor	Series Present Worth Factor	
N	(*F*/*P*,*r*,*N*)	(*P*/*F*,*r*,*N*)	(*A*/*F*,*r*,*N*)	(*F*/*A*,*r*,*N*)	(*A*/*P*,*r*,*N*)	(*P*/*A*,*r*,*N*)	(*A*/*G*,*r*,*N*)
1	1.2840	0.77880	1.0000	1.0000	1.2840	0.77880	0.00000
2	1.6487	0.60653	0.43782	2.2840	0.72185	1.38533	0.43782
3	2.1170	0.47237	0.25428	3.9327	0.53830	1.85770	0.83505
4	2.7183	0.36788	0.16530	6.0497	0.44932	2.22558	1.19290
5	3.4903	0.28650	0.11405	8.7680	0.39808	2.51208	1.51306
6	4.4817	0.22313	0.08158	12.2584	0.36560	2.73521	1.79751
7	5.7546	0.17377	0.05974	16.7401	0.34376	2.90899	2.04855
8	7.3891	0.13534	0.04445	22.4947	0.32848	3.04432	2.26867
9	9.4877	0.10540	0.03346	29.8837	0.31749	3.14972	2.46046
10	12.1825	0.08208	0.02540	39.3715	0.30942	3.23181	2.62656
11	15.6426	0.06393	0.01940	51.5539	0.30342	3.29573	2.76958
12	20.0855	0.04979	0.01488	67.1966	0.29891	3.34552	2.89206
13	25.7903	0.03877	0.01146	87.2821	0.29548	3.38429	2.99641
14	33.1155	0.03020	0.00884	113.072	0.29287	3.41449	3.08488
15	42.5211	0.02352	0.00684	146.188	0.29087	3.43801	3.15955
16	54.5982	0.01832	0.00530	188.709	0.28932	3.45633	3.22229
17	70.1054	0.01426	0.00411	243.307	0.28814	3.47059	3.27481
18	90.0171	0.01111	0.00319	313.413	0.28722	3.48170	3.31860
19	115.584	0.00865	0.00248	403.430	0.28650	3.49035	3.35499
20	148.413	0.00674	0.00193	519.014	0.28595	3.49709	3.38514
21	190.566	0.00525	0.00150	667.427	0.28552	3.50234	3.41003
22	244.692	0.00409	0.00117	857.993	0.28519	3.50642	3.43053
23	314.191	0.00318	0.00091	1102.69	0.28493	3.50961	3.44737
24	403.429	0.00248	0.00071	1416.88	0.28473	3.51208	3.46117
25	518.013	0.00193	0.00055	1820.30	0.28457	3.51401	3.47246
26	665.142	0.00150	0.00043	2338.32	0.28445	3.51552	3.48166
27	854.059	0.00117	0.00033	3003.46	0.28436	3.51669	3.48916
28	1096.63	0.00091	0.00026	3857.52	0.28428	3.51760	3.49526
29	1408.10	0.00071	0.00020	4954.15	0.28423	3.51831	3.50020
30	1808.04	0.00055	0.00016	6362.26	0.28418	3.51886	3.50421
31	2321.57	0.00043	0.00012	8170.30	0.28415	3.51930	3.50745
32	2980.96	0.00034	0.00010	10 492.0	0.28412	3.51963	3.51007
33	3827.63	0.00026	0.00007	13 473.0	0.28410	3.51989	3.51219
34	4914.77	0.00020	0.00006	17 300.0	0.28408	3.52010	3.51389
35	6310.69	0.00016	0.00005	22 215.0	0.28407	3.52025	3.51526

APPENDIX C

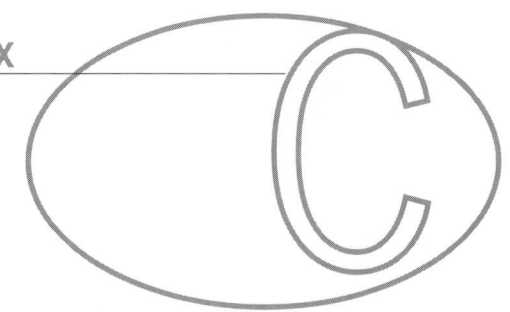

Compound Interest Factors for Continuous Compounding, Continuous Compounding Periods

r = 1% **Continuous Compounding, Continuous Compounding Periods**

T	Sinking Fund Factor (A/F,r,T)	Uniform Series Factor (F/A,r,T)	Capital Recovery Factor (A/P,r,T)	Series Present Worth Factor (P/A,r,T)
1	0.99501	1.0050	1.0050	0.99502
2	0.49502	2.0201	0.50502	1.9801
3	0.32836	3.0455	0.33836	2.9554
4	0.24503	4.0811	0.25503	3.9211
5	0.19504	5.1271	0.20504	4.8771
6	0.16172	6.1837	0.17172	5.8235
7	0.13792	7.2508	0.14792	6.7606
8	0.12007	8.3287	0.13007	7.6884
9	0.10619	9.4174	0.11619	8.6069
10	0.09508	10.5171	0.10508	9.5163
11	0.08600	11.6278	0.09600	10.4166
12	0.07843	12.7497	0.08843	11.3080
13	0.07203	13.8828	0.08203	12.1905
14	0.06655	15.0274	0.07655	13.0642
15	0.06179	16.1834	0.07179	13.9292
16	0.05763	17.3511	0.06763	14.7856
17	0.05397	18.5305	0.06397	15.6335
18	0.05071	19.7217	0.06071	16.4730
19	0.04779	20.9250	0.05779	17.3041
20	0.04517	22.1403	0.05517	18.1269
21	0.04279	23.3678	0.05279	18.9416
22	0.04064	24.6077	0.05064	19.7481
23	0.03867	25.8600	0.04867	20.5466
24	0.03687	27.1249	0.04687	21.3372
25	0.03521	28.4025	0.04521	22.1199
26	0.03368	29.6930	0.04368	22.8948
27	0.03226	30.9964	0.04226	23.6621
28	0.03095	32.3130	0.04095	24.4216
29	0.02972	33.6427	0.03972	25.1736
30	0.02858	34.9859	0.03858	25.9182
31	0.02752	36.3425	0.03752	26.6553
32	0.02652	37.7128	0.03652	27.3851
33	0.02558	39.0968	0.03558	28.1076
34	0.02469	40.4948	0.03469	28.8230
35	0.02386	41.9068	0.03386	29.5312
40	0.02033	49.1825	0.03033	32.9680
45	0.01760	56.8312	0.02760	36.2372
50	0.01541	64.8721	0.02541	39.3469
55	0.01364	73.3253	0.02364	42.3050
60	0.01216	82.2119	0.02216	45.1188
65	0.01092	91.5541	0.02092	47.7954
70	0.00986	101.375	0.01986	50.3415
75	0.00895	111.700	0.01895	52.7633
80	0.00816	122.554	0.01816	55.0671
85	0.00746	133.965	0.01746	57.2585
90	0.00685	145.960	0.01685	59.3430
95	0.00631	158.571	0.01631	61.3259
100	0.00582	171.828	0.01582	63.2121

$r = 2\%$ **Continuous Compounding, Continuous Compounding Periods**

	Sinking Fund Factor	Uniform Series Factor	Capital Recovery Factor	Series Present Worth Factor
T	$(A/F,r,T)$	$(F/A,r,T)$	$(A/P,r,T)$	$(P/A,r,T)$
1	0.99003	1.0101	1.0100	0.99007
2	0.49007	2.0405	0.51007	1.9605
3	0.32343	3.0918	0.34343	2.9118
4	0.24013	4.1644	0.26013	3.8442
5	0.19017	5.2585	0.21017	4.7581
6	0.15687	6.3748	0.17687	5.6540
7	0.13309	7.5137	0.15309	6.5321
8	0.11527	8.6755	0.13527	7.3928
9	0.10141	9.8609	0.12141	8.2365
10	0.09033	11.0701	0.11033	9.0635
11	0.08128	12.3038	0.10128	9.8741
12	0.07373	13.5625	0.09373	10.6686
13	0.06736	14.8465	0.08736	11.4474
14	0.06189	16.1565	0.08189	12.2108
15	0.05717	17.4929	0.07717	12.9591
16	0.05303	18.8564	0.07303	13.6925
17	0.04939	20.2474	0.06939	14.4115
18	0.04615	21.6665	0.06615	15.1162
19	0.04326	23.1142	0.06326	15.8069
20	0.04066	24.5912	0.06066	16.4840
21	0.03832	26.0981	0.05832	17.1477
22	0.03619	27.6354	0.05619	17.7982
23	0.03424	29.2037	0.05424	18.4358
24	0.03246	30.8037	0.05246	19.0608
25	0.03083	32.4361	0.05083	19.6735
26	0.02932	34.1014	0.04932	20.2740
27	0.02793	35.8003	0.04793	20.8626
28	0.02664	37.5336	0.04664	21.4395
29	0.02544	39.3019	0.04544	22.0051
30	0.02433	41.1059	0.04433	22.5594
31	0.02328	42.9464	0.04328	23.1028
32	0.02231	44.8240	0.04231	23.6354
33	0.02140	46.7396	0.04140	24.1574
34	0.02054	48.6939	0.04054	24.6692
35	0.01973	50.6876	0.03973	25.1707
40	0.01632	61.2770	0.03632	27.5336
45	0.01370	72.9802	0.03370	29.6715
50	0.01164	85.9141	0.03164	31.6060
55	0.00998	100.208	0.02998	33.3564
60	0.00862	116.006	0.02862	34.9403
65	0.00749	133.465	0.02749	36.3734
70	0.00655	152.760	0.02655	37.6702
75	0.00574	174.084	0.02574	38.8435
80	0.00506	197.652	0.02506	39.9052
85	0.00447	223.697	0.02447	40.8658
90	0.00396	252.482	0.02396	41.7351
95	0.00352	284.295	0.02352	42.5216
100	0.00313	319.453	0.02313	43.2332

$r = 3\%$ **Continuous Compounding, Continuous Compounding Periods**

	Sinking Fund Factor	Uniform Series Factor	Capital Recovery Factor	Series Present Worth Factor
T	$(A/F,r,T)$	$(F/A,r,T)$	$(A/P,r,T)$	$(P/A,r,T)$
1	0.98507	1.0152	1.0151	0.98515
2	0.48515	2.0612	0.51515	1.9412
3	0.31856	3.1391	0.34856	2.8690
4	0.23530	4.2499	0.26530	3.7693
5	0.18537	5.3945	0.21537	4.6431
6	0.15212	6.5739	0.18212	5.4910
7	0.12838	7.7893	0.15838	6.3139
8	0.11060	9.0416	0.14060	7.1124
9	0.09679	10.3321	0.12679	7.8874
10	0.08575	11.6620	0.11575	8.6394
11	0.07673	13.0323	0.10673	9.3692
12	0.06923	14.4443	0.09923	10.0775
13	0.06290	15.8994	0.09290	10.7648
14	0.05748	17.3987	0.08748	11.4318
15	0.05279	18.9437	0.08279	12.0791
16	0.04870	20.5358	0.07870	12.7072
17	0.04509	22.1764	0.07509	13.3168
18	0.04190	23.8669	0.07190	13.9084
19	0.03905	25.6089	0.06905	14.4825
20	0.03649	27.4040	0.06649	15.0396
21	0.03418	29.2537	0.06418	15.5803
22	0.03209	31.1597	0.06209	16.1050
23	0.03019	33.1239	0.06019	16.6141
24	0.02845	35.1478	0.05845	17.1083
25	0.02686	37.2333	0.05686	17.5878
26	0.02539	39.3824	0.05539	18.0531
27	0.02404	41.5969	0.05404	18.5047
28	0.02279	43.8789	0.05279	18.9430
29	0.02163	46.2304	0.05163	19.3683
30	0.02055	48.6534	0.05055	19.7810
31	0.01955	51.1503	0.04955	20.1815
32	0.01861	53.7232	0.04861	20.5702
33	0.01774	56.3745	0.04774	20.9474
34	0.01692	59.1065	0.04692	21.3135
35	0.01615	61.9217	0.04615	21.6687
40	0.01293	77.3372	0.04293	23.2935
45	0.01050	95.2475	0.04050	24.6920
50	0.00862	116.056	0.03862	25.8957
55	0.00713	140.233	0.03713	26.9317
60	0.00594	168.322	0.03594	27.8234
65	0.00498	200.956	0.03498	28.5909
70	0.00419	238.872	0.03419	29.2515
75	0.00353	282.925	0.03353	29.8200
80	0.00299	334.106	0.03299	30.3094
85	0.00254	393.570	0.03254	30.7306
90	0.00216	462.658	0.03216	31.0931
95	0.00184	542.926	0.03184	31.4052
100	0.00157	636.185	0.03157	31.6738

$r = 4\%$ **Continuous Compounding, Continuous Compounding Periods**

	Sinking Fund Factor	Uniform Series Factor	Capital Recovery Factor	Series Present Worth Factor
T	(*A*/*F*,*r*,*T*)	(*F*/*A*,*r*,*T*)	(*A*/*P*,*r*,*T*)	(*P*/*A*,*r*,*T*)
1	0.98013	1.0203	1.0201	0.98026
2	0.48027	2.0822	0.52027	1.9221
3	0.31373	3.1874	0.35373	2.8270
4	0.23053	4.3378	0.27053	3.6964
5	0.18067	5.5351	0.22067	4.5317
6	0.14747	6.7812	0.18747	5.3343
7	0.12379	8.0782	0.16379	6.1054
8	0.10606	9.4282	0.14606	6.8463
9	0.09231	10.8332	0.13231	7.5581
10	0.08133	12.2956	0.12133	8.2420
11	0.07237	13.8177	0.11237	8.8991
12	0.06493	15.4019	0.10493	9.5304
13	0.05865	17.0507	0.09865	10.1370
14	0.05329	18.7668	0.09329	10.7198
15	0.04865	20.5530	0.08865	11.2797
16	0.04462	22.4120	0.08462	11.8177
17	0.04107	24.3469	0.08107	12.3346
18	0.03794	26.3608	0.07794	12.8312
19	0.03514	28.4569	0.07514	13.3083
20	0.03264	30.6385	0.07264	13.7668
21	0.03039	32.9092	0.07039	14.2072
22	0.02835	35.2725	0.06835	14.6304
23	0.02650	37.7323	0.06650	15.0370
24	0.02482	40.2924	0.06482	15.4277
25	0.02328	42.9570	0.06328	15.8030
26	0.02187	45.7304	0.06187	16.1636
27	0.02057	48.6170	0.06057	16.5101
28	0.01937	51.6214	0.05937	16.8430
29	0.01827	54.7483	0.05827	17.1628
30	0.01724	58.0029	0.05724	17.4701
31	0.01629	61.3903	0.05629	17.7654
32	0.01540	64.9160	0.05540	18.0491
33	0.01458	68.5855	0.05458	18.3216
34	0.01381	72.4048	0.05381	18.5835
35	0.01309	76.3800	0.05309	18.8351
40	0.01012	98.8258	0.05012	19.9526
45	0.00792	126.241	0.04792	20.8675
50	0.00626	159.726	0.04626	21.6166
55	0.00498	200.625	0.04498	22.2299
60	0.00399	250.579	0.04399	22.7321
65	0.00321	311.593	0.04321	23.1432
70	0.00259	386.116	0.04259	23.4797
75	0.00210	477.138	0.04210	23.7553
80	0.00170	588.313	0.04170	23.9809
85	0.00138	724.103	0.04138	24.1657
90	0.00112	889.956	0.04112	24.3169
95	0.00092	1092.53	0.04092	24.4407
100	0.00075	1339.95	0.04075	24.5421

r = 5% **Continuous Compounding, Continuous Compounding Periods**

	Sinking Fund Factor	Uniform Series Factor	Capital Recovery Factor	Series Present Worth Factor
T	(*A*/*F*,*r*,*T*)	(*F*/*A*,*r*,*T*)	(*A*/*P*,*r*,*T*)	(*P*/*A*,*r*,*T*)
1	0.97521	1.0254	1.0252	0.97541
2	0.47542	2.1034	0.52542	1.9033
3	0.30896	3.2367	0.35896	2.7858
4	0.22583	4.4281	0.27583	3.6254
5	0.17604	5.6805	0.22604	4.4240
6	0.14291	6.9972	0.19291	5.1836
7	0.11931	8.3814	0.16931	5.9062
8	0.10166	9.8365	0.15166	6.5936
9	0.08798	11.3662	0.13798	7.2474
10	0.07707	12.9744	0.12707	7.8694
11	0.06819	14.6651	0.11819	8.4610
12	0.06082	16.4424	0.11082	9.0238
13	0.05461	18.3108	0.10461	9.5591
14	0.04932	20.2751	0.09932	10.0683
15	0.04476	22.3400	0.09476	10.5527
16	0.04080	24.5108	0.09080	11.0134
17	0.03732	26.7929	0.08732	11.4517
18	0.03426	29.1921	0.08426	11.8686
19	0.03153	31.7142	0.08153	12.2652
20	0.02910	34.3656	0.07910	12.6424
21	0.02692	37.1530	0.07692	13.0012
22	0.02495	40.0833	0.07495	13.3426
23	0.02317	43.1639	0.07317	13.6673
24	0.02155	46.4023	0.07155	13.9761
25	0.02008	49.8069	0.07008	14.2699
26	0.01873	53.3859	0.06873	14.5494
27	0.01750	57.1485	0.06750	14.8152
28	0.01637	61.1040	0.06637	15.0681
29	0.01532	65.2623	0.06532	15.3086
30	0.01436	69.6338	0.06436	15.5374
31	0.01347	74.2294	0.06347	15.7550
32	0.01265	79.0606	0.06265	15.9621
33	0.01189	84.1396	0.06189	16.1590
34	0.01118	89.4789	0.06118	16.3463
35	0.01052	95.0921	0.06052	16.5245
40	0.00783	127.781	0.05783	17.2933
45	0.00589	169.755	0.05589	17.8920
50	0.00447	223.650	0.05447	18.3583
55	0.00341	292.853	0.05341	18.7214
60	0.00262	381.711	0.05262	19.0043
65	0.00202	495.807	0.05202	19.2245
70	0.00156	642.309	0.05156	19.3961
75	0.00120	830.422	0.05120	19.5296
80	0.00093	1071.96	0.05093	19.6337
85	0.00072	1382.11	0.05072	19.7147
90	0.00056	1780.34	0.05056	19.7778
95	0.00044	2291.69	0.05044	19.8270
100	0.00034	2948.26	0.05034	19.8652

$r = 6\%$ **Continuous Compounding, Continuous Compounding Periods**

	Sinking Fund Factor	Uniform Series Factor	Capital Recovery Factor	Series Present Worth Factor
T	(*A*/*F*,*r*,*T*)	(*F*/*A*,*r*,*T*)	(*A*/*P*,*r*,*T*)	(*P*/*A*,*r*,*T*)
1	0.97030	1.0306	1.0303	0.97059
2	0.47060	2.1249	0.53060	1.8847
3	0.30423	3.2870	0.36423	2.7455
4	0.22120	4.5208	0.28120	3.5562
5	0.17150	5.8310	0.23150	4.3197
6	0.13846	7.2222	0.19846	5.0387
7	0.11495	8.6994	0.17495	5.7159
8	0.09739	10.2679	0.15739	6.3536
9	0.08380	11.9334	0.14380	6.9542
10	0.07298	13.7020	0.13298	7.5198
11	0.06419	15.5799	0.12419	8.0525
12	0.05690	17.5739	0.11690	8.5541
13	0.05078	19.6912	0.11078	9.0266
14	0.04558	21.9394	0.10558	9.4715
15	0.04111	24.3267	0.10111	9.8905
16	0.03723	26.8616	0.09723	10.2851
17	0.03384	29.5532	0.09384	10.6568
18	0.03085	32.4113	0.09085	11.0067
19	0.02821	35.4461	0.08821	11.3363
20	0.02586	38.6686	0.08586	11.6468
21	0.02376	42.0904	0.08376	11.9391
22	0.02187	45.7237	0.08187	12.2144
23	0.02017	49.5817	0.08017	12.4737
24	0.01863	53.6783	0.07863	12.7179
25	0.01723	58.0282	0.07723	12.9478
26	0.01596	62.6470	0.07596	13.1644
27	0.01480	67.5515	0.07480	13.3684
28	0.01374	72.7593	0.07374	13.5604
29	0.01277	78.2891	0.07277	13.7413
30	0.01188	84.1608	0.07188	13.9117
31	0.01106	90.3956	0.07106	14.0721
32	0.01031	97.0160	0.07031	14.2232
33	0.00961	104.046	0.06961	14.3655
34	0.00897	111.510	0.06897	14.4995
35	0.00837	119.436	0.06837	14.6257
40	0.00599	167.053	0.06599	15.1547
45	0.00432	231.329	0.06432	15.5466
50	0.00314	318.092	0.06314	15.8369
55	0.00230	435.211	0.06230	16.0519
60	0.00169	593.304	0.06169	16.2113
65	0.00124	806.707	0.06124	16.3293
70	0.00091	1094.77	0.06091	16.4167
75	0.00067	1483.62	0.06067	16.4815
80	0.00050	2008.51	0.06050	16.5295
85	0.00037	2717.03	0.06037	16.5651
90	0.00027	3673.44	0.06027	16.5914
95	0.00020	4964.46	0.06020	16.6109
100	0.00015	6707.15	0.06015	16.6254

r = 7% **Continuous Compounding, Continuous Compounding Periods**

	Sinking Fund Factor	Uniform Series Factor	Capital Recovery Factor	Series Present Worth Factor
T	(*A*/*F*,*r*,*T*)	(*F*/*A*,*r*,*T*)	(*A*/*P*,*r*,*T*)	(*P*/*A*,*r*,*T*)
1	0.96541	1.0358	1.0354	0.96580
2	0.46582	2.1468	0.53582	1.8663
3	0.29956	3.3383	0.36956	2.7059
4	0.21663	4.6161	0.28663	3.4888
5	0.16704	5.9867	0.23704	4.2187
6	0.13411	7.4566	0.20411	4.8993
7	0.11070	9.0331	0.18070	5.5339
8	0.09325	10.7239	0.16325	6.1256
9	0.07976	12.5373	0.14976	6.6773
10	0.06905	14.4822	0.13905	7.1916
11	0.06036	16.5681	0.13036	7.6712
12	0.05318	18.8052	0.12318	8.1184
13	0.04716	21.2046	0.11716	8.5354
14	0.04206	23.7779	0.11206	8.9241
15	0.03768	26.5379	0.10768	9.2866
16	0.03390	29.4979	0.10390	9.6246
17	0.03061	32.6726	0.10061	9.9397
18	0.02772	36.0774	0.09772	10.2335
19	0.02517	39.7292	0.09517	10.5075
20	0.02291	43.6457	0.09291	10.7629
21	0.02090	47.8462	0.09090	11.0011
22	0.01910	52.3513	0.08910	11.2231
23	0.01749	57.1830	0.08749	11.4302
24	0.01603	62.3651	0.08603	11.6232
25	0.01472	67.9229	0.08472	11.8032
26	0.01353	73.8837	0.08353	11.9711
27	0.01246	80.2767	0.08246	12.1275
28	0.01148	87.1332	0.08148	12.2735
29	0.01058	94.4869	0.08058	12.4095
30	0.00977	102.374	0.07977	12.5363
31	0.00902	110.833	0.07902	12.6546
32	0.00834	119.905	0.07834	12.7649
33	0.00771	129.635	0.07771	12.8677
34	0.00714	140.070	0.07714	12.9636
35	0.00661	151.262	0.07661	13.0529
40	0.00453	220.638	0.07453	13.4170
45	0.00313	319.087	0.07313	13.6735
50	0.00218	458.792	0.07218	13.8543
55	0.00152	657.044	0.07152	13.9817
60	0.00107	938.376	0.07107	14.0715
65	0.00075	1337.61	0.07075	14.1348
70	0.00053	1904.14	0.07053	14.1793
75	0.00037	2708.09	0.07037	14.2107
80	0.00026	3848.95	0.07026	14.2329
85	0.00018	5467.90	0.07018	14.2485
90	0.00013	7765.31	0.07013	14.2595
95	0.00009	11 025.0	0.07009	14.2672
100	0.00006	15 652.0	0.07006	14.2727

r = 8% **Continuous Compounding, Continuous Compounding Periods**

T	Sinking Fund Factor (A/F,r,T)	Uniform Series Factor (F/A,r,T)	Capital Recovery Factor (A/P,r,T)	Series Present Worth Factor (P/A,r,T)
1	0.96053	1.0411	1.0405	0.96105
2	0.46107	2.1689	0.54107	1.8482
3	0.29493	3.3906	0.37493	2.6672
4	0.21213	4.7141	0.29213	3.4231
5	0.16266	6.1478	0.24266	4.1210
6	0.12985	7.7009	0.20985	4.7652
7	0.10657	9.3834	0.18657	5.3599
8	0.08924	11.2060	0.16924	5.9088
9	0.07587	13.1804	0.15587	6.4156
10	0.06528	15.3193	0.14528	6.8834
11	0.05670	17.6362	0.13670	7.3152
12	0.04964	20.1462	0.12964	7.7138
13	0.04373	22.8652	0.12373	8.0818
14	0.03874	25.8107	0.11874	8.4215
15	0.03448	29.0015	0.11448	8.7351
16	0.03081	32.4580	0.11081	9.0245
17	0.02762	36.2024	0.10762	9.2917
18	0.02484	40.2587	0.10484	9.5384
19	0.02240	44.6528	0.10240	9.7661
20	0.02024	49.4129	0.10024	9.9763
21	0.01833	54.5694	0.09833	10.1703
22	0.01662	60.1555	0.09662	10.3494
23	0.01510	66.2067	0.09510	10.5148
24	0.01374	72.7620	0.09374	10.6674
25	0.01252	79.8632	0.09252	10.8083
26	0.01142	87.5559	0.09142	10.9384
27	0.01043	95.8892	0.09043	11.0584
28	0.00953	104.917	0.08953	11.1693
29	0.00872	114.696	0.08872	11.2716
30	0.00798	125.290	0.08798	11.3660
31	0.00731	136.766	0.08731	11.4532
32	0.00670	149.198	0.08670	11.5337
33	0.00615	162.665	0.08615	11.6080
34	0.00564	177.254	0.08564	11.6766
35	0.00518	193.058	0.08518	11.7399
40	0.00340	294.157	0.08340	11.9905
45	0.00225	444.978	0.08225	12.1585
50	0.00149	669.977	0.08149	12.2711
55	0.00099	1005.64	0.08099	12.3465
60	0.00066	1506.38	0.08066	12.3971
65	0.00044	2253.40	0.08044	12.4310
70	0.00030	3367.83	0.08030	12.4538
75	0.00020	5030.36	0.08020	12.4690
80	0.00013	7510.56	0.08013	12.4792
85	0.00009	11 211.0	0.08009	12.4861
90	0.00006	16 730.0	0.08006	12.4907
95	0.00004	24 965.0	0.08004	12.4937
100	0.00003	37 249.0	0.08003	12.4958

r = 9% **Continuous Compounding, Continuous Compounding Periods**

	Sinking Fund Factor	Uniform Series Factor	Capital Recovery Factor	Series Present Worth Factor
T	(*A*/*F*,*r*,*T*)	(*F*/*A*,*r*,*T*)	(*A*/*P*,*r*,*T*)	(*P*/*A*,*r*,*T*)
1	0.95567	1.0464	1.0457	0.95632
2	0.45635	2.1913	0.54635	1.8303
3	0.29036	3.4440	0.38036	2.6291
4	0.20769	4.8148	0.29769	3.3592
5	0.15836	6.3146	0.24836	4.0264
6	0.12570	7.9556	0.21570	4.6361
7	0.10255	9.7512	0.19255	5.1934
8	0.08535	11.7159	0.17535	5.7028
9	0.07212	13.8656	0.16212	6.1682
10	0.06166	16.2178	0.15166	6.5937
11	0.05322	18.7915	0.14322	6.9825
12	0.04628	21.6076	0.13628	7.3378
13	0.04050	24.6888	0.13050	7.6626
14	0.03564	28.0602	0.12564	7.9594
15	0.03150	31.7492	0.12150	8.2307
16	0.02794	35.7855	0.11794	8.4786
17	0.02487	40.2020	0.11487	8.7052
18	0.02221	45.0343	0.11221	8.9122
19	0.01987	50.3218	0.10987	9.1015
20	0.01782	56.1072	0.10782	9.2745
21	0.01602	62.4374	0.10602	9.4325
22	0.01442	69.3638	0.10442	9.5770
23	0.01300	76.9425	0.10300	9.7090
24	0.01173	85.2349	0.10173	9.8297
25	0.01060	94.3082	0.10060	9.9400
26	0.00959	104.236	0.09959	10.0408
27	0.00869	115.099	0.09869	10.1329
28	0.00787	126.984	0.09787	10.2171
29	0.00714	139.989	0.09714	10.2941
30	0.00648	154.219	0.09648	10.3644
31	0.00589	169.789	0.09589	10.4287
32	0.00535	186.825	0.09535	10.4874
33	0.00487	205.466	0.09487	10.5411
34	0.00443	225.862	0.09443	10.5901
35	0.00403	248.178	0.09403	10.6350
40	0.00253	395.536	0.09253	10.8075
45	0.00160	626.638	0.09160	10.9175
50	0.00101	989.079	0.09101	10.9877
55	0.00064	1557.50	0.09064	11.0324
60	0.00041	2448.96	0.09041	11.0609
65	0.00026	3847.05	0.09026	11.0791
70	0.00017	6039.69	0.09017	11.0907
75	0.00011	9478.43	0.09011	11.0981
80	0.00007	14 871.0	0.09007	11.1028
85	0.00004	23 329.0	0.09004	11.1058
90	0.00003	36 594.0	0.09003	11.1077
95	0.00002	57 397.0	0.09002	11.1090
100	0.00001	90 023.0	0.09001	11.1097

$r = 10\%$ **Continuous Compounding, Continuous Compounding Periods**

	Sinking Fund Factor	Uniform Series Factor	Capital Recovery Factor	Series Present Worth Factor
T	(*A*/*F*,*r*,*T*)	(*F*/*A*,*r*,*T*)	(*A*/*P*,*r*,*T*)	(*P*/*A*,*r*,*T*)
1	0.95083	1.0517	1.0508	0.95163
2	0.45167	2.2140	0.55167	1.8127
3	0.28583	3.4986	0.38583	2.5918
4	0.20332	4.9182	0.30332	3.2968
5	0.15415	6.4872	0.25415	3.9347
6	0.12164	8.2212	0.22164	4.5119
7	0.09864	10.1375	0.19864	5.0341
8	0.08160	12.2554	0.18160	5.5067
9	0.06851	14.5960	0.16851	5.9343
10	0.05820	17.1828	0.15820	6.3212
11	0.04990	20.0417	0.14990	6.6713
12	0.04310	23.2012	0.14310	6.9881
13	0.03746	26.6930	0.13746	7.2747
14	0.03273	30.5520	0.13273	7.5340
15	0.02872	34.8169	0.12872	7.7687
16	0.02530	39.5303	0.12530	7.9810
17	0.02235	44.7395	0.12235	8.1732
18	0.01980	50.4965	0.11980	8.3470
19	0.01759	56.8589	0.11759	8.5043
20	0.01565	63.8906	0.11565	8.6466
21	0.01395	71.6617	0.11395	8.7754
22	0.01246	80.2501	0.11246	8.8920
23	0.01114	89.7418	0.11114	8.9974
24	0.00998	100.232	0.10998	9.0928
25	0.00894	111.825	0.10894	9.1792
26	0.00802	124.637	0.10802	9.2573
27	0.00720	138.797	0.10720	9.3279
28	0.00647	154.446	0.10647	9.3919
29	0.00582	171.741	0.10582	9.4498
30	0.00524	190.855	0.10524	9.5021
31	0.00472	211.980	0.10472	9.5495
32	0.00425	235.325	0.10425	9.5924
33	0.00383	261.126	0.10383	9.6312
34	0.00345	289.641	0.10345	9.6663
35	0.00311	321.155	0.10311	9.6980
40	0.00187	535.982	0.10187	9.8168
45	0.00112	890.171	0.10112	9.8889
50	0.00068	1474.13	0.10068	9.9326
55	0.00041	2436.92	0.10041	9.9591
60	0.00025	4024.29	0.10025	9.9752
65	0.00015	6641.42	0.10015	9.9850
70	0.00009	10 956.0	0.10009	9.9909
75	0.00006	18 070.0	0.10006	9.9945
80	0.00003	29 800.0	0.10003	9.9966
85	0.00002	49 138.0	0.10002	9.9980
90	0.00001	81 021.0	0.10001	9.9988
95	0.00001	133 587.0	0.10001	9.9993
100	0.00000	220 255.0	0.10000	9.9995

$r = 11\%$ **Continuous Compounding, Continuous Compounding Periods**

	Sinking Fund Factor	Uniform Series Factor	Capital Recovery Factor	Series Present Worth Factor
T	(*A*/*F*,*r*,*T*)	(*F*/*A*,*r*,*T*)	(*A*/*P*,*r*,*T*)	(*P*/*A*,*r*,*T*)
1	0.94601	1.0571	1.0560	0.94696
2	0.44702	2.2371	0.55702	1.7953
3	0.28135	3.5543	0.39135	2.5552
4	0.19902	5.0246	0.30902	3.2360
5	0.15002	6.6659	0.26002	3.8459
6	0.11767	8.4981	0.22767	4.3923
7	0.09485	10.5433	0.20485	4.8817
8	0.07796	12.8264	0.18796	5.3202
9	0.06504	15.3749	0.17504	5.7129
10	0.05489	18.2197	0.16489	6.0648
11	0.04674	21.3953	0.15674	6.3800
12	0.04010	24.9402	0.15010	6.6624
13	0.03461	28.8973	0.14461	6.9154
14	0.03002	33.3145	0.14002	7.1420
15	0.02615	38.2453	0.13615	7.3450
16	0.02286	43.7494	0.13286	7.5269
17	0.02004	49.8936	0.13004	7.6898
18	0.01762	56.7522	0.12762	7.8357
19	0.01553	64.4083	0.12553	7.9665
20	0.01371	72.9547	0.12371	8.0836
21	0.01212	82.4948	0.12212	8.1885
22	0.01074	93.1442	0.12074	8.2825
23	0.00952	105.032	0.11952	8.3667
24	0.00845	118.302	0.11845	8.4422
25	0.00751	133.115	0.11751	8.5097
26	0.00668	149.650	0.11668	8.5703
27	0.00595	168.108	0.11595	8.6245
28	0.00530	188.713	0.11530	8.6731
29	0.00472	211.713	0.11472	8.7166
30	0.00421	237.388	0.11421	8.7556
31	0.00376	266.048	0.11376	8.7905
32	0.00336	298.040	0.11336	8.8218
33	0.00300	333.753	0.11300	8.8499
34	0.00268	373.618	0.11268	8.8750
35	0.00239	418.119	0.11239	8.8975
40	0.00137	731.372	0.11137	8.9793
45	0.00078	1274.32	0.11078	9.0265
50	0.00045	2215.38	0.11045	9.0538
55	0.00026	3846.48	0.11026	9.0695
60	0.00015	6 674.0	0.11015	9.0785
65	0.00009	11 574.0	0.11009	9.0838
70	0.00005	20 067.0	0.11005	9.0868
75	0.00003	34 788.0	0.11003	9.0885
80	0.00002	60302.218	0.11002	9.0895
85	0.00001	104525.668	0.11001	9.0901
90	0.00001	181176.095	0.11001	9.0905
95	0.00000	314030.679	0.11000	9.0906
100	0.00000	544301.288	0.11000	9.0908

r = 12% **Continuous Compounding, Continuous Compounding Periods**

	Sinking Fund Factor	Uniform Series Factor	Capital Recovery Factor	Series Present Worth Factor
T	(*A*/*F*,*r*,*T*)	(*F*/*A*,*r*,*T*)	(*A*/*P*,*r*,*T*)	(*P*/*A*,*r*,*T*)
1	0.94120	1.0625	1.0612	0.94233
2	0.44240	2.2604	0.56240	1.7781
3	0.27693	3.6111	0.39693	2.5194
4	0.19478	5.1340	0.31478	3.1768
5	0.14596	6.8510	0.26596	3.7599
6	0.11381	8.7869	0.23381	4.2771
7	0.09116	10.9697	0.21116	4.7357
8	0.07446	13.4308	0.19446	5.1426
9	0.06171	16.2057	0.18171	5.5034
10	0.05172	19.3343	0.17172	5.8234
11	0.04374	22.8618	0.16374	6.1072
12	0.03726	26.8391	0.15726	6.3589
13	0.03192	31.3235	0.15192	6.5822
14	0.02749	36.3796	0.14749	6.7802
15	0.02376	42.0804	0.14376	6.9558
16	0.02062	48.5080	0.14062	7.1116
17	0.01794	55.7551	0.13794	7.2498
18	0.01564	63.9261	0.13564	7.3723
19	0.01367	73.1390	0.13367	7.4810
20	0.01197	83.5265	0.13197	7.5774
21	0.01050	95.2383	0.13050	7.6628
22	0.00922	108.443	0.12922	7.7387
23	0.00811	123.332	0.12811	7.8059
24	0.00714	140.119	0.12714	7.8655
25	0.00629	159.046	0.12629	7.9184
26	0.00554	180.386	0.12554	7.9654
27	0.00489	204.448	0.12489	8.0070
28	0.00432	231.577	0.12432	8.0439
29	0.00381	262.164	0.12381	8.0766
30	0.00337	296.652	0.12337	8.1056
31	0.00298	335.537	0.12298	8.1314
32	0.00264	379.379	0.12264	8.1542
33	0.00233	428.811	0.12233	8.1745
34	0.00206	484.546	0.12206	8.1924
35	0.00183	547.386	0.12183	8.2084
40	0.00100	1004.25	0.12100	8.2648
45	0.00054	1836.72	0.12054	8.2957
50	0.00030	3353.57	0.12030	8.3127
55	0.00016	6117.46	0.12016	8.3220
60	0.00009	11 154.0	0.12009	8.3271
65	0.00005	20 330.0	0.12005	8.3299
70	0.00003	37 051.0	0.12003	8.3315
75	0.00001	67 517.0	0.12001	8.3323
80	0.00001	123 032.0	0.12001	8.3328
85	0.00000	224 185.0	0.12000	8.3330
90	0.00000	408 498.0	0.12000	8.3332
95	0.00000	744 339.0	0.12000	8.3332
100	0.00000	1 356 282.0	0.12000	8.3333

$r = 13\%$ **Continuous Compounding, Continuous Compounding Periods**

	Sinking Fund Factor	Uniform Series Factor	Capital Recovery Factor	Series Present Worth Factor
T	*(A/F,r,T)*	*(F/A,r,T)*	*(A/P,r,T)*	*(P/A,r,T)*
1	0.93641	1.0679	1.0664	0.93773
2	0.43781	2.2841	0.56781	1.7611
3	0.27255	3.6691	0.40255	2.4842
4	0.19061	5.2464	0.32061	3.1191
5	0.14199	7.0426	0.27199	3.6766
6	0.11003	9.0882	0.24003	4.1661
7	0.08758	11.4179	0.21758	4.5960
8	0.07107	14.0709	0.20107	4.9734
9	0.05851	17.0923	0.18851	5.3049
10	0.04870	20.5331	0.17870	5.5959
11	0.04090	24.4515	0.17090	5.8515
12	0.03459	28.9140	0.16459	6.0759
13	0.02942	33.9960	0.15942	6.2729
14	0.02514	39.7835	0.15514	6.4460
15	0.02156	46.3745	0.15156	6.5979
16	0.01856	53.8805	0.14856	6.7313
17	0.01602	62.4286	0.14602	6.8485
18	0.01386	72.1634	0.14386	6.9513
19	0.01201	83.2496	0.14201	7.0417
20	0.01043	95.8749	0.14043	7.1210
21	0.00907	110.253	0.13907	7.1906
22	0.00790	126.627	0.13790	7.2518
23	0.00688	145.274	0.13688	7.3055
24	0.00601	166.511	0.13601	7.3526
25	0.00524	190.695	0.13524	7.3940
26	0.00458	218.237	0.13458	7.4304
27	0.00401	249.602	0.13401	7.4623
28	0.00350	285.322	0.13350	7.4904
29	0.00307	326.000	0.13307	7.5150
30	0.00269	372.327	0.13269	7.5366
31	0.00235	425.084	0.13235	7.5556
32	0.00206	485.166	0.13206	7.5722
33	0.00181	553.588	0.13181	7.5869
34	0.00158	631.510	0.13158	7.5997
35	0.00139	720.249	0.13139	7.6110
40	0.00072	1386.71	0.13072	7.6499
45	0.00038	2663.34	0.13038	7.6702
50	0.00020	5108.78	0.13020	7.6807
55	0.00010	9793.12	0.13010	7.6863
60	0.00005	18 766.0	0.13005	7.6892
65	0.00003	35 954.0	0.13003	7.6907
70	0.00001	68 879.0	0.13001	7.6914
75	0.00001	131 948.0	0.13001	7.6919
80	0.00000	252 759.0	0.13000	7.6921
85	0.00000	484 177.0	0.13000	7.6922
90	0.00000	927 467.0	0.13000	7.6922
95	0.00000	1 776 608.0	0.13000	7.6923
100	0.00000	3 403 172.0	0.13000	7.6923

r = 14% **Continuous Compounding, Continuous Compounding Periods**

T	Sinking Fund Factor (A/F,r,T)	Uniform Series Factor (F/A,r,T)	Capital Recovery Factor (A/P,r,T)	Series Present Worth Factor (P/A,r,T)
1	0.93163	1.0734	1.0716	0.93316
2	0.43326	2.3081	0.57326	1.7444
3	0.26822	3.7283	0.40822	2.4497
4	0.18650	5.3619	0.32650	3.0628
5	0.13810	7.2411	0.27810	3.5958
6	0.10635	9.4026	0.24635	4.0592
7	0.08411	11.8890	0.22411	4.4621
8	0.06780	14.7490	0.20780	4.8123
9	0.05544	18.0387	0.19544	5.1168
10	0.04582	21.8229	0.18582	5.3815
11	0.03820	26.1756	0.17820	5.6116
12	0.03207	31.1825	0.17207	5.8116
13	0.02707	36.9418	0.16707	5.9855
14	0.02295	43.5666	0.16295	6.1367
15	0.01954	51.1869	0.15954	6.2682
16	0.01668	59.9524	0.15668	6.3824
17	0.01428	70.0350	0.15428	6.4818
18	0.01225	81.6328	0.15225	6.5681
19	0.01053	94.9735	0.15053	6.6432
20	0.00906	110.319	0.14906	6.7085
21	0.00781	127.970	0.14781	6.7652
22	0.00674	148.274	0.14674	6.8146
23	0.00583	171.629	0.14583	6.8575
24	0.00504	198.494	0.14504	6.8947
25	0.00436	229.396	0.14436	6.9272
26	0.00377	264.942	0.14377	6.9553
27	0.00327	305.829	0.14327	6.9798
28	0.00283	352.860	0.14283	7.0011
29	0.00246	406.959	0.14246	7.0196
30	0.00213	469.188	0.14213	7.0357
31	0.00185	540.768	0.14185	7.0497
32	0.00160	623.105	0.14160	7.0619
33	0.00139	717.815	0.14139	7.0725
34	0.00121	826.757	0.14121	7.0817
35	0.00105	952.070	0.14105	7.0897
40	0.00052	1924.47	0.14052	7.1164
45	0.00026	3882.66	0.14026	7.1297
50	0.00013	7825.95	0.14013	7.1363
55	0.00006	15 767.0	0.14006	7.1396
60	0.00003	31 758.0	0.14003	7.1413
65	0.00002	63 959.0	0.14002	7.1421
70	0.00001	128 805.0	0.14001	7.1425
75	0.00000	259 389.0	0.14000	7.1427
80	0.00000	522 353.0	0.14000	7.1428
85	0.00000	1 051 897.0	0.14000	7.1428
90	0.00000	2 118 268.0	0.14000	7.1428
95	0.00000	4 265 676.0	0.14000	7.1428
100	0.00000	8 590 023.0	0.14000	7.1429

$r = 15\%$ **Continuous Compounding, Continuous Compounding Periods**

	Sinking Fund Factor	Uniform Series Factor	Capital Recovery Factor	Series Present Worth Factor
T	(*A/F,r,T*)	(*F/A,r,T*)	(*A/P,r,T*)	(*P/A,r,T*)
1	0.92687	1.0789	1.0769	0.92861
2	0.42874	2.3324	0.57874	1.7279
3	0.26394	3.7887	0.41394	2.4158
4	0.18246	5.4808	0.33246	3.0079
5	0.13429	7.4467	0.28429	3.5176
6	0.10277	9.7307	0.25277	3.9562
7	0.08075	12.3843	0.23075	4.3337
8	0.06465	15.4674	0.21465	4.6587
9	0.05249	19.0495	0.20249	4.9384
10	0.04308	23.2113	0.19308	5.1791
11	0.03566	28.0465	0.18566	5.3863
12	0.02971	33.6643	0.17971	5.5647
13	0.02488	40.1913	0.17488	5.7182
14	0.02093	47.7745	0.17093	5.8503
15	0.01767	56.5849	0.16767	5.9640
16	0.01497	66.8212	0.16497	6.0619
17	0.01270	78.7140	0.16270	6.1461
18	0.01081	92.5315	0.16081	6.2186
19	0.00921	108.585	0.15921	6.2810
20	0.00786	127.237	0.15786	6.3348
21	0.00672	148.907	0.15672	6.3810
22	0.00574	174.084	0.15574	6.4208
23	0.00492	203.336	0.15492	6.4550
24	0.00421	237.322	0.15421	6.4845
25	0.00361	276.807	0.15361	6.5099
26	0.00310	322.683	0.15310	6.5317
27	0.00266	375.983	0.15266	6.5505
28	0.00228	437.909	0.15228	6.5667
29	0.00196	509.856	0.15196	6.5806
30	0.00169	593.448	0.15169	6.5926
31	0.00145	690.567	0.15145	6.6029
32	0.00124	803.403	0.15124	6.6118
33	0.00107	934.500	0.15107	6.6194
34	0.00092	1086.81	0.15092	6.6260
35	0.00079	1263.78	0.15079	6.6317
40	0.00037	2682.86	0.15037	6.6501
45	0.00018	5687.06	0.15018	6.6589
50	0.00008	12 047.0	0.15008	6.6630
55	0.00004	25 511.0	0.15004	6.6649
60	0.00002	54 014.0	0.15002	6.6658
65	0.00001	114 355.0	0.15001	6.6663
70	0.00000	242 097.0	0.15000	6.6665
75	0.00000	512 526.0	0.15000	6.6666

r = 20% **Continuous Compounding, Continuous Compounding Periods**

	Sinking Fund Factor	Uniform Series Factor	Capital Recovery Factor	Series Present Worth Factor
T	(*A*/*F*,*r*,*T*)	(*F*/*A*,*r*,*T*)	(*A*/*P*,*r*,*T*)	(*P*/*A*,*r*,*T*)
1	0.90333	1.1070	1.1033	0.90635
2	0.40665	2.4591	0.60665	1.6484
3	0.24327	4.1106	0.44327	2.2559
4	0.16319	6.1277	0.36319	2.7534
5	0.11640	8.5914	0.31640	3.1606
6	0.08620	11.6006	0.28620	3.4940
7	0.06546	15.2760	0.26546	3.7670
8	0.05059	19.7652	0.25059	3.9905
9	0.03961	25.2482	0.23961	4.1735
10	0.03130	31.9453	0.23130	4.3233
11	0.02492	40.1251	0.22492	4.4460
12	0.01995	50.1159	0.21995	4.5464
13	0.01605	62.3187	0.21605	4.6286
14	0.01295	77.2232	0.21295	4.6959
15	0.01048	95.4277	0.21048	4.7511
16	0.00850	117.663	0.20850	4.7962
17	0.00691	144.821	0.20691	4.8331
18	0.00562	177.991	0.20562	4.8634
19	0.00458	218.506	0.20458	4.8881
20	0.00373	267.991	0.20373	4.9084
21	0.00304	328.432	0.20304	4.9250
22	0.00249	402.254	0.20249	4.9386
23	0.00203	492.422	0.20203	4.9497
24	0.00166	602.552	0.20166	4.9589
25	0.00136	737.066	0.20136	4.9663
26	0.00111	901.361	0.20111	4.9724
27	0.00091	1102.03	0.20091	4.9774
28	0.00074	1347.13	0.20074	4.9815
29	0.00061	1646.50	0.20061	4.9849
30	0.00050	2012.14	0.20050	4.9876
31	0.00041	2458.75	0.20041	4.9899
32	0.00033	3004.23	0.20033	4.9917
33	0.00027	3670.48	0.20027	4.9932
34	0.00022	4484.24	0.20022	4.9944
35	0.00018	5478.17	0.20018	4.9954
40	0.00007	14 900.0	0.20007	4.9983
45	0.00002	40 510.0	0.20002	4.9994
50	0.00001	110 127.0	0.20001	4.9998
55	0.00000	299 366.0	0.20000	4.9999
60	0.00000	813 769.0	0.20000	5.0000
65	0.00000	2 212 062.0	0.20000	5.0000
70	0.00000	6 013 016.0	0.20000	5.0000
75	0.00000	16 345 082.0	0.20000	5.0000

r = 25% **Continuous Compounding, Continuous Compounding Periods**

	Sinking Fund Factor	Uniform Series Factor	Capital Recovery Factor	Series Present Worth Factor
T	(*A*/*F*,*r*,*T*)	(*F*/*A*,*r*,*T*)	(*A*/*P*,*r*,*T*)	(*P*/*A*,*r*,*T*)
1	0.88020	1.1361	1.1302	0.88480
2	0.38537	2.5949	0.63537	1.5739
3	0.22381	4.4680	0.47381	2.1105
4	0.14549	6.8731	0.39549	2.5285
5	0.10039	9.9614	0.35039	2.8540
6	0.07180	13.9268	0.32180	3.1075
7	0.05258	19.0184	0.30258	3.3049
8	0.03913	25.5562	0.28913	3.4587
9	0.02945	33.9509	0.27945	3.5784
10	0.02236	44.7300	0.27236	3.6717
11	0.01707	58.5705	0.26707	3.7443
12	0.01310	76.3421	0.26310	3.8009
13	0.01008	99.1614	0.26008	3.8449
14	0.00778	128.462	0.25778	3.8792
15	0.00602	166.084	0.25602	3.9059
16	0.00466	214.393	0.25466	3.9267
17	0.00362	276.422	0.25362	3.9429
18	0.00281	356.069	0.25281	3.9556
19	0.00218	458.337	0.25218	3.9654
20	0.00170	589.653	0.25170	3.9730
21	0.00132	758.265	0.25132	3.9790
22	0.00103	974.768	0.25103	3.9837
23	0.00080	1252.76	0.25080	3.9873
24	0.00062	1609.72	0.25062	3.9901
25	0.00048	2068.05	0.25048	3.9923
26	0.00038	2656.57	0.25038	3.9940
27	0.00029	3412.24	0.25029	3.9953
28	0.00023	4382.53	0.25023	3.9964
29	0.00018	5628.42	0.25018	3.9972
30	0.00014	7228.17	0.25014	3.9978
31	0.00011	9282.29	0.25011	3.9983
32	0.00008	11 920.0	0.25008	3.9987
33	0.00007	15 307.0	0.25007	3.9990
34	0.00005	19 655.0	0.25005	3.9992
35	0.00004	25 239.0	0.25004	3.9994
40	0.00001	88 102.0	0.25001	3.9998
45	0.00000	307 516.0	0.25000	3.9999
50	0.00000	1 073 345.0	0.25000	4.0000
55	0.00000	3 746 353.0	0.25000	4.0000

APPENDIX D

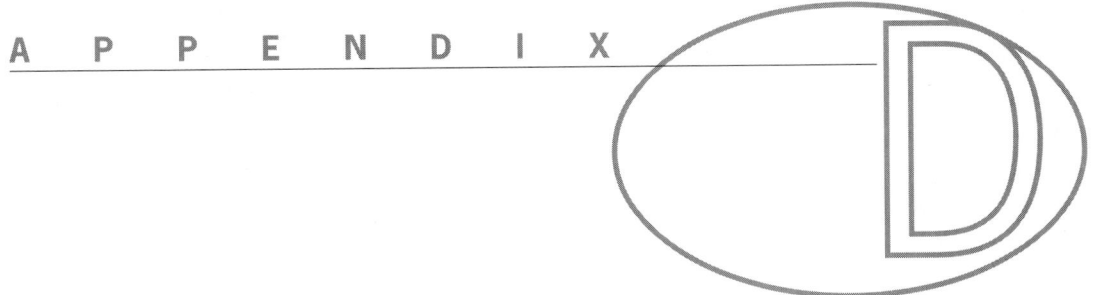

Answers to Selected Problems

CHAPTER 1

CHAPTER 2

2.1	$120
2.3	$5000
2.5	8%
2.7 (a)	$1210
2.9	18.75%
2.11	$29 719
2.13 (a)	5 years
2.15 (a)	$6728
2.17 (a)	26.6%
2.19	5%
2.21	$665 270
2.23	Brand 2
2.25	$i_{e \text{ continuous}} = 8.32\%$
2.27	$i_{e \text{ weekly}} = 5.65\%$
2.29	0.5%
2.31	$2140
2.37	Decisional equivalence holds
2.43 (b)	Lost $60 by locking in

CHAPTER 3

3.3	$317.22
3.5	20.3%
3.7	$74 790
3.9	$18 466 per year
3.11	$94.13
3.13	$2664
3.15	11.7%
3.17	5.8 years
3.19 (b)	$26.44
3.21	162.5 MW
3.23	$3.98
3.25	$34 616
3.27	$3339
3.31	$3 600 000
3.33	$8 013 000
3.35	$74 500
3.37	No
3.39	$86 million
3.41	$122 316
3.43	27 months
3.45	$21 098
3.47	$3 086 200
3.49	$257 143
3.51	No more than $504
3.53	Up to $4587
3A.1	$1701
3A.3	$8353

CHAPTER 4

4.1	BC and CD
4.5	A, AB, BC, ABC, and BCD
4.7 (a)	−$2164
4.9 (a)	18%
4.11	Second offer, PW = $137 000
4.13	Earthen dam, PW = $396 000
4.15	No
4.17	Hydraulic press, AW = $24 716
4.19	Plastic liner, PW = $1 100 000
4.21	XJ3, PW = −$6565

4.23	No
4.25	3.6 years
4.27	20%
4.29	$134
4.31 (a)	T, PW = $96 664
4.33	T, AW = $26 154
4.35 (a)	Only B
4.39	Landfill site, AW = $125 351
4.43	Curtains, payback period = 1.67 years

CHAPTER 5

5.1 (a)	9.2%
5.3	12.4%
5.5	B and D
5.7 (b)	7.6%
5.9	New machine
5.11	21.7%
5.13	3.9%
5.15	Approximate ERR > 30%
5.17 (c)	Clip Job
5.19	Used refrigerator
5.21 (b)	2.46%
5.23 (b)	2
5.25	E
5.27	$119
5.29	A and C
5.31	B
5.33	Payback period
5.35	IRR or present worth
5.37	Payback period
5.39	Annual worth
5A.1 (d)	12.6%, No
5A.3 (e)	21.3%, No

CHAPTER 6

6.1 (a)	Functional loss
6.3	Use-related loss
6.5	Time-related loss
6.7 (b)	$7714
6.9 (b)	$5806
6.13	12%
6.15	A, PW = −$22 321
6.17	$22 059, $14 706, $7353, 0
6.19	$4471
6.21	$1834
6.25	Current assets, fixed assets, depreciation, income taxes, net income (operations), and net income
6.27	Equity ratio = 0.296

CHAPTER 7

7.5 (c)	6 years
7.7	Every 3 years
7.11 (d)	Move robot immediately
7.13 (a)	14 years
7.15 (b)	No
7.17 (c)	Immediately
7.19 (a)	10 feasible alternatives
7.21 (a)	Additional 5 years
7.23	Every 8 years
7.25	No
7.27 (a)	Less than $10 344
7.29	Overhaul in 5 years
7.31 (a)	8 feasible alternatives

7.33 (a) A2 12 years, B2 12 years

7.35 (a) A4 9 years, B4 15 years

7.37 $184 397

CHAPTER 8

8.1 (a) Class 10, 30%

8.3 $IRR_{before\text{-}tax} = 18.5\%$

8.5 $19 250

8.7 $IRR_{after\text{-}tax} = 9.3\%$

8.9 $26 779

8.11 $51 929

8.15 6.23%, 6.72%

8.17 Leasing

8.19 $10 242

8.21 $280 588

8.23 Annual cost of $1647

8.25 $IRR_{after\text{-}tax} = 3.7\%$, $MARR_{after\text{-}tax} = 6.0\%$

8.27 $4189

8.29 $10 737 for A, $6040 for B

8.31 Alternative 2

8.33 T, PW = $9974

8.35 20.7%, 15.1%, 11.7%

8.37 T

8.39 $IRR_{after\text{-}tax} = 8.5\%$

CHAPTER 9

9.1 (a) Actual

9.3 (a) $292

9.5 (a) 14.5%

9.7 (b) −2.9%

9.9 67 years

9.11 14.3%

9.13 $14 683

9.15 11 216 million rubles

9.17 (a) 7.75%

9.21 −$125 532

9.23 (a) 16%

9.25 (b) $243 547

9.27 (c) 2.13%

9.29 (b) −$18 801, $998

CHAPTER 10

10.1 1.55, 2

10.3 (c) 0

10.5 (a) $1599 per day

10.11 $212 500

10.15 (a) $1 201 200 per year

10.17 (d) $1333 per day

10.19 (a) River route

10.23 (b) 1.07

10.25 (a) 1.08, 1.1

10.27 (a) 0.75

10.29 (a) Tennis courts, BCR = 1.18

10.31 Yes, BCR = 1.66

10.37 (c) $34 510

10.39 (a) 2 lanes

CHAPTER 11

11.3 (a) Break-even analysis for multiple projects

11.7 Most sensitive to the first cost

11.9 (a) Most sensitive to the first cost

11.11 (a) 179 deliveries per month

11.13 (a) \$141 045, outside of the likely values

11.15 (b) 47 375 boards per year

11.17 \$17 535

11.19 (a) \$4146

11.19 (b) Break-even operating hours = 671

11.21 Break-even maintenance cost \$61.50

11.23 Break-even interest rate 11.2%

11.25 Lease if tax rate is below 27%

CHAPTER 12

12.1 \$855

12.3 X1000

12.5 \$23 367

12.7 \$9074

12.9 E(buy ticket) = 150

12.11 (c) E(send stock) = 4000

12.13 E(combined rate) = 15.9%

12.15 E(partnership) = \$113 750

12.17 (a) AW(optimistic, public accept) = \$925 000

12.17 (b) E(new product) ≅ \$107 975

12.21 (c) Average PW ≅ −\$160 000

12.23 Average PW ≅ −\$5600

12.25 Average PW ≅ −\$20 000

12.27 Average PW ≅ −\$15 700

CHAPTER 13

13.1 4, 5, 6, 7, 8, and 10

13.3 (b) L4

13.5 (a)
$$\begin{bmatrix} 1 & 1 & 5 & \frac{1}{3} \\ 1 & 1 & \frac{1}{3} & 2 \\ \frac{1}{5} & 3 & 1 & 2 \\ 3 & \frac{1}{2} & \frac{1}{2} & 1 \end{bmatrix}$$

13.7 (a) 1, 3, 7, 8

13.9 D, 57.5

13.11
$$\begin{bmatrix} 1 & 1 & \frac{1}{3} & 5 & \frac{1}{5} \\ 1 & 1 & \frac{1}{3} & 5 & \frac{1}{5} \\ 3 & 3 & 1 & 5 & \frac{1}{3} \\ \frac{1}{5} & \frac{1}{5} & \frac{1}{5} & 1 & \frac{1}{5} \\ 5 & 5 & 3 & 5 & 1 \end{bmatrix}$$

13.17 $[0.13 \quad 0.13 \quad 0.56 \quad 0.13 \quad 0.05]^T$

13.23 C

13.25 (b) $[0.05 \quad 0.17 \quad 0.17 \quad 0.61]^T$

13.29
$$\begin{bmatrix} 0.05 & 0.61 & 0.14 & 0.44 \\ 0.17 & 0.26 & 0.26 & 0.07 \\ 0.17 & 0.08 & 0.46 & 0.05 \\ 0.61 & 0.05 & 0.14 & 0.44 \end{bmatrix}$$

13.31 $[0.24 \quad 0.20 \quad 0.17 \quad 0.40]^T$

13.A1 CR = 0.1343

13.A3 CR = 0.0393

Glossary

acid-test ratio: The ratio of quick assets to current liabilities. Quick assets are cash, accounts receivable, and marketable securities — those current assets considered to be highly *liquid*. The acid-test ratio is also known as the quick ratio.

actual dollars: Monetary units at the time of payment.

actual interest rate: The stated, or observed, interest rate based on actual dollars. If the real interest rate is i' and the inflation rate is f, the actual interest rate i is found by:

$$i = i' + f + i'f$$

actual internal rate of return, IRR_A: The internal rate of return on a project based on actual dollar cash flows associated with the project; also the real internal rate of return which has been adjusted upwards to include the effect of inflation.

actual MARR: The minimum acceptable rate of return for *actual dollar* cash flows. It is the real MARR adjusted upwards for inflation.

amortization period: The duration over which a loan is calculated to be repaid.

analytic hierarchy process (AHP): A multi-attribute utility theory (MAUT) approach used for large, complex decisions, which provides a method for establishing the criteria weights.

annual worth method: Comparing alternatives by converting all cash flows to a uniform series, i.e., an annuity.

annuity: A series of uniform-sized receipts or disbursements that start at the end of the first period and continue over a number, N, of regularly spaced time intervals.

annuity due: An annuity whose first of N receipts or disbursements is immediate, at time 0, rather than at the end of the first period.

arithmetic gradient series: A series of receipts or disbursements that start at the end of the first period and then increase by a constant amount from period to period.

arithmetic gradient to annuity conversion factor: Denoted by $(A/G,i,N)$, gives the value of an annuity, A, that is equivalent to an arithmetic gradient series where the constant increase in receipts or disbursements is G per period, the interest rate is i, and the number of periods is N.

asset-management ratios: Financial ratios that assess how efficiently a firm is using its assets. Asset management ratios are also known as efficiency ratios. Inventory turnover is an example.

assets: The economic resources owned by an enterprise.

backward induction: See **rollback procedure**

balance sheet: A financial statement which gives a snapshot of an enterprise's financial position at a particular point in time, normally the last day of an accounting period.

base period: A particular date associated with *real dollars* that is used as a reference point for price changes; also the period from which the expenditure shares are calculated in a Laspeyres price index.

base year: The year on which real dollars are based.

benefit-cost ratio: The ratio of the present worth (or annual worth) of benefits to the present worth(or annual worth) of costs. That is,

$$\text{BCR} = \frac{\text{PW(benefits)}}{\text{PW(costs)}}$$

bond: An investment that provides an annuity and a future value in return for a cost today. It has a "par" or "face" value, which is the amount for which it can be redeemed after a certain period of time. It also has a "coupon rate," meaning that the bearer is paid an annuity, usually semi-annually, calculated as a percentage of the face value.

book value: The depreciated value of an asset for accounting purposes, as calculated with a depreciation model.

break-even analysis: The process of varying a parameter of a problem and determining what parameter value causes the performance measure to reach some threshold or "break-even" value.

capacity: The ability to produce, often measured in units of production per time period.

capital cost: The depreciation expense incurred by the difference between what is paid for the assets required for a particular capacity and what the assets could be resold for some time after purchase.

capital cost allowance (CCA): The maximum depreciation expense allowed for tax purposes on all assets belonging to an asset class.

capital cost allowance (CCA) asset class: A categorization of assets for which a specified CCA rate is used to compute CCA. Numerous CCA asset classes exist in the CCA system.

capital cost allowance (CCA) rate: The maximum depreciation rate allowed for assets in a designated asset class within the CCA system.

capital cost allowance (CCA) system: The system established by the Canadian government whereby the amount and timing of depreciation expenses on capital assets is controlled.

capital cost tax factor (CCTF): A value that summarizes the effect of the future benefit of tax savings due to the CCA. It allows analysts to take these benefits into account when calculating the value of an asset. The *new* CCTF takes into account the "half-year rule" where only half of the capital cost of an asset can be used to calculate the CCA in the first year. This rule came into effect in November 1981. The *old* CCTF takes into account the entire amount of a capital expense in one year.

capital expense: The expenditure associated with the purchase of a long-term depreciable asset.

capital recovery factor: Denoted by $(A/P,i,N)$, gives the value, A, of the periodic payments or receipts that is equivalent to a present amount, P, when the interest rate is i and the number of periods is N.

capitalized value: The present worth of an infinitely long uniform series of cash flows.

cash flow diagram: A chart that summarizes the timing and magnitude of cash flows as they occur over time. The X axis represents time, measured in periods, and the Y axis represents the size and direction of the cash flows. Individual cash flows are indicated by arrows pointing up (positive cash flows, or receipts) or down (negative cash flows, or disbursements).

challenger: A potential replacement for an existing asset. See **defender**.

comparison methods: Methods of evaluating and comparing projects, such as present worth, annual worth, payback, and IRR.

compound amount factor: Denoted by $(F/P,i,N)$, gives the future amount, F, that is equivalent to a present amount, P, when the interest rate is i and the number of periods is N.

compound interest: The standard method of computing interest where interest accumulated in one interest period is added to the principal amount used to calculate interest in the next period.

compound interest factors: Functions that define the mathematical equivalence of certain common cash flow patterns.

compounding period: The interest period used with the compound interest method of computing interest.

consistency ratio: A measure of the consistency of the reported comparisons in a PCM. The consistency ratio ranges from 0 (perfect consistency) to 1 (no consistency). A consistency ratio of 0.1 or less is considered acceptable.

constant dollars: See **real dollars**.

consumer price index (CPI): The CPI relates the average price of a standard set of goods and services in some base period to the average price of the same set of goods and services in another period. Currently, Statistics Canada uses a base year of 1992 for the CPI.

continuous compounding: Compounding of interest which occurs continuously over time, i.e., as the length of the compounding period tends toward zero.

continuous models: Models that assume all cash flows and all compounding of cash flows occur continuously over time.

cost of capital: The minimum rate of return required to induce investors to invest in a business.

cost principle of accounting: A principle of accounting which states that assets are to be valued on the basis of their cost as opposed to market or other values.

current assets: Cash and other assets that could be converted to cash within a relatively short period of time, usually a year or less.

current dollars: See **actual dollars**.

current liabilities: Liabilities that are due within some short period of time, usually a year or less.

current ratio: The ratio of all current assets to all current liabilities. It is also known as the working capital ratio.

debt management ratio: See **leverage ratio**.

decision matrix: A multi-attribute utility theory (MAUT) method in which the rows of a matrix represent criteria, and the columns alternatives. There is an extra column for the weights of the criteria. The cells of the matrix (other than the criteria weights) contain an evaluation of the alternatives.

decisional equivalence: Decisional equivalence is a consequence of indifference on the part of a decision maker among available choices.

decision tree: A graphical representation of the logical structure of a decision problem in terms of a sequence of decisions and chance events.

declining-balance method of depreciation: A method of modelling depreciation where the loss in value of an asset in a period is assumed to be a constant proportion of the asset's current value.

defender: An existing asset being assessed for possible replacement. See **challenger**.

deflation: The decrease, over time, in average prices. It can also be described as the increase in the purchasing power of money over time.

depreciation: The loss in value of a capital asset.

debt-management ratio: See **leverage ratio**.

discrete models: Models that assume all cash flows and all compounding of cash flows occur at the ends of conventionally defined periods like months or years.

dominated: See **inefficient**.

economic life: The service life of an asset that minimizes its average cost of use.

effective interest rate: The actual but not usually stated interest rate, found by converting a given interest rate (with an arbitrary compounding period, normally less than a year) to an equivalent interest rate, with a one-year compounding period.

efficiency: A multi-criterion decision-making (MCDM) method which permits the identification of a subset of superior alternatives.

efficiency ratios: See **asset-management ratios**.

efficient: An alternative is **efficient** if no other alternative is valued as at least equal to it for all criteria and preferred to it for at least one.

engineering economics: Science that deals with techniques of quantitative analysis useful for selecting a preferable alternative from several technically viable ones.

equity ratio: A financial ratio which is the ratio of total owners' equity to total assets. The smaller this ratio is, the more dependent the firm is on debt for its operations and the higher are the risks the company faces.

equivalence: A condition that exists when the value of a cost at one time is equivalent to the value of the related benefit at a different time.

equivalent annual cost (EAC): An annuity that is mathematically equivalent to a generally more complicated set of cash flows.

expected value: A summary statistic of a random variable which gives the mean or average value.

expensed: Term applied to an asset with a CCA rate of 100%. For all intents and purposes, this is the same as treating the cost of the asset as an operating cost rather than a capital cost.

external rate of return (ERR): The rate of return on a project where any cash flows that are not invested in the project are assumed to earn interest at a predetermined rate (such as the MARR).

extraordinary item: A gain or loss which does not typically result from a company's normal business activities and is therefore not a recurring item.

financial accounting: The process recording and organizing the financial data of a business. The data cover both flows over time, like revenues and expenses, and levels, like an enterprise's resources and the claims on those resources, at a given date.

financial analysis: Comparison of a firm's financial ratios with ratios computed for the same firm from previous financial statements and with industry standard ratios.

financial ratios: Ratios between key amounts taken from the financial statements of a firm. They give an analyst a framework for answering questions about the firm's liquidity, asset management, leverage, and profitability.

future worth: See the definition of **interest rate**.

future worth method: Comparing alternatives by taking all cash flows to future worth.

geometric gradient series: A set of disbursements or receipts that change by a constant *proportion* from one period to the next in a sequence of periods.

geometric gradient to present worth conversion factor: Denoted by $(P/A,g,i,N)$, gives the present worth, P, that is equivalent to a geometric gradient series where the base receipt or disbursement is A, and where the rate of growth is g, the interest rate is i, and the number of periods is N.

goal: The decision to be made in AHP.

growth adjusted interest rate, i°:

$$i^\circ = \frac{1+i}{1+g} - 1 \text{ so that } \frac{1}{1+i^\circ} = \frac{1+i}{1+g}$$

where i is the interest rate and g is the growth rate. The growth adjusted interest rate is used in computing the geometric gradient to present worth conversion factor.

income statement: A financial statement which summarizes an enterprise's revenues and expenses over a specified accounting period.

independent projects: Two projects are independent if the expected costs and the expected benefits of each of the projects do not depend on whether or not the other one is chosen.

inefficient: An alternative that is not efficient.

inflation: The increase, over time, in average prices of goods and services. It can also be described as the decrease in the purchasing power of money over time.

inflation rate: The rate of increase in average prices of goods and services over a specified time period, usually a year; also, the rate of decrease in purchasing power of money over a specified time period, usually a year.

installation costs: Costs of acquiring capacity (excluding the purchase cost) which may include disruption of production, training of workers, and perhaps a reorganization of other production.

interest: The compensation for giving up the use of money.

interest period: The base unit of time over which an interest rate is quoted. The interest period is referred to as the compounding period when compound interest is used.

interest rate: If the right to P at the beginning of a time period exchanges for the right to F at the end of the period, where $F = P(1+i)$, i is the interest rate per time period. In this definition, P is called the *present worth* of F, and F is called the *future worth* of P.

internal rate of return (IRR): That interest rate, i^*, such that, when all cash flows associated with a project are discounted at i^*, the present worth of the cash inflows equals the present worth of the cash outflows.

inventory-turnover ratio: A financial ratio that captures the number of times that a firm's inventories are replaced (or turned over) per year. It provides a measure of whether the firm has more or less inventory than normal.

Laspeyres price index: A commonly used price index which measures weighted average changes in prices of a set of goods and services over time as compared with the prices in a base period. The weights are the expenditure shares in the base period. The weights are then converted to percentages by multiplying by 100.

leverage ratios: Financial ratios that capture the extent to which a firm relies on debt for its operations. These are also known as debt-management ratios. The equity ratio is an example of a leverage ratio.

liabilities: Claims, other than those of the owners, on a business's assets.

liquidity ratio: A financial ratio that evaluates the ability of a business to meet its current liability obligations. The current ratio and quick ratio are two examples of liquidity ratios.

long-term assets: Assets that are not expected to be converted to cash in the short term, usually taken to be one year.

long-term liabilities: Liabilities that are not expected to draw on the business's current assets.

management accounting: The process of analyzing and recording the costs and benefits of the various activities of an enterprise. The goal of management accounting is to provide managers with information to help in decision making.

market: A group of buyers and sellers linked by trade in a particular product or service.

market equivalence: The ability to exchange one cash flow for another at zero cost.

market failure: Condition in which output or consumption decisions are made in which aggregate benefits to all persons who benefit from the decision are less than aggregate costs imposed on persons who bear costs that result from the decision.

market value: Usually taken as the actual value an asset can be sold for in an open market.

mathematical equivalence: An equivalence of cash flows due to the mathematical relationship between time and money.

minimum acceptable rate of return (MARR): An interest rate that must be earned for any project to be accepted.

modified benefit-cost ratio: The ratio of the present worth (or annual worth) of benefits minus the present worth (or annual worth) of operating costs to the present worth (or annual worth) of capital costs, that is,

$$\text{BCRM} = \frac{\text{PW(benefits)} - \text{PW(operating costs)}}{\text{PW(capital costs)}}$$

Monte Carlo simulation: A procedure that constructs the probability distribution of an outcome performance measure of a project by repeatedly sampling from the input random variable probability distributions.

multi-attribute utility theory (MAUT): An MCDM approach in which criteria weights and preference measures are combined mathematically to determine a best alternative.

multi-criterion decision making (MCDM): Methods that explicitly take into account multiple criteria and guide a decision maker to superior or optimal choices.

mutually exclusive projects: Projects are mutually exclusive if, in the process of choosing one, all the other alternatives are excluded.

net cash flow: The difference between cash inflows and outflows for the period. The net cash flow, A_t, is given by $A_t = R_t - D_t$, where R_t is cash inflow in period t, and D_t is cash disbursed in period t.

net-profit ratio: See **return on total assets ratio**.

nominal dollars: See **actual dollars**.

nominal interest rate: The conventional method of stating the annual interest rate. It is calculated by multiplying the interest rate per compounding period by the number of compounding periods per year.

owners' equity: The interest of the owner or owners of a firm in its assets.

pairwise comparison: An evaluation of the importance or preference of a pair of criteria (or alternatives), based on an AHP value scale.

pairwise comparison matrix (PCM): A device for storing pairwise comparison evaluations.

par value: The price per share set by a firm at the time the shares are originally issued.

payback period: The period of time it takes for an investment to be recouped when the interest rate is assumed to be zero.

payback period method: A method used for comparing alternatives by comparing the periods of time required for the investments to pay for themselves.

performance measures: Calculated values that allow conclusions to be drawn from data.

present worth: See the definition of **interest rate**.

present worth factor: Denoted by $(P/F,i,N)$, gives the present amount, P, that is equivalent to a future amount, F, when the interest rate is i and the number of periods is N.

present worth method: Comparing alternatives by taking all cash flows to present worth.

price index: A number, usually a percentage, that relates prices of a given set of goods and services in some period, t_1, to the prices of the same set of goods and service in another period, t_0.

principal eigenvector: The eigenvector associated with the largest eigenvalue, λ_{max}, of a PCM.

priority weights: Weights calculated for alternatives by normalizing the elements of a PCM and averaging the row entries.

probability: The limit of a long-run proportion or relative frequency; see Close-Up 12.1 for other views of probability.

profitability ratios: Financial ratios that give evidence of how productively assets have been employed in producing a profit. Return on total assets (or net-profit ratio) is an example of a profitability ratio.

project: A term used throughout this text to mean "investment opportunity."

project balance: If a project has a sequence of net cash flows $A_0, A_1, A_2, \ldots, A_T$, and the interest rate is i', there are $T + 1$ project balances, $B_0, B_1, \ldots, B_T$, one at the end of each period t, $t = 0,1, \ldots, T$. A project balance, B_t, is the cumulative future value of all cash flows, up to the end of period t, compounded at the rate, i'.

quick ratio: See **acid-test ratio**.

random variable: A variable that takes on one or more values with probabilities given by its probability distribution.

real dollars: Monetary units of constant purchasing power.

real interest rate: The interest rate, i', is the interest rate that would yield the same number of real dollars in the absence of inflation as the actual interest rate yields in the presence of inflation at the rate f. It is given by

$$i' = \frac{(1+i)}{(1+f)} - 1$$

real internal rate of return: The internal rate of return on a project based on real dollar cash flows associated with the project.

real MARR: The minimum acceptable rate of return when cash flows are expressed in real, or constant, dollars.

recovery period: The designated service life for depreciation calculation purposes in U.S. tax law.

related but not mutually exclusive projects: For pairs of projects in this category, the expected costs and benefits of one project depend on whether the other one is chosen.

repeated lives: Used for comparing alternatives with different service lives, based on the assumption that alternatives can be repeated in the future, with the same costs and benefits, as often as necessary. The life of each alternative is repeated until a common total time period is reached for all alternatives.

retained earnings: The cumulative sum of earnings from normal operations, in addition to gains (or losses) from transactions such as the sale of plant assets or investments that have been reinvested in the business, i.e., not paid out as dividends.

retire: To remove an asset from use without replacement.

return-on-total-assets ratio: A financial ratio that captures how productively assets have been employed in producing a profit. It is also known as the net-profit ratio.

rollback procedure: A procedure in decision tree analysis that computes an expected value at each chance node and selects a preferred alternative at each decision node; also known as backward induction.

salvage value: Either the actual value of an asset at the end of its useful life (when it is sold), or an estimate of the salvage value calculated using a depreciation model.

scenario analysis: The process of examining the consequences of several possible sets of variable values associated with a project.

scrap value: Either the actual value of an asset at the end of its physical life (when it is broken up for the material value of its parts), or an estimate of the scrap value calculated using a depreciation model.

sensitivity analysis: Methods that assess the sensitivity of an economic measure to uncertainties in estimates in the various parameters of a problem.

sensitivity graph: A graph of the changes in a performance measure, holding all other variables fixed.

series present worth factor: Denoted by (*P*/*A*,*i*,*N*), gives the present amount, *P*, that is equivalent to an annuity, *A*, when the interest rate is *i* and the number of periods is *N*.

simple interest: A method of computing interest where interest earned during an interest period is not added to the principal amount used to calculate interest in the next period. Simple interest is rarely used, except as a method of calculating approximate interest.

simple investment: A project that consists of one or more cash outflows at the beginning, followed only by one or more cash inflows.

sinking fund: Interest bearing account into which regular deposits are made in order to accumulate some amount.

sinking fund factor: Denoted by (*A*/*F*,*i*,*N*), gives the size, *A*, of a repeated receipt or disbursement that is equivalent to a future amount, *F*, when the interest rate is *i* and the number of periods is *N*.

specialist company: A firm that concentrates on manufactuing a limited range of very specialized products.

statement of changes in financial position: An accounting report which shows how much cash was generated by a company's operation and by other sources of financing during an accounting period.

straight-line method of depreciation: A method of modelling depreciation which assumes that the rate of loss in value of an asset is constant over its useful life.

study period: A period of time used to compare alternatives.

sunk costs: Costs that were incurred in the past and are no longer relevant in replacement decisions.

term: The duration over which a loan agreement is valid.

trend analysis: A form of financial analysis which traces the financial ratios of a firm over several accounting periods.

undepreciated capital cost (UCC): The remaining book value of assets subject to depreciation for taxation purposes. For any given year, the UCC balance can be calculated as follows:

$$\text{UCC}_{\text{opening}} + \text{additions} - \text{disposals} - \text{CCA} = \text{UCC}_{\text{ending}}$$

uniform series compound amount factor: Denoted by (*F*/*A*,*i*,*N*), gives the future value, *F*, that is equivalent to a series of equal-sized receipts or disbursements, *A*, when the interest rate is *i* and the number of periods is *N*.

weighted average cost of capital: A weighted average of the costs of borrowing and of selling shares. The weights are the fractions of total capital that come from the different sources.

working capital: The difference between total current assets and total current liabilities.

working capital ratio: See **current ratio**.

Index

N

O

P

Q

R

List of Formulas

After-tax IRR:

$$\text{IRR}_{\text{after-tax}} \cong \text{IRR}_{\text{before-tax}} \times (1 - t)$$

After-tax MARR:

$$\text{MARR}_{\text{after-tax}} \cong \text{MARR}_{\text{before-tax}} \times (1 - t)$$

Benefit-cost ratio:

$$\text{BCR} = \frac{\text{PW (benefits)}}{\text{PW(costs)}}$$

Book Value, Declining Balance:

$$\text{BV}_{db}(n) = P(1 - d)^n$$

Book Value, Straight Line:

$$BV_{sl}(n) = P - n\left(\frac{P - S}{N}\right)$$

Capital Cost Tax Factor (new):

$$\text{CCTF}_{\text{new}} = 1 - \frac{td\left(1 + \frac{i}{2}\right)}{(i + d)(1 + i)}$$

Capital Cost Tax Factor (old):

$$\text{CCTF}_{\text{old}} = 1 - \frac{td}{(i + d)}$$

Capitalized Value:

$$P = \frac{A}{i}$$

Capital Recovery Formula:

$$A = (P - S)(A/P, i, N) + Si$$

Compound Interest:

$$F = P(1 + i)^N$$

Compound Interest Factors:

- **Compound Amount Factor**

$$(F/P, i, N) = (1 + i)^N$$

- **Present Worth Factor**

$$(P/F, i, N) = \frac{1}{(1 + i)^N}$$

- **Sinking Fund Factor**

$$(A/F, i, N) = \frac{i}{(1 + i)^N - 1}$$

- **Uniform Series Compound Amount Factor**

$$(F/A, i, N) = \frac{(1 + i)^N - 1}{i}$$

- **Capital Recovery Factor**

$$(A/P, i, N) = \frac{i(1 + i)^N}{(1 + i)^N - 1}$$

- **Series Present Worth Factor**

$$(P/A, i, N) = \frac{(1 + i)^N - 1}{i(1 + i)^N}$$

- **Arithmetic Gradient to Annuity Conversion Factor**

$$(A/G, i, N) = \frac{1}{i} - \frac{N}{(1 + i)^N - 1}$$

- **Geometric Gradient Series to Present Worth Factor**

$$(P/A, g, i, N) = \frac{(P/A, i^\circ, N)}{1 + g}$$

$$(P/A, g, i, N) = \left[\frac{(1 + i^\circ)^N - 1}{i^\circ(1 + i^\circ)^N}\right]\frac{1}{1 + g}$$

Depreciation Amount, Straight Line:

$$D_{sl}(n) = \frac{P - S}{N}$$

Depreciation Amount, Declining Balance:

$$D_{db}(n) = BV_{db}(n - 1) \times d$$